The University of Chicago School Mathematics Project

Functions, Statistics, and Trigonometry

Teacher's Edition

Authors

Rheta N. Rubenstein
James E. Schultz
Sharon L. Senk
Margaret Hackworth
John W. McConnell
Steven S. Viktora
Dora Aksoy
James Flanders
Barry Kissane
Zalman Usiskin

Technology requirements:
Students should have access to computers equipped with
graphing software, a statistics package, and BASIC. Graphics calculators with
statistical capabilities will suffice for most (but not all) questions.
Student access to technology is needed for homework and tests.

Teachers should have access to comparable technology for
in-class demonstrations. (See pages T40-T42.)

About the Cover
If one obtains data of the position of a point on a vibrating reed, the ordered pairs
(time, distance from the point at rest) can be modeled by a function whose graph is a sine wave.
Thus this picture, in which statistics lead to a function that is trigonometric,
involves the three major themes of this book.

Scott, Foresman
Editorial Offices: Glenview, Illinois   Regional Offices: Sunnyvale, California •
Atlanta, Georgia • Glenview, Illinois • Oakland, New Jersey • Dallas, Texas

Acknowledgments

Authors

Rheta N. Rubenstein
Associate Professor of Education, University of Windsor, Windsor, Ontario (formerly) Mathematics Department Head, Renaissance H.S., Detroit, MI

James E. Schultz
Associate Professor of Mathematics, Ohio State University, Columbus

Sharon L. Senk
Associate Professor of Mathematics, Michigan State University, East Lansing

Margaret Hackworth
Mathematics Supervisor, Pinellas County Schools, Largo, FL

John W. McConnell
Instructional Supervisor of Mathematics, Glenbrook South H.S., Glenview, IL

Steven S. Viktora
Chairman, Mathematics Department, New Trier H.S., Winnetka, IL

Dora Aksoy
UCSMP

James Flanders
Assistant Professor of Mathematics and Statistics, Western Michigan University, Kalamazoo

Barry Kissane
Associate Professor of Education, Murdoch University, Perth, Western Australia

Zalman Usiskin
Professor of Education, The University of Chicago

UCSMP Production and Evaluation

Series Editors: Zalman Usiskin, Sharon L. Senk

Managing Editor: Daniel Hirschhorn

Technical Coordinator: Susan Chang

Director of Evaluation: Catherine Sarther (Mount Mary College, Milwaukee, WI)

We wish to acknowledge the generous support of the **Amoco Foundation** and the **Carnegie Corporation of New York** in helping to make it possible for these materials to be developed and tested.

ISBN: 0-673-33239-x

Contents · Teacher's Edition

UCSMP Helps You Update Your Curriculum and Better Prepare Your Students!

As reports from national commissions have shown, students currently are not learning enough mathematics, and the curriculum has not kept pace with changes in mathematics and its applications.

In response to these problems, UCSMP has developed a complete program for grades 7-12 that upgrades the school mathematics experience for the average student. The usual four-year high-school mathematics content — and much more — is spread out over six years. The result is that students learn more mathematics and they are better prepared for the variety of mathematics they will encounter in their future mathematics courses and in life.

In addition, UCSMP helps students view their study of mathematics as worthwhile, as full of interesting information, as related to almost every endeavor. With applications as a hallmark of all UCSMP materials, students no longer ask, "How does this topic apply to the world I know?"

For a complete description of the series, see pages **T19-T52** at the back of the **Teacher's Edition.**

In short, UCSMP...

> **"**This is the most interesting math book I've ever used. This book makes me think. **"**
>
> **Functions, Statistics, and Trigonometry** student, Falmouth High School, ME

- **Offers a wide variety of content and applications.**
- **Prepares students to use mathematics effectively in today's world.**
- **Develops independent thinking and learning.**
- **Helps students improve their performance.**
- **Provides the practical support you need.**

Imagine using a text that has been developed as part of a coherent 7-12 curriculum design, one that has been tested on a large scale *before* publication, and most important, a text that has bolstered students' mathematical abilities, which is reflected in test scores. Read on to find out how UCSMP has done that, and more.

For a detailed discussion of the development and testing of *Functions, Statistics, and Trigonometry*, see pages *T45-T49*.

Years of field-testing and perfecting have brought impressive results!

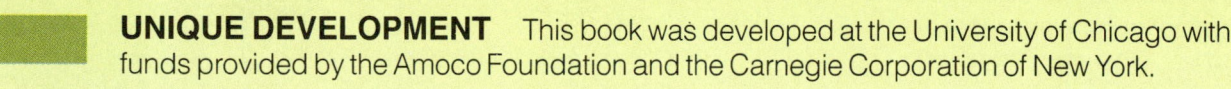

UNIQUE DEVELOPMENT This book was developed at the University of Chicago with funds provided by the Amoco Foundation and the Carnegie Corporation of New York.

PLANNING Initial planning was done with input from professors, classroom teachers, school administrators, and district and state supervisors of mathematics, with attention to the recommendations by national commissions and international studies. The UCSMP secondary curriculum is the first full mathematics curriculum to implement the recommendations of the NCTM Standards committees.

AUTHORSHIP Authors were chosen for expertise in the relevant areas of school mathematics and for classroom experience. All of the authors of *Functions, Statistics, and Trigonometry* are experienced teachers and are active in the field of mathematics education. Several have taught statistics at the high-school level.

FIELD-TESTING AND EVALUATION Pilot testing began with teaching by the initial team of authors and revising based on their firsthand experiences. Then, there was evaluation based on local studies, and the materials were revised again. Further evaluation and revisions were based on national studies. And finally, there were the **Scott, Foresman** enhancements, such as color photography, about 600 blackline masters, and software specifically designed for this course, to better meet the needs of teachers and students.

Offers a Wide Variety of Content and Applications

Wider Scope

Functions, Statistics, and Trigonometry was developed to attract and keep students in mathematics. The course integrates ideas of functions and trigonometry usually studied at this level, with the statistics and data analysis necessary to function successfully in the worlds of today and tomorrow.

FUNCTIONS are widely recognized as a major unifying theme in mathematics. **Functions, Statistics, and Trigonometry** builds that understanding with real-world problems, establishing a firm foundation for future work in mathematics courses and in applications the students will meet in their lives.

LESSON 4-3

Exponential Functions

Consider a population of bacteria that doubles in number every hour. Suppose there are 10 bacteria initially. The table shows the population P at the end of each of the first 4 hours.

A graph of the data is below. The population P is a function of the number of hours h. If N is that function, then $P = N(h) = 10 \cdot 2^h$ because h indicates the number of times the population has doubled.

Hours passed	Population
0	10
1	20
2	40
3	80
4	160

What can be said about the points between those on the left graph? Because time is a continuous variable, after half an hour you would expect there to be $N\left(\frac{1}{2}\right) = 10 \cdot 2^{1/2}$ bacteria. This is $10\sqrt{2} \approx 14.14$ bacteria, which should be truncated to 14 since it isn't meaningful to consider fractions of bacteria. Similarly, other rational periods of time can be interpreted to give a more complete graph, like the one at the right above.

Now consider times between rational values of h, such as $\sqrt{2}$, $\sqrt{3}$, or π hours. To evaluate irrational powers of numbers, such as $2^{\sqrt{2}}$, $2^{\sqrt{3}}$, or 2^{π}, proximate the value of the exponent. For example, $2^{\sqrt{3}}$ can be estimated f approximations to $\sqrt{3}$.

Because $1 < \sqrt{3} < 2$, then $2^1 < 2^{\sqrt{3}} < 2^2$, so $2^1 < 2^{\sqrt{3}} < 4$.

Because $1.7 < \sqrt{3} < 1.8$, then $2^{1.7} < 2^{\sqrt{3}} < 2^{1.8}$, so $3.25 < 2^{\sqrt{3}} < 3.4$

Because $1.73 < \sqrt{3} < 1.74$, then $2^{1.73} < 2^{\sqrt{3}} < 2^{1.74}$, so $3.32 < 2^{\sqrt{3}} <$

You can get a good approximation to $2^{\sqrt{3}}$ by using your calculator's ering key, with 2 as the base and $\sqrt{3}$ as the exponent. To the neare thousandth, $2^{\sqrt{3}} = 3.322$.

LESSON 4-3 Exponential Functions

STATISTICS play a major role in today's world. Collecting, analyzing, processing, and displaying data are ever increasing needs for today's students as they move into the work force and into society. ***Functions, Statistics, and Trigonometry*** provides an important study of data analysis and statistics that all students need. Statistics and algebraic concepts are integrated as students display, describe, transform, interpret, and model numerical data.

TRIGONOMETRY is fundamental to many real-world problems. In ***Functions, Statistics, and Trigonometry,*** traditional topics of trigonometry are integrated with matrix representations and exploration of real-world phenomena using the trigonometric functions.

LESSON 5-6

Graphs of the Sine, Cosine, and Tangent Functions

Consider the sine function, $f(\theta) = \sin \theta$, with θ measured in radians. This is the function that maps each real number θ to the y-coordinate of the image of $(1, 0)$ under a rotation of θ.

The sine function is positive for $0 < \theta < \pi$ and negative for $\pi < \theta < 2\pi$.

The maximum value is 1 for $\theta = \frac{\pi}{2}$

and the minimum is -1 for $\theta = \frac{3\pi}{2}$.

Other values of this function are given in the table below.

θ	0	$\frac{\pi}{6}$	$\frac{\pi}{4}$	$\frac{\pi}{3}$	$\frac{\pi}{2}$	$\frac{2\pi}{3}$	$\frac{3\pi}{4}$	$\frac{5\pi}{6}$	π	$\frac{7\pi}{6}$	$\frac{5\pi}{4}$	$\frac{4\pi}{3}$	$\frac{3\pi}{2}$	$\frac{5\pi}{3}$	$\frac{7\pi}{4}$	$\frac{11\pi}{6}$	2π
$\sin \theta^*$	0	0.5	0.707	0.866	1	0.866	0.707	0.5	0	-0.5	-0.707	-0.866	-1	-0.866	-0.707	-0.5	0
$\sin \theta^{**}$	0	$\frac{1}{2}$	$\frac{\sqrt{2}}{2}$	$\frac{\sqrt{3}}{2}$	1	$\frac{\sqrt{3}}{2}$	$\frac{\sqrt{2}}{2}$	$\frac{1}{2}$	0	$-\frac{1}{2}$	$-\frac{\sqrt{2}}{2}$	$-\frac{\sqrt{3}}{2}$	-1	$-\frac{\sqrt{3}}{2}$	$-\frac{\sqrt{2}}{2}$	$-\frac{1}{2}$	0

* Some entries are approximations.
**exact values

The graph of all points on the sine function: $\theta \to \sin \theta$, for $0 \leq \theta \leq 2\pi$, is shown here.

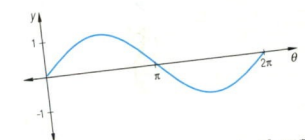

The image of $(1, 0)$ under a rotation of θ repeats itself every 2π radians. As a result, the y-coordinates in the ordered pairs of the function $f(\theta) = \sin \theta$ repeat every 2π. Thus, the table and graph above can be easily extended both to the right and left without calculating any new sine values. A more extended graph of the function $f(\theta) = \sin \theta$ appears below.

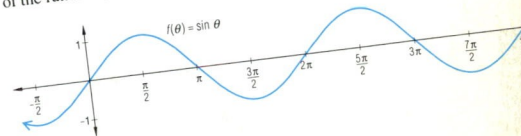

LESSON 10-1

Binomial Probability Distributions

Recall that a binomial probability experiment is one in which there are a fixed number of independent trials, and on each trial there are only two possible outcomes (with fixed probabilities) that can be called *success* and *failure*. In Lesson 8-9, it was proved that in a binomial experiment with n trials, each having probability of success p and probability of failure $q = 1 - p$, then the probability of exactly k successes is

$$_nC_k \cdot p^k q^{n-k}.$$

For instance, studies of long sequences of free throws by basketball players have found no evidence that successive shots are dependent. Thus it is reasonable to consider successive free-throw attempts as independent trials of a binomial experiment in which making a free throw is considered a success, and missing a free throw is considered a failure. Then, the distribution of the numbers k of free throws made in n attempts by a player with a known free-throw success percentage p yields a *binomial probability distribution*.

Example 1 Assume a basketball player makes 75% of free-throw attempts and that the attempts are independent of each other.

a. Determine and graph the probability distribution for the number of free throws made in 8 attempts.
b. In one game this player shoots 8 free throws and makes only 4 of them. Is it unusual for the player to shoot so far below average?

Solution
a. Let x be the number of successes in 8 trials. Then x can be any whole number between 0 and 8, inclusive. The probability p of success is given to be 0.75. By the Binomial Probability Theorem,
$$P(x) = {}_8C_x \cdot (.75)^x (.25)^{8-x}.$$
Evaluating that expression for each possible value of x generates the numbers in the table below. The data are graphed in the histogram below.

x	$P(x)$		
0	$1 \cdot$	$(.25)^8$	≈ 0.000
1	$8 \cdot (.75)$	$(.25)^7 \approx$	0.000
2	$28 \cdot (.75)^2$	$(.25)^6 \approx$	0.004
3	$56 \cdot (.75)^3$	$(.25)^5 \approx$	0.023
4	$70 \cdot (.75)^4$	$(.25)^4 \approx$	0.087
5	$56 \cdot (.75)^5$	$(.25)^3 \approx$	0.208
6	$28 \cdot (.75)^6$	$(.25)^2 \approx$	0.311
7	$8 \cdot (.75)^7$	$(.25) \approx$	0.267
8	$1 \cdot (.75)^8$		0.100

Prepares Students to Use Mathematics Effectively in Today's World

"The projects pulled together the ideas we learned and made us apply them in real life. It is the first time I ever had to do this, and I loved it."

Functions, Statistics, and Trigonometry student, Davis High School, Yakima, WA

Technology

The availability of scientific calculators, graphics calculators, and computers makes it possible for students to work with functions and statistics at a level never before possible.

With technology, a broader range of both pure and applied problems are available to students. It is also possible to create a classroom environment that encourages exploration and experimentation.

Evidence shows that appropriate use of technology enhances student understanding and problem-solving skills.

Definition

The **natural logarithm** of x is the logarithm of x to the base e, written **ln x**. That is, for all x, ln x = log_e x.

The function with equation $y = \ln x$ is the **natural logarithm function**. Note that $y = \ln x$ if and only if $e^y = x$.

The graphs of the exponential function and its inverse, the natural logarithm function, are shown together at the right. These were produced on a function grapher. As with all inverse functions, the graphs are reflection images of each other over the line $y = x$.

On calculators, the key for natural logarithms is usually marked [ln] or [ln x], while [log] is used to refer to common logarithms. Since logarithmic and exponential functions are inverses of each other, you may find them related on your calculator, possibly through [INV] or [2nd] keys. Natural logarithms are often used in models of situations. Example 2 illustrates one such situation.

Example 2 Under certain conditions, the height h in feet above sea level can be approximated from the atmospheric pressure P in pounds per squar[e] inch (psi) at that level using the equation
$$h = \frac{\ln P - \ln 14.7}{-0.000039}.$$
If human blood, at 0.9 psi, "boils" at body temperature, at what heig[ht] above sea level will blood boil in an unpressurized cabin?

Solution Substitute $P = 0.9$, and evaluate with a calculator.
$$h = \frac{\ln 0.9 - \ln 14.7}{-0.000039} \approx \frac{-0.10536 - 2.6878}{-0.000039} \approx 71,621$$

Without a pressurized cabin, heights greater than 72,000 feet ab[ove sea] level would be fatal.

Logarithms can be found to any base b, $b > 0$, $b \neq 1$. But only three b[ases] (for common logarithms), e (for natural logarithms), and 2 (due to it[s relation] to half-life and doubling) are generally used in practical applications[.]

Solution

a. Draw a scatter plot and obtain the correlation from a statistics package.

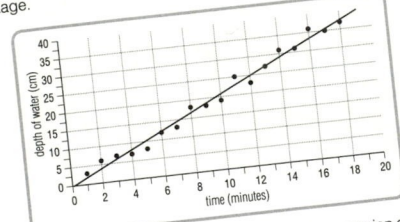

b. The statistics package we used calculated a regression equation $y = 2.103x - .2026$ and $r^2 = 0.983$. So $r = \pm\sqrt{0.983} \approx \pm 0.99$. Becau[se] the relation between the variables is positive (y increases as x increases), we conclude that $r = 0.99$.
c. A correlation of 0.99, indicates a very strong positive relation betw[een] the two variables. A line is a good model for these data, as is clear fr[om] the closeness of the points to the line of best fit. It seems likely that th[e] water was running into the tub at a rather steady rate, that the bathtu[b] had a fairly regular shape, and that the boy's measurements were reasonably careful.

Example 2 verifies that the correlation coefficient r is unaffected if you sw[ap] the variables on the *x*- and *y*-axes.

Example 2 Use the same data as for Example 1, but plot Time vs. Depth. Find th[e] regression line and the correlation coefficient.

Solution

Reading Lessons

These provide opportunities for students to learn mathematics without the pressure of having to master the content. These special lessons, appearing at the end of each chapter and indicated with a blue band at the top of the page, include historical applications, such as the question of authorship of *The Federalist Papers,* and discussion of the mathematics behind recent developments, such as fractals and calculators.

Projects

Projects allow students to become engaged in extended tasks. Students may collect and analyze data or research mathematical ideas. Students are responsible for reporting the results of their work. One set of projects is included at the end of each chapter. These projects offer a wonderful opportunity for students to work cooperatively in small groups.

LESSON

2-8

The Men's Mile Record

In 1976, the International Athletic Federation announced that it would only recognize track records in meters—with one exception. The exception is the one-mile run. The mile run received this special treatment because it has long had a special mystique about it which has captured the interest even of those not normally interested in athletics.

Perhaps this was best illustrated in 1954 when the Englishman Roger Bannister became the first man in recorded history to run a mile in less than four minutes. This athletic feat was headline news around the world, with many commentators regarding the event as "breaking through a barrier." Bannister's run broke the nine-year-old record of the Swede Gunder Hägg, and this long gap between records had encouraged many to think of a sub-four-minute mile as an impenetrable barrier, representing the limit of human potential. With hindsight, of course, we see that this is not the case, and that the barrier was both temporary and psychological.

The table on the next page lists the world record holders for the men's one-mile run since 1875 and the year in which each record was set. The first entry in the table indicates that in 1875 Englishman Walter Slade ran the mile in 4:24.5 minutes, 24.5 seconds, a new world record time. This is written as 4:24.5. Nowadays, this time is often bettered by high school runners and the women's world record is lower.

The current record (as of 1990) of 3:46.32, set in 1985 by another Englishman, Steve Cram, indicates how much progress has been made over the intervening 110 years. In a present day world-class field, Walter Slade, the record holder in 1875, would be beaten by well over an eighth of a mile! Has this progress been steady or erratic? Should we expect the record to continue to fall or have we now almost reached the 'impenetrable barrier'? To answer such questions, a statistical analysis of the data is useful.

A scatter plot is a useful first step in modeling the men's mile record. Here the data from the table are graphed with the record (in seconds) on the vertical axis and the associated year on the horizontal axis.

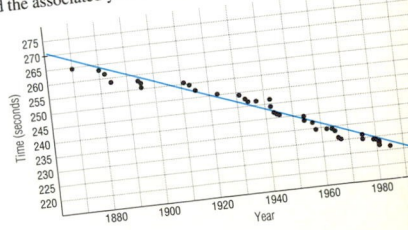

Projects

1. Find a long tape measure and a means of measuring angles. Use trigonometry to measure some "inaccessible" distances of interest to you in your environment. Distances are inaccessible if you are not able, or not permitted, to measure them directly. Some examples of inaccessible distances include: the width of a river or lake, the length of a rail tunnel through a hill, the height of a tall building, and the height of a tower surrounded by a security wall. You may want to measure each distance and angle several times to maximize your accuracy.

2. Running tracks may consist of two straight sections and two curves, which are semicircular in shape. You may have noticed that athletics tracks have "staggered" starting positions for most races, especially those over 100 m.
   a. In which direction do the athletes run on the track shown at the right? Why are the starting

3. Surveyors and cartographers use trigonometry extensively to locate points on the earth and to represent points on maps. In particular, they use a network of fixed *bench marks* to locate property lines and other landmarks.
   a. Find out how trigonometric points and bench marks are determined.
   b. Find some trigonometric points or bench marks in your neighborhood.
   c. Describe how these locations are used to locate other places.

4. Use the Law of Sines to prove the following.
   a. $\dfrac{a - c}{c} = \dfrac{\sin A - \sin C}{\sin C}$
   b. $\dfrac{a}{b} = \dfrac{\sin A}{\sin B}$
   c. $\dfrac{b + c}{b - c} = \dfrac{\sin B + \sin C}{\sin B - \sin C}$
   d. The bisector of an interior angle of a triangle divides the opposite side into parts whose ratio is equal to the ratio of the sides adjacent to the angle bisected. That is, if $\overline{AD}$ bisects $\angle BAC$, then $\dfrac{x}{y} = \dfrac{c}{b}$.

Develops Independent Thinking and Learning

"I liked the fact that FST challenged me. It used every kind of mathematics which I had previously studied and helped me in my other mathematics classes. FST taught me to think mathematically!"

Functions, Statistics, and Trigonometry student, Renaissance High School, Detroit, MI

Reading

Because explanations are clear and well written, students move away from a passive learning stance. By reading the text and using it as a reference tool, students develop the independence they will need in future mathematics classes and in their lives.

Because students can read and understand the text, teachers have the freedom to teach in a variety of ways, to use the class time more creatively, to spend more time on open-ended exploration, and to discuss alternate strategies for solving problems. When students read, the learning environment is enriched.

Problem Solving

In every lesson, students use a variety of problem-solving strategies. The combination of real-world problems, data analysis, and calculator and computer technology, all combine to create a setting that models future problem solving that students will encounter in and after mathematics classrooms.

LESSON

2-2

Linear Models

A **linear function** is a function which can be described by an equation of the form $y = mx + b$, where m and b are constants. Recall that the graph of every function of this form is a line with slope m and y-intercept b. Both the domain and range of the general linear function are the set of all real numbers.

Some situations can be modeled exactly by linear functions. In other situations, the variables of interest may not take on all real number values, but a linear function still provides a convenient model. In either case, linear models can be used to predict values of either the dependent or independent variable.

Example 1

A firm specializes in sanding and varnishing hardwood floors. The cost C for this service is $50 setup charge plus $2 per square foot for materials and labor. Suppose A is the area of the floor in square feet.
a. Find an equation to represent C as a function of A.
b. Draw a graph of the relation between C and A.
c. What will the firm charge to sand and varnish the floor of a room 12 feet by 11 feet?
d. What is the largest area that can be treated for $275?

Solution
a. The cost C is $50 plus $2 for each of the A square feet. Constant increase situations such as this are modeled by linear functions.

$$C = 50 + 2A$$

b. The graph is a line with C-intercept 50 and slope equal to 2.

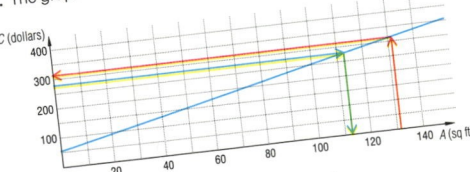

c. The area of the room is $A = 12 \cdot 11 = 132$ sq ft. Substitute 132 for A.
$$C = 50 + 2A$$
$$C = 50 + 2(132)$$
$$= 314$$

The cost will be $314
d. In the equation $C = 50 + 2A$, substitute 275 for C and solve for A.
$$275 = 50 + 2A$$
$$2A = 225$$
$$A = 112.5$$
So the largest area treated for $275 is 112.5 sq ft.

86

T10

Four Kinds of Questions

in each lesson engage students in a
variety of activities designed to enhance
learning and performance.

COVERING THE READING   questions examine what students
have read in the lesson.

APPLYING THE MATHEMATICS   questions offer a broad range
of applications and extensions of the lesson concepts.

REVIEW   questions, keyed to past lessons, help students maintain
and improve performance on important skills and concepts, and preview
ideas to prepare students for topics that will be studied later.

EXPLORATION   questions further extend the ideas of the lesson by
including applications, generalizations, open-ended experiments, or
research.

Questions

Covering the Reading

In 1–3, convert to degrees.
1. $\frac{1}{3}$ revolution counterclockwise
2. $\frac{3}{4}$ revolution clockwise
3. 0.43 revolution counterclockwise
4. Express the latitude of Mobile, Alabama, 30°42′ N, in decimal degrees.
5. At the left is a graph of a unit circle.
   a. Give the length of $\overarc{AC}$.
   b. What is the magnitude in radians of the rotation that maps $\overrightarrow{OA}$ to $\overrightarrow{OC}$?
6. An angle whose measure is given as 3 is assumed to be measured in __?__.

In 7 and 8, a. draw a circle with radius 1. On this circle, heavily shade an arc
with the given length. b. Give the measure in degrees of the central angle of
this arc.
7. $\frac{\pi}{3}$
8. $\frac{3\pi}{2}$
9. Convert −225° to radians exactly without a calculator.
10. Convert 7π to degrees exactly without a calculator.

In 11 and 12, convert to radians. Round your answer to the nearest
thousandth.
11. 370°
12. −40°

In 13 and 14, convert to decimal degrees. Round your answer to the nearest
thousandth.
13. −4.2 radians
14. −4.2π radians

Applying the Mathematics

15. An angle whose measure is $\frac{\pi}{2}$ is about __?__ times as large as an angle
whose measure is $\frac{\pi^{\circ}}{2}$.
16. Order from smallest to largest: 1 revolution, 1 degree, 1 grad,
1 radian.
17. Express $22\frac{3}{8}$ degrees as an angle measure in degrees, minutes, and
seconds.
18. An LP record is played at $33\frac{1}{3}$ revolutions per minute (rpm). Through
how many degrees will a point on the edge of an LP move in three
seconds?

In 19 and 20, use the fact that the planet Jupiter rotates on its axis at a rate of
approximately 0.6334 radians per hour.
19. Is this faster or slower than the Earth's rate of rotation?
20. What is the approximate length of the Jovian day (the time it takes Jupiter to make a complete revolution)?

21. Convert x radians to degrees.

Review

22. What translation will map the graph of $y = x^2$ onto the graph of
$y = (x + 7)^2 - 1$? *(Lesson 3-2)*
23. Find an equation for the parabola that contains (−2, 5), (4, 11), and
(10, 53). *(Lesson 2-7)*
24. *Multiple choice.* Which of the following values of r^2 indicates the
strongest linear relationship between two variables? *(Lesson 2-5)*
   (a) 0.90   (b) 0.50   (c) 0.09   (d) −0.95
25. The following instructions are taken from a spaghetti packet. Explain
what is wrong with the instructions (assuming all portions are equal).
(Previous course)

A perfect portion measures up every time!

HERE'S HOW IT WORKS:

1. **GRASP** a handful of spaghetti.

2. **PLACE** it on **END** in one of
the portion circles on the box. (For
instance, when the circle of spaghetti
matches the 2-portion circle, you know it's
the right amount to cook for two people.)

26. The circle at the right has diameter 18 m. Find
   a. its circumference.
   b. its area. *(Previous course)*

18 m

27. Use the circle at the left. Find a. the length of $\overarc{AB}$ and b. the
shaded region. *(Previous course)*

Exploration

In 28 consult an encyclopedia or book on the history of mathematics
28. What systems of angle measurement were used by ancient cultures
other than the Babylonians?

Helps Students Improve Their Performance

Progress Self-Test

This provides the opportunity for feedback and correction — before students are tested formally. The Student Edition contains full solutions for all of these questions to allow for accurate self-evaluation.

Chapter Review

The main objectives of the chapter are organized into sections corresponding to the four main types of understanding this book promotes: Skills, Properties, Uses, and Representations. Thus, the Chapter Review extends a multi-dimensional approach to understanding, offering a broader perspective that helps students put everything in place.

SKILLS include simple and complicated procedures for getting answers. The emphasis is on how to carry out algorithms.

PROPERTIES cover the mathematical *justifications for* procedures and other theory. To fully understand ideas, students must answer the common question, "But why does it work that way?"

USES include real-world applications of the mathematics. To effectively apply what they learn, students must know when different models or techniques are relevant.

REPRESENTATIONS cover ways to picture what is being studied. Visual images range from graphs of functions to effective displays of statistical data.

CHAPTER 3

Progress Self-Test

Take this test as you would take a test in class. You will need an automatic grapher. Then check your work with the solutions at the back of the book.

In 1 and 2, the translation $(x, y) \to (x - 1, y + 5)$ is applied to the function $y = 3x^2$.

1. Write an equation of the image.

2. What are the coordinates of the vertex of the image?

In 3 and 4, suppose that $y = 12x + x^2 - x^3$, which is graphed below, is transformed under $S: (x, y) \to (-x, 2y)$.

3. Sketch a graph of the image.

4. Give an equation for the graph of the image.

In 5 and 6, use the given data. To make the calculations easier, an entomologist subtracted 20 g from the mass of each of 170 beetle specimens she had collected. She found the following statistics for the weights of her specimens after this transformation:
mean: 7.9 g   median: 7.4 g   maximum: 11.8 g
minimum: 3.4 g   standard deviation: 1.3 g.

5. For the actual beetle masses, give
  a. the median   **b.** the range
  c. the variance.

6. To compare her results with those of other entomologists, she converted some statistics to ounces (1 g ≈ 0.0353 oz). Give the mean and standard deviation of her sample in ounces.

In 7 and 8, suppose $\sum_{i=1}^{5} g_i = 2$. Evaluate the expression.

7. $\sum_{i=1}^{5} (g_i + 6)$ **8.** $\sum_{i=1}^{5} 4g_i$

In 9–11, consider $a(x) = 18 - 3x$ and $b(y) = 4x^2$.

9. Evaluate $(b \circ a)(2)$.   **10.** Find $a(b(x))$.

11. Give the domain of $a \circ b$.

In 12–14, consider $f(x) = \dfrac{2}{x-3}$.

12. Find an equation for f^{-1}, the inverse of f.

13. Graph f and f^{-1} with an automatic grapher. Sketch or print a hardcopy of the graphs.

14. Is the inverse of f a function? If the inverse is a function, prove it. If it is not, explain why.

In 15 and 16, the function $h(t) = -16t^2 + 60t + 20$ describes the height $h(t)$ in feet above the ground of an object t sec after being thrown straight up at 60 ft/sec from 20 ft above the ground.

15. Use an automatic grapher to graph this function. Sketch or print a hard copy of the graph that shows all its key features.

16. Estimate to the nearest integer the time at which the object hits the ground.

17. Prove that the function $f(x) = 4x^3 - 2x$ is an odd function.

In 18 and 19, consider the graph below of the function g, which is the image of $y = |x|$ under a scale change.

18. Find an equation for g.

19. Describe the symmetries of g.

In 20 and 21, *true* or *false*.

20. If each element of a data set is multiplied by k, then the variance of the new data set is k times the variance of the original data set.

21. Consider two functions f and g. If f is the image of g under a translation, then the graph of f is congruent to the graph of g.

196

T12

Quizzes and Test Masters

Quizzes and Test Masters in the Teacher's Resource File offer further help to assess mastery. You choose the test format that best suits your needs.

Quiz and Test Writer

Quiz and Test Writer provides computer-generated quizzes and tests for additional assessment flexibility.

CHAPTER 3

Chapter Review

Questions on **SPUR** Objectives

SPUR stands for **S**kills, **P**roperties, **U**ses, and **R**epresentations.
The Chapter Review questions are grouped according to the SPUR Objectives for this chapter.

SKILLS deal with the procedures used to get answers.

■ **Objective A:** *Find composites of functions.*
(Lesson 3-7)
In 1 and 2, let $f(t) = 6t - 2$ and $g(t) = t - t^2$.
1. Evaluate **a.** $f(g(-1))$ **b.** $g(f(-1))$.
2. Find a formula for each composite.
   a. $f(g(t))$ **b.** $g(f(t))$

In 3 and 4, consider the function f mapping A to B, and g mapping B to C.

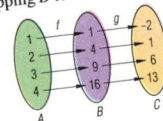

3. Evaluate $g(f(4))$.
4. Evaluate $(g \circ f)(2)$.

In 5 and 6, let $m(x) = \frac{3}{x}$ and $n(x) = x + 5$.
5. Evaluate
   a. $(m \circ n)(-6)$ **b.** $(n \circ m)(-6)$.
6. Find a formula for each function.
   a. $m \circ n$ **b.** $n \circ m$

■ **Objective B:** *Find inverses of functions.*
(Lesson 3-8)
In 7–10, for each function, **a.** describe the inverse using a set of ordered pairs or an equation.
b. State whether the inverse is a function.
7. $\{(3, 7), (4, 8), (5, 9), (6, 10), (7, 11)\}$
8. $y = 2x + 7$ 9. $f(x) = |x|$ 10. $g(x) = \frac{2}{x+1}$

PROPERTIES deal with the principles behind the mathematics.

■ **Objective C:** *Use the Graph Translation Theorem or the Graph Scale Change Theorem to find transformation images. (Lessons 3-2, 3-5)*
11. In an equation for a function or relation, if x is replaced by $x - h$ and y is replaced by $y - k$, how is the graph of the image related to the graph of the preimage?
12. In an equation for a function or relation, if x is replaced by $\frac{x}{a}$ and y by $\frac{y}{b}$, how is the graph of the resulting equation related to the graph of the original?

In 13 and 14, *multiple choice*.
13. Which translation has the effect of moving each point 3 units down and 8 units to the right?
    (a) $T(x, y) = (x - 3, y + 8)$
    (b) $T(x, y) = (x + 8, y - 3)$
    (c) $T(x, y) = (x - 8, y + 3)$
    (d) $T(x, y) = (x - 8, y - 3)$
14. Which scale change has the effect on a graph of scaling horizontally by a factor of 12 and vertically by a factor of $\frac{1}{5}$?
    (a) $S(x, y) = (12x, 5y)$ (b) $S(x, y) = \left(\frac{x}{12}, 5y\right)$
    (c) $S(x, y) = \left(12x, \frac{y}{5}\right)$ (d) $S(x, y) = \left(\frac{x}{12}, \frac{y}{5}\right)$

In 38–40, **a.** determine if the function is odd, even, or neither. **b.** If the function is odd or even, prove it. If it is neither, give a counterexample.
39. $s(t) = 5t^2 - t^4$
38. $f(x) = 8x^3$
40. $g(m) = |3m - 4|$

■ **Objective H:** *Identify properties of composites or inverses. (Lessons 3-7, 3-8)*
41. If $f(x) = \sqrt{x}$ and $g(x) = x + 10$, what is the domain of $f(g(x))$?
    If two relations are inverses of each other, what transformation maps one to the other?

In 43 and 44, *multiple choice*.
43. If (p, q) is a point on the graph of a relation, what point must be on its inverse?
    (a) $(-p, q)$ (b) $(p, -q)$
    (c) $(-p, -q)$ (d) (q, p)
44. Which condition is sufficient for concluding that f and g are inverses of each other?
    (a) $f(g(x)) = g(f(x))$ for all x in the domains of f and g.
    (b) $f(g(x)) = x$ for all x in the domain of g and $g(f(x)) = x$ for all x in the domain of f.
    (c) f is always positive and g is always negative.
    (d) f and g each pass the vertical line test.

[USE]S deal with applications of mathematics in real situations.

■ **Objective I:** *Use translations or scale changes to analyze data. (Lessons 3-3, 3-6)*
Use a translation to find mentally the average of these bowling scores:
102, 112, 110, 106, 115.
   a. Identify the transformation used to scale the scores below.

original scores	scaled scores	frequency
3	15	1
4	20	3
5	25	2
7	35	7
8	40	2

   b. Find the mode, mean, and median of the original data.
   c. Find the mode, mean, and median of the scaled scores.
   d. What property of data transformations is shown in **b** and **c**?

47. A swimmer training for the World Championships recorded the following times for the 800-m freestyle event (in minutes:seconds).

8:28.71	8:31.06	8:30.14
8:31.02	8:29.91	8:27.88
8:28.13	8:29.37	8:30.00
8:28.65	8:29.50	8:31.36

Using a translation to simplify the computation, find the mean and standard deviation of her times.

48. For a sample of a certain butterfly species, a scientist measured their lengths and found a mean length of 1.76 inches with a range of 3.02 inches and a standard deviation of 0.53 inches. If the data are converted to millimeters, give the following statistics:
    a. mean   **b.** range   **c.** variance.

[REP]RESENTATIONS deal with pictures, graphs, or objects that illustrate concepts.

■ **Objective J:** *Apply the Graph Translation Theorem or the Graph Scale Change Theorem to sketch or identify graphs. (Lessons 3-2, 3-5)*
   Sketch the graphs of $y = |x - 5| + 7$ and its parent function on the same set of axes.

50. For $k(x) = 5 + \frac{1}{x+6}$,
    a. sketch a graph of k,
    b. give the equations of the asymptotes,
    c. give the coordinates of the intercepts,
    d. identify the parent equation and the transformation that maps the parent function to $y = k(x)$.

Provides the Practical Support You Need

Continual involvement of teachers and instructional supervisors — in planning, writing, rewriting, and evaluating — has made this program convenient and adaptable to your needs.

Before each chapter you'll find the following:

 OBJECTIVES are letter-coded and keyed to Progress Self-Test, Chapter Review, Lesson Masters, and Chapter Tests, Forms A and B — showing a direct correspondence between what is taught and what is tested.

 The **OVERVIEW** anticipates and addresses your needs for the upcoming chapter.

The **PERSPECTIVES** are a unique feature that provides the rationale for the inclusion of topics or approaches, provides mathematical background, and makes connections within UCSMP materials. This is interesting information you'll really use.

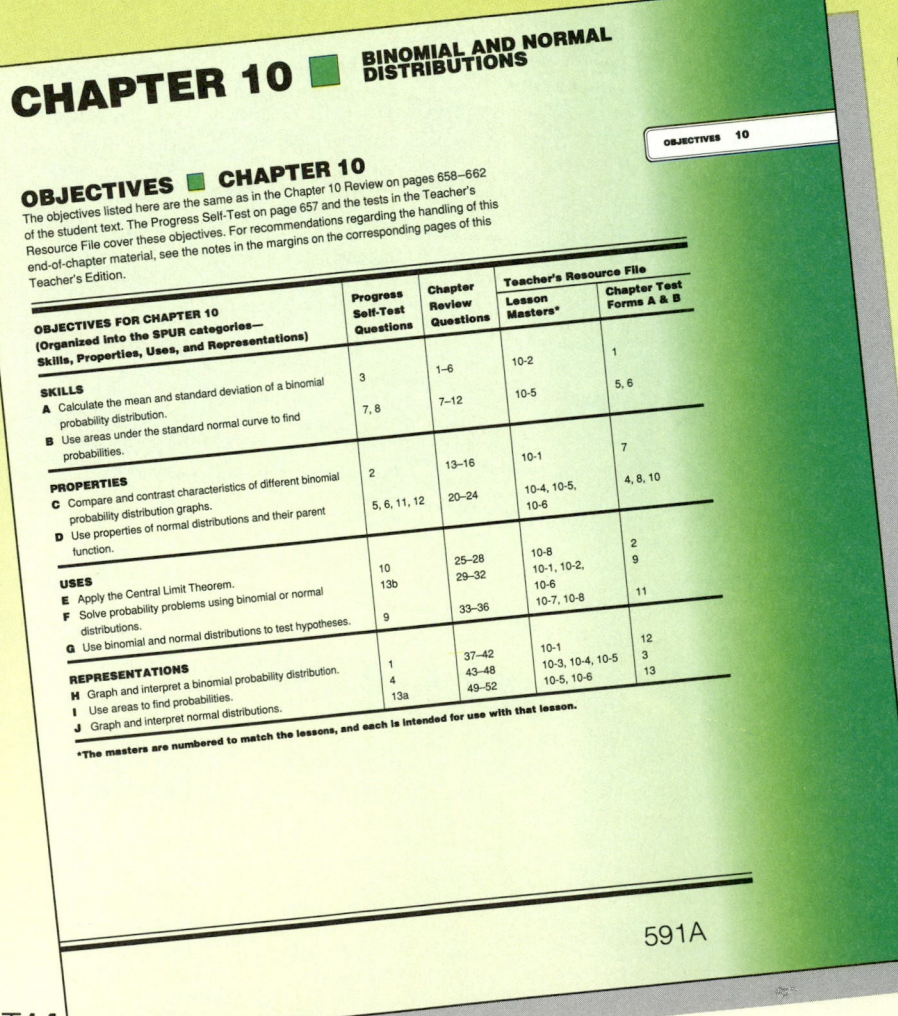

CHAPTER 10 ■ BINOMIAL AND NORMAL DISTRIBUTIONS

OBJECTIVES 10

OBJECTIVES ■ CHAPTER 10

The objectives listed here are the same as in the Chapter 10 Review on pages 658–662 of the student text. The Progress Self-Test on page 657 and the tests in the Teacher's Resource File cover these objectives. For recommendations regarding the handling of this end-of-chapter material, see the notes in the margins on the corresponding pages of this Teacher's Edition.

OBJECTIVES FOR CHAPTER 10— (Organized into the SPUR categories— Skills, Properties, Uses, and Representations)	Progress Self-Test Questions	Chapter Review Questions	Teacher's Resource File		Chapter Test Forms A & B
			Lesson Masters*		
SKILLS					
A Calculate the mean and standard deviation of a binomial probability distribution.	3	1–6	10-2		1
B Use areas under the standard normal curve to find probabilities.	7, 8	7–12	10-5		5, 6
PROPERTIES					
C Compare and contrast characteristics of different binomial probability distribution graphs.	2	13–16	10-1		7
D Use properties of normal distributions and their parent function.	5, 6, 11, 12	20–24	10-4, 10-5, 10-6		4, 8, 10
USES					
E Apply the Central Limit Theorem.	10 13b	25–28 29–32	10-8 10-1, 10-2, 10-6		2 9
F Solve probability problems using binomial or normal distributions.	9	33–36	10-7, 10-8		11
G Use binomial and normal distributions to test hypotheses.					
REPRESENTATIONS					
H Graph and interpret a binomial probability distribution.	1	37–42	10-1		12
I Use areas to find probabilities.	4	43–48	10-3, 10-4, 10-5		3
J Graph and interpret normal distributions.	13a	49–52	10-5, 10-6		13

*The masters are numbered to match the lessons, and each is intended for use with that lesson.

591A

OVERVIEW 10

OVERVIEW ■ CHAPTER 10

This chapter constitutes the last extended discussion of statistics in this book. (The sole later mention of statistics is in conjunction with the study of ellipses in Chapter 12.) The goal of the chapter is to discuss the normal distribution and the use of statistics in hypothesis testing.

The shape of histograms of binomial distributions with large numbers of trials leads into the study of the normal curve and its properties. Using the techniques of Chapter 4, students standardize the normal curve, and use a table of probabilities to estimate the area of regions under the curve. These probabilities are then used in conjunction with the Central Limit Theorem and the Law of Large Numbers to calculate confidence intervals for particular occurrences of events in continuous data sets.

The chapter easily divides into three parts with three lessons in each. The first part begins with binomial probability distributions. Lesson 10-1 presents an analysis of one experiment having a variable probability and a fixed number of trials, and a second with a fixed probability and a variable number of trials. In Lesson 10-2, the formulas for the mean and standard deviation of such distributions are discussed. Lesson 10-3 introduces the idea of representing probability as an area between a curve and the x-axis.

The next three lessons provide an extended introduction to the normal distribution. In Lesson 10-4, the function $f(x) = e^{-x^2}$, the parent of the normal curves, is discussed. In Lesson 10-5, the parent normal function is scaled so that the area between the image and the x-axis

is 1, yieldi... distribution... use a table... mate area... under the... the skills ... use to fin... happenin... uted dat... the use ... approxim... which be...

The l... chapte... The ide... in Less... Centra... mathe... normal... eses, ... close... using... hypo... hazar... test...

PERSPECTIVES 10

PERSPECTIVES ■ CHAPTER 10

The Perspectives provide the rationales for the inclusion of topics or app... some mathematical background, and make connections with other less... UCSMP texts.

10-1
BINOMIAL PROBABILITY DISTRIBUTIONS
We call the function that maps the ordered triple (k, n, p) onto $_nC_k p^k (1-p)^{n-k}$ the binomial probability distribution $B(k, n, p)$.

In this lesson students explore the values and graphs of distributions for various values of p and n. In particular, we are interested in what happens as n gets larger, because the distribution approaches a normal distribution.

10-2
MEAN AND STANDARD DEVIATION OF A BINOMIAL DISTRIBUTION
The mean of a distribution is the mean value you would expect to get as the number of trials increases without bound; for this reason it is also called the expected value of the distribution. This mean can be calculated by finding the sum of each value multiplied by its probability. The letter μ is used to indicate that this is a theoretical mean and not a sample mean. Students are also given a general formula for the variance and the square root of the variance is the standard deviation. These formulas are then applied to binomial distributions.

591B

T14

The **DAILY PACING CHART** shows you at a glance two alternate ways to pace the chapter.

TESTING OPTIONS list the chapter quizzes and tests for ease of planning.

Professional Sourcebook for UCSMP

Also at your fingertips is a wealth of information that includes valuable material on topics ranging from research to review. See pages **T19-T52** at the back of the **Teacher's Edition.**

---

mal curve. Then we scale the triangle using the Area Scale Change Theorem so that its area is 1. This same idea will be applied to the parent normal curve in Lesson 10-5.

10-4
THE PARENT OF THE NORMAL CURVE
The parent of the normal curve is the function $f(x) = e^{-x^2}$. It has a bell shape but the area under it is not 1. Students are expected to be able to identify the critical points of the curve, including the maximum and inflection points, and to describe its curvature and asymptotic behavior. Students are not expected to calculate inflection can be at ± 1, as this requires calculus.

10-5
THE STANDARD NORMAL DISTRIBUTION
The parent normal curve function cannot be a probability distribution because the area under it is not 1. By doing an appropriate vertical and horizontal scale change, the area under the curve can be made equal to 1 and the x-coordinates of the points of inflection can be at ± 1. The significance of this is that the function becomes a probability distribution with mean 0 and standard deviation 1.

The normal distribution is important because so many natural phenomena seem to be normally distributed. Normal distributions are close approximations to binomial distributions, and the distributions of sample means are normal. The last of these is discussed in Lesson 10-8. The normal distribution is so important that tables listing areas under this curve have been compiled to indicate the probability that an event can happen.

10-6
OTHER NORMAL DISTRIBUTIONS
When a random variable is normally distributed, it is rather unlikely that its mean is 0 and its standard deviation is 1. However, if its mean is m and its standard deviation is s,

then the transformation $x \to \dfrac{x - m}{s}$

will transform it into a random variable whose distribution is standard normal. Then tables can be used to find probabilities, and the results can be applied to the given distribution. This lesson considers some distributions that are assumed to be normal and uses tables to find probabilities that the random variable lies in a particular interval.

10-7
USING PROBABILITY TO MAKE JUDGMENTS
Probabilities indicate only the likelihood that an event may occur under certain conditions or assumptions. When a probability is low, but not zero, we consider the event unlikely, but not impossible.

This lesson is devoted to inferential statistics, the use of statistics to make judgments. With actual data, if the probability of the results we find is very low, we may decide that our assumptions are wrong about the situation. These ideas, along with those of the null hypothesis, alternative hypothesis, and significance level, form the basis of hypothesis testing.

10-8
SAMPLING DISTRIBUTIONS AND THE CENTRAL LIMIT THEOREM
The question of likelihood of many kinds of events depends on the size of the sample. It may not be unusual for one student to miss a particular problem, but it could be quite unlikely for many students to miss that problem. The question of

the effect of sample size on the distribution of means of a sample is covered in the Central Limit Theorem (CLT): If samples of size n are taken from a population with mean μ and standard deviation σ, then as n increases, the sample means approach a normal distribution with mean μ and standard deviation $\dfrac{\sigma}{\sqrt{n}}$. The power of this theorem is that it applies to means of samples of any size over 30. The proof of this theorem is beyond the scope of this course, so in this lesson, computer simulations are exhibited to give credibility to the theorem.

10-9
CONFIDENCE AND CAUTIONS IN REASONING
The ideas from the previous two lessons are applied here to discuss the fundamental question of how large a sample needs to be in order to have a particular level of confidence in the results. The Central Limit Theorem indicates what will happen with samples of a particular size; the notion of a particular likelihood being a test value is extended here to the idea of confidence interval. The surprise is that the accuracy of a random sample has nothing to do with the size of the population. A random sample of 1200 taken from a population of 240,000,000 can be treated with the same confidence as one of 1200 from a population of 2400.

591C

DAILY PACING CHART ■ CHAPTER 10

For optimal student performance, the Progress Self-Test and the Chapter Review should be covered. (See *General Teaching Suggestions: Mastery* on page T35 of this Teacher's Edition.) By following the pace of the Full Course given here, students can complete 12 or 13 chapters of the book by the end of the year. Students following the pace of the Minimal Course spend more time to complete lessons that are combined with quizzes and to consider the Chapter Review. When all lessons are covered in order (the recommendation of the authors), students in the Minimal Course can cover all but the last two chapters. For more information on pacing, see *General Teaching Suggestions: Pace* on page T34 of this Teacher's Edition.

DAY	MINIMAL COURSE	FULL COURSE
1	10-1	10-1
2	10-2	10-2
3	10-3	10-3
4	Quiz (TRF); Start 10-4	Quiz (TRF); 10-4
5	Finish 10-4	10-5
6	10-5	10-6
7	10-6	Quiz (TRF); 10-7
8	Quiz (TRF); Start 10-7	10-8
9	Finish 10-7	10-9
10	10-8	Progress Self-Test
11	10-9	Chapter Review
12	Progress Self-Test	Chapter Test (TRF)
13	Chapter Review	
14	Chapter Review	
15	Chapter Test (TRF)	

TESTING OPTIONS
■ Quiz for Lessons 10-1 through 10-3   ■ Chapter 10 Test, Form A   ■ Chapter 10 Test, Cumulati
■ Quiz for Lessons 10-4 through 10-6   ■ Chapter 10 Test, Form B

A Quiz and Test Writer is available for generating additional questions, additional quizzes, or additional forms of the Chapter Test.

PROVIDING FOR INDIVIDUAL DIFFERENCES
The student text has been written for, and tested with, average students. It has also been used successfully with students who are better prepared as well as with students who are poorly prepared.

The Lesson Notes often include Error Analysis and Alternate Approach features to help you assist those students who need more help. Students of all abilities often learn from their peers and may benefit from small group work referenced as appropriate throughout the Notes. A blackline Lesson Master (in the Teacher's Resource File), keyed to the chapter objectives, is available for each lesson to provide more practice. (However, we advise you to use these masters sparingly; again refer to the suggestions on pace on page T34.)

Extension activities are provided in the Lesson Notes for use with students who need to be challenged even more than the requirements of completing the Full Course provide.

591D

T15

Super Teaching Support

*"With **Functions, Statistics, and Trigonometry,** I feel better about the way I am preparing my students than I have ever felt before. I know that their understanding is broader and deeper than that of students in any other course at this level. "*

William Kring, Teacher, Davis High School, Yakima, WA

1 **RESOURCES** save you time by coordinating all of the ancillaries with the lesson.

2 **OBJECTIVES** are letter-coded for easy reference.

3 **TEACHING NOTES** provide everything you need for lesson planning. These include suggestions for the examples, dialogue to stimulate higher-order thinking, and more.

4 **MAKING CONNECTIONS** helps you connect present content and ideas to material covered in an earlier or later lesson, chapter, or text.

5 **ALTERNATE APPROACH** offers a different strategy for presenting the lesson.

6 **TECHNOLOGY** suggests ways to involve students with computers and graphics calculators. Combine this with the Computer Masters and the new Scott, Foresman software for valuable and interesting activities.

7 **ERROR ANALYSIS** pinpoints typical student errors and provides remediation techniques.

8 **ADDITIONAL EXAMPLES** provide examples parallel to those in the text for added flexibility.

9 **NOTES ON QUESTIONS** highlight important aspects of questions and provide helpful suggestions to enhance learning.

10 **MORE PRACTICE** lists the Lesson Master which can be used for additional practice.

11 **EXTENSION** offers high-interest activities for all students, as well as enrichment activities for students needing additional challenge. These well-liked activities provide ideas for technology, careers, connections with other fields of study, and additional applications.

12 **SMALL GROUP WORK** includes activities and projects suitable for groups of students.

13 **EVALUATION** tells what you need to know about the quizzes and tests. It also provides **Alternative Assessment** suggestions to encourage different formats, such as oral presentation and cooperative learning.

14 **ADDITIONAL ANSWERS** locates answers to questions in the lesson. Shorter answers are overprinted in red near the questions.

15 **LESSON MASTERS** are pictured where you need them for your convenience.

LESSON 7-3

1

RESOURCES
- Lesson Master 7-3
- Quiz for Lessons 7-1 through 7-3
- Computer Master 18

2

OBJECTIVES

B Compute probabilities.
C Find the number of ways of selecting or arranging objects.
D Evaluate expressions using factorials.
H Use probability to solve problems.
I Use Counting Principles to find the number of ways of arranging objects.

3

TEACHING NOTES

4

Making Connections
We treat the terms *arrangement* and *permutation* as synonyms. The latter word will be introduced in Lesson 7-5. Not until the symbol $_nP_r$ is used is there a need to explain what the "P" denotes.

5

6

Alternate Approach: Technology   There are ways of dealing with factorials higher than a student's calculator will estimate. In later lessons, students will learn to write the factors to divide out common factors when possible. Also, higher factorials can be approxi-

T16

mated using Stirling's formula $n! \approx \sqrt{2n\pi}\left(\dfrac{n}{e}\right)n$, where e is the base of natural logarithms, or with a simple program like the following one.

```
20   INPUT N
30   F = 1
40   FOR X = 1 TO N
50     F = F*X
60   NEXT X
70   PRINT N;"! = "; F
80   END
```

Error Analysis Suggest to students to first decide whether the given situation is one with replacement or without, and also to show their calculations with their answers.

ADDITIONAL EXAMPLES

1. A quiz has 10 questions, each of which can be answered "always," "sometimes," or "never." If you guess on each question, what is the probability of answering all questions correctly?

$\left(\dfrac{1}{3}\right)^{10} \approx .000017$

2. How many license plates are possible with 4 letters from A to Z (except O and I) followed by two digits from 0 through 9?

$24^4 \cdot 10^2 = 33{,}177{,}600$

3. How many orders of finish are possible in a 7-horse race?

$7! = 5040$

NOTES ON QUESTIONS
Making Connections for Question 2: The word *sample* here suggests a subset, signalling that the order of selection of the boy or girl does not matter. If the sample did not have to have two children of different genders, then the answer is $26 \cdot 25$ when the order makes a difference (for instance, a winner and a runner-up) and $26 \cdot \dfrac{25}{2}$ when the order does not matter. This is the distinction between permutations ... to be dis-

---

MORE PRACTICE
For more questions on SPUR Objectives, use *Lesson Master 7-3*, shown on page 413.

EXTENSION
Recently, Marilyn vos Savant posed a probability problem in her "Ask Marilyn" column in *Parade Magazine*. Judging from the mail she received, the problem confounded many people, including several mathematicians. You may wish to have your students try to solve that problem.

The problem: Suppose you are on a game show, and the host gives you a choice of three doors. Behind one door is a valuable prize. Behind each of the others is a goat. You pick one door, say Door 2. The host, who knows what is behind each door, then opens a door to reveal a goat. The host now gives you a choice: Do you want with your choice or do you want to switch to the other unopened door? Is it to your advantage to switch? Explain.

Answer: You should switch. Your probability of winning if you do not switch is $\frac{1}{3}$. Your probability of winning if you switch is $\frac{2}{3}$.

Rationale: Suppose these are the doors.

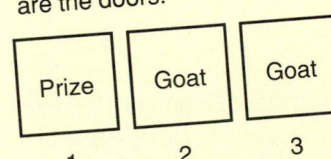

Prize	Goat	Goat
1	2	3

If you pick Door 1 and switch, you will lose. If you pick Door 2 and switch, you win. If you pick Door 3 and switch, you win. Hence, when you switch, you win 2 out of 3 times. You might have students work in **small groups** to imitate the game, using cards labeled P,

---

PROJECTS
Projects 2 and 4 are related to the content of this lesson.

EVALUATION
A quiz covering Lessons 7-1 through 7-3 is provided in the Teacher's Resource File.

ADDITIONAL ANSWERS
3.a.

Components Designed for Ease of Teaching

Student Edition
full color

Teacher's Edition
annotated with margin notes

Teacher's Resource File
About 600 blackline masters
   cover your every classroom need!

Quizzes and Test Masters
Quizzes (at least one per chapter)
Chapter Tests, Forms A and B (parallel forms)
Chapter Tests, Cumulative Form
Comprehensive Tests (four per text, including
   Final Exam, primarily multiple choice)

Lesson Masters
(one or two pages per lesson)

Computer Masters

Answer Masters
(provide answers for questions in student text;
   oversized type to enable display in class,
   allowing students to grade their own work)

Teaching Aid Masters
(masters for overhead transparencies; forms,
   charts, and graphs from the P.E.; coordinate
   grids; and more)

Additional Ancillaries
Solution Manual
Computer Software (for Macintosh and IBM)
   GraphExplorer
   StatExplorer
   Quiz and Test Writer

The University of Chicago School Mathematics Project

Functions, Statistics, and Trigonometry

Authors

Rheta N. Rubenstein
James E. Schultz
Sharon L. Senk
Margaret Hackworth
John W. McConnell
Steven S. Viktora
Dora Aksoy
James Flanders
Barry Kissane
Zalman Usiskin

Technology requirements:
Students should have access to computers equipped with
graphing software, a statistics package, and BASIC. Graphics calculators with
statistical capabilities will suffice for most (but not all) questions.
Student access to technology is needed for homework and tests.

About the Cover
If one obtains data of the position of a point on a vibrating reed, the ordered pairs
(time, distance from the point at rest) can be modeled by a function whose graph is a sine wave.
Thus this picture, in which statistics lead to a function that is trigonometric,
involves the three major themes of this book.

Scott, Foresman
Editorial Offices: Glenview, Illinois   Regional Offices: Sunnyvale, California •
Atlanta, Georgia • Glenview, Illinois • Oakland, New Jersey • Dallas, Texas

Acknowledgments

Authors

Rheta N. Rubenstein
*Associate Professor of Education, University of Windsor, Windsor, Ontario
(formerly) Mathematics Department Head, Renaissance H.S., Detroit, MI*

James E. Schultz
Associate Professor of Mathematics, Ohio State University, Columbus

Sharon L. Senk
Associate Professor of Mathematics, Michigan State University, East Lansing

Margaret Hackworth
Mathematics Supervisor, Pinellas County Schools, Largo, FL

John W. McConnell
Instructional Supervisor of Mathematics, Glenbrook South H.S., Glenview, IL

Steven S. Viktora
Chairman, Mathematics Department, New Trier H.S., Winnetka, IL

Dora Aksoy
UCSMP

James Flanders
*Assistant Professor of Mathematics and Statistics, Western Michigan University,
Kalamazoo*

Barry Kissane
Associate Professor of Education, Murdoch University, Perth, Western Australia

Zalman Usiskin
Professor of Education, The University of Chicago

UCSMP Production and Evaluation

Series Editors: Zalman Usiskin, Sharon L. Senk

Managing Editor: Daniel Hirschhorn

Technical Coordinator: Susan Chang

Director of Evaluation: Catherine Sarther (Mount Mary College, Milwaukee, WI)

We wish to acknowledge the generous support of the **Amoco Foundation** and the
Carnegie Corporation of New York in helping to make it possible for these
materials to be developed and tested.

It takes many people to put together a project of this kind and we cannot thank them all by name. We wish particularly to acknowledge Carol Siegel, who coordinated the printing of pilot and formative editions of these materials and their use in schools; Peter Bryant, Maura Byrne, Dan Caplinger, Janine Crawley, Lewis Garvin, Kurt Hackemer, Maryann Kannappan, Mary Lappan, Teresa Manst, Yuri Mishina, Lee Resta, Lorena Shih, Vicki Ritter, and David Wrisley of our technical staff; and editorial assistants Matt Ashley, Laura Gerbec, Jon Golub, Eric Kolaczyk, Ben Krug, Kevin Leuthold, John McNamara, Robert Schade, and Matt Solit.

We wish to acknowledge and give thanks to the following teachers who taught preliminary versions of this text, participated in the field testing or formative evaluations, and contributed ideas to help to improve this text.

Todd Biederwolf
M. L. King High School
Detroit, Mi

David Case
Woodward High School
Cincinnati, OH

Leslie Chew
Kenwood Academy
Chicago Public Schools

Sharon Llewellyn
Renaissance High School
Detroit, MI

Gerald Pillsbury
Brentwood School
Los Angeles, CA

Ray Thompson
Thornton Fractional High School South
Lansing, IL

Alan Bunner
The Culver Academies
Culver, IN

Joseph Chamberlin
The Culver Academies
Culver, IN

Raymond Heintz
Thornton Fractional High School North
Calumet City, IL

Cheryl Murphy
Newark High School
Newark, Ohio

George Pryjma
Niles Township High School North
Skokie, IL

David Williams
Southwestern High School
Detroit, MI

We also wish to express our thanks and appreciation to the many other schools and students who have used earlier versions of these materials.

UCSMP

Functions, Statistics, and Trigonometry

The University of Chicago School Mathematics Project (UCSMP) is a long-term project designed to improve school mathematics in grades K-12. UCSMP began in 1983 with a six-year grant from the Amoco Foundation, whose support continued in 1989 with a grant through 1994. Additional funding has come from the Ford Motor Company, the Carnegie Corporation of New York, the National Science Foundation, the General Electric Foundation, GTE, Citibank/Citicorp, and the Exxon Education Foundation.

The project is centered in the Departments of Education and Mathematics of the University of Chicago, and has the following components and directors:

Resources	Izaak Wirszup, Professor Emeritus of Mathematics
Primary Materials	Max Bell, Professor of Education
Elementary Teacher Development	Sheila Sconiers, Research Associate in Education
Secondary	Sharon L. Senk, Associate Professor of Mathematics, Michigan State University Zalman Usiskin, Professor of Education
Evaluation	Larry Hedges, Professor of Education

From 1983-1987, the director of UCSMP was Paul Sally, Professor of Mathematics. Since 1987, the director has been Zalman Usiskin.

The text *Functions, Statistics, and Trigonometry* was developed by the Secondary Component (grades 7-12) of the project, and constitutes the fifth year in a six-year mathematics curriculum devised by that component. As texts in this curriculum completed their multi-stage testing cycle, they were published by Scott, Foresman. A list of the six texts follows.

Transition Mathematics
Algebra
Geometry
Advanced Algebra
Functions, Statistics, and Trigonometry
Precalculus and Discrete Mathematics

A first draft of this course, then titled *Functions and Statistics with Computers*, was begun in the summer of 1986, and completed and edited during the 1986-87 school year. It was piloted in three schools during 1986-87. After a major revision, a second pilot edition, entitled *Functions, Statistics, and Trigonometry with Computers*, was used in six schools during 1987-88. Further changes were made and a field trial edition was given a formal test in 1988-89 in seven schools. In this and all other studies, some students had previous UCSMP courses, some had not. This Scott, Foresman edition is based on improvements suggested by that testing, by the authors and editors, and by some of the many teacher and student users of earlier editions.

Comments about these materials are welcomed. Address queries to Secondary Mathematics Product Manager, Scott, Foresman, 1900 East Lake Avenue, Glenview, Illinois 60025, or to UCSMP, The University of Chicago, 5835 S. Kimbark, Chicago, IL 60637.

This book differs from other books at this level in six major ways. First, it has **wider scope**. It gives strong attention to statistics as well as to the ideas of functions and trigonometry normally found at this level. Statistics is important not only for the consumer but also for the prospective college student; as many college majors require statistics as require calculus. Functions are important throughout mathematics and its applications. Trigonometry is essential for the sciences and engineering. Putting these ideas together has enabled the authors to integrate and intertwine them.

Second, this book requires **computer access**. Computers are used throughout the book to promote a student's ability to visualize functions; to explore relations between equations and their graphs; to simulate experiments; to generate data; to analyze data; and to develop the concept of limit. It is expected that a demonstration computer is available in class whenever it is needed. To use the computer most effectively, it is assumed that a class can meet in a computer room when needed and that every student can have access for homework and tests when necessary. This book is not geared to a particular type of hardware; any microcomputer with the ability to support graphing software, a statistics package, and the BASIC language may be used. A graphics calculator with statistical capabilities will suffice for most (but not all) questions.

Third, **reading and problem solving** are emphasized throughout. Students can and should be expected to read this book. The explanations were written for students and tested with them. The first set of questions in each lesson, called "Covering the Reading," guides students through the reading and checks their knowledge of critical words, rules, explanations, and examples. A second set of questions, called "Applying the Mathematics," extends students' understanding of the principles and applications of the lesson. Like skills, problem solving must be practiced; when practiced it becomes far less difficult. To further widen students' horizons, "Exploration" questions are provided in every lesson. Projects, detailed at the end of each chapter, are larger-scale assignments, and develop persistence and group problem-solving skills.

Fourth, there is a **reality orientation** towards both the selection of content and the methods taught the student for working out problems. Functions, statistics, and trigonometry are all areas of mathematics that arose from real-world considerations and they remain rich in applications. Each type of function is studied in detail for its applications to real world problems. Real data are used for many of the statistics examples. The variety of content of this book permits lessons on strategies for solving problems to be embedded in application settings. The reality orientation extends also to the methods assumed in solving the problems: students are assumed to have scientific calculators at all times.

Fifth, **four dimensions of understanding** are emphasized: skill in carrying out various algorithms; developing and using mathematical properties and relationships; applying mathematics in realistic situations; and representing or picturing mathematical concepts. We call this the SPUR approach: **S**kills, **P**roperties, **U**ses, **R**epresentations. With the SPUR approach, concepts are discussed in a rich environment that enables more students to be reached.

Sixth, the **instructional format** is designed to maximize the acquisition of skills and concepts. The book is organized around lessons usually meant to be covered in one day. The lessons have been sequenced into carefully constructed chapters which combine gradual practice with techniques to achieve mastery and retention. Concepts introduced in a lesson are reinforced through "Review" questions in the immediately succeeding lessons. This gives students several nights to learn and practice important concepts. At the end of each chapter, a modified mastery learning scheme is used to solidify acquisition of concepts from the chapter so that they may be applied later with confidence. It is critical that the end-of-chapter content be covered. To maintain skills, important ideas are reviewed in later chapters.

CONTENTS

Purposes of this book

For many people, functions are the most important content in all of high school mathematics. In your earlier work, you should have studied linear, quadratic, exponential, and logarithmic functions, and perhaps also trigonometric and circular functions. In this book, you will review and extend ideas about these functions. Many of the extensions are done with the aid of a computer or a calculator with graphics and statistical capabilities, for this technology has changed the ways in which people deal with the mathematical ideas in this course.

The ability of computers to store and analyze information has made statistics an increasingly important subject to know. Statistics are used by people who work in government or journalism, who have to make decisions in business, by people who need to analyze or interpret the results of medical or psychological studies, and by people who wish simply to understand the world. The field of statistics is relatively new; even bar and circle graphs were unknown until about 200 years ago, and much of statistics has been developed in this century.

If you further study in any area in which mathematics is encountered, it is likely that you will need the mathematics presented in this book. A thorough knowledge of functions and trigonometry is needed for calculus, an area of mathematics that is fundamental in engineering and in the physical sciences. Statistics is required for any who major in the social sciences or business. Even if you never take a course in mathematics itself in college, you are likely to encounter many of the ideas you see here in other courses and in your daily life.

Another purpose of this book is to review and bring together, in a cohesive way, what you have learned in previous courses. Functions, statistics, and trigonometry cover a wide range of topics. You may find topics close together in this book that you encountered in different chapters of books in previous years or that you studied even in different years. Mathematics is a unified discipline in the sense that what is learned in one area can be applied in all other areas, and we wish you to have that spirit of mathematics.

Materials needed

In addition to lined and unlined notebook paper, pencils, and erasers you typically use when doing mathematics, you will need graph paper and a scientific calculator at all times. You will also need the capability to automatically graph functions and calculate statistics. This can be accomplished with a computer that has graphing and statistics software, or with a calculator that has a small screen on which graphs can be shown and that has built-in functions for calculating statistics involving one or two variables. Many assignments will take you quite a bit longer to complete and some may even be inaccessible if you do not have such time-saving and accurate technology.

Organization of this book

We want you to read this book. The reading in each lesson is designed to introduce, explain, and relate the important ideas of the lesson. Read slowly and carefully; if there are calculations or graphs, verify them as you read. Consequently, it is best to have technology nearby. Use this book's glossary and index if you encounter terms with which you are not familiar. Keeping a dictionary handy is also a good idea.

The Questions following each lesson are designed to get you to think about these ideas and are of four types. Questions "Covering the Reading" follow the vocabulary and examples of the lesson. Questions "Applying the Mathematics" may require you to extend the content of the lesson or deal with the content in an unfamiliar way. Do not expect to be able to do all the questions right away. If your first attack on a problem is unsuccessful, try a different approach. If you can, go away from the problem and come back to it a little later. "Review" questions deal with ideas from previous lessons, chapters, and sometimes even previous courses. "Exploration" questions extend the content of the lesson and deal with situations that usually have many possible ways of being approached. Answers to most of the odd-numbered questions are found in the Selected Answers section in the back of the book.

People who work with mathematics for their living routinely deal with problem situations that take more than a single evening to study and resolve. At the end of each chapter, there is a page or two of project ideas for extended work. Some projects ask you to gather data and look closely at everyday situations; others involve exploring certain mathematical ideas in more depth than usual; still others require that you use a computer to solve mathematical problems that would otherwise be inaccessible. Some projects may result in work suitable for exhibits or science fairs.

Also at the end of each chapter are a variety of features designed to help you master the material of that chapter. A chapter Summary reviews the major ideas in the chapter. It is followed by a list of the vocabulary and symbols you are expected to know. Then there is a Progress Self-test for the chapter. We strongly recommend that you take this test and check your work with the solutions given in the back of the book. Finally, there is a Chapter Review, a set of review questions organized by the objectives of the chapter. You should do these also and check your answers to all odd-numbered questions with those found in the back of the book.

The authors have tried to make this book easy to understand and interesting to read. But we know that learning new content can at times be difficult. You can make your job easier if you ask friends or your teacher for help when ideas are not clear. We hope you enjoy this book and wish you much success.

OBJECTIVES ■ CHAPTER 1

The objectives listed here are the same as those listed in the Chapter 1 Review on pages 72–78 of the student text. The Progress Self-Test on pages 69–71 and the tests in the Teacher's Resource File cover these objectives. For recommendations regarding the handling of this end-of-chapter material, see the notes in the margins on the corresponding pages of this Teacher's Edition.

OBJECTIVES FOR CHAPTER 1 (Organized into the SPUR categories— Skills, Properties, Uses, and Representations)	Progress Self-Test Questions	Chapter Review Questions	Teacher's Resource File	
			Lesson Masters*	Chapter Test Forms A & B
SKILLS				
A Calculate measures of center and spread for data sets.	15, 21–24	1–8	1-3, 1-4, 1-5, 1-8	5b,c,d, 7b,c, 9b
PROPERTIES				
B Use Σ-notation to represent a sum, mean, variance, or standard deviation.	14	9–12	1-4, 1-8	5a
C Describe relations between measures of center or measures of spread.	16	13–16	1-4, 1-5, 1-6, 1-8	6
USES				
D Use samples to make inferences about populations.	4, 5	17–20	1-1	1
E Determine relationships and interpret data presented in a table.	1–3	21–26	1-1	2–4, 9a
F Use statistics to describe data sets or to compare or contrast data sets.	13	27–31	1-3, 1-4, 1-5, 1-6, 1-7, 1-8	7d,f
REPRESENTATIONS				
G Read and interpret bar graphs, circle graphs, or coordinate graphs.	6–9	32–35	1-2, 1-3	8, 10
H Read and interpret box plots.	17–20	36–41	1-5	9d
I Read and interpret histograms.	10–12	42–48	1-6	7a,e
J Draw graphs to display data.	25	49–54	1-2, 1-3, 1-5, 1-6, 1-7	9c

*The masters are numbered to match the lessons, and each is intended for use with that lesson.

OVERVIEW ■ CHAPTER 1

Before proceeding with this chapter, read the Professional Sourcebook on pages T19–52 of this book. Make sure that you have the requisite hardware and are familiar with the statistics software that you will need beginning with Lesson 1-7. (See the following notes regarding Lesson 1-7.) Also, you will probably want to read the introduction for an overview of the UCSMP program and its scope. This will help you know what your students are expected to know before this course and what they are expected to learn in it.

Make certain that students have read the "To the Student" pages and know what supplies they are expected to have for this class.

If students have not had previous UCSMP courses, it is appropriate to point out the four types of questions in each lesson, to describe the end-of-chapter material, and to indicate the location of the glossary, index, and answers to odd-numbered questions.

A successful implementation of the first chapter of a textbook sets the stage for an exciting and fruitful year for students. In addition to its mathematical objectives, this chapter has the following objectives:
a. To begin with content that is quite different from that of students' previous mathematics courses, thus requiring the students to read at the level that is necessary for success in this book;

b. To help students review important skills that will be needed in later chapters;
c. To introduce students to some of the technology that will be used in this course;
d. To give students experience with questions that have many possible answers and questions that require students to use their judgment in selecting the most appropriate answer from several choices.

Newspapers, magazines, and televised programs use graphs, tables, charts, and other displays to summarize data—often with the intention of influencing opinions. Students need to be able to read and critically analyze such displays in order to make judgments and form opinions of their own. The first two lessons of the chapter focus on techniques for accurately reading and interpreting complex displays and graphs. Students also need to be able to construct accurate displays to communicate information to others—these are topics of Lessons 1-4 and 1-5.

The statistics presented in this chapter are primarily descriptive. They are also primarily univariate; that is, they deal with one variable, not with relations between variables. Measures of central tendency (mean, median, and mode), and measures of distribution (range, interquartile range, variance, and standard deviation) are

developed in Lessons 1-3 through 1-8. The computer is introduced in Lesson 1-7. Students are expected to become familiar with the machine and use a statistics software package to draw histograms and scatter plots and to obtain the summary statistics mentioned above. In Chapter 2, statistics packages are used to find equations for lines of best fit and correlations between two variables.

In this chapter there is a systematic review of linear equations in one and two variables. This is in anticipation of Chapter 2, which begins with a study of linear modeling. The review occurs in the text, in Questions Covering the Reading, and in Applying the Mathematics as well as in the Review Questions.

It is important that you keep close to a lesson-per-day pace in this chapter. You will find that the review format and the use of computers and calculators will enable students to master essential skills and apply them in context. There are two occasions in this chapter which may require an extra day. These are the first class session, which is often spent on organizational matters, and the computer-oriented Lesson 1-7. If you think students require more time to master a concept, it is more appropriate to spend an extra day or two at the end of the chapter than during a lesson.

PERSPECTIVES ■ CHAPTER 1

The Perspectives provide the rationales for the inclusion of topics or approaches, give some mathematical background, and make connections with other lessons and with other UCSMP texts.

1-1
COLLECTING DATA
This lesson initiates a central theme of this course—the use of statistics to make inferences. It introduces the important distinction between a population and a sample and applies this idea to the capture-recapture technique for estimating wildlife populations.

1-2
TABLES AND GRAPHS
In this chapter, data are organized using tables, graphs (circle, bar, line, histogram), and plots (scatter, stem-and-leaf, box). This lesson focuses on tables, circle graphs, and bar graphs. The examples tend to be more complex than students would have encountered in earlier courses.

1-3
OTHER DISPLAYS
The displays encountered in this lesson are scatter plots, their related line graphs, and stem-and-leaf plots. The scatter plots involve time-series data, and this data is used in calculating average rates of change, in anticipation of the further review of slope in Chapter 2.

1-4
MEASURES OF CENTER
The traditional measures of center (central tendency), the mean, the median, and the mode, are reviewed here. Summation notation is introduced. Older statistics books referred to the mean, median, and mode as measures of *central tendency;* many newer books refer to these as measures of *center.* We use the latter term because it is simpler.

1-5
QUARTILES, PERCENTILES AND BOX PLOTS
Box plots (sometimes called *box-and-whiskers plots*) are displays of data based on quartiles and percentiles. They can be used to show differences between groups of data. The interquartile range, a measure of spread based on percentiles, is also discussed.

1-6
HISTOGRAMS
A histogram is a special type of bar graph used to display frequency or relative frequency distributions. Since histograms can distort the appearance of data, this lesson discusses both the use and misuse of histograms.

1-7
USING A STATISTICS PACKAGE
This is the first lesson to introduce students to a computer statistical package. From this lesson on, computers and appropriate software should be available to students.

Most statistics packages (that is, software that organizes, analyzes, and displays data) enable data to be displayed in most, if not all, of the ways mentioned in Lessons 1-1 through 1-6. This entire lesson is devoted to an orientation to whatever statistics package or packages you intend to use in this course. Students are expected to spend class time for this lesson in a computer lab.

1-8
VARIANCE AND STANDARD DEVIATION
The variance of a set of data and its square root, called the standard deviation, are measures of spread based on the mean. (The interquartile range is a measure of spread related to the median.) This lesson provides practice in calculating these statistics.

1-9
WHO WROTE *THE FEDERALIST* PAPERS?
The last lesson in each chapter is what UCSMP calls a "reading lesson." The contents of the lesson are not intended to be mastered and the ideas in the lesson are not included in any of the tests provided with this book. However, this lesson should not be skipped. First, it provides an opportunity for a non-threatening look at extensions of the content of the chapter. Second, it provides an opportunity for an extra day on Review Questions in preparation for the chapter test.

The methods used by the statisticians who studied the authorship of *The Federalist* papers are surprisingly simple. Yet they can be and have been applied to resolve the authorship of other disputed works. Furthermore, study of this lesson can be connected with virtually any course in United States history.

DAILY PACING CHART ■ CHAPTER 1

Every chapter of UCSMP *Functions, Statistics, and Trigonometry* includes lessons, a Progress Self-Test, and a Chapter Review. For optimal student performance, the Progress Self-Test and the Chapter Review should be covered. (See *General Teaching Suggestions: Mastery* on page T35 of this Teacher's Edition.)

By following the pace of the Full Course given here, students can complete 12 or 13 chapters of the book by the end of the year. Students following the pace of the Minimal Course spend more time to complete lessons that are combined with quizzes and to consider the Chapter Review. When all lessons are covered in order (the recommendation of the authors), students in the Minimal Course can cover all but the last two chapters. For more information on pacing, see *General Teaching Suggestions: Pace* on page T34 of this Teacher's Edition.

DAY	MINIMAL COURSE	FULL COURSE
1	1-1	1-1
2	1-2	1-2
3	1-3	1-3
4	Quiz (TRF); Start 1-4	Quiz (TRF); 1-4
5	Finish 1-4	1-5
6	1-5	1-6
7	1-6	Quiz (TRF); 1-7
8	Quiz (TRF); Start 1-7	1-8
9	Finish 1-7	1-9
10	1-8	Progress Self-Test
11	1-9	Chapter Review
12	Progress Self-Test	Chapter Test (TRF)
13	Chapter Review	
14	Chapter Review	
15	Chapter Test (TRF)	

TESTING OPTIONS

■ Quiz for Lessons 1-1 through 1-3   ■ Chapter 1 Test, Form A
■ Quiz for Lessons 1-4 through 1-6   ■ Chapter 1 Test, Form B

A Quiz and Test Writer is available for generating additional questions, additional quizzes, or additional forms of the Chapter Test.

PROVIDING FOR INDIVIDUAL DIFFERENCES

The student text has been written for, and tested with, average students. It has also been used successfully with students who are well prepared as well as with students who are poorly prepared.

The Lesson Notes often include Error Analysis and Alternate Approach features to help you assist those students who need more help. Students of all abilities often learn from their peers and may benefit from small group work referenced as appropriate throughout the Notes. A blackline Lesson Master (in the Teacher's Resource File), keyed to the chapter objectives, is available for each lesson to provide more practice. (However, we advise you to use these masters sparingly; again refer to the suggestions on pace on page T34.)

Extension activities are provided in the Lesson Notes for use with students who need additional challenges.

Making Sense of Data

UNITED KINGDOM
30.0%

ASIA
25.5%

AUSTRALIA
24.2%

EUROPE
24.1%

17.9%

UNITED STATES
37.2%

20.9%

17.2%

MEN

CARIBBEAN
29.2%

AFRICA
21.6%

WOMEN
29.5%

24.6%

15.8%

14.7%

17.5%

17.7%

LATIN AMERICA

Bars show percent of people who could **not** smell androstenone. Red bars represent women, blue bars represent men.

In 1986, 1.5 million people in the United States and abroad participated in a landmark study conducted by the National Geographic Society and the Monell Chemical Senses Center in Philadelphia. It was the first comprehensive study of smell conducted in the world, and it involved the largest scientific sampling of its kind. A study of this magnitude could only be undertaken in the computer age; analyzing returns by hand would have been impossible. Even doing a preliminary analysis of about 26,000 randomly selected returns from the United States and about 100,000 returns from abroad took a great deal of effort.

In addition to the result depicted above, researchers also found that women not only think they can smell more accurately than men, they generally can; in general, there is little change in the ability to smell until about age 60 or 70; and nearly two in three people have suffered a smell loss at some time.

Statistics is the branch of mathematics dealing with the collection, organization, analysis, and interpretation of information, usually numerical information, called **data**. (Data is the plural of *datum*, the Latin word for fact.) In this chapter you will encounter some of the fundamental concepts of statistics. You will also learn how to use the statistical features found on calculators and computers to help organize, summarize, and analyze data.

LESSON 1-1

RESOURCES
■ Lesson Master 1-1

OBJECTIVES

Letter codes refer to the SPUR Objectives on page 3A.
D Use samples to make inferences about populations.
E Determine relationships and interpret data.

TEACHING NOTES

We strongly recommend not discussing this lesson until students have had an opportunity to read it and try the questions on their own. Reading mathematics may be a new expectation for some students. To read well, a student must read carefully, watching for important terms and symbols. Students should read with a pencil and paper. Reading is an active, not passive, process.

In **Example 2,** an inference about a population is drawn from a sample by using a proportion. Students may be familiar with the "proportion" reasoning required to answer the question, *If 40 of 200 dentists surveyed recommended Glimmer Toothpaste, how many dentists out of 3000 would recommend Glimmer?* Most students can answer *600* without too much effort. The idea that the ratio in the sample carries over to the ratio in the population is easily accepted. However, this intuition does not necessarily make a capture-recapture problem easy. You may find it helps some students to set up the problem as a proportion as shown on page 5.

Collecting Data

When planning to collect data you must decide what type of data to collect and from whom to collect it. The first issue involves *variables*, the second involves *populations* and *samples*.

In statistics, a **variable** is a characteristic of a person or thing which can be classified, counted, ordered, or measured. For instance, some variables that describe students in your class are gender, religion, number of siblings, rank in class, height, and family income. Some variables describing a country are population, area, number of tons of steel produced, and infant mortality rate.

The set of all individuals or objects you want to study is called the **population** for that study. If you can not or do not collect data from the entire population, but study only a part of it, that part actually studied is called a **sample**. A sample is a subset of the population.

Sometimes, for reasons such as fairness or legal requirements, the entire population must be studied. For instance, to be fair the president of a club might want to get opinions from every member. Gathering facts or opinions through an interview or questionnaire is called a **survey**. The U.S. Constitution requires that every ten years an enumeration, or census, or survey of the entire population of the United States be taken.

Other times, for reasons such as cost, safety, or preservation of a product, it is preferable to take a sample. For instance, it might be too expensive to ask all owners of a particular make of truck whether they are pleased with the product. So the manufacturer will survey a sample of the owners. To evaluate the taste of a new delivery of apples, a grocer who tasted every apple (the population) would destroy the product! So the grocer will taste one or two of the apples (a sample).

You should be able to distinguish between populations and samples and identify the variables being studied.

■ ■ ■ ■ ■ ■ ■ ■

Example 1   A medical laboratory technician counts the number of white blood cells in a patient's blood. Identify the variable, population, and sample.

Solution   The variable is the number of white blood cells. The population is all the patient's blood; and the sample is the part of the patient's blood withdrawn.

Note that in Example 1, if the population (the entire blood supply) had been studied (withdrawn), the patient would have died.

When samples are taken **randomly**, that is, in a way so that every member of the population has an equal chance of being chosen, data from the sample can be used to estimate information about the population.

4

For instance, consider the question of estimating the size of a population of a certain type of animal. This is a question of importance to ecologists, conservationists, biologists, food producers, and often to citizens interested in animals. But how can you count birds or whales or buffalo or rabbits? Each species of animal presents its own problems for counting. Because direct counting is difficult or impossible, indirect methods are used to estimate population size.

Biologists sometimes use a method called the **capture-recapture method**. In this method, certain animals are captured, tagged, and released. Later, a sample is captured. The estimate is based on the assumption that when the sample is chosen randomly, the ratio of tagged animals to the number of animals caught is nearly the same as the percentage of tagged animals in the entire population. That is,

$$\frac{\text{number of tagged animals in sample}}{\text{number of animals in sample}} = \frac{\text{number of tagged animals in population}}{P}$$

where P is the unknown number of animals in the population.

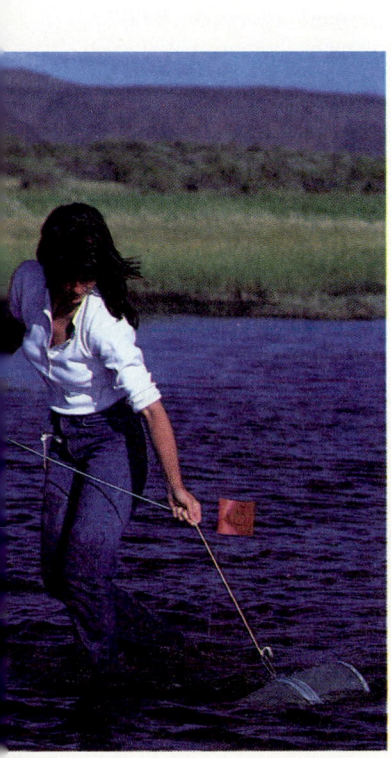

Example 2   From a variety of locations in a lake a biologist catches 80 fish, then tags and returns them to the lake. A week later 60 fish are caught from the same locations. Exactly 12 of the 60 have tags. Assuming that both catches are taken randomly, estimate the number of fish in the entire lake.

Solution   The biologist found that $\frac{12}{60}$ of the sample had tags. Thus, it is reasonable to estimate that $\frac{12}{60}$ of *all* the fish in the lake were tagged. Let P be the population of fish in the lake. Then $\frac{12}{60} = \frac{80}{P}$. So $P = 400$. There are about 400 fish in the lake.

Check   If there are 400 fish in the lake, then $\frac{80}{400}$ or $\frac{1}{5}$ were tagged. This equals $\frac{12}{60}$, the ratio of tagged fish found in the sample.

How reliable is the estimate? The capture-recapture technique gives a good estimate of a population provided the capture locations are representative of all possible habitats of the animal, no unusual events (such as chemicals deposited in the lake) occurred during the week, and tagging the animals did not affect their chance of being recaptured.

A good report of a well-conducted study describes *what was studied* (the variable or questions), *who was studied* (the sample or population), and *what was found* (a summary of the data or answers to the questions). The description of the sample should indicate how many people were involved, how they were chosen, and some background information on them. Unfortunately, at times survey data are reported without indicating much about the sample. If a sample is not chosen randomly from the population, the data from the sample may not apply to the population.

LESSON 1-1   Collecting Data   5

After reading this lesson, students should be able to discuss potential benefits and limitations of using a sample to answer questions about a population.

Alternate Approach   Students may not have a solid understanding of the concept of random. *Random* does not mean haphazard but is defined in terms of probability. Randomness is studied in some depth in Chapters 7 and 10, so there is no need to dwell on this concept here. A simple classroom activity you could try would be to ask each student to write four digits (0, 1, 2, . . . , 9) down at "random" on four little scraps of paper. Tell students they may use a digit more than once. Have the students sort these scraps of paper and count the number of zeros, ones, and so on. Ask them what we should expect if the numbers were truly chosen at random. If you try this activity without discussing the meaning of random first, the class results will probably contain fewer zeros than 10% and more sevens than 10%. (Groups of people tend not to choose numbers at random.) Did students choose more odds than evens? Ask students to think of methods of choosing digits at random. Some may be familiar with the random number generator on their calculator.

Alternatively, you might introduce capture-recapture experiments by demonstrating a simulation as suggested in Project 4 on pages 66 and 67.

ADDITIONAL EXAMPLES
1. A physician takes a biopsy of a suspicious growth to check for malignancy. Identify the variable, population, and sample.
The variable is the state (malignant, nonmalignant) of the growth. The population is the entire growth. The sample is the small portion taken for the biopsy.

Consider the following article which appeared in *USA Today* in June, 1986.

Family meals losing out

By Nanci Hellmich
USA TODAY

Fewer USA families are sitting down together for meals, says a new survey of 2,077 women.

Among families:
- 43 percent eat breakfast separately.
- 29 percent eat dinner separately.
- 20 eat together only on weekends.

The reason: With more women working, it's difficult for families to find time to be together. says Mark Clements, who did the survey for Conde Nast Publications.

Among the women:
- 34 percent say they eat only when hungry.
- 72 percent skip meals occasionally.
- 44 percent eat more now than three years ago.
- 75 percent do most of the cooking.

As you read the article above you might wonder, who are these 2,077 women who were asked about their eating habits? Do you think these families are representative of all U.S. families?

If the 2,077 women were women at suburban supermarkets between 9:00 A.M. and 11:00 A.M., for instance, then the sample was not representative of all women. Women who work during those hours could not have been included in the sample and so the sample was not random. In that case we would call the sample **biased**. Because this article reports nothing about how the sample was chosen, the reader cannot be sure how well these results reflect the population of all families in the United States.

Questions

Covering the Reading

These questions check your understanding of the reading. If you cannot answer a question, you should go back to the reading to help you find an answer.

In 1–3, define each term.
1. variable **characteristic of a person or thing which can be classified, counted, ordered, or measured**
2. population **set of individuals or objects studied**
3. statistics **the branch of mathematics dealing with the collection, organization, analysis, and interpretation of data**
4. Give three examples of variables which were not mentioned in the lesson that might be collected for a person. **sample: weight, favorite color, age**

6

In 5 and 6, identify: **a.** the population; **b.** the sample; and **c.** the variable of interest.   **See margin.**

5. A pastry inspector counts the number of raisins per cookie in 10 oatmeal-raisin cookies in a batch fresh out of the oven.

6. In order to learn the TV habits of all Carlton students, those students entering the north entrance of the school between 7:30 A.M. and 7:45 A.M. are asked which TV programs they watched last night.

7. Refer to the National Geographic Smell Study described in the chapter opener.
   a. Describe the three samples mentioned.   **See margin.**
   b. What population was studied?   **the entire world population**

8. Give three characteristics of a situation which might force a person to study a sample rather than a population.   **See margin.**

9. What is a random sample?   **a sample in which every member of the population under study has an equal chance of being selected**

10. In a capture-recapture study, suppose 60 deer are tagged. On the recapture, 52 deer are caught of which 9 are found to have been tagged. Estimate the number of deer in the forest.   **about 350 deer**

11. An ecologist is trying to determine how many turtles live in a bog. Suppose 30 turtles from various locations are captured and tagged. Days later 35 turtles are captured, 17 of which have tags. Estimate the number of turtles in the bog.   **about 62 turtles**

In 12–14, refer to the article about family meals at the end of the lesson.

12. *Multiple choice.* About what fraction of the families eat together only on weekends?

   (a) $\frac{1}{4}$ (b) $\frac{1}{5}$ (c) $\frac{1}{3}$ (d) $\frac{2}{5}$ (e) $\frac{3}{10}$ **(b)**

13. a. What percent of women reported that they do most of the cooking?
   b. How many women responded that they do most of the cooking?
   a. 75%   b. about 1558 women

14. Explain why a sample taken from female nursing home residents would probably not be representative of the population of all U.S. females.
   sample: because such a sample would not include women who work or cook.

Applying the Mathematics

These questions extend the content of the lesson. You should take your time, study the examples and explanations in the lesson, and try a variety of methods. Check your answers to odd-numbered questions with the ones in the back of the book.

15. The final count in the 1980 U.S. census was 226,545,805 people. The general population is estimated to have been under-counted by one percent. That is, only 99 percent of the people were counted.
   a. Estimate the actual 1980 population.   **228,834,146 people**
   b. What are some factors which might have caused the population to be under-counted?   **sample: Some people did not respond to the census. Homeless people were not counted.**

LESSON 1-1 Collecting Data   **7**

ADDITIONAL ANSWERS
7.a. 1.5 million people from the U.S. and abroad; about 26,000 people from the U.S.; about 100,000 people from abroad

8. sample: (1) too expensive to test the entire population; (2) if the population is studied, it will be destroyed (such as biting into apples); (3) the population does not stay in one place (such as a population of animals)

7

16. The following passage appeared in a story in the May 21, 1990, edition of the *Chicago Tribune:*

> Non-smokers who live with smokers have a 20 percent to 30 percent higher risk of dying from heart disease than do other non-smokers, a researcher said Sunday. "Passive smoking causes heart disease," said Stanton Glantz of the University of California, San Francisco. Glantz is a researcher and statistician who conducts research in cardiology. "There are now 11 studies of the effects of passive smoking on heart disease deaths" in non-smokers, he said. "All but one [study] the tobacco industry funded show an increased risk."

How might the study funded by the tobacco industry be different from the other ten studies? **sample: The sample selected by the tobacco industry may not have been representative of the population.**

17. A student is investigating the greatest number of pieces into which *n* straight cuts can divide a circular region. The sample below is used.

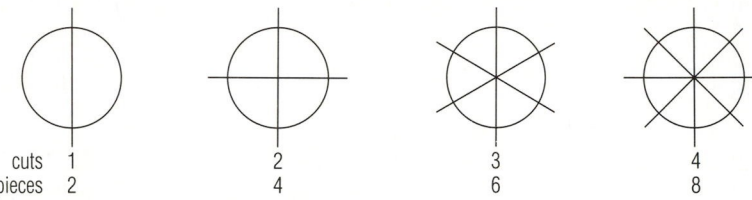

From these data the student concludes that *n* cuts divide a circle into at most 2*n* pieces.

   a. What is wrong with the student's reasoning?

   b. What is the maximum number of pieces into which a circular region can be cut with 4 straight cuts? **11 pieces**

   a) The student did not use a representative sample.

Review. *Each lesson in this book contains review questions to practice using ideas you have studied earlier. In this lesson all review questions cover ideas from previous courses.*

18. *Skill sequence.* Solve.

   a. $7x = 140$ **20**   **b.** $.7x = 140$ **200**

   c. $140x = 7$ $\frac{1}{20}$ **d.** $140x = .7$ $\frac{1}{200}$

19. In 1986 it was estimated that about 19,000,000 people in Nigeria were of the Animist religion; they accounted for 18% of the total population. Estimate the total population of Nigeria in 1986. **about 106,000,000 people**

20. **a.** The population of Austin, Texas, increased by 48.89% between 1970 and 1980. If the 1970 population was 360,463, what was the 1980 population? **about 536,700 people**

   b. The population of Daytona Beach, Florida, increased by 52.67% between 1970 and 1980. If the 1980 population was 258,762, what was the 1970 population? **about 169,500 people**

8

21. *Multiple choice.*
Which segment has
the greatest slope?
(a)

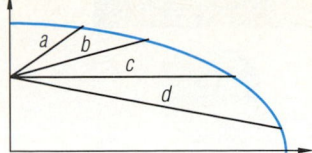

22. Find the slope of the line through (3, 7) and (2, -1). **8**

In 23 and 24, consider a circle
with central angles as shown at
the right.

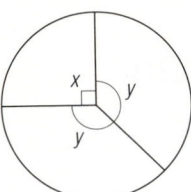

23. Find the measure in
degrees.
a. *x* **90°**   **b.** *y* **135°**

24. What percent of the area of
the circle is the area of each
sector?
x: 25%; y: 37.5%

Exploration

*Exploration questions often require that you use reference books or other
sources. Frequently they have many possible answers.*

25. a. Ask a sample of students in your school about their families' eating
habits.
 i. How many eat breakfast separately?
 ii. How many eat dinner separately?
 iii. How many eat together only on weekends?
b. How representative do you think your sample is of the following populations?
 i. families of all students in the school
 ii. all families in the United States
 iii. all families in the Soviet Union
c. How many students do you think you would need in order to have a
sample that is representative of all families of students in the United
States?

26. Use whatever references you can find to estimate the current (this year)
population of the county or parish in which your school is located.

25,26) Answers will vary.

NAME _____

LESSON **MASTER 1-1**
QUESTIONS ON **SPUR** OBJECTIVES

■**USES** *Objective D (See pages 72–78 for objectives.)*
In 1–7, use this table from *The Statistical Abstract of the United States 1990.*

Travel by U.S. Residents by Selected Trip Characteristics: 1983 to 1988 (in Millions)

Characteristic	Trips						Person-Trips[1]					
	1983	1984	1985	1986	1987	1988	1983	1984	1985	1986	1987	1988
Total	540.9	528.2	558.4	592.3	636.3	656.1	1057.8	1012.0	1077.6	1121.5	1191.1	1232.5
Purpose												
Visit friends and relatives	181.9	180.5	206.8	214.9	210.5	213.8	384.6	384.6	430.8	442.5	437.1	425.0
Other pleasure	189.6	170.5	177.6	200.3	234.4	241.5	401.4	349.7	376.0	418.9	464.6	502.1
Business or convention	103.6	114.5	133.3	140.0	157.5	155.6	146.4	156.1	185.2	189.0	206.2	211.7
Other	65.8	62.7	40.7	37.1	33.9	45.1	125.4	121.6	85.6	71.1	83.2	92.9
Mode of transport												
Auto, truck, recreation vehicle	396.1	380.9	376.1	405.6	433.7	472.6	839.8	794.0	797.7	832.1	889.4	952.2
Airplane	110.2	118.8	140.5	143.4	160.7	154.6	174.2	174.5	217.3	231.0	239.9	238.0
Other	26.6	28.5	41.8	43.3	41.9	28.9	43.8	43.5	62.6	58.4	61.8	41.7
Vacation trip	307.8	333.3	339.8	354.3	366.2	396.2	642.3	689.8	728.7	752.4	775.2	830.9
Weekend trip	225.5	211.4	224.0	252.0	274.4	271.7	485.0	434.9	470.1	513.5	559.4	560.0

[1] A count of times each person (child or adult) goes on a trip.
Source: US Travel Data Center, Washington, DC, *National Travel Survey*, annual. (Copyright.)

1. Explain the information conveyed by the number 214.9 in the fourth column
("1986") under "Trips."
 214.9 million trips were made in 1986 by U.S. residents
 with the purpose of visiting friends and relatives.

2. Which numbers in the fourth column total 592.3?
 Two sets: Numbers under "Purpose," and numbers
 under "Mode of Transport"

3. How many trips were taken by airplane in 1987? **160.7 million**

4. How many people took trips by airplane in 1987? **239.9 million**

5. *Multiple choice*   If *p* is the number of people who took
trips and *t* is the number of trips, which equation could
express the relationship between *p* and *t*?   **(b)**
(a) $p = 200t + 1.6$ (b) $p = 1.6t + 200$
(c) $t = 200p + 1.6$ (d) $t = 1.6p + 200$

6. Is the ratio of person-trips to trips higher for the auto, truck, recreation
vehicle category or the airplane category? Give a reason why this happens.
 auto, truck, recreational vehicle; Families or groups
 of friends usually drive rather than fly.

7. For what other purpose might people travel besides those listed in the table?
 sample: medical reasons, collecting data for studies

Functions, Statistics, and Trigonometry © Scott, Foresman and Company   **1**

NAME _____
Lesson MASTER 1-1 (page 2)

■**USES**   *Objective E*
In 8 and 9, identify a. the population; b. the sample; and c. the variable.

8. Before cleaning a sofa with a new cleaning solution, a man cleans a 5 cm by
5 cm section which is not visible.
 a. **cloth surface of sofa**
 b. **the 5 cm × 5 cm section**
 c. **degree of cleaning**

9. In July 1990 the city of Chicago announced that it would set up roadside
checkpoints to stop motorists at random in order to check for drunken drivers.
 a. **motorists in Chicago**
 b. **drivers stopped at random**
 c. **level of alcohol in the motorist's blood**

In 10 and 11, determine why a sample is used rather than making a survey of
the total population.

10. The Nielsen television rating service determines the U.S. television ratings
with a sample of about 1200 homes.
 too many homes in the population

11. A fireworks company tests some of the sparklers that it manufactures.
 The company cannot use all its product in testing.

12. A group of people were trying to estimate the
number of Great Northern beans in a one-pound
bag. They withdrew 50 beans and replaced them
with pinto beans. They mixed the beans well.
They withdrew 120 beans of which 4 were pinto
beans. About how many Great Northern beans
were in the original one-pound bag?   **1500**

13. In order to estimate the number of blue gills in a
small lake, a biologist captured and carefully
tagged 85 of the fish. She then released them.
One week later she caught 122 blue gills, of
which 18 were tagged. About how many blue
gills were in the lake?   **576 (or 580)**

14. In 1988 it was estimated that about 6.5 million
Ghanaians could read English. If the literacy rate
(in English) was about 45%, estimate the
population of Ghana in 1988.   **about 14.4
million**

2   *Functions, Statistics, and Trigonometry* © Scott, Foresman and Company

LESSON 1-2

RESOURCES
- Lesson Master 1-2
- Teaching Aid 1: Table and graph for **Example 1**
- Teaching Aid 2: Tables for **Questions 9 and 11–13**
- Teaching Aid 3: Table and graph for **Additional Examples 1 and 2**
- Protractor

OBJECTIVES

Letter codes refer to the SPUR Objectives on page 3A.

G Read and interpret bar graphs or circle graphs.

J Draw graphs to display data.

TEACHING NOTES

This lesson emphasizes the details of reading and analyzing tables at a level required for success in business or in college. Tell your students to expect to have to use some effort in answering the questions. These questions require careful attention to details displayed in each graph.

Again, we recommend against "teaching" this lesson before the students have read it for themselves and answered the questions. Students should practice reading mathematics in order to use mathematics outside of school.

Teaching Aids 1 and 2 can be used to facilitate the discussions of **Example 1** and **Questions 9 and 11–13** or as work sheets or note sheets when answering the questions.

LESSON

1-2

Tables and Graphs

To make sense of data, it helps to organize the data in tables or graphs. In this lesson you will study the use of tables, circle graphs, and bar graphs.

Consider the question "How much does it cost to raise a child?" The U.S. Department of Agriculture, in a study of Midwestern families with no more than 5 children living in homes with both a mother and father, collected the following data:

Area	Food	Clothing	Housing	Medical Care	Education	Transportation	All Other	Total
Estimated average cost of raising a child from birth to age 18 at a moderate cost level in Midwestern urban and rural areas in 1986								
Expenditure in dollars								
Urban	21,032	6,260	30,896	5,796	1,824	13,972	12,448	92,228
Rural	19,487	5,836	28,998	5,292	1,824	13,428	11,050	85,915
Expenditure as a percent								
Urban	22.8	6.8	33.5	6.3	2.0	15.1	13.5	100.0
Rural	22.7	6.8	33.8	6.2	2.1	15.6	12.9	100.0

When reading a table you should ask yourself:

1. What is being presented?

To answer this question, identify the variables. These are often named as labels of the rows and columns. Then determine the units of the numbers. If percents are given identify the base. For this table there are nine variables. One variable is area, and the other eight are the total costs of raising a child from birth to age 18 for food, clothing, housing, medical care, education, transportation, other types of expenses, and the sum of all the expenses. The costs in each category are expressed both in dollars and as a percent of the overall total.

2. Are the data trustworthy?

To answer this question, some things to consider are the data source, the accuracy of the data, and the time when the data were collected. Ideally, the data source should be given, allowing you to verify the data if you want. For the preceding table, the source is a government agency we consider reputable. Data should be reported as accurately as possible. In this case, data are reported to the nearest dollar or tenth of a percent. This seems too accurate. Rounding to the nearest hundred dollars might have been more reasonable. Knowing when the data were collected helps you decide what conclusions to make. Because the data are from 1986, due to inflation the present costs are probably somewhat higher than those reported.

3. What conclusions can you draw from the data?

Look for patterns and try to form reasonable generalizations. Following are some reasonable conclusions from these data.

10

According to estimates of the U.S. Department of Agriculture, in 1986 the average cost of raising a child in a Midwestern, urban area from birth to age 18 was about $92,000. The cost of raising a child in a Midwestern rural area was about $6,000 lower. Housing and food together made up the largest part (about 56%) of the total cost in each area.

Data consisting of a sum and its component parts are often displayed in a *circle graph* (sometimes called a *pie chart*). To make a circle graph you can calculate the measure of a central angle corresponding to each component part of the sum. Recall that one revolution is 360°.

■ ■ ■ ■ ■ ■ ■ ■ ■ ■ ■

Example 1 Draw a circle graph showing the data for urban areas from the preceding table.

Solution First, calculate the measure of the central angle corresponding to the percent of each type of expense. For instance, 22.8% of the total cost is for food. Thus, for the sector representing food, the measure x of the central angle should be 22.8% of 360°:

$$x = (.228)(360°) \approx 82°.$$

The results for all types of expenses (rounded to the nearest degree) are given in the table below at the left. Second, draw a circle and sectors with central angles having these measures. Label the sectors with the type of expense and the percent of expense it represents. This is done at the right below.

Type of expense	Percent	Angle Measure
food	22.8	82
clothing	6.8	24
housing	33.5	121
medical care	6.3	23
education	2.0	7
transportation	15.1	54
all other	13.5	49

Estimated average cost of raising a child from birth to age 18 in a midwestern urban area by type of expense in 1986

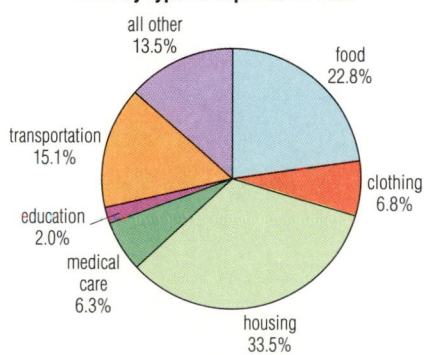

Notice that in the above circle, the areas of the sectors are proportional to the percentages. That is, the housing sector's area is 33.5% of the area of the circle, and so on, as it should be. This is because you are viewing the circle from directly above its center. But notice what happens if the circle is viewed from a position not directly above its center. From this perspective, the circle appears as an ellipse, and the ratios of areas are changed. The nearer sectors occupy more area. Housing takes up over 40% of the area of this ellipse. The graph is said to be *distorted*; some would call it misleading.

housing
33.5%

To picture data from two or more samples, sometimes separate graphs are made. For instance, to compare and contrast the expenses of families in urban and rural areas, you could make two circle graphs side by side.

LESSON 1-2 Tables and Graphs   11

ADDITIONAL EXAMPLES
Note: The data used in Additional Example 1 is taken from a table that is fully reproduced on Teaching Aid 3. If possible when discussing this example, use the teaching aid so that students get the full benefit of interpreting the data.

You may want to use one or both of the Additional Examples (with Teaching Aid 3) as **small group work.** Several Additional Examples in subsequent lessons may also be suitable as small group work. Additional Example 1 can be extended by asking the same questions for the other years.

1.

Employment Status (in 1000s)	Graduates 1988	Dropouts 1988
Civilian population	**8,999**	**4,231**
In labor force	7,638	2,763
Percent of population	84.9	65.3
Employed	6,932	2,235
Percent of labor force	90.8	80.9
Unemployed	706	528
Percent of labor force	9.2	19.1
Not in labor force	1,361	1,468
Percent of Population	15.1	34.7

a. In 1988, how many high school graduates were employed or seeking employment?
7,638,000

b. What does 19.1 indicate?
Sample: Of the dropouts that were in the labor force, 19.1% of them were unemployed.

c. In 1988, how many high school dropouts were employed? Unemployed but able to work? Unemployed and unavailable to work?
2,235,000; 528,000; 1,468,000

11

d. Calculate the percentage of 1988 dropouts for each category in part **c.** Convert these percentages to degrees and draw a circle graph showing the three regions.
percentages: 53, 12, 35; degrees: 191, 43, 126

Employment Status of Persons Who Dropped Out of High School in 1988

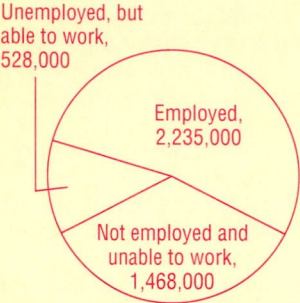

Unemployed, but able to work, 528,000

Employed, 2,235,000

Not employed and unable to work, 1,468,000

e. Compare the job outlook for high school graduates to that of dropouts.
Sample: The job outlook for high school graduates is about 10 percent better than that of the dropouts and the percentage of the graduates unavailable to work is about half of the percentage of dropouts unavailable to work.

2.

Percent of 11th Grade Students, by Amount of Day Spent on Homework and by Race/Ethnicity: 1983–84

☐ White ■ Black ▨ Hispanic
Note: Percents may not total 100 percent due to rounding.
*None assigned or didn't do

Another way to compare and contrast data from various samples is to use a *bar graph*. One axis of a bar graph labels categories or variables; the other is a numerical scale typically with counts or percents. A well-made bar graph identifies the categories being described, labels the numerical scale in equal intervals, uses bars of equal widths for each category, and provides a legend, or explanation of the data, if data from more than one sample or population is being pictured. In order to portray relations between data accurately, numerical scales usually should begin at 0.

Note how both numerical and visual approaches can be used to answer questions about data given in graphs.

Example 2 Consider the graph below.

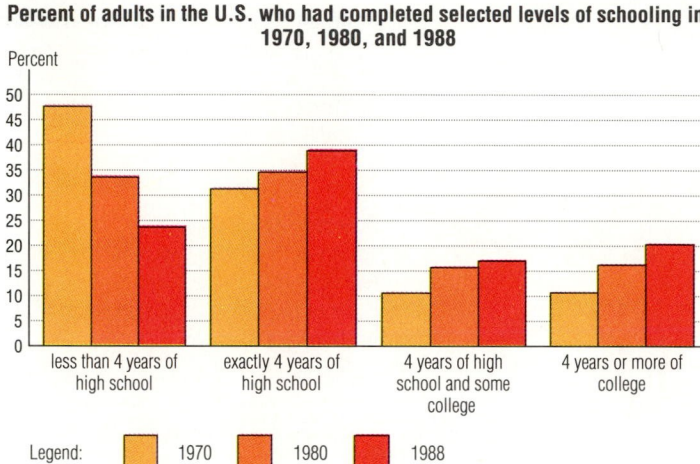

Percent of adults in the U.S. who had completed selected levels of schooling in 1970, 1980, and 1988

Legend: ☐ 1970 ■ 1980 ■ 1988

a. About what percent of adults in 1970 had completed four years of high school or more?
b. Had a greater percent of adults completed high school in 1980 or 1988?

Solution

a. Estimate the percent of adults in 1970 who completed exactly 4 years of high school, 4 years of high school and some college, and 4 years or more of college. Find the sum. 31 + 11 + 11 = 53. Thus about 53% of adults in 1970 had completed 4 years of high school or more.
b. In each of the three categories showing at least 4 years of high school education the bar for 1988 is higher than the bar for 1980. Thus the percent of adults completing high school in 1988 was higher than the percent of adults in 1980.

Check

a. About 47% of adults completed less than 4 years of high school. So, about 100% − 47% = 53% completed 4 or more years. Taking into account round-off effects, this agrees with the earlier percentage.
b. About 24% of adults reported having not completed high school in 1988, about 34% in 1980. So a greater percent had completed high school in 1988.

12

Today you see bar graphs everywhere. You may not realize that bar graphs were invented! The first bar graph appeared in the 1786 book *The Commercial and Political Atlas*, by William Playfair (1759–1823). Playfair was a pioneer in the use of graphical displays and wrote extensively about them.

Questions

Covering the Reading

In 1–6, refer to the table on Estimated Average Cost of Raising a Child.

1. What is the difference between the cost of clothing for a child in an urban area and a child in a rural area?   **$424**

2. *True* or *false*.   When raising a child, parents in rural areas spend more on transportation than parents in urban areas.   **False**

3. *True* or *false*. The percent of the cost of raising a child spent for clothing is equal in urban and rural areas.   **True**

4. What percent of an urban family's cost of raising a child is spent on food, clothing, or housing?   **63.1%**

5. For each group—urban and rural—expenses for housing are about how many times expenses for clothing?   **about 5 times**

6. Draw a circle graph showing the cost of raising a child in a rural Midwestern area.   **See margin.**

7. Refer to the bar graph showing school completion by adults.
   a. About what percent of adults in 1988 had completed four years of high school or more?   **about 75-80%**
   b. Would an adult in 1970, 1980, or 1988 be more likely to have attended college?   **1988**
   c. Justify your answer to part **b**.   **See margin.**

8. About how many years ago were bar graphs invented, and by whom?   **about 200 years ago by William Playfair**

Applying the Mathematics

9. The table at the right gives the numbers of accidental deaths (in thousands) by principal types of accident. The data came from the National Safety Council.

Cause	1985	1986	1987	1988
Motor Vehicles	45.9	48.3	48.7	49.0
Falls	12.0	11.3	11.3	12.0
Drowning	5.3	5.6	5.3	5.0
Fires, Burns	4.9	4.6	4.8	5.0
Poison (solid, liquid)	4.1	4.4	4.4	5.3
Ingestion of food, objects	3.6	3.5	3.2	3.6
Firearms	1.6	1.6	1.4	1.4
Poison (gas)	1.1	1.0	1.0	1.0

   a. How many people died in 1988 from motor vehicle accidents?
   b. Over the four years shown, how many people died as a result of firearm accidents?   **(a) about 49,000; (b) about 6000**
   c. *True* or *false*.   In 1986, 11.2% of accidental deaths were the results of falls.   **False**
   d. Assuming that the number of deaths from accidents not listed is negligible, what percent of accidental deaths in 1987 resulted from poisonings?   **about 6.7%**

(Source: National Assessment of Educational Progress (NAEP), "Homework: What Do National Assessment Results Tell Us?" 1986 from *Youth Indicators 1988— Trends in the Well-Being of American Youth*, p. 70.)

a. About what percent of white 11th grade students do at least one hour of homework?
40%

b. For which group of 11th grade students (white, black, or Hispanic) was the percent of students who did at least some homework the largest?
black; white 66%, black 71%, Hispanic 60%

NOTES ON QUESTIONS
Questions 13–16 and 24: These questions require students to write about mathematics. Writing should go hand in hand with learning and understanding mathematics. Require your students to write in complete sentences. Answers should vary. For each of these questions, ask volunteers to read what they wrote and then ask for comments. Discuss whether the conclusions are found in the data or involve interpretations or opinion. **Questions 13, 14a, and 14b** ask students to draw conclusions from only the data presented; **Questions 14c and 15b** ask them to make statements that go beyond the data. **Questions 14 and 15** could be assigned as **small group work.**

ADDITIONAL ANSWERS
6. See Additional Answers in the back of this book.

7.c. Both bars for those having attended college ("4 years of high school and some college" and "4 years or more of college") are higher for 1988 than for 1980 or 1970.

10. Use the bar graph below representing advertising expenditures in billions of dollars for 1985. Source: *The 1987 Information Please Almanac*.

a. What does the 25.2 represent?   **See margin.**
b. How much money was spent on advertising in 1985?
c. *True* or *false*.   Newspapers and television together get more than half of all advertising expenditures.   **False**
d. What percent of advertising expenditures goes into direct mail?
b) about $94.7 billion; (d) about 16.4%

In 11–13, use the table below showing the number and percent of families classified by income level and education completed by the householder .

Percent distribution of families by income level and education of householder										
Highest level of education completed by householder	Number of families (1,000)	Percent distribution of families by income level								Median income (dollars)
		Under $5,000	$5,000-$9,999	$10,000-$14,999	$15,000-$24,999	$25,000-$34,999	$35,000-$49,999	$50,000-$74,999	$75,000 and over	
Elementary School:										
Less than 8 years	5,245	8.9	22.7	20.6	22.9	12.9	7.6	3.5	0.9	14,859
8 years	3,462	7.0	14.9	18.2	28.6	15.4	10.1	4.5	1.3	17,920
High School:										
1-3 years	8,005	7.5	13.0	13.9	24.9	17.4	13.9	7.2	2.2	20,372
4 years	23,630	3.9	6.1	8.8	20.8	21.0	22.6	13.1	3.7	29,450
College:										
1-3 years	10,720	2.3	3.8	6.0	16.5	19.0	25.9	19.2	7.3	35,907
4 years or more	13,478	0.8	1.2	2.4	9.0	12.5	24.0	27.5	22.6	48,126

Source: U.S. Bureau of the Census, *Current Population Reports,* series P-60, No. 162, and unpublished data.

11. Suppose "poverty" is defined to be "living in a family with an annual income under $10,000."
a. What percent of the families whose householder completed less than 8 years of education live in poverty?   **≈ 31.6%**
b. What percent of the families whose householder completed 1–3 years of college live in poverty?   **≈ 6.1%**
c. How many families whose householder completed exactly 8 years of education live in poverty?   **about 760,000**
d. How many families whose householder completed 4 years of college or more do not live in poverty?   **about 13,200,000**

12. Make a bar graph or circle graph representing the percent of families in each income range for the following level of education completed by the householder.
a. 1–3 years of high school   **b.** 4 years or more of college
See margin.
13. Write a short paragraph drawing some conclusions from the above data.
See margin.

14

In 14 and 15, refer to the bar graph below on cases of selected diseases reported in the United States between 1950 and 1990.

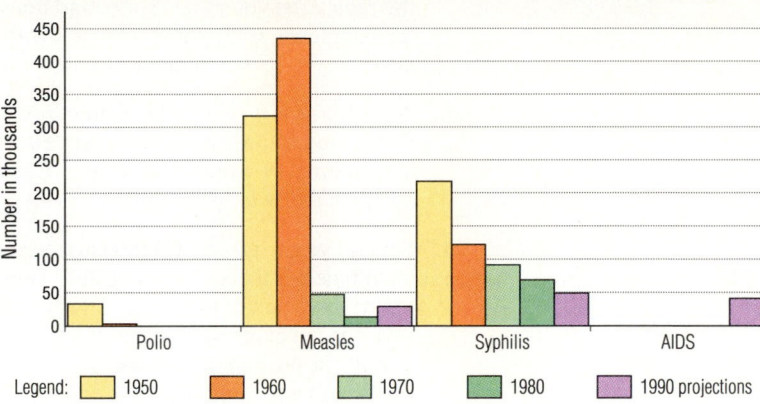

Total Number of Reported Cases of Selected Diseases between 1950 and 1990

Legend: 1950   1960   1970   1980   1990 projections

Note: Polio data for 1970, 1980, and 1990 are too small to display graphically.

14. a. Describe the pattern of syphilis cases between 1950 and 1990.
   b. Write a sentence or two about the trend in the incidence of measles between 1950 and 1990.
   c. Why do you think there is only one bar for AIDS?
   See margin.

15. Read parts **a** and **b** before doing part **a**.   **See margin.**
   a. Copy the graph, adding the following data for the total numbers of reported cases of gonorrhea: 1950, 286,700; 1960, 258,900; 1970, 600,100; 1980, 1,004,000; 1990, 734,000.
   b. Extend the entire graph to what you predict it would look like if data for the year 2000 were entered, and give reasons for your predictions.

16. Why might the bar graph at the right be misleading?
sample: Since the scale for the vertical axis does not start from zero, you may get the misleading impression that the Wacky Widget Company is doubling its profits every year.

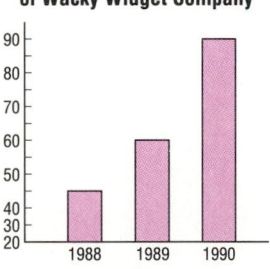

Annual Profit (in thousands of dollars) of Wacky Widget Company

Food
100°

17. Data are sometimes displayed in three-dimensional figures like the "cylinder graph" on the left. Suppose this cylinder has a radius of 5 cm and is 2 cm tall.
   a. What percent of the total volume does the slice of "Food" take up?
   b. If housing, taking up 15% of a family's budget, is to be shown on the same cylinder, what should be the angle cut off by its slice?
   a) 27.8%; b) 54°

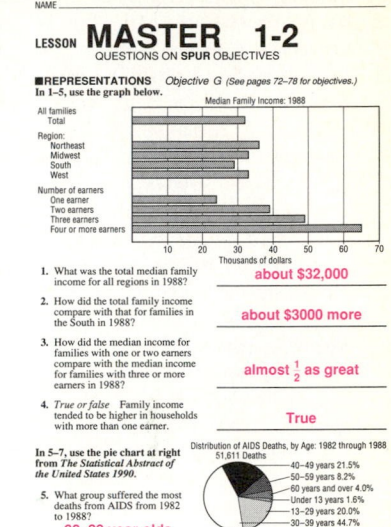

Review

18. Complete the analogy. Set is to subset as population is to ___?___ . *(Lesson 1-1)* **sample**

19. Suppose a pollster wants to send a questionnaire to a random sample of the registered voters in a state, and does so by sending the form only to the people whose last name begins with the letter M. Why isn't this a good way to pick a random sample? *(Lesson 1-1)* **Sample: Not everyone has an equal chance of being selected.**

20. A zoologist is trying to determine how many ducks live in a swamp. Five ducks are captured randomly, tagged, and released. Two weeks later 30 are captured of which 4 have tags. Estimate the size of the population of ducks in the swamp. *(Lesson 1-1)* **about 38 ducks**

21. Several years ago, ABC television asked viewers to call area code 900 telephone numbers to register their support or opposition to the United States' invasion of Grenada. Eighty percent of the callers supported the invasion, in contrast to the 30 percent support found in a survey of randomly chosen people. *(Lesson 1-1)*
a. What factors might account for such different percentages?
b. Which survey was likely to be more representative of the entire U.S. population at that time?
a) See margin. b) the random survey

22. a. Find five ordered pairs (x, y) such that $y = 3x - 4$.   **See margin.**
b. Graph these ordered pairs. They should lie on a line.   **See margin.**
c. What is the slope of this line?   **3**
d. What is its y-intercept? *(Previous course)*   **−4**

23. In 1980 the population of San Antonio, Texas, was about 786,000. In 1988 it was about 941,000. What was the average annual growth in population during this period?   *(Previous course)* **about 19,000 people per year**

Exploration

24. Find an example of a circle graph or bar graph in a newspaper, magazine, or other publication. Is the graph distorted? What conclusions can you draw from the data?   **Answers will vary.**

16

16

1-3

Other Displays

William Playfair, who was mentioned in Lesson 1-2, was the first to display changes in a variable over time. These data are called **time-series data** and sometimes can be represented well in a bar graph. (See, for example, the graph for questions 14 and 15 in Lesson 1-2.) However, drawing many bars can be tedious; and sometimes bar graphs can give false impressions.

With time-series data there are two variables, and so another way to display them is with a coordinate graph. If the points are plotted but not connected, the graph is called a **scatter plot**. If the data points are connected with line segments, the graph is called a **line graph**.

Example 1 Consider the population of Boston between 1850 and 1988.

Year	1850	1900	1950	1960	1970	1980	1988
Population (in thousands)	137	561	801	697	641	563	578

Source: U.S. Bureau of the Census

Show these data **a.** in a bar graph; **b.** in a scatter plot; and **c.** in a line graph.

Solution

a.

b.

c.

LESSON 1-3

RESOURCES
- Lesson Master 1-3
- Coordinate graph paper
- Quiz for Lessons 1-1 through 1-3
- Teaching Aid 4: Graphs for **Examples 1 and 2**
- Teaching Aid 5: Stemplots for **Example 3** and **Question 9**
- Teaching Aid 6: Tables and graph for **Questions 10–12 and 17–19**

OBJECTIVES

Letter codes refer to the SPUR Objectives on page 3A.
A Calculate measures of spread for data sets.
F Use statistics to describe data sets or to compare or contrast data sets.
G Read and interpret bar graphs or coordinate graphs.
J Draw graphs to display data.

TEACHING NOTES

Making Connections
This lesson combines statistics with traditional precalculus notions. The thorough treatment of functions begins in Chapter 2.
Examples 1 and 2 focus on the vocabulary of graphing and the interpretation of graphs of situations where time is the independent variable. Students should be able to interpret a graph in the context of the situation that the graph illustrates. In particular, they should interpret the slope of a segment between two points on a graph as an average rate of change.

The idea of graphing the status of one variable versus time is a major technique in scientific investigations and is a skill in which students are often deficient. (However, in our studies students in earlier UCSMP courses have consistently outperformed comparable students in ideas related to graphing.)

Be careful in your use of the terms *maximum* and *minimum* with respect to graphs. In the examples and questions the terms are used to identify attributes of the dependent variable, and so are features measured vertically on a graph. This will also be the case in the study of functions, where the maximum and minimum values in the range of a function are of interest. But any variable may have extreme values and these may be represented horizontally. For example, the minimum value of time t in **Example 1** is 1850. For such values it is better to use terms like *earliest* or *latest,* rather than *minimum* or *maximum.*

Example 3 shows a stemplot. Stemplots can show the data in the order the data were recorded, as in the stemplot before **Example 3,** or in numerical order as in **Example 3.** Both are commonly accepted.

A stemplot is like a bar graph. This can be clarified for your students by rotating the stemplot 90° counterclockwise so the leaves extend vertically.

Colleagues who teach science or social studies may have additional data you can use for projects or review.

ADDITIONAL EXAMPLES
1. Consider the prices of gold in dollars/fine oz between 1975 and 1988.

Year	1975	1980	1985
Price	161	613	318

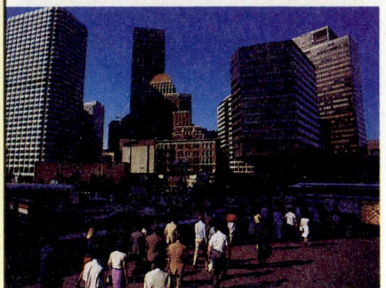

Read each graph in Example 1 from left to right. Observe that the population increased between 1850 and 1950, decreased between 1950 and 1980, and increased a little between 1980 and 1988. But because the intervals on the bar graph do not represent equal intervals of time, you cannot get an accurate impression of how rapidly the population changed in Boston. In contrast, on the scatter plot and on the coordinate graph, both the horizontal and vertical axes are scaled evenly, so you can estimate more accurately the rate of change of population.

Recall that the **average rate of change** between two points on the coordinate plane is the slope of the segment joining them. If the points are (x_1, y_1) and (x_2, y_2), the slope m of the line through these points is given by

$$m = \frac{y_2 - y_1}{x_2 - x_1}.$$

Recall also that when a graph slants up as you read from left to right, its slope is positive; when a graph slants down as you read from left to right, its slope is negative; and when a graph is horizontal, its slope is 0. If a graph has positive slope on some interval, the graph is **increasing** on that interval; if the graph has negative slope on some interval, it is **decreasing** on that interval; and if the graph has zero slope, it is **constant** on that interval.

Example 2   Calculate the average rate of change in the population of Boston in the time interval
a. between 1850 and 1900;
b. between 1950 and 1960.

Solution
a. On this interval the endpoints of the graph are (1850, 137) and (1900, 561). Thus, the average rate of change of the population is

$$\frac{y_2 - y_1}{x_2 - x_1} = \frac{561 - 137}{1900 - 1850} = \frac{424}{50} = 8.48.$$

The unit for the numerator is thousands of people; for the denominator it is years. So between 1850 and 1900 the population increased by an average of 8480 people per year.
b. Calculate the slope between (1950, 801) and (1960, 697).

$$\frac{697 - 801}{1960 - 1950} = \frac{-104}{10} = -10.4$$

Between 1950 and 1960 the rate of change was -10,400 people per year. The population decreased by an average of 10,400 people per year.

Notice that in Example 2, the absolute value of the answer in **b** is greater than the absolute value of the answer in **a**. This indicates that between 1950 and 1960, the population changed more per year than it did between 1850 and 1900. The coordinate graph of Example 1 gives an accurate impression of this, but the bar graph gives a false impression.

18

Some methods of displaying data are quite old. Coordinate graphs were developed by René Descartes and Pierre Fermat in the 16th century. Other methods are quite new. In the last half of the 20th century several new ways to represent data have been invented. Here is an example of a *stem-and-leaf diagram* or *stemplot* invented by Professor John Tukey of Princeton University in the 1960s.

Consider the following 24 scores from a math quiz.

| 75 | 32 | 80 | 95 | 62 | 75 | 98 | 93 | 84 | 87 | 94 | 85 |
| 70 | 39 | 84 | 78 | 98 | 78 | 90 | 68 | 75 | 82 | 76 | 85 |

In this form it is difficult to see patterns in these data. Below at the right the same data have been recorded in a stem-and-leaf diagram.

In this stemplot the stem is the number of 10s in the score. Each leaf is the last digit of a single score. For instance, the circled leaf represents the score of 68.

Stems	Leaves
3	2 9
4	
5	
6	2 ⑧
7	5 5 0 8 8 5 6
8	0 4 7 5 4 2 5
9	5 8 3 4 8 0

Note that the stem-and-leaf diagram is something like a bar graph, because the stems are like categories and the number of leaves for that stem is the number of grades in that category. However, unlike a bar graph, in the stemplot the individual data values are not lost.

From the stemplot you can see clearly the **maximum**, the highest score (98; there are two of them), and the **minimum**, the lowest score (32). Thus you can calculate the **range** of the data, that is, the difference between the highest and lowest scores. Note also that most scores *cluster*, or bunch up, in the 70s, 80s, or 90s, indicating overall that the class performed quite well in the quiz, but that there are two **outliers**, that is, scores which are very different from all the rest.

You can compare two related sets of data in a **back-to-back stemplot,** as shown in Example 3. In a back-to-back stemplot the stem is written in the center of the display, with one set of leaves to the right of the stem and another set of leaves to the left.

■ ■ ■ ■ ■ ■ ■ ■

Example 3 The quiz grades in Theresa Chair's two sections of geometry are shown below. To help compare the two data sets, the leaves for each stem are written in order (from the center out).

a. Find the range of scores in each class.
b. How many students in each class took the quiz?
c. How many students scored in the 80s?
d. Which scores appear to be outliers in each class?

1st Period		3rd Period
	3	2 9
0	4	
8 5 3	5	
9 6 5 2 0	6	8
5 5 2 0	7	0 5 8 8 8 9
5 2 1	8	0 2 4 5 5 5 6
	9	0 3 5 8
0 0 0	10	

Year	1986	1987	1988
Price	368	448	438

(Source: *Statistical Abstract of the United States 1990*)

Show these data in a

a. bar graph.

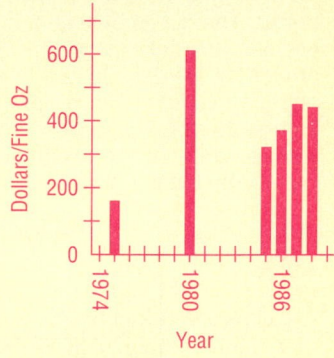

Prices of Gold in Dollars/Fine Oz Between 1975 and 1988

b. scatter plot.

Prices of Gold in Dollars/Fine Oz Between 1975 and 1988

c. line or coordinate graph.

Prices of Gold in Dollars/Fine Oz Between 1975 and 1988

Solution

a. In the 1st period class the highest score is 100, and the lowest is 40; so the range is $100 - 40 = 60$. In the 3rd period class the range is $98 - 32 = 66$.

b. Each leaf represents the score of a single student. So count the number of leaves on each side of the stem. There were 19 students who took the quiz in 1st period and 20 in 3rd period.

c. Count the number of leaves to the left and right of the stem representing 80. In the 1st period there are three scores in the 80s (85, 82, 81) and in the 3rd period there are seven (80, 82, 84, three scores of 85, and 86), for a total of 10 scores in the 80s.

d. The three scores of 100 in the 1st period, and the two in the 30s (32 and 39) in the 3rd period appear to be outliers. These scores are quite different from others in the same class. Scores in the 1st period cluster in the 50s, 60s, 70s, and 80s while those in 3rd period cluster in the 70s, 80s, and 90s.

Questions

Covering the Reading

1. A single word meaning the same thing as *average rate of change* between two points is __?__.   **slope**

In 2 and 3, refer to the data on the population of Boston given in the lesson.

2. a. Find the average rate of change in population between 1900 and 1950.   **about 4800 people per year**
   b. Find the average rate of change in population between 1980 and 1988.   **about 1875 people per year**
   c. During which of the intervals—1850 to 1900, 1900 to 1950, or 1980 to 1988—did the population of Boston increase most rapidly?
   The most rapid increase was from 1850 to 1900.

3. State one ten-year interval in which the average rate of change of population was negative.   **sample: 1960 to 1970**

4. *True* or *false*.   Stem-and-leaf diagrams were invented before the 20th century.   **False**

5. Consider a set of data. The smallest value is called the __a.__ and the largest is called the __b.__. The difference of the two is the __c.__.
See margin.

6. Use the stemplot at the right. The data represent scores on a 100-point quiz.

```
6 | 0 3
7 | 8 9
8 | 4 9
9 | 2 3 7 8
```

   a. What are the values of the stems?   **60, 70, 80, and 90**
   b. What score does 8|4 represent?   **84**
   c. What score is the maximum?   **98**
   d. What is the range of scores?   **38**

7. What advantage does a stem-and-leaf plot have over a bar graph?
sample: The individual data values are not lost.

20

---

2. Calculate the average rate of change in the price of gold in the time interval

a. between 1985 and 1987.
65 dollars/fine oz

b. between 1980 and 1985.
-59 dollars/fine oz

3. The ages of the Wimbledon tennis champions in the men's and women's singles from 1970 to 1990 are shown below. The dot (•) between two stems breaks the stem preceding the dot into two equal parts; for instance, 20–24 and 25–29. The leaves were entered in chronological order from left to right.

```
     Men |   | Women
         | 1 |
      87 | • | 999
412242432101 | 2 | 1120
  597576 | • | 7898656789
       1 | 3 | 1103
         | • |
         | 4 |
```

a. Find the range of ages for the men and for the women.
men 14, women 14

b. How many women were from 25 to 29 years old when they won the championship?
10

c. How old is the youngest person to win Wimbledon in this time period? the oldest?
17, 33

d. Are there any outliers? If so, what ages are they?
No

e. *True* or *false*. The male winners tend to be repeat winners more often than the female winners.
False

f. *True* or *false*. Men tend to win the championship at earlier ages than women.
True

NOTES ON QUESTIONS
Making Connections for Questions 9 and 21:
Students may note that Babe Ruth's overall home run record was superior to that of Roger Maris. Some may use

8. Use the following scores of last year's Advanced Algebra students on their midyear exam.

```
68  86  65  68  78  92
86  84  98  87  52  67
94  92  74  66  86  84
78  92  66  83  94  44
```

a. Make a stem-and-leaf diagram of the data.   **See margin.**
b. Find the range of scores.   **54**

9. Consider the back-to-back stemplot at the right. The data represent the number of home runs per season hit by Babe Ruth and Roger Maris. Data for Ruth are for each year he played with the New York Yankees. Data for Maris are for the 10 years he played in the American League. Ruth held the record for the number of home runs in one season until 1961 when Maris broke that record.

Maris		Ruth
8	0	
6 4 3	1	
8 6 3	2	2 5
9 3	3	4 5
	4	1 1 6 6 6 7 9
	5	4 4 9
1	6	0

a. What does the entry 8 | 0 represent?   **8 home runs in a season by Maris**
b. How many years did Ruth play for the Yankees?   **15 years**
c. What is the greatest number of home runs each hit in a single season during his career?   **Ruth hit 60; Maris hit 61.**
d. Which number appears to be an outlier?   **Maris's 61**

Babe Ruth

Applying the Mathematics

In 10–12, consider the table and graph below showing expenditures for magazine advertising from 1985 to 1988.

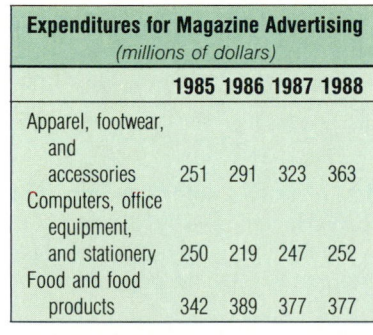

Expenditures for Magazine Advertising
(millions of dollars)

	1985	1986	1987	1988
Apparel, footwear, and accessories	251	291	323	363
Computers, office equipment, and stationery	250	219	247	252
Food and food products	342	389	377	377

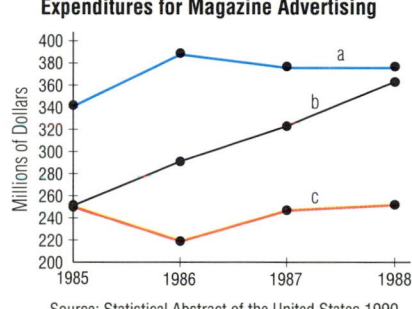

Expenditures for Magazine Advertising

Source: Statistical Abstract of the United States 1990

10. Match each of graphs a, b, and c with its expenditures category.
See margin.

11. *Multiple choice.* During which of these intervals were the expenditures for computers, office equipment, and stationery decreasing?
(a) 1985 to 1986   (b) 1986 to 1987
(c) 1987 to 1988   (d) none of the above   **(a)**

12. a. Find the average rate of change of advertising expenditures for food and food products from 1985 to 1988.   **11.7 million dollars per year**
b. Suppose that you expect expenditures for 1988 to 1991 to follow the same trend as from 1985 to 1988. Which category would you predict to have the greatest expenditures in 1991? Justify your answer.
See margin.

LESSON 1-3   Other Displays   **21**

words such as *average, mean,* or *median,* thus providing a natural introduction to Lesson 1-4. It is difficult to compare players from different eras because the baseball itself changes over time. From 1961, when Maris and Mickey Mantle hit over 50 home runs, until 1990 when Cecil Fielder hit 51 home runs, no one had hit over 50 home runs, perhaps due to the use of a "less lively" ball. The data for these questions, found in Teaching Aid 5, is also used in Question 13 of Lesson 1-4.

Questions 10–12: The table and graph can be found on Teaching Aid 6.

Making Connections for for Question 16: Parts **a** and **b** should be seen as identical, as are parts **c** and **d** (except for the choice of variable). The solution to **c** and **d** is 10 less than the solution to **a** and **b**. This presages a discussion of these ideas in Chapter 11.

ADDITIONAL ANSWERS
5.a. minimum

b. maximum

c. range

8.a.

4	4
5	2
6	5 6 6 7 8 8
7	4 8 8
8	3 4 4 6 6 6 7
9	2 2 2 4 4 8

10.a. Food and food products

b. Apparel, footwear, and accessories

c. Computers, office equipment, and stationery

12.b. See margin, page 22.

FOLLOW-UP

MORE PRACTICE
For more questions on SPUR Objectives, use *Lesson Master 1-3,* shown on page 23.

EVALUATION
A quiz covering Lessons 1-1 through 1-3 is provided in the Teacher's Resource File.

Alternative Assessment
You may wish to prepare an informal questionnaire which elicits students' feelings about their study of mathematics thus far. You might also ask them to describe what mathematics is. Use the questionnaire sometime early in the course. Then periodically help students realize that they are expanding their conceptions of what mathematics is.

ADDITIONAL ANSWERS
12.b. sample: apparel, because they have been steadily increasing and will probably exceed the expenditures of food and food products by 1991

14.b. Sample: The number of leaves on the tree is always a whole number. A line graph would give the misleading impression that fractional numbers of leaves could be on the tree.

c. After leaves began to appear in March, the number of leaves increased each month, reaching a maximum in July. After July, the number of leaves decreased each month, until there were no leaves left in the middle of October.

22

13. Use the stem-and-leaf plot for the Top Twenty Money Making Movies in 1984. Each stem represents $10,000,000. For instance, 9 | 5 represents revenue of $95,000,000. (All values have been truncated to millions.)

7	0 4 6 8 8 9
8	2 2 6 9
9	5 6
10	9
11	5
12	7 9
13	
14	1
15	
16	5
17	
18	
19	3
20	9

a. The largest money maker in 1984 was *E.T., the Extra-Terrestrial.* About how much money did *E.T.* earn that year?

b. The underlined leaf represents *Gone With the Wind,* a movie first issued in 1939. About how much money did it earn in 1984?

c. How many movies made from 80 to 100 million dollars?   **6 movies**

a) $209 million; b) $76 million

14. Refer to the scatter plot below.

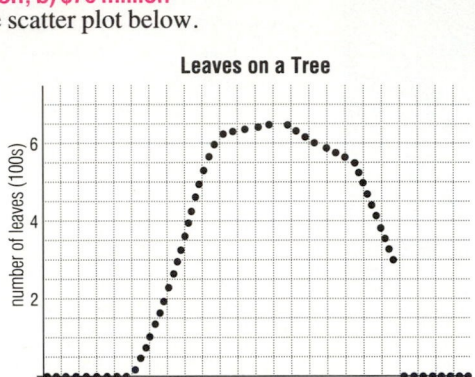

a. Estimate the maximum number of leaves on the tree at one time.

b. Why is a scatter plot more appropriate than a line graph for this situation?

c. Describe what happened to the number of leaves as the months passed.

a) about 640; b,c) See margin.

Review

15. A medical researcher administers an experimental drug to 20 persons with AIDS. He monitors the number of times they caught an infection. Identify the following for this study. *(Lesson 1-1)*   **See margin.**

a. the sample   **b.** the population   **c.** the variable

16. *Skill sequence.*   Solve. *(Previous course)*

a. $\frac{x}{8} = 16$ **128**   **b.** $0.125x = 16$ **128**

c. $\frac{x + 10}{8} = 16$ **118**   **d.** $\frac{1}{8}(n + 10) = 16$ **118**

In 17–19, use the following table of estimated costs of owning and operating an automobile. The costs listed are estimates over a 12-year life span of a vehicle (1984), or a 10-year life span (1970-81).   *(Lesson 1-2, Previous course)*

22

Item	Total Cost ($1000)				Item	Total Cost ($1000)			
	1970	1976	1981	1984		1970	1976	1981	1984
LARGE SIZE 4-DOOR SEDAN					COMPACT CAR				
Total	11.9	17.9	32.0	36.8	Total	(NA)	14.6	25.7	28.0
Costs excluding taxes	10.5	16.4	30.2	34.2	Costs excluding taxes	(NA)	13.5	24.3	26.1
Depreciation	3.2	4.9	9.2	11.6	Depreciation	(NA)	3.8	7.1	8.8
Repairs and maintenance	1.5	3.7	6.2	6.2	Repairs and maintenance	(NA)	3.0	5.2	4.7
Replacement tires	.4	.4	.9	.8	Replacement tires	(NA)	.4	.6	.5
Accessories	0.0	.1	.2	.2	Accessories	(NA)	.1	.2	.2
Gasoline	1.7	3.2	8.6	8.2	Gasoline	(NA)	2.3	6.1	5.4
Oil	.2	.2	.2	.2	Oil	(NA)	.2	.2	.2
Insurance	1.7	1.7	4.0	5.9	Insurance	(NA)	1.6	4.0	5.2
Garaging, parking, etc.	1.8	2.2	.9	1.1	Garaging, parking, etc.	(NA)	2.1	.9	1.1
Taxes and fees	1.4	1.6	1.8	2.6	Taxes and fees	(NA)	1.1	1.4	1.9

Source: *Statistical Abstract of the United States, 1987.*

17. What does **Total Cost** (*$1000*) mean? **The various costs of owning and operating a car are listed in thousands of dollars.**

18. For each category of automobile describe the trends in costs for:
a. repairs and maintenance; and **b.** gasoline. **See margin.**

19. *Multiple choice.* In general, which is the best description of how the total cost ℓ of a large sedan compares to the total cost c of a compact car?
(a) $c \approx 1.2\ell$ (b) $\ell \approx 1.2c$ (c) $\ell \approx c$ (d) $\ell \approx c + 3000$ **(b)**

20. Refer to the circle graph below. *(Lesson 1-2)*
a. What is wrong with the graph? **See margin.**
b. About how many M&Ms sold every day are brown or tan?
about 35 million

Mouth-Watering Colors

More than 100 million M&M plain and peanut chocolate candies are sold every day. The colors:

Green 10%
Orange 15%
Tan 5%
Red 20%
Brown 30%
Yellow 20%

Source: M&M/Mars

Exploration

21. Refer to Question 9. **a,b) Answers may vary.**
a. Which of the two baseball players do you think was a better home-run hitter? Why?
b. Other than number of home runs hit per year, what are some other variables you might consider to determine who was the greatest major league home-run hitter of all time?

LESSON 1-3 Other Displays **23**

OBJECTIVES

Letter codes refer to the SPUR Objectives on page 3A.
A Calculate measures of center for data sets.
B Use Σ-notation to represent a sum or mean.
C Describe relations between measures of center.
F Use statistics to describe data sets or to compare or contrast data sets.

TEACHING NOTES

For many students this lesson is a review of the measures of center. What is probably not review is the use of summation (Σ) notation. The notation is used throughout the text so emphasize it at this time. Students should be able to work with and evaluate simple expressions written with summation notation.

Making Connections In this lesson the index i is used only as a subscript. Later in the course the index will be used as a variable in expressions such as

$$\sum_{i=0}^{100} i.$$

LESSON

1-4

Measures of Center

Consider again the quiz grades from Ms. Chair's geometry class.

1st Period		3rd Period
	3	2 9
0	4	
8 5 3	5	
9 6 5 2 0	6	8
5 5 2 0	7	0 5 8 8 8 9
5 2 1	8	0 2 4 5 5 5 6
	9	0 3 5 8
0 0 0	10	

As noted earlier each group has some extreme scores. Is there a "typical" score, a single score that can represent the whole class? To describe typical values in a data set statisticians use **measures of center**, sometimes called **measures of central tendency**. The two most common of these measures are the *mean* and *median*.

The **mean** of a data set is simply the arithmetic average, that is, the sum of the data divided by the number of items in the data set. To calculate the mean for T. Chair's 1st period class, divide the total number of points the class earned by the number of students in the class.

$$\frac{40 + 53 + 55 + 58 + 60 + 62 + 65 + 66 + 69 + 70 + 72 + 2(75) + 81 + 82 + 85 + 3(100)}{19}$$

$$= \frac{1368}{19} = 72$$

In the 3rd period class the students earned a total of 1560 points. So the mean score in 3rd period is

$$\frac{1560}{20} = 78.$$

One way to interpret the mean is to consider it as the amount each person in the class would get if all the points earned by the class were shared equally. For instance, in the 3rd period class there were enough points scored so that each student could have scored 78.

The **median** is the middle value of a set of data placed in increasing order. When the data set has an even number of elements, the median is the average of the two middle values.

■ ■ ■ ■ ■ ■ ■ ■

Example 1 Find the median score for each of Theresa Chair's classes.

Solution There are 19 scores for 1st period, so the middle score is the 10th value as you count from either the maximum or minimum. The median in the 1st period is 70. In the 3rd period there are 20 scores. So the median is the average of the two middle scores (the 10th and 11th).

$$\frac{80 + 82}{2} = \frac{162}{2} = 81$$

So the median in 3rd period is 81.

24

In general, the students in 3rd period scored better than the students in 1st period. In 1st period the mean and median are 72 and 70, respectively. In 3rd period they are 78 and 81, respectively.

The mean and median values of a data set are not always close together, as the next example shows.

Example 2   The Wacky Widget Company has 15 employees. The jobs, the number of people having each job, and the annual salary for each job are given below.

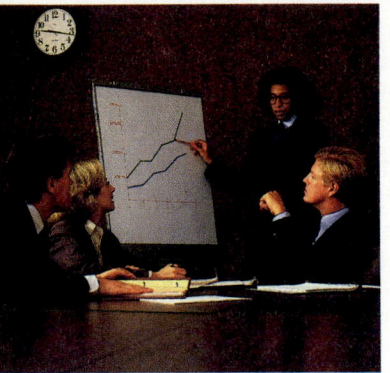

Job	Number having job	Annual salary
president	1	200,000
vice president	1	50,000
supervisor	2	25,000
sales representative	4	21,000
warehouse worker	2	15,000
custodian	2	15,000
clerical worker	3	12,000

a. Find the mean salary.
b. Find the median salary.
c. Why do you suppose most employees were upset by a recent newspaper headline reporting "Average worker at Wacky Widget making $32,000"?

Solution

a. The mean (in thousands of dollars) is

$$\frac{200 + 50 + 2(25) + 4(21) + 4(15) + 3(12)}{15} = \frac{480}{15} = 32.$$

So the mean salary is $32,000.
b. The median in a set of 15 numbers is the 8th number. The 8th salary is $21,000, the salary of a sales representative.
c. Even though the newspaper accurately reported the mean salary, 13 of the 15 employed earn far less than the mean. In this case the typical salary at Wacky Widget Company is better represented by the median.

A third measure is sometimes reported as a typical measure. The **mode** of a data set is the most common item in that set. Unlike the mean and median, the mode is always a member of the data set. In T. Chair's geometry classes, the mode in 1st period is 100: it occurred three times. In the 3rd period there were two modes: 78 and 85. Even though in the 3rd period the modes are fairly close to the mean and median, in the 1st period the mode is an extreme value. For this reason, although in many books the mode is considered a measure of central tendency, and in this book we sometimes use the mode, we do not consider it a measure of the center of a data set.

Technology The ability to do sums is basic to spreadsheet software and a notation related to summation notation is used. If a_i is the element in the ith row of column a, then with some spreadsheets the sum of a_3 to a_{10} (for example) is designated as SUM(A3:A10).

Error Analysis Emphasize that when finding the median the data must first be *ranked* from least to greatest or vice versa.

ADDITIONAL EXAMPLES
1. Find the median age of the women winning Wimbledon singles championships from 1970 to 1990.

```
1
.   9 9 9
2   1 1 2 0
.   7 8 9 8 6 5 6 7 8 9
3   1 1 0 3
.
```
27

2. The chart below gives information about the height of the varsity basketball team at Hillsboro High School.

Number of players	Height in inches
1	67
3	69
3	70
2	71
2	72
1	74
1	79

a. How many members does the team have?
13

b. Find the mean height.
71 in.

c. Find the median height.
70 in.

d. Which number, the mean or the median, best describes the "average" height of a member of the team? Why?
median; the mean is higher because of the outlier of 79.

Each of the mean, median, and mode has merits in different situations. A retailer stocking T-shirts likes to know the modal size. A teacher trying to see quickly how his or her class performed on an exam would probably look for the median score. A city council member budgeting the local income tax may want to know both the mean and median family income. A statistician frequently uses the mean because of its theoretical properties.

Calculating a mean involves finding a sum, and sums are so basic to mathematics that a shorthand notation for representing numbers and sums is commonly used. This notation is also used in developing formal definitions of the mean and other statistical measures.

Refer again to the scores of T. Chair's 1st period geometry students. Let f_i (read f-sub-i) be the score of the ith student in the 1st period. Here the scores are ranked from lowest to highest, f_1 is the lowest, f_2 the next lowest, and so on, but in general the data do not have to be in any special order.

The sum $f_1 + f_2 + f_3 + \ldots + f_{19}$ can be written $\sum_{i=1}^{19} f_i$, read, ''The sum of the f-sub-i's as i goes from 1 to 19.'' The symbol Σ is the capital Greek letter *sigma*; it corresponds to the English letter S for sum. This notation is called **summation notation** or **sigma-notation** or **Σ-notation**.

To find the total number of points earned by the top five scorers in 1st period, you must sum the 16th to the 20th score, $f_{16} + f_{17} + f_{18} + f_{19} + f_{20}$. The sum is
$$\sum_{i=16}^{20} f_i = 82 + 85 + 100 + 100 + 100 = 467.$$

In general, the expression $\sum_{i=a}^{b} x_i$ is read ''the sum of the x-sub-i's as i goes from a to b.'' It represents the sum of the set of x-values from x_a to x_b. The number i is called an **index** because it indicates the position of a number in an ordered list.

Note that the mean score in 1st period is

$$\frac{f_1 + f_2 + f_3 + \ldots + f_{19}}{19}, \text{ which in } \Sigma\text{-notation is } \frac{\sum_{i=1}^{19} f_i}{19}.$$

Definition

Let $\{x_1, x_2, \ldots, x_n\}$ be a set of n numbers. Then the **mean** $\bar{x}$ of the data set is

$$\bar{x} = \frac{\sum_{i=1}^{n} x_i}{n} = \frac{1}{n} \sum_{i=1}^{n} x_i.$$

Example 3

An apartment building has 200 apartments. Let p_i = the number of pets in the ith apartment.

a. What does $\sum_{i=1}^{200} p_i$ represent?

b. Use Σ-notation to express the mean number of pets per apartment.

Solution

a. $\sum_{i=1}^{200} p_i = p_1 + p_2 + p_3 + \ldots + p_{200}$. This represents the total number of pets in the 200 apartments.

b. To find the mean, divide the total number of pets by the number of apartments.

$$\bar{p} = \frac{\sum_{i=1}^{200} p_i}{200} \text{ or } \frac{1}{200}\sum_{i=1}^{200} p_i$$

Questions

Covering the Reading

1. Name two measures commonly used to describe the center of a data set.   **mean, median**

In 2–4, define each term.

2. mean   **The mean is the arithmetic average or the sum of the data divided by the number of items in the data set.**

3. median   **See margin.**

4. mode   **The mode is the value or values that appear most often.**

In 5 and 6, *true* or *false*. If true, explain why. If false, give a counterexample.   **See margin.**

5. The mean of a data set is always greater than its median.   **See margin.**

6. The mean of a data set is sometimes not a member of the set.   **See margin.**

In 7 and 8, refer to Example 2.

7. In the computation of the mean, why is 25 multiplied by 2, and 21 multiplied by 4?   **There are 2 supervisors with a salary of $25,000 and 4 sales representatives earning $21,000.**

8. Consider the salaries of all employees except the president.
  a. Find the mean, median, and modal salaries.
  b. Compare your answers in part **a** to those in the solution to Example 2. By how much has each measure of center changed?
  c. In general, is the mean or the median more affected by extreme values?   **the mean**
  a) The mean salary is $20,000; the median salary is $18,000; there are two modal salaries, $21,000 and $15,000.
  b) change in mean = $12,000; change in median = $3000

In 9–12, consider the following set of ages of students in a college seminar.
$$x_1 = 21, x_2 = 27, x_3 = 18, x_4 = 19, x_5 = 19, x_6 = 20, x_7 = 20$$

9. Calculate:

   a. $\displaystyle\sum_{i=1}^{5} x_i$ **104**   **b.** $\displaystyle\sum_{i=3}^{6} x_i.$ **76**

10. Write an expression using Σ-notation to represent the sum of
    the ages of the people in the seminar. $\displaystyle\sum_{i=1}^{7} x_i$

11. What measure of center does $\bar{x}$ represent? **the mean**

12. *Multiple choice.* Which of the following expresses the mean of the
    set?

    (a) $\displaystyle\sum_{i=1}^{7} x_i$ (b) $\displaystyle\sum_{i=19}^{27} x_i$ (c) $\dfrac{1}{7}\displaystyle\sum_{i=1}^{7} x_i$ (d) $\dfrac{1}{7}\displaystyle\sum_{i=19}^{27} x_i$ **(c)**

13. Consider again the data for number of
    home runs hit by Roger Maris and
    Babe Ruth.
    a. Calculate the mean and median for
    Maris.
    b. Calculate the mean and median for
    Ruth.
    c. Use measures of center to state rea-
    sons why one of these men was a
    better home-run hitter than the
    other.
    See margin.

Maris		Ruth
8	0	
6 4 3	1	
8 6 3	2	2 5
9 3	3	4 5
	4	1 1 6 6 6 7 9
	5	4 4 9
1	6	0

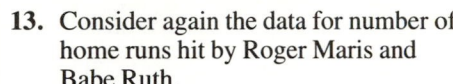
Roger Maris

Applying the Mathematics

14. Janine's scores on her first four science exams give her an average
    (mean) of 88. The maximum score on each test is 100.
    a. What does she need on the fifth exam to have an overall average of
    exactly 90?   **98**
    b. What is the highest average she could have at the end of six exams?
    92

15. According to the Department of Commerce, the mean and median price
    of new houses sold in the United States in mid-1988 were $141,200 and
    $117,800. Which is likely to be the mean, and which the median? Ex-
    plain your answer.   **See margin.**

16. Make up a data set consisting of five temperatures for which the mean is
    positive and the median negative.
    sample: -5°, -3°, -2°, 6°, 9°; mean = 1°; median = -2°

17. When data are non-numerical but can be ordered, such as letter grades,
    the mean cannot be calculated; but both the median and mode can be
    found. Last year in Mr. Flag's history course the final grades were 8 As,
    6 Bs, 3 Cs, 3 Ds, and 1 F.
    a. What was the median grade?   **B**
    b. What was the modal grade?   **A**
    c. Why doesn't it make sense to find the mean grade?
    d. If the number of grades is even, sometimes you can find the median
    and sometimes you cannot. Explain each situation.
    c,d) See margin.

28

In 18–20, refer to the table below. Let the index i be the number of the region, 1 to 6, and let e_i, x_i, m_i, and p_i represent the value of exports in 1987, exports in 1988, imports in 1987, and imports in 1988, respectively.

Exports from the United States, and Imports into the U.S.				
(in millions of dollars)				
Region	Exports		Imports	
	1987	1988	1987	1988
1. Western Hemisphere	94,795	113,155	120,605	135,135
2. Western Europe	69,718	87,995	99,934	105,035
3. Eastern Europe (including USSR)	2,200	3,650	2,118	2,385
4. Asia	73,268	99,705	184,195	200,360
5. Oceania	6,526	8,242	4,550	5,299
6. Africa	6,283	7,431	12,680	11,710

18. *Multiple choice.* Which expression represents the total value of exports from the United States in these six regions during 1987?

(a) $\sum_{i=1}^{6} e_i$ (b) $\sum_{i=1}^{6} x_i$ (c) $\sum_{i=1}^{6} m_i$ (d) $\sum_{i=1}^{6} p_i$ **(a)**

19. **a.** Write an expression using Σ-notation representing the total value of all imports from these regions in 1987. $\sum_{i=1}^{6} m_i$

   b. Evaluate the expression in part **a.** **$424,082,000,000**

20. **a.** Evaluate **See margin.**

   i. $\sum_{i=1}^{6} p_i$ **ii.** $\sum_{i=1}^{6} x_i$ **iii.** $\sum_{i=1}^{6} p_i - \sum_{i=1}^{6} x_i$

   b. What quantity does $\sum_{i=1}^{6} p_i - \sum_{i=1}^{6} x_i$ represent? **See margin.**

   c. *True* or *false.* $\sum_{i=1}^{6} p_i - \sum_{i=1}^{6} x_i = \sum_{i=1}^{6} (p_i - x_i)$. Justify your answer.

   See margin.

21. A line has a slope of 5 and contains the point (2, 3). Name two other points on the line. *(Previous course)* **sample: (1, -2) and (4, 13)**

22. *Skill sequence.* Find the area of each triangle. *(Previous course)*

a.

b.

c.

42 units²   **$\frac{1}{2}$ hz units²**   **2m² units²**

NAME _____

LESSON **MASTER 1-4**
QUESTIONS ON **SPUR** OBJECTIVES

■SKILLS *Objective A (See pages 72–78 for objectives.)*

1. Use the stemplot   Mathematics Class Enrollment   a. **23.4**
   at the right. Find   0 | 9
   a. the mean,   1 | 2 4 4 5 7 8 9   b. **26**
   b. the median, and   2 | 0 4 6 6 7 7 7 8 9 9
   c. the mode.   3 | 2 3 3 6   c. **27**

2. Jerry Attic was a dedicated golfer. His scores over a two-month period were 93, 85, 85, 103, 97, 87, 88, 86, 94, 99, 101, 85, and 89. What would Jerry have to shoot in his next game to lower his mean score to 90?
   Mean: 91.7 (or 92), Median: 89, Mode: 85, 68

3. In an algebra class of 24 students the mean grade on a test was 82; in another algebra class with 30 students the mean grade was 74. What is the combined mean of the two classes?   **77.6**

■PROPERTIES *Objective B*
In 4 and 5, suppose that x_i equals the number of loaves of bread sold on the ith day by the corner grocery store.

$x_1 = 15, x_2 = 30, x_3 = 29, x_4 = 24, x_5 = 31, x_6 = 12, x_7 = 34, x_8 = 26$
$x_9 = 24, x_{10} = 17, x_{11} = 28, x_{12} = 33, x_{13} = 41, x_{14} = 29$

4. Write an expression using Σ which indicates the total number of loaves of bread sold. $\sum_{i=1}^{14} x_i$

5. Find

   a. $\sum_{i=1}^{7} x_i$ b. $\sum_{i=8}^{14} x_i$ c. $\sum_{i=1}^{4} x_i^2$

   175   **198**   **2542**

In 6 and 7, refer to the table below. Let i be the number of the region (1 to 4). Let a_i be the total number of households with motor vehicles in region i. Let b_i, c_i, d_i, and e_i represent the number of vehicles from model years 1973 and earlier, 1974–1978, 1979–1982, and 1983 and later, respectively.

Source: The Statistical Abstract of the United States 1990

Household Motor Vehicle Transportation, By Region and Vehicle Model Year: 1985 (in millions)					
Region	Households with motor vehicles	1973 and earlier	1974 to 1978	1979 to 1982	1983 and later
Northeast	15.0	3.0	8.1	7.6	6.1
Midwest	19.5	6.4	11.4	9.0	7.0
South	27.0	9.2	15.0	13.8	10.5
West	16.1	8.3	8.9	7.3	5.1

Functions, Statistics, and Trigonometry © Scott, Foresman and Company   **7**

NAME _____
Lesson MASTER 1-4 (page 2)

6. *Multiple choice* Which expression represents the total number of households with motor vehicles?   **(a)**

   (a) $\sum_{i=1}^{4} a_i$ (b) $\sum_{i=1}^{4} b_i$ (c) $\sum_{i=1}^{4} c_i$ (d) $\sum_{i=1}^{4} d_i$ (e) $\sum_{i=1}^{4} e_i$

7. **a.** Write an expression using Σ-notation representing the total number of vehicles from 1973 and earlier. $\sum_{i=1}^{4} b_i$

   b. Evaluate the expression in part a.   **26.9 million**

■PROPERTIES *Objective C*
8. The weekly salaries of workers in a small factory are $250, $250, $375, $400, $400, $400, $400, $425, $500, and $900. If the top salary is increased to $1000, how will that affect the

   a. mean?   **It will go up by $10.**   b. median?   **Median will not change.**

9. *True or false* The mean of a data set is always between the minimum and maximum values of the set.   **True**

10. *True or false* The median of a data set is always in the data set.   **False**

■USES *Objective F*
11. Use the table below. On the average which league has bigger ball parks? Use a measure of center to justify your answer.

   The National League has bigger ball parks. The mean capacity is 53,199, as opposed to 51,838 for the American League.

8   *Functions, Statistics, and Trigonometry © Scott, Foresman and Company*

23. The graph below shows the number of heliports in operation in the USA in recent years. (Source: *Statistical Abstract of the United States, 1987.*)
(Lesson 1-3)

a. Between what two years is the slope of the graph smallest? **1983–84**
b. What was the approximate annual rate of increase of the number of heliports between 1970 and 1985? **about 160 per year**
c. Use your answer to part **b** to predict the number of heliports in operation in 1986. **about 3360 heliports**
d. Why would it be unwise to use these data to predict the number of heliports in operation in the year 2010? **Sample: the demand for heliports might change drastically over the next 20 years.**

24. What might be misleading about the graph shown below? *(Lesson 1-1)* **See margin.**

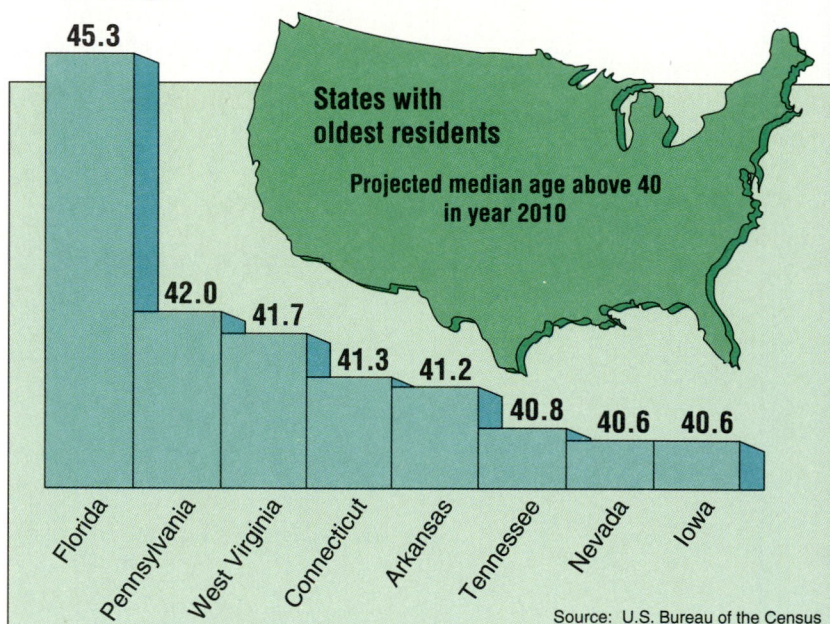

States with oldest residents

Projected median age above 40 in year 2010

Source: U.S. Bureau of the Census

Exploration

25. a. Find a set of five numbers whose mean is 10, whose median is 15, and whose mode is 30. **sample: -20, -5, 15, 30, 30**
   b. Given that $x < y < z$, find a set of five numbers whose mean is x, whose median is y, and whose mode is z.
   sample: 2.5x − y − z, 2.5x − z, y, z, z

Quartiles, Percentiles, and Box Plots

When describing data it is useful to describe both central values and how much the data are spread out from the center. The mean and median are the most common measures of center. Common measures of spread are: *range*, which was defined in Lesson 1-3; *standard deviation* and *variance*, which you will study in Lesson 1-8; and *interquartile range* which you will study in this lesson.

Consider the question, "How much does it cost to go to college at a public institution?" Here are some data from 1989 on average charges for in-state tuition, room, and board for public colleges by state. [Source: *College 'Scope*, Citicorp] Numbers have been rounded to the nearest $10.

Basic Yearly Student Charges for Room, Board, and Tuition in Public Colleges by State U.S. Average $4210					
Rank	State	Cost	Rank	State	Cost
1.	Arkansas	$2980	26.	Massachusetts	$4120
2.	South Dakota	3110	26.	Indiana	4120
3.	Oklahoma	3210	28.	West Virginia	4140
4.	North Dakota	3240	29.	Colorado	4150
5.	New Mexico	3270	30.	South Carolina	4160
6.	Mississippi	3300	31.	Iowa	4220
7.	North Carolina	3350	32.	Minnesota	4260
8.	Nebraska	3360	33.	Oregon	4390
9.	Alaska	3400	34.	Delaware	4400
10.	Idaho	3450	35.	Hawaii	4510
11.	Tennessee	3480	36.	New Hampshire	4570
12.	Alabama	3510	37.	Washington	4580
13.	Kansas	3720	38.	New York	4630
14.	Texas	3740	39.	Maine	4730
15.	Kentucky	3760	40.	Ohio	4790
16.	Louisiana	3780	41.	Pennsylvania	4810
17.	Georgia	3870	42.	California	4830
18.	Wisconsin	3940	43.	Connecticut	4970
19.	Wyoming	3960	44.	Michigan	5020
20.	Missouri	4000	45.	Illinois	5100
21.	Montana	4020	46.	New Jersey	5340
22.	Nevada	4060	47.	Maryland	5600
22.	Arizona	4060	48.	Virginia	5890
24.	Utah	4070	49.	Rhode Island	6040
25.	Florida	4090	50.	Vermont	6530

In this table, the states are not listed alphabetically, but rather they are **rank-ordered**. The data are sequenced in ascending order, that is, from lowest to highest. Putting data in order of ranking is a helpful first step in organizing information. Notice that in this ranking some ranks appear more than once. For example, Arizona and Nevada are tied for 22nd position. Then Utah, the next state, is 24th.

LESSON 1-5

RESOURCES
- ■ Lesson Master 1-5
- ■ Coordinate graph paper
- ■ Teaching Aid 8: Table and box plots for **Examples 1–3**
- ■ Teaching Aid 9: Box plots for **Example 4, Questions 10 and 12,** and Additional Example 4

OBJECTIVES

Letter codes refer to the SPUR Objectives on page 3A.
A Calculate measures of center and spread for data sets.
C Describe relations between measures of center and measures of spread.
F Use statistics to describe data sets or to compare or contrast data sets.
H Read and interpret box plots.
J Draw graphs to display data.

TEACHING NOTES

Notice that the U.S. average in the table on page 31 is the mean of the averages for the states. Quite a different U.S. average might be computed if each state's average were weighted by the number of students in the state.

What have been called measures of *dispersion* are now called measures of *spread* by many statisticians. The *standard deviation* (discussed in Lesson 1-8) is a measure of spread associated with the mean. The *interquartile range,* introduced in this lesson, is a measure of spread associated with percentiles, and thus with the median.

There are two commonly used definitions for the *p*th percentile in a set. Consider the following set of 5 scores: 11, 12, 14, 19, 19. Under our definition on page 33, the score 11 is at the 20th percentile, because 20% of the scores are less than *or equal* to 11. The number 12 is at the 40th percentile, 14 is at the 60th percentile, and 19 is at the 100th percentile. One tends not to think of any other percentiles for this data set. This is the definition which agrees with the origin of percentile, in that, in the set of integers from 1 to 100, the integer *n* is at the *n*th percentile. With this definition, there must be more than 200 scores for a score to be at the 0th percentile (rounded to the nearest percent).

Sometimes scores equal to the value being considered are not counted in determining its percentile. (See **Question 14.**) If this were the case, then in the above set, 11 is at the 0th percentile, 12 is at the 20th percentile, 14 is at the 40th percentile, and 19 is at the 60th percentile. With this definition, no score can be at the 100th percentile. In the set of integers from 1 to 100, the integer *n* is at the (*n* − 1)st percentile. This is the definition used in many standardized tests.

Although the median is commonly thought of as the 50th percentile, it cannot be that (even rounded to the nearest percent) if the number of data in the set is small with an odd number of elements. However, it is always as close to the 50th percentile as any element of the data set.

When students are constructing their own box plots, encourage them to use graph paper. It simplifies the construction and usually yields greater accuracy.

Students may be more convinced of the value of box plots if you make one using test scores from past classes,

First you can find the two measures of center. Under the title, the mean total cost, $4120, is reported. (The calculated value, $4212.60, is rounded to the same degree of accuracy as the actual data.) The rank order makes it easy to determine the median total cost. It is the mean of the 25th and 26th entries. So the median is $\frac{4090 + 4120}{2} = \$4105$.

So for the question "How much did it cost to go to college in 1989 at a public institution?" you could answer that the median total cost was $4105 or the mean total cost was $4210.

Now we examine the spread, because the measures of center do not indicate in any way how the costs vary from state to state. One measure of spread is the range, which is $6530 − $2980, or $3550. The spread of a data set can also be identified by giving several single scores called *quartiles*. **Quartiles** are so named because they are values which divide an ordered set into four subsets of approximately equal size. To calculate the quartile scores, first order the set and locate its median. By definition, the **second quartile** (also called the **middle quartile**) is the median. The **first quartile** (or the **lower quartile**) is the median of the numbers below the location of the median. The **third** (or **upper**) **quartile** is the median of the numbers above that location.

■ ■ ■ ■ ■ ■ ■ ■

Example 1 For the set of college costs, find

a. the lower quartile,

b. the upper quartile.

Solution

a. The median was calculated earlier. It is $4105 and its location is between the 25th and 26th entries. Thus there are 25 entries below the median. (They are in the left half of the preceding table.) Thus the number in the 13th position, $3720, the cost in Kansas, is the lower quartile.

b. Similarly, the upper quartile is the 13th value above the median. The total cost for the 38th-ranked state is $4630 for New York.

The difference between the third quartile and the first quartile is called the **interquartile range**. It gives a measure of the spread around the center of the data. Specifically, it tells a range in which you will find the middle 50% of the data. For the preceding data the interquartile range is $4630 − 3720 = \$910$.

The quartiles together with the minimum and maximum of the data set provide a **five-number summary** of the data.

A **box plot**, or **box-and-whiskers plot**, is a visual representation of the five-number summary of a data set. Box plots were invented in the 1970s by John Tukey. (Lesson 1-3 described another of his inventions, stemplots.) A box plot is constructed as follows:

1. Draw a number line including the minimum and maximum data values.
2. Draw a rectangle with opposite sides at the lower and upper quartiles of the data. (Sometimes these segments are called "hinges.")
3. In the box, draw a segment parallel to the hinges at the median.
4. Draw segments from the midpoints of the hinges to the minimum and maximum values. (These segments are called "whiskers.")

32

Below is a box plot for the college cost data.

Total Cost for College (in 100s of dollars)

Box plots convey a great amount of information. From this box plot, you can see that there are as many colleges with costs within the small interval $3720 to $4105 as there are in the larger interval $4630 to $6530.

Each of the five numbers (the minimum, first quartile, median, third quartile, and maximum) is often associated with a *percentile*.

> **Definition**
>
> The *p*th **percentile** of a set of numbers is a value in the set such that *p* percent of the numbers are less than or equal to that value.

By definition, every value in a data set is less than or equal to the maximum, so the maximum value is at the 100th percentile. For the college costs, $\frac{1}{50} = 2\%$ of the numbers are at or below the minimum value, so the minimum value for those data is at the 2nd percentile. The cost of college in Vermont is at the 100th percentile. The median is often called the 50th percentile and the lower and upper quartiles are called the 25th and 75th percentiles, respectively. In practice, however, the quartiles may not correspond *exactly* to these percentiles.

Example 2 For the set of college costs, find
  a. the percentile rank of $3720 (the first quartile)
  b. the college cost at the 85th percentile.

Solution

a. There are 13 costs less than or equal to $3720, and $\frac{13}{50} = .26 = 26\%$.

So $3720 is the 26th percentile.
b. 85% of 50 is $(.85)(50) = 42.5$. The rankings are whole numbers, so there is no such ranking as 42.5. Typically when dealing with percentiles, people round up to the next integer. So the 85th percentile corresponds with the 43rd cost, $4970, the cost in Connecticut.

The lengths of the whiskers depend on percentiles. Some people end whiskers at the 5th and 95th percentiles; others end them at the 10th and 90th percentiles. Still others use a procedure based on the interquartile range to determine outliers, and then draw whiskers to represent all values except the outliers.

or past ACT or SAT results. The scores are then represented as absolute measures (the number or percent correct) and relative measures (the percentile). You may choose to present box plots of quiz and test results for this course throughout the year. A good project later in the year would be to make a box plot of class or school scores on college entrance exams and compare them to the national averages.

Making Connections In Lesson 1-7, students will use a computer to construct box plots. Statistical packages vary in how they handle outliers and the upper and lower quartiles. Check yours to see how it operates. You may want to demonstrate the box-plot feature of your package during this lesson in preparation for Lesson 1-7.

ADDITIONAL EXAMPLES
In **1–3,** use the following scores of a mathematics class on the midyear exam.

43	68	73	78	80	88	92
52	70	74	78	82	89	93
65	70	75	78	85	90	94
66	71	75	78	87	90	94
67	72	76	79	87	90	98

1. Find the first quartile, median, and third quartile.
71, 78, 89

2. Find **a.** the percentile rank of 89 **b.** the test score at the 20th percentile.
a. 77th percentile; b. 70

3. Use the $1.5 \times$ IQR criterion to determine if there are any outliers.
43 is an outlier.
Note: You may want to have your students construct a box plot using the data above.

4. Use the following box plots which show the number of games won by the 25 leading pitchers in baseball's National League (NL) and American League (AL) in 1989. The statistics are from *Information Please Almanac Atlas and Yearbook 1990*.

Number of Games Won by
25 Leading Pitchers NL and AL 1986

a. What was the range of numbers of games won by National League pitchers?
16

b. Three quarters of the American League leading pitchers won no more than _____ games.
7

c. The maximum for the National League is the same as a(n) _____ for the American League.
outlier

d. Generally speaking, in which league were the leading pitchers more successful? Explain.
NL; Sample: Except for a few outliers, the NL had a higher median and upper quartile.

NOTES ON QUESTIONS
Question 10: The scale on the number line is of test scores, which are not necessarily percentiles. Emphasize that the scale on any box plot has the same units as the original data. You might ask students how many scores are included in the plot for **Question 10.** The correct answer is that it cannot be determined.

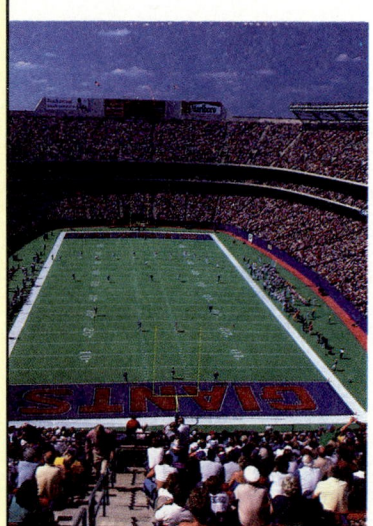
Giants Stadium

34

The following criterion for determining outliers is used frequently.
Let IQR be the interquartile range.
Add $1.5 \times$ IQR to the third quartile. Any value above that sum is an outlier.
Subtract $1.5 \times$ IQR from the first quartile. Any value below that difference is an outlier.

Example 3 Consider the college costs presented earlier. Use the $1.5 \times$ IQR criterion to determine if there are any outliers.

Solution IQR = \$910; so $1.5 \times$ IQR = $1.5 \times 910 = \$1365$.
The cost at the third quartile is \$4630; adding $1.5 \times$ IQR gives
$$4630 + 1365 = 5995.$$
So any cost greater than 5995 is an outlier. Similarly, any cost \$1365 lower than the first quartile is an outlier.
$$3720 - 1365 = 2355$$
Thus the only outliers are \$6040 and \$6530, the costs of college in Rhode Island and Vermont.

When the $1.5 \times$ IQR criterion is used to determine outliers, the outliers are often marked as separate points on the box plot, as in the following version of a box plot of college costs.

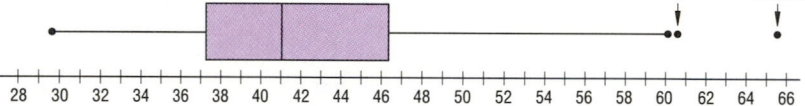

Box plots are valuable in comparing data sets.

Example 4 The National Football League is separated into two parts—the American Football Conference (AFC) and the National Football Conference (NFC). Here are separate box plots of the capacities of the football stadiums used by the AFC and NFC.

Capacities of Stadiums (in 10,000s)

Use the plots to answer the following.
a. What is the median capacity in each conference?
b. What is the size of the largest stadium in each conference?
c. About what percent of the stadiums in the AFC hold fewer than 60,000 people?
d. On the whole, which conference has larger stadiums?

Solution

a. Look at the vertical segment inside each box. The median capacity is about 70,000 in the AFC and about 63,000 in the NFC.
b. The maximum is at the endpoint of the whisker on the right. For the AFC it is about 92,000; for the NFC, about 80,000.
c. In the box plot for the AFC, 60,000 indicates the lower quartile, so about 25% of the stadiums in the AFC hold fewer than 60,000 people.
d. On the whole, AFC stadiums are larger than NFC stadiums. The median size in the AFC is close to the 75th percentile in the NFC.

Questions

Covering the Reading

In 1–5, refer to the data on page 31 concerning annual cost of going to college.

1. What cost is at the 34th percentile?   **$3870**

2. What cost is at the 98th percentile?   **$6040**

3. At what percentile is a cost of $4000?   **40th percentile**

4. At what percentile is a cost of $4830?   **84th percentile**

5. At what percentile is the cost of college in Illinois?   **90th percentile**

6. The 25th percentile is often called the ___?___ quartile.   **first or lower**

7. What percentile corresponds to the third quartile?   **75th percentile**

8. What percentile corresponds to the maximum score?   **100th percentile**

9. a. What numbers are reported in a five-number summary?
b. Which give(s) the center?   **the median**
c. Which give(s) information about the spread in the middle of the data set?   **the first and third quartile**
d. Which give(s) information about extreme values?   **max. and min.**
a) maximum, minimum, and the 3 quartiles

10. Use the following box plot of student test scores on last year's advanced algebra mid-year exam.

a. What is the median score?   **83**
b. What is the interquartile range?   **29**
c. What percent of the students scored between 62 and 91?   **50%**
d. What is the interval of scores of students who ranked below the lower quartile?   **48 to 62**

Error Analysis for Question 11: Some people (and some statistics packages) remove the median from a data set with an odd number of elements before they find the upper and lower quartiles. We do not do this. For **Question 11,** if the median is removed, then the upper quartile is the average of the 6th and 7th ranked costs (corresponding to $475) and the lower quartile is the average of the 19th and 20th ranked movies ($185). Check your software for how this is handled, and reassure students who delete the median that their answers are valid. As with most statistical measures and displays, the difference in the techniques is less noticeable for large data sets than for small ones.

Questions 12 and 13: Point out the value of box plots for making comparisons.

Making Connections for Question 17: Students may convert to inches (6′6″ = 78″) or to fractions of a foot (6′6″ = 6.5′) or transform the data using deviations from 6′. In the latter case, Stretch is -2″, Lanky is 0″, Skyler and Knees are +2″, and so on. Then computations can be done in inches. This anticipates the concept of deviation from the mean which is covered in Lesson 1-7, and translations of data, covered in Lesson 3-3.

Question 21: Part **a** of this question is appropriate as a **small-group** activity.

MORE PRACTICE
For more questions on SPUR
Objectives, use *Lesson Master 1-5*, shown on page 37.

EXTENSION
You might obtain a district report of SAT or ACT scores from some previous year (without names, of course) and discuss the relationships between your district percentiles and the national percentiles.

ADDITIONAL ANSWERS
11.a. median: $31; lower quartile: $18.50; upper quartile: $47.50

b. The values $97 and $113 are outliers.

c.

Amount of money spent at grocery store

10 20 30 40 50 60 70 80 90 100 110
dollars

d. Sample: About half the values are clustered in the relatively small interval from $18.50 to $47.50. The spread between the extreme values is almost 4 times the spread between the first and third quartiles. Also the spread among the "big spenders" is less than that among those who spend less.

13.a.

Division A
Division B
Division C

10 20 30 40 50 60 70
Annual salary in 1000s of dollars

Applying the Mathematics

11. At the right is a stem plot of the amount of money spent by 25 shoppers at a grocery store. The stem is in $10 units.
   a. Find the median, lower quartile, and upper quartile.
   b. Use the $1.5 \times IQR$ criterion to determine which, if any, values are outliers.
   c. Construct a box plot using Xs for outliers.
   d. Write several sentences about the center, spread, and extreme values of this data set.
   See margin.

0	3 8
1	0 1 7 8 9
2	0 0 3 6 8
3	1 3 4 7
4	2 5 5
5	0
6	5
7	2 6
8	
9	7
10	
11	3

12. The box plots below represent the number of hours per week students used computers at three levels of schools in 1985. Endpoints of whiskers correspond to the 10th and 90th percentiles.

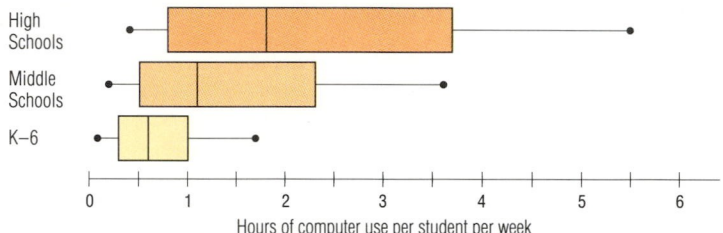

Adapted from data of the Center for Social Organization of Schools, Johns Hopkins University (1985)

True or *false*.
   a. The number of hours per week K–6 students use computers varies from nearly none in some schools to about 1.8 in others.   **True**
   b. In 1985, in about half of the U.S. high schools, the average student used computers from about 0.8 hour to about 3.5 hours per week.
   c. The median number of hours per week that computers are used decreases as students go through school.   **False**
   d. Some high school students use computers three times as many hours per week as others.   **True**
   b) True

13. Use the five-number summaries given below for the annual salaries in three divisions of a company. The minimum, first quartile, median, third quartile, and maximum are abbreviated min, Q_1, M, Q_3, and max, respectively. Data are in 1000s of dollars.

Division	min	Q_1	M	Q_3	max
A	18	28	34	40	57
B	23	35	43	48	59
C	19	24	27	30	43

   a. Make three box plots to represent this information graphically.
   b. *True* or *false*.   Nearly all of the salaries in Division C are in the middle 50% of salaries in Division B.   **False**
   c. Overall, which division has the lowest salaries?   **Division C**
   a) See margin.

14. Some books and standardized tests use a different definition of percentile. They say that the pth percentile of a set of numbers is a value in the set such that p percent of the numbers are less than that value. According to this definition,
   a. what percentile is the minimum value in a data set?   **0th percentile**
   b. Can any value ever be at the 100th percentile?   **No**

15. A data set has 100 elements $a_1, a_2, \ldots, a_{100}$. Write an expression in Σ-notation to represent the mean of the set.   *(Lesson 1-4)*   **See margin.**

16. Suppose $x_1 = 2$, $x_2 = 7$, and $x_3 = 4$. Evaluate each expression.   *(Lesson 1-4)*

   a. $\displaystyle\sum_{i=1}^{3} x_i$ **13**   **b.** $\displaystyle\left(\sum_{i=1}^{3} x_i\right)^2$ **169**   **c.** $\displaystyle\sum_{i=1}^{3} x_i^2$ **69**

17. Here are the heights of four members of the Reach High basketball team.
   Stretch: 5'10"   Skyler: 6'2"   Lanky: 6'   Knees: 6'2"

   They have a choice of adding Marvin (6'1") for his mobility or Harry (7'1") for his height. Calculate the mean and median for the heights of
   a. the current team;   **mean = 6' 1/2"; median = 6' 1"**
   b. the current team plus Marvin;   **mean = 6' 3/5"; median = 6' 1"**
   c. the current team plus Harry.   **mean = 6' 3"; median = 6' 2"**
   d. Which statistic in **c** do you think better describes the team, and why?
   e. In general, which of the mean and median is less affected by extreme values?   *(Lesson 1-4)*   **median**
   d) See margin.

In 18 and 19, state **a.** the slope; **b.** an equation for the line.   *(Previous course)*

18.   a) **- 1**
   b) **sample: y = - x**

19. 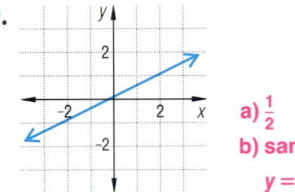   a) $\frac{1}{2}$
   b) **sample:**
   $y = \frac{x}{2}$

20. *Skill sequence.*   Suppose $x \neq 0$. Simplify each expression.

   a. $\dfrac{13x - x}{2x}$ **6**   **b.** $\dfrac{20x + 16x}{2x + 4x}$ *(Previous course)*   **6**

21. **a.** Collect the heights of a sample of 50 people in your school.
   b. Make a box plot of the data.
   c. Estimate any outliers visually from the box plot.
   d. Identify outliers using the $1.5 \times$ IQR method.
   e. Is your sample representative of any larger population? Explain.
   Answers will vary.

22. Find a report on your school that uses percentiles. Describe the way percentiles are calculated and what they tell about you and your classmates.   **Answers will vary.**

23. Refer to Example 4. What accounts for the fact that AFC stadiums are generally larger than NFC stadiums? **sample: AFC stadiums are newer.**

Margin notes (right column):

15. sample: $\dfrac{1}{100}\displaystyle\sum_{i=1}^{100} a_i$

17.d. sample: median, because four of the five team members are shorter than 6'3", the mean

---

NAME _____

LESSON MASTER   1-5
QUESTIONS ON **SPUR** OBJECTIVES

■**SKILLS** *Objective A* (See pages 72–78 for objectives.)
In 1 and 2, use the stemplot at the right.

1. Identify the **a.** first quartile,   Mathematics Class Enrollment
   b. the third quartile, and **c.**   0 | 9
   the interquartile range.   1 | 2 4 4 5 7 8 9 9
                              2 | 0 4 6 6 7 7 7 7 8 8 9
                              3 | 2 3 3 6
   a. _____ **17.5**   b. _____ **28**   c. _____ **10.5**

2. What class enrollment is at the
   a. 12th percentile? _____ **14**   **b.** 80th percentile? _____ **28**

■**PROPERTIES** *Objective C*
3. What percentile corresponds to the median of a data set?   **50th**

4. *True or false* The third quartile is the median of the data above the median.   **True**

■**USES** *Objective F*
5. The stemplot below gives the number of World Series appearances of major league franchises through 1989.

American League	National League
9 9 7 5 4 3 2 1 0	1 3 4 4 7 8
3 1	0 5 6 8
2	
3 3	

   a. How many World Series have there been?   **86**
   b. Find the five-number summaries for each league.
   American: (1, 3, 6, 9, 33); National: (1, 4, 7.5, 15, 18)
   c. Are there any outliers using the $1.5 \times$ IQR criterion?
   33 is an outlier ($1.5 \times$ IQR = 9)

NAME _____
Lesson MASTER 1-5 (page 2)

   d. Describe similarities and differences between the two groups.
   sample: ten teams in both; typical number of appearances less than 10 in the American, more like 10 in the National; the American League dominated by one.

■**REPRESENTATIONS** *Objective H*
6. Refer to the box plot below which represents the length (in years) English monarchs were on the throne (through George VI).

   0 10 20 30 40 50 60 70
Lengths of Reigns of English Rulers (in Years)
Source: *Information Please Almanac Atlas and Yearbook 1990*

   a. What is the median length for the reign of an English monarch?   **about 20 years**
   b. What percent of the monarchs were on the throne between 9.5 and 34 years?   **50**

■**REPRESENTATIONS** *Objective J*
7. Draw box plots to illustrate the data of problem 5 of this Lesson Master.

   National League

   American League

   0 5 10 15 20 25 30 35

LESSON 1-6

Histograms

A **histogram** is a special type of bar graph. It breaks the range of values of a numerical variable into non-overlapping intervals of equal width, and displays the number of values that fall into each interval. These numbers are sometimes called **frequencies**. For this reason a histogram representing actual counts is sometimes called a **frequency distribution**, and one representing the counts as parts of a total is called a **relative frequency distribution**.

To make a histogram, first, organize the data into non-overlapping intervals of equal width. Choosing the width of the interval is a matter of judgment; there is usually not a single best size. Too few intervals will lump all the data together; too many will result in only a few numbers in each one.

Second, count the number of observations per interval and record the results in a **frequency table**, that is, a table that gives the frequency or relative frequency for each of the intervals created.

Finally, draw the histogram. First mark the endpoints of the intervals on a horizontal axis and a scale for the frequencies on a vertical axis. Then draw a bar to represent the frequency in each interval. Unlike other bar graphs, because histograms often represent continuous variables they are drawn with no horizontal space between bars (unless an interval is empty, in which case its bar has height 0).

Example 1 Consider the data on college costs from Lesson 1-5. Display these data in a histogram.

Solution The range of costs is 6530 − 2980 = 3550. This is about 3500. So intervals of size 500 seem reasonable. Rather than start the intervals at the minimum, we begin at the greatest multiple of 500 less than the minimum. Thus our first interval goes from 2500 to 3000; the second from 3000 to 3500. A refinement is needed because 3000 would be in two intervals. So we made an arbitrary decision: the intervals are 2500–2999, 3000–3499, 3500–3999, and so on. The frequency table is below on the left, and the corresponding histogram below on the right.

Total Cost ($)	Frequency
2500–2999	1
3000–3499	10
3500–3999	8
4000–4499	15
4500–4999	9
5000–5499	3
5500–5999	2
6000–6499	1
6500–6999	1

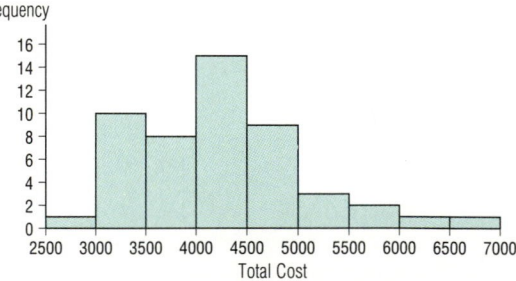

38

In Example 1, the histogram contains 9 intervals ranging from 2500 to 7000, each with width 500. In general, if the histogram ranges from a to b, and there are n intervals, then the width of each interval is $\frac{b-a}{n}$. If the first interval starts with the value a, then the endpoints of the intervals are the following:

A well made histogram, like a box plot, gives some information about the center and spread of the data. For instance, like the box plot of the college cost data drawn in Lesson 1-5, the histogram above suggests that in most states annual costs at public colleges are around $4000, but that there are some that charge less than $3000, and some that charge more than $6000.

However, unlike a box plot, histograms seldom indicate the median or any other exact value. For instance, the histogram above shows that the highest cost occurs in just one state, and that the cost is between $6500 and $7000. However, there is no way to determine where in that interval the cost falls.

Furthermore, a poor choice of intervals on a histogram can mask the spread of the data. For instance, the histogram at the right has too few intervals to convey the college cost data effectively.

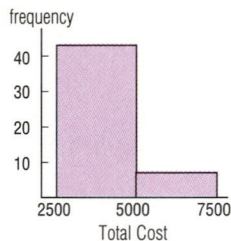

You should be able to gain information from histograms.

Example 2

The histogram below shows the results of all students in Ms. Chair's school on last year's final exam in geometry. The intervals include the right endpoints but not the left. For instance, the left two intervals are $20 < x \leq 30$ and $30 < x \leq 40$.

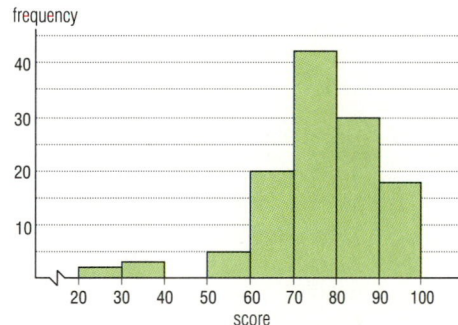

a. About how many students got scores between 60 and 70 including 70?
b. About how many students took the final exam?
c. In which 10-point interval does the median probably fall?

Two aspects of this lesson make it appropriate for this level. First is the use of algebra to obtain the endpoints of the intervals. You may write each as a single fraction as follows:

$$a + \frac{b-a}{n} = \frac{(n-1)a+b}{n}$$
$$a + 2\frac{b-a}{n} = \frac{(n-2)a+2b}{n}$$
$$\vdots$$
$$a + (n-1)\frac{b-a}{n} = \frac{a+(n-1)b}{n}.$$

The right sides of these equations show that each endpoint of a bar in the histogram is a weighted average of the endpoints of the range.

Making Connections
Rewriting the left side of the first equation was covered in Question 19 of Lesson 1-8.

The second aspect of this lesson that makes it suitable for an upper level high school course is the use of unequal intervals in histograms. This may cause appropriate emphasis, but it also distorts data.

Error Analysis Students may not distinguish between a frequency distribution and a relative frequency distribution. The values of a frequency distribution are nonnegative integers. The values of a relative frequency distribution are numbers between and including 0 and 1. These are typically represented as decimals to the same decimal place, fractions with the same denominator, or percents. The graph preceding **Example 3** uses decimals; Additional Example 3 which follows uses fractions. Emphasize that there is no single "best" set of intervals for making a histogram.

Alternate Approach:
Technology You may use your statistics package to demonstrate multiple views of a data set. This will illustrate the point, and it will ease students' use of the statistical package in the next lesson.

1. Consider the test score data from Additional Examples 1–3 in Lesson 1-5:

```
43 68 73 78 80 88 92
52 70 74 78 82 89 93
65 70 75 78 85 90 94
66 71 75 78 87 90 94
67 72 76 79 87 90 98
```

Display these data in a histogram.

Scores of a Mathmatics Class on Their Midyear Exam

2. The histogram below shows the results of all students in Ms. Chair's class on a chapter test. The intervals include the left endpoint, but not the right. For instance, the left interval is $40 \le x < 50$.

Test Results of All Students in Ms. Chair's Class

a. Why is it not possible to tell how many students are in the class?

The value for each bar shows only the ratio of students in each category to the total number of students.

b. If there are 22 students in the class, how many students received grades greater than or equal to 80 and less than 90?

6

Solution
a. Read the height of the bar between 60 and 70. About 20 students got scores in this interval.
b. Total the heights of the bars. Note that for four of the intervals you must estimate the heights. Reading the intervals from left to right, an estimate is

$$2 + 3 + 0 + 5 + 20 + 42 + 30 + 18 = 120.$$

So about 120 students took the geometry exam.
c. When the 120 scores are in order, the median is between the 60th and 61st scores. From the estimates in part **b** we conclude that there are about 30 scores below 70, and about 48 between 70 and 80. So the 60th and 61st scores are between 70 and 80, and so the median must be between 70 and 80.

Below on the left the frequencies on the geometry final exam from Example 2 have been converted to **relative frequencies** by dividing each frequency by 120, the estimated total number of students. Below at the right is a histogram based on the relative frequencies.

Score	frequency	relative frequency
$20 < x \le 30$	2	.02
$30 < x \le 40$	3	.02
$40 < x \le 50$	0	0
$50 < x \le 60$	5	.04
$60 < x \le 70$	20	.17
$70 < x \le 80$	42	.35
$80 < x \le 90$	30	.25
$90 < x \le 100$	18	.15
total	120	1.00

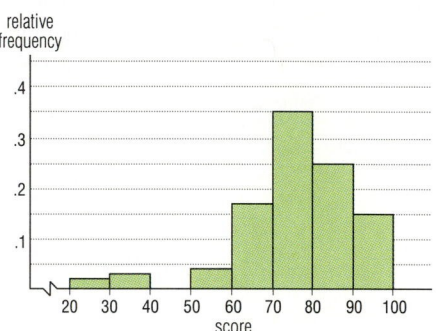

By adjusting the scales on the y-axis, the histogram based on relative frequencies can look congruent to the histogram based on counts.

Example 3 Use the table of relative frequencies.
a. About what percent of the students scored between 70 and 80?
b. About what percent of the students scored less than or equal to 50 on the exam?

Solution
a. .35 is 35%. So about 35% of the students scored between 70 and 80.
b. $.02 + .02 + 0 = .04$. So about 4% scored at or below 50.

Histograms can be distorted if the intervals are not of equal width. Below at the left is a table whose data are in the histogram at the right. These data are rounded from the U.S. Bureau of the Census estimates of the late 1980s.

Age	Projection (1000s)
under 5	16,900
5–17	48,800
18–24	25,200
25–44	81,200
45–64	60,500
over 65	34,900

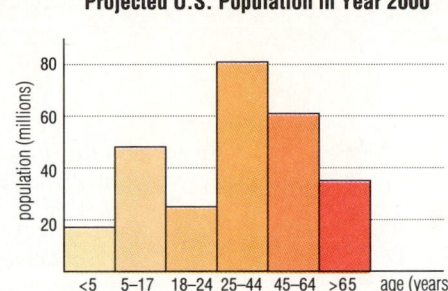

Projected U.S. Population in Year 2000

The histogram gives the impression that there are the fewest number of people at the ages from birth to 5, but the bar is shortest because there are only 5 years (0, 1, 2, 3, 4) represented. That is also the reason the bar for 18–24 is so short. This histogram would be clearer if there were equal intervals, for example: < 20, 21–40, 41–60, 61–80, > 80.

Questions

Covering the Reading

1. What is the difference between frequency and relative frequency?
   See margin.
2. Suppose the interval of SAT scores from 200 to 800 is split into 15 intervals of equal width.
   a. What will be the width of each interval? **40**
   b. What will be the middle interval? **480 to 520**

3. Suppose the interval from x to y is split into n equal parts.
   a. What will be the width of each part? **$(y - x)/n$**
   b. What will be the endpoints of the left interval? **x and $x + (y - x)/n$**

In 4 and 5, consider the data on costs of attending college from Lesson 1-5 and Example 1 of this lesson.

4. a. Which interval has the highest frequency? **$4000 to $4499**
   b. In which interval is the median? **$4000 to $4499**

5. a. Make another histogram of the same data using intervals of size 300 beginning at 2700. **See margin.**
   b. Mark the mean and median on the histogram. **See margin.**

6. Consider Ms. Chair's geometry students from Examples 2 and 3.
   a. What percent of students scored above 80? **40%**
   b. In which interval is the 25th percentile? **$60 < x \le 70$**

7. Use the projected population data in this lesson.
   a. About how many 23-year olds are projected for the year 2000?
   b. Which age group is expected to have the most people, 5–24, 25–44, 45–64, or 65–84? **25–44**
   a) cannot determine from the histogram

LESSON 1-6 Histograms **41**

---

c. In which 10-point interval does the median fall? Justify your answer.
$70 \le x < 80$; since 5.5/11 = 50%, we can total the relative frequencies from either end to find that 5.5/11, and hence the median, would fall in the interval $70 \le x < 80$.

3. Use this table of relative frequencies.

Score	Relative frequency
$40 \le x < 50$	$\frac{2}{15}$
$50 \le x < 60$	$\frac{3}{15}$
$60 \le x < 70$	0
$70 \le x < 80$	$\frac{4}{15}$
$80 \le x < 90$	$\frac{1}{15}$
$90 \le x < 100$	$\frac{5}{15}$

a. About what percent of the students had scores in the interval 80–89?
7%

b. About what percent of the students passed (scores greater than or equal to 60)?
67%

mean = $4210
median = $4105

NOTES AND QUESTIONS

Questions 8 and 9: Ask students how many representatives your state has. The numbers may have changed since 1988–90 because they are modified based on the most recent census.

Question 13: This question may prompt a lively classroom discussion. There should be no disagreement about x; it is the mode because its bar is the tallest. The median can be calculated by counting. If students are skeptical, choose scales for both axes. For example, the horizontal axis could be $0 \le x < 5, 5 \le x < 10, \ldots,$ $95 \le x < 100$. Estimate the heights of the bars and calculate the median.

ADDITIONAL ANSWERS
9.a.

Number of Representatives	Frequency
1 to 5	22
6 to 10	17
11 to 15	3
16 to 20	2
21 to 25	3
26 to 30	1
31 to 35	1
36 to 40	0
41 to 45	1

b.

Applying the Mathematics

In 8 and 9, use this information. The number of representatives each state has in the U.S. Congress depends on the population of the state. Below are the numbers of representatives in each state in the 100th Congress (1988–90).

7	1	5	4	45	6	6	1	19	10
2	2	22	10	6	5	7	8	2	8
11	18	8	5	9	2	3	2	2	14
3	34	11	1	21	6	5	23	2	6
1	7	27	3	1	10	8	4	9	1

8. Identify the **a.** minimum, **b.** maximum, and **c.** range of number of representatives per state.   **a) 1; b) 45; c) 44**

9. **a.** Make a frequency table showing the distribution of the number of representatives of states. The first few lines are given below.

Number of Representatives	frequency
1 to 5	?
6 to 10	?
11 to 15	?

   b. Draw a histogram to represent these data.   **See margin.**

10. Refer to the frequency distribution below of the scores of students taking the Latin Achievement Test of the College Entrance Examination Board in 1985.

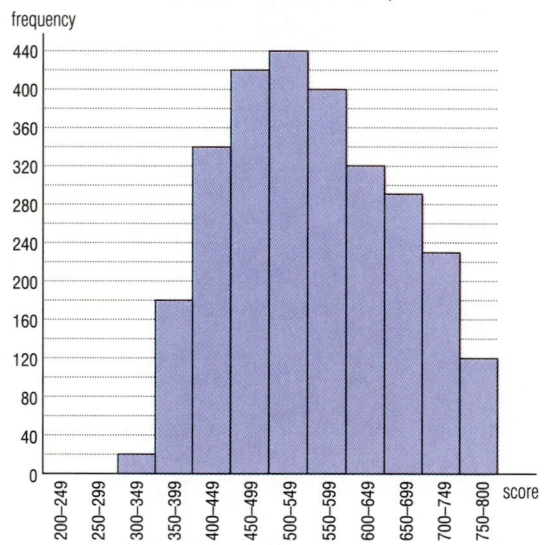

College Board Latin Achievement Test Scores, 1985

   a. About how many students scored between 400 and 499?   **about 760**
   b. What score interval was the mode?   **500–549**
   c. About how many students took the Latin exam in 1985? (Read the graph to the nearest 10.)   **about 2760**
   d. What percent of the students scored 700 or better?   **about 13%**
   e. Estimate the location of the median score.   **about 550**

42

11. The following item appeared on a questionnaire. Indicate any flaws that exist in the intervals used.   **See margin.**

> Indicate how many years you have been with this company.
>
> _____ 1–10 years
> _____ 10–20 years
> _____ 21–39 years
> _____ more than 40 years

11. Sample: The intervals are not of uniform length; the first two intervals overlap at 10; the last two intervals exclude 40.

12. A physician obtained cholesterol levels from a sample of patients. The relative frequency distribution is given in the table below.

Cholesterol level	relative frequency
170–179	.12
180–189	.16
190–199	.25
200–209	.15
210–219	.10
220–229	.07
230–239	.06
240–249	.05
250–259	.00
260–269	.04

12.c.

a. A cholesterol level under 200 is desirable. What percent of this sample has cholesterol level in the desirable range?   **53%**

b. A cholesterol level of 240 to 260 is considered by some to put the patient at moderately high risk for heart attack. What percent of the patients are in this group? **5%**

c. Draw a histogram for these data.   **See margin.**

13. On the diagram below, x, y, and z represent the mean, median, and mode, but not necessarily in that order. Which is which?

x = mode, y = mean, z = median

In 14 and 15, data are being reported about consecutive integer values. Such data can be represented with a *dot frequency diagram* or a histogram in which the bars are centered over the individual values.

14. *Multiple choice.*   Below is the distribution of heights of 24 people with the same occupation. Which is the most likely occupation of this group?

(a) dancers   (b) basketball players
(c) jockeys   (d) circus performers   **(b)**

15. The histogram below shows the responses of 100 people to the question "How many pairs of shoes do you own?"

Let i be the number of pairs of shoes owned and f_i = the frequency of that number.

a. What value of i is the mode? **3**   b. Estimate f_5. **13**

c. Evaluate $\sum_{i=1}^{11} f_i$. **100**

d. What does the quantity $\dfrac{\sum_{i=1}^{11}(if_i)}{\sum_{i=1}^{11}f_i}$ represent? **the mean number of pairs of shoes owned**

16. a. Between 12 noon and 12 midnight, how many times do the minute and hour hands of a clock point in the same direction? **11 times**

b. The intervals between the times in part **a** are of equal length. Use this information to calculate the first time after 1 P.M. when the hands point in the same direction. **1:05:27**

Review

17. The graph below shows the number of U.S. farms from 1975 to 1989, according to *The Statistical Abstract of the United States, 1990.*

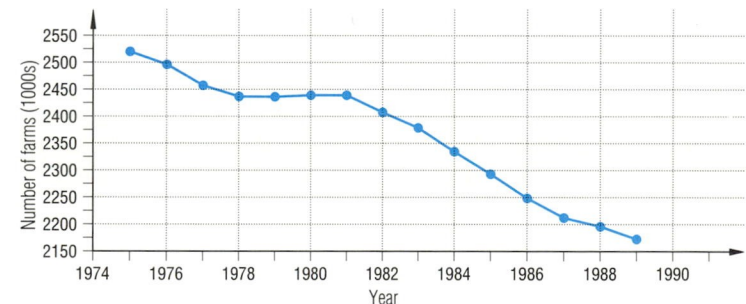

a. U.S. farms occupy a total of about one billion acres. This value has remained fairly constant in the recent past. How has the average farm size changed in recent years? **It has steadily increased.**

b. Use both the graph and the data in **a** to estimate the average farm size for each of 1975 and 1989. **1975: ≈ 400 acres; 1985: ≈ 440 acres**

c. Estimate the average rate of change in the number of U.S. farms between 1984 and 1989. *(Lessons 1-4, 1-3)* **about 30,000 farms/year**

44

18. Use the table below on the SAT-Math scores of 119 students in the University of Chicago Laboratory Schools in 1988. *(Lesson 1-5, 1-4)*

score	n	%
750–800	6	5
700–740	15	13
650–690	19	16
600–640	24	20
550–590	16	13
500–540	18	15
450–490	10	8
400–440	4	3
350–390	5	4
300–340	2	2
250–290	0	0
200–240	0	0
TOTAL	119	100

a. *True* or *false*.   There is at least one student who scored 800 on the exam.   **False**

b. What is the lowest possible score in the 70th percentile?   **650**

c. Determine the smallest and largest possible values for the median of scores.   **smallest 600, largest 640**

19. The mean height of the 25 students in Mr. Kolowski's 3rd-grade class is 121 cm, and the mean height of the 20 students in Miss Jackson's 2nd-grade class is 116 cm. Find the mean height of the combined group of 45 students. *(Lesson 1-4)*   **about 119 cm**

20. Let $p_1 = 2$, $p_2 = 3$, $p_3 = 5$, $p_4 = 7$, $p_5 = 11$. Show that
$$\sum_{i=1}^{5}(p_i^2) \neq \left(\sum_{i=1}^{5}p_i\right)^2 . \ \textit{(Lesson 1-4)} \quad \sum_{i=1}^{5}(p_i^2) = 208 \neq 784 = \left(\sum_{i=1}^{5}p_i\right)^2$$

21. Give an equation of the line through $(-1, -7)$ and $(5, -8)$.
(Previous course)   **sample: $x + 6y = -43$**

22. Evaluate $2\pi\sqrt{\dfrac{(1.3 + 1.7)^2}{1.3^2 + 1.7^2}}$ to the nearest thousandth.   *(Previous course)*
8.808

Exploration

23. a. Locate a histogram from a publication.
b. Decide whether the histogram distorts or accurately represents the data.
c. Write a short paragraph summarizing what it shows.
Answers will vary.

LESSON MASTER 1-6

■PROPERTIES *Objective C (See pages 72–78 for objectives.)*

1. The table at the right gives the relative frequency of arrests by age group in the U.S. in 1988.

Age	Relative Frequency
Under 15	.05
15–19	.21
20–24	.21
25–29	.18
30–34	.14
35–39	.09
40–44	.05
45–49	.03
50–54	.02
55 and over	.03

Source: Information Please Almanac: Atlas and Yearbook 1990

a. *True or false* About 90% of those arrested are between the ages of 15 and 54.
True

b. Which age intervals accounted for the largest frequency of arrests?
15–19 and 20–24

■USES *Objective F*

2. Below are the numbers of delegates to the 1988 Democratic National Convention from the 50 states, the District of Columbia, "Democrats Abroad," American Samoa, Guam, Puerto Rico, and the Virgin Islands.

65	17	43	48	368	55	63	19	154	94	28	24	200	91
61	45	65	76	29	84	119	162	91	47	88	28	30	23
22	125	30	292	95	22	183	56	54	202	28	53	20	84
212	28	20	86	77	47	91	18	24	22	12	10	61	11

a. Complete the frequency distribution table below for these data.

Number of Delegates	0–39	40–79	80–119	120–159	160–199	200–239	240–279	280–319	320–359	360–399
Frequency	21	16	10	2	2	3	0	1	0	1

b. Find the mean. **75.0**   **c.** Find the median. **54.5**

■REPRESENTATIONS *Objective I*

3. Below are the 1950 and the projected 2075 age distributions of the population in the U.S.

(Source: U.S. Census Bureau)

Functions, Statistics, and Trigonometry © Scott, Foresman and Company   **11**

Lesson MASTER 1-6 (page 2)

a. The total population in 1950 was about 151.1 million. In what age group was the median age?   **30–39**

b. How does the relative number of children (ages 0–9) in 1950 compare with the relative number of children projected in 2075?   **1950: 19%; 2075: 11%**

c. In what ways are the population distributions different?
sample: even distribution in a wider range of ages (0–69) in 2075 than in 1950 (10–59). Children make up the most crowded age group in 1950; 30–39-year bracket is the most crowded in 2075.

■REPRESENTATIONS *Objective J*

4. The following table gives the number of deaths due to heart disease and cancer in the U.S. in 1986.

Age	Number of Deaths (in 1000s) Due to Heart Disease	Cancer
15–24	1.1	2.1
25–34	3.7	5.6
35–44	12.4	15.0
45–54	33.0	37.8
55–64	94.3	98.9
65–74	180.8	146.8
75–84	238.0	116.6

Source: Statistical Abstract of the United States 1990

a. Draw a histogram for the 2nd column and a histogram for the 3rd column.

c. Discuss the shapes of these distributions, including measures of center.
sample: The first one is increasing; mean/median are high, probably 65–74. The second one is more evenly distributed over 55–84-year age range, although still heavily skewed for later ages. Mean/median is probably 55–64.

12   *Functions, Statistics, and Trigonometry © Scott, Foresman and Company*

RESOURCES
■ Lesson Master 1-7
▣ Computer Master 1

OBJECTIVES

Letter codes refer to the SPUR Objectives on page 3A.

F Use statistics to describe data sets or to compare or contrast data sets.

J Draw graphs to display data.

TEACHING NOTES

Technology There is no better way to prepare for a class session on computers than to have done the examples and problems yourself on the machine. Experiment with available options, make mistakes so that you learn how to recover, and generally enjoy the power of the tools.

Small Group Work Encourage students to work on computers in pairs or small groups. All students must get appropriate experience at this point in the course. Ensure that every student in a group has an opportunity to use the keyboard.

In almost every class there is a student who is quite proficient at computers. (A survey taken in 1989 found that about 25% of all households in the U.S. had personal computers.) Students who are very proficient at computers can help others when problems arise. Things work more smoothly when there are several people able to answer students' questions.

You and your students may need to refer to hardware and software manuals. Keep them accessible.

LESSON

1-7

Using a Statistics Package

It can be very tedious and time-consuming (as you may have already realized!) to organize and analyze large sets of data. When calculating statistics with paper and pencil or even with a scientific calculator the possibility of making errors is great. To analyze data quickly and accurately, a computer is indispensable. Nowadays many calculators have the power of small computers to organize, analyze, and display data.

The purpose of this lesson is to familiarize you with features common to most *statistics packages* for computers or calculators with statistical features. A **statistics package** is a collection of programs for organizing, analyzing, and displaying data.

Virtually all statistics packages have three types of statistical features: data entry and editing, calculating various statistics, and displaying data in tables or graphs. Some also allow you to get **hard copy** of your work. That is, when a printer is connected to your computer it will print out data, text, or displays you have created on the screen.

Sometimes there is a data set stored already in the package or on your system. Such a set is called a **demonstration file**. You can use the demonstration file to practice other features without having to enter data directly. Of course, you must learn how to enter your own data. Most times you will enter data directly from a keyboard or keypad; but sometimes your teacher may prepare *files* for you to use. A **file** is a program or a set of data organized on a disk. A file is accessed by its **file name**. To **edit** a file means to change values (perhaps to correct a typing error) or to add or delete data set values.

President	Age
1 T. Roosevelt	42
2 Taft	51
3 Wilson	56
4 Harding	55
5 Coolidge	51
6 Hoover	54
7 F. Roosevelt	51
8 Truman	60
9 Eisenhower	62
10 Kennedy	43
11 Johnson	55
12 Nixon	56
13 Ford	61
14 Carter	52
15 Reagan	69
16 Bush	64

All statistics packages and statistical calculators will compute measures of center. Most do minimum and maximum values; some will calculate percentiles and other measures.

For instance, we used a statistics package to enter the names of the 20th-century presidents of the United States and their ages at inauguration. The data are shown at the left.

When the appropriate keys were pressed, the package automatically calculated certain statistics for the data. Some should be familiar to you: *mean* in the top row; *minimum, maximum,* and *range* in the bottom row. Three other statistics are obvious: *count* refers to the number of entries, *sum* is the total of the ages, and *missing* indicates incomplete data. Others, such as standard deviation (Std. Dev.) and variance will be studied in the next lesson. Still others, such as the sum of the squares of the data (Sum Squared) will not be used much in this course, nor will "standard error" and "coefficient of variance."

46

x₁: Age

Mean:	Std. Dev.:	Std. Error:	Variance:	Coef. Var.:	Count:
55.125	7.177	1.779	50.65	12.91	16

Minimum:	Maximum:	Range:	Sum:	Sum Squared:	No. Missing:
42	69	27	882	49380	0

The most powerful packages create all the statistical displays you have learned thus far. Other computer software and graphing calculators have more limited graphics options. The statistics package we used for this lesson can make scatter plots, bar graphs, pie charts (circle graphs), line graphs, and box plots.

Most packages have some **default settings**, that is, settings which are set automatically and stay that way until the user changes them.

Example Show the data on the inaugural ages of 20th-century U.S. presidents in
a. a box plot,   **b.** a histogram.

Solution

a. On the package used in this lesson you would choose "Box Plot." The plot is shown below. (This package presents the box plot using a vertical number line; in Lesson 1-5 the number line was horizontal.) The manual for the software explains that the ends of the whiskers are drawn to the numbers at the 10th and 90th percentiles, rather than to the extremes. Any data points outside this range, in this case the 42, 64, and 69, are shown as small circles.

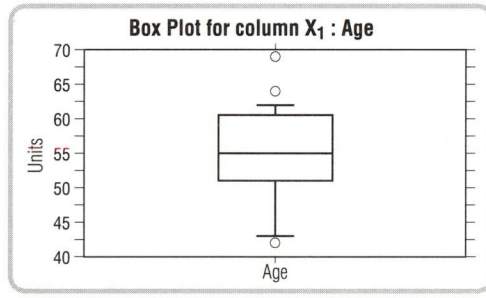

b. The histogram at the left below was produced with the default settings. Because of the scale used, it is hard to interpret the meaning of the histogram. The histogram below on the right is drawn by the same software after specifying 6 intervals of width 5 starting at age 40.

LESSON 1-7   *Using a Statistics Package*   **47**

Some things you should think about as preparation for using computers include the following.
a. Get to know the person in charge of the computer lab. Give that person a copy of the software and the text.
b. Startup and shutdown procedures for systems depend on the lab environment. Do your machines have hard disks? Are they networked together? Are you working off a minicomputer or a mainframe?
c. Save data sets in the event you need to use them again. Encourage students to save data on their own diskettes.
d. Have a plan for managing the use of printers. Students do not need to print everything that appears on the screen. However, when hard copy is required, there is the potential for everyone wanting access at the same time.

Computer programs vary in the way they handle outliers. In some cases, outliers are not displayed at all. The extremes in a box plot may represent the extremes of the data, as the ones in the lesson do, or they may represent the 5th and 95th, or 10th and 90th, percentiles. Other programs use more complex formulas.

Your software probably has default settings for the histogram which divides the interval from a to b into n boxes. This setting may be difficult to interpret for a given example if $\frac{b-a}{n}$ is not a multiple of a convenient power of 10.

If your students have calculators with statistical capabilities, they should learn how to use the stat (statistical) mode. This is especially important for students who do not have access to computers. They can use the data about the ages at the time of inauguration of 20th century U.S. Presidents and compare their results with those given in the lesson. Discuss appropriate uses of technology. If the data set is small, a calcu-

47

Yearly Average Daily Temperature (°F) for Selected Cities

Yearly average daily temperature (°F)

In part **b** of the Example, you must still figure out if each interval contains its left or right endpoint. Your manual should contain such information; and in many cases you can figure this out from the original data. For instance, the data set for the presidents has seven ages between 50 and 55: 51, 51, 51, 52, 54, 55, and 55. If the bar between 50 and 55 contained the right endpoint but not the left (that is, $50 < x \le 55$), the bar would have to show a frequency (count) of 7. But the bar is only 5 units high, so the interval must contain its left endpoint, that is, $50 \le x < 55$.

Notice that both of the axes in the histograms are labeled. Some calculator and computer screens are so small that axes are not labeled. In general, because hardware and software vary considerably, consult manuals and talk with your teacher and friends for specific details about your system.

Questions

Covering the Reading

In 1–3, tell what the term means with respect to a statistics package or statistical calculator.   **See margin.**

1. demonstration file   **2.** hard copy

3. Name three features found in virtually every statistics package.

In 4 and 5, *true* or *false*.

4. Some computers and calculators do not label axes in statistical graphs.
   True

5. All statistics packages can draw box plots.   **False**

6–12. Correct answers will vary depending on statistics package used.

Applying the Mathematics

6. Which of these statistics does your statistics package calculate?
   a. mean   **b.** median   **c.** mode
   d. minimum   **e.** maximum   **f.** range

7. Which of these graphs will your statistics package make?
   a. bar graph   **b.** circle graph   **c.** box plot
   d. histogram   **e.** scatter plot   **f.** line graph

Does your statistics software have some sample or demonstration files? If so, do questions 8–12, otherwise, skip to Question 13.

In 8–12, open the demonstration file and choose a variable. Do the following exercises.

8. Sort the variable values in ascending order. If possible, obtain a printout of the screen.

9. For the chosen variable, find the
   a. mean,   **b.** median,   **c.** mode.

10. Create a histogram of the variable. Obtain a printout of the histogram appearing on the screen.

48

11. Use the same variable and construct a box plot.
    a. Obtain a hard copy of the box plot.
    b. Give the upper quartile.
    c. Give the lower quartile.
    d. Does your program indicate outliers? If so, what are they?

12. A scatter plot requires pairs of data. Choose two variables you think might be related. Use one for the horizontal or *x*-axis and the other for the vertical or *y*-axis. Graph the scatter plot and if possible obtain a print-out of the screen. In what way, if any, do the variables seem related?

13. Use the data about the ages of 20th-century U.S. presidents at the time of inauguration. **See margin.**
    a. Enter the data.
    b. Calculate and record all the measures your statistics package can do for these data, and which you have studied.
    c. Make at least one of the following: box plot or histogram.

14. Below are the amounts of snowfall in inches in Syracuse, NY, from the winter of 1942–43 through the winter of 1989–90. (Source: National Weather Service: Syracuse, NY.) To determine the year, read the table from left to right and from top down. That is, in 1942–43 it snowed 76.5 inches; in 1943–44 it snowed 66.5 inches; and so on.

76.5	66.5	128.7	67.8	110.6	75.5
76.6	118.0	92.8	100.5	77.5	85.9
101.4	146.8	76.1	141.1	137.2	134.8
130.5	77.3	116.5	83.8	97.3	118.8
83.0	81.2	97.9	125.5	157.2	133.7
81.2	123.2	105.5	95.8	145.0	161.2
118.5	93.4	79.0	123.1	66.0	113.6
116.4	104.9	93.5	111.4	97.8	162.0

*Inauguration of
Theodore Roosevelt*

    a. Enter the data, check it for accuracy, and edit as necessary before going on.
    b. Calculate the mean annual snowfall in Syracuse during this 48-year period. **106.4 inches**
    c. Make a graph of your choice of the data. **See margin.**
    d. Write a few sentences describing how much it typically snows in Syracuse in the winter and how the amount of snowfall varies. **See margin.**

Review

15. Consider the data set $\{c, -3c, 7c, 2c, 6c, -c\}$, $c > 0$. Find the
    a. range **10c**   b. median **1.5c**   c. mean. *(Lesson 1-4)* **2c**

16. *Multiple choice.* Which expression is used to calculate the mean $\overline{X}$ of n data values? *(Lesson 1-4)*

    (a) $\frac{1}{\overline{X}}\sum_{i=1}^{n} X_i$ (b) $\frac{1}{n}\sum_{i=1}^{n} X_{n-1}$ (c) $\frac{1}{n}\sum_{i=1}^{n} X_i$ (d) $\frac{1}{X_i}\sum_{i=1}^{n} n$ **(c)**

17. In a capture-recapture experiment to determine the number of elk in a national park, a biologist captured and marked 50 elk and returned them to the park. One week later, 80 elk were captured of which 24 were marked. Give an estimate for the number of elk in the park. *(Lesson 1-1)*
    about 167 elk

LESSON 1-7 *Using a Statistics Package* **49**

Snowfall in Syracuse, NY
(1942 – 1990)

18. Suppose 10 students each tossed a coin repeatedly and recorded the number of heads which occurred.
    a. If the mean number of heads obtained was 18.2, find the total number of times the coin came up heads.   **182 times**
    b. What is a good guess for the total number of coin tosses for all 10 students? *(Lesson 1-4)*   **360 tosses**

In 19 and 20, refer to the circle graphs below which describe the variability in the ratio of students per microcomputer in U.S. public schools in 1989. A ratio of students to micros in the range from 1:1 to 29:1 is considered acceptable by many experts. (Many people would say that 1:1 is ideal.) In 1989 there were about 15,400 public senior high schools and 12,500 public junior high schools. *(Lesson 1-2)*

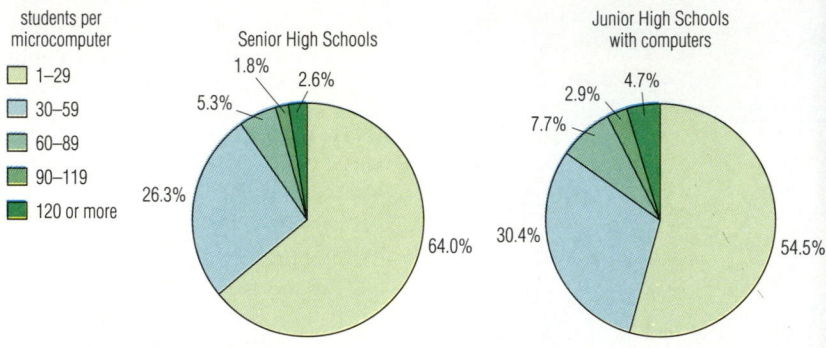

Source: Quality Education Data, Inc., 1989

19. About how many senior high schools had acceptable levels of microcomputers in 1989?   **about 9,860**

20. About how many junior highs did not have acceptable levels of available microcomputers?   **about 5,690**

21. The initial temperature of an apartment at 5:00 P.M. was 86 when the air conditioner was turned on. At 8:30 P.M. the temperature was 78. What was the average rate of change in temperature?   *(Lesson 1-3)*
    about -2.3° per hour

22. *Skill sequence.*   Solve. *(Previous course)*
    a. $3p - 33 = 30$ b. $3k - 33 = 30k$ c. $3n^2 - 33 = 30n$
    $p = 21$ $k = -11/9$ $n = 11$ or $n = -1$

Exploration

23. Consult the National Weather Service in your area. Get some data on amounts of snowfall for the past 10 to 15 years. Calculate the same statistical measures and make the same type of graph as you made in Question 14. Write a few sentences comparing and contrasting the snowfall in your area with the snowfall in Syracuse. (If it seldom snows where you live, choose data from someplace where it does snow.)
    Answers will vary.

1-8

Variance and Standard Deviation

Consider the dot frequency distributions below with the heights of the ten players on two women's college basketball teams.

Macaws

Sweet Dreams

Note that each team has a mean height of 74". In this way the heights are quite similar. However, although the ranges are equal ($79 - 69 = 10$ in each case), the spread of the heights is different in each distribution. The heights of the Macaws seem more spread out than those of the Sweet Dreams.

The interquartile range, which uses the median as a measure of center, is one *measure of spread*. Two other measures of spread, which describe how spread out the scores are in relation to the mean, are the *variance* and *standard deviation*. Both the variance and standard deviation are calculated from the **deviation**, or difference of each data value from the mean. The variance is roughly the average of the squared deviations. The **standard deviation** is the square root of the variance. Because each of these measures is based on the mean, they are used only when it makes sense to calculate a mean.

Here is an algorithm for calculating the variance and standard deviation for a data set with *n* numbers.

1. Calculate the mean of the data.
2. Find the deviation (difference) of each value from the mean.
3. Square each deviation and add the squares.
4. Divide the sum of squared deviations by $n - 1$. This is the variance.
5. Take the square root of the variance. This is the standard deviation.

RESOURCES
■ Lesson Master 1-8

OBJECTIVES

Letter codes refer to the SPUR Objectives on page 3A.
A Calculate measures of center and spread for data sets.
B Use Σ-notation to represent a sum, mean, variance, or standard deviation.
C Describe relations between measures of center or measures of spread.
F Use statistics to describe data sets or to compare or contrast data sets.

TEACHING NOTES

Making Connections
After this lesson students will have studied three measures of center (mean, median, and mode) and four measures of spread (range, interquartile range, variance, and standard deviation).

It is to a student's advantage to associate the steps in the algorithm on page 51 with the order of operations in the formula on page 52. Spend time discussing deviation from the mean and the effect of squaring those deviations. It is important for students to understand that variance is always greater than or equal to zero.

On page 52, the overbar is used in $14.\overline{2}$ to denote a repeating digit and in $\overline{x}$ to denote the mean. Students should realize that the uses are unrelated.

Example 1 Find the variance and standard deviation for the heights of the Macaws.

Solution
To find the variance and standard deviation by hand it often helps to organize the work in columns.
(1) Write the heights in a column, as shown below. Find the mean by adding these numbers and dividing by 10.
(2) In a second column, record the result of subtracting the mean from each score. Some deviations are positive; others, negative or zero.
(3) Square each deviation; record each result and the sum in a third column.
(4) Divide the sum in column 3 by n − 1, in this case 9, to get the variance.
(5) Find the square root of the variance to get the standard deviation. The results of these steps are shown below.

Height	Deviation	Square of deviation
69	−5	25
70	−4	16
70	−4	16
72	−2	4
73	−1	1
74	0	0
77	3	9
78	4	16
78	4	16
79	5	25
Total 740		128

The mean is $\frac{740}{10} = 74$ in. The variance is $\frac{128}{9} = 14.\overline{2}$ in.^{2} and the standard deviation is $\sqrt{14.\overline{2}} \approx 3.77$ in.

Notice that the unit for the variance is the square of the unit for the original variable, but the unit for the standard deviation is the same as that of the original variable.

The variance and standard deviation can be described using Σ-notation. Recall that for a data set with numbers $\{x_1, x_2, x_3, \ldots, x_n\}$, the mean $\bar{x}$ is

$$\bar{x} = \frac{\sum_{i=1}^{n} x_i}{n}.$$

For the same set of data, each deviation from the mean can be written as $x_i - \bar{x}$, and the square of the deviation as $(x_i - \bar{x})^2$. The variance and standard deviation can then be written as shown in the following definitions.

52

Definitions

Let $\bar{x}$ be the mean of the set $\{x_1, x_2, ..., x_n\}$. Then the **variance** s^2 and **standard deviation** s are given by

$$s^2 = \frac{\sum_{i=1}^{n}(x_i - \bar{x})^2}{n-1} \quad \text{and} \quad s = \sqrt{s^2} = \sqrt{\frac{\sum_{i=1}^{n}(x_i - \bar{x})^2}{n-1}}.$$

Example 2   Use the definition above to calculate s^2 and s for the heights of the Sweet Dreams.

Solution   Follow the same steps used in Example 1. In the table below, column i at the left labels the individual whose height is used. The symbols x_i, $x_i - \bar{x}$, and $(x_i - \bar{x})^2$ represent the height, deviation from the mean, and the square of the deviation, respectively.

i	x_i	$x_i - \bar{x}$	$(x_i - \bar{x})^2$
1	69	-5	25
2	73	-1	1
3	74	0	0
4	74	0	0
5	74	0	0
6	74	0	0
7	74	0	0
8	74	0	0
9	75	1	1
10	79	5	25

Total: $\quad \sum_{i=1}^{10} x_i = 740 \qquad\qquad \sum_{i=1}^{10}(x_i - \bar{x})^2 = 52$

The mean: $\bar{x} = \dfrac{\sum_{i=1}^{10} x_i}{10} = \dfrac{740}{10} = 74$ inches.

The variance: $s^2 = \dfrac{\sum_{i=1}^{n}(x_i - \bar{x})^2}{n-1} = \dfrac{52}{10-1} = 5.\overline{7}$ square inches.

The standard deviation: $s = \sqrt{\dfrac{\sum_{i=1}^{n}(x_i - \bar{x})^2}{n-1}} = \sqrt{5.\overline{7}} \approx 2.40$ inches.

Because the deviations are squared, values of data farther from the mean contribute more to the variance than values close to the mean. The heights of the Macaws differ more from the mean than the heights of the Sweet Dreams. As a result the variance and standard deviation for the data about the Macaws are greater than the corresponding statistics for the data about the Sweet Dreams. In general, groups with most data close to the mean have smaller standard deviations than do groups with most data far from the mean.

Below are three relative frequency histograms, each displaying a data set with mean 15 but different standard deviations. Each bar represents the percent of observations falling within an interval containing its left endpoint. For example, in histogram C the tallest bar contains values x such that $2.5 \leq x < 7.5$. The three histograms illustrate the fact that the more the data vary from the mean, the larger the standard deviation.

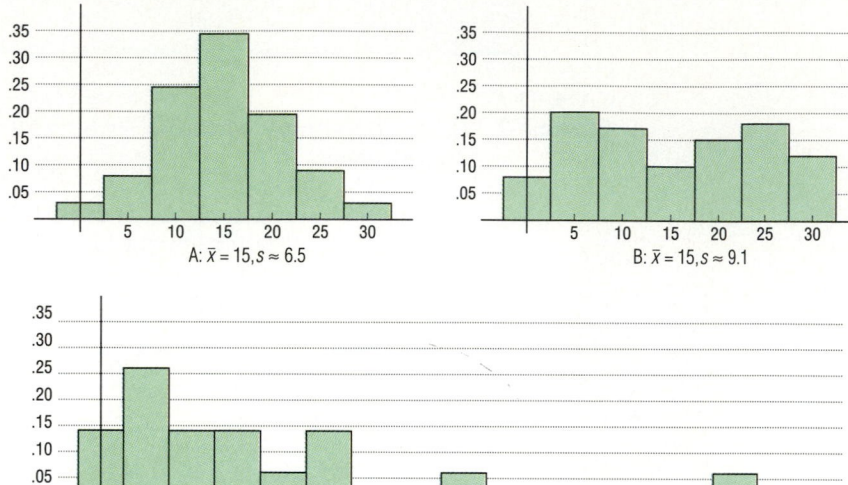

A: $\bar{x} = 15, s \approx 6.5$

B: $\bar{x} = 15, s \approx 9.1$

C: $\bar{x} = 15, s \approx 16.2$

In some books, $s^2 = \dfrac{\displaystyle\sum_{i=1}^{n}(x_i - \bar{x})^2}{n-1}$ is called the *sample variance*, and the symbol σ^2 (σ is the lower case Greek letter sigma) is used to represent the *population variance*: $\sigma^2 = \dfrac{\displaystyle\sum_{i=1}^{n}(x_i - \bar{x})^2}{n}$. Note that the denominator in the population variance is n, rather than $n-1$. Then the symbol σ is used to represent the population standard deviation: $\sigma = \sqrt{\sigma^2} = \sqrt{\dfrac{\displaystyle\sum_{i=1}^{n}(x_i - \bar{x})^2}{n}}$. So some statistics packages have two sets of symbols: s^2 and s, and σ^2 and σ. Other calculators and programs use only the formulas for σ and σ^2. Because we deal mostly with samples in this book, we use the formulas for sample variance and standard deviation, unless stated otherwise. Check your calculator or statistics software using the following set of test scores.

$$90, 80, 75, 68, 100, 92, 85, 82$$

You should find that the mean is 84. If your statistics package gives a standard deviation of about 10.1, it is using a formula equivalent to the one used in this lesson. If it gives about 9.4, it is using a formula equivalent to the population standard deviation, that is, a denominator of n instead of $n-1$.

ADDITIONAL ANSWERS

6. $\sigma^2 = \dfrac{\sum\limits_{i=1}^{n}(x_i - \overline{x})^2}{n}$

Covering the Reading

1. State whether the following are measures of center or measures of spread.
   a. mean   **center**
   b. range   **spread**
   c. standard deviation   **spread**
   d. median   **center**
   e. variance   **spread**
   f. interquartile range   **spread**

2. *Multiple choice.*   The standard deviation of a set of scores is
   (a) the sum of the deviation scores.
   (b) the difference between the highest and lowest scores.
   (c) the score that occurs with the greatest frequency.
   (d) the average spread of the scores with respect to the mean.
   (e) none of (a)–(d)   **(e)**

3. For the Macaws and Sweet Dreams, give the difference of
   a. the means   **0 inches**
   b. the ranges   **0 inches**
   c. the standard deviations.   **about 1.37 inches**

4. Suppose you know the distance in miles each student in a class lives from school. For this data set, state the units for the
   a. mean   **miles**
   b. range   **miles**
   c. variance   **square miles**
   d. standard deviation.   **miles**

5. Find the standard deviation of each data set.
   a. 4, 7, 11, 13, 15   **about 4.5**
   b. 8, 9, 10, 11, 12   **about 1.6**

6. Give the formula for the population variance σ used in some programs.
   See margin.

7. The mean of a group of high school students on the SAT (Scholastic Aptitude Test) Verbal section is 453.
   a. Jerry has a deviation from the mean of -23. What is his SAT Verbal score?   **430**
   b. Sam has a squared deviation of 289. What is his SAT Verbal score?   **470 or 436**
   c. Whose score would contribute more to the variance, Jerry's or Sam's?   **Jerry's**

8. Suppose two samples have the same mean, but different standard deviations s_1 and s_2, with $s_1 < s_2$. Which sample will show more variability?
   the second sample

Applying the Mathematics

9. Perry found the variance of a data set to be -27. Why must his answer be wrong?   **Variance is the sum of squared deviations divided by $n - 1$, so it must be nonnegative.**

10. a. Consider the weights in kilograms of a group of teenagers. If the standard deviation is 8.2 kg, what is the variance?
    b. If the variance is 20 kg^2, what is the standard deviation?
    a) 67.24 kg^2; b) about 4.5 kg

11. Use the following hypothetical frequency distributions for ACT (American College Test) scores.

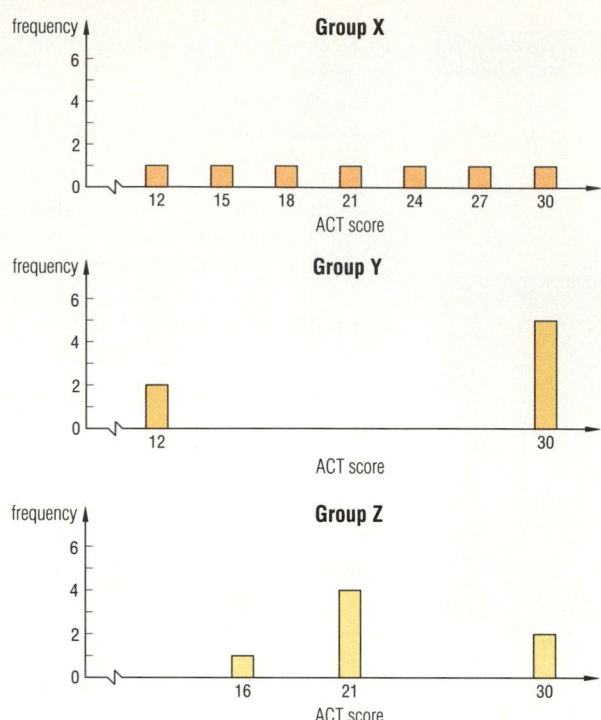

a. Match each group with its best description.
   i. consistently near the mean **Group Z**
   ii. very widely spread   **Group Y**
   iii. evenly distributed   **Group X**
b. Without calculating, tell which group's ACT scores have the greatest standard deviation and which have the smallest.
c. Verify your answer to part **b** with calculations.
   b) See margin. c) $s_X \approx 6.5$, $s_Y \approx 8.8$, $s_Z \approx 5.2$

12. For a data set, suppose the sum of squared deviations about the mean is 500. Find the value of $\sqrt{\dfrac{500}{n}}$ and $\sqrt{\dfrac{500}{n-1}}$ for the following values of n.
   a. 5   **See margin.**   b. 50   **See margin.**   c. 500   **See margin.**
   d. *True* or *false*.   For large values of n, dividing by $n-1$ gives nearly the same result as dividing by n.   **True**

13. In 1988, there were more than 1.1 million students in the United States who took the SAT. On the mathematics section $\bar{x} = 476$ and $s = 120$. Students receive scores rounded to the nearest 10. What is the interval of student scores that lie within one standard deviation of the mean?
   360 to 600

For 14 and 15, use the following data.

Time (in seconds) for 20 sixth graders to run 200 meters									
70	80	80	85	90	100	100	100	100	100
100	105	105	105	120	130	130	130	140	150

14. Find a group of five running times whose standard deviation is as small as possible.   **all 5 times of 100 seconds**

15. Find a group of four running times whose standard deviation is larger than 25 seconds. Compute the standard deviation.   **sample: 70 seconds, 80 seconds, 140 seconds, 150 seconds; $s \approx 40.8$ seconds**

16. *Multiple choice.*   A class of students is said to be *homogeneous* if the students in the class are very much alike on some measure. Here are four classes of students who were tested on a 20-point spelling test. Which class is the most likely to be homogeneous with respect to spelling?
(a) $n = 20$ $\overline{X} = 15.3$ $s = 2.5$
(b) $n = 25$ $\overline{X} = 12.1$ $s = 5.4$
(c) $n = 18$ $\overline{X} = 11.3$ $s = 7.9$
(d) $n = 30$ $\overline{X} = 10.4$ $s = 3.2$ **(a)**

17. Anita computed $\sum_{i=1}^{15}(X_i - \overline{X})^2$ to be 850 for a set of data. Find, if possible,
   a. the number of elements in the data set   **15**
   b. the mean   **not possible**
   c. the variance   **about 60.7**
   d. the standard deviation.   **about 7.8**

Review

18. *Multiple choice.* $\sum_{i=1}^{n} X_i$ equals

   (a) $\overline{X}$ (b) $\dfrac{\overline{X}}{n}$ (c) $\dfrac{n}{X}$ (d) none of (a)–(c).   *(Lesson 1-4)*   **(d)**

19. If the interval from a to b is split into n equal parts, the leftmost part is from a to $a + \dfrac{b-a}{n}$. Write the expression $a + \dfrac{b-a}{n}$ as a single fraction with denominator n. *(Previous course)*   **$\dfrac{(n-1)\,a + b}{n}$**

20. The *1990 Information Please Almanac* reports on the universities with the greatest numbers of library books. Below is a list of the numbers of books in the collections of universities with more than 3 million books. *(Lessons 1-6, 1-1)*

Institution	Volumes	Institution	Volumes
Columbia	5,740,832	U of Calif., Berkeley	7,190,821
Cornell	4,924,421	U of Calif., L. A.	5,812,163
Duke	3,668,935	U of Chicago	4,970,720
Harvard	11,496,906	U of Illinois	7,377,051
Indiana	4,011,675	U of Michigan	6,133,171
Michigan State	3,301,739	U of Minnesota	4,473,262
North Carolina	3,520,273	U of Pennsylvania	3,499,741
Northwestern	3,346,817	U of Texas	5,888,776
Ohio State	4,254,266	U of Virginia	3,003,066
Princeton	4,070,827	U of Washington	4,764,341
Rutgers	3,065,533	U of Wisconsin	4,804,386
Stanford	5,740,162	Yale	8,538,156
U of Arizona	3,329,146		

   a. Suggest a criticism of the level of accuracy of these data.
   b. Describe these data in a histogram of your choosing.
   See margin.

---

b. sample:

Number of Volumes (in millions of books)

NAME _____

LESSON **MASTER   1-8**
QUESTIONS ON **SPUR** OBJECTIVES

■**SKILLS** *Objective A (See pages 72–78 for objectives.)*
1. Find the standard deviation of each data set.

   a. 3, 7, 0, 6, 0, 4, 5, 2, 5, 7   b. 43, 47, 50, 56, 60, 64, 65, 72, 75, 87
   2.60   **13.65**

2. a. Calculate the mean and standard deviation for the set 5, 5, 6, 7, 9, 10, 12, 12.   **mean: 8.25, s.d.: 2.92**

   b. Make up another data set of eight numbers using only 5, 6, 7, 9, 10, and 12 which has the same mean but a smaller standard deviation.
   sample: 5, 6, 7, 7, 9, 10, 10, 12 (mean: 8.25, s.d.: 2.22)

■**PROPERTIES** *Objective B*
3. *Multiple choice*   Which of the following is a correct formula for finding the standard deviation of the data set $a_1, a_2, a_3, ..., a_{20}$?   **(d)**

   (a) $\sqrt{\dfrac{\sum_{i=1}^{20}(a_i - a_{20})^2}{19}}$ (b) $\sqrt{\dfrac{(\sum_{i=1}^{20}(a_i - \bar{a}))^2}{19}}$

   (c) $\sqrt{\dfrac{\sum_{i=1}^{20}(a_i^2 - \bar{a}^2)}{19}}$ (d) $\sqrt{\dfrac{\sum_{i=1}^{20}(a_i - \bar{a})^2}{19}}$

■**PROPERTIES** *Objective C*
4. *Multiple choice*   If y is the variance of a set of n numbers, which of these is the standard deviation of the set?   **(c)**

   (a) $\sqrt{\dfrac{y}{n}}$ (b) $\sqrt{\dfrac{y}{n-1}}$ (c) $\sqrt{y}$ (d) y^2

5. If the standard deviation of a data set is 4.9, what is the variance?   **24.01 (or 24)**

NAME _____
Lesson MASTER 1-8 (page 2)

■**USES** *Objective F*
6. The average monthly precipitation, in inches, for Juneau is given below.

   3.7   3.7   3.3   2.9   3.4   3.0
   4.1   5.0   6.4   7.7   5.2   4.7
   Source: *The World Almanac and Book of Facts 1989*

   a. Find the mean and standard deviation of these data.   **mean 4.43   s.d.: 1.46**

   b. How many of these data are within one standard deviation of the mean?   **9**

7. a. The heights, in feet, of the 20 tallest buildings in the U.S. are given below.

   1454   1377   1250   1136   1127   1046   1002
   985   952   915   900   859   858   853
   851   850   841   813   808   792
   Source: *Information Please Almanac Atlas and Yearbook 1990*

   Find the standard deviation of these heights.   **193.1**

   b. The heights, in feet, of the 21st to 41st tallest buildings in the U.S. are given below.

   790   779   775   775   774   765   754
   750   750   747   744   743   730   725
   724   714   710   708   707   705   700
   Source: *Information Please Almanac: Atlas and Yearbook 1990*

   Find the standard deviation of these heights.   **28.1**

   c. Which set of heights is more variable based on the answer to parts **a** and **b**? Explain why this seems reasonable.
   The 20 tallest buildings have a standard deviation about 7 times as much as the 21st to 41st tallest buildings.

MORE PRACTICE

For more questions on SPUR Objectives, use *Lesson Master 1-8,* shown on page 57.

EXTENSION

It is sometimes possible to find the interquartile range, but not be able to determine the variance and the standard deviation. This is because the former is based on the median; whereas, the latter are determined by the mean. The following data set is an example that satisfies these conditions.

Ages of children	Frequency
1	2
2	1
4	2
5	3
6	2
over 6	1

The median is 5; the IQR is 4. The mean, variance and standard deviation cannot be determined because the age of the child over 6 is not known. Provide the above set of data to students and have them discover which measures of center and spread can or cannot be found.

ADDITIONAL ANSWERS

21.a. sample:

b. The first shows more variability.

22.b. sample:

---

21. Two data sets of heights of people each have minimum = 50″, median = 67″, and maximum = 80″. One has interquartile range = 15″; the other has IQR = 10″.
   a. Draw possible box plots for each data set.   **See margin.**
   b. Which data set shows more variability? *(Lessons 1-5, 1-4)*   **See margin.**

In 22 and 23, the percent of Advanced Placement Examinations in Mathematics or Computer Science taken by women is given below. Source: The College Board, *AP Yearbook, 1989.*

Year	1974	1979	1984	1989
Percent Women	26	32	35	36

22. a. *Multiple choice.*   Which of the following would be an appropriate graph for representing these data?
   (i)   box plot   (ii)   circle graph
   (iii) histogram   (iv) line graph   **(iv)**
   b. Draw such a graph.   *(Lessons 1-6, 1-5, 1-3, 1-2)*   **See margin.**

23. The total number of students taking AP Exams in Mathematics or Computer Science was about 44,000 in 1984 and about 89,000 in 1989.
   a. What was the average annual increase in the number of women taking AP Exams in these areas during this period?
   b. What was the average annual increase in the number of men taking these exams in this period? *(Lesson 1-3)*
   a) about 3300 women per year; b) about 5700 men per year

Exploration

P. L. Chebychev

24. The Russian mathematician P. L. Chebychev proved a remarkable theorem which implies that in *any* data set, if p is the portion of data that lies within k standard deviations to either side of the mean, then

$$p \geq 1 - \frac{1}{k^2}.$$

   a. According to Chebychev, what percent of a data set must lie within 2 standard deviations of the mean?   **75%**
   b. What percent must lie within 3 standard deviations?   **about 89%**
   c. Test Chebychev's theorem on a data set of your choice.
   Answers will vary.

1-9

Who Wrote *The Federalist* Papers?

James Madison

Alexander Hamilton

Once data are collected and organized they can be used to help answer many questions, sometimes in surprising ways. In fact an analysis of frequency distributions helped decide the authorship of some famous documents in U.S. history, *The Federalist* papers.

The Federalist papers were written between 1787 and 1788 under the pen name "Publius" to persuade the citizens of the State of New York to ratify the Constitution. Of the 85 *Federalist* papers, 14 were known to be written by James Madison, 51 by Alexander Hamilton, and 5 by John Jay. Of the remaining 15, three were joint works and 12 were called "disputed" because historians were unsure whether they were written by Hamilton or Madison. The dispute could not be settled by comparing the ideas in the papers, because at that time the philosophies of the men were similar.

In an attempt to identify which man had authored each of the disputed papers, in the 1960s two statisticians, Frederick Mosteller of Harvard University and David Wallace of The University of Chicago, used computers to count the occurrence of key words in documents by the men. They first examined documents other than *The Federalist* papers known to have been authored by the two men: 48 papers written by Hamilton and 50 papers by Madison. Key words were chosen so that they did not reflect either author's writing style. The words were also chosen to be independent of the context of the paper being studied. For this particular study, the words chosen were 'by,' 'from,' and 'to.'

Then Mosteller and Wallace divided the counts of these key words by the number of words in the document. For example, if the key word 'by' occurred 32 times in a paper of 2075 words, the rate of occurrence was $32/2075 = 0.0154$. They reported each relative frequency as the rate per 1000:

$$(0.0154) \cdot 1000 = 15.4 \text{ occurrences per 1000 words.}$$

LESSON 1-9 Who Wrote *The Federalist* Papers? **59**

LESSON 1-9

RESOURCES
■ Teaching Aid 12: Table and graph for this lesson

OBJECTIVES

There are no SPUR Objectives for any reading lesson.

TEACHING NOTES

At the end of each chapter is a lesson we often call a "reading lesson." The goal is to provide a reading experience like the ones students might have in other courses, in college, in their careers, or simply as educated citizens. This lesson differs from other lessons in the chapter in that it does not have questions Applying the Mathematics; there are no objectives relating to it in the SPUR Review; there is no Lesson Master; and it is not tested in any chapter tests or cumulative tests from the Teacher's Resource File. However, we do not consider it as an optional lesson, and you may find ideas from this lesson mentioned later in the book. For further information see page T37 in the Professional Sourcebook.

The following table shows the results of their counts of the key words 'by,' 'from,' and 'to' in the papers of known authorship. Take a few minutes to read the table before going on.

Frequency Distribution of Rate per Thousand Words in 48 Hamilton and 50 Madison Papers of the words 'by,' 'from,' and 'to'								
'by'			**'from'**			**'to'**		
Rate	H	M	Rate	H	M	Rate	H	M
1–3*	2		1–3*	3	3	23–26*		3
3–5	7		3–5	15	19	26–29	2	2
5–7	12	5	5–7	21	17	29-32	2	11
7–9	18	7	7–9	9	6	32–35	4	11
9–11	4	8	9–11		1	35–38	6	7
11–13	5	16	11–13		3	38–41	10	7
13–15		6	13–15		1	41–44	8	6
15–17		5				44–47	10	1
17–19		3				47–50	3	2
						50–53	1	
						53–56	1	
						56-59	1	
Totals	48	50	Totals	48	50	Totals	48	50

Source: Frederick Mosteller and David L. Wallace (1964), *Inference and Disputed Authorship: The Federalist*, Reading, MA: Addison-Wesley.
*Each interval excludes its upper end point. Thus a paper with a rate of exactly 3 per 1000 words would appear in the count for the 3–5 interval.

Notice that there are three pairs of frequency distributions within the table, one for each of the words 'by,' 'from,' and 'to.' Under each word there are three columns, one for the rate and one for each author (H for Hamilton, M for Madison.) The asterisk (*) by the first rate for each word directs you to the footnote. The footnote explains that a rate of 1–3 means the rate was between 1 and 3 per thousand in this sense:

$$\text{If } r \text{ is the rate per thousand, then } 1 \le r < 3.$$

The numbers under H and M are the frequencies for that rate. For example, under 'by,' the 2 in the first H column means that two of Hamilton's papers used 'by' between 1 and 3 times per thousand. The last row shows the totals.

Overall, in the papers of known authorship, the word 'by' is used much more frequently by Madison. The rate of its occurrence seems to distinguish the two authors. The word 'to' seems more used by Hamilton. In contrast, the use of the word 'from' does not distinguish one man's writing from the other's.

Based on this information, Mosteller and Wallace compared the use of the word 'by' in the disputed *Federalist* papers to its use in the papers known to be authored by Hamilton and Madison. On the next page are relative frequency distributions comparing the use of the word 'by' in the Hamilton papers, the Madison papers, and the disputed papers.

The horizontal axes represent the rates per 1000 words. The vertical axes represent the fraction of papers in each group with the given frequency. For instance, the tallest column on Madison's graph extends to .32. This represents the 16 papers by Madison of the 50 studied which used 'by' between 11 and 13 times per thousand words.

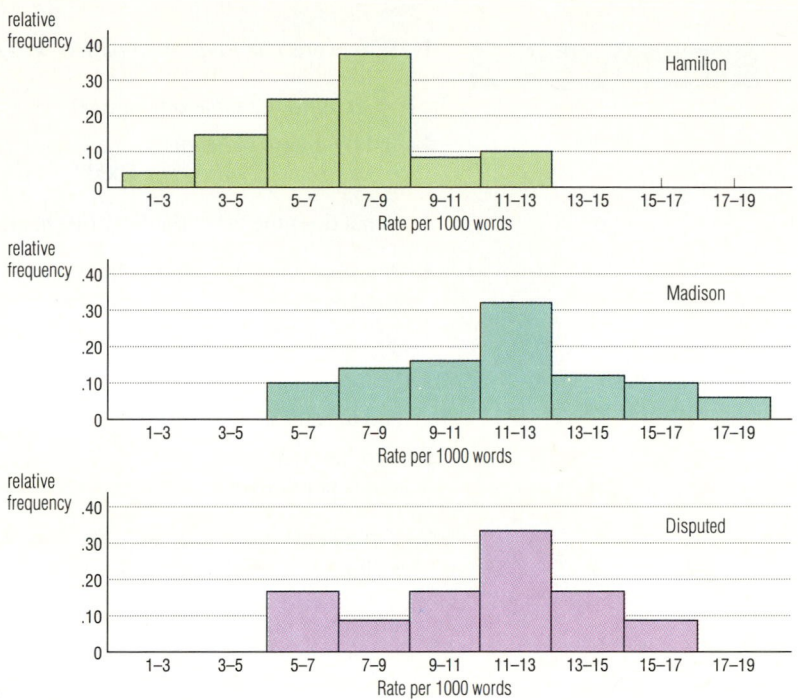

Distributions of rates of occurrence of the word 'by'

The shape of the graph for the disputed papers is clearly more like the shape of Madison's than of Hamilton's. The median rate of occurrence of the word 'by' is 11–13 for Madison's and for the disputed papers, and the median rate is only 7–9 for Hamilton's papers. This evidence suggests that the disputed papers are Madison's. In fact, the full research study demonstrated that it is extremely likely that Madison authored 11 of the 12 disputed papers and probably the 12th as well.

Beware! Statistical reasoning is *inferential*, not deductive. That is, statistical research does not *prove* findings with certainty as in geometry or algebra. Instead, they report what is very *likely*, and give a measure for a level of confidence in that conclusion. For example, Mosteller and Wallace reported that for the *most* disputed paper, their evidence yields odds of 80 to 1 that the author was Madison. Here is what 80 to 1 means: If 81 analyses were done in the same way as theirs on comparable documents, the findings would be in favor of Madison in 80 cases and against him in 1 case. They called these odds strong but not overwhelming. The odds were much higher, and considered to be overwhelming, for the other 11 papers.

Applications of mathematics to history, such as this authorship dispute, are becoming more common. In fact, this new discipline has a name, *cliometrics*, in honor of the Greek muse of history, Clio.

Questions

Covering the Reading

1. Why were *The Federalist* papers written?
   to persuade the citizens of New York to ratify the Constitution

In 2–4, refer to the table on page 60.

2. In how many of Madison's papers was the word 'from' used at a rate between 3 and 5 words per 1000?   **19**

3. What does the 50 in the lower right corner represent?
   50 of Madison's papers were studied.

4. *Multiple choice.*   On the 'to' chart in the first column, a rate per thousand of 38–41 means the rate r is in which interval?
   (a) $38 < r < 41$ (b) $38 < r \le 41$
   (c) $38 \le r \le 41$ (d) $38 \le r < 41$ **(d)**

In 5 and 6, refer to the histograms on page 61.

5. In what percent of the disputed papers was 'by' used between 5 and 7 times per thousand?   **about 16%**

6. What is the mode interval of the frequency distribution of Hamilton's use of the word 'by'?   **7 to 9 times per thousand**

7. *Multiple choice.*   Which best expresses one of the populations and sample sets used by Mosteller and Wallace?
   (a) A sample of 48 papers by Hamilton was used to represent the population of all of Hamilton's writing.
   (b) A sample of 50 papers by Madison was used to represent the population of disputed documents.
   (c) A sample of 12 disputed papers was used to represent the population of all of Madison's writing.   **(a)**

8. *True* or *false.*   Statistical reasoning proves its findings with absolute certainty.   **False**

9. What is the name of the discipline which applies mathematics to history?   **cliometrics**

Review

In the last lesson of each chapter there are no Applying the Mathematics questions. Instead, to help you prepare for the chapter test, there are usually more Review questions than in other lessons.

10. Complete the analogy: Interquartile range is to median as standard deviation is to __?__.   *(Lesson 1-8, 1-5)*   **mean**

11. **a.** Calculate the mean and standard deviation for the set 2, 2, 3, 4, 4, 5, 6, 6.   **mean = 4; standard deviation ≈ 1.6**
    b. Make up another data set of eight numbers using only 2, 3, 4, 5, or 6 which has the same mean but a smaller standard deviation.
    (Lesson 1-8)   **sample: 2, 3, 3, 3, 4, 5, 6, 6**

62

12. Imagine a shoe store with 20 salespersons. Suppose that on a particular day Mr. Webb sold 30 pairs of shoes and that this was the average number sold for all salespersons. Miss Feet sold 8 pairs less than the average and Ms. Slipper sold 40 pairs of shoes.
  a. Which salesperson had sales which contribute the most to variance among salespersons? **Ms. Slipper**
  b. Which salesperson had sales which contribute the least to variance among salespersons? *(Lesson 1-8)* **Mr. Webb**

13. Suppose a data set consists of weights in pounds of hogs shown at a state fair. State the unit for the
  a. mean **lb**   **b.** range **lb**
  c. variance **lb²**   **d.** standard deviation. *(Lesson 1-8, 1-4)* **lb**

In 14 and 15, use your statistics package or calculator as needed.

The table below gives average monthly temperature (*T*) and average monthly precipitation (*P*) for Jakarta, Indonesia and Sydney, Australia from *The Weather Handbook*.

		J	F	M	A	M	J	J	A	S	O	N	D
Jakarta	*T* (°F)	84	84	86	87	87	87	87	87	88	87	86	85
	P (in.)	12	12	8	6	5	4	3	2	3	4	6	8
Sydney	*T* (°F)	78	78	76	71	66	61	60	63	67	71	74	77
	P (in.)	4	4	5	5	5	5	5	3	3	3	3	3

14. a. Find the median, first, and third quartiles of the temperature data of each city. **See margin.**
  b. Use the interquartile range to decide which city shows less variability in temperature during the year. **Jakarta**
  c. Make two box plots using the same scale to illustrate the temperature data. *(Lesson 1-5, 1-4)* **See margin.**

Jakarta, Indonesia

Sydney, Australia

15. a. Find the mean and standard deviation of the precipitation data for each city. **See margin.**
  b. Use the standard deviation to decide which city shows less variability in rainfall during the year. **Sydney**
  c. Make two histograms using the same scale to illustrate the data in precipitation. *(Lesson 1-8, 1-6, 1-4)* **See margin.**

16. Show that $\frac{1}{8} \sum\limits_{i=1}^{8} x_i = \sum\limits_{i=1}^{8} \frac{x_i}{8}$. *(Lesson 1-4)* **See margin.**

LESSON 1-9 Who Wrote *The Federalist* Papers? **63**

Top: Katherine Hepburn and Henry Fonda
Bottom: Marlee Matlin

17. Use the box plot of the ages at which people received Academy Awards (''Oscars'') between 1928 and 1989 for best performance by an actor or actress in a leading role.

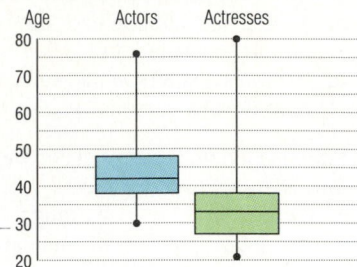

a. Estimate the median age of the men who won Oscars for best performance in a leading role. *(Lessons 1-5, 1-4)* **about 42 years old**

b. The youngest and oldest men to receive Oscars for best performance during this period were Marlon Brando at age 30 for *On the Waterfront*, and Henry Fonda at age 76 for *On Golden Pond*. What is the range of ages for winning actors? *(Lessons 1-5, 1-3)* **46 years**

c. The five-number summary for the data on actresses is: 21, 27, 33, 38, 80. The youngest and oldest actresses to receive Oscars during this period were Marlee Matlin in 1986 at age 21, and Jessica Tandy in 1989 at age 80. Which, if either of these, is an outlier using the $1.5 \times \text{IQR}$ method? *(Lesson 1-5)* **Age 80 is an outlier.**

d. Do gender differences exist in the ages at which men and women win Oscars? **Yes**

e. Write a short paragraph to justify your answer in part **d**. **See margin.** *(Lessons 1-5, 1-4)*

18. The circle graphs below are from a study of about 6700 entering freshmen conducted at Ohio State University in the fall of 1986. The graph on the left shows the college placement of students who had less than four years of college preparatory mathematics (CPM); the one on the right shows the placement of students who had four or more years of CPM. *(Lesson 1-2)*

Mathematics Placement Level by Number of Years of College Preparatory Mathematics

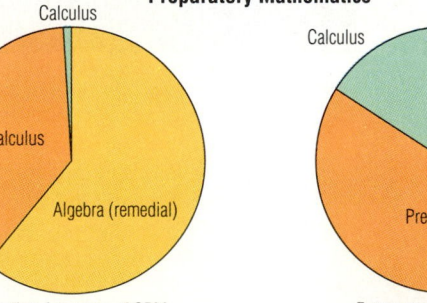

Less than four years of CPM

Four or more years of CPM

64

NOTES ON QUESTIONS
**Small Group Work for
Question 23:** You might
want students to work
together in groups for this
question.

a. Into what mathematics course were students who took less than four years of college preparatory mathematics most often placed?

b. *True* or *false.* Most students who took at least four years of college preparatory math in high school were placed into precalculus.

c. Estimate the percent of students with four or more years of CPM who were placed into calculus. (Hint: Use a protractor.)

a) algebra (remedial); b) True; c) about 17%

19. If the interval from x to y is split into n equal parts, there are $n - 1$ endpoints between x and y.

23.a.,d.
Average Monthly Temperature

● Jakarta temp
○ Sydney temp

a. Give the coordinates of the first, second, and third endpoints.

b. Give the coordinates of the last and next-to-last endpoints.

c. Write the coordinates in parts **a** and **b** as single fractions with denominator n.   *(Lesson 1-6, Previous course)*

See margin.

20. What is the area of an $n°$ sector of a circle with radius 2"? **$(n\pi)/90$ in.^{2}**
(Lesson 1-2, Previous course)

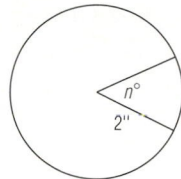

**b. The temperature
decreases and then
increases in a periodic
pattern.**

**c. changes in seasons at
more southerly latitudes**

21. Consider the line $y + 5 = 2(x + 4)$. *(Previous course)*

a. Show that $y = 2x + 3$ is an equivalent sentence.

b. What are the slope and y-intercept of the line?

c. Graph the line.

a) See margin. b) slope = 2, y-intercept = 3; c) See margin.

**e. The temperature in
Jakarta is relatively
constant because Jakarta
is very close to the
equator.**

22. Find an equation of the line passing through $(4, 0)$ and $(-2, 12)$.
(Previous course)   **sample: $y = -2x + 8$**

Exploration

23. Refer to the data for Questions 14 and 15.   **See margin.**

a. Make a scatterplot of month versus temperature in Sydney. Plot the months on the horizontal axis. Repeat the months to show a two-year cycle.

b. Describe the pattern(s) you observe.

c. What physical factors account for these patterns?

d. Plot the temperatures for Jakarta on the same scatter plot.

e. What physical factor(s) account for the relatively constant temperature in Jakarta?

Projects

PROJECTS

The projects provide an opportunity for extended work on topics related to the content of the chapter. See page T37 for more information on the purposes and management of projects.

Solutions to the projects are found at the end of the Additional Answers section at the back of the book.

You may wish to allow students to work in pairs or **small groups** on these projects. Do not expect all students to attempt all the projects. Several projects are suggested so that students may choose ones which particularly interest them.

Detection of misleading graphs or displays **(Project 1)** may be an activity for the class to engage in for the entire year. A bulletin board or class scrapbook could be made of these graphs, along with critiques of each item. Similarly, examples of excellent graphics may be collected and displayed with analyses of their virtues.

Random phenomena of the types suggested for **Project 3** are dealt with analytically in Chapters 7 and 10, when the binomial probability distribution is discussed.

Project 5 follows up the reading lesson (1-9). Statistical inferences like those suggested in Project 5 are also studied in Chapter 10.

Projects provide an opportunity for you to apply the concepts of this chapter to more extended questions and situations. You should plan to spend more time on a project than on typical questions.

1. Collect at least six published graphs or tables. Analyze each display as follows.
   a. Provide the source of the display (who drew or published the graph or table) and the source of the data (who collected the data).
   b. Evaluate how trustworthy the data are. Were they fairly gathered? Are they timely?
   c. Draw at least two conclusions from each display. Explain your reasoning.
   d. Critique the display. That is, describe its strengths and weaknesses. Is it appealing? Accurate? Are there any errors? Is it ambiguous? If there are many weaknesses, create a new display which avoids the original difficulties.

2. What cars are most popular in your community?
   a. Go to a large parking lot (near a shopping center, office building, or school) and classify at least 60 cars by the following criteria: style (van, truck, limousine, sports car, and so on); color; year; and manufacturer.
   b. Report the results with at least 3 tables or displays.
   c. Write a short paragraph summarizing and interpreting your findings.
   d. Describe the differences that might have resulted if you had collected your data at a different location (senior citizens center, executive office garage, or used car lot), or different time (church lot on Sunday, movies on budget night).

3. a. Perform one of the experiments listed below.
      i. Toss a group of ten pennies 100 times. Count the number of heads in each group toss.
      ii. Toss six dice 100 times. Count the number of evens in each set.
      iii. Toss three dice 100 times. Count the number of times each sum from 3 to 18 appears.
   b. Calculate the mean $\bar{x}$ and standard deviation s for the data from each experiment.
   c. Make at least 3 histograms using different-sized intervals on the horizontal axis (for example, by 1s, by 5s, by 10s). Mark $\bar{x}$ on each axis. Describe the different impressions resulting from the various choices of intervals.

4. Conduct a simulation of a capture-recapture experiment. Use congruent objects of two different colors; for example, two colors of beans or marbles.
   a. Place a large, unknown number of items of one color in an opaque container (for example, 2 lb of red beans in a 3 lb coffee can). This is the population for the experiment.
   b. Select a sample (grab a handful) and 'tag' the initial sample by replacing each item with an item of the other color. (That is, replace the red beans by white beans.)
   c. Mix the items well.
   d. Select a second sample, and from it estimate the size of the entire population.
   e. Repeat the experiment 6 to 10 times using different-sized samples.
   f. From **a–e,** what would you estimate the actual size of the population to be?

g. Prepare a table or graph of predicted population size as a function of sample size.

h. Count the items in the population. (That is, count all the beans in the container.) How accurate were your predictions?

i. Write several paragraphs summarizing and interpreting your findings.

5. Obtain three samples of text; one of them some essays you have written recently and two of them chosen from the list below:
   > a book intended for beginning readers;
   > some pages from *The Complete Works of William Shakespeare;*
   > yesterday's newspaper;
   > the first chapter of *Functions, Statistics, and Trigonometry*;
   > another school textbook you are now using.

   Choose a common word of interest and count the number of occurrences of the word in your three samples. Make a relative frequency graph of your results. Present your results in a display to highlight the similarities and differences between texts. Can you conclude that your writing style is closer to one or the other styles? Why or why not?

6. Obtain a copy of either a recent edition of *The Statistical Abstract of the United States* or an almanac. Pick some data that you find interesting or surprising (or both) and that are presented in a table. Design a suitable poster, at least 24″ x 30″ in dimensions, that interprets the data in the table, including some displays to support your interpretation. Make sure that you choose a suitable headline for your poster that will attract people's attention.

7. A preliminary report on the study of smell described on page 3 was in *National Geographic* in October, 1987. Since then many other articles have appeared with reports on further analyses of the original data, and on more recent studies of smell. Prepare a report on what is known about smell, and why this research is important.

8. Compile a database of information about the members of your mathematics class.
   a. Ask each person to complete an information sheet for the following data. Where appropriate, use metric measurements (that is, height and lengths in centimeters).

gender	foot length
age	grade in school
height	circumference of wrist
pulse rate	circumference of neck
eye color	arm span (fingertip to fingertip)
hand span	number of siblings

   b. Construct a computer file with the class database. The database will be used again in later chapters.
   c. Choose at least three variables. Use a statistics package to display and summarize the data. For each variable decide which type of display is most appropriate (box plots, frequency distributions, circle graphs, scatter plots, line graphs, or bar charts). Whenever appropriate, calculate statistics such as the mean, median, standard deviation, quartiles, percentiles, and range.
   d. Write a short paragraph describing a "typical" student in your class in terms of the variables you analyzed. For numerical variables, this will involve interpreting both the center and spread of the distributions.

Project 8, the construction of a class database, is particularly suited to **small-group work,** but may be undertaken as a whole class activity. You should feel free to change the list of attributes in Project 8. We have intentionally included some area measures in anticipation of exploring nonlinear relationships. We have avoided more sensitive variables, such as weight, family income, and parental data. You are in the best position to decide on the merits of particular variables for your class. In Chapter 2, students study the relationships between variables, and the database of **Project 8** will supply information about many different variables.

Summary

Statistics is the branch of mathematics dealing with collecting, organizing, displaying, analyzing, and interpreting numerical information. Samples are frequently used to study the characteristics of large populations. It is important that a sample is truly representative of its population, so that statistical findings can be generalized. Random selection is a common method used to form representative samples and avoid biases.

Data can be organized and displayed in several ways including tables, graphs, and stem-and-leaf diagrams. Pie charts, bar graphs, line graphs, scatter plots, box plots, histograms, and relative frequency histograms are the kinds of graphs studied in this chapter.

Summary statistics help give a quick impression of a data set and allow for comparison of sets. Statistics such as the mean and median (and sometimes the mode) are measures of center. Measures of spread include the range, interquartile range, variance, and standard deviation. The five-number summary, which includes quartiles and extremes, is another way to summarize the center and spread in a data set. The symbol for summation, Σ (sigma), provides a short way to express formulas for means, variances, and standard deviations.

Data sets are frequently large and unwieldly to deal with by hand, so computer assistance is necessary. A statistics package is a sophisticated computer program used to perform statistical tasks very quickly. It is possible to enter, edit, store, and retrieve data files on the computer, which then can be analyzed numerically and graphically.

Statistics can be used to determine the likelihood that patterns in observed data are similar to each other. For example, when the authorship of a work is disputed, statisticians have analyzed frequency distributions of certain words in known works by the possible authors in order to arrive at a likely determination of the author.

Vocabulary

For the starred (*) terms you should be able to give a definition of the term.
For the other terms, you should be able to give a general description and a specific example of each.

Lesson 1-1
statistics
*data, variable
*population, *sample
survey
random, randomly
capture-recapture
   method
biased

Lesson 1-2
table
circle graph, pie chart
bar graph

Lesson 1-3
time-series data
coordinate graph
scatter plot, line graph
*average rate of change
*slope
increasing, decreasing
stem-and-leaf diagram
stemplot
*maximum, *minimum
*range, outlier
back-to-back stemplot

Lesson 1-4
measures of center
measures of central
   tendency
*mean, *$\bar{x}$
*median, *mode
Σ (sigma)
summation notation
sigma-notation
Σ-notation

Lesson 1-5
rank ordered
*quartile
first (lower) quartile
second (middle)
   quartile
third (upper) quartile
interquartile range
   (IQR)
five-number summary
box plot
box-and-whiskers plot
*percentile
$1.5 \times$ IQR method

Lesson 1-6
histogram
*frequency
*relative frequency
frequency distribution
relative frequency
   distribution
frequency table

Lesson 1-7
statistics package
hard copy
demonstration file
   file, file name
edit, default settings

Lesson 1-8
measures of spread
deviation from
   the mean
*variance s^2
*standard deviation, s
sample variance
sample standard
   deviation
population
   variance, σ^2
population standard
   deviation, σ

68

Progress Self-Test

See margin for answers not shown below.

Take this test as you would take a test in class. You will need graph paper and a calculator. You may want to use a statistics package. Then check the test yourself using the solutions at the back of the book.

In 1–3, refer to the following table from *The Statistical Abstract of the United States 1990*.

NO. **191.** PERSONS WITH ACTIVITY LIMITATION, BY SELECTED CHRONIC CONDITIONS: 1980 AND 1985

[Covers civilian noninstitutional population. Conditions classified according to ninth revision of International Classification of Diseases. Based on National Health Interview Survey; see Appendix iii. See headnote, table 182]

CONDITION	TOTAL[1]	AGE			SEX		RACE		FAMILY INCOME		
		Under 45 years	45–64 years	65 years and over	Male	Female	White	Black	Under $20,000	$20,000 to $34,999	$35,000 and over
1980											
Persons with limitation (mil.)	**31.4**	**10.2**	**10.4**	**10.8**	**15.5**	**15.9**	**27.1**	**3.9**			
Percent limited by—											
Heart conditions.	16.4	4.3	19.7	24.5	18.0	14.7	16.6	15.2			
Arthritis and rheumatism.	17.5	5.4	20.0	26.5	11.0	23.7	17.4	18.3			
Hypertension[2]	9.9	3.3	13.2	13.1	7.7	12.1	8.6	19.2	(NA)	(NA)	(NA)
Impairment of back/spine	9.2	13.9	9.9	4.0	9.1	9.2	9.4	7.7			
Impairment of lower extremities and hips.	8.0	10.7	7.2	6.4	9.0	7.1	8.1	7.2			
Percent of all persons with—											
No activity limitation.	85.6	93.2	76.1	54.8	85.3	85.9	85.6	84.9			
Activity limitation	14.4	6.8	23.9	45.2	14.7	14.1	14.4	15.1			
In major activity.	10.9	4.2	18.8	39.0	11.2	10.6	10.8	12.2			
1985											
Persons with limitation (mil.)	**32.7**	**11.6**	**10.4**	**10.7**	**15.3**	**17.4**	**28.0**	**4.1**	**16.6**	**7.1**	**4.7**
Percent limited by—											
Heart conditions.	17.4	4.7	21.5	27.1	18.2	16.7	17.5	16.9	21.5	14.4	14.8
Arthritis and rheumatism.	18.9	5.4	22.8	29.7	12.4	24.6	18.9	20.3	25.0	14.3	13.5
Hypertension[2]	10.5	2.9	15.2	14.2	7.9	12.8	9.0	21.3	15.1	7.0	5.3
Impairment of back/spine	9.2	12.5	10.4	4.4	8.9	9.4	9.4	7.0	9.1	11.0	10.6
Impairment of lower extremities and hips.	8.9	10.7	8.2	7.8	9.4	8.5	9.0	8.1	10.1	8.5	9.6
Percent of all persons with—											
No activity limitation.	86.0	92.8	76.6	60.4	86.4	85.6	85.9	85.5	79.7	89.2	91.9
Activity limitation	14.0	7.2	23.4	39.6	13.6	14.4	14.1	14.5	20.3	10.8	8.1
In major activity.	9.5	4.9	17.5	24.1	9.7	9.4	9.4	11.1	14.2	7.3	5.1

NA Not Available.   [1]Includes persons with unknown family income and other races, not shown separately.   [2]Covers all cases of hypertension, regardless of conditions.

Source: U.S. National Center for Health Statistics, *Vital and Health Statistics*, series 10, and unpublished data.

Note: numbers in the ''Percent limited by'' categories do not add to 100% because ailments other than those listed are not included in this summary.

1. What information is conveyed by the number 15.9 in the first line of the table?
2. How many persons over 64 years old were limited by heart conditions in 1980?
3. Which chronic condition most frequently limited activity by individuals 45–64 years of age as reported in 1985?

   1) 15,900,000 females had activity limitations due to chronic medical conditions.
   2) 24.5 million
   3) arthritis and rheumatism

In 4 and 5, to estimate the size of a flock of wild geese, an ornithologist captured and tagged twelve geese. She released them, and the following week captured seventeen geese from the same flock.

4. If three geese in her second sample were tagged when they were caught, estimate the size of the flock.   **about 68 geese**
5. State an assumption used in drawing your conclusion in Question 4.

ADDITIONAL ANSWERS
9. Sample: At 4 A.M. to 5
A.M., the apartment has
had all night to cool off.
The apartment heats up
during the day. The very
sharp fall at 5 P.M.
probably corresponds to
the cooling system being
activated when the
temperature gets too high.

In 6–9, use the graph below which shows the temperature in degrees Celsius of an apartment on a warm summer day.

Apartment Temperature on a Summer Day

6. Estimate the temperature at 2 P.M.   **about 29°C**
7. Find the average rate of change of temperature between 6 A.M. and 4 P.M.   **1°C per hour**
8. What are the times of the day when the apartment is the coolest? hottest?
9. Give a plausible explanation for the extreme temperatures at these times.
   8) coolest 4 A.M.–5 A.M.; warmest 5 P.M.

In 10–12, Jill tossed eight pennies a number of times, and constructed the following frequency distribution of the number of heads observed in each group toss.

Number of heads

10. Find the median number of heads in the tosses.
11. In Jill's experiment, what percentile rank should be assigned to the event of getting three heads in a toss of eight coins?
12. Explain how to convert the graph to display a relative frequency distribution.
10) 4 heads; 11) the 40th percentile;
12) Convert each frequency to a relative frequency by dividing by 20, the total number of observations.

13. Each of the histograms below shows the frequency of monthly expenses for entertainment of two groups of students. The bars represent 10s of dollars and each interval includes only the left endpoint. Thus the bar on the left indicates expense x, such that $10 \leq x < 20$. If each of the distributions below has the same mean, which has the greater standard deviation?   **(b)**

(a)

(b)

In 14–16, to monitor office long-distance calls, Sheila checked her company's telephone account, which listed the length in minutes of each call. Here is the data for September 11th:

$X_1 = 1$ $X_2 = 1$ $X_3 = 2$ $X_4 = 1$
$X_5 = 12$ $X_6 = 2$ $X_7 = 1$ $X_8 = 1$

14. Find $\sum_{i=1}^{8} X_i$.   **21**

15. Find the mean length of the calls for that day.
16. Explain why so few calls exceed the mean in length.
15) 2.625 minutes; 16) Only one call exceeded the length of the mean because that one call was an isolated outlier, far from all the other data items, which pulled the mean up much higher than it would have been otherwise.

70

25. sample:

In 17–20, use the following box plots which show the ages at inauguration of the last 20 presidents and vice presidents of the United States.

17. What was the range of ages for vice presidents? **32 years**
18. About three-quarters of the last 20 presidents were older than __?__ years when they were inaugurated. **51**
19. *Multiple choice.*   The median inaugural age for presidents falls in which part of the distribution of vice-presidential ages?
    (a) below the first quartile
    (b) between the median and the 25th percentile
    (c) between the median and the 75th percentile
    (d) above the upper quartile **(c)**
20. Determine the outlier values in both data sets using the $1.5 \times$ IQR method.
    presidents: 69
    vice presidents: there are no outliers

In 21–25, refer to the following table describing the estimated crude petroleum production in millions of 42-gallon barrels in 1988 by the members of the Organization of Petroleum Exporting Countries (OPEC). (Source: *Information Please Almanac*, 1989.)

Algeria	244
Ecuador	113
Gabon	64
Iran	661
Iraq	975
Indonesia	402
Kuwait	402
Libya	365
Nigeria	475
Qatar	146
Saudi Arabia	1,570
United Arab Emirates	329
Venezuela	578

21. Find the lower and upper quartiles.
22. Which production figures seem to be outliers?
23. Find the mode of these production figures.
24. Find the standard deviation of these production figures. $s = \sqrt{166{,}514} \approx 408$ **(million barrels)**
25. Draw a histogram with five intervals to display the information in the table.

21) lower quartile = 195; upper quartile = 619.5;
22) 64 and 1,570; 23) mode = 402

Chapter Review

Questions on SPUR Objectives

See margin for answers not shown below.

SPUR stands for **S**kills, **P**roperties, **U**ses, and **R**epresentations.
The Chapter Review questions are grouped according to the SPUR Objectives for this chapter.

SKILLS deal with the procedures used to get answers.

■ **Objective A:** *Calculate measures of center and spread for data sets. (Lessons 1-3, 1-4, 1-5, 1-8)*

In 1 and 2, use the data set below.

$$2, 4, 6, 7, 7, 8, 8, 9, 9, 10$$

1. Find the **a.** range **b.** median.   **a) 8; b) 7.5**
2. Find the **a.** mean **b.** variance **c.** standard deviation.   **a) 7; b) 6; c) about 2.4**

In 3–6, use the scores in the stemplot below. The stem represents units.

```
 4 | 3 7
 5 | 0 6
 6 | 0 4 5
 7 | 2 5
 8 | 7 7 8
 9 | 2 5
10 | 0
```

3. How many scores are given?   **15**
4. Identify the **a.** minimum **b.** maximum **c.** range.   **a) 4.3; b) 10.0; c) 5.7**
5. Which score is the mode?   **8.7**
6. Identify the score at the **a.** median **b.** first quartile **c.** third quartile.   **a) 7.2; b) 5.6; c) 8.8**

7. A bowler has scores of

$$132, 181, 150, 97, \text{ and } 165.$$

   a. What is the mean score?   **145; b) 175**
   b. What score would the bowler need on the next game to bring the average up to 150?

8. In one geometry class with 20 students the mean grade on an exam was 73; in another class with 25 students the mean grade was 81. What is the combined mean of the two classes?   **$77\frac{4}{9}$**

PROPERTIES deal with the principles behind the mathematics.

■ **Objective B:** *Use Σ-notation to represent a sum, mean, variance, or standard deviation.*
(Lessons 1-4, 1-8)

In 9–11, suppose g_i equals the number of points Vicki scored in the ith basketball game so far this season, and

$g_1 = 14, g_2 = 12, g_3 = 18, g_4 = 14,$
$g_5 = 18, g_6 = 19, g_7 = 27, g_8 = 16,$
$g_9 = 12, g_{10} = 19, g_{11} = 26, g_{12} = 18.$

9. Write an expression using Σ which indicates the total number of points Vicki has scored so far.

$$\sum_{i=1}^{12} g_i$$

10. Find   **a.** $\displaystyle\sum_{i=1}^{6} g_i$ **95**   **b.** $\displaystyle\sum_{i=9}^{12} g_i$.   **75**

11. *Multiple choice.* Which expression represents the mean number of points Vicki scored per game?

(a) $\displaystyle\frac{\sum_{i=1}^{11} g_i}{11}$ (b) $\displaystyle\frac{\sum_{i=1}^{12} g_i}{12}$

(c) $\displaystyle\frac{\sum_{i=12}^{1} g_i}{12}$ (d) $\displaystyle\frac{\sum_{i=9}^{12} g_i^2}{12}$ **(b)**

12. *Multiple choice.* Which of the following is a correct formula for finding the standard deviation of the data set $a_1, a_2, a_3, a_4, \ldots, a_{12}$?

(a) $\sqrt{\dfrac{\sum_{i=1}^{12} (a_i - a_{12})^2}{11}}$

(b) $\sqrt{\dfrac{\sum_{i=1}^{12} (a_i - \bar{a})^2}{11}}$

(c) $\sqrt{\dfrac{\sum_{i=1}^{12} a_i - \bar{a}^2}{11}}$

(d) $\sqrt{\dfrac{\sum_{i=1}^{12} a_i^2 - \bar{a}^2}{11}}$ **(b)**

■ **Objective C**: *Describe relations between measures of center or measures of spread.*
(Lessons 1-4, 1-5, 1-6, 1-8)

In 13 and 14, *multiple choice.*

13. The definition of standard deviation uses which of the following?
    (a) mean   (b) median
    (c) mode   (d) range   **(a)**

14. If the standard deviation of a set of n numbers is y, which of these is the variance of the set?

    (a) $\sqrt{y}$ (b) y^2 (c) $\dfrac{y}{n-1}$ (d) $\dfrac{y}{n}$ **(b)**

15. Which is generally affected more by extreme values in the data set—the mean or the median?   **the mean**

16. a. *True* or *false.* The mean and median of a data set can never be equal.   **False**
    b. If your answer in part **a** is true, explain why. If false, give a counterexample.
    sample data set: 1, 2, 3, 4, 5;
    mean = median = 3

USES deal with applications of mathematics in real situations.

■ **Objective D**: *Use samples to make inferences about populations.* *(Lesson 1-1)*

17. Give two reasons for using a sample rather than a population.

18. The Mayor's office of a small city wants to estimate the teenage unemployment rate in the city. Criticize the following means of choosing a sample for this purpose:
    a. going to the local arcade between 9 A.M. and 4 P.M. and interviewing teens;
    b. surveying people who look like teens at the city's shopping malls after 6 P.M.

19. To study a species of fox, a team captures and tags 15 foxes from a large forest and then releases them. The following week, they capture 20 foxes and find 4 of them tagged.
    a. Estimate the number of foxes in the forest.
    b. Give two assumptions necessary for answering part **a**.
    a) about 75 foxes

20. Consider the paragraphs in the Chapter Summary on page 68. Count the number of words in each one.
    a. Calculate the average number of words in these paragraphs.   **about 60 words/para.**
    b. Give several reasons why the sample you used in part **a** may not be a good one to determine the average number of words in a paragraph of this book.

use the answers provided in the book as a study aid. For further information see pages T38–39.

ADDITIONAL ANSWERS
17. Sample answer:
(1) Studying the entire population may be prohibitively expensive.
(2) Members of the population actually studied may be destroyed in the process.

18.a. Sample answer: This is not random. Teens who are employed are not more likely to be found at the arcade during working hours.

b. Sample answer: This is not random. Teens who are employed are more likely to be found at the malls after 6 P.M. than those who are unemployed.

19.b. Sample answer: (1) The foxes captured in each sample are representative of the entire population of foxes. (2) When the samples have been chosen randomly, the percentage of tagged animals among the animals caught in the second sample is nearly the same as the percentage of tagged animals in the entire population.

20.b. Sample answer: The sample was selected according to an arbitrary criterion, rather than randomly; paragraphs apart from those in the Chapter 1 Summary had no chance to be included. The paragraphs in the Chapter 1 Summary are not necessarily representative of those in the rest of the book. Furthermore, the sample chosen was too small— just a few paragraphs— to be reliable.

■ **Objective E:** *Determine relationships and interpret data presented in a table.* (*Lesson 1-1*)

In 21–26, use the table below, from *The Statistical Abstract of the United States, 1990*, providing details of people moving households between 1987 and 1988.

AGE AND REGION	Total (1,000)	Non-movers (same house)	Movers (different house in U.S.) Total	Same county	Different county Total	Same State	Different State	Movers from abroad
Total	**237,382**	**81.9**	**17.6**	**11.3**	**6.4**	**3.6**	**2.7**	**.5**
1–4 years old	14,531	73.1	26.3	17.4	9.0	4.7	4.3	.6
5–9 years old	17,880	79.8	19.7	13.4	6.3	3.3	2.9	.5
10–14 years old	16,520	84.2	15.3	10.4	4.9	2.9	2.0	.5
15–19 years old	17,996	82.3	17.2	11.0	6.1	3.5	2.7	.5
20–24 years old	18,839	63.6	35.2	22.3	12.9	7.4	5.5	1.2
25–29 years old	21,524	67.4	31.8	20.5	11.3	6.5	4.8	.8
30–44 years old	56,128	81.5	17.9	11.3	6.6	3.8	2.8	.5
45–54 years old	23,795	89.5	10.2	6.3	3.9	2.3	1.6	.3
55–64 years old	21,642	92.8	7.1	3.9	3.1	1.9	1.2	.1
65–74 years old	17,472	95.0	4.9	2.7	2.2	1.2	1.0	.1
75 years old and over	11,055	95.3	4.7	2.9	1.7	1.3	.4	.1
Northeast	49,026	88.1	11.4	7.3	4.1	2.4	1.7	.5
Midwest	58,222	83.1	16.6	11.0	5.7	3.5	2.2	.3
South	81,255	79.6	19.9	11.9	8.0	4.4	3.6	.5
West	48,880	77.9	21.3	14.5	6.8	3.8	2.9	.8

NO. **25.** MOBILITY STATUS OF THE POPULATION, BY SELECTED CHARACTERISTICS: 1987–1988

Source: U.S. Bureau of the Census, *Current Population Reports*, series P-20, forthcoming report.

21. Which numbers in the table total 237,382?

22. Explain the meaning of the number 7.4 in the sixth column (''same state'').

23. Give a reason why the sum of the numbers 49,026, 58,222, 81,255, and 48,880 in the first column is 237,383 and not 237,382.

24. How many people 55–64 years old moved to a different county within their own state in 1987–1988?   **about 411,198 people**

25. How many people 20–44 years old moved to the United States in 1987–1988?

26. Which of the age groups shown was most mobile in the period shown?

21) the first column from 14,531 through 11,055
22) Between 1987 and 1988, 7.4% of the 20–24 year old segment moved to a different county within the same state.
23) There is cumulative rounding error because each figure is rounded.
25) about 678,900 people
26) the 20–24 year old age group

■ **Objective F:** *Use statistics to describe data sets or to compare or contrast data sets.*
(*Lessons 1-3, 1-4, 1-5, 1-6, 1-7, 1-8*)

27. Each of the members of a Girl Scout troop sold cookies as part of a fundraising effort. The number of boxes of cookies sold is given below.

63	78	102	69	42	174	81
73	82	94	92	79	62	68
71	73	74	69	11	88	80
63	74	69	71	77	70	93
87	67	77	77	62	85	176

a. Find the mean and standard deviation of these sales figures.   **mean ≈79.2; s.d. ≈28.6**

b. Which sales seem to be outliers to these data?   **11, 42, 174, and 176**

c. Remove the outliers, and find the mean and standard deviation of the remaining sales figures.   **mean ≈ 76.5; s.d. ≈ 10.2**

d. Write several sentences describing the sales of this troop. Include notions of center and spread of the distribution. Comment on extreme cases.

74

28. Refer to the table below which gives the median sales price of existing single family homes during 1989 for 14 large metropolitan statistical areas. Prices are in 1000s of dollars.

Boston	186.2
Chicago	105.0
Cleveland	76.2
Dallas	92.4
Detroit	73.1
Houston	68.7
Los Angeles	218.0
Miami	88.1
New York	186.3
Philadelphia	108.8
San Francisco	265.7
Seattle	109.1
St. Louis	75.3
Washington D.C.	140.5

a. Compute the five-number summary for these data.

b. What conclusions can you make from these data regarding the price of housing in the largest metropolitan areas in the United States?

29. Use the data below for normal daily maximum temperatures by month in Juneau, Alaska and Minneapolis-St. Paul, Minnesota. Data have been rounded to the nearest degree Fahrenheit.

Month	Juneau, Alaska	Minneapolis-St. Paul, Minnesota
January	27	20
February	34	26
March	37	38
April	47	56
May	55	69
June	61	79
July	64	83
August	63	81
September	56	71
October	47	60
November	38	41
December	32	27

a. Which city has the higher summer temperatures?   **Minneapolis-St. Paul**

b. Which has the lower winter temperatures?

c. On the average, which city has a higher temperature? Use a measure of center to justify your answer.

d. On the average, which city shows greater variability in temperature? Use a measure of spread to justify your answer.

30. In a botany experiment, Lana recorded the number of days it took for each of ten plants to flower from the time the buds were first visible. She obtained the following data:

13, 15, 12, 10, 17, 18, 8, 10, 13, 14.

a. For these data find the mean, variance, and standard deviation.

b. In an earlier experiment, Lana found that, when fertilizer was applied, the number of days before plants flowered had mean 11 and standard deviation 1.5. What seems to be the effect of the fertilizer?

31. The stemplot below gives the scores of a sample of 25 male and 25 female high school students on a test which measures study habits.

a. Find the five-number summaries for each group.

b. Are there any outliers using the $1.5 \times$ IQR criterion?

c. Describe similarities and differences between the two groups.

Male		Female
5 0	7	
8	8	
2 1	9	0
4 9 8	10	1 3 9 9
9 5 5 4 3	11	5 8
8 6	12	6 6 9
5 2	13	7 7 8
7 6 0	14	0 2 5 9 9
5 1	15	2 4 4
9	16	5 8
	17	8
8 0	18	
	19	
	20	0

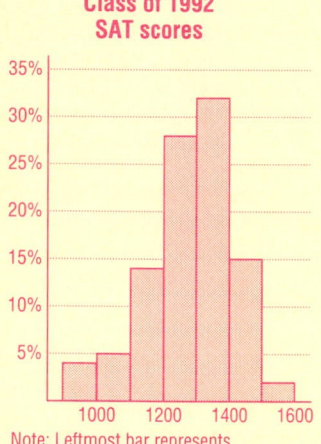
REPRESENTATIONS deal with pictures, graphs, or objects that illustrate concepts.

■ **Objective G:** *Read and interpret bar graphs, circle graphs, or coordinate graphs.*
(*Lessons 1-2, 1-3*)

32. Use the displays at the right from a 1985 atlas.
   a. Describe the main difference between the populations of the U.S.A. and Guatemala at age 0–9.
   b. What factors might account for such different shapes of the age structure of these two countries?

Age Structure of Population

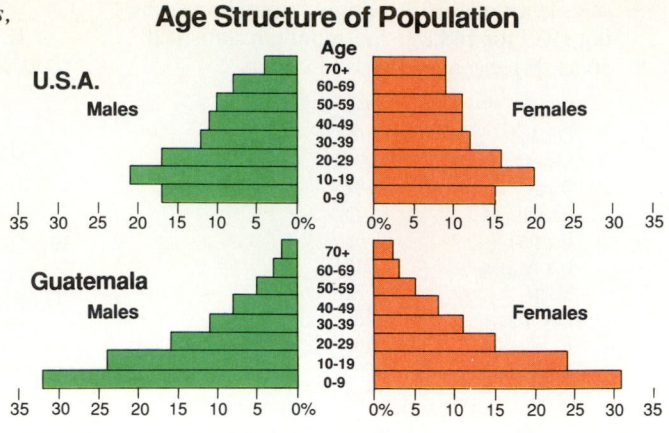

The horizontal bars in these diagrams represent the percentage of the male population and the percentage of the female population in the age group shown.

33. Use the figures below, taken from p. 324 of L. Wolken and J. Glocker's *Invitation to Economics* (second edition), published by Scott Foresman.

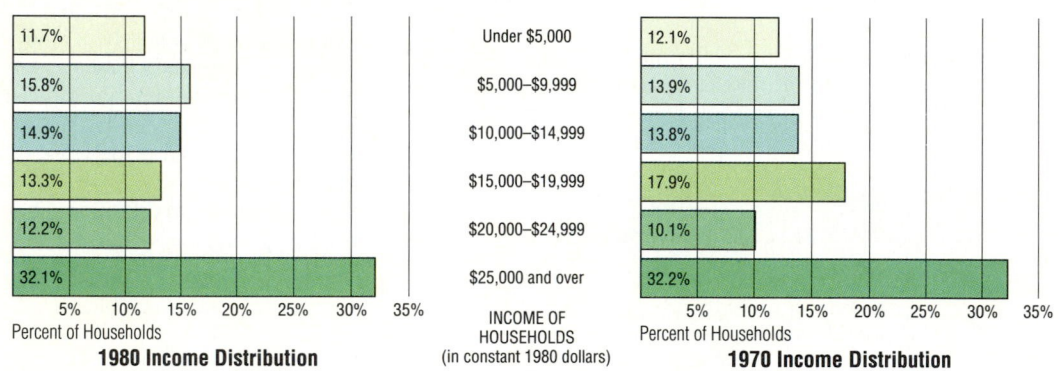

   a. What percent of households had an income below ten thousand dollars in 1980? **27.5%**
   b. In which income level was there the greatest change in the percent of households from 1970 to 1980? **in the $15,000–$19,999 level**

34. Refer to the circle graph at the right which shows who paid for the research and development done in the U.S. in 1988. If the total amount of research funded was about $126,115 million, about how much was funded by:
   a. the Federal government? **about $60,500 million**
   b. industry? **about $60,500 million**

Source of Funds

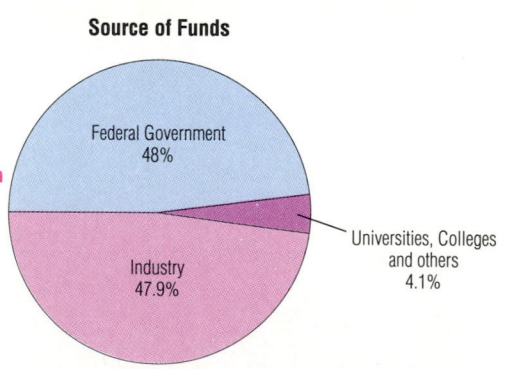

35. Use the graph at the right from p. 291 of *Drive Right* (eighth edition), by Johnson *et al,* (Scott, Foresman, 1987). The graph shows the blood alcohol concentration of a person attending a party in hours after stopping drinking.

a. For how long after the person stopped drinking did the blood alcohol concentration continue to rise?   **≈ 1.5 hours**

b. The text says "Alcohol is absorbed into the body very quickly, but it is very slow to leave." What features of the graph are consistent with this statement?   **the steepness of the slope up and shallowness of the slope down**

c. The legal limit for blood alcohol level in some states is 0.10%. If the person started drinking at 9 P.M., during which hours was the person legally not permitted to drive?   **about 11:30 P.M. to 4:30 A.M.**

d. To estimate a formula for blood alcohol consumption, a police officer assumed the graph was linear between (5, 0.15) and (13, 0). Find the slope of the line joining those two points, and state the unit in which it is measured.   **slope = − .15/8 ≈ − .02 percent blood-alcohol concentration per hour**

■ **Objective H:** *Read and interpret box plots. (Lesson 1-5)*

In 36–41, refer to the box plots at the right which represent the 1987 incomes of householders from ages 15 to 44 in the U.S.A., based on estimates from data collected by the Bureau of the Census. The endpoints of each box plot represent the 10th and the 90th percentiles.

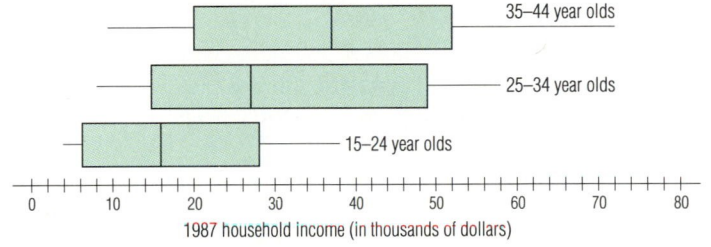

36. What was the median household income for 25–34 year olds?   **about $27,000**

37. What income was exceeded by only 25% of 35–44 year old householders?   **about $52,000**

38. What was the lower quartile of the income for 15–24 year old householders?   **about $6,000**

39. Only 10% of 15–24 year old householders in 1987 had an income of more than ___?___.
about $38,000

40. The middle 50% of 25–34 year old householders in 1987 had incomes between ___?___ and ___?___.

41. Give a reason for using 10th and 90th percentiles rather than the minimum and maximum points on these box plots.
40) about $15,000; about $49,000
41) Sample: Scores below the 10th percentile and above the 90th percentile are not very representative of the population.

51. sample:

Boxes sold

52. sample:

Median home prices ($1000s)

53. sample:

Daily Maximum Temperatures Juneau, Alaska

temperature (°F)

Daily Maximum Temperatures Minneapolis-St. Paul, Minnesota

temperature (°F)

■ **Objective I:** *Read and interpret histograms.*
(Lesson 1-6)

In 42–45, use the following data. Zoe counted the number of students waiting in a line to be served at the school cafeteria at five-minute intervals over her lunch hour. Below is the frequency distribution of her results.

42. How many students were in the longest line recorded?   **11 students**
43. Find the **a.** mode **b.** median **c.** range.
44. What percent of the waiting lines was composed of nine or more students? $33\frac{1}{3}\%$
45. Give a reason for preferring a relative frequency distribution over a frequency distribution.
43a) 7; b) 8; c) 5

In 46–48, consider the histogram below showing the percentages of the total U.S. population in 1988 in age categories between 25 and 74. The source is *The Statistical Abstract of the United States, 1990.*

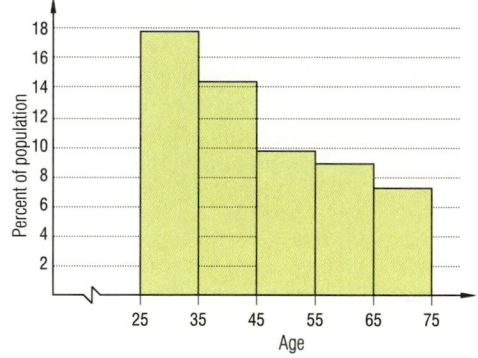

46. The population of the United States was about 246 million in 1988. Estimate the number of people between 65 and 74 years old in that year.   **about 17 million people**
47. Approximately what percentage of the total U.S. population were between 25 and 74 years old in 1988?   **about 57%**
48. *True* or *false.*   The histogram shows that in 1998 almost 18% of the U.S. population will be between 35 and 44 years old.   **False**

■ **Objective J:** *Draw graphs to display data.*
(Lessons 1-2, 1-3, 1-5, 1-6, 1-7)

In 49 and 50, draw a graph to display the data. The data are from the 870 students who entered the University of Chicago as the class of 1992.

49. Of these, 42% came from the Midwest, 24% from Mid-Atlantic states, 10% from New England, 8% from the South, 4% from the Southwest, 10% from the West, and 2% from foreign countries.
50. Their combined SAT scores (SAT Math and SAT Verbal) were as follows.

1500–1600	2%
1400–1499	15%
1300–1399	32%
1200–1299	28%
1100–1199	14%
1000–1099	5%
below 1000	4%

51. Refer to the data on cookie sales given for Question 27. Display these data in a histogram.
52. Refer to the data on home sales in Question 28. Display the data in a box plot.
53. Refer to the monthly temperatures for Juneau and Minneapolis-St. Paul given for Question 29. Illustrate the data and justify your responses to the questions by drawing two histograms.
54. Refer to the test data in Question 31. Draw box plots to illustrate your responses to those questions.

78

CHAPTER 2 ■ FUNCTIONS AND MODELS

OBJECTIVES ■ CHAPTER 2

The objectives listed here are the same as in the Chapter 2 Review on pages 138–142 of the student text. The Progress Self-Test on pages 136–137 and the tests in the Teacher's Resource File cover these objectives. For recommendations regarding the handling of this end-of-chapter material, see the notes in the margins on the corresponding pages of this Teacher's Edition.

OBJECTIVES FOR CHAPTER 2 (Organized into the SPUR categories— Skills, Properties, Uses, and Representations)	Progress Self-Test Questions	Chapter Review Questions	Teacher's Resource File	
			Lesson Masters*	Chapter Test Forms A & B
SKILLS				
A Evaluate functions using Euler's notation.	3a, 4a	1–6	2-1, 2-4	2a
B Find linear or quadratic functions which contain specified points.	11a	7–10	2-2, 2-7	3, 7
PROPERTIES				
C Identify the variables, domain, and range of functions.	2, 3b	11–20	2-1, 2-2, 2-4, 2-6	2b
D Identify properties of the line of best fit and of the correlation coefficient.	6, 10, 12d	21–28	2-3, 2-5	4, 10, 11e
USES				
E Find and interpret linear models.	12a,b,c,e,f	29–33	2-2, 2-3	5, 11a,b,c
F Use step functions to model situations.	5, 9	34–38	2-4	9, 13
G Find and interpret quadratic models.	8, 11b	39–41	2-6, 2-7	8, 12
REPRESENTATIONS				
H Graph linear, quadratic, and step functions.	4b	42–45	2-2, 2-4, 2-6	2c
I Interpret properties of relations from graphs.	1	46–53	2-1, 2-4, 2-6	1
J Use scatter plots to draw conclusions about models for data.	7	54–65	2-2, 2-3, 2-4 2-5, 2-7	6, 11d

*The masters are numbered to match the lessons, and each is intended for use with that lesson.

OVERVIEW ■ CHAPTER 2

A model of a real situation uses mathematical terminology, operations, and concepts to describe relationships among variables. Mathematical models are powerful tools in the sciences because they can be used to make predictions or can be linked to other mathematical models as part of a larger theoretical structure.

Models can emerge from theory or from observations. In this chapter we use the latter to build models using the line, the parabola, and stepwise graphs. The idea of modeling is introduced visually. The first linear models result from an eyeball fit to data. Immediately thereafter we move to the idea of a best-fitting line. This requires technology, either in the form of computer software or a statistical calculator. The goal here is to analyze the line which is computed, not the method of computation. Correlation is introduced as a measure of how well a linear model fits a data set.

Functions are a major tool in model-building, so this chapter reviews the language, symbolism, and properties of functions. Key ideas of domain and range are critical in judging the validity of a model and setting limits on viewing windows. Quadratic functions are used to describe certain curvilinear patterns. Computation of a fitted parabola is done via systems of equations and computer. Integer-valued functions emerge as models in cost and counting problems.

Often students are asked to critique the use of a model. What assumptions or simplifications are needed for the model? Should all the data points be used, or are some outliers? Can the model be used to predict events, through interpolation or extrapolation, or is it best treated as a good description of observed data? These questions are important facets of intelligent use of mathematical models.

The topics of this chapter may seem lengthy at first glance, but in fact the lessons are rather easily accomplished, each in one day.

PERSPECTIVES ■ CHAPTER 2

The Perspectives provide the rationales for the inclusion of topics or approaches, give some mathematical background, and make connections with other lessons and to other UCSMP texts.

2-1

THE LANGUAGE OF FUNCTIONS

It is assumed that students have studied functions in a previous course or courses, and have likely been exposed to a wide range of models for thinking about functions. This lesson gives two equivalent definitions for functions (as a set of ordered pairs and as a correspondence), discusses the basic properties and language of functions, and relates functions to modeling. Three representations of functions are discussed: ordered pairs, graphs, and rules.

2-2

LINEAR MODELS

In teaching modeling, it is often instructive to distinguish three types of models: the exact model, where the mathematics gives the precise value (as is often found in dealing with money); the almost exact theory-based model (such as in describing the path of a projectile); and what might be called the impressionistic model, where there is little basis to predict the model, but the data fit it (such as using lines of fit for test scores).

This lesson includes one example of each type. It begins with a constant increase cost situation that reviews finding the slope, *y*-intercept, and other points of a linear function. A second example, in theory, is of a constant rate situation but, in practice, is only close. The final example is impressionistic. In the last two situations, the data are graphed in a scatter plot whose points lie close to a line. A ruler or piece of string can be used to establish a line. By using the point-slope or slope-intercept form for an equation of a line, an equation for a line of fit can be found.

2-3

THE LINE OF BEST FIT

This lesson follows naturally from Lesson 2-2. Instead of eyeballing a line of fit, the lesson discusses the statistical line of best fit, which is the line that minimizes the sum of squares of the deviations from the points to the line and which is called the least squares or regression line. Students should be able to use a statistical package to draw a line of best fit on a scatter plot of data and use the computer-generated equation of that line to make predictions about the population which the line models.

You may notice that though we define the word *regression,* we do not use it to any great extent in the lesson, preferring rather to use *best fit* consistently. This is a choice for linguistic reasons. It is possible that students have encountered the word *regression* in other contexts, and some of these do not agree with the way it is used by statisticians. Most of the uses of *regression* impart the notion of going backward, which is the generic definition of the word.

In statistics *regression* refers to a measure of the extent to which two variables increase together linearly, or how one decreases as the other increases linearly. The idea was first described by Sir Francis Galton in 1877 to explain results in an experiment with sweet peas. Offspring were normally distributed about a population average, not their parents' average. (Normal distributions are studied in Chapter 10.) Galton called this "reversion," or going back to the population mean. In 1885 he used the word *regression* in describing a similar phenomenon among heights of children in relation to their parents' heights. This particular example is found in Chapter 12.

2-4
STEP FUNCTIONS
Some models are almost linear, but a more in-depth look indicates that they are not continuous but jump. For instance, the cost of a phone call might jump by 20¢ each minute. A linear model might be exact for every whole number of minutes but be off in between. A step function is needed to model this situation.

This lesson introduces the concept and symbols of the greatest integer function $f(x) = \lfloor x \rfloor$, which is quite useful in obtaining exact models for the phone call situation and others like it. Students should be able to evaluate the function for any x, read a graph of f or other step functions, and analyze the use of the function in practical situations.

The greatest integer function is a *rounding down* function; it rounds each real number down to the nearest integer. There is obviously a counterpart, the *rounding up* function. Until a generation ago, the notation [ ] was universally used for the greatest integer function. However, in order to distinguish between the rounding down and rounding up functions, new notations were invented and they were given new names: the *floor function* $\lfloor\ \rfloor$ (for the greatest integer or rounding down function) and *ceiling*

function $\lceil\ \rceil$ (for the rounding up function). Many books use the older notation [ ] for the greatest integer function. Students are likely to see either notation in calculator and computer manuals.

The greatest integer function is utilized in many applications done by computers since computers can deal with only a finite number of points between two points on the number line, and so cannot give exact graphs or data for continuous functions. In discrete mathematics, the greatest integer function is called the floor function.

Statistics provides a natural link between the continuous and the discrete. Most collections of data are sets of discrete elements, and so are examples of discrete sets. Yet we have consistently been fitting continuous models to those discrete sets in an effort to find reasonably *simple* descriptions of the relationships between variables.

2-5
CORRELATION
When is a line a good fit to data? The correlation coefficient helps answer this question. It is important for students to know that most statistical packages can produce a slope and an intercept for a line of best fit for a set of data, but the resulting line might be worthless as a model. The correlation coefficient gives an indication of the strength of a linear relationship. You should emphasize that the correlation coefficient has finite range ($-1 \le r \le 1$) and that both ends of the range indicate strong correlation while the middle indicates weak or no linear correlation.

As with the line of best fit, the goal is to introduce the notion of correlation as a measure of linearity, and show some understanding of what the sign and magnitude of r tell about a data set. The goal is not the hand calculation of correlation, but students should be able to use a statistics package to find r and interpret the result.

2-6
QUADRATIC MODELS
This lesson introduces data sets in which the variables are obviously *not* linearly related, but are curvilinear. The focus is on quadratic modeling, so quadratic equations and the use of the quadratic formula are reviewed. It is assumed that students are quite familiar with quadratic functions.

This lesson reviews the general form $y = ax^2 + bx + c$, its graph, and the quadratic formula. Students should use technology to determine the best fitting parabola. If you don't have the technology, discuss the two applications of quadratics in the text. The first (gravity) is an almost-exact, theory-based model. The second (electric production) gives a situation which leads to an impressionistic model.

2-7
FINDING QUADRATIC MODELS
Even without the technology to perform polynomial regressions, students can "eyeball" a polynomial to fit data. This lesson addresses the general problem of fitting a line to two points and extends it to fitting a parabola to three noncollinear points. In Lesson 9-2, this skill is extended to polynomials of higher degrees. The computation of the coefficients of the quadratic function requires the solution of a 3×3 system of equations.

2-8
THE MEN'S MILE RECORD
This reading lesson discusses in depth the application of both linear modeling and correlation to the men's mile record, the most famous record in running (the only nonmetric distance for which official world records are still kept). Students should be able to understand the arguments and analyze the extent to which the model is a valid one.

DAILY PACING CHART ■ CHAPTER 2

For optimal student performance, the Progress Self-Test and the Chapter Review should be covered. (See *General Teaching Suggestions: Mastery* on page T35 of this Teacher's Edition.) By following the pace of the Full Course given here, students can complete 12 or 13 chapters of the book by the end of the year. Students following the pace of the Minimal Course spend more time to complete lessons that are combined with quizzes and to consider the Chapter Review. When all lessons are covered in order (the recommendation of the authors), students in the Minimal Course can cover all but the last two chapters. For more information on pacing, see *General Teaching Suggestions: Pace* on page T34 of this Teacher's Edition.

DAY	MINIMAL COURSE	FULL COURSE
1	2-1	2-1
2	2-2	2-2
3	2-3	2-3
4	Quiz (TRF); Start 2-4	Quiz (TRF); 2-4
5	Finish 2-4	2-5
6	2-5	2-6
7	2-6	Quiz (TRF); 2-7
8	Quiz (TRF); Start 2-7	2-8
9	Finish 2-7	Progress Self-Test
10	2-8	Chapter Review
11	Progress Self-Test	Chapter Test (TRF)
12	Chapter Review	
13	Chapter Review	
14	Chapter Test (TRF)	

TESTING OPTIONS

■ Quiz for Lessons 2-1 through 2-3   ■ Chapter 2 Test, Form A   ■ Chapter 2 Test, Cumulative Form
■ Quiz for Lessons 2-4 through 2-6   ■ Chapter 2 Test, Form B

A Quiz and Test Writer is available for generating additional questions, additional quizzes, or additional forms of the Chapter Test.

PROVIDING FOR INDIVIDUAL DIFFERENCES

The student text has been written for, and tested with, average students. It has also been used successfully with students who are better prepared as well as with students who are poorly prepared.

The Lesson Notes often include Error Analysis and Alternate Approach features to help you assist those students who need more help. Students of all abilities often learn from their peers and may benefit from small group work referenced as appropriate throughout the Notes. A blackline Lesson Master (in the Teacher's Resource File), keyed to the chapter objectives, is available for each lesson to provide more practice. (However, we advise you to use these masters sparingly; again refer to the suggestions on pace on page T34.)

Extension activities are provided in the Lesson Notes for use with students who need to be challenged even more than the requirements of completing the Full Course provide.

Functions and Models

RESOURCES
■ Teaching Aid 13: Table and graph for page 79

We recommend 11–14 days be spent on this chapter: 8–10 days on the lessons and quizzes; 1 day for the Progress Self-Test; 1 or 2 days for the Chapter Review; and 1 day for a Chapter Test. (See the Daily Pacing Chart on page 79D.) Do not spend more than 14 days on this chapter.

USING PAGE 79
The quadratic model shown on the opening page of Chapter 2 was generated using a statistics package. As was the case for the opener to Chapter 1, the accuracy of the coefficients of the terms in the model affects the degree of precision to which the model can be used as a predictor of mileage flown in a given year. If you use a different statistics package to generate a quadratic model for these data, you may get slightly different results.

A teaching aid is provided to facilitate discussion of this example.

Passenger Miles for U.S. Intercity Air Traffic			
Year	Miles (billions)	Year	Miles (billions)
1945	3	1979	210
1950	10	1980	204
1955	23	1981	201
1960	32	1982	214
1965	54	1983	232
1970	109	1985	263
1975	136	1986	293
1976	150	1987	322
1978	189		

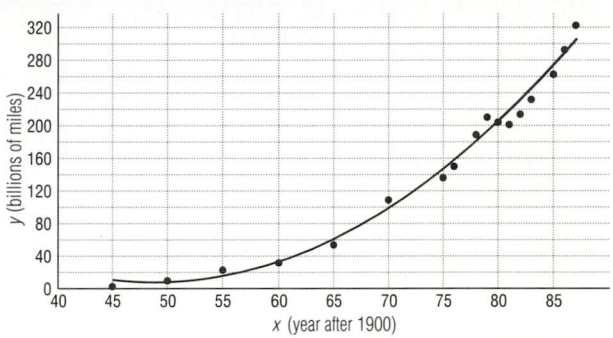

In Chapter 1 you learned to use measures of center and spread to describe data involving a single variable, that is, **univariate** data. In this chapter, you will learn how to describe data involving relations between two variables, or **bivariate** data. For example, the data from the 1988/89 *FAA Aviation Forecasts* show the number of passenger miles for intercity air traffic in the U.S. in various years. These data, pictured in the scatter plot, show that there has been a rather consistent increase in the number of air passenger miles flown annually since 1945, though not every year has more miles than the previous year.

To describe relations more fully, mathematicians seek *models* for data. A **mathematical model** is a mathematical description of a real situation, usually involving some simplification and assumptions concerning the situation. Such models not only describe relations between data, they also can be used to make estimates and predictions.

A function such as a linear or a quadratic function can be a good model. For instance, if x is the year after 1900 and y the number of billions of passenger miles, then the air traffic data can be modeled by the quadratic function $y = 0.203x^2 - 19.788x + 490.043$, which is graphed on the scatter plot above. Although the data points do not fall exactly on the graph, they are rather close. With such a model, you might reasonably predict the number of air passenger miles for 1977 or 1989, even though this information was not provided in the FAA publication.

RESOURCES
■ Lesson Master 2-1

OBJECTIVES

A Evaluate functions using Euler's notation.
C Identify the variables, domain, and range of functions.
I Interpret properties of relations from graphs.

TEACHING NOTES

Of the two definitions of *function* on page 80, the first is static: a set of ordered pairs in which each first element is paired with exactly one second element. The second definition is dynamic: a correspondence which makes use of one value to obtain another value. Both aspects of function are important and both are used throughout this book.

The terms *independent* and *dependent* are related to model-building. Determining which of two variables should be represented on the *x*-axis is not automatic. The concepts of domain and range take on new meaning in applications. In some instances linear and quadratic functions are used to model real world phenomena which do not have continuous domains and/or ranges. While the model might be accurate (and simpler), students need to understand that there may be restrictions on the domain and range. For example, a model relating the independent variable number of dosages of a medication to a

LESSON

2-1

The Language of Functions

The chapter opener lists many ordered pairs of numbers. The first number is a year; the second is the number of passenger miles flown in the U.S. that year. Any set of ordered pairs is a relation. In many contexts, the second number in each ordered pair depends in some way on the first number. For this reason, the first variable in a relation is called the *independent variable* and the second variable is called the *dependent variable*.

Two definitions of function are commonly used in mathematics. They are equivalent but they stress different aspects of functions. One definition is as a special type of relation. This is useful for graphing.

> **Definition**
>
> A **function** is a set of ordered pairs in which each first element is paired with exactly one second element.

In the list of ordered pairs of a function, the set of first elements is the **domain** of the function. The set of second elements is the **range**. The domain consists of all allowable values of the independent variable; the range is the set of possible values for the dependent variable.

The other definition stresses the independent-dependent variable idea.

> **Definition**
>
> A **function** is a correspondence between two sets A and B in which each element of A corresponds to exactly one element of B.

The domain is the set A; the range is the set B.

For the air traffic data we say that the number of passenger miles is a function of the year. The domain is the set of all years in which there has been commercial air traffic in the United States; the range is the set of numbers of passenger miles flown in those years.

In this course you may assume that the domain of a function is the set of all real numbers for which the function is defined, unless some other domain is explicitly stated.

Example 1 A laundry service normally charges $1.00 per shirt for laundering. Customers get a $1.00 discount if they have 6 or more shirts laundered.

a. Which is true, "the cost C is a function of the number n of shirts" or "the number n of shirts is a function of the cost C"?

80

b. Identify the independent and dependent variables.

c. State the domain and range of the function.

Solution

a. Because there is exactly one cost C for a given number of shirts, the cost is a function of the number of shirts. The cost of laundering either 5 shirts or 6 shirts is the same, so the number of shirts is not a function of the cost.

b. Because C depends on n, n is the independent variable and C is the dependent variable.

c. The domain is the set of all values for n. So the domain is the set of nonnegative integers. The range is the set of all possible values for C. Any whole-number cost is possible, so the range is also the set of nonnegative integers.

Functions can be represented in many ways; among the most frequently used are: ordered pairs in tables or lists, rules expressed in words or symbols, and coordinate graphs. You should know how to recognize functions expressed in each of these forms and how to convert from one form to another.

Example 2   Consider again the cost C of laundering n shirts, which is $1.00 per shirt with a $1.00 discount if 6 or more shirts are laundered.

a. List three ordered pairs in the function.

b. Write an equation to express the relation between C and n.

c. Graph the function.

Solution

a. Pick any three values of n in the domain of the function, and calculate C. We chose 5, 6, and 9. For 5 shirts the cost is $1.00 per shirt, or $5.00. For 6 and 9 shirts, the cost is $1.00 per shirt with a $1.00 discount. So, the cost for 6 shirts is $5.00 and the cost for 9 shirts is $8.00. Three ordered pairs in this function are (5, 5), (6, 5), and (9, 8).

b. For 5 or fewer shirts, $C = n$. For 6 or more shirts $C = n - 1$. This can be written as

$$C = \begin{cases} n & \text{if } n \leq 5 \\ n-1 & \text{if } n > 5 \end{cases}.$$

c. A graph is drawn below.

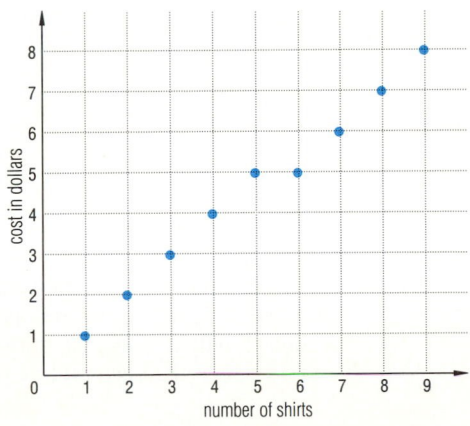

---

dependent variable of cure rates might be linear, but the domain would certainly not include negative real numbers.

It is best not to overemphasize restrictions on the domain and range. The point the students should get is that mathematical models are tools in the hands of the people who create them, and the tool varies depending on the needs and desires of the creator.

Students may consider the questions posed in **Example 1** as trivial, but do not underestimate their value. They introduce important vocabulary and make the point that the formal language of functions is related to realistic applications.

Error Analysis   When the argument of a function is an expression, as in **Example 4b** and **Question 5,** students are very prone to misapplying the distributive property. While there are exceptions, in general $f(a + b) \neq f(a) + f(b)$. Use additional examples as necessary. This issue will arise again in the discussions of transformations and composition of functions in Chapter 3.

An important exercise for students is to state in words the rule of a function. For example, $f(x) = \dfrac{1}{x + 2}$ can be read as "add 2 to the argument then take the reciprocal." In constrast, $g(x) = \dfrac{1}{x} + 2$ means that the reciprocal of the argument is taken first, then 2 is added to the result.

ADDITIONAL EXAMPLES

1. The cost of renting a car is $35 per day plus $.40/mile for mileage over 100 miles.

a. Which is true, "cost is a function of miles driven" or "miles driven is a function of cost"?

Cost is a function of miles driven.

81

Example 3 Find the range of the function whose graph is at the right. A rule for the function is $y = -x^2 + 3$.

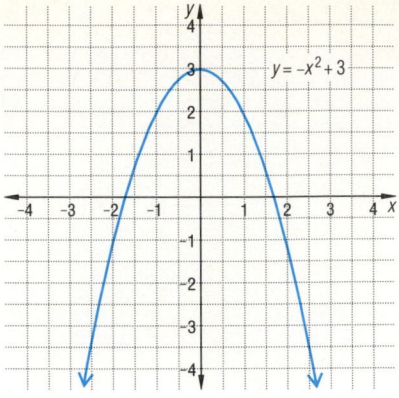

Solution The graph is a parabola, with maximum point (0, 3). From the graph, you can see that the values of the function are all real numbers less than or equal to 3. So the range is $\{y: y \leq 3\}$.

Check Since $x^2 \geq 0$ for all real x, $-x^2 \leq 0$. So $-x^2 + 3 \leq 3$. This checks with the graph.

Rules for functions are sometimes written using *function notation*. The symbol $f(x)$, which is read "f of x," was invented by the mathematician Leonhard Euler (pronounced "oiler") in the 18th century. It indicates the value of the dependent variable when the independent variable is x. Euler's notation is particularly useful when evaluating two or more functions at specific values of the independent variable.

Example 4 Let $f(x) = x^2 + 3$ and $g(x) = (x + 3)^2$.
a. Evaluate $f(5)$ and $g(5)$.
b. Does $f(2 + 5) = f(2) + f(5)$?

Solution
a. The rule for f says "to the square of x add 3." So
$$f(5) = 5^2 + 3 = 28.$$
The rule for g says "to x add 3, then square the result." So
$$g(5) = (5 + 3)^2 = 8^2 = 64.$$
b. $f(2 + 5) = f(7) = 7^2 + 3 = 52$
$f(2) + f(5) = (2^2 + 3) + (5^2 + 3) = 7 + 28 = 35$
So $f(2 + 5) \neq f(2) + f(5)$.

Example 4 illustrates that, in general, $f(a + b) \neq f(a) + f(b)$. That is, there is no distributive property for function notation.

In a function, there is only one member of the range paired with each member of the domain. So if you draw any vertical line through the graph of a function, it will intersect the graph at no more than one point. This is often referred to as the **vertical line test** for determining functions. You can see

82

how this works on the two following graphs. Both show relations, but only the one at the left represents a relation in which y is a function of x.

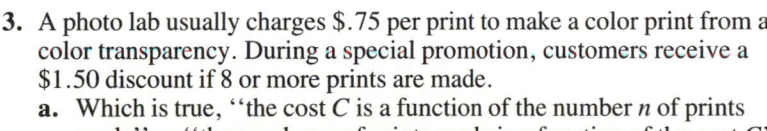

y is a function of x—any vertical line intersects the graph no more than once.

y is not a function of x—some vertical lines intersect the graph in two places.

The three ways of looking at functions—as sets of ordered pairs, as graphs, or by their rules—highlight different aspects of functions. These three ways are united when we think of a function as a set of ordered pairs of the form $(x, f(x))$ which can be graphed in a coordinate plane.

Questions

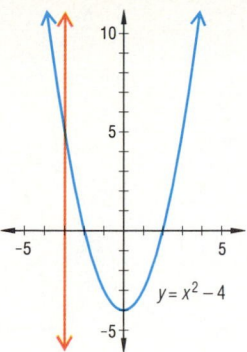

Covering the Reading

In 1 and 2, give a definition.

1. function **See margin.**

2. domain **See margin.**

3. A photo lab usually charges \$.75 per print to make a color print from a color transparency. During a special promotion, customers receive a \$1.50 discount if 8 or more prints are made.
   a. Which is true, "the cost C is a function of the number n of prints made" or "the number n of prints made is a function of the cost C"?
   b. What is the cost of making 10 color prints from transparencies?
   c. Write an equation expressing the relation between C and n.
   d. Graph the relation in part c for $n \le 12$.
      a) C is a function of n. b) \$6.00; c,d)   **See margin.**

4. Consider the relationship $x = f(t)$. Identify the letter which represents each of the following.
   a. the function   *f*
   b. the dependent variable   *x*
   c. the independent variable   *t*

5. Let $f(x) = 4x - 1$. Evaluate.
   a. $f(6)$ **23**   b. $f(1 + 4)$ **19**   c. $f(1) + f(4)$ **18**

6. Let $g(x) = x^2 - x$. Evaluate.
   a. $g(-7)$ **56**   b. $3 \cdot g(10)$ **270**   c. $g(3 \cdot 10)$ **870**

NOTES ON QUESTIONS
Question 16: Technically, the range is discrete, consisting of nonnegative real numbers to two decimal places, but for simplicity, we often think of the range as simply the nonnegative real numbers. Part **d** continues the review of slope begun in Chapter 1, and is a key part to this question. Students need to become accustomed to finding real world interpretations of key features of functions. Students with previous UCSMP experience should be at an advantage here.

Question 17: This is the first of several questions in Lessons 2-1 through 2-5 reviewing quadratic functions in preparation for Lesson 2-6. Because of the scale on the graph, students may not be able to read the range exactly. However, from the symmetry of the graph they know that the minimum value of the range is $f(4)$ and this can be calculated from the formula.

ADDITIONAL ANSWERS
1. A function is a set of ordered pairs in which each first element is paired with exactly one second element; or, a function is a correspondence between two sets A and B in which each element of A corresponds to exactly one element of B.

2. The domain of a function is the set of all allowable values of the independent variable.

3.c. $c = \begin{cases} .75n & \text{if } n < 8 \\ .75n - 1.50 & \text{if } n \ge 8 \end{cases}$

d.

NOTES ON QUESTIONS

Question 23: Use this question as a check that students have mastered the solution of 2 × 2 systems of linear equations. There are other questions on 2 × 2 systems in Lesson 2-2 and some 3 × 3 systems to solve in Lessons 2-5 and 2-6. Mastery in solving 2 × 2 and 3 × 3 linear systems is assumed in Lesson 2-7.

ADDITIONAL ANSWERS
11.a.,b., 12.a.,b., 13.a.,b., 14.a.,b., 15.c.,d., 17a.,b., 20.a., 22.a.,b.,c. See Additional Answers in the back of this book.

FOLLOW-UP

MORE PRACTICE
For more questions on SPUR Objectives, use *Lesson Master 2-1,* shown on page 85.

EXTENSION
Technology Technological support for the second definition of function on page 80 comes from the metaphor of a calculator as a "function machine." A calculator provides a hands-on version of the metaphor. Below is a list of similar characteristics to share with your class.

Calculator
1. number entered (input)
2. output of a calculation on display
3. inputs which do not lead to an error message
4. all possible numerical outputs
5. each input has exactly one output
6. input not in domain generates an error message

In 7–10, **a.** tell whether the graph shows a relation and **b.** tell whether the graph shows y as a function of x.

7.

a) Yes; b) No

8.

a) Yes; b) Yes

9.

a) Yes; b) No

10.

a) Yes; b) Yes

Applying the Mathematics

In 11–14, consider the relation defined by the sentence. **a.** Sketch a graph. **b.** Tell if the relation is a function. If so, give its domain and range.

11. $z(x) = x + 3$ **See margin.**

12. $y = |x|$ **See margin.**

13. $y > x$ **See margin.**

14. $h(x) = \sqrt{x}$ **See margin.**

15. At a sale, all clothing prices are reduced by 20%.
　a. Write an equation for the sale price P of clothing originally marked at d dollars.
　b. Identify the domain and range of this function.
　c. List three ordered pairs in this function.
　d. Draw a graph of the function.   **a) $P = 0.8d$; b) domain = range = the set of all nonnegative real numbers; c,d) See margin.**

16. The graph at the right represents the cost of renting a car for a day as a function of the number of miles driven.
　a. Write two ordered pairs which are part of this function.
　b. What is the range of the function?
　c. Find the slope of the line.   **0.08**
　d. What does the slope represent?

Car Rental Cost

　a) sample: (0, 20), (500, 60); b) {$C : C \geq 20$};
　d) the rental cost (in dollars) per mile

84

17. At the right is a graph of $f(t) = t^2 - 8t + 9$.

a. State the domain and range of f.

b. State two ordered pairs which are part of this function.
See margin.

In 18 and 19, give an example of a function satisfying the given condition. Then give its domain and a rule.

18. The range is {0}.
sample: $f(x) = 0$; **domain is the set of all real numbers**

19. The range is the set of all negative integers.
sample: $y = -x$; **domain is the set of all positive integers**

20. Suppose two dice are rolled a fixed number of times. If x is the sum of the values shown on top of the dice, let $f(x)$ be the number of times the sum x appears.

a. What is the domain of f? **See margin.**

b. If the dice are rolled 100 times, what are the possible values for $f(x)$? **range = the set of all integers from 0 to 100**

Review

21. Without graphing, tell whether the point (32, 98) is on the line with equation $y = 3x + 4$. Justify your answer. *(Previous course)*
No; $3 \cdot 32 + 4 = 100$

22. a. State the slope-intercept form of the equation of a line.
b. State the point-slope form of the equation of a line.
c. Write an equation for the line which has slope -3 and passes through the point (2, 7). *(Previous course)* **See margin.**

23. Solve the system $\begin{cases} a + b = 11 \\ 4a + 2b = 25 \end{cases}$. *(Previous course)* $a = \frac{3}{2}, b = \frac{19}{2}$

24. *Skill sequence.* Match the equation to the shape of its graph. *(Previous course)*

a. $y = \dfrac{3}{x}$ **II**   I. parabola

b. $y = 3x$ **III**   II. hyperbola

c. $y = \dfrac{x}{3}$ **III**   III. line

d. $y = 3x^2$ **I**

Exploration

25. Describe the sales tax where you live **a.** in words, **b.** with a graph, **c.** with an equation or equations relating it to an item's price.
Answers will vary.

(dice illustration)

Function
1. argument or independent variable x
2. value of the function or dependent variable $f(x)$
3. domain of the function f
4. range of the function f
5. each first element is paired with exactly 1 second element
6. argument not in domain means the function is not defined for that argument

NAME _____

LESSON **MASTER 2-1**
QUESTIONS ON **SPUR** OBJECTIVES

■**PROPERTIES** *Objective A (See pages 138–142 for objectives.)*
In 1 and 2, let $f(x) = 3x^2 - 1$.

1. Evaluate
   a. $f(1)$. **2**   **b.** $f(4)$. **47**   **c.** $f(5)$. **74**

2. *True or false*. Justify your answer.
   a. $f(1) + f(4) = f(1 + 4)$ **b.** $f(1 + 4) = f(1) \cdot f(4)$
   a. **False; $f(1) + f(4) = 2 + 47 = 49$; $f(1 + 4) = f(5) = 74$; $49 \neq 74$**
   b. **False; $f(1 + 4) = f(5) = 74$; $f(1) \cdot f(4) = 2 \cdot 47 = 94$; $74 \neq 94$**

3. Let $g(x) = |x| - x$. *True or false*.
   a. $g(4) = g(-4)$ **False**   **b.** $g(4) = -g(-4)$ **False**

4. Suppose $h(x) = 4^x$. Evaluate
   a. $h(3) + h(-3)$. **$\frac{4097}{64}$**   **b.** $\dfrac{h(3) - h(1)}{3 - 1}$. **30**

■**PROPERTIES** *Objective C*
In 5–8, state a. the domain and b. the range.

5. $y = |x + 1|$ **a.** **all reals**   **b.** **all nonnegative reals**
6. $g(x) = x^2 + 5$ **a.** **all reals**   **b.** **{y: $y \geq 5$}**
7. $f(x) = -\frac{1}{3}x - 7$ **a.** **all reals**   **b.** **all reals**
8. $y = \dfrac{5}{x^2}$ **a.** **all nonzero reals**   **b.** **all positive reals**

■**REPRESENTATIONS** *Objective I*
In 9–12, state whether the graph shows a function.

9.   10.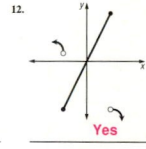

    No   **Yes**

Functions, Statistics, and Trigonometry © Scott, Foresman and Company   **17**

NAME _____
Lesson MASTER 2-1 (page 2)

11.   12.

     No   **Yes**

In 13–16, for each relation determine the a. domain and b. range.

13.   14.

   a. **all reals**   **a.** **all reals**
   b. **{-1, 0, 1}**   **b.** **all reals**

15.   16.

   a. **{x: $-4 \leq x \leq 4$}**   **a.** **all reals**
   b. **{y: $-4 \leq y \leq 4$}**   **b.** **{y: $y \geq 2$ or $y \leq -2$}**

18   *Functions, Statistics, and Trigonometry © Scott, Foresman and Company*

85

LESSON 2-2

RESOURCES
- Lesson Master 2-2
- Teaching Aid 14: Table and graph for **Example 1a** and **Question 3**
- Teaching Aid 15: Table and graph for **Example 2** and **Question 4**
- Teaching Aid 16: Table and graph for **Questions 9 and 12**

OBJECTIVES

B Find linear functions which contain specified points.
C Identify the variables, domain, and range of a function.
E Find and interpret linear models.
H Graph linear functions.
J Use scatter plots to draw conclusions about models for data.

TEACHING NOTES

This lesson introduces the modeling of situations involving a constant rate of change. Students should be able to (1) "fit" (by eye) a line to a scatter plot, (2) find an equation of the form $y = mx + b$ for the line (in this equation m is the rate of change and b is the initial value of y), and (3) use their model to predict, by interpolation or extrapolation, values of y for different values of x. They should also be able to discuss whether their predictions are appropriate.

Most of the data sets here are close to linear to facilitate modeling with a line. Emphasize that the slope is a rate of change. Having reviewed slope in Chapter 1, students should be quite familiar with the idea that the subtractions in the numerator and denomi-

LESSON

2-2

Linear Models

A **linear function** is a function which can be described by an equation of the form $y = mx + b$, where m and b are constants. Recall that the graph of every function of this form is a line with slope m and y-intercept b. Both the domain and range of the general linear function are the set of all real numbers.

Some situations can be modeled exactly by linear functions. In other situations, the variables of interest may not take on all real number values, but a linear function still provides a convenient model. In either case, linear models can be used to predict values of either the dependent or independent variable.

Example 1 A firm specializes in sanding and varnishing hardwood floors. The cost C for this service is $50 setup charge plus $2 per square foot for materials and labor. Suppose A is the area of the floor in square feet.

a. Find an equation to represent C as a function of A.
b. Draw a graph of the relation between C and A.
c. What will the firm charge to sand and varnish the floor of a room 12 feet by 11 feet?
d. What is the largest area that can be treated for $275?

Solution

a. The cost C is $50 plus $2 for each of the A square feet. Constant increase situations such as this are modeled by linear functions.

$$C = 50 + 2A$$

b. The graph is a line with C-intercept 50 and slope equal to 2.

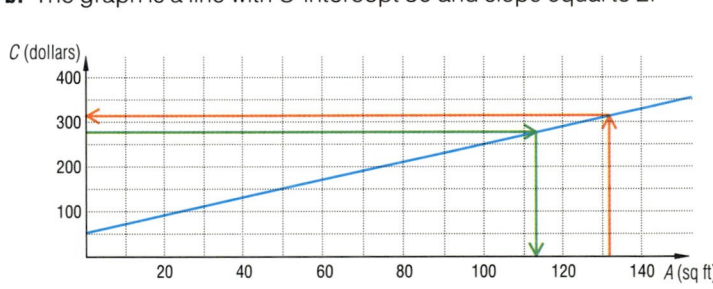

c. The area of the room is $A = 12 \cdot 11 = 132$ sq ft.
Substitute 132 for A. $C = 50 + 2A$
$$C = 50 + 2(132)$$
$$= 314$$
The cost will be $314
d. In the equation $C = 50 + 2A$, substitute 275 for C and solve for A.
$$275 = 50 + 2A$$
$$2A = 225$$
$$A = 112.5$$
So the largest area treated for $275 is 112.5 sq ft.

86

Check

c. Read across to $A = 132$, up to the line $C = 50 + 2A$, and over to the C-axis. A cost of slightly more than $300 seems reasonable.

d. Read up to $C = 275$, across to the line $C = 50 + 2A$, and down to the A-axis. An area of just over 110 square feet is found.

In some situations the relation between two variables may be expected to be approximately, but not exactly, linear. For instance, Carla thought that her rate of jogging was fairly constant, and *assumed* that the small deviations from a constant rate were the result of slight fluctuations in her speed and inaccuracies in timing. Data from some recent laps are given below.

Number of laps	2	4	6	8	10	12	15
Time (minutes and seconds)	3:15	6:46	10:20	13:25	16:45	20:10	25:45
Time (seconds)	195	406	620	805	1005	1210	1545
Average rate (secs/lap)	98	101	103	101	100	101	103

The bottom row in the table is found by dividing the third row by the first. The average rate is the mean number of seconds per lap. The data showed Carla that she jogged at an average rate of about 1 minute 40 seconds or $1\frac{2}{3}$ minutes per lap. Thus, if T stands for the time (in minutes) to complete L laps, a model for Carla's jogging can be expressed by the function

$$T = \frac{5}{3}L.$$

Carla could check her model making a scatter plot of the data and observing that the data points almost, but not quite, lie on the graph of the line $T = \frac{5}{3}L$. We say that this model is a good fit for these data.

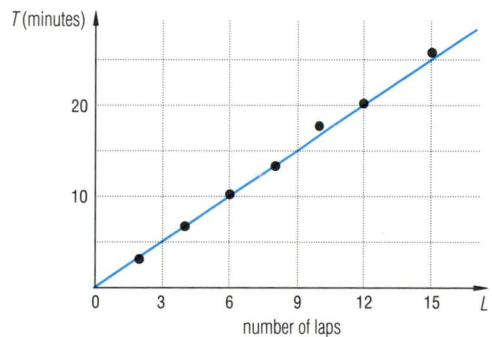

Even when nothing in the situation suggests that there might be a linear relation between two variables, sometimes such relations appear.

Consider the following situation. A consumer panel purchased a number of sticks of sugarless bubble gum and evaluated their quality, based upon taste and the length of time the flavor lasted. Each stick was rated on a 0 to 10 scale. The average price of each stick (in cents) was also recorded. The data are listed in the table on the next page.

---

nator of the slope formula calculate the changes, while the division gives the rate. The slope tells how fast the dependent variable changes as the independent variable increases by one unit. If the increase is constant, then a linear model is appropriate. (Here again students with previous UCSMP experience should have an advantage.)

Example 1 reviews finding the equation of a line when its slope and intercept are known. You can also use the equation to discuss the appropriateness of the model. In practice a floor finisher would probably round area to the nearest square foot, and, of course, cost would be rounded to at least the nearest cent. So a more accurate mathematical model would be a step function (see Lesson 2-4). In contrast, because the finisher would most likely round to the nearest square foot before using *any* model to calculate the price, the linear one is perfectly accurate on that restricted domain, and it is simpler. Similar restrictions would apply to estimating area for given costs, as in part **d** of the example—always round the answer up to the nearest whole foot.

Example 2 was deliberately designed to use integral values in the domain and range. You may want to mention that (11, 5) appears to be an outlier. Clearly, Chomp is not a "best buy." A teaching aid is provided to assist with discussion of this example.

Pointing out the meaning of the prefixes may help students remember the difference between interpolation and extrapolation: *inter* = between; *extra* = outside.

Alternate Approach: Technology You may want to do the Extension activity which follows as a classroom demonstration.

Brand	Cost/Stick	Rating
Chomp	11¢	5
Wow	7	5
Slosh	12	8
Yummo	10	7
Orsum	9	6
Blowup	15	9
Slurp	13	8
Yes!	6	5

The scatter plot above on the right shows that, with the exception of the point (11, 5), all the data seem to lie close to a straight line.

One way of finding a linear model for a data set is to make a scatter plot and to place a transparent ruler on the plot and move it until its edge seems to give a line close to the data. For a good linear model about half the data points are on either side of the line and most of the points are close to the line. A useful linear model need not pass through any of the data points exactly.

Example 2 **a.** Find a linear model for the relation between price and quality of bubble gum.
b. Use the model to predict the rating of gum which costs 4¢/stick.

Solution

a. Find an equation for the line above. Let *x* = the price in cents, and *y* = the quality. Use (15, 9) and (6, 5) to find the slope *m*.

$$m = \frac{y_2 - y_1}{x_2 - x_1} = \frac{9 - 5}{15 - 6} = \frac{4}{9}$$

Substitute $m = \frac{4}{9}$ and (6, 5) in the point-slope equation of a line.

$$y - 5 = \frac{4}{9}(x - 6)$$

Solve for *y*. $\qquad y = \frac{4}{9}x + \frac{7}{3}$

b. To predict quality from price, substitute *x* = 4 and solve for *y*.

$$y = \frac{4}{9}(4) + \frac{7}{3} = \frac{37}{9} \approx 4.1$$

A rating of about 4 is predicted.

Check

b. Plot (4, 4.1) on the graph. The point seems to follow the trend for the relation between quality and price.

The function $y = \frac{4}{9}x + \frac{7}{3}$ is a good model for the relation between price and rating of quality except for the Chomp brand. It reflects the general trend that for every increase of 1 cent in price (one unit of *x*) there is an increase of about $\frac{4}{9}$ point in quality rating (one unit of *y*).

88

This model can be used to make predictions between known values of data, such as to predict what the quality rating would be for a price of 8¢/stick. Prediction like this is called **interpolation**. Prediction made beyond the known values of the data, such as that made in Example 2, is called **extrapolation**. Extrapolation is often more hazardous than interpolation, because it depends on an assumption that a relationship will continue past the known data. For instance, for a price of 20¢ per stick, the model $y = \frac{4}{9}x + \frac{7}{3}$ predicts a rating of $\frac{4}{9}(20) + \frac{7}{3} = \frac{101}{9} \approx 11.2$. But the rating scale only goes up to 10! So the model cannot possibly hold for large values of x.

Questions

Covering the Reading

1. Define a linear function.   **See margin.**

2. Refer to Example 1.
   a. *True* or *false*. C is a linear function of A.   **True**
   b. What will the firm charge to sand and varnish a floor measuring 18 ft by 10 ft?   **$410**
   c. What size floor can be sanded and varnished for $400?
      175 square feet

3. Refer to Carla's jogging table and graph.
   a. About how far can Carla jog in a quarter of an hour?
   b. What is the slope of the line drawn on the scatter plot?   **5/3**
   c. What physical quantity does the slope represent?
   d. By how much does Carla's actual jogging time for 12 laps differ from what is predicted by the linear model?   **a) about 9 laps; c) the average time (in minutes) per lap; d) 10 seconds**

4. Consider the model $y = \frac{4}{9}x + \frac{7}{3}$ for the relation between price and quality of chewing gum.
   a. What does the slope mean?   **See margin.**
   b. Use this equation to predict the likely rating of gum which costs 8¢ a stick.   **about 6**
   c. According to this equation, what would a chewing gum with a rating of 3 be likely to cost?   **1.5¢ a stick**
   d. What is the largest possible domain for which the model can hold?
      0 < x ≤ 69/4

5. Prediction between known values of a data set is called __a__; prediction beyond known values is called __b__.
   a) interpolation; b) extrapolation

6. The thickness of a book is determined by the thickness of its cover and its pages. The front and back covers of a book like this one are each about 3 mm thick and every 25 pages adds an additional millimeter to the thickness. Let n be the number of pages and T be its thickness.
   a. How thick would a 600-page book be?   **30 mm**
   b. If total thickness is to be under 48 mm, what is the largest number of pages that can be in the book?   **1049 pages**
   c. Give an equation relating n to T.   **T = (1/25)n + 6**
   d. Graph the relation between n and T.   **See margin.**

LESSON 2-2   Linear Models   89

Small Group Work and Technology for Question 8: Students have to draw a line that fits the data, and not all students will have the same answer. It is important for students to realize that there is no single right answer. If you do not do the **small group** activity suggested under Extension, you may want to redo **Question 8** using technology.

Making Connections for Question 9: Discuss the limitations of the linear model. For instance, a line that fits reasonably well has the equation $y = 0.17x + 70$. But if that is the case, when $x = 85$, then $y = 84.5$, which means that a person at the age of 85 would have an expectation of living a negative number of years. Thus either the domain of the model needs to be restricted or a nonlinear model should be used. Students should save their work and answer to use in Question 10 of the next lesson, students are asked to use a statistical package to find the line of best fit for the same data. A teaching aid with these data is provided.

Question 12: A teaching aid is provided. Note that Euler's function notation can be used even when no algebraic formula is available.

FOLLOW-UP

MORE PRACTICE
For more questions on SPUR Objectives, use *Lesson Master 2-2,* shown on page 91.

Small Group Work After discussing the Questions, you may want to do the following activity. Ask students whether they expect the relation between wrist size and neck size to be linear. Why? Then divide the class into groups. Give each group some string and a meter stick. Have them measure the circumferences in centimeters of the wrist and the neck of each student in the group. Graph the ordered pairs (wrist, neck) on a single sheet of acetate. Use a transparent ruler, and "eyeball" a line of fit. Find the slope, *y*-intercept, and equation of the line. Discuss the relation between the two variables and what these characteristics mean. Then enter the data in a statistical package and show students how to find the line of "best fit." Compare and contrast the values of the slope and *y*-intercept determined by the package to those of the line found by eye-balling. You may also want to compare your data with that of Example 3 in Lesson 2-3.

Be sure to emphasize that a linear model does not have to pass through any of the data points. Students tend to want to make the line go through a data point in the "middle" of the set. Yet, in one sense, that thinking is on the right track. As is pointed out in the next lesson, the line of best fit does contain the center of gravity of the data set. But the center of gravity does not have to be one of the points.

As you discuss where to place the ruler for your linear model mention that in addition to wanting about half the points on either side of the line, you do not want a situation like the following:

This has all the points at both ends below the line and would best be illustrated with a curvilinear model:

---

7. *Multiple choice.* For which of the following scatter plots would a linear model be least suitable? **(b)**

8. The table below gives the percent of married women, with children aged 6 to 17, in the labor force in the United States, from 1950 to 1987.

Year	1950	1960	1970	1975	1980	1985	1987
Percent in labor force	28.3	39.0	49.2	52.3	61.7	67.8	70.6

(Source: U.S. Department of Commerce, Bureau of the Census)

a. Draw a scatter plot of these data with Year on the horizontal axis.
b. Draw a line which seems to fit the data. **a, b) See margin.**
c. Find an equation for your line. **sample: $P = 1.14Y - 2195$**
d. According to your answer in part **c**, about what percent of married women, with children aged 6 to 17, were in the labor force in the year: **i.** 1982? **ii.** 1900? **See margin.**
e. According to the model, what percent of women, with children aged 6 to 17, will be in the labor force in the year 2000? **85%**
f. Over what domain do you think your model accurately describes the relation between participation in the labor force and year? **sample: 1940 to 2000**

9. The following table lists estimated life expectancy for people of different ages.

Current age	Expected life span	Current age	Expected life span	Current age	Expected life span
1	75.1	30	76.3	60	80.2
5	75.2	35	76.6	65	81.7
10	75.3	40	77.0	70	83.5
15	75.4	45	77.5	75	85.7
20	75.7	50	78.1	80	88.1
25	75.7	55	79.0	85	91.1

a. Make a scatter plot of the data. Use current age as the independent variable. (Hint: Scale the vertical axis from 72 to 92 with 1-yr units.)
b. Draw a line which seems to fit the data. **a, b) See margin.**
c. Find an equation for your line. **sample: $y = 0.16x + 72.6$**
d. According to your answer in part **c**, what is a person's life expectancy at birth? **72.6 years**
e. *Multiple choice.* What does the slope of your line represent?
  (i) expected months of life span gained per year of age
  (ii) expected months of life span lost per year of age
  (iii) expected years of life span gained per year of age
  (iv) expected years of life span lost per year of age **(iii)**
f. Calculate the life expectancy of a person age 15 according to your model. By how much does it differ from the corresponding value in the table? **75 years; 0.4 years**

90

g. At what period in life, if any, is a person's life span greater than that predicted by your model?　**0 to 20 years, 70 to 85 years**

10. Let $f(x) = -3x + 5$.　*(Lesson 2-1)*
  a. Evaluate $f(-3)$.　**14**　 **b.** Evaluate $2f(-1.5)$.　**19**
  c. *True* or *false*.　$f(1 + 6) = f(1) + f(6)$. Justify your answer.　**See margin.**

11. Let $g(x) = \frac{1}{2}x^2$. *(Lesson 2-1, Previous course)*

  a. Draw a graph of $y = g(x)$.　 **b.** State the domain and range of g.
  See margin.

12. The graph at the right is from *Weather on the Planets*, by George Ohring. It shows how temperature is a function of latitude on Earth and Mars when it is spring in one hemisphere on each planet. Let L be the latitude on each planet, $E(L)$ be the average temperature on the Earth, and $M(L)$ be the average temperature on Mars at latitude L.

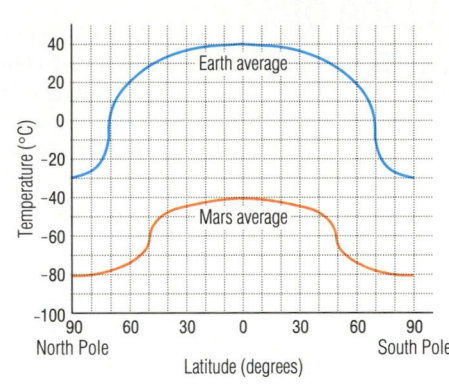

Martian surface in the Chryse area photographed by NASA's Lander 1

  a. Estimate $E(60)$.　**about 20°C**
  b. Estimate $E(0) - M(0)$, and state what quantity this expression represents.
  c. State whether L is the dependent or independent variable.
  d. Which planet is colder? Justify your answer.　*(Lesson 2-1)*
  b) See margin. c) independent; d) See margin.

In 13 and 14, solve the system.　*(Previous course)*

13. $\begin{cases} 10a + b = 13 \\ 30a + b = 25 \end{cases}$
  $a = \frac{3}{5}, b = 7$

14. $\begin{cases} 3x - 2y = -17 \\ 5x + y = 2 \end{cases}$
  $x = -1, y = 7$

15. According to the Federal Bureau of Investigation, 17,545 people were murdered in the U.S. in 1980. For 10,296 of these, the murder weapon was a gun. What percent of 1980 murders were committed with a gun?
(Previous course)　**about 58.7%**

16. Pick some task that is at least one minute long and that can be repeated many times without stopping, like Carla's jogging task in this lesson, or walking from one place to another, or playing a piece with a musical instrument, or singing a song. **a.** Do this task at least 10 times, trying to make the duration of each task as similar as you can. **b.** Keep track of the cumulative times (to the nearest second) and make a table like that on page 87. **c.** Graph the points in your table and find an equation for a line which closely fits the data. **d.** Use this information to rate your ability to keep the times consistent.　**Answers will vary.**

ADDITIONAL ANSWERS
8.a.,b.,d., 9.a.,b., 10.c., 11.a.,b., 12.b.,d. See Additional Answers in the back of this book.

RESOURCES
■ Lesson Master 2-3
■ Quiz for Lessons 2-1
  through 2-3
■ Teaching Aid 17: Graphs
  for pages 92 and 93
■ Teaching Aid 18: Graph for
  Example 3
▣ Computer Master 2

OBJECTIVES

D Identify properties of the
  line of best fit.
E Find and interpret linear
  models.
J Use scatter plots to draw
  conclusions about models
  for data.

TEACHING NOTES

Error Analysis Make
sure students understand the
terms *observed values, predicted values,* and *errors.*
The line of best fit minimizes
the sum of the squared
errors. The definition of *error*
is important; it is the *vertical*
difference between the data
point and the corresponding
point on the line of best fit.
Students sometimes confuse
error with the perpendicular
distance to the line. The following diagram clarifies this
misconception.

The vertical distance is always larger than the perpendicular distance. A teaching
aid, containing the first four
graphs of the lesson, is
provided.

92

LESSON

2-3

The Line of Best Fit

In the previous lesson, we found a linear model from a scatter plot by drawing
a line close to all the data points. In this lesson we present a procedure for
constructing a linear model which fits a set of data points better than any other
line. This line is called the *line of best fit* or *regression line*.

Data collected from sources such as experiments and surveys are called
observed values. The points predicted by the linear model are called
predicted values. The line of best fit is found by minimizing the **errors**
or **deviations** in the predictions, which are the differences between observed
and expected values of the dependent variable.

Consider a very simple case with
three data points (2, 5), (6, 3),
and (10, 7). (In practice it
would not be wise to try to build
a mathematical model, linear or
otherwise, for such a small data
set. This example is used only
to illustrate a procedure.) Start
with a scatter plot, like the one
on the right.

Here are two possible linear models, *A* and *B*, for these data:

To decide which of these linear models fits the data better, examine the errors
in each case. Here we define "error" to be "observed *y* − predicted *y*." Each
error is the directed length of a vertical segment on the graph, and can be positive or negative.

| | line *A* | | | line *B* | | |
x	observed *y*	predicted *y*	error	observed *y*	predicted *y*	error
2	5	6	−1	5	5	0
6	3	5	−2	3	5	−2
10	7	4	3	7	5	2

Notice that the sum of the errors for each model is zero, and thus the mean of
the errors is zero for both models. Thus, a comparison of the mean of the errors is not sufficient to tell which of *A* or *B* is a better model. A better comparison uses the *sums of the squares of the errors* instead. The smaller the sum,
the better the line fits the data.

For line A, the sum of the squares of the errors is $(-1)^2 + (-2)^2 + (3)^2 = 14$. For line B, the sum of the squares of the errors is $(0)^2 + (2)^2 + (-2)^2 = 8$. So line B fits the data better than line A, according to the criterion of least sum of squares.

The **line of best fit** is the line with the smallest value for the sum of the squares of the errors. For this reason, the process of finding the line of best fit is sometimes called the **method of least squares**.

A method for finding the line of best fit was published by the French mathematician Adrien Legendre in 1805. Using calculus, Legendre found formulas for the slope m and y-intercept b of the line of best fit for any set of data. The formulas are rather complicated.

Fortunately, statistics packages and some calculators have been programmed to calculate b and m quickly, so you don't have to calculate them by hand. Statistics packages that find the line of best fit usually call this line either "Least Squares Fit" or "Regression Line." With a statistics package, for the three data points shown earlier we find the slope m and the y-intercept b of the line of best fit are $m = 0.25$ and $b = 3.5$. So an equation of the line of best fit is $y = 0.25\,x + 3.5$.

The line for $y = 0.25\,x + 3.5$ is graphed at the right with the original data.

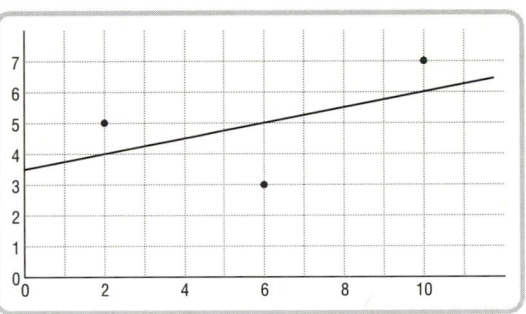

Example 1 Find the sum of the squares of the errors for the regression line $y = 0.25x + 3.5$.

Solution First find the predicted y-values for the three known x-values.
At $x = 2$, $y = 0.25(2) + 3.5 = 4$.
At $x = 6$, $y = 0.25(6) + 3.5 = 5$.
At $x = 10$, $y = 0.25(10) + 3.5 = 6$.
The observed values are as follows.
At $x = 2$, $y = 5$.
At $x = 6$, $y = 3$.
At $x = 10$, $y = 7$.
Thus the sum of the squares of the errors for the regression line is:
$(5 - 4)^2 + (3 - 5)^2 + (7 - 6)^2 = 1^2 + (-2)^2 + 1^2 = 6$.

Observe that this measure of error is indeed less than that of either of the other two lines.

Although you will be using a computer to find the line of best fit for a set of data, there is always one point on the line you can determine by hand. That point is the **center of gravity** of the data. The coordinates of the center of gravity are the mean of the observed x-values and the mean of the observed y-values.

Example 2 Find the center of gravity of the original three data points, and verify that it is on the line of best fit, $y = 0.25x + 3.5$.

Solution The mean of the x-coordinates is $\frac{2+6+10}{3} = \frac{18}{3} = 6$. The mean of the y-coordinates is $\frac{5+3+7}{3} = \frac{15}{3} = 5$. So, the center of gravity is the point (6, 5). Substituting (6, 5) into the equation for the line of best fit gives $5 = .25(6) + 3.5$ or $5 = 1.5 + 3.5$, which is true.

Notice that lines A and B also contain the center of gravity of the data. The center of gravity alone is not sufficient to determine the line of best fit.

Example 3 There are several theories about relations between measurements of different parts of the body (called *anthropometrics*). One theory states that there is a linear relationship between the wrist circumference (C_w) and the neck circumference (C_n). To study this theory, the data below were gathered from seventeen persons.

C_w in cm	C_n in cm	C_w in cm	C_n in cm
14.4	31.0	16.3	36.4
14.8	33.5	16.7	37.3
15.0	33.0	16.9	35.3
15.4	35.2	17.1	36.6
15.7	34.2	17.2	38.5
16.1	33.7	17.6	39.8
16.2	32.2	17.8	38.3
16.2	35.5	17.9	37.1
16.2	35.8		

Notice that the wrist circumference of 16.2 cm is matched with three different neck circumferences (32.2, 35.5, and 35.8) in the data above. This leads to three y-coordinates corresponding to the same x-coordinate. Thus the wrist vs. neck circumference data is not a function. However, we use a function, the line of best fit, to model it. The discrepancy is tolerable, and the equation of the line allows for predictions.

a. Use a computer to find the line of best fit for predicting neck size y (in cm) from wrist size x (in cm).
b. What is the likely neck size of a person with wrist circumference 16 cm?
c. What is the likely neck size of a person with wrist circumference 6.5 in.?

94

Solution

a. Enter the data and follow the recommended procedure for the statistics package you use. The line of best fit shown below on the scatter plot is from our package. An equation for the line is $y = 1.92x + 4.23$.

b. Substitute $x = 16$.
$y = 1.92(16) + 4.23 = 34.95$
A neck size of 35.0 cm is predicted.

c. First convert 6.5 inches to centimeters:
$$x = 6.5 \text{ in.} \times 2.54 \tfrac{\text{cm}}{\text{in.}} = 16.51 \text{ cm}$$
$$y = 1.92(16.51) + 4.23 \approx 35.93.$$
A neck size of about 35.9 cm (14.1 inches) is predicted.

Questions

Covering the Reading

1. For a data set the line of best fit is the line in which the sum of the squares of the differences between observed and **a** values is **b**.
   a) predicted; b) smallest
2. *True* or *false*.   Another name for the line of best fit is regression line.
   True

In 3–5, a student fitted the line ℓ to the data points $(2, 1)$, $(4, 5)$, and $(6, 3)$.

3. **a.** What is the observed value of y at $x = 2$?   **1**
   b. What is the predicted value of y at $x = 2$?   **3**
   c. Find the error of each of the three points from line ℓ.
   d. Find the sum of the squares of the deviations of the three points from line ℓ.   **7.69**
   c) -2, 1.2, -1.
4. **a.** Find the center of gravity of the data.   **(4, 3)**
   b. Is the center of gravity on line ℓ?   **No**
   c. What tells you that line ℓ cannot be the line of best fit for the data?
   The center of gravity is not on line ℓ.

Regarding (2), if the slope of the line of best fit is calculated to too few decimal places, any prediction is likely to be poor. The level of accuracy of the values of m and b affects all statistics that depend on those values. In general, it is better not to round until the end of all computations.

ADDITIONAL ANSWERS
5.b.

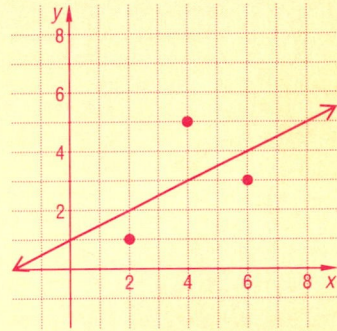

d. The center of gravity is on this line, and the sum of the squares of the deviations is smaller than in Question 4.

7.a.,b.,d.

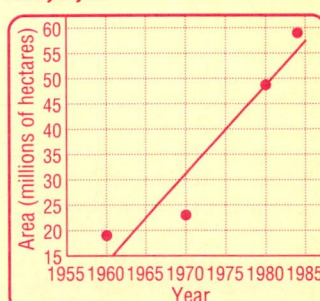

b. $y = 29$

d. The equation in part c is a better model, since it contains the center of gravity of the data and has a smaller total for the sum of the squares of the errors.

9.a.

10.a.

**c. The length of the
winning long jump
increases very slightly
(about 1.8 cm) from one
year to the next.**

**e. The center is
(1941, 7.8), which the
regression line contains.**

**f. (1896, 6.34) and
(1968, 8.90) appear to be
outliers and are likely the
result of, respectively, a
weak field of competitors
and personal excellence.**

12.a.

Applying the Mathematics

5. Using a computer program which finds regression lines, the student read the following output: M = 0.5   B = 1
   a. What is an equation of the line of best fit?   **y = 0.5x + 1**
   b. Graph the data and the line of best fit.   **See margin.**
   c. Find the sum of the squares of the deviations.   **6**
   d. How do you know that this line is a better fit than line ℓ?
      See margin.

6. Verify that the center of gravity of the (wrist, neck) data in Example 3 is on the line of best fit.
   The center is (16.32, 35.49); 35.49 ≈ 1.92(16.32) + 4.23.

7. Flooding in India is increasing because of deforestation. Here is the area of land (in millions of hectares) that has been subject to flooding.

Year	1960	1970	1980	1984
Area	19	23	49	59

(Source: Centre for Science and Environment, New Delhi, India, 1987)

   a. Graph these points.   **a, b) See margin.**
   b. Estimate and graph a line of best fit for these points.
   c. A computer program for a line of best fit gives the following output:
      M = 1.7320   B = -3381
      Write an equation of this line.   **y = 1.7320x − 3381**
   d. Graph the line in part **c.** Which is a better model of the data—your equation in part **b** or the equation in part **c**?   **See margin.**

8. Refer to the data on wrist and neck sizes in Example 3. Let x = neck circumference and y = wrist circumference.
   a. Find the line of best fit for predicting wrist size from neck size.
   b. What is the predicted wrist size of a person with a neck circumference of 37.5 cm?   **about 17 cm**
   c. What is the predicted wrist size of a person with a neck circumference of 17 in.?   **about 19.1 cm or 7.5 in.**
   a) sample: y = 0.368x + 3.247

9. The following table lists again the estimates of life expectancy for people of different ages.

Current age	Expected life span	Current age	Expected life span	Current age	Expected life span
1	75.1	30	76.3	60	80.2
5	75.2	35	76.6	65	81.7
10	75.3	40	77.0	70	83.5
15	75.4	45	77.5	75	85.7
20	75.7	50	78.1	80	88.1
25	75.7	55	79.0	85	91.1

   a. Make a scatter plot of the data. Use current age as the independent variable.   **See margin.**
   b. Find an equation of the line of best fit.   **y = 0.165x + 72.3**
   c. Calculate the life expectancy of a person age 15 according to your model. By how much does it differ from the corresponding value in the table?   **74.8 years; 0.6 years**
   d. Compare your answers to parts **b** and **c** to your answers to the corresponding parts of Question 9 of Lesson 2-2. Which model in general produces the least amount of error?
      The model calculated here produces the least amount of error.

10. The following table shows the winning jumps in the men's long jump event at the Olympic games.

Carl Lewis

Year	Gold Medalist	Jump
1896	Ellery Clark, United States	6.34 m
1900	Alvin Kraenzlein, United States	7.19 m
1904	Myer Prinstein, United States	7.34 m
1908	Francis Irons, United States	7.48 m
1912	Albert Gutterson, United States	7.60 m
1920	William Pettersson, Sweden	7.15 m
1924	DeHart Hubbard, United States	7.45 m
1928	Edward B. Hamm, United States	7.74 m
1932	Edward Gordon, United States	7.64 m
1936	Jesse Owens, United States	8.06 m
1948	Willie Steele, United States	7.82 m
1952	Jerome Biffle, United States	7.57 m
1956	Gregory C. Bell, United States	7.83 m
1960	Ralph H. Boston, United States	8.12 m
1964	Lynn Davies, Great Britain	8.07 m
1968	Robert Beamon, United States	8.90 m
1972	Randy Williams, United States	8.24 m
1976	Arnie Robinson, United States	8.35 m
1980	Lutz Dombrowski, East Germany	8.54 m
1984	Carl Lewis, United States	8.54 m

a. Make a scatter plot of these data.   **See margin.**
b. Find the line of best fit predicting the winning jump for a given year.
c. What does the slope of the line tell you about the average rate of change in the length of the winning long jump?   **See margin.**
d. Use the line of best fit to predict the winning jump for the Seoul Olympics in 1988, and the error in the prediction (the actual jump by Carl Lewis was about 8.72 m).   **8.88 m; an error of 0.16 m**
e. Calculate the center of gravity of the data, and verify that the regression line passes through this point.
f. Which data points seem to be outliers here? Why are they so far from the linear model?

b) $y = 0.018x - 26.9$; e,f) See margin.

Review

11. Suppose $Q(x) = \sqrt{x + 7}$.
  a. Find $Q(2)$.   **3**
  b. What is the domain of Q? $\{x : x \geq -7\}$
  c. Give the range of Q.   *(Lesson 2-1)*   **range = the set of all nonnegative real numbers**

In 12 and 13, **a.** sketch a graph and **b.** identify the domain and range.
(Lesson 2-1, Previous course)   **See margin.**

12. $y = \dfrac{12}{x}$

13. $y = \dfrac{12}{x^2}$

14. To develop a roll of film with up to 36 exposures, a photo lab charges $1.49 per roll plus $.15 per print. Let n = the number of prints developed and $f(n)$ = the cost of developing a roll of film with n prints.
  a. Write a formula for $f(n)$. $f(n) = 1.49 + .15n$
  b. What is the domain of f?   **all integers n such that $0 < n \leq 36$**
  c. Find the charge for developing a roll of film and making 20 prints.
  d. How many prints were made if the cost of developing and printing was $6.29? *(Lesson 2-2)*   c) $4.49; d) 32

NAME _____

LESSON **MASTER 2-3**
QUESTIONS ON **SPUR** OBJECTIVES

■**PROPERTIES** *Objective D (See pages 138–142 for objectives.)*
In 1 and 2, *true* or *false*. Consider the line k of best fit for a data set.

1. Line k passes through the center of gravity of the data set.   **True**

2. The sum of the squares of the deviations of observed values from predicted values is greater with k than with any other line.   **False**

■**USES** *Objective E*

3. The table contains data from the years 1950 to 1985 for the number of U.S. families with no children under the age of 18 living at home. *(Source: Youth Indicators 1988)*

Year (after 1900)	50	55	60	65	70	75	80	85
Number of families (1000s)	20,267	23,181	25,662	27,073	28,666	30,057	30,517	31,112

  a. Find an equation of the line of best fit for the data set. $y = 303x + 6596$
  b. Use this equation to predict the likely number of families with no children younger than 18 living at home in 1990.   **33,866,000**
  c. Over what domain would you expect this model to hold?   **sample: $35 \leq x \leq 100$**

4. The table below contains information about the record times of the women's 100-meter run from 1911 to 1974. *(Source: Progression of World Best Performances and Official IAAF World Records, 1987)*

Year	Winner	Time	Year	Winner	Time
1911	Siina Simila, Finland	18.8	1935	Helen Stephens, U.S.A.	11.6
1913	Nina Popova, USSR	13.1	1936	Helen Stephens, U.S.A.	11.5
1920	Marie Kiessling, Germany	13.0	1952	Marjorie Jackson, Australia	11.4
1921	Marie Kiessling, Germany	12.8	1955	Shirley Strickland, Australia	11.3
1925	Emmy Haux, Germany	12.7	1961	Wilma Rudolph, U.S.A.	11.2
1925	Helene Schmidt, Germany	12.4	1965	Irena Kirszenstein, Poland	11.1
1928	Myrtle Cook, Canada	12.0	1968	Wyonia Tyus, U.S.A.	11.0
1932	Tollien Schuurman, Holland	11.9	1973	Renate Stecher, East Germany	10.9
1933	Stanislawa Walasiewicz, Poland	11.8	1973	Renate Stecher, East Germany	10.8
1934	Stanislawa Walasiewicz, Poland	11.7			

NAME _____
Lesson MASTER 2-3 (page 2)

  a. Plot these data, using the horizontal axis for the year (after 1900) the record was set.
  b. Find the line of best fit, letting x be the year since 1990. $y = -.038x + 13.41$
  c. Use the line of best fit to predict the record for this event in 1988.   **10.07 seconds**
  d. Use the line of best fit to predict when the record for this event will be 10 s.   **in 1990**

■**REPRESENTATIONS** *Objective J*
In 5 and 6, a student fit a line ℓ to the data points (0, 5), (2, 2.5), (4, 5), and (6, 2), as shown at right.

5. a. What is the observed value of y at $x = 4$?   **5**
  b. What is the predicted value of y at $x = 4$?   **3**
  c. Find the error of each of the four points from line ℓ.   **-1, -2, 2, .5**
  d. Find the sum of the squares of the deviations of the four points from line ℓ.   **9.25**

6. Using technology, the student found that an equation of the line of best fit is $y = -0.325x + 4.6$.
  a. Graph the four data points and the line of best fit.
  b. Find the sum of the squares of the deviations.   **5.575**
  c. Verify that the center of gravity is on this line.   **center of gravity: (3, 3.625); $3.625 = -0.325(3) + 4.6$**
  d. How do you know that this line is a better fit than line ℓ?   **Sum of the squares of the deviations is much smaller for this line than it is for ℓ.**

NOTES ON QUESTIONS
Question 19: These data are from the Second International Mathematics Study of 1981–82. The comparison is quite misleading because virtually all students take mathematics at 8th grade and quite a smaller percentage (16% at the time of that study, but higher now) take mathematics in 12th grade.

FOLLOW-UP

MORE PRACTICE
For more questions on SPUR Objectives, use *Lesson Master 2-3,* shown on page 97.

EXTENSION
Alternate Approach
Our approach is to present the reasoning which leads to the formulas, and allow students to accept the computer's calculations as accurate. However, you may wish at least to present the formulas to your students.

Let X_i and Y_i represent the observed values of the two variables, having means $\bar{x}$ and $\bar{y}$ respectively. If $x_i = X_i - \bar{x}$ and $y_i = Y_i - \bar{y}$ are the deviations of the data from their means, then the slope m and y-intercept b of the line of best fit are:

$$b = \frac{\sum\limits_{i=1}^{n} x_i y_i}{\sum\limits_{i=1}^{n} x_i^2} \text{ and } m = \bar{y} - b\bar{x}.$$

The formula for b comes from using calculus to minimize a generalized sum of squares developed the same way as the examples in the lesson.

PROJECTS
The projects for Chapter 2 are on pages 133 and 134. All are related to the content of this lesson and could be started at this point.

EVALUATION
A quiz covering Lessons 2-1 through 2-3 is provided in the Teacher's Resource File.

---

15. Which, interpolation or extrapolation, generally gives a more accurate prediction? *(Lesson 2-2)*   **interpolation**

16. Suppose $f(x) = x^2 - 9$. For what value(s) of x does $f(x) = 0$? *(Previous course)*   **-3, 3**

In 17 and 18, solve. *(Previous course)*

17. $a^3 = 4^6$ **16**

18. $(-2)^n \cdot (-2)^5 = (-2)^{18}$ **13**

19. The pie charts below display the attitudes of eighth- and twelfth-grade mathematics students about mathematics, as found in an international study. Which of the following statements is *true* based solely on the information in the two charts?

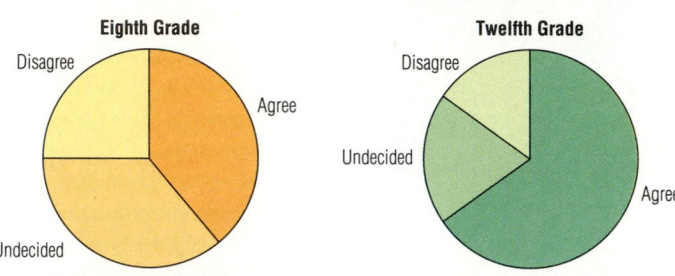

I would like to work at a job that lets me use Mathematics.

Eighth Grade — Disagree, Agree, Undecided

Twelfth Grade — Disagree, Undecided, Agree

(a) A larger percent of mathematics students in the twelfth grade than in eighth grade have an opinion about working at a job which requires use of mathematics.

(b) About 40% of mathematics students in the eighth grade seem inclined to pursue a mathematical orientation in their future careers.

(c) A larger proportion of mathematics students in the twelfth grade are looking for jobs. *(Lesson 1-2)*   **a, b**

Exploration

20. Legendre, who developed the method of least squares, was a mathematician of great breadth and originality. In addition to his work in calculus and statistics, he contributed to number theory and geometry. Find out more about his contributions to mathematics.   **Sample: Some of Legendre's most significant contributions resulted from his study of gravitational attraction in a branch of applied mathematics called potential theory. From this work came a class of functions known as Legendre polynomials.**

Legendre's contributions in number theory were fundamental; the Legendre symbol is named after him. He also formulated a version of Euclidean geometry upon which many modern versions are based.

98

Step Functions

What function models the following situation?

> You have *a* dollars available to spend on cassettes each of which costs $8.99. How many cassettes *n* can you buy?

You know from your study of algebra that if you buy *n* cassettes, the cost is $8.99n$.

From this you might conclude that $a = 8.99n$, so $n = \dfrac{a}{8.99}$.

But this conclusion does not take into account the fact that *n* can only be a positive integer. For instance, if you have $30 to spend, you cannot buy $\dfrac{30}{8.99} \approx 3.3$ cassettes. You can buy only 3.

The table below shows some pairs of amounts of money available and the number of cassettes you can buy for that amount. These data are also shown in a scatter plot below.

Amount available in dollars	5	10	15	20	25	30	35	40	45	50
Number of cassettes	0	1	1	2	2	3	3	4	5	5

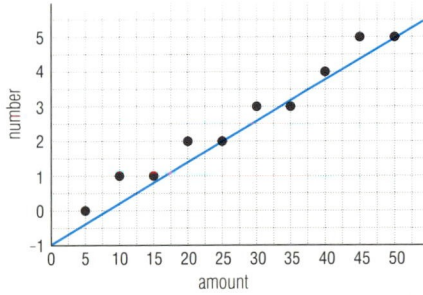

Clearly, these data show a linear trend. But neither the simple linear model you get from elementary algebra techniques,

$$n = \frac{a}{8.99},$$

nor the line of best fit shown above,

$$n = .112a - .467,$$

gotten from a statistical package, is as precise as can be.

An exact model which describes the preceding situation can be found in the family of functions we call *step functions*. Step functions are functions whose graphs look like steps. Many step functions are based on the *greatest integer function*.

RESOURCES
■ Lesson Master 2-4
■ Teaching Aid 18: Graph for **Example 2**

OBJECTIVES

A Evaluate functions using Euler's notation.
C Identify the variables, domain, and range of functions.
F Use step functions to model situations.
H Graph step functions.
I Interpret properties of relations from graphs.
J Use scatter plots to draw conclusions about models for data.

TEACHING NOTES

Error Analysis Some students have a hard time evaluating the floor and ceiling functions when the value of the independent variable is negative, as in **Example 1a.** Using number lines can help to clarify the situation. The direction of the greatest integer function is always to the left; $\lfloor x \rfloor$ less than or equal to *x*; for example, $\lfloor -4.3 \rfloor = -5$. This is easily seen in terms of rounding down.

If students have a greatest integer key INT on their calculator, have them test it for correct output of the function for both positive and negative arguments. Most versions of BASIC have an INT command meaning "greatest integer" function. Some computer languages and calculators use INT to mean "integer part of" or "truncate." If a machine gives -7 for INT(-6.7), then INT means greatest integer. If it gives -6, then INT means truncate.

Definition

The **greatest integer function** is the function f defined on all real numbers such that $f(x)$ is the greatest integer less than or equal to x.

The greatest integer function is also called the **rounding-down** or the **floor function**. The greatest integer less than or equal to x is often denoted by the symbol $\lfloor x \rfloor$. In some computer and calculator languages, $\lfloor x \rfloor$ is represented by INT(x).

Example 1   Evaluate **a.** $\lfloor -2.2 \rfloor$ **b.** $\left\lfloor \dfrac{29}{10} \right\rfloor$ **c.** INT (-1)

Solution   Each expression involves the greatest integer or rounding-down function.

a. $\lfloor -2.2 \rfloor = -3$ because the greatest integer less than or equal to -2.2 is -3.

b. $\left\lfloor \dfrac{29}{10} \right\rfloor = 2$ because rounding down 2.9 gives 2.

c. INT $(-1) = -1$.

Example 2   **a.** Draw a graph of $f(x) = \lfloor x \rfloor$.
   b. State the domain and range of f.

Solution

a. For any integer x, the greatest integer less than or equal to x is x. For any real number that is not an integer, we round down to find $\lfloor x \rfloor$. The graph is drawn at the right. Note that the symbol ●—○ indicates that a segment includes a left endpoint, but not a right one. You should check the values for Example 1 by locating them on the graph. They are $(-2.2, -3)$, $\left(\dfrac{29}{10}, 2 \right)$, and $(-1, -1)$.

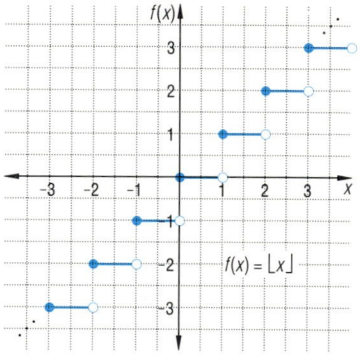

$f(x) = \lfloor x \rfloor$

b. The domain is the set of all real numbers. The range is the set of integers.

A function is **continuous** if its graph can be drawn without lifting the pencil off the paper and **discontinuous** if it cannot be drawn that way. Observe that the greatest integer function is a discontinuous function. At each integer in the domain of f there is a **point of discontinuity**.

100

Example 3 **a.** Find an exact formula for the number of cassettes n you can purchase at $8.99 each if you have a dollars.
b. Identify the points of discontinuity of this function.

Solution

a. The strategy used to find n is to divide the amount available by $8.99 and round down. Thus

$$n = \left\lfloor \frac{a}{8.99} \right\rfloor.$$

b. The function is discontinuous at each point where $\frac{a}{8.99}$ is an integer. Some points of discontinuity occur when $a = 8.99, 17.98, 26.97,\ldots$.

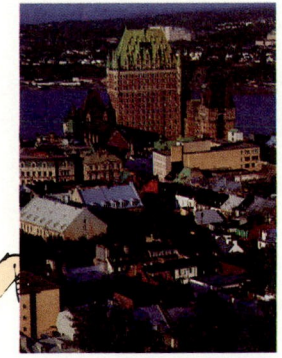
Quebec City, Quebec, Canada

The function which pairs each number x with the smallest integer greater than or equal to x is called the **rounding up** or **ceiling function**. The symbol $\lceil x \rceil$ is often used to denote the smallest integer greater than or equal to x. For instances, $\lceil 4.1 \rceil = 5$ and $\lceil 4.9 \rceil = 5$. The ceiling function also provides exact models for some situations which are approximately linear.

For example, with one long distance phone company, to call Quebec City from Detroit on a Sunday in 1990, it costs 33¢ for the first minute and 32¢ for each additional minute or fraction thereof.

Let m = the number of minutes of the call, and c = the cost of the call. A linear model which approximates this situation is
$$c_1 = .33 + .32(m - 1).$$
An exact model for this situation is
$$c_2 = .33 + .32 \lceil m - 1 \rceil.$$

Example 4 Calculate the cost of a call lasting 25 minutes 30 seconds to Quebec City from Detroit on a Sunday.

Solution 25 minutes 30 seconds is 25.5 minutes. Substitute $m = 25.5$ into the exact model.
$$c = .33 + .32 \lceil 25.5 - 1 \rceil$$
$$= .33 + .32 \lceil 24.5 \rceil$$
$$= .33 + .32 (25)$$
$$= .33 + 8.00$$
$$c = 8.33$$
It will cost $8.33 for the call.

Detroit, Michigan

The floor and ceiling functions are often used in computing. Their applications include rounding and testing for divisibility. For example, the BASIC program below prints out the factors of a number.

```
10   REM FINDING FACTORS
20   INPUT "NUMBER";N
30   FOR I = 1 TO N
40     X = N/I
50     IF X = INT(X) THEN PRINT I
60   NEXT I
70   END
```

The decision-making line of this program is line 50. If X = INT (X), then X must be an integer, indicating that I is a factor of N.

Questions

Covering the Reading

In 1–4, evaluate.

1. $\lfloor 25.9 \rfloor$ **25**   **2.** $\lfloor -25.9 \rfloor$ **-26**   **3.** INT(7/8)   **0**   **4.** $\lfloor 16 \rfloor$ **16**

5. For what values of x does $\lfloor x \rfloor = x$?   **all integers**

6. *Multiple choice.*   Refer to the graph at the left of the greatest integer function. What part of the domain of the function does this portion of the graph represent?
(a) the set of numbers between 0 and 1
(b) the set of numbers between 0 and 1 including 0
(c) the set of numbers between 0 and 1 including 1
(d) the set of numbers from 0 to 1   **(b)**

7. Let $f(a) = \frac{a}{8.99}$, $g(a) = \left\lfloor \frac{a}{8.99} \right\rfloor$, and $h(a) = .112a - .467$.
a. Which function gives the exact number of cassettes costing $8.99 which can be purchased for *a* dollars?   **g**
b. Calculate $f(40)$, $g(40)$, and $h(40)$.   **$f(40) \approx 4.45, g(40) = 4, h(40) = 4.013$**
c. When $a = 40$, for which function is the difference between actual and predicted number of cassettes the greatest?   **f**

8. Suppose you can get one free Fresh Fizz with every four coupons.
a. How many Fresh Fizzes can you get with 11 coupons?   **2**
b. *Multiple choice.*   How many Fresh Fizzes can you get with *c* coupons?
(i) $\lfloor c \rfloor$ (ii) $\left\lfloor \frac{c}{4} \right\rfloor$ (iii) $\lfloor 4 \rfloor$ (iv) $\lfloor 4c \rfloor$ **ii**

9. A school bus holds 40 students. Suppose *n* students are going on a field trip.
a. Write a formula for $f(n)$, the number of buses needed for the trip, using the ceiling function.   **$f(n) = \left\lceil \frac{n}{40} \right\rceil$**
b. Verify that the function $g(n) = -\left\lfloor \frac{-n}{40} \right\rfloor$ also gives the number of buses needed.   **See margin.**

102

102

10. Let $f(x) = \lceil x \rceil$, the ceiling function.
    a. State the domain and range of f.   **See margin.**
    b. Draw a graph of $y = f(x)$ for $-3 \leq x \leq 4$.   **See margin.**
    c. State three values of x at which f is discontinuous.
       sample: -2, 0, 2

Applying the Mathematics

11. *Multiple choice.*   Until 1975, the U.S. Post Office charged equally for every ounce mailed. On Sept. 14, 1975, the rate was changed to 10¢ for up to one ounce and 9¢ for each additional ounce or fraction of an ounce. Which formula gives the cost of mailing a letter of x ounces at these rates?
    (a) $c(x) = 9 - 10 \lfloor 1 - x \rfloor$
    (b) $c(x) = 10 - 9 \lfloor x - 1 \rfloor$
    (c) $c(x) = 10 - 9 \lfloor 1 - x \rfloor$
    (d) $c(x) = 9 - 10 \lfloor x - 1 \rfloor$ **(c)**

12. Some airlines provide an Airfone® service, for passengers to make telephone calls from the air. Here is a graph of the cost of making calls of various duration.

    Airfone® Costs

    a. How much would a 10-minute phone call cost?   **$15.75**
    b. Is the cost per minute lower for several short calls or one longer call of the same total duration?   **one longer call**
    c. Express the cost, C, in dollars as a function of the length of call, L, in minutes.   **See margin.**

13. A saleswoman earns a $135 bonus for every $1000 worth of merchandise she sells.
    a. What bonus would she earn for selling $2700 worth of merchandise?
    b. Write an expression using either the floor or the ceiling function for the bonus she will earn when she sells M dollars worth of merchandise.   **a) $270; b) $135 \left\lfloor \dfrac{M}{1000} \right\rfloor$**

14. Run the program given in the lesson to find the factors of
    a. 105   b. 197   c. 7128.
    a) 1, 3, 5, 7, 15, 21, 35, 105; b) 1, 197; c) See margin.

15. a. Complete the table below.

X	X + 0.5	INT(X + 0.5)
12.4	12.9	12
12.7	13.2	13
4.49	4.99	4
5.50	6.00	6

    b. Describe the output of INT(X + 0.5) when any real number is input for X.   **The output is X rounded to the nearest integer; halves are rounded up.**

16. a. Describe in words the function $F(x) = x - \lfloor x \rfloor$, $x > 0$.
    b. Draw a graph of this function.
    See margin.

LESSON 2-4   Step Functions   **103**

FOLLOW-UP

MORE PRACTICE
For more questions on SPUR Objectives, use *Lesson Master 2-4,* shown on page 103.

EXTENSIONS
1. If there are stairs in your school, measure the depth of each stair and the horizontal distance from one stair to the next. Next pick a point at the bottom of the stairs to be the origin, and select horizontal and vertical lines through this point to be the coordinate axes. Determine an equation for the stairs.

2. Technology Students with programming experience could modify the BASIC program on page 102 to print only the prime factors of a whole number input by the user.

Sample:

```
10  REM FINDING
    PRIME FACTORS
20  INPUT "NUMBER";
    N
30  FOR I = 1 TO
    INT(SQR(N))
40  X = N/I
50  IF X<>INT(X)
    GOTO 130
60  C = 0
70  FOR J = 1 TO X
80  P = X/J
90  IF P = INT(P) THEN
    LET C = C + 1
100 NEXT J
110 IF C = 2 THEN
    PRINT X
120 NEXT I
130 END
```

ADDITIONAL ANSWERS
18.a., 21. See Additional Answers in the back of this book.

20.a. first: mean = 8.41, s.d. = 1.05; second: mean = 6.54, s.d. = 1.04

b. The first astronomer's data set has a larger mean than the second, but both astronomers are equally consistent.

Review

17. Tell whether the relation graphed is a function.   *(Lesson 2-1)*

a.

function

b.

function

c.

function

d.

not a function

18. Let $f(x) = 2x^2 + 5x$. *(Lesson 2-1, Previous course)*
   a. Graph $y = f(x)$ for $-3 \leq x \leq 3$.   **See margin.**
   b. For what values of x does $f(x) = 0$?   **0, -5/2**

19. Here are the world records for the women's 4×100 m track relay since 1922. (Source: International Athletic Foundation.)

Time (seconds)	Country	Year	Time (seconds)	Country	Year
53.2	Czechoslovakia	1922	43.9	United States	1964
51.4	Great Britain	1922	43.6	USSR	1968
50.4	Germany	1926	43.4	United States	1968
49.7	Germany	1928	42.8	United States	1968
48.4	Canada	1932	42.8	West Germany	1972
46.9	United States	1932	42.8	East Germany	1973
46.4	Germany	1936	42.6	East Germany	1974
46.1	Australia	1952	42.51	East Germany	1974
45.9	Germany	1952	42.50	East Germany	1976
45.6	USSR	1953	42.27	East Germany	1978
45.2	USSR	1956	42.09	East Germany	1979
45.1	Germany	1956	41.85	East Germany	1980
44.9	Australia	1956	41.60	East Germany	1980
44.5	Australia	1956	41.53	East Germany	1983
44.4	United States	1960	41.37	East Germany	1985
44.3	United States	1961			

   a. Find the line of best fit for predicting world records for this event.
   b. Predict the year in which the 40-second barrier will be broken for the first time. *(Lesson 2-3)*
   a) sample: $y = -6.106t + 2232.55$; b) 1988

20. Two astronomers, each at a different observatory in the same city, checked the angular separation (in degrees) of two newly discovered stars. Each measured the separation ten times. Here are their results.

First astronomer	8.03	7.56	8.12	6.88	10.06	8.64	9.35	7.15	9.53	8.82
Second astronomer	8.51	7.37	7.32	6.43	4.92	6.55	6.91	5.52	5.72	6.17

   a. Find the means and standard deviations of their measurements.
   b. Describe any differences between their results. Which astronomer is more consistent? *(Lesson 1-8)*
   See margin.

Exploration

21. Find the current charges for postage for different classes of mail. Can any of the postal rates be modeled by step functions? If so, which ones?
See margin.

104

2-5

Correlation

The scatter plot and line of best fit drawn below show how average annual precipitation is related to average annual temperature in 50 cities in the United States.

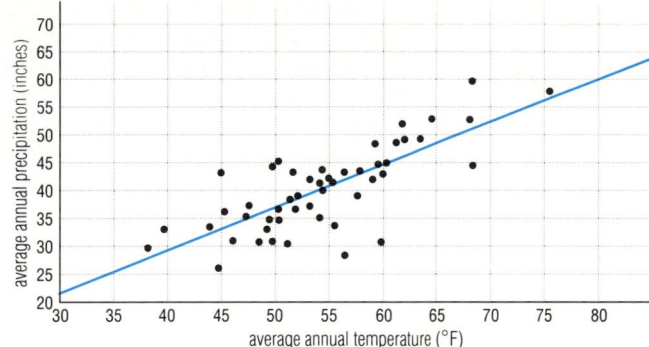

Many of the data points are close to the line of best fit. Overall, a line is a reasonably good model for these data. But how good is "reasonably good"?

To measure the strength of the linear relation between two variables, a measure called the **correlation coefficient** is used. The correlation coefficient, often denoted by the letter r, was first defined by the English statistician Karl Pearson (1857-1936).

Because the procedure for calculating the correlation coefficient is tedious, you are not expected in this course to calculate r by hand. The purpose of this lesson is to help you interpret a value of r when a computer or calculator calculates it for you.

The correlation coefficient is a number between -1 and 1. For the weather data above, $r \approx 0.76$. Some data sets and the corresponding values of r are shown in the scatter plots below.

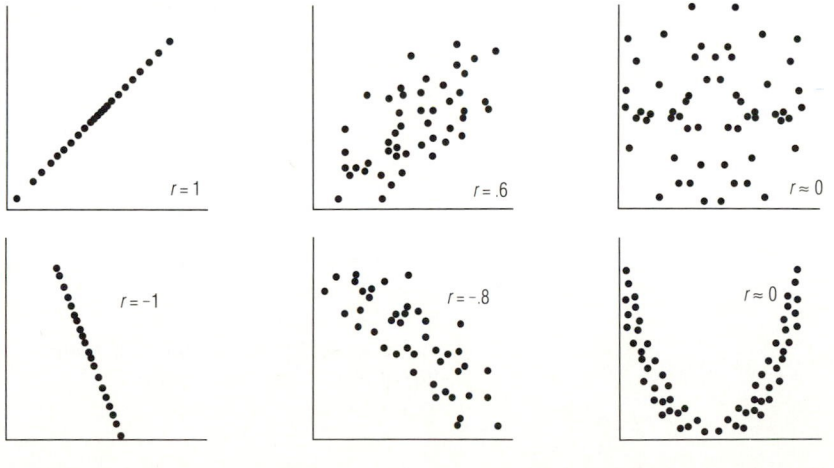

OBJECTIVES

D Identify properties of the line of best fit and of the correlation coefficient.

J Use scatter plots to draw conclusions about models for data.

TEACHING NOTES

Recall that, if the independent and dependent variables are interchanged, the lines of best fit are not necessarily the same. First, error is always attributed to the dependent variable, so the best-fit technique will usually minimize a different set of numbers. Second, the slope is a rate depending on units of measure of the variables. For instance, in **Example 1,** the slope of the best fitting line is inches/minute; in **Example 2,** minutes/inch. In contrast, the correlation coefficient is measure-free and hence gives the same information no matter which variable is chosen as the independent variable. A teaching aid with the table and graph of **Example 1** is available.

In general, the sign of r indicates the *direction* of the relation between the variables, and its magnitude indicates *the strength* of the relation. Positive values of r indicate a *positive relation* between the variables. That is, larger values of one variable are associated with larger values of the other. Negative values of r indicate a *negative relation* between the variables. That is, larger values of one variable are associated with smaller values of the other.

The extreme values of 1 and -1 indicate a perfect linear relation. That is, all data points lie on a line. Thus $r = \pm 1$ is sometimes called a **perfect correlation**. Perfect correlations are rare. A relation for which most of the data fall close to a line is called *strong*. A *weak* relation is one for which, although a linear trend can be seen, many points are not very close to the line. A correlation close or equal to 0 indicates that the variables are not related by a linear model. Note, however, that as indicated in the *curvilinear* relation on the previous page, if $r = 0$ the variables might be strongly related in some other way. The number line below summarizes these relations.

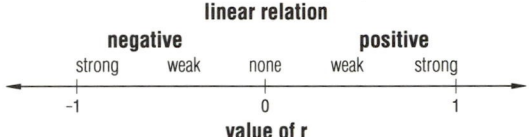

There are no strict rules about how large a correlation must be to be considered strong. In some cases $|r| = 0.5$ is considered fairly strong, and in others it might be considered moderate or weak.

Some statistics packages give values of r^2 rather than of r. This is because r^2 is used in advanced statistical techniques. You can calculate $|r|$ by taking the square root of r^2 and determine the sign by observing the direction of the relation in the scatter plot.

■ ■ ■ ■ ■ ■ ■■

Example 1   A boy measured the depth of the water in a bathtub at one-minute intervals after the faucet was turned on. His data are below.

Time (minutes)	**Depth** (cm)	**Time** (minutes)	**Depth** (cm)
1	3	10	19
2	6	11	25
3	7	12	23
4	7	13	27
5	8	14	31
6	12	15	31
7	13	16	36
8	18	17	35
9	18	18	37

a. Use a statistics package to draw a scatter plot of the data and a line of best fit.
b. Have the package calculate r.
c. Describe the relationship between time and depth, and comment on the quality of the linear model.

Solution

a. Draw a scatter plot and obtain the correlation from a statistics package.

b. The statistics package we used calculated a regression equation of $y = 2.103x - .2026$ and $r^2 = 0.983$. So $r = \pm\sqrt{0.983} \approx \pm 0.99$. Because the relation between the variables is positive (y increases as x increases), we conclude that $r = 0.99$.

c. A correlation of 0.99, indicates a very strong positive relation between the two variables. A line is a good model for these data, as is clear from the closeness of the points to the line of best fit. It seems likely that the water was running into the tub at a rather steady rate, that the bathtub had a fairly regular shape, and that the boy's measurements were reasonably careful.

Example 2 verifies that the correlation coefficient r is unaffected if you switch the variables on the x- and y-axes.

■ ■ ■ ■ ■ ■ ■ ■ ■

Example 2 Use the same data as for Example 1, but plot Time vs. Depth. Find the regression line and the correlation coefficient.

Solution

Our statistics package gives the regression equation $y = .4668x + .2684$ and $r^2 = 0.983$. Again r is positive, so $r = \sqrt{0.983} \approx 0.99$.

In general, as illustrated in Examples 1 and 2, when two variables are plotted first as (x, y) and then as (y, x), the lines of best fit are different but the correlation coefficients are the same.

b. Comment on the quality of the linear model.
With a correlation so near to 1, the linear model is a very good fit.

2. The world population (in millions) over the past 350 years is estimated as follows:

Year	Population
1650	510
1700	625
1750	710
1800	910
1850	1130
1900	1600
1950	2510

a. Find an equation for the line of best fit for these data.
$P \approx 5.97Y - 9619$

b. Use the line to estimate the world population in the year 2000.
2339

c. What is the linear correlation between year and population?
0.91

d. Use your answer to part **c** to judge the accuracy of the answer to part **b.** Does the actual world population support your judgment?
Given the high correlation, the answer should be accurate, but the actual world population does not fit this value because during the last 100 years there has been an extraordinary change in the pattern of population growth. There are not enough points from recent years in this analysis to reflect that change. (This again points out the dangers of extrapolation.)

It is important to note that while r provides a mathematical measure of *linearity*, it does not provide information about *cause and effect*. It is up to the people who analyze and interpret the data to determine why two variables might be related. For instance, there is a large positive correlation between shoe size and reading level of children. But this does not mean that learning to read better causes your feet to grow or that wearing bigger shoes improves your reading. The correlation is large because each variable is related to age. Older children generally have both larger feet and higher reading skills than younger children.

This idea is sometimes summarized as *correlation does not mean causation*.

Questions

Covering the Reading

In 1–3, *true* or *false*.

1. A correlation coefficient measures the strength of the linear relation between two variables.   **True**

2. If the line of best fit for a data set has positive slope, the correlation between the two variables is positive.   **True**

3. If the slope of the line of best fit to a data set is 2, the correlation coefficient is 2.   **False**

In 4–9, match the scatter plot with the best description.
   (a) strong negative correlation   (b) weak negative correlation
   (c) strong positive correlation   (d) correlation approximately zero

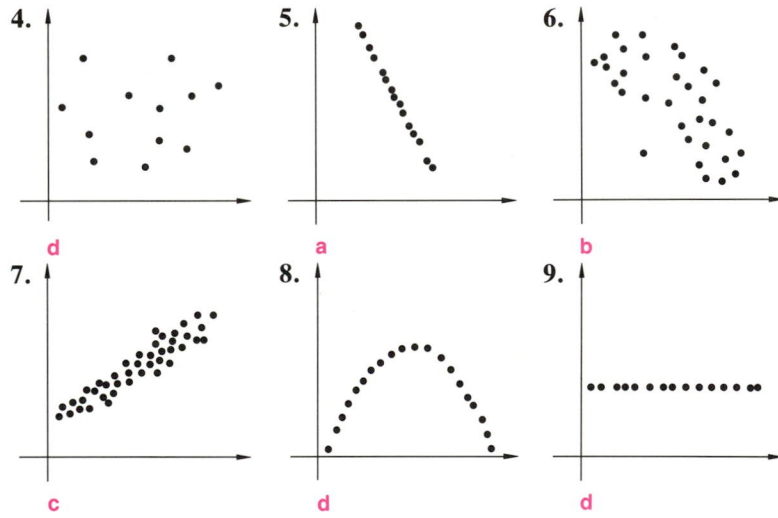

10. Draw a scatter plot for a data set showing perfect positive correlation.
   See margin.
11. Refer to the scatter plot showing the relation between temperature and precipitation. Suppose temperature were plotted on the vertical axis and precipitation on the horizontal axis. Give the correlation coefficient.
   ≈ 0.76
12. Suppose for some data set $r^2 = .64$. Find all possible values of r.
   0.8, ⁻0.8

108

13. A person measured the relationship between the duration and cost of various phone calls and obtained $r^2 = 0.77$. Find the correlation between the two variables.   **about 0.88**

Applying the Mathematics

In 14–16, use the data below, which show 1988 American League Baseball averages. (The batting average is the relative frequency of hits—the higher the average the better. Earned run average is a measure of pitching—the lower the average the better.)

Nolan Ryan, Texas Rangers

Team	Team Batting Average	Team Earned Run Average	Proportion of Games Won
Boston	.283	3.97	.543
Minnesota	.274	3.93	.562
Toronto	.268	3.80	.537
Oakland	.263	3.44	.642
New York	.263	4.24	.528
California	.261	4.32	.463
Cleveland	.261	4.16	.481
Kansas City	.259	3.65	.522
Seattle	.257	4.15	.422
Milwaukee	.257	3.45	.537
Texas	.252	4.05	.435
Detroit	.250	3.71	.543
Chicago	.244	4.12	.441
Baltimore	.238	4.54	.335

14. a. Without doing any calculation estimate whether the correlation between team batting average and proportion of games won is positive, negative, or about zero. Write a sentence explaining your prediction.   **See margin.**

  b. Use a statistics package to find the correlation between team batting average and proportion of games won.   **0.65**

  c. Do your results in **b** verify your prediction in **a**? Why or why not?

  d. Find r^2 for the relation between the proportion of games won and team batting average.   **c) Yes, but the correlation is not as strong as one might expect.   d) 0.42**

15. a. Repeat Question 14 for the relation between the team earned run averages and proportion of games won.   **See margin.**

  b. Which of the averages—team batting average or earned run average—is more highly correlated with the proportion of games won?
  Earned run average, since the correlation coefficient is greater.

16. With advanced statistics it can be shown that if s_x and s_y are the standard deviations for the observed x_i and y_i values, and r is the correlation coefficient, then the slope m of the regression line of y on x (y is the dependent variable) is $m = r \cdot \dfrac{s_y}{s_x}$. Refer to the data used in Examples 1 and 2.

  a. Calculate the standard deviations s_d and s_t for depth and time, respectively.   **$s_d = 11.33, s_t = 5.34$**

  b. Evaluate $r \cdot \dfrac{s_d}{s_t}$. Which slope does it equal?   **2.10; slope of the regression line of depth on time**

  c. Evaluate $r \cdot \dfrac{s_t}{s_d}$. Which slope does it equal?   **See margin.**

19.a.

20.a.

b.

d. Sample: The correlation may be low because the sample was so small that individual factors (such as body weight, fatigue, and so on) may have influenced the results.

e. Sample: There appears to be a positive correlation between drug dosage and reaction time, but this may be attributable to other factors.

17. The bathtub referred to in Example 1 is 38 cm deep. Find the correlation between the distance between the water level and the top of the tub and the time for which the faucet was running.   **-0.99**

18. There is a large positive correlation between the number *n* of firefighters at a fire and the dollar value *D* of the fire damage.
   a. Does this mean that firefighters cause the damage?   **no**
   b. Name a variable to which both *n* and *D* are related.
   sample: the size of the fire

In 19 and 20, use the data below which show the reaction times (*t*) of a group of ten people administered various dosages (*d*) of a drug. Data were collected to decide whether people should be advised about possible dangers of taking the drug before driving a car. Reaction time was measured as the average time for a person to respond to a red light over several trials.

Dosage *d* (mg)	Reaction time *t* (sec)
85	0.5
89	0.6
90	0.2
95	1.2
95	1.6
103	0.6
107	1.0
110	1.8
111	1.0
115	1.5

19. a. Using pencil and paper make a scatter plot of the relation between dosage and reaction time with dosage on the horizontal axis.
   b. Estimate the correlation coefficient. Does a linear model seem appropriate?
   a) See margin.   b) Sample: 0.6; a linear model seems appropriate.

20. Enter the data into a statistics package.
   a. Plot *d* on the horizontal axis and *t* on the vertical. Find the line of best fit and the correlation coefficient.   **See margin.**
   b. Plot *t* on the horizontal axis and *d* on the vertical. Find the line of best fit and the correlation coefficient.   **See margin.**
   c. Which measure, slope or correlation coefficient, does not change when the variables on the horizontal and vertical axes are switched?
   d. Suggest a reason for the low correlation between the two variables.
   e. Write a sentence describing the relationship between drug dosage and reaction time.
   c) correlation coefficient; d, e) See margin.

Review

21. Refer to the data for Questions 19 and 20.
   a. Find the mean dosage administered.   **100 mg**
   b. Find the mean reaction time.   **1.0 sec**
   c. How are answers to parts **a** and **b** related to the lines of best fit found in Question 20?   *(Lessons 2-3, 1-4)*   **See margin.**

22. Refer to the scatter plot at the beginning of the lesson showing the relation between temperature and precipitation. An equation for the line of best fit for predicting rainfall from temperature is $y = 0.766x - 1.225$.
   a. *True* or *false.*   In general, the hotter the city, the less precipitation there is.   **False**

110

b. The slope of this line indicates that, for each increase of 1° F in average annual temperature, the average annual precipitation ___?___.

c. Use this equation to predict the annual rainfall for Sault St. Marie, MI, which has an average annual temperature of 39.7°.

d. If the average annual precipitation is 55 inches, what is the expected average annual temperature? **about 73.4°F**

e. Wichita, KS has average annual temperature of 56.4°F and average annual precipitation of 28.6″. Determine the difference between the observed and predicted amount of precipitation for this outlier.
(Lessons 1-3, 2-2, 2-3) **about ⁻13.4 inches**

b) increases about 0.766 inches; c) about 29.2 inches

23. *Multiple choice.* In 1990, the cost for first-class postage was 25¢ for up to an ounce and 20¢ for each additional ounce or fraction thereof. Which is the correct graph for this relation? *(Lesson 2-4)*

(a)

(b)

(c)

(d) 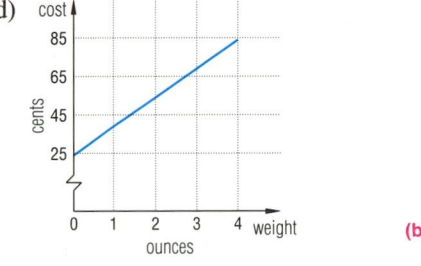 **(b)**

24. A taxi charges $1.50 for the first mile plus $.12 for each additional tenth of a mile or fraction thereof. Let $f(m)$ = the charge for a trip of m miles.
a. Evaluate $f(5)$. **$6.30** **b.** Evaluate $f(7.8)$. **$9.66**
c. Find a formula for f. *(Lesson 2-2, 2-4)*
sample: $f(m) = 1.50 + .12\lceil 10(m-1)\rceil$

25. Solve the system. *(Previous course)*
$$\begin{cases} a + b + c = 8 \\ 2a - b - c = 1 \\ 3b - 10c = 1 \end{cases} \quad a = 3, b = \frac{51}{13}, c = \frac{14}{13}$$

Exploration

26. Are heights and weights strongly correlated? Use a sample of heights and weights from at least 10 people to explore this question.
Answers will vary.

...viation scores can be related to concepts in physics. Imagine the scores to be on a balance beam with its fulcrum in the middle. In a weak correlation, the deviation scores are either very close to the fulcrum or on opposite sides of it; in a strong correlation, the deviation scores are consistently together—farther from the fulcrum—so they apply a large torque or moment to the beam. For this reason, the Pearson correlation coefficient is sometimes called the *product-moment correlation coefficient.*

PROJECTS
Project 2 on page 133 is related to this lesson.

ADDITIONAL ANSWERS
19.a., 20.a.,b.,d.,e. See margin, page 109.

21.c. The point (100, 1) is the center of gravity for the (dosage, reaction time) data; it lies on the line of best fit found in 20a. The point (1, 100) is the center of gravity for the (reaction time, dosage) data; it lies on the line of best fit found in 20b.

OBJECTIVES

C Identify the variables, domain, and range of functions.
G Find and interpret quadratic models.
H Graph quadratic functions.
I Interpret properties of relations from graphs.

TEACHING NOTES

Making Connections
Students will study how to generate quadratic models in Lesson 2-7. This lesson reviews the algebra of quadratic equations and the properties of parabolas. Have students memorize the quadratic formula if they have not done so by this time. Do not spend time having students draw careful graphs of quadratics; a sketch is all that is expected here. Do make certain, however, that students are fluent in reading the coordinates of vertices and intercepts from a given graph.

Also make certain that students are able to analyze $y = ax^2 + bx + c$ and describe generally how a, b, and c affect the graph of the parabola. For example, they should have a clear picture in their mind of $y = x^2$, $y = 2x^2$, and $y = -x^2$, and should know that nonzero values of b or c indi-

---

LESSON

2-6

Quadratic Models

In each of the three scatter plots below the correlation is approximately 0. So there is no good-fitting line.

However, as noted in the previous lesson, and as illustrated by the graphs in the center and on the right above, even though the correlation is 0, there may be a *nonlinear* relation between the variables. The points on the center plot seem to lie near a U-shaped curve; the ones on the right show a regular repetitive pattern.

There are many types of nonlinear models. Examples include polynomial models, exponential models, and trigonometric models. You will learn about all of these later in this course. In this lesson, we focus on **quadratic models**, that is, on models based on quadratic functions. Recall that a quadratic function is of the form

$$ax^2 + bx + c,$$

where $a \neq 0$. Recall also that graphs of quadratic functions are parabolas. If $a < 0$, the parabola has a maximum point; if $a > 0$, the parabola has a minimum point.

The domain of a quadratic function is the set of all real numbers. When $a < 0$ the range is the set of all real numbers less than or equal to the maximum value; when $a > 0$ the range is the set of all real numbers greater than or equal to the minimum value. The y-intercept is the y-coordinate of the point where $x = 0$.

$$f(0) = a \cdot 0^2 + b \cdot 0 + c = c$$

So c is the y-intercept. The x-intercepts are the x-coordinates of the points where $y = 0$, found by solving the quadratic equation $ax^2 + bx + c = 0$. From the quadratic formula, the x-intercepts are when

$$x = \frac{-b \pm \sqrt{b^2 - 4ac}}{2a}.$$

112

Example 1 Consider the function $f(x) = 2x^2 - 3x - 1$.

a. Find the y- and x-intercepts.
b. Sketch a graph of $y = f(x)$.

Solution

a. The y-intercept is -1, because $f(0) = 2(0)^2 - 3(0) - 1 = -1$. To find the x-intercepts, substitute 0 for y, and solve the resulting quadratic equation.

$$2x^2 - 3x - 1 = 0$$

So
$$x = \frac{3 \pm \sqrt{9 - 4(2)(-1)}}{2(2)}$$

$$= \frac{3 \pm \sqrt{17}}{4}.$$

To the nearest hundredth, $x = 1.78$ or $x = -0.28$. The intercepts give three points on the graph: $(0, -1)$, $(1.78, 0)$, and $(-0.28, 0)$.

b. Because the coefficient of x^2 is positive, the parabola has a minimum point. Plot the intercepts. Find some additional points.

For $x = 1$, $y = 2(1)^2 - 3(1) - 1 = -2$.
For $x = 3$, $y = 2(3)^2 - 3(3) - 1 = 8$.
For $x = -2$, $y = 2(-2)^2 - 3(-2) - 1 = 13$.

Thus $(1, -2)$, $(3, 8)$, and $(-2, 13)$ are also on the graph. Plot these and draw the parabola. A graph is shown below.

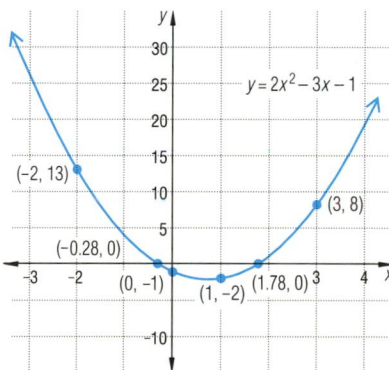

$y = 2x^2 - 3x - 1$

Some physical situations are modeled by quadratic functions. Among the most well known is a famous function from physics. In the 17th century, Isaac Newton discovered that the height h of an object at time t after it has been thrown upward with an initial velocity v_0 from an initial height h_0 satisfies the formula

$$h = -\frac{1}{2}gt^2 + v_0 t + h_0,$$

where g is the **acceleration due to gravity**. Recall that velocity is the rate of change of distance with respect to time; it is measured in units such as

Isaac Newton

cate that the parabola does not have a vertex at the origin. (In Lesson 3-1 use of an automatic grapher is introduced. Some teachers prefer to introduce the grapher here.) Do not feel compelled to discuss all the transformation images of $y = x^2$, nor the alternative form $y = a(x - h)^2 + k$. These topics are covered thoroughly in Chapter 3.

It is important that students understand why a linear fit is not appropriate for the peak power problem. Not only do the data appear to fall on a curve, but if a line were used all the data points on the far left and far right would fall above the line. This kind of pattern often indicates that a curvilinear model is more appropriate.

Emphasize the paragraph after **Example 3** which points out the danger of extrapolating from quadratic models. The model is reasonable for only a small range of temperatures on either end of those given in the data set. The daily low and high are unlikely to fall below 65°F or above 105°F.

Alternate Approach If your statistics package gives quadratic models, you may want to enter the peak power load data and compare and contrast its output from that given in the text.

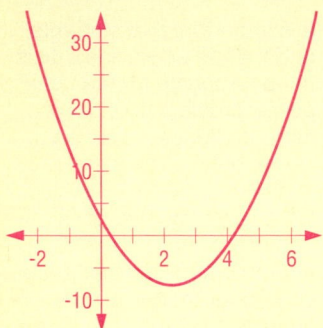
miles per hour or meters per second. Acceleration is the rate at which velocity changes, so it is measured in units such as miles per hour per hour or meters per second^2. Near the surface of the earth g is approximately 32 ft/sec^2 or 9.8 m/sec^2.

■ ■ ■ ■ ■ ■ ■ ■

Example 2   Suppose a ball is thrown upward from a height of 15 m with an initial velocity of 20 m/sec.

a. Find a model for the relation between height h and time t after the ball is released.
b. How high is the ball after 2 seconds?
c. At what time is the ball 25 m off the ground?

Solution

a. The conditions satisfy Newton's equation. Here $v_0 = 20$ m/sec and $h_0 = 20$ m, so we use $g = 9.8$ m/sec^2.

$$h = \frac{1}{2} \cdot -9.8t^2 + 20t + 15$$

b. We are given $t = 2$, and asked to find h. So

$$h = \frac{1}{2} \cdot -9.8(2)^2 + 20(2) + 15.$$

$$h = 35.4 \text{ m}.$$

After 2 seconds the ball is about 35.4 m above the ground.

c. We are given $h = 25$, and asked to find t. So

$$25 = \frac{1}{2} \cdot -9.8t^2 + 20t + 15$$

$$0 = \frac{1}{2} \cdot -9.8t^2 + 20t - 10.$$

Solve for t. By the quadratic formula,

$$t = \frac{-20 \pm \sqrt{20^2 - 4(-4.9)(-10)}}{2(-4.9)}$$

$$= \frac{-20 \pm \sqrt{204}}{-9.8}$$

$$\approx \frac{-20 \pm 14.28}{-9.8}$$

$$t \approx .6 \text{ or } t \approx 3.5.$$

There are two times at which the ball is 25 m off the ground—on the way up, about .6 seconds after being thrown, and on the way down, about 3.5 seconds after being thrown.

Many curvilinear relations that occur in the everyday world can be approximated rather well by a quadratic model.

Consider the following situation. Utility companies need to be able to predict the peak power load to be able to handle operations efficiently. The peak power load is the maximum amount of power that must be generated

114

each day in order to meet demand. The data below give the daily high temperature and the peak power load for a 21-day period one summer in one city.

Temperature (°F)	Peak load (megawatts)	Temperature (°F)	Peak load (megawatts)
94	135.0	98	150.1
96	131.7	100	157.9
95	140.7	102	175.6
88	115.4	103	198.5
84	113.4	106	225.2
90	123.0	105	205.3
97	143.2	92	130.0
71	95.0	92	130.0
67	101.6	85	112.4
79	105.2	89	113.5
87	112.7	74	103.9

A scatter plot of the data shows that a linear model may not be a good fit. Although the correlation coefficient $r \approx .862$ is reasonably high, visually the data seem to fall on a curve which increases more rapidly as temperature rises. Thus the power company hypothesizes that a quadratic model might be a better fit. They think that a one-unit increase in temperature from 99° to 100°F might result in a bigger increase in the demand for power than an increase from 79° to 80° F. With a statistics package they find that a model for these data is $y = 0.131x^2 - 19.986x + 861.09$ where $x =$ the temperature in degrees Fahrenheit and $y =$ the peak load in megawatts.

Question 1: In part **a** the data appear to be sinusoidal; in part **c** a quadratic model might be preferable to a linear model.

Questions 3, 6, and 7: Unlike some texts, we do not control quadratics so that almost all solutions are rational numbers. When dealing with real data, solutions with rational numbers rarely occur. Fortunately, calculators do arithmetic with decimal approximations as easily as they do arithmetic with whole numbers.

Questions 8–10: Teaching Aid 21 may be used.

Example 3 Use the quadratic model on page 115 to predict:

a. the peak power when the high temperature is 93°,

b. the temperature at which the peak power load is expected to reach 250 megawatts, the plant's maximum capacity.

Solution

a. Substitute $x = 93$ in the model:

$$y = 0.131x^2 - 19.986x + 861.09$$
$$= 0.131(93)^2 - 19.986(93) + 861.09$$
$$= 135.411.$$

So the model predicts that when the temperature hits 93° F, the peak power load will be about 135 megawatts.

b. We want the value of x for which $y = 250$. Substitute $y = 250$ into the given equation.

$$0.131x^2 - 19.986x + 861.09 = 250$$

So $\quad 0.131x^2 - 19.986x + 611.09 = 0.$

Use the quadratic formula, with $a = 0.131$, $b = -19.986$, and $c = 611.09$.

$$x = \frac{19.986 \pm \sqrt{(-19.986)^2 - 4(0.131)(611.09)}}{2(0.131)}$$

$$= \frac{19.986 \pm 8.901}{0.262}$$

So $x \approx 42.3$ or $x \approx 110.3$.

From the graph, it is clear that the larger value is needed. So according to this model the power plant's capacity of 250 megawatts is likely to be needed when the temperature hits 110° F. This is close enough to recent high temperatures for the utility company to seek additional energy sources.

Check

a. (93, 135) appears to be close to the graph of the model.

b. Substitute $x = 110$ in the model.

$y = 0.131(110)^2 - 19.986(110) + 861.09 \approx 247.73$. It checks.

Notice that in Example 3 the model also predicts that peak power load will also reach 250 megawatts when the temperature is only 42°F. This illustrates again the danger of extrapolation. It also suggests that a quadratic model may be inappropriate. In fact, an exponential model of the type you will study in Chapter 4 may be more appropriate. The quadratic model here is useful for interpolation and for extrapolation over limited portions of the domain.

In Lesson 2-7, you will study how to find equations for quadratic models like those shown in this lesson.

Questions

Covering the Reading

1. *Multiple choice.* For which of the following scatter plots does a nonlinear model for the data seem to be most necessary? **(a)**

(a)

(b)

(c)

(d)

2. What is the general form of a quadratic function? $y = ax^2 + bx + c$

3. Consider the function $f(x) = 2x^2 - x - 4$.
   a. Give the y-intercept. **−4**
   b. Give the x-intercept(s). $(1 + \sqrt{33})/4 \approx 1.69, (1 - \sqrt{33})/4 \approx -1.19$
   c. Sketch a graph. **See margin.**

4. Which of the following graphs have a maximum point?
   (a) $y = 8x^2 - 3x - 7$ (b) $y = 2x + 4x^2$
   (c) $y = 6 - 2x^2$ (d) $y = 177 + 5x - x^2$ **c, d**

5. At a certain moment, a ball is falling with velocity −65 ft/sec. At what rate is its velocity changing? **32 ft/sec²**

6. Refer to the situation in Example 2.
   a. How high is the ball after 1.5 seconds? **about 34 m**
   b. Find t so that $h = 0$. **about 4.7 seconds**
   c. Over what domain does the quadratic model hold? **0 ≤ t ≤ 4.7**

7. Suppose a ball is thrown upward at a velocity of 44 ft/sec from a cliff 200 feet above a dry riverbed.
   a. Write an equation for the height h (in feet above the riverbed) of the ball after t seconds. $h = -16t^2 + 44t + 200$
   b. Predict the height of the ball after 3 seconds. **188 feet**
   c. At what time will the ball hit the riverbed?
      about 5.2 seconds after it is thrown

In 8–10, consider the quadratic model for peak power load as a function of temperature in Example 3.

8. a. Predict the peak power load at 80°F. **about 100.6 megawatts**
   b. What kind of prediction (extrapolation or interpolation) is made in part **a**? **interpolation**

9. The utility company is considering a proposal to expand capacity to 300 megawatts. What temperature is likely to cause this power load? What factors other than temperature might influence the utility company's decision to expand its capacity? **about 115°F; sample: population growth, expanded area covered**

10. a. Find the y-intercept of the graph of the model. **865.557**
    b. Interpret the y-intercept in terms of peak load. **See margin.**
    c. Is this extrapolation too far beyond the domain? **Yes**

LESSON 2-6 Quadratic Models **117**

ADDITIONAL ANSWERS
3.c.

10.b. Sample: The figure is an extrapolation which suggests a tremendous amount of power is needed at low temperature. This result is counterintuitive and underscores the limits of extrapolation.

11.e.

12.a.

b.

Applying the Mathematics

11. A piece of an artery or a vein is approximately the shape of a cylinder. The French physiologist and physician Jean Louis Poiseuille (1799-1869) discovered experimentally that the velocity v at which blood travels through the arteries or veins is a function of the distance r of the blood from the axis of symmetry of the cylinder. Specifically, for a wide arterial capillary the following formula might apply:
$$v = 1.185 - (185 \times 10^4)r^2,$$
where r is measured in cm, and v in cm/sec.
  a. Find the velocity of blood traveling on the axis of symmetry of the capillary.   **1.185 cm/sec**
  b. Find the velocity of blood traveling 6×10^{-4} cm from the axis of symmetry.   **0.519 cm/sec**
  c. According to this model, where in the capillary is the velocity of the blood 0?   **8×10^{-4} cm from the axis of symmetry**
  d. For this application, what is the domain of the function?
  e. Sketch a graph of this function.
  d) $0 \leq r \leq 8 \times 10^{-4}$ cm; e) See margin.

In 12 and 13, to test the hypothesis that underinflated or overinflated tires can increase tire wear, new tires of the same type were tested for wear at different pressures. The results are shown in the table below.

12. Make a scatter plot of the data.
  a. If you were given data only for $x = 29$ to $x = 33$, what kind of model would best fit the data? Find an equation for such a model and graph it on your scatter plot.
  b. If you were given data only for $x = 33$ to $x = 36$, what kind of model would best fit the data? Find an equation for such a model, and graph it on your scatter plot.   **a) a linear model; $y = 2.7x - 51.9$; b) a linear model; $y = -3x + 136$**

x Pressure (in psi)	y Mileage (in thousands)
29	25
30	30
31	33
32	35
33	36
34	35
35	32
36	27

13. If you use all the data, a quadratic model explains the relation between x and y quite well.
  a. Draw another smooth curve through the data points to approximate a parabola.   **See margin.**
  b. Estimate the coordinates of the vertex of your parabola.
  c. In the equation $y = ax^2 + bx + c$ for your parabola, is a positive or negative?   **b) (33, 36); c) negative**

14. *Multiple choice.* On which of the following graphs do the slopes of segments joining points on the curve measure velocity?   **(a)**

           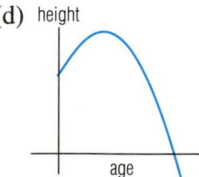

(a) distance / time   (b) weight / height   (c) time / distance   (d) height / age

15. Biologists measure the diversity of plant and animal species to study the environment. Below are data on bird species diversity (BSD), foliage height diversity (FHD), and plant species diversity (PSD) from a study conducted at sites in Pennsylvania, Vermont, and Maryland. (Source: Kolmes and Mitchell, *Information Theory and Biological Diversity. UMAP Journal*, Spring, 1990)

Killdeer and Red-Headed Woodpecker

Site	BSD	FHD	PSD
A	0.639	0.043	0.972
B	1.266	0.448	1.911
C	2.265	0.745	2.344
D	2.403	0.943	1.768
E	1.721	0.731	1.372
F	2.739	1.009	2.503
G	1.332	0.577	1.367
H	2.285	0.859	1.776
I	2.277	1.021	2.464
J	2.127	0.825	2.176
K	2.567	1.093	2.816

a. Find the correlation between BSD and FHD.   **0.95**
b. Find the correlation between BSD and PSD.   **0.81**
c. *True* or *false*.   From these data, birds appear to be selecting habitats more on the basis of foliage height diversity than on the species of the plants producing the different heights. *(Lesson 2-5)*   **True**

16. Evaluate $\lfloor 5.9 \rfloor - \lceil 5.9 \rceil$.   *(Lesson 2-4)*   **−1**

17. After a long hot spell, a swimming pool contains only 45 thousand liters of water. At noon, an inlet valve is opened, letting 40 liters of water per minute into the pool. Assume no water evaporates.
a. Give a formula for the volume V liters of water in the pool m minutes after noon.   **$V = 40m + 45{,}000$**
b. What will be the volume of the pool at 1:45 P.M.?
c. When will the pool reach its maximum capacity of 55 thousand liters? *(Lesson 2-2)*   **b) 49,200 liters; c) 4:10 P.M.**

18. Solve the system. *(Previous course)*
$$\begin{cases} 5a + 2b + c = 3 \\ a - b - c = 11 \\ 3a + b - c = 9 \end{cases}$$ **$a = 3, b = \text{-}4, c = \text{-}4$**

19. In a capture-recapture experiment in a lake in central New York, 213 largemouth bass were caught, their fins marked, and then released. Several months later, 104 bass were caught; among these 13 had marked fins.
a. Estimate the number of largemouth bass in the lake.
b. State several assumptions about the bass and the lake that justify your estimate in part **a**.   *(Lesson 1-1)*
a) 1700 bass; b) See margin.

20. What techniques do people use to measure the height of objects in air, such as the ball in Example 2?   **Answers will vary.**

LESSON 2-6   Quadratic Models   **119**

OBJECTIVES

B Find quadratic functions which contain specified points.

G Find and interpret quadratic models.

J Use scatter plots to draw conclusions about models for data.

TEACHING NOTES

In this lesson students continue to develop models to fit data. First, they review how to generate a linear model given two points, and then they learn how to fit a quadratic model using three noncollinear points. To do this students need to be able to set up and solve systems of equations in two or three unknowns (for the linear and quadratic models, respectively).

Making Connections
The purpose of **Example 1** is the same as the purpose of Example 2 in Lesson 2-2, namely, to find an equation for the line through two points. In Lesson 2-2, the slope m was calculated and then substituted into one of the equations to find b. Here, we use systems and the language of functions because this process generalizes more easily to finding an equation for a quadratic through three points.

LESSON

2-7

Finding Quadratic Models

As you have seen, both linear and nonlinear models may be used to describe relationships in data. The same general method used to find linear models can be applied to find an equation for a nonlinear model. First, draw a curve that seems to fit the data well. Then find an equation for that curve. A technique using systems of equations can be applied to find either linear or nonlinear models. First, we show how this technique works for finding linear models.

Here are some data for the average price in dollars paid to farmers per 100 lb of lamb or turkey since 1930. (Source: The U.S Agriculture Department.) Notice that both prices increase substantially over that time. A scatter plot and a line approximating the data are shown below.

Year	Price of lamb	Price of turkey
1930	$7.76	$20.20
1940	8.10	15.20
1950	25.10	32.80
1960	17.90	25.40
1970	26.40	22.60
1975	42.10	34.80
1980	63.60	41.30

An equation for any nonvertical line is $f(x) = mx + b$. The function is determined when the values for the coefficients m and b are known. Each point $(x, f(x))$ on the line satisfies the equation for f. If you substitute two points on the line into the equation for f, you get a system of two equations in two unknowns. Solving this system gives values for m and b, and thus, an equation for the line.

■ ■ ■ ■ ■ ■ ■ ■

Example 1 Find an equation of the form $f(x) = mx + b$ for the line drawn above.

Solution Choose two points on the line. We use (10, 20) and (60, 40). Substitute in $f(x) = mx + b$.

$$f(10) = m \cdot 10 + b = 20$$
$$f(60) = m \cdot 60 + b = 40$$

So the system for m and b is: $\begin{cases} 10m + b = 20 \\ 60m + b = 40 \end{cases}$.

To solve this system, subtract the first equation from the second and solve for m.

$$50m = 20$$
$$m = 0.4$$

Substitute $m = 0.4$ in $10m + b = 20$, and solve for b.

$$10(0.4) + b = 20$$
$$b = 16$$

120

The solution to the system is $m = 0.4$ and $b = 16$, so an equation for the linear function is $f(x) = 0.4x + 16$.

Check   Check that the two points are on the line.
$f(10) = 0.4(10) + 16 = 20$ and
$f(60) = 0.4(60) + 16 = 40$. It checks.

This procedure of constructing and solving a system of equations can also be used to find quadratic models. Recall that, in general, a quadratic function has an equation of the form

$$f(x) = ax^2 + bx + c.$$

To determine a quadratic model, you must find the coefficients, a, b, and c. Recall also that three equations are necessary to find three unknowns. That is, three points are needed to determine a quadratic function, while only two points were needed for a linear function. You can construct three equations by selecting any three points $(x, f(x))$ on the curve, and then substituting their coordinates into the quadratic function. Each point will give you a single equation for the coefficients, a, b, and c. Solving these three equations will give you values for a, b, and c.

To illustrate how to do this, consider the following data collected by a veterinarian working for a large pig cooperative interested in increasing the weight of its pigs. Twenty four randomly selected pigs were each given a daily dosage of a food supplement. Three pigs each received the same dosage, and their percent weight gain was averaged. The table below shows the average percent weight gain in one month for pigs in relation to the dosage.

Dosage (in pellets)	0	1	2	3	4	5	6	7
Percent weight gain	10	13	21	24	22	20	16	13

A scatter plot of the relation between dosage and percent weight gain shows a quadratic trend. To find a quadratic function to model these data, draw a parabola which passes through the data set and read the coordinates of three points on the curve.

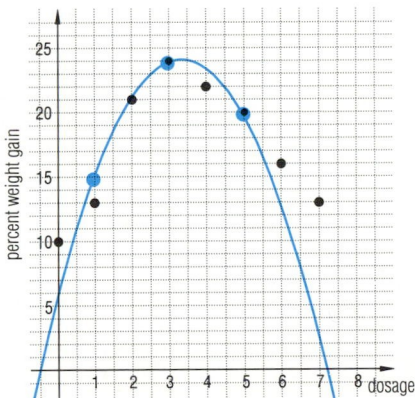

The method used in this lesson is sometimes called *parabolic* or *quadratic interpolation* (see Project 5 on page 134). In Lesson 9-2, this method is extended to fitting polynomials of any degree to data.

Error Analysis Students are sometimes unclear about why three points, and so three equations and three unknowns, are needed to determine a quadratic function. Point out that two points do not suffice. For example, $(0, 0)$ and $(1, 1)$ are on both $y = x^2$ and $y = -(x - 1)^2 + 1$, but the parabolas are not the same.

In **Example 2** the three points chosen are equally spaced along the independent variable (the x-axis). Any three points would do, but equally-spaced points save some work. Notice that after c was eliminated, the two resulting equations in a and b had opposite coefficients. This sets up a simple 2×2 system to solve. To increase the chance of finding a good-fitting model, students should be encouraged to choose three points which are spread out. If a student had chosen $(0, 10)$, $(1, 13)$, and $(2, 21)$, the first condition would have been met, but not the second. The resulting system would be easy to solve but the model would not fit the data as well.

Alternate Approach
You may want to demonstrate the use of an automatic grapher to plot the equations you generate and to verify that the points used to generate the equation are on the graph. Students will be expected to use an automatic grapher beginning with Lesson 3-1.

When working with $y = ax^2 + bx + c$, students are accustomed to knowing a, b, and c and finding values of x and y. As a result, some students will have trouble going from known xs and ys to finding a, b, and c. This understanding is part of a mature view of functions, so it should be emphasized.

ADDITIONAL EXAMPLE
(This example covers the content of both **Examples 1 and 2** in the lesson.) The ten tallest buildings in Minneapolis (as of 1990) are listed below together with their heights (in feet) and the number of stories in each.

Building	Height	Stories
IDS Center	775	57
Norwest	772	57
Multifoods Tower	668	52
Piper Jaffray Tower	579	42
Pillsbury Center	561	42
Opus	503	36
Plaza VII	475	36
Lincoln Centre	454	32
Foshay Tower	447	32
Marriott Hotel	440	32

a. Graph these points.

b. Fit a linear model to these data by eyeballing the data and solving a system of two equations in two unknowns.
Sample: If the points (460, 33) and (740, 56) are chosen, then an approximate equation for the line is $y = .0821x - 4.8$.

Example 2 Find an equation for a quadratic model for the weight-gain data pictured on page 121.

Solution We choose three points that are equally spaced on the x-axis, such as (1, 15), (3, 24), and (5, 20). Substitute the coordinates of each point into the equation $f(x) = ax^2 + bx + c$ to get a system of three equations for a, b, and c.

$$f(x) = ax^2 + bx + c$$
$$f(1) = 15 = a + b + c$$
$$f(3) = 24 = 9a + 3b + c$$
$$f(5) = 20 = 25a + 5b + c$$

Now solve the system. To eliminate c subtract the first equation from the second, and the second from the third.

$$9 = 8a + 2b$$
$$-4 = 16a + 2b$$

Subtract these two equations to eliminate b.

$$13 = -8a$$

So,
$$a = -\frac{13}{8} = -1.625.$$

Substitute into $9 = 8a + 2b$ to get b.

$$9 = 8(-1.625) + 2b$$
$$b = 11$$

Substitute in $15 = a + b + c$ to get c.

$$15 = -1.625 + 11 + c$$
$$c = 5.625$$

So, a quadratic function modeling these data is given by

$$f(x) = -1.625x^2 + 11x + 5.625.$$

Had we chosen three points not equally spaced on the x-axis, we might have had to perform more operations to find the coefficients of the model, but the procedure would have been the same.

Just as we did for linear models, we can study the fit of a quadratic model to the data by evaluating the function at various data points. The table below shows several actual and predicted values associated with the data for the pigs studied in Example 2.

Dosage	Actual percent weight gained	Predicted percent weight gain	Error (actual – predicted)
0	10	5.625	4.375
1	13	15	-2
3	24	24	0
5	20	20	0
6	16	13.125	2.875

Note that the error is 0 for two of the points in the original data set used to find the equation of the model. Because most of the other errors are rather small, the quadratic model is a reasonable fit.

Although the quadratic model is useful for interpolation, you must be very careful when extrapolating from nonlinear models. The quadratic model in

122

Example 2 is a good fit on the domain, $0 \le x \le 7$, but for $x < 0$ the model makes no sense and for $x > 7$ we have no data to be sure of the fit.

As shown in Example 1, two points determine a unique line, or linear model. Example 2 shows that three noncollinear points determine a unique quadratic model. These results can be generalized. You need at least four points to determine the four coefficients necessary for a **cubic model**:

$$g(x) = ax^3 + bx^2 + cx + d.$$

Five points can determine a quartic (4th degree) model, and so on. Such polynomial models can always be found by solving systems of equations. In general, a polynomial of degree n requires at least $n + 1$ points, which lead to a system of $n + 1$ equations in $n + 1$ unknowns. In practice, when modeling real data, it is rarely necessary to use polynomial models of degree greater than 2.

The more powerful statistics packages can determine the quadratic model of best fit. For instance, for the data in Example 2 the best fitting quadratic model has equation:

$$y = -1.116x^2 + 7.925x + 8.668.$$

The question of how the quadratic model of best fit for a data set is determined is beyond the scope of this course.

Questions

Covering the Reading

In 1 and 2, refer to the data on prices of lamb and turkey at the beginning of the lesson.

1. **a.** Use the point-slope method to find an equation of the line containing $(10, 20)$ and $(60, 40)$.   **Sample: $y - 20 = 2/5(x - 10)$**
   b. Why must your equation be equivalent to the one in Example 1?
   See margin.
2. **a.** Use the 1930 data, $(7.76, 20.20)$, and the 1980 data, $(63.60, 41.30)$, to construct a system of equations for the coefficients of a linear model.
   b. Solve your system in **a** and explain why you get a different linear model from the one in the text.
   See margin.

In 3–6, refer to Example 2.

3. Show that $f(5) = 20$.   **$f(5) = -1.625(5)^2 + 11(5) + 5.625 = 20$**

4. **a.** Use the model to predict the percent weight gain in one month of pigs fed 4 pellets daily.   **about 24%**
   b. Is the prediction in part **a** extrapolation or is it interpolation?
   interpolation
5. **a.** Use the model to predict the percent of weight gain in one month of pigs fed 10 pellets daily.   **about -47%**
   b. Is the prediction in part **a** extrapolation or is it interpolation?
   c. Explain why the prediction in part **a** may be unreasonable.
   b) extrapolation; c) See margin.

LESSON 2-7   Finding Quadratic Models   **123**

c. Fit a quadratic model to the data.
Sample: If the points (470, 35), (620, 47), and (770, 57) are used, then an approximate equation for the parabola is $y = -0.0000444x^2 + 0.1284x - 15.6$

d. Which model seems to fit better?
Both fit well. The quadratic model fits better (in the sense that the sum of the squared deviations from the given points to the model is less) but the linear model is simpler. The linear model suggests that there is about .082 stories per foot, which equals 12 feet per story.

NOTES ON QUESTIONS
Making Connections for Question 1: This question asks students to use the method of Lesson 2-2 to confirm the answer to **Example 1.** Students may wonder why a new method is needed. As noted in the lesson, explain that the method of this lesson is more widely applicable.

ADDITIONAL ANSWERS
1.b. because it was constructed using the same two points, and two points determine a line uniquely

2.a. $7.76m + b = 20.2$
$63.6m + b = 41.3$

b. $m \approx 0.38$; $b \approx 17.3$; $y = .38x + 17.27$; the points used in this construction did not lie exactly on the line constructed in the text.

5.c. It is unlikely that giving the pigs an unusually large dose of food supplement will make them lose half of their body weight.

NOTES ON QUESTIONS
Technology for Question 9: Stopping distance involves reaction time and time to decelerate. Reaction distance is a linear function of speed and the distance it takes to decelerate is (in theory) quadratic. Thus one should expect a quadratic model to fit quite well, but a third-degree polynomial model fits better. You can save time by giving students the answer to part **f** and asking them to do part **g.** If your statistics package does polynomial regression, you might want to determine the equations of the best fitting quadratic and polynomial models and compare them to the answers to parts **b, c,** and **f.** Whether or not your statistics package does nonlinear regression, use an automatic grapher to plot the equations found in **b, c,** and **f** on the same screen. Point out that each passes through some of the data points but none passes through all.

Small Group Work for Question 20: Groups of students can share interesting data sets. You may wish to save some of the sets on disk for practice or projects.

FOLLOW-UP

MORE PRACTICE
For more questions on SPUR Objectives, use *Lesson Master 2-7,* shown on page 125.

EXTENSION
Finding the maximum number of pieces into which a circular region (two-dimensional) can be cut is discussed in **Question 8.** You may wish to pose the three-dimensional problem of cutting a cylinder (pizza, cake, wheel of cheese, and so on) with n cuts. To simplify the problem, require that all cuts be either parallel or perpendicular to the table. (for $n > 0$, $f(n) = n^2 - n + 2$)

6. Graham chose the points (1, 13), (3, 24), and (6, 16) to find a quadratic model for the data on pigs.
   a. Write the three equations obtained from these points.
   b. Solve the system of equations from part **a**.
   c. Compare your solutions to those in Example 2.
   d. Which system, the one in part **a** or the one in Example 2, was easier to solve?
   See margin.
7. Find an equation for the quadratic function containing the three points (1, 4), (2, 11), and (3, 20). $f(x) = x^2 + 4x - 1$

Applying the Mathematics

8. A pizza is cut by a number of straight cuts as shown below. The table shows the largest number of pieces $f(n)$ into which it is cut by n cuts.

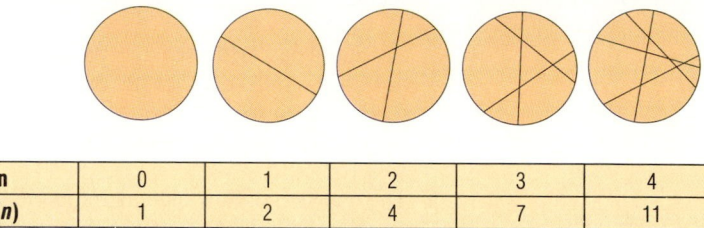

n	0	1	2	3	4
f(n)	1	2	4	7	11

   a. Find a quadratic model for these data. $f(n) = 0.5n^2 + 0.5n + 1$
   b. Use your model to find the greatest number of pizza pieces produced by five straight cuts. Check your answer by drawing a diagram.   **See margin.**
9. Based on tests made by the Bureau of Public Roads, here are the distances (in feet) it takes to stop in minimum time under emergency conditions. Reaction time is considered to be 0.75 second.

mph	10	20	30	40	50	60	70
Stopping distance	19	42	73	116	173	248	343

(Source: American Automobile Association; quoted in *The Man-Made World*, p.15.)

   a. Construct a scatter plot for these data.   **See margin.**
   b. Find a quadratic model for these data. Use the data for 10, 30, and 50 mph. $y = 0.058x^2 + 0.4x + 9.25$
   c. Find another quadratic model for the data, using 10, 40, and 70 mph.
   d. Check the predictions for 20 and 60 mph for each of the two models in parts **b** and **c**. Which model results in the smaller sum of squared errors for these data?   **See margin.**
   e. If your statistics package can fit polynomial models to data, find an equation for the best-fitting quadratic model. What stopping distance does it predict for a car traveling 60 mph?
   f. A third degree polynomial will fit four points exactly. Find a third degree polynomial that fits the above data for 10, 30, 50, and 70 mph.
   g. Does your cubic model from part **f** fit the data better than the quadratic models?   **Yes**
   c) $y = 0.072x^2 - 0.38x + 15.6$;
   e) $f(x) = 0.072x^2 - 0.49x + 19.714$; $f(60) \approx 249.5$;
   f) $f(x) = 0.0005x^3 + 0.013x^2 + 1.55x + 1.75$

124

10. Recall that the height of an object thrown upwards with an initial velocity of v_0 from a point h_0 meters above ground is given by $h(t) = -4.9t^2 + v_0t + h_0$, where t is time in seconds. Consider an individual drop of water coming out of a fountain 5 m above ground level which spurts water up with a velocity of 30 m/sec. *(Lesson 2-6)*

  a. How high is the drop after 3 seconds?   **50.9 m**

  b. When will the drop be back to the level of the fountain?

  c. If there is a little wind, the drop will not travel along a vertical line but its time in the air will not be affected. If this is the case, determine how long it would take the drop of water to hit the sidewalk.
  b) about 6.12 seconds after it is spurted; c) about 6.3 seconds

In 11–13, *true* or *false*. *(Lesson 2-5)*

11. If two variables are not linearly related, the correlation between them is near zero.   **True**

12. If the correlation between two variables is near zero, they are not linearly related.   **True**

13. If the correlation between two variables is near zero they might be related by some nonlinear relation.   **True**

14. For what values of x does $\lfloor x \rfloor = \lceil x \rceil$?   *(Lesson 2-4)*
   when x is an integer

15. One airline offers its frequent flyers a free trip for every 20,000 miles flown. Give a formula for the number n of free trips earned by flying m miles.   *(Lesson 2-4)* $n = \left\lfloor \dfrac{m}{20{,}000} \right\rfloor$

In 16 and 17, tell whether or not the relation is a function.   *(Lesson 2-1)*

16. $\{(4, 5), (8, -7), (-5, -5), (0, 9), (9, 9)\}$ **function**

17. $\{(5, 4), (-7, 8), (-5, -5), (9, 0), (9, 9)\}$ **not a function**

18. Give the domain and range of the function $f(t) = 2^t$.   *(Lesson 2-1)*
   See margin.

19. During February, Mrs. Winston recorded the number of absentees from her class of twenty-five students for each of 20 successive school days. Here are her data.   *(Lessons 1-4, 1-6)*

$$0, 2, 0, 0, 1, 0, 1, 1, 1, 2, 5, 8, 5, 3, 2, 1, 0, 0, 0, 0$$

  a. Give the range, mean, median, and mode of the number of daily absentees.   **range = 8; mean = 1.6; median = 1; mode = 0**

  b. Draw a histogram of the results.   **See margin.**

  c. Locate the mean, median, and mode on the graph.   **See margin.**

20. Find a data set that interests you. Does either a linear or quadratic model fit the data? If so, use the techniques described in this lesson to find an equation for the model. If not, what other patterns do you find in your data?   **Answers will vary.**

PROJECTS
Projects 4 and 5 on pages 133 and 134 are related to this lesson.

ADDITIONAL ANSWERS
6.a.,b.,c.,d., 8.b., 9.a.,d., 18., 19.b.,c. See Additional Answers in the back of this book.

RESOURCES
- Teaching Aid 13: Table for **Questions 15 and 16**
- Teaching Aid 21: Scatter plots and graph for the men's mile record

TEACHING NOTES

Any of a large number of data sets might have been picked as a vehicle for this reading lesson. This particular one was chosen for the following reasons: (1) It is famous. (There are only a few graphs in the entire 1990 World Almanac, but one of them is a graph of the evolution of the record since 1954. It is famous because the 4-minute mile was once considered an impenetrable barrier, because records exist for it longer than for almost any other event in sports, because the distance in the customary system is so simple, and because the distance (between a sprint and a marathon) represents an interesting question for the ultimate speed of humans. If the record is broken, you will be able to extend the data. (2) It is surprising that over a period of 130 years a line fits the data points so well. As far as we know, the first appearance of the *graph* of these data was in 1974 in a test version of *Algebra Through Applications,* a precursor to UCSMP *Algebra*). (3) Since a line cannot theoretically be a good long-term model for these data (every line with negative slope has to intersect the *x*-axis, but the mile can never be run in zero time), the example points out the need for a more sophisticated model. Even the

The Men's Mile Record

In 1976, the International Athletic Federation announced that it would only recognize track records in meters—with one exception. The exception is the one-mile run. The mile run received this special treatment because it has long had a special mystique about it which has captured the interest even of those not normally interested in athletics.

Perhaps this was best illustrated in 1954 when the Englishman Roger Bannister became the first man in recorded history to run a mile in less than four minutes. This athletic feat was headline news around the world, with many commentators regarding the event as "breaking through a barrier." Bannister's run broke the nine-year-old record of the Swede Gunder Hägg, and this long gap between records had encouraged many to think of a sub-four-minute mile as an impenetrable barrier, representing the limit of human potential. With hindsight, of course, we see that this is not the case, and that the barrier was both temporary and psychological.

The table on the next page lists the world record holders for the men's one mile run since 1875 and the year in which each record was set. The first entry in the table indicates that in 1875 Englishman Walter Slade ran the mile in 4 minutes, 24.5 seconds, a new world record time. This is written as 4:24.5. Nowadays, this time is often bettered by high school runners and the women's world record is lower.

The current record (as of 1990) of 3:46.32, set in 1985 by another Englishman, Steve Cram, indicates how much progress has been made over the intervening 110 years. In a present day world-class field, Walter Slade, the record holder in 1875, would be beaten by well over an eighth of a mile! Has this progress been steady or erratic? Should we expect the record to continue to fall or have we now almost reached the 'impenetrable barrier'? To answer such questions, a statistical analysis of the data is useful.

A scatter plot is a useful first step in modeling the men's mile record. Below, the data from the table are graphed with the record (in seconds) on the vertical axis and the associated year on the horizontal axis.

126

Roger Bannister upon breaking the four-minute barrier

The History of the World Record for the Men's Mile*				
Time	**Athlete**	**Country**	**Year**	**Location**
4:24.5	Walter Slade	England	1875	London, England
4:23.2	Walter George	England	1880	London, England
4:21.4	Walter George	England	1882	London, England
4:18.4	Walter George	England	1884	Birmingham, England
4:18.2	Fred Bacon	Scotland	1894	Edinburgh, Scotland
4:17.0	Fred Bacon	Scotland	1895	London, England
4:15.6	Thomas Conneff	United States	1895	Travers Island, New York
4:15.4	John Paul Jones	United States	1911	Cambridge, Massachusetts
4:14.4	John Paul Jones	United States	1913	Cambridge, Massachusetts
4:12.6	Norman Taber	United States	1915	Cambridge, Massachusetts
4:10.4	Paavo Nurmi	Finland	1923	Stockholm, Sweden
4:09.2	Jules Ladoumegue	France	1931	Paris, France
4:07.6	Jack Lovelock	New Zealand	1933	Princeton, New Jersey
4:06.8	Glenn Cunningham	United States	1934	Princeton, New Jersey
4:06.4	Sydney Wooderson	England	1937	London, England
4:06.2	Gunder Hägg	Sweden	1942	Göteborg, Sweden
4:06.2	Arne Andersson	Sweden	1942	Stockholm, Sweden
4:04.6	Gunder Hägg	Sweden	1942	Stockholm, Sweden
4:02.6	Arne Andersson	Sweden	1943	Göteborg, Sweden
4:01.6	Arne Andersson	Sweden	1944	Malmö, Sweden
4:01.4	Gunder Hägg	Sweden	1945	Malmö, Sweden
3:59.4	Roger Bannister	England	1954	Oxford, England
3:58.0	John Landy	Australia	1954	Turku, Finland
3:57.2	Derek Ibbotson	England	1957	London, England
3:54.5	Herb Elliot	Australia	1958	Dublin, Ireland
3:54.4	Peter Snell	New Zealand	1962	Wanganui, New Zealand
3:54.1	Peter Snell	New Zealand	1964	Auckland, New Zealand
3:53.6	Michel Jazy	France	1965	Rennes, France
3:51.3	Jim Ryun	United States	1966	Berkeley, California
3:51.1	Jim Ryun	United States	1967	Bakersfield, California
3:51.0	Filbert Bayi	Tanzania	1975	Kingston, Jamaica
3:49.4	John Walker	New Zealand	1975	Göteborg, Sweden
3:49.0	Sebastian Coe	England	1979	Oslo, Norway
3:48.8	Steve Ovett	England	1980	Oslo, Norway
3:48.53	Sebastian Coe	England	1981	Zurich, Switzerland
3:48.40	Steve Ovett	England	1981	Koblenz, West Germany
3:47.33	Sebastian Coe	England	1981	Brussels, Belgium
3:46.32	Steve Cram	England	1985	Oslo, Norway

*Source: 1990 Information Please Almanac

The data seem to conform rather closely to a linear model. The line of best fit for predicting world record time y from the year x is
$$y = 914.156 - 0.346x.$$

Because r^2 is about .972 and the slope of the regression line is negative, the correlation r between record time and year is about -0.99, indicating that a line is a remarkably good model for these data.

quadratic model used in the lesson is insufficient. (4) Still, the linear model has been a very good predictor. (In *Algebra Through Applications*, the record was predicted to be 3:46 in 1985. At the time of the prediction, the record was 3:51.1. In that book, a prediction of 3:40 is given for the year 2000.)

An interesting sidelight is that Roger Bannister, who broke the 4-minute barrier in 1954, was a physician, and he correctly believed the barrier was psychological, not physiological.

The conversion from customary units to metric is necessary if one wishes to compare the mile record with records for other distances. World records are not kept for any other of our customary units. Here is one way to convert Sebastian Coe's 800-meter record to miles per hour.

$$\frac{800 \text{ m}}{1 \text{ min } 41.8 \text{ s}} = \frac{800 \text{ m}}{101.8 \text{ s}} =$$
$$\frac{800 \text{ m}}{101.8 \text{ s}} \cdot \frac{60 \text{ s}}{1 \text{ min}} \cdot \frac{60 \text{ min}}{1 \text{ h}} \cdot$$
$$\frac{1 \text{ km}}{1000 \text{ m}} \approx \frac{28.3 \text{ km}}{h} =$$
$$\frac{28.3 \text{ km}}{h} \cdot \frac{0.6214 \text{ mi}}{1 \text{ km}} \approx \frac{17.6 \text{ mi}}{h}$$

Throughout this book, there are problems that require conversion of units, and a discussion of this process will be helpful.

Project 1 on page 133 asks students to compare these data with those for women at the same distance.

If the trend evident in the scatter plot continues, we might use the line of best fit to extrapolate from the data and to predict likely future changes in the world record. For example, to predict the men's mile record in the year 2000, substitute $x = 2000$ in the above equation:

$$y = 914.156 - 0.346(2000) \approx 222.16.$$

So in the year 2000 we anticipate a record time of 222.16 seconds, or 3 minutes 42.16 seconds. Such a prediction seems plausible.

The slope of the line of best fit tells how much improvement might be expected each year. The negative slope indicates that the world record is expected to get smaller each year, while the numerical size suggests that it should decrease by about 0.35 seconds each year.

Extrapolation into the far future, however, is a different matter. The line of best fit crosses the x-axis, suggesting that in the future a man will complete the one-mile run in no time at all, and shortly afterwards the record will be lowered to negative time! Consequently, a linear model is inappropriate to describe the trend over a very long time. A likely shape of the longer term trend for the mile record is shown here.

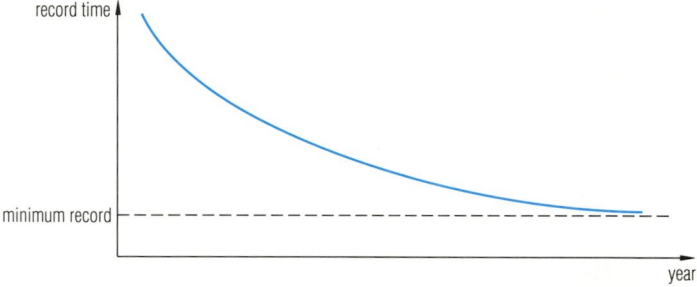

However, in the short term, a line is a good fit because we are seeing only part of this larger curve. A continually decreasing curve like this one will appear always approximately linear if you zoom in on it closely enough.

The dotted line, a horizontal *asymptote* to the curve, represents a minimum time for running a mile that humans are unable to better. The minimum time is related to fundamental characteristics of human beings, such as lung capacity, the ability of the blood to use oxygen, and respiration rate. No one knows what that minimum is for the mile run.

One possible way of estimating this minimum for the mile is to extrapolate from a nonlinear model. For instance, assume that a quadratic model is a useful approximation to the mile record data.

To estimate an equation for a quadratic model, it may be wise to use only more recent data, for instance, since 1930. Before that time world records were measured with less sophisticated timing equipment, and thus might be less reliable. Recently, times have been measured to the hundredth of a second.

128

Also, before 1930, it might be argued that athletes were more likely to be amateurs, and now the best athletes are almost full-time professionals trained by experts on exercise, diet, and biomechanics. The scatter plot below shows a quadratic curve of good fit through the data beginning with the year 1930.

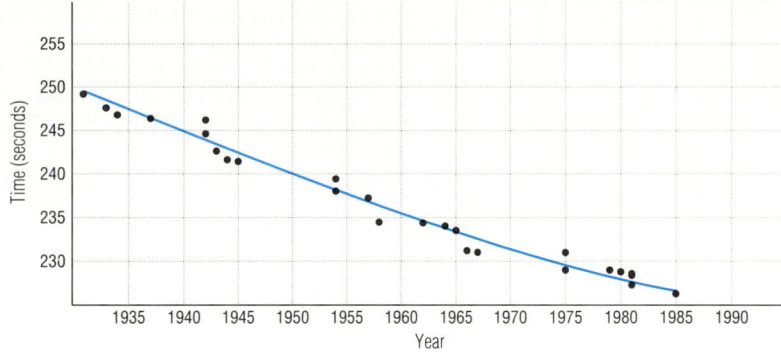

An equation of a quadratic model given by our statistics package is
$$y = 11170.0396 - 10.745375x + 0.00263596x^2.$$

Using a function grapher, this curve can be graphed and its relative minimum point found to be about (2038, 219.3). So, this quadratic model predicts that the world record for the men's one-mile run will not go below about 3 minutes 39.3 seconds, and that this will be reached around 2038.

As with the linear model, we know that the quadratic model cannot be relied upon for extrapolation too far beyond the data. Otherwise, the world record will rise after 2038! Also, the model depends critically on the data, and the choice of 1930 as a cut-off was essentially arbitrary.

Sebastian Coe

Another approach to the problem is to study the average speed of runners, rather than the time taken to run a certain distance. For example, the 1985 world record for the men's 800 meter run was 1:41.8, set by Sebastian Coe in 1981. Coe's average speed for this run was

$$\frac{800}{1000} \cdot \frac{3600}{101.8} \text{ kilometers per hour} = 28.3 \text{ kph.}$$

Because there are 0.6214 miles in a kilometer, this speed is $28.3 \cdot 0.6214 \approx 17.6$ miles per hour. We might guess that the ultimate world record for the mile could not be run much faster than this speed (the best in 1985 for about half a mile), and compute the world record accordingly. Thus, if a person ran at a speed of 17.6 miles per hour for a whole mile, not merely for 800 m, the time would be $\frac{1}{17.6}$ hours or about 3 minutes 25 seconds. This prediction is somewhat lower than the prediction from the quadratic model.

Only time will tell whether such predictions as these will seem pessimistic, optimistic, or about right.

Questions

Covering the Reading

1. Convert Steve Cram's 1985 world record of 3:46.32 into seconds.
   226.32 seconds
2. Which world record for the mile stood for the longest period of time?
   Thomas Conneff's 4:15.6 (1895-1911)
3. Suppose that Walter Slade and Steve Cram ran their world-record times for the mile run in the same race and that each ran at a constant speed. By how many yards would Cram beat Slade?
   about 254 yards

In 4–7, use the line of best fit for the men's mile record data.

4. Predict the mile record for each year.
   a. 1960 **3:55.97**   **b.** 2020 **3:35.21**   **c.** 3000 **-2:03.87**
5. Predict when the first $3\frac{1}{2}$-minute mile will be run by a man.
   in the year 2035
6. On average, how much has the world record decreased each year?
   0.346 seconds
7. The equation of the regression line is given to considerable accuracy.
   a. Give a reason for this level of accuracy.   **See margin.**
   b. Use the less-accurately stated equation $y = 914.2 - 0.35x$ for predicting the record for 1960, 2020, and 3000.
   c. Compare your answers to part **b** and Question 4.   **See margin.**
   b) 1960 3:48.2; 2020 3:27.2; 3000 -2:15.8
8. **a.** Use the quadratic model given in the lesson to predict the world record for the men's one-mile run in the year 2000.   **3:43.1**
   b. By how much does this prediction differ from the record time predicted by the line of best fit?   **≈ .944 second**

In 9 and 10, use the fact that Sebastian Coe also holds the current world record for the 1000 m run. He set the record of 2:12.40 in 1981.

9. Convert this achievement into a speed expressed in
   a. kilometers per hour   **b.** miles per hour.
      about 27.2 kph   **about 16.9 mph**
10. If someone ran a mile at the same speed at which Coe ran his record-setting 1000 m, what would be his time?   **3:33.05**

Review

In 11–14, match the scatter plot with the most likely correlation coefficient r.
(a) $r = -.8$ (b) $r = 0$ (c) $r = .7$ *(Lesson 2-5)*

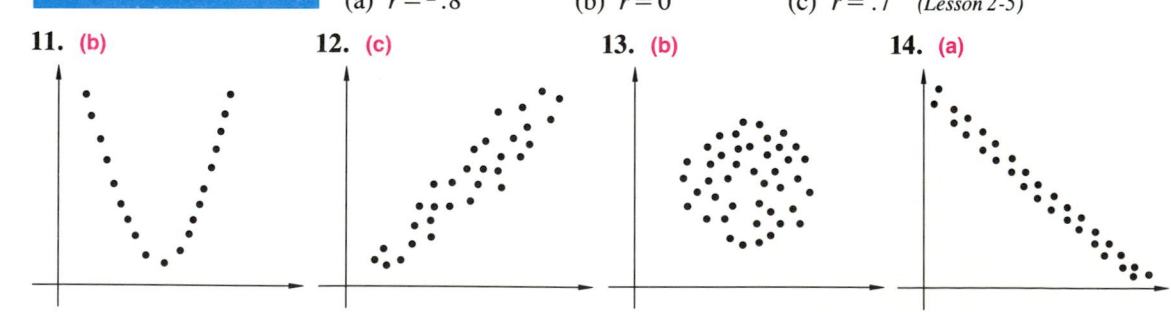

11. (b)   **12.** (c)   **13.** (b)   **14.** (a)

130

Year	Billions of miles
1945	3
1950	10
1955	23
1960	32
1965	54
1970	109
1975	136
1976	150
1978	189
1979	210
1980	204
1981	201
1982	214
1983	232
1985	263
1986	293
1987	322

In 15 and 16, use the data at the left (which were presented also in the chapter opener) showing the number of passenger miles for intercity traffic in the United States as a function of year.

15. Let $y =$ the year and $f(y) =$ the number of passenger miles.
   a. Evaluate $f(1980)$. **b.** Solve $f(y) = 150,000,000$.
   204 billion **1976**

16. a. Enter the data on a statistics package.
   b. Find the line of best fit for the data. **$y = 7.405x - 14,452.8$**
   c. Find the correlation between year and number of passenger miles.
   d. Use the line of best fit to predict the number of air passenger miles flown in 1965. **about 98 billion miles**
   e. For the linear model find the deviation of the predicted from the observed value for 1965. **about 44 billion miles**
   f. Use the quadratic model for these data given in the chapter opener to predict the number of air passenger miles in 1965.
   g. For the quadratic model find the deviation of the predicted from the observed value for 1965. *(Lessons 2-1, 2-3, 2-5, 2-6)*
   c) 0.954; f, g) See margin.

17. The graph at the right is from one of the earliest reports linking smoking to lung cancer. The line through the data is the line of best fit.
   a. Identify the independent and dependent variables.
   b. In which countries was the male death rate for lung cancer less than that predicted by the line of best fit?
   c. In which countries was the male death rate for lung cancer greater than that predicted by the line of best fit?
   d. In which country was the deviation of predicted death rate from observed death rate the greatest?
   e. Why do you think the authors chose to compare cigarette consumption and death rate 20 years apart, rather than in the same year? *(Lessons 1-2, 2-1, 2-3)*
   See margin.

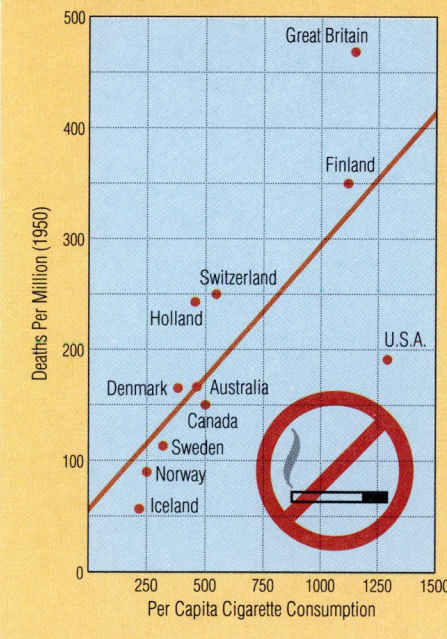

Crude Male Death Rate For Lung Cancer In 1950 and Per Capita Consumption Of Cigarettes In 1930 In Various Countries.

Report of the Advisory Committee to the Surgeon General, *Smoking and Health* (Washington, D.C., 1964), p. 176; based on R. Doll, "Etiology of Lung Cancer," *Advances in Cancer Research*, 3 (1955), 1-50.

In 18 and 19, state the domain and range of each function. *(Lessons 2-1, 2-4)*

18. $y = |x|$ **See margin.** **19.** $f(a) = \lfloor a \rfloor$ **See margin.**

LESSON 2-8 The Men's Mile Record 131

e. Sample: Because the effects of cigarette smoking on lung cancer incidence are not immediately noticeable, someone who began smoking in 1930 would probably not die from lung cancer in the same year.

18. The domain is the set of all real numbers; the range is the set of all nonnegative real numbers.

19. The domain is the set of all real numbers; the range is the set of all integers.

FOLLOW-UP

Solutions to the projects are found at the end of the Additional Answers section at the back of the book.

PROJECTS
Projects 1 and 3 on page 133 are related to the content of this lesson. Project 3 is suitable for **small groups.**

In 20 and 21, identify the error made in the use of the correlation coefficient. *(Lesson 2-5)*

20. We found a high correlation ($r = 1.89$) between students' SAT scores and their grade-point average. $-1 \le r \le 1$; r cannot equal 1.89

21. The correlation between family income and amount spent by the family on vacations is $r = 0.68$ dollars.   r has no units

22. The table below lists the end-of-quarter interest rates on 3-month certificates of deposit during a four-year period from mid-1979 to mid-1983 (a period during which interest rates fluctuated greatly). It also gives the end-of-quarter values of Standard and Poor's 500 Stock Composite Average for the same period.

Year	Quarter	Interest rate	Standard and Poor's 500
1979	3	11.89	109.32
	4	13.43	107.94
1980	1	17.57	104.69
	2	8.49	114.24
	3	11.29	125.46
	4	18.65	135.76
1981	1	14.43	136.06
	2	16.90	131.21
	3	16.84	116.18
	4	12.49	122.55
1982	1	14.21	111.96
	2	14.46	109.61
	3	10.66	120.42
	4	8.66	135.28
1983	1	8.69	152.96
	2	9.20	168.11

a. Find the correlation coefficient between interest rate and the S & P 500 average.   -0.365
b. Write a sentence describing in words the relation between these two variables. *(Lesson 2-5)*   See margin.

23. A manufacturer of copying machines is interested in improving its customer support services. As a result, a sample of 40 customers was surveyed to determine the amount of downtime (in hours) they had experienced in the previous month. The results are below.

2	6	5	6	11	19	28	4	37	0
14	5	3	8	2	1	12	7	21	0
4	0	5	7	10	9	12	8	18	9
24	16	7	1	54	9	8	14	39	40

a. Construct a box plot of these data.   See margin.
b. How many customers are having excessive downtime using the $1.5 \times$ IQR criterion? *(Lesson 1-5)*   4 customers

Exploration

24. Contact the athletic department of your school for records in the boys' or girls' mile run. Plot the records over time. Do the data from your school show trends similar to the world records?   Answers will vary.

CHAPTER 2
Projects

1. How does the women's mile record compare to the men's?
   a. Consult an almanac or other reference to obtain the world record times for the women's mile run.
   b. Enter the women's records into a statistical package.
   c. Do these data show a linear trend? If so, find an equation for the line of best fit for the data.
   d. Use the equation in part **c** and the regression line given in Lesson 2-7 to predict the year in which a woman will break the existing men's record in the mile run. Is the year near enough in the future for you to have faith in this extrapolation?
   e. Write a brief report comparing and contrasting the performance of males and females in the mile run.

2. Use the database constructed as a project at the end of Chapter 1.
   a. Some people claim that most people are squares. That is, their height and arm span are approximately equal.
      i. If this conjecture were true, what would you expect a scatter plot of height vs. arm span to resemble?
      ii. Make a scatter plot of height and arm span for students in your class, and find the line of best fit. Are most people in your class squares?
      iii. Are there any outliers? If so, who are they? Remove these data from the set and recalculate the regression line.
   b. Examine some other pairs of variables to find a pair with a high positive correlation.
      i. Find an equation of the line of best fit for predicting one variable from the other.
      ii. Reverse the order of the variables in **i** (that is, make the former independent variable the new dependent variable). Find an equation of the line of best fit. Use your line of best fit to predict one variable from the other for some friends who are not in your class. How accurate is your model?

   c. Which variables, if any, show a negative correlation? Find an appropriate model for one such relation.
   d. Write a brief report describing which variables seem most strongly related in your class.

3. Much evidence has been gathered in recent years regarding the dangers of smoking.
   a. Find data showing the relation between measures of smoking and measures of health. Some examples might be the percent of American adults who smoke in relation to year, the percent of teenage girls who smoke in relation to year, or the per capita cigarette consumption in relation to lung cancer or throat cancer rates.
   b. Display the data in tables and graphs.
   c. Find some mathematical models for your data and use them to make some predictions.
   d. Use your models to support the argument that smoking is a health hazard.
   e. What arguments might the tobacco industry present to counter your arguments in part **d**?

4. Consult an almanac, encyclopedia, or other reference to obtain the world record at various points in time for a particular sports event that interests you. Try to obtain as complete data as possible. Do not restrict yourself to Olympic results (which occur only every four years).
   a. Plot your data on a scatter plot with year on the horizontal axis.
   b. Decide whether a linear or quadratic model suits your data best.
   c. Draw a line or parabola to fit your data.
   d. Select some points on the line or parabola to find an equation for it.
   e. Are there any outliers in your data? If there are, give reasons for them.
   f. Predict the likely world records in your event over the next decade.
   g. Predict when the world record in your event is likely to reach a significant milestone.

PROJECTS

The projects provide an opportunity for extended work on topics related to the content of the chapter. (See pages 66 and T37 for more information on the purposes of projects.) Some or all of these projects may be suitable for **small group work.**

Solutions to the projects are found at the end of the Additional Answers section at the back of the book.

You might expect the following relationships for the pairs of variables suggested in **Project 2.**
1. Height and arm span are linearly related—and in fact the quotient of these is nearly equal for most people.
2. Age and length of hand span have a linear relationship for young children, and little or no relationship for people over the age of 15.
3. The area of a foot should be proportional to the square of the length of a foot.
4. Wrist circumference and pulse rate are unrelated.

This project can be started any time after Lesson 2-2 and returned to as students learn about new models.

In part **a** of **Project 2,** there is a reference to discarding outliers. You might have a discussion about whether statisticians should discard data that do not fit a given model. There is no clear answer to the question. In some contexts discarding data is warranted (for example, data improperly entered into a computer). In other contexts it is not (most randomly generated samples should not be tampered with). At any rate, good statistical practice dictates that the statistician explain any tampering with a data set.

A good source of information for **Project 3** are the publications of the Office of the U.S. Surgeon General available through the U.S. Government Printing Office. Your school nurse and the Physical Education Department should have other references.

Students who do **Project 4** might be interested in also making models for Olympic results in the women's javelin, and comparing their Olympic and world record models.

Besides the sources stated in **Project 4** of Chapter 2, another source for the world records for the women's javelin is the book *Progression of World Best Performances and IAAF Approved World Records,* researched and compiled by Ekkehard zur Megede and published by the International Amateur Athletic Federation, 1987.

The normal distribution function $N(x)$ referred to in **Project 5** is discussed in Chapter 10. The value of the function is the area to the left of x under the standard normal curve

$$f(x) = \frac{1}{\sqrt{2\pi}}\, e^{-\frac{x^2}{2}}.$$

The graph of f is bell-shaped as shown below, and the total area under f is 1.

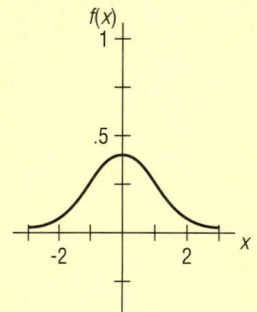

We get the following quadratic model for $N(x)$ using the points (0, .5), (1, .8413), and (2, .9772) from the table on page 134. (Note that this is not the equation of a parabola which approximates the normal curve, but rather of a function that approximates the area under that curve.)
$N(x) = -0.1027x^2 + 0.4440x + 0.5$
Your students who do this project should get a quadratic

---

5. The following table gives some values for the *normal distribution function N* which you will use later in this course. (It is not necessary to know at this time what the numbers represent.)

x	N(x)	x	N(x)
0.0	.5000	1.2	.8849
0.2	.5793	1.4	.9192
0.4	.6554	1.6	.9452
0.6	.7257	1.8	.9641
0.8	.7881	2.0	.9772
1.0	.8413		

To find values for $N(x)$ for values of x between 0 and 2 that are not given in the table, interpolation is necessary. *Linear interpolation* models the data with a linear function, while *quadratic interpolation* models the data with a quadratic function.
  a. Plot these data on a scatter plot.
  b. Draw a line to model the data. Select two points on the line to find the equation of the linear model. (Choose one point with a small *x*-coordinate and one with a large *x*-coordinate.)
  c. Draw a parabola to model the data. Select three points on the curve and find an equation of the quadratic model. (Choose one point with a small *x*-coordinate, one with a mid-range *x*-coordinate, and one with a large *x*-coordinate.)
  d. Compare your linear and quadratic models by using them to predict the values of $N(x)$ for some values not in the table. Try at least $x = 0.3, 0.72, 1.55,$ and 1.812.
  e. Ask your teacher for a larger table of values for $N(x)$ and check how much better your quadratic model fits the table than does your linear model.

6. In the next column are the latitude (in degrees North) and the average daily maximum temperature in April for various cities in North America.
  a. Convert each latitude to decimal notation. For instance, 35°26′ is read 35 degrees, 26 minutes. There are 60 minutes in a degree.
    So $35°26′ = 35\frac{26}{60} \approx 35.42°$.
  b. Use a computer package to draw a scatter plot with latitude on the *x*-axis.
  c. Find an equation of the line of best fit for these data.

Place	Latitude	Temperature (°F)
Acapulco, Mexico	16°51′	87
Bakersfield, CA	35°26′	73
Caribou, ME	46°52′	50
Charleston, SC	32°54′	74
Chicago, IL	41°59′	55
Dallas-Ft. Worth, TX	32°54′	75
Denver, CO	39°46′	54
Duluth, MN	46°50′	52
Great Falls, MT	47°29′	56
Juneau, AK	58°18′	39
Kansas City, MO	39°19′	59
Los Angeles, CA	33°56′	69
Mexico City, Mexico	19°25′	78
Miami, FL	25°49′	81
New Orleans, LO	29°59′	77
New York, NY	40°47′	60
Ottawa, Canada	45°26′	51
Phoenix, AZ	33°26′	83
Quebec, Canada	46°48′	45
Salt Lake City, UT	40°47′	58
San Francisco, CA	37°37′	65
Seattle, WA	47°27′	56
Vancouver, Canada	49°18′	58
Washington, DC	38°51′	64

d. Interpret the sign and magnitude of the slope of the regression line.
e. Over what domain do you expect the regression line to fit the data well?
f. Predict the average daily April maximum temperature for these cities: Detroit, MI 42°22′N; Tampa, FL 27°49′N
g. The actual average daily high temperature in April is 47°F for Detroit and 72°F for Tampa. Find the percent error for each of your predicted values in part **d**.
h. Which cities appear to be outliers? Give plausible reasons why these cities might have a different relation between latitude and temperature than others.
i. What would you expect a linear model for the relation between latitude in South America and temperature to be? Why? Find some data on cities in South America and test your conjecture.
j. Find the latitudes and average daily maximum temperatures in cities in other parts of the world, such as in Africa, Asia, or Europe. Explain any big differences between the regression lines you find for these areas and those found in parts **c** and **i**.

Summary

Any set of ordered pairs is a relation. Functions are particular kinds of relations—those for which each first element has exactly one second element. Another way to view a function is a rule of correspondence between two sets A and B, which relates each element of A to exactly one element of B. The set of first elements or the set A is the domain, and the set of second elements is the range of a function.

Bivariate data can often be well-represented by linear models—linear functions which are reasonable simplifications of the observed relation. Scatter plots are used to determine the type of the relation and the feasibility of a linear model. A linear model can be approximated by drawing a line close to all the data points. However, there exists a line that causes the total squared error between observed and predicted values to be a minimum. This line is called the line of best fit or the regression line. Nowadays, statistical packages and some calculators are capable of calculating the slope-intercept form of this line. Using an equation of the regression line, predictions can be made for unobserved values of the dependent and independent variables.

Some data have y-values that are constant over an interval of x-values and then jump to another level. These can be modeled by step functions such as the greatest integer function. Step functions have points of discontinuity.

The strength of a linear relation between two variables is measured by the correlation coefficient r. The sign of the correlation coefficient indicates the direction of the relation between the variables, and its magnitude indicates the extent of the linearity. Although perfect correlations ($r = \pm 1$) are rare, an r with an absolute value close to 1 indicates a strong linear relation. The world record for running the mile has a very strong linear correlation with the year. A strong linear relation, however, does not necessarily imply that one of the variables causes the other. An r-value close to zero indicates that the variables are not related linearly.

Sometimes nonlinear functions model a data set more closely than a line. Polynomial models were discussed in this chapter. An algebraic method to determine the coefficients of a quadratic model was developed. This method can be generalized for higher-order polynomial models. Some statistics software is capable of determining formulas for polynomial models.

When making predictions with any kind of mathematical model, interpolation is always safer than extrapolation. Linear or quadratic models can produce very reasonable predictions for intervals over which the original data were observed. However, one has to use caution outside the domain of observed values.

Vocabulary

For the starred (*) terms you should be able to give a definition of the term.
For the other terms you should be able to give a general description and a specific example of each.

Lesson 2-1
univariate data
bivariate data
mathematical model
relation
*independent variable
*dependent variable
*function, *domain
*range, vertical line test

Lesson 2-2
*linear function
interpolation
extrapolation

Lesson 2-3
*line of best fit
regression line
observed value
predicted value
errors in prediction
deviation
method of least squares
center of gravity

Lesson 2-4
step function
*greatest integer function
rounding-down function
floor function, continuous
discontinuous
point of discontinuity
rounding-up function
ceiling function

Lesson 2-5
*correlation coefficient
perfect correlation

Lesson 2-6
nonlinear model
*quadratic model
acceleration due to gravity

Lesson 2-7
cubic model
quartic model
polynomial model

---

with coefficients close to the ones above. A standard normal distribution table is provided in Appendix D, page 844. Use it to respond to students who ask you about part **e** of the project.

SUMMARY

The Summary gives an overview of the entire chapter and provides an opportunity for students to consider the material as a whole. Thus, the Summary can be used to help students relate the various concepts presented in the chapter.

VOCABULARY

Terms, symbols, and properties are listed by lesson to provide a checklist of concepts a student must know.

Progress Self-Test

ADDITIONAL ANSWERS
4.b.

See margin for answers not shown below.

Take this test as you would take a test in class. You will need graph paper and a calculator. You may want to use a statistics package. Then check the test yourself using the solutions at the back of the book.

In 1 and 2, **a.** state if the relation represents a function. If it does not, give a reason for your answer. **b.** Determine the domain and the range of the relation.

1.

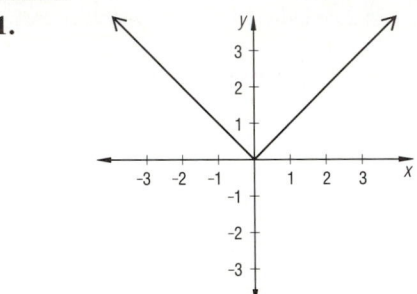

2. $\{(3, 4), (4, 9), (6, 12), (0, -1), (4, 7)\}$

3. Let $k(x) = x^2 + 7$.
 a. Evaluate $k(-3)$.   **16**
 b. Find the range of k.   **$\{x : x \geq 7\}$**

4. Suppose $f(x) = \lceil x \rceil + 1$.
 a. Evaluate $f(6\pi)$.   **20**
 b. Graph $y = f(x)$ for $-3 \leq x \leq 3$.

5. A company operates a shuttle service from the airport to downtown. The company uses shuttle buses that hold a maximum of thirty passengers.
 a. How many buses are needed to hold 131 passengers?   **5**
 b. Write a formula for the number n of buses needed to transport p passengers.

6. For a set of data, the line of best fit has equation $y = 3.7x - 14.1$ and $r^2 = 0.37$. What is the correlation between x and y?   **0.61**

1 a) The relation is a function.
 b) The domain of the function is the set of all real numbers; the range is the set of all nonnegative real numbers.
2 a) the relation is not a function, since it contains the points (4, 9) and (4, 7).
 b) domain: {0, 3, 4, 6}; range: {-1, 4, 7, 9, 12}
5 b) sample: $n = \left\lceil \dfrac{p}{30} \right\rceil$

In 7 and 8, use the following data showing the average number of miles traveled by all U.S. passenger vehicles per gallon of fuel consumed (commonly known as **mpg**).

Year	mpg	Year	mpg
1970	13.57	1981	15.54
1972	13.69	1982	16.33
1974	13.43	1983	16.65
1976	13.72	1984	16.68
1978	14.06	1985	17.80
1980	15.15	1986	18.24

7. a. Make a scatter plot of these data.
 b. Use one sentence to describe the relationship between year and vehicle efficiency in miles per gallon.
 c. Explain why a linear model may not be appropriate here.

8. A quadratic model for the number of miles per gallon m as a function of the year y (after 1900) is $m = 0.025y^2 - 3.67y + 145.84$. Use this model to predict the average mpg for passenger vehicles in 1990.   **18.04 mpg**

In 9 and 10, *multiple choice*.

9. An electrical consultant charges $60 for the first hour of her time, and $45 for each hour or part hour afterwards. Which formula represents her charge C in dollars as a function of the time h in hours?
 (a) $C = 45 - 60 \lfloor h - 1 \rfloor$
 (b) $C = 45 - 60 \lfloor 1 - h \rfloor$
 (c) $C = 60 - 45 \lfloor h - 1 \rfloor$
 (d) $C = 60 - 45 \lfloor 1 - h \rfloor$ **(d)**

10. Each number below is the correlation coefficient between two variables. Which indicates the strongest linear relation?
 (a) 0.64   (b) 0.09
 (c) -0.5   (d) -.82   **(d)**

7 b) Sample: Vehicle efficiency increases slightly between 1974 and 1978, then begins to increase more rapidly over time.
 c) Sample: The rate of increase in vehicle efficiency (which corresponds to the slope of the linear model) is not constant.

11. A ball is thrown off the top of a high building. The table below shows the height *h* in feet of the ball above ground level *t* seconds after being thrown.

t	1	2	3	5	6
h	299	311	291	155	39

   a. Construct and solve a system of equations to find a quadratic model for these data.
   b. Use the model from part **a** to estimate when the ball will be 80 feet above ground level.

12. The following table lists the results for the men's 800-meter run at the modern Olympic Games.

Year	Runner	Time (sec)
1896	Edwin Flack, Australia	131.0
1900	Alfred Tysoe, Great Britain	121.4
1904	James Lightbody, United States	116.0
1908	Mel Sheppard, United States	112.8
1912	Ted Meredith, United States	111.9
1920	Albert Hill, Great Britain	113.4
1924	Douglas Lowe, Great Britain	112.4
1928	Douglas Lowe, Great Britain	111.8
1932	Thomas Hampson, Great Britain	109.8
1936	John Woodruff, United States	112.9
1948	Malvin Whitfield, United States	109.2
1952	Malvin Whitfield, United States	109.2
1956	Tom Courtney, United States	107.7
1960	Peter Snell, New Zealand	106.3
1964	Peter Snell, New Zealand	105.1
1968	Ralph Doubell, Australia	104.3
1972	David Wottle, United States	105.9
1976	Alberto Juantorena, Cuba	103.5
1980	Steve Ovett, Great Britain	105.4
1984	Joaquin Cruz, Brazil	103.0

   a. Find an equation for the line of best fit for predicting the winning time from the year.
   b. Use the equation to predict the winning time in the Olympic Games in the year 2000.
   c. According to the equation, when will an Olympic Games men's 800-meter race be run in less than 1 minute 40 seconds? (Remember that the Olympic Games only take place every four years.)

   d. What is the correlation *r* between year and winning time?
   e. Paul Ereng of Kenya won the 800-m race in the 1988 Summer Olympic Games with a time of 1:43.45. What would be the error in your prediction if you used the linear model for these data?
   f. Identify the prediction made in part **c** as an example of interpolation or extrapolation.

11 a) sample: $299 = a + b + c$
$291 = 9a + 3b + c$
$39 = 36a + 6b + c$
so $h(t) = {}^-16t^2 + 60t + 255$
b) about 5.7 seconds after it is thrown

12 a) sample: $y = .202x + 503$
b) 99 seconds
c) 1996
d) about $^-0.87$
e) about 2.0 seconds
f) extrapolation, since our known data only goes up to 1984

CHAPTER REVIEW

The main objectives for the chapter are organized here into sections corresponding to the types of understanding this book promotes: Skills, Properties, Uses, and Representations (called SPUR objectives).

USING THE CHAPTER REVIEW

Students should be able to answer questions like these with about 85% accuracy by the end of the chapter. (See pages 72 and T38–39 for more information.)

Chapter Review

Questions on **SPUR** Objectives

SPUR stands for **S**kills, **P**roperties, **U**ses, and **R**epresentations.
The Chapter Review questions are grouped according to the SPUR Objectives for this chapter.

SKILLS deal with the procedures used to get answers.

■ **Objective A:** *Evaluate functions using f(x) notation.* (Lessons 2-1, 2-4)

In 1 and 2, let $f(x) = 3^x$.

1. Evaluate
   a. $f(1)$ **3**   **b.** $f(-1)$ $\frac{1}{3}$

2. *True* or *false*. Justify your answer.
   a. $f(2) + f(-2) = 0$ **b.** $f(2) \cdot f(3) = f(5)$

3. Let $g(x) = x^2 + 3$.
   a. Evaluate $g(-1)$. **4**
   b. Does $g(4) - g(2) = g(2)$? Justify your answer.

4. Let $h(x) = |x - 5|$. *True* or *false*.
   a. $h(3) = h(-3)$ **False**   **b.** $-h(3) = -h(7)$ **True**

5. Suppose $p(x) = \lfloor x \rfloor$. Evaluate.
   a. $p(12.9)$ **12**   **b.** $p(-5)$ **-5**
   c. $p(\pi)$ **3**   **d.** $p(-3.5)$ **-4**

6. Find the value of each expression.
   a. $\lceil -2.9 \rceil$ **-2**   **b.** $\lfloor \sqrt{75} \rfloor$ **9**

■ **Objective B:** *Find linear or quadratic functions which contain specified points.* (Lessons 2-2, 2-7)

7. Show that the points $(0, 1)$, $(2, -1)$, and $(3, 4)$ lie on the graph of $p(t) = 2t^2 - 5t + 1$.

8. Find an equation for the quadratic function through the points $(1, 7)$, $(4, 5)$, and $(5, 2)$.

9. The points $(0, .33)$, $(1, .6)$, $(5, 1.68)$, and $(11, 3.3)$ lie on a line.
   a. How many points are necessary to determine an equation of the line containing these points? **two**
   b. Write an equation in slope-intercept form of the line. **y = 0.27x + 0.33**

10. The four points, $(0, -5)$, $(2, 9)$, $(4, 47)$, and $(6, 109)$ lie on a parabola.
    a. How many points are necessary to determine the coefficients of the quadratic model containing these four points? **3**
    b. Write a system of equations to find the coefficients.
    c. Solve the system in part **b**.

PROPERTIES deal with the principles behind the mathematics.

■ **Objective C:** *Identify the variables, domain, and range of functions.* (Lessons 2-1, 2-2, 2-4, 2-6)

In 11–16, state **a.** the domain and **b.** the range.

11. $y = \lceil x \rceil$ 12. $r(x) = \sqrt{x}$

13. $j(x) = \dfrac{2}{-x}$ 14. $f(t) = 2t^2 - 18$

15. $y = mx + b$, $m \neq 0$

16. $\{(-2, 2), (-1, 2), (0, 2), (2, 2), (5, 2)\}$

In 17 and 18, identify the **a.** independent variable and **b.** dependent variable.

17. $3x^2 = y$ 18. $f(t) = |t|$

In 19 and 20, identify the values at which the function is not continuous.

19. $f(x) = \dfrac{48}{x^2}$ 20. $y = \lfloor x \rfloor$

■ **Objective D:** *Identify properties of the line of best fit and of the correlation coefficient.* (Lessons 2-3, 2-5)

21. Explain what is meant by a correlation of 0.95.

22. A statistics package gives values of r^2, where r is the correlation coefficient between two variables. If $r^2 = .54$, what are all possible values of r? **0.73, -0.73**

23. For a set of data, the line of best fit is given by $y = 6.2 - 1.7x$ and $r^2 = 0.80$. What is the correlation coefficient? ≈ -0.89

In 24–26, *true* or *false*.

24. If the sum of the errors calculated using a linear model for a given data set is zero, then the line of best fit goes through all the data points. **False**

25. The line of best fit always contains the center of gravity of a data set. **True**

26. The correlation coefficient r is measured in the same unit as the slope of the regression line. **False**

27. *Multiple choice.* An r value of .23 indicates
(a) a strong positive relation
(b) a weak positive relation
(c) a weak negative relation
(d) a strong negative relation. **(b)**

28. What value of r indicates a perfect negative relation? **-1.0**

USES deal with applications of mathematics in real situations.

■ **Objective E:** *Find and interpret linear models.*
(Lessons 2-2, 2-3)

29. The table contains data for total U.S. coal production in tens of millions of short tons from 1973 to 1982.

Year (after 1900)	Production
73	59.9
74	61.0
75	65.5
76	68.5
77	69.7
78	67.0
79	78.1
80	83.0
81	82.4
82	83.8

One linear model for these data is:
$P = 2.9Y - 152.9$
where P is the production (in tens of million of tons), and Y is the number of years after 1900.
a. Plot the data and the model on the same coordinate system.
b. Which data points appear to be outliers?
c. Use the model to predict the likely coal production in 1987. **994 million tons**
d. The actual coal production in 1987 was 917 million tons. How big is the prediction error in part **c**? **77 million tons**
e. When would you expect coal production to reach 945 million tons annually? **1985**
f. Give a reason why this model should not be used to predict coal production in the year 2010.

30. A rental car company charges $35.99 per day plus $.18 per mile for a full-size car.
a. Find an equation for the cost C of renting a car for one day and driving m miles.
b. How far could a person drive in one day without spending over $100 on rental fees?

31. An archaeologist counted the flintstones and charred bones at several sites, and obtained the following data:

Number of flintstones	18	53	23	8	47	16	3	81	55	37
Number of bones	2	4	2	0	6	0	1	9	5	4

a. Find the correlation between the number of flintstones and the number of bones. **0.95**
b. Does it seem appropriate to use a linear function to model these data? Justify your answer.
c. The archaeologist claimed that the data prove that the flintstones were used to light fires that charred the bones. Criticize this claim.

30 a) $C = 35.99 + 0.18\,m$; b) 355.6 mi

In 32 and 33, use the following table listing the heights and shoe sizes for 13 men in an office. Shoe sizes are either whole numbers or half sizes.

Height (in.)	Shoe size	Height (in.)	Shoe size
70	$10\frac{1}{2}$	71	12
73	$9\frac{1}{2}$	69	9
68	7	66	$8\frac{1}{2}$
69	10	71	9
72	10	70	10
68	9	73	$11\frac{1}{2}$
74	12		

32. a. Make a scatter plot of the data by hand. Plot height on the horizontal axis.
   b. Draw a line that fits the data.
   c. Determine an equation for your line.
   d. Interpret the slope and y-intercept of your line.
   e. According to your model, what size shoe would you expect a man to wear if his height is **i.** 72 inches? **ii.** 65 inches?

33. a. Find an equation for the line of best fit for these data. $y = 0.426x - 20.13$
   b. Use the line of best fit to predict the expected shoe size of a man 6′2″ tall.
   c. Over what domain would you expect this model to hold?   **b) 11½; c) between about 62 and 78 inches**

■ **Objective F:** *Use step functions to model situations.* *(Lesson 2-4)*

34. *Multiple choice.* Which of the following gives the century number c from the year y?

   (a) $c = \left\lfloor \dfrac{y}{100} \right\rfloor$ (b) $c = \left\lfloor \dfrac{y}{100} \right\rfloor + 1$

   (c) $c = \lfloor 100y \rfloor$ (d) $c = \left\lfloor \dfrac{y}{100} \right\rfloor + 100$ **(b)**

35. A cereal manufacturer offers a free pound of bananas for every six coupons cut from their boxes of breakfast cereal. If each box has one coupon, write a formula for p, the number of pounds of bananas an individual can receive, as a function of b, the number of boxes of cereal purchased. $p = \left\lfloor \dfrac{b}{6} \right\rfloor$

36. A salesperson earns a $167.50 bonus for every $1000 worth of merchandise sold. Write a formula for the amount of bonus earned from selling k dollars worth of merchandise.

37. *Multiple choice.* In 1990, the cost of phoning Los Angeles from Chicago between 8 A.M. and 5 P.M. Monday through Friday using one phone company was 33¢ for the first minute and 22¢ for each extra minute or part thereof. Which of the following gives the cost d in dollars as a function of the time t in minutes?
   (a) $d = 0.33 + 0.22 \lfloor 1 - t \rfloor$
   (b) $d = 0.33 - 0.22 \lfloor 1 - t \rfloor$
   (c) $d = 0.22 + 0.33 \lfloor 1 - t \rfloor$
   (d) $d = 0.22 - 0.33 \lfloor 1 - t \rfloor$ **(b)**

38. Refer to the data in Question 38. Find a formula using the ceiling function for d as a function of t. $d = 0.33 + 0.22 \lceil t - 1 \rceil$

■ **Objective G:** *Find and interpret quadratic models.* *(Lessons 2-6, 2-7)*

39. A toy rocket launched off a cliff over the sea follows a quadratic model for height H feet above sea level as a function of time t seconds after launch: $H = 310 + 110t - 16t^2$.
   a. Predict the height of the rocket 5 seconds after launch.   **460 feet above sea level**
   b. How long after launching will the rocket hit the sea?   **about 9 seconds**

40. The Illinois Department of Conservation has published information concerning the growth rate of largemouth bass. The average length of the fish is 9 inches at 2 years of age, 11.6 inches at 3 years, 17.4 inches at 6 years, and 20.7 inches at 10 years.
   a. Construct a scatter plot of these data.
   b. Give a reason for *not* expecting age and length to be linearly related.
   c. Use some of the data to construct a suitable polynomial model relating length to age.
   d. Use your model from **c** to predict the length of a largemouth bass that is 8 years old.
   e. Use your model to give the likely age of a largemouth bass measuring 13.5 inches in length.   **3.8 years of age**

41. The wind chill measures the rate of heat loss from body surfaces. It is commonly used as an index of how cold it feels when the wind is blowing on a cold day. The following data give the wind chills for actual temperatures of 30°F at various wind speeds. Wind speeds greater than 45 mph have little additional chilling effect. (Source: National Weather Service, NOAA, U.S. Commerce Department).

Wind speed (mph)	5	10	15	20	25	30	35	40	45
Wind chill (°F)	27	16	9	4	1	-2	-4	-5	-6

a. Identify the independent and dependent variables in this case.
b. Construct a scatter plot of these data.
c. Find a suitable quadratic model for the data.

a) **Temperature and wind speed are independent variables; wind chill is dependent.**
c) **sample: .025x² − 1.993x + 34.69**

REPRESENTATIONS deal with pictures, graphs, or objects that illustrate concepts.

■ **Objective H:** *Graph linear, quadratic, and step functions.* (Lessons 2-2, 2-4, 2-6)

In 42–45, sketch a graph of each function on the domain $-4 \le x \le 4$.

42. $f(x) = .5x + 3$

43. $y = 2x^2 - 18x$

44. $y = \lfloor x \rfloor$

45. $y = 3 + 2\lceil x - 1 \rceil$

■ **Objective I:** *Interpret properties of relations from graphs.* (Lessons 2-1, 2-4, 2-6)

In 46–49, state whether the graph represents a function.

46.

not a function

47.

function

48.

function

49.
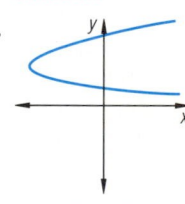
not a function

In 50–53, determine the domain and range of the relation.

50.

51.

52.

53.
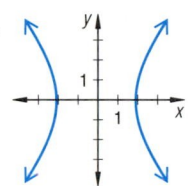

50) domain: {x: -2 ≤ x < 3}; range: {-2, -1, 0, 1, 2}
51) domain: the set of all real numbers; range: {x: x ≤ 3}
52) domain: the set of all real numbers; range: {x: x ≤ 2}
53) domain: {x: x ≤ -2 or x ≥ 2}; range: the set of all real numbers

---

40.a.

b. Sample: Like most other animals, the bass probably reach a mature stage and stop growing.

c. sample: y = -0.16x² + 3.38x + 2.9

d. 19.7 inches

41.b.

42.

43., 44., 45. See Additional Answers in the back of this book.

■ **Objective J:** *Use scatter plots to draw conclusions about models for data.*
(Lessons 2-2, 2-3, 2-4, 2-5, 2-7)

In 54-57, for each scatter plot, **a**. determine whether a linear or a curvilinear model is more suitable, and **b**. state whether the correlation coefficient is likely to be positive, negative, or approximately zero.

54.

55.

56.

57.
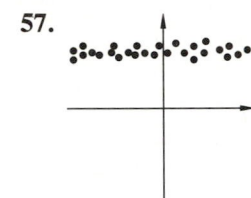

In 58–61, use the display below showing the line of best fit through data relating the number of hours spent studying for a test and the mark on the test for a group of students.

$$y = 1.887x + 60.762, \ r^2 = 0.51$$

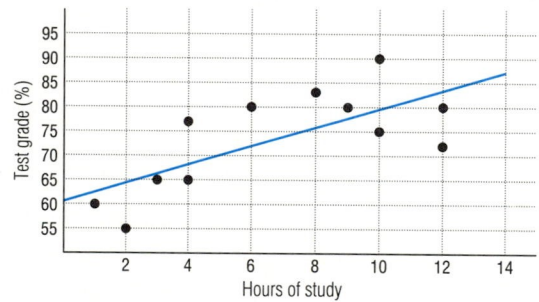

58. How many students were in the group?
59. For the student who studied six hours,
   a. what is the observed y-value?
   b. what is the predicted y-value?
   c. what is the error of prediction?
   58) 12 students
   59) a. 80 percent; b. ≈ 72 percent
   c. ≈ 8 percent

60. For these data, $\bar{x} = 6.75$ and $\bar{y} = 73.5$.
   a. What is the center of gravity of the data?
   b. Verify that the line of best fit contains the center of gravity.
61. What is the correlation between the number of hours studied and the test grade?   **0.71**

In 62 and 63, use the display below showing a quadratic model fitted to the hours of study vs. test grade data above.

$$y = -.455x^2 + 7.979x + 46.775$$

62. For the student who studied six hours,
   a. what is the observed y-value?
   b. what is the predicted y-value?
   c. what is the error of prediction?
63. According to the quadratic model, what is the amount of time a student should study in order to achieve a maximum score? Explain why the existence of such a point might make sense.

In 64 and 65, consider the graphs of both the linear and the quadratic functions fitted on the hours of study vs. grade data.

64. One of the students who studied ten hours seems to be an outlier. Which model comes closer to the student's actual performance?
65. Which seems to be a better model? Why?
62) a. 80 percent; b. ≈ 78 percent; c. ≈ 2 percent
64) the quadratic model

CHAPTER 3 ■ TRANSFORMATIONS OF FUNCTIONS AND DATA

OBJECTIVES ■ CHAPTER 3

The objectives listed here are the same as in the Chapter 3 Review on pages 197–200 of the student text. The Progress Self-Test on page 196 and the tests in the Teacher's Resource File cover these objectives. For recommendations regarding the handling of this end-of-chapter material, see the notes in the margins on the corresponding pages of this Teacher's Edition.

OBJECTIVES   3

OBJECTIVES FOR CHAPTER 3 (Organized into the SPUR categories— Skills, Properties, Uses, and Representations)	Progress Self-Test Questions	Chapter Review Questions	Teacher's Resource File	
			Lesson Masters*	Chapter Test Forms A & B
SKILLS				
A Find composites of functions.	9–11	1–6	3-7	17–19
B Find inverses of functions.	12–14	7–10	3-8	13, 14
PROPERTIES				
C Use the Graph Transition Theorem or the Graph Scale Change Theorem to find transformation images.	1, 2, 4, 20	11–20	3-2, 3-5	1, 4, 11, 12
D Describe the effects of translations or scale changes on functions and their graphs.	21	21–24	3-2, 3-5	2
E Use summation properties to evaluate and rewrite expressions.	7, 8	25–28	3-3, 3-6	10
F Describe the effects of translations or scale changes on measures of center or spread.	5	29–34	3-3, 3-6	8, 9
G Describe the symmetries of graphs.	17, 19	35–40	3-4	6, 7, 16
H Identify properties of composites or inverses.	11, 14	41–44	3-7, 3-8	15, 20, 21
USES				
I Use translations or scale changes to analyze data.	6	45–48	3-3, 3-6	24
REPRESENTATIONS				
J Apply the Graph Translation Theorem or the Graph Scale Change Theorem to make or identify graphs.	15, 16	49–56	3-2, 3-5	3
K Use an automatic grapher to draw the graph of a function and read key values.	3, 18	57–61	3-1	5
L Graph inverses of functions.	13	62–65	3-8	22, 23

*The masters are numbered to match the lessons, and each is intended for use with that lesson.

143A

OVERVIEW ■ CHAPTER 3

This chapter discusses a topic common to statistics and to the study of functions, namely the transformation of a set of data or a set of points into a more convenient form to be studied. We can make observations from this easier representation and apply them to the information which was not as conducive to study.

Two transformations are considered in this chapter: translations (adding a constant to all values of a variable) and scale changes (multiplying each value by a constant).

What we know about the graph of a particular function or relation in its most convenient placement (perhaps symmetric to an axis) can be applied to transformation images of that graph. What we know about the equation of a relation can be applied to determine an equation for the transformation image.

An automatic grapher, which

may be new to many students, is introduced in Lesson 3-1. Students first learn to draw graphs of what should be familiar functions so that they can concentrate on using the features of their graphers properly and efficiently.

There is also some review of function notation and the algebraic manipulation of equations of functions. If the graphs of functions appear to be symmetric over the y-axis or through the origin (see Appendix A), then students are asked in Lesson 3-4 to prove algebraically that the function is even or odd, respectively.

The primary objectives of Lessons 3-2 and 3-5 are for students to gain an understanding of how transformations affect important features of graphs and how those effects are represented in the changes in equations for the relations. The objectives in Lessons 3-3 and 3-6 are

for students to transfer this understanding to data sets and to predict the effects of the same kinds of transformations on measures of center and spread. In Lessons 3-8 and 3-9, the techniques of graphing and proof are applied to study inverses of functions and to show that all parabolas are similar.

Students who have studied UCSMP materials before will be familiar with transformations, and will likely be able to maintain a 1 day per lesson pace throughout the chapter, completing it in 12 days with the test. The applications of transformations to data sets will be the only completely new content. Students with no previous experience with an automatic grapher and no previous work with transformations will need extra days to complete the chapter.

PERSPECTIVES ■ CHAPTER 3

The Perspectives provide the rationales for the inclusion of topics or approaches, give some mathematical background, and make connections with other lessons and to other UCSMP texts.

3-1

USING AN AUTOMATIC GRAPHER

The term *automatic grapher* refers either to a calculator that has graphing capability or to a computer with graphing software.

This lesson introduces an automatic grapher as a tool for discovering important attributes of functions. Students should be able to use the automatic grapher to display functions, change viewing windows, and read coordinates of points from the display.

For students who have used such technology previously, this is a review lesson, easily covered in a day following the Chapter 2 test. For students new to the technology, an extra day will be needed.

3-2

THE GRAPH TRANSLATION THEOREM

A translation is the mathematical name for a slide. Translations are caused by adding a constant to a variable; the resulting graph is an image congruent to the original.

The Graph Translation Theorem is one of two theorems about graphing studied in this chapter. It is a powerful theorem, as it is applicable to all relations. In later chapters, it will be applied to the phase shift of trigonometric functions and to the graphs of conics not in standard position.

3-3

TRANSLATIONS OF DATA

This lesson follows the lesson on translating graphs in order to em-

phasize the parallel ideas of translating graphs and translating data. The effects of adding a constant to each data value on the displays of those values, on measures of center, and on measures of spread are investigated.

Many people apply the skill of translating to find a mean in day-to-day routines without being aware that they are doing so. It is intuitively sensible that averaging 105, 108, and 103 can be done mentally by averaging 5, 8, and 3, and then adding 100. This lesson shows that such a strategy works for all data sets. The same strategy may be used to calculate mode, median, variance, and standard deviation of data sets.

3-4
SYMMETRIES OF GRAPHS

The function $y = ax^n$ is symmetric with respect to the y-axis when n is even and is symmetric with respect to the origin when n is odd. In the first case, the function is an even function, and in the latter case it is an odd function.

This lesson gives the algebraic definition for even functions (for all x, $f(-x) = f(x)$) and odd functions (for all x, $f(-x) = -f(x)$). It provides more practice with a function grapher and asks students to prove whether a function has the property of being an odd or an even function.

3-5
THE GRAPH SCALE CHANGE THEOREM

The Graph Scale Change Theorem is the second of the theorems about graphing which apply to all relations. The development in this lesson parallels the one in Lesson 3-2 of the Graph Translation Theorem. Just as subtracting h from x added h to the x-coordinates of points which satisfy a relation, so dividing x by a multiplies the x-coordinates by a.

The Graph Scale Change Theorem is also very powerful. This theorem will be applied in later chapters to determine the amplitude and period of trigonometric functions and to the graphs of ellipses. Students should be able to sketch graphs of equations which are scale changes of parent functions and write equations for scale change images.

3-6
SCALE CHANGES OF DATA

In the same way that Lesson 3-2 on translating data followed Lesson 3-1 on translating graphs, this lesson on scaling data immediately follows the lesson on scaling graphs. Students will see scaling graphs and scaling data as parallel concepts.

This lesson investigates the effects of scaling data on the display, the measures of center, and the measures of spread. A scale change by a of a data set multiplies the range and measures of central tendency by a, the standard deviation by $|a|$, and the variance by a^2.

It is not uncommon to combine scale changes and translations. This is first done in this book in Lesson 5-3 in connection with graphs of circular functions and is done in connection with the standardization of data in Chapter 10.

3-7
COMPOSITION OF FUNCTIONS

We place composition of functions here for two reasons. First, it provides a mathematical basis for the discussion of inverse functions and relations in Lesson 3-8 (which in turn provides language to discuss exponential and logarithmic functions in Chapter 4). Second, the idea of composition of *transformations* is employed in Chapter 5 in the discussion of linear changes, that is, transformations which map (x, y) onto $(ax + by + c, dx + ey + f)$.

It is natural to think of combining two transformations into one; already there has been the combination of a horizontal and a vertical translation into one two-dimensional translation. Do the same for scale changes. Students who have had UCSMP *Geometry* and UCSMP *Advanced Algebra* will be familiar with this idea as it relates to transformations. But in this lesson, the discussion centers around functions of a single variable.

This lesson formalizes the concept of composition of functions by defining composition and introducing the ∘ symbol. This lesson may be review for many students.

3-8
INVERSE FUNCTIONS

Like most college texts, we define the inverse of a function as the relation in which the components of all ordered pairs of the function are switched. Under this definition, every function has an inverse, but of course not all inverses are functions.

This lesson begins with that definition, shows that the graph of an inverse is a reflection of the original function over the line $y = x$, and presents the Inverse Function Theorem as an algebraic tool used to prove that two functions are inverses of each other. Students should be able to write an equation for the inverse of a given function, graph one function given the graph of the other, and use the Inverse Function Theorem.

3-9
ARE ALL PARABOLAS SIMILAR?

This reading lesson applies the transformations of this chapter to investigate whether all parabolas are similar. Changes are used to prove that all parabolas with equations of the form $y = ax^2$ are similar to the graph of $y = x^2$.

Translations then show that the graph of $y = ax^2 + bx + c$ is congruent to the graph of $y = ax^2$. From this it follows that all parabolas are similar; that is, they all have the same shape. The reason they do not all look similar is that they are unbounded figures and we see more of some parabolas than others; the more we see of a parabola in a given space, the thinner it looks.

DAILY PACING CHART ■ CHAPTER 3

For optimal student performance, the Progress Self-Test and the Chapter Review should be covered. (See *General Teaching Suggestions: Mastery* on page T35 of this Teacher's Edition.) By following the pace of the Full Course given here, students can complete 12 or 13 chapters of the book by the end of the year. Students following the pace of the Minimal Course spend more time to complete lessons that are combined with quizzes and to consider the Chapter Review. When all lessons are covered in order (the recommendation of the authors), students in the Minimal Course can cover all but the last two chapters. For more information on pacing, see *General Teaching Suggestions: Pace* on page T34 of this Teacher's Edition.

DAY	MINIMAL COURSE	FULL COURSE
1	3-1	3-1
2	3-1	3-2
3	3-2	3-3
4	3-3	Quiz (TRF); 3-4
5	Quiz (TRF); Start 3-4	3-5
6	Finish 3-4	3-6
7	3-5	Quiz (TRF); 3-7
8	3-6	3-8
9	Quiz (TRF); Start 3-7	3-9
10	Finish 3-7	Progress Self-Test
11	3-8	Chapter Review
12	3-9	Chapter Test (TRF)
13	Progress Self-Test	
14	Chapter Review	
15	Chapter Review	
16	Chapter Test (TRF)	

TESTING OPTIONS
■ Quiz for Lessons 3-1 through 3-3   ■ Chapter 3 Test, Form A   ■ Chapter 3 Test, Cumulative Form
■ Quiz for Lessons 3-4 through 3-6   ■ Chapter 3 Test, Form B   ■ Comprehensive Test, Chapters 1–3

A Quiz and Test writer is available for generating additional questions, additional quizzes, or additional forms of the Chapter Test.

PROVIDING FOR INDIVIDUAL DIFFERENCES
The student text has been written for, and tested with, average students. It has also been used successfully with students who are well prepared as well as with students who are poorly prepared.

   The Lesson Notes often include Error Analysis and Alternate Approach features to help you assist those students who need more help. Students of all abilities often learn from their peers and may benefit from small group work referenced as appropriate throughout the Notes. A blackline Lesson Master (in the Teacher's Resource File), keyed to the chapter objectives, is available for each lesson to provide more practice. (However, we advise you to use these masters sparingly; again refer to the suggestions on pace on page T34.)

   Extension activities are provided in the Lesson Notes for use with students who need to be challenged even more than the requirements of completing the Full Course provide.

Transformations of Functions and Data

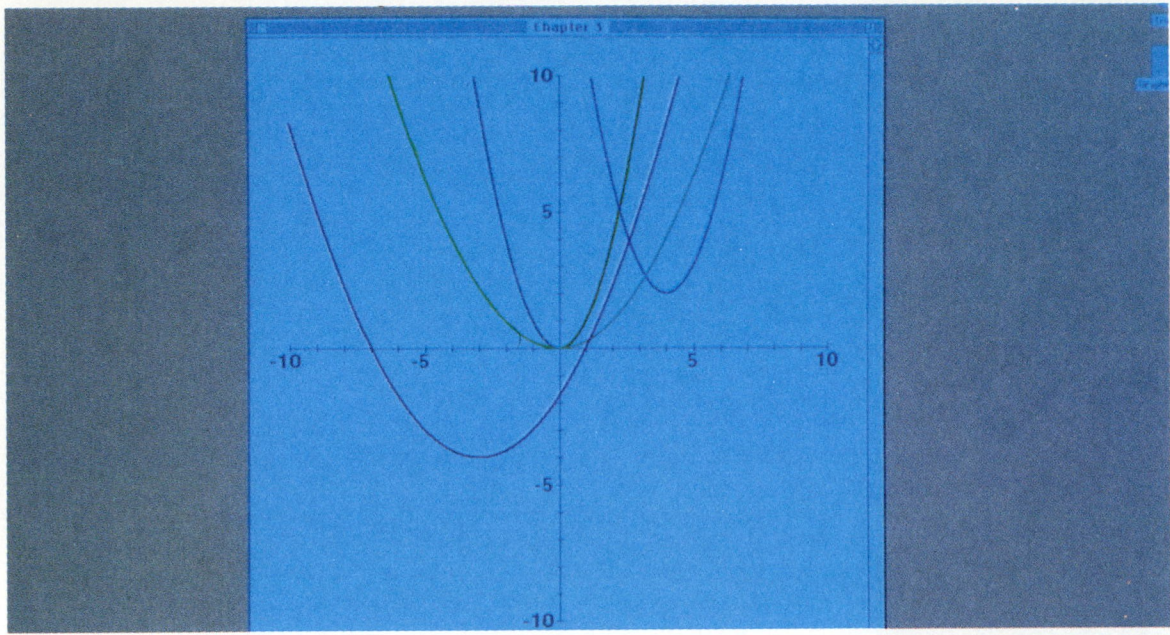

A **transformation** is a one-to-one correspondence between sets of points. Two important transformations are *translations* and *scale changes*.

The picture above shows the *images* of the graph of the parabola $y = x^2$ under various transformations, including a translation, a scale change, and *composites* of translations and scale changes.

Transformations are often described by algebraic formulas. In this chapter you will investigate algebraic descriptions of translations and scale changes and study the effects of such transformations on functions, on statistical measures, and on properties of those functions and statistics.

We recommend 12–16 days be spent on this chapter: 9–12 days on the lessons and quizzes, with the extra days for students for whom automatic graphers or transformations are new; 1 day for the Progress Self-Test; 1 or 2 days for the Chapter Review; and 1 day for a Chapter Test. (See the Daily Pacing Chart on page 143D.) Do not spend more than 16 days on this chapter.

USING PAGE 145

You might use the following introduction for this chapter. Remind students that parabolas extend without end, so a graph can only show a part of a parabola. Then ask: Which of these parabolas do you think are congruent? Which are similar? Which are neither congruent nor similar? Explore their definitions of congruence and similarity. The following general definitions are found in UCSMP materials: two figures are congruent if and only if one can be mapped onto the other by a composite of reflections, rotations, and translations; two figures are similar if and only if one can be mapped onto the other by a composite of those transformations and size changes. Do these definitions change any students' views regarding which parabolas are congruent and which are similar? Conclude by telling students they will learn the answers in this chapter. (Of course, they will learn a lot of other things, too.) (There are two pairs of congruent parabolas; they all are similar.)

This chapter contains much content that will likely be review for students. It might be worthwhile to determine how much is review, so that your time can be put to the best possible use.

OBJECTIVE

K Use an automatic grapher
to draw the graph of a function and read key values.

TEACHING NOTES

This lesson is designed to
allow students to become familiar with the basic features
of an automatic grapher, so
the Examples and most of
the Questions are devoted to
developing the skills of displaying a graph on the computer monitor or calculator
screen, changing the size of
the viewing window to see important features of the graph,
and getting a hard copy of the
graph, either with a printer or
by hand-sketching. See
pages T40–42 and T52 for
more information on computers and graphing calculators.

Making Connections
The functions used to practice these skills are related to
the parent functions dis-
cussed in the lesson. Familiarity with all the graphs is not
assumed, though we expect
that students have seen the
parent graphs before. You
may want to refer students to
Appendix A, Graphs of Parent Functions, and find the coordinates of a point or two on
each graph to make certain
that they make a connection
between the equations and
the curves.

A parent function is the simplest form of the equation for
a certain type of function.

LESSON

3-1

Using an Automatic Grapher

Graphing functions by plotting points by hand is often very tedious. By using
an **automatic grapher**, a calculator or computer software that draws the
graph of a relation, you can draw and analyze graphs of functions more
easily.

Five important aspects of using an automatic grapher are:
   (a) entering a function or relation
   (b) graphing the function
   (c) reading points and scales on the screen
   (d) changing the scales on the axes
   (e) printing a hard copy of the graph (if possible).

On many automatic graphers the relation to be graphed must be entered as a
formula for y or $f(x)$ in terms of x. Thus,
   $y = 7x$ and $f(x) = 3^x - 2$ can be entered directly
but $x = 4y$ and $x + y = 10$ may not be accepted in that form.

On many automatic graphers you must use the keys *, /, and ^ to indicate
multiplication, division, and powering, respectively. Because automatic
graphers differ, you should check with your teacher, your classmates, or the
grapher's manual for the particular operation symbols available to you.

The **viewing window** or **viewing rectangle** displays the graph of the function
for a specific domain and range. For some graphers you must specify a
domain and range; others select a **default** domain and range if you do not.

For instance, when the function $F(x) = 12 + 4x - \frac{1}{2}x^2$ was graphed on the default window $-8 \leq x \leq 8$, $-10 \leq y \leq 10$ of one grapher, the graph looked like
the one below.

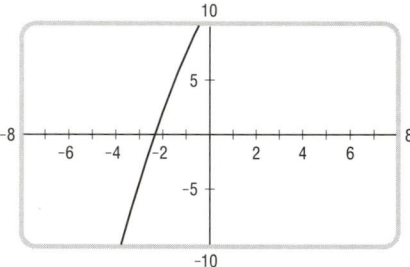

The function has the general form $f(x) = ax^2 + bx + c$, so you might predict
that the curve would be a parabola. But some of its key features, namely its
vertex and other x-intercept, are not visible. In such cases you will want to
change the size of the viewing window. Depending on your automatic grapher, you either change or input new values for the domain or range.

144

Example 1 Estimate to the nearest integer the coordinates of the vertex and the x-intercepts of the parabola $F(x) = 12 + 4x - \frac{1}{2}x^2$.

Solution In the default window, the y-coordinate of the vertex of the parabola appears to be greater than 10. There also appears to be an x-intercept to the right of 8. So change the domain and range accordingly. After some experimenting with different values, we settled on a domain of $-5 \le x \le 12$ and a range of $-20 \le y \le 22$. The graph is shown below.

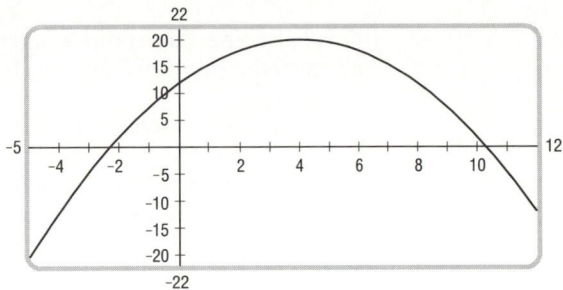

It appears that the vertex is near (4, 20). One x-intercept is between -3 and -2, and the other is between 10 and 11. To the nearest integer the x-intercepts are approximately -2 and 10. (In a later lesson you will learn how the grapher can help you approximate coordinates to greater accuracy.)

Check The x-intercepts occur when $F(x) = 0$. Solve the equation $0 = 12 + 4x - \frac{1}{2}x^2$. Use the quadratic formula to find $x = \frac{8 \pm \sqrt{160}}{2} = 4 \pm \sqrt{40}$. That is, $x \approx -2.3$ or $x \approx 10.3$.

Predicting the shape of a graph given its equation, and finding an equation for a relation given its graph, are important mathematical skills. To develop these skills, you will find it helpful to look for "resemblances" among graphs of "families" of functions. As in human families, families of functions often have a *parent* from which other related functions can be derived.

The parent of all parabolas is the function $f(x) = x^2$. Six other important parent functions are shown below and on the next page.

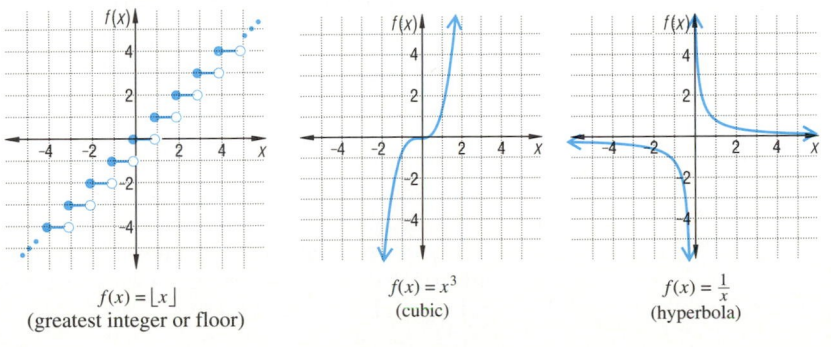

$f(x) = \lfloor x \rfloor$
(greatest integer or floor)

$f(x) = x^3$
(cubic)

$f(x) = \frac{1}{x}$
(hyperbola)

This lesson considers four parent functions: $y = x^2$; $y = |x|$, $y = \sqrt{x}$, and $y = \frac{1}{x}$. A function which stems from any parent function is referred to as an "offspring."

Become familiar with the automatic graphers your students have before class so that you may be better prepared to answer their questions.

Small Group Work You can work through the Examples during class with individual students or small groups of students using automatic graphers. If small groups are used, students should alternate turns at the keyboard or calculator. Identify students who have automatic grapher experience, as they can be valuable assistants to you in answering technical questions of other students. Most students with graphics calculators probably will not have the capability to print hard copies of their graphs, so encourage them to make neat, accurate sketches when necessary.

Error Analysis Emphasize that sketches should include all important features. For example, a quadratic function should "appear" to be symmetric. A sketch of a quadratic function such as this does not make clear whether the student understands symmetry. Also, all local maximum and minimum

points of polynomial functions should be included in sketches of the function.

Most automatic graphers have a *trace* capability, in which a cursor moves along the graph, giving coordinates of points as it progresses.

For **Example 1,** some students may know that the x-coordinate of the vertex of the graph of $y = ax^2 + bx + c$ is $-\frac{b}{2a}$. Here, $-\frac{4}{2(-\frac{1}{2})} = 4$, and substituting this value for x yields $y = 20$.

You might use a calculator to verify that $F(4)$ is a maximum.

The trace function can be used to check the result of **Example 1.** If you use this function, note that on many graphers the points traced are not necessarily those with convenient coordinates. The vertex even may not be graphed.

The zoom function on automatic graphers can enlarge a part of a graph for closer study. The teaching aid for **Example 2** enables you to discuss this, but the best effect is achieved by projecting a calculator or computer output directly onto a screen. One caution is that students may forget that they are using the zoom feature, and lose sight of the overall picture.

Some function graphers and graphics calculators have a feature that prints the coordinates of any point on a graph indicated by the user. Therefore, if a student reports that the vertex of a parabola is (3.0213, 3.9874), there is reason to suspect that the student has not thought about the situation.

Using a function grapher makes it possible to graph many more functions in a lesson than you are probably used to, so students may have many more questions. Having students sketch graphs by hand is not emphasized, but sometimes it is helpful to draw the curve first and then fit the axes to the graph.

Technology We use BASIC notation for entering functions because at this writing the majority of function

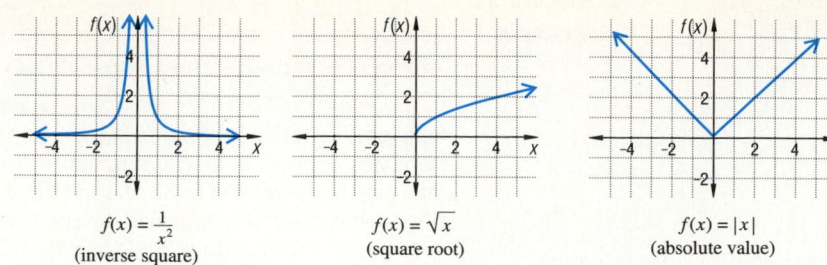

$f(x) = \frac{1}{x^2}$
(inverse square)

$f(x) = \sqrt{x}$
(square root)

$f(x) = |x|$
(absolute value)

Graphs of other parent functions are given in Appendix A. Refer to the graphs above or the Appendix until you become familiar with their shapes.

Most automatic graphers can plot more than one graph in a single window. If your grapher plots in color you can distinguish graphs by their colors. On a monochrome screen you will need to depend more on your knowledge of mathematics to distinguish the graphs.

Example 2
a. Graph $f(x) = \sqrt{ax}$ for $a = \frac{1}{2}$, 1, 2, and 3.

b. What happens to the graph as a increases?

Solution

a. You are asked to graph $y = \sqrt{\frac{1}{2}x}$, $y = \sqrt{x}$, $y = \sqrt{2x}$, and $y = \sqrt{3x}$.

Enter one equation at a time and do not clear the screen until all four graphs have been drawn. For the window $-1 \le x \le 4$, $-1 \le y \le 4$, our grapher produced the following display.

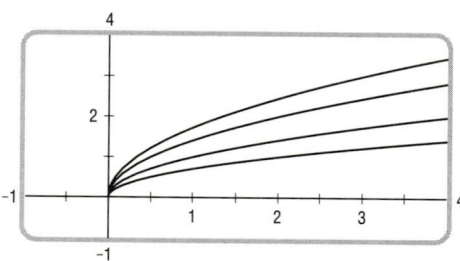

b. As the value of a increases the graph is higher. For a given positive x-value, the function with the least y-value is $y = \sqrt{\frac{1}{2}x}$; the one with the greatest y-value is $y = \sqrt{3x}$.

Some graphers have a *zoom* feature similar to those found on cameras. This feature enables you to change the window without retyping intervals for x or y. For instance, suppose in Example 2 above that you want to study the behavior of the square root functions near the origin. If you zoom by a factor of 10 around the origin, then the viewing rectangle becomes $-.1 \le x \le .4$, $-.1 \le y \le .4$. The result follows.

146

Again each graph shows a square root function. The function with the greatest y-value for a given x is still $y = \sqrt{3x}$; and the one with the smallest y-value is still $y = \sqrt{\frac{1}{2}\,x}$.

In general, none of the mathematical properties of a function change when you change the viewing rectangle. However, sometimes a change in size of a viewing rectangle may give you a different visual impression of the graph. For example, the graph of $y = \sqrt{x}$ is certainly not linear, but the portion in the window $10.1 \le x \le 10.2$, $0 \le y \le 5$ may seem linear.

You should become familiar with how your automatic grapher handles discontinuities of the parent functions pictured earlier in this lesson. The function $y = \lfloor x \rfloor$ is discontinuous at each integer, because the graph jumps at those points. The hyperbola and inverse square curves are discontinuous at $x = 0$ because $f(x)$ has a zero denominator for $x = 0$. Recall that an **asymptote** is a line that the graph of a function $y = f(x)$ approaches as the variable x approaches a fixed value or increases or decreases without bound. Each of the graphs of $f(x) = \frac{1}{x}$ and of $f(x) = \frac{1}{x^2}$ has a vertical and a horizontal asymptote.

■ ■ ■ ■ ■ ■ ■ ■ ■

Example 3
a. Graph $y = \frac{1}{x-6}$ and its parent $y = \frac{1}{x}$ on the same set of axes.
b. How is the graph of $y = \frac{1}{x-6}$ related to the graph of its parent?

Solution **a.** To enter the function $y = \frac{1}{x-6}$ you will need to use parentheses. That is, enter $y = 1/(x-6)$. The graph of $y = \frac{1}{x}$ is in black; the graph of $y = \frac{1}{x-6}$ is in blue.

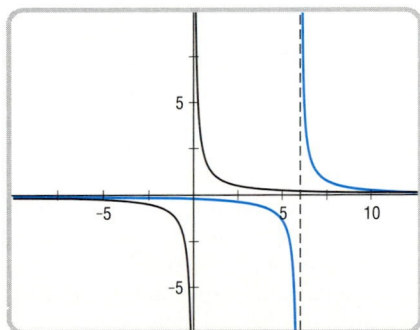

graphers require it. Some function plotters allow the user to enter functions in virtually the same form as is written on paper, including exponents and fractions. At the other extreme, some of the older function plotters require the user to change the DEF FN line of the program by hand, as students did in the programs of Chapter 2.

ADDITIONAL EXAMPLES
1. Sketch the graph of $y = 10 - x^2$, naming key points on it.

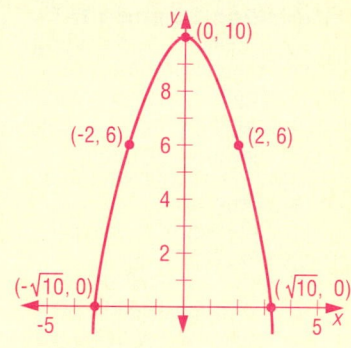

2. Sketch the graph of $f(x) = 4x^4 - 10x^3$.

NOTES ON QUESTIONS
Question 5: Ask students for the y-intercept. (-24) Do not accept a value like -23.9849 from the graph; they should use the given formula.

Technology for Question 6: Although it is common to speak of the graph of $f(x) = 2x^2$ as "thinner" than the graph of $h(x) = 0.5x^2$, in Lesson 3-9, all parabolas will be proven similar. Here, more of the graph of f is being seen than the graph of h. Zoom in on the graph of f and it will look wider.

148

b. The graph of $y = \dfrac{1}{x-6}$ appears to be congruent to the graph of its parent. It seems to be the image of $y = \dfrac{1}{x}$ under a translation of 6 units to the right. Whereas the parent function has a vertical asymptote at $x = 0$, the function $y = \dfrac{1}{x-6}$ has an asymptote (shown as a dashed vertical line on our grapher) at $x = 6$.

If your grapher cannot produce hard copy you must transfer sufficient information from the screen to your *hand sketch* of the graphs. A sketch generally should meet the following criteria:
  the axes are labeled;
  enough values are shown on each axis to show the scales of the axes and the domain and range for the portion sketched;
  the shapes are appropriate (*e.g.* ⋃ rather than ⋁ for a parabola);
  the intercepts are approximately correct;
  the points where the function is discontinuous are properly indicated.

Questions

Covering the Reading

1. a. What is a default window?
   b. If your grapher has a default window, state its dimensions.
   a) See margin. b) Answers will vary depending upon the grapher used.

In 2 and 3, part of a window is shown, in which a graph is intersecting the x-axis. Approximate the x-intercept(s) to some reasonable level of accuracy.

2.

3.

sample: $x \approx -7$ and $x \approx 7$ **sample:** $x \approx -40$, $x \approx 100$, and $x \approx 240$

4. a. Describe two ways a function might be discontinuous.
   b. Give an example of a parent function to illustrate each type in **a**. **See margin.**

5. Consider the function $f(x) = x^2 - 2x - 24$.
   a. Use an automatic grapher to graph the function. **See margin.**
   b. Estimate to the nearest integer the x-intercepts of the graph. **-4 and 6**
   c. Solve $f(x) = 0$ to verify your estimate in **b**. **$x = -4$ or $x = 6$**
   d. Estimate the coordinates of the vertex of the parabola. **(1, -25)**

6. Consider $f(x) = 2x^2$, $g(x) = 0.5x^2$, and $h(x) = -x^2$.
   a. Graph all three functions simultaneously on an automatic grapher.
   b. Graph $y = x^2$ on the same set of axes. Describe how each graph in part **a** compares to the graph of the parent function. **See margin.**

In 7 and 8, plot each pair of functions on the same set of axes.
a. Describe the relation between each pair of graphs. **b.** Sketch the graphs on paper or print hard copy. (Check with your teacher or a manual about how to enter square root and absolute value functions.) **See margin.**

7. $y = \sqrt{x}$, $y = \sqrt{x-5}$ **8.** $y = |x|$, $y = |4x|$

Applying the Mathematics

9. **a.** On one set of axes, graph $f(x) = \frac{1}{x}$, $g(x) = \frac{1}{x+6}$, and $h(x) = \frac{1}{x} + 6$.
   b. Give an equation of the vertical asymptotes of each curve.
   c. How is each of g and h related to f?
   a, c) See margin. b) for $f(x)$, $x = 0$; for $g(x)$, $x = -6$; for $h(x)$, $x = 0$

In 10 and 11, **a.** give the size of a suitable viewing window on your automatic grapher so that the vertex and x-intercepts of the graph of the function are shown. **b.** Sketch a graph of the function.

10. $y = (x-3)^2 - 4$ **See margin.**   11. $y = |x+10| - 15$ **See margin.**

12. **a.** Graph the function $f(x) = x^3 - x$ on the given windows.
   i. $-1 \le x \le 1$, $-1 \le y \le 1$
   ii. $-5 \le x \le 5$, $-5 \le y \le 5$
   iii. $-10 \le x \le 10$, $-10 \le y \le 10$
   iv. $-100 \le x \le 100$, $-100 \le y \le 100$ **See margin.**
   b. Which window provides the most useful graph?   **Answers may vary.**

13. **a.** On a single set of axes, graph $f(x) = \frac{1}{x^2}$ and $g(x) = \frac{4}{x^2}$.
   b. *True* or *false*. For all x in the domains of f and g, $f(x) < g(x)$.
   c. Justify your response to part **b** both algebraically (by using the formulas) and geometrically (by using the graphs).
   a) See margin; b) True; c) See margin.

14. **a.** Graph $y = x - \lfloor x \rfloor$.   **See margin.**
   b. At what values of x is the function discontinuous?   **at all integers**

Review

15. The number of fish caught by various competitors in a catch-and-release contest are listed in the table below.

Number of fish	0	1	2	3	6	9
Number of competitors	4	7	13	8	2	1

   a. How many competitors were there?   **35 competitors**
   b. Find the mean, median, and mode of the numbers of fish caught by the competitors.   **mean ≈ 2.23; median = 2; mode = 2**
   c. What is the standard deviation of the numbers of fish caught by the competitors?   **1.8**
   d. Display the data with a histogram.   *(Lessons 1-4, 1-6, 1-8)*
   See margin.

16. **a.** Draw $\triangle ABC$ with $A = (3, 5)$, $B = (-4, -2)$, and $C = (2, -2)$.
   b. On the same axes, draw $\triangle A'B'C'$, the triangle determined by new points whose x-coordinates are 6 more than the corresponding coordinates in $\triangle ABC$.
   c. How are $\triangle ABC$ and $\triangle A'B'C'$ related?   *(Previous course)*
   See margin.

17. **a.** Draw parallelogram $PQRS$ with $P = (0, 0)$, $Q = (0, -4)$, $R = (-3, -6)$, and $S = (-3, -2)$.
   b. Draw $P'Q'R'S'$, the image of $PQRS$ under the transformation which maps (x, y) to $(x+5, y-2)$.
   c. How are $PQRS$ and $P'Q'R'S'$ related?   *(Previous course)*
   See margin.

b. The graph of $f(x) = 2x^2$ is a horizontal shrinking of the graph of $y = x^2$; the graph of $g(x) = 0.5x^2$ is a horizontal stretch of the graph of $y = x^2$; the graph of $h(x) = -x^2$ is a reflection of the graph of $y = x^2$ over the x-axis.

7.a. The graph of $y = \sqrt{x} - 5$ is the image of the graph of $y = \sqrt{x}$ under a translation 5 units to the right.

b.

8.a.,b., 9.a.,c., 10.a. See margin, page 150.

10.b., 11.a.,b., 12.a., 13.a.,c., 14.a., 15.d., 16.a.,b.,c., 17.a.,b.,c. See Additional Answers in the back of this book.

In 18 and 19, use the following data about U.S. education. *(Source: USA Statistics in Brief, 1988)*

EDUCATION	Unit	1970	1980	1984	1985	1986
School enrollment	Mil.	60.4	58.6	58.9	59.8	60.1
Elementary (grades K-8)	Mil.	37.1	30.6	30.3	30.7	31.1
Secondary (grades 9-12)	Mil.	14.7	14.6	13.9	14.1	14.0
Higher education	Mil.	7.4	11.4	12.3	12.5	12.4
School expenditures	$Bil.	75.7	182.8	247.2	266.2	282.1
Elementary and secondary	$Bil.	48.2	112.3	148.9	160.8	170.0
Public	$Bil.	45.5	104.1	136.5	147.6	156.0

18. Give a reason why the figures for "school enrollment" are not simply a sum of the figures for "Elementary," "Secondary," and "Higher education." *(Lesson 1-2)*   **See below.**

19. **a.** Calculate the expenditures per student in 1970 and in 1986.
    b. Did the average expenditure per student increase or decrease between these years? *(Lesson 1-4)*   **a) 1970: about $1250 per student; 1986: about $4690 per student; b) increase**

20. Write without an exponent: $\dfrac{14.2 \cdot 10^5}{10^{-2}}$. *(Previous course)*   **142,000,000**

Exploration

21. **a.** Using only the parent functions $y = x^2$, $y = x^3$, $y = \sqrt{x}$, $y = |x|$, $y = \lfloor x \rfloor$, $y = \dfrac{1}{x}$, and $y = \dfrac{1}{x^2}$, and the operation of addition, create some new functions. For instance, $y = \sqrt{x} + |x|$.
    b. What is the most interesting graph you can make from adding two of these parents? Why do you think it is interesting?
    a) sample: $y = x^2 + \lfloor x \rfloor$, **b) See below.**

18) School enrollment today includes more than the traditional elementary, secondary, and higher education categories. It also includes categories such as pre-school and vocational school.
21b) sample: $y = \lfloor x \rfloor + |x|$; Functions defined by adding the greatest integer function to other functions often have unusual graphs. However, many graphers always plot continuous graphs, whereas graphs containing the greatest integer function should have discontinuities.

3-2

The Graph Translation Theorem

Even when an automatic grapher is used to sketch the graph of a relation, it is helpful to know what kind of graph to expect. In this lesson, you will see how to predict the shape and location of some graphs in relation to the graphs of their parents.

For instance, below on the left are graphs of the parent $y = |x|$ (in black) and $y = |x - 4|$ (in blue). Below at the right are graphs of $y = \sqrt{x}$ and $y = \sqrt{x} + 3$.

The graph of $y = |x - 4|$ is the *image* of the graph of $y = |x|$ under the *translation* of 4 units to the right. Similarly, the graph of $y = \sqrt{x} + 3$ is the image of the graph of $y = \sqrt{x}$ under the translation of 3 units up.

Translations are mappings. Specifically, the mapping which adds 4 to the x-coordinate of each point in the plane is a horizontal translation of 4 units. This translation can be written as $(x, y) \rightarrow (x + 4, y)$, which is read "the ordered pair (x, y) is mapped onto the ordered pair $(x + 4, y)$." If T is such a translation, it can be written $T(x, y) = (x + 4, y)$ or $T: (x, y) \rightarrow (x + 4, y)$. Similarly, the mapping that adds 3 to the y-coordinate of each point in a plane is a vertical translation. It can be written as $T(x, y) = (x, y + 3)$ or $T:(x, y) \rightarrow (x, y + 3)$.

The general translation of the plane translates figures horizontally and vertically at the same time.

Definition

The **translation** of h units horizontally and k units vertically is the transformation that maps each point (x, y) to $(x + h, y + k)$.

When $h > 0$, each image point is to the right of its preimage point, so the translation is to the right; when $h < 0$, the translation is to the left; when $h = 0$, there is no horizontal change. Similarly, the translation is up for $k > 0$; down for $k < 0$; and there is no vertical change for $k = 0$.

For instance, if $T(x, y) = (x - 1, y + 6)$, then $T(4, 2) = (4 - 1, 2 + 6) = (3, 8)$. T translates $(4, 2)$ 1 unit to the left and 6 units up.

LESSON 3-2 The Graph Translation Theorem   **151**

RESOURCES
- Lesson Master 3-2
- Teaching Aid 23: Graphs for **Examples 2 and 3**
- Teaching Aid 24: Graphs for **Question 17**
- Computer Master 6

OBJECTIVES

C Use the Graph Translation Theorem to find transformation images.
D Describe the effects of translations on functions and their graphs.
J Apply the Graph Translation Theorem to make or identify graphs.

TEACHING NOTES

Students who studied UCSMP *Advanced Algebra* should be familiar with translations and the Graph Translation Theorem.

This lesson may require 2 days if students have not had experience with translations.

Students may be unsure that replacing x by $x - h$ translates a graph to the right rather than the left. You might have them draw two graphs on the same axes to see that it is true.

Technology A computer master is provided if you wish to spend more time on this topic.

At this point you should insist that your students at least know the parent graphs for $y = x^2$, $y = |x|$ and $y = \frac{1}{x}$.

They should be able to focus on issues having to do with

Example 1   The graph of $y = x^2$ is shown at the
right, together with its image under
a translation T. The point on the
image that corresponds to the
preimage vertex (0, 0) on is (2, -5).

a. Find a rule for the translation T.
b. Find the image of (3, 9) under T.

Solution

a. $T(0, 0) = (2, -5)$, so the second graph has been obtained from the
graph of $y = x^2$ by a translation 2 units to the right and 5 units down.
Thus, the rule for T is $T(x, y) = (x + 2, y - 5)$.
b. $T(3, 9) = (3 + 2, 9 - 5) = (5, 4)$.

There is a direct relationship between replacing a variable expression in an
equation and finding the image of a graph under a transformation.

Consider again the graphs of $y = |x|$ and
$y = |x - 4|$, drawn at the right. As the arrow from the point (8, 8) to (12, 8) indi-
cates, the graph of $y = |x - 4|$ can be
obtained from the graph of $y = |x|$ by
the translation of 4 units to the right, or
$(x, y) \rightarrow (x + 4, y)$. Note that adding 4 to
each x-coordinate corresponds to replacing x by $x - 4$ in the equation of the
preimage.

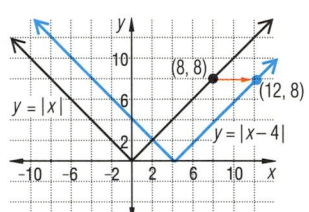

Similarly, graphs of the equations
$y = x^2$ and $y = (x - 2)^2 - 5$ are drawn
at the right. Notice that the graph of
$y = (x - 2)^2 - 5$ is the graph of the image of $y = x^2$ under the translation that
maps (x, y) to $(x + 2, y - 5)$. Notice
that $y = (x - 2)^2 - 5$ is equivalent to
$y + 5 = (x - 2)^2$. So the translation
$(x, y) \rightarrow (x + 2, y - 5)$ has yielded the
same graph as the graph found by replacing x by $x - 2$ and y by $y + 5$ in the
equation of the preimage.

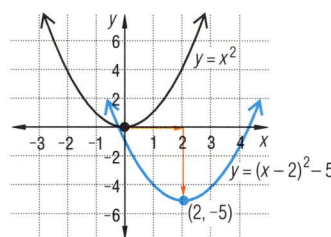

This leads to an important generalization.

> ### Graph Translation Theorem
>
> In a relation described by a sentence in x and y, the following two
> processes yield the same graph:
> (1) replacing x by $x - h$ and y by $y - k$ in the sentence;
> (2) applying the translation (x, y) to $(x + h, y + k)$ to the graph of the
> original relation.

152

Under the translation $T(x, y) = (x + h, y + k)$, an equation of the image of $y = f(x)$ is $y - k = f(x - h)$.

The Graph Translation Theorem can be applied to sketch a graph if an equation is given and to write an equation if a graph is given.

Example 2   Sketch a graph of $y = 1 + \sqrt{x + 2}$.

Solution   Start with the graph of the parent function $y = \sqrt{x}$. Rewrite the sentence to see the replacements in relation to the function $y = \sqrt{x}$.
$$y - 1 = \sqrt{x + 2} = \sqrt{x - -2}$$
In the equation $y = \sqrt{x}$, y has been replaced by $y - 1$ and x has been replaced by $x + 2$. By the Graph Translation Theorem, the graph of $y - 1 = \sqrt{x + 2}$ is the image of the graph of $y = \sqrt{x}$ under the translation $T(x, y) = (x - 2, y + 1)$. Therefore, its graph is translated 2 units to the left and 1 unit up from the graph of $y = \sqrt{x}$. The graphs are drawn below.

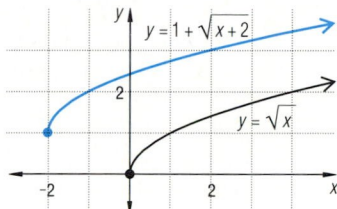

Check   Check the image of a point. The endpoint of the graph has been translated from $(0, 0)$ to $(-2, 1)$. Does $(-2, 1)$ work in the equation $y = 1 + \sqrt{x + 2}$? Does $1 = 1 + \sqrt{-2 + 2}$? Yes.

Example 3   On the right are graphs of the function $y = C(x) = x^2$ and its image $y = D(x)$ under the translation $(x, y) \rightarrow (x + 5, y - 4)$. Find an equation for the image.

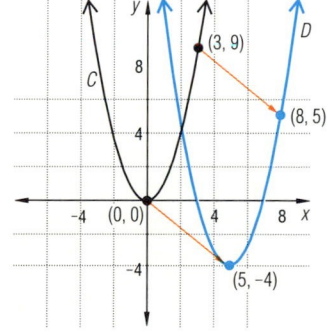

Solution   In $y = x^2$, the equation of the preimage, replace x by $x - 5$ and y by $y - -4 = y + 4$. Thus, an equation of the image is $y + 4 = (x - 5)^2$ or equivalently, $y = (x - 5)^2 - 4$.

Check $(0, 0)$ and $(3, 9)$, which are on $y = x^2$, have the images $(5, -4)$ and $(8, 5)$, respectively, under this translation. Do these images satisfy $y = (x - 5)^2 - 4$? Yes.

ADDITIONAL ANSWERS
7.b. $7 - 5 = |-8 + 6|$
$2 = 2$
$6 - 5 = |-7 + 6|$
$1 = 1$
$5 - 5 = |-6 + 6|$
$0 = 0$

9.b.

10.a. $g(x) = (x - 5)^2 - 3 = x^2 - 10x + 22$

b.

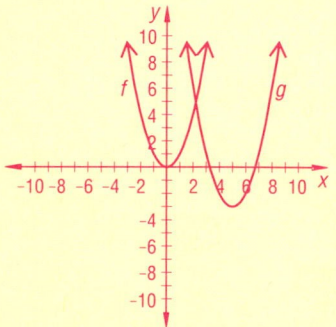

FOLLOW-UP

MORE PRACTICE
For more questions on SPUR Objectives, use *Lesson Master 3-2,* shown on page 155.

EXTENSION
You might wish to extend **Question 23** and algebraically prove that a line and its translation image are parallel. This activity may be suitable for **small group work**.

ADDITIONAL ANSWERS
7.b., 9.b., 10.a.,b. See margin, page 153.

11.b.

12.b.

19.a $\overline{x} = \dfrac{\sum\limits_{i=1}^{n} x_i}{n}$

b. $\sqrt{\dfrac{\sum\limits_{i=1}^{n} (x_i - \overline{x})^2}{n-1}}$

23.b. The image of y = 3x – 5 under T(x, y) = (x + h, y + 3h) is y = 3x – 5.

Covering the Reading

In 1–3, find the image of each point under the given translation.
1. (3, 5), vertical by $^-6$ **(3, -1)**   2. ($^-4$, $^-7$), horizontal by 3   **(-1, -7)**
3. (r, v), horizontal by a and vertical by b.   **(r + a, v + b)**

In 4 and 5, find the image of each point under $T(x, y) = (x - 2, y + 5)$.
4. (-2, 2)   **(-4, 7)**   5. (p, q) **(p – 2, q + 5)**

6. *Multiple choice.*   Which translation has the effect of sliding a graph 6 units down and 7 units to the left?
   (a) $T(x, y) = (x - 6, y - 7)$ (b) $T(x, y) = (x + 7, y + 6)$
   (c) $T(x, y) = (x - 7, y + 6)$ (d) $T(x, y) = (x - 7, y - 6)$ **(d)**

7. Suppose $y = |x|$ and $T(x, y) = (x - 6, y + 5)$.
   a. Find the images of (-2, 2), (-1, 1), and (0, 0).
   b. Verify that all three images satisfy $y - 5 = |x + 6|$.
   See margin. a) T(-2, 2) = (-8, 7); T(-1, 1) = (-7, 6); T(0, 0) = (-6, 5)
8. Suppose that under some translation, $T(1, 3)$ is mapped to $(7, 0)$.
   a. Find a formula for $T(x, y)$.   **b.** What is $T(12, 40)$?
   T(x, y) = (x + 6, y – 3)   **(18, 37)**

In 9 and 10, the graph of the given function f is translated 5 units to the right and 3 units down. **a.** Find an equation of its image g. **b.** Sketch graphs of f and g on the same set of axes.

9. $f(x) = |x|$ 10. $f(x) = x^2$ **See margin.**
   g(x) = |x – 5| – 3; b) See margin.

In 11 and 12, **a.** state a rule for a translation that maps the graph of f onto the graph of g. **b.** Graph f and g on the same set of axes.   **See margin for graphs.**

11. $f(x) = \sqrt{x}$, $g(x) = \sqrt{x + 3}$ 12. $f(x) = |x|$, $g(x) = |x - 1| - 2$
    a) (x, y) → (x – 3, y)   **a) (x, y) → (x + 1, y – 2)**

In 13 and 14, a translation image of $y = x^2$ is shown. Find an equation for the image.

13.

y = x² – 2

14.

y = (x – 3)² – 2

Applying the Mathematics

15. The equation $x^2 + y^2 = 16$ describes a circle of radius 4 with center at the origin. Describe the graph of the relation $(x + 7)^2 + (y - 6)^2 = 16$.
    a circle of radius 4 with center at (-7, 6)
16. *Multiple choice.*   A parabola has its vertex at $(2, ^-3)$. Which of the following may be an equation for the parabola?
    (a) $f(x) = (x + 2)^2 + 3$ (b) $f(x) = (x + 2)^2 - 3$
    (c) $f(x) = (x - 2)^2 - 3$ (d) $f(x) = (x - 2)^2 + 3$ **(c)**

154

17. *Multiple choice.* Zephyr Yacht Sales formerly gave its sales people 6% commission on their total sales each month. Now they have added a base salary of $300 per month to the commission. Which pair of graphs represents the former pay policy along with the new one? **(a)**

(a)

(b)

(c)

18. Find an equation for the translation image of $y = \dfrac{1}{x}$ pictured at the right.

$y = \dfrac{1}{x+2} - 1$

Review

19. Consider the data set $\{x_1, x_2, x_3, \ldots, x_n\}$.
   a. Write an expression using Σ-notation for the mean $\bar{x}$ of the set.
   b. Write an expression using Σ-notation for the standard deviation s of the set. *(Lessons 1-4, 1-8)* **See margin.**

Monthly Electric Costs

In 20–22, use the graph at the left, which shows residential electric rates in a big city. *(Previous Course, Lesson 1-3)*

20. What is the minimum monthly fee paid by a customer even if no electricity is used? **≈ $9.92**

21. *Multiple choice.* Which is a unit for the slope between two points on this graph?
   (a) dollars
   (b) kilowatt hours
   (c) dollars per kilowatt hour
   (d) kilowatt hours per dollar **(c)**

22. *True* or *false.* The cost per kilowatt hour declines after a customer uses 400 kilowatt hours. **True**

Exploration

23. a. Consider the linear equation $f(x) = 3x - 5$. Find an equation for the image of f under the following transformations.
   i. $T(x, y) = (x + 1, y + 3)$ $y - 3 = 3(x - 1) - 5$ or $y = 3x - 5$
   ii. $T(x, y) = (x + 2, y + 6)$ $y - 6 = 3(x - 2) - 5$ or $y = 3x - 5$
   iii. $T(x, y) = (x - 4, y - 12)$ $y + 12 = 3(x + 4) - 5$ or $y = 3x - 5$
   b. Make a conjecture based on the results of part **a**. **See margin.**
   c. Prove your conjecture in **b**. **See margin.**
   d. Generalize this problem to any line of the form $y = mx + b$.
   The image of $y = mx + b$ under $T(x, y) = (x + h, y + mh)$, is $y = mx + b$.

LESSON 3-3

Translations of Data

A data set can be transformed just as a set of points can. Numbers are sometimes transformed to make them more convenient for calculating statistics. Consider the times taken by a male world-class swimmer for the 400-m individual medley event. Typical times are 4 minutes, 20.36 seconds (written 4:20.36) or 4 minutes, 21.81 seconds (4:21.81). Swimmers and expert commentators will commonly describe these times as "20.36" and "21.81" respectively, assuming that a person listening will know the meaning. The data have been transformed by subtracting 4 minutes. The same term is used as for geometric points; such a transformation of data is called a **translation**.

A **translation** of a set of data $\{x_1, x_2, \ldots, x_n\}$ is a transformation that maps each x_i to $x_i + h$, where h is some constant. This can be expressed as

$$T : x \rightarrow x + h \text{ or } T(x) = x + h.$$

The number $x + h$ or the point it represents is called the **image** of x. In the situation above the transformation mapping each original swimming time x (in minutes) onto its image is $T(x) = x - 4$.

As might be expected, translations have an effect on displays and the measures of center and spread of data.

Example 1 Suppose that a fast food restaurant employs 11 part-time workers, and that their earnings for a week are (in dollars):

$$40, 46, 47, 48, 50, 50, 52, 52, 52, 53, 60.$$

Suppose also that every worker is given an end-of-the-year bonus of $25 on his or her final check.

a. Find the amount received by each employee for the last check.
b. Compare the distributions of employee pay prior to and then including the bonus amount.
c. Describe the effect of the bonus on the median, range, mean, and standard deviation of the set of earnings.

Solution

a. The translation applied here is $x \rightarrow x + 25$. The adjusted earnings on the last check are   65, 71, 72, 73, 75, 75, 77, 77, 77, 78, 85.
b. Graphing the two data sets on the same axes shows that when each element of the data set has been translated 25 units, the new frequency distribution is the image, 25 units to the right, of the original frequency distribution.

c. For the original data, the median is 50; the range is $60 - 40 = 20$, and the mean is 50. For the transformed data, the median is 75; the range is $85 - 65 = 20$; and the mean is 75.

The standard deviations for the two data sets are calculated below. In the table, x_i represents the original earnings of the ith person and x_i' represents his or her earnings after the bonus.

Original Earnings x_i	Deviation $x_i - \bar{x}$	Square of deviation $(x_i - \bar{x})^2$	Earnings After Bonus x_i'	Deviation $x_i' - \bar{x}'$	Square of deviation $(x_i' - \bar{x}')^2$
40	-10	100	65	-10	100
46	-4	16	71	-4	16
47	-3	9	72	-3	9
48	-2	4	73	-2	4
50	0	0	75	0	0
50	0	0	75	0	0
52	2	4	77	2	4
52	2	4	77	2	4
52	2	4	77	2	4
53	3	9	78	3	9
60	10	100	85	10	100
Totals 550		250	825		250

The variance of the original data is $\dfrac{\sum\limits_{i=1}^{11}(x_i - \bar{x})^2}{10} = \dfrac{250}{10} = 25.$

So, its standard deviation is $\sqrt{25} = 5$. Similarly, the

variance of the translated data is $\dfrac{\sum\limits_{i=1}^{11}(x_i' - \bar{x}')^2}{10} = \dfrac{250}{10} = 25.$ So the

standard deviation of the translated data is also $\sqrt{25}$ or 5. Thus, adding \$25 to each individual's earnings increased the mean and median each by \$25, but had no effect on the range and standard deviation.

You can generalize the results in Example 1 by examining what happens to the measures of center and spread of a data set $\{x_1, x_2, \ldots, x_n\}$ under a translation by h. Under such a translation, each value x_i is mapped to $x_i + h$. To find the mean of the image set, you must find

$$\frac{\sum\limits_{i=1}^{n}(x_i + h)}{n}.$$

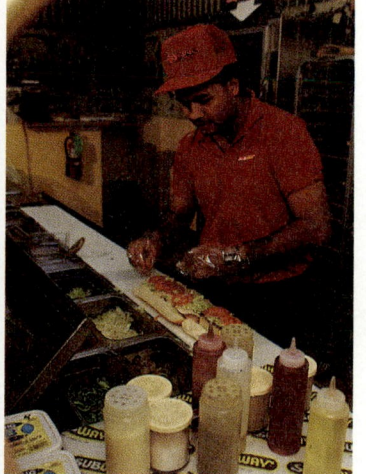

using either their computer statistics packages or calculators with statistical capabilities. Most statistics packages have the capability to perform transformations on data sets and students should learn how theirs works.

ADDITIONAL EXAMPLES
1. Ten students in a class earned scores of 93, 95, 91, 96, 88, 90, 93, 95, 80, and 100 on a test. Translate the data by using a constant of -100 to mentally find the mean of these scores.
92.1

2. A worker records the time it takes to get from home to the parking lot of the factory and finds a mean time of 20.6 minutes with a standard deviation of 3.5 minutes. If it consistently takes 5 minutes to get from the parking lot to the worker's place in the factory, find the mean and standard deviation of the time it takes the worker to get from home to that place in the factory.
25.6; 3.5

Error Analysis For **Example 2,** students who resort to paper and pencil computation or who give an answer of 8.5 for the standard deviation have not grasped the theorem on page 158.

NOTES ON QUESTIONS
Question 10: This question can be done by either computing $(-2 + 8) + (7 + 8) + (2.5 + 8) + (1.3 + 8) + (9.2 + 8)$ or by computing $(-2 + 7 + 2.5 + 1.3 + 9.2) + 5(8)$. The generalization is found in **Question 15a.**

NOTES ON QUESTIONS
Question 13: You might have students compute the standard deviation for each set of scores.

Making Connections for Question 16: This question applies the Graph Translation Theorem and reviews the line of best fit. A statistics package will be helpful. The equation of the line of best fit

By the definition of Σ, that expression represents
$$\frac{(x_1 + h) + (x_2 + h) + (x_3 + h) + \ldots + (x_n + h)}{n}.$$

Using the associative and commutative properties of addition, rewrite the expression as
$$\frac{(x_1 + x_2 + x_3 + \ldots + x_n) + \overbrace{(h + h + h + \ldots + h)}^{n \text{ terms}}}{n} = \frac{\left(\sum_{i=1}^{n} x_i\right) + nh}{n}$$

$$= \frac{\sum_{i=1}^{n} x_i}{n} + h = \overline{x} + h.$$

This proves that, under a translation by h, the mean of the image set of data is h units more than the mean of the original set of data. It also can be shown that after a translation of h units, the median and mode of the image are also increased by h units.

Theorem

Adding h to each number in a data set adds h to each of the mean, median, and mode.

What happens to the measures of spread of a data set after a translation by h? The minimum m is mapped to $m + h$; and the maximum M is mapped to $M + h$. So the range of the translated data is
$$(M + h) − (m + h) = M − m,$$
which is the range of the original data.

Similarly, in the calculation of the variance and standard deviation, a translation by h maps each value x_i to $x_i + h$. By the theorem above, the mean $\overline{x}$ is mapped to $\overline{x} + h$. So, each new deviation equals $(x_i + h) − (\overline{x} + h) = x_i − \overline{x}$, which is the original deviation. Because each individual deviation stays the same under a translation, the variance and standard deviation also stay the same under a translation.

Theorem

Adding h to each number in a data set does not change the range, interquartile range, variance, and standard deviation of the data.

Because the measures of spread of a data set do not vary under a translation, they are said to be **invariant** under translation.

The preceding theorems can be used to simplify calculations of measures of center or spread.

158

■ ■ ■ ■ ■ ■ ■ ■ ■

Example 2   A swimmer recorded his last ten times for the 400-m individual medley in minutes and seconds:

4:21.81	4:20.36	4:22.06	4:22.16	4:21.77
4:21.62	4:20.97	4:20.80	4:20.50	4:20.38.

Find the mean and standard deviation of these times.

Solution   To simplify calculations, subtract 4 minutes from each time. That is, use the seconds only. The translated data set is

21.81, 20.36, ... , 20.38.

For the translated data set find the mean and standard deviation, either by hand or with a computer or calculator. The mean is $\frac{212.43}{10}$, or about 21.24 seconds. The sum of the squared deviations is 4.5971, so the variance is $\frac{4.5971}{9} \approx .5108$ and the standard deviation is about 0.71 seconds. So, by the theorem above, for the original data the mean is 4:21.24, while the standard deviation is 0.71 second.

Questions

Covering the Reading

1. A transformation that maps a number x to $x + h$ is called a(n) ___?___ .
   translation

2. Suppose $T : x \rightarrow x - 4$, where x is a time in minutes. Evaluate each expression.
   a. $T(4.9)$ **0.9 minutes**   **b.** $T(4$ minutes, 38 seconds$)$ **38 seconds**

In 3–8, refer to the data on earnings in Example 1. Suppose that, instead of a bonus being added, $2 was deducted from each worker's earnings to pay for a staff holiday party. Give the following measures for the adjusted earnings.

3. range   **20**   4. mode   **50**   5. median   **48**

6. mean   **48**   7. variance   **25**   8. standard deviation   **5**

9. Name three statistical measures that are invariant under a translation.
   Samples: variance, standard deviation, interquartile range, and range

In 10 and 11, suppose $x_1 = -2$, $x_2 = 7$, $x_3 = 2.5$, $x_4 = 1.3$, and $x_5 = 9.2$. Evaluate the given expression.

10. $\sum_{i=1}^{5} (x_i + 8)$ **58**   11. $\sum_{i=1}^{5} x_i + 8$ **26**

12. On a set of test scores for a class of n students, let $\bar{x}$ be the mean and s the standard deviation. Later, every score is increased by b bonus points.
   a. What is the mean of the image scores? $\bar{x} + b$
   b. What is the standard deviation of the image scores? s

LESSON 3-3   Translations of Data   **159**

It is translated left 100 units.

g. mean height = 15.5 cm; mean weight = 1.08 kg; standard deviation = 9.01 cm; standard deviation = 3.53 kg

h.

It is translated down 20 units and left 100 units.

FOLLOW-UP

MORE PRACTICE
For more questions on SPUR
Objectives, use *Lesson Mas-
ter 3-3,* shown on page 161.

EVALUATION
A quiz covering Lessons 3-1
through 3-3 is provided in the
Teacher's Resource File.

ADDITIONAL ANSWERS
13.a.,c., 15.a.,b., 16.b.,c.,
e.,g.,h., See margin,
pages 158–159.
18.b.

19.a.,b.

20.a.,b.

23.a. *x*-intercepts:
-3 + √5 ≈ -0.8, -3 – √5 ≈ -5.2;
y-intercept: 4

b.

Applying the Mathematics

13. Consider the two frequency distributions below.

Original scores	Frequency	Transformed scores	Frequency
6	2	17	2
10	3	21	3
12	1	23	1
16	2	27	2
22	2	33	2

  a. Make a dot frequency diagram showing the two sets of scores.
  b. Identify the transformation used to get the transformed scores.
  c. Find the range, mode, mean, and median for each set of scores.
  a,c) See margin. b) $x \rightarrow x + 11$

14. Jim's score on a standardized test placed him at the 71st percentile. If each test score was decreased by 7, give Jim's percentile rank in the new distribution, if possible.   **71st**

15. Let $\{x_1, x_2, x_3, \ldots, x_n\}$ be a data set, and k a constant.
  a. Prove that $\sum_{i=1}^{n} (x_i + a) = \sum_{i=1}^{n} x_i + na$.

  b. Prove that $\sum_{i=1}^{n} (x_i - \bar{x}) = 0$, where $\bar{x}$ refers to the mean of the set.

  (Hint: Apply the result of part **a**.)   **See margin.**

16. Consider the following data which give the height h (in cm) and weight w (in kg) of the students in a first-grade class.

h	112	106	114	109	110	122	129	112	117	126	128	101
w	20	18	19	17	20	24	25	21	26	23	25	15

  a. Find the mean height and mean weight. Enter these data into a computer file. (You will use the data again in later lessons, so save the data file if you can.)   **mean height = 115.5 cm; mean weight = 21.1 kg**
  b. Find the standard deviation of the heights and of the weights.
  c. Draw a scatter plot with h on the horizontal axis.   **See margin.**
  d. Find the line of best fit for predicting w from h. $w = .342h - 18.43$
  e. Reduce each height by 100 cm. (Your statistical package may be able to do this for you.) Draw a new scatter plot. How is this scatter plot different from that in part **c**?   **See margin.**
  f. Without using a computer, write an equation of the line of best fit for part **e**. Check using the computer. $w = .342h + 15.78$
  g. Reduce each height by 100 cm and each weight by 20 kg. Find the mean and standard deviation of the transformed variables.
  h. Draw a new scatter plot. How is this scatter plot different from that in part **a**.   **See margin.**
  i. Without using a computer write down an equation for the line of best fit for part **e**. Check using the computer. $w = .342h - 4.22$
  b,g) See margin.

17. Generalize the result of Question 16. If bivariate data are translated, which (if any) of the following are invariant?
  (a) means of the two variables
  (b) standard deviations of the two variables
  (c) equation of the line of best fit
  (d) slope of the line of best fit
  (e) correlation between the two variables   **(b), (d), (e)**

160

Review

18. Suppose the translation $T: (x, y) \rightarrow (x + 6, y + 5)$ is applied to the graph of the function $y = \frac{1}{x^2}$.

   a. Find an equation for the image. $y - 5 = 1/(x - 6)^2$ or $y = 1/(x - 6)^2 + 5$
   b. Sketch graphs of the preimage and image using an automatic grapher.
   (Lessons 3-1, 3-2)   **See margin.**

In 19 and 20,
   a. sketch a graph of the function by hand;
   b. check your prediction with an automatic grapher;
   c. state the transformation that maps the parent function to the given function.   *(Lesson 3-2)*

19. $y = |x - 7| + 10$ **a,b) See margin.**   **20.** $y = \lfloor x \rfloor + 1$ **a,b) See margin.**
   c) $(x, y) \rightarrow (x + 7, y + 10)$ **c) $(x, y) \rightarrow (x, y + 1)$**

In 21 and 22, write an equation for the given graph. (Hint: the parents are the absolute value and cubic functions, respectively.)   *(Lesson 3-2)*

21.

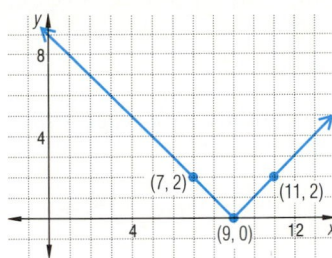

$y = |x - 9|$

22.

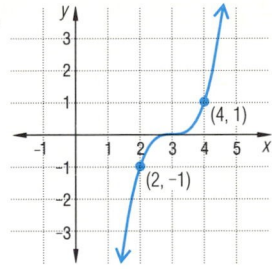

$y = (x - 3)^3$

23. Suppose $f(x) = x^2 + 6x + 4$.
   a. Give the intercepts of $y = f(x)$.
   b. Check your work by making a graph with an automatic grapher.
   (Lessons 2-6, 3-1)
   See margin.

24. Use the boxplot below to estimate the value of each measure.

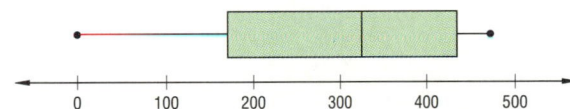

   a. median   **≈ 325**
   b. interquartile range   *(Lessons 1-4, 1-5)*   **≈ 265**

25. *Skill sequence.*   Simplify each expression.   *(Previous course)*

   a. $\dfrac{x}{\frac{1}{5}}$ **5x**   **b.** $\dfrac{4x}{\frac{1}{9}}$ **36x**   **c.** $\left(\dfrac{x}{\frac{1}{6}}\right)^2$ **$36x^2$**

Exploration

26. Give an example of some situation other than swimming events where data are often translated, and explain why the translated data are used.
   samples: stock market prices (for efficiency, the first digit of the price is often dropped on tickers or in quoting prices); times for marathons (hours are dropped when calling out finishing times because it is assumed people know them)

NAME _____

LESSON **MASTER 3-3**
QUESTIONS ON **SPUR** OBJECTIVES

■**PROPERTIES**   *Objective E (See pages 197–200 for objectives.)*
1. Given that $a_1 = 1$, $a_2 = 1$, $a_3 = 2$, $a_4 = 3$, $a_5 = 5$, and $a_6 = 8$, evaluate.

   a. $\sum_{i=1}^{6}(a_i + 1)$ **26**

   b. $\sum_{i=1}^{6} a_i + 1$ **21**

   c. $\sum_{i=1}^{6} a_i^2$ **104**

2. Show that $\sum_{i=1}^{n}(5 + x_i) = \sum_{i=1}^{n} x_i + 5n$ for all n.

Sample: $\sum_{i=1}^{n}(5 + x_i) = (5 + x_1) + (5 + x_2) + \ldots + (5 + x_n)$.

As addition is commutative and associative, rearranging,
$= (x_1 + x_2 + \ldots + x_n) + \underbrace{(5 + 5 + \ldots + 5)}_{n \text{ times}}$
$= \sum_{i=1}^{n} x_i + 5n.$

■**PROPERTIES**   *Objective F*
3. *True or false.* Under a translation, the mean of a data set stays the same. Justify your answer.
False. Suppose k is added to each of the n elements of a data set.
Generalizing Question 2, $\sum_{i=1}^{n}(k + x_i) = \sum_{i=1}^{n} x_i + k_n$. As the mean
of the new data set is $\dfrac{\sum_{i=1}^{n}(k + x_i)}{n} = \dfrac{\sum_{i=1}^{n}(k_i + x_n)}{n} = \dfrac{\sum_{i=1}^{n} x_i}{n} + k$,
this is k different from the mean of the original set.

Functions, Statistics, and Trigonometry © Scott, Foresman and Company   **33**

NAME _____
Lesson MASTER 3-3 (page 2)

4. Suppose 13 is subtracted from each element in a data set. Describe the effect of this transformation on each measure.
   a. median   **13 less than the original median**
   b. mode   **13 less than the original mode**
   c. variance   **does not change**
   d. standard deviation   **does not change**

■**USES**   *Objective I*
5. Use a translation to calculate mentally the average number of pages in a chapter of a book if the individual chapters have 85, 78, 75, 82, and 85 pages. Indicate how you did this.
   The average is 81. Subtract 80 from each value. Find
   the average $\left(\frac{5 + -2 + -5 + 2 + 5}{5} = 1\right)$. This must be 80 less
   than the original average.

6. A 10 k track runner finished his 11 performances of the season in the following times (minutes:seconds):

8:32.11	8:20.00	8:12.55
8:30.02	8:29.99	8:14.88
8:28.91	8:15.32	8:20.00
8:25.07	8:12.04	

Find **a.** the mean and **b.** the standard deviation of his times. [Hint: Use a translation to simplify the computation.]
   a. __**8:21.90**__   **b.** __**0:7.61 (that is, 7.61 seconds)**__

34   *Functions, Statistics, and Trigonometry* © Scott, Foresman and Company

LESSON

3-4

Symmetries of Graphs

Many graphs have *symmetries*. For instance, the graphs of $f(x) = |x|$ and $g(x) = \frac{1}{x^2}$ are *symmetric to the y-axis*. That is, they coincide with their reflection images over the line.

A graph is **symmetric with respect to the y-axis** if, for every point (x, y) on the graph, then so is $(-x, y)$.

Any function whose graph is symmetric with respect to the *y*-axis is called an *even function*.

Definition

A function is an **even function** if and only if for all values of *x* in its domain, $f(-x) = f(x)$.

Looking at the graph of a function often gives a good idea of whether or not the graph is symmetric over the *y*-axis. However, to be sure your impression is accurate, apply the definition above to prove that a function is even.

■ ■ ■ ■ ■ ■ ■ ■

Example 1 Prove that $f(x) = |x|$ is an even function.

> **Solution** To show that $f(x) = |x|$ is an even function, show that $f(-x) = f(x)$ for all *x*. By the definition of *f*, $f(-x) = |-x|$.
> By the definition of absolute value, $|-x| = |x|$.
> Since $|x| = f(x)$, $f(-x) = f(x)$.

Similarly, if for each point (x, y) on a graph there is a corresponding point $(x, -y)$ on the graph, the graph is said to be **symmetric with respect to the x-axis**. The graph of $x = y^2$ at the right is symmetric over the *x*-axis. Note, however, that this graph does not represent a function.

162

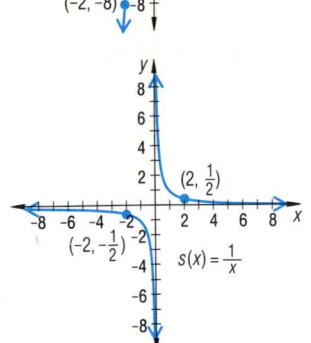

In general, unless a function is a subset of the x-axis, its graph cannot be symmetric with respect to the x-axis. If it were, it would have two points with the same x-coordinate, which would violate the definition of function.

Consider the functions $h(x) = x^3$ and $s(x) = \frac{1}{x}$, graphed at the left. These functions are not symmetric to the y-axis or the x-axis, but they do have a certain "balance." If you rotate each graph 180 degrees around the origin, it maps onto itself. Such graphs are said to be *symmetric to the origin*. A graph is **symmetric to the origin** if and only if for every (x, y) on the graph, then so is $(-x, -y)$.

A function whose graph is symmetric to the origin is called an *odd function*.

> **Definition**
>
> A function f is an **odd function** if and only if for all values of x in its domain, $f(-x) = -f(x)$.

If you are not sure if a function is even, odd, or neither, then an automatic grapher may help you decide.

▪ ▪ ▪ ▪ ▪ ▪ ▪ ▪

Example 2   Use an automatic grapher to see if the function $f(x) = x^3 - x$ appears to be odd, even, or neither. If it appears to be even or odd, prove it.

Solution   A graph of f is shown at the right.
It appears to be symmetric to the origin, so f seems to be an odd function. To prove this, substitute $-x$ for x in the formula for f and simplify.

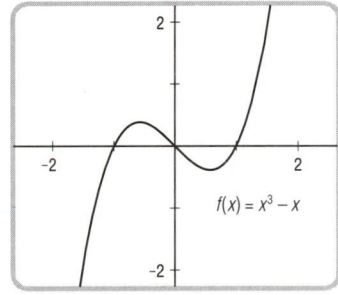

$$f(-x) = (-x)^3 - (-x)$$
$$= -x^3 + x$$
$$= -(x^3 - x)$$
$$= -f(x)$$

Thus for all x, $f(-x) = -f(x)$.
So f is an odd function.

Even and odd functions get their names from the **power functions**; that is, from functions of the form $f(x) = x^n$, where n is a positive integer. All power functions with even exponents, for instance $y = x^2$, are even functions. All power functions with odd exponents, for instance $y = x^3$, are odd functions. However, as shown in Example 1, in the Questions, and later in the graphs of trigonometric functions, not all even functions involve even numbers and not all odd functions involve odd numbers.

LESSON 3-4   Symmetries of Graphs   **163**

Error Analysis   Some students think a function is even or odd when they manage to provide a numerical example that meets the condition. The difference between a specific example and proving something true for all values in the domain may need clarification.

Students may have difficulty with the notation in the definitions of symmetry and of even and odd functions. When confronted with x and $-x$ in the same setting, students tend to think that x is always positive and $-x$ is always negative. To counter this, you might say "opposite x-values give the same y-values in even functions." Keep appealing to symmetry: if (x, y) and $(-x, y)$ are on a graph, the points are reflection images of each other over the y-axis.

Making Connections
Some students might not see symmetry to the origin as a 180° rotation. They may, however, see that an odd function maps onto itself if it is reflected over the x-axis and then reflected over the y-axis (or vice versa). This has been discussed in previous UCSMP courses and comes up again in Chapter 11.

ADDITIONAL EXAMPLES
1. Prove that the squaring function is an even function.
The squaring function s has equation $s(x) = x^2$; For all x, $(-x)^2 = x^2$, thus x, $s(-x) = s(x)$, by definition and so the function is even.

2. Use an automatic grapher to conjecture whether the function $f(x) = 2x^3$ is even, odd, or neither. Then prove your conjecture.
odd; for all x, $f(-x) = 2(-x)^3 = -2x^3 = -f(x)$

3. Does the graph of $g(x) = 5x^3 - 10x^2$ appear to have line symmetry or point symmetry?

Reflection symmetries with respect to the *x*-axis, *y*-axis, and origin are special cases of more general symmetries.

A graph is said to be **reflection-symmetric** if the graph can be mapped onto itself by a reflection over some line ℓ. The reflecting line ℓ is called the **axis** or **line of symmetry** of the graph. Similarly, a graph is said to be **rotation-symmetric** or to have **symmetry to point P** if the graph can be mapped to itself under a rotation of 180° around P. The point P is called a **center of rotation**.

Example 3 Consider the function $f(x) = (x + 3)^2 - 10$. Which, if either, of line symmetry or point symmetry does f have?

Solution Sketch a graph of the function.
The graph of $f(x) = (x + 3)^2 - 10$, shown at the left, appears to have line symmetry. It is reflection-symmetric over the line $y = -3$, which is the vertical line through the vertex.

Notice that the function f in Example 3 can be rewritten as $y + 10 = (x + 3)^2$, so its graph is the image of $y = x^2$ under the translation $(x, y) \rightarrow (x - 3, y - 10)$; and the line of symmetry of f is the image of the line of symmetry of the parent function under the same translation.

In general, if f is a function and each point (x, y) on it is mapped to $(x + h, y + k)$, then the graph of the image is congruent to the graph of the preimage, and all properties associated with its shape are invariant. Specifically, lines of symmetry map to lines of symmetry, maxima to maxima, minima to minima, vertices to vertices, and symmetry points to symmetry points.

Example 4 Consider the function $y = F(x) = \dfrac{1}{x - 5} - 4$.

a. Give equations for the asymptotes of its graph.
b. Describe any lines or points of symmetry.

Solution

a. Rewriting F as $y + 4 = \dfrac{1}{x - 5}$, shows that, by the Graph Translation Theorem, F is the image of $y = \dfrac{1}{x}$ under the translation $T(x, y) = (x + 5, y - 4)$. The graph of the parent function has asymptotes $x = 0$ and $y = 0$. Each asymptote of the parent is translated 5 units to the right and 4 units down. So the asymptotes of F are $x = 5$ and $y = -4$.

b. Sketch a graph of $y = F(x)$.
The graph at the left appears to have rotation symmetry around $(5, -4)$, the point of intersection of the asymptotes. Indeed, $(5, -4)$ is the translation image of $(0, 0)$, the center of rotation for $y = \dfrac{1}{x}$, under $T : (x, y) \rightarrow (x + 5, y - 4)$.

164

Examples 3 and 4 are instances of the following generalization.

> **Theorem**
>
> Consider the translation T that maps (x, y) to $(x + h, y + k)$. If f is an even function, then $T(f)$ is reflection-symmetric with respect to the line $x = h$. If f is an odd function, then $T(f)$ is rotation-symmetric to the point (h, k).

Questions

Covering the Reading

1. For each type of function in the left column name two properties from the right column.
   - **a.** even function
   - **b.** odd function
   - **i.** The graph is symmetric through the origin.
   - **ii.** The graph is symmetric over the y-axis.
   - **iii.** If (x, y) is in the function, so is $(-x, y)$.
   - **iv.** If (x, y) is in the function, so is $(-x, -y)$.

   a) ii, iii; b) i, iv

2. Suppose z is a relation which includes the point $(-3, 5)$. What other point must be included in the relation if z is
   - **a.** odd? **b.** even? **c.** symmetric over the x-axis?

   (3, -5)   **(3, 5)**   **(-3, -5)**

3. Prove that $q(x) = x^7$ is an odd function. **See margin.**

In 4 and 5, **a.** tell if the function is odd, even, or neither. **b.** If either odd or even, prove it.

4. $f(x) = x$ **a) odd; b) See margin.**   5. $g(x) = x^3 - 5$ **a) neither**

6. Which of the following functions are power functions?
   $a(x) = 2, \quad b(x) = x^2, \quad c(x) = 2^x, \quad d(x) = x^{11}, \quad e(x) = x^{-1}$ **b, d**

7. Give a counterexample to prove that $f(t) = t^3 - 2$ is not an even function.
   sample: $f(2) = 2^3 - 2 = 6, f(-2) = (-2)^3 - 2 = -10$

Applying the Mathematics

In 8–10, use an automatic grapher to graph the given functions on the same coordinate axes.
   - **a.** Sketch or print a hard copy of the graphs of the functions.
   - **b.** Determine if each function is odd, even, or neither. (It is not necessary to prove them odd or even.)
   - **c.** Identify which function is the parent, and describe the relationship between the graph of the parent and the other graph(s).
   - **d.** Give equations for all axes of symmetry for reflection-symmetry; give coordinates for all centers of rotation for point symmetry.

8. $g(x) = x^3, h(x) = -x^3, j(x) = x^3 + 2$ **See margin.**

9. $k(x) = |x|, m(x) = |x| - 2, n(x) = |x + 4| - 2$ **See margin.**

10. $q(x) = \frac{1}{x}, r(x) = \frac{1}{x + 5}, s(x) = \frac{1}{x} + 3$ **See margin.**

11. If $f(x) = -g(x)$ for all values of x, how are their graphs related?
    The graph of g is a reflection image of the graph of f over the x-axis.

LESSON 3-4 Symmetries of Graphs   165

9.a.

b. k and m are even; n is neither.

c. k is the parent; m is a translation of k by 2 units down; n is a translation of k by 4 units to the left and 2 units down.

d. The axis of symmetry for k and m is $x = 0$; for n the axis of symmetry is $x = -4$.

10.a.

10.b.,c.,d. See margin, page 166.

EXTENSION
Prove that a function is both even and odd if and only if the domain of the function contains $-x$ whenever it contains x and if the function only has zero values. (To be odd and even, one must have $f(-x) = -x$ and $f(-x) = x$ for all x. So $-x = x$, and so $x = 0$, from which $f(-x) = f(x) = 0$ for all x.)

ADDITIONAL ANSWERS
10.b. *q* is odd; *r* and *s* are neither.

c. *q* is the parent; *r* is a translation of *q* by 5 units to the left; *s* is a translation of *q* by 3 units up.

d. *q* has point symmetry around (0, 0); *r* has point symmetry around (-5, 0); and *s* has point symmetry around (0, 3).

17.a. about $8 billion

b. about $2 billion

18.a. about $9 billion

b. the increase in total home video spending between 1983 and 1990

20.a.,b.

U.S. Home Video Industry

Videocassette recorders
Ownership in millions of households

Source: Motion Picture Association of America, Paul Kagan Associates

Home video spending
In billions of dollars, estimates

In 12 and 13, if $x_1 = 3$, $x_2 = -8$, $x_3 = 9$, and $x_4 = -1$, evaluate the expression. *(Lesson 3-3)*

12. $\sum_{i=1}^{4} x_i + 3$ **6**

13. $\sum_{i=1}^{4} (x_i + 3)$ **15**

14. The upper and lower quartiles of a set of 30 test scores are 62 and 87, respectively. What are the new values for the upper and lower quartiles if, because of a marking error,
  a. each score is increased 3 points?   **65 and 90**
  b. each score is increased k points?   *(Lesson 3-3)*   **$62 + k$ and $87 + k$**

15. *Skill sequence.* Solve for x.   *(Previous course)*
  a. $x^2 - 9 = 0$ **b.** $x^3 - 9x = 0$ **c.** $x^3 - 5x = 0$
  a) $x = 3$ or $x = -3$;   **b) $x = 3$ or $x = 0$ or $x = -3$;**   **c) $x = \sqrt{5}$ or $x = 0$ or $x = -\sqrt{5}$**

In 16–18, refer to the graph at the left.   *(Lesson 1-2, 1-3, 2-1)*

16. a. About how many more households in the U.S. had VCRs in 1989 than in 1983?   **about 54 million**
  b. By about what percent did VCR ownership increase in households between 1983 and 1989?   **about 775%**

17. Of the amount estimated to be spent by households on videos in 1990, about how much was estimated for each of the following?
  a. rentals   **b.** sales   **See margin.**

18. Suppose $T(y)$ represents the total home video spending in year y.
  a. Estimate $T(1990) - T(1983)$.
  b. What does the answer to part **a** represent?   **See margin.**

19. Find the image of the point $(-1, 2)$ under the given scale change transformation.   *(Previous course)*
  a. $S(x, y) = (x, 4y)$ **b.** $S(x, y) = (2x, -3y)$ **c.** $S(x, y) = (-x, y)$
  $(-1, 8)$ **$(-2, -6)$**   **$(1, 2)$**

20. a. Draw rectangle $ABCD$ with $A = (-2, 2)$, $B = (8, 2)$, $C = (8, -4)$, and $D = (-2, -4)$.   **See margin.**
  b. Draw $A'B'C'D'$, the image under the transformation that maps (x, y) to $\left(\frac{1}{2}x, 2y\right)$.   **See margin.**
  c. *True* or *false*. $ABCD$ and $A'B'C'D'$ are congruent.   *(Previous course)*   **False**

21. a. Can the graph of a function be mapped onto itself by a rotation whose magnitude is not 180° or 360°? If so, give some examples. If not, explain why not.
  b. Repeat Question 21a for graphs of relations.

---

21. a) Yes; for example, the function {(1, 3), (-3, 1), (-1, -3), (3, -1)} is mapped onto itself by a rotation of 90° and of 270°. **b)** Yes; for example, any regular polygon whose center is the origin, or any circle whose center is the origin is mapped onto itself by a rotation.

166

LESSON 3-5

The Graph Scale Change Theorem

Consider the function $y = x^3 - 3x$. You know that adding constants to x or y will translate the graph. What is the effect of multiplying or dividing x or y by a constant? Consider replacing y by $\frac{y}{2}$. Then, for the new function, $\frac{y}{2} = x^3 - 3x$, which can be written $y = 2x^3 - 6x$. Call the new function g. Below, the graphs of f and g are plotted on the same coordinate system.

Notice the two graphs are *not* congruent. Each point (x, y) on f has been mapped to a point on g with the same x-coordinate, but with twice the y-coordinate. For instance, the y-value of the relative maximum point $(-1, 4)$ on g is twice that of the y-value of the relative maximum point $(-1, 2)$ on f. Similarly, the relative minimum of g is twice as far below the x-axis as that of f. We say that g is a vertical stretch of f, or g is the image of f under a **vertical scale change**.

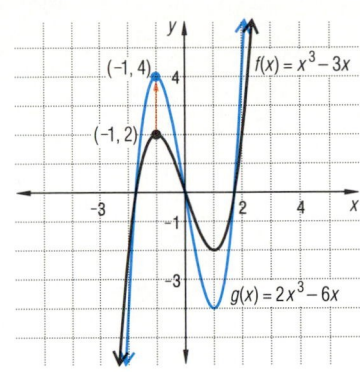

What happens if x is replaced by $\frac{x}{2}$? Replacing y by $\frac{y}{2}$ resulted in a vertical stretch, so you might conjecture that the new graph will be stretched horizontally. This is the case. Below are graphs of $f(x) = x^3 - 3x$ and a new function h, where $h(x) = \left(\frac{x}{2}\right)^3 - 3\left(\frac{x}{2}\right)$.

The relative maximum and minimum of h is each twice as far from the y-axis as its corresponding point of f. Similarly, the x-intercepts of the image are also twice as far from the y-axis as are the x-intercepts of the preimage.

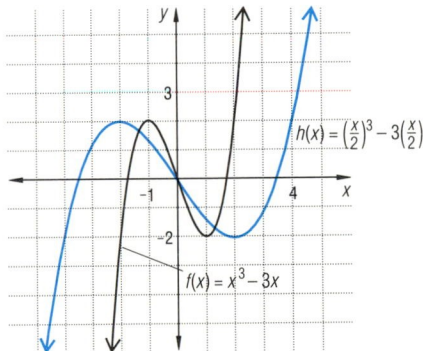

The graph of h is the image of the graph of f under a **horizontal scale change** of magnitude 2. That is, each point on h is the image of a point on f under the mapping $(x, y) \rightarrow (2x, y)$.

RESOURCES
- Lesson Master 3-5
- Teaching Aid 27: Graphs of scale changes of $y = x^3 - 3x$
- Teaching Aid 28: Graphs for **Examples 1 and 2**
- Teaching Aid 29: Graphs for **Questions 13–16**
- Computer Master 8

OBJECTIVES

C Use the Graph Scale Change Theorem to find transformation images.

D Describe the effects of scale changes on functions and their graphs.

J Apply the Graph Scale Change Theorem to make or identify graphs.

TEACHING NOTES

Technology Students may think that replacing x by $\frac{x}{a}$ in an equation should "shrink" the graph of the equation when $a > 1$. Use an automatic grapher to verify the graphs on the first page of the lesson, and to illustrate that the result is really a stretch.

The completed graphs are available on a teaching aid. A computer master is also provided should you wish to spend more time on this topic.

Emphasize that scale change images are generally *not* congruent to preimages and that the limitations of automatic graphers do *not* account for differences in shape. However, in the special case where $a = b$, a size change

167

In general, a **scale change** centered at the origin with **horizontal scale factor** $a \neq 0$ and **vertical scale factor** $b \neq 0$ is a transformation that maps (x, y) to (ax, by). The scale change S can be written

$$S: (x, y) \to (ax, by) \text{ or } S(x, y) = (ax, by).$$

When the scale factors a and b are equal, the scale change is called a **size change**. Notice that in the preceding instances, replacing x by $\frac{x}{2}$ in an equation for a function results in the scale change $S: (x, y) \to (2x, y)$; and replacing y by $\frac{y}{2}$ leads to a scale change $S: (x, y) \to (x, 2y)$. These results generalize.

Graph Scale Change Theorem

In a relation described by a sentence in x and y, the following two processes yield the same graph:

(1) replacing x by $\frac{x}{a}$ and y by $\frac{y}{b}$ in the sentence;

(2) applying the scale change $(x, y) \to (ax, by)$ to the graph of the original relation.

Under the scale change $S: (x, y) \to (ax, by)$, an equation for the image of $y = f(x)$ is $\frac{y}{b} = f\left(\frac{x}{a}\right)$. Notice that multiplication in the transformation corresponds to division in the equation of the image. This is similar to the Graph Translation Theorem in Lesson 3-2 for $y = f(x)$, where addition in the translation $(x, y) \to (x + h, y + k)$ corresponded with subtraction in the image equation $y - k = f(x - h)$.

■ ■ ■ ■ ■ ■ ■ ■

Example 1 Sketch the graph of $y = \lfloor 0.5\,x \rfloor$.

Solution The graph of $y = \lfloor 0.5x \rfloor$ is a scale change image of the parent graph $y = \lfloor x \rfloor$, shown below at the left. To find the scale factors, rewrite the given equation $y = \lfloor 0.5x \rfloor$ as $y = \left\lfloor \frac{x}{2} \right\rfloor$. By the Graph Scale Change Theorem, the result of replacing x by $\frac{x}{2}$ is the scale change that maps (x, y) to $(2x, y)$, so the parent graph is stretched horizontally by a factor of 2. The graph of $y = \left\lfloor \frac{x}{2} \right\rfloor$ is below at the right.

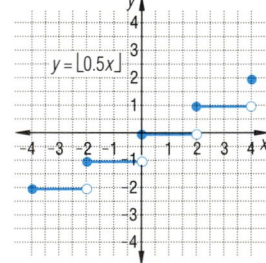

168

Check

Check that a point satisfying the equation is on the graph. For example, suppose $x = 3$. Then $y = \lfloor 0.5(3) \rfloor = \lfloor 1.5 \rfloor = 1$. Is the point $(3, 1)$ on the graph? Yes.

Example 2   Sketch the graph of $y = 1.5 \lfloor 0.5x \rfloor$.

Solution   Rewrite the given equation to see the scale change factors.

$$\frac{y}{1.5} = \left\lfloor \frac{x}{2} \right\rfloor .$$

As in Example 1, the parent $y = \lfloor x \rfloor$ is stretched horizontally by a factor of 2. Also, the graph is stretched vertically by a factor of 1.5 . The graph is shown at the right.

Check

Check a point. If $x = 2$, then $y = 1.5 \lfloor 0.5 \cdot 2 \rfloor = 1.5 \lfloor 1 \rfloor = 1.5$. The point $(2, 1.5)$ satisfies the equation. Is it on the graph? Yes.

When a scale factor for the x- or y-coordinate is negative, the graph is *reflected* over an axis. To see this, consider the horizontal and vertical scale changes by -1,

$$S_1 : (x, y) \rightarrow (-x, y) \text{ and } S_2 : (x, y) \rightarrow (x, -y).$$

In S_1, each x-value is replaced by its opposite, which produces a reflection over the y-axis. Similarly, in S_2, replacing y by $-y$ produces a reflection over the x-axis.

Example 3   **a.** Sketch the image of the graph of $y = x^2$ under $S(x, y) = (x, -3y)$.
b. Give an equation for the image.

Solution

a. The graph of $y = x^2$ is shown below in black. S is a vertical scale change by -3. The effect of the 3 is to stretch the graph vertically by a factor of 3. The effect of the negative sign is to reflect it about the x-axis. The resulting image is shown below in blue.

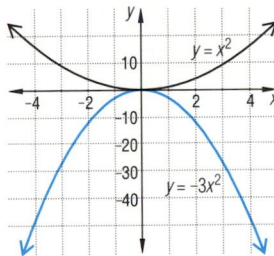

b. In the parent equation $y = x^2$, replace y by $\frac{y}{-3}$. An equation of the image is $\frac{y}{-3} = x^2$, which may be solved for y: $y = -3x^2$.

---

b. Give an equation for the image.
$y = -\frac{1}{8} x^3$

4. The line $41x - 29y = 700$ contains the points $(39, 31)$ and $(10, -10)$. Use this information to obtain two points on the line with equation $20.5x - 87y = 700$.
The second line is the image of the first under the scale change $S(x, y) = \left(2x, \frac{y}{3} \right)$, so two points on that line are $\left(78, \frac{31}{3} \right)$ and $\left(20, \frac{-10}{3} \right)$.

NOTES ON QUESTIONS
Making Connections for Question 8b: The horizontal scale factor of $\frac{1}{3}$ can be explained by realizing that $3x = \frac{x}{\frac{1}{3}}$. This same kind of rewriting will be applied in Chapter 6 in the study of periods of trigonometric functions.

Questions 13–16: A teaching aid is provided with the graphs for these questions.

Making Connections for Question 16: There are infinitely many size changes that map f onto g. In general, the parabola $y = x^2$ can be mapped onto the parabola $y = ax^2$ by any size change S of the form $S(x, y) = (kx, ak^2y)$. In Lesson 3-9, we choose $k = \frac{1}{a}$ and thus show the parabolas are similar.

Making Connections for Question 24: This transformation is studied in more detail in Chapter 11.

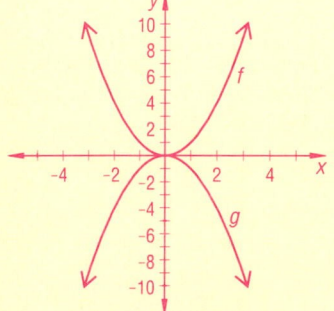
Unlike a translation, a scale change does not preserve distance; that is, a preimage segment does not necessarily have the same length as its image. However, the following is true.

> **Theorem**
>
> Under a scale change, the intercepts of a graph are mapped to the intercepts of the image of the graph.

Proof Consider the graph of $y = f(x)$ and the scale change $S:(x, y) \rightarrow (ax, by)$, $a \neq 0$, $b \neq 0$. An x-intercept is the x-coordinate of a point for which the y-coordinate is zero. Let r be an x-intercept of f. So $(r, 0)$ is a point on the graph of f. The image of the x-intercept is $S(r, 0) = (a \cdot r, b \cdot 0) = (ar, 0)$, which is a point on the x-axis, and thus $a \cdot r$ is an x-intercept of the image of f.

Similarly, if t is the y-intercept of f, then $(0, t)$ is on the graph of f. The image of the y-intercept is $S(0, t) = (a \cdot 0, b \cdot t) = (0, bt)$, which is a point on the y-axis, and thus $b \cdot t$ is the corresponding y-intercept of the image of f.

Example 4 **a.** Find the intercepts of the graph of $y = x^2 - 8$.

b. Use the intercepts in **a** to find the intercepts of $\frac{y}{2} = (3x)^2 - 8$.

Solution

a. The x-intercepts occur where $y = 0$. So if $0 = x^2 - 8$, then $x^2 = 8$ and $x = \pm\sqrt{8}$. The y-intercepts occur where $x = 0$; then $y = 0^2 - 8 = -8$. Thus, for $y = x^2 - 8$, the x-intercepts are $\pm\sqrt{8}$; the y-intercept is -8.

b. The graph of $\frac{y}{2} = (3x)^2 - 8$ is the image of the graph of $y = x^2 - 8$ under $S(x, y) = (\frac{x}{3}, 2y)$. By the theorem above, the images of the intercepts in part **a** under this scale change are the intercepts of the image.

$$S(0, -8) = (0, -16)$$
$$S(\sqrt{8}, 0) = (\frac{\sqrt{8}}{3}, 0)$$
$$S(-\sqrt{8}, 0) = (\frac{-\sqrt{8}}{3}, 0).$$

So -16 is the y-intercept and $\pm\frac{\sqrt{8}}{3}$ are the x-intercepts for $\frac{y}{2} = (3x)^2 - 8$.

170

Check

a. Use an automatic grapher, and estimate the coordinates of the points of intersection of $y = x^2 - 8$ with each axis.

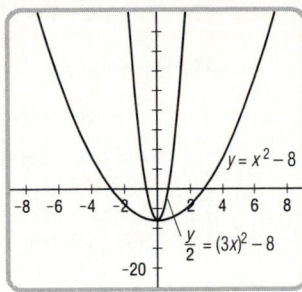

Using a zoom feature, the y-intercept is -8 and the x-intercepts are $\pm 2.83 \approx \pm\sqrt{8}$.

b. Rewrite $\frac{y}{2} = (3x)^2 - 8$ as $g(x) = y = 2((3x)^2 - 8)$. Then $g\left(\frac{\sqrt{8}}{3}\right) =$

$2\left(\left(3 \cdot \frac{\sqrt{8}}{3}\right)^2 - 8\right) = 2((\sqrt{8})^2 - 8) = 0$, as required for an x-intercept.

Questions

Covering the Reading

1. Under a scale change with horizontal factor a and vertical factor b, the image of (x, y) is ___?___.   **(ax, by)**

2. Suppose $y = f(x)$. If S maps each point (x, y) in the plane to $(3x, 4y)$, the image of $y = f(x)$ is ___?___. $y = 4f\left(\frac{x}{3}\right)$

3. Consider the parabola $y = x^2$. Let $S(x, y) = \left(\frac{x}{2}, 3y\right)$.

   a. Find images of $(-3, 9)$, $(0, 0)$, and $\left(\frac{1}{2}, \frac{1}{4}\right)$ under S.

   b. Show that the images in part **a** are on the graph of $\frac{y}{3} = (2x)^2$.
   See margin.

4. Describe the effect of the scale change $S: (x, y) \rightarrow (-2x, 2y)$ on a graph.   **The graph is stretched horizontally and vertically by a factor of 2, and reflected over the y-axis.**

5. *True* or *false*. Under a scale change, the graph of the preimage and image are congruent.   **False**

6. Refer to the Graph Scale Change Theorem.
   a. Explain the reason for the restrictions $a \neq 0$ and $b \neq 0$.
   b. If $a = b$, what special name is given to the transformation?
   a) See margin. b) size change

7. State whether the property is invariant under a scale change.
   a. number of x-intercepts   **Yes**   **b.** values of x-intercepts   **No**
   c. number of y-intercepts   **Yes**   **d.** values of y-intercepts   **No**

8. Sketch a graph.   **See margin.**
   a. $y = 3\lfloor x \rfloor$ **b.** $y = \lfloor 3x \rfloor$

---

d. $g(x) = -x^2$

10.a.,b.

c. The image is the pre-image stretched horizontally by a factor of 2 and vertically by a factor of $\frac{1}{3}$.

d. $g(x) = \frac{1}{3}\left\lfloor \frac{x}{2} \right\rfloor$

11.a. x-intercepts of $y = x^2 - 16$ and of $\frac{y}{4} = x^2 - 16$:

-4 and 4

b.

In 9 and 10, **a.** sketch f; **b.** sketch the image g of f under the given scale
change; **c.** describe how the image is related to the preimage; **d.** give an
equation for the image. **See margin.**

9. $f(x) = x^2$, $S: (x, y) \rightarrow (x, -y)$ **10.** $f(x) = \lfloor x \rfloor$, $S: (x, y) \rightarrow \left(2x, \frac{1}{3}y\right)$

11. a. Find the x- and y-intercepts of the graphs of $y = x^2 - 16$ and
$\dfrac{y}{4} = x^2 - 16$.

b. Sketch the graphs on the same set of axes.
See margin.

12. Give another name for the horizontal scale change of magnitude -1.
reflection over the y-axis

Applying the Mathematics

In 13 and 14, identify a scale change that maps f to g.

13.

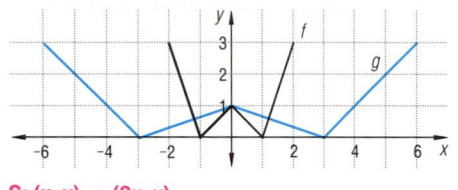

S: $(x, y) \rightarrow (3x, y)$

14.

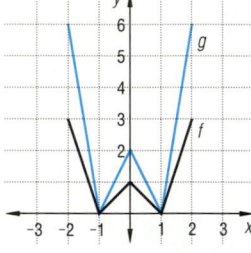

S: $(x, y) \rightarrow (x, 2y)$

15. The graph of f shown
at the right has the
following features.
x-intercepts: 1, 4
y-intercept: 1
maximum: $(-2, 3)$
minimum: $(2, -1)$

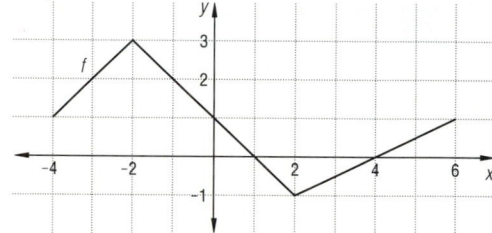

a. Graph the image
of f under
$S(x, y) = \left(3x, \frac{1}{2}y\right)$. **See margin.**

b. Find the coordinates of the named feature of the image of f.
i. x-intercepts **(3, 0) and (12,0)** **ii.** y-intercept **(0, 1/2)**
iii. maximum **(-6, 3/2)** **iv.** minimum **(6, -(1/2))**

16. Scale changes on parabolas can be visualized as either horizontal or vertical scale changes. Refer to the graph of the parabolas below.

a. Write equations for f
and g.
b. Under what horizontal
scale change is g the
image of f?
c. Under what vertical
scale change is g the
image of f?
See margin.

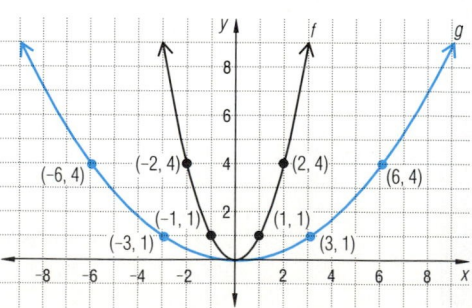

172

17. The formula $s = \sqrt{24d}$ can be used to estimate the pre-brake speed s in mph of a vehicle which skidded d feet on a dry concrete road before coming to a stop.

   a. This formula can be considered a formula for a vertical scale change image of the function $s = \sqrt{d}$ by what factor? $\sqrt{24} = 2\sqrt{6}$

   b. About how fast was a car going which skidded 27 ft before stopping?

   c. About how far would a car skid if its brakes locked at 55 mph?

   d. A similar formula, $s = 2\sqrt{5d}$, is used to estimate the speed of a car based on skid marks on a wet concrete road. What scale change is used to obtain the formula for a wet surface from that for a dry surface?   b) about 25.5 mph; c) about 126 feet; d) $S(s, d) = \left(s, \frac{6}{5}d\right)$

Review

18. Consider the function $y = \frac{1}{x^2}$ and the translation $T(x, y) = (x - 3, y + 1)$.
(Lessons 3-1, 3-2, 3-4)

   a. Sketch the image of f under this translation.   **See margin.**

   b. Find the x-intercepts of the image.   **none**

   c. Find the y-intercept of the image.   **10/9**

   d. Find equations for any asymptotes of the image. $x = -3; y = 1$

   e. Describe the symmetries of the graph of the image.
   The graph is symmetric with respect to the line $x = -3$.

In 19 and 20, **a.** identify the function as odd, even, or neither; **b.** if the function is odd or even, prove it. *(Lesson 3-4)*

19. $f(x) = x - 8$ **neither**   **20.** $g(x) = 3x^4 + 37$ **a) even;**
                                                                      b) See margin.

21. A teacher recorded the number of days each child in a class was absent from school for a whole year. The results are summarized in the table below. *(Lessons 1-4, 1-8)*

# days absent	0	1	2	3	8	21	46
# of children	4	7	4	5	2	1	1

   a. How many children are in the class?   **24**

   b. Find the mean, median, and mode for number of days absent.

   c. Which of the measures in part **b** represents the year's absences least well, and why?   **See below.**

   d. Find the standard deviation of the number of days absent.   **9.8 days**
   b) mean ≈ 4.7; median = 2; mode = 1

22. *Skill sequence.* Simplify. *(Previous course.)*

   a. $\dfrac{1}{\frac{1}{5}}$ **5**   **b.** $\dfrac{1}{\frac{1}{n}}$ **n**   **c.** $\dfrac{1}{\frac{1}{n+3}}$ **n + 3**

23. *Skill sequence.* Use the Power of a Product Property to rewrite the expression. *(Previous course.)*

   a. $(5y)^3$ $125y^3$ **b.** $(xy)^3$ x^3y^3 **c.** $(xy)^p$ $x^p y^p$

Exploration

24. Begin with a graph of a relation of your own choice. Replace x by y and replace y by $-x$; and graph the new relation. Repeat this with different relations until you can determine what transformation these substitutions cause.   **This transformation is a clockwise rotation of 90°.**

21 c) Sample: the mean; the mean is affected by the outlier score, 46. Only 4 students were absent more than 4 days.

20.b. We need to show $g(-x) = g(x)$ for all x.
$$g(-x) = (-x)^4 + 37 =$$
$$[(-1)(x)]^4 + 37 =$$
$$(-1)^4(x)^4 + 37 = x^4 + 37$$
$$= g(x)$$

NAME _____

LESSON **MASTER 3-5**
QUESTIONS ON **SPUR** OBJECTIVES

■**PROPERTIES** *Objective C (See pages 197–200 for objectives.)*
1. *Multiple choice.* Which scale change has the effect of shrinking a graph horizontally by a factor of 4 and stretching it vertically by a factor of 3?

   (a) $S(x, y) \rightarrow (\frac{x}{4}, 3y)$

   (b) $S(x, y) \rightarrow (4x, \frac{y}{3})$

   (c) $S(x, y) \rightarrow (4x, 3y)$

   (d) $S(x, y) \rightarrow (\frac{x}{3}, 4y)$ **(a)**

2. Find an equation for the image of the graph of $y = x^3$ under the transformation $(x, y) \rightarrow (-2x, \frac{y}{3})$.   **sample: $y = -\frac{3}{8}x^3$**

3. What transformation maps the graph of $y = \frac{1}{x}$ onto the graph of $y = \frac{1}{5x^2}$?
   sample: $(x, y) \rightarrow (x, \frac{y}{5})$

■**PROPERTIES** *Objective D*
4. *True or false.* Under a size change, a graph is stretched or shrunk by the same factor vertically and horizontally.   **True**

5. Suppose the graph of a function g has x-intercepts $(-2, 0)$ and $(7, 0)$. Give the coordinates of the new x-intercepts if a horizontal stretch of 3 and a vertical stretch of 5 is applied to the graph of g.
   (-6, 0) and (21, 0)

36   *Functions, Statistics, and Trigonometry © Scott, Foresman and Company*

NAME _____
Lesson MASTER 3-5 (page 2)

■**REPRESENTATIONS** *Objective J*
6. a. Sketch graphs of $f(x = \lceil x \rceil$ and $g(x) = 2\lceil 0.2x \rceil$ on the same set of axes.

   b. *True or false.* g is the image of f under $(x, y) \rightarrow (0.2x, \frac{y}{2})$.   **False**

   c. *True or false.* g is the image of f under $(x, y) \rightarrow (5x, 2y)$.   **True**

In 7 and 8, the graph is a scale change image of the graph of $y = x^2$. Write an equation for the graph.

7.   **8.**

sample: $y = 5x^2$ **sample: $y = \frac{-x^2}{2}$**

RESOURCES
- Lesson Master 3-6
- Quiz for Lessons 3-4 through 3-6
- Teaching Aid 30: Table and box plots for **Example 1**
- Computer Master 8

OBJECTIVES

E Use the summation properties to evaluate and rewrite expressions.

F Describe the effects of scale changes on measures of center or spread.

I Use scale changes to analyze data.

TEACHING NOTES

Students may be unfamiliar with the Consumer Price Index, a number which attempts to model the cost of living, that is, the cost of maintaining a household. When the CPI (based on 1982–84) is 159 it means that something that cost $100 in 1982–84 would cost approximately $159 now.

Small Group Work The cost of goods varies from place to place and changes over time, of course, so you might wish to have small groups of students research more recent figures. If they consult a source with information prior to 1988, then the CPI values will be based on 1967 costs and will be quite different.

You might have students verify the percent (about 59%) stated in the first paragraph. The percent of increase from $287.45 to $456.23 can be calculated as follows:

LESSON 3-6

Scale Changes of Data

Scale changes can also be applied to data sets. Scale changes, like translations of data, also affect the measures derived from the data. Consider, for example, the Consumer Price Index (CPI), a government cost-of-living measure. To calculate the CPI, the cost of a specified set of goods is totaled, say $287.45, at a baseline time period (currently 1982–84). This total is then scaled down to 100. If the same set of goods costs $456.23 some years later, the CPI at that later time is calculated from the proportion:

$$\frac{456.23}{287.45} = \frac{CPI}{100} \quad \text{or} \quad CPI \approx 159.$$

Rescaling on a basis of 100 makes it easy to find the percent increase in prices: 59% since the baseline year.

A **scale change** of a set of data $\{x_1, x_2, \ldots, x_n\}$ is a transformation that maps each x_i to ax_i, where a is a nonzero constant. That is, S is a scale change if and only if

$$S : x \rightarrow ax, \text{ or } S(x) = ax.$$

The number a is called the **scale factor** of the scale change. The number ax or the point it represents again is called the **image** of x. In the situation above, the 1982–84 cost x of a set of goods is mapped to the later CPI value via the following scale change:

$$S : x \rightarrow 1.59x, \text{ or } S(x) \approx 1.59x.$$

When a scale change is applied to a data set, the process is usually called **scaling** or **rescaling**.

Example 1

Consider again the data from Lesson 3-3 for the weekly earnings (in dollars) of eleven employees:

$$40, 46, 47, 48, 50, 50, 52, 52, 52, 53, 60.$$

Suppose that each employee works only half a week.

a. Find the scaled weekly earnings (in dollars).
b. Compare the median, mean, range, and standard deviation of the original and scaled data.

Solution

a. Apply the transformation $x \rightarrow \frac{1}{2}x$ to each value. The scaled earnings in dollars are

$$20, 23, 23.5, 24, 25, 25, 26, 26, 26, 26.5, 30.$$

b. In Example 1 of Lesson 3-3, these measures for the original data were calculated: median = 50, range = 20, mean = 50, and standard deviation = 5.

174

For the scaled data, the median is 25, the range is $30 - 20 = 10$, and the mean is $\frac{275}{11} = 25$. The standard deviation, as shown in the following calculations, is 2.5.

Scaled Earnings x_i	Deviation $x_i - \bar{x}$	Square of deviation $(x_i - \bar{x})^2$
20	-5	25
23	-2	4
23.5	-1.5	2.25
24	-1	1
25	0	0
25	0	0
26	1	1
26	1	1
26	1	1
26.5	1.5	2.25
30	5	25
Totals 275.0		62.5

So, $\quad s^2 = \dfrac{\sum\limits_{i=1}^{11} (x_i - \bar{x})^2}{10} = \dfrac{62.5}{10} = 6.25.$

Thus, $\quad s = \sqrt{6.25} = 2.5.$

Notice that the median, range, mean, and standard deviation of the scaled data is $\frac{1}{2}$ of the corresponding measure for the original data.

Box plots of the scaled earnings and original earnings illustrate the effects of scaling on both measures of center and spread.

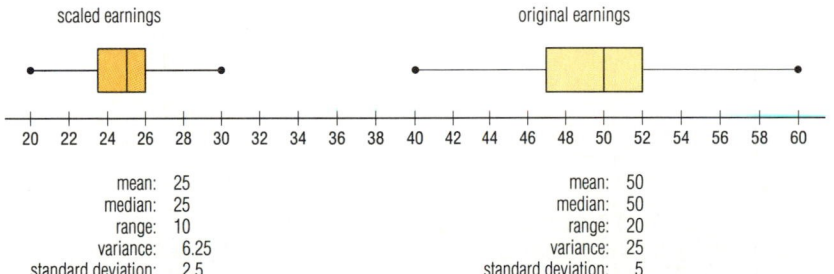

scaled earnings

mean:	25
median:	25
range:	10
variance:	6.25
standard deviation:	2.5

original earnings

mean:	50
median:	50
range:	20
variance:	25
standard deviation:	5

The mode of a data set is the point of maximum frequency. Under a scale change the frequencies do not change, so the mode of the original data is mapped to the mode of the image data. For the data above, the image mode, 26, is half of the original mode, 52.

To describe in general the effect of a scale change on statistical measures for a data set, represent the set as $\{x_1, x_2, x_3, \ldots, x_n\}$. Under a scale change that multiplies each value by a scale factor a, the image data set is
$$\{ax_1, ax_2, ax_3, \ldots, ax_n\}.$$

$\dfrac{456.23 - 287.45}{287.45} \approx 58.72\%$

A teaching aid is provided showing the box plots for **Example 1.** It should be evident that scaling data by a factor of 1/2 affects the display and the mean, median, mode, range, and interquartile range by the scale factor of 1/2.

In the theorem preceding **Example 2,** the standard deviation is multiplied by $|a|$. Since a could be negative, but the standard deviation must be positive, finding the absolute value is necessary.

The presentation in this lesson is similar to that of Lesson 3-3 and uses the same data. Assist students to work through the mathematical symbols in the proof of the theorem on page 176, as students need to be comfortable using sigma notation.

Error Analysis The word "rescaling" is used before **Example 3.** "Scaling" and "rescaling" are synonyms because all data is in some sense scaled to begin with. Make sure students do not misinterpret "rescaling" to mean "scaled twice."

Technology A computer master is provided that offers practice with both translations and scale changes of data.

ADDITIONAL EXAMPLES
1. The teachers in a school have a mean salary of $30,000 with a standard deviation of $4000. If each teacher is given a 5% raise, what will be their new mean salary, and what will be the new standard deviation?
$31,500; $4200

2. To convert from miles to kilometers, you can multiply the number of miles by 1.61. Suppose data is collected about the number of miles that cars can go on a tank of gas. What will be the effect of changing from miles to kilometers on **a.** the median of the data; **b.** the variance of

the data; and **c.** the standard deviation of the data?
a. The median will be multiplied by 1.61.
b. The variance will be multiplied by 1.61².
c. The standard deviation will be multiplied by 1.61.

NOTES ON QUESTIONS
Error Analysis for Question 9b: Some students may calculate the statistical measures for both sets, but only one set of scores needs to be calculated. The measures for the scaled scores can be found using the theorem in this lesson.

Making Connections for Question 11: This question parallels Question 16 in Lesson 3-3. The equation in part **d** can be found from the answer to part **b** by replacing h with $2.54h$:
$w + 18.43 = .342(2.54h)$
$w = .869h - 18.43$
The equation in part **f** can be derived from the equation in part **d** by replacing w with $.454w$:
$.454w = .869h - 18.43$
$w = 1.914h - 40.59$

ADDITIONAL ANSWERS
1. A scale change of a data set is a transformation that maps each element x_i of the data set to ax_i, where a is a non-zero constant.

9.a. S: $x_i \to \frac{3}{2} x_i$

b.

original scores	scaled scores
range = 16	range = 24
mode = 10	mode = 15
mean = 13	mean = $\frac{39}{2}$
median = 11	median = $\frac{33}{2}$

c.

original scores

scaled scores

To find the mean $\bar{x}$ of the image data, you need to evaluate $\dfrac{\displaystyle\sum_{i=1}^{n} (ax_i)}{n}$.

By definition of Σ, that expression represents $\dfrac{ax_1 + ax_2 + ax_3 + \ldots + ax_n}{n}$.

Rewrite the expression as $\dfrac{a(x_1 + x_2 + x_3 + \ldots + x_n)}{n}$, which can be represented as $a\left[\dfrac{\displaystyle\sum_{i=1}^{n} x_i}{n}\right]$. Thus, by definition of the mean, $\overline{x'} = a\bar{x}$.

This proves that, under a scale change, the mean of a set of data is mapped to the mean of the image set of data. A similar result holds for the mode and median.

> **Theorem**
>
> Multiplying each element of a data set by the factor a multiplies each of the mode, mean, and median by the factor a.

What happens to measures of spread, such as the range, variance, and standard deviation, when data are scaled?

Consider the data set $\{x_1, x_2, x_3, \ldots, x_n\}$ and its image under a scale change of magnitude a, $\{ax_1, ax_2, ax_3, \ldots, ax_n\}$. The mean of this data set is $a\bar{x}$ where $\bar{x}$ is the mean of the original data set. So the variance of the image data is given by

$$\frac{\displaystyle\sum_{i=1}^{n} (ax_i - a\bar{x})^2}{n-1} = \frac{\displaystyle\sum_{i=1}^{n} [a(x_i - \bar{x})]^2}{n-1} \qquad \text{Distributive Property}$$

$$= \frac{\displaystyle\sum_{i=1}^{n} [a^2(x_i - \bar{x})^2]}{n-1} \qquad \text{Power of a Product Property}$$

As in the derivation of the mean of the image data above, it can be shown that

$$\sum_{i=1}^{n} [a^2(x_i - \bar{x})^2] = a^2 \sum_{i=1}^{n} (x_i - \bar{x})^2.$$

Hence the variance of the image data is given by

$$\frac{a^2 \displaystyle\sum_{i=1}^{n} (x_i - \bar{x})^2}{n-1} = a^2 \left[\frac{\displaystyle\sum_{i=1}^{n} (x_i - \bar{x})^2}{n-1}\right] = a^2 s^2,$$

where s^2 represents the variance of the original data set. To get the standard deviation, take the square root. Thus the standard deviation of the image data is $|a| \cdot s$, which is $|a|$ times the standard deviation of the original data set.

176

Theorem

If each element of a data set is multiplied by a, then the variance is a^2 times the original variance, the standard deviation is $|a|$ times the original standard deviation, and the range is $|a|$ times the original range.

Example 2 Consider the data on weekly earnings in Example 1. Find the mean, variance, and standard deviation of the earnings if each person works for three weeks.

Solution For the data shown, the mean is 50, the variance is 25, and the standard deviation is 5. If each person works three weeks, this can be considered as scaling by a factor $a = 3$. Thus, the mean and standard deviation will triple and the variance will be multiplied by $3^2 = 9$. So, for the image data, the mean is $3(50) = 150$; the variance is $(3)^2 (25) = 225$; and the standard deviation is $3(5) = 15$.

Check Calculate the statistics directly from the three-week earnings.

A common reason for rescaling data involves changes of units, as the next example shows.

Example 3 In recent years an obstetrician recorded the birth weight (in pounds) of the babies she delivered. She found that the distribution had a mean of 6.75 lb and a standard deviation of 1.14 lb. Find the mean and standard deviation of the distribution of birth weights in kilograms.

Solution There are 0.454 kg per pound. So, the data need to be scaled by 0.454. Thus, the mean is $(0.454) \cdot (6.75) = 3.06$ kg and the standard deviation is $(0.454) \cdot (1.14) = 0.52$ kg .

Questions

Covering the Reading

1. Define: scale change of a set of data.   **See margin.**

In 2 and 3, use the fact that in 1990 the CPI was 131.6.

2. On the average, what was the percentage increase of costs of goods from 1982–84 to 1990?   **31.6%**

3. Approximately what was the 1990 cost of a set of items costing \$577 in 1982–84?   **about \$759**

4. Refer to the original data on earnings of workers in Example 1. Consider their earnings if they are paid every four weeks. For the four-week earnings, find the
    a. range;   **80**   b. mode;   **208**   c. median;   **200**
    d. mean;   **200**   e. variance;   **400**   f. standard deviation.   **20**

LESSON 3-6   Scale Changes of Data   **177**

177

Applying the Mathematics

5. Suppose $Y_1 = 6$, $Y_2 = 0$, $Y_3 = -4$, $Y_4 = 7$, $Y_5 = -6$, $Y_6 = 2$. Evaluate the given expression. **a.** $\sum\limits_{i=1}^{6} 10Y_i$ **50** **b.** $\sum\limits_{i=1}^{6} kY_i$ **5k** **c.** $\sum\limits_{i=1}^{6}\left(\dfrac{Y_i}{m}\right)$ $\dfrac{5}{m}$

6. Let $\bar{x}$ = the mean and s = the standard deviation of scores on a test for a class of n students. Suppose everyone's score is multiplied by r. For the image scores find the
   a. mean; $r\bar{x}$ **b.** variance; r^2s^2 **c.** standard deviation. $|r|s$

7. As part of a study of the yield of orange trees, a biologist weighed a sample of sixty oranges, and found a mean weight of 0.32 pounds and a range of 0.21 pounds. Find the mean and range of the weights in grams.
   mean ≈ 145 grams, range ≈ 95 grams

8. Let x_i be an individual student's score on a test with M points possible. Write a rule which can be used by a teacher to convert each student's score to a percent. $S: x_i \to \dfrac{100}{M}x_i$

9. Refer to the frequency distributions below.

original scores	frequency		scaled scores	frequency
6	2		9	2
10	3		15	3
12	1		18	1
16	2		24	2
22	2		33	2

   a. Write a formula for the transformation used to get the scaled scores.
   b. Find the range, mode, mean, and median for each set of scores.
   c. Make two dot frequency diagrams of the scores. Compare and contrast the distributions.
   See margin.

10. Let M represent the maximum value of a data set and let m represent the minimum value.
    a. Write an expression for the range of the data set. $M - m$
    b. After a scale change by b, what are the maximum and minimum values of the image data set? **See margin.**
    c. Write and simplify an expression for the range of the scaled data.
    $|bM - bm|$

11. Consider the following data which give the height h in cm and weight w in kg of the students in a first-grade class:

h	112	106	114	109	110	122	129	112	117	126	128	101
w	20	18	19	17	20	24	25	21	26	23	25	15

    a. Enter these data into a computer file. (You may have already done so for Question 16 of Lesson 3-3.) Draw a scatter plot with h on the horizontal axis. **See margin.**
    b. Find the line of best fit for predicting w from h. $w = .342h - 18.43$
    c. Use your computer statistics package to convert the height to inches (1 in. = 2.54 cm). Draw a new scatter plot. How is the scatter plot different from that in part **a**? **See margin.**
    d. Without using a computer, write an equation of the line of best fit for part **c**. Check using the computer. $w = .869h - 18.43$
    e. Convert height to inches and kilograms to pounds (1 lb = 0.454 kg). Draw a new scatter plot. How is the scatter plot different from that in part **a**? **See margin.**

178

f. Without using a computer, write the equation of the line of best fit for part **e.** Check using the computer.   *w = 1.914h − 40.59*

12. Generalize the result of Question 11. If bivariate data are rescaled, which (if any) of the following are invariant?
   a. means of the two variables   **b.** variances of the two variables
   c. equation of the line of best fit   **d.** slope of the line of best fit
   e. correlation between the two variables   **e**

13. If each element of a data set is divided by the standard deviation of the data set, prove that the variance and standard deviation of the image set is each equal to 1.   **See margin.**

Review

14. Consider the functions $f(x) = \lceil x \rceil$ and $g(x) = \lceil 3x \rceil$.
   a. Identify a scale change that maps f to g.   **S: (x, y) → (x/3, y)**
   b. Identify a scale change that maps g to f.   **T: (x, y) → (3x, y)**
   c. Sketch graphs of f and g on two different sets of axes.   *(Lesson 3-5)*
   See margin.

In 15–18, match the graph to the function.   *(Lesson 3-2)*

(a)   (b)   (c)   (d)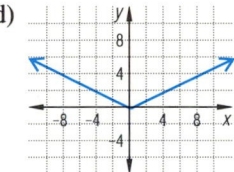

15. $f(x) = |x + 3|$ **(a)**   **16.** $g(x) = |2x|$ **(b)**

17. $h(x) = \frac{1}{2}|x|$ **(d)**   **18.** $j(x) = |x − 3|$ **(c)**

19. Refer to the functions in Questions 15–18. Name all that are even functions.   *(Lesson 3-4)*   **functions g and h**

20. A certain hyperbola is a translation image of the parent $y = \frac{1}{x}$ and has asymptotes $x = 1$ and $y = -7$. Give an equation of the hyperbola.
   (Lesson 3-2) $y = \frac{1}{x-1} - 7$

In 21 and 22, are the two triangles similar? Justify your answers.
(Previous course)

21.

   See margin.

22.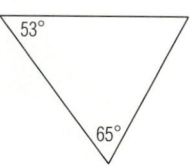

   See margin.

23. *Skill sequence.* Let $f(x) = x^2 + 7x$. Evaluate the expression.   *(Lesson 2-1)*
   a. $f(3)$ **30**   **b.** $f(a)$ **a² + 7a**   **c.** $f(a − 4)$ **a² − a − 12**

Exploration

24. What goods and services are used to calculate the Consumer Price Index?   **Goods from each of the following categories: food, clothing, housing, fuels, transportation, medical care**

LESSON 3-6   Scale Changes of Data   **179**

179

PROJECTS
The projects for Chapter 3 are described on page 194. Project 2 is related to the content of this lesson.

EVALUATION
A quiz covering Lessons 3-4 through 3-6 is provided in the Teacher's Resource File.

Alternate Assessment
Students can write a paragraph explaining how measures of center (mean, median, and mode) and measures of spread (range and standard deviation) are affected by translating a set of data and by scaling a set of data. Optionally, they can write additional paragraphs explaining why these measures are affected by these kinds of transformations.

21. Yes; each side of the second triangle has a length 2.5 times the corresponding side of the first triangle.

22. Yes; corresponding angles have the same measure.

NAME _____

OBJECTIVES

A Find composites of functions.
H Identify properties of composites.

TEACHING NOTES

There is no standard way to read $f \circ g$. You can read $f \circ g$ as "g followed by f" or "the composite of f with g." In some contexts one of these readings will be more natural than the other; for instance, in the Smith-Lopez family example, the language of composites is better than the language of "followed by" because in $f \circ m$, f actually precedes m in time.

Another way is to read both $(f \circ g)(x)$ and $f(g(x))$ as "f of g of x," and think of them as applying g, then f. Students who are uncomfortable with what seems to be a reversal of order could be reminded that we always work with nested parentheses from the inside out.

The nonnumerical example of composition using the Smith and Lopez families emphasizes the idea of a function of a function. You might wish to avoid extending the example to students' families, as in real life the delineation of people and their roles may not be as clear, or the same for everyone. This discussion could be uncomfortable for some students.

LESSON

3-7

Composition of Functions

Many people are interested in family trees, which show their ancestors and relatives. Here is part of Deena Smith's family tree. (For simplicity, "family" in this lesson refers to biological parents and children.) It shows that Aileen Long Smith and Tony Smith are Deena's parents.

Mary Long   **Des Long**

Merle Smith   **Cecil Smith**

Aileen Long

Tony Smith

Deena Smith

Functions can be used to describe the relationships between pairs of members of this tree. For example, suppose m is the "mother" function defined by

$$m(x) \text{ is the mother of } x,$$

and f is the "father" function defined by

$$f(x) \text{ is the father of } x.$$

Then m(Deena) = Aileen, m(Tony) = Merle, f(Deena) = Tony, and so on.

In studying family trees, as in studies of other functions, functions can be combined, so that the value of one function becomes the argument of another. For example, if the mother function is applied, followed by the father function, we have m(Deena) = Aileen,

and f(Aileen) = Des.

In words, the father of the mother of Deena is Des. This combination can be written

$$f(m(\text{Deena})) = f(\text{Aileen})$$
$$= \text{Des.}$$

We say that functions m and f have been *composed* to make a new function, which can be called the "maternal grandfather" function.

> **Definition**
>
> Suppose f and g are functions. The **composite** of f with g, written $f \circ g$, is the function defined by
>
> $$(f \circ g)(x) = f(g(x)).$$
>
> The domain of $f \circ g$ is the set of values of x in the domain of g for which $g(x)$ is in the domain of f.

Using the symbol for composition, $(f \circ m)(\text{Deena}) = f(m(\text{Deena}))$
$$= f(\text{Aileen})$$
$$= \text{Des.}$$

180

Example 1 Consider Deena Smith's family tree. What biological relation does the composite function $m \circ f$ represent?

Solution Use the definition. Note the order of the two functions. The composite $m \circ f$ is the function defined as $m(f(x))$, the mother of the father of x. This is the paternal grandmother function.

Check Consider a specific case.
$$(m \circ f)(\text{Deena}) = m(f(\text{Deena}))$$
$$= m(\text{Tony})$$
$$= \text{Merle}$$
Merle is Deena's father's mother.

Notice that each of $f \circ m$ and $m \circ f$ is a function, since there is a single output associated with each input.

Also notice that the two functions $f \circ m$ and $m \circ f$ are *different* functions. The range of $f \circ m$ contains only men, while the range of $m \circ f$ contains only women. This illustrates that, in general, $f \circ m \neq m \circ f$, or that *composition of functions is not commutative*. The next two examples also illustrate this important fact.

Example 2 For $f(x) = x^2 + x$ and $g(x) = x - 5$, evaluate
a. $f(g(7))$ **b.** $g(f(7))$.

Solution

a. To evaluate $f(g(7))$ first evaluate $g(7)$. $g(7) = 7 - 5 = 2$
Then use this output as the input to f. So $f(g(7)) = f(2)$
$$= 2^2 + 2 = 6.$$

b. $g(f(7)) = g(7^2 + 7)$
$$= g(56) = 56 - 5 = 51$$

To graph composites of functions, it is tedious to evaluate each point, as in Example 2. An alternative is to find a single formula for the composite.

Example 3 Let $f(x) = x^2 + x$ and $g(x) = x - 5$.
a. Derive a formula for each of $f(g(x))$ and $g(f(x))$.
b. Verify that $f \circ g \neq g \circ f$ by graphing.

Solution

a. The function g maps x to $x - 5$. So substitute $x - 5$ for $g(x)$.
$$f(g(x)) = f(x - 5)$$
Now use $x - 5$ as the input for function f.
$$f(g(x)) = f(x - 5) = (x - 5)^2 + (x - 5)$$
$$= x^2 - 10x + 25 + (x - 5)$$
So $f(g(x)) = x^2 - 9x + 20$.
Similarly, $g(f(x)) = g(x^2 + x) = (x^2 + x) - 5$.
So $g(f(x)) = x^2 + x - 5$.

LESSON 3-7 Composition of Functions   **181**

You might note that we could not define a function b for "brother of" or s for "sister of" because a function must have a unique value. This could be solved by defining "oldest brother of" or "youngest sister of," but that does not have the same simplicity as mother or father.

Examples 2 and 3 demonstrate the calculation of values and formulas for composite functions. Two other important concepts are that the composition of functions is not commutative as shown in **Examples 1 and 3** and that the domain of the composite function is not necessarily the same as the domain of either of its component functions as shown in **Example 4.**

Technology A computer master is provided should you wish to spend more time on this topic.

ADDITIONAL EXAMPLES
1. Refer to the Smith-Lopez family tree. Evaluate $f(f(\text{Deena Smith}))$ and describe the function $f \circ f$.
Cecil Smith; paternal grandfather of

2. Let $f(x) = x^2$ and $g(x) = \frac{1}{3x + 1}$. Evaluate.
a. $f(g(4))$ **b.** $g(f(4))$
c. $(f \circ g)(4)$
$\frac{1}{169}; \frac{1}{49}; \frac{1}{169}$

3. For the functions of Additional Example 2, **a.** find $f(g(x))$ and $g(f(x))$ **b.** verify that $f \circ g \neq g \circ f$ by graphing.
a. $f(g(x)) = \frac{1}{(3x + 1)^2}$;
$g(f(x)) = \frac{1}{3x^2 + 1}$.

b. The graphs have only two points in common, when $x = 0$ and when $x = -1$.

4. Give the domains of the functions *f*, *g*, and *f ∘ g* in Additional Example 2.

domain of *f* = set of all real numbers; domain of *g* = set of all real numbers except -$\frac{1}{3}$; domain of *f ∘ g* = set of all real numbers except -$\frac{1}{3}$

NOTES ON QUESTIONS
Error Analysis for Question 3: Ask if $f(g(x)) = g(f(x))$. Some may think that $f(g(x)) = g(f(x))$ because it is true sometimes. The problem is one of quantification: while it may be true that *there exists an x* for which $f(g(x)) = g(f(x))$, it is not true here that *for all x, $f(g(x)) = g(f(x))$*.

Questions 4 and 5: It sometimes helps to have descriptive names for functions. Here *m* is the "adding 8" function and *n* is the "dividing into 4" function. The noncommutativity of the application of these functions is sometimes more evident when these names are used.

Questions 6 and 7: The note for **Question 3** applies here as well. **Question 6** is false, though in **Question 7** it appears that composition of functions is commutative. Reiterate that it is only one specific numerical example, which is not sufficient to infer *true* for all *t*.

Question 11: Our experience is that even most adults do not realize that either the discount or the tax could be taken first. An interesting extension of this question is to consider whether the order of computing the discount and tax makes a difference to the seller, who has to pay the tax, or to the government. Which order would each prefer? Why? (seller: discount first because the seller would receive 0.9*x* and the government 0.045*x*; government: tax first because the government would receive 0.05*x* and the seller 0.895*x*)

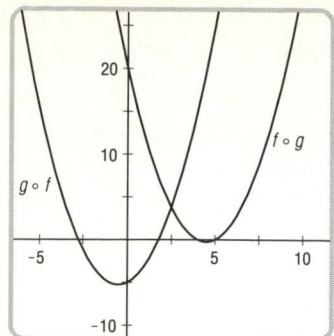

b. The two functions, *f ∘ g* and *g ∘ f*, are graphed at the left. Clearly *f ∘ g ≠ g ∘ f*, as the graph of each is a different parabola.

Check Evaluate each function at *x* = 7, and compare with the results of Example 2. $f(g(7)) = 7^2 - 9 \cdot 7 + 20 = 6$; $g(f(7)) = 7^2 + 7 - 5 = 51$. The results agree with those in Example 2; $f(g(7)) \neq g(f(7))$.

Notice in Example 3 that although *f ∘ g* and *g ∘ f* are *not* the same function, the graphs intersect so there is at least one value for *x* at which the functions have the same value. This is the *x*-value at that point of intersection.

The domain of a composite function is not necessarily the domain of either of its component functions, as Example 4 illustrates.

Example 4   Let $F(t) = 13 - t$ and $G(t) = \frac{2}{t+1}$. Find the domain of *G ∘ F*.

Solution 1   The domain of *F* is the reals. The domain of *G ∘ F* is the set of all *t* in the domain of *F* for which *F*(*t*) is in the domain of *G*. Because *G* is not defined when the denominator *t* + 1 equals 0, the domain of *G* is the set of reals except -1. So the domain of *G ∘ F* excludes the point(s) for which *F*(*t*) = -1. This is where $13 - t = -1$ or $t = 14$. So the domain of *G ∘ F* is the reals except 14 and -1.

Solution 1   Find a formula for *G ∘ F* and analyze its domain.
$$(G \circ F)(t) = G(F(t)) = G(13 - t)$$
$$= \frac{2}{13 - t + 1}$$
$$= \frac{2}{14 - t}$$
This function is defined for all reals except when the denominator is 0. The domain of *G ∘ F* is the reals except *t* = 14. However, *F* is not defined when *t* = -1. So neither -1 nor 14 is the domain of *G ∘ F*.

Check   Draw a graph of $y = (G \circ F)(x) = \frac{2}{14 - x}$. You will see that *x* = 14 is a vertical asymptote of *G ∘ F*.

Questions

In 1 and 2, consider Deena Smith's family tree.
1. What biological relation does the composite *m ∘ m* represent? **maternal grandmother**
2. Suppose Deena marries Harper Scott and they have a son, Joe.
   a. Evaluate $(f \circ m)$(Joe). **Tony Smith**
   b. Explain why $(f \circ m)$(Joe) $\neq$ $(m \circ f)$(Joe). **See margin.**

3. Refer to Examples 2 and 3. Verify the following.
   a. $f(g(0)) \neq g(f(0))$ b. $f(g(2.5)) = g(f(2.5))$ **See margin.**

182

In 4 and 5, consider $m(x) = x + 8$ and $n(x) = \frac{4}{x}$.

4. **a.** Evaluate $m(n(7))$.   **60/7**   **b.** Find a formula for $(m \circ n)(x)$.
   c. State the domain of $m \circ n$.   **b) $(m \circ n)(x) = 4/x + 8$**
   the set of all nonzero real numbers
5. **a.** Evaluate $n(m(7))$.   **4/15**   **b.** Find a formula for $(n \circ m)(x)$.
   c. State the domain of $n \circ m$.   **the set of all real numbers except -8.**
   b) $(n \circ m)(x) = \dfrac{4}{x+8}$
6. *True* or *false*. Composition of functions is commutative.   **False**

In 7 and 8, consider $g(t) = t^2$ and $h(t) = 2t + 1$.

7. Evaluate: **a.** $g(h(-2))$ **9**   **b.** $h(g(-2))$.   **9**

8. **a.** Find a formula for each of $(g \circ h)(t)$ and $(h \circ g)(t)$.
   b. Show that $g \circ h \neq h \circ g$ by graphing each composite function on the same set of axes.
   See margin.

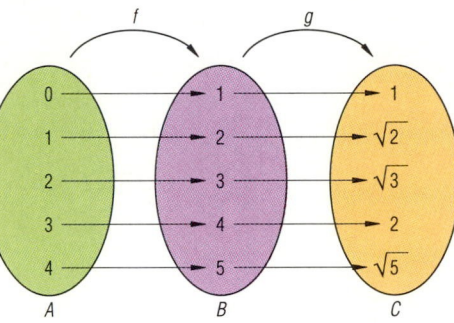

Applying the Mathematics

9. Consider the sets A, B, and C below.
   a. Evaluate $g(f(3))$.   **2**
   b. The composite $g \circ f$ maps 4 to what number?   **$\sqrt{5}$**
   c. If the rules defining the mappings are $f(x) = x + 1$ and $g(x) = \sqrt{x}$, write a formula for $g(f(x))$.
   d. If the domain of f is extended to the set of all reals, what is the domain of $g \circ f$?   **$\{x : x \geq -1\}$**
   c) $g(f(x)) = \sqrt{x + 1}$

10. Consider a discount function $D(x) = 0.9x$ and a rebate function $R(x) = x - 100$, where x is the price of electronic equipment.
    a. Explain why it is appropriate to call D a discount function and R a rebate function.   **See margin.**
    b. A stereo system normally sells for \$1200. Evaluate $D(R(1200))$ and $R(D(1200))$.   **$D(R(1200)) = 990$; $R(D(1200)) = 980$**
    c. Refer to your answer to **b**. If you are buying a stereo for \$1200, is it better to apply the discount after the rebate or before the rebate?
    d. Prove that, in general, $D \circ R \neq R \circ D$.   **c) before; d) $(D \circ R)(x) = D(x - 100) = 0.9(x - 100) = 0.9x - 90$; $(R \circ D)(x) = R(0.9x) = 0.9x - 100$**

11. Consider a 10% discount function $D(x) = 0.9x$ and a 5% total-with-tax function $T(x) = 1.05x$.
    a. If you buy an item with a list price of x dollars, what will it cost you after this discount and tax?   **.945x dollars**
    b. If you have the choice, which is better to take first, the discount or the tax?   **It doesn't matter; both have the same effect.**

12. Consider the functions $k(x) = 2x - 7$, $m(x) = 5x + 12$, and $n(x) = 8 - x$.
    a. Find $k \circ m$ and $m \circ k$.   **b.** Find $m \circ n$ and $n \circ m$.
    c. Prove that the composite of any two linear functions is a linear function. (Hint: let $f(x) = ax + b$ and $g(x) = cx + d$.)
    See margin.

ADDITIONAL ANSWERS

2.b. $(f \circ m)$(Joe) = f(Joe's mother) = Megan's father while $(m \circ f)$(Joe) = m(Joe's father) = Harper's mother

3.a. f(g(0)) = f(0 − 5) = f(-5) = (-5)² + (-5) = 25 − 5 = 20; g(f(0)) = g((0)² + 0) = g(0) = 0 − 5 = -5

b. f(g(2.5)) = f(2.5 − 5) = f(-2.5) = (-2.5)² + (-2.5) = 6.25 − 2.5 = 3.75; g(f(2.5)) = g((2.5)² + 2.5) = g(8.75) = 8.75 − 5 = 3.75

8.a.,b., 12.c. See Additional Answers in the back of this book.

10.a. D reduces the price you must pay by an amount proportional to the regular price. R reduces the price you must pay by a fixed amount, regardless of the regular price.

12.a. $(k \circ m)(x) = 10x + 17$; $(m \circ k)(x) = 10x − 23$

b. $(m \circ n)(x) = 52 − 5x$; $(n \circ m)(x) = -4 − 5x$

NAME _____

LESSON **MASTER   3-7**
QUESTIONS ON **SPUR** OBJECTIVES

■**SKILLS**   *Objective A (See pages 197–200 for objectives.)*
In 1 and 2, let $f(x) = \frac{4}{x}$, and $g(x) = x^2 - 2$.

1. Evaluate
   a. $f(g(2))$.   **2**   **b.** $g(f(-1))$.   **14**   **c.** $(f \circ g)(0)$.   **-2**

2. Find a formula for each function.
   a. $f(g(x))$ $\dfrac{4}{x^2 - 2}$
   b. $g(f(x))$ $\dfrac{16}{x^2} - 2 = \dfrac{16 - 2x^2}{x^2}$
   c. $(g \circ g)(x)$ $(x^2 - 2)^2 - 2 = x^4 - 4x^2 + 2$

3. Consider the function m mapping A to B and n mapping B to C. Evaluate

   a. $m(n(0))$.   **2**   **b.** $n(m(0))$.   **2**   **c.** $(m \circ n)(3)$.   **0**

■**PROPERTIES**   *Objective H*
4. *True or false*. $f \circ g$ and $g \circ f$ are the same function if and only if f and g are the same function.   **False**

5. If $p(x) = x^2 - 3$ and $q(x) = \sqrt{x}$, determine the domain of each function.
   a. $p(q(x))$ **all nonnegative reals**
   b. $q(p(x))$ **$\{x : |x| \geq \sqrt{3}\}$**

Functions, Statistics, and Trigonometry © Scott, Foresman and Company

39

184

Making Connections for Questions 14 and 15:
You might ask what the affect will be if 8 is subtracted from each data value, and then the values are tripled; and what it will be if the values are tripled and then 8 is subtracted from each value. This combines the ideas of transformation and composition, as will be done in Chapter 10.

Making Connections for Question 23: This question reviews a skill needed for the determination of inverse functions in the next lesson.

FOLLOW-UP

MORE PRACTICE
For more questions on SPUR Objectives, use *Lesson Master 3-7,* shown on page 183.

EXTENSION
Choose two functions *f* and *g*. Does the composite value ever equal the product of the values? That is, does $f(g(x)) = f(x) \cdot g(x)$? Does the composite value ever equal the sum of the values? That is, does $f \circ g(a + b) = f(a) + g(b)$? The answers will depend on which functions you choose, but these values are seldom the same.

ADDITIONAL ANSWERS
17. Neither; if *f* were odd, then $f(-x) = -f(x)$ for all *x*. If *f* were even, then $f(-x) = f(x)$ for all *x*. However, $f(-9) = 0$. Since $f(9) \neq 0$, *f* is neither even nor odd.

23.a. $y = \dfrac{12}{x}$ for $x \neq 0$

b. $y = \dfrac{10}{x} + 1$ for $x \neq 0$

c. $y = \dfrac{4}{x}$ for $x \neq 0$

d. $y = \dfrac{1}{x + 2}$ for $x \neq -2$

24.a.,b. See Additional Answers in the back of this book.

---

13. Is the composite of two quadratic functions a quadratic function? Justify your answer. **No, for example, let $f(x) = 2x^2$ and $g(x) = x^2$. Then $f(g(x)) = f(x^2) = 2x^4$, which is not a quadratic function.**

In 14 and 15, suppose a scientist has collected some data with mean 3.92 and variance 1.69. What will be the effect of each transformation on the mean and standard deviation of those data? *(Lessons 3-3, 3-6)*

14. Subtract 8 from each data value. **The mean will go down by 8. The standard deviation will not be affected.**

15. Triple each data value.
The mean and the standard deviation will be tripled.

In 16 and 17, suppose $f(x) = x^2 + 9x$. *(Lessons 2-6, 3-4)*

16. Solve $f(x) = 0$. **$x = 0$ or $x = -9$**

17. Is *f* odd, even, or neither? Justify your answer. **See margin.**

In 18 and 19, find an equation for the image of $f(x) = x^2$ under the transformation. *(Lessons 3-2, 3-5)*

18. $(x, y) \rightarrow \left(\dfrac{x}{3}, -y\right)$ **$y = -9x^2$** 19. $(x, y) \rightarrow (x + r, y - s)$
$y = (x - r)^2 - s$

20. A grocery shop is giving away one free piece of an oven-proof dinnerware set for every \$8 worth of grocery purchase. Give a formula for the number *n* of free pieces earned by spending *d* dollars. *(Lesson 2-4)* **$n = \left\lfloor \dfrac{d}{8} \right\rfloor$**

21. Tell whether the graph represents a function. *(Lesson 2-1)*

a.

b.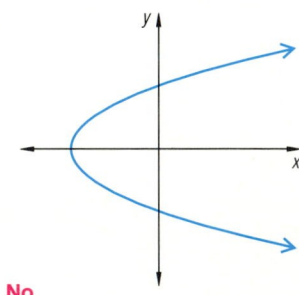

Yes   **No**

22. *Skill sequence.* Simplify. *(Previous course)*
a. $2^8 \cdot 2^5$ **2^{13}** b. $(2^3)^5$ **2^{15}** c. $(b^x)^y$ **b^{xy}**

23. *Skill sequence.* Solve for *y*. *(Previous course)*
a. $xy = 12$ b. $xy = 10 + x$ c. $x = \dfrac{4}{y}$ d. $x + 2 = \dfrac{1}{y}$
See margin.

24. a. Draw $\triangle ABC$ with $A = (2, 7)$, $B = (2, 4)$, and $C = (-2, 4)$.
b. Let *r* be a transformation that maps each point (x, y) to (y, x). Draw $\triangle A'B'C' = r(\triangle ABC)$.
c. What type of transformation is *r*? *(Previous course)*
a,b) See margin. c) reflection

25. Find a function *f* such that $f(f(x)) = x$ for all *x* in its domain.
samples: $f(x) = x$, $f(x) = \dfrac{1}{x}$

184

Inverse Functions

Recall that by definition a function is a set of ordered pairs in which each first element is paired with exactly one second element. If you switch coordinates in each pair, the resulting set of ordered pairs is called the **inverse of the function**.

Example 1 Let $F = \{(-3, 9), (-2, 4), (-1, 1), (0, 0), (1, 1), (2, 4), (3, 9)\}$. Find the inverse of F.

Solution Let G be the inverse of F. The ordered pairs in G are found by switching the x- and y-coordinates of each pair in F.
$G = \{(9, -3), (4, -2), (1, -1), (0, 0), (1, 1), (4, 2), (9, 3)\}$

Note that G contains $(9, -3)$ and $(9, 3)$, so G is not a function.

If the original function is described by an equation, then switching the variables in the equation gives an equation for the inverse.

Example 2 **a.** Give an equation for the inverse of the function $y = x^2$.
b. Sketch a graph of $y = x^2$ and its inverse on the same set of axes.
c. Is the inverse a function?

Solution
a. To form the inverse, switch x and y. The inverse of $y = x^2$ is $x = y^2$.
b.

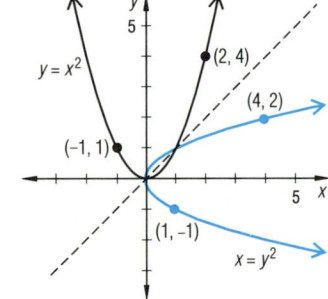

c. The graph of $x = y^2$ fails the vertical line test, so the inverse is not a function.

Note that in Example 2, the graphs of $y = x^2$ and $x = y^2$ are reflection images of each other over the line $y = x$. This is due to the fact that the mapping $(x, y) \rightarrow (y, x)$ is the reflection over the line $y = x$. Thus, the graphs of any function and its inverse are reflection images of each other over the line $y = x$.

RESOURCES
■ Lesson Master 3-8
■ Teaching Aid 31: Graphs for **Examples 2 and 3**

OBJECTIVES

B Find inverses of functions.
H Identify properties of inverses.
L Graph inverses of functions.

TEACHING NOTES

Students should be able to think of inverse functions in two ways. One is that *they undo each other.* For example, the halving function $f(x) = x/2$ undoes the doubling function $g(x) = 2x$. This way leads to the notion that the composite of two inverse functions is the identity function.

The other way of thinking of inverse functions is that the elements of *their ordered pairs* (their domain and range) *are switched.* This way is most useful in comparing the graphs of a function and its inverse, and in finding an equation for the inverse of a function.

Examples 2 and 3 provide opportunities to point out that if the graphs of a function and its inverse intersect then the intersections must lie on the line $y = x$. A teaching aid with these graphs is provided.

Making Connections In this book, the notation f^{-1} is reserved only for inverses which are functions. This is not the case in some other books, and if you wish to use the notation more generally, be certain to discuss it with your students.

186

· · · · · · ■ ■

Example 3 Consider the function $y = \frac{1}{x} + 2$.

a. Give an equation for its inverse.

b. Graph $y = \frac{1}{x} + 2$ and its inverse on the same set of axes.

c. Is the inverse a function?

Solution

a. To form the inverse of $y = \frac{1}{x} + 2$, switch x and y. So an equation of the inverse is $x = \frac{1}{y} + 2$.

Solve for y. $x - 2 = \frac{1}{y}$

$$y = \frac{1}{x - 2}$$

b. The graphs drawn by an automatic grapher are shown at the right. To check, note that each branch of the inverse is the image of one of its branches under a reflection over $y = x$.

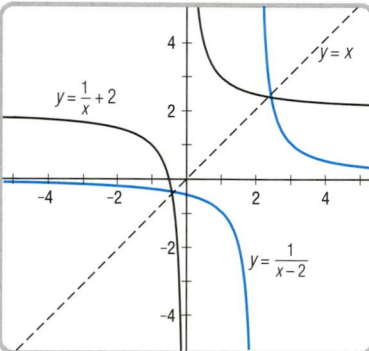

c. The graph of $y = \frac{1}{x - 2}$ passes the vertical line test, and so it is a function.

Note from the preceding examples that sometimes the inverse of a function is a function, and sometimes it is not. When the inverse of a function f *is* a function, it is denoted by the symbol f^{-1}, read "f inverse." So the inverse of $f(x) = \frac{1}{x} + 2$ in Example 3 can be written $f^{-1}(x) = \frac{1}{x - 2}$. Note also that f^{-1} does not denote the reciprocal of f.

Because the inverse of a function is found by switching the x- and y-coordinates, the domain and range of the inverse are found by switching the domain and range of the original function. Thus, the domain of f^{-1} is the range of f, and the range of f^{-1} is the domain of f. Hence, if f is a function whose inverse is also a function, then $f(f^{-1}(x))$ and $f^{-1}(f(x))$ can be calculated for all x in the domains of both functions.

Note that for the function in Example 3,

$$f(f^{-1}(3)) = f\left(\frac{1}{3 - 2}\right) = f(1) = \frac{1}{1} + 2 = 3.$$

Also, $f^{-1}(f(3)) = f^{-1}\left(\frac{1}{3} + 2\right) = f^{-1}\left(\frac{7}{3}\right) = \frac{1}{\frac{7}{3} - 2} = \frac{1}{\frac{1}{3}} = 3.$

Example 4 illustrates that $f \circ f^{-1}$ and $f^{-1} \circ f$ are always equal.

■ ■ ■ ■ ■ ■ ■

Example 4   Verify that if $f(x) = \frac{1}{x} + 2$ and $f^{-1}(x) = \frac{1}{x-2}$, then $f(f^{-1}(x)) = x$ for all $x \neq 2$ and $f^{-1}(f(x)) = x$ for all $x \neq 0$.

Solution

First find $f(f^{-1}(x))$:

$$f(f^{-1}(x)) = f\left(\frac{1}{x-2}\right) \text{ for } x \neq 2$$

$$= \frac{1}{\frac{1}{x-2}} + 2$$

$$= x - 2 + 2$$

$$= x.$$

Now find $f^{-1}(f(x))$:

$$f^{-1}(f(x)) = f^{-1}\left(\frac{1}{x} + 2\right) \text{ for } x \neq 0$$

$$= \frac{1}{\left(\frac{1}{x} + 2\right) - 2}$$

$$= \frac{1}{\frac{1}{x}}$$

$$= x.$$

This characteristic property of the composition of inverses is an instance of the following theorem.

Inverse Function Theorem

Given any two functions f and g, then f and g are inverse functions if and only if $f(g(x)) = x$ for all x in the domain of g, and $g(f(x)) = x$ for all x in the domain of f.

When f and g are inverse functions, $f = g^{-1}$ and $g = f^{-1}$. The theorem states that $f = g^{-1}$ if and only if $f \circ g$ and $g \circ f$ are the function $I(x) = x$, a function which is called an **identity function**. This theorem enables you to test if two functions are inverse functions even if you have not derived one from the other.

■ ■ ■ ■ ■ ■ ■ ■

Example 5   **a.** Use the Inverse Function Theorem to show that

$f(x) = 3x + 4$ and $g(x) = \frac{1}{3}x - 4$ are *not* inverses.

b. Verify your result in **a** on an automatic grapher.

Solution

a. Show that at least one of $f(g(x)) = x$ or $g(f(x)) = x$ fails to hold.

First, $f(g(x)) = f\left(\frac{1}{3}x - 4\right)$

$$= 3\left(\frac{1}{3}x - 4\right) + 4$$

$$= x - 12 + 4$$

$$= x - 8.$$

No further work is necessary.

b. The graphs of f and g are shown at the right. Note that neither is the reflection image of the other over $y = x$.

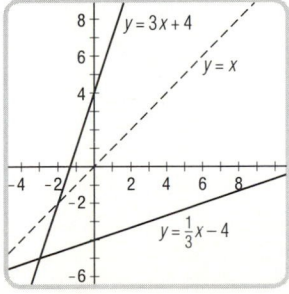

In **2** and **3,** give an equation for the inverse of the function and tell whether the inverse is a function.

2. $f(x) = 6x + 5$

$f^{-1}(x) = \frac{1}{6}x - \frac{5}{6}$; **yes**

3. $y = \dfrac{4}{3x - 1}$

$y = \dfrac{4 + x}{3x}$; **yes**

4. Verify that the functions of **Question 2** are inverses by taking the composites of the functions in each direction.

$f(f^{-1}(x)) = f\left(\frac{1}{6}x - \frac{5}{6}\right) =$

$6\left(\frac{1}{6}x - \frac{5}{6}\right) + 5 = x;$

$f^{-1}(f(x)) = f^{-1}(6x + 5) =$

$\frac{1}{6}(6x + 5) - \frac{5}{6} = x$

5. Explain why the functions $f(x) = x^2$ and $g(x) = x^{-2}$ are not inverses.

One explanation is that $f(g(x)) = f(x^{-2}) = (x^{-2})^2$ $= x^{-4}$, and since $x^{-4} \neq x$ for all x, f and g are not inverses. Alternatively, graph f and g. The graph of f is a parabola; the graph of g is an inverse square curve; clearly they are not reflection images of one another over the line $y = x$.

NOTES ON QUESTIONS
Error Analysis for Question 4: This is an opportunity for students to check how their automatic graphers interpret the expression $-x^2$. If it happens that the graph of $f(x) = -x^2$ opens up, then they will need to enter $f(x) = -(x^2)$ to get the correct graph for the function.

Questions 13 and 14:
These questions are paradigms for applications of inverse functions; if a function translates from one system to another, then its inverse translates back.

Questions

Covering the Reading

1. Define: inverse of a function.   **See margin.**

2. Give an example of a function whose inverse is *not* a function.
   sample: $f(x) = x^2$

3. Let $f = \{(5, 4), (6, 6), (7, 8), (8, 10)\}$.
   a. Find g, the inverse of f.   **{(4, 5), (6, 6), (8, 7), (10, 8)}**
   b. Graph f and g on the same set of axes.   **See margin.**
   c. What transformation maps f to g?   **reflection over y = x**
   d. What transformation maps g to f?   **reflection over y = x**

4. a. Let $f(x) = -x^2$. Graph f and its inverse on the same set of axes.
   b. Is the inverse a function?   **No**
   a) See margin.

In 5–8, **a.** give an equation for the inverse of the function. **b.** Is the inverse a function?   **See margin.**

5. $f(x) = 3x + 6$ **6.** $f(x) = x$ **7.** $y = x^3$ **8.** $y = \sqrt{x}$

9. If $h(-1) = 4$, what is $h^{-1}(4)$?   **-1**

In 10–12, **a.** determine if the functions are inverses by calculating $g \circ f$ and $f \circ g$; **b.** check your conclusion in part **a** on an automatic grapher.

10. $f(x) = x + 7$;
   $g(x) = x - 7$
   See margin.

11. $f(x) = 2x + 1$;
   $g(x) = \frac{1}{2}x - 1$

12. $f(x) = \frac{2}{x} - 5$;
   $g(x) = \frac{2}{x + 5}$

Applying the Mathematics

13. At one point in the summer of 1990, one U.S. dollar was worth 2670 Mexican pesos. Let $M(x)$ be the amount in pesos of an item priced at x U.S. dollars and $U(x)$ be the amount in dollars of an item priced at x Mexican pesos. [Note: M gives Mexican values; U gives U.S. values.]
   a. Write expressions for $M(x)$ and $U(x)$.   **M(x) = 2670x; U(x) = x/2670**
   b. What was the U.S. price of an item going for 20,000 pesos?   **$7.49**
   c. Are M and U inverses of each other?   **Yes**

14. A rule for converting from degrees Fahrenheit to degrees Celsius is "subtract 32, then multiply by $\frac{5}{9}$."
   a. Determine the rule for converting Celsius to Fahrenheit.   **$F = \frac{9}{5}C + 32$**
   b. Do the two rules represent inverse functions?   **Yes**

In 15 and 16, for each graph part **a,** sketch the graph of the inverse of the function; **b.** state whether or not the inverse is a function.

15.

16.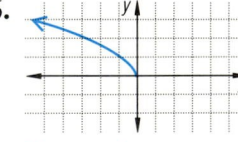

See margin.   **See margin.**

17. For $h(p) = \frac{1}{p}$, show that $h(p) = h^{-1}(p)$ for all $p \neq 0$.   **See margin.**

"Let's go over to Celsius' place. I hear it's only 36° over there."

188

18. Let $f(x) = |x|$. **a.** Determine the inverse of f. **b.** Is the relation in part **a** a function? Justify your answer.
a) $|y| = x$; b) No. For example, the points (1, 1) and (1, -1) are on the graph.

19. **a.** Let $f(x) = mx + b$, where $m \neq 0$. Find a formula for $f^{-1}(x)$.
b. *True* or *false*. The inverse of every linear function is a linear function. If *true*, explain why; if *false*, give a counterexample.
a) $f^{-1}(x) = \frac{x-b}{m}$; b) See margin.

Review

20. A clockmaker makes pendulums for grandfather clocks from a mass of metal. She finds that the length L of the pendulum (in centimeters) is a linear function of its mass m of metal (in grams):

$L = f(m) = \frac{m}{10} - 30$. The time t (in seconds) for the pendulum to swing across and back once is a function of L, $t = g(L) = 2\pi\sqrt{\frac{L}{980}}$.

a. Evaluate $g(f(950))$.   **about 1.6**
b. For a pendulum made from 950 grams of metal, give the time it will take to swing across and back once.   **about 1.6 seconds**
c. Find $(g \circ f)(m)$.   **See margin.**
d. Give the domain of $g \circ f$.   **all real numbers m such that $m \geq 300$**
e. What relationship is expressed by $g \circ f$? *(Lesson 3-7)*   **See margin.**

21. Consider the transformation $T(x, y) = (3x, y - 2)$.   **See margin.**
a. Describe the effects of T on the graph of a function.
b. Graph the image of $y = \frac{1}{3}x + 2$ under T. *(Lesson 3-5)*

22. A Mexican student who wants to attend a university in the USA took the itemized costs (tuition, room, board) from the university's catalog and multiplied each by 2670, the cost of one U.S. dollar in pesos. What kind of a transformation did the student apply to the data? *(Lesson 3-6)*   **a scale change**

23. **a.** Prove that $p(t) = 5 - |t|$ is an even function.
b. Verify the result using a function grapher. *(Lessons 3-1, 3-4)*
a) $p(-t) = 5 - |-t| = 5 - |t| = p(t)$; b) See margin.

24. **a.** Construct a scatter plot to display the data in the following table.

Number of black elected officials (total of local, state, and federal levels)								
Year	1972	1974	1976	1978	1980	1982	1984	1986	1988
Total	2300	3000	4000	4500	5000	5200	5900	6400	6800

Source: *Statistical Abstract of the United States, 1989*

b. Find the line of best fit for these data.   **$t = 272.5y - 534761$**
c. Use the linear model to predict the number of black elected officials in the year 1996.   **about 9150**
d. What factors might influence the accuracy of your prediction? *(Lessons 1-3, 3-3)*
a,d) See margin.

25. **a.** Rewrite $y = (x - 8)^2 + 9$ in the form $y = ax^2 + bx + c$.
b. Rewrite $y = x^2 + 12x - 3$ in the form $y = a(x - h)^2 + k$.
(Previous course)   **a) $y = x^2 - 16x + 73$; b) $y = (x + 6)^2 - 39$**

Exploration

26. Find two functions f and g such that $f(g(x)) = x$ but $g(f(x)) \neq x$.
sample: $f(x) = x^2$, $g(x) = \sqrt{x}$; $f(g(x)) = x$, for all nonnegative x; $g(f(x)) = |x|$

Harold Washington, Mayor of Chicago, 1983-1987

FOLLOW-UP

MORE PRACTICE
For more questions on SPUR Objectives, use *Lesson Master 3-8*, shown below.

EXTENSION
An approximate way to convert from degrees Fahrenheit to degrees Celsius is "subtract 30, divide by 2." Compare this guide with the rule in **Question 14**.
a. For what range(s) of temperatures is this approximate guide reasonably accurate? (sample: $0 \leq F \leq 100$; $-18 \leq C \leq 38$)
b. For what range(s) of temperatures is this approximate guide rather inaccurate?(sample: $F < 0$ or $F > 100$; $C < -18$ or $C > 38$)
c. Answer parts **a** and **b** for the inverse approximate guide for converting from degrees Celsius to degrees Fahrenheit. (sample: **a.** $32 \leq F \leq 68$; $0 \leq C \leq 20$; **b.** $F < 32$ or $F > 68$; $C < 0$ or $C > 20$)

NAME _____

LESSON **MASTER   3-8**
QUESTIONS ON **SPUR** OBJECTIVES

■**SKILLS**   *Objective B (See pages 197–200 for objectives.)*
In 1 and 2, **a.** find an equation for the inverse of the function, and **b.** state whether the inverse is a function.

1. $g(x) = -\frac{2}{3x}$ **a.** ___$y = -\frac{2}{3}x$___   2. $y = x^2 - 5$ **a.** ___$y = \pm\sqrt{x + 5}$___
   b. ___Yes___   **b.** ___No___

■**PROPERTIES**   *Objective H*
3. If $p(1) = 0.25$, $p(2) = 0.53$, and $p(3) = 1.01$, give the coordinates of three points on the graph of the inverse of p.   **(0.25, 1), (0.53, 2), and (1.01, 3)**
4. *True or false.* If $f(5) = f(-5)$, then the inverse of f is not a function. Justify your answer.
True. Sample: If $f(5) = f(-5) = c$, then (5, c) and (-5, c) are in f. Then (c, 5) and (c, -5) are in the inverse relation of f. Since c is matched with two different second elements, this relation is not a function.
5. Describe two functions f and g such that $f(g(x)) = g(f(x))$ for all x in the domains of f and g, but f and g are not inverses of each other.
Sample: $f(x) = 1$ and $g(x) = 1$

■**REPRESENTATIONS**   *Objective L*
In 6 and 7, graph on the same set of axes.
6. The function in Question 1 and its inverse
7. The function in Question 2 and its inverse

The two graphs coincide.

40   *Functions, Statistics, and Trigonometry © Scott, Foresman and Company*

RESOURCES
■ Teaching Aid 32: Graphs of parabolas

TEACHING NOTES

Similarity of curves and un-bounded figures may be a new idea to many students. Their definition of similarity may only extend to figures with proportional sides and congruent angles or figures with the same shape. This definition, though widely used, is not all-encompassing. A more mathematical definition of similarity is "There is a composite of reflections and size changes that maps one figure onto the other." Students who have had UCSMP Geometry will be familiar with the above mathematical definition.

Error Analysis If students are puzzled by the idea, you might begin by asking if the two circles are similar:

Students will readily agree. Are they still similar when their positions are changed? (Yes)

How about now? (Yes)

LESSON

Are All Parabolas Similar?

Informally, two figures are similar if they have the same shape but not necessarily the same size. All circles are similar, all squares are similar, but not all triangles are similar. Examples are shown below.

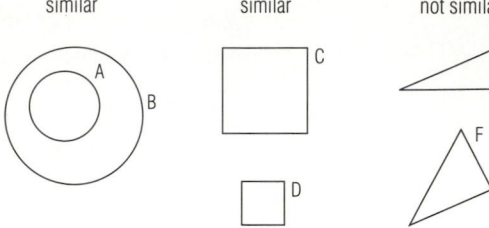

similar   similar   not similar

What about parabolas? Are they all similar, like circles and squares? Or are there different shapes for parabolas, as there are for triangles?

The answer to this question needs a more precise definition of similar. Recall the formal definition of *similar* from geometry: two figures are similar if and only if one is the image of the other under a composite of reflections, rotations, translations (all of which are called *isometries*) and size changes. So figures related by a size change (as in circles A and B above) are similar. In fact, any figures related by a size change followed by a reflection, rotation, or translation are similar. Triangles E and F above do not satisfy this definition, so they are not similar figures.

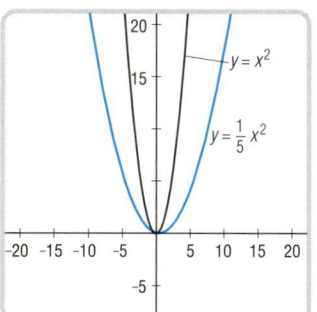

Now consider two parabolas, for instance, those with equations $y = x^2$ and $y = \frac{1}{5}x^2$ (shown at the left). Some people look at these and think, "the shapes are different, so the parabolas are not similar." You will see in this lesson that this conclusion is incorrect.

Recall from Lesson 3-5 that a size change is a scale change with horizontal and vertical scale factors the same: $S:(x,y) \rightarrow (kx, ky)$.

To map $y = x^2$ to $y = \frac{1}{5}x^2$ replace x by $\frac{x}{5}$ and y by $\frac{y}{5}$. Then rewrite:

$$\frac{y}{5} = \left(\frac{x}{5}\right)^2 = \frac{x^2}{25}$$

$$y = \frac{1}{5}x^2.$$

This shows that the graph of $y = \frac{1}{5}x^2$ is the image of the graph of $y = x^2$ under the size change $S:(x,y) \rightarrow (5x, 5y)$. Because the graphs of $y = \frac{1}{5}x^2$ and $y = x^2$ are related by a size change, they are similar.

190

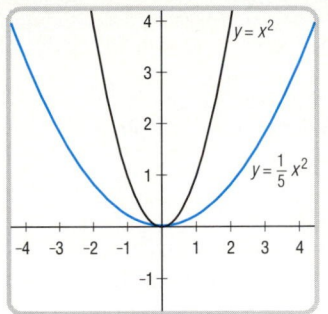

To illustrate that the two parabolas are similar, zoom in on the graph on the previous page using a zoom factor of 5. The result is shown at the left.

Compare this pair of graphs with those on the previous page. The graph of $y = x^2$ after zooming in looks congruent to that of $y = \frac{1}{5}x^2$ before zooming in. By comparing the two sets of graphs, you can see that the two parabolas are similar, with the graph for $y = \frac{1}{5}x^2$ merely "larger" than that of $y = x^2$.

Another way of thinking about this is to regard the graph of $y = x^2$ as the shape of a needle-point under a microscope. The graph of $y = \frac{1}{5}x^2$ is merely a magnified version of the same needle-point shape.

This idea generalizes. Using the same argument with a replacing $\frac{1}{5}$, it can be shown that the image of any parabola $y = ax^2$ is the parabola $y = x^2$ under a size change. Consider the mapping $S:(x, y) \rightarrow \left(\frac{x}{a}, \frac{y}{a}\right)$. Begin with
$$y = x^2.$$

The image of the parent function is found by replacing y with ay and x with ax and then solving for y:
$$ay = (ax)^2 = a^2x^2$$
$$y = ax^2.$$

Thus, for all $a \neq 0$, the graph of $y = ax^2$ is similar to the graph of $y = x^2$.

For any curve, a translation image of that curve is congruent to the preimage. So, mapping each point (x, y) on $y = ax^2$ to $(x + h, y + k)$ yields the congruent image $y - k = a(x - h)^2$ or $y = a(x - h)^2 + k$. Thus $y = a(x - h)^2 + k$, which is the composite of a size change and a translation, is also similar to $y = x^2$.

Further, for any parabola of the form $y = ax^2 + bx + c$, you can complete the square on $ax^2 + bx + c$, and write an equation for that parabola in the form $y = a(x - h)^2 + k$.

Thus, any parabola with a vertical axis of symmetry is similar to the parabola $y = x^2$. Since any parabola can be rotated onto one with a vertical axis, and rotations preserve shape, we have proved the following remarkable result.

Theorem

All parabolas are similar.

Just as there is only one circular shape and one square shape, there is only one parabolic shape!

Are these parabolas similar?

Some students might realize that we are looking at only a small portion of a function that is increasing without bound. If we draw a "window" on $y = x^2$ above,

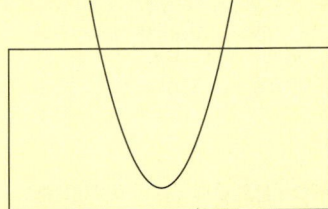

the part inside the rectangle appears to be wider than the corresponding portion of the graph of $y = 2x^2$. Try graphing $y = x^2$ for $-2 \leq x \leq 2$ and $-1 \leq y \leq 4$ and $y = 2x^2$ for $-1 \leq x \leq 1$ and $-.5 \leq y \leq 2$.

Alternate Approach Another approach for students still unsure is to put a coordinate grid transparency on an overhead projector about 3 or 4 feet from the blackboard. Place a transparency with a parabola on it (a teaching aid is provided) on top of the coordinate grid so that the parabola is the graph of $y = x^2$. Move the overhead further away from the blackboard; until the parabola shows the graph of $y = \frac{1}{2}x^2$ on the coordinate grid.

ADDITIONAL EXAMPLES
1. a. Graph $y = x^3$ and $y = ax^3$ for several values of a. Is the graph of $y = x^3$ similar to the graph of $y = ax^3$?
Yes

b. Graph $y = x^n$ and $y = ax^n$ for several values of a. Is the graph of $y = x^n$ similar to the graph of $y = ax^n$?
Yes

192

2. Graph $\dfrac{x^2}{9} + \dfrac{y^2}{4} = 1$ and $\dfrac{x^2}{16} + \dfrac{y^2}{4} = 1$. Use your results to determine if all ellipses are similar.
No

NOTES ON QUESTIONS
Questions 6 and 7: The point is subtle: There are many transformations that can map one curve onto another (or, for that matter, any figure onto any other). If only one of them is a similarity transformation, then the figures are similar. Our intuition tends to see the stretches that map parabolas onto each other, not the size changes, which is another reason why our first thought is that parabolas are not similar.

NOTES ON QUESTIONS
Making Connections for Question 22: In general, two hyperbolas are similar if and only if the vertical angles formed by their asymptotes and in whose interiors the branches of the hyperbolas lie are congruent. All hyperbolas with equations of the form $y = \dfrac{k}{x}$ have horizontal and vertical asymptotes, so the angles are all right angles. These ideas are discussed in more detail in Chapter 12.

ADDITIONAL ANSWERS
4.b.

Questions

Covering the Reading

In 1 and 2, **a.** give the image of the point under the size change $S:(x, y) \rightarrow (5x, 5y)$; **b.** show that the preimage point is on the graph of $y = x^2$ and the image point is on the graph of $y = \frac{1}{5}x^2$.

1. $(0, 0)$
a) **(0, 0); b) $0 = 0^2$; $0 = \frac{1}{5}(0)^2$**

2. $(2, 4)$
a) **(10, 20); b) $4 = 2^2$; $20 = \frac{1}{5}(10)^2$**

3. a. Give a size change that maps the graph of $y = 5x^2$ onto that of $y = \frac{1}{5}x^2$. **$S(x, y) \rightarrow (25x, 25y)$**

   b. *True* or *false*. The graphs of $y = 5x^2$ and $y = \frac{1}{5}x^2$ are similar.
   True

4. Consider $f(x) = x^2$, $g(x) = 2x^2$, $h(x) = 2(x + 5)^2 - 35$, and $j(x) = 2x^2 + 20x + 15$.
   a. Without graphing, name:
   i. two functions whose graphs are identical. **h, j**
   ii. all graphs congruent to $y = g(x)$. **graphs of h and j**
   iii. all graphs similar to $y = f(x)$. **graphs of g, h, and j**
   b. Verify your answer by using an automatic grapher and plotting all curves on the same set of axes. **See margin.**

5. To show that the graph of $y = 2x^2 - 8x + 13$ is similar to that of $y = x^2$, you need to show that the graph of $y = 2x^2 - 8x + 13$ is a translation image of $y = 2x^2$.
   a. By completing the square, rewrite $y = 2x^2 - 8x + 13$ in the form $y = 2(x - h)^2 + k$. **$y = 2(x - 2)^2 + 5$**
   b. Use your result in **a** to give the translation mapping $y = 2x^2$ onto $y = 2x^2 - 8x + 13$. **$(x, y) \rightarrow (x + 2, y + 5)$**
   c. What size change maps $y = x^2$ onto $y = 2x^2$? **$(x, y) \rightarrow (\frac{1}{2}x, \frac{1}{2}y)$**

In 6 and 7, recall that $y = x^2$ can be transformed to $y = \frac{1}{5}x^2$ by the size change $S:(x, y) \rightarrow (5x, 5y)$.

6. Show that $S:(x, y) \rightarrow (2x, 0.8y)$ also transforms $y = x^2$ onto $y = \frac{1}{5}x^2$.
   See margin.

7. a. Give an equation of the image of $y = x^2$ under the scale change $S:(x, y) \rightarrow (ax, by)$. **$y = \frac{b}{a^2}x^2$**
   b. What conditions must a and b satisfy if the image of $y = x^2$ under $S:(x, y) \rightarrow (ax, by)$ is $y = \frac{1}{5}x^2$? **$\frac{b}{a^2} = \frac{1}{5}$**
   c. Use your result from part **b** to find yet another pair of values a and b for which the image of $y = x^2$ under $S:(x, y) \rightarrow (ax, by)$ is $y = \frac{1}{5}x^2$. **sample: $a = \sqrt{5}$, $b = 1$**

Review

In 8 and 9, let $f(x) = x^3$ and $g(x) = x^{-3}$. *True* or *false*. *(Lessons 3-7, 3-8)*

8. $f(g(x)) = g(f(x))$ **True** **9.** f and g are inverses of each other. **False**

10. If $f: x \rightarrow ax$, $g: x \rightarrow x - h$, and $h: x \rightarrow x^2$, composing the functions in which order yields the function $t: x \rightarrow a(x - h)^2$? *(Lesson 3-8)* **$f \circ h \circ g$**

192

11. Evaluate $r(s(3))$ when $s(n) = |3n - 2|$ and $r(n) = \lfloor n - \frac{1}{2} \rfloor$.
(Lessons 2-4, 3-8)   **6**

12. *True* or *false*.   The inverse of an even function is always a function.
(Lessons 2-1, 3-4, 3-8)   **False**

13. a. Identify all symmetries of the curve $f(x) = 2(x - 6)^2 + 8$.
   b. *True* or *false*. f is an even function. *(Lesson 3-4)*   **False**
   a) symmetric with respect to the line $x = 6$

14. *Multiple choice.*   The graph of which relation has rotation symmetry?

   (a) $y = |x|$ (b) $y = x^2$ (c) $y = x^3$ (d) $y = \frac{1}{x^2}$ *(Lesson 3-4)*
   (c)

15. *Multiple choice.*   A transformed set of data has a variance twice that of
   the original set. How were the data transformed?
   (a) translated by 2   (b) multiplied by $\sqrt{2}$
   (c) multiplied by 2   (d) multiplied by 4 *(Lesson 3-6)*   **(b)**

16. In the USA, rainfall is measured in inches. In
   most other countries it is measured in millimeters. (There are exactly 25.4 mm in an inch.)
   Change the data at the right for Kansas City,
   Missouri, to mm to conform to the rest of a
   meteorological handbook. *(Lesson 3-6)*
   mean ≈ 5.84 mm; range ≈ 40.89 mm; variance ≈ 58 mm²

April daily rainfall
mean: 0.23 in.
range: 1.61 in.
variance: 0.09 in.^{2}

In 17 and 18, refer to the graphs at the left. Identify a translation or scale
change that maps the first function to the second. *(Lessons 3-2, 3-5)*

17. a. f to g **b.** g to f
   See margin.

18. a. f to h **b.** h to f
   See margin.

19. For a certain set of data, the line of best fit is given as $y = 131 - 4.6x$,
   with $r^2 = 0.61$. What is the correlation coefficient between variables x
   and y? *(Lesson 2-5)*   **-0.78**

20. Rewrite $x_1y_1 + x_2y_2 + x_3y_3 + \ldots + x_ny_n$ using Σ notation. *(Lesson 1-4)*
   See margin.

21. *Skill sequence.* Simplify.   *(Previous course)*

   a. $\dfrac{2^{15}}{2^{10}}$ $2^5 = 32$ **b.** $\left(\dfrac{x^7}{x^2}\right)^3$ x^{15} **c.** $\left(\dfrac{6x^5}{2x}\right)^p$ $3^p x^{4p}$

Exploration

22. a. Are all hyperbolas with equations of the form $y = \frac{k}{x}$ similar?
   b. Are all inverse square curves $y = \frac{k}{x^2}$ similar?

   a) See margin. b) Yes; apply $S(x, y) = \left(\sqrt[3]{k}x, \sqrt[3]{k}y\right)$ to $y = \frac{1}{x^2}$.

Margin notes:

6. An equation for the image of $y = x^2$ under S is given by substituting $\frac{x}{2}$ for x and $\frac{y}{0.8}$, or $\frac{5y}{4}$ for y.

$$\frac{5y}{4} = \left(\frac{x}{2}\right)^2$$
$$\frac{5y}{4} = \frac{x^2}{4}$$
$$y = \frac{1}{5}x^2$$

17.a. $(x, y) \rightarrow \left(\frac{2}{3}x, \frac{1}{2}y\right)$

b. $(x, y) \rightarrow \left(\frac{3}{2}x, 2y\right)$

18.a. $(x, y) \rightarrow (x + 4, y + 3)$

b. $(x, y) \rightarrow (x - 4, y - 3)$

20. $\sum\limits_{i=1}^{n} x_iy_i$

22.a. Yes; apply $S(x, y) = (\sqrt{k}x, \sqrt{k}y)$ to $y = \frac{1}{x}$ if $k > 0$.
If $k < 0$, apply $S(x, y) = (\sqrt{-k}x, \sqrt{-k}y)$.

Projects

The projects provide an opportunity for extended work on topics related to the content of the chapter. (See pages 66 and T37 for more information on the purposes of projects. Some or all of these projects may be suitable for **small group work.**

Solutions to the projects are found at the end of the Additional Answers section at the back of the book.

The goal of **Project 1** is for students to realize that equation coefficients can determine parabolas in much the same way as they determine lines. All of the parabolas in parts **a–d** are congruent. Congruent parabolas with a particular point in common, or with the same symmetry line, might be considered as part of one family of parabolas.

There are many possible responses to the questions in **Project 2.** Ask students who do this project to write their results as they would a science experiment: write a brief overview, explain their choices of variables to study, present their conclusions in words and with graphs, and write a conclusion.

The answer to part **b** of **Project 3** should include a proof of the result.

Most almanacs have perpetual calendars that enable answers to parts **5a–c** of **Project 4** to be checked. These calendars do not go back before 1700 because in western countries the calendar was changed from the Julian to the Gregorian calendar in the years 1582–1752.

1. Use an automatic grapher to explore the location of the vertices of parabolas of the form $y = ax^2 + bx + c$ when a and c are constant and b varies.
   a. Draw a parabola for $y = x^2 + 8x + 5$. Draw five more parabolas with the same values of a and c, but different values of b.
   b. What point do all the graphs have in common? Justify your answer.
   c. Find the vertex of each parabola you drew. On what type of curve do all vertices lie?
   d. Draw 5 to 10 more parabolas with the same values of a and c as in part **a**. Does your answer to part **c** still hold? Can you justify your conclusion?
   e. Repeat parts **a–d** using parabolas in which a is a constant different from 1.
   f. Write a report summarizing your findings.

2. Use the class database constructed as a project at the end of Chapter 1. Consider the variables for which measurements are involved.
   a. Convert these measurements to other units (*e.g.*, change centimeters to inches). Do not remove the original data. Rather, apply transformations to create new data sets.
   b. Compare the descriptive statistics of both the transformed data and the original data to confirm the results of Lessons 3-3 and 3-6.
   c. In addition, examine relations between pairs of transformed variables (*e.g.* between arm span and foot length). What statistics are invariant under these transformations?

3. The set of all scale changes under composition forms a mathematical structure called a *commutative group* with these properties:
   i. **Closure.** If S_1 and S_2 are scale changes, then so is $S_2 \circ S_1$.
   ii. **Commutativity.** $S_1 \circ S_2 = S_2 \circ S_1$
   iii. **Associativity.** $(S_3 \circ S_2) \circ S_1 = S_3 \circ (S_2 \circ S_1)$
   iv. **Identity.** There is a scale change I such that $S \circ I = I \circ S = S$.
   v. **Inverse.** For every S there is an inverse S^{-1} such that $S \circ S^{-1} = S^{-1} \circ S = I$.

Each of these properties can be proven. For instance, for property **ii** let $S_1(x, y) = (ax, by)$ and $S_2(x, y) = (cx, dy)$. Then
$$(S_1 \circ S_2)(x) = S_1[S_2(x, y)] = S_1(cx, dy)$$
$$= (acx, bdy);$$
$$(S_2 \circ S_1)(x) = S_2[S_1(x, y)] = S_2(ax, by)$$
$$= (cax, dby).$$
Because the real numbers are commutative and associative, $acx = cax$ and $bdy = dby$.
So $S_1 \circ S_2 = S_2 \circ S_1$.
   a. Prove the remaining properties for scale changes.
   b. Do translations form a commutative group under composition?

4. Repeated use of the greatest integer function can enable you to find the day of the week for any given date. Consider a particular date. Follow this algorithm.
   1. Let d be the date of the month.
   2. Let m be the number of the month in the year, with January regarded as month 13 and February as month 14 of the previous year. The other months are numbered 3 through 12 as usual.
   3. Let y be the year.
   4. Compute $W = d + 2m + \left\lfloor \frac{3(m + 1)}{5} \right\rfloor + y + \left\lfloor \frac{y}{4} \right\rfloor - \left\lfloor \frac{y}{100} \right\rfloor + \left\lfloor \frac{y}{400} \right\rfloor + 2$.
   5. Then the *remainder*, when W is divided by seven, is the day of the week, according to the codes below.

1	for	Sunday	5	for	Thursday
2	for	Monday	6	for	Friday
3	for	Tuesday	0	for	Saturday
4	for	Wednesday			

   a. As a check, use the algorithm above to calculate what day of the week is today.
   b. Calculate the day of the week on which you were born.
   c. Calculate the day of the week on which some famous event in history since 1700 occurred.
   d. Explain the significance of $\left\lfloor \frac{y}{4} \right\rfloor$, $\left\lfloor \frac{y}{100} \right\rfloor$, and $\left\lfloor \frac{y}{400} \right\rfloor$ in the function.

Summary

Function equations, graphs, and data can be transformed in similar ways. Two such transformations—translations and scale changes—were studied in this chapter. When translated or scaled, graphs of functions bear resemblances to the graphs of parent functions. Translations slide graphs, and scale changes stretch or shrink them, horizontally and vertically. So some features of graphs can be predicted if just the equation is given, and equations can be written if just the graph is given. Some symmetries of graphs of functions can also be predicted from their equations, such as the symmetries of odd and even functions.

Under a translation, measures of center for a data set are translated, while measures of spread are unaffected. When data are scaled, however, measures of both center and spread are also scaled.

An automatic grapher is a valuable tool for getting accurate graphs of functions quickly and for estimating coordinates of points such as intercepts, intersections, maxima, and minima. Changing the viewing window and zooming in are useful features for analyzing and describing graphs. By viewing more than one function in the same window, you can see the effects of various transformations.

Figures and their translation images are congruent. Figures and their scale change images are similar if the horizontal and vertical scale factors are the same. From this, it can be proved that *all* parabolas are similar.

Composition of functions can be performed by letting one function, say g, operate on the outputs of another, f, for those outputs in g's domain. The composite of g with f is denoted by $(g \circ f)(x)$. In general, composition is not commutative.

The inverse of a function f can be obtained by switching xs and ys in the defining equation or switching x- and y-coordinates in the set of ordered pairs. When the inverse is a function, it is denoted by f^{-1}. The graphs of a function and its inverse are reflection images of each other with respect to the line $y = x$. Another characteristic property of inverse functions is that the composite of f and f^{-1} is the identity function; that is, $f(f^{-1}(x)) = f^{-1}(f(x)) = x$.

Vocabulary

For the starred (*) items you should be able to give a definition of the term.
For the other items you should be able to give a general description and a specific example of each.

Lesson 3-1
*transformation
automatic grapher
default
viewing window
viewing rectangle
parent
zoom
asymptote

Lesson 3-2
*translation (of a graph)
Graph Translation Theorem

Lesson 3-3
*translation (of data)
image (of a data value)
invariant

Lesson 3-4
*symmetric with respect to the
  y-axis
*even function
*symmetric with respect to the
  x-axis
*symmetric to the origin
*odd function
power function
reflection-symmetric
axis of symmetry
*line of symmetry
rotation-symmetric
symmetry to point P
*center of rotation

Lesson 3-5
vertical scale change
horizontal scale change
horizontal scale factor

vertical scale factor
*scale change (of a graph)
Graph Scale Change Theorem
size change

Lesson 3-6
scale change (of data)
scale factor
magnitude of scale change
scaling, rescaling

Lesson 3-7
*composition (of functions)
composite

Lesson 3-8
*inverse of a function, f^{-1}
Inverse Function Theorem
*identity function

PROGRESS SELF-TEST

The Progress Self-Test provides the opportunity for feedback and correction. The Chapter Review provides additional opportunities for practice.

USING THE PROGRESS SELF-TEST
Assign the Progress Self-Test as a one-night assignment. (See pages 69 and T35 for more information.)

ADDITIONAL ANSWERS
1. $y = 3x^2 + 6x + 8$

3.

12. $f^{-1}(x) = \dfrac{2}{x} + 3$

13.

14. Yes; each x-value has exactly one y-value.

15.

17. We need to show that
$f(-x) = -f(x)$ **for all** x**.**
$f(-x) = 4(-x)^3 - 2(-x) =$
$-4x^3 + 2x = -(4x^3 - 2x) = -f(x)$

19. g **is symmetric with respect to the** y**-axis.**

Progress Self-Test

See margin for answers not shown below.
Take this test as you would take a test in class. You will need an automatic grapher. Then check your work with the solutions at the back of the book.

In 1 and 2, the translation $(x, y) \to (x - 1, y + 5)$ is applied to the function $y = 3x^2$.

1. Write an equation of the image.

2. What are the coordinates of the vertex of the image? **(−1, 5)**

In 3 and 4, suppose that $y = 12x + x^2 - x^3$, which is graphed below, is transformed under $S: (x, y) \to (-x, 2y)$.

3. Sketch a graph of the image.

4. Give an equation for the graph of the image.
   $y = -24x + 2x^2 + 2x^3$

In 5 and 6, use the given data. To make the calculations easier, an entomologist subtracted 20 g from the mass of each of 170 beetle specimens she had collected. She found the following statistics for the weights of her specimens after this transformation:

mean: 7.9 g   median: 7.4 g   maximum: 11.8 g
minimum: 3.4 g   standard deviation: 1.3 g.

5. For the actual beetle masses, give
   a. the median   **27.4 g**   b. the range   **8.4 g**
   c. the variance.   **1.69 g²**

6. To compare her results with those of other entomologists, she converted some statistics to ounces (1 g ≈ 0.0353 oz). Give the mean and standard deviation of her sample in ounces.
   mean ≈ 0.98 oz; s. d. ≈ 0.046 oz

In 7 and 8, suppose $\sum\limits_{i=1}^{5} g_i = 2$. Evaluate the expression.

7. $\sum\limits_{i=1}^{5} (g_i + 6)$ **32**

8. $\sum\limits_{i=1}^{5} 4g_i$ **8**

In 9–11, consider $a(x) = 18 - 3x$ and $b(x) = 4x^2$.

9. Evaluate $(b \circ a)(2)$.   10. Find $a(b(x))$.
                                     $18 - 12x^2$
11. Give the domain of $a \circ b$.
    9) 576; 11) the set of all real numbers

In 12–14, consider $f(x) = \dfrac{2}{x - 3}$.

12. Find an equation for f^{-1}, the inverse of f.

13. Graph f and f^{-1} with an automatic grapher. Sketch or print a hardcopy of the graphs.

14. Is the inverse of f a function? If the inverse is a function, prove it. If it is not, explain why.

In 15 and 16, the function $h(t) = -16t^2 + 60t + 20$ describes the height $h(t)$ in feet above the ground of an object t sec after being thrown straight up at 60 ft/sec from 20 ft above the ground.

15. Use an automatic grapher to graph this function. Sketch or print a hard copy of the graph that shows all its key features.

16. Estimate to the nearest integer the time at which the object hits the ground.   **t = 4**

17. Prove that the function $f(x) = 4x^3 - 2x$ is an odd function.

In 18 and 19, consider the graph below of the function g, which is the image of $y = |x|$ under a scale change.

18. Find an equation for g.   **g(x) = 3|x|**

19. Describe the symmetries of g.

In 20 and 21, *true* or *false*.

20. If each element of a data set is multiplied by k, then the variance of the new data set is k times the variance of the original data set.   **False**

21. Consider two functions f and g. If f is the image of g under a translation, then the graph of f is congruent to the graph of g.   **True**

Chapter Review

Questions on **SPUR** Objectives

SPUR stands for **S**kills, **P**roperties, **U**ses, and **R**epresentations.
The Chapter Review questions are grouped according to the SPUR Objectives for this chapter.
See margin for answers not shown below.

SKILLS deal with the procedures used to get answers.

■ **Objective A:** *Find composites of functions.*
(Lesson 3-7)
In 1 and 2, let $f(t) = 6t - 2$ and $g(t) = t - t^2$.
1. Evaluate **a.** $f(g(-1))$ **-14** **b.** $g(f(-1))$.   **-72**
2. Find a formula for each composite.
   a. $f(g(t))$ **b.** $g(f(t))$

In 3 and 4, consider the function f mapping A to B, and g mapping B to C.

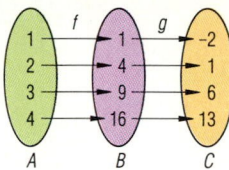

3. Evaluate $g(f(4))$.
   13
4. Evaluate $(g \circ f)(2)$.
   1

In 5 and 6, let $m(x) = \frac{3}{x}$ and $n(x) = x + 5$.
5. Evaluate
   a. $(m \circ n)(-6)$ **-3** **b.** $(n \circ m)(-6)$. **$4\frac{1}{2}$**
6. Find a formula for each function.
   a. $m \circ n$ **b.** $n \circ m$

■ **Objective B:** *Find inverses of functions.*
(Lesson 3-8)
In 7–10, for each function, **a.** describe the inverse using a set of ordered pairs or an equation.
b. State whether the inverse is a function.
7. $\{(3, 7), (4, 8), (5, 9), (6, 10), (7, 11)\}$
8. $y = 2x + 7$ 9. $f(x) = |x|$ 10. $g(x) = \frac{2}{x+1}$

PROPERTIES deal with the principles behind the mathematics.

■ **Objective C:** *Use the Graph Translation Theorem or the Graph Scale Change Theorem to find transformation images.* *(Lessons 3-2, 3-5)*
11. In an equation for a function or relation, if x is replaced by $x - h$ and y is replaced by $y - k$, how is the graph of the image related to the graph of the preimage?
12. In an equation for a function or relation, if x is replaced by $\frac{x}{a}$ and y by $\frac{y}{b}$, how is the graph of the resulting equation related to the graph of the original?

In 13 and 14, *multiple choice*.
13. Which translation has the effect of moving each point 3 units down and 8 units to the right?
    (a) $T(x, y) = (x - 3, y + 8)$
    (b) $T(x, y) = (x + 8, y - 3)$
    (c) $T(x, y) = (x - 8, y + 3)$
    (d) $T(x, y) = (x - 8, y - 3)$ **(b)**
14. Which scale change has the effect on a graph of scaling horizontally by a factor of 12 and vertically by a factor of $\frac{1}{5}$?   **(c)**
    (a) $S(x, y) = (12x, 5y)$ (b) $S(x, y) = \left(\frac{x}{12}, 5y\right)$
    (c) $S(x, y) = \left(12x, \frac{y}{5}\right)$ (d) $S(x, y) = \left(\frac{x}{12}, \frac{y}{5}\right)$

RESOURCES
■ Chapter 3 Test, Form A
■ Chapter 3 Test, Form B
■ Chapter 3 Test, Cumulative Form
■ Comprehensive Test, Chapters 1–3
■ Quiz and Test Writer

CHAPTER REVIEW

The main objectives for the chapter are organized here into sections corresponding to the types of understanding this book promotes: Skills, Properties, Uses, and Representations (called SPUR objectives).

USING THE CHAPTER REVIEW
Students should be able to answer questions like these with about 85% accuracy by the end of the chapter. (See pages 72 and T38–39 for more information.)

ADDITIONAL ANSWERS
2.a. $f(g(t)) = -6t^2 + 6t - 2$

b. $g(f(t)) = -36t^2 + 30t - 6$

6.a. $(m \circ n)(x) = \frac{3}{x + 5}$

b. $(n \circ m)(x) = \frac{3}{x} + 5$

7.a. $\{(7, 3), (8, 4), (9, 5), (10, 6), (11, 7)\}$

b. Yes

8.a. $y = \frac{x - 7}{2}$

b. Yes

9.a. $x = |y|$

b. No

10.a.,b., 11., 12. See margin, page 198.

10.a. $y = \frac{2}{x} - 1$

b. Yes

11. The image is translated *b* units to the right and *k* units up.

12. The image is stretched by a factor of *a* horizontally and by a factor of *b* vertically.

26. True. $\sum_{i=1}^{n} (x_i - 7) =$
$(x_1 - 7) + (x_2 - 7) + \ldots +$
$(x_n - 7) = (x_1 + x_2 + \ldots$
$+ x_n) - (n)(7) =$
$\left(\sum_{i=1}^{n} x_i \right) - 7n$

28. $\sum_{i=1}^{n} (x_i - y_i) =$
$(x_1 - y_1) + (x_2 - y_2) + \ldots +$
$(x_n - y_n) = (x_1 - x_2 + \ldots + x_n)$
$+ (-y_1 - y_2 - \ldots - y_n) =$
$x_1 + x_2 + \ldots + x_n) -$
$(y_1 + y_2 + \ldots + y_n) =$
$\sum_{i=1}^{n} x_i - \sum_{i=1}^{n} y_i$

29. The mean goes up by 10.

30. The standard deviation is unchanged.

31. The median is multiplied by *k*.

32. The variance is multiplied by k^2.

33. The standard deviation is unchanged.

34. The standard deviation is divided by 2.

38.a. odd

b. We want to show that $f(-x) = -f(x)$ for all x
$f(-x) = 8(-x)^3 = -8x^3 = -f(x)$

39.a. even

b. We want to show that $f(-t) = f(t)$ for all t. $f(-t) =$
$5(-t)^2 - (-t)^4 = 5t^2 - t^4 = f(t)$

40.a. neither

In 15 and 16, find an equation for the image of $y = x^2$ under the given transformation.

15. $(x, y) \rightarrow (x + 7, y - 3)$ **$y = x^2 - 14x + 46$**

16. $(x, y) \rightarrow \left(2x, \frac{y}{3} \right)$ **$y = x^2/12$**

In 17 and 18, suppose $f(x) = |x|$. Find an equation for the image of f under the transformation.

17. $S(x, y) = \left(\frac{x}{4}, 5y \right)$ **18.** $T(x, y) = (x + 1, y)$
   $y = 5|4x|$ **$y = |x - 1|$**

19. What transformation maps the graph of $y = \sqrt{x}$ onto the graph of $y = \sqrt{10x}$? **$(x, y) \rightarrow (x/10, y)$**

20. What transformation maps the graph of $y = \frac{1}{x}$ onto the graph of $y = \frac{1}{x - 8} + 9$?

   $(x, y) \rightarrow (x + 8, y + 9)$

■ **Objective D:** *Describe the effects of translations and scale changes on functions and their graphs.* (*Lessons 3-2, 3-5*)

In 21 and 22, *true* or *false*.

21. Under a translation, asymptotes are mapped to asymptotes. **True**

22. Under a scale change, y-intercepts are mapped to y-intercepts. **True**

23. Under which transformation—translation or scale change—is the number of x-intercepts invariant? **scale change**

24. Which scale change has the effect of reflecting a graph over the y-axis? **$(x, y) \rightarrow (-x, y)$**

■ **Objective E:** *Use summation properties to evaluate and rewrite expressions.* (*Lessons 3-3, 3-6*)

25. Suppose $\sum_{i=1}^{5} a_i = 8$. Evaluate the given expression.

a. $\sum_{i=1}^{5} (a_i + 2)$ **18**   **b.** $\sum_{i=1}^{5} a_i + 2$ **10**

c. $\sum_{i=1}^{5} 7a_i$ **56**   **d.** $\sum_{i=1}^{5} ka_i$ **8k**

26. *True* or *false*. $\sum_{i=1}^{n} (x_i - 7) = \sum_{i=1}^{n} x_i - 7n$
   Justify your answer.

27. Find k so that $\sum_{i=1}^{n} \left(\frac{x_i}{2} \right) = k \sum_{i=1}^{n} x_i$. **$k = \frac{1}{2}$**

28. Prove that $\sum_{i=1}^{n} (x_i - y_i) = \sum_{i=1}^{n} x_i - \sum_{i=1}^{n} y_i$.

■ **Objective F:** *Describe the effects of translations or scale changes on measures of center or spread.* (*Lessons 3-3, 3-6*)

In 29 and 30, suppose 10 is added to each element in a data set. Describe the effect of this transformation on each measure.

29. mean   **30.** standard deviation

In 31 and 32, if each element in a data set is multiplied by k, how is the following measure affected?

31. median   **32.** variance

In 33 and 34, what is the effect of the transformation on the standard deviation of a set of data?

33. subtracting 11 from each element

34. dividing each element by 2

■ **Objective G:** *Describe the symmetries of graphs.* (*Lesson 3-4*)

In 35 and 36, *true* or *false*.

35. If a function can be mapped to itself under a rotation of 180° around the origin, then it is an odd function. **True**

36. If a function is symmetric over the line $x = 10$, then it is an even function. **False**

37. Classify the function as odd, even, or neither.

a.
even

b.
odd

c.
even

d.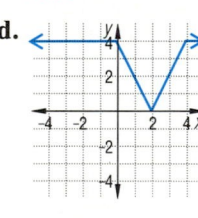
neither

In 38–40, **a.** determine if the function is odd, even, or neither. **b.** If the function is odd or even, prove it. If it is neither, give a counterexample.

38. $f(x) = 8x^3$ **39.** $s(t) = 5t^2 - t^4$

40. $g(m) = |3m - 4|$

■ **Objective H:** *Identify properties of composites or inverses.* (Lessons 3-7, 3-8)

41. If $f(x) = \sqrt{x}$ and $g(x) = x + 10$, what is the domain of $f(g(x))$?   {x : x ≥ -10}

42. If two relations are inverses of each other, what transformation maps one to the other?
reflection over the line y = x

In 43 and 44, *multiple choice*.

43. If (p, q) is a point on the graph of a relation, what point must be on its inverse?
(a) $(-p, q)$ (b) $(p, -q)$
(c) $(-p, -q)$ (d) (q, p) **(d)**

44. Which condition is sufficient for concluding that f and g are inverses of each other?
(a) $f(g(x)) = g(f(x))$ for all x in the domains of f and g.
(b) $f(g(x)) = x$ for all x in the domain of g and $g(f(x)) = x$ for all x in the domain of f.
(c) f is always positive and g is always negative.
(d) f and g each pass the vertical line test.   **(b)**

USES deal with applications of mathematics in real situations.

■ **Objective I:** *Use translations or scale changes to analyze data.* (Lessons 3-3, 3-6)

45. Use a translation to find mentally the average of these bowling scores:
102, 112, 110, 106, 115.   **109**

46. a. Identify the transformation used to scale the scores below.   s → 5s

original scores	scaled scores	frequency
3	15	1
4	20	3
5	25	2
7	35	7
8	40	2

b. Find the mode, mean, and median of the original data.
c. Find the mode, mean, and median of the scaled scores.
d. What property of data transformations is shown in **b** and **c**?

47. A swimmer training for the World Championships recorded the following times for the 800-m freestyle event (in minutes:seconds).

8:28.71	8:31.06	8:30.14
8:31.02	8:29.91	8:27.88
8:28.13	8:29.37	8:30.00
8:28.65	8:29.50	8:31.36

Using a translation to simplify the computation, find the mean and standard deviation of her times.

48. For a sample of a certain butterfly species, a scientist measured their lengths and found a mean length of 1.76 inches with a range of 3.02 inches and a standard deviation of 0.53 inches. If the data are converted to millimeters, give the following statistics:
a. mean   **b.** range   **c.** variance.
   44.70 mm   76.71 mm   ≈ 181 mm²

REPRESENTATIONS deal with pictures, graphs, or objects that illustrate concepts.

■ **Objective J:** *Apply the Graph Translation Theorem or the Graph Scale Change Theorem to make or identify graphs.* (Lessons 3-2, 3-5)

49. Sketch the graphs of $y = |x - 5| + 7$ and its parent function on the same set of axes.

50. For $k(x) = 5 + \dfrac{1}{x+6}$,
a. sketch a graph of k,
b. give the equations of the asymptotes,
c. give the coordinates of the intercepts,
d. identify the parent equation and the transformation that maps the parent function to $y = k(x)$.

---

b. g(1) = 1, g(-1) = 7

46.b. mode = 7; mean = 6; median = 7

c. mode = 35; mean = 30; median = 35

d. Multiplying each element of a data set by the factor *a* multiplies each of the mode, mean, and median by a factor of *a*.

47. mean = 8:29.64; standard deviation = 1.15 seconds

49.

50.a.

b. x = -6, y = 5

c. x-intercept: $\left(-\frac{31}{5}, 0\right)$;
y-intercept: $\left(0, \frac{31}{6}\right)$

d. parent equation: $y = \frac{1}{x}$;
transformation: $(x, y) \rightarrow (x - 6, y + 5)$

51. a. Sketch a graph of $y = 3\lfloor 0.4x \rfloor$.
  b. Name three x-values at which the function is discontinuous.

52. a. Graph $f(x) = x^3$ and $g(x) = 2x^3$ on the same set of axes.
  b. *True* or *false*. g is the image of f under the transformation $(x, y) \rightarrow (2x, y)$. **False**

In 53–56, the graph is a translation or scale change image of the given parent function. Write an equation for the graph.

53. parent: $y = x^2$
$y = (x + 2)^2 + 5$

54. parent: $y = |x|$
$y = |x + 3| - 4$

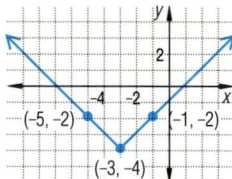

55. parent: $y = \sqrt{x}$
$y = \sqrt{2x}$

56. parent: $y = \dfrac{1}{x^2}$

$y = \dfrac{1}{(x - 4)^2} - 2$

■ **Objective K:** *Use an automatic grapher to draw the graph of a function and read key values.*
(Lesson 3-1)

57. Sketch the graph of $R(x) = -0.8x^2 - 4.3x + 1.8$ on the domain $-10 \le x \le 10$.

58. The graph below was produced by an automatic grapher. The viewing window is defined by $-4 \le x \le 4$ and $-2 \le y \le 18$. Give the approximate coordinates of
  a. the x-intercept(s)   **(2.2, 0), (-3.8, 0)**
  b. the y-intercept(s)   **(0, 16)**
  c. the maximum point.   **(-0.8, 17)**

In 59–61, **a.** draw a graph using an automatic grapher. **b.** State the domain and range of the function.

59. $y = \dfrac{1}{x + 5}$ **60.** $g(x) = \sqrt{x - 5}$

61. $f(x) = 3(x - 8)^2 + 10$

■ **Objective L:** *Graph inverses of functions.*
(Lesson 3-8)

In 62 and 63, **a.** graph the function and its inverse. **b.** Determine if the inverse is a function. **c.** Give an equation for the inverse.

62. $h(x) = \dfrac{3}{x}$ **63.** $y = x^2$

In 64 and 65, two graphs f and g are drawn. Tell whether or not f and g are inverses of each other.

64.

No

65.

Yes

OBJECTIVES ■ CHAPTER 4

The objectives listed here are the same as in the Chapter 4 Review on pages 263–266 of the student text. The Progress Self-Test on page 262 and the tests in the Teacher's Resource File cover these objectives. For recommendations regarding the handling of this end-of-chapter material, see the notes in the margins on the corresponding pages of this Teacher's Edition.

OBJECTIVES FOR CHAPTER 4 (Organized into the SPUR categories—Skills, Properties, Uses, and Representations)	Progress Self-Test Questions	Chapter Review Questions	Teacher's Resource File	
			Lesson Masters*	Chapter Test Forms A & B
SKILLS				
A Evaluate $b^{m/n}$ for $b > 0$.	1–3	1–12	4-1, 4-2	1–3
B Solve exponential equations.	8, 9	13–16	4-8	11, 12, 24
C Evaluate logarithms.	4–6, 12	17–25	4-5, 4-6, 4-7	4–8, 22
PROPERTIES				
D Describe properties of exponential and logarithmic functions.	7, 13	26–32	4-3, 4-4, 4-5, 4-6	9, 10
E Use properties of logarithms.	14, 15	33–43	4-7, 4-8, 4-9	17, 20b
USES				
F Use rational exponents to model situations.	17	44–47	4-1, 4-2	18
G Use exponential models to solve problems.	10, 11	48–57	4-3, 4-4, 4-9	19
H Use logarithmic functions to model data.	20	58–61	4-6, 4-9	23
REPRESENTATIONS				
I Graph exponential and logarithmic functions.	16	62–68	4-3, 4-5, 4-6	20a, 21
J Interpret graphs of exponential and logarithmic functions.	18, 19	69, 70	4-3, 4-5, 4-6	13–16, 20c

*The masters are numbered to match the lessons, and each is intended for use with that lesson.

OVERVIEW ■ CHAPTER 4

Rational power functions, exponential functions, and logarithmic functions are all based on the equation $b^r = p$ (base raised to exponent equals power), where $b > 0$ and $b \neq 1$ except in special cases. The differences are in whether the base or the exponent is kept constant, and which of the other two is the domain variable (x) and which is the range variable (y).

$$x^r = y \quad r\text{th power}$$
$$y^r = x \quad r\text{th root}$$
$$(y = x^{1/r})$$
$$b^x = y \quad \text{exponential base } b$$
$$b^y = x \quad \text{logarithmic base } b$$
$$(y = \log_b x)$$

The study of these functions includes the approximation of values using a calculator, the computation of exact values by inspection or pru-dent use of properties, and the application of these skills to solve exponential or logarithmic equations. The graphs of the functions are analyzed using a function grapher, and the statistical methods of previous chapters are extended to model data which appears to be exponential or logarithmic in nature.

The first two lessons of the chapter review the meaning and notation of rational roots. While the meaning is clear, there is not a consistent use of the symbols among mathematics textbooks or even among mathematics educators. This inconsistency is reflected in the fact that calculators interpret the value of $x^{p/q}$ differently. This issue is discussed in the notes for Lesson 4-1.

Lessons 4-3 and 4-4 involve the finding and use of exponential models. As with linear and quadratic modeling, the approach is to take an equation which is a generic model and, by substituting known values, to solve a system of equations to find a specific equation.

Lessons 4-5 through 4-7 discuss common logarithms, natural logarithms, and the properties they share. These properties are applied to the solving of equations of the form $a^x = b$ in Lesson 4-8 and to the rewriting of $a^x = b$ as a linear relation between $\log a$ and $\log b$ in Lesson 4-9.

The chapter concludes with a description of how logarithms are used in the pre-programming of scientific calculators to approximate powers of any positive number.

PERSPECTIVES ■ CHAPTER 4

The Perspectives provide the rationales for the inclusion of topics or approaches, give some mathematical background, and make connections with other lessons and to other UCSMP texts.

4-1

NTH ROOT FUNCTIONS

An nth root of a number x is a solution r to the equation $r^n = x$. This lesson discusses only the real nth roots of real numbers; when $x \geq 0$ these can be found by examining the graph of the nth root function $g(x) = x^{1/n}$.

The language of nth roots is often misunderstood. The phrase "nth root of x" can apply to any x, real or complex. The symbol $x^{1/n}$ is used only when x is a nonnegative real number. The radical notation $\sqrt[n]{x}$ means $x^{1/n}$ when x is nonnegative but also has a use when x is negative and n is an odd positive integer greater than 1.

This lesson reviews the notation and evaluation of real nth roots, the use of nth roots in solving equations of the form $ax^n = b$, the graphing of nth root functions, and some applications.

4-2

RATIONAL POWER FUNCTIONS

When x is positive and m and n are integers, the mth power of $x^{1/n}$ is the rational power $x^{m/n}$, and the function with equation $y = x^{m/n}$ is a rational power function. Noninteger rational powers are not defined when the base is not positive because there is no way to do this and still keep the general properties of powers such as $(x^m)^n = x^{mn}$.

This lesson applies the general properties of powers to the evaluation of rational powers, the solving of equations of the form $ax^r = b$ when r is rational, the graphing of rational power functions, and some applications.

4-3

EXPONENTIAL FUNCTIONS

When b is positive and x is irrational, the real power b^x can be estimated by calculating or estimating b^r for rational estimates r of x. This property enables the defining of exponential functions $f(x) = ab^x$ with positive base $b \neq 1$ and $a \neq 0$, where x may be any real number. Properties and graphs of exponential functions are described in this lesson, with applications to growth situations ($b > 1$) and decay situations ($b < 1$) also included.

4-4

FINDING EXPONENTIAL MODELS

This lesson concentrates on constructing mathematical models of situations which are exponential in nature. Given two points of such a model, students are instructed to find an equation of the form $y = ab^x$ by solving a system of two equations with two unknowns. The well-known examples of population growth and half-life of radioactive elements are presented to model exponential growth and decay.

4-5

LOGARITHMIC FUNCTIONS

We use the term *logarithm function base b* to apply to the function that maps x onto $\log_b x$. The term *logarithmic function* applies to any off-spring of the logarithm functions.

This lesson concentrates on the common logarithm function, that is, the logarithm function base 10. It reviews the calculation of values of these functions and their application to the solving of equations involving real powers of 10. This is a two-day lesson if students have not encountered logarithms previously.

4-6

e AND NATURAL LOGARITHMS

The constant e is generated through continuous compounding of an investment, as the limit of $\left(1 + \frac{1}{n}\right)^n$ as $n \to \infty$. This continuous compounding is generalized in the Continuous Change Model for exponential growth or decay. The exponential function $f(x) = e^x$ is the limit of functions of the form $f(x) = \left(1 + \frac{1}{n}\right)^{nx}$ as $n \to \infty$. The function is applied to the situation of continuous growth or decay, and its inverse is defined as the natural logarithm function. Students are asked to evaluate expressions involving e or the natural logarithm, and use the Continuous Change Model to solve problems.

4-7

PROPERTIES OF LOGARITHMS

Historically, logarithms arose as important aids in computation. The speed with which their use spread indicates their utility; Napier's work was done around 1609, and by 1616 Kepler had used logarithms to help in his computations of planetary orbits.

The four properties of logarithms are derived here from the properties of powers. These properties enable exponential equations to be rewritten as linear equations involving logarithms.

4-8

SOLVING EXPONENTIAL EQUATIONS

Until this lesson, students have been able to solve equations of the form $b^x = a$ only when a and b are easily seen to be powers of the same number. Now their knowledge of logarithms enables them to solve this equation for any positive values of a and b. To elaborate, in Lesson 4-4, students learned that a solution to $b^x = a$ (when a and b are both positive and $b \neq 1$) is $x = \log_b a$. In Lesson 4-7, they saw that $b^x = a$ is equivalent to $x = \frac{\log a}{\log b}$. Thus they already have two ways of solving exponential equations. From equating the two solutions, the Change of Base Theorem $\left(\log_b a = \frac{\log_c a}{\log_c b}\right)$ appears. This makes possible the calculation of logarithms of numbers with any base c.

4-9

EXPONENTIAL AND LOGARITHMIC MODELING

This lesson provides an informal introduction to the idea of log-linear modeling. Students should be able to use the properties of logarithms to rewrite exponential equations as linear equations with logarithmic co-efficients, and vice versa. Instead of first trying to fit an exponential or a logarithmic curve directly to a scatter plot, one of the variables is transformed by taking logs of all the values of the appropriate variable. Then a statistics package can be used to find a line of best fit for the transformed data.

4-10

THE SCIENTIFIC CALCULATOR

This reading lesson describes how scientific calculators may approximate values of the exponential and natural logarithm functions, and how the properties of logarithms may be used to increase calculation speed. How calculators may be programmed to do this is illustrated using BASIC computer programs.

DAILY PACING CHART ■ CHAPTER 4

For optimal student performance, the Progress Self-Test and the Chapter Review should be covered. (See *General Teaching Suggestions: Mastery* on page T35 of this Teacher's Edition.) By following the pace of the Full Course given here, students can complete 12 or 13 chapters of the book by the end of the year. Students following the pace of the Minimal Course spend more time to complete lessons that are combined with quizzes and to consider the Chapter Review. When all lessons are covered in order (the recommendation of the authors), students in the Minimal Course can cover all but the last two chapters. For more information on pacing, see *General Teaching Suggestions: Pace* on page T34 of this Teacher's Edition.

DAY	MINIMAL COURSE	FULL COURSE
1	4-1	4-1
2	4-2	4-2
3	4-3	4-3
4	Quiz (TRF); Start 4-4	Quiz (TRF); 4-4
5	Finish 4-4	4-5
6	4-5	4-6
7	4-6	4-7
8	4-7	Quiz (TRF); 4-8
9	Quiz (TRF); Start 4-8	4-9
10	Finish 4-8	4-10
11	4-9	Progress Self-Test
12	4-10	Chapter Review
13	Progress Self-Test	Chapter Test (TRF)
14	Chapter Review	
15	Chapter Review	
16	Chapter Test (TRF)	

TESTING OPTIONS

■ Quiz for Lessons 4-1 through 4-3 ■ Chapter 4 Test, Form A ■ Chapter 4 Test, Cumulative Form
■ Quiz for Lessons 4-4 through 4-7 ■ Chapter 4 Test, Form B

A Quiz and Test Writer is available for generating additional questions, additional quizzes, or additional forms of the Chapter Test.

PROVIDING FOR INDIVIDUAL DIFFERENCES

The student text has been written for, and tested with, average students. It has also been used successfully with students who are better prepared as well as with students who are poorly prepared.

The Lesson Notes often include Error Analysis and Alternate Approach features to help you assist those students who need more help. Students of all abilities often learn from their peers and may benefit from small group work referenced as appropriate throughout the Notes. A blackline Lesson Master (in the Teacher's Resource File), keyed to the chapter objectives, is available for each lesson to provide more practice. (However, we advise you to use these masters sparingly; again refer to the suggestions on pace on page T34.)

Extension activities are provided in the Lesson Notes for use with students who need to be challenged even more than the requirements of completing the Full Course provide.

Power, Exponential, and Logarithmic Functions

The 1923 earthquake in Japan caused extensive damage.

Power functions, exponential functions, and logarithmic functions are used for modeling real-life situations and solving problems like those below.

For a ship, the speed s in knots varies directly with the seventh root of the horsepower p of the engines. Write a formula for s as a function of p.

An earthquake in Japan in 1923 measured 8.9 on the Richter scale. The earthquake that occurred in and around San Francisco on October 17, 1989, registered 7.1. How many times more intense was the Japanese earthquake?

In 1988, the population of India was about 816.8 million, with an average annual growth rate of 2.1%. If this growth rate continues, when will India's population reach one billion?

Carbon-14 has a half-life of about 5730 years. A human thigh bone is found with 41% of the original quantity of carbon-14 remaining. How old is the bone?

These three important families of functions share the characteristic that properties of the functions in them are derived from properties of powers, that is, expressions of the form x^y.

CHAPTER 4

For students who have previously studied similar content, particularly for those who have studied UCSMP *Advanced Algebra,* this chapter should be done at a pace of one lesson a day. For students who have not studied logarithms before or for whom the applications are new, we recommend that you still try to do 3 or 4 lessons at about a lesson-a-day pace, then pause for a day of review as necessary. You can use questions from the Chapter Review for additional practice on those days. Assuming 10 days for the lessons, 0–3 review days, 1 day for the Progress Self-Test, 1–2 days for the Chapter Review, and one day for testing, this chapter will take 13–17 days.

USING PAGE 201
Four applications are given. For each question, you might ask how many students think they could answer it even before studying the chapter. This will give you an opportunity to assess the familiarity of students with the material and pace the chapter accordingly.

Making Connections
The first question is repeated as Question 14 in Lesson 4-1, so it will be on the assignment. A question similar to the Richter scale question is Question 12 in Lesson 4-7. Population growth and half-life are discussed in Lesson 4-4.

LESSON 4-1

RESOURCES
■ Lesson Master 4-1

OBJECTIVES

A Evaluate $b^{m/n}$ for $b > 0$.
F Use rational exponents to model situations.

TEACHING NOTES

There are two major issues associated with the definition and use of nth roots. The first is the distinction between the English phrase "nth root of," the radical notation $\sqrt[n]{x}$, and the power notation $x^{1/n}$. *No two of these mean the same thing.*

Remind students that when x is positive and n is even, the symbol $\sqrt[n]{x}$ stands for just one of the nth roots of a real number, but the equation $r^n = x$ has two real solutions, denoted $\pm\sqrt[n]{x}$. For example, $\sqrt[4]{81} = 3$, while $x^4 = 81$ implies that $x = \pm\sqrt[4]{81} = \pm 3$.

The second issue is more subtle, and has to do with the value of $\sqrt[n]{x}$ when n is odd and x is negative. By the definition of nth root on page 202, -2 is a cube root of -8. The difficulty comes with the introduction of the notations for nth roots: $\sqrt[n]{x}$ and $x^{1/n}$. With respect to rational exponents in general, it is necessary that the power be defined only for nonnegative x. Unfortunately, the necessity is not apparent until the definitions in this lesson are expanded to include rational exponents having numerators other than 1, and that is not discussed until Lesson 4-2.

*n*th Root Functions

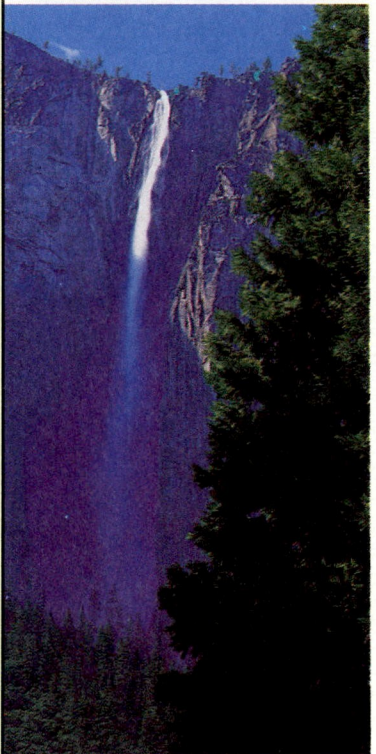

Ribbon Falls, Yosemite Valley

Square roots arise naturally out of real situations. For example, a body in free fall travels about d feet in t seconds, where $d = 16t^2$. To calculate how long it takes water to fall the 1612-foot height of Ribbon Falls in California's Yosemite National Park, substitute 1612 for d and solve for t.

$$1612 = 16t^2$$
$$t^2 = 100.75$$
$$t = \sqrt{100.75} \text{ or } t = -\sqrt{100.75}$$

Since $t > 0$, $t \approx 10$ seconds.

Finding the square root is the inverse operation of squaring.

Similarly, finding the cube root is the inverse of cubing. For instance,

5 is a cube root of 125 because $5^3 = 125$;
-6 is a cube root of -216 because $(-6)^3 = -216$.

Square roots and cube roots are instances of a more general pattern.

Definition

Let n be an integer with $n \geq 2$. r is an **nth root** of x if and only if $r^n = x$.

If n is odd, x has exactly one real nth root. For example, for $n = 5$, $3^5 = 243$ while $(-3)^5 = -243$, so 3 is the only 5th root of 243. Similarly, $-\frac{1}{3}$ is the 5th root of $-\frac{1}{243}$. If n is even and x is positive, x has two real nth roots. For example, for $n = 4$, $3^4 = (-3)^4 = 81$, so both 3 and -3 are 4th roots of 81. Also, those two 4th roots, 3 and -3, are opposites of each other. If n is even and x is negative, x has no real nth roots. For example, for $n = 4$, there is no real number r for which $r^4 = -81$ because the 4th power of a real number cannot be negative.

There are no special names for nth roots other than square roots (when $n = 2$) and cube roots (when $n = 3$).

■ ■ ■ ■ ■ ■ ■ ■

Example 1   Show that 6 and -6 are fourth roots of 1296.

Solution   6 is a fourth root of 1296 because $6^4 = 6 \cdot 6 \cdot 6 \cdot 6 = 1296$.
-6 is a fourth root of 1296 because $(-6)^4 = 1296$.

202

Recall some useful vocabulary: in the expression r^n, r is the *base* and n is the *exponent*; the entire expression r^n is called a *power*. The nth root of a nonnegative number can be expressed as a power using the exponent $\frac{1}{n}$. To do so, recall the Power of a Power Property: for any nonnegative base x and real exponents m and n,

$$(x^m)^n = x^{mn}.$$

Thus, because

$$(8^{1/3})^3 = 8^{3/3} = 8^1 = 8,$$

$8^{1/3}$ is a cube root of 8.

To ensure that every power with exponent $\frac{1}{n}$ has a unique value, the use of non-integer powers is restricted to positive bases.

An nth root can also be expressed using the **radical** symbol $\sqrt{}$. The square root of 2 can be written as $\sqrt{2}$, the cube root of 2 is $\sqrt[3]{2}$, and the nth root of a positive number x is $\sqrt[n]{x}$. That is, $\sqrt[n]{x} = x^{1/n}$.

Definitions

When $x \geq 0$, and n is an integer, $n \geq 2$,
$\sqrt[n]{x} = x^{1/n} =$ the positive nth root of x.

When $x < 0$, and n is an odd integer, $n \geq 3$,
$\sqrt[n]{x} =$ the real nth root of x.

Note that the symbol $x^{1/n}$ is not used when $x < 0$. Also, the definition does not include the case of $x < 0$ and n even because, for example, $\sqrt[4]{-81}$ is not a real number.

■ ■ ■ ■ ■ ■ ■ ■ ■

Example 2   Evaluate each expression.

   a. $1296^{1/4}$ **b.** $\sqrt[5]{-32}$

Solution

a. From Example 1, you know that both 6 and -6 are 4th roots of 1296. By definition, $1296^{1/4}$ represents the positive 4th root, so $1296^{1/4} = 6$.
b. $\sqrt[5]{-32} = -2$, because $(-2)^5 = -32$.

All scientific calculators give accurate estimates of nth roots of positive numbers. On some you can calculate $\sqrt[n]{x}$ using the key sequence $x\,\boxed{\sqrt[n]{x}}\,n$; on others you use the key sequence $\boxed{\text{INV}}\,\boxed{y^x}$ or $\boxed{\text{2nd f}}\,\boxed{y^x}$; and on still others you enter $\boxed{y^x}\,\boxed{(}\,1\,\boxed{\div}\,n\,\boxed{)}$ to evaluate nth roots. Test your calculator with the value $\sqrt[4]{33} = 2.3968$ (rounded to 4 decimal places). Some calculators will

An example of the difficulty follows: Assume that $(-8)^{1/3} = -2$. Then because $\frac{1}{3} = \frac{2}{6}$, we ought to be able to write $(-8)^{2/6} = -2$. Then, if properties held as usual, $(-8)^{2/6} = ((-8)^2)^{1/6} = 64^{1/6} = 2$, not -2. So $(-8)^{1/3}$ cannot be defined and still keep the integrity of the transitive property of equality and the power properties of algebra.

For these reasons, some calculators give an error message when a negative base is input with a fractional exponent. Unfortunately, some recent calculators actually *give* answers for odd rational powers of negative bases, and one that we tested gave -2 for $(-8)^{2/6}$.

Another possible issue concerns $x^{1/n}$ for $x = 0$. Many calculators will give you 0 for $0^{1/3}$. However, older calculators may give an error message. This is because they compute y^x as $e^{x \ln y}$ (as explained in Lesson 4-10) and $e^{1/3 \ln 0}$ leads to an error as $\ln 0$ is undefined. We have defined $0^n = 0$ for $n > 0$.

Regarding **Example 3,** be aware that some automatic graphers graph $y = x^{1/3}$ for all real x and do not use the restriction $x \geq 0$. That is, they graph $y = \sqrt[3]{x}$ instead.

Making Connections
Students know that calculators give approximate values of nth roots but may be bothered if their calculator does not give integer answers when they exist, such as a calculator displaying 2.999999 as the fifth root of 243. Mention that Lesson 4-10 will discuss how calculators do the arithmetic of powers. On some calculators, to find an nth root of x, a more accurate answer may be found by entering $x\,\boxed{\sqrt[n]{x,}}\,n$ than by entering $x\,\boxed{x^y}\,(1/n)$. This is because the calculator stores more digits than it displays.

1. a. Show that 1/2 is a cube root of 1/8.

$\left(\frac{1}{2}\right)^3 = \left(\frac{1}{2}\right) \cdot \left(\frac{1}{2}\right) \cdot \left(\frac{1}{2}\right) = \frac{1}{8}$

b. Show that -1/2 is not a cube root of 1/8.

The cube of $-\frac{1}{2}$ is negative; it cannot be a cube root of a positive number.

2. Evaluate: **a.** $\left(\frac{1}{27}\right)^{1/3}$

b. $\sqrt[4]{81}$.

a. $\frac{1}{3}$; b. 3

3. a. Graph the functions $f(x) = x^4$ and $g(x) = x^{1/4}$.

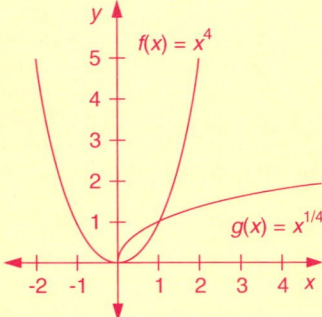

b. On what domain are functions with these formulas inverses?

{x: x ≥ 0}

4. Give a formula for the radius r of a sphere with surface area S.

$S = 4\pi r^2$, so $r = \sqrt{\dfrac{S}{4\pi}}$

not calculate powers or roots of negative numbers. On such calculators, to evaluate nth roots of negative numbers you will have to do the calculations using positive numbers, and adjust the sign of the answer as necessary.

Taking the nth power and taking the nth root of a number are *inverse operations*. Each "undoes" the result of the other. For instance, if you start with the number 10, raise it to the 4th power (to get $10^4 = 10,000$), and then take the 4th root (to get $\sqrt[4]{10,000}$), you end up with the original number, 10. Similarly, as Example 3 shows, the functions

$$f(x) = x^n \quad \text{and} \quad g(x) = x^{1/n}$$

are inverses of each other as long as $x^{1/n}$ is defined and represents a function.

Example 3   **a.** Use an automatic grapher to plot $f(x) = x^3$ and $g(x) = x^{1/3}$ on the same set of axes.
   b. For what values of x is $f(x) = g^{-1}(x)$? Justify your answer.

Solution

a. Below are the graphs produced by one grapher. The graph of $f(x) = x^3$ is black; the graph of $g(x) = x^{1/3}$ is in orange.

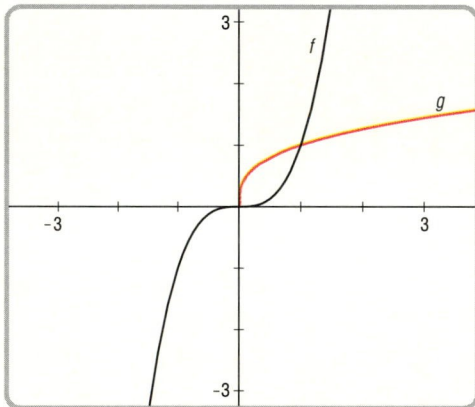

b. g is not defined when $x<0$. So g can only be the inverse of f if the domain of f is limited to nonnegative real numbers. So suppose $x \ge 0$. Then

$$f(g(x)) = f\left(x^{1/3}\right) = \left(x^{1/3}\right)^3 = x$$

and $g(f(x)) = g\left(x^3\right) = \left(x^3\right)^{1/3} = x$.

Thus, $f = g^{-1}$ and $g = f^{-1}$ for the domain $\{x : x \ge 0\}$.

The patterns in Example 3 generalize.

Theorem

Let n be an integer, $n \ge 2$. On the domain $\{x : x \ge 0\}$, the functions $f(x) = x^n$ and $g(x) = x^{1/n}$ are inverse functions.

204

Functions with equations of the form $y = x^{1/n}$, where n is an integer, $n \geq 2$, are called **nth root functions**. Because $x^{1/n}$ is defined only for $x \geq 0$, the domain of all these functions is the set of nonnegative real numbers. The range is also the set of nonnegative reals. Some nth root functions are graphed below.

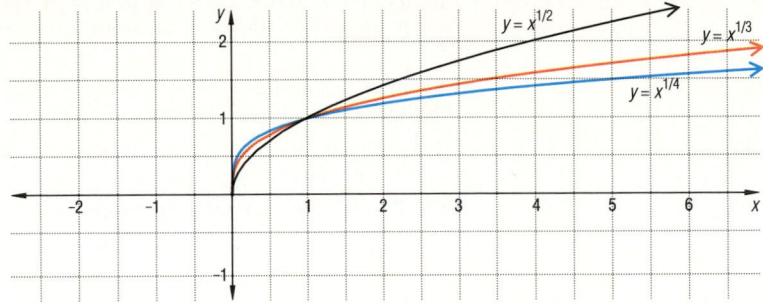

Some practical situations can be modeled using nth root functions.

■ ■ ■ ■ ■ ■ ■

Example 4 Spherical ball bearings are made by dropping a volume of molten metal down a shaft. Express the radius r in mm of an individual ball bearing as an nth root function of its volume V in mm^3.

Solution The volume V of a sphere with radius r is given by

$$V = \frac{4}{3}\pi r^3.$$

Solve for r.
$$r^3 = \frac{3V}{4\pi}$$

$$r = \sqrt[3]{\frac{3V}{4\pi}}.$$

So, the radius is given by $r = f(V) = \left(\frac{3V}{4\pi}\right)^{1/3}$.

Check Substitute a value for V, say $V = 2$, in the formula for r. Then $f(2) = \left(\frac{3 \cdot 2}{4\pi}\right)^{1/3} \approx 0.782$. Now check that these values work in the original formula: $V = \frac{4}{3} \cdot \pi \cdot (0.782)^3 \approx 2$. It does.

Questions

1. Show that 7 is a fifth root of 16,807. $7^5 = 16,807$

2. **a.** Show that 3.5 is a fourth root of 150.0625 . $(3.5)^4 = 150.0625$
   b. Give the other real fourth root of 150.0625.   **-3.5**

In 3 and 4, a number x and a value of n are given. **a.** Give the number of real nth roots x has. **b.** Find all real nth roots of x.

3. $x = 4096, n = 4$ **a) 2; b) -8 and 8**   4. $x = 243, n = 5$ **a) 1; b) 3**

In 5–7, evaluate without a calculator.

5. $1000^{1/3}$ **10**   6. $\sqrt[4]{625}$ **5**   7. $\sqrt[3]{-216}$ **-6**

8. For what values of x and n are $x^{1/n}$ and $\sqrt[n]{x}$ equal?
   $x \geq 0$ and n an integer with $n \geq 2$

In 9 and 10, *true* or *false*.

9. $36^{1/2} = -6$ **False**   10. $(125)^{1/3} > \sqrt[3]{-125}$ **True**

11. Let $f(x) = x^4$ and $g(x) = x^{1/4}$.
    a. Plot $y = f(x)$ and $y = g(x)$ on the same set of axes.   **See margin.**
    b. How can the domains of f and g be restricted so that f and g are inverse functions?   **The domains of f and g can be restricted to the set of all nonnegative real numbers.**

In 12 and 13, *true* or *false*.

12. The domain of each nth root function is the set of real numbers.   **False**

13. The graph of every nth root function passes through the point $(1, 1)$.
    True

14. It has been found that the speed s in knots of a ship is a function of the horsepower p developed by its engines, with $s = 6.5p^{1/7}$. How fast will a ship travel with engines producing 650 horsepower?   **about 16.4 knots**

15. The ancient Greeks wanted to construct a cube that would have twice the volume of a cube at the altar at Delos. If the given cube had side of length one unit, to the nearest thousandth what should be the length of a side of the constructed cube?   **1.260 units**

In 16 and 17, use an automatic grapher.

16. **a.** Plot $y = x^{1/2}$ and $y = -x^{1/2}$ on one set of axes.   **See margin.**
    b. What single equation describes the union of these two graphs?   **$x = y^2$**

17. **a.** Plot $f(x) = \sqrt[5]{x}$ and $g(x) = \sqrt[5]{x+2} + 3$ on one set of axes.
    b. What transformation maps the graph of f to the graph of g?
    c. What transformation maps the graph of g to the graph of f?
    a) See margin. b) $(x, y) \rightarrow (x - 2, y + 3)$; c) $(x, y) \rightarrow (x + 2, y - 3)$

18. The volume V of a cone with radius r and height equal to the radius is given by $V = \frac{1}{3}\pi r^3$.
    a. Give a formula for the radius in terms of V.   **$r = (3V/\pi)^{1/3}$**
    b. A cone with equal radius and height has volume 60 mm^3. Use the formula in part **a** to find its height.   **about 3.9 mm**

206

19. Determine whether the functions $f(x) = \frac{2}{x}$ and $g(x) = \frac{x}{2}$ are inverse functions. Justify your answer. *(Lesson 3-8)* **Not inverses; (2, 1) is on graph of $f(x)$, but (1, 2) is not on graph of $g(x)$.**

20. *Multiple choice.* In 1990, a first class airmail letter to Australia cost 45¢ per half ounce or fraction thereof to mail up to 2 ounces. Which formula describes the cost to send a letter weighing w ounces for $0 < w \leq 2$? *(Lesson 2-4)*
(a) $45 - 45\lceil 1 - 2w \rceil$ (b) $45\lceil w \rceil$
(c) $45\lceil 2w \rceil$ (d) $45 - 45\lceil 0.5 - w \rceil$ **(c)**

In 21 and 22, suppose a real estate analyst wants to estimate the average price of a three-bedroom home in a city. She picks a 30 square-block region on one side of a city and calculates the average price of three-bedroom homes.

21. a. Name the population. **3-bedroom homes in the city**
b. Name the sample. *(Lesson 1-1)*
3-bedroom homes in a 30-block region on one side of the city
22. a. Why might the sample not be representative of the population?
b. Suggest a better way for choosing a representative sample.
(Lesson 1-1) **a) See margin. b) sample: The analyst could have randomly selected blocks throughout the city.**

23. *Skill sequence.* Solve. *(Previous course)*
a. $3^z = 81$ **z = 4** **b.** $3^{y-1} = 81$ **y = 5**

24. Here are results on a computer analysis relating the independent variable, year of graduation, and dependent variable, percent of high school graduates reporting having ever smoked a cigarette.

Year	% reporting smoking
1975	73.6
1978	75.3
1981	71.0
1983	70.6
1985	68.8
1987	67.2

$\bar{x} = 1981.5$ $\bar{y} = 71.1$
$s_x = 4.46$ $s_y = 2.99$
Source: *1989 World Almanac*, p. 215

a. The correlation is -0.92. Is it reasonable to assume a line would be a good model of data? **Yes**
b. Give a point which must be on the line of best fit. **(1981.5, 71.1)**
c. The line of best fit is $y = -0.61457x + 1288.9$. What is the predicted value for 1995? *(Lessons 2-3, 2-5)* **62.8%**

25. Here are data on standardized algebra achievement scores for ten students: 15, 16, 16, 19, 21, 25, 25, 26, 29, 35. Compute the mean and standard deviation and compare this group to the reported national mean of 21 and standard deviation of 5. *(Lesson 1-8)* **mean = 22.7; standard deviation = 6.5. The mean and s.d. are both higher than the national figures.**

26. Consider functions of the form $y = x^{1/n}$. Theoretically, there is no largest n possible. Experiment with your automatic grapher. Plot $y = x^{1/n}$ for some large values of n. What is the largest value of n your machine can handle before memory or screen resolution is overloaded? **Answers will vary.**

FOLLOW-UP

MORE PRACTICE
For more questions on SPUR Objectives, use *Lesson Master 4-1*, shown below.

NOTES ON QUESTIONS
Making Connections for Question 23: These equations are review for Lesson 4-3. It is expected that students will solve part **a** by trial and error, or by recognizing that $81 = 3^4$ and equating exponents, and then use that solution to solve part **b**

Question 24: You might ask whether the year causes reduced smoking. Also, point out that the data spans a 12-year period of time. Is it reasonable to ask about the predicted value of the function for 1995?

Small Group Work for Question 26: Students can work in pairs or small groups to answer this question. One student can enter data while others keep a record and help analyze results.

NAME _____

LESSON **MASTER 4-1**
QUESTIONS ON **SPUR** OBJECTIVES

■**SKILLS** *Objective A (See pages 263–266 for objectives.)*
1. *Multiple choice.* Which expression is not equal to 4? **(b)**
(a) $\sqrt[3]{64}$ (b) $8^{\frac{1}{2}}$
(c) $256^{\frac{1}{4}}$ (d) $\sqrt{16}$

In 2–5, assume each variable represents positive numbers. Rewrite using a radical sign.
2. $x^{\frac{1}{7}}$ $\sqrt[7]{x}$ 3. $(3y)^{\frac{1}{2}}$ $\sqrt{3y}$
4. $(z^5)^{\frac{1}{3}}$ $\sqrt[3]{z^5}$ or $2\sqrt[3]{z^5}$ 5. $(w^2)^{\frac{1}{2}}$ $\sqrt{w^2}$

In 6–9, rewrite without a radical sign. Assume $a > 0$ and $c > 0$.
6. $\sqrt[4]{c}$ $c^{\frac{1}{4}}$ 7. $\sqrt[5]{32c^2}$ $2c^{\frac{2}{5}}$
8. $\sqrt[3]{a}$ $a^{\frac{1}{3}}$ 9. $\sqrt[6]{64c^{1200}}$ $2^{\frac{1}{6}}c^{100}$

In 10–14, evaluate without using a calculator.
10. $81^{\frac{1}{4}}$ **3** 11. $\sqrt[3]{343}$ **7**
12. $\sqrt{10000}$ **100** 13. $128^{\frac{1}{7}}$ **2**
14. Order from smallest to largest: 2^5, 5^2, $2^{\frac{1}{5}}$, 2^5, $5^{\frac{1}{2}}$, $2^{\frac{1}{5}}$, $2^{\frac{1}{2}}$, $5^{\frac{1}{2}}$, 5^2, 2^5

■**USES** *Objective F*
15. **a.** A cube has edges of length e. Express e in terms of the surface area A of the cube.
b. If the surface area of a cube is 96 cm², what is the cube's edge length?
a. $e = \sqrt{\frac{A}{6}}$ b. **4 cm**

16. Students traced different sized circular objects to draw a model of a planetary system. Express the radius r of a circular orbit as a function of its area A.
$r = \sqrt{\frac{A}{\pi}}$

Functions, Statistics, and Trigonometry © Scott, Foresman and Company **41**

OBJECTIVES

A Evaluate $b^{m/n}$ for $b > 0$.
F Use rational exponents to model situations.

TEACHING NOTES

You may want to review the distinction between the words *power* and *exponent*. In the power x^y, y is the exponent. We say that x^y is the yth power of x, so this confirms that x^y is the power.

Making Connections In the equation $T(p) = 10p^{0.585}$ in paragraph one on page 208, the exponent 0.585 is the solution to $2^t = 1.5$. See Lesson 4-8, page 245, Question 14.

Example 1 points out that students can evaluate expressions with rational exponents mentally. Taking the root first is generally a good idea, as this allows one to work with a smaller number.

Technology In the check for **Example 1,** notice that $\approx$ is used rather than =. With a calculator we first changed 3/7 to a decimal and wrote down the 0.4285714 on the display. Then we calculated $128^{0.4285714}$ and got 7.99999989. The reason $128^{0.4285714}$ gives an approximate value is because 3/7 is not exactly 0.4285714. However, this calculator stores more digits than it can display. When we calculated 3/7, stored it in memory, and

4-2

Rational Power Functions

According to a certain microwave cookbook, whenever you double the amount being cooked, the cooking time should be multiplied by 1.5. For example, if it takes 10 minutes to cook one portion, it takes 15 minutes to cook two portions and 22.5 minutes to cook four portions. Using techniques you will learn in Lesson 4-8, it can be shown that if it takes 10 minutes to cook one portion then the approximate time $T(p)$ it takes to cook p portions is given by

$$T(p) = 10p^{0.585}.$$

What is the time needed to cook 5 portions? From the formula,

$$T(5) = 10 \cdot 5^{0.585}.$$

Use a calculator to get $5^{0.585} \approx 2.56$. So $T(5) \approx 10(2.56) = 25.6$ minutes.

But what does it *mean* to find the 0.585 power of a number? As in the case for the exponent $\frac{1}{n}$, the properties of powers indicate how rational exponents can be interpreted.

For instance, suppose $x > 0$. Then,

$$x^{\frac{3}{5}} = x^{\frac{1}{5} \cdot 3} \qquad \text{Definition of division}$$

$$= \left(x^{\frac{1}{5}}\right)^3 \qquad \text{Power of a Power Property}$$

$$= \left(\sqrt[5]{x}\right)^3. \qquad \text{Definition of } \sqrt[n]{x}$$

Similarly, $\quad x^{\frac{3}{5}} = x^{3 \cdot \frac{1}{5}} \qquad \text{Definition of division}$

$$= \left(x^3\right)^{\frac{1}{5}} \qquad \text{Power of a Power Property}$$

$$= \sqrt[5]{x^3}. \qquad \text{Definition of } \sqrt[n]{x}$$

These arguments can be generalized.

> **Rational Exponent Theorem**
>
> For all positive integers m and n, and base $x > 0$,
>
> $x^{\frac{m}{n}} = \left(x^{\frac{1}{n}}\right)^m = \left(\sqrt[n]{x}\right)^m$, the mth power of the nth root of x and
>
> $x^{\frac{m}{n}} = \left(x^m\right)^{\frac{1}{n}} = \left(\sqrt[n]{x^m}\right)$, the nth root of the mth power of x.

208

Thus by writing $5^{0.585}$ as $5^{\frac{585}{1000}} = 5^{\frac{119}{200}}$, $5^{0.585}$ can be interpreted either as the 200th root of the 119th power of 5 or as the 119th power of the 200th root of 5. You would not compute $5^{0.585}$ by hand, but some rational powers can be computed by hand.

- - - - - - - - -

Example 1 Evaluate $128^{\frac{3}{7}}$.

Solution 1 Find the 3rd power of the 7th root of 128:
$$128^{\frac{3}{7}} = \left(\sqrt[7]{128}\right)^3 = 2^3 = 8.$$

Solution 2 Find the 7th root of the 3rd power of 128:
$$128^{\frac{3}{7}} = \sqrt[7]{128^3} = \sqrt[7]{2,097,152} = 8.$$

Check Use a calculator. Key in $128\ \boxed{y^x}\ \boxed{(}\ 3\ \boxed{\div}\ 7\ \boxed{)}\ \boxed{=}$ or its equivalent.

Notice in Example 1 that Solution 1 involves much smaller numbers than Solution 2. Generally, it is easier to take roots before powering when evaluating rational exponents. Notice also that while $128^{\frac{3}{7}}$ is an integer, most numbers raised to rational exponents can only be approximated by a decimal. For example, the value of $132^{\frac{2}{5}}$ to three decimal places is 7.051.

Recall the following general properties for powers using real number exponents. From these properties you can determine properties of zero and negative exponents.

Postulates for Powers

For any positive real numbers x and y and real numbers m and n:

Product of Powers Property	$x^m \cdot x^n = x^{m+n}$
Power of a Product Property	$(xy)^n = x^n y^n$
Quotient of Powers Property	$\dfrac{x^m}{x^n} = x^{m-n}\quad (x \neq 0)$
Power of a Quotient Property	$\left(\dfrac{x}{y}\right)^n = \dfrac{x^n}{y^n}\quad (y \neq 0).$

These properties can be used to rewrite numbers with zero as an exponent. For instance,
$$\frac{2^5}{2^5} = 2^{5-5} = 2^0 \text{ by the Quotient of Powers Property.}$$

But $\dfrac{2^5}{2^5} = 1$ because any nonzero number divided by itself is 1. So $2^0 = 1$.

then did the powering using the stored exponent, we got 8.0000000. Perhaps the best way to do this is to key in

$128\ \boxed{y^x}\ \boxed{(}\ 3\ \boxed{\div}\ 7\ \boxed{)}$.

Continue helping your students find more efficient and accurate ways to do these calculations. When using a calculator, use all of its power.

Alternate Approach Another example that illustrates the Negative Exponent Theorem is the following: $\dfrac{x^5}{x^7} = x^{-2}$ by the Quotient of Powers Property. But $\dfrac{x^5}{x^7} = \dfrac{x^5}{x^5} \cdot \dfrac{1}{x^2}$ by the Product of Powers Property, so $x^{-2} = \dfrac{1}{x^2}$.

The Negative Exponent Theorem may appear cumbersome as stated. You could separate it into two parts to emphasize the conditions to your students.
1. For $x > 0$ and n a real number, $x^{-n} = \dfrac{1}{x^n}$. This allows you to deal with an expression such as $(17.2)^{-2/3}$.
2. For $x \neq 0$, and n an integer, $x^{-n} = \dfrac{1}{x^n}$. This allows you to include negative bases, as long as the exponent is integral, such as $(-17.2)^{-3}$.

When solving a problem like that in **Example 2,** have students parse out the exponent: The negative in the exponent means reciprocal; the denominator means 3rd root; the numerator means 5th power.

You should remind your students that $x \geq 0$ in the definition of a rational power function, at the top of page 211. This guarantees that the graph of the function will only be in the first quadrant. The graphs following this definition are shown on a teaching aid.

In general, using the Quotient of Powers Property, when $x > 0$, $\frac{x^n}{x^n} = x^{n-n} = x^0$. But $x^n \neq 0$, so $\frac{x^n}{x^n} = 1$. So $x^0 = 1$. This proves the following result.

Zero Exponent Theorem

If b is any nonzero real number, $b^0 = 1$.

Properties of negative exponents can also be deduced from the Postulates for Powers. Consider 2^{-7}. What does this mean? By the Product of Powers Property,

$$
\begin{aligned}
2^7 \cdot 2^{-7} &= 2^{7 + (-7)} \\
&= 2^0 \\
&= 1. \qquad \text{Zero Exponent Theorem}
\end{aligned}
$$

Dividing each side of the equation by 2^7 gives

$$2^{-7} = \frac{1}{2^7}.$$

This result is generalized as the following theorem.

Negative Exponent Theorem

For $x > 0$ and n a real number, or for $x \neq 0$ and n an integer,

$$x^{-n} = \frac{1}{x^n}.$$

So x^n and x^{-n} are reciprocals. For example, 9^{-2} and 9^2 are reciprocals:

$$9^{-2} = \frac{1}{9^2} = \frac{1}{81},$$

$$9^2 = \frac{1}{9^{-2}} = \frac{1}{\frac{1}{81}} = 81.$$

▪ ▪ ▪ ▪ ▪ ▪ ▪ ▪ ▪

Example 2 Evaluate $27^{-\frac{5}{3}}$.

Solution $27^{-\frac{5}{3}} = \dfrac{1}{27^{\frac{5}{3}}} = \dfrac{1}{(\sqrt[3]{27})^5} = \dfrac{1}{3^5} = \dfrac{1}{243}.$

Check Use your calculator, with 27 as the base and $-\frac{5}{3}$ as the exponent. Key in 27 $\boxed{y^x}$ $\boxed{(}$ 5 $\boxed{\div}$ 3 $\boxed{\pm}$ $\boxed{)}$ $\boxed{=}$ or its equivalent. Our display shows 0.00411523. On the same calculator, we get the same display for $\frac{1}{243}$.

Any function f with equation of the form $f(x) = x^{\frac{m}{n}}$, where m and n are non-zero integers, is a **rational power function**.

Both the domain and range of a rational power function are the set of nonnegative real numbers. Graphs of rational power functions can be drawn using an automatic grapher or by plotting a few points by hand and sketching a curve through them. Here is a table of values and a sketch for $y = x^{\frac{3}{2}}$. For comparison, the graphs of $y = x$ and $y = x^2$ are also sketched.

$y = x$	$y = x^{\frac{3}{2}}$	$y = x^2$
0	0	0
.5	0.4	0.25
1	1	1
1.5	1.8	2.25
2	2.8	4
2.5	4.0	6.25
3	5.2	9
3.5	6.5	12.25
4	8	16

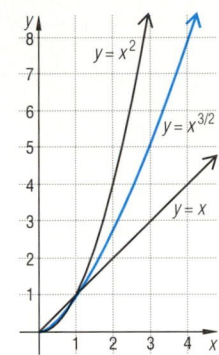

Notice that for all $x \neq 1$, the graph of $y = x^{\frac{3}{2}}$ lies between the graphs of $y = x^1$ and $y = x^2$. This is as expected, for $\frac{3}{2}$ is between 1 and 2.

■ ■ ■ ■ ■ ■ ■ ■ ■

Example 3   Let $f(x) = x^{\frac{5}{3}}$.

    a. Find an equation for the inverse of f.
    b. Is the inverse a function?
    c. Use an automatic grapher to plot f and its inverse g.

Solution

a. The inverse of $y = x^{\frac{5}{3}}$ is $x = y^{\frac{5}{3}}$. To solve for y, raise each side to the $\frac{3}{5}$ power. $x^{\frac{3}{5}} = \left(y^{\frac{5}{3}}\right)^{\frac{3}{5}} = y^1$. So the inverse of f is $y = x^{\frac{3}{5}}$.

b. For the inverse $y = x^{\frac{3}{5}}$, to each $x \geq 0$ there corresponds a unique value of y. So the inverse is a function.
c. The graphs are produced at the right.

Check   Note that the graphs of f and g are reflection images over the line $y = x$, verifying that f and g are inverses. The graph of g passes the vertical line test, verifying that g is a function.

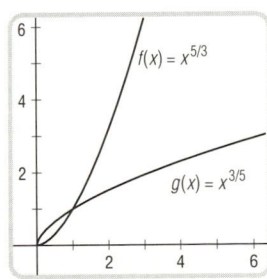

LESSON 4-2   Rational Power Functions   211

b. Find the surface area of this cylinder to the nearest square centimeter if its volume is 1 liter.
Substituting 1000 cm³ for V in the formula for S.A. in part a, S.A. ≈ 369 cm².
Note: For all cylinders with volume 1 liter, this cylinder has the smallest surface area.

NOTES ON QUESTIONS
Error Analysis for Question 1: We use 64 as a base because both the cube root and square root come out even. If students switch these, they will answer 16 for part **b** and 1/512 for part **d**.

Questions 3, 9, 16, and 17: These involve exponents which are decimals rather than fractions. They are important because many students incorrectly think that a decimal power has different properties than a fractional power. Students will work more with decimal exponents in Lesson 4-3.

Question 15: If you did not address the issue during the discussion of Lesson 4-1, do this question in class and make sure students understand that the ambiguity here exemplifies the reason we do not allow negative bases with rational exponents.

Question 16: Because rational powers are usually calculated by considering numerator and denominator separately, students do not realize the continuity of powers. The two functions here have graphs that are relatively close to each other because 2.2 and 2 are close. You might graph $h(x) = x^3$ to show that the graph of g is between the graphs of f and h.

ADDITIONAL ANSWERS
11.a.,b.

12.a.

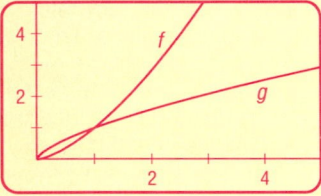

b. They are reflections of each other over the line y = x. Thus, they are inverses. This can be verified by computing
$f(g(x)) = f(x^{3/2}) = (x^{3/2})^{2/3} = x$ **and** $g(f(x)) = g(x^{2/3}) = (x^{2/3})^{3/2} = x$**.**

16.a.

When a formula contains integral powers, it is often useful to rewrite the formula as a rational power function, as the next example illustrates.

Example 4   **a.** Express the surface area A of a cube as a function of its volume V.
b. Find the volume of a cube with surface area 93 square inches.

Solution
a. Let s be the length of one edge of the cube. The surface area is $A = 6s^2$. Because $V = s^3$, each edge of the cube has length $s = \sqrt[3]{V}$. So the area of each face is $s^2 = \left(\sqrt[3]{V}\right)^2 = V^{\frac{2}{3}}$. Thus $A = 6V^{\frac{2}{3}}$.
b. Substitute 93 for A in the formula found in **a** and solve for V.

$$6V^{\frac{2}{3}} = 93$$
$$V^{\frac{2}{3}} = 15.5$$

Raise each side to the $\frac{3}{2}$ power. $\left(V^{\frac{2}{3}}\right)^{\frac{3}{2}} = (15.5)^{\frac{3}{2}}$
$$V \approx 61.024$$

A cube with a surface area of 93 square inches has a volume of about 61 cubic inches.

Check   Substitute 61 for V in the formula. $A = 6 \cdot (61)^{\frac{2}{3}} \approx 92.98$; that checks. (Substituting 61.024 for V yields 93.000.)

Questions

Covering the Reading

1. Evaluate each number without a calculator.
   a. $64^{1/3}$ **4**   **b.** $64^{3/2}$ **512**   **c.** $64^{-1/2}$ $\frac{1}{8}$ **d.** $64^{-2/3}$ $\frac{1}{16}$

In 2–4, suppose the variables represent positive numbers. Write each expression using fractional exponents.
   2. $\left(\sqrt[3]{r}\right)^5$ $r^{5/3}$ **3.** $k^{-0.3}$ $k^{-3/10}$ **4.** $\sqrt[4]{t^9}$ $t^{9/4}$

In 5 and 6, write each expression using a radical sign. Assume $x > 0$ and $y > 0$.
   5. $x^{7/8}$ $(\sqrt[8]{x})^7$ or $\sqrt[8]{x^7}$ **6.** $y^{-8/5}$ $(\sqrt[5]{y})^{-8}$ or $\sqrt[5]{y^{-8}}$

In 7 and 8, give the exact value without using a calculator.
   7. $16^{-3/4}$ $\frac{1}{8}$ **8.** $125^{-5/3}$ $\frac{1}{3125}$

9. Without using a calculator, decide which is larger, $0.1^{0.3}$ or $0.1^{-0.3}$. Check with a calculator.   **$0.1^{-0.3}$**

10. Refer to the microwave cooking time information at the start of this lesson. Use two different methods to find the approximate time needed to cook six portions.   **about 28.5 minutes**

212

11. **a.** Make a table of values and sketch a graph of $y = x^{5/2}$ for $0 \le x \le 3$ and $0 \le y \le 3^{5/2}$.
   b. Plot the inverse of this function on the same set of axes.
   c. Write an equation for the inverse.   **a,b) See margin. c) $y = x^{2/5}$**

12. **a.** Use an automatic grapher to plot $f(x) = x^{2/3}$ and $g(x) = x^{3/2}$ on the same set of axes.
   b. How are the graphs related? Justify your answer.   **a,b) See margin.**

13. **a.** Express the volume of a cube as a function of its surface area.
   b. Find the volume of a cube whose surface area is 34.56 cm².
      a) $V = (A/6)^{3/2}$; b) 13.824 cm³

Applying the Mathematics

14. The earth's atmospheric pressure decreases as you ascend from the surface. It can be shown that at altitude h kilometers $(0 < h < 80)$, the pressure P in Newtons per square centimeter (N/cm²) is approximately given by the formula $P = 10.13e^{-0.116h}$. Give the approximate pressure at
   a. 30 km above the earth   **0.31 N/cm²**
   b. 50 km above the earth   **0.031 N/cm²**
   c. ground level.   **10.13 N/cm²**

15. This question shows why rational exponents are not used with negative bases.
   a. If $(-125)^{1/3}$ is the cube root of -125, then $(-125)^{1/3}$ is equal to ___?___.
   b. According to the Rational Exponent Theorem, $(-125)^{2/6}$ equals $((-125)^2)^{1/6}$ which equals ___?___.
   c. Are your answers to **a** and **b** the same?   **a) -5; b) 5; c) No**

In 16 and 17, use an automatic grapher to plot each pair of functions on the same axes.

16. **a.** Sketch graphs of $f(x) = x^2$ and $g(x) = x^{2.2}$ for $x \ge 0$.   **See margin.**
   b. Where do the graphs intersect?   **(0, 0) and (1, 1)**
   c. For what values of x is $g(x) > f(x)$?   **$x > 1$**
   d. For what values of x is $g(x) < f(x)$?   **$0 < x < 1$**

17. **a.** Sketch graphs of $f(x) = x^{0.4}$ and $g(x) = x^{-0.4}$ for $x \ge 0$.   **See margin.**
   b. *True* or *false*. The graphs are reflection images of each other over the line $y = x$.   **False**

18. Solve $7^x = 20$ for x to the nearest tenth using trial and error and a calculator.   **1.5**

Review

19. Order from smallest to largest.   *(Lesson 4-1, Previous course)*
   $$3^4, 3^{-4}, 4^{-3}, 3^{1/4}, \frac{3}{4}$$ **$3^{-4}, 4^{-3}, \frac{3}{4}, 3^{1/4}, 3^4$**

20. For what values of x is $\sqrt[n]{x} > x$?   *(Lesson 4-1)*   **$0 < x < 1$**

21. Consider the function $f(x) = \sqrt[3]{x} - 2$.
   a. Give an equation for the inverse of f.   **$f^{-1}(x) = (x + 2)^3$**
   b. Sketch f and its inverse on the same set of axes.   **See margin.**
   c. What is the largest domain on which both f and its inverse can be defined so each is a function?   *(Lessons 4-1, 3-8)*   **the set of all real numbers**

LESSON 4-2   Rational Power Functions   **213**

21.b.

22. The keys of a piano are tuned to have frequencies in cycles per second (cps) so that each note (white *and* black) has a frequency $2^{1/12}$ times that of the previous note. Often, the A below middle C is tuned to a frequency of 220 cps.
    a. Find the frequency of middle C. **about 262 cps**
    b. Find the frequency of the A above middle C, which is one octave (12 notes) higher than the A below middle C. *(Lesson 4-1)* **440 cps**

23. Use the table below from *The Statistical Abstract of the United States, 1990.*

Cotton-Acreage, Production, and Value, by State: 1986 to 1988

STATE	Acreage harvested (1,000)			Yield per acre (lb.)			Production (1,000 bales)			Farm price per lb. (cents)			Farm value (mil. dol.)		
	1986	1987	1988	1986	1987	1988	1986	1987	1988	1986	1987	1988	1986	1987	1988
U.S.	8,468	10,035	11,891	552	706	623	9,731	14,760	15,446	52.4	64.3	56.4	2,449	4,555	4,057
AL	313	333	355	506	572	514	330	397	380	52.1	64.8	51.3	83	123	94
AZ	323	380	477	1,224	1,342	1,127	823	1,062	1,120	60.6	73.4	68.7	239	374	365
AR	480	550	675	602	786	747	602	901	1,050	49.7	63.5	51.7	144	275	261
CA	990	1,141	1,337	1,088	1,258	1,024	2,245	2,991	2,853	59.1	69.6	63.6	637	999	863
GA	195	245	315	455	662	564	185	338	370	58.3	61.8	53.5	52	100	95
LA	570	600	645	567	782	707	673	977	950	49.8	63.2	52.0	161	296	237
MO	160	189	237	588	838	628	196	330	310	51.5	66.0	51.6	48	105	77
MS	1,000	1,010	1,190	571	829	738	1,190	1,745	1,830	50.9	63.6	51.0	291	533	448
OK	350	400	400	288	415	348	210	346	290	43.4	58.4	42.8	44	97	60
TN	335	435	535	567	700	529	396	634	590	49.0	63.0	51.2	93	192	145
TX	3,476	4,431	5,340	356	508	473	2,576	4,686	5,260	46.8	60.1	51.5	579	1,351	1,253

  a. For Tennessee, in 1986, explain the relationship between the numbers 335, 567, 49.0, and 93.
  b. Which was generally the worst of the three seasons shown for cotton production? Justify your answer. *(Lesson 1-2)*
  a,b) See margin.

Exploration

24. The expression 0^0 is called an *indeterminate form* because there are several values which might be reasonable for 0^0.
    a. Compute the sequence $0^2, 0^1, 0^{1/2}, 0^{1/4}, 0^{1/8}, 0^{1/16}, \ldots$. What does this imply as a reasonable value for 0^0? **0, 0, 0, ...; $0^0 = 0$**
    b. Compute the sequence $2^0, 1^0, \left(\frac{1}{2}\right)^0, \left(\frac{1}{4}\right)^0, \left(\frac{1}{8}\right)^0, \left(\frac{1}{16}\right)^0, \ldots$. What does this imply as a reasonable value for 0^0? **1, 1, 1, ...; $0^0 = 1$**
    c. Compute the sequence $2^2, 1^1, \left(\frac{1}{2}\right)^{1/2}, \left(\frac{1}{4}\right)^{1/4}, \left(\frac{1}{8}\right)^{1/8}, \ldots$. What does this sequence imply? **4, 1, .71, .71, .77, ...; $0^0 = 1$**
    d. Tell what value you get for 0^0 when you use the $\boxed{y^x}$ key on different calculators. **Answers depend on the calculator used. Most will display an error message.**

LESSON 4-3

Exponential Functions

RESOURCES
- Lesson Master 4-3
- Quiz for Lessons 4-1 through 4-3
- Teaching Aid 34: Graph of $y = 2^x$ and $y = \left(\frac{1}{2}\right)^x$
- Computer Master 10

Consider a population of bacteria that doubles in number every hour. Suppose there are 10 bacteria initially. The table shows the population P at the end of each of the first 4 hours.

A graph of the data is below. The population P is a function of the number of hours h. If N is that function, then $P = N(h) = 10 \cdot 2^h$ because h indicates the number of times the population has doubled.

Hours passed	Population
0	10
1	20
2	40
3	80
4	160

What can be said about the points between those on the left graph? Because time is a continuous variable, after half an hour you would expect there to be $N\left(\frac{1}{2}\right) = 10 \cdot 2^{1/2}$ bacteria. This is $10\sqrt{2} \approx 14.14$ bacteria, which should be truncated to 14 since it isn't meaningful to consider fractions of bacteria. Similarly, other rational periods of time can be interpreted to give a more complete graph, like the one at the right above.

Now consider times between rational values of h, such as $\sqrt{2}$, $\sqrt{3}$, or π hours. To evaluate irrational powers of numbers, such as $2^{\sqrt{2}}$, $2^{\sqrt{3}}$, or 2^{π}, approximate the value of the exponent. For example, $2^{\sqrt{3}}$ can be estimated from approximations to $\sqrt{3}$.

Because $1 < \sqrt{3} < 2$, then $2^1 < 2^{\sqrt{3}} < 2^2$, so $2^1 < 2^{\sqrt{3}} < 4$.

Because $1.7 < \sqrt{3} < 1.8$, then $2^{1.7} < 2^{\sqrt{3}} < 2^{1.8}$, so $3.25 < 2^{\sqrt{3}} < 3.49$.

Because $1.73 < \sqrt{3} < 1.74$, then $2^{1.73} < 2^{\sqrt{3}} < 2^{1.74}$, so $3.32 < 2^{\sqrt{3}} < 3.35$.

You can get a good approximation to $2^{\sqrt{3}}$ by using your calculator's powering key, with 2 as the base and $\sqrt{3}$ as the exponent. To the nearest thousandth, $2^{\sqrt{3}} = 3.322$.

LESSON 4-3 Exponential Functions   215

OBJECTIVES

D Describe properties of exponential functions.
G Use exponential models to solve problems.
I Graph exponential functions.
J Interpret graphs of exponential functions.

TEACHING NOTES

The exponential model of the bacteria population which begins this lesson is another example of applying a continuous model to a discrete situation—bacteria populations are whole numbers only. This motivates the question asked in the second paragraph. What can be said about the points in between? In answering this question, we use the term *truncated*. A mathematical definition of truncated is $y = \lfloor x \rfloor$; that is, any decimal portion of x is ignored.

You might ask students why certain restrictions are needed to define an exponential function. (If the coefficient a or the base b in $y = ab^x$ is zero, $f(x) = 0$ for all x. If the base $b = 1$, then $f(x) = a$ for all x. These are constant functions, not exponential functions.) **Question 14** addresses the matter of $b = 1$ from a graphical perspective.

If you think about $P = 10 \cdot 2^h$ out of the context of the bacteria situation, then
h can be any positive real number. The function mapping h to P is an example
of an exponential function.

> **Definition**
>
> An **exponential function with base *b*** is a function with a formula of the form
> $f(x) = ab^x$, where $a \neq 0$, $b > 0$, and $b \neq 1$.

$P = 10 \cdot 2^h$ is related to the parent exponential function whose equation is
$P = 2^h$. The domain of these and all exponential functions is the set of all real
numbers, and their graphs are continuous curves. These two functions are
graphed below, with $P = 2^h$ in black and $P = 10 \cdot 2^h$ in orange.

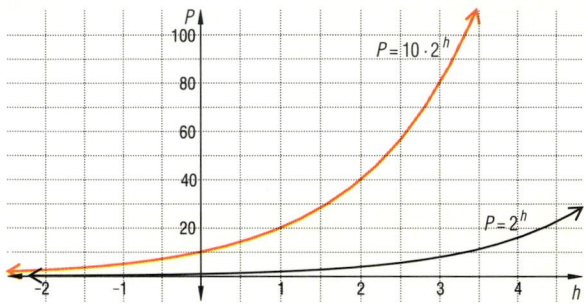

Exponential functions occur in nature when a quantity changes by a constant
factor during a given time period. Specifically, if a growth rate is a constant
percent of the population, the relation between time and population can be
described using an exponential function. In the bacteria growth above, the
population doubles or grows 200 percent every hour. So the base of the
function is 2.

Example

In 1985, the population of New Zealand was 3,295,000, with an average
annual growth rate of 1.4%. Assume that this growth rate continues
indefinitely.
a. Estimate the population of New Zealand in each of 1986, 1987, and
1988.
b. Express the population P as a function of n, the number of years after
1985.
c. Estimate New Zealand's population in the year 2000.

Solution   A growth rate of 1.4% means the new population is 1.014
times the old.
a. In 1986: $P = (3,295,000)(1.014) \approx 3,341,000$.
    In 1987: $P = (3,295,000)(1.014)^2 \approx 3,388,000$.
    In 1988: $P = (3,295,000)(1.014)^3 \approx 3,435,000$.
b. Generalize the pattern in **a.** In n years after 1985, the population will
have increased by 1.4% n times. That is,
$$P = f(n) = (3,295,000)(1.014)^n.$$
c. The year 2000 is 15 years after 1985, so $n = 15$.
$$P = f(15) = (3,295,000)(1.014)^{15} \approx 4,059,000$$
So under the assumption that the growth rate remains at 1.4%, the
population in year 2000 will be about 4,059,000.

216

The exponential function in the Example arises from assuming that the population growth rate is constant. In fact, there are many reasons why the growth rate may change: such as a trend toward smaller families, or changes in health care or nutrition resulting in a change in death rates. It is unwise to extrapolate growth rates further than a few years. In fact, by 1988 the growth rate in New Zealand had changed considerably, and its population was still only about 3,300,000.

The graph of any exponential function $f(x) = ab^x$ with base $b > 1$ and $a > 0$ is called an **exponential growth curve**. Here are the graphs of two such curves, with $f(x) = 3^x$ in black and $g(x) = 10^x$ in orange.

Based on the graphs above and your knowledge of powers, note the following properties of all exponential functions.

(1) The domain is the real numbers.
(2) The range is the set of positive real numbers. That is, $f(x)$ is *strictly positive* because $ab^x > 0$ for all real x.
(3) Because the range is the set of positive real numbers, every positive real number can be expressed as some power of b.
(4) The graph contains the point $(0, 1)$; it does not intersect the x-axis.
(5) The function is *strictly increasing* because as x values increase, corresponding y values increase.
(6) As x gets larger, $f(x)$ grows without bound.
(7) As x gets smaller, $f(x)$ is always positive but it approaches zero. (For example, on the graph of $f(x) = 3^x$, even for negative values as far from zero as -20, 3^x is still positive but very small. In this case, 3^{-20} represents the reciprocal of 3^{20}, or about 0.00000000029.) The graph has the x-axis as an asymptote.

However, when $b < 1$ (but still positive), the graph of the exponential function looks different. Consider the graphs of $g(x) = (0.5)^x$ and $f(x) = 2^x$.

5.b.

9. Sample: Exponential functions with bases in the range $0 < b < 1$ are always decreasing. As x gets smaller, $f(x)$ grows without bound. As x gets larger, $f(x)$ remains positive but approaches zero. Exponential functions with $b > 1$ are always increasing. As x gets larger, $f(x)$ grows without bound. As x gets smaller, $f(x)$ remains positive but approaches zero.

10.c.

MORE PRACTICE
For more questions on SPUR
Objectives, use *Lesson Mas-
ter 4-3,* shown on page 219.

EXTENSION
Small Group Work Obviously, $x^y = y^x$ when $x = y > 0$.
But it is also the case that
$2^4 = 4^2$. Are there any other
pairs of values for x and y
that satisfy this equation?
(Yes, for each value of $x \neq 1$,
there are two values of y that
work, one of them x and the
other not equal to x.) You
might have students get together in small groups to find,
with automatic graphers, solutions to this equation for various values of x both greater
than and less than 1. Collect
the work of the groups to obtain an approximate graph of
all solutions to this equation.
(The graph is the union of the
part of the line $y = x$ in the
first quadrant and a curve re-
sembling one branch of a hyperbola.)

EVALUATION
A quiz covering Lessons 4-1
through 4-3 is provided in the
Teacher's Resource File.

ADDITIONAL ANSWERS
**4.b., 5.b., 9., 10.c. See
margin, page 217.**

11.a.

12.a.

Notice that the graph of $g(x) = (0.5)^x$ is a reflection image of $f(x) = 2^x$ over the
y-axis. This is not so surprising when you recall two simple ideas.
   (1) $.5 = 2^{-1}$, so $(0.5)^x = (2^{-1})^x = 2^{-x}$.
   (2) Reflection over the y-axis occurs when x is replaced by $-x$ in an
        equation.

Graphs of exponential functions with base b such that $0 < b < 1$ are called
exponential decay curves. The word "decay" comes from the fact that these
exponential functions model situations—such as radioactive decay—in which
a quantity is diminishing at a constant rate.

Look again at the graphs of $g(x) = (0.5)^x$ and $f(x) = 2^x$. In general, the graph
of $g(x) = b^x$ for $0 < b < 1$ has many of the properties of $f(x) = b^x$ for $b > 1$. In
both cases, the domain is the reals and the range is the positive reals; and each
includes the intercept $(0, 1)$. When $0 < b < 1$, $g(x) = b^x$ is *strictly decreasing*.
As x increases, b^x is positive, but approaches zero; the x-axis is an asymptote.
As x decreases, the function increases without bound. So the end behavior of
an exponential decay function is reversed from that of an exponential growth
function.

Questions

The Great Wall of China

Covering the Reading

1. In the bacteria population in the lesson, how many bacteria will there be
   after 3.5 hours? **about 113**

2. Approximate $2^{\sqrt{5}}$ to the nearest hundredth. **4.71**

3. Determine if the function is an exponential function.
   a. $k(m) = 4^m$ **b.** $s(t) = 6$ **c.** $j(z) = z^2$ **d.** $p(x) = (0.6)^x$
   Yes   **No**   **No**   **Yes**

In 4 and 5, **a.** solve using a calculator and trial and error; **b.** show the solution by referring to the graph of $f(x) = 3^x$ or $g(x) = 10^x$.

4. $3^x = 4.76$ 5. $10^x = 3.5$
   a) $\approx$ 1.42; b) See margin.   **a) $\approx$ 0.54; b) See margin.**

6. The most populous country in the world is the People's Republic of
   China. In 1987 its population was estimated to be about 1,062,000,000;
   and the average annual growth rate was about 0.9%. Assuming this rate
   remains unchanged,
   a. find the population in 1988 and 1989;
   b. express the population P as a function of n, the number of years after
        1987; **$P(n) = 1{,}062{,}000{,}000 (1.009)^n$**
   c. predict the population of China in 1995. **about 1,141,000,000**
   a) 1988: about 1,072,000,000; 1989: about 1,081,000,000

In 7 and 8, *true* or *false*.

7. All exponential functions contain the point $(0, 1)$. **False**

8. The graph of every function of the form $y = b^x$ where $b \neq 0$ or $b \neq 1$ is always above the x-axis. **True**

9. State three ways in which exponential functions $y = b^x$ with bases b in the
   interval $0 < b < 1$ differ from exponential functions with $b > 1$.
   See margin.

218

10. Consider $f(x) = 4^x$ and $g(x) = 5^x$.
  a. Without graphing, which function has greater values when $x > 0$?
  b. Without graphing, which function has greater values when $x < 0$?
  c. Check your answers to **a** and **b** by graphing f and g on the same set of axes.   **a) g; b) f; c) See margin.**

11. a. Graph $f(x) = 3^x$ and $g(x) = \left(\frac{1}{3}\right)^x$ on one set of axes scaled in 0.25 units for $-3 < x < 3$.   **See margin.**
  b. Rewrite the equation for g using a negative exponent.   **g(x) = 3⁻ˣ**
  c. What transformation maps the graph of f onto the graph of g?
    (x, y) → (⁻x, y)

12. a. Graph $f(x) = 5^x$ and $g(x) = x^5$ on the same set of axes.
  b. As x increases, between which two consecutive integers does $g(x)$ first exceed $f(x)$?   **between 1 and 2**
  c. Estimate to the nearest tenth all values of x where $f(x) = g(x)$.
    a) See margin. c) 1.8 and 5.0

13. A certain substance decays so that each year 90% of the previous year's material is still present.
  a. After 2 years how much of an initial 2 kilograms of material would remain?   **1.62 kg**
  b. After n years how much of an initial 2 kilograms of material would remain?   **2(0.9)ⁿ kg**
  c. *True or false.*   After six years, more than half the material will have decayed.   **False**

14. a. Graph $y = b^x$ for $b = 1$.   **See margin.**
  b. Would you call the graph in part **a** an exponential growth curve? Why or why not?   **No; There is no growth, the function is constant.**

In 15–18, simplify without a calculator.   *(Lesson 4-2, 4-1)*

15. $32^{3/5}$ **8**   **16.** $\left(\frac{1}{81}\right)^{-1/4}$ **3**   **17.** $\sqrt[5]{32^4}$ **16**   **18.** $81^{-3/4}$ $\frac{1}{27}$

19. A conservation group estimated that the population of an endangered species was decreasing according to the formula $P = 2700(81)^{-t/8}$ where t is the number of years after 1990.
  a. Without using a calculator, find the population in 1992.   **900**
  b. Find the population in 1994.   **300**
  c. Rewrite the formula without negative or rational exponents.
    (Lesson 4-2) $P = 2700/\sqrt{3^t}$

In 20 and 21, let $f(t) = 3t + 13$ and $g(t) = t^2 + 2t$.   *(Lesson 3-7)*

20. *True or false.* $f(g(-3)) = g(f(-3))$ **False**

21. a. Find a formula for $f \circ g$.   **(f ∘ g)(t) = 3t² + 6t + 13**
  b. Give the domain and range of $f \circ g$.
    domain = the set of all real numbers; range = {x: x ≥ 10}

22. What factor other than a change in birth rate or death rate can affect the growth rate of a country?   **Immigration and emigration; annexation or other border changes**

14.a.

■**PROPERTIES** *Objective D (See pages 263–266 for objectives.)*
In 1–5, *true or false*. Consider the exponential function $f(x) = ab^x$ with $b > 0$, $b \neq 1$, and $a > 0$.

1. The graph of $f(x)$ never crosses the y-axis.   **False**
2. When $a = 1$, $f(x)$ always passes through the point $(0, 1)$.   **True**
3. When $b > 1$, $f(x)$ is strictly increasing.   **True**
4. The range of $f(x)$ is the set of all real numbers.   **False**
5. The function $g(x) = 3x^3$ is not an example of an exponential function.   **True**

■**USES** *Objective G*
6. *Multiple choice.* The population of India was about 820 million in 1988 and was increasing at a rate of about 2.1% per year. Which of the following expressions best estimates the population P of India (in millions), t years after 1988?

(a) $P = 820.021^t$ (b) $P = (820 \cdot 1.021)^t$
(c) $P = 820(1.021)^t$ (d) $P = 820(0.021)^t$ **(c)**

■**REPRESENTATIONS** *Objective I*
7. **a.** Graph $g(x) = 6^x$ and $h(x) = 8^x$.
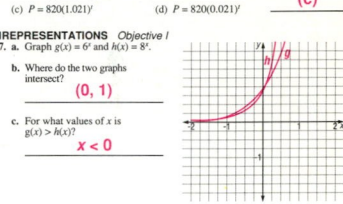
  b. Where do the two graphs intersect?   **(0, 1)**
  c. For what values of x is $g(x) > h(x)$?   **x < 0**

8. **a.** Graph $f(x) = 2^{-x}$ when $-2 \le x \le 2$.

  b. Use the graph to estimate $\sqrt{2}$ and $\frac{1}{\sqrt{2}}$ to the nearest tenth.   **1.4**   **.7**

■**REPRESENTATIONS** *Objective J*
9. Consider the graph $f(x) = am^{-x}$ and $g(x) = bn^{-x}$.

  a. Determine a and b.   **a = b = 3**
  b. For what values of x is $g(x) \le f(x)$?   **x ≤ 0**
  c. For what values of x is $g(x) > f(x)$?   **x > 0**
  d. If both m and n are positive, which is the larger value?   **m**

TEACHING NOTES

Making Connections
Because it takes three non-collinear points to determine a circle, and three to determine a quadratic function, students may be surprised to read that an exponential model is determined by just two points. The theoretical justification for this does not come until Lesson 4-9, when the exponential equation is seen as equivalent to a linear equation in the logs of the variables. The method for finding the model for Example 1 in Lesson 2-7 was to subtract the equations. Here a variable is eliminated by dividing the equations in the system:

$$\frac{108}{74} = \frac{ab^5}{ab^3}$$
$$1.459 \approx b^2$$
$$1.208 \approx b$$

Alternate Approach Students may question whether **Example 1** could be done more efficiently by redefining the independent variable t to be the number of hours after you begin counting the bacteria. Then the two known points are $(0, 74)$ and $(2, 108)$. The equation of the model is $f(t) = 74(1.208)^t$. This approach is usually easier because you already know $a = 74$ and the student can merely substitute $(2, 108)$ into $f(t) = 74b^t$ to find b.

Finding Exponential Models

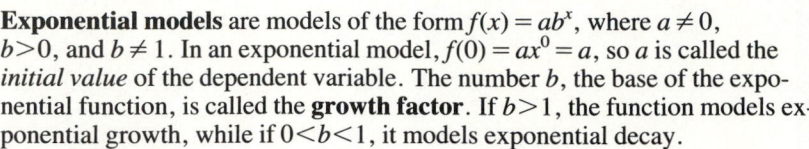

Exponential models are models of the form $f(x) = ab^x$, where $a \neq 0$, $b > 0$, and $b \neq 1$. In an exponential model, $f(0) = ax^0 = a$, so a is called the *initial value* of the dependent variable. The number b, the base of the exponential function, is called the **growth factor**. If $b > 1$, the function models exponential growth, while if $0 < b < 1$, it models exponential decay.

As in a linear model, once the two values a and b are known, the model is specified completely and can be used to make predictions. To determine the constants a and b in the exponential model, you set up and solve a system of two equations.

Example 1

In a laboratory experiment on the growth of insects, there were 74 insects three days after the beginning of the experiment and 108 after an additional two days. Assume that the insect population grows exponentially. Find
a. an exponential model for the population;
b. the initial number of insects;
c. the number of insects 6.5 days after the beginning of the experiment.

Solution
a. An exponential model for the number of insects $f(t)$ after t days is $f(t) = ab^t$. You are given $f(3) = 74$ and $f(5) = 108$. Substitute these values in the equation to get a system.

$$74 = ab^3$$
$$108 = ab^5$$

Divide the second equation by the first: $\frac{108}{74} = b^2$. Because b must be positive in an exponential model, $b = \sqrt{\frac{108}{74}} \approx 1.208$. To find a, substitute this value of b into one of the equations involving a and b. Using the first equation,

$$74 \approx a(1.208)^3$$
$$a \approx \frac{74}{1.763} \approx 41.970.$$

So an exponential model is $f(t) = 41.97 \cdot 1.208^t$.
b. Initially, $t = 0$. Use the equation from part **a.**

$$f(0) = 41.97 \cdot (1.208)^0 = 41.97$$

So about 42 insects were present initially.
Another reasonable exponential model is $f(t) = 42 \cdot (1.208)^t$.
c. Substitute $t = 6.5$. $f(6.5) = 42 \cdot (1.208)^{6.5} \approx 143.44$. So about 143 insects were present 6.5 days after the experiment began.

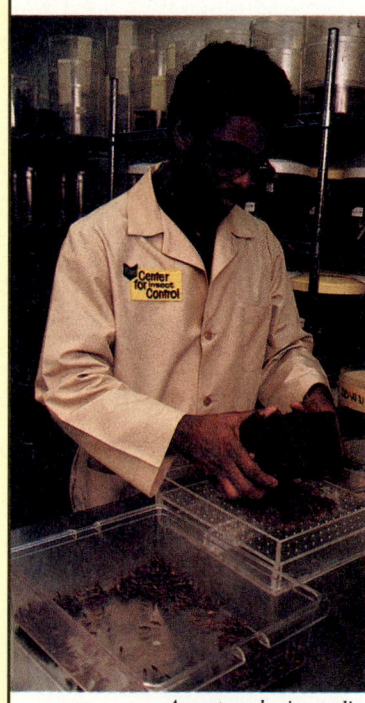

An entomologist studies insect populations.

220

Check Draw a graph of the function $f(t) = 42 \cdot (1.208)^t$. The output from an automatic grapher is shown at the left.

This graph passes through the points (3, 74) and (5, 108). The answers to **b** and **c** seem reasonable, since the graph passes near the points (0, 42) and (6.5, 144).

In the model $f(t) = 42(1.208)^t$, the growth factor 1.208 indicates that at the end of each day there are about 1.208 times as many insects as at the beginning of that day. An equivalent statement is that the growth rate is about 20.8% per day.

Situations involving exponential growth or decay often contain information about the **doubling time** or the **half-life** of a quantity, that is, about how long it takes a quantity to double or to decay to half its original amount.

Example 2 The half-life of a certain radioactive substance is 40 days. If 10 grams of the substance are present initially, how much of the substance will be present in 90 days?

Solution Let $f(t)$ be the amount of the substance present t days after the radioactive decay starts. The function f is exponential, so a model for it has the form $f(t) = ab^t$. We need to evaluate $f(90)$. First find the values of a and b. Given is $f(0) = 10$, so $a = 10$. By the definition of half-life,

$$f(40) = \frac{1}{2} \cdot 10 = 5, \text{ so}$$

$$5 = 10b^{40}.$$

Solve for b.
$$.5 = b^{40}$$

$$b = (.5)^{1/40} \approx .9828$$

So a model is
$$f(t) = 10(.9828)^t.$$

Thus
$$f(90) \approx 10(.9828)^{90} \approx 2.1.$$

So after 90 days there will be about 2.1 grams of the radioactive substance left.

Check Make a table using the given information about the initial amount and the half-life.

Number of half-life periods	0	1	2	3
Number of days after decay begins	0	40	80	120
Amount present (grams)	10	5	2.5	1.25

The amount (2.1 g) found after 90 days (2.25 half-life periods) seems reasonable.

Students should realize that the equations in the text, $A(t) = P(1 + y)^t$ and $A(n) = A_0(0.5)^n$ are simply adaptations of $f(x) = ab^x$. The $(1 + y)$ in $A(t) = P(1 + y)^t$ is b. You add 1 to y in order to "preserve" your initial investment. In the other equation, $A(n) = A_0(0.5)^n$, every time n increases by 1 the quantity is halved. In other words every time the half-life period goes by, half of the quantity remains.

Students who have used UCSMP materials before should be very familiar with the compound interest formula $A(t) = P(1 + y)^t$. If it is new material for your students, you might want to consider the compound interest situation modeled by $A(t) = 1500(1.055)^t$.
a. What is the principal amount in this situation? ($1500)
b. What is the annual interest rate? (5.5%)
c. How much will be in the account after 15 years? ($3348.71)

Alternate Approach In **Example 3,** instead of the exponent representing the number of 5700-year periods, you could let n be the actual number of years since decay began and use the model $A(n) = A_0(0.5)^{n/5700}$, which can be rewritten as
$$A(n) = A_0(0.5^{1/5700})^n$$
$$= A_0(0.999878402)^n.$$

Another possibility is to substitute the two points $(0, A_0)$ and $\left(5700, \frac{1}{2}A_0\right)$ into the general form of an exponential equation, which also leads to $A(n) = A_0(.999878402)^n$. In this form, it is evident that over 99.9% of the carbon-14 is retained from one year to the next. However, it masks the notion of halving, and involves an awkward decimal.

You may wish to use a statistics package to demonstrate the examples in this lesson.

If the half-life of a substance is known, and if n represents the number of half-life periods, then it can be shown that the amount of a radioactive substance is multiplied by 0.5 each time n increases by 1. Thus a general exponential model for radioactive decay is $A(n) = A_0(0.5)^n$, where A_0 is the initial amount present and $A(n)$ is the amount present after n half-life periods.

■ ■ ■ ■ ■ ■ ■ ■ ■

Example 3   The half-life of radioactive carbon-14 is 5700 years. That is, the amount of carbon-14 in a nonliving object decays to one-half its previous amount about every 5700 years. Let A_0 be the original amount of carbon-14 in a substance. Let $A(n)$ be the amount left after n 5700-year periods. About what percent of the original amount of carbon-14 would you expect to find in a substance after 2000 years?

Solution   Use the radioactive decay model $A(n) = A(0.5)^n$. 2000 years is $\frac{2000}{5700} = \frac{20}{57}$ of a 5700-year period. So evaluate the function A when $n = \frac{20}{57}$.

$$A\left(\frac{20}{57}\right) = A(0.5)^{20/57} \approx A(0.5)^{0.351} \approx 0.78A$$

After 2000 years, about 78% of the carbon-14 remains.

Check   Since 2000 years is less than half of the half-life, you expect more than half of the material to be left. It checks.

Another well-known exponential model is the one for compound interest. Suppose you deposit P dollars in an account which pays an annual yield y, compounded yearly. Then at the end of every year, provided you leave the interest in the account, your balance is multiplied by $(1 + y)$. After t years, your balance $A(t)$ is given by the exponential model, $A(t) = P(1 + y)^t$.

Questions

Covering the Reading

In 1 and 2, consider the exponential model $f(x) = ab^x$.

1. *True* or *false*.   The initial value of the model is $f(1)$.   **False**

2. If $0 < b < 1$, what type of exponential model is f?
exponential decay model

3. For the exponential model $B = 37(1.32)^x$, give
a. the initial value;   **37**
b. the growth factor.   **1.32**

4. Use the following data from a bacterial growth experiment.

Hours passed (h)	0	1	2	3	4	5
Population (p)	200	600	1,800	5,400	16,200	48,600

a. Every time one additional hour passes, what happens to the population?   **It triples.**
b. Give an exponential model for these data.   **$P = 200(3)^h$**
c. Estimate the population size after 3 hours and 30 minutes have passed.   **about 9,350**

222

5. Suppose an exponential model of the form $f(t) = ab^t$ contains the two points (3, 20) and (5, 40).

  a. Substitute the two points into the model to get a system of equations for a and b.   **$20 = ab^3$; $40 = ab^5$**
  b. Solve the system to yield an equation for the model.   **$f(t) = \sqrt{50}(\sqrt{2})^t$**
  c. Give the initial value for the model.   **$\sqrt{50}$**
  d. What is the growth factor for the model?   **$\sqrt{2}$**

6. In a study of the change in an insect population, there were about 170 insects four weeks after the study began, and about 320 after two more weeks. Assuming an exponential model of growth, estimate

  a. the initial number of insects,   **48**
  b. the number of insects five weeks after the study began.
    233

7. Radium has a half-life of 1620 years.

  a. Give an exponential model for the amount of radium left from an initial mass m as a function of time t.   **$A(t) = m(0.5)^{t/1620}$**
  b. If the initial mass of some radium was 3 kg, how much would you expect to find after 4000 years?   **about 0.54 kg**

In 8 and 9, the half-life of the carbon isotope C^{14} is about 5700 years.

8. About how many years does it take 1000 grams of this substance to decay to 500 grams?   **5700 years**

9. About what percent of the original amount would you expect to find after 1000 years?   **about 89%**

10. Philip T. Rich deposits $500 in an account paying 6.25% annual yield. If he leaves the account untouched, how much will the investment be worth at the end of three years?   **about $600**

Applying the Mathematics

In 11 and 12, an exponential model for the population P (in millions) of Indonesia during the 1980s is given by $P = 148.7(1.021)^y$ where y stands for the number of years after 1980.

11. What is the annual growth rate of Indonesia's population?   **2.1%**

12. Use an automatic grapher to estimate when the population reached 175 million.   **about 1988**

13. A tour guide noticed that the larger the size of the group, the more time it took to assemble everyone for an event. The guide timed people and collected the following data.

Number of people	2	3	4	5	6	7	8	9	10
Minutes to assemble	2	2.6	3.4	4.4	5.7	7.4	9.7	12.5	16.3

  a. Graph the data.   **See margin.**
  b. Assuming a linear model, draw a line which appears to fit the data and write its equation.   **sample: $y = 2x - 3$ (on graph)**
  c. Assuming an exponential model, use two data points to find values for a and b in $f(x) = ab^x$.
  d. Which of the two models, linear or exponential, seems to fit the data better? Why?
    c) sample: (using (4, 3.4) and (7,7.4)); $f(x) = 1.2(1.3)^x$; d) the exponential model, because the errors are much smaller than the errors using the linear model

LESSON 4-4 Finding Exponential Models   **223**

---

14. A mosquito population doubles every 15 days. If the population is initially 2000, what will it be t days later? **$2000(2)^{t/15}$**

15. When a certain drug enters the bloodstream, its potency decreases exponentially with a half-life of 3 days. If the initial amount of the drug present is A_0, how much will be present 30 days later? **about 0.001 A_0**

Review

16. If the inflation rate was 7% when the cartoon below was published in 1977, what was the hourly wage rate? *(Lesson 4-3)* **about $5.53**

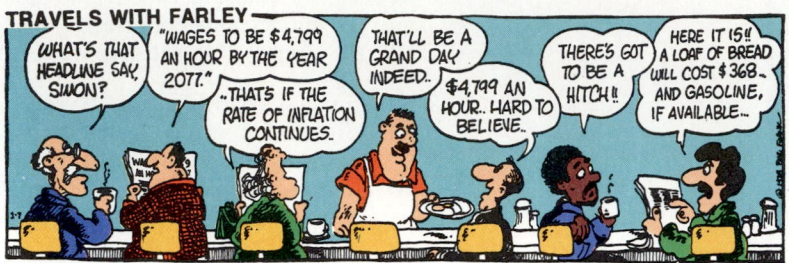

17. Consider the function $k(x) = 3(1.3)^x$. Give
   a. the domain; **the set of all real numbers**
   b. the range; **the set of all positive real numbers**
   c. equations for any asymptotes. *(Lesson 4-3)* **$y = 0$**

18. a. Give a formula for the surface area of a sphere with volume V.
   b. Find the surface area of a sphere with volume 1 m³. *(Lesson 4-2)*
   a) $A = (36\pi V^2)^{1/3}$; b) $\sqrt[3]{36\pi} \approx 4.84$ m²

19. Evaluate without using a calculator: $1024^{-0.4}$. *(Lesson 4-2)* **$\frac{1}{16}$**

20. Write $\sqrt[7]{64}$ as a power of 4. *(Lesson 4-2)* **$4^{3/7}$**

21. Let $f(x) = x^2 + 3$.
   a. Sketch f and its inverse on the same set of axes. **See margin.**
   b. Find an equation for the inverse. **$x = y^2 + 3$**
   c. Is the inverse of f a function? *(Lessons 3-8, 2-6)* **No**

22. Determine whether the graph represents a function. *(Lesson 2-1)*

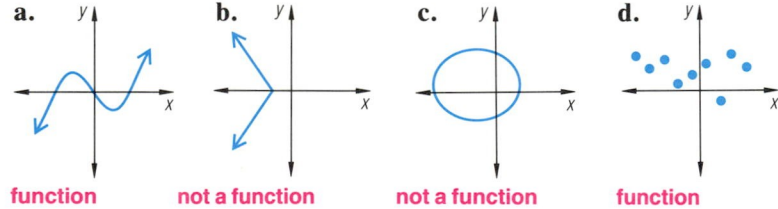

   a. **function**   b. **not a function**   c. **not a function**   d. **function**

Exploration

23. Find out how archaeologists use carbon-14 to date such items as skeletons or fossils. **Answers will vary.**

224

224

4-5

Logarithmic Functions

Napier's Rods or "Bones" were a device to facilitate multiplication. Invented in 1617 by John Napier, the Rods were popular for more than fifty years.

Recall that if y is a function of x, the inverse of the function can be found by switching x and y. The inverses of exponential functions are especially important. Consider the exponential function $y = b^x$ with base $b > 0$, $b \neq 1$. Its inverse is $x = b^y$. Solving that equation for y involves *logarithms*.

Definition

Let $b > 0$ and $b \neq 1$. Then y is the **logarithm of x to the base b**, written $y = \log_b x$, if and only if
$$b^y = x.$$

The word "logarithm" was first used by John Napier (1550–1617) of Scotland, who is usually credited as being the inventor of logarithms because of his 1614 brochure, *Mirifici logarithmorum canonis descriptio* (A Description of the Wonderful Law of Logarithms). It is often abbreviated as "log" in both spoken and written mathematics.

To find the log of a number with a particular base, you sometimes can use the definition directly. First you need to write the number as a power; that is, as a base raised to an exponent.

Example 1 Evaluate:

a. $\log_5 125$
b. $\log_5 \sqrt{5}$
c. $\log_5 0.04$.

Solution Because each base is 5, write each number as a power of 5. Then apply the definition of logarithm: the exponent is the logarithm.

a. $125 = 5^3$. So $\log_5 125 = 3$.

b. $\sqrt{5} = 5^{1/2}$. So $\log_5 \sqrt{5} = \frac{1}{2}$.

c. $0.04 = \frac{1}{25} = \frac{1}{5^2} = 5^{-2}$. So $\log_5 0.04 = -2$.

In Example 2, you must solve an equation before applying the definition of logarithm.

LESSON 4-5 Logarithmic Functions **225**

LESSON 4-5

RESOURCES
■ Lesson Master 4-5

OBJECTIVES

C Evaluate logarithms.
D Describe properties of logarithmic functions.
I Graph logarithmic functions.
J Interpret graphs of logarithmic functions.

TEACHING NOTES

Making Connections
The roots of Napier's word *logarithm* are *log,* meaning "word, discourse, or study of," and *arithmorum,* meaning "of arithmetic." Students are invited to explore more of the historical uses of logarithms in Project 4, which involves slide rules.

Although this lesson assumes no mastery of logarithms, we expect that students have some familiarity with them. Thus, ideas are presented at a rather quick pace. If students have not studied logarithms, be sure they understand the pattern below.

This	Means that
base^{exponent} = power	exponent = log of power, base b
$b^t = p$	$t = \log_b p$
$10^2 = 100$	$2 = \log_{10} 100$

Alternate Approach
Another way students can approach **Example 1** is to rewrite $y = \log_b x$ in the form $b^y = x$. So $5^y = 125$, and

Example 2   Evaluate $\log_4 8$.

Solution   From the definition, $\log_4 8$ is the power of 4 that equals 8. So, write 8 as a power of 4.
$$8 = 4^y$$
To solve this equation, express each side as a power to the same base; in this case, 2.

$$2^3 = \left(2^2\right)^y \qquad \text{Substitution}$$
$$2^3 = 2^{2y} \qquad \text{Power of a Power Property}$$

The exponents must be the same if the two expressions are equal. So,
$$3 = 2y$$
$$y = \frac{3}{2}.$$
Thus
$$\log_4 8 = \frac{3}{2}.$$

Check   Does $4^{3/2} = 8$? $4^{3/2} = (2^2)^{3/2} = 2^3 = 8$.   It checks.

Unlike Examples 1 and 2, most logarithms cannot be evaluated directly. Nowadays, a calculator or computer is used to give approximations. This is a recent development; as late as the 1960s, tables of logarithms were needed.

Scientific calculators have a key marked $\boxed{\log}$ which calculates values of logarithms to base 10. These are sometimes called **common logarithms**. Check on your calculator that $\log_{10} 100 = 2$ and that $\log_{10} 7 = 0.845$, to three decimal places. (The $\boxed{\ln}$ key calculates $\log_e x$, which you will study in the next lesson. Those are usually called *natural logarithms*.) The first table of common logarithms was produced by Henry Briggs (1561–1631) of England, who worked with Napier. In 1624 he published *Arithmetica logarithmica*, containing common logarithms to fourteen decimal places.

Example 3   Solve $10^x = 4$ to the nearest tenth.

Solution   Use the definition of a logarithm to rewrite the equation as $x = \log_{10} 4$. Evaluate $\log_{10} 4$ with a calculator.
$$\log_{10} 4 \approx 0.60206 \approx 0.6$$

Check   Use your calculator to verify that $10^{0.6} \approx 3.98 \approx 4$.

The exponential functions $f(x) = 2^x$, $g(x) = 3^x$, and $h(x) = 4^x$ are shown below in black. For $f(x) = 2^x$, its inverse $x = 2^y$ or $y = \log_2 x$ is shown in blue. The inverse passes the vertical line test, so it is a function and can be labeled $f^{-1}(x) = \log_2(x)$. g^{-1} and h^{-1} are also graphed below. In general, the function which maps *x* to $\log_b x$ is the **logarithmic function base *b***. This function is the inverse of the exponential function with base *b*. That is,
if
$$f(x) = b^x,$$
then
$$f^{-1}(x) = \log_b x.$$

226

Like the family of exponential functions, the graphs of the logarithmic functions also form a *family* of curves with a characteristic shape and similar properties. From the graphs of the various logarithmic functions above, you can see that for any logarithmic function where $b > 1$:

(1) The domain is the positive real numbers.
(2) The range is all real numbers.
(3) The graph contains $(1, 0)$; that is, for any base b, $\log_b 1 = 0$.
(4) The function is strictly increasing.
(5) The end behavior, as x gets larger, is to increase without bound.
(6) As x gets smaller and approaches 0, the values of the function are negative with larger and larger absolute values.
(7) The line $x = 0$ (the y-axis) is an asymptote of the graph.

The fourth property listed above, that logarithmic functions are strictly increasing, was also true for exponential functions with base $b > 1$. That is, if $x_2 > x_1$, then $\log_b x_2 > \log_b x_1$. This property allows you to deduce the solution to logarithmic equations, because if $\log_a x = \log_a y$, then it must be true that $x = y$.

For another result of the "strictly increasing" property, note that $\log_{10} 100 = 2$, $\log_{10} 1000 = 3$, and so on. For example, since 826 is between 10^2 and 10^3, $\log_{10} 826$ is between 2 and 3. In fact, all 3-digit integers have common logarithms beginning with 2.

n	$\log_{10} n$
$1 = 10^0$	0
$10 = 10^1$	1
$100 = 10^2$	2
$1000 = 10^3$	3
$10,000 = 10^4$	4

In general, the common logarithm of numbers between successive powers of 10 is a number between successive integers. This property can be used in reverse. A number with logarithm between 3 and 4 must be between 1000 and 10,000.

Caution: Common logarithms are often *written* without indicating the base 10. In this book, when you see an expression like log 8.7, you can assume it means $\log_{10} 8.7$. However, in some computer programs log refers to natural logarithms.

You can use the inverse relationship between exponents and logs, expressed in the definition of logarithm, to find easily a number with a given logarithm.

students are not asked to graph logarithm functions in this lesson. If you wish to demonstrate the family of functions in the lesson using a function grapher, define F(X) = LOG(X)/LOG(B) to graph $f(x) = \log_b x$. Also, if your students are using automatic function graphers, have them graph $y = \log x$ to see whether the base is 10 or e.

ADDITIONAL EXAMPLES
1. Evaluate **a.** $\log_6 \frac{1}{6}$
b. $\log_6 36$ **c.** $\log_6 \sqrt[5]{36}$.
a. -1; b. 2; c. $\frac{2}{5}$

2. Evaluate $\log_9 243$.
$\frac{5}{2}$

3. Solve $10^y = 73$ to the nearest hundredth.
1.86

4. Solve $\log t = 2.9$.
$10^{2.9} \approx 794.3$

Example 4   Solve $\log x = 3.724$.

Solution   Because no base is given, assume the base is 10. Apply the definition.

$$\log_{10} x = 3.724 \text{ if and only if } x = 10^{3.724}.$$

Use a calculator.

$$x \approx 5296.63$$

Check 1   Estimate. Since $3 < \log x < 4$, x must be a number between 1,000 and 10,000. It checks.

Check 2   Does $\log 5296.63 \approx 3.724$? It does.

Questions

Covering the Reading

1. *Multiple choice.*   Which statement is read "p is the logarithm of q base s''?
   (a) $\log_q p = s$ (b) $\log_p q = s$ (c) $p = \log_q s$ (d) $p = \log_s q$ **(d)**

2. Write $2^7 = 128$ using logs.   **$\log_2 128 = 7$**

3. Write $\log_6 216 = 3$ using exponents.   **$6^3 = 216$**

4. *True* or *false*. The function with equation $y = \log_4 x$ is the inverse of the function with equation $y = 4^x$.   **True**

In 5–8, evaluate.

5. $\log_4 256$ **4**   6. $\log_7 1$ **0**   7. $\log_5 0.2$ **-1**   8. $\log_9 27$ **$\frac{3}{2}$**

In 9–11, evaluate without a calculator.

9. $\log 1000$ **3**   10. $\log 1$ **0**   11. $\log 0.01$ **-2**

In 12–14, solve.

12. $10^x = 8$ **≈ 0.9**   13. $\log_{10} x = 3.121$ 14. $\log_{10} x = -1.4$
                                                   ≈ 1321   **≈ 0.04**

15. Consider the function $f(x) = \log_b x$ where $b > 1$.
    a. State the domain of f.   **the set of all positive real numbers**
    b. *True* or *false*. The graph contains the point $(1, 0)$.   **True**
    c. Which axis is an asymptote of the graph?   **the y-axis**

16. *Multiple choice.* If $\log_{10} p = 1.6274$, then p must be between which of the following? (Do not use a calculator.)
    (a) 0 and 1   (b) 1 and 2   (c) 1 and 10   (d) 10 and 100   **(d)**

Applying the Mathematics

In 17 and 18, use this information. The pH level of blood having bicarbonate concentration b and carbonic acid concentration c is given by
$\text{pH} = 6.1 + \log\left(\frac{b}{c}\right)$. Find the pH level of blood when:

17. $b = 32, c = 3$ **about 7.1**   18. $b = 20c$ **about 7.4**

228

19. *Multiple choice.* In which interval is $\log_7 62$?
    (a) $0 \le x < 1$ (b) $1 \le x < 2$ (c) $2 \le x < 3$ (d) $8 \le x < 9$ **(c)**

In 20–22, use the measure of egg quality called the "Haugh unit," given in the *U.S. Egg and Poultry Magazine* of 1937. The number of Haugh units of an egg is given by $100 \log (H - \frac{1}{100} \sqrt{32.2} \, (30w^{0.37} - 100) + 1.9)$ where w is the weight of the egg in grams and H is the height of the albumen in millimeters when the egg is broken onto a flat surface.

20. Find the number of Haugh units of an egg for which $w = 60$ and $H = 5$.
    about 68.4 Haugh units
21. An egg weighing 53.2 g has albumen height 5.5 mm. Find the number of Haugh units. **about 75.3 Haugh units**

22. For two eggs of the same weight, which has the larger number of Haugh units, that with the larger albumen height or that with the smaller albumen height? **the egg with the larger albumen height**

23. Solve for x: $\log_7 (4x - 3) = \log_7 5$. **2**

Review

24. Suppose a function with rule $f(t) = ab^t$ contains the two points $(5, 50)$ and $(7, 100)$. Find a and b. *(Lesson 4-4)* **$a \approx 8.84$, $b \approx 1.414$**

25. The population P of Bolivia in 1988 was estimated as 6,900,000 with an average annual growth rate of 2.6%. Assume that this growth rate continues.
    a. Express P as a function of the number of years after 1988.
    b. Estimate Bolivia's population in the year 2000. *(Lesson 4-3)*
    a) $P = 6,900,000(1.026)^y$; b) about 9,400,000 people

In 26 and 27, use the fact that in 1980 there were about 14.6 million Americans of Hispanic origin. According to some estimates, the Hispanic population in the U.S. will double in about 28 years. Assume that the doubling time remains constant indefinitely.

26. In what year will the Hispanic population reach 8 times its 1980 size?
    2064
27. Find an exponential model for the growth of the Hispanic American population. *(Lesson 4-4)* **$P(t) = 14.6(2)^{t/28}$ where $P(t)$ is in millions of people and t is years after 1980.**

28. Give the exact value of $512^{-2/3}$. *(Lesson 4-2)* **$\frac{1}{64}$**

29. The power P of a radio signal varies inversely as the square of the distance from the transmitter: $P = \frac{k}{d^2}$. Assume that all variables represent positive real numbers.
    a. Solve for d, writing your solution with radicals. **$d = \sqrt{k/P}$**
    b. Write your solution without radical notation. *(Lesson 4-1)* **$d = (k/P)^{1/2}$**

30. The twenty-seven golfers in a tournament had scores which averaged six strokes above par, with a standard deviation of 3.1 strokes. If par is 72, what were the actual mean and standard deviation? *(Lesson 3-3)*
    mean = 78 strokes; standard deviation = 3.1 strokes

Exploration

31. Find out what prompted Napier and Briggs to invent logarithms.
    See margin.

LESSON 4-5 Logarithmic Functions **229**

e and Natural Logarithms

Examine the banking advertisement at the left carefully. Notice that a distinction is made between the interest *rate* and the annual *yield*. Here is how the yield is calculated from the rate. The ad states that interest compounds monthly, so on an investment of P dollars between $9 and $25,000, the bank uses the *prorated* monthly interest rate of $\frac{0.0575}{12}$. After one month, the total interest earned is $P \cdot \frac{0.0575}{12}$. So after one month, the total in the account would be $P + P \cdot \frac{0.0575}{12} = P\left(1 + \frac{0.0575}{12}\right)$. Thus, the monthly scale factor is $\left(1 + \frac{0.0575}{12}\right)$. If no deposits or withdrawals are made, then the balance after one year is $P\left(1 + \frac{0.0575}{12}\right)^{12} = 1.05904P$, indicating the effective annual "yield" of 5.904% stated in the advertisement.

In general, when the principal P (the initial amount) is invested at an annual interest rate r compounded n times per year, the interest rate in each period is $\frac{r}{n}$ and the number of compoundings in one year is n. The balance A at the end of one year is $P\left(1 + \frac{r}{n}\right)^n$ and the yield is $\left(1 + \frac{r}{n}\right)^n - 1$. At the end of t years, the amount A the investment is worth is given by $A = P\left(1 + \frac{r}{n}\right)^{nt}$.

Consider another case. Suppose you invest $1 for 1 year (365 days) at a huge 100% annual interest rate. What is your balance after 1 year? It depends on the compounding schedule the bank uses, as the following table shows.

Compounding Schedule	n	$P\left(1+\frac{r}{n}\right)^n$	Balance in $ after 1 year
annually	1	$1(1 + 1)^1$	2.00
semi-annually	2	$1\left(1 + \frac{1}{2}\right)^2$	2.25
quarterly	4	$1\left(1 + \frac{1}{4}\right)^4$	2.44141
monthly	12	$1\left(1 + \frac{1}{12}\right)^{12}$	2.61304
daily	365	$1\left(1 + \frac{1}{365}\right)^{365}$	2.71457
hourly	8760	$1\left(1 + \frac{1}{8760}\right)^{8760}$	2.71813
every second	31,536,000	$1\left(1 + \frac{1}{31,536,000}\right)^{31,536,000}$	2.71828

As you might expect, the balance after one year increases with greater numbers of compounding periods. However, that balance does not increase

230

without bound, so you do *not* earn an infinite sum of money. As the number n of compounding periods increases, the balances approach a limit of $2.71828\ldots$, the number called e.

Definition

$$e = \lim_{n \to \infty} \left(1 + \frac{1}{n}\right)^n$$

The definition is read "e equals the limit of $\left(1 + \frac{1}{n}\right)^n$ as n goes to infinity."
The number e is named after the mathematician Leonhard Euler (1707–1783), who discovered its importance. The number e is irrational; its decimal expansion, like that of π, is known to many places; to thirteen decimal places,

$$e \approx 2.7182818284590\ldots .$$

The function $y = e^x$ is of the form $y = ab^x$, so it is an exponential function. It has so many applications that it is built into scientific calculators and computers, and is often referred to as *the* exponential function. Some scientific calculators have an $\boxed{e^x}$ key for evaluating the exponential function. On others, you may need to press $\boxed{\text{INV}}$, $\boxed{\text{2nd}}$, or $\boxed{\text{f}}$ first to access the exponential function. To test your calculator, confirm that $e^{.08} \approx 1.083287$.

The function $y = e^x$ is a limit of functions, which means its graph is the limiting shape for a family of graphs. Consider the exponential functions $f_n(x) = \left(1 + \frac{1}{n}\right)^{nx} = \left[\left(1 + \frac{1}{n}\right)^n\right]^x$ which correspond to compounding n times a year for x years. For $x > 0$, their graphs become steeper as n increases. However, they do not become so steep as to approach a vertical line. Their limit is the function with equation $f(x) = e^x$.

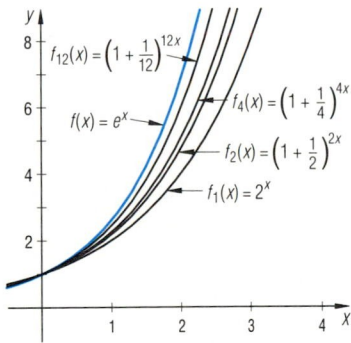

It can also be shown that for large n, $\left(1 + \frac{r}{n}\right)^n \approx e^r$, or, more precisely,

$$\lim_{n \to \infty} \left(1 + \frac{r}{n}\right)^n = e^r.$$

The Continuous Change Formula is sometimes called "pert." It can be derived from the formula for periodic growth by taking the limit as n goes to infinity. With your best students you might go through this argument.

$$A = P\left(1 + \frac{r}{n}\right)^{nt}$$

$$\lim_{n \to \infty} A = \lim_{n \to \infty} P\left(1 + \frac{r}{n}\right)^{nt}$$

Factor out P, which is not involved in the limit.

$$= P \lim_{n \to \infty} \left[\left(1 + \frac{1}{n/r}\right)^n\right]^t$$

Rewrite the exponent.

$$= P \lim_{n \to \infty} \left[\left(1 + \frac{1}{n/r}\right)^{n/r \cdot r}\right]^t$$

$$= P\left[\lim_{n \to \infty} \left(1 + \frac{1}{n/r}\right)^{n/r \cdot r}\right]^t$$

Use the definition of e and replace A by $A(t)$ to emphasize that A depends on t.
$$A(t) = Pe^{rt}$$

Making Connections   A "trademark" of situations which are appropriately modeled with the Continuous Change Model is that the rate of change of the amount (of money, population, substance, and so on) is proportional to the amount. In calculus terms, these are situations where $\frac{dy}{dt} = ky$, where t stands for time and k is a constant.

As in Lesson 4-5, we can find the inverse. If $f(x) = y = e^x$, then $\log_e y = x$ or $\ln y = x$. The inverse of f is found by switching variables: $\ln x = y$ or $f^{-1}(x) = \ln x$.

Technology   On some calculators e may have to be displayed by the key sequence:

1 $\boxed{\text{INV}}$ $\boxed{\text{LN}}$. This works

because $\boxed{\text{INV}}$ $\boxed{\text{LN}}$ is $\boxed{e^x}$ and $e^1 = e$.

Ask your students to find ln 0 and ln (-2). Why are both of these undefined?

ADDITIONAL EXAMPLES

1. Suppose $10,000 is put into a 5-year certificate of deposit that pays 7.35% interest compounded continuously.

a. What is the balance at the end of this period?
$10{,}000e^{0.3675} \approx \$14{,}441.20$

b. How does this compare with the balance if the interest were compounded annually?
It is about $185 greater.

2. Consider the region bounded by the graphs of $y = 1/x$, the x-axis, the line $x = a$, and the line $x = b$, where $x > 0$. Using calculus, it can be proven that the area of that region is $\ln b - \ln a$. What then is the area bounded by the graph of $y = 1/x$, the x-axis, the line $x = 1$, and the line $x = 7$?
$\ln 7 - \ln 1 \approx 1.946$

Recall from the beginning of this lesson that $\left(1 + \frac{r}{n}\right)^{n} - 1$ is the yield. The limit indicates that, as the number of compounding periods increases, the yield is $e^{r} - 1$. When this yield is used to calculate interest, the interest is said to be *compounded continuously*.

For example, if $500 is put in a savings account at an 8% *rate* compounded continuously, the *yield* is $e^{.08} - 1 \approx 1.083287\ldots - 1 \approx 8.3287\%$.

After one year the account would grow to
$$500 + 500(e^{.08} - 1) = 500(e^{.08}) \approx \$541.64.$$

After t years the amount would be $500(e^{.08})^{t} = 500e^{.08t}$. This is an instance of the following general formula.

Continuous Change Formula

If an initial quantity P grows or decays continuously at an annual rate r, the amount $A(t)$ after t years is given by $A(t) = Pe^{rt}$.

Example 1 Suppose $500 is invested in an account paying an 8% annual interest rate.

a. What is the balance after 4.5 years if interest is compounded continuously?

b. How does this balance compare with quarterly compounding over 4.5 years?

Solution

a. Use the Continuous Change Model $A(t) = Pe^{rt}$ where P is 500, r is 0.08, and t is 4.5.
$$A(4.5) = 500e^{0.08(4.5)} = 500e^{0.36} \approx 500(1.433329)$$
The balance after 4.5 years is $716.66.

b. Use the formula $A = P\left(1 + \frac{r}{n}\right)^{nt}$ where $P = 500$, $r = 0.08$, and $t = 4.5$.
$$A = 500\left(1 + \frac{0.08}{4}\right)^{4 \cdot 4.5} = 500(1.02)^{18} = \$714.12$$
Continuous compounding earns $2.54 more in this case.

Check You expect continuous compounding to give a slightly greater balance. It does.

Many situations in the everyday world involve continuous change, or close approximations to it. Examples are radioactive decay and most natural growth processes. The growth of large human populations, presented in the chapter opener or as in Question 15, is approximately continuous.

Like all other exponential functions, $f(x) = e^{x}$ has an inverse which is a logarithmic function, $f^{-1}(x) = \log_{e} x$. The logarithmic function to the base e is so important that it is given a special name and symbol.

232

Definition

The **natural logarithm** of x is the logarithm of x to the base e, written **ln x**. That is, for all x, $\ln x = \log_e x$.

The function with equation $y = \ln x$ is the **natural logarithm function**. Note that $y = \ln x$ if and only if $e^y = x$.

The graphs of the exponential function and its inverse, the natural logarithm function, are shown together at the right. These were produced on a function grapher. As with all inverse functions, the graphs are reflection images of each other over the line $y = x$.

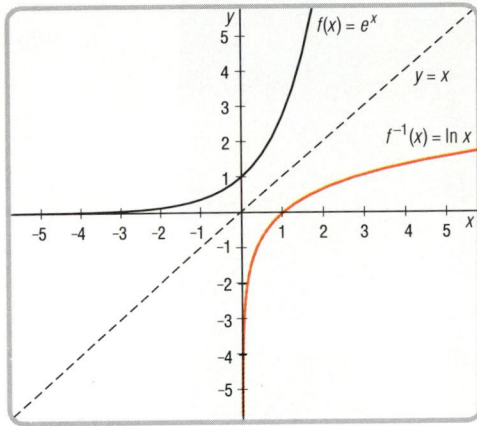

On calculators, the key for natural logarithms is usually marked $\boxed{\text{ln}}$ or $\boxed{\text{ln }x}$, while $\boxed{\text{log}}$ is used to refer to common logarithms. Since logarithmic and exponential functions are inverses of each other, you may find them related on your calculator, possibly through $\boxed{\text{INV}}$ or $\boxed{\text{2nd}}$ keys. Natural logarithms are often used in models of situations. Example 2 illustrates one such situation.

Example 2 Under certain conditions, the height h in feet above sea level can be approximated from the atmospheric pressure P in pounds per square inch (psi) at that level using the equation

$$h = \frac{\ln P - \ln 14.7}{-0.000039}.$$

If human blood, at 0.9 psi, "boils" at body temperature, at what height above sea level will blood boil in an unpressurized cabin?

Solution Substitute $P = 0.9$, and evaluate with a calculator.

$$h = \frac{\ln 0.9 - \ln 14.7}{-0.000039} \approx \frac{-0.10536 - 2.6878}{-0.000039} \approx 71{,}621$$

Without a pressurized cabin, heights greater than 72,000 feet above sea level would be fatal.

Logarithms can be found to any base b, $b > 0$, $b \neq 1$. But only three bases, 10 (for common logarithms), e (for natural logarithms), and 2 (due to its relation to half-life and doubling) are generally used in practical applications.

LESSON 4-6   *e* and Natural Logarithms   **233**

Questions

Covering the Reading

1. a. Evaluate $\left(1 + \frac{1}{n}\right)^n$ for the given value of n.
   i. 1000   **2.716924**   **ii.** 1,000,000   **2.718281**
b. *True* or *false.* $\lim\limits_{n\to\infty} \left(1 + \frac{1}{n}\right)^n = e$ **True**

2. a. Evaluate.   **i.** $\left(1 + \frac{5}{1000}\right)^{1000}$ **ii.** $\left(1 + \frac{5}{1000000}\right)^{1000000}$
b. What is $\lim\limits_{n\to\infty} \left(1 + \frac{5}{n}\right)^n$?   **a) i. 146.5756; ii. 148.4113; b) e^5**

3. Suppose \$800 is invested in an account paying 7% annual interest for 4 years. Give the balance if interest is
a. compounded continuously.   **b.** compounded monthly.
\$1058.50   **\$1057.64**

In 4–7, evaluate.

4. $e^{1.64}$ **$\approx$ 5.16**   **5.** $e^{-0.8}$ **$\approx$ 0.45**   **6.** $\ln 5.16$ **$\approx$ 1.64**   **7.** $\ln(0.45)$ **$\approx$ -0.8**

In 8–11, evaluate without a calculator.

8. e^0 **1**   **9.** $\ln 1$ **0**   **10.** $\ln e$ **1**   **11.** $\ln e^5$ **5**

12. Refer to Example 2. A hiker measured atmospheric pressure to be about 10 psi. Estimate the altitude of the hiker.   **about 9900 feet**

Applying the Mathematics

13. a. Answer without calculating. Which is larger, $\log 17$ or $\ln 17$?
b. Check your answer by evaluating each expression on a calculator.
a) $\ln 17$; b) $\log 17 \approx 1.2$, $\ln 17 \approx 2.8$

14. It can be shown that the velocity V reached by a rocket when all its propellant is burned is given by $V = c \ln R$ where c is the exhaust velocity of the engine and R is the ratio $\dfrac{\text{takeoff weight}}{\text{burnout weight}}$ for the rocket. (The burnout weight is the takeoff weight minus the weight of the fuel.) Space Shuttle engines produce exhaust velocities of 4.6 kilometers per second. If the ratio R for a Space Shuttle is 3.4, what velocity can it reach from its own engine?   **$\approx$ 5.6 km/sec**

15. a. The population of New Zealand in 1985 was 3,295,000 with an average growth rate of 1.4%. If this growth rate continues, use the Continuous Change Model to predict New Zealand's population in 2000.   **about 4,065,000**
b. Compare your answer with the Example in Lesson 4-3. Explain any similarities or differences.   **This answer is slightly higher; Lesson 4-3 does not assume continuous growth.**

16. a. Use an automatic grapher to graph the functions
$$f(x) = e^{3x}, \quad g(x) = \ln 3x, \quad \text{and} \quad h(x) = \frac{1}{3}\ln x.$$ **See margin.**
b. Which, g or h, is the inverse of f?   **h**

17. a. Evaluate $\ln(e^{6.2})$ without using a calculator.   **b.** Evaluate $e^{\ln 8.1}$.
c. Generalize the results of parts **a** and **b**.   **d.** Predict the shape of the graph of $f(x) = e^{\ln x}$. Check using an automatic grapher.
a) 6.2; b) 8.1; c) $\ln(e^a) = a$; $e^{\ln a} = a$; d) the part of the line $y = x$ in the first quadrant

234

In 18 and 19, a particular satellite has a radioisotope power supply, with power output given by the exponential model $P = 60e^{-t/300}$, where t is the time in days and P is the power output in watts.

18. What is the power output at the end of its first year?   **about 17.8 watts**

19. When the power output drops below 8 watts, there will be insufficient power to operate the satellite. Use an automatic grapher to decide how long the satellite will remain operable.   **about 604 days**

Review

In 20 and 21, give an equation for the inverse of the function. *(Lesson 4-5)*

20. $f(x) = 7^x$ $f^{-1}(x) = \log_7 x$ **21.** $g(x) = \log_6 x$ $g^{-1}(x) = 6^x$

22. If $3^{2.3} = 12.514$, find $\log_3 12.514$. *(Lesson 4-5)*   **2.3**

23. If $\log_8 27.8576 = 1.6$, calculate $8^{1.6}$. *(Lesson 4-5)*   **27.8576**

24. Evaluate $\log_{10} \sqrt[3]{10}$ without a calculator. *(Lesson 4-5)* $\frac{1}{3}$

25. Which symbol, $=$, $>$, or $<$, makes $900^{3/5}$ __?__ $(\sqrt[3]{900})^5$ true? *(Lesson 4-2)* $<$

26. The value of a used car sometimes is modeled by $V = Cd^t$ where V is the current value in dollars, C is the original cost, and t is the number of years after the car is new. One 1987 model sold for $8100 in 1990 and $6910 in 1991.
  a. Find values for C and d.   **b.** Give the original cost. *(Lesson 4-4)*
     $C \approx 13{,}050; d \approx 0.853$ **$13,050**

27. On the PSAT (Preliminary Scholastic Aptitude Test), scores are reported in a range of 20–80. To estimate SAT score from PSAT, you should multiply by 10. A group of 35 students had an average of 491 and standard deviation of 103 on the SAT. Assuming that their SAT scores were comparable, what would you expect as mean and standard deviation of the same group's PSAT scores? *(Lesson 3-6)*
  mean = 49.1; standard deviation = 10.3

Exploration

28. The number e^x equals the infinite sum $e^x = 1 + \frac{x}{1!} + \frac{x^2}{2!} + \frac{x^3}{3!} + \frac{x^4}{4!} + \dots$

where $4! = 4 \cdot 3 \cdot 2 \cdot 1$, $3! = 3 \cdot 2 \cdot 1$, etc.
  a. Evaluate e^1 using the five terms shown in the sum above. How close is your answer to the value in the text?   **2.70833; within .01 of e**
  b. Compute $e^{0.15}$ in using the five terms in the formula above. Compare to a calculator value.   **1.1618336; within .0000007 of e$^{.15}$**
  c. How many terms of the sum need to be included to get e accurate to the thousandths place?   **7 terms (up through $\frac{x^6}{6!}$)**

29. a. Use an automatic grapher to plot $y = e^x$ and $y = 1 + x$ on the same set of axes. When does $e^x = 1 + x$? Find some numbers so that $e^x \approx 1 + x$.
  b. Explain why for some values of r, the functions $y = a(1 + r)^x$ and $y = ae^{rx}$ are approximately equal.
  See margin.

LESSON 4-7

RESOURCES
■ Lesson Master 4-7
■ Quiz for Lessons 4-4 through 4-7
■ Teaching Aid 36: Theorems about logarithms

OBJECTIVES

C Evaluate logarithms.
E Use properties of logarithms.

TEACHING NOTES

The four theorems of this lesson are summarized on a teaching aid. Since logs are exponents, students should expect logs and exponents to have similar properties. Here are the properties of exponents that correspond to each of the theorems in this lesson, stated in informal language:
Logarithm of 1:
  The zero power of any number is 1.
Logarithm of a Product:
  To multiply two powers with the same base, add exponents.
Logarithm of a Quotient:
  To divide two powers with the same base, subtract exponents.
Logarithm of a Power:
  The power of a power is found by multiplying the exponents.

Technology For most computations, students these days do not need to use properties of logarithms. Most calculations can be done immediately using a calculator. **Example 4,** however, shows that knowing the properties can be helpful even when using a calculator. But this need is lessening; there

---

LESSON

Properties of Logarithms

Recall that any positive number except 1 can be the base b of a logarithm, and that $x = \log_b m$ if and only if $b^x = m$. This definition and the properties of powers, listed in Lessons 4-1 and 4-2, allow you to prove corresponding properties of logs.

For example, because $b^0 = 1$ for any nonzero b, then by the definition of log, $\log_b 1 = 0$. In words, the logarithm of 1 to any base is zero. That is, the graph of $f(x) = \log_b x$ contains $(1, 0)$ for any base b.

> **Theorem (Logarithm of 1)**
>
> For any base b, $\log_b 1 = 0$.

Suppose you know two positive numbers, x and y, and their logarithms to a certain base b:

$$\log_b x = m \quad \text{and} \quad \log_b y = n.$$

To find the logarithm of the product xy, first express xy as a power of that base b.

$x = b^m$	definition of $\log_b$
$y = b^n$	definition of $\log_b$
$xy = b^m \cdot b^n$	Multiplication Property of Equality
$xy = b^{m+n}$	Product of Powers Property

Therefore, $\log_b (xy) = m + n$.   definition of $\log_b$

This proves the following theorem.

> **Theorem (Logarithm of a Product)**
>
> For any base b and for any positive real numbers x and y,
>
> $$\log_b(xy) = \log_b x + \log_b y.$$

In words, the log of a product is the sum of the logs of its factors. The Logarithm of a Product Theorem can also be used to simplify sums of logs.

Example 1   Evaluate $\log 250 + \log 4$ without a calculator.

Solution $\log 250 + \log 4 = \log(250 \cdot 4) = \log 1000 = 3$

Check   Use a calculator. $\log 250 \approx 2.3979$ and $\log 4 \approx 0.6021$, so $\log 250 + \log 4 \approx 2.3979 + 0.6021 = 3.0000$.

236

236

The log of a quotient $\frac{x}{y}$ is found using the corresponding power property $\frac{b^m}{b^n} = b^{m-n}$.

Theorem (Logarithm of a Quotient)

For any base b and for any positive real numbers x and y,

$$\log_b\left(\frac{x}{y}\right) = \log_b x - \log_b y.$$

Similarly, the log of a power is derived from the Power of a Power Property: $(b^m)^n = b^{mn}$.

Theorem (Logarithm of a Power)

For any base b, any positive real number x and any real number p,

$$\log_b(x^p) = p\log_b x.$$

You are asked to prove the two theorems above in the Questions. The theorems on these two pages can be used to solve or rewrite *logarithmic equations*.

Example 2   If $t \geq 0$ and $\ln P = \frac{1}{2}\ln t - \ln 6$, write an expression for P without logarithms.

Solution

$$\ln P = \frac{1}{2}\ln t - \ln 6$$
$$= \ln t^{1/2} - \ln 6 \quad \text{Log of a Power Theorem}$$
$$= \ln\left(\frac{t^{1/2}}{6}\right) \quad \text{Log of a Quotient Theorem}$$

Then $P = \dfrac{t^{1/2}}{6}.$ If $\log_b x = \log_b y$, then $x = y$.

The properties of logarithms can also be used to rewrite *exponential* equations. This skill is needed when you wish to find linear models for the logarithms of variables. You will see how to do this in Lesson 4-9.

Example 3   Given the exponential model $y = ab^x$, find a linear model for $\ln y$.

Solution   Take the natural log of each side.
$$\ln y = \ln(ab^x)$$
$$= \ln a + \ln b^x \quad \text{Log of a Product Theorem}$$
So $\ln y = (\ln b)x + \ln a \quad \text{Log of a Power Theorem}$

If a and b are constant, this is a linear model for $\ln y$ with slope $\ln b$ and y-intercept $\ln a$.

---

are calculators that can evaluate the number in **Example 4** directly.

You can also use **Example 4** to highlight a benefit of using common logarithms. Although it is as easy to find $\ln 2^{365} \approx 252.9987$ as it is to find $\log 2^{365} \approx 109.8759$, the latter result allows you to infer immediately that 2^{365} is a 110-digit number. **Question 11** also addresses this benefit.

If any students have begun Project 4, ask them to demonstrate how to do the examples above using a slide rule.

ADDITIONAL EXAMPLES
1. Simplify without a calculator:
a. $\log_6 2 + \log_6 3$
b. $\log_5 200 - \log_5 8$
c. $\log 85 - \log 17 + \frac{1}{2}\log 400$.

a. 1; b. 2; c. 2

2. Rewrite without logarithms:

a. $\ln x = \frac{1}{3}\ln a$

x = $\sqrt[3]{a}$

b. $\log x = \log a - 4\log b$

x = $\dfrac{a}{b^4}$

3. Write a linear equation equivalent to the exponential equation $A = Pe^{rt}$.
ln A = rt + ln P

4. Which number is larger, 400^{500} or 500^{400}?

400^{500} has 1302 digits; 500^{400} has 1080 digits, so the first number is larger.

NOTES ON QUESTIONS
Making Connections for Question 8: The formula used is Newton's universal law of gravitation, also known as the Inverse Square Law. F is the force (attraction) between two bodies, G is the universal gravitational constant, m and M are the masses of the two bodies, and d is the distance between them.

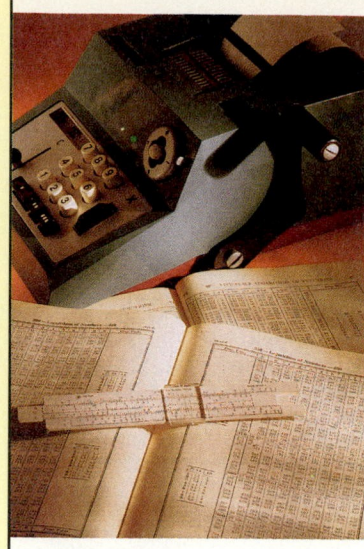

Because it is easier to add than it is to multiply, logs were often used in the past to evaluate powers, roots, products, and quotients. The invention of logarithms by Napier and the subsequent publication of tables of logarithms made numerical calculations much easier in areas such as astronomy, trade, navigation, engineering, and warfare. The great French mathematician Pierre-Simon Laplace (1749–1827) even suggested that the invention of logarithms doubled the life of astronomers, because it reduced the labors of calculation.

The slide rule was invented in 1622 by the Englishman William Oughtred (1574–1660); it used logarithms to determine lengths, and allowed approximate calculations to be performed mechanically. This technology persisted until the invention of the electronic calculator in the 1970s. Now the slide rule, like the hand-cranked adding machine, is a museum piece.

However, some numbers are too large even for many calculators or computers to evaluate. For example, if a coin is tossed every day of a (non-leap) year, and the result ''heads'' or ''tails'' recorded, there are 2^{365} possible outcomes. If you try to evaluate 2^{365} using the powering key, an 'overflow' error will occur on any machine limited to numbers less than 10^{100} (that is, numbers with fewer than 100 digits). However, as shown in Example 4, the Logarithm of a Power Theorem can be used to approximate this large power.

Example 4 Evaluate 2^{365}.

Solution Let $x = 2^{365}$.
Take the log of each side.

$$\log x = \log (2^{365})$$
$$\log x = 365 \log 2 \qquad \text{Log of a Power Theorem}$$
$$\log x \approx 109.87595$$
$$x = 10^{109.87595} \qquad \text{definition of log}$$
$$= 10^{109} \cdot 10^{0.87595} \qquad \text{Product of Powers Property}$$
$$= 10^{109} \cdot 7.515$$

So 2^{365} is about $7.515 \cdot 10^{109}$, a 110-digit number!

Questions

Covering the Reading

In 1–3, evaluate without using a calculator.
1. $\log 5 + \log 2$ **1**   2. $\log 100^6$ **12**   3. $\log 2000 - \log 2$ **3**

4. **a.** Prove the Logarithm of a Quotient Theorem.
   b. Prove the Logarithm of a Power Theorem.   **a,b) See margin.**

In 5–7, use the facts that $\log_6 11 \approx 1.3383$ and $\log_6 4 \approx 0.7737$ to evaluate.
5. $\log_6 44$ **2.112**   6. $\log_6 \left(\dfrac{11}{4}\right)$ **0.5646**   7. $\log_6 \sqrt[3]{121}$ **0.8922**

238

8. Given $F = \dfrac{G \cdot m \cdot M}{d^2}$, give $\ln F$ in terms of $\ln G$, $\ln m$, $\ln M$, and $\ln d$.
$\ln F = \ln G + \ln m + \ln M - 2 \ln d$

9. If $\log a = 3 \log b - 2 \log c$, find an expression for a that does not involve logarithms. **$a = b^3/c^2$**

10. Given $y = 700(1.05)^x$, write a linear function for $\ln y$.
$\ln y = \ln 700 + x \ln 1.05$

11. a. Evaluate 13^{100} by each of the following methods.
 i. using the powering key on your calculator
 ii. using common logs   **i) sample: overflow error; ii) 2.48 x 10^{111}**
 iii. using natural logs   **1.64 × e^{256}**
 b. Which of the methods in part **a** is best? Why?
 c. How many digits does 13^{100} have?   **112**
 b) ii; i gives an overflow error with most calculators and iii gives an answer that is more difficult to interpret.

Applying the Mathematics

12. The Richter scale for measuring the intensity of earthquakes has been used since its invention in 1932 by Charles F. Richter (1901–85). The scale has the property that its values are common logarithms. The Japanese earthquake of 1923 is estimated to have measured 8.9 on the Richter scale. The 1906 San Francisco earthquake is estimated to have been 8.3. If $\log x = 8.9$ and $\log y = 8.3$, x is how many times as big as y? (This gives you an idea of how much more intense the Japanese earthquake was.)   **about 4 times as great**

13. Prove that $\log_b \left(\dfrac{1}{n}\right) = -\log_b n$.   **See margin.**

14. Let $f(x) = \log (3x)$, $g(x) = \log 3 + \log x$, and $h(x) = (\log 3)(\log x)$.
 a. Use an automatic grapher to plot all functions on the same set of axes.   **See margin.**
 b. Which two of the functions are identical?   **f and g**
 c. What theorem in this lesson justifies your response to part **b**?
 Logarithm of a Product Theorem

In 15–17, *decibels* (dB) are used to measure the intensity of a sound according to the formula:
$$dB = 10 \log (P \cdot 10^{16})$$
where P is the power of the sound in watts/cm^2.

15. A soft whisper has power $P = 10^{-15}$ watts/cm^2. Find its dB level.   **10 dB**

16. Thunder has power $P = 10^{-5}$ watts/cm^2. Find its dB level.   **110 dB**

17. Compare the decibels for P and $10P$.
 The decibel level for 10P is 10 more than the decibel level for P.

Review

18. The population of the People's Democratic Republic of Yemen (formerly known as South Yemen) was estimated to be 2,488,000 in 1989, with an average annual growth rate of 2.8%.
 a. Use the Continuous Change Formula, $A(t) = Pe^{rt}$ where A is the population after t years from 1989 and P is the 1989 population, to find an equation for Yemen's population in future years.
 b. On what assumptions do this formula rest?
 c. Use the equation to estimate the 1995 population of Yemen.
 (Lesson 4-6)   **a) $A(t) = 2,488,000e^{.028t}$; b) Population growth is constant and continuous. c) about 2,943,000**

Margin notes (right column):

$b^m = x^p$ **def. of logarithm**

$b^{m/p} = x$ **take 1/p power of both sides**

$\dfrac{m}{p} = \log_b x$ **def. of logarithm**

$m = p \cdot \log_b x$ **mult. by p**

So,
$\log_b x^p = p \cdot \log_b x$ **substitution**

If $p = 0$, then
$\log_b x^0 = \log_b 1 = 0$
$= 0 \log_b x.$

13. $\log_b \dfrac{1}{n} =$
$\log_b 1 - \log_b n =$
$0 - \log_b n$
Therefore, $\log_b \dfrac{1}{n} = -\log_b n$.

14.a.

239

MORE PRACTICE
For more questions on SPUR Objectives, use *Lesson Master 4-7,* shown on page 239.

PROJECTS
The projects for Chapter 4 are described on pages 259–260. Projects 3 and 4 are related to the content of this lesson.

EVALUATION
A quiz covering Lessons 4-4 through 4-7 is provided in the Teacher's Resource File.

19. If t is the number of years and r is the annual rate, match the interest formulas with their compounding periods. *(Lesson 4-6)*

   a. $A(t) = Pe^{rt}$ **iii**   **i.**   annual compounding

   b. $A(t) = P\left(1 + \dfrac{r}{n}\right)^{nt}$ **ii**   **ii.** periodic compounding

   c. $A(t) = P(1 + r)^t$ **i**   **iii.** continuous compounding

In 20 and 21, evaluate. *(Lesson 4-5)*

20. $\log_2 32$ **5**   21. $\ln e^5$ **5**

22. *Multiple choice.* Which of the following is closest to $\log_4 120$?
   (a) 3   (b) 4   (c) 5   (d) 30 *(Lesson 4-5)*   **(a)**

23. When air is let out of a balloon, the amount of air in the balloon is thought to decay exponentially, so that the volume V at time t seconds is given by $V = Ab^t$. A balloon is inflated to a volume of 5 liters and then released. Its volume after 4 seconds is 0.5 liters. *(Lesson 4-4)*
   a. Calculate values for A and b.   **$A = 5, b \approx 0.56$**
   b. How much air is left in the balloon 2 seconds after being released?
     $\approx$ 1.58 liters

In 24 and 25, *multiple choice.* Below are graphs of two exponential functions $A = kb^t$ with $k > 0$. Match the graph to the value of b. *(Lesson 4-3)*

24.   25.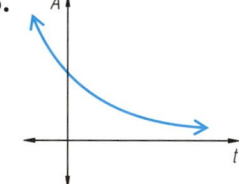

   (a) $b < 0$ (b) $0 < b < 1$ (c) $b = 1$ (d) $b > 1$
  24) (d); 25) (b)

26. Find an equation for the translation image of $y = \dfrac{5}{x}$ shown at the right. *(Lesson 3-2)*
  $y = \dfrac{5}{x - 10} - 5$

27. Given $\log 2 \approx 0.3010$, $\log 3 \approx 0.4771$, and $\log 7 \approx 0.8451$, it is possible to use the properties of logs to find the common logs of most positive integers less than or equal to 40.
   a. Use these values to find
     i. $\log 5$ **ii.** $\log 27$ **iii.** $\log 28$.
   b. Give five numbers less than 50 whose logarithms *cannot* be found from $\log 2$, $\log 3$, and $\log 7$.   **11, 13, 17, 19, 23, 29, 31, 37, 41, 43, 47**
  a) i) log 5 = log 10 − log 2 ≈ 1 − .3010 = .6990; ii) log 27 = 3log 3 ≈ 1.4313;
  iii) log 28 = 2log 2 + log 7 ≈ 2(.3010) + .8451 = 1.4471

Solving Exponential Equations

You already know how to solve equations that contain constant exponents. For example, $x^3 = 15$ can be solved by taking the cube root of each side. However, when an equation such as $3^x = 15$ has a variable as the exponent, you need new techniques.

An equation with a variable exponent is called an **exponential equation**. Approximate solutions to exponential equations may be found by using guess-and-check or graphing strategies. For instance, $3^x = 15$ must have a solution between 2 and 3 because $3^2 = 9$ and $3^3 = 27$. A first guess might be $x = 2.4$; a calculator shows that $3^{2.4} \approx 13.97$, a little low. A check of $x = 2.5$ shows that $3^{2.5} \approx 15.59$, a little high. Thus $2.4 < x < 2.5$. Further guesses and checks can be used to narrow the interval that contains the solution, thus increasing the level of accuracy of the answer.

To solve the equation $3^x = 15$ graphically, you can find the x-coordinate of the point on the graph of $y = 3^x$ that has y-coordinate 15. At the right is a graph of $y = 3^x$.

From the graph you can also see that when $y = 15$, $x \approx 2.5$.

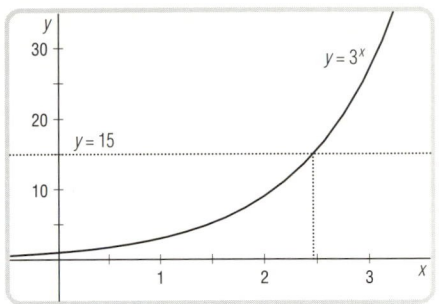

By changing the size of the viewing rectangle on an automatic grapher or by reading the coordinates of a point on the curve marked by a cursor, you can increase the accuracy of your estimated solution.

The graph of $y = 3^x$ on the window $2 \le x \le 3$, $10 \le y \le 20$ is at the right. The coordinates of the cursor suggest that the solution to $3^x = 15$, to three decimal places, is $x = 2.465$.

(2.465, 15.000)

Any equation of the form $b^x = a$, where $b > 0$ and $b \neq 1$, can be solved by using logarithms. If you take the logarithm to any base of each side of an exponential equation, the exponents become factors in a linear equation.

RESOURCES
■ Lesson Master 4-8

OBJECTIVES

B Solve exponential equations.
E Use properties of logarithms.

TEACHING NOTES

The first step of the algorithm for solving $a = b^x$ is to take a logarithm of each side of the equation. Point out that the logarithms must be defined, so a and b need to be positive.

The Change of Base Theorem is necessary in order to use many automatic graphers and calculators to find logarithms to bases other than 10 or e. Students can practice this skill in **Question 11.**

The paragraphs following **Example 2** are important. Students will develop an appreciation for natural logs when they see that natural logs can be easier to use than common logs.

Making Connections As Project 2 on page 259 points out, there is a corresponding change of base theorem for exponential expressions: $a^x = b^{x \log_b a}$. This is not given a special role in most math books because the base b is almost always chosen to be the number e. For example, 8^x can be converted to base e as follows: $8 = e^{\ln 8}$ so $8^x = (e^{\ln 8})^x = e^{x \ln 8}$. Similarly, you can write a continuous change equation with a base other than e: $e^{0.0575 t} \approx (1.05919)^t$.

Example 1　Solve $3^x = 15$ to five decimal places.

Solution　Take logs of both sides. Common logs are convenient.

$$\log(3^x) = \log 15$$
$$x \log 3 = \log 15 \qquad \text{Log of a Power}$$
$$x = \frac{\log 15}{\log 3}$$

So
$$x \approx 2.46497.$$

Check　Use your calculator.　$3^{2.46497} \approx 14.999942 \approx 15$

In general, to solve $b^x = a$ for x using an arbitrary base c, take the logarithm of each side of the equation to that base.

$$\log_c b^x = \log_c a$$

Use the Logarithm of a Power Theorem:　$x \log_c b = \log_c a$.

Divide each side by $\log_c b$ to solve for x:　$x = \frac{\log_c a}{\log_c b}$.

So the solution to $b^x = a$ is $x = \frac{\log_c a}{\log_c b}$. However, the original equation is equivalent to $x = \log_b a$. This proves a theorem about the quotient of logs with the same base.

Theorem (Change of Base)

For all values of a, b, and c for which the logarithms exist:
$$\log_b a = \frac{\log_c a}{\log_c b}.$$

Example 2　Evaluate $\log_4 128$.

Solution　Use the Change of Base Theorem. You may choose any base for c. Choose $c = 10$ in order to use common logarithms on your calculator.

$$\log_4 128 = \frac{\log 128}{\log 4} = 3.5$$

Check 1　This solution can be checked exactly.
$$4^{3.5} = 4^{7/2} = (\sqrt{4})^7 = 2^7 = 128$$

Check 2　Use natural logarithms.
$$\log_4 128 = \frac{\ln 128}{\ln 4} = 3.5$$

242

In a problem like that in Example 2, use your calculator so you do not have to round your numbers until the final step. That reduces the risk of introducing round-off error.

Although any base for logarithms can be used to solve an exponential equation, sometimes one base is preferred over others. For example, the Continuous Change Formula from Lesson 4-6 gives rise to equations involving e. When you take logarithms, $\ln e = \log_e e = 1$, so using natural logarithms makes the solution easier.

Example 3 The 1989 population of Mexico was estimated at 87 million. Its annual growth rate is 2.4%. Estimate when the population will reach 100 million.

Solution Assume the growth rate stays constant. Use the Continuous Change Formula

$$A(t) = Pe^{rt}$$

where t represents the number of years after 1989. Then $A(t) = 100$ million, $P = 87$ million, and $r = 0.024$. You need to solve

$$100 = 87 \, e^{0.024t} \text{ for } t.$$
$$\frac{100}{87} = e^{0.024t}$$

Take the natural log of each side.

$$\ln\left(\frac{100}{87}\right) = \ln\left(e^{0.024t}\right)$$

$$\ln\left(\frac{100}{87}\right) = 0.024t \ln e$$
$$= 0.024t \, (1)$$
$$= 0.024t$$

So
$$t = \frac{\ln\left(\frac{100}{87}\right)}{0.024} \approx 5.803.$$

So if the growth rate remains constant, the population is expected to reach 100 million in about 6 years from 1989, or around the year 1995.

The assumption of continuous change would not be a good one over a long period of time for a country like Mexico, but it is reasonable over a short term like the six years.

An assumption of continuous change is frequently appropriate in physical situations, like those of decay.

Question 8: Some people have serious reservations regarding the use of the decay of radioactive elements to date objects. In particular, such dating shows that the world is quite a bit older than people of certain religious groups believe. However, the reservations tend to be about the use of such dating for objects far older than paintings, since carbon dating has been shown to be very accurate for examples like these. See also **Questions 13 and 25.**

Question 10: If any student has a relative or friend that works in a bank or deals with investments, have the student ask that acquaintance whether he or she is familiar with the rule of 72. It is a very useful rule, surprisingly accurate over the entire range of percents (2% to 20%) usually found in investments.

Question 11c: It is doubtful that a student will see the relationship from the answers; the relationship should be seen from the calculation that led to the answers.

Example 4 Assuming carbon-14 has a half-life of 5730 years, give the approximate age of a skull found by an archaeologist if the skull has 63% of its original carbon-14 concentration.

Solution Use the Continuous Change Formula, $A(t) = Pe^{rt}$ with $P = 1$. The information about half-life implies that $A(5730) = 0.5$. So you can determine r. Substitute: $0.5 = 1 \cdot e^{5730r}$. Take the natural log of each side:

$$\ln 0.5 = 5730r.$$

So
$$r = \frac{\ln 0.5}{5730} \approx -0.000121.$$

Now, substitute for $A(t)$, P, and r in the Continuous Change Formula and solve the resulting equation for t.

$$0.63 = 1 \cdot e^{-0.000121t}$$

Take natural logs: $\ln 0.63 = -0.000121t$
$$t \approx 3818.$$

So the skull is about 3800 years old.

Check More than half the carbon-14 was left, so the age is expected to be less than half the half-life. It checks.

Notice in Example 4 that the rate of growth is negative, indicating decay. The technique of radiocarbon dating, and dating by other radioactive substances, is very important to archaeologists and anthropologists interested in early civilizations. The results are only approximations, however, since they depend on an assumption that the decay rates are constant. Although most scientists believe carbon-14 is an accurate dating technique for about ten half-life periods (up to about fifty thousand years), there is some question of its accuracy for over twenty half-life periods (more than hundreds of thousands of years).

Archaeologists and physical anthropologists examine human fossils to determine dating, gender, age, and cause of death. (Photo courtesy of West Virginia University Photography Dept.)

Questions

Covering the Reading

1. Solve the equation $3^x = 15$ using natural logarithms. Check your solution against that in Example 1. **$x \approx 2.465$; It checks.**

2. Consider the equation $5^x = 15$.
   a. Use guess-and-check to find the solution rounded to the nearest tenth.
   b. Draw a graph that justifies your answer to **a**. **See margin.**
   c. Solve the equation using logarithms. **See margin.**
   a) 1.7

In 3 and 4, solve and check.

3. $7^x = 30$ **$x \approx 1.75$; $7^{1.75} \approx 30$**

4. $6^z = 0.4$ **$z \approx -0.51$; $6^{-0.51} \approx 0.4$**

In 5 and 6, evaluate.

5. $\log_6 20$ **≈ 1.67**

6. $\log_{12} 7$ **≈ 0.783**

7. In 1987, the population of the United States was estimated at 244 million, with an annual growth rate of 0.7%. If this rate continues, estimate when the population will reach 300 million. **2017**

244

8. An art dealer in 1988 used carbon-14 dating to determine whether a painting was likely to have been painted by the great Italian artist and scientist, Leonardo da Vinci (1452–1519). A specimen of paint was found to have 96% of the original amount of carbon-14. Is it plausible that the painting could be a da Vinci? Justify your answer.   **See margin.**

Applying the Mathematics

9. If you invest money at 7% interest compounded continuously, how long will it take for your money to double?   **about 9.9 years**

10. a. Generalize the answer to Question 9. That is, give a rule for the length of time t it takes to double an investment at interest rate $r\%$ assuming continuous compounding. $t = (\ln 2)/(0.01r)$
b. A rule of thumb used by some people to determine the length of time t in years it would take to double their money at $r\%$ interest is called the *Rule of 72*. By this rule, the time t is estimated to be $t = \frac{72}{r}$. To see how this compares with your rule for part **a**, graph the two rules on an automatic grapher simultaneously.   **See margin.**
c. State how well the rule of thumb approximates the continuous compounding case.   **The Rule of 72 is a very close approximation.**

11. a. Find $\log_6 7776$.   **5**
b. Find $\log_{7776} 6$.   **1/5**
c. What is the relationship of the answers to **a** and **b**?   **They are reciprocals.**
d. Use the Change of Base Theorem to explain the relationship in **c**.
   d) $\log_6 7776 = \log 7776/\log 6$; $\log_{7776} 6 = \log 6/\log 7776$
12. In BASIC, LOG(X) means $\ln x$. There is no built-in function for finding common logs. Write a BASIC function for finding common logarithms. $\dfrac{\text{LOG (X)}}{\text{LOG (10)}}$

13. *Encyclopædia Britannica* (Fifteenth Edition, Volume 19, 1985, p. 783) gives the following formula for estimating the age t years of a specimen using the original number N_0 of radioactive atoms in the sample, the number N of radioactive atoms in the sample today, and the half-life $t_{0.5}$ of the substance.

$$t = \frac{t_{0.5}}{0.693} \ln\left(\frac{N_0}{N}\right)$$

a. Check that this formula works for Example 4 by replacing $\frac{N_0}{N}$ with $\frac{P}{A(t)}$.
b. Prove that the formula is correct.   **a,b) See margin.**

14. In Lesson 4-2, you were given the equation $T(p) = 10p^{0.585}$. Assuming $T(p) = ap^b$, derive that equation from the data given on page 208.
See margin.

Review

In 15–17, evaluate without a calculator. *(Lessons 4-7, 4-6, 4-5)*
15. $\ln e$ **1**   **16.** $\log 0.001$ **-3**   **17.** $\ln e^{0.03}$ **0.03**

18. Use the laws of logarithms to express $\log N$ in terms of $\log A$ and $\log B$ for $N = \sqrt{\frac{A}{B^3}}$. *(Lesson 4-7)* $\log N = \frac{1}{2}(\log A - 3\log B)$

13.b. Use the Continuous Change Formula.
$A(t) = Pe^{rt}$.
From the half-life of the substance,
$$A(t_{0.5}) = \frac{1}{2}P$$
$$\text{So, } \frac{1}{2}P = Pe^{rt_{0.5}}$$
$$\frac{1}{2} = e^{rt_{0.5}}$$
$$\ln\frac{1}{2} = \ln(e^{rt_{0.5}})$$
$$-0.693 = rt_{0.5}$$
$$r = \frac{-0.693}{t_{0.5}}.$$

Substituting into the original formula,
$$A(t) = Pe^{(-0.693/t_{0.5})\cdot t}$$
$$\frac{1}{e^{(-0.693/t_{0.5})\cdot t}} = \frac{P}{A(t)}$$
$$e^{(-0.693/t_{0.5})\cdot t} = \frac{P}{A(t)}$$
$$\ln(e^{(-0.693/t_{0.5})\cdot t}) = \ln\left(\frac{P}{A(t)}\right)$$
$$\frac{0.693}{t_{0.5}}\cdot t = \ln\left(\frac{P}{A(t)}\right)$$
$$t = \frac{t_{0.5}}{0.693}\cdot \ln\left(\frac{P}{A(t)}\right)$$

14. See margin, p. 246.

245

MORE PRACTICE
For more questions on SPUR Objectives, use *Lesson Master 4-8,* shown on page 245.

PROJECTS
Project 2 is related to the content of this lesson.

ADDITIONAL ANSWERS
14. Let $T(p) = Ap^b$. Since $T(1) = 10$, $A(1)^b = 10$, and so $A = 10$. Now solve $T(2) = 10 \cdot 2^b = 15$ for b.
$10 \cdot 2^b = 15$
$2^b = 1.5$
$b = \dfrac{\log 1.5}{\log 2} \approx 0.585$
So, $T(p) = 10p^{0.585}$.

22.b. Sample: Positive correlation means for an increase in age, the fish, in general, gets longer.

23.b. For every increase of one year in age, there is an increase, on average, of 2.22 inches in length.

19. *Multiple choice.* Which function could produce the graph at the left? Assume $a > 0$, $b > 0$, and $b \neq 1$. *(Lesson 4-6)*
(a) $y = ab^x$ (b) $y = -ab^x$
(c) $y = a \log_b x$ (d) $y = -a \log_b x$ **(c)**

In 20 and 21, use this situation from psychology. Learning and forgetting are often modeled with logarithmic functions. In one experiment, subjects studied nonsense syllables (like "gpl") and were asked to recall them after t seconds. A model for remembering was found to be:

$$P = 92 - 25 \ln t \quad \text{for } t \geq 1$$

where P is the percent of students who remembered a syllable after t seconds.

20. **a.** What percent of students remembered after 1 second?   **92%**
     b. What percent remembered after 10 seconds?   **about 34%**

21. Use an automatic grapher to give the approximate time after which only half of the students remembered a syllable. *(Lesson 4-6)*   **about 5 seconds**

In 22 and 23, use the typical lengths of Illinois channel catfish of various ages given below. (Source: Illinois Department of Conservation)

Age (years)	1	2	3	4	5	6	7	8	9	10
Length (inches)	6.4	9.6	12.6	14.3	16.7	18.5	21.0	22.6	25.6	26.6

22. **a.** Find the correlation r between the ages of the fish and their lengths.
     b. Interpret the sign of the correlation r. *(Lesson 2-6)*
     a) ≈ 0.997; b) See margin.
23. **a.** Find the equation of the line of best fit for predicting length from age.   **$y = 2.22x + 5.2$**
     b. Interpret the slope of the line.   **See margin.**
     c. Use the line to predict the length of a twelve-year-old channel catfish in Illinois.   **≈ 31.8 inches**
     d. Suggest a reason for being cautious about your prediction in **c**.
     (Lesson 2-5)   **The growth of the catfish might tend to slow as the fish ages.**

24. *Multiple choice.* Which of the following is an equation for the graph of the image of the parabola $y = x^2$ under $S(x, y) = \left(\frac{1}{2}x, 7y\right)$? *(Lesson 3-5)*

(a) $\frac{y}{2} = 49x^2$ (b) $2y = \frac{x^2}{49}$ (c) $\frac{y}{7} = 4x^2$ (d) $y - 7 = \left(x - \frac{1}{2}\right)^2$
(c)

Exploration

25. Has the world population growth rate been constant in this century? Find data to back up your answer.

Here are data from 4 years. Let $x = $ year-1900.

Year	Population (billions)	x
1900	1.6	0
1950	2.5	50
1970	3.6	70
1990	5.2	90

If the growth rate has been constant, then $P = Ab^x$. Using the data from 1900 and 1950, $A = 1.6$ and $b \approx 1.009$. Then $P = 1.6(1.009)^{70}$ predicts a population of 3 billion for 1970, and $P = 1.6(1.009)^{90}$ predicts a population of 3.6 billion for 1990. The growth rate has not been constant; it has increased markedly.

246

4-9

Exponential and Logarithmic Modeling

RESOURCES
- Lesson Master 4-9
- Teaching Aid 37: Detergent data for **Questions 1–4**
- Teaching Aid 38: Word recall data for the lesson and **Questions 3 and 9**
- Computer Master 12

OBJECTIVES

E Use properties of logarithms.
G Use exponential models to solve problems.
H Use logarithmic functions to model data.

TEACHING NOTES

The lesson focuses on two types of log-linear modeling. In both the goal is to represent the data on a scatter plot which appears linear. In the first type, a log transformation is applied to the dependent variable A in $A = kb^t$, and results in the model, $\log A = \log k + t \log b$. In the second variation, the independent variable t undergoes the log transformation and results in the general model $P = a \log t + b$.

Two teaching aids are provided. The first shows the original and transformed data and graphs for the detergent experiment. The second shows comparable information for the recall-of-words experiment.

Because of the introductory nature of the lesson we do *not* discuss two important issues of log-linear modeling in the text, but some students may express curiosity about these issues. One is the issue of how to determine the

---

Linear and quadratic models can be used to describe many data sets. For other data, however, exponential or logarithmic models seem to be more appropriate. In this lesson you will study how to determine whether either an exponential or a logarithmic model seems appropriate for a given situation.

For example, consider an experiment concerned with the spread of liquid detergent film on water. A small amount of detergent was dropped onto a still tank of water, and the area A of water (in square centimeters) covered by a thin film of detergent at various times t (in seconds) was recorded. (A video camera was used to record the actual spread of the detergent, and the areas were calculated.) Below, the data are given in the table and plotted by a statistics package.

t (sec)	A (cm²)
0	1.9
1	2.8
2	3.6
3	4.5
4	6.3
5	8.3
6	10.5
7	13.8
8	18.6
9	26.8
10	31.7

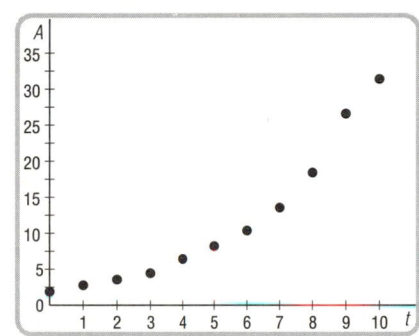

The shape of the plot suggests that a linear model is inappropriate. An appropriate nonlinear model may be an exponential of the form

$$A = kb^t,$$

with k and b constants. For the detergent film data, it seems clear that the data represent an exponential growth situation, so $b > 1$.

In Example 3 of Lesson 4-7, you saw that an exponential equation can be changed to an equivalent linear equation by taking logs of each side.

Logarithms to any base can be used. If you take the common log of each side of the equation $A = kb^t$, you get:

$$\log A = \log k + t \log b$$
$$= (\log b)t + \log k.$$

Since t is the independent variable and $\log A$ the dependent variable, this is a linear model for $\log A$ in terms of t with $\log k$ and $\log b$ as constants.

LESSON 4-9 *Exponential and Logarithmic Modeling*   **247** best among many choices of models for curvilinear data. So far in this book, the choices include quadratic (Lesson 2-6), direct exponential (Lesson 4-4), and now log-linear models. **Question 21** hints at one approach to choosing a best model, but often the decision depends on the particular context of the situation being modeled. For instance, growth or decay might influence the decision to try an exponential rather than a quadratic model. An exponential model should be considered when the data seems to fit a curve which is concave up and appears to have a horizontal asymptote. (For example, data modeling the temperature of coffee as it cools over time.) Similarly, a logarithmic model might be appropriate when the data seems to fit a curve which is concave down and has a vertical asymptote. It is even possible that different models are more appropriate over different parts of the domain of the independent variable. Encourage students to experiment with different models—the computer makes such experimentation easier.

The other issue not presented in the lesson involves numerical accuracy. We have mentioned before that in models where year is a variable, accuracy is increased by coding the year instead of using the entire four-digit number. In **Questions 11 and 13,** for example, you might suggest that students use "years since 1980" and "years since 1790," respectively, as the independent variable. This coding increases the accuracy of the model from the tens place to the tenths place.

The recoding is especially necessary when your statistics package provides coefficients for a line of best fit rounded to the nearest thousandth. Problems of insufficient accuracy are compounded in log-linear models, particularly when the equation for the model is rewritten

248

To see whether such a model is a good fit for the detergent film data, the logarithms of the areas were obtained and graphed against the times, as shown below.

t	log A
0	0.28
1	0.45
2	0.56
3	0.65
4	0.80
5	0.92
6	1.02
7	1.14
8	1.27
9	1.43
10	1.50

The variables log A and t seem to be linearly related. The data are close to the line of best fit drawn by a computer program. The regression equation it found was

$$\log A = 0.121t + 0.303.$$

Example 1 Rewrite the relationship $\log A = 0.121t + 0.303$ to give A as a function of t.

Solution Given

$$\log A = 0.121t + 0.303.$$

Then
$$A = 10^{0.121t + 0.303} \qquad \text{definition of base 10 logarithm}$$
$$= 10^{0.121t} \cdot 10^{0.303} \qquad \text{Power of a Product Property}$$
$$= (10^{0.121})^t \cdot 10^{0.303}. \qquad \text{Power of a Power Property}$$

So
$$A = 2.01(1.32)^t. \qquad \text{Calculation of rational powers and Commutative Prop. of Mult.}$$

Check Check some values of t.

When $t = 0$, $A = 2.01(1.32)^0 = 2.0$, while the recorded value was 1.9.
When $t = 5$, $A = 2.01(1.32)^5 \approx 8.1$, while the recorded value was 8.3.
When $t = 9$, $A = 2.01(1.32)^9 \approx 24.5$, while the recorded value was 26.8.

The model seems to fit the data well.

To study the relationship between A and t for the detergent film data, the first step was to replace each A value by $\log A$. In that step, the variable A was *transformed* into the new variable $\log A$. This is an example of a **logarithmic transformation**.

You can see from the graphs that in this case the logarithmic transformation produces a scattergram that is roughly linear while the raw data is definitely nonlinear. The correlation coefficient for the transformed data, $r = .998$, confirms directly that the transformed data are linear, and indirectly that the original exponential model is appropriate.

Since only a few values were involved, the logarithmic transformation was computed with a calculator. With a large data set, it is easier to transform variables by computer rather than by hand. Study your statistics package to find out if and how your software does a logarithmic transformation.

Now consider the following data and scatter plot from an experiment on the effect of practice time t in seconds on the percent correct recall P of unfamiliar words. For instance, when people were given 5 seconds to concentrate on the words, they recalled 53% of the meanings correctly. The relationship between P and t is clearly nonlinear.

Time practicing in (t) seconds	% recalled correctly (P)
1	30
2	43
5	53
10	65
15	74
20	76
30	85

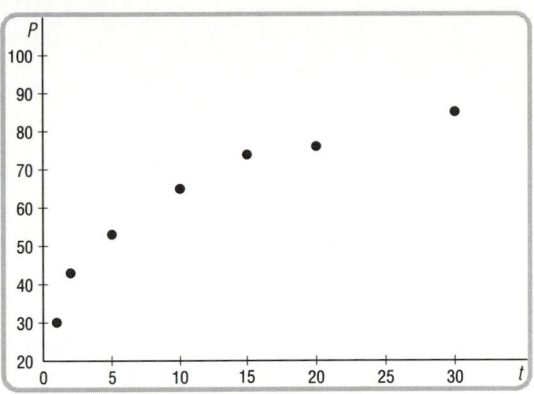

The graph suggests that an appropriate model may be logarithmic of the form
$$P = a \log t + b$$
for some positive a and b.

To test whether this model is appropriate for the practice time data, first find the logarithms of the values of t. Any base is suitable. Then plot P against the logarithm of t.

t	ln t	P
1	0	30
2	.70	43
5	1.61	53
10	2.30	65
15	2.71	74
20	3.00	76
30	3.40	85

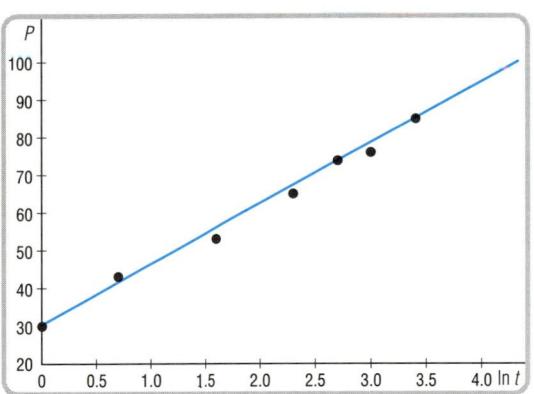

The effect of the transformation is to show a near linear relationship between P and $\ln t$, as the scatter plot shows. So, a logarithmic model is appropriate for these data. A statistics package, with variables P and $\ln t$, calculates an equation for the line of best fit:
$$P = 15.7 \ln t + 30.1.$$

in exponential form, as in **Example 2.**

Because we are not attempting to make an exhaustive study of log-linear modeling in this book, we have not introduced the name "log-linear" to students. The name derives from the fact that first the data is operated on by a log function, then the resulting data is modeled linearly to give a linear function of those log values. It is an example of composition of functions.

Technology We recommend that before class you (or a student assistant) enter the data for the Examples into a computer for demonstration purposes. At that time, check to see if your statistics package can make log transformations of data as well as find a line of best fit of the transformed data—otherwise logs need to be calculated and then entered into the computer.

Making Connections
One of the greatest examples of modeling of all time, Kepler's finding that the time the planets go around the sun is related to their distance from the sun, used logarithms. The details are discussed in Chapter 12.

ADDITIONAL EXAMPLES
1. Rewrite $G = 36(7)^x$ as a linear model of log G in terms of x.
log G = log 36 + (log 7)x

2. Solve ln $R = 9x - 5.52$ for R.
$$R = e^{(9x - 5.52)} = \frac{e^{9x}}{e^{5.52}} \approx .004e^{9x}.$$

3. Referring to **Example 2** in the lesson, estimate the amount of practice time needed to exceed 98% recall.
Solve 98 = 15.7 ln t + 30.1; $t \approx$ 76 seconds

NOTES ON QUESTIONS
Question 10: When the ratio of the largest to the smallest value of the dependent variable is great, taking the logarithms of that variable may be appropriate. Here that ratio is almost 25.

ADDITIONAL ANSWERS
10.a.

b.

year (t) after 1980	log n
2	2.64
3	3.15
4	3.50
5	3.80
6	4.03

c.,d.

e. log n = 0.343t + 2.052, where t is year after 1980.

f. $10^{\log n} = 10^{0.343t} + 2.052$
$n = 10^{0.343t} \cdot 10^{2.052}$
$n \approx 112.7 \, (10^{0.343})^t$
$n \approx 112.7 \, (2.03)^t$

11.c. Sample: the Civil War and World War I; less immigration during this period

---

■ ■ ■ ■ ■ ■ ■ ■ ■

Example 2   Assuming that $P = 15.7 \ln t + 30.1$ is an appropriate model for the word recall experiment, approximate the amount of practice time needed to exceed 95% recall.

Solution   Substitute 95 for P and solve for t.

$$95 = 15.7 \ln t + 30.1$$
$$64.9 = 15.7 \ln t$$
$$\ln t \approx 4.133758$$

So by definition of $\ln t$, $t \approx e^{4.133758}$
$$\approx 62.4.$$

The model predicts that people will exceed 95% accuracy after about 63 seconds of practice.

Check   Extend the line on the graph. When $P = 95$, $\ln t$ is a little more than 4. So $\ln t \approx 4$, or $t \approx e^4 \approx 55$, consistent with the solution.

As with all real data, it is important to keep equations and predictions like those in Examples 1 and 2 in perspective. For instance, in Example 2 perhaps it is not reasonable to assume most people will ever achieve 95% recall of the words. It may be that after a certain number of seconds of practice, boredom becomes a factor and percent recall actually declines. However, for practice times within the range observed, the model seems to be quite accurate. As always, it is usually more defensible to interpolate than to extrapolate.

Also, other nonlinear models may also describe the data well. In particular, a quadratic model for A as a function of t in Example 1 might be appropriate. A decision to prefer an exponential or logarithmic function over other possible models is often made because one model is predicted by, or consistent with, other theories or hypotheses.

Questions

Covering the Reading

In 1–4, refer to the detergent film experiment.

1. What are the values of $\log k$ and $\log b$ in the equation $\log A = \log k + t \log b$?   **log k = 0.303, log b = 0.121**

2. Verify that $(t, \log A) = (5, 0.92)$ is close to the line of best fit.
   0.121(5) + 0.303 = 0.908

3. After 12 seconds, what area do you expect the film to have?
   about 57 cm²

4. What area of film would you expect 8.6 seconds after the experiment begins?   **about 22.1 cm²**

5. **a.** *True* or *false*.   If $0 < b$ and $b \neq 1$, $y = ab^x$ if and only if $\log y = \log a + x \log b$.   **True**
   b. A logarithmic transformation maps an exponential equation to what type of equation?   **linear**

250

6. Rewrite the equation $A = 12(3)^t$ in the form $\log A = mt + b$, where m and b are constants. **log A = 0.477t + 1.08**

7. Rewrite the equation $\log C = 1.2r + 3.3$ in the form $C = kb^r$ (where k and b are constants). **C = 1995.3 · 15.8^r**

In 8 and 9, refer to the word recall experiment.

8. According to the model, what is the expected percent recall after 4 seconds of practice? **about 51.9%**

9. How much time should a subject expect to practice in order to achieve 90% recall? **about 45 sec.**

10. Consider the following data on the number of AIDS cases reported in the U.S. by state health departments between 1982 and 1986.

Year (t)	1982	1983	1984	1985	1986
Number of cases (n)	434	1416	3196	6242	10,620

a. Draw a scatter plot of the data with year (t) on the horizontal axis.
b. Find the log of n for each value above.
c. Draw a scatter plot of log n vs. t.
d. Draw a line that seems to fit the data in c.
e. Find an equation for the line in part d.
f. Show that the equation in part e is equivalent to an exponential equation. **a)–f) See margin.**
g. Use the models in e and f to predict the number of AIDS cases in 1988. (The actual number of AIDS cases reported by state health departments in 1988 was 15,463.) **about 62,500**
h. Over what domain do you think your models in e and f will hold? **sample: between 1982 and 1986**

11. Some historians have said that on the basis of census data up to 1860, President Lincoln estimated that the U.S. population by 1930 would be over 250 million. Here are the census figures from 1790 to 1860.

Census Year	Population
1790	3,929,214
1800	5,308,483
1810	7,239,881
1820	9,638,453
1830	12,866,020
1840	17,069,453
1850	23,191,876
1860	31,444,321

a. Find a linear relationship between log P, the logarithm of population, and y, the year after 1800. Use y as the independent variable.
b. Does your model predict the same 1930 U.S. population as Lincoln did? **No, it predicts slightly below 250 million.**
c. The actual population of the U.S. in 1930 was about 122.8 million. What factors might account for any difference between your prediction and the actual population? **See margin.**
d. Convert the equation in part a to an equivalent exponential equation.

a) log P = 0.01282y + 6.7255; d) P = (5,314,960) 10$^{0.01282y}$ or 5,314,960 (1.30)^y

NAME _____

LESSON **MASTER   4-9**
QUESTIONS ON **SPUR** OBJECTIVES

■**PROPERTIES**   Objective E (See pages 263–266 for objectives.)
1. True or false. The equation $y = ab^{kx}$ is equivalent to $\log y = mx + \log a$, where $a > 0$, $b > 0$, $b \neq 1$, and m is a real constant.   **True**

2. Rewrite the equation $T = 5.3e^t$ in the form $\log T = mt + b$, where m and b are constants.
   log T = (log e)t + log 5.3

3. Rewrite the equation $\ln R = -9.86t - 2.30$ in the form $R = Ie^{rt}$, where I and r are constants.
   R = .1e$^{-9.86t}$

■**USES**   Objectives G and H
4. Use the compound interest formula $A = P\left(1 + \frac{r}{n}\right)^{nt}$ to decide how long it would take $1000 to double at 9% interest compounded quarterly.
   about 7 years and 3 quarters, or 7.79 years

5. Suppose bacteria of a certain type were found on Mars and could grow from 80 to 164 in number in 3 hours. Find k for the growth formula $y = ne^{kt}$.
   k ≈ .239/hour

52   Functions, Statistics, and Trigonometry © Scott, Foresman and Company

FOLLOW-UP

MORE PRACTICE
For more questions on SPUR
Objectives, use *Lesson Master 4-9,* shown on page 251.

EXTENSION
1. Question 17 can be generalized. How long does it take for an investment at $r\%$ continuously compounded to be multiplied by k?

$$\left(\dfrac{\ln k}{.01r}\right)$$

2. Small Group Work
Students can research the population of a few local communities as determined by the U.S. Census since 1960 and determine a growth model for each.

PROJECTS
The projects for Chapter 4 are described on pages 259–260. Project 1 is related to the content of this lesson.

12. Statisticians often use r^2 to decide which of two models is a better fit for a data set. Refer to the detergent-film data.
    a. Use a statistics package to fit a linear model to the data.
    b. Calculate r^2 for the data.   **a) $A = 2.844t - 2.509$; b) .876.**
    c. Which is larger, the r^2 for the (t, A) data or the r^2 for the $(t, \log A)$ data? Which model, the linear model or the exponential model, fits the original data better?   **r^2 for the $(t, \log A)$ data; the exponential model**

13. Refer to the recall data.
    a. Use your statistics package to find a model for the data in the form $P = a \log t + b$.   **$P = 36.2 \log t + 30.1$**
    b. Use your model in **a** to predict the amount of practice needed to achieve 95% recall.   **about 62 seconds**
    c. Compare your answer in **b** to that in Example 2. Does the answer seem to depend on whether common or natural logarithms are used?   **It is the same answer. The answer doesn't seem to depend on whether common or natural logs are used.**

Review

In 14 and 15, solve and check. *(Lesson 4-8)*

14. $6^x = 70$
    $x \approx 2.37$; Check: $6^{2.37} \approx 70$

15. $\log_5 376 = t$
    $t \approx 3.684$; Check: $5^{3.684} \approx 376$

16. Assuming carbon-14 has a half-life of 5730 years, give the approximate age of a wooden spear found by an anthropologist, if the spear has 34% of the original carbon-14 concentration. *(Lesson 4-8)*   **about 8900 years**

17. Find a formula for the length of time it takes to *triple* an investment under continuous compounding at $r\%$ interest. *(Lesson 4-8)*   **$t = \dfrac{\ln 3}{.01r}$**

18. The graph of $y = \log_2 x$ is scaled by $S(x, y) = \left(x, \dfrac{1}{\log_2 3} y\right)$.
    a. Use the Graph Scale Change Theorem to write an equation for the image as a quotient of logarithms.   **$y = \log_2 x/\log_2 3$**
    b. Use the Change of Base Theorem to write the equation in **a** as a single logarithm function. *(Lessons 4-8, 3-5)*   **$y = \log_3 x$**

19. *True* or *false*. $\log\left(\dfrac{5}{3}\right)^2 = 2 \log 5 - \log 3$ *(Lesson 4-7)*   **False**

20. If d dollars are invested at a rate of $r\%$ per annum, compounded continuously, give a formula for the value of the investment after t years. *(Lesson 4-6)*   **$A(t) = de^{(rt/100)}$**

21. Refer to the detergent-film data in the lesson.   *(Lesson 2-7)*
    a. Fit a quadratic equation to these data, using the points $(1, 2.8)$, $(5, 8.3)$, and $(9, 26.8)$.   **$y = 0.40625x^2 - 1.0625x + 3.45625$**
    b. What values for A does your model from part **a** predict for $t = 0$, 8, and 10?   **$(0, 3.45625)$; $(8, 20.95625)$; $(10, 33.45625)$**
    c. Which model—the exponential or quadratic—fits the data better?   **exponential**

Exploration

22. Collect data on the number of AIDS cases reported in recent years for your city or state. Which type of function—linear, exponential, or logarithmic—best models the data?   **Answers may vary.**

4-10

The Scientific Calculator

Hand-held calculators are a recent invention. Before the 1970s, the only electronic calculators available were the size of typewriters, were limited to performing the four basic operations, and were rather expensive—as much as three or four hundred dollars for a machine with no more capability than what a $5 machine will do today. Today's scientific calculators have many more built-in functions.

How does a scientific calculator work? How can the operations of addition, subtraction, multiplication, and division be used to find powers and logarithms quickly and accurately? Ideas from this chapter provide some answers to the questions.

Of course, different calculators work in different ways, and each inner working is usually a closely-guarded secret of the manufacturer. So this lesson will explore how a calculator *might* be programmed to evaluate powers, exponentials, and logs. In practice, slightly different methods may be used, but they tend to apply the same underlying mathematical principles.

The exponential function plays a central role. It has so many applications that it is built into scientific calculators and computers. Calculators and computers find values for the exponential function by approximating the sum of several terms of a series,

$$e^x = 1 + \frac{x}{1} + \frac{x^2}{2\cdot1} + \frac{x^3}{3\cdot2\cdot1} + \frac{x^4}{4\cdot3\cdot2\cdot1} + \dots .$$

For instance,

$$e^2 = 1 + \frac{2}{1} + \frac{2^2}{2\cdot1} + \frac{2^3}{3\cdot2\cdot1} + \frac{2^4}{4\cdot3\cdot2\cdot1} + \dots .$$

Notice that it is necessary to use only the operations of addition, multiplication, and division to evaluate the terms of this series. A machine can be programmed to perform these operations almost instantly. The denominators of successive terms of this series become large very quickly, while the corresponding numerators do not. So the value of later terms is very small, and the sum of the terms approaches a constant number rather quickly. A good approximation of any power of e can be found with just a few terms. To illustrate this, consider the BASIC program below which prints out sums of the first n terms of this series for $n = 1$ to 20.

In this program, SUM stores the current sum of successive terms. The most important lines are lines 60 and 70. Line 60 adds the next TERM to give a new SUM. Line 70 computes the new TERM by multiplying the previous TERM by X and dividing by the next integer N.

```
5    REM SERIES FOR E^X
10   INPUT "POWER OF E?"; X
20   LET SUM = 0
30   LET TERM = 1
40   PRINT "N", "SUM OF N TERMS"
50   FOR N = 1 TO 20
60      LET SUM = SUM + TERM
70      LET TERM = TERM * X/N
80      PRINT N, SUM
90   NEXT N
100  END
```

TEACHING NOTES

Point out to your students that these algorithms are valid and accurate, but are not necessarily the ones used in a given calculator. The actual algorithms used are often trade secrets, that is, information known only to the calculator manufacturer. For this reason, students may discover minor differences in rounding and calculation time between different calculator brands. Also, calculators approximate values to more digits than are displayed in order to give more accurate values.

We do not give a formal definition of the sum of a convergent infinite series at this time, but students tend to have little difficulty with the intuitive idea. Students do already have some experience with the notion of convergence from their study of functions which have graphs with asymptotes. This lesson should further contribute to building their intuition on the subject.

When the BASIC program was run, the sum stabilized after 25 terms. Point out that the number of terms required for the series to stabilize will vary with the computer and software used.

Technology You might wish to demonstrate the successive approximations of power series using an automatic grapher. On the same axes, in order, graph

$y = 1 + x,\ y = 1 + x + \dfrac{x^2}{2}$,

$y = 1 + x + \dfrac{x^2}{2} + \dfrac{x^3}{6}$,

$y = 1 + x + \dfrac{x^2}{2} + \dfrac{x^3}{6} + \dfrac{x^4}{24}$, and, finally, $y = e^x$. Do a similar demonstration for $\ln(1 + x)$, and point out how the polynomials move away from $\ln(1+x)$ outside $-1 \geq x \geq 1$.

Another example that may interest your students uses the following recursive algorithm, based on Newton's method of approximation, to approximate $\sqrt{x}$ with $a_n \geq 0$ and $n \geq 0$.

$$\begin{cases} a_0 = \dfrac{x}{2}; \\ a_n = 0.5\left(a_{n-1} + \dfrac{x}{a_{n-1}}\right), \text{ for } n > 1. \end{cases}$$

Generation of this sequence enables one to find square roots with a basic four-function calculator.

POWER OF E? 0.7

N	SUM OF N TERMS
1	1
2	1.7
3	1.945
4	2.002167
5	2.012171
6	2.013571
7	2.013735
8	2.013751
9	2.013753
10	2.013753
.	.
.	.
.	.
20	2.013753

Some lines of output when this computer program was run with $X = 0.7$ are shown to the left.

Notice that the first line of the table gives the first term in the series for $e^{0.7}$, the second line gives the sum of the first two terms, the third the sum of the first three terms, and so on. For all lines of output from the 9th through the 20th, the value of SUM printed was 2.013753.

A scientific calculator does not use BASIC, but the ideas behind it are the same as those above. Most scientific calculators have an $\boxed{e^x}$ key or an $\boxed{\text{INV}}\boxed{\text{In}}$-combination of keys for evaluating the exponential function. The effect of the key(s) is to activate a program like the one above, usually taking only the first few terms. The program stops when a good approximation is found; that is, when successive terms do not differ by more than the internal accuracy of the calculator (usually one or two digits more than the display). The series stabilizes so quickly that you hardly notice the time involved.

Using a calculator, $e^{0.7} \approx 2.013753$ to six decimal places.

The natural logarithm function can also be evaluated on machines by using a series, but the series is valid only for small values of the variable:

$$\ln(1+x) = x - \frac{x^2}{2} + \frac{x^3}{3} - \frac{x^4}{4} + \frac{x^5}{5} - \ldots, \text{ if } -1 < x \leq 1.$$

Below is a computer program in BASIC to print the first n terms of this series for $n = 1$ to 100.

```
10    REM SERIES FOR LN(1+X)
20    INPUT "TYPE X, FOR -1<X≤1"; X
30    LET SUM = 0
40    LET TERM = X
50    PRINT "N", "SUM OF N TERMS"
60    PRINT
70    FOR N = 1 TO 100
80        LET SUM = SUM + TERM/N
90        LET TERM = -X * TERM
100       PRINT N, SUM
110   NEXT N
120   END
```

When the program was run with $X = 0.5$, the SUM stabilized at $\ln 1.5 \approx 0.405465108$ after only 25 terms of the series were evaluated.

When $x = 1$ in the series for $\ln(1+x)$,

$$\ln(1+1) = \ln 2 = 1 - \frac{1}{2} + \frac{1}{3} - \frac{1}{4} + \frac{1}{5} - \ldots.$$

254

This series for ln 2 stabilizes very slowly, unfortunately, so that a very large number of terms is needed to get reasonable accuracy. For example, to stabilize at ln 2 = 0.693 required over 3400 terms. Because of time constraints, a program like the one above in a calculator will efficiently evaluate ln $(1 + x)$ for only a limited range of values of x.

What if you need to find ln 100 ? The series cannot be used directly, since it is valid only for $-1 < x \le 1$. The calculator uses the properties of logarithms to find an equivalent expression to ln 100.

From the properties of exponents and logarithms, for all $x > 0$,

$$\ln \frac{1}{x} = \ln x^{-1} \quad \text{log of a quotient}$$

$$= -\ln x. \quad \text{log of a power}$$

So

$$\ln x = -\ln \frac{1}{x}.$$

Let $x = 100$:

$$\ln 100 = -\ln \frac{1}{100}$$

$$= -\ln \left(1 - \frac{99}{100}\right)$$

$$= -\ln \left(1 + -\frac{99}{100}\right).$$

So ln 100 can be evaluated by running the program for LN(1 + X) with
$X = -\frac{99}{100}$, then taking the opposite.

This series stabilizes quite slowly. Find the sum of 1000 terms by changing line 70 to read 70 FOR N = 1 TO 1000.
The result is $\ln \left(1 - \frac{99}{100}\right) \approx -4.60516625$,
so then $\ln 100 = -\ln \left(1 - \frac{99}{100}\right) \approx 4.60516625$.
Since ln 100 = 4.6051702 (to seven decimal places), the first thousand terms stabilize the series only to four decimal places. More efficient (that is, faster) series are almost certainly used in calculators.

It is easy to obtain common logarithms on a calculator once natural logarithms have been found. Using the Change of Base Theorem:

$$\log_{10} x = \frac{\ln x}{\ln 10},$$

so

$$\log x \approx 0.434294 \cdot \ln x.$$

For example, if ln $x = 7.5958899$,
then log $x = (.434294)(7.5958899) \approx 3.2988494$.
(To check: $x = 1990$, and log $x = 3.2988531$, which agrees to five decimal places.)

To get powers and roots on a calculator, the exponential function can be used in a clever way, because *any* exponential function can be written as an exponential function with base e. For example, if $y = 8^x$, then $\ln y = x \ln 8$. So by the definition of natural logarithm, $y = e^{x \ln 8}$. Thus for all x, $8^x = e^{x \ln 8}$. To find $8^{3.1}$, for example, the calculator evaluates

$$8^{3.1} = e^{3.1 \ln 8}.$$

A series is used to find $\ln 8$, a multiplication gives $3.1 \ln 8$, and another series gives $e^{3.1 \ln 8}$. So

$$8^{3.1} \approx e^{3.1(2.0794)} \approx e^{6.4463} \approx 630.346.$$

By a similar argument, you can show that in general, $a^x = e^{x \ln a}$. So a powering key can be programmed to find a^x by calculating $e^{x \ln a}$. A series is used to find $\ln a$, a multiplication gives $x \ln a$, and another series is used to find $e^{x \ln a}$. This accounts for the fact that, on some calculators, the powering key works more slowly than other keys. A powering key can also be used to find a square root by raising a number to the $\frac{1}{2}$ power, although an alternative, faster method is usually used.

The invention of logarithms in the seventeenth century revolutionized calculation. Slide rules were a mechanical implementation of this mathematical theory. Similarly, the extraordinary power and speed of the scientific calculator can be explained partly by the theory of exponential and logarithmic functions that you have seen in this chapter; the "mechanical implementation" of that theory, however, depends on the high technology world of electronics and the silicon chip, which allows these fairly old mathematical ideas to be used in new and dazzling ways.

This silicon chip contains the equivalent of about 350,000 transistors.

Questions

Covering the Reading

1. Refer to the series for e^x on the first page of the lesson. Give the next two terms of the series. **$x^5/5\cdot4\cdot3\cdot2\cdot1$, $x^6/6\cdot5\cdot4\cdot3\cdot2\cdot1$**

2. **a.** Run the computer program for e^x to evaluate e^{10} and e^2.
   b. Which stabilizes more quickly?
   a) $e^{10} \approx 21950.38$; $e^2 \approx 7.389057$; b) e^2

3. Give a series for $\ln 0.9$. **$\ln(0.9) = (-0.1) - \frac{(-0.1)^2}{2} + \frac{(-0.1)^3}{3} - \frac{(-0.1)^4}{4} + \frac{(-0.1)^5}{5} - \cdots$**

In 4 and 5, run the computer program for $\ln(1 + x)$ in the lesson to evaluate the expression correct to five decimal places.

4. $\ln 1.7$ **0.53062**   5. $\ln 62$ **4.12713**

6. Why is the series for $\ln(1 + x)$ not valid for $x = -1$?
   because ln 0 is undefined

7. Evaluate $11^{2.6}$ on your calculator using $\boxed{\ln}$ and $\boxed{e^x}$ keys, but not the powering key. (Check your result with the powering key.)
   $11^{2.6} = e^{\ln 11^{2.6}} = e^{2.6\ln 11} \approx 510.1$

8. **a.** Write $f(x) = 6^x$ as an exponential function with base e.   **$6^x = e^{x \ln 6}$**
   b. Use your answer to part **a** find $6^{0.3}$ without using the powering key.
   $6^{0.3} = e^{0.3 \ln 6} \approx 1.712$

9. The table below lists the estimates of expected lifespan for people of different ages.

Current age (years)	Expected Lifespan (years)	Current age (years)	Expected Lifespan (years)
1	75.1	45	77.5
5	75.2	50	78.1
10	75.3	55	79.0
15	75.4	60	80.2
20	75.7	65	81.7
25	75.7	70	83.5
30	76.3	75	85.7
35	76.6	80	88.1
40	77.0	85	91.1

a. Draw a scatter plot of current age x versus expected lifespan y.
b. Which of the two functions, $y = ab^x$ or $y = a \log x$, seems to fit these data better? **$y = ab^x$**
c. Determine an equation of one of the forms in part **b** to fit these data.
d. For the data above, a statistics package calculated the equation $y = 0.165x + 72.283$ for the line of best fit. Which of the models, the linear model or the model from part **c**, predicts the expected lifespan of a 40-year old with the greatest accuracy? **the model in part c**
e. Comment on the suitability of the models in parts **c** and **d** for predicting the expected lifespan of a 90-year old. *(Lessons 4-9, 2-3)*
   a,e) See margin. c) $y = 72.6 (1.002)^x$

10. It has been suggested that the time taken t (seconds) for a rowing shell to cover 2000 m is related to the number x of oarsmen in the boat, according to the power function

$$t = ax^n.$$

Here are the winning times for various events at the 1980 Olympics rowing regatta.

x	1	2	4	8
t	429.6	408.0	368.2	349.1

a. Take logs of each side of the equation for t above.
b. Graph $\log t$ (vertical axis) against $\log x$ (horizontal axis) for the data. **See margin.**
c. Find an equation for the line of best fit for the graph in part **b**.
d. Interpret the slope of your line in part **c**. **See margin.**
e. Give a model of the form $t = ax^n$ for the 1980 Olympics data.
   (Lessons 4-9, 4-2) **a) $\log t = \log (ax^n) = \log a + n\log x$;**
   c) $\log t = -0.105\log x + 2.635$; e) $t = 432x^{-0.105}$

In 11 and 12, solve. *(Lessons 4-8, 4-1)*

11. $8^x = 20$ **$x \approx 1.44$**

12. $3y^5 = 105$ **$y \approx 2.036$**

13. Use the Change of Base Theorem to graph $f(x) = \log_2 x$ on a function grapher. Print out or sketch the result. *(Lesson 4-8)* **See margin.**

EXTENSION
Use the series from this lesson to estimate various values of e^x and ln x without using the keys for them on the calculator. For what values of x is an approximation to e^x found within .01 after only 4 terms of the series?
$(-.7245 < x < .6756)$

14. Judge which of a linear model, exponential model, or logarithmic model would be most appropriate for each scattergram. *(Lessons 4-9, 4-4, 2-2)*

a.   **b.**   **c.**   **d.**

exponential   **linear**   **linear**   **logarithmic**

In 15 and 16, write without logs.

15. $\log 5 + \log 20$ **2**

16. $\log_c \frac{b}{c} + \log_c \frac{c^2}{b}$ *(Lesson 4-7)*   **1**

17. **a.** Use a function grapher to draw the inverse g^{-1} of $g(x) = 5^x$.
    b. Find $g^{-1}(10)$. *(Lessons 4-6, 4-5)*
    a) See margin. b) ≈ 1.43

18. Describe the transformation which takes the graph of $y = x^2$ to the graph of $y = (x + 8)^2 - 11$. *(Lesson 3-2)*   **$(x, y) \rightarrow (x - 8, y - 11)$**

19. The sector ABO is cut from a circle with radius 11 inches. The central angle is 100°. *(Previous course)*
    a. What is the area of the sector?   **$605\pi/18$ in.^{2} ≈ 105.6 in.^{2}**
    b. What is the length, in inches, of $\overset{\frown}{AB}$?   **$55\pi/9$ in. ≈ 19.2 in.**

20. The shadow of a tree is 40 feet when the shadow of a 6 foot forest ranger is 7 feet. Estimate the height of the tree.
(Previous course)
about 34 feet

Exploration

21. In Question 29 of Lesson 4-6, it was noted that $e^x \approx 1 + x$ for small values of x. Use the series on the first page of this lesson to explain why this approximation works.
Using the series, $|e^x - (1 + x)| = \frac{x^2}{2} + \frac{x^3}{3 \cdot 2} + \frac{x^4}{4 \cdot 3 \cdot 2} + \dots$. When $x \approx 0$, the terms of the series get very small very quickly and the sum remains small.

258

CHAPTER 4

Projects

RESOURCES
■ Teaching Aid 39: Semi-log paper
■ Teaching Aid 40: Log-log paper

PROJECTS

The projects provide an opportunity for extended work on topics related to the content of the chapter. (See pages 66 and T37 for more information on the purposes of projects.

Solutions to the projects are found at the end of the Additional Answers section at the back of the book.

Project 1 shows how log-linear models can be graphed by first applying the log transformation to regular graph paper, then plotting the untransformed data. Semi-log and log-log paper are needed for the assignment. We have included a teaching aid for each type of paper for you to photocopy and distribute, or students can purchase the paper by the sheet at most office supply stores.

Using semi-log or log-log paper requires great care when labeling the scales.

In parts **a** and **b** of **Project 1,** students should get the graphs of lines. The second line in **a** is a vertical translation image by log 3 of the first line because log 3(1.6)^x = log 3 + x log 1.6. So the lines are parallel, a fact which may surprise your students. The data collected and studied in part **c** could be interesting because some states are currently growing at a slower rate than in the past. Therefore the graphs your students make should not be linear, and the population model should not be exponential.

1. The graph above shows the U.S. population from 1790 to 1860 on special paper called *semi-log* graph paper. The scale on the *y*-axis is graduated in proportion with the logarithms of the numbers represented. The *x*-axis is scaled uniformly, which is why the paper is only "semi" logarithmic. With such paper, you do not need a calculator or computer to do a logarithmic transformation. You merely graph the populations directly (watch the scale carefully; the halfway mark between vertical measures is the geometric mean of those values, not the arithmetic mean.) Similarly, you can read values directly from either scale. Extending the line in this case, the predicted 1930 population is about 250 million.

   a. Obtain some semi-log graph paper. Notice carefully the numbers on the vertical axis. Label the vertical axis from 1 to 100 and the horizontal axis from 1 to 10. Graph the function $f(x) = 1.6^x$. What shape is your graph? Graph $g(x) = 3(1.6)^x$ on the same set of axes. What kind of transformation maps f to g? Explain why.

   b. Graph other exponential functions on semi-log paper, for example $k(x) = e^x$, $m(x) = e^{2x}$, and $n(x) = 1.8^x$. Describe the graph of any exponential function that is plotted on semi-log paper.

   c. Obtain some population data from several years for your state. Plot the data on semi-log paper. Estimate the population in your state in 1990, and check it against 1990 census data. Explain any deviations from a linear model for the log of the population.

   d. Log-log paper has logarithmic scales on both the *x*- and *y*-axes. Use log-log paper to graph $f(x) = x^3$, $g(x) = x^{1.5}$, and $h(x) = x^{-2.2}$ on the same set of axes. What do you notice? What is true about the graph of any function of the form $y = x^m$ when it is plotted on log-log paper?

2. Just as logarithms can be converted from one base to another using the Change of Base Theorem, so can exponential functions be converted from one base to another.

   a. Use a function grapher to graph $f(x) = 8^x$ and $g(x) = e^{2.07944x}$ on the same graph. What do you notice?

   b. Graph $f(x) = 8^x$ and $h(x) = 6^{1.16056x}$ on the same graph. What do you notice?

   c. Prove that $8^x = e^{x \ln 8}$ by starting with $y = 8^x$. Use this result to account for what you observed about the graphs in **a**.

   d. Explain your observations in **b**.

   e. Rewrite $f(x) = 8^x$ as an exponential function with base 3.

   f. Any exponential function of the form $f(x) = ab^x$ can be rewritten as an exponential function of the form $g(x) = ke^{ax}$. Develop a method of doing this. (Notice that this means that you never need to find any powers except powers of e. The power key on your calculator is unnecessary, provided you have an e^x key!)

CHAPTER 4 Projects **259**

3. One of the more influential documents in the recent history of the world was the small booklet published in 1803 by the Englishman, Thomas R. Malthus (1766–1834), *An Essay of the Principle of Population as it affects the Future Improvement of Society, with Remarks on the Speculations of Mr. Godwin, M. Condorcet and other Writers: A View of its Past and Present Effects on Human Happiness with an Inquiry into our Prospects Respecting the Future Removal or Mitigation of the Evils Which it Occasions.* As the (lengthy) title suggests, Malthus was concerned about the growth of human populations and the consequences of allowing such growth to go unchecked. His main claim was that populations grow exponentially, while the food supply does not. Consequently he claimed that populations will expand to the limits of subsistence, and will be held there by famine, war, and ill-health. According to Malthus, "vice" (e.g., infanticide), "misery," and "self-restraint" would check population growth.

   a. Consult an encyclopedia or other source to find out more about Malthus. Why did he write his book? What solutions did he propose? How did people of his era view his book? What are the current views?

   b. Malthus assumed that, at best, the food supply increases by a constant amount each year. With the benefit of hindsight, criticize this assumption.

   c. Here are some data on recent world population trends taken from the United Nation's *Demographic Yearbook,* 1980, and the 1990 *Information Please Almanac.* The figures are in millions, estimated at the middle of the years shown.

	Asia	Europe	Africa	South America	USSR	North America	Oceania
1950	1380	392	219	164	180	166	13
1955	1514	408	244	187	196	182	14
1960	1683	425	275	215	214	199	16
1965	1878	445	311	247	231	214	18
1970	2091	460	354	283	244	226	19
1975	2319	474	406	323	254	236	21
1980	2558	484	472	365	267	246	23
1985	2831	492	566	410	278	264	24

   In recent years, has population growth in various parts of the world followed an exponential trend? Has total world population increased exponentially since 1950? Justify your response with graphs and equations.

4. Lesson 4-7 mentions that logarithms and the slide rule were once used to simplify complex calculations. Do one of the following.

   a. Find an advanced algebra or trigonometry book published before 1965. Find out how logarithms were used. Write a description of what you find.

   b. Locate a slide rule and instruction book. Find out how to multiply, divide, and take powers of numbers. Demonstrate this skill to your class.

   c. Before the slide rule was invented, John Napier invented a calculation aid now called "Napier's Bones." Find a description and make a set. Show how they were used for computations.

Summary

In this chapter, you studied power functions, which have the general form $f(x) = x^k$, $x > 0$, $k \neq 0$; exponential functions to base b, general form $f(x) = ab^x$; and logarithmic functions, base b, with general form $f(x) = \log_b x$, $x > 0$. Exponential and logarithmic functions are defined only if b is a positive number different than 1.

In an exponential function the exponent can assume any real value. An exponent that is the reciprocal of a positive integer n is equivalent to an nth root:
$$f(x) = x^{1/n} = \sqrt[n]{x} \text{ for } x > 0.$$
Any rational number can be used as an exponent: $x^{p/q} = \sqrt[q]{x^p} = (\sqrt[q]{x})^p$ for $x \geq 0$.

Exponential functions are useful in modeling situations in which a quantity changes by a constant factor in a given time period. These situations include exponential growth (when the base $b > 1$) and exponential decay ($b < 1$). Just as two points determine a linear function, two data points are sufficient to determine an exponential function.

Logarithmic functions are inverses of exponential functions. Although any positive base other than one is possible, common logarithms (base 10) are especially convenient because the decimal number system also has base 10, so $\log 100 = 2$, $\log 1000 = 3$, and so on. Natural logarithms (base e) are often used because they are easily calculated from series. The Continuous Change Model $A(t) = Pe^{rt}$, with initial value P and annual growth rate r, gives the amount $A(t)$ of a substance after t years when growth or decay is continuous. Several natural phenomena follow this model.

Laws of exponents can be used to prove theorems about logs of products, quotients, and powers, and these in turn can be used to solve exponential equations that you could previously only solve through trial and error. The Change of Base Theorem allows you to convert logarithms from one base to another.

Finally, you saw how logarithmic functions and transformations can be used to model real situations. When a relationship between two variables is curvilinear, a log transformation of one (or both) variables sometimes allows you to find a line of best fit, and consequently an exponential or logarithmic relationship.

Vocabulary

For the starred (*) terms you should be able to give a definition of the term.
For the other items you should be able to give a general description and a specific example of each.

Lesson 4-1
radical
nth root, $\sqrt[n]{x}$, $x^{1/n}$
nth root functions

Lesson 4-2
rational exponent, $x^{m/n}$, $\sqrt[n]{x^m}$, $(\sqrt[n]{x})^m$, x^n
rational power functions

Lesson 4-3
*exponential function
exponential growth curve
exponential decay curve

Lesson 4-4
*exponential model
initial value
growth factor
half-life, doubling time

Lesson 4-5
*logarithm, base, $\log_b x$
*common logarithm
logarithmic function

Lesson 4-6
interest rate, yield
*e
continuous compounding
*natural logarithm, ln
*Continuous Change Formula

Lesson 4-7
*Logarithm of 1 Theorem
*Logarithm of a Product Theorem
*Logarithm of a Quotient Theorem
*Logarithm of a Power Theorem

Lesson 4-8
exponential equation
*Change of Base Theorem

Lesson 4-9
logarithmic transformation

CHAPTER 4 Summary and Vocabulary   **261**

The Progress Self-Test provides the opportunity for feedback and correction. The Chapter Review provides additional opportunities for practice.

USING THE PROGRESS SELF-TEST
Assign the Progress Self-Test as a one-night assignment. (See pages 69 and T35 for more information.)

ADDITIONAL ANSWERS
7. domain = range = the set of all nonnegative real numbers

16.a.,b.

Progress Self-Test

See margin for answers not shown below.

Take this test as you would take a test in class. You may want to use a scientific calculator and an automatic grapher. Then check the test yourself using the solutions at the back of the book.

In 1–6, evaluate or simplify without a calculator.

1. $64^{2/3}$ **16**

2. $\sqrt[5]{x^{10}y^5}$ for $x>0$, $y>0$ **x^2y**

3. $\left(\frac{9}{25}\right)^{-1/2}$ **$\frac{5}{3}$**

4. $\log 1$ **0**

5. $\ln e^2$ **2**

6. $\log_3 27$ **3**

7. State the domain and range of $f(x) = 3x^{1/4}$.

8. Solve: $7^x = 2401$. **4**

9. Solve $5^x = 47$ to the nearest thousandth. **2.392**

10. The U.S. population in 1960 was about $180 \cdot 10^6$, and in 1970 was about $209 \cdot 10^6$. Assuming it is growing exponentially, predict the 1990 population. **about $282 \cdot 10^6$**

11. Strontium-90 has a half-life of 25 years. How much will be left of 10 grams of strontium-90 after 30 years? **about 4.4 grams**

12. Approximate $\log_5 16$ to the nearest thousandth. **1.723**

13. Consider $g(t) = 3^t$.
a. What is g^{-1}? **$g^{-1}(t) = \log_3 t$** **1.9**
b. Approximate $g^{-1}(8)$ to the nearest tenth.

14. Rewrite without logarithms:
$\frac{1}{2}\log a + \log b = \frac{1}{3}\log c$. **$a^{1/2}b = c^{1/3}$**

15. *True* or *false*. $10^{\log x} = e^{\ln x}$ for any positive number x. **True**

16. a. Graph $f(x) = 8^x$.
b. On the same set of axes, graph f^{-1}, the inverse of f.

17. Using a certain predictive model, the number L of a softball team's losses varies inversely as the square of the number of runs R scored in a season.

$$L = \frac{k}{R^2}$$

Write R as a function of L. **$R = \sqrt{\frac{k}{L}}$**

In 18 and 19, consider the graphs below.
(a)   (b)

(c)   (d)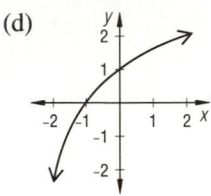

18. Name all which show logarithmic growth. **(b), (d)**

19. Which could be the graph of $y = e^x$? **(c)**

20. At the beginning of an experiment, a cup of coffee at 100° C is in a 20° C room. At one-minute intervals, the temperature of the coffee was measured, and the difference in temperature y from room temperature recorded. Finally, the natural logarithms of temperature differences were plotted against time t, and the following line fitted to the data: $\ln y = \ln 80 + t \ln (0.875)$.
a. Rewrite the equation giving y as a function of t. **$y = 80(0.875)^t$**
b. What was the temperature difference at $t = 8$ minutes? **$\approx 27.5°$**

Chapter Review

Questions on SPUR Objectives

RESOURCES
- Chapter 4 Test, Form A
- Chapter 4 Test, Form B
- Chapter 4 Test, Cumulative Form
- Quiz and Test Writer

SPUR stands for **S**kills, **P**roperties, **U**ses, and **R**epresentations.
The Chapter Review questions are grouped according to the SPUR Objectives for this chapter.
See margin for answers not shown below.

CHAPTER REVIEW

The main objectives for the chapter are organized here into sections corresponding to the four main types of understanding this book promotes: Skills, Properties, Uses, and Representations. We call these the SPUR objectives.

USING THE CHAPTER REVIEW
Students should be able to answer questions like these with about 85% accuracy by the end of the chapter. (See pages 72 and T38–39 for more information.)

SKILLS deal with the procedures used to get answers.

■ **Objective A:** *Evaluate $b^{m/n}$ for $b>0$.*
(Lessons 4-1, 4-2)

For 1–4, evaluate without a calculator.

1. $125^{1/3}$ **5**
2. $125^{5/3}$ **3125**
3. $125^{-2/3}$ **$\frac{1}{25}$**
4. $\sqrt[3]{-125}$ **-5**

In 5 and 6, assume a and t are positive. Rewrite using a radical sign.

5. $a^{2/3}$ **$\sqrt[3]{a^2}$ or $(\sqrt[3]{a})_2$**
6. $t^{-3/7}$ **$\frac{1}{\sqrt[7]{t^3}}$ or $\frac{1}{(\sqrt[7]{t})^3}$**

In 7 and 8, assume all variables are positive. Write without a radical sign.

7. $\sqrt[5]{st^7}$ **$s^{1/5}t^{7/5}$**
8. $\frac{1}{\sqrt[6]{a^2b^3}}$ **$a^{-1/3}b^{-1/2}$**

In 9 and 10, without a calculator estimate the number.

9. $8^{1.96}$ **about 60**
10. $26^{-0.34}$ **$27^{-\frac{1}{3}} = \frac{1}{3}$**

In 11 and 12, insert the appropriate symbol ($>$, $<$, $=$).

11. $(10{,}000)^{3/4} \underline{\ ?\ } (1000)^{4/3}$ **$<$**
12. $\sqrt[10]{(-2)^{10}} \underline{\ ?\ } -2$ **$>$**

■ **Objective B:** *Solve exponential equations.*
(Lesson 4-8)

In 13–16, solve.

13. $e^m = 8$ **≈ 2.08**
14. $5^n = 9$ **≈ 1.37**
15. $11^m \geq 32$ **$m \geq 1.45$**
16. $15(1.08)^n = 40$ **$n \approx 12.7$**

■ **Objective C:** *Evaluate logarithms.*
(Lessons 4-5, 4-6, 4-7)

In 17–22, evaluate without using a calculator.

17. $\log_{10} 10{,}000$ **4**
18. $\log_3 \left(\frac{1}{27}\right)$ **-3**
19. $\ln e^{-1.73}$ **-1.73**
20. $\log_2 2048$ **11**
21. $\log_a (\sqrt[3]{a^4})$ **$\frac{4}{3}$**
22. $e^{\ln 5.3}$ **5.3**

23. Order from smallest to largest, without using a calculator:
$$\log_3 5, \ln 5, \log_2 5, \log 5.$$

In 24 and 25, evaluate.

24. $\log_{12} \pi$ **≈ 0.461**
25. $\log_6 73$ **≈ 2.39**

23) $\log 5, \log_3 5, \log_2 5$

PROPERTIES deal with the principles behind the mathematics.

■ **Objective D:** *Describe properties of exponential and logarithmic functions.* *(Lessons 4-3, 4-4, 4-5, 4-6)*

In 26–28, consider the exponential function $f(t) = ab^t$ with $b>0$, $b \neq 1$, and $a>0$.

26. What are
   a. the domain and **all real numbers**
   b. the range of f? **all real numbers >0**

27. Under what condition is f
   a. strictly increasing? **$b>1$**
   b. strictly decreasing? **$0<b<1$**

28. Give equations for any asymptotes of f. **$y=0$**

29. Define: e.

30. Give the point(s) of intersection of the graphs of $a(x) = 7^x$ and $b(x) = 9^x$. **(0, 1)**

29) $e = \lim\limits_{n \to \infty} \left(1 + \frac{1}{n}\right)^n$

---

31. Without graphing, decide which of the functions $k(t) = (0.2)(1.3)^t$ and $m(t) = (1.3)(0.2)^t$ models exponential growth and which models exponential decay.

32. *Multiple choice.* Which of the following is the inverse of $y = 3^x$?
   (a) $y = 3 \log x$ (b) $y = \log_3 x$
   (c) $y = x^3$ (d) $y = \sqrt[3]{x}$ **(b)**

■ **Objective E:** *Use properties of logarithms.*
(Lessons 4-7, 4-8, 4-9)

33. State **a.** the domain and **b.** the range of the function $y = \ln x$.

In 34–36, use the fact that $\log_8 5 \approx 0.774$ and $\log_8 12 \approx 1.195$ to evaluate the expression.

34. $\log_8 60$ $\approx$ **1.969**   **35.** $\log_8 \left(\frac{25}{12}\right)$ $\approx$ **0.353**

36. $\log_8 \sqrt[5]{12}$ $\approx$ **0.239**

37. If $\log y = 3 \log x + \log 17$, give an expression for y that does not involve logarithms. $y = 17x^3$

38. If $A = \frac{C^3}{D}$, express $\log A$ in terms of $\log C$ and $\log D$. $\log A = 3\log C - \log D$

39. Rewrite $y = 8(5)^x$ as a linear model for $\ln y$ in terms of x. $\ln y = \ln 8 + (\ln 5)x$

40. 18^{150} is an n-digit number. Find n. **189**

In 41 and 42, *true* or *false*.

41. $\log_{1/2} 8 = \log_2 \frac{1}{8}$ **True**

42. For $x > 0$, $\log_{10} x = (\log_{10} e)(\ln x)$. **True**

43. Prove that for $a, x, y, z > 0$,
$\log_a \left(\frac{x}{y}\right) + \log_a \left(\frac{y}{z}\right) + \log_a \left(\frac{z}{x}\right) = 0.$

USES deal with applications of mathematics in real situations.

■ **Objective F:** *Use rational exponents to model situations.* (Lessons 4-1, 4-2)

44. One analysis of the rise of AIDS in the American population modeled the cube root of cases N as a function of time T in years: $N^{1/3} = kT$ where k is a constant. Write N as a function of T. $N = k^3 T^3$

45. The weight W of an object varies inversely as the square of its distance from the center of the earth: $W = \frac{k}{d^2}$. Write d as a function of W. $d = \sqrt{\frac{k}{W}}$

46. $T(d) = \left(\frac{1}{2}\right)^{d/10} T_0$ gives the toxicity of sewage after d days of treatment. T_0 is the initial value of toxicity.
   a. Evaluate $T(4)$. $T(4) \approx 0.76T_0$
   b. What value of d would give a toxicity of 10% of T_0? $d \approx 33$ **days**
   c. A chemical accident causes the toxicity to rise to 5 times the safe level. How many days will it take to restore the quality of water to a safe level? **24 days**

47. An estimate for the population of the earth P in billions is
$P = 4(2)^{(Y - 1975)/35}$
where Y is the year.
   a. Evaluate P for $Y = 1975$. $P = 4$
   b. Estimate P in 1999. **about 6.4**
   c. When will the 1990 population be doubled? **2025**

■ **Objective G:** *Use exponential models to solve problems.* (Lessons 4-3, 4-4, 4-9)

48. A bacteria population was counted every day for a week with the following results:

Day (*d*)	1	2	3	4	5	6	7
Population (*P*)	6	12	24	48	96	192	384

Give an exponential model for P. $P = 3 \cdot 2^d$

49. Find an equation for the exponential curve passing through $(2, 21)$ and $(5, 64)$.
$y \approx 9.99(1.45)^x$

264

50. The intensity of sunlight at points below the ocean is thought to decrease exponentially with the depth of the water. Suppose the intensity on the surface is 100 units, and the intensity at a point 1 meter below the surface is 32.5 units.
   a. Give an exponential model for the intensity I as a function of depth d. $I = 100(.325)^d$
   b. Find the intensity at a depth of 0.5 meters.
   c. If special equipment is needed for divers to see when the intensity drops below 0.2 units, at what depth is this equipment needed?

51. *Multiple choice.* The population of Burma was about 41 million in 1988 and was increasing at a rate of 2.1% per annum. Which of the following models the population P (in millions) of Burma y years after 1988?
   (a) $P = 41 \cdot (1.021)^y$ (b) $P = 41 \cdot (0.021)^y$
   (c) $P = (41 \cdot 1.021)^y$ (d) $P = 41.021^y$ **(a)**

52. The population of Tanzania in 1987 was about 24.3 million, with an annual growth rate of 3.5%. If the population is assumed to change continuously,
   a. give a model for the population n years after 1987. $P = (24.3)e^{0.035n}$, **in millions**
   b. Predict the population in 2000.
   c. Predict when the population will reach 30 million. **b)** ≈ **38.3 million; c) 1993**

In 53 and 54, Benjamin Franklin in 1790 established a trust of $8000. He specified that his investment should compound for 200 years, at the end of which time the funds should be split evenly between the cities of Boston and Philadelphia, and used for loans ''to young apprentices like himself.'' Franklin anticipated the trust would be worth $20.3 million after 200 years.

53. Assuming the trust funds were compounded annually, what annual yield did Franklin project? **4.0%**

54. The actual monies in the fund in 1990 amounted to $6.5 million. What was the actual annual yield? **3.4%**

55. A certain radioactive substance has a half-life of 4 hours. Let A be the original amount of the substance, and L the amount left after h hours.
   a. Give an exponential model for L in terms of A and h. $L = A(.841)^h$ or $L = Ae^{-.25\ln(0.5)h}$
   b. What percent of the original amount of the substance will remain after 7 hours? ≈ **30%**

56. In 1811 an earthquake estimated as high as 8.8 on the Richter scale occurred near New Madrid, Missouri. The quake was so large, and disturbed such a wide area, that it caused church bells 1000 miles away in Boston to ring. In October, 1989, an earthquake measuring 6.9 on the Richter scale occured in the San Francisco area, killing 270. How many dead might there have been if the 1989 earthquake were as severe as the 1811? Assume deaths are proportional to the intensity of the earthquake. **about 21,450**

57. The formula $\log w = -2.866 + 2.722 \log h$ estimates the normal weight w in pounds of a girl h inches tall. Estimate the normal weight of a 55"-tall girl. **about 74 pounds**

■ **Objective H:** *Use logarithmic functions to model data.* (Lessons 4-6, 4-9)

58. A line of best fit for $\log b$ in terms of c is given by $\log b = 0.38c - 0.9$. Rewrite the equation to give b as a function of c. $b = 0.126(10)^{0.38c}$

59. In 1619, Kepler observed that the length a of the semi-major axis of a planet's elliptical orbit was directly related to the time T it took for the planet to complete a revolution around the sun. When T is measured in days, and a in millions of kilometers, the logarithms of the data for the first six planets (Mercury, Venus, Earth, Mars, Jupiter, and Saturn) produce almost collinear points. The equation for the line is

$$\ln T = \frac{3}{2} \ln a - 1.72.$$

Write T as a function of a. (This is Kepler's Third Law of Planetary Motion.) $T = (.179)a^{3/2}$

60.a.,c.,d., 62.a. See
margin, page 264.

63.a.

65. They are reflections of
each other over the line
$y = x$.

66.

67. b.

60. When Julie began to practice bowling, she recorded her average score s after w weeks. The
results are shown below.

Average score (s)	75	94	103	106	118	124	131	133
Week (w)	1	3	5	7	15	20	25	30

  a. Draw a scatter plot of s against w.
  b. Which of the following is the most likely
    model of the relationship between s and w
    for constants a and b?
    (i) $s = aw + b$ (ii) $s = ab^w$
    (iii) $s = a \ln w + b$ **(iii)**
  c. Draw a scatter plot of s against $\ln w$.
  d. Use a statistics package to find a line of best
    fit for the scatter plot in **c.**
  e. Use the relationship in **d** to predict Julie's
    average bowling score after a year of prac-
    tice, if it continues to improve in a similar
    way to the first 30 weeks. **≈ 140**

61. In the manufacture of glass, the hardening process is called *gelling*. Below is a graph showing
the time it takes glass to gel at temperatures
ranging from 100° to 300°C. On the graph, the
scale of the time t has undergone a logarithmic
transformation.
  a. *Multiple choice.* If the curve was drawn for
    the data without any logarithmic transforma-
    tion, the graph would
    (i) still decrease but much more rapidly
    (ii) still decrease but not as rapidly
    (iii) increase at about the same rate
    (iv) not change. **a) (i); b) ≈ 30 min**
  b. Estimate the gel time of glass at 150° C.

REPRESENTATIONS deal with pictures, graphs, or objects that illustrate concepts.

■ **Objective I:** *Graph exponential and logarithmic
functions. (Lessons 4-3, 4-5, 4-6)*

62. **a.** Graph $g(x) = 3^x$ and $h(x) = 3^{2x}$.
  b. Which function has greater values when
    $x > 0$? **h**
  c. Which function has greater values when
    $x < 0$? **g**

63. **a.** Graph $f(x) = 10^x$ for $0 \le x \le 2$.
  b. Use the graph in **a** to estimate $\sqrt{10}$ and $10^{1.6}$
    to the nearest tenth. **3.2; 39.8**

64. Select the two functions below which are reflection images of each other over the y-axis.
  (a) $a(t) = 5^x$ (b) $b(t) = x^5$
  (c) $c(t) = x^{-5}$ (d) $d(t) = 5^{-x}$ **(a) and (d)**

65. How is the graph of $f(t) = \log_3 t$ related to the
graph of $g(t) = 3^t$?

66. Sketch the graph of $y = \log_5 x$ for
$0 < x \le 125$.

67. **a.** Predict the shape of the graph of
    $f(x) = \ln e^x$. **linear**
  b. Check using a function grapher.

68. What point does the family of graphs of
$y = \log_b x$ have in common? **(1, 0)**

■ **Objective J:** *Interpret graphs of exponential and
logarithmic functions (Lessons 4-3, 4-5, 4-6)*

In 69 and 70, consider the graphs of (a)–(d) below.

(a)   (b)

(c)   (d)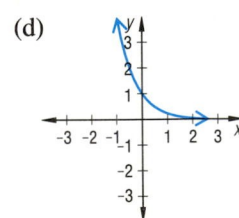

69. Name all that show exponential growth. **(a)**

70. Which represent(s) the function
$y = \log(x - a)$, where $a > 0$? **(c)**

OBJECTIVES ■ CHAPTER 5

The objectives listed here are the same as in the Chapter 5 Review on pages 334–338 of the student text. The Progress Self-Test on pages 332–333 and the tests in the Teacher's Resource File cover these objectives. For recommendations regarding the handling of this end-of-chapter material, see the notes in the margins on the corresponding pages of this Teacher's Edition.

OBJECTIVES FOR CHAPTER 4 (Organized into the SPUR categories— Skills, Properties, Uses, and Representations)	Progress Self-Test Questions	Chapter Review Questions	Teacher's Resource File Lesson Masters	Chapter Test Forms A & B
SKILLS				
A Convert between degrees, radians, and revolutions.	1, 2	1–14	5-1	1, 2
B Find sines, cosines, and tangents of acute angles.	6, 8, 11, 12, 18, 20	15–24	5-3, 5-5	5–7
C Evaluate circular functions outside the first quadrant.	7, 19	25–36	5-4, 5-5	16, 17
D Use trigonometry to find sides or angles in triangles.	13	37–44	5-3, 5-8, 5-9	9, 10, 19, 20
E Find circular arc length and areas of sectors.	14	45–48	5-2	3, 4
PROPERTIES				
F State properties of the sine, cosine, and tangent functions.	9	49–58	5-6, 5-7	15
G Use theorems about sines, cosines, and tangents.	14–17	59–68	5-6, 5-7, 5-8, 5-9	12–14
USES				
H Solve problems involving lengths of arcs or areas of sectors.	3–5	69, 70	5-2	8
I Solve problems using trigonometric ratios in right triangles.	22–23	71–74	5-3	11
J Solve problems involving laws of sines and cosines.	24–25	75–79	5-8, 5-9	18
REPRESENTATIONS				
K Represent sines, cosines, and tangents in the coordinate plane.	10	80–86	5-3, 5-4	21a
L Draw or interpret graphs of the parent sine, cosine, and tangent functions.	21	87–90	5-6	21b

*The masters are numbered to match the lessons, and each is intended for use with that lesson.

OVERVIEW ■ CHAPTER 5

This chapter begins the *Functions, Statistics and Trigonometry* presentation of trigonometry. Students first review the measures of angles, arc lengths and sectors of circles, and then are introduced to the definitions of the sine, cosine, and tangent functions. The definitions are presented as ratios in right triangles in Lesson 5-3, and in terms of rotations of the point (1, 0) around the unit circle in Lesson 5-4. With the second definition, the periodicity of the functions becomes evident, they can be graphed, and their most common properties derived. The last three lessons of the chapter return to trigonometry in its literal sense, the study of triangle measurement.

The chapter makes heavy use of the calculator to solve problems, so there is as much or more emphasis on students being able to approximate angle measures or trigonometric functions as there is on working with exact values. Yet, whenever possible, formulas and identities are proved as theorems. This is intended to provide students not with simply a formula to use, but with a method for reconstructing a formula they may have forgotten.

PERSPECTIVES ■ CHAPTER 5

The Perspectives provide the rationales for the inclusion of topics or approaches, give some mathematical background, and make connections with other lessons and to other UCSMP texts.

5-1
MEASURES OF ANGLES AND ROTATIONS
This lesson mentions four different ways of measuring angles: degrees, revolutions, radians, and gradients (grads). Students should be able to convert between fractions of a revolution, degrees, and radians. They should also be able to convert from degree, minute, second measure to decimal degrees for calculator use. (The grad is used in some engineering applications, but it is not used in this course beyond this lesson.)

5-2
LENGTHS OF ARCS AND AREAS OF SECTORS
In the first lesson, students learned to convert between degree measure and radian measure. This lesson illustrates why radian measure is often preferable to degree measure. Some formulas are simpler when radians are used.

5-3
TRIGONOMETRIC RATIOS OF ACUTE ANGLES
The definitions of sine, cosine, and tangent are presented as ratios of sides of a right triangle, which we hope is review for most students. Students are expected to apply the ratios and use calculators appropriately. As an application, the slope of a line is shown to be the tangent of the angle formed by the line and the positive direction of the *x*-axis.

5-4
THE SINE, COSINE, AND TANGENT FUNCTIONS
In some contexts the sine, cosine, and tangent functions are called *trigonometric* functions only when their arguments θ are such that $0° < \theta < 180°$, that is, when the arguments can be measures of angles in triangles. When the domain extends beyond that interval, these functions are called *circular* functions, reflecting a definition dependent on the unit circle. We do not adhere to such strict linguistics, because this is not the language used by mathematicians. Instead we have chosen to let the context of the discussion determine the terms we use. Because most of the applications in this chapter involve triangles, the chapter is called Trigonometric Functions.

The right triangle definitions of the trigonometric functions do not apply to right or obtuse angles, nor to real numbers, so over the centuries mathematicians have developed a number of ways of extending the ratio idea, each involving a rectangular coordinate system. The traditional way is by means of *reference triangles;* then one still speaks of sines, cosines, and tangents of angles regardless of the measures of such angles. A second way, the *wrapping function* approach, was developed in the 1940s and first appeared in high school books in the newer mathematics curricula of the late 1950s and early 1960s. In this approach, the trigonometric functions become functions of real numbers; the disadvantage is that the "wrapping" of a real line about a unit circle is a completely new idea that is being used to introduce another new idea.

The method we use is about a century old. It has the advantages of both the other methods; it is related to geometry and it allows the argument θ to be any real number. In this approach, (cos θ, sin θ) is defined to be the image of (1, 0) under a rotation of magnitude θ around the origin. We choose this definition for its elegance and the ease with which many properties of sine and cosine can be developed. The tangent is defined as the ratio of sine to cosine.

All of these approaches result in equivalent mathematics, so one object of the lesson is to point out that this new way of treating the cosine and sine is consistent with the old way.

5-5

EXACT VALUES OF TRIGONOMETRIC FUNCTIONS

This lesson applies the definition of sine and cosine to obtain values for the sine, cosine, and tangent of multiples of $\pi/6$ (30°) or $\pi/4$ (45°). These values provide (1) key points in the behavior of these functions, (2) a basis from which to graph them, and (3) the means to check some of the identities that relate the functions.

5-6

GRAPHS OF THE SINE, COSINE, AND TANGENT FUNCTIONS

Until this lesson, sin θ, cos θ, and tan θ have been treated as numbers. In this lesson, the corresponding functions
sine: $\theta \to \sin \theta$,
cosine: $\theta \to \cos \theta$, and
tangent: $\theta \to \tan \theta$
are graphed and analyzed for their domain, range, relative minima or maxima (if any), intercepts, end behavior, and periodicity.

5-7

PROPERTIES OF SINES, COSINES, AND TANGENTS

This lesson discusses several theorems relating values of the sine, cosine, and tangent functions. From the definition of the cosine and sine as coordinates of a particular point on the unit circle follows immediately the Pythagorean Identity $\cos^2\theta + \sin^2\theta = 1$. Then, from the symmetry of the circle, the sines and cosines of $-\theta$, $\pi - \theta$, and $\frac{\pi}{2} - \theta$ can be found in terms of sin θ and cos θ. By division, the tangents of these numbers can be found in terms of tan θ. The graphs of these

functions, together with calculator or hand calculation of values, provide verification for the theorems deduced.

With the exception of the Pythagorean Identity, all the theorems of this lesson are based on the geometric properties of points on the unit circle.

5-8

THE LAW OF COSINES

In geometry, students learn that two triangles are congruent if certain sets of corresponding parts are congruent: SAS (two sides and the included angle); SSS (three sides); ASA (two angles and the included side); AAS (two angles and a non-included side); and, if they have studied from UCSMP *Geometry*, SSA (two sides and the angle opposite the longer side). Another way of interpreting these theorems is that they tell when the measures of all unknown parts of the triangle are uniquely determined. It is natural to want to determine these parts, and the Law of Cosines enables this to be done for the SAS and SSS conditions. In this lesson, the Law of Cosines is derived, applied, and proved using the definitions of cosine and sine and the distance formula.

5-9

THE LAW OF SINES

When the AAS or ASA conditions are satisfied, the Law of Sines enables the determination of the unique measures of the other two sides. When the SSA condition is given, the Law of Sines enables determination of a second angle, which with the Law of Cosines allows the third side to be determined. When the SSA condition is satisfied but the given angle is not opposite the larger side, there may be no value or two values possible for the third side. All these are considered in this lesson.

5-10

FROM WASHINGTON TO BEIJING

A spherical triangle is a triangle whose sides are arcs of great circles on a sphere. Historically, the trigonometry of spherical triangles came very early due to its applications in astronomy; some people feel that it predated the trigonometry of plane triangles. In this lesson, the spherical Law of Cosines is stated (but not proved) and applied to the problem of finding the shortest route between two points on the surface of the earth.

DAILY PACING CHART ■ CHAPTER 5

For optimal student performance, the Progress Self-Test and the Chapter Review should be covered. (See *General Teaching Suggestions: Mastery* on page T35 of this Teacher's Edition.) By following the pace of the Full Course given here, students can complete 12 or 13 chapters of the book by the end of the year. Students following the pace of the Minimal Course spend more time to complete lessons that are combined with quizzes and to consider the Chapter Review. When all lessons are covered in order (the recommendation of the authors), students in the Minimal Course can cover all but the last two chapters. For more information on pacing, see *General Teaching Suggestions: Pace* on page T34 of this Teacher's Edition.

DAY	MINIMAL COURSE	FULL COURSE
1	5-1	5-1
2	5-2	5-2
3	5-3	5-3
4	Quiz (TRF); Start 5-4	Quiz (TRF); 5-4
5	Finish 5-4	5-5
6	5-5	5-6
7	5-6	5-7
8	5-7	Quiz (TRF); 5-8
9	Quiz (TRF); Start 5-8	5-9
10	Finish 5-8	5-10
11	5-9	Progress Self-Test
12	5-10	Chapter Review
13	Progress Self-Test	Chapter Test (TRF)
14	Chapter Review	
15	Chapter Review	
16	Chapter Test (TRF)	

TESTING OPTIONS

■ Quiz for Lessons 5-1 through 5-3   ■ Chapter 5 Test, Form A   ■ Chapter 5 Test, Cumulative Form
■ Quiz for Lessons 5-4 through 5-7   ■ Chapter 5 Test, Form B

A Quiz and Test Writer is available for generating additional questions, additional quizzes, or additional forms of the Chapter Test.

PROVIDING FOR INDIVIDUAL DIFFERENCES

The student text has been written for, and tested with, average students. It has also been used successfully with students who are better prepared as well as with students who are poorly prepared.

The Lesson Notes often include Error Analysis and Alternate Approach features to help you assist those students who need more help. Students of all abilities often learn from their peers and may benefit from small group work referenced as appropriate throughout the Notes. A blackline Lesson Master (in the Teacher's Resource File), keyed to the chapter objectives, is available for each lesson to provide more practice. (However, we advise you to use these masters sparingly; again refer to the suggestions on pace on page T34.)

Extension activities are provided in the Lesson Notes for use with students who need to be challenged even more than the requirements of completing the Full Course provide.

Trigonometric Functions

Early astronomers used various forms of the astrolabe, Ptolemy's quadrant, and Jacob's staff for measuring triangles to the stars.

We recommend 13–16 days be spent on this chapter: 10–12 days on the lessons and quizzes, 1 day for the Progress Self-Test, 1 or 2 days for the Chapter Review, and 1 day for a Chapter Test. (See Daily Pacing Chart on page 267D.) This pace assumes that students have studied trigonometry before. Add two or three days if this is not the case.

USING PAGE 267
This is an appropriate time to assess how much your students know about the content of the chapter. In some physics classes, right triangle trigonometry is discussed quite early and some of your students may have had this content before now. You might direct students attention to the list of chapter contents printed on page 267, and ask for a show of hands to indicate familiarity with the various topics listed.

The word *trigonometry* is derived from Greek words meaning "triangle measurement." The origins of trigonometry have been traced back to the Egyptians of the 13th century B.C., whose tables of shadow lengths correspond to today's tangent and cotangent functions. The Babylonians and Greeks used trigonometry to study the heavens. Travelers of recent centuries used trigonometry to navigate, and physicists use trigonometry to study sound, light, and other waves. In this chapter, you will extend your previous study of right-triangle trigonometry to solve problems like the following.

What is the diameter of the moon?

How far is it from Washington, D.C. to Beijing, China?

What area is swept clean by a windshield wiper?

You will also study properties of functions based on the trigonometric ratios.

Earthrise from Moon surface

OBJECTIVE

A Convert between degrees, radians, and revolutions.

TEACHING NOTES

A teaching aid is provided to assist students with the conversions considered in this lesson.

Many books state that latitude is measured as the angle between the radius from the center of the earth to a point on its surface and the radius that intersects the equator on the same meridian as the point. Due to the oblateness of Earth, this is only an estimate of the latitude. That is why the phrase *approximated by* is used in **Example 1.**

Students should be able to estimate the size of an angle in radian measure. They should think of 1 radian as a little less than 60°.

Radian measure is needed for virtually all real-world applications involving trigonometric functions. Lesson 5-2 shows students how some formulas are simpler when written in radians. The radian measure of an angle is a real number equal to the length of the corresponding *arc on a unit circle*, and is unit free. At this point we say π *radians* in order to emphasize the origin of the measure, though, later in the text we will drop the word *radians*, and when an

LESSON

5-1

Measures of Angles and Rotations

Landsat view of Central Park in New York City

Recall that an **angle** is the union of two rays (its **sides**) with the same endpoint (its **vertex**).

An angle can be thought of as having been generated by rotating a ray either counterclockwise or clockwise around its endpoint from one position to another. For instance, in ∠AQB at the right, you can think of $\overrightarrow{QB}$ as the image of $\overrightarrow{QA}$ under a counterclockwise rotation with center Q.

The *measure* of an angle is a number that represents the size and direction of rotation used to generate the angle. In trigonometry, angles generated by counterclockwise rotations are measured with positive numbers. Those generated by clockwise rotations are measured with negative numbers.

You probably first learned to measure angles in *degrees*. Angle AQB above has measure about 40°. If $\overrightarrow{QB}$ is considered as the initial side and $\overrightarrow{QA}$ is its image under a clockwise rotation, then the measure of ∠BQA is about -40°.

Another unit for measuring rotations is the **revolution**, related to degrees by the conversion formula:

$$1 \text{ revolution counterclockwise} = 360°.$$

By multiplying both sides of the conversion formula by the same number, you can convert from revolutions to degrees and vice versa. For instance, an angle generated by $\frac{1}{8}$ of a revolution counterclockwise has degree measure of 45° because

$$\frac{1}{8} (1 \text{ revolution counterclockwise}) = \frac{1}{8} (360°) = 45°.$$

Similarly,

$$\frac{1}{12} (1 \text{ revolution clockwise}) = \frac{1}{12} (-360°) = -30°.$$

Why should there be 360° in a complete revolution? It is thought that the ancient Babylonians first introduced the division of a circle into 360 parts as a measure of angle size. This method was probably developed through their study of astronomy, in connection with the construction of a calendar to aid in their yearly planting and harvesting.

268

The Babylonians used a number system with base 60, called the **sexagesimal system**. Their use of this system explains some other traditional measures of parts of angles. A degree is divided into sixty **minutes**. A minute is divided into sixty **seconds**. Minutes are represented by a single tick mark and seconds by a double tick mark. For instance, 5 degrees, 22 minutes, 30 seconds is written as 5° 22′ 30″, and represents $5 + \frac{22}{60} + \frac{30}{3600}$ degrees. It is often inconvenient to work with such numbers, and decimal degrees are more commonly used these days:

$$5° \ 22′ \ 30″ = 5 + \frac{22}{60} + \frac{30}{3600}$$
$$= 5 + .3\overline{6} + .008\overline{3}$$
$$= 5.375.$$

Some scientific calculators have a procedure for converting between degrees, minutes, and seconds and decimal degrees, sometimes using a key labeled [DMS]. You should check your calculator to see how it works. This conversion is helpful since the sexagesimal measures are still commonly used in geography, as in Example 1.

Example 1 The latitude of Central Park in New York City is 40° 47′ N, approximated by the measure of the angle in the drawing at the left. Give the measure of this angle as a decimal number of degrees.

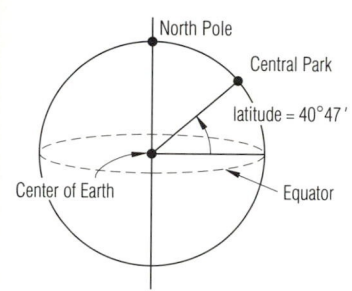

Solution As a fraction, 40° 47′ is $40\frac{47}{60}$ degrees. As a decimal, this is about 40.783°.

Another important unit for measuring angles or magnitudes of rotation is the **radian**. Radians have been in use for only about 100 years. You will see later in this chapter that there are some mathematical reasons for preferring their use over degrees. They are also very important in the study of calculus and other advanced mathematics. Radian measure is based on the following idea.

Consider a circle O with radius 1, called a **unit circle**, as pictured below at the left. Rotate $\overrightarrow{OA}$ to $\overrightarrow{OP}$ to create $\angle AOP$. The radian measure of the central $\angle AOP$ is defined as the length of $\overset{\frown}{AP}$. Thus, an angle whose measure is 1 radian cuts off an arc on a unit circle of length 1. At the right below, $m\angle BOQ = 1$ radian. Note that the length of $\overset{\frown}{BQ}$ is equal to a radius.

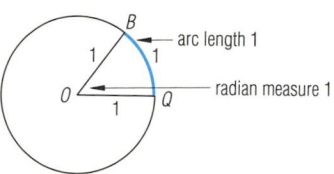

angle is measured in degrees the degree sign will always be indicated, as in sin A = sin $\frac{\pi}{6}$ = sin 30°. The next lesson will also give students practice working with radians.

Making Connections
Some science classes and earlier UCSMP courses cover the idea of multiplying by a conversion factor to change units. A conversion factor is a fraction whose numerator and denominator are equal quantities but with different units. This is what is done in **Example 2,** where the conversion factor is

$\frac{180°}{\pi \text{ radians}}$. In **Example 3,** degrees are converted to radians by multiplying by $\frac{\pi \text{ radians}}{180°}$.

ADDITIONAL EXAMPLES
1. Mecca, a city in Saudi Arabia, is at latitude 21°27′N and longitude 39°49′E. Change these measures into their decimal equivalents.
21.45°N, 39.82°E

2. Convert 31.4159 radians to degrees.
1800°

3. Convert 42° to radians **a.** exactly **b.** approximately.
a. $\frac{7}{30}\pi$; b. ≈ .73 radians

Radian measure is also used to measure the magnitude of a rotation. For instance, in the figure above at the left, if $\overrightarrow{OA}$ were rotated through one complete revolution or 360°, it would determine an arc whose length is the circumference of the unit circle: 2π. Similarly, a 180° rotation on a unit circle determines an arc length of π, and a 90° rotation determines an arc length of $\frac{\pi}{2}$.

360° rotation
arc length 2π

180° rotation
arc length π

90° rotation
arc length $\frac{\pi}{2}$

Therefore, degrees and radians are related by the following conversion formulas:

$$360° = 2\pi \text{ radians}$$
$$180° = \pi \text{ radians}$$
$$90° = \frac{\pi}{2} \text{ radians.}$$

Many situations involve angles with measures which are multiples of 30° and 45°. Using the conversions

$$360° = 1 \text{ revolution} = 2\pi \text{ radians}$$

and dividing by 12 and 8, respectively, you can derive the following:

$$30° = \frac{1}{12} \text{ revolution} = \frac{\pi}{6} \text{ radians}$$
$$45° = \frac{1}{8} \text{ revolution} = \frac{\pi}{4} \text{ radians.}$$

The following table lists a set of equivalent measures that result directly from the basic relationships above:

Degrees	0°	30°	45°	60°	90°	120°	135°	150°	180°	360°
Radians	0	$\frac{\pi}{6}$	$\frac{\pi}{4}$	$\frac{\pi}{3}$	$\frac{\pi}{2}$	$\frac{2\pi}{3}$	$\frac{3\pi}{4}$	$\frac{5\pi}{6}$	π	2π
Revolutions	0	$\frac{1}{12}$	$\frac{1}{8}$	$\frac{1}{6}$	$\frac{1}{4}$	$\frac{1}{3}$	$\frac{3}{8}$	$\frac{5}{12}$	$\frac{1}{2}$	1

Because 180° = π radians, you can use the conversion factors $\dfrac{180°}{\pi \text{ radians}}$ or $\dfrac{\pi \text{ radians}}{180°}$ to convert between radian and degree measure.

270

Example 2 Convert 1 radian to degrees.

Solution Use the conversion factor $\frac{180°}{\pi \text{ radians}}$.

So
$$1 \text{ radian} = 1 \text{ radian} \cdot \frac{180°}{\pi \text{ radians}}$$
$$= \frac{180°}{\pi}$$
$$\approx 57.3°$$

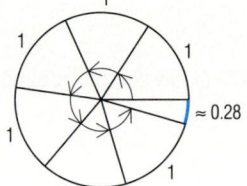

Check The circumference of a unit circle is $2\pi \approx 6.28$, so there are about $6\frac{1}{4}$ radians in one revolution. So one radian is slightly less than $\frac{1}{6}$ revolution. $\frac{1}{6}$ revolution would have measure 60°, so one radian should be slightly less than 60°. It checks.

Example 3 Convert 100° to radians **a.** exactly; **b.** approximately.

Solution

a. $100° = 100° \cdot \frac{\pi \text{ radians}}{180°} = \frac{100}{180}\pi = \frac{5}{9}\pi$

b. Use a calculator to get a decimal approximation for $\frac{5\pi}{9}$.

$$\frac{5\pi}{9} \approx 1.7453$$

So 100° is about 1.7 radians.

Check

a. $\frac{3\pi}{9} < \frac{5\pi}{9} < \frac{6\pi}{9}$, so using the table of degrees, radians, and revolutions on page 270, $60° < \frac{5\pi}{9} < 120°$. So $\frac{5\pi}{9} = 100°$ checks.

b. One radian is a bit less than 60°. So 100° should be less than 2 radians. It checks.

There are other ways of measuring sizes of angles. A **grad** (an abbreviation for **gradient**) is defined as one hundredth of a right angle. So there are 100 grads in 90 degrees and 400 grads in one revolution.

Most scientific calculators allow you to specify whether angles are measured in degrees, radians, or grads, often by using a key marked $\boxed{\text{DRG}}$. Some even allow you to make conversions between radians and degrees. Check your calculator to see if it has these keys and learn how to use them.

Caution: It is customary to omit the word *radians* when giving the radian measure of an angle. For example, we may write " $\frac{\pi}{6}$ " instead of " $\frac{\pi}{6}$ radians" for the radian measure of an angle of 30°. For this reason, you should always indicate degree measure with the ° symbol to distinguish degree measures from radian measures.

Questions 11–14: Encourage students to estimate the answers before using their calculators. Anticipating answers is always a good safeguard against incorrect use of a calculator.

Making Connections for Questions 26 and 27:
These questions are important review for the next lesson.

ADDITIONAL ANSWERS
7.a.

8.a.

25. The area of circle 2 is not twice the area of circle 1. When measuring the amount of spaghetti needed for two portions with these instructions, the result will not be equal to twice the amount needed for one portion.

28. Sample: The Chinese used a system of angle measurement in which a circle was divided into the same number of degrees as a solar year has days.

Questions

Covering the Reading

In 1–3, convert to degrees.
1. $\frac{1}{3}$ revolution counterclockwise **120°**
2. $\frac{3}{4}$ revolution clockwise **⁻270°**
3. 0.43 revolution counterclockwise **154.8°**
4. Express the latitude of Mobile, Alabama, 30°42′ N, in decimal degrees. **30.7° N**
5. At the left is a graph of a unit circle.
   a. Give the length of $\overset{\frown}{AC}$. **π units**
   b. What is the magnitude in radians of the rotation that maps $\overrightarrow{OA}$ to $\overrightarrow{OC}$? **π radians**
6. An angle whose measure is given as 3 is assumed to be measured in __?__. **radians**

In 7 and 8, **a.** draw a circle with radius 1. On this circle, heavily shade an arc with the given length. **b.** Give the measure in degrees of the central angle of this arc.

7. $\frac{\pi}{3}$ **a) See margin. b) 60°**   8. $\frac{3\pi}{2}$ **a) See margin. b) 270°**

9. Convert ⁻225° to radians exactly without a calculator. **$\frac{-5\pi}{4}$**

10. Convert 7π to degrees exactly without a calculator. **1260°**

In 11 and 12, convert to radians. Round your answer to the nearest thousandth.
11. 370°   **6.458 radians**   12. ⁻40°   **⁻0.698 radians**

In 13 and 14, convert to decimal degrees. Round your answer to the nearest thousandth.
13. ⁻4.2 radians   **⁻240.642°**   14. ⁻4.2π radians   **⁻756.000°**

Applying the Mathematics

15. An angle whose measure is $\frac{\pi}{2}$ is about __?__ times as large as an angle whose measure is $\frac{\pi}{2}^{\circ}$.   **57**

16. Order from smallest to largest: 1 revolution, 1 degree, 1 grad, 1 radian.   **1 grad, 1 degree, 1 radian, 1 revolution**

17. Express $22\frac{3}{8}$ degrees as an angle measure in degrees, minutes, and seconds.   **22° 22′ 30″**

18. An LP record is played at $33\frac{1}{3}$ revolutions per minute (rpm). Through how many degrees will a point on the edge of an LP move in three seconds?   **600°**

272

In 19 and 20, use the fact that the planet Jupiter rotates on its axis at a rate of approximately 0.6334 radians per hour.

19. Is this faster or slower than the Earth's rate of rotation?   **faster**

20. What is the approximate length of the Jovian day (the time it takes Jupiter to make a complete revolution)?   **about 9.9 hours**

21. Convert x radians to degrees.   **x radians $\cdot \dfrac{180°}{\pi \text{ radians}} = \dfrac{180x}{\pi}$ degrees**

Review

22. What translation will map the graph of $y = x^2$ onto the graph of $y = (x + 7)^2 - 1$?   *(Lesson 3-2)*   **$(x, y) \rightarrow (x - 7, y - 1)$**

23. Find an equation for the parabola that contains $(-2, 5)$, $(4, 11)$, and $(10, 53)$.   *(Lesson 2-7)*   **$y = 1/2 \, x^2 + 3$**

24. *Multiple choice.*   Which of the following values of r^2 indicates the strongest linear relationship between two variables?   *(Lesson 2-5)*
(a) 0.90   (b) 0.50   (c) 0.09   (d) -0.95 **(a)**

25. The following instructions are taken from a spaghetti packet. Explain what is wrong with the instructions (assuming all portions are equal).
(Previous course)   **See margin.**

A perfect portion measures up every time!

HERE'S HOW IT WORKS:

1. GRASP a handful of spaghetti.

2. PLACE it on **END** in one of the portion circles on the box. (For instance, when the circle of spaghetti matches the 2-portion circle, you know it's the right amount to cook for two people.)

26. The circle at the right has diameter 18 m. Find
   a. its circumference.   **18π m**
   b. its area.   *(Previous course)*   **81π m^2**

18 m

27. Use the circle at the left. Find **a.** the length of $\overset{\frown}{AB}$ and **b.** the area of the shaded region.   *(Previous course)*   **a) 5π cm; b) 25π cm^2**

Exploration

In 28 consult an encyclopedia or book on the history of mathematics.

28. What systems of angle measurement were used by ancient civilizations other than the Babylonians?   **See margin.**

NAME _____

LESSON MASTER 5-1
QUESTIONS ON **SPUR** OBJECTIVES

■**SKILLS** *Objective A (See pages 334–338 for objectives.)*
In 1 and 2, convert to a. degrees and b. radians.

1. $\frac{1}{12}$ revolution counterclockwise   2. $\frac{4}{5}$ revolution clockwise
  a. _____ **30°**   a. _____ **-288°**
  b. _____ **$\frac{\pi}{6}$**   b. _____ **$\frac{-8\pi}{5}$**

In 3 and 4, convert to radians without using a calculator.
3. 45° _____ **$\frac{\pi}{4}$**   4. 210° _____ **$\frac{7\pi}{6}$**

In 5 and 6, convert to degrees without using a calculator.
5. $-\frac{5\pi}{6}$ radians _____ **-150°**   6. $\frac{\pi}{5}$ radians _____ **36°**

In 7 and 8, convert to radians, correct to the nearest thousandth.
7. 112° _____ **1.955**   8. -150° _____ **-2.618**

In 9 and 10, convert to degrees, correct to the nearest hundredth.
9. 6 radians _____ **343.77°**   10. -2.5 radians _____ **-143.24°**

In 11 and 12, tell how many revolutions are represented by each rotation.
11. $\frac{5\pi}{4}$ radians _____ **$\frac{5}{8}$**   12. 800° _____ **$\frac{20}{9} = 2.2$**

Functions, Statistics, and Trigonometry © Scott, Foresman and Company   **53**

OBJECTIVES

E Find circular arc length and areas of sectors.

H Solve problems involving lengths of arcs or areas of sectors.

TEACHING NOTES

Making Connections
The formulas in this lesson are not the only formulas for which radians are simpler than degrees. The infinite series that are used to calculate sines and cosines are elegant in radians and quite complicated in degrees:

$$\sin x = x - \frac{x^3}{3!} + \frac{x^5}{5!} - \frac{x^7}{7!} + \cdots;$$

$$\cos x = 1 - \frac{x^2}{2!} + \frac{x^4}{4!} - \frac{x^6}{6!} + \cdots.$$

The idea is quite analogous to the advantages of using base e to base 10 for logarithms, as students saw in Chapter 4.

To find the length of an arc on a circle we take a fraction of the circumference. Students should understand these formulas well enough so that if they forget them they can reconstruct them. Notice that if θ is replaced by 2π (the entire circumference) in each formula, then the familiar formulas for circumference and area of the entire circle appear. That is, in both formulas, one is dealing with $\frac{\theta}{2\pi}$ of the circle, so if one multiplies by $\frac{2\pi}{\theta}$, the original formula results.

LESSON

5-2

Lengths of Arcs and Areas of Sectors

You may have studied angles and rotations for years without ever using radians. You may be wondering why radians are used and if they are ever needed. One advantage of radians over degrees is that certain formulas are simpler when written with radians.

Example 1 Find the length of an arc of a 75° central angle in a circle of radius 6 ft.

Solution The 75° arc is $\frac{75}{360}$ of the circumference of the circle. The circumference has length $2\pi r$, or $2\pi \cdot 6$ ft. So the length of the arc is

$$\frac{75}{360} \cdot 2\pi \cdot 6 \text{ ft}$$

which simplifies to $\frac{5}{2}\pi$ ft exactly, or 7.85 ft approximately.

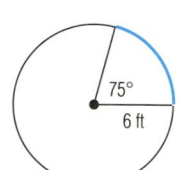

In general, the length s of an arc of a central angle of $\theta°$ in a circle of radius r is given by the formula $s = \frac{\theta°}{360°} \cdot 2\pi r$. But notice how much simpler the formula is if the central angle is measured in radians.

Circular Arc Length Formula

If s is the length of the arc of a central angle of θ radians in a circle of radius r, then $s = r\theta$.

Proof The central angle is $\frac{\theta}{2\pi}$ of a revolution. So the length of the arc is $\frac{\theta}{2\pi}$ of the circumference. The circumference of the circle is $2\pi r$.

Thus $s = \frac{\theta}{2\pi} \cdot 2\pi r$.

So $s = r\theta$.

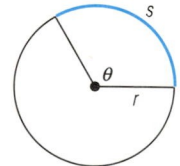

Had the measure of the central angle in Example 1 been given as $\frac{5}{12}\pi$ radians instead of 75°, the Circular Arc Length Formula would give the arc length immediately as $6 \cdot \frac{5}{12}\pi$, or $\frac{5}{2}\pi$, which equals the result in Example 1.

A **disk** is the union of a circle and its interior. A **sector of a circle** is that part of a disk that is on or in the interior of a central angle. The sector at the right below is formed by a central angle of measure θ.

disk

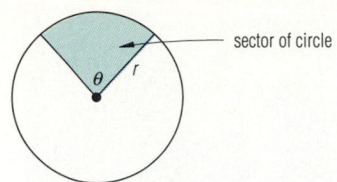

sector of circle

The area of a sector is $\frac{\theta°}{360°}$ of the area of a circle.

$$A = \frac{\theta°}{360°}\pi r^2$$

If θ is given in radians, the formula is simpler.

> ### Circular Sector Area Formula
>
> If A is the area of the sector formed by a central angle of θ radians in a circle of radius r, then $A = \frac{1}{2}r^2\theta$.

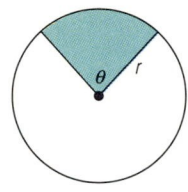

Proof The proof is very similar to the previous proof. The area A of the sector is $\frac{\theta}{2\pi}$ of the area of the circle. The area of the circle is πr^2.

Thus $\qquad\qquad\qquad A = \frac{\theta}{2\pi}\cdot\pi r^2$

$$A = \frac{1}{2}r^2\theta.$$

▪ ▪ ▪ ▪ ▪ ▪ ▪ ▪

Example 2 Find the area of the sector of a circle of radius 5 cm if the central angle of the sector is $\frac{2\pi}{3}$.

Solution Since the angle is given in radians, use the Circular Sector Area Formula with $r = 5$, $\theta = \frac{2\pi}{3}$.

$$A = \frac{1}{2}r^2\theta = \frac{1}{2}\cdot 5^2\cdot\frac{2\pi}{3} = \frac{25\pi}{3}\text{ cm}^2$$

Check Convert $\frac{2\pi}{3}$ to degrees. $\frac{2\pi}{3} = \frac{2}{3}\cdot 180° = 120°$. The area of the sector is $\frac{120}{360}$ or $\frac{1}{3}$ of the area of the circle. The area of the circle is $\pi r^2 = 25\pi$ cm^2. So it checks.

Caution: It is essential that the central angle be in radians in order to apply the Circular Sector Area Formula. The solution for Example 2 illustrates this.

Since θ is one of many Greek letters used in mathematics, you might like to display the entire Greek alphabet somewhere in the classroom. The alphabet is listed below and also provided on a teaching aid. In this book, the letters alpha, beta, theta, pi, and phi are used for angles or arc measures; mu, rho, and lowercase sigma appear in statistics; and uppercase sigma appears in sums.

A	α	alpha
B	β	beta
Γ	γ	gamma
Δ	δ	delta
E	ε	epsilon
Z	ζ	zeta
H	η	eta
Θ	θ	theta
I	ι	iota
K	κ	kappa
Λ	λ	lambda
M	μ	mu
N	ν	nu
Ξ	ξ	xi
O	o	omicron
Π	π	pi
P	ρ	rho
Σ	σ	sigma
T	τ	tau
Y	υ	upsilon
Φ	φ	phi
X	χ	chi
Ψ	ψ	psi
Ω	ω	omega

ADDITIONAL EXAMPLES
1. Find the length of an arc of a central angle with measure $\pi/3$ in a circle with radius 5 cm.
5π/3 cm ≈ 5.24 cm

Error Analysis To reemphasize the importance of using estimates to check answers and, thereby, avoid errors, you might consider the following check for Additional Example 1: Since $\pi/3$ is one-sixth of the way around the circle, the chord of this arc equals the radius. The length of the arc itself should be slightly more.

2. Find the area of a 27° sector of a circle with radius 10 feet.
7.5π ft² ≈ 23.56 ft²

3. Assume the earth travels around the sun in a circular orbit with radius 93,000,000 miles. How far does the earth travel in a single day? (Use 365.25 days for the length of a year.)
186,000,000π/365.25 ≈ 1,600,000 miles

NOTES ON QUESTIONS
Question 9: This question can be done by using conversion factors, as follows:

$$\frac{1 \text{ day}}{24 \text{ hr}} \cdot \frac{1 \text{ hr}}{60 \text{ min}} \cdot 2\pi r =$$

$$\frac{1 \text{ day}}{24 \text{ hr}} \cdot \frac{1 \text{ hr}}{60 \text{ min}} \cdot 2\pi$$

$$\frac{500 \text{ m}}{1 \text{ day}} \cdot \frac{100 \text{ cm}}{1 \text{ m}} \approx \frac{218 \text{ cm}}{1 \text{ min}} \cdot$$

Making Connections for Question 10: Problems like these can help improve the geography knowledge of students. You may wish to discuss the following: Since the answer is about 326 miles, why does one airline report the distance as 338 miles and a Rand McNally Road Atlas say the distance is 410 miles? (The air distance between cities is normally the great circle distance between airports. Road distance is almost never great circle distance since roads follow paths determined by topographic features.)

Question 13: Like **Question 9,** this question offers an opportunity for students to convert units. This is an important skill for students to master.

Question 14: This question was posed on the first page of this chapter (page 267).

Example 3 A water irrigation arm 500 m long rotates around a pivot P once every day. How much area is irrigated every hour?

Solution 1 In a day the irrigation arm covers πr^2 square meters, where $r = 500$. In an hour it covers $\frac{1}{24}$ of that.

$$\frac{1}{24} \cdot \pi \cdot 500^2 = \frac{250000}{24}\pi$$
$$\approx 32{,}700 \text{ square meters}$$

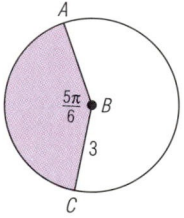

500 m

Solution 2 Use the Circular Sector Area Formula. In a day the arm rotates 1 revolution, or 2π radians. So in an hour it rotates $\frac{2\pi}{24}$ radians. Thus $A = \frac{1}{2}r^2\theta = \frac{1}{2} \cdot 500^2 \cdot \frac{2\pi}{24} = \frac{250000}{24}\pi$ square meters.

Questions

Covering the Reading

In 1 and 2, use the circle below, in which $m\angle ABC = \frac{5\pi}{6}$ radians.

$\frac{5\pi}{6}$ B

3

A

C

1. Compute the area of sector ABC. $\frac{15\pi}{4}$ **square units**

2. Compute the length of $\overset{\frown}{AC}$. $\frac{5\pi}{2}$ **units**

In 3 and 4, consider a circle with radius 8 cm and a central angle of $\frac{3\pi}{4}$.

3. Find the length of the arc cut off by this angle. **6π cm**

4. Find the area of the sector determined by this angle. **24π cm²**

In 5 and 6, repeat Questions 3 and 4 if the central angle has measure 48°.
See below.

7. Refer to Example 1. Find the area of the sector exactly. $\frac{15\pi}{2}$ **ft²**

5) $\frac{32}{15}\pi$ **cm; 6)** $\frac{128}{15}\pi$ **cm²**

276

8. Refer to Example 2. Find the length of the arc of the sector to the nearest tenth of a millimeter. **104.7 mm**

9. Refer to Example 3. To the nearest centimeter, how far does the tip of the arm travel in one minute? **218 cm**

Applying the Mathematics

10. The diagram at the right shows a cross section of the earth. G represents Grand Rapids, MI (longitude 85°40′ W and latitude 42°58′ N) and L represents Louisville, KY (85°46′ W, 38°15′ N). Because their longitudes are close, assume that Grand Rapids is directly north of Louisville. If the radius of the earth is about 3,960 miles, estimate the air distance from Grand Rapids to Louisville. *(Lesson 5-2)* **about 326 miles**

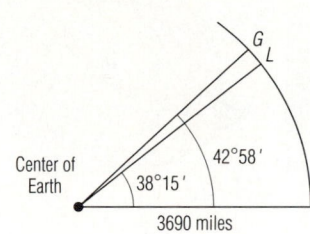

Diagram not to scale

11. Below, the circle $x^2 + y^2 = 25$ and the line $y = x$ are graphed. Find the area of the shaded sector. **25π/8 square units**

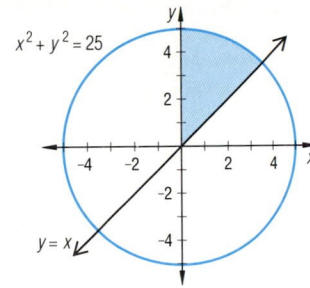

12. If a sector in a circle with a central angle of $\frac{\pi}{6}$ has an area of 3π square meters, what is the radius of the circle? **6 m**

13. Suppose you ride a bike with 22″ wheels (in diameter) so that the wheels make 150 revolutions per minute.
    a. Find the number of inches traveled during each revolution. **22π in.**
    b. How many inches are traveled each minute? **3300π inches**
    c. Use your answer from part **b** to find the speed, in miles per hour, that you are traveling. (Hint: Write the units in each step as you multiply by appropriate conversion factors. Cancel units until mph remains.)
    about 9.8 miles per hour

14. The windshield wiper at the back of a hatchback has an 18″ blade mounted on a 10″ arm as shown at the right. If the wiper turns through an angle of 110°, what area is swept clean?
    209π in.²

LESSON 5-2   Lengths of Arcs and Areas of Sectors   277

NOTES ON QUESTIONS
Error Analysis for Question 19: If a student answers 162°, that student has forgotten that the hour hand moves in the 7 minutes.

Making Connections for Questions 21–23: These questions provide an important review of concepts used in the next lesson.

FOLLOW-UP

MORE PRACTICE
For more questions on SPUR Objectives, use *Lesson Master 5-2*, shown on page 277.

EXTENSION
You might like to "localize" **Question 10** to find the air distance from your location to another location at about the same longitude. The problem of finding the distance from your city to another city at the same latitude is more difficult and is discussed in Lesson 5-10.

PROJECTS
The projects for Chapter 5 are described on pages 329–330. Project 2 is related to the content of this lesson.

Review

In 15 and 16, convert to radians. *(Lesson 5-1)*

15. 45° $\frac{\pi}{4}$ **radians**

16. $^-240°$ $\frac{-4\pi}{3}$ **radians**

In 17 and 18, convert to degrees. *(Lesson 5-1)*

17. $-\frac{\pi}{6}$ radians   **$^-30°$**

18. $\frac{3\pi}{2}$ radians   **270°**

19. If the time is exactly 8:07, what is the measure of $\angle ABC$ on the clock face at the left? *(Lesson 5-1)*   **48°**

20. Find m$\angle E$. *(Previous course)*   **86° 54'**

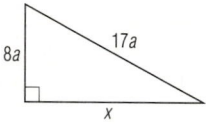

21. Solve for x exactly. *(Lesson 5-1, Previous course)*   **15a**

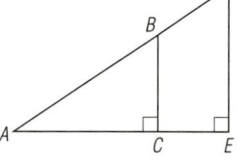

22. In the figure at the right, $\overline{BC}$ and $\overline{DE}$ are perpendicular to $\overline{AE}$. Complete to make a true sentence. *(Previous course)*

   a. $\triangle ABC \sim$ ___?___   **$\triangle ADE$**

   b. $\frac{BC}{AB} =$ ___?___   **$\frac{DE}{AD}$**

23. In right triangle PQR, $PQ = 11$ and m$\angle Q = 30$. Find the lengths of

   a. $\overline{PR}$ **$\frac{11}{2}$**

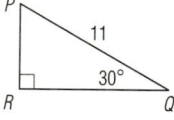

   b. $\overline{QR}$. *(Previous course)*   **$\frac{11\sqrt{3}}{2}$**

Exploration

24. The German astronomer Johannes Kepler (1571–1630) formulated three laws of planetary motion. Find out what these laws were and how they used elliptical arcs and/or sectors.   **Law of Orbits: All planets move in elliptical orbits having the sun as one focus. Law of Areas: A line joining any planet to the sun sweeps out equal areas in equal times. Law of Periods: The square of the period of any planet is proportional to the cube of the planet's mean distance from the sun.**

278

LESSON

5-3

Trigonometric Ratios of Acute Angles

The ancient Greeks, notably Hipparchus (c. 140 B.C.), used ratios of the lengths of sides of right triangles to solve practical problems of surveying and astronomy. Here are three such **trigonometric ratios**.

> **Definition**
>
> Let θ be an acute angle in a right triangle. Then
>
> $\sin \theta = \dfrac{\text{side opposite } \theta}{\text{hypotenuse}}$,
>
> $\cos \theta = \dfrac{\text{side adjacent to } \theta}{\text{hypotenuse}}$,
>
> $\tan \theta = \dfrac{\text{side opposite } \theta}{\text{side adjacent to } \theta}$.

For example, in the triangle at the right,

$$\sin A = \frac{5}{13}, \quad \sin B = \frac{12}{13},$$

$$\cos A = \frac{12}{13}, \quad \cos B = \frac{5}{13},$$

and $\quad \tan A = \frac{5}{12}, \quad \tan B = \frac{12}{5}.$

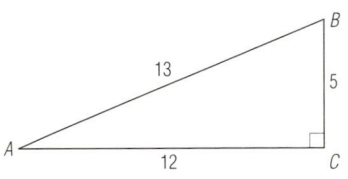

The abbreviations *sin*, *cos*, and *tan* stand for the words **sine**, **cosine**, and **tangent**, respectively. Georg Joachim Rhaeticus (1514–1576) is credited with giving the trigonometric ratios these names.

Although these ratios derive from a particular $\triangle ABC$, in fact the same ratios would be obtained in *any* right triangle with an angle congruent to $\angle A$. This follows from the properties of similar triangles. At the right, $\triangle ABC \sim \triangle ADE$ because of the AA Similarity Theorem.

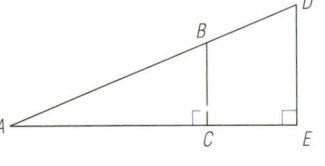

Using $\triangle ABC$, $\cos A = \dfrac{AC}{AB}$; using $\triangle ADE$, $\cos A = \dfrac{AE}{AD}$.

Because the two triangles are similar, the ratios of corresponding sides are equal, so $\dfrac{AC}{AB} = \dfrac{AE}{AD}$, and the two values for $\cos A$ are equal. In general, the value of $\cos A$ depends on the measure of $\angle A$, not on any specific right triangle containing $\angle A$.

> **OBJECTIVES**

B Find sines, cosines, and tangents of acute angles.

D Use trigonometry to find sides or angles in triangles.

I Solve problems using trigonometric ratios in right triangles.

K Represent sines, cosines, and tangents in the coordinate plane.

> **TEACHING NOTES**

We assume that students have encountered trigonometric ratios in a previous course. If not, this may be a two-day lesson. We suggest covering through **Example 2** and assigning **Questions 1–8, 13, 15, and 19–26** on the first day.

For students who have never studied these ratios, you may need to note that $\angle \theta$ in the beginning of the lesson has two sides, but one of these sides has a special name, the hypotenuse. The other side is called the adjacent side.

Error Analysis You will find it necessary to remind students often to check the mode (degree, radian, or grad) to which their calculator is set. Many calculators automatically reset to the degree mode when the "clear" key is pressed, but may not reset when the "clear entry" key is pressed. Even experienced users have made the mistake of using the wrong key on many occasions. The

For most acute angles, trigonometric ratios are not simple fractions and so they are usually approximated. Before the 1970s, decimal approximations of sines, cosines, and tangents were found in tables. Today, calculators give these approximations.

Check to see how to use your calculator to obtain or estimate these trigonometric ratios. First, set the calculator to degrees or radians, whichever is needed. Then, check to see whether you enter the name of the ratio or the angle measure first. To evaluate sin 30° on some calculators you enter ⌊sin⌋ 30 ⌊=⌋; on others you enter 30 ⌊sin⌋. The correct value of sin 30° is 0.5.

■ ■ ■ ■ ■ ■ ■ ■

Example 1 Evaluate **a.** cos 80° **b.** tan 71° 30′ **c.** sin $\frac{\pi}{3}$.

Solution Use a calculator. For parts **a** and **b** set the calculator to degrees. Answers below are rounded to the nearest thousandth.
a. cos 80° ≈ 0.174
b. tan 71° 30′ = tan 71.5° ≈ 2.989
c. Since no degree sign is indicated, the angle measure is given in radians. Set the calculator to radians. sin $\frac{\pi}{3}$ ≈ .866

In many applications you know only the length of one side and the measure of one acute angle in a right triangle. By choosing the appropriate trigonometric ratio, you can determine the length of either other side.

■ ■ ■ ■ ■ ■ ■ ■

Example 2 Suppose that the distance along the constant slope from the top bank of a river to the edge of the water is 23.6 m, and a surveyor found the land slopes downward at an angle of 26° 45′. Find d, the horizontal distance from the top of the bank to the river's edge.

Solution Given here is the measure of an acute angle and the length of the hypotenuse in a right triangle. You must find the side adjacent to the acute angle. Use the cosine ratio, and solve for d.

$$\cos 26° 45' = \frac{d}{23.6}$$
$$d = (23.6)(\cos 26° 45')$$
$$d = (23.6)(\cos 26.75°)$$
$$d \approx 21.1$$

So the horizontal distance from the bank of the river to the edge of the water is about 21.1 meters.

280

Check Use a simpler triangle with measurements close to those given. Use 30° for the angle and 24 for the hypotenuse. In any 30-60-90 triangle, the ratio of sides is $1:\sqrt{3}:2$, so the side opposite the 30° angle is $\frac{1}{2} \cdot 24 = 12$ and the side adjacent is $12\sqrt{3} \approx 20.8$. This is very close to the obtained value. It checks.

If you know lengths of at least two sides of a right triangle, you can use trigonometry to find the measures of the acute angles of the triangle.

Example 3　One portion of a wide staircase has 6 steps. Each step goes up 18 cm for every change of 28 cm horizontally. The sides of the staircase are to be trimmed with a triangular oak panel. At what angle *B* should the oak be cut?

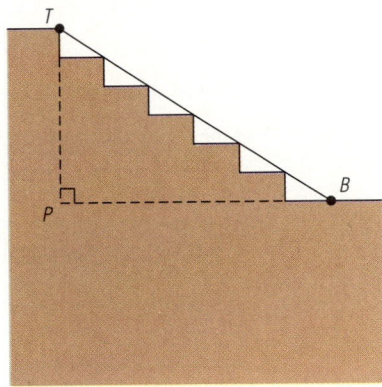

Solution　From the top of the stairs *T*, draw a vertical segment $\overline{TP}$ to the horizontal floor.

$$TP = 6 \cdot 18 \text{ cm} = 108 \text{ cm}$$
and
$$PB = 6 \cdot 28 \text{ cm} = 168 \text{ cm}.$$

This information gives the lengths of the sides adjacent to and opposite $\angle B$. So use the tangent ratio.

$$\tan B = \frac{TP}{PB}$$
$$= \frac{108}{168}$$
$$\approx .643$$

Thus *B* is the acute angle whose tangent is about 0.643. To find *B* with your calculator, press .643 [INV] or [2nd] and then [tan]. You should find $m\angle B \approx 32.7°$

Check 1　If $\triangle TPB$ were isosceles, then $m\angle B$ would be 45°. Because $TP < PB$, $m\angle B$ should be less than 45°. So it checks.

Check 2　$m\angle B$ is close to 30°, so *TP* should be close to $\frac{1}{2}TB$. $TB = \sqrt{108^2 + 168^2} \approx 200$ cm, and $TP = 108$. This is close enough.

3. To the nearest tenth of a degree, give the measures of the angles in a right triangle whose sides measure 20, 21, and 29.
46.4°, 43.6°, 90°

4. Find the measure of the acute angle formed by the line $y = 3x + 1$ and the *x*-axis.
≈ 1.25 radians or ≈ 71.57°

NOTES ON QUESTIONS
Error Analysis for Questions 5–7: The following answers indicate that the calculator has been used in the wrong mode. For **Question 5,** 0.97; for **Question 6,** 1.00 (from 0.9997); for **Question 7,** 1.20.

Question 13: This question can be used to introduce the decimal approximation 0.707 to $\frac{\sqrt{2}}{2}$, an approximation that students should learn.

Making Connections for Question 14: This may be the first time students have seen a formula for the area of a triangle in terms of two sides and the included angle. This is a good time to review the conditions for triangle congruence and to point out that these conditions determine the area of a triangle as well. The formula in this question involves the SAS condition. Heron's Formula, which students may have seen in their geometry course, is the formula for the SSS condition.

There is a surprising link between slopes of lines and the tangent of an angle, illustrated in the diagram below.

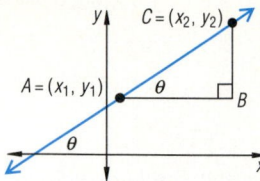

Recall that the slope of the line through (x_1, y_1) and (x_2, y_2) is

$$m = \frac{y_2 - y_1}{x_2 - x_1}.$$

Assume $m \geq 0$. Then, in $\triangle ABC$, $y_2 - y_1$ is the length of $\overline{CB}$ and $x_2 - x_1$ is the length of $\overline{AB}$.

So
$$m = \frac{CB}{AB}.$$

Let θ be the measure of $\angle CAB$. Then $\frac{CB}{AB} = \tan \theta$, so $m = \tan \theta$. For positive slopes, this proves the following theorem. A proof for negative slopes is similar.

Tangent-Slope Theorem

Let m be the slope of a line that forms an acute angle θ with the positive x-axis. Then, $m = \tan \theta$.

A practical use of this theorem is shown in Example 4, where an approximation to an equation of a linear model is obtained quickly, without the use of a computer statistics package.

■ ■ ■ ■ ■ ■ ■ ■■

Example 4 The scatter plot below shows an approximately linear relationship between two variables. A close-fitting line has been drawn. The angle θ is measured to be about 22°. Give an equation for the line.

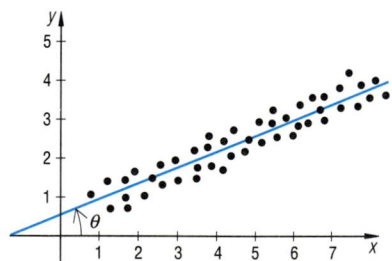

Solution   The line has y-intercept about 0.6 and slope
$$m \approx \tan 22°.$$
Use the slope-intercept form of a linear equation:
$$y \approx (\tan 22°) \, x + 0.6$$
or
$$y \approx 0.4x + 0.6.$$

Questions

Covering the Reading

In 1–4, use △ABC at the right. Give exact values for each.

1. sin A $\frac{5}{7}$
2. cos A $\frac{2\sqrt{6}}{7}$
3. tan B $\frac{2\sqrt{6}}{5}$
4. sin B $\frac{2\sqrt{6}}{7}$

In 5–7, use a calculator to give approximations to the nearest hundredth.

5. sin 83°   **0.99**
6. cos $\frac{5\pi}{12}$ **0.26**
7. tan 16° 35′   **0.30**

8. Find DE and EF to the nearest hundredth.
   DE ≈ 15.36 units; EF ≈ 14.14 units

In 9 and 10, give an equation of the line which

9. contains (0, 4) and is inclined at 30° to the positive x-axis.
   y = (tan 30°) x + 4 ≈ 0.577x + 4
10. has y-intercept of ⁻2 and is inclined at 48° to the positive x-axis.
    y = (tan 48°) x − 2 ≈ 1.11x − 2

In 11 and 12, find m∠A to the nearest tenth of a degree.

11.

    ≈ 46.4°

12.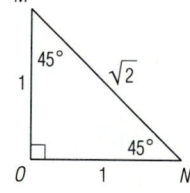

    ≈ 48.6°

Applying the Mathematics

13. Recall the isosceles right triangle (45-45-90 triangle). The one at the right has legs of length 1, so the Pythagorean Theorem gives $MN = \sqrt{2}$.

    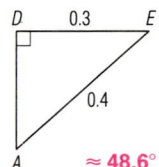

    a. Give exact values of sin M, cos M, and tan M.
    b. Use a calculator to verify that your values in part **a** are correct.   **a,b) See margin.**

14. a. Prove that the area of △ABC with sides a and b and included acute angle θ is $\frac{1}{2}ab \sin\theta$. (Hint: Draw the altitude from B.)   **See margin.**

    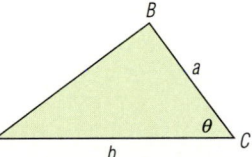

    b. Use part **a** to find the area of △ABC where a = 7, b = 12, and m∠C = 52°.
    about 33 square units

LESSON 5-3   Trigonometric Ratios of Acute Angles   **283**

15. The sails of the windmill illustrated at the right measure 44 feet across from tip to tip. The pivot O is 27 feet off the ground. How far off the ground is point P after the sail rotates 70° from the horizontal as shown?
about 47.7 ft

16. A ship has maintained a bearing of 47°. (A *bearing* is the angle measured clockwise from due north.)
   a. How far has the ship sailed if it is 200 miles north of its original position? **about 293 miles**
   b. How far east of its original position is the ship? **about 214 miles**
   c. If the ship's average speed is 12 miles per hour, for how long has it been sailing? **about 24 hours**

17. From a distance, a surveyor measures the *elevation* of Chicago's Sears Tower (the world's tallest building; height 1454 feet) to be about 6°.

   a. Assuming the measurement of 6° is correct, how far is the surveyor from the building, to the nearest 100 feet? **13,800 feet**
   b. If the elevation is only correct to the nearest degree (that is, the elevation is 6° ± 1°), give the maximum and minimum values for the surveyor's distance from Sears Tower.
   maximum ≈ 16,600 feet; minimum ≈ 11,800 feet

18. Find the area of the parallelogram at the right, to the nearest tenth of a square centimeter. **277.7 cm²**

Review

19. In the unit circle O, $\overarc{PQ}$ has length 1.3. *(Lessons 5-2, 5-1)*
   a. Give the radian measure of $\angle POQ$.
   b. Give the degree measure of $\angle POQ$.
   c. Give the length of major arc $\overarc{PRQ}$.
   d. Give the area of the sector of $\overarc{PRQ}$.
   a) 1.3 radians; b) about 74.5°;
   c) $2\pi - 1.3$ units; d) $\pi - 0.65$ square units

284

20. A *nautical mile* is a unit of distance frequently used in navigation. It is defined as the length of an arc $\overset{\frown}{AB}$ on the circumference of the earth with a central angle of one minute. If the diameter of the earth is 7927 miles, find the length of a nautical mile to the nearest ten feet. (The common navigational measure of speed, the *knot*, is defined as one nautical mile per hour.)   *(Lesson 5-2)*   **6090 feet**

In 21–23, if (1, 0) is rotated the given amount about (0, 0), in which quadrant (I, II, III, or IV) will its image lie?   *(Lesson 5-1)*

21. $-30°$ **IV**   **22.** $\frac{2\pi}{3}$ **II**   **23.** $539°$ **II**

24. a. What is an equation of the circle with center at the origin and radius 1? $x^2 + y^2 = 1$
   b. *P* is a point on the circle in part **a**. Its *x*-coordinate is $-\frac{1}{2}$. Find two possibilities for its *y*-coordinate.   *(Previous course)*
   $-\sqrt{3}/2$ **or** $\sqrt{3}/2$

In 25 and 26, give the image of (m, n) under the transformation.
(Previous course)

25. reflection over the *y*-axis
   ($-m, n$)
26. reflection over the *x*-axis
   ($m, -n$)

Exploration

27. For small central angles, the length of the angle's arc and the length of the arc's chord are approximately equal.

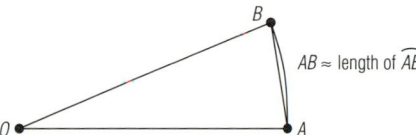

$AB \approx$ length of $\overset{\frown}{AB}$

Astronomical measurements are commonly made by measuring the small angle formed by the extremities of a distant object. The moon covers an angle of 31 minutes when it is about 239,000 miles away.
   a. Use this information to estimate the diameter of the moon.

   b. Compare your estimate to the value found in an almanac or other reference book.
   a) about 2200 miles
   b) Answers may vary; the two values should be close.

LESSON 5-4

RESOURCES
- Lesson Master 5-4
- Teaching Aid 43: Definitions of sine and cosine
- Computer Master 13

OBJECTIVES

C Evaluate circular functions outside the first quadrant.
K Represent sines, cosines, and tangents in the coordinate plane.

TEACHING NOTES

Some people distinguish between trigonometric and circular functions according to whether the argument is measured in degrees or radians, respectively. There is no compelling reason for this distinction, so we do not make it. Calculators can be set to work with either degrees or radians, so the difference in application depends on the context of the problem and the user's preference. Students should be aware that many computer languages and all the automatic graphers we have examined assume radian measure.

Alternate Approach To get the definition across, you might draw a large unit circle and point to (1, 0). Now trace the circle slowly counterclockwise. Every so often, stop and ask how many degrees and how many radians θ your finger has traced through and ask the approximate coordinates of the point. This gives an estimate of cos θ and sin θ for that value of θ. Be certain to stop at the points (0, 1), (-1, 0), and (0, -1). Now trace

---

LESSON

The Sine, Cosine, and Tangent Functions

For the sine, cosine, and tangent, the definitions in terms of ratios of sides of right triangles are limited to positive acute angles; that is, to angles θ in the interval $0° < \theta < 90°$ or $0 < \theta < \frac{\pi}{2}$ radians.

The trigonometric functions can be applied to more than positive acute angles by considering rotations and the coordinates of points on the particular unit circle that has its center at the origin.

The point (1, 0) is on this circle. Consider its image under a rotation of 70° about the origin $O = (0, 0)$. What are the coordinates of the image?

Let $P = (x, y)$ be the image. P is on this circle because $OP = 1$. To find the values of x and y, construct a right triangle OPQ as shown at the right. Now $\cos 70° = \frac{x}{1}$ and $\sin 70° = \frac{y}{1}$. Thus exact values for the coordinates are $x = \cos 70°$ and $y = \sin 70°$. A calculator shows $x \approx 0.342$ and $y \approx 0.940$. So $P = (\cos 70°, \sin 70°) \approx (0.342, 0.940)$.

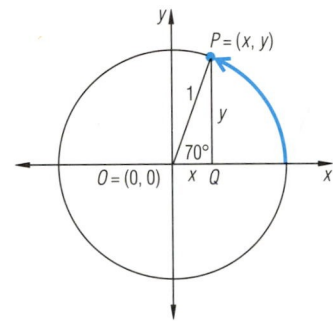

By generalizing the ideas from the preceding instance to allow θ to be the magnitude of *any* rotation on the unit circle, the sine, cosine, and tangent can be defined as functions of *any* real number. (The definition for tangent follows Example 2.)

Definitions

For all real numbers θ, **cos θ** and **sin θ** are the first and second coordinates of the image of the point (1, 0) under a rotation of magnitude θ about the origin.

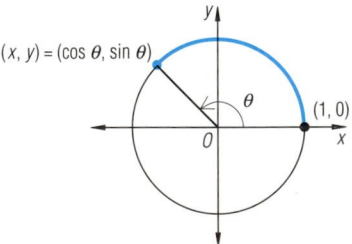

286

286

It is easy to find values of the cosine and sine when θ is a multiple of 90° or $\frac{\pi}{2}$ radians.

■ ■ ■ ■ ■ ■ ■

Example 1 Find: **a.** cos 180° **b.** sin 180°.

> **Solution** The rotation of 180° maps (1, 0) onto (-1, 0).
> By definition, (cos 180°, sin 180°) = (-1, 0).
> So cos 180° = -1 and sin 180° = 0.
> The radian equivalent to 180° is π radians,
> so cos π = -1 and sin π = 0.

Cosines and sines of other multiples of 90° or $\frac{\pi}{2}$ radians are shown on the unit circle below.

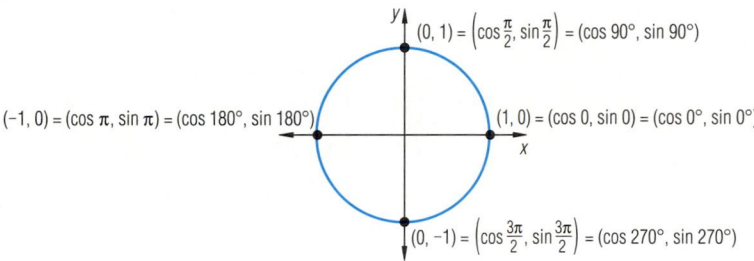

You can approximate other values of sin θ or cos θ using a scientific calculator.

■ ■ ■ ■ ■ ■ ■

Example 2 (1, 0) is rotated 305° about the origin. Find the coordinates of its image rounded to the nearest thousandth.

> **Solution** By definition, the image of (1, 0) after a rotation of 305° is (cos 305°, sin 305°). Set your calculator to degree mode and calculate.
> cos 305° ≈ 0.574 and sin 305° ≈ -0.819.
> So the image of (1, 0) under a rotation of 305°
> is approximately (0.574, -0.819).
>
> **Check** The image lies
> in the fourth quadrant,
> closer to the y-axis than
> the x-axis. So the
> coordinates seem
> correct.

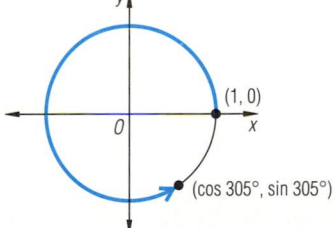

clockwise and do the same thing for negative values of θ. You might then use the teaching aid, which gives the definitions of sine and cosine as well as some of the key points.

ADDITIONAL EXAMPLES
1. Find **a.** cos 270° **b.** sin 270°.
a. 0; b. -1

2. Estimate, to the nearest thousandth, the coordinates of the image of (1, 0) under a rotation of -10π/9.
(-.940, .342)

3. Estimate, to the nearest thousandth:
a. cos 500° **b.** cos 500.
a. -0.766; b. -0.884

To find tan θ on the unit circle, consider first the case when θ is in the first quadrant. Then, because $\qquad$ $\tan \theta = \dfrac{\text{side opposite } \theta}{\text{side adjacent to } \theta}$,

$$\tan \theta = \frac{y}{x}.$$

But $y = \sin \theta$ and $x = \cos \theta$, so $\qquad \tan \theta = \dfrac{\sin \theta}{\cos \theta}$.

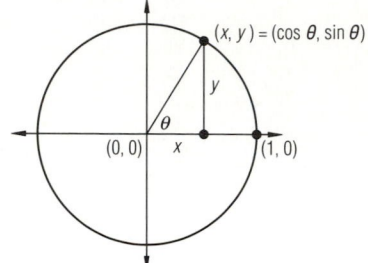

Extending this idea to any quadrant gives a definition of tan θ for other real numbers θ.

> **Definition**
>
> For all real numbers θ, provided $\cos \theta \neq 0$, $\tan \boldsymbol{\theta} = \dfrac{\sin \theta}{\cos \theta}$.

When $\cos \theta$ *does* equal zero, which occurs at any odd multiple of $\frac{\pi}{2}$ or $90°$, then tan θ is *undefined*.

You can determine whether the value of a trigonometric function is positive or negative by using coordinate geometry and the unit circle. The cosine is positive when the image of point $(1, 0)$ is in the first or fourth quadrant. The sine is positive when the image is in the first or second quadrant. The tangent is positive when the sine and cosine have the same sign (first and third quadrants); tan θ is negative when sin θ and cos θ have opposite signs (second and fourth quadrants). The following table summarizes this information for values of θ between 0 and 2π.

magnitude of rotation	quadrant of image	cosine	sine	tangent
$0 < \theta < \frac{\pi}{2}$	first	+	+	+
$\frac{\pi}{2} < \theta < \pi$	second	−	+	−
$\pi < \theta < \frac{3\pi}{2}$	third	−	−	+
$\frac{3\pi}{2} < \theta < 2\pi$	fourth	+	−	−

Example 3 Find **a.** sin(-7) **b.** cos(-7) **c.** tan(-7). Write your answer rounded to the nearest thousandth.

Solution

a. No unit is given, which means the angle measure is given in radians. Set your calculator to radian mode. Use the appropriate key sequence for your calculator to find sin(-7). You should see -0.6569... displayed. So, to the nearest thousandth, sin(-7) ≈ -0.657.
b. Repeat part **a** for the cosine: cos(-7) ≈ 0.754.
c. tan(-7) ≈ -0.871

Check

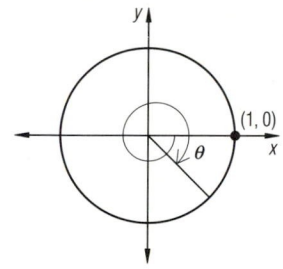

a. Imagine a unit circle. The minus sign indicates a clockwise rotation. Because 7 is a little more than 2π, a rotation of a little more than one revolution is involved. The image of (1, 0) is thus in the fourth quadrant. So sin(-7) should be negative, consistent with the calculated value.
b. If θ is in the fourth quadrant, the value of cos θ is positive. It checks.
c. $\dfrac{\sin(-7)}{\cos(-7)} \approx \dfrac{-0.657}{0.754} \approx -0.871$. It checks.

As you have seen, there is a unique value of sin θ associated with each value of θ. So, the ordered pairs $(\theta, \sin\theta)$ define a function, called the **sine function**. In symbols, $f(\theta) = \sin\theta$.

Similarly, the **cosine function** and the **tangent function** are defined respectively by the ordered pairs $(\theta, \cos\theta)$ and $(\theta, \tan\theta)$. In symbols, $g(\theta) = \cos\theta$ and $h(\theta) = \tan\theta$. Unless specified otherwise, it is assumed that θ is measured in radians for each of these three functions. These functions are known as **trigonometric functions**. Because they are defined in terms of the unit circle, they are also called **circular functions**.

The domain of the cosine and sine functions is the set of all possible magnitudes of rotations, that is, the set of all real numbers. The range of these functions is the set of all possible x-coordinates (for the cosine) and the set of all possible y-coordinates (for the sine) of points on the unit circle. So the range of the cosine and sine functions is the set of real numbers from -1 to 1. The domain of the tangent function is the set of all values of θ for which cos $\theta \neq 0$, that is, the set of all real numbers except odd multiples of $\frac{\pi}{2}$. Its range is the set of all possible slopes of lines, that is, the set of all real numbers.

In BASIC, the circular functions are built in and are abbreviated SIN, COS, and TAN, respectively. Their arguments are assumed to be real numbers in radians rather than degrees. To compute the values of the circular functions when the argument is in degrees, a conversion factor must be introduced. The following BASIC statement defines a function S(X) which computes the approximate sine of X, where X is measured in degrees.

DEF FN S(X) = SIN(X * 3.14159265/180)

Questions

Covering the Reading

In 1–3, approximate to the nearest thousandth.

1. $\sin 260°$ **-0.985**
2. $\cos\left(-\dfrac{10\pi}{7}\right)$ **-0.223**
3. $\tan\left(-\dfrac{10\pi}{7}\right)$ **-4.381**

4. Find the image of $(1, 0)$ under a rotation of $200°$ about the origin.
   about (-0.94, -0.34)

In 5–7, give exact values.

5. $\sin\left(-\dfrac{3\pi}{2}\right)$ **1**
6. $\cos\left(-\dfrac{3\pi}{2}\right)$ **0**
7. $\tan\left(-\dfrac{3\pi}{2}\right)$ **undefined**

8. Give two values of θ, in radians, for which $\tan \theta$ is undefined.
   sample: -3π/2 and π/2

In 9–11, use the figure at the right. Which point is the image of $(1, 0)$ under the rotation with the given magnitude?

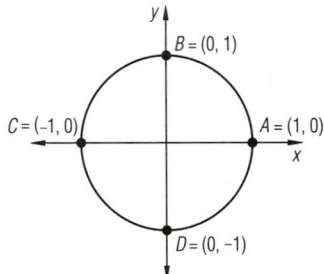

9. $\dfrac{9\pi}{2}$ **B**

10. -13π **C**

11. $-\dfrac{17\pi}{2}$ **D**

In 12 and 13, let P be the image of $(1, 0)$ under a rotation of θ around the origin.

12. If P is in the second quadrant, state the sign of the following.
   a. $\cos \theta$ **negative**   **b.** $\tan \theta$ **negative**

13. If $\sin \theta < 0$ and $\cos \theta < 0$, in what quadrant is P? **III**

In 14–16, state **a.** the domain, and **b.** the range.

14. $f(\theta) = \sin \theta$
   a) the set of all real numbers
   b) {y : -1 ≤ y ≤ 1}

15. $g(\theta) = \cos \theta$ **See below.**

16. $h(\theta) = \tan \theta$ **See below.**

17. Find three values of θ for which $\cos \theta = 1$.   **sample: θ = 0, 2π, -8π**

Applying the Mathematics

In 18 and 19, draw a unit circle on graph paper. Use the graph paper to approximate the values to the nearest tenth. Check using a calculator.

18. $\sin 155°$ **0.4**

19. $\cos{-35°}$ **0.8**

20. For what values of θ between 0 and 2π is $f(\theta) = \sin \theta$ positive?
   0 < θ < π

21. As θ increases from 0 to $\dfrac{\pi}{2}$, tell whether $y = \cos \theta$ increases or decreases. **decreases**

15 a) the set of all real numbers
   b) {y: -1 ≤ y ≤ 1}
16 a) the set of all real numbers except odd multiples of π/2
   b) the set of all real numbers

22. This question relates the tangent function to the use of the word *tangent* in geometry. At the right, line ℓ is a *tangent* to the unit circle at $(1, 0)$, and $\overrightarrow{OP}$ intersects ℓ at Q.

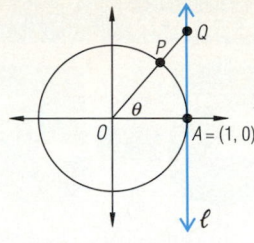

a. Prove that $QA = \tan \theta$, for $0 < \theta < \frac{\pi}{2}$.

b. Draw a diagram like that at the right for the case of $\frac{\pi}{2} < \theta < \pi$. Explain how to find $\tan \theta$ from your diagram.   **See margin.**

a) $\tan \theta = \frac{QA}{OA} = \frac{QA}{1} = QA$

Review

23. A fishing boat sails at a steady speed of 13 mph on a bearing of 67° (from north). If the boat set sail at 3:00 A.M., describe its location east and north of its port at 7:00 A.M.   *(Lesson 5-3)*
about 48 miles east and 20 miles north of its port

24. *Multiple choice.*   A 30′ ladder used by firefighters is safe only when it leans against a building at a 75° angle or less to the ground. What is the maximum height in feet on a building it can reach?   *(Lesson 5-3)*

(a) $30 \cos 75°$ (b) $30 \sin 75°$ (c) $\frac{30}{\cos 75°}$ (d) $\frac{30}{\sin 75°}$ **(b)**

25. A crank is used to pull a bucket out of a well. The rope holding the bucket wraps around a 3″ diameter rod. Assuming that the rope fits snugly on the rod as the bucket is pulled up, estimate how many times you would have to rotate the crank to raise the bucket 15 feet.
(Lesson 5-2)   **about 19 times**

In 26–28, *skill sequence*. Find the exact solution.   *(Lesson 4-8)*

26. $5^x = 3125$ **5**   **27.** $e^x = 3125$ **ln 3125**   **28.** $4^x = 3125$
$\log_4 3125$

In 29 and 30, define.   *(Previous course)*

29. supplementary angles
two angles whose sum is 180°

30. complementary angles
two angles whose sum is 90°

Exploration

31. a. To the nearest hundredth, $\sin .01 = .01$. For what values of θ between 0 and $\frac{\pi}{2}$ is θ within 0.01 of $\sin \theta$?

b. $\cos \frac{\pi}{4} = \sin \frac{\pi}{4}$. For what values of θ between 0 and $\frac{\pi}{2}$ is $\cos \theta = \sin \theta$, if both values are rounded to the nearest hundredth?
a) $0 \le \theta \le 0.39249$; b) $0.78243 \le \theta \le 0.78837$

FOLLOW-UP

MORE PRACTICE
For more questions on SPUR Objectives, use *Lesson Master 5-4*, shown below.

PROJECTS
The projects for Chapter 5 are described on pages 329–330. Project 5 is related to the content of this lesson.

ADDITIONAL ANSWERS
22.b.

NAME _____

LESSON **MASTER   5-4**
QUESTIONS ON **SPUR** OBJECTIVES

■**SKILLS** *Objective C (See pages 334–338 for objectives.)*

1. Find

  a. cos 120°.   **-.5**   **b.** sin 180°.   **0**

2. The point (1, 0) is rotated 300° about the origin. Find the coordinates of its image correct to three decimal places.   **(.5, -.866)**

3. Find to the nearest hundredth.

  a. sin (-12°)   **-.21**   **b.** cos (-95°)   **-.09**   **c.** tan (-7)   **-.87**

In 4–6, approximate to the nearest thousandth.

4. sin 215°   **-.574**

5. tan $\left(\frac{-5\pi}{4}\right)$ **-1**

6. cos $\left(\frac{10\pi}{7}\right)$ **-.223**

7. Find all values of θ between 0° and 360° for which tan $\theta = 2.5$.   **68.2°, 248.2°**

■**REPRESENTATIONS** *Objective K*
In 8–11, refer to the figure. Which point is the image of (1, 0) under the rotation with the given magnitude?

8. $\frac{7\pi}{2}$ **D**

9. -2π **A**

10. $\frac{-3\pi}{2}$ **B**

11. 9π **C**

Functions, Statistics, and Trigonometry © Scott, Foresman and Company   **57**

LESSON 5-5

OBJECTIVES

B Find sines, cosines, and tangents of acute angles.
C Evaluate circular functions outside the first quadrant.

TEACHING NOTES

A beneficial class exercise might be to review the values of the sines and cosines labeled on the large circle shown on page 295. Consider both degree and radian measures for θ. A teaching aid is provided.

Students might ask if there are other values of θ for which $\sin \theta$, $\cos \theta$, and $\tan \theta$ can be found exactly. Under the conditions that θ be a rational multiple of π or 180°, and that the trigonometric values of θ are rational numbers or rational multiples of simple radicals, only the values given on page 295 qualify. However, if the conditions are that θ be a whole number multiple of $\frac{\pi}{180}$ or 1° and that the trigonometric values of θ be in terms of rational numbers and simple radicals, then any multiple of $\frac{\pi}{60}$ or 3° can be found using identities to be derived in Chapter 11.

Throughout this lesson we adhere to the common convention of rationalizing the denominator of expressions such as $\frac{1}{\sqrt{3}}$, but this is not to be taken as a recommendation that students be required to

LESSON

5-5

Exact Values of Trigonometric Functions

For most angles, the sine, cosine, and tangent are not known exactly, but must be approximated. However, you already know exact values of the sine, cosine, and tangent of multiples of 180°, and of the sine and cosine of multiples of 90°. By applying some geometry you can also obtain exact values of the trigonometric functions for 30°, 45°, 60°, and their multiples.

Consider first the 45-45-90 triangle shown below.

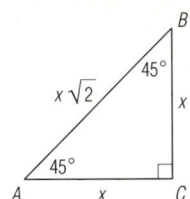

If $AC = BC = x$, then by the Pythagorean Theorem, $AB = x\sqrt{2}$. So by definition:

$$\sin 45° = \frac{\text{side opposite 45° angle}}{\text{hypotenuse}} = \frac{x}{x\sqrt{2}} = \frac{1}{\sqrt{2}} = \frac{\sqrt{2}}{2}$$

$$\cos 45° = \frac{\text{side adjacent to 45° angle}}{\text{hypotenuse}} = \frac{x}{x\sqrt{2}} = \frac{1}{\sqrt{2}} = \frac{\sqrt{2}}{2}$$

$$\tan 45° = \frac{\text{side opposite 45° angle}}{\text{side adjacent to 45° angle}} = \frac{x}{x} = 1.$$

In radian measure, $\qquad 45° = \frac{\pi}{4}.$

Thus, $\qquad \sin \frac{\pi}{4} = \cos \frac{\pi}{4} = \frac{\sqrt{2}}{2}$

$$\tan \frac{\pi}{4} = 1.$$

Now consider the 30-60-90 triangle. Recall from geometry that if $FE = x$, then $DF = 2x$. By the Pythagorean Theorem, $DE = x\sqrt{3}$. So

$$\sin 30° = \cos 60° = \frac{FE}{DF} = \frac{x}{2x} = \frac{1}{2}$$

$$\cos 30° = \sin 60° = \frac{DE}{DF} = \frac{x\sqrt{3}}{2x} = \frac{\sqrt{3}}{2}$$

$$\tan 30° = \frac{FE}{DE} = \frac{x}{x\sqrt{3}} = \frac{1}{\sqrt{3}} = \frac{\sqrt{3}}{3}$$

$$\tan 60° = \frac{DE}{FE} = \frac{x\sqrt{3}}{x} = \sqrt{3}.$$

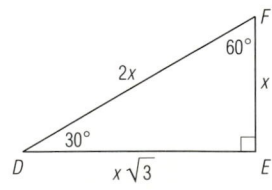

292

In radian measure, $30° = \frac{\pi}{6}$ and $60° = \frac{\pi}{3}$,

so $\qquad \sin \frac{\pi}{6} = \cos \frac{\pi}{3} = \frac{1}{2}$

$\qquad\qquad \cos \frac{\pi}{6} = \sin \frac{\pi}{3} = \frac{\sqrt{3}}{2}$

$\qquad\qquad \tan \frac{\pi}{6} = \frac{\sqrt{3}}{3}$

$\qquad\qquad \tan \frac{\pi}{3} = \sqrt{3}.$

You should memorize these values, as you will very likely be expected to know them on standardized tests or in future mathematics courses.

Using the definitions of the trigonometric functions that arise from rotations of $(1, 0)$ on the unit circle, you can find exact values of the circular functions for all integral multiples of $\frac{\pi}{6}$, $\frac{\pi}{4}$, and $\frac{\pi}{3}$.

Example 1 Find $\sin \frac{7\pi}{6}$, $\cos \frac{7\pi}{6}$, and $\tan \frac{7\pi}{6}$.

Solution $\frac{7\pi}{6}$ is greater than π and less than $\frac{3\pi}{2}$, so $\left(\cos \frac{7\pi}{6}, \sin \frac{7\pi}{6}\right)$ is a point in the third quadrant. It is the image of $\left(\cos \frac{\pi}{6}, \sin \frac{\pi}{6}\right)$ under a rotation of π around the origin.

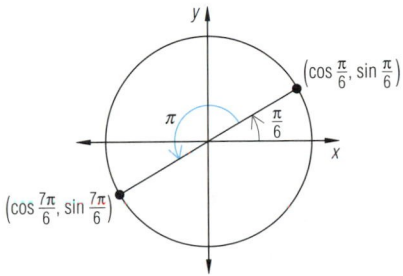

So if $\left(\cos \frac{\pi}{6}, \sin \frac{\pi}{6}\right) = (a, b)$, then $\left(\cos \frac{7\pi}{6}, \sin \frac{7\pi}{6}\right) = (-a, -b)$.

You know $\sin \frac{\pi}{6} = \frac{1}{2}$ and $\cos \frac{\pi}{6} = \frac{\sqrt{3}}{2}$.

So $\qquad \sin \frac{7\pi}{6} = -\frac{1}{2}$

and $\qquad \cos \frac{7\pi}{6} = -\frac{\sqrt{3}}{2}.$

Then $\qquad \tan \frac{7\pi}{6} = \dfrac{\sin \frac{7\pi}{6}}{\cos \frac{7\pi}{6}} = \dfrac{-\frac{1}{2}}{-\frac{\sqrt{3}}{2}} = \dfrac{1}{\sqrt{3}} = \dfrac{\sqrt{3}}{3}.$

do so. Historically denominators were rationalized because it was easier and more accurate to find a decimal approximation for $\frac{\sqrt{3}}{3}$ than for $\frac{1}{\sqrt{3}}$. Students should know the two expressions are equivalent, but in today's world of hand-held calculators the ease and greater accuracy arguments are no longer as valid.

ADDITIONAL EXAMPLES

1. Find $\sin \frac{-2\pi}{3}$, $\cos \frac{-2\pi}{3}$, and $\tan \frac{-2\pi}{3}$.

$-\frac{\sqrt{3}}{2}$, $-\frac{1}{2}$, $\sqrt{3}$

2. Find $\sin \frac{3\pi}{4}$, $\cos \frac{3\pi}{4}$, and $\tan \frac{3\pi}{4}$.

$\frac{\sqrt{2}}{2}$, $-\frac{\sqrt{2}}{2}$, -1

3. Use the formula $A = \frac{1}{2} ab \sin \theta$ to find the area of a regular octagon inscribed in a circle of radius 10. (The formula was proved in Question 14, page 283.)

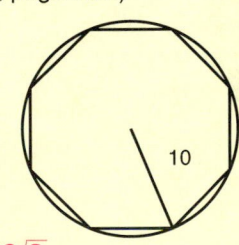

200√2

Notice that the rotation of $\frac{7\pi}{6}$ with center at the origin also has magnitude $-\frac{5\pi}{6}$, so you can use Example 1 to find that $\sin\left(-\frac{5\pi}{6}\right) = -\frac{1}{2}$, $\cos\left(-\frac{5\pi}{6}\right) = \frac{-\sqrt{3}}{2}$, and $\tan\left(-\frac{5\pi}{6}\right) = \frac{\sqrt{3}}{3}$.

The quarter circle in Quadrant II is the image of the quarter circle in Quadrant I under reflection about the y-axis. This idea is used in the next example.

■ ■ ■ ■ ■ ■ ■ ■ ■

Example 2 Find $\cos\frac{2\pi}{3}$, $\sin\frac{2\pi}{3}$, and $\tan\frac{2\pi}{3}$.

Solution Use a unit circle and locate P', the image of $(1, 0)$ under a rotation of $\frac{2\pi}{3}$. $P' = \left(\cos\frac{2\pi}{3}, \sin\frac{2\pi}{3}\right)$ is also the image of $P = \left(\cos\frac{\pi}{3}, \sin\frac{\pi}{3}\right)$ under a reflection over the y-axis.

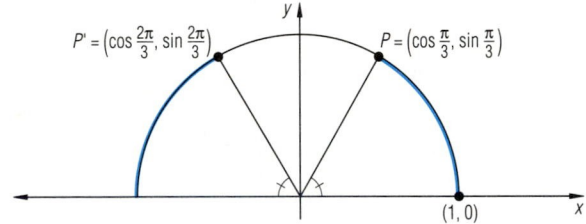

So the x-coordinates of P and P' are opposites and the y-coordinates are equal.

$$\cos\frac{2\pi}{3} = -\cos\frac{\pi}{3} \qquad \text{and} \qquad \sin\frac{2\pi}{3} = \sin\frac{\pi}{3}$$
$$= -\frac{1}{2} \qquad\qquad\qquad\qquad = \frac{\sqrt{3}}{2}.$$

Thus, $\tan\frac{2\pi}{3} = \dfrac{\sin\frac{2\pi}{3}}{\cos\frac{2\pi}{3}} = \dfrac{\frac{\sqrt{3}}{2}}{-\frac{1}{2}} = -\sqrt{3}.$

Check Use a calculator set to radian mode.

$\cos\frac{2\pi}{3} = -0.5 = -\frac{1}{2}$, $\sin\frac{2\pi}{3} \approx 0.866 \approx \frac{\sqrt{3}}{2}$, and $\tan\frac{2\pi}{3} \approx -1.732 \approx -\sqrt{3}$.

On the unit circle following are the images of $(1, 0)$ under all rotations of integral multiples of $\frac{\pi}{6}$ or $\frac{\pi}{4}$. You should be able to give exact values of the sine, cosine, and tangent functions for all values of θ pictured. Cosines and sines of such values are found by relating them to one of the points in the first quadrant or on the positive axes. The tangents are found using the ratios of sines to cosines. For negative multiples, simply locate the positive counterpart.

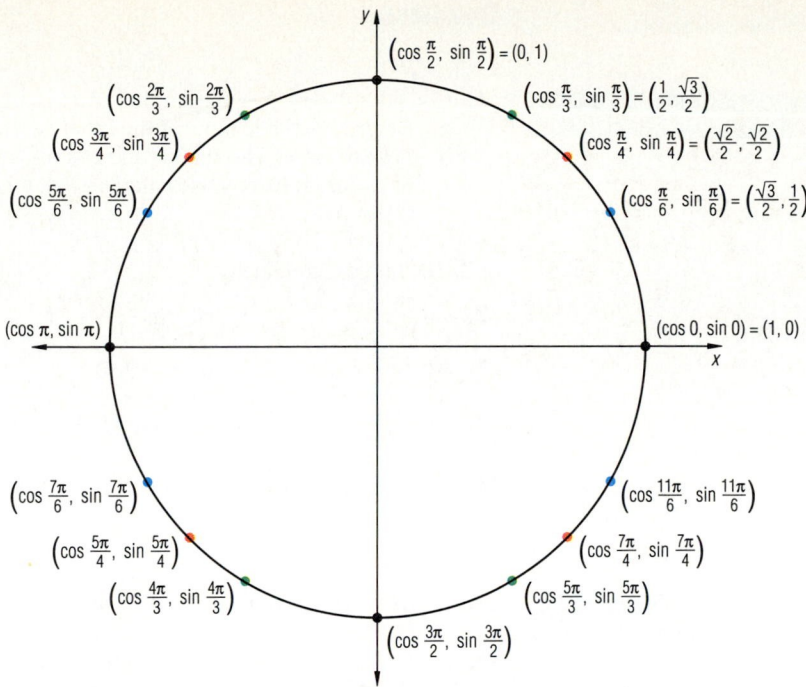

Because you can find exact values of certain angles, you can also obtain exact areas of certain polygons involving those angles.

Example 3   A regular hexagon has sides of length 16. Find its exact area in simplest form.

Solution   Each angle in a regular hexagon has measure 120°. Thus the area A of the hexagon is equal to the area of 6 equilateral triangles of side 16, or of twelve 30-60-90 triangles with hypotenuse 16.

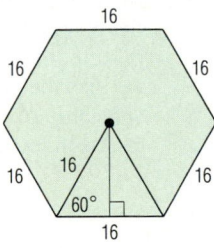

Thus

$$A = 12\left(\frac{1}{2} \cdot 8 \cdot 16 \sin 60°\right)$$

$$= 768 \cdot \frac{\sqrt{3}}{2}$$

$$= 384 \sqrt{3}.$$

Check   The area of a circle circumscribed about the hexagon is 256π. Is $256\pi > 384\sqrt{3}$? $804 > 665$, so it checks.

Questions

Covering the Reading

1. Consider the 30-60-90 triangle $\triangle ABC$ with the short leg $AB = 1$.
   a. Determine the lengths of the other two sides. $\sqrt{3}$ **and 2**
   b. *True* or *false*. The sin 30° and cos 60° calculated using $\triangle ABC$ would
      be different from those calculated at the beginning of the lesson by a
      factor of x. **False**

In 2–10, find exact values.

2. $\sin \frac{5\pi}{4}$ $-\frac{\sqrt{2}}{2}$

3. $\cos \frac{5\pi}{4}$ $-\frac{\sqrt{2}}{2}$

4. $\tan \frac{5\pi}{4}$ **1**

5. $\sin 330°$ $-\frac{1}{2}$

6. $\cos 330°$ $\frac{\sqrt{3}}{2}$

7. $\tan 330°$ $-\frac{\sqrt{3}}{3}$

8. $\sin \left(\frac{17\pi}{6} \right)$ $\frac{1}{2}$

9. $\cos(-60°)$ $\frac{1}{2}$

10. $\tan(-405°)$ **-1**

11. Find the exact area in simplest form of a hexagon with sides of length
    10 cm. $150\sqrt{3}$ **cm²**

Applying the Mathematics

12. Find two values of θ between $-\frac{\pi}{2}$ and $\frac{\pi}{2}$ for which $\cos \theta = \frac{1}{2}$.
    $-\pi/3$ **and** $\pi/3$

13. a. Find four values of θ between -2π and 2π such that $\cos \theta = \sin \theta$.
    b. What is the value of $\tan \theta$ for each value of θ in part **a**? **1**
    a) $\theta = -7\pi/4, -3\pi/4, \pi/4, 5\pi/4$

In 14 and 15, consider the drawing at the left of a cube with side length
5 meters. $\overline{BE}$ and $\overline{AB}$ are diagonals of two faces of the cube. $\overline{DB}$ is a diagonal
of the cube. (Note that m$\angle ACB$ = m$\angle DEB$ = 90°.)

14. Find $\sin \theta$ exactly. $\frac{\sqrt{2}}{2}$ 15. Find $\cos \phi$ exactly. $\sqrt{\frac{2}{3}}$

16. Use the formula $A = \frac{1}{2} ab \sin \theta$ (from Question 14, Lesson 5-3) to
    find the exact area in simplest form of a regular dodecagon (12-sided
    polygon) inscribed in a circle with radius 9. **243 square units**

17. Prove that the y-axis is
    the perpendicular
    bisector of $\overline{AB}$. (This
    verifies the use of
    reflection over the
    y-axis in Example 2.)
    See margin.

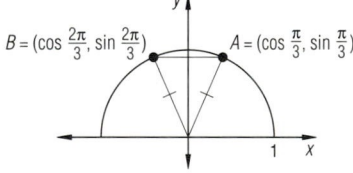

Review

18. a. Evaluate sin 111° to four decimal places. **0.9336**
    b. Find x such that $\sin x = \sin 111°$ and $0 < x < 90°$. *(Lesson 5-4)* **69°**

19. a. Prove that $\cos \theta \cdot \tan \theta = \sin \theta$ for all $\cos \theta \neq 0$.
    b. Why is it impossible to have $\cos \theta = 0$ in part **a**? *(Lesson 5-4)*
    a) $\cos \theta \cdot \tan \theta = \cos \theta \cdot \frac{\sin \theta}{\cos \theta}$ if $\cos \theta \neq 0$
    $= \frac{\cos \theta}{\cos \theta} \cdot \sin \theta = \sin \theta$
    b) Tan θ would be undefined.

296

20. The volume V of a right circular cone of height h and radius r is
$$V = \frac{1}{3}\pi r^2 h.$$

Consider a cross section of a cone shown at the right. Find a formula for the volume of a cone in terms of r and θ. *(Lesson 5-3)* $\quad V = \frac{1}{3}\pi r^3 \tan\theta$

21. To measure the distance across a lake for the purpose of planning a new freeway bridge, surveyors placed poles at D, E, and F so that $\overline{ED} \perp \overline{EF}$, and $ED = 100$ m. With a transit they found $m\angle EDF = 32°\,45'$. Find the distance d across the lake. *(Lesson 5-3)* **about 64.3 m**

22. Identify all expressions equal to the area of $\triangle XYZ$. *(Lesson 5-3)*

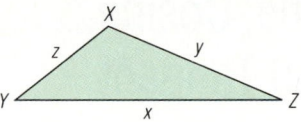

(a) $xy \sin Z$ (b) $\frac{1}{2}xy \sin Z$

(c) $\frac{1}{2}z \sin X$ (d) $xz \cos Y$ **(b)**

23. Given that the equatorial diameter of the earth is about 7927 miles and the earth rotates on its axis once in approximately 23 hours 56 minutes, estimate the speed $\left(\text{in } \frac{\text{miles}}{\text{hr}}\right)$ at which a point on the equator is rotating around the earth's center. *(Lesson 5-2)* **about 1041 $\frac{\text{miles}}{\text{hour}}$**

24. Here are the latitudes of four cities. Order the cities in distance from the equator, from closest to farthest. *(Lesson 5-1)*

Vladivostok, U.S.S.R.	43°10′N
Hobart, Australia	42°52′S
Syracuse, NY	43°2′N
Detroit, MI	42°20′N

Detroit, Hobart, Syracuse, Vladivostok

25. The graph of $y = e^x$ is scaled by $S(x, y) = \left(\frac{1}{2}x, \frac{1}{3}y\right)$. Write an equation for the image graph. *(Lessons 3-5, 4-6)* $\quad y = \frac{e^{2x}}{3}$

26. Find $f(3t)$ if $f(t) = t^3 + t + 4$. *(Lesson 2-1)* $27t^3 + 3t + 4$

27. Find the distance between $(3, 4)$ and $(10, 15)$. *(Previous course)* $\quad \sqrt{170}$

Exploration

In 28 and 29, look in other math books or develop the formulas yourself.

28. Find a formula for the area of a regular polygon of n sides which is inscribed in a circle with radius r. $A = \frac{1}{2}nr^2 \sin\left(\frac{2\pi}{n}\right)$ or $nr^2 \sin\frac{\pi}{n}\cos\frac{\pi}{n}$

29. Find a formula for the area of a regular polygon of n sides with a side of length s. $A = \dfrac{ns^2}{4\tan\left(\frac{\pi}{n}\right)}$

LESSON 5-6

RESOURCES
- Lesson Master 5-6
- Teaching Aid 45: Table and graph of the sine function
- Teaching Aid 46: Table and graph of the tangent function
- Computer Master 14

OBJECTIVES

F State properties of the sine, cosine, and tangent functions.

G Use theorems about sines, cosines, and tangents.

L Draw or interpret graphs of the parent sine, cosine, and tangent functions.

TEACHING NOTES

In the tables on pages 298 and 300, sin θ and tan θ are given in terms of both decimal approximations and exact values. Students should be comfortable with both expressions. The graphs of sin θ and tan θ are also shown on these pages.

Point out the scale on the θ-axis, which we have marked in multiples of $\frac{\pi}{2}$. We know of at least one automatic grapher which allows the user to choose such a scale. Most automatic graphers, however, label the axis in decimal radians, which serves to hide the relationship to π. Students may find it valuable to label the θ-axis with both exact values and decimal approximations. Teaching aids are provided.

LESSON

Graphs of the Sine, Cosine, and Tangent Functions

Consider the sine function, $f(\theta) = \sin \theta$, with θ measured in radians. This is the function that maps each real number θ to the y-coordinate of the image of $(1, 0)$ under a rotation of θ.

The sine function is positive for $0 < \theta < \pi$ and negative for $\pi < \theta < 2\pi$.

The maximum value is 1 for $\theta = \frac{\pi}{2}$

and the minimum is -1 for $\theta = \frac{3\pi}{2}$.

Other values of this function are given in the table below.

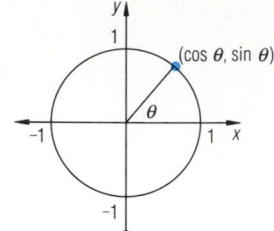

θ	0	$\frac{\pi}{6}$	$\frac{\pi}{4}$	$\frac{\pi}{3}$	$\frac{\pi}{2}$	$\frac{2\pi}{3}$	$\frac{3\pi}{4}$	$\frac{5\pi}{6}$	π	$\frac{7\pi}{6}$	$\frac{5\pi}{4}$	$\frac{4\pi}{3}$	$\frac{3\pi}{2}$	$\frac{5\pi}{3}$	$\frac{7\pi}{4}$	$\frac{11\pi}{6}$	2π
$\sin \theta$*	0	0.5	0.707	0.866	1	0.866	0.707	0.5	0	-0.5	-0.707	-0.866	-1	-0.866	-0.707	-0.5	0
$\sin \theta$**	0	$\frac{1}{2}$	$\frac{\sqrt{2}}{2}$	$\frac{\sqrt{3}}{2}$	1	$\frac{\sqrt{3}}{2}$	$\frac{\sqrt{2}}{2}$	$\frac{1}{2}$	0	$-\frac{1}{2}$	$-\frac{\sqrt{2}}{2}$	$-\frac{\sqrt{3}}{2}$	-1	$-\frac{\sqrt{3}}{2}$	$-\frac{\sqrt{2}}{2}$	$-\frac{1}{2}$	0

* Some entries are approximations.
**exact values

The graph of all points on the sine function: $\theta \to \sin \theta$, for $0 \le \theta \le 2\pi$, is shown here.

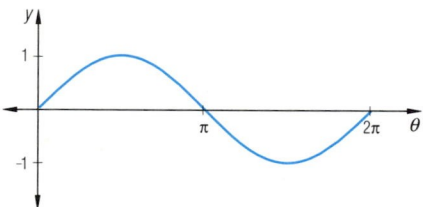

The image of $(1, 0)$ under a rotation of θ repeats itself every 2π radians. As a result, the y-coordinates in the ordered pairs of the function $f(\theta) = \sin \theta$ repeat every 2π. Thus, the table and graph above can be easily extended both to the right and left without calculating any new sine values. A more extended graph of the function $f(\theta) = \sin \theta$ appears below.

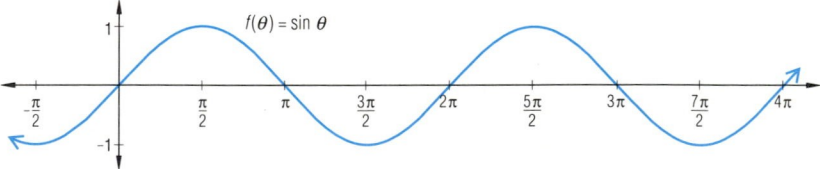

298

The graph for $0 \leq \theta \leq 2\pi$ shows one **cycle** of the graph of the sine function. The graph of the entire sine function has infinitely many cycles.

The sine function is an example of a *periodic function*.

Definition

A function f is **periodic** if and only if there is a positive real number p such that $f(x + p) = f(x)$ for all x. The smallest such positive number is the **period** of the function.

The period of the sine function is 2π, because 2π is the smallest positive number such that $\sin(x + 2\pi) = \sin x$ for all x.

The domain of the sine function is the set of all possible values of θ, the set of all real numbers. The range of the function is the set of all possible values of $\sin \theta$, that is, the set of y-coordinates on the unit circle, $\{y: -1 \leq y \leq 1\}$.

The y-intercept of the sine function is 0 and the x-intercepts are integral multiples of π: ... , $-2\pi, -\pi, 0, \pi, 2\pi, 3\pi, 4\pi, ...$.

The cosine function has characteristics that correspond to those of the sine function.

▪ ▪ ▪ ▪ ▪ ▪ ▪ ▪ ▪

Example 1   Consider the cosine function $g: \theta \to \cos \theta$.
  a. Give the domain and range of g.
  b. What are the maximum and minimum values of g?
  c. What is its period?
  d. Display a graph of g.

Solution

a. Because θ can be any real number, the domain is the set of all real numbers. On the unit circle, it is clear that the cosine function takes all values between -1 and 1. So the range is $\{y: -1 \leq y \leq 1\}$.
b. The maximum value is 1 and the minimum value is -1.
c. The image of $(1,0)$ is the same after each complete revolution of a point around the unit circle. So the period is 2π.
d. The graph below is for the domain $\{\theta: -5 \leq \theta \leq 10\}$. This is enough to show several cycles of the cosine function.

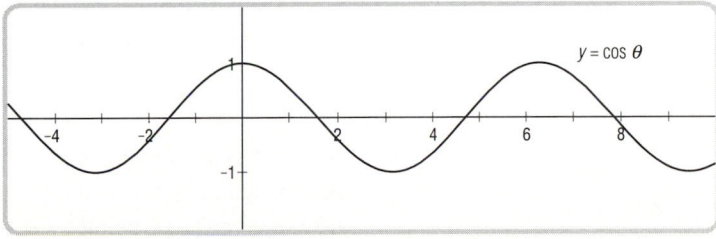

Check   The graph is consistent with **a, b,** and **c**.

Circular functions are useful for modeling situations having periodic behavior, and often angles and rotations are not explicitly involved. For this reason, radian measure is used almost exclusively in this chapter.

Though we use θ for the independent variable, many automatic graphers will use x. Students should realize that $g(\theta) = \cos \theta$ defines the same function as $f(x) = \cos x$.

Example 1 indicates the major properties of sine waves that students should know at this time: the domain and range, maximum and minimum values, the period, and the graph. **Example 2** anticipates the properties of sines and cosines to be encountered in the next lesson. **Example 3** provides more practice in the solving of simple trigonometric equations.

ADDITIONAL EXAMPLES
1. The graph of the cosine function is a translation image of the graph of the sine function. Give an equation for the translation.
sample: $T(x, y) = (x + \frac{\pi}{2}, y)$

2. Given $\tan 50° \approx 1.192$, find three other values of x with $\tan x \approx 1.192$.
samples: 230°, 410°, -130°

3. Find three solutions (in radians) to the equation $\sin \theta = \frac{1}{4}$.
samples: $\theta \approx .253$, $\theta \approx 6.536$, $\theta \approx 2.889$

NOTES ON QUESTIONS

Question 4: You might wish to emphasize that students should complete this question carefully. For part **b** encourage students to mark both the θ axis and the cos θ axis with exact and decimal approximations.

Question 13: You might pose the following questions: Find all solutions to sin $x = 3$. (There are none.) Why? (3 is not in the range of the sine function; 3 is not a second co-ordinate of a point on the unit circle.)

Making Connections for Question 14: Many periodic functions arise from natural phenomena, of which the graph of blood pressure is only one. All of them can be estimated by sums of sine and cosine functions. This idea, part of what is called Fourier analysis, is a topic of a reading lesson in *Precalculus and Discrete Mathematics*.

Error Analysis for Question 16: Students sometimes think the step function is periodic because the function exhibits a predictable pattern at periodic intervals. Being periodic is more than just a repeated pattern—it is also repeated values. A step function does not meet this second requirement.

Making Connections for Question 17: This is the first time in this book that students have been asked to graph an offspring to the parent trigonometric functions. This idea is studied in detail in Chapter 6.

ADDITIONAL ANSWERS
1.a.

The y-intercept of the cosine function is 1 and the x-intercepts are odd multiples of $\frac{\pi}{2}$: $\ldots, \frac{-5\pi}{2}, \frac{-3\pi}{2}, \frac{-\pi}{2}, \frac{\pi}{2}, \frac{3\pi}{2}, \frac{5\pi}{2}, \ldots$. Note that the graph of the cosine function is congruent to that of the sine function. In fact, it can be shown that the graph of $g(\theta) = \cos\theta$ is a translation image of the graph of $f(\theta) = \sin\theta$.

Now we turn to the tangent function. From the values of sine and cosine for $0 \le \theta \le 2\pi$ and the definition $\tan\theta = \frac{\sin\theta}{\cos\theta}$, a table of values for the tangent can be generated.

θ	0	$\frac{\pi}{6}$	$\frac{\pi}{4}$	$\frac{\pi}{3}$	$\frac{\pi}{2}$	$\frac{2\pi}{3}$	$\frac{3\pi}{4}$	$\frac{5\pi}{6}$	π	$\frac{7\pi}{6}$	$\frac{5\pi}{4}$	$\frac{4\pi}{3}$	$\frac{3\pi}{2}$	$\frac{5\pi}{3}$	$\frac{7\pi}{4}$	$\frac{11\pi}{6}$	2π
tan θ*	0	0.577	1	1.732	—	-1.732	-1	-0.577	0	0.577	-1	1.732	—	-1.732	-1	-0.577	0
tan θ**	0	$\frac{1}{\sqrt{3}}$	1	$\sqrt{3}$	—	$-\sqrt{3}$	-1	$-\frac{1}{\sqrt{3}}$	0	$\frac{1}{\sqrt{3}}$	-1	$\sqrt{3}$	—	$-\sqrt{3}$	-1	$-\frac{1}{\sqrt{3}}$	0

*Some entries are approximations.
**exact values

The table illustrates certain properties of tangents which follow immediately from the definition of tan θ.
(1) When sin θ and cos θ are both positive or both negative, tan θ is positive. When only one of sin θ or cos θ is negative, tan θ is negative.
(2) When sin $\theta = 0$, tan $\theta = 0$.
(3) When cos $\theta = 0$, tan θ is undefined.

Here is a graph of $h(\theta) = \tan\theta$.

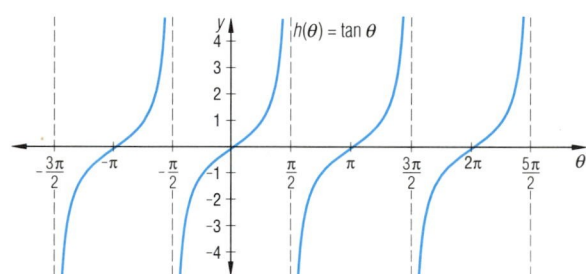

The graph exhibits a variety of ways in which the tangent function differs from the sine and cosine functions.
(1) The period of the tangent function is π.
(2) The domain is the set of all real numbers except odd multiples of $\frac{\pi}{2}$; that is, all x such that $x \ne \frac{\pi}{2} + n\pi$ for all integers n.
(3) The range of the tangent function is the set all real numbers.
(4) There are no relative minimum or maximum values.
(5) The tangent function has an x-intercept at multiples of π; that is, at $n\pi$ for all integers n.
(6) When cos θ is close to 0, sin θ is close to either 1 or -1 and tan θ becomes large in magnitude. So the lines with equations $\theta = \frac{\pi}{2} + n\pi$, where n is an integer, are vertical asymptotes to the graph. The tangent function is not continuous at all values of θ for which cos $\theta = 0$.

300

300

Most automatic graphers have the tangent function built into them. Check one available to you to see how it handles the asymptotes of $y = \tan \theta$.

The periodic nature of the sine, cosine, and tangent functions is summarized in the following theorem.

> **Periodicity Theorem**
>
> For every θ, and for every integer n:
> $$\sin (\theta + 2\pi n) = \sin \theta,$$
> $$\cos (\theta + 2\pi n) = \cos \theta,$$
> and $$\tan (\theta + \pi n) = \tan \theta.$$

The above statement of the Periodicity Theorem expresses the values of θ in radians. In degrees, $\cos (\theta + n \cdot 360°) = \cos \theta$, $\sin (\theta + n \cdot 360°) = \sin \theta$, and $\tan (\theta + n \cdot 180°) = \tan \theta$ for every θ and for every integer n.

Example 2 Given that $\cos 2.300 \approx -0.666$, find three other values of θ with $\cos \theta \approx -0.666$.

Solution The cosine function has a period of 2π, so some other solutions can be found by adding or subtracting multiples of $2\pi \approx 6.283$ to 2.300. For instance,
$$2.300 + 2\pi \approx 8.583,$$
$$2.300 + 4\pi \approx 14.866,$$
$$2.300 - 2\pi \approx -3.983.$$

Check Find $\cos (8.583)$, $\cos (14.866)$, and $\cos (-3.983)$ on your calculator. In each case, the value is about -0.666.

Example 3 Find three solutions to the equation $\tan \theta = 4$.

Solution Set your calculator to radians and find one value of θ whose tangent is 4. (A possible key sequence is 4 [INV] [TAN].) To the nearest thousandth, $\tan 1.326 \approx 4$. The period of the tangent function is π. So other solutions can be found by adding or subtracting multiples of π. For instance,
$$1.326 + \pi \approx 4.467,$$
and $$1.326 - \pi \approx -1.816$$
are two other solutions.

Check 1 Find $\tan (1.326)$, $\tan (4.467)$, and $\tan (-1.1816)$ on your calculator. Each is approximately equal to 4.

Check 2 Draw a graph of $y = \tan x$ on an automatic grapher. Trace with a cursor or zoom to show that the points (1.326, 4), (4.767, 4), and (-1.816, 4) are on the graph.

b.

4. a.

θ	y	y
0	1	1
$\frac{\pi}{6}$	$\frac{\sqrt{3}}{2}$	0.866
$\frac{\pi}{4}$	$\frac{\sqrt{2}}{2}$	0.707
$\frac{\pi}{3}$	$\frac{1}{2}$	0.5
$\frac{\pi}{2}$	0	0
$\frac{2\pi}{3}$	$-\frac{1}{2}$	-0.5
$\frac{3\pi}{4}$	$-\frac{\sqrt{2}}{2}$	-0.707
$\frac{5\pi}{6}$	$-\frac{\sqrt{3}}{2}$	-0.866
π	-1	-1
$\frac{7\pi}{6}$	$-\frac{\sqrt{3}}{2}$	-0.866
$\frac{5\pi}{4}$	$-\frac{\sqrt{2}}{2}$	-0.707
$\frac{4\pi}{3}$	$-\frac{1}{2}$	-0.5
$\frac{3\pi}{2}$	0	0
$\frac{5\pi}{3}$	$\frac{1}{2}$	0.5
$\frac{7\pi}{4}$	$\frac{\sqrt{2}}{2}$	0.707
$\frac{11\pi}{6}$	$\frac{\sqrt{3}}{2}$	0.866

4.b.

c.

6.a.

b. $-2\pi < x < -\dfrac{3\pi}{2}$,

$-\pi < x < -\dfrac{\pi}{2}$, $0 < x < \dfrac{\pi}{2}$,

$\pi < x < \dfrac{3\pi}{2}$, $2\pi < x < \dfrac{5\pi}{2}$,

$3\pi < x < -\dfrac{7\pi}{2}$

17.a.,b.

c. The graph of f has the same period as $y = \sin \theta$, but its range is $\{y: -2 \le y \le 2\}$. The graph of g has the same range as $y = \sin \theta$, but its period is π.

Questions

Covering the Reading

1. Sketch a graph of $f(\theta) = \sin \theta$, for
   a. $-2\pi \le \theta \le 4\pi$.   **b.** $0° \le \theta \le 1080°$.
   a,b) See margin.

2. Consider $f(\theta) = \sin \theta$ for $-2\pi \le \theta \le 4\pi$. Give the interval(s) in which the function is
   a. positive.   **b.** negative.
   $-2\pi < \theta < -\pi, \, 0 < \theta < \pi, \, 2\pi < \theta < 3\pi$ $-\pi < \theta < 0, \, \pi < \theta < 2\pi, \, 3\pi < \theta < 4\pi$

3. One solution to the equation $\sin \theta = 0.85$ is $\theta \approx 1.016$. Find three other solutions to this equation.
   sample: $1.016 + 2\pi \approx 7.299, \, 1.016 - 4\pi \approx -11.550, \, 1.016 + 10\pi \approx 32.432$

4. **a.** Make a table of values for $y = \cos \theta$, using $\theta = 0, \dfrac{\pi}{6}, \dfrac{\pi}{4}, \dfrac{\pi}{3}, \dfrac{\pi}{2}$, and their positive integral multiples less than 2π.
   b. Plot $y = \cos \theta$ for $0 \le \theta \le 2\pi$ on graph paper.
   c. Graph $y = \cos \theta$ for $-2\pi \le \theta \le 4\pi$ on an automatic grapher.
   d. How many cycles of the curve in part **b** appear in the graph in part **c**?
   a,b,c) See margin. d) 3 cycles

5. Find three solutions of the equation $\cos \theta = 0.315$.
   sample: $\theta \approx 1.250, \, \theta \approx 1.250 + 2\pi, \, \theta \approx 1.250 - 4\pi$

6. **a.** Graph $y = \tan \theta$ on an automatic grapher over the interval $-2\pi \le \theta \le 4\pi$.   **See margin.**
   b. On which intervals in $-2\pi \le x \le 4\pi$ is $\tan x$ positive?   **See margin.**

In 7–11, name the function(s)—sine, cosine, or tangent—for which the property is true.

7. The domain is the set of all real numbers.   **sine, cosine**

8. The range is the set of all real numbers.   **tangent**

9. The function contains the point $\left(\dfrac{\pi}{2}, 1\right)$.   **sine**

10. The function contains the point $\left(\dfrac{\pi}{4}, 1\right)$.   **tangent**

11. The graph is symmetric to the y-axis.   **cosine**

In 12 and 13, find all solutions in the interval $0 \le x \le 4\pi$ to the nearest 0.001.

12. $\tan x = -0.3$
    $\approx 2.850, \approx 5.992, \approx 9.133, \approx 12.275$

13. $\tan x = 3$
    $\approx 1.249, \approx 4.391, \approx 7.532, \approx 10.674$

Applying the Mathematics

14. The graph at the right gives normal blood pressure ranges. The changes in pressure from systolic (from heart to arteries) to diastolic (from veins to heart) create the pulse. For this graph find
    a. the range $\{y: 80 \le y \le 120\}$
    b. the maximum and minimum values
    c. the period.   **0.8 seconds**
    b) maximum = 120; minimum = 80

302

15. Describe the translation with smallest positive magnitude that maps the graph of $f(x) = \sin x$ onto that of $g(x) = \cos x$. $(x, y) \rightarrow (x - \frac{\pi}{2}, y)$

16. Is the function $f(x) = \lfloor x \rfloor$ periodic? Why or why not?   **No. There is no number p such that $f(x + p) = f(x)$ for all values of x.**

17. **a.** Graph $f(\theta) = 2 \sin \theta$ using an automatic grapher.
    b. On the same set of axes, graph $g(\theta) = \sin 2\theta$.
    c. Compare and contrast the graphs of f and θ to the graph of $y = \sin \theta$.
    See margin.

Review

In 18 and 19, A is a point on the unit circle with center O at the origin. Find the coordinates of A for the given value of θ. *(Lessons 5-5, 5-4)*

18.

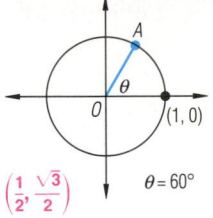

$\left(\frac{1}{2}, \frac{\sqrt{3}}{2} \right)$ $\theta = 60°$

19.

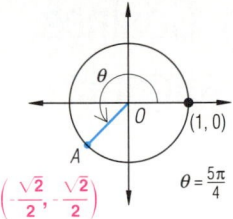

$\left(-\frac{\sqrt{2}}{2}, -\frac{\sqrt{2}}{2} \right)$ $\theta = \frac{5\pi}{4}$

In 20 and 21, give the exact value without using a calculator.   *(Lesson 5-5)*

20. $\cos \frac{3\pi}{4}$ $-\frac{\sqrt{2}}{2}$

21. $\sin \frac{11\pi}{6}$ $-\frac{1}{2}$

22. A small drive wheel A, 2 cm in diameter, is connected with a pulley to a large wheel B, 13 cm in diameter. How many revolutions of A are needed to produce a $\frac{\pi}{2}$ radian rotation in B? *(Lesson 5-2)* $1\frac{5}{8}$ **revolutions**

In 23 and 24, give an equation for the image of the graph of the given equation under the given scale change. *(Lesson 3-5)*

23. $y = 2\sqrt{x} - 3$, for $S(x, y) = \left(2x, \frac{1}{3}y \right)$ $y = \frac{\sqrt{2x}}{3} - 1$

24. $y = e^{x - 4}$, for $S(x, y) = (1300x, 0.12y)$ $y = 0.12e^{x/1300 - 4}$

Exploration

25. The number of hours of daylight in cities in the United States varies over the year. Use an almanac to find the daylight hours for a city for one day each month. Graph this data using day of year as x-value and number of daylight hours as y-value. What kind of function seems to model these data?   **sine function**

FOLLOW-UP

MORE PRACTICE
For more questions on SPUR Objectives, use *Lesson Master 5-6,* shown below.

EXTENSION
This is an extension of **Question 16.** Ask students whether the following graph shows a periodic function. (Yes) Is it a step function? (No)

NAME _____

LESSON **MASTER   5-6**
QUESTIONS ON **SPUR** OBJECTIVES

■**PROPERTIES** *Objective F (See pages 334–338 for objectives.)*

1. For what values of θ in $0 \le \theta \le 4\pi$ is $\cos \theta = 0$? $\frac{\pi}{2}, \frac{3\pi}{2}, \frac{5\pi}{2}, \frac{7\pi}{2}$

In 2 and 3, *true* or *false*, given $f(x) = \sin x$.

2. f is an even function.   **False**   3. The minimum value of f is -1.   **True**

4. *Multiple choice.* For what values of θ between 0 and 2π is $\sin \theta > 0$ and $\cos \theta < 0$?
   (a) $0 < \theta < \frac{\pi}{2}$ (b) $\frac{\pi}{2} < \theta < \pi$
   (c) $\pi < \theta < \frac{3\pi}{2}$ (d) $\frac{3\pi}{2} < \theta < 2\pi$ **(b)**

■**PROPERTIES** *Objective G*

5. For what values of x between 0 and 2π are both $\sin x$ and $\tan x$ negative? $\frac{3\pi}{2} < x < 2\pi$

■**REPRESENTATIONS** *Objective L*

6. Consider the function $b(x) = \cos x$, where x is measured in radians.
   a. Sketch a graph of b for $-\pi \le x \le 2\pi$.
   b. What is the period of b? 2π

7. Let $g(\theta) = \tan \theta$, where θ is in degrees.
   a. Sketch a graph of $y = g(\theta)$ for $-180° \le \theta \le 360°$.
   b. State its period. $180°$

Functions, Statistics, and Trigonometry © Scott, Foresman and Company   **59**

Properties of Sines, Cosines, and Tangents

Some properties of trigonometric functions follow rather quickly from their definitions. Four properties are derived in this lesson.

Remember that $(\cos \theta, \sin \theta)$ is defined to be the image of the point $(1, 0)$ under a rotation of θ around the origin.

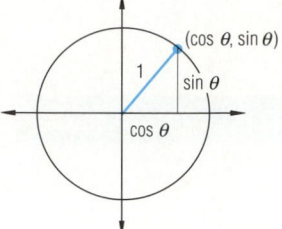

An equation for the unit circle is $x^2 + y^2 = 1$. Because, for every θ, the point $(\cos \theta, \sin \theta)$ is on the unit circle, the coordinates must satisfy the equation of the circle. Thus every number θ satisfies the equation $(\cos \theta)^2 + (\sin \theta)^2 = 1$. It is customary to write $\cos^2 \theta$ and $\sin^2 \theta$ for $(\cos \theta)^2$ and $(\sin \theta)^2$, respectively. This argument thus proves a theorem called the Pythagorean Identity.

Pythagorean Identity

For every θ, $\cos^2 \theta + \sin^2 \theta = 1$.

An **identity** is an equation that is true for all values of the variable(s) for which the expressions are defined. The name of the above identity comes from the Pythagorean Theorem, because in the first quadrant, as shown above, $\cos \theta$ and $\sin \theta$ are the sides of a right triangle with hypotenuse 1. Among other things, the Pythagorean Identity enables you to obtain one of $\cos \theta$ or $\sin \theta$ if you know the other.

Example 1 If $\cos \theta = $, find $\sin \theta$.

Solution Substitute into the Pythagorean Identity.

$$\left(\frac{3}{5}\right)^2 + \sin^2 \theta = 1$$

$$\frac{9}{25} + \sin^2 \theta = 1$$

$$\sin^2 \theta = \frac{16}{25}$$

$$\sin \theta = \pm \frac{4}{5}$$

There are two possible values of $\sin \theta$, $\frac{4}{5}$ or $-\frac{4}{5}$.

304

Check 1 Refer to the unit circle. There are two points whose x-coordinate ($\cos\theta$) is $\frac{3}{5}$. One is in the first quadrant, in which case the y-coordinate ($\sin\theta$) is $\frac{4}{5}$. The other is in the fourth quadrant, where the sine is $-\frac{4}{5}$.

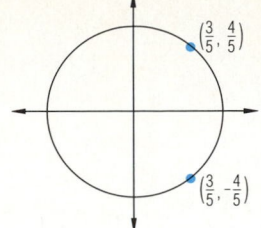

Check 2 Use your calculator to find the acute angle θ whose cosine is 0.6. In degrees, $\theta \approx 53.16°$. Now find the sine of this angle. $\sin 53.16° \approx 0.8$. (In radians, $\theta \approx 0.927$ and again $\sin 0.927 \approx 0.8$.) So the positive value for $\sin\theta$ checks.

A second property relates a trigonometric function of θ and the same function of $-\theta$. In the unit circle below, let P be the image of $(1, 0)$ under a rotation of θ and Q be its image under a rotation of $-\theta$. P and Q are reflection images of each other over the x-axis, so their x-coordinates (cosines) are equal, but their y-coordinates (sines) are opposites. It follows that the ratios of the y-coordinate to the x-coordinate (tangents) are opposites.

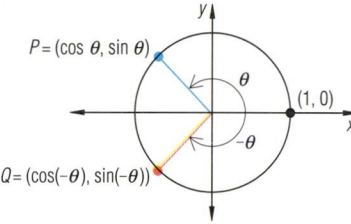

This proves the following.

Opposites Theorem

For every θ, $\quad \sin(-\theta) = -\sin\theta$,
$\cos(-\theta) = \cos\theta$, and
$\tan(-\theta) = -\tan\theta$.

Recall that an odd function is one for which $f(-x) = -f(x)$ for all x and an even function is one for which $f(-x) = f(x)$ for all x. Thus, the Opposites Theorem can be stated using the language of functions: the sine and tangent functions are odd functions, and the cosine function is an even function. The sine and tangent functions are symmetric to the origin; the cosine function is symmetric to the y-axis. You can verify this by looking back at the graphs of these functions in Lesson 5-6.

---

can also use symmetry to support the Complements Theorem as follows:

The coordinates of R are related to the coordinates of S. $R = (\cos 16°, \sin 16°)$ and $S = (\cos 74°, \sin 74°)$. But, by reflecting R over $y = x$, we can conclude that $S = (\sin 16°, \cos 16°)$. Thus, $\cos 74° = \sin 16°$ and $\sin 74° = \cos 16°$. You might mention that, before the days of calculators, when everyone used tables to find values of trigonometric functions, the Complements Property made it possible for the table to include only values of θ such that $0 \le \theta \le \frac{\pi}{4}$ or $0 \le \theta \le 45°$, rather than for values for the entire first quadrant.

If you discuss **Examples 2 and 4** in class, refer to the unit circle and use symmetry to verify the results. This approach is helpful for students who are visual learners.

ADDITIONAL EXAMPLES
1. If $\sin\theta = \frac{1}{3}$ find $\cos\theta$.

$\frac{2\sqrt{2}}{3}$ or $-\frac{2\sqrt{2}}{3}$

2. If $\sin x = 0.681$, find $\sin(-x)$ and $\sin(\pi - x)$.

-0.681, 0.681
3. Explain graphically why $\sin(\pi - \theta) = \sin\theta$ for all θ.

Graphing each function, $y = \sin(\pi - \theta)$ and $y = \sin\theta$ have graphs that coincide. On the unit circle, the point $(\cos(\pi - \theta), \sin(\pi - \theta))$ is the reflection image of $(\cos\theta, \sin\theta)$ over the y-axis. Since the two points have the same second coordinate, $\sin(\pi - \theta) = \sin\theta$.

Question 2: Students should realize that there are two possible values of sin u as stated in this question. Later, this will not usually be stated in the question.

Questions 15–17: Before the days of automatic graphers, it would be unreasonable to have all these questions in the same lesson because they would be so time consuming.

Question 18: The given equation *is* true for multiples of π. This gives you a chance to reinforce the fact that for a statement to be an identity it must be true for *all* values of the variables for which the expressions are defined. You might want to have your students use an automatic grapher and graph $y = \sin (\pi - \theta)$ and $y = \sin \pi - \sin \theta$ and determine for themselves if there are any values of θ for which the statement is valid.

Question 27: This question utilizes the first few terms of the infinite series for sin x.

The third property relates trigonometric functions of θ and $\pi - \theta$. Let $P = (\cos \theta, \sin \theta)$, and let Q be the reflection image of P over the y-axis, as in the diagram at the right. Q and P have opposite first coordinates and equal second coordinates. Recall from geometry that reflections preserve angle measure, so

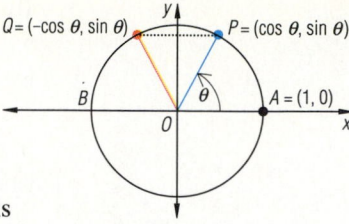

$$m\angle QOB = \theta.$$
Also, $$m\angle AOB = \pi.$$
Thus, $$m\angle AOQ = m\angle AOB - m\angle QOB$$
$$= \pi - \theta.$$

So, by the definitions of cosine and sine in Lesson 5-4,
$$Q = (\cos (\pi - \theta), \sin (\pi - \theta)).$$

Because Q is the image of P under a reflection over the y-axis, it is also true that $Q = (\text{-}\cos \theta, \sin \theta)$. Thus,
$$(\cos (\pi - \theta), \sin (\pi - \theta)) = (\text{-}\cos \theta, \sin \theta).$$
Therefore $$\cos (\pi - \theta) = \text{-}\cos \theta,$$
and $$\sin (\pi - \theta) = \sin \theta.$$
Dividing the second quantity by the first gives
$$\tan (\pi - \theta) = \text{-}\tan \theta.$$

This completes the proof of the following theorem.

Supplements Theorem

For every θ, $\quad \sin (\pi - \theta) = \sin \theta,$
$\qquad\qquad\quad \cos (\pi - \theta) = \text{-}\cos \theta,$
and $\qquad\qquad \tan (\pi - \theta) = \text{-}\tan \theta$

When this theorem is written using degree measures it may be more obvious why it is called the Supplements Theorem:

$$\sin (180° - \theta) = \sin \theta;$$
$$\cos (180° - \theta) = \text{-}\cos \theta; \text{ and}$$
$$\tan (180° - \theta) = \text{-}\tan \theta.$$

In words, supplementary angles have the same sine, opposite cosines, and opposite tangents.

Example 2 Suppose $\cos \theta = 0.15$. Evaluate without using a calculator.

a. $\cos (\text{-}\theta)$ **b.** $\cos (\pi - \theta)$

Solution

a. By the Opposites Theorem, $\cos (\text{-}\theta) = \cos \theta = 0.15$.
b. By the Supplements Theorem, $\cos (\pi - \theta) = \text{-}\cos \theta = \text{-}0.15$.

306

The fourth property relates trigonometric function values of θ and $\frac{\pi}{2} - \theta$.

Example 3 Verify graphically that for all θ, $\sin\left(\frac{\pi}{2} - \theta\right) = \cos\theta$.

Solution Graph $y = \sin\left(\frac{\pi}{2} - x\right)$ and $y = \cos x$. If graphed on the same window they will coincide for all x. Below are separate graphs of the two functions, showing that they are equal for all x.

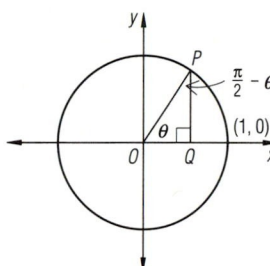

To prove the property of Example 3 for values of θ between 0 and $\frac{\pi}{2}$, consider the unit circle at the left. In right $\triangle PQO$, $\angle POQ$ and $\angle OPQ$ are complementary. So if $m\angle POQ = \theta$, then $m\angle OPQ = \frac{\pi}{2} - \theta$.

$$\sin\theta = \frac{PQ}{OP} = \cos\left(\frac{\pi}{2} - \theta\right)$$

$$\cos\theta = \frac{OQ}{OP} = \sin\left(\frac{\pi}{2} - \theta\right)$$

These equations yield the following identity which also holds for all other real values of θ.

Complements Theorem

For every θ, $\sin\left(\frac{\pi}{2} - \theta\right) = \cos\theta$,

and $\cos\left(\frac{\pi}{2} - \theta\right) = \sin\theta$.

In degree notation, the Complements Theorem states that
$$\sin(90° - \theta) = \cos\theta \text{ and}$$
$$\cos(90° - \theta) = \sin\theta.$$
You can verify the Supplements and Complements Theorems with your calculator. For instance, in radian mode, check that $\sin(\pi - 2) = \sin 2 \approx 0.909$. In degree mode, check that $\cos 16° = \sin(90° - 16°) = \sin 74° \approx 0.961$.

3.

15.a.

b. The two graphs are symmetric to each other over the y-axis and over the x-axis.

16.a.

b. The two graphs are symmetric to each other over the x-axis.

Example 4 Given $\sin \theta = 0.875$ and $-\frac{\pi}{2} < \theta < 0$, find
  a. $\sin (-\theta)$; **b.** $\sin (\pi - \theta)$; **c.** $\cos \left(\frac{\pi}{2} - \theta \right)$.

Solution
  a. Use the Opposites Theorem: $\sin (-\theta) = \text{-} \sin \theta = 0.875$.
  b. Use the Supplements Theorem: $\sin (\pi - \theta) = \sin \theta = 0.875$.
  c. Use the Complements Theorem: $\cos \left(\frac{\pi}{2} - \theta \right) = \sin \theta = 0.875$.
  Since these identities hold for all values of θ, the restriction $-\frac{\pi}{2} < \theta < 0$ has no effect on the solutions.

Questions

Covering the Reading

1. *True* or *false*. For all θ, $\cos^2 \theta = 1 - \sin^2 \theta$. **True**

2. If $\cos \theta = \frac{24}{25}$, what are two possible values of $\sin \theta$? $\pm \frac{7}{25}$

3. Draw a unit circle and an angle θ to illustrate $\cos(40°) = \cos(\text{-}40°)$. **See margin.**

4. *True* or *false*. $f(\theta) = \sin \theta$ is an even function. **False**

In 5 and 6, $\sin \theta = \frac{3}{7}$ and $\cos \theta > 0$. Evaluate without using a calculator.

5. $\sin (\text{-}\theta)$ $\text{-}\frac{3}{7}$ 6. $\cos (\theta + 2\pi)$ $\frac{2\sqrt{10}}{7}$

In 7 and 8, $\cos \theta = \frac{2}{9}$ and $0 < \theta < \frac{\pi}{2}$. Evaluate without using a calculator.

7. $\cos (180° - \theta)$ $\text{-}\frac{2}{9}$ 8. $\sin (90° - \theta)$ $\frac{2}{9}$

In 9 and 10, if $\tan \theta = 0.375$ and $0 < \theta < \frac{\pi}{2}$, calculate

9. $\tan (\text{-}\theta)$ $\text{-}0.375$ 10. $\tan (\pi - \theta)$. $\text{-}0.375$

In 11–14, **a.** *true* or *false*. **b.** If true, what theorem supports your answer?

11. $\sin (\text{-}120°) = \sin (120°)$
  a) False

12. $\cos \left(\frac{\pi}{6} \right) = \cos \left(\frac{5\pi}{6} \right)$
  a) False

13. $\tan \left(-\frac{5\pi}{4} \right) = \text{-}\tan \left(\frac{5\pi}{4} \right)$
  a) True; b) Opposites Theorem

14. $\tan 425° = \tan (\text{-}245°)$
  a) False

Applying the Mathematics

In 15–17, use an automatic grapher.

15. **a.** Plot $y = \tan x$ and $y = \tan(\text{-}x)$ on the same set of axes.
  b. Describe the relation between the two graphs.
  c. What theorem in this lesson is illustrated by the graphs in part **a**?
  a,b) See margin. c) tan(-θ) = -tan θ (Opposites Theorem)

16. **a.** Plot $y = \cos x$ and $y = \cos(\pi - x)$ on the same set of axes.
  b. Describe the relation between the two graphs.
  c. What theorem in this lesson is illustrated by the graphs in part **a**?
  a,b) See margin. c) cos(π - x) = -cos x (Supplements Theorem)

17. **a.** Plot $y = \sin x$ and $y = \cos \left(\frac{\pi}{2} - x \right)$ on the same set of axes.
  b. Describe the relation between the two graphs.
  c. What theorem in this lesson is illustrated by the graphs in part **a**?
  a,b) See margin. c) cos $\left(\frac{\pi}{2} - x \right)$ = sin x (Complements Theorem)

18. Give a counterexample to prove that $\sin(\pi - \theta) = \sin\pi - \sin\theta$ is *not* an identity.   **See margin.**

In 19 and 20, simplify.

19. $\cos x + \cos(\pi - x)$ **0**

20. $\sin\theta + \sin(\pi - \theta)$ **2sin θ**

Review

21. Find two values of θ (in radians) for which $\sin\theta = .1999$. *(Lesson 5-6)*
   sample: θ ≈ 0.2013 and θ ≈ 2.9403

22. The function pictured below, $y = f(x)$, has values of 1 at even integral values of x; 0 at odd integral values; and segments connect the points at successive integral values of x.

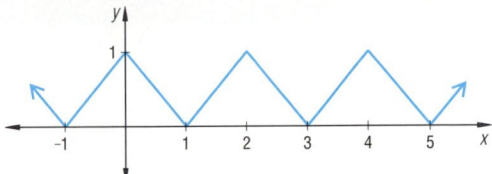

   a. What is the period of this function?   **2**

   b. Sketch the graph of $y = f\left(x - \dfrac{1}{2}\right)$.   *(Lessons 5-6, 3-2)*   **See margin.**

In 23 and 24, give exact values.

23. $\cos\left(-\dfrac{\pi}{6}\right)$ $\dfrac{\sqrt{3}}{2}$

24. $\sin\left(\dfrac{2\pi}{3}\right)$ *(Lesson 5-5)* $\dfrac{\sqrt{3}}{2}$

25. a. Calculate the coordinates of the image of $(1, 0)$ under a rotation about $(0, 0)$ of $-137°$.   **about (-0.7314, -0.6820)**

   b. Find $\tan(-137°)$ using your answer to part **a**. *(Lesson 5-4)*
   ≈ 0.9325

26. A kite is attached to 300 ft of string, which makes a 42° angle to the horizon. The end of the string is being held 4 ft above ground.

   a. About how high is the kite?   **≈ 205 ft**

   b. What assumption did you need to make about the string to answer part **a**?   **The string is straight.**

   c. The assumption in part **b** is in fact not true. How does this affect your answer to part **a**? *(Lesson 5-3)*   **Since the string would actually be curved, the kite might be lower than 205 ft.**

Exploration

27. For part of its domain the sine function can be approximated by the function $f(x) = x - \dfrac{1}{6}x^3 + \dfrac{1}{120}x^5$, where x is in radians.

   a. Use your function grapher to compare the graph of $y = \sin x$ to the graph of $y = x - \dfrac{1}{6}x^3 + \dfrac{1}{120}x^5$. For what interval of x-values are the two graphs close to each other?   **sample: -2 ≤ x ≤ 2**

   b. Compare the graph of $y = x - \dfrac{1}{6}x^3 + \dfrac{1}{120}x^5 - \dfrac{1}{5040}x^7$ to the graph of $y = \sin x$. For what interval of x-values are the two graphs close to each other?   **sample: -3 ≤ x ≤ 3**

FOLLOW-UP

MORE PRACTICE
For more questions on SPUR Objectives, use *Lesson Master 5-7*, shown below.

EXTENSION
Small Group Work   The hyperbolic sine, hyperbolic cosine, and hyperbolic tangent are functions which are defined similarly to their corresponding trigonometric functions, except the hyperbola $x^2 - y^2 = 1$ in place of the unit circle. You might have small groups of students try to derive properties of the hyperbolic functions similar to the Pythagorean Identity and Opposites Theorem of this lesson. ($\cosh^2\theta - \sinh^2\theta = 1$; $\sinh(-\theta) = -\sinh\theta$; $\cosh(-\theta) = \cosh\theta$.)

PROJECTS
The projects for Chapter 5 are described on pages 329–330. Project 5 is related to the content of this lesson.

EVALUATION
A quiz covering Lessons 5-4 through 5-7 is provided in the Teacher's Resource File.

NAME _____

OBJECTIVES

D Use trigonometry to find sides or angles in triangles.
G Use theorems about sines, cosines, and tangents.
J Solve problems involving laws of sines and cosines.

TEACHING NOTES

This lesson and the following one on the Law of Sines round out the applications of trigonometric functions to solving triangle problems. Most students find both lessons relatively easy, perhaps because the applications of these theorems are rather straightforward. This will give you a chance to review some of the more difficult problems of previous lessons in this chapter.

The proof of the Law of Cosines is an elegant one in its use of transformations to justify the coordinates of *A*. (A teaching aid with the diagram needed for the proof is provided.) Many books require the construction of a perpendicular from *A* to the *x*-axis and the use of right-triangle definitions of sine and cosine to find the coordinates. Students in this course should be able to accept the scale change (more precisely a *size* change) that maps point *D* to point *A*. You might want to mention briefly that in this proof, we have derived rectangular coordinates for a point with polar coordinates.

The Law of Cosines

Baseball diamond at Wrigley Field, Chicago, Illinois

The bases on a baseball diamond are vertices of a square of side 90 ft. From the Pythagorean Theorem, you know that the distance from second base (*S*) to home plate (*H*) is

$$SH = \sqrt{90^2 + 90^2}$$
$$= \sqrt{16200}$$
$$\approx 127 \text{ ft.}$$

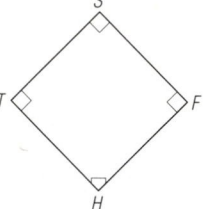

So, a ball thrown from second base to home plate would travel a distance along the ground of about 127 ft.

You did not need trigonometry to find *SH*. But now, suppose a player is 10 feet behind second base. How far does the player have to throw to get the ball to first base (*F*)? It is possible to find this distance by using a powerful theorem called the *Law of Cosines*.

In the statement of the Law of Cosines, we use the standard procedure of indicating a segment length in a triangle by the small letter corresponding to the capital letter of the opposite vertex.

Theorem (Law of Cosines)

In any $\triangle ABC$, $c^2 = a^2 + b^2 - 2ab \cos C$.

Notice the power of the Law of Cosines. Given two sides and the included angle of a triangle (the SAS condition), you can find the third side. Before proving this theorem, here is how to use it to find out how far the player has to throw the baseball.

310

Example 1 A player is 10 feet behind second base. How far is it from that position to first base?

Solution In $\triangle PSF$, $PS = 10$, $SF = 90$, and $m\angle PSF = 135°$. Since two sides and the included angle are given, the Law of Cosines can be used. For this situation,
$$PF^2 = PS^2 + SF^2 - 2 \cdot PS \cdot SF \cdot \cos S$$
$$= 10^2 + 90^2 - 2 \cdot 10 \cdot 90 \cdot \cos 135°$$
$$\approx 8200 - 1800 \cdot (-.707)$$
$$\approx 9473.$$
So $PF \approx \sqrt{9473} \approx 97$ feet.

Check Since $\angle PSF$ is obtuse, PF should be the longest side of $\triangle PSF$ and, by the Triangle Inequality, $PF < 100$. So the answer is reasonable.

Here is a proof of the Law of Cosines.

Proof Impose a coordinate plane on $\triangle ABC$ so that $C = (0, 0)$ and $B = (a, 0)$. Let D be the intersection of $\overrightarrow{AC}$ and the unit circle. Then $D = (\cos C, \sin C)$. Since $AC = b$, A can be considered the image of D under the scale change $(x, y) \rightarrow (bx, by)$, so $A = (b \cos C, b \sin C)$.

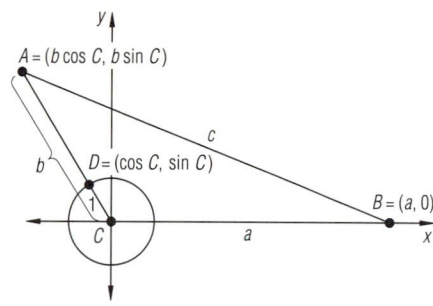

By the distance formula, $c = \sqrt{(b \cos C - a)^2 + (b \sin C - 0)^2}$.
The rest of the proof involves rewriting the equation so that it has the form in which it appears in the theorem.

Square both sides. $\qquad c^2 = (b \cos C - a)^2 + (b \sin C - 0)^2$

Expand the binomials. $\qquad c^2 = b^2\cos^2 C - 2ab \cos C + a^2 + b^2 \sin^2 C$

Apply the commutative
property of addition. $\qquad c^2 = a^2 + b^2 \sin^2 C + b^2 \cos^2 C - 2ab \cos C$

Factor. $\qquad c^2 = a^2 + b^2(\sin^2 C + \cos^2 C) - 2ab \cos C$

Use the Pythagorean
Identity. $\qquad c^2 = a^2 + b^2 - 2ab \cos C$

Example 1 shows how the Law of Cosines can be used given SAS. The Law of Cosines can also be used to find the measure of any angle of a triangle if the lengths of all three sides are known (the SSS condition).

Emphasize that the Law of Cosines is applicable to *any* triangle whereas the Pythagorean Theorem can only be used in a right triangle.

Error Analysis The most common error in applying the Law of Cosines is to add instead of subtract (or vice versa), particularly when $\cos C$ is negative. Also watch for an order of operations error, subtracting $2ab$ before multiplying by $\cos C$.

If students ask why one would want to find $m\angle C$ in **Example 2,** you might challenge them to find a reason. Perhaps the construction crew for the walkways in the drawing needs to know how to set up guidelines for the bulldozers to follow, or the building of a small wall around the plot requires creating an appropriately shaped cornerstone.

The last paragraph of the lesson should be qualified. The first *known* appearance of the Law of Cosines is in Euclid's *Elements.*

ADDITIONAL EXAMPLES
1. A parallelogram has a 70° angle and sides of length 10 in. and 15 in. What is the length of its longer diagonal?
$\sqrt{325-300 \cos 110°} \approx$ 20.68 in.

2. To the nearest degree, find the measure of the smallest angle in a triangle whose sides have lengths 4 cm, 5 cm, and 6 cm.
$\cos C = .75$, $C \approx 41°$

■ ■ ■ ■ ■ ■

Example 2   A triangular plot of land is to be used as a garden. The measures of the sides of the plot are $AB = 8$ m, $BC = 12$ m, and $AC = 18$ m. Find the measure of **a.** ∠C and **b.** ∠B.

Solution

a. You are given $a = 12$, $b = 18$, and $c = 8$. You need to find m∠C.

By the Law of Cosines, $c^2 = a^2 + b^2 - 2ab\cos C$.
Substitute. $8^2 = 12^2 + 18^2 - 2(12)(18)\cos C$
Solve. $-404 = -432\cos C$
    $.935 \approx \cos C$
    $m\angle C \approx 20.7°$

b.
By the Law of Cosines, $b^2 = a^2 + c^2 - 2ac\cos B$.
Substitute. $18^2 = 12^2 + 8^2 - 2(12)(8)\cos B$
Solve. $116 = -192\cos B$
    $-0.604 \approx \cos B$
    $m\angle B \approx 127.2°$

Check   The smallest angle of the triangle should be opposite the shortest side. The largest angle should be opposite the longest side. $m\angle A \approx 180° - 127.2° - 20.7° = 32.1°$. So, ∠B is the largest and ∠C is the smallest; they are opposite the largest and smallest sides, respectively.

The first appearance of the Law of Cosines is in Book II of Euclid's *Elements*, written around 300 B.C. It can be considered as a generalization of the Pythagorean Theorem. Notice that if m∠C = 90° in △ABC, then
$$c^2 = a^2 + b^2 - 2ab\cos C.$$
But cos 90° = 0, so $c^2 = a^2 + b^2$.

Questions

Covering the Reading

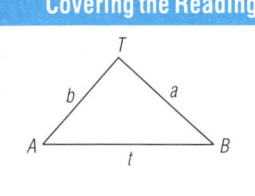

1. Refer to the baseball example at the start of the lesson. Suppose that an infielder is standing 20 feet behind second base. To the nearest foot, how far is the player from third base?   ≈ **105 ft**

2. Refer to the triangle at the left.
   a. Write a formula for *t* in terms of ∠T, *a*, and *b*.
   b. Suppose $a = 6$, $b = 5$, and m∠T = 120°. Find the exact length *t*.
   a) $t = \sqrt{a^2 + b^2 - 2ab \cdot \cos T}$; b) $\sqrt{91}$

312

In 3 and 4, refer to the proof of the Law of Cosines.

3. **a.** Find the slope of $\overline{CD}$. **tan C**
   b. Find the slope of $\overline{CA}$. **tan C**
   c. Use the results of parts **a** and **b** to show that A, D, and C are collinear. **See margin.**

4. Use the distance formula to show that $AC = b$. **See margin.**

5. Refer to Example 2. Use the Law of Cosines to find m∠A. **≈ 32.1°**

6. In △XYZ, $x = 10$, $y = 15$, and $z = 20$. Find
   a. m∠Z **≈ 104.5°**   **b.** m∠X. **≈ 29.0°**

In 7–10, *multiple choice*. Which formula can be used to find the required measure directly?

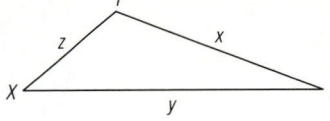

(a) $x^2 = y^2 + z^2 - 2yz \cos X$
(b) $y^2 = x^2 + z^2 - 2xz \cos Y$
(c) $z^2 = x^2 + y^2 - 2xy \cos Z$
(d) none of (a)–(c)

7. Given: x, y, z.
   Find: m∠X. **(a)**

8. Given: x, y, m∠X.
   Find: z. **(a)**

9. Given: x, y, m∠Z.
   Find: z. **(c)**

10. Given: x, y, m∠Z.
    Find: m∠X. **(d)**

Applying the Mathematics

11. Find the length of a diagonal of a regular pentagon which has a side of length 10. **≈ 16.2**

12. The sides of a triangle are of lengths 5.9 m, 2.8 m, and 5.9 m. Find the measure of the smallest angle. **≈ 27.45°**

13. Prove that there is no triangle with sides 5 cm, 9 cm, and 15 cm using
    a. the Law of Cosines;   **b.** the Triangle Inequality. **See margin.**

14. A sign at a mountain overlook tells a hiker (H) that she is 5.9 km from a microwave tower (M) and 7.8 km from the highest visible peak (P). She estimates that the angle between M and P from her position to be 40°. Find the horizontal distance between M and P.
    ≈ 5.0 km

Review

15. Name a property or theorem that justifies the statement $\sin \frac{\pi}{12} = \sin \frac{25\pi}{12}$.
    (Lesson 5-7) **Periodicity Theorem**

16. At the right the circle with radius r has central angle θ measured in radians. Prove that the area A of the shaded region is $A = \frac{1}{2}r^2(\theta - \sin \theta)$.
    (Lessons 5-3, 5-2)

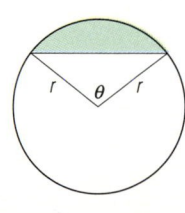

Area = (Area of sector) − (Area of triangle)
$$= \frac{1}{2}r^2\,\theta - \frac{1}{2}r^2 \sin \theta$$
$$= \frac{1}{2}r^2\,(\theta - \sin \theta)$$

LESSON 5-8 The Law of Cosines   **313**

In 17 and 18, find all solutions between 0° and 360°. *(Lessons 5-7, 5-5)*

17. $\cos \theta = \dfrac{\sqrt{3}}{2}$ **30°, 330°**   **18.** $\sin \theta = 0.4567$ $\approx$ **27.17°, $\approx$ 152.83°**

In 19–21, state exact values. *(Lesson 5-5)*

19. $\sin 30°$ $\dfrac{1}{2}$ **20.** $\cos \dfrac{2\pi}{3}$ $-\dfrac{1}{2}$ **21.** $\tan(-120°)$ $\sqrt{3}$

In 22 and 23, find all solutions to the equation with $0 \le \theta \le 2\pi$. *(Lessons 5-5, 5-4)*

22. $\cos \theta = .5$ $\dfrac{\pi}{3}, \dfrac{5\pi}{3}$ **23.** $\tan \theta = 1$ $\dfrac{\pi}{4}, \dfrac{5\pi}{4}$

24. A plane flying at 32,000 ft starts a steady descent 100 miles from an airport A. At what angle θ does the plane descend? *(Lesson 5-3)* $\approx$ **3.5°**

25. Find the area of $\triangle ABC$ with sides and angles as shown at the right. *(Lesson 5-3)* $\approx$ **10.34 square units**

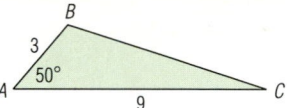

Exploration

26. Let a, b, and c be the sides of a triangle, and suppose that c is the longest side.
  a. If $c^2 > a^2 + b^2$, what is true about $\triangle ABC$?
  b. If $c^2 < a^2 + b^2$, what can you conclude? $\triangle ABC$ **is acute.**
  c. Prove your conjectures in parts **b** and **c**.
  a) $\triangle ABC$ **is obtuse, with obtuse $\angle C$. c) See margin.**

314

314

LESSON 5-9

The Law of Sines

In Question 15 of Lesson 5-3, you were asked to prove that the area K of a triangle is equal to one half the product of the length of any two sides and the sine of their included angle. So in $\triangle ABC$,

$$K = \tfrac{1}{2}bc \sin A,$$

and $K = \tfrac{1}{2}ac \sin B,$

and $K = \tfrac{1}{2}ab \sin C.$

Because the area of a given triangle is constant,

$$\tfrac{1}{2}bc \sin A = \tfrac{1}{2}ac \sin B = \tfrac{1}{2}ab \sin C.$$

From these equations comes an important and quite elegant result.

Multiply by 2.

$$bc \sin A = ac \sin B = ab \sin C$$

Divide by abc.

$$\frac{bc \sin A}{abc} = \frac{ac \sin B}{abc} = \frac{ab \sin C}{abc}$$

Simplify.

$$\frac{\sin A}{a} = \frac{\sin B}{b} = \frac{\sin C}{c}$$

This argument proves that in any triangle, the ratio of the sine of an angle to the length of the opposite side is constant. This important property is called the *Law of Sines*.

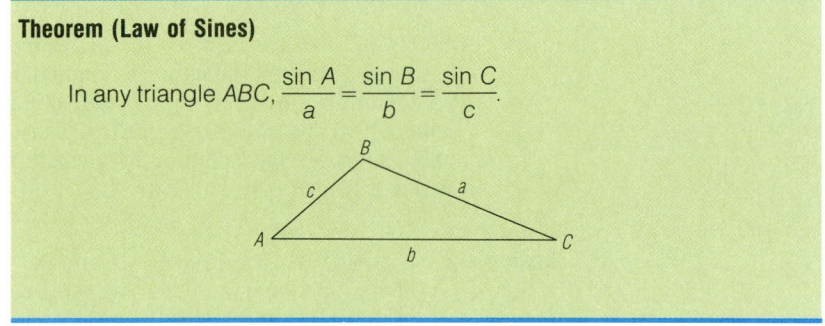

Theorem (Law of Sines)

In any triangle ABC, $\dfrac{\sin A}{a} = \dfrac{\sin B}{b} = \dfrac{\sin C}{c}$.

You can use the Law of Sines to find the length of a second side of a triangle given the measures of one side and two angles (the ASA or AAS conditions).

LESSON 5-9 The Law of Sines **315**

LESSON 5-9

RESOURCES
■ Lesson Master 5-9

OBJECTIVES

D Use trigonometry to find sides or angles in triangles.
G Use theorems about sines, cosines, and tangents.
J Solve problems involving laws of sines and cosines.

TEACHING NOTES

Before beginning this lesson, you might list the conditions that lead to triangle congruence and associate them with the corresponding theorem, as in the following table.

Condition	Theorem
SAS	Law of Cosines
SSS	Law of Cosines
AAS	Law of Sines
ASA	Law of Sines
SsA	Law of Sines
SSA	Law of Sines (possibly ambiguous)

Because of the potential for ambiguity, many geometry courses dismiss SSA triangle congruence as a theorem, so your students may not accept the possible unique solutions to a SSA problem using the Law of Sines. This will be reinforced by **Example 2,** which shows two different solutions to such a problem. In **Example 3,** when the side opposite the given angle is longer than the other given side, there is a unique triangle and no ambiguity is possible. This situation is labeled SsA and is a triangle congruence theorem, where S means the longer of the two sides with lengths S and s.

There are three other SSA possibilities that can occur:
1. The side opposite the given angle can be too short to even create a triangle. For example, $m\angle A = 30°$, $b = 8$ and $a = 3$. Students will get an error message on their calculators because $\sin B > 1$.
2. The side opposite the given angle can be exactly long enough to create a right triangle. For example, $m\angle A = 30°$, $b = 8$, and $a = 4$. There is a unique triangle in this case.
3. The two given sides can be equal. Again, there is a unique triangle.

We want students to realize that in a SSA case if the side opposite the given angle is longer than the other given side there is one triangle; otherwise there may be more than one triangle. The fact that there can be 0, 1, or 2 solutions should not be treated as any more unusual than the fact that a quadratic equation can have 0, 1, or 2 real solutions.

In doing the questions, encourage students to draw an accurate diagram (if none is given) as they start a question. This should help them to select an appropriate method to solve the problem. A diagram should also help students to recognize when they are dealing with an ambiguous case.

Example 1   A camper wants to find the distance f across a lake as shown. He measures the distance from the flagpole to the edge of the headquarters and measures the angles as indicated. Find f.

Solution   Use the Triangle-Sum Theorem to find the measure of $\angle B$.
$m\angle B = 180° - (110° + 40°) = 30°$
Use the Law of Sines to find f.
$$\frac{\sin F}{f} = \frac{\sin B}{b}$$
$$\frac{\sin 40°}{f} = \frac{\sin 30°}{400}$$
$$f \sin 30° = 400(\sin 40°)$$
$$f = \frac{400(\sin 40°)}{\sin 30°}$$
$$f \approx 514$$
It is about 514 meters across the lake from the beach house to headquarters.

Check   Recall that in a triangle the longer sides are opposite larger angles. The 514 m side is opposite the 40° angle, and the smaller 400 m side is opposite the (smaller) 30° angle, as it should be.

The Law of Sines can also be used to determine the measure of a second angle of a triangle when two sides and a nonincluded angle are known. This is the SSA condition. However, as you studied in geometry, in general SSA is not a condition that guarantees congruence. Thus, when the Law of Sines is used in an SSA situation, two solutions may result. Such an *ambiguous case* is illustrated in Example 2.

Example 2   In $\triangle ABC$, $m\angle A = 30°$, $AB = 8$, and $BC = 5$. Find the two possible values of $m\angle C$ (to the nearest degree) and sketch the triangles determined by these values.

Solution   Make a rough sketch. Use the Law of Sines.
$$\frac{\sin 30°}{5} = \frac{\sin C}{8}$$
Solve for $\sin C$. $\sin C = \dfrac{8 \sin 30°}{5}$
$$\sin C = 0.8$$

316

A calculator shows m∠C ≈ 53°. The triangle determined by this value is shown below.

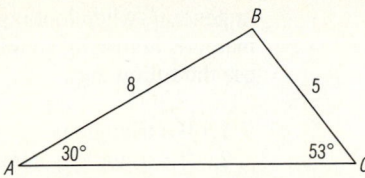

However, from the Supplements Theorem, a second angle with sine equal to 0.8 is 180° − 53° ≈ 127°. The triangle determined by this value of C is shown below.

When the side opposite the given angle is larger than the other given side, the SSA condition leads to a unique triangle. This situation is then called the SsA condition.

Example 3   In △XYZ, x = 10, z = 3, and m∠X = 42.5°. Find m∠Z to the nearest tenth of a degree.

Solution   From the Law of Sines,
$$\frac{\sin X}{x} = \frac{\sin Z}{z}.$$

So
$$\frac{\sin 42.5°}{10} = \frac{\sin Z}{3}$$

$$\sin Z = \frac{3 \sin 42.5°}{10}$$

$$\sin Z \approx 0.20268.$$

A calculator shows that
$$m∠Z \approx 11.7°.$$

From the Supplements Theorem,
$$\sin 11.7° = \sin(180° − 11.7°) = \sin 168.3°.$$

But if 168.3° were a solution to this problem, then the sum of the angles of △XYZ would be more than 180°. So m∠Z = 11.7° is the only possible solution.

As you will see in the questions, there are also SSA cases when no triangle is possible.

In general, when looking for missing angle or side measures in triangles, try methods involving simpler computations first. If these methods do not work, use the following.

1. If a triangle is a right triangle, use right triangle trigonometric ratios.
2. If a triangle is not a right triangle, consider the Law of Sines. It is useful for the ASA, AAS, and SSA conditions.
3. If the Law of Sines is not helpful, consider the Law of Cosines. The Law of Cosines is most directly applicable to SAS and SSS conditions.

Questions

Covering the Reading

1. In any triangle, the ratios of the sine of the angles to __?__ are equal.
   the lengths of the sides opposite them
2. Tell if the statement is equivalent to the Law of Sines for $\triangle ABC$.
   a. $\frac{A}{a} = \frac{B}{b} = \frac{C}{c}$ **No**

   b. $\frac{a}{\sin A} = \frac{b}{\sin B} = \frac{c}{\sin C}$ **Yes**

   c. $\frac{a}{b} = \frac{\sin A}{\sin B}, \frac{b}{c} = \frac{\sin B}{\sin C}, \frac{a}{c} = \frac{\sin A}{\sin C}$ **Yes**

   d. $ab \sin C = bc \sin A = ac \sin B$ **Yes**

In 3 and 4, use the Law of Sines to find x.

3.
   ≈ **7.3 units**

4. ≈ **12.2**

5. In $\triangle ABC$, suppose m$\angle C = 45°$, m$\angle A = 30°$, and $c = 12$. Find the lengths of the other two sides. $a = 6\sqrt{2}, b = 6 + 6\sqrt{3} \approx 16.4$

6. In $\triangle XYZ$, $x = 14$, $z = 21$, and m$\angle X = 40°$.
   a. Find the two possible measures of $\angle Z$. ≈ **74.6°**, ≈ **105.4°**
   b. Draw two noncongruent triangles satisfying the conditions of this question. **See margin.**

7. The Costas want to check the survey of a plot of land they are thinking of buying. The plot is triangular with one side on the lakefront. The other two sides are 320 ft and 452 ft long. The angle between the lakeshore and the 320 ft side measures 53°.
   a. Find the measure of the angle between the lakeshore and the 452 ft side. ≈ **34.4°**
   b. About how many feet of lake frontage does the plot have? ≈ **565 ft**

318

In 8 and 9, *true* or *false*.

8. When given SSA conditions, it is possible for two noncongruent triangles to be determined. **True**

9. When given ASA conditions, it is possible for two noncongruent triangles to be determined. **False**

In 10–15, *multiple choice*. A triangle is given with an unknown side or angle measure *x*. Which strategy for finding *x* is computationally the easiest?
   (a) definition of right triangle trigonometric ratios
   (b) Law of Sines
   (c) Law of Cosines
Do not solve for *x*.

10.

(c)

11.

(a)

12.

(b)

13.

(b)

14.

(c)

15.
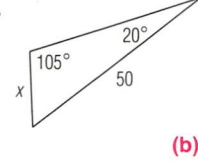
(b)

Applying the Mathematics

16. A piston driven by a radial arm is shown below. At the time that the radial arm is 80° above horizontal, what is *x*? **≈ 9.02 cm**

17. To find the height of a mountain, surveyors often find the angle of elevation to the top from two points at the same altitude a fixed distance apart. Suppose that the angles of elevation from two points 500 meters apart are 35°20′ and 25°46′. How high is the mountain above the altitude of the two points? **756 m**

18. Use the Law of Sines to prove that the base angles of an isosceles triangle are congruent. **See margin.**

19. Find the length of the third side in the triangle of Question 13. **≈ 14.2**

LESSON 5-9   The Law of Sines   **319**

$$\frac{\sin B}{AC} = \frac{\sin C}{AC}$$ **Substitution**
sin B = sin C multiplication by AC
Therefore, m∠B = m∠C or m∠B = 180° − m∠C. The second equation implies m∠B + m∠C = 180°, which is impossible. So, m∠B = m∠C.

NAME _____

LESSON **MASTER 5-9**
QUESTIONS ON **SPUR** OBJECTIVES

■**SKILLS** *Objective D (See pages 334–338 for objectives.)*
In 1 and 2, use the Law of Sines to find *x*.

1. ≈ 7.66 2. ≈ 5.21

3. In △ABC, suppose m∠C = 40°, m∠A = 30° and c = 15. Find the lengths of the two other sides. *a* = 11.67, *b* = 21.93

4. In △XYZ, x = 15, z = 22, and m∠X = 35°.
   a. Find the two possible measures of ∠Z. ≈ 57.27° and ≈ 122.73°
   b. Draw two noncongruent triangles satisfying the conditions of this question.

Triangle 1   Triangle 2

■**PROPERTIES** *Objective G*

5. For which of the following triangles is the Law of Sines the easiest way of finding *x*? **(b)**

Functions, Statistics, and Trigonometry © Scott, Foresman and Company 63

NAME _____
Lesson MASTER 5-9 (page 2)

6. How many possible values exist for m∠Z in triangle XYZ below? **1**

7. *True* or *false*. The Law of Sines can be used to prove that the base angles of an isosceles triangle are congruent. **True**

■**USES** *Objective J*

8. To find the height of a building, first find the angle of elevation to the top of the building from two points a fixed distance apart. Suppose that the angles of elevation from two points 100 meters apart are 20° and 15°. Find the height of the building to the nearest meter. **102 meters**

64 *Functions, Statistics, and Trigonometry* © Scott, Foresman and Company

320

Review

20. In $\triangle ABC$, if $AB = 4$, $AC = 5$, and m$\angle A = 110°$, find BC. (*Lesson 5-8*)
≈ 7.4

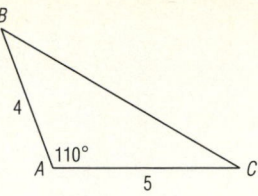

21. *Multiple choice.* Which function is graphed below? (*Lesson 5-6*)

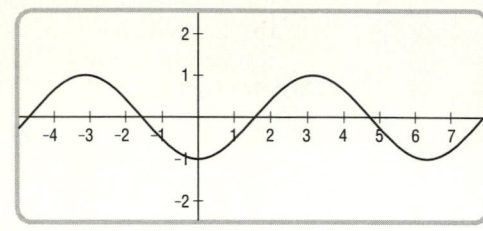

(a) $y = \cos(-x)$ (b) $y = \cos(x)$ (c) $y = -\cos(x)$ (d) $y = \sin(-x)$
(c)

22. Give equations for all vertical asymptotes of $y = \tan \theta$ in the domain $0 < \theta < 4\pi$. (*Lesson 5-6*) $\theta = \frac{\pi}{2}, \theta = \frac{3\pi}{2}, \theta = \frac{5\pi}{2},$ and $\theta = \frac{7\pi}{2}$

23. The function $y = f(x)$, graphed in black below, is periodic. (*Lesson 5-6*)

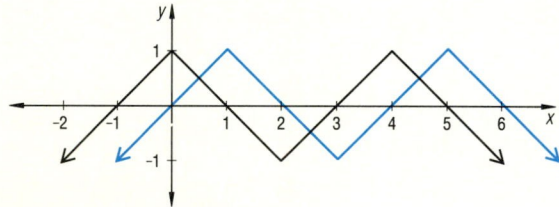

a. What is its period?   4
b. The graph is shifted right one unit to give the graph in blue. What is the period of the image graph?   4

In 24–26, find exact values without a calculator. (*Lesson 5-5*)

24. $\cos 150°$ $-\frac{\sqrt{3}}{2}$ 25. $\sin\left(-\frac{\pi}{6}\right)$ $-\frac{1}{2}$ 26. $\tan \frac{7\pi}{4}$ -1

Exploration

27. Consider $\triangle WXY$, in which $WX = 10$ and m$\angle W = 30°$.
   a. Find m$\angle Y$ for each of the following possible values of XY.
      i. $XY = 3$ ii. $XY = 5$
      iii. $XY = 6$ iv. $XY = 1$

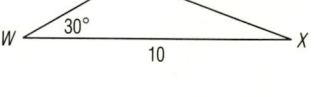

   b. What values of XY will yield the following number of solutions for m$\angle Y$?
      i.   exactly two   ii.   exactly one   iii.   none
      a) i) no solution; ii) 90°; iii) about 56.4° or about 123.6°; iv) no solution
      b) i) 5<XY<10; ii) XY = 5 or XY ≥ 10; iii) XY<5

5-10

From Washington to Beijing

Wang FuJing Street in downtown Beijing, China

Some passengers flying from Washington, D.C. to Beijing, China were surprised to learn about halfway through their journey that they were passing over part of Alaska. Neither Washington nor Beijing is that far north, so why were they going this way? In this lesson we use ideas from the plane trigonometry you have studied in this chapter to find distances on a sphere. First we review some geography and define "shortest distances" on a sphere. Then we use these ideas to find the shortest distance between two cities having the same longitude or latitude. Finally, we calculate the shortest distance between Washington, D.C. and Beijing to illustrate how you can use trigonometry to find the shortest distance between any two cities on Earth.

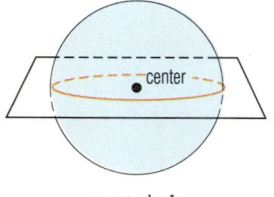

great circle

If a plane contains a point in the interior of a sphere, then the intersection of the plane and the sphere is a circle. A **great circle** of a sphere is the intersection of a sphere and a plane containing the center of the sphere. This circle has the same center as the sphere. Thus, a great circle has the same radius and circumference as the sphere. Great circles are important because the shortest distance between two points on a sphere is measured along the great circle containing them.

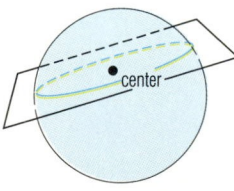

not a great circle

On Earth, which is approximately a sphere of radius 3960 mi, the equator is a great circle. There are infinitely many great circles containing the north pole *N* and the south pole *S*. Each semicircle with endpoints *N* and *S* is called a "line" of longitude, or **meridian**. The meridian through Greenwich, England is called the **Greenwich meridian** or **prime meridian**. *Longitudes* are measured using angles east or west of Greenwich, and so all longitudes are between 0° and 180°. In the figure below, the longitude of *A* is θ. Because *A* is east of Greenwich, θ measures longitude east. The meridian which is 180°W (or 180°E) is called the **International Date Line**.

RESOURCES
- Teaching Aid 49: Longitudes and Latitudes
- Teaching Aid 50: Washington, D.C. and Beijing

TEACHING NOTES

This lesson is challenging, not simply because spherical trigonometry is probably new to students, but because the reading requires many references to complicated drawings. Students need to know some elementary concepts of geography before they can use the Spherical Law of Cosines to find the shortest distances between any two points on Earth.

We suggest that when discussing this lesson in class you have a globe available and a piece of string that you can place on the globe to show great circle routes. (If the string is fixed at two points and not allowed to be slack, and nothing is in its way, it will trace the approximate great circle route.) You will also need an atlas in the classroom so you can look up the latitude and longitude of various cities. A road atlas is a nice source of data for comparing shortest distances to distances via roads in the United States and Canada; an almanac may give air distances between major cities in the world.

Error Analysis Some students have a weak understanding of geography. Years spent looking at flat maps of a spherical world have reinforced the incorrect notion that the shortest distance between cities with the same latitudes lies along that same line of latitude.

Notice what students must be able to do to find the distance between any two points on the earth: (1) From the latitudes, obtain the degree measures of two sides of the spherical triangles as determined by each location and the North Pole. These sides are along meridians. (2) From the longitudes, obtain the degree measures of two sides of the spherical triangle as determined by each location and the North Pole. These sides are along meridians. (2) From the longitudes, obtain the degree measure of the angle between these meridians. (3) Since (1) and (2) give two sides and an included angle, the Spherical Law of Cosines can be applied to obtain the degree measure of the third side of the triangle, which is the shortest arc. (4) Translate that degree measure into distance by determining what part of the circumference of a great circle it is. The last part is exactly what students did in Lesson 5-2 to find lengths of arcs of circles.

Latitudes measure the extent to which a point is north or south of the equator. They are approximately determined by the angle subtended at the center of the Earth by an arc on a line of longitude; so all latitudes are between 0° and 90°. In the diagram at the right, the latitude of P is θ. Because P is north of the equator, θ measures latitude north. The equator itself can be considered at 0°N latitude or 0°S latitude. Notice that a "line" of latitude is a circle, but, except for the equator, lines of latitude are *not* great circles.

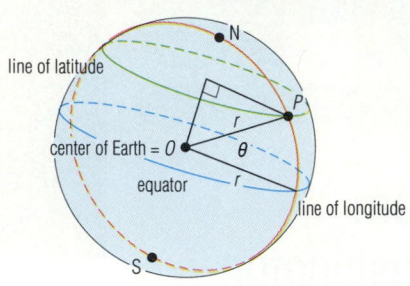

The position of any point on Earth can be determined by its longitude and latitude. In the figure at the right, W represents Washington, D.C. Its location is specified by 77°0'W, 38°55'N. This means that Washington, D.C. is on a line of longitude 77°0' west of the Greenwich meridian and on a line of latitude 38°55' north of the equator.

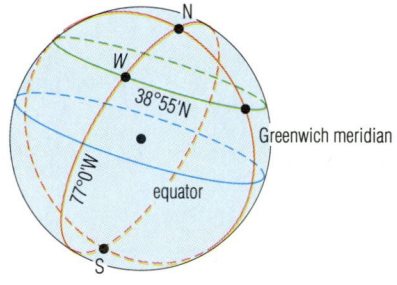

What is the shortest distance from the North Pole to Washington? On any sphere the shortest distance between two points is the length of the minor arc of the great circle connecting them. So the shortest way to go from the North Pole to Washington is to go along the 77°0'W meridian.

As you can see from the diagram at the right, because m∠DOE = 38°55', m∠NOD = 51°5'.

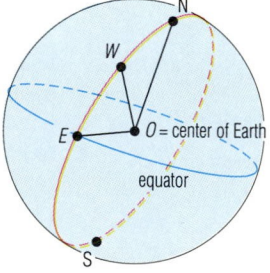

So, the length of $\widehat{NW}$ is $\frac{51°5'}{360°} \cdot 2 \cdot \pi \cdot 3960$ mi ≈ 3530 mi.

Washington, D.C. is about 3530 miles from the North Pole.

In this way you can find the shortest distance between any two places on the Earth that are on the same line of longitude. For instance, Jackson, Mississippi (90°12′W, 32°22′N) and St. Louis, Missouri (90°12′W, 38°35′N) are on the same meridian. They are points J and L in the figure at the right. Because they are on the same great circle, the shortest distance between them is the length of $\widehat{JL}$ on the longitude line.

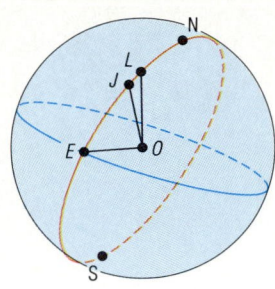

$$m\angle JOL = m\angle LOE - m\angle JOE$$
$$= 38°35′ - 32°22′$$
$$= 6°13′$$

So, the distance between the cities is

$$\frac{6°13′}{360°} \cdot 2 \cdot \pi \cdot 3960 \text{ mi} \approx 430 \text{ mi.}$$

However, the *shortest* distance between cities with the same latitude cannot be found in this way. Consider Ankara, Turkey (32°55′E, 39°55′N) and Beijing, China (116°25′E, 39°55′N). Many people think that the shortest distance between Ankara and Beijing is along the line of 39°55′N latitude, but this is not the case because lines of latitude are not great circles.

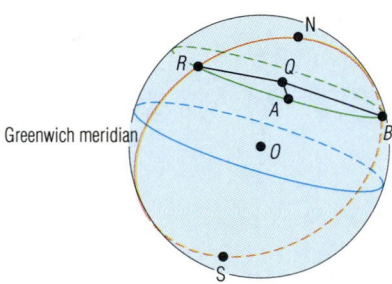

Greenwich meridian

To show this, first find the distance between A and B along the 39°55′N line of latitude. Let R be the point on the Greenwich meridian at latitude 39°55′. If Q is the center of the circle that is this "line" of latitude,

$$m\widehat{AB} = m\angle AQB. \text{ So,}$$
$$m\widehat{AB} = m\angle RQB - m\angle RQA$$
$$= 116°25′ - 32°55′$$
$$= 83°30′.$$

To find the length of $\overparen{AB}$, you need to know the radius of circle Q.

One radius of this circle is $\overline{RQ}$.

Because $\overline{RQ} \parallel \overline{EO}$, $m\angle QRO = 39°55'$. Also,

$$\frac{RQ}{RO} = \cos \angle QRO.$$

Hence, $RQ = RO \cdot \cos \angle QRO.$

But RO is the radius of the earth. Therefore,
$$RQ \approx 3960 \cdot \cos 39°55'$$
$$\approx 3040 \text{ mi.}$$

Hence the distance between Ankara and Beijing along the line of latitude is about

$$\frac{83°30'}{360°} \cdot 2 \cdot \pi \cdot 3040 \text{ mi} \approx 4430 \text{ mi.}$$

Now compare this distance with the great circle distance between Ankara and Beijing. This uses *spherical triangles*, that is, triangles whose sides are arcs of great circles, and a *Spherical Law of Cosines,* which is presented here without proof.

Spherical Law of Cosines

If ABC is a spherical triangle with sides a, b, and c, then
$\cos c = \cos a \cos b + \sin a \sin b \cos C$.

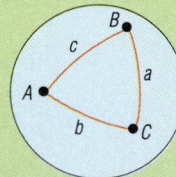

You can use the Spherical Law of Cosines to find the great circle distance between Ankara and Beijing. The key is to let Ankara (A) and Beijing (B) be two vertices of the spherical triangle and to let one of the poles be the third vertex of the spherical triangle. In this case, the North Pole is used. We wish to find the length of n, the great circle arc from A to B. From the Spherical Law of Cosines:

$$\cos n = \cos a \cos b + \sin a \sin b \cos N.$$

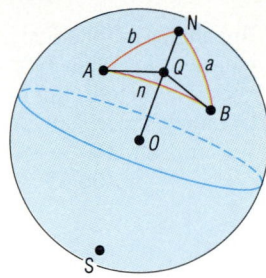

$m\angle N$ is the same as $m\angle AQB$ that was found before in calculating the distance from Ankara to Beijing along the line of latitude. Angle measures a and b are easy to find.

$$b = 90° - \text{latitude of Ankara}$$
$$= 90° - 39°55'$$
$$= 50°5'$$

Because Ankara and Beijing have the same latitude, $a = 50°5'$ also. Now substitute into the Spherical Law of Cosines.

$$\cos n = \cos 50°5' \cos 50°5' + \sin 50°5' \sin 50°5' \cos 83°30'$$
$$\approx .4783$$
$$n \approx 61.4°$$

So, the length of the great circle through Ankara and Beijing is about

$$\frac{61.4°}{360°} \cdot 2 \cdot \pi \cdot 3960 \text{ mi} \approx 4240 \text{ mi}.$$

Notice that this is about 190 miles shorter than the path from A to B along the line of latitude.

The most general and most common problem of this type is to find the shortest distance between two cities not on the same latitude or longitude, for instance, Washington (W) and Beijing (B).

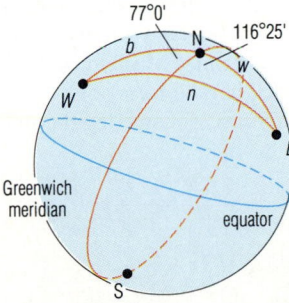

To use the Spherical Law of Cosines, we need to find b, w, and m$\angle N$, where b is the great circle arc measure from Washington to the North Pole (N) and w is the great circle arc measure from Beijing to the North Pole. Earlier we found that $w = 50°5'$ and $b = 51°5'$. Finding m$\angle N$ is more complicated. You might think that

$$m\angle N = 77°0' + 116°25'$$
$$= 193°25'.$$

But angles in spherical triangles, like those in plane triangles, must have measures less than 180°. This means that the shortest great circle arc from Washington to Beijing crosses the International Date Line instead of the Greenwich meridian, and

$$m\angle N = 360° - 193°25'$$
$$= 166°35'.$$

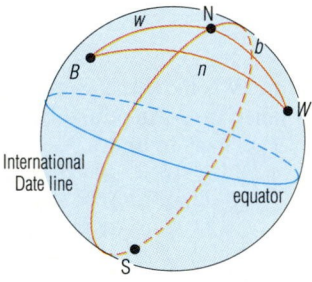

Now substitute into the Spherical Law of Cosines.

$$\cos n = \cos b \cos w + \sin b \sin w \cos N$$
$$= (\cos 51°5')(\cos 50°5') + (\sin 51°5')(\sin 50°5')(\cos 166°35')$$
$$\approx -.1777$$
$$n \approx 100.2°$$

So the shortest distance from Washington to Beijing is about

$$\frac{100.2°}{360°} \cdot 2 \cdot \pi \cdot 3960 \text{ mi} \approx 6930 \text{ mi}.$$

Problems involving distances in navigation and astronomy have been important for millennia and led to the development of trigonometry. Spherical trigonometry, in fact, developed in ancient Greece before plane trigonometry. Euclid knew some of its fundamentals, and by the time of Menelaus (about 100 A.D.), Greek trigonometry reached its peak. Nasir-Eddin (1201–1274), an Arabian mathematican, systematized both plane and spherical trigonometry, but his work was unknown in Europe until the middle fifteenth century. Johannes Muller (1436–1476), also known as Regiomontanus, presented the Spherical Law of Cosines given in this lesson and also a *Spherical Law of Sines*, an amazing theorem that is left for you to find as an exploration.

Questions

Covering the Reading

In 1 and 2, *multiple choice*. Tell whether the figure is (a) always, (b) never, or (c) sometimes (but not always) a great circle.

1. line of longitude **(a)**

2. line of latitude **(c)**

In 3 and 4, state the common name of the meridian.

3. 0°W **Greenwich meridian or prime meridian**

4. 180°E **International Date Line**

5. What is the name of the great circle which is 0°N latitude? **the equator**

6. Find the distance from Kinshasa, Zaire (15°17′E, 4°18′S) to the South Pole. **≈ 5923 miles**

7. Find the distance between Fresno, California (119°48′W, 36°44′N) and Reno, Nevada (119°49′W, 39°30′N). Assume that the cities are on the same meridian. **≈ 190 miles**

8. *True* or *false*. An arc of a line of latitude (other than the equator) can be the side of a spherical triangle. **False**

9. Use the distances calculated in the text from Ankara to Beijing. If an airplane flies at 550 mph along the line of latitude instead of the great circle arc, about how much longer will the flight take? **about 20 minutes**

In 10 and 11, consider Chicago, Illinois (88°W, 42°N),
Providence, RI (71°W, 42°N), and Rome, Italy (12°E, 42°N).

10. a. Find the distance from Chicago to Providence along the line of latitude. **≈ 873 miles**
b. Find the great circle distance from Chicago to Providence.
c. How much longer is the line of latitude distance? **≈ 87 miles**
d. To the nearest percent, how much longer is the line of latitude distance? **≈ 10%**
b) ≈ 785 miles

11. a. Find the distance from Chicago to Rome along the line of latitude.
b. Find the great circle distance from Chicago to Rome. **≈ 4796 mi**
c. How much longer is the line of latitude distance? **≈ 340 miles**
d. To the nearest percent, how much longer is the line of latitude distance? **≈ 7%**
a) ≈ 5136 miles

12. Find the great circle distance between Prague, Czechoslovakia (14°26′E, 50°5′N) and Rio de Janeiro, Brazil (43°12′W, 22°57′S). **≈ 6137 miles**

Review

In 13–15, suppose $\cos \theta = .4$. *(Lesson 5-7)*

13. Use the Pythagorean Identity to find two values for $\sin \theta$. **≈ ±.917**

14. Find $\cos(\theta + 2880°)$. **.4**

15. Find $\cos(\pi - \theta)$. **−.4**

NOTES ON QUESTIONS
Questions 10–12: If students remark about the complexity of the calculations in these questions, you might remind them of the "good old days" (before 1970) when the sines and cosines had to be found in tables and the multiplications had to be either done by hand or with logarithms.

16. Consider the triangle CUB, where m$\angle C = 84°$, $c = 17$, and $u = 11$. Find the measure of $\angle U$. *(Lesson 5-9)* ≈ **40°**

17. To draw a map, a cartographer needed to find the distances between point Z across the lake and each of points X and Y on another side. The cartographer found $XY \approx 0.3$ miles, m$\angle X \approx 50°$, and m$\angle Y \approx 100°$. Find the distances from X to Z and from Y to Z. *(Lesson 5-9)*
XZ ≈ .59, YZ ≈ .46

18. A satellite traveling in a circular orbit 1600 km above the earth is due to pass directly over a tracking station at noon. Assume that the satellite takes two hours to make an orbit and that the radius of the earth is 6400 km. If the tracking antenna is aimed 30° above the horizon (that is, m$\angle HPS = 30$), at what time will the satellite pass through the beam of the antenna? *(Lesson 5-9)*
about 12:05:23 P.M.

19. In $\triangle PAM$ below, find AM. *(Lesson 5-8)*
≈ **8.04**

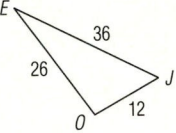

20. In $\triangle JOE$ above, find m$\angle O$ and redraw $\triangle JOE$ more accurately. *(Lesson 5-8)* ≈ **139.7°**

21. Give the domain and range of the cosine function. *(Lesson 5-6)*
domain: the set of all real numbers; range: $\{y : -1 \leq y \leq 1\}$

22. Sketch a graph of $y = \tan \theta$ for $-\pi < \theta < \pi$. *(Lesson 5-6)*   **See margin.**

23. Let $f(x) = 2\pi x$. Let $g(x) = x^2$.
    a. Write $f(g(x))$.   b. Write $g(f(x))$. *(Lesson 3-7)*
    $f(g(x)) = 2\pi x^2$ **$g(f(x)) = 4\pi^2 x^2$**

24. The graph of $y = x^3$ is transformed by $S(x, y) \rightarrow (ax, by)$. What is an equation of the image? *(Lesson 3-6)*   **Sample: $y = \dfrac{b}{a^3} x^3$**

25. What transformation maps the graph of $y = \ln x$ to the graph of $y = \ln(x + 2) - 5$? *(Lesson 3-2)*   **$(x, y) \rightarrow (x - 2, y - 5)$**

Exploration

26. Consult an encyclopedia or a book on spherical trigonometry. What is the Spherical Law of Sines?   **In a spherical triangle with sides a, b, and c and opposite angles A, B, and C, $\dfrac{\sin a}{\sin A} = \dfrac{\sin b}{\sin B} = \dfrac{\sin c}{\sin C}$.**

328

Projects

1. Find a long tape measure and a means of measuring angles. Use trigonometry to measure some "inaccessible" distances of interest to you in your environment. Distances are inaccessible if you are not able, or not permitted, to measure them directly. Some examples of inaccessible distances include: the width of a river or lake, the length of a rail tunnel through a hill, the height of a tall building, and the height of a tower surrounded by a security wall. You may want to measure each distance and angle several times to maximize your accuracy.

2. Running tracks may consist of two straight sections and two curves, which are semicircular in shape. You may have noticed that athletics tracks have "staggered" starting positions for most races, especially those over 100 m.

   a. In which direction do the athletes run on the track shown at the right? Why are the starting marks staggered?
   b. Design a track with six lanes, each 2 m wide, that allows for a 200 m race. Decide where the starting positions should be for each lane. Give precise directions that a sports official could easily follow.
   c. Use your track from part **b**. Where should the starting positions be for a 400 m race?

3. Surveyors and cartographers use trigonometry extensively to locate points on the earth and to represent points on maps. In particular, they use a network of fixed *bench marks* to locate property lines and other landmarks.
   a. Find out how trigonometric points and bench marks are determined.
   b. Find some trigonometric points or bench marks in your neighborhood.
   c. Describe how these locations are used to locate other places.

4. Use the Law of Sines to prove the following.
   a. $\dfrac{a-c}{c} = \dfrac{\sin A - \sin C}{\sin C}$
   b. $\dfrac{a}{b} = \dfrac{\sin A}{\sin B}$
   c. $\dfrac{b+c}{b-c} = \dfrac{\sin B + \sin C}{\sin B - \sin C}$
   d. The bisector of an interior angle of a triangle divides the opposite side into parts whose ratio is equal to the ratio of the sides adjacent to the angle bisected. That is, if $\overrightarrow{AD}$ bisects $\angle BAC$, then $\dfrac{x}{y} = \dfrac{c}{b}$.

   e. In an acute scalene triangle, the largest angle is opposite the longest side.
   f. A triangle is equilateral if and only if it is equiangular.

The projects provide an opportuntiy for extended work on topics related to the content of the chapter. (See pages 66 and T37 for more information on the purposes of projects.) Some or all of these projects may be suitable for **small group work.**

Solutions to the projects are found at the end of the Additional Answers section at the back of the book.

Project 1 provides students with an opportunity to use trigonometry outdoors and thus to see it in a local and practical perspective. We recommend that students attempt this project in small groups of two or three, and that they plan on equipment needs before taking to the field. If more than one group attempts the project, have them compare results and methods after they finish.

Project 2 does not require work outside, but if a track is available it could be used by students to check their solutions to parts **b** and **c.**

Project 3 is similar to Project 1.

Projects 4 and 5 provide opportunities for students who want to prove some theorems or invent their own formal mathematics.

5. As their name suggests, the circular functions are defined in terms of a unit circle. To understand the significance of the circle, imagine using a different shape centered at the origin O and passing through $A = (1, 0)$. Below are three possible alternatives: two squares (figures A and B) and a regular octagon (figure C).

In each case, you could define the functions *side, coside,* and *tide* (to distinguish them from sine, cosine, and tangent) as follows. For any rotation θ in radians, let $P(\theta)$ be the point on the figure such that m$\angle POA = \theta$. Then define
$$P = (\text{coside } \theta, \text{ side } \theta),$$
and tide $\theta = \dfrac{\text{side } \theta}{\text{coside } \theta}$ if coside $\theta \neq 0$.

a. Choose either Figure A or Figure B. Draw (separate) graphs of $s(x) = \text{side } x$, $c(x) = \text{coside } x$, and $t(x) = \text{tide } x$. Compare and contrast these with the graphs of the circular functions.

b. Investigate theorems of the side, coside, and tide functions analogous to those studied in this chapter, such as the Pythagorean Identity, the Opposites Theorem, or the Supplement Theorem. Prove your results.

c. Choose one of the two figures you *didn't* choose in part **a** above. Repeat steps **a** and **b** and compare and contrast your results.

Figure A

Figure B

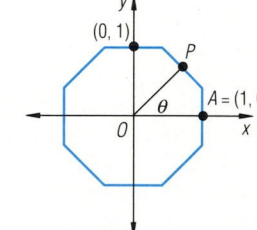

Figure C

Summary

Various units are used to measure angles and rotations. Degrees are in common practical use, while radians are preferred in many cases because of their mathematical convenience. Degrees are divided into minutes and seconds, although it is usually more convenient to use decimal degrees to deal with parts of a degree. By convention, measures of counterclockwise rotations are positive and of clockwise rotations are negative.

When the central angle θ of a circle of radius r is measured in radians, the associated arc length is $s = r\theta$ and the area of the sector is $A = \frac{1}{2}r^2\theta$.

For acute angles, the trigonometric ratios sine, cosine, and tangent can be defined using ratios of side lengths of right triangles. These ratios can be evaluated exactly for some angles. If a line of slope m on the coordinate plane intersects the x-axis at an angle of θ, then $m = \tan\theta$.

The definitions of the trigonometric functions over the set of real numbers use the unit circle. The coordinates of the image of the point $(1, 0)$ under a rotation θ are $(\cos\theta, \sin\theta)$. For all θ except integral multiples of $\frac{\pi}{2}$, $\tan\theta = \frac{\sin\theta}{\cos\theta}$. Following this definition, $\sin\theta$ and $\cos\theta$ both have the same sign when the image of $(1, 0)$ is in the first or third quadrant, and opposite signs in the second and fourth quadrants. The periodic nature of trigonometric functions, and hence the Periodicity Theorem, is a consequence of the values of sine and cosine repeating with each rotation around the unit circle. The Pythagorean Identity, $\sin^2\theta + \cos^2\theta = 1$, is a consequence of the Pythagorean Theorem applied in the first quadrant. The Opposites and Supplements Theorems are derived from reflections of points over the x- and y-axes, while the Complements Theorem is a consequence of the definitions of sine and cosine in a right triangle.

Trigonometry can be used to find unknown measures in all triangles. The Law of Cosines is helpful when you have the SAS or the SSS condition. The Law of Sines is helpful when you have the ASA or AAS condition. In the case of SSA, zero, one, or two triangles may result.

Spherical trigonometry deals with figures whose sides are arcs of great circles on a sphere. There are Laws of Sines and Cosines in spherical trigonometry, and the latter can be used to find the shortest distance between two points on Earth. The result that the shortest distance always is along an arc of a great circle is important in air and ship navigation.

Vocabulary

For the starred (*) items you should be able to give a definition of the terms.
For the other items you should be able to give a general description and a specific example of each.

Lesson 5-1
trigonometry
*angle, *sides (of an angle)
*vertex (of an angle)
revolution, sexagesimal
degree, minute, second
*radian, unit circle
gradient (grad)

Lesson 5-2
Circular Arc Length Formula
disk, sector of a circle
Circular Sector Area Formula

Lesson 5-3
*trigonometric ratios
sine, cosine, tangent in terms
   of right triangles
sin, cos, tan
Tangent-Slope Theorem

Lesson 5-4
sine, cosine, tangent functions
   in terms of unit circles
trigonometric functions
circular functions

Lesson 5-6
cycle
*periodic, *period
Periodicity Theorem

Lesson 5-7
*identity, Pythagorean Identity
Opposites Theorem
Supplements Theorem
Complements Theorem

Lesson 5-8
Law of Cosines

Lesson 5-9
Law of Sines, ambiguous case

Lesson 5-10
great circle
meridian, Greenwich meridian,
   prime meridian
longitude, International Date
   Line
latitude, spherical triangles
Spherical Laws of Cosines,
   Sines

Progress Self-Test

See margin for answers not shown below.

Take this test as you would take a test in class. You will need graph paper and a calculator. Then check the test yourself using the solutions at the back of the book.

In 1 and 2, give your answer to the nearest hundredth.

1. Convert 1.4 radians to degrees. ≈ **80.21°**

2. Convert 67 degrees to radians. ≈ **1.17**

3. The city of Adelaide, Australia has latitude 34°55′S and longitude 138°43′E. Assuming the Earth is a sphere of radius 6400 km, find how far Adelaide is from the South Pole. ≈ **6150 km**

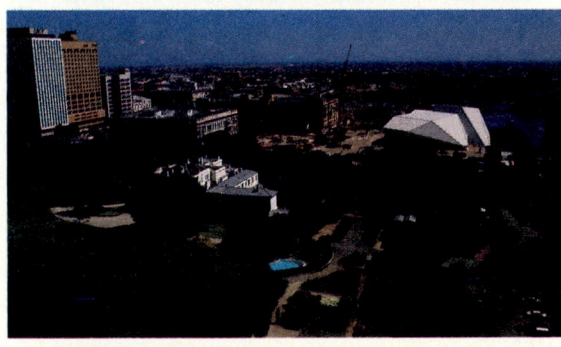

Adelaide, Australia

4. Find the area of a sector of a circle whose radius is 12 cm and whose central angle is $\frac{5\pi}{4}$ radians. **90π cm² ≈ 282.7 cm²**

5. A sprinkler of length 20 m rotates through an angle of 40°, as shown below. What area is traversed by the sprinkler? $\frac{400\pi}{9}$ **m² ≈ 139.6 m²**

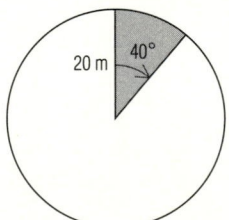

In 6–8, give the value to the nearest thousandth.

6. sin 47° **.731** 7. tan (−3) **.143** 8. cos 16°35′ **.958**

9. State **a.** the domain and **b.** the range of the cosine function.

10. On the unit circle at the right, give the coordinates of P. $\left(\cos\frac{4\pi}{5}, \sin\frac{4\pi}{5}\right)$

In 11–13, consider △JBM below.

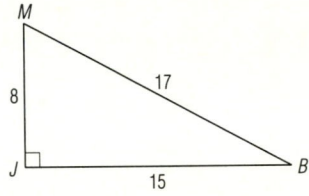

11. Find sin M exactly. $\frac{15}{17}$

12. Find tan B exactly. $\frac{8}{15}$

13. Find m∠B to the nearest tenth of a degree. **28.1°**

In 14–18, evaluate if $\sin\theta = 0.80$ and $0 < \theta < \frac{\pi}{2}$.

14. $\sin(\theta - 2\pi)$ **.80** 15. $\cos\left(\frac{\pi}{2} - \theta\right)$ **.80**

16. $\sin(\theta - \pi)$ **-.80** 17. $\cos\theta$ **.60**

18. θ ≈ **.93 (radian)**

In 19 and 20, give exact values.

19. $\sin\frac{3\pi}{4}$ $\frac{\sqrt{2}}{2}$ 20. $\tan\frac{\pi}{3}$ $\sqrt{3}$

9 a) the set of real numbers
  b) -1 ≤ y ≤ 1

21. Consider the function $f(x) = \cos(2\pi - x)$.
   a. Draw the graph of f for $-2\pi \le x \le 2\pi$.
   b. State the period of f.   **b) 2π; c) cos x**
   c. f is the same as one of the parent trigonometric functions. Which one is it?

22. A ladder 9 m long is placed on a wall so that its base is 2 m from the wall. Find, to the nearest degree, the angle the ladder makes with the ground.   **≈ 77°**

23. A certain variety of wheat forms a mound in the shape of a cone as shown below. If the wheat is dry and allowed to fall naturally, then $m\angle\theta = 41°$. How high is the mound when its diameter is 12 feet?   **≈ 5.2 ft**

24. Find x in the triangle below.   **2.0 units**

25. A team of surveyors wishes to find the distance between a public marina and a private pier. They make the measurements shown in the picture below. What is the distance between the marina and the pier?   **≈ 247.3 m**

CHAPTER REVIEW

The main objectives for the chapter are organized here into sections corresponding to the types of understanding this book promotes: Skills, Properties, Uses, and Representations (called SPUR objectives).

USING THE CHAPTER REVIEW
Students should be able to answer questions like these with about 85% accuracy by the end of the chapter. (See pages 72 and T38−39 for more information.)

Chapter Review

Questions on **SPUR** Objectives

SPUR stands for **S**kills, **P**roperties, **U**ses, and **R**epresentations.
The Chapter Review questions are grouped according to the SPUR Objectives for this chapter.
See margin for answers not shown below.

SKILLS deal with the procedures used to get answers.

■ **Objective A:** *Convert between degrees, radians, and revolutions.* (Lesson 5-1)

In 1 and 2, convert to **a.** degrees **b.** radians.

1. $\frac{1}{5}$ revolution counterclockwise   **a) 72°; b)** $\frac{2\pi}{5}$

2. $\frac{2}{3}$ revolution clockwise   **a) -240°; b)** $-\frac{4}{3}\pi$

In 3 and 4, convert to radians without a calculator.

3. 30° $\frac{\pi}{6}$ **4.** 135° $\frac{3\pi}{4}$

In 5 and 6, convert to degrees without a calculator.

5. $\frac{7\pi}{12}$ radians   **105°**   **6.** $-\frac{\pi}{8}$ radians   **-22.5°**

In 7 and 8, convert to radians, to the nearest thousandth.

7. -313°   **-5.463**   **8.** 48°   **0.838**

In 9 and 10, convert to degrees, to the nearest hundredth.

9. 2 radians   **114.59°**   **10.** -8.6 radians   **-492.74°**

In 11 and 12, tell how many revolutions are represented by each rotation.

11. $\frac{3\pi}{4}$ radians $\frac{3}{8}$ **12.** 990° $2\frac{3}{4}$

13. Convert 16° 14′ 30″ to decimal degrees, to the nearest thousandth.   **16.242°**

14. Convert 103.42° to degrees, minutes, and seconds.   **103° 25′ 12″**

■ **Objective B:** *Find sines, cosines, and tangents of acute angles.* (Lessons 5-3, 5-5)

15. Refer to triangle *TRY*. Find
a. sin *T*   **24/25**
b. sin *Y*   **7/25**
c. cos *T*.   **7/25**

In 16−18, give exact values.

16. $\sin \frac{\pi}{3}$ $\frac{\sqrt{3}}{2}$ **17.** $\cos \frac{\pi}{4}$ $\frac{\sqrt{2}}{2}$ **18.** $\tan \frac{\pi}{6}$ $\frac{\sqrt{3}}{3}$

In 19−21, approximate to the nearest thousandth.

19. tan 1.1   **1.965**   **20.** sin 16° 40′   **.287**
21. cos 82.13°   **.137**

In 22−24, find θ, where $0 < \theta < \frac{\pi}{2}$, to the nearest hundredth.

22. sin θ = 0.32   **.33**   **23.** tan θ = 0.5   **.46**

24. Find the exact radian measure such that $\sin \theta = \frac{\sqrt{3}}{2}$. $\frac{\pi}{3}$

■ **Objective C:** *Evaluate circular functions outside the first quadrant.* (Lessons 5-4, 5-5)

In 25−28, evaluate to the nearest hundredth.

25. sin 3   **.14**   **26.** cos (-4.2)   **-.49**
27. tan (-151°)   **.55**   **28.** sin 643°   **-.97**

In 29−32, give exact values.

29. $\cos \frac{5\pi}{4}$ $-\frac{\sqrt{2}}{2}$ **30.** sin 210° $-\frac{1}{2}$

31. tan (-45°)   **-1**   **32.** $\sin \frac{9\pi}{2}$ **1**

33. The point $(1, 0)$ is rotated $112°$ clockwise about the origin. Find the coordinates of its image, rounded to the nearest hundredth. **($-.37, .93$)**

34. To the nearest hundredth, find all solutions of $\tan \theta = 3$ between 0 and 2π.   **$\approx 1.25, \approx 4.39$**

35. Find four values of θ, between $0°$ and $720°$, for which $\cos \theta = 0.214$. **$77.6°, 282.4°, 437.6°, 642.4°$**

36. Solve $\sin x = -\frac{1}{2}$ exactly in the interval $-\frac{\pi}{2} \le x \le \frac{5\pi}{2}$. $-\frac{\pi}{6}, \frac{7\pi}{6}, \frac{11\pi}{6}$

 Objective D: *Use trigonometry to find sides or angles in triangles. (Lessons 5-3, 5-8, 5-9)*

In 37–40, find x.

37.

≈ 9.33

38.

≈ 39.44

39.

$\approx 82.82°$

40.
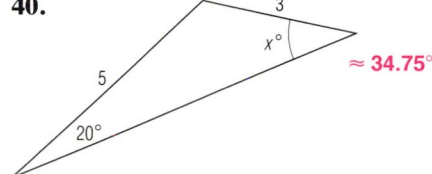
$\approx 34.75°$

41. Find to the nearest tenth of a degree the size of the angles of a 5-12-13 triangle.   **$22.6°, 67.4°, 90°$**

42. In $\triangle XYZ$, $m\angle X = 48°$, $m\angle Y = 68°$, and $y = 10$. Find x to the nearest tenth.   **8.0 units**

43. Calculate the area of the triangle.   **≈ 20.67 cm²**

44. Find the area of the parallelogram below.   **≈ 179.76 ft²**

 Objective E: *Find circular arc length and areas of sectors. (Lesson 5-2)*

45. Find the length of the arc of a $120°$ central angle in a circle with radius 9 inches.   **6π in.**

46. The arc of a central angle of $\frac{\pi}{6}$ radians has a length of 3π feet. Find the radius of the circle.   **18 ft**

47. Find the area of the sector whose central angle is $\frac{\pi}{6}$ radians in a circle whose radius is 10 meters. $\frac{25\pi}{3}$ **m²**

48. The area of a sector in a circle of radius 6 inches is 48 square inches. Find the measure of the central angle to the nearest tenth of a degree.   **152.8°**

PROPERTIES deal with the principles behind the mathematics.

 Objective F: *State properties of the sine, cosine, and tangent functions. (Lessons 5-6, 5-7)*

49. State **a.** the domain and **b.** the range of the sine function.

50. For what values of θ is $\tan \theta$ undefined?
odd multiples of $\pi/2$

In 51 and 52, for $f(x) = \cos(x)$, *true* or *false*.

51. f is an even function.   **True**

52. The maximum value of f is 1.   **True**
49 a) the set of all real numbers
    b) $\{y : -1 \le y \le 1\}$

53. *Multiple choice.* For what values of θ between 0 and 2π is $\sin \theta < 0$ and $\cos \theta > 0$?

(a) $0 < \theta < \frac{\pi}{2}$

(b) $\frac{\pi}{2} < \theta < \pi$

(c) $\pi < \theta < \frac{3\pi}{2}$

(d) $\frac{3\pi}{2} < \theta < 2\pi$ **(d)**

ADDITIONAL ANSWERS

65. $-\sin\left(\dfrac{\pi}{2} - \theta\right) = -\cos\theta$

   Complements Theorem

$= \cos(\pi - \theta)$

   Supplements Theorem

66. A formula for the area of a triangle, given two sides *a* and *b* and the included angle *C*, is $A = \dfrac{1}{2}ab\sin C$. Since sin 70° = sin 110°, the two triangles must have the same area.

67. In the Law of Cosines, $c^2 = a^2 + b^2 - 2ab\cos C$. If $C = 90°$, then cos *C* = 0 and $c^2 = a^2 + b^2$.

72.a. ≈ **13.8 mi**

c. ≈ **4.56 hours after 1 P.M., at 5:33 and 36 seconds; around 5:30 P.M.**

54. In what interval(s) between 0 and 2π are both the cosine and tangent functions negative? $\dfrac{\pi}{2} < \theta < \pi$

55. *Multiple choice.* The statement that for all θ, $\sin(\pi - \theta) = \sin\theta$ is called
   (a) The Supplements Theorem
   (b) The Complements Theorem
   (c) The Opposites Theorem
   (d) The Periodicity Theorem. **(a)**

In 56–58, *true* or *false*. For all θ:

56. $\cos(\theta + 3\pi) = \cos\theta$ **False**

57. $\sin(\theta + 6\pi) = \sin\theta$ **True**

58. $\cos^2\theta + \sin^2\theta = \tan^2\theta$. **False**

■ **Objective G:** *Use theorems about sines, cosines, and tangents. (Lessons 5-6, 5-7, 5-8, 5-9)*

59. If $\cos\theta = \dfrac{1}{4}$, without using a calculator find all possible values of:
   a. $\sin\theta$;   **b.** $\tan\theta$. **a)** $\pm\dfrac{\sqrt{15}}{4}$; **b)** $\pm\sqrt{15}$

60. Given that $\tan 2.45 \approx -0.828$, find three other values of θ for which $\tan\theta \approx -0.828$.
   sample: $-2\pi + 2.45 \approx -3.83$; $-\pi + 2.45 \approx -.69$; $\pi + 2.45 \approx 5.59$

In 61–64, given $\sin 28° \approx 0.469$, without using a calculator, find

61. $\cos 62°$ ≈ **0.469**   **62.** $\sin 152°$ ≈ **.469**

63. $\sin(-152°)$ ≈ **-.469**   **64.** $\sin 1108°$. ≈ **.469**

65. Use theorems of sines and cosines to prove that $-\sin\left(\dfrac{\pi}{2} - \theta\right) = \cos(\pi - \theta)$.

66. Explain why the two triangles below have the same area.

67. Explain how the Pythagorean Theorem follows from the Law of Cosines.

68. In $\triangle EFG$, m$\angle E = 35°$, $EF = 4$, and $FG = 9$. Tom claims that there is exactly one triangle satisfying these conditions. Karen claims that there are two. Who is correct?
Tom (This is the SsA condition.)

USES deal with applications of mathematics in real situations.

■ **Objective H:** *Solve problems involving lengths of arcs or areas of sectors. (Lesson 5-2)*

69. A radar screen represents a circle of radius 40 miles. If the arm shown below makes 25 revolutions per minute, what area is mapped by the radar in one second? ≈ **2094 mi²**

70. A theatre is planned as shown below. The internal radius of the building is $AO = 26$ m and m$\angle AOB = 160°$. If the stage area has a radius of 5 m, find the area of the seating section.
$\dfrac{868\pi}{3}$ m² ≈ **909 m²**

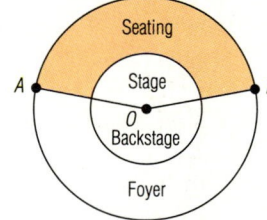

336

■ Objective I: *Solve problems using trigonometric ratios in right triangles.* *(Lesson 5-3)*

71. A ladder against a wall makes a 60° angle with the ground. If the base of the ladder is 8 feet from the wall, find the length of the ladder. **16 ft**

72. Refer to the diagram below. A tanker sails due west from port at 1 P.M. at a steady speed. At 2:15 P.M., the bearing of a lighthouse from the ship is 37° (from North).

   a. If the lighthouse is 18.3 miles due north of the port, how far out to sea is the ship?
   b. Find the speed of the ship. **≈ 11 mph**
   c. At what time will the lighthouse lie on a bearing of 70° from the ship, assuming that the speed of the ship is constant?

73. A school building casts a shadow 18 m long when the elevation of the sun is 20° as shown below. How high is the top of the roof if $AB = 16$ m? **≈ 9.5 m**

74. Guy wires 60 feet long are to be used to steady a radio tower 50 feet tall. What angle θ will the wires make with the ground? **≈ 56.4°**

■ Objective J: *Solve problems using the laws of sines and cosines.* *(Lessons 5-8, 5-9)*

75. An airport controller notes from radar that one jet 12° west of south and 20 miles from the airport is flying toward a private plane at the same altitude which is 27 miles directly south of the airport.
   a. How far apart are the planes? **≈ 8.52 mi**
   b. If the jet is travelling at 470 mph and the private plane at 180 mph, and they do not change course, how long will it be before they crash? **.013 hr ≈ 47 seconds**

76. Find the length of the longest diagonal of a parallelogram with one angle of 39° and sides of 8.2 cm and 10.1 cm. **≈ 17.3 cm**

77. A team of surveyors measuring from A to B across a pond positions a vertical stake at S, a point from which they can measure distances to A and B across dry ground. The measures are shown in the diagram. Find AB. **≈ 105.2 m**

EVALUATION
Three tests are provided for this chapter in the Teacher's Resource File. Chapter 5 Test, Forms A and B, cover just Chapter 5. The third test is Chapter 5 Test, Cumulative Form. About 50% of this test covers Chapter 5, 25% covers Chapter 4, and 25% covers earlier chapters. For information on grading, see *General Teaching Suggestions: Grading* on page T43 in this Teacher's Edition.

ASSIGNMENT RECOMMENDATION
We strongly recommend that you assign the reading and some questions from Lesson 6-1 for homework the evening of the test.

If you do not give assignments on the days of tests, you may cover one less *chapter* over the course of the year.

78. Forest lookout stations at A and B are 10 miles apart. The rangers in A spot a fire 10° east of north. They call the rangers at B, who locate the fire as 30° north of west with respect to B.
   a. If B is directly northeast of A, find the distance of the fire from A.
   b. Find the distance of the fire from B.
   a) ≈ **10.3 mi;**
   b) ≈ **6.1 mi**

79. In the diagram below, $\overline{AB}$ and $\overline{CD}$ are parallel chords in $\odot O$. $\odot O$ has diameter 40 mm and $AB = CD = 35$ mm.
   a. Find θ to the nearest tenth of a degree.
   b. Find the shaded area.   ≈ **743.1 mm²**
   a) 122.1°

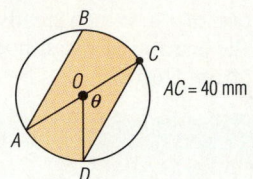

AC = 40 mm

REPRESENTATIONS deal with pictures, graphs, or objects that illustrate concepts.

■ **Objective K:** *Represent sines, cosines, and tangents in the coordinate plane.* *(Lessons 5-3, 5-4)*

In 80–83, refer to the unit circle below. Which letter could represent the value given?

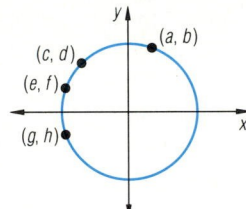

80. $\sin 70°$ **(b)**
81. $\cos (-160°)$ **(g)**
82. $\cos 135°$ **(c)**
83. $\sin 920°$ **(h)**

84. Draw a picture using a unit circle to show why the cosine function is an even function.

85. Give an equation for the line through $(-2, 3)$ that intersects the positive x-axis at an angle of 20°.

86. In the diagram below, find θ to the nearest tenth of a degree.   **18.4°**

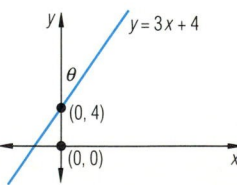

85) $y = (\tan 20°)x + (3 + 2 \tan 20°) \approx .364x + 3.728$

■ **Objective L:** *Draw or interpret graphs of the sine, cosine, and tangent functions.*
(Lesson 5-6)

87. Consider the function $f(x) = \sin x$, where x is measured in radians.
   a. Sketch a graph of f without using an automatic grapher.
   b. What is the period of f?   **2π**

88. Let $g(\theta) = \tan \theta$, where θ is in degrees.
   a. Sketch a graph of $y = g(\theta)$.
   b. State its period.   **180°**
   c. Write equations for two of the asymptotes of g.   **sample: $\theta = 90°$, $\theta = 270°$**

89. Consider the graphs of $S(x) = \sin x$ and $C(x) = \cos x$. What translation maps
   a. S to C?
   b. C to S?

90. a. Use an automatic grapher to draw
$$y = \cos (x)$$
$$y = \cos (-x)$$
and $$y = \cos (\pi - x)$$
on the same set of axes.
   b. Describe the relations among the curves.
   c. What theorems do the relations between these three graphs represent?

89. a) $T(x, y) \rightarrow (x + \frac{\pi}{2}, y)$

b) $T(x, y) \rightarrow (x - \frac{\pi}{2}, y)$

338

CHAPTER 6 ■ GRAPHS OF CIRCULAR FUNCTIONS

OBJECTIVES ■ CHAPTER 6

The objectives listed here are the same as in the Chapter 6 Review on pages 393–396 of the student text. The Progress Self-Test on page 392 and the tests in the Teacher's Resource File cover these objectives. For recommendations regarding the handling of this end-of-chapter material, see the notes in the margins on the corresponding pages of this Teacher's Edition.

OBJECTIVES FOR CHAPTER 6 (Organized into the SPUR categories— Skills, Properties, Uses, and Representations)	Progress Self-Test Questions	Chapter Review Questions	Teacher's Resource File	
			Lesson Masters*	Chapter Test Forms A & B
SKILLS				
A Evaluate inverse trigonometric functions.	5, 6	1–8	6-5	8, 9
B Evaluate reciprocal functions.	7, 8	9–14	6-7	10, 11
C Solve equations involving circular functions.	11	15–24	6-6, 6-7	13–16
PROPERTIES				
D Identify the amplitude, period, frequency, phase shift, and other properties of circular functions.	1–3	25–32	6-1, 6-2, 6-3	1–5, 18
E State properties of inverse trigonometric functions.	13, 14	33–36	6-5	19
F State properties of the reciprocal trigonometric functions.	15b	37–40	6-7	20
USES				
G Model phenomena with transformation images of circular functions.	10, 12, 16	41–46	6-1, 6-2, 6-3, 6-4	7
REPRESENTATIONS				
H Graph transformation images of circular functions.	9b	47–52	6-1, 6-2, 6-3, 6-6	17
I State equations for graphs of circular functions.	4, 9a	53–60	6-1, 6-2, 6-3	12
J Graph or identify graphs of inverse trigonometric functions or reciprocal trigonometric functions.	15a,c	61–64	6-5, 6-7	6

*The masters are numbered to match the lessons, and each is intended for use with that lesson.

OVERVIEW ■ CHAPTER 6

Chapter 5 may be viewed as the development of the basics of the trigonometric functions. This chapter builds on those basics. The graphs of the sine, cosine, and tangent are extended via scale changes and translations to obtain equations for the graphs of any sine wave. The inverses of the functions, which students have already encountered in solving equations, are analyzed. The "other" three trigonometric functions (secant, cosecant, and cotangent) are defined. And, in the last lesson of the chapter, linear combinations of sines and cosines are shown to be capable, perhaps unexpectedly, of approximating square waves. Thus, this chapter is a natural extension of Chapter 5.

The flow of this chapter from preceding chapters is signalled by the relative lack of new concepts. Almost every idea in this chapter has been studied in previous chapters in some way, though the ideas may have specific names as they apply to trigonometric functions. For instance, the effects of changes in *amplitude*, *phase shift*, and *period* on sine and cosine functions are special cases of the scale changes and translations studied in Chapter 3. Modeling using trigonometric functions is an extension of similar ideas found in Chapters 2 and 4. And of course the trigonometric functions themselves build on the function concept introduced in Chapter 2.

Periodic phenomena are often modeled by trigonometric functions, and a variety of such phenomena appear in this chapter. In the opening lesson, the terminology of sine waves is introduced and frequency is discussed. In the next lesson, the phase shift relating voltage and current flow appears. In the next two lessons, equations for modeling sine waves are found, with applications to rotating wheels, hours of daylight, temperatures, and orbits. Equations involving these models are solved via the inverse functions and then using the three reciprocal functions cotangent, secant, and cosecant. Alternating current is described in the reading lesson. Most of these applications do not involve triangles, and for this reason radian measure dominates this chapter, and we use the term *circular functions* to describe the content.

PERSPECTIVES ■ CHAPTER 6

The Perspectives provide the rationales for the inclusion of topics or approaches, give some mathematical background, and make connections with other lessons and to other UCSMP texts.

6-1

SCALE CHANGE IMAGES OF CIRCULAR FUNCTIONS

This lesson covers the graphs of $y = b \sin \frac{x}{a}$ and $y = b \cos \frac{x}{a}$, where a and b are nonzero real constants. These graphs are all sine waves, and their shapes are predictable as transformation images of the parent functions $y = \sin x$ and $y = \cos x$. By comparing them with the parent functions, the amplitude, period, and frequency of these sine waves can rather easily be found. Students are expected to check their predictions using an automatic grapher.

6-2

TRANSLATION IMAGES OF CIRCULAR FUNCTIONS

When applied to sine waves, a horizontal translation is called *phase shift*. This name comes from the fact that sine waves picture the flow of current through a circuit. This lesson applies horizontal and vertical translations to the graphs of the circular functions.

6-3

LINEAR CHANGES OF CIRCULAR FUNCTIONS

A linear change is a type of affine transformation. In general, an *affine transformation* in two dimensions is a mapping of the form $(x, y) \rightarrow (ax + by + c, dx + ey + f)$, where $ab \neq 0$ and $cd \neq 0$. If $c = f = 0$, then the affine transformation is called a *linear transformation*. That is, a linear transformation has the form $(x, y) \rightarrow (ax + by, cx + dy)$. Linear transformations in dimension n can be represented by $n \times n$ matrices. We have chosen the term *linear change* to discuss those affine transformations in which $b = d = 0$.

When a scale change $(x, y) \rightarrow (ax, by)$ and a translation $(x, y) \rightarrow (x + h, y + k)$ are applied to a graph in that order, the effect is to apply the *linear change* $(x, y) \rightarrow (ax + h, by + k)$ to the graph. A sine wave can be most efficiently defined as the image of $y = \sin x$ under a linear change.

The linear change described above can be effected by substituting $\frac{x - a}{h}$ for x and $\frac{y - b}{k}$ for y. The effects of these substitutions on the amplitude, phase shift, and period of sine functions are all easily obtained from properties of the linear change. They are the focus of this lesson.

6-4

MODELING WITH CIRCULAR FUNCTIONS

In the previous lesson, an equation for a sine wave is given and the task is to determine various properties of the wave. The task is reversed in this lesson. Here data are given from which a wave can be graphed and the student is asked to determine an equation. The two examples may be new to students: the length of daylight as a function of the time of the year, and the distance of an artificial satellite from the equator is a function of the amount of time it has been in orbit.

6-5

INVERSE CIRCULAR FUNCTIONS

In one sense this lesson merely formalizes the mathematics that students encountered first in Chapter 5. Students are not expected to master the graphs of these functions, but are expected to be able to use them to solve equations and model situations.

The equations to be solved here have solutions between 0 and 2π (equivalently, between 0 and 360°) and are of the form $\sin x = k$, $\cos x = k$, or $\tan x = k$.

6-6

SOLVING TRIGONOMETRIC EQUATIONS

This lesson extends the previous lesson in two ways. Now we want to find all real solutions to a particular equation, which because of periodicity, typically means there are infinitely many solutions. And the equations include quadratic equations in a trigonometric function, where it is useful to think of the trigonometric expression as a *chunk*, that is, as a single entity.

6-7

THE SECANT, COSECANT, AND COTANGENT FUNCTIONS

This lesson introduces students to the reciprocal circular functions: secant (reciprocal of cosine), cosecant (reciprocal of sine), and cotangent (reciprocal of tangent). These functions are presented in the same manner as the functions sine, cosine, and tangent were presented in Chapter 5. They are used in right-triangle situations, and then analyzed as functions. Students are expected to be able to give domain, range, maximum and minimum points, and graphs of the parent reciprocal functions.

6-8

ALTERNATING CURRENT AND SQUARE WAVES

A Fourier series is a series of the form

$$\frac{1}{2} a_0 + \sum_{n=1}^{\infty} (a_n \cos nx + b_n \sin nx).$$

This series is periodic with period 2π. If two functions f and $|f|$ have integrals on the interval $-\pi \leq x \leq \pi$, and f is periodic with period 2π, then under a variety of conditions it can be approximated by a Fourier series. These conditions are not particularly restrictive, and in this lesson a simple example is given of how Fourier series can approximate the square wave found in alternating current.

DAILY PACING CHART ■ CHAPTER 6

For optimal student performance, the Progress Self-Test and the Chapter Review should be covered. (See *General Teaching Suggestions: Mastery* on page T35 of this Teacher's Edition.) By following the pace of the Full Course given here, students can complete 12 or 13 chapters of the book by the end of the year. Students following the pace of the Minimal Course spend more time to complete lessons that are combined with quizzes and to consider the Chapter Review. When all lessons are covered in order (the recommendation of the authors), students in the Minimal Course can cover all but the last two chapters. For more information on pacing, see *General Teaching Suggestions: Pace* on page T34 of this Teacher's Edition.

DAY	MINIMAL COURSE	FULL COURSE
1	6-1	6-1
2	6-2	6-2
3	6-3	6-3
4	Quiz (TRF); Start 6-4	Quiz (TRF); 6-4
5	Finish 6-4	6-5
6	6-5	6-6
7	6-6	Quiz (TRF); 6-7
8	Quiz (TRF); Start 6-7	6-8
9	Finish 6-7	Progress Self-Test
10	6-8	Chapter Review
11	Progress Self-Test	Chapter Test (TRF)
12	Chapter Review	
13	Chapter Review	
14	Chapter Test (TRF)	

TESTING OPTIONS

- ■ Quiz for Lessons 6-1 through 6-3
- ■ Quiz for Lessons 6-4 through 6-6
- ■ Chapter 6 Test, Form A
- ■ Chapter 6 Test, Form B
- ■ Chapter 6 Test, Cumulative Form
- ■ Comprehensive Test, Chapters 1–6

A Quiz and Test Writer is available for generating additional questions, additional quizzes, or additional forms of the Chapter Test.

PROVIDING FOR INDIVIDUAL DIFFERENCES

The student text has been written for, and tested with, average students. It has also been used successfully with students who are better prepared as well as with students who are poorly prepared.

The Lesson Notes often include Error Analysis and Alternate Approach features to help you assist those students who need more help. Students of all abilities often learn from their peers and may benefit from small group work referenced as appropriate throughout the Notes. A blackline Lesson Master (in the Teacher's Resource File), keyed to the chapter objectives, is available for each lesson to provide more practice. (However, we advise you to use these masters sparingly; again refer to the suggestions on pace on page T34.)

Extension activities are provided in the Lesson Notes for use with students who need to be challenged even more than the requirements of completing the Full Course provide.

Graphs of Circular Functions

Top left: Sound wave produced by a violin
Below left: Sound wave produced by a clarinet
Below right: Sound wave of a pure tone

Sound is produced by fluctuations in the pressure of the air. Different kinds of fluctuations cause us to hear different kinds of sounds. Variations in air pressure can be picked up by a microphone, and can be pictured by an *oscilloscope* as a graph of air pressure versus time.

In the pictures above, different sounds produced by various musical instruments are displayed on an oscilloscope. From such graphs you can see why sound is said to travel in waves.

A *pure tone* is a tone in which air pressure varies *sinusoidally* with time; that is, as a *sine wave*. For most people, sounds in which the air pressure rises and falls almost periodically are considered to be quite pleasant.

Pure tones seldom occur in nature. They can be produced, however, by certain tuning forks and electronic music synthesizers. Mathematically and physiologically, pure tones or sine waves are the foundation of all musical sound. An analysis of musical sound depends on knowledge of trigonometric functions, particularly of their graphical properties.

In this chapter you will study the effects of transformations on the graphs of the parent sine, cosine, and tangent functions, and you will learn methods for solving trigonometric equations. You will also learn how graphs of the trigonometric functions are used to model sound, electricity, and other periodic phenomena.

CHAPTER 6

This chapter should take 8–10 days for the lessons and quizzes, a day for the Progress Self-Test, 1 or 2 days for the Chapter Review, and 1 day for testing. The total is 11–14 days. (See the Daily Pacing Chart on page 339D.) Even if you decide to spend more time on projects, do not spend more than three weeks on this chapter.

USING PAGE 339
Science departments in most high schools have an oscilloscope. Bringing one to class is a good way to start the chapter. Vary the pitch and volume of a tone to produce changes in the period and amplitude of the tone. Point out that the first three lessons of this chapter deal with the mathematics of these changes in sine waves. Then combine tones to produce more complicated waves. The last lesson of this chapter deals with the kind of combination that is displayed.

OBJECTIVES

D Identify the amplitude, period, frequency, phase shift, and other properties of circular functions.

G Model phenomena with transformation images of circular functions.

H Graph transformation images of circular functions.

I State equations for graphs of circular functions.

TEACHING NOTES

You may be accustomed to considering functions of the form $y = A \sin bx$ and $y = A \cos bx$. We use the forms

$$y = b \sin \frac{x}{a} \text{ and } y = b \cos \frac{x}{a}$$

so that students can easily apply the Graph Scale Change Theorem. That is, the forms we use are equivalent to $\frac{y}{b} = \sin \frac{x}{a}$ and

$\frac{y}{b} = \cos \frac{x}{a}$. Because these are only special cases of something students already know, this lesson is easy for most students.

Technology The actual process of graphing these sine waves can be quite tedious. Some sketching of sine waves by hand is useful but time-consuming. Capitalize on the speed and accuracy of automatic graphers. Especially valuable for students is seeing an image graphed with its preimage, as in **Examples 1–3.** If your computer has a color monitor the graphics can be impres-

LESSON
6-1

Scale Change Images of Circular Functions

These two electronic music synthesizers are differently programmed, then played simultaneously.

340

As was remarked on the previous page, pure sound tones travel in sine waves.

> **Definition**
>
> A **sine wave** is the image of the graph of the sine or cosine function under a composite of translations and scale changes.

The higher pitched the sound, the shorter the *period* of its wave. The louder the sound, the greater the *amplitude* of its wave. The **amplitude** of a sine wave is one half the difference between the maximum and minimum values attained by the function.

Variations in period or amplitude can be derived from horizontal or vertical scale changes, respectively, of the parent cosine and sine functions. Equations for and graphs of such functions can be derived from the Graph Scale Change Theorem.

Example 1　　**a.** Find an equation for the image of $y = \cos x$ under the scale change $(x, y) \rightarrow (x, 3y)$.
b. Graph the parent and the image on the same set of axes.
c. State the amplitude of each function.

Solution

a. By the Graph Scale Change Theorem, an equation of the image is

$\frac{y}{3} = \cos x$ or equivalently $y = 3 \cos x$.

b. The graph of $y = \cos x$ is in black and the graph of its scale change image is in blue.

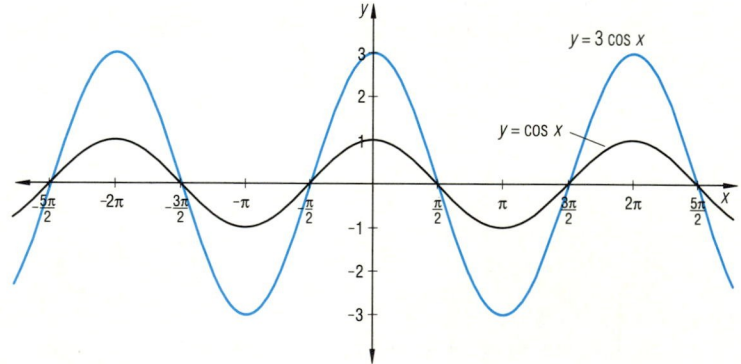

c. The maximum value of $y = \cos x$ is 1; its minimum value is -1. So its amplitude is $\frac{1}{2}(1--1) = \frac{1}{2} \cdot 2 = 1$. The extreme values of $y = 3 \cos x$ are 3 and -3, so its amplitude is $\frac{1}{2}(3--3) = \frac{1}{2} \cdot 6 = 3$.

Notice that the scale change of Example 1 is a vertical scale change. If the graphs in Example 1 represent sound waves, then the equation $y = 3 \cos x$ represents a sound 3 times as loud as that represented by $y = \cos x$. Vertical scale changes affect the amplitude of a wave. Horizontal scale changes affect the period.

■ ■ ■ ■ ■ ■ ■ ■

Example 2　**a.** Sketch a graph of $y = \sin 4x$.
b. How do its amplitude and period compare to that of its parent $y = \sin x$?

Solution

a. Compare the given function to $y = \sin x$ by rewriting the equation $y = \sin 4x$ as $y = \sin\left(\dfrac{x}{\frac{1}{4}}\right)$. From the Graph Scale Change Theorem, the graph of $y = \sin 4x$ is the image of $y = \sin x$ under a horizontal scale change with magnitude $\frac{1}{4}$. Below are the parent graph (black) and the image graph (blue).

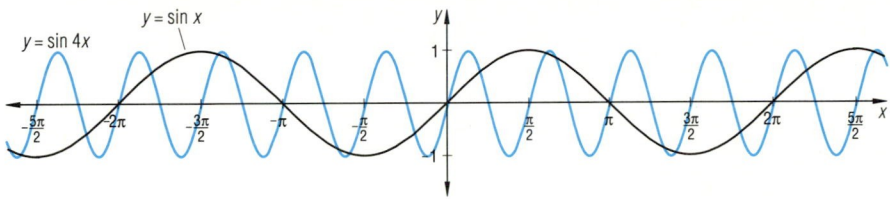

b. $y = \sin 4x$ has the same amplitude as $y = \sin x$, but the period of $y = \sin 4x$ is one-fourth of the period of $y = \sin x$. That is, the period of $y = \sin 4x$ is $\frac{2\pi}{4} = \frac{\pi}{2}$.

Note that the graph of $y = \sin 4x$ completes four cycles for every one completed by $y = \sin x$. In terms of sound waves, $y = \sin 4x$ can be interpreted as modeling a sound having four times the *frequency* of the original wave $y = \sin x$. The **frequency** is the reciprocal of the period and represents the number of cycles per unit of time. Whereas the original graph completes one cycle in 2π seconds, the image graph completes four cycles in the same amount of time. If these graphs represent sound waves, then doubling the frequency results in a pitch one octave higher (recall Question 22 from Lesson 4-2), so the graph of $y = \sin 4x$ represents a pitch two octaves higher than the pitch of $y = \sin x$.

The procession of examples in this lesson is smooth. **Example 1** asks students to graph a wave in which there has only been a change in amplitude (a vertical scale change). In **Example 2,** only the period is changed (a horizontal scale change). In **Example 3** both are changed. In **Example 4** the problem is reversed; given the graph, students are asked to find an equation. As an alternative to showing the graphs with an automatic grapher, we provide a teaching aid with the graphs for **Examples 3 and 4.**

Error Analysis　The most difficult part of the algebra in this lesson is realizing that

$\sin(kx)$ is equal to $\sin\left(\dfrac{x}{\frac{1}{k}}\right)$

and thus that replacing x by kx means a horizontal scale change of magnitude $1/k$, not k. Although this idea has been discussed earlier, in Chapter 3, some students may still get confused. Such confusion is natural and should be expected.

ADDITIONAL EXAMPLES
1. a. Graph $y = 5 \sin x$ for $-2\pi \le x \le 2\pi$.

b. What is the amplitude of this function?
5

2. a. Graph $y = \cos 6x$ for $-2\pi \le x \le 2\pi$.

b. How do the amplitude and period of this function compare with that of $y = \cos x$?

amplitude: same as $y = \cos x$; period: $\frac{1}{6}$ of $y = \cos x$

3. Consider the graph of $y = \frac{1}{4} \sin 2x$. **a.** Give its amplitude. **b.** Give its period. **c.** Give its frequency.

a. $\frac{1}{4}$; b. π; c. $\frac{1}{\pi}$

4. The graph below shows $y = \cos x$ and an image of $y = \cos x$ under a scale change. Find an equation for the image.

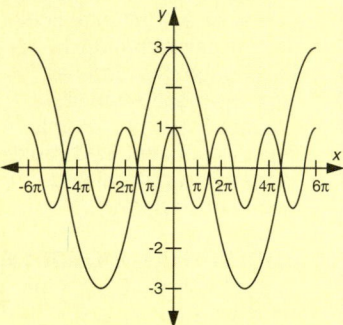

$y = 3 \cos \frac{x}{3}$

Example 3 Consider the graph of the function with equation $y = -5\cos\left(\frac{x}{2}\right)$.

a. Identify its amplitude.
b. Find its period.
c. Find its frequency.
d. Check your answers to parts **a** and **b** using an automatic grapher.

Solution Divide both sides of the equation by -5 to put the function rule in a form that can be analyzed using the Graph Scale Change Theorem.

$$-\frac{y}{5} = \cos\left(\frac{x}{2}\right)$$

a. In the equation $y = \cos x$ of the parent function, y has been replaced by $-\frac{y}{5}$. Thus, the graph of the parent is stretched by a factor of -5 in the vertical direction. In particular, this means that the maximum and minimum values of the parent graph are multiplied by -5. Hence the given function has amplitude $\frac{1}{2}(5 - -5) = 5$.

b. In the equation $y = \cos x$, x has been replaced by $\frac{x}{2}$, indicating a horizontal stretch factor of 2. Thus, the period 2π of the parent graph is also stretched by a factor of 2. So the function $y = -5\cos\left(\frac{x}{2}\right)$ has a period of $2 \cdot 2\pi = 4\pi$.

c. The period of $y = -5\cos\left(\frac{x}{2}\right)$ is 4π. So, its frequency is $\frac{1}{4\pi}$.

d. Graphs of $y = -5\cos\left(\frac{x}{2}\right)$ and $y = \cos x$ are shown below. You can verify by inspection that the amplitude and period found above are correct. (Remember, $4\pi \approx 12.6$.)

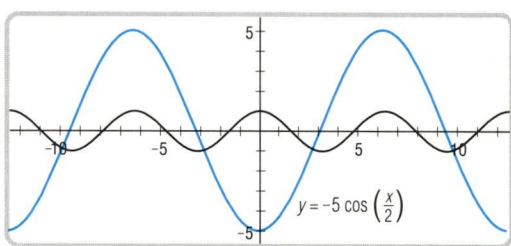

$$y = -5 \cos\left(\frac{x}{2}\right)$$

In general, from the Graph Scale Change Theorem, the functions defined by

$$y = b \sin\left(\frac{x}{a}\right) \text{ and } y = b \cos\left(\frac{x}{a}\right),$$

where $a \ne 0$ and $b \ne 0$, are images of the parent functions

$$y = \sin x \text{ and } y = \cos x$$

under the scale change that maps (x, y) to (ax, by). The theorem below indicates the relationship of the constants to the properties of the graph.

342

Theorem

The functions defined by $y = b \sin\left(\dfrac{x}{a}\right)$ and $y = b \cos\left(\dfrac{x}{a}\right)$ have amplitude $|b|$ and period $2\pi |a|$.

A proof of the theorem is just a generalization of the reasoning in parts **a** and **b** of Example 3.

Example 4 The graph below shows an image of $y = \sin x$ under a scale change. Find an equation for f.

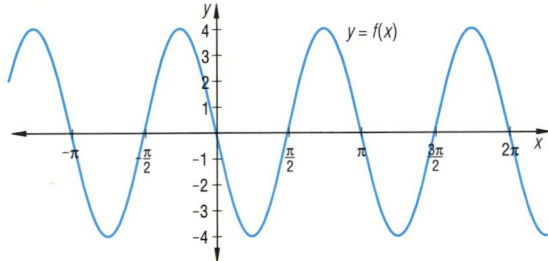

Solution An equation for f is of the form $y = b \sin\left(\dfrac{x}{a}\right)$. From the graph, the amplitude is 4 and the period is π. So $|b| = 4$ and $2\pi|a| = \pi$.

Thus $b = 4$ or -4 and $a = \dfrac{1}{2}$ or $-\dfrac{1}{2}$. Consider the four possibilities.

$$y = 4 \sin(2x)$$
$$y = 4 \sin(-2x)$$
$$y = -4 \sin(2x)$$
$$y = -4 \sin(-2x)$$

Since $\sin(-x) = -\sin x$, the first and fourth equations are equivalent, as are the second and third. Substituting a few specific points shows that either $y = 4 \sin(-2x)$ or $y = -4 \sin(2x)$ could be an equation for the given function.

Check Use an automatic grapher to graph either equation $y = 4 \sin(-2x)$ or $y = -4 \sin(2x)$.

Questions

Covering the Reading

1. a. Find an equation of the image of the function $y = \sin x$ under the transformation $(x, y) \rightarrow \left(4x, \dfrac{y}{3}\right)$. **$y = \dfrac{1}{3} \sin \dfrac{x}{4}$**

b. Graph the parent and the image when $-4\pi \le x \le 8\pi$ on the same set of axes. **See margin.**

c. Find the amplitude of the image. **$\dfrac{1}{3}$**

d. Find the period of the image. **8π**

LESSON 6-1 Scale Changes of Circular Functions **343**

mine the amplitude and period. Now fit the scale on the axes to the given function. Any of the equations in this lesson can be graphed by this process; this is why the underlying theorem is called the Graph Scale Change Theorem.

Making Connections for Question 10: This is a modeling question in the sense that the graph is given first and then an equation must be determined. Situations like this will be studied further in Lesson 6-4.

ADDITIONAL ANSWERS

1.b.

2.d.

5.b.

6. sample:

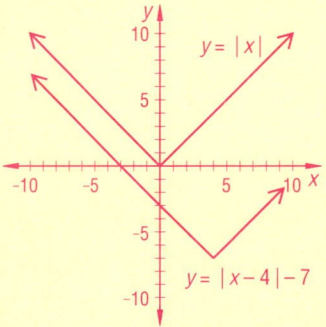
2. Consider the function $y = \frac{1}{2} \cos x$.

   a. *True* or *false*. The graph of this function is a sine wave.   **True**

   b. What is its period?   **2π**

   c. What is its amplitude?   **$\frac{1}{2}$**

   d. Sketch a graph of $y = \frac{1}{2} \cos x$ and $y = \cos x$ on the same set of axes for $-\pi \le x \le 2\pi$.   **See margin.**

3. The graph of $y = 4 \sin 2x$ is sketched below.

   a. Identify the amplitude.   **4**

   b. Give the period.   **π**

   c. If this graph represents a sound wave, then it is **i.** times as loud and has **ii.** times the frequency of the parent sound wave.   **i) 4; ii) 2**

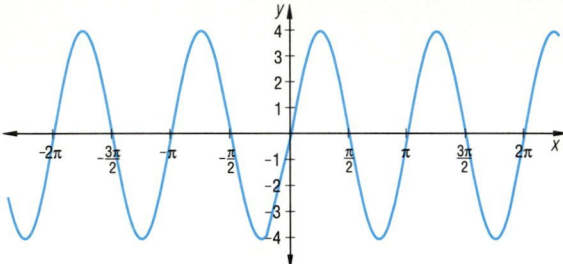

4. If one tone has a frequency of 440 cycles per second, and a second has a frequency of 880 cycles per second, which has the higher pitch and how much higher is that pitch?   **The second has a higher pitch by one octave.**

5. a. Give the period and amplitude of $y = \frac{1}{3} \sin 5x$.   **period $= \frac{2\pi}{5}$; ampl. $= \frac{1}{3}$**

   b. Check using an automatic grapher.   **See margin.**

6. Sketch one complete cycle of $8y = \cos \frac{x}{12}$ by hand, and label the zeros of the function.   **See margin.**

In 7 and 8, *multiple choice*.

7. Which equation could yield the graph below?

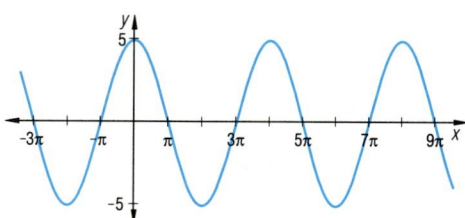

   (a) $y = 5 \sin 2x$ (b) $y = 5 \cos 2x$ (c) $y = 5 \sin \frac{x}{2}$ (d) $y = 5 \cos \frac{x}{2}$
   (d)

8. A sound wave whose parent graph is $y = \sin x$ has five times the frequency, and is four times as loud, as the parent. What is a possible equation for this sound wave?

   (a) $y = 5 \sin 4x$ (b) $y = 4 \sin 5x$ (c) $y = 4 \sin \frac{1}{5}x$ (d) $y = \frac{1}{4} \sin \frac{1}{5}x$
   (b)

344

9. Consider the function g with the equation $g(x) = \sin\left(-\frac{x}{3}\right)$.
    a. Use the Opposites Theorem to rewrite the equation for g without a negative argument. $g(x) = -\sin\left(\frac{x}{3}\right)$
    b. Give the period and the amplitude of g. **period $= 6\pi$; amplitude $= 1$**
    c. Sketch a graph of g on the domain $-3\pi \le x \le 6\pi$. **See margin.**

10. Residential electricity is called AC for "alternating current," because the direction of current flow alternates through a circuit. The current (measured in amperes) is a sine function of time. The graph below models an AC situation.

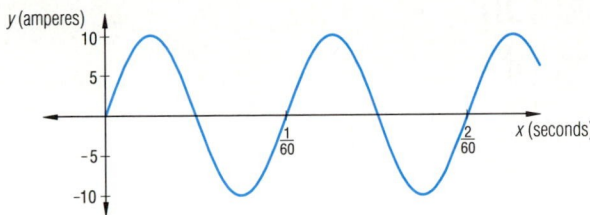

    a. Write an equation for current (amperes) as a function of time (seconds). $y = 10 \sin(120\pi x)$
    b. Find the current produced at 0.04 seconds. **about 5.9 amperes**

11. Consider the functions $f(x) = \tan x$, $g(x) = \tan 2x$, and $h(x) = 2\tan x$.
    a. Plot graphs of these functions on the same set of axes, when $-\frac{\pi}{2} \le x \le \frac{5\pi}{2}$. **See margin.**
    b. What transformation maps f to g? $(x, y) \rightarrow (x/2, y)$
    c. What transformation maps f to h? $(x, y) \rightarrow (x, 2y)$
    d. Which two functions have the same period? **f and h**
    e. Why is the amplitude of each function undefined? **None of the functions has a maximum or minimum.**

12. Consider $f(x) = |x|$ and the transformation $T: (x, y) \rightarrow (x + 4, y - 7)$.
    a. Find an equation for the image of f under T. $y = |x - 4| - 7$
    b. On one set of axes, sketch a graph of f and its image under T.
    (Lesson 3-2) **See margin.**

13. A periodic function is graphed at the left. Identify
    a. the domain and range; **domain $= \{$reals$\}$; range $= \{y: 1 \le y \le 3\}$**
    b. any maximum or minimum values; **max $= 3$, min $= 1$**
    c. the period. *(Lessons 5-6, 2-1)* **2**

14. State the Graph Translation Theorem. *(Lesson 3-2)* **See margin.**

In 15 and 16, do not use a calculator. *True* or *false*. *(Lessons 4-7, 4-2)*

15. $\left(\frac{9}{16}\right)^{3/2} = \frac{27}{64}$ **True**

16. $\frac{\log 15}{\log 3} = \log 5$ **False**

17. Other than pitch and intensity, what other characteristics do sounds have? **Some characteristics are related to pitch (pureness of tone, overtones, Doppler effect) or to intensity (damping, threshhold); also, noise and echoes.**

---

14. In an equation for a relation, if x is replaced by $x - h$ and y by $y - k$, the graph of the resulting equation is the image of the graph of the original equation under the translation which maps (x, y) to $(x + h, y + k)$.

NAME _____

■**PROPERTIES** *Objective D (See pages 393–396 for objectives.)*

1. a. Find an equation for the image of the graph of $y = \cos x$ under the transformation $(x, y) \rightarrow \left(2x, \frac{y}{4}\right)$. $y = \frac{1}{4}\cos\left(\frac{x}{2}\right)$

   b. Find the amplitude of the image. $\frac{1}{4}$

   c. State the period of the image. 4π

2. Consider the function $y = \frac{1}{2}\sin x$.
   a. What is its period? 2π
   b. What is its amplitude? $\frac{1}{2}$
   c. *True or false.* The graph of $y = \frac{1}{2}\sin x$ makes twice as many cycles from 0 to 2π as the graph of $y = \sin x$. **False**

In 3–5, give a. the period and b. the amplitude of each function.

3. $y = \sin\left(\frac{x}{3}\right)$ a. 6π b. 1

4. $y = 4\cos(2x)$ a. π b. 4

5. $\frac{y}{3} = \frac{\sin x}{2}$ a. 2π b. $\frac{3}{2}$

■**USES** *Objective G*

6. A certain sound wave has equation $y = 30\cos 10\pi t$. Give an equation of a sound wave with half the frequency and twice the loudness of this one. $y = 60\cos 5\pi t$

7. One tone has a frequency of 440 cycles per second, and a second has a frequency of 220 cycles per second. Which has the higher pitch, and by how much is it higher? The tone with the frequency of 440 cycles per second has a pitch one octave higher than the pitch of the other tone.

NAME _____
Lesson MASTER 6-1 (page 2)

8. A sound wave whose parent graph is $y = 2\sin x$ has four times the frequency and half the loudness of the parent. What is a possible equation for this sound wave?
   (a) $y = 4\sin\frac{1}{2}x$ (b) $y = \sin 4x$
   (c) $y = \sin\frac{1}{4}x$ (d) $y = 8\sin\frac{1}{4}x$ **(b)**

■**REPRESENTATIONS** *Objective H*
In 9 and 10, sketch one cycle of the graph without using an automatic grapher.

9. $y = 2\sin 2x$

10. $y = 4\cos\frac{\pi}{2}x$

■**REPRESENTATIONS** *Objective I*

11. *Multiple choice.* Which is the equation corresponding to the following graph?

   (a) $2y = \sin\frac{\pi}{2}x$ (b) $y = 2\sin x$
   (c) $y = 2\sin\frac{1}{4}x$ (d) $y = \frac{1}{2}\sin 4x$ **(b)**

OBJECTIVES

D Identify the amplitude, period, frequency, phase shift, and other properties of circular functions.

G Model phenomena with transformation images of circular functions.

H Graph transformation images of circular functions.

I State equations for graphs of circular functions.

TEACHING NOTES

This lesson extends the transformations of graphs of circular functions to include translations. Students should be able to predict the effect of translations on the position of sine waves in the coordinate plane, and to check their predictions using an automatic grapher. This is easy for most students.

There are two basic reasons for the skill in this lesson. First, in order to model the sine waves that occur in real situations, it is necessary to be able to position the wave anywhere in the coordinate plane. As the name *phase shift* indicates, the particular application of sine waves to alternating current is quite important. Second, translations illustrate and confirm certain properties of circular functions. For instance, the property that $\cos\left(x - \frac{\pi}{2}\right) =$ $\sin x$ for all x is confirmed by noting that a shift of the

Translation Images of Circular Functions

Below are graphs of the parent sine function and its image under the translation $(x, y) \rightarrow (x, y + 2)$. According to the Graph Translation Theorem, an equation for the image is $y - 2 = \sin x$ or $y = \sin x + 2$.

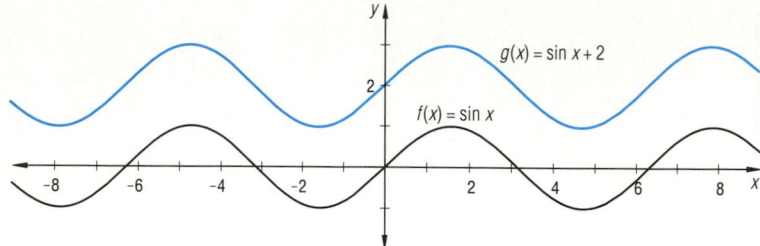

The maximum and minimum values of the sine function are 1 and -1, respectively. So the maximum and minimum values of $y = \sin x + 2$ are 3 and 1, respectively. Thus the amplitude of $y = \sin x + 2$ is $\frac{1}{2}(3 - 1) = \frac{1}{2} \cdot 2 = 1$, the amplitude of the parent sine function.

Similarly, the period of $y = \sin x + 2$ equals 2π, the period of the parent sine function.

In general, if two curves are translation images of each other, then they are congruent. Thus, translation of a sine wave preserves both its amplitude and its period.

Below are graphs of $f(x) = \sin x$ and its image under a translation two units to the right. By the Graph Translation Theorem, an equation for the image is $t(x) = \sin(x - 2)$. The graph of t is congruent to the graphs of f and g.

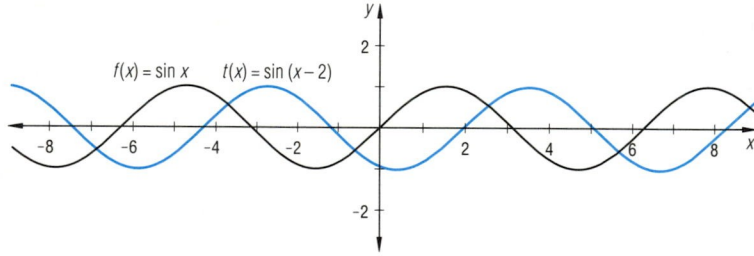

Because circular functions model many natural phenomena such as sound and electricity, horizontal translations of them have a special name—*phase shift*. In general, the **phase shift** of a sine wave is the least positive or the greatest negative horizontal translation that maps the graph of $y = \cos x$ or $y = \sin x$ onto the given wave. The phase shift of the function $y = \sin(x - h)$ is h units, if $-2\pi < h < 2\pi$. In the case above, the phase shift of the function t from the sine function is 2.

346

Example 1 Consider the function $f(x) = \sin\left(x + \frac{\pi}{3}\right)$.

a. Identify the phase shift. **b.** Sketch two cycles of the curve.

Solution

a. The graph of f is the image of $y = \sin x$ under a horizontal translation of $-\frac{\pi}{3}$ units. Thus the phase shift is $-\frac{\pi}{3}$.

b. Lightly draw two cycles of the parent graph.

Translate each zero, maximum, and minimum point $\frac{\pi}{3}$ units to the left, and sketch a graph of $f(x) = \sin\left(x + \frac{\pi}{3}\right)$ through the image points. The result, in color below, is two cycles of f.

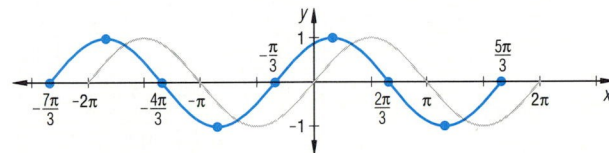

Label the intercepts.

Even when using a calculator or computer to sketch graphs, it is worth your time to anticipate their features.

Example 2 Consider the function defined by $y = \cos(x - \pi) + 1$.
a. Compare and contrast the graph of the function with the graph of its parent.
b. Graph the function on an automatic grapher.

Solution

a. Rewrite the sentence to fit the Graph Translation Theorem.
$$y - 1 = \cos(x - \pi)$$
This function is the image of the parent $y = \cos x$ under a translation right π units and up 1 unit. Both the parent and image have amplitude 1 and period 2π.
b. Part of the graph of $y = \cos(x - \pi) + 1$ is shown below.

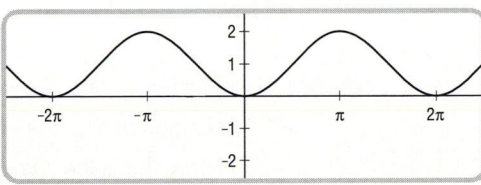

(The check follows on the next page.)

graph of $y = \cos x$ to the left by $\frac{\pi}{2}$ units yields the graph of $y = \sin x$.

Making Connections
The general formulas for $\cos(a - b)$ and $\sin(a - b)$ also could be used to confirm properties of phase shifts; these are discussed in Chapter 11.

Technology As in the previous lesson, it is valuable for your students to verify the graphs in the examples during class discussion. Seeing the graphs drawn dynamically on an automatic grapher helps emphasize which graph is the image and which is the preimage. Again, color highlights this distinction even more vividly.

We do not expect students to become conversant in the language of voltage, current, and inductance. The examples are included to show students the origin of the term *phase shift* and to point out the importance of sine waves in applications.

ADDITIONAL EXAMPLES
1. a. Graph two cycles of the function $y = \cos\left(x - \frac{\pi}{2}\right)$.

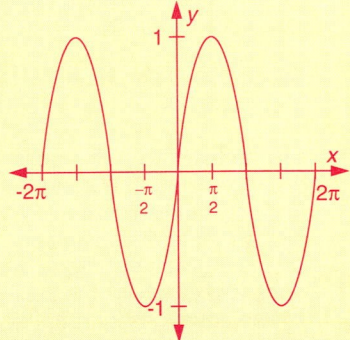

b. What is the phase shift of this function from $y = \cos x$?
$\frac{\pi}{2}$

c. What is the phase shift of this function from $y = \sin x$?
0; it is the same function.

2. Compare the graphs of $y = \sin x$ and $y = \sin\left(x + \frac{5\pi}{6}\right) + 2$.

Check Verify several points on the graph.

Try $(\pi, 2)$.   Does $2 = \cos (\pi - \pi) + 1$?
                 Yes, $2 = \cos 0 + 1$. It checks.

Try $(0, 0)$.   Does $0 = \cos (0 - \pi) + 1$?
                       $0 = \cos (-\pi) + 1$?
                 Yes, $0 = -1 + 1$. It checks.

In the above example, part of the translation of the parent $y = \cos x$ is "up 1 unit." This is often called a *vertical shift* of 1.

From an analysis of the phase shift you can determine an equation for a translation image of any of the parent circular functions.

Example 3   The graph below is a translation image of the graph of $y = \tan x$. Find an equation for it.

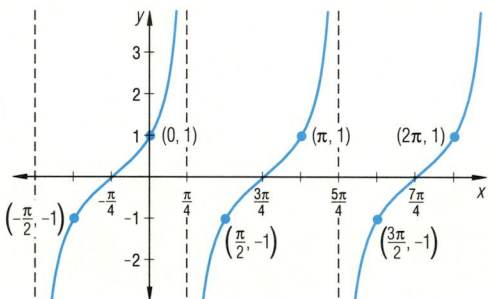

Solution   Inspecting the zeros and asymptotes, the graph is the image of the graph of $y = \tan x$ under a translation of $\frac{\pi}{4}$ to the left. By the Graph Translation Theorem, an equation for the graph is $y = \tan\left(x + \frac{\pi}{4}\right)$.

Check   Substitute some points on the graph. Try $\left(\frac{3\pi}{4}, 0\right)$. Does $0 = \tan\left(\frac{3\pi}{4} + \frac{\pi}{4}\right)$? Does $0 = \tan \pi$? Yes. At $x = \frac{\pi}{4}$, $\tan\left(\frac{\pi}{4} + \frac{\pi}{4}\right) = \tan \frac{\pi}{2}$, which is undefined. It checks.

Electrical engineers have long been concerned with phase shifts. In an alternating current circuit, for example, the voltage and the current flow are sinusoidal functions of time. If these sine waves coincide, then they are said to be *in phase*. If the current flow lags behind the voltage, then the circuit is *out-of-phase* and an *inductance* is created. Inductance helps to keep current flow stable.

Example 4   Maximum inductance in an alternating current occurs when the current flow lags behind the voltage by $\frac{\pi}{2}$. Assume that the current and the voltage have the same amplitude and period, and that the voltage is given by $y = \sin x$.

348

In a situation of maximum inductance, find an equation for the current, and sketch the two sine waves.

Solution Maximum inductance occurs when the current has phase shift $\frac{\pi}{2}$. So an equation for the current is $y = \sin\left(x - \frac{\pi}{2}\right)$. Both waves are graphed at the right.

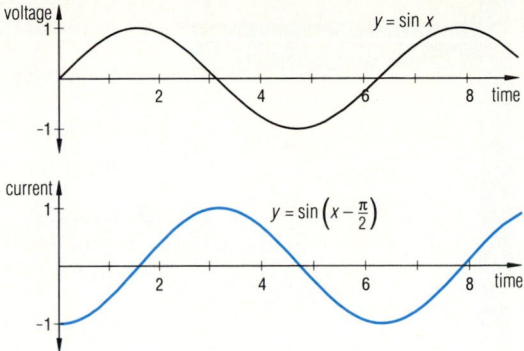

Because the goal in Example 4 was simply to illustrate maximum inductance, any sine wave could have been used. Below are graphs of $y = \cos x$ (black) and $y = \cos\left(x - \frac{\pi}{2}\right)$ (blue), which also are out of phase by $\frac{\pi}{2}$, the voltage wave leading the current wave.

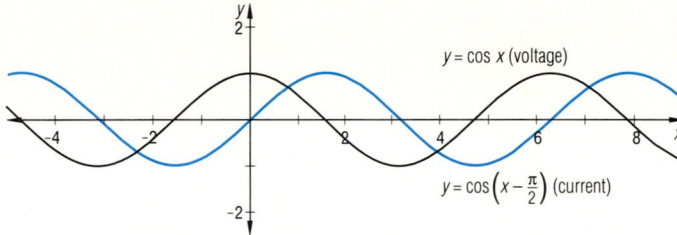

Examine all four graphs above. Notice that the graph of $y = \cos\left(x - \frac{\pi}{2}\right)$ coincides with the graph of $y = \sin x$. This suggests an identity similar to the Complements Theorem, namely that for all real numbers x,

$$\cos\left(x - \frac{\pi}{2}\right) = \sin x.$$

This statement can be proved using the Opposites and Complements Theorems.

$\cos\left(x - \frac{\pi}{2}\right) = \cos\left(-\left(x - \frac{\pi}{2}\right)\right)$ $\cos(-x) = \cos x$

$\qquad\qquad = \cos\left(\frac{\pi}{2} - x\right)$ Distributive and Commutative Properties

$\qquad\qquad = \sin x$ Complements Theorem

Many other properties may be suggested to you by translating sine waves using a function grapher, perhaps including some not covered in this text.

Making Connections for Question 21: This skill sequence is practice for the idea of thinking of trigonometric expressions as chunks, needed in the equation-solving of Lesson 6-6.

ADDITIONAL ANSWERS

1.b.

2. The graph of the function $y = \sin\left(x - \frac{\pi}{4}\right) - 1$ is the image of $y = \sin x$ under a translation $\frac{\pi}{4}$ units to the right and one unit down.

3.

Questions

Covering the Reading

1. Consider the function $y = \cos\left(x + \dfrac{\pi}{3}\right)$. **a.** Identify the phase shift. $-\dfrac{\pi}{3}$
   b. Sketch two cycles of the curve. **See margin.**

2. Compare and contrast the graph of the function $y = \sin\left(x - \dfrac{\pi}{4}\right) - 1$ and the graph of its parent. **See margin.**

3. Sketch by hand the graph of $y = \cos\left(x - \dfrac{\pi}{3}\right)$ for $-2\pi \le x \le 2\pi$.
   See margin.

In 4–7, match the equation with its graph.

4. $y = \sin\left(x + \dfrac{\pi}{3}\right)$ **(c)**

5. $y = \sin\left(x - \dfrac{\pi}{2}\right)$ **(b)**

6. $y = \cos x + 1$ **(a)**

7. $y = \sin x - 1$ **(d)**

(a)

(b)

(c)

(d)

(e)
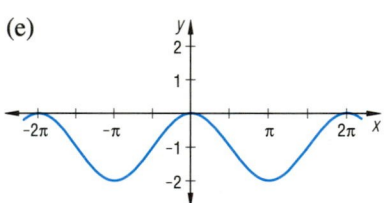

8. Sketch by hand two cycles of $y = \sin(x + \pi) + 2$. **See margin.**

9. The graph at the right is a translation image of the graph of $y = \tan x$. Find an equation for the graph.

   sample: $y = \tan\left(x - \dfrac{\pi}{2}\right) + 1$

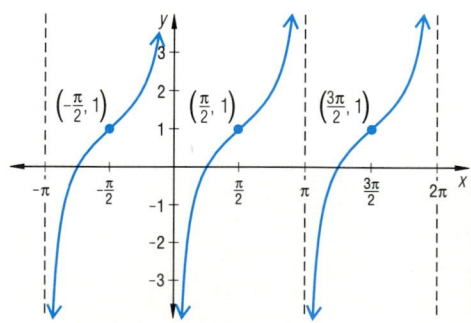

350

350

Applying the Mathematics

In 10 and 11, write an equation for the function that is graphed, using the sine function as the parent function.

10.

sample: $y = \sin\left(x - \frac{\pi}{3}\right)$

11.

sample: $y = \sin x + 2$

12. Maximum *capacitance* occurs in an alternating current circuit when the current flow leads the voltage by $\frac{\pi}{2}$. If the voltage wave is modeled by $y = 2 \cos x$, give an equation for the current flow for maximum capacitance. **$y = 2\cos(x + \pi/2)$**

13. Write four equations for the sine wave shown, two using sine and two using cosine.

samples: $y = -\sin x$; $y = -\sin(x - 2\pi)$; $y = \cos\left(x + \frac{\pi}{2}\right)$; $y = -\cos\left(x - \frac{\pi}{2}\right)$

14. *True* or *false*. $\sin\left(x - \frac{\pi}{2}\right) = -\cos x$. If true, prove using appropriate theorems; if false, give a counterexample. **True; See margin.**

15. a. Sketch or print a hardcopy of the graph of $f(\alpha) = \tan\left(\alpha - \frac{\pi}{6}\right)$ on the interval $-\frac{\pi}{2} \le \alpha \le 2\pi$. **See margin.**

  b. Give equations for the asymptotes of the function on this interval.
  $\alpha = -\frac{\pi}{3}$, $\alpha = \frac{2\pi}{3}$, $\alpha = \frac{5\pi}{3}$

Review

16. Consider the sine wave $f(x) = 2 \sin(3x)$. *(Lesson 6-1)*
  a. Graph f and its parent on the same set of axes for $-\pi \le x \le 2\pi$.
  b. Identify the amplitude and frequency of f.
  a) See margin. b) amplitude = 2; frequency = $(2\pi)/3$

17. a. Graph $y = \cos 3x$ and $y = 3 \cos x$ on the same axes. **See margin.**
  b. Approximate to the nearest hundredth at least one value of x between 0 and 2π where $\cos 3x = 3 \cos x$. *(Lesson 6-1)* **≈ 1.57**

In 18 and 19, give exact values. *(Lesson 5-4)*

18. $\sin\left(-\frac{2\pi}{3}\right)$ **$-\frac{\sqrt{3}}{2}$**

19. $\cos\frac{9\pi}{4}$ **$\frac{\sqrt{2}}{2}$**

20. Given $g(x) = |x|$ and $p(x) = 1 - x^2$, let $c(x) = g(p(x))$.
  a. Write an expression for c. **$c(x) = |1 - x^2|$**
  b. State the domain and range of c. *(Lesson 2-8)* **domain = the set of all real numbers; range = the set of all positive real numbers**

21. *Skill sequence*. Expand. *(Previous course)*
  a. $(3x - 2y)^2$ **b.** $[3(x - 5) - 2(y + 2)]^2$ **c.** $[3 \sin x - 2 \cos x]^2$
  a) $9x^2 - 12xy + 4y^2$; b) $9x^2 - 12xy + 4y^2 - 114x + 76y + 361$
  c) $9\sin^2 x - 12(\sin x)(\cos x) + 4\cos^2 x$

Exploration

22. Find out how phase shift is interpreted in the study of sound.
  Answers will vary; "acoustic impedance" is related to the effectiveness of an instrument as a transmitter of sound.

FOLLOW-UP

MORE PRACTICE
For more questions on SPUR Objectives, use *Lesson Master 6-2*, shown below.

NAME _____

LESSON **MASTER 6-2**
QUESTIONS ON **SPUR** OBJECTIVES

■**PROPERTIES** *Objective D (See pages 393–396 for objectives.)*

1. Consider the graph of the function $f(x) = \sin\left(x - \frac{\pi}{2}\right)$.
  What is its phase shift from the parent sine curve? $\frac{\pi}{2}$

2. What is the phase shift of the graph of $y = 2\cos(x + \pi) - 1$ from $y = \cos x$? $-\pi$

■**USES** *Objective G*

3. If a voltage flow in an alternating current circuit is modeled by $y = 2 \sin x$, state a function for the current that has the same amplitude and period, but has phase shift such that maximum inductance occurs.
  $y = 2\sin\left(x - \frac{\pi}{2}\right)$

■**REPRESENTATIONS** *Objective H*
In 4–7, sketch a graph of the function.

4. $y = \cos(x - \pi)$

5. $f(t) = 2 \sin(t - 2\pi)$

6. $f(x) = \cos\left(x - \frac{\pi}{2}\right) - 1$

7. $y = \tan\left(x + \frac{\pi}{6}\right) - 1$

Functions, Statistics, and Trigonometry © Scott, Foresman and Company   **67**

NAME _____
Lesson MASTER 6-2 (page 2)

■**REPRESENTATIONS** *Objective I*

8. a. Write an equation for the image of $y = \cos x$ under a phase shift of $\frac{-\pi}{3}$.
  $y = \cos\left(x + \frac{\pi}{3}\right)$

  b. Graph the image in part a.
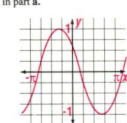

9. Write an equation for the translation image of $y = \tan x$ that is graphed below.

  $y = \tan\left(x + \frac{\pi}{4}\right)$

68   *Functions, Statistics, and Trigonometry © Scott, Foresman and Company*

LESSON 6-3

RESOURCES
- Lesson Master 6-3
- Quiz for Lessons 6-1 through 6-3
- Teaching Aid 53: Graphs for linear changes of a parabola
- Computer Master 16

OBJECTIVES

D Identify the amplitude, period, frequency, phase shift, and other properties of circular functions.

G Model phenomena with transformation images of circular functions.

H Graph transformation images of circular functions.

I State equations for graphs of circular functions.

TEACHING NOTES

Making Connections
The functions studied in this lesson are often seen in the form $y = A \sin (Bx + C) + D$ or $y = A \sin B (x - C) + D$. We have chosen the form $\dfrac{y - k}{b} = \sin \dfrac{x - h}{a}$ for two reasons. First, this form is a natural extension of the work we have done with graphs of all functions (and relations). The parabola with which the lesson begins points this out. (A teaching aid is provided for further discussion.) Second, the expressions $\dfrac{y - k}{b}$ and $\dfrac{x - h}{a}$ are of the same form as the expression $\dfrac{x - \mu}{\sigma}$ used in statistics to standardize data so that the mean is 0 and standard deviation is 1. This idea, to be applied also in

LESSON

6-3

Linear Changes of Circular Functions

In Lessons 6-1 and 6-2 you learned how, in circular functions, scale changes affect amplitude and frequency and how translations introduce phase shifts. In this lesson you will learn how composites of scale changes and translations affect circular functions. First, consider how such a composite affects a familiar non-periodic function. Let S be the scale change $S:(x, y) = (3x, 2y)$; let T be the translation $T(x, y) = (x - 4, y + 1)$. If S is applied to the parent parabola $y = x^2$, and then T to the result, the graphs below are produced.

Parent S (Parent) $T \circ S$ (Parent)

The images of the parent function $y = x^2$ under S and $T \circ S$ can be described algebraically. By the Graph Scale Change Theorem, the image of $y = x^2$ under the scale change $(x, y) \rightarrow (3x, 2y)$ is

$$\frac{y}{2} = \left(\frac{x}{3}\right)^2.$$

Thus the image of $y = x^2$ under the composite $T \circ S$ is

$$\frac{y - 1}{2} = \left(\frac{x + 4}{3}\right)^2.$$

The composite $T \circ S$ is an example of a *linear change*. It is called a linear change because when $T \circ S$ is expressed as a single transformation, linear expressions arise. Specifically,

$$T \circ S(x, y) = T(3x, 2y)$$
$$= (3x - 4, 2y + 1).$$

So $T \circ S$ maps (x, y) to $(3x - 4, 2y + 1)$, and $3x - 4$ and $2y + 1$ are linear expressions.

In general, a **linear change** is the composite of scale changes and translations. Suppose S is a scale change and T a translation with

$$S(x, y) = (ax, by), \text{ where } a \neq 0 \text{ and } b \neq 0,$$
$$\text{and } T(x, y) = (x + h, y + k).$$

Then the linear change $\quad T \circ S(x, y) = T(S(x, y))$

$$= T(ax, by)$$
$$= (ax + h, by + k).$$

Thus $T \circ S$ maps (x, y) to $(ax + h, by + k)$.

To see how such a linear change affects the equation of a relation, consider the image (x', y') of the point (x, y) under the linear change.

352

Then $\qquad x' = ax + h \qquad$ and $\qquad y' = by + k.$

So $\qquad \dfrac{x' - h}{a} = x \qquad$ and $\qquad \dfrac{y' - k}{b} = y.$

This justifies the following theorem.

Graph Standardization Theorem

In a relation described by a sentence in x and y, the following yield the same graph:
(1) applying the scale change $(x, y) \rightarrow (ax, by)$ where $a \neq 0$ and $b \neq 0$, followed by applying the translation $(x, y) \rightarrow (x + h, y + k)$;
(2) applying the linear change $(x, y) \rightarrow (ax + h, by + k)$ to the graph of the original sentence;
(3) replacing x by $\dfrac{x - h}{a}$ and y by $\dfrac{y - k}{b}$ in the sentence for the relation.

The word "standardization" is used in the Graph Standardization Theorem for two reasons. First, linear changes are often used to transform data into "standard" scores; you will study this application in Chapter 10. Second, such linear changes are often used to generate what are called "standard forms" of equations for many curves.

Example 1   **a.** Use the Graph Standardization Theorem to describe the graph of
$$y = \cos\left(\dfrac{x + \pi}{3}\right).$$
   b. Check part **a** using an automatic grapher.

Solution
a. The given equation results from replacing x by $\frac{x + \pi}{3}$ in the equation $y = \cos x$. Thus it is the image of $y = \cos x$ under the scale change $(x, y) \rightarrow (3x, y)$ followed by the translation $(x, y) \rightarrow (x - \pi, y)$. This transformation stretches horizontally by a factor of 3, then translates π units to the left. The parent and image under the scale change are in black and blue; the image under the composite is in orange.

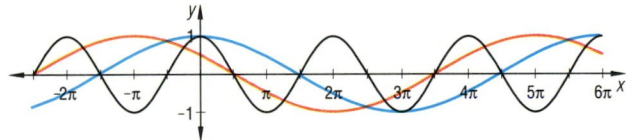

b. A graph of $y = \cos\left(\frac{x + \pi}{3}\right)$ on the domain $-8 \leq x \leq 27$ is below.

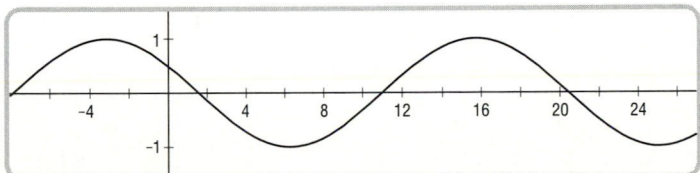

Chapter 10, indicates why we called the theorem of this lesson the Graph Standardization Theorem.

In order to use the valuable notion of a phase shift, under which all points on a graph are translated a fixed amount horizontally, the shape of the graph must be transformed first. Thus, on page 352, the scale changes of the parent function come before the translation T. If this is not done, then the intersection of the graph and the y-axis is the *only* point which truly undergoes a phase shift of $\dfrac{\pi}{3}$; all other points are shifted by $\dfrac{\pi}{3}$ modified by the horizontal scale change of 3.

Technology   Few would graph **Example 2** by hand. The goal of the lesson is for students to check what is drawn with an automatic grapher.

You might center class discussion around **Example 3**. Here the origin of the coordinate system is ground level under the center of the wheel. Concentrate on students' understanding of the relationships between the equation and the graph, not on the origin of the equation. Project 6 on page 389 develops the equation in detail.

The model in **Example 3** is a natural extension of the definition of $(\cos \theta, \sin \theta)$ as a rotation image of $(1, 0)$ about the origin. Now the circle has radius 20 and is not centered at the origin (the vertical shift), the amount of revolution is a function of time (the scale change), and the boarding is not at a point on the same horizontal line as the center of the wheel (the horizontal shift).

ADDITIONAL EXAMPLES
1. a. Describe the graph of
$$y = \sin 2\left(x - \dfrac{\pi}{4}\right).$$

From the graphs in Example 1 you can see that the amplitude of
$y = \cos\left(\frac{x+\pi}{3}\right)$ is 1. You can also ascertain that its period is 6π, and its phase
shift is π.

The theorem below, which follows directly from the Graph Standardization
Theorem, indicates the relationship of the constants in the equation to the
properties of the graph.

> **Theorem**
>
> The graphs of the functions with equations
> $$\frac{y-k}{b} = \sin\left(\frac{x-h}{a}\right) \text{ and } \frac{y-k}{b} = \cos\left(\frac{x-h}{a}\right), a \neq 0, b \neq 0,$$
> have: amplitude $|b|$,
> period $2\pi|a|$,
> phase shift h from the parent functions,
> and vertical shift k.

Example 2   Consider $y = 2\sin(3x + \pi)$.
a. Draw a graph using an automatic grapher.
b. Describe the graph as the image of $y = \sin x$ under a composite of
transformations.
c. Determine the amplitude, period, and phase shift of the graph.

Solution
a. Here is a graph of $y = 2\sin(3x + \pi)$.

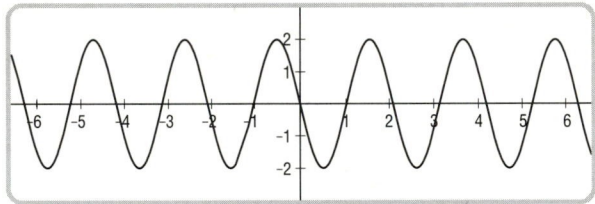

b. Rewrite the equation as $\frac{y}{2} = \sin(3x + \pi) = \sin\left(3\left(x + \frac{\pi}{3}\right)\right)$
$$= \sin\left(\frac{x + \frac{\pi}{3}}{\frac{1}{3}}\right).$$

The graph is the image of $y = \sin x$ under a horizontal scale change with
a factor of $\frac{1}{3}$, a vertical scale change with a factor of 2, and a translation
$\frac{\pi}{3}$ to the left.

c. The amplitude of the sine wave is 2. The period is $\frac{2\pi}{3}$, one-third that of
$y = \sin x$. The phase shift is $-\frac{\pi}{3}$.

354

Check for part c   The extreme values of the function are 2 and -2, so the amplitude is $\frac{1}{2}(2--2) = \frac{1}{2}(4) = 2$. The interval $-\frac{\pi}{3}$ to $\frac{\pi}{3}$ contains a complete wave, so the period is $\frac{2\pi}{3}$. The point (0, 0) on $y = \sin x$ has been translated to $\left(-\frac{\pi}{3}, 0\right)$, so the phase shift is $-\frac{\pi}{3}$.

Circular functions of the form studied in this lesson naturally arise from *circular motion*, which describes a point moving around a circle at uniform speed.

Example 3   Suppose the top of a Ferris wheel 40 feet in diameter is 45 feet above the ground, and it takes 10 seconds to reach the top from the boarding point. If the wheel turns at a uniform speed of 2 revolutions per minute, it can be shown that the equation $d = 25 + 20 \cos\left(\frac{\pi}{15}(t - 10)\right)$ gives the distance d feet a person is above the ground at time t seconds. (You may wish to consult the Chapter 6 Projects for more information.) Graph and analyze this function.

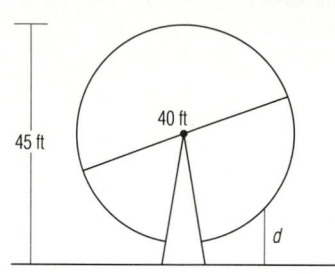

Solution   For such a complicated function use an automatic grapher. A graph is shown below.

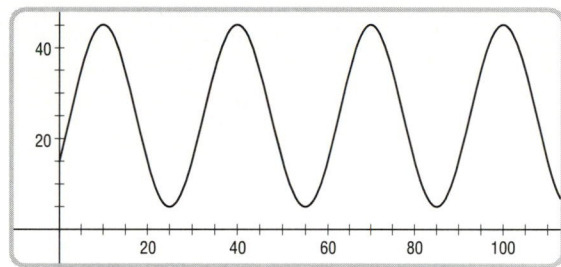

To analyze the function, rewrite its equation in the form

$$\frac{y - k}{b} = \cos\left(\frac{x - h}{a}\right).$$

$$\frac{d - 25}{20} = \cos\left(\frac{\pi}{15}(t - 10)\right)$$

$$= \cos\left(\frac{t - 10}{\frac{15}{\pi}}\right)$$

From this form you can deduce that the following transformations have been applied to $y = \cos x$ to get the given function: a horizontal stretch by a factor of $\frac{15}{\pi}$ and a vertical stretch by a factor of 20, followed by a translation of 10 to the right and 25 up. Thus, the period of f is $\frac{15}{\pi}$ times that of $\cos x$ or $2\pi \cdot \frac{15}{\pi} = 30$. The amplitude of f is 20, the phase shift is 10, and the vertical shift is 25. Note also that the domain is the set of nonnegative numbers, because negative time is not reasonable in this situation.

4.d.

5.a. amplitude = 2; period = 2; phase shift = 0; vertical shift = 1

b.

6.a. amplitude = 6; period = π; phase shift = $-\frac{3}{2}$; vertical shift = -8

b.

11.b.

FOLLOW-UP

MORE PRACTICE
For more questions on SPUR
Objectives, use *Lesson Master 6-3*, shown on page 357.

EXTENSION
Ask students to separate the
linear changes in this lesson
into the scale changes and
translations that make them
up. For instance, the linear
change of Question 6 is the
composite of the scale
change $S(x, y) = \left(\frac{1}{2}x, 6y\right)$ and
the translation $T(x, y) =$
$\left(x - \frac{3}{2}, y - 8\right)$.

PROJECTS
Project 6 is related to the content of this lesson.

EVALUATION
A quiz covering Lessons 6-1
through 6-3 is provided in the
Teacher's Resource File.

ADDITIONAL ANSWERS
3.a., b., 4.d., 5.a., b., 6.a.,
b., 11.b. See margin,
pages 354–355.

12.b.

13. $(S \cdot T)(x, y)$
  $= S(x + h, y + k)$
  $= (a(x + h), b(y + k))$
 $(T \cdot S)(x, y) = T(ax, by)$
  $= (ax + h, by + k)$
  $\neq (a(x + h), b(y + k))$

15.a.

Questions

Covering the Reading

1. A linear change is a composite of what two types of transformations?
   scale changes and translations
2. *True* or *false*. A linear change is a transformation that maps (x, y) to
   $(ax + h, by + k)$ where a, b, h, and k are constants and $a \neq 0$ and $b \neq 0$.
   True
3. Consider the function $y = \sin\left(\frac{x + \pi}{2}\right)$.
   a. Sketch a graph of the function for $-2\pi \leq x \leq 4\pi$.
   b. State the amplitude, period, and phase shift of the graph.
   c. The graph is the image of $y = \sin x$ under the composite of what two
      transformations? $(x, y) \rightarrow (2x, y)$ followed by $(x, y) \rightarrow (x - \pi, y)$
   a,b) See margin.
4. Consider the graph of $y = 3\cos\left(\frac{x - \pi}{4}\right)$.
   a. State its amplitude.   **3**
   b. What is its period?   **8π**
   c. Find the number of units the graph is translated horizontally from the
      graph of $y = \cos\frac{x}{4}$.   **π**
   d. Sketch the graph on the interval $-4\pi \leq x \leq 8\pi$.   **See margin.**

In 5 and 6, **a.** give the amplitude, period, phase shift, and vertical shift of the
graph. **b.** Sketch the graph.

5. $y = 2\cos \pi x + 1$
   See margin.
6. $s = -8 + 6\sin(2t + 3)$
   See margin.

7. Refer to Example 3. How many seconds after boarding does a person
   first get to be 40 feet above the ground?   **≈ 6.5 seconds**

Applying the Mathematics

In 8 and 9, write an equation for a function whose graph will have the given
characteristics.

8. parent $y = \cos x$, phase shift $\frac{\pi}{2}$, period π, amplitude 2   **$y = 2\cos 2(x - \pi/2)$**

9. parent $y = \sin x$, phase shift π, period $\frac{\pi}{2}$, amplitude 4   **$y = 4\sin 4(x - \pi)$**

10. Write an equation for the cosine curve below (Hint: the function has
    period 2π).

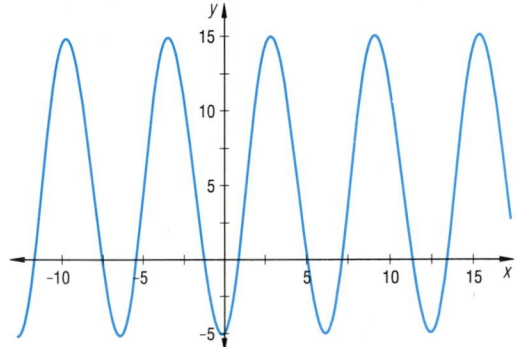

$y = 10\cos(x - \pi) + 5$

356

11. Suppose $f(x) = \tan\left(\frac{x-\pi}{3}\right)$.

    a. Describe a transformation that maps $y = \tan x$ to $y = f(x)$.

    b. Draw a graph of $y = f(x)$.   **See margin.**

    c. State the period and phase shift of the graph of $f(x)$.

     a) $(x, y) \rightarrow (3x + \pi, y)$; c) period $= 3\pi$; phase shift $= \pi$

12. Let $S = \left(2x, \frac{y}{3}\right)$ and $T = (x + 5, y - 1)$.

    a. Find an equation for the image of $y = x^2$ under $T \circ S$.

    b. Graph the image of $y = x^2$ under $T \circ S$.   **See margin.**

    c. The composite $T \circ S$ maps (x, y) to what point?   **$\left(2x + 5, \frac{y}{3} - 1\right)$**

    a) $3(y - 1) = ((x - 5)/2)^2$

13. Let $S(x, y) = (ax, by)$ and $T(x, y) = (x + h, y + k)$.
True or *false*. $(S \circ T)(x, y) = (T \circ S)(x, y)$. Justify your answer.
False. See margin.

14. Find a scale change S and a transformation T such that $y = \sin\left(\frac{x + \pi}{2}\right)$ is the image of $y = \sin x$ under $S \circ T$.   **$S(x, y) = (2x, y)$; $T(x, y) = (x - \pi, y)$**

Review

15. a. Graph $y = \cos\frac{x}{2}$ and $y = \sin 3x$ on $0 \le x \le \pi$.   **See margin.**

    b. Give approximate solutions correct to the nearest hundredth to
$\cos\frac{x}{2} = \sin 3x$ for $0 \le x \le \pi$. *(Lesson 6-1)*   **0.45, 0.63, 2.24, 3.14**

16. Find all x between 0 and 4π such that $\tan x = \sqrt{3}$. *(Lesson 5-5)*
  $\dfrac{\pi}{3}, \dfrac{4\pi}{3}, \dfrac{7\pi}{3}, \dfrac{10\pi}{3}$

In 17 and 18, **a.** find $f(g(x))$ and $g(f(x))$. **b.** Are f and g inverses of each other? *(Lessons 4-6, 4-1, 3-8)*

17. $f(x) = x^3 - 1$; $g(x) = x^{1/3} + 1$
a) $f(g(x)) = (x^{1/3} + 1)^3 - 1$ or $x + 3x^{2/3} + 3x^{1/3}$; $g(f(x)) = (x^3 - 1)^{1/3} + 1$; b) No

18. $f(x) = e^{x/2}$; $g(x) = \ln x^2$ **a) $f(g(x)) = x$; $g(f(x)) = x$; b) Yes**

19. The height h (in feet) above the ground of a model rocket at time t (in seconds) is given by $h = -32t^2 + 160t + 192$. *(Lessons 3-1, 2-6)*

    a. Graph this function.   **See margin.**

    b. Find the maximum height of the rocket.   **392 feet**

    c. When will the rocket hit the ground?   **after 6 seconds**

Exploration

20. a. Draw a unit circle.

    b. Draw its image under the scale change $(x, y) \rightarrow (3x, 2y)$.

    c. Draw the image of the figure produced in part **b** under the translation $(x, y) \rightarrow (x - 4, y + 1)$.

    d. Choose another scale change and translation, and draw the image of the unit circle under the composite of these transformations.

    e. Make a conjecture: what type of figure is the image of a circle under a linear change?

    a–c) See margin. d) Answers may vary. e) an ellipse

NAME _____

■**PROPERTIES** *Objective D (See pages 393–396 for objectives.)*

1. Consider the function $y = \frac{1}{3}\cos\left(\frac{x+\pi}{2}\right)$. State

  a. the amplitude.   **b.** the period.   **c.** the phase shift.

    $\frac{1}{3}$ 4π $-\pi$

2. Consider $y = 3\sin(2x - \pi) + 2$. Determine

  a. the amplitude.   **b.** the period.   **c.** the phase shift.

    3   π $\frac{\pi}{2}$

3. Suppose the transformation $(x, y) \rightarrow (3x + 2, \frac{y}{2} - 2)$ is applied to the graph of $y = \cos x$.

  a. State an equation for the image. $y = \frac{1}{2}\cos\left(\frac{x-2}{3}\right) - 2$

  b. Find the amplitude, period, phase shift, and vertical shift of the image.
amplitude $= \frac{1}{2}$, period $= 6\pi$, phase shift $= 2$, vertical shift $= -2$

4. Let $S(x, y) = \left(\frac{x}{2}, -y\right)$ and $T(x, y) = (x - 4, 2y)$. Find the image of $y = \cos x$ under the composite transformation $T \circ S$. $y = -2\cos(2x + 8)$

In 5 and 6, write a function whose graph will have the given characteristics.

5. parent $y = \cos x$, phase shift $\frac{\pi}{3}$, period 2π, amplitude 4. $y = 4\cos\left(x - \frac{\pi}{3}\right)$

6. parent $y = \sin x$, phase shift π, period $\frac{\pi}{4}$, amplitude 2. $y = 2\sin(8(x - \pi))$

■**USES** *Objective G*

7. A simple pendulum is shown below. The angular displacement from vertical (in radians) as a function of time t in seconds is given by $f(t) = \frac{3}{5}\sin\left(3t + \frac{\pi}{2}\right)$.

  a. What is the angular displacement for $t = 0$? $\frac{3}{5}$

  b. What is the frequency of f? $\frac{3}{2}\pi$ **swings/sec.**

  c. How long will it take for the pendulum to make 3 complete swings? $\frac{2}{\pi}$ **seconds**

NAME _____
Lesson MASTER 6-3 (page 2)

8. The figure at the right shows a Ferris wheel rotating at 50 revolutions per hour. The distance d of point P from the ground as a function of time t in hours can be modeled by a sine wave equation of the form
$$\frac{d-k}{b} = \frac{\sin(t-h)}{a}$$

  a. What is the amplitude of the sine wave?   **20**

  b. What is the period of the sine wave? $\frac{1}{50}$ **of a hour, or 72 sec**

  c. When does point P reach its highest point, if it is at the bottom of the Ferris wheel at time $t = 0$?   **in 36 seconds**

■**REPRESENTATIONS** *Objective H*
In 9–11, sketch a graph of the function.

9. $\frac{y-1}{3} = \cos\left(x - \frac{\pi}{2}\right)$ **10.** $y = 6 - 5\cos(2x + 1)$ **11.** $\frac{y+2}{2} = \sin\left(\frac{1}{2}x + 2\pi\right)$

■**REPRESENTATIONS** *Objective I* $y = 2\cos\left(x - \frac{\pi}{4}\right)$

12. Give an equation for the cosine wave below.

Modeling with Circular Functions

Seattle, Washington

TEACHING NOTES

Emphasize the order of the steps in the process used in the examples. First the amplitude and period are determined to find the magnitudes *a* and *b* of the scale changes inherent in the situation. Then the translation(s) of the scaled graph are determined. This order allows any horizontal translation to be treated as a phase shift.

Any situation that can be modeled by a sine function can be modeled by a cosine function, and vice versa. The choice of which to use can be subtle or even arbitrary. For the Seattle daylight situation in **Example 1,** a sine model is chosen because the "starting" point of the wave is on the *y*-axis. A cosine model would require both vertical and horizontal translations, so the sine model is simpler. In the satellite situation of **Example 2,** however, either model requires vertical and horizontal translations. We chose a cosine model to balance the use of functions in the lesson. **Questions 3 and 7** ask for second models of these situations using the complementary function. A teaching aid is provided with the graphs arising from each of these examples.

Many phenomena are periodic. For instance, the graph below gives the number of hours *y* of daylight for Seattle, WA, as a function of the number *x* of days after March 21.

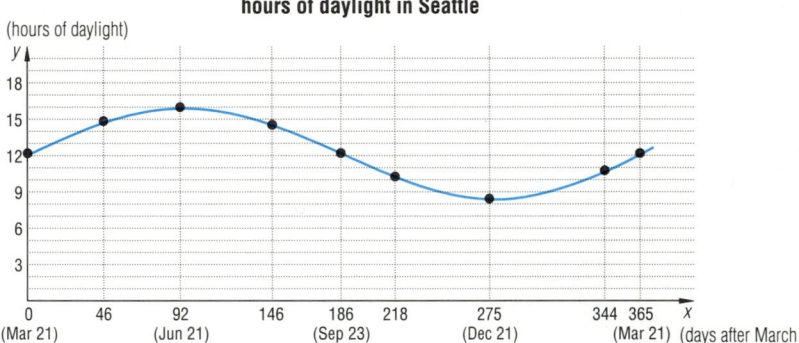

hours of daylight in Seattle

March 21 is the vernal equinox, the date on which night and day are approximately of equal length. From March 21 until June 21 the number of hours of daylight in Seattle increases to a maximum of about 16. From June 21 until December 21 the number of hours of daylight decreases, reaching 12 again around September 21 (the autumnal equinox) and a minimum of about 8.5 on December 21. Then the number of hours of daylight starts to increase again. The intermediate days on the graph are values for May 6 ($x = 46$), August 14 ($x = 146$), October 25 ($x = 218$), and February 28 ($x = 344$). A sine (or cosine) wave appears to be an appropriate model for the relation between *x* and *y*.

Because the graph above appears to be the image of the parent function $y = \sin x$ under the composite of scale changes and translations, an equation of the form

$$\frac{y - k}{b} = \sin\left(\frac{x - h}{a}\right)$$

can be used as a model for this situation. Each of the constants *a*, *b*, *h*, and *k* can be determined from the graph.

358

The maximum number of hours of daylight in Seattle is 16 and the minimum is 8.5. The amplitude is half the range of hours, so

$$b = \frac{1}{2}(16 - 8.5) = 3.75.$$

Ignoring leap year, the period is 365 days.
Thus,

$$2\pi a = 365$$

so

$$a = \frac{365}{2\pi}.$$

Thus so far, the equation is $\dfrac{y - k}{3.75} = \sin\left(\dfrac{x - h}{\frac{365}{2\pi}}\right).$

Finally, compare a graph of $\dfrac{y}{3.75} = \sin\left(\dfrac{x}{\frac{365}{2\pi}}\right)$ with that of the graph of the

number of hours of daylight in Seattle. Notice that a vertical translation maps $(0, 0)$ onto $(0, 12.25)$. Thus, the translation

$$(x, y) \rightarrow (x, y + 12.25)$$

produces the image from the parent. This implies that $h = 0$ and $k = 12.25$ in the model. So the final equation

$$\frac{y - 12.25}{3.75} = \sin\left(\frac{x}{\frac{365}{2\pi}}\right)$$

models the relation between x, the number of days after March 21, and y, the number of hours of daylight.

To check this equation, substitute values of x, calculate y, and verify that the points are on the graph. For instance, if $x = 275$ (which corresponds to December 21),

$$\frac{y - 12.25}{3.75} = \sin\left(\frac{275}{\frac{365}{2\pi}}\right)$$

$$y = 3.75 \sin\left(\frac{2\pi(275)}{365}\right) + 12.25$$

$$\approx 8.5,$$

and $(275, 8.5)$ is on the graph.

The final equation can be used to predict the number of hours of daylight for any specified date of the year. For instance, to predict the number of hours of daylight in Seattle on July 30, note that July 30 is the 131st day after March 21. So evaluate y when $x = 131$.

$$\frac{y - 12.25}{3.75} = \sin\left(\frac{131}{\frac{365}{2\pi}}\right)$$

$$y = 3.75 \sin\left(\frac{2\pi \cdot 131}{365}\right) + 12.25$$

$$y \approx 15.2 \text{ hours}$$

You can also transform the equation so that the independent variable represents the day of the year beginning with January 1 instead of the day after March 21. Example 1 shows how to do this.

Making Connections
The sine wave model illustrates several phenomena associated with the number of hours of daylight: The daylight periods are all about the same length around June 21st, when they are longest. They are also all about the same length around December 21st, when they are shortest. The daylight period changes most around the equinoxes, March 21st and September 21st. These ideas, which involve rate of change, are studied in more detail in *Precalculus and Discrete Mathematics*.

ADDITIONAL EXAMPLES
1. In Houston, New Orleans, and Jacksonville, all of which are at about 30°N latitude, the longest daylight period (June 21) is about 14 hours 5 minutes (845 minutes) long, and the shortest daylight period (December 21) is about 10 hours 13 minutes (613 minutes) long. The vernal equinox (March 21) has about 12 hours 9 minutes (729 minutes) of daylight. Assuming a sine wave model for the length of daylight, find an equation which gives the number of minutes of daylight in these three cities as a function of the day of the year.
The amplitude is 116 minutes, the period is 365, so $a = \dfrac{365}{2\pi}$, the vertical translation has magnitude 729 minutes, and the horizontal translation is of magnitude 80 (as in Example 1). Thus an equation is $\dfrac{y - 729}{116} =$

$\sin\left(\dfrac{x - 80}{\frac{365}{2\pi}}\right).$

Note that these cities are closer to the equator than Seattle, so the number of hours of daylight does not vary as much—the amplitude of the sine curve is smaller.

2. In the situation of **Example 2,** how far north (or south) of the equator is the satellite 45 minutes after launch?

$y = 4500 \cos \frac{\pi}{60}(45 - 15)$

= 0, so the satellite is directly above the equator.

NOTES ON QUESTIONS
Error Analysis for Question 3: Students sometimes think that on new questions they cannot use answers previously obtained. The hint suggests that they should use a previously known piece of information.

Questions 3 and 7: These questions show that the same data can be described either by a sine function or a cosine function.

In general, replace x by $x - \frac{\pi}{2}$ to change from sine to cosine, and replace x by $x + \frac{\pi}{2}$ to change from cosine to sine.

Question 9: Because temperatures are not solely determined by the latitude of a location, they do not provide as good a fit to a sine wave as length of daylight.

Small Group Work for Question 20: Small groups of students may select locations of interest to them. One group should try to find data for your location (if it is not already mentioned in this lesson). Your local weather bureau has such information.

■ ■ ■ ■ ■ ■ ■

Example 1   **a.** Find an equation which gives the number of hours of daylight in Seattle as a function of the day of the year.
b. Use it to confirm the number of hours of daylight for July 30.

Solution

a. March 21 is the 80th day of the year, so replace x by $x - 80$ in the "final equation" above. This yields the equation

$$\frac{y - 12.25}{3.75} = \sin\left(\frac{x - 80}{\frac{365}{2\pi}}\right)$$

where x now represents the day of the year.

b. July 30 is the 211th day of the year. So,

$$\frac{y - 12.25}{3.75} = \sin\left(\frac{211 - 80}{\frac{365}{2\pi}}\right)$$

$$y = 3.75 \sin\left(\frac{2\pi \cdot 131}{365}\right) + 12.25 \approx 15.2 \text{ hours,}$$

the same value predicted by the original model.

The graphs of the two equations are shown below,

$$y = 3.75 \sin \frac{2\pi x}{365} + 12.25 \text{ in black}$$

and

$$y = 3.75 \sin\left(\frac{2\pi(x - 80)}{365}\right) + 12.25 \text{ in blue.}$$

hours of daylight in Seattle

Note that as is predicted by the Graph Standardization Theorem, the graph of the second equation shows a phase shift of 80 units (days) from the first.

■ ■ ■ ■ ■ ■ ■ ■

Example 2   A satellite is launched from Cape Canaveral into an orbit which goes alternately north and south of the equator. Its distance from the equator over time can be approximated by a sine wave. Suppose that the satellite is y kilometers north of the equator at time x minutes. (When the satellite is south of the equator, $y < 0$.) It reaches 4500 km, its farthest point north of the equator, 15 minutes after the launch. Half an orbit later it is 4500 km south of the equator, its farthest point south. Each orbit takes 2 hours.

a. Find an equation using the cosine function which models the distance of the satellite from the equator.
b. How far away from the equator is the satellite 1 hour after the launch?

360

360

Solution

a. Sketch a graph with x in minutes, y in kilometers. You are given that the point (15, 4500) is on the graph. Also, a full orbit takes 120 minutes, so half an orbit takes 60 minutes. Thus, the satellite will be 4500 km south of the equator 60 min after it is 4500 km north of the equator. So (75, -4500) is on the graph.

To model this situation with a cosine function, use $\dfrac{y-k}{b} = \cos\left(\dfrac{x-h}{a}\right)$

and follow the steps used in the Seattle daylight problem. First find the amplitude b.

$$b = \tfrac{1}{2}(4500 - (-4500)) = 4500$$

Next calculate a. The period (one full orbit) is 120 minutes, so

$$2\pi a = 120$$

and

$$a = \frac{60}{\pi}.$$

Finally, compare the graph of the satellite's position to the graph of $\dfrac{y}{4500} = \cos\left(\dfrac{x}{\frac{60}{\pi}}\right)$. The graph above is the image of the graph of

$\dfrac{y}{4500} = \cos\left(\dfrac{x}{\frac{60}{\pi}}\right)$ under the translation $(x, y) \rightarrow (x + 15, y)$. So to get an

equation for the position of the satellite relative to the equator, replace x

by $x - 15$ in the equation $\dfrac{y}{4500} = \cos\left(\dfrac{x}{\frac{60}{\pi}}\right)$. The desired equation is

$$\frac{y}{4500} = \cos\left(\frac{x-15}{\frac{60}{\pi}}\right)$$

or

$$\frac{y}{4500} = \cos\left(\frac{\pi}{60}(x-15)\right).$$

b. One hour after launch, $x = 60$.

So

$$\frac{y}{4500} = \cos\left(\frac{\pi}{60}(60 - 15)\right)$$

$$y = 4500\cos\left(\frac{3\pi}{4}\right)$$

$$y \approx -3180.$$

The satellite will be about 3180 km south of the equator after one hour.

Check The point (60, -3180) appears to be on the graph, and so part **b** and the graph serve as checks for the model.

Landsat satellite

9.a.,b.

10.a.,b.

c. sample: $\dfrac{y - 12.5}{7} =$

$\cos\left(\dfrac{2\pi(x - 172)}{365}\right)$

11.a. $\frac{1}{3}$, **b.** 10π, **c.** 0, **d.** 0

e.

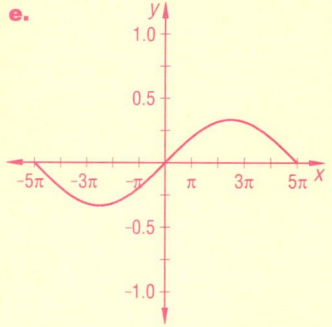

12.a. 1.5, **b.** 5π, **c.** 2.5, **d.** 1
e.

Questions

Covering the Reading

In 1–3, refer to the two equations for the Seattle daylight problem given in this lesson.

1. Verify that both equations predict about 16 hours of daylight on June 21. **See margin.**

2. How many hours of daylight do the models predict for July 4? **about 15.9 hours**

3. **a.** Model this situation with an equation of the form $\dfrac{y - k}{b} = \cos\left(\dfrac{x - h}{a}\right)$. [Hint: Do you need to start from scratch?]
   b. Use this model to confirm your prediction in Question 2.
   a) See margin. b) about 15.9 hours

In 4–8, refer to Example 2.

4. Verify that the satellite is directly over the equator $1\frac{3}{4}$ hours into its flight. **See margin.**

5. How far north of the equator is Cape Canaveral? **≈ 3180 km**

6. Without using a calculator, about how far from the equator is the satellite after 3 hours? **≈ 3180 km**

7. Write an equation in the form $\dfrac{y - k}{b} = \sin\left(\dfrac{x - h}{a}\right)$ to model the satellite's position relative to the equator. **sample:** $\dfrac{y}{4500} = \sin\left(\dfrac{x + 15}{60/\pi}\right)$

8. Another satellite launched from Cape Canaveral has an orbit of 4 hours. Its farthest distances north and south are 5200 km, which occur 20 minutes and $2\frac{1}{3}$ hours after launch, respectively.
   a. Write an equation to describe the distance y of the satellite from the equator x hours after launch.
   b. How far south of the equator will this satellite be after 1 hour?
   a) $\dfrac{y}{5200} = \cos\left(\dfrac{x - 20}{120/\pi}\right)$;   **b) 2600 km**

Applying the Mathematics

9. Average monthly temperatures in degrees Fahrenheit for Chicago for the years 1951–81 are given below.

Month	Jan	Feb	Mar	Apr	May	Jun	Jul	Aug	Sep	Oct	Nov	Dec
Avg. temp (°F)	21.4	26.0	36.0	48.8	59.1	68.6	73.0	71.9	64.7	53.5	39.8	27.7

   a. Plot these data on a scattergram. Plot x = the month after January on the horizontal axis, and y = the temperature on the vertical axis.
   b. Sketch a good-looking sine curve to fit the data. **a,b) See margin.**
   c. What is the period of this function? **12**
   d. Estimate the amplitude a of your curve to the nearest degree.
   e. *Multiple choice.* Which of these four models is the best for these data?

   (i) $\dfrac{y}{a} = \cos\left(\dfrac{\pi x}{6}\right)$ (ii) $\dfrac{y - 46}{a} = \cos\left(\dfrac{\pi x}{6}\right)$

   (iii) $\dfrac{y - 46}{-a} = \cos\left(\dfrac{\pi x}{6}\right)$ (iv) $\dfrac{y - 26}{a} = \sin\left(\dfrac{\pi x}{6}\right)$ **(iii)**

   f. Find another equation equivalent to your answer to part **e** that also describes these data. **sample:** $\dfrac{y - 46}{-26} = \sin\left(\dfrac{\pi(x + 3)}{6}\right)$

   d) sample: about 26 degrees

362

10. Listed below are the hours of daylight in Anchorage, AK, for ten days of the year.

Date	Jan 1	Feb 28	Mar 21	Apr 27	May 6	June 21	Aug 14	Sep 23	Oct 25	Dec 21
Hours	5.59	10.23	12.38	15.91	16.71	19.40	15.93	12.16	9.14	5.44

  a. Sketch a scatterplot of the data indicating dates as days of the year.
  b. Using paper and pencil, fit a sine wave to the scatterplot.
  c. Determine an equation of a cosine function which would model the data.
  d. Using the model in part **c**, estimate the hours of daylight on July 3 in Anchorage.   **a,b,c) See margin. d) sample: about 19.4 hours**

Review

In 11 and 12, for the graph give the **a.** amplitude, **b.** period, **c.** phase shift, and **d.** vertical shift, and **e.** draw the graph on an interval that shows at least one complete wave.   *(Lessons 6-3, 6-1)*

11. $3y = \sin\left(\frac{x}{5}\right)$ **See margin.**   **12.** $\frac{y-1}{1.5} = \cos\left(\frac{x-2.5}{2.5}\right)$ **See margin.**

In 13 and 14, the graph is a translation image of a parent trigonometric function. Find an equation for the graph.   *(Lesson 6-2)*

13.

14.

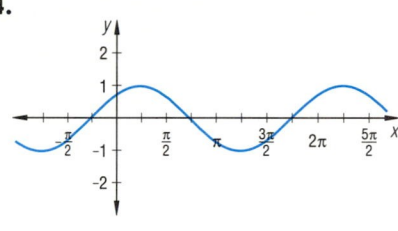

sample: $y = \tan\left(x + \frac{\pi}{3}\right)$ sample: $y = \cos\left(x - \frac{\pi}{4}\right)$

15. Solve $\sin c = 0$, for $0 \le c \le 6\pi$.   *(Lesson 5-6)*   **$0, \pi, 2\pi, 3\pi, 4\pi, 5\pi,$ or 6π**

In 16 and 17, solve to the nearest tenth.   *(Lessons 4-6, 4-3)*

16. $0.3^a = 9$ **-1.8**   **17.** $\ln x = -0.11$ **0.9**

18. *Skill sequence.* Factor.   *(Previous course)*
  a. $k^2 - 1$ **b.** $25t^2 - 9$ **c.** $(p + 3)^2 - 16$
  (k + 1)(k − 1)   **(5t + 3)(5t − 3)**   **(p + 7)(p − 1)**

19. Draw a scatterplot with 15 points (x, y) which show a weak, negative correlation between x and y.   *(Lesson 3-6)*   **sample: See margin.**

Exploration

20. a. How is the latitude of a city mathematically related to the number of hours of daylight at that city?
  b. Collect some data for a city other than Seattle and Anchorage and test your hypothesis in part **a**.   **Answers may vary.**
  a) Sample: the further from the equator, the greater the difference in number of hours of daylight for summer and winter.

6-5

Inverse Circular Functions

Many situations arise where the measure of an angle is unknown, and circular functions allow you to find it. For instance, suppose a plane flying at an altitude of 5.5 miles begins its descent 90 miles from an airport. The angle of the path of descent can be found using trigonometry.

The sides adjacent to θ and opposite θ are known, as the diagram above indicates. So, $\tan \theta = \frac{5.5}{90} \approx 0.0611$.

To find an angle whose tangent is about .0611, the INV or 2nd key on a calculator can be used. The key sequence $5.5 \div 90 = $ INV tan results in a displayed value of about 3.497. So the angle of descent is about 3.5°.

The solution to the problem above required the *inverse* of the tangent function, which is preprogrammed into scientific calculators. In symbols, the inverse of the tangent function is tan^{-1}. You can write $\theta = \tan^{-1} 0.0611$, which is read "$\theta$ is the angle whose tangent is 0.0611."

In this lesson, properties of the inverses of the sine, cosine, and tangent functions are examined. Are the inverses of the parent trigonometric functions actually functions? You can answer this question by examining the graphs of their inverses. Recall that the graph of the inverse of a function is obtained by reflecting its graph over the line $y = x$. Below, the graphs of the parent functions are in black, and the graphs of their inverses are in blue.

By the vertical line test, none of the inverses is a function. In order to define inverses which are functions, the domains of the parent sine, cosine, and tangent functions must be restricted. The following criteria are generally used to choose an appropriate domain for each inverse function.

364

1. The domain should include the angles between 0 and $\frac{\pi}{2}$, because these are the measures of the acute angles in a right triangle.
2. On the restricted domain, the function should take on all values in its range.
3. If possible, the function should be continuous on the restricted domain.

These criteria lead to the following choices for restrictions on the domains of the cosine, sine, and tangent functions.

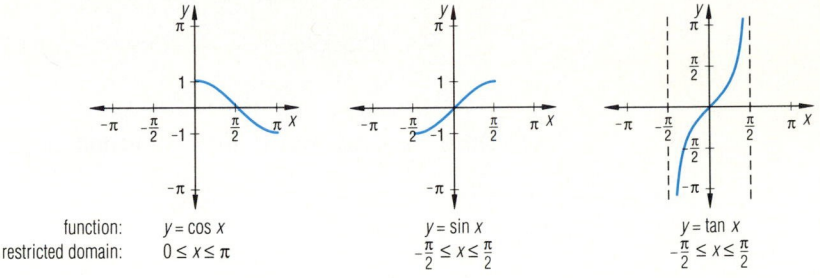

function:	$y = \cos x$	$y = \sin x$	$y = \tan x$
restricted domain:	$0 \le x \le \pi$	$-\frac{\pi}{2} \le x \le \frac{\pi}{2}$	$-\frac{\pi}{2} \le x \le \frac{\pi}{2}$

If each of the preceding graphs is reflected over the line $y = x$, the resulting graph represents a function. These functions are called *inverse trigonometric functions* or *inverse circular functions*. They are denoted cos^{-1}, sin^{-1}, and tan^{-1} and called the *inverse cosine*, *inverse sine*, and *inverse tangent* functions, respectively.

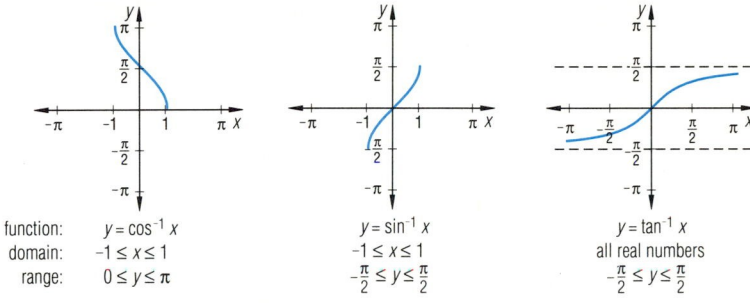

function:	$y = \cos^{-1} x$	$y = \sin^{-1} x$	$y = \tan^{-1} x$
domain:	$-1 \le x \le 1$	$-1 \le x \le 1$	all real numbers
range:	$0 \le y \le \pi$	$-\frac{\pi}{2} \le y \le \frac{\pi}{2}$	$-\frac{\pi}{2} \le y \le \frac{\pi}{2}$

These ideas are formalized in the following definitions.

Definitions

The **inverse cosine**, **inverse sine**, and **inverse tangent** functions, denoted by **$\cos^{-1}x$**, **$\sin^{-1}x$**, and **$\tan^{-1}x$**, respectively, are defined as follows:

$y = \cos^{-1}x$ if and only if $x = \cos y$ and $0 \le y \le \pi$;

$y = \sin^{-1}x$ if and only if $x = \sin y$ and $-\frac{\pi}{2} \le y \le \frac{\pi}{2}$;

$y = \tan^{-1}x$ if and only if $x = \tan y$ and $-\frac{\pi}{2} < y < \frac{\pi}{2}$.

The notations **Arccos**, **Arcsin**, and **Arctan** are sometimes used in place of $\cos^{-1}$, $\sin^{-1}$, and $\tan^{-1}$.

LESSON 6-5 Inverse Circular Functions **365**

are negative numbers x for which $\cos x = -0.1$.

With the definition of inverse used in this book, every function has an inverse. In some places the inverses of sin, cos, and so on, are called arcsin, arccos, and so on, using a small a. Then the inverses with restricted domains are called Arcsin, Arccos, and so on, using capital A. But even this use is not consistent, and definitions vary among mathematical dictionaries.

Because $\sin^2 x$ means $(\sin x)^2$ and $\sin^3 x$ means $(\sin x)^3$, it is very possible that a person could interpret $\sin^{-1} x$ to mean $(\sin x)^{-1}$. This is the reason that the names Arcsin, Arccos, and Arctan remain in use.

Error Analysis The most difficult part of dealing with inverses is reversing one's thinking. Emphasize that $\sin^{-1} x$, $\cos^{-1} x$, and $\tan^{-1} x$ are real numbers representing angles or magnitudes of rotation. Use the definition to rewrite an expression. For example, if $x = \sin^{-1}\frac{1}{2}$, then $\sin x = \frac{1}{2}$. That is, go back to the original function if students have difficulty.

A teaching aid with the graphs of the inverse trigonometric functions is provided. Another teaching aid is provided with the rather complicated figure of **Example 4.**

Some values of inverse circular functions can be found exactly without using a calculator.

Example 1 Evaluate $\sin^{-1}\left(\frac{1}{2}\right)$.

Solution If $y = \sin^{-1}\left(\frac{1}{2}\right)$, then by definition of $\sin^{-1}$, y is the unique number in the interval $-\frac{\pi}{2} \leq y \leq \frac{\pi}{2}$ whose sine is $\frac{1}{2}$. Because $\sin\frac{\pi}{6} = \frac{1}{2}$ and $\frac{\pi}{6}$ is in that interval, $y = \frac{\pi}{6}$. Thus $\sin^{-1}\left(\frac{1}{2}\right) = \frac{\pi}{6}$.

Other values of inverse circular functions require a calculator.

Example 2 Evaluate Arctan(-3.7).

Solution Apply the definition of the inverse tangent function:
$y = \text{Arctan}(-3.7)$ if and only if $\tan y = -3.7$ and $-\frac{\pi}{2} < y < \frac{\pi}{2}$. Set the calculator in radian mode, and use the key sequence

$$3.7 \boxed{+/-}\boxed{\text{INV}}\boxed{\tan}$$

or its equivalent. The calculator should display -1.3068 radians. Consequently, Arctan(-3.7) ≈ -1.3068.

Check Evaluate $\tan(-1.3068)$ with a calculator in radian mode.

Key in $1.3068\boxed{+/-}\boxed{\tan}$ or its equivalent.

This results in a display of -3.6995, which rounds to -3.7. It checks.

Inverse trigonometric functions are often used to model situations in which the dependent variable is an angle or a rotation.

Example 3 An 18-ft ladder leans against a building as shown at the left.
a. Express y, the measure of the angle the ladder makes with the ground, as a function of x, the height of the top of the ladder.
b. What are the domain and range of this function?

Solution
a. From the diagram you can see that
$$\sin y = \frac{x}{18}.$$
So
$$y = \sin^{-1}\left(\frac{x}{18}\right).$$

366

b. x is the height of the top of the ladder so the domain, allowing the ladder to be completely horizontal or vertical, is $\{x: 0 \le x \le 18\}$. y is the measure of the angle made with the ground, so the range is $\left\{ y: 0 \le y \le \frac{\pi}{2} \right\}$.

Check   Graph the function $y = \sin^{-1}\left(\frac{x}{18}\right)$. By the Graph Scale Change Theorem, you should expect to see the image of $y = \sin^{-1} x$ under the scale change $(x, y) \rightarrow (18x, y)$. Below is the graph produced on an automatic grapher using the window $-20 \le x \le 20, -2 \le y \le 2$.

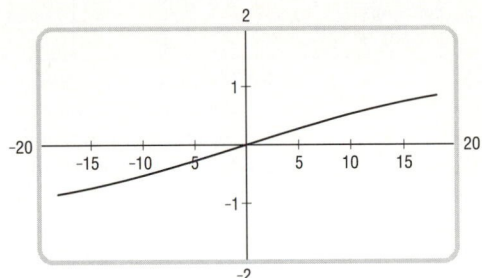

The graph shows that for the domain $\{x: 0 \le x \le 18\}$, the range is $\left\{ y: 0 \le y \le \frac{\pi}{2} \right\}$. It checks.

Example 4   The Landsat 2 satellite can only see a portion of the earth's surface (bounded by a *horizon circle* as shown below at the left) at any given time. Imagine looking at a cross section of the satellite and the earth, as shown below at the right. Point C is the center of the earth, H is a point on the horizon circle, and S is the location of the satellite. Let $\alpha = m\angle HCS$. Let r be the radius of the earth and h be the distance in km of the satellite above the earth. $\angle CHS$ is a right angle.

a. Write a formula giving α in terms of r and h.
b. The radius of the earth is about 6378 km. To the nearest tenth of a degree, what is α when Landsat 2 is 956 km above the earth?

  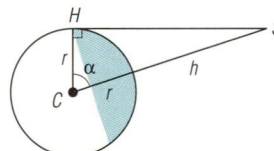

Solution

a. Given are the hypotenuse of $\triangle CHS$ and the side adjacent to α. Use the cosine function.

$$\cos \alpha = \frac{r}{r+h}$$

Solve for α.

$$\alpha = \cos^{-1}\left(\frac{r}{r+h}\right)$$

b. For $r = 6378$ and $h = 916$,

$$\alpha = \cos^{-1}\left(\frac{6378}{6378 + 956}\right)$$
$$\approx \cos^{-1}(0.869648)$$
$$\approx 29.58°.$$

So

$$\alpha = 29.6°.$$

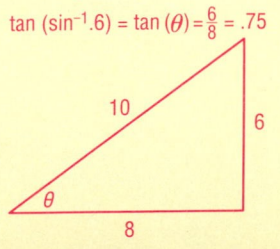
Questions

Covering the Reading

1. How is the expression "$\theta = \sin^{-1} k$" read?
   Theta is the angle whose sine is k.
2. **a.** Complete the table of values below.

points on $y = \cos x$	$(0, ?)$	$\left(\frac{\pi}{6}, ?\right)$	$\left(\frac{\pi}{4}, ?\right)$	$\left(\frac{\pi}{3}, ?\right)$	$\left(\frac{\pi}{2}, ?\right)$	$\left(\frac{2\pi}{3}, ?\right)$	$\left(\frac{3\pi}{4}, ?\right)$	$\left(\frac{5\pi}{6}, ?\right)$	$(\pi, ?)$
	1	$\sqrt{3}/2$	$1/\sqrt{2}$	$1/2$	0	$-1/2$	$-1/\sqrt{2}$	$-\sqrt{3}/2$	-1
corresponding point on $y = \cos^{-1}x$	$(1, 0)$	$\left(\frac{\sqrt{3}}{2}, \frac{\pi}{6}\right)$	$\left(\frac{1}{\sqrt{2}}, \frac{\pi}{4}\right)$	$\left(\frac{1}{2}, \frac{\pi}{3}\right)$	$\left(0, \frac{\pi}{2}\right)$	$\left(-\frac{1}{2}, \frac{2\pi}{3}\right)$	$\left(-\frac{1}{\sqrt{2}}, \frac{3\pi}{4}\right)$	$\left(-\frac{\sqrt{3}}{2}, \frac{5\pi}{6}\right)$	$(-1, \pi)$

   b. Graph $y = \cos x$ and $y = \cos^{-1} x$ on the same coordinate system. Use the same scale on each axis. **See margin.**
   c. State the domain and range of the function $y = \cos^{-1} x$.
   d. What transformation maps the graph of $y = \cos x$ onto the graph of $y = \cos^{-1}x$? **$(x, y) \rightarrow (y, x)$**
   c) domain $= \{x : -1 \le x \le 1\}$; range $= \{y : 0 \le y \le \pi\}$

In 3–5, find the exact measure of the angle in **a.** radians and **b.** degrees.

3. $\cos^{-1}\left(\frac{\sqrt{3}}{2}\right)$ **a)** $\frac{\pi}{6}$; **b)** $30°$

4. Arctan (1)   **a)** $\frac{\pi}{4}$; **b)** $45°$

5. $\sin^{-1}\left(-\frac{\sqrt{2}}{2}\right)$ **a)** $-\frac{\pi}{4}$; **b)** $-45°$

In 6–8, use a calculator to compute the radian measure of the angle.

6. $\sin^{-1}\left(\frac{\sqrt{3}}{3}\right)$ ≈ 0.615

7. Arccos ($-.9$) ≈ 2.691

8. $\tan^{-1}(-4.88)$ ≈ -1.369

9. *True* or *false*. For all n and θ:
   a. if $\theta = \sin^{-1} n$, then $n = \sin \theta$.   **True**
   b. if $n = \sin \theta$, then $\theta = \sin^{-1} n$.   **False**

10. A radar tracking station is located 2 miles from a rocket launching pad. If a rocket is launched straight upward, express the angle of elevation θ of the rocket from the tracking station as a function of the altitude a (in miles) of the rocket. (Assume there are no effects due to the rotation of the earth.) $\theta = \tan^{-1}(a/2)$

Applying the Mathematics

11. Refer to Example 4.   **11 a)** $\angle HSC = \sin^{-1}\left(\frac{r}{r+h}\right)$
    a. Give a formula for m$\angle HSC$ in terms of r and h. (m$\angle HSC$ is called the *angular separation of the horizon*.)
    b. What is the angular separation for Landsat 2 when it is 900 km from earth? $\approx 61.2°$
    c. Give a relationship between cos α and sin($\angle HSC$).   **cos α = sin($\angle HSC$)**

12. **a.** Evaluate tan(sin⁻¹ .6) on your calculator.   **0.75**
    b. Draw an appropriate triangle to show how the answer to part **a** could have been found without a calculator.   **See margin.**
    c. Evaluate $\tan\left(\sin^{-1}\frac{b}{c}\right)$, where $b \neq c$, $c \neq 0$. $\tan\left(\sin^{-1}\frac{b}{c}\right) = \frac{b}{\sqrt{c^2 - b^2}}$
    d. Does it matter if the angle is measured in degrees or in radians?   **No**

368

13. A rectangular picture 3 ft high is hung on a vertical wall so that the bottom edge is at your eye level. Your "view" of this picture is determined by the angle y at your eye cut off by the top and bottom edges of the picture.

   a. Write an equation for y as a function of the distance x between your eye and the wall. $y = \tan^{-1}(3/x)$

   b. Suppose now that the rectangular picture is hung on a vertical wall so that the bottom edge is 1 ft above your eye level. Express y as a function of x. $y = \tan^{-1}\left(\frac{4}{x}\right) - \tan^{-1}\left(\frac{1}{x}\right)$

14. Compute without using your calculator.

   a. $\sin\left(\sin^{-1}\left(-\dfrac{\sqrt{2}}{2}\right)\right)$ $\dfrac{-\sqrt{2}}{2}$ **b.** $\cos\,(\text{Arccos}\,(.6))$ 0.6

15. Use the definition of the inverse trigonometric functions to prove that $\sin(\sin^{-1} x) = x$ for all x such that $-1 \le x \le 1$.   See margin.

16. The equation $E = 3 \cos (60\pi t)$ describes the electrical voltage E in a circuit at a time t.

   a. Solve for t in terms of E. $t = \dfrac{\cos^{-1}\left(\frac{E}{3}\right)}{60\pi}$

   b. How is the graph of t as a function of E related to the graph of the parent inverse function $y = \cos^{-1} x$?   It is the image of the parent function under S: $(x,y) \rightarrow \left(3x, \frac{y}{60\pi}\right)$.

Review

In 17 and 18, a function is given. Determine its **a.** domain; **b.** range; **c.** period; and **d.** if appropriate, its amplitude. *(Lessons 6-3, 6-2)*

17. $y = \tan\left(x + \dfrac{\pi}{4}\right)$ See margin.   **18.** $y = 2 \sin (4x) - 5$ See margin.

19. Find the area of the shaded sector in circle O below. *(Lesson 5-2)*

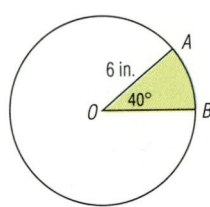

4π in.^{2}

In 20–22, solve for x. *(Previous course)*

20. $x^2 + 6x + 5 = 0$ **21.** $2x^2 = 1 - x$ **22.** $2x^2 + 6x = 10$

$-1, -5$ $\dfrac{1}{2}, -1$ $\dfrac{-3 \pm \sqrt{29}}{2}$

---

23. Use the following data from *The World Almanac* on the number of accidental deaths by gas leaks in the U.S. for each month in 1980.

Month (x)	Jan	Feb	Mar	Apr	May	Jun	Jul	Aug	Sep	Oct	Nov	Dec
Number (y)	158	139	113	84	58	68	57	51	55	119	124	216

In the scatter plot below, January 1980 is represented by 1, February 1980 by 2, and so on.

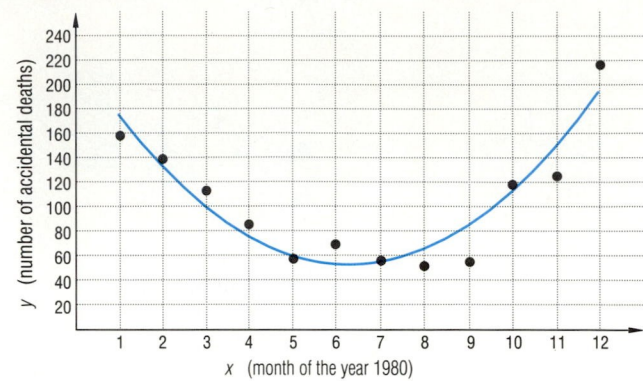

A statistics package fit the quadratic model $y = 4.4x^2 - 55.5x + 227.6$ to the data.
a. What number of accidental deaths is predicted by this model for March, 1980?   **about 101**
b. What meaning would $x = 14$ have in this context?   **February 1981**
c. Use the model to predict the number of accidental deaths for $x = 14$.
d. Explain why a periodic function is probably a better model for these data than a quadratic function.   **See margin.**
e. Estimate the amplitude and period of a sine wave that fits these data.
(Lessons 6-4, 6-1, 2-7)
c) 313; e) sample: amplitude ≈ 80; period ≈ 12 months

24. Which of the following are equal to $\sin\theta$ for all θ? *(Lesson 5-7)*
a. $\cos\left(\frac{\pi}{2} - \theta\right)$ b. $\sin(-\theta)$ c. $\sin(\pi - \theta)$ d. $\sin(2\pi + \theta)$
a, c, d

25. How many solutions between 0 and 2π are there to $\tan x = -\sqrt{3}$? *(Lesson 5-5)*   **2**

Exploration

26. Consider the functions $f(x) = \sin(\sin^{-1} x)$ and $g(x) = \sin^{-1}(\sin x)$.
a. What is the largest domain on which these functions are defined?
b. Graph each function using an automatic grapher.   **See margin.**
c. Compare and contrast the two graphs, and explain any discrepancies between your answers to parts **a** and **b**.
a) domain of $f = \{x : -1 \le x \le 1\}$; domain of g = the set of all real numbers;
c) The graph of f is a segment; the graph of g is an infinite "sawtooth."

370

Solving Trigonometric Equations

The following are examples of *trigonometric equations*.

$$40 = 52 \cos x \qquad 34 = 53 \sin \alpha - 160 \qquad 15 = 25 \cos(3t + 20)$$

When solving trigonometric equations, a fundamental issue is the domain of the variable.

Three domains commonly arise:
1. the restricted domains of the parent functions studied in the previous lesson,
2. an interval equal in size to the period of the parent function,
3. the set of all real numbers for which the function is defined.

The solution that arises from considering all real values of the variable is called the **general** solution to the trigonometric equation. To solve trigonometric equations, you need to use the inverse circular functions, the theorems studied in Chapter 5, and the period of the function involved.

■ ■ ■ ■ ■ ■ ■ ■■■

Example 1   Consider the equation $\sin \theta = 0.8$.

a. Find the solution between $-\frac{\pi}{2}$ and $\frac{\pi}{2}$.

b. Solve the equation when $0 \le \theta \le 2\pi$.

c. Find the general solution.

Round all solutions to the nearest thousandth.

Solution

a. By definition of the inverse sine function, a solution is

$$\theta = \sin^{-1} 0.8.$$

Set a calculator to radian mode, and key in

$$.8 \;\boxed{\text{INV}}\;\boxed{\text{sin}}$$

or an equivalent key sequence. The display shows

$$\theta \approx 0.927.$$

On the unit circle, this is the first-quadrant solution.

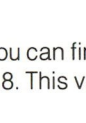

b. From the Supplements Theorem you can find another value of θ between 0 and 2π for which $\sin \theta = 0.8$. This value is the second-quadrant solution shown above.

$$\theta \approx \pi - 0.927$$
$$\approx 2.214$$

In the third and fourth quadrants, $\sin \theta \le 0$, so there are no other solutions and the solution set for $\sin \theta = 0.8$ with $0 \le \theta \le 2\pi$ is $\{0.927, 2.214\}$.

---

between 0 and 2π because the line $y = k$ intersects the graph of the sine and cosine functions in at most 2 points on this interval. Similarly, tan $x = k$ has exactly one

solution between $-\dfrac{\pi}{2}$ and $\dfrac{\pi}{2}$

because $y = k$ intersects the graph of the tangent function exactly once. Thus the key to finding all solutions is to find all the solutions in one cycle and then add integer multiples of the period to describe all other solutions.

Students should recall from previous courses that odd numbers are those of the form $2n + 1$ and even numbers are those of the form $2n$, where n is any integer. Then it is a small step to the idea that $2nx$ stands for an even multiple of x and $(2n + 1)x$ stands for an odd multiple of x. If n is allowed to be negative (which is perfectly permissible), then one does not need the symbol $\pm$ before $2nx$ or $(2n + 1)x$; the symbol + suffices.

Making Connections
The word *chunk* in **Example 3** will be familiar to students who have studied from UCSMP *Algebra*. Chunking is the process of thinking of a mathematical expression as a single entity. It is a fundamental thinking process. We chunk letters when we read words, we chunk numerals when we read multidigit numbers, we chunk when we think of $\dfrac{x - h}{a}$ as a single number, and so on.

ADDITIONAL EXAMPLES
1. Consider the equation cos $x = 0.456$.

a. Find the solution between 0 and π.
≈ **1.097**

b. Find all solutions between 0 and 2π.
≈ **2π – 1.097 ≈ 5.186 as well as 1.097**

c. Describe all real solutions.
$x \approx 1.097 + 2n\pi$ or $x \approx$ 5.186 + $2n\pi$, where n is any integer.

c. The general solution follows from part **b** and from the Periodicity Theorem. Adding or subtracting multiples of 2π to the two solutions in part **b** generates all possible solutions to sin $\theta = 0.8$.

$$
\begin{array}{ll}
\vdots & \vdots \\
\theta \approx 0.927 - 4\pi & \theta \approx 2.214 - 4\pi \\
\theta \approx 0.927 - 2\pi & \theta \approx 2.214 - 2\pi \\
\theta \approx 0.927 & \theta \approx 2.214 \\
\theta \approx 0.927 + 2\pi & \theta \approx 2.214 + 2\pi \\
\vdots & \vdots
\end{array}
$$

Thus, the general solution to sin $\theta = 0.8$ is

$$\theta \approx 0.927 + 2\pi n \quad \text{or} \quad \theta \approx 2.214 + 2\pi n, \text{ where } n \text{ is any integer.}$$

Check On the same set of axes, graph $y = \sin x$ and $y = 0.8$.
Each point of intersection of the sine wave and the line corresponds to a solution to the equation sin $x = 0.8$.

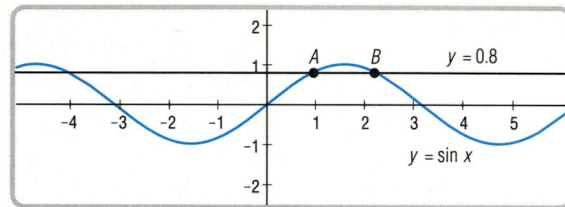

Point A represents the solution found in part **a**. Points A and B represent the two solutions found in part **b**. The infinite number of solutions found in part **c** are represented by the set of all points of intersection.

To solve more complex equations involving circular functions, find a solution using an inverse trigonometric function, and then use other properties of functions, particularly the Periodicity Theorem, to find other solutions.

■ ■ ■ ■ ■ ■ ■ ■ ■ ■

Example 2 The voltage E volts in a circuit after t seconds ($t > 0$) is given by
$$E = 12 \cos 2\pi t.$$
To the nearest .01 second, at which times in the first three seconds does $E = 10$?

Solution If $\qquad\qquad 10 = 12 \cos 2\pi t,$

then $\qquad\qquad \cos 2\pi t = \dfrac{10}{12}.$

Think of $2\pi t$ as a "chunk" representing the argument θ and solve for $2\pi t$. A calculator in radian mode gives the first-quadrant solution.

$$\theta = 2\pi t = \cos^{-1}\left(\dfrac{10}{12}\right)$$
$$\approx 0.5857$$

Another solution arises from the Opposites Theorem. So
$$\theta \approx -0.5857 \text{ also.}$$

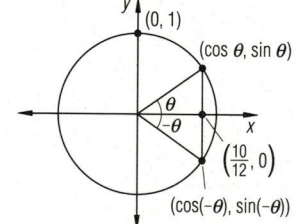

Because the period of the cosine function is 2π, if multiples of 2π are added to or subtracted from ± 0.5857, other solutions are generated. So the general solution is $2\pi t \approx 2n\pi \pm 0.5857$, for n an integer. Dividing each side by 2π gives

$$t \approx n \pm 0.0932.$$

Examining different values of n shows that there are six values of t,

$$1 - 0.0932, \ 2 - 0.0932, \ 3 - 0.0932, \ 0 + 0.0932, \ 1 + 0.0932, \ 2 + 0.0932,$$

or

$$0.9068, \qquad 1.9068, \qquad 2.9068, \qquad 0.0932, \qquad 1.0932, \qquad 2.0932,$$

in the interval $0 \le t \le 3$. E equals 10 volts after approximately 0.09, 0.91, 1.09, 1.91, 2.09, and 2.91 seconds.

Check 1 Verify that $E \approx 10$ for all six values of t. For instance for the first two values, $12 \cos (2\pi(0.09)) \approx 10$, and $12 \cos (2\pi(0.91)) \approx 10$.

Check 2 The function $E = 12 \cos 2\pi t$ has period equal to $\frac{2\pi}{2\pi} = 1$.

Consistent with the period, the three solutions 0.09, 1.09, and 2.09 represent differences of one second, as do the three solutions 0.91, 1.91, and 2.91.

In some trigonometric equations, the value of the circular function is itself a chunk. Example 3 illustrates a trigonometric equation that has quadratic form.

■ ■ ■ ■ ■ ■ ■ ■ ■■

Example 3 Suppose $2 \sin^2 \theta = 1 - \sin \theta$.
  a. Find all solutions such that $0 \le \theta \le 2\pi$.
  b. Find the general solution.

Solution

a. Think of $\sin \theta$ as a chunk. The equation is quadratic with $\sin \theta$ as the variable. Rewrite the equation with zero on one side.

$$2 \sin^2 \theta + \sin \theta - 1 = 0$$

Solve by factoring or by using the quadratic formula.

$$(2 \sin \theta - 1)(\sin \theta + 1) = 0$$

So $\qquad 2 \sin \theta - 1 = 0 \qquad$ or $\quad \sin \theta + 1 = 0$

$$\sin \theta = \tfrac{1}{2} \qquad \text{or} \qquad \sin \theta = -1.$$

Using the inverse sine function there are two solutions:

$$\theta = \sin^{-1}\!\left(\tfrac{1}{2}\right) \qquad \text{or} \qquad \theta = \sin^{-1}(-1).$$

So $\qquad\qquad\qquad \theta = \tfrac{\pi}{6} \qquad$ or $\qquad \theta = -\tfrac{\pi}{2}.$

By the Supplements Theorem, $\sin \theta = \sin(\pi - \theta)$. Thus $\pi - \tfrac{\pi}{6} = \tfrac{5\pi}{6}$ and $\pi - \left(-\tfrac{\pi}{2}\right) = \tfrac{3\pi}{2}$ also satisfy the given equation. Even though $-\tfrac{\pi}{2}$ is not in the requested interval, $\pi - \left(-\tfrac{\pi}{2}\right) = \tfrac{3\pi}{2}$ is. So the solutions to $2 \sin^2 \theta = 1 - \sin \theta$ in the interval $0 \le \theta \le 2\pi$ are $\tfrac{\pi}{6}, \tfrac{5\pi}{6}$, and $\tfrac{3\pi}{2}$.

b. By the Periodicity Theorem, adding or subtracting multiples of 2π does not change the value of the sine. So the general solution is

$$\theta = \tfrac{\pi}{6} \pm 2n\pi, \ \tfrac{5\pi}{6} \pm 2n\pi, \ \text{or} \ \tfrac{3\pi}{2} \pm 2n\pi, \text{ for any integer } n.$$

LESSON 6-6 Solving Trigonometric Equations **373**

2. Consider the equation

$$\frac{y - 12.25}{3.75} = \sin\!\left(\frac{x - 80}{\frac{365}{2\pi}}\right) \text{ for}$$

the number of hours of daylight y in the xth day of the year. Approximately when is a daylight period 11 hours long in Seattle? [Hint:

$$\sin\!\left(\frac{2\pi(x - 80)}{365}\right) = -.333. \ldots$$

The sine function has negative values for both the third and fourth quadrants. So

$$\left(\frac{2\pi(x - 80)}{365}\right) \approx -.34 \text{ or } 2.80.$$

This accounts for the two solutions.]
on about the 60th and 283rd day of each year

3. Solve for x (in degrees): $3 \cos^2 x + 4 \cos x + 1 = 0$.
$x \approx 109° + 360°n$; 251° + 360°n; or 180° + 360°n

NOTES ON QUESTIONS
Questions 7 and 13: To make the chunk explicit, you might have students substitute y for cos x in **Question 7** and substitute y for sin θ in **Question 13.**

Question 20: You may want to discourage the use of an automatic grapher on this question.

ADDITIONAL ANSWERS
2.b.

9.b.

Sample: The graphs of $y = 3 \cos \theta$ **and** $y = 7$ **do not intersect, so there is no solution to** $3 \cos \theta = 7$.

17.b.

Check 1 Substitute values in the original equation. For example, does $2 \sin^2 \frac{\pi}{6} = 1 - \sin \frac{\pi}{6}$? Does $2\left(\frac{1}{2}\right)^2 = 1 - \frac{1}{2}$? Yes.

Check 2 Graph $y = 2 \sin^2 x$ and $y = 1 - \sin x$ on the same set of axes. The x-coordinates of the points of intersection are the solutions to the given equation. Below is the output from an automatic grapher.

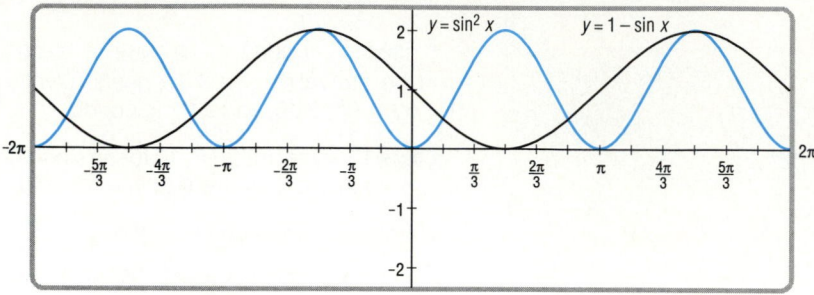

The graph shows that there are three points of intersection in the interval $0 \leq x \leq 2\pi$, as found in part **b**.

Questions

Covering the Reading

1. Consider the equation $5 \cos \theta = 2$.
   a. Find the solution(s) between 0 and π. ≈ 1.159
   b. Find all solutions between 0 and 2π. ≈ 1.159, ≈ 5.124
   c. Give a general solution to the equation.
      $\theta \approx 1.159 + 2\pi n$ or $\theta \approx 5.124 + 2\pi n$ for any integer n
2. Refer to Example 2.
   a. How many solutions are there to the equation $12 \cos 2\pi t = 10$ when $0 < t < 5$?   **10**
   b. Draw a pair of graphs to justify your answer to part **a**.   **See margin.**
   c. Give all solutions to $12 \cos 2\pi t = 10$ when $3 < t < 5$.
      ≈ 3.093, ≈ 3.907, ≈ 4.093, ≈ 4.907
3. Solve $\cos \theta = 0.34$ to the nearest degree for the domain indicated.
   a. $\{\theta: 0° \leq \theta° < 90°\}$ **70°**   b. $\{\theta: 0° \leq \theta° < 360°\}$ **70°, 290°**
   c. $\{\theta: 360° \leq \theta° < 1080°\}$ **430°, 650°, 790°, 1010°**

4. The equation $4 \sin \theta + 3 = 0$ has solutions 3.99 and 5.44 on the interval $0 \leq \theta \leq 2\pi$. Give a general solution to the equation.
   $\theta \approx 3.99 + 2n\pi$ or $\theta \approx 5.44 + 2n\pi$, for any integer n

In 5 and 6, solve for $0 \leq \theta < 2\pi$.

5. $5 \cos \theta + 1 = 0$ 6. $5 \sin^2\theta - 1 = 0$
   ≈ 1.772, ≈ 4.511 ≈ 0.464, ≈ 2.678, ≈ 3.605, ≈ 5.820

7. For the equation $2 \cos^2 x - 3 \cos x + 1 = 0$,
   a. solve when $0 \leq x < 2\pi$;   b. give a general solution.
   a) $0, \frac{\pi}{3}, \frac{5\pi}{3}$; b) $x = 2n\pi$ or $x = \frac{\pi}{3} + 2n\pi$ or $x = \frac{5\pi}{3} + 2n\pi$, for any integer n

Applying the Mathematics

8. Solve $5 \sin (2x + 1) = 1$ when $0 < x < 2\pi$.
   ≈ 0.970, ≈ 2.743, ≈ 4.112, ≈ 5.884
9. a. Give a general solution of $3 \cos \theta = 7$.   **no solution**
   b. Use a graph to explain your answer to part **a**.   **See margin.**

374

10. *True* or *false*. For $|a| \leq 1$, there are exactly two solutions to the equation $\cos \theta = a$ for $0 \leq \theta < 2\pi$.   **False**

11. Under what conditions will the equation $a \cos \theta = b$ have:
    a. zero solutions; **b.** one solution; **c.** two solutions;
    on the interval $0 \leq \theta < 2\pi$?   **a) $\left|\frac{b}{a}\right| > 1$; b) $\left|\frac{b}{a}\right| = 1$; c) $\left|\frac{b}{a}\right| < 1$**

In 12 and 13, solve on the interval $0 \leq \theta \leq 2\pi$.

12. $\tan \theta - \sqrt{3} = 2 \tan \theta$ **$\frac{2\pi}{3}, \frac{5\pi}{3}$**   13. $\cos^2 \theta + \sin \theta + 1 = 0$ **$\frac{3\pi}{2}$**

In 14 and 15, use the following information. When an object is thrown at an angle of α to the horizontal and with a velocity of v ft/sec, its horizontal and vertical displacements as a function of time t are given by $f(t) = vt \cos \alpha$ and $g(t) = vt \sin \alpha - 16t^2$, respectively.

14. A baseball was thrown with a velocity of 53 ft/sec, and was 34 ft above ground after 2 seconds. What was its initial angle?   **about 68°**

15. At what angle should you set a tennis practice machine so that balls it throws with a velocity of 75 ft/sec are 30 ft away from it after 0.5 seconds?   **about 37°**

Review

16. Give an exact value for $\text{Arcsin}\left(\frac{\sqrt{3}}{2}\right)$. *(Lesson 6-5)*   **$\frac{\pi}{3}$**

17. **a.** What is the domain of $y = \tan^{-1} x$?   **the set of all real numbers**
    b. Sketch graphs of $y = \tan x$ and $y = \tan^{-1} x$ on the same set of axes.
    (Lesson 6-5)   **See margin.**

18. Find the maximum and minimum values of the function $g(t) = 13 \sin(2t + 0.7)$ to the nearest tenth on the interval $0 \leq t \leq \pi$.
    (Lesson 6-3)   **maximum = 13; minimum = -13**

19. Write an equation for the sine curve below. *(Lesson 6-1)*

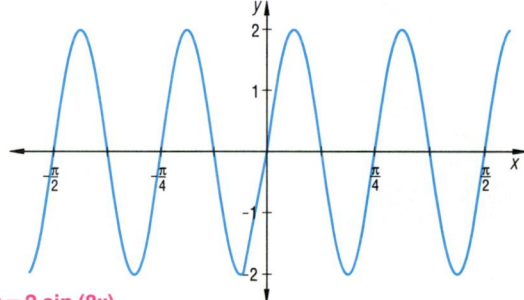

Sample: y = 2 sin (8x)

20. Suppose $f(x) = x^3$ for $x \geq 0$.
    a. Find a formula for $g(x) = \frac{1}{f(x)}$.   **$g(x) = x^{-3}$, for all $x > 0$**
    b. Graph $y = g(x)$.   **See margin.**
    c. Find a formula for $h(x) = f^{-1}(x)$.   **$h(x) = x^{1/3}$ for all $x > 0$**
    d. Graph $y = h(x)$.   **See margin.**
    e. Which of g or h has a graph that is the reflection image of the graph of f over the line $y = x$? *(Lessons 4-1, 3-8, 3-1)*   **h**

LESSON 6-6 Solving Trigonometric Equations   **375**

Exploration

21. A portion of a roller-coaster track is to be sinusoidal as illustrated below. The high and low points of the track are separated by 60 meters horizontally and 35 meters vertically. The low point is 5 meters below ground level. *(Lesson 6-4)*

a. Write an equation for the height y in meters of a point on this part of the track at a horizontal distance x meters from the high point.

b. The contractor building the roller coaster is preparing to cut the timbers to be set vertically every 5 meters starting at the foot of the high point. How many timbers are needed?   **10**

c. The following program calculates lengths of vertical timbers spaced 5 meters apart and prints the results.
  i.   Fill in the blanks in the program.
  ii. Run the program to verify your answers to part **b**.   **See margin.** Give the lengths of the vertical timbers to the nearest hundredth.

```
10 REM ROLLER-COASTER TIMBERS
20 DEF FN F(X) = _____    1.75*(cos(3.14159*X/60)) + 12.5
30 PRINT "TIMBER NO.", "LENGTH"
40 FOR X = 0 TO __ STEP 5    45
50    PRINT X/5 + 1, FN F(X)
60 NEXT X
70 END
```

a) sample: $y = 17.5 \cos\left(\pi x/60\right) + 12.5$

22. Wheelchair ramps are commonly required to have angles of elevation between about 2.9° and 4.8°. Find the angle of elevation of a ramp 16 feet long with a rise of 15 inches. Is it within the specified range? *(Lesson 5-3)*   **≈ 4.48°; It is within the specified range.**

23. Order from smallest to largest without using a calculator. *(Lessons 4-6, 4-5)*
$\log_{10} 8, \quad \log_2 8, \quad \log_6 8, \quad \ln 8$ **$\log_{10} 8$, $\log_6 8$, $\ln 8$, $\log_2 8$**

24. a. How many solutions are there to the given equation on the interval $0 \le \theta < 2\pi$?
  i. $\cos \theta = 0.3$ **2**   **ii.** $\cos 2\theta = 0.3$ **4**
  iii. $\cos 3\theta = 0.3$ **6**   **iv.** $\cos 4\theta = 0.3$ **8**
  b. Generalize your results in part **a**.
  c. How many solutions are there to the equation $\sin n\theta = a$, where $|a| < 1$ and n is a positive integer, on the interval $0 \le \theta < 2\pi$?   **2n**
  b) If n is a nonnegative integer, $\cos n\theta = 0.3$ has 2n solutions on $0 \le \theta < 2\pi$.

LESSON 6-7

The Secant, Cosecant, and Cotangent Functions

You have seen how tan θ can be defined as the ratio of sin θ to cos θ. Three other common circular functions, *secant*, *cosecant*, and *cotangent*, can also be defined in terms of the sine and cosine. As with the definition of tangent, division by zero must be avoided.

Definitions

Let θ be any real number. Then

secant of θ = **sec** θ = $\dfrac{1}{\cos \theta}$, for $\cos \theta \neq 0$;

cosecant of θ = **csc** θ = $\dfrac{1}{\sin \theta}$, for $\sin \theta \neq 0$;

cotangent of θ = **cot** θ = $\dfrac{\cos \theta}{\sin \theta}$, for $\sin \theta \neq 0$.

Notice that $\dfrac{\cos \theta}{\sin \theta} = \dfrac{1}{\frac{\sin \theta}{\cos \theta}}$, except when sin $\theta = 0$ or cos $\theta = 0$. So

$\cot \theta = \dfrac{1}{\tan \theta}$ except when cot $\theta = 0$ or tan $\theta = 0$. Because each of the secant, cosecant, and cotangent functions can be expressed as the reciprocal of a parent trigonometric function, these functions are sometimes called **reciprocal trigonometric functions**.

Example 1 Find sec $\frac{7\pi}{6}$: **a.** exactly; and **b.** approximately, using a calculator.

Solution

a. By definition of secant, sec $\dfrac{7\pi}{6} = \dfrac{1}{\cos \frac{7\pi}{6}} = \dfrac{1}{-\frac{\sqrt{3}}{2}} = \dfrac{-2}{\sqrt{3}} = \dfrac{-2\sqrt{3}}{3}$.

b. Because there is no $\boxed{\text{sec}}$ key on calculators, you need to use the reciprocal key, $\boxed{1/x}$. On some calculators you may use the key sequence (in radian mode) 7 $\boxed{\text{X}}$ $\boxed{\pi}$ $\boxed{\div}$ 6 $\boxed{=}$ $\boxed{\cos}$ $\boxed{1/x}$, which displays -1.155... .

Check $\dfrac{-2\sqrt{3}}{3} \approx -1.155$, so it checks.

For values of θ between 0 and $\frac{\pi}{2}$, values of the reciprocal trigonometric functions can be expressed in terms of the sides of a right triangle.

LESSON 6-7 The Secant, Cosecant, and Cotangent Functions **377**

Theorem

Given a right triangle with angle θ. Then

$$\sec \theta = \frac{\text{hypotenuse}}{\text{side adjacent to } \theta}$$

$$\csc \theta = \frac{\text{hypotenuse}}{\text{side opposite } \theta}$$

$$\cot \theta = \frac{\text{side adjacent to } \theta}{\text{side opposite } \theta}.$$

The proof of this theorem follows immediately from the definitions of the trigonometric ratios given in Lesson 5-3 and the property that $\dfrac{1}{\frac{a}{b}} = \dfrac{b}{a}$.

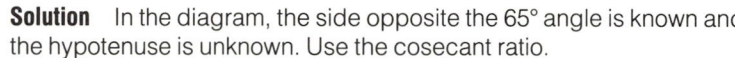

Example 2 A safe angle for a fire ladder is 65° to the ground. How long should an adjustable-height ladder be to reach 19 feet up the side of a building at that safe angle?

Solution In the diagram, the side opposite the 65° angle is known and the hypotenuse is unknown. Use the cosecant ratio.

$$\csc 65° = \frac{d}{19}$$

$$d = 19 \csc 65°$$

Set a calculator to degree mode. On some calculators you may use the key sequence 19 $\boxed{\times}$ $\boxed{(}$ 65 $\boxed{\sin}$ $\boxed{1/x}$ $\boxed{)}$ $\boxed{=}$, which displays 20.9642. The ladder needs to be about 21 feet long.

Check Check using the sine ratio. Is $\frac{19}{21} \approx \sin 65°$?

Yes, because $\frac{19}{21} \approx .905$ and $\sin 65° \approx .906$.

Because the expressions $\sec \theta$ and $\cos \theta$ are reciprocals, you can use the properties of the cosine function to determine characteristics of the secant function. These characteristics enable you to graph the function $y = \sec x$.

Function Properties	Graph Properties
When $\cos x$ is positive, $\sec x$ is positive. When $\cos x$ is negative, $\sec x$ is negative.	The graphs of $y = \cos x$ and $y = \sec x$ are on the same side of the x-axis.
When $\cos x = 0$, $\sec x$ is undefined.	There is a vertical asymptote of $y = \sec x$ when $\cos x = 0$.
Because $|\cos x| \leq 1$ for all x, $|\sec x| \geq 1$. The smaller $|\cos x|$ is, the larger $|\sec x|$ is.	The closer the graph of $y = \cos x$ is to the x-axis, the farther the graph of $y = \sec x$ is.
$\sec x = \cos x$ when $\cos x = \pm 1$.	The graphs of $y = \cos x$ and $y = \sec x$ intersect when $\cos x = \pm 1$.

378

These properties are exhibited in the graphs below. The graph of $y = \cos x$ is in black; the graph of $y = \sec x$ is in blue.

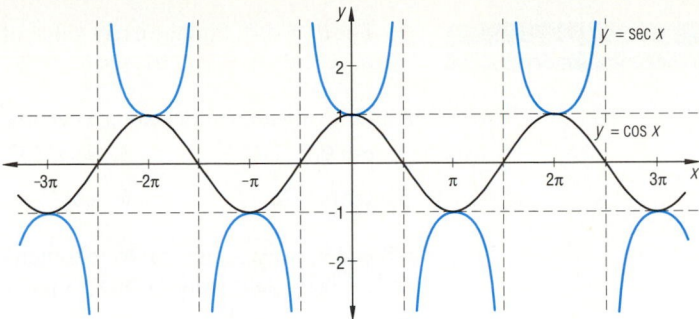

Example 3 Consider the graph of the function $y = \sec x$.
a. Identify the domain and range.
b. Find the period.
c. Identify any minimum or maximum values.

Solution

a. The domain consists of all real numbers except $x = \frac{\pi}{2} + n\pi$, for all integers n. From both the definition and the graph above, you can see that the range is $\{y: y \leq -1 \text{ or } y \geq 1\}$.
b. The period of $y = \cos x$ is 2π, so the graph of $y = \sec x$ also has period 2π.
c. There are no maximum or minimum values of $\sec x$. However, 1 is a relative minimum and -1 is a relative maximum of the secant function over certain intervals.

In Question 14 you are asked to analyze the graph of $y = \csc x$ in relation to the graph of $y = \sin x$.

The graph of $y = \cot x$ is shown below in blue. It is a reflection image, over any vertical line $x = \frac{\pi}{4} + n\pi$ where n is an integer, of the graph of $y = \tan x$, which is drawn in black.

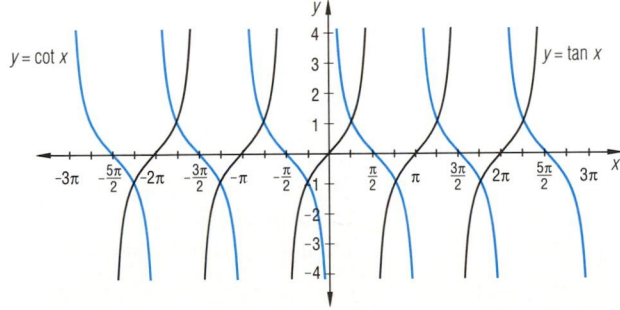

---

of the polygon has measure θ, what is the ratio of R to r in terms of θ?

$$\frac{R}{r} = \csc\left(\frac{\theta}{2}\right)$$

3. Consider the function $y = \csc x$.

a. What is its period?

2π

b. Identify any relative maximum or relative minimum values.

relative maximum values are -1, when $x = \frac{3\pi}{2} + 2n\pi$ for any integer n; relative minimum values are 1, when $x = \frac{\pi}{2} + 2n\pi$, for any integer n.

NOTES ON QUESTIONS
Questions 8 and 9:
These questions may be solved using either a reciprocal function or one of the parent functions. Students can use one as a check of the other.

Questions 16–18: Perhaps the most important generalization students can make from this lesson is that some properties of a function transfer to its reciprocal function. Among these properties are its period and its evenness or oddness.

Question 24: Because of the lesson in which this review question is included, some students may notice that $\theta = \sec^{-1}\left(\frac{300}{x}\right)$ even though we have not defined inverse functions for the secant, cosecant, and cotangent. You may wish to follow this up (see the Extension below).

ADDITIONAL ANSWERS
14.a.

379

Questions

Covering the Reading

1. For $\theta = 45°$, find the exact value of
   a. $\csc \theta$ $\sqrt{2}$ **b.** $\sec \theta$ $\sqrt{2}$ **c.** $\cot \theta$.   **1**

In 2–7, evaluate without using a calculator.

2. $\csc 90°$ **1**

3. $\sec 150°$ $\frac{-2\sqrt{3}}{3}$

4. $\cot(-135°)$ **1**

5. $\sec\left(-\frac{11\pi}{6}\right)$ $\frac{2\sqrt{3}}{3}$

6. $\cot \frac{17\pi}{3}$ $-\frac{\sqrt{3}}{3}$

7. $\csc\left(-\frac{5\pi}{2}\right)$ **-1**

In 8 and 9, use a reciprocal trigonometric function to find x to the nearest
tenth. Check your work by using a parent trigonometric function.

8.

9.

2.4 in.   **14.9 cm**

10. *Multiple choice.* The base of a ladder is placed so that it makes an angle
    of 56° with the ground. If it reaches 40 ft up the side of a building, how
    long is the ladder?
    (a) $40 \sin 56°$ (b) $40 \cos 56°$ (c) $40 \sec 56°$ (d) $40 \csc 56°$ **(d)**

In 11 and 12, in the interval $0 \leq x < 2\pi$, identify
a. all values at which the function is undefined; **b.** all x-intercepts.

11. $y = \sec x$ **a)** $\frac{\pi}{2}, \frac{3\pi}{2}$; **b) none**

12. $y = \cot x$ **a) 0, π; b)** $\frac{\pi}{2}$ **and** $\frac{3\pi}{2}$

Applying the Mathematics

13. Consider the triangle at the right. Express
    each of the following in terms of x, y, or z.
    a. $\csc \theta$ **z/x**   **b.** $\cot \theta$ **y/x**
    c. $\sec(90° - \theta)$ **z/x**   **d.** $\cot(90° - \theta)$ **x/y**
    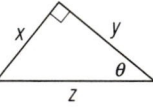

14. Let $f(x) = \sin x$ and $g(x) = \csc x$.
    a. Use an automatic grapher to graph both functions on the same set of
    axes.   **See margin.**
    b. *True or false.* The graphs of $y = f(x)$ and $y = g(x)$ are on the same
    side of the x-axis.   **True**
    c. Give the domain and range of g.   **See margin.**
    d. State equations of all asymptotes to the graph of $y = g(x)$.
    e. For what values of x between 0 and 2π does $f(x) = g(x)$?   **$\pi/2$, $3\pi/2$**
    d) $x = n\pi$ for any integer n

15. A guy wire is attached to an
    electrical pole 5 feet from the
    top of the pole. The wire
    makes a 28° angle with the
    pole and is anchored to the
    ground 14 feet from the base
    of the pole. How tall is
    the pole?   **about 31 feet tall**

380

In 16 and 17, **a.** sketch the function using an automatic grapher. **b.** Give the period of the function.

16. $y = \sec 6x$ **a) See margin. b) π/3**   **17.** $y = 3 \csc 2x$ **a) See margin. b) π**

18. Prove that $y = \cot x$ is an odd function.   **See margin.**

Review

19. Consider the equation $2 \sin^2 \theta - \sin \theta - 3 = 0$.
   a. Find the solution(s) between 0 and 2π.   **3π/2**
   b. Find the general solution.   *(Lesson 6-6)* $\frac{3\pi}{2} + 2n\pi$ **for any integer** n

20. In Lesson 6-4, Example 1, the equation
$$\frac{y - 12.25}{3.75} = \sin\left(\frac{x - 80}{\frac{365}{2\pi}}\right)$$
was shown to give y, the number of hours of daylight in Seattle, as a function of x, the day of the year.
   a. Find two days during the year when this function predicts 9 hours of daylight.   **Jan. 19 and Nov. 19**
   b. Between what two dates can you expect no more than 9 hours of daylight in Seattle?   *(Lesson 6-6)*   **between Nov. 20 and Jan 19.**

21. Given $2 \cos\left(\frac{\pi t}{2}\right) + 1 = 0$.
   a. Solve for t in the interval $-2\pi \le t \le 2\pi$. $\frac{-16}{3}, \frac{-8}{3}, \frac{-4}{3}, \frac{4}{3}, \frac{8}{3}, \frac{16}{3}$
   b. On the same set of axes, graph $y = 2 \cos\left(\frac{\pi t}{2}\right)$ and $y = -1$; and verify your answer in part **a**.   *(Lesson 6-6)*   **See margin.**

In 22 and 23, evaluate in degrees.   *(Lesson 6-5)*

22. $\tan^{-1} 1$ **45°**   **23.** $\cos^{-1}(-0.5)$ **120°**

24. A ship travels 300 km along a bearing of $\theta°$. Give an equation for the bearing as a function of x, the distance north of the original position.
   (Lesson 6-5) $\theta = \cos^{-1}\left(\frac{x}{300}\right)$

25. Find the area of the trapezoid shown at the right.   *(Lesson 5-3)*

   about 238 m²

26. a. *True* or *false*. The graphs of $e(x) = 6^x$ and $p(x) = x^6$ are reflection images of each other over the y-axis.   **False**
   b. Use an automatic grapher to justify your answer in **a**.   *(Lesson 4-3)*
   See margin.

27. Evaluate $\sum_{k=1}^{10} k$.   *(Lesson 1-4)*   **55**

Exploration

28. Analyze the graph of $\frac{y - k}{b} = \sec\left(\frac{x - h}{a}\right)$.   **See margin.**

16.a.

17.a.

18. cot (-x)

$$= \frac{\cos(-x)}{\sin(-x)} \quad \text{(def. of cot.)}$$

$$= -\frac{\cos x}{\sin x} \quad \text{(Opp. Thm.)}$$

$$= -\cot x \quad \text{(def. of cot.)}$$

21.b.

26., 28. See Additional Answers in the back of this book.

TEACHING NOTES

This reading lesson explores an application of the circular functions to the study of electricity. The questions are not difficult and they provide a review of periodicity, of summation notation, and of the use of automatic graphers. We hope that students will be inspired to create wave forms of their own after studying square waves.

Background for the mathematics of this lesson is found in the Perspectives for this chapter (page 339C).

As in a number of previous lessons, students are asked to accept the notion of a series without a formal definition. Again, we expect that the context of the lesson will make such a definition unnecessary. The idea of a limit is also introduced informally along with the notation $\lim_{n \to \infty}$.

Students will study infinite series in detail in Chapter 8.

Technology Graphing the functions in this lesson is time consuming, even with a computer, so be prepared for periods of waiting. If you choose to do this in class, you may wish to go over questions while the computer does its work. A computer is also needed to run the BASIC program on page 385 and for **Question 5.**

Alternating Current and Square Waves

Electrical current can be generated by rotating a rectangular wire through the magnetic field created by two opposite poles of a magnet, as shown above. The strength of the current (measured in amperes) depends on the position of the wire. In the diagram above, the wire lies in a plane perpendicular to the magnetic field. In this position the current is zero. As the wire rotates, the point on the top at position A moves toward position B, and the current increases, attaining its maximum when the plane of the wire is parallel to the magnetic field. As the wire continues to rotate, the current decreases, becoming zero when the wire has made a half turn (that is, when the point formerly at A is at position C). The minimum current occurs when the top of the wire has moved to position D. As it rotates from position D, the current increases and reaches zero again when one complete revolution has been completed. If the wire rotates at a constant speed, the relationship between current and time is sinusoidal as shown in the graph below.

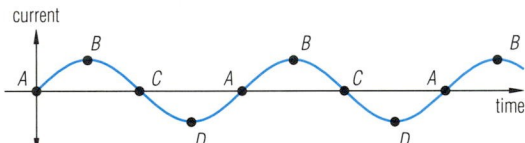

Current like this is called *alternating current* because positive and negative current flow in opposite directions. In practice, the current flow changes direction so often (usually 60 changes per second) that you cannot detect it happening. Electrical appliances are designed to work smoothly on such changes in direction and magnitude.

In some practical situations such as television, radar, and high-frequency telecommunications, a steadier current source is needed. These currents can be produced by combining different sinusoidal currents.

382

In the 18th century the French mathematician Jean Baptiste Fourier proved that any periodic function can be approximated by the sum of sine functions. Consider the problem of combining sine functions to approximate a square wave such as that below. (A square wave, of course, is not a function because of its vertical segments.) Such a wave would produce a current with a steady magnitude, alternating only in direction of flow.

Amazingly, such a square wave can be approximated using sums of sine waves whose equations follow a simple pattern. Consider the two functions

$$f(x) = \sin x$$

and

$$g(x) = \frac{1}{3} \sin 3x.$$

The amplitude of g is one-third the amplitude of f, and the frequency of g is three times that of f. Shown below are the graphs of the two functions, together with the graph of their sum,

$$s(x) = f(x) + g(x) = \sin x + \frac{1}{3} \sin 3x.$$

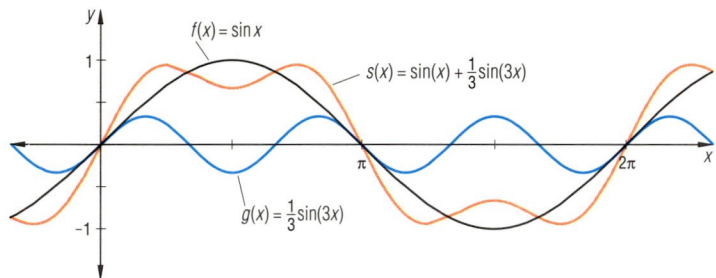

Notice that the function s is periodic. The graph of the sum of f and g looks a little like a square wave.

To produce an even better approximation to a square wave, other functions with higher frequencies can be added. For example, consider the function below. It is the sum of a *fundamental function*, $f(x) = \sin x$, together with two functions each having a frequency which is an odd multiple of the fundamental, and an amplitude which is the reciprocal of its frequency. These functions are called *harmonics*.

$$F_5(x) = \sin x + \frac{1}{3} \sin 3x + \frac{1}{5} \sin 5x$$

$$\uparrow \qquad \uparrow \qquad \uparrow$$

fundamental   3rd harmonic   5th harmonic

From the graph, you can see that F_5 is closer to a square wave than the previous graph.

Better approximations to a square wave come from adding more odd harmonics. The following is part of the graph of

$$F_{19}(x) = \sin x + \frac{1}{3}\sin 3x + \frac{1}{5}\sin 5x + \ldots + \frac{1}{19}\sin 19x.$$

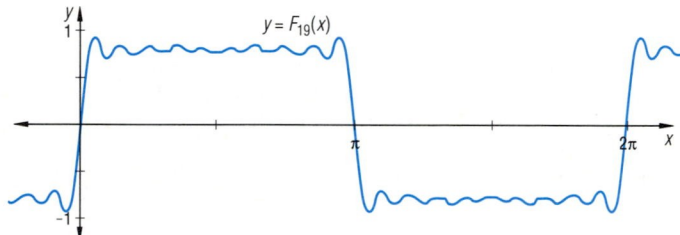

Below is the graph of $F_{43}(x) = \sin x + \frac{1}{3}\sin 3x + \ldots + \frac{1}{43}\sin 43x$ on an interval including $0 \le x \le 2\pi$.

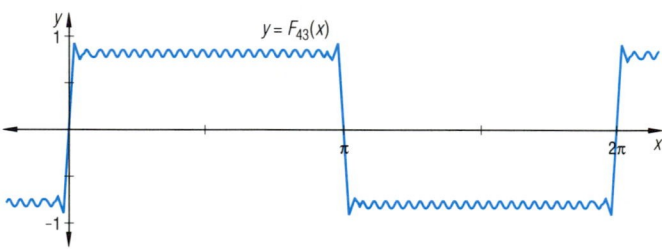

By changing the window on an automatic grapher, you can see how well F_{43} approximates a square wave.

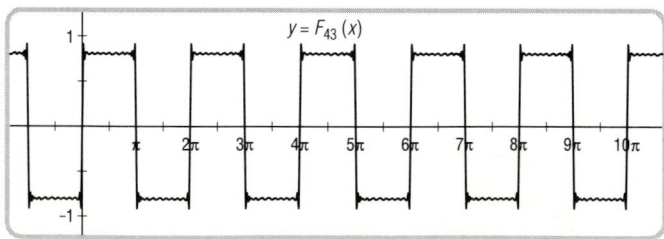

384

The more harmonics added, the better is the approximation to a square wave. As the number of harmonics approaches infinity, the limiting function F_∞ is a square wave. Using the symbol $\lim_{n \to \infty}$ to mean the limit as n approaches infinity,

$$F_\infty(x) = \lim_{n \to \infty}\left[\sum_{k=1}^{n}\left(\frac{1}{2k-1}\right)\sin(2k-1)x\right].$$

The limiting process can be checked using a numerical method. Here is a computer program to find the sum of the first N terms of the series

$$F_N(x) = \sin x + \frac{1}{3}\sin 3x + \frac{1}{5}\sin 5x + \ldots + \frac{1}{2N-1}\sin(2N-1)x$$

on the interval $0 \le x \le 6.5$.

```
10  INPUT "NUMBER OF TERMS"; N
20  FOR X = 0 TO 6.5 STEP 0.5
30    SUM = 0
40    FOR K = 1 TO (2 * N − 1) STEP 2
50    SUM = SUM + (1/K) * SIN(K * X)
60    NEXT K
70    PRINT X, SUM
80  NEXT X
90  END
```

Running this program for $N = 1000$ gave us the following results.

```
0       0
.5      .7851049
1       .7855077
1.5     .7856423
2       .785598
2.5     .7853331
3       .7837934
3.5     -.7847828
4       -.7853751
4.5     -.7855993
5       -.7856467
5.5     -.7854977
6       -.7848293
6.5     .784242
```

These values indicate a good approximation to a square wave, since the value of the function, truncated to two decimal places, is one of only two values, each the opposite of the other.

NOTES ON QUESTIONS
Technology for Question 5: To calculate just one value, change line 20 to INPUT X.

Question 17: You may need to remind students that these are inverse functions, not reciprocal functions.

FOLLOW-UP

PROJECTS
Projects 3, 4, and 5 are related to the content of this lesson.

ADDITIONAL ANSWERS
4.b.

9.b. sec (-x)

$= \dfrac{1}{\cos(-x)}$ (def. of sec.)

$= \dfrac{1}{\cos x}$ (Opp. Thm.)

$= \sec x$ (def. of sec.)

10.

14.b.

Questions

In 1–3, give the period of the function.

1. $F_1(x) = \sin x$ **2π**

2. $F_3(x) = \sin x + \frac{1}{3}\sin 3x$ **2π**

3. $F_5(x) = \sin x + \frac{1}{3}\sin 3x + \frac{1}{5}\sin 5x$ **2π**

4. a. Expand $F(x) = \displaystyle\sum_{k=1}^{6}\left(\frac{1}{2k-1}\right)\sin(2k-1)x$.

b. Graph $y = F(x)$ for the interval $-\pi \le x \le 2\pi$.   **See margin.**

c. State the period of F.   **2π**

d. Approximately what shape is the wave that results?   **square wave**

a) $\sin x + \frac{1}{3}\sin 3x + \frac{1}{5}\sin 5x + \frac{1}{7}\sin 7x + \frac{1}{9}\sin 9x + \frac{1}{11}\sin 11x$

5. Use the computer program in the text to find the value of

$$F(x) = \sum_{k=1}^{1000}\left(\frac{1}{2k-1}\right)\sin(2k-1)x \text{ to the nearest ten-thousandth.}$$

a. $F(1.5)$ **b.** $F(5.9)$ **c.** $F(6.3)$ **d.** $F(-0.6)$

a) .7856; b) −.7847; c) .7939; d) −.7850

In 6–8, find the exact value.   *(Lesson 6-7)*

6. $\sec 30°$ $\dfrac{2\sqrt{3}}{3}$ **7.** $\csc \dfrac{7\pi}{6}$ **−2**   **8.** $\tan 540°$ **0**

In 9–11, consider the function $y = \sec x$.   *(Lessons 6-7, 3-4)*

9. a. Is the function odd, even, or neither?   **even**

b. Justify your answer.   **See margin.**

10. Sketch a graph of the function when $-\pi \le x \le 2\pi$.   **See margin.**

11. State its range. $\{y: y \le -1 \text{ or } y \ge 1\}$

12. Solve $8 - 5\cos(2x - 4) = 12$ when $0 \le x < 2\pi$.   *(Lesson 6-6)*
$\approx 0.11, \approx 0.75, \approx 3.25, \approx 3.89$

13. Solve $5\sin x - 1 = \dfrac{1}{\sin x}$ when $-\pi \le x < \pi$.   *(Lesson 6-6)*
$\approx -2.78, \approx -0.37, \approx 0.59, \approx 2.55$

14. Consider $f(x) = \frac{1}{3}\cos(x - \pi)$.

a. Describe a transformation that maps the graph of $g(x) = \cos x$ onto the graph of f.

b. Sketch the graph of f when $-2\pi \le x \le 2\pi$.   *(Lesson 6-3)*

a) $(x, y) \to \left(x + \pi, \frac{1}{3}y\right)$; b) See margin.

386

In 15 and 16, a gear with a 4 cm radius rotates counterclockwise at a rate of 120 revolutions per minute. The gear starts with the tooth labeled A, level with the center of the wheel, as shown at the left. *(Lessons 6-5, 6-1)*

15. a. Write an equation to give the vertical distance y that point A is above or below its starting position at time t.
   b. How far above or below the starting position will point A be after 5 minutes?   **0 cm**
   a) Sample: $y = 4 \sin (240 \pi t)$ where t is time in minutes.

16. Write an equation for the vertical distance h that point A is above or below point B at time t. $h = 4 + 4\sin(240\pi t)$

17. a. Sketch graphs of $f(x) = \cot x$ and $g(x) = \cot^{-1} x$ on the same set of axes.
   b. State the domain and range of each function.   *(Lesson 6-5)*
   a,b) See margin.

18. What is the period of $y = \tan 8x$?   *(Lesson 6-1)* $\pi/8$

19. Three sound waves have displacements at t seconds given by
$$d_1(t) = 8 \sin 12\pi t,$$
$$d_2(t) = 6 \sin 8\pi t,$$
$$\text{and } d_3(t) = 7 \sin 15\pi t. \quad \textit{(Lesson 6-1)}$$
   a. Rank the corresponding sounds from loudest to softest.
   b. Rank the corresponding sounds from highest to lowest pitch.
   a) d_1, d_3, d_2, b) d_3, d_1, d_2

20. *Skill sequence.* Find the standard deviation of the data.   *(Lesson 1-7)*
   a. 2, 2, 2, 2, 2, 2, 2, 2, 2   **0**
   b. $a, a, a, a, a, a, a, a, a$ **0**
   c. 2, 2, 2, 1, 4, 1, 2, 2, 2 $\dfrac{\sqrt{3}}{2}$

Exploration

21. a. Draw graphs of
$$G_1(x) = \cos x,$$
$$G_2(x) = \cos x + \frac{1}{3} \cos 3x, \text{ and}$$
$$G_3(x) = \cos x + \frac{1}{3} \cos 3x + \frac{1}{5} \cos 5x. \quad \textbf{See margin.}$$
   b. Predict the period and shape of $G_4(x)$.   **2π; Answers may vary.**
   c. Is $G_4(x)$ an odd function, an even function, or neither?   **even**
   d. Predict the period and shape of $G_n(x)$ for large n.
   e. Check your predictions in **d** with an automatic grapher.   **See margin.**
   d) The period is 2π. The graph looks like a succession of sharp peaks alternately pointing upward and downward.

17.a.

b. $f(x)$: domain = the set of all real numbers x such that $x \neq n\pi + \dfrac{\pi}{2}$ for any integer n, range = the set of all real numbers; $g(x)$: domain = the set of all real numbers, range =
$$\left(y\text{: } -\frac{\pi}{2} < y < \frac{\pi}{2} \right)$$

21.a.

G_1

G_2

G_3

e. A graph of $G_{15}(x)$ on an automatic grapher seems to confirm the prediction.

Projects

The projects provide an opportunity for extended work on topics related to the content of the chapter. (See pages 66 and T37 for more information on the purposes of projects.) Some or all of these projects may be suitable for **small group work.**

Solutions to the projects are found at the end of the Additional Answers section at the back of the book.

Project 1 is good review of Lesson 6-7 and makes use of the statistics package on your computer. Though they should expect the time of sunset to be periodic, the sinusoidal nature may surprise students even though the length of the daylight period is sinusoidal.

Project 2 illustrates the diversity of situations from which circular functions arise. *Proper safety precautions should be exercised by students using razor blades for the project.* If your candles don't cut cleanly, you can substitute sausage, hotdogs or some other cylindrical object.

The point of **Project 3** will not be made unless the student uses values of p and q with $\frac{p}{q}$ closer to 1 than is true with the given values.

Project 4 examines Gregory's series for arctan, which was of considerable importance in early attempts to find decimal approximations to π. A computer is needed to deal with the series efficiently. Students may like to consult some library books for this project, such as Peter Beckman's *The History of π* or Howard

1. From an almanac or other reference, find the times of sunrise and sunset for various dates in a particular city in a particular year.
   a. On the same set of axes, make scatter plots of date versus time of sunrise, and date versus time of sunset. Sketch curves of good fit through the data. Find an equation to model each set of data.
   b. How would Daylight Savings Time affect a graph of time of sunset or sunrise?
   c. From the equations of sunset and sunrise in relation to date, how could you determine an equation for the number of hours of daylight in relation to the date?
   d. How does the latitude or longitude of a city affect the time of sunset or sunrise? Collect other data and supply theoretical evidence to support your answer.

2. a. Cut out a strip of paper about 10 cm long and 3 cm wide. Wind it tightly around a small "birthday cake" candle and, with a razor blade, carefully cut through the paper and candle at a slant, as shown below.

   i. Look at the cross section of the candle that you have cut. What curve do you see? The length of one axis of this curve was determined by the diameter of the candle. Which one? What determined the length of the other axis of the curve?

   ii. Unwind the paper. What curve results? What property of the curve was determined by the size of the candle? What property of the curve was determined by the slant of the cut?
   b. Repeat the steps in part **a** using candles of different sizes, and cuts at different slants. Generalize your observations.

3. a. Explore the patterns that result when two sine functions with the same amplitude but different periods that are "nearly equal" are added. For instance, graph
   $$y = \sin 5x + \sin 6x,$$
   $$y = \sin 9x + \sin 8x,$$
   $$y = \sin 10x + \sin 12x,$$
   and other functions of the form
   $$y = \sin px + \sin qx, \text{ where } \frac{p}{q} \approx 1.$$
   b. Find out how such functions help explain the variations in intensity heard in some music. (These variations are called "beats," not to be confused with the "beat" of a percussion instrument.)

4. a. Graph the functions
   $$S_1(x) = \sin x,$$
   $$S_2(x) = \sin x + \frac{1}{2} \sin 2x,$$
   $$S_3(x) = \sin x + \frac{1}{2} \sin 2x + \frac{1}{3} \sin 3x, \text{ and}$$
   $$S_{10}(x) = \sum_{i=1}^{10} \frac{\sin ix}{i}.$$
   Describe the patterns you observe in general shape, amplitude, and period. Predict the behavior of the function $S_n(x) = \sum_{i=1}^{n} \frac{\sin ix}{i}$ for large n. Test your conjectures using an automatic grapher.
   b. Examine the sum $T_n(x) = \sum_{i=1}^{n} \frac{\cos ix}{i}$ for various values of n. Compare and contrast these functions with those produced in part **a**.

388

5. Consider functions of the form
   $$f(x) = a \sin x + b \cos x.$$
   Here are some examples:
   $a = 1, b = 1$: $f(x) = \sin x + \cos x$;
   $a = \sqrt{3}, b = 1$: $f(x) = \sqrt{3} \sin x + \cos x$;
   $a = 1, b = -1$: $f(x) = \sin x - \cos x$;
   $a = 5, b = 12$: $f(x) = 5 \sin x + 12 \cos x$.

   a. Can you predict the graphs of such functions from the values of a and b? Graph the four examples above. Then choose some other values for a and b. What shapes are the graphs?

   b. Can you predict the amplitude of a graph before you draw it? For example, what is the amplitude of $f(x) = 3 \sin x + 4 \cos x$? Check your prediction using an automatic grapher. What is the period of $f(x) = a \sin x + b \cos x$?

   c. What is the phase shift of $f(x) = a \sin x + b \cos x$? Predict the phase shift of $g(x) = \sin x + \sqrt{3} \cos x$.
   Check your prediction using the function grapher.

   d. For $a > 0$ and $b > 0$, the amplitude and the phase shift are related to a right triangle with legs of lengths a and b. Find out what the relationship is.

   e. How are the graphs of $y = a \sin x - b \cos x$ and $y = a \sin x + b \cos x$ related if a and b are positive? How does your response to the question change if a and b are not both positive?

6. Recall the ferris wheel problem in Lesson 6-3. The top of the wheel is 45 feet from the ground, the wheel is 40 feet in diameter, and is traveling at a uniform speed of 2 revolutions per minute. After first boarding the wheel, it takes 10 seconds to reach the top. An equation for the height $f(t)$ off the ground of a person t seconds after boarding is:

   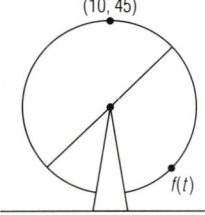

   (10, 45)

   $f(t)$

   $$f(t) = 25 + 20 \cos\left(\frac{\pi}{15}(t - 10)\right).$$

   This project is intended to show you how the function was derived. The method used is based on the one used in Lesson 6-4.

a. You don't know the height off the ground of the boarding point. In other words, you don't know $f(0)$. You do know what $f(t)$ is when $t = 10$: $f(10) = 45$ feet. So build the function relative to $t = 10$ or $t - 10 = 0$. The wheel is turning at a rate of 2 revolutions per minute, or $\frac{4\pi \text{ radians}}{60 \text{ seconds}} = \frac{\pi}{15}$ radians per second. At this rate, the wheel turns through $\frac{\pi}{15}(t - 10)$ radians in $t - 10$ seconds. This expression allows you to determine where the boarding point was, measured in radians from the top of the wheel. Find that measure.

b. By design in part a, $\frac{\pi}{15}(t - 10) = 0$ when $t = 10$. At this time, the value of the function should be a maximum. The circular function which has a maximum at 0 is $y = \cos x$, so $\cos\left(\frac{\pi}{15}(t - 10)\right)$ has a maximum at $t = 10$.
   Sketch $f(t) = \cos\left(\frac{\pi}{15}(t - 10)\right)$ on an automatic grapher to verify this. Find the next value of t at which this function has a maximum.

c. The center of the wheel is 25 feet above the ground, and a person's position relative to that point varies between 20 feet above the center and 20 feet below the center. That is, the amplitude of the cosine curve modeling this situation is 20, and the curve is the image of $y = \cos x$ under a vertical translation of 25. This leads to the final form of the function:

   $$f(x) = 25 + 20 \cos\left(\frac{\pi}{15}(t - 10)\right).$$

   Check this by verifying that 25 seconds after boarding, a person is at the bottom of the wheel (10 seconds to reach the top and 15 more to reach the bottom).

d. Gather information about the size and velocity of a Ferris wheel at an amusement park you know. Develop an equation that models the height the person is off the ground as a function of the time t after the person boards the Ferris wheel.

Eves' *An Introduction to the History of Mathematics.*

Project 5 deals with the rather surprising fact that the "sum" of two sine waves of the same period is also a sine wave. Further, the amplitude, period, and frequency of the sum can be readily predicted from the original functions.

7. Find out about *atonal* music. In what parts of the world does it have the longest heritage? What are some of the instruments on which it is played? What are some mathematical properties of this type of music?

8. The Scottish mathematician James Gregory (1638–1675) found the following series for Arctan x:

$$\text{Arctan } x = x - \frac{x^3}{3} + \frac{x^5}{5} - \frac{x^7}{7} + \ldots . \{x: -1 \le x \le 1\}$$

Here is a BASIC program to approximate Arctan x using the first 100 terms of Gregory's series:

```
10    INPUT "A NUMBER BETWEEN -1 AND 1";X
20    SUM = X
30    TERM = X
40    PRINT 1,SUM
50    FOR N = 3 TO 199 STEP 2
60        TERM = -X * X *(N – 2)/N * TERM
70        SUM = SUM + TERM
80        PRINT N,SUM
90    NEXT N
100   END
```

a. Explain the function of lines 30 and 60 of this program.

b. RUN the program with $x = 1$ to find an approximation for $\frac{\pi}{4}$. RUN the program with some other value of x $\{x: -1 \le x \le 1\}$, and compare your output with the value of Arctan x given by your calculator. Check that the formula does *not* work for $|x| > 1$.

c. Modify the program to evaluate more terms of the series. Use your modification to evaluate Arctan 1. Compare the computer's output to the exact value of Arctan 1.

d. The astronomer Abraham Sharp (1651–1742) used this series to find $\text{Arctan}\frac{\sqrt{3}}{3}$ to approximate π to 72 decimal places. Use the computer program to find $\text{Arctan }\frac{\sqrt{3}}{3}$. Compare the rate of convergence with that for Arctan 1 obtained in **c**. (Notice that the computer is able to use only a limited accuracy for calculation of this kind. Special programs are needed to evaluate to more than about 12-decimal-place accuracy.)

e. Read about these and other uses of Gregory's Arctan series to evaluate π in Peter Beckmann's *A History of* π or Howard Eves' *An Introduction to the History of Mathematics*.

Summary

The images of graphs of sine and cosine functions under a composite of translations and scale changes are called sine waves. These are useful for describing real-world periodic phenomena such as alternating electrical currents, sound waves, and daylight hours over a year. In fact, the graph of any periodic function can be closely approximated by the graph of a sum of a finite number of sine functions.

Starting with the graphs of parent circular functions, the Graph Scale Change Theorem, the Graph Translation Theorem, and the Graph Standardization Theorem can be used to predict the shapes of sine waves. The graph of the equation

$$\frac{y-k}{b} = \sin\left(\frac{x-h}{a}\right) \text{ or } \frac{y-k}{b} = \cos\left(\frac{x-h}{a}\right), \text{ with}$$

$a \neq 0$, has amplitude $|b|$, period $2\pi|a|$, horizontal shift h, and vertical shift k. Horizontal translation magnitudes are often called phase shifts.

The inverses of the sine, cosine, and tangent functions are not functions. However, if the domains of the parent functions (and equivalently, the range of their inverses) are restricted as noted below, their inverses *are* functions.

$y = \cos^{-1} x = \text{Arccos } x$ is defined for $0 \leq y \leq \pi$;

$y = \sin^{-1} x = \text{Arcsin } x$ is defined for $\frac{-\pi}{2} \leq y \leq \frac{\pi}{2}$;

$y = \tan^{-1} x = \text{Arctan } x$ is defined for $\frac{-\pi}{2} < y < \frac{\pi}{2}$.

A calculator typically gives values of these functions only when the input is in these intervals.

Equations involving the circular functions are called trigonometric equations. Solving trigonometric equations requires the use of the inverses of the circular functions. This leads to solutions in a restricted domain. Due to the periodic nature of circular functions, a general solution can be obtained by using the period of the parent function.

Secant, cosecant, and cotangent are reciprocals of the cosine, sine, and tangent functions, respectively. They are convenient in solving certain trigonometric equations. As there are no specific keys on calculators for these functions, their values can be calculated by using the $\boxed{1/x}$ key.

Vocabulary

For the starred (*) items you should be able to give a definition of the term.
For the other terms you should be able to give a general description and a specific example of each.

Lesson 6-1
sinusoidal
*sine wave
*amplitude
*frequency

Lesson 6-2
*phase shift
vertical shift
in phase
out of phase

Lesson 6-3
linear change
Graph Standardization Theorem

Lesson 6-5
inverse circular functions,
inverse trigonometric functions
*cos^{-1} (Arccos)
*sin^{-1} (Arcsin)
*tan^{-1} (Arctan)

Lesson 6-6
trigonometric equation
general solution

Lesson 6-7
reciprocal trigonometric functions
*secant (sec)
*cosecant (csc)
*cotangent (cot)

Lesson 6-8
fundamental function
harmonics

Progress Self-Test

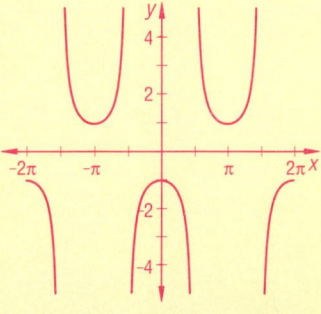
See margin for answers not shown below.

Take this test as you would take a test in class. You
may want to use an automatic grapher. Then check
the test yourself using the solutions at the back of the
book.

1. Give **a.** the domain **b.** the range **c.** the amplitude and **d.** the period of $y = \frac{1}{2}\sin\left(\frac{1}{2}x\right)$.

2. Consider $f(x) = \sin\left(x - \frac{\pi}{6}\right)$.   **a)** $(x, y) \to \left(x + \frac{\pi}{6}, y\right)$
   a. What transformation maps the graph of the
   parent sine function to the graph of f?
   b. Give the phase shift of the graph of f. $\frac{\pi}{6}$

3. Consider $y = \cos(3x + \pi)$.
   a. Graph the function.   **b.** State its period.
   c. State its phase shift.   **b)** $2\pi/3$; **c)** $-\pi/3$
   d. Estimate to tenths the solution(s) in
   $0 \le x \le \frac{\pi}{4}$ to $\cos(3x + \pi) = 0.5$.   **0.7**

4. *Multiple choice.* Which is the graph of
   $y = \cos 2x + 1$?

   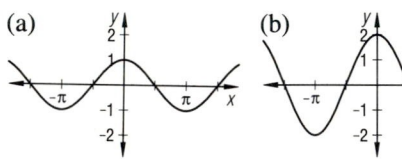

   (c)

   In 5–8, evaluate without a calculator.
5. $\cos^{-1}\left(\frac{1}{2}\right)$ $\frac{\pi}{3}$
6. $\text{Arctan}\left(\tan\frac{\pi}{4}\right)$ $\frac{\pi}{4}$
7. $\cot\left(-\frac{\pi}{6}\right)$ $-\sqrt{3}$
8. $\sec 405°$ $\sqrt{2}$

9. **a.** Write an equation for a function whose
   parent is $y = \cos x$ with phase shift $-\pi$ and
   period $\frac{\pi}{2}$. $y = \cos(4(x + \pi))$
   b. Graph the function in part **a.**

10. The voltage E in volts in a circuit after t seconds
    ($t > 0$) is given by $E = 14\cos 5\pi t$. After how
    many seconds is the voltage first equal to
    12 volts?   **about 0.034 second**

11. Give a general solution to $\sin^2\theta - \frac{1}{4} = 0$.

12. The hum you hear on a radio when it is not func-
    tioning properly may be a sinusoidal sound
    wave with a frequency of 60 vibrations per sec-
    ond. If the amplitude of this wave is 0.1 and the
    displacement is 0 at time $t = 0$, write an equation
    for the displacement y as a function of time t.
    $y = 0.1\sin(120\pi t)$

    In 13 and 14, *true or false*.
13. The function $y = \sin^{-1}\theta$ is defined for
    $0 \le \theta \le 2\pi$.   **False**
14. If $\tan\theta = 2$, then $\theta = \tan^{-1}2$.   **False**

15. Consider the function $y = \sec(x - \pi)$.
    a. Sketch a graph of the function.
    b. State the period of the graph. 2π
    c. State equations of two of the asymptotes.

16. A sine wave model
    for the average
    temperature T for
    Grand Rapids,
    Michigan, during
    month n can be
    found using the data
    shown in the graph
    at the right.
    a. What is the
    amplitude of
    the wave?   **24.5**
    b. What is the period of the wave?   **12 months**
    c. Write an equation of the wave in the form
    $\frac{T - h}{a} = \cos\left(\frac{n - k}{b}\right)$.
    d. Rewrite the function in **c** in the form
    $T = a\cos(bn + c) + d$.
    e. Use the model and the values from parts **a**
    and **b** to estimate the average temperature for
    Grand Rapids in February.   **about 25°F**

    c) $\frac{T - 46.5}{24.5} = \cos\left(\frac{n - 6}{6/\pi}\right)$
    d) $T = 24.5\cos\left(\frac{\pi n}{6} - \pi\right) + 46.5$

**Average Monthly Temperature
for Grand Rapids, Michigan**

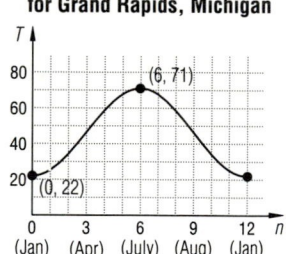

Chapter Review

Questions on **SPUR** Objectives

SPUR stands for **S**kills, **P**roperties, **U**ses, and **R**epresentations.
The Chapter Review questions are grouped according to the SPUR Objectives for this chapter.
See margin for answers not shown below.

SKILLS deal with the procedures used to get answers.

Objective A: *Evaluate inverse trigonometric functions.* *(Lesson 6-5)*

1–3, evaluate without a calculator. State angle measures in degrees.

1. $\sin^{-1}\left(\frac{1}{2}\right)$ **30°**   2. Arctan (1)   **45°**

3. $\cos^{-1}\left(-\frac{\sqrt{2}}{2}\right)$ **135°**

4–6, give the value to the nearest tenth of a radian.

4. $\tan^{-1}(10)$ **1.5**   5. Arcsin (.1895)   **0.2**

6. $\cos^{-1}(-0.8753)$ **2.6**

7 and 8, evaluate without a calculator.

7. $\cos\text{Arccos}\left(\frac{4}{5}\right)$ **$\frac{4}{5}$**   8. $\sin^{-1}\left(\sin\left(\frac{3\pi}{4}\right)\right)$ **$\frac{\pi}{4}$**

Objective B: *Evaluate reciprocal functions.* *(Lesson 6-7)*

9–11, give exact values.

9. $\csc\frac{\pi}{2}$ **1**   10. $\cot\left(-\frac{\pi}{4}\right)$ **-1**   11. $\sec 390°$ **$\frac{2\sqrt{3}}{3}$**

12–14, evaluate to the nearest hundredth.

12. $\sec 28°$ **1.13**   13. $\csc 3.7$ **-1.89**   14. $\cot 237°$ **0.65**

Objective C: *Solve equations involving circular functions.* *(Lessons 6-6, 6-7)*

In 15–17, give the number of solutions to the equation if $0 \le x \le 2\pi$.

15. $8\sin(x+3) = 5$ **2**

16. $5\sin(x+3) = 8$ **0**

17. $7\cos(3x+2) = 5$ **6**

In 18 and 19, solve when $0 \le \theta \le 2\pi$.

18. $\tan\theta = -0.4$ $\theta \approx$ **2.76 or 5.90**

19. $12\cos^2\theta = 29\cos\theta - 15$ $\theta \approx$ **0.72 or 5.56**

In 20–23, give a general solution.

20. $3\tan\theta - 5 = 0$

21. $\frac{1}{4}\sin(2x+1) = \frac{1}{9}$

22. $\sin^2\theta + 3\sin\theta + 1 = 0$

23. $\sec\theta = 4$

24. Consider the equation $\cos 2x = \frac{1}{2} - x$.
   a. How many solutions are there when $0 \le x \le 2\pi$?   **2**
   b. Round the smallest solution in $0 \le x \le 2\pi$ to the nearest hundredth.   **1.12**

PROPERTIES deal with the principles behind the mathematics.

Objective D: *Identify the amplitude, period, frequency, phase shift, and other properties of circular functions.* *(Lessons 6-1, 6-2, 6-3)*

25–28, give, if it exists, **a.** the period, **b.** the amplitude, **c.** the phase shift.

25. $\frac{y}{5} = \sin\frac{x}{2}$ 26. $y = 2\cos(3\pi x)$

27. $y = 2\cos\left(x - \frac{\pi}{3}\right)$ 28. $h(\theta) = \frac{1}{2}\tan 2\theta$

29. Suppose the transformation
   $(x, y) \rightarrow (2x + 1, \frac{y}{3} - 1)$ is applied to the graph
   of the function $y = \sin x$.
   a. State an equation for the image.
   b. Find the amplitude, period, phase shift, and vertical shift of the image.   **1/3; 4π; 1; -1**

   a) sample: $3y + 3 = \sin\left(\frac{x-1}{2}\right)$

CHAPTER REVIEW

The main objectives for the chapter are organized here into sections corresponding to the types of understanding this book promotes: Skills, Properties, Uses, and Representations (called SPUR objectives).

USING THE CHAPTER REVIEW
Students should be able to answer questions like these with about 85% accuracy by the end of the chapter. (See pages 72 and T38–39 for more information.)

ADDITIONAL ANSWERS
20. $\theta \approx 1.03 + n\pi$ for any integer n

21. $x \approx 2.87 + n\pi$ or $x \approx 0.84 + n\pi$ for any integer n

22. $\theta \approx 5.89 + 2n\pi$ or $\theta \approx 3.53 + 2n\pi$ for any integer n

23. $\theta \approx 1.32 + 2n\pi$ or $\theta \approx 4.97 + 2n\pi$ for any integer n

25.a. 4π, b. 5, c. 0

26.a. $\frac{2}{3}$, b. 2, c. 0

27.a. 2π, b. 2, c. $\frac{\pi}{3}$

28.a. $\frac{\pi}{2}$

b. does not exist

c. 0

30. Let $S(x, y) = \left(\frac{x}{3}, -2y\right)$ and $T(x, y) = (x + 6, y)$.

  a. Find the image of the graph of $y = \cos x$ under the composite transformation $T \circ S$.

  b. Find a rule for the single transformation that maps the graph of $y = \cos x$ to the function in part **a**. **See below.**

31. For the graph of $y = -4 \cos \frac{x}{3}$, state

  a. the amplitude **4**

  b. the period **6π**

  c. the frequency. **$\frac{1}{6\pi}$**

32. State **a.** the maximum and **b.** the minimum value of the function $f(t) = 10 + 5 \sin 2t$.
  a) 15; b) 5

■ **Objective E:** *Apply properties of inverse trigonometric functions.* *(Lesson 6-5)*

33. For what values of θ is the following statement true? If $k = \sin \theta$, then $\theta = \sin^{-1} k$.
  $-\pi/2 \le \theta \le \pi/2$

30 a) sample: $y = -2 \cos(3x - 18)$

  b) $(x, y) \to \left(\frac{x}{3} + 6, -2y\right)$

34. *Multiple choice.* If $\sin 5x = a$ and $-\frac{\pi}{10} \le x \le \frac{\pi}{10}$, then x equals

  (a) $5 \sin^{-1} a$ (b) $\sin^{-1} \frac{a}{5}$

  (c) $\sin^{-1} 5a$ (d) $\frac{1}{5} \sin^{-1} a$. **(d)**

35. State **a.** the domain and **b.** the range of $y = \cos^{-1} x$. **a) $\{x: -1 \le x \le 1\}$; b) $\{y: 0 \le y \le \pi\}$**

36. *True* or *false*. The function $y = \text{Arctan } x$ has period π. **False**

■ **Objective F:** *Apply properties of the reciprocal trigonometric functions.* *(Lesson 6-7)*

In 37 and 38, consider the function $f(x) = \csc x$.

37. What is the period of f? **2π**

38. For what values is f undefined?
  for $x = n\pi$ for any integer n

In 39 and 40, *true* or *false*.

39. $\sec \theta = \frac{1}{\csc \theta}$, for all θ. **False**

40. The function $y = \cot x$ is an even function.
  False

USES deal with applications of mathematics in real situations.

■ **Objective G:** *Analyze phenomena described by sine waves.* *(Lessons 6-1, 6-2, 6-3, 6-4)*

41. An alternating current I in amps of a circuit at time t in seconds is given by the formula $I = 40 \cos 60\pi t$. **a) max: 40 amps; min: -40 amps**

  a. Find the maximum and minimum currents.

  b. How many times per second does the current reach its maximum? **30**

42. The voltage V in volts in a circuit after t seconds $(t > 0)$ is given by $V = 120 \cos 60\pi t$.

  a. Find the first time the voltage is 100.

  b. Find all times at which the voltage is maximized.

  a) after about 0.0031 seconds

  b) $t = n/30$ seconds for any integer n.

43. A certain sound wave has equation $y = 60 \cos 20\pi t$. Give an equation of a sound wave with twice the frequency and that is half as loud as this one. **$y = 30 \cos 40\pi t$**

44. A simple pendulum is shown at the right. The angular displacement from vertical (in radians) as a function of time t in seconds is given by

$f(t) = \frac{1}{2} \sin\left(2t + \frac{\pi}{2}\right)$.

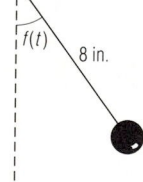

  a. What is the initial angular displacement?

  b. What is the frequency of f? **$1/\pi$**

  c. How long will it take for the pendulum to make 5 complete swings? **5π seconds, or about 15.7 seconds**

  a) 1/2 radian

45. The figure below shows a waterwheel rotating at 4 revolutions per minute. The distance d of point P from the surface of the water as a function of time t in seconds can be modeled by a sine wave with equation of the form $\dfrac{d-k}{b} = \sin\left(\dfrac{t-h}{a}\right)$.

a. What is the amplitude of the wave? **10**
b. What is the period of the wave? **15**
c. If point P emerges from the water just as you start a stopwatch, write an equation relating d and t.
d. Approximately when does point P first reach its highest point?

c) sample: $\dfrac{d-7}{10} = \sin\left(\dfrac{t-1.851}{\frac{7.5}{\pi}}\right)$

d) after about 5.6 seconds

46. The following table gives the percent of the United States budget spent on defense in the years between 1974 and 1989. (Source: *Statistical Abstract of the United States*, 1990.)

Year (since 1970)	Percent of budget	Year (since 1970)	Percent of budget
4	30.9	12	24.8
5	27.3	13	26.0
6	25.2	14	26.8
7	24.8	15	27.1
8	23.8	16	28.1
9	24.1	17	29.2
10	23.5	18	28.8
11	23.6	19	27.7

a. Draw a scattergram of the data with "year since 1970" as the independent variable. Sketch a sine wave which models the data.
b. What is the amplitude of the sine wave in part **a**? **sample: about 3.5**
c. What is the period of the sine wave in part **a**? **sample: about 13**
d. Supply arguments why a sine wave may or may not be a good model for long-term description of the percent of the U.S. budget spent on defense. **Answers may vary.**

REPRESENTATIONS deal with pictures, graphs, or objects that illustrate concepts.

■ **Objective H:** *Graph transformation images of graphs of circular functions.*
(Lessons 6-1, 6-2, 6-3, 6-6)

In 47 and 48, sketch one cycle of the graph without using an automatic grapher.

47. $y = \frac{1}{2}\sin\frac{1}{2}x$ **48.** $y = 8\cos\pi x$

49. Draw a graph to illustrate that $\tan 2\theta = \cos\theta$ has three solutions between 0 and π.

In 50–52, **a.** sketch a graph of the function. **b.** State the period and maximum value.

50. $\dfrac{y-1}{2} = \cos(x-\pi)$ **b) period 2π; maximum 3**

51. $f(t) = 7\sin(6t - 2\pi)$ **52.** $y = 6 - 5\cos(4x+2)$
b) period $\pi/3$; max 7   **b) period $\pi/2$; max 11**

■ **Objective I:** *State equations for graphs of circular functions.* *(Lessons 6-1, 6-2, 6-3)*

53. a. Write an equation for the image of the graph of $y = \sin x$ under a phase shift of $\frac{\pi}{6}$. $y = \sin\left(x - \frac{\pi}{6}\right)$
b. Check by graphing.

54. a. Write an equation for the image of the graph of $y = \cos x$ under a phase shift of $-\frac{4\pi}{3}$. $y = \left(x + \frac{4\pi}{3}\right)$
b. Check by graphing.

51.a.

52.a.

53.b.

54.b.

61.

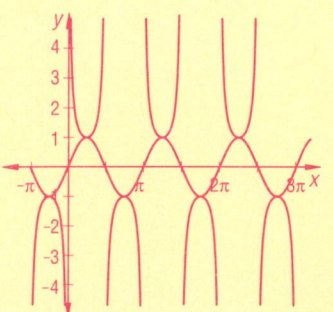

In 55–58, match each graph with its equation.

55. $y = \sin\left(x - \frac{\pi}{3}\right)$ **(b)**

56. $y = 1 + \sin x$ **(a)**

57. $y = \cos\left(x - \frac{\pi}{2}\right)$ **(d)**

58. $y = \cos x - 1$ **(c)**

(a)

(b)

(c)

(d)

59. Give an equation for the sine wave below.

60. Find an equation for the graph at the right which is a translation image of the graph of $y = \tan x$.
sample: $y = \tan (x + \pi/6)$

59) sample: $\frac{y-3}{4} = \sin (2\pi x)$

■ **Objective J:** *Graph or identify graphs of inverse trigonometric functions or reciprocal trigonometric functions.* (*Lessons 6-5, 6-7*)

In 61 and 62, graph on the same set of axes.

61. $y = \cos x$ for $0 \le x \le \pi$ and $y = \cos^{-1} x$.

62. $y = \sin x$ and $y = \frac{1}{\sin x}$ for $-\pi \le x \le 3\pi$.

63. *True* or *false*. The graph of the equation $y = \cot x$ has the same asymptotes as the graph of $y = \tan x$. **False**

64. *Multiple choice.* Which function is graphed below?
(a) $y = \csc x$ (b) $y = \sec x$
(c) $y = \sin x$ (d) $y = \sin^{-1} x$ **(d)**

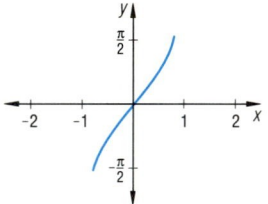

CHAPTER 7 ■ PROBABILITY AND SIMULATION

OBJECTIVES ■ CHAPTER 7

The objectives listed here are the same as in the Chapter 7 Review on pages 451–454 of the student text. The Progress Self-Test on page 450 and the tests in the Teacher's Resource File cover these objectives. For recommendations regarding the handling of this end-of-chapter material, see the notes in the margins on the corresponding pages of this Teacher's Edition.

OBJECTIVES FOR CHAPTER 7 (Organized into the SPUR categories— Skills, Properties, Uses, and Representations)	Progress Self-Test Questions	Chapter Review Questions	Teacher's Resource File Lesson Masters*	Teacher's Resource File Chapter Test Forms A & B
SKILLS				
A List sample spaces and events for probabilistic experiments.	1a	1–6	7-1	1a
B Compute probabilities.	1b, 2, 3	7–12	7-1, 7-2, 7-3	1b, 2, 3
C Find the number of ways of selecting or arranging objects.	4	13–16	7-3, 7-5	4
D Evaluate expressions using factorials.	5, 6	17–24	7-3, 7-5	5, 6
PROPERTIES				
E State properties of probabilities.	7	25–30	7-1, 7-2, 7-7	7, 15b,c
F Determine whether events are mutually exclusive, independent, or complementary.	8	31–38	7-2, 7-4	8
G Solve equations using factorials.	9	39–42	7-5	9
USES				
H Use probability to solve problems.	10	43–46	7-1, 7-2, 7-3, 7-4	10
I Use Counting Principles to find the number of ways of arranging objects.	11, 12	47–52	7-3, 7-5	11
REPRESENTATIONS				
J Construct, graph, and interpret probability distributions.	13	53–59	7-7	12, 14
K Use computers to generate pseudo-random numbers.	14	60–63	7-8	13, 15a
L Use computers for counting.	15	64–68	7-6	15d

*The masters are numbered to match the lessons, and each is intended for use with that lesson.

OVERVIEW ■ CHAPTER 7

This chapter has as its main themes probability, counting, and simulation. The intent is to review certain concepts, tackle some important types of questions of counting and probability, and prepare students for the study of binomial and normal distributions in Chapters 8 and 10.

A typical sequence in chapters that discuss counting and probability is to begin with counting, discuss permutations and combinations, and then cover probability. Instead, we begin with probability, as it provides the rationale for many of the counting situations to follow. We also delay the discussion of combinations for the next chapter, so students do not confuse them with permutations.

In Lessons 7-2 and 7-3 we present the basic properties of counting, which should be familiar to most students, but may be new to some even at this level. These counting techniques enable the theory of independent events to be verified easily in Lesson 7-4 with examples.

The Multiplication Counting Principle flows naturally into the discussion of permutations in Lesson 7-5. The calculation of permutations is applied directly in Lesson 7-6, where computer searches are discussed and analyzed.

Lesson 7-7 serves as a review of much of the previous lessons, putting together the calculations of probability with the study of probability distributions and the calculation of expected value.

The last two lessons introduce probability simulations. Such simulations require the concept of randomness, which is studied in some detail in Lesson 7-8. A particular type of simulation, the Monte Carlo simulation, is discussed in Lesson 7-9.

Functions underlie much of the discussion. The notations $P(\ )$ and $N(\ )$ are used from the first lesson on, even though probability distributions are not explicitly discussed until the seventh lesson. Multiple representations of distributions are used throughout the chapter, including graphs, tables, and equations.

If your students have not had much background in probability you may need to spend an extra day in Lessons 7-1 though 7-4, and use Project 1 in class.

PERSPECTIVES ■ CHAPTER 7

The Perspectives provide the rationales for the inclusion of topics or approaches, give some mathematical background, and make connections with other lessons and to other UCSMP texts.

7-1

FUNDAMENTAL PROPERTIES OF PROBABILITY

The language of set theory is particularly appropriate for studying probability. The sample space is the set of possible outcomes of an experiment; an event is a subset of the sample space.

In this lesson, probabilities are calculated for events in which the sample space is small and finite and the outcomes are equally likely. Similarities are drawn between the ideas of relative frequency and probability, and the basic properties of a probability function P defined on the events and outcomes of a sample space are established.

7-2

ADDITION COUNTING PRINCIPLES

This lesson presents the Addition Counting Principle and uses it to develop theorems for finding probabilities of unions of events (mutually exclusive or not), and probabilities of complementary events. Students should be able to compute the number of elements in a union of sets, and apply the theorems appropriately.

7-3

MULTIPLICATION COUNTING PRINCIPLE

If the first component of an ordered pair comes from a set with x elements and the second component comes from a set with y elements, then there are xy different ordered pairs possible. This is the Multiplication Counting Principle. This principle is further applied to counting arrangements with replacement of selections and without replacement. Students may already be familiar with the Multiplication Counting Principle from UCSMP's *Transition Mathematics* or *Algebra*.

7-4

INDEPENDENT EVENTS

We define the events *A* and *B* to be independent if and only if $P(A \cap B) = P(A) \cdot P(B)$. There is another definition of independence based on conditional probability which we use in UCSMP *Algebra*, but the definition we use in this book directly relates to the Multiplication Counting Principle and lends itself to binomial experiments. In this lesson, students are asked to determine whether events are independent, and to calculate probabilities if independence is known.

7-5

PERMUTATIONS

Students will have had practice with factorials and expressions containing factorials before this lesson, so the difficulty of learning the notation *and* the concept of permutations at the same time is avoided. In this lesson, the most general type of permutation without replacement is covered, and the familiar formulas are presented.

7-6

COMPUTERS AND COUNTING

This lesson uses BASIC programs to count and list all permutations of possible outcomes. Students should be able to understand the programs and modify them to solve a variety of problems. They should also be able to explain how nested loops work in a program. Such programs are needed when doing some types of computer searches. This lesson also provides students with more exposure to permutation problems.

7-7

PROBABILITY DISTRIBUTIONS

A probability distribution is a function which pairs the mutually exclusive outcomes of a sample space (the domain) with their probabilities of occurrence. When statisticians search for mathematical models to describe patterns in experimental data, they assume or test the existence of an underlying probability distribution based on calculations or expectations of how the data are expected to behave.

Probability distributions are theoretical, whereas relative frequency distributions (that one often sees in reports and other writing) are based on actual data. The mean of the relative frequency distribution has as its counterpart the expected value of the probability distribution.

This discussion of probability distributions could have been presented after Lesson 7-2, but we chose to provide students with more time to become comfortable with probability concepts before connections with functions were made.

7-8

RANDOM NUMBERS

No method for actually generating random numbers is known. In fact, it is not known if it is theoretically possible to generate such numbers. Until recent times, if random numbers were desired, one would look in a table for them. (That is hardly random, since two different people might use identical numbers.) In this lesson, students use the computer to generate pseudo-random numbers, that is, numbers that behave much like random numbers. They are shown how to simulate experiments by generating random numbers to fall within specific intervals.

7-9

MONTE CARLO METHODS

A Monte Carlo simulation is simply the use of random numbers to simulate an event. Monte Carlo methods are an outgrowth of the ability of computers to perform calculations very quickly. However, in this lesson, random number tables are used instead of computers.

Two different types of simulations are described. The first is to determine the number of reservations an airline company should accept if it knows that there are likely to be a certain percentage of no-shows. The second is the determination of the area in a region bounded by two lines and a parabola.

This reading lesson is designed to show the power of simulation. In both the reservation situation and the finding of the area under a parabola, simulation substitutes for an exact answer whose derivation requires college level mathematics.

DAILY PACING CHART ■ CHAPTER 7

For optimal student performance, the Progress Self-Test and the Chapter Review should be covered. (See *General Teaching Suggestions: Mastery* on page T35 of this Teacher's Edition.) By following the pace of the Full Course given here, students can complete 12 or 13 chapters of the book by the end of the year. Students following the pace of the Minimal Course spend more time to complete lessons that are combined with quizzes and to consider the Chapter Review. When all lessons are covered in order (the recommendation of the authors), students in the Minimal Course can cover all but the last two chapters. For more information on pacing, see *General Teaching Suggestions: Pace* on page T34 of this Teacher's Edition.

DAY	MINIMAL COURSE	FULL COURSE
1	7-1	7-1
2	7-2	7-2
3	7-3	7-3
4	Quiz (TRF); Start 7-4	Quiz (TRF); 7-4
5	Finish 7-4	7-5
6	7-5	7-6
7	7-6	Quiz (TRF); 7-7
8	Quiz (TRF); Start 7-7	7-8
9	Finish 7-7	7-9
10	7-8	Progress Self-Test
11	7-9	Chapter Review
12	Progress Self-Test	Chapter Test (TRF)
13	Chapter Review	
14	Chapter Review	
15	Chapter Test (TRF)	

TESTING OPTIONS

■ Quiz for Lessons 7-1 through 7-3   ■ Chapter 7 Test, Form A   ■ Chapter 7 Test, Cumulative Form
■ Quiz for Lessons 7-4 through 7-6   ■ Chapter 7 Test, Form B

A Quiz and Test Writer is available for generating additional questions, additional quizzes, or additional forms of the Chapter Test.

PROVIDING FOR INDIVIDUAL DIFFERENCES

The student text has been written for, and tested with, average students. It has also been used successfully with students who are better prepared as well as with students who are poorly prepared.

The Lesson Notes often include Error Analysis and Alternate Approach features to help you assist those students who need more help. Students of all abilities often learn from their peers and may benefit from small group work referenced as appropriate throughout the Notes. A blackline Lesson Master (in the Teacher's Resource File), keyed to the chapter objectives, is available for each lesson to provide more practice. (However, we advise you to use these masters sparingly; again refer to the suggestions on pace on page T34.)

Extension activities are provided in the Lesson Notes for use with students who need to be challenged even more than the requirements of completing the Full Course provide.

Probability and Simulation

Weather forecasts are frequently given in terms of probability.

Many situations in life involve uncertainty. Will it rain on the day of the class picnic? Will there be enough people at a student council meeting to have a quorum? Can you expect to win a lottery once in your life if you play each week? Answers to these questions involve *probabilities*, numbers which indicate the measure of certainty of an event.

The mathematics of probability theory has as its foundation a series of letters between the French mathematicians Pascal and Fermat in 1654 concerning dice games. Nowadays, probability theory has many applications beyond the analysis of games of chance, including setting insurance rates, weather forecasting, and political polling.

The simplest of probability questions can be answered by direct counting. In more complicated situations, theorems are applied to calculate the number of ways in which things can happen. Sometimes the situations can be modeled with a picture, and probabilities calculated by comparing lengths of segments or areas of regions. Probabilities arising from still more complex situations, such as those involving weather forecasting, cannot be calculated directly, but they can be estimated from *simulations* of the situation.

In this chapter, you will study a broad range of situations involving probability, from counting to simulation.

We recommend that 12–15 days be spent on this chapter: 9–11 days on the lessons and quizzes; 1 day for the Progress Self-Test; 1 or 2 days for the Chapter Review; and 1 day for a Chapter Test. (See the Daily Pacing Chart on page 397D.) Therefore, this chapter should be done at the usual lesson per day pace, with possibly an extra day or two for students who have never studied counting or probability.

USING PAGE 397
Four areas of application of probability are given on the page; you might ask students if they can think of others. The point to make is that probabilities occur in many different activities and are important in these activities.

LESSON 7-1

RESOURCES
■ Lesson Master 7-1
■ Teaching Aid 60: Sample space for tossing two dice

OBJECTIVES

A List sample spaces and events for probabilistic experiments.
B Compute probabilities.
E State properties of probabilities.
H Use probability to solve problems.

TEACHING NOTES

The sample space of an experiment is not always unique. For instance, in **Example 1b,** one could make the case that the sample space consists of the 36 ordered pairs (1, 1) through (6, 6). However, since the sum of the numbers is recorded and the dice are not distinguishable, we chose to consider the 11 sums as the sample space. However, in **Example 2,** the sample space is viewed as having 36 outcomes, because it is easier to calculate the probability when the outcomes are equally likely. A teaching aid is provided so that you do not have to list the 36 possibilities.

Example 3 can be used to illustrate the difference between a relative frequency and a probability. Here we *assume* that the births are equally likely even though it is known that the relative frequencies are significantly different from 1/2. Relative frequencies are ratios calculated from data, whereas probabilities are values (hopefully close to relative frequencies) that are either

LESSON 7-1

Fundamental Properties of Probability

Probability is the branch of mathematics that models uncertainty. More specifically, the *probability of an event* is a number from 0 to 1 that measures the certainty or uncertainty of the event. A probability of 0 signifies an impossible event; a probability of 1 indicates an event is certain to happen. Probabilities may be written in any way that numbers are written. An event which is expected to happen one-quarter of the time has a probability which can be written as $\frac{1}{4}$, or $\frac{2}{8}$, or .25, or 25% and so on.

Probabilities are determined in a variety of ways. When tossing a coin, you might assume that heads and tails are equally likely. From this assumption, it follows that the probability of a head is $\frac{1}{2}$ and the probability of a tail is $\frac{1}{2}$. In 2500 tosses of the coin you would expect about 1250 heads. If you do not wish to toss the coin so many times, you might simulate the tossing by computer. In contrast, if you have some reason to expect that the coin is unbalanced, you might actually toss the coin a large number of times and take the long-term relative frequency of heads as an estimate of the probability. For instance, if it comes up heads 1205 times in 2500 tosses, then the relative frequency of heads is $\frac{1205}{2500} = 48.2\%$, and you might pick 48% as the probability of heads for that coin.

In the tossing or simulation to calculate a probability, each toss is called an **experiment** or **trial**. The relative frequency above is the result of repeating an experiment 2500 times. A result of an experiment is called an **outcome**.

If a single coin is tossed, there are only two possible outcomes for each trial: heads (H), or tails (T). For that experiment, the *sample space* is the set {H, T}.

> **Definition**
>
> The set of all possible outcomes of an experiment is the **sample space** for the experiment.

Example 1

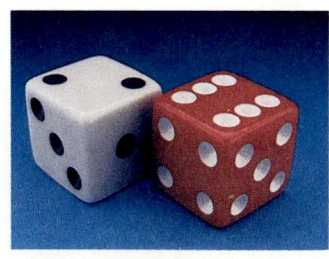

Determine the sample space for each experiment.

a. A die is rolled and the number on the top face is recorded.
b. A pair of dice are rolled and the sum of the numbers on the top faces is recorded.

Solution

a. All the integers from 1 to 6 are possible. So $S = \{1, 2, 3, 4, 5, 6\}$.
b. The smallest sum is $1 + 1 = 2$; the largest is $6 + 6 = 12$; and all the integers in-between are possible. So $S = \{2, 3, 4, ..., 11, 12\}$.

398

An **event** is a subset of the sample space for an experiment. For instance, for the experiment in Example 1, part **a**, one event is {2, 4, 6}, which can be described as "rolling an even number."

If a sample space has n possible outcomes, all of which are equally likely, then the experiment is called **fair** or **unbiased** and each outcome has probability $\frac{1}{n}$. In a fair experiment, the probability of any event is related in a simple way to the number of outcomes in that event:

> ### Definition
>
> Let E be an event in a finite sample space S. If each outcome in S is equally likely, then the **probability of the event E occurring**, denoted **$P(E)$**, is given by
>
> $$P(E) = \frac{\text{number of outcomes in the event}}{\text{number of outcomes in the sample space}}.$$

If $N(E)$ denotes the number of elements in set E, then $P(E) = \frac{N(E)}{N(S)}$. For the event "rolling an even number with a fair die," $E = \{2, 4, 6\}$, $N(E) = 3$, and $S = \{1, 2, 3, 4, 5, 6\}$. So $N(S) = 6$ and $P(E) = \frac{3}{6} = \frac{1}{2}$. You would expect to roll an even number half the time.

Note that in order to apply this definition, the outcomes in the sample space must be *equally likely*. You need to pay attention to this requirement when you set up or interpret sample spaces. For example, if you toss two fair coins, and record all possible orders of heads and tails, there are four equally likely outcomes in the sample space: HH, HT, TH, and TT.

Thus, $P(\text{HH}) = P(\text{HT}) = P(\text{TH}) = P(\text{TT}) = \frac{1}{4}$.

Now suppose you are interested in the number of *heads* that come up when you toss two fair coins. Then the sample space is
{2 heads, 1 head, 0 heads}.
There are three outcomes, but they are not equally likely because the event "one head" has two outcomes, HT and TH.

$$P(1 \text{ head}) = \frac{2}{4} = \frac{1}{2}, \text{ whereas } P(2 \text{ heads}) = P(0 \text{ heads}) = \frac{1}{4}.$$

hypothesized on the basis of an educated guess, calculated on the basis of some assumptions such as randomness or equal likelihood, or estimated from known relative frequencies.

The distinction between a relative frequency and a probability may be represented graphically. Suppose a coin is flipped 1, 2, 3, . . . , n times and $P(n)$ represents the relative frequency of heads. The graph of the pairs $(n, P(n))$ should theoretically approach (from both directions) the line $y = \frac{1}{2}$ as an asymptote.

Small Group Work If your students have not studied probability before, you might wish to separate the class into groups to do Project 1 on page 448. Regarding the Properties of Probabilities on page 401, (ii) and (iii) could both be written as "if and only if" statements. The question is whether one wishes to allow impossible outcomes in the sample space. For instance, one could ask, "What is the probability of getting 13 as the sum of numbers in the tosses of two dice?" While 13 is not a possible outcome, it could be given as an outcome in the sample space. Then we would have $P(\{13\}) = 0$, but {13} is not the null set.

Making Connections In function terms, the Properties of Probabilities are properties of the domain and range of a probability function P. The fact that the independent variable in this function (an event) may be the occurrence of one *or more* of the outcomes listed in the domain makes a probability function more complicated than most of the functions your students have studied up to this point. For this reason we do not discuss P as a function in this lesson, but use the notation in anticipation of discussing *probability distribution* functions in Lesson 7-7.

1. A test is given with five possible grades: A, B, C, D, and F.

a. Give the sample space for the experiment.
{A, B, C, D, F}

b. Identify the event "passing."
{A, B, C, D}

2. Suppose two fair dice are rolled. What sum has the highest probability of occurring? What is its probability?
7; $\frac{1}{6}$

3. If births of boys and girls are assumed equally likely, what is the probability that a family with four children has all girls?
$\frac{1}{16}$

NOTES ON QUESTIONS
Error Analysis for Questions 1–3: If students have missed any of these, you may want to clarify the definitions of the terms. A useful geometric analogy is to view the sample space as the entire space, an outcome as a point, and an event as a figure.

Question 11: This question asks for the probability of the complement of the event {A, B, O}. This calculation is formalized in Lesson 7-2.

Consider rolling two dice: one white, one red. There are 6 ways that each die might land. So there are $6 \cdot 6 = 36$ ways the two dice can come up. Each is equally likely if both dice are fair. All the possible outcomes of this experiment are shown below.

Sample Space for Tossing Two Dice

Example 2   Suppose two fair dice are rolled. Find P(sum of dice is 10).

Solution   The event "sum of dice is 10" has 3 outcomes, circled above. The sample space has 36 outcomes.

So P(Sum of 10) $= \frac{\text{number of ways to get 10}}{\text{total number of outcomes}} = \frac{3}{36} = \frac{1}{12}$.

Example 3   A researcher is studying the number of boys and girls in families with three children. Assume that the birth of a boy or a girl is equally likely. Find the probability that a family of three children has exactly one boy.

Solution   Let B represent a boy, G a girl, and a triple of letters the sexes of the children from oldest to youngest. List the outcomes. Then an appropriate sample space (listed alphabetically) is
{BBB, BBG, BGB, BGG, GBB, GBG, GGB, GGG}.
Three outcomes have exactly one boy. Thus, the probability of exactly one boy in a three-child family is $\frac{3}{8}$. So P(one boy) $= \frac{3}{8}$.

Actually, more boys than girls are born. In the U.S., the relative frequency of the birth of a boy in a single birth is closer to 52%. So the assumption in Example 3 is a little off. Later in this chapter you will see how to calculate probabilities when the outcomes are not equally likely.

Relative frequencies and probabilities have important similarities and differences. For example, both yield values between 0 and 1 inclusive. However, the meanings of the two values differ. A relative frequency of 0 means an event *has not occurred*; for example, the relative frequency of snow on July 4th in Florida since 1900 is 0. In comparison, a probability of 0 means the event is *impossible*. For example, the probability of tossing a sum of 17 with two normal dice is 0. Similarly, a relative frequency of 1 means that an event *has occurred* in each known trial, while a probability of 1 means that an event *must always occur*. For example, the probability is 1 that the toss of a die gives a number less than 7. Notice that this event consists of the whole sample space of the experiment. These properties are summarized on the next page.

400

Properties of Probabilities

Let S be the sample space associated with an experiment, and let $P(E)$ be the probability of an event in S. Then:
(i) $0 \leq P(E) \leq 1$.
(ii) If $E = S$, then $P(E) \equiv 1$.
(iii) If $E = \emptyset$, then $P(E) = 0$.

Questions

Covering the Reading

1. In 1988, 593,000 of the 2,138,000 men and women in the USA armed forces were in the Navy. What was the relative frequency of Navy personnel in the armed forces in 1988? **about 27.7%**

2. To test a new package design, a carton of a dozen eggs is dropped from a height of 18 inches. The number of broken eggs is counted. Determine a sample space for the experiment. **{0, 1, 2, ..., 12}**

3. Let A be an event. If $P(A) = .25$ and there are 200 equally likely outcomes in the sample space, how many outcomes are in A? **50**

In 4–6, suppose two fair dice are rolled as in Example 2. For the named event, **a.** list its outcomes and **b.** give its probability.

4. sum is 11
   a) {(5, 6), (6, 5)}; b) 1/18

5. both dice show the same number ("doubles")
   a) {(1, 1), (2, 2), (3, 3), (4, 4), (5, 5), (6, 6)}; b) 1/6

6. at least one die shows a 3
   a) {(3, 1), (3, 2), (3, 3), (3, 4), (3, 5), (3, 6), (1, 3), (2, 3), (4, 3), (5, 3), (6, 3)}; b) $\frac{11}{36}$

7. Find the probability that a family of three children has exactly two girls. $\frac{3}{8}$

8. Consider the experiment of tossing 3 fair coins.
   a. Give a sample space for the experiment.
   b. Find P(exactly 2 heads). $\frac{3}{8}$
   c. *True* or *false*. P(3 heads) $= P$(0 heads) **True**
   d. Find P(at least 2 heads). $\frac{4}{8} = \frac{1}{2}$
   a) {HHH, HHT, HTH, HTT, THH, THT, TTH, TTT}

9. State the difference between a probability of zero and a relative frequency of zero. **sample: A probability of 0 means an event is impossible; a relative frequency of 0 means an event has never happened.**

10. Give an example different from the one in the lesson of an event A such that $P(A) = 1$.
    sample: the probability that today will be either sunny or not sunny

Applying the Mathematics

11. All human blood can be typed as one of O, A, B, or AB, but the distribution of the blood type varies with race. Among black Americans 49% have type O, 27% have type A, and 20% have type B. What is the probability that a black American has type AB blood? **4%**

12. Consider a fair tetrahedral die with faces numbered 1, 3, 5, 7. Two of these dice are tossed. Find:
    a. P(sum is even) **1**   **b.** P(sum is odd) **0**   **c.** P(sum is 8). $\frac{1}{4}$

LESSON 7-1 Fundamental Properties of Probability   **401**

NOTES ON QUESTIONS
Questions 16 and 17:
The representation of odds is tricky. In general, if an event has probability $\frac{m}{n}$ for occurring, then the odds in favor of that event are $m:(n-m)$ and the odds against that event are $n:(n-m)$. For instance, if you think the odds are 2 to 1 that a team will win a particular game, then the probability the team will win is $\frac{2}{3}$ and the probability the team will lose is $\frac{1}{3}$.

FOLLOW-UP

MORE PRACTICE
For more questions on SPUR Objectives, use *Lesson Master 7-1*, shown on page 401.

EXTENSION
Is it possible to number the faces of two fair dice so that each of the sums from 2 to 12 is equally likely? Allow the possibility that two faces on the same die could have the same number. (No, there are 11 sums; each cannot occur the same number of times in 36 tosses. However, it is possible to have each of the sums from 1 to 12 be equally likely—number one die with 3 sides of 1 and 3 sides of 2; number the faces of a second die 0, 2, 4, 6, 8, and 10.)

PROJECTS
The projects for Chapter 7 are described on page 448. Project 1 is related to the content of this lesson.

In 13 and 14, two bags each contain five slips of paper on which one of the angle measures 0°, 30°, 45°, 60°, and 90° are written. One slip is drawn from each bag. Let a and b be the measures from bags 1 and 2, respectively.

13. Find $P(a + b \geq 90°)$ $\frac{3}{5}$ **14.** Find $P(\sin(a + b) \leq \sin 90°)$.   **1**

15. A baseball player with a batting average of .318 for the first 85 at-bats gets up to bat for the 86th time.
   a. Estimate the probability that this batter makes a hit.   ≈ 0.318
   b. Suppose the batter makes a hit. What is the new batting average?   .326

In 16 and 17, if the *odds against* an event are x to y, the probability of the event is $\frac{y}{x+y}$.

16. If the odds against winning are 5 to 1, what is the probability of winning?   1/6

17. Suppose the probability of having exactly one boy in a family of three children is $\frac{3}{8}$. What are the odds against having exactly one boy in a family of three children?   **5 to 3**

18. The definition of probabilities can be extended to apply to infinite sample spaces. If every outcome in an infinite sample space is equally likely, and E is an event in that space, then

$$P(E) = \frac{\text{measure of event}}{\text{measure of sample space}},$$

where the measure may be length, area, volume, etc.
   a. Due to a storm, the electricity went off at 2:13 and so both hands of the clock stopped between 2 and 3. If all times are equally likely, what is the probability that both hands would be between the same two consecutive numbers on the face of the clock?
   b. Think of the minute hand as a spinner. If the clock face is fair and the minute hand spun, what is the probability it lands between 6 and 8?   a) $\frac{1}{12}$;   b) $\frac{1}{6}$

19. Solve. $\log_3 9^x = 27$ *(Lesson 4-7)*   **13.5**   **Review**

20. If the half-life of a drug in the body is 10 hours, what percent of the drug remains after   **a.** 10 hours;   **b.** 24 hours?   *(Lesson 4-4)*
a) 50%; b) ≈ 18.95%

21. Give the negation of the statement "A coin tossed 4 times showed at least one head."   *(Previous course)*   A coin tossed 4 times showed no heads.

22. From the town dock and another point 100 m apart, Mr. Wu sights his cabin and takes angle measures as indicated on the drawing. Find the distance from the cabin to the dock.   *(Lesson 5-9)*   about 1.3 km

133°
44°
100 m

23. Simulate the experiment of Example 3 by tossing three coins. Call one side of each coin a boy and the other side a girl. Repeat the experiment a large number of times. How close is your relative frequency of "one boy" to the probability calculated in the example?   **Exploration**
Answers will vary, but the relative frequency should be close to 37.5%.

402

Addition Counting Principles

Because probabilities are often calculated by dividing one count by another, being able to count elements of sets is an important skill. This and the next lesson cover the fundamental principles of counting.

Recall that $A \cup B$, the **union** of sets A and B, contains all elements that are either in A or in B or in both. If A and B have no elements in common, they are called **disjoint** or **mutually exclusive**. The first counting principle is a basic model for addition.

The Addition Counting Principle (Mutually Exclusive Form)

If two finite sets A and B are mutually exclusive, then
$N(A \cup B) = N(A) + N(B)$.

Now think of A and B as events in a sample space S. Divide both sides of the addition counting principle formula by $N(S)$.

$$\frac{N(A \cup B)}{N(S)} = \frac{N(A)}{N(S)} + \frac{N(B)}{N(S)}$$

A basic theorem about probability is obtained.

Theorem (Probability of the Union of Mutually Exclusive Events)

If A and B are mutually exclusive events in the same finite sample space, then $P(A \cup B) = P(A) + P(B)$.

Example 1   If two fair dice are tossed, what is the probability of a sum of 7 or 11?

Solution 1   Let A = "tossing a 7" and B = "tossing an 11". You might wish to refer to the list of outcomes in the previous lesson. There are 6 outcomes in A and 2 outcomes in B. So there are 8 outcomes in $A \cup B$. Because there are 36 outcomes in the sample space S,

$$P(A \cup B) = \frac{N(A \cup B)}{N(S)} = \frac{8}{36} \approx 0.22.$$

Solution 2   Define A, B, and S as in the first solution:
$P(A) = \frac{N(A)}{N(S)} = \frac{6}{36}$ and $P(B) = \frac{N(B)}{N(S)} = \frac{2}{36}$. Because A and B are mutually

exclusive, $P(A \cup B) = P(A) + P(B) = \frac{2}{36} + \frac{6}{36} = \frac{8}{36}$.

OBJECTIVES

B Compute probabilities.
E State properties of probabilities.
F Determine whether events are mutually exclusive, independent, or complementary.
H Use probability to solve problems.

TEACHING NOTES

In words, the Addition Counting Principle says that to total the number of things in two sets, add the number in each set and subtract the number you counted twice. If the events are mutually exclusive, simply add the number of items in each set. This principle may be rather elementary, but it does help to deduce the more sophisticated properties of probability.

Error Analysis   While the distinction between mutually exclusive events and not mutually exclusive events is easy to make in the examples of this lesson, it seems that many people do not make the distinction so easily. For example, the number of people reported as having learning disorders or physical disabilities, or being afflicted with a disease is sometimes overestimated because the estimator does not take into account the possibility that the same individual could have more than one disorder.

In applying the preceding theorem, it is necessary that the events be in the same sample space. After all, the probability of tossing a head with a fair coin is 0.5 and the probability that a boy is born is 0.52, but finding the probability of tossing heads or having a boy by adding .50 + .52 would give you a number greater than 1, which is impossible.

Events which are not mutually exclusive overlap. Recall that the **intersection** $A \cap B$ of two sets A and B is the set of elements in both A and B.

Consider the following situation. Participants at a two-day conference could register for only one of the days or both. There were 231 participants on Friday; 252 on Saturday. The total number of people who registered for the conference was 350. Because $231 + 252 \neq 350$, there must have been people who attended both days.

Symbolically, if A is the set of Friday attendees and B is the set of Saturday attendees, then $N(A) = 231$, $N(B) = 252$, and $N(A \cap B) = 350$. The overlap $A \cap B$ between the participants on Friday and Saturday is shown in the *Venn diagram* to the right.

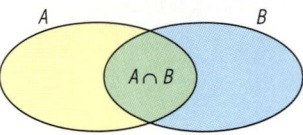

Let f be the number who attended only on Friday, s be the number who attended only on Saturday, and y be the number who attended on both Friday and Saturday.

So
$$N(A) = 231 = f + y,$$
$$N(B) = 252 = y + s,$$
and
$$N(A \cup B) = 350 = f + y + s.$$

To solve this system of three equations for y, add the first two equations and subtract the third.

$$231 + 252 - 350 = f + 2y + s - (f + y + s)$$
$$133 = y$$

Thus the number of participants who registered for both Friday and Saturday, $N(A \cap B)$, is 133. This is an instance of the more general relationship
$$N(A \cap B) = N(A) + N(B) - N(A \cup B).$$
This property is usually written solved for $N(A \cup B)$.

The Addition Counting Principle (General Form)

For any finite sets A and B,
$$N(A \cup B) = N(A) + N(B) - N(A \cap B).$$

If A and B are two events in the same sample space S, then dividing both sides by $N(S)$ yields $\frac{N(A \cup B)}{N(S)} = \frac{N(A)}{N(S)} + \frac{N(B)}{N(S)} - \frac{N(A \cap B)}{N(S)}$. The fractions all stand for probabilities, identified in the next theorem.

404

Theorem (Probability of a Union)

If A and B are any events in the same finite sample space, then
$P(A \text{ or } B) = P(A \cup B) = P(A) + P(B) - P(A \cap B)$.

Example 2 The name of a participant in the conference described previously is drawn from a hat for a door prize. What is the probability that this person attended on both days?

Solution 1 Use the definition of probability. Since 133 of the 350 participants attended both days,

$P(\text{winner attended both days}) = \frac{133}{350} = 0.38$.

Solution 2 $P(A \cap B)$ is desired, so use the Probability of a Union Theorem. $P(A) = \frac{231}{350}$ and $P(B) = \frac{252}{350}$. Because everyone who attended the conference attended on either Friday or Saturday $P(A \cup B) = 1 = \frac{350}{350}$.
By the theorem,

$$P(A \cup B) = P(A) + P(B) - P(A \cap B).$$

$$\frac{350}{350} = \frac{231}{350} + \frac{252}{350} - P(A \cap B)$$

$$-\frac{133}{350} = -P(A \cap B)$$

So $\qquad P(A \cap B) = \frac{133}{350}$.

Example 3 A pair of dice is thrown. What is the probability that the dice show doubles or a sum over 7?

Solution Use the sample space shown on page 400 for a pair of dice. Find the probability of each event, then compute the probability of their union using the theorem above.

$P(\text{doubles}) = \frac{6}{36}$

$P(\text{sum over } 7) = \frac{15}{36}$

$P(\text{doubles } and \text{ sum over } 7) = \frac{3}{36}$ (double 4s, 5s, or 6s)

So, $P(\text{doubles } or \text{ sum over } 7) = P(\text{doubles}) + P(\text{sum over } 7) - P(\text{both})$

$$= \frac{6}{36} + \frac{15}{36} - \frac{3}{36}$$

$$= \frac{18}{36} = \frac{1}{2}.$$

Notice that the calculation of the probability for the union of mutually exclusive events, discussed at the beginning of this lesson, is a special case of the Probability of a Union Theorem. When A and B are mutually exclusive events, $A \cap B = \emptyset$ and so $P(A \cap B) = 0$. Then

$$P(A \cup B) = P(A) + P(B) - P(A \cap B)$$

reduces to $\qquad P(A \cup B) = P(A) + P(B)$.

Questions 1 and 2: You
might ask: When will
$N(X \cup Y) = N(X \cup Y)$?
(when $X = Y$) When will
$N(X \cup Y) = N(X)$? (when Y
is a subset of X) When will
$N(X \cap Y) = N(X)$? (when X is
a subset of Y)

Questions 7 and 12:
The events in **Question 7**
could be the answer for
Question 12.

Question 16: There
seems to be a pattern after
parts **a** and **b**. That pattern
even holds for divisibility by 5
and 6, but it does not hold for
7. That is, the probability that
a random number from 1 to
300 is divisible by 7 is not $\frac{1}{7}$
but $\frac{42}{300}$; in general the probability that a number from 1 to
m is divisible by n is $\dfrac{\left\lfloor \frac{m}{n} \right\rfloor}{m}$.

ADDITIONAL ANSWERS
14. Sample: *A* and not *A*
are mutually exclusive
and exhaust the entire
sample space. Therefore,
$1 = P(A \cup \text{not } A) = $
$P(A) + P(\text{not } A)$. Hence,
$P(\text{not } A) = 1 - P(A)$.

In Example 4, the events are both mutually exclusive and exhaust the entire sample space. Such events are called **complementary events**. The complement of an event A is called **not A**.

- - - - - - - - -

Example 4   Two dice are tossed. Find the probability of each event.

a. Their sum is seven.   **b.** Their sum is *not* seven.

Solution

a. Use the diagram of the two-dice experiment on page 400. There are 6 possibilities, all on the diagonal from lower left to upper right.

So $P(\text{sum is } 7) = \frac{6}{36} = \frac{1}{6}$.

b. You could count again. Or you could recognize that "sum is 7" and "sum is not 7" are mutually exclusive events that exhaust the entire sample space. Then

$$P(\text{sum is 7 } or \text{ sum is not 7}) = P(\text{sum is 7}) + P(\text{sum is not 7}).$$
So, $1 = P(\text{sum is 7}) + P(\text{sum is not 7})$.
Thus $P(\text{sum is not 7}) = 1 - P(\text{sum is 7})$

$$= 1 - \frac{1}{6}.$$

Thus the probability that the sum of two dice is not seven is $\frac{5}{6}$.

Check   For part **b** there are 30 possibilities, so $P(\text{not } 7) = \frac{30}{36} = \frac{5}{6}$.

The same reasoning as in Example 4 proves the following general theorem.

Theorem (Probability of Complements)

If A is any event, then $P(\text{not } A) = 1 - P(A)$.

Questions

Covering the Reading

In 1 and 2, suppose two dice are rolled. Use the sample space on page 400. **a.** State whether the events X and Y are mutually exclusive. **b.** Find $N(X \cup Y)$. **c.** Find $N(X \cap Y)$.

1. X = The first die shows 6. Y = The sum of the dice is 10.
   a) No; b) 8; c) 1

2. X = The first die shows 5. Y = The sum of the dice is 3.
   a) Yes; b) 8; c) 0

In 3 and 4, refer again to the sample space for tossing two fair dice on page 400.

3. What is the probability that white shows an even number or red shows a multiple of 3? $\frac{2}{3}$

4. What is the probability that at least one of the dice shows a 4? $\frac{11}{36}$

406

5. In the situation used in Example 2, find the probability that the winner attended only on Saturday.   **119/350**

6. The probability that a manufactured computer chip is usable is 0.993. What is the probability that it is not usable?   **0.007**

7. The probability of Carol winning the election for Student Government presidency is 0.4 and the probability of Carl winning is 0.5.
   a. What is the probability that Carol or Carl will win?   **0.9**
   b. What is the probability that Carl will not win?   **0.5**
   c. What is the probability that neither Carl nor Carol will win?   **0.1**

In 8–10, assume A and B are events in the same sample space. Under what conditions are the following true?

8. $P(A \text{ or } B) = P(A) + P(B) - P(A \cap B)$ **all conditions**

9. $P(A \text{ or } B) = P(A) + P(B)$ **A and B are mutually exclusive events**

10. $P(\text{not } A) = 1 - P(A)$ **all conditions**

11. *True* or *false*. Complementary events are a special case of mutually exclusive events.   **True**

Applying the Mathematics

12. Give an example of two mutually exclusive events that are not complementary.
    sample: The temperature is above 90°; the temperature is below 40°.

13. Name a sport in which "win the game" and "lose the game" are not complementary events.   **sample: hockey**

14. Write an argument explaining why $P(\text{not } A) = 1 - P(A)$.   **See margin.**

15. The manager of a little league baseball team contacted the local meteorological office regarding the likely weather for the opening day of the season, and was given the table of probabilities below.

	Windspeed (mph)		
	less than 10	10–30	more than 30
Sunny	0.30	0.16	0.08
Partly cloudy	0.14	0.08	0.06
Overcast	0.09	0.06	0.03

   a. What is the probability that opening day will be sunny?   **0.54**
   b. What is the probability that the windspeed will be 30 mph or less?
   c. What is the probability that the weather will be overcast or the windspeed will be greater than 30 mph?   **0.32**
   d. What is the probability that the weather will not be overcast and the windspeed will be less than 10 mph?   **b) 0.83; d) 0.44**

16. A whole number between 1 and 300 (inclusive) is chosen at random. Find the probability that the number is
    a. divisible by 3 $\frac{1}{3}$ b. divisible by 4 $\frac{1}{4}$ c. divisible by 3 or by 4. $\frac{1}{2}$

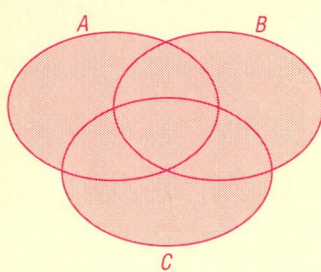
17. Consider the dart board pictured here. The radii of the concentric circles are 2″, 6″, 10″, and 11″. You get 10, 20, 30, 40, or 50 points if your dart lands in the regions as indicated. Suppose you always hit the square when you throw, you are equally likely to hit any point on the square, and you cannot land on a line. Determine the probability that your dart scores

  a. over 30 points   ≈ **0.162**
  b. under 30 points.   ≈ **0.630**

Review

18. A basketball player shoots two free throws. Each is either a basket or a miss. Determine the sample space for each experiment under each circumstance.
  a. You record the result of each shot in order.   **{BB, BM, MB, MM}**
  b. You count the number of baskets made.   *(Lesson 7-1)*   **{0, 1, 2}**

19. The table below gives the probability that a randomly chosen M & M® peanut candy is a particular color.

color	Brown	Red	Yellow	Green	Orange
probability	0.3	x	x	x	0.1

  If no other color is possible, find the value of x.   *(Lesson 7-1)*   **0.2**

20. If a whole number from 1 to 50 is picked at random, what is the probability that it is a perfect square?   *(Lesson 7-1)*   **7/50**

21. Solve: $7 = \log\left(\frac{1}{H}\right)$.   *(Lesson 4-6)*   **$10^{-7}$**

22. Give an equation for the parabola graphed at the right.   *(Lessons 3-2, 3-5)*
   $y = -x^2 + 4x$

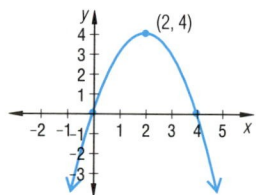

23. Solve: $t^2 - 100t = 21$.   *(Previous course)*   **$50 \pm \sqrt{2521}$**

Exploration

24. a. Draw a Venn diagram for $A \cup B \cup C$.   **See margin.**
   b. Extend the Probability of a Union Theorem to cover any 3 events in the same sample space. That is, give a formula for $P(A \cup B \cup C)$.
   c. Give an example of the use of the formula you find in part **b**.
   b) $P(A \cup B \cup C) = P(A) + P(B) + P(C) - P(A \cap B) - P(A \cap C) - P(B \cap C) + P(A \cap B \cap C)$

c) Sample: Let S = {students in your high school}, A = {students enrolled in an English course}, {B = students enrolled in a math course}, C = {students enrolled in a social studies course}. $P(A \cup B \cup C)$ is the probability that a student in your school is enrolled in an English, math, or social studies course.

7-3

Multiplication Counting Principles

CHEESE WITH PEPPERONI

THIN CRUST | THICK CRUST | STUFFED

CHEESE WITHOUT PEPPERONI

Pete Seria decided to offer a special on his famous pizza pies. He limited the special to cheese pizzas with or without pepperoni and with a choice of thin crust, thick crust, or stuffed. Pete wondered how many different versions of pizza were possible, so he sketched this tree diagram.

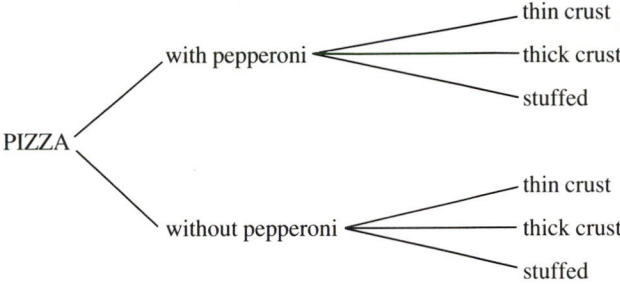

Counting the paths, Pete saw there were only six different possible pizzas, and thought that offer wasn't attractive enough. So he decided to advertise that each pizza came in one of four sizes: individual, small, medium and large. Rather than continuing with the tree (since he'd run out of room on his paper) Pete noticed that for each of 2 choices about pepperoni there were 3 choices of crust and $2 \cdot 3 = 6$ possible pizzas. And for each of these there were 4 possible sizes; so that gave $6 \cdot 4 = 24$ total choices.

Pete did not have to diagram or list all the possible choices because he knew a fundamental use of multiplication.

The Multiplication Counting Principle

Let A and B be any finite sets. The number of ways to choose one element from A and then one element from B is $N(A) \cdot N(B)$.

The Multiplication Counting Principle has an obvious extension to choices made from more than two sets. The number of ways to choose one element from set A_1, then one element from set A_2, …, and then one element from set A_k is $N(A_1) \cdot N(A_2) \cdot \ldots \cdot N(A_k)$.

LESSON 7-3

RESOURCES
- Lesson Master 7-3
- Quiz for Lessons 7-1 through 7-3
- Computer Master 18

OBJECTIVES

B Compute probabilities.
C Find the number of ways of selecting or arranging objects.
D Evaluate expressions using factorials.
H Use probability to solve problems.
I Use Counting Principles to find the number of ways of arranging objects.

TEACHING NOTES

Tree diagrams are useful to show visually all possible arrangements and the logic supporting the Multiplication Counting Principle. They become less practical, however, as the number of arrangements increases. Tree diagrams can help to represent situations of conditional probability, as is done in UCSMP's *Precalculus and Discrete Mathematics*.

Making Connections
We treat the terms *arrangement* and *permutation* as synonyms. The latter word will be introduced in Lesson 7-5. Not until the symbol $_nP_r$ is used is there a need to explain what the "P" denotes.

Alternate Approach: Technology There are ways of dealing with factorials higher than a student's calculator will estimate. In later lessons, students will learn to write the factors to divide out common factors when possible. Also, higher factorials can be approxi-

Example 1

a. How many ways are there of answering a test having 5 true-false questions?
b. If you guess on each question, what is the probability of answering all questions correctly?

Solution

a. There are two ways to answer each of the five questions. Think of each answer as a choice from a set with 2 elements in it, T or F.
$A_1 = \{T,F\}$, $A_2 = \{T,F\}$, $A_3 = \{T,F\}$, $A_4 = \{T,F\}$, and $A_5 = \{T,F\}$.
The number of ways to answer the test equals
$N(A_1) \cdot N(A_2) \cdot N(A_3) \cdot N(A_4) \cdot N(A_5) = 2 \cdot 2 \cdot 2 \cdot 2 \cdot 2 = 2^5 = 32$.

b. When you guess, all outcomes are equally likely. One of those 32 outcomes is "all answers correct," so the probability of answering all questions correctly is $\dfrac{1}{32}$, or about 3%.

Check

a. Below is a list of the 16 possible ways to answer the 5 questions if the first is answered T.

TTTTT	TTFTT	TFTTT	TFFTT
TTTTF	TTFTF	TFTTF	TFFTF
TTTFT	TTFFT	TFTFT	TFFFT
TTTFF	TTFFF	TFTFF	TFFFF

In the Questions you are asked to list the other 16.

© 1968 United Feature Syndicate, Inc.

Reprinted by permission of UFS, Inc.

The situation in Example 1 is an instance of making repeated choices *with replacement* from a set, here the set $\{T,F\}$. That is, you may give the same answer as many times as you wish. Each way of answering the 5 questions gives rise to an **arrangement** of the symbols. Two choices for each of 5 questions gives 2^5 arrangements. In general:

Theorem (Selections with Replacement)

Let S be a set with n elements. Then there are n^k possible arrangements of k elements from S *with replacement*.

Proof Use the Multiplication Counting Principle. Here $N(S) = n$, so the number of possible ways to choose one element from S, each of k times, is $\underbrace{N(S) \cdot N(S) \cdot ... \cdot N(S)}_{k \text{ times}} = \underbrace{n \cdot n \cdot ... \cdot n}_{k \text{ factors}} = n^k$.

410

Example 2   The mathematics section of some versions of the PSAT (Preliminary Scholastic Aptitude Test) have had 30 multiple-choice questions with 5 options each, and 20 multiple-choice questions with 4 options each. How many ways are there to answer this section?

Solution   First, figure out how many ways there are of answering each part of the test. Applying the Selections with Replacement Theorem, 5 choices for 30 questions gives 5^{30} possible arrangements on the first part, and 4 choices for 20 questions gives 4^{20} arrangements on the second part. Now, apply the Multiplication Counting Principle to the two parts. There are $5^{30} \cdot 4^{20}$, or about 10^{33} different ways of answering these versions of the PSAT mathematics section.

Example 3 illustrates that when repeated choices from a set are made *without replacement*, the Multiplication Counting Principle can still be applied.

Example 3   Candace decides to rank order the five colleges to which she plans to apply. How many rankings can she make?

Solution   For her first choice there are 5 possibilities. For second choice there are 4 remaining colleges. For third choice there remain 3 colleges. There are 2 colleges left for the fourth choice and 1 school is left for fifth. Altogether there are $5 \cdot 4 \cdot 3 \cdot 2 \cdot 1 = 120$ possible rankings.

The answer to Example 3 is often denoted as 5!, read "*five factorial.*"

Definition

For n a positive integer, **n factorial** is the product of the positive integers from 1 to n. In symbols,
$$n! = n \cdot (n-1) \cdot (n-2) \cdot (n-3) \cdot \ldots \cdot 3 \cdot 2 \cdot 1.$$

Many scientific calculators have a factorial key [n!] or [x!] on them. Check yours and find out how to use it.

The result of Example 3 can easily be generalized.

Theorem (Selections without Replacement)

Let S be a set with n elements. Then there are $n!$ possible arrangements of the n elements *without replacement*.

Making Connections for Question 13: This is an important question to discuss, for two reasons. First, it generalizes the procedures of many previous questions. Second, it prepares students for the formula for permutations to be seen in Lesson 7-5.

Question 19: This question assumes that each book has either a chapter on photography or one on painting. That assumption seems reasonable given the placement of the books in the store.

Technology for Question 24: The maximum factorial calculators can handle and the number of digits shown can vary. One calculator gives $1.7112 \cdot 10^{98}$ for 69! and an error message for 70!. Another shows $1.711224524 \cdot 10^{98}$, more digits but no higher factorial. A third one gives $5.17346099264 \cdot 10^{499}$ for 253!, and its memory is exceeded for 254!.

ADDITIONAL ANSWERS
3.a.

12.a. {(1, 1), (1, 2), (1, 3), (1, 4), (1, 5), (1, 6), (2, 1), (2, 2), (2, 3), (2, 4), (2, 5), (2, 6), (3, 1), (3, 2), (3, 3), (3, 4), (3, 5), (3, 6), (4, 1), (4, 2), (4, 3), (4, 4), (4, 5), (4, 6), (5, 1), (5, 2), (5, 3), (5, 4), (5, 5), (5, 6), (6, 1), (6, 2), (6, 3), (6, 4), (6, 5), (6, 6)}

MORE PRACTICE

For more questions on SPUR Objectives, use *Lesson Master 7-3*, shown on page 413.

EXTENSION

Recently, Marilyn vos Savant posed a probability problem in her "Ask Marilyn" column in *Parade Magazine*. Judging from the mail she received, the problem confounded many people, including several mathematicians. You may wish to have your students try to solve that problem.

The problem: Suppose you are on a game show, and the host gives you a choice of three doors. Behind one door is a valuable prize. Behind each of the others is a goat. You pick one door, say Door 2. The host, who knows what is behind each door, then opens a door to reveal a goat. The host now gives you a choice: Do you want to stay with your choice or do you want to switch to the other unopened door? Is it to your advantage to switch? Explain.

Answer: You should switch. Your probability of winning if you do not switch is $\frac{1}{3}$. Your probability of winning if you switch is $\frac{2}{3}$.

Rationale: Suppose these are the doors.

Prize	Goat	Goat
1	2	3

If you pick Door 1 and switch, you will lose. If you pick Door 2 and switch, you win. If you pick Door 3 and switch, you win. Hence, when you switch, you win 2 out of 3 times. You might have students work in **small groups** to imitate the game, using cards labeled *P*, G_1, and G_2.

Covering the Reading

1. If each of Pete Seria's pizzas came in only 3 sizes, how many different specials would he have had?   **18**

2. A sample of two children (one boy and one girl) is to be chosen from a class of 12 boys and 14 girls. In how many ways can the sample be chosen?   **168**

3. A diner serves a bargain breakfast which includes eggs (over-easy, poached, or scrambled); pancakes or toast; and juice (orange, tomato, or grapefruit). A breakfast must include one selection from each category.
   a. Draw a tree diagram showing the different possible breakfasts.
   b. How many possible breakfasts are there?   **a) See margin. b) 18**

4. Refer to Example 1, part **a**. Write the 16 ways to answer the test not given in the solution.   **FTTTT; FTFTT; FFTTT; FFFTT; FTTTF; FTFTF; FFTTF; FFFTF; FTTFT; FTFFT; FFTFT; FFFFT; FTTFF; FTFFF; FFTFF; FFFFF**

5. A test has 10 true-false questions. If you know 3 and guess on 7, what is the probability that you will get them all correct?   **1/128 ≈ .0078**

6. How many ways are there of answering a test (assuming you answer all items) if the test has 10 multiple choice questions each with 4 choices?
   $4^{10} = 1{,}048{,}576$

7. How many ways are there of answering a test with 10 true-false and 20 multiple choice questions, each with 5 choices, if you must answer each question? $2^{10} \cdot 5^{20} \approx 9.8 \times 10^{16}$

8. Refer to Example 2. The PSAT mathematics section has a penalty for guessing, so if you don't know an answer you should leave the question blank. The four-option multiple choice questions therefore really have five possible answers: A, B, C, D, or blank; similarly, the five-option questions have six possible answers. Write an expression for the number of ways of answering the PSAT mathematics section. $6^{30} \cdot 5^{20} \approx 2 \cdot 10^{37}$

9. Evaluate 6! without using a calculator.   **720**

10. Evaluate 12! using a calculator.   **479,001,600**

Applying the Mathematics

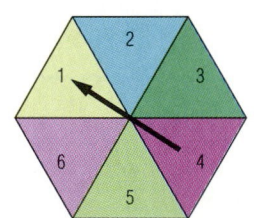

11. In how many ways can the batting order of a 10-person softball team be set? $10! = 3{,}628{,}800$

12. The spinner at the left has 6 congruent regions.
    a. If it is spun twice, list the possible outcomes.   **See margin.**
    b. If it is spun 10 times, how many possible outcomes are there?
    c. If the spinner is fair, what is the probability of getting ten 1s in a row?   **b) $6^{10} = 60{,}466{,}176$; c) $(1/6)^{10} \approx 1.7 \times 10^{-8}$**

13. If there are *n* television programs on at a particular time, how many possible rankings are there of three of these? $n(n-1)(n-2)$

14. Solve $\frac{8!}{p} = 6!$.   **56**

412

15. A 10-speed bicycle gets its 10 speeds by selecting one from each of two sets of gears. The front set has two gears. How many gears does the rear set have?   **5**

16. Consider $\frac{n!}{r!} = 72$, where n and r are positive integers.

   a. To find all possible solutions (n, r), how many different pairs of numbers must you check if $10 \geq n \geq 6$ and $8 \geq r \geq 5$?   **20**

   b. Solve $\frac{n!}{r!} = 72$, where n and r satisfy the conditions of part **a.**
   $n = 9, r = 7$

17. In Example 2, a calculator was used to get $5^{30} \cdot 4^{20} \approx 10^{33}$. Use the facts that $4^{20} = 2^{40}$ and $5^{30} \cdot 2^{30} = 10^{30}$ to derive the same result without a calculator.
   $5^{30} \cdot 4^{20} = \frac{10^{30}}{2^{30}} \cdot 2^{40} = 2^{10} \cdot 10^{30} = 1,024 \cdot 10^{30} \approx 10^3 \cdot 10^{30} = 10^{33}$

Review

18. A record store receives a new shipment of albums. Of these, r are rock, c are country, and b are both rock and country. How many records are in the shipment?   *(Lesson 7-2)*   **$r + c - b$**

19. There are 37 books on a bookstore shelf labeled ''Photography & Painting.'' 18 books contain chapters on photography, and 23 books have chapters on painting. What is the probability that a randomly selected book has chapters both on photography and painting?
   (Lesson 7-2)   **$\frac{4}{37} \approx 0.11$**

20. **a.** What is the probability of rolling a sum of 10 with two fair dice?
   b. What are the odds against rolling a sum of 10 with two fair dice?
   (Lesson 7-1)   **a) $\frac{1}{12} \approx 0.08$; b) 11 to 1**

In 21 and 22, give exact values without using a calculator.   *(Lesson 6-5)*

21. $\tan^{-1} 1$ **$\frac{\pi}{4}$**   **22.** $\cos^{-1}\left(\frac{-\sqrt{2}}{2}\right)$ **$\frac{3\pi}{4}$**

23. Refer to the graph below.   *(Lessons 2-1, 5-6)*
   a. What are the domain and range of f?
   b. Give the maximum and minimum values of f.
   c. What is the period of f?   **4**

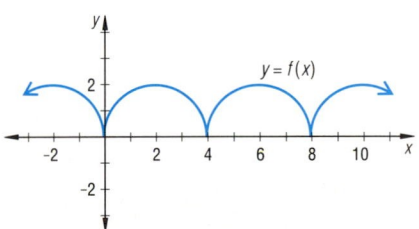

a) **domain = the set of all real numbers; range = $\{y: 0 \leq y \leq 2\}$**
b) **maximum value = 2; minimum value = 0**

Exploration

24. **a.** What is the largest factorial your calculator can display without going into scientific notation?   **13! on many calculators**
   b. What is the largest factorial your calculator can estimate in scientific notation?   **69! on many calculators**

PROJECTS
Projects 2 and 4 are related to the content of this lesson.

EVALUATION
A quiz covering Lessons 7-1 through 7-3 is provided in the Teacher's Resource File.

ADDITIONAL ANSWERS
3.a., 12.a. See margin, page 411.

NAME _____

LESSON **MASTER** **7-3**
QUESTIONS ON **SPUR** OBJECTIVES

■**SKILLS**   *Objective B (See pages 451–454 for objectives.)*

1. Jamie has 3 black pairs, 2 brown pairs, and one gold pair of socks in one drawer. If she randomly picks a pair of socks from the drawer, what is the probability that she will pick a brown pair?   **$\frac{1}{3}$**

2. A green die and an orange die are tossed. What is the probability that the green die shows a three and the blue die shows a number greater than three?   **$\frac{1}{12}$**

■**SKILLS**   *Objective C*

3. Suppose a state forms its license plates with three letters followed by four digits from 0 to 9, and that the first letter must be either A, B, or C. How many different license plates are possible?   **20,280,000**

4. How many code symbols can be formed using 4 out of the 5 letters V, W, X, Y, Z if the letters

   a. are not repeated?   **120**

   b. can be repeated?   **625**

   c. are not repeated and the code must begin with Y?   **24**

   d. are not repeated and the code must end with YZ?   **6**

■**SKILLS**   *Objective D*

5. Evaluate each of the following.
   a. $7! - 5!$ **4920**   **b.** $4! + 2!$ **26**   **c.** $\frac{8!}{4!}$ **1680**

Functions, Statistics, and Trigonometry © Scott, Foresman and Company   **77**

NAME _____
Lesson MASTER 7-3 (page 2)

6. A funhouse at an amusement park has 13 doors inside. Once a person has gone through a door, the person cannot go through it again. How many ways are there to go through the funhouse?   **13!**

■**USES**   *Objective I*

7. A certain new car is available with a choice of four exterior colors, three interior colors, two types of trim, and three types of engines. How many different sets of colors and engines are available?   **72**

8. How many ways are there of answering a test (assuming one answers all items) if the test has 50 multiple-choice items, each with 4 choices?   **4^{50}**

78   *Functions, Statistics, and Trigonometry © Scott, Foresman and Company*

RESOURCES
■ Lesson Master 7-4
■ Teaching Aid 61: Sample spaces with and without replacement

OBJECTIVES

F Determine whether events are mutually exclusive, independent, or complementary.
H Use probability to solve problems.

TEACHING NOTES

Words used in mathematics generally have meanings which are more precise than their definitions in everyday usage, as with *independent*. A person is considered to act independently when the person does not worry about what others do or think; two events are independent if the probability of one does not influence the probability of the other. Thus events from repeated trials with replacement can be independent because you do not know what transpired, but events from repeated trials without replacement are not independent because the first event affects what is possible for the second. A teaching aid with the two sock sample spaces is given to facilitate discussing the differences in the sample spaces.

In **Example 1,** the events are independent because the second spin is not affected by what happened on the first. However, in **Example 2,** event *A* affects event *B*. For instance, if the first spin was a 6, then event *B* cannot occur; if the first spin was a 3, event *A* cannot occur, but *B* can. In informal language, we

LESSON

7-4

Independent Events

When you flip a fair coin twice, the result of the first flip has no bearing on the result of the second flip. This is sometimes expressed by saying "the coin has no memory." Thus *if a coin is fair*, even if 10 heads have occurred in a row, the probability of heads on the 11th toss is still $\frac{1}{2}$. The tosses of the coin are called *independent events* because the results of one toss do not affect the results of any other toss.

Similarly, selections with replacement are considered to be independent selections because later selections "do not remember" what happened with earlier selections. For instance, suppose that six socks are in a drawer, 4 of them orange, 2 black, labeled **1 2 3 4 1 2**. If a first sock is blindly taken, put back, and then a second sock is blindly taken, what is the probability that both are black?

Let *A* be the event that the first sock is black, and *B* be the event that the second sock is black. Thus $A \cap B$ is the event that both socks are black. Since in each selection there are 2 black socks among the 6, $P(A) = \frac{2}{6}$ and $P(B) = \frac{2}{6}$. To calculate $P(A \cap B)$, a rectangular array can be drawn as was done for the dice in Lesson 7-1. Each pair in the array is an equally likely outcome in the sample space for "selecting two socks, with replacement." For instance, **4 1** means that the orange sock #4 is the first sock and black sock #1 is the second sock.

Sample Space with Replacement					
1 1	1 2	1 3	1 4	1 1	1 2
2 1	2 2	2 3	2 4	2 1	2 2
3 1	3 2	3 3	3 4	3 1	3 2
4 1	4 2	4 3	4 4	4 1	4 2
1 1	1 2	1 3	1 4	1 1	1 2
2 1	2 2	2 3	2 4	2 1	2 2

In 4 of the 36 pairs, both socks are black. So $P(A \cap B) = \frac{4}{36}$. This is the product of $P(A)$ and $P(B)$.

More generally, if S_1 is the sample space for the first selection and S_2 is the sample space for the second selection, and the selections are independent, then an array formed like the one above will have $N(S_1) \cdot N(S_2)$ elements, of which $N(A) \cdot N(B)$ are both black. So

$$P(A \cap B) = \frac{N(A) \cdot N(B)}{N(S_1) \cdot N(S_2)} = \frac{N(A)}{N(S_1)} \cdot \frac{N(B)}{N(S_2)} = P(A) \cdot P(B).$$

This relationship between the probabilities of *A*, *B*, and $A \cap B$ is taken as the definition of independent events.

414

Definition

Events A and B are **independent events** if and only if
$$P(A \cap B) = P(A) \cdot P(B).$$

Example 1
The circular area around a fair spinner is divided into six equal parts and numbered, as shown at the left. Consider spinning it twice (the spinner cannot stop on a boundary line). Define two events as follows.

    A: the first spin stops on an even number;
    B: the second spin stops on a multiple of 3.
Decide whether events A and B are independent.

Solution
  A and B are independent if $P(A \cap B) = P(A) \cdot P(B)$.
$$P(A) = \frac{3}{6} = \frac{1}{2}$$
$$P(B) = \frac{2}{6} = \frac{1}{3}$$

$P(A \cap B)$ can be obtained from the sample space illustrated below. The six circled points represent those for which both A and B are true.

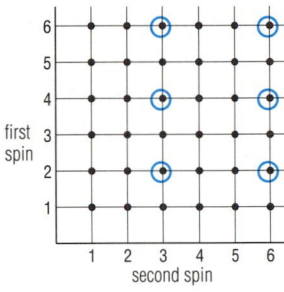

So $P(A \cap B) = \frac{6}{36} = \frac{1}{6}$. Because $P(A) \cdot P(B) = \frac{1}{2} \cdot \frac{1}{3} = \frac{1}{6} = P(A \cap B)$, events A and B are independent.

Check Since the spinner does not have a memory, the outcome on the first spin does not influence the second spin.

Now examine what happens if two socks are blindly taken *without replacement* from a drawer with 4 orange socks and 2 black socks. $P(A)$, the probability that the first sock is black, is still $\frac{2}{6}$. But $P(B)$, the probability that the second is black, depends on whether the first one taken was black or was orange. If the first one taken was black, then there are 4 orange socks and 1 black sock left, so the probability that the second is black is $\frac{1}{5}$. If the first one taken was orange, then there are 3 orange and 2 black socks left, so the probability that the second sock is black is $\frac{2}{5}$. Thus the value of $P(B)$ depends on the outcome of event A. For this reason, selections without replacement are not independent events; rather, they are **dependent events**.

---

would say the occurrence of B depends on what happens in A. The mathematical language formalizes the informal English.

Example 3 points out that many confuse independent events with mutually exclusive events. Your primary objective should be to make certain students understand the difference in practice, not just in theory. As is indicated by the engine-failure example, the dependence of events may not be obvious at first glance. The mathematical definition of independence provides a test.

To elaborate further on **Example 3,** the events might not be independent if, for example, the same mechanic worked on all three engines. However, if three *different* mechanics each worked on one of the engines, then the events would be independent, because whether one mechanic forgets the seals has no effect on whether the second or third mechanic will also forget.

ADDITIONAL EXAMPLES
1. Imagine that a fair coin is tossed 4 times. Let A = getting all heads and B = getting all tails. Are A and B independent or dependent?

$P(A \cap B) = 0$, so $P(A \cap B) \neq P(A) \cdot P(B)$ and the events are dependent.

2. Two normal dice are tossed. Let C = the sum is 7 and D = the first die shows a 5. Are C and D dependent or independent?
$P(C) = \frac{6}{36}$, $P(D) = \frac{1}{6}$, and
$P(C \cap D) = \frac{1}{36}$, so
$P(C) \cdot P(D) = P(C \cap D)$, so the events are independent.

This result may come as a surprise, because regardless of what the first die shows, there is still a 1/6 chance of getting a 7 for the sum. This is not true for any other sum.

It is possible to calculate $P(A \cap B)$ even when A and B are dependent. Here is a listing for the sock experiment without replacement.

Sample Space without Replacement				
1 2	1 3	1 4	1 1	1 2
2 1	2 3	2 4	2 1	2 2
3 1	3 2	3 4	3 1	3 2
4 1	4 2	4 3	4 1	4 2
1 1	1 2	1 3	1 4	1 2
2 1	2 2	2 4	2 4	2 1

These outcomes consist of all those in the independent case except those with the same sock selected twice. Thus there are only 30 outcomes. In only two are both socks black, so $P(A \cap B) = \frac{2}{30} = \frac{1}{15}$.

Example 2 Consider the spinner in Example 1 and the events
A: the first spin shows less than 3;
B: the sum of the spins is less than 5.
Decide whether events A and B are independent.

Solution Check to see if $P(A \cap B) = P(A) \cdot P(B)$.
A has outcomes {1, 2}, so $P(A) = \frac{2}{6} = \frac{1}{3}$.
B has outcomes {(1, 1), (1, 2), (1, 3), (2, 1), (2, 2), (3, 1)}, so
$$P(B) = \frac{6}{36} = \frac{1}{6}.$$
The five circled points below represent $A \cap B$ in the sample space for spinning twice.

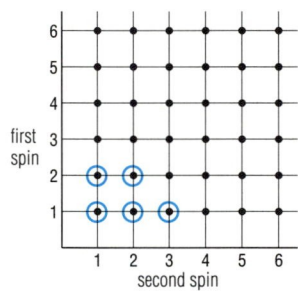

So, $P(A \cap B) = \frac{5}{36}$.

In this case, $P(A) \cdot P(B) = \frac{1}{3} \cdot \frac{1}{6} \neq P(A \cap B)$.

So, the events A and B are *not* independent.

Check The outcome of event A is part of the sum in event B. It can be expected that these events are dependent.

The notion of independence can be extended to more than two events. Consider this true story.

In May, 1983, an airline jet carrying 172 people between Miami and Nassau lost its engine oil, power, and 12,000 feet of altitude over the Atlantic Ocean before a safe recovery was made. When all three engines' warning lights

416

indicating low oil pressure all lit up at nearly the same time, the crew's initial reaction was that something was wrong with the warning system, not the oil pressure. As one person stated in *The Miami Herald* of May 5, 1983,

> "They considered the possibility of a malfunction in the indication system because it's such an unusual thing to see all three with low pressure indications. The odds are so great that you won't get three indications like this. The odds are way out of sight, so the first thing you'd suspect is a problem with the indication system."

Example 3   Aviation records show that for the most common engine on a Boeing 727 airliner, there is an average of 0.04 "inflight shutdowns" per 1000 hours of running time. So the probability of an engine's failure in a particular hour is about 0.00004 . Suppose the failures of the three engines were independent. What is the probability of three engines failing in the same hour?

Solution   Let A, B, and C be the engines. If they are independent events,

$$P(\text{A, B and C fail}) = P(\text{A fails}) \cdot P(\text{B fails}) \cdot P(\text{C fails})$$
$$= (0.00004) \cdot (0.00004) \cdot (0.00004)$$
$$= 6.4 \times 10^{-14}.$$

The number 6.4×10^{-14} is the sort of "out of sight" probability the writer of the article mentioned. This probability could be interpreted that *if* failures were independent, then the reciprocal of the probability above yields the result that about once in every 16,000,000,000,000 hours of flight would all three engines fail simultaneously.

After the incident it was discovered that a mechanic doing routine maintenance work on the plane had failed to install six tiny rubber seals on three engines' oil plugs. The gaps this error created allowed all the oil to leak out when the engines were fired up.

The crew members assumed that oil pressure problems in the three engines were independent events. Had this in fact been the case they would have been correct in their assumption that the probability of all three engines failing simultaneously was extremely small. But the three failures were all due to one cause, a mechanic's error. So the failures were dependent events, and the event was not as unlikely as the crew thought.

Questions

Covering the Reading

1. A drawer contains 5 socks, 3 red and 2 blue.
   a. If you pick a sock blindly, replace it, and pick another, what is the probability that both socks are red?   **9/25 = .36**
   b. If you pick two socks blindly without replacing the first, what is the probability that both are red?   **6/20 = .3**
   c. If $F =$ the first sock is red, and $S =$ the second sock is red, in which of parts **a** and **b** are F and S independent events?   **a**

LESSON 7-4   Independent Events   **417**

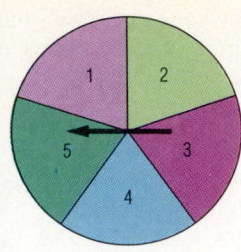

In 2 and 3, suppose you have a spinner with 5 congruent areas and a die, both fair. Decide whether the two events are independent by finding $P(A)$, $P(B)$, and $P(A \cap B)$, and checking whether the multiplication rule holds.

2. A: the spinner shows 4; B: the die shows 4.   **See margin.**

3. A: the spinner shows 4; B: the sum of the spinner and die is over 6.
No, they are not independent; $P(A) = 1/5$, $P(B) = 1/2$, $P(A \cap B) = 2/15 \neq 1/5 \cdot 1/2$.

4. If events A, B, and C are independent, then $P(A \cap B \cap C) = \underline{?}$.
$P(A) \cdot P(B) \cdot P(C)$

In 5 and 6, refer to the reporting of the airline incident. Suppose the failure rate per hour of each engine is 0.0005, and assume failures of the engines are independent events.

5. What is the probability that all three engines fail in the same hour?
$1.25 \cdot 10^{-10}$

6. What is the probability that none of the engines fail?   **≈ 0.9985**

7. Two fair dice are tossed. Use the definition of independence to classify the following pairs of events as independent, mutually exclusive, or neither.
a. One die shows 2. Another die shows 3.   **independent**
b. One die shows 2. The same die does not show 3.   **neither**
c. One die shows 2. The same die shows 3.   **mutually exclusive**

8. A fair coin is tossed three times. Consider the events:
  A: at most one head occurs;
  B: heads and tails each occur at least once.
Are A and B independent? Justify your answer using the definition of independence.   **Yes; $P(A) = \frac{1}{2}$, $P(B) = \frac{3}{4}$, $P(A \cap B) = \frac{3}{8} = \frac{1}{2} \cdot \frac{3}{4}$.**

9. The information booklet about a wildlife reservation park states that the probability of observing an eagle on a given day is .21, and the probability for a hawk is .17. The booklet also indicates that, based on records, 2% of the visitors observe both of these prey birds on the same day. In this park, are observing an eagle and observing a hawk independent events? Explain your answer.
No; $P(Eagle) \cdot P(Hawk) = .21 \cdot .17 = .0357 \neq .02$.

10. A motorist has done some observation during her commuting on one section of the road with 5 traffic lights. Her chance of being stopped by any one of them is 25%, while her chance of being stopped by all five of them is 1%. Show that the timings of the lights are not independent.
See margin.

11. An auto manufacturer buys plastic molded parts from two companies: Ace and Best. Here is the frequency distribution of parts by quality that the auto manufacturer has received.

	Ace	Best
excellent	60	24
acceptable	272	20
unacceptable	16	8

a. What is the probability that a randomly selected part is from Ace?
b. What is the probability that a randomly selected part is excellent?
c. What is the probability that a randomly selected part is both from Ace and excellent?   **0.15**
d. Are the quality of parts and their sources independent?   **no**
a) 0.87; b) 0.21

418

In 12–14, consider Chuck's Restaurant, which has the following items on its menu. *(Lesson 7-3)*

Entrées	Beverages	Desserts
Hamburger	Coffee	Cheese Cake
Shrimp Sandwich	Tea	Apple Pie
Grilled Chicken	Milk	Brownie
Vegetarian Delight	Orange Juice	Frozen Yogurt

12. a. If you order milk, an entrée, and a dessert, how many possible orders are there?   **16**
   b. Show them in a tree diagram.   **See margin.**

13. How many ways are there to order an entrée, beverage, and dessert?   **64**

14. If you cannot eat either hamburger or chicken, how many ways are there to order an entree, beverage, and dessert?   **32**

15. A family of five is arranging itself in a line for a photo. How many possible arrangements are there? *(Lesson 7-3)*   **5! = 120**

In 16–18, evaluate. *(Lesson 7-3)*

16. $6 \cdot 5!$ **720**   **17.** $\frac{10!}{9!}$ **10**   **18.** $\frac{95!}{93!}$ **95 · 94 = 8930**

19. *True* or *false*. If A and B are mutually exclusive events, then $P(A \cap B) = P(A) + P(B)$. *(Lesson 7-2)*   **False**

20. There are nine felt-tip and nine ball-point pens in a box, three red, three blue, and three black of each kind. A pen is selected at random from the box. What is the probability that it is black or has a ball-point end? *(Lessons 7-1, 7-2)* $\frac{2}{3}$

21. a. Draw $f(x) = \cos x$ and $g(x) = \sin\left(x + \frac{5\pi}{2}\right)$, when $-\pi \le x \le 2\pi$, on the same axes.   **See margin.**
   b. *True* or *false*. For all x, $\sin\left(x + \frac{5\pi}{2}\right) = \cos x$.   *(Lesson 6-2)*   **True**

22. Give an equation for the quadratic function that contains the three points $(-2, 3)$, $(3, 8)$, and $(5, 38)$.   *(Lesson 2-7)* $y = 2x^2 - x - 7$

23. *Skill sequence*. If $G(r) = r^2 - 6r - 11$, find
   a. $G(2t)$ **b.** $G(t + 7)$.   **c.** $G(2t + 7)$.   *(Lesson 2-1)*
   $4t^2 - 12t - 11$ $t^2 + 8t - 4$ $4t^2 + 16t - 4$

24. Explain in words why, if two events can happen and they are mutually exclusive, then they cannot be independent. Include an example with your explanation.
   Sample: If events can happen, then they have positive probabilities. If they are independent, then the probability of both happening is the product of these probabilities, a positive number. But if they are mutually exclusive, the probability of both happening is 0, by definition of "mutually exclusive." So they cannot be both independent and mutually exclusive. For instance, the birth of twins and the birth of triplets cannot happen at the same time; they are mutually exclusive. Therefore, the birth of twins and the birth of triplets are not independent events.

LESSON 7-4 Independent Events   **419**

FOLLOW-UP

MORE PRACTICE
For more questions on SPUR Objectives, use *Lesson Master 7-4*, shown below.

21.a.

NAME

LESSON **MASTER   7-4**
QUESTIONS ON **SPUR** OBJECTIVES

■**PROPERTIES**   *Objective F (See pages 451–454 for objectives.)*
In 1–4, decide whether the two events are independent by finding $P(A)$, $P(B)$, and $P(A \cap B)$, and check whether the multiplication rule holds.

1. Two cards are picked at random without replacement from a standard deck. A = the first card is black and B = the second card is red.   **Not independent**

2. Two dice are rolled. A = the first die shows a 6 and B = the second die shows a number less than 5.   **Independent**

3. Two dice are rolled. A = the first die shows a 4 and B = the sum is less than 8.   **Not independent**

4. In a family of three children, A = the first child is a boy and B = at least two girls are in the family.   **Not independent**

■**USES**   *Objective H*

5. A police officer gave the following tickets one month. No person received more than one ticket.

Driver's Age	No Seat Belt	Speeding
16–20	8	20
21–35	10	35
Over 35	58	13

   a. What is the probability that a randomly chosen person received a ticket for no seat belt? $\frac{19}{63}$
   b. What is the probability that a randomly chosen person is 21–35 years old? $\frac{5}{16}$
   c. Are the age of a person being 21–35 and receiving a ticket for no seat belt independent?   **No**

6. The chance that Marge will buy a hamburger is .25; the chance that she will buy milk is .7. If the events are independent, what is the probability she will buy both?   **.175**

Functions, Statistics, and Trigonometry © Scott, Foresman and Company

79

RESOURCES
■ Lesson Master 7-5

OBJECTIVES

C Find the number of ways of selecting or arranging objects.
D Evaluate expressions using factorials.
G Solve equations using factorials.
I Use Counting Principles to find the number of ways of arranging objects.

TEACHING NOTES

In permutation problems, all of the objects may be arranged, as with the basketball teams, or just some of the objects may be arranged, as in **Example 1,** where not all of the seven horses can finish first, second, or third.

Point out that in all the counting problems of this lesson, *the order of the elements being counted matters.* If the 4, 6, and 1 horses finished first, second, and third, this is different than if the order of finish had been 6, 4, and 1. If we were counting combinations, these would not be distinct. This idea is the difference between permutations and combinations, which is the topic of Lesson 8-6.

Note that the Permutation Theorem is a special case of the formula for $_nP_r$. Let $r = n$ in that theorem and the formula becomes $_nP_n = n(n-1)(n-2) \cdot \ldots \cdot (n-n+1)$. Notice that each factor is one less than the preceding, and the last factor is 1, which indicates that $_nP_n = n!$.

LESSON
7-5

Permutations

Montana St.	Montana St.	Montana St.	Montana St.	Montana St.
Boise State	Boise State	Boise State	Boise State	Boise State
Montana	Montana	Montana	Montana	Montana
Idaho	Idaho	Idaho	Idaho	Idaho
Nevada-Reno	Nevada-Reno	Nevada-Reno	Nevada-Reno	Nevada-Reno
N. Arizona	N. Arizona	Idaho State	Idaho State	Weber State
Idaho State	Weber State	N. Arizona	Weber State	N. Arizona
Weber State	Idaho State	Weber State	N. Arizona	Idaho State

In 1990, the Big Sky Basketball Conference had eight teams: Montana State, Boise State, Montana, Idaho, Nevada-Reno, Northern Arizona, Idaho State, and Weber State. Assuming no ties, how many different possible standings could result?

The answer is given by the Selections without Replacement Theorem from Lesson 7-3; for the eight teams, the number of different orders of finish is
$$8! = 8 \cdot 7 \cdot 6 \cdot 5 \cdot 4 \cdot 3 \cdot 2 \cdot 1 = 40,320.$$

Each different arrangement of a set of objects is called a **permutation**. Six of the 40,320 possible permutations of the teams in the Big Sky Conference are listed above.

In the language of permutations, the Selections without Replacement Theorem can be restated as a Permutation Theorem.

Permutation Theorem

There are $n!$ permutations of n different elements.

Similar reasoning can be used when not all of the n original objects are selected for arrangement.

Example 1

Seven horses are in a race at Churchill Downs. You want to predict which horse will finish first, which second, and which third. How many different predictions are possible?

Solution Unlike the standings situation, only three of the seven horses are selected. Use the Multiplication Counting Principle.

	first	second	third
Any of the 7 horses can be first.	7		
For each of these, 6 horses can be second.	7 ·	6	
For each of these, 5 horses can be third.	7 ·	6 ·	5

So there are $7 \cdot 6 \cdot 5 = 210$ possible predictions. If you picked the horses blindly, the probability of predicting the first, second, and third place finishers correctly is $\frac{1}{210}$.

420

When the order of the objects is taken into account, the number of ways of arranging 3 objects out of a set of 7 objects is referred to as the number of permutations of 7 objects taken 3 at a time. In general, the number of permutations of **n objects taken r at a time,** denoted $_nP_r$, can be calculated by multiplying n by the consecutive integers less than itself, until r factors have been taken. The second factor is $n - 1$, the third factor is $n - 2$, and so on. So the rth factor is $n - (r - 1)$ or $n - r + 1$.

Theorem (Formula for $_nP_r$)

The number of permutations of n objects taken r at a time is
$$_nP_r = n(n - 1)(n - 2) \cdot \ldots \cdot (n - r + 1).$$

Notice that $_7P_3 = 7 \cdot 6 \cdot 5 = \dfrac{7 \cdot 6 \cdot 5 \cdot 4 \cdot 3 \cdot 2 \cdot 1}{4 \cdot 3 \cdot 2 \cdot 1} = \dfrac{7!}{4!}$, and the denominator is determined by n and r: $4 = 7 - 3$. In a similar way, any product of consecutive integers can be written as a quotient of factorials. This provides an alternate way of calculating $_nP_r$.

Theorem (Alternate Formula for $_nP_r$)

$$_nP_r = \frac{n!}{(n - r)!}.$$

Example 2 How many different five-letter words can be formed from the word BIRTHDAY? (The words do not have to make sense in English or any other language).

Solution 1 One such word is BIRTH. Other words are RABID and YADIR. The question asks for the number of permutations of 8 objects taken 5 at a time. Use the first formula for $_nP_r$.

$$_8P_5 = \underbrace{8 \cdot 7 \cdot 6 \cdot 5 \cdot 4}_{5 \text{ factors}} = 6720 \text{ words}$$

Solution 2 Use the Alternate Formula for $_nP_r$.
$$_8P_5 = \frac{8!}{(8 - 5)!} = \frac{8!}{3!} = 6720 \text{ words}$$

When all n objects are selected, the alternate formula gives $_nP_n = \dfrac{n!}{(n - n)!} = \dfrac{n!}{0!}$. Because $_nP_n$ should be $n!$, the quantity $0!$ is defined to equal 1.

Definition

$$0! = 1$$

Technology There are many situations in which the number of permutations is very large and calculators are essential. In many cases some arithmetic rewriting can speed calculation, and in some cases such rewriting is necessary. For example, $_{100}P_3$ cannot be computed on most calculators in the form $\frac{100!}{97!}$, because both 100! and 97! are too large. However, the alternative $_{100}P_3 = 100 \cdot 99 \cdot 98$ is easily computed.

Alternate Approach We define $0! = 1$ because this makes it consistent with the formula for $_nP_r$. This value for $0!$ is also the only value that satisfies the criterion that $n! = n(n - 1)!$ for $n = 1$. Specifically, one would need $1! = 1 \cdot 0!$, from which $1 = 1 \cdot 0!$, from which $0! = 1$. It is also the only value that works in formulas for $_nC_r$.

Some students may enjoy trying to find 5-letter permutations from BIRTHDAY that are words in the English language. Here are some that are not mentioned on page 421: BRAID, THIRD, DIARY, DAIRY, HAIRY, HABIT.

ADDITIONAL EXAMPLES
1. In a 4-person relay race, the fastest runner typically is the last person to run for the relay team. If that place is set, how many different orders are there for the other runners?
3! = 6

2. How many different four-letter words can be formed from the word EQUATIONS? (The words do not have to make sense.)
$_9P_4$ = 3024

3. Solve: $_xP_{10} = 20(_xP_8)$.
$\dfrac{x!}{(x - 10)!} = \dfrac{20x!}{(x - 8)!}$, so
$20(x - 10)! = (x - 8)!$, from which $20 = (x - 8)(x - 9)$. Solving the quadratic yields $x = 4$ or $x = 13$, of which only $x = 13$ is in the domain of the original equation.

NOTES ON QUESTIONS
Question 3a: You might ask how you know when to stop writing factors. (In the writing of $_nP_r$ as a product of factors, there are r factors.)

Question 14: Stress that students should try to answer the question without calculating the four values. It is reasonably obvious that $_{17}P_{15}$ is less than $_{17}P_{16}$ (each begins with the factor 17, the first has 15 factors, the second 16), and that $_{17}P_{16}$ is less than $_{18}P_{16}$ (each has 16 factors but the latter begins with the higher number). Similarly, $_{18}P_{15}$ is between $_{17}P_{15}$ and $_{18}P_{16}$. So the question is, which is larger, $_{18}P_{15}$ or $_{17}P_{16}$? Write out the factors of each and find the common factors to determine this.

Making Connections for Question 15: The skill of rewriting factorials is needed in proofs, particularly those involving combinations in the next chapter.

Example 3 Solve $_nP_5 = 7 \cdot {}_nP_4$.

Solution 1 Use the Formula for $_nP_r$.
$$n(n-1)(n-2)(n-3)(n-4) = 7n(n-1)(n-2)(n-3)$$
Divide each side of the equation by $n(n-1)(n-2)(n-3)$.
So $$n - 4 = 7$$
$$n = 11.$$

Solution 2 Use the Alternate Formula for $_nP_r$.
Rewrite. $$\frac{n!}{(n-5)!} = 7\frac{n!}{(n-4)!}$$
Divide by $n!$ and multiply means by extremes.
$$(n-4)! = 7(n-5)!$$
A key step is to use the fact that $n! = n(n-1)!$ or, more generally,
$(n-a)! = (n-a)(n-a-1)!$
Rewrite. $$(n-4)(n-5)! = 7(n-5)!$$
$$n - 4 = 7$$
$$n = 11$$

Check $_{11}P_5 = \frac{11!}{6!}$ and $7 \cdot {}_{11}P_4 = 7 \cdot \frac{11!}{7!} = \frac{7}{7} \cdot \frac{11!}{6!}$.
So $\quad {}_{11}P_5 = 7 \cdot {}_{11}P_4$.

Questions

Covering the Reading

1. List all the permutations of the letters in USA.
   USA, UAS, SUA, SAU, AUS, ASU

2. Seven band members line up to take a picture. How many permutations are possible? **7! = 5040**

3. Write $_{11}P_3$ as
   a. a product of integers.   **b.** a ratio of two factorials. $\frac{11!}{8!}$
   $11 \cdot 10 \cdot 9$

In 4 and 5, evaluate.

4. $_{12}P_5$ **95,040**   5. $_7P_2$ **42**

6. Refer to the 8 teams in the Big Sky basketball conference. In how many ways can the first 5 positions in the standings be filled? **6720**

7. **a.** How many permutations consisting of two letters each can be formed from the letters of UCSMP? **20**
   b. List them all. **UC, US, UM, UP, CU, CS, CM, CP, SU, SC, SM, SP, MU, MC, MS, MP, PU, PC, PS, PM**

In 8 and 9, **a.** evaluate. **b.** In terms of choosing items from a set, explain your answer.

8. $_nP_1$ **a) n; b) See below.**   9. $_nP_n$ **a) n!; b) See below.**

10. Solve for n: $_nP_3 = 5 \cdot {}_nP_2$. **7**

8b) If n objects are being taken one at a time there are n possibilities, one for each object.
9b) If you take all the objects from a set of n objects, your selection can be any of the n! permutations of the n objects.

422

11. **a.** How many ID numbers are there consisting of a permutation of four of the digits 1, 2, 3, 4, 5 and 6?   **360**
    b. If one of the ID numbers is chosen at random, what is the probability that it is 3416? $\frac{1}{360} \approx$ **0.003**

12. An exhibition hall has eight doors. Through how many pairs of doors can you enter and leave the hall if you must enter and leave in different ways?   **56**

13. Consider the word WOMBAT.
    a. How many permutations of the letters of the word are there?   **720**
    b. How many permutations of the letters in the word end in a vowel?
    c. How many permutations of the letters in the word do not end in a vowel?   **b) 240; c) 480**

The wombat is a burrowing marsupial native to Australia.

14. Place in increasing numerical order.
    $_{17}P_{15},\ _{18}P_{16},\ _{17}P_{16},\ _{18}P_{15}$ $_{17}P_{15},\ _{17}P_{16},\ _{18}P_{15},\ _{18}P_{16}$

15. **a.** Show that $_6P_4 = 6 \cdot _5P_3$. $_6P_4 = \frac{6!}{(6-4)!} = \frac{6!}{2!} = 6 \cdot \frac{5!}{2!} = 6 \cdot \frac{5!}{(5-3)!} = 6 \cdot _5P_3$
    b. Prove that $_nP_r = n \cdot _{n-1}P_{r-1}$ for all integers n and r with $0 \le r \le n$.
    $_nP_r = \frac{n!}{(n-r)!} = n \cdot \frac{(n-1)!}{(n-r)!} = n \cdot \frac{(n-1)!}{(n-1)-(r-1)} = n \cdot _{n-1}P_{r-1}$

16. Each row of an aircraft has three seats on each side of the aisle. In how many different ways can a woman, her husband and four children occupy a row of seats, if the two parents sit next to the aisle?   **48**

17. A bag contains 5 black, 4 white, and 3 green marbles. These marbles are drawn in succession, each marble being replaced before the next one is drawn. What is the probability of drawing a black, then a white, and then a green marble?   *(Lessons 7-1, 7-4)* $\frac{5}{144} \approx$ **0.03**

18. On January 8, 1989, a two-engine plane crashed in England after both engines shut down. Reporters quoted experts as saying that the probability of both engines failing were 10^{-6}. Assuming independence of engine failure,
    a. what is the probability of one engine failing? 10^{-3}
    b. Are the engines on this plane more or less reliable than the ones described in Lesson 7-4?   **less**
    c. Give a reason why the engine failures may *not* have been independent events.   *(Lesson 7-4)*   **sample: A system upon which both engines depend, such as the electrical system, may have failed.**

In 19–21, identify the events as independent, mutually exclusive, complementary, or none of these. (There may be more than one correct answer.) *(Lessons 7-1, 7-2, 7-4)*

19. One student is chosen from a class, then another student is chosen.   **none**

20. Fumbling in the dark, a person blindly tries one key from the key ring, drops the ring, then blindly tries a key again.   **independent**

21. Kevin is on time for class; Kevin is late for class.   **mutually exclusive and complementary**

FOLLOW-UP

MORE PRACTICE
For more questions on SPUR Objectives, use *Lesson Master 7-5*, shown on page 423.

PROJECTS
Projects 2 and 4 are related to the content of this lesson.

ADDITIONAL ANSWERS
23.a.

● February   ○ July

---

22. When one coin is tossed there are two possible outcomes—heads or tails. How many outcomes are possible when the following numbers of coins are tossed? *(Lesson 7-3)*
    a. 2 **4**   b. 5 **32**   c. 8 **256**   d. n 2^n

23. The following data show the average daily household TV use in hours and minutes in the United States, according to the *1989 Information Please Almanac*. *(Lessons 1-2, 2-3)*

Year	February	July
1975-76	6:49	5:33
1979-80	7:22	5:48
1980-81	7:23	6:00
1981-82	7:22	6:09
1982-83	7:33	6:23
1983-84	7:38	6:28
1984-85	7:49	6:36
1985-86	7:48	6:37
1986-87	7:35	6:32

   a. Plot these data on a suitable graph to show the trends for TV viewing time.   **a) See margin. b–c) See below.**
   b. Find equations for the lines of best fit for each month.
   c. Compare the slopes and y-intercepts of the lines of best fit. Interpret the differences.
   d. Use the lines of best fit to estimate the average daily viewing time for 1990-91 for each of February and July.   **about 8:09 for February, about 7:08 for July**

24. *Skill sequence.* Find the perimeter of the polygon.   *(Previous course)*

   a.
   56

   b.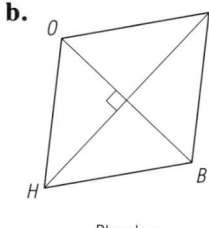
   Rhombus
   $HM = 8$, $OB = 6$
   20

   c.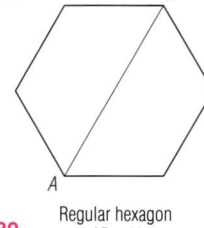
   Regular hexagon
   $AB = 10$
   30

Exploration

25. $7 \cdot 6 \cdot 5 = 15 \cdot 14$. Find two other sets of consecutive integers greater than one whose products are equal.
   sample: $7 \cdot 8 \cdot 9 \cdot 10 = 2 \cdot 3 \cdot 4 \cdot 5 \cdot 6 \cdot 7$ and $8 \cdot 9 \cdot 10 = 2 \cdot 3 \cdot 4 \cdot 5 \cdot 6$

---

23 b) For July: $y = 4.884x + 386.9$, where x = year after 1970 (for example, $x = 6$ for 1975–76) and y = minutes. **For February:** $y = 6.489x + 291.8$, where x = year after 1970 and y = minutes.

c) Sample: The slope for July is higher, but the y-intercept for February is higher. Thus, people used to watch more TV in February, and still do, but the gap is narrowing at a rate of about 1.6 minutes per year.

Computers and Counting

Some calculators and computers have special keys or function statements for calculating the number of permutations of *n* things taken *r* at a time. If your computer does not have such a feature, you can write short programs to calculate factorials or to calculate or list permutations.

The BASIC program below calculates *n*! for any positive integer *n* input by the user (within the bounds of the computer's ability to do arithmetic).

```
10   REM FACTORIAL PROGRAM
20   INPUT "A POSITIVE INTEGER "; N
30   FACT = N
40   FOR I = N – 1 TO 1 STEP -1
50     FACT = I * FACT
60   NEXT I
70   PRINT N; "! = "; FACT
80   END
```

Notice that this is a recursive program. The computer first stores the original value of *n* in a location called FACT. Then each time it executes the loop in lines 40 to 60 it multiplies the current value of FACT by the next smaller integer.

■ ■ ■ ■ ■ ■ ■ ■ ■ ■

Example 1 Modify the program above to calculate the number of permutations of *n* objects taken *r* at a time for integers *n* and *r*, $r \le n$, input by the user.

Solution The $_nP_r$ Theorem indicates that you must tell the computer to multiply $n \cdot (n-1) \ldots (n-r+1)$. So, add a new line 25 to prompt for the input of *r*, edit line 40 to stop when $n - r + 1$ is reached, and edit the print statement in line 70. The new program reads as follows.

```
10   REM PERMUTATIONS PROGRAM
20   INPUT "HOW MANY OBJECTS?"; N
25   INPUT "HOW MANY AT A TIME?"; R
30   FACT = N
40   FOR I = N – 1 TO N – R + 1 STEP -1
50     FACT = I * FACT
60   NEXT I
70   PRINT N; "P"; R; " = " ; FACT
80   END
```

Some problems can be solved fairly quickly by inspecting a complete list of outcomes. For example, in Lesson 7-1 you listed the complete 36-outcome sample space for two dice. For three dice, the sample space would have $6^3 = 216$ outcomes; listing these by hand would be time consuming. A computer program with *nested loops* provides an efficient way of getting a complete list involving two or more variables.

LESSON 7-6 Computers and Counting **425**

LESSON 7-6

RESOURCES
■ Lesson Master 7-6
■ Quiz for Lessons 7-4 through 7-6
■ Teaching Aid 62: Programs for **Questions 3, 4, 7, and 8**

OBJECTIVE

L Use computers for counting.

TEACHING NOTES

Technology We continue our practice of indenting the interior of loops (the lines between FOR . . . and NEXT . . .) when writing computer programs and recommend that your students do so as well. Unfortunately, some versions of BASIC do not display programs exactly as they are entered, so the helpful indentation does not appear on the screen when the program is listed.

When one loop is nested in another, as in the program on page 426, we indent again. Stress that an inner loop must be completely contained within an outer loop in order for it to be nested.

In the nested loops that students encounter, the number of repetitions of a loop is known in advance. Sample sizes are usually fixed in experiment design, and even simulation programs usually generate a given number of events. Thus the use of the computer to count is somewhat artificial. The computer's unique power is in its ability to generate lists and process information as it runs through them, as in the problem of determining those three-digit

Consider the following situation.
When you take a true-false test, you
receive 1 point for every question that
you get right and zero points for each
one you get wrong. The BASIC program at the right generates all possible ways to answer a three-question
test, and calculates the score for each
pattern. Read the program, paying
particular attention to lines 30 to 170.

```
10   REM TEST SCORE PATTERNS
20   PRINT "PATTERN","SCORE"
30   FOR Q1 = 0 TO 1
40    FOR Q2 = 0 TO 1
50     FOR Q3 = 0 TO 1
100      SC = Q1 + Q2 + Q3
110      PRINT Q1;Q2;Q3,SC
150     NEXT Q3
160    NEXT Q2
170  NEXT Q1
200  END
```

There are three loops in the program. The outer loop in lines 30 and 170 generates the score Q1 on the first question; the middle loop in lines 40 and 160
generates the score on the second question; and the inner loop in lines 50 and
150 generates the score Q3 on the third question. The score (SC) is calculated
in line 100. Line 110 prints the response patterns and their corresponding
scores.

When you run the program it produces the output at the right.

By the Multiplication Counting Principle, you could have predicted that
there are $2 \cdot 2 \cdot 2 = 8$ possible response patterns. From the output you
can also tell that there is one way to
score 0, three ways to score 1, three
ways to score 2, and one way to score 3.

PATTERN	SCORE
000	0
001	1
010	1
011	2
100	1
101	2
110	2
111	3

Example 2
a. Rewrite the program to show all possible patterns and scores for a
four question true-false test.
b. Predict the number of different patterns your program for part **a** will
produce.
c. Run the program and give the score distribution.
d. If a person were to answer such a test by blind guessing, find the
probability of getting at least 3 items right.

Solution

a. Insert another loop to generate the possible response on the fourth
question.
```
60  FOR Q4 = 0 TO 1
140  NEXT Q4
```
Change the formula for the score and change the PRINT statement to
take into account the score on the fourth item. That is, edit lines 100
and 110 as follows.
```
100  SC = Q1 + Q2 + Q3 + Q4
110  PRINT Q1;Q2;Q3;Q4,SC
```

b. By the Multiplication Counting Principle, the output will have
$2 \times 2 \times 2 \times 2 = 16$ lines.

426

c. The revised program produces the output at the right. There is one score of 0, four scores of 1, six scores of 2, four scores of 3, and one score of 4.

d. There are 5 ways of scoring at least 3 points. The sample space has 16 equally likely outcomes. So the probability of getting at least 3 items correct by blind guessing is $\frac{5}{16}$ or about 31%.

PATTERN	SCORE
0000	0
0001	1
0010	1
0011	2
0100	1
0101	2
0110	2
0111	3
1000	1
1001	2
1010	2
1011	3
1100	2
1101	3
1110	3
1111	4

3. Write a program to search all two-digit numbers to determine whether any are equal to the sum of the squares of their digits.

```
10  FOR T = 1 TO 9
20    FOR U = 0 TO 9
30      LET NUM = 10*T + U
40      IF NUM = T*T + U*U
          THEN PRINT T;U
50    NEXT U
60  NEXT T
70  END
```

Some algebra problems can also be solved by making a complete list of all possibilities. Here is an example.

Which 3-digit numbers equal the sum of the cubes of their digits?

One such number is 407 because $4^3 + 0^3 + 7^3 = 64 + 0 + 343 = 407$. Are there any other numbers like this?

To solve this problem algebraically, you might start by letting h represent the hundreds digit, t the tens digit, and u the units digit of a 3-digit number. Then the value of the number is $100h + 10t + u$.

The sum of the cubes of the digits is $h^3 + t^3 + u^3$.

So you need to find all positive integers with $100h + 10t + u = h^3 + t^3 + u^3$.

Normally you need at least three equations to solve for three unknowns. Yet there is only one equation here. Although it is possible to solve this problem algebraically, the solution is tedious.

It is much easier to use a computer to check the sum of the cubes for all possible three-digit numbers. The program below searches a complete list of numbers with $1 \le h \le 9$, $0 \le t \le 9$, and $0 \le u \le 9$:

```
10  FOR H = 1 TO 9
20    FOR T = 0 TO 9
30      FOR U = 0 TO 9
40        LET NUM = 100*H + 10*T + U
50        IF NUM = H*H*H + T*T*T + U*U*U THEN PRINT H;T;U
60      NEXT U
70    NEXT T
80  NEXT H
90  END
```

The program has three loops, one each for the hundreds, tens, and units digits. The units loop is nested within the tens loop, which is nested within the hundreds loop. Using the Multiplication Counting Principle, you can see that lines 40 and 50 are executed $9 \cdot 10 \cdot 10 = 900$ times, once for each integer from 100 to 999 inclusive. So the program conducts an exhaustive search of all three-digit numbers.

When the program is RUN, the output is

153
370
371
407

These are the only 3-digit numbers with the property that the number equals the sum of the cubes of its digits.

Computer-generated lists and computer searches generally take much less time than making lists or counting by hand. However, you should consider the time it takes to write the program.

A disadvantage of solving problems with computers is that it is often difficult to check the computer's output, which may be thousands of pages long. However, computer searches are now common in applications such as scheduling routes for business people who travel frequently, or routing telephone calls between exchanges.

Questions

Covering the Reading

1. If a user inputs the number 6 in the FACTORIAL program, what will be printed when the program is run?   **6! = 720**

2. Consider the PERMUTATIONS program of Example 1.
   a. What is the output if the input is $n = 8$ and $r = 3$?   **8P6 = 336**
   b. What is the output if the input is $n = 5$ and $r = 0$?   **5P0 = 5**

In 3 and 4, how many lines of output are generated by the computer program segment?

3.
```
10   FOR X = 1 TO 25
20     FOR Y = 1 TO 30
30       PRINT X,Y,X^2 + Y^2
40     NEXT Y
50   NEXT X    750
```

4.
```
10   FOR R = 1 TO 6
20     FOR S = 1 TO 6
30       FOR T = 1 TO 4
40         PRINT R + S + T
50       NEXT T
60     NEXT S
70   NEXT R    144
```

428

5. Refer to the program for test score patterns in Example 2.
  a. Explain how to modify it to produce all possible outcomes for a true-false test with 5 questions.   **See margin.**
  b. How many lines should it print?   **32 lines of scores**
  c. Run the program, and give the score distribution.   **See margin.**
  d. What is the probability of getting at least 80% of the items correct on such a test by blind guessing?   **6/32 = .1875**

6. Show that 153, 370, 371, and 407 are equal to the sum of the cubes of their digits.   **See margin.**

7. Use the program below, which searches for four-digit numbers equal to the sum of the fourths power of their digits:

```
10   FOR TH = 1 TO 9
20     FOR H = 0 TO 9
30       FOR T = 0 TO 9
40         FOR U = 0 TO 9
50           LET NUM = 1000*TH + 100*H + 10*T + U
60           IF NUM = TH*TH*TH*TH + H*H*H*H +
              T*T*T*T + U*U*U*U THEN PRINT TH;H;T;U
70         NEXT U
80       NEXT T
90     NEXT H
100 NEXT TH
110 END
```

  a. How many loops does the program have?   **4**
  b. How many times is line 50 executed?   **9000**
  c. How many numbers does the program search?   **9000**
  d. Enter the program in a computer and run it. Are there any four-digit numbers equal to the sum of the fourth powers of their digits? If so, what are they?   **Yes; 1634, 8208, 9474**

8. Refer to this program for finding the capacities of cylinders with radii the integers from 1 to 5 inches and heights the integers from 1 to 4 inches.

```
10 FOR H = 1 TO 4                    b)  1   1   3.141593
20   FOR R = 1 TO 5                       1   2   12.56637
30     LET CAP = 3.141593 * R * R * H     1   3   28.27434
40     PRINT H,R,CAP
50   NEXT R
60 NEXT H
70 END
```

  a. How many lines of output will this program have when run?   **20**
  b. Give the first three lines of output for the program.   **See above.**
  c. Give the last line of output.   **4   5   314.1593**

9. a. Write a computer program and use it to find a square and a cube whose sum is 2222. (Use A*A for the square and B*B*B for the cube, rather than A^2 and B^3, respectively.) $5^2 + 13^3 = 2222$
  b. Modify the computer program from part **a** to find a square and a cube whose sum is 5442. $23^2 + 17^3 = 5442$

10. a. Write a computer program to list all the outcomes of tossing 3 dice.
  b. Refine your program to calculate the probability that the sum of the numbers on the dice is equal to or greater than 15.   **a,b) See margin.**

---

6. $1^3 + 5^3 + 3^3 =$
$1 + 125 + 27 = 153$;
$3^3 + 7^3 + 0^3 = 27 + 343 + 0$
$= 370$; $3^3 + 7^3 + 1^3 =$
$27 + 343 + 1 = 371$;
$4^3 + 0^3 + 7^3 = 64 + 0 + 343$
$= 407$

10.a.
```
10   FOR I = 1 to 6
20     FOR J = 1 to 6
30       FOR K = 1 to 6
40         PRINT I; " + ";J;
            " + ";K; " = "; I+J+K
50       NEXT K
60     NEXT J
70   NEXT I
80 END
```

b. Add the following:
```
5    N=0
75   PRINT N/216
```
and change line 40 to :
```
40   IF I+J+K > 14 THEN
       N=N+1
```
The probability is about .09.

429

Review

11. a. In how many ways can a club of 23 people line up for a photo?
   b. In how many ways can four of these people be chosen to be president, vice-president, secretary and treasurer of the club?   *(Lesson 7-5)*
   a) $23! \approx 2.60 \cdot 10^{22}$; b) $\frac{23!}{19!} = 212,5$

12. Suppose you plan to visit the six large islands in the state of Hawaii: Hawaii, Kauai, Lanai, Maui, Molokai, and Oahu. If you are to visit each island once, in how many different orders might you travel?
(Lesson 7-5)   **720**

13. Solve for n: $17! = n \cdot 16!$ *(Lesson 7-5)*   **17**

14. Solve for n and r: ${}_nP_r = 600$ *(Lesson 7-5)* $n = 25, r = 2$ or $n = 600, r = 1$

15. The wheel on the TV program *Wheel of Fortune* is divided equally into 24 compartments with one labeled "Lose a Turn." Find the probability that "Lose a Turn" will be under the pointer on
   a. a single spin $\frac{1}{24} \approx .042$
   b. two successive spins $\frac{1}{24^2} \approx .0017$
   c. three successive spins.   *(Lesson 7-4)* $\frac{1}{24^3} \approx 7.2 \times 10^{-5}$

16. Suppose a basketball player has a free throw percentage of .75. What is the probability that the player will not make the next free throw?   *(Lesson 7-2)*   **0.25**

In 17 and 18, *true* or *false*.   For all real numbers θ,

17. $\sin \theta = \cos \left(\frac{\pi}{2} - \theta \right)$.
   True

18. $\cot \theta = \tan \left(\frac{\pi}{2} - \theta \right)$.   *(Lessons 5-7, 6-7)*
   True

19. The doubling-time of a certain population is 10 years. Find the percent of increase in the population after
   a. 5 years
      about 41%
   b. 24 years.   *(Lesson 4-4)*
      about 428%

Exploration

20. Make up a problem whose answer you do not know that can be solved by making a computer generated list or doing a computer search. Solve your problem.   **Answers will vary.**

21. a. What happens if you input numbers other than positive integers greater than 1 in the FACTORIAL program on page 425?
   b. Modify the program so that it can calculate 0!
   a) If $n \leq 1$, the computer will print the value of $n(n-1)$, giving incorrect values for both 0! and 1!. If $n > 1$ but n is not an integer, the computer will print the value of $n(n-1)(n-2) \dots (n-k)$ where k is the largest positive number such that $n - k > 1$.
   b) Change line 30 to 30 FACT = 1. Change line 40 to 40 for I = 1 TO N.

430

7-7

Probability Distributions

In Lesson 7-1, the 36 possible outcomes for the tossing of two dice are listed. If the dice are fair, then the probabilities of obtaining each possible sum are given in the table below. Check that the sum of the probabilities is 1, so no possible outcome has been excluded.

x (Sum of dice)	2	3	4	5	6	7	8	9	10	11	12
P(x) (Probability of Outcome)	$\frac{1}{36}$	$\frac{2}{36}$	$\frac{3}{36}$	$\frac{4}{36}$	$\frac{5}{36}$	$\frac{6}{36}$	$\frac{5}{36}$	$\frac{4}{36}$	$\frac{3}{36}$	$\frac{2}{36}$	$\frac{1}{36}$

In this situation, if x = the sum of the numbers on the top faces of two dice, then x is a *random variable*, and the pairs of numbers $(x, P(x))$ in the table above represent a *probability distribution*. In general, a **random variable** is a variable whose values are numbers determined by the outcome of an experiment. A **probability distribution** is a function which maps each value of a random variable onto its probability.

The probability distribution P for the sum of two dice is graphed at the right. From both the table and the graph you can see that the domain of P is the set of integers between 2 and 12, inclusive; the range is the set

$$\left\{\frac{1}{36}, \frac{2}{36}, \frac{3}{36}, \frac{4}{36}, \frac{5}{36}, \frac{6}{36}\right\}.$$

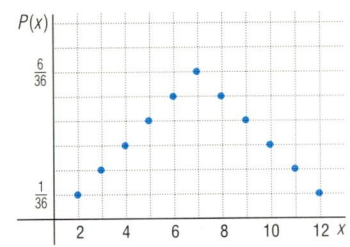

Now compare this probability distribution with the distribution of relative frequencies you might get if you actually tossed two dice. In an experiment, two dice were tossed 78 times, and the sum of the numbers on the top faces was recorded. Below is a table of each sum and its relative frequency.

x (Sum of dice)	2	3	4	5	6	7	8	9	10	11	12
R(x) (Relative Frequency)	$\frac{2}{78}$	$\frac{4}{78}$	$\frac{6}{78}$	$\frac{6}{78}$	$\frac{13}{78}$	$\frac{12}{78}$	$\frac{14}{78}$	$\frac{11}{78}$	$\frac{5}{78}$	$\frac{2}{78}$	$\frac{3}{78}$

At the right is a graph of the relative frequencies. The graph of the relative frequencies resembles the graph of the probability distribution. Each graph shows that for $x = 2$ or 12, both $P(x)$ and $R(x)$ are small; and for $x = 6$, 7, or 8, both $P(x)$ and $R(x)$ are relatively large.

RESOURCES
■ Lesson Master 7-7
■ Teaching Aid 63: Tables and graphs for page 431

OBJECTIVES

E State properties of probabilities.
J Construct, graph, and interpret probability distributions.

TEACHING NOTES

Our use of the term *random variable* to refer to the domain variable of a probability distribution does not represent uniform usage. Often, the probability distribution itself is called a random variable. That is, you may see in other books that a random variable is not a variable, but a function. That unusual use of the word *variable* can cause confusion.

A teaching aid of the distributions on page 431 is provided to help illustrate that a relative frequency distribution will tend not to have the regularity of the corresponding probability distribution. For the dice example, the text explains that $\bar{x} \approx 7.06$ and $\mu = 7$.

Error Analysis In **Example 1b** the percent error between $\bar{x}$ and μ is less than 1%. A quick reading may lead students to believe that the percent error is 86%, not 0.86%.

Emphasize that the domain of a probability distribution function consists of *all* the mutually exclusive outcomes of an experiment.

The distributions can be compared further by calculating their means. To find the mean sum $\bar{x}$ resulting from the experiment, add all 78 outcomes and divide by 78. There were two 2s, four 3s, six 4s, etc., so the total is
$$2 \cdot 2 + 3 \cdot 4 + 4 \cdot 6 + \ldots + 11 \cdot 2 + 12 \cdot 3 = 551.$$
Then the mean sum is
$$\bar{x} = \frac{551}{78} \approx 7.06.$$
Alternatively, you could divide each outcome by 78 before adding. Then the mean sum is
$$\bar{x} = 2 \cdot \frac{2}{78} + 3 \cdot \frac{4}{78} + 4 \cdot \frac{6}{78} + 5 \cdot \frac{6}{78} + 6 \cdot \frac{13}{78} + 7 \cdot \frac{12}{78} + 8 \cdot \frac{14}{78}$$
$$+ 9 \cdot \frac{11}{78} + 10 \cdot \frac{5}{78} + 11 \cdot \frac{2}{78} + 12 \cdot \frac{3}{78}$$
$$= \frac{551}{78} \approx 7.06.$$
Think of the 11 possible values of the random variable x as $x_1, x_2, \ldots, x_{11}$. This way of calculating the mean generalizes to the following formula for computing the mean $\bar{x}$ of n numbers x_i, if the relative frequency $R(x_i)$ of each x_i is known.
$$\bar{x} = x_1 \cdot R(x_1) + x_2 \cdot R(x_2) + \ldots + x_n \cdot R(x_n) = \sum_{i=1}^{n} (x_i \cdot R(x_i))$$
The *mean* or *expected value of a probability distribution* is found by using the probability $P(x_i)$ instead of the relative frequency $R(x_i)$. The expected value of the probability distribution is usually denoted by μ, the lower case Greek letter mu (pronounced "mew").

Definition

Let $\{(x_1, P(x_1)), (x_2, P(x_2)), \ldots, (x_n, P(x_n))\}$ be a probability distribution. The **mean** or **expected value** μ of the distribution is
$$\mu = \sum_{i=1}^{n} (x_i \cdot P(x_i)).$$

The mean of a probability distribution is also called the mean or expected value of the random variable. It indicates the average value you may expect for the random variable as an experiment is repeated. As with a probability for an event, an expected value for a variable is a fixed theoretical value which is approximated but not necessarily obtained every time an experiment is repeated.

Example 1 Consider the experiment of rolling two dice and adding the numbers on the top faces.

a. Calculate the expected value of the probability distribution.
b. What is the percent error between the mean sum $\bar{x}$ of the relative frequencies, calculated above, and the expected value?

Solution

a. Refer to the table of values of x_i and $P(x_i)$ at the beginning of the lesson.

432

The expected value $\sum\limits_{i=1}^{11} (x_i \cdot P(x_i))$

$= 2 \cdot \frac{1}{36} + 3 \cdot \frac{2}{36} + 4 \cdot \frac{3}{36} + 5 \cdot \frac{4}{36} + 6 \cdot \frac{5}{36} + 7 \cdot \frac{6}{36} + 8 \cdot \frac{5}{36}$

$\quad + 9 \cdot \frac{4}{36} + 10 \cdot \frac{3}{36} + 11 \cdot \frac{2}{36} + 12 \cdot \frac{1}{36}$

$= \frac{2}{36} + \frac{6}{36} + \frac{12}{36} + \frac{20}{36} + \frac{30}{36} + \frac{42}{36} + \frac{40}{36} + \frac{36}{36} + \frac{30}{36} + \frac{22}{36} + \frac{12}{36}$

$= 7.$

b. The mean sum for the experiment at the beginning of the lesson is $\frac{551}{78} \approx 7.06$. This differs by only 0.06 from the expected value, so the percent error from the expected value is $\frac{0.06}{7} \approx 0.0086 = 0.86\%$, less than 1%.

When asked to give a probability distribution, you are usually expected to define a random variable, give a table or rule for the probability function, and show a graph. Frequently, a probability distribution is shown as a histogram. It is common in such cases to center each bar over the individual value of the random variable. The height of each bar is the corresponding probability.

Example 2 Consider a family with 3 children. Assume that births of boys and girls are equally likely. Define the random variable of the experiment to be the number of girls.

a. Find the probability for each value of the random variable.
b. Construct a histogram of the probability distribution.
c. Find the expected value of the distribution.

Solution The sample space consists of 8 outcomes, the ordered triples BBB, BBG, BGB, BGG, GBB, GBG, GGB, and GGG, so the random variable takes on the values $x = 0, 1, 2, 3$.

a. Use the definition to calculate the probability of each event.

x_i = number of girls	0	1	2	3
$P(x_i)$	$\frac{1}{8}$	$\frac{3}{8}$	$\frac{3}{8}$	$\frac{1}{8}$

b.
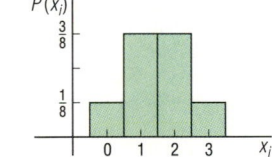

c. Use the definition of expected value.

$$\mu = \sum_{i=1}^{3} (x_i \cdot P(x_i)) = 0 \cdot \frac{1}{8} + 1 \cdot \frac{3}{8} + 2 \cdot \frac{3}{8} + 3 \cdot \frac{1}{8}$$

$$= \frac{12}{8}$$

$$= 1.5 \text{ girls}$$

Check For part **c**: if the births of boys and girls are equally likely, they each happen half of the time. In a family of three children, $\frac{1}{2}$ of 3 *is* 1.5, the expected value calculated in part **c**.

b.

c. 2.5

NOTES ON QUESTIONS
Making Connections for Question 7: This question extends **Example 2** to 4 children; Additional Example 2 deals with 5 children. All of these are useful preparation for the binomial distributions to be discussed in Chapter 10.

Question 10c: The Teaching Notes for this lesson include a derivation of an absolute value function that contains all the points.

Making Connections for Question 12: The probability distribution can be pictured by referring to the array of 36 outcomes shown in Lesson 7-1 and looking at the diagonals with slope 1. All outcomes on the same diagonal have the same difference.

Question 13: For many people, the loss of $.20 per ticket on the average (calculated from the expected value) is not enough to deter them from the slim chance that they could win. Yet the more tickets one buys, the more one is expected to lose.

Question 15: This absolute value model generalizes to other binomial distributions only in the sense that a distribution with $2n$ points can be modeled by the absolute value of a polynomial function of degree $n - 1$. Here $n = 2$.

In the preceding histogram, each bar has an area of $1 \cdot P(x_i) = P(x_i)$. So the area of each bar is the probability of the event to which it corresponds; and thus the sum of the areas of the bars is 1. This property is true for *all* probability distributions, and will be applied in Chapter 10.

Questions

Covering the Reading

1. What is a random variable?   **a variable whose values are numbers determined by the outcome of an experiment**
2. What is a probability distribution?
   a function which maps each value of a random variable onto its probability
3. *True* or *false*.   In a probability distribution, each element in the range must be a number between 0 and 1, inclusive.   **True**

In 4 and 5, explain why the given table does not define a probability distribution.

4.
x	3	4	5	6	7
$f(x)$	$\frac{1}{16}$	$\frac{11}{16}$	$-\frac{1}{16}$	$\frac{2}{16}$	$\frac{3}{16}$

$-\frac{1}{16}$ **is not a probability; it is not between 0 and 1.**

5.
x	0	1	2	3
$f(x)$	$\frac{1}{4}$	$\frac{2}{4}$	$\frac{3}{4}$	$\frac{1}{4}$

The sum of the probabilities is not 1.

6. Find the expected value of the following probability distribution.

x	5	6	7	8	9	10
$P(x)$	$\frac{1}{15}$	$\frac{2}{15}$	$\frac{3}{15}$	$\frac{4}{15}$	$\frac{4}{15}$	$\frac{1}{15}$

$\frac{116}{15} = 7.7\overline{3}$

7. Consider a family with four children.   **See margin.**
   a. List all possible outcomes in the sample space, including birth orders.
   b. Let x = the number of girls in the family. Assuming boys and girls are equally likely, list the pairs in the probability distribution for this experiment.
   c. Make a histogram of the probability distribution.
   d. Find the expected value of x.

8. What is the total area of the bars in a histogram for a probability distribution?   **1 square unit**

9. *True* or *false*.   The mean of a probability distribution is always a value of the random variable of the experiment.   **False**

Applying the Mathematics

10. Consider the graph of the probability distribution associated with rolling two dice and recording the sum.
    a. Superimpose two lines which contain the points of the graph.
    b. Show that the area enclosed by the two lines in part **a** and the x-axis is 1.
    c. Determine the equations for the two lines in part **a**.

    a) See margin. b) It is a triangle with base 12 units and height $\frac{6}{36}$ units. The area is $\frac{1}{2}bh = \frac{1}{2} \cdot 12 \cdot \frac{6}{36} = 1$. c) $y = \frac{1}{36}x - \frac{1}{36}$ and $y = -\frac{1}{36}x + \frac{13}{36}$

434

In 11 and 12, an experiment involves rolling two dice and recording the absolute value of the difference between the numbers showing on the two dice. For example, if you roll a 4 and a 6, the outcome is 2.

11. The following table of relative frequencies was formed after 360 trials.

Difference	0	1	2	3	4	5
Relative frequency	$\frac{62}{360}$	$\frac{98}{360}$	$\frac{77}{360}$	$\frac{60}{360}$	$\frac{38}{360}$	$\frac{25}{360}$

    a. Make a scatterplot of these data. **See margin.**
    b. Find the mean difference. ≈ 1.97

12. Let X = the absolute value of the difference between the numbers showing on two dice.
    a. Make a table of values of the probability distribution for X.
    b. Make a scatter plot of the distribution.
    c. Find the expected value of X. **a,b) See margin. c)** ≈ 1.94

13. In a lottery, the *value* of a ticket is a random variable, defined to be the amount of money you win less the cost of playing. Suppose that in a lottery for charity with 225 tickets, each ticket costs $1. First prize is $50, second prize is $30, and third prize is $20. Then the possible values of the random variable are $49, $29, $19, and $-1.
    a. Why is one of the values negative?
    b. The probability of winning first prize is $\frac{1}{225}$. The same probability holds for second and third prizes. Find the probability of winning nothing.
    c. Find the expected value of a ticket. **a) It represents a loss; it is negative because $1 must be paid for the ticket. b) 222/225; c)** $\approx -56¢$

14. An ecologist collected the data shown in the table below on the life span of a species of deer. Based on this sample, what is the expected lifespan of this species? **about 4.6 years**

Age at death (years)	1	2	3	4	5	6	7	8
Number	2	30	86	132	173	77	40	10

15. The dot frequency graph below is of the probability function in Example 2. The graph of $f(x) = -\frac{1}{4}\left|x - \frac{3}{2}\right| + \frac{1}{2}$ is superimposed.
    a. Verify that each of the four points $(x_i, P(x_i))$ is on the graph of f.
        See margin.

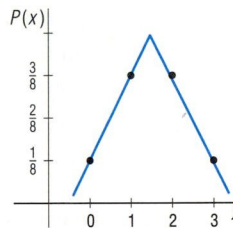

    b. Show that the equation for f is the image of $y = |x|$ under the linear change $L(x, y) = \left(x + \frac{3}{2}, -\frac{1}{4}y + \frac{1}{2}\right)$. **See margin.**

15.a.

$f(0) = \frac{-1}{4}\left|0 - \frac{3}{2}\right| + \frac{1}{2} = \frac{1}{8}$;

$f(1) = \frac{-1}{4}\left|1 - \frac{3}{2}\right| + \frac{1}{2} = \frac{3}{8}$;

$f(2) = \frac{-1}{4}\left|2 - \frac{3}{2}\right| + \frac{1}{2} = \frac{3}{8}$;

$f(3) = \frac{-1}{4}\left|3 - \frac{3}{2}\right| + \frac{1}{2} = \frac{1}{8}$;

b. Substituting $x - \frac{3}{2}$ **for** x

and $\dfrac{y - \frac{1}{2}}{\frac{-1}{4}}$ **for** y **into** $y = |x|$

results in $\dfrac{y - \frac{1}{2}}{\frac{-1}{4}} = \left|x - \frac{3}{2}\right|$ **or**

$y = \frac{-1}{4}\left|x - \frac{3}{2}\right| + \frac{1}{2}$

MORE PRACTICE
For more questions on SPUR
Objectives, use *Lesson Master 7-7*, shown on page 435.

EXTENSION
Small Group Work You may wish to divide the class into groups and give each group some dice. Have them toss the dice 36 times and find the mean and percent error from the expected value if the dice were fair. The question of when the difference between the observed mean and expected value is large enough is basic to statistics, and you might discuss what difference seems sufficient to the students to cause them to believe the dice are not fair.

ADDITIONAL ANSWERS
22.c.

Review

16. Describe the output for the following program. *(Lesson 7-6)*
```
10   FOR I = 1 TO 9
20     FOR J = 0 TO 9
30       PRINT I, J
40     NEXT J
50   NEXT I
60   END
```
It prints the integers from 10 to 99.

17. Consider eight athletes getting ready to run a 100-meter-dash heat. *(Lessons 7-3, 7-5)*
    a. In how many ways can the runners finish the race? **$8! = 40,320$**
    b. If three of the athletes are clearly superior to the rest and are sure to be the top three, in how many ways can the race end? **$3!5! = 720$**

18. A kindergarten teacher has 20 felt-tip pens, which are evenly divided among the colors yellow, green, red, blue, and black. Find the probability that the first six pens chosen by the children have the following characteristics. *(Lessons 7-1, 7-3)*
    a. All pens are of different colors. **0**
    b. Three red pens are chosen, then three blue. **$1/48450 \approx 2.1 \cdot 10^{-5}$**

19. The 1990 *Statistical Abstract of the United States* states that in 1988 82,231,200 households owned a camera and 70,867,368 households owned or rented video equipment. The total number of households that owned a camera or owned or rented video equipment was 89,479,000.
    a. How many households in 1988 both owned a camera and owned or rented video equipment? **63,619,568**
    b. If there were 91,066,000 households in the U.S. in 1988, determine the probability that a randomly picked household owned a camera but did not own or rent a video equipment? **$\approx 20\%$**

20. In a study of the change of a bacteria population, there were about 130,000 bacteria three hours after the beginning of the experiment and about 290,000 two hours after that. Assuming an exponential model of growth, find
    a. the initial number of bacteria, **about 39,000**
    b. the number of bacteria six hours after the experiment began.
    (Lesson 4-4) **about 430,000**

21. A class of 27 students took a test, with mean score 57.6 and standard deviation 11.3. Suppose everyone's score is multiplied by 2. For the image scores find
    a. the mean; **115.2**   b. the variance. *(Lesson 3-6)* **510.76**

22. Consider the graph of the equation $y = 3x^2 + 6x + 1$. *(Lesson 2-1)*
    a. Give the x-intercept(s).
    b. Give the y-intercept. **1**
    c. Sketch the graph. **See margin.**
    a) $\dfrac{-3 + \sqrt{6}}{3} \approx -0.18$ and $\dfrac{-3 - \sqrt{6}}{3} \approx -1.8$

Exploration

23. a. Find the life expectancy for a person of your age and gender.
    b. Find out how life expectancies are determined.
    a,b) Answers will vary.

Random Numbers

The probability distribution for rolling one die is easy to draw and to remember. It is shown below both as a scatter plot and a histogram.

Probability Distribution for a Fair Die

This distribution can be described by the rule $P(n) = \frac{1}{6}$ for each n in the domain $\{1, 2, 3, 4, 5, 6\}$. The probability that each face will land up is $\frac{1}{6}$. Any function with an equation $f(x) = k$, where k is a fixed value, is a **constant function**. So the function with equation $P(n) = \frac{1}{6}$ is a constant function. A distribution that may be represented by a constant function is called a **uniform distribution**. "Uniform" should suggest to you the same idea as "constant," the same everywhere, unchanging.

Many natural phenomena have approximately uniform distributions. For example, in the first few minutes of a rainfall, if you look at several equal sized slabs of cement, the distribution, or number of drops of rain on each slab, is nearly uniform. In practice, randomness can only be imagined or assumed, and is only approximated by real phenomena.

If a list contains a uniformly distributed set of numbers, and the position of each number is independent of the others, the list is called a **set of random numbers**. In a set of **random digits**, each digit has the same probability of occurring, each pair of consecutive digits has the same probability of occurring, each trio of consecutive numbers has the same probability of occurring, and so on. A fair die is a device for generating random numbers from $\{1, 2, 3, 4, 5, 6\}$, and for this reason dice are used for playing games in which you want everyone to have the same chance.

However, there is no known way to generate a set of truly random numbers. All that computers or calculators can do is generate numbers that are close to random. For this reason, the numbers generated are called *pseudo-random*. ("Pseudo" means false or pretended.)

In BASIC, a uniform distribution of pseudo-random numbers with decimal values between 0 and 1 is obtained using the RND function. In some versions of BASIC, this function always has an argument, usually 1. In other versions the argument may be optional. The computer program at the top of page 438 will print ten random numbers, with a different set printed each time it is run.

RESOURCES
■ Lesson Master 7-8
▯ Computer Master 19

OBJECTIVE

K Use computers to generate pseudo-random numbers.

TEACHING NOTES

In order to simulate an experiment in which there is an underlying (assumed) probability distribution, since the domain variable in that distribution is a random variable, the domain values must be found by some random process. A spinner, die, or coin could be used to generate the domain values, but here we focus on the use of random numbers. The reason is that such numbers can simulate whatever probabilities we need. For instance, if we want to simulate the days of the week occurring randomly, we would need 7 outcomes that occur with equal frequency. This is difficult with a coin or die and few spinners have 7 sectors, but a BASIC function could be defined to do it. (See Additional Example 3.)

In the programs for **Examples 1 and 2,** the < sign is used in line 30 instead of the ≤ sign. Point out that the way the RND(1) function works, numbers can be divided into 100 groups by the first two digits, .00 to .99. So in **Example 1,** 85% or 85 of these 100 groups are in the closed interval that starts with .00 and ends with .84. The remaining 15% are in the closed interval from .85 through .99. So < is the proper symbol.

```
10  FOR I = 1 TO 10
20    PRINT RND(1)
30  NEXT I
40  END
```

In some versions of BASIC it is necessary to *seed* the random number generator first, in order to avoid getting the same numbers every time RND is used. In this case, add the line

```
5  RANDOMIZE
```

to the program above, and provide a proper seed value when the computer asks

```
RANDOM NUMBER SEED (-32768 to 32767)?
```

The programs in the rest of this lesson assume that a seed is either not needed or has already been given.

Random or pseudo-random numbers can be used to represent outcomes in experiments which depend on chance. Data for a real event that have been obtained without actual observation of the event are called **simulated** data. Simulation is a powerful tool for studying random phenomena.

Example 1 A new operation to restore eyesight is thought to be successful 85% of the time, when performed well on appropriate patients. Give a BASIC program that will generate data to simulate the results of 60 operations.

Solution You want a program to indicate a successful operation 85% of the time. The value of RND(1) will be less than .85 for 85% of the time, on average, since RND has a uniform distribution. So generate a random number x between 0 and 1. If $x < .85$, print the word SUCCESS, otherwise print the word FAILURE. The IF...THEN...ELSE statement is useful here. Repeat this process a total of 60 times, once for each operation.

An appropriate program is

```
10  FOR I = 1 TO 60
20    LET X = RND(1)
30    IF X < .85 THEN PRINT "SUCCESS " ELSE PRINT "FAILURE"
40  NEXT I
50  END
```

Each eye operation in Example 1 can be considered an experiment with two outcomes: success and failure. The probability distribution is given by $P(\text{success}) = .85$ and $P(\text{failure}) = .15$. If you assign the value 1 to success and 0 to failure, you can calculate the expected value, or the mean of the distribution.

$$\mu = 1 \cdot .85 + 0 \cdot .15 = .85$$

Not surprisingly, $\mu = .85$ is consistent with the fact that over many repetitions of the experiment (that is, the operation), 85% of them can be expected to result in success.

438

Example 2 Write a BASIC program to simulate selecting five samples of 16 corn stalks each from a field which is believed to be 25% damaged because of a drought.

Solution We call the outcomes Healthy and Damaged, with $P(H) = .75$ and $P(D) = .25$. Let a generated random number less than .75 indicate the selection of a healthy stalk. The program is similar to that in Example 1, except that it repeats the sampling loop 5 times and uses a probability of .75 in line 30.

```
5  FOR SAMPLE = 1 TO 5
10   FOR I = 1 TO 16
20     LET S = RND(1)
30     IF S < .75 THEN PRINT "H"; ELSE PRINT "D";
40   NEXT I
50     PRINT
60   NEXT SAMPLE
70  END
```

Check An actual run of the program produced the output at the left: The first sample has two damaged stalks, indicating $\frac{2}{16}$, or 12.5% damage. The second and third samples indicate 31.25% damage, and the last two samples show 25% damage. The program seems to be generating random samples that simulate the drought damage.

```
HDHHHHHHHHHHHHHDH
HHHHHDDHDHHDDHHH
HDHHHDHHDHHHDDHH
HHHDDHHHDHHHDHHH
HHHHHHHHDDDHHHHD
```

The range of the RND function in BASIC is $0 \le x < 1$. This is a convenient range if the function is used to simulate an event with a certain probability, as in Examples 1 and 2. If a different range is needed, you can use the greatest integer function to generate numbers within an appropriate range.

Suppose for instance, you want to select numbers randomly from the set $\{1, 2, 3\}$. Because the RND function has a uniform distribution, the value of RND(1) will lie in each of the equal-sized intervals $0 \le x < \frac{1}{3}$, $\frac{1}{3} \le x < \frac{2}{3}$, and $\frac{2}{3} \le x < 1$ one third of the time. So 3 * RND(1) will lie in each of the intervals $0 \le x < 1$, $1 \le x < 2$, and $2 \le x < 3$ one-third of the time. Now apply the greatest integer function. INT(3 * RND(1)) will generate each of the integers 0, 1, and 2 one-third of the time. Finally, INT(3 * RND(1)) + 1 will generate the desired integers 1, 2, and 3 with the appropriate frequencies. The following table summarizes this reasoning.

Function	Range	Graph
RND(1)	$\{x: 0 \le x < 1\}$	
3*RND(1)	$\{x: 0 \le x < 3\}$	
INT(3*RND(1))	$\{0, 1, 2\}$	
INT(3*RND(1))+1	$\{1, 2, 3\}$	

The procedure is easy to generalize to any set of the form $\{1, 2, \ldots, n\}$.

NOTES ON QUESTIONS
Error Analysis for Question 4: Ask for a variety of answers. Check that students do not associate random with "haphazard." A random phenomenon must have some aspect of equal probabilities within it.

Example 3 Define a BASIC function appropriate for simulating the throw of a fair die.

Solution The range of the function must be the set of equally likely outcomes {1, 2, 3, 4, 5, 6}.

Start with RND(1).

Multiply by 6 to get 6 * RND(1).

Take the greatest integer: INT(6*RND(1)).

Add 1: INT(6*RND(1)) + 1.

An appropriate BASIC function is
DEF FN RN(X) = INT(6*RND(1)) + 1

There are many other ways of generating pseudo-random numbers. For example, some computers have a continuously-running digital clock. Some digits from the clock, such as the thousandths place of the seconds, can be used to generate psuedo-random numbers based on the specific instant the number is requested.

Some calculators have pseudo-random number generators built into them. A key labelled RAN#, or its equivalent, gives a different pseudo-random number in the interval $0 \le x < 1$ each time it is pressed. Check your calculator to see if it has such a facility and, if it does, find out how to use it.

Questions

Covering the Reading

In 1 and 2, *multiple choice*.

1. Which of the following is a constant function?
   (a) $f(x) = x$ (b) $x = 0.5$
   (c) $f(x) = 0.5$ (d) $f(x) = 0.5x$ **(c)**

2. Which is a graph of a constant function?

   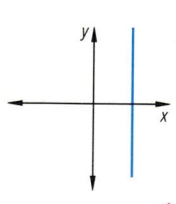

   (a)   (b)   (c)   (d)

   (c)

3. What characterizes a set of random numbers?
   Each number in the set has an equal probability of occurring.

4. Name one natural phenomenon that is random, or nearly so.
   sample: the day of the week when the most rain falls

440

5. a. How many pseudo-random numbers will the following program print?   **41**

```
10  FOR K = 10 TO 50
20    PRINT RND(1)
30  NEXT K
40  END
```

b. Give the interval from which the numbers are chosen. $0 \le x < 1$

c. Run the program twice and describe the output.
Each output consists of 41 pseudo-random numbers between 0 and 1.

6. Modify the program given in the solution to Example 1 to simulate 200 operations.   **Change line 10 to: 10 FOR I = 1 TO 200**

7. Given the BASIC program in the solution of Example 2.
   a. Run the program and determine the percentage of damage after 5 samples.   **The answer should be about 25%.**
   b. Adjust line 5 so that the program simulates selecting 20 samples.
   c. Run the adjusted program and calculate the percentage of damage. Is it closer to 25% than in part **a**?
   b) Change line 5 to: 5 FOR SAMPLE = 1 TO 20; c) It usually will be.

8. a. What is the probability that a number less than 0.4 is generated by RND(1)?   **0.4**
   b. What is the probability that a number between 0.4 and 0.7 is generated by RND(1)?   **0.3**

In 9–12, give the set of possible values of the BASIC expression.

9. 12*RND(1)   **{0 ≤ x < 12}**

10. INT(2*RND(1))   **{0, 1}**

11. INT(RND(1)) + 1   **{1}**

12. INT(2*RND(1)) + 1   **{1, 2}**

13. Define a BASIC function that will give a uniform distribution of the integers 5, 6, 7, 8, 9, 10.   **DEF FN RN(X) = INT(6* RND(1)) + 5**

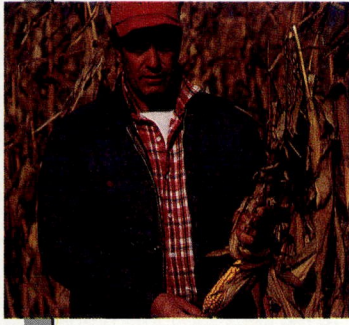

14. Refer to the situation in Example 2. Define randomly picking a corn stalk from the field as an experiment.
    a. What are the possible outcomes?   **Healthy, Damaged**
    b. Make a table of the probability distribution.   **See below.**
    c. Assigning 1 to a healthy stalk and 0 to a damaged one, calculate the expected value of the outcome.   **.75**
    d. Interpret the result in part **c** in terms of the samples of 16 stalks collected in the example.   **sample: One would expect, on average, 12 of 16 stalks to be healthy.**

    b)

H	D
$\frac{3}{4}$	$\frac{1}{4}$

15. Consider the following program.

```
10  FOR I = 1 TO 10
20    FOR J = 1 TO 4
30      PRINT INT(6*RND(1)) + 1;
40    NEXT J
50    PRINT
60  NEXT I
```

a. How many random numbers will be printed by the program?   **40**
b. For which of the following simulations would the program be useful?
   (a) 4 sets of 10 dice throws   (b) 10 sets of 4 dice throws
   (c) 40 sets of 6 dice throws   (d) none of (a)–(c)   **(a) or (b)**

MORE PRACTICE
For more questions on SPUR
Objectives, use *Lesson Master 7-8,* shown on page 441.

EXTENSION
Small Group Work Students can work in groups and invent a simulation like that in **Example 1.** Then they should carry out their simulation on a computer and write up the results they get.

PROJECTS
Project 3 is related to the content of this lesson.

ADDITIONAL ANSWERS
17.b.

16. Given the BASIC program:

```
10  FOR I = 1 TO 10
20    LET W = RND(1)
30    LET X = 5*W
40    LET Y = INT(X)
50    LET Z = Y + 2
60    PRINT X, Y, Z, W
70  NEXT I
```

a) **X is a random number in the interval $0 \le X < 5$. Y is a random integer between 0 and 4, inclusive. Z is a random integer between 2 and 6, inclusive.**

a. Run the program. Describe the possible values of X, Y, and Z.
b. Express Z as a function of W.
   b) **$Z = INT(5* W) + 2$**

Review

17. The distribution of family sizes in a small village is given in the table below.

S = family size	2	3	4	5	6	7	8
$P(S)$	$\frac{2}{87}$	$\frac{9}{87}$	$\frac{19}{87}$	$\frac{31}{87}$	$\frac{15}{87}$	$\frac{8}{87}$	$\frac{3}{87}$

a. Why can this be considered as a probability distribution?
b. Graph the distribution. **See margin.**
c. What is the average size of a family in this village? *(Lesson 7-7)*
   a) **All values for $P(S)$ are between 0 and 1; their sum is 1. c) $432/87 \approx 5$**

18. a. Using a fair coin, what is the expected number of heads in five tosses? **2.5**
b. When five fair coins are tossed simultaneously 150 times, what is the total expected number of heads? *(Lesson 7-7)* **375**

19. a. How many four-letter permutations can be made using the letters in the word CHAPTER? **840**
b. List five of these permutations which have a meaning in English.
   (Lesson 7-5) **sample: PART, CHAP, TRAP, RATE, TEAR**

20. a. Show that ${}_8P_5 = 56 \cdot {}_6P_3$. ${}_8P_5 = \frac{8!}{(8-5)!} = 8 \cdot 7 \cdot \frac{6!}{3!} = 56 \cdot {}_6P_3$
b. Prove that ${}_nP_r = (n^2 - n) \cdot {}_{n-2}P_{r-2}$. *(Lesson 7-5)* **See below.**

21. A person recorded the following scores on an electronic game: 810, 670, 630, 820, 710, 590, 6450, 920, 610, 770.
a. Find the mean of these scores. **1298**
b. Find the median score. **740**
c. Which is the better descriptive measure here? **median**
d. Find the standard deviation of the scores. **about 1813**
e. How does the score of 6450 affect the size of the standard deviation of these scores? *(Lessons 1-4, 1-8)* **very much; if that score is removed, the standard deviation is about 112.**

Exploration

22. Simulate 500 tosses of two dice, recording the sum. **Answers will vary.**
a. How close are your relative frequencies to the corresponding probabilities if the dice are fair?
b. Do you think your simulation is a good one?

20b) ${}_nP_r = \frac{n!}{(n-r)!} = \frac{n \cdot (n-1) \cdot (n-2)!}{(n-r)!} = n(n-1) \frac{(n-2)!}{[(n-2)-(r-2)]!} = (n^2 - n) {}_{n-2}P_{r-2}$

442

Monte Carlo Methods

The first means of generating random numbers used equipment similar to that in gambling casinos. As a result, the methods of using random numbers to simulate events are called **Monte Carlo methods**, named after the well-known Monte Carlo casino in the principality of Monaco.

Consider a particular case to illustrate a Monte Carlo method, the situation that occurs with airplane reservations. It is a common practice for airline companies to accept more reservations than there are available seats, to compensate for the fact that many people do not show up for their scheduled flights. If the airline accepts too many reservations, it is likely that some passengers will not be able to travel, despite having reserved seats. If they accept too few reservations, many flights will depart with empty seats. Neither of these is desirable from the company's viewpoint.

Suppose that an airline is initiating a new service using a small commuter aircraft with a 24-seat capacity. How many reservations should it accept for each flight?

The company could solve this problem by experimenting with accepting different numbers of reservations. However, at the beginning they may find this too expensive if too many planes depart with empty seats. If too many planes are overbooked, the company may lose many future customers. So, a Monte Carlo method of studying the problem is a good idea here.

Suppose, based on past records, that the probability that a person with a reservation shows up for the flight is 0.9. Further, suppose each passenger acts independently, that is, without regard to what other passengers do.

A means of simulating events with probability 0.9 is needed. This can be done with a computer program, like that discussed in Lesson 7-8, or a table of random (actually pseudo-random) numbers. A table of random numbers is a big list of the digits 0 through 9, each of which is about as likely to occur in any location as any other digit. In other words, the distribution of these digits is approximately uniform with

$$P(0) \approx P(1) \approx P(2) \approx \ldots \approx P(9) \approx .1.$$

Here is a portion of a table of random numbers.

24130	48360	22527	97265	76393	64216	79309	30624	36168	03785
61637	57039	97581	83716	65606	12197	79210	69071	10084	27512
77565	34094	29939	69526	36927	37889	74103	65611	29875	29856
89482	37071	19973	36710	48081	78772	33135	10851	27655	18215
12597	74528	23682	13825	24746	02688	93681	01001	54092	33362
94904	31273	04146	56831	31008	16510	64722	20183	10263	27593

Once you start using a random number table by randomly choosing an entry in the table, you can choose a second random digit by selecting the digit next to it—left, right, up, down, or diagonally.

Strings of digits selected from a random number table can be used to simulate an experiment by coding the events (the probabilities of which are known)

RESOURCES
- Teaching Aid 64: Table of random numbers
- Teaching Aid 65: Table of simulated reservations
- Teaching Aid 66: Simulation of area under a parabola

TEACHING NOTES

If an airline were to accept only 200 reservations on a jet seating 200 passengers, and 95% of the people showed up, there would be 10 empty seats. It is worth the risk of not satisfying a passenger with a reservation to have the opportunity to satisfy 10 more customers.

Hotels use the same idea in booking more rooms than they have. This seems to cause problems at conventions of mathematics teachers; mathematics teachers seem to carry through on their reservations at a much higher rate than the general public!

The use of Monte Carlo methods to estimate area is perhaps conceptually easier to understand than its use to simulate airplane reservations, even though in this situation, probability is calculated from a sample space with an infinite number of points. If calculus were not available, this might be the most common way to calculate area.

with the occurrence of selected digits (also a known probability). Each of the following six trials was generated by selecting 24 numbers, the appearance of a '0' being coded as a no-show. (A no-show could be coded with any other digit because each one has a 0.1 probability of appearance, the same as that of a passenger with a reservation failing to show up for a flight.)

Trial 1:	71194	18738	44013	48840	6321
Trial 2:	94595	56869	69014	60045	1842
Trial 3:	57740	84378	25331	12566	5867
Trial 4:	38867	62300	08158	17983	1643
Trial 5:	56865	05859	90106	31595	7154
Trial 6:	18663	72695	52180	20847	1223

The results, if only 24 reservations are accepted, are shown at the right.

Trial	Arrived	Not Arrived (no-shows)
1	22	2
2	21	3
3	23	1
4	21	3
5	21	3
6	22	2

This suggests that the airline might reasonably accept 25 reservations and seldom be overbooked. But how often is seldom?

The table of random numbers on page 443 can be used to determine the number of reservations needed to fill the plane. Each trial ends when there are 24 shows.

Trial	Simulated Reservations (0 means no-show)						Reservations Needed
Trial 1	24130	48360	22527	97265	76393	6	26
Trial 2	42167	93093	06243	61680	37856	16	27
Trial 3	37570	39975	81837	16656	06121	9	26
Trial 4	77921	06907	11008	42751	27756	534	28
Trial 5	09429	93969	52636	92737	88974		25
Trial 6	10365	61129	87529	85689	48237		25
Trial 7	07119	97336	71048	08178	77233	13	27
Trial 8	51085	12765	51821	51259	77452		25
Trial 9	82368	21382	52474	60268	89368		25
Trial 10	10100	15409	23336	25490	43127	30414	30

Accepting 25 reservations never gets the airline into trouble; in fact, at least 25 are always needed to fill the plane. But accepting 26 reservations in this simulation would result in people with reservations yet no seat about 40% of the time.

By using a computer, thousands of trials can be run and the probability of people with no seat estimated rather closely. The company then has the information it needs to decide how many reservations to accept. This same idea is used by hotels in deciding how many reservations they can accept.

Monte Carlo methods were pioneered by John von Neumann [1903–1957], one of the inventors of modern day computer programming. These methods can be utilized in a wide variety of problems, including some purely mathematical ones.

444

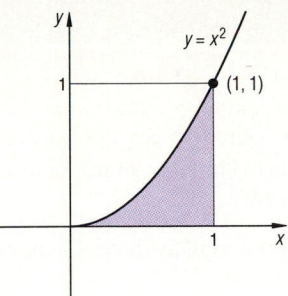

One important problem of this type is to find the area enclosed by curves on the coordinate plane. The great mathematician Archimedes (287–212 B.C.) was the first to find the area of the region under the parabola $y = x^2$ between $x = 0$ and $x = 1$.

At first, this may not look like a problem which can be approached by a Monte Carlo simulation. However, it can be done! The idea is that if a point is randomly selected in the unit-square region bounded by $(0, 0)$, $(0, 1)$, $(1, 1)$, and $(1, 0)$, then the probability it lies in the shaded region is the ratio of the shaded area to the area of the square:

$$P(\text{point is in the shaded region}) = \frac{\text{shaded area}}{\text{area of square}}.$$

To conduct the simulation, select a large number n of points at random in the unit-square region, and test each to see whether or not it lies under the parabola. If c of them do, a good estimate of P(point is in the shaded area) is $\frac{c}{n}$.

Since the area of the square is known to be 1, the shaded area under the parabola is easily estimated.

At the right is a BASIC program which carries out these steps.

```
10   REM MONTE CARLO AREA
20   RANDOMIZE
30   INPUT "HOW MANY POINTS"; N
40   C = 0
50   FOR P = 1 TO N
60     X = RND: Y = RND
70     IF Y < X^2 THEN C = C+1
80   NEXT P
90   AREA = C/N
100  PRINT "THE APPROXIMATE AREA IS", AREA
110  END
```

The program randomly picks N points (X, Y) in the square and counts the number C of these that satisfy $Y < X^2$. The probability estimate is then C/N. One result of running the program three times with N = 1000 (thus randomly choosing 1000 points in the square) was the approximations .333, .343, and .319 for the area under the curve. The mean of these values is .332 . Thus, a good estimate for the area is about $\frac{1}{3}$.

Using calculus, it can be shown that the area under the parabola $y = x^2$ between $x = 0$ and $x = 1$ is exactly $\frac{1}{3}$. The simulation is quite accurate.

In any Monte Carlo simulation study, you should conduct a number of experiments to get a sense of the accuracy of your results. For the case of the area under the parabola, the three estimates were very close to each other when 1000 points were chosen, so it was not necessary to conduct more experiments. This will not always be the case. If you choose a smaller number of points, you will usually find that answers on successive trials are not as close to each other.

Question 17: One place to "find" three tables of random numbers is to generate them by computer. Discuss any samples thought to be defective, and suggest that students take second samples from suspect sources in order to critique their own conclusions from the first sample.

FOLLOW-UP

EXTENSION
An enterprising student or students might call a local hotel to determine how many reservations (if any) the hotel will take over its capacity. It is possible that such information is considered confidential and would not be divulged. Even so, the number of hotel rooms could be found, and students might ask what is the percentage of people with reservations who actually use them. From that data, **small groups** of students can do simulations that could be used to advise the hotel regarding the number of reservations it should accept. This would be a very appropriate project.

EVALUATION
Alternative Assessment
Have students write a couple of paragraphs summarizing the major points of this lesson. Ask them to incorporate their opinions concerning the use of Monte Carlo techniques.

PROJECTS
Project 3 is related to the content of this lesson.

Questions

Covering the Reading

1. Here is a line from a random number table:
   77503   82629   35404   44646   02544   54765   96252   41533.
   Starting at 77503, use these random numbers to simulate 24 reservations for a 24-seat flight. How many persons arrived?   **21**

2. Repeat Question 1 using 4 instead of 0 to represent a no-show. Comment on any differences.   **18 people arrived; see margin**.

In 3 and 4, refer to the 10-trial reservation simulation.

3. If the airline allowed 26 reservations per flight, how many flights would be overbooked?   **4**

4. **a.** Use a random number table to simulate ten more flights. How many reservations would have been necessary to fill the plane on each flight?
   b. Compare your answer to part **a** with that in the text. Comment on any differences.   **a,b) Answers will vary.**

A number of assumptions about passenger behavior were necessary in the airline reservations example. In 5 and 6, give an example of how the assumption listed might not hold.   **See margin.**

5. Flights at different times of the day are considered the same.

6. Passengers act independently of each other.

In 7–10, suppose that the MONTE CARLO AREA program is run to find the area under the parabola $y = x^2$ between $x = 0$ and $x = 1$. What is the approximation produced if the following points are chosen?

7. $(0.3, 0.8)$, $(0.6, 0.2)$, $(0.4, 0.1)$, $(0.9, 0.5)$ **0.75**

8. $(0.16, 0.74)$, $(0.77, 0.68)$, $(0.88, 0.96)$, $(0.04, 0.71)$, $(0.85, 0.37)$, $(0.2, 0.01)$ **≈ 0.33**

9.

   ≈ 0.27

10.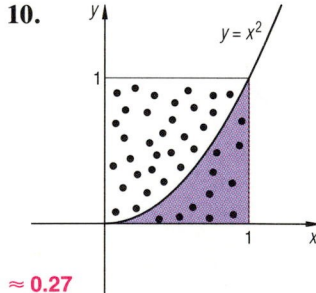

    ≈ 0.27

11. Adapt the MONTE CARLO AREA program to estimate the area under the parabola $y = x^2$ between $x = 0$ and $x = 2$ using 1000 points.
    Change line 60 to 60 X = 2*RND(1):Y = 2*RND(1) and change line 90 to 90 AREA = (C/N)*4. The actual area is $\frac{8}{3}$.

446

Review

12. Assume births of boys and girls are equally likely in families with 4 children. Define an experiment where the outcome is the number of girls.
    a. Find the probability for each individual outcome and make a table.
    b. Construct a scattergram of the probability distribution.
    c. Find the expected value of the distribution.   *(Lessons 7-1, 7-7)*
    a, b) See margin. c) 2

13. Solve $_nP_2 = 182$.   *(Lesson 7-6)*   **n = 14**

14. Consider the experiment of tossing a fair coin five times.
    a. Determine the probability of at least one tails.   **31/32 ≈ .97**
    b. Determine the probability of exactly two heads.   **10/32 ≈ .31**
    c. *True* or *false*. The two events in parts **a** and **b** are independent. Justify your answer.   *(Lessons 7-1, 7-2, 7-3, 7-5)*   **See below.**

15. Refer to the BASIC program below.   *(Lesson 7-4)*

```
10  FOR R = 5 TO 12
20    FOR H = 20 TO 30
30      LET V = 3.141593*R*R*H/3
40      PRINT R, H, V
50    NEXT H
60  NEXT R
70  END
```

    a. How many lines of output would a run of this program have?   **88**
    b. Based on the calculation in line 30, for what is this program used?
    to find the volume of cones of different sizes

16. According to *The Statistical Abstract of the United States, 1987*, the number of passports (in thousands) issued in the U.S. were as follows:

Year 19—	70	73	74	75	76	77	78
Number	2219	2729	2415	2334	2817	3107	3234
Year 19—	79	80	81	82	83	84	85
Number	3170	3020	3222	3764	4122	4718	4968

    a. Construct a scatter plot of these data.   **See margin.**
    b. Does a linear model seem appropriate for the data?   **yes**
    c. Find the equation of the line of best fit for predicting the number of passports from the year.   **sample: y = 172.8x − 10,270**
    d. Predict the likely number of passports to be issued in the year 1995.
    e. What is the correlation between number of passports issued and the year?   *(Lessons 1-2, 2-2, 2-3, 2-5)*   **d) about 6150; e) about .91**

Exploration

17. Find three different tables of random numbers.   **Answers will vary.**
    a. Select three samples of 100 consecutive digits (horizontally, vertically, or diagonally) in each table. Calculate the relative frequency of the digits 0 through 9 in each sample.
    b. Discuss the differences and similarities between the frequencies you calculated in part **a**.

14c) False. The probability of having exactly two heads and at least one tails is $\frac{10}{32}$. So $P(A \cap B) = \frac{10}{32}$. Since $P(A) \cdot P(B) = \frac{31}{32} \cdot \frac{10}{32} \neq \frac{10}{32}$, the events are not independent.

LESSON 7-9   Monte Carlo Methods   **447**

Projects

1. Obtain a pair of dice distinguishable from each other (for instance, two dice of different sizes or colors).
   a. Throw the dice fifty times, recording the results of each die on each throw.
   b. Construct a relative frequency distribution for each of the two dice separately.
   c. Construct a relative frequency distribution for the sum of the two dice.
   d. Repeat steps **a** to **c** several times. Calculate the total relative frequency distribution for the sum of the two dice.
   e. Describe how close the relative frequencies of occurrence of various sums are to their probabilities for a larger number of tosses. Do your dice seem to be fair?

2. a. The Grump family, consisting of two parents and two children, are about to sit down together at a circular table with four seats. In how many different ways can they do this if rotation images like the following are regarded as the same?

   b. Generalize your answer to part **a**. That is, in how many different ways can 5, 6, 7, ... , n people be seated at a circular table? Make drawings and write mathematical arguments to support your answers.

3. Investigate the following situation. If r and s are numbers between 0 and 9 inclusive, what is the probability P that the function $f(x) = x^2 + rx + s$ has real roots?
   a. Consider the case where r and s are integers, and conduct an experiment. Randomly choose 20 pairs (r, s) of integers between 0 and 9, calculate the relative frequency of the number of functions having real roots, and estimate P. Repeat the experiment and revise your estimate as necessary.
   b. Calculate the probability of having real roots if r and s are integers between 0 and 9. [Hint: there are 100 outcomes in the sample space.]
   c. Assume r and s are real numbers between 0 and 9. Modify the computer program given in Lesson 7-9 to estimate the probability that f has real roots.
   d. Repeat parts **b** and **c** allowing r and s to be between -10 and 10. Compare the probabilities you get with those found in parts **b** and **c**.

4. a. Calculate the probability that n people have n different birthdates, for $n = 2, 3, 4,$ and k.
   b. What is the smallest value of k for which this probability is less than .5? (This is known as the **birthday problem**.)
   c. Test the result of part **b** on a group of people or with a set of birthdays of famous people.

5. Statistics and probability are used in determining insurance premiums (the amounts that people pay for insurance). Pick a type of insurance (e.g., auto or life).
   a. Find out all the variables that affect the premiums you would have to pay if you wanted this type of insurance and what it would cost you to obtain such insurance.
   b. Write an essay summarizing how insurance companies use statistics and probability to determine these rates.

448

Summary

Probabilities of events in finite sample spaces can be calculated using counting and an approach similar to calculating relative frequencies. If all outcomes in a sample space are equally likely, the probability of an event is the ratio of the number of individual outcomes making up the event to the number of outcomes in the sample space. The probability of the union of two events A and B satisfies $P(A \cup B) = P(A) + P(B) - P(A \cap B)$. If A and B are mutually exclusive, $P(A \cup B) = P(A) + P(B)$. If A and B are complementary, $P(B) = 1 - P(A)$.

The number of ways to choose one element from each of two sets A and B is $N(A) \cdot N(B)$. This leads to the definition that A and B are independent events (the occurrence of one does not change the probability for the other), if and only if $P(A \cap B) = P(A) \cdot P(B)$.

From the Multiplication Counting Principle, the number of arrangements of n different items is $n!$. The number of arrangements of r of n items *with* replacement is n^r. The number of arrangements of r items out of a given set of n *without* replacement (called a permutation of r out of n things) $= {}_nP_r = n(n-1) \cdot \ldots \cdot (n-r+1) = \dfrac{n!}{(n-r)!}$.

A probability distribution is a function which maps each value of a random variable (determined by the outcome of an experiment) onto its probability. The probability distribution for an experiment with a finite number of outcomes can be represented by a table, a scattergram, or a histogram. The mean, or the expected value, of a probability distribution is the sum of the products of each possible outcome with its respective probability; that is,

$$\mu = \sum_{i=1}^{n} (x_i \cdot P(x_i)).$$

The probability distribution of an experiment is uniform when all the outcomes have the same probability. Randomness can be approximated manually (such as by throwing dice) or using strings of computer-generated pseudo-random numbers. By proper coding of experiments or events with random numbers, Monte Carlo methods enable simulation of real life situations such as the number of people who will appear at an event, or estimations in mathematical situations such as the area under a curve.

Vocabulary

For the starred (*) terms you should be able to give a definition of the term.
For the other terms you should be able to give a general description and a specific example of each.

Lesson 7-1
trial, experiment
outcome of an experiment
*sample space
fair, unbiased
*probability of an event
equally likely
Properties of Probabilities

Lesson 7-2
union of sets, disjoint sets
*mutually exclusive events
Addition Counting Principle
  (for mutually exclusive sets;
  general form)
Probability of a Union Theorem
  (for mutually exclusive sets;
  general form)
complementary events, not A
Probability of Complements
  Theorem

Lesson 7-3
Multiplication Counting Principle
Selections with Replacement
  Theorem
*n factorial, $n!$
Selections without Replacement
  Theorem

Lesson 7-4
*independent events
dependent events

Lesson 7-5
*permutation
Permutation Theorem
*permutations of n objects taken r at
  a time, ${}_nP_r$
Formula for ${}_nP_r$ Theorem
Alternate Formula for ${}_nP_r$ Theorem

Lesson 7-7
random variable
probability distribution
*mean of a probability distribution
*expected value of a probability distribution

Lesson 7-8
constant function
*uniform distribution
at random, randomly
*random numbers
pseudo-random numbers
RND(1), RND
simulated

Lesson 7-9
Monte Carlo method

The Progress Self-Test provides the opportunity for feedback and correction. The Chapter Review provides additional opportunities for practice.

USING THE PROGRESS SELF-TEST
Assign the Progress Self-Test as a one-night assignment. (See pages 69 and T35 for more information.)

ADDITIONAL ANSWERS
1.a. {(1, 1), (1, 2), (1, 3), (2, 1), (2, 2), (2, 3), (3, 1), (3, 2), (3, 3)}

13.a. Each value of P(x) satisfies 0 ≤ P(x) ≤ 1 and their sum is 1.

b.

Progress Self-Test

See margin for answers not shown below.

Take this test as you would take a test in class. You may want to use a calculator and a computer. Then check the test yourself using the solutions at the back of the book.

1. Assume that each of the two spinners shown below is equally likely to land in each of the three regions.

  a. Give the sample space for the result of spinning *both* spinners.
  b. What is the probability that the sum of the two spinners is greater than 4?   **3/9 ≈ 0.33**

2. When two fair dice of different colors are rolled, what is the probability that the red die will show an odd number, and the green die will show an even number? $\frac{1}{4} = 0.25$

3. Suppose $P(A \cap B) = 0.5$, $P(A) = 0.6$, and $P(B) = 0.8$. Find $P(A \cup B)$.   **0.9**

4. a. How many permutations of the letters in the word MASTER are there?   **720**
   b. How many of these permutations begin with A and end with T?   **24**

In 5 and 6, evaluate.

5. $\frac{9!}{3!}$ **60,480**   **6.** $_5P_3$ **60**

7. *True* or *false*. If an event A contains all the possible outcomes of an experiment, then $P(A) = 1$.   **True**

8. Consider the events of scoring under 50 and scoring between 50 and 100, inclusive, on a test. Determine if the events are mutually exclusive, complementary, or both, if the maximum possible score on the test is
   a. 100   **both**   **b.** 120   **mutually exclusive**

9. Solve. $_nP_4 = 56\,_nP_2$ **10**

10. A consumer protection group reports that 25% of 5-lb bags of sugar of a certain brand are underweight. Three bags of sugar are selected at random. Assuming that the report is correct, what is the probability that
   a. all three bags are underweight;   **1/64 ≈ .016**
   b. no bag is underweight?   **27/64 ≈ .42**

11. If you have three pairs of jeans, two pairs of sneakers, and five sweatshirts, how many different outfits consisting of jeans, a sweatshirt, and a pair of sneakers can you make?   **30**

12. A test contains 10 true-false questions and 5 multiple choice questions, each with four choices. Assuming a student answers all questions, how many different answer sheets are possible?   **1,048,576**

13. a. Verify that the table below shows a probability distribution.

x	1	2	3	4	5
P(x)	.03	.47	.22	.05	.23

   b. Graph the probability function in part **a**.
   c. Calculate the mean of the distribution.   **2.98**

14. A computer program is to be written for simulating birthday months. Write a BASIC statement which will generate a uniform distribution of the integers 1, 2, 3, …, 12.   **INT (12*RND(1)) + 1**

15. Consider the program below, written to print a chart of the volumes of cylinders with radii and heights in inches.

```
10  REM VOLUMES OF CYLINDERS
20  REM RAD = RADIUS, HT = HEIGHT
30  FOR RAD = 4 TO 6 STEP 0.5
40    FOR HT = 8 TO 10 STEP 0.5
50      VOL = 3.1416 * RAD * RAD * HT
60      PRINT RAD, HT, VOL
70    NEXT HT
80  NEXT RAD
90  END
```

   a. For how many different radii is the volume calculated?   **5**
   b. Give the first two lines of output.
   c. Give the last line of output.
   d. How many lines of output will be printed?
      b) 4   8   402.1248
         4   8.5   427.2576
      c) 6   10   1130.976
      d) 25

450

Chapter Review

Questions on **SPUR** Objectives

SPUR stands for **S**kills, **P**roperties, **U**ses, and **R**epresentations.
The Chapter Review questions are grouped according to the SPUR Objectives for this chapter.
See margin for answers not shown below.

SKILLS deal with the procedures used to get answers.

■ **Objective A:** *List sample spaces and events for probabilistic experiments.* (*Lesson 7-1*)

In 1–3, consider the experiment of tossing three different coins.

1. Write the sample space for the experiment.
2. List the outcomes in the event "at least two tails show up." **{HTT, THT, TTH, TTT}**
3. *True or false.* The event "no tails show up" consists of a single outcome. **True**

In 4–6, assume that a right-hand page of this book is picked at random and the page number is recorded.

4. What is the sample space for this experiment?
5. Write the set of outcomes in the event "the page is the first page of a chapter."
6. *True or false.* The event "the page has an odd number" is the empty set. **False**

■ **Objective B:** *Compute probabilities.*
(*Lessons 7-1, 7-2, 7-3*)

In 7 and 8, consider rolling two dice, and recording the numbers on the top faces. Find each probability.

7. P(each die is even) **1/4 = .25**
8. P(the sum is even) **1/2 = .5**

9. Consider the experiment of tossing a coin and a standard die. The coin is marked "1" (heads) and "2" (tails). Find the probability that the sum is less than 5. $\frac{5}{12} \approx$ **.417**

In 10 and 11, let x be a randomly selected number from $\{1, 2, 3, 5, 8, 13, 21, 34, 55, 89\}$. Calculate the probability.

10. $P(x$ is even *or* $x < 2^5)$ $\frac{8}{10} =$ **.8**
11. $P(x$ is even *and* $x < 2^5)$ $\frac{2}{10} =$ **.2**
12. When two fair dice are tossed, what is the probability the first is a 3 or the second is odd? **21/36 ≈ .583**

■ **Objective C:** *Find the number of ways of selecting or arranging objects.* (*Lessons 7-3, 7-5*)

13. A coin is tossed seven times. How many possible outcomes are there? **128**
14. List the permutations of the digits 4, 0, and 7. **407, 470, 047, 074, 740, 704**
15. a. How many permutations of the letters of the word NUMBER are there? **720**
    b. How many of these end in BER? **6**
16. a. How many permutations consisting of four letters each can be formed from the letters of DINOSAUR? **1680**
    b. How many of the permutations in **a** end in R? **210**
    c. How many of the permutations in **a** start with D and end in R? **30**

CHAPTER REVIEW

The main objectives for the chapter are organized here into sections corresponding to the types of understanding this book promotes: Skills, Properties, Uses, and Representations (called SPUR objectives).

USING THE CHAPTER REVIEW
Students should be able to answer questions like these with about 85% accuracy by the end of the chapter. (See pages 72 and T38–39 for more information.)

ADDITIONAL ANSWERS
1. **{HHH, HHT, HTH, HTT, THH, THT, TTH, TTT}**

4. **{1, 3, 5, . . ., last odd page in book}**

5. **{3, 79, 143, 201, 267, 339, 397, 455, 523, 591, 663, 727, 779}**

■ **Objective D:** *Evaluate expressions using factorials.* (Lessons 7-3, 7-5)

In 17–19, evaluate without using a calculator.

17. $7!$ **5040**   **18.** $0!$ **1**   **19.** $\frac{21!}{20!}$ **21**

20. Write $\frac{16!}{12!}$ as a product of consecutive integers.
16 · 15 · 14 · 13

21. Evaluate. $\frac{120!}{116!}$ **197,149,680**

22. *True* or *false*. $\frac{12!}{6!6!} = \frac{12!}{8!4!}$ **False**

In 23 and 24, evaluate.

23. $_{10}P_4$ **5040**   **24.** $_5P_5$ **120**

PROPERTIES deal with the principles behind the mathematics.

■ **Objective E:** *State properties of probabilities.* (Lessons 7-1, 7-2, 7-7)

25. If $P(A) = .23$ in an experiment, determine $P(\text{not } A)$. **0.77**

26. Explain why $P(E) = 1.5$ cannot be a correct statement for any event E. **The probability of an event is always between 0 and 1, inclusive.**

In 27–30, *true* or *false*.

27. If A and B are mutually exclusive events, and $P(A) = .7$, then $P(B) = .3$. **False**

28. If A and B are complementary events and $P(A) = k$, then $P(B) = 1 - k$. **True**

29. If A and B are independent events, then $P(A \text{ or } B) = 1$. **False**

30. If the sample space of A is ø, then $P(A) = 0$. **True**

■ **Objective F:** *Determine whether events are mutually exclusive, independent, or complementary.* (Lessons 7-2, 7-4)

In 31 and 32, determine if the pair of events are mutually exclusive.

31. scoring an 80 on a test and scoring a 95 on the same test **mutually exclusive**

32. throwing a sum of 9 on two dice and throwing a 2 on at least one die **mutually exclusive**

33. *True* or *false*. Selecting a king from a deck of cards and then picking a king from the remaining cards are two independent events. **False**

34. Determine if finding a TV viewer to be an adult and finding a TV viewer to be a teenager are complementary events. Explain your reasoning. **No, a viewer could be a child.**

In 35–38, information is given about the probabilities of events A and B. Deduce which (if any) of the terms below apply to the events.

(a) mutually exclusive
(b) complementary
(c) independent

35. $P(A) = .5, P(B) = .2, P(A \cup B) = .6$ **(c)**
36. $P(A) = .5, P(B) = .2, P(A \cup B) = .7$ **(a)**
37. $P(A) = .4, P(B) = .6, P(A \cup B) = 1$ **(a, b)**
38. $P(A) = .33, P(B) = .3, P(A \cup B) = .099$ **(c)**

■ **Objective G:** *Solve equations using factorials.* (Lesson 7-5)

In 39–42, solve.

39. $\frac{x!}{56} = 6!$ **8**   **40.** $\frac{t!}{(t-1)!} = 19$ **19**

41. $_nP_5 = 12 \cdot _nP_4$ **16**   **42.** $_5P_c = _6P_3$ **4 or 5**

■ Objective H: *Calculate probabilities in real situations.* Lessons 7-1, 7-2, 7-4)

In 43 and 44, consider the following situation. An ornithologist feeds a special nutrient to 26 of the 257 pelicans in a bird sanctuary and tags them. A week later, she captures a pelican in the sanctuary. Assume the special nutrient does not affect the behavior of the birds.

43. What is the probability that the pelican is a tagged one?   **26/257 ≈ .1012**

44. If the pelican is not tagged, what is the probability that the next one she catches will be tagged if she
   a. releases the first bird?   **26/257 ≈ .1012**
   b. does not release the first bird?
        26/256 ≈ .1016

In 45 and 46, consider a business which needs computer diskettes. Of the two independent suppliers they usually use, U has a .4 probability and C has a .7 probability of filling any given order in three days. They order diskettes from both suppliers.

45. What is the probability that both U and C will fill the order in 3 days?   **0.28**

46. What is the probability that the order will be filled by one of the suppliers in 3 days?   **0.82**

■ Objective I: *Use counting principles to find the number of ways of arranging objects.*
(Lessons 7-3, 7-5)

47. How many different ways can a student answer 20 multiple choice questions, each of which has four choices? $4^{20} \approx 1.1 \times 10^{12}$

48. Susan has a choice of n math classes, five science classes and two history classes. In how many ways can she select one of each of the three kinds of classes?   **10n**

49. In how many ways can the starting five on a basketball team line up?   **120**

50. A committee of two students is to be chosen from 340 juniors and 330 seniors. In how many ways can the committee be chosen to include both a junior and a senior?   **112,200**

51. How many automobile license plates consisting of two letters followed by four digits are there if repetitions of letters or digits are allowed?

52. A car seats five people, two in front and three in back.
   a. In how many ways can a family of five be seated in the car?   **120**
   b. In how many ways can a family of five be seated in the car if only two of the family members have a driver's license?   **48**

51) 6,760,000

■ Objective J: *Construct, graph, and interpret probability distributions.* (Lesson 7-7)

In 53–55, a fair die was formed from a cube. The faces contain 0, 1, 4, 9, 16, and 25 dots, respectively. The die is tossed one time, and the top face recorded.

53. Show the probability distribution as a table.
54. Graph the distribution as a histogram.
55. Find the mean of the probability distribution.
     9.16

56. Tell why this table does *not* show a probability distribution.

x	-1	0	1	4
$P(x)$	$\frac{1}{2}$	$\frac{1}{4}$	$\frac{1}{8}$	$\frac{1}{5}$

57. Construct a graph of the probability distribution for the function P where $P(s)$ is the probability of getting a maximum score of s when two dice are rolled.

56. The sum of values of $P(x)$ is not 1.

ADDITIONAL ANSWERS
53.

Face	0	1	4	9	16	25
Prob.	$\frac{1}{6}$	$\frac{1}{6}$	$\frac{1}{6}$	$\frac{1}{6}$	$\frac{1}{6}$	$\frac{1}{6}$

54.

57.

58. In a lottery, 120 tickets are sold at $1 each. First prize is $50 and second prize is $20. Find the expected value of a ticket.   **about -42¢**

x = profit in dollars	-1	19	49
$P(x)$	$\frac{118}{120}$	$\frac{1}{120}$	$\frac{1}{120}$

59. The clerk in charge of textbooks took a sample of books that were returned at the end of the year and judged whether they were reusable. Here is the distribution of books that were not reusable.

x = age of book in years	1	2	3	4	5	6
Number not reusable	3	10	10	12	20	30

On the basis of these data, what is the expected lifespan of a book?   **about 4.5 years**

■ **Objective K:** *Use computers to generate pseudo-random numbers.* (Lesson 7-8)

60. Give the range of the BASIC function $J(x)$ defined below: **{1, 2, 3, 4, 5, 6, 7, 8, 9, 10}**

```
DEF FN J(X) = INT (10*RND(1)) + 1.
```

61. Define a BASIC function to generate at random a member of the set
$$\{1, 2, 3, \dots, 289\}.$$
DEF FN RN(X) = INT(289*RND(1)) + 1

In 62 and 63, consider the program

```
10  FOR M = 1 TO 50
20      FOR N = 1 TO 6
30          PRINT INT(6*RND(1)) + 1;
40      NEXT N
50      PRINT
60  NEXT M
70  END
```

62. a. How many random numbers will be generated by the program?
   b. What does line 50 do?

63. *Multiple choice.* For which of the following would the program be useful?
   (a) simulating 6 sets of 50 throws of a die
   (b) simulating 50 throws of a die
   (c) simulating 50 sets of 6 throws of a die
   (d) simulating the results of 6 spins of a spinner with 50 spaces.   **(c)**

62 a) 300
   b) It gives a blank line after every six lines of output.

■ **Objective L:** *Use computers for counting.* (Lesson 7-6)

In 64–66, refer to the program below, which prints the length of the hypotenuse of right triangles with legs of integral lengths from 7 to 16 inches:

```
10  FOR L1 = 7 TO 16
20      FOR L2 = 7 TO 16
30          LET HYP = SQR(L1*L1 + L2*L2)
40          PRINT L1,L2,HYP
50      NEXT L2
60  NEXT L1
70  END
```

64. How many loops are in the program?   **2**
65. When the program is run, how many lines of output will there be?   **100**
66. If lines 10 and 20 are changed to

```
10  FOR L1 = 7 TO 16 STEP 0.5
20  FOR L2 = 7 TO 16 STEP 0.5
```

   a. How many lines of output will there be?
   b. Describe what this modified program is doing.

In 67 and 68, consider the BASIC program below.

```
10  FOR I = 5 TO 9
20      FOR J = 0 TO 9
30          FOR K = 0 TO 9
40              PRINT I,J,K
50          NEXT K
60      NEXT J
70  NEXT I
80  END
```

67. How many lines of output will there be when the program is run?   **500**
68. Describe the output.   **Every integer between 500 and 999**

66 a) 361
   b) It prints the lengths of the hypotenuse of right triangles with legs whose lengths are in the set {7, 7.5, 8, 8.5, ..., 15, 15.5, 16}.

CHAPTER 8 ■ SEQUENCES, SERIES, AND COMBINATIONS

OBJECTIVES ■ CHAPTER 8

The objectives listed here are the same as in the Chapter 8 Review on pages 519–522 of the student text. The Progress Self-Test on page 517 and the tests in the Teacher's Resource File cover these objectives. For recommendations regarding the handling of this end-of-chapter material, see the notes in the margins on the corresponding pages of this Teacher's Edition.

OBJECTIVES FOR CHAPTER 8 (Organized into the SPUR categories— Skills, Properties, Uses, and Representations)	Progress Self-Test Questions	Chapter Review Questions	Teacher's Resource File	
			Lesson Masters*	Chapter Test Forms A & B
SKILLS				
A Use explicit or recursive formulas to determine terms of sequences.	1, 2	1–6	8-1	1, 5a
B Find the nth term of an arithmetic or geometric sequence.	3, 4b	7–10	8-1	3
C Find the sum of the terms of a sequence.	9, 10	11–16	8-3, 8-4	7
D Expand binomials.	13a, 14	17–22	8-8	8
PROPERTIES				
E Classify a sequence as arithmetic, geometric, or neither.	4a	23–30	8-1	2
F Decide whether a sequence has a limit, and if it does, determine the limit.	6	31–38	8-2	5b
G Tell whether an infinite series converges. If it does, give the limit.	8	39–44	8-5	6, 11
H Prove identities involving combinations.	12	45–48	8-6	13
USES				
I Use sequences and series to solve problems.	5, 7, 11	49–54	8-1, 8-2, 8-3, 8-4, 8-5	4, 9, 10
J Use combinations to compute the number of ways of arranging or selecting objects.	15	55–58	8-6	16
K Determine probabilities in situations involving binomial experiments.	16, 17	59–63	8-9	12
REPRESENTATIONS				
L Locate numerical properties represented by the patterns in Pascal's Triangle.	18	64–67	8-7	14, 15

*The masters are numbered to match the lessons, and each is intended for use with that lesson.

OVERVIEW ■ CHAPTER 8

Sequences are fundamental to mathematics; they occur whenever we wish to study the change in a quantity over time, to examine a pattern, or simply to list some data. Your students will have studied sequences in some form or another; the differences will lie in the formality with which they have studied the content and the variety of sequences they have encountered.

This chapter begins by reviewing the basic notation and properties of arithmetic and geometric sequences, which are defined simply as functions having as domain the set of integers greater than a given integer. Without formally defining limit, students study the limiting behavior of sequences in order to determine whether the sequences converge or diverge. Available analytic techniques and a computer are used to

make judgments. We consider an intuitive background in the study of limits to be essential for the more formal treatment of the concept later in calculus.

Series are indicated sums of terms of sequences. Thus any summation Σ is a series. However, most of the important applications of series occur when the terms are not data from an experiment but are values of a function whose formula is known.

Series are the primary topic of Lessons 8-3 through 8-5. Classic algorithms for finding the sum S_n of arithmetic and geometric series are presented and used to prove theorems for finding S_n explicitly. The theorems, in turn, provide formulas for quick solutions to applications involving series.

The last half of the chapter deals with combinations and their application to the powers of binomials and to binomial probabilities. First, counting problems leading to combinations are considered. Pascal's Triangle is introduced as a way of representing combinations. Then it is shown that Pascal's Triangle also contains the coefficients of the powers of the binomial $a + b$. The reading lesson explains how these ideas are applied in an area that is very important to all of business and industry, quality control.

Throughout the chapter students need scientific calculators. Calculators having keys for calculating combinations will come in handy. Computers are needed to demonstrate how sequences and finite series approach limits and to simulate some of the binomial distributions in the last two lessons.

PERSPECTIVES ■ CHAPTER 8

The Perspectives provide the rationales for the inclusion of topics or approaches, give some mathematical background, and make connections with other lessons and to other UCSMP texts.

8-1

FORMULAS FOR SEQUENCES

This lesson focuses on arithmetic and geometric sequences in preparation for Lessons 8-3 and 8-4. The terms *arithmetic progression* and *geometric progression* were once commonly used for these sequences; those terms are generally not used these days because they mistakenly suggest that these sequences are qualitatively different from other sequences.

Recursively, an arithmetic sequence is any sequence in which, for all n, the $(n + 1)$st term is found by adding a fixed constant d (for constant difference) to the nth term. Because sequences are functions, we use one letter to name them. If the terms of an arithmetic sequence are $a_1, a_2, a_3, \ldots$, then an explicit formula is $a_n = a_1 + (n - 1)d$, and we call the sequence a.

Geometric sequences are to arithmetic sequences as multiplication is to addition. A geometric sequence is a sequence in which, for all n, the $(n + 1)$st term is found by multiplying the nth term by a constant. If g is a geometric sequence, then an explicit formula is $g_n = g_1 r^{n-1}$, and we call r the constant ratio.

These sequences are important because many applications of linear and exponential functions can evolve into applications of sequences.

8-2

LIMITS OF SEQUENCES

Treated as a function, a sequence $a_1, a_2, a_3, \ldots, a_n$, has a limit if its graph has a horizontal asymptote as $n \to \infty$. Arithmetic sequences do not have a limit unless the constant difference is 0, in which case they are constant. Geometric sequences

have the limit 0 if the constant ratio r satisfies $|r| < 1$. However, other sequences have interesting limits. In this lesson, the notion of a limit is introduced intuitively, and students see how computer programs enable a person to make educated guesses as to the limiting behavior of a sequence. From the fact that

$$\lim_{n \to \infty} \frac{1}{n} = 0,$$ fundamental properties

of limits, and some algebraic manipulation, students can deduce limits of sequences that have rational formulas. In *Precalculus and Discrete Mathematics*, the ideas are carried further in a detailed discussion of the end behavior of rational functions.

8-3

ARITHMETIC SERIES

A series is the sum of a finite number of terms of a sequence; an arithmetic series is derived from an arithmetic sequence. The two common formulas for finding the sum of arithmetic series are given here. They serve as an introduction to the greater variety of series that will be seen in the next two lessons.

Although simple computer programs to add successive terms of an arithmetic sequence might be written, we have avoided doing so. It is usually much more efficient to use the algebraic formulas and a calculator than to use a computer program to deal with these situations, especially since no equations beyond linear ones are involved.

8-4
GEOMETRIC SERIES
This lesson is to geometric series as the previous lesson was to arithmetic series. Given appropriate information about a geometric sequence, students should be able to find the sum of the first n terms using a formula. If the constant ratio $r > 1$, then as n increases, both sequences and series can grow quite large rather quickly. The case where $|r| < 1$ leads into the next lesson.

8-5
INFINITE SERIES
This lesson concentrates primarily on infinite geometric series because they can be studied analytically using the theorem from Lesson 8-4. In a geometric series, if the constant ratio $|r| < 1$, then as n increases, the sequence approaches 0 and the corresponding series also seems to approach a limit. The expression for that limit $\left(\dfrac{g_1}{1 - r}\right)$ is wonderfully simple.

The limits of other convergent infinite series are not easy to obtain but can be estimated with a computer.

8-6
COMBINATIONS
A combination is a subset of a set, and counting these subsets is surprisingly useful. The first task of this lesson is to exhibit situations where combinations are desired. Then, a formula for the number of combinations of n things taken r at a time is derived.

The different common notations that exist for the number of combinations of n things taken r at a time, $_nC_r$, $C(n, r)$, and $\binom{n}{r}$ are an indication of the importance and breadth of this concept. We use the $_nC_r$ notation most often in this book. The $C(n, r)$ notation, which we do not use in this book, emphasizes that combinations are functions of two variables. The $\binom{n}{r}$ notation is perhaps the most commonly used notation in advanced courses, but students often confuse it with fractions.

8-7
PASCAL'S TRIANGLE
There are a number of ways to define Pascal's Triangle. For young children, it is usually defined as a triangular array in which the terms of each succeeding row are sums of terms of the preceding row. Formally, that means that Pascal's Triangle has been defined recursively as a two-dimensional sequence. In algebra, it is often defined as an array of binomial coefficients. Here we define it as an array of combinations and point out its recursive nature. In Lesson 8-8 the connection is made with binomial coefficients.

8-8
THE BINOMIAL THEOREM
It is one of the wonderful interrelationships in mathematics that the combinations of n things taken r at a time are the entries on the nth row of Pascal's Triangle and the coefficients in the expansion of $(x + y)^n$. In this lesson, the connections are explained, and the Binomial Theorem is used to solve certain counting problems in preparation for the next lesson.

8-9
BINOMIAL PROBABILITIES
A binomial experiment has a fixed number of independent trials, each with two possible outcomes A and B that have fixed probabilities p and $q = 1 - p$ of occurrence, respectively. From this definition, it can be deduced that the probability that A occurs k times in n trials is $_nC_k \cdot p^k q^{1-k}$. This lesson shows why this formula works and applies it to a variety of situations.

The mathematics here is powerful and widely applicable. Later, in Chapter 10, the probabilities of k occurrences for fixed values of n and p are considered together in the formation of binomial distribution functions. These functions picture Pascal's Triangle, and their limits as $n \to \infty$ are normal distributions.

8-10
QUALITY CONTROL BY SAMPLING
As with all reading lessons, this one is intended to shed some light into rather sophisticated uses of the mathematics students have studied in this chapter. This lesson presents a way that binomial probabilities can be used by manufacturers to study issues of quality control of their products. The lesson begins by considering a plan to determine whether a box of nails has too many defectives. Students are shown how to test the plan for effectiveness. The computation is tedious even with a calculator, but fortunately the key elements can be approximated by binomial probabilities.

DAILY PACING CHART ■ CHAPTER 8

For optimal student performance, the Progress Self-Test and the Chapter Review should be covered. (See *General Teaching Suggestions: Mastery* on page T35 of this Teacher's Edition.) By following the pace of the Full Course given here, students can complete 12 or 13 chapters of the book by the end of the year. Students following the pace of the Minimal Course spend more time to complete lessons that are combined with quizzes and to consider the Chapter Review. When all lessons are covered in order (the recommendation of the authors), students in the Minimal Course can cover all but the last two chapters. For more information on pacing, see *General Teaching Suggestions: Pace* on page T34 of this Teacher's Edition.

DAY	MINIMAL COURSE	FULL COURSE
1	8-1	8-1
2	8-2	8-2
3	8-3	8-3
4	Quiz (TRF); Start 8-4	Quiz (TRF); 8-4
5	Finish 8-4	8-5
6	8-5	8-6
7	8-6	8-7
8	8-7	Quiz (TRF); 8-8
9	Quiz (TRF); Start 8-8	8-9
10	Finish 8-8	8-10
11	8-9	Progress Self-Test
12	8-10	Chapter Review
13	Progress Self-Test	Chapter Test (TRF)
14	Chapter Review	
15	Chapter Review	
16	Chapter Test (TRF)	

TESTING OPTIONS

■ Quiz for Lessons 8-1 through 8-3   ■ Chapter 8 Test, Form A   ■ Chapter 8 Test, Cumulative Form
■ Quiz for Lessons 8-4 through 8-7   ■ Chapter 8 Test, Form B

A Quiz and Test Writer is available for generating additional questions, additional quizzes, or additional forms of the Chapter Test.

PROVIDING FOR INDIVIDUAL DIFFERENCES

The student text has been written for, and tested with, average students. It has also been used successfully with students who are better prepared as well as with students who are poorly prepared.

The Lesson Notes often include Error Analysis and Alternate Approach features to help you assist those students who need more help. Students of all abilities often learn from their peers and may benefit from small group work referenced as appropriate throughout the Notes. A blackline Lesson Master (in the Teacher's Resource File), keyed to the chapter objectives, is available for each lesson to provide more practice. (However, we advise you to use these masters sparingly; again refer to the suggestions on pace on page T34.)

Extension activities are provided in the Lesson Notes for use with students who need to be challenged even more than the requirements of completing the Full Course provide.

Sequences, Series, and Combinations

(Inset: cystic fibrosis cells)

Although few hereditary diseases can be cured at present, tests have been developed to detect carriers of many genetic disorders. By assessing the probabilities that various genetic diseases will arise in their family, couples can make informed decisions concerning their future.

For instance, suppose that Mr. and Mrs. Washington each had siblings with cystic fibrosis, a disease now known to be hereditary. If the Washingtons have two children, what is the probability that one will have the disease? This situation can be analyzed as a *binomial experiment*, a type of situation which you will study in this chapter.

Probabilities in binomial experiments can be calculated from just a few pieces of information. In the case of cystic fibrosis, it is known through observation that about two out of three people who have siblings with cystic fibrosis are carriers of the disease. Also, the probability that two carriers of the disease have an afflicted child is about 25%. From this information, it can be found that if Mr. and Mrs. Washington have two children, then the probability that at least one has the disease is just over 20%. (Of course, if detection tests show that either Mr. or Mrs. Washington is not a carrier, the probability is drastically reduced.)

Binomial experiments such as this one often appear in business, politics, and sports as well as medicine, and they require a knowledge of counting techniques. In this chapter, these techniques are developed through the study of sequences, series, and combinations.

CHAPTER 8

We recommend 13–16 days be spent on this chapter: 10–12 days on the lessons and quizzes, 1 day for the Progress Self-Test, 1 or 2 days for the Chapter Review, and 1 day for a Chapter Test. (See the Daily Pacing Chart on page 455D.) If your students have not previously studied sequences, then you will need to spend at least one extra day.

USING PAGE 455

If your students studied probability in their biology classes in connection with genetics, they were applying the binomial theorem and binomial probabilities. Specifically, if two organisms each with a dominant gene *G* and a recessive gene *g* mate, then their offspring will have a 1/4 probability of having each of the following genes: *GG*, *Gg*, *gG*, and *gg*. An offspring with gene type *GG* does not exhibit the recessive trait and cannot pass it on to the next generation. Offspring with gene type *Gg*, or *gG*, are "carriers" of the recessive trait (they do not exhibit it but can pass it on), and an offspring with gene type *gg* exhibits the recessive trait. These possibilities are like the terms in the expansion of $(G + g)^2$.

Ask students how many of them studied this in biology. Do they know of anyone with cystic fibrosis? Do they know other genetic disorders or conditions in humans? (Some are hemophilia, albinism, muscular dystrophy, and color blindness.)

RESOURCES
■ Lesson Master 8-1

OBJECTIVES

A Use explicit or recursive formulas to determine terms of sequences.
B Find the nth term of an arithmetic or geometric sequence.
E Classify a sequence as arithmetic, geometric, or neither.
I Use sequences to solve problems.

TEACHING NOTES

The lesson assumes that students have studied arithmetic and geometric sequences before. If your students have no background with sequences or recursion, plan on spending two days on this lesson.

Emphasize the difference between a recursive formula for a sequence and an explicit formula. In a recursive formula, every term but the first (or first few) is found by using previous terms. In an explicit formula, any term can be found without using the previous ones.

Our definition allows sequences to begin at any integer k because it is convenient at times to start a sequence with a 0th term. (This happens with the sequence of rows in Pascal's Triangle.) Sometimes it is helpful to begin a sequence with a year. However, unless otherwise indicated, sequences defined in terms of k begin with $k = 1$.

Formulas for Sequences

Consider rectangular arrays of dots with n columns and $n + 1$ rows. Let R_n be the number of dots in the nth rectangular array.

$R_1 = 2$　　$R_2 = 6$　　$R_3 = 12$　　$R_4 = 20$　　$R_5 = 30$

In general, the nth array has n dots in each of $n + 1$ rows, so $R_n = n(n + 1)$.

$R_n = n(n + 1)$ is a formula that pairs each positive integer n with the corresponding number R_n. Such a correspondence is a type of function called a *sequence*.

> **Definition**
>
> A **sequence** is a function whose domain is a set of consecutive integers greater than or equal to k.

Unless otherwise indicated, $k = 1$ for a sequence. Each element in the range of a sequence is called a **term** of the sequence. The corresponding positive integer in the domain is its **position** in the sequence. Often, the letter n is used to denote the position of a term, and a subscripted variable such as R_n or a_n is used to denote the term itself.

A formula such as $R_n = n(n + 1)$, which determines the nth term of a sequence directly from n, is called an **explicit formula**. In contrast, a **recursive formula** is one in which the first term or first few terms are given, and then the nth term is expressed in terms of the preceding term(s).

For the sequence R above, notice that each term after the first is the sum of the previous one and some even number.

$$R_2 = \ 6 = \ 2 + 4 \ = R_1 + 4$$
$$R_3 = 12 = \ 6 + 6 \ = R_2 + 6$$
$$R_4 = 20 = 12 + 8 \ = R_3 + 8$$
$$R_5 = 30 = 20 + 10 = R_4 + 10$$

Notice also that in each case the even number added is twice the subscript of R_n. That is $R_n = R_{n-1} + 2n$. So a recursive formula for this sequence is

$$\begin{cases} R_1 = 2 \\ R_n = R_{n-1} + 2n, \text{ for } n > 1. \end{cases}$$

The pattern in the recursive formula can be pictured.

R_1　　$R_2 = R_1 + 2 \cdot 2$　　$R_3 = R_2 + 2 \cdot 3$　　$R_4 = R_3 + 2 \cdot 4$　　$R_5 = R_4 + 2 \cdot 5$

456

Example 1
a. Calculate R_6 using the explicit formula.
b. Calculate R_6 using the recursive formula.
c. Find R_{14} using either formula.

Solution

a. Using the explicit formula,
$$R_6 = 6(6+1) = 6 \cdot 7 = 42.$$

b. Using the recursive formula,
$$R_6 = R_{6-1} + 2 \cdot 6$$
$$= R_5 + 12$$
$$= 30 + 12$$
$$= 42.$$

c. From the explicit formula you know that
$$R_{14} = 14(14+1)$$
$$= 14 \cdot 15 = 210.$$

Check

a and b. Draw an array with 6 columns and 7 rows. As shown at the right, $R_6 = 42$.

c. Find the first 13 terms: 2, 6, 12, 20, 30, 42, 56, 72, 90, 110, 132, 156, 182. So $R_{13} = 182$. Now use the recursive definition. $R_{14} = R_{13} + 2 \cdot 14$. So $R_{14} = 182 + 28 = 210$. It checks.

R_6

Two important types of sequences are *arithmetic* and *geometric* sequences. An **arithmetic sequence** is one in which the difference between consecutive terms is constant. For example, the sequence with first term equal to -7 and a constant difference of 3 is an arithmetic sequence whose first six terms are

$$-7, -4, -1, 2, 5, 8.$$

Each term beyond the first is three more than the previous term. Each term is also equal to -7 plus a number of 3s. For instance, if the nth term of this sequence is called a_n, then

$$a_n = -7 + 3 \cdot (n-1).$$

The preceding pattern generalizes so that any arithmetic sequence can be generated both explicitly and recursively.

Theorem

If n is a positive integer and a_1 and d are constants, then the formulas
$$a_n = a_1 + (n-1)d$$

and $\begin{cases} a_1 \\ a_n = a_{n-1} + d, \, n > 1 \end{cases}$

generate the terms of the arithmetic sequence with first term a_1 and constant difference d.

Error Analysis   In **Example 2,** students may not immediately recognize that the sequence is arithmetic because it is not written in the form of the explicit definition in the theorem preceding the example. Point out that a_1 and d in the theorem are constants (similar to m and b in the slope-intercept form of a line), and so a_n is a linear function of n.

In **Example 4,** you might start the sequence with $g_0 = $ \$12,800. Then g_n stands for its value n years after purchase. (As is, $g_n + 1$ stands for this value, since the sequence begins with g_1.)

Technology   Although computers can be used to determine and print terms of arithmetic sequences, it is usually more efficient to use a calculator and the formula for the nth term. This is especially the case if a program must first be written. The emphasis in this course is on using a computer as a tool when it is convenient or necessary to do so, rather than using it just because it is available. On the other hand, terms of geometric sequences are more difficult to compute, and the program GEOMETRIC SEQUENCE GENERATOR, which follows **Example 4,** can come in handy.

Making Connections   Both **Examples 3 and 4** can be discussed as applications of exponential modeling. You may wish to relate them to Lesson 4-4.

1. What is the 9th term of the sequence 2, 4, 6, 8, . . . ? Did you use an explicit formula or a recursive formula to get this term?
If students repeatedly added 2 to get the term 18, then they are using recursive thinking. If they doubled 9 to get 18, they are using an explicit formula.

Point out that without being given one type of formula or the other, you cannot be certain what the 9th term is. For instance, this sequence might be 2, 4, 6, 8, 0, 2, 4, 6, 8, 0, . . ., which is the sequence of last digits of the even positive integers; in this case the 9th term is 8.

2. Give a recursive formula for the increasing sequence of positive odd numbers.
$$\begin{cases} a_1 = 1 \\ a_n = a_{n-1} + 2 \text{ for } n > 1 \end{cases}$$

3. What term is 344 in the arithmetic sequence 8, 15, 22, 29, . . . ?
49th

4. Suppose a model train collection valued now at $2000 increases 5% in value each year.

a. Give its value next year.
$2100

b. Give the value of the collection n years from now.
$2000(1.05)^n$

c. What will its value be 10 years from now?
about $3300

The explicit formula in the theorem shows that arithmetic sequences are linear functions of n. Given such a function as an explicit formula, you can find a recursive formula.

Example 2 Give a recursive formula for the sequence whose nth term is $a_n = 27 - 4n$.

Solution The formula for a_n is linear, so from the previous theorem it generates an arithmetic sequence.
The first term is $a_1 = 27 - 4(1) = 23$.
The second term is $a_2 = 27 - 4(2) = 19$.
The constant difference is $d = a_2 - a_1 = 19 - 23 = -4$.
So a recursive definition is
$$\begin{cases} a_1 = 23 \\ a_n = a_{n-1} - 4, \, n > 1. \end{cases}$$

Check Find the first few terms, using both the explicit and recursive definitions. For both methods, you get 23, 19, 15, 11, 7, ...

In an explicit formula for an arithmetic sequence, if any three of the numbers d, n, a_1, and a_n are known, the fourth can always be found.

Example 3 What is the position of 127 in the arithmetic sequence below?
$$16, 19, 22, ..., 127, ...$$

Solution The constant difference d is 3, $a_1 = 16$, and $a_n = 127$. Substitute into the explicit formula from the preceding theorem and solve for n.
$$a_n = a_1 + (n - 1)d$$
$$127 = 16 + (n - 1)3$$
$$127 = 3n + 13$$
$$n = 38$$

Check Find the 38th term. $a_{38} = 16 + (38 - 1)3 = 127$. It checks.

A **geometric sequence** is one in which the ratio between consecutive terms is constant. For instance, in the geometric sequence
$$3, \frac{3}{2}, \frac{3}{4}, \frac{3}{8}, ...,$$
the constant ratio is $\frac{1}{2}$. Below at the left is an explicit formula for the nth term of this sequence; at the right is a recursive formula.

$$g_n = 3\left(\frac{1}{2}\right)^{n-1} \qquad\qquad \begin{cases} g_1 = 3 \\ g_n = \frac{1}{2} g_{n-1}, \, n > 1 \end{cases}$$

Generalizing for any geometric sequence gives the following.

458

Question 14: On a calculator, the exact answer for $3(1.1)^5$ is 4.83153, which is far too accurate for the situation.

Question 16: Students might find this formula two ways. One way is to see that the rectangles are 1×3, 2×4, 3×5, and 4×6. The pattern is then rather easy to generalize. Another way is to see that the sequence 3, 8, 15, 24, . . . is quadratic and to fit a quadratic model to the terms using ideas from Lesson 2-6. That is, this function contains (1, 3), (2, 8), (3, 15), (4, 24), ...

Theorem

If n is a positive integer and g_1 and r are constants, then the formulas
$$g_n = g_1 r^{n-1}$$

and
$$\begin{cases} g_1 \\ g_n = rg_{n-1}, \ n > 1 \end{cases}$$
generate the terms of the geometric sequence with first term g_1 and constant ratio r.

Notice that the explicit formula in the theorem shows that geometric sequences are exponential functions of n.

Example 4

A particular car depreciates 25% in value each year. Suppose the original cost is $12,800 .

a. Find the value of the car in its second year.
b. Write an explicit formula for the value of the car in its nth year.
c. In how many years will the car be worth about $1000?

Solution

a. Let g_n = the amount the car is worth in dollars in year n. So the initial amount $g_1 = 12800$. After 1 year it is worth 75% of the previous amount, so $g_2 = 12800(.75) = 9600$.
b. The situation generates a geometric sequence. Find g_n when $g_1 = 12800$ and $r = .75$.
$$g_n = 12800(0.75)^{n-1}$$
c. Find n so that $g_n = 1000$.
Substitute.
$$12800(.75)^{n-1} = 1000$$
$$(.75)^{n-1} = 0.078125$$
Take the logarithm of each side.
$$(n-1) \log .75 = \log 0.078125$$
$$n - 1 \approx 8.86$$
$$n \approx 9.86$$
That is, in about ten years, the car will be worth about $1000.

Both explicit and recursive formulas can be used in computer programs to generate sequences. For instance, the following BASIC program generates the first N terms of any geometric sequence using an explicit formula.

```
10  REM GEOMETRIC SEQUENCE GENERATOR
20  INPUT "ENTER THE FIRST TERM, CONSTANT RATIO,
    AND NUMBER OF TERMS"; FIRST, RATIO, N
30  FOR K = 1 TO N
40  TERM = FIRST * RATIO ^ (K-1)
50  PRINT K, TERM
60  NEXT K
70  END
```

LESSON 8-1 Formulas for Sequences **459**

Below is the output from a computer when the user input 12800, .75, and 10 for FIRST, RATIO, and N, respectively.

1	12800
2	9600
3	7200
4	5400
5	4050
6	3037.5
7	2278.125
8	1708.594
9	1281.445
10	961.0840

Note that the 9th and 10th terms of the sequence provide a check to the work in part **c** of Example 4. In the Questions you are asked to write a program that uses a recursive formula.

Explicit formulas are useful because they allow you to calculate values directly. There are, however, several reasons for recursive formulas. First, sometimes an explicit formula cannot be found. Second, some biological processes, such as the genetic instructions for the spiral growth of certain shells, work recursively. Third, computer programs often run faster using recursive rather than explicit formulas, particularly when the number of terms generated is large.

Wenteltrap Shell

Questions

Covering the Reading

In 1 and 2, refer to the sequence R in the lesson.

1. What is the 7th term?   **56**

2. Find R_{30}.   **930**

In 3 and 4, *true* or *false*.

3. The domain of every sequence is the set of positive integers.   **False**

4. The set of terms of a sequence is its range.   **True**

5. If t_n is a term in a sequence, what is the following term? t_{n+1}

In 6–8, **a.** give the first 3 terms of the sequence. **b.** Identify the sequence as arithmetic or geometric, or neither.

6. $\begin{cases} C_1 = 2 \\ C_n = C_{n-1} - 2, \, n > 1 \end{cases}$ **7.** $b_n = \dfrac{n(n+1)}{2}$ **8.** $d_n = -(1.05)^{n-1}$

a) 2, 0, -2; b) arithmetic   a) 1, 3, 6; b) neither

8a) -1, -1.05, -1.1025; b) geometric

9. Find the first term and the constant difference of the arithmetic sequence defined by $k_n = \dfrac{4-n}{2}$. $k_1 = \frac{3}{2}, d = -\frac{1}{2}$

10. What is the position of 3282 in the arithmetic sequence 6, 18, 30, … , 3282,…?   **274**

460

11. Consider the sequence generated by the following.

$$\begin{cases} t_1 = 18; \\ t_n = \frac{1}{2}t_{n-1}, \ n > 1. \end{cases}$$

    a. Write the first four terms of the sequence.   **18, 9, 9/2, 9/4**
    b. Write an explicit formula for t_n. $t_n = 18(1/2)^{n-1}$
    c. Find t_{20} using either the explicit or recursive formula.
       $18/2^{19} \approx 3.4 \times 10^{-5}$

12. Suppose a car bought for \$15,000 depreciates 20% per year.
    a. Find the value of the car in its 3rd year.   **\$9600**
    b. In what year will the value of the car first fall below \$5000?
       in the 6th year

Applying the Mathematics

13. How many terms are in the geometric sequence 6, 12, 24, ... , 768?   **8**

14. Several long distance runners are on a special exercise program. On the first day, three miles are to be run, and on each successive day of the program 10% more miles are to be run than on the previous day. How far must they run on the sixth day?   **$3 \cdot 1.10^5 \approx 4.8$ miles**

In 15 and 16, consider the following sequence of dots in rectangular arrays.

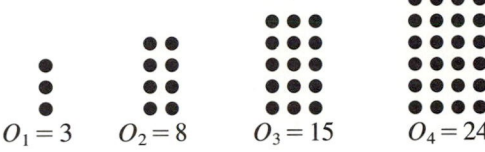

$O_1 = 3$ $O_2 = 8$ $O_3 = 15$ $O_4 = 24$

15. *Multiple choice.* Recall that n is the index of O_n. Which of the following describes those rectangles?
    (a) The width and length equal the index.
    (b) The width equals the index and the length is one more than the index.
    (c) The width equals the index and the length is two more than the index.   **(c)**

16. **a.** Write an explicit formula for O_n. $O_n = n(n + 2)$
    b. What is O_{100}?   **10,200**

17. A female bee has both female and male parents; but a male bee has only a female parent.

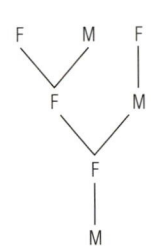

    a. At the left is part of the ancestral family tree of a male bee. The number of bees in consecutive generations are 1, 1, 2, 3. Continue the ancestral family tree for the male bee for three more generations.
    b. Complete the recursive formula for the number of bees in the nth generation of the male's ancestral family tree.

$$\begin{cases} b_1 = 1 \\ b_2 = 1 \\ b_n = \underline{\ ?\ } \end{cases}$$ $b_n = b_{n-1} + b_{n-2}$ **for** $n \ge 3$

    c. Make an ancestral family tree for 3 generations of a female bee.
    d. Write a recursive formula for the number of bees in the nth generation of the female's ancestral family tree.
    a, c, d) See margin.

---

b. Sample:

```
10  REM GEOMETRIC
      SEQUENCE
      GENERATOR,
      RECURSIVE FORMULA
20  INPUT "ENTER THE
      FIRST TERM,
      CONSTANT RATIO,
      AND NUMBER OF
      TERMS"; FIRST,
      RATIO, N
30  LET TERM = FIRST
40  FOR I = 1 TO N
50  PRINT I, TERM
60  TERM = TERM * RATIO
70  NEXT I
80  END
```

NAME _____

LESSON **MASTER 8-1**
QUESTIONS ON **SPUR** OBJECTIVES

■SKILLS   *Objective A*   *(See pages 519–522 for objectives.)*
In 1 and 2, give the first 3 terms of the sequence.   **3.03, 3.0603, 3.090903**

1. $g_n = 3(1.01)^n$

2. $\begin{cases} A_1 = 3 \\ A_n = A_{n-1} + 4n; \ n > 1 \end{cases}$ **3, 11, 23**

3. Give an explicit formula for the sequence defined by $\begin{cases} b_1 = 8 \\ b_n = b_{n-1} + 4; \ n > 1 \end{cases}$ **sample: $b_n = 4 + 4n$**

■SKILLS   *Objective B*

4. Find the 20th term of the arithmetic sequence 2, 11, ...   **173**

5. Find the 8th term of the geometric sequence $3, \frac{3}{2}, \frac{3}{4}, ...$ **$\frac{3}{128}$**

■PROPERTIES   *Objective E*
In 6–8, identify each sequence as arithmetic, geometric, or neither.

6. $A_n = 14(0.98)^{n-1}$ **geometric**

7. 0, 4, 10, 18, 28, ...   **neither**

8. $\begin{cases} b_1 = x + 2y \\ b_n = b_{n-1} - (x - y) \end{cases}$ for a fixed x and y **arithmetic**

■USES   *Objective I*

9. The cost of buying shoes went up about 6% a year during a 5-year span. At this rate, about what would a \$30 pair of shoes cost 15 years from now?   **\$35.73**

10. Suppose the cost of a home in New Orleans went down by 1% in a year. If that rate holds, what would a \$75,000 home cost ten years from now?   **\$67,829**

MORE PRACTICE
For more questions on SPUR Objectives, use *Lesson Master 8-1*, shown on page 461.

EXTENSION
Are there any right triangles whose sides are three consecutive terms of a geometric sequence? (Yes; the constant ratio is $\frac{1+\sqrt{5}}{2}$.)

PROJECTS
The projects for Chapter 8 are described on pages 514–515. Project 1 is related to the content of this lesson.

ADDITIONAL ANSWERS
18.a.,b. See margin, pages 460–461.

19.a. Sample:
```
10  REM ARITHMETIC
    SEQUENCE
    GENERATOR
20  INPUT "ENTER THE
    FIRST TERM,
    CONSTANT
    DIFFERENCE, AND
    NUMBER OF TERMS";
    FIRST, DIFF, N
30  FOR K = 1 TO N
40    TERM = FIRST +
      DIFF * (K – 1)
50    PRINT K, TERM
60  NEXT K
70  END
```

b.
1	-97
2	-93
3	-89
4	-85
5	-81
6	-77
7	-73
8	-69
9	-65
10	-61
11	-57
12	-53
13	-49
14	-45
15	-41
16	-37
17	-33
18	-29
19	-25
20	-21

18. **a.** Use the GEOMETRIC SEQUENCE GENERATOR program to list the first twenty terms of the geometric sequence 36, 24, 16, … .
    b. Write a program that uses a recursive formula for the sequence in part **a**.   **a, b) See margin.**

19. **a.** Modify the GEOMETRIC SEQUENCE GENERATOR program to generate terms of an arithmetic sequence.   **See margin.**
    b. Use your program to print the first 20 terms of the arithmetic sequence with $a_1 = -97$ and $d = 4$.   **See margin.**
    c. Which term of this sequence is the number -1?   **25th**

Review

20. Refer to the graph below.   *(Lesson 6-2)*

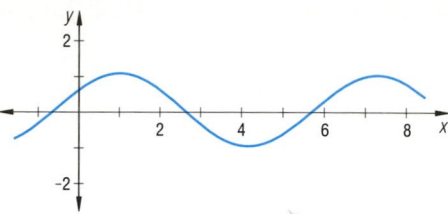

Multiple choice. Which can be an equation for the graph?

(a) $y = \cos\left(x - \frac{\pi}{6}\right)$ (b) $y = \cos\left(x + \frac{\pi}{6}\right)$

(c) $y = \sin\left(x - \frac{\pi}{6}\right)$ (d) $y = \sin\left(x + \frac{\pi}{6}\right)$ **(d)**

In 21 and 22, find the length x.   *(Lessons 5-8, 5-9)*

21.
    ≈ 19.7

22.
    ≈ 25.3

23. Evaluate $\frac{2(2^{13} - 1)}{2 - 1}$.   *(Previous course)*   **16,382**

24. *Skill sequence.* Simplify.   *(Previous course)*

    a. $\dfrac{\frac{2}{x}}{\frac{3}{x}}$ **$\frac{2}{3}$**

    b. $\dfrac{\frac{1}{n} - 4}{\frac{3}{n} + 6}$ **$\frac{1-4n}{3+6n}$**

Exploration

25. Consider the right triangle with sides 3, 4, and 5. Notice that the lengths of the sides form an arithmetic sequence.
    a. Determine three other right triangles with side lengths in arithmetic sequence.   **samples: {6, 8, 10}, {9, 12, 15}, {12, 16, 20}**
    b. Let the sides of a right triangle be a, $a + d$, and $a + 2d$ units long. Use the Pythagorean Theorem to determine a relation between a and d.
    c. Based on the result in part **b**, make a general statement about all the right triangles whose sides form an arithmetic sequence.
    b) $(a + 2d)^2 = a^2 + (a + d)^2$; $(a - 3d)(a + d) = 0$, so $a = 3d$ ($a \neq -d$)
    c) If $a = 3d$, the 3 sides are {$3d$, $4d$, $5d$} for any number $d > 0$. So all the

right triangles whose sides form an arithmetic sequence are similar to the 3, 4, 5 right triangle.

462

462

Limits of Sequences

Consider the following four sequences and their graphs.

$$a_n = -\frac{n}{4}: \quad -\frac{1}{4}, -\frac{1}{2}, -\frac{3}{4}, -1, -\frac{5}{4}, \ldots \qquad b_n = \frac{n^2}{4}: \quad \frac{1}{4}, 1, \frac{9}{4}, 4, \frac{25}{4}, \ldots$$

$$c_n = \frac{2n+1}{n}: \quad \frac{3}{1}, \frac{5}{2}, \frac{7}{3}, \frac{9}{4}, \frac{11}{5}, \ldots \qquad d_n = \frac{(-1)^n}{n}: \quad -1, \frac{1}{2}, -\frac{1}{3}, \frac{1}{4}, -\frac{1}{5}, \ldots$$

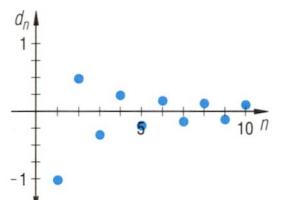

There are important differences and similarities among the four sequences listed and graphed above. The sequence with terms $a_n = -\frac{n}{4}$ is an arithmetic sequence whose terms decrease steadily as n increases. Each term is $\frac{1}{4}$ less than its predecessor; the points all lie on a line. The terms of $b_n = \frac{n^2}{4}$ increase as n increases; all points of the graph lie on a parabola. Sequence a has a maximum value but no minimum value. Sequence b has a minimum value but no maximum value. As n increases, the terms in the former decrease without bound and those in the latter increase without bound.

In contrast, as n increases, the terms of the sequence $c_n = \frac{2n+1}{n}$ get closer and closer to 2. For example, $c_{100} = \frac{201}{100} = 2.01$, $c_{1000} = \frac{2001}{1000} = 2.001$, and $c_{10,000} = \frac{20,001}{10,000} = 2.0001$. We say that this sequence has a **limit** or **limiting value** of 2, written $\lim_{n \to \infty} c_n = 2$. This sentence is commonly read, "the limit of c_n, as n approaches infinity, is 2." The symbol ∞ represents infinity. This limit is represented on the graph of c with a horizontal asymptote at $y = 2$.

LESSON 8-2   Limits of Sequences   **463**

LESSON 8-2

RESOURCES
■ Lesson Master 8-2
■ Teaching Aid 67: Graphs of sequences from page 463

OBJECTIVES

F Decide whether a sequence has a limit, and if it does, determine the limit.
I Use sequences to solve problems.

TEACHING NOTES

This lesson informally presents the limit of a sequence and the ideas of convergent and divergent sequences. Students should be able to graph sequences in the coordinate plane and decide whether a sequence seems to have a limit either by using a computer program or by using algebra and applying properties of limits, including the theorem that $\lim_{n \to \infty} \frac{1}{n} = 0$.

A major purpose of this lesson is to provide students with background experience before they encounter a more formal treatment of limits in a later course. Thus we do not give a formal "epsilon-delta" (ε-δ) definition of a limit. Likewise, we have avoided an explicit definition of ∞. Still, you might give a verbal description of these ideas: The limit of a sequence is L if and only if, no matter how close to L you decide to be, after a certain point in the sequence all terms are that close to L. The limit of a sequence is ∞ if and only if no matter how large a number you pick, after a certain point in the sequence all terms of the sequence are larger than that number.

463

The sequence with terms $d_n = \frac{(-1)^n}{n}$ is like the sequence with terms c_n in that as n increases, the value of $\frac{(-1)^n}{n}$ gets closer and closer to a number. For example, $d_{100} = \frac{(-1)^{100}}{100} = 0.01$ and $d_{10,000} = \frac{(-1)^{10,000}}{10,000} = 0.0001$. Although the terms alternate between positive and negative values, this sequence also has a limit. The limit is 0, written $\lim_{n \to \infty} d_n = 0$. The graph has $y = 0$ as its horizontal asymptote.

Because neither sequence a nor b approaches a constant value as n increases, neither sequence has a limit as n approaches infinity. Another way of saying this is, "the limit does not exist." This is consistent with the lack of horizontal asymptotes to the graphs of a and b.

The following terms are used to describe the behavior of sequences with respect to limits.

> **Definition**
>
> A sequence which has a finite limit L is said to be **convergent** or to **converge to L**. A sequence which does not have a finite limit is said to be **divergent** or to **diverge**.

Sequences a and b are divergent. Sequences c and d are convergent.

Sometimes you can use a computer to help decide whether a sequence is convergent or divergent. For example, the computer program below prints out successive terms of the sequence

$$\frac{1}{2}, \frac{2}{3}, \frac{3}{4}, \dots, \frac{n}{n+1}, \dots.$$

```
 5   REM SEQUENCE BY INFINITE LOOP
10   LET N = 1
20   PRINT N, N/(N + 1)
30   LET N = N + 1
40   GOTO 20
```

The "infinite loop" in this computer program is in lines 20, 30, and 40. Statement 20 prints n and $t_n = \frac{n}{n+1}$. Statement 30 increases n by 1. Statement 40 returns to statement 20 to print the next position and term of the sequence. The computer will run the program until it overflows, malfunctions, or you stop it. For most computers there are ways to interrupt an infinite loop temporarily or stop it permanently. Check how your machine works. On some machines, you may have to press the CONTROL key simultaneously with another key. Or your machine may have a BREAK key which serves the same purpose.

Example 1 Is the sequence defined by $t_n = \dfrac{n}{n+1}$ convergent or divergent? If it is convergent, give its limit.

Solution Run the computer program above. Here are some lines of output of this program:

1	0.5
2	0.666667
3	0.75
⋮	⋮
19	0.95
20	0.952381
⋮	⋮
99	0.99
100	0.990099
⋮	⋮
399	0.9975

As n increases, the terms of the sequence appear to approach 1. So it seems that the sequence is convergent, and that

$$\lim_{n\to\infty} t_n = \lim_{n\to\infty} \frac{n}{n+1} = 1.$$

By changing line 20 of the preceding SEQUENCE BY INFINITE LOOP program you can check whether any sequence generated by an explicit formula appears to have a limit.

Example 2 Find $\lim\limits_{n\to\infty} \dfrac{(-1)^n\, n}{2n-1}$.

Solution Modify the formula in statement 20 of the preceding computer program. Here is some output:

```
10  LET N = 1
20  PRINT N, (-1)^N*N/(2*N − 1)
30  LET N = N + 1
40  GOTO 20
```

1	−1
2	0.6666667
3	−0.6
4	0.5714286
⋮	⋮
49	−0.5051546
50	0.5050505
⋮	⋮
499	−0.5005015
500	0.5005005

Successive terms seem to approach 0.5 and −0.5, alternately. So it appears that $\lim\limits_{n\to\infty} \dfrac{(-1)^n\, n}{2n-1}$ does not exist and the sequence generated by $t_n = \dfrac{(-1)^n\, n}{2n-1}$ is divergent.

Making Connections In *Precalculus and Discrete Mathematics*, students are shown how to determine convergence (and limits) or divergence of all the types of sequences represented in this lesson.

ADDITIONAL EXAMPLES

1. Is the sequence defined by $s_n = \dfrac{n^2}{2n+100}$ convergent or divergent? If it is convergent, give its limit.
divergent

2. Find $\lim\limits_{n\to\infty} \dfrac{(-1)^n}{n^2}$
0

3. Find $\lim\limits_{n\to\infty} \dfrac{6n+4}{5-3n}$
-2

Technology for Question 3c: This sequence is a subset of the rational function $f(x) = \dfrac{3x+2}{2x-1}$. However, the sequence exhibits much simpler behavior than that function; it has no zeros and no vertical asymptotes. If students use an automatic grapher to graph the sequence and they are unable to restrict the domain, they will have to ignore all the extra points the grapher gives them.

Questions 6 and 7: Note that the computer does not provide proofs, but we know enough about the expressions in the numerator and denominator of these fractions to believe that the behavior of the sequence will not be unusual. This gives us confidence in what the computer signals.

Making Connections for Question 11: This question can be answered by using the techniques of the next lesson. In fact, students may have used that technique in the past to rewrite repeating decimals as fractions. If they do not know the technique, we recommend putting off instructions until the next lesson. At this point appeal to students' memories and common sense: where have they seen this decimal before? Suggest that they use a calculator and trial and error if necessary.

Question 14c: This kind of thinking is needed in formal approaches to the idea of limit.

Making Connections for Question 16: Students encountered this sequence in Lesson 4-6. The generalization is in **Question 25.**

Making Connections for Question 24: This is important review for Lesson 8-3.

In this book, only an intuitive idea of the limit of a sequence is introduced. You will see a more precise definition in later courses. Without a formal definition of limit, you cannot prove statements about them. But two particular theorems are useful to know, even without formal proofs.

Theorem

$$\lim_{n \to \infty} \frac{1}{n} = 0$$

Notice that each term of the sequence $1, \frac{1}{2}, \frac{1}{3}, \frac{1}{4}, \ldots, \frac{1}{n}, \ldots$ is smaller than the preceding term and that each term is positive. For very large values of n, $\frac{1}{n}$ is a very small positive number. So the theorem seems reasonable.

The following theorem can also be used to help find limits.

Theorem

If $\lim_{n \to \infty} a_n$ and $\lim_{n \to \infty} b_n$ exist and c is a constant, then:

(1) $\lim_{n \to \infty} c = c$

(2) $\lim_{n \to \infty} (a_n + b_n) = \lim_{n \to \infty} a_n + \lim_{n \to \infty} b_n$

(3) $\lim_{n \to \infty} (a_n b_n) = (\lim_{n \to \infty} a_n)(\lim_{n \to \infty} b_n)$

(4) $\lim_{n \to \infty} (c \, a_n) = c \lim_{n \to \infty} a_n$.

In words, the parts of the preceding theorem say that the limit of a constant sequence is that constant, the limit of a sum is the sum of the individual limits, the limit of a product is the product of the individual limits, and the limit of a constant times the terms of a sequence equals the product of the constant and the limit of the sequence.

For instance, to find the limit of the sequence

$$\frac{3}{1}, \frac{5}{2}, \frac{7}{3}, \frac{9}{4}, \ldots, \frac{2n+1}{n}, \ldots$$

you need to find out if $\lim_{n \to \infty} \frac{2n+1}{n}$ exists. Rewrite $\frac{2n+1}{n}$ as a sum of fractions and simplify.

$$\frac{2n+1}{n} = \frac{2n}{n} + \frac{1}{n}$$
$$= 2 + \frac{1}{n}$$

So
$$\lim_{n \to \infty} \frac{2n+1}{n} = \lim_{n \to \infty} \left(2 + \frac{1}{n}\right)$$
$$= \lim_{n \to \infty} 2 + \lim_{n \to \infty} \frac{1}{n} = 2 + 0 = 2.$$

The limit of the sequence is 2. You can use a computer to check this result.

466

Here is another example of finding a limit by rewriting an expression in terms of $\frac{1}{n}$ and constants.

Example 3 Find $\lim_{n\to\infty} \frac{2-n}{4-3n}$.

Solution Rewrite the expression after dividing numerator and denominator by n.

$$\lim_{n\to\infty} \frac{2-n}{4-3n} = \lim_{n\to\infty} \frac{\frac{2}{n}-1}{\frac{4}{n}-3}$$

$$= \lim_{n\to\infty} \frac{2\left(\frac{1}{n}\right)-1}{4\left(\frac{1}{n}\right)-3}$$

$$= \frac{\lim_{n\to\infty} 2\left(\frac{1}{n}\right)-1}{\lim_{n\to\infty} 4\left(\frac{1}{n}\right)-3}$$

Use the results that $\lim_{n\to\infty} \frac{1}{n} = 0$ and $\lim_{n\to\infty} c = c$.

So the limit is $\frac{2(0)-1}{4(0)-3} = \frac{-1}{-3} = \frac{1}{3}$.

Check Substitute some large values for n. For instance, if $n = 1000$,

$$\frac{2-n}{4-3n} = \frac{2-1000}{4-3000} = \frac{-998}{-2996} \approx 0.3331 \approx \frac{1}{3}.$$

Questions

Covering the Reading

In 1 and 2, *true* or *false*.

1. If each term in a sequence is less than the preceding term, the sequence has a limit.   **False**

2. If a sequence has a limit, then the graph of the sequence has a horizontal asymptote.   **True**

3. Consider the sequence $\frac{5}{1}, \frac{8}{3}, \frac{11}{5}, \frac{14}{7}, \ldots$, where $a_n = \frac{3n+2}{2n-1}$.
   a. Find the tenth term of the sequence.   **32/19 ≈ 1.684**
   b. Evaluate a_{100} and a_{1000} to three decimal places.
   c. Graph the sequence.   **See margin.**
   d. Does the graph appear to have a horizontal asymptote? If so, what is its equation?   **Yes; y = 3/2**
   e. Does the sequence appear to have a limit? If so, what is it?
      b) $a_{100} \approx 1.158$, $a_{1000} \approx 1.502$; e) Yes; $\lim_{n\to\infty} a_n = 3/2$

4. a. Graph the sequence with nth term $\frac{3n-7}{4n}$. **See margin.**

b. Is the sequence convergent or divergent? If it is convergent, give its limit. **convergent; 3/4**

5. Use the computer program below to study the sequence

$$\frac{4}{2}, \frac{7}{5}, \frac{12}{10}, \frac{19}{17}, \cdots, \frac{n^2+3}{n^2+1}, \cdots.$$

```
10  LET N = 1
20  PRINT N, (N*N + 3)/(N*N + 1)
30  LET N = N + 1
40  GOTO 20
```
converges; $\lim\limits_{n\to\infty} \frac{n^2+3}{n^2+1} = 1$

Does the sequence converge or diverge? If it converges, state its limit.

In 6 and 7, use computer programs to decide whether the sequence is convergent or divergent. If it is convergent, give its limit.

6. $-1, 1, \frac{3}{7}, \frac{2}{7}, \cdots, \frac{n}{n^2-2}, \cdots$
convergent; 0

7. $0, \frac{6}{4}, \frac{24}{9}, \frac{120}{25}, \cdots, \frac{n^3-n}{n^2}, \cdots$
divergent

In 8–11, find the limit without a computer.

8. $\lim\limits_{n\to\infty} \frac{3}{n}$ **0**

9. $\lim\limits_{n\to\infty} \frac{11-5n}{6n+23}$ $-\frac{5}{6}$

10. The constant sequence $8, 8, 8, 8, \ldots$. **8**

11. The sequence formed by an increasing number of 3s: $0.3, 0.33, 0.333, 0.3333, \ldots$. **1/3**

Applying the Mathematics

In 12 and 13, consider the computer program at the right.

```
10  LET N = 1
20  PRINT N, SQR(N)/(3 + SQR(N))
30  LET N = N + 1
40  GOTO 20
```

12. Find the limit of the sequence. **1**

13. Change statement 30 to 30 LET N = N + 10. What is the effect of this change on the program output? **The output stabilizes after fewer lines of output.**

14. Let $t_n = \frac{24{,}000}{n}$.

a. Find $t_1, t_2, t_3, t_4, t_{1000},$ and $t_{10{,}000}$. **See margin.**

b. This sequence has a limit L. Use the theorems in the lesson to determine L. **See below.**

c. Find a number x such that for all $n > x$, t_n is within 0.1 of L. **sample: $x = 240{,}000$**

15. In parts **a–d**, tell whether the limit exists. If it does, state it.

a. $\lim\limits_{n\to\infty}\left(\frac{2}{3}\right)^n$ **b.** $\lim\limits_{n\to\infty}\left(\frac{3}{2}\right)^n$ **c.** $\lim\limits_{n\to\infty}(-4)^n$ **d.** $\lim\limits_{n\to\infty}(-0.99)^n$

e. Make a conjecture about $\lim\limits_{n\to\infty} r^n$.

**a) exists, 0; b) does not exist; c) does not exist; d) exists, 0;
e) $\lim\limits_{n\to\infty} r^n = 0$ if $|r| < 1$ and $\lim\limits_{n\to\infty} r^n$ does not exist if $|r| > 1$.**

14b) $\lim\limits_{n\to\infty} t_n = \lim\limits_{n\to\infty} \frac{24{,}000}{n} = 24{,}000 \lim\limits_{n\to\infty} \frac{1}{n} = 24{,}000 \cdot 0 = 0$

468

16. Some important mathematical numbers are defined as the limit of a sequence. One example is e, defined as the limit of the following sequence as n approaches infinity:

$$\left(1 + \frac{1}{1}\right)^1, \left(1 + \frac{1}{2}\right)^2, \left(1 + \frac{1}{3}\right)^3, \dots, \left(1 + \frac{1}{n}\right)^n, \dots.$$

Use a computer program to find the value of e to five decimal places.
$e \approx 2.71828$

17. a. Find $\lim\limits_{n \to \infty} \left(\frac{1}{n} - 5\right)$. **-5**

b. Give an equation of the horizontal asymptote of $y = \frac{1}{x} - 5$. **$y = -5$**

c. Why are the answers to **a** and **b** similar?
The horizontal asymptote of the equation $y = a/x + b$ is the limit, as $n \to \infty$, of $a/n + b$.

Review

18. For the geometric sequence $100, -50, 25, -12.5, \dots$, write:
a. an explicit formula,   **b.** a recursive formula. *(Lesson 8-1)*
$g_n = 100(-0.5)^{n-1}$ **$g_1 = 100; g_{n+1} = (-0.5)g_n, n \geq 1$**

19. For the arithmetic sequence: $3x + 2y, 4x + y, 5x, \dots$, find:
a. the constant difference,   **b.** the 50th term. *(Lesson 8-1)*
$x - y$ **$52x - 47y$**

20. How many terms are there in the arithmetic sequence $10, 15, 20, \dots, 105$? *(Lesson 8-1)* **20**

21. Consider the geometric sequence with terms $5, 15, 45, 135, 405, \dots$.
a. Find a formula for t_n, the nth term. **$t_n = 5(3)^{n-1}$**
b. Write the log of the first five terms in the geometric sequence.
c. Prove that the numbers found in part **b** form an arithmetic sequence.
(Lessons 4-5, 8-1) **b) .699, 1.176, 1.653, 2.130, 2.607; c) See margin.**

22. A number between 1 and 1000 (inclusive) is chosen at random. What is the probability that it is divisible by 7 or 11? *(Lesson 7-2)* **0.22**

23. Find the other measures of the sides and angles of two noncongruent triangles in which $a = 20$, $m\angle C = 25°40'$, and $c = 10$.
(Lesson 5-9) **$m\angle A = 119°58'$, $m\angle B = 34°21'$, and $b \approx 13$; or $m\angle A = 60°01'$, $m\angle B = 94°19'$, $b \approx 23$**

24. *Skill sequence.* Represent using summation notation.
a. $a_1 + a_2 + a_3 + a_4$ **b.** $b_2 + b_3 + b_4 + \dots + b_{10}$
c. $m_1 + m_2 + m_3 + \dots + m_n$ *(Lesson 1-4)*

a) $\sum\limits_{i=1}^{4} a_i$; b) $\sum\limits_{i=2}^{10} b_i$; c) $\sum\limits_{i=1}^{n} m_i$

Exploration

25. Consider the sequence whose terms are defined by $\left(1 + \frac{r}{n}\right)^n$, where r is a constant.
a. Use a computer or calculator to determine the limit of the sequence when $r = 1, 3$, and 8.
b. Each of the limits in **a** is related to a power of e. Determine the relation and generalize what you observe.

a) $\lim\limits_{n \to \infty} \left(1 + \frac{1}{n}\right)^n \approx 2.718$, $\lim\limits_{n \to \infty} \left(1 + \frac{3}{n}\right)^n \approx 20.086$, $\lim\limits_{n \to \infty} \left(1 + \frac{8}{n}\right)^n \approx 2980.958$

b) $\lim\limits_{n \to \infty} \left(1 + \frac{1}{n}\right)^n = e$, $\lim\limits_{n \to \infty} \left(1 + \frac{3}{n}\right)^n = e^3$, $\lim\limits_{n \to \infty} \left(1 + \frac{8}{n}\right)^n = e^8$. In general, $\lim\limits_{n \to \infty} \left(1 + \frac{x}{n}\right)^n = e^x$.

FOLLOW-UP

MORE PRACTICE
For more questions on SPUR Objectives, use *Lesson Master 8-2*, shown below.

PROJECTS
Projects 2 and 5 are related to the content of this lesson.

NAME _____

LESSON **MASTER 8-2**
QUESTIONS ON **SPUR** OBJECTIVES

■**PROPERTIES** *Objective F (See pages 519–522 for objectives.)*
In 1–8, decide whether the sequence is convergent or divergent. If it is convergent, give its limit.

1. $3, \frac{5}{4}, 1, \dots, \frac{2n+1}{3n-2}$ — **convergent; $\frac{2}{3}$**

2. $2, 4, 8, 16, \dots$ — **divergent**

3. nth term $= \frac{n^2 + 2}{n}$ — **divergent**

4. $\frac{1}{2}, \frac{2}{9}, \frac{3}{28}, \dots, \frac{n}{n^3 + 1}$ — **convergent; 0**

5. $\frac{2}{3}, \frac{8}{9}, \frac{32}{27}, \dots, \frac{2(\frac{4}{3})^{n-1}}{}$ — **divergent**

6. nth term $= \frac{2 - 3n}{5n}$ — **convergent; $-\frac{3}{5}$**

7. $5, -\frac{5}{2}, \frac{5}{4}, -\frac{5}{8}, \dots, 5(-\frac{1}{2})^{n-1}$ — **convergent; 0**

8. nth term $= \frac{3n^2 + 5n}{7n^2}$ — **convergent; $\frac{3}{7}$**

■**USES** *Objective I*

9. A certain town had a population of 50,000 in 1990. If the growth rate remains at a constant -3%, what will the population be in the year 2000? — **36,871**

10. Emilio is selling cookies for his club. Suppose at each stop he makes, he sells 2 more boxes of cookies than he sold at the last stop. If he sold one box of cookies at his first stop, how many boxes did he sell at his 12th stop? — **23**

OBJECTIVES

C Find the sum of the terms
of a sequence.
I Use series to solve
problems.

TEACHING NOTES

The distinction between
sequences and *series* is a difficult one for many students.
One reason for the difficulty
is that subscripts are found in
both. The second reason is
that out of any numerical
sequence comes a sequence
of series! You may wish to
write the following to emphasize that a series is a sum:

$S_1 = a_1$
$S_2 = a_1 + a_2$
$S_3 = a_1 + a_2 + a_3$

. . .

$$S_n = \sum_{i=1}^{n} a_n$$

Making Connections
The story about Gauss is a
classic, but if your students
covered Chapter 13 of
UCSMP *Advanced Algebra*,
the story will be familiar to
them.

Error Analysis The steps
on page 471 for proving the
theorem on 472 should not
be difficult for students with
prior experience with
sequences and their notation.
If students have not had sufficient experience, ensure that
students understand Gauss'
steps for the specific instance
of finding S_{100} before explaining the generalization.

LESSON 8-3

Arithmetic Series

A child builds a figure with colored blocks in stages as illustrated above.
The number of blocks set down in each stage is a term in the arithmetic
sequence of odd numbers 1, 3, 5, ... The final pattern consists of a 6×6
square of 36 blocks. So the number of blocks in the sixth pattern is the sum of
the first six odd numbers.

$$1 + 3 + 5 + \ldots + 11 = 36$$

If a_n is the number of blocks added in the nth stage and S_n is the total number
of blocks in the nth figure, the total S_6 can be written using summation notation.

$$S_6 = a_1 + a_2 + a_3 + \ldots + a_6 = \sum_{n=1}^{6} a_n$$

S_6 is the value of a *series*. In general, a **series** is an indicated sum of terms of a
sequence. If the number of terms added is infinite, the resulting series is an
infinite series. If a finite number of terms are added, the resulting series is
called a **finite series**. Suppose the sum of the first n terms of a sequence is
denoted by S_n. If $a_1, a_2, a_3, \ldots, a_n$ are terms in an arithmetic sequence, then

$$S_n = a_1 + a_2 + a_3 + \ldots + a_n = \sum_{i=1}^{n} a_i$$

is called a **finite arithmetic series**, or simply, an **arithmetic series**.

An explicit formula for the nth odd number is $a_n = 2n - 1$, so the finite arithmetic series from the pattern of child's blocks can also be written as

$$\sum_{n=1}^{6} (2n - 1).$$

To evaluate this expression, replace n with each of the integers 1 through 6,
and add the resulting terms.

$$\sum_{i=1}^{6} (2n - 1) = [2(1) - 1] + [2(2) - 1] + [2(3) - 1] + [2(4) - 1] + [2(5) - 1] + [2(6) - 1]$$

$$= 1 + 3 + 5 + 7 + 9 + 11 = 36$$

Note the importance of the parentheses in $\sum_{n=1}^{6} (2n - 1)$. In the expression

$\sum_{n=1}^{6} 2n - 1$, the 1 is not part of the summation, and a different sum is
indicated.

470

Example 1 Evaluate $\displaystyle\sum_{n=1}^{6} 2n - 1$.

Solution Each of the six terms is of the form $2n$. Add them and then subtract one from the result.

$$\sum_{n=1}^{6} 2n - 1 = \left(\sum_{n=1}^{6} 2n\right) - 1$$
$$= [2(1) + 2(2) + 2(3) + 2(4) + 2(5) + 2(6)] - 1$$
$$= (2 + 4 + 6 + 8 + 10 + 12) - 1$$
$$= 41$$

So $\displaystyle\sum_{n=1}^{6} (2n - 1)$ and $\displaystyle\sum_{n=1}^{6} 2n - 1$ have different values.

To find the sum of the terms of a finite arithmetic sequence, one method is to add every term. Another method was discovered by the great mathematician Karl Friedrich Gauss (1777–1855). Part of the folklore of mathematics is that he discovered this method while he was in the third grade. He was asked to add the first 100 integers. Here is what he did.

Karl Friedrich Gauss as a youth

Suppose S_{100} represents the sum of the first 100 integers.

Then $S_{100} = \displaystyle\sum_{n=1}^{100} n$

or $\quad S_{100} = 1 + 2 + 3 + \ldots + 99 + 100$.
He rewrote the series beginning with the last term.
$\quad S_{100} = 100 + 99 + 98 + \ldots + 2 + 1$

Then he added the two equations, term by term.
$\quad 2S_{100} = 101 + 101 + 101 + \ldots + 101 + 101$

The right side of the equation has 100 terms, each equal to 101. Thus
$\quad 2S_{100} = 100(101)$.

He divided each side of the equation by 2.

$$S_{100} = \frac{100\,(101)}{2} = 5050.$$

It is further said that Gauss wrote nothing but the answer of 5050 on his slate, having done all the work in his head!

Gauss's strategy can be used to find a formula for the value S_n of any arithmetic series. First, write S_n starting with the first term a_1 and successively add the constant difference d. Second, write S_n starting with the last term a_n and successively subtract the constant difference d.

$$S_n = a_1 + (a_1 + d) + (a_1 + 2d) + \ldots + [a_1 + (n - 1)d]$$
$$S_n = a_n + (a_n - d) + (a_n - 2d) + \ldots + [a_n - (n - 1)d]$$

Now add the two preceding equations, term by term.

$$2S_n = \underbrace{(a_1 + a_n) + (a_1 + a_n) + (a_1 + a_n) + \ldots + (a_1 + a_n)}_{n \text{ terms}}$$

The right side has n terms each equal to $a_1 + a_n$.
So $\qquad\qquad\qquad\qquad 2S_n = n(a_1 + a_n)$.

Dividing by 2: $\qquad\qquad\qquad S_n = \dfrac{n}{2}(a_1 + a_n)$.

LESSON 8-3 Arithmetic Series **471**

1. Evaluate: **a.** $\sum_{i=1}^{8} (n^2 - 1)$

b. $\sum_{i=1}^{8} (n^2) - 1$.

a. 196; b. 203

2. A packer had to fill 100 boxes identically with machine tools. The packer filled the first box in 13 minutes, but got faster by the same amount each time as time went on. If he filled the last box in 8 minutes, what was the total time that it took to fill the 100 boxes?
1050 minutes = 17.5 hours

3. Vertical lengths of wood 12 inches wide are used to support stairs leading to the second story of an apartment building. If there are 12 stairs, each stair is 10 inches higher than the previous stair, and the first stair is 10 inches higher than the floor, what is the total length of wood needed?
780 inches = 65 feet

4. A new business decides to rank its 9 employees by how well they work and pay them amounts that are in arithmetic sequence with a constant difference of $500 a year. If the total amount paid the employees is to be $250,000, what will the employees make per year?
from $25,778 to $29,778

This formula for S_n is useful if you know the first and nth terms of the series. If you do not know the nth term, you can find it using $a_n = a_1 + (n-1)d$ from Lesson 8-1. This leads to the second formula in the theorem below, which you are asked to derive in the questions.

Theorem

The sum $S_n = a_1 + a_2 + \ldots + a_n$ of an arithmetic series with first term a_1 and constant difference d is given by

$$S_n = \frac{n}{2}(a_1 + a_n)$$

or

$$S_n = \frac{n}{2}[2a_1 + (n-1)d].$$

Example 2

A student borrowed $4000 for college expenses. The loan was to be repaid over a 100-month period, with monthly payments as follows:
$$\$60.00, \$59.80, \$59.60, \ldots, \$40.20.$$
How much did the student pay over the life of the loan?

Solution Find the sum $60.00 + 59.80 + 59.60 + \ldots + 40.20$. Because the terms show a constant difference ($d = -0.20$), this is an arithmetic series with $a_1 = 60.00$, $n = 100$, and $a_{100} = 40.20$. Substitute into the formula

$$S_n = \frac{n}{2}(a_1 + a_n).$$

So

$$S_{100} = \frac{100}{2}(60.00 + 40.20)$$

$$= 5010.$$

The student paid back a total of $5010.

Example 3

In training for a marathon, an athlete runs 7500 meters on the first day, 8000 meters the next day, 8500 meters the third day, and each day thereafter runs 500 m more than on the previous day. How far will the athlete have run in all at the end of thirty days?

Solution The distances form an arithmetic sequence, with $a_1 = 7500$ and $d = 500$. Because the final term is not known, use the second formula in the theorem.

$$S_n = \frac{30}{2}[2a_1 + (30-1)d]$$

$$= \frac{30}{2}[2(7500) + (30-1)500]$$

$$= 442,500 \text{ m}$$

The athlete will have run 442.5 km in thirty days.

The formulas for S_n are also useful if S_n is known and you must find one of a_1, a_n, n, or d.

Example 4

A woman is building a log playhouse for her children, with the design at the left for part of the roof. How should she cut up a ten-foot log to get the five logs needed, if each log is to be eight inches longer than the one before it?

Solution The lengths of the logs form an arithmetic sequence with $n = 5$, $d = 8$, and $S_5 = 120$. If she knew a_1, she would know how to cut the logs. Substitute into the formula for S_n and solve for a_1.

$$S_5 = \frac{5}{2}[2a_1 + (5-1)8]$$

$$120 = \frac{5}{2}(2a_1 + 32)$$

$$120 = 5a_1 + 80$$

Thus $a_1 = 8$.

She should cut the top log 8 inches long. The other logs should be 16, 24, 32, and 40 inches long.

Check $8 + 16 + 24 + 32 + 40 = 120$

Questions

Covering the Reading

1. What is the main difference between a sequence and a series?
A series is an indicated sum of terms of a sequence.

In 2–5, evaluate.

2. $\sum_{n=1}^{5} (2n + 1)$ **35**

3. $\sum_{n=1}^{5} 2n + 1$ **31**

4. $\sum_{n=1}^{5} (2^n + 1)$ **67**

5. $\sum_{n=1}^{5} 2^n + 1$ **63**

6. *Multiple choice.* Refer to the child's pattern of blocks at the start of the lesson. Which expression represents the number of blue, green, and red blocks in the 6×6 square?

(a) $\sum_{n=1}^{3} (2n - 1)$

(b) $\sum_{n=1}^{3} (4n - 1)$

(c) $\sum_{n=1}^{3} (4n - 3)$

(d) $\sum_{n=0}^{3} (2n + 1)$ **(c)**

7. Find the sum of the first one thousand positive integers.   **500,500**

8. A woman borrowed $6000 for 5 years. Her monthly payments were $145, $144.25, $143.50, $142.75, …, $100.75.
   a. How much did she pay over the life of the loan?   **$7372.50**
   b. How much interest did she pay on this loan?   **$1372.50**

9. Refer to Example 3. How far will the marathoner have run after six weeks? **745.5 km**

10. Refer to Example 4. How should a 12-foot log be cut to get five logs whose lengths form an arithmetic sequence with constant difference equal to 8 in.? **12.8 in., 20.8 in., 28.8 in., 36.8 in., 44.8 in.**

Applying the Mathematics

In 11 and 12, evaluate.

11. $\sum_{p=1}^{7} (-1)^p p$ **-4**

12. $\sum_{p=1}^{7} (p)^p$ **873,612**

In 13 and 14, suppose a display of cans in a supermarket is built with one can on top, two cans in the next row, and one more can in each succeeding row.

13. If there are 12 rows of cans, how many cans are in the display? **78**

14. If there are 171 cans in the display, how many cans are in the bottom row? **18**

15. As part of a sales promotion, a large flag is to be made by attaching strips of plastic ribbon three inches wide to a pole. If the first strip is six inches long and each successive strip is four inches longer than the previous one, how many strips can be cut from a roll of plastic ribbon 966 inches long? **21**

16. Write an expression for the sum of the first k terms of an arithmetic sequence with first term f and kth term m. $S_k = \frac{k}{2}(f + m)$

17. Use the formulas $S_n = \frac{n}{2}(a_1 + a_n)$ and $a_n = a_1 + (n-1)d$ to prove that the sum of the first n terms of an arithmetic series with first term a_1 and constant difference d can also be written $S_n = \frac{n}{2}[2a_1 + (n-1)d]$.

$S_n = \frac{n}{2}[a_1 + a_n] = \frac{n}{2}[a_1 + (a_1 + (n-1)d)] = \frac{n}{2}[2a_1 + (n-1)d]$

Review

In 18 and 19, use a computer program to decide whether the sequence is convergent or divergent. If it is convergent, give its limit. *(Lesson 8-2)*

18. $\frac{8}{3}, \frac{11}{5}, \frac{16}{35}, \ldots, \frac{n^2 + 7}{4n^2 - 1}, \ldots$ **convergent; $\frac{1}{4}$**

19. $\frac{8}{3}, \frac{11}{7}, \frac{16}{11}, \ldots, \frac{n^2 + 7}{4n - 1}, \ldots$ **divergent**

20. Evaluate $\lim_{n \to \infty} \frac{7n + 4}{11n - 4}$. *(Lesson 8-2)* **$\frac{7}{11}$**

474

21. Let $t_n = \cos\left(\dfrac{\pi}{6}n\right)$.

    a. Write exact values for the first five terms of the sequence generated by this formula.   **$\sqrt{3}/2, 1/2, 0, -1/2, -\sqrt{3}/2$**

    b. State whether the sequence is arithmetic, geometric, or neither.

    c. Does $\lim\limits_{n\to\infty} t_n$ exist? If so, what is it? *(Lessons 5-5, 8-1, 8-2)*

      b) neither; c) does not exist

22. Life scientists sometimes test animals' responses to doses of drugs or nutrients in amounts that form a geometric sequence. To test the content of vitamin A in carrots, pieces are fed to rats. Six dosage levels are arranged in a geometric sequence. If the lowest dose is 20 mg and the highest is 200 mg, find the other four doses. *(Lesson 8-1)*   **See below.**

23. According to *The 1987 Information Please Almanac*, the population of Mexico was about 83.5 million in 1986, and the annual growth rate was 2.5%.

    a. Use this information to write a recursive formula for the sequence of annual Mexican populations.   **See below.**

    b. Estimate the population of Mexico in the year 2000.   **$\approx$ 118 million**

    c. Give two reasons why the annual growth rate may *not*, in fact, remain constant. *(Lesson 8-1)*   **If any of the birth, death, immigration, or emigration rates change, the growth rate may change.**

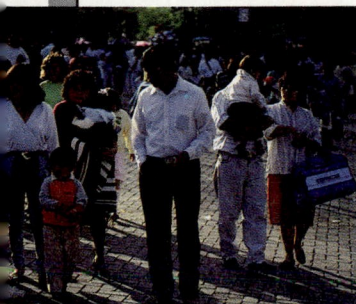

Chapultepec Park, Mexico City

24. Solve $3^x = 120$. *(Lesson 4-8)*

    $\dfrac{\log 120}{\log 3} \approx 4.358$

Exploration

25. Consider the arithmetic sequence of positive odd integers,
$$1, 3, 5, 7, \ldots ,$$
and the new sequence formed by adding consecutive terms of the sequence.

$$S_1 = \sum_{n=1}^{1} a_n = 1$$

$$S_2 = \sum_{n=1}^{2} a_n = 1 + 3 = 4$$

$$S_3 = \sum_{n=1}^{3} a_n = 1 + 3 + 5 = 9$$

$$\vdots$$

    a. Determine S_4, S_5, and S_6.   **$S_4 = 16, S_5 = 25, S_6 = 36$**

    b. Based on the results in **a**, write a rule for the sum of the first n odd integers, S_n, different from the two general formulas given in the lesson for any arithmetic series.   **$S_n = n^2$**

    c. Show algebraically that your rule in **b** can be derived from the general formula $S_n = \dfrac{n}{2}(a_1 + a_n)$.   **Starting with $S_n = \dfrac{n}{2}(a_1 + a_n)$, substitute**

      $a_1 = 1$ and $a_n = 2n - 1$. So $S_n = \dfrac{n}{2}(1 + 2n - 1) = \dfrac{n}{2}(2n) = n^2$.

22) $20(10)^{1/5} \approx 32$ **mg,** $20(10)^{2/5} \approx 50$ **mg,**
$20(10)^{3/5} \approx 80$ mg, $20(10)^{4/5} \approx 126$ mg

23a) $\begin{cases} P_{1986} = 83{,}500{,}000 \\ P_{n+1} = 1.025\, P_n \text{ for } n \geq 1986 \end{cases}$

MORE PRACTICE
For more questions on SPUR Objectives, use *Lesson Master 8-3*, shown below.

EVALUATION
A quiz covering Lessons 8-1 through 8-3 is provided in the Teacher's Resource File.

NAME _____

LESSON **MASTER 8-3**
QUESTIONS ON **SPUR** OBJECTIVES

■**SKILLS** *Objective C (See pages 519–522 for objectives.)*
In 1–8, evaluate.

1. $\displaystyle\sum_{n=1}^{5}(3n+2)$ **55**

2. $\displaystyle\sum_{n=1}^{5} 3n + 2$ **47**

3. $\displaystyle\sum_{i=1}^{4}(7-3i)$ **-2**

4. $\displaystyle\sum_{k=1}^{4}(10-7k)$ **-30**

5. $\displaystyle\sum_{n=1}^{5} 3^n + 1$ **364**

6. $\displaystyle\sum_{n=1}^{5}(3^n+1)$ **368**

7. $\displaystyle\sum_{n=1}^{200}(n+1)$ **20,100**

8. $\displaystyle\sum_{n=1}^{300} 2n$ **90,300**

■**USES** *Objective I*

9. Bobbi borrowed some money from her mother. She returned the money in payments of $10.80, $10.40, $10.00, and so on. Her last payment was $1.20. If she was charged no interest, what was the amount she had borrowed from her mother?   **$150.00**

10. Walt collected 2 aluminum cans the first day, 5 aluminum cans the second day, 8 aluminum cans the third day, and so on, until he had collected a total of 1365 aluminum cans. How many days did it take him to collect the aluminum cans?   **30**

C Find the sum of the terms of a sequence.
I Use series to solve problems.

This lesson begins with a classic story that has a geometric sequence embedded in it. The total number of grains of wheat is astonishingly huge, yet students may not be able to appreciate its magnitude. Therefore, you might wish to discuss **Question 24** during class.

You might compare the opening geometric series with an arithmetic series. Suppose, instead, that 1 grain of wheat was put on the first square, 2 grains on the second, 3 on the third, and so on, with n grains on the nth square. Then there would be a total of only 2080 grains on the entire chessboard!

Error Analysis You may find that even students who have had little trouble with summation notation get a bit confused when the index is in the exponent, as in

$S_n = \sum_{i=1}^{n} a_1 r^{i-1}$. For this

reason, **Example 1** and **Questions 5 and 6** are important. For additional practice with the notation, you might want to have students write out the 8 terms of the sum in **Example 2.**

8-4

Geometric Series

There is a story about the origin of the game of chess. The king of Persia, after learning how to play, offered the game's inventor a reward. Clearing a chessboard, the inventor asked for one single grain of wheat on the first square, twice that on the second square, twice that again on the third, twice that again on the fourth, and so on for the whole board. The king, ready to give jewelry, gold, and other riches, thought this was a modest request, easily granted. Was it?

The terms of the request are $2^0, 2^1, 2^2, \ldots, 2^{63}$. They form a geometric sequence with $g_1 = 1$ and $r = 2$, and so the nth term is $g_n = 2^{n-1}$. Representing the total award for the entire chessboard by S_{64},

$$S_{64} = \sum_{n=1}^{64} 2^{n-1} = 1 + 2 + 4 + 8 + \ldots + 2^{63}.$$

S_{64} is the value of a *finite geometric series*, analogous to the finite arithmetic series of the previous lesson.

To evaluate S_{64} without adding every term, you can use an approach similar to the one used for arithmetic series. First, write the series and then multiply each side of the equation by the constant ratio, 2.

$$S_{64} = 1 + 2 + 4 + \ldots + 2^{62} + 2^{63}$$
$$2S_{64} = \quad\ \ 2 + 4 + \ldots + 2^{62} + 2^{63} + 2^{64}$$

Then, subtract the second equation from the first, term by term.

$$S_{64} - 2S_{64} = 1 + (2 - 2) + (4 - 4) + \ldots + (2^{62} - 2^{62}) + (2^{63} - 2^{63}) - 2^{64}$$

Simplify.

$$-S_{64} = 1 - 2^{64}$$
So $\qquad S_{64} = 2^{64} - 1.$

Thus the total number of grains of wheat on the chessboard would be $2^{64} - 1$, or about 1.85×10^{19}. If you assume that each grain can be approximated by a 4 mm × 1 mm × 1 mm rectangular box, the total volume of wheat would be about 74 cubic kilometers, many many times the amount of wheat in the world! It would have been a very large reward!

In general, a **geometric series** is an indicated sum of terms of a geometric sequence. If the first n terms of the sequence with first term g_1 and constant ratio r are added, the sum S_n is called the ***n*th partial sum** and is written

$$S_n = g_1 + g_1 r + g_1 r^2 + \ldots + g_1 r^{n-1}$$

$$= \sum_{i=1}^{n} g_1 r^{i-1}.$$

476

The procedure used to find S_{64} can be generalized.
$$S_n = g_1 + g_1 r + g_1 r^2 + \ldots + g_1 r^{n-1}$$

Multiply S_n by r.
$$r S_n = \quad g_1 r + g_1 r^2 + \ldots + g_1 r^{n-1} + g_1 r^n$$

Subtract the preceding equations.
$$S_n - r S_n = g_1 - g_1 r^n$$

Factor each side.
$$(1 - r) S_n = g_1 (1 - r^n)$$

Divide each side by $1 - r$.
$$S_n = \frac{g_1(1 - r^n)}{1 - r}$$

This proves the following theorem.

Theorem

The sum $S_n = g_1 + g_2 + \ldots + g_n$ of a geometric series with first term g_1 and constant ratio r is given by
$$S_n = \frac{g_1(1 - r^n)}{1 - r}.$$

Example 1 Evaluate $\displaystyle\sum_{i=1}^{7} 18\left(\frac{1}{3}\right)^{i-1}$.

Solution This is a geometric series with $g_1 = 18$, $r = \frac{1}{3}$, and $n = 7$. Use the theorem above.
$$S_n = \frac{g_1(1 - r^n)}{1 - r}$$

Substituting, $S_7 = \dfrac{18\left(1 - \left(\frac{1}{3}\right)^7\right)}{1 - \frac{1}{3}} = \dfrac{18\left(1 - \frac{1}{2187}\right)}{\frac{2}{3}}$

$$= 27\left(1 - \frac{1}{2187}\right)$$
$$= 27 - \frac{1}{81}$$
$$= 26\frac{80}{81}.$$

Check Evaluate each term and add them.
$$18 + 6 + 2 + \frac{2}{3} + \frac{2}{9} + \frac{2}{27} + \frac{2}{81} = 26\frac{80}{81}$$

If you multiply the numerator and denominator of the formula for the sum of a geometric series by -1, you get an alternative formula useful for situations where $r > 1$.
$$S_n = \frac{g_1(r^n - 1)}{(r - 1)}$$

Because students know how to solve exponential equations, the computer program on page 478 is not necessary for solving a problem like **Example 3.** Still, the list of values of the geometric series shows the growth in these values better than a simple computation of the answer.

ADDITIONAL EXAMPLES

1. Evaluate $\displaystyle\sum_{i=1}^{6} 10(0.75)^{i-1}$.

32.88085938

2. In a set of 10 nested dolls (each doll fits inside another doll), each doll is $\frac{5}{6}$ the height of the next taller one. If the height of the first doll is 15 cm, what is the total height of the dolls?

$90\left(1 - \left(\frac{5}{6}\right)^{10}\right) \approx 75$ cm

3. Suppose you have two children who marry and each of them has two children. Each of these offspring has two children, and so on. If all these progeny marry but none marry each other, and all have two children, in how many generations will you have a thousand descendants?

9

Example 2

The set of a music show includes a backdrop in a design of nested triangles; the first four are shown at the left. The middle triangle is the first built, having a perimeter of 0.55 meters. Each successive triangle has perimeter twice the previous one. What is the sum of the perimeters of the first eight nested triangles?

Solution   The answer is the value of the finite geometric series where $g_1 = 0.55$, $r = 2$, and $n = 8$.

$$S_n = \sum_{n=1}^{8} 0.55(2)^{n-1}$$

Use the alternative formula for S_n.

$$S_n = \frac{g_1(r^n - 1)}{r - 1}$$
$$= \frac{0.55(2^8 - 1)}{2 - 1}$$
$$= 0.55(256 - 1)$$
$$= 140.25$$

The sum of the perimeters is 140.25 meters.

A computer can be used to evaluate a geometric series. For example, consider the series which gives the total number of natural ancestors that you have: 2 parents, 4 grandparents, 8 great-grandparents, and so on. Assuming that no one appears twice in your ancestor tree, in the past n generations you have S_n ancestors, where $S_n = 2 + 4 + 8 + \ldots + 2^n$.

Here is a computer program which finds the sum by generating each term (line 40) and keeping a running total (line 50).

```
10   SUM = 0
20   INPUT "HOW MANY GENERATIONS"; N
30   FOR I = 1 TO N
40     TERM = 2^I
50     SUM = SUM + TERM
60     PRINT I, SUM
70   NEXT I
80   END
```

Running the program with N = 5 gives the following output.

```
1     2
2     6
3     14
4     30
5     62
```

478

Example 3 How many generations must you go back before you have a million ancestors, assuming that no one appears twice?

Solution Solve $$1,000,000 = \frac{2(2^n - 1)}{2 - 1}.$$

$$1,000,000 = 2(2^n - 1)$$
$$500,000 = 2^n - 1$$
$$2^n = 500,001$$
$$n = \frac{\ln 500,001}{\ln 2}$$
$$\approx 18.9$$

So, assuming no one appears twice on your ancestor tree, if you go back 19 generations, you will have had a million ancestors. (It is very likely that many people have appeared twice on your tree.)

Check Running the preceding computer program with $n = 20$ gives the following.

⋮	⋮
17	262142
18	524286
19	1048574
20	2097150

So it checks.

Questions

In 1 and 2, refer to the chessboard story at the beginning of the lesson.

1. Explain why the number of grains on the squares of the chessboard forms a geometric sequence. **The ratio between consecutive terms is constant (in this case 2).**

2. *True* or *false*. The total number of grains on the first 63 squares of the chessboard is about half that on all 64 squares. **True**

3. *Multiple choice.* The expression $\sum_{i=1}^{n} a_1 r^{i-1}$ equals

(a) $\dfrac{a_1(1 - r)^n}{1 - r}$ (b) $\dfrac{a_1(1 - r^n)}{1 - r}$ (c) $\dfrac{a_1(1 - r)^n}{r - 1}$ (d) $\dfrac{a_1(1 - r^n)}{r - 1}$.

(b)

In 4–6, find the sum.

4. the first six terms of the geometric sequence with first term -2 and constant ratio 3 **-728**

5. $\displaystyle\sum_{n=1}^{8} 3(0.5)^{n-1}$ **≈ 5.98**

6. $\displaystyle\sum_{i=1}^{20} 10(-2)^{i+1}$ **-13,981,000**

LESSON 8-4 Geometric Series **479**

7. As first prize winner in a lottery, you are offered a million dollars in cash or a prize consisting of one cent on July 1, two cents on July 2, four cents on July 3, and so on, with the amount doubling each day until the end of July. Which prize is more valuable?
   the latter prize (it is worth $21,474,836.47)

8. Refer to Example 2.
   a. What is the sum of the perimeters of the first 7 triangles?   **69.85 m**
   b. Use your answer in part **a** to give the perimeter of the eighth triangle.
   70.4 m

9. Consider the geometric sequence 32, 24, 18, … .
   a. Use logarithms to find how many terms must be added to give a sum of more than 127.   **17**
   b. Check with a computer or a calculator. $S_{17} = 127.03783$

Applying the Mathematics

10. Consider the expression $p + mp + m^2p + \ldots + m^{k-1}p$.
    a. Rewrite the expression using Σ-notation.
    b. Write the sum as a fraction.
    a) $\displaystyle\sum_{i=1}^{k} pm^{i-1}$; b) $\dfrac{p(1-m^k)}{1-m}$

In 11 and 12, find the sum of the geometric series.

11. $3200 + 800 + 200 + \ldots + 3.125$ 12. $8 + {}^{-}12 + \ldots + {}^{-}60.75$
    4265.625   **${}^{-}33.25$**

13. a. Give an example of a geometric sequence with common ratio one.
    b. Explain why the formula for finding the sum of the first n terms of this sequence is not appropriate.
    c. Give the sum of the first n terms of your answer to part **a**.   **n; ng**
    a) See margin. b) The denominator of the sum would be zero.

14. Find an explicit formula for the nth term of a geometric sequence with common ratio 1.5 for which the sum of the first four terms is 65.
    $g_n = 8(1.5)^{n-1}$

15. A superball bounces to three-quarters of the height from which it falls. What is the total distance traversed by a superball dropped vertically from 10 feet above the ground to the moment when it hits the ground for the eighth time? (The actual bounce would be closer to vertical. The bounces are shown spread out to help you visualize the problem.)   **≈ 62 ft**

16. Here is a computer program to evaluate a geometric series.
    a. If a number greater than or equal to 3 is input, give the first 3 lines of output. (You do *not* need to run the program.)
    b. If N = 15 is input, what is the output?
    a,b) See margin.

```
10  SUM = 0
20  INPUT "NO. OF TERMS"; N
30  FOR I = 1 TO N
40    TERM = 4 * 3 ^ I
50    SUM = SUM + TERM
60    PRINT I, SUM
70  NEXT I
80  END
```

480

17. a. Use the computer program applied in Example 3 to estimate the number of ancestors you had during the last 50 generations.
　b. Explain why your answer in part **a** is so far off.
　a) ≈ 2.25×10^{15} people; b) Not that many people have existed; the assumption that no one appears twice is incorrect.

Review

18. A new graduate accepts a job as a data processing clerk at a starting salary of $13,500 per annum, with an annual increment of $460. If he stays in the job, how much will he earn in total after ten years? *(Lesson 8-3)*
　$155,700

19. Let $t_n = \dfrac{2^n}{2^n - 1}$.　a) 2, $\dfrac{4}{3}$, $\dfrac{8}{7}$, $\dfrac{16}{15}$, $\dfrac{32}{31}$; b) Yes; the limit is one.
　a. Write the first five terms of the sequence generated by t_n.
　b. Does $\lim\limits_{n \to \infty} t_n$ exist? If so, what is it? *(Lessons 8-1, 8-2)*

Ireland

20. Some countries have flags consisting of three stripes of different colors. The flags of Ireland and the Ivory Coast use the same colors, but in a different order. How many different flags can be designed consisting of three congruent stripes, either horizontal or vertical, where one is orange, one is white, and one is green? (Ignore the size of the flag and other markings.) *(Lesson 7-6)*　**12**

Ivory Coast

21. Consider tossing a fair coin four times.
　a. How many elements are there in the sample space?　**16**
　b. *True* or *false*. $P(2 \text{ heads}) = \dfrac{1}{2}$. Explain your answer. *(Lesson 7-4)*
　See below.

22. Evaluate without a calculator: $\dfrac{373!}{372!}$. *(Lesson 7-3)*　**373**

23. Consider $f(x) = \dfrac{1}{x + 8} + 5$.
　a. What transformation maps the graph of $y = \dfrac{1}{x}$ to the graph of $y = f(x)$?
　b. Give equations for the asymptotes of the graph of $y = f(x)$.
　(Lesson 3-3)　a) $(x, y) \to (x - 8, y + 5)$; b) $x = -8, y = 5$

Exploration

24. Refer to the chessboard story at the start of this lesson.
　a. Express the volume for a single grain of wheat in terms of cubic kilometers, and explain how the volume 74 km³ was obtained.
　b. If 74 cubic kilometers of wheat could be spread evenly over the area of the 50 states, how deep would the wheat be?
　a) The volume of a single grain is about 4 mm × 1 mm × 1 mm; in terms of km³ that is $(4 \times 10^{-6})(10^{-6})(10^{-6}) = 4 \times 10^{-18}$ km³. The volume of the grain is $(1.85 \times 10^{19}\text{ grains})(4 \times 10^{-18}\text{ km}^3/\text{grain}) = 7.4 \times 10 = 74$ km³.

　b) The area of the United States is about 9.52×10^6 km², so the wheat would be about 7.8 mm deep.
　21 b) False; of the 16 equally likely elements in the sample space, 6 have exactly 2 heads. So $P(2 \text{ heads}) = \dfrac{6}{16} \neq \dfrac{1}{6}$.

FOLLOW-UP

MORE PRACTICE
For more questions on SPUR Objectives, use *Lesson Master 8-4*, shown below.

EXTENSION
Small Group Work　The following problem introduces a geometric sequence requiring difficult computation. You might have small groups work on the solution. The earth's population, which passed 5 billion in 1987, has been growing at a rate at which it doubles about every 35 years. If this continued, how long would it be before there was a person for every square foot on the planet? For simplicity of calculation, you might suggest that students assume the population will be 6 billion in the year 2000 and that the population will continue to double every 40 years. Then assume the earth is a sphere with a radius of about 4000 miles. (about the year 2800)

Infinite Series

Fleance, a jumping flea, is at the center of a circular ring of radius 1 yd. Suppose the flea jumps along a radius toward the ring, but that each jump is half the previous jump (as Fleance gets tired). If the first jump is 18 in. long, will it ever reach the ring? If so, how many jumps will it take to reach the ring?

The jumps form a geometric sequence

$$\frac{1}{2}, \frac{1}{4}, \frac{1}{8}, \frac{1}{16}, \ldots.$$

The total distance jumped equals $\frac{1}{2} + \frac{1}{4} + \frac{1}{8} + \ldots$.

The table below shows the total distances S_n after various numbers of jumps.

Number of jumps	Total Distance Jumped	
1	$S_1 = \frac{1}{2}$	$= \frac{1}{2}$
2	$S_2 = \frac{1}{2} + \frac{1}{4}$	$= \frac{3}{4}$
3	$S_3 = \frac{1}{2} + \frac{1}{4} + \frac{1}{8}$	$= \frac{7}{8}$
4	$S_4 = \frac{1}{2} + \frac{1}{4} + \frac{1}{8} + \frac{1}{16}$	$= \frac{15}{16}$
.	.	
.	.	
n	$S_n = \frac{1}{2} + \frac{1}{4} + \ldots + \frac{1}{2^n}$	

The total distances $S_1, S_2, S_3, S_4, \ldots, S_n$ form a new sequence, $\frac{1}{2}, \frac{3}{4}, \frac{7}{8}, \frac{15}{16}, \ldots$, the **sequence of partial sums** of the infinite series $\frac{1}{2} + \frac{1}{4} + \frac{1}{8} + \ldots$. That is, the nth term of this new sequence is the sum of the first n terms of the original sequence. In the case of the flea jumps, the original sequence is geometric with first term $\frac{1}{2}$ and common ratio $\frac{1}{2}$. So you can find S_n using the theorem proved in Lesson 8-4 for the sum of a geometric series.

$$S_n = \frac{\frac{1}{2}\left(1 - \left(\frac{1}{2}\right)^n\right)}{1 - \frac{1}{2}}$$

$$= 1 - \left(\frac{1}{2}\right)^n$$

To find out the number of jumps needed to reach the edge of the ring, you need to know the smallest positive integer n such that $S_n \geq 1$.

482

Substitute $1 - \left(\frac{1}{2}\right)^n$ for S_n and solve.

If
$$1 - \left(\frac{1}{2}\right)^n \geq 1,$$

then
$$-\left(\frac{1}{2}\right)^n \geq 0.$$

Notice that the left side of the inequality is always negative. So the sentence $S_n \geq 1$ has no real solution. This means that Fleance never reaches the ring.

You can use a computer to study $\lim\limits_{n\to\infty} S_n$. The program below prints out values of the partial sums of the series modeling Fleance's progress.

```
10  PRINT "NO. OF JUMPS","DIST. JUMPED"
20  N = 1
30  S = 1 - (0.5)^N
40  PRINT N,S
50  N = N + 1
60  GOTO 30
```

Here are some lines of output from this program:

NO. OF JUMPS	DIST. JUMPED
1	0.5
2	0.75
3	0.875
⋮	⋮
6	0.984375
7	0.9921875
8	0.9960938
⋮	⋮
14	0.999939
15	0.9999695
⋮	⋮
23	0.9999999
24	0.9999999
25	1
26	1

From this output you can see that although the flea never actually gets to the ring, it gets close enough for all practical purposes. Note that after 7 jumps, it is within 0.01 yd of the ring; after 14 jumps it is within 0.0001 yd, and after 23 jumps it is within 0.0000001 yd of the ring. After 25 or more jumps, it is even closer than this, and the computer rounds off the total to 1, correct to the 7-decimal place accuracy of the output. Thus the sequence

$$\frac{1}{2}, \frac{3}{4}, \frac{7}{8}, \frac{15}{16}, \dots$$

of partial sums of the infinite series $\frac{1}{2} + \frac{1}{4} + \frac{1}{8} + \frac{1}{16} + \dots$ converges to 1.

That is,
$$\lim_{n\to\infty} \sum_{i=1}^{n} \frac{1}{2^i} = 1.$$

Thus, the limit of the sequence of total distances traveled by Fleance in a horizontal direction is 1 yd.

The computer is a brilliant tool for helping to decide whether infinite series seem to diverge or not, and to what limit they converge if that is the case. It helps a great deal to work through an example or two, as is done on page 483 and in **Questions 9 and 23.** We have provided a teaching aid with the data on the jumping flea, so that you can compare values found from the formula with values found by computer.

Error Analysis   Writing out the partial sums will be helpful for students who may still be confusing sequences and series. You can use **Question 3.**

Note in **Example 1** that the terms of the sequence do not approach 0, so the series will diverge. However, the converse of this is false, as the harmonic series discussed in the last paragraph of the lesson shows. Since this is not at all obvious to students, it is important that **Question 9** be discussed.

At this point in their mathematical careers, since students do not have calculus, the computer is indispensable for analyzing infinite series which are not geometric. The mathematics needed to analyze even the simple series in **Example 3** is very sophisticated. In fact, it took the mathematician Euler to prove that $\sum\limits_{i=1}^{\infty} \frac{1}{n^2} = \frac{\pi^2}{6}$. Even after 4100 terms, the estimate may not be close enough to convince students of this, because to six decimal places, $\frac{\pi^2}{6} \approx 1.644934$.

Using summation notation, an infinite series is expressed as

$$\sum_{i=1}^{\infty} a_i = a_1 + a_2 + a_3 + \ldots .$$

As you saw in the flea example, an important question about infinite series is whether the sum of the series exists. And if the sum does exist, what is it? These questions can be answered by looking at the sequence of partial sums of the series,

$$S_1, S_2, S_3, \ldots, S_n.$$

If this sequence converges, then its limit, written S_∞, is called *the sum of the series*.

Definition

The **sum S_∞ of an infinite series** $\sum\limits_{i=1}^{\infty} a_i$ is the limit of the sequence of partial sums S_n of the series, provided the limit exists and is finite. In symbols,

if $\qquad S_n = \sum\limits_{i=1}^{n} a_i,$

then $\qquad S_\infty = \sum\limits_{i=1}^{\infty} a_i = \lim\limits_{n \to \infty} S_n = \lim\limits_{n \to \infty} \sum\limits_{i=1}^{n} a_i.$

If $\lim\limits_{n \to \infty} S_n$ exists, the series is called *convergent* and its sum is S_∞. If the limit does not exist, the series is *divergent*.

Example 1 Consider the infinite geometric series
$$\frac{1}{2} + \frac{3}{2} + \frac{9}{2} + \frac{27}{2} + \frac{81}{2} + \ldots + \frac{3^{n-1}}{2} + \ldots .$$
Find the first five partial sums. Does the sequence of partial sums seem to converge? If so, what is the sum of the series?

Solution
$$S_1 = \frac{1}{2}$$

$$S_2 = \frac{1}{2} + \frac{3}{2} = 2$$

$$S_3 = \frac{1}{2} + \frac{3}{2} + \frac{9}{2} = \frac{13}{2} = 6.5$$

$$S_4 = \frac{1}{2} + \frac{3}{2} + \frac{9}{2} + \frac{27}{2} = \frac{40}{2} = 20$$

$$S_5 = \frac{1}{2} + \frac{3}{2} + \frac{9}{2} + \frac{27}{2} + \frac{81}{2} = \frac{121}{2} = 60.5$$

The sequence of partial sums appears to increase without a limit. This infinite series seems to diverge.

484

To examine infinite geometric series in general, consider the geometric sequence,

$$g_1, g_1r, g_1r^2, \ldots, g_1r^{n-1}, \ldots .$$

For large values of n, $|g_1r^{n-1}|$ is large if $|r| > 1$, and $|g_1r^{n-1}|$ is small if $|r| < 1$. This suggests that the convergence or divergence of the geometric series depends on the magnitude of the common ratio r. The sum of the first n terms of the geometric sequence with first term g_1 and common ratio r is

$$S_n = \frac{g_1(1 - r^n)}{1 - r} ,$$

or

$$S_n = \frac{g_1 - g_1r^n}{1 - r} .$$

For $|r| < 1$, the term g_1r^n is very close to zero for very large values of n. In fact, it can be proved that

$$\lim_{n \to \infty} g_1r^n = 0 \text{ for } |r| < 1.$$

So

$$\lim_{n \to \infty} S_n = \lim_{n \to \infty} \left(\frac{g_1 - g_1r^n}{1 - r} \right)$$

$$= \frac{g_1 - 0}{1 - r} ;$$

and

$$S_\infty = \frac{g_1}{1 - r} .$$

When $|r| > 1$, the term g_1r^n is farther from zero for every successive value of n, and the infinite series is divergent. For $|r| = 1$, the sequence is either $g_1, g_1, g_1, \ldots$, for which $S_n = ng_1$, or the sequence is $g_1, {}^-g_1, g_1, {}^-g_1, \ldots$, for which S_n is either 0 or g_1; in each case, the series is divergent. The following theorem summarizes these results.

Theorem

Consider the infinite geometric series
$g_1 + g_1r + g_1r^2 + \ldots + g_1r^{n-1} + \ldots$, with $g_1 \neq 0$.

a. If $|r| < 1$, the series converges and $S_\infty = \dfrac{g_1}{1 - r}$.

b. If $|r| \geq 1$, the series diverges.

Note that in the series modeling the jumps of a flea, $g_1 = \frac{1}{2}$ and $r = \frac{1}{2}$, so the series converges and $S_\infty = \dfrac{\frac{1}{2}}{1 - \frac{1}{2}} = 1$. In Example 1, $g_1 = \frac{1}{2}$ and $r = 3$, so the series diverges.

Example 2

Because of air resistance, the length of each swing of a certain pendulum is 95% of the length of the previous swing. If the first swing has length 40 cm, find the total length the pendulum will swing before coming to rest.

Solution Successive lengths of swing, measured in cm, form a geometric sequence.

$$40, 40(0.95), 40(0.95)^2, \ldots$$

You want the sum of the series

$$40 + 40(0.95) + 40(0.95)^2 + \ldots ,$$

where $g_1 = 40$ and $r = 0.95$. Use the theorem above. Because $|r| < 1$, the sum exists.

$$S_\infty = \frac{g_1}{1-r}$$
$$= \frac{40}{1 - 0.95}$$
$$= 800 \text{ cm}$$

The pendulum swings through 800 cm, or 8 m.

To analyze infinite nongeometric series thoroughly, more advanced mathematics is needed. However, properties of nongeometric series often can be studied with a computer or calculator.

Example 3

With advanced mathematics it can be proved that the infinite series

$$\frac{1}{1^2} + \frac{1}{2^2} + \frac{1}{3^2} + \frac{1}{4^2} + \ldots$$

is convergent. Use a computer to approximate the limit as accurately as possible.

Solution Use this computer program to print the sequence of partial sums.

```
10  LET SUM = 0
20  LET N = 1
30  SUM = SUM + 1/(N * N)
40  PRINT N, SUM
50  LET N = N + 1
60  GOTO 30
```

Run the program, and leave it running until at least ten consecutive terms are the same. Our computer gives a sum of 1.644725 for $N = 4100$.

The **harmonic series** $\frac{1}{1} + \frac{1}{2} + \frac{1}{3} + \ldots + \frac{1}{n} + \ldots$ is the sum of the reciprocals of the natural numbers. Although the partial sums of the harmonic series change slowly for large values of n, they do continue to grow as n gets larger and can be made larger than any specified number. That is, the series is divergent, and has no sum. Because the partial sums sometimes change slowly, you should be careful and not leap too quickly to the conclusion that a series has converged just from examining the partial sums.

486

In 1 and 2, refer to the series modeling the flea's jump.

1. How far has the flea gone after 6 jumps?   **63/64 yd**

2. How close to the edge of the ring is the flea after 100 jumps?
   $1/2^{100} \approx 8 \times 10^{-31}$ yd

3. Consider $\frac{1}{3} + \frac{1}{9} + \frac{1}{27} + \ldots$.   **a) $\frac{1}{3}, \frac{4}{9}, \frac{13}{27}$**
   a. Write the first three terms of the sequence of partial sums of this series.
   b. Write an explicit formula for the nth partial sum.   **$\frac{1 - (1/3)^n}{2}$**
   c. Does the series converge? If so, what is its sum?
      Yes; 1/2

In 4–6, state whether or not the geometric series is convergent. If the series is convergent, give its sum.

4. $3 - \frac{9}{2} + \frac{27}{4} - \frac{81}{8} + \ldots$
   not convergent

5. $5 + 4 + 3.2 + 2.56 + \ldots$
   convergent; 25

6. $16 - 12 + 9 - 6.75 + \ldots$
   convergent; 64/7 ≈ 9.14

7. Refer to Example 2. How far would the pendulum travel if the length of each swing is only 85% of the length of the previous swing?   **$266.\bar{6}$ cm**

8. Refer to the computer program in Example 3. Modify it to determine the sum of $\frac{1}{1^4} + \frac{1}{2^4} + \frac{1}{3^4} + \ldots$ to the nearest millionth.
   change line 30: 30 SUM = SUM + 1/(N*N*N*N); SUM ≈ 1.082323

9. a. Write the first five terms of the harmonic series.
   b. Use a computer to find how many terms of the harmonic series must be added before the sum exceeds each of the following
      i. 3   **11**   ii. 5   **83**
   c. *True* or *false*.   The harmonic series is divergent.   **True**
   a) $\frac{1}{1} + \frac{1}{2} + \frac{1}{3} + \frac{1}{4} + \frac{1}{5}$

10. Consider a pile driver driving a 3-meter pile into the ground. The first hit drives the pile in 50 cm, the second hit drives the pile in 40 cm further, and successive distances driven form a geometric sequence.
    a. How far will the pile be driven into the ground if the pile driver is allowed to run forever?   **250 cm**
    b. What percent of the total distance driven is reached after 20 hits?
       b) ≈ 98.8%

11. a. Under what condition(s) will the following geometric series converge? $b + \frac{b^2}{4} + \frac{b^3}{16} + \frac{b^4}{64} + \ldots$ **$|b| < 4$**

    b. Under the condition(s) in part a, what is $\lim_{n \to \infty} \sum_{i=1}^{n} \frac{b^n}{4^{n-1}}$?   **$\frac{4b}{4-b}$**

12. Find $\sum_{i=1}^{\infty} a_i$ for the sequence defined by
    $$\begin{cases} a_1 = 3; \\ a_n = (0.6)a_{n-1}, \ n \geq 2. \end{cases}$$ **7.5**

NAME _____

LESSON **MASTER   8-5**
QUESTIONS ON **SPUR** OBJECTIVES

■PROPERTIES   *Objective G (See pages 519–522 for objectives.)*
In 1–8, state whether the geometric series is convergent. If the series is convergent, give its sum.

1. $15 + 10 + \frac{20}{3} + \frac{40}{9} + \ldots$ **convergent; 45**

2. $\frac{1}{2} - \frac{5}{8} + \frac{25}{32} - \frac{125}{128} + \ldots$ **not convergent**

3. $\frac{1}{2} - \frac{2}{5} + \frac{8}{25} - \frac{32}{125} + \ldots$ **convergent; $\frac{5}{18}$**

4. $\frac{1}{6} + \frac{1}{3} + \frac{2}{3} + \ldots$ **not convergent**

5. $270 - 300 + \frac{1000}{3} - \ldots$ **not convergent**

6. $500 + 50 + 5 + \ldots$ **convergent; $\frac{5000}{9}$**

7. $\frac{5}{6} - \frac{5}{8} + \frac{15}{32} - \ldots$ **convergent; $\frac{10}{21}$**

8. $200 + 150 + \frac{225}{2} + \ldots$ **convergent; 800**

9. Use a computer to conjecture whether the following series is convergent. If it seems to be convergent, give what seems to be its limit.
   $\frac{1}{1^4} + \frac{1}{2^4} + \frac{1}{3^4} + \frac{1}{4^4} + \ldots$ **1.1175**

■USES   *Objective I*

10. A rubber ball dropped 12 feet rebounds on each bounce to $\frac{2}{3}$ of the distance from which it fell. How far will it travel before coming to rest?   **60 ft**

88   Functions, Statistics, and Trigonometry © Scott, Foresman and Company

13. A rubber ball dropped 60 feet rebounds on each bounce to $\frac{3}{4}$ of the distance from which it fell. How far will it travel before coming to rest? **420 ft**

14. Find the sum of the series
$$S = 1 + 3x + 5x^2 + 7x^3 + \dots , |x| < 1.$$
(Hint: Consider $S - Sx$). $\frac{1+x}{(1-x)^2}$

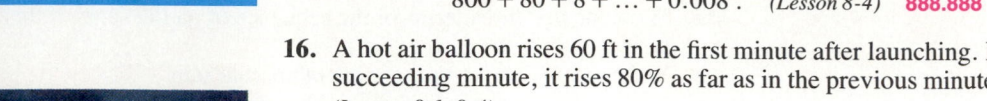

Review

15. Evaluate the sum of this geometric series:
$$800 + 80 + 8 + \dots + 0.008 .$$ *(Lesson 8-4)* **888.888**

16. A hot air balloon rises 60 ft in the first minute after launching. In each succeeding minute, it rises 80% as far as in the previous minute. *(Lessons 8-1, 8-4)*
    a. How far does it rise during the sixth minute after launching? $\approx$ **19.7 ft**
    b. How far will it have risen after 10 minutes? $\approx$ **268 ft**
    c. How long will it take to reach a height of 260 feet? $\approx$ **9 minutes**

17. Prove that the sum of the integral powers of 2 from 2^0 to 2^{n-1} is $2^n - 1$. *(Lesson 8-4)* $2^0 + 2^1 + \dots + 2^{n-1} = 1 + 2^1 + \dots + 2^{n-1} = \frac{2^n - 1}{2 - 1} = 2^n - 1$

In 18 and 19, evaluate. *(Lessons 8-3, 8-4)*

18. $\sum_{k=1}^{6} 3^k$ **1092**

19. $\sum_{k=1}^{6} (3 - k)$ **-3**

20. Find the first positive term of the arithmetic sequence $-101, -97, -93, \dots$. *(Lesson 8-3)* **3**

21. *Skill sequence.* *(Lesson 7-6)*
    a. How many permutations of the letters of the word FLEAS are there?
    b. How many of the permutations in part **a** begin with the letter A? **24**
    c. How many of them begin with A and end with S? **a) 120; c) 6**

22. *True* or *false.* For $n \geq 1$, $n! = n \cdot (n-1)!$ *(Previous course)* **True**

Exploration

23. Consider the following function defined by an infinite series.
$$s(x) = x - \frac{x^3}{3!} + \frac{x^5}{5!} - \dots + \frac{(-1)^{n-1} x^{2n-1}}{(2n-1)!} + \dots$$
    a. Approximate $s(1)$ and $s\left(\frac{\pi}{2}\right)$ using the first five terms in the definition. **0.841471; $s(\pi/2) \approx$ 1.0000035**
    b. Write a computer program which accepts as input x and n, and produces as output $s(x)$ approximated by the nth partial sum of the series. **See margin.**
    c. This series converges to $\sin(x)$ for any real value of x. Use the computer program you wrote in part **b** to determine how many terms of the series are needed to obtain an accuracy of two, three, and ten digits for $-\pi \leq x \leq \pi$. **For 2 digits, 6 terms are needed; for 3 digits, 7 terms are needed; and for 10 digits, 12 terms are needed.**

488

LESSON
8-6

Combinations

Sequences and series can often be used to solve counting problems. Consider the following situation. Six friends, June, Kevin, Lewis, Maria, Nelson, and Olivia, leave a restaurant, one at a time. Each person says good-bye to each of the others with a handshake. If the order in which the people part does not matter, how many partings take place?

To answer this question you might consider one person at a time. Suppose June starts by shaking hands with each of the others. So, June says good-bye to each of Kevin, Lewis, Maria, Nelson, and Olivia. Now Kevin and June have already said good-bye to each other, so Kevin shakes hands with each of Lewis, Maria, Nelson, and Olivia. The following pattern develops:

    J   parts with K, L, M, N, and O;
    K   parts with L, M, N, and O;
    L   parts with M, N, and O;
    M   parts with N and O;
and N   parts with O.

Thus, the number of partings that takes place is

$$5 + 4 + 3 + 2 + 1 = 15.$$

The number of partings can also be determined using the Multiplication Counting Principle. A parting occurs when two of the six people meet to hug or shake hands. There are six choices for the first person in the pair and five choices for the second, so there are $6 \cdot 5 = 30$ ordered pairs of people. This set of thirty pairs are the permutations of 2 objects chosen from 6. But the 30 pairs contain JK *and* KJ, NO *and* ON, and so on. Thus, the actual number of partings is only $\frac{1}{2}$ of 30, or 15.

The methods used to solve the preceding problem can be generalized.

■ ■ ■ ■ ■ ■ ■ ■ ■

Example 1   Suppose that n people meet at a wedding, and each person greets each of the others with a hug. How many hugs take place?

Solution 1   Consider the number of people each person hugs. No person greets himself or herself. So the first person hugs $n - 1$ people; the second person, having already greeted the first, hugs each of the $n - 2$ other people; the third person then hugs each of the $n - 3$ others, etc. So the number of hugs is $(n - 1) + (n - 2) + (n - 3) + ... + 1$, an arithmetic series with first term $n - 1$, last term 1, and $n - 1$ terms. So the sum is $\left(\frac{n-1}{2}\right)[(n-1) + 1] = \left(\frac{n-1}{2}\right)n$.

Solution 2   There are $n(n - 1)$ pairs of two people chosen from n people. In this set each hug is counted twice—once as AB, another time as BA. So the number of unordered pairs, corresponding to the number of hugs, is $\frac{n(n-1)}{2}$.

LESSON 8-6 Combinations   **489**

LESSON 8-6

RESOURCES
■ Lesson Master 8-6

OBJECTIVES

H Prove identities involving combinations.
J Use combinations to compute the number of ways of arranging or selecting objects.

TEACHING NOTES

Combinations and permutations are quite different. Emphasize: *Permutations are arrangements, and combinations are subsets.* In permutations, order matters; in combinations, order does not. But they are both names given to objects in counting problems, and since $_nP_r$ is used to derive the formula for $_nC_r$, it is difficult to cover combinations without bringing permutations into the discussion.

Alternate Approach   It is useful to compare similar combination and permutation problems, as is done on page 490. You might have students write all the permutations of the 5 letters A, B, C, D, and E taken 3 at a time. It takes a while but is worth the effort. Then group the 60 permutations by the letters in them: ABC, ABD, ABE, ACD, ACE, ADE, BCD, BCE, BDE, CDE. Since these are the combinations, it should then become clear that $_5P_3 = 3! \cdot _5C_3$.

As with permutations, calculating combinations involving large numbers may require students to do some arithmetic before they can use their calculators. Have students

The preceding problems involve *combinations*. A **combination** is a collection of objects in which the order of the objects does not matter.

▪ ▪ ▪ ▪ ▪ ▪ ▪ ▪ ■ ■

Example 2 *Multiple choice.* Which of the following situations involves combinations?
(a) Twenty students are semi-finalists for three scholarships—one for $1500, one for $1000, and one for $500. In how many different ways can the scholarships be awarded?
(b) Twenty students are semi-finalists for three $1000 scholarships. In how many different ways can the scholarships be awarded?

Solution For choice (a), the amounts awarded are different, so the order of the winning students matters. For choice (b), the scholarships are equal, so the order of the winning students does not matter. Choice (a) involves permutations; choice (b) involves combinations.

The answer to the question in situation (a) of Example 2 is the number of permutations of 20 things taken 3 at a time. In contrast, the answer to situation (b) is called the number of *combinations* of 20 people taken 3 at a time, written $_{20}C_3$. To calculate $_{20}C_3$, note that for any one of these combinations of 3 students, say A, B, and C, there are $3! = 6$ possible permutations:

$$\text{A B C,}$$
$$\text{A C B,}$$
$$\text{B A C,}$$
$$\text{B C A,}$$
$$\text{C A B,}$$
$$\text{and C B A.}$$

Thus $_{20}C_3 = 3! \, _{20}P_3$. So $_{20}C_3 = \frac{_{20}P_3}{3!} = \frac{6840}{6} = 1140$. So there are 1140 different ways to award three $1000 scholarships when choosing from 20 students.

In general, the **number of combinations of n things taken r at a time** is written $_nC_r$ or $\binom{n}{r}$. Some people read this as "n choose r." A formula for $_nC_r$ can be derived from the formula for $_nP_r$ by generalizing the previous paragraph. Each combination of r objects can be arranged in $r!$ ways. So $r! \, _nC_r = \, _nP_r$.

Thus $\qquad\qquad _nC_r = \, _nP_r \cdot \frac{1}{r!}$.

Substitute from the $_nP_r$ Calculation Theorem.

$$_nC_r = \frac{n!}{(n-r)!} \cdot \frac{1}{r!}$$

So $\qquad\qquad _nC_r = \frac{n!}{(n-r)!r!}$.

490

$_nC_r$ Calculation Theorem

For all whole numbers n and r, with $r \leq n$,

$$_nC_r = \frac{n!}{(n-r)!\, r!}.$$

Example 3 A menu at a Chinese restaurant contains 48 main dishes. A group of friends decides to order 6 different dishes. In how many different ways can they do this?

Solution There are 48 dishes, from which 6 are to be chosen. Since the order in which the dishes are chosen does not matter, you want $_{48}C_6$. Use the $_nC_r$ Calculation Theorem.

$$_{48}C_6 = \frac{48!}{42!\, 6!}$$

A calculator gives $\quad _{48}C_6 = 12{,}271{,}512.$

There are over 12 million different meals possible!

Sometimes the number resulting from computing a combination is too large for a calculator's memory. Or the display may only give you an approximate answer, because scientific notation has been used. In such cases, it may be necessary to do some calculation by hand first, as Example 4 shows.

Example 4 A football team with a roster of 60 players randomly selects 7 players from the roster for drug testing. What is the probability that its star quarterback and its star middle linebacker are both selected?

Solution There are $_{60}C_7$ combinations of 7 players that could be picked. To determine how many combinations include the two star players, note that each such combination includes 5 players from the remaining 58 on the roster. So there are $_{58}C_5$ combinations of 7 players from 60 that include the star quarterback and star middle linebacker. Thus the probability that these two stars are selected, assuming randomness, is

$$\frac{_{58}C_5}{_{60}C_7} = \frac{\dfrac{58!}{53!5!}}{\dfrac{60!}{53!7!}} = \frac{58!7!}{60!5!} = \frac{7 \cdot 6}{60 \cdot 59} = 0.012\ldots.$$

Questions

Covering the Reading

1. Suppose the 26 students in a class each shake hands once with each other.
   a. Use an arithmetic series to calculate the number of handshakes that take place.
   b. *True* or *false*.   The number of handshakes that takes place is $_{26}C_2$.
   a) $25 + 24 + 23 + \ldots + 1 = \frac{25}{2}(25 + 1) = 325$; b) True

LESSON 8-6 Combinations   **491**

ADDITIONAL EXAMPLES

1. Twenty distinct points are chosen on a circle.

a. How many segments are there with these points as endpoints?
190

b. How many triangles are there with these points as vertices?
1140

c. How many quadrilaterals are there with these points as vertices?
4845

2. Which is the permutation problem, which the combination problem?

a. You have six colors to choose from and you wish to choose three for a flag. How many choices of colors are possible?

b. A flag with stripes of three different colors can use any of six colors. How many flags are possible?
a is the combination problem; b is the permutation problem.

3. At a restaurant you can order pizza with any of 9 different toppings. How many different pizzas are possible with 3 of those toppings?
84

4. Five students from your class are randomly picked to be interviewed by a local newspaper reporter who is interested in how you are using calculators and computers in school. If your class contains 20 students, what is the probability that neither you nor your best friend in the class will be picked?
$\frac{_{18}C_5}{_{20}C_5} = \frac{21}{38} \approx .55$

You may wish to change the number 20 to the number of students in your class.

491

2. Match each item on the left with one on the right.
   a. combination **(ii)**   (i) an arrangement of objects in order
   b. permutation **(i)**   (ii) a selection of objects without regard to order

3. a. How many combinations of the letters UCSMP taken two at a time are possible? **10**
   b. List them all. **UC CS SM MP US CM SP UM CP UP**

4. For $r > 1$, which is larger, ${}_nC_r$ or ${}_nP_r$? Justify your answer.
   Sample: ${}_nP_r = \dfrac{n!}{(n-r)!} > \dfrac{n!}{(n-r)!\,r!} = {}_nC_r$

In 5–7, evaluate.

5. ${}_{18}C_3$ **816**   6. ${}_{14}C_1$ **14**   7. ${}_{103}C_0$ **1**

8. In how many ways can a person order 8 different dishes from a list of 48?
   377,348,994

9. Suppose 13 players from a track squad of 52 are chosen for drug testing.
   a. What is the number of combinations of players that could be chosen?
   b. How many of these combinations include the star male and female sprinters? **a) ${}_{52}C_{13} = 635,013,559,600$; b) ${}_{50}C_{11} = 37,353,738,800$**

Applying the Mathematics

10. A lottery that requires guessing 6 numbers out of 50, in any order, sells $1 tickets. If you buy 12,000,000 different tickets, are you sure to win? Justify your answer. **See margin.**

11. A poker deck has 52 different cards: 13 each of clubs, diamonds, hearts, and spades.
    a. How many different five-card poker hands are possible?
    b. How many of the hands in part **a** contain only spades?
    c. What is the probability of getting a poker hand that is all spades?
    a) ${}_{52}C_5 = 2,598,960$ hands; b) ${}_{13}C_5 = 1287$ hands; c) See margin.

In 12–14, evaluate the expression.

12. ${}_nC_n$ **1**   13. ${}_nC_1$ **n**   14. ${}_nC_0$ **1**

15. Six points are in a plane so that no three are collinear, as shown on the right.
    a. How many triangles can be formed having these points as vertices? **20 (two are shown)**
    b. Generalize your results in **a** to the case of n noncollinear points. ${}_nC_3 = \dfrac{n!}{3!(n-3)!}$

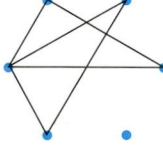

16. Each year a company holds a drawing in which 3 of its 100 employees get free vacations to Tahiti. What is the probability that the 3 employees chosen are the three bosses? $\dfrac{1}{{}_{100}C_3} = \dfrac{1}{161,700} \approx .000006$

17. a. Evaluate each expression.
    i. ${}_9C_4$ **126** **ii.** ${}_9C_5$ **126** **iii.** ${}_8C_2$ **28** **iv.** ${}_8C_6$ **28**
    b. Find a value of k other than 3 for which ${}_{10}C_3 = {}_{10}C_k$. **k = 7**
    c. Generalize the pattern observed in parts **a** and **b**. ${}_nC_r = {}_nC_{n-r}$
    d. Prove that your generalization is true.
    ${}_nC_r = \dfrac{n!}{(n-r)!\,r!} = \dfrac{n!}{(n-(n-r))!\,(n-r)!} = {}_nC_{n-r}$

492

18. In Example 3, the 48 dishes are equally divided into seafood, meat, and vegetable. How many different meals are possible if the friends want two dishes from each group? **1,728,000**

Review

19. a. Determine the sum of the infinite geometric series

$$1 - \frac{1}{4} + \frac{1}{16} - \frac{1}{64} + \ldots + \left(-\frac{1}{4}\right)^{n-1} + \ldots \quad \textbf{.8}$$

b. Based on the result in part **a**, determine the sum $\sum_{i=1}^{\infty} \left[b \left(-\frac{1}{4}\right)^{i-1} \right]$, where b is any real number. *(Lessons 8-2, 8-5)* **.8b**

20. a. Evaluate, giving your answer as a fraction.

$$\frac{3\left(1 - \left(\frac{1}{3}\right)^6\right)}{1 - \frac{1}{3}} \quad \frac{\textbf{364}}{\textbf{81}}$$

b. Write a geometric series having the value in part **a** as a sum. *(Lesson 8-4)* **3 + 1 + 1/3 + 1/9 + 1/27 + 1/81**

21. Suppose you deposit $50 one year on the 15th of January in an account which pays 5.75% annual interest compounded monthly. On the first of each month following you also deposit $50. If interest is paid on the 15th of the month for the money in the account on the previous day, how much will you have in the account on January 1 of the following year? *(Lesson 8-4)* **$668.77**

22. Solve $6 \cdot {}_nP_3 = {}_9P_4$. *(Lesson 7-5)* **9**

23. A test has 10 multiple-choice questions with five choices, 15 true-false questions, and 5 questions which require matching from a list of 6 choices. In how many ways can you answer:
a. if you must answer each question; $\mathbf{5^{10} \cdot 2^{15} \cdot 6! = 2.304 \times 10^{14}}$
b. if you must answer the matching questions, but you have the option of leaving some true-false or multiple choice questions blank?
(Lesson 7-3) $\mathbf{6^{10} \cdot 3^{15} \cdot 6! \approx 6.2469 \times 10^{17}}$

Exploration

24. In professional baseball, basketball, and ice hockey, the championship is determined by two teams playing a best-four-games-out-of-seven series. Call the two teams X and Y.
a. In how many different ways can the series occur if team X wins? For instance, two different 6-game series are XXYXYX and XYYXXX. (Hint: Determine the different number of 4-game series, 5-game series, 6-game series, and 7-game series.) **35**
b. There is a ${}_nC_r$ with n and r both less than 10 which equals the answer for part **a**. Find n and r. **n = 7, r = 4 or n = 7, r = 3**
c. Explain why the combination of part **b** solves the problem of part **a**.
${}_7C_4$ gives the number of ways X can win exactly 4 out of 7 games. A series of less than 7 games can be thought of as a 7-game series in which X loses all games after its fourth victory. Therefore ${}_7C_4$ counts all the possible series.

FOLLOW-UP

MORE PRACTICE
For more questions on SPUR Objectives, use *Lesson Master 8-6*, shown below.

PROJECTS
Project 6 on page 515 is related to the content of this lesson.

OBJECTIVE

L Locate numerical properties represented by the patterns in Pascal's Triangle.

TEACHING NOTES

Point out that Pascal's Triangle displays a set of numbers, just as a graph does. (For this reason, the objective relating to Pascal's Triangle is included under Representations in the Chapter Review.) In this lesson and the next students study the two most important (and quite different) meanings for the numbers in the triangle.

On page 494, rows 0–6 of Pascal's Triangle are given. In **Example 1,** row 7 is begun and it is completed in **Example 4.** Row 8 is found in **Question 7.** Rows 9, 10, and 11 are found in the Review Questions of the next lesson. Students should be able to generate the first ten rows quickly by hand. That is, they should know these numbers well enough not to have to painstakingly add to get the new rows. Therefore, you may wish to write the triangle by having students call out numbers one by one as quickly as they can until those rows are filled.

LESSON

8-7

Pascal's Triangle

For a given value of n, you can calculate ${}_nC_0, {}_nC_1, {}_nC_2, \ldots, {}_nC_n$, a total of $n + 1$ calculations. For instance, for $n = 4$, the 5 values are

$${}_4C_0 = 1 \qquad {}_4C_1 = 4 \qquad {}_4C_2 = 6 \qquad {}_4C_3 = 4 \qquad {}_4C_4 = 1.$$

When the values of ${}_nC_r$ are recorded systematically, a beautiful pattern emerges. Below, the values of ${}_nC_r$ with values of n and r from 0 to 6 are arranged in an array in the form of a right triangle. In the Western world this array is called **Pascal's Triangle**, after Blaise Pascal [1623–1662], a French mathematician and philosopher.

		0	1	2	3	4	5	6
	0	1						
	1	1	1					
	2	1	2	1				
n	3	1	3	3	1			
	4	1	4	6	4	1		
	5	1	5	10	10	5	1	
	6	1	6	15	20	15	6	1

(column header row labeled r)

The isosceles triangle form of Pascal's Triangle is shown below. Notice that the numbering starts with row 0 at the top.

```
row 0 →                    1
row 1 →                 1     1
row 2 →              1     2     1
row 3 →           1     3     3     1
row 4 →        1     4     6     4     1
row 5 →     1     5    10    10     5     1
row 6 →  1     6    15    20    15     6     1
         ⋮
```

Entries in Pascal's Triangle can be identified by row number (n) and column number (r), and so the terms can be considered as a two-dimensional sequence. The following definition provides an explicit formula for the terms in the nth row, where n can be any whole number.

Definition

Let n and r be nonnegative integers with $r \leq n$. The **$(r + 1)$st term in row n of Pascal's Triangle** is ${}_nC_r$.

494

Example 1 Find the first four terms in row 7 of Pascal's Triangle.

Solution By the definition, the first four terms of row 7 are
$$_7C_0, \, _7C_1, \, _7C_2, \text{ and } _7C_3.$$

These are $\quad \dfrac{7!}{7!0!}, \dfrac{7!}{6!1!}, \dfrac{7!}{5!2!}, \text{ and } \dfrac{7!}{4!3!},$

or 1, 7, 21, and 35.

Look closely at Pascal's Triangle. In its rows and columns and along the diagonals are many types of sequences, and many interesting patterns. Here are some properties which appear to be true for every row of Pascal's Triangle. The properties are described both in words and in symbols.

1. The first and last terms in each row are ones.
   That is, for each whole number n, $_nC_0 = _nC_n = 1$.
2. The second and next-to-last terms in the nth row equal n.
   For each whole number n, $_nC_1 = _nC_{n-1} = n$.
3. Each row is symmetric.
   For any pair of whole numbers n and r $(r \leq n)$, $_nC_r = _nC_{n-r}$.
4. The sum of the terms in row n is 2^n.
   For any pair of whole numbers
   $$n \text{ and } r \, (r \leq n), \, \sum_{r=0}^{n} {}_nC_r = 2^n.$$

You are asked to verify Property 1 in the Questions. Properties 2 and 3 are proved in Example 2 and 3, respectively. You are asked to prove Property 4 in the next lesson.

Example 2 Prove that the second and next-to-last term in the nth row of Pascal's triangle is n.

Solution For any row n, the second entry is $_nC_1$.
$$_nC_1 = \frac{n!}{(n-1)!1!} = \frac{n!}{(n-1)!} = \frac{n(n-1)!}{(n-1)!} = n$$
Similarly, the next-to-last entry in the nth row is $_nC_{n-1}$.
$$_nC_{n-1} = \frac{n!}{(n-(n-1))!(n-1)!} = \frac{n!}{(n-n+1)!(n-1)!} = \frac{n!}{1!(n-1)!} = n$$

So $_nC_1 = _nC_{n-1} = n$.

Example 3 Prove that for whole numbers n and r, where $r \leq n$, $_nC_r = _nC_{n-r}$.

Solution Use the $_nC_r$ Calculation Theorem.
$$_nC_r = \frac{n!}{(n-r)!r!}$$
Also, $_nC_{n-r} = \dfrac{n!}{[n-(n-r)]!(n-r)!} = \dfrac{n!}{[n-n+r]!(n-r)!} = \dfrac{n!}{r!(n-r)!}$

So for whole numbers n and r, where $r \leq n$, $_nC_r = _nC_{n-r}$.

In class discussion, you may wish to assess students' understanding of, and ability to use, the five properties of Pascal's Triangle discussed on pages 495–496 and their ability to relate these meaningfully to properties of combinations. (For example, saying that $_nC_0 = 1$ does not explain Property 1 meaningfully; the student should be able to explain why: there is only one subset of 0 elements in a set of n elements. That subset is the null set.)

Making Connections
Students should also be able to prove the properties of Pascal's Triangle. We provide proofs for some and ask students for proofs of others. Here is where the proofs are found in this lesson and the next.

Property	Proof Location
1	Question 5
2	Example 2
3	Example 3
4	Lesson 8-8 (Question 15)
5	Question 14

The proofs of Properties 1–3 and 5 use the definition of the triangle in terms of $_nC_r$. The proof for Property 4 utilizes the Binomial Theorem and is therefore delayed until Lesson 8-8. However, we ask students to verify particular instances of Property 4 in **Questions 7b and 8c.**

ADDITIONAL EXAMPLES
1. Use the formula for $_nC_r$ to find the first four terms in row 9 of Pascal's Triangle.
$_9C_0 = 1, \, _9C_1 = 9, \, _9C_2 = 36, \, _9C_3 = 84$

2. Give a polynomial formula for the third term in the nth row of Pascal's Triangle.
$$\frac{n(n-1)}{2} = \frac{1}{2}n^2 - \frac{1}{2}n$$

(More generally, the rth term in the nth row has a polynomial formula of degree $r-1$.)

495

3. Prove that for all positive integers n and r with $r \le n$,

$$_nC_{r+1} = \frac{n-r}{r+1} \, _nC_r$$

$$_nC_{r+1} = \frac{n!}{(r+1)!(n-r-1)!}$$

$$= \frac{n!(n-r)}{(r+1)r!(n-r)!} = \frac{n-r}{r+1} \, _nC_r$$

4. Here are the first 6 terms of row 11 of Pascal's Triangle: 1, 11, 55, 165, 330, 462. Use them to construct row 12.
1, 12, 66, 220, 495, 792, 924, 792, 495, 220, 66, 12, 1

Error Analysis Since the rows are symmetric, the last six terms of row 11 are the same numbers in reverse order. After noting that 1 is the first and last term of row 12, add consecutive terms to get the others.

NOTES ON QUESTIONS
Making Connections for Question 12: This is the lead-in to Lesson 8-8.

Question 13: In recent years new properties of Pascal's Triangle have been discovered (some by students). A teaching aid with the first 12 rows of Pascal's Triangle is provided.

Another property, not so easily seen but easily checked, is the following.

5. From row 1 (next to the top) on, the sum of two consecutive terms equals the term immediately below and between them. For example, the 4 and 6 in row 4, generated by $_4C_1$ and $_4C_2$, add to 10, which is the entry just below them generated by $_5C_2$. (This and another instance are highlighted by the small triangles in the following figure.) In general, for whole numbers n and r with $1 \le r \le n$, $_nC_{r-1} + _nC_r = _{n+1}C_r$.

Pascal's Triangle, with the shaded hexagons representing odd numbers.

The proof of property 5 is left as a question. Properties 1, 3, and 5 together provide a way to generate Pascal's Triangle recursively.

Example 4   Use Properties 1, 3, and 5 to construct row 7 in Pascal's Triangle.

Solution   Start by copying row 6. For row 7, Property 1 states that its first and last entries are 1. To find the second, third, and fourth entries in row 7, add consecutive pairs of terms in row 6 as described in Property 5; the results are $1 + 6 = 7$, $6 + 15 = 21$, and $15 + 20 = 35$. Use the symmetry property (or continue adding pairs of terms of row 6) to find the rest of the terms. Below is the result.

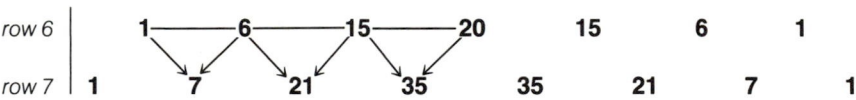

Check 1   Do the entries equal, in order, $_7C_0, _7C_1, _7C_2, \dots$? In Example 1, it was shown that $_7C_0 = 1$, $_7C_1 = 7$, $_7C_2 = 21$, and $_7C_3 = 35$.

Check 2   Use Property 4.
Does $1 + 7 + 21 + 35 + 35 + 21 + 7 + 1 = 128 = 2^7$? Yes.

Pascal's Triangle was known to mathematicians long before Pascal studied it. For example, the Persian mathematician and poet Omar Khayyam [c. 1048–1122] used it around the year 1100. It was also known by Chinese mathematicians by at least the fourteenth century. But it was Pascal who wrote extensively about the triangular arrays of numbers and its properties in a 1653 publication, *Treatise on the Arithmetic Triangle*. It is for this reason that the triangle has his name.

496

Questions

Covering the Reading

1. Refer to row 6 of Pascal's Triangle.
   a. How many terms are in the row?   **7**
   b. The third term is $_6C_r$. Find r.   **2**
   c. The middle term is $_6C_s$. Find s.   **4**
   d. What is the sum of the numbers in this row?   **64**
   e. Express your answer to part **d** as a power of 2.   **2^6**

In 2–4, match the English description of the pattern in Pascal's Triangle to the description using combination notation.
   (a) For each n, $_nC_1 = {_n}C_{n-1}$
   (b) For each n and each r $(r \le n)$, $_nC_r = {_n}C_{n-r}$
   (c) For each n, $_nC_0 = {_n}C_n = 1$

2. Each row in the isosceles triangle is symmetric to a vertical line.   **(b)**

3. The second and next-to-last entries in each row are equal.   **(a)**

4. The first and last entries in a row are 1.   **(c)**

5. Prove that for all whole numbers n, $_nC_0 = \dfrac{n!}{0!(n-0)!} = \dfrac{n!}{(n-n)!n!} = {_n}C_n$
   $_nC_0 = {_n}C_n = 1$.

6. Verify the property $_{n+1}C_r = {_n}C_{r-1} + {_n}C_r$ for $n = 7$, $r = 3$.   **See margin.**

7. a. Give the entries in row 8 of Pascal's Triangle.
   b. Check your answer to **a** by showing the entries add to 2^8.
      a,b) $1 + 8 + 28 + 56 + 70 + 56 + 28 + 8 + 1 = 256 = 2^8$

8. The first eight entries of row 14 of Pascal's Triangle are
   1, 14, 91, 364, 1001, 2002, 3003, 3432.
   a. How many other entries are there in row 14?   **7**
   b. List the remaining entries.   **3003, 2002, 1001, 364, 91, 14, 1**
   c. What is the sum of the entries in row 14?   **$2^{14} = 16,384$**
   d. List the entries in row 15.   **1, 15, 105, 455, 1365, 3003, 5005, 6435,
      6435, 5005, 3003, 1365, 455, 105, 15, 1**

9. *True* or *false*.   Blaise Pascal was the first person to study the triangle that now bears his name.   **False**

Applying the Mathematics

10. Find the first three numbers in the 100th row of Pascal's Triangle.
    1, 100, 4950

11. What are the last two terms in the row of Pascal's Triangle whose terms add to 2^{27}?   **27, 1**

12. a. Expand the following.
    i. $(x+y)^1$ **$x+y$**   ii. $(x+y)^2$ iii. $(x+y)^3$
    b. Relate the coefficients of the results in part **a** to Pascal's Triangle.
       ii) $x^2 + 2xy + y^2$; iii) $x^3 + 3x^2y + 3xy^2 + y^3$; b) See margin.

13. Find another pattern in Pascal's Triangle, not mentioned in this lesson.
    See margin.

14. Prove Property 5 in the lesson which states that any entry in the Pascal's Triangle that has two entries directly above it is the sum of those two entries. That is, show that $_{n+1}C_r = {_n}C_{r-1} + {_n}C_r$, where $1 \le r \le n$.
    See margin.

12.b. The coefficients of $(x + y)^n$ are the entries in the nth row of Pascal's Triangle.

13. sample: the "hockey stick" pattern. In any shape as shown below, with any number of terms in the "handle" but only one in the "blade," the sum of the terms in the handle is equal to the term in the blade. For example, $1 + 3 + 6 + 10 = 20$.

14. $_nC_{r-1} + {_n}C_r =$
$$\frac{n!}{(n-r+1)!(r-1)!} + \frac{n!}{(n-r)!r!}$$
$$= \frac{r \cdot n!}{(n-r+1)!r!} + \frac{n! \cdot (n-r+1)}{(n-r+1)!r!}$$
$$= \frac{n!(r+n-r+1)}{(n-r+1)!r!} = \frac{(n+1)!}{(n+1-r)!r!}$$
$$= {_{n+1}}C_r$$

FOLLOW-UP

MORE PRACTICE
For more questions on SPUR
Objectives, use *Lesson Master 8-7*, shown on page 497.

EXTENSION
To find more patterns in Pascal's Triangle, students could transform the triangle and look for patterns in the image. One such transformation is to rewrite an entry as 0 if the entry is even, 1 if it is odd. The first ten rows of the image are shown here, but you may wish to encourage students to write 20 rows to see the patterns.

```
          1
         1 1
        1 0 1
       1 1 1 1
      1 0 0 0 1
     1 1 0 0 1 1
    1 0 1 0 1 0 1
   1 1 1 1 1 1 1 1
  1 0 0 0 0 0 0 0 1
 1 1 0 0 0 0 0 0 1 1
1 0 1 0 0 0 0 0 1 0 1
```

A similar transformation is to replace each element in the triangle with its remainder when divided by 3. Still another exploration could use a completely different triangle: For instance, what patterns can be found in the familiar multiplication table when it is represented as a triangle? Here are the first 8 rows.

```
           1
          2 2
         3 4 3
        4 6 6 4
       5 8 9 8 5
      6 10 12 12 10 6
     7 12 15 16 15 12 7
    8 14 18 20 20 18 14 8
```

EVALUATION
A quiz covering Lessons 8-4 through 8-7 is provided in the Teacher's Resource File.

Review

15. A soccer league which has grown to include 22 teams is to be broken into two 11-team divisions. In how many ways can this be done? *(Lesson 8-6)*
705,432

16. Consider a collection of Russian dolls with descending sizes so that they fit into each other. Assume that each doll has $\frac{3}{4}$ the height of the next bigger one.
  a. What is the ratio of the volumes of two consecutive dolls? **27/64**
  b. If the biggest doll requires a wooden piece of 1400 cm³, how much wood would be needed to carve out
    i. 10 dolls; **ii.** infinitely many of them? *(Lessons 8-4, 8-5)*
    ≈ 2421 cm³ **≈ 2422 cm³**

In 17 and 18, a restaurant offers cheese pizza with thick or thin crust, and with or without any of the following toppings: anchovies, mushrooms, onions, peppers, pepperoni, sausage. *(Lessons 7-3, 8-6)*

17. How many different kinds of pizza can be made? **128**

18. How many thin crust pizzas can be made with exactly two toppings? **15**

19. Find x if the sequence whose first three terms are $4, x, \frac{3}{2}x$ is
  a. arithmetic; **8** **b.** geometric. *(Lesson 8-1)* **6**

```
10   FOR I = 5 TO 6
20     FOR J = 1 TO 2
30       FOR K = 1 TO 2
40         PRINT I;J;K
50       NEXT K
60     NEXT J
70   NEXT I
80   END
```

20. Give the output of the program at the left. *(Lesson 7-4)*
511, 512, 521, 522, 611, 612, 621, 622

21. *Skill sequence.* Find the exact volume. *(Previous course)*

  a. rectangular solid
  b. right circular cylinder
  c. 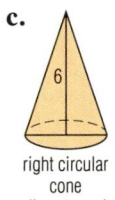 right circular cone diameter = 4

  96 units³   **24π units³**   **8π units³**

Exploration

22. a. Use the definition $\begin{bmatrix} n \\ k \end{bmatrix} = \dfrac{1}{(n+1)\binom{n}{k}} = \dfrac{k!(n-k)!}{(n+1)!}$ to evaluate each term in the first four rows of the following triangle.

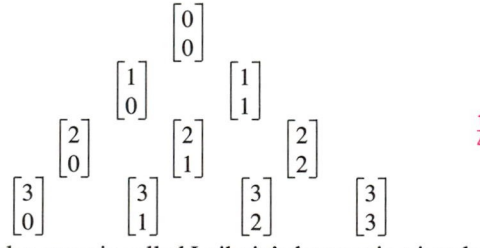

  b. This triangular array is called Leibniz's harmonic triangle. Where in the triangle does the harmonic sequence appear?
  c. Describe some other pattern in this triangle.
    b) The first entries in the rows, or the last entries in the rows.
    c) sample: The sum of any two adjacent terms is equal to the term directly above the two terms. For example, 1/4 + 1/12 = 1/3.

LESSON 8-8

The Binomial Theorem

You saw in the previous lesson that Pascal's Triangle arises from evaluating $_nC_r$ for whole numbers n and r, $r \le n$. This lesson examines the connections between $(x + y)^n$ and Pascal's Triangle, as well as some applications of those connections.

To begin, find the *series expansion* of $(x + y)^n$ for $n = 0$, 1, and 2.
$$(x + y)^0 = 1$$
$$(x + y)^1 = 1x + 1y$$
$$(x + y)^2 = 1x^2 + 2xy + 1y^2$$

Notice that the coefficients of the terms in each expansion are the entries in the corresponding row of Pascal's Triangle. This pattern continues to hold. To see why, consider $(x + y)^3 = (x + y)(x + y)(x + y)$. You can find this product by multiplying each x and y term of the first factor by an x or y term of each of the other factors, and then adding those partial products. In all, eight different partial products are computed and then added. These are highlighted in color below.

$$(x + y)\,(x + y)\,(x + y) \rightarrow xxx \rightarrow x^3$$
$$(x + y)\,(x + y)\,(x + y) \rightarrow xxy \rightarrow x^2y$$
$$(x + y)\,(x + y)\,(x + y) \rightarrow xyx \rightarrow x^2y$$
$$(x + y)\,(x + y)\,(x + y) \rightarrow xyy \rightarrow xy^2$$
$$(x + y)\,(x + y)\,(x + y) \rightarrow yxx \rightarrow x^2y$$
$$(x + y)\,(x + y)\,(x + y) \rightarrow yxy \rightarrow xy^2$$
$$(x + y)\,(x + y)\,(x + y) \rightarrow yyx \rightarrow xy^2$$
$$(x + y)\,(x + y)\,(x + y) \rightarrow yyy \rightarrow y^3$$

Thus, $(x + y)^3 = 1x^3 + 3x^2y + 3xy^2 + 1y^3$.

Now examine the 8 partial products. The result x^3 occurs when x is used as a factor three times and y is used as a factor zero times. There are three y terms from which to choose (one in each of the factors); the number of ways to choose zero of them is $_3C_0 = 1$, so x^3 occurs in $_3C_0 = 1$ way. That is, x^3 occurs once so its coefficient in the series is 1.

The product x^2y occurs when x is chosen from two of the factors and y from one. There are three y terms from which to choose, so this can be done in $_3C_1 = 3$ ways. (Equivalently, there are two x terms, and $_3C_2 = {_3C_1} = 3$.) So x^2y occurs three times and its coefficient in the series is 3. Similarly, xy^2 occurs when x is chosen from one factor and y from two. There are three y terms from which to choose so this can be done in $_3C_2 = 3$ ways. So the coefficient of xy^2 in the series is 3.

Finally, y^3 occurs when all three y terms are chosen, and this occurs in $_3C_3 = 1$ way. Thus
$$(x + y)^3 = x^3 \quad + \quad 3x^2y \quad + \quad 3xy^2 \quad + \quad y^3,$$
but in the language of combinations,
$$(x + y)^3 = {_3C_0}x^3 + {_3C_1}x^2y + {_3C_2}xy^2 + {_3C_3}y^3.$$

RESOURCES
■ Lesson Master 8-8
■ Teaching Aid 70:
Expansion of $(x + y)^3$

OBJECTIVE

D Expand binomials.

TEACHING NOTES

This should not prove to be a difficult lesson for students, given their extensive practice with combinations and Pascal's Triangle during the last couple of days. However, the notations used in the two forms of the theorem may be intimidating, so sufficient time should be spent on verifying expansions of $(x + y)^n$ for particular values of n. A teaching aid for $n = 3$, duplicating what is on page 499, is provided. **Questions 1–7** can be used to guide class discussion.

Making Connections A proof of the Binomial Theorem today requires mathematical induction. Omar Khayyam's proof would not be considered a rigorous proof under today's standards. We leave the proof for *Precalculus and Discrete Mathematics*. Still, the connection should be strongly made with combinations.

Error Analysis The algebraic manipulation in expansions like **Example 2** is not easy for most students; it requires very careful writing. Encourage students to be systematic, to write down the form of the Binomial Theorem for the particular value of n if needed, and then, directly under that form, do the substitutions. Also encourage checks.

The application of the Binomial Theorem to the coin-tossing problem in Example 3 is quite elegant in that the powers of the variables H and T serve double duty as indices. For example, H^3T^2 represents an instance of getting 3 heads and 2 tails in a flip of five coins. Order does not matter, so the coefficient $_5C_2$ counts the number of possible combinations of 3 heads and 2 tails.

As is shown in the solution to **Example 3,** the compact notation allows for a quick solution to the problem of how many tosses have *at least* two heads. Students generally have difficulty with these types of problems, but they will see others in the next lesson.

ADDITIONAL EXAMPLES
1. Give the coefficient of a^9b^2 in $(a+b)^{11}$.
55

2. a. Expand $(4x+5y)^5$.
**$1024x^5 + 6400x^4y +$
$16{,}000x^3y^2 + 20{,}000x^2y^3 +$
$12{,}500xy^4 + 3125y^5$**

b. Expand $(4x^2+5y)^5$.
**$1024x^{10} + 6400x^8y +$
$16{,}000x^6y^2 + 20{,}000x^4y^3 +$
$12{,}500x^2y^4 + 3125y^5$**

3. A coin is flipped 6 times. How many of the possible outcomes have at least 3 heads?
42

Similarly,
$$(x+y)^4 = (x+y)(x+y)(x+y)(x+y)$$
$$= {_4C_0}x^4 + {_4C_1}x^3y + {_4C_2}x^2y^2 + {_4C_3}xy^3 + {_4C_4}y^4$$
$$= x^4 + 4x^3y + 6x^2y^2 + 4xy^3 + y^4.$$

In general, the series expansion of $(x+y)^n$ has $_nC_0x^n = x^n$ as its first term and $_nC_ny^n = y^n$ as its last. The second term is $_nC_1x^{n-1}y = nx^{n-1}y$, and the second from the last is $_nC_{n-1}xy^{n-1} = nxy^{n-1}$. The sum of the exponents in each term is n, and the coefficient of $x^{n-k}y^k$ is $_nC_k$. These ideas are generalized in the following important theorem.

Binomial Theorem

For any nonnegative integer n,
$$(x+y)^n = {_nC_0}x^n + {_nC_1}x^{n-1}y + {_nC_2}x^{n-2}y^2 + \ldots + {_nC_k}x^{n-k}y^k + \ldots + {_nC_n}y^n$$
$$= \sum_{k=0}^n {_nC_k}x^{n-k}y^k.$$

The oldest known proof of the Binomial Theorem is by Omar Khayyam. Because of their application in this theorem, the combinations $_nC_k$ are sometimes called **binomial coefficients**.

The Binomial Theorem can be used to expand any binomial.

Example 1 Find the coefficient of x^2y^3 in $(x+y)^5$.

Solution From the five factors of $(x+y)^5$, y is to be chosen three times and x twice. This can be done in $_5C_3 = 10$ ways, so the coefficient of x^2y^3 is 10.

Check $(x+y)^5 = x^5 + 5x^4y + 10x^3y^2 + \mathbf{10x^2y^3} + 5xy^4 + y^5$

Example 2 Expand $(2v-3)^4$.

Solution Use the Binomial Theorem with $n=4$.
$$(x+y)^4 = x^4 + 4x^3y + 6x^2y^2 + 4xy^3 + y^4$$
Substitute $2v$ for x and -3 for y.
$$(2v-3)^4 = (2v)^4 + 4(2v)^3(-3) + 6(2v)^2(-3)^2 + 4(2v)(-3)^3 + (-3)^4$$
$$= 16v^4 + 4(8)(-3)v^3 + 6(4)(9)v^2 + 4(2)(-27)v + 81$$
$$= 16v^4 - 96v^3 + 216v^2 - 216v + 81$$

Check Let $v=2$. The power $(2v-3)^4 = (2\cdot2-3)^4 = 1^4 = 1$.
For $v=2$, the series expansion has the value
$$16(2^4) - 96(2^3) + 216(2^2) - 216(2) + 81 = 256 - 768 + 864 - 432 + 81$$
$$= 1.$$
It checks.

The Binomial Theorem can also be used to solve counting problems.

■ ■ ■ ■ ■ ■ ■ ■ ■■

Example 3 A coin is flipped five times. How many of the possible outcomes have at least two heads?

Solution Substitute H for x, T for y, and 5 for n in the Binomial Theorem.
$$(H + T)^5 = {}_5C_0H^5 + {}_5C_1H^4T + {}_5C_2H^3T^2 + {}_5C_3H^2T^3 + {}_5C_4HT^4 + {}_5C_5T^5$$
$$= \mathbf{1}\,H^5 + \mathbf{5}\,H^4T + \mathbf{10}\,H^3T^2 + \mathbf{10}\,H^2T^3 + \mathbf{5}\,HT^4 + \mathbf{1}\,T^5$$

The coefficients correspond to the number of different ways of obtaining 5 heads, 4 heads and 1 tail, ... , and 5 tails, respectively. Thus 5 heads can occur in **1** way, 4 heads and 1 tail can occur in **5** ways, and so on. The number of ways in which at least two heads occur is therefore
$$\mathbf{1 + 5 + 10 + 10} = 26.$$

Check Of the total $2^5 = 32$ outcomes, 1 has no heads and 5 have one head, so 6 have fewer than 2 heads. That leaves $32 - 6 = 26$ outcomes with at least 2 heads. It checks.

Questions

Covering the Reading

1. **a.** In the product $(x + y)(x + y)(x + y)$ what is the coefficient of x^2y?
   b. Why can this coefficient be derived from a combination problem?
   a) 3; b) See margin.
2. Write the series expansion of $(a + b)^5$.
   $a^5 + 5a^4b + 10a^3b^2 + 10a^2b^3 + 5ab^4 + b^5$
3. What is the coefficient of x^2y^4 in the expansion of $(x + y)^6$? **15**

In 4–7, expand.

4. $(v + 2)^3$
   $v^3 + 6v^2 + 12v + 8$
5. $(x - 1)^6$
   $x^6 - 6x^5 + 15x^4 - 20x^3 + 15x^2 - 6x + 1$
6. $(1 - 2q)^5$
   $1 - 10q + 40q^2 - 80q^3 + 80q^4 - 32q^5$
7. $(10p + 2)^3$
   $1000p^3 + 600p^2 + 120p + 8$

8. How many of the possible outcomes of 5 tosses of a coin have
   a. exactly 3 tails? **10**   **b.** at least 3 tails? **16**

Applying the Mathematics

In 9–11, *true or false*.   Suppose kx^py^q is a term in the series expansion of $(x + y)^n$.

9. $p + q = n$ **True**   10. $k = {}_nC_q$ **True**   11. $k = {}_nC_p$ **True**

12. Expand $(x^3 - y^2)^4$. $x^{12} - 4x^9y^2 + 6x^6y^4 - 4x^3y^6 + y^8$

13. Mr. and Mrs. Ippy are planning to have four children. Assuming that boys and girls are equally likely, what is the probability they will have at least two girls?   **11/16 = 68.75%**

14. Rewrite $\displaystyle\sum_{i=0}^{7}\binom{7}{i}a^{7-i}5^i$ in the form $(x + y)^n$.   **$(a + 5)^7$**

NOTES ON QUESTIONS
Question 8: Students can list how many possible outcomes of 5 tosses have 0 heads, 1 head, 2 heads, 3 heads, 4 heads, and 5 heads. Make the row in Pascal's Triangle explicit. Then the question is easy to answer.

Questions 9–11: Students often guess when they do not know. You might suggest the following method of solution: Pick a value of n, say 3. Then let $p = 2$, $q = 1$. This forces $k = 3$. Now check the various equations.

Question 12: You might suggest that students begin by expanding $(a - b)^4$. Then substitute x^3 for a and y^2 for b.

ADDITIONAL ANSWERS
1.b. The coefficient is the number of combinations of 3 x-terms taken two at a time for 3 y-terms taken one at a time.

NOTES ON QUESTIONS
Question 15: This proof is so elegant that students do not think it is a proof. Point out that not all proofs have to be long!

Question 24: This question presents a use of the Binomial Theorem which was much more important in the days before calculators. Students who are interested might be encouraged to practice the skill. It is possible to do such approximations mentally.

FOLLOW-UP

MORE PRACTICE
For more questions on SPUR Objectives, use *Lesson Master 8-8*, shown on page 501.

PROJECTS
Project 7 on page 515 is related to the content of this lesson.

15. Use the Binomial Theorem to show that $\sum_{r=0}^{n} {}_nC_r = 2^n$.   **See below.**

Hint: Let $2^n = (1 + 1)^n$. (This is property 4 from Lesson 8-7.)

16. Alternately add and subtract the entries in the rows of Pascal's triangle as follows.

$$a_1 = 1 - 1$$
$$a_2 = 1 - 2 + 1$$
$$a_3 = 1 - 3 + 3 - 1$$
$$a_4 = 1 - 4 + 6 - 4 + 1$$

Give an explicit formula for a_n.   **$a_n = 0$ for all n**

15) $2^n = (1 + 1)^n$
$$= \sum_{r=0}^{n} {}_nC_r(1)^{n-r}(1)^r$$
$$= \sum_{r=0}^{n} {}_nC_r$$

Review

In 17 and 18, refer to the following numbers, which are the first six terms of row 10 of Pascal's Triangle.   1   10   45   120   210   252   *(Lesson 8-7)*

17. Which term represents ${}_{10}C_4$?   **210**

18. **a.** Write the rest of row 10.   **b.** Write row 11 in full.
    c. Write row 9 in full.   **1, 9, 36, 84, 126, 126, 84, 36, 9, 1**
    a) 210, 120, 45, 10, 1; b) 1, 11, 55, 165, 330, 462, 462, 330, 165, 55, 11, 1

In 19 and 20, suppose that the names of four people in a baking competition are to be chosen from among a list of 100 finalists.
a. Tell whether the situation represents a combination or permutation.
b. State the number of possible outcomes.   *(Lessons 7-6, 8-6)*

19. The first receives a $10,000 prize, the second receives $5,000, the third $1,000, and the fourth $500.   **a) permutation; b) 94,109,400**

20. All four people receive identical prizes: a trip to Paris, France.
    a) combination; b) 3,921,225

21. Three integers between 1 and 100 (inclusive) are chosen at random. What is the probability that they are three consecutive integers?
    (Lessons 7-1, 8-6)   **≈ 0.000606**

22. **a.** Under what conditions does the following infinite series have a sum?

$$\sum_{n=2}^{\infty} xy^n = xy^2 + xy^3 + xy^4 + xy^5 + \ldots$$ **$x = 0$ or $|y| < 1$**

   b. What is that sum, if it exists?   *(Lesson 8-5)* $\frac{xy^2}{1-y}$

23. Area codes are three digit numbers. The first digit may not be a 0 or 1. The second digit must be a 0 or 1. The third digit may not be a 0 or 1. How many area codes are possible with these restrictions?   *(Lesson 7-3)*
    128

Exploration

24. **a.** Expand 1.001^5 using $(1 + .001)^5$.
    b. How many terms are needed to get an answer accurate to the nearest thousandth?   **2**
    c. Use your results from parts **a** and **b** to estimate 1.003^8 to the nearest millionth.   **1.024254**
    d. Give the complete decimal expansion of 1.003^8.
    e. How close is your calculator value of 1.003^8 to the value of your answer in part **d**?   **Sample: The calculator gives 1.024253518.**
    a) $1 + 5 \times 10^{-3} + 10 \times 10^{-6} + 10 \times 10^{-9} + 5 \times 10^{-12} + 1 \times 10^{-15}$;
    d) 1.024253517683628429502561

LESSON 8-9

Binomial Probabilities

Suppose that the probability of recovering from a certain kind of cancer after a certain treatment is known. If four patients with this disease are selected randomly and their medical condition is observed for a fixed period of time after the treatment, what is the probability that two of these patients recover?

This situation is called a *binomial experiment* because, as you will learn later in this lesson, a formula for calculating the probabilities associated with such a situation is related to the Binomial Theorem. A **binomial experiment** has the following features:
1. There are repeated situations, called *trials*.
2. There are only two possible outcomes, often called *success* and *failure*, for each trial.
3. The trials are independent.
4. Each trial has the same probability of success.
5. The experiment has a fixed number of trials.

In the situation described above, each trial is one individual's battle with cancer. Recovering from the disease within the fixed time period is called a success S, and not recovering is called a failure F. The trials are independent because no person's individual health influences another's.

Listed below are all the possible outcomes for four trials.

Outcomes for 4 Trials in a Binomial Experiment				
exactly 0 successes	exactly 1 success	exactly 2 successes	exactly 3 successes	exactly 4 successes
FFFF	FFFS	FFSS	FSSS	SSSS
	FFSF	FSFS	SFSS	
	FSFF	FSSF	SSFS	
	SFFF	SFFS	SSSF	
		SFSF		
		SSFF		

Notice that the number of ways to have exactly k successes among the 4 trials is ${}_4C_k$. That is, the number of outcomes in the columns above, 1, 4, 6, 4, 1, are precisely the numbers from row 4 of Pascal's Triangle.

The chart should look familiar. It has 16 outcomes similar to tossing four fair coins. However, unlike the situation in which fair coins are tossed, the preceding sixteen outcomes may not be equally likely because the probability of success and the probability of failure may not be equal.

Example 1 Suppose that the probability of recovery within one year after chemotherapy is 0.7 for patients with a certain cancer.

a. Find the probability that exactly two of the four patients currently being monitored recover within one year after chemotherapy.
b. Find the probability that two or three patients recover.

LESSON 8-9 *Binomial Probabilities* **503**

OBJECTIVE

K Determine probabilities in situations involving binomial experiments.

TEACHING NOTES

The power of the Binomial Theorem combined with the applicability of combinations enables probabilities to be calculated for events students would not expect to be amenable to mathematical treatment. The outcomes do not have to be equally likely for their probabilities to be calculated.

A possible lesson plan follows: Begin with **Questions 1–5,** which relate to the discussion on page 503. State the theorem on page 504. Then go through **Example 1,** followed by **Questions 6 and 7.** Finally, go through **Example 2** and **Questions 8 and 9. Example 3** wraps this all up and is a prelude to Chapter 10, where binomial probability distributions are discussed in more detail.

1. The probability of getting a sum of 7 in a toss of two fair dice is known to be $\frac{1}{6}$.

a. What is the probability of getting exactly two 7s in 5 tosses?

≈ **0.16**

b. What is the probability of getting at least two 7s in 5 tosses?

≈ **0.20**

(See Additional Example 3 below.)

2. What is the probability of getting at least 8 of 10 questions on a test correct, if you feel you have an 80% chance of answering each individual question correctly?

≈ **.68**

3. Determine the probability distribution from the binomial experiment of Additional Example 1.

Let x be the number of 7s in 5 tosses of two fair dice.

x	$P(x)$
0	.4019
1	.4019
2	.1608
3	.0322
4	.0032
5	.0001

NOTES ON QUESTIONS
Questions 1–10: These questions can be used to guide the class discussion. See Teaching Notes.

NOTES ON QUESTIONS
Questions 10–12: Mastery is not expected at this time, but it is important for students to be familiar with the idea of distributions, and it helps in putting the individual answers to other questions in perspective.

Making Connections for Question 15: This question is similar to the airline reservation situation studied in Lesson 7-9. One might wonder how many small cars should a company have in order to be able to satisfy these customers? Car rental

Solution

a. For each patient, $P(\text{success}) = 0.7$; so $P(\text{failure}) = 1 - 0.7 = 0.3$. As listed in the column labeled *exactly 2 successes*, there are 6 combinations in which exactly 2 of the 4 patients recover. The probability of each of these combinations is $(0.7)^2(0.3)^2$. So

$$P(\text{exactly 2 recoveries in 4 trials}) = 6(0.7)^2(0.3)^2$$
$$= 6(0.0441)$$
$$= 0.2646.$$

b. Because the events are mutually exclusive,
$P(\text{2 or 3 recoveries}) = P(\text{exactly 2 recoveries}) + P(\text{exactly 3 recoveries})$.
From part **a**, $P(\text{exactly 2 recoveries}) = 0.2646$.
There are 4 ways to have exactly 3 recoveries. Each has probability $(0.3)(0.7)^3$. So

$$P(\text{exactly 3 recoveries}) = 4(0.7)(0.3)^3 = 0.4116.$$
Thus, $P(\text{2 or 3 recoveries})$ $= 0.2646 + 0.4116 = 0.6762.$

The results of Example 1 can be generalized. In a binomial experiment, if the probability of success is p, then the probability of failure is $q = 1 - p$. So the probability of exactly 2 successes in 4 trials is $_4C_2 p^2 q^2 = 6p^2q^2$, and the probability of exactly 3 successes in 4 trials is $_4C_3 p^3 q = 4p^3q$.

These ideas can be generalized still further.

Theorem

Suppose that in a binomial experiment with n trials the probability of success is p in each trial, and the probability of failure is q, where $q = 1 - p$. Then
$$P(\text{exactly } k \text{ successes}) = {_nC_k} \cdot p^k q^{n-k}.$$

Example 2 A quiz has 8 multiple choice questions, each with 4 alternatives. If a student guesses randomly on every question, what is the probability of getting 5 or more correct?

Solution This is a binomial experiment. Answering each question is a trial, so there are $n = 8$ trials. The probability of success (getting a single question correct) is 0.25; the probability of failure is 0.75 . Getting 5 or more correct is equivalent to the mutually exclusive events of getting exactly 5 or exactly 6 or exactly 7 or exactly 8 correct. So the desired probability is

$P(\text{5 or better})$
$= P(\text{exactly 5}) + P(\text{exactly 6}) + P(\text{exactly 7}) + P(\text{exactly 8})$
$= {_8C_5}\,(.25)^5(.75)^3 + {_8C_6}\,(.25)^6(.75)^2 + {_8C_7}\,(.25)^7(.75) + {_8C_8}(.25)^8$
$\approx 56(.000412) + 28(.000137) + 8(.000046) + 1(.000015)$
$\approx .027298.$

So, sheer random guessing on an 8-item multiple choice quiz yields a probability less than 3% of getting 5 or more items correct. (Chances usually improve substantially with study!)

504

The probability distribution generated from the probability of x successes in a binomial experiment is called a **binomial probability distribution**. Here is an example.

■ ■ ■ ■ ■ ■ ■ ■ ■

Example 3   In a binomial experiment, suppose that the probability of success is 0.7. Determine and graph the probability distribution for the number of successes in 5 trials.

Solution   Let x be the number of successes. Then x can be any integer from 0 to 5, inclusive. Let $P(x) =$ the probability of exactly x successes. By the preceding theorem, $P(x) = {}_5C_x p^x q^{5-x}$, where $p = 0.7$, and $q = 1 - p = 0.3$. Evaluate $P(x)$ for each possible value of x. The probability distribution is represented below in a table and a graph.

x	$P(x)$	
0	$q^5 = (.3)^5$	$= .00243$
1	$5pq^4 = 5(.7)(.3)^4$	$= .02835$
2	$10p^2q^3 = 10(.7)^2(.3)^3$	$= .1323$
3	$10p^3q^2 = 10(.7)^3(.3)^2$	$= .3087$
4	$5p^4q = 5(.7)^4(.3)$	$= .36015$
5	$p^5 = (.7)^5$	$= .16807$

Check   The sum of the probabilities should be 1. You should check this.

Binomial probability distributions are discussed further in Chapter 10.

Questions

Covering the Reading

1. State five characteristics of a binomial experiment.   **See margin.**

In 2–5, state whether or not the experiment is a binomial experiment. If not, identify the missing property or properties.

2. A die is rolled 7 times and the number of times a 3 shows up is recorded.
is a binomial experiment

3. Four cards are selected from a standard deck (of 52 cards) without replacement. The number of aces is recorded.
No, it does not satisfy property 3.

4. A bag contains 2 red, 3 green and 5 blue marbles. The bag is shaken, a marble is selected, its color is recorded, and it is replaced. This is repeated five times.   **No, it does not satisfy property 2.**

5. 100 people are selected at random and asked if they voted in the last presidential election   **is a binomial experiment**

In 6 and 7, suppose as in Example 1 that the probability of recovering from a certain cancer is 0.7.

6. What is the probability that all 4 of the 4 patients will recover?   **≈ 0.24**

7. What is the probability that at least 2 of the 4 will recover?   **≈ 0.92**

LESSON 8-9   Binomial Probabilities   **505**

companies avoid these kinds of problems by keeping more big cars on hand than they need. Then, when a renter comes in and desires a small car, if they have no small car, they offer the bigger car at no extra cost. Airlines cannot do this since everyone comes at once for a particular flight, but sometimes they upgrade their best customers to first class.

Question 19:   You may wish to adapt this question to your state lottery. For example, in Illinois, you must pick 6 numbers correctly out of 54. But you get to pick two sets of 6 numbers on each $1 ticket. What is the probability of winning on one $1 ticket? $\left(\frac{1}{12,913,582.5} \approx 7.74 \times 10^{-8}\right)$

Question 20:   If you are not in Hawaii, do your students know any Hawaiian words or names? The most common are the song *Aloha oe* ("Greetings to you"), *hula*, *poi*, *luau*, *Honolulu*, and the names of the major islands: *Oahu*, *Maui*, *Kauai*, *Hawaii*, *Lanaii*, and *Molokai*. These use all the letters.

ADDITIONAL ANSWERS
1. (1) There are repeated situations, called trials. (2) There are only two possible outcomes, usually called success and failure. (3) The trials are independent. (4) Each trial has the same probability of success. (5) The experiment has a fixed number of trials.

8. Refer to Example 2.
   a. Where in the solution is the requirement of independent events applied?   **when the binomial formula is used**
   b. Where in the solution is the fact that events are mutually exclusive applied?   **when the probabilities of 5 correct, 6 correct, 7 correct, and 8 correct are added**

9. Suppose a multiple choice quiz has 10 questions, each with 5 alternatives. If a student answers by random guessing, what is the probability that at least 7 questions are answered correctly?   **≈ 0.00086**

10. In a binomial experiment, suppose that the probability of success is 0.4. Use a table and graph to represent the probability distribution for the number of successes in 6 trials.   **See margin.**

11. In the distribution of Example 3, give the
   a. mode   **4**   **b.** expected value.   **3.5**

12. Consider tossing a fair coin 6 times.
   a. Construct a probability distribution for this experiment, where x is the number of heads that occurs.   **See margin.**
   b. Graph the probability distribution.   **See margin.**
   c. Find the probability of getting 2 or 3 heads.   **.546875**
   d. Find the probability of getting at least 1 head.   **.984375**

13. Refer to the drawing at the right. On any one spin, the spinner is equally likely to land in any of the six regions.
   a. What is the probability of the spinner landing in a shaded region?   **1/3**
   b. Suppose a trial is a single spin, and that "success" means the spinner lands in a shaded region. Complete the following probability distribution for the number of successes in 4 spins.

Probability distribution of exactly x successes in 4 spins

x	0	1	2	3	4
$P(x)$	.198	.395	.29	.099	012

   c. Graph the distribution in part **b**.   **See margin.**
   d. Find P(at least 3 successes).   **0.111**
   e. Find P(at least 1 success).   **.802**

14. A baseball player has a batting average of .250. This can be interpreted to mean that the probability of a hit is $.250 = \frac{1}{4}$. Some people think this means that in 4 times at bat, the batter is *sure* to get a hit. Give the probability that this batter gets
   a. exactly 1 hit in 4 times at bat,   **27/64 ≈ .42**
   b. at least 1 hit in 4 times at bat.   **175/256 ≈ .68**

15. Suppose that in a binomial experiment, P(success on one trial) $= a$.
   a. Express P(failure on one trial) in terms of a.   **$1 - a$**
   b. Find P(success on exactly 3 of 7 trials).   **$35(1 - a)^4 a^3$**

506

16. A car rental company finds that, on the average, 86% of renters want small cars, and the remainder want big cars. If 10 people at random come to rent cars, find the probability that
  a. exactly 7 of them want small cars,   ≈ 0.115
  b. no more than 3 of them want big cars.   ≈ .96
$$18) \ 8\sin^3\theta - 12\sin^2\theta\cos\theta + 6\sin\theta\cos^2\theta - \cos^3\theta$$

In 17 and 18, expand.   *(Lessons 7-1, 7-5, 8-8)*

17. $(a + 2b)^6$ **18.** $(2\sin\theta - \cos\theta)^3$ **See above.**
$a^6 + 12a^5b + 60a^4b^2 + 160a^3b^3 + 240a^2b^4 + 192ab^5 + 64b^6$

19. In a state lottery, you must pick 6 numbers correctly (in any order) out of 55. What is the probability of winning with one ticket?   *(Lesson 8-6)*
  1 in $_{55}C_6 = 1/28,989,675 \approx 3.4 \times 10^{-8}$

20. Transliterated into English, the Hawaiian language has only the following letters: A, E, H, I, K, L, M, N, O, P, U, W. Every Hawaiian word and syllable ends with a vowel, and some words have no consonants. Two consonants never occur without a vowel between them. How many four-letter words are possible?   *(Lessons 7-3, 7-5)*
  2375

21. Below are results from a survey reported in *The Wall Street Journal*, Dec. 2, 1986.   *(Lessons 7-1, 7-2)*
  a. Are all five categories mutually exclusive? Explain your answer.
  b. Determine the probability that a randomly selected customer eats candy at least once a week.   a) See margin. b) 56%

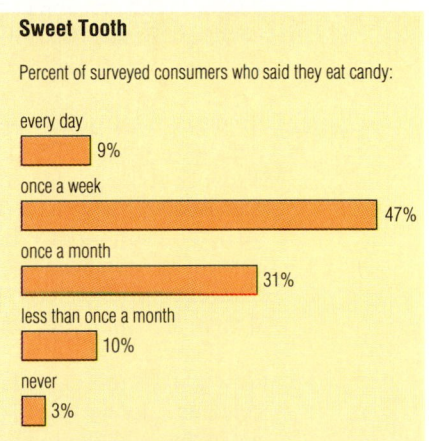

Sweet Tooth

Percent of surveyed consumers who said they eat candy:

every day — 9%
once a week — 47%
once a month — 31%
less than once a month — 10%
never — 3%

Source: National Confectioners Association

22. *Skill sequence.* Solve for x.   *(Lesson 4-8)*
  a. $.2^x = .05$ ≈ 1.86
  b. $1 - .4^x = .95$ ≈ 3.27
  c. $1 - p^x = a, p > 0, a < 1$ $\dfrac{\log(1-a)}{\log p}$

23. *True* or *false.*   The lines with equations $2x - 7y = 8$ and $7x + 2y = 1$ are perpendicular.   *(Previous course)*   **True**

24. Binomial trials are sometimes called *Bernoulli trials*.   **See margin.**
  a. Who was Bernoulli?
  b. What work of Bernoulli is related to binomial probabilities?

Translation from Hawaiian: "one, three, five, seven, to answer briefly"

MORE PRACTICE
For more questions on SPUR Objectives, use *Lesson Master 8-9*, shown below.

24.a. The Bernoulli family made significant contributions to mathematics in the seventeenth century. No less than eight brilliant mathematicians were produced by the family. Jakob Bernoulli (1654–1705) made important contributions in probability theory. His brother, Johann (1667–1748), did important work in the theory of functions. He coined the term *functio*, the root of the present term *function*, in 1714.

b. Jakob Bernoulli wrote *Ars Conjectardi*, the first significant book on probability. In this work, the theorem which is in this lesson appeared for the first time; it sometimes is still called Bernoulli's theorem.

NAME _____

LESSON
8-10

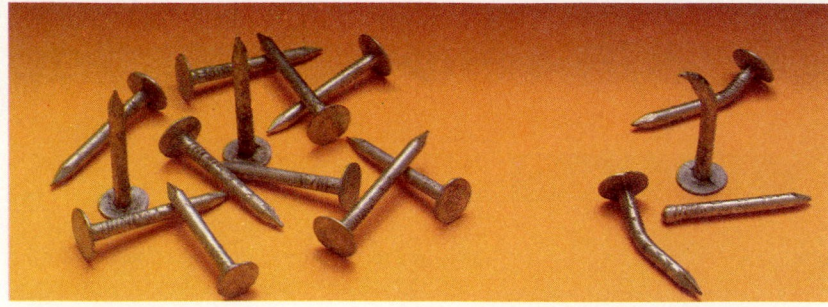

Quality Control by Sampling

Suppose a carpenter buys a box of nails for fastening shingles to a roof. The nails should have large flat heads and straight shafts so that they will firmly hold the shingles to the roof. But sometimes defective nails are produced. A nail may have no head, or it may be bent.

When items like nails are manufactured in large quantities, a few defective items are expected to occur. How should the manufacturer or the consumer decide whether a box has too many defective nails?

The box could be emptied and each nail inspected for defects. But this is time-consuming. A more efficient procedure for assessing quality involves taking a random sample of the nails, and seeing how many defective nails it contains. On the basis of the sample, a decision is made about the acceptability of the whole box.

To monitor the quality of a product, a manufacturer often sets up a sampling and decision plan. This is a procedure for deciding whether to accept or reject a supply of the product based on a sample. For example, suppose you (as the manufacturer or as the buyer) design the following sampling and decision plan. You will take a random sample of 20 nails from a box of 100. On the basis of the number of defectives in the sample, you either "accept" or "reject" the whole box. If there are 0 or 1 defective nails you accept the box. If there are 2 or more defectives you reject it.

How good is this plan? For instance, suppose that a box of 100 nails actually contains 2 defective nails. (Of course, you would not know this, or you would not bother with the sampling process!) How likely is the box to be accepted or rejected? To answer this question, consider all the possibilities. The sample of 20 nails may contain (a) only acceptable nails; (b) exactly 1 of the 2 defective nails; or (c) the 2 defective nails. The probability of each of these occurrences can be found using techniques you have studied. For example,

$$P\begin{pmatrix}\text{0 defective in sample} \\ \text{if 2 defectives in box}\end{pmatrix} = \frac{\text{number of samples with 0 defective}}{\text{possible number of samples}}.$$

A sample of 20 with 0 defective nails consists entirely of nails taken from the 98 acceptable nails in the box. The number of such samples is the number of ways of choosing 20 objects from 98. That is,

$$\text{number of samples with 0 defective} = {}_{98}C_{20}.$$
$$\text{The total possible number of samples} = {}_{100}C_{20}.$$

So, $P(\text{0 defective}) = \dfrac{{}_{98}C_{20}}{{}_{100}C_{20}}.$

508

Similarly,

$$P \begin{pmatrix} \text{1 defective in sample} \\ \text{if 2 defectives in box} \end{pmatrix} = \frac{\text{number of samples with 1 defective}}{\text{possible number of samples}}.$$

A sample of 20 with 1 defective consists of 19 nails taken from the 98 acceptable nails and 1 taken from the 2 defective nails. The number of ways that *both* these events can occur can be found using the Multiplication Counting Principle.

Number of samples with 1 defective $= {}_{98}C_{19} \cdot {}_{2}C_{1}$.

So,

$$P(\text{1 defective}) = \frac{{}_{98}C_{19} \cdot {}_{2}C_{1}}{{}_{100}C_{20}}.$$

Similarly,

$$P(\text{2 defective}) = \frac{{}_{98}C_{18} \cdot {}_{2}C_{2}}{{}_{100}C_{20}}.$$

Large factorials such as 100! or 98! cannot be computed on most calculators, so it is necessary to cancel before performing the arithmetic. For example,

$$P(\text{2 defective}) = \frac{{}_{98}C_{18} \cdot {}_{2}C_{2}}{{}_{100}C_{20}}$$

$$= \frac{\frac{98!}{80! \ 18!} \cdot \frac{2!}{0! \ 2!}}{\frac{100!}{80! \ 20!}}$$

$$= \frac{98! \ 2! \ 80! \ 20!}{80! \ 18! \ 0! \ 2! \ 100!}$$

$$= \frac{98! \ 20!}{100! \ 18!}$$

$$= \frac{98! \ (20) \ (19) \ 18!}{(100) \ (99) \ 98! \ 18!}$$

$$= \frac{20 \cdot 19}{100 \cdot 99}$$

$$\approx 0.0383.$$

Similarly, $P(\text{1 defective}) \approx 0.3232$ and $P(\text{0 defective}) \approx 0.6384$.

Notice that because the three events, "0 defective," "1 defective," and "2 defective," are mutually exclusive and are the only three possibilities, their probabilities add to 1.

$$P(0) + P(1) + P(2) \approx 0.6384 + 0.3232 + 0.0383 \approx 1$$

Returning to your decision plan, the probability that the box is accepted is

$$P(\text{0 defective}) + P(\text{1 defective}) = 0.6384 + 0.3232$$
$$\approx 0.9616.$$

So, with this sampling plan, about 96.2% of the boxes with exactly 2 defective nails will be accepted. This seems reasonable. From the perspective of the customer, two defective nails per box is probably tolerable. From the perspective of a manufacturer, only about 4 of such boxes will be rejected.

But what happens when this sampling plan is used with boxes of nails containing more than 2 defective nails? According to the sampling plan, if a box contains 3 defective nails, it will be judged acceptable if the random sample

The mathematics can be summarized as follows: The basic question is: What is the probability that the population will be acceptable in quality even though it does not satisfy the criterion? First you establish a quality criterion. (In the case of nails in this lesson, it begins as 2 defectives in a box.) Then you establish a size of sample. (For the nails, the sample is 20 nails taken from a box.) In the discussion of pages 508–510, we find that the sample is not sufficiently large, or the criterion sufficiently strict, to reject enough boxes. If there is a box with 2 or more defective ones, there is too high a probability that it would not be detected in the sample.

The discussion on page 511 can serve as motivation for studying polynomials in Chapter 9, for here we find that dealing with a polynomial of degree 20 saves time in this situation.

We recommend that after the lesson is summarized, the questions be considered in order.

of 20 nails contains either 0 or 1 defective nail. The probability of this compound event of mutually exclusive outcomes is

$$P(0 \text{ or } 1 \text{ defective}) = P(0 \text{ defective}) + P(1 \text{ defective})$$

$$= \frac{_{97}C_{20}}{_{100}C_{20}} + \frac{_{97}C_{19} \cdot _{3}C_{1}}{_{100}C_{20}}$$

$$\approx 0.8989.$$

The first four lines of the table below summarize the results obtained above. The rest of the table was constructed using similar analysis to calculate the probability of acceptance of a box with a specified number of defective nails among the 100 nails.

Actual number of defective nails out of 100	Probability of acceptance
0	1
1	1
2	0.9616
3	0.8989
4	0.8224
5	0.7395
6	0.6554
8	0.4972
10	0.3630
15	0.1453
20	0.0498

The graph of these data, shown below, is called an *operating characteristic curve*. Both the table and the graph suggest that this sampling plan does not effectively identify boxes of poor quality. For example, with as many as 10 defective nails in a box of 100, there is still a large probability (about 36%) that the box will be accepted. In other words, the sampling plan only identifies about two-thirds of boxes that are as seriously deficient as this. A better sampling plan is needed.

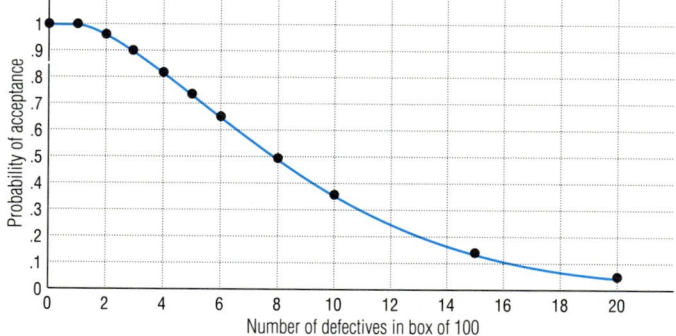

To find a better sampling plan, quality control engineers study how different criteria for the number of defective nails regarded as acceptable, and the size of the samples drawn, affect the probability of acceptance. However, the calculations can be tedious, as you can see from the above examples, and the operating characteristic curve may be difficult to construct.

Fortunately, for certain sample sizes (no more than 20% of the whole box) the binomial probability distribution can be used to approximate the operating characteristic curve. If the proportion of defectives in a box is p, and a sample of size n is taken, then the number of defective nails in the sample is approximately binomially distributed. So $P(k)$, the probability that there are exactly k defective nails in a sample, is given by

$$P(k) \approx {}_nC_k \cdot p^k(1-p)^{n-k}.$$

Specifically, if the proportion of defective nails in a box is p, a sample of 20 is taken, and the box is acceptable if less than two defective nails are found, this formula predicts that the probability of acceptance is

$$P(0 \text{ or } 1) \approx {}_{20}C_0 \cdot p^0 \cdot (1-p)^{20} + {}_{20}C_1 p^1(1-p)^{19}$$
$$= (1-p)^{20} + 20p(1-p)^{19}.$$

Consider the case of a box of 100 nails with 2 defective nails. Then $p = 0.02$ and $1 - p = 0.98$. So the probability of acceptance predicted by the binomial approximation is

$$P(0 \text{ or } 1) \approx 0.98^{20} + 20(0.02)(0.98)^{19}$$
$$\approx 0.940.$$

This is a little less than the actual value of 0.9616, derived before. If the number of defective nails is actually 10 (so $p = 0.10$), then the probability of acceptance is predicted to be

$$P(0 \text{ or } 1) \approx 0.9^{20} + 20(0.1)(0.9)^{19}$$
$$\approx 0.3917,$$

which is a little more than the actual value of 0.3630 in the previous table, but still a good approximation.

So for the values $p = 0.02$ and $p = 0.10$, the function

$$f(p) = (1-p)^{20} + 20p(1-p)^{19}$$

predicts values close to the actual probabilities calculated earlier. In general, the function f, which is graphed below, is a rather close match to the operating characteristic curve graphed on page 510.

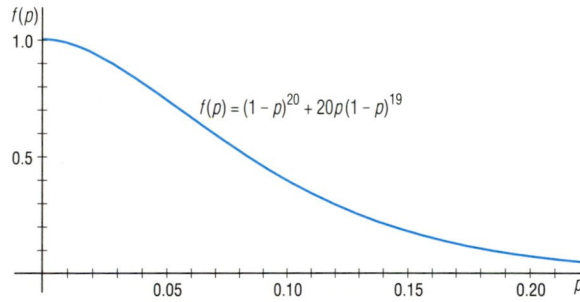

Furthermore, $f(p)$ is easy to evaluate and independent of the size of the box, provided the sample is not too large in relation to the box. So the same formula and graph can be used to study samples of 20 nails from a box of 1000 as easily as samples of 20 from a box of 100. Similar functions can be used to study other sampling and decision plans. For instance, in the Questions you are asked to study the plan to sample 15 nails and to reject the box if there are any defective nails in the sample.

Questions

In 1–3, refer to the sampling and decision plan in the lesson, for which a box of 100 nails is accepted if no more than 1 of a sample of 20 is defective.

1. If a box actually contains 1 defective, explain why the probability of accepting it is 1. **No sample can ever have more than (the) one defective nail, so all boxes will be accepted.**
2. If a box actually contains 4 defective nails,
   a. find the probability that a sample of 20 will contain
      i.   no defective nails; **≈ .4033**   ii. 1 defective nail; **≈ .4191**
      iii. more than 1 defective nail. **≈ .1776**
   b. What is the probability of accepting such a box under this sampling plan? Check your result with the table in the text. **.8224**

3. At what point is the probability of acceptance 0; that is, when is it impossible to pick an acceptable sample? [Hint: Consider the total number of defective nails in the box.] **when there are 82 or more defective nails in the box**

In 4 and 5, refer to the first operating characteristic curve in the lesson.

4. If a box of 100 nails contains 12 defective nails, what is the probability that it will be accepted? **≈ .25**

5. *True* or *false*. About half of boxes of 100 nails that have 8 defective nails will be accepted under this sampling plan. **True**

6. Consider the binomial approximation model of the operating characteristic curve of the lesson, $f(p) = (1-p)^{20} + 20p(1-p)^{19}$.
   a. Evaluate $f(0.05)$. **$f(.05) \approx .7358$**
   b. Describe in words what the probability $f(0.05)$ approximates.
   c. What is the difference between the actual probability and $f(0.05)$?
      b) See margin. c) ≈ .0036

In 7 and 8, consider the following sampling plan to test boxes of 100 nails, some of which are defective. *Using a sample of size 15, accept the whole batch if there are no defective nails.*

7. Suppose a particular box contains 3 defective nails.
   a. What is the probability that the first nail examined is defective? **.03**
   b. Find the actual probability that the box will be accepted.
   c. Use a binomial approximation to estimate the probability that boxes like this will be accepted. **≈ .6333**
      b) ≈ .6108
8. a. Use a binomial probability approximation to find a function to approximate probabilities of acceptance for boxes with x defective nails. **$f(x) = (1 - x/100)^{15}$**
   b. Use the formula in part **a** to estimate the probability that a box of 100 nails with 10 defectives will be judged acceptable. **≈ .2059**
   c. Graph this function. **See margin.**
   d. Find a whole number x such that a box with x defectives has a probability of about 0.5 of being accepted. **4 or 5**

512

Review

In 9 and 10, use the Binomial Theorem.   *(Lesson 8-8)*

9. Expand $(x^3 - y^2)^6$ and simplify.
$(x^3 - y^2)^6 = x^{18} - 6x^{15}y^2 + 15x^{12}y^4 - 20x^9y^6 + 15x^6y^8 - 6x^3y^{10} + y^{12}$

10. Give the coefficient of the term of the series expansion $(2b - c^2)^5$ that contains b^4c^2.   **-80**

11. The average woman in Kenya bears 8 children. If the probability of the birth of a boy is 0.52, find the probability that a woman with 8 children has

   a. 4 boys and 4 girls,   **≈ .2717**

   b. more boys than girls.   *(Lesson 8-9)*   **≈ .408**

12. *True* or *false*.   Flipping a coin 15 times and recording the number of heads is a binomial experiment.   *(Lesson 8-9)*   **True**

13. Solve $_{n+5}C_1 = {}_nC_2$.   *(Lesson 8-6)*   **5**

In 14 and 15, does the sequence have a limit as $n \to \infty$? If so, state the number L to which the sequence converges.   *(Lesson 8-2)*

14. $t_n = 8 + \dfrac{(-1)^n}{n}$ **yes; 8**

15. $v_n = \dfrac{100n}{n+1}$ **yes; 100**

16. *Multiple choice.*   A textbook is chosen at random at the bookstore of a high school. Let A be the event "the price of the book is over \$20," and let B be the event "the price of the book is over \$25." The events A and B are

   (a) mutually exclusive   (b) complementary

   (c) independent   (d) none of the above.   *(Lessons 7-2, 7-4)*
   (d)

17. Given $P(E \cup F) = 0.57$, $P(E) = 0.45$, and $P(F) = 0.23$, find $P(E \cap F)$.
   (Lesson 7-2)   **0.11**

18. *Skill sequence.* Solve for $0 < \theta < \pi$.

   a. $8 \cos \theta = 5$ **≈ 0.896**

   b. $8 \cos^2 \theta = 5$ **≈ 0.659 or ≈2.483**

   c. $8 \cos^2 \theta = 5 \cos \theta$ *(Lessons 5-7, 6-6)*   **≈ 0.896 or π/2**

Exploration

19. Create and defend your own sampling strategy to determine if a box of 100 nails is acceptable.
   Answers will vary.

Kenyan family

Projects

The projects provide an opportunity for extended work on topics related to the content of the chapter. (See pages 66 and T37 for more information on the purposes of projects.) Some or all of these projects may be suitable for **small group work.**

Solutions to the projects are found at the end of the Additional Answers section at the back of this book.

Project 1 has more meaning if the student who attempts it is aware that there is no polynomial formula that will generate all the primes.

Projects 2 and 4 involve geometric sequences and series.

Project 3 has students explore series based on the harmonic series that was first introduced in Lesson 8-5.

Project 5 presents a recursively defined sequence of rational numbers. The sequence converges quite rapidly to 1.732051, which students may recognize as an approximation for $\sqrt{3}$. The result for part **c** is independent of the starting fraction. Part **d** suggests that a generalization may be possible.

Project 6 is included because of the growing number of states in the U.S. which support lotteries. We include the caveat about purchasing tickets because we do not wish to encourage experimental approaches to the project. If anything, the results of doing the project should discourage participation in lotteries.

Material for **Project 7** can be found in most calculus books.

1. In 1934 a young Indian student named Sundaram devised the following "sieve" to identify some prime numbers. Here is the beginning of his table—each row (or column) of which is an arithmetic sequence.

4	7	10	13	16	19	22	…
7	12	17	22	27	32	37	…
10	17	24	31	38	45	52	…
13	22	31	40	49	58	67	…
16	27	38	49	60	71	82	…
·	·	·	·	·	·	·	…

Sundaram's table has the following properties: If N occurs in the table, then $2N + 1$ is not a prime number; if N does not occur in the table, then $2N + 1$ is a prime number.

a. Verify the first property for three numbers in the table and the second property for three numbers that are not (and never will be) in the table.

b. Identify the common difference d for each row. What is the pattern in these common differences as you read down the table?

c. Write an explicit formula for the kth term in the nth row. Use this result to prove that if N occurs in the table, then $2N + 1$ is composite.

d. Prove that if $2N + 1$ is not prime, then N is in the table. (Note that this is equivalent to the second property stated above.)

e. Can all prime numbers be found by using Sundaram's table? Why or why not?

2. The concept of infinity has intrigued people throughout history. In the last 100 years several artists and mathematicians have created spectacular works of art based on the infinite repetition of geometric forms. The Dutch artist Maurits Escher is probably most noted for his *tessellations*, which are tilings of the plane with congruent objects. Tessellations can be thought of as providing a finite pattern for covering the infinite plane. In other works of art, such as *Square Limit*, Escher represented infinite divisions within a finite space. Investigate Escher's use of art in the representation of infinity.

3. If an infinite series containing only positive terms is to converge, it is necessary that the terms approach 0 as n approaches infinity. However, this is not sufficient; the text points out that the harmonic series diverges even though the terms approach 0.

a. Explain why the following series diverges.

$$A = 1 + \frac{1}{2} + \underbrace{\frac{1}{4} + \frac{1}{4}}_{2 \text{ terms}} + \underbrace{\frac{1}{8} + \frac{1}{8} + \frac{1}{8} + \frac{1}{8}}_{4 \text{ terms}} +$$

$$\underbrace{\frac{1}{16} + \frac{1}{16} + \frac{1}{16} + \frac{1}{16} + \frac{1}{16} + \frac{1}{16} + \frac{1}{16} + \frac{1}{16}}_{8 \text{ terms}} + \cdots$$

b. Show that each term of the harmonic series
$$H = 1 + \frac{1}{2} + \frac{1}{3} + \frac{1}{4} + \frac{1}{5} + \frac{1}{6} + \cdots$$
is greater than or equal to the corresponding term of series A. Explain how this result shows that the harmonic series diverges.

c. The following variation of the harmonic series converges.
$$B = 1 - \frac{1}{2} + \frac{1}{3} - \frac{1}{4} + \frac{1}{5} - \frac{1}{6} + \frac{1}{7} - \frac{1}{8} + \cdots$$
$$+ (-1)^{n+1} \frac{1}{n} + \cdots .$$
Use a computer to approximate B to six decimal places.

d. If every term of the harmonic series which contains a 9 is deleted (such as $\frac{1}{9}$, $\frac{1}{95}$, $\frac{1}{409}$, etc.), the resulting series converges. Use a computer to approximate the sum of this series to four decimal places.

4. Many interesting curves can be created by recursive definitions. For instance, begin with a square with a given side, say _____. At each stage, replace each segment with the shape shown at the right. The original square and subsequent stages leading to a limit curve sometimes called the *dragon curve* are shown at the top of the next page.

514

a. Determine the perimeter, area and dimension of the dragon curve.

b. Find out about other curves that are defined recursively. Some famous ones are called snowflake curves and space-filling curves.

c. Design your own recursive curve.

5. Consider the sequence defined by

$$t_n = \frac{a_n}{b_n}$$

where $\begin{cases} a_1 = 2; \\ b_1 = 1; \\ a_n = a_{n-1} + 3 \cdot b_{n-1}, n > 1; \\ b_n = a_{n-1} + b_{n-1}, n > 1. \end{cases}$

So $t_1 = \frac{a_1}{b_1} = \frac{2}{1}$,

$$t_2 = \frac{a_2}{b_2} = \frac{a_1 + 3b_2}{a_1 + b_1} = \frac{2 + 3}{2 + 1} = \frac{5}{3},$$

and $t_3 = \frac{a_3}{b_3} = \frac{a_2 + 3b_2}{a_2 + b_2} = \frac{5 + 3 \cdot 3}{5 + 3} = \frac{14}{8}$.

a. Evaluate the next six terms of this sequence. Does the sequence appear to converge? If so, what appears to be the limit?

b. Use the computer program below to examine the convergence of the sequence.

```
10   A = 2
20   B = 1
30   N = 1
40   PRINT "N", "A", "B", "TERM"
50   TERM = A/B
60   PRINT N, A, B, TERM
70   OLDB = B
80   OLDA = A
90   A = OLDA + 3*OLDB
100  B = OLDA + OLDB
110  N = N + 1
120  GOTO 50
```

c. Investigate the sequence defined as above except that $a_1 = 5$ instead of $a_1 = 2$. Compare and contrast your results with those in parts **a** and **b**.

d. Investigate the sequence defined by $t_n = \frac{a_n}{b_n}$,

where $\begin{cases} a_1 = 2; \\ b_1 = 1; \\ a_n = a_{n-1} + 5 \cdot b_{n-1}, n > 1; \\ b_n = a_{n-1} + b_{n-1}, n > 1. \end{cases}$

Compare and contrast your results with those in **a**, **b**, and **c**.

6. a. Obtain the rules, entry sheets, and an information sheet for a lottery. Use this information to find the number of different entries possible for each game, the probability that you will win first prize with a single entry, and the probability of winning *any* prize with a single entry. Would you advise someone to buy lottery tickets? Why or why not?

b. Repeat the steps in part **a** for another lottery. In which lottery does the person have a better chance of winning? Suggest some reasons why a particular game might be played.

(Note: Do not attempt to buy a ticket in a state lottery without first checking that it is legal for you to do so. All states restrict sales to persons over a certain age.)

7. Isaac Newton generalized the Binomial Theorem to rational exponents. That is, he derived series expansions for such expressions as $(x + y)^{-3}$, $(x + y)^{2/3}$, and $(x + y)^{-5/6}$. What did Newton find? What are the first four terms of the series expansions of the binomials above? How can this extended Binomial Theorem be used to aid in calculations?

SUMMARY

The Summary gives an overview of the entire chapter and provides an opportunity for students to consider the material as a whole. Thus, the Summary can be used to help students relate the various concepts presented in the chapter.

VOCABULARY

Terms, symbols, and properties are listed by lesson to provide a checklist of concepts a student must know.

Summary

A sequence is a function whose domain is the set of consecutive integers greater than or equal to k. An explicit formula for a sequence allows you to find any specified term directly. A recursive formula describes each term in terms of previous ones and is often more efficient for generating the terms.

Two important kinds of sequences are arithmetic and geometric sequences. In an arithmetic sequence, the difference between successive terms is a constant, the constant difference. In a geometric sequence, the ratio of successive terms is a constant, the constant ratio. If an arithmetic sequence has first term a_1 and constant difference d, the nth term $a_n = a_1 + (n-1)d$. If a geometric sequence has first term g_1 and constant ratio r, the nth term $g_n = g_1 r^{n-1}$.

Some infinite sequences converge to a limit while others diverge. A computer is useful for deciding whether or not a sequence is convergent, but it is not an infallible tool. Limits of some sequences can be evaluated algebraically by applying theorems about limits.

A series is the sum of the terms of a sequence. Explicit formulas for arithmetic and geometric series exist. The sum S_n of the first n terms of an arithmetic sequence is given by $S_n = \frac{n}{2}[2a_1 + (n-1)d] = \frac{n}{2}(a_1 + a_n)$. For a geometric series, $S_n = \frac{g_1(1-r^n)}{1-r}$. To examine infinite series,

it is useful to form the sequence S_1, S_2 ... of partial sums. An infinite series may have a limit; some do not. For an infinite geometric series with first term g and constant ratio r, the limit exists when $|r| < 1$; $S_\infty = \frac{g_1}{1-r}$.

The number of ways to select r unordered items from a set of n elements is $_nC_r = \frac{_nP_r}{r!} = \frac{n!}{(n-r)!r!}$. Each selection is called a combination.

A famous configuration of combinations is called Pascal's Triangle. Each row of Pascal's Triangle gives coefficients of terms in the expanded form of the power of a binomial. In particular, for all positive integers n, $(x+y)^n = {}_nC_0 x^n + {}_nC_1 x^{n-1}y + {}_nC_2 x^{n-2}y^2 + \ldots + {}_nC_n y^n$.

A binomial experiment has a fixed number of trials, each with only two possible outcomes (success and failure), the probabilities of which are fixed and which sum to 1. In a binomial experiment with n trials and probability of success p, the probability of exactly k successes is $_nC_k p^k (1-p)^{n-k}$. This is the $(k+1)$st term in the expansion of $(x+y)^n$, where $x = p$ and $y = 1-p$.

A binomial probability distribution underlies some of the theory of quality control. The distribution of the number of defective items in a large population of a product can be predicted from the observed number of defective items in a sample.

Vocabulary

For the starred (*) terms you should be able to give a definition of the term.
For the other terms you should be able to give a general description and a specific example of each.

Lesson 8-1
*sequence, term, position
explicit formula, recursive formula
*arithmetic sequence
*geometric sequence

Lesson 8-2
limit of a sequence, limiting value
infinity, ∞, *divergent
*convergent, converge to L
infinite loop (in a computer program)

Lesson 8-3
*series, *arithmetic series
infinite series, finite series

Lesson 8-4
*geometric series, nth partial sum

Lesson 8-5
sequence of partial sums
sum of an infinite series, S_∞
harmonic series

Lesson 8-6
*combination, combinations of
  n things taken r at a time
$_nC_r$, $\binom{n}{r}$
$_nC_r$ Calculation Theorem

Lesson 8-7
Pascal's Triangle

Lesson 8-8
series expansion of $(x+y)^n$
Binomial Theorem
binomial coefficients

Lesson 8-9
*binomial experiment
binomial probability distribution

Lesson 8-10
operating characteristic curve

Progress Self-Test

The Progress Self-Test provides an opportunity for feedback and correction. The Chapter Review provides additional opportunities for practice.

See margin for answers not shown below.

Take this test as you would take a test in class. You will need a computer. Then check the test yourself using the solutions at the back of the book.

In 1 and 2, consider this computer program that generates terms in a sequence.

```
10  PRINT "N", "G(N)"
20  G = 0
30  FOR N = 1 TO 10
40      G = G + N ^ 2
50      PRINT N, G
60  NEXT N
70  END
```

1. Is the formula for the sequence explicit or recursive?

2. Find the first 4 terms of the sequence.
1) recursive; 2) 1, 5, 14, 30

3. The first term of an arithmetic sequence is -12 and the constant difference is d. Find the 15th term. $-12 + 14d$

4. Consider the sequence $9, 3, 1, \frac{1}{3}, \dots, \frac{1}{243}$.
 a. Is the sequence arithmetic, geometric, or neither?   **geometric**
 b. Find the 7th term of the sequence.   **1/81**

5. An employee begins a job paying $19,000 per year with the guarantee of a 5% increase in salary each year.
 a. What is the employee's salary during the fourth year?   **$21,994.88**
 b. What is the employee's salary during the nth year?   **$19,000(1.05)$^{n-1}$**

6. Consider the sequence given by
$$\begin{cases} a_1 = 3; \\ a_n = a_{n-1} - \frac{1}{2}n, \ n \geq 2. \end{cases}$$
 a. Sketch a graph of this sequence.
 b. Is the sequence convergent? If so, find its limit.   **no**

7. On each swing, a certain pendulum swings 70% of the length of its previous swing. How far will the pendulum swing before coming to rest if its first swing is 75 cm?   **250 cm**

8. a. Complete the following computer program to help determine
$$\sum_{i=1}^{\infty} \frac{3}{i^3} = \lim_{n \to \infty} \sum_{i=1}^{n} \frac{3}{i^3}.$$
```
10  LET SUM = 0
20  LET I = 1
30  SUM = SUM +  i.
40  PRINT I,  ii.
50  I = I + 1
60  GOTO 30   i) 3/(I*I*I); ii) SUM
```
 b. Use your program from part **a** to find
$$\lim_{n \to \infty} \sum_{i=1}^{n} \frac{3}{i^3} \text{ correct to three decimal places.}$$ **3.606**

9. Find the sum of all the integers n from 100 to 200.   **15,150**

10. Find $\sum_{i=1}^{10} 7\left(\frac{3}{4}\right)^i$ to the nearest ten-thousandth.   **19.8174**

11. Lillian has begun running every day for exercise. The first day she ran 0.5 miles. The 12th day she ran 3.25 miles. If she ran a constant distance farther each day than on the previous day, how many total miles will she have run in all 12 days combined?   **22.5 miles**

12. Prove that for all integers n greater than or equal to 1,
$$_nC_2 = {}_nC_{n-2} = \frac{n(n-1)}{2}.$$

13. a. Expand $(x + y)^3$. $x^3 + 3x^2y + 3xy^2 + y^3$
 b. Explain how your answer to **a** is related to Pascal's Triangle.   **The coefficients are the terms in row 3 of Pascal's Triangle**

In 14 and 15, *multiple choice*.

14. Consider $(2c - b)^{10}$. The first term in its binomial expansion is $(2c)^{10}$. What is the 4th term?
 (a) $(2c)^7$
 (b) $(2c)^7(-b)^3$
 (c) $10 \cdot 9 \cdot 8(2c)^7(-b)^3$
 (d) $\frac{10 \cdot 9 \cdot 8}{3!}(2c)^7(-b)^3$ **(d)**

ADDITIONAL ANSWERS
6.a.

12. $_nC_2 = \dfrac{n!}{(n-2)!2!} =$
$\dfrac{n(n-1)(n-2)!}{(n-2)!2} = \dfrac{n(n-1)}{2}$
$_nC_{n-2} = \dfrac{n!}{(n-n+2)!(n-2)!}$
$= \dfrac{n!}{2!(n-2)!} = \dfrac{n(n-1)}{2}$
So $_nC_2 = {}_nC_{n-2} = \dfrac{n(n-1)}{2}$.

15. In a talent contest with 50 finalists (one from each state), the top three winners will be given identical prizes. How many different sets of possible winners are there?

(a) 3^{50} (b) 50^3

(c) $50 \cdot 49 \cdot 48$ (d) $\dfrac{50 \cdot 49 \cdot 48}{3!}$ **(d)**

16. Suppose the following computer program is used to simulate binomial events.

```
10  FOR K = 1 TO 80
20  X = RND(1)
30  IF X < 0.63 THEN PRINT
    "YES" ELSE PRINT "NO"
40  NEXT K
```

a. What is the probability of success for this binomial event? **.63**

b. If the first three random numbers generated by the computer were 0.162387, 0.818193, and 0.905168, give the first three lines of the output of the program. **YES, NO, NO**

17. Suppose the probability that a randomly selected heart transplant patient will survive more than one year is 0.86. Find the probability that at least 4 of 5 randomly selected heart patients will survive for longer than a year. **≈ .853**

18. Show three places in Pascal's Triangle where the following property is displayed.

$$_nC_0 + {_nC_2} + {_nC_4} + \ldots + {_nC_{n-1}} =$$
$$_nC_1 + {_nC_3} + {_nC_5} + \ldots + {_nC_n},$$

where n is an odd integer.

Sample—row 1: 1 = 1; row 3: 1 + 3 = 3 + 1; row 5: 1 + 5 + 10 = 10 + 5 + 1

Chapter Review

Questions on **SPUR** Objectives

SPUR stands for **S**kills, **P**roperties, **U**ses, and **R**epresentations.
The Chapter Review questions are grouped according to the SPUR Objectives for this chapter.
See margin for answers not shown below.

SKILLS deal with the procedures used to get answers.

■ **Objective A:** *Use explicit or recursive formulas to determine terms of sequences. (Lesson 8-1)*

In 1 and 2, write the first 5 terms of the sequence.

1. $R_n = n^2 - n$ **0, 2, 6, 12, 20**

2. $\begin{cases} B_1 = -6; \\ B_n = B_{n-1} + 3n, \ n \geq 2 \end{cases}$ **-6, 0, 9, 21, 36**

In 3 and 4, consider the following program.

```
10   PRINT "N", "A(N)"
20   DEF FN F(N) = 1/N
30   FOR N = 1 TO 30
40       PRINT N, (FN F(N))^2
50   NEXT N
60   END
```

3. Determine the first and last lines of output when this program is run.   **1 1; 30 0.00111111**

4. Write an explicit formula for the *n*th term of the sequence this program produces. $t_n = \left(\frac{1}{n}\right)^2$

5. Give an explicit formula for the sequence defined by
$\begin{cases} k_1 = 22{,}000 \\ k_n = 0.8k_{n-1}, \ n \geq 2. \end{cases}$ $k_n = 22{,}000(.8)^{n-1}$

6. Give a recursive formula for the sequence defined by $A_n = 32 - 5n$. $\begin{cases} A_1 = 27 \\ A_n = A_{n-1} - 5 \quad \text{for } n > 1 \end{cases}$

■ **Objective B:** *Find the nth term of an arithmetic or geometric sequence. (Lesson 8-1)*

In 7 and 8, read the computer program and give the 100th line of output. (It should not be necessary to RUN the program.)

7.
```
10   LET TERM1 = 5000
20   FOR I = 1 TO 1000
30       TERM = TERM1 − 7*I
40       PRINT I, TERM
50   NEXT I
60   END        100   4300
```

8.
```
10   LET TERM = 7
20   FOR I = 1 TO 1000
30       TERM = 1.01*TERM
40       PRINT I, TERM
50   NEXT I
60   END        100   18.933697
```

9. Find the 50th term of the arithmetic sequence 84, 67, 50, 33, … .   **-749**

10. Find the 12th term of the geometric sequence 24, -84, 294, … . $24(-3.5)^{11} \approx$ **-23,171,797.77**

■ **Objective C:** *Find the sum of the terms of a sequence. (Lessons 8-3, 8-4)*

In 11–14, evaluate the arithmetic or geometric series.

11. $103 + 120 + 137 + 154 + \ldots + 290$ **2358**

12. $(5u - v) + (4u + v) + (3u + 3v) + \ldots + (23v - 7u)$ **143v − 13u**

13. $1 + 3 + 9 + 27 + \ldots + 3^{14}$ **7,174,453**

14. $x + x^2y + x^3y^2 + \ldots + x^{21}y^{20}$ $\frac{x - x^{22}y^{21}}{1 - xy}$

15. Evaluate $\sum_{i=1}^{5} (i^2 - i)$.   **40**

16. Evaluate $\sum_{n=1}^{20} 10(0.6)^{n-1}$. $\approx$ **24.999**

CHAPTER REVIEW

The main objectives for the chapter are organized here into sections corresponding to the types of understanding this book promotes: Skills, Properties, Uses, and Representations (called SPUR objectives).

USING THE CHAPTER REVIEW

Students should be able to answer questions like these with about 85% accuracy by the end of the chapter. (See pages 72 and T38–39 for more information.)

■ **Objective D:** *Expand binomials. (Lesson 8-8)*

In 17–19, expand.

17. $(a+b)^2$ $\quad a^2 + 2ab^2 + b^2$

18. $(x+y)^6$ $\quad x^6 + 6x^5y + 15x^4y^2 + 20x^3y^3 + 15x^2y^4 + 6xy^5 + y^6$

19. $(2x-5)^4$ $\quad 16x^4 - 160x^3 + 600x^2 - 1000x + 625$

In 20 and 21, *true* or *false*.

20. The first term of the binomial expansion of $(5p+2q)^{17}$ is $5p^{17}$. **False**

21. The second term of the binomial expansion of $(x-y)^{12}$ is $_{12}C_2(x)^{10}(-y)^2$. **False**

22. Find the 10th term in the expansion of $(p+q)^{10}$. **$10pq^9$**

PROPERTIES deal with the principles behind the mathematics.

■ **Objective E:** *Classify a sequence as arithmetic, geometric, or neither. (Lesson 8-1)*

In 23–28, classify the sequence as possibly being arithmetic, geometric, or neither.

23. $13, 24, 35, 46, \ldots$ **arithmetic**

24. $44, 50, 57, 65, \ldots$ **neither**

25. $3u, u+v, 2v-u, 3v-3u, \ldots$ **arithmetic**

26. $\begin{cases} a_1 = -2 \\ a_n = (a_{n-1})^2 + 1, \, n>1 \end{cases}$ **neither**

27. $t_n = -8(-0.7)^n$ **geometric**

28. $\begin{cases} a_1 = x \\ a_n = y - a_{n-1}, \, n \geq 2 \end{cases}$ **neither**

In 29 and 30, the computer program generates a sequence. Classify the sequence as arithmetic, geometric or neither.

29.
```
10  A = 17
20  N = 1
30  PRINT N, A
40  N = N + 1
50  A = A + 14
60  GOTO 30
```
arithmetic

30.
```
10  A = -8
20  N = 1
30  PRINT N, A
40  N = N + 1
50  A = A + 7*N
60  GOTO 30
```
neither

■ **Objective F:** *Decide whether a sequence has a limit, and if it does, determine the limit. (Lesson 8-2)*

In 31–34, tell whether a limit of the arithmetic or geometric sequence exists. If it does, state the limiting value.

31. $4, 6, 9, 13.5, 20.25, \ldots$ **limit does not exist**

32. $\begin{cases} a_1 = 4 \\ a_n = a_{n-1} \end{cases}$ **Yes; 4**

33. $80, 60, 45, 33.75, \ldots$ **Yes; 0**

34. $80, 60, 40, 20, \ldots$ **limit does not exist**

In 35 and 36, find the limit of the sequence if it exists.

35. $h_n = \dfrac{7}{n}$ **yes; 0**

36. $\begin{cases} v_1 = 1 \\ v_n = (-1)^n \cdot v_{n-1} \text{ for } n>1 \end{cases}$ **limit does not exist**

37. Use the computer program below, which prints terms of the sequence

$$-\frac{19}{69}, -\frac{22}{67}, -\frac{25}{65}, \ldots, \frac{3n+16}{2n-71}, \ldots.$$

```
10  LET N = 1
20  PRINT N, (3*N + 16)/(2*N - 71)
30  LET N = N + 1
40  GOTO 20
```

a. Does the sequence converge? **a) Yes; b) 1.5**

b. If the sequence is convergent, give its limit. If it is divergent, explain how you can tell.

38. Use a computer program to decide whether the sequence below is convergent or divergent. If the sequence converges, give its limit as a fraction in lowest terms.

$1, \dfrac{11}{16}, \dfrac{16}{24}, \dfrac{33}{22}, \ldots, \dfrac{n^2+7}{8n}, \ldots$ **divergent**

■ **Objective G:** *Tell whether an infinite series converges. If it does, give the limit. (Lesson 8-5)*

In 39–41, for the geometric series shown, state whether or not the series is convergent. If the series is convergent, give its sum.

39. $8 - \dfrac{40}{3} + \dfrac{200}{9} - \dfrac{1000}{27} + \ldots$ **not convergent**

40. $7 + 5.6 + 4.48 + \ldots$ **convergent; 35**

41. $a - \dfrac{a}{5} + \dfrac{a}{25} - \dfrac{a}{125} + \ldots$ **$\dfrac{5}{6}a$**

In 42 and 43, use a computer to conjecture whether the series is convergent. If it seems to be convergent, give what seems to be its limit.

42. $1 - \frac{1}{2} + \frac{1}{3} - \frac{1}{4} + \frac{1}{5} - \ldots + \frac{(-1)^{n+1}}{n} + \ldots$ **≈ .69**

43. $\frac{1}{1^5} + \frac{1}{2^5} + \frac{1}{3^5} + \frac{1}{4^5} + \ldots + \frac{1}{n^5} + \ldots$ **≈ 1.0369**

44. a. Under what condition does the following geometric series have a value?
$a^3b^2 + a^3b^4 + a^3b^6 + a^3b^8 + \ldots$ **$b^2 < 1$**
b. Give an expression for the sum, when it exists. **$\frac{a^3b^2}{1-b^2}$**

■ **Objective H:** *Prove identities involving combinations.* *(Lesson 8-6)*

In 45–47, prove the given identity.

45. $_nC_r = {_nC_{n-r}}$ for all positive integers n and r, $r \le n$.

46. $_nC_1 = {_nC_{n-1}} = n$

47. $_nC_r = {_{n-1}C_r} + {_{n-1}C_{r-1}}$

48. *True* or *false*. For all positive integers n and r such that $r \le n$, $_nP_r > {_nC_r}$. **False**

45) $_nC_r = \frac{n!}{(n-r)!r!} = \frac{n!}{(n-r)!(n-(n-r))!} = {_nC_{n-r}}$

USES deal with applications of mathematics in real situations.

■ **Objective I:** *Use sequences and series to solve problems.* *(Lessons 8-1, 8-2, 8-3, 8-4, 8-5)*

49. One sunflower produces 500 seeds. Each seed produces a flower which produces 500 seeds, and so on. Let a_n = the number of seeds after n generations. Write a formula for a_n that is
a. explicit; **b.** recursive.
c. Assuming no losses, how many sunflower seeds are produced at the end of five generations, starting with a first generation of 100 sunflowers?

50. In a set of steps leading to a ship, the third step is 60 inches above the water and the fifth step is 71 inches above the water. Assuming all steps have equal rise, how far is the ninth step above the water? **93 in.**

51. In mid-1986, the population of Indonesia was about 168 million, with an average annual growth rate of 2.1%. Estimate the population in the middle of the year 2000, assuming that the population grows geometrically and the growth rate does not change. **≈ 224.7 million**

52. The coach of a soccer team rewards the players on the team by putting stars on jerseys, and increases by 2 stars every time the team wins. The first win gave 3 stars.
a. Determine how many times the team has won if their jerseys have 17 stars on them? **8**
b. Determine the least number of wins needed in order to have 100 stars on the jersey. **10**

53. In a certain housing complex, rents are increased by 4% every year. Consider a family which has been living at this complex for 15 years and assume that they used to pay $200 per month in the first year.
a. Determine the monthly rent they are paying this year. **$364.33 per month**
b. At the end of this year, how much will the family have paid during their 15 years of residency? [Hint: consider the total rent the family has paid in each complete year.] **≈ $48,057**

54. The tracks on an LP record can be approximated as concentric circles (rather than a continuous spiral) with a difference of .3 mm between the radii of each consecutive circle. Consider a record on which the grooves start 15 cm away from the center and end 3 cm away from it. What is the distance covered by the needle of a record player in one playing of this LP?
7218π cm $\approx$ 22,676 cm

■ **Objective J:** *Use combinations to compute the number of ways of arranging or selecting objects.* (Lesson 8-6)

55. A committee of three is to be chosen from a faculty of 13 math teachers. How many ways of choosing the committee are there? **286**

56. A DJ wants to select eight hits from the top 40 to play in the next half hour. In how many ways can she do this? **76,904,685**

57. There are 20 members of a tennis club. If each must play the others exactly once in a tennis match, how many games must be played? **190**

58. A euchre hand consists of five cards from a deck of 32 cards. Half the cards are red; half, black.
    a. How many different euchre hands are there? **201,376**
    b. How many euchre hands contain only red cards? **4368**

■ **Objective K:** *Determine probabilities in situations involving binomial experiments.* (Lesson 8-9)

In 59 and 60, the McDonnells know that each of their 5 children likes roughly 80% of the suppers they serve. Assume that the children's likes and dislikes are independent.

59. What is the probability that all 5 children will like a given supper? **.32768**

60. What is the probability that at least one person will dislike supper? **.67232**

61. If Mr. and Mrs. Washington both have siblings with cystic fibrosis, the probability that any given child born to them has the disease is $\frac{1}{9}$. If the family will have three children, what is the probability that at least one will have cystic fibrosis? **$\approx$ 30%**

62. Twenty percent of young trees die during their first winter. Walter, a landscape contractor, is to install 10 trees at the front of an auditorium. He installs 11 to be safe. What is the probability that at least 10 trees will survive? **$\approx$ 32%**

63. What is the probability that a student will get a perfect score by random guessing on an exam with 17 multiple-choice questions, each of which has four options? **$1/4^{17} \approx 5.8 \times 10^{-11}$**

REPRESENTATIONS deal with pictures, graphs, or objects that illustrate concepts.

■ **Objective L:** *Locate numerical properties represented by the patterns in Pascal's Triangle.* (Lesson 8-7)

In 64–67, show three places in Pascal's Triangle where the given property is represented.

64. $_nC_r = {_nC_{n-r}}$

65. $_nC_r = {_{n-1}C_r} + {_{n-1}C_{r-1}}$

66. $\sum_{i=0}^{n} [(-1)^i \cdot i \cdot {_nC_i}] = 0$

67. $\sum_{i=0}^{n} (i \cdot {_nC_i}) = n \cdot 2^{n-1}$

64) True in every row; each is symmetric.
65) Every element off the border; each is the sum of the two elements above it.
66) True in every row.
67) True in every row.

OBJECTIVES ■ CHAPTER 9

The objectives listed here are the same as in the Chapter 9 Review on pages 587–590 of the student text. The Progress Self-Test on page 586 and the tests in the Teacher's Resource File cover these objectives. For recommendations regarding the handling of this end-of-chapter material, see the notes in the margins on the corresponding pages of this Teacher's Edition.

OBJECTIVES FOR CHAPTER 9 (Organized into the SPUR categories— Skills, Properties, Uses, and Representations)	Progress Self-Test Questions	Chapter Review Questions	Teacher's Resource File	
			Lesson Masters*	Chapter Test Forms A & B
SKILLS				
A Use finite differences and systems of equations to determine an equation for a polynomial function from data points.	1	1–4	9-2	1, 9
B Calculate or approximate zeros and relative extrema of polynomial functions.	2	5–8	9-3	12
C Divide polynomials.	3	9–12	9-4	7
D Factor polynomials and solve polynomial equations using the Factor Theorem, sums or differences of powers, grouping terms, or trial and error.	4, 5	13–20	9-5, 9-6, 9-8, 9-9	10, 11
E Perform operations with complex numbers.	7	21–26	9-6	8
PROPERTIES				
F Apply the vocabulary of polynomials.	8–10	27–30	9-1, 9-2, 9-3	5
G Apply the Remainder Theorem, Factor Theorem, and Factor-Solution-Intercept Equivalence Theorem.	6	31–34	9-4, 9-5	2, 14
H Apply the Fundamental Theorem of Algebra and Conjugate Zeros Theorem.	11, 12	35–38	9-7	13b
USES				
I Construct and interpret polynomials that model real world situations.	13	39–43	9-1, 9-2	3
REPRESENTATIONS				
J Represent two- or three-dimensional figures with polynomials.	14, 16	44–46	9-1, 9-2	13a
K Relate properties of polynomial functions and their graphs.	15	47–53	9-3, 9-5, 9-7	4, 6

*The masters are numbered to match the lessons, and each is intended for use with that lesson.

OVERVIEW ■ CHAPTER 9

Polynomial functions are important for a variety of reasons. Any finite set of points with no two on the same vertical line can be modeled by a polynomial function. Even a function that is not a polynomial function, if it satisfies certain broad conditions, can be approximated by a polynomial function. Linear, quadratic, and power functions, important in their own right, are all classes of polynomial functions.

This chapter has three themes. Modeling is the topic of the first three lessons. Applications to annuities and volume are used to introduce the terminology and basic properties of polynomials in Lesson

9-1. Lesson 9-2 covers the fitting of polynomial functions to a finite number of data points. Lesson 9-3 discusses the various shapes of polynomial function graphs, thus providing some intuition for the types of data that might be modeled by these functions.

The second theme is factoring a polynomial. This is found in Lessons 9-4 and 9-5, in which the general theory is established through the Remainder and Factor Theorems, and again in Lessons 9-8 and 9-9, in which techniques for factoring are described and practiced.

The third theme is finding the zeros of polynomial functions. The

major result is the Fundamental Theorem of Algebra, given in Lesson 9-7. For this, complex numbers are reviewed in Lesson 9-6. The relationships between the zeros and the coefficients of a polynomial are discussed in Lesson 9-10.

Before automatic graphers, much of the theory of polynomial functions was employed to determine the properties of a graph. Even though automatic graphers do now exist, there are still many reasons to learn about polynomials. And, since the mathematical theory is exquisite, there are some interesting and unexpected results to encounter along the way.

PERSPECTIVES ■ CHAPTER 9

The Perspectives provide the rationales for the inclusion of topics or approaches, give some mathematical background, and make connections with other lessons and to other UCSMP texts.

9-1
POLYNOMIAL MODELS
The application in this lesson to annuities is very useful. If you want a loan of a particular amount of money and need to pay it back in a certain amount of time, you can determine the equal monthly payments from the interest rate.

For any number of compounding periods greater than 2, the polynomial model will have a degree higher than 2. This application should be familiar to students who have studied UCSMP *Algebra* or *Advanced Algebra*.

9-2
FINDING POLYNOMIAL MODELS
All continuous functions on closed intervals and many other functions can be approximated by polynomial functions to as close a degree of accuracy as desired (a theorem known as the Weierstrass Approximation Theorem, beyond the scope of this course but the subject of a

reading lesson in *Precalculus and Discrete Mathematics*). Thus there is never a question whether given data points can be approximated by a polynomial model. The only question is how high a degree you wish to allow the polynomial to have.

In Chapter 2, students were shown how to determine linear and quadratic models by solving a system of equations. This lesson extends the idea to polynomials of higher degree. First is the determination of the degree of the polynomial that fits given data. This is done by taking successive differences of values of the polynomial until a constant is reached. Then, once the degree n is established, solving a system of $n + 1$ equations in $n + 1$ unknowns determines the coefficients. Unlike some of the models of earlier chapters, the models of this and the previous lesson are exact models; they fit the given data points exactly.

The Polynomial Difference Theorem is a theorem of discrete mathe-

matics. Its continuous analog is that the nth derivative of a polynomial function is a constant nonzero function if and only if the polynomial is of degree n.

Students who have studied UCSMP *Advanced Algebra* should be familiar with the content of this lesson.

9-3
GRAPHS OF POLYNOMIAL FUNCTIONS
There are many ways to find zeros of polynomial functions. This lesson concentrates on two of the ways that have been influenced by technology: the use of programs to generate tables of values and the use of graphs. In each the idea is to find an interval in which a zero is located, and then to consider smaller and smaller subintervals that contain the zero until the desired accuracy is reached. This lesson cannot be done without appropriate technology.

9-4

DIVISION AND THE REMAINDER THEOREM

Since we are going to study the zeros of polynomials in some detail, the ability to find quotients is quite useful. This lesson covers polynomial long division, which will be new to UCSMP students, and proves the Remainder Theorem. If the dividend is $p(x)$ and the divisor is of the form $x - c$, this provides students with a relatively quick way to check their work; the remainder should be $p(c)$.

Precalculus and Discrete Mathematics studies the analogies between integer and polynomial division.

9-5

THE FACTOR THEOREM

A number c is a solution to a polynomial equation $p(x) = 0$ if and only if $x - c$ is a factor of $p(x)$. This statement, known as the Factor Theorem, enables a factor $x - c$ to be found from a zero of a polynomial $p(x)$. By division, a polynomial function $q(x) = \dfrac{p(x)}{x - c}$, whose degree is one less, can be examined for its zeros. Thus the Factor Theorem is a powerful tool for finding more zeros of a polynomial, given that you can find one zero. In this lesson, the Factor Theorem is proven and applied.

In contrast to the delay of polynomial long division, UCSMP texts cover the Factor Theorem early. A Factor Theorem for Quadratics is given in UCSMP *Algebra* as a powerful way to factor any quadratic. In UCSMP *Advanced Algebra*, the Factor Theorem is proved using the Graph Translation Theorem.

9-6

COMPLEX NUMBERS

A complex number is defined here to be a number of the form $a + bi$, where a and b are real and $i^2 = -1$. We assume that complex numbers satisfy the field properties; this enables them to be operated with as polynomials are, and so the rules for operating with them become theorems rather than definitions.

This lesson reviews the four fundamental operations with complex numbers and shows students how to rewrite the nonreal solutions to quadratic equations in $a + bi$ form.

The geometry of these numbers (graphing and interpretations of the operations) can be found in Lesson 13-5.

Students who have studied UCSMP *Advanced Algebra* have seen this content. If your students have never studied complex numbers, then plan to spend more than one day on this lesson—perhaps three days on this and Lesson 9-7 together.

9-7

THE FUNDAMENTAL THEOREM OF ALGEBRA

Complex numbers arose as numbers that could represent solutions to quadratic equations. Then it was found that these same numbers sufficed to solve all cubic and quartic equations, even those with nonreal coefficients. Still, the discovery by Gauss in 1797 that all polynomial equations with real or nonreal coefficients have a complex solution was an extraordinary breakthrough. After all, if you need new numbers to solve all quadratics with real coefficients, is it not natural to believe you would need still different new numbers to solve some equations of higher degree?

This lesson begins with a brief history of that theorem of Gauss, now known as the Fundamental Theorem of Algebra. From it the lesson goes on to prove (with the aid of the Factor and Remainder Theorems) that every nonzero polynomial of degree n has at most n zeros and, if multiplicities are counted, exactly n zeros. Furthermore, if the polynomial has real coefficients, the complex zeros come in conjugate pairs. These theorems, put together, enable all the zeros of many polynomial functions to be determined exactly.

9-8

FACTORING SUMS AND DIFFERENCES OF POWERS

If $x^n = a^n$, then $x^n - a^n = 0$, and solutions can be found by factoring $x^n - a^n$. In this lesson, the Sums and Differences of Powers are factored, and students learn the factorizations for the sum and difference of two cubes.

9-9

ADVANCED FACTORING TECHNIQUES

The "advanced" factoring techniques studied here are found in many first-year algebra books. They involve more than the direct application of the Factor Theorem, learned formulas, or the distributive property, in its simplest form.

In this lesson, students are asked to factor expressions of the form $ax + bx + ay + by$ and $ax^2 + bxy + cy^2$. In Chapter 12, students will apply these factorizations to graph relations of the form $(ax + by + c)(dx + ey + f) = 0$, which are degenerate conic sections.

9-10

ROOTS AND COEFFICIENTS OF POLYNOMIALS

Many students will already be familiar with the fact that if r and s are the roots of a quadratic equation $ax^2 + bx + c = 0$, then $r + s = \dfrac{-b}{a}$ and $rs = \dfrac{c}{a}$. This lesson discusses the extension of that result to cubic equations and then further to all polynomial equations. The latter extension, which involves the idea of combinations from Chapter 8, enables equations to be formed rather easily with any given solutions.

DAILY PACING CHART ■ CHAPTER 9

For optimal student performance, the Progress Self-Test and the Chapter Review should be covered. (See *General Teaching Suggestions: Mastery* on page T35 of this Teacher's Edition.) By following the pace of the Full Course given here, students can complete 12 or 13 chapters of the book by the end of the year. Students following the pace of the Minimal Course spend more time to complete lessons that are combined with quizzes and to consider the Chapter Review. When all lessons are covered in order (the recommendation of the authors), students in the Minimal Course can cover all but the last two chapters. For more information on pacing, see *General Teaching Suggestions: Pace* on page T34 of this Teacher's Edition.

DAY	MINIMAL COURSE	FULL COURSE
1	9-1	9-1
2	9-2	9-2
3	9-3	9-3
4	Quiz (TRF); Start 9-4	Quiz (TRF); 9-4
5	Finish 9-4	9-5
6	9-5	9-6
7	9-6	Quiz (TRF); 9-7
8	Quiz (TRF); Start 9-7	9-8
9	Finish 9-7	9-9
10	9-8	9-10
11	9-9	Progress Self-Test
12	9-10	Chapter Review
13	Progress Self-Test	Chapter Test (TRF)
14	Chapter Review	
15	Chapter Review	
16	Chapter Test (TRF)	

TESTING OPTIONS

■ Quiz for Lessons 9-1 through 9-3   ■ Chapter 9 Test, Form A   ■ Chapter 9 Test, Cumulative Form
■ Quiz for Lessons 9-4 through 9-6   ■ Chapter 9 Test, Form B   ■ Comprehensive Test, Chapters 1–9

A Quiz and Test Writer is available for generating additional questions, additional quizzes, or additional forms of the Chapter Test.

PROVIDING FOR INDIVIDUAL DIFFERENCES

The student text has been written for, and tested with, average students. It has also been used successfully with students who are better prepared as well as with students who are poorly prepared.

The Lesson Notes often include Error Analysis and Alternate Approach features to help you assist those students who need more help. Students of all abilities often learn from their peers and may benefit from small group work referenced as appropriate throughout the Notes. A blackline Lesson Master (in the Teacher's Resource File), keyed to the chapter objectives, is available for each lesson to provide more practice. (However, we advise you to use these masters sparingly; again refer to the suggestions on pace on page T34.)

Extension activities are provided in the Lesson Notes for use with students who need to be challenged even more than the requirements of completing the Full Course provide.

Polynomial Functions

A polynomial function like P_4 could be used to model the shape of the nose of this aircraft.

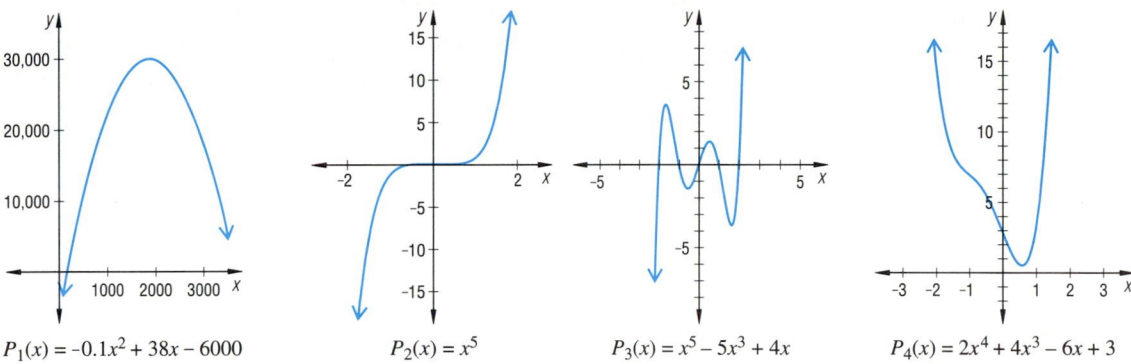

$P_1(x) = -0.1x^2 + 38x - 6000$ $\qquad$ $P_2(x) = x^5$ $\qquad$ $P_3(x) = x^5 - 5x^3 + 4x$ $\qquad$ $P_4(x) = 2x^4 + 4x^3 - 6x + 3$

A *polynomial* in one variable is a sum of multiples of nonnegative integer powers of that variable. The quadratic function P_1 and power function P_2 graphed above are examples of *polynomial functions*.

Other polynomial functions such as P_3 and P_4 exhibit more complicated behavior.

Despite their variety, all polynomial functions share many important properties. The theory underlying these properties involves the complex numbers, division, and factoring.

The variety of polynomial functions makes them candidates for modeling many real situations, some of which they model exactly and others which they approximate. In this chapter you will see applications of polynomial functions and their graphs in farming and economics and other fields, as well as see some of the beautiful mathematical results that apply to these functions.

CHAPTER 9

This chapter can be covered at a lesson-per-day pace, particularly if your students have had any experience with polynomial functions. Since Lessons 9-2, 9-3, 9-6, 9-9, and 9-10 are relatively short, some flexibility can be exercised in allocating the time needed to cover them. Allowing 10–12 days for the lessons and quizzes, 1 day for the Progress Self-Test, 1 or 2 days for the Chapter Review, and 1 day for the Chapter Test, this is a 13–16 day chapter.

USING PAGE 523
Alternative Assessment
Engage students in an informal question-and-answer session to find out what they know about factoring, about complex numbers, about operations with polynomials (including long division), and about graphing functions of the type found on this page. Almost all students should have familiarity with some of these ideas. This will enable you to determine in advance where students may have trouble and where you should plan to focus your time.

CHAPTER 9 Polynomial Functions **523**

OBJECTIVES

F Apply the vocabulary of polynomials.

I Construct and interpret polynomials that model real world situations.

J Represent two- or three-dimensional figures with polynomials.

TEACHING NOTES

A possible way to introduce this lesson follows: Go through the example on page 524 line by line, asking students after each sentence if they follow what was being done. Continue with the discussion on page 525. Note that, although M (the monthly payment) can be written as a quotient of polynomials, if you know the interest rate you can determine M by substitution, as is done there. Or, if you know how much you are able to pay per month, you could determine the maximum interest rate you could afford.

If the amount of money is a constant M and is multiplied by the scale factor x in that interval, then the total amount paid after $n + 1$ payments has grown to $Mx^n + Mx^{n-1} + Mx^{n-2} + \ldots + Mx + M$. If the amount paid at the time of payment k is not necessarily constant but called a_{n-k}, then the total amount paid after $n + 1$ payments is $a_nx^n + a_{n-1}x^{n-1} + a_{n-2}x^{n-2} + \ldots + a_1x + a_0$. This is the general polynomial in x of degree n. This description demonstrates

Polynomial Models

Stuart Dent wanted to buy a used car but did not have enough cash on hand. A bank agreed to lend him $3000 at a 12% annual interest rate to be repaid by equal monthly installments of $M. How do you determine M so that the loan is completely paid after 36 equal installments? If there were no interest, M would be $\frac{$3000}{36}$, or about $83.33. But since there is interest, the situation is more complicated.

Here is an analysis of the situation. On most loans an interest rate of $r\%$ per year means that $\frac{r}{12}\%$ of the unpaid balance is charged each month. So Stu will be charged $\frac{12\%}{12} = 1\% = .01$ times the unpaid balance each month. The rest of his monthly payment will be subtracted from the amount still to be paid on the principal.

Let B_i be the balance after i months. Then $B_0 = 3000$. Let r be the monthly interest rate. Here r is $\frac{1}{12}$ of the yearly rate, or 1%, but it is easier to do the analysis in general than to do it for the specific case.

The balance after 1 month is the original principal + interest − amount paid. That is,
$$B_1 = 3000 + 3000r - M = 3000(1 + r) - M.$$
Similarly, $\quad B_2 = B_1 + B_1r - M = B_1(1 + r) - M.$
In general, $\quad B_{i+1} = B_i + B_ir - M = B_i(1 + r) - M$ for all $i \geq 1$.

It is convenient to let $x = 1 + r$, since that expression appears so often. By repeated substitution,
$$B_1 = 3000x - M$$
$$B_2 = B_1x - M = (3000x - M)x - M = 3000x^2 - Mx - M$$
$$B_3 = B_2x - M = (3000x^2 - Mx - M)x - M = 3000x^3 - Mx^2 - Mx - M.$$

Continuing this process a total of 36 times,
$$B_{36} = 3000x^{36} - Mx^{35} - Mx^{34} - \ldots - Mx - M.$$
The expression for B_{36} is an instance of a *polynomial in the variable x*.

Definition

A **polynomial in x** is an expression of the form
$$a_nx^n + a_{n-1}x^{n-1} + a_{n-2}x^{n-2} + \ldots + a_1x + a_0$$
where n is a nonnegative integer and $a_n \neq 0$.

524

Note that the variable of a polynomial cannot have negative or noninteger exponents. The number n is the **degree** of the polynomial and the numbers $a_n, a_{n-1}, a_{n-2}, \ldots, a_0$ are its **coefficients**. The number a_n is called the **leading coefficient** of the polynomial. For example, in the polynomial obtained in analyzing Stuart's loan repayment schedule, the degree is 36 and the leading coefficient is 3000.

Notice that except for the leading coefficient 3000, all of the coefficients of the polynomial for B_{36} are equal. Since B_{36} is the balance after 36 months, $B_{36} = 0$. Thus

$$0 = 3000x^{36} - M(x^{35} + x^{34} + \ldots + x + 1).$$

The expression in parentheses is a geometric series, so it can be rewritten.

$$0 = 3000x^{36} - M\frac{x^{36} - 1}{x - 1}$$

This equation can be solved for M.

$$M = 3000x^{36} \cdot \frac{x - 1}{x^{36} - 1}$$

Now recall that $x = 1 + r = 1 + .01 = 1.01$. Substituting for x,
$$M \approx \$99.64293\ldots .$$

Because the amount Stu pays each month is constant, in total he would probably be required to pay 36 installments of $99.65, or $3587.40.

Polynomials arise in other situations that you have studied before.

■ ■ ■ ■ ■ ■ ■ ■

Example 1   Tamra is saving her summer earnings for college. The table below shows the amount of money earned each summer.

After grade	earned
8	$ 600
9	$ 900
10	$1100
11	$1500
12	$1600

At the end of each summer she put her money in a savings account with an annual yield of 7%. How much will be in her account when she goes to college if no additional money is added or withdrawn, and the interest rate remains constant?

Solution   Recall the compound interest formula, $A = P(1 + r)^t$, which gives the value of P dollars invested at an annual interest rate r after t years. The money Tamra put in the bank after grade 8 earns interest for 4 years. It is worth $600(1.07)^4$ when Tamra enters college. Similarly, the amount saved at the end of grade 9 is worth $900(1.07)^3$. Adding the values from each summer gives the total amount in the bank account.

$$600(1.07)^4 + 900(1.07)^3 + 1100(1.07)^2 + 1500(1.07) + 1600$$

from summer after grade 8	from summer after grade 9	from summer after grade 10	from summer after grade 11	from summer after grade 12

Evaluating this expression shows that Tamra will have about $6353.

that annuities are a powerful enough application to cover polynomials of any degree.

In **Example 3,** note that each term of the answer has degree 3. This should be expected, since the answer stands for volume, and the dimension of volume is 3.

Students may wonder if there are special names for polynomials with four terms, five terms, and so on. There are no such names in common use, though quadrinomials (4 terms) can be found in some older texts.

ADDITIONAL EXAMPLES
1. a. Suppose you can earn $\frac{1}{2}$% interest compounded each month (6% per year), paid at the end of the month. If you deposit $100 at the beginning of each month from January through December, how much will be in the account on January 1st of the next year?
$100(1.005^{12} + 1.005^{11} + \ldots + 1.005 + 1) =$
$100\left(\frac{1.005^{13} - 1}{0.05}\right) \approx \$1339.72$

b. Generalize part **a** for any monthly interest rate r.
$100\left(\frac{(1 + r)^{13} - 1}{r}\right)$

2. Evaluate $-x^3 + 2x^5 - x^7$ when $x = -3$.
1728

3. The radius R of a circle is increased by r.

a. Express the area of the enlarged circle as a polynomial in r and R.
$\pi R^2 + 2\pi Rr + \pi r^2$

b. Give the degree of the polynomial.
2, which is to be expected since the expression represents area

LESSON 9-1   Polynomial Models   525

The amount that Tamra has when she enters college depends on the interest rate she can get. If her annual yield is r and again $x = 1 + r$, her savings $A(x)$ can be considered as a *polynomial function of x*.

$$A(x) = 600x^4 + 900x^3 + 1100x^2 + 1500x + 1600$$

In general, a **polynomial function** is a function whose rule can be written as a polynomial.

In the function A above, the coefficients of the terms of the polynomial represent the amount invested each successive year. The exponents represent the number of years each amount is on deposit. The value $A(x)$ of the function is the amount Tamra would have if the annual interest rate were $x - 1$.

Example 2   Consider the preceding function A. Evaluate $A(1.085)$, and state what it represents.

Solution   Substitute $x = 1.085$ in the formula for $A(x)$.

$$A(1.085) = 600(1.085)^4 + 900(1.085)^3 + 1100(1.085)^2 + 1500(1.085) + 1600$$
$$\approx \$6503.52$$

This represents the amount Tamra's investment would be worth at the end of the summer after completing grade 12, if deposited in an account paying $1.085 - 1 = .085 = 8.5\%$ interest, compounded annually.

Recall that a **monomial** is a polynomial with one term; a **binomial** is a polynomial with two terms; and a **trinomial** is a polynomial with three terms.

Monomials, binomials, and trinomials may be of any degree. For instance, the trinomial $4x^2 - 5x + 1$ is of degree 2, while $a^{10} - a^{20} + a^{30}$ is of degree 30.

Polynomials may also involve several variables. For instance, each of the following is a *polynomial in x and y*.

$$x^2y^3 - 3y^3 + 2x^2 - 6$$
$$x^3 + x^2y^2 + y^5$$

The **degree of a polynomial in more than one variable** is the largest sum of the exponents of the variables in any term. Each of the preceding polynomials in x and y has degree 5.

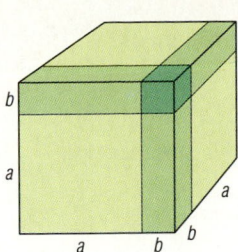

Example 3

a. Express the volume V of a cube with sides of length $a + b$ as a polynomial function in terms of a and b.
b. State the degree of the polynomial.

Solution

a. $V = (a + b)^3$. Use the Binomial Theorem to rewrite the right side.
$V = a^3 + 3a^2b + 3ab^2 + b^3$
b. The sum of the exponents of the variables in each term is 3, so V has degree 3.

Check As illustrated in the figure at the left, the volume of the cube equals the sum of the volumes of 8 rectangular solids. In the top layer are two boxes with volumes equal to b^2a, and one each of volumes b^3 and a^2b. In the bottom layer two of the visible boxes have volume a^2b, and the third has volume b^2a. The box hidden from view has volume a^3. So the total volume is
$$V = (2b^2a + b^3 + a^2b) + (2a^2b + b^2a + a^3)$$
$$= b^3 + 3b^2a + 3a^2b + a^3.$$

Questions

Covering the Reading

1. Consider the polynomial $3p^2 - 5p^6$.
   a. What is its degree? **6**   **b.** What is its leading coefficient? **-5**
   c. *True* or *false*. The polynomial is a binomial. **True**

2. Let $p(x) = 6x^8 + 12x^5 - 4x^2 + 9$. If $p(x)$ is of degree n, state the value of each of the following.
   a. n **8**   **b.** a_n **6**   **c.** a_0 **9**   **d.** a_2 **-4**   **e.** a_{n-1} **0**

3. Determine whether each expression can be written as a polynomial.
   a. $3x^2$ **Yes**   **b.** 5^x **No**   **c.** $2x - 5$ **Yes**
   d. $x^{1/2}$ **No**   **e.** $\dfrac{x^2 + x}{2}$ **Yes**   **f.** $\dfrac{x^2 + x}{x - 1}$ **No**

In 4 and 5, write the general form.

4. a 4th degree polynomial in y $a_4y^4 + a_3y^3 + a_2y^2 + a_1y + a_0$

5. an nth degree polynomial in x $a_nx^n + a_{n-1}x^{n-1} + \ldots + a_1x + a_0$

In 6–8, refer to the description of Stuart Dent's car loan.

6. What does the polynomial $3000x^{36} - Mx^{35} - Mx^{34} - \ldots - Mx - M$ represent?   **the balance after 36 months**

7. How much interest will Stu pay on the $3000 loan during the three years?   **$587.40**

8. Suppose Stu is given the option of borrowing $3000 for 2 years at 12% annual interest. What will be his monthly payment on this loan?
   $141.22

Making Connections for Question 3: Technically, students do not know enough mathematics to show that the expressions in parts **b** and **d** cannot be written as polynomials. So this question has to be answered on faith; if it does not look as if it can be placed in polynomial form relatively easily, then it probably does not represent a polynomial. The division of polynomials in Lesson 9-4 can be used to show that the expression of part **f** cannot be written as a polynomial.

Questions 19 and 20: These questions review ideas that are needed for Lesson 9-2.

Making Connections for Questions 24 and 25: These questions review a skill that will be particularly useful in Lesson 9-9, but at this time point out that the process of multiplication and the process of factoring undo each other.

Question 27: Bank loans historically have lower interest rates than auto dealer loans, but special incentives can change that.

9. Refer to Example 1. Suppose Tamra can invest her earnings at 7.5% annual yield. Determine her savings when she enters college. **$6,403.04**

10. Shamika saved her earnings for several summers just as Tamra did. A polynomial for her savings is
$$S(x) = 1000x^4 + 1250x^3 + 1300x^2 + 1400.$$
    a. What did Shamika deposit in the bank the first summer she saved?
    b. One summer Shamika did not save any money. Which summer was this? **summer after grade 11**
    c. Evaluate $S(1.0575)$ and explain what it represents.
       a) $1000; c) See margin.

11. Give the degree of the polynomial $7x^3y + 8y^2x^9$. **11**

12. a. Find a polynomial for V, the volume of a cube with sides of length $x + 2y$. **$V = (x + 2y)^3 = x^3 + 6x^2y + 12xy^2 + 8y^3$**
    b. State the degree of the polynomial. **3**

Applying the Mathematics

In 13 and 14, give an example, if possible.

13. a binomial in three variables, with degree 5 **sample: $x^3yz + xyz$**

14. a trinomial in two variables, of degree 4 **sample: $x^2y^2 + x^3y + xy$**

15. The expression $\sum_{i=0}^{n} c_i x^i$ is a polynomial. Give **a.** its leading coefficient and **b.** its degree. **a) c_n; b) n**

16. A box is constructed from a rectangular piece of cardboard 60 cm by 80 cm. Squares with edges of length x are cut from each corner and the edges are folded to make an open box as shown.
    a. Find a polynomial function for the volume of the resulting box in terms of x. **$V(x) = x(80 - 2x)(60 - 2x) = 4x^3 - 280x^2 + 4800x$**
    b. Graph the function using an automatic grapher. **See margin.**
    c. Find the value of x (to the nearest cm) that gives the box with the greatest volume. **$x = 11$ cm**

In 17–19, recall that the **square numbers**, 1, 4, 9, 16, 25, …, are the values of the function $s(n) = n^2$ when n is a positive integer. The **triangular numbers** $t(n)$ are the numbers of dots in shapes like those pictured below.

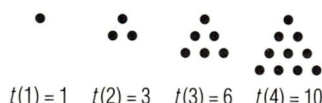

$t(1) = 1 \quad t(2) = 3 \quad t(3) = 6 \quad t(4) = 10$

17. *True* or *false*. $s(n)$ is a polynomial function of n. **True**

18. Let $p(n) = t(n) + t(n - 1)$. Use the pictures above to show that $p(n)$ is a polynomial function of n. **For all n, $t(n) + t(n - 1) = s(n)$. Therefore, $p(n) = t(n) + t(n - 1) = n^2$ is a polynomial function of n.**

528

19. There exists a quadratic function which produces the triangular numbers. Use three triangular numbers to determine this function. *(Lesson 2-7)*

$$t(n) = \frac{n(n+1)}{2} = \frac{1}{2}n^2 + \frac{1}{2}n$$

20. Solve the system.

$$\begin{cases} x + y + z = 0 \\ 9x + 3y + z = 12 \\ x - y + z = -4 \end{cases}$$ *(Lesson 2-7)* **(x, y, z) = (1, 2, -3)**

In 21–23, use the following information. In a large city school district, Board of Education policy states that students must maintain a C average and have no failing grades to participate in extracurricular activities. The boxplots below represent the percentages of students at 49 city high schools declared ineligible in five activities. *(Lesson 1-5)*

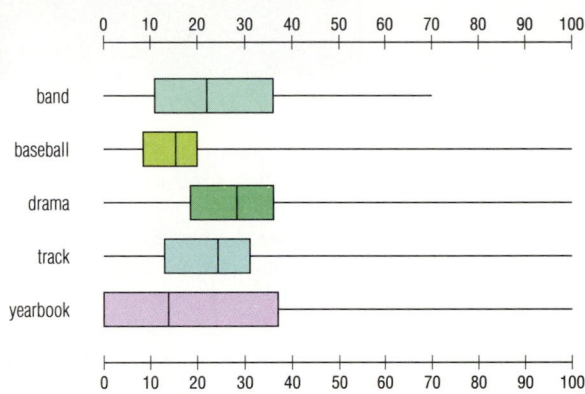

21. Which activity has the lowest median rate of ineligibility? **yearbook**

22. *True* or *false*. Students participating in athletic events have higher rates of ineligibility than students participating in other activities. **False**

23. Write a paragraph or two comparing and contrasting the rates of ineligibility in these five activities. **See margin.**

In 24 and 25, expand. *(Previous course)*

24. $(x + 3)(3x + 5)$
$3x^2 + 14x + 15$

25. $(x - 2y)(2x + 8y)$
$2x^2 + 4xy - 16y^2$

26. Draw an example of each geometric solid.
a. a regular tetrahedron
b. a pyramid with a square base *(Previous course)* **a,b) See margin.**

27. a. Find out the current rates on auto loans from an auto dealer.
b. Are the rates different depending on how long the loan is? If so, which have the lower rates, shorter loans or longer loans?
c. Determine what a bank in your area offers for an auto loan. Does the bank give a better rate than an auto dealer? **a,b,c) Answers will vary.**

FOLLOW-UP

MORE PRACTICE
For more questions on SPUR Objectives, use *Lesson Master 9-1*, shown below.

EXTENSION
Tables of annuities enable the reader to more easily answer the question that begins this lesson. A reasonable project for this chapter is to find such tables and explain how they are developed.

26.b.

TEACHING NOTES

Part of the proof of the Polynomial Difference Theorem is quite symbolic. If you think your students are ready for it, you may want to explain it. Suppose there exists a polynomial function $x \rightarrow P(x)$ of degree n that fits $n + 1$ points $(x_1, y_1), (x_2, y_2), \ldots (x_{n+1}, y_{n+1})$, forming an arithmetic sequence with constant difference d. Then for all k, the difference in values

$$y_{k+1} - y_k = P(x_{k+1}) - P(x_k)$$
$$= P(x_k + d) - P(x_k)$$
$$= \sum_{i=0}^{n} a_i(x_k + d)^i - \sum_{i=0}^{n} a_i(x_k)^i.$$

Now, by the Binomial Theorem, each element of the left summation can be expanded, and when it is, its only term of degree n will equal the corresponding term that appears in the right summation. Since they are subtracted, these terms of degree n will not appear in the difference. Thus the differences can be fit by a polynomial of degree $n - 1$.

Fruit is sometimes displayed in layers of triangular patterns as shown above. The result is a *tetrahedral* pattern.

Below are the first four tetrahedral numbers and their geometric representations.

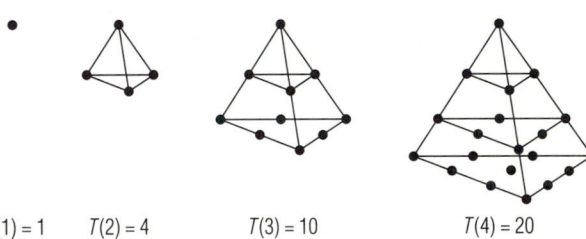

$T(1) = 1 \qquad T(2) = 4 \qquad T(3) = 10 \qquad T(4) = 20$

The 5th tetrahedral number is the sum of the 4th tetrahedral number and the 5th triangular number. So

$$T(5) = 20 + 15 = 35.$$

Thus if $T(x)$ represents the xth tetrahedral number and $t(x)$ is the xth triangular number, a recursive formula for the xth tetrahedral number is

$$\begin{cases} T(1) = 1 \\ T(x) = T(x - 1) + t(x) \quad \text{for } x > 1. \end{cases}$$

Suppose that for the 25th anniversary of a store the owner wants to arrange apples in a tetrahedral stack 25 rows high. She would like to find an explicit formula for predicting the number of apples in a stack with x rows.

She begins by graphing x versus $T(x)$ to see what kind of curve they suggest. At the left is her graph.

This graph is clearly not linear. Is it a parabola? Is it part of a higher degree polynomial? Or is it a part of an exponential curve?

530

The family of polynomial functions in one variable has a special property that enables you to predict from a table of values whether or not a function is polynomial, and if so, the degree of the polynomial. To examine this property consider first some polynomials of small degree evaluated at consecutive integers. For instance, consider the first degree polynomial $f(x) = 5x + 3$.

Observe that the differences between consecutive values of $f(x)$, calculated as right minus left are constant.

x	0	1	2	3	4	5
$f(x) = 5x + 3$	3	8	13	18	23	28
		5	5	5	5	5

Note also that this constant difference is equal to the slope of the function $f(x) = 5x + 3$.

For second degree polynomial functions, that is, for quadratic functions, the differences between consecutive values are not equal, but differences of these differences are constant. For instance, if $f(x) = x^2 + x + 3$ is evaluated at the integers from -1 to 5 inclusive, the following results.

x	-1	0	1	2	3	4	5
$f(x) = x^2 + x + 3$	3	3	5	9	15	23	33
1st differences		0	2	4	6	8	10
2nd differences			2	2	2	2	2

Consider now a cubic polynomial, say $f(x) = 2x^3 + x - 10$. For the integers between 0 and 6 inclusive, the following differences occur.

x	0	1	2	3	4	5	6
$f(x) = 2x^3 + x - 10$	-10	-7	8	47	122	245	428
1st differences		3	15	39	75	123	183
2nd differences			12	24	36	48	60
3rd differences				12	12	12	12

In this case the 3rd differences are constant. Note also that 4th differences and beyond will all equal zero.

Each of the preceding examples is an instance of the following theorem. Its proof is beyond the scope of this book and is therefore omitted.

Polynomial Difference Theorem

The function $y = f(x)$ is a polynomial function of degree n if and only if, for any set of x-values that form an arithmetic sequence, the nth differences of corresponding y-values are equal and nonzero.

The Polynomial Difference Theorem tells you that if a function is polynomial of degree n, that it will yield constant nth differences. Also, it tells you that if a function produces constant differences (from an arithmetic sequence of x-values) then the function is polynomial. Moreover, the degree of the polynomial is the n corresponding to the nth differences that are equal.

Emphasize that the system on page 532 and, more generally, the system formed by substituting in the manner described is always solvable by subtracting consecutive equations. You may wish to point out that, for the example on that page, the constants that appear in these equations are, in fact, the first 4 values of the function T in the consecutive terms, the first 3 first differences (3, 6, and 10), the first 2 second differences (3 and 4), and the first 3rd difference (1). A similar pattern always occurs.

ADDITIONAL EXAMPLE
There exists a polynomial formula of degree 4 for the sum of the cubes of the integers from 1 to n. Find that formula.
Let $P(n) = an^4 + bn^3 + cn^2 + dn + e$ be the polynomial. Use $P(1) = 1$, $P(2) = 9$, $P(3) = 36$, $P(4) = 100$, and $P(5) = 225$ as the points to fit. The system to be solved is $225 = 625a + 125b + 25c + 5d + e$; $100 = 256a + 64b + 16c + 4d + e$; $36 = 81a + 27b + 9c + 3d + e$; $9 = 16a + 8b + 4c + 2d + e$; $1 = a + b + c + d + e$.
$P(n) = \frac{1}{4}n^4 + \frac{1}{2}n^3 + \frac{1}{4}n^2$.

NOTES ON QUESTIONS
Questions 6 and 7:
Point out that although a polynomial function does not fit the data, there are other functions that do. In **Question 6,** $y = \sqrt[3]{x}$, the inverse of the polynomial function $y = x^3$ when $x \geq 0$. In **Question 7,** $y = 2^x$. This is an exponential function; for such functions, each row of consecutive differences is a multiple of the first row. (The continuous analog is that the derivative of an exponential function is a multiple of that function.)

NOTES ON QUESTIONS
Questions 18 and 19:
Students are not expected to expand the binomial powers but to use the Binomial Theorem.

Making Connections for Questions 21 and 22:
These questions continue the review of factoring in anticipation of later lessons in this chapter.

ADDITIONAL ANSWERS
1.a. $f(-1) = 3$, $f(0) = 5$,
$f(1) = 3$, $f(2) = -3$, $f(3) = -13$, $f(4) = -27$, $f(5) = -45$, $f(6) = -67$, $f(7) = -93$

b. 1st differences: 2, -2, -6, -10, -14, -18, -22, -26; 2nd differences: -4, -4, -4, -4, -4, -4, -4

2.a. $f(-1) = -2$, $f(0) = 1$, $f(1) = 2$, $f(2) = 7$, $f(3) = 22$, $f(4) = 53$, $f(5) = 106$, $f(6) = 187$, $f(7) = 302$

b. 1st differences: 3, 1, 5, 15, 31, 53, 81, 115; 2nd differences: -2, 4, 10, 16, 22, 28, 34; 3rd differences: 6, 6, 6, 6, 6, 6

Example Consider the tetrahedral numbers described on the first page of this lesson. Can they be described by a polynomial? If so, what is the degree of the polynomial?

Solution Construct a table of consecutive tetrahedral numbers, and take differences between consecutive terms.

x	1	2	3	4	5	6
$T(x)$	1	4	10	20	35	56
1st differences		3	6	10	15	21
2nd differences			3	4	5	6
3rd differences				1	1	1

Notice that the third differences appear to be constant. Thus, from the Polynomial Difference Theorem, the function $x \to T(x)$ is a polynomial function of degree 3.

In order to find an explicit formula for this cubic polynomial, you can expand the technique used in Lesson 2-7 for finding quadratic models. You know the formula for $T(x)$ is of the form
$$T(x) = ax^3 + bx^2 + cx + d.$$
You must find four coefficients a, b, c, and d in order to determine $T(x)$. So use any four data points in the table of tetrahedral numbers, and substitute them into the cubic form above. Using $x = 4$, 3, 2, and 1 gives the following four equations:

$$T(4): \ 20 = 64a + 16b + 4c + d$$
$$T(3): \ 10 = 27a + \ 9b + 3c + d$$
$$T(2): \ \ 4 = \ 8a + \ 4b + 2c + d$$
$$T(1): \ \ 1 = \ \ a + \ \ b + \ c + d$$

Solve this system. To eliminate d, subtract each successive pair of equations above. This gives the following system.
$$10 = 37a + 7b + c$$
$$6 = 19a + 5b + c$$
$$3 = \ 7a + 3b + c$$

Repeating the same procedure with the sentences above eliminates c.
$$4 = 18a + 2b$$
$$3 = 12a + 2b$$

Subtracting these two equations gives $1 = 6a$; so $a = \frac{1}{6}$.

To find the other coefficients, substitute back.
$$3 = 12a + 2b, \text{ so when } a = \frac{1}{6},$$
$$3 = 12\left(\frac{1}{6}\right) + 2b. \text{ Thus, } b = \frac{1}{2}.$$

Similarly, $3 = 7a + 3b + c$, so when $a = \frac{1}{6}$ and $b = \frac{1}{2}$,
$$3 = 7\left(\frac{1}{6}\right) + 3\left(\frac{1}{2}\right) + c. \text{ Thus, } c = \frac{1}{3}.$$

Also, $1 = a + b + c + d$, so when $a = \frac{1}{6}$, $b = \frac{1}{2}$, and $c = \frac{1}{3}$, then $d = 0$.

532

The resulting polynomial formula for the xth tetrahedral number, then, is

$$T(x) = \frac{1}{6}x^3 + \frac{1}{2}x^2 + \frac{1}{3}x.$$

By substituting 1 and 5 for x, you can check that this formula is correct.

$$T(1) = \frac{1}{6} \quad + \frac{1}{2} \quad + \frac{1}{3} \quad = 1$$

$$T(5) = \frac{1}{6}(125) + \frac{1}{2}(25) + \frac{1}{3}(5) = 35$$

Hence to find the number of apples needed in a display 25 rows high, let $x = 25$. Then $T(x) = T(25) = 2925$. So almost 3000 apples are needed for the anniversary display.

Notice that to find the cubic model for the sequence of tetrahedral numbers, four noncollinear data points were used. Five noncollinear points are needed to determine a quartic (4th degree) model. In general, to find the equation for a polynomial of degree n you need $n + 1$ points, no 3 collinear, no 4 on the same parabola, and so on, which lead to a system of $n + 1$ equations in $n + 1$ unknowns.

Questions

Covering the Reading

In 1 and 2, a formula for a function f is given. **a.** Evaluate f for integers between -1 and 7, inclusive. **b.** Take differences between consecutive values until a nonzero constant is found.

1. $f(x) = -2x^2 + 5$ **See margin.**   **2.** $f(x) = x^3 - x^2 + x + 1$ **See margin.**

In 3 and 4, refer to the Polynomial Difference Theorem.

3. If the y-values are all equal to 7 for the 4th differences of consecutive integral x-values, what is the degree of the polynomial?   **4**

4. The theorem applies to differences taken from what sorts of x-values?
those that form an arithmetic sequence

In 5–7, use the data listed in each table.
a. Determine if y is a polynomial function of x of degree less than 6.
b. If the function is polynomial, find its degree.

5.

x	1	2	3	4	5	6	7	8	9
y	2	12	36	80	150	252	392	576	810

a) Yes; b) 3

6.

x	1	8	27	64	125	216	343
y	1	2	3	4	5	6	7

See below.

7.

x	1	3	5	7	9	11	13
y	2	8	32	128	512	2048	8192

a) No

6a) It cannot be determined. The sequence of x-values is not an arithmetic sequence.

LESSON 9-2 Finding Polynomial Models   **533**

533

8. Show how to find the 7th tetrahedral number using each of the following.
   a. adding the 7th triangular number to the 6th tetrahedral number
   b. using the explicit formula for $T(x)$ developed in the lesson
   a) $T(7) = T(6) + t(7) = 56 + 28 = 84$; b) See margin.

9. Consider the function f described by the data points below.

x	0	1	2	3	4	5
$f(x)$	0	5	14	33	68	125

   a. Show that f is polynomial, and find its degree.   **See margin.**
   b. Determine an equation for f.   **$f(x) = x^3 - x^2 + 5x$**

Applying the Mathematics

10. Consider the sequence determined by the following:
    $$\begin{cases} t_1 = 17 \\ t_n = t_{n-1} - 5 \text{ for } n > 1. \end{cases}$$
    a. List the first six terms of the sequence.   **17, 12, 7, 2, -3, -8**
    b. Tell whether the sequence can be described explicitly by a polynomial.   **Yes**
    c. If the answer to part **b** is yes, find the polynomial.   **$t(x) = -5x + 22$**

11. Square numbers of items can be stacked into pyramids, just as triangular numbers can.   **a) 30, 55, 91**
    a. The first 3 square-pyramid numbers are 1, $1 + 4 = 5$, and $1 + 4 + 9 = 14$. Find the next three square-pyramid numbers.
    b. Determine an explicit polynomial for these numbers.
    c. If apples are to be stacked in a square base pyramid display with 12 layers, how many apples are needed?
    b) $f(x) = \frac{1}{3}x^3 + \frac{1}{2}x^2 + \frac{1}{6}x$; c) $f(12) = 650$

12. Suppose $f(x) = mx + b$.
    a. Calculate $f(5) - f(4)$.   **m**
    b. Evaluate $f(x+1) - f(x)$.   **m**
    c. Describe in words what the difference in part **b** represents.
    See below.

13. a. If $f(x) = ax^3 + bx^2 + cx + d$, find $f(1), f(2), f(3), f(4), f(5)$, and $f(6)$.
    b. Prove that the third differences of these values are constant.
    a,b) See margin.

14. Refer again to the tetrahedral numbers in the lesson. Suppose you try to model the data with a quadratic function.
    a. Use $T(1)$, $T(2)$, and $T(3)$ to find a quadratic model for the data, and verify that this formula works for the first three numbers in the sequence.
    b. Verify that this quadratic model does not predict the 4th tetrahedral number in the sequence.   **$f(4) = 24 - 6 + 1 = 19 \neq T(4)$**
    c. Graph the quadratic function from part **a** and the cubic model developed in the lesson on the same set of axes. Explain why both models fit the data at $x = 1, 2,$ and 3.   **See margin.**
    a) $f(x) = \frac{3}{2}x^2 - \frac{3}{2}x + 1$; $f(1) = 1, f(2) = 4, f(3) = 10$

Review

15. The geometric series $c_1 + c_1r + c_1r^2 + \ldots + c_1r^{n-1}$ is a polynomial in the variable __**a.**__ of degree __**b.**__.   *(Lesson 9-1)*   **a) r; b) $n-1$**

12c) The change in a linear function is a constant, the slope of the function.

534

16. Huana Brian Rich started saving on his 14th birthday. On that day he set aside $100. On each successive birthday he has added to his account twice the amount that he deposited the previous birthday. His money is invested at 9% compounded annually and he makes no additional deposits or withdrawals.
   a. How much money will he have in his account after making the deposit on his 21st birthday? **$27,912.91**
   b. Write a polynomial in x to represent Huana's savings, where $x = 1 + \frac{r}{100}$ if the account pays r% annual interest. **See below.**
   c. What is the degree of the polynomial in part **b**? **7**
   d. What is the leading coefficient in the polynomial of part **b** and what does it represent? *(Lesson 9-1)*
   $100; the amount that has earned interest for the longest time

17. Rose Gardner is expanding her garden. The original dimensions of the rectangular plot were x by y. She is now adding h meters to each of the length and width. *(Lesson 9-1)*
   a. Write a product of two binomials to express the area of her new garden. **$(x + h)(y + h)$**
   b. Expand the product. **$xy + xh + yh + h^2$**
   c. Make a diagram of the original plot and the expansion. Identify each of the areas represented by the terms in part **b**. **See margin.**

In 18 and 19, tell the degree of the polynomial. *(Lesson 9-1)*

18. $(a - 5)(a - 2)^3$ **4**   19. $(x^2 - 5y)^3$ **6**

20. Consider the quadratic function $f(x) = 2x^2 + 9x - 5$.
   a. Identify its y-intercept. **-5**
   b. Identify its x-intercepts. **-5, 1/2**
   c. Identify its line of symmetry. **$x = -9/4$**
   d. Sketch the graph from the information above. **See margin.**
   e. Use the results of part **b** to factor $f(x)$. **$f(x) = (2x - 1)(x + 5)$**
   (Lesson 2-6, Previous course)

In 21 and 22, factor each expression. *(Previous course)*

21. $x^2 - 10x + 25$ **$(x - 5)^2$**   22. $x^2 - 11x + 30$ **$(x - 5)(x - 6)$**

Exploration

23. A complete set of dominoes of order n consists of t tiles with two spaces filled according to the following criterion:
   Each whole number less than or equal to n (including the blank, representing 0) is paired with each other number exactly once.
   Note that this means that each number is paired with itself exactly once.
   a. Draw a complete set of dominoes of order 3. **See margin.**
   b. Make a table of the number of dominoes in a complete set for orders 1, 2, 3, 4, 5, and 6.
   c. Determine a formula for t in terms of n. **$t(n) = \frac{1}{2}n^2 + \frac{3}{2}n + 1$**

b)
n	1	2	3	4	5	6
t	3	6	10	15	21	28

16b) $f(x) = 100x^7 + 200x^6 + 400x^5 + 800x^4 + 1600x^3 + 3200x^2 + 6400x + 12800$

LESSON 9-2 Finding Polynomial Models **535**

NAME _____

LESSON **MASTER** **9-2**
QUESTIONS ON **SPUR** OBJECTIVES

■**SKILLS** *Objective A (See pages 587–590 for objectives.)*
In 1 and 2, use the data listed in each table.
a. Can y be a polynomial function of x of degree less than 5?
b. If so, find an equation for it.

1.
x	1	2	3	4	5	6
y	-1	4	11	20	31	44

a. ____ **Yes**
b. **$y = x^2 + 2x - 4$**

2.
x	1	2	3	4	5	6
y	5	17	47	107	209	365

a. ____ **Yes**
b. **$y = 2x^3 - 3x^2 + 7x - 1$**

■**PROPERTIES** *Objective F*

3. Consider the function f described by the data points below. Show that f is a polynomial, and find its degree.

x	0	1	2	3	4	5	6
y	-8	-3	0	55	288	897	2152

Differences:

1st:	5	3	55	233	609	1255
2nd:		-2	52	178	376	646
3rd:			54	126	198	270
4th:				72	72	72
5th:					0	0
6th:						

degree: ____ **4**

94   *Functions, Statistics, and Trigonometry* © Scott, Foresman and Company

NAME _____
Lesson MASTER 9-2 (page 2)

■**USES** *Objective I*

4. Soup cans are to be displayed so that the top layer has 2 cans, the second layer has 6 cans, the third layer has 12 cans, the fourth layer has 20 cans, the fifth layer has 30 cans, and so on. Determine an explicit polynomial $f(x)$ for the number of cans in the xth layer. **$f(x) = x^2 + x$**

■**REPRESENTATIONS** *Objective J*

5. The lengths (measured in feet) of the sides of a right triangle, shown below, form an arithmetic sequence.

Determine the area of the triangle in square feet. **54 ft^2**

Functions, Statistics, and Trigonometry © Scott, Foresman and Company   95

LESSON 9-3

RESOURCES
- Lesson Master 9-3
- Quiz for Lessons 9-1 through 9-3
- Teaching Aid 71: Program TABULATE
- Computer Master 22

OBJECTIVES

B Calculate or approximate zeros and relative extrema of polynomial functions.
F Apply the vocabulary of polynomials.
K Relate properties of polynomial functions and their graphs.

TEACHING NOTES

This is an appropriate lesson to take place in a computer lab. Students can enter the program TABULATE and use it to verify **Example 2.** A teaching aid is provided with this program and with the output from that example. Then students can use this program to answer questions on the Lesson Master or to do the Additional Examples.

Technology Technology is used here in two ways: first, to generate a table of values; second, to graph. Each of these ways can be utilized to find a zero of a polynomial function and to find coordinates of relative minimum and relative maximum points. The software and calculators available today have made these tasks more or less routine.

Graphs of Polynomial Functions

You are familiar with graphs of polynomial functions of degree 1 or 2. You know that the graph of every 1st degree polynomial function is a line, and the graph of every quadratic function is a parabola. The graphs of higher degree polynomial functions do not have special names; nor do all polynomial functions of the same degree have graphs of the same shape. However, graphs of higher degree polynomial functions do show a certain regularity, and like linear and quadratic functions, higher degree polynomial functions have graphs that can be described using key points and intervals. The key points are x-intercepts and relative maxima or relative minima.

Consider the graphs of the cubic polynomial functions

$$f(x) = x^3 + 2,$$
$$g(x) = -x^3 + 5x + 2.$$

$f(x) = x^3 + 2$

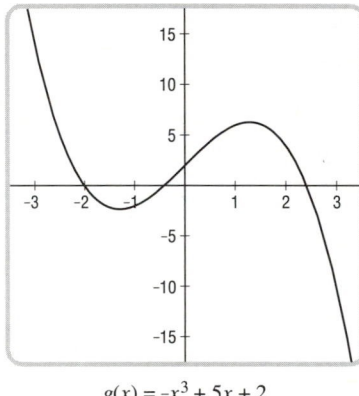

$g(x) = -x^3 + 5x + 2$

Notice that each graph above is continuous; that is, it has no gaps or jumps. Every polynomial function

$$p(x) = a_n x^n + a_{n-1} x^{n-1} + \ldots + a_1 x + a_0$$

is continuous over the real numbers, because for any small change in the x-value, there is a relatively small change in the value of $p(x)$.

Notice, however, that the two graphs above are quite different. The graph of f appears to have no relative extreme points, whereas the graph of g has at least one relative minimum and one relative maximum. Also, the graph of f appears to have only one x-intercept, whereas the graph of g seems to have three x-intercepts.

536

Another view of each of these graphs, produced using a zoom factor of 2, supports these observations.

$f(x) = x^3 + 2$

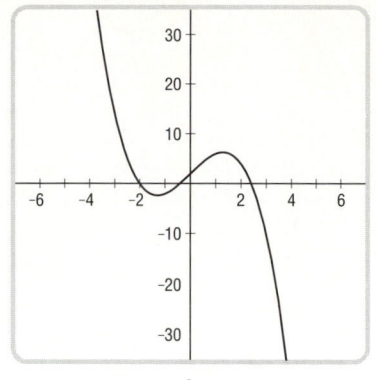

$g(x) = -x^3 + 5x + 2$

It turns out that there are two types of graphs of cubic polynomial functions. The two graphs above exemplify the two types. As you will learn later in this chapter, the numbers of zeros and relative extrema of a polynomial function are related to the degree of the polynomial.

To find the x-intercepts (the zeros) of a polynomial function

$$p(x) = a_n x^n + a_{n-1} x^{n-1} + \ldots + a_1 x + a_0,$$

you must find the values of x such that $p(x) = 0$. We call these the **zeros of the polynomial $p(x)$**.

You already know how to find exact zeros of linear or quadratic polynomial functions. There are no formulas for finding zeros of polynomial functions with degrees higher than 4 (and the formulas for the cubic and quartic functions are quite complicated). To solve $p(x) = 0$ for $n \geq 2$, there are five basic techniques.

1. Trial-and-error.
2. Draw a graph.
3. Make a table of values.
4. Factor the polynomial.
5. Apply properties of transformations.

Factoring polynomials is studied later in this chapter, and the use of transformations to solve equations was studied in Chapter 3. You are already familiar with trial-and-error as a problem-solving technique. The rest of this lesson concentrates on graphs and tables.

Our use of the term *zeros of a polynomial* is an expansion of the term *zero of a function*. Purists may feel that only functions have zeros, but mathematicians often speak of zeros of expressions.

Making Connections
On page 537 students are told that they will learn about the numbers of zeros and relative extrema of a polynomial function of degree *n*. They will learn specifically about the number of zeros in Lesson 9-7. The number of zeros indicates how many times a horizontal line can intersect the graph. Thus it also indicates the number of "wiggles" the graph can have, but that is not explicitly stated. In general, a polynomial function of degree *n* can have at most *n* different zeros and at most $n-1$ relative maxima or minima.

In **Example 1,** two of the zeros of the function *g* are found. In **Question 3,** students are asked to find the third zero.

1. Graph the function $p(x) = -x^3 + 4x + 3$. From the graph you should be able to find one of the zeros exactly. Find the other two to the nearest tenth.

-1 (exactly), -1.3, 2.3

2. Use the program TABULATE to approximate (to the nearest hundredth):

a. the smallest zero of $p(x) = -x^3 + 4x + 3$;

-1.30

b. the largest zero of this function.

2.30

You may point out that later in the chapter the students will be able to find all these zeros exactly.

3. Use a table of values to approximate the coordinates of the relative maximum point (to the nearest hundredth) of the function p from Additional Examples 1 and 2.

(1.15, 6.08)

To find these coordinates exactly requires calculus.

Example 1 Use the graphs above to find, to the nearest tenth, the two x-intercepts of $g(x) = -x^3 + 5x + 2$ that are negative.

Solution From the graph, there are two places to the left of the y-axis at which the graph crosses the x-axis. One zero of g appears to be $x = -2$. This is easily verified: $g(-2) = -(-2)^3 + 5(-2) + 2 = 0$.

Another zero appears to be between -1 and 0. Use your automatic grapher to zoom, rescale, or create a window around this region to get a better view of the zero. For instance, at the right is the graph of g in the window $-1 < x < 0$ and $-2.5 < y < 2.5$. This indicates that the second zero is near $x = -0.4$. This too can be checked:

$$g(-0.4) = -(-0.4)^3 + 5(-0.4) + 2$$
$$= 0.064$$
$$\approx 0.$$

To get a better approximation you could zoom or rescale to show even smaller intervals for x and y or you can use a table of values. Many automatic graphers also produce tables of values. If yours does not, you can use a computer program to do so. Here is a BASIC program called TABULATE which prints out values of the function from Example 1.

```
10    REM PROGRAM TABULATE
20    DEF FN G(X) = -X^3 + 5*X + 2
30    INPUT "LEFT ENDPOINT"; L
40    INPUT "RIGHT ENDPOINT"; R
50    INPUT "INCREMENT"; I
60    PRINT
70    PRINT "X", "G(X)"
80    FOR X = L TO R STEP I
90      PRINT X, FN G(X)
100   NEXT X
110   END
```

In line 20, the function is defined. From lines 30 to 50, the program asks the user to determine the domain values to be used by typing in values for the left and right endpoints (L and R, respectively) of an interval and for the size of an increment (I). The loop from lines 80 to 100 then prints out an ordered list of values of x and associated values of the function defined in line 20.

To find the zero for g between -1 and 0, run TABULATE with $L = -1$, $R = 0$, and $I = 0.1$. The following shows a computer printout.

538

```
LEFT ENDPOINT? -1
RIGHT ENDPOINT? 0
INCREMENT? .1

X      G(X)
-1     -2
-.9    -1.771
-.8    -1.488
-.7    -1.157
-.6    -.784
-.5    -.375
-.4    .064
-.3    .527
-.2    1.008
-.1    1.501
0      2
```

Notice that the values of the function change from negative to positive as x goes from -0.5 to -0.4. Because polynomial functions are continuous, somewhere in-between -0.5 and -0.4 the value of the function must be zero. A better approximation to the zero can be found by using the program with successively smaller intervals and steps. The example below shows how.

Example 2   Use **TABULATE** to approximate to the nearest hundredth the zero of $g(x) = -x^3 + 5x + 2$ in the interval -0.5 to -0.4.

Solution   First run the program with -0.5 and -0.4 as endpoints and 0.01 as the increment. The output below at the left shows that $g(-0.42)$ is negative and $g(-0.41)$ is positive. So the value of the function must be zero between $x = -.42$ and $x = -.41$. Run the program again with $-.42$ and $-.41$ as endpoints and increments of 0.001. The output is shown at the right below.

```
LEFT  ENDPOINT? -.5        LEFT  ENDPOINT? -.42
RIGHT ENDPOINT? -.4        RIGHT ENDPOINT? -.41
INCREMENT? .01             INCREMENT? .001

X       G(X)               X       G(X)
-.5     -.375              -.42    -.025912
-.49    -.332351           -.419   -.02144
-.48    -.289408           -.418   -.016966
-.47    -.246177           -.417   -.012488
-.46    -.202664           -.416   -.008009
-.45    -.158875           -.415   -.003527
-.44    -.114816           -.414   .000958
-.43    -.070493           -.413   .005445
-.42    -.025912           -.412   .009934
-.41    .018921            -.411   .014426
-.4     .064               -.41    .018921
```

From the output, you can see that $g(-.415) < 0$ and $g(-.414) > 0$. Rounding to the nearest hundredth, a zero is at -0.41.

Question 3: The other zeros have been found in **Example 1** of the lesson.

Question 5: From the given data, differences can be taken to determine the equation of the polynomial function. You might have students try to determine the equation. ($f(x) = -2x^4 + 6x^2 + 3x + 3$)

Making Connections for Question 9: The students know more powerful techniques for finding the maximum point and zeros of the function h, namely, using the quadratic formula. These techniques can be used to check (or be checked with) the graphing or table-listing techniques of this lesson.

Questions 10–13: The "advanced mathematics" needed to prove this is covered in *Precalculus and Discrete Mathematics*. As $x \to \infty$ and as $x \to -\infty$, the end behavior of a given 4th degree polynomial function is the same. This can happen if there is one change in direction, as in the function f of **Question 10.** More generally, at any relative maximum or minimum point the function changes direction, so there must be an odd number of changes of direction.

Making Connections for Questions 18 and 19: These questions continue the review of polynomial operations. **Question 18** is particularly important for the next lesson.

In the questions you are asked to find the third zero of $g(x) = -x^3 + 5x + 2$, which seems to be close to 2.4, more accurately.

Both graphs and tables of values can be used to find relative extrema of polynomials. From the second pair of graphs in this lesson, notice that the graph of g has a low point between -2 and 0. By zooming or entering the coordinates of a new window you can show that, to the nearest tenth in this interval, the smallest value of y (about -2.3) occurs when x is approximately -1.3.

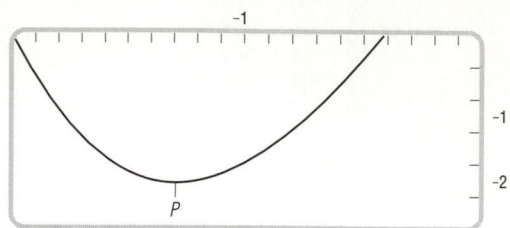

Thus, the relative minimum point P is near $(-1.3, -2.3)$. (With the TABULATE program and some automatic graphers, the coordinates can be found more precisely; with calculus, they can be determined exactly.)

Example 3 Use a table of values to approximate the relative maximum of $g(x) = -x^3 + 5x + 2$.

Solution From the initial graph of g in this lesson you can see that g has a relative maximum between $x = 1$ and $x = 2$. Use TABULATE with an interval between 1 and 2 and an increment of 0.1 to produce the table below.

```
LEFT ENDPOINT? 1
RIGHT ENDPOINT? 2
INCREMENT? .1


X       G(X)
1       6
1.1     6.169
1.2     6.272
1.3     6.303
1.4     6.256
1.5     6.125
1.6     5.904
1.7     5.587
1.8     5.168
1.9     4.641
2       4
```

Observe that in the G(X) column the maximum value is 6.303. This occurs when $x = 1.3$. Thus, an estimate for the relative maximum point is $(1.3, 6.3)$.

540

Now that all the key points of g are well approximated, you can easily identify some intervals of special interest.

For example, you can find the intervals where the function is positive (or negative) using the graph and the zeros already identified. Where is g positive? Look for the parts of the graph above the x-axis. One interval is between the zeros near $-.4$ and 2.4. Another is for all the values of x to the left of $x = -2$.

So $g(x) > 0$ when $x < -2$ or approximately when $-0.4 < x < 2.4$.

Similarly, g is negative when the graph is below the x-axis. That is,

$g(x) < 0$ about when $-2 < x < -0.4$ or about when $x > 2.4$.

You may also ask: when is g increasing? In other words, as you read the values of g going from left to right across the graph, where do the y-values climb? This time you can use the relative extrema you have already found. Starting at the far left, g values are falling until they reach the relative minimum near $x = -1.3$. Then they increase until they reach the relative maximum near $x = 1.3$. From that point on, they decrease. This can be summarized formally as

g is increasing about when $-1.3 < x < 1.3$;
g is decreasing about when $x < -1.3$ or $x > 1.3$.

In the questions you will analyze other polynomials. To use TABULATE for other functions, change line 20 to apply to the function of interest to you.

Questions

Covering the Reading

In 1 and 2, *true* or *false*.
1. The graph of a cubic polynomial function may have three real zeros.
   True
2. The graphs of all cubic polynomial functions have the same shape.
   False
3. Find the third zero of $g(x) = -x^3 + 5x + 2$, correct to the nearest hundredth, using **a.** an automatic grapher; **b.** a table of values.
   a) 2.41; b) 2.41
4. Does $f(x) = x^3 + 2$ have any relative extrema? If so, state the coordinates of each one.   **No**

5. The table at the right gives values of a polynomial function.
   a. Plot these points and sketch a graph of $y = f(x)$.   **See margin.**
   b. Between which pairs of consecutive integers must the zeros of f occur?
   -2 **and** -1, -1 **and** 0, 1 **and** 2

x	f(x)
-2	7
-1	-1
0	3
1	7
2	-1
3	-33

ADDITIONAL ANSWERS
5.a.

6.a.

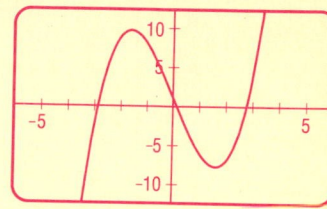

7.a. $V(x) = x(100 - 2x)(70 - 2x) = 4x^3 - 340x^2 + 7000x$

b.

c. $x < 0$ **or** $35 < x < 50$; x **cannot have a value in these intervals because the volume must be a positive quantity. (Actually** $0 < x < 35$ **since each of the three dimensions must be positive.)**

10.a.

11.a.

12.a.

13.a.

Applying the Mathematics

6. Use the function $G(x) = x^3 - 8x + 1$.
   a. Sketch a graph of G using an automatic grapher.   **See margin.**
   b. Identify and verify the y-intercept of G.   **0; $1 = 0^3 - 8 \cdot 0 + 1$**
   c. Estimate the zeros of G to the nearest tenth.   **-2.9, 0.1, 2.8**
   d. Estimate the coordinates of the relative extrema of G to the nearest hundredth.   **(-1.63, 9.71), (1.63, -7.71)**
   e. In what intervals is G positive? Negative?
   f. In what intervals is G increasing? Decreasing?   **G is increasing about when $x < -1.63$ or $x > 1.63$. G is decreasing about when $-1.63 < x < 1.63$.**
   e) $G(x)$ is positive about when $-2.9 < x < 0.1$ or $x > 2.8$.
   $G(x)$ is negative about when $x < -2.9$ or $0.1 < x < 2.8$.

7. An open box is constructed by cutting square corners with side x cm from a sheet of cardboard 100 cm by 70 cm, as shown below.

   a. Write a polynomial function for the volume of the box.
   b. Use an automatic grapher to graph the function.
   c. When is the function negative? What does this tell you?
   d. Find the coordinates of the relative maximum point of the function.
   e. For what size square cut does the maximum volume occur?
   f. What is the maximum volume the box can have?   **$\approx 42,400$ cm^3**
   a,b,c) See margin. d) $\approx$ (13.5, 42,400); e) $x \approx 13.5$ cm

8. Match each function with a characteristic.
   a. $A(x) = 4 - x^2 - x$ **II**   I.   no relative extrema
   b. $B(x) = (x - 2)^3$ **I**   II.   one relative maximum
   c. $C(x) = 7x - 9$ **I**   III.   one relative minimum
   d. $D(x) = (x - 5)^2 - 6$ **III**   IV.   both a relative maximum and a
   e. $E(x) = x^3 - 5x$ **IV**   relative minimum

9. A missile is fired from a nozzle 10 ft above the ground with an initial velocity of 500 ft per sec and follows a projectile path. Its height (vertical displacement) $h(t)$ at time t in seconds is given by $h(t) = -16t^2 + 500t + 10$.
   a. After how many seconds will the missile be 3000 ft above the ground?   **≈ 8 sec and ≈ 23 sec**
   b. What will be the maximum height of the missile?   **≈ 3900 ft**
   c. How many seconds after launching will the missile hit the ground?   **≈ 31 sec**

In 10–13, it can be proved using advanced mathematics that every quartic has either one or three relative extrema. For each of the following:
a. sketch a graph;
b. tell how many relative maxima and relative minima the function has.

10. $f(x) = x^4$ **11.** $g(t) = (t - 3)^4 + 2$

12. $h(x) = x^4 - x^2$ **13.** $j(x) = x^4 - 5x^3 - 1$

10a, 11a, 12a, 13a) See margin.
10b) 1 minimum; 11b) 1 minimum; 12b) 2 minima, 1 maximum; 13b) 1 minimum

542

14. Find a polynomial model for the following data points. *(Lesson 9-2)*

x	0	1	2	3	4	5
y	0	0	8	54	192	500

$y = x^4 - x^3$

15. You invest $750 at the ends of three consecutive years in an account paying $r\%$ annual yield. Your money is compounded annually and none is added or withdrawn.
  a. Write a polynomial function describing the amount in the bank at the end of the third year. $f(x) = 750x^2 + 750x + 750$
  b. Determine the amount if $r = 7.8\%$. *(Lesson 9-1)* **$2430.06**

16. a. In how many ways can you choose
  i. 5 objects out of 6? **6**
  ii. 99 objects out of 100? **100**
  iii. 364 objects out of 365? **365**
  b. Generalize the results of part **a** and prove your answer. *(Lesson 8-6)*
  See below.

17. The population of Turkey in 1988 was about 52.9 million, with an annual growth rate of 2.8%. Assume the population change is continuous and maintains the same rate.
  a. Give a model for the population n years after 1988. $P(n) = 52.9(1.028)^n$
  b. Predict the population in 2000. $P(12) = 73.7$ million
  c. Predict when the population will reach 100 million. *(Lesson 4-9)*
  in the year 2011

18. *Skill sequence.* Multiply and simplify.
  a. $(x + 3)(4x)$ $4x^2 + 12x$
  b. $(x + 3)(4x - 5)$ $4x^2 + 7x - 15$
  c. $(x + 3)(2x^2 + 4x - 5)$ *(Previous course)* $2x^3 + 10x^2 + 7x - 15$

19. *Skill sequence.* Solve.
  a. $x(x - 7) = 0$ **0, 7**
  b. $(y + 9)(3y - 7) = 0$ **-9, 7/3**
  c. $(z - 5)(z + 8)(z + 15) = 0$ *(Previous course)* **5, -8, -15**

Fish market on the Golden Horn, Istanbul, Turkey

20. Create some quintic (5th degree) polynomials of your own. Find functions fitting these criteria.
  a. a quintic with exactly one real zero **sample: $x^5 - 1$**
  b. a quintic with exactly three real zeros
  c. a quintic with exactly five real zeros
  b) sample: $x(x - 1)(x - 2)(x^2 + 1) = x^5 - 3x^4 + 3x^3 - 3x^2 + 2x$
  c) sample: $x(x - 1)(x - 2)(x + 1)(x + 2) = x^5 - 5x^3 + 4x$

16b) There are n ways to choose $n - 1$ objects from a collection of n objects. *Proof*: Choosing $n - 1$ objects from n objects is the same as selecting 1 object which will not be chosen. There are n ways of doing that.

LESSON 9-3 Graphs of Polynomial Functions **543**

OBJECTIVES

C Divide polynomials.
G Apply the Remainder Theorem.

TEACHING NOTES

It is useful to point out the similarities between polynomial division and integer division. The quotient "may come out even" (as with the example on page 544 and **Example 1**). Then, $\frac{dividend}{divisor}$ = quotient, or dividend = divisor · quotient. For instance, $\frac{54}{9}$ = 6, so 54 = 9 · 6.

The other possibility is that there is a remainder, as in **Examples 2 and 3.** Then, $\frac{dividend}{divisor}$ = quotient + $\frac{remainder}{divisor}$. The remainder is of smaller degree than the divisor; otherwise it could be divided further. When this form is written with multiplication, dividend = divisor · quotient + remainder. This latter form is the form used in the Remainder Theorem. For integers, $\frac{55}{9}$ = 6 + $\frac{1}{9}$, so 55 = 9 · 6 + 1.

A teaching aid with three of the completed long division problems of this lesson is provided to facilitate making these points.

Some books write the quotient as far to the left as possible above the dividend. We put the terms of the quotient above the corresponding

LESSON
9-4

Division and the Remainder Theorem

Pictured here is the first printed example of long division
(53497 ÷ 83, with the answer $644\frac{45}{83}$), from an arithmetic
by Philippi Calandri, published in 1491.

In earlier courses you learned how to add, subtract, and multiply polynomials. Now you will learn how to divide them. The procedure is similar to dividing integers and relies on the same inverse relationship between multiplication and division.

For instance, because $5 \cdot 13 = 65$, from the definition of division you may conclude that $\frac{65}{13} = 5$. Recall that 65 is called the *dividend*, 13 is called the *divisor*, and 5 is the *quotient*. Similarly for polynomials, because $(x + 3)(x + 5) = x^2 + 8x + 15$, you may conclude that $\frac{x^2 + 8x + 15}{x + 5} = x + 3$, provided $x \neq {}^-5$.

If you do not recognize the factorization, then you can use long division. The procedure of long division with polynomials is illustrated by the division of $6x^2 + 13x - 5$ by $3x - 1$ below.

Step 1: Look at the first terms of both the dividend and divisor.

$$
\begin{array}{r}
2x \\
3x - 1 \overline{)6x^2 + 13x - 5} \\
\underline{6x^2 - 2x} \\
15x - 5
\end{array}
$$

Think: $3x\overline{)6x^2} = 2x$. This is the first term in the quotient; write it above the dividend. Now multiply $3x - 1$ by $2x$ and subtract the result from $6x^2 + 13x - 5$.

Step 2: Look at the first term of the divisor and the new dividend $15x - 5$.

$$
\begin{array}{r}
2x + 5 \\
3x - 1 \overline{)6x^2 + 13x - 5} \\
\underline{6x^2 - 2x} \\
15x - 5 \\
\underline{15x - 5} \\
0
\end{array}
$$

Think: $3x\overline{)15x} = 5$. This is the second term in the quotient; write it above the dividend to the right of $2x$. Multiply $3x - 1$ by 5 and subtract the result from $15x - 5$. Since 0 is left, the division is finished and there is no remainder.

544

If some of the coefficients in the dividend polynomial are zero, you need to fill in all the missing powers of the variable in the dividend, using zero coefficients. Example 1 demonstrates this and also shows two ways to check a long division answer.

Example 1 Divide $5x^3 - 2x - 36$ by $x - 2$.

Solution The coefficient of x^2 in the polynomial is zero. Insert $0x^2$ so that all powers of x appear in the dividend.

$$
\begin{array}{r}
5x^2 + 10x + 18 \\
x - 2 \overline{\smash{)}5x^3 + 0x^2 - 2x - 36} \\
\underline{5x^3 - 10x^2} \\
10x^2 - 2x - 36 \\
\underline{10x^2 - 20x} \\
18x - 36 \\
\underline{18x - 36} \\
0
\end{array}
$$

Think: $x \overline{\smash{)}5x^3} = 5x^2$.
Multiply $x - 2$ by $5x^2$ and subtract.
Think: $x \overline{\smash{)}10x^2} = 10x$.
Multiply $x - 2$ by $10x$ and subtract.
Think: $x \overline{\smash{)}18x} = 18$.
Multiply $x - 2$ by 18; subtract to get 0.

Check 1 Multiply quotient by divisor and use the distributive property to simplify.
$$
\begin{aligned}
(x - 2)(5x^2 + 10x + 18) &= x(5x^2 + 10x + 18) - 2(5x^2 + 10x + 18) \\
&= 5x^3 + 10x^2 + 18x - 10x^2 - 20x - 36 \\
&= 5x^3 - 2x - 36
\end{aligned}
$$

Check 2 Graph $f(x) = \dfrac{5x^3 - 2x - 36}{x - 2}$ and $g(x) = 5x^2 + 10x + 18$ on the same set of axes. The graphs should coincide except when $x = 2$. The graphs drawn by an automatic grapher on the window $-5 \le x \le 5$, $0 \le y \le 50$ are shown below.

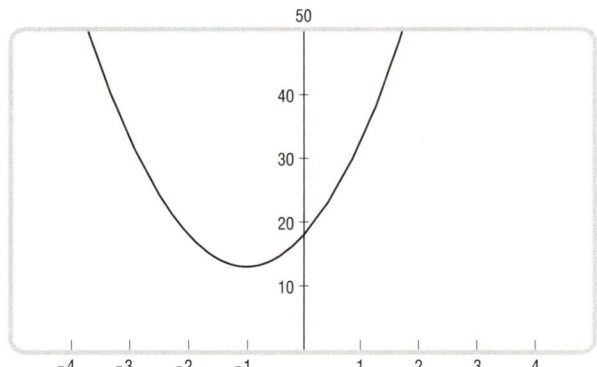

As in division of integers, not all polynomial division problems ''come out even.'' When this happens, there is a remainder. The remainder is a polynomial whose degree is always less than the divisor. An instance of this is given in Example 2.

terms of the dividend because that is the way it is done with integer long division.

Error Analysis In the long division, students should subtract from the entire dividend. Some books subtract just from the corresponding terms. We prefer our method because no dividend terms get lost.

Error Analysis Some texts instruct students to change the signs of the subtrahend in the subtraction of polynomials and then add. We encourage students to do this mentally. If students visibly change the signs, they may not be able to determine what they wrote before the sign change and it will be very difficult for them to check their work for errors.

As with all paper and pencil algorithms, proficiency takes practice. This practice is distributed through the rest of the chapter. But the algorithm is meaningless unless the student can check answers.

Example 2 Divide $2x^4 - 8x^3 + 12$ by $x^2 - 2$.

Solution

$$
\begin{array}{r}
2x^2 - 8x + 4 \\
x^2 - 2\overline{\smash{)}2x^4 - 8x^3 + 0x^2 + 0x + 12} \\
\underline{2x^4 - 4x^2} \\
-8x^3 + 4x^2 + 0x + 12 \\
\underline{-8x^3 + 16x} \\
4x^2 - 16x + 12 \\
\underline{4x^2 - 8} \\
-16x + 20
\end{array}
$$

Since the degree of $-16x + 20$ is less than that of $x^2 - 2$, the division is complete.

The quotient is $2x^2 - 8x + 4$ with remainder $-16x + 20$.

Example 2 can be checked by using this relationship:

$$\text{quotient} \cdot \text{divisor} + \text{remainder} = \text{dividend}.$$
$$(2x^2 - 8x + 4)(x^2 - 2) + (-16x + 20)$$
$$= (2x^4 - 8x^3 + 16x - 8) + (-16x + 20) = 2x^4 - 8x^3 + 12$$

As a further check you may substitute a value for x and verify the last equation.

In general, when a polynomial $f(x)$ is divided by a polynomial $d(x)$, it produces a quotient $q(x)$ and remainder $r(x)$. Either $r(x) = 0$ or $r(x)$ has degree less than the degree of the divisor $d(x)$. In symbols,

$$f(x) = q(x) \cdot d(x) + r(x)$$
$$\text{dividend} = \text{quotient} \cdot \text{divisor} + \text{remainder}.$$

Consider the special case when the divisor $d(x)$ is linear (its degree is one) and of the form $x - c$. For the remainder to be of a lower degree than the divisor, it must be a constant, possibly zero. Then

$$f(x) = q(x)(x - c) + r(x), \text{ where } r(x) \text{ is a constant.}$$

This equation is true for all x. In particular, let $x = c$. Then

$$f(c) = q(c)(c - c) + r(c).$$

This can be simplified.

$$f(c) = q(c) \cdot 0 + r(c)$$
$$f(c) = r(c)$$

This says that the value of the polynomial at $x = c$ is precisely the remainder when f is divided by $x - c$. These steps prove the Remainder Theorem.

546

Remainder Theorem

If a polynomial $f(x)$ is divided by $x - c$, then the remainder is $f(c)$.

■ ■ ■ ■ ■ ■ ■ ■ ■

Example 3   Find the remainder when $7x^5 - x^3 + 1$ is divided by $x - 2$.

> **Solution**   Use the Remainder Theorem. The value of the polynomial at $x = 2$,
>
> $$7(2)^5 - (2)^3 + 1 = 217,$$
>
> is the remainder when $7x^5 - x^3 + 1$ is divided by $x - 2$.
>
> **Check**   Use polynomial long division.

$$
\begin{array}{r}
7x^4 + 14x^3 + 27x^2 + 54x + 108 \\
x - 2 \overline{)\ 7x^5 \quad\quad - \ x^3 \quad\quad\quad\quad + \ 1\ } \\
\underline{7x^5 - 14x^4} \\
14x^4 - \ x^3 \quad\quad\quad\quad + \ 1 \\
\underline{14x^4 - 28x^3} \\
27x^3 \quad\quad\quad\quad + \ 1 \\
\underline{27x^3 - 54x^2} \\
54x^2 \quad + \ 1 \\
\underline{54x^2 - 108x} \\
108x + \ 1 \\
\underline{108x - 216} \\
217
\end{array}
$$

Notice that the Remainder Theorem has limitations:
    a. It only applies when dividing by a linear factor.
    b. It does not produce a quotient.

Despite its limitations, the Remainder Theorem has a very useful consequence, the *Factor Theorem*, which you will study in Lesson 9-5.

Questions

Covering the Reading

In 1–3, find the quotient when the first polynomial is divided by the second.
  1. $3x^2 - x - 10, \ x - 2$ $3x + 5$

  2. $x^4 - 9x^3 + 19x^2 - 2x - 3, \ x - 3$ $x^3 - 6x^2 + x + 1$

  3. $6z^4 - 3z^3 - 2z - 8, \ 2z^2 + 2$ $3z^2 - \frac{3}{2}z - 3$

  4. a. Verify that $(x - 3)(x^2 + 5x + 1) = x^3 + 2x^2 - 14x - 3$.
      b. Write as a polynomial in simplest terms.
        i. $\dfrac{x^3 + 2x^2 - 14x - 3}{x - 3}$ $x^2 + 5x + 1$ **ii.** $\dfrac{x^3 + 2x^2 - 14x - 3}{x^2 + 5x + 1}$ $x - 3$

a) $(x - 3)(x^2 + 5x + 1) = x^3 - 3x^2 + 5x^2 - 15x + x - 3 = x^3 + 2x^2 - 14x - 3$

LESSON 9-4   Division and the Remainder Theorem   **547**

NOTES ON QUESTIONS
Error Analysis for Question 3:   Some students may stop when a fraction appears in the quotient. You may have to tell them to continue. Additional Example 2 is another quotient of this type.

Question 4:   Part **a** should be applied to make part **b** easier.

NOTES ON QUESTIONS
Question 13:   The graph of the function g will look identical to the graph of $f(x) = x - 4$ except at the single point $(4, 0)$, which is on the graph of f but not on the graph of g.

Making Connections for Question 19:   This question could be used as a lead-in to the next lesson. (Notice that we expect UCSMP students to have seen the Factor Theorem in a previous course.)

Making Connections for Question 20:   This question anticipates work in Lesson 9-8.

ADDITIONAL ANSWERS
11.a. $q(x) = a + 5$, $r(x) = 5a + 15$

b. Degree of remainder is 1. Degree of divisor is 2.

c. $(a^2 - 5)(a + 5) + (5a + 15) = (a^3 + 5a^2 - 5a - 25) + (5a + 15) = a^3 + 5a^2 - 10$

12.a. $q(x) = 3b^2 + 16$, $r(x) = 21b^2 - 27b + 162$

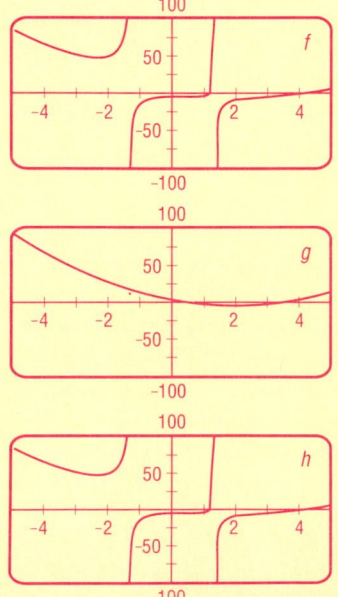
548

5. In a division problem, state a relationship between the dividend, divisor, quotient, and remainder.   **dividend = (divisor)(quotient) + remainder**

6. If $f(x) = d(x) \cdot q(x) + r(x)$, $f(x) = x^2 + 3x + 7$, and $d(x) = x + 2$, find possible polynomials for $q(x)$ and $r(x)$.   **$q(x) = x + 1$, $r(x) = 5$**

In 7 and 8, **a.** use the Remainder Theorem to find the remainder when the first polynomial is divided by the second. **b.** Check by dividing.

7. $y^3 + 4y^2 - 9y - 40$, $y - 3$
   a) -4; b) $y^3 + 4y^2 - 9y - 40 = (y - 3)(y^2 + 7y + 12) - 4$

8. $2x^3 - 11x^2 - 2x + 5$, $x - 5$
   a) -30; b) $2x^3 - 11x^2 - 2x + 5 = (x - 5)(2x^2 - x - 7) - 30$

9. Suppose you know that when $f(x)$ is divided by $x - 2$, the remainder is zero. What is $f(2)$?   **0**

Applying the Mathematics

10. Let $f(x) = x^5 - 32$ and $g(x) = x - 2$.
    a. Express $\dfrac{f(x)}{g(x)}$ as a polynomial in simplest terms.
    b. The expression in part **a** is equal to a geometric series. Identify the first term and the common ratio of the series.
    a) $x^4 + 2x^3 + 4x^2 + 8x + 16$; b) first term: x^4; common ratio: $2/x$

In 11 and 12, **a.** find the quotient and remainder when the first polynomial is divided by the second. **b.** Show that the degree of the remainder is less than the degree of the divisor. **c.** Check by multiplying the divisor by the quotient and then adding the remainder.

11. $a^3 + 5a^2 - 10$, $a^2 - 5$ **See margin.**

12. $3b^5 + 16b^3 - 27b + 50$, $b^3 - 7$ **See margin.**

13. **a.** *True* or *false*.
    For all $x \neq 4$, $g(x) = \dfrac{x^3 - 12x^2 + 48x - 64}{x^2 - 8x + 16}$ is a linear function.   **True**
    b. Justify your answer to part **a** using a graph.   **See margin.**
    c. Justify your answer to part **a** using polynomial long division.
    $x^3 - 12x^2 + 48x - 64 = (x^2 - 8x + 16)(x - 4)$

14. What is the quotient if $x^3 + 2x^2y - 2xy^2 - y^3$ is divided by $x - y$?
    $x^2 + 3xy + y^2$

Review

15. Let $h(x) = x^4 + 3x^3 - 11x^2 - 3x + 10$.
    a. Graph $y = h(x)$ on a window that shows all zeros and relative extrema.   **See margin.**
    b. To the nearest hundredth, estimate the endpoints of the intervals where h is increasing, and the intervals where h is decreasing.
    c. Identify the intervals where h is positive, and the intervals where it is negative.   *(Lesson 9-3)*
    **b) increasing: $-3.69 < x < -0.13$; $x > 1.57$; decreasing: $x < -3.69$; $-0.13 < x < 1.57$
    c) positive: $x < -5$; $-1 < x < 1$; $x > 2$; negative: $-5 < x < -1$; $1 < x < 2$**

16. The total daily revenue from a hot dog stand at a ball park is given by the equation $R(x) = x \cdot p(x)$, where x is the number of hot dogs sold and $p(x)$ is the price (in cents) of one hot dog. The price of a hot dog is given by the equation $p(x) = 250 + .002x - .00001x^2$. **a) $R(x) = 250x + .002x^2 - .00001x^3$**

  a. Find a polynomial expression in simplest terms equal to $R(x)$.

  b. Draw a graph of R on a window that shows all relative extrema and x-intercepts for $x \geq 0$. [Hint: you will need values of x and y larger than 1000.] **See margin.**

  c. Determine the number of hot dogs to be sold to produce the maximum daily revenue. **≈ 3000**

  d. What is the maximum daily revenue possible? **$\approx \$500,000$**

  e. What is the largest domain on which this model could hold?

  (Lesson 9-3) **$0 \leq x \leq 5100$**

17. Consider the table of values below.

x	-1	0	1	2	3	4
y	3	2	3	12	35	78

  a. Find a polynomial function to fit these data. **$f(x) = x^3 + x^2 - x + 2$**

  b. Find, to the nearest tenth, all zeros of this function. **-2**

  c. Find, to the nearest tenth, all relative extrema of the function.

  (Lessons 9-2, 9-3) **(-1, 3.0), (.3, 1.8)**

18. Let $g(x) = (2ax - 3b)^7$, where a and b are constants. *(Lessons 9-1, 8-8)*

  a. What is the degree of g? **7**

  b. What is the leading coefficient of g? **$128a^7$**

  c. What is the constant term of g? **$-2187b^7$**

19. Consider $f(x) = 6x^2 + 13x - 5$.

  a. Solve $f(x) = 0$. **$x = -5/2$ or $x = 1/3$**

  b. Factor $f(x)$. **$6x^2 + 13x - 5 = (3x - 1)(2x + 5)$**

  c. State a relationship between the results to parts **a** and **b**.

  (Previous course) **$(x - a)$ is a factor $\Leftrightarrow f(a) = 0$.**

Exploration

20. a. Divide each polynomial by $x - y$. **See below.**

  i. $x^3 - y^3$ **ii.** $x^5 - y^5$

  iii. $x^7 - y^7$ **iv.** $x^9 - y^9$

  b. Generalize the pattern above. $\dfrac{x^n - y^n}{x - y} = \sum\limits_{i=0}^{n-1} x^{n-1-i} y^i$

21. a. Refer to Example 2, use an automatic grapher to plot on the same set of axes

$$f(x) = \frac{2x^4 - 8x^3 + 12}{x^2 - 2}$$

$$g(x) = 2x^2 - 8x + 4$$

and $\quad h(x) = 2x^2 - 8x + 4 + \dfrac{-16x + 20}{x^2 - 2}$.

  b. Write several sentences comparing and contrasting the graphs.
  See margin.

20a) i) $x^2 + xy + y^2$

ii) $x^4 + x^3y + x^2y^2 + xy^3 + y^4$
iii) $x^6 + x^5y + x^4y^2 + x^3y^3 + x^2y^4 + xy^5 + y^6$
iv) $x^8 + x^7y + x^6y^2 + x^5y^3 + x^4y^4 + x^3y^5 + x^2y^6 + xy^7 + y^8$

LESSON 9-4   Division and the Remainder Theorem   **549**

FOLLOW-UP

MORE PRACTICE
For more questions on SPUR Objectives, use *Lesson Master 9-4*, shown below.

PROJECTS
Project 4 on page 584 is related to the content of this lesson.

21.b. The graphs of f and h are identical. Both have a vertical asymptote at $x = \sqrt{2}$. The graph of g is a parabola with line of symmetry $x = 2$. As x increases, points on the graph of f get closer and closer to the graph of g.

NAME _____

LESSON **MASTER 9-4**
QUESTIONS ON **SPUR** OBJECTIVES

■SKILLS *Objective C (See pages 587–590 for objectives.)*
In 1–10, determine a. the quotient and b. the remainder when the first polynomial is divided by the second.

		a.		b.	
1.	$x^3 + 2x^2 - 5x + 6$; $x + 4$	a.	$x^2 - 2x + 3$	b.	-6
2.	$2x^3 - 3x^2 + 6x - 7$; $2x + 1$	a.	$2x^2 - 4x + 8$	b.	-11
3.	$4x^3 - 2x^2 + 8x - 10$; $x^2 + 1$	a.	$4x - 2$	b.	$4x - 8$
4.	$6x^4 - x^3 + x^2 + 2x - 1$; $2x + 3$	a.	$3x^3 - 5x^2 + 8x - 11$	b.	32
5.	$x^4 + x^2 - 4x + 6$; $x - 1$	a.	$x^3 + x^2 + 2x - 2$	b.	4
6.	$x^5 - 3x^4 + 2x^3 - x^2 + x + 1$; $x^2 + 1$	a.	$x^3 - 3x^2 + x + 2$	b.	-1
7.	$x^5 + 2x^4 - 3x^3 + x^2 - 2x + 1$; $x^2 - x + 1$	a.	$x^3 + 3x^2 - x - 3$	b.	$-4x + 4$
8.	$9x^6 - 3x^4 + 18x^2 - 3$; $3x^2 - 3$	a.	$3x^4 + 2x^2 + 8$	b.	21

■PROPERTIES *Objective G*
In 11–15, use the Remainder Theorem to find the remainder when the first polynomial is divided by the second.

9.	$x^3 - 2x^2 + 7x - 1$; $x - 1$	5
10.	$2x^3 + 6x^2 - 5x + 6$; $x + 3$	21
11.	$x^4 - 3x^3 + 5x^2 - 8x + 1$; $x + 1$	18
12.	$x^4 - 4x^3 + 6x^2 - 10x + 4$; $x + 2$	96

97

RESOURCES
■ Lesson Master 9-5
▪ Computer Master 23

OBJECTIVES

D Factor polynomials and solve polynomial equations using the Factor Theorem.
G Apply the Remainder Theorem, Factor Theorem, and Factor-Solution-Intercept Equivalence Theorem.
K Relate properties of polynomial functions and their graphs.

TEACHING NOTES

When used with reference to an equation, the words *solution* and *root* are synonyms. A solution to a polynomial equation $p(x) = 0$ is called a *zero* of the polynomial, a *zero* of the polynomial function p, and is an *x-intercept* of its graph. For this reason, when $p(c) = 0$ for some number c, that number may be called a solution to the equation, a root of the equation, a zero of the polynomial, a zero of the polynomial function, or an *x*-intercept of the graph.

The Factor-Solution-Intercept Equivalence Theorem includes five equivalent statements. The middle three are simply the same statement in different language and thus a consequence of definitions. The first is equivalent to these three due to the Factor Theorem. The last is equivalent to these three because of the Remainder Theorem.

The pattern of the Factor Theorem is quite simple and easily learned when the zero is an integer. Both **Examples 1 and 2** involve rational

LESSON
9-5

The Factor Theorem

Consider the following polynomial functions. Each is shown in standard and factored form, its zeros are listed, and it is graphed. What pattern between the zeros and factors do you find?

$q(x) = x^2 - 3x - 10$
$\qquad = (x + 2)(x - 5)$

zeros of q: $-2, 5$

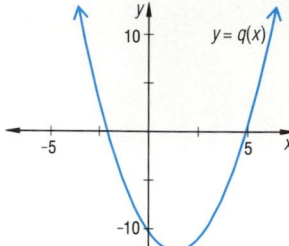

$r(x) = x^3 + x^2 - 12x$
$\qquad = x(x + 4)(x - 3)$

zeros of r: $-4, 0, 3$

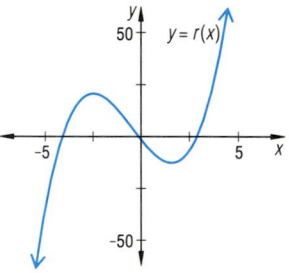

$s(x) = 2x^4 - 5x^3 - 57x^2 + 90x$
$\qquad = x(x + 5)(x - 6)(2x - 3)$

zeros of s: $-5, 0, \frac{3}{2}, 6$

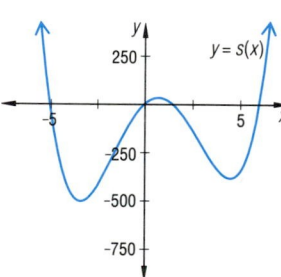

The relations between the expanded form of the polynomials and the zeros are subtle. However, when looking at the *factored* forms and the zeros a pattern emerges: each zero c of the polynomial appears to correspond with a linear factor $(x - c)$.

This result holds for all polynomials and can be proved using the Remainder Theorem from Lesson 9-4.

> **Factor Theorem**
>
> For a polynomial $f(x)$, a number c is a solution to $f(x) = 0$ if and only if $(x - c)$ is a factor of f.

550

Proof The theorem says "if and only if," so two statements must be proved:

(1) If $(x - c)$ is given to be a factor of f, then c is a solution to $f(x) = 0$.

(2) If c is a solution to $f(x) = 0$, then $(x - c)$ is a factor of f.

In (1), $(x - c)$ is given to be a factor of f. Then, by the definition of a factor, when f is divided by $(x - c)$, the remainder is zero. So, $f(x) = (x - c)q(x)$. In particular, at $x = c$, $f(x) = f(c) = (c - c)q(c) = 0$. So $f(c) = 0$; that is, c is a solution to $f(x) = 0$.

In (2), c is given to be a solution to $f(x) = 0$. Then, by the definition of a solution, $f(c) = 0$. From the Remainder Theorem, you know that when $f(x)$ is divided by $(x - c)$, the remainder is $f(c)$, which is zero, thus making $(x - c)$ a factor of f.

Graphically, a solution to $f(x) = 0$ is an x-intercept of the graph. Putting this fact together with the Remainder and Factor theorems produces the following result.

Factor-Solution-Intercept Equivalence Theorem

For any polynomial function f, the following are equivalent:
$(x - c)$ is a factor of f;
$f(c) = 0$;
c is an x-intercept of the graph of $y = f(x)$;
c is a zero of f;
the remainder when f is divided by $(x - c)$ is 0.

Example 1 Factor $g(x) = 6x^3 - 25x^2 - 31x + 30$.

Solution Begin with a graph to see if any zeros are obvious.

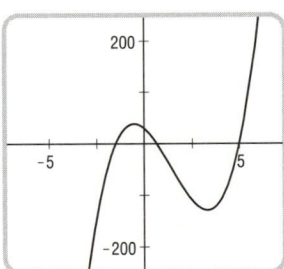

There appears to be a zero at $x = 5$. Verify by substitution:
$g(5) = 6(5)^3 - 25(5)^2 - 31(5) + 30 = 0$.
Thus one linear factor is $(x - 5)$. Divide $6x^3 - 25x^2 - 31x + 30$ by this factor to find another factor.

zeros, and students should realize that if $\frac{a}{b}$ is a zero to a polynomial equation with integer coefficients, not only is $\left(x - \frac{a}{b}\right)$ a factor of the polynomial, but so is $(bx - a)$. Still, students may not realize the generality of the theorem, so you may wish to mention zeros that are not rational. (See the Extension for this lesson.) In the next lesson, the Factor Theorem will be applied to zeros that are not real numbers.

If you discuss **Example 1,** point out that it is possible that one of the other zeros (-3/2 or 2/3) might have been seen first from the graph. If so, the division would have resulted in a quadratic whose solutions are the other two zeros. There is an analogy with the factorization of integers into primes; regardless of which factor is found first, the ultimate factorization has the same factors.

$$\begin{array}{r} 6x^2 + 5x - 6 \\ x - 5\overline{)6x^3 - 25x^2 - 31x + 30} \\ \underline{6x^3 - 30x^2} \\ 5x^2 - 31x + 30 \\ \underline{5x^2 - 25x} \\ -6x + 30 \\ \underline{-6x + 30} \\ 0 \end{array}$$

The quotient $6x^2 + 5x - 6$ is a second factor. This polynomial can be factored as a product of binomials.
$$6x^2 + 5x - 6 = (2x + 3)(3x - 2)$$
So
$$g(x) = (x - 5)(6x^2 + 5x - 6)$$
$$= (x - 5)(2x + 3)(3x - 2).$$

Check

1. Multiply the factors.
$$\begin{aligned} (x - 5)(2x + 3)(3x - 2) &= (2x^2 - 7x - 15)(3x - 2) \\ &= 6x^3 - 21x^2 - 45x - 4x^2 + 14x + 30 \\ &= 6x^3 - 25x^2 - 31x + 30 \\ &= g(x); \text{ it checks.} \end{aligned}$$
2. Check the zeros of g. In the solution above, $g(5)$ was shown to equal 0. By the Factor Theorem, the factor $(2x + 3)$ corresponds to a zero of $-\frac{3}{2}$, and the factor $(3x - 2)$ corresponds to a zero of $\frac{2}{3}$. These appear to be the x-intercepts on the graph. Substitute to show that $g\left(\frac{2}{3}\right) = 0$ and $g\left(-\frac{3}{2}\right) = 0$.

The Factor Theorem can also be used to find equations for polynomial functions, given their zeros.

■ ■ ■ ■ ■ ■ ■ ■ ■

Example 2   Find an equation for a polynomial function with zeros -1, $\frac{4}{5}$, and $-\frac{8}{3}$.

Solution   Each zero indicates a factor of the polynomial. Call the polynomial $p(x)$. Then $p(x)$ has factors $(x + 1)$, $\left(x - \frac{4}{5}\right)$, and $\left(x + \frac{8}{3}\right)$. It may have other factors as well. So $p(x) = k(x + 1)\left(x - \frac{4}{5}\right)\left(x + \frac{8}{3}\right)$ where k may be any nonzero constant or a polynomial in x. One possible answer is found by letting $k = 1$. Then $p(x) = (x + 1)\left(x - \frac{4}{5}\right)\left(x + \frac{8}{3}\right)$.

Check   Substitute -1 for x. Is $p(-1) = 0$?
$$p(-1) = k(-1 + 1)\left(-1 - \frac{4}{5}\right)\left(-1 + \frac{8}{3}\right)$$
The second factor is zero, so the product is zero. Similarly, $p\left(\frac{4}{5}\right) = 0$ and $p\left(-\frac{8}{3}\right) = 0$.

552

Notice that the degree of $p(x)$ in Example 2 is at least 3. However, the degree of k is not known, so the degree of $p(x)$ cannot be determined, nor can the coefficients of its terms. Graphs of many polynomial functions go through the points $(-1, 0)$, $\left(\frac{4}{5}, 0\right)$, and $\left(-\frac{8}{3}, 0\right)$. Below are graphs of three different polynomial functions, two of degree 3 and one of degree 4, all with zeros of -1, $\frac{4}{5}$, and $-\frac{8}{3}$.

$f(x) = (x+1)\left(x-\frac{4}{5}\right)\left(x+\frac{8}{3}\right)$

$g(x) = 5(x+1)\left(x-\frac{4}{5}\right)\left(x+\frac{8}{3}\right)$

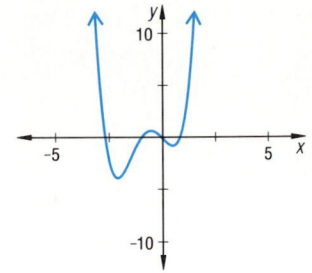

$h(x) = x\,(x+1)\left(x-\frac{4}{5}\right)\left(x+\frac{8}{3}\right)$

Questions

Covering the Reading

1. State the Factor Theorem.   **For a polynomial $f(x)$, a number c is a solution to $f(x) = 0$ if and only if $(x - c)$ is a factor of f.**

2. If $f(x) = x^3 - 3x^2 + 6x - 18$ has a factor of $(x - 3)$, what must be true about $f(3)$?   **$f(3) = 0$**

In 3 and 4, one fact is stated about a polynomial. State at least two other conclusions you can draw.

3. The number -1 is a solution to $g(x) = x^4 - x^3 + 3x^2 + 3x - 2 = 0$.
See margin.

4. The graph of $h(x) = x^6 + x^5 - 2x^4$ crosses the x-axis at $x = -2$.
See margin.

In 5 and 6, **a.** find a zero of the polynomial function by graphing. **b.** Use this information to factor the polynomial into linear factors.

5. $r(x) = 7x^3 - 22x^2 - 67x + 10$ **a) See margin. b) $r(x) = (x+2)(x-5)(7x-1)$**

6. $t(x) = 3x^3 + 20x^2 - 108x - 80$ **a) See margin. b) $t(x) = (x+10)(3x+2)(x-4)$**

7. Corina factored $g(x) = 3x^3 + 4x^2 - 17x - 6$ into $(x+3)(x-2)(3x+1)$.
   a. Graph $y = g(x)$ and $f(x) = (x+3)(x-2)(3x+1)$ on the same set of axes.   **See margin.**
   b. Is Corina's factorization correct?   **Yes**

8. Find an equation for a polynomial function with zeros equal to -5, $\frac{9}{2}$, and $\frac{7}{3}$.   **sample: $f(x) = (x+5)\left(x-\frac{9}{2}\right)\left(x-\frac{7}{3}\right) = x^3 - \frac{11}{6}x^2 - \frac{71}{3}x + \frac{105}{2}$**

7.a.

20.c. $p(5) = 35$, $p(6) = 51$

MORE PRACTICE
For more questions on SPUR
Objectives, use *Lesson Mas-
ter 9-5,* shown on page 555.

EXTENSION
The quadratic equation
$x^2 + x - 1 = 0$ has solutions
$x = \dfrac{-1 \pm \sqrt{5}}{2}$. Thus, by the

Factor Theorem, $x^2 + x - 1 =$
$\left(x - \dfrac{-1 + \sqrt{5}}{2}\right)\left(x - \dfrac{-1 - \sqrt{5}}{2}\right) =$
$\left(x + \dfrac{1}{2} - \dfrac{\sqrt{5}}{2}\right) \cdot \left(x + \dfrac{1}{2} + \dfrac{\sqrt{5}}{2}\right).$
Have students use this idea
to make up and factor other
quadratics whose solutions
are not rational.

ADDITIONAL ANSWERS
**20.c. See margin,
page 553.**

22.a.

Applying the Mathematics

9. Without dividing, choose two factors of $x^3 + 6x^2 + 3x - 10$ from the four possibilities $(x + 2)$, $(x - 2)$, $(x + 1)$, and $(x - 1)$.   **(x + 2), (x − 1)**

In 10 and 11, find the other solutions to the equation.

10. $4x^3 + 20x^2 - 68x - 84 = 0$; -1 is one solution.   **-7, 3**

11. $2p^4 + 13p^3 + 12p^2 - 17p - 10 = 0$; 1 and $-\frac{1}{2}$ are solutions.   **-2, -5**

In 12 and 13, a graph of a polynomial function with integer zeros is given. Find an equation of the given degree for the graph.   **See below.**

12. degree 3

13. degree 4

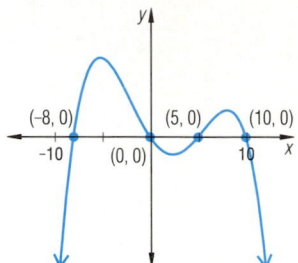

Review

14. Complete. If a polynomial $g(x)$ is divided by __?__, then the remainder is $g(a)$.   *(Lesson 9-4)*   **(x − a)**

15. Suppose $m(x) = x^5 - x^3 + 3x$. Without dividing, find the remainder when $m(x)$ is divided by $x - 2$.   *(Lesson 9-4)*   **m(2) = 30**

16. *Multiple choice.* A polynomial $q(x)$ is divided by $x - 3$ and the remainder is 7. Which of the following points must be on the graph of $q(x)$?
   (Lesson 9-4)
   (a) $(3, 7)$ (b) $(7, 3)$
   (c) $(3, 0)$ (d) $(7, 0)$
   (e) none of these   **(a)**

In 17 and 18, find the quotient when the first polynomial is divided by the second.   *(Lesson 9-4)*

17. $x^5 - 243$, $x - 3$
    $x^4 + 3x^3 + 9x^2 + 27x + 81$

18. $y^5 + 2y^4 - 7y^3 - 14y^2$, $y^2 - 7$
    $y^3 + 2y^2$

19. At the right is a graph of a polynomial function $y = f(x)$.   *(Lesson 9-3)*
   a. Identify the zeros of f.   **m, p, r**
   b. Identify the relative extrema.
   c. Where is f increasing?
   d. Where is f decreasing?   **$n < x < q$**
   e. Where is f positive?
   f. Where is f negative?
      b) (n, f(n)), (q, f(q)); c) x<n, x>q; e) m<x<p, x>r; f) x<m, p<x<r

   12) f(x) = (x + 3)(x − 2)(x − 5) = x³ − 4x² − 11x + 30
   13) f(x) = (x + 8)x(x − 5)(x − 10) = x⁴ − 7x³ − 70x² + 400x

20. Pentagonal numbers $p(n)$ are so-called because they are the numbers of dots in arrays like those pictured below. *(Lessons 2-7, 9-2)*

$p(1) = 1$ $p(2) = 5$ $p(3) = 12$ $p(4) = 22$

a. Show that there is a polynomial of degree 2 which generates the pentagonal numbers. **1st differences: 4, 7, 10; 2nd differences: 3, 3**

b. Determine the quadratic polynomial $p(n)$.

c. Use the polynomial of part **b** to calculate $p(5)$ and $p(6)$, and verify your answer geometrically. **b) $p(n) = \frac{3}{2}n^2 - \frac{1}{2}n$; c) See margin.**

21. At some point during the men's final match in the 1990 U.S. Open Tennis tournament, Pete Sampras had the following statistics.

First service in: 53%

First service points won: 83%

Second service points won: 48%

Assuming that he had not double faulted (missed both services) up to that point, what was the probability that, next time he served, Sampras

a. got in his first service; **.53**

b. won the point on his first service; **.83**

c. won the point on his second service; **.48**

d. lost the point? **.0901 + .2444 ≈ .33**

e. What percentage of the points that he had served on had Pete Sampras won up to that point? *(Lessons 7-1, 7-4)* **≈ .67**

Pete Sampras

Exploration

22. a. Graph
$$y = x + 2$$
$$y = x^2 + 4$$
$$y = x^3 + 8$$
$$y = x^4 + 16$$

and $y = x^n + 2^n$ for two other values of n of your choice.

b. For what values of n does $y = x^n + 2^n$ have real zeros? For these values of n, what binomial is a factor of $x^n + 2^n$?

a) See margin. b) odd values of n; a factor is $(x + 2)$

22.a.

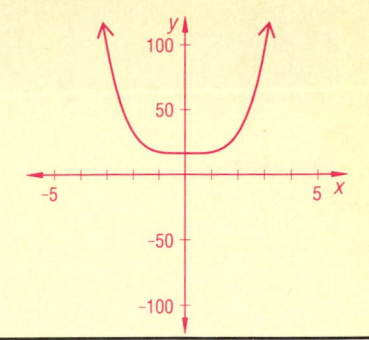

NAME _____

LESSON MASTER   9-5
QUESTIONS ON **SPUR** OBJECTIVES

■SKILLS *Objective D (See pages 587–590 for objectives.)*
In 1 and 2, find the zeros of each polynomial function by graphing. Use this information to factor the polynomial into linear factors.

1. $f(x) = x^3 + 2x^2 - x - 2$

$f(x) = (x + 1)(x - 1)(x + 2)$

2. $f(x) = 2x^3 + 3x^2 - 8x + 3$

$f(x) = (x - 1)(x + 3)(2x - 1)$

■PROPERTIES *Objective G*
In 3–8, let $f(x) = 2x^4 - 5x^3 - 23x^2 + 38x + 24$. Write *true* or *false* for each.

3. $f(x)$ has a zero at $x = -3$, so $(x - 3)$ is a factor of $f(x)$. — **False**

4. $x = 4$ is a zero of the function, so the graph crosses the x-axis at $x = 4$. — **True**

5. $f\left(-\frac{1}{2}\right) = 0$, so $f(x)$ crosses the y-axis at $y = -\frac{1}{2}$. — **False**

6. 2 is an x-intercept of the graph of $y = f(x)$, so the remainder when f is divided by $(x - 2)$ is 0. — **True**

7. $(x - 4)$ is a factor of $f(x)$, so $f(4) = 0$. — **True**

8. If $-\frac{1}{2}$ is a zero of $f(x)$, then $\left(x + \frac{1}{2}\right)$ is a factor of $f(x)$. — **True**

NAMF _____
Lesson MASTER 9-5 (page 2)

■REPRESENTATIONS *Objective K*
In 9 and 10, tell how many real zeros each function has.

9.

3

10.

4

LESSON 9-6

OBJECTIVES

D Factor polynomials and solve polynomial equations using the Factor Theorem or trial and error.
E Perform operations with complex numbers.

TEACHING NOTES

Some books define the imaginary part of $a + bi$ to be bi, and not b. We view that as an inappropriate definition; as ordered pairs in rectangular coordinates, $a + bi = (a, b)$ and no i is needed in the description of the complex number.

Examples 1 and 5 stress the definition of i as $\sqrt{-1}$.
Examples 2 and 4 give straightforward computations with complex numbers.
Example 3 shows that polynomials that cannot be factored into polynomials with real coefficients can be factored into polynomials with nonreal coefficients.

You could add a box for "nonreal numbers" to the chart at the bottom of page 559, between the boxes for complex numbers and imaginary numbers. Project 5 on page 584 discusses two types of irrational numbers; these could also be added to the chart.

LESSON

Complex Numbers

Consider the polynomial $f(x) = x^2 + 1$. Its zeros are the solutions to $x^2 + 1 = 0$ or, more simply, $x^2 = -1$. Does this last sentence have any solutions?

That depends on the domain of available numbers. If x is a real number, $x^2 = -1$ has no solution, since the square of any real number is either positive or zero.

The graph at the right verifies that $f(x) = x^2 + 1$ has no real zeros, because it does not cross the x-axis.

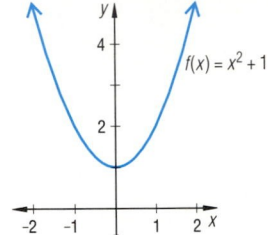

However, it is possible to define solutions to $x^2 + 1 = 0$ that are not real numbers. Since $x^2 = -1$, these solutions are called *the square roots of negative 1*. One of these solutions is denoted as $\sqrt{-1}$. It is customary also to call this number i.

> **Definition**
>
> $i = \sqrt{-1}$

The number i and its nonzero, real-number multiples are called **imaginary numbers** because when they were first used, mathematicians had no concrete way of representing them. They are, nevertheless, a reality in mathematics, and many important applications have been found for them.

Clearly, $i^2 = -1$. Note that if we assume some properties of real numbers can be extended to i, then it is also true that

$$(-i)^2 = (-1)(i)(-1)(i) = (-1)(-1)(i)(i) = (1)(i)^2 = -1.$$

Thus, in the domain of imaginary numbers, the original polynomial $f(x) = x^2 + 1$ has two zeros, i and $-i$. You can verify this by applying the quadratic formula to the equation $x^2 + 1 = 0$.

If operations with imaginary numbers are assumed to satisfy the commutative and associative properties of multiplication, the square root of any negative number can be expressed as a multiple of i. Example 1 gives an instance.

556

■ ■ ■ ■ ■ ■ ■ ■

Example 1   Show that a square root of -5 is $i\sqrt{5}$.

Solution   Multiply $i\sqrt{5}$ by itself.
$$i\sqrt{5} \cdot i\sqrt{5} = i \cdot i \cdot \sqrt{5}\sqrt{5} = i^2 \cdot 5 = -1 \cdot 5 = -5$$

In general, if k is a positive real number, then $\sqrt{-k} = i\sqrt{k}$, so the square root of any negative number is the product of i and a real number. This means that all equations of the form $x^2 + k = 0$ where $k > 0$ have two solutions, $i\sqrt{k}$ and $-i\sqrt{k}$.

When an imaginary number is added to a real number, the sum is called a **complex number**.

Definitions

A **complex number** is a number of the form $a + bi$, where a and b are real numbers and $i = \sqrt{-1}$. The number a is the **real part** and b is the **imaginary part** of $a + bi$.

Complex numbers have many similarities to binomials. For instance, two complex numbers $a + bi$ and $c + di$ are **equal** if and only if their real parts are equal and their imaginary parts are equal. That is, $a + bi = c + di$ if and only if $a = c$ and $b = d$. Furthermore, the arithmetic of complex numbers is similar to the arithmetic of binomials.

■ ■ ■ ■ ■ ■ ■ ■

Example 2   Let $z = 4 + 5i$ and $w = 2 - 3i$. Express each of the following in $a + bi$ form:

a. $z + w$ **b.** $z - w$ **c.** zw **d.** $\frac{z}{2}$.

Solution **a.** $z + w = (4 + 5i) + (2 - 3i)$
$$= (4 + 2) + (5i - 3i)$$
$$= 6 + 2i$$

b. $z - w = (4 + 5i) - (2 - 3i)$
$$= 4 + 5i\ - 2 + 3i$$
$$= (4 - 2) + (5i + 3i)$$
$$= 2 + 8i$$

c. $zw = (4 + 5i) \cdot (2 - 3i)$
$$= (4)(2) - (4)(3i) + (5i)(2) - (5i)(3i)$$
$$= 8 - 12i + 10i - 15i^2$$
$$= 8 - 2i - 15\,(i^2)$$
$$= 8 - 2i - 15(-1) \qquad \text{Use } i^2 = -1.$$
$$= 23 - 2i$$

d. $\dfrac{z}{2} = \dfrac{4 + 5i}{2}$
$$= \frac{4}{2} + \frac{5}{2}i$$
$$= 2 + \frac{5}{2}i$$

1. Write the square roots of -3.

a. without using i **b.** using i

a. $\sqrt{-3}$, $-\sqrt{-3}$; **b.** $i\sqrt{3}$, $-i\sqrt{3}$

2. Let $w = 1 - 2i$ and $z = 6 + 9i$. Express in $a + bi$ form.

a. $w + z$ **b.** $w - 3z$ **c.** wz
d. z^2
a. 7 + 7i; **b.** -17 − 29i;
c. 24 − 3i; **d.** -45 + 108i

3. Factor $x^2 + 9$ over the set of polynomials with complex coefficients.
$(x + 3i)(x - 3i)$

4. Express $\dfrac{2 + 3i}{9 - 7i}$ in $a + bi$ form.
$\dfrac{-3}{130} + \dfrac{41}{130}i$

5. Solve $2x^2 + 5x + 4 = 0$ over the set of complex numbers and put all nonreal solutions in $a + bi$ form.
$\dfrac{-5}{4} + \dfrac{\sqrt{7}}{4}i$, $\dfrac{-5}{4} - \dfrac{\sqrt{7}}{4}i$

The example illustrates the following theorem:

It is not necessary to memorize this theorem. Simply remember to operate as you do with binomials: combine like terms when adding and use the distributive property when multiplying.

Just as finding solutions to equations depends on the domain of allowable numbers, so does factoring. With the introduction of complex numbers the sum of two squares can be factored.

Proof Multiply the factors, using the distributive property.
$$(a + bi)(a - bi) = a^2 - abi + abi - (b^2 i^2)$$
$$= a^2 - b^2(-1)$$
$$= a^2 + b^2$$

Complex number pairs of the form $a + bi$ and $a - bi$ are called **complex conjugates**. Each is the complex conjugate of the other. As the preceding proof confirms, the product of complex conjugates is a real number.

In the previous lesson, the Factor Theorem stated that if c is a solution of $f(x) = 0$, then $(x - c)$ must be a factor of $f(x)$. The theorem is still true when c is a complex number. For instance, i and $-i$ are solutions of the equation $x^2 + 1 = 0$, and $x^2 + 1$ factors into $(x - i)(x + i)$.

Example 3 Show that $(x - i)$ and $(x + i)$ are factors of $x^2 + 1$.

Solution 1 Multiply. $(x - i)(x + i) = x^2 + ix - ix - i^2$
$$= x^2 - i^2$$
$$= x^2 + 1$$

Solution 2 Notice that $x + i$ and $x - i$ are complex conjugates. By the previous theorem,
$$(x + i)(x - i) = x^2 + 1^2$$
$$= x^2 + 1.$$

Often a domain is given for factoring. $x^2 - 9$ can be factored over the set of polynomials with integer coefficients: $x^2 - 9 = (x + 3)(x - 3)$. $x^2 - 3$ can be factored over the set of polynomials with real coefficients into $(x - \sqrt{3})(x + \sqrt{3})$. However, $x^2 + 3$ cannot be factored over this set; it is only factorable over the set of polynomials with complex number coefficients, into $(x - i\sqrt{3})(x + i\sqrt{3})$.

Complex conjugates are useful in performing division of complex numbers.

Example 4   Express $\dfrac{3 - 4i}{6 + i}$ in $a + bi$ form.

Solution   When the denominator is multiplied by its complex conjugate, the result is a real number. So multiply numerator and denominator by the conjugate of the denominator.

$$\frac{3 + 4i}{6 + i} = \frac{3 - 4i}{6 + i} \cdot \frac{6 - i}{6 - i} = \frac{14 - 27i}{36 + 1} = \frac{14}{37} - \frac{27}{37}i$$

Check   Multiply the quotient by the divisor.

$$\left(\frac{14}{37} - \frac{27}{37}i\right)(6 + i) = \frac{1}{37}(14 - 27i)(6 + i)$$

$$= \frac{1}{37}(84 + 14i - 162i - 27i^2)$$

$$= \frac{1}{37}(111 - 148i) = 3 - 4i$$

Complex numbers make possible the solution of all quadratic equations.

Example 5   Solve $x^2 - 4x + 13 = 0$.

Solution   Use the quadratic formula.

$$x = \frac{4 \pm \sqrt{16 - 4(1)(13)}}{2(1)} = \frac{4 \pm \sqrt{-36}}{2} = \frac{4 \pm 6i}{2} = 2 \pm 3i$$

Note that the two solutions are complex conjugates. In general, if a quadratic equation with real coefficients has a negative discriminant, then the two solutions are complex conjugates of each other.

For the complex number $a + bi$, if $a = 0$, then $a + bi = 0 + bi = bi$, so every imaginary number is also complex. Similarly, for any real number a, $a = a + 0i$. Thus, every real number is also complex. The diagram at the right shows the way that many types of numbers are related.

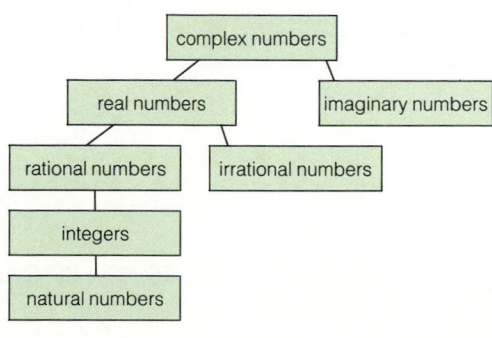

NOTES ON QUESTIONS
Question 10:   Some mathematicians would answer "false" to this question. They would argue that the field of real numbers is isomorphic to a subfield of the complex numbers, which means that real numbers can be treated as if they are complex numbers, but that the objects are different. The answer depends on how the complex numbers are defined and how rigorous one wishes to be. The interpretation we take here is that the field of complex numbers is given and that the real numbers constitute a subfield of the field of complex numbers.

NOTES ON QUESTIONS
Making Connections for Questions 20 and 21: A generalization of this result will be found in the next lesson. Whenever a polynomial equation has real coefficients, any nonreal solutions come in complex conjugate pairs.

Question 25: Some people prefer to write the i to the left of the radical sign so that it does not get put under the radical sign by mistake. Others like to put it to the right so that the $a + bi$ form is clearly shown.

ADDITIONAL ANSWERS
22.

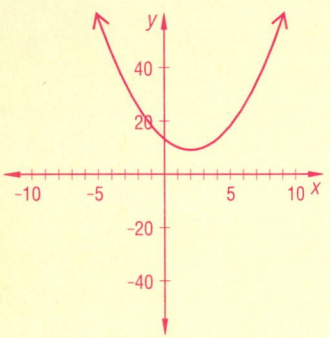

The graph does not cross the x-axis.

31.b.

Covering the Reading

In 1–6, simplify.

1. $\sqrt{-25}$ **5i**

2. $\sqrt{-8}$ **2i√2**

3. $(7i)^2$ **-49**

4. i^3 **-i**

5. i^4 **1**

6. $(-i)^6$ **-1**

7. Show that $i\sqrt{7}$ and $-i\sqrt{7}$ are each square roots of -7.
   $(i\sqrt{7})^2 = i^2 \cdot 7 = -1 \cdot 7 = -7; (-i\sqrt{7})^2 = (-i)^2 \cdot 7 = -1 \cdot 7 = -7$

8. A complex number is a number of the form __**a.**__ where a and b are __**b.**__ numbers.   **a) a + bi; b) real**

9. When are two complex numbers equal?   **$a + bi = c + di \Leftrightarrow a = c$ and $b = d$**

10. *True* or *false*.   Every real number is a complex number.   **True**

11. Give the conjugate.
    a. $2 - 3i$ **2 + 3i**
    b. $6i + 5$ **-6i + 5**

12. Factor $x^2 - 3$ over the reals.   **(x − √3)(x + √3)**

In 13–17, write in $a + bi$ form.

13. $(4 + 8i) + (-3 - 9i)$ **1 − i**

14. $5(2 - 3i)$ **10 − 15i**

15. $(2 - 3i)(4 + 7i)$ **29 + 2i**

16. $\dfrac{3 + i}{2 - i}$ **1 + i**

17. $\dfrac{5 - 3i}{1 + 2i}$ **$-\frac{1}{5} - \frac{13}{5}i$**

In 18 and 19, **a.** factor over the set of polynomials with complex coefficients; **b.** check by multiplying.

18. $x^2 + 25$ **a) (x + 5i)(x − 5i);**
    b) $x^2 + 5ix − 5ix − 25i^2 = x^2 + 25$

19. $9z^2 + 18$ **See below.**

In 20 and 21, **a.** solve and express the solutions in $a + bi$ form.
b. What relation do the solutions of each sentence have to one another?

20. $x^2 - 2x + 26 = 0$ **a) 1 + 5i, 1 − 5i**
    b) They are complex conjugates.

21. $9x^2 + 229 = 12x$ **See below.**

22. Refer to Example 5. Graph $f(x) = x^2 - 4x + 13$. What properties does the graph of a quadratic function have if its roots are complex?   **See margin.**

21a) $\frac{2}{3} + 5i, \frac{2}{3} - 5i$; b) **They are complex conjugates.**

Applying the Mathematics

23. Find the 20th term of the arithmetic sequence
    $2, 5 + i, 8 + 2i, 11 + 3i, \dots$.   **59 + 19i**

24. Let $f(x) = x^2 - 4x + 5$.
    a. Evaluate $f(2 + i)$.   **f(2 + i) = 0**
    b. Evaluate $f(2 - i)$.   **f(2 − i) = 0**
    c. Use the results of parts **a** and **b** to factor f over the complex numbers.   **$x^2 − 4x + 5 = (x − (2 + i))(x − (2 − i)) = (x − 2 − i)(x − 2 + i)$**

25. Use the Quadratic Formula to factor $x^2 - 2x + 3$ over the complex numbers.   **$x^2 − 2x + 3 = (x − 1 − \sqrt{2}i)(x − 1 + \sqrt{2}i)$**

19a) $9z^2 + 18 = 9(z^2 + 2) = 9(z - i\sqrt{2})(z + i\sqrt{2})$
   b) $9(z - i\sqrt{2})(z + i\sqrt{2}) = 9(z^2 + zi\sqrt{2} - zi\sqrt{2} - 2i^2)$
   $= 9(z^2 + 2)$
   $= 9z^2 + 18$

560

26. Let $z = \frac{1}{2} + \frac{\sqrt{3}}{2}i$.

   a. Calculate z^3 and write the result in $a + bi$ form.   **$-1 + 0i$**

   b. Let w be the complex conjugate of z. Calculate w^3 and write it in $a + bi$ form.

   c. Both z and w are cube roots of __?__ .   **-1**

   d. Find another cube root of the answer to part **c**.   **-1 itself**

     b) $w = \frac{1}{2} - \frac{\sqrt{3}}{2}i$; $w^3 = -1 + 0i$

Review

27. According to the Factor Theorem, if $g(x)$ is a polynomial, a number b is a solution to $g(x) = 0$ if and only if __?__ . *(Lesson 9-5)*
$(x - b)$ is a factor of $g(x)$

28. $h(x)$ is a polynomial function whose graph intersects the x-axis at $(-2, 0)$. State at least two conclusions you can make from this information.
(Lesson 9-5)   **Samples: $h(-2) = 0$; -2 is a zero of $h(x)$; $(x + 2)$ is a factor of $h(x)$; the remainder when $h(x)$ is divided by $(x + 2)$ is 0.**

29. Identify three different polynomial functions with zeros at 2, 4, and $-\frac{6}{5}$.

(Lesson 9-5)   **All are of the form $p(x) = k(x)(x - 2)(x - 4)(x + 6/5)$, where $k(x)$ is a polynomial or nonzero constant function.**

30. A wooden cube with edge of length 10 cm has a square hole with edge of length x cm bored through from top to bottom. Give the surface area of the new shape.   **$600 - 2x^2 + 40x$**

(Lesson 9-1, Previous course)

10 cm

31. The following table gives the numbers a particular spinner can land on, and the probability that each number appears.

Number x_i	1	2	3	4	5	6	7	8
$P(x_i)$	$\frac{1}{24}$	$\frac{1}{12}$	$\frac{1}{8}$	$\frac{1}{4}$	$\frac{5}{24}$	$\frac{1}{6}$	$\frac{1}{12}$	$\frac{1}{24}$

   a. Verify that P satisfies the two conditions for a probability distribution.   **$P(x_i) \geq 0$ for all x_i and $\sum_{i=1}^{8} P(x_i) = 1$.**

   b. Draw a histogram of the distribution.   **See margin.**

   c. Calculate the mean of the distribution   **$109/24 \approx 4.542$**

   d. Explain what the result in part **c** indicates in terms of the spinner.
(Lesson 7-7)   **Over a long period of time, the average of the numbers the spinner lands on will approach 109/24.**

Exploration

32. Explore patterns in powers of i.

   a. Predict the value of each of i^{1992}, i^{1993}, and i^{2000}.

   b. Describe in words or in symbols how to evaluate a large power of i.

a) $i^{1992} = 1$; $i^{1993} = i$; $i^{2000} = 1$; b) To find i^m, find the remainder when m is divided by 4. If the remainder is 0, then $i^m = 1$. If the remainder is 1, then $i^m = i$. If the remainder is 2, then $i^m = -1$. If the remainder is 3, then $i^m = -i$.

FOLLOW-UP

MORE PRACTICE
For more questions on SPUR Objectives, use *Lesson Master 9-6*, shown below.

EXTENSION
A student interested in other "numbers" that are not real (nor complex for that matter) might research quaternions, and write a brief report. (A quaternion has the form $a + bi + cj + dk$, where a, b, c, and d are real numbers and $i^2 = y^2 = k^2 = ijk = -1$. Quaternions were developed in the nineteenth century by the Irish mathematician Sir William Rowan Hamilton.)

EVALUATION
A quiz covering Lessons 9-4 through 9-6 is provided in the Teacher's Resource File.

Alternative Assessment
Students can write a brief essay explaining the diagram on page 554 in words and giving one or two reasons why these ideas are important.

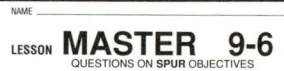

NAME _____

LESSON **MASTER** **9-6**
QUESTIONS ON **SPUR** OBJECTIVES

■**SKILLS** *Objective D (See pages 587–590 for objectives.)*
In 1–7, factor over the set of polynomials with coefficients in the indicated domain.

1. $6v^2 + 7v - 20$; integers   **$(2v + 5)(3v - 4)$**

2. $w^3 - 4w^2 - 3w + 12$; reals   **$(w - \sqrt{3})(w + \sqrt{3})(w - 4)$**

3. $b^3 - 4b^2 - 4b + 1$; integers   **$(b + 1)(b^2 - 5b + 1)$**

4. $81x^4 - y^4$; reals   **$(9x^2 + y^2)(3x - y)(3x + y)$**

5. $c^2 - 14c + 53$; complex numbers   **$(c - 7 + 2i)(c - 7 - 2i)$**

6. $21p^2 + 32p - 5$; integers   **$(3p + 5)(7p - 1)$**

7. $25n^2 + 9$; complex numbers   **$(5n - 3i)(5n + 3i)$**

■**SKILLS** *Objective E*
In 8 and 9, for each problem find a. $z + w$, b. $z - w$, and c. zw.

8. $z = 3 + 2i$, $w = 4 - 5i$
  a. **$7 - 3i$**
  b. **$-1 + 7i$**
  c. **$22 - 7i$**

9. $z = -1 + 6i$, $w = 3 - 7i$
  a. **$2 - i$**
  b. **$-4 + 13i$**
  c. **$39 + 25i$**

In 10 and 11, express each in the form $a + bi$.

10. $\frac{2 + 3i}{6 - i}$ **$\frac{9}{37} + \frac{20}{37}i$**

11. $\frac{1 - i}{2 + 3i}$ **$-\frac{1}{13} - \frac{5}{13}i$**

12. Determine the sum $\sum_{k=1}^{8} k \cdot i^k$ where $i = \sqrt{-1}$.   **$4 - 4i$**

RESOURCES
■ Lesson Master 9-7
■ Teaching Aid 73: Graphs for **Example 2** and **Question 19**

OBJECTIVES

H Apply the Fundamental Theorem of Algebra and Conjugate Zeros Theorem.

K Relate properties of polynomial functions and their graphs.

TEACHING NOTES

Error Analysis Some students may not realize that the Fundamental Theorem of Algebra applies to polynomials with real number coefficients. Stress that real numbers are complex numbers.

Alternative Assessment
To probe whether students understand what is meant by multiplicity, you might ask for a quadratic equation that has a zero with multiplicity 2. Such a quadratic is equivalent to an equation of the form $(x - c)^2 = 0$. In previous courses, c might have been called a "double root" to this equation.

The theorem proved at the bottom of page 564 can be generalized. If p and q are different polynomial functions of degrees n and m, with $n \geq m$, then their graphs can have at most n points in common. Consider the polynomial $p(x) - q(x)$. It has degree at most n. Thus the equation $p(x) - q(x) = 0$ has at most n solutions. Thus $p(x) = q(x)$ for at most n values of x.

LESSON

9-7

The Fundamental Theorem of Algebra

As you know, the solutions to any quadratic equation can be found using the quadratic formula. The graph of a quadratic function intersects the x-axis at two points when there are two real zeros, in exactly one point when there is a single real solution, and does not intersect the x-axis when its zeros are complex. As the drawings below indicate, the graph of a quadratic function, its discriminant, its zeros, and its factors are related.

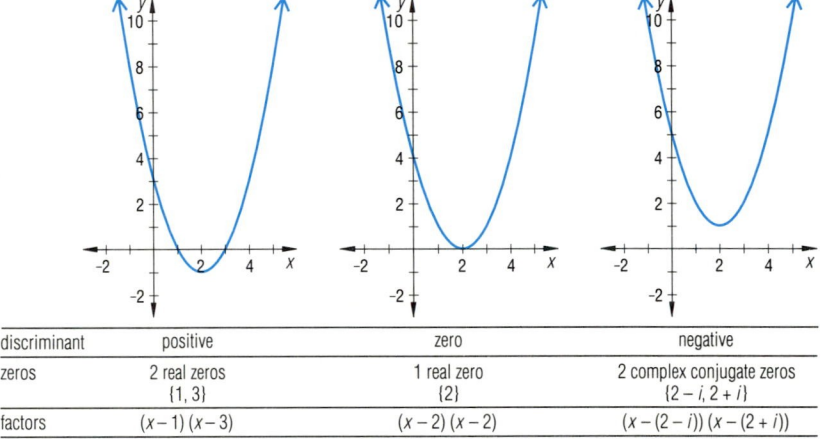

discriminant	positive	zero	negative
zeros	2 real zeros $\{1, 3\}$	1 real zero $\{2\}$	2 complex conjugate zeros $\{2 - i, 2 + i\}$
factors	$(x - 1)(x - 3)$	$(x - 2)(x - 2)$	$(x - (2 - i))(x - (2 + i))$

In general, consider the quadratic function $g(x) = ax^2 + bx + c$ with real coefficients. Then if the discriminant d is positive, g has two real zeros. If d is zero, g has one real zero, and if d is negative, g has two complex conjugate zeros.

You may wonder how zeros and graphs of higher degree polynomial functions are related. For example, are there formulas like the quadratic formula to provide zeros to all higher degree polynomials? Can all polynomials be decomposed into linear factors as quadratics can? What is the greatest number of zeros a polynomial can have? These questions intrigued mathematicians for many years. In this lesson these questions will be answered, and the implications of these issues for graphs of polynomial functions will also be discussed.

An Arabian mathematician, Al-Khowarizmi, is believed to have first discovered the quadratic formula in about 825 A.D. His use of the formula was limited to what are now called *real solutions*. Europeans first learned of his discovery in the year 1202 when it was translated into Latin by Fibonacci, the Italian mathematician also known for the sequence with his name.

562

*Niccolo Tartaglia
(woodcut, 1546)*

For centuries mathematicians sought general formulas, like the quadratic formula, for finding zeros of higher degree polynomials. Several Italian mathematicians of the 16th century made progress with cubics. The works of Scipione del Ferro (1465–1526) and of Niccolo Tartaglia (1500–1557) were published by Girolamo Cardano (1501–1576) in 1545 in his treatise on algebra, *Ars Magna* (''Great Art''). In that book, formulas for zeros for classes of cubics are given and complex numbers are recognized as legitimate solutions to equations. Shortly after, Ludovico Ferrari (1522–1565), a student of Cardano, found a method for finding exact zeros of any polynomial of degree 4. In all these discoveries, no new numbers were needed beyond the complex numbers. The search for a formula for zeros to all quintics and beyond continued.

In 1797, the following key result connecting the previous investigations was discovered by Gauss when he was 18 years old. It made him famous among mathematicians.

Fundamental Theorem of Algebra

If $p(x)$ is any polynomial of degree $n \geq 1$ with complex coefficients, then $p(x)$ has at least one complex zero.

This theorem, whose proof is beyond the scope of this book, is remarkable because:
  –it refers to all polynomials;
  –it tells the complex nature of at least one zero; and
  –it leads to another simple, yet powerful result.

Theorem

A polynomial of degree n has at most n zeros.

Proof Let $p(x)$ be a polynomial of degree n. By the Fundamental Theorem of Algebra, $p(x)$ has at least one complex zero, call it c. Then, by the Factor and Remainder Theorems, when $p(x)$ is divided by $(x - c)$ the quotient, $q(x)$, is a polynomial of degree $n - 1$. Now begin again. $q(x)$ has at least one complex zero, so divide $q(x)$ by the factor associated with that zero to get a quotient $r(x)$ of degree $n - 2$, and so on. Each division reduces the degree of the previous polynomial by 1, so the process of repeated division can have at most n steps, each providing one zero.

Thus another important consequence of the Fundamental Theorem of Algebra is that regardless of the degree of the polynomial, each of its zeros is a complex number; no new types of numbers are needed!

Of course, some of the zeros at the various divisions may be the same. For example, the quadratic $g(x)$ at the beginning of this lesson factors to $(x - 2)^2$. It is said to have a *double zero*. The **multiplicity of a zero** r for a polynomial is the highest power of $(x - r)$ that appears as a factor of that polynomial.

LESSON 9-7 The Fundamental Theorem of Algebra **563**

In **Example 2,** if a line intersected the graph of the polynomial function $y = f(x)$ more than 5 times, then f would have degree more than 5. A teaching aid is provided.

You may wish to emphasize the steps in the solution to **Example 3** as follows: We know, because the polynomial is a cubic, that there are exactly 3 zeros, counting multiplicities. In part **a**, the solution $3i$ is confirmed. In part **b**, the Conjugate Zeros Theorem gives us a $-3i$ solution. By the Factor Theorem, we know two of the three linear factors of the polynomial. Finally, using long division, the third factor is found, from which the third zero is identified.

ADDITIONAL EXAMPLES

1. Find all zeros of $p(x) = 4x^{32} + x^{30}$, and give the multiplicity of each zero.
$p(x) = x^{30}(2x + i)(2x - i)$, so $p(x)$ has 32 zeros, 0 with multiplicity 30, and $\frac{i}{2}$ and $\frac{i}{2}$ with multiplicity 1

2. The graph of a polynomial function $y = f(x)$ has 4 points in common with the x-axis and 6 points in common with the line $y = 5$. What can be said about the degree of $f(x)$? **It is at least 6.**

3. Let $p(x) = x^4 - 2x^3 + 6x^2 - 2x + 5$.

a. Verify that i is a zero of $p(x)$.
$i^4 - 2i^3 + 6i^2 - 2i + 5 = 1 + 2i - 6 - 2i + 5 = 0$

b. Verify that $(1 + 2i)$ is a zero of $p(x)$.
$(1 + 2i)^4 - 2(1 + 2i)^3 + 6(1 + 2i)^2 - 2(1 + 2i) + 5 = -7 - 24i - (-22 - 4i) + (-18 + 24i) - (2 + 4i) + 5 = 0$.

c. Find all other zeros of $p(x)$.
The two other zeros are $-i$ and $1 - 2i$.

■ ■ ■ ■ ■ ■ ■ ■ ■

Example 1   Find the zeros of $p(x) = x^3 - 6x^2 + 9x$.

> **Solution**   You can factor out an x immediately, to get
> $$p(x) = x(x^2 - 6x + 9).$$
> The quadratic factor is a perfect square trinomial. Thus
> $$p(x) = x(x - 3)^2.$$
> This means $p(x)$ has three zeros: the zero 0 has multiplicity 1; the zero 3 has multiplicity 2.

In Example 1 the polynomial function p has degree 3. Notice that it has exactly 3 complex zeros, all of which were found in the solution. The zeros (all of which are real in this case) are indicated in the graph at the right. So there is no need to look at a larger domain to find other zeros.

In general, the idea of multiplicities of zeros and the fact that no new numbers are needed to factor polynomials of any degree implies the following result.

Number of Zeros of a Polynomial Theorem

A polynomial of degree $n \geq 1$ with complex coefficients has exactly n complex zeros, if multiplicities are counted.

From the graph of a polynomial you can determine information about its degree. Not only does the Number of Zeros Theorem dictate the maximum number of intersections of a polynomial graph and the x-axis, it also determines the number of intersections the graph may have with any horizontal line.

Theorem

Let $p(x)$ be a polynomial of degree $n \geq 1$ with real coefficients. Then the graph of $p(x)$ can cross any horizontal line $y = d$ at most n times.

Proof   Let $p(x)$ be a polynomial with degree $n \geq 1$ with real coefficients. The points of intersection of the graph of $y = p(x)$ and the horizontal line $y = d$ are the solutions of the equation $p(x) = d$. This equation is equivalent to $g(x) = p(x) - d = 0$. The degree of $g(x)$ is the same as the degree of $p(x)$ because the two polynomials differ only by a constant. Thus, $g(x)$ has at most n zeros. So the graph of $p(x)$ has at most n intersections with $y = d$.

564

■ ■ ■ ■ ■ ■ ■ ■

Example 2   A polynomial function $y = f(x)$ is graphed below. What is the lowest possible degree of $f(x)$?

Solution   The polynomial crosses the horizontal line $y = -2$ five times. So the degree of $f(x)$ must be at least 5.

When one (or more) of the coefficients of the polynomial is nonreal, the solutions cannot be pictured on a standard graph. However, the Number of Zeros of a Polynomial Theorem can still be applied. For instance, the polynomial $g(x) = -3x^5 - ix$ has degree 5, so it has 5 zeros.

When a polynomial has real coefficients, then its nonreal zeros always come in complex conjugate pairs. The proof of this theorem is long, so we omit it.

Conjugate Zeros Theorem

Let $p(x) = a_n x^n + a_{n-1}x^{n-1} + \dots + a_1 x + a_0$, where $a_n, a_{n-1}, \dots a_1, a_0$ are all real numbers, and $a_n \neq 0$. If $z = a + bi$ is a zero of $p(x)$, then the complex conjugate of z, $a - bi$, is also a zero of $p(x)$.

■ ■ ■ ■ ■ ■ ■ ■

Example 3   Let $p(x) = 2x^3 - x^2 + 18x - 9$.
   a. Verify that $3i$ is a zero of $p(x)$.
   b. Find the remaining zeros of $p(x)$ and their multiplicities.

Solution

a. $\begin{aligned} p(3i) &= 2(3i)^3 - (3i)^2 + 18\,(3i) - 9 \\ &= 54i^3 \quad - 9i^2 \quad + 54i \quad - 9 \\ &= -54i \quad + 9 \quad + 54i \quad - 9 \\ &= 0 \end{aligned}$

b. Because $3i$ is a zero, then, by the Conjugate Zeros Theorem, so is its conjugate $-3i$. The Factor Theorem implies that $(x - 3i)$ and $(x + 3i)$ are factors of $p(x)$. Thus their product $(x - 3i) \cdot (x + 3i) = x^2 + 9$ is a factor of $p(x)$. Divide $p(x)$ by $x^2 + 9$ to find another factor:

$$
\begin{array}{r}
2x - 1 \\
x^2 + 9 \overline{\smash{\big)}\, 2x^3 - x^2 + 18x - 9} \\
\underline{2x^3 + 18x } \\
-x^2 - 9 \\
\underline{-x^2 - 9} \\
0
\end{array}
$$

Thus $p(x) = (x^2 + 9)(2x - 1)$. So the zeros of $p(x)$ are $3i$, $-3i$, and $\frac{1}{2}$.

Check In part **a**, $p(3i)$ was shown to equal 0. Similarly,
$$p(-3i) = 2(-3i)^3 - (-3i)^2 + 18(-3i) - 9$$
$$= 54i + 9 - 54i - 9 = 0.$$
Also, $p\left(\frac{1}{2}\right) = 2\left(\frac{1}{2}\right)^3 - \left(\frac{1}{2}\right)^2 + 18\left(\frac{1}{2}\right) - 9$
$$= \frac{1}{4} - \frac{1}{4} + 9 - 9 = 0.$$

Neils Abel

The question of finding a formula for exact zeros to all polynomials was not settled until the early 19th century. In 1824, a Norwegian mathematician, Neils Abel (1802–1829), wrote a conclusive proof that it is impossible to construct a general formula for zeros of any polynomial beyond degree 4. Abel's work had several effects. First, the theory he developed contributed to the foundation of another advanced branch of modern mathematics, *group theory*. Second, rather than searching for exact zeros to polynomials, mathematicians knew they had to rely on approximation techniques. These are studied in another branch of advanced mathematics called *numerical analysis*.

Questions

Covering the Reading

1. When and by whom was the quadratic formula first discovered? **about 825 A.D., by Al-Khowarizmi**
2. Name three mathematicians who contributed to the analysis of zeros of cubic or quartic polynomials. **Scipione del Ferro, Niccolo Tartaglia, Girolamo Cardano**
3. State the Fundamental Theorem of Algebra. **See below.**

In 4–9, *true* or *false*.

4. Every polynomial has at least one real zero. **False**

5. A polynomial of degree n has at most n real zeros. **True**

6. A cubic may have 4 zeros. **False**

7. A polynomial with zeros 2, 3, and -1 could be of degree greater than three. **True**

8. All polynomials, regardless of degree, have zeros that are complex. **True**

9. Suppose $p(x)$ is a polynomial with real coefficients. If $2 + 3i$ is a zero of $p(x)$, then $2 - 3i$ is a zero of $p(x)$. **True**

10. Consider $q(x) = (x - 5)^2(x + 2)^3$.
    a. Identify the zeros of $q(x)$ and give their multiplicities.
    b. Verify your answer to part **a** by graphing $y = q(x)$. **See margin.**
    a) 5 (multiplicity 2), -2 (multiplicity 3)

3) If $p(x)$ is any polynomial of degree $n \geq 1$ with complex coefficients, then $p(x)$ has at least one complex zero.

566

566

11. Find all zeros of $p(x) = x^5 - 16x$. **0, ±2, ±2i**

12. **a.** How many zeros does $r(x) = 3x^{11} - ix$ have? **11**
    b. Find one real zero. **0**

13. Consider the polynomial $p(x) = x^4 + 2x^3 + 11x^2 + 2x + 10$.
    a. Verify that i is a zero of $p(x)$. **See margin.**
    b. Find the remaining zeros of $p(x)$ and their multiplicities.
    $-i, -1 + 3i, -1 - 3i$ (all of multiplicity 1)

14. Suppose the graph at the right represents the polynomial function $y = g(x)$. What is the lowest possible degree for $g(x)$? **4**

Applying the Mathematics

15. The zeros of $x^n - 1$ are called the *nth roots of unity*. Find the fourth roots of unity. **±1, ±i**

16. Suppose $p(x)$ is a polynomial with real coefficients and $p(4 + 9i) = 0$. What is $p(4 - 9i)$? **0**

17. Find a polynomial $p(x)$ with real coefficients, leading coefficient 1, and of the lowest degree possible that has the two zeros 3 and $1 - 2i$.
    $p(x) = (x - 3)(x - 1 + 2i)(x - 1 - 2i) = x^3 - 5x^2 + 11x - 15$

18. **a.** Find the zeros of $t(x) = 2x^2 + ix + 3$. **$i, -\frac{3}{2}i$**
    b. The zeros are not complex conjugates. Explain why this does not contradict the Conjugate Zeros Theorem.
    The coefficients of $t(x)$ are not all real numbers.

19. Tell whether each of the following could or could not be part of the graph of a fourth degree polynomial function. Explain your decision for each one. **See margin.**

    a.   **b.**   **c.**

20. The curve pictured below is the graph of a polynomial function p of degree 5 with real coefficients. Copy the figure and insert a horizontal axis so that each condition is satisfied. **See margin.**
    a. p has one real root.
    b. p has five real roots.

$y = p(x)$

20.a.

b.

MORE PRACTICE
For more questions on SPUR
Objectives, use *Lesson Master 9-7*, shown on page 567.

PROJECTS
Project 6 on page 584 is
related to the content of
this lesson.

Review

21. Let $z = 3 - 8i$. Write in $a + bi$ form.   *(Lesson 9-6)*
  a. w, the complex conjugate of z **3 + 8i**
  b. $w + z$ **6 + 0i**   **c.** $w - z$ **0 + 16i**   **d.** wz **73 + 0i**   **e.** $\frac{w}{z}$ $-\frac{55}{73} + \frac{48}{73}i$

In 22–24, factor the polynomial into linear factors over the complex numbers.   *(Lesson 9-6)*

22. $3p^2 - 4$ **23.** $4p^2 - 9$ **24.** $4p^2 + 9$
  $(\sqrt{3}\,p + 2)(\sqrt{3}\,p - 2)$ $(2p + 3)(2p - 3)$ $(2p + 3i)(2p - 3i)$

25. How can you check a polynomial division problem when the remainder is not zero?   *(Lesson 9-4)*   **Multiply the quotient and divisor, then add the remainder. The result should equal the dividend.**

26. Approximate, to the nearest tenth, all the real zeros of the polynomial $p(x) = x^5 - 4x^3 + 8x^2 - 12$.   *(Lesson 9-3)*   **-2.5, -1.0, 1.5**

27. A silo is in the shape of a cylinder with a hemispherical top. The height of the cylinder is 7 m.   *(Lesson 9-1)*
  a. Write a formula for the polynomial function V which expresses the volume $V(r)$ of the silo as a function of r, its radius. $V(r) = \frac{2}{3}\pi r^3 + 7\pi r^2$
  b. What is the degree of V?   **3**
  c. What is the leading coefficient of V? $\frac{2}{3}\pi$
  d. Suppose the radius may be anywhere from 3 m to 5 m. What are the maximum and minimum capacities of the silo, assuming the entire space can be filled?   **maximum ≈ 812 m³; minimum ≈ 254 m³**

28. The government of a certain country decides to increase the salaries of all government employees by 5%. Disregarding all other factors (such as promotions, resignations, or new hirings), how will this raise affect
  a. the average salary of government employees nationally;
  b. the standard deviation of salaries?   *(Lesson 3-6)*   **a,b) The average salary and the standard deviation of the salaries will each increase 5%.**

In 29 and 30, *true* or *false*.   *(Lesson 3-4)*

29. The graph of every odd function passes through the origin.   **True**

30. If z is a zero of an odd function f, then $-z$ is also a zero of f.   **True**

Exploration

31. Write your nine-digit social security number. (Make up one if you do not have such a number.) Use the digits in order as the coefficients of a polynomial of degree 8 with alternating signs. For instance, if your social security number is 369–46–4564, your polynomial is
$$y = 3x^8 - 6x^7 + 9x^6 - 4x^5 + 6x^4 - 4x^3 + 5x^2 - 6x + 4.$$
  a. Graph your social security number polynomial. Tell how many real zeros it has.   **Answers will vary.**
  b. If the nine digits of a social security number may be any one of 0 through 9, and the system of alternating signs of coefficients is used to create a polynomial, what is the least number m of real zeros the polynomial can have? What is the greatest number M of real zeros it may have? Can a social security polynomial have any number of real zeros between m and M? Why or why not?   **See margin.**

568

LESSON 9-8

Factoring Sums and Differences of Powers

What are the cube roots of 125? Any cube root of 125 must be a solution to $x^3 = 125$. According to the extension of the Fundamental Theorem of Algebra, $x^3 - 125$ must have three complex zeros. The easiest one to find is 5, since $125 = 5^3$. Is 5 the only one, with multiplicity three? A quick check shows that $(x - 5)^3$ is *not* equal to $x^3 - 125$. A different approach is needed.

■ ■ ■ ■ ■ ■ ■ ■

Example 1   Find the cube roots of 125 other than 5.

Solution   Since $(x - 5)$ is a factor of $x^3 - 125$, divide to get the other one.

$$
\begin{array}{r}
x^2 + 5x + 25 \\
x - 5 \overline{)\,x^3 + 0x^2 + 0x - 125} \\
\underline{x^3 - 5x^2} \\
5x^2 + 0x \\
\underline{5x^2 - 25x} \\
25x - 125 \\
\underline{25x - 125} \\
0
\end{array}
$$

Thus $x^3 - 125 = (x - 5)(x^2 + 5x + 25)$. Now use the quadratic formula to find the zeros of the quadratic factor.

$$x = \frac{-5 \pm \sqrt{5^2 - 4 \cdot 25}}{2} = \frac{-5 \pm \sqrt{-75}}{2} = \frac{-5 \pm 5\sqrt{3}\,i}{2}$$

The other cube roots of 125 are $-\frac{5}{2} + \frac{5\sqrt{3}}{2}i$ and $-\frac{5}{2} - \frac{5\sqrt{3}}{2}i$.

Check $\left(-\frac{5}{2} + \frac{5\sqrt{3}}{2}i\right)^3 = \left(-\frac{5}{2} + \frac{5\sqrt{3}}{2}i\right)\left(-\frac{5}{2} + \frac{5\sqrt{3}}{2}i\right)^2$

$= \left(-\frac{5}{2} + \frac{5\sqrt{3}}{2}i\right)\left(\frac{25}{4} - \frac{25\sqrt{3}}{2}i - \frac{75}{4}\right)$

$= \left(-\frac{5}{2} + \frac{5\sqrt{3}}{2}i\right)\left(-\frac{25}{2} - \frac{25\sqrt{3}}{2}i\right)$

$= \frac{125}{4} + \frac{125\sqrt{3}}{4}i - \frac{125\sqrt{3}}{4}i + \frac{375}{4}$

$= \frac{500}{4}$

$= 125$

A check for $\left(-\frac{5}{2} - \frac{5\sqrt{3}}{2}i\right)$ is asked for in Question 1.

OBJECTIVE

D Factor polynomials and solve polynomial equations using sums or differences of powers.

TEACHING NOTES

As you discuss **Example 1,** you might note that it is possible to find all the cube roots and fourth roots of real numbers by factoring, but not higher roots.

Alternate Approach   For questions like that in **Example 2,** many teachers like to have students first write down the general formula they will use. $x^3 + y^3 = (x + y)(x^2 - xy + y^2)$. Then substitute $2a$ for x and $3b$ for y directly underneath this formula. This has the advantage of recording the general theorem.

Checking the result of **Example 3** in class may enable students to see the pattern well enough so that they could reproduce the factorization even if they forgot it. You might note that the polynomial $a^4 - a^3b + a^2b^2 - ab^3 + b^4$ can be factored into two quadratic polynomials with real coefficients. (This can be proven from the Conjugate Zeros Theorem.) However, there is no factorization into polynomials with rational coefficients.

In the solution of Example 1, the difference of two cubes $x^3 - 125$ is factored into two polynomials with real coefficients, one of which is $x - 5$. More generally, consider the problem of factoring $x^3 - y^3$, the difference of *any* two cubes. Notice that $x = y$ is clearly a solution to $x^3 - y^3 = 0$, so $x - y$ is one factor. Divide $x^3 - y^3$ by $x - y$ to find the other factor.

$$
\begin{array}{r}
x^2 + \quad xy + y^2 \\
x - y \overline{)x^3 + 0x^2y + 0xy^2 - y^3} \\
\underline{x^3 - \quad x^2y} \\
x^2y \\
\underline{x^2y - \quad xy^2} \\
xy^2 - y^3 \\
\underline{xy^2 - y^3} \\
0
\end{array}
$$

Thus $x^3 - y^3 = (x - y)(x^2 + xy + y^2)$.

This proves the second part of the following theorem. You are asked to prove the first part in the Questions.

Sums and Differences of Cubes Theorem

For all x and y,
$$x^3 + y^3 = (x + y)(x^2 - xy + y^2)$$
$$x^3 - y^3 = (x - y)(x^2 + xy + y^2).$$

Neither one of the quadratic factors in the Sums and Differences of Cubes Theorem can be factored further over the set of polynomials with real coefficients.

▪ ▪ ▪ ▪ ▪ ▪ ▪ ▪ ▪

Example 2 Factor $8a^3 + 27b^6$.

Solution This is the instance of the sum of two cubes $x^3 + y^3$ where $x = 2a$ and $y = 3b^2$. Apply the previous theorem.
$$8a^3 + 27b^6 = (2a)^3 + (3b^2)^3$$
$$= (2a + 3b^2)((2a)^2 - (2a)(3b^2) + (3b^2)^2)$$
$$= (2a + 3b^2)(4a^2 - 6ab^2 + 9b^4)$$

The method used to find factors of sums and differences of cubes generalizes to all odd powers.

Sums and Differences of Odd Powers Theorem

For all x and y and for all odd positive integers n,
$$x^n + y^n = (x + y)(x^{n-1} - x^{n-2}y + x^{n-3}y^2 - \ldots - xy^{n-2} + y^{n-1})$$
$$x^n - y^n = (x - y)(x^{n-1} + x^{n-2}y + x^{n-3}y^2 + \ldots + xy^{n-2} + y^{n-1}).$$

570

You will verify the theorem for specific values of n in the Questions. Example 3 applies it to a sum of fifth powers.

■ ■ ■ ■ ■ ■ ■ ■

Example 3 Factor $a^5 + b^5$.

Solution Since $a^5 + b^5$ is a sum, $a + b$ is a factor, and in the second factor the signs alternate.

Thus, $a^5 + b^5 = (a + b)(a^4 - a^3b + a^2b^2 - ab^3 + b^4)$.

A similar theorem for the factorization of sums and differences of even powers does not exist. If n is a positive even integer, then $x^n + y^n$ does not have a linear factor with real coefficients. To factor the difference of two powers $x^n - y^n$ for n even, consider the even power as the square of some lower power, and reduce the problem to the difference of two squares. Example 4 illustrates a specific case.

■ ■ ■ ■ ■ ■ ■ ■

Example 4 Factor $x^6 - 64$ completely over the set of polynomials with real coefficients.

Solution $\quad x^6 = (x^3)^2$, and $64 = 2^6 = (2^3)^2$.
So $\qquad\qquad x^6 - 64 = (x^3)^2 - (2^3)^2$
$\qquad\qquad\qquad\qquad = (x^3 - 2^3)(x^3 + 2^3)$.

Each of the factors is a sum or difference of cubes, so they can be factored further. Hence,
$$x^6 - 64 = (x - 2)(x^2 + 2x + 4)(x + 2)(x^2 - 2x + 4).$$
Since $x^2 + 2x + 4$ and $x^2 - 2x + 4$ each has discriminant -12, they cannot be factored into polynomials with real coefficients. So this factorization is complete.

Questions

Covering the Reading

1. Show that $-\frac{5}{2} - \frac{5\sqrt{3}}{2}i$ is a cube root of 125.   **See margin.**

In 2 and 3, **a.** find the real zeros of the function by factoring; **b.** verify your result by drawing a graph.

2. $f(x) = 2x^3 - 32x$ **See margin.**   3. $g(x) = x^3 - 64$ **See margin.**

4. **a.** Show that $x - 3$ is a factor of $x^3 - 27$ by using long division.
    b. Check your answer by multiplying $x - 3$ by the quotient.
    c. Show that the quotient is not factorable over the set of polynomials with real coefficients by calculating its discriminant.
    d. Determine the three cube roots of 27.
    a) See margin. b) $(x - 3)(x^2 + 3x + 9) = x^3 - 27$; c) $b^2 - 4ac = -27$
    d) $3, -\frac{3}{2} + \frac{3\sqrt{3}}{2}i, -\frac{3}{2} - \frac{3\sqrt{3}}{2}i$

LESSON 9-8 Factoring Sums and Differences of Powers   571

NOTES ON QUESTIONS
Question 8: Because this is a proof, students should be ready to give justifications. The basic justification here is the Distributive Property; other justifications are the laws of exponents ($x^m \cdot x^n = x^{m+n}$) and the commutativity and associativity of addition and multiplication.

Error Analysis for Questions 9 and 10: Some students may need to write 32 as 2^5 and 10,000,000 as 10^7.

Question 27g: The generalization is sometimes called the Rational Root Theorem, when applied to solving polynomial equations.

ADDITIONAL ANSWERS
1. See Additional Answers in the back of this book.

2.a. $2x^3 - 32x = 2x(x^2 - 16) = 2x(x - 4)(x + 4)$; zeros are 0 and ± 4

b.

3.a. $x^3 - 64 = (x - 4)(x^2 + 4x + 16)$; real zero is $x = 4$

b.

4.a. See Additional Answers in the back of this book.

In 5–7, **a.** describe the polynomial as a difference of squares, sum of squares, difference of cubes, or sum of cubes; **b.** factor.

5. $a^2 - 64x^2$
See margin.

6. $125x^3 + 1$
See margin.

7. $27 - 64a^3$
See margin.

8. Prove: For all x and y,
$$x^3 + y^3 = (x + y)(x^2 - xy + y^2).$$
$(x + y)(x^2 - xy + y^2) = x^3 + x^2y - x^2y - xy^2 + xy^2 + y^3 = x^3 + y^3$

In 9 and 10, **a.** factor the expression using the Sums and Differences of Odd Powers Theorem; **b.** justify your answer using either multiplication, division, or a graph.

9. $x^5 - 32$ See margin.

10. $x^7 + 10{,}000{,}000$ See margin.

In 11 and 12, *true* or *false*. Justify your answer.

11. $(x + y)$ is a factor of $x^4 + y^4$.
False. Sample: if you divide $x^4 + y^4$ by $x + y$, the remainder is $2y^4$.

12. $(a - b)$ is a factor of $a^n - b^n$, if n is a positive odd integer.
True, by the Sums and Differences of Odd Powers Theorem

Applying the Mathematics

In 13–15, factor completely over the set of polynomials with integer coefficients.

13. $28x^2y^2 - 7x^4$
$7x^2(2y + x)(2y - x)$

14. $x^4 - y^4$
$(x^2 + y^2)(x + y)(x - y)$

15. $8x^3y^3 + 343z^3$
See margin.

16. Refer to Example 4.
  a. Factor $x^6 - 64$ as a difference of cubes.
  b. Verify that the result you get in part **a** can be factored further to get the solution shown in Example 4. See margin.
  a) $x^6 - 64 = (x^2)^3 - (4)^3 = (x^2 - 4)(x^4 + 4x^2 + 16)$

17. Consider $p(x) = x^5 - 5x^3 + 4x$.
  a. Find the zeros of the polynomial by factoring.
  b. Verify your results in part **a** by graphing the function p. See margin.
  a) $x^5 - 5x^3 + 4x = x(x + 2)(x - 2)(x + 1)(x - 1)$, so the zeros are $x = 0, \pm 1, \pm 2$.

In 18 and 19, **a.** factor the given binomial over integers using the Sums and Differences of Odd Powers Theorem; **b.** verify the factorization by multiplying.

18. $x^7 + y^7$ See margin.

19. $t^9 - 512$ See margin.

Review

20. *True* or *false*. The Fundamental Theorem of Algebra guarantees that the equation $4x^4 + 16x^3 + 5x^2 + 25 = 0$ has at least one real solution. *(Lesson 9-7)* **False**

21. *True* or *false*. The polynomial $c(x) = x^3 - 1271x^2 + 1273x - 1272$ has at least one real zero. *(Lesson 9-7)* **True**

22. What is the minimum degree of the polynomial function graphed at the right? *(Lesson 9-7)*
5

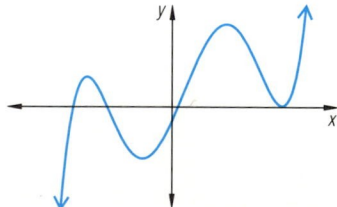

572

23. Suppose that $(x - 3)$ is a factor of the polynomial $f(x)$. Determine which of the following statements must be true.
  a. 3 is a solution to the equation $f(x) = 0$.   **True**
  b. $^-3$ is a solution to the equation $f(x) = 0$.   **False**
  c. The graph of $f(x)$ crosses the x-axis at $(3, 0)$.   **True**
  d. When $f(x)$ is divided by $(x - 3)$, the remainder is zero.   **True**
  e. $f(3) = 0$ *(Lesson 9-5)*   **True**

24. Suppose each face of an 8 cm by 10 cm by 12 cm block of wood is planed by x cm.   *(Lessons 9-1, 9-3)*
  a. Write a polynomial expression equal to the volume of the resulting block. $V(x) = (8 - 2x)(10 - 2x)(12 - 2x) = {^-8}x^3 + 120x^2 - 592x + 960$
  b. Find the value of x (to the nearest millimeter) which will leave the block with half the original volume.   **1 cm**

25. Consider the functions $f(x) = x^3 - 27x$ and $g(x) = (x - 2)^3 - 27(x - 2)$.
  (Lessons 3-3, 3-4, 9-3)   **a) x-intercepts: 0, $3\sqrt{3}$, $^-3\sqrt{3}$; y-intercept: 0**
  a. Find the x- and y-intercepts of f.
  b. State whether f is odd, even, or neither.   **odd**
  c. Find the x- and y-intercepts of g.
  d. State whether g is odd, even, or neither.   **neither**
  c) x-intercepts: 2, $2 + 3\sqrt{3}$, $2 - 3\sqrt{3}$; y-intercept: 46

26. Suppose that 36% of the labor force of a certain country are women. A company hires twelve new workers, and three of them are women.
  a. Determine the probability that this would occur based on the national statistics.
  b. What is the probability that fewer than four of the new workers are women?   *(Lesson 8-9)*   **a)** $\binom{12}{3}(.36)^3(.64)^9 \approx 18\%$; **b)** $\approx 32\%$

Exploration

27. Consider the polynomial which when factored is
  $(3x - 5)(2x + 9)(2x + 7)$.
  a. When it is expanded, what will the leading coefficient be?   **12**
  b. When it is expanded, what will the constant term be?   **$^-315$**
  c. Write the zeros of the polynomial as simple fractions.   **5/3, $^-9/2$, $^-7/2$**
  d. Ignoring the sign, what is the relationship of the numerators of the zeros to the linear factors?
  e. Ignoring the sign, what is the relationship of the denominators of the zeros to the linear factors?
  f. Suppose a polynomial is $(ax + b)(cx + d)(ex + f)(gx + h)$. Tell what the constant term will be, what the leading coefficient will be, and give the relationship of these products to the coefficients in the original polynomial in expanded form.
  g. Generalize parts **a** to **f**.   **If a simple fraction a/b is a zero of a polynomial P with integer coefficients, then a is a factor of the constant term of P and b is a factor of the leading coefficient;**
  d) The numerators are the constant terms in the factors.

e) The denominators are the coefficients of x in the factors.
f) constant term: $bdfh$; leading coefficient: $aceg$. The constant term is the product of the constant terms of the factors. The leading coefficient is the product of the leading coefficients of the factors.

FOLLOW-UP

MORE PRACTICE
For more questions on SPUR Objectives, use *Lesson Master 9-8*, shown below.

EXTENSION
You might wish to expand on **Question 27.** According to the generalization found there, what rational numbers are possible solutions to $12x^3 - 29x^2 + 8x + 4 = 0$? (all rational numbers that can be written as $\frac{a}{b}$, where a is a factor of 4 and b is a factor of 12) Search these numbers to find the solutions. $(-\frac{2}{3}, -\frac{1}{4}, 2)$

19.a. $(t - 2)(t^8 + 2t^7 + 4t^6 + 8t^5 + 16t^4 + 32t^3 + 64t^2 + 128t + 256)$

b. $(t - 2)(t^8 + 2t^7 + 4t^6 + 8t^5 + 16t^4 + 32t^3 + 64t^2 + 128t + 256) = t^9 + 2t^8 + 4t^7 + 8t^6 + 16t^5 + 32t^4 + 64t^3 + 128t^2 + 256t - 2t^8 - 4t^7 - 8t^6 - 16t^5 - 32t^4 - 64t^3 - 128t^2 - 256t - 512 = t^9 - 512$

NAME _____

LESSON **MASTER   9-8**
QUESTIONS ON **SPUR** OBJECTIVES

■**SKILLS** *Objective D (See pages 587–590 for objectives.)*
  1. Find the three cube roots of 64.   **4, $^-2 + 2\sqrt{3}i$, $^-2 - 2\sqrt{3}i$**
  2. Find the four fourth roots of 81.   **3, $^-3$, $3i$, ^-3i**

In 3–8, factor completely over the set of polynomials with integer coefficients.
  3. $m^4 - n^2$ **$(m^2 + n)(m^2 - n)$**
  4. $56x^3y^4 - 14x^2y^2 - 7x$ **$7x(2xy^2 - 1)(4xy^2 + 1)$**
  5. $6x^3 - x^2 - 15x$ **$x(2x + 3)(3x - 5)$**
  6. $b^5 - 10b^3 + 9b$ **$b(b + 1)(b - 1)(b + 3)(b - 3)$**
  7. $v^2 - u^2$ **$(v + u)(v - u)$**
  8. $c^3d - c^2d - 2cd$ **$cd(c + 1)(c - 2)$**

In 9–12, factor over polynomials with integer coefficients using the sum or difference of squares, or the sum or difference of cubes.
  9. $216n^3 + 125$ **$(6n + 5)(36n^2 - 30n + 25)$**
  10. $n^2 + 64m^2$ **$(n + 8m)(n - 8m)$**
  11. $81t^4 - 16s^4$ **$(9t^2 + 4s^2)(3t + 2s)(3t - 2s)$**
  12. $64v^6 - 729$ **$(2v + 3)(2v - 3)(4v^2 - 6v + 9)(4v^2 + 6v + 9)$**

In 13 and 14, factor the expression using the Sums and Differences of Odd Powers Theorem.
  13. $32f^5 + 243$ **$(2f + 3)(16f^4 - 24f^3 + 36f^2 - 54f + 81)$**
  14. $78,125 - 128g^7$ **$(5 - 2g)(15,625 + 6250g + 2500g^2 + 1000g^3 + 400g^4 + 160g^5 + 64g^6)$**

D Factor polynomials and solve polynomial equations using the grouping of terms.

You might wish to summarize the various kinds of factoring that students are expected to know. Before this course, students should have become familiar with the following types of factoring:
   common monomial;
   difference of squares;
   perfect squares; and
   quadratic trinomials.
From earlier lessons in this chapter, they should be familiar with:
   use of the Factor Theorem;
   sum of nth powers; and
   difference of nth powers.
In this lesson they will learn another technique, grouping, which they may have seen before.

These are not the only kinds of factoring techniques. For instance, though many students are taught that a sum of squares can never be factored over polynomials with real coefficients, it is the case that $x^4 + 4 = (x^2 - 2x + 2)(x^2 + 2x + 2)$, and more generally, $x^4 + \dfrac{b^4}{4} = \left(x^2 - bx + \dfrac{b^2}{2}\right) \cdot \left(x^2 + bx + \dfrac{b^2}{2}\right)$, and if b is even, only integers are involved in the factoring.

The strategy used in **Example 1** is quite powerful. Find two terms that have a common factor and factor it out;

Advanced Factoring Techniques

In previous lessons, polynomials were factored by using graphs, recognizing special patterns (sums and differences), and using the quadratic formula. In this lesson, you will utilize chunking and another factoring technique called *grouping*. Grouping is used to factor polynomials which contain groups of terms with common factors, usually monomials. It involves only the repeated application of the distributive property. Example 1 gives an instance of this technique.

Example 1   Factor $x^3 + 2x^2 - 9x - 18$.

Solution   Observe that the first two terms have a common factor of x^2 and the last two terms are each divisible by -9. Grouping these terms yields
$$x^3 + 2x^2 - 9x - 18 = x^2(x + 2) - 9(x + 2).$$

Note that this shows that another common factor is $(x + 2)$. Now apply the distributive property.
$$= (x^2 - 9)(x + 2)$$

Finally, factor the difference of squares.
$$= (x - 3)(x + 3)(x + 2)$$

Check 1   Multiply the factors. You are asked to do this in the questions.

Check 2   Draw graphs of $y = x^3 + 2x^2 - 9x - 18$ and $y = (x - 3)(x + 3)(x + 2)$. If the factorization is correct, the two graphs will coincide. At the right is the output from an automatic grapher. The two graphs are identical. Also, the graphs have zeros at -3, -2, and 3, as predicted by the factored form.

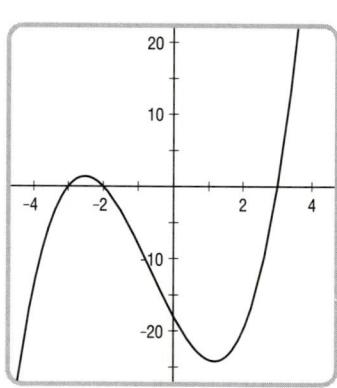

Grouping can be applied to factor the general trinomial $ax^2 + bx + c$. If there exist two numbers n_1 and n_2 such that $n_1 n_2 = ac$ and $n_1 + n_2 = b$, then the middle term bx can be split up into $n_1 x + n_2 x$ (since $n_1 + n_2 = b$). This new polynomial, $ax^2 + n_1 x + n_2 x + c$, is always factorable by grouping. Example 2 illustrates this.

574

Example 2   Factor $6x^2 - 13x + 5$.

Solution   Here $a = 6$, $b = -13$, and $c = 5$. Are there two numbers whose product is 30 and whose sum is -13? Mentally you may realize that the numbers are -3 and -10. So rewrite the given polynomial as
$$6x^2 - 3x - 10x + 5.$$
Now group:
$$= (6x^2 - 3x) + (-10x + 5)$$
$$= 3x(2x - 1) - 5(2x - 1)$$
$$= (3x - 5)(2x - 1).$$

Check   Multiply: $(3x - 5)(2x - 1) = 6x^2 - 3x - 10x + 5$
$$= 6x^2 - 13x + 5.$$

Combining grouping and chunking can allow the factoring of polynomials with more than one variable. When a polynomial in two variables is equal to 0, as in Example 3, grouping terms may help you draw the graph of the relation.

Example 3   Graph the set of ordered pairs (x, y) satisfying $y^2 - xy + 5x - 5y = 0$.

Solution   The form of the equation makes it difficult to evaluate numbers and plot by hand; and many automatic graphers cannot accept an equation if it is not solved for y. Grouping with repeated applications of the Distributive Property yields the following.
$$0 = y^2 - xy + 5x - 5y$$
$$= y(y - x) + 5(x - y)$$
$$= y(y - x) - 5(y - x)$$
$$= (y - 5)(y - x)$$
So the original equation is true if and only if $y - 5 = 0$ or $y - x = 0$; that is, if $y = 5$ or $y = x$. The graphs of these equations are lines. Thus the graph of the original equation is the union of these two lines, as shown below.

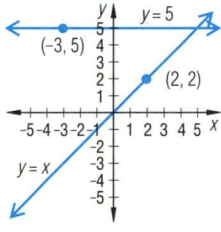

Check   Pick any point on the line $y = x$, say $(2, 2)$. Substitute in the original equation. Does $2^2 - 2 \cdot 2 + 5 \cdot 2 - 5 \cdot 2 = 0$? Yes. It checks. Similarly, any point on the line $y = 5$, say $(-3, 5)$, checks:
$$5^2 - (-3)(5) + 5(-3) - 5 \cdot 5 = 0.$$

then look to see if the other factor is present in the other two terms. When four terms are the product of two binomials, as in **Example 1,** there are always two ways to group. The other way is given as **Question 1.**

Some students may have been taught to factor quadratic trinomials using the method of **Example 2.** Most students, however, are taught to use trial and error and have not seen this method.

Alternate Approach
Another way to check both **Examples 1 and 2** is to substitute values for the variables in the given polynomial and in the factored answer.

The set of ordered pairs in **Example 3** is an example of a quadratic relation whose graph is a degenerate hyperbola. These graphs are discussed further in Chapter 12.

ADDITIONAL EXAMPLES
1. Factor $6a^2b + 2ac + 3ab^2 + bc$.
$(3ab + c)(2a + b)$

2. Factor $4x^2 + 4x - 15$ by grouping and check your answer.
$4x^2 + 4x - 15 = 4x^2 + 10x - 6x - 15 = 2x(2x + 5) - 3(2x + 5) = (2x - 3)(2x + 5)$; check by letting $x = 2$: $4(2)^2 + 4(2) - 15 = 9$ and $(2 \cdot 2 - 3)(2 \cdot 2 + 5) = 9$.

3. Describe the set of ordered pairs (x, y) satisfying $x^3 + 3x^2 - yx - 3y = 0$.
the union of the set of ordered pairs satisfying $x + 3 = 0$ and the set of ordered pairs satisfying $x^2 - y = 0$. The first is the vertical line through $(-3, 0)$; the second is the parent parabola.

NOTES ON QUESTIONS
Questions 6 and 11:
Students may be able to generalize the idea of these questions and **Example 2.** Suppose an equation is desired whose graph is the union of the graphs of two

575

relations. You can rewrite the equations for those relations so that the right sides are 0. Then multiply the nonzero sides to get an equation for the desired relation.

Question 18: Students should use each part as an aid in doing the next part.

Question 23: The Rational Roots Theorem from the exploration question of Lesson 9-8 can be applied here.

ADDITIONAL ANSWERS

1. $(x^3 - 9x) + (2x^2 - 18) =$
$x(x^2 - 9) + 2(x^2 - 9) =$
$(x + 2)(x^2 - 9) =$
$(x + 2)(x - 3)(x + 3)$

2. $(x + 2)(x - 3)(x + 3) =$
$(x + 2)(x^2 - 9) =$
$x^3 + 3x^2 - x - 3x - 6x -$
$18 = x^3 + 2x^2 - 9x - 18$

3. $x^3 + 3x^2 - 4x - 12 =$
$x^2(x + 3) - 4(x + 3) =$
$(x^2 - 4)(x + 3) =$
$(x - 2)(x + 2)(x + 3)$

4. $12x^2 + 8x + 1 =$
$12x^2 + 6x + 2x + 1 =$
$6x(2x + 1) + 1(2x + 1) =$
$(6x + 1)(2x + 1)$

5. $2x^2 - 9x - 5 =$
$2x^7 - 10x + 1x - 5 =$
$2x(x - 5) + 1(x - 5) =$
$(2x + 1)(x - 5)$

6.a.

7.a. $2x^3 - 5x^2 + 6x - 15 =$
$x^2(2x - 5) + 3(2x - 5) =$
$(x^2 + 3)(2x - 5)$, so the zeros of f are $\pm i\sqrt{3}$, $\frac{5}{2}$

Covering the Reading

In 1 and 2, refer to Example 1.

1. Factor the polynomial by rewriting it as $(x^3 - 9x) + (2x^2 - 18)$ and then applying the distributive property twice.   **See margin.**

2. Verify that the product $(x - 3)(x + 3)(x + 2)$ equals $x^3 + 2x - 9x - 18$.   **See margin.**

3. Factor the polynomial $x^3 + 3x^2 - 4x - 12$ by grouping the first and second terms and the third and fourth terms and then applying the distributive property twice.   **See margin.**

In 4 and 5, use grouping to factor the following trinomials.

4. $12x^2 + 8x + 1$ **See margin.**   **5.** $2x^2 - 9x - 5$ **See margin.**

6. a. Graph the set of ordered pairs (x, y) satisfying
$y^2x - x^3 - y^3 + x^2y = 0$.   **See margin.**
  b. Describe the graph in words.
  It is the union of the two lines $y = x$ and $y = -x$.

Applying the Mathematics

7. Find the zeros of the function $f(x) = 2x^3 - 5x^2 + 6x - 15$:
  a. by grouping and factoring,
  b. by drawing a graph.   **a, b) See margin.**

In 8 and 9, factor.

8. $ax - bx - by + ay$ **9.** $2x^2 + xy - 2xz - yz$
(x + y)(a − b)   **(x − z)(2x + y)**

10. Draw a graph of $\{ (x, y): -x^2 + xy + 2y^2 = 0 \}$.
  See margin.

11. Find an equation describing the graph of the union of the two relations $y = x^2$ and $x = y^2$.   **$xy - y^3 - x^3 + x^2y^2 = 0$**

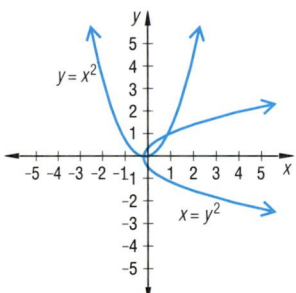

12. Let $f(x) = x^2 + 6x + 8$. Simplify $\dfrac{f(x) - f(b)}{x - b}$.   **$x + b + 6$**

In 13–16, factor over the set of polynomials with integer coefficients. *(Lesson 9-8)*

13. $x^3 - 8$ **(x − 2)(x² + 2x + 4)**

14. $t^5 + u^5$
(t + u)(t⁴ − t³u + t²u² − tu³ + u⁴)

15. $36x^6 - y^2z^4$
(6x³ − yz²)(6x³ + yz²)

16. $x^6y^3 - 27z^6$
(x²y − 3z²)(x⁴y² + 3x²yz² + 9z⁴)

17. Determine the three cube roots of 10. *(Lesson 9-8)* **See below.**

18. *Skill sequence.* Find all real solutions. *(Lessons 6-6, 9-6, 9-7)*
 a. $2x^3 - x^2 = 0$ **0 (multiplicity 2), 1/2**
 b. $2x^3 - x^2 - 2x + 1 = 0$ **1, -1, 1/2**
 c. $2\sin^3\theta - \sin^2\theta - 2\sin\theta + 1 = 0$ **See below.**

19. A propane gas tank is in the shape of a cylinder with hemispheres on each end, as shown at the left. Suppose that the total length of the tank is 10 ft. *(Lessons 9-1, 9-3)*
 a. Find the volume $V(r)$ of the tank in terms of the radius r.
 b. Draw a graph of the function $y = V(r)$ over the largest domain over which the function is defined.
 c. What is the largest volume such a tank can have? **b) See margin.**

20. A meteorologist has predicted that the probability of rain in each of the two days of a weekend is 0.6. Code a set of random digits to represent the weather on both days of the weekend. *(Lesson 7-8)* **See below.**

21. a. How many permutations consisting of four letters each can be formed from the letters of DINOSAUR? **8!/4! = 1680**
 b. How many of the permutations in part **a** end in R? *(Lesson 7-5)*
 7 · 6 · 5 = 210

22. Suppose a binomial probability is given by $_nC_rp^rq^{n-r} = 45p^r\left(\frac{1}{9}\right)^{10-r}$
 a. How many trials are there? **10**
 b. How many successes are there? **r**
 c. What is the probability of success for each trial? *(Lesson 8-9)* **8/9**

23. Consider the function $f(x) = 12x^4 - 8x^3 - 27x^2 + 18x$.
 a. Graph the function on the domain $-4 \le x \le 4$. Use a range that allows you to see all four zeros and three relative extrema.
 b. Estimate the zeros to the nearest tenth.
 c. Use factoring to determine the zeros exactly.
 a) See margin. b) -1.5, 0, 0.7, 1.5
 c) f(x) = x(2x + 3)(3x − 2)(2x − 3), so the zeros are 0, $-\frac{3}{2}$, $\frac{2}{3}$, $\frac{3}{2}$.

17) $\sqrt[3]{10}$, $-\dfrac{\sqrt[3]{10}}{2} \pm \dfrac{\sqrt{3}\,\sqrt[3]{10}}{2}i$

18c) $\dfrac{\pi}{2} + \pi n, \dfrac{\pi}{6} + 2\pi n, \dfrac{5\pi}{6} + 2\pi n$ for all integers n

19a) $V(r) = \pi r^2(10 - 2r) + \dfrac{4}{3}\pi r^3$; c) $\dfrac{500\pi}{3} \approx 524\ \text{ft}^3$

20) Let the digits 0, 1, 2, 3, 4, 5 denote rain on one day and let the digits 6, 7, 8, 9 denote no rain. The weather on the weekend is represented by choosing 2 random digits.

MORE PRACTICE
For more questions on SPUR Objectives, use *Lesson Master 9-9*, shown below.

7.b.

10.

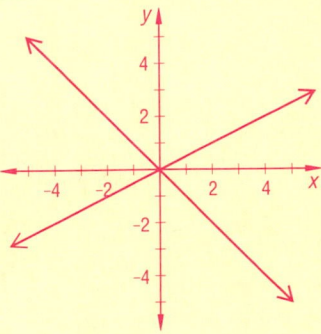

19.b., 23.a. See Additional Answers in the back of this book.

NAME _____

LESSON **MASTER** **9-9**
QUESTIONS ON **SPUR** OBJECTIVES

■**SKILLS** *Objective D (See pages 587–590 for objectives.)*
In 1–4, factor each expression.

1. $6x^2 + 13x - 5$ — **(2x + 5)(3x − 1)**
2. $12a^2 - 5a - 2$ — **(4a + 1)(3a − 2)**
3. $12x^2 + 28x + 15$ — **(6x + 5)(2x + 3)**
4. $x^3 - 4x^2 - x + 4$ — **(x − 1)(x + 1)(x − 4)**

In 5–8, solve.

5. $20y^2 = y + 63$ — **y = -1.75 or y = 1.8**
6. $43x = 18x^2 - 5$ — **x = -$\frac{1}{9}$ or x = $\frac{5}{2}$**
7. $15(z^2 - 3z) = 5z^2 - 2z + 9$ — **z = 4.5 or z = 0.2**
8. $x^3 + 4x^2 - x - 4 = 0$ — **x = -1, x = 1, or x = -4**

In 9 and 10, use factoring to find the zeros of the polynomial.

9. $f(x) = 2x^3 + 3x^2 - 8x - 12$ — **-2, 2, $\frac{-3}{2}$**
10. $g(x) = x^3 - 2x^2 - x + 2$ — **1, -1, 2**

In 11–14, factor.

11. $2x^2 + 2xy + 3x + 3y$ — **(2x + 3)(x + y)**
12. $8x^2 - 3yw - 2xw + 12xy$ — **(2x + 3y)(4x − w)**
13. $xv + 2xy - 3yv - 6y^2$ — **(x − 3y)(v + 2y)**
14. $21x - 28xy + 3w - 4yw$ — **(3 − 4y)(7x + w)**

In 15–17, solve for x.

15. $3x^2 - 5xy - 2y^2 = 0$ — **x = 2y or x = $\frac{-4}{3}$**
16. $10x^2 + 15xz = 12x + 18z$ — **x = $\frac{6}{5}$ or x = $\frac{-3}{2}z$**
17. $15x^2 - 10x = 18ax - 12a$ — **x = $\frac{6a}{5}$ or x = $\frac{2}{3}$**

103

9-10

Roots and Coefficients of Polynomials

In mathematics, it is common to switch the given information with what is to be found and thus invent a new problem. For instance, instead of using the lengths of two sides of a rectangle to find its area, you could be given its area and search for possible lengths of sides. Or, instead of being given x and asked to find tan x, you could be given tan x and asked to find x. As you saw in the previous lessons of this chapter, instead of multiplying binomials to obtain a polynomial, a common problem is to be given the polynomial and asked to find its factors.

To reverse the process of solving equations, you could begin with the solutions and, from them, find the equation. Of course, more than one equation could have the same solution. For instance, if $x = 1$ is the solution, any of the following could be an equation, and you could find many more.

$$\log x = 0 \qquad 2^x = 2 \qquad 3x^2 - 6x + 8 = 5 \qquad x - 1 = 0$$

However, if the equation must be a polynomial equation in standard form $a_nx^n + a_{n-1}x^{n-1} + \ldots + a_1x + a_0 = 0$, then the choice is more restricted. Only the two right equations above are polynomial equations, and only the rightmost is in standard form. If the multiplicities of solutions are to be considered as 1 unless otherwise indicated, then only the equations of the form $ax - a = 0$, where a is any nonzero complex number, would have only the solution $x = 1$. Finally, if the leading coefficient must be 1, then only the equation $x - 1 = 0$ satisfies all these criteria and has only the solution $x = 1$.

What polynomial equation in standard form, and with leading coefficient 1, has the solutions $\frac{2}{3}$ and -5? Using the Factor Theorem, the polynomial can be found by multiplying $\left(x - \frac{2}{3}\right)(x - -5)$ to obtain the left side of the desired equation, $x^2 + \frac{13}{3}x - \frac{10}{3} = 0$.

In general, if a polynomial equation in standard form has two roots r_1 and r_2, then the polynomial is found by multiplying $(x - r_1)(x - r_2)$, and the equation can be written $x^2 - (r_1 + r_2)x + r_1r_2 = 0$. This argument proves the following theorem.

Theorem

For the quadratic equation $x^2 + bx + c = 0$, the sum of the roots is $-b$ and the product of the roots is c.

For example, to find two numbers whose sum is 10 and whose product is 40, you need only solve the quadratic equation $x^2 - 10x + 40 = 0$. The solutions are the desired numbers. This exact problem was used by Girolamo Cardano in 1533 to introduce complex numbers for the first time.

If a polynomial equation has three roots (counting multiplicities), then you know the polynomial is of degree 3. If the roots are r_1, r_2, and r_3, then the polynomial is

$$(x - r_1)(x - r_2)(x - r_3).$$

Girolamo Cardano (woodcut, 1539)

To write the polynomial in standard form, multiply the three binomials. The result is the left side of the equation

$$x^3 - (r_1 + r_2 + r_3)x^2 + (r_1r_2 + r_1r_3 + r_2r_3)x - r_1r_2r_3 = 0.$$

Notice how the multiplication of the three binomials creates this polynomial. The coefficient of x^3 is 1 because there is only one way to multiply the three xs from the binomials. To explain the x^2 term, notice that there are three ways to choose two of the xs, and in each case one of the r_i must be the third factor. This creates the x^2 term. For the x term, there are three ways to choose the x, but the other two factors in each term must be pairs of r_i. And then, choosing all three r_i from each binomial can be done in only one way. (You are asked to confirm this result in the Questions.) The alternating subtraction and addition occurs because each root is subtracted in the binomial, and so when there are an even number of roots multiplied, the subtraction is converted to an addition.

Theorem

For the cubic equation $x^3 + bx^2 + cx + d = 0$, the sum of the roots is $-b$, the sum of the products of the roots two at a time is c, and the product of the three roots is $-d$.

The theorems for quadratic and cubic equations can be generalized to polynomial equations of any degree. If a polynomial equation $p(x) = 0$ has n roots $r_1, r_2, \ldots, r_n$, then the polynomial $p(x)$ is

$$(x - r_1)(x - r_2)(x - r_3) \ldots (x - r_n).$$

The key to understanding the generalization is to see the multiplication of the binomials $(x - r_i)$ as a problem in combinations. Each product contributing to the power x^k is a result of k times choosing x and $n - k$ times choosing one of the r_i. So there are $_nC_k$ products with the power x^k. The coefficient of x^k is the sum of all these products, multiplied by -1 if an odd number of r_i have been chosen.

The second idea in this lesson is the writing of an equation for the union of two relations. This idea is quite powerful. For instance, if you want an equation for the union of $y = \sin x$ and $y = \sin(x - \pi)$, rewrite each equation so that one side is zero and multiply the nonzero sides. The result is $(y - \sin x)(y - \sin(x - \pi)) = 0$; the graph looks like a row of clams.

Point out to students that this is the idea that was used in Lesson 9-9 to create equations for the union of two lines, but here the equations may be of relations other than polynomials.

For instance, to find a polynomial equation with roots -8, 9, 2, and 5, you could multiply
$$(x + 8)(x - 9)(x - 2)(x - 5)$$
and set the product equal to zero. You could also determine the coefficients by taking the products of the roots one at a time, two at a time, three at a time, and four at a time. The coefficient of x^4 is 1. Products one at a time are just the terms, and the opposite of their sum is the coefficient of x^3,
$$-8 + 9 + 2 + 5 = 8.$$

There are $_4C_2 = 6$ products two at a time. Their sum gives the coefficient of x^2.
$$-8 \cdot 9 + -8 \cdot 2 + -8 \cdot 5 + 9 \cdot 2 + 9 \cdot 5 + 2 \cdot 5 = -55$$

There are $_4C_3 = 4$ products three at a time. The opposite of their sum gives the coefficient of x.
$$-8 \cdot 9 \cdot 2 + -8 \cdot 9 \cdot 5 + -8 \cdot 2 \cdot 5 + 9 \cdot 2 \cdot 5 = -494$$

There is $_4C_4 = 1$ product four at a time. This is the product of the 4 roots, and it gives the constant term, the coefficient of x^0.
$$-8 \cdot 9 \cdot 2 \cdot 5 = -720$$

A polynomial equation with roots -8, 9, 2, and 5 is therefore
$$x^4 - 8x^3 - 55x^2 + 494x - 720 = 0.$$

Finding equations with given solutions sometimes arises when working with databases. In databases, people or other variables are typically categorized in various ways, and the categories are identified by numbers. For instance, in a database for students in a high school, one variable might be the section of the community in which a student lives. Name this variable SECT and suppose it has the 8 integer values 1 to 8, each identifying a particular section of the community. Suppose that a survey is to be done of how students get to school, and further suppose sections 3, 5, and 6 are very close to the school. To choose students from these sections, you could use the following instruction.

    IF SECT = 3 OR SECT = 5 OR SECT = 6, THEN...

The three equations and two OR statements could be replaced by a single equation.

    IF (SECT − 3)*(SECT − 5)*(SECT − 6) = 0, THEN...

580

Multiplying the three binomials does not result in a shorter form of the line of the instruction. To see this, replace SECT by x and use the results of the previous theorem to rewrite the polynomial in standard form.

$$(x-3)(x-5)(x-6) = x^3 - (3+5+6)x^2 + (3 \cdot 5 + 5 \cdot 6 + 3 \cdot 6)x - 3 \cdot 5 \cdot 6$$
$$= x^3 - 14x^2 + 63x - 90$$

Since computers are often preprogrammed to use logarithms to calculate powers, and because that arithmetic involves approximations, it is better to put the final polynomial in nested form.

$$= ((x - 14)x + 63)x - 90$$

The replacement of x by SECT in the decision line gives the following.

IF ((SECT − 14)*SECT + 63)*SECT − 90 = 0, THEN...

This is no shorter than the first line, but it involves no OR statements and fewer mathematical operations. For large programs, the result of the replacement as a savings of computer time and memory could be substantial.

Questions

1. Give four equations different from those in this lesson that have the solution $x = 1$. Try to make your equations not look much alike.
   samples: $x^4 - 1 = 0$; $x^4 - x^3 + 4x^2 - 3x - 1 = 0$; $5^{x-1} = 1$; $\log(2x - 1) = 0$

2. A quadratic equation has solutions $-\frac{3}{4}$ and 6. If the leading coefficient of the quadratic is 1, write the equation in standard form. $x^2 - \frac{21}{4}x - \frac{9}{2} = 0$

3. Solve Cardano's problem. $5 \pm \sqrt{15}\,i$

4. The sum of two numbers is 17 and their product is 100. What are the numbers? $\frac{17}{2} \pm \frac{\sqrt{111}}{2}\,i$

5. Find a polynomial equation with the three solutions 1, 2, and 3.
   sample: $x^3 - 6x^2 + 11x - 6 = 0$

6. A polynomial equation in x has the five solutions -10, -5, 0, 5, and 10. The leading coefficient of the polynomial is 1.
   a. Give the coefficient of x^4.   0
   b. Give the coefficient of x^3.   -125
   c. Give a possible equation. $x^5 - 125x^3 + 2500x = 0$

7. Consider the equation $x^5 - 3x^2 + 60 = 0$.
   a. What is the sum of the roots of this equation?   0
   b. What is the product of the roots of this equation?   -60

8. In instructions to find a file in a database, how could the following OR statement be replaced by a statement with one equation and no OR?
   IF SECT = 7 OR SECT = 2 OR SECT = 6, THEN...
   IF ((SECT − 15) * SECT + 68) * SECT − 84 = 0, THEN...

Review

9. Find the zeros of $p(x) = x^5 - x^3 + x^2 - 1$ by grouping. Describe any multiple zeros. *(Lessons 9-7, 9-9)* **-1 (multiplicity 2), $\frac{1}{2} \pm \frac{\sqrt{3}}{2}i$, 1**

10. Factor the polynomial $t^3 - 1$ over the set of
    a. integers,   b. reals,   c. complex numbers. *(Lessons 9-6, 9-8)*
    a) $(t-1)(t^2 + t + 1)$; b) $(t-1)(t^2 + t + 1)$; c) See below.

11. The first two terms of a geometric sequence are $6i$ and -12. Find
    a. the common ratio,   b. the next two terms,   c. the 100th term.
    (Lesson 8-4) **2i**   **24i, 48**   **$3 \cdot 2^{100}$**

12. Sketch the graph of a fourth degree polynomial satisfying the given condition. **See margin.**
    a. four real roots, one of multiplicity two
    b. no real roots   *(Lesson 9-7)*

13. When a polynomial $g(x)$ is divided by $3x + 4$, the remainder is 0. Make a conclusion about the equation $g(x) = 0$. *(Lessons 9-4, 9-5)*
    -4/3 is a solution.

14. a. Find a general equation for a cubic function with zeros at -1, 0, and 2.
    b. Use the result in part **a** to find an equation for the cubic function passing through the points $(-1, 0)$, $(0, 0)$, $(1, -1)$, and $(2, 0)$.
    (Lessons 9-2, 9-5) **a) $ax^3 - ax^2 - 2ax = 0$; b) $p(x) = \frac{1}{2}x^3 - \frac{1}{2}x^2 - x$**

15. When a polynomial $p(x)$ is divided by $x - 12$, the quotient is $x^3 + 5$ and the remainder is -1. Determine $p(x)$. *(Lesson 9-4)*
    $p(x) = x^4 - 12x^3 + 5x - 61$

16. Consider a (closed) rectangular box with dimensions x, $x + 3$, and $x - 5$. Write a polynomial in expanded form for
    a. $V(x)$, the volume of the box.   **$V(x) = x^3 - 2x^2 - 15x$**
    b. $S(x)$, the surface area of the box.   **$S(x) = 6x^2 - 8x - 30$**
    c. If x is measured in cm, in what unit is each of $V(x)$ and $S(x)$ measured? *(Lesson 9-1)* **$V(x)$ is in cm³; $S(x)$ is in cm².**

17. About 20.8% of the U.S. population in 1988 was between the ages 40 and 60. Determine the probability that the first four people called in a national telephone survey in 1988 were all between the ages 40 and 60. *(Lessons 7-3, 7-4)* **$\approx (.208)^4 \approx .2\%$**

18. The 12 oz packages (in glass jars) of a certain brand of peanuts are found to weigh 24.5 ounces on the average and the weights have a standard deviation of 0.45 oz. The company decides to use plastic jars which weigh 8 ounces less than the glass ones. Determine the mean and standard deviation of the weight of the new packages. *(Lesson 3-3)*
    mean = 16.5 oz; standard deviation = 0.45 oz

Exploration

19. Look into a database program and explain how it sets up files and how it enables the user to make choices from those files. **Answers will vary.**

10c) $(t-1)(t + \frac{1}{2} + \frac{\sqrt{3}}{2}i)(t + \frac{1}{2} - \frac{\sqrt{3}}{2}i)$

582

Projects

1. On some tests you are asked to "complete the pattern," or to find the next term in a sequence such as

$$1, 4, 9, \ldots .$$

These are misleading questions, because there are always many justifiable answers to such problems. In fact, for any n points, if you add one more point of your own, you can find a polynomial of degree n to model the data. For example, below is a table of differences for the pattern begun above.

Only one second difference, 2, is known.

Now make up any value you like for the third difference, 6 for instance, and work backward from that 3rd difference to the next y-value.

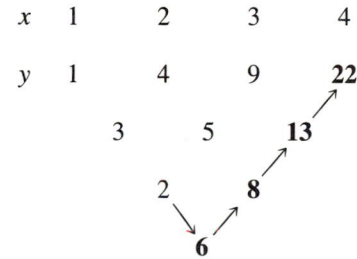

  a. Use the method of finite differences to find a polynomial for these data. Check that all four values fit the polynomial.

  b. Now go back to the original data and choose a different value for the third difference, and work backwards to find the y-value associated with $x = 4$. Use finite differences to find a polynomial different from that in part **a** that models the original data.

  c. Find a value for y when $x = 4$ that could be justified by a quadratic model.

  d. If n data points are given, what is the degree of the polynomial needed to fit all of the data? What do the results of your investigation in parts **a–c** tell you about the number of polynomials of higher degree that might fit the data?

2. The graph of every quadratic function $f(x) = ax^2 + bx + c$ is reflection-symmetric to the line $x = -\frac{b}{2a}$. The purpose of this project is to explore the symmetry of the graph of the general cubic function $f(x) = ax^3 + bx^2 + cx + d$.

  a. Graph $y = x^3$ and four other cubic polynomials. Describe any line or point symmetry you observe.

  b. Examine specifically several polynomials with equations of the form $f(x) = ax^3 + px$. Prove that each of these is an odd function, and explain how your proof shows that the graph of $f(x) = ax^3 + px$ is point-symmetric to the origin. That is, $(0, 0)$ is a center of symmetry for the graph of f.

  c. Now, examine graphs of equations of the form $y = ax^3 + px + q$. What is the center of symmetry for these curves?

  d. Consider equations of the form
$$f(x) = a(x - h)^3 + p(x - h) + q.$$
What are the coordinates of the center of symmetry for these curves?

  e. When an equation of the form
$$f(x) = a(x - h)^3 + p(x - h) + q$$
is expanded and like terms are combined, explain why it must be of the form
$$f(x) = ax^3 + bx^2 + cx + d.$$
By equating coefficients of terms of equal degree, prove that $h = -\frac{b}{3a}$. Use this result to find a center of symmetry for the graph of
$$f(x) = ax^3 + bx^2 + cx + d.$$

The projects provide an opportunity for extended work on topics related to the content of the chapter. (See pages 66 and T37 for more information on the purposes of projects.)

Solutions to the projects are found at the end of the Additional Answers section at the back of this book.

Project 1 allows for a great deal of freedom. Encourage students to create patterns that start simple but do not remain that way. For instance, 2, 4, 6, 13, ...

Project 2 explains why the graphs of cubics can be classified into two types.

For **Project 3,** you might have students look for patterns in Pascal's Triangle other than those mentioned in this chapter and include them in the writeup.

Regarding **Project 4,** synthetic division is part of the subject matter of the next UCSMP text, *Precalculus and Discrete Mathematics.*

Many books on the history of mathematics contain the information needed for **Projects 5 and 6.**

3. Examine each diagonal of Pascal's triangle. These are the sequences of numbers on a line slanted at 60° to the horizontal. Explain why the nth term of each diagonal can be represented by a polynomial. Find examples of patterns in this triangle that can be described by linear, quadratic, cubic, quartic, and quintic polynomials.

4. *Synthetic division* is another way to perform long division of a polynomial by a linear factor. Find a book that explains this procedure. Teach yourself to do it. Prove why it works. Figure out how to use the Remainder Theorem in conjunction with it.

5. *Algebraic numbers* and *transcendental numbers* are two types of real numbers. Two transcendental numbers you have studied are π and e. What are the distinguishing characteristics of transcendental or algebraic numbers? How are they related to the rational and irrational numbers? Where are they used? What led to their discovery?

6. Like Abel, Evariste Galois contributed important results to the study of polynomials at an early age. And also like Abel, Galois died at an early age. Find out more about the mathematical work of these two men and the tragic circumstances that led to their early deaths.

Summary

A polynomial in x of degree n is an expression of the form $a_n x^n + a_{n-1} x^{n-1} + \ldots + a_1 x + a_0$, where $a_n \neq 0$. Polynomials model long-term loan repayments, savings accounts with periodic deposits, volumes and surface areas of three dimensional figures, and so forth.

For any nth degree polynomial, the nth differences of y-values corresponding to any arithmetic sequence of x-values are equal and nonzero. If a polynomial of degree n fits a given set of data, then $n + 1$ points give rise to $n + 1$ equations that determine its coefficients.

The graphs of polynomial functions of a given degree n share certain characteristics. To approximate or determine specific values for a given polynomial (such as extrema, zeros, or intercepts), automatic graphers or computer programs that tabulate values of the function over specified intervals can be used.

The algorithm for dividing polynomials is very similar to that for dividing integers. If there is a remainder, its degree is always less than the degree of the dividend. Thus, when a polynomial $f(x)$ is divided by the linear factor $(x - c)$, the remainder is $f(c)$. Consequently, c is a solution to $f(x) = 0$ if and only if $(x - c)$ is a factor of $f(x)$. Thus, monomial factors of polynomial functions can be used to determine x-intercepts of their graphs, or to find solutions to polynomial equations. Conversely, formulas for polynomial functions always can be constructed if their zeros are known.

Imaginary numbers are all nonzero, real-number multiples of $i = \sqrt{-1}$. Complex numbers are sums of imaginary and real numbers. Operations with complex numbers are similar to those with polynomials. The Fundamental Theorem of Algebra and its consequences guarantee that every polynomial of degree $n \geq 1$ with complex coefficients has exactly n complex zeros, if multiplicities are counted.

Though factoring polynomials is helpful, no general formula exists for finding exact solutions for all polynomial equations. All sums and differences of the same odd power can be factored, as can differences of even powers. Another technique is the grouping of terms with common factors and then using the distributive property.

Coefficients of a polynomial are directly determined by its roots. If the leading coefficient is 1, the next coefficient is the opposite of the sum of the roots, the next one is the sum of the products of the roots taken two at a time, and so on.

Vocabulary

For the starred (*) terms you should be able to give a definition of the term.
For the other terms you should be able to give a general description and a specific example of each.

Lesson 9-1
*polynomial in x
degree of a polynomial
coefficients, leading coefficient
polynomial function
degree of a polynomial in more
than one variable
square numbers, triangular numbers

Lesson 9-2
tetrahedral numbers
Polynomial Difference Theorem

Lesson 9-3
TABULATE program
zero of a polynomial

Lesson 9-4
dividend, divisor, quotient,
remainder
Remainder Theorem

Lesson 9-5
Factor Theorem
Factor-Solution-Intercept
Equivalence Theorem
pentagonal numbers

Lesson 9-6
*imaginary numbers, i
*complex numbers, $a + bi$
*real part, imaginary part of a
complex number
equal complex numbers
*complex conjugates

Lesson 9-7
Fundamental Theorem of Algebra
multiplicity of a zero
Number of Zeros of a Polynomial
Theorem
Conjugate Zeros Theorem

Lesson 9-8
Sums and Differences of Cubes
Theorem
Sums and Differences of Odd
Powers Theorem

Lesson 9-9
factoring by grouping

Lesson 9-10
Roots and Coefficients of
Polynomials Theorem

Progress Self-Test

PROGRESS SELF-TEST

The Progress Self-Test provides an opportunity for feedback and correction. The Chapter Review provides additional opportunities for practice.

USING THE PROGRESS SELF-TEST
Assign the Progress Self-Test as a one-night assignment. (See pages 69 and T35 for more information.)

ADDITIONAL ANSWERS
3. $q(x) = 3x^3 - 13x^2 + 34x - 82$; $r(x) = 154$

5. $(t^2 + 10y)(t^4 - 10t^2y + 100y^2)$

11. It must have 7 zeros (counting multiplicity). Since complex zeros come in conjugate pairs, and the polynomial has an odd number of zeros, at least one of the roots must be real.

See margin for answers not shown below.

Take this test as you would take a test in class. Then check the test yourself using the solutions at the back of the book.

1. y is a polynomial function of x. Using the points below, find an equation of least degree for the polynomial. $y = 3x^2 - 8x + 1$

x	1	2	3	4	5	6	7
y	-4	-3	4	17	36	61	92

2. Approximate to the nearest hundredth the smallest positive x-intercept of the graph of $y = x^5 - x^3 + 3x^2 - 9$. **1.42**

3. Determine the quotient and remainder when $3x^4 - 7x^3 + 8x^2 - 14x - 10$ is divided by $x + 2$.

4. Factor $9x^2 - 3x - 2$. **$(3x - 2)(3x + 1)$**

5. Factor $t^6 + 1000y^3$ over the set of polynomials with integer coefficients.

6. *Multiple choice.* When $f(x) = x^8 - 7x^6 + 6x^3 - 4x^2 + 12x - 8$, then $f(1) = 0$. Which is a factor of $f(x)$?
   (a) $x - 1$ (b) $x + 1$
   (c) $x - 2$ (d) $x + 2$ **(a)**

7. Let $z = 2 + 5i$ and $w = 3 - i$. Express each of the following in $a + bi$ form.

   a. $2z + w$ **$7 + 9i$**   **b.** $\dfrac{z}{w}$ **$\dfrac{1}{10} + \dfrac{17}{10}i$**

In 8–10, use the polynomial
$$g(x) = (x - 8)^3(2x + 3)^2(5x - 1).$$

8. Which zero of g has multiplicity two?   **-3/2**

9. What is the degree of g?   **6**

10. How many x-intercepts does the graph of $g(x)$ have?   **3**

11. Explain why a polynomial of degree seven with real coefficients must have at least one real zero.

12. Solve $x^4 + 11x^2 + 10 = 0$ completely, given that one of the solutions is i.   **$x = i, -i, \sqrt{10}i, -\sqrt{10}i$**

13. The prices for a thin crust spinach pizza at Sophia's pizzeria are listed below.

10"	12"	14"
$5.95	$6.95	$8.50

If pricing is based on a quadratic function containing these three points, what should a 16" pizza cost (to the nearest penny)?   **$10.66**

14. In a right triangle the sides are as given below.

Find x to the nearest hundredth.   **1.27**

15. Part of the graph of a polynomial function t is shown below.

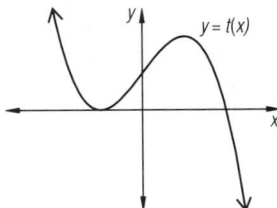

a. At least how many real zeros (including multiplicities) does t have?   **3**
b. What is the lowest degree the polynomial $t(x)$ must have?   **3**

16. A 25-meter-high cylindrical grain silo is to have 35 cm of insulation between two layers of sheet metal, as shown below.

Express the volume V of insulation material as a polynomial function of r, the radius of the outer layer of the silo.   **$V(r) = \dfrac{35}{2}\pi r - \dfrac{49\pi}{16}$**

586

Chapter Review

Questions on **SPUR** Objectives

RESOURCES
- Chapter 9 Test, Form A
- Chapter 9 Test, Form B
- Chapter 9 Test, Cumulative Form
- Comprehensive Test, Chapters 1–9
- Quiz and Test Writer

SPUR stands for **S**kills, **P**roperties, **U**ses, and **R**epresentations.
The Chapter Review questions are grouped according to the SPUR Objectives for this chapter.
See margin for answers not shown below.

SKILLS deal with the procedures used to get answers.

■ **Objective A:** *Use finite differences and systems of equations to determine an equation for a polynomial function from data points. (Lesson 9-2)*

1. In a set of data points, the x-values form an arithmetic sequence. If the 5th differences of y-values for the consecutive x-values are the first set of differences equal to 0,
   a. what is the degree of the polynomial which can be used to model this data?   **4**
   b. How many data points are necessary to determine an equation for this polynomial?   **5**

In 2 and 3, use the data listed in the table. **a.** Determine if y is a polynomial function of x of degree less than 5. **b.** If so, find an equation for it.

2.
x	1	2	3	4	5	6	7
y	6	13	26	45	70	101	138

3.
x	2	4	6	8	10	12	14
y	4	16	64	256	1024	4096	16384	**No**

2a) Yes: b) $y = 3x^2 - 2x + 5$

4. The following are the first six triangular numbers, t_n.
   $t_1 = 1, t_2 = 3, t_3 = 6, t_4 = 10, t_5 = 15, t_6 = 21$
   a. Determine a polynomial formula for t_n.
   b. Calculate t_{100}.   **a) $t_n = \frac{1}{2}n^2 + \frac{1}{2}n$; b) 5050**

■ **Objective B:** *Calculate or approximate zeros and relative extrema of polynomial functions. (Lesson 9-3)*

In 5–7, consider the function $g(t) = t^3 - 2t^2 - 4t + 3$.

5. Determine, to the nearest hundredth, a negative zero of g using the program TABULATE.   **-1.62**

6. Determine, to the nearest tenth, a local maximum of g by graphing.   **(-.7, 4.5)**

7. $g(t) = (t - 3)(t^2 + t - 1)$. Explain why $t = 3$ must be a t-intercept of the graph of g.

8. Approximate, to the nearest hundredth, the largest x-intercept of the graph of $y = x^3 - x^2 - 10x + 10$.   **3.16**

■ **Objective C:** *Divide polynomials. (Lesson 9-4)*

In 9–12, determine the quotient and remainder when the first polynomial is divided by the second.

9. $2t^5 - 3t^3 + t^2 + 11, \ t^2 + 3$

10. $t^6, t^3 - 1$ **quotient: $t^3 + 1$; remainder: 1**

11. $x^6 - 2x^5 + x^4 - 2x^3 + 2x^2 - x + 1, \ x^3 - 1$

12. $x^3 + 2x^2y - wxy^2 - y^3, \ x - y$

■ **Objective D:** *Factor polynomials and solve polynomial equations using the Factor Theorem, sums or differences of powers, grouping terms, or trial and error. (Lessons 9-5, 9-6, 9-8, 9-9)*

13. *True or false.*   If $(x - \sqrt{5})$ is a factor of $x^4 - 2x^2 - 15$, then $x = \sqrt{5}$ is a solution for $x^4 - 2x^2 - 15 = 0$.   **True**

14. Solve $x^3 - x^2 - 4x + 4 = 0$.   **-2, 1, 2**

In 15–20, factor over the set of polynomials with coefficients in the indicated domain.

15. $243t^5 - u^5$; integers

16. $x^5 + x^2 - 2x^3 - 2$; reals

17. $z^3 - 8$; complex numbers

18. $8x^2 + 10x + 3$; integers   **$(4x + 3)(2x + 1)$**

19. $20x^2 - 8x - 1$; integers   **$(10x + 1)(2x - 1)$**

20. $t^5 - t^4 - 16t + 16$; integers

15) $(3t - u)(81t^4 + 27t^3u + 9t^2u^2 + 3tu^3 + u^4)$
16) $(x + \sqrt{2})(x - \sqrt{2})(x + 1)(x^2 - x + 1)$
17) $(z - 2)(z + 1 - \sqrt{3}i)(z + 1 + \sqrt{3}i)$
20) $(t - 2)(t + 2)(t^2 + 4)(t - 1)$

CHAPTER REVIEW

The main objectives for the chapter are organized here into sections corresponding to the types of understanding this book promotes: Skills, Properties, Uses, and Representations (called SPUR objectives).

USING THE CHAPTER REVIEW
Students should be able to answer questions like these with about 85% accuracy by the end of the chapter. (See pages 72 and T38–39 for more information.)

ADDITIONAL ANSWERS
7. By the Factor-Solution-Intercept Equivalence Theorem, (t – 3) is a factor of $g(t)$ if and only if 3 is a t-intercept.

9. quotient: $2t^3 - 9t + 1$; remainder: $27t + 8$

11. quotient: $x^3 - 2x^2 + x - 1$; remainder: 0

12. quotient: $x^2 + 3xy + (3 - w)y^2$; remainder: $(2 - w)y^3$

■ **Objective E:** *Perform operations with complex numbers. (Lesson 9-6)*

21. Let $z = 3 + 2i$ and $w = 5 - 4i$. Express each of the following in $a + bi$ form.
   a. $z + w$ **8 − 2i**   **b.** $z - w$ **−2 + 6i**
   c. zw **23 − 2i**   **d.** $\frac{z}{4}$ $\frac{3}{4} + \frac{1}{2}i$

22. Repeat Question 21 if $z = 1 + 6i$ and $w = -2 - 3i$.
   a) −1 + 3i; b) 3 + 9i; c) 16 − 15i; d) $\frac{1}{4} + \frac{3}{2}i$

23. Express $\frac{6+i}{3-4i}$ in $a + bi$ form. $\frac{14}{25} + \frac{27}{25}i$

24. Express $\frac{2+2i}{-8+3i}$ in $a + bi$ form. $-\frac{10}{73} - \frac{22}{73}i$

25. Determine the sum $\sum\limits_{k=1}^{100} i^k$, where $i = \sqrt{-1}$. **0**

26. Show that a cube root of 64 is $-2 + 2\sqrt{3}\,i$.
   $(-2 + 2\sqrt{3}\,i)^3 = (-2 + 2\sqrt{3}\,i)(-8 - 8\sqrt{3}\,i) = 64$

PROPERTIES deal with the principles behind the mathematics.

■ **Objective F:** *Apply the vocabulary of polynomials. (Lessons 9-1, 9-2, 9-3)*

27. Given the polynomial $3x^7 + x^5 - 2x^3 + 6x^2 - 127$, indicate
   a. the degree of the polynomial, **7**
   b. the leading coefficient, **3**
   c. the coefficient of x^4, **0**
   d. the constant term, **−127**
   e. the number of zeros this polynomial must have. **7 (counting multiplicities)**

28. Write a cubic polynomial with two terms and a leading coefficient of 4. **sample: $4x^3 + 1$**

29. *True* or *false*. If a polynomial $p(x)$ has a relative minimum at $(5, 1)$, then its graph never crosses the x-axis. **False**

30. Consider the polynomial $m(z) = (z - 3)(z + 2)^2(z - 1)^3$. Determine the zero(s) with multiplicity **a.** one, **b.** two, **c.** three. **a) 3; b) −2; c) 1**

■ **Objective G:** *Apply the Remainder Theorem, Factor Theorem, and Factor-Solution-Intercept Equivalence Theorem. (Lessons 9-4, 9-5)*

In 31–33, $f(x)$ is a polynomial function. *True* or *false*.

31. If the remainder is 6 when $f(x)$ is divided by $(x - 1)$, then $f(6) = 1$. **False**

32. If the remainder is 0 when $f(x)$ is divided by $(x^2 - 9)$, then $f(3) = 0$. **True**

33. If the remainder is 0 when $f(x)$ is divided by $(x^2 - 5)$, then $f(x) = (x^2 - 5)p(x)$, where $p(x)$ is another polynomial. **True**

34. Given that $g(x) = (x + 2)^2(3x - 4)(5x + 1)$, how many x-intercepts does the graph of $g(x)$ have? **3**

■ **Objective H:** *Apply the Fundamental Theorem of Algebra and Conjugate Zeros Theorem. (Lesson 9-7)*

35. **a.** Show that every real number is equal to its complex conjugate.
   b. *True* or *false*. Every odd-degree polynomial with real coefficients must have at least one real zero. Explain your answer.

36. If the graph of a fourth degree polynomial does not cross the horizontal axis, what conclusions can be made about its zeros? **They are not real.**

37. Solve $x^4 + 5x^2 - 36 = 0$ completely, given that one of the solutions is $3i$. **3i, −3i, 2, −2**

38. *True* or *false*. If $2 + i$ is a zero of $z^3 + 3z^2 - 23z + 35$, then $1 + i$ cannot be another zero of it. Explain your answer.
   True; $2 + i$ and $2 - i$ are zeros since zeros come in conjugate pairs. If $1 + i$ were a zero, then $1 - i$ would also have to be a zero. But, since the polynomial is a cubic it only has three zeros. The third zero must be real.

■ **Objective I:** *Construct and interpret polynomials that model real situations.* *(Lessons 9-1, 9-2)*

39. Mehmet asks the payment office at his job to directly deposit $40 of each salary check into a special bank account where interest is calculated monthly. He keeps this account (and his job) for a whole year without any other transactions. Assume that he opened the account with an initial amount of $100, and deposits take effect on the last business day of each month. Let $x = 1 + r$, where r is the monthly interest rate for this account.
  a. Write a polynomial $M(x)$ which gives the balance in Mehmet's account at the end of the year.
  b. Calculate how much Mehmet's account would have at the end of the year if the monthly interest rate is .5625%. ≈ **$602**

40. A rectangular horse-training area is to be fenced with 1000 ft of wire leaving a 20-ft opening for a gate, as shown below.

20 ft

The area of the enclosed region is given by $w\ell$, and the amount of available wire indicates that $2\ell + 2w - 20 = 1000$.
  a. Express ℓ in terms of w and use this to write a polynomial $A(w)$ which gives the area of the training pen.
  b. Use the polynomial $A(w)$ in part **a** to determine the area of the training pen when its width is
    i. 100 feet   **41,000 ft²**
    ii. 250 feet   **65,000 ft²**
    iii. 300 feet.   **63,000 ft²**
  c. Approximate the dimensions of the pen with the largest possible area. ℓ = **255 ft, w = 255 ft**
  a) $\ell = 510 - w$; $A(w) = w(510 - w) = 510w - w^2$

41. The total daily revenue from a yogurt stand is given by the equation $R(c) = c \cdot p(c)$, where c is the number of cups of yogurt sold and $p(c)$ is the price of one cup of yogurt. The price of one cup is given by the equation
$$p(c) = 2.50 + .001c - .00002c^2.$$
  a. Find a polynomial expression in simplest terms equal to $R(c)$.
  b. Use an automatic grapher to approximate the number of cups of yogurt that would produce the maximum daily revenue.   ≈ **220**
  c. What is the maximum daily revenue possible?   **$385**
  a) $R(c) = 2.50c + .001c^2 - .00002c^3$

In 42 and 43, a confectioner is designing a new fruit bit-and-nut treat that looks like an ice cream cone with a hemisphere of goodies on top.

The height of the cone is h cm, and the radius is r cm. The entire shape is to be filled with the fruit bits and nuts.

42. Write a formula for the volume of the treat in terms of r and h. $V = \frac{1}{3}\pi r^2 h + \frac{2}{3}\pi r^3$

43. a. If the height of the cone is fixed to be 6 cm, write a polynomial in r for the volume of fruit bits and nuts needed.
  b. Identify the degree of the polynomial and tell why it is a reasonable degree for the situation.
  c. What is the leading coefficient of the polynomial?
  a) $V(r) = 2\pi r^2 + \frac{2}{3}\pi r^3$

  b) **3; volume is a 3-dimensional quantity.**

  c) $\frac{2}{3}\pi$

REPRESENTATIONS deal with pictures, graphs, or objects that illustrate concepts.

■ **Objective J:** *Represent two- or three-dimensional figures with polynomials.* (*Lessons 9-1, 9-2*)

44. A cone is inscribed in a sphere of radius 5 in., as shown below. Let x be the distance from the center of the sphere to the base of the cone.

   a. Express the volume of the cone as a polynomial function of x.
   b. What is the degree of this polynomial? **3**
   c. Find the value of x which makes the volume of the cone 27π. **4**

45. A cube with side s is expanded by adding h cm to each edge as shown. $s^3 + 3s^2h + 3sh^2 + h^3$

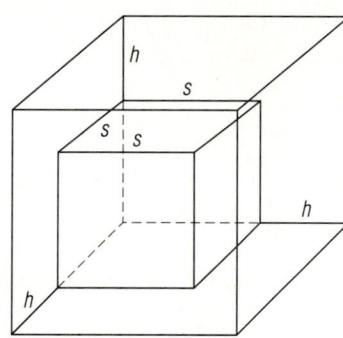

Write the volume of the new figure as a polynomial in expanded form.

46. The lengths (measured in cm) of the sides of a right triangle, shown below, form an arithmetic sequence.

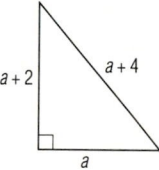

Determine the area of the triangle in cm^2.
24 cm^2

■ **Objective K:** *Relate properties of polynomial functions and their graphs.* (*Lessons 9-3, 9-5, 9-7*)

In 47–50, **a.** graph the polynomial function using an automatic grapher. **b.** Estimate any relative extrema. **c.** Estimate all real zeros.

47. $y = -2x^2 + 20x - 42$
48. $y = x^3 + 6x^2 + 11x + 6$
49. $y = 3x^4 + 25x^3 - 53x^2 - 54x + 72$
50. $y = x^5 - x^3 + x^2 - 1$

In 51–53, consider the part of the graph of $f(x) = .1x^5 - 2x^3 + x^2 - 27$ shown below.

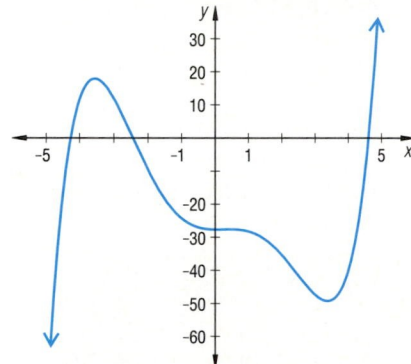

a) 3; b) f has 2 nonreal zeros.

51. a. How many real zeros does f seem to have?
   b. The number of real zeros and the degree of f do not match. Give a possible explanation.

52. *True* or *false*. The equation
$$.1x^5 - 2x^3 + x^2 - 27 = 0$$
has exactly three solutions. Explain your answer. If the statement is *false*, suggest a change to make it *true*.

53. Explain why the provided graph of f probably contains all the essential information about the function.

52) False. The equation has exactly three *real* solutions.
53) Since f has only three real zeros, it does not cross the x-axis again. Thus there are no more maxima, minima, or intercepts.

590

CHAPTER 10 ■ BINOMIAL AND NORMAL DISTRIBUTIONS

OBJECTIVES ■ CHAPTER 10

The objectives listed here are the same as in the Chapter 10 Review on pages 658–662 of the student text. The Progress Self-Test on page 657 and the tests in the Teacher's Resource File cover these objectives. For recommendations regarding the handling of this end-of-chapter material, see the notes in the margins on the corresponding pages of this Teacher's Edition.

OBJECTIVES FOR CHAPTER 10 (Organized into the SPUR categories— Skills, Properties, Uses, and Representations)	Progress Self-Test Questions	Chapter Review Questions	Teacher's Resource File Lesson Masters*	Teacher's Resource File Chapter Test Forms A & B
SKILLS				
A Calculate the mean and standard deviation of a binomial probability distribution.	3	1–6	10-2	1
B Use areas under the standard normal curve to find probabilities.	7, 8	7–12	10-5	5, 6
PROPERTIES				
C Compare and contrast characteristics of different binomial probability distribution graphs.	2	13–16	10-1	7
D Use properties of normal distributions and their parent function.	5, 6, 11, 12	20–24	10-4, 10-5, 10-6	4, 8, 10
USES				
E Apply the Central Limit Theorem.	10	25–28	10-8	2
F Solve probability problems using binomial or normal distributions.	13b	29–32	10-1, 10-2, 10-6	9
G Use binomial and normal distributions to test hypotheses.	9	33–36	10-7, 10-8	11
REPRESENTATIONS				
H Graph and interpret a binomial probability distribution.	1	37–42	10-1	12
I Use areas to find probabilities.	4	43–48	10-3, 10-4, 10-5	3
J Graph and interpret normal distributions.	13a	49–52	10-5, 10-6	13

*The masters are numbered to match the lessons, and each is intended for use with that lesson.

591A

OVERVIEW ■ CHAPTER 10

This chapter constitutes the last extended discussion of statistics in this book. (The sole later mention of statistics is in conjunction with the study of ellipses in Chapter 12.) The goal of the chapter is to discuss the normal distribution and the use of statistics in hypothesis testing.

The shape of histograms of binomial distributions with large numbers of trials leads into the study of the normal curve and its properties. Using the techniques of Chapter 4, students standardize the normal curve, and use a table of probabilities to estimate the area of regions under the curve. These probabilities are then used in conjunction with the Central Limit Theorem and the Law of Large Numbers to calculate confidence intervals for particular occurrences of events in continuous data sets.

The chapter easily divides into three parts with three lessons in each. The first part begins with binomial probability distributions. Lesson 10-1 presents an analysis of one experiment having a variable probability and a fixed number of trials, and a second with a fixed probability and a variable number of trials. In Lesson 10-2, the formulas for the mean and standard deviation of such distributions are discussed. Lesson 10-3 introduces the idea of representing probability as an area between a curve and the x-axis.

The next three lessons provide an extended introduction to the normal distribution. In Lesson 10-4, the function $f(x) = e^{-x^2}$, the parent of the normal curves, is discussed. In Lesson 10-5, the parent normal function is scaled so that the area between the image and the x-axis

is 1, yielding the standard normal distribution. Students are asked to use a table of probabilities to estimate areas of specific regions under the curve. In Lesson 10-6, the skills of Lesson 10-5 are put to use to find probabilities of events happening in any normally distributed data set. Also included here is the use of the normal distribution to approximate a binomial distribution, which brings the discussion full circle.

The last three lessons of the chapter discuss hypothesis testing. The idea of hypotheses is put forth in Lesson 10-7. In Lesson 10-8, the Central Limit Theorem, the main mathematical tool in the use of the normal distribution in testing hypotheses, is discussed. The chapter closes with a reading lesson on using confidence intervals for hypothesis testing and the potential hazards facing the user of such testing.

PERSPECTIVES ■ CHAPTER 10

The Perspectives provide the rationales for the inclusion of topics or approaches, give some mathematical background, and make connections with other lessons and to other UCSMP texts.

10-1
BINOMIAL PROBABILITY DISTRIBUTIONS

We call the function that maps the ordered triple (k, n, p) onto $_nC_kp^k(1 - p)^{n-k}$ the binomial probability distribution $B(k, n, p)$.

In this lesson students explore the values and graphs of distributions for various values of p and n. In particular, we are interested in what happens as n gets larger, because the distribution approaches a normal distribution.

10-2
MEAN AND STANDARD DEVIATION OF A BINOMIAL DISTRIBUTION

The mean of a distribution is the mean value you would expect to get as the number of trials increases without bound; for this reason it is also called the expected value of the distribution. This mean can be calculated by finding the sum of each value multiplied by its probability. The letter μ is used to indicate that this is a theoretical mean and not a sample mean. Students are also given a general formula for the variance and, the square root of the variance is the standard deviation. These formulas are then applied to binomial distributions.

10-3
REPRESENTING PROBABILITIES BY AREAS

The movement from the discrete (binomial) to the continuous (normal) requires the consideration of probabilities that are not calculated by counting. This lesson considers probability as a ratio of areas—often called geometric probability, which students may have seen in UCSMP *Algebra* or *Geometry*. To interpret the area under the standard normal curve as a probability, students calculate the probability that a randomly selected point is in a particular region of a triangle graphed in the first quadrant. We use the triangle because it is a fair approximation to the shape of a normal curve. Then we scale the triangle using the Area Scale Change Theorem so that its area is 1. This same idea will be applied to the parent normal curve in Lesson 10-5.

10-4
THE PARENT OF THE NORMAL CURVE
The parent of the normal curve is the function $f(x) = e^{-x^2}$. It has a bell shape but the area under it is not 1. Students are expected to be able to identify the critical points of the curve, including the maximum and inflection points, and to describe its curvature and asymptotic behavior. Students are not expected to calculate inflection points, as this requires calculus.

10-5
THE STANDARD NORMAL DISTRIBUTION
The parent normal curve function cannot be a probability distribution because the area under it is not 1. By doing an appropriate vertical and horizontal scale change, the area under the curve can be made equal to 1 and the x-coordinates of the points of inflection can be at ±1. The significance of this is that the function becomes a probability distribution with mean 0 and standard deviation 1.

The normal distribution is important because so many natural phenomena seem to be normally distributed. Normal distributions are close approximations to binomial distributions, and the distributions of sample means are normal. The last of these is discussed in Lesson 10-8. The normal distribution is so important that tables listing areas under this curve have been compiled to indicate the probability that an event can happen.

10-6
OTHER NORMAL DISTRIBUTIONS
When a random variable is normally distributed, it is rather unlikely that its mean is 0 and its standard deviation is 1. However, if its mean is m and its standard deviation s, then the transformation $x \rightarrow \dfrac{x - m}{s}$ will transform it into a random variable whose distribution is standard normal. Then tables can be used to find probabilities, and the results can be applied to the given distribution. This lesson considers some distributions that are assumed to be normal and uses tables to find probabilities that the random variable lies in a particular interval.

10-7
USING PROBABILITY TO MAKE JUDGMENTS
Probabilities indicate only the likelihood that an event may occur under certain conditions or assumptions. When a probability is low, but not zero, we consider the event unlikely, but not impossible.

This lesson is devoted to inferential statistics, the use of statistics to make judgments. With actual data, if the probability of the results we find is very low, we may decide that our assumptions are wrong about the situation. These ideas, along with those of the null hypothesis, alternative hypothesis, and significance level, form the basis of hypothesis testing.

10-8
SAMPLING DISTRIBUTIONS AND THE CENTRAL LIMIT THEOREM
The question of likelihood of many kinds of events depends on the size of the sample. It may not be unusual for one student to miss a particular problem, but it could be quite unlikely for many students to miss that problem. The question of

the effect of sample size on the distribution of means of a sample is covered in the Central Limit Theorem (CLT): If samples of size n are taken from a population with mean μ and standard deviation σ, then as n increases, the sample means approach a normal distribution with mean μ and standard deviation $\dfrac{\sigma}{\sqrt{n}}$. The power of this theorem is that it applies to means of samples of any size over 30. The proof of this theorem is beyond the scope of this course, so in this lesson, computer simulations are exhibited to give credibility to the theorem.

10-9
CONFIDENCE AND CAUTIONS IN REASONING
The ideas from the previous two lessons are applied here to discuss the fundamental question of how large a sample needs to be in order to have a particular level of confidence in the results. The Central Limit Theorem indicates what will happen with samples of a particular size; the notion of a particular likelihood being a test value is extended here to the idea of confidence interval. The surprise is that the accuracy of a random sample has nothing to do with the size of the population. A random sample of 1200 taken from a population of 240,000,000 can be treated with the same confidence as one of 1200 from a population of 2400.

DAILY PACING CHART ■ CHAPTER 10

For optimal student performance, the Progress Self-Test and the Chapter Review should be covered. (See *General Teaching Suggestions: Mastery* on page T35 of this Teacher's Edition.) By following the pace of the Full Course given here, students can complete 12 or 13 chapters of the book by the end of the year. Students following the pace of the Minimal Course spend more time to complete lessons that are combined with quizzes and to consider the Chapter Review. When all lessons are covered in order (the recommendation of the authors), students in the Minimal Course can cover all but the last two chapters. For more information on pacing, see *General Teaching Suggestions: Pace* on page T34 of this Teacher's Edition.

DAY	MINIMAL COURSE	FULL COURSE
1	10-1	10-1
2	10-2	10-2
3	10-3	10-3
4	Quiz (TRF); Start 10-4	Quiz (TRF); 10-4
5	Finish 10-4	10-5
6	10-5	10-6
7	10-6	Quiz (TRF); 10-7
8	Quiz (TRF); Start 10-7	10-8
9	Finish 10-7	10-9
10	10-8	Progress Self-Test
11	10-9	Chapter Review
12	Progress Self-Test	Chapter Test (TRF)
13	Chapter Review	
14	Chapter Review	
15	Chapter Test (TRF)	

TESTING OPTIONS

■ Quiz for Lessons 10-1 through 10-3   ■ Chapter 10 Test, Form A   ■ Chapter 10 Test, Cumulative Form
■ Quiz for Lessons 10-4 through 10-6   ■ Chapter 10 Test, Form B

A Quiz and Test Writer is available for generating additional questions, additional quizzes, or additional forms of the Chapter Test.

PROVIDING FOR INDIVIDUAL DIFFERENCES

The student text has been written for, and tested with, average students. It has also been used successfully with students who are better prepared as well as with students who are poorly prepared.

The Lesson Notes often include Error Analysis and Alternate Approach features to help you assist those students who need more help. Students of all abilities often learn from their peers and may benefit from small group work referenced as appropriate throughout the Notes. A blackline Lesson Master (in the Teacher's Resource File), keyed to the chapter objectives, is available for each lesson to provide more practice. (However, we advise you to use these masters sparingly; again refer to the suggestions on pace on page T34.)

Extension activities are provided in the Lesson Notes for use with students who need to be challenged even more than the requirements of completing the Full Course provide.

Binomial and Normal Distributions

We recommend that this chapter be covered in 12–15 days: 9–11 days for the lessons and quizzes, 1 day for the Progress Self-Test, 1 or 2 days for the Chapter Review, and 1 day for the Chapter Test.

USING PAGE 591
You might have your students measure their heights, rounding to the nearest 2 inches. Then get a distribution for the class. Discuss how the distribution might change if all other students from the same grade level in the school were measured. This will help to introduce distributions and the accuracy of samples, two ideas from the chapter.

Below are two distributions related to heights of people. The graph on the left shows the distribution of heights (in inches) of the players in the National Basketball Association during the 1990-91 season. The graph on the right shows the relative frequency distribution of heights of American adults, with the relative frequency distributions for males and females superimposed.

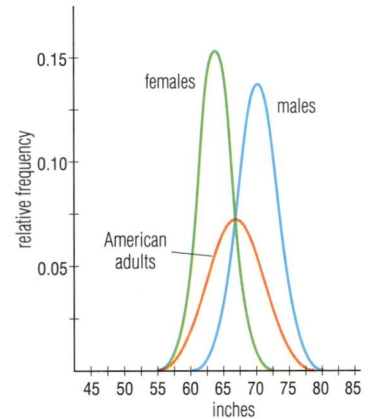

The graph on the left approximates the graph of a *binomial probability distribution*. The graph on the right is a bell-shaped curve called a *normal curve*, picturing a probability distribution called a *normal distribution*. Because of both their frequent occurrence in our world, and their theoretical properties, binomial and normal distributions are very important functions in mathematics, and are often employed to make or to support decisions in situations involving uncertainty.

CHAPTER 10 Binomial and Normal Distributions   591

Binomial Probability Distributions

Recall that a binomial probability experiment is one in which there are a fixed number of independent trials, and on each trial there are only two possible outcomes (with fixed probabilities) that can be called *success* and *failure*. In Lesson 8-9, it was proved that in a binomial experiment with *n* trials, each having probability of success *p* and probability of failure $q = 1 - p$, then the probability of exactly *k* successes is

$$_nC_k \cdot p^k q^{n-k}.$$

For instance, studies of long sequences of free throws by basketball players have found no evidence that successive shots are dependent. Thus it is reasonable to consider successive free-throw attempts as independent trials of a binomial experiment in which making a free throw is considered a success, and missing a free throw is considered a failure. Then, the distribution of the numbers *k* of free throws made in *n* attempts by a player with a known free-throw success percentage *p* yields a *binomial probability distribution*.

Example 1 Assume a basketball player makes 75% of free-throw attempts and that the attempts are independent of each other.

a. Determine and graph the probability distribution for the number of free throws made in 8 attempts.

b. In one game this player shoots 8 free throws and makes only 4 of them. Is it unusual for the player to shoot so far below average?

Solution

a. Let *x* be the number of successes in 8 trials. Then *x* can be any whole number between 0 and 8, inclusive. The probability *p* of success is given to be 0.75. By the Binomial Probability Theorem,

$$P(x) = {_8C_x} \cdot (.75)^x (.25)^{8-x}.$$

Evaluating that expression for each possible value of *x* generates the numbers in the table below. The data are graphed in the histogram below.

x	P(x)				
0	1	·		$(.25)^8$	≈ 0.000
1	8	·	(.75) ·	$(.25)^7$	≈ 0.000
2	28	·	$(.75)^2$ ·	$(.25)^6$	≈ 0.004
3	56	·	$(.75)^3$ ·	$(.25)^5$	≈ 0.023
4	70	·	$(.75)^4$ ·	$(.25)^4$	≈ 0.087
5	56	·	$(.75)^5$ ·	$(.25)^3$	≈ 0.208
6	28	·	$(.75)^6$ ·	$(.25)^2$	≈ 0.311
7	8	·	$(.75)^7$ ·	$(.25)$	≈ 0.267
8	1	·	$(.75)^8$		≈ 0.100

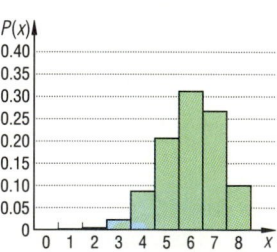

592

b. From the table, the probability of making exactly 4 free throws is about .09. Thus, making only 4 of 8 free throws is not very unusual for a basketball player with a free throw percentage of 75%. In fact, for such a player, the probability of making 4 or fewer free throws in any 8 successive attempts is

$$\sum_{x=0}^{4} P(x) \approx 0.11.$$

That is, about 11% of the time a player with a 75% free throw average will make no more than 4 of 8 free throws.

Check The 9 probabilities in part **a** should add to 1, which is the case.

As you know, if n is large, the value of $_nC_k$ may overflow the memory of your calculator for some values of k. Also, the calculation of many values of a binomial probability distribution can be tedious. Often a computer is a useful tool for overcoming these difficulties. Some statistics packages have built-in routines for evaluating binomial probabilities. Check your class package to see if it has such a feature. If it does, you will need to learn how to use that feature. If it does not, you can write a program to evaluate the binomial probabilities or use the following BASIC program, BINOMIAL GENERATOR.

```
5     REM CALCULATING BINOMIAL PROBABILITIES
10    INPUT "NUMBER OF TRIALS"; N
20    INPUT "PROBABILITY OF SUCCESS"; P
30    X = N
40    INPUT "NUMBER OF SUCCESSES"; K
50    C = 1
60    FOR I = 1 TO K
70      C = C * X/I
80      X = X − 1
90    NEXT I
100   PROB = C * P^K * (1 − P)^(N − K)
110   PRINT "PROBABILITY OF EXACTLY ";K;" SUCCESSES"
120   PRINT "IN ";N;" TRIALS IS ";PROB
130   INPUT "TYPE 1 TO CONTINUE, 0 TO STOP"; AGAIN
140   IF AGAIN = 1 THEN GO TO 30
150   END
```

It is useful to study this program carefully. The loop in statements 60 to 90 uses recursion to calculate the binomial coefficient, because

$$C = \frac{n(n-1)(n-2)(n-3) \ldots (n-k+1)}{1 \cdot 2 \cdot 3 \cdot 4 \cdot \ldots \cdot k} = \frac{n!}{k!(n-k)!} = {}_nC_k.$$

Then statement 100 evaluates $_nC_k \cdot p^k (1-p)^{n-k}$, the probability of exactly k successes in n binomial trials. Statements 130 and 140 allow another binomial probability to be evaluated for a different value of k without entering n and p again.

LESSON 10-1 Binomial Probability Distributions   593

---

aid is provided. How $_nC_k$ is calculated in the loop in lines 60 to 90 is not obvious. Use **Question 2** to guide class inspection of the program.

In **Example 2,** note that although the probability of getting exactly 25 heads in 50 tosses of a fair coin is only about 0.11, the probability of getting near 25 heads is far greater.

A teaching aid is provided to facilitate discussion of the graphs and patterns in **Example 3.**

Question 19 is time-consuming, though not especially difficult. If you choose to do this question, plan to add a half day to the lesson.

ADDITIONAL EXAMPLES
1. A baseball player has a .300 batting average.

a. If, in fact, the player has a .300 probability of getting a hit, determine the probability distribution for the number of hits in 5 at-bats in a game.

x	P(x)
0	0.168
1	0.360
2	0.309
3	0.132
4	0.028
5	0.002

b. How unusual is it for this batter to make 3 or more hits in a game?
The probability is about 0.162, about 1 in 6, not particularly unusual.

2. A coin that is biased so that heads occurs 60% of the time is tossed 50 times by someone who does not know it is biased. What is the probability that between 23 and 27 heads occurs, so that the person is by mistake rather sure the coin is fair?
≈ .0153 + .0259 + .0405 + .0584 + .0778 = 0.2179; this points out the need for a larger sample of tosses

3. a. Draw graphs of the binomial probability distributions with $p = 0.6$, when $n = 2, 5, 10,$ and 50.

$B(k, 2, 0.6)$

$B(k, 5, 0.6)$

$B(k, 10, 0.6)$

$B(k, 50, 0.6)$

b. Describe the effect of increasing n on the graph.
As n increases, the graph increasingly resembles a bell-shaped curve, the distribution spreads out and flattens, and near the mode it looks more and more symmetric.

594

Example 2 A fair coin is tossed 50 times. What is the probability of getting

a. exactly 25 heads;

b. from 30 to 35 heads?

Solution

a. The answer is $_{50}C_{25}(0.5)^{25}(1 - 0.5)^{25}$. This is very difficult to evaluate on some calculators. With a statistics package or the program above, with inputs $N = 50$, $P = 0.5$ and $K = 25$, the probability is found to be about 0.11.

b. Again, $N = 50$ and $P = 0.5$. Six probabilities must be calculated, for $K = 30, 31, 32, 33, 34,$ and 35. Output from our computer shows the following probabilities.

K	30	31	32	33	34	35
$P(K$ successes in N trials)	.0419	.0270	.0160	.0087	.0044	.0020

The individual probabilities are mutually exclusive, so the probability of getting from 30 to 35 heads is the sum
.0419 + .0270 + .0160 + .0087 + .0044 + .0020 .
Thus P(from 30 to 35 heads) $\approx$ .1000, or 10%.

The expression $_nC_kp^k(1 - p)^{n-k}$ has three variables: k, n, and p. If all three are allowed to vary, a function $B(k, n, p)$ of three variables is created. This function is called a **binomial distribution** function. Then $B(k, n, p)$ represents the probability of getting exactly k successes in n binomial trials, each of which has probability p of success. The numbers n and p are called *characteristics* of the binomial distribution.

Graphing all values of such a function would require four dimensions! However, if two of the three variables are held constant, then the function can be graphed in a 2-dimensional coordinate plane with the third variable on the horizontal axis and $B(k, n, p)$ on the vertical axis.

For instance, if the number of trials is fixed at 8, and a particular value of p is chosen, the graph of $B(k, 8, p)$ can be determined for the nine possible values of k. Below are graphs showing the distributions for the probability of success on six such binomial experiments with probabilities of success ranging from $p = 0.05$ to $p = 0.95$.

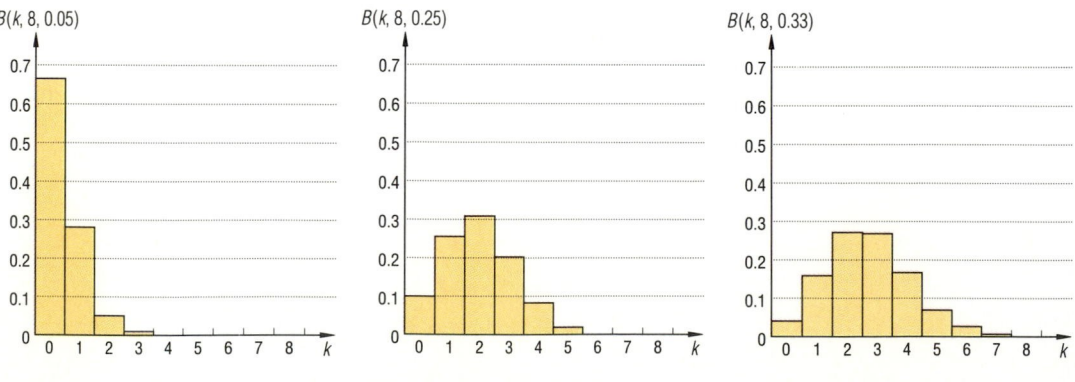

$B(k, 8, 0.05)$ $B(k, 8, 0.25)$ $B(k, 8, 0.33)$

$B(k, 8, 0.67)$ $B(k, 8, 0.75)$ $B(k, 8, 0.95)$

Notice that the graph for $B(k, 8, 0.75)$ represents the situation in Example 1, showing the probability of k successes in 8 free throws from a player with a free-throw success rate of 75%. The other graphs could represent similar situations with other success rates.

Notice the patterns in the graphs above:
1. As p increases, the mode of the distribution (the maximum value of the probability function) increases along the k-axis. This should agree with your intuition that as success becomes more likely, the most likely number of successes should increase.
2. When two values of p total 1, for instance $p = 0.05$ and $p = 0.95$, the corresponding pair of graphs are reflection images of each other. This is not surprising when you consider that a probability of success of 0.05 is equivalent to a probability of failure of 0.95.

From the preceding observations, you might expect that when $n = 8$ and $p = 0.5$, the binomial distribution is symmetric to the line $k = 4$, the mode of the distribution. As shown in the graph below, this is indeed the case.

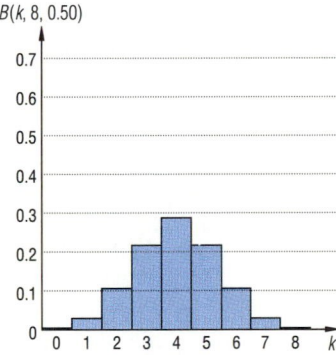

$B(k, 8, 0.50)$

Using the same k- and B-axes, it is also useful to study the graphs of binomial distributions for fixed p and various values of n, the number of trials.

Question 4: You might ask why the statement is true and then ask for a generalization. (It is true because if the probability that something can happen is 0.75, then the probability it cannot happen is 0.25; so the distribution of x successes is the same as the distribution of $8 - x$ failures. A generalization is that the distributions of $B(k, n, p)$ and $B(k, n, q)$ are reflection images of each other if $p + q = 1$.)

Question 12: The Wheel of Fortune spinning wheel has 24 sections, 21 with prizes, 1 "lose a turn," 1 "bankrupt," and 1 "free spin." You might ask: If each section is equally likely, what is the probability of spinning 10 times without going bankrupt? $\left(\left(\frac{23}{24}\right)^{10} \approx .65\right)$

Question 14: The distribution here is a multinomial probability distribution.

ADDITIONAL ANSWERS
1.a.

x	P(x)
0	.01680
1	.08958
2	.20902
3	.27869
4	.23224
5	.12386
6	.04129
7	.00786
8	.00066

Example 3 **a.** Draw graphs of the binomial probability distributions with $p = 0.5$, when $n = 2, 6, 10,$ and 50.

b. Describe the effect of increasing n on the distribution $B(k, n, 0.5)$.

Solution

a. The data for the graphs determined by $n = 2, 6,$ and 10 can be generated by hand. But for the case of $n = 50$, use a statistics package or the program **BINOMIAL GENERATOR**. The graphs follow.

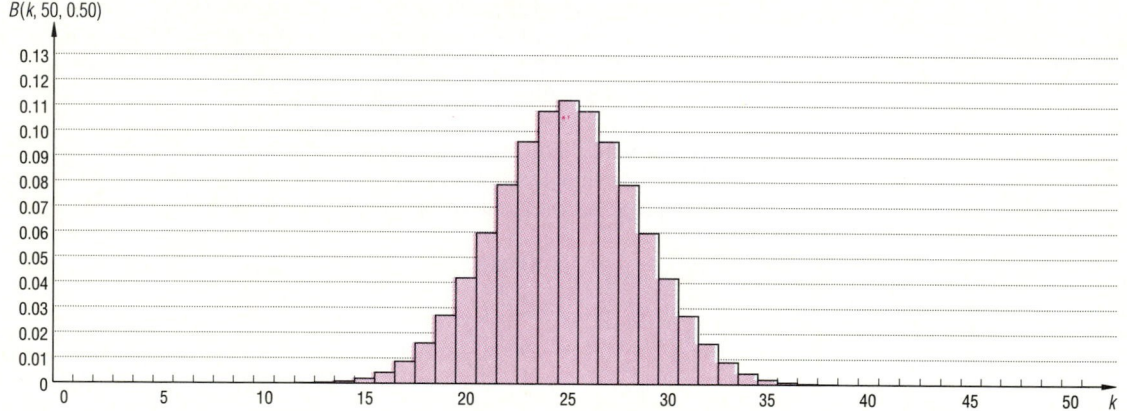

b. When $p = 0.5$, as n increases, the following patterns can be observed:

1. The mode of the distribution increases along the k-axis. That is, the more trials, the greater the most likely value.
2. The graph increasingly resembles a bell-shaped curve.
3. The distribution "spreads out."
4. The maximum value in the range, which is the maximum probability, decreases. That is, the distribution "flattens."

In the next lesson you will see theorems that confirm these four and the two previous patterns.

596

Questions

Covering the Reading

1. Assume a basketball player makes 40% of free-throw shots taken and that these are independent.
   a. Determine and graph the probability distribution for the number of free throws made in 8 attempts.   **See margin.**
   b. What is the probability that the player will make at least 6 of the next 8 shots?   **≈ 5%**
   c. Compare and contrast the graph in part **a** to the one drawn in Example 1.   **Sample: The mode in part a is less than the mode in Example 1; this graph is a little more spread out**.

2. Suppose you want to calculate the probability of getting exactly 5 successes in 8 binomial trials, where the probability of success in each trial is 0.25, and you use the BINOMIAL GENERATOR program given in the lesson.
   a. What are the values of N, P, and K?   **N = 8, P = .25, K = 5**
   b. What are the values of C and X after each run of the loop in lines 60–90?   **See margin.**

3. The function $B(k, n, p)$ represents the probability of getting exactly __**a.**__ successes in __**b.**__ trials of a __**c.**__ experiment in which the probability of success on each trial is __**d.**__.   **a) k; b) n; c) binomial; d) p**

In 4–7, consider binomial probability distributions $B(k, n, p)$.   *True* or *false*.

4. For $n = 8$, the graphs of $B(k, n, 0.75)$ and $B(k, n, 0.25)$ are reflection images of each other.   **True**

5. For a fixed value of n, as p increases, the mode of the distribution increases along the k-axis.   **True**

6. For $p = 0.5$, the graph is symmetric over the line $k = \dfrac{n}{p}$.   **False**

7. For $p = 0.5$, as n increases, the probability of getting the modal number of successes increases.   **False**

In 8 and 9, use a statistics package or data produced by the BINOMIAL GENERATOR program.

8. a. Graph the probability distributions $B(k, n, p)$ for $n = 7$ and $p = 0.8, 0.6, 0.5, 0.4, 0.2$.   **a) See margin. b) p = .8 and .2; p = .6 and .4**
   b. Which graphs are reflection images of each other?
   c. Describe the effect on the graph of decreasing p.
      The mode decreases; the graph bunches up nearer the y-axis.

9. a. Graph the probability distributions $B(k, n, p)$ for $p = 0.3$ and $n = 5, 10, 25,$ and 100.   **See margin.**
   b. Describe the effect on this distribution of fixing p and increasing n.
      The mode increases; the graph approximates a bell-shaped curve; the distribution "spreads out" and "flattens."

Applying the Mathematics

10. A multiple choice test has 15 questions each with 5 alternatives. Assume that a student answers all questions by random guessing. Determine the probability that the student gets
    a. no more than 5 right answers;   **≈ .939**
    b. exactly 5 right answers;   **≈ .103**
    c. more than 5 right answers.   **≈ .061**

LESSON 10-1   Binomial Probability Distributions   **597**

8.a. (continued)

$B(k, 7, 0.4)$

$B(k, 7, 0.2)$

9.a.

$B(k, 5, 0.3)$

$B(k, 10, 0.3)$

**9.a. (continued)
See Additional Answers in the back of this book.**

598

MORE PRACTICE
For more questions on SPUR Objectives, use *Lesson Master 10-1*, shown on page 599.

PROJECTS
The projects for Chapter 10 are described on pages 654–655. Projects 1 and 2 are related to the content of this lesson.

ADDITIONAL ANSWERS
11.b. Sample: It seems effective because the probability of all 5 people recovering without an effective drug is only about .03.

d. Sample: Again, the drug seems to be effective. The second experiment is more conclusive because the result of 18 people out of 25 recovering is less likely to happen by chance (.022) than the result of 5 out of 5 people recovering (.031).

12.b. Sample: The mode would be shifted to the right and the distribution would have more spread.

15. Sample:
```
10  N = 20: S = 0:
    M = 0: L = 0
20  FOR I = 1 TO N
30  P = RND(1)
40  IF P < .35 THEN
    S = S + 1
50  IF P >= .35 AND
    P < .85 THEN
    M = M + 1
60  IF P = .85 THEN
    L = L + 1
70  NEXT I
80  PRINT S; "SMALL
    WATERMELONS"
90  PRINT M; "MEDIUM
    WATERMELONS"
100 PRINT L; "LARGE
    WATERMELONS"
110 END
```

16.a.

x	P(x)
1	.125
2	.375
4	.375
8	.125

11. Suppose that a certain disease has a 50% recovery rate.
   a. What is the probability that five of five people with the disease will recover within a year?   **.03125**
   b. In one experiment, a drug is tested on 5 people and all 5 recover within a year. Considering your answer to part **a**, comment on the effectiveness of the drug.   **See margin.**
   c. In a second experiment, 25 people get the drug and 18 recover within a year. Find the probability that 18 or more people recover even if the drug has no effect on the illness.   **≈ .022**
   d. Consider your answers to parts **b** and **c**. Does the drug appear to be effective? Which experiment might be considered more conclusive and why?   **See margin.**

12. A wheel used on a certain TV game show has 19 sections. Of these, 9 sections have big prizes, 9 sections have small prizes, and one section throws the player out of the game. The scatter plot below shows the probability distribution of winning *k* big prizes in 50 spins of this wheel.

$B\left(k, 50, \frac{9}{19}\right)$

   a. Are the chances for a player to win a big prize about the same as those for other outcomes? What feature of the graph supports your answer?
   b. If the wheel is spun more than 50 times, how would you expect the graph to change?   **See margin.**
   a) Yes, the graph is nearly symmetric to the line $k = 25$ so P(winning large prize) ≈ .5.

Review

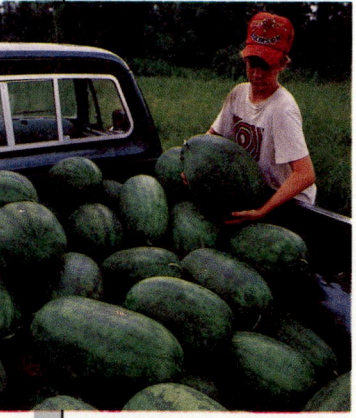

In 13 and 14, determine if the given situation is a binomial experiment. *(Lesson 8-9)*

13. At a highway weigh station, all trucks are stopped and weighed. Trucks over 40 tons are not allowed to continue traveling on that highway. Statistics at this weigh station show that about 2% of trucks are rerouted. It is desired to determine the probability that more than 3 of the first 50 trucks weighed on a particular day will be rerouted.   **Yes**

14. 100 watermelons are being emptied from a truck at a farmer's market. Each one is weighed and put into one of three piles (small, medium, and large). In the past years, this farm has produced about 35% small, 50% medium, and 15% large watermelons. It is desired to determine the probability that this will again be the distribution today.   **No**

15. Write a computer program which would use the RND function to model the situation in Question 14 for a selection of 20 watermelons. How many small, medium, and large watermelons do you end up with? *(Lesson 7-8)* **See margin.**

16. Three fair coins are tossed. Heads is given a value of 2, tails 1.
  a. Find the probability distribution for the product of the values of the three outcomes.
  b. What is the mean of the distribution? *(Lesson 7-7)* **3.375**

In 17 and 18, find θ. *(Lessons 5-8, 5-9)*

a)

x	1	2	4	8
P(x)	.125	.375	.375	.125

17.

W — 15 — Y
θ
8
20
X

≈ 117.5°

18.

A — — D
12   12
50°   C — 15
B

θ

≈ 22°

Exploration

19. A *cumulative binomial probability table* gives, for fixed n and p, the probabilities for k or fewer successes. Below is a table of cumulative binomial probabilities for $n = 7$. The number in row k and column p represents the probability of no more than k successes in 7 binomial trials, each of which has probability p. For instance, the probability of no more than 2 successes in 7 binomial trials, each of which has probability of success equal to 0.8, is about 0.0047. That is,
$$B(0, 7, 0.8) + B(1, 7, 0.8) + B(2, 7, 0.8) \approx .0047.$$

$n = 7$

k	p: 0.01	0.05	0.1	0.2	0.3	0.4	0.5	0.6	0.7	0.8	0.9	0.95	0.99
0	.9321	.6983	.4783	.2097	.0824	.0280	.0078	.0016	.0002	.0000	.0000	.0000	.0000
1	.9980	.9556	.8503	.5767	.3294	.1586	.0625	.0188	.0038	.0004	.0000	.0000	.0000
2	1.0000	.9962	.9743	.8520	.6471	.4199	.2266	.0963	.0288	.0047	.0002	.0000	.0000
3	1.0000	.9998	.9973	.9667	.8740	.7102	.5000	.2898	.1260	.0333	.0027	.0002	.0000
4	1.0000	1.0000	.9998	.9953	.9712	.9037	.7734	.5801	.3529	.1480	.0257	.0038	.0000
5	1.0000	1.0000	1.0000	.9996	.9962	.9812	.9375	.8414	.6706	.4233	.1497	.0444	.0020
6	1.0000	1.0000	1.0000	1.0000	.9998	.9984	.9922	.9720	.9176	.7903	.5217	.3017	.0679

  a. Use the table to determine the probability of five or fewer successes in a binomial experiment with 7 trials, if in each trial the probability of success is 0.8. Compare this answer to the one you get by applying directly the method learned in Chapter 8.
  b. Cumulative binomial probability tables can also be used to calculate probabilities for single events. Figure out how to use a cumulative binomial table to determine the probability of exactly 2 successes in 7 trials, each with 0.4 probability of success. Verify your answer by direct calculation using the Binomial Probability Theorem.
  a) .4233 from the table; .4232832 from direct calculation
  b) P(exactly 2 successes) = P(at least 2) − P(at least 1) = .4199 − .1586 = .2613; P(exactly 2 successes) = $_7C_2$ (.4)^2(.6)^5 = .2612736; they agree.

TEACHING NOTES

The two theorems of this lesson provide a way to quickly calculate the mean μ and standard deviation σ of a binomial probability distribution. The first two examples of the lesson justify that they are true for $n = 3$.

At the bottom of page 601, the formula for the variance of a set of n numbers $x_1, x_2, ..., x_n$ is written in two ways. Here are the missing steps, with the index of the summation always from $i = 1$ to $i = n$.

$$\frac{\sum (x_i - \bar{x})^2}{n}$$

$$= \frac{\sum x_i^2 - 2x\sum x_i + \sum (\bar{x})^2}{n}$$

$$= \frac{\sum x_i^2 - 2\frac{\sum x_i}{n}\sum x_i + n(\bar{x})^2}{n}$$

$$= \frac{\sum x_i^2}{n} - 2\bar{x}^2 + \bar{x}^2$$

$$= \frac{\sum x_i^2}{n} - \bar{x}^2.$$

600

Mean and Standard Deviation of a Binomial Distribution

In the previous lesson you studied the shapes of the graphs of binomial probability distributions produced under two conditions: (1) fixing the number n of trials, and allowing p, the probability of success on an individual trial, to vary; or (2) fixing the value of p, and allowing n to vary. In particular you saw that if p is fixed, then as n increases, the mode of the distribution increases and the distribution spreads out. In this lesson we examine the effect of n and p on other measures of center and spread, namely, the mean and standard deviation. You will see that despite the variety of shapes binomial distributions can have, both of these statistics can be calculated using surprisingly simple formulas.

Consider a basketball player with an 80% free-throw average. Experience and intuition might suggest that in the next five shots the most likely number of successes is $5(0.8) = 4$ free throws. This is exactly what is found by probability theory.

In a binomial experiment with 5 trials and a 0.8 probability of success on each trial, the values of the probability distribution $P(x) = B(x, 5, 0.8)$ are as follows.

x = number of successes	5	4	3	2	1	0
P(x)	$(.8)^5$	$5(.8)^4(.2)$	$10(.8)^3(.2)^2$	$10(.8)^2(.2)^3$	$5(.8)(.2)^4$	$(.2)^5$

Recall that the mean (or expected value) of a probability distribution equals the sum of the products of each possible value of a random variable and its probability. Here

$$\text{mean} = \mu = \sum_{x=0}^{5} x \cdot P(x)$$

$$= 5(.8)^5 + 4 \cdot 5(.8)^4(.2) + 3 \cdot 10(.8)^3(.2)^2 + 2 \cdot 10(.8)^2(.2)^3 + 1 \cdot 5(.8)(.2)^4 + 0 \cdot (.2)^5$$
$$= 1.6384 + \quad 1.6384 \quad + \quad 0.6144 \quad + \quad 0.1024 \quad + \quad 0.0064 \quad + \quad 0$$
$$= 4.0.$$

This confirms that our intuition is correct.

The following example shows how the mean of a binomial distribution varies with p, the probability of success of a single event.

■ ■ ■ ■ ■ ■ ■ ■

Example 1   A dart player has probability p of hitting the bull's-eye with a single dart, and all attempts are independent. Find the expected number of bull's-eyes the player will hit with three attempts.

Solution   This is a binomial experiment with $n = 3$ and probability of success p. Letting $1 - p = q$, then the probability distribution for the random variable *number of bull's-eyes* is shown below.

number of bull's-eyes	3	2	1	0
probability	p^3	$3p^2q$	$3pq^2$	q^3

Let μ be the mean of this distribution. Then by the definition of the mean of a probability distribution (Lesson 7-7),

$$\mu = 3(p^3) + 2(3p^2q) + 1(3pq^2) + 0(q^3)$$
$$= 3p^3 + 6p^2q + 3pq^2$$
$$= 3p(p^2 + 2pq + q^2)$$
$$= 3p(p + q)^2.$$

But $p + q = 1$,
so $\mu = 3p$.
So the player can expect to hit $3p$ bull's-eyes with three shots.

Check   Substitute some values for p. If $p = 0$, $3p = 0$. If $p = 0.5$, $3p = 1.5$. If $p = 1$, $3p = 3$. Each of these expected values agrees with intuition.

The result in Example 1 generalizes.

Theorem (Mean of a Binomial Distribution)

The mean μ of a binomial distribution with n trials and probability p of success on each trial is given by $\mu = np$.

The variance and standard deviation of a probability distribution $P(x) = B(x, n, p)$ can also be calculated directly from n and p. Recall that the population variance σ^2 of n numbers $x_1, x_2, \ldots, x_n$ equals

$$\frac{\sum_{i=1}^{n} (x_i - \bar{x})^2}{n},$$

where $\bar{x}$ is the mean of the numbers. This can be rewritten as

$$\sigma^2 = \frac{\sum_{i=1}^{n} x_i^2}{n} - \bar{x}^2.$$

The analogous formula for the **variance σ^2 of a probability distribution** with n outcomes and mean μ is

$$\sigma^2 = \left(\sum_{i=1}^{n} (x_i^2 \cdot P(x_i)) \right) - \mu^2.$$

ADDITIONAL EXAMPLES
1. Suppose a dart player has probability p of hitting the bull's-eye with a single dart, and all attempts are independent. Prove that the expected number of bull's-eyes the player will hit in two attempts is $2p$.
Let $q = 1 - p$. By the definition of the mean of a probability distribution, $\mu = 2(p^2) + 1(2pq) + 0(q^2) = 2p^2 + 2pq = 2p(p + q) = 2p$.

2. Prove that the variance and standard deviation for the distribution generated by the dart player in Additional Example 1 are $2pq$ and $\sqrt{2pq}$.
By the definition of the variance of a probability distribution, $\sigma^2 = 2^2(p^2) + 1^2(2pq) + 0^2(q^2) - \mu^2 = 4p^2 + 2pq - (2p)^2 = 2pq$. From this, $\sigma = \sqrt{2pq}$.

3. In the binomial distribution of tossing a fair coin 250 times and counting the number of heads, what is the mean and standard deviation?
mean = 125; standard deviation = $\sqrt{62.5} \approx 7.9$

Example 2 Find the variance and standard deviation for the distribution generated by the darts player in Example 1.

Solution Use the table of values of the probability distribution given in the solution of Example 1, and the result that $\mu = 3p$. By definition,
$$\sigma^2 = [3^2 \cdot p^3 + 2^2 \cdot (3p^2q) + 1^2 \cdot (3pq^2) + 0^2 \cdot q^3] - (3p)^2.$$
Expand and rearrange terms to simplify.
$$\sigma^2 = 9p^3 + 12p^2q + 3pq^2 - 9p^2$$
Rewrite $12p^2q$ as $9p^2q + 3p^2q$.
$$\sigma^2 = 9p^3 + 9p^2q + 3p^2q + 3pq^2 - 9p^2$$
$$= 9p^2(p + q) + 3pq(p + q) - 9p^2$$
$$= 9p^2 + 3pq - 9p^2 \qquad\qquad \text{(since } p + q = 1)$$
Thus $\sigma^2 = 3pq$.

As always, the standard deviation is the square root of the variance.
$$\sigma = \sqrt{3pq}$$

The result in Example 2 also generalizes.

Theorem (Variance and Standard Deviation of a Binomial Distribution)

In a binomial distribution with n trials, each with probability p of success and probability q of failure, the variance $\sigma^2 = npq$ and the standard deviation $\sigma = \sqrt{npq}$.

Example 3 The probability that a newborn child will be female is about $p = 0.48$. If a large hospital expects to have 2500 births in a given year, find

a. the expected number of females born;
b. the expected standard deviation of the number of female children.

Solution If x is the number of female children in 2500 births, then $P(x) = B(x, n, p) = B(x, 2500, 0.48)$. By the previous theorems,
a. the mean $\mu = 2500(0.48) = 1200$;
b. the standard deviation $\sigma = \sqrt{2500(0.48)(0.52)} = \sqrt{624} \approx 25$.

Check It is tedious to calculate by hand a table of values of the probability distribution. Below is a graph based on the output generated by a statistics package.

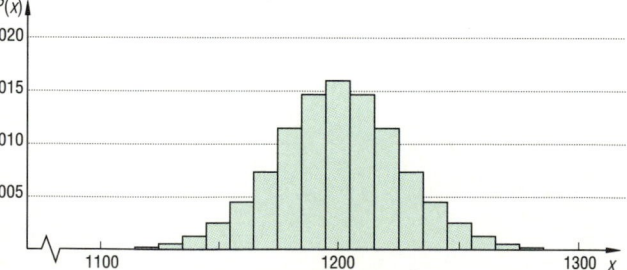

From the graph you can see that the mean appears to be about 1200, and that the other values that are likely to occur are within 50 units, or two standard deviations, of the mean.

Questions

Covering the Reading

In 1 and 2, find **a.** the mean, **b.** the variance, and **c.** the standard deviation of the binomial distribution related to an experiment with the given characteristics.

1. $n = 10$, $p = .3$
   a) 3; b) 2.10; c) ≈1.45

2. $n = 50$, $p = 0.75$
   a) 37.5; b) 9.375; c) ≈ 3.06

3. Suppose a dart player has probability $p = \frac{2}{3}$ of hitting a bull's-eye with a single dart. In 100 attempts, find
   a. the expected number of bull's-eyes; $66.\overline{6}$
   b. the standard deviation of the distribution. $\sqrt{200}/3 \approx 4.7$

4. Refer to Example 3. For the same hospital and year, determine
   a. the expected number of males;   **1300**
   b. the standard deviation of the number of males.   **≈ 25**

Applying the Mathematics

5. Suppose that both parents in a family with 4 children carry genes for blood types A and B. It is known that the blood types of their children are then independent and that each child has probability $\frac{1}{4}$ of having blood type A. Let $x =$ the number among the 4 children in the family who have blood type A.
   a. Construct a table for the probability distribution $P(x)$.   **See margin.**
   b. Draw a histogram of the distribution.   **See margin.**
   c. Find the mean number of children with blood type A and mark the location of the mean on the histogram.   **1; See margin.**
   d. Find the standard deviation of x. $\sqrt{3}/2 \approx .87$

6. A baseball player has a batting average a, where $0 \le a \le 1$. Let x equal the number of hits made in the next 25 times at bat, and consider the probability distribution $P(x)$. For this distribution, find
   a. the mean;   **25a**   **b.** the variance. $25a(1-a)$ or $25a - 25a^2$

7. The results of a recent survey indicate that 40% of male voters support the use of federal funds for child care. A "concerned father" decides to conduct his own survey with 200 randomly selected male voters. Assume that the result of the large-scale survey is accurate. Determine the number of voters in the sample who can be expected to support the legislation in two different ways:
   a. by calculating 40% of the number in the sample;   **40% · 200 = 80**
   b. by using the characteristics of the binomial experiment.
      $\mu = np = 200 \cdot 0.4 = 80.$

NOTES ON QUESTIONS
Question 5: You might ask how many students know their blood type. It is a useful bit of information to know.

Question 6: Try some values to see if the answers are reasonable. For instance, when the batting average $a =$ .250, the mean is 25(.250) = 6.25, so the player should expect 6 to 7 hits, with a standard deviation of $\sqrt{25(.250)(.750)} \approx 2.2$.

ADDITIONAL ANSWERS
5.a.

x	P(x)
0	.316
1	.422
2	.211
3	.047
4	.004

b.,c.

8. How many six-sided dice must be tossed if the expected number of 1s showing is to be 4? **24**

9. If, for some binomial probability distribution, $\mu = 45$ and $\sigma = 6$, find n and p. **$n = 225$, $p = .2$**

10. Consider binomial distributions with $p = 0.8$ and various numbers of trials.
    a. Calculate the mean and standard deviation of the distribution for each of the following numbers of trials.
       i. 10   ii. 100
       iii. 50   iv. 1000   **See margin.**
    b. *True* or *false*.
       i. For a fixed value of p, the mean of a binomial distribution is directly proportional to n, the number of trials.   **True**
       ii. For a fixed value of p, the standard deviation of a binomial distribution is directly proportional to n, the number of trials.
           False

11. According to a study by a federal agency, the probability is about 0.2 that a polygraph (lie detector) test given to a truthful person suggests that the person is deceptive. (That is, in about 20% of cases a truthful person will be labeled deceptive.) Suppose that a firm gives polygraph tests to 20 job applicants. Suppose also that all 20 answer truthfully.
    a. What is the probability that the polygraph tests show that at least one person is deceptive?   **about .988**
    b. What is the expected number of these 20 truthful persons who will be classified as deceptive? What is the standard deviation of the number expected to be classified as deceptive?   **$\mu = 4$, $\sigma = \sqrt{3.2} \approx 1.8$**
    c. What is the probability that the number classified as deceptive is less than the mean?   **≈ 41%**

Review

12. Customs officers have the following policy for accepting shipments of packaged dried fruit from another country. Select a sample of 8 boxes and check. If two or more are found substandard, reject the entire shipment.   *(Lesson 10-1)*
    a. Determine and graph the probability distribution of the number of shipments rejected out of the next 10 if 5% of the dried-fruit production of the country is defective.
    b. What is the probability that at least 8 of the next 10 shipments of dried fruit will be accepted?   **about .981**
    c. How would the graph of the probability distribution change if the dried-fruit production of the exporting country is
       i. 10% defective?   ii. 20% defective?
    d. How would the graph of the probability distribution change if
       i. 15 or ii. 20 boxes are sampled instead of 8?   **a, c, d) See margin.**

13. Approximate all real solutions to $x^3 + 2x - 1 = 0$ to the nearest tenth.   *(Lessons 3-1, 9-3)*   **0.5**

14. Two fair coins are tossed and a fair die is rolled. Find the probability of getting two heads and rolling a number greater than 4.   *(Lessons 7-3, 7-4)*   **1/12**

15. Suppose that 3% of a city's 327,000 water patrons have complaints about polluted water. Of these, 25% are from the residential area next to the chemical plant in town. There are 15,840 water patrons in this area.
  a. Determine the probability that a water patron is from the area near the chemical plant.
  b. Determine the probability that a water patron is from the area near the plant and has polluted water.
  c. Show that living near the chemical plant and having polluted water are *not* independent events.   *(Lessons 7-1, 7-4)*
  a) ≈ .048; b) ≈ .0075; c) See below.

In 16 and 17, solve exactly.   *(Lessons 4-7, 4-8)*

16. $e^y = 198$ **y = ln 198**   **17.** $\log(m^2 + 10) = 3$ **±3√110**

18. a. Graph $f(x) = e^x$ and $g(x) = e^{-x}$ on the same set of axes.   **See margin.**
  b. What transformation maps the graph of f to the graph of g?
  reflection over the y-axis
  c. What is $\lim\limits_{x \to \infty} g(x)$?   **0**
  d. What is $\lim\limits_{x \to -\infty} f(x)$?   *(Lessons 4-6, 8-2)*   **0**

19. A company compares the amount of time devoted each week to volunteer work by its executives to the amount of time devoted weekly to volunteer work by all its employees.
  a. What is the population?   **all the employees**
  b. What is the sample?   *(Lesson 1-1)*   **the executives**

Exploration

20. a. Find a dart board and, using area, compute the geometric probability p that a dart that hits the target at random is a bull's-eye.
  b. Compute answers for Examples 1 and 2 of this lesson using the value of p found in part **a**.   **a,b) Answers will vary.**

15c) *P*(living near the plant and having polluted water) ≈ .0075;
P(living near the plant) · *P*(having polluted water) = (.048)(.03)
= .00144
Since *P*(near plant and polluted water) ≠ *P*(near plant) · *P*(polluted water), the events are not independent.

FOLLOW-UP

MORE PRACTICE
For more questions on SPUR Objectives, use *Lesson Master 10-2*, shown below.

EXTENSION
Small Group Work   Here is a general proof of the Mean of a Binomial Distribution Theorem.

$$\mu = \sum_{k=0}^{n} k\binom{n}{k}p^k(1-p)^{n-k}$$

$$= \sum_{k=0}^{n} \frac{kn!}{k!(n-k)!} p^k(1-p)^{n-k}$$

$$= \sum_{k=0}^{n} \frac{n(n-1)!}{(k-1)!(n-k)!} pp^{k-1}(1-p)^{n-k}$$

$$= np\sum_{k=1}^{n} \binom{n-1}{k-1}p^{k-1}(1-p)^{n-k}$$

$$= np(p + (1-p))^{n-1}$$

$$= np$$

The general formula for the variance of a binomial probability distribution is far more difficult to deal with, but can be derived in the same way.

LESSON

10-3

Representing Probabilities by Areas

In Lessons 10-1 and 10-2, you may have been surprised to see *discrete* probability functions graphed as histograms. You may have wondered why a scattergram was not used. The reason for using histograms is that the bars show area, and these areas equal the probabilities being graphed. Here is another example. Suppose 60% of American adults favor raising taxes to fund child care, and it is assumed that each adult acts independently of others. If 5 American adults are chosen at random, let x be the number who favor raising taxes to support child care, and let $P(x)$ be the probability that exactly x favor such an act. Then the probability distribution $P(x) = B(x, n, p) = B(x, 5, 0.6)$ is binomial with six points. The coordinates of these points are given in the table below; they are represented by the six bars in the histogram.

x	P(x)
0	.010
1	.077
2	.230
3	.346
4	.259
5	.078

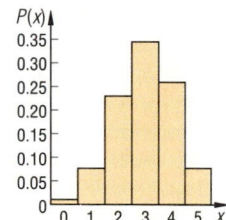

The base of each bar of the histogram is 1 unit and the height of each bar is $P(x)$, so the area of each bar is $1 \cdot P(x) = P(x)$. For instance, the area of the bar at $x = 4$ is $P(4) \approx .259$. Because P is a probability distribution, the sum of the probabilities is 1. So the total area of the bars is also 1. Probabilities for any interval of outcomes can be found by summing the areas of the bars in the range. For example, the probability that 1, 2, or 3 of 5 adults surveyed favor increasing taxes to support child care equals $P(1) + P(2) + P(3)$. That is, $P(1 \leq x \leq 3) = .077 + .230 + .346 = .653$, which is the sum of the areas of the rectangles over $x = 1$, $x = 2$, and $x = 3$.

Probabilities may also arise from a study of *continuous* random variables. In such cases, it is common to determine probabilities by finding the *area under a curve*—meaning the area between the curve, vertical lines at the endpoints of an interval, and the horizontal axis. Example 1 involves finding the area of a triangular region under a curve where the total area is not 1.

606

Example 1

Suppose that a point is selected at random in the triangular region OMN, where $O = (0, 0)$, $M = (2, 3)$, and $N = (4, 0)$.

a. Find $P(x<1)$, the probability that the x-coordinate of the point is less than 1.

b. Find $P(x = 2)$, the probability that the x-coordinate of the point is 2.

c. Find w so that $P(x<w)$, the probability that the x-coordinate of the point is less than w, is 0.32.

Solution First note that

$$\text{area}(\triangle OMN) = \tfrac{1}{2}bh = \tfrac{1}{2} \cdot 4 \cdot 3 = 6.$$

a. The points with x-coordinate less than 1 lie in the shaded region at the right bounded by the x-axis, the line $x = 1$, and $\overline{OM}$. Because the point is selected at random from $\triangle OMN$,

$$P(x<1) = \frac{\text{shaded area}}{\text{area of } \triangle OMN}$$

$$= \frac{0.75}{6}$$

$$= 0.125.$$

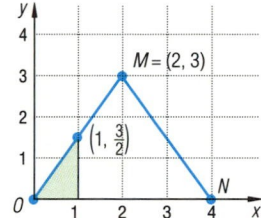

b. The points with x-coordinate 2 lie on the vertical line $x = 2$. This line has width 0; hence it has no area. Thus,

$$P(x = 2) = \frac{0}{\text{area of } \triangle OMN} = 0.$$

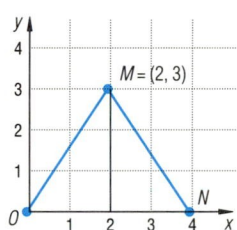

c. Because 0.32 is about one-third, the points with x-coordinate less than w lie in a region which covers about one-third the total area of $\triangle OMN$. So the vertical line $x = w$ for the shaded region is somewhere between $x = 1$ and $x = 2$. The segment $\overline{OM}$ lies on the line with equation $y = \tfrac{3}{2}x$. Thus, as shown below, the region has vertices $(0, 0)$, $(w, 0)$, and $(w, \tfrac{3}{2}w)$.

$$P(x<w) = \frac{\text{shaded area}}{\text{area of } \triangle OMN}$$

$$0.32 = \frac{\tfrac{1}{2}(w)\left(\tfrac{3}{2}w\right)}{6}$$

$$2.56 = w^2$$

Since $w>0$,

$$w = 1.6.$$

Error Analysis   In part **b** it may be confusing to students that the probability of an x-coordinate being equal to 2 is zero, even though $x = 2$ is not impossible. Remind them that $x = 2$ represents one coordinate of a point, that a point has no dimension and, therefore, no area.

ADDITIONAL EXAMPLES

1. A point is selected at random from inside the trapezoid with vertices $(0, 0)$, $(1, 3)$, $(6, 3)$, and $(7, 0)$.

a. What is the probability that its x-coordinate is equal to 2.5?
0

b. Find the probability that its x-coordinate is less than 2.5.
$\frac{1}{3}$

c. Give the value of w so that $P(x < w) = .9$
w = 5.9

2. What scale change can be applied to the trapezoid of Additional Example 1 so that the area of the image is equal to 1?
Any transformation of the form $(x, y) \rightarrow (ax, by)$ with $ab = \frac{1}{18}$.

In each part of Example 1, the area of the shaded region was divided by 6, the area under the curve, to obtain the probability that a point in the triangle is in the shaded region. Obviously, calculation would be easier if the area under a curve were 1. In that case, the probability of an event is simply the area of the region representing that event. You can then use probability and area interchangeably. Fortunately, it is always possible to make the area of a region equal to 1 by applying a suitable scale change. For instance, the triangle in Example 1 has area 6, so apply the vertical scale change

$$S:(x, y) \rightarrow \left(x, \frac{y}{6}\right)$$

to construct a triangle with area 1. The image of $\triangle OMN$ under this transformation is shown in the figure below. The area of the image $\triangle O'M'N'$ is

$$\frac{1}{2}bh = \frac{1}{2}(4)\left(\frac{1}{2}\right) = 1.$$

Area Scale Change Theorem

If the scale change
$$S:(x, y) \rightarrow (ax, by)$$
is applied to a region, then the area of the image is $|ab|$ times the area of the preimage.

Proof The area of any region can be approximated to as great an accuracy as you wish by adding the areas of squares or parts of squares. Start with a unit square with horizontal and vertical sides.

Apply $S_{a,b}(x, y) \rightarrow (ax, by)$ to this square. The image is a rectangle with vertices as shown below.

The area of the image rectangle is the product of the lengths of its sides.

$$\text{Area} = |(ah + a) - ah| \cdot |(bk + b) - bk|$$
$$= |a| \cdot |b|$$
$$= |ab|$$

Thus the area of the unit square is multiplied by $|ab|$. Since the area of any region is the sum of partial or whole unit squares, when $S_{a,b}$ is applied, the area of the image is multiplied by $|ab|$.

Example 2   Let $\triangle O'M'N'$ be the image of $\triangle OMN$ of Example 1 under the scale change

$$S: (x, y) \rightarrow \left(x, \frac{y}{6}\right).$$

Determine the probability that the x-coordinate of a randomly selected point in $\triangle O'M'N'$ is less than 1.

Solution   Here is a drawing of the image, the same drawing as on the previous page, but magnified.

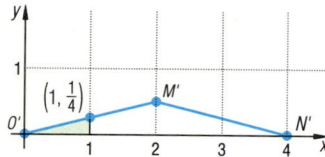

Under this scale change,

$$\left(1, \frac{3}{2}\right) \rightarrow \left(1, \frac{1}{4}\right).$$

Thus, because the area of $\triangle O'M'N' = 1$, the event $x < 1$ is shown by the shaded area above.

So $P(x<1) = $ area of shaded region $= 0.125$.

Check   The answer is the probability calculated in Example 1, part **a**.

Notice that in the figure of Example 2, only the y-axis was rescaled. This keeps the random variable the same as in the original problem.

Questions

NOTES ON QUESTIONS
Questions 1–8: We recommend going through these questions in order.

Covering the Reading

1. Refer to the binomial probability distribution with $p = 0.6$ and $n = 5$ graphed in this lesson.
   a. What is the total area of the bars for which $x \neq 0$?   ≈ **.990 units²**
   b. What is the probability that $x \neq 0$?   ≈ **.990**

In 2–4, suppose a point is selected at random in the triangular region shown below.

2. Find the probability that the *x*-coordinate of the point is less than 1.5. **≈ .28**

3. Find the probability that the *x*-coordinate of the point is 2. **0**

4. Find *w* so that $P(x<w)$, the probability that the *x*-coordinate of the point is less than *w*, is 0.32. **1.6**

5. If the scale change $S_{-1,3}$ is applied to a figure, what happens to the area of that figure? **It triples.**

6. *True* or *false*. The area of any triangle remains the same under the scale change
$$S: (x, y) \rightarrow \left(6x, \frac{y}{6}\right). \quad \textbf{True}$$

7. Refer to $\triangle OCD$ above.
 a. Draw $\triangle O'C'D'$, its image under the scale change
$$S: (x, y) \rightarrow \left(x, \frac{y}{10}\right). \quad \textbf{See margin.}$$
 b. Show that the area of the image is 1 using both the area formula for a triangle and the Area Scale Change Theorem. **See margin.**
 c. Determine the probability that the *x*-coordinate of a randomly selected point in $\triangle O'C'D'$ is less than 1.5. **≈ .28**

8. *Multiple choice.* Let *x* be a continous random variable, and let $P(x)$ be the probability that *x* occurs. Match each shaded region with the correct probability.

 a.

 b.

 c.

   (i) $P(x<c)$

   (ii) $P(x>c)$

   (iii) $P(c<x<d)$

(iv)   (iv) $P(x<d)$

   (v) $P(x>d)$

(iii)

(ii)

In 9–11, refer to the triangles from Examples 1 and 2. Suppose a point is selected randomly from the interior of $\triangle OMN$.

9. **a.** Determine the probability that the x-coordinate of the point is less than 3. **.875**
   b. Use the image of $\triangle OMN$ under the scale change

$$S: (x, y) \rightarrow \left(x, \frac{y}{6}\right).$$

   to verify your answer in part **a**. **See margin.**

10. Determine the probability that the y-coordinate of the point is greater than 2. **1/9**

11. Find w so that the probability is .75 that the x-coordinate is less than w.
    $4 - \sqrt{2}$

grade equivalent	percent
2.0– 2.9	0.5
3.0– 3.9	3.5
4.0– 4.9	5.2
5.0– 5.9	18.4
6.0– 6.9	23.8
7.0– 7.9	22.8
8.0– 8.9	16.4
9.0– 9.9	7.3
10.0–10.9	1.5
11.0–11.9	0.5
12.0–12.9	0.1

12. The table at the left summarizes the mathematics achievement scores (as grade equivalents) of a large group of students entering seventh grade.
    a. Draw a histogram of these data such that the area of the histogram equals 1. **a,c) See margin.**
    b. If a student is selected at random from this group of seventh graders, what is the probability that the student's mathematics achievement is at least one year above grade level? **25.8%**
    c. Shade the region of the histogram corresponding to part **b**.

13. A dart is thrown at a circle with radius four inches lying entirely within a square with side ten inches. Assuming that the dart is equally likely to land on all points inside the square and that the dart lands within the square, find the probability that the dart lands
    a. in the circle; **≈ .503**
    b. outside the circle; **≈ .497**
    c. on the circle. **0**

14. A circular archery target 122 cm in diameter consists of a bull's-eye of diameter 12.2 cm, surrounded by nine evenly spaced concentric circles. Assume that an arrow that hits a target from 90 meters away is equally likely to land anywhere on the target. Find the probability that the arrow
    a. hits the bull's-eye; **.01**
    b. lands somewhere in the outermost 4 circles; **.64**
    c. lands within the fifth ring in from the outer edge. **.11**

15. What vertical scale change can be applied to the semicircular region at the right so that the area of the image is 1?

    sample: $(x, y) \rightarrow \left(x, \frac{2y}{\pi}\right)$

23.a.

NAME _____

LESSON MASTER 10-3
QUESTIONS ON **SPUR** OBJECTIVES

■**REPRESENTATIONS** *Objective I (See pages 658–662 for objectives.)*

1. In a certain game, the number of points earned depends on which ring into which you can throw a ball, as shown. If the rings are equidistant and assuming that all throws landed in a ring, what is the probability of getting 50 points on a single throw?

$\frac{1}{16}$

In 2 and 3, a point is randomly selected from the triangular region shown below. The nonhorizontal sides are given by line segments AB and BC.

2. Calculate the area of the region. **6**

3. Determine the probability that the x-coordinate of the selected point is:
   a. less than 1. $\frac{1}{8} = .125$
   b. between 2 and 3. $\frac{3}{8} = .375$

In 4 and 5, a point is randomly selected from the triangular region shown on the right. The nonhorizontal sides are given by line segments CD and DE.

4. Find the value of k such that the probability that the x-coordinate of the selected point is less than k. $P(x < k) = .2$. **≈ 1.90**

5. Find the area of triangle CDE if the region is transformed by a size change of $\frac{2}{3}$ so that $S(x, y) = (\frac{2}{3}x, \frac{2}{3}y)$. **≈ 5.33**

Functions, Statistics, and Trigonometry © Scott, Foresman and Company

107

MORE PRACTICE
For more questions on SPUR
Objectives, use *Lesson Master 10-3*, shown on page 611.

EXTENSION
Ask students to try to prove the Area Scale Change Theorem for triangles with horizontal bases. (Take a triangle with vertices (x_1, y_1), $(x_1 + s, y_1)$, and $(x_2, y_1 + h)$. Its area is $\frac{1}{2}|h| \cdot |s|$. Under the scale change $(x, y) \rightarrow (ax, by)$, the image triangle has vertices (ax_1, by_1), $(ax_1 + as, by_1)$, and $(ax_2, by_1 + bh)$. The area of the image is consequently $\frac{1}{2}|as| \cdot |bh|$, which is $|ab|$ times the area of the preimage.)

PROJECTS
Project 7 on page 655 is related to the content of this lesson.

EVALUATION
A quiz covering Lessons 10-1 through 10-3 is provided in the Teacher's Resource File.

ADDITIONAL ANSWERS
23.a. See margin, page 611.

b.

c.

16. Determine the mean and standard deviation of a binomial distribution where $n = 25$ and $p = .47$. *(Lesson 10-2)*   **mean = 11.75; s.d. ≈ 2.495**

17. Consider a multiple-choice test of 20 questions, each of which has four choices.
    a. Find the expected number of questions a student guessing on all questions will get right.   **5**
    b. Find the standard deviation for the corresponding binomial distribution. *(Lessons 10-1, 10-2)* $\sqrt{3.75} \approx$ **1.94**

18. *True* or *false*.   The graph of a binomial distribution with a fixed probability of success ''spreads out'' as the number of trials is increased. *(Lessons 10-1, 10-2)*   **True**

19. Factor $2a^5 + 486a^5b^5c^{10}$ over the set of polynomials with integer coefficients. *(Lesson 9-8)* $2a^5(1 + 243\,b^5c^{10})$

20. There are eleven girls in Carla's Girl Scout troop and thirteen in her second-grade class. Altogether, there are twenty girls in the two groups. How many girls besides Carla are in both the Girl Scout troop and the class? *(Lesson 7-2)*   **3 besides Carla**

21. Give a counterexample to show that it is not always true that $\tan x \neq \tan(-x)$. *(Lesson 5-7)*   **For x = 0, tan 0 = 0 = tan(-0).**

22. The graph below represents an EKG reading of the electrical impulses from the heart (abnormal in this case). Each large square represents 0.2 seconds. Find the period of this heartbeat pattern. *(Lesson 5-6)*
    ≈ 0.36 seconds

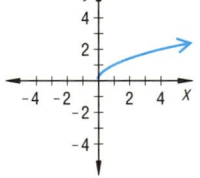

23. Refer to the graph of $y = \sqrt{x}$ at the right. Sketch the graph of
    a. $y = \sqrt{x} + 2$;
    b. $y = -\sqrt{x} + 2$;
    c. $y = \sqrt{x + 2}$. *(Lessons 3-2, 3-5)*
    a-c) See margin.

24. A student has test scores of 90, 72, 85, and 93. What score must the student get on a fifth test to have a 92 average? *(Lesson 1-4)*   **120**

25. A stick of length one meter is broken into two pieces at a randomly selected point. Find the probability that the difference in the lengths of the pieces is less than 10 centimeters.   **1/10**

612

10-4

The Parent of the Normal Curve

You have seen that when the probability p of success is fixed, then as the number n of trials increases, the graph of a binomial probability distribution approaches a bell-shaped curve. For instance, below is a set of binomial probability distributions for $p = .8$ showing the change in the shape of the distribution as n takes on the values of 5, 20, 50, and 100.

It can be shown using ideas of calculus that as n increases, the values of the binomial distribution, which is a function of a discrete random variable, approach those of a continuous function called a *normal distribution*. Many data sets of natural phenomena have approximately normal distributions. For instance, the heights of one thousand randomly chosen junior girls in a selected school district were measured in inches. The graph of the relative frequency distribution of heights looked like the graph below.

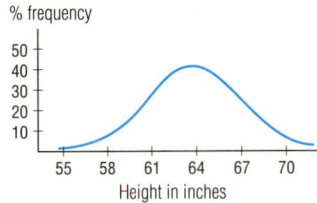

Other examples of data which have approximately normal distributions are the amounts of annual rainfall in a single city over a long period of time and the weights of individual fruits from a particular orchard in a good year.

The graph of a normal distribution is called a **normal curve**. Normal curves were first studied by Gauss and Pierre Simon de Laplace (1749–1827) in the early part of the 19th century. The parent of the normal curves is the graph of $f(x) = e^{-x^2}$, where e is the base of natural logarithms.

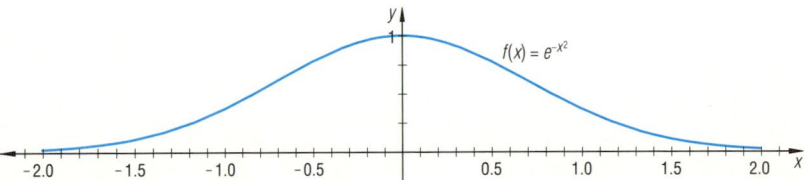

The graph for the girls' heights is an image of the parent normal curve under linear changes. The parent normal curve and images of it under linear changes are often called **bell-shaped curves.**

LESSON 10-4 The Parent of the Normal Curve   613

LESSON 10-4

RESOURCES
- Lesson Master 10-4
- Teaching Aid 77: Graph of $f(x) = e^{-x^2}$

OBJECTIVES

D Use properties of normal distributions and their parent function.
I Use areas to find probabilities.

TEACHING NOTES

The questions in this lesson require the MONTE CARLO AREA program from Lesson 7-9.

The graphs of binomial distributions are used to motivate this lesson, but of course the normal curve itself is continuous. You may wish to use a function grapher for demonstration purposes, or you can use the teaching aid provided. The asymptotic behavior can be investigated by zooming in on the graph for large positive and negative values of x. Have students verify that the maximum point is (0, 1), and that the inflection points (±0.71, 0.61) are on the curve.

Making Connections
Students do not need to know the exact values of x or y at the inflection points, and it requires calculus to find them. (They are points where the second derivative is 0.) The inflection points are included in this lesson because they are used in the transformation of the parent normal function to the standard normal function in the next lesson, and they are 1 standard deviation away from the mean.

The following properties are evident from the graph of $f(x) = e^{-x^2}$. (You should confirm them using an automatic grapher.) First, (0, 1) is the maximum point, and the y-intercept of the curve is 1. Second, the function is symmetric to the y-axis. This can be confirmed by noting that

$$f(-x) = e^{-(-x)^2} = e^{-x^2} = f(x).$$

Third, the x-axis is an asymptote to the graph of $f(x) = e^{-x^2}$ at both ends. This can be confirmed both visually and analytically. As $|x|$ gets larger, so does x^2. Raising e to the opposite of those large positive values gives the reciprocals of e to those large values; the results are smaller and smaller positive numbers.

A fourth property of the graph of $y = e^{-x^2}$ is the way its curvature changes. Notice that near the y-intercept the graph is curved downward, called *concave down*. Farther away from the y-axis, the graph is curved upward, called *concave up*.

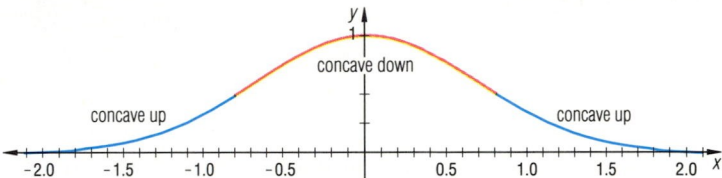

At two points, the graph changes its direction of concavity. These points are called **inflection points**. The inflection points for $y = e^{-x^2}$ occur when $x = \pm \dfrac{1}{\sqrt{2}} \approx \pm 0.71$; they have the same y-value, $y \approx 0.61$. Another window for this function in the first quadrant is shown below, with a vertical line through the inflection point.

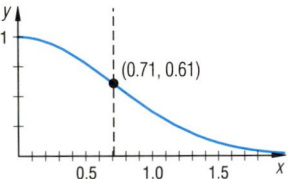

Fifth, it can be shown that the area under the parent of the normal curve is finite even though the graph never touches the x-axis. Finding the exact value of the area requires calculus, but it can be approximated using a computer and Monte Carlo methods.

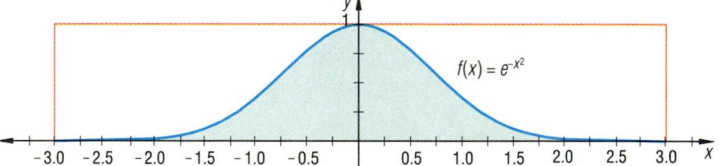

614

Imagine a rectangle containing the part of $f(x) = e^{-x^2}$ between $x = -3$ and $x = 3$ and between $y = 0$ and $y = 1$, as shown at the bottom of page 614. Randomly pick n points in the rectangle and count the number m which are under the curve. Then an approximation to the area under the curve is $\frac{m}{n}$ times the area of the rectangle. We modified the program MONTE CARLO AREA from Lesson 7-9 and ran it ten times, picking $n = 3000$ points from the rectangle with $-3 \leq x \leq 3$ and $0 \leq y \leq 1$. The number of points under f generated by the computer in the ten runs were 908, 928, 914, 843, 891, 879, 885, 897, 900, and 842. The mean of these ten numbers is 888.7. So $\frac{m}{n} \approx 0.296$, giving an area under the curve of about 1.777, which is very close to 1.768, the value (to thousandths) for the area under $f(x) = e^{-x^2}$ that can be found using calculus.

Using calculus, it can be shown that the area between the complete graph of $f(x) = e^{-x^2}$ and the x-axis equals $\sqrt{\pi} \approx 1.7725$. Clearly, not much area is added to the region under $f(x) = e^{-x^2}$ when $|x| > 3$. In fact, because $\frac{1.768}{\sqrt{\pi}} \approx 0.997$, about 99.7% of the area under $f(x) = e^{-x^2}$ is between $x = -3$ and $x = 3$.

Caution: In general, even if the graph of a function is above the x-axis and approaches it asymptotically, the area under the curve may be infinite. For example, the area under the curve $y = \frac{1}{x}$ for $x > 1$ is infinite.

Questions

Covering the Reading

1. *True* or *false.*   Given a binomial probability distribution where the probability of success on a single trial is 0.8, then as the number of trials increases the graph of the distribution approaches a bell-shaped curve.
   True

2. **a.** Using a calculator or computer, evaluate $y = e^{-x^2}$ with x increasing by steps of 0.3 from -1.5 to 1.5.
   b. Plot these values and connect them with a smooth curve.
   a,b) See margin.

3. Give an equation of the indicated feature of the graph of $y = e^{-x^2}$.
   a. line of symmetry   **x = 0**   **b.** asymptote(s)   **y = 0**

4. For the graph of $y = e^{-x^2}$, give the coordinates of all
   a. inflection points;   **b.** relative extrema.
   (.71, 61), (-.71, 61)   **(0, 1)**

In 5 and 6, *true* or *false*.

5. The graph of $y = e^{-x^2}$ is concave up in the interval $x \leq -\frac{1}{\sqrt{2}}$.   **True**

6. The parent of the normal curve is an even function.
   True

NOTES ON QUESTIONS
Questions 2–4:   These questions are made easier because the function is an even function.

ADDITIONAL ANSWERS

2.a.

x	y
-.15	.105
-1.2	.237
-.9	.445
-.6	.698
-.3	.914
0	1
.3	.914
.6	.698
.9	.445
1.2	.237
1.5	.105

b.

7. Consider the distribution of heights of junior girls shown in this lesson. It has been found that the median height of junior girls is 163 cm, and 22% of them have a height greater than 169 cm. Estimate the probability that a randomly chosen junior girl will have height h in cm as follows.
   a. $h > 163$ **.50**   b. $h < 169$ **.78**   c. $163 < h < 169$ **.28**

8. What percent of the area between $f(x) = e^{-x^2}$ and the x-axis is between $x = -2$ and $x = 2$?   **≈ 99.5%**

In 9–11, consider the equation $g(x) = e^{-x^2/2}$.

9. a. Without graphing, describe the symmetries of g.
   b. Without graphing, state equations for all asymptotes of g.   **y = 0**
   c. Use an automatic grapher to draw the graph of $y = g(x)$.
   d. Estimate the coordinates of the inflection points of g, correct to the nearest hundredth.   **(1, .61), (-1, .61)**
   a) reflection-symmetric to the y-axis; c) See margin.

10. Modify the MONTE CARLO AREA program to estimate the area between the graph of g and the x-axis when $-3 \leq x \leq 3$.   **See margin.**

11. Consider the graph of g and the graph of its parent function $f(x) = e^{-x^2}$. Identify a transformation that maps the graph of f to the graph of g.
    $(x, y) \rightarrow (\sqrt{2}\,x, y)$

In 12 and 13, let $f(x) = e^{-x^2}$, $g(x) = 2^{-x^2}$, and $h(x) = 3^{-x^2}$.

12. a. What type of transformation maps the graph of f to the graph of g? (Hint: $2 \approx e^{0.693}$.)   **a) $(x, y) \rightarrow (1.2x, y)$; b) $(x, y) \rightarrow (.95x, y)$**
    b. What type of transformation maps the graph of f to the graph of h?

13. Use the MONTE CARLO AREA program to estimate the area between the graph of g and the x-axis when $-3 \leq x \leq 3$.   **The area is about 2.13.**

14. Refer to $\triangle OQR$ below. A point inside the triangle is chosen at random.
    a. Determine the probability that the x-coordinate of the point is between 1 and 2.   **3/8 = .375**
    b. Determine the value of b such that the probability is 0.4 that the x-coordinate of the point is less than b.   **$\sqrt{3.2} \approx 1.79$**
    c. Draw the image of the triangle under the scale change
    $$S: (x, y) \rightarrow \left(x, \frac{1}{4}y\right).$$ **See margin.**
    d. What is the area of the image?   *(Lesson 10-3)*   **2 units²**

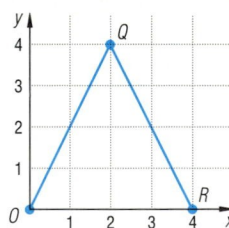

616

15. The figure at the right is the graph of a uniform (constant) probability distribution f. Specifically, $f(x) = 0.1$ for $0 \le x \le 10$, and $f(x) = 0$ elsewhere.

a. Suppose a point is selected randomly from the rectangular region. What is the probability that its x-coordinate is between 1.5 and 8.5? **.7**

b. If 5 points are selected randomly from the rectangular region, what is the probability that all 5 will have x-coordinates between 1.5 and 8.5? *(Lessons 7-4, 10-3)* $(.7)^5 \approx .168$

16. Express the quotient $\dfrac{x^5 - 32}{x - 2}$ as a polynomial. *(Lesson 9-4)*
$x^4 + 2x^3 + 4x^2 + 8x + 16$

17. Expand $(3x - 5)^4$ using the Binomial Theorem. *(Lesson 8-8)*
$81x^4 - 540x^3 + 1350x^2 - 1500x + 625$

18. Consider two events A and B in the same sample space. *True* or *false*. Justify your answer.
a. If $P(A) + P(B) = 1$, then A and B are complementary events.
b. If A and B are complementary events, then $P(A) + P(B) = 1$.
(Lesson 7-2) **a) False, A and B might not be mutually exclusive.**
b) True, by definition of complementary events

19. A school compares its students' performance on a college-entrance examination to the performance of students in the nation.
a. What is the population?
b. What is the sample? *(Lesson 1-1)*
a) all students in the nation who take the test; b) students in the school who take the test

Exploration

20. Find out some things about the mathematician Pierre Simon de Laplace.
Answers will vary.

21. a. Modify the MONTE CARLO AREA program from Lesson 7-9 to estimate the area under $y = 1/x$ for $1 \le x \le n$ and for $n = 10, 20, 50, 100,$ and 1000.

b. Explain how your answers to part **a** support the statement in the final sentence of this lesson.   **Answers will vary.**

21a) Sample: Insert line 25 INPUT "UPPER ENDPOINT OF INTERVAL "; B
change lines 60, 70, 90 to
60 X = B*RND + 1 : Y = RND
70 IF Y < 1/X THEN C = C + 1
90 AREA = (B − 1)*C/N

FOLLOW-UP

MORE PRACTICE
For more questions on SPUR Objectives, use *Lesson Master 10-4*, shown below.

EXTENSION
As mentioned on page 613, normal curves were first studied extensively by Gauss and Laplace. But the normal distribution was probably first found by Abraham DeMoivre in 1733 and later rediscovered by Gauss and Laplace. Interested students can research who discovered the normal distribution and when.

NAME _____

LESSON **MASTER   10-4**
QUESTIONS ON **SPUR** OBJECTIVES

■REPRESENTATIONS   *Objective I (See pages 658–662 for objectives.)*

1. *True* or *false*   In a binomial probability distribution, the probability of success on a single trial must be greater than 0.5 in order for the graph to approach a bell-shaped curve as the number of trials increases.   **False**

In 2–9, consider the function with equation $y = e^{-x^2}$.

2. Is the function odd, even, or neither?   **Even**

3. Give all intercepts.   **y-intercept is 1. (0, 1)**

4. Give the coordinates of all relative extrema.   **(0, 1)**

5. Give equations for all asymptotes.   **y = 0**

6. Is the area between the graph of $y = e^{-x^2}$ and the x-axis finite or infinite?   **Finite**

7. *True* or *false*   Over 99% of the area between the curve $y = e^{-x^2}$ and the x-axis is between −3 and 3.   **True**

8. How many inflection points does the graph have?   **2**

9. The graph of $y = e^{-x^2}$ is concave up for $-\frac{1}{\sqrt{2}} \le x \le \frac{1}{\sqrt{2}}$. Use this information to give coordinates of the inflection point(s). $\left(-\frac{1}{\sqrt{2}}, \frac{1}{\sqrt{e}}\right), \left(\frac{1}{\sqrt{2}}, \frac{1}{\sqrt{e}}\right)$

108   *Functions, Statistics, and Trigonometry © Scott, Foresman and Company*

617

OBJECTIVES

B Use areas under the standard normal curve to find probabilities.
D Use properties of normal distributions and their parent function.
I Use areas to find probabilities.
J Graph and interpret normal distributions.

TEACHING NOTES

One of the two main ideas in this lesson is that the standard normal distribution is a probability distribution. Students should be able to calculate probabilities of the form $P(z < a)$, $P(z < a)$, and $P(a < z < b)$ for any a and b between -3.09 and 3.09. A teaching aid is provided for **Example 2.** Refer students to the table in Appendix D on page 844. The application of these skills in realistic contexts begins in Lesson 10-6.

The second main idea in this lesson is the reasoning that leads to the scale changes of the parent normal distribution to make it the standard normal distribution with the special properties that distribution has. Of the six properties listed on page 619, students will have to accept properties (1), (2), and (4), though they can verify (2) and (4) with an automatic grapher. Furthermore, they

618

LESSON

10-5

The Standard Normal Distribution

An important offspring of the parent function $y = e^{-x^2}$ is $f(z) = \dfrac{1}{\sqrt{2\pi}}e^{-\frac{z^2}{2}}$, whose graph is the bell-shaped curve shown below.

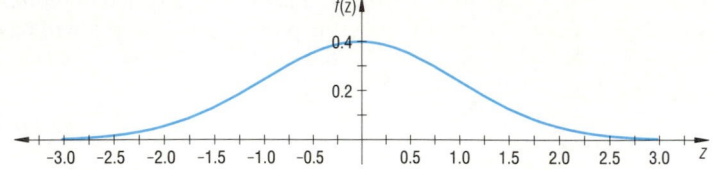

This graph is known as the **standard normal curve**. It represents a probability distribution called the **standard normal distribution**. Because of its special properties (listed below), it is the most important probability distribution in the entire field of probability and statistics.

The transformation which maps the graph of $y = e^{-x^2}$ onto the graph of f is the scale change $(x,y) \to \left(\dfrac{x}{\sqrt{2}}, \dfrac{y}{\sqrt{2\pi}}\right)$. This scale change is the composite of two scale changes. The first ensures that inflection points of the curve are at $x = \pm 1$; the second ensures that the area under the standard normal curve is 1. Here is how.

The first scale change is $S_1: (x, y) \to (\sqrt{2}x, y)$; an equation of the image of $y = e^{-x^2}$ transformed by S_1 is

$$y = e^{-\left(\frac{x}{\sqrt{2}}\right)^2},$$

which simplifies to
$$y = e^{-\frac{x^2}{2}}.$$

Recall that the inflection points of $y = e^{-x^2}$ are the points where $x = \pm\dfrac{1}{\sqrt{2}}$, so the points of inflection of the image curve occur at $x = \pm 1$.

The parent and image under S_1 are graphed below.

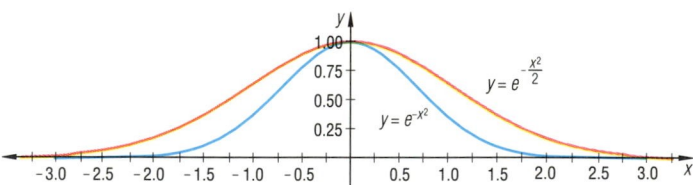

Recall that the area under $y = e^{-x^2}$ is $\sqrt{\pi}$. By the Area Scale Change Theorem, the area under $y = e^{-\frac{x^2}{2}}$, the image of S_1, is $\sqrt{2}\,\sqrt{\pi} = \sqrt{2\pi}$.

Now to find the second scale change S_2, so that the area under $y = e^{-\frac{x^2}{2}}$, transformed by S_2, is 1, use the scale change

$$S_2: (x, y) \rightarrow \left(x, \frac{y}{\sqrt{2\pi}} \right).$$

An equation of the image of $y = e^{-\frac{x^2}{2}}$ under S_2 is

$$\sqrt{2\pi}\, y = e^{-\frac{x^2}{2}}.$$

Solve for y to obtain an equation of the standard normal curve:

$$y = \frac{1}{\sqrt{2\pi}} e^{-\frac{x^2}{2}}.$$

A graph of this curve and its preimage is shown below.

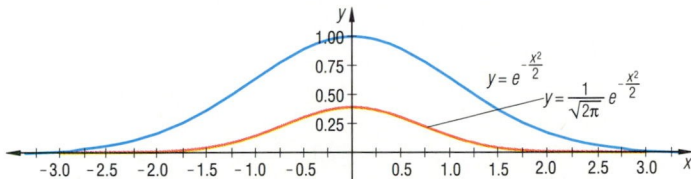

In discussions of the normal distribution, sometimes the horizontal axis is labeled as the z-axis. (This is customary for the standard normal curve; the next lesson introduces "z-scores.") Using z for the horizontal axis, the standard normal distribution has the following properties.

(1) The area under the curve and above the z-axis is 1.
(2) It is concave down for $-1 < z < 1$, and concave up for $|z| > 1$; its inflection points occur when $z = \pm 1$.
(3) It is an even function, and so is symmetric to the line $z = 0$.
(4) Its maximum value is $f(0) = \dfrac{1}{\sqrt{2\pi}} \approx .3989$.
(5) $f(z) > 0$ for all real numbers.
(6) As $z \rightarrow \infty$ or $z \rightarrow -\infty$, $f(z) \rightarrow 0$.

Since the standard normal curve represents a probability distribution, probabilities for intervals of z-values are areas under the curve. Specifically, the shaded area below represents the probability that z is less than zero. Because of properties (1) and (3) above, $P(z < 0) = 0.5$ and $P(z > 0) = 0.5$.

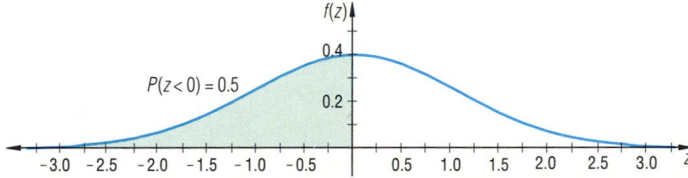

need to accept that one standard deviation from the mean are the points of inflection. This property is discussed on pages 618–619 before **Example 3.**

Technology   If students have a calculator which gives normal distribution values, we suggest that they check how the machine works. You may wish to make a note of students having this capability and to familiarize yourself with the mechanics so that you can assist those students.

ADDITIONAL EXAMPLES
1. Find the probability that a randomly chosen observation from a standard normal distribution is less than 0.96.
.8315 ≈ .83

2. Find the probability that an observation from a normally distributed random variable falls within the given range.

a. between 0 and 0.96
≈ .3315

b. greater than 0.96
≈ .1685

c. less than -0.96
≈ .1685

d. between -0.96 and 0.96
≈ .6630

3. What percent of the data are within 1.5 standard deviations of the mean? (Use Appendix D.)
≈ 87%

The standard normal distribution is so important that probabilities derived from it have been calculated and recorded in tables. Below is part of the Standard Normal Distribution Table given in Appendix D. It gives the area under the standard normal curve to the left of a given positive number a. The table does not give values for $a > 3.09$ because there is very little area under the standard normal curve to the right of $z = 3.09$.

					Standard Normal Distribution Table $P(z<a)$ for $a \geq 0$					
a	**0**	**1**	**2**	**3**	**4**	**5**	**6**	**7**	**8**	**9**
0.0	.5000	.5040	.5080	.5120	.5160	.5199	.5239	.5279	.5319	.5359
.										
.										
.										
0.6	.7257	.7291	.7324	.7357	.7389	.7422	.7454	.7486	.7517	.7549
0.7	.7580	.7611	.7642	.7673	.7704	.7734	.7764	.7794	.7823	.7852
0.8	.7881	.7910	.7939	.7967	.7995	.8023	.8051	.8078	.8106	.8133
0.9	.8159	.8186	.8212	.8238	.8264	.8289	.8315	.8340	.8365	.8389
1.0	.8413	.8438	.8461	.8485	.8508	.8531	.8554	.8577	.8599	.8621
.										
.										
.										
1.7	.9554	.9564	.9573	.9582	.9591	.9599	.9608	.9616	.9625	.9633
.										
.										
.										
2.9	.9981	.9982	.9982	.9983	.9984	.9984	.9985	.9985	.9986	.9986
3.0	.9987	.9987	.9987	.9988	.9988	.9989	.9989	.9989	.9990	.9990

■ ■ ■ ■ ■ ■ ■

Example 1 Find the probability that a randomly chosen observation from a standard normal distribution is less than 0.85.

Solution Imagine or sketch a standard normal curve. The desired probability is the area under the curve to the left of the line $z = 0.85$.

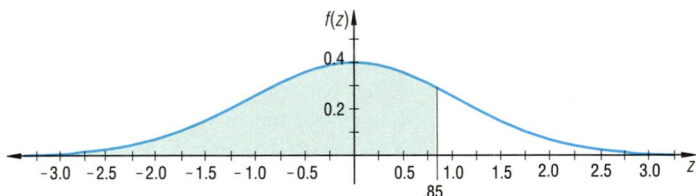

Use the table directly. Read down the a-column until you get to 0.8. Go across this row until you get to the column headed by 5. The entry there is .8023. So

$$P(z<0.85) \approx 0.8023.$$

By adding and subtracting areas and using symmetry properties of the standard normal distribution, the Standard Normal Distribution Table can be used to calculate probabilities that are not of the form $P(z<a)$ with $a \geq 0$.

Example 2 Find the probability that a randomly chosen observation from a standard normal distribution is

a. between 0 and 0.85;

b. greater than 0.85;

c. less than -1.73;

d. greater than -1.73.

Solution

a. $P(0<z<0.85)$ is the area under the normal curve between $z = 0$ and $z = 0.85$. This is the difference of two areas.

$P(0<z<0.85) = P(z<0.85) - P(z \leq 0)$

From Example 1, $P(z<0.85) \approx 0.8023$. From the symmetry of f, $P(z \leq 0) = 0.5$. So

$$P(0<z<0.85) \approx 0.8023 - 0.5000$$
$$= 0.3023.$$

b. $P(z>0.85)$ equals the area of the shaded region below.

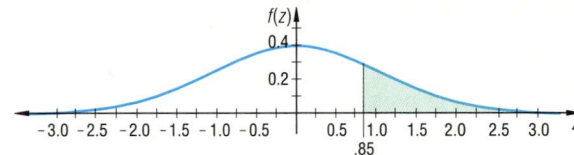

To find the area under the standard normal curve to the *right* of a given value, subtract the area to its left from the total area under the curve of 1.

$$P(z>0.85) = 1 - P(z<0.85)$$
$$\approx 1 - 0.8023$$
$$= 0.1977$$

c. To find $P(z< -1.73)$, use symmetry.

$$P(z< -1.73) = P(z>1.73)$$

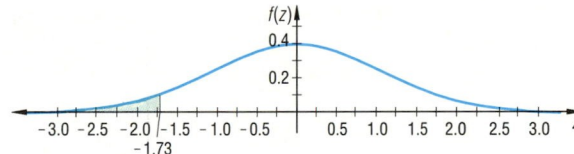

Now find $P(z>1.73)$ as in part **b**.

$$P(z>1.73) = 1 - P(z<1.73)$$
$$\approx 1 - .9582$$
$$= 0.0418$$

So $P(z< -1.73) \approx 0.0478$.

d. $P(z > -1.73)$ is the area of the shaded region below.

You can subtract the area found in part **c** from the total area.
$$P(z > -1.73) = 1 - P(z < -1.73)$$
$$\approx 1 - 0.0418$$
$$= 0.9582$$

Or you can use symmetry.
$$P(z > -1.73) = P(z < 1.73)$$
$$\approx 0.9582$$

Note that the probability for any particular value, such as $P(z = a)$, is zero by the area interpretation. For this reason, if z is normally distributed, it is always the case that $P(z < a) = P(z \leq a)$.

Some calculators or statistical packages provide values from the normal distribution table, and some provide related values. For example, some give 0.8413 in response to $P(z < 1)$. However, others give 0.3413, which is the area between scores of 0 and 1, rather than the area to the left of 1; and still others give 0.1587, the area to the right of 1. Check that you know how to use properly the technology available to you.

When the z-values are considered as data points, the properties of the standard normal distribution imply the data have a mean of 0 and a variance, thus a standard deviation, equal to 1.

Example 3 derives another property of the standard normal distribution.

Example 3   What percent of the data in a standard normal distribution are within one standard deviation of the mean?

Solution   Sketch the standard normal curve.

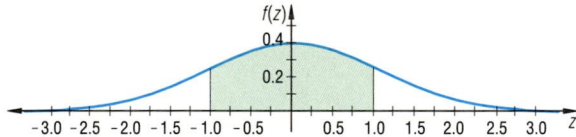

Being within one standard deviation of the mean means that $-1 < z < 1$.
So
$$P(-1 < z < 1) = 2 \cdot P(0 < z < 1)$$
$$= 2 \left[P(z < 1) - P(z < 0) \right]$$
$$\approx 2(0.8413 - 0.5)$$
$$= 0.6826.$$
Thus, about 68% of normally distributed data are within one standard deviation of the mean.

Check   The shaded area seems to be about $\frac{2}{3}$ of the area under the curve.

622

Questions

NOTES ON QUESTIONS
Question 9: A more formal way of putting this is: About two-thirds of the time, a normally distributed random variable will have a value within one standard deviation from the mean.

Questions 11, 15–18: These require the table in Appendix D.

ADDITIONAL ANSWERS
1.a.

Covering the Reading

1. **a.** Sketch a graph of the standard normal curve. **See margin.**
   b. State an equation for the standard normal curve. $f(z) = \frac{1}{\sqrt{2\pi}} e^{\frac{-z^2}{2}}$

2. For the standard normal curve, state the
   a. coordinates of all relative extrema; $(0, 1/\sqrt{2\pi}) \approx (0, .399)$
   b. equation(s) of the line(s) of symmetry; $z = 0$
   c. equation(s) of the asymptotes. $y = 0$

3. Evaluate, given the fact that in a standard normal distribution,
   $$P(z < 1.6) \approx .9452.$$

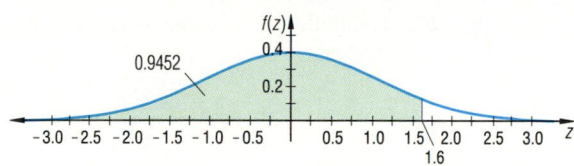

   a. $P(z > 1.6)$ **.0548**   **b.** $P(0 < z < 1.6)$ **.4452**
   c. $P(|z| < 1.6)$ **.8904**   **d.** $P(z > -1.6)$ **.9452**

In 4–7, evaluate using the Normal Distribution Table in Appendix D.

4. $P(z < 1.88)$ **.9699**   5. $P(z > 0.07)$ **.4721**

6. $P(-1.5 < z < 1.3)$ **.8364**   7. $P(z > -2)$ **.9772**

8. State **a.** the mean, and **b.** the standard deviation of the standard normal distribution. **a) 0; b) 1**

In 9 and 10, *true* or *false*.

9. About two-thirds of the observations from a standard normal distribution are within one standard deviation of the mean. **True**

10. The area between the standard normal curve and the z-axis is 1. **True**

11. Show that about 95% of the values in the standard normal distribution are within two standard deviations of the mean.
    $P(-2 < z < 2) = 2P(0 < z < 2) \approx 2(.9772 - .5) = 2(.4772) = .9544$

12. About what percent of the data in a standard normal distribution are within three standard deviations of the mean? $\approx$ **99.7%**

13. Give a reason why the Normal Distribution Table does not list probabilities for z larger than 3.09. **Only a tiny percent of scores are beyond 3.09 standard deviations from the mean.**

14. How do the areas under the parent $y = e^{-x^2}$ and the standard normal curve compare? **The area under $y = e^{-x^2}$ is $\sqrt{\pi}$; the area under the standard normal curve is 1.**

Applying the Mathematics

In 15 and 16, find the value of c satisfying the equation.

15. $P(z<c)=0.975$ ≈ 1.96

16. $P(z\geq c)=0.4522$ $\approx .12$

17. Which value of the random variable in a standard normal distribution is exceeded by about 75% of the distribution? $\approx -.67$

18. Complete: 90% of the observations of a standard normal distribution fall within ___?___ standard deviations of the mean. ≈ 1.65

19. Use the MONTE CARLO AREA program to approximate the area under the standard normal curve on the following intervals.
  a. $-2<x<2$ $\approx .954$ **b.** $-3<x<3$ $\approx .997$

Review

20. The table below summarizes the SAT-Math scores of a large group of students.

score	percent
750-800	8
700-740	23
650-690	21
600-640	17
550-590	8
500-540	13
450-490	6
400-440	4
350-390	2

  a. Construct a histogram of these data.   **See margin.**
  b. If a student is selected at random from this group, what is the probability that the student's SAT-Math score is 600 or more? $\approx .69$
  c. Shade the region of the histogram corresponding to part **b**.
  (Lesson 10-3)   **See margin.**

21. If for some binomial probability distribution $\mu = 20$ and $\sigma = 2.3$, find n and p.   *(Lesson 10-2)*   **n = 27, p = .7355**

22. Let $f(x) = 2x^2 + 3x + 4$. Evaluate $\dfrac{f(x) - f(h)}{x - h}$.   *(Lesson 9-9)*   **2x + 3 + 2h**

23. A box has length 4 inches, width 3 inches, and height 2 inches. Another box with twice the volume has dimensions $4 + x$, $3 + x$, and $2 + x$ inches. Approximate the value of x to the nearest 0.1 inch.   *(Lessons 9-1, 9-3)*
  x ≈ .7 inch

24. In a certain city, 60% of the citizens are in favor of a school bond rate increase. A sample of ten citizens is taken at random. What is the probability that fewer than half of them will be in favor of the increase?
  (Lesson 8-9) $\approx .166$

25. In how many ways can a basketball coach select her starting team of five from a squad of twelve?   *(Lesson 8-6)*   **792**

26. A small airline company has the following number of flights on any given day between five cities in a state and the company's main operational city (the hub).

City	Number of flights	Distance to the hub (in miles)
I	3	150
II	2	75
III	5	210
IV	8	240
V	2	95

a. Determine the mean and standard deviation of the number of miles for flights with this airline. **$\bar{x} = 188$ miles, s.d. $= 61.2$**

b. Without reentering data, determine what the new mean and standard deviation of flight mileage would be if
   i. the distances to the hub are measured in kilometers (1 mile = 1.61 km); **$\bar{x} \approx 303$ km, s.d. ≈ 99 km**
   ii. an extra 20 miles is added to each distance (so flying time calculations can take into account the time spent in queue before landings). *(Lessons 3-3, 3-6)* **$\bar{x} = 208$, s.d. $= 61.2$ miles**

In 27 and 28, **a.** find the exact zeros of the function. **b.** check your answer by graphing the function and locating zeros correct to the nearest hundredth. *(Lessons 2-1, 4-5, 9-5)*

27. $s(x) = 3 \ln x - 5$

a) $e^{\frac{5}{3}}$; b) See margin.

28. $p(x) = x^3 - 3x^2 + x$

a) $0, \dfrac{3 \pm \sqrt{5}}{2}$; b) See margin.

Exploration

29. a. Examine the differences between successive values in the first row of the Normal Distribution Table. Use these to estimate the following values.
   i. $P(z < 0.035)$ **.5140**   **ii.** $P(z < 0.068)$ **.5271**

b. Why are the differences referred to in part **a** almost constant?

c. Why are differences between successive values in other rows of the table not constant?

b) The graph is nearly a straight line for the interval $0 < z < .09$. c) The graph is not close to linear for the other rows of the table.

MORE PRACTICE
For more questions on SPUR Objectives, use *Lesson Master 10-5*, shown below.

PROJECTS
Project 2 on page 654 is related to the content of this lesson.

NAME _____

LESSON **MASTER 10-5**
QUESTIONS ON **SPUR** OBJECTIVES

■**SKILLS** *Objective B (See pages 658–662 for objectives.)*
In 1 and 2, use the table in Appendix D.

1. Determine the area under the standard normal curve from $z = 1.74$ to $z = 3.08$.   **.0399**

2. Evaluate $P(.65 < z < 2.91)$.   **.2559**

■**PROPERTIES** *Objective D*

3. Use the program MONTE CARLO AREA to estimate the area under the curve defined by
$f(z) = \dfrac{1}{\sqrt{2\pi}} e^{-\frac{z^2}{2}}$ between $z = -0.32$ and $z = 0.32$.   **sample: .248**

■**REPRESENTATIONS** *Objective I*

4. The shaded area under the standard normal curve is about .6826. Determine
   a. $P(-1 < z < 1)$.   **.6826**
   b. $P(0 < z < 1)$.   **.3413**
   c. $P(z < 1)$.   **.8413**

■**REPRESENTATIONS** *Objective J*

5. Which outcome value in a standard normal distribution is exceeded by three-fourths of the distribution?   **-.67**

6. What percent of values in a standard normal distribution lie between 1 and 2 standard deviations above the mean?   **≈ 13.6%**

Functions, Statistics, and Trigonometry © Scott, Foresman and Company   **109**

625

In the previous lesson you learned to calculate probabilities by finding areas under the standard normal curve

$$f(x) = \frac{1}{\sqrt{2\pi}} e^{-\frac{x^2}{2}} .$$

Any composite of translations and scale changes of the standard normal curve produces an image that is also a normal curve. In particular, if m and s are constants and $s \neq 0$, it can be shown that the graph of the equation

$$f(x) = \frac{1}{s\sqrt{2\pi}} e^{-\frac{1}{2}\left(\frac{x-m}{s}\right)^2}$$

is a normal curve with area under the curve also equal to 1. Thus, such curves can represent probability distributions.

Below are three different normal curves generated by using three different pairs of values of m and s.

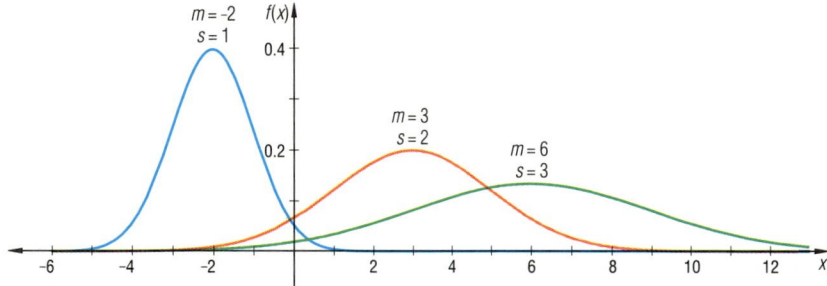

Notice that the value of m in the equation indicates the line of symmetry of the normal curve, and s is a measure of the spread of the curve. In particular, it can be proved that if x represents a continuous random variable and $f(x)$ its probability, then m is the mean and median of the probability distribution, and s is its standard deviation.

As in the case of the standard normal distribution, answers to probability questions about other normal curves can be found by calculating areas. However, in order to be able to use the Standard Normal Distribution Table as in Lesson 10-5, you must first transform the given normal curve into the standard normal curve.

As an example of how this is done, consider the fact that the heights of adult men in the United States are approximately normally distributed with a mean of 70 inches and a standard deviation of 3 inches. Thus, the relative frequency distribution of the heights is approximated by the following graph.

626

$$f(x) = \frac{1}{3\sqrt{2\pi}} \, e^{-\frac{1}{2}\left(\frac{x-70}{3}\right)^2}$$

Recall that if each number in a data set is translated by a constant, the mean of the distribution is also translated by that constant. In particular, if $m = 70$ is subtracted from each height in the preceding distribution, the mean of the resulting distribution is $70 - 70 = 0$. This translation has no effect on the standard deviation of the resulting distribution.

If the translated heights are each multiplied by $\frac{1}{3}$, the standard deviation of the resulting distribution is $\frac{1}{3} \cdot 3 = 1$. Thus, the linear change $x \rightarrow \frac{x-70}{3}$ maps the normal distribution above onto a distribution with mean 0 and standard deviation 1.

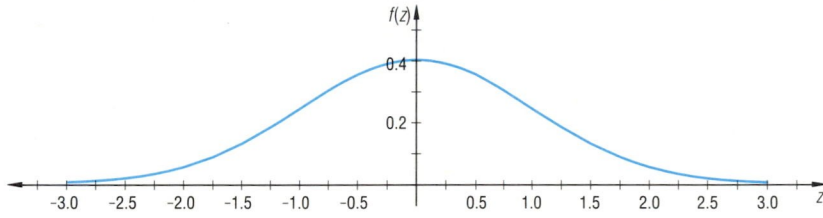

With the Graph Standardization Theorem you can show that this distribution can be transformed into the standard normal curve.

The preceding analysis generalizes to the following theorem.

> **Theorem**
>
> If a variable x has a normal distribution with mean m and standard deviation s, then the image variable
>
> $$z = \frac{x-m}{s}$$
>
> has the standard normal distribution.

The value of the image variable z is called the **standard score** or **z-score** for the value x. The process of getting z-values from an original data set by applying the transformation

$$x \rightarrow \frac{x-m}{s}$$

is often referred to as **standardizing** the variable. By standardizing the values of normal distributions, you can answer many probability questions.

similar to **Example 1,** so you may choose to do this question as part of your class discussion.

The normal approximation to a binomial distribution allows students to estimate probabilities in binomial situations with comparative ease. The steps are shown in **Example 3:** Determine n, p, and q from the given information; then calculate np and $\sqrt{npq}$ to obtain the mean and standard deviation you want for the normal distribution. Next, calculate z scores for the data using $\frac{x - np}{\sqrt{npq}}$, and find the corresponding probabilities from tables for the z-scores.

LESSON 10-6 Other Normal Distributions   **627**

Example 1   An adult American male is selected randomly. What is the probability that the man is shorter than 6 feet tall?

Solution   Because the height x of adult American males has approximately a normal distribution with a mean of 70 inches and standard deviation of 3 inches, the variable $z = \frac{x-70}{3}$ has approximately a standard normal distribution. The z-score associated with $x = 72$ is $z = \frac{72-70}{3} = \frac{2}{3} \approx 0.67$. So a height of 72 inches is $\frac{2}{3}$ of a standard deviation above the mean of 70 inches, as illustrated below.

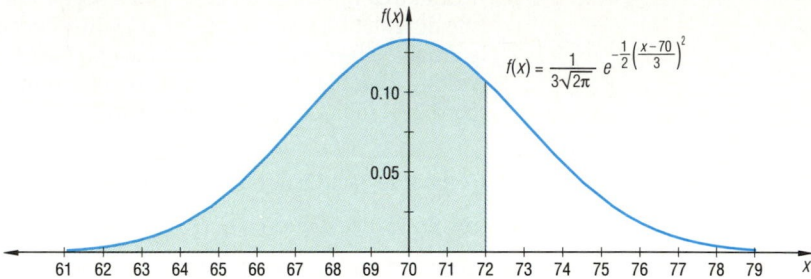

$$f(x) = \frac{1}{3\sqrt{2\pi}} e^{-\frac{1}{2}\left(\frac{x-70}{3}\right)^2}$$

From the table of values for the standard normal distribution, the proportion of men shorter than 72 inches tall is
$$P(x<72) = P(z<0.67)$$
$$\approx 0.7486.$$
So about 75% of adult men in the U.S. are shorter than 72 inches tall. The probability of a randomly selected male being shorter than 6 feet tall is about 0.75.

Example 2   The refills for a particular mechanical pencil are supposed to be 0.5 mm in diameter. Refills below 0.485 mm in diameter do not stay in the pencil, while those above 0.520 mm do not fit in the pencil at all. If a firm makes refills with diameters whose differences from the correct size are normally distributed with mean 0.5 mm and standard deviation .01 mm, find the probability that a randomly chosen refill will fit.

Solution   Let $x =$ the diameter of the refill in mm. The task is to find $P(0.485<x<0.520)$, where x is normally distributed with $m = 0.5$ and $s = 0.01$. Change to standard scores using
$$z = \frac{x-0.5}{0.01}$$
The z-score z_1 associated with $x = 0.485$ is
$$z_1 = \frac{0.485-0.5}{0.01}$$
$$= -1.5.$$
The z-score z_2 associated with $x = 0.520$ is
$$z_2 = \frac{0.520-0.5}{0.01}$$
$$= 2.$$

628

According to the preceding theorem, z_1 and z_2 are scores from the standard normal distribution. The range of standardized scores between z_1 and z_2 is shown below.

Use the table of values for the standard normal distribution.

$$P(0.485 < x < 0.520) = P(-1.5 < z < 2.0) = P(z < 2.0) - P(z < -1.5)$$
$$= P(z < 2.0) - [1 - P(z < 1.5)]$$
$$\approx 0.9772 - (1 - 0.9332)$$
$$= 0.9104$$

The probability that a refill will fit is about 0.91. That is, about 91% of the refills will fit the pencil, 9% will not.

The answer to Example 2 is low; the manufacturer should probably adjust the machinery to get more accurate thicknesses.

The normal distribution is related to many other probability distributions. In particular, as you have seen earlier in this chapter, some binomial distributions are approximately bell-shaped.

For instance, the binomial probability distribution with $n = 100$ and $p = 0.2$ has mean $\mu = 100(0.2) = 20$ and standard deviation
$$\sigma = \sqrt{npq} = \sqrt{100(0.2)(0.8)} = \sqrt{16} = 4.$$
Below is the probability histogram for this binomial distribution, with the curve for the normal distribution with $m = 20$ and $s = 4$ superimposed.

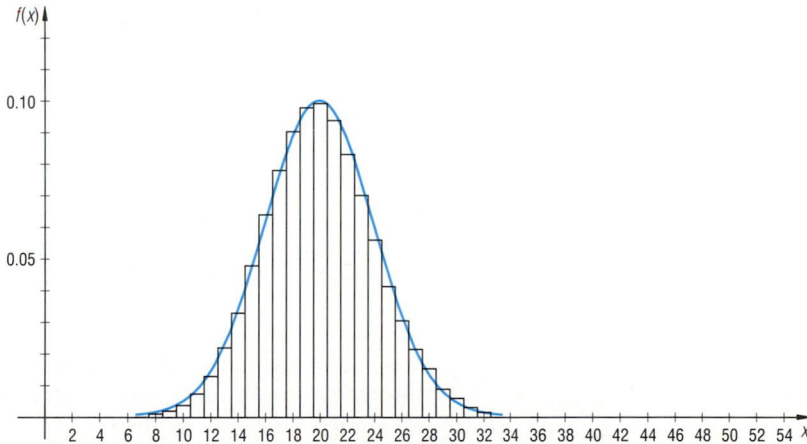

For these values of n and p the normal distribution appears to approximate the binomial distribution quite well.

In general, if n is quite large, or when p is close to 0.5, a normal curve approximates a binomial distribution quite well. If n is small and p is near 0 or 1, the binomial is not well approximated by a normal distribution. With $q = 1 - p$, a rule of thumb is that a binomial distribution can be approximated by a normal distribution with mean np and standard deviation $\sqrt{npq}$ provided np and nq are each ≥ 5.

The graphs of binomial distributions in Lessons 10-1 and 10-2 provide evidence for this approximation. Example 3 shows how to apply it.

▪ ▪ ▪ ▪ ▪ ▪ ▪ ▪ ▪

Example 3 In the past, 85% of ticketed passengers show up for a regularly scheduled Chicago–New York flight with a capacity of 240. The airline company accepts 270 reservations. Estimate the probability that a flight will be overbooked.

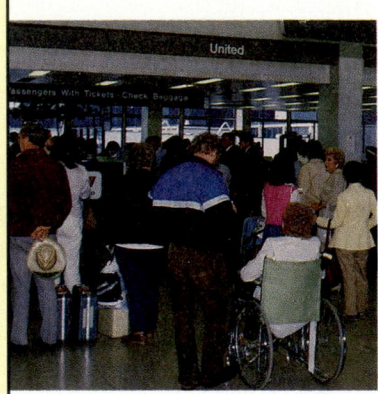

Solution Because each passenger either arrives or doesn't arrive, and all passengers are assumed to act independently, the number of passengers arriving is binomially distributed. The distribution has $n = 270$, $p = 0.85$, and $q = 0.15$. Because both $np = 229.5$ and $nq = 40.5$ are greater than 5, the normal approximation to the binomial distribution can be used. The appropriate normal distribution has mean $np = 229.5$ and standard deviation $\sqrt{npq} \approx 5.87$. The flight is overbooked when the number n of passengers is greater than 240. Thus, you must calculate $P(n > 240)$, which is represented by the area shaded below.

$$P(n > 240) \approx P\left(z > \frac{240 - 229.5}{5.87}\right)$$
$$\approx P(z > 1.79)$$
$$\approx 1 - 0.9633$$
$$= 0.0367$$

So the flight will be overbooked about 3.7% of the times it departs. If the flight operates once a day except on Sundays, company officials should expect it to be overbooked about once a month.

Note how much simpler the calculation is for the normal approximation than is the calculation for the exact answer using the binomial distribution. The exact answer would require calculating a sum of 30 terms:

$$_{270}C_{241}(.85)^{241}(.15)^{29} + {}_{270}C_{242}(.85)^{242}(.15)^{28} + \ldots + {}_{270}C_{270}(.85)^{270}(.15)^{0}.$$

If you need very accurate values for binomial probabilities, and a normal distribution approximation is not appropriate, try to use a statistical package or a computer program for the calculations.

Questions

Covering the Reading

In 1–3, consider equations of the form $f(x) = \dfrac{1}{s\sqrt{2\pi}} e^{-\frac{1}{2}\left(\frac{x-m}{s}\right)^2}$, where m and s are constants and $s \neq 0$.

1. What name is given to the graph of such an equation? **normal curve**

2. If x represents a random variable, and $f(x)$ its probability, identify the statistical measure represented by the constant.
   a. m **mean**
   b. s **standard deviation**

3. Sketch by hand a graph of $y = f(x)$ where $m = 6$ and $s = 2$.
   See margin.

4. Suppose a variable x is normally distributed with mean 10 and standard deviation 3. If the transformation $x \to \dfrac{x - 10}{3}$ is applied to each data point, what type of distribution results?   **standard normal**

In 5–7, use the fact that the heights of adult men in the United States are approximately normally distributed with mean 70″ and standard deviation 3″.

5. What proportion of men are shorter than five feet, nine inches tall? $\approx$ **.37**

6. What is the median height?   **70″**

7. How many of a group of 1200 male employees of a company can the basketball coach of the company team expect to be taller than six feet, six inches? $\approx$ **4 or 5**

8. Refer to Example 2. What is the probability a randomly chosen refill will be
   a. too large;   **.0228**
   b. too small?   **.0668**

9. Suppose that the lifetime t (in hours) of a torch battery is normally distributed with mean 100 hours and standard deviation 4 hours.
   a. Transform t to a variable z following the standard normal distribution.
   b. What are the mean and standard deviation of the distribution of z-scores?   **a)** $z = \dfrac{t - 100}{4}$ **; b) mean = 0; s.d. = 1**

10. Refer to Example 3. Find the probability that the flight will be overbooked if 280 reservations are accepted. $\approx$ **.37**

11. Under what conditions may a binomial distribution be approximated by a normal distribution?   **if np and nq are each ≥ 5**

12. A binomial distribution with $p = 0.15$ and $n = 80$ can be approximated by the normal distribution with mean __**a.**__ and standard deviation __**b.**__.
    a) 12; b) $\sqrt{10.2} \approx 3.19$

13. State one advantage of using a normal approximation for a binomial distribution.   **Calculations are simpler.**

NOTES ON QUESTIONS
Questions 14 and 15:
Point out that it is very possible that the distribution is not normal. There are other distributions studied in statistics that might apply here. Statisticians would take the data and see what kind of distribution gives a good fit before making an assumption like the one used here.

Question 16: In contrast to **Questions 14 and 15,** these test scores are scaled so that a normal distribution fits the test scores distribution well.

ADDITIONAL ANSWERS
17.a. $f(x) =$

$$\frac{1}{2\sqrt{2\pi}} e^{-\frac{1}{2}\left(\frac{x-8}{2}\right)^2}$$

b.,c.

d. $g(x) =$

$$\frac{1}{6\sqrt{2\pi}} e^{-\frac{1}{2}\left(\frac{x-8}{6}\right)^2}$$

e. The maximum height of the first curve, ≈ .20, is about 3 times as great as the maximum height of the second curve, ≈ .066.

f. The area under each is 1.

Applying the Mathematics

In 14 and 15, assume that the time for a certain surgical incision to heal is normally distributed with a mean of 150 hours and a standard deviation of 20 hours.

14. What proportion of incisions should heal within a week?   **.8159**

15. If patients must stay in the hospital until their wounds heal, what is the probability that a randomly chosen patient will stay ten or more days? **almost 0**

16. In March 1990, the scaled scores on the SAT verbal section had mean 424 and standard deviation 111. Assume that the scaled scores are normally distributed. What is the probability that a randomly selected student had an SAT verbal score between 500 and 600?   **≈ .19**

17. **a.** Write an equation for a normal distribution with mean 8 and standard deviation 2.
   b. Graph the equation using an automatic grapher.
   c. On the same set of axes, graph a distribution with the same mean and three times the standard deviation.
   d. Write an equation for the curve drawn in part **c**.
   e. How do the maximum heights of the two curves compare?
   f. How do the areas under the two curves compare?
   See margin.

Review

18. Suppose the random variable z has a standard normal distribution. Determine
   a. $P(z \geq 2.67)$;   **.0038**
   b. a number n such that $P(1 < z \leq n) = .1549$.   **$n = 2.67$**
   (Lesson 10-5)

19. Complete: In a standard normal distribution, 99% of the observations are within __?__ standard deviations of the mean.   *(Lesson 10-5)*   **2.58**

20. *True* or *false*.   In an experiment for which the frequencies of outcomes is approximated by $f(x) = ke^{-x^2}$, the mean of the distribution of outcomes x is 0.   *(Lesson 10-4)*   **True**

21. The graph of $f(x) = \frac{1}{\sqrt{2\pi}} e^{-\frac{x^2}{2}}$ undergoes the scale change

$$S:(x, y) \rightarrow \left(\sqrt{2}x, \frac{1}{\sqrt{2\pi}} y\right).$$

   Determine the area between the image curve and the x-axis.
   (Lesson 10-4)   **$1/\sqrt{\pi}$ unit² ≈ .56 unit²**

22. How many ways are there of selecting a committee of five faculty and two students from a group of 20 faculty and 300 students?   *(Lesson 8-6)*
   (15,504)(44,850) = 695,354,400

23. A population of 300 viral cells grows by a factor of 2.4 every hour. Express the population p as a function of the number h of hours passed.   *(Lesson 4-4)*   **$p = 300 \cdot (2.4)^h$**

632

In 24 and 25, suppose a spacecraft system is composed of three independent subsystems X, Y, and Z, and the probability that these will fail during a mission is 0.002, 0.006, and 0.003, respectively. *(Lessons 7-2, 7-3, 7-4)*

24. If the three are connected in series, as below, a failure in any one of the three will lead to a failure in the whole system.

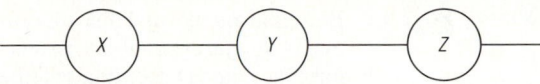

What is the probability that the system will be reliable—that is, it will not fail? **≈ .989**

25. Suppose the subsystems are connected as shown at the right. In this case, both X and one of Y and Z must be reliable for the whole system to be reliable. What is the probability that this whole system will not fail? **≈ .998**

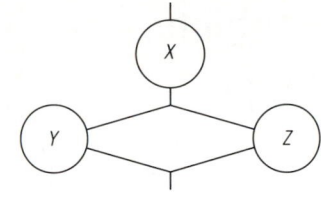

26. Refer to the function graphed at the right.
   a. Sketch the inverse of f.
   b. Find $f^{-1}(-1)$. *(Lesson 6-5)*
a) See margin. b) π

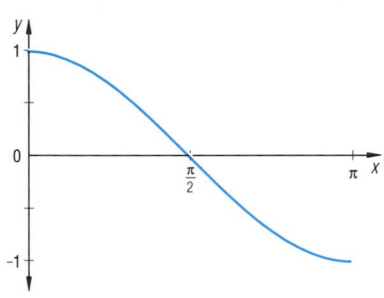

27. *Skill sequence.* Each circle has radius 10 m. Find the shaded area. *(Lesson 5-2, Previous course)*

a.

50π ≈ 157 m²

b.
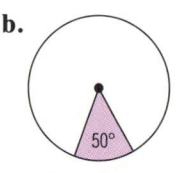
$\frac{125}{9}\pi \approx 43.6$ m²

c.
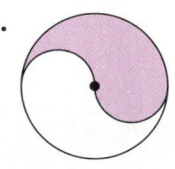
50π ≈ 157 m²

Exploration

28. Some books give the following rule of thumb for finding the standard deviation of a normally distributed variable:

$$s \approx \frac{\text{range}}{6}.$$

Explain why this rule of thumb works.

Sample: This rule works because for a normally distributed variable, virtually all data values fall within 3 standard deviations of the mean. Therefore the complete range is equal to about 6 standard deviations, from 3 below the mean to 3 above the mean.

FOLLOW-UP

MORE PRACTICE
For more questions on SPUR Objectives, use *Lesson Master 10-6*, shown below.

PROJECTS
Projects 2, 4, and 7 on pages 654–655 are related to the content of this lesson.

EVALUATION
A quiz covering Lessons 10-4 through 10-6 is provided in the Teacher's Resource File.

26.a.

NAME _____

LESSON **MASTER 10-6**
QUESTIONS ON SPUR OBJECTIVES

■**PROPERTIES** *Objective D (See pages 658–662 for objectives.)*

1. a. Write an equation for a normal distribution with $\mu = 40$ and $\sigma = 4$. $f(x) = \frac{1}{4\sqrt{2\pi}} e^{-\frac{1}{2}\left(\frac{x-40}{4}\right)^2}$

  b. What is the maximum value of the function in part a?   ≈ .0997

2. What is the median value in a normal distribution with mean 15 and standard deviation 1.6?   15

■**USES** *Objective F*
In 3 and 4, assume that the lifespan of a certain type of light bulb is normally distributed with a mean of 200 hours and a standard deviation of 15 hours.

3. What is the probability that a light bulb of this type will last up to 220 hours?   ≈ .91

4. What is the probability that a light bulb of this type will last no more than 188 hours?   ≈ .21

■**REPRESENTATIONS** *Objective J*

5. Use an automatic grapher to sketch the graph of a normal distribution with $m = 10$ and $s = 3$.

6. About 68% of the area under the curve (and above the x-axis) in Question 5 is between what values of x?   7 < x < 13

110   *Functions, Statistics, and Trigonometry © Scott, Foresman and Company*

OBJECTIVE

G Use binomial and normal distributions to test hypotheses.

TEACHING NOTES

Example 1 is of a type students have seen before, but the new idea is the situation in which it is presented. Before **Example 1** we do not know whether the coin is fair, so we do **Example 1** as a hypothetical case: If the coin were fair, what would happen? Then, after the example, we question that hypothesis because the probability of the event is so small.

It will be helpful to go through the logic with students and connect the reasoning with the steps in **Example 2:** An event occurs. We make assumptions such as randomness or normal distribution (Step 1) in order to determine the probability that the event occurred (Step 3). We assign a significance level (Step 2) below which we consider the event to be unusual. If the probability for the event is below the significance level, then we reject the hypothesis that our assumptions were correct (Step 4).

LESSON

10-7

Using Probability to Make Judgments

Binomial and normal distributions are often used to help people make judgments or *inferences* about issues. The two most important uses of statistical inference are (1) deciding whether data support a particular statement, and (2) estimating a characteristic of a population from what is known about a sample. The former is studied in this lesson; the latter in the following two lessons.

Consider the following case. In 1986, two candidates for a political office in Illinois got exactly the same number of votes. Following state law, the county clerk tossed a coin to determine the winner. Clearly, before flipping the coin the people involved needed to know that the coin used to pick the winner was fair.

Assume as a test of fairness you toss a coin 40 times. If it comes up heads 29 times, is it a fair coin?

Some people are surprised to learn that mathematics cannot be used to provide a definitive "yes-or-no" answer to this question. Instead, mathematics provides only information about the probability of fairness.

Here the expected number of heads from a fair coin is $40 \cdot \frac{1}{2} = 20$. So a reasonable question to ask is: if a fair coin is tossed 40 times, what is the probability that the number of heads differs by 9 or more from the expected value of 20 heads?

Example 1

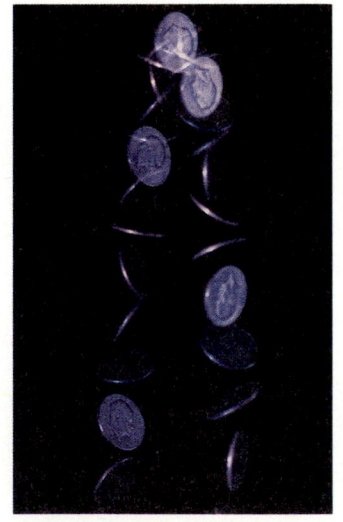

Calculate the probability that in 40 tosses of a fair coin, the number of heads is either less than or equal to 11 or greater than or equal to 29.

Solution Let x = the number of heads in 40 tosses of a fair coin. Because of the symmetry of the binomial distribution with $p = 0.5$, $P(x \le 11 \text{ or } x \ge 29) = 2P(x \ge 29)$. Because both np and nq are greater than 5, a normal approximation can be used to estimate the probability. Note that $\mu = np = 20$ and $\sigma = \sqrt{npq} = \sqrt{10}$. So the variable $z = \frac{x - 20}{\sqrt{10}}$ has a standard normal distribution. The z-score of interest here is $z = \frac{29 - 20}{\sqrt{10}} \approx 2.85$. The Standard Normal Distribution Table indicates that

$$P(z < 2.85) \approx 0.9978.$$
So $$P(z \ge 2.85) \approx 1 - 0.9978$$
$$= 0.0022.$$
Thus $$P(x \le 11 \text{ or } x \ge 29) \approx 2(0.0022)$$
$$= 0.0044.$$

So when a fair coin is tossed 40 times, the probability of getting any number of heads that is 9 or more away from the expected value of 20 is about 0.0044; that is less than 1 chance in 200.

634

Is a 1-in-200 chance a small enough probability to conclude that the coin is not fair? That depends on the situation. Here the candidates might wish to see the coin tossed more times before agreeing to its use.

Researchers in different disciplines and for different sorts of studies choose different values for the probability levels they consider minimally acceptable for making judgments. Common values are 0.10, 0.05, and 0.01. Such a value is called a **significance level** for an experiment. For example, to say that an experiment is *significant at the 0.05 level* means that the probability of the given results, or more extreme ones, being strictly due to chance is less than 0.05.

Significance levels are sometimes set before an experiment to avoid debate after the fact. For example, if the significance level 0.01 had been agreed to before testing the coin, and 29 heads appeared in 40 tosses, both candidates would have had to reject the assumption that the coin is fair.

Such reasoning is an example of statistical *hypothesis testing*. Sometimes hypothesis tests are expressed in terms of a *null hypothesis* and an *alternative hypothesis*.

Consider again the example of flipping a coin 40 times to determine if it was fair. A coin was hypothesized as being unbiased. Such an assumption is called a null hypothesis, abbreviated H_0 and pronounced "H-nought." The word "null" suggests that the coin is not out of the ordinary. Usually, a **null hypothesis** is a statement of "no effect" or "no difference." A null hypothesis is usually contrasted with an **alternative hypothesis** H_1 which identifies an alternative conclusion about the same event. An alternative hypothesis is usually the complement of the null hypothesis. For the coin flip, the null hypothesis and an alternative hypothesis are

H_0: the probability that the coin lands heads is 0.5.

H_1: the probability that the coin lands heads is not 0.5.

Another alternative hypothesis is

H_2: the probability that the coin lands heads is greater than 0.5.

The choice of null and alternative hypotheses is subjective. But once the choices are made, probability theory and statistical inference are used to determine which of the two hypotheses is more reasonable.

■ ■ ■ ■ ■ ■ ■ ■

Example 2   A small office has 3 male and 5 female employees. Of 3 people promoted, only 1 is a female. From this information, can you conclude that the office discriminates against females?

Solution
Step 1: State the hypotheses.
   H_0: The probability of being promoted was equal for each person.
   H_1: The probability of being promoted was not equal for each person.

Step 2: Set a significance level.
   We use 0.05, a common level.

The traditional notation of stating null and alternative hypotheses is included here not just for formality, but because students who continue studying statistics are likely to see it. In practice, however, more and more statistical analyses are being written with more prose and less technical symbolism.

Alternative Assessment
Have your students write out some problems both ways—with symbols and with words. Their words will provide some insight into how well they are understanding the concepts.

Students may be surprised by the conclusion in **Example 2.** You may wish to poll the class before doing the example to see what their intuitive feelings are about whether the office discriminates against females.

Step 3: Calculate probabilities.

What is the probability that when 3 people are randomly selected from the 8 employees, no more than 1 is female? That is, find the probability of choosing 2 males and 1 female, or choosing 3 males from this office, based completely on chance.

P(2 male, 1 female) =

$$\frac{\text{number of choices of 2 males from 3 and 1 female from 5}}{\text{number of choices of 3 people from 8}}$$

$$= \frac{{}_3C_2 \cdot {}_5C_1}{{}_8C_3} = \frac{3 \cdot 5}{56} = \frac{15}{56}$$

P(3 male) $\quad = \frac{{}_3C_3 \cdot {}_5C_0}{{}_8C_3} = \frac{1 \cdot 1}{56} = \frac{1}{56}$

So the total probability of 1 or no females being promoted based on random selection is

$$\frac{16}{56} \approx 0.29.$$

Step 4: Draw conclusions.

Because $0.29 \geq 0.05$, the evidence does not contradict the null hypothesis. If the promotions happened by chance alone, there is a 29% probability that exactly one female would be promoted, so from this information alone a person is not justified in claiming gender discrimination.

Note carefully the conclusions drawn in Examples 1 and 2. Neither the null nor the alternative hypothesis is ever proven exactly. Instead, one or the other is supported by a probability level. Unlike deductive proof, hypothesis testing merely lets you be more or less confident in your conclusions—never certain.

Questions

Covering the Reading

1. Suppose a fair coin is tossed 50 times, and 30 heads occur.
   a. Find the following probabilities.
      i. $P(H = 30)$ ≈ .042
      ii. $P(H \leq 30)$ ≈ .9207
   b. If a 0.05 significance level is used to test the null hypothesis that the coin is fair, should the null hypothesis be accepted? **Yes**

2. What does it mean to say an experiment is significant at the 0.01 level?
   The probability that the results are due to chance is less than .01.

In 3 and 4, consider an office in which five of its employees (10 females and 10 males) receive large bonuses. Test the hypothesis that the office discriminates (assuming all employees have an equal chance of receiving bonuses under the stated conditions). Use the 0.05 significance level.

3. All five people getting bonuses are female. P(5 female) = .016; using the .05 level, the evidence supports the contention that the office discriminates.

4. One of the five people getting a bonus is female.
   P(1 female, 4 male) = .135; using the .05 level, the evidence does not support the contention that the office discriminates.

636

In 5 and 6, state an appropriate null hypothesis H_0 and an alternative hypothesis H_1.

5. A new car averages 30 miles per gallon on the highway. The owner switches to a new motor oil that claims to increase gas mileage. After driving 2000 miles with the new oil, the owner wants to determine if gas mileage has actually increased. **H_0: Gas mileage is not increased with the new oil. H_1: Gas mileage is increased with the new oil.**

6. A bag of kitty litter is supposed to contain 10 lbs. A consumer group thinks that the manufacturer is consistently filling the bags with less than the labeled amount. They weigh some bags.
H_0: The mean weight is 10 lb. H_1: The mean weight is less than 10 lb.

Applying the Mathematics

7. A quality control inspector in a factory accepts a shipment of parts from a supplier if she finds no more than 5 faulty parts in a random selection of 100.
   a. Is it possible for her to accept a shipment that has more than 5% bad parts? **Yes**
   b. Is it possible for her to reject a shipment that has fewer than 5% bad parts? **Yes**

8. An audio cassette manufacturing company claims that no more than $\frac{1}{2}\%$ of their cassettes are defective. A consumer advocate tests 2000 of the company's cassettes and finds 16 defective ones. Using the 1% significance level, is the advocate justified in declaring the company fraudulent? **No; P(16 or more defective) $\approx .0287 > .01$.**

9. A baseball player had a career batting average of .221 through last season and has made 11 hits in 30 times at bat so far this season. Using a 0.05 significance level, is the batter doing better this season? **Yes; P(11 or more hits) $\approx .027 < .05$.**

10. Researchers testing new drugs often start with the assumption (null hypothesis) that the drugs do not work, and then try to disprove their assumption. Consider an illness which has a natural recovery rate of about 35%. Suppose 183 patients suffering from the illness are treated with a new drug developed to cure that illness. Determine the smallest number of recoveries necessary for the company to conclude with 95% confidence that the drug is effective. **75**

Review

11. In any normal distribution, what percent of the area under the curve is within two standard deviations of the mean? *(Lesson 10-6)*
0.9544 or about 95%

12. Suppose that an airline knows that when given the choice of chicken Kiev or pasta primavera, about 65% of airline passengers prefer chicken and 35% prefer pasta. Suppose also that the capacity of an aircraft is 240 passengers. If the airline wants to be able to meet everyone's choice of meal on at least 95% of the flights, what is
   a. the minimum number of chicken dinners and **169**
   b. the minimum number of pasta dinners they should load on the aircraft? *(Lesson 10-6)* **97**

13. The upper 1% of the area under the standard normal curve falls above what z-score? *(Lesson 10-5)* **2.33**

LESSON 10-7 *Using Probability to Make Judgments* **637**

MORE PRACTICE
For more questions on SPUR Objectives, use *Lesson Master 10-7*, shown on page 637.

EXTENSION
Small Group Work
Groups of students can write problems from their own experience that are similar to **Questions 7–10.**

PROJECTS
Projects 4 and 6 on page 655 are related to the content of this lesson.

14. Write a BASIC statement which will generate random integers between 3 and 18, inclusive. *(Lesson 7-8)*   **INT (RND(1)*16) + 3**

15. In a lottery, the net gain on each ticket is the difference between the cost of the ticket and the amount won with that ticket. Let the net gain for an entrant on a ticket be x dollars. Here is a probability distribution for a state lottery which has a first prize of $1,000,000, five second prizes of $100,000 each, and in which 3,000,000 tickets are bought.

x	-1	99,999	999,999	
$P(x)$	$\frac{2,999,994}{3,000,000}$	$\frac{5}{3,000,000}$	$\frac{1}{3,000,000}$	*(Lesson 7-7)*

   a. What is the cost of a ticket?   **1 dollar**
   b. Verify that this is a probability distribution.
   c. Find the expected value of a ticket.   **negative 50 cents**
   d. What would you expect to win if you bought 10 tickets?
   e. How much will the state make if 5,000,000 tickets are sold?
   b) Σ P(x), each ≥ 0, is 1. d) negative 5 dollars; e) $3.5 million

16. Of a penny, nickel, dime, one dollar bill, and five dollar bill, one coin and one bill are selected at random.
   a. Identify the sample space for this experiment.
   b. Find the probability that the total amount of money is more than $1.05. *(Lesson 7-1)*   **2/3**
   a) {P\$, P5, N\$, N5, D\$, D5}

In 17–19, consider $\triangle FUR$. *True* or *false*.

17. If $\angle F$ is a right angle, then $\sin F = \frac{u}{r}$. *(Lesson 5-3)*   **False**

18. $\frac{\sin F}{f} = \frac{\sin R}{r} = \frac{\sin U}{u}$ *(Lesson 5-9)*   **True**

19. $f^2 = r^2 + u^2 + 2ur \cos F$ *(Lesson 5-8)*   **False**

20. In the table at the right, a manufacturer of potato chips reported producing the number of bags per day (BPD) of potato chips.

Year	BPD
1980	200
1982	1000
1984	3000
1989	80000

   a. Which of the following—linear, quadratic, or exponential—seems to be the best model for this firm's growth in production?   **exponential**
   b. Find an equation that relates BPD to the year after 1980.
   c. Use the model from part **b** to predict the following:
      i. the BPD in 1992   **≈ 500,000**
      ii. the year in which BPD first exceeds 1,000,000.   **1994**
      (Lessons 2-2, 2-6, 4-9)   **b) sample: BPD = 225 · 1.9ˣ**

Exploration

21. Flip a coin 100 times and record the number of heads that come up. Test the hypothesis that the coin you used is fair.   **Answers will vary.**

638

10-8

Sampling Distributions and the Central Limit Theorem

As you know, it is often too costly, too difficult, or even impossible to study an entire population. Usually a sample is taken. Then inferences about the population are made from the sample data. However, even if a random sample is taken, the characteristics of the sample are not likely to be identical to the corresponding characteristics of the population. For instance, the means of various samples from the same population are likely to differ somewhat from each other, and from the population mean. Fortunately, if samples are selected randomly, probability and statistics can be used to predict, within preset significance levels, characteristics of the population.

In order to understand how measures of a sample are related to the corresponding measures of the population from which it is taken, it is helpful to study how various samples drawn from the same population vary.

Consider a simple case: the uniform distribution of random digits from 0 through 9. Using the definition of the mean of a probability distribution, the population mean is $\mu = \frac{1}{10}(0 + 1 + 2 + 3 + 4 + 5 + 6 + 7 + 8 + 9) = 4.5$. If repeated samples of these random digits are taken, and the mean of each sample is calculated, what are some of the characteristics of the distribution of sample means?

▪ ▪ ▪ ▪ ▪ ▪ ▪ ▪ ▪▪

Example 1 Choose 14 samples, each of size 5, from the uniform distribution of integers between 0 and 9. Find

a. the mean of each sample;

b. the mean of the distribution of sample means.

Solution Use a computer, calculator, or table of random numbers to simulate selecting samples of size 5 from a uniform distribution of integers between 0 and 9. For instance, in the table of random numbers in Appendix C choose any row, and read off blocks of 5 digits at a time. We use row 27.

$$29676 \quad 20591 \quad 68086 \ldots$$

a. The mean of the first sample is $\frac{2+9+6+7+6}{5} = 6$.

The means of the 14 samples are

6, 3.4, 5.6, 3.4, 4, 4.6, 7.6, 4.6, 5.8, 4.6, 5.4, 3, 4.4, 3.4.

b. The mean of the 14 means above is 4.7.

Note that in Example 1 above, two types of means were calculated—the means of individual samples, and the mean of these means. We have used $\bar{x}$ to stand for the mean of a sample. We shall use M to stand for the mean of a set of sample means. As you will see later, under certain conditions both $\bar{x}$ and M approximate the population mean μ.

The probability distribution for random digits from 0 to 9 is shown at the left below. At the right is the distribution of Example 1's 14 means. When the samples are all the same size, such a distribution of means is called a *sampling distribution*. Note that although the original population has a uniform distribution, the sample means do not. However, the population mean $\mu = 4.5$, and the mean of the distribution of sample means $M = 4.7$, are almost equal.

Below is a relative frequency distribution of sample means from 1000 random samples of size 5 from the digits between 0 and 9. Note that the distribution of sample means is clearly not uniform. It approximates a bell-shaped curve, with the mean M of the distribution about 4.5 .

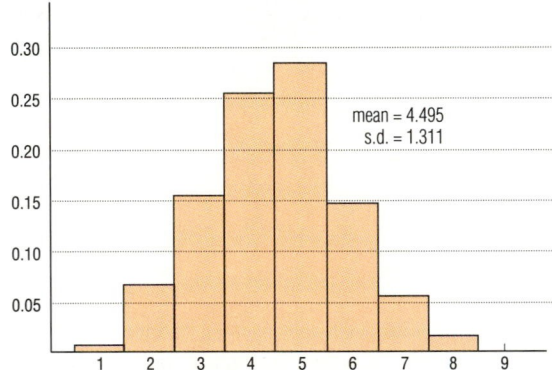

The work with samples drawn from a uniform distribution can be generalized to populations with other distributions. If a large number of random samples of sufficient and equal size are drawn from *any* population, the distribution of sample means approaches a normal distribution, and the mean M of the sample means approaches the mean μ of the population.

As the size n of the individual samples increases, another pattern emerges.

Example 2 Simulate drawing the following samples from the uniform distribution of digits from 0 to 9. Find the mean of each sample, and the mean and standard deviation of each distribution of sample means.
  a. 1000 samples of size 15
  b. 1000 samples of size 100

640

Solution Simulations of this size are best done by computer or programmable calculator. Below is the output from a statistics package.

a. 1000 sample means ($n = 15$) gave $M = 4.494$.

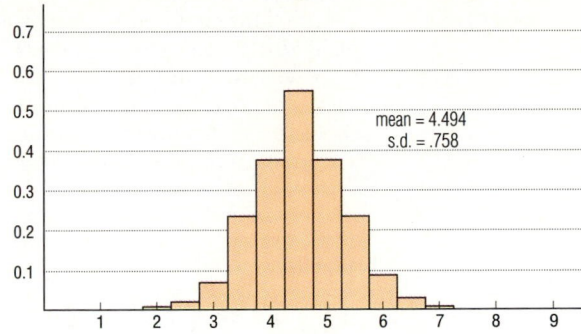

mean = 4.494
s.d. = .758

b. 1000 sample means ($n = 100$) gave $M = 4.5031$.

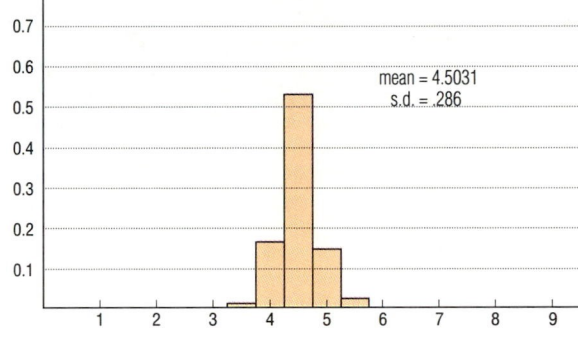

mean = 4.5031
s.d. = .286

Note that as the size n of the sample increases, M does not vary much, but the standard deviation s of the distribution of sample means decreases. Surprisingly, the relations between n, s, and the population standard deviation σ, first proved by Laplace, are quite simple. They are summarized in the *Central Limit Theorem*.

> **The Central Limit Theorem (CLT)**
>
> Suppose a large number of random samples of size n are selected from any population with mean μ and standard deviation σ.
> Then, as n increases,
> 1. the mean M of the distribution of sample means approaches μ;
> 2. the standard deviation s of the distribution of sample means approaches $\frac{\sigma}{\sqrt{n}}$; and
> 3. the distribution of sample means approaches a normal curve.

641

The Central Limit Theorem is *central* because it deals with the mean of a population, a measure of its *center*. It is a *limit* theorem because it deals with what can be expected as samples of larger and larger size are involved. Note that the Central Limit Theorem states that even if the underlying population distribution is not normal, the means of samples of sufficient size will be normally distributed. However, the Central Limit Theorem does not specify exactly what that sufficient size is. Fortunately, a corollary to the Central Limit Theorem provides a rule of thumb for determining how large the samples should be so that the distribution of sample means is close enough to normal. The proof of this corollary is beyond the scope of this course.

Corollary

Consider a population with mean μ and standard deviation σ from which random samples of size n are taken. Then the distribution of sample means is approximately normal with mean $M \approx \mu$ and standard deviation $s \approx \dfrac{\sigma}{\sqrt{n}}$ whenever one of the following occurs:

a. the population itself is normally distributed, or

b. the sample size $n \geq 30$.

The Central Limit Theorem and its corollary allow you to predict answers about the sample means for "large" samples even when the population is not normal.

Example 3 The graph below represents the probability distribution for the variable x, the amount of time needed by a technician to perform routine maintenance on a copy machine. Suppose this population distribution is known to have $\mu = 1$ hour and $\sigma = 1$ hour. If the company services 50 such copying machines, what is the probability that on repeated service calls the average maintenance time for the 50 machines exceeds 45 minutes?

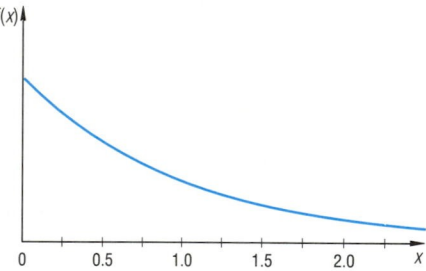

Solution Let $\bar{x}$ equal the mean maintenance time on 50 copy machines. This can be considered as the mean of the maintenance times of a sample of 50 randomly selected machines. The Central Limit Theorem and its corollary imply that $\bar{x}$ is approximately normally distributed with mean $M = 1$ and standard deviation $\frac{\sigma}{\sqrt{n}} = \frac{1}{\sqrt{50}} \approx 0.14$. Thus, the variable $z = \frac{\bar{x} - 1}{0.14}$ has a standard normal distribution. Because 45 minutes is 0.75 hours, the probability that the average maintenance time exceeds 45 minutes is

$$P(\bar{x} > 0.75) \approx P\left(\frac{\bar{x} - 1}{0.14} > \frac{0.75 - 1}{0.14}\right)$$
$$\approx P(z > -1.77)$$
$$= P(z < 1.77)$$
$$\approx 0.9616.$$

Check In a standard normal distribution, about 95% of the area under the curve is within 2 standard deviations of the mean 0. It seems reasonable that about 96% of the area is to the right of $z > -1.77$.

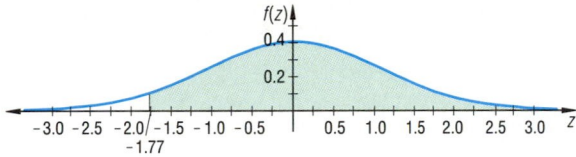

Notice that in Example 3 the mean μ and standard deviation σ of the population were known, and from them questions about the sample mean $\bar{x}$ and its probability distribution were answered. In Lesson 10-9 you will read about predicting μ and σ from $\bar{x}$ and s.

Questions

Covering the Reading

In 1 and 2, consider the uniform distribution of integers from 0 to 9.
  a. Choose random samples of the indicated size.
  b. Find the mean of each sample.
  c. Plot the distribution of sample means.
  d. Find the mean of the distribution of sample means.

1. 10 random samples each of size 5   **a-d) Answers may vary.**

2. 100 random samples each of size 5   **a-d) Answers may vary.**

NOTES ON QUESTIONS
Question 1: This question can be done by hand.

Technology for Question 2: A computer or programmable calculator is needed for that many random samples.

LESSON 10-8 Sampling Distributions and the Central Limit Theorem   **643**

In 3–5, consider a population from which a very large number of random samples is taken. *True* or *false*.

3. The mean of the sample means approximates the population mean.
   True

4. The standard deviation of the sample means approximates the standard deviation of the population.   **False**

5. The shape of the distribution of the sample means is the same as the shape of the population distribution. **False**

6. To what kinds of samples do the Central Limit Theorem and its corollary apply?   **random samples**

7. The distribution of heights of average young adult women in the United States is approximately normal with $\mu = 63.7$ inches and $\sigma = 2.6$ inches. A medical study records the means of the heights of random samples of 100 of these women. Estimate the mean and standard deviation of the numbers recorded.   **mean ≈ 63.7 in., s.d. ≈ 0.26 in.**

In 8 and 9, refer to Example 3.

8. Give **a.** the mean, **b.** the standard deviation and **c.** the shape of the probability distribution for the average amount of time needed by a technician to perform maintenance on a random sample of 50 such copy machines.   **a) 1 hour or 60 minutes; b) 0.14 hours or 8.4 minutes; c) approximately normal**

9. Suppose another firm records data for maintenance on a random sample of 100 such machines. Find the probability that the average maintenance time for these machines exceeds 45 minutes.   **≈ .9938**

Applying the Mathematics

In 10–12, the relative frequency distributions of 100 sample means from the same population are given. Match each graph to the most likely sample size:
   (a) 2   (b) 15   (c) 40

10.

(c)

11.

(a)

12.

(b)

13. Refer to the graphs in Questions 10–12 above. Was the population from which the samples were drawn normal? How can you tell?
No; the distribution for small samples is skewed.

14. What advantage do larger sample sizes give in estimating population means? **The standard deviation for the distribution of sample means is smaller, giving a more precise estimate.**

In 15 and 16, in a recent year the scores of students on the ACT college entrance exam had a normal distribution with $\mu = 18.6$ and $\sigma = 5.9$.

15. What is the probability that a randomly chosen student from the population who took the exam has a score of 22 or higher? **≈ .28**

16. The average score of the 84 students at Acme High who took the ACT that year was $\bar{x} = 20.6$. What is the probability that the mean score for 84 students selected randomly from all who took the test nationally is 20.6 or higher? **almost zero**

Review

17. A researcher is testing for color preference of babies. There are ten balls of identical size in a box: five are red, five are blue. A baby is observed picking up four balls from the box. The baby chooses three red balls and a blue one.
  a. State a null hypothesis H_0 and an alternative hypothesis H_1 for this experiment.
  b. Determine which hypothesis should be accepted at the .10 significance level. *(Lesson 10-7)* **a,b) See margin.**

18. In 1987 the average annual pay for a city employee (excluding the education sector) in the U.S. was about $25,500, with a standard deviation of $1500. A citizen's group in one city claims the bus drivers are overpaid compared to other city employees in the U.S. To support this claim, they point to a bus driver who makes $28,500 annually. Assuming that salaries are normally distributed, is the group justified to make the statement at the significance level of
  a. 0.05; **Yes**   **b.** 0.01? *(Lesson 10-7)* **No**
  $P(\text{salary} > 28{,}500) \approx .0228$

19. An electrical company estimates the lifetime of an automatic dishwasher to be approximately normally distributed with mean ten years and standard deviation ten months, and guarantees to make spare parts available to the original purchaser within twelve years of the purchase date. What proportion of dishwashers will need no repairs before the guarantee expires? *(Lesson 10-6)* **≈ .0082 or about 0.8%**

NOTES ON QUESTIONS
Question 15: Have students compare this question with **Question 16.**

Question 16: The mean of 20.6 is over 3 standard deviations above what would be expected for the mean of a sample of this size taken from the entire ACT population. The probability of this occurring is less than .01. Thus we might infer that the students from Acme High were better-scoring students than a similar group chosen randomly from the national population.

ADDITIONAL ANSWERS
17.a. H_0: The baby has no color preference between red and blue. H_1: The baby prefers red to blue.

b. P(3 or 4 red balls) ≈ .26, so we cannot reject H_0. There is no evidence to support a claim of preference for red balls.

ADDITIONAL ANSWERS
23.a.,c.

b. sample: y = -1.476x² + 20.81x + 4.29

d. Sample: Weather is generally periodic; a sine curve is more appropriate.

The Performing Arts Center, Nashville

20. Consider the standard normal distribution for variable z. Determine the probability that z is between -1.65 and 1.65. *(Lesson 10-5)* $\approx .901$

21. *True* or *false*.   Given any normally distributed variable, its mean, median, and mode are equal. *(Lesson 10-4)* **True**

22. Use grouping to factor $6a^2 - 10a + 3ab - 5b$. *(Lesson 9-9)*
(2a + b)(3a − 5)

23. The following are average monthly temperatures in °F for Nashville, TN.

Jan.	Feb.	Mar.	Apr.	May	June	July	Aug.	Sept.	Oct.	Nov.	Dec.
37	40	49	60	68	76	79	78	72	60	49	41

a. Draw a scatter plot of this data, using x = month of the year.
b. Pick three points and determine a quadratic function to model the data.
c. Graph the function you found in part **b** on the same axes with the scatter plot.
d. Explain why a quadratic function is probably not the best choice for modeling data on weather. *(Lessons 1-3, 2-7, 6-4)* **See margin.**

24. Solve $z = \dfrac{x - m}{s}$ for x. *(Previous course)* **x = zs + m**

Exploration

25. Using this text, randomly choose a page and then 20 consecutive sentences on that page. Count the number of words in each sentence. Then calculate the mean of these counts. Repeat this process ten times, with different pages.
a. Determine the mean M and standard deviation s of the ten means.
b. Based on the Central Limit Theorem, what can you predict about the mean and standard deviation of the distribution of the number of words in the sentences in this book? **a,b,d) Answers may vary.**
c. State one assumption which allows the use of the CLT in making the prediction in part **b**.
d. Discuss how different (if at all) the results would be for parts **a** and **b** if you used a
   i. physics textbook; **ii.** novel; **iii.** computer software manual.
   c) The number of words per sentence is normally distributed.

646

10-9

Confidence and Cautions in Statistical Reasoning

Consider the following situation. In Ohio in 1989–90, the scores of the seniors taking the mathematics section of the Scholastic Aptitude Test (the SAT-M) had a mean of 499 and a standard deviation of 119. Only about 23.5% of seniors in that state took the SAT-M that year, and there are two main reasons why this group is not a random sample of all seniors—it is self-selected, and virtually all are college-bound.

Suppose that you want to know how *all* high school seniors in Ohio might do on the SAT-M. After considerable effort and expense, you give the SAT-M to a random sample of 400 high school seniors and find that the mean of this sample is 440. What can you say about the mean on the SAT-M if all seniors in Ohio took this test? That is, what does your sample mean predict about the population mean?

Clearly, another sample of 400 students would be likely not to yield a mean of 440 again. By the Central Limit Theorem, in repeated samples of size 400, the sample means are approximately normally distributed with mean $M \approx \mu$ and standard deviation $s \approx \dfrac{\sigma}{\sqrt{n}}$, where μ and σ are the mean and standard deviation of the population you are trying to describe.

With no other information available, the "best guess" is that the population mean is the sample mean and that the variation among the SAT-M scores of the entire population of seniors in the state of Ohio is equal to the variation among the SAT-M scores of the seniors who took the test in 1990 (an unrealistic assumption, but one that you make, because without some value for σ the mathematics is a bit more complicated). That is, assume $\sigma = 119$. Thus, in repeated samples of 400 randomly chosen students the distribution should be approximately normal with $M \approx \mu \approx 440$ and $s \approx \dfrac{\sigma}{\sqrt{n}} = \dfrac{119}{\sqrt{400}} = \dfrac{119}{20} \approx 6$.

With this information and assumptions, you can make inferences about μ, the mean of the population. The reasoning goes as follows. You know that in a normal distribution about 68% of the data fall within one standard deviation of the mean. Thus the probability is about 0.68 that a sample mean $\bar{x}$ will be within 6 points of the population mean 440. So you can say that in about 68% of random samples of size 400 from the population of seniors, the mean will fall between $440 - 6$ and $440 + 6$.

The interval $440 - 6 \le \mu \le 440 + 6$, or $434 \le \mu \le 446$, is called the *68% confidence interval* for the population mean μ. The quantity ± 6 is called the *margin of error* for the confidence interval.

There have been books written on the misuse of statistics; perhaps the most famous is Darrell Huff's *How to Lie with Statistics*. We have purposefully not gone into great detail on this subject because we believe that if students understand the proper use of statistics, then they will tend to notice the misuses without instruction.

One usually tries to keep the confidence interval as small as possible, but to make it smaller requires a larger sample size. Because the sample size *n* varies inversely with the square of the width *A* of the confidence interval (see page 649), to halve the width requires four times the sample size. Thus if a sample of 1200 people gives an accuracy of about 3% with a confidence of 95%, then a sample of 4800 people would be required to give an accuracy of 1.5% with the same confidence.

Jerzy Neyman

Confidence intervals were invented in 1937 by Jerzy Neyman (1884–1981), a Polish mathematician who for many years was Professor of Mathematics at the University of California at Berkeley. Statisticians have invented techniques for finding confidence intervals for many different parameters based on various assumptions. But all confidence intervals share two properties: there is an interval of possible values constructed from the sample data, and a confidence level gives the probability that the method produces an interval which contains the true value of the parameter.

As is true when testing hypotheses, the user begins by choosing a confidence level. Usually it is at least 90%, and 95% and 99% are commonly used. For a 95% level for μ, you know that 95% of the sample means fall within two standard deviations of the population mean; for this situation, between 428 and 452. You can report with 95% confidence that the mean score of the population of seniors in Ohio on the SAT-M is between 428 and 452. Similarly, since 99% of the sample means fall within three standard deviations, the interval for a 99% level is $422 \leq \mu \leq 458$.

Compare these results with those found earlier. Notice that except for rounding, each interval is centered on $\bar{x} = 440$, the sample mean.

Also, as the level of confidence increased, so did the width of the confidence interval. This should make sense, because if you want to place a higher probability on your estimate, you need a wider estimate to "cover your bases."

What does this mean if you would like a "finer tuning" or closer estimate of the mean? You have two choices: choose a lower confidence level or use a larger sample. Generally, people who use statistics do not want to use a confidence level below 90%, so in order to get within a decreased margin of error, they must increase the sample size.

Surprisingly, it is relatively simple to figure out how large a sample size is needed to ensure a certain margin of error at a desired confidence level. Suppose you wanted to estimate the mean score of all seniors in Ohio on the SAT-M with 99% confidence to within 10 points. How large a sample size is necessary?

You know that the standard score

$$z = \frac{\bar{x} - \mu}{\frac{\sigma}{\sqrt{n}}},$$

648

and that for the confidence level to be 99%, z must equal 2.57. You want $|\bar{x} - \mu| = 10$. If you assume as we did earlier that $\sigma = 119$, you must find n so that

$$2.57 = \frac{10}{\frac{119}{\sqrt{n}}} \cdot$$

Rewriting the right side gives

$$2.57 = \frac{10\sqrt{n}}{119} \cdot$$

Solving this for n,

$$\frac{(2.57)\,(119)}{10} = \sqrt{n}\,,$$

or $\qquad\qquad n \approx 935.$

That is, to estimate the population to within 10 points at the 99% confidence level, you would have to select a random sample of nearly 1000 students. You as an experimenter would need to decide whether the increase in precision of measurement is worth the increased cost of selecting and testing this much larger sample.

In general, to find the sample size needed so that you can estimate a mean μ for a population with known standard deviation σ at a given confidence level and within accuracy $A = |\bar{x} - \mu|$, you must first find the z-score corresponding to that confidence level. Then you must solve the following equation for n:

$$z = \frac{A}{\frac{\mu}{\sqrt{n}}} \cdot$$

This gives $\qquad\qquad n = \left(\frac{z\sigma}{A}\right)^2 \cdot$

Notice that n varies directly with the square of z and inversely with the square of A. A result is that as the desired accuracy gets smaller, the needed sample size increases quite rapidly. Consequently, getting a high level of confidence and small margin of error simultaneously can result in the need for very large sample sizes, and these, in turn, can be very costly. Even in statistics, there is no such thing as a free lunch!

You should note that just as medications carry warnings about their potential harmful effects, if used incorrectly, the techniques described in this lesson to make inferences about populations must be applied with caution. First, the data collected must be from random samples of the population. The methods used here to determine a confidence interval for μ do not apply to other types of samples such as volunteers or similar "convenient" samples. Second, as a practical matter, if the sample size is relatively small, outliers can have a large effect on the confidence interval. You should always look for outliers in a sample, and try to correct them or justify their removal before computing a sample mean.

Error Analysis for Questions 9 and 10: Before an election, it is not uncommon for newspapers or TV stations to conduct polls using 500 or 800 people. These results are then generalized to apply to an election which may involve more than 100 times as many people. It is natural that people are skeptical that this can be done; they do not know enough about statistics to see why it can work. They often, however, do know enough to know the possible pitfalls, such as a nonrandom sample, a biased interviewer, and so on.

Third, the techniques developed here require that you know the standard deviation σ of the population. In most surveys this is unrealistic. There are ways to estimate σ from the sample standard deviation s and to construct confidence intervals based on s, but they require other distributions than the binomial or normal. Fourth, the margin of error in a confidence interval covers only random sampling error. That is, the margin of error describes the typical error expected because of chance variation in random selection. It does not describe errors arising from inaccurate data collection or data entry. Fancy formulas can never compensate for sloppy data. Finally, you should understand what statistical confidence *does not* say. If you say that you are 95% confident that the mean SAT-M score of seniors in a state is between 428 and 452, this means that to calculate these numbers you used techniques that give correct results in 95% of all possible random samples. It *does not* say that the probability is 0.95 that the true population mean is between 428 and 452. The true population mean either is or is not in the interval between 428 and 452. As always, realized events have probabilities zero or one; all other probability values are for future or hypothetical events.

Questions

Covering the Reading

In 1 and 2, refer to the SAT-M data in the text with samples of size 400. For each confidence level, state **a.** the confidence interval and **b.** the margin of error.

1. 68% level   **a) $434 \leq \mu \leq 446$; b) ± 6**   2. 95% level   **a) $428 \leq \mu \leq 452$; b) ± 12**

3. Find a 90% confidence interval for the SAT-M scores of the seniors in the state of Ohio. Assume, as in the lesson, that $\bar{x} = 440$, $\sigma = 119$, and $n = 400$.   **$430 \leq \mu \leq 450$**

In 4 and 5, *true* or *false*.

4. In most statistical applications a confidence level of at least 90% is used. **True**
5. As the confidence level increases, the width of the confidence interval increases. **True**

6. Below are the 90% and 95% confidence intervals for a population mean μ. Which is which?   **a is 90%, b is 95%**

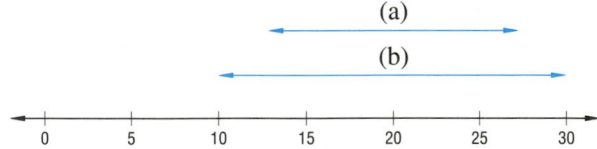

7. Identify two ways to decrease the range of a confidence interval.
lower the confidence level; increase the sample size

650

8. Refer again to the SAT-M data in the lesson. Suppose you wanted to estimate μ with 95% confidence to within 10 points. How large a sample would be needed?   **544**

9. When opinion polls are conducted, the results are often reported as a percent in favor of something $\pm$ a margin of error as a percent. In most polls it is standard practice, unless stated otherwise, to report the margin of error for a 95% confidence interval.

   a. Suppose you read that 28% of Americans recently surveyed felt that homelessness was the most serious problem in the U.S., and that the poll has a margin of error of $\pm 3\%$. What is the 95% confidence interval for the percent of adults who think that homelessness is the most serious problem in the U.S.?   **25% to 31%**

   b. Can you be certain that the true population percent falls within the interval in part **a**?   **No**

10. A pollster wants accuracy to within 2% in a close election between candidates A and B, with a 95% confidence interval. Assuming a 50% probability of a person preferring candidate A, then $\sigma = \frac{1}{2}$. If so, how many people need to be polled?   **about 2400**

11. The test for cholesterol level is not perfectly precise. Moreover, level of blood cholesterol varies from day to day. Suppose that repeated measurements for an individual on different days vary normally with $\sigma = 5$ units. On a single test, Alisa's cholesterol level is reported to be 180.

   a. Find a 95% confidence level for her mean cholesterol level.

   b. A person with a cholesterol level of about 200 is considered moderately at-risk for a stroke or heart attack, and if the person is either overweight or smokes the risk is even higher. Should Alisa be concerned about the results of her cholesterol test?   **No**

   a) $170 \le \mu \le 190$

Review

In 12 and 13, refer to the Central Limit Theorem. *True* or *false*.   *(Lesson 10-8)*

12. As more and more samples are selected, the mean of the sample means approaches the population mean.   **True**

13. As more and more samples are selected, the standard deviation of the sample means approaches the standard deviation of the population.   **False**

In 14 and 15, consider the set of two-digit numbers from 10 to 99.
(Lessons 7-7, 10-8)

14. Suppose one number is chosen at random.

   a. Name the type of probability distribution that results.

   b. Find the mean of this distribution.   **a) uniform distribution; b) 54.5**

15. Suppose random samples of size 40 are selected, and their means calculated.

   a. What type of distribution is the distribution of sample means?

   b. How does the mean of the distribution in part **a** compare to the mean of the distribution in Question 14?   **They are approximately equal.**

   a) approximately normal

PROJECTS
Projects 5 and 6 on page 655 are related to the content of this lesson.

EXTENSION
In the early part of this century, William S. Gasset, who was an Irish statistician, published many papers on interpreting data obtained by sampling. Gasset wrote under the pen name "Student." He developed procedures for determining probabilities from small samples when the standard deviation of the population is not known. The probability distribution is not a normal distribution and, in honor of Gasset's pen name, is called Student's *t*-distribution. Students may be interested in finding out who Student was and what Student's *t*-distribution is.

16. *Multiple choice.* Which of the following questions does a hypothesis test answer?
  a. Is the sample random?
  b. Is the experiment properly designed?
  c. Is the result due to chance?
  d. Is the result important? *(Lesson 10-7)* **(c)**

In 17–19, consider a standard normal distribution of a variable z. Suppose it is known that the probability that z is less than a is p. Express each of the following in terms of p. *(Lesson 10-5)*

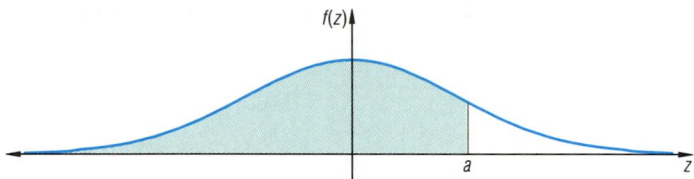

17. $P(z \geq a)$ **1 − p** **18.** $P(-a \leq z \leq a)$ **2p − 1** **19.** $P(z < -a)$ **1 − p**

20. The height of a certain species of plant is known to be normally distributed with mean 24″ and standard deviation 2″. If a nursery grows 3000 plants of this species in preparation for Mother's Day, how many should they expect to be shorter than 18″ tall (and consequently, too short to sell as top-quality)? *(Lesson 10-6)* **about 4**

In 21–24, *multiple choice.* What equation corresponds to the graph?
(Lessons 4-6, 10-4, 10-5, 10-6)

  (a) $y = e^{-x}$

  (b) $y = e^{-x^2}$

  (c) $y = \dfrac{1}{\sqrt{2\pi}} e^{-\frac{x^2}{2}}$

  (d) $y = \dfrac{1}{\sqrt{2\pi}} e^{-\frac{1}{2}(x-1)^2}$

21.

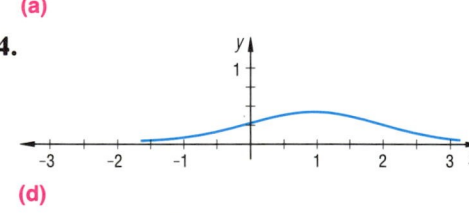

(b)

22.

(a)

23.

(c)

24.

(d)

In 25 and 26, **a.** evaluate the expression, and **b.** identify the result as an arithmetic series, geometric series, or neither. *(Lessons 8-3, 8-4)*

25. $\sum_{i=1}^{7} (2i - 1)$ **a) 49; b) arithmetic**   **26.** $\sum_{k=4}^{6} 2^k$ **a) 112; b) geometric**

27. a. How many permutations consisting of four letters each can be formed from the letters of the word CONFIDE? **840**

 b. How many of the permutations in **a** begin with a vowel? **360**
 (Lesson 7-5)

28. The land area of the United States is 3,615,105 square miles. The area of Colorado is 104,247 square miles. What is the probability that a randomly chosen point of land in the U.S. is

 a. in Colorado? **≈ .029**   **b.** not in Colorado? *(Lesson 7-1)* **≈ .971**

Exploration

29. What size sample is used in ratings of TV programs? What is the accuracy of these ratings? What is the confidence level?
Answers will vary.

Projects

The projects provide an opportunity for extended work on topics related to the content of the chapter. (See pages 66 and T37 for more information on the purposes of projects.) Some or all of these may be suitable for **small group work.**

Solutions to the projects are found at the end of the Additional Answers section at the back of this book.

Project 1 is appropriate for students who are adept at working with tools.

Project 6 could not be more open-ended, yet it may be the one we recommend most. If students have some motivating concern, then what better time to tackle the problem with the mathematics they have just been studying? Consider allowing individuals, small groups, or the class as a whole to do this project.

1. In the early 1870s the English physician, explorer, and scientist, Sir Francis Galton, designed an apparatus he called a *quincunx* to illustrate binomial experiments. Sometimes the quincunx is called a Galton board. The original quincunx had a glass face and a funnel at the top. Small balls were poured through the funnel and cascaded through an array of pins. Each ball struck on a pin at each level; theoretically, it then had an equal probability of falling to the right or left. The balls collected in compartments at the bottom. Turning the quincunx upside down sent all the balls back to its original position. Galton's original quincunx still survives in England (see photo below); large replicas exist in many science museums, and smaller ones can be purchased from science suppliers.

Galton's original quincunx, made in 1873

   a. Obtain an example, or build your own Galton board, with at least ten rows of pins. Release the balls and observe the distribution. Repeat this experiment 20–30 times and describe any trends.
   b. What binomial probability distribution does your quincunx represent? Discuss the relation between your observations in part **a** and this probability distribution.

2. The curves in the graph below are *percentile curves* or *cumulating percentage curves*. Each ordered pair (x, y) represents the percent y of the data in the set that are less than x. For instance, in the data set pictured by these curves, about 60% of the boys and about 18% of the girls had scores less than 35 on a test of coordination.

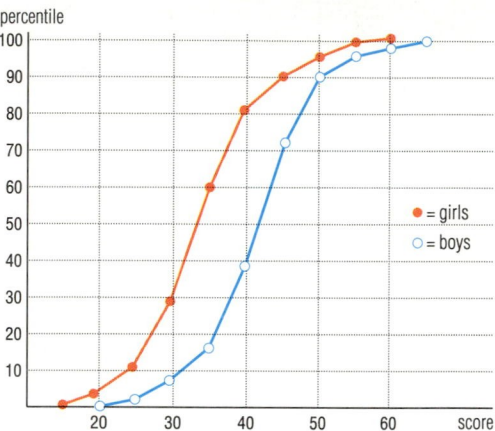

   a. Obtain examples of data sets, each with at least 25 numbers, but different distributions. For instance, collect examples of data sets with distributions that are approximately uniform, binomial, exponential, and normal. Calculate percentile ranks for the scores in each data set and plot the corresponding percentile curve. Compare and contrast the shapes of the percentile curves.
   b. Galton (see Project 1) called such a percentile curve an *ogive* (pronounced "oh-jive"), based on an architectural term. Find pictures of ogives in architecture books and compare and contrast them to the curves you drew in part **a**.

3. Refer to Project 8 of Chapter 1 (*p.* 67) and the related projects in Chapters 2 and 3.
   a. Determine a way to choose a random sample of at least 30 students in your school. Survey that sample with the same variables as in the earlier projects.

b. Use displays and descriptive statistics to describe a typical student in the random sample.

c. Make inferences to describe a typical student in your school.

d. Compare your results from part **b** above to the results you obtained in earlier chapters. How typical is your class of the school as a whole?

4. When scores on two forms of the same test have different distributions, z-scores can be used to compare individual performances on the two forms. For instance, suppose on Form 1 of a test, $\bar{x} = 80$ and $s = 4$, and on Form 2, $\bar{x} = 70$ and $s = 8$. Then a score of 78 on Form 1 corresponds to $z = \frac{78 - 80}{4} = -0.5$; whereas, a score of 78 on Form 2 corresponds to $z = \frac{78 - 70}{8} = 1$. So a score of 78 on Form 2 is substantially higher with respect to the group who took the test than a score of 78 on Form 1. Consider a test such as the SAT or ACT that is given frequently.

a. Find the mean and standard deviation for a given national sample (say, all seniors who took the test in the U.S.) during the past 10 years.

b. Find the mean and standard deviation for your state and your school over the same period.

c. Use standard scores to compare and contrast the performances of the three groups over time. Suggest explanations for any trends you observe.

5. a. Describe a way to pick sentences at random from this chapter, and pick 30 such sentences. Count the number of occurrences of the letter e in each sentence.

b. Use your data from part **a** to determine a 95% confidence interval for the percent of e occurrences in a mathematics book.

c. Repeat parts **a** and **b** using 30 sentences randomly chosen from another book.

d. Compare your answers to parts **b** and **c**.

6. Consider a problem that concerns you: crime, drug abuse, air pollution, poverty, etc. Design a study to investigate some aspect of the problem. Use descriptive and inferential statistics, as appropriate, to report the results of your study.

7. a. Use a computer, calculator, or table of random numbers to choose two random *digits* between 0 and 9, inclusive. Then calculate their sum. Repeat this until you have 50 such random sums. For instance, your first ten sums might be

14   10   5   3   15   3   6   7   7   12.

Let $s =$ the sum of the two random digits. Then s can be any integer between 0 and 18, inclusive. Find $P(s \leq n)$ for every integer from 0 to 18. Describe the probability distribution with a table of values and with a graph.

b. Modify part **a** to choose two random *numbers* between 0 and 9, inclusive. That is, consider the analogous *continuous* random variable s. To calculate $P(s \leq 5)$ in this case it will help to think geometrically. For instance, to calculate $P(s \leq 5)$ and $P(s \leq 12)$ you will need to consider the areas of the figures below.

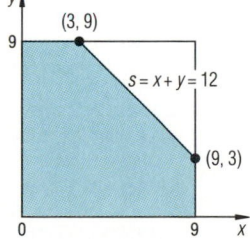

Use areas to find $P(s \leq n)$ for all integers n between 0 and 18. Again describe the probability distribution with a table and with a graph.

c. Compare and contrast the distributions $(n, P(s \leq n))$ in parts **a** and **b**.

Summary

Binomial probabilities are used to analyze two-outcome events of repeated independent trials. The binomial distribution function $B(k, n, p)$ represents the probability of getting exactly k successes in n binomial trials, each of which has probability p of success. By fixing two of the parameters, two-dimensional graphs of this function can be obtained. Usually computer programs are used to calculate the necessary binomial probabilities.

The mean μ and standard deviation σ of a binomial probability distribution are given by $\mu = np$ and $\sigma = \sqrt{npq}$. For large n, graphs of binomial probability distributions resemble a bell-shaped curve.

A commonly observed continuous distribution is the normal distribution, characterized by bell-shaped distribution curves. Many naturally occurring phenomena can be approximated by normal distributions. The parent of all these curves is given by $f(x) = e^{-x^2}$. The standard normal distribution has a mean of 0 and a standard deviation of 1; its equation is $f(x) = \dfrac{1}{\sqrt{2\pi}} e^{-\frac{x^2}{2}}$. Areas under this or any other bell-shaped curve can be calculated using calculus or closely approximated using computer programs or tables. Probabilities of events in any normal distribution can be estimated by first standardizing the variable with z-scores and then using tabulated values for the standard normal distribution.

Binomial and normal distributions are often used to make judgments or inferences about issues. Assuming a binomial or normal distribution, one can determine how likely it is that a certain observed event happened by chance. Then, based on predefined significance levels, one can decide if the calculated probability contradicts the null hypothesis that the event *is* a reasonable outcome with the assumed distribution. If so, you accept the alternative hypothesis that the assumed distribution is not a proper one. If the data fail to contradict the null hypothesis, then your conclusion is to accept it.

The Central Limit Theorem states that if a large number of random samples of size n are selected from any population with mean μ and standard deviation σ, then as n increases, the distribution of sample means has the following properties: (1) its mean M approaches μ; (2) its standard deviation s approaches $\dfrac{\sigma}{\sqrt{n}}$; and (3) it approaches a normal distribution.

The theorem can be applied if the original population is itself normally distributed or if the sample size $n \geq 30$.

Based on the relations above, interval estimates can be formed for the mean of a population using sample results. The intervals may be narrow or wide depending on the preset confidence level and the size of the sample. These inference techniques must be applied with caution, paying attention to proper sampling, unbiased experiment conditions, and satisfaction of the necessary conditions in the theorems.

Vocabulary

For the starred (*) items you should be able to give a definition of the term.
For the other terms you should be able to give a general description and a specific example of each.

Lesson 10-1
* binomial probability
    distribution
BINOMIAL GENERATOR
    program
binomial distribution
    function, $B(k, n, p)$
characteristics of a binomial
    distribution
cumulative binomial
    probability table

Lesson 10-2
Mean of a Binomial
    Distribution Theorem
variance of a probability
    distribution
Variance and Standard
    Deviation of a Binomial
    Distribution Theorem

Lesson 10-3
discrete, continuous
    situations
area under a curve
Area Scale Change Theorem

Lesson 10-4
* normal distribution
normal curve
bell-shaped curves
concave down, concave up
inflection points

Lesson 10-5
* standard normal curve
* standard normal distribution

Lesson 10-6
standard score, z-score
standardizing a variable

Lesson 10-7
statistical inference
significance levels
hypothesis testing
null hypothesis, H_0
alternative hypothesis, H_1

Lesson 10-8
sampling distribution
mean of a sampling
    distribution, M
Central Limit Theorem

Lesson 10-9
confidence interval
margin of error

656

Progress Self-Test

Take this test as you would take a test in class. You will need Appendix D and either the BASIC program BINOMIAL GENERATOR or a statistics package that generates binomial probabilities. Then check the test yourself using the solutions at the back of the book.

1. Suppose the probability that a randomly selected patient with a particular heart condition will survive more than one year is 0.89. Draw a histogram of the binomial probability distribution that x of 10 heart patients will survive a year or more.

2. Describe how the domain, range, and the shape of the binomial distribution in Question 1 would change if 100 patients are studied.

3. A binomial probability distribution $B(k, n, p)$ has $p = 0.2$ and $n = 60$. What are the mean μ and standard deviation σ of the distribution?

4. What is the probability of randomly picking a point from the area outside the square and inside the circle below? $1 - \frac{2}{\pi} \approx .36$

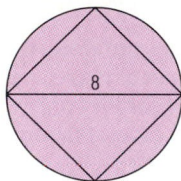

In 5 and 6, consider $f(x) = \frac{1}{2} e^{-x^2}$.

5. The graph of f is symmetric over what line?

6. *True* or *false*.   The equation for f is an equation for the standard normal curve.   **False**

In 7 and 8, find the probability that a randomly chosen observation z from a standard normal distribution is

7. less than 1.3;   **.9032**

8. between -1.5 and 1.5 .   **.8664**

2) The domain would increase; the range would "flatten" or decrease; the shape would approach a normal distribution.

3) $\mu = 12$, $\sigma = \sqrt{9.6} \approx 3.1$

5) $x = 0$

9. In 1987 the average annual salary for workers in the construction industry was about $20,100, with a standard deviation of $2100. A survey of the 50 members of a local carpenters' union finds that their mean annual salary is $18,200.
   a. State H_0, a null hypothesis, and H_1, an alternative hypothesis, that the carpenters could use to test that their mean salary is lower than the national average for their industry.
   b. Test the union's hypothesis at the 0.05 level.

10. *True* or *false*.   The Central Limit Theorem is true only for large samples ($n \geq 30$) from a population.   **False**

11. Complete: 95% of all observations on a normal distribution fall within __?__ standard deviations of the mean.   **1.96**

12. If a variable x has a normal distribution with mean m and standard deviation s, then the transformed variable $z = \dfrac{x - m}{s}$ has what distribution?   **standard normal**

13. A survey of the students at Peaks High School shows that they travel an average of 2.15 miles to school every day with a standard deviation of 0.36 miles. Assuming a normal distribution, determine the probability that a randomly picked student at Peaks High travels over 3 miles to school.   **.0091; just under 1%**

657

The main objectives for the chapter are organized here into sections corresponding to the types of understanding this book promotes: Skills, Properties, Uses, and Representations (called SPUR objectives).

USING THE CHAPTER REVIEW
Students should be able to answer questions like these with about 85% accuracy by the end of the chapter. (See pages 72 and T38–39 for more information.)

CHAPTER 10

Chapter Review

Questions on **SPUR** Objectives

SPUR stands for **S**kills, **P**roperties, **U**ses, and **R**epresentations.
The Chapter Review questions are grouped according to the SPUR Objectives for this chapter.
See margin for answers not shown below.

SKILLS deal with the procedures used to get answers.

■ **Objective A:** *Calculate the mean and standard deviation of a binomial probability distribution.* (*Lesson 10-2*)

In 1 and 2, find the mean and standard deviation for the distribution.

1. a binomial distribution with $p = .6$, $n = 60$
2. a binomial distribution with $q = .3$, $n = 100$

In 3 and 4, consider a test with 20 items.

3. If all items are True-False and students guess randomly, what is the expected mean and standard deviation? $\mu = 10$, $\sigma = \sqrt{5} \approx 2.24$

4. If all items are multiple choice with four options and students guess randomly, what is the expected mean and standard deviation?

5. How many questions with five options should there be on a multiple choice test to have an expected mean of 20 when students guess randomly? **100**

6. If $\mu = 158$ and $\sigma = 5$ for a binomial probability distribution, determine n and p (with three-digit accuracy). $n = 188$, $p = .842$

1) $\mu = 36$, $\sigma = \sqrt{14.4} \approx 3.8$
2) $\mu = 30$, $\sigma = \sqrt{21} \approx 4.58$
4) $\mu = 5$, $\sigma = \sqrt{3.75} \approx 1.94$

■ **Objective B:** *Use areas under the standard normal curve to find probabilities.* (*Lesson 10-5*)

In Questions 7–11, use the table in Appendix D.

7. **a.** Determine the area under the standard normal curve from $z = -1.28$ to $z = 1.28$.
   b. What is the probability that a random variable z with standard normal distribution takes on a value between -1.28 and 1.28?

In 8–10, evaluate the given probability.

8. $P(z<2)$ 9. $P(1<z<2)$ 10. $P(z<-2.34)$

11. Verify the statement that about 99.7% of the area under the standard normal curve is between $z = -3$ and $z = 3$.

12. Suppose that the variable z is normally distributed with a mean of 0 and standard deviation 1, and that the probability that z is between $-a$ and a is p. Express each of the following in terms of p.

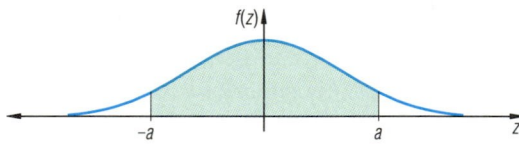

a. $P(0<z<a)$ **b.** $P(z \geq a)$ **c.** $P(|z|>a)$

$\frac{p}{2}$ $\frac{1-p}{2}$ $1-p$

7 a) .7994 units²; b) .7994
8) .9772
9) .1359
10) .0096
11) $P(-3<z<3) = 2P(0 \leq z<3) = 2(.9987 - .5) = 2(.4987) = .9974 \approx 99.7\%$

PROPERTIES deal with the principles behind the mathematics.

■ **Objective C:** *Compare and contrast characteristics of different binomial probability distribution graphs.* (Lesson 10-1)

In 13–15, consider binomial probability distributions described by $B(k, n, p)$. *True* or *false*.

13. The graph of any binomial distribution is symmetric to the line $k = m$, the mode of the distribution. **False**

14. With fixed p, as n increases, the graph of a binomial distribution ''spreads out.'' **True**

15. With fixed n, as p increases, the mode of a binomial distribution moves to the right along the k-axis. **True**

16. **a.** Describe the relation between the graphs of $B(k, 10, .45)$ and $B(k, 10, .55)$.
    b. Generalize the property in part **a** for any n and p.

17. **a.** Show that $f(x) = e^{-x^2}$ is an even function.
    b. What kind of symmetry does this imply for the graph of any normal distribution with mean μ and standard deviation σ?

16a) They are reflection images of each other over the vertical line $k = 5$.
 b) Given n and p, the graphs of $B(k, n, p)$ and $B(k, n, 1 - p)$ are reflections of each other over the vertical line $k = \frac{n}{2}$.

17a) $f(-x) = e^{-(-x)^2} = e^{-x^2} = f(x)$
 b) reflection symmetry over the y-axis.

■ **Objective D:** *Apply properties of normal distributions and their parent function.*
(Lessons 10-4, 10-5, 10-6)

In 18–20, *true* or *false*.

18. The graph of $y = e^{-x^2}$ never intersects the x-axis. **True**

19. The area between the graph of $y = e^{-x^2}$ and the x-axis is infinite. **False**

20. The area between the standard normal curve and the x-axis is 1. **True**

21. Use the program MONTE CARLO AREA to estimate the area under the curve defined by $f(x) = e^{-x^2}$ between its inflection points, $(-0.71, 0.61)$ and $(0.71, 0.61)$. **≈ 1.21**

22. Given that the equation which defines the standard normal curve is $f(z) = \dfrac{1}{\sqrt{2\pi}} e^{-\frac{1}{2}z^2}$, approximate the maximum value of this function to two decimal places. **.40**

23. **a.** Write an equation for the normal distribution with $\mu = 2$ and $\sigma = 1$.
    b. What is the maximum value of the function in part **a**?

24. What is the median value in a normal distribution with mean μ and standard deviation σ? **μ**

23a) $f(x) = \dfrac{1}{\sqrt{2\pi}} e^{-\frac{1}{2}(x - 2)^2}$; b) $\dfrac{1}{\sqrt{2\pi}} \approx .40$

USES deal with applications of mathematics in real situations.

■ **Objective E:** *Apply the Central Limit Theorem.*
(Lesson 10-8)

In 25 and 26, *true* or *false*.

25. If the distribution of population data is not known to be normal, one needs to pick large samples ($n \geq 30$) to apply the Central Limit Theorem. **True**

26. Given a population with mean μ and standard deviation σ, the standard deviation of the means of k samples of size n from this population is approximately $\dfrac{\sigma}{\sqrt{k}}$. **False**

27. Suppose that the weights of full-grown oranges in a citrus grove are normally distributed with mean 7.3 ounces and standard deviation 2.1 ounces. Determine the probability that 50 oranges randomly picked from this grove have a mean weight greater than 10.5 ounces.

28. A highway patrol group decides to check the results they have read in a report which claimed that the speed of cars in their region of the highway has a mean of 63 and standard deviation of 15 miles per hour. They track 100 cars every day for a month and record the average speed observed at the end of each day. If the numbers in the report are correct, describe the distribution of the data collected by this group.

27) almost zero; 28) normally distributed, mean 63, s.d. 1.5

■ **Objective F:** *Solve probability problems using binomial or normal distributions.*
(Lessons 10-1, 10-2, 10-6)

29. 20% of the deer in a wildlife preservation park are believed to have a certain antler infection. A zoologist randomly captures 15 deer and checks their antlers. Assuming the prediction is accurate, determine the probability that more than three of the deer will have infected antlers.

30. An allergy test requires that small samples of chemicals are injected under the skin of the patient. Individuals sensitive to any one of the chemicals have a reaction time that is approximately normally distributed with a mean of 11 hours and a standard deviation of 3 hours. How often can one expect a reaction time of longer than 18 hours with such individuals?

31. A new test of extroversion-introversion was administered to a large group of adults, and found to yield scores that were approximately normally distributed with a mean of 35 and a standard deviation of 8. For the instruction manual, various scores are to be interpreted as in the table below. Determine the highest and the lowest scores for each of these five groups.

highly extroverted	top 5%	≥ 49
extroverted	next to top 15%	42–48
average	middle 60%	29–41
introverted	next to bottom 15%	22–28
highly introverted	bottom 5%	≤ 21

32. Based on factory and field tests, a car company determines that the trouble-free mileage for their new model is normally distributed with a mean of 31,600 miles and a standard deviation of 16,200 miles. The car is marketed with a 40,000-mile guarantee. Determine the percentage of cars of this make the company should expect to repair under the guarantee before 40,000 miles. **.52%**

■ **Objective G:** *Use binomial and normal distributions to test hypotheses.* (*Lessons 10-7, 10-8*)

In 33 and 34, a machine is supposed to produce steel rods with a diameter of 1.25 cm. A study of 200 randomly chosen pieces produced by the machine shows that the mean diameter is 1.14 cm and the sample standard deviation is 0.05 cm. The owner of the machine is interested in finding out if the settings on the machine are truly off.

33. *Multiple choice.* To answer this question, the most appropriate set of hypotheses is
(a) H_0: Diameters of rods produced by the machine are 1.14 cm.
H_1: Diameters of rods produced by the machine are over 1.14 cm.
(b) H_0: Diameters of rods produced by the machine are 1.14 cm.
H_1: Diameters of rods produced by the machine are 1.25 cm.
(c) H_0: Diameters of rods produced by the machine are 1.25 cm.
H_1: Diameters of rods produced by the machine are under 1.25 cm. **(c)**

34. Answer the owner's question at the 0.05 significance level.

35. The weight of the checked luggage allowed for one person on a domestic flight is 20 kg. For flight safety purposes, one airline randomly samples 450 of its customers and studies the weights of their checked luggage. The mean weight is found to be 27.8 kg and the sample standard deviation to be 5.3 kg. Can the airline officers claim that passengers are checking in luggage heavier than 20 kilograms? Use a 0.01 significance level.

36. Refer to the highway patrol group in Question 28. Suppose that, based on the total of 3000 cars they track in a month, they calculate an average speed of 58.75 mph. Can they claim that the results in the report are wrong using a 0.10 significance level?

660

■ **Objective H:** *Graph and interpret a binomial probability distribution.* (Lesson 10-9)

37. Tell why this table does *not* show a probability distribution.

x	−1	0	1	4
$P(x)$	$\frac{1}{2}$	$\frac{1}{4}$	$\frac{1}{8}$	$\frac{1}{5}$

38. Assuming that the births of boys and girls are equally likely, the probability of x boys in a family with 4 children is given below.

x = number of boys	0	1	2	3	4
$P(x)$	$\frac{1}{16}$	$\frac{1}{4}$	$\frac{3}{8}$	$\frac{1}{4}$	$\frac{1}{16}$

a. Finish the table of probabilities.
b. Graph the distribution as a histogram.

39. Let x be the number of items a person gets right (by guessing) on a 3-item test. Then the probability $P(x)$ of getting x items correct is
$$P(x) = {}_3C_x\left(\frac{1}{3}\right)^x\left(\frac{2}{3}\right)^{3-x}$$
a. Construct a table of probabilities for $x = 0, 1, 2, 3$.
b. Graph the probability distribution as a scatter plot.

40. Construct a graph of the probability distribution for the function P where $P(x)$ is the probability of getting k heads in 5 tosses of a coin that is biased with $P(\text{heads}) = .6$.

41. The graph of a binomial probability distribution is given below.

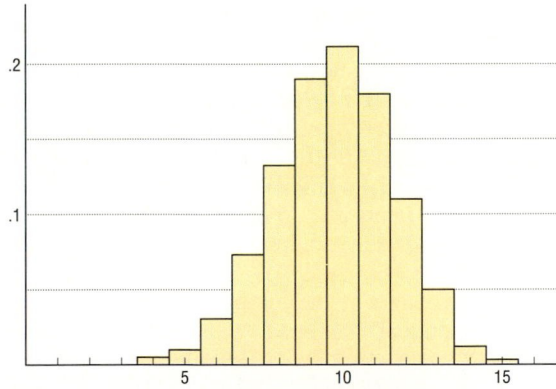

a. Approximate the mean of the distribution. **10**
b. Estimate $P(3 < k \leq 7)$.   **0.25**

42. Give two reasons why the following graph *cannot* represent a binomial probability distribution.

■ **Objective I:** *Use areas to find probabilities.* (Lessons 10-3, 10-4, 10-5)

In 43 and 44, refer to the sonar screen below, which is on board a marine research ship looking for dolphins. The screen has a radius of 4 inches and the lines indicating depth (in feet) are equidistant from each other. Assume that this is a blind search; that is, the crew has no idea about where the dolphins are.

43. What is the probability that the first image will indicate a depth no more than 100 ft?   **1/9**

44. *True* or *false*.   The probability that the image appears right at the center is 0. Explain your answer.   **True; the area of a point is zero, and ratios of areas represent probability values.**

40.

42. Samples: it is continuous and not discrete; $P(x) > 0$ for arbitrarily large values of x.

In 45–47, a point is randomly selected from the triangular region shown below. The non-horizontal sides are given by $y = \frac{5}{3}x$ and $y = -\frac{5}{2}x + \frac{25}{2}$.

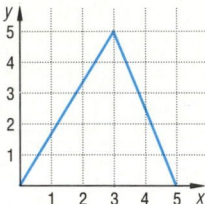

45. a. Calculate the area of the region.   **12.5**
  b. Determine the probability that the x-coordinate of the selected point is
    i. less than 3;   **.6**
    ii. between 2 and 4.   **.6̄3**

46. Find the value of k such that $P(x<k) = .5$.

47. *True* or *false*.   If the triangular region is transformed by a size change of $\frac{1}{2}$ so that $S(x, y) = \left(\frac{1}{2}x, \frac{1}{2}y\right)$, the probabilities calculated in Question 45 (with x-coordinate scaled accordingly) are divided by 4. Explain your answer.

48. The shaded area under the standard normal curve is about .9544. Determine
  a. $P(-2<z<2)$ **b.** $P(0<z<2)$ **c.** $P(z\le 2)$.

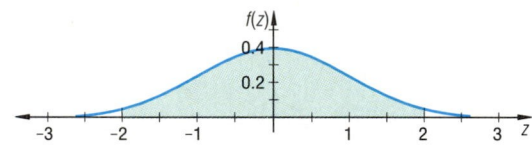

46) $k = \sqrt{7.5} \approx 2.74$
47) False; each area is divided by 4, so the ratios do not change.
48a) .9544
  b) .4772
  c) .9772

■ **Objective J:** *Graph and interpret normal distributions.* *(Lessons 10-5, 10-6)*

49. Below are the graphs of three normal distributions. Identify
  a. the standard normal distribution   **ii**
  b. the distribution with $\sigma = 2$ **iii**
  c. the distribution with $\sigma = \frac{1}{2}$.   **i**

50. Use an automatic grapher to graph a normal distribution with $m = 4$ and $s = 2$.

51. Consider a normal distribution with mean m and standard deviation s. Explain why the curve below cannot represent the distribution.

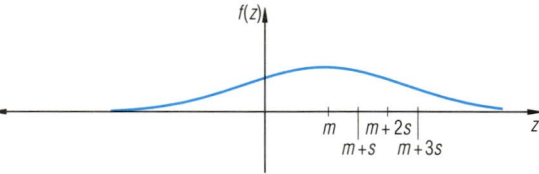

52. a. Graph the curve defined by $f(z) = e^{-z^2}$, the "parent" of the normal curves.
  b. The graph in part **a** is symmetric to the line $z = 0$, and about 68% of the area under it is between $z = -0.71$ and $z = 0.71$. On the same axes, draw the curve which represents the normal distribution with mean 0 and standard deviation 0.71.
  c. *True* or *false*.   About 68% of the area under the new curve is also between $z = -0.71$ and $z = 0.71$.   **True**
  d. Explain why the curves in **a** and **b** are not identical.   **Sample: The curve in part a is not a normal distribution; the area beneath it is not 1. The curve in part b is a normal distribution.**

51) Sample: Much less than 99.7% of the area lies within 3 standard deviations of the mean.

662

CHAPTER 11 ■ MATRICES AND TRIGONOMETRY

OBJECTIVES ■ CHAPTER 11

The objectives listed here are the same as in the Chapter 11 Review on pages 723–726 of the student text. The Progress Self-Test on page 722 and the tests in the Teacher's Resource File cover these objectives. For recommendations regarding the handling of this end-of-chapter material, see the notes in the margins on the corresponding pages of this Teacher's Edition.

OBJECTIVES FOR CHAPTER 11 (Organized into the SPUR categories— Skills, Properties, Uses, and Representations)	Progress Self-Test Questions	Chapter Review Questions	Teacher's Resource File	
			Lesson Masters*	Chapter Test Forms A & B
SKILLS				
A Multiply matrices, where possible.	2, 4	1–4	11-1	1
B Use matrices to solve systems of equations.	18	5–9	11-7	2
PROPERTIES				
C Apply properties of matrices and matrix multiplication.	5	10–13	11-1, 11-3, 11-7	12
D Find the inverse of a 2×2 matrix.	9b, 16, 17	14–17	11-7	3, 4
E Apply the Addition and Double Angle Identities.	13–15	18–27	11-5, 11-6	7, 8, 13
USES				
F Use a matrix to organize information.	1, 3	28–34	11-1, 11-2	5
REPRESENTATIONS				
G Represent reflections, rotations, scale changes, and size changes as matrices.	9a, 10	35–43	11-2, 11-3, 11-4	9, 10
H Represent composites of transformations as matrix products.	11	44–47	11-3, 11-4	11
I Find the image of a figure under a transformation using a matrix.	6–8, 12	48–51	11-2, 11-3, 11-4	6

*The masters are numbered to match the lessons, and each is intended for use with that lesson.

OVERVIEW ■ CHAPTER 11

The use of matrices in algebra, geometry, discrete mathematics, finance, computing, biology, differential equations, linear algebra, and in statistics make it necessary for today's students to become familiar with them.

Different aspects of matrices are studied in three UCSMP texts. The common use of matrices to solve systems of equations is covered in UCSMP *Advanced Algebra*. In *Precalculus and Discrete Mathematics*, the use of matrices in representing graphs, and in describing transition states is studied.

This chapter reviews multiplication of matrices and their use to represent transformations, and then uses matrices to develop the formulas for cos $(\alpha + \beta)$ and sin $(\alpha + \beta)$. Various applications of these formulas and of the connection between

transformations and matrices close out the chapter.

Lesson 11-1 presents the language, symbols, and rules for multiplying matrices. The next three lessons then apply matrix multiplication to generate geometric transformations. In Lesson 11-2, students learn about matrices for some of the transformations they studied in Chapter 3. These include matrices for reflection through either axis or the line $y = x$, size changes of magnitude k, and scale changes $S(x, y) = (ax, by)$. Lesson 11-3 develops matrices for rotations through angles of 90°, 180°, or 270° around the origin.

Lesson 11-4 applies the trigonometry of Chapter 5 to produce a matrix for any rotation about the origin. The general rotation matrix is then applied in Lesson 11-5 to prove the Addition Identities for

Cosine and Sine. These identities are then used in Lesson 11-6 to prove the Double Angle Identities. The emphasis on these identities is their practical utility—students are asked to calculate accurately and decide which formula is appropriate in a given situation.

In Lesson 11-7, the chapter further connects matrices with transformations in the finding of inverses of 2×2 matrices. Lesson 11-8 describes how matrices can be used in computer programs to manipulate graphics.

The trigonometric identities in this chapter are all covered in *Precalculus and Discrete Mathematics*. If your students will be taking that course, then you do not need to spend much time on the proofs of the identities and can concentrate more on the matrix representations and arithmetic calculations.

PERSPECTIVES ■ CHAPTER 11

The Perspectives provide the rationales for the inclusion of topics or approaches, give some mathematical background, and make connections with other lessons and to other UCSMP texts.

11-1

MATRIX MULTIPLICATION

Matrix multiplication is an easy operation to learn if one begins with multiplication of one row (on the left) by one column (on the right). This immediately indicates when the rows and columns will be compatible. Then, if a matrix with m rows is multiplied by a compatible matrix with n columns, there are mn possible products and the only question is where to put them. The obvious answer: put the product of the ith row and the jth column in that place in the product.

We do not attempt to teach matrix addition. In part this is because matrix addition is done in an element-by-element mode (that is, corresponding elements are

added), and this easily leads students into expecting that multiplication is also element-by-element. So by beginning their study of matrices with multiplication, students do not need to "unlearn" an inappropriate algorithm.

11-2

MATRICES FOR TRANSFORMATIONS

This lesson uses matrix multiplication to represent transformations in the coordinate plane. The general idea is this: The product of a 2×2 matrix $\begin{bmatrix} a & b \\ c & d \end{bmatrix}$ and a 2×1 matrix $\begin{bmatrix} x \\ y \end{bmatrix}$ is the 2×1 matrix $\begin{bmatrix} ax + by \\ cx + dy \end{bmatrix}$. Thus, if the image of

(x, y) under a transformation is given by a formula of the form $(ax + by, cx + dy)$, then we can think of (x, y) as represented by the matrix $\begin{bmatrix} x \\ y \end{bmatrix}$ and the transformation as represented by the matrix $\begin{bmatrix} a & b \\ c & d \end{bmatrix}$.

This lesson does *not* attempt to answer the question "If I have a transformation, how do I find its matrix?". That question is discussed in Lesson 11-4. Rather, we expect students to see how matrix representations of reflections and scale changes can be used to transform figures. All the matrices for transformations in this lesson have dimensions 2×2.

11-3

MATRICES FOR COMPOSITES OF TRANSFORMATIONS

UCSMP *Geometry* students, in their study of congruence, learn a Two-Reflection Theorem: the composite of two reflections over intersecting lines is a rotation whose center is the point of intersection of the lines and whose magnitude is twice the acute angle between the lines measured from the first line to the second. Here that theorem is recalled and verified, but not deduced, for the special case where the first line is the x-axis and the second is the line $y = x$. Students know a formula for each of these reflections and so can obtain a formula for the composite.

But this composite of transformations is represented by the product of the matrices for the the reflections. Then, products of matrices for 90° rotations lead to matrices for rotations of 180° and 270°. Thus, by the end of this lesson, students have a matrix for any rotation about the origin whose magnitude is a multiple of 90°.

To help in remembering all of these matrices, it is proved that if a 2×2 matrix exists for a transformation, then the image of (1, 0) is the first column of the matrix and the image of (0, 1) is the second column.

11-4

THE GENERAL ROTATION MATRIX

In this lesson, we assume that there is a 2×2 matrix for R_θ, the rotation of magnitude θ about the origin. From this assumption and the theorem of the previous lesson,

the matrix $\begin{bmatrix} \cos \theta & -\sin \theta \\ \sin \theta & \cos \theta \end{bmatrix}$ for R_θ

is derived and applied. The derivation is quite simple: the first column is the image of (1, 0) under this rotation (from the definition of cosine and the sine). The second column is shown to be the image of the first under a rotation of 90°.

Any transformation of the plane that maps the origin to itself and

maps lines onto lines will have a 2×2 matrix.

11-5

IDENTITIES FOR cos $(\alpha + \beta)$ AND sin $(\alpha + \beta)$

The image of (1, 0) under the rotation R_β is $(\cos \beta, \sin \beta)$. The image of that point under R_α is $(\cos (\alpha + \beta), \sin (\alpha + \beta))$. When this is done with matrices, we simply multiply:

$$\begin{bmatrix} \cos \alpha & -\sin \alpha \\ \sin \alpha & \cos \alpha \end{bmatrix} \begin{bmatrix} \cos \beta \\ \sin \beta \end{bmatrix}.$$

This gives the familiar formulas for $\cos (\alpha + \beta)$ and $\sin (\alpha + \beta)$ both in one line.

This derivation seems to have been discovered in the first decade of this century. The "wrapping function" approach found in some textbooks is newer, having been invented in the 1940s. In the wrapping function approach $\cos (\alpha - \beta)$ is found by equating distances, and $\cos (\alpha + \beta)$ is found, then $\sin (\alpha - \beta)$ is found by using the identity $\sin x = \cos (\pi - x)$. In *Precalculus and Discrete Mathematics* we combine the transformation definitions of cosine and sine with the equating of distances.

11-6

IDENTITIES FOR cos 2θ AND sin 2θ

Regardless of the approach used to find the identities in the last lesson, by letting θ stand for both α and β, these identities are derived quite easily. They are called "double angle" identities because of their historical origin, not because α and β must be angles. Some teachers prefer to call them "double argument" identities.

11-7

INVERSES OF 2×2 MATRICES

Let T be the set of all transformations with 2×2 matrices and let M be the set of the corresponding matrices. Then the algebraic structures T with composition and M with multiplication are isomorphic;

that is, their results correspond. Two examples of this isomorphism that students have seen previously are (1) if M_1 is the matrix for T_1 and M_2 is the matrix for T_2, then $M_2 \cdot M_1$ is the matrix for $T_2 \circ T_1$; and (2) the

identity matrix $\begin{bmatrix} 1 & 0 \\ 0 & 1 \end{bmatrix}$ is the

matrix for the identity transformation. Here we continue the correspondence; inverse matrices correspond to inverse transformations.

We also present an application of inverses of matrices to cryptography to show the usefulness of inverses of matrices outside the algebraic context.

11-8

MATRICES IN COMPUTER GRAPHICS

The importance of matrices in virtually every field of study and every industry has caused computer designers to incorporate matrix operations into their computer hardware. For three-dimensional graphics, the techniques outlined in this lesson are typical of "graphics engines" currently on the market.

There are two new mathematical concepts introduced here. The first is the use of 3×3 matrices to represent transformations that map (x, y) onto $(ax + by + e, cx + dy + f)$. The

idea is to represent (x, y) as $\begin{bmatrix} x \\ y \\ 1 \end{bmatrix}$

and the transformation as

$\begin{bmatrix} a & b & e \\ c & d & f \\ 0 & 0 & 1 \end{bmatrix}$. This makes it possible

to find matrices for all isometries and similarity transformations in the plane, though that is not done here.

The second idea is the rotation about a point other than the origin. The idea is to translate the point to the origin; rotate there (since a formula is known for that), and then translate back.

The lesson ends with a discussion of the problem of picturing three-dimensional objects on a two-dimensional computer screen via the use of world coordinates.

DAILY PACING CHART ■ CHAPTER 11

For optimal student performance, the Progress Self-Test and the Chapter Review should be covered. (See *General Teaching Suggestions: Mastery* on page T35 of this Teacher's Edition.) By following the pace of the Full Course given here, students can complete 12 or 13 chapters of the book by the end of the year. Students following the pace of the Minimal Course spend more time to complete lessons that are combined with quizzes and to consider the Chapter Review. When all lessons are covered in order (the recommendation of the authors), students in the Minimal Course can cover all but the last two chapters. For more information on pacing, see *General Teaching Suggestions: Pace* on page T34 of this Teacher's Edition.

DAY	MINIMAL COURSE	FULL COURSE
1	11-1	11-1
2	11-2	11-2
3	11-3	11-3
4	Quiz (TRF); Start 11-4	Quiz (TRF); 11-4
5	Finish 11-4	11-5
6	11-5	11-6
7	11-6	Quiz (TRF); 11-7
8	11-7	11-8
9	Quiz (TRF); Start 11-8	Progress Self-Test
10	Finish 11-8	Chapter Review
11	Progress Self-Test	Chapter Test (TRF)
12	Chapter Review	
13	Chapter Review	
14	Chapter Test (TRF)	

TESTING OPTIONS

■ Quiz for Lessons 11-1 through 11-3   ■ Chapter 11 Test, Form A   ■ Chapter 11 Test, Cumulative Form
■ Quiz for Lessons 11-4 through 11-7   ■ Chapter 11 Test, Form B

A Quiz and Test Writer is available for generating additional questions, additional quizzes, or additional forms of the Chapter Test.

PROVIDING FOR INDIVIDUAL DIFFERENCES

The student text has been written for, and tested with, average students. It has also been used successfully with students who are better prepared as well as with students who are poorly prepared.

The Lesson Notes often include Error Analysis and Alternate Approach features to help you assist those students who need more help. Students of all abilities often learn from their peers and may benefit from small group work referenced as appropriate throughout the Notes. A blackline Lesson Master (in the Teacher's Resource File), keyed to the chapter objectives, is available for each lesson to provide more practice. (However, we advise you to use these masters sparingly; again refer to the suggestions on pace on page T34.)

Extension activities are provided in the Lesson Notes for use with students who need to be challenged even more than the requirements of completing the Full Course provide.

Matrices and Trigonometry

This chapter should take 11–14 days: 8 – 10 days for the lessons and quizzes, 1 day for the Progress Self-Test, 1 or 2 days for the Chapter Review, and 1 day for a Chapter Test.

USING PAGE 663
The drawings of the monkeys on this page of the chapter were made using a computer program with a built-in rotation feature. One monkey was drawn and the program used a version of the techniques described in this chapter to reproduce the other images. The differences you see between the monkeys are due to computer round-off error.

Alternative Assessment
Matrices are taught even in some junior high classes. Informally assess your students' knowledge of matrices. If they remember matrix multiplication and have had matrices for transformations, the first three lessons of this chapter will be easy, and you may be able to do them in two days.

Have you ever wondered how computer games produce animated images? A standard programming technique uses matrices and trigonometry. A *matrix* (plural, *matrices*) is a rectangular arrangement of objects.

For instance, to generate an animation with cartoon figures, a single figure might be described by a set of points. These data are stored in a matrix in which each column represents the *x*- and *y*-coordinates. Then the figure is changed by transformations such as scale changes, translations, or rotations. These transformations can also be stored as matrices. Multiplying the matrix representing a figure by a matrix representing a transformation gives a matrix for the image of the original figure. For example, the following represents a transformation that translates the figure and rotates it 22.5° clockwise.

$$\begin{bmatrix} .9239 & .3827 & 82.016 \\ -.3827 & .9239 & 37.462 \\ 0 & 0 & 1 \end{bmatrix}$$

In this chapter you will learn matrices for several transformations, including a rotation of any magnitude around the origin. You will also apply matrices to derive some trigonometric identities, and to illustrate some elementary computer graphics techniques.

TEACHING NOTES

Emphasize that matrices with numerical elements are a type of shorthand—they represent ideas in a rectangular array. Newspapers contain many examples of matrices (**Question 26**). Students may be familiar with the phrase "dot-matrix printer" or the matrix of the screen of a computer monitor or television. These are more general uses of the word *matrix* to mean "any rectangular array."

Error Analysis A common error among beginning students of matrix multiplication is to get the matrices in the wrong order. To help avoid this mistake, emphasize the arrows in the dimensions shown at the bottom of page 665 and the shading in the row-column pattern in the solution to **Example 2.** Encouraging this attention to setting up a multiplication will also help students correctly translate applications into matrix form more quickly.

Even when students learn what sizes of matrices *may* be multiplied, it may take them a while to learn that for most square matrices *A* and *B*, $AB \neq BA$. One notable exception is that $AI = IA$, where *I* is the appropriate

LESSON

11-1

Matrix Multiplication

The Math Team holds an autumn and a spring fund-raiser. Below are the numbers of each item sold at each sale last year.

	trail treats	carob chews	fruit clusters	nut bars
autumn	40	100	0	40
spring	75	108	80	65

The numbers are presented in a **matrix**, a rectangular array typically enclosed by square brackets. Each object in a matrix is called an **element**. The matrix above has 2 rows (labeled autumn and spring) and 4 columns (one for each item). The labels are not part of the matrix. The matrix above is said to have *dimensions* 2 by 4, written 2×4. In general, a matrix with m rows and n columns has **dimensions $m \times n$**. It is sometimes called an $m \times n$ matrix. Each element may be identified by giving its row and column. For instance, the element in row 2, column 1 is 75.

■ ■ ■ ■ ■ ■ ■ ■

Example 1 The prices for trail treats, carob chews, fruit clusters, and nut bars are 1.00, 1.00, 0.50, and 1.50 dollars, respectively. Present these values in a matrix.

Solution There are two possible matrices. One is a 4×1 matrix. It has 4 rows and 1 column, and is shown below.

$$\begin{bmatrix} 1.00 \\ 1.00 \\ 0.50 \\ 1.50 \end{bmatrix}$$

The other is a 1×4 matrix, and is shown below.

$$\begin{bmatrix} 1.00 & 1.00 & 0.50 & 1.50 \end{bmatrix}$$

To calculate the total amount of money (the *revenue*) received from each fund-raiser, the club treasurer multiplies each quantity sold by its price, and then adds the results. For the autumn sale the revenue was $200 because
$$40(1.00) + 100(1.00) + 0(0.50) + 40(1.50) = 200.$$
This can be considered as the product of a matrix of quantities and a matrix of prices.

$$\begin{bmatrix} 40 & 100 & 0 & 40 \end{bmatrix} \cdot \begin{bmatrix} 1.00 \\ 1.00 \\ 0.50 \\ 1.50 \end{bmatrix} = \begin{bmatrix} 200.00 \end{bmatrix}$$

664

The first matrix, the quantity matrix, is called a **row matrix** because it consists of a single row. The second matrix, the price matrix, is a **column matrix** because it consists of a single column.

Similarly, the revenue received in the spring sale was $320.50.

$$75(1.00) + 108(1.00) + 80(.50) + 65(1.50) = 320.50$$

This can be represented as the following product of matrices.

$$\begin{bmatrix} 75 & 108 & 80 & 65 \end{bmatrix} \cdot \begin{bmatrix} 1.00 \\ 1.00 \\ 0.50 \\ 1.50 \end{bmatrix} = \begin{bmatrix} 320.50 \end{bmatrix}$$

The two sets of products can be combined in a single product by multiplying the original 2×4 matrix for quantities by the 4×1 price matrix from Example 1. If Q is the matrix of quantities sold, and U is the matrix of the unit prices, then $QU = R$ represents the revenue.

$$\begin{array}{c} \\ \text{autumn} \\ \text{spring} \end{array} \begin{array}{cccc} \text{trail} & \text{carob} & \text{fruit} & \text{nut} \\ \text{treats} & \text{chews} & \text{clusters} & \text{bars} \\ \begin{bmatrix} 40 & 100 & 0 & 40 \\ 75 & 108 & 80 & 65 \end{bmatrix} \end{array} \cdot \begin{bmatrix} 1.00 \\ 1.00 \\ 0.50 \\ 1.50 \end{bmatrix} \begin{array}{l} \text{trail treats} \\ \text{carob chews} \\ \text{fruit clusters} \\ \text{nut bars} \end{array}$$

In general, matrix multiplication is done using rows from the left matrix and columns from the right matrix. Multiply the first element in the row by the first element in the column, the second element in the row by the second element in the column, and so on. Finally, add the resulting products. The shading below shows the first row times the first column, which gave a total of $200.00.

$$\begin{bmatrix} 40 & 100 & 0 & 40 \\ 75 & 108 & 80 & 65 \end{bmatrix} \cdot \begin{bmatrix} 1.00 \\ 1.00 \\ 0.50 \\ 1.50 \end{bmatrix}$$

Similarly, we saw that the second row of Q times the column of U gave $320.50. These two results may be represented in a 2×1 matrix.

$$Q \cdot U = R = \begin{bmatrix} 200.00 \\ 320.50 \end{bmatrix}$$

Notice how the dimensions of the original matrices relate to the product matrix:
The product of a 2×4 matrix and a 4×1 matrix is a 2×1 matrix.

(must be equal)

In general, the product of two matrices A and B exists if and only if the number of columns of A equals the number of rows of B.

identity matrix. However, we recommend that, at this point, the properties of matrices should not be over-emphasized. The focus should remain on the meaning and skills of matrix multiplication.

In the definition of matrix multiplication, the notation c_{ij} for an element of matrix C may take some time for students to understand.

Technology Some calculators can perform operations with matrices. If your students have such a calculator, you may wish to require that they also have their owner's handbooks with them when they ask you questions about using this technology.

Definition

Suppose A is an $m \times n$ matrix and B is an $n \times p$ matrix. Then the **product matrix** $A \cdot B$ is an $m \times p$ matrix whose element in row i and column j is the sum of the products of elements in row i of A and corresponding elements in column j of B.

As usual, often we write AB for the product $A \cdot B$.

Example 2 Consider $A = \begin{bmatrix} 3 & 0 \\ 1 & 2 \end{bmatrix}$ and $B = \begin{bmatrix} 4 & 3 & 2 \\ 1 & 0 & 5 \end{bmatrix}$. Find, if it exists,

a. AB;   **b.** BA.

Solution

a. By definition, the product of a 2×2 by a 2×3 matrix is a 2×3 matrix. This means you should expect to find 6 elements. The product of row 1 of A and column 1 of B is

$$3 \cdot 4 + 0 \cdot 1 = 12.$$

Write this in the 1st row and 1st column of the product matrix.

$$\begin{bmatrix} 3 & 0 \\ 1 & 2 \end{bmatrix} \cdot \begin{bmatrix} 4 & 3 & 2 \\ 1 & 0 & 5 \end{bmatrix} = \begin{bmatrix} 12 & - & - \\ - & - & - \end{bmatrix}$$

The product of row 1 of A and column 2 of B is $3 \cdot 3 + 0 \cdot 0 = 9$. This is the element in row 1 and column 2 of the product.

$$\begin{bmatrix} 3 & 0 \\ 1 & 2 \end{bmatrix} \cdot \begin{bmatrix} 4 & 3 & 2 \\ 1 & 0 & 5 \end{bmatrix} = \begin{bmatrix} 12 & 9 & - \\ - & - & - \end{bmatrix}$$

The other four elements of the product matrix are found by using the same row ▢ by column ▢ pattern. For instance, the element in the 2nd row, 3rd column of the product is found by multiplying the 2nd row of A by the 3rd column of B, shown here along with the final result.

$$\begin{bmatrix} 3 & 0 \\ 1 & 2 \end{bmatrix} \cdot \begin{bmatrix} 4 & 3 & 2 \\ 1 & 0 & 5 \end{bmatrix} = \begin{bmatrix} 12 & 9 & 6 \\ 6 & 3 & 12 \end{bmatrix}$$

b. To calculate the element in the 1st row, 1st column of BA, you would have to multiply the shaded numbers in pairs.

$$\begin{bmatrix} 4 & 3 & 2 \\ 1 & 0 & 5 \end{bmatrix} \cdot \begin{bmatrix} 3 & 0 \\ 1 & 2 \end{bmatrix}$$

The row by column multiplication cannot be done. The product BA does not exist.

666

As Example 2 illustrates, matrix multiplication is not commutative. However, matrix multiplication *is* associative. That is, for matrices A, B, C where multiplication exists, $(AB)C = A(BC)$. In Example 3 one product of three matrices is calculated. In the questions you will verify associativity for this case.

Example 3 Suppose that the Math Team mentioned earlier makes a profit of 40% of its sales in the autumn and 50% in the spring. Let $P = [.4 \quad .5]$ represent the profit percents. Calculate $P(QU)$, and describe what this product represents.

Solution Earlier we calculated

$$QU = \begin{bmatrix} 40 & 100 & 0 & 40 \\ 75 & 108 & 80 & 65 \end{bmatrix} \cdot \begin{bmatrix} 1.00 \\ 1.00 \\ 0.50 \\ 1.50 \end{bmatrix} = \begin{bmatrix} 200.00 \\ 320.50 \end{bmatrix}$$

Thus
$$P(QU) = [.4 \quad .5] \cdot \begin{bmatrix} 200.00 \\ 320.50 \end{bmatrix}$$
$$= [80 + 160.25]$$
$$= [240.25].$$

The product represents the total amount of profit earned by the Math Team's two fund-raisers.

Questions

Covering the Reading

1. What is a matrix? **a rectangular array of numbers**

In 2–5, use the matrix at the right which gives the number of members in the House of Representatives by gender for the 98th–101st Congresses.

$$\begin{array}{c} & \text{male} \quad \text{female} \\ \begin{array}{c} 1983 \\ 1985 \\ 1987 \\ 1989 \end{array} & \begin{bmatrix} 413 & 21 \\ 412 & 22 \\ 412 & 23 \\ 408 & 25 \end{bmatrix} \end{array}$$

2. What are the dimensions of this matrix?
   4 × 2
3. What is the element in row 3, column 2?   **23**

4. If a_{ij} represents the element in row i and column j, what is a_{21}?   **412**

5. What does the sum of the numbers in each row represent?
   total number of male and female representatives

6. If A and B are matrices, under what circumstances does the product AB exist?   **The number of columns in A equals the number of rows in B.**

In 7 and 8, give the dimensions of AB.

7. A is 3×2, B is 2×4 **3 × 4**   8. A is 5×1, B is 1×7 **5 × 7**

In 9–12, multiply.

Members of the 101st Congress taking the oath of office

9. $[8 \quad 2] \cdot \begin{bmatrix} 9 \\ 1 \end{bmatrix}$ **[74]**

10. $\begin{bmatrix} 7 & 3 & 1 \\ 0 & 4 & 2 \end{bmatrix} \cdot \begin{bmatrix} -1 \\ 1 \\ -2 \end{bmatrix}$ $\begin{bmatrix} \mathbf{-6} \\ \mathbf{0} \end{bmatrix}$

11. $\begin{bmatrix} a & b \\ c & d \end{bmatrix} \cdot \begin{bmatrix} x \\ y \end{bmatrix}$ $\begin{bmatrix} \mathbf{ax + by} \\ \mathbf{cx + dy} \end{bmatrix}$

12. $\begin{bmatrix} 1 & 0 \\ 2 & 3 \end{bmatrix} \cdot \begin{bmatrix} -2 & 3 \\ 1 & -4 \end{bmatrix}$ $\begin{bmatrix} \mathbf{-2} & \mathbf{3} \\ \mathbf{-1} & \mathbf{-6} \end{bmatrix}$

LESSON 11-1 *Matrix Multiplication*   **667**

NOTES ON QUESTIONS
Error Analysis for Questions 9–12: Before doing each multiplication, it is helpful to ask students to give the dimensions of the product. Then put in blanks in each place in the product matrix and fill each blank by multiplying a row by a column. Pick the blank before picking the row and column so that students see that for element a_{ij}, row i and column j are to be multiplied.

Question 14: Many students do not understand associativity of an operation until they see it with matrix multiplication. They have a way of adding or multiplying three numbers at a time, but they have no algorithm for multiplying three matrices. Thus they are forced to take a product like PQU and do one of the multiplications first. Explain that, because of associativity, no parentheses are needed in a product like PQU.

13. Give the dimensions of two matrices which cannot be multiplied.
    sample: 3 × 2 and 3 × 4
14. Refer to the Math Team fund-raiser described in the lesson. In Example 3 the matrix $P(QU)$ was calculated. Calculate $(PQ)U$. That is, find the product matrix PQ, then multiply this result by U. Verify that $P(QU) = (PQ)U$. **See margin.**

In 15 and 16, the matrix N gives the number of tickets sold for a Children's Theater performance. The matrix C gives the unit cost in dollars for each ticket.

$$N = \begin{array}{r} \\ \text{Weekday matinee} \\ \text{Weekend matinee} \\ \text{Weekend evening} \end{array} \overset{\text{Adults Children}}{\begin{bmatrix} 250 & 340 \\ 273 & 320 \\ 170 & 405 \end{bmatrix}} \qquad C = \begin{bmatrix} 6.00 \\ 4.00 \end{bmatrix} \begin{array}{l} \text{adults} \\ \text{children} \end{array}$$

a) $\begin{bmatrix} 2860 \\ 2918 \\ 2640 \end{bmatrix}$

15. **a.** Find NC.
    b. What were the theater's receipts for the weekday performance?
    $2860
16. A portion of the receipts for each performance goes to a children's health charity: 50% for the weekday performance and 40% for each of the weekend performances. Let $P = [.50 \quad .40 \quad .40]$. Find the total contribution to charity for all three performances. **$3653.20**

In 17 and 18, solve for the variables.

17. $\begin{bmatrix} 3 & 4 \\ 1 & 2 \end{bmatrix} \cdot \begin{bmatrix} x \\ 1 \end{bmatrix} = \begin{bmatrix} -2 \\ 0 \end{bmatrix}$
    x = -2

18. $\begin{bmatrix} 2 & a \\ 3 & b \end{bmatrix} \cdot \begin{bmatrix} 5 \\ 6 \end{bmatrix} = \begin{bmatrix} 7 \\ 8 \end{bmatrix}$
    a = -1/2, b = -7/6

19. For square matrices, $M^2 = M \cdot M$. Let $M = \begin{bmatrix} 0 & -2 \\ -2 & 0 \end{bmatrix}$.
    a. Calculate M^2. **a)** $\begin{bmatrix} 4 & 0 \\ 0 & 4 \end{bmatrix}$; **b)** $\begin{bmatrix} 0 & -8 \\ -8 & 0 \end{bmatrix}$
    b. Calculate M^3.

20. The diagram below shows the major highways connecting four cities.

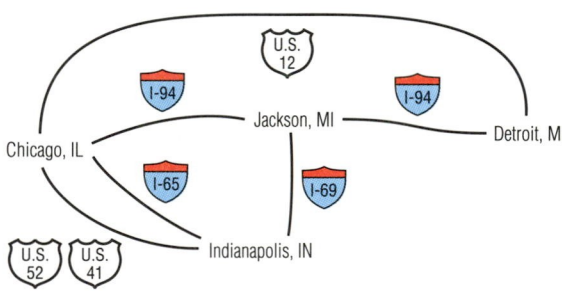

$$\begin{array}{r} \\ \text{Chicago} \\ \text{Jackson} \\ \text{Detroit} \\ \text{Indianapolis} \end{array} \overset{\text{Chicago Jackson Detroit Indianapolis}}{\begin{bmatrix} 0 & 1 & 1 & 2 \\ 1 & 0 & 1 & 1 \\ 1 & 1 & 0 & 0 \\ 2 & 1 & 0 & 0 \end{bmatrix}}$$

a. Code the number of direct routes (not through any other city on the diagram) between each pair of cities into a matrix as begun above. (Note that U.S. 41–U.S. 52 is considered to be one road.)
b. Multiply the matrix from part **a** by itself and interpret what it signifies. **a,b) See below.**

21. Five pairs each of red, blue, yellow, and green socks are in a drawer. Four pairs are selected randomly and not replaced. Let b = number of blue pairs selected. *(Lesson 7-7)*
a. List all possible values of b. **{0, 1, 2, 3, 4}**
b. Make a probability distribution table for the random variable b.
c. Construct a histogram of the probability distribution.
b,c) See margin.

22. Let (x, y) be the image of the point $(1, 0)$ under a rotation of magnitude θ about the origin. Then $x = $ __**a.**__ and $y = $ __**b.**__ . *(Lesson 5-4)*
a) cos θ; b) sin θ

23. $f(x) = x^2 + 2$, $g(x) = 3x - 1$. Find a rule for $f(g(x))$. *(Lesson 3-7)*
$f(g(x)) = 9x^2 - 6x + 3$

24. a. Draw $\triangle SAD$, where $S = (3, 7)$, $A = (0, 5)$, and $D = (2, -3)$.
b. Draw $\triangle S'A'D'$, the reflection image of $\triangle SAD$ over the y-axis.
c. If r_y represents reflection over the y-axis, then r_y: $(x, y) \rightarrow$ __?__.
(Lesson 3-4) **a,b) See margin. c) (-x, y)**

25. Match each transformation with the *best* description.
(Lessons 3-2, 3-4, 3-5, Previous course)
a. $M(x, y) = (x + 3, y - 2)$ **v**   (i) reflection over the x-axis
b. $N(x, y) = (x, -y)$ **i**   (ii) reflection over the line $y = x$
c. $P(x, y) = (y, x)$ **ii**   (iii) scale change
d. $Q(x, y) = \left(\dfrac{x}{3}, \dfrac{y}{3}\right)$ **iv**   (iv) size change
e. $V(x, y) = (0.1x, 10y)$ **iii**   (v) translation

26. Matrices frequently appear in the business and sports sections of newspapers.
a. Find an example of a matrix in a newspaper.
b. State its dimensions.
c. Describe what each row and column represents.
a-c) Answers may vary.

20a) See matrix on page 668.

b)
```
     C  J  D  I
  C [ 6  3  1  1 ]
  J [ 3  3  1  2 ]
  D [ 1  1  2  3 ]
  I [ 1  1  3  5 ]
```

The matrix represents the number of ways of traveling from one city to another by traveling through exactly one other city.

21.b.

b	0	1
$P(b)$	.282	.470

2	3	4
.217	.031	.001

c.

24.a.,b.

OBJECTIVES

F Use a matrix to organize information.

G Represent reflections, scale changes, and size changes as matrices.

I Find the image of a figure under a transformation using a matrix.

TEACHING NOTES

A possible way to introduce this lesson follows: Begin with **Questions 1 and 2.** Then review the geometric meaning of the transformations with a figure such as the parabola with equation $y = x^2$. What is an equation for its image under each of the reflections in the theorem on page 671? ($y = -x^2$, $y = x^2$, $x = y^2$) Next review the algebraic formulas for these reflections. Then do **Question 3,** showing how matrix multiplication gives the same image as the algebraic formula.

In covering the theorem on page 671, note that since $(0, 0)$ is its own image under every transformation with a 2×2 matrix, only reflections over lines containing the origin can have 2×2 matrices. The matrices here represent reflections over all the symmetry lines of the square with vertices $(1, 1)$, $(1, -1)$, $(-1, -1)$, and $(-1, 1)$ except the line $y = -x$. That is given as Additional Example 1, which follows.

LESSON

11-2

Matrices for Transformations

Matrices can represent geometric figures as well as numerical data. To do so, let the point (a, b) be written as the matrix $\begin{bmatrix} a \\ b \end{bmatrix}$. Such a 2×1 matrix is sometimes called a **point matrix**. Then an n-gon is represented by a $2 \times n$ matrix in which the columns are the coordinates of consecutive vertices. For instance, $\triangle QRS$ at the right can be represented by the matrix

$$\begin{bmatrix} -1 & 3 & 3 \\ -1 & -1 & 2 \end{bmatrix}.$$
$$\begin{array}{ccc} \uparrow & \uparrow & \uparrow \\ Q & R & S \end{array}$$

If named $\triangle SRQ$, the triangle is represented by

$$\begin{bmatrix} 3 & 3 & -1 \\ 2 & -1 & -1 \end{bmatrix}.$$

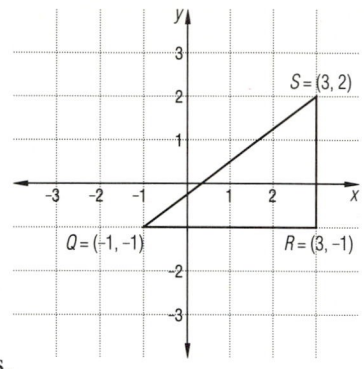

Matrices may also represent transformations.

■ ■ ■ ■ ■ ■ ■ ■

Example 1   **a.** Multiply the matrix for $\triangle QRS$ above by the matrix $\begin{bmatrix} 1 & 0 \\ 0 & -1 \end{bmatrix}$ and graph the resulting image, $\triangle Q'R'S'$.

b. Describe the transformation represented by the matrix $\begin{bmatrix} 1 & 0 \\ 0 & -1 \end{bmatrix}$.

Solution

a. To multiply, the matrix for $\triangle QRS$ must be on the right.

$$\begin{bmatrix} 1 & 0 \\ 0 & -1 \end{bmatrix} \cdot \begin{bmatrix} -1 & 3 & 3 \\ -1 & -1 & 2 \end{bmatrix} = \begin{bmatrix} -1 & 3 & 3 \\ 1 & 1 & -2 \end{bmatrix}$$
$$\begin{array}{cccccc} \uparrow & \uparrow & \uparrow & \uparrow & \uparrow & \uparrow \\ Q & R & S & Q' & R' & S' \end{array}$$

b. $\triangle Q'R'S'$ is the image of $\triangle QRS$ under reflection over the x-axis.

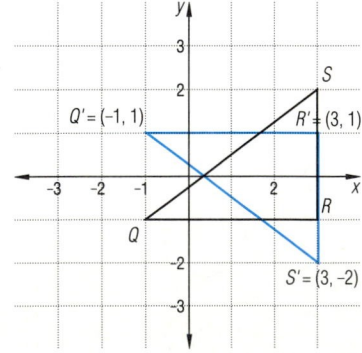

670

In general, multiplying $\begin{bmatrix} 1 & 0 \\ 0 & -1 \end{bmatrix}$ by any point matrix $\begin{bmatrix} x \\ y \end{bmatrix}$ gives

$$\begin{bmatrix} 1 & 0 \\ 0 & -1 \end{bmatrix} \cdot \begin{bmatrix} x \\ y \end{bmatrix} = \begin{bmatrix} x \\ -y \end{bmatrix}.$$

This means that multiplication by $\begin{bmatrix} 1 & 0 \\ 0 & -1 \end{bmatrix}$ on the left maps (x, y) to $(x, -y)$. You should recognize this transformation as r_x, the reflection over the x-axis.

The above result can be generalized. Let F be a matrix for a geometric figure. Whenever multiplication by F of matrix M produces a matrix for the image of F under a transformation T, then M is called the **matrix representing the transformation T**, or the matrix for T.

Below are matrices for r_x, two other reflections, and the image of $\triangle QRS$ under each. Note that in each matrix the first column is the image of $(1, 0)$ and the second column is the image of $(0, 1)$ under that transformation. You should verify that the matrices are correct.

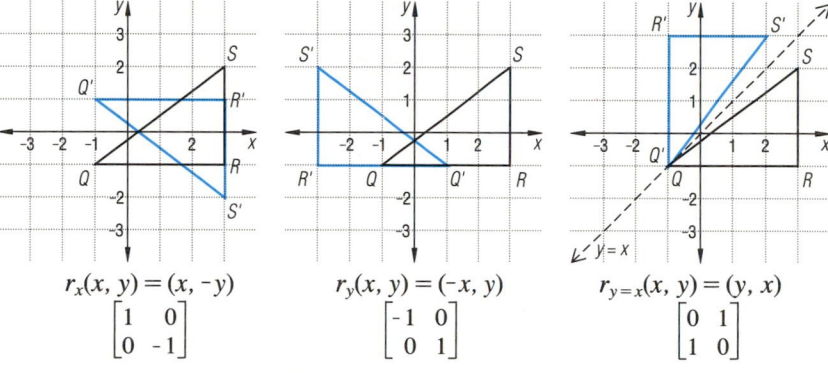

$r_x(x, y) = (x, -y)$ $r_y(x, y) = (-x, y)$ $r_{y=x}(x, y) = (y, x)$

$\begin{bmatrix} 1 & 0 \\ 0 & -1 \end{bmatrix}$ $\begin{bmatrix} -1 & 0 \\ 0 & 1 \end{bmatrix}$ $\begin{bmatrix} 0 & 1 \\ 1 & 0 \end{bmatrix}$

Theorem (Matrices for Reflections)

$\begin{bmatrix} 1 & 0 \\ 0 & -1 \end{bmatrix}$ is the matrix for r_x, reflection over the x-axis.

$\begin{bmatrix} -1 & 0 \\ 0 & 1 \end{bmatrix}$ is the matrix for r_y, reflection over the y-axis.

$\begin{bmatrix} 0 & 1 \\ 1 & 0 \end{bmatrix}$ is the matrix for $r_{y=x}$, reflection over the line $y = x$.

You can demonstrate scale changes using the matrices of **Examples 2 and 3** (a teaching aid is provided), or you can ask what is an equation for the image of $y = x^2$ under the transformations of **Questions 5–8**.

Making Connections
Students may wonder why they need matrices to do transformations that they could do geometrically by plotting points and algebraically with formulas. Point out the following: (1) matrices give an *arithmetic* way of doing these transformations; (2) one of the advantages of matrices will appear when transformations need to be composed, an application they will see in the next lesson; (3) a second advantage of matrices is their ability to represent some more complicated algebraic formulas, which they will see in Lesson 11-4.

A bonus of studying this application of matrices is that prior to this point, many students will have thought of transformations as objects that connect preimages to images, not as objects that exist in their own right. Matrices make the latter conception automatic and give students new insight into the transformations they used in previous chapters.

1. Multiply $\begin{bmatrix} -1 & 3 & 3 \\ -1 & -1 & 2 \end{bmatrix}$, the matrix for $\triangle QRS$, by the matrix $\begin{bmatrix} 0 & -1 \\ -1 & 0 \end{bmatrix}$.

a. What is the product matrix?
$\begin{bmatrix} 1 & 1 & -2 \\ 1 & -3 & -3 \end{bmatrix}$

b. What transformation does the matrix seem to represent?
reflection over the line
y = -x

2. Multiply the matrix
$\begin{bmatrix} 0 & 5 & 10 & 10 \\ 0 & 5 & 15 & 20 \end{bmatrix}$ by $\begin{bmatrix} .2 & 0 \\ 0 & .2 \end{bmatrix}$.

a. What is the product matrix?
$\begin{bmatrix} 0 & 1 & 2 & 2 \\ 0 & 1 & 3 & 4 \end{bmatrix}$

b. Interpret the multiplication geometrically.
a size change of
magnitude .2 on the
quadrilateral with vertices
(0, 0), (5, 5), (10, 15), and
(10, 20)

3. Apply $\begin{bmatrix} -4 & 0 \\ 0 & 3 \end{bmatrix}$ to the quadrilateral given in Additional Example 2.

a. What is the image?
$\begin{bmatrix} 0 & -20 & -40 & -40 \\ 0 & 15 & 45 & 60 \end{bmatrix}$

b. Describe the transformation.
a transformation that
stretches the figure 4
times horizontally,
reflects it over the y-axis,
and stretches it 3 times
vertically

What transformation does the matrix $\begin{bmatrix} 1 & 0 \\ 0 & 1 \end{bmatrix}$ represent?

Since $\begin{bmatrix} 1 & 0 \\ 0 & 1 \end{bmatrix} \cdot \begin{bmatrix} x \\ y \end{bmatrix} = \begin{bmatrix} x \\ y \end{bmatrix}$, the transformation represented by this matrix maps any point (x,y) to itself, and is the *identity transformation*.

Thus, the matrix $\begin{bmatrix} 1 & 0 \\ 0 & 1 \end{bmatrix}$ is called the **2 × 2 identity matrix.** Note that while matrix multiplication is not commutative in general, the identity matrix gives the same result whether multiplied on the left or on the right.

Size changes may be represented by matrices.

Example 2   **a.** Multiply $\begin{bmatrix} 3 & 0 \\ 0 & 3 \end{bmatrix}$ by the matrix for $\triangle JKL$, when $J = (-1, 2)$, $K = (1, 3)$, and $L = (1, -1)$.
b. Graph the preimage and image.

c. What transformation does $\begin{bmatrix} 3 & 0 \\ 0 & 3 \end{bmatrix}$ represent?

Solution
a. Write the matrix for $\triangle JKL$ to the right of the given matrix.
$$\begin{bmatrix} 3 & 0 \\ 0 & 3 \end{bmatrix} \cdot \begin{bmatrix} -1 & 1 & 1 \\ 2 & 3 & -1 \end{bmatrix} = \begin{bmatrix} -3 & 3 & 3 \\ 6 & 9 & -3 \end{bmatrix}$$
b.

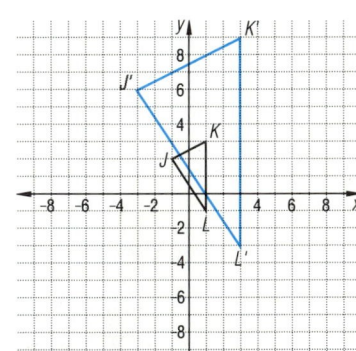

c. The matrix $\begin{bmatrix} 3 & 0 \\ 0 & 3 \end{bmatrix}$ represents a size change of magnitude 3.

672

In general, a size change of magnitude k for $k \neq 0$ is represented by the matrix $\begin{bmatrix} k & 0 \\ 0 & k \end{bmatrix}$. More generally, a matrix of the form $\begin{bmatrix} a & 0 \\ 0 & b \end{bmatrix}$ with $a \neq 0$, $b \neq 0$ represents the scale change $S(x, y) = (ax, by)$. This is the transformation that stretches a preimage horizontally by a and vertically by b and is symbolized by $S_{a,b}$.

Example 3
a. Apply $\begin{bmatrix} 3 & 0 \\ 0 & 2 \end{bmatrix}$ to the figure at the right.
b. Identify the transformation.

Solution

a. Here is the product that represents the image of the square.

$$\begin{bmatrix} 3 & 0 \\ 0 & 2 \end{bmatrix} \cdot \begin{bmatrix} 4 & 4 & -2 & -2 \\ 4 & -2 & -2 & 4 \end{bmatrix} = \begin{bmatrix} 12 & 12 & -6 & -6 \\ 8 & -4 & -4 & 8 \end{bmatrix}$$

The images of the interior figures are found with the following products.

segment
$$\begin{bmatrix} 3 & 0 \\ 0 & 2 \end{bmatrix} \cdot \begin{bmatrix} -1 & 0 \\ 1 & 2 \end{bmatrix} = \begin{bmatrix} -3 & 0 \\ 2 & 4 \end{bmatrix}$$

rectangle
$$\begin{bmatrix} 3 & 0 \\ 0 & 2 \end{bmatrix} \cdot \begin{bmatrix} 1 & 1 & 3 & 3 \\ 1 & 2 & 2 & 1 \end{bmatrix} = \begin{bmatrix} 3 & 3 & 9 & 9 \\ 2 & 4 & 4 & 2 \end{bmatrix}$$

triangle
$$\begin{bmatrix} 3 & 0 \\ 0 & 2 \end{bmatrix} \cdot \begin{bmatrix} -1 & 3 & 2 \\ -1 & 0 & -1 \end{bmatrix} = \begin{bmatrix} -3 & 9 & 6 \\ -2 & 0 & -2 \end{bmatrix}$$

b. Graph the image.

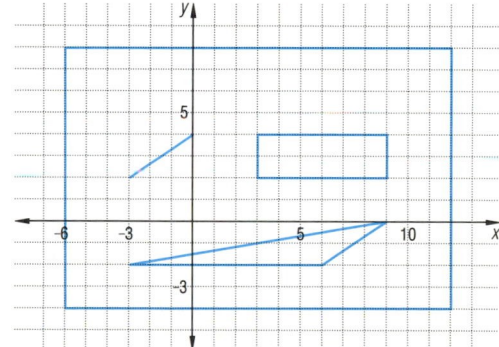

Notice that the image is a distortion, not a simple enlargement of the original figure. Every part of the original figure is stretched three times horizontally and twice vertically. The transformation is the scale change $S_{3,2}$.

Questions

Covering the Reading

1. *Multiple choice.* Which matrix represents the point $(5, -2)$?

   (a) $[5 \ -2]$ (b) $[-2 \ 5]$ (c) $\begin{bmatrix} 5 \\ -2 \end{bmatrix}$ (d) $\begin{bmatrix} -2 \\ 5 \end{bmatrix}$ **(c)**

2. Refer to the figure at the right.
   a. Write $\triangle XYZ$ as a matrix. $\begin{bmatrix} 4 & 2 & -2 \\ -2 & 3 & 2 \end{bmatrix}$
   b. Multiply the matrix for $\triangle XYZ$
      by $\begin{bmatrix} -1 & 0 \\ 0 & 1 \end{bmatrix}$. $\begin{bmatrix} -4 & -2 & 2 \\ -2 & 3 & 2 \end{bmatrix}$
   c. The matrix given in part **b** represents
      what transformation?
      a reflection over the y-axis

3. Match the matrix with its transformation.

   a. $\begin{bmatrix} -1 & 0 \\ 0 & 1 \end{bmatrix}$ **iii**   (i) identity

   b. $\begin{bmatrix} 0 & -1 \\ -1 & 0 \end{bmatrix}$ **v**   (ii) reflection over x-axis

   c. $\begin{bmatrix} 1 & 0 \\ 0 & -1 \end{bmatrix}$ **ii**   (iii) reflection over y-axis

   d. $\begin{bmatrix} 1 & 0 \\ 0 & 1 \end{bmatrix}$ **i**   (iv) reflection over $y = x$

   e. $\begin{bmatrix} 0 & 1 \\ 1 & 0 \end{bmatrix}$ **iv**   (v) none of these

4. Verify that $\begin{bmatrix} a & b \\ c & d \end{bmatrix}\begin{bmatrix} 1 & 0 \\ 0 & 1 \end{bmatrix} = \begin{bmatrix} 1 & 0 \\ 0 & 1 \end{bmatrix}\begin{bmatrix} a & b \\ c & d \end{bmatrix}$.   **See margin.**

In 5 and 6, use the trapezoid *TRZD* below.

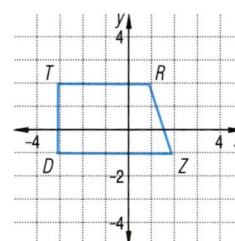

a. Find the matrix for the image of *TRZD* under the transformation.
b. Describe the transformation.

5. $\begin{bmatrix} 1.5 & 0 \\ 0 & 1.5 \end{bmatrix}$ **See margin.**   6. $\begin{bmatrix} 2 & 0 \\ 0 & 3 \end{bmatrix}$ **See margin.**

In 7 and 8, state a matrix for the transformation.

7. size change of magnitude a $\begin{bmatrix} a & 0 \\ 0 & a \end{bmatrix}$

8. scale change $S_{a,b}$ $\begin{bmatrix} a & 0 \\ 0 & b \end{bmatrix}$

674

9. Prove that $\begin{bmatrix} -1 & 0 \\ 0 & 1 \end{bmatrix}$ is the matrix for r_y, reflection over the y-axis.
   See margin.

10. **a.** Write the matrix for $r_{y=x}$. **a)** $\begin{bmatrix} 0 & 1 \\ 1 & 0 \end{bmatrix}$; **b)** $\begin{bmatrix} 1 & 0 \\ 0 & 1 \end{bmatrix}$
    b. Square the matrix.
    c. What transformation does the matrix found in part **b** represent?
    the identity

11. **a.** Find the image of $\triangle ABC = \begin{bmatrix} 2 & 3 & 0 \\ -1 & 2 & 4 \end{bmatrix}$

    under the transformation T with matrix $\begin{bmatrix} -1 & 0 \\ 0 & -1 \end{bmatrix}$. $\begin{bmatrix} -2 & -3 & 0 \\ 1 & -2 & -4 \end{bmatrix}$
    b. Describe T. **It is a rotation of 180° around the origin.**

In 12 and 13, consider the drawing at the right of the face of a cat Felix.

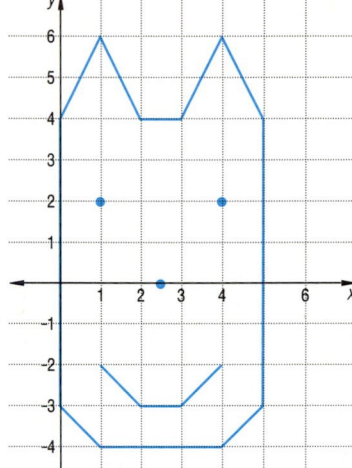

12. Write a matrix F to represent the outline of Felix's face.
    See margin.
13. Find a matrix for a transformation whose product with F produces the desired image.
    a. a face similar to Felix's with four times the area
    b. a face that is half as tall and twice as wide as Felix's
    c. a face that is longer and thinner than Felix's
    See margin.

14. Below are matrices for three collinear points A, B, and C, and a matrix T for a transformation:
    $$A = \begin{bmatrix} 1 \\ 4 \end{bmatrix}, B = \begin{bmatrix} 0 \\ 2 \end{bmatrix}, C = \begin{bmatrix} -2 \\ -2 \end{bmatrix}; T = \begin{bmatrix} 2 & 7 \\ -1 & -4 \end{bmatrix}.$$
    a. Find the images of the points under the transformation. **See below.**
    b. Are the images collinear? **yes**
    c. Does the transformation preserve distance? **no**

15. The matrix $\begin{bmatrix} \dfrac{\sqrt{3}}{2} & \dfrac{1}{2} \\ \dfrac{1}{2} & -\dfrac{\sqrt{3}}{2} \end{bmatrix}$ is associated with the reflection over a line ℓ which

    contains the origin. What is the measure of the acute angle formed by ℓ and the x-axis? (You may need to experiment with graph paper and a protractor.) **$\pi/12$**

14a) $A' = \begin{bmatrix} 30 \\ -17 \end{bmatrix}, B' = \begin{bmatrix} 14 \\ -8 \end{bmatrix}, C' = \begin{bmatrix} -18 \\ 10 \end{bmatrix}$

19.b. The first row represents production costs in Oakland in thousands of dollars for each of the years 1988, 1989, and 1990. The second row represents the same numbers for Charleston.

c. No. A 3 × 3 matrix cannot be multiplied by a 2 × 3 matrix, because the number of columns in the first matrix must correspond to the number of rows in the second.

20.a. $\begin{bmatrix} -7 & -1 \\ -14 & -10 \end{bmatrix}$

b. $\begin{bmatrix} -7 & -1 \\ -14 & -10 \end{bmatrix}$

25.a.

16. Matrix X has dimensions 3×5 and matrix Y has dimensions 5×4.
   a. What are the dimensions of XY?   **3 × 4**
   b. How many elements are in XY?   *(Lesson 11-1)*   **12**

In 17 and 18, multiply if the product exists.   *(Lesson 11-1)*

17. $\begin{bmatrix} 2 & 3 \\ 0 & 4 \end{bmatrix}\begin{bmatrix} x \\ y \end{bmatrix}$ $\begin{bmatrix} 2x + 3y \\ 4y \end{bmatrix}$

18. $\begin{bmatrix} 3 & 0 & 1 \\ 2 & -1 & -2 \end{bmatrix} \cdot \begin{bmatrix} 4 & 0 \\ 1 & -3 \end{bmatrix}$
The product does not exist.

19. A clothing manufacturer has factories in Oakland, CA, and Charleston, SC. The quantities (in thousands) of each of three products manufactured are given in the production matrix P below. The costs in dollars for producing each item during three years are given in the cost matrix C below.

	Coats	Pants	Shirts
Oakland	10	10	22
Charleston	20	15	0

$= P$,

	1988	1989	1990
Coats	30	30	32
Pants	5	7	8
Shirts	2	3	3.5

$= C$

   a. Calculate PC.
   b. Interpret PC by telling what each element represents.
   c. Does CP exist? Why or why not?   *(Lesson 11-1)*
   See margin.

20. Consider $X = \begin{bmatrix} 1 & 2 \\ 3 & -1 \end{bmatrix}$, $Y = \begin{bmatrix} 4 & 3 \\ 0 & -1 \end{bmatrix}$, and $Z = \begin{bmatrix} -2 & 0 \\ 1 & -1 \end{bmatrix}$.
   a. Find $(XY)Z$.   **b.** Find $X(YZ)$.
   c. What property of matrix multiplication do the results of parts **a** and **b** illustrate?   *(Lesson 11-1)*   **the associative property**
   a, b) See margin.

21. The height of a certain flower species is known to be normally distributed with mean 23 inches and standard deviation 2 inches. If a nursery grows 3000 of this plant in preparation for Mother's Day, about what percent should they expect to be shorter than 18 inches?   *(Lesson 10-6)*
   about .62 percent

In 22 and 23, solve for θ.   *(Lessons 5-7, 6-7)*

22. $\cos(-\theta) = -\cos\theta$
$\theta = \frac{\pi}{2} + k\pi$ for all integers k

23. $\cot\theta = \dfrac{1}{\tan\theta}$
all reals except $\theta = k\pi/2$ for all integers k

24. Let $k(x) = \sqrt{x}$ and $n(x) = 3x + 1$. *True* or *false*.
   $(k \circ n)x = (n \circ k)(x)$.
   Justify your answer.   *(Lesson 3-7)*
   False. $(k \circ n)(x) = k(n(x)) = k(3x + 1) = \sqrt{3x + 1}$
   $(n \circ k)(x) = n(k(x)) = n(\sqrt{x}) = 3\sqrt{x} + 1$

25. A cube in a three-dimensional coordinate system can be represented by the matrix C at the right.

$C = \begin{bmatrix} 1 & 1 & 1 & -1 & -1 & -1 & 1 & -1 \\ 1 & 1 & -1 & 1 & -1 & 1 & -1 & -1 \\ 1 & -1 & 1 & 1 & 1 & -1 & -1 & -1 \end{bmatrix}$

   a. Graph the cube.
   b. Let T be the transformation represented by the matrix
   $M = \begin{bmatrix} 2 & 0 & 0 \\ 0 & 2 & 0 \\ 0 & 0 & 2 \end{bmatrix}$. Calculate and graph MC.
   c. Describe the transformation T.
   See margin.

Review

Exploration

11-3

Matrices for Composites of Transformations

**REFLECT
ROTATE**

As you know, when two transformations are composed, the result is another transformation. Such composites can be represented using matrices.

Example 1   Consider $\triangle ABC = \begin{bmatrix} 2 & 6 & 6 \\ 1 & 1 & 3 \end{bmatrix}$.

   a. Find a matrix for $\triangle A'B'C'$, the image of $\triangle ABC$ under a reflection over the *x*-axis.

   b. Reflect the image over $y = x$ to obtain a matrix for $\triangle A''B''C''$.

Solution

a. The matrix for r_x is $\begin{bmatrix} 1 & 0 \\ 0 & -1 \end{bmatrix}$.

So $\triangle A'B'C'$ is represented by the matrix

$$\begin{bmatrix} 1 & 0 \\ 0 & -1 \end{bmatrix} \cdot \begin{bmatrix} 2 & 6 & 6 \\ 1 & 1 & 3 \end{bmatrix} = \begin{bmatrix} 2 & 6 & 6 \\ -1 & -1 & -3 \end{bmatrix}.$$

b. The matrix for $r_{y=x}$ is $\begin{bmatrix} 0 & 1 \\ 1 & 0 \end{bmatrix}$. Apply this matrix to the matrix for

$\triangle A'B'C'$ to get a matrix for $\triangle A''B''C''$.

$$\begin{bmatrix} 0 & 1 \\ 1 & 0 \end{bmatrix} \cdot \begin{bmatrix} 2 & 6 & 6 \\ -1 & -1 & -3 \end{bmatrix} = \begin{bmatrix} -1 & -1 & -3 \\ 2 & 6 & 6 \end{bmatrix}.$$

At the right is a graph of $\triangle ABC$ and its images.

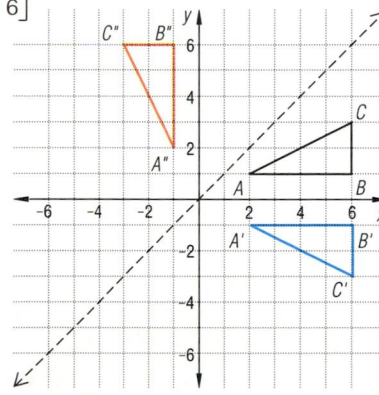

OBJECTIVES

C Apply properties of matrices and matrix multiplication.
G Represent rotations as matrices.
H Represent composites of transformations as matrix products.
I Find the image of a figure under a transformation using a matrix.

TEACHING NOTES

Error Analysis   As with the notation for composition of functions, the matrices in a representation of a composite transformation are written from right to left. Although the notation for composite functions, using parentheses, makes this right-to-left process relatively obvious, the straightforward notation of matrix multiplication hides it altogether. So students may need to be continually reminded that the order of transformations represented in a matrix multiplication is right to left.

Some students may confuse the symbol ∘ for composition with the symbol · for multiplication. We feel that the distinction should be carefully maintained. However, students may start to call the matrices by the names for the transformations. That is, they may use R to name both the transformation (the rotation of 90° about the origin) and the matrix $\begin{bmatrix} 0 & -1 \\ 1 & 0 \end{bmatrix}$. We feel that this double meaning is acceptable. As long as students know the difference between a matrix and a transformation, its much simpler and quite convenient to call the matrix by its transformation name.

In **Example 1,** we use an equal sign between the notations for a matrix and a triangle. This may bother some students. Point out that this is the customary use of the equal sign to indicate that two symbols name the same object. Obviously the triangle (a set of points) and the matrix (a rectangular array of numbers) are not the same thing. But then, neither are the symbols 1.5 and $\frac{3}{2}$, yet we would write $1.5 = \frac{3}{2}$.

A teaching aid is provided with the graph of **Example 1.** You may wish to have students redraw in the line $y = x$ on the graph to help emphasize the reflection of $\triangle A'B'C'$ to its image $\triangle A''B''C''$.

Making Connections
The general rotation matrix is presented in Lesson 7-4, and R_{90} is needed for its development. Students should know the matrix for R_{90}.
  In UCSMP *Advanced Algebra*, students will have seen R_{90} used to show that two oblique lines in the coordinate plane are perpendicular if and only if the product of their slopes is -1. The idea is that a rotation of 90° geometrically maps a line onto a line perpendicular to it. Algebrai-

678

In Example 1, $\triangle A''B''C''$ is the image of $\triangle ABC$ under the *composite* of the reflections $r_{y=x}$ and r_x. That is, $\triangle A''B''C'' = r_{y=x}(r_x(\triangle ABC))$. (Note that the first transformation is r_x; the second, $r_{y=x}$.) To find a single matrix for the composite $r_{y=x} \circ r_x$, notice that the matrix for $\triangle A''B''C''$ came from the product

$$\begin{bmatrix} 0 & 1 \\ 1 & 0 \end{bmatrix} \cdot \left(\begin{bmatrix} 1 & 0 \\ 0 & -1 \end{bmatrix} \cdot \begin{bmatrix} 2 & 6 & 6 \\ 1 & 1 & 3 \end{bmatrix} \right).$$

Because matrix multiplication is associative, the preceding expression may be rewritten as

$$\left(\begin{bmatrix} 1 & 0 \\ 0 & 1 \end{bmatrix} \cdot \begin{bmatrix} 1 & 0 \\ 0 & -1 \end{bmatrix} \right) \cdot \begin{bmatrix} 2 & 6 & 6 \\ 1 & 1 & 3 \end{bmatrix}.$$

So multiplying the matrices for $r_{y=x}$ and r_x, in that order, gives the matrix for the composite $r_{y=x} \circ r_x$.

$$\begin{bmatrix} 0 & 1 \\ 1 & 0 \end{bmatrix} \cdot \begin{bmatrix} 1 & 0 \\ 0 & -1 \end{bmatrix} = \begin{bmatrix} 0 & -1 \\ 1 & 0 \end{bmatrix}.$$

Compare the preimage and the final image in Example 1. A single transformation that maps $\triangle ABC$ to $\triangle A''B''C''$ is R_{90}, the *rotation* of 90° counterclockwise with the origin as its center. (Note: we usually omit the degree symbol in the subscript for rotations.)
So $R_{90} = r_{y=x} \circ r_x$.

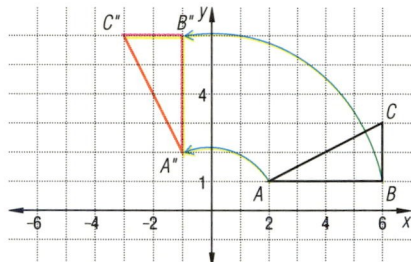

The results of Example 1 can be generalized:

Theorem

If M is the matrix associated with a transformation t, and N is the matrix associated with a transformation u, then NM is the matrix associated with the transformation $u \circ t$.

■ ■ ■ ■ ■ ■ ■

Example 2   Use the fact that a rotation of 180° can be considered as a 90° rotation followed by another 90° rotation to find a matrix for R_{180}.

Solution $R_{180} = R_{90} \circ R_{90}$. Thus the product of two matrices for R_{90} equals the matrix for R_{180}.

$$\begin{bmatrix} 0 & -1 \\ 1 & 0 \end{bmatrix} \cdot \begin{bmatrix} 0 & -1 \\ 1 & 0 \end{bmatrix} = \begin{bmatrix} -1 & 0 \\ 0 & -1 \end{bmatrix}$$

So $\begin{bmatrix} -1 & 0 \\ 0 & -1 \end{bmatrix}$ is a matrix for R_{180}.

Check Apply this matrix to △*ABC* of Example 1.

$$\begin{bmatrix} -1 & 0 \\ 0 & -1 \end{bmatrix} \cdot \begin{bmatrix} 2 & 6 & 6 \\ 1 & 1 & 1 \end{bmatrix} = \begin{bmatrix} -2 & -6 & -6 \\ -1 & -1 & -3 \end{bmatrix}$$

The resulting matrix should represent the image of △*ABC* under a rotation of 180° around the origin. The graph below illustrates that it does.

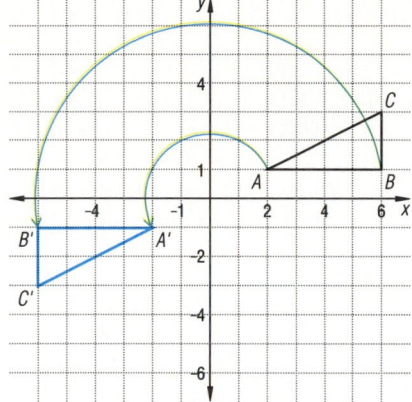

Similarly, a matrix for R_{270} can be found by noting that $R_{270} = R_{180} \circ R_{90}$. In the questions you are asked to prove that R_{270} is represented by $\begin{bmatrix} 0 & 1 \\ -1 & 0 \end{bmatrix}$.

The following theorem summarizes these results.

Theorem: Matrices for Rotations

$\begin{bmatrix} 0 & -1 \\ 1 & 0 \end{bmatrix}$ is the matrix for R_{90}, rotation of 90° around the origin.

$\begin{bmatrix} -1 & 0 \\ 0 & -1 \end{bmatrix}$ is the matrix for R_{180}, rotation of 180° around the origin.

$\begin{bmatrix} 0 & 1 \\ -1 & 0 \end{bmatrix}$ is the matrix for R_{270}, rotation of 270° around the origin.

You may wonder how you will ever remember matrices for each of the transformations you have studied. The key is a simple but beautiful result: if a transformation can be represented by a 2 × 2 matrix, then the first column of the matrix is the image under that transformation of (1, 0), and the second column of the matrix is the image of (0, 1).

cally, the coordinates are switched and one of them is changed to its opposite. If this algebra is done with the slope formula, then the numerator and denominator are switched, and one of them is changed to its opposite. Thus the slope of the image is the opposite of the reciprocal of the slope of the preimage.

Stress the general property of rotations: $R_\alpha \circ R_\beta = R_{\alpha+\beta}$. This property is applied in the next lesson.

Emphasize the use of the Matrix Basis Theorem to facilitate remembering matrices for transformations. We recommend that you go over its proof, but do not expect students to understand it. The proof is so short that many students do not see what has occurred.

For instance, suppose you forget the matrix for R_{90}. Visualize the 90° rotation of (1, 0) and (0, 1). Note that $R_{90}(1, 0) = (0, 1)$ and $R_{90}(0, 1) = (-1, 0)$. So the matrix for R_{90} is

image of image of
(1, 0) (0, 1)
↓ ↓
$$\begin{bmatrix} 0 & -1 \\ 1 & 0 \end{bmatrix}.$$

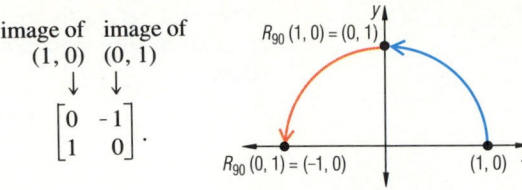

This relation is called the Matrix Basis Theorem because the matrix is *based* on the images of the points (1, 0) and (0, 1).

Matrix Basis Theorem

Suppose t is a transformation represented by a 2×2 matrix. If $t(1, 0) = (x_1, y_1)$ and $t(0, 1) = (x_2, y_2)$, then t has the matrix
$$\begin{bmatrix} x_1 & x_2 \\ y_1 & y_2 \end{bmatrix}.$$

Proof Let M be the 2×2 matrix for t. Because $t(1, 0) = (x_1, y_1)$ and $t(0, 1) = (x_2, y_2)$, the image matrix is $\begin{bmatrix} x_1 & x_2 \\ y_1 & y_2 \end{bmatrix}$, so

$$M \cdot \begin{bmatrix} 1 & 0 \\ 0 & 1 \end{bmatrix} = \begin{bmatrix} x_1 & x_2 \\ y_1 & y_2 \end{bmatrix}.$$

    ↑ ↑   ↑ ↑
1st 2nd image image
point point of 1st of 2nd
       point point

But $\begin{bmatrix} 1 & 0 \\ 0 & 1 \end{bmatrix}$ is the identity matrix for multiplication, so $M = \begin{bmatrix} x_1 & x_2 \\ y_1 & y_2 \end{bmatrix}$.

■ ■ ■ ■ ■ ■ ■ ■

Example 3 Use the Matrix Basis Theorem to verify that $\begin{bmatrix} -1 & 0 \\ 0 & -1 \end{bmatrix}$ represents the transformation R_{180}.

Solution Under a 180° rotation, the image of (1, 0) is (-1, 0) and the image of (0, 1) is (0, -1). Thus, by the Matrix Basis Theorem,

$$R_{180} = \begin{bmatrix} -1 & 0 \\ 0 & -1 \end{bmatrix}.$$

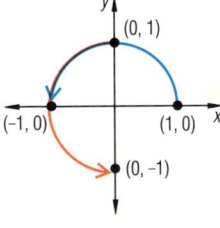

Check $\begin{bmatrix} -1 & 0 \\ 0 & -1 \end{bmatrix} \begin{bmatrix} x \\ y \end{bmatrix} = \begin{bmatrix} -x \\ -y \end{bmatrix}$,
and $R_{180}(x, y) = (-x, -y)$. It checks.

680

Questions

Covering the Reading

1. In the composite $r_{y=x} \circ r_x$, which reflection is done first? $\quad r_x$

2. What is a single transformation for $r_{y=x} \circ r_x$? $\quad \begin{bmatrix} 0 & -1 \\ 1 & 0 \end{bmatrix}$

3. What is a matrix for R_{90}? $\quad \begin{bmatrix} 0 & -1 \\ 1 & 0 \end{bmatrix}$

4. Refer to Example 1.
   a. Reflect $\triangle ABC$ over the x-axis. Then reflect its image $\triangle A'B'C'$ over the y-axis to get a second image $\triangle A''B''C''$. **See margin.**
   b. What transformation can take you directly from $\triangle ABC$ to $\triangle A''B''C''$? $\quad R_{180}$
   c. The composite $r_y \circ r_x$ is represented by
   $\begin{bmatrix} -1 & 0 \\ 0 & 1 \end{bmatrix} \cdot \begin{bmatrix} 1 & 0 \\ 0 & -1 \end{bmatrix}$. Find this product. $\quad \begin{bmatrix} -1 & 0 \\ 0 & -1 \end{bmatrix}$
   d. What single transformation is represented by the product in part c? $\quad R_{180}$

5. Refer to the figures at the right.
   a. What single transformation maps $\triangle RQS$ to $\triangle R'Q'S'$? $\quad R_{90}$
   b. Write a matrix for that transformation. $\quad \begin{bmatrix} 0 & -1 \\ 1 & 0 \end{bmatrix}$
   c. Use matrices to verify your answers in parts a and b.
   $\begin{bmatrix} 0 & -1 \\ 1 & 0 \end{bmatrix} \begin{bmatrix} 4 & 1 & 1 \\ 6 & -2 & 4 \end{bmatrix} = \begin{bmatrix} -6 & 2 & -4 \\ 4 & 1 & 1 \end{bmatrix}$

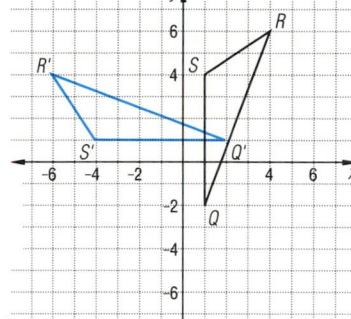

6. a. Derive the matrix for R_{270} by multiplying matrices for R_{90} and R_{180}.
   b. Does the order in which you multiply the matrices matter? Explain. **See below.**

7. Consider $\triangle DEF$ represented by the matrix $\begin{bmatrix} 0 & 6 & 6 \\ 0 & 0 & 2 \end{bmatrix}$.
   a. Find the reflection image of $\triangle DEF$ over $y = x$.
   b. Transform the image in part a by a scale change, $S_{2,0.6}$.
   c. Give a matrix for a single transformation for the composite $S_{2,0.6} \circ r_{y=x}$.

a) $\begin{bmatrix} 0 & 0 & 2 \\ 0 & 6 & 6 \end{bmatrix}$; b) $\begin{bmatrix} 0 & 0 & 4 \\ 0 & 3.6 & 3.6 \end{bmatrix}$; c) $\begin{bmatrix} 0 & 2 \\ .6 & 0 \end{bmatrix}$

6a) $\begin{bmatrix} 0 & -1 \\ 1 & 0 \end{bmatrix} \begin{bmatrix} -1 & 0 \\ 0 & -1 \end{bmatrix} = \begin{bmatrix} 0 & 1 \\ -1 & 0 \end{bmatrix}$

b) No. A rotation of 270° can be considered as a rotation of 180° followed by a rotation of 90°, or vice-versa.

NOTES ON QUESTIONS
Question 6: Students can be asked to generalize this property. They should generalize both the commutativity and the fact that it applies to all rotations about the same point. But it does not apply in general to rotations with different centers.

Questions 8–10: Students should become comfortable with these questions, since the Matrix Basis Theorem is applied in the next lesson.

Error Analysis for Questions 11–14: Students who are having difficulty should identify which transformation is to be done first. Stress that the order in which the matrices are written is the same as the order in which the transformations are written.

Question 16: Although this question may seem difficult, the process actually becomes quite easy for students after they have seen one or two cases.

Making Connections for Question 17: This is preparation for Lesson 11-7.

ADDITIONAL ANSWERS
4.a. $\triangle A'B'C' =$
$\begin{bmatrix} 2 & 6 & 6 \\ -1 & -1 & -3 \end{bmatrix}$;
$\triangle A''B''C'' =$
$\begin{bmatrix} -2 & -6 & -6 \\ -1 & -1 & -3 \end{bmatrix}$;

682

NOTES ON QUESTIONS

Question 22: You might use this question as a lead-in to Lesson 11-4. You could say: "You know the image of $(1, 0)$ under R_{30}. Now the question is: How can you find the image of all other points under this transformation? Read the next lesson to find out."

ADDITIONAL ANSWERS

11. $\begin{bmatrix} -14 & 9 & -89 \\ -14 & 3 & -53 \end{bmatrix}$

12. $\begin{bmatrix} 28 & 30 & -110 \\ 0 & 3 & -18 \end{bmatrix}$

15.b.

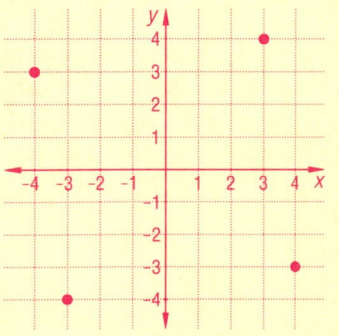

In 8–10, for each transformation give   **See below.**
a. the image of $(1, 0)$;
b. the image of $(0, 1)$;
c. the 2×2 matrix for the transformation derived from the Matrix Basis Theorem.

8. R_{-90} 9. $r_{x=-y}$ 10. $r_{x=y}$

Applying the Mathematics

In 11 and 12, a transformation t has matrix $\begin{bmatrix} 2 & 5 \\ 1 & 3 \end{bmatrix}$ and a transformation u has matrix $\begin{bmatrix} 4 & 2 \\ -1 & 3 \end{bmatrix}$. Let $\triangle ABC$ be represented by $\begin{bmatrix} 8 & 3 & 2 \\ -2 & 0 & -5 \end{bmatrix}$. Calculate and graph.

11. $(t \circ u)(\triangle ABC)$ **See margin.**

12. $(u \circ t)(\triangle ABC)$.   **See margin.**   13) $R_{90} \circ R_{270} = \begin{bmatrix} 0 & -1 \\ 1 & 0 \end{bmatrix}\begin{bmatrix} 0 & 1 \\ -1 & 0 \end{bmatrix} = \begin{bmatrix} 1 & 0 \\ 0 & 1 \end{bmatrix}$

13. Use matrices to verify that $R_{90} \circ R_{270}$ is the identity transformation.

14. a. Calculate a matrix for $r_x \circ r_{y=x}$.   a) $r_x \circ r_{y=x} = \begin{bmatrix} 1 & 0 \\ 0 & -1 \end{bmatrix}\begin{bmatrix} 0 & 1 \\ 1 & 0 \end{bmatrix} = \begin{bmatrix} 0 & 1 \\ -1 & 0 \end{bmatrix}$
    b. Describe the composite transformation.
    R_{270}

15. a. Find the image of $(3, 4)$ under each of R_{90}, R_{180}, and R_{270}.
    b. Graph $(3, 4)$ and its three images from part **a**.   **See margin.**
    c. The four points are vertices of what kind of polygon?   **square**
    d. Find the center and radius of a circle that passes through these four points.   a) $(-4, 3)$, $(-3, -4)$, $(4, -3)$; d) center $(0, 0)$, radius 5

16. Find a 2×2 matrix for t if t maps (x, y) onto $(3x - y, x + y)$.
    $\begin{bmatrix} 3 & -1 \\ 1 & 1 \end{bmatrix}$

Review

17. Fill in the blanks.   *(Lesson 11-2)*
    $\begin{bmatrix} \underline{a.} & \underline{b.} \\ \underline{c.} & \underline{d.} \end{bmatrix} \cdot \begin{bmatrix} 1 & 0 \\ 0 & 1 \end{bmatrix} = \begin{bmatrix} w & x \\ y & z \end{bmatrix}$ a) w; b) x; c) y; d) z

18. Write the matrix for the size change of magnitude 6.   *(Lesson 11-2)* $\begin{bmatrix} 6 & 0 \\ 0 & 6 \end{bmatrix}$

19. Find x and y so that the following is true.   *(Lesson 11-1)*
    $\begin{bmatrix} 1 & -1 \\ x & 3 \end{bmatrix} \cdot \begin{bmatrix} 2 & y \\ 0 & 1 \end{bmatrix} = \begin{bmatrix} 2 & 6 \\ -4 & -11 \end{bmatrix}$ $x = -2, y = 7$

20. The point $(t, 0.8)$ is in the second quadrant and on the unit circle. Find t.   *(Lesson 5-4)* $-.6$

8a) $(0, -1)$; b) $(1, 0)$; c) $\begin{bmatrix} 0 & 1 \\ -1 & 0 \end{bmatrix}$; 9a) $(0, -1)$; b) $(-1, 0)$; c) $\begin{bmatrix} 0 & -1 \\ -1 & 0 \end{bmatrix}$

10a) $(0, 1)$; b) $(1, 0)$; c) $\begin{bmatrix} 0 & 1 \\ 1 & 0 \end{bmatrix}$

21. A child's swing is mounted 12 ft off the ground. If the ropes supporting the swing are 10'6'' long, and the swing rotates through a vertical angle of 40°, as shown at the right, how high off the ground is the bottom of the swing in the position shown? *(Lesson 5-3)*   ≈ **4 feet**

22. *Multiple choice.* The point $(1, 0)$ is rotated 30° counterclockwise around the origin. Which statement is *false*?   *(Lessons 5-2, 5-3, 5-4)*
(a) Its image is $(\cos 30°, \sin 30°)$.
(b) Its image is $\left(\dfrac{\sqrt{3}}{2}, \dfrac{1}{2}\right)$.
(c) The arc length from $(1, 0)$ to the image is $\dfrac{\pi}{6}$.
(d) The arc length from $(1, 0)$ to the image is $\tan 30°$.   **(d)**

23. **a.** Graph $f(\theta) = 2 \cos \theta$ for $-\pi \le \theta \le 3\pi$ using an automatic grapher.
b. On the same set of axes, graph $g(\theta) = \cos 2\theta$.
c. Based on the graphs, determine a transformation S which would map the graph of f onto the graph of g.   *(Lessons 3-5, 6-1)*   **$(x, y) \to (x/2, y/2)$**
a,b) See margin.

In 24 and 25, simplify.   *(Previous course)*

24. $\dfrac{\sqrt{2}}{2}\left(-\dfrac{\sqrt{2}}{2}\right)$ $-\dfrac{1}{2}$

25. $\dfrac{3}{2} \cdot \dfrac{\sqrt{3}}{2} + \dfrac{\sqrt{3}}{2}$ $\dfrac{5\sqrt{3}}{4}$

Exploration

26. Write a Matrix Basis Theorem for three dimensions. Verify your theorem with an example.

Sample: Suppose t is a transformation represented by a 3×3 matrix. If $t(1, 0, 0) = (x_1, y_1, z_1)$, $t(0, 1, 0) = (x_2, y_2, z_2)$, and $t(0, 0, 1) = (x_3, y_3, z_3)$, then t has the matrix

$$\begin{bmatrix} x_1 & x_2 & x_3 \\ y_1 & y_2 & y_3 \\ z_1 & z_2 & z_3 \end{bmatrix}.$$

Example: Let t by the reflection over the xy-plane; that is, $t(x, y, z) = (x, y, -z)$ for all points (x, y, z). Then $t(1, 0, 0) = (1, 0, 0)$, $t(0, 1, 0) = (0, 1, 0)$, and $t(0, 0, 1) = (0, 0, -1)$. The matrix of t is

$$\begin{bmatrix} 1 & 0 & 0 \\ 0 & 1 & 0 \\ 0 & 0 & -1 \end{bmatrix}.$$

To verify that this represents t, compute the product

$$\begin{bmatrix} 1 & 0 & 0 \\ 0 & 1 & 0 \\ 0 & 0 & -1 \end{bmatrix} \cdot \begin{bmatrix} x \\ y \\ z \end{bmatrix} = \begin{bmatrix} x \\ y \\ -z \end{bmatrix}.$$

MORE PRACTICE
For more questions on SPUR Objectives, use *Lesson Master 11-3*, shown below.

PROJECTS
Project 1 on page 718 is related to the content of this lesson.

EVALUATION
A quiz covering Lessons 11-1 through 11-3 is provided in the Teacher's Resource File.

23.a.,b

NAME _____

OBJECTIVES

G Represent rotations as matrices.

H Represent composites of transformations as matrix products.

I Find the image of a figure under a transformation using a matrix.

TEACHING NOTES

The proof of the Rotation Matrix Theorem is important because understanding it will save time in the next lesson. A teaching aid with the graph is provided.

In **Example 1,** we use the equal sign between a name for a transformation and its matrix. Again the idea is that these are two names for the same transformation, not that they are the same object.
You might note that, if 0.707 is used as an approximation to $\frac{\sqrt{2}}{2}$, then Check 1 yields the matrix

$$\begin{bmatrix} 0 & .{-}999698 \\ .999698 & 0 \end{bmatrix}$$

Remind students that if they are unsure about a matrix for a transformation, a specific example like that in Check 2 can be very helpful.

Example 2 shows how computers can simulate a spinning wheel. A teaching aid is provided to help in explanation.

The General Rotation Matrix

You have seen that size changes with center at (0, 0), scale changes, reflections over the *x*- and *y*-axes and the line *y* = *x*, rotations of 90°, 180°, and 270°, and several other transformations can all be represented by 2 × 2 matrices. For what other transformations do 2 × 2 matrices exist?

The answer, perhaps a surprise, is that *all* rotations around (0, 0) can be represented by 2 × 2 matrices. To develop a general rotation matrix requires the Matrix Basis Theorem and some trigonometry.

Let θ be the magnitude of a rotation around the origin. In this discussion, θ is in degrees. First, note that composition of rotations with the same center is commutative, so $R_\theta \circ R_{90} = R_{90} \circ R_\theta$.

Consider the images A' and B' of the points $A = (1, 0)$ and $B = (0, 1)$. By definition of the cosine and sine,
$A' = R_\theta(1, 0) = (\cos \theta, \sin \theta)$.
Since $B' = R_\theta(B)$ and $B = R_{90}(A)$, then $B' = (R_\theta \circ R_{90})(A) = (R_{90} \circ R_\theta)(A) = R_{90}(A')$. Thus, if $B' = (x, y)$, then

$$\begin{bmatrix} x \\ y \end{bmatrix} = \begin{bmatrix} 0 & -1 \\ 1 & 0 \end{bmatrix} \cdot \begin{bmatrix} \cos \theta \\ \sin \theta \end{bmatrix} = \begin{bmatrix} -\sin \theta \\ \cos \theta \end{bmatrix}.$$

$$B' = \quad R_{90} \quad \cdot \quad (A')$$

Thus, $B' = R_\theta(0, 1) = (-\sin \theta, \cos \theta)$.

From the images of (1, 0) and (0, 1) and the Matrix Basis Theorem, the matrix for R_θ follows.

Rotation Matrix Theorem

The matrix for R_θ, the rotation of magnitude θ about the origin, is
$$\begin{bmatrix} \cos \theta & -\sin \theta \\ \sin \theta & \cos \theta \end{bmatrix}.$$

Proof The first column is $R_\theta(1, 0) = (\cos \theta, \sin \theta)$. The second column is $R_\theta(0, 1) = (-\sin \theta, \cos \theta)$.

684

In the questions you are asked to verify that the matrices for R_{90}, R_{180}, and R_{270} are special cases of this theorem.

■ ■ ■ ■ ■ ■ ■ ■ ■ ■

Example 1   Find a 2 × 2 rotation matrix for R_{45}.

Solution   Use the Rotation Matrix Theorem.

$$R_{45} = \begin{bmatrix} \cos 45° & -\sin 45° \\ \sin 45° & \cos 45° \end{bmatrix} = \begin{bmatrix} \frac{\sqrt{2}}{2} & -\frac{\sqrt{2}}{2} \\ \frac{\sqrt{2}}{2} & \frac{\sqrt{2}}{2} \end{bmatrix}$$

Check 1   Using $R_{45} \circ R_{45} = R_{90}$, is the square of the matrix for R_{45} equal to the matrix for R_{90}?

$$\begin{bmatrix} \frac{\sqrt{2}}{2} & -\frac{\sqrt{2}}{2} \\ \frac{\sqrt{2}}{2} & \frac{\sqrt{2}}{2} \end{bmatrix} \cdot \begin{bmatrix} \frac{\sqrt{2}}{2} & -\frac{\sqrt{2}}{2} \\ \frac{\sqrt{2}}{2} & \frac{\sqrt{2}}{2} \end{bmatrix} = \begin{bmatrix} 0 & -1 \\ 1 & 0 \end{bmatrix}$$

Yes, it checks.

Check 2   Triangle ABC below is represented by $\begin{bmatrix} 1 & 2 & 2 \\ 3 & 3 & 6 \end{bmatrix}$. Its image, $\triangle A'B'C'$, is represented by the product

$$\begin{bmatrix} \frac{\sqrt{2}}{2} & -\frac{\sqrt{2}}{2} \\ \frac{\sqrt{2}}{2} & \frac{\sqrt{2}}{2} \end{bmatrix} \cdot \begin{bmatrix} 1 & 2 & 2 \\ 3 & 3 & 6 \end{bmatrix} = \begin{bmatrix} -\sqrt{2} & -\frac{\sqrt{2}}{2} & -2\sqrt{2} \\ 2\sqrt{2} & \frac{5\sqrt{2}}{2} & 4\sqrt{2} \end{bmatrix}.$$

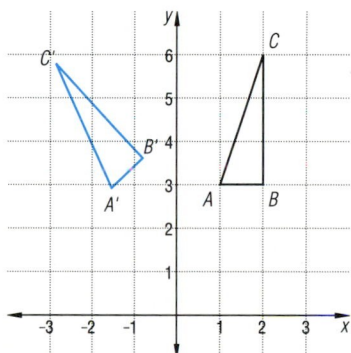

You are asked to verify that $AO = A'O$ and $m\angle AOA' = 45°$ in the Questions.

Computer animators use matrices to produce rotation images. Points are identified as ordered pairs on a coordinate grid. To rotate a set of points around the origin, a matrix for the rotation is multiplied by a matrix for the points.

ADDITIONAL EXAMPLES
1. a. Give a matrix for R_{-45}

$$\begin{bmatrix} -\frac{\sqrt{2}}{2} & \frac{\sqrt{2}}{2} \\ -\frac{\sqrt{2}}{2} & -\frac{\sqrt{2}}{2} \end{bmatrix}$$

b. Check the answer by multiplying that matrix by the matrix for R_{45}.

The product of the matrices is $\begin{bmatrix} 1 & 0 \\ 0 & 1 \end{bmatrix}$, **the matrix for R_0.**

2. In the computer game of **Example 2** in the lesson, what are the coordinates of the endpoints of the letter T if the wheel is stopped after 1.2 seconds?

$$\begin{bmatrix} \cos 1.2\pi & -\sin 1.2\pi \\ \sin 1.2\pi & \cos 1.2\pi \end{bmatrix} \cdot$$

$$\begin{bmatrix} 1.7 & 1.8 & 4.0 \\ 4.8 & 8.0 & 7.0 \end{bmatrix} \approx$$

$$\begin{bmatrix} 1.4 & 3.2 & 0.9 \\ -4.9 & -7.5 & -8.0 \end{bmatrix}$$

$$\begin{bmatrix} \frac{\sqrt{3}}{2} & -\frac{1}{2} \\ \frac{1}{2} & \frac{\sqrt{3}}{2} \end{bmatrix} \begin{bmatrix} 1 & 2 & 2 \\ 3 & 3 & 6 \end{bmatrix} =$$

$$\begin{bmatrix} \frac{\sqrt{3}-3}{2} & \frac{2\sqrt{3}-3}{2} & \frac{2\sqrt{3}-6}{2} \\ \frac{3\sqrt{3}+1}{2} & \frac{2\sqrt{3}+3}{2} & \frac{6\sqrt{3}+2}{2} \end{bmatrix}$$

5. $R_{90} =$
$$\begin{bmatrix} \cos 90 & -\sin 90 \\ \sin 90 & \cos 90 \end{bmatrix} =$$
$$\begin{bmatrix} 0 & -1 \\ 1 & 0 \end{bmatrix}$$

6. $R_{180} =$
$$\begin{bmatrix} \cos 180 & -\sin 180 \\ \sin 180 & \cos 180 \end{bmatrix} =$$
$$\begin{bmatrix} -1 & 0 \\ 0 & -1 \end{bmatrix}$$

7. $R_{270} =$
$$\begin{bmatrix} \cos 270 & -\sin 270 \\ \sin 270 & \cos 270 \end{bmatrix} =$$
$$\begin{bmatrix} 0 & 1 \\ -1 & 0 \end{bmatrix}$$

8.a. $A = (1, 3)$,
$A' = (-\sqrt{2}, 2\sqrt{2})$;
$AO =$
$\sqrt{(1-0)^2 + (3-0)^2}$
$= \sqrt{1+9} = \sqrt{10}$;
$A'O =$
$\sqrt{(-\sqrt{2}-0)^2 + (2\sqrt{2}-0)^2}$
$= \sqrt{2+8} = \sqrt{10}$

b. Let Q be the point
$(3, 0)$. Then $m\angle A'OA =$
$m\angle A'OQ - m\angle AOQ$,
$m\angle A'OQ = \tan^{-1}\left(\frac{2\sqrt{2}}{-\sqrt{2}}\right) +$
$180° \approx 116.6°$ and
$m\angle AOQ = \tan^{-1}\left(\frac{3}{1}\right) \approx$
$71.6°$. So $m\angle A'OA \approx 116.6°$
$- 71.6° = 45°$.

Example 2 In a computer game, the wheel below (pictured on a coordinate grid) spins counterclockwise around the origin at the rate of π radians per second. The **T** (in the word **TO**) has endpoints at (-0.8, 5.3), (-2.3, 7.5), and (-0.3, 7.8). Approximate the coordinates of the endpoints of **T** if someone playing the game hits a key to stop the wheel after 5.43 seconds.

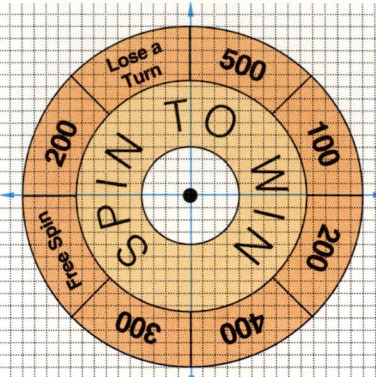

Solution In 5.43 seconds, the wheel covers 5.43π radians. As any multiple of 2π revolutions preserves the image, it is the last 1.43π of a revolution that should be considered.
The matrix for $R_{1.43\pi}$ is

$$\begin{bmatrix} \cos 1.43\pi & -\sin 1.43\pi \\ \sin 1.43\pi & \cos 1.43\pi \end{bmatrix} \approx \begin{bmatrix} -.218 & .976 \\ -.976 & -.218 \end{bmatrix}.$$

Multiply the matrix for the preimage by this matrix.

$$\begin{bmatrix} -.218 & .976 \\ -.976 & -.218 \end{bmatrix} \cdot \begin{bmatrix} -0.8 & -2.3 & -0.3 \\ 5.3 & 7.5 & 7.8 \end{bmatrix} \approx \begin{bmatrix} 5.3 & 7.8 & 7.7 \\ -0.4 & 0.6 & -1.4 \end{bmatrix}$$

Thus, the coordinates of the endpoints of **T** in its new position are (5.3, -0.4), (7.8, 0.6), and (7.7, -1.4).

Check 1.43π radians is about $\frac{7}{10}$ of a full revolution, or about 260°. The image is shown below.

686

Questions

Covering the Reading

1. What is the matrix for R_t, the rotation of magnitude t about the origin?
   See below.

2. **a.** Give the 2×2 matrix for R_{30}.   **See below.**
   b. Check your result by multiplying the matrix for R_{30} by the matrix for the triangle in the *check* of Example 1, and plotting the preimage and image on the same set of axes.   **See margin.**

3. If $\sin 27° \approx 0.454$ and $\cos 27° \approx 0.891$, what is an approximate matrix for R_{27}?   **See below.**

4. If $\sin(-12°) = -0.208$ and $\cos(-12°) = 0.978$, what is an approximate matrix for R_{-12}? $\begin{bmatrix} 0.978 & 0.208 \\ -0.208 & 0.978 \end{bmatrix}$

In 5–7, verify the Rotation Matrix Theorem for these rotations.

5. R_{90} 6. R_{180} 7. R_{270}
   5–7) See margin.

8. In Example 1, verify:
   a. $AO = A'O$;   **b.** $m\angle AOA' = 45°$.   **See margin.**

9. Refer to Example 2. The endpoints of the I in "SPIN" are given by the matrix $\begin{bmatrix} -5.3 & -7.8 \\ 0.8 & 1.2 \end{bmatrix}$. Approximate the coordinates of the I after it is turned 3.2π radians counterclockwise. $\begin{bmatrix} 4.8 & 7.0 \\ 2.5 & 3.6 \end{bmatrix}$

Applying the Mathematics

10. **a.** Use the Rotation Matrix Theorem to find a matrix for R_{60}.
    b. Verify your result in part **a** by showing $(R_{60})^3 = R_{180}$.   **See margin.**

11. The 30-60-90 triangle shown at the right may be represented by the matrix $\begin{bmatrix} 0 & 8 & 6 \\ 0 & 0 & 2\sqrt{3} \end{bmatrix}$.
    a. Find a matrix for its image under R_{210}. (Use exact values.)
    b. Graph the image.
       See margin.

    a) $\begin{bmatrix} 0 & -4\sqrt{3} & -2\sqrt{3} \\ 0 & -4 & -6 \end{bmatrix}$

1) $\begin{bmatrix} \cos t & -\sin t \\ \sin t & \cos t \end{bmatrix}$; 2a) $\begin{bmatrix} \dfrac{\sqrt{3}}{2} & -\dfrac{1}{2} \\ \dfrac{1}{2} & \dfrac{\sqrt{3}}{2} \end{bmatrix}$; 3) $\begin{bmatrix} 0.891 & -0.454 \\ 0.454 & 0.891 \end{bmatrix}$

10.a. $R_{60} = \begin{bmatrix} \cos 60 & -\sin 60 \\ \sin 60 & \cos 60 \end{bmatrix} = \begin{bmatrix} \dfrac{1}{2} & -\dfrac{\sqrt{3}}{2} \\ \dfrac{\sqrt{3}}{2} & \dfrac{1}{2} \end{bmatrix}$

b. $(R_{60})^3 = \begin{bmatrix} \dfrac{1}{2} & -\dfrac{\sqrt{3}}{2} \\ \dfrac{\sqrt{3}}{2} & \dfrac{1}{2} \end{bmatrix} \cdot \begin{bmatrix} \dfrac{1}{2} & -\dfrac{\sqrt{3}}{2} \\ \dfrac{\sqrt{3}}{2} & \dfrac{1}{2} \end{bmatrix} \cdot \begin{bmatrix} \dfrac{1}{2} & -\dfrac{\sqrt{3}}{2} \\ \dfrac{\sqrt{3}}{2} & \dfrac{1}{2} \end{bmatrix} =$

$\begin{bmatrix} -\dfrac{1}{2} & -\dfrac{\sqrt{3}}{2} \\ \dfrac{\sqrt{3}}{2} & -\dfrac{1}{2} \end{bmatrix} \cdot \begin{bmatrix} \dfrac{1}{2} & -\dfrac{\sqrt{3}}{2} \\ \dfrac{\sqrt{3}}{2} & \dfrac{1}{2} \end{bmatrix} =$

$\begin{bmatrix} -1 & 0 \\ 0 & -1 \end{bmatrix} = R_{180}$

11.b.

12. A student is designing a club logo to be produced on a computer screen. The outline is a regular pentagon inscribed in a circle with radius 10. One vertex is at (0, 10). Find (to the nearest thousandth) the coordinates of the other vertices.

(-9.511, 3.090)
(-5.878, -8.090)
(5.878, -8.090)
(9.511, 3.090)

13. **a.** Show that a matrix for $R_{-\theta}$, a clockwise rotation of θ around the origin, is $\begin{bmatrix} \cos\theta & \sin\theta \\ -\sin\theta & \cos\theta \end{bmatrix}$.   **See margin.**

    b. Use part **a** to determine the matrix for a clockwise rotation of 90° around the origin. $\begin{bmatrix} \cos 90° & \sin 90° \\ -\sin 90° & \cos 90° \end{bmatrix} = \begin{bmatrix} 0 & 1 \\ -1 & 0 \end{bmatrix}$

Review

14. Use the Matrix Basis Theorem to determine the matrix for the reflection over the line $y = -x$.   *(Lesson 11-3)*
    See below.

15. Is the transformation the identity transformation; that is, does it map every figure to itself?   *(Lesson 11-2)*
    a. rotation of 360°   **yes**
    b. reflection over the *y*-axis   **no**
    c. size change of magnitude 1   **yes**
    d. size change of magnitude -1   **no**

In 16–18, for a standard normal distribution find each of the following. *(Lesson 10-5)*

16. the mean   **0**

17. the standard deviation   **1**

18. the area between the curve and the *x*-axis   **1**

19. **a.** Determine an equation for a sine curve *s* which has an amplitude of 3 and a phase shift of $\frac{\pi}{4}$ from the parent sine curve.

    b. Explain why the point $\left(\frac{\pi}{2}, \frac{3\sqrt{2}}{2}\right)$ should be on *s*.   *(Lesson 6-2)*

    a) $y = 3\sin\left(x - \frac{\pi}{4}\right)$; **b)** The graph of $y = 3\sin\left(x - \frac{\pi}{4}\right)$ is the image of the graph of $y = \sin x$ under $(x, y) \to \left(x + \frac{\pi}{4}, 3y\right)$. The transformation takes the point $\left(\frac{\pi}{4}, \frac{\sqrt{2}}{2}\right)$, which is on the curve $y = \sin x$, and maps it to $\left(\frac{\pi}{2}, \frac{3\sqrt{2}}{2}\right)$.

    14) $t(1, 0) = (0, -1)$ and $t(0, 1) = (-1, 0)$ so, $r_{y=-x} = \begin{bmatrix} 0 & -1 \\ -1 & 0 \end{bmatrix}$.

688

In 20–22, *true* or *false*. *(Lessons 5-3, 5-5, 5-7)*

20. $\cos(30° + 60°) = \cos 30° + \cos 60°$ **False**

21. $\sin(90°) = 2 \sin 45°$ **False**

22. $\cos \frac{\pi}{3} + \sin \frac{\pi}{3} = 1$ **False**

23. Consider the points $(1, 0)$, $(-1, 4)$, and $(-3, 16)$. *(Lessons 2-7, 9-2)*
    a. Find an equation for the quadratic function that passes through these points. $y = x^2 - 2x + 1$
    b. Find an equation for a cubic function that passes through these points.   **sample:** $y = x^3 + 4x^2 - 3x - 2$

In 24–26, *multiple choice*. The data set shows a correlation coefficient r that is
    (a) strongly positive
    (b) strongly negative
    (c) approximately zero.   *(Lesson 2-5)*

24.   **25.** 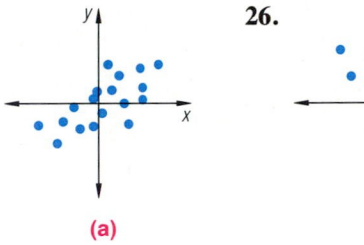   **26.**

         (c)   **(a)**   **(b)**

27. Suppose a print shop states that they charge $65 to print 500 posters and $125 to print 1000 posters. Assume that the cost c (in dollars) is linearly related to the number n of copies made.
    a. Find an equation relating c to n. $c = .12n + 5$
    b. How much should the printer charge to print 2000 posters?
    (Lesson 2-2)   **$245**

Exploration

28. The *determinant* of the matrix $M = \begin{bmatrix} a & b \\ c & d \end{bmatrix}$ is $ad - bc$.

    a. Multiply the matrix M by $S = \begin{bmatrix} 0 & 1 & 1 & 0 \\ 0 & 0 & 1 & 1 \end{bmatrix}$.
    b. How is the area of the figure MS related to the determinant of M?
    a) $\begin{bmatrix} 0 & a & a+b & b \\ 0 & c & c+d & d \end{bmatrix}$; b) **The area of MS is the absolute value of the determinant of M.**

MORE PRACTICE
For more questions on SPUR Objectives, use *Lesson Master 11-4*, shown below.

PROJECTS
Project 2 on page 718 is related to the content of this lesson.

LESSON

11-5

Identities for cos $(\alpha + \beta)$ and sin $(\alpha + \beta)$

Matrices are used to transform images displayed on graphics workstations, such as the one shown here.

Suppose that a computer animator wishes to rotate the point $P = (\cos \alpha, \sin \alpha)$ counterclockwise $\beta°$ around the origin. The image of P will be Q, where $Q = (\cos(\alpha + \beta), \sin(\alpha + \beta))$, as illustrated here.

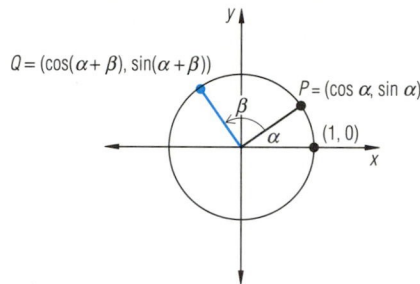

As you know, in general

$$\cos(\alpha + \beta) \neq \cos \alpha + \cos \beta$$

and

$$\sin(\alpha + \beta) \neq \sin \alpha + \sin \beta.$$

For instance, when $\alpha = 45°$ and $\beta = 30°$,

$$\sin (45° + 30°) = \sin 75° \approx 0.966,$$

but

$$\sin 45° + \sin 30° \approx 0.707 + 0.5$$
$$\approx 1.207.$$

Still, the coordinates of Q *can* be stated in terms of cosines and sines of α and β. The idea is to use rotation matrices to compute $(R_\alpha \circ R_\beta) (1, 0)$ (or, because composition of rotations with the same center is commutative, $(R_\beta \circ R_\alpha) (1, 0)$).

Theorem (Addition Identities for the Cosine and Sine)

For all real numbers α and β,
$$\cos(\alpha + \beta) = \cos \alpha \cos \beta - \sin \alpha \sin \beta$$
$$\sin(\alpha + \beta) = \sin \alpha \cos \beta + \cos \alpha \sin \beta.$$

690

Proof By definition of the sine and cosine functions,
$$(\cos(\alpha + \beta),\ \sin(\alpha + \beta)) = R_{\alpha+\beta}\,(1,\ 0).$$

But it is also true that $\quad R_{\alpha+\beta} = R_\alpha \circ R_\beta.$
Thus, $\quad (\cos(\alpha + \beta),\ \sin(\alpha + \beta)) = (R_\alpha \circ R_\beta)\,(1,\ 0)$
$$= R_\alpha(R_\beta(1,\ 0))$$
$$= R_\alpha(\cos\beta,\ \sin\beta).$$

Now translate this into matrices.

$$\begin{bmatrix} \cos(\alpha + \beta) \\ \sin(\alpha + \beta) \end{bmatrix} = \begin{bmatrix} \cos\alpha & -\sin\alpha \\ \sin\alpha & \cos\alpha \end{bmatrix} \begin{bmatrix} \cos\beta \\ \sin\beta \end{bmatrix}$$

$$= \begin{bmatrix} \cos\alpha\cos\beta - \sin\alpha\sin\beta \\ \sin\alpha\cos\beta + \cos\alpha\sin\beta \end{bmatrix}$$

Two matrices are equal if and only if their corresponding elements are equal, so

$$\cos(\alpha + \beta) = \cos\alpha\cos\beta - \sin\alpha\sin\beta$$
and $\quad\quad\quad\quad\quad \sin(\alpha + \beta) = \sin\alpha\cos\beta + \cos\alpha\sin\beta.$

The preceding identities are very useful in mathematics. You should either memorize them or be able to reconstruct them.

Using the exact values of cosine and sine of 30°, 45°, 60°, and 90°, these formulas lead to the exact values for the sine and cosine of many other angles.

▪ ▪ ▪ ▪ ▪ ▪ ▪ ▪ ▪ ▪

Example 1 Find an exact value for sin 75°.

Solution Let $\alpha = 45°$ and $\beta = 30°$. Then

$$\sin 75° = \sin(45° + 30°)$$
$$= \sin 45°\cos 30° + \cos 45°\sin 30°$$
$$= \frac{\sqrt{2}}{2} \cdot \frac{\sqrt{3}}{2} + \frac{\sqrt{2}}{2} \cdot \frac{1}{2} = \frac{\sqrt{6} + \sqrt{2}}{4}.$$

Check Evaluate $\dfrac{\sqrt{6} + \sqrt{2}}{4}$ and sin 75° on your calculator. You should get approximately 0.966 both times.

It is also possible to derive formulas for $\sin(\alpha - \beta)$ and $\cos(\alpha - \beta)$. Example 2 shows how to do this for $\sin(\alpha - \beta)$. You are asked to verify that $\cos(\alpha - \beta) = \cos\alpha\cos\beta + \sin\alpha\sin\beta$ in the Questions.

▪ ▪ ▪ ▪ ▪ ▪ ▪ ▪ ▪ ▪

Example 2 Derive a formula for $\sin(\alpha - \beta)$ in terms of sines and cosines of α and β.

Solution Rewrite $\alpha - \beta$ as $\alpha + (-\beta)$. Then

$\sin(\alpha - \beta) = \sin(\alpha + (-\beta))$
$\quad\quad\quad\quad = \sin\alpha\cos(-\beta) + \cos\alpha\sin(-\beta)$ Addition Identity for the Sine
$\quad\quad\quad\quad = \sin\alpha\cos\beta + \cos\alpha\,(-\sin\beta)$ Opposites Theorems
$\quad\quad\quad\quad = \sin\alpha\cos\beta - \cos\alpha\sin\beta.$ Simplify

---

1. Use the sine of a sum to find an exact value for sin 105°

sin (60° + 45°) =
sin 60° cos 45° +
cos 60° sin 45° =

$\left(\dfrac{\sqrt{3}}{2} \cdot \dfrac{\sqrt{2}}{2} + \dfrac{1}{2} \cdot \dfrac{\sqrt{2}}{2}\right) =$

$\dfrac{\sqrt{6} + \sqrt{2}}{4}$

2. Derive a formula for cos (30° + x) in terms of cos x and sin x.

cos (30° + x) =
cos 30° cos x – sin 30° sin x

$= \dfrac{\sqrt{3}}{2}\cos x + \dfrac{1}{2}\sin x$

3. Prove the Complements Theorem for the sine function.

$\sin\left(\dfrac{\pi}{2} - \theta\right)$

$= \sin\dfrac{\pi}{2}\cos\theta - \cos\dfrac{\pi}{2}\sin\theta$

$= 1 \cdot \cos\theta - 0 \cdot \sin\theta$

$= \cos\theta$

The formulas for $\cos(\alpha + \beta)$ and $\sin(\alpha + \beta)$ provide another way to prove some theorems about circular functions, such as the Supplements and Complements theorems in Lesson 5-7.

▪ ▪ ▪ ▪ ▪ ▪ ▪ ▪

Example 3 Prove the Supplements Theorem for the sine function:
$\sin(\pi - \theta) = \sin \theta$ for all real numbers θ.

Solution From the identity in Example 2,
$$\sin(\pi - \theta) = \sin \pi \cos \theta - \cos \pi \sin \theta$$
$$= 0 \cdot \cos \theta - (-1)\sin \theta$$
$$= \sin \theta.$$

Questions

Covering the Reading

1. In the proof of the theorem in the lesson it is stated that $(R_\alpha \circ R_\beta)(1, 0) = R_\alpha(\cos \beta, \sin \beta)$. Explain why this is true. **See margin.**

2. Consider the statement: *For all α and β,*
$cos(\alpha + \beta) = cos \, \alpha + cos \, \beta$.
   a. Give one pair of values for α and β for which the statement is true.
   b. Give a counterexample to the statement.
   a) sample: $\alpha = \frac{\pi}{2}$, $\beta = \frac{3\pi}{4}$; b) See margin.

In 3 and 4, simplify without using a calculator.

3. $\sin 75° \cos 15° + \cos 75° \sin 15°$ $\sin(75° + 15°) = \sin(90°) = 1$

4. $\cos \frac{11\pi}{12} \cos \frac{7\pi}{12} - \sin \frac{11\pi}{12} \cos \frac{7\pi}{12}$ $\cos\left(\frac{11\pi}{12} + \frac{7\pi}{12}\right) = \cos\left(\frac{3\pi}{2}\right) = 0$

In 5 and 6, give an exact value.

5. $\cos 75°$ $\frac{\sqrt{6} - \sqrt{2}}{4}$ 6. $\sin 15°$ $\frac{\sqrt{6} - \sqrt{2}}{4}$

In 7–9, use the formula for $\cos(\alpha + \beta)$ to prove the following.

7. $\cos(\alpha - \beta) = \cos \alpha \cos \beta + \sin \alpha \sin \beta$ **See margin.**

8. Supplements Theorem for the cosine function
$\cos(\pi - x) = \cos \pi \cos x + \sin \pi \sin x = -\cos x$

9. Complements Theorem for the cosine function
$\cos\left(\frac{\pi}{2} - x\right) = \cos \frac{\pi}{2} \cos x + \sin \frac{\pi}{2} \sin x = \sin x$

Applying the Mathematics

In 10–12, give exact values.

10. $\sin 255°$ $\frac{-\sqrt{2} - \sqrt{6}}{4}$ 11. $\cos\left(-\frac{13\pi}{12}\right)$ $\frac{-\sqrt{2} - \sqrt{6}}{4}$

12. $\tan 75°$ $2 + \sqrt{3}$

13. Give an exact matrix for $R_{5\pi/12}$. (Hint: $\frac{5\pi}{12} = \frac{3\pi}{12} + \frac{2\pi}{12}$.)

$$\begin{bmatrix} \frac{\sqrt{6} - \sqrt{2}}{4} & \frac{-\sqrt{6} - \sqrt{2}}{4} \\ \frac{\sqrt{6} + \sqrt{2}}{4} & \frac{\sqrt{6} - \sqrt{2}}{4} \end{bmatrix}$$

692

14. *True* or *false*.   Exact values of $\sin\theta$ and $\cos\theta$ can be found for all integral multiples of $\frac{\pi}{12}$ between 0 and 2π. Justify your answer. **See margin.**

15. a. Without graphing, predict which two of the following functions are identical.

$$f(x) = 2\,\sin\!\left(x + \frac{\pi}{6}\right)$$

$$g(x) = 2\,\sin x + 2\,\sin\frac{\pi}{6}$$

$$h(x) = \sqrt{3}\,\sin x + \cos x \quad \textbf{\textit{f and h}}$$

b. Check your prediction by graphing $y = f(x)$, $y = g(x)$, and $y = h(x)$ on the same set of axes.   **See margin.**

c. Explain your results using a theorem in this lesson.   **See margin.**

16. Use the identity for $\sin(\alpha + \beta)$ to show that $\sin(x + 2\pi n) = \sin x$ for all integers n.   **See margin.**

17. Prove that $\sin\!\left(x - \frac{\pi}{2}\right) = -\cos x$.   **See margin.**

18. Write the matrix for $R_{\alpha+\beta}$ in terms of sines and cosines of α and β. **See below.**

19. $\triangle PQR$ has acute angles P and Q with $\cos P = \frac{1}{2}$ and $\cos Q = \frac{1}{3}$. Find:

   a. $\cos(P + Q)$ **$(1-2\sqrt{6})/6$**   **b.** $\cos R$.   **$(2\sqrt{6}-1)/6$**

18) $R_{\alpha+\beta} = \begin{bmatrix} \cos\alpha\cos\beta - \sin\alpha\sin\beta & -\sin\alpha\cos\beta - \cos\alpha\sin\beta \\ \sin\alpha\cos\beta + \cos\alpha\sin\beta & \cos\alpha\cos\beta - \sin\alpha\sin\beta \end{bmatrix}$

Review

20. Determine the coordinates of the image of point $(2, 5)$ after it is rotated counterclockwise around the origin by $42°$.   *(Lesson 11-4)* $\approx(-1.86, 5.05)$

In 21 and 22, consider the polygon represented by the matrix

$$P = \begin{bmatrix} 1 & 6 & 8 & 3 \\ 1 & 1 & 4 & 4 \end{bmatrix}.$$ **21b) It is a scale change: $(x, y) \rightarrow (2x, -y)$**

21. a. Calculate $P' = \begin{bmatrix} 2 & 0 \\ 0 & -1 \end{bmatrix} \cdot P$. $\begin{bmatrix} 2 & 12 & 16 & 6 \\ -1 & -1 & -4 & -4 \end{bmatrix}$

b. Describe the transformation mapping P onto P'.   *(Lessons 11-2, 11-3)*

22. a. Calculate $P'' = \begin{bmatrix} \cos 45° & -\sin 45° \\ \sin 45° & \cos 45° \end{bmatrix} \cdot P$.   **See margin.**

b. Describe the transformation mapping P onto P''.   *(Lessons 11-2, 11-4)*
   It is a counterclockwise rotation of 45°.

23. Kevin, Laura, and Sergio work part-time after school. The matrix R below gives the hourly payment each one earns. The matrix H shows the number of hours each worked on the days of a certain week.   *(Lesson 11-1)*

$$R = \begin{bmatrix} 5.25 \\ 5.40 \\ 5.50 \end{bmatrix} \begin{matrix} \text{Kevin} \\ \text{Laura} \\ \text{Sergio} \end{matrix}$$

	Kevin	Laura	Sergio
Monday	3	4	4
Tuesday	5	5	4
Wednesday	2	4	4
Thursday	4	2	4
Friday	3	0	4

$= H$

How many hours did Laura work in this week?   **15**
On which day was the combined time of the three the least?   **Friday**
Which matrix, HR or RH, gives the total money earned by Kevin, Laura, and Sergio during this week?   **HR**
How much did the three of them earn on Thursday?   **$53.80**

Sidebar (right column):

15.b.

c. $f(x) = 2\,\sin\!\left(x + \dfrac{\pi}{6}\right) =$

$2\left(\sin x\,\cos\dfrac{\pi}{6} + \cos x\,\sin\dfrac{\pi}{6}\right)$

$= 2\left(\dfrac{\sqrt{3}}{2}\,\sin x + \dfrac{1}{2}\,\cos x\right) =$

$\sqrt{3}\,\sin x + \cos x = h(x)$

16., 17. See Additional Answers in the back of this book.

22.a.

$$\begin{bmatrix} 0 & \dfrac{5\sqrt{2}}{2} & 2\sqrt{2} & -\dfrac{\sqrt{2}}{2} \\ \sqrt{2} & \dfrac{7\sqrt{2}}{2} & 6\sqrt{2} & \dfrac{7\sqrt{2}}{2} \end{bmatrix}$$

EXTENSION
Alternate Approach An alternate proof of the Addition Identities follows. Since $R_{\alpha+\beta} = R_\alpha \circ R_\beta$,

$$\begin{bmatrix} \cos(\alpha+\beta) & \sin(\alpha+\beta) \\ \sin(\alpha+\beta) & \cos(\alpha+\beta) \end{bmatrix}$$

$$= \begin{bmatrix} \cos\alpha & -\sin\alpha \\ \sin\alpha & \cos\alpha \end{bmatrix} \cdot$$

$$\begin{bmatrix} \cos\beta & -\sin\beta \\ \sin\beta & \cos\beta \end{bmatrix} =$$

$$\begin{bmatrix} \cos\alpha\cos\beta - \sin\alpha\sin\beta \\ \sin\alpha\cos\beta + \cos\alpha\sin\beta \end{bmatrix}$$

$$\begin{bmatrix} -\sin\alpha\cos\beta - \cos\alpha\sin\beta \\ \cos\alpha\cos\beta - \sin\alpha\sin\beta \end{bmatrix}$$

and corresponding elements of matrices are equal. Although this proof is even more elegant than the proof given in the lesson, we have found that students are more satisfied with the one we give there.

Small Group Work Students working together can use the alternate approach given above to prove theorems in this lesson. For instance, an alternate approach for **Example 3** is to use the matrix form of the transformation identity $R_{\pi-\theta} = R_\pi \circ R_{-\theta}$

ADDITIONAL ANSWERS
25.c. True. Since g is at least of degree 4, and since the tails of the graph both point in the same direction, the degree of g must be an even number greater than 2. Since nonreal roots come in pairs, there must be an even number of nonreal roots and thus an even number of real roots. So, there cannot be only 3 real roots; there must be at least 4.

694

---

24. Solve $x^4 - 2x^3 - 13x^2 - 4x - 30 = 0$ completely, given that one of the solutions is $\sqrt{2}\,i$. *(Lesson 9-7)* $x = i\sqrt{2}, x = -i\sqrt{2}, x = -3, \text{ or } x = 5$

25. Consider the graph of a polynomial function g pictured below.

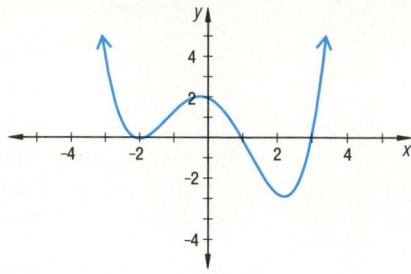

a. How many real zeros does g seem to have? **3**
b. What is the smallest possible degree g can have? **4**
c. *True* or *false*. The factorization of g over the set of polynomials with real number coefficients has at least four factors. Explain your answer. *(Lessons 9-5, 9-6)* **See margin.**

26. Give an exact value.

a. $\sec\frac{\pi}{6}$ $\frac{2\sqrt{3}}{3}$ **b.** $\csc\frac{\pi}{4}$ $\sqrt{2}$ **c.** $\cot\frac{\pi}{3}$ *(Lesson 6-7)* $\frac{\sqrt{3}}{3}$

27. *Multiple choice*. Which correlation coefficient indicates the strongest linear relationship? *(Lesson 2-5)*
(a) 0.43 (b) 0.16 (c) -0.79 (d) 0.1 **(c)**

Exploration

28. a. Use the fact that $\tan(\alpha+\beta) = \dfrac{\sin(\alpha+\beta)}{\cos(\alpha+\beta)}$ to derive a formula for $\tan(\alpha+\beta)$ in terms of tangents of α and β. (Hint: divide both numerator and denominator of the fraction by $\cos\alpha\cos\beta$.)
b. Check your answer to part **a** by using some values of tangents known to you.

a) $\tan(\alpha+\beta) = \dfrac{\sin(\alpha+\beta)}{\cos(\alpha+\beta)} = \dfrac{\sin\alpha\cos\beta + \cos\alpha\sin\beta}{\cos\alpha\cos\beta - \sin\alpha\sin\beta}$

$$= \frac{\dfrac{\sin\alpha\cos\beta}{\cos\alpha\cos\beta} + \dfrac{\cos\alpha\sin\beta}{\cos\alpha\cos\beta}}{\dfrac{\cos\alpha\cos\beta}{\cos\alpha\cos\beta} - \dfrac{\sin\alpha\sin\beta}{\cos\alpha\cos\beta}}$$

$$= \frac{\dfrac{\sin\alpha}{\cos\alpha} + \dfrac{\sin\beta}{\cos\beta}}{1 - \dfrac{\sin\alpha}{\cos\alpha} \cdot \dfrac{\sin\beta}{\cos\beta}}$$

$$= \frac{\tan\alpha + \tan\beta}{1 - \tan\alpha\tan\beta}$$

b) Sample: If $\alpha = 30°$ and $\beta = 45°$, then $\tan(30° + 45°) = \tan 75° \approx 3.732$, and
$\dfrac{\tan 30° + \tan 45°}{1 - \tan 30° \tan 45°} \approx \dfrac{.577 + 1}{1 - (.577)(1)} \approx 3.732$.

694

11-6

Identities for cos 2θ and sin 2θ

Ignoring air resistance and wind, an object thrown, kicked, or hit into the air will travel in a parabolic trajectory. For example, if a golf ball is driven off the ground with velocity v m/sec at an initial angle of θ degrees to the ground, the horizontal distance d that the ball travels is given by $d = \dfrac{v^2 \sin 2\theta}{g}$, where g is the acceleration due to gravity.

Thus, a golf ball hit off a tee at 45 meters per second at an initial angle of 30° to the ground will travel about

$$\frac{\left(45 \; \frac{m}{sec}\right)^2 \cdot \sin(2 \cdot 30°)}{9.81 \; \frac{m}{sec^2}} \approx 179 \text{ m.}$$

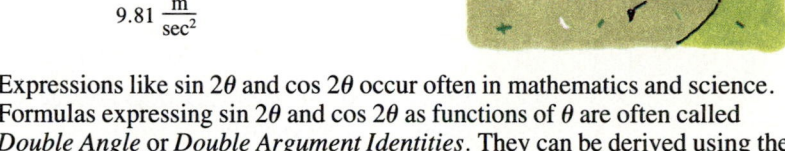

Expressions like $\sin 2\theta$ and $\cos 2\theta$ occur often in mathematics and science. Formulas expressing $\sin 2\theta$ and $\cos 2\theta$ as functions of θ are often called *Double Angle* or *Double Argument Identities*. They can be derived using the Addition Identities for the sine and cosine functions.

Theorem: Double Angle Identities

For all real numbers θ,
$$\sin 2\theta = 2 \sin \theta \cos \theta;$$
and
$$\cos 2\theta = \cos^2\theta - \sin^2\theta$$
$$= 2 \cos^2\theta - 1$$
$$= 1 - 2 \sin^2\theta.$$

Proof Set $\alpha = \theta$ and $\beta = \theta$ in the formulas for $\sin(\alpha + \beta)$ and $\cos(\alpha + \beta)$.

Then
$$\sin 2\theta = \sin(\theta + \theta)$$
$$= \sin \theta \cos \theta + \cos \theta \sin \theta$$
$$= 2 \sin \theta \cos \theta.$$

Similarly,
$$\cos 2\theta = \cos(\theta + \theta)$$
$$= \cos \theta \cos \theta - \sin \theta \sin \theta$$
$$= \cos^2\theta - \sin^2\theta.$$

Substituting $\sin^2\theta = 1 - \cos^2\theta$ from the Pythagorean Identity yields the second form of the Double Angle Identity for the cosine function.
$$\cos 2\theta = \cos^2\theta - (1 - \cos^2\theta)$$
$$= \cos^2\theta - 1 + \cos^2\theta$$
$$= 2 \cos^2\theta - 1$$

You are asked to derive the third form of the Double Angle Identity for the cosine function in the Questions.

This is a straightforward lesson. However, even students who know that $\sin(\alpha + \beta) \neq \sin \alpha + \sin \beta$ may wrongly believe that $\sin 2x = 2 \sin x$. They are readily convinced by going through a special case such as that given in **Example 1** or Additional Example 1 below.

The major reasons for having double angle identities are (1) the existence of certain formulas that involve expressions with them, as in the first paragraph of this lesson, (2) the ability to deduce formulas for $\cos(n\theta)$ and $\sin(n\theta)$ from them, as **Example 2** exemplifies, and (3) to use them to deduce the half-angle formulas given in **Questions 18 and 19.** Today, the last reason is not particularly important, as values of any of the trigonometric functions can be found to as much accuracy as one wishes. Still, in theory, by halving and adding as appropriate, you can with these identities get as close as you wish to the value of these functions for any argument.

Graphically, the Double Angle Identities mean that the graphs of $y = \sin 2x$ and $y = 2 \sin x \cos x$ coincide, as do the graphs of $y = \cos 2x$ and $y = \cos^2 x - \sin^2 x$. The graphs of the sine functions are shown below. You are asked to verify the identities for cos $2x$ graphically in the Questions.

$y = \sin 2x$

$y = 2 \sin x \cos x$

Since the Double Angle Identities are true for all θ, you can use a specific value of θ to check them.

Example 1 For $\theta = \frac{\pi}{6}$, verify the identity for sin 2θ.

Solution When $\theta = \frac{\pi}{6}$, sin $2\theta = 2 \sin \theta \cos \theta$ becomes

$\sin \left(2 \cdot \frac{\pi}{6}\right) = 2 \sin \frac{\pi}{6} \cos \frac{\pi}{6}$.

Is this true?

$$\sin \left(2 \cdot \frac{\pi}{6}\right) = \sin \frac{\pi}{3} = \frac{\sqrt{3}}{2},$$

and $\qquad 2 \sin \frac{\pi}{6} \cos \frac{\pi}{6} = 2 \cdot \frac{1}{2} \cdot \frac{\sqrt{3}}{2} = \frac{\sqrt{3}}{2}$ also.

The identity is verified.

From the identities for cos 2θ and sin 2θ, identities for cos 4θ and sin 4θ can be found.

Example 2 Express cos 4θ as a function of cos θ only.

Solution First, express cos 4θ as a function of cos 2θ.
$$\cos 4\theta = \cos 2(2\theta)$$
$$= 2 \cos^2(2\theta) - 1$$
Now use the Double Angle Identity again.
$$\cos 4\theta = 2(2 \cos^2\theta - 1)^2 - 1$$
$$= 2(4 \cos^4\theta - 4 \cos^2\theta + 1) - 1$$
$$= 8 \cos^4\theta - 8 \cos^2\theta + 2 - 1$$
$$= 8 \cos^4\theta - 8 \cos^2\theta + 1$$

696

Check Substitute a value for θ for which the values of $\cos \theta$ and $\cos 4\theta$ are known. We use $\theta = 45°$.

$$\cos 4\theta = \cos (4 \cdot 45°)$$
$$= \cos 180°$$
$$= -1$$

$$8 \cos^4\theta - 8 \cos^2\theta + 1 = 8\left(\frac{\sqrt{2}}{2}\right)^4 - 8\left(\frac{\sqrt{2}}{2}\right)^2 + 1$$
$$= 8\left(\frac{4}{16}\right) - 8\left(\frac{2}{4}\right) + 1$$
$$= -1$$

Questions

Covering the Reading

In 1 and 2, use the formula $d = \dfrac{v^2 \sin 2\theta}{g}$.

1. How far will a soccer ball travel with respect to the ground if you kick it at $34 \frac{m}{sec}$ at an initial angle of $45°$ to the ground?   **about 118 m**

2. How far will a tennis ball travel horizontally if thrown at $25 \frac{m}{sec}$ at an angle of $90°$ to the ground?   **0 m**

3. Derive the Double Angle Identity $\cos 2\theta = 1 - 2 \sin^2\theta$.   **See margin.**

4. Find a value of θ for which $\cos 2\theta \neq 2 \cos \theta$.
   sample: $\theta = 30°$

5. **a.** Graph on the same set of axes
$$f(x) = \cos^2 x - \sin^2 x$$
$$g(x) = 2\cos^2 x - 1$$
   and
$$h(x) = 1 - 2 \sin^2 x.$$ **See margin.**
   b. Find another equation whose graph coincides with those of f, g, and h.   **sample: $R(x) = \cos(2x)$**

6. Use $\theta = 60°$ and a formula for $\cos 2\theta$ to find an exact value for $\cos 120°$
   $\cos 120° = \cos(2 \cdot 60°) = \cos^2 60° - \sin^2 60° = \left(\frac{1}{2}\right)^2 - \left(\frac{\sqrt{3}}{2}\right)^2 = \frac{1}{4} - \frac{3}{4} = \frac{-1}{2}$

In 7–10, write each expression as the sine or cosine of a single argument.

7. $2 \cos^2 25° - 1$ **$\cos 50°$**

8. $2 \sin 35° \cos 35°$ **$\sin 70°$**

9. $1 - 2 \sin^2 \frac{3\pi}{8}$ **$\cos \frac{3\pi}{4}$**

10. $\cos^2 \frac{4\pi}{9} - \sin^2 \frac{4\pi}{9}$ **$\cos \frac{8\pi}{9}$**

11. Express $\sin 4\theta$ as a function of $\cos \theta$ and $\sin \theta$.
   $\sin 4\theta = 4 \sin \theta \cos^3 \theta - 4 \sin^3 \theta \cos \theta$

Applying the Mathematics

12. Suppose a golf ball can be hit at a velocity of $45 \frac{m}{sec}$. Use the formula
$$d = \frac{v^2 \sin 2\theta}{g}.$$
   a. At what angle should the golfer hit the ball in order to have it go the maximum distance?   **45°**
   b. What horizontal distance will the ball then travel?   **about 207 m**

LESSON 11-6 Identities for $\cos 2\theta$ and $\sin 2\theta$ **697**

Questions 18 and 19:
These identities are not necessary for the development of important trigonometric concepts. Their primary use is in formulas for integration of trigonometric functions in calculus. Most people do not commit these formulas to memory; they can always be looked up.

ADDITIONAL ANSWERS

3. $\cos 2\theta = \cos (\theta + \theta) =$
$\cos \theta \cos \theta - \sin \theta \sin \theta =$
$\cos^2 \theta - \sin^2 \theta =$
$(1 - \sin^2 \theta) - \sin^2 \theta =$
$1 - 2 \sin^2 \theta$

5.a.

16.b. $r_L =$
$$\begin{bmatrix} \cos (2 \cdot 45°) & \sin (2 \cdot 45°) \\ \sin (2 \cdot 45°) & -\cos (2 \cdot 45°) \end{bmatrix}$$
$$= \begin{bmatrix} \cos 90° & \sin 90° \\ \sin 90° & -\cos 90° \end{bmatrix}$$
$$= \begin{bmatrix} 0 & 1 \\ 1 & 0 \end{bmatrix} = r_y = x$$

17.a. $r_L = \begin{bmatrix} \frac{1}{2} & \frac{\sqrt{3}}{2} \\ \frac{\sqrt{3}}{2} & -\frac{1}{2} \end{bmatrix}$

b. $r_L \begin{bmatrix} 1 \\ 0 \end{bmatrix} = \begin{bmatrix} \frac{1}{2} \\ \frac{\sqrt{3}}{2} \end{bmatrix}$

MORE PRACTICE
For more questions on SPUR Objectives, use *Lesson Master 11-6*, shown on page 699.

EXTENSION
Alternate Approach
Students can prove the double angle identities by using the transformation identity $R_{2\theta} = R_\theta \circ R_\theta$ and the alternate approach in the Extension of Lesson 11-5.
Proof:

$$\begin{bmatrix} \cos 2\theta & -\sin 2\theta \\ \sin 2\theta & \cos 2\theta \end{bmatrix} =$$

$$\begin{bmatrix} \cos \theta & -\sin \theta \\ \sin \theta & \cos \theta \end{bmatrix} \cdot$$

$$\begin{bmatrix} \cos \theta & -\sin \theta \\ \sin \theta & \cos \theta \end{bmatrix} =$$

$$\begin{bmatrix} \cos\theta\cos\theta - \sin\theta\sin\theta \\ \sin\theta\cos\theta + \cos\theta\sin\theta \end{bmatrix}$$

$$\begin{bmatrix} -\sin\theta\cos\theta - \cos\theta\sin\theta \\ \cos\theta\cos\theta - \sin\theta\sin\theta \end{bmatrix}$$

$$=$$

$$\begin{bmatrix} \cos^2\theta - \sin^2\theta & -2\sin\theta\cos\theta \\ 2\sin\theta\cos\theta & \cos^2\theta - \sin^2\theta \end{bmatrix}$$

and corresponding elements of equal matrices are equal.

PROJECTS
Project 3 on pages 718 and 719 is related to the content of this lesson.

ADDITIONAL ANSWERS
16.b., 17.a.,b. See margin, page 697.

18.a.,b., 19.a.,b. See Additional Answers in the back of this book.

26.b. For this distribution, $P(z < 233) \approx 0.2$ and $P(z < 288) \approx .92$. Therefore $P(233 < z < 288) \approx .92 - .02 = .90$.

c. Part a gives an interval centered at the mean in which 90% of all births occur. In contrast, the interval in part b is centered below the mean. The interval in part b must be wider than the interval in part a because gestation periods at the lowest extreme of the

13. If $\angle A$ is acute and $\sin A = \frac{4}{5}$, find exact values for each of the following:
    a. $\cos A$; **3/5**   b. $\sin 2A$. **24/25**

14. Suppose $\cos A = \frac{1}{4}$. Find $\cos 2A$, if
    a. $0° < A < 90°$; **-7/8**   b. $270° < A < 360°$. **-7/8**

In 15–17, consider a line L which passes through the origin at an angle θ with the positive x-axis. Then the matrix $\begin{bmatrix} \cos 2\theta & \sin 2\theta \\ \sin 2\theta & -\cos 2\theta \end{bmatrix}$ represents r_L, reflection over L.

15. Suppose that L is the x-axis. Verify that the matrix for r_L equals that for r_x. $r_L = \begin{bmatrix} \cos(2 \cdot 0°) & \sin(2 \cdot 0°) \\ \sin(2 \cdot 0°) & -\cos(2 \cdot 0°) \end{bmatrix} = \begin{bmatrix} 1 & 0 \\ 0 & -1 \end{bmatrix} = r_x$

16. Suppose that L is the line $y = x$.
    a. What is θ in this case? $\theta = 45°$
    b. Verify that $r_L = r_{y=x}$. **See margin.**

17. a. If L is the line $y = \frac{1}{\sqrt{3}}x$, find a matrix of exact values for r_L.
    b. Use your matrix from part a to find the image of $(1, 0)$ when reflected over $y = \frac{1}{\sqrt{3}}x$. Check your answer with a drawing. **See margin.**

In 18 and 19, the double angle formulas are employed to derive **half angle formulas**. Consider the equations
$$\cos 2\theta = \cos^2\theta - \sin^2\theta$$
and
$$1 = \cos^2\theta + \sin^2\theta.$$

18. a. Add the equations and show that $\cos \theta = \pm\sqrt{\frac{1+\cos 2\theta}{2}}$. (The sign used is determined by the quadrant in which θ lies.)
    b. Use the formula in part a to show that $\cos \frac{\pi}{8} = \frac{\sqrt{2+\sqrt{2}}}{2}$. **See margin.**

19. a. Show that $\sin \theta = \pm\sqrt{\frac{1-\cos 2\theta}{2}}$. **See margin.** (The sign used is determined by the quadrant in which θ lies.)
    b. Find an exact value for $\sin \frac{\pi}{8}$.

20. Write $\cos(3\theta)$ as a function of $\sin\theta$ and $\cos\theta$. $\cos(3\theta) = \cos^3\theta - 3\sin^2\theta\cos\theta$

• Review

21. Find an exact value for $\sin 195°$. *(Lesson 11-5)* $\frac{\sqrt{2}-\sqrt{6}}{4}$

In 22 and 23, *multiple choice*.
22. For all real numbers α and β, $\cos\alpha\cos\beta - \sin\alpha\sin\beta =$
    (a) $\cos(\alpha + \beta)$ (b) $\sin(\alpha + \beta)$
    (c) $\cos(\alpha - \beta)$ (d) $\sin(\alpha - \beta)$. *(Lesson 11-5)* **(a)**

698

23. The matrix $\begin{bmatrix} 1 & 0 \\ 0 & -1 \end{bmatrix}$ represents

    (a) reflection over the *x*-axis   (b) reflection over the *y*-axis

    (c) reflection over the line $y = x$ (d) none of (a) – (c).   *(Lesson 11-2)*

    (a)

24. A transformation which can be represented by a matrix maps a square with vertices at $(0, 0)$, $(0, 1)$, $(1, 1)$, and $(1, 0)$ to a parallelogram with corresponding vertices at $(0, 0)$, $(2, 3)$, $(-2, 4)$, and $(-4, 1)$. What is the matrix for the transformation?   *(Lesson 11-4)* $\begin{bmatrix} -4 & 2 \\ 1 & 3 \end{bmatrix}$

25. If $\begin{bmatrix} a & -4 \\ -7 & b \end{bmatrix}\begin{bmatrix} 3 & 4 \\ 7 & 9 \end{bmatrix} = \begin{bmatrix} -1 & 0 \\ 0 & -1 \end{bmatrix}$, find *a* and *b*.   *(Lesson 11-1)*   **a = 9, b = 3**

26. The *gestation period* (time from conception to birth) in humans has an approximately normal distribution with a mean of 266 days and a standard deviation of 16 days.

    a. Determine an interval for the number of days after conception when 90% of births occur.   **from 239 to 293 days**

    b. Show that the following statement is true: 90% of births occur from 233 to 288 days after conception.

    c. Compare and contrast the results in parts **a** and **b**. Explain any discrepancies.   *(Lesson 10-6)*   **b,c) See margin.**

Husbands and wives at a Lamaze class

27. Given that $a = 2 + i$ and $b = 1 - 2i$, calculate

    a. $a^2 - b^2$ **6 + 8i**   **b.** $\dfrac{a}{b}$.   *(Lesson 9-6)*   **i**

28. Suppose that a test of anxiety is given to representative samples of people of ages 10, 15, 20, 40, and 60, and that the respective means of anxiety levels on the test are 40, 70, 65, 60, and 50.   *(Lessons 1-1, 3-6)*

    a. Draw a scatter plot of these data using age as the independent variable.   **See margin.**

    b. *Multiple choice.* The correlation between age and anxiety level in the samples is

        (i) strongly positive;   (ii) strongly negative;   (iii) nearly zero.

        (iii)

Exploration

29. You have seen derivations of the following identities:

$$\cos 2\theta = 2\cos^2\theta - 1$$
$$\cos 4\theta = 8\cos^4\theta - 8\cos^2\theta + 1.$$

Can $\cos n\theta$, where *n* is a positive integer, always be expressed as a polynomial in $\cos\theta$? If it can be, what can you say about the degree and leading coefficient of the polynomial? Justify your answers.
See margin.

---

interval from 233 to 288 are less likely than gestation periods at the extremes of the interval 239 to 293.

28.a.

29. Yes; as a polynomial in cos θ, cos nθ has degree n and its leading coefficient is 2^{n-1}. From a table of trigonometric identities, cos nθ = 2 cos (n – 1)θ cos θ – cos (n – 2)θ for all integers n ≥ 2. So if n ≥ 3, cos (n – 1)θ is a polynomial in cos θ of degree n – 1 with leading coefficient 2^{n-2}. Therefore, cos nθ = 2 cos (n – 1)θ cos θ – cos(n – 2)θ has degree n and leading coefficient 2^{n-1}.

LESSON

11-7

Inverses of 2 x 2 Matrices

You have seen that a R_θ, a rotation of magnitude θ around the origin, can be represented by the matrix $\begin{bmatrix} \cos\theta & -\sin\theta \\ \sin\theta & \cos\theta \end{bmatrix}$.

So using the Opposites Theorems, a matrix for $R_{-\theta}$ is $\begin{bmatrix} \cos(-\theta) & -\sin(-\theta) \\ \sin(-\theta) & \cos(-\theta) \end{bmatrix} = \begin{bmatrix} \cos\theta & \sin\theta \\ -\sin\theta & \cos\theta \end{bmatrix}$

Thus, the composite transformation $R_{-\theta} \circ R_\theta$ can be represented by the product

$$\begin{bmatrix} \cos\theta & \sin\theta \\ -\sin\theta & \cos\theta \end{bmatrix} \begin{bmatrix} \cos\theta & -\sin\theta \\ \sin\theta & \cos\theta \end{bmatrix}$$

$$= \begin{bmatrix} \cos^2\theta + \sin^2\theta & -\cos\theta\sin\theta + \sin\theta\cos\theta \\ -\sin\theta\cos\theta + \cos\theta\sin\theta & \sin^2\theta + \cos^2\theta \end{bmatrix}$$

$$= \begin{bmatrix} 1 & 0 \\ 0 & 1 \end{bmatrix},$$

which is the 2×2 identity matrix. You can verify that $R_\theta \circ R_{-\theta}$, the composite in the reverse order, also is the identity transformation.

Because each composite $R_{-\theta} \circ R_\theta$ and $R_\theta \circ R_{-\theta}$ is the identity, the transformation $R_{-\theta}$ undoes the transformation R_θ. Thus, R_θ and $R_{-\theta}$ are examples of *inverse transformations*. Their associated matrices are called *inverse matrices*.

Definitions

Two transformations S and T are **inverse transformations** if and only if both $S \circ T$ and $T \circ S$ map any point onto itself.

Two 2×2 matrices are **inverse matrices** if and only if their product is the 2×2 identity matrix for matrix multiplication.

700

In general, only square matrices can have inverses because both products MN and NM must exist.

In symbols, if M and N are both 2×2 matrices, then M and N are inverses if and only if $MN = NM = \begin{bmatrix} 1 & 0 \\ 0 & 1 \end{bmatrix}$. The inverse of matrix M is written M^{-1}.

Some inverses are easy to find. Recall that $\begin{bmatrix} 3 & 0 \\ 0 & 5 \end{bmatrix}$ is the matrix for the scale change

$$S: (x, y) \rightarrow (3x, 5y).$$

To undo the effect of this scale change, you can apply the scale change

$$S: (x, y) \rightarrow \left(\tfrac{1}{3}x, \tfrac{1}{5}y\right),$$

whose associated matrix is $\begin{bmatrix} \frac{1}{3} & 0 \\ 0 & \frac{1}{5} \end{bmatrix}$. This suggests that the inverse of the

matrix $\begin{bmatrix} 3 & 0 \\ 0 & 5 \end{bmatrix}$ is $\begin{bmatrix} \frac{1}{3} & 0 \\ 0 & \frac{1}{5} \end{bmatrix}$. This can be verified.

Example 1 Show that $\begin{bmatrix} 3 & 0 \\ 0 & 5 \end{bmatrix}$ and $\begin{bmatrix} \frac{1}{3} & 0 \\ 0 & \frac{1}{5} \end{bmatrix}$ are inverses.

Solution Because matrix multiplication is not generally commutative, check both products.

$$\begin{bmatrix} 3 & 0 \\ 0 & 5 \end{bmatrix} \begin{bmatrix} \frac{1}{3} & 0 \\ 0 & \frac{1}{5} \end{bmatrix} = \begin{bmatrix} 1 & 0 \\ 0 & 1 \end{bmatrix}$$

$$\begin{bmatrix} \frac{1}{3} & 0 \\ 0 & \frac{1}{5} \end{bmatrix} \begin{bmatrix} 3 & 0 \\ 0 & 5 \end{bmatrix} = \begin{bmatrix} 1 & 0 \\ 0 & 1 \end{bmatrix}$$

Other inverses may not be so obvious.

Example 2 Find the inverse of $\begin{bmatrix} 3 & 7 \\ 2 & 5 \end{bmatrix}$.

Solution As the inverse of a 2×2 matrix is also 2×2, you must find a matrix of the form $\begin{bmatrix} w & x \\ y & z \end{bmatrix}$ such that

$$\begin{bmatrix} 3 & 7 \\ 2 & 5 \end{bmatrix}\begin{bmatrix} w & x \\ y & z \end{bmatrix} = \begin{bmatrix} w & x \\ y & z \end{bmatrix}\begin{bmatrix} 3 & 7 \\ 2 & 5 \end{bmatrix} = \begin{bmatrix} 1 & 0 \\ 0 & 1 \end{bmatrix}.$$

$$\begin{bmatrix} 3 & 7 \\ 2 & 5 \end{bmatrix}\begin{bmatrix} w & x \\ y & z \end{bmatrix} = \begin{bmatrix} 3w+7y & 3x+7z \\ 2w+5y & 2x+5z \end{bmatrix}$$

So $3w + 7y = 1$,
$3x + 7z = 0$,
$2w + 5y = 0$,
and $2x + 5z = 1$.

To solve this system, note that the first and third equations use the same two variables. Multiply the first equation by 2 and the third by -3 and add the resulting sentences. This yields $y = -2$. Substituting this value into the first equation gives $w = 5$. Similarly, multiplying the second equation by 2 and the fourth by -3 and adding the results yields $z = 3$. Substituting this value into the second equation gives $x = -7$.

So $\begin{bmatrix} w & x \\ y & z \end{bmatrix}$, the inverse of $\begin{bmatrix} 3 & 7 \\ 2 & 5 \end{bmatrix}$, is $\begin{bmatrix} 5 & -7 \\ -2 & 3 \end{bmatrix}$.

Check $\begin{bmatrix} 3 & 7 \\ 2 & 5 \end{bmatrix} \cdot \begin{bmatrix} 5 & -7 \\ -2 & 3 \end{bmatrix} = \begin{bmatrix} 15-14 & -21+21 \\ 10-10 & -14+15 \end{bmatrix} = \begin{bmatrix} 1 & 0 \\ 0 & 1 \end{bmatrix}$

$\begin{bmatrix} 5 & -7 \\ -2 & 3 \end{bmatrix} \cdot \begin{bmatrix} 3 & 7 \\ 2 & 5 \end{bmatrix} = \begin{bmatrix} 15-14 & 35-35 \\ -6+6 & -14+15 \end{bmatrix} = \begin{bmatrix} 1 & 0 \\ 0 & 1 \end{bmatrix}$

If the inverse of a 2×2 matrix exists, you can find it by generalizing the procedure used in Example 2, or you can use the following theorem.

Theorem (Inverse of a 2 × 2 Matrix)

If $ad - bc \neq 0$, then

$$\begin{bmatrix} a & b \\ c & d \end{bmatrix}^{-1} = \begin{bmatrix} \dfrac{d}{ad-bc} & \dfrac{-b}{ad-bc} \\ \dfrac{-c}{ad-bc} & \dfrac{a}{ad-bc} \end{bmatrix}.$$

702

Proof It must be shown that the product of the matrices, in either order, is the identity matrix.

$$\begin{bmatrix} a & b \\ c & d \end{bmatrix} \begin{bmatrix} \dfrac{d}{ad-bc} & \dfrac{-b}{ad-bc} \\ \dfrac{-c}{ad-bc} & \dfrac{a}{ad-bc} \end{bmatrix} = \begin{bmatrix} \dfrac{ad}{ad-bc} - \dfrac{bc}{ad-bc} & \dfrac{-ab}{ad-bc} + \dfrac{ab}{ad-bc} \\ \dfrac{cd}{ad-bc} - \dfrac{cd}{ad-bc} & \dfrac{-bc}{ad-bc} + \dfrac{ad}{ad-bc} \end{bmatrix}$$

$$= \begin{bmatrix} \dfrac{ad-bc}{ad-bc} & \dfrac{0}{ad-bc} \\ \dfrac{0}{ad-bc} & \dfrac{ad-bc}{ad-bc} \end{bmatrix}$$

$$= \begin{bmatrix} 1 & 0 \\ 0 & 1 \end{bmatrix}$$

You are asked to verify the multiplication in the other order in the Questions.

Example 3 Use the Inverse of a 2 × 2 Matrix Theorem to find $\begin{bmatrix} 2 & -3 \\ 6 & -1 \end{bmatrix}^{-1}$.

Solution In $\begin{bmatrix} 2 & -3 \\ 6 & -1 \end{bmatrix}$, $a = 2$, $b = -3$, $c = 6$, and $d = -1$. So,

$$ad - bc = (2)(-1) - (-3)(6) = 16.$$

Therefore, $\begin{bmatrix} 2 & -3 \\ 6 & -1 \end{bmatrix}^{-1} = \begin{bmatrix} -\dfrac{1}{16} & \dfrac{3}{16} \\ -\dfrac{6}{16} & \dfrac{2}{16} \end{bmatrix}$.

Check

$$\begin{bmatrix} 2 & -3 \\ 6 & -1 \end{bmatrix} \begin{bmatrix} -\dfrac{1}{16} & \dfrac{3}{16} \\ -\dfrac{6}{16} & \dfrac{2}{16} \end{bmatrix} = \begin{bmatrix} -\dfrac{2}{16} + \dfrac{18}{16} & \dfrac{6}{16} - \dfrac{6}{16} \\ -\dfrac{6}{16} + \dfrac{6}{16} & \dfrac{18}{16} - \dfrac{2}{16} \end{bmatrix} = \begin{bmatrix} 1 & 0 \\ 0 & 1 \end{bmatrix}$$

$$\begin{bmatrix} -\dfrac{1}{16} & \dfrac{3}{16} \\ -\dfrac{6}{16} & \dfrac{2}{16} \end{bmatrix} \begin{bmatrix} 2 & -3 \\ 6 & -1 \end{bmatrix} = \begin{bmatrix} -\dfrac{2}{16} + \dfrac{18}{16} & \dfrac{3}{16} - \dfrac{3}{16} \\ -\dfrac{12}{16} + \dfrac{12}{16} & \dfrac{18}{16} - \dfrac{2}{16} \end{bmatrix} = \begin{bmatrix} 1 & 0 \\ 0 & 1 \end{bmatrix}$$

ADDITIONAL EXAMPLES

1. Show that $\begin{bmatrix} 6 & 11 \\ 2 & 4 \end{bmatrix}$ and $\begin{bmatrix} 2 & -5.5 \\ -1 & 3 \end{bmatrix}$ are inverses of each other.
The product in both orders is $\begin{bmatrix} 1 & 0 \\ 0 & 1 \end{bmatrix}$.

2. Find the inverse of $\begin{bmatrix} 1 & 5 \\ -3 & 4 \end{bmatrix}$.
$\begin{bmatrix} \dfrac{4}{19} & -\dfrac{5}{19} \\ \dfrac{3}{19} & \dfrac{1}{19} \end{bmatrix}$

3. Show that $\begin{bmatrix} 3 & -15 \\ -4 & 20 \end{bmatrix}$ has no inverse.
Its determinant
$3 \cdot 20 - (-15)(-4) = 0$.

We have mentioned earlier that only *square matrices* can have inverses. However, not all square matrices have inverses. If $M = \begin{bmatrix} a & b \\ c & d \end{bmatrix}$ and $ad - bc = 0$, then $\frac{1}{ad-bc}$ is undefined and M^{-1} cannot exist. The expression $ad - bc$ is called the **determinant** of the matrix because it can be used to determine whether the matrix has an inverse. The word *determinant* is abbreviated as **det**, and we write

$$\det \begin{bmatrix} a & b \\ c & d \end{bmatrix} = ad - bc.$$

Thus, if $M = \begin{bmatrix} a & b \\ c & d \end{bmatrix}$ is a matrix whose determinant is nonzero, the preceding theorem implies that $M^{-1} = \begin{bmatrix} \dfrac{d}{\det M} & \dfrac{-b}{\det M} \\ \dfrac{-c}{\det M} & \dfrac{a}{\det M} \end{bmatrix}$.

Matrices and their inverses have applications in many areas besides transformations. In 1929–31, a mathematician named Lester Hill devised a way to use matrices to encode messages. He assigned every integer to a letter according to the following scheme based on arithmetic sequences.

```
A:   ...,   -51,   -25,    1,    27,   ...,   26n + 1,    ...
B:   ...,   -50,   -24,    2,    28,   ...,   26n + 2,    ...
C:   ...,   -49,   -23,    3,    29,   ...,   26n + 3,    ...
D:   ...,   -48,   -22,    4,    30,   ...,   26n + 4,    ...
.
.
.
X:   ...,   -28,    -2,   24,    50,   ...,   26n + 24,   ...
Y:   ...,   -27,    -1,   25,    51,   ...,   26n + 25,   ...
Z:   ...,   -26,     0,   26,    52,   ...,   26n + 26,   ...
```

To code or encipher a message like MATH TEAM, first put the letters into 2×2 matrices, four at a time like this:

$$\begin{bmatrix} M & A \\ T & H \end{bmatrix} \text{ and } \begin{bmatrix} T & E \\ A & M \end{bmatrix}.$$

Then, using the above coding scheme, find an integer corresponding to each letter. Using M = 13, A = 1, T = 20, H = 8, and E = 5, code the messages as

$$\begin{bmatrix} 13 & 1 \\ 20 & 8 \end{bmatrix} \text{ and } \begin{bmatrix} 20 & 5 \\ 1 & 13 \end{bmatrix}.$$

Next, multiply each 2×2 matrix on the left by a *coding matrix* with determinant ± 1, such as $\begin{bmatrix} 2 & 1 \\ 1 & 0 \end{bmatrix}$.

This gives
$$\begin{bmatrix} 2 & 1 \\ 1 & 0 \end{bmatrix}\begin{bmatrix} 13 & 1 \\ 20 & 8 \end{bmatrix} = \begin{bmatrix} 46 & 10 \\ 13 & 1 \end{bmatrix}$$

and
$$\begin{bmatrix} 2 & 1 \\ 1 & 0 \end{bmatrix}\begin{bmatrix} 20 & 5 \\ 1 & 13 \end{bmatrix} = \begin{bmatrix} 41 & 23 \\ 20 & 5 \end{bmatrix}.$$

Finally, change back to letters, giving
$$\begin{bmatrix} T & J \\ M & A \end{bmatrix} \text{ and } \begin{bmatrix} O & W \\ T & E \end{bmatrix}.$$

The coded message is then TJMAOWTE. Notice that the A in MATH is coded as J; and the A in team is coded as T. This makes the code difficult to break.

To decode or decipher a message such as TJMAOWTE, first break up the letters into groups of four:
$$\begin{bmatrix} T & J \\ M & A \end{bmatrix} \text{ and } \begin{bmatrix} O & W \\ T & E \end{bmatrix}.$$

Then replace each letter by an appropriate integer:
$$\begin{bmatrix} 20 & 10 \\ 13 & 1 \end{bmatrix} \text{ and } \begin{bmatrix} 15 & 23 \\ 20 & 5 \end{bmatrix}.$$

The decoding matrix is the *inverse* of the coding matrix. Multiply each matrix on the left by this matrix. By the theorem of this lesson,
$$\begin{bmatrix} 2 & 1 \\ 1 & 0 \end{bmatrix}^{-1} = \begin{bmatrix} 0 & 1 \\ 1 & -2 \end{bmatrix}.$$

The products are
$$\begin{bmatrix} 0 & 1 \\ 1 & -2 \end{bmatrix}\begin{bmatrix} 20 & 10 \\ 13 & 1 \end{bmatrix} = \begin{bmatrix} 13 & 1 \\ -6 & 8 \end{bmatrix}$$

and
$$\begin{bmatrix} 0 & 1 \\ 1 & -2 \end{bmatrix}\begin{bmatrix} 15 & 23 \\ 20 & 5 \end{bmatrix} = \begin{bmatrix} 20 & 5 \\ -25 & 13 \end{bmatrix}.$$

Replace each integer by its corresponding letter.
$$\begin{bmatrix} M & A \\ T & H \end{bmatrix} \text{ and } \begin{bmatrix} T & E \\ A & M \end{bmatrix}$$

Write the message: MATH TEAM.

Inverses exist for square matrices with dimensions higher than 2×2, but the most practical way to find them is by using computer software packages or sophisticated calculators which find inverses of matrices.

Question 2: Point out the use of the Pythagorean Identity $\sin^2 \theta + \cos^2 \theta = 1$.

Question 3: This question shows that, at times, the inverse of a matrix can be found without a formula or solving an equation.

Question 16: In general, a linear system of the form
$$\begin{cases} ax + by = c \\ dx + ey = f \end{cases}$$
can be written as $AX = B$,

where $A = \begin{bmatrix} a & b \\ d & e \end{bmatrix}$,

$X = \begin{bmatrix} x \\ y \end{bmatrix}$, and $B = \begin{bmatrix} c \\ f \end{bmatrix}$.

When A has an inverse, $X = A^{-1}B$. If A does not have an inverse, then there is no unique solution.

ADDITIONAL ANSWERS
1. See Additional Answers in the back of this book.

2.a.

det $\begin{bmatrix} \cos \theta & -\sin \theta \\ \sin \theta & \cos \theta \end{bmatrix}$

$= \cos^2 \theta + \sin^2 \theta = 1,$

so $\begin{bmatrix} \cos \theta & -\sin \theta \\ \sin \theta & \cos \theta \end{bmatrix} =$

$\begin{bmatrix} \cos \theta & \sin \theta \\ -\sin \theta & \cos \theta \end{bmatrix}.$

4., 16.c. See Additional Answers in the back of this book.

Questions

Covering the Reading

In 1 and 2, refer to the matrices for R_θ and $R_{-\theta}$.

1. Verify that $R_\theta \circ R_{-\theta} = \begin{bmatrix} 1 & 0 \\ 0 & 1 \end{bmatrix}$.   **See margin.**

2. a. Use the Inverse of a 2×2 Matrix Theorem to find $\begin{bmatrix} \cos \theta & -\sin \theta \\ \sin \theta & \cos \theta \end{bmatrix}^{-1}$.
   b. *True* or *false*. $R_\theta^{-1} = R_{-\theta}$ **a) See margin. b) True**

3. a. Write a matrix for the size change $S(x, y) = (4x, 4y)$.   **a–d) See below.**
   b. What transformation T undoes S?
   c. Write a matrix for T.
   d. *True* or *false*.   The matrices for S and T are inverses. Justify your answer.

4. Complete the proof of the Inverse of a 2×2 Matrix Theorem.
   See margin.

In 5 and 6, find the inverse of the matrix if it exists.

5. $\begin{bmatrix} -3 & 7 \\ 6 & -14 \end{bmatrix}$ **The inverse does not exist.** **6.** $\begin{bmatrix} 8 & 3 \\ 2 & 7 \end{bmatrix}$ $\begin{bmatrix} .14 & -.06 \\ -.04 & .16 \end{bmatrix}$

7. Find det $\begin{bmatrix} 3 & -12 \\ 6 & 5 \end{bmatrix}$.   **87**

8. In order for the inverse of a 2×2 matrix to exist, what must be true about its determinant?   **It must be nonzero.**

In 9 and 10, explain why the matrix does not have an inverse.

9. $\begin{bmatrix} 4 & 2 \\ 2 & 1 \end{bmatrix}$ **Its determinant is zero.**   **10.** $\begin{bmatrix} 1 & 2 & 3 \\ 4 & 5 & 6 \end{bmatrix}$ **It is not square.**

In 11 and 12, refer to the coding scheme derived by Lester Hill. Use the coding matrix $\begin{bmatrix} 2 & 1 \\ 3 & 2 \end{bmatrix}$.

11. a. Give a number matrix for the word CUBS.   **a–c) See below.**
   b. Multiply this matrix on the left by the coding matrix.
   c. Write the coded message.

12. Decode the message UTWXJVSORAAB.   **SPINACH PIZZA**

3a) $S = \begin{bmatrix} 4 & 0 \\ 0 & 4 \end{bmatrix}$; b) a size change $T(x, y) = \left(\frac{1}{4}x, \frac{1}{4}y\right)$; c) $T = \begin{bmatrix} .25 & 0 \\ 0 & .25 \end{bmatrix}$;

d) True. $ST = \begin{bmatrix} 1 & 0 \\ 0 & 1 \end{bmatrix} = TS$.

11a) $\begin{bmatrix} 3 & 21 \\ 2 & 19 \end{bmatrix}$; b) $\begin{bmatrix} 2 & 1 \\ 3 & 2 \end{bmatrix} \cdot \begin{bmatrix} 3 & 21 \\ 2 & 19 \end{bmatrix} = \begin{bmatrix} 8 & 61 \\ 13 & 101 \end{bmatrix}$; c) HIMW

13. a. *True* or *false*. $\det \begin{bmatrix} 90 & -75 \\ 63 & 112 \end{bmatrix} = -\det \begin{bmatrix} 63 & 112 \\ 90 & -75 \end{bmatrix}$ **True**

   b. Generalize the result of part **a**: $\det \begin{bmatrix} W & X \\ Y & Z \end{bmatrix} = \underline{\quad?\quad}$. $-\det \begin{bmatrix} Y & Z \\ W & X \end{bmatrix}$

14. Find the determinant for the matrix representing R_θ.   **1**

15. Let $A = \begin{bmatrix} 2 & 5 \\ 1 & 4 \end{bmatrix}$ and $B = \begin{bmatrix} -1 & -2 \\ 3 & 4 \end{bmatrix}$.   **See below.**

   a. Find AB.   **b.** Find $(AB)^{-1}$.
   c. Find A^{-1}.   **d.** Find B^{-1}.
   e. Find $A^{-1}B^{-1}$ and $B^{-1}A^{-1}$.
   f. Generalize from the results of parts **b** and **e**.

16. Consider the following system.
$$\begin{cases} x + 2y = 11 \\ 3x - y = 12 \end{cases}$$

   a. Multiply $\begin{bmatrix} 1 & 2 \\ 3 & -1 \end{bmatrix}\begin{bmatrix} x \\ y \end{bmatrix}$. (The answer means that it is possible to represent the left side of the system by a matrix equation.) $\begin{bmatrix} x + 2y \\ 3x - y \end{bmatrix}$

   b. Find $\begin{bmatrix} 1 & 2 \\ 3 & -1 \end{bmatrix}^{-1}$. $\begin{bmatrix} 1/7 & 2/7 \\ 3/7 & -1/7 \end{bmatrix}$

   c. Multiply, on the left, both sides of the equation $\begin{bmatrix} 1 & 2 \\ 3 & -1 \end{bmatrix}\begin{bmatrix} x \\ y \end{bmatrix} = \begin{bmatrix} 11 \\ 12 \end{bmatrix}$ by the answer to part **b**.   **See margin.**

   d. What is the solution to the system of equations? $x = 5, y = 3$

   e. Use the method of parts **a–d** to solve the following system.
$$\begin{cases} -8x - 3y = 10 \\ 4x + 3y = 5 \end{cases}$$ $x = -\frac{15}{4}, y = \frac{20}{3}$

17. Simplify $\sin(a + b) + \sin(a - b)$. *(Lesson 11-5)*   **2 sin a cos b**

18. Consider the graph of the function $y = 6 \sin x \cos x$.
   a. What is its amplitude?   **3**
   b. What is its period?
   c. Use parts **a** and **b** to write another equation for the function.
   d. Use a Double Angle Identity to rewrite the formula for the function and verify your answers in parts **a** and **b**. *(Lessons 6-1, 11-6)*
   c) $y = 3\sin(2x)$; d) $6 \sin x \cos x = 3(2 \sin x \cos x) = 3\sin(2x)$

In 19 and 20, rewrite each expression as the cosine or sine of a single argument. *(Lessons 11-5, 11-6)*

19. $\cos^2 \frac{\pi}{30} - \sin^2 \frac{\pi}{30}$ $\cos\frac{\pi}{15}$

20. $\cos 2\theta \cos \theta - \sin 2\theta \sin \theta$ $\cos 3\theta$

15a) $\begin{bmatrix} 13 & 16 \\ 11 & 14 \end{bmatrix}$; b) $\begin{bmatrix} 7/3 & -8/3 \\ -11/6 & 13/6 \end{bmatrix}$; c) $\begin{bmatrix} 4/3 & -5/3 \\ -1/3 & 2/3 \end{bmatrix}$; d) $\begin{bmatrix} 2 & 1 \\ -1.5 & -.5 \end{bmatrix}$;

e) $A^{-1}B^{-1} = \begin{bmatrix} 31/6 & 13/6 \\ -5/3 & -2/3 \end{bmatrix}$, $B^{-1}A^{-1} \begin{bmatrix} 7/3 & -8/3 \\ -11/6 & 13/6 \end{bmatrix}$; f) $(AB)^{-1} = B^{-1}A^{-1}$

MORE PRACTICE
For more questions on SPUR Objectives, use *Lesson Master 11-7*, shown below.

EXTENSION
Have students prove that, where *I* is the identify matrix, $MN = I$ implies that $NM = I$.

Let $I = \begin{bmatrix} 1 & 0 \\ 0 & 1 \end{bmatrix}$ and suppose $MN = I$. Then
$(MN)M = IM$
$(MN)M = M$
$M(NM) = M$
$M^{-1}M(NM) = M^{-1}M$
$I(NM) = I$
$NM = I.$

PROJECTS
Projects 4 and 5 on pages 719 and 720 are related to the content of this lesson.

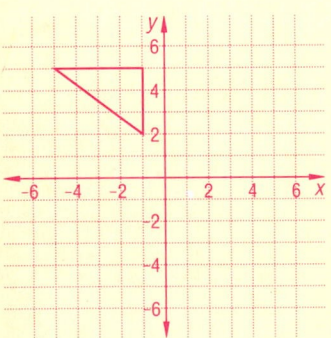
In 21 and 22, let $M_1 = \begin{bmatrix} \cos a & -\sin a \\ \sin a & \cos a \end{bmatrix}$ and $M_2 = \begin{bmatrix} 1 & 0 \\ 0 & -1 \end{bmatrix}$.

21. What transformation is represented by each matrix?
  a. M_1 **R_a**
  b. M_2 *(Lessons 11-2, 11-4)* **r_x**

22. *True* or *false.* $M_1 M_2 = M_2 M_1$. Justify your answer. *(Lessons 11-1, 11-2)*
  See below.

23. Let $A = (2, -1)$, $B = (5, -1)$, and $C = (5, -5)$.
  a. Write a matrix M representing $\triangle ABC$. $\begin{bmatrix} 2 & 5 & 5 \\ -1 & -1 & -5 \end{bmatrix}$
  b. Draw $(r_x \circ R_{90})(\triangle ABC)$. **See margin.**
  c. Write a matrix M' representing $(r_x \circ R_{90})(\triangle ABC)$. $\begin{bmatrix} 1 & 1 & 5 \\ -2 & -5 & -5 \end{bmatrix}$
  d. Draw $(R_{90} \circ r_x)(\triangle ABC)$. **See margin.**
  e. Write a matrix representing $(R_{90} \circ r_x)(\triangle ABC)$. $\begin{bmatrix} -1 & -1 & -5 \\ 2 & 5 & 5 \end{bmatrix}$
  (Lessons 7-2, 11-3, Previous course)

24. Find all three solutions to the equation $x^3 = 343$. *(Lesson 9-8)*
  See below.

In 25 and 26, give all real solution(s) **a.** exactly; **b.** to the nearest hundredth. *(Lessons 4-1, 4-8)*

25. $x^{2/3} = 81$ **a) 729; b) 729.00**

26. $3^x = 20$ **a) $\log_3 20$; b) 2.73**

Exploration

27. Figure out a way to use matrices to code or decode messages whose lengths are not multiples of 4. **Answers will vary.**

22) False. $\begin{bmatrix} \cos a & -\sin a \\ \sin a & \cos a \end{bmatrix} \begin{bmatrix} 1 & 0 \\ 0 & -1 \end{bmatrix} = \begin{bmatrix} \cos a & \sin a \\ \sin a & -\cos a \end{bmatrix}$;

$\begin{bmatrix} 1 & 0 \\ 0 & -1 \end{bmatrix} \begin{bmatrix} \cos a & -\sin a \\ \sin a & \cos a \end{bmatrix} = \begin{bmatrix} \cos a & -\sin a \\ -\sin a & -\cos a \end{bmatrix}$

24) $7; -\frac{7}{2} + \frac{7\sqrt{3}}{2}i; -\frac{7}{2} - \frac{7\sqrt{3}}{2}i$

708

11-8

Matrices in Computer Graphics

Computer simulation of a space shuttle landing (Courtesy Evans & Sutherland)

The four pictures of the space shuttle above are a small part of a computer simulation of take-offs and landings. For the human eye to "see" smooth motion, at least 15 pictures of the jet must appear each second, each one showing the new position of the jet, its shadow, and any changes in background. The screen that showed this simulation has about 1024×1024 pixels. So to change every pixel 15 times a second requires $15 \times 1024 \times 1024 \approx 15,700,000$ *point computations*, which are instructions for individual pixels. Each point computation requires a large number of multiplications and additions.

How does computer graphics software perform and illustrate these operations fast enough to fool the human eye? One way is to make sure the hardware operates as quickly as possible. Another is to use mathematics to make the point computations for each picture as efficient as possible. Matrices play a role in accomplishing both. This lesson shows how ideas from this chapter form the foundation of computer graphics.

TEACHING NOTES

The purpose of this lesson is to introduce students to the mathematics of computer animation, without attempting to teach the topic for mastery. Therefore, Questions Covering the Reading are included, but there are no Questions Applying the Mathematics.

The four-step process for producing three-dimensional graphics which is described on pages 712 and 713 is an over-simplification of the process actually used. It does, however, relate the major steps in the process to matrix multiplication.

Making Connections
To review Lesson 11-7, you might wish to discuss the inverse $(T_{a,b})^{-1} = T_{-a,-b}$ of a translation $T_{a,b}$. In matrix form,

$$(T_{a,b})^{-1} = \begin{bmatrix} 1 & 0 & a \\ 0 & 1 & b \\ 0 & 0 & 1 \end{bmatrix}^{-1} = \begin{bmatrix} 1 & 0 & -a \\ 0 & 1 & -b \\ 0 & 0 & 1 \end{bmatrix}.$$

You can check by multiplying that $(T_{a,b})^{-1} \circ T_{a,b} = I$, where here I is the identity for multiplication of 3×3 matrices.

LESSON 11-8   Matrices in Computer Graphics   **709**

On the first page of the chapter, you saw various images of a cartoon character. Suppose you want to change the figure from the position below on the left so that it is at the position on the right and is at an angle of 22.5° clockwise from vertical.

Every point of the figure, its eyes, nose, hands, feet, etc., must be transformed in a consistent manner. To produce such an image, graphics programmers generate a point matrix for the figure, just as you have done for simple geometric figures in this chapter, and then multiply that matrix by a matrix representing the transformation.

Three steps describe the transformation of the cartoon figure. Take any point of the picture, say a key on the calculator at (50, 70).
1. Translate the key to the origin. Call this transformation T_1.
2. Rotate the image about the origin 22.5° clockwise. This is $R_{-22.5}$.
3. Translate the rotated figure so that the key is at (155, 83). Call this second translation T_2.

If you can write each transformation as a matrix, then you can find a single matrix which represents the composite $T_2 \circ R_{-22.5} \circ T_1$. (Recall that multiplication of matrices is associative, so grouping parentheses are not needed.)

Unfortunately, a translation cannot be represented by a 2×2 matrix. This is because any 2×2 matrix maps the origin into the origin,

$$\begin{bmatrix} a & b \\ c & d \end{bmatrix} \begin{bmatrix} 0 \\ 0 \end{bmatrix} = \begin{bmatrix} 0 \\ 0 \end{bmatrix},$$

whereas under a nonzero translation the image of the origin is a different point.

Computer graphics programs get around this dilemma by expressing matrices for points, translations, and rotations in *homogeneous form*. The **homogeneous form** for any point (x, y) is $(x, y, 1)$. The key of the calculator has homogeneous coordinates (50, 70, 1), and the top of the figure's head is (25, 108, 1). Thus, the origin is (0, 0, 1) in homogeneous form.

710

You may think that appending a 1 to the coordinates of every point doesn't gain much information. However, it allows you to write every translation of the plane as a 3×3 matrix. Here is how.

The translation h units horizontally and k units vertically shifts each point according to the rule $T: (x, y) \rightarrow (x + h, y + k)$. In matrix form

$$T: \begin{bmatrix} x \\ y \end{bmatrix} \rightarrow \begin{bmatrix} x + h \\ y + k \end{bmatrix}.$$

This translation can be expressed using homogeneous coordinates with the

matrix $\begin{bmatrix} 1 & 0 & h \\ 0 & 1 & k \\ 0 & 0 & 1 \end{bmatrix}$, because $\begin{bmatrix} 1 & 0 & h \\ 0 & 1 & k \\ 0 & 0 & 1 \end{bmatrix} \begin{bmatrix} x \\ y \\ 1 \end{bmatrix} = \begin{bmatrix} x + h \\ y + k \\ 1 \end{bmatrix}$.

In particular, the image of the origin is $\begin{bmatrix} 1 & 0 & h \\ 0 & 1 & k \\ 0 & 0 & 1 \end{bmatrix} \begin{bmatrix} 0 \\ 0 \\ 1 \end{bmatrix} = \begin{bmatrix} h \\ k \\ 1 \end{bmatrix}$, as needed.

Now you can find matrices for the translations applied to the cartoon figure. T_1, the translation to the origin requires a shift of 50 left and 70 down. Thus

$$T_1 = \begin{bmatrix} 1 & 0 & -50 \\ 0 & 1 & -70 \\ 0 & 0 & 1 \end{bmatrix}.$$

T_2, the other translation, is 155 right and 83 up. So

$$T_2 = \begin{bmatrix} 1 & 0 & 155 \\ 0 & 1 & 83 \\ 0 & 0 & 1 \end{bmatrix}.$$

Now the rotation matrix must be made into a 3×3 matrix. In general, the

2×2 transformation matrix $\begin{bmatrix} a & b \\ c & d \end{bmatrix}$ has homogeneous form $\begin{bmatrix} a & b & 0 \\ c & d & 0 \\ 0 & 0 & 1 \end{bmatrix}$. So the

homogeneous form for a reflection over the x-axis, $\begin{bmatrix} 1 & 0 \\ 0 & -1 \end{bmatrix}$, is $\begin{bmatrix} 1 & 0 & 0 \\ 0 & -1 & 0 \\ 0 & 0 & 1 \end{bmatrix}$.

$R_{-22.5}$, the rotation matrix for the figure, is

$$R_{-22.5} = \begin{bmatrix} \cos(-22.5) & -\sin(-22.5) & 0 \\ \sin(-22.5) & \cos(-22.5) & 0 \\ 0 & 0 & 1 \end{bmatrix}.$$

The composite which represents the motion of the figure is

$$T_2 \qquad \circ \qquad R_{-22.5} \qquad \circ \qquad T_1.$$

So

$$\begin{bmatrix} 1 & 0 & 155 \\ 0 & 1 & 83 \\ 0 & 0 & 1 \end{bmatrix} \begin{bmatrix} \cos(-22.5) & -\sin(-22.5) & 0 \\ \sin(-22.5) & \cos(-22.5) & 0 \\ 0 & 0 & 1 \end{bmatrix} \begin{bmatrix} 1 & 0 & -50 \\ 0 & 1 & -70 \\ 0 & 0 & 1 \end{bmatrix}$$

$$\approx \left(\begin{bmatrix} 1 & 0 & 155 \\ 0 & 1 & 83 \\ 0 & 0 & 1 \end{bmatrix} \begin{bmatrix} .9239 & .3827 & 0 \\ -.3827 & .9239 & 0 \\ 0 & 0 & 1 \end{bmatrix} \right) \begin{bmatrix} 1 & 0 & -50 \\ 0 & 1 & -70 \\ 0 & 0 & 1 \end{bmatrix}$$

$$\approx \begin{bmatrix} .9239 & .3827 & 155 \\ -.3827 & .9239 & 83 \\ 0 & 0 & 1 \end{bmatrix} \begin{bmatrix} 1 & 0 & -50 \\ 0 & 1 & -70 \\ 0 & 0 & 1 \end{bmatrix}$$

$$\approx \begin{bmatrix} .9239 & .3827 & 82.016 \\ -.3827 & .9239 & 37.462 \\ 0 & 0 & 1 \end{bmatrix}.$$

This checks for the calculator key:

$$\begin{bmatrix} .9239 & .3827 & 82.016 \\ -.3827 & .9239 & 37.462 \\ 0 & 0 & 1 \end{bmatrix} \begin{bmatrix} 50 \\ 70 \\ 1 \end{bmatrix} = \begin{bmatrix} 155 \\ 83 \\ 1 \end{bmatrix}.$$

To find the image of any point P on the figure, then, you would multiply the 3×3 matrix for $T_2 \circ R_{-22.5} \circ T_1$ by the point matrix for P in homogeneous form. Notice that the difficulty of the multiplication is not greatly increased by going from 2×2 form to homogeneous form. The last row of all 3×3 homogeneous form matrices is $[0 \quad 0 \quad 1]$, so that row contributes very little to the complexity of calculations.

When computer scientists wish to picture three-dimensional objects on a two-dimensional computer screen, they generalize the preceding process. First, the object must be oriented with respect to a coordinate system called **world coordinates**. World coordinates for an object maintain a record of where the object is in space. Just as 3×3 matrices are used for homogeneous coordinates of 2×2 matrices, 4×4 matrices are used to transform the position and orientation of any 3-D object. The point (x, y, z) in 3-space has homogeneous coordinates $(x, y, z, 1)$, and a translation of 5 units in x, -3 units in y, and

2.5 units in z is represented by the matrix $\begin{bmatrix} 1 & 0 & 0 & 5 \\ 0 & 1 & 0 & -3 \\ 0 & 0 & 1 & 2.5 \\ 0 & 0 & 0 & 1 \end{bmatrix}.$

712

Second, the computer scientist must determine the viewing point. Where should the viewer be stationed? This is a separate issue from computing world coordinates. A viewer on the right side of the classroom has a different view of your teacher than a person on the left side. But the teacher's world coordinates are the same for both viewing positions. Once the viewing position is determined, a third step is that the view must be restricted, or clipped, to fit into the boundary of a computer screen, as though the computer screen were a window. For the last step, the projection of coordinates must be accomplished. Where on the 2D screen will the 3D points appear?

All of these steps can be performed with 4×4 matrices. Here is a summary.

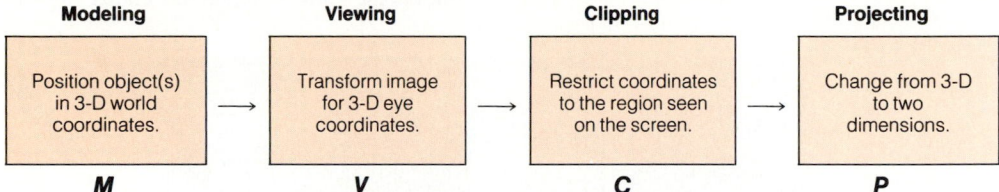

Modeling	Viewing	Clipping	Projecting
Position object(s) in 3-D world coordinates.	Transform image for 3-D eye coordinates.	Restrict coordinates to the region seen on the screen.	Change from 3-D to two dimensions.
M	V	C	P

The matrix product $P \cdot C \cdot V \cdot M$ is applied to points in the world coordinates to give points on the screen. (Note the order of the matrices—M represents the first-applied transformation.) Each frame of an animated sequence has a different 4-matrix product.

The complexity of three-dimensional graphics makes it difficult to do high-quality work on personal computers. The multiplications needed to perform the transformations take a long time. However, supercomputers and high-quality graphics workstations do matrix operations quickly. *Vector processors* are designed to multiply a row times a column in one step. On supercomputers these vector processors vastly shorten the time needed for each 4-matrix product. On graphics workstations, more than one processor might be devoted to the matrix multiplications. Here is how the work can be shared among processors. Consider the following multiplication.

$$\begin{bmatrix} 3 & -1 & 2 & 7 \\ 5 & 8 & 4 & 3 \\ 2 & 9 & -6 & 4 \\ 0 & 0 & 0 & 1 \end{bmatrix} \begin{bmatrix} 33 \\ 27 \\ 11 \\ 1 \end{bmatrix} = \begin{bmatrix} x' \\ y' \\ z' \\ 1 \end{bmatrix}$$

A human or microcomputer would have to perform 12 multiplications and 9 additions to find x', y', and z'. If there were three processors sharing the task, each could have 4 multiplications and 3 additions. For example, the first processor could find $x' = 3 \cdot 33 + -1 \cdot 27 + 2 \cdot 11 + 7 \cdot 1$. The second processor could find y', and the third, z'. So three processors could cut the computational time to $\frac{1}{3}$ that of a single processor. If the processors were vector processors, they could further reduce the time to that of a single step.

High-powered computers for graphics are organized to perform matrix multiplications quickly so that the motion that is represented appears smooth and natural. When combined with shadows and light reflections, also computable with matrices, you can get the realism which fascinates viewers, as in the picture below.

Scene from the movie "Star Wars"

Questions

Covering the Reading

1. A commercial film projects 24 frames a second; a computer simulation has 5000×5000 pixels for each frame. How many point calculations must be performed per second for computer simulation of the commercial film? **6×10^8**

2. Write the homogeneous form of $(3, -7)$. **$(3, -7, 1)$**

3. Write the nonhomogeneous form of $(2, 0, -1, 1)$. **$(2, 0, -1)$**

4. Write the homogeneous form of the matrix for R_{30}, using exact values. **See margin.**

5. Give the 2×2 matrix which has homogeneous form
$$\begin{bmatrix} -1 & 0 & 0 \\ 0 & -1 & 0 \\ 0 & 0 & 1 \end{bmatrix}.$$ **$\begin{bmatrix} -1 & 0 \\ 0 & -1 \end{bmatrix}$**

6. What is the homogeneous form of the 2×2 identity matrix? **$\begin{bmatrix} 1 & 0 & 0 \\ 0 & 1 & 0 \\ 0 & 0 & 1 \end{bmatrix}$**

7. Find the image of $(25, 108)$, the top of the cartoon figure's head, under the composite transformation in the lesson. **$\approx (146.4, 127.7)$**

8. Give a 4×4 matrix which represents a translation of -4 units in x, 0.3 units in y, and 0 units in z. **See margin.**

9. Why are 3 processors, not 4, used to speed the multiplication of 4×4 matrices representing 3-D graphics? **See margin.**

10. Prove that if A and B are 3×3 matrices in homogeneous form, then so is the product AB. **See margin.**

11. A microcomputer with a single processor runs a 3-D computer game with 10 frames per second. In an analysis of the computation steps, the designers found that 60% of the time the processor was doing computations involved in the matrix operations of 3-D graphics and 40% in running the video display. If they replace the single processor with a graphics "engine" that has three processors to split up the matrix multiplications and if they don't change the speed of the video display, how many frames per second can they expect? **no more than about 16.7 frames per second**

Question 26: A projection matrix does not represent a transformation because it is not one-to-one.

ADDITIONAL ANSWERS

13.a. $\begin{bmatrix} .985 & .174 \\ -.174 & .985 \end{bmatrix}$

b. $R_{10} = \begin{bmatrix} .985 & -.174 \\ .174 & .985 \end{bmatrix}$

15. If T_1 and T_2 are scale changes with T_1:

$(x, y) \rightarrow \left(3x, \frac{1}{2}y\right)$ and

T_2: $(x, y) \rightarrow \left(\frac{1}{3}x, 2y\right)$,

then $T_1 \circ T_2 = I$, where I is the identity transformation.

26. This matrix represents the projection of a 3D object onto the xy-plane. It makes the z-coordinate 0 because the third row of the matrix is composed of all zeros.

Review

12. Let $A = \begin{bmatrix} 1 & 0 \\ 0 & -1 \end{bmatrix}$. a) r_x; b) $\begin{bmatrix} 1 & 0 \\ 0 & -1 \end{bmatrix}$; c) r_x
   a. What transformation does A represent?
   b. Find A^{-1}.
   c. What transformation does A^{-1} represent? *(Lessons 11-2, 11-7)*

13. a. Give the entries of the matrix for R_{-10}, rounded to the nearest thousandth.
    b. What matrix is the inverse of the matrix in part **a**? *(Lessons 11-4, 11-7)*
    See margin.

14. Give an example of a 2×2 matrix with a zero determinant. *(Lesson 11-7)* sample: $\begin{bmatrix} 4 & 8 \\ -3 & -6 \end{bmatrix}$

15. State a corresponding sentence involving transformations.

$$\begin{bmatrix} 3 & 0 \\ 0 & \frac{1}{2} \end{bmatrix} \cdot \begin{bmatrix} \frac{1}{3} & 0 \\ 0 & 2 \end{bmatrix} = \begin{bmatrix} 1 & 0 \\ 0 & 1 \end{bmatrix}$$ *(Lessons 11-2, 11-7)* **See margin.**

16. Give a formula for $\sin 3\theta$ in terms of $\sin \theta$. *(Lesson 11-6)*
   $\sin 3\theta = 3 \sin \theta - 4 \sin^3 \theta$

17. Use a Double Angle Identity to show that the graph of the function $f(x) = 10 \cos^2 x - 5$ has amplitude 5. *(Lesson 11-6)*
   $f(x) = 5(2 \cos^2 x - 1) = 5 \cos 2x$, which has amplitude 5.

18. Give an exact value. *(Lesson 11-5)*
   a. $\cos 255°$ $\frac{\sqrt{2} - \sqrt{6}}{4}$ b. $\sin \frac{29\pi}{2}$ 1

19. Suppose that a triangle is represented by the matrix $\begin{bmatrix} a & c & e \\ b & d & f \end{bmatrix}$. Tell what the following product represents.
   $$\begin{bmatrix} \cos 25° & -\sin 25° \\ \sin 25° & \cos 25° \end{bmatrix} \cdot \begin{bmatrix} a & c & e \\ b & d & f \end{bmatrix}$$ *(Lesson 11-4)*
   a counterclockwise rotation of 25°

20. A matrix is used to transform a figure. $(1, 0)$ has image $(3, -1)$; $(0, 1)$ has image $(5, 4)$. *(Lesson 11-3)*
   a. What is the matrix? $\begin{bmatrix} 3 & 5 \\ -1 & 4 \end{bmatrix}$
   b. What is the image of $(0, 0)$? **(0, 0)**
   c. What is the image of $(1, 1)$? **(8, 3)**

21. Multiply. $\begin{bmatrix} 1 & 2 & 3 \\ 4 & 5 & 6 \\ 7 & 8 & 9 \end{bmatrix} \cdot \begin{bmatrix} -1 & -2 \\ -3 & -4 \\ -5 & -6 \end{bmatrix}$ *(Lesson 11-1)* $\begin{bmatrix} -22 & -28 \\ -49 & -64 \\ -76 & -100 \end{bmatrix}$

22. Consider two thin lenses with focal lengths f_1 and f_2 at a distance d apart.

716

Physicists and engineers studying optics have shown that the product

$$\begin{bmatrix} 1 & 0 \\ -\frac{1}{f_1} & 1 \end{bmatrix} \begin{bmatrix} 1 & d \\ 0 & 1 \end{bmatrix} \begin{bmatrix} 1 & 0 \\ -\frac{1}{f_2} & 1 \end{bmatrix}$$

describes the image of light passing through this system.   *(Lesson 11-1)*

a. Calculate this product for the special case when the lenses touch each other, that is, $d = 0$.

b. Calculate this product for the general case given above.
See below.

In 23 and 24, the graph below represents the standard normal distribution for the random variable x. Find the number a satisfying the following.
(Lesson 10-5)

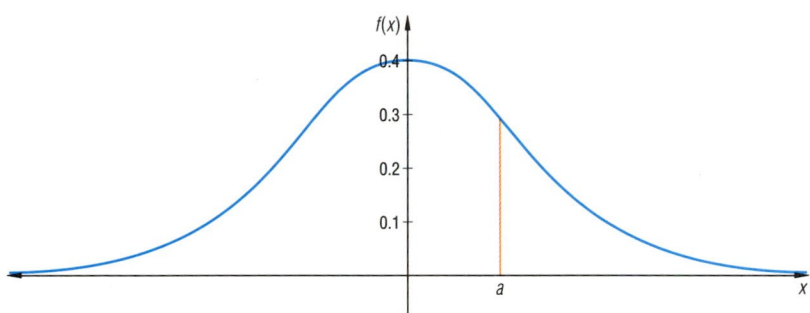

23. $P(x < a) = 0.75$ **about 0.67**   **24.** $P(x \geq a) = 0.38$ **about 0.31**

Exploration

25. Estimate your world coordinates in math class with relation to the front door of your house or apartment. Let x be feet north; y be feet east; and z be feet up. (If you are 2 miles west of your house, for instance, you would have a y-coordinate of $-2 \cdot 5280 = -10560$. If you are on the third floor of your school and your front door is at the level of the school entrance, $z \approx 24$ for you.)   **Answers will vary.**

26. One kind of projection matrix is $\begin{bmatrix} 1 & 0 & 0 & 0 \\ 0 & 1 & 0 & 0 \\ 0 & 0 & 0 & 0 \\ 0 & 0 & 0 & 1 \end{bmatrix}$. What do you suppose this does to coordinates of a 3-D object? Why?
See margin.

22a) $\begin{bmatrix} 1 & 0 \\ \frac{-1}{f_1} - \frac{1}{f_2} & 1 \end{bmatrix}$; b) $\begin{bmatrix} 1 - \frac{d}{f_2} & d \\ \frac{d - f_1 - f_2}{f_1 f_2} & 1 - \frac{d}{f_1} \end{bmatrix}$

FOLLOW-UP

EXTENSION
Small Group Work
Groups of students might want to use the ideas of this lesson to find a 3×3 matrix for every reflection and every rotation. The 3-step process on page 710 shows how to do this. For reflections, translate a point on the reflecting line to the origin, next reflect over the line using the formula in Questions 15–17 of Lesson 11-6, and then translate back to the original point. For rotations, translate the center to the origin, rotate using the Rotation Matrix Theorem, and then translate back.

PROJECTS
Project 1 on page 718 is related to the content of this lesson.

Projects

1. Write a computer program to display a figure and its images under some of the transformations described in this chapter. Include the images under a reflection, rotation, translation, and scale change.

2. In this chapter you learned how to represent rotations in the plane with 2×2 matrices. Find some books which describe how to represent three-dimensional rotations by matrices. Prepare a report on your findings.

3. The formula $d = \dfrac{v^2 \sin 2\theta}{9.8}$, for the distance d in meters at which an object projected from ground level with an initial velocity v (in meters per second) and initial angle θ to the horizontal will land, only applies if the points of release and landing are at the same horizontal level. This assumption is often not satisfied. For instance, in the shot-put an athlete propels a shot, a heavy metal sphere, and tries to maximize the horizontal distance it travels. A shot-putter releases the shot at about shoulder level, not at ground level.

Mary Harrington

It can be shown that when the shot is released at a height h meters above the ground with velocity v meters per second at angle θ with the horizontal, its horizontal displacement d is given by

$$d = \frac{v^2 \sin 2\theta}{19.6} + \frac{v^2 \cos \theta}{9.8} \sqrt{\sin^2 \theta + \frac{19.6h}{v^2}}.$$

When the release velocity and release height are given, d is solely a function of θ.

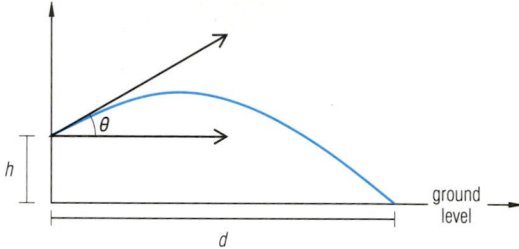

a. World-class shot-putters can easily achieve a release velocity of 13 meters per second. Suppose for such a shot-putter $h = 1.90$m. Modify the program TABULATE from Lesson 9-3 or use other table-generating software to construct a table of values for d as a function of θ, for angles between 20° and 50° in increments of 5°. Using smaller increments, find the angle θ at which the shot should be released so as to achieve the maximum value of d. According to this model, what is the maximum distance possible for this shot-putter?

b. Consider a shorter athlete who releases the shot from a height of 1.7m at the same velocity. What is the maximum distance this shot-putter can attain?

c. How do the data in parts **a** and **b** above compare to records in the shot-put for your school? for the Olympics?

718

4. It is possible to express the complex number $a + bi$ as a 2×2 matrix of the form $\begin{bmatrix} a & -b \\ b & a \end{bmatrix}$, where a and b are real numbers.

a. Express each complex number below in matrix form.
   i. $2 + 3i$
   ii. -1
   iii. $-4 + 8i$
   iv. i

b. Each 2×2 matrix below represents a complex number. Express it in $a + bi$ form.
   i. $\begin{bmatrix} 3 & -7 \\ 7 & 3 \end{bmatrix}$
   ii. $\begin{bmatrix} 0 & 4 \\ -4 & 0 \end{bmatrix}$

c. Under this representation, multiplication of complex numbers corresponds to matrix multiplication. For example, the matrix representation of $(1 + 4i)(-2 + i)$ is
$$\begin{bmatrix} 1 & -4 \\ 4 & 1 \end{bmatrix} \cdot \begin{bmatrix} -2 & -1 \\ 1 & -2 \end{bmatrix} = \begin{bmatrix} -6 & 7 \\ -7 & -6 \end{bmatrix}.$$
Use the following steps to verify that this is true for any two complex numbers $z = a + bi$ and $w = c + di$.
   i. Represent z and w in matrix form.
   ii. Express $z \cdot w$ in matrix form.
   iii. Find $z \cdot w$ in $a + bi$ form.
   iv. Compare your results from **ii** and **iii**.

d. Matrices of the same dimension are added by adding the corresponding elements. For example,
$$\begin{bmatrix} 1 & 3 \\ -2 & 4 \end{bmatrix} + \begin{bmatrix} 6 & 5 \\ 8 & 0 \end{bmatrix} = \begin{bmatrix} 7 & 8 \\ 6 & 4 \end{bmatrix}.$$
Verify that under matrix representation of complex numbers, matrix addition corresponds to addition of complex numbers in $a + bi$ form.

e. For a complex number $a + bi$, its absolute value $|a + bi| = \sqrt{a^2 + b^2}$. How is this related to the determinant of the matrix representation of $a + bi$?

f. Which matrices represent complex numbers with absolute value 1?

5. a. Two 3×3 matrices M and N are inverse matrices if and only if their products MN and NM are the 3×3 identity matrix. Verify that
$$\begin{bmatrix} 2 & 1 & 1 \\ 1 & 4 & -3 \\ -1 & -3 & 3 \end{bmatrix}^{-1} = \begin{bmatrix} \frac{3}{7} & -\frac{6}{7} & -1 \\ 0 & 1 & 1 \\ \frac{1}{7} & \frac{5}{7} & 1 \end{bmatrix}.$$

b. Consider the system
$$\begin{aligned} 2x + y + z &= 0 \\ x + 4y - 3z &= 5 \\ -x - 3y + 3z &= 7. \end{aligned}$$
Verify that you can represent this system with the matrix equation
$$\begin{bmatrix} 2 & 1 & 1 \\ 1 & 4 & -3 \\ 1 & -3 & 3 \end{bmatrix} \cdot \begin{bmatrix} x \\ y \\ z \end{bmatrix} = \begin{bmatrix} 0 \\ 5 \\ 7 \end{bmatrix}.$$

c. Multiply both sides of the matrix representation for the system by the inverse matrix in part **a**. What is the solution to the system?

d. The BASIC program below (called INVERTER) computes the inverse of a 3×3 matrix M. When you input the rows of M be sure to separate the elements of each row by commas. Verify that the program is correct by using it to compute the inverse of $\begin{bmatrix} 2 & 1 & 1 \\ 1 & 4 & -3 \\ -1 & -3 & 3 \end{bmatrix}$, and by comparing your result to the matrix in part **a**.

e. Use the program INVERTER to solve the system
$$\begin{aligned} 3x + y + z &= 4 \\ 2x - 2y - z &= 2 \\ 3x - y - z &= 1 . \end{aligned}$$
Check your results by substitution.

```
10    REM PROGRAM INVERTER
20    INPUT "FIRST ROW OF MATRIX M"; M(1,1), M(1,2), M(1,3)
30    INPUT "SECOND ROW OF MATRIX M"; M(2,1), M(2,2), M(2,3)
40    INPUT "THIRD ROW OF MATRIX M"; M(3,1), M(3,2), M(3,3)
50    LET N(1,1) = M(2,2)*M(3,3)  −  M(2,3)*M(3,2)
60    LET N(1,2) = M(2,1)*M(3,3)  −  M(2,3)*M(3,1)
70    LET N(1,3) = M(2,1)*M(3,2)  −  M(2,2)*M(3,1)
80    LET N(2,1) = M(1,2)*M(3,3)  −  M(1,3)*M(3,2)
90    LET N(2,2) = M(1,1)*M(3,3)  −  M(1,3)*M(3,1)
100   LET N(2,3) = M(1,1)*M(3,2)  −  M(1,2)*M(3,1)
110   LET N(3,1) = M(1,2)*M(2,3)  −  M(1,3)*M(2,2)
120   LET N(3,2) = M(1,1)*M(2,3)  −  M(1,3)*M(2,1)
130   LET N(3,3) = M(1,1)*M(2,2)  −  M(1,2)*M(2,1)
140   LET D = M(1,1)*N(1,1)  −  M(1,2)*N(1,2) + M(1,3)*N(1,3)
150   FOR I = 1 TO 3
160   PRINT ((-1)^(1 + I))*N(1,I)/D,  ((-1)^(2 + I))*N(2,I)/D,  ((-1)^(3 + I))*N(3,I)/D
170   NEXT I
180   END
```

Summary

An $m \times n$ matrix is a rectangular array of elements arranged in m rows and n columns. If $m = n$, the matrix is a square matrix. The elements of all matrices in this chapter are real numbers.

If A and B are matrices, the product $C = AB$ exists if the number of columns of A equals the number of rows of B. The element in the ith row and jth column of C is the term-by-term product of the ith row of A and the jth column of B. Matrix multiplication has many applications. If the elements of A represent coefficients, the elements of B are variables, and C is a column matrix of constants, then $AB = C$ represents a linear system of equations. If A contains numbers of items and B represents corresponding unit values of the items, then AB gives total values of the items.

Some transformations can be represented by 2×2 matrices. Any n-gon can be represented by a $2 \times n$ matrix whose columns are its vertices. If M represents a transformation T and P represents a polygon, then MP represents $T(P)$. This enables transformations to be done using matrices. If M_1 and M_2 are matrices for transformations T_1 and T_2, then $M_1 M_2$ represents $T_1 \circ T_2$. Thus composites of transformations can also be done with matrices. Lifelike computer animations can be performed by multiplying a matrix for a set of points for a figure by successive matrices representing the transformations desired of the animated figure.

Some of the transformations for which 2×2 matrices exist are: reflections over the x-axis, y-axis, and line $x = y$; all rotations with center at the origin; all size changes with center $(0,0)$; and all scale changes of the form $(x,y) \rightarrow (ax,by)$. The Matrix Basis Theorem provides an easy way to recall the matrix for a given transformation; you need only know the images of $(1,0)$ and $(0,1)$ under that transformation.

If a matrix exists for a transformation, then the multiplicative inverse of that matrix is the matrix for the inverse of the transformation. Matrices and their inverses can be used to solve systems and to make and break codes.

From the matrix for a rotation, formulas for $\cos(x + y)$ and $\sin(x + y)$ can be rather quickly derived. From these formulas, formulas for $\cos(x - y)$, $\sin(x - y)$, $\cos 2x$, and $\sin 2x$ follow. These identities enable exact values for certain cosines and sines to be determined and to express multi-argument trigonometric functions in terms of single arguments.

Vocabulary

For the starred (*) items you should be able to give a definition of the term.
For the other terms you should be able to give a general description and a specific example of each.

Lesson 11-1
matrix, matrices
element
* dimensions of a matrix, $m \times n$
row matrix
column matrix
matrix multiplication
product matrix

Lesson 11-2
point matrix
matrix representing a transformation
Matrices for Reflections Theorem
2×2 Identity Matrix

Lesson 11-3
Matrices for Rotations Theorem
Matrix Basis Theorem

Lesson 11-4
Rotation Matrix Theorem

Lesson 11-5
Addition Identities for the Cosine and
  Sine

Lesson 11-6
Double Angle Identities

Lesson 11-7
* inverse transformations
* inverse matrices
Inverse of a 2×2 Matrix Theorem
* square matrices
determinant, det
coding matrix

Lesson 11-8
homogeneous form
world coordinates
vector processor

Progress Self-Test

See margin for answers not shown below.

Take this test as you would take a test in class. You
will need graph paper and a calculator. Then check
the test yourself using the solutions at the back of the
book.

In 1–3, use the following matrices. P gives the number of registered voters (in millions) in each part of a
state. V represents the percents of registered voters in
those areas by their indications of how they would
vote in an election.

$$\begin{matrix} \text{Urban} \\ \text{Rural} \\ \text{Suburban} \end{matrix} \begin{bmatrix} 3.4 \\ 0.6 \\ 2.1 \end{bmatrix} = P$$

$$\begin{matrix} & \text{Urban} & \text{Rural} & \text{Suburban} \end{matrix}$$
$$\begin{matrix} \text{Democrat} \\ \text{Republican} \\ \text{Independent} \end{matrix} \begin{bmatrix} .50 & .47 & .29 \\ .29 & .41 & .58 \\ .21 & .12 & .13 \end{bmatrix} = V$$

1. How many people in the state are expected to
   vote Republican? **2.45 million**

2. Find VP.

3. What does the matrix VP represent?

4. Multiply $\begin{bmatrix} 5 & -1 \\ 2 & 0 \end{bmatrix} \begin{bmatrix} 3 & 8 \\ -1 & 1 \end{bmatrix}$. $\begin{bmatrix} 16 & 39 \\ 6 & 16 \end{bmatrix}$

5. M is a 3×4 matrix, N is a $4 \times t$ matrix, and Z is
   a 3×5 matrix. If $MN = Z$, find the value of t. **5**

In 6–8 use the figure below.

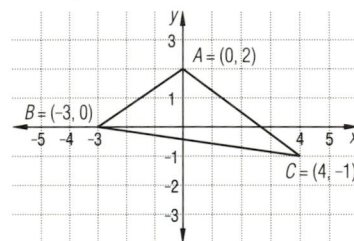

6. Write a matrix M for $\triangle ABC$.

7. Find XM, where $X = \begin{bmatrix} -1 & 0 \\ 0 & 2 \end{bmatrix}$.

8. Graph the image represented by XM.

6) $\begin{bmatrix} 0 & -3 & 4 \\ 2 & 0 & -1 \end{bmatrix}$; 7) $\begin{bmatrix} 0 & 3 & -4 \\ 4 & 0 & -2 \end{bmatrix}$

9. a. Give a matrix representing a reflection over
      the line $y = x$.
   b. Use the matrix in part **a** to show that the
      transformation $r_{y=x}$ is its own inverse.

10. *Multiple choice.* Which matrix represents R_{30}?

(a) $\begin{bmatrix} \frac{\sqrt{3}}{2} & \frac{1}{2} \\ \frac{1}{2} & \sqrt{3} \end{bmatrix}$ (b) $\begin{bmatrix} \frac{\sqrt{3}}{2} & -\frac{1}{2} \\ \frac{1}{2} & \frac{\sqrt{3}}{2} \end{bmatrix}$

(c) $\begin{bmatrix} -\frac{\sqrt{3}}{2} & \frac{1}{2} \\ -\frac{1}{2} & -\frac{\sqrt{3}}{2} \end{bmatrix}$ (d) $\begin{bmatrix} \frac{1}{2} & -\frac{\sqrt{3}}{2} \\ \frac{\sqrt{3}}{2} & \frac{1}{2} \end{bmatrix}$

(e) $\begin{bmatrix} \frac{1}{2} & \frac{\sqrt{3}}{2} \\ \frac{\sqrt{3}}{2} & \frac{1}{2} \end{bmatrix}$ **(b)**   **11a) r_x; b) $\begin{bmatrix} -1 & 0 \\ 0 & 1 \end{bmatrix}$**

11. A figure is transformed by $R_{180} \circ r_x$.
    a. What transformation is done first?
    b. What single matrix represents the
       composite?

12. What is the image of $\triangle PQR$, where $P = (0, 0)$,
    $Q = (1, 0)$, and $R = (0, 1)$, under a rotation of
    magnitude θ?

13. Write as the sine or cosine of a single argument:
    $\sin 83° \cos 42° - \cos 83° \sin 42°$. **sin 41°**

14. Using an identity for $\cos(\alpha + \beta)$, simplify
    $\cos(\pi + \theta)$.

15. If $\sin A = 0.40$ and A is acute, find $\sin 2A$.

16. Find the inverse of $C = \begin{bmatrix} 1 & -1 \\ 3 & 5 \end{bmatrix}$. $\begin{bmatrix} 5/8 & 1/8 \\ -3/8 & 1/8 \end{bmatrix}$

17. Show that $G = \begin{bmatrix} 3x & 6 \\ 5x & 10 \end{bmatrix}$ has no inverse.

18. a. Write a matrix equation to represent the
       system.
       $$\begin{cases} 3x - y = 6 \\ 5x - 2y = 11 \end{cases}$$
    b. Solve the equation in part **a**.

17) **det $G = (3x)(10) - (5x)(6) = 30x - 30x = 0$.
A matrix with a zero determinant has no inverse.**

18a) $\begin{bmatrix} 3 & -1 \\ 5 & -2 \end{bmatrix} \begin{bmatrix} x \\ y \end{bmatrix} = \begin{bmatrix} 6 \\ 11 \end{bmatrix}$; b) $\begin{bmatrix} x \\ y \end{bmatrix} = \begin{bmatrix} 1 \\ -3 \end{bmatrix}$

722

Chapter Review

Questions on **SPUR** Objectives

SPUR stands for **S**kills, **P**roperties, **U**ses, and **R**epresentations.
The Chapter Review questions are grouped according to the SPUR Objectives for this chapter.
See margin for answers not shown below.

SKILLS deal with the procedures used to get answers.

■ **Objective A:** *Multiply matrices, where possible.*
(Lesson 11-1)

In 1–4, use the following matrices.

$$A = \begin{bmatrix} \frac{1}{2} & 3 \\ 5 & -1 \end{bmatrix} \qquad B = \begin{bmatrix} 1 & 5 & -1 \\ -1 & 0 & 1 \end{bmatrix}$$

$$C = \begin{bmatrix} x \\ 3 \end{bmatrix} \qquad D = \begin{bmatrix} 1 & -7 \end{bmatrix}$$

1. *Multiple choice.* Which of the following products is not possible?
   (a) AC (b) CD
   (c) BA (d) A^2C **(c)**

2. Find AB.

3. Find DC.   **[x − 21]**

4. Find A^2. $\begin{bmatrix} 61/4 & -3/2 \\ -5/2 & 16 \end{bmatrix}$

■ **Objective B:** *Use matrices to solve systems of equations.* *(Lesson 11-7)*

5. Solve for x: $\begin{bmatrix} 1 & 2 \\ -1 & 1 \end{bmatrix}\begin{bmatrix} x \\ 1 \end{bmatrix} = \begin{bmatrix} 7 \\ -4 \end{bmatrix}$.   **5**

6. Consider the following system.
$$\begin{cases} x + 3y = 11 \\ 2x - y = 8 \end{cases}$$
   a. Represent the system by a matrix equation.
   b. Find the inverse of the coefficients matrix in part **a**.
   c. Multiply on the left both sides of the matrix equation by the answer in part **b**.
   d. What is the solution to the system of equations?   **x = 5 and y = 2**

In 7 and 8, solve the given system of equations using matrices.

7. $\begin{cases} 5x + 2y = 6 \\ 10x - 10y = -9 \end{cases}$ **x = $\frac{3}{5}$ and y = $\frac{3}{2}$**

8. $\begin{cases} 8x + 3y = 10 \\ x + 8y = -14 \end{cases}$ **x = 2 and y = -2**

9. Solve the system $\begin{cases} x + 2y + 3z = 10 \\ 3x + y + 2z = 13 \\ 2x + 3y + z = 13 \end{cases}$,

   given that $\begin{bmatrix} 1 & 2 & 3 \\ 3 & 1 & 2 \\ 2 & 3 & 1 \end{bmatrix}^{-1} = \begin{bmatrix} -\frac{5}{18} & \frac{7}{18} & \frac{1}{18} \\ \frac{1}{18} & -\frac{5}{18} & \frac{7}{18} \\ \frac{7}{18} & \frac{1}{18} & -\frac{5}{18} \end{bmatrix}$.

   x = 3, y = 2, z = 1

PROPERTIES deal with the principles behind the mathematics.

■ **Objective C:** *Apply properties of matrices and matrix multiplication.* *(Lessons 11-1, 11-3, 11-7)*

In 10 and 11, the matrix M has dimensions 4×2. The matrix P has dimensions 4×5. Give the dimensions of N if:

10. $MN = P$ **2 × 5**

11. $PN = M$.   **5 × 2**

12. *Multiple choice.* Which of the following statements about all 2×2 matrices R, S, and T is *false*?
    (a) RS is a 2×2 matrix.
    (b) $(RS)T = R(ST)$
    (c) There is a matrix I such that $RI = IR = R$.
    (d) $RS = SR$ **(d)**

13. Name three types of transformations that can be represented by 2×2 matrices.

CHAPTER REVIEW

The main objectives for the chapter are organized here into sections corresponding to the types of understanding this book promotes: Skills, Properties, Uses, and Representations (called SPUR objectives).

USING THE CHAPTER REVIEW
Students should be able to answer questions like these with about 85% accuracy by the end of the chapter. (See pages 72 and T38–39 for more information.)

ADDITIONAL ANSWERS

2. $\begin{bmatrix} -\frac{5}{2} & \frac{5}{2} & \frac{5}{2} \\ 6 & 25 & -6 \end{bmatrix}$

6.a. $\begin{bmatrix} 1 & 3 \\ 2 & -1 \end{bmatrix}\begin{bmatrix} x \\ y \end{bmatrix} = \begin{bmatrix} 11 \\ 8 \end{bmatrix}$

b. $\begin{bmatrix} \frac{1}{7} & \frac{3}{7} \\ \frac{2}{7} & -\frac{1}{7} \end{bmatrix}$

c. $\begin{bmatrix} \frac{1}{7} & \frac{3}{7} \\ \frac{2}{7} & -\frac{1}{7} \end{bmatrix}\begin{bmatrix} 1 & 3 \\ 2 & -1 \end{bmatrix}\begin{bmatrix} x \\ y \end{bmatrix}$
$= \begin{bmatrix} \frac{1}{7} & \frac{3}{7} \\ \frac{2}{7} & -\frac{1}{7} \end{bmatrix}\begin{bmatrix} 11 \\ 8 \end{bmatrix}$;
$\begin{bmatrix} x \\ y \end{bmatrix} = \begin{bmatrix} 5 \\ 2 \end{bmatrix}$

13. sample: a rotation about the origin, a reflection over a line containing the origin, and the identity transformation

■ **Objective D:** *Find the inverse of a 2 × 2 matrix.*
(Lesson 11-7)

In 14 and 15, find the inverse of the given matrix.

14. $\begin{bmatrix} 2 & 1 \\ -5 & 4 \end{bmatrix}$

15. $\begin{bmatrix} \cos 40° & -\sin 40° \\ \sin 40° & \cos 40° \end{bmatrix}$

$\begin{bmatrix} \cos 40° & \sin 40° \\ -\sin 40° & \cos 40° \end{bmatrix}$

16. Find $\begin{bmatrix} \frac{2}{9} & \frac{1}{9} \\ -\frac{1}{3} & \frac{4}{3} \end{bmatrix}^{-1}$. $\begin{bmatrix} 4 & -1/3 \\ 1 & 2/3 \end{bmatrix}$

17. Tell which *two* of the following matrices have no inverse.

(a) $\begin{bmatrix} 0 & 0 \\ 0 & 0 \end{bmatrix}$

(b) $\begin{bmatrix} 1 & 0 \\ 0 & 1 \end{bmatrix}$

(c) $\begin{bmatrix} 0 & 1 \\ 1 & 0 \end{bmatrix}$

(d) $\begin{bmatrix} -2 & 4 \\ -6 & 8 \end{bmatrix}$

(e) $\begin{bmatrix} -2 & 4 \\ -1 & 2 \end{bmatrix}$ (a) and (e)

■ **Objective E:** *Apply the Addition and Double Angle identities.* *(Lessons 11-5, 11-6)*

In 18 and 19, state the exact value.

18. $\sin 105°$ $\frac{\sqrt{6} + \sqrt{2}}{4}$

19. $\cos \frac{5\pi}{12}$ $\frac{\sqrt{6} - \sqrt{2}}{4}$

In 20 and 21, A and B are acute angles with $\sin A = .8$ and $\cos B = .5$. Find:

20. $\cos (A - B)$
    $\approx .99$

21. $\sin 2A$.
    .96

In 22 and 23, simplify.

22. $\cos (\pi - \theta)$ $-\cos \theta$

23. $\sin \left(\theta - \frac{\pi}{2} \right)$
    $-\cos \theta$

24. Give a formula for $\cos 4\theta$ in terms of $\cos \theta$ only.
    $\cos 4\theta = 8\cos^4 \theta - 8\cos^2 \theta + 1$

In 25–27, write as the sine or cosine of a single argument.

25. $2 \sin A \cos A$ $\sin(2A)$

26. $\cos \frac{\pi}{5} \cos \frac{\pi}{3} + \sin \frac{\pi}{5} \sin \frac{\pi}{3}$ $\cos \left(\frac{2\pi}{15} \right)$

27. $2 \cos^2 25° - 1$ $\cos 50°$

USES deal with applications of mathematics in real situations.

■ **Objective F:** *Use a matrix to organize information.* *(Lessons 11-1, 11-2)*

In 28–31, use the following matrix of Foreign Exchange rates.

	United Kingdom (pound)	Canada (dollar)	Japan (yen)	United States (dollar)
1970	2.40	0.96	0.0028	1.00
1975	2.22	0.98	0.0034	1.00
1980	2.21	0.84	0.0040	1.00
1985	1.15	0.76	0.0040	1.00
1990	1.6!	0.86	0.0068	1.00

$= F$

28. Describe the information given by each row of F.

29. How many U.S. dollars were 1000 Japanese yen worth in 1990?   **$6.80**

30. If a tourist entered the U.S. in 1990 with 150 pounds U.K., 212 dollars Canadian, and 100 dollars U.S., what is the U.S. equivalent of that cash?   **$523.82**

31. *True* or *false*. To compare the total cash value of the tourist in Question 30 at five-year intervals between 1970 and 1990, one would look at the product matrix FC, where

$$C = \begin{bmatrix} 150 \\ 212 \\ 0 \\ 100 \end{bmatrix}$$ **True**

In 32–34, use the following production matrix P and cost matrix C.

$$P = \begin{array}{c} \\ \\ \\ \\ \\ \end{array} \begin{array}{cc} \text{Melons} & \text{Lettuce} \\ & \text{(head)} \end{array}$$

$$P = \begin{bmatrix} 1000 & 10{,}000 \\ 2000 & 10{,}000 \\ 500 & 20{,}000 \\ 500 & 5{,}000 \end{bmatrix} \begin{array}{l} \text{Farm } A \\ \text{Farm } B \\ \text{Farm } C \\ \text{Farm } D \end{array}$$

$$\begin{array}{cc} \text{Cost to} & \text{Cost to} \\ \text{produce} & \text{consumer} \end{array}$$

$$C = \begin{bmatrix} .45 & .90 \\ .65 & 1.00 \end{bmatrix} \begin{array}{l} \text{Melons} \\ \text{Lettuce} \end{array}$$

32. a. Calculate PC.
b. What does PC represent?

33. a. Find the total cost to produce melons and lettuce on Farm C.   **$13,225**
b. What row and/or column contains this cost in PC?   **3rd row, 1st column**

34. Find the total cost to the consumer of the melons and lettuce produced on Farm A.   **$10,900**

32a) $\begin{bmatrix} 6950 & 10900 \\ 7400 & 11800 \\ 13225 & 20450 \\ 3475 & 5450 \end{bmatrix}$ **b) It is the total cost of production (first column) and total cost to the consumer (second column) resulting from the produce from the farm (the rows).**

REPRESENTATIONS deal with pictures, graphs, or objects that illustrate concepts.

■ **Objective G:** *Represent reflections, rotations, scale changes, and size changes as matrices.*
(Lessons 11-2, 11-3, 11-4)

35. Write the matrix representing r_x, reflection over the x-axis. $\begin{bmatrix} 1 & 0 \\ 0 & -1 \end{bmatrix}$

36. A graph is scaled by shrinking it vertically $\frac{1}{3}$. Write the corresponding matrix. $\begin{bmatrix} 1 & 0 \\ 0 & 1/3 \end{bmatrix}$

In 37 and 38, describe the transformation represented by the matrix.

37. $\begin{bmatrix} 11 & 0 \\ 0 & 11 \end{bmatrix}$ **A size change by a factor of 11.**

38. $\begin{bmatrix} 0 & 1 \\ 1 & 0 \end{bmatrix}$ $r_{y=x}$

39. Write the matrix which represents a 90° rotation counterclockwise about the origin.

40. Tell what rotation is represented by $\begin{bmatrix} -1 & 0 \\ 0 & -1 \end{bmatrix}$. R_{180}

41. Give the matrix for R_{32}.

42. Give the matrix with exact values for R_{120}.

43. A transformation which can be represented by a matrix takes $(0, 0)$ to $(0, 0)$, $(1, 0)$ to $(5, 0)$, and $(0, 1)$ to $(0, -1)$.
a. What is the matrix for the transformation?
b. What is the image of $(1, 1)$?   **$(5, -1)$**

a) $\begin{bmatrix} 5 & 0 \\ 0 & -1 \end{bmatrix}$

■ **Objective H:** *Represent composites of transformations as matrix products.*
(Lessons 11-3, 11-4)

44. A figure is rotated 90° counterclockwise, then reflected over the y-axis.
a. Write a matrix product to represent the composite transformation.
b. Compute the matrix product from part **a**.
c. What single transformation is equivalent to the two given transformations? $r_{y=x}$

45. Tell, in words, what composition of transformations is represented by
$$\begin{bmatrix} \cos 32° & -\sin 32° \\ \sin 32° & \cos 32° \end{bmatrix} \cdot \begin{bmatrix} 0 & 1 \\ 1 & 0 \end{bmatrix} \cdot \begin{bmatrix} 0 & -1 \\ 1 & 0 \end{bmatrix}.$$

46. The matrix representing R_{135} is
$$A = \begin{bmatrix} -\dfrac{\sqrt{2}}{2} & -\dfrac{\sqrt{2}}{2} \\ \dfrac{\sqrt{2}}{2} & -\dfrac{\sqrt{2}}{2} \end{bmatrix}.$$
a. Compute A^2.
b. What transformation does A^2 represent?

47. Let Y be the matrix representing r_y.
a. Compute Y^3.
b. What transformation does Y^3 represent?

46a) $\begin{bmatrix} 0 & 1 \\ -1 & 0 \end{bmatrix}$; **b) R_{-90}**

47a) $\begin{bmatrix} -1 & 0 \\ 0 & 1 \end{bmatrix}$; **b) r_y**

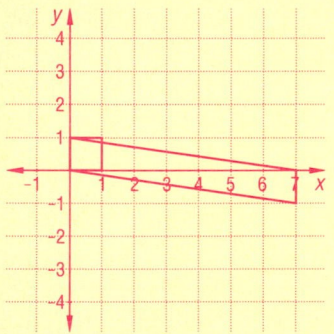
■ **Objective I:** *Find the image of a figure under a transformation using a matrix.*
(Lessons 11-2, 11-3, 11-4)

48. Refer to △*ABC* below.

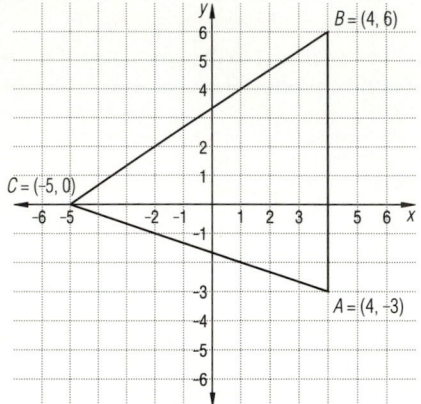

a. Find a matrix for the image △*A′B′C′* under the transformation represented by $\begin{bmatrix} 1 & 1 \\ 0 & -1 \end{bmatrix}$.

b. Draw the image.

a) $\begin{bmatrix} 1 & 10 & -5 \\ 3 & -6 & 0 \end{bmatrix}$

49. a. Find a matrix for the vertices of the image of the square having opposite vertices $(0, 0)$ and $(1, 1)$ under the transformation with matrix $\begin{bmatrix} 7 & 0 \\ -1 & 1 \end{bmatrix}$. **sample:** $\begin{bmatrix} 0 & 0 & 7 & 7 \\ 0 & 1 & 0 & -1 \end{bmatrix}$

b. Draw the square and its image on the same axes.

50. Let $A = (7, 0)$, $B = (0, 2)$, $C = (-1, -1)$.
a. Describe $(r_x \circ R_{90})(\triangle ABC)$ as the product of three matrices.
b. Find a single matrix for the image triangle.

51. A two-dimensional figure has been rotated counterclockwise around the origin by 75°. Write the matrix for the transformation which would return the figure to its original position.

50a) $\begin{bmatrix} 1 & 0 \\ 0 & -1 \end{bmatrix} \begin{bmatrix} 0 & -1 \\ 1 & 0 \end{bmatrix} \begin{bmatrix} 7 & 0 & -1 \\ 0 & 2 & -1 \end{bmatrix}$

b) $\begin{bmatrix} 0 & -2 & 1 \\ -7 & 0 & 1 \end{bmatrix}$

51) $R_{-75} = \begin{bmatrix} \cos(-75°) & -\sin(-75°) \\ \sin(-75°) & \cos(-75°) \end{bmatrix}$

$= \begin{bmatrix} \cos 75° & \sin 75° \\ -\sin 75° & \cos 75° \end{bmatrix}$

726

CHAPTER 12 ■ QUADRATIC RELATIONS

OBJECTIVES ■ CHAPTER 12

The objectives listed here are the same as in the Chapter 12 Review on pages 776–778 of the student text. The Progress Self-Test on page 775 and the tests in the Teacher's Resource File cover these objectives. For recommendations regarding the handling of this end-of-chapter material, see the notes in the margins on the corresponding pages of this Teacher's Edition.

OBJECTIVES FOR CHAPTER 12 (Organized into the SPUR categories— Skills, Properties, Uses, and Representations)	Progress Self-Test Questions	Chapter Review Questions	Teacher's Resource File	
			Lesson Masters*	Chapter Test Forms A & B
SKILLS				
A Use properties of ellipses and hyperbolas to write equations describing them.	5, 8	1–4	12-2, 12-3	6, 9
B Find equations for rotation images of figures.	9, 11	5–10	12-4	10
C Rewrite equations of conic sections in standard form of a quadratic relation in two variables.	14	11–14	12-5	5
PROPERTIES				
D State and apply properties of ellipses and hyperbolas to draw and describe them.	6, 7, 10	15–20	12-1, 12-3, 12-4	2, 7
E Describe the intersections of a plane and a cone of 2 nappes.	1	21–24	12-1, 12-5	1
USES				
F Determine information about elliptical orbits.	15	25, 26	12-2	8
REPRESENTATIONS				
G Graph or identify graphs of ellipses and hyperbolas with equations in their standard form.	2, 3	27–34	12-2, 12-3	4
H Graph transformation images of ellipses and hyperbolas.	4	35–38	12-2, 12-3, 12-5	3
I Describe graphs of quadratic equations.	12, 13	39–42	12-5	11

*The masters are numbered to match the lessons, and each is intended for use with that lesson.

OVERVIEW ■ CHAPTER 12

This short chapter brings together ideas from a variety of sources to examine the quadratic relations (those with equations of the form $Ax^2 + Bxy + Cy^2 + Dx + Ey + F = 0$). The theoretical results are among the most beautiful in all of elementary mathematics, and the applications include two of the most famous examples of mathematical modeling of all time, Kepler's modeling of the orbits of planets and Galton's modeling of the relationship between the heights of parents and their children.

In comparison with the linear relations, the quadratic relations are more varied, as might be expected. Their graphs may be parabolas, ellipses (including circles), hyperbolas, two lines, a single line, a single point, or even the empty set. Parabolas, ellipses, and hyperbolas constitute the conic sections, and often a chapter like this one is called Conic Sections. Since students have already studied parabolas, the first attention in the chapter is given to ellipses.

The first question we address is the connection between the three-dimensional geometry of the cone that gives rise to a conic section and the equation for the two-dimensional section. The movement from the cone to the geometric two-dimensional definition of an ellipse is done in Lesson 12-1. In Lesson 12-2 the steps from the geometry to the algebra are completed. With the use of translations, we arrive at an equation for any ellipse with horizontal and vertical lines of symmetry. Discussions of planetary orbits are interspersed throughout these lessons.

Despite the fact that hyperbolas look quite different from ellipses, their geometric and algebraic properties are quite analogous. And so what took two lessons for the ellipse is presented in Lesson 12-3 for the hyperbola.

Lesson 12-4 discusses how to rotate any relation about the origin. Galton's data serve as an introduction. Because rotations do not change the degree of any relation

(the substitutions for x and for y are linear combinations of x and y), this establishes that all conic sections have equations of the second degree.

In Lesson 12-5, the problem is reversed, and we show that every quadratic relation is either a conic section or the so-called degenerate conic section (two lines, a single point, or the null set).

Each conic section has special reflection properties. Lesson 12-6 examines these.

For some classes, this will be the last chapter of the book that is covered. For some students, it may be their last mathematics in high school. It is a fitting close, for the chapter combines algebra, two- and three-dimensional geometry, trigonometry, relations and functions, and statistics, and includes some beautiful proofs and wonderful applications. Thus not only are all the themes of this course included, but the most important themes of all of high school mathematics are juxtaposed.

PERSPECTIVES ■ CHAPTER 12

The Perspectives provide the rationales for the inclusion of topics or approaches, give some mathematical background, and make connections with other lessons and to other UCSMP texts.

12-1

THE GEOMETRY OF THE ELLIPSE

There are two common ways to characterize ellipses. One is three-dimensional, as the intersection of a plane and a single nappe of a cone. The other is the two-dimensional locus definition as the set of points in a plane such that the sum of the distances of each point from two given points is a constant. Often books show the three-dimensional definition, the far more intuitive one, but do not connect it with the two-dimensional form that leads to equations. Shown in this lesson is the gorgeous connection between these definitions discov-

ered by the French mathematician Dandelin less than 200 years ago. This enables us to be confident that the locus definition indeed describes the same ellipses that the conic sections describe.

From the locus definition, without benefit of equations in x and y, it can be shown that every ellipse has two symmetry lines. The significance of this is that symmetry is a geometric property; we do not want students to feel that they have to have an equation for the curve to know of its symmetry.

One of the problems of this lesson is to reconstruct the data that led Kepler to discover that the ratio

of the cube of the radius of a planetary orbit to the square of the time of revolution is a constant. It was the first example of logarithmic modeling.

12-2

THE ALGEBRA OF THE ELLIPSE

By conveniently locating the foci at $(c, 0)$ and $(-c, 0)$, so that the lines of symmetry of an ellipse are the axes, the equation $\dfrac{x^2}{a^2} + \dfrac{y^2}{b^2} = 1$ for an ellipse can be derived. Students who have studied Chapter 12 of UCSMP *Advanced Algebra* will have studied this content.

In the above equation, $a > b$. If the foci are on the y-axis at $(0, c)$ and $(0, -c)$, then an equation of the same form arises but $b > a$. (If $a = b$; then the circle with center $(0, 0)$ and radius a results.) In both cases, the relationship between $a, b,$ and c is quite simple: $c^2 = |a^2 - b^2|$.

Here $2a$ is the length of the major axis, $2b$ the length of the minor axis, and $2c$ the distance between the foci. In the questions, the eccentricity $\frac{c}{a}$ of an ellipse is introduced. The relationships between these quantities enable information to be deduced about orbits of planets.

12-3

THE HYPERBOLA

The hyperbola is formed by a plane intersecting *both* nappes of the cone and not being parallel to an edge. The locus is the set of points in a plane such that the *absolute difference* of the distances of each point from two given points is a constant.

The hyperbola has the same two symmetry lines as the ellipse (namely, the line through its foci and the perpendicular bisector of the segment whose endpoints are the foci), and the proof is virtually identical. So is the derivation of the following equation for the hyperbola with foci $(c, 0)$ and $(-c, 0)$: $\frac{x^2}{a^2} - \frac{y^2}{b^2} = 1$. If the foci are on the y-axis at $(0, c)$ and $(0, -c)$, an equation is $\frac{y^2}{a^2} - \frac{x^2}{b^2} = 1$. In both cases, $c^2 = a^2 + b^2$.

The hyperbola has one major feature that distinguishes it from the ellipse, its asymptotes. The classic transformation strategy is utilized to deal with these; we consider the simplest hyperbola with an equation of one of the above forms, $x^2 - y^2 = 1$. (It is akin to the unit circle.) We find its asymptotes. Then we scale change it—and along with it, its asymptotes—to the other hyperbolas of that form.

12-4

ROTATING RELATIONS

A generation ago, axes would be translated and rotated. Now it is considered far easier to translate and rotate the graphs of relations. The Graph Rotation Theorem of this lesson is another in the mold of the Graph Translation Theorem in Lesson 3-2 and the Graph Scale Change Theorem in Lesson 3-5. It enables a sentence for the rotation image of any graph to be found if you know a sentence for the preimage.

Thus, to find an equation for any conic section, one determines how much their axes are tilted to the horizontal and vertical, rotates a curve in standard position that amount, and then translates it. The algebraic manipulation is still quite formidable, so the examples given in this lesson are simple. Rotating the hyperbola $x^2 - y^2 = 1$ an amount $\frac{\pi}{4}$ about the origin gives a hyperbola of the form $xy = k$ as its image. Thus the hyperbolas that students have seen in other contexts are in fact congruent to the hyperbolas that have been studied in the previous lesson.

12-5

THE GENERAL QUADRATIC

Any conic section in the plane can be mapped by a rotation onto a conic section with horizontal and/or vertical symmetry. A further translation can get a key point (normally a center or vertex of the conic) to be at the origin. The logic of this lesson is as follows. First we note that if $B = 0$ in the general quadratic $Ax^2 + Bxy + Cy^2 + Dx + Ey + F = 0$, then one can complete the square to get equations for ellipses, hyperbolas, and parabolas with horizontal or vertical symmetry lines, and which type we get can thus be determined. Unless we get a degenerate conic, we find that the graph is an ellipse, parabola, or hyperbola, depending on whether AC is positive, zero, or negative.

To consider the cases when $B \neq 0$, we determine the magnitude θ of a rotation under which there will be no xy term in an equation for the image of the quadratic relation above. This magnitude is such that $\tan 2\theta = \frac{B}{C - A}$.

Next is an amazing fact: under a rotation or a translation, the value of $B^2 - 4AC$ in $Ax^2 + Bxy + Cy^2 + Dx + Ey + F = 0$ is the same as the value of $B'^2 - 4A'C'$ in its image $A'x^2 + B'xy + C'y^2 + D'x + E'y + F' = 0$. Thus, if $B' = 0$, $B^2 - 4AC = -4A'C'$. Thus the sign of $B^2 - 4AC$ is the opposite of the sign of $A'C'$, and we know what conics are determined by any given value of $A'C'$. Thus, unless we get a degenerate conic, the graph of the general quadratic relation is an ellipse, parabola, or hyperbola, depending on whether $B^2 - 4AC$ is negative, zero, or positive, the signs opposite those found when considering just AC.

12-6

ITS ALL DONE WITH MIRRORS

The conics have applications to mirrors and to orbits. The mirror applications are discussed in this lesson. (The orbit applications, first deduced by Newton, require calculus.) It is proved that ellipses possess the following reflection property: an object travelling from one focus of an ellipse will bounce off of the ellipse in the direction of the other focus. This explains the phenomena of whispering galleries. The corresponding property for a parabola, that an object traveling from the focus of a parabola will bounce off the parabola in a direction that is parallel to the parabola's axis, is mentioned but not proved. That property is the reason that parabolic mirrors are used in headlights and in receiving dishes for communication and telescopes.

727C

DAILY PACING CHART ■ CHAPTER 12

For optimal student performance, the Progress Self-Test and the Chapter Review should be covered. (See *General Teaching Suggestions: Mastery* on page T35 of this Teacher's Edition.) By following the pace of the Full Course given here, students can complete 12 or 13 chapters of the book by the end of the year. Students following the pace of the Minimal Course spend more time to complete lessons that are combined with quizzes and to consider the Chapter Review. When all lessons are covered in order (the recommendation of the authors), students in the Minimal Course can cover all but the last two chapters. For more information on pacing, see *General Teaching Suggestions: Pace* on page T34 of this Teacher's Edition.

DAY	MINIMAL COURSE	FULL COURSE
1	12-1	12-1
2	12-2	12-2
3	Start 12-3	Start 12-3
4	Finish 12-3	Finish 12-3
5	12-4	12-4
6	Quiz (TRF); Start 12-5	Quiz (TRF); 12-5
7	Finish 12-5	12-6
8	12-6	Progress Self-Test
9	Progress Self-Test	Chapter Review
10	Chapter Review	Chapter Test (TRF)
11	Chapter Review	
12	Chapter Test (TRF)	

TESTING OPTIONS

■ Quiz for Lessons 12-1 through 12-4   ■ Chapter 12 Test, Form A   ■ Chapter 12 Test, Cumulative Form
■ Chapter 12 Test, Form B

A Quiz and Test Writer is available for generating additional questions, additional quizzes, or additional forms of the Chapter Test.

PROVIDING FOR INDIVIDUAL DIFFERENCES

The student text has been written for, and tested with, average students. It has also been used successfully with students who are better prepared as well as with students who are poorly prepared.

The Lesson Notes often include Error Analysis and Alternate Approach features to help you assist those students who need more help. Students of all abilities often learn from their peers and may benefit from small group work referenced as appropriate throughout the Notes. A blackline Lesson Master (in the Teacher's Resource File), keyed to the chapter objectives, is available for each lesson to provide more practice. (However, we advise you to use these masters sparingly; again refer to the suggestions on pace on page T34.)

Extension activities are provided in the Lesson Notes for use with students who need to be challenged even more than the requirements of completing the Full Course provide.

Quadratic Relations

Johannes Kepler discussing planetary motion with Emperor Rudolph II

In the 3rd century B.C., only about 25 years after the appearance of Euclid's *Elements,* another Greek mathematician, Appollonius, wrote a set of books entitled *Conic Sections,* dealing with the curves formed by intersecting a cone with a plane: ellipses, parabolas, and hyperbolas.

Appollonius did not consider any applications of these curves, but 1800 years later, in the 9-year period from 1609 to 1618, the German scholar Johannes Kepler performed one of the greatest examples of mathematical modeling of all time. Until Kepler, almost all people thought that the planets moved in circles (either around the earth or the sun) and that the stars were fixed on a celestial sphere. Using the measurements of the positions of the planets obtained by his teacher, the Danish astronomer Tycho Brahe, Kepler found that the orbits of the planets around the sun were ellipses, and that these ellipses satisfied several simple but not obvious properties.

At about the same time, Descartes applied his new coordinates (the rectangular coordinates you know very well) to the geometry of the conic sections and found that all conics could be described by equations of the form $Ax^2 + Bxy + Cy^2 + Dx + Ey + F = 0$. Because the highest term in this equation has degree 2, the sets of ordered pairs (x, y) satisfying it are called *quadratic relations*.

This chapter explores the relationships between the equations of quadratic relations and their graphs, the conic sections. Along the way, you will encounter a variety of applications of these curves.

CHAPTER 12 Quadratic Relations **727**

CHAPTER 12

This chapter should be done in 10–12 days: 7 or 8 days for the lessons and quiz, 1 day for the Progress Self-Test, 1 or 2 days for the Chapter Review, and 1 day for a Chapter Test. Note that Lesson 12-3 is a two-day lesson. (See Daily Pacing Chart on page 727D.)

**USING PAGE 727
Alternative Assessment**
As with most chapters, this is a time to probe what your students know about the content of the chapter. Have they studied equations for ellipses? Do they know any equations for hyperbolas? (Of course they should know some equations for parabolas.) Point out how any equations they give are special kinds of quadratic relations.

LESSON 12-1

OBJECTIVES

D State and apply properties of ellipses to draw and describe them.
E Describe the intersections of a plane and a cone of 2 nappes.

TEACHING NOTES

The purpose of this lesson is to connect the three-dimensional and two-dimensional characterizations of ellipses. How do we know that the intersection of the cone in the left drawing on page 728 is the same ellipse as that given by the definition on page 730? We could get a three-dimensional equation for the cone, intersect it with a plane, and then get a two-dimensional curve and somehow prove it is an ellipse. But it is easier to use the elegant method of Dandelin.

Like all elegant methods, this one requires that the explanation be read slowly. You may wish to go through page 729 with your students line by line. Plastic models of Dandelins spheres have been available for some time, and such a model would be an excellent visual aid if it is possible for you to locate one.

LESSON 12-1

The Geometry of the Ellipse

The "cone" that gives rise to the conic sections is not the same as the cones studied in geometry. The conic section cone is formed by rotating one of two intersecting lines about the other. The fixed line is called the **axis** of the cone; the point of intersection of these lines is the cone's **vertex**. The conic section cone has two parts, called **nappes**, one on each side of its vertex, as shown in each of the three figures below.

In general, a **cross section** or **section** of a 3-dimensional figure is the intersection of a plane with that figure. The three figures below are the sections of a cone formed when a plane intersects the cone but does not contain the cone's vertex.

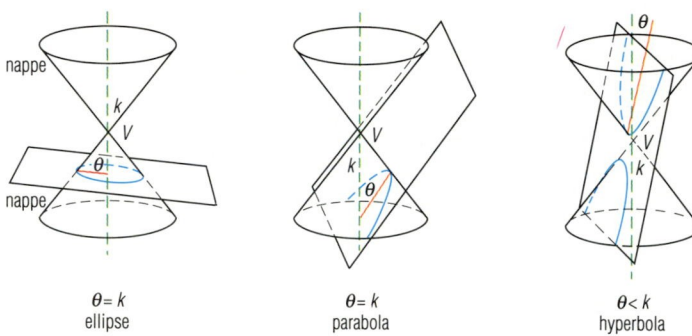

| $\theta = k$ | $\theta = k$ | $\theta < k$ |
| ellipse | parabola | hyperbola |

Because drawings can only show a part of a cone, it may not be obvious that a fixed cone intersects every plane in space. Let k be the measure of the angle between the lines that determined the cone (that is, between its axis and the rotating line). Let θ be the measure of the smallest angle between the cone's axis and the plane. (This angle cannot be obtuse, so $\theta \le 90°$.) If $\theta > k$, the plane intersects only one nappe of the cone and the section is an *ellipse*. If $\theta = k$, the plane is parallel to an edge of the cone, so again the plane intersects only one nappe, and the section is a *parabola*. If $\theta < k$, the plane intersects both nappes of the cone, and a *hyperbola* is formed.

The three-dimensional geometric descriptions of conic sections in the previous paragraph are like those used by Appollonius to define the three types of conic sections. The French mathematician Germinal Pierre Dandelin (1794–1847) gave an elegant way to obtain a two-dimensional algebraic characterization for one of the conic sections, the ellipse, from the above three-dimensional description. Dandelin's method involves spheres that are nested inside the cone and applies the following properties.

728

1. The intersection of a cone and a sphere nested in the cone is a circle (in blue at the right) whose points are all the same distance from the vertex of the cone.

2. The intersection of a sphere and a plane tangent to the sphere is a single point. (Think of a hard ball resting on a flat surface; in theory, there is one point in common.)

3. The lengths of all tangents from a given point to a given sphere are equal.

Now consider an ellipse formed by a plane M intersecting a cone as drawn at the right. There are only two spheres that can be nested in the cone and that are tangent to the plane M. One sphere S_1 is between the plane and the vertex. A larger sphere S_2 is on the other side of M. Let F_1 and F_2 be the points of tangency of the spheres with the plane and let C_1 and C_2 be the circles of intersection of the spheres and the cone. Notice that the cone and plane determine an ellipse; then, all of these named spheres, circles, and points are determined by the ellipse and are fixed.

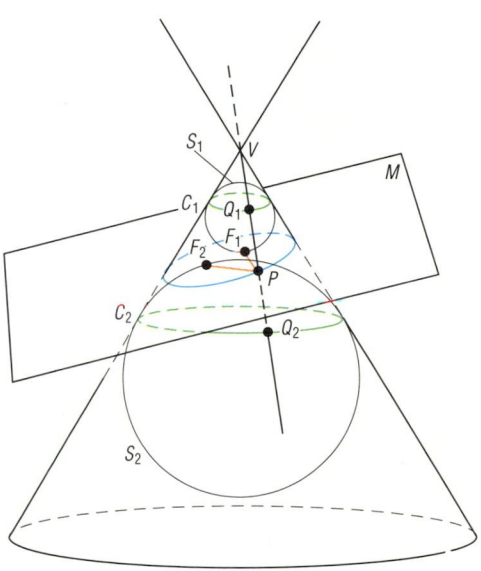

What we want is a property that relates any point on the ellipse to the fixed points F_1 and F_2. Let P be any point on the ellipse. P is on an edge $\overleftrightarrow{PV}$ of the cone and that edge intersects the circles C_1 and C_2 at Q_1 and Q_2, respectively. Notice that Q_1Q_2 is the distance along the cone between the circles, and since VQ_2 and VQ_1 are constant differences (property 3 above),

$$Q_1Q_2 = VQ_2 - VQ_1 = \text{a constant.}$$

So the distance between the circles along any edge of the cone is the same.

Emphasize the logic: We prove that the intersection of the cone and the plane has the property described in the definition on page 730. Then we are justified in defining an ellipse as a curve that has that property. Refer to the drawing on page 729 (for which there is a teaching aid). The idea of the proof is that $PF_1 + PF_2$ equals the distance (as measured along an edge of the cone) between the circles of intersection of the spheres and the cone. Since the circles are parallel, that distance is constant.

Some students may have learned that the ellipse with equation $\frac{x^2}{a^2} + \frac{y^2}{b^2} = 1$ is symmetric to the x- and y-axes by noting that if (x, y) is on the graph, so are $(x, -y)$ and $(-x, y)$. This algebraic proof has two weaknesses: It does not apply to all ellipses and it uses algebra on a property that is fundamentally geometric. The proof given on page 730 relies on the fact that reflections preserve distance, a property that should be known to students who have studied UCSMP *Geometry* or UCSMP *Advanced Algebra*.

ADDITIONAL EXAMPLE
Small Group Work
Groups of students can shine a flashlight on a flat wall. They should turn the flashlight so the outline is a hyperbola, an ellipse, and a parabola.
If the outline lies entirely on the wall, the outline is an ellipse. If an edge of the beam is parallel to the wall, then the outline is a parabola. If the flashlight is turned still further, the outline is a hyperbola.

Ellipse overprinted on a photograph of an egg. Only the orange segment is an axis of symmetry of the cross-section of the egg.

Now notice that both $\overline{PF_1}$ and $\overline{PQ_1}$ are tangents to sphere S_1 from P, so $PF_1 = PQ_1$. Similarly, both $\overline{PF_2}$ and $\overline{PQ_2}$ are tangents to sphere S_2, so $PF_2 = PQ_2$. Thus $PF_1 + PF_2 = PQ_1 + PQ_2 = Q_1Q_2$ (since P is on $\overline{Q_1Q_2}$), which is a constant. That means that the sum of the distances from any point P on the ellipse to the points F_1 and F_2 is constant. This property is normally taken as the two-dimensional definition of an ellipse with *foci* F_1 and F_2 and a given *focal constant*. (The word *foci* [pronounced "foe sigh"] is the plural of *focus*.)

Definition

Let F_1 and F_2 be two given points in a plane and k be a positive real number with $k > F_1F_2$. Then the **ellipse with foci F_1 and F_2 and focal constant k** is the set of all points P in the plane which satisfy

$$PF_1 + PF_2 = k.$$

This definition can be used to draw an ellipse. Put two pins into a piece of paper at the foci. Tie a string between the pins so that the string is not tight and the length of string is the focal constant. Then, keeping the string taut, trace out a curve with a pencil; that curve is an ellipse.

As the two foci of an ellipse move near each other, the ellipse looks more like a circle. When the foci F_1 and F_2 are the same point F, then the ellipse is the set of points P such that $PF + PF = k$, or $PF = \frac{k}{2}$, which *is* a circle.

Some people call an ellipse egg-shaped, and indeed the word "oval" comes from the Latin word for egg, "ovum." But the cross sections of many eggs have only one line of symmetry. Eggs are often more pointed at one end than the other, while every ellipse has two symmetry lines.

Theorem

An ellipse with foci F_1 and F_2 is reflection-symmetric to $\overleftrightarrow{F_1F_2}$ and to the perpendicular bisector of $\overline{F_1F_2}$.

Proof Let F_1 and F_2 be the foci of an ellipse with focal constant k. Then for any point P on the ellipse, $PF_1 + PF_2 = k$. To show that the ellipse is reflection-symmetric to $m = \overleftrightarrow{F_1F_2}$, it must be shown that the ellipse coincides with its reflection image over that line. Let $P' = r_m(P)$. Because F_1 and F_2 are on the reflecting line, $r_m(F_1) = F_1$ and $r_m(F_2) = F_2$. Because reflections preserve distance, $P'F_1 = PF_1$ and $P'F_2 = PF_2$. So for all P, $P'F_1 + P'F_2 = PF_1 + PF_2 = k$. Thus P' is on the ellipse. So the image of each point on the ellipse is also on the ellipse.

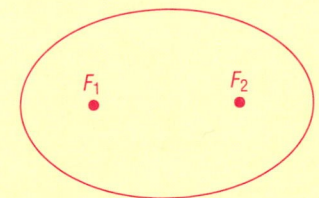

To show that the image is the entire ellipse, an indirect argument is used. Suppose a point Q were on the original ellipse but not on its reflection image. Then its image, $r_m(Q)$, would be on the original ellipse (using the argument above). But $r_m(r_m(Q)) = Q$. So Q is on the image, and thus all points of the ellipse are on the image.

A similar argument can be written to show that the perpendicular bisector of $\overline{F_1F_2}$ is also a symmetry line for the ellipse.

One discovery by Kepler about planetary orbits was that a planet does not go around the sun at a constant speed, but moves at a varying rate so that a segment from the sun to the planet sweeps out equal areas in equal times. This is now known as Kepler's second law. Another discovery, known as his first law, is that the orbit of each planet is an ellipse with the sun at one focus. His third discovery, called his third law, relates the planet's average distance from the sun to its period, the length of time it takes to make one revolution. You are asked to investigate Kepler's third law in the Questions.

Questions

Covering the Reading

1. What is the difference between the cones studied in geometry and the cones that give rise to the conic sections?   **See margin.**

2. Describe how a plane intersects a given cone to result in the given figure.
   a. parabola $\theta = k$ **b.** ellipse $\theta > k$
   c. hyperbola $\theta < k$ **d.** exactly one point   **at the vertex of the cone**
   e. no points   **not possible**

3. **a.** If a sphere is *nested* in a cone, how are these figures related?
   b. Draw a picture of a sphere nested in a cone.   **a,b) See margin.**

4. Give a definition for *ellipse*:
   a. in three dimensions;   **b.** in two dimensions.   **a,b) See margin.**

5. Suppose the intersection of a cone and a plane is an ellipse.
   a. Describe the location of the Dandelin spheres for this ellipse.
   b. How are the foci of the ellipse related to the location of the Dandelin spheres?   **a) The smaller of the spheres is between the plane and the vertex of the cone, tangent to the plane at one focus of the ellipse. The plane is between the vertex and the larger of the spheres; it is tangent to the sphere at the other focus. b) Each Dandelin sphere is tangent to the plane. The foci are the points of tangency.**

LESSON 12-1   The Geometry of the Ellipse   **731**

3.a. The center of the sphere is on the axis of the cone. The intersection of the cone and the sphere is a circle whose points are all the same distance from the vertex of the cone.

3.b.

4.a. The intersection of a cone and a plane which intersects only one nappe of the cone.

b. Sample: The collection of all the points on a plane the sum of whose distances to two fixed points on the same plane is a constant.

9. The ellipse should have the shape of the one below.

In 6 and 7, *true* or *false*.

6. Every ellipse has at least two symmetry lines.   **True**

7. A circle is a special type of ellipse.   **True**

8. Apply Kepler's second law to make a statement about the planet Earth.   **Sample: Earth does not move at a constant speed in its orbit about the sun.**

Applying the Mathematics

In 9 and 10, use pins (or tacks) and string to draw the ellipse satisfying the conditions.

9. The distance between the foci is 3 inches and the focal constant is 4 inches.   **See margin.**

10. The foci are the same point and the focal constant is 5 centimeters. **The ellipse should be a circle of radius 2.5 centimeters.**

11. Prove that the perpendicular bisector of the segment joining the foci of an ellipse is a symmetry line for the ellipse.   **See margin.**

12. In Kepler's day, the orbits of the six known planets around the sun were thought to be circular, and the radii of these circles were calculated in **astronomical units (a.u.)**, where 1 astronomical unit was the radius of Earth's orbit. Here are the values that Kepler had for these radii and for the periods of the planets.

Planet	Radius r (a.u.)	Period t (days)	r^3/t^2
Mercury	0.389	87.77	7.64×10^{-6}
Venus	0.724	224.70	7.52×10^{-6}
Earth	1.000	365.25	7.50×10^{-6}
Mars	1.524	686.98	7.50×10^{-6}
Jupiter	5.200	4,332.62	7.49×10^{-6}
Saturn	9.510	10,759.20	7.43×10^{-6}

Kepler discovered that $\dfrac{r^3}{t^2}$ is nearly a constant. Verify Kepler's calcula-

tions. What is the mean value that you get for $\dfrac{r^3}{t^2}$?
The mean value for $r^3/t^2 \approx 7.51 \times 10^{-6}$

13. Suppose the two foci of an ellipse are 1 meter apart.
a. What is the smallest possible focal constant for this ellipse?
b. As the focal constant gets larger, what happens to the shape of the ellipse?
c. Is there a largest possible focal constant for this ellipse? If so what is it?   **a) There is no smallest focal constant, but it must be larger than 1 meter. b) As the focal constant gets larger, the ellipse looks more and more like a circle. c) See below.**

Review

14. Give an exact value for $\sin \dfrac{\pi}{12}$.   *(Lesson 11-5)* $\dfrac{\sqrt{6} - \sqrt{2}}{4}$

15. a. Write a matrix for the composite transformation $r_{y=x} \circ R_{90}$.
b. Find a matrix for the image of $\triangle ABC = \begin{bmatrix} 2 & 4 & 5 \\ 1 & 3 & 1 \end{bmatrix}$ under the transformation in part **a**.
c. Sketch $\triangle ABC$ and its image on the same set of axes.   *(Lesson 11-3)*
See margin.

13c) There is no largest focal constant; it can be arbitrarily large as long as it is larger than 1 meter.

16. *Multiple choice.* Suppose $u^2 - v^2 = w^2$. Under which of these conditions is w a real number? *(Lesson 9-6)*

I. $u > v > 0$ II. $u = v$ III. $u < v < 0$

(a) I and II only   (b) II and III only

(c) I and III only   (d) I, II, and III   **(d)**

17. Determine the sum of the first 100 multiples of 3. [Hint: any multiple of 3 can be written in the form $3k$, where k is an integer.] *(Lesson 8-3)*
15,150

18. The following are the average monthly temperatures in °F for Huron, South Dakota. **a,b) See margin. c) See below.**

Jan	Feb	Mar	Apr	May	June	July	Aug	Sep	Oct	Nov	Dec
11	18	29	46	57	68	74	72	61	49	32	19

    a. Draw a scattergram of the data.

    b. Sketch a sine curve to fit the data.

    c. Determine an equation for the curve in part **b** to model the data.

    (Lessons 1-3, 6-4)

In 19 and 20, expand the binomials and combine like terms. *(Previous course)*

19. $(a + b)^2 - (a - b)^2$ **4ab**   **20.** $(2p - \sqrt{d})^2 - d^2$
                                                $4p^2 - 4p\sqrt{d} + d - d^2$

21. Find the distance between the points (a, b) and (x, y). *(Previous course)*
$\sqrt{(a-x)^2 + (b-y)^2}$

Exploration

22. The mathematician Hypatia wrote one of her books, *On the Conics of Apollonius*, around 400 A.D. Find out something else about this mathematician. **sample: Hypatia, an Alexandrian mathematician of note, was the daughter of Theon of Alexandria, a Greek geometrician. She also wrote commentaries on the works of Diophantus.**

18c) sample: $t = 31.5\sin\left(\frac{\pi}{6}m - \frac{\pi}{2}\right) + 42.5$, where t = average monthly temperature in °F and m = months after January.

FOLLOW-UP

MORE PRACTICE
For more questions on SPUR Objectives, use *Lesson Master 12-1*, shown below.

PROJECTS
The projects for Chapter 12 are described on pages 771–773. Projects 1, 2, 3, and 5 are related to the content of this lesson.

EXTENSION
Small Group Work A group of students may be interested in constructing a model of Dandelin's spheres. About the only materials needed would be spheres of two sizes and transparent plastic material for constructing a cone and an intersecting plane.

NAME _____

LESSON **MASTER 12-1**
QUESTIONS ON **SPUR** OBJECTIVES

■**PROPERTIES** *Objective D (See pages 776–778 for objectives.)*
In 1 and 2, use pins or tacks and string.

1. Draw an ellipse in which the distance between the foci is 1 inch and the focal constant is 1.5 inches.

2. Show all lines of symmetry for the figure drawn in Exercise 1.

■**PROPERTIES** *Objective E*

3. Draw a plane intersecting a cone where the section is an ellipse.

4. If the focal constant of an ellipse is 3 cm, what can be said about the distance between the foci?   **It is less than 3 cm.**

5. Name the figure formed when a plane intersects a cone as described.

  a. The measure of the smallest angle between the cone's axis and the plane equals the measure of the angle between the lines that determine the cone.   **parabola**

  b. The measure of the smallest angle between the cone's axis and the plane is greater than the measure of the angle between the lines that determine the cone.   **hyperbola**

120   *Functions, Statistics, and Trigonometry © Scott, Foresman and Company*

LESSON

12-2

The Algebra of the Ellipse

From the two-dimensional definition of an ellipse in Lesson 12-1, an equation for any ellipse in the plane can be found. If the foci are $F_1 = (a, b)$ and $F_2 = (c, d)$ and the focal constant is k, then any point $P = (x, y)$ on the ellipse must satisfy

$$PF_1 \quad + \quad PF_2 \quad = k$$

or $\sqrt{(x-a)^2 + (y-b)^2} + \sqrt{(x-c)^2 + (y-d)^2} = k.$

This is an equation for the ellipse. But this equation is rather unwieldy and it does not lend itself to graphing even with an automatic grapher. A more common approach is to begin with an ellipse that is symmetric to the x- and y-axes, find its equation, and then transform that equation as needed to find equations for other ellipses. This is the approach we will take.

To simplify the algebraic manipulations, the focal constant is called $2a$, and the foci are on the x-axis, symmetric to the origin. So let $F_1 = (-c, 0)$, $F_2 = (c, 0)$, and $P = (x, y)$; note that $c > 0$. The following eleven steps are numbered for reference.

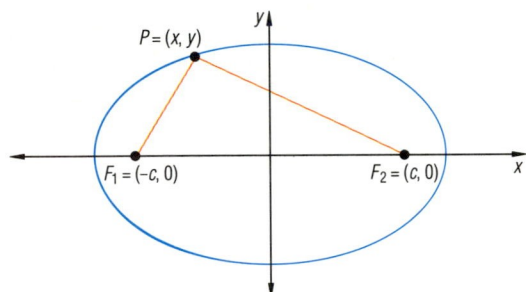

1. By the definition of an ellipse,
$$PF_1 + PF_2 = 2a.$$
Using the Distance Formula, this becomes
$$\sqrt{(x+c)^2 + y^2} + \sqrt{(x-c)^2 + y^2} = 2a.$$

2. Subtract one of the square roots from both sides.
$$\sqrt{(x-c)^2 + y^2} = 2a - \sqrt{(x+c)^2 + y^2}$$

3. Square both sides (the right side is like a binomial).
$$(x-c)^2 + y^2 = 4a^2 - 4a\sqrt{(x+c)^2 + y^2} + (x+c)^2 + y^2$$

4. Expand the binomials and do appropriate subtractions.
$$-2cx = 4a^2 - 4a\sqrt{(x+c)^2 + y^2} + 2cx$$

5. Use the Addition Property of Equality and rearrange terms.
$$4a\sqrt{(x+c)^2 + y^2} = 4a^2 + 4cx$$

6. Multiply both sides by $\frac{1}{4}$.
$$a\sqrt{(x+c)^2 + y^2} = a^2 + cx$$

734

7. Square both sides a second time.
$$a^2[(x+c)^2 + y^2] = a^4 + 2a^2cx + c^2x^2$$
8. Expand the binomial and subtract $2a^2cx$ from both sides.
$$a^2x^2 + a^2c^2 + a^2y^2 = a^4 + c^2x^2$$
9. Subtract a^2c^2 and c^2x^2 from both sides, then factor.
$$(a^2 - c^2)x^2 + a^2y^2 = a^2(a^2 - c^2)$$
10. Since $c > 0$, then $F_1F_2 = 2c$; and since $2a > F_1F_2$, then $2a > 2c$. So $a > c > 0$. Thus $a^2 > c^2$ and $a^2 - c^2$ is positive. So $a^2 - c^2$ can be considered as the square of some real number, say b. Now let $a^2 - c^2 = b^2$ and substitute.
$$b^2x^2 + a^2y^2 = a^2b^2$$
11. Dividing both sides by a^2b^2, where $b^2 = a^2 - c^2$, gives
$$\frac{x^2}{a^2} + \frac{y^2}{b^2} = 1.$$

This argument yields the *standard form* for an equation of this ellipse.

> **Theorem (Equation for an Ellipse)**
>
> The ellipse with foci $(c, 0)$ and $(-c, 0)$ and focal constant $2a$ has equation $\dfrac{x^2}{a^2} + \dfrac{y^2}{b^2} = 1$, where $b^2 = a^2 - c^2$.

By substitution into the equation, it is easy to verify that the points $(a, 0)$, $(-a, 0)$, $(0, b)$, and $(0, -b)$ are on this ellipse. Notice that $|x|$ cannot be greater than a, nor can $|y|$ be greater than b; otherwise the left side would be greater than 1. This shows that those four points are extreme points for the ellipse; the four points help to graph it.

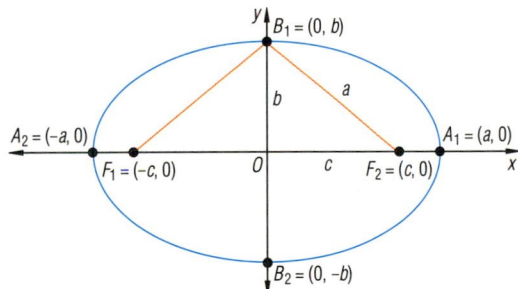

Because $2a$ is the focal constant, $B_1F_1 + B_1F_2 = 2a$. Also, $B_1F_1 = B_1F_2$, so each of these distances is a. You can see in the drawing that $b^2 + c^2 = a^2$, as was used in step 10 of the argument yielding the equation. Thus $a > b$.

The segments $\overline{A_1A_2}$ and $\overline{B_1B_2}$ and their lengths are called the **major** and **minor** axes of the ellipse, respectively. The major axis contains the foci and is always longer. (If $a = b$, the ellipse is a circle.) The two axes lie on the symmetry lines and intersect at the **center** of the ellipse. The previous diagram illustrates the following theorem.

Note that the theorem on page 736 applies both to ellipses with foci $(c, 0)$ and $(-c, 0)$ and to ellipses with foci $(0, c)$ and $(0, -c)$. It is for this reason that we use the terms *horizontal axis* and *vertical axis* rather than *major axis* and *minor axis*. The major axis is always the longer axis; it is the axis that contains the foci.

Error Analysis   Some students will be confused by the theorem on page 736 because on the previous page the focal constant is $2a$. They are not wrong. In every ellipse, the focal constant equals the length of the major axis.

The equation for the translation image of an ellipse in standard form (middle of page 737) is important for students to know, but it is simply an application of the Graph Translation Theorem.

Theorem

In the ellipse with equation $\frac{x^2}{a^2} + \frac{y^2}{b^2} = 1$,

$2a$ is the length of the horizontal axis,
$2b$ is the length of the vertical axis, and, for $c^2 = |a^2 - b^2|$,
$2c$ is the distance between the foci.

If $a \geq b$ as in the proof of the equation for an ellipse, then $(c, 0)$ and $(-c, 0)$ are the foci, the focal constant is $2a$, and $c^2 = a^2 - b^2$. However, if $a < b$, then the major axis of the ellipse is vertical. In this case the foci are $(0, c)$ and $(0, -c)$, the focal constant is $2b$, and $c^2 = b^2 - a^2$.

Example 1 illustrates both cases.

Example 1 Determine the foci of the ellipse and sketch its graph.

a. $\dfrac{x^2}{36} + \dfrac{y^2}{9} = 1$

b. $\dfrac{x^2}{9} + \dfrac{y^2}{36} = 1$

Solution

a. For this ellipse, $a = 6$ and $b = 3$. The foci are $(c, 0)$ and $(-c, 0)$, where $c^2 = a^2 - b^2 = 36 - 9 = 27$. So $c = \sqrt{27}$. So $(\sqrt{27}, 0) \approx (5.2, 0)$ and $(-\sqrt{27}, 0) \approx (-5.2, 0)$ are the foci. The extreme points for this ellipse are $(6, 0)$, $(-6, 0)$, $(0, 3)$, and $(0, -3)$. This enables the ellipse to be sketched rather easily.

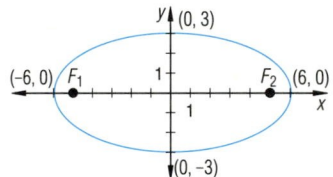

b. The ellipse in part **b** is the reflection image of the ellipse in part **a** over the line $y = x$. So its foci are $(0, \sqrt{27})$ and $(0, -\sqrt{27})$. The extreme points are $(3, 0)$, $(0, -3)$, $(0, 6)$, and $(0, -6)$. Its graph is sketched below.

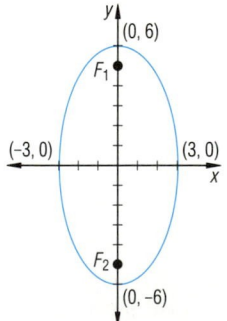

736

You can check each graph in Example 1 by finding a pair of numbers satisfying the equation and showing that the corresponding point is on the graph. You can check the foci showing that the sum of the distances from this point to the two foci is 12.

The equation of an ellipse does not represent a function, so some automatic graphers have special routines to graph ellipses. Typically, such graphers allow you to plot ellipses centered at the origin, and also plot their translation images. On some of these graphers you must enter the constants a and b in the standard form of the equation for an ellipse.

Below are the graphs of $\dfrac{x^2}{49} + \dfrac{y^2}{16} = 1$ and $\dfrac{(x-5)^2}{49} + \dfrac{(y+2)^2}{16} = 1$ as produced by an automatic grapher. Notice that each has major axis of length $2 \cdot 7 = 14$ and minor axis of $2 \cdot 4 = 8$.

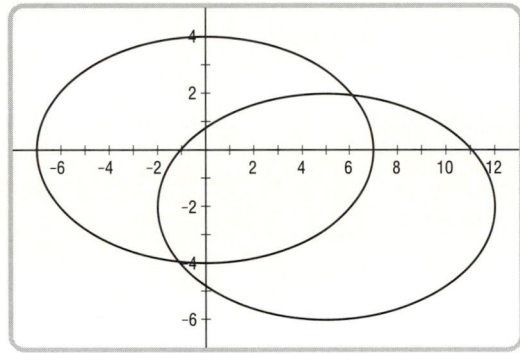

In general, because translations preserve distance, the graph of the ellipse with equation

$$\frac{(x-h)^2}{a^2} + \frac{(y-k)^2}{b^2} = 1$$

has center (h, k), horizontal axis of length $2a$, and vertical axis of length $2b$.

The theorems about ellipses can be used to answer questions about the orbits of planets.

Example 2   The closest the earth gets to the sun is approximately 91.4 million miles, and the farthest is 94.6 million miles. Given that the orbit of the earth is an ellipse with the sun at one focus, how far from the sun is the other focus?

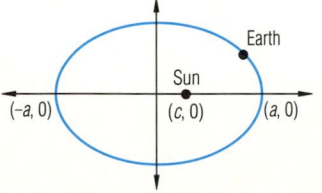

Solution   A picture, like the exaggerated one at the left, helps. Think of the center of the orbit as being the origin and place the sun at $(c, 0)$, where the unit is one million miles. The maximum and minimum distances from the sun will be achieved when the earth is intersecting the x-axis. From the given information, $a - c = 91.4$ and $c - (-a) = 94.6$. This is a system of two linear equations in two variables, easily solved. Adding these equations, $2a = 186.0$ so $a = 93.0$. Thus $c = 1.6$. So the second focus is $(-1.6, 0)$ and is 3.2 million miles from the sun.

The sun itself has a radius of about 433,000 miles, so the second focus is about 2.8 million miles from the surface of the sun.

Not long after Kepler, Isaac Newton (1642–1727) *deduced* that the orbits of the planets were ellipses from assumed principles of force and mass. Actually, Newton worked backward from Kepler's third law to discover these principles, and then worked forward to apply them to describe how the planets moved. The mathematics of his time was insufficient to deal with these ideas, so Newton developed *calculus* to solve the problem. This is another one of the great mathematical achievements of all time. Newton understood the generality of his results, that any object or body moves around a second body in an elliptical orbit. The orbits of the moon, artificial satellites around the earth, and of comets around the sun obey these laws; the orbits are always ellipses.

Questions

Covering the Reading

1. Use the distance formula to write an equation for the ellipse with foci (11, 6) and (-2, 7) and focal distance 15. (The equation does not have to be in standard form.) $\sqrt{(x-11)^2+(y-6)^2}+\sqrt{(x+2)^2+(y-7)^2}=15$

2. **a.** Find an equation in standard form for the ellipse with foci (5, 0), and (-5, 0) and focal constant 26 by going through the steps that led to the first theorem of this lesson. $(x^2/169)+(y^2/144)=1$
   b. Give the focal distance, length of major axis, and length of minor axis for this ellipse.
   focal distance = 10, major axis = 26, minor axis = 24

In 3 and 4, **a.** sketch a graph of the ellipse, **b.** determine its foci, and **c.** determine the length of the major axis.

3. $\dfrac{x^2}{25}+\dfrac{y^2}{16}=1$ **See margin.**

4. $\dfrac{x^2}{16}+\dfrac{y^2}{25}=1$ **See margin.**

In 5 and 6, consider the ellipse with equation $\dfrac{x^2}{a^2}+\dfrac{y^2}{b^2}=1$.

5. Identify
   a. its center,   **b.** its endpoints of the horizontal axis,
   c. the length of its horizontal axis.
   a) (0, 0); b) (a, 0) and (-a, 0); c) 2a

6. If the horizontal axis is the major axis, what is true about a and b?
   a > b

7. Consider the ellipse with equation $\dfrac{(x-1)^2}{36}+\dfrac{(y+7)^2}{9}=1$.
   a. Explain how its graph is related to the graph of one of the ellipses in Example 1.   **a) The graph is the translation image of the graph in Example 1a under T(x, y) → (x + 1, y − 7).**
   b. Graph this ellipse.   **b) See margin.**

8. Consider the ellipse with equation $\dfrac{(x-h)^2}{a^2}+\dfrac{(y-k)^2}{b^2}=1$.
   a. What are the coordinates of its center?   **(h, k)**
   b. What is the length of the vertical axis?   **2b**
   c. How is this ellipse related to the ellipse with equation $\dfrac{x^2}{a^2}+\dfrac{y^2}{b^2}=1$?
   It is the translation image under T(x, y) → (x + h, y + k).

738

9. The closest distance the planet Pluto comes to the sun is about 2756 million miles, and its farthest distance from the sun is about 4551 million miles. In fact, it gets "inside" Neptune's orbit and until 1998 will be closer to the sun than Neptune. Is the second focus of Pluto's orbit nearer or farther from the sun than the earth? **farther**

10. Find an equation for the ellipse pictured at the right.
$\frac{x^2}{9} + y^2 = 1$

Voyager 2 view of Neptune

11. a. Give an equation for the ellipse whose center is the origin, whose horizontal axis has length 20, and whose vertical axis has length 14.
    b. Give an equation for the image of the ellipse in part **a** under the translation $(x, y) \rightarrow (x + 3, y - 6)$.   **a)** $\frac{x^2}{100} + \frac{y^2}{49} = 1$; **b)** $\frac{(x-3)^2}{100} + \frac{(y+6)^2}{49} = 1$

12. Explain why the set of points (x, y) satisfying $2x^2 + 3y^2 = 12$ is an ellipse.   **See margin.**

13. If the curve with equation $\frac{x^2}{25} + \frac{y^2}{25} = 1$ is an ellipse, state the length of the major and minor axes. Justify your answer.   **See margin.**

In 14–17, the **eccentricity** of an ellipse is the ratio of the distance between the foci to the length of its major axis.

14. a. What is the eccentricity of the ellipse with equation $\frac{x^2}{a^2} + \frac{y^2}{b^2} = 1$?

    (Assume $a \geq b$; give a formula in terms of a, b, and/or c.)
    b. Is the eccentricity of a long, thin ellipse greater or less than the eccentricity of an ellipse that is more like a circle?   **greater**
    a) Eccentricity $= c/a$, where $c^2 = a^2 - b^2$.

15. a. What is the eccentricity of a circle?   **0**
    b. What are the largest and smallest possible values of the eccentricity?
    $0 \leq$ eccentricity < 1

16. Refer to Question 9. Pluto has the most eccentric orbit of the planets. What is the eccentricity of Pluto's orbit?   **≈ 0.2457**

17. Even if an automatic grapher does not have a special feature for graphing conics, you can still graph an ellipse by following these steps.

    Step 1: Solve $\frac{x^2}{a^2} + \frac{y^2}{b^2} = 1$ for y.

    Step 2: Graph one of the solutions.
    Step 3: Graph the other solution. The union of the graphs is the ellipse.

    a. Graph the ellipse with equation $\frac{x^2}{4} + \frac{y^2}{9} = 1$ on an automatic grapher.

    b. Give the coordinates of four points on the ellipse other than the endpoints of the axes.
    a) **See margin. b) sample:** $\left(1, \frac{3\sqrt{3}}{2}\right), \left(1, -\frac{3\sqrt{3}}{2}\right), \left(-1, \frac{3\sqrt{3}}{2}\right), \left(-1, -\frac{3\sqrt{3}}{2}\right)$

NOTES ON QUESTIONS
Question 19: This question is meant to clarify the difference between the terms *focal distance* and *focal constant*.

Making Connections for Question 21: This is an important review for Lesson 12-4.

FOLLOW-UP

MORE PRACTICE
For more questions on SPUR Objectives, use *Lesson Master 12-2*, shown on page 739.

PROJECTS
Projects 3 and 5 on pages 771 and 772 are related to the content of this lesson.

ADDITIONAL ANSWERS
19.

20.c.

Review

18. Describe the different kinds of intersections formed by a conic section cone and a plane which contains the vertex and some other points of the cone. *(Lesson 12-1)* **The intersection can be a single line or a pair of intersecting lines through the vertex.**

19. Draw an ellipse in which the distance between the foci is 2 inches and the focal constant is 5 inches. *(Lesson 12-1)* **See margin.**

20. Consider the equation $\cos 2\theta = \sin^2 \theta - \cos^2 \theta$.
    a. Show that the equation holds for $\theta = \dfrac{\pi}{4}$. **See below.**
    b. Give an example to show that the equation is *not true* for all values of θ. **See below.**
    c. Sketch the graphs of $f(\theta) = \cos 2\theta$ and $f(\theta) = \sin^2 \theta - \cos^2 \theta$ on the same set of axes to explain the results in parts **a** and **b**. **See margin.**
    d. Replace one side of the equation with another expression so that the resulting sentence is an identity. **Sample: $\cos 2\theta = \cos^2 \theta - \sin^2 \theta$**
    (Lessons 5-5, 5-6, 11-6)

21. Given that $R_\theta = \begin{bmatrix} -.5736 & -.8192 \\ .8192 & -.5736 \end{bmatrix}$, estimate θ if $0° < \theta < 360°$. *(Lesson 11-4)*
    ≈ 125°

22. Write a matrix which represents a similarity transformation that results in an image whose area is five times the area of the preimage. *(Previous course, Lesson 11-2)* **See below.**

23. A certain space probe can transmit 500 megabytes of electronic information (1 megabyte $= 2^{20}$ bytes) during the first year of its mission. Due to decrease in energy sources and increasing distance, the transmission drops by 10% every year. Determine how many megabytes of information this probe will have transmitted:
    a. in 15 years; **about 3971 megabytes**
    b. if it operates "forever." *(Lessons 8-4, 8-5)* **5000 megabytes**

Exploration

24. Although the earth is closer to the sun at some times than others, it is not this aspect of the earth's orbit that determines whether it is winter or summer. (It could not be, for when it is winter in one hemisphere, it is summer in the other.) What aspect of the earth's orbit causes the seasonal changes? **the inclination of the earth's axis of rotation**

20a) $\cos(2 \cdot \frac{\pi}{4}) = \cos\left(\frac{\pi}{2}\right) = 0$; $\sin^2\left(\frac{\pi}{4}\right) - \cos^2\left(\frac{\pi}{4}\right) = \left(\frac{\sqrt{2}}{2}\right)^2 - \left(\frac{\sqrt{2}}{2}\right)^2 = 0$

b) Sample: Let $\theta = \frac{\pi}{3}$. Then $\cos 2\theta = -\frac{1}{2}$, but $\sin^2 \theta - \cos^2 \theta = \left(\frac{\sqrt{3}}{2}\right)^2 - \left(\frac{1}{2}\right)^2 = \frac{3}{4} - \frac{1}{4} = \frac{1}{2}$.

22) Sample: $\begin{bmatrix} \sqrt{5} & 0 \\ 0 & \sqrt{5} \end{bmatrix}$

740

The Hyperbola

LESSON 12-3

RESOURCES
- Lesson Master 12-3
- Teaching Aid 89: Dandelin spheres for the hyperbola
- Computer Master 29

It is not known for certain how or why the dinosaurs became extinct, but about 65 million years ago, after nearly 140 million years of roaming the earth, they suddenly disappeared. One theory is that a star came close to the sun and spewed dangerous radiation, and the dinosaurs, being so large, could not recover from it. If that star swerved past the sun, not to return, then its path relative to the sun would not be an ellipse. In the neighborhood of the sun, it can be shown that its path would be one branch of a hyperbola.

In Lesson 12-1 you saw that the hyperbola is one type of conic section; this makes it a relative of the ellipse. That certain orbits are hyperbolas suggests other similarities between these two quite different-looking figures. In fact, much of what you have seen in the previous two lessons can be adapted for the hyperbola.

Here is Dandelin's way of relating the definition of a hyperbola as a conic section to a two-dimensional definition in terms of foci and a focal constant.

Consider a hyperbola formed by a plane M intersecting both nappes of a cone, as shown at the right. Two spheres are tangent to the cone and to the plane M, one in each nappe. Let F_1 and F_2 be the points of tangency of the spheres with the plane, and let C_1 and C_2 be the circles of intersection of the spheres and the cone. The planes of C_1 and C_2 are each perpendicular to the axis of the cone, so C_1 and C_2 are parallel.

There is a property that relates any point P on the hyperbola to the fixed points F_1 and F_2. Suppose P is on the nappe of the cone containing sphere S_1. P is on an edge $\overleftrightarrow{PV}$ of the cone and that edge intersects the circles at Q_1 and Q_2. Since Q_1Q_2 is the distance along the edge between the circles, and the circles lie in parallel planes, Q_1Q_2 is a constant, regardless of the position of P.

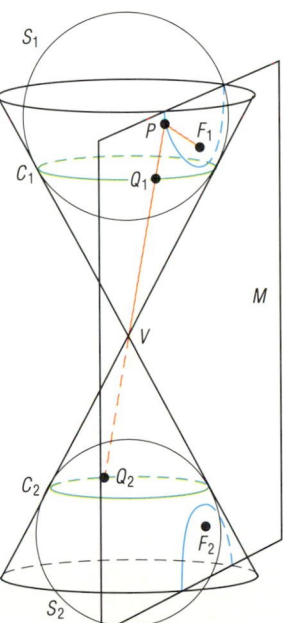

LESSON 12-3   The Hyperbola   **741**

742

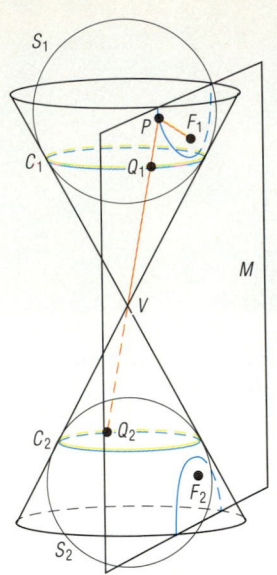

Now notice that $\overline{PF_1}$ and $\overline{PQ_1}$ are tangents to sphere S_1 from point P, and $\overline{PF_2}$ and $\overline{PQ_2}$ are tangents from P to sphere S_2, so $PF_1 = PQ_1$ and $PF_2 = PQ_2$. So $PF_2 - PF_1 = PQ_2 - PQ_1 = Q_1Q_2$, a constant. That means that the difference $PF_2 - PF_1$ is a constant.

If P were located on the other nappe of the cone, then the argument above would result in the difference $PF_1 - PF_2 = PQ_1 - PQ_2 = Q_1Q_2$, the same constant.

Thus either $PF_2 - PF_1 = Q_1Q_2$ or $PF_1 - PF_2 = Q_1Q_2$. These two equations are equivalent to the single equation $|PF_2 - PF_1| = Q_1Q_2$. That is, the absolute value of the difference of the distances from any point P on the hyperbola to F_1 and F_2 is a constant. This algebraic property is normally taken as the two-dimensional definition of a hyperbola.

Definition

Let F_1 and F_2 be two given points in a plane and k be a positive real number with $k < F_1F_2$. Then the **hyperbola with foci F_1 and F_2 and focal constant k** is the set of all points P in the plane which satisfy
$$|PF_1 - PF_2| = k.$$

For instance, here is a hyperbola with foci F_1 and F_2 that are 14 units apart, and $k = 10$. Notice that, unlike the ellipse, this hyperbola is unbounded; that is, a point can be further than any specified distance from either focus and still be 10 units closer to one focus than to the other.

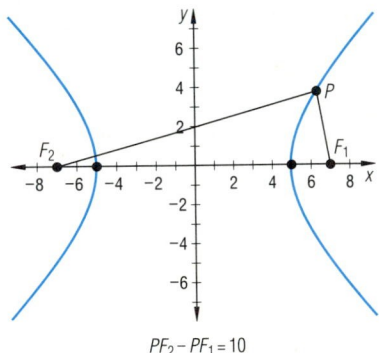

$$PF_2 - PF_1 = 10$$

Hyperbolas possess the same symmetries as ellipses.

Theorem

A hyperbola with foci F_1 and F_2 is reflection-symmetric to $\overleftrightarrow{F_1F_2}$ and to the perpendicular bisector of $\overline{F_1F_2}$.

Proof The proof follows the ideas of the corresponding proof for the ellipse found in Lesson 12-1 and is left to you.

From the two-dimensional definition of a hyperbola, an equation for any hyperbola in the plane can be found, again by a method similar to that used in Lesson 12-2 for the ellipse. If the foci are $F_1 = (a, b)$ and $F_2 = (c, d)$ and the focal constant is k, then any point $P = (x, y)$ on the hyperbola must satisfy the equation

$$|PF_1 - PF_2| = k$$

or

$$\sqrt{(x-a)^2 + (y-b)^2} - \sqrt{(x-c)^2 + (y-d)^2} = \pm k.$$

This is an equation for the hyperbola. Again it is unwieldy and it does not lend itself to graphing even with an automatic grapher. What we will do is to begin with a hyperbola whose foci are particularly well chosen and find an equation for that hyperbola.

Theorem (Equation for a Hyperbola)

The hyperbola with foci $(c, 0)$ and $(-c, 0)$ and focal constant $2a$ has equation $\dfrac{x^2}{a^2} - \dfrac{y^2}{b^2} = 1$, where $b^2 = c^2 - a^2$.

Proof The proof is identical to the proof of the equation for an ellipse in standard form, with two exceptions. First, by the definition of hyperbola,

$$|PF_1 - PF_2| = 2a.$$

With $P = (x, y)$, $F_1 = (c, 0)$, and $F_2 = (-c, 0)$, this equation is equivalent to

$$\sqrt{(x-c)^2 + y^2} - \sqrt{(x--c)^2 + y^2} = \pm 2a.$$

Now the steps are identical to those in steps 2–9 in the proof of the theorem for the ellipse. A difference comes in step 10, because $c > a > 0$, so $c^2 > a^2$, and we let $b^2 = c^2 - a^2$. This accounts for the minus sign in the equation for the hyperbola (step 11) where there is a plus sign for the ellipse.

The equation $\dfrac{x^2}{a^2} - \dfrac{y^2}{b^2} = 1$ is said to be the **standard form** for an equation of this hyperbola. The simplest equation of this form is $x^2 - y^2 = 1$, which occurs when $a = b = 1$. Then, since $b^2 = c^2 - a^2$, $1 = c^2 - 1$, so $c = \pm\sqrt{2}$. Thus the hyperbola with foci $(\sqrt{2}, 0)$ and $(-\sqrt{2}, 0)$ and focal constant 2 is the set of points (x, y) that satisfy the equation $x^2 - y^2 = 1$. Every other hyperbola of the form $\dfrac{x^2}{a^2} - \dfrac{y^2}{b^2} = 1$ is a scale transformation image of this hyperbola. So it helps to study the hyperbola $x^2 - y^2 = 1$ in some detail.

Students may wonder why we did not begin with the unit circle and stretch it to get equations for ellipses. The reason is that typically the foci of a conic do not stretch into the foci of its image. A circle has only one focus (its center) and that focus could never be mapped onto the two foci of any noncircular ellipse.

However, the asymptotes do stretch with the hyperbola itself, so the stretch argument works quite well in finding equations for asymptotes.

In the **Example,** because the equation of part **b** is identical to the equation of part **a** except with x and y switched, it should be obvious that each is the reflection image of the other over the line $y = x$. Note that the asymptotes also are reflection images of each other.

ADDITIONAL EXAMPLE
Name the vertices and asymptotes of each hyperbola.

a. $\dfrac{x^2}{100} - \dfrac{y^2}{36} = 1$.
vertices (10, 0) and (-10, 0)

b. $\dfrac{y^2}{100} - \dfrac{x^2}{36} = 1$.
asymptotes $\dfrac{x}{10} - \dfrac{y}{6} = 0$
and $\dfrac{x}{10} + \dfrac{y}{6} = 0$, or
equivalently, $x = \dfrac{5}{3} y$ and
$x = -\dfrac{5}{3} y$

Here is a table of some values. Notice that it contains the points $(1, 0)$, and $(-1, 0)$. Other points can be found by solving for y. Since $y^2 = x^2 - 1$, $y = \pm\sqrt{x^2 - 1}$. The symmetry of the hyperbola to the line through its foci (the x-axis) and to the perpendicular bisector of the segment connecting its foci (the y-axis) enables points to be found in the other quadrants. This hyperbola is graphed below.

x	$y = \pm\sqrt{x^2 - 1}$
1	0
2	$\pm\sqrt{3} \approx \pm 1.7$
3	$\pm\sqrt{8} \approx \pm 2.8$
4	$\pm\sqrt{15} \approx \pm 3.9$
5	$\pm\sqrt{24} \approx \pm 4.9$
6	$\pm\sqrt{35} \approx \pm 5.9$
-1	0
-2	$\pm\sqrt{3} \approx \pm 1.7$
-3	$\pm\sqrt{8} \approx \pm 2.8$
-4	$\pm\sqrt{15} \approx \pm 3.9$
-5	$\pm\sqrt{24} \approx \pm 4.9$

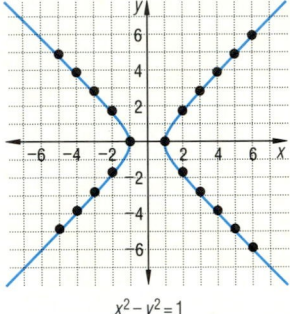

$x^2 - y^2 = 1$

Unique among the conic sections, every hyperbola has **asymptotes**, lines it approaches as x gets farther from the foci. The asymptotes for the hyperbola $x^2 - y^2 = 1$ are $y = x$ and $y = -x$. This is because as x gets larger, $\sqrt{x^2 - 1}$ becomes closer and closer to $\sqrt{x^2}$, which is $|x|$. So y gets closer and closer to $\pm x$. You can verify this by examining the above table. Even for as small a value as $x = 6$, $y \approx \pm 5.92$, which is close to $\pm x$. Alternately, you can also use an automatic grapher to investigate the asymptotes of this hyperbola. Below are two views of the hyperbola $x^2 - y^2 = 1$ and its asymptotes. The one on the left has the window $-4 \leq x \leq 4$, $-4 \leq y \leq 4$. The one on the right shows the window $-20 \leq x \leq 20$, $-20 \leq y \leq 20$.

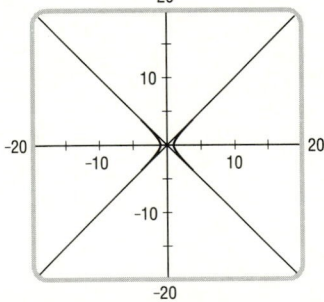

Note that as you look at larger values of $|x|$ in each window, the hyperbola gets closer to its asymptote.

It is important to realize, however, that the hyperbola never reaches its asymptotes. If you zoom or rescale around a point that appears to be on $y = \pm x$, you will see that there is always some distance between the graphs. For instance, below is a view of the graphs of $x^2 - y^2 = 1$ and $y = x$ near the point (15, 15). The graph of $y = x$ definitely does not coincide with the graph of $x^2 - y^2 = 1$.

Graphs of other hyperbolas can be determined by looking at scale changes of this parent. The scale transformation $(x, y) \rightarrow (ax, by)$ maps the graph of $x^2 - y^2 = 1$ onto the graph of $\dfrac{x^2}{a^2} - \dfrac{y^2}{b^2} = 1$. The asymptotes $y = \pm x$ are mapped onto the lines $\dfrac{y}{b} = \pm \dfrac{x}{a}$ or, solving for y, $y = \pm \dfrac{b}{a} x$. With this information, the hyperbola with equation in standard form $\dfrac{x^2}{a^2} - \dfrac{y^2}{b^2} = 1$ can be graphed.

By substitution into the equation, you can verify that $(a, 0)$ and $(-a, 0)$ are on this hyperbola. These are the **vertices** of the hyperbola. When $|x| < a$, then y^2 is negative, which is impossible. Thus the hyperbola contains no points with x-coordinate between $-a$ and a.

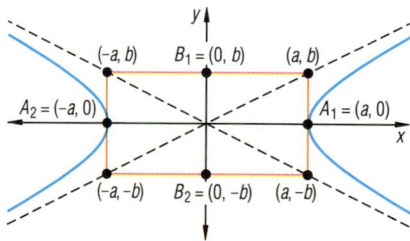

Both asymptotes contain the origin. One asymptote goes through the points (a, b) and $(-a, -b)$; the other goes through $(a, -b)$ and $(-a, b)$. Drawing the rectangle with these four points as vertices, and the lines determined by its diagonals, as done above at the right, helps to position the hyperbola.

Notice that since $b^2 = c^2 - a^2$, then $a^2 + b^2 = c^2$, and so the sides and diagonals of the guiding rectangle have lengths $2a$, $2b$, and $2c$.

The segments $\overline{A_1A_2}$ and $\overline{B_1B_2}$ of the guiding rectangle, and their lengths, are the **axes** of the hyperbola. The two axes lie on the symmetry lines and intersect at the **center** of the hyperbola. Thus there is a theorem for hyperbolas just as there was for ellipses.

> **Theorem**
>
> In the hyperbola with equation $\frac{x^2}{a^2} - \frac{y^2}{b^2} = 1$,
>
> $2a$ is the length of the horizontal axis,
> $2b$ is the length of the vertical axis, and
> $2c$ is the distance between the foci, where $c^2 = a^2 + b^2$.
>
> The coordinates of the vertices are $(a, 0)$ and $(-a, 0)$, and the foci are $(c, 0)$ and $(-c, 0)$.

If x and y are switched in the equation for a hyperbola, the general form becomes $\frac{y^2}{a^2} - \frac{x^2}{b^2} = 1$. Its graph is the reflection image of the graph of $\frac{x^2}{a^2} - \frac{y^2}{b^2} = 1$ over the line $y = x$, and the foci of the image hyperbola are on the y-axis. In this case, $2a$ is the length of the vertical axis, and $2b$ is the length of the horizontal axis.

■ ■ ■ ■ ■ ■ ■ ■

Example Sketch a graph of the hyperbola, and determine equations of its asymptotes.

a. $\frac{x^2}{9} - \frac{y^2}{16} = 1$

b. $\frac{y^2}{9} - \frac{x^2}{16} = 1$

Solution

a. In this hyperbola the foci are on the x-axis, $a = 3$ and $b = 4$. Thus the vertices of the hyperbola are $(3, 0)$ and $(-3, 0)$. To sketch the hyperbola, draw a rectangle centered at the origin with horizontal axis of length $2 \cdot 3 = 6$, and vertical axis of length $2 \cdot 4 = 8$. The diagonals of this rectangle determine the asymptotes. They have equations $\frac{y}{4} = \pm\frac{x}{3}$ or $y = \pm\frac{4}{3}x$. The hyperbola has vertices $(\pm 3, 0)$ and is bounded by these asymptotes. Below at the left is a sketch.

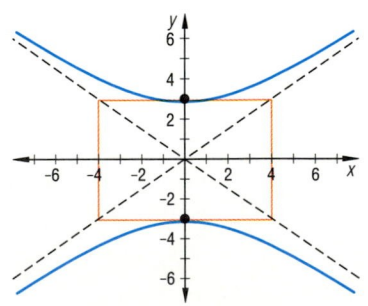

746

b. In this hyperbola, $a = 3$ and $b = 4$. The vertices of this hyperbola are $(0, 3)$ and $(0, -3)$. The asymptotes are also $\frac{y}{3} = \pm \frac{x}{4}$. However, the foci and vertices are on the y-axis, and the vertices are $(0, 3)$ and $(0, -3)$. A graph is above at the right.

Check
a. Analyze the equation for values excluded from the domain or range.

In part **a**, $\frac{x^2}{9} - 1 = \frac{y^2}{16}$ or $y = \pm \sqrt{16\left(\frac{x^2}{9} - 1\right)}$. The expression under the

radical sign is a real number if and only if $\frac{x^2}{9} - 1 \geq 0$ or $x^2 \geq 9$. That is, all

points on the hyperbola must have $|x| \geq 3$. This agrees with the sketch we drew.
b. You are asked to check the work for part **b** in the Questions.

Hyperbolas are used to locate objects that emit sound waves. The idea comes directly from the definition of hyperbola. Suppose a whale at an unknown point W emits a sound. Let A and B be locations of two underwater devices that can receive the sound, and suppose A and B are 10,000 feet apart. Suppose a sound from the whale is received 0.5 second later at point A than at point B. The speed of sound in water is known to be about 5000 feet per second. So the difference of the distances WA and WB is 2500 feet. Thus the position of the whale must be on a hyperbola with foci A and B and with focal constant 2500 feet.

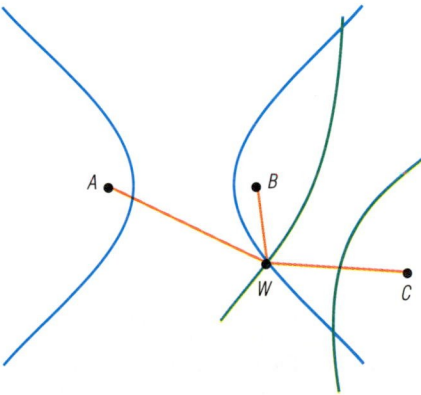

If another device receives the same sound at point C and the time is recorded, then the position of the whale can be located on two other hyperbolas (one with foci A and C, the other with foci B and C). The whale's position at the time of emitting the sound, which is the solution to this system of two equations in two variables, can be located quite precisely.

10. Let $F_1 = (\sqrt{2}, 0)$, $F_2 = (-\sqrt{2}, 0)$, and $P = (1, 0)$. We have to show $|PF_1 - PF_2| = 2$, or equivalently, $(PF_1 - PF_2)^2 = 4$. $(PF_1 - PF_2)^2 = $

$\left(\sqrt{(1 - \sqrt{2})^2 + (0 - 0)^2} - \sqrt{(1 + \sqrt{2})^2 + (0 - 0)^2}\right)^2 = $

$\left(\sqrt{3 - 2\sqrt{2}} - \sqrt{3 + 2\sqrt{2}}\right)^2 = 3 - 2\sqrt{2} - $

$\sqrt{(3 - 2\sqrt{2})(3 + 2\sqrt{2})} + 3 + 2\sqrt{2} = 6 - 2\sqrt{9 - 8} = 4$

11. Let $A = (5000, 0)$ and $B = (-5000, 0)$. An equation for the hyperbola is
$\frac{x^2}{1,562,500} - \frac{y^2}{23,437,500} = 1.$

13.c.

16.c.

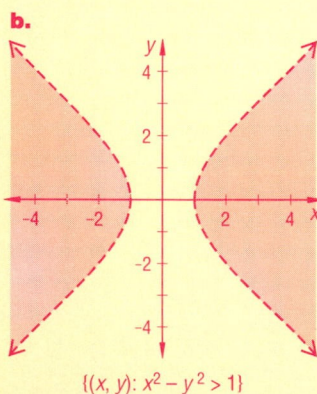
Questions

Covering the Reading

1. *True* or *False*.   The paths of some bodies in space may be hyperbolic in shape.   **True**

2. Suppose ℓ and m are intersecting lines and m is rotated around ℓ to form a cone. Now a plane P intersects the cone and does not contain ℓ or m. Which position of P gives a hyperbola, which an ellipse, which a parabola?
   a. P is parallel to m.   **parabola**
   b. P is perpendicular to ℓ.   **ellipse**
   c. P is parallel to ℓ.   **hyperbola**

3. a. Draw a plane intersecting a cone to form a hyperbola.   **See margin.**
   b. Draw the Dandelin spheres for this hyperbola.   **See margin.**
   c. How are the Dandelin spheres related to the foci of the hyperbola?
      The spheres are tangent to the plane at the foci of the hyperbola.

4. If $|m - n| = k > 0$, what are the possible values of $m - n$?
   $m - n = k$ **or** $m - n = {}^-k$

5. Draw two points F_1 and F_2, 2 units apart.
   a. Draw six points that are 1 unit closer to F_1 than to F_2.   **See margin.**
   b. *True* or *false*.   All six points must lie on the same branch of a hyperbola with foci F_1 and F_2.   **True**

6. Consider the hyperbola $\dfrac{x^2}{25} - \dfrac{y^2}{49} = 1$.   **See margin.**
   a. Find the coordinates of six points on the graph.
   b. Give equations for its asymptotes.
   c. Use the results of parts **a** and **b** to sketch a graph of the hyperbola.

7. a. Graph the hyperbola with equation $\dfrac{x^2}{16} - \dfrac{y^2}{9} = 1$.   **See margin.**
   b. Graph the hyperbola with equation $\dfrac{y^2}{16} - \dfrac{x^2}{9} = 1$.   **See margin.**
   c. How are the graphs in part **a** and **b** related?
      They are reflection images of each other over the line $y = x$.

8. Check part **b** of the Example of this lesson.   **See margin.**

Applying the Mathematics

9. Prove that every hyperbola is symmetric to the line containing its foci.
   See margin.

10. Use the definition of hyperbola to verify that the point $(1, 0)$ is on the hyperbola with foci $(\sqrt{2}, 0)$ and $(-\sqrt{2}, 0)$ and focal constant 2.
   See margin.

11. Find an equation for the hyperbola on which the whale of this lesson lies. (Hint: Let A and B be points on the x-axis, 10,000 units apart.)
   See margin.

12. a. Explain why the graph of $x^2 - 2y^2 = 2$ is a hyperbola.
    b. Find the foci and focal constant of this hyperbola.
       a) Dividing both sides of the equation $x^2 - 2y^2 = 2$ by 2 gives the equivalent equation $x^2/2 - y^2 = 1$. This is the standard form of the equation for a hyperbola with $a = \sqrt{2}$ and $b = 1$. b) foci: $(1, 0)$, $({}^-1, 0)$; focal constant: $2\sqrt{2}$

748

13. Consider the hyperbola with equation $(x + 2)^2 - (y + 3)^2 = 1$.
   a. Name its vertices. **(-1, -3), (-3, -3)**
   b. Give equations for its asymptotes. **$y = x - 1, y = -x - 5$**
   c. Sketch a graph of the hyperbola. **See margin.**

14. Consider the hyperbola with equation $\dfrac{(x - h)^2}{a^2} - \dfrac{(y - k)^2}{b^2} = 1$. Identify

   a. its center **(h, k)**   **b.** its vertices **(a + h, k), (-a + h, k)**
   c. its foci   **d.** its asymptotes.
   c) $(\sqrt{a^2 + b^2} + h, k), (-\sqrt{a^2 + b^2} + h, k)$; d) $\frac{y - k}{b} = \frac{x - h}{a}, \frac{y - k}{b} = \frac{-(x - h)}{a}$

Review

15. **a.** What is an equation for the ellipse shown at the right? **$(x^2/4) + (y^2/9) = 1$**
   b. Where are its foci? *(Lesson 12-2)*
      $(0, \sqrt{5}), (0, -\sqrt{5})$

16. Consider the ellipse with equation $\dfrac{(x + 2)^2}{25} + \dfrac{(y - 5)^2}{4} = 1$.
   a. Determine the center and the foci of this ellipse.
   b. How long are its major and minor axes? **major axis: 10; minor axis: 4**
   c. Graph this ellipse. *(Lesson 12-2)* **See margin.**
   a) center: (-2, 5); foci: $(\sqrt{21} - 2, 5), (-\sqrt{21} - 2, 5)$

17. Give the matrix for the rotation of the given magnitude around the origin. *(Lesson 11-4)*

   a. $\dfrac{3\pi}{2}$ $\begin{bmatrix} 0 & 1 \\ -1 & 0 \end{bmatrix}$ **b.** θ $\begin{bmatrix} \cos\theta & -\sin\theta \\ \sin\theta & \cos\theta \end{bmatrix}$ **c.** $-\theta$ $\begin{bmatrix} \cos\theta & \sin\theta \\ -\sin\theta & \cos\theta \end{bmatrix}$

18. The operational lifespan of a certain brand of toner for laser printers is known to be normally distributed with mean 4500 pages of text and standard deviation 500 pages. If a printing office is buying 50 of these toners, how many of them should they expect to need replacement after fewer than 4000 pages? *(Lesson 10-6)* **about 8**

19. **a.** Use the Binomial Theorem to write out the terms of the following powers of $(1 + 1)$. **See margin.**
      i. $(1 + 1)^0$ **ii.** $(1 + 1)^1$
      iii. $(1 + 1)^2$ **iv.** $(1 + 1)^3$
   b. *True* or *false*.   The binomial expansion of $(1 + 1)^n$ for any n gives the elements in the nth row of Pascal's Triangle. Justify your answer. *(Lessons 8-6, 8-7, 8-8)* **See margin.**

In 20 and 21, state the degree of the polynomial. *(Lesson 9-1)*

20. $3x^2 + 6x^4 - 7x^5$ **5**   **21.** $x^4 + x^2y^5 + y^6$ **7**

22. Simplify. $(\sqrt{2}\,a + b)^2 + (\sqrt{2}\,a - b)^2$ *(Previous course)* **$4a^2 + 2b^2$**

Exploration

23. **a.** Describe the graph of the set of points (x, y) satisfying $x^2 - y^2 < 1$.
   b. Describe the graph of the set of points (x, y) satisfying $x^2 - y^2 > 1$.
   c. Generalize parts **a** and **b** to apply to any hyperbola in standard form.
      a–c) See margin.

FOLLOW-UP

MORE PRACTICE
For more questions on SPUR Objectives, use *Lesson Master 12-3*, shown below.

EXTENSION
Small Group Work If students did the Extension for Lesson 12-1, they might adapt their model of Dandelin's spheres for the hyperbola.

PROJECTS
Projects 1 through 4 on pages 771 and 772 are related to the content of this lesson.

23.c. Given a hyperbola $\dfrac{x^2}{a^2} - \dfrac{y^2}{b^2} = 1$, the set of points satisfying $\dfrac{x^2}{a^2} - \dfrac{y^2}{b^2} < 1$ is the set of points between the branches of the hyperbola. The set of points satisfying $\dfrac{x^2}{a^2} - \dfrac{y^2}{b^2} > 1$ is the set of points outside the branches of the hyperbola.

NAME _____

LESSON MASTER 12-3
QUESTIONS ON SPUR OBJECTIVES

■SKILLS *Objective A (See pages 776–778 for objectives.)*

1. Find an equation for the hyperbola with foci $(0, 3)$ and $(0, -3)$ and focal constant 4. $\dfrac{y^2}{4} - \dfrac{x^2}{5} = 1$

2. Give equations for the asymptotes of the hyperbola $\dfrac{x^2}{25} - \dfrac{y^2}{9} = 1$. $y = \pm\dfrac{3}{5}x$

■PROPERTIES *Objective D*

3. Draw a hyperbola in which the distance between the foci is 1.5 inches and the focal constant is .5 inch.

■REPRESENTATIONS *Objective G*
In 4 and 5, graph each equation.

4. $\dfrac{x^2}{25} - \dfrac{y^2}{4} = 1$

5. $\dfrac{y^2}{16} - \dfrac{x^2}{49} = 1$

■REPRESENTATIONS *Objective H*
In 6 and 7, graph each equation.

6. $\dfrac{(x - 3)^2}{16} - \dfrac{(y - 1)^2}{25} = 1$

7. $y^2 - \dfrac{(x + 2)^2}{9} = 1$

122   *Functions, Statistics, and Trigonometry* © Scott, Foresman and Company

749

LESSON 12-4

RESOURCES
- ■ Lesson Master 12-4
- ■ Quiz for Lessons 12-1 through 12-4
- ■ Teaching Aid 90: Graphs for **Example 3**

OBJECTIVES

B Find equations for rotation images of figures.
D State and apply properties of ellipses and hyperbolas to draw and describe them.

TEACHING NOTES

This lesson has more reading than most, so it is a good idea to summarize it. There are five parts: (1) the description of Galton's statistical work, which shows a reason for rotating ellipses; (2) a proof and an example of the Graph Rotation Theorem; (3) a second example of that theorem, showing that relations with equations of the form $xy = k$ are hyperbolas; (4) a verbal argument explaining how any isometry in the plane can be performed on a relation, and why the equation for the image will have the same degree as the pre-image; and (5) an example of performing an isometry on a parabola. We suggest that you begin with the questions and discuss the parts of the lesson as needed.

LESSON

12-4

Rotating Relations

Francis Galton (1882)

In Lesson 12-1, it was noted that Kepler's discovery that the planets go around the sun in elliptical orbits is one of the great examples of mathematical modeling of all time. Another brilliant example of mathematical modeling, which also involved ellipses, occurred only a little over a century ago, in 1886. It came as the result of the work of Francis Galton (1822–1911), an English scientist and statistician.

Galton was attempting to show the influences of heredity on height. He collected the heights of 928 adult children from 205 families and tabulated their heights against the mean height of their respective parents (which he called the *midparent*). He then grouped the data as shown in the graph below. He discovered that, in the grouped data, the sets of entries with equal probability of occurrence lay on a series of similar ellipses with the same center and whose axes were at the same angle of inclination.

750

He showed his findings to a mathematician, J. Hamilton Dickson, who proved that Galton's discovery followed from the assumption that these heights were related by a bivariate normal distribution. The ellipses are the contour lines of equal probabilities, as shown here.

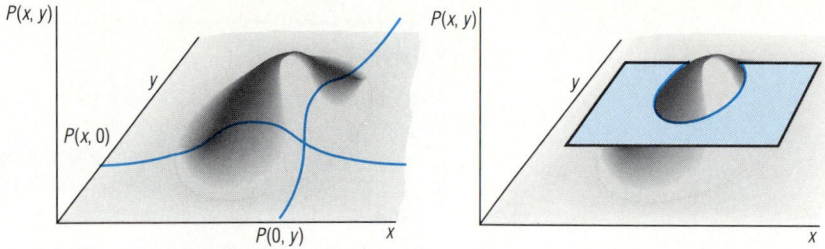

This work of Galton led to the ideas of correlation and variance and their connection with regression and the line of best fit. While those details are beyond the scope of this course, you can determine the equations of Dalton's inclined ellipses—that is, ellipses that are rotated so that their axes are no longer horizontal or vertical—like the one pictured above.

Recall that if the point (x, y) is rotated through magnitude θ about the origin, then its image (x', y') can be found by matrix multiplication.

$$\begin{bmatrix} x' \\ y' \end{bmatrix} = \begin{bmatrix} \cos \theta & -\sin \theta \\ \sin \theta & \cos \theta \end{bmatrix} \begin{bmatrix} x \\ y \end{bmatrix}$$

The inverse of the matrix for R_θ is the matrix for $R_{-\theta}$. Multiply both sides of the preceding matrix equation by that inverse on the left to obtain an equation for (x, y) in terms of (x', y').

$$\begin{bmatrix} \cos \theta & \sin \theta \\ -\sin \theta & \cos \theta \end{bmatrix} \begin{bmatrix} x' \\ y' \end{bmatrix} = \begin{bmatrix} \cos \theta & \sin \theta \\ -\sin \theta & \cos \theta \end{bmatrix} \begin{bmatrix} \cos \theta & -\sin \theta \\ \sin \theta & \cos \theta \end{bmatrix} \begin{bmatrix} x \\ y \end{bmatrix}$$

$$= \begin{bmatrix} 1 & 0 \\ 0 & 1 \end{bmatrix} \begin{bmatrix} x \\ y \end{bmatrix}$$

$$= \begin{bmatrix} x \\ y \end{bmatrix}$$

Now multiply the two matrices on the left side to get

$$\begin{bmatrix} x'\cos \theta + y'\sin \theta \\ -x'\sin \theta + y'\cos \theta \end{bmatrix} = \begin{bmatrix} x \\ y \end{bmatrix}.$$

Equating these matrices gives the following.

$$x = x'\cos \theta + y'\sin \theta$$
$$y = -x'\sin \theta + y'\cos \theta$$

Error Analysis Students may not understand how Galton's table on page 750 relates to the contour graphs on page 751. They should think of Galton's table as being in the xy-plane of a three-dimensional coordinate system. Then think of the frequencies as being the values of z for each pair (x, y). The values of z will be highest in the middle and taper off in all directions. If heights are normally distributed for each weight, and vice versa, then the bivariate normal distribution is the theoretical distribution that occurs. The cross-sections could be circles if x and y were converted into standard scores, but when they are not standardized the cross-sections are most likely to be noncircular ellipses.

The proof of the Graph Rotation Theorem follows the same pattern as its counterparts, the Graph Translation Theorem and the Graph Scale Change Theorem. Begin with the formula for the image under the transformation that gives x' and y' in terms of x and y. Then solve the system (here represented by matrices) for x and y in terms of x' and y'. Make those substitutions for x and y to rotate the given graph.

Students may not realize that the degree of xy is 2. Mention that the equations for the images in **Examples 1 and 2** are all of second degree. This is critical knowledge for the analyses of the next lesson, in which all quadratic relations are considered.

In **Example 2,** students may wonder what hyperbola would have given $xy = 1$. The answer is $x^2 - y^2 = 2$. See Additional Example 2, which follows.

Now, if a sentence involving x and y is known, then the above expressions can be substituted for x and y and the result will be a sentence involving x' and y', where the new sentence describes the rotation image. Once the sentence for the image is found, the primes in x' and y' are removed so that the resulting sentence is in the variables x and y. The result is the following theorem.

Graph Rotation Theorem

In a relation described by a sentence in x and y, the following two processes yield the same graph:
1. replacing x by $x \cos \theta + y \sin \theta$ and y by $-x \sin \theta + y \cos \theta$;
2. applying the rotation of magnitude θ about the origin to the graph of the original equation.

Example 1 Find an equation for the image of the ellipse $\frac{x^2}{4} + \frac{y^2}{9} = 1$ under a rotation of 30° about the origin.

Solution To simplify computation, first multiply both sides of the given equation by 36. The equation becomes $9x^2 + 4y^2 = 36$. Now, to find an equation for the image, use part (1) of the Graph Rotation Theorem.

Replace x by $x \cos 30° + y \sin 30°$, which is $\frac{\sqrt{3}}{2}x + \frac{1}{2}y$. Replace y by $-x \sin 30° + y \cos 30°$, which is $-\frac{1}{2}x + \frac{\sqrt{3}}{2}y$.

$$9\left(\frac{\sqrt{3}}{2}x + \frac{1}{2}y\right)^2 + 4\left(-\frac{1}{2}x + \frac{\sqrt{3}}{2}y\right)^2 = 36$$

$$\frac{9}{4}(\sqrt{3}x + y)^2 + (-x + \sqrt{3}y)^2 = 36$$

$$\frac{9}{4}(3x^2 + 2\sqrt{3}xy + y^2) + (x^2 - 2\sqrt{3}xy + 3y^2) = 36$$

$$\frac{31}{4}x^2 + \frac{5}{2}\sqrt{3}xy + \frac{21}{4}y^2 = 36$$

For a simpler expression, multiply both sides by 4.
$$31x^2 + 10\sqrt{3}xy + 21y^2 = 144$$

Check Find a point on the preimage. Check that the coordinates of its image under R_{30} satisfy the equation. We take (2, 0). Its image under a rotation of 30° is

$$\begin{bmatrix} \cos 30° & -\sin 30° \\ \sin 30° & \cos 30° \end{bmatrix}\begin{bmatrix} 2 \\ 0 \end{bmatrix} = \begin{bmatrix} \frac{\sqrt{3}}{2} & -\frac{1}{2} \\ \frac{1}{2} & \frac{\sqrt{3}}{2} \end{bmatrix}\begin{bmatrix} 2 \\ 0 \end{bmatrix} = \begin{bmatrix} \sqrt{3} \\ 1 \end{bmatrix}.$$

Is $(\sqrt{3}, 1)$ a solution to the equation found for the image? Substitute $\sqrt{3}$ for x and 1 for y.

Does $31(\sqrt{3})^2 + 10\sqrt{3}(\sqrt{3})(1) + 21(1)^2 = 144$?
Does $93 + 30 + 21 = 144$? Yes, it checks.

752

Some automatic graphers allow you to graph any equation of the form $Ax^2 + Bxy + Cy^2 + Dx + Ey + F = 0$. If you have access to such a grapher you can also check the solution to Example 1 by graphing. Below are the graphs of $9x^2 + 4y^2 - 36 = 0$ and $31x^2 + 10\sqrt{3}\,xy + 21y^2 - 144 = 0$ produced by such a grapher.

 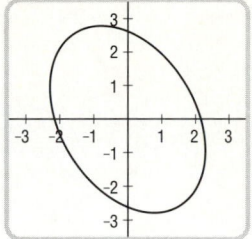

The second appears to be the image of the first under a rotation of 30° around the origin.

In the solution of Example 1 there appears an xy term. This is typical of quadratic relations that are not symmetric to the axes. Example 2 shows that rotation of the parent hyperbola can result in a hyperbola whose equation contains *only* an xy term.

Example 2   Rotate the hyperbola $x^2 - y^2 = 1$ by a magnitude of $-\frac{\pi}{4}$ about the origin.

Solution   Use the Graph Rotation Theorem. Replace x by $x\cos\left(-\frac{\pi}{4}\right) + y\sin\left(-\frac{\pi}{4}\right)$; replace y by $-x\sin\left(-\frac{\pi}{4}\right) + y\cos\left(-\frac{\pi}{4}\right)$. Note that $\sin\left(-\frac{\pi}{4}\right) = -\frac{\sqrt{2}}{2}$ and $\cos\left(-\frac{\pi}{4}\right) = \frac{\sqrt{2}}{2}$, so an equation for the image is

$$\left(\frac{\sqrt{2}}{2}x - \frac{\sqrt{2}}{2}y\right)^2 - \left(\frac{\sqrt{2}}{2}x + \frac{\sqrt{2}}{2}y\right)^2 = 1.$$

Since $\left(\frac{\sqrt{2}}{2}\right)^2 = \frac{1}{2}$, the computation is easier than in Example 1.

$$\frac{1}{2}x^2 - xy + \frac{1}{2}y^2 - \left(\frac{1}{2}x^2 + xy + \frac{1}{2}y^2\right) = 1$$
$$-2xy = 1$$
$$xy = -\frac{1}{2}$$

Check   The graph of $xy = -\frac{1}{2}$ is known to be a hyperbola whose graph lies entirely in the 2nd and 4th quadrants and whose asymptotes are the x- and y-axes. This is what would be expected from rotating the hyperbola $x^2 - y^2 = 1$ by a magnitude of $\frac{\pi}{4}$ clockwise.

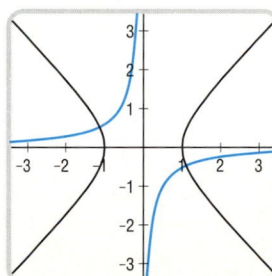

When the asymptotes of a hyperbola are perpendicular, the hyperbola is called a **rectangular hyperbola**. The preimage and image hyperbolas in Example 2 are both rectangular hyperbolas. Also, each is a rotation image of the other so, they are congruent. Now if a size change of magnitude $\sqrt{2}$ is applied to the hyperbola $xy = -\frac{1}{2}$ (that is, if $(x, y) \rightarrow (\sqrt{2}x, \sqrt{2}y)$), the result is a third hyperbola, $\frac{x}{\sqrt{2}} \cdot \frac{y}{\sqrt{2}} = -\frac{1}{2}$, which simplifies to $xy = -1$. When this hyperbola is reflected over the x-axis, its image is a fourth hyperbola $xy = 1$. All of these transformations give rise to similar figures, so this reasoning shows that the hyperbolas $x^2 - y^2 = 1$ and $xy = 1$ are similar. More generally, all rectangular hyperbolas are similar. Still more generally, two hyperbolas are similar if and only if the angles between their asymptotes that include them are congruent.

Notice that in Examples 1 and 2, both the given equations and the equations of the images are polynomials of degree 2. Suppose a relation is described by a sentence involving a polynomial, as in these examples. In previous chapters you learned how to find an equation for the translation image of any relation: replace x by $x - h$ and y by $y - k$. You also know that you can reflect any relation over the line $y = x$ by switching x and y. These substitutions, like the substitution in the Graph Rotation Theorem, never change the degree of the polynomial, because x and y are each replaced by some linear expression in x and/or y. As a result, if you begin with a polynomial of a particular degree and apply any composite of three types of transformations—a translation, a reflection over the line $y = x$, or a rotation around the origin—you will end up with a polynomial of the same degree.

An **isometry** is a composite of translations, rotations, and reflections. It can be proved that if all translations and rotations are possible, then any isometry needs a maximum of one reflection. Specifically, only reflection over the line $y = x$ is needed. Some isometries need no reflections; for example, rotations of magnitude θ around a point (a, b) can be accomplished by translating the figure using $T_{-a,-b}$, rotating θ about the origin, then translating the rotated image using $T_{a,b}$. Thus the Graph Translation Theorem and the Graph Rotation Theorem, together with reflection over the line $y = x$, give the means to perform any isometry in the plane on any relation.

What has been exemplified for ellipses and hyperbolas can also be done for parabolas. Any parabola is the image of the parabola $y = ax^2$ under a composite of translations and rotations, so any parabola will have an equation involving a polynomial of degree 2.

754

Example 3   Find an equation for the image of the parabola $y = x^2$ when rotated 20° about the point (5, 8).

Solution   Following the discussion above, first translate the parabola using $T_{-5,-8}$. (This maps (5, 8) onto the origin.)

$$y + 8 = (x + 5)^2$$

Then, obtain an equation for the image after a rotation of 20° about the origin.

Replace x by $x \cos 20° + y \sin 20° \approx .94x + .34y$, and
y by $-x \sin 20° + y \cos 20° \approx -.34x + .94y$.

$$-.34x + .94y + 8 \approx (.94x + .34y + 5)^2$$

Now, translate the rotated image using $T_{5,8}$.

$$-.34(x - 5) + .94(y - 8) + 8 \approx (.94(x - 5) + .34(y - 8) + 5)^2$$

This simplifies to

$$.88x^2 + .64xy + .12y^2 - 4.21x - 2.59y + 3.68 \approx 0.$$

The images of the original parabola after each of the three transformations are graphed below.

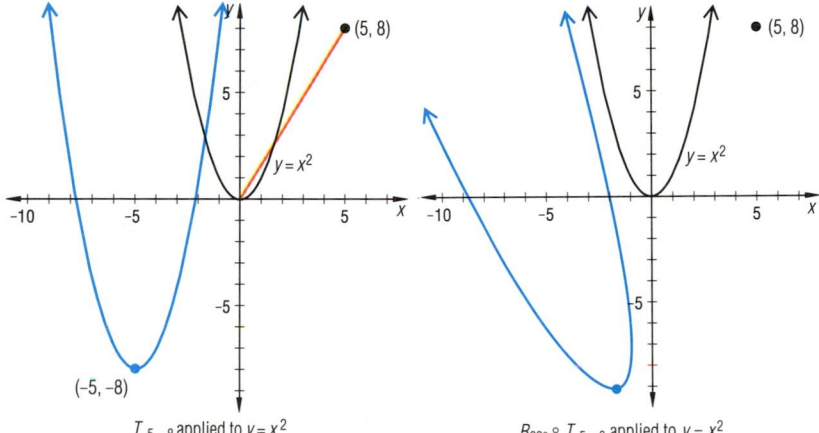

$T_{-5, -8}$ applied to $y = x^2$ $R_{20°} \circ T_{-5, -8}$ applied to $y = x^2$

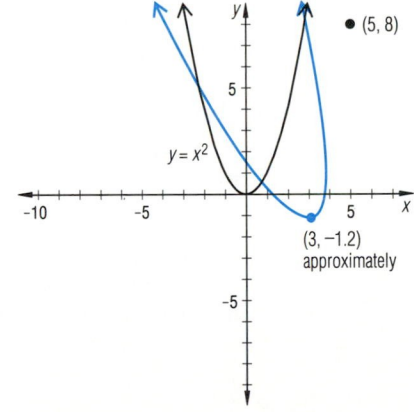

$(T_{5, 8} \circ (R_{20°} \circ T_{-5, -8})) =$ rotation of 20° about (5, 8)

NOTES ON QUESTIONS
Making Connections for
Question 4:   Notice that the equation in the answer looks like it might be an equation for a hyperbola or ellipse; in the next lesson students will see how to determine the type of conic from its equation.

Question 6:   A way of showing that these hyperbolas are not congruent is to note that the distance between the vertices of $x^2 - y^2 = 1$ is 2, while the distance between the vertices of $xy = 1$ is $2\sqrt{2}$. Since isometries preserve distance, these hyperbolas cannot be related by an isometry.

Question 8:   Students should examine the particular relation that is given and not go blindly into substitutions. The image of the circle is the circle itself. If you would like to emphasize this point, you might forge ahead with the substitution as if you were a student who did not sufficiently think about the problem. Then substitute $x \cos 40° + y \sin 40°$ for x and $-x \sin 40° + y \cos 40°$ for y. Show that it all works out at the end. You could even do it in general, with θ in place of 40°, or have students do this.

Question 9:   First students need to find an equation for this ellipse as if it were untilted. Then they should rotate that ellipse 30°.

756

Question 11: This question generalizes Additional Example 2.

Making Connections for Question 20: Completing the square is needed for the next lesson.

ADDITIONAL ANSWERS
7. The equation of the image of the ellipse is found by substituting linear combinations of x and y for x and y. Since the original is of degree 2, the resulting equation also has degree 2.

9.a. sample: $139x^2 - 78\sqrt{3}\,xy + 217y^2 = 6400$

b. major axis: $y = \dfrac{\sqrt{3}}{3}\,x$;

minor axis: $y = -\dfrac{\sqrt{3}}{3}\,x$

10. An equation for the x-axis is $y = 0$. By the Graph Rotation Theorem, the image of the equation $-x\sin\theta + y\cos\theta = 0$. Solving for y yields $y = x\tan\theta$.

12.a. sample: $-\dfrac{3}{4}x^2 - \dfrac{\sqrt{3}}{2}xy - \dfrac{1}{4}y^2 + \dfrac{3\sqrt{3}+5}{2}x + \dfrac{3\sqrt{3}+3}{2}y - \dfrac{18\sqrt{3}+17}{4} = 0$

b.

13.a.

Questions

Covering the Reading

1. What situation was Galton trying to model when he found that ellipses describe points with equal probability of occurrence? **the influence of heredity on height**

2. If a figure F is the image of a figure G under R_θ, then G is the image of F under what transformation? **$R_{-\theta}$**

In 3 and 4, find an equation for the image of the given figure under the given rotation.

3. $\dfrac{x^2}{25} + \dfrac{y^2}{16} = 1;\ R_{60°}$

4. $y = x^2 + 4;\ R_{\pi/4}$

sample: $91x^2 - 18\sqrt{3}\,xy + 73y^2 = 1600$;   **sample: $-x^2 - 2xy - y^2 - \sqrt{2}x + \sqrt{2}y = 8$**

5. Tell whether or not the hyperbola is a rectangular hyperbola.
   a. $xy = -2$ **rectangular**   b. $x^2 - y^2 = 8$ **rectangular**
   c. $y^2 - x^2 = 8$ **rectangular**   d. $\dfrac{x^2}{4} - \dfrac{y^2}{9} = 1$ **not rectangular**

6. *Multiple choice.* The hyperbolas $x^2 - y^2 = 1$ and $xy = 1$ are
   (a) congruent
   (b) similar but not congruent
   (c) neither similar nor congruent. **(b)**

7. Explain why, if an ellipse is rotated, its image must have an equation of degree 2. **See margin.**

8. Find an equation for the image of $x^2 + y^2 = 10$ under a rotation of 40° around the origin without using the Graph Rotation Theorem. **$x^2 + y^2 = 10$**

Applying the Mathematics

9. Galton's ellipse, pictured on page 750, has a major axis of length about 8 and a minor axis of length about 5, and the major axis makes an angle of about 30° to the x-axis. **See margin.**
   a. Find an equation for this ellipse.
   b. Find equations for the lines containing its major and minor axes.

10. Suppose the x-axis is rotated θ about the origin. Prove that an equation for its image is $y = x\tan\theta$. (This provides another way of showing that the slope of a line that makes an angle of θ measured counterclockwise from the positive x-axis is $\tan\theta$.) **See margin.**

11. Suppose the hyperbola $xy = k$ is rotated $\dfrac{\pi}{4}$ about the origin.
    a. What is an equation for its image? **sample: $y^2 - x^2 = 2k$**
    b. *True* or *false*. Both $xy = k$ and its rotation image in part **a** are rectangular hyperbolas. **True**

12. a. Find an equation for the image of the parabola $y = (x - 2)^2 + 3$ when rotated 30° about the point (2, 3).
    b. Check your work by graphing the image and preimage with an automatic grapher. **a,b) See margin.**

13. Consider the hyperbola with equation $\frac{x^2}{36} - y^2 = 1$.
  a. Sketch its graph.   **See margin.**
  b. Give equations for its asymptotes.
  c. State an equation for the image of this curve under reflection over the line $y = x$.   *(Lesson 12-3)*   **b)** $y = \frac{1}{6}x, y = -\frac{1}{6}x$; **c)** $\frac{y^2}{36} - x^2 = 1$

14. Find an equation for the ellipse with foci $(0, 3)$ and $(0, -3)$ and focal constant 7.   *(Lesson 12-2)*   **sample:** $\frac{4}{13}x^2 + \frac{4}{49}y^2 = 1$

15. What is an identity?   *(Lesson 5-7)*
  an equation which is true for all values in the domain of the variables

In 16–19, *multiple choice.*   Which expression below equals the given expression?   *(Lessons 5-7, 11-6)*
  (a) 1   (b) $\sin \theta$ (c) $\sin 2\theta$ (d) $\cos 2\theta$

16. $\cos^2 \theta - \sin^2 \theta$ **(d)**   **17.** $\cos^2 \theta + \sin^2 \theta$ **(a)**

18. $2 \sin \theta \cos \theta$ **(c)**   **19.** $1 - 2 \sin^2 \theta$ **(d)**

20. *Skill sequence.*   Complete the square.   *(Previous course)*
  a. $x^2 + 10x + \underline{\;?\;}$ **25**
  b. $x^2 + bx + \underline{\;?\;}$ **$b^2/4$**
  c. $2x^2 + 7x + \underline{\;?\;}$ **49/8**
  d. $ax^2 + bx + \underline{\;?\;}$ **$b^2/4a$**

21. Ask each of about 25 students for the height of his or her biological parents. Graph the ordered pairs of the form (height of student, midparent height).
  a. Find the center of gravity (mean student height, mean midparent height) of this bivariate distribution.
  b. Find an equation for the line of best fit.
  c. The line from part **b** estimates the major axis of the ellipses which contain points of equal probability in the distribution. Derive an equation for the ellipse that seems to contain about $\frac{2}{3}$ of the data points.   **Answers will vary.**

FOLLOW-UP

MORE PRACTICE
For more questions on SPUR Objectives, use *Lesson Master 12-4*, shown below.

EXTENSION
Students might be interested in finding out more about Sir Francis Galton. They could write a brief report.

EVALUATION
A quiz covering Lessons 12-1 through 12-4 is provided in the Teacher's Resource File.

Alternative Assessment
If this is the last chapter that you will cover in the course, you may wish to once again use an informal questionnaire of the kind suggested with Lesson 1-3. (See page 22.) You and your students might find it interesting to compare their earlier responses, at the beginning of the course, with their responses now, at the conclusion of the course.

NAME _____

LESSON **MASTER 12-4**
QUESTIONS ON **SPUR** OBJECTIVES

■**SKILLS** *Objective B (See pages 776–778 for objectives.)*
In 1–8, find an equation for the image of the given figure under the given rotation.

1. $\frac{x^2}{16} - \frac{y^2}{9} = 1$; $R_{90°}$ $-39x^2 + 50\sqrt{3}xy + 11y^2 = 576$

2. $25x^2 - y^2 = 1$; $R_{45°}$ $12x^2 + 26xy + 12y^2 = 1$

3. $y = x^2 + 2$; $R_{\frac{\pi}{3}}$ $-2\sqrt{3}x + 2y = x^2 + 2\sqrt{3}xy + 3y^2 + 8$

4. $x^2 + y^2 = 49$; $R_{\frac{\pi}{2}}$ $x^2 + y^2 = 49$

5. $\frac{x^2}{25} + \frac{y^2}{64} = 1$; $R_{\frac{\pi}{4}}$ $217x^2 + 78\sqrt{3}xy + 139y^2 = 6400$

6. $\frac{x^2}{64} - \frac{y^2}{16} = 1$; $R_{\frac{\pi}{2}}$ $y^2 - 4x^2 = 64$

7. $y = x^2 + 2x$; $R_{45°}$ $-3\sqrt{2}x - \sqrt{2}y = x^2 + 2xy + y^2$

8. $x^2 + y^2 = 9$; $R_{180°}$ $x^2 + y^2 = 9$

9. Find an equation for the image of the parabola $y = (x + 1)^2 - 3$ when rotated 30° about the point (-1, -3).
    $-2x + 2\sqrt{3}y = 3x^2 + 2\sqrt{3}xy = y^2$

10. Find an equation for the image of the ellipse $\frac{x^2}{4} - \frac{y^2}{9} = 1$ when rotated 45° about the point (1, 1).
    $\frac{13}{2}x^2 + 5xy - 5\sqrt{2}x - 13\sqrt{2}y + \frac{13}{2}y^2 = 23$

■**PROPERTIES** *Objective D*
In 11 and 12, tell whether the hyperbola is a rectangular hyperbola.

11. $xy = 25$ **Yes**

12. $\frac{x^2}{25} - \frac{y^2}{16} = 1$ **No**

Functions, Statistics, and Trigonometry © Scott, Foresman and Company   **123**

RESOURCES
- Lesson Master 12-5
- Teaching Aid 91: Degenerate conics
- Computer Master 30

OBJECTIVES

C Rewrite equations of conic sections in standard form of a quadratic relation in two variables.

E Describe the intersections of a plane and a cone of 2 nappes.

H Graph transformation images of ellipses and hyperbolas.

I Describe graphs of quadratic equations.

TEACHING NOTES

After students have read the lesson, you may wish to summarize as follows: Put a table on the board with the following six columns: equation, type of conic, $A, B, C,$ $B^2 - 4AC$. Then use a variety of equations known to the students, for instance: $y = 3x^2$, parabola, 3, 0, 0, 0; $x^2 + y^2 = 9$, circle, 1, 0, 1, -4; $\frac{x^2}{4} + \frac{y^2}{9} = 1$, ellipse, $\frac{1}{4}, 0, \frac{1}{9}, -\frac{1}{9}$; $x^2 - y^2 = 1$, hyperbola, 1, 0, -1, 4; $xy = 100$, hyperbola, 0, 1, 0, 1. Point out that the type of conic and whether the value of $B^2 - 4AC$ is positive, negative, or zero corresponds exactly as predicted by the last theorem of the lesson. The goal of the entire lesson is to prove that rather amazing theorem. (An outline of the development is found in the Perspectives for this lesson on page 727C.)

Notice in **Example 1** that the solution stops when the equation is in a form that

---

You have learned that any conic section has an equation that is a polynomial of degree 2. That is, every conic section has an equation of the form
$$Ax^2 + Bxy + Cy^2 + Dx + Ey + F = 0,$$
where at least one of A, B, and C is not zero. This is the **standard form of a quadratic relation in two variables.** Thus every conic section is described by a quadratic relation.

In this lesson, the converse question is discussed. Does every quadratic relation describe a conic? The answer is Yes and No! Here's an explanation.

First, notice that if $B = 0$, then there is no xy term in the equation. The equation has the form
$$Ax^2 + Cy^2 + Dx + Ey + F = 0$$
which, by reordering the terms, can be rewritten as
$$Ax^2 + Dx + Cy^2 + Ey + F = 0.$$
In this form, the equation is set up for completing the square. First factor out A and C from the x- and y-terms, respectively.
$$A(x^2 + \frac{D}{A}x + \quad) + C(y^2 + \frac{E}{C}y + \quad) + F = 0$$

Then complete the squares and add the appropriate constants,
$-F, A \cdot \dfrac{D^2}{4A^2}$, and $C \cdot \dfrac{E^2}{4C^2}$, to each side, and simplify.

■ ■ ■ ■ ■ ■ ■ ■

Example 1 Show that the graph of $3x^2 - y^2 + 6x + 7y - 12 = 0$ is a hyperbola.

Solution Reorder the terms.
$$3x^2 + 6x - y^2 + 7y - 12 = 0$$
Factor.
$$3(x^2 + 2x + \quad) - (y^2 - 7y + \quad) - 12 = 0$$
Complete the square. Be careful to add the same amounts to both sides and watch out for the numbers that have been factored out of the parentheses.
$$3(x^2 + 2x + 1) - (y^2 - 7y + 12.25) - 12 = 0 + 3 - 12.25$$
Rewrite the binomials as squares and write a single constant on the right side.
$$3(x + 1)^2 - (y - 3.5)^2 = 2.75$$
This is actually far enough to see what is going on. By dividing both sides by 2.75, an equation of the form
$$\frac{(x + 1)^2}{a^2} - \frac{(y - 3.5)^2}{b^2} = 1$$
is obtained. This is an equation for a hyperbola with center at (-1, 3.5).

Check Use an automatic grapher that graphs equations of the form $Ax^2 + Bxy + Cy^2 + Dx + Ey + F = 0$. Input $A = 3$, $B = 0$, $C = -1$, $D = 6$, $E = 7$, and $F = -12$. At the right is the output from such a grapher.

The graph is a hyperbola with $x = -1$ and $y = 3.5$ as axes of symmetry.

Example 1 illustrates the graph of a standard form of a quadratic relation in two variables where $B = 0$ and A and C have opposite signs. For such values of A, B, and C, the graph is typically a hyperbola. If A and C have the same sign, then it is possible that no points may satisfy the equation. For example, there are no ordered pairs of real numbers (x, y) which satisfy $3x^2 + 4y^2 + 5 = 0$. (Note that for all real x and y, the left side of the equation is positive.) But if A and C are both positive and more than one point satisfies the equation, then the graph of the quadratic relation is an ellipse. (The special case of a circle occurs when $A = C$.) If one of A or C is zero, and there are points which satisfy the equation, then the graph is typically a parabola.

Notice the word "typically" occurs in the preceding paragraph twice. The reason is that, on occasion, the expression
$$Ax^2 + Cy^2 + Dx + Ey + F$$
is factorable into the product of two linear expressions. For instance, the left side of
$$9x^2 - 4y^2 + 3x + 22y - 30 = 0$$
is factorable, so the equation can be rewritten as
$$(3x + 2y - 5)(3x - 2y + 6) = 0.$$
But if the product of two numbers is zero, then one or the other (or both) is zero. So
$$3x + 2y - 5 = 0 \text{ or } 3x - 2y + 6 = 0.$$
These are equations of lines. So a point satisfies the first quadratic relation if and only if it is on one or both of the two lines.

Geometrically, this situation can still be interpreted as a conic section, that is, as the intersection of a cone and a plane, but the plane contains the vertex of the cone. These are called **degenerate conic sections**. There are several degenerate conic sections possible.

a single point
(degenerate ellipse)

a line
(degenerate parabola)

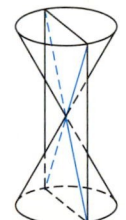

two intersecting lines
(degenerate hyperbola)

allows it to be recognized as a hyperbola.

In the determination of the angle in **Example 2,** if θ is a solution, then so is $\theta + \dfrac{\pi}{2}$. A proof is simple: $\tan 2\left(\theta + \dfrac{\pi}{2}\right) = \tan(2\theta + \pi) = \tan 2\theta$ since the period of the tangent function is π.

Geometrically, it is easy to see when a conic section is degenerate; the intersecting plane contains the vertex of the cone. The drawings on page 759 show the possibilities (a teaching aid contains them, also). However, the algebraic counterpart is not so simple, and students are likely to ask how you can determine whether an equation graphs as a degenerate conic or some other conic. One way to decide is to graph the conic. But this may not be easy to do if the available technology does not accept conics in standard form. Algebraically, you can try to solve for one variable in terms of the other and try to find five points on the conic. If it is impossible to find five points, then the conic either consists of a single point or it is the null set. If three of the five points are on the same line, then the degenerate conic consists of intersecting or parallel lines (use the $B^2 - 4AC$ criterion to determine which) because no line can intersect an ellipse, hyperbola, or parabola in more than two points.

Students may ask if it is coincidence that the discriminant $b^2 - 4ac$ for a quadratic has the same form as the expression $B^2 - 4AC$ used to discriminate conics. A reasonable response is that it seems to be coincidence, but perhaps someone will find in the future that there is a connection, similar to the way that Dandelin connected the two-dimensional and three-dimensional characterizations of conics.

This information is summarized in the following theorem.

> **Theorem**
>
> The graph of $Ax^2 + Cy^2 + Dx + Ey + F = 0$ is
> 1. an ellipse, a point, or the null set, if $AC > 0$;
> 2. a parabola, a single line, or the null set, if $AC = 0$;
> 3. a hyperbola or a pair of intersecting lines, if $AC < 0$.

Now you see why we wrote "Yes and No" to answer the question of whether the graph of a quadratic relation is always a conic section.

There still remains the possibility that B is not zero; that is, there is an xy term. In the last lesson, you saw that this can occur by rotating a conic section so that its axes are no longer horizontal or vertical. Consequently, in order to classify such a conic section, you must rotate it back in order to apply the above theorem. The argument is long, but read carefully and try to follow it.

Begin with the general quadratic relation
$$Ax^2 + Bxy + Cy^2 + Dx + Ey + F = 0.$$
We need to find θ so that under a rotation of θ about the origin, the image has no xy term, so substitute $x \cos \theta + y \sin \theta$ for x, and $-x \sin \theta + y \cos \theta$ for y. The expressions become rather complicated, but you do not have to worry about the substitutions for x and y in Dx and Ey because they do not yield a term in xy.

$$A(x \cos \theta + y \sin \theta)^2 + B(x \cos \theta + y \sin \theta)(-x \sin \theta + y \cos \theta)$$
$$+ C(-x \sin \theta + y \cos \theta)^2 + \ldots + F = 0$$

We are looking for the xy-coefficient of the image, so consider only the terms in xy that will result from the squaring and multiplication of the binomials.

$$A(2xy \cos \theta \sin \theta) + B(xy \cos^2\theta - xy \sin^2\theta) + C(-2xy \sin \theta \cos \theta) + \ldots + F = 0$$

Factoring out xy from each of these terms gives
$$xy\,[2A \cos \theta \sin \theta + B(\cos^2 \theta - \sin^2 \theta) - 2C \sin \theta \cos \theta] + \ldots + F = 0.$$

Thus the coefficient of the new xy term is
$$B' = 2A \cos \theta \sin \theta + B(\cos^2 \theta - \sin^2 \theta) - 2C \sin \theta \cos \theta.$$

Notice that each expression in θ is related to a double-angle formula, so
$$B' = A \sin 2\theta + B \cos 2\theta - C \sin 2\theta$$
or $\qquad B' = (A - C) \sin 2\theta + B \cos 2\theta.$

Remember that you want a value of θ so that $B' = 0$. So consider this equation.
$$0 = (A - C) \sin 2\theta + B \cos 2\theta$$

This is equivalent to
$$B \cos 2\theta = (C - A) \sin 2\theta.$$

760

There are now two possibilities. If $C = A$, then solve $B \cos 2\theta = 0$ to find the magnitude of the rotation θ under which the image of the original quadratic has no xy term. That equation always has the solution $2\theta = \frac{\pi}{2}$, or $\theta = \frac{\pi}{4}$. If $C \neq A$, then neither B nor $C - A$ is 0 and so neither $\sin 2\theta$ nor $\cos 2\theta$ is 0. Then divide by $(C - A) \cos 2\theta$ to get

$$\frac{B}{C - A} = \frac{\sin 2\theta}{\cos 2\theta};$$

that is,

$$\frac{B}{C - A} = \tan 2\theta.$$

This can be solved for 2θ to find the value of θ that will rotate the quadratic relation into one that has no xy term.

■ ■ ■ ■ ■ ■ ■ ■

Example 2   What rotation will cause the image of the graph of
$$x^2 + 3xy + 4y^2 + 2x - y + 5 = 0$$
to have an equation with no xy term?

Solution   Here $A = 1$, $B = -3$, and $C = 4$. $C \neq A$, so solve $\tan 2\theta = \frac{3}{4 - 1} = 1$. The solution is $2\theta = \tan^{-1} 1 = \frac{\pi}{4}$ and so $\theta = \frac{\pi}{8}$.

A rotation of magnitude $\frac{\pi}{8}$ will cause the image to have no xy term.

Finding an equation in standard form for the image in Example 2 first requires finding $\cos \theta$ and $\sin \theta$ knowing $\tan 2\theta$, then doing the substitutions as in Lesson 12-4. The amount of algebraic manipulation required is considerable but necessary if the aim is to describe the specific shape and position of the image. Fortunately, if all that is desired is to know whether the quadratic relation is a parabola, ellipse, or hyperbola, there is a criterion that does not require finding this equation. It is based on the following theorem: under a rotation, the value of the discriminant $B^2 - 4AC$ does not change.

Theorem

Let S be the set of ordered pairs (x, y) satisfying
$$Ax^2 + Bxy + Cy^2 + Dx + Ey + F = 0,$$
and let $S' = R_\theta (S)$. Furthermore, suppose that
$$A'x^2 + B'xy + C'y^2 + D'x + E'y + F' = 0$$
is an equation for S'. Then $B^2 - 4AC = (B')^2 - 4A'C'$.

The proof for this theorem follows on the next page.

NOTES ON QUESTIONS
Error Analysis for Question 6:   Some students may answer that this is an ellipse. It is, more specifically, a degenerate ellipse, one whose focal constant and focal distance are both 0.

Questions 9 and 10:
Any integer multiple of $\frac{\pi}{2}$ can be added to the magnitude.

Proof The proof requires quite a bit of algebraic manipulation. It is begun here, and you are asked to fill in some other details as one of the questions. To perform the rotation R_θ, $x \cos \theta + y \sin \theta$ is substituted for x in the equation for S, and $-x \sin \theta + y \cos \theta$ is substituted for y. After some algebraic manipulation, the result is the equation for S'. It can be shown that

$$A' = A \cos^2\theta - B \cos \theta \sin \theta + C \sin^2\theta,$$
$$B' = 2(A - C) \sin \theta \cos \theta + B(\cos^2\theta - \sin^2\theta) \quad \text{(as stated on page 760),}$$

and $C' = A \sin^2\theta + B \cos \theta \sin \theta + C \cos^2\theta.$

The next step is to find $(B')^2 - 4A'C'$. This too requires quite a bit of manipulation. To shorten the writing, let $s = \sin \theta$ and $t = \cos \theta$. Then

$$A' = At^2 - Bts + Cs^2,$$
$$B' = 2(A - C)ts + B(t^2 - s^2),$$

and $C' = As^2 + Bts + Ct^2.$

Now calculate $B'^2 - 4A'C'$ and use the fact that $s^2 + t^2 = 1$ to finish the proof.

The above theorem holds for *any* rotation, so it certainly holds for the rotation for which $B' = 0$. Thus for the equation of the rotation image that has no xy term, $B^2 - 4AC = (B')^2 - 4A'C' = -4A'C'$. But, from the first theorem of this lesson, the type of conic section is determined by the sign of $A'C'$. The sign of $-4A'C'$ is just the opposite. Putting this all together, the following theorem is obtained.

Theorem

Let S be the set of ordered pairs (x, y) satisfying
$$Ax^2 + Bxy + Cy^2 + Dx + Ey + F = 0.$$
Then:

if $B^2 - 4AC = 0$, then S is a parabola, two parallel lines, a line, or the null set.

if $B^2 - 4AC < 0$, then S is an ellipse, a single point, or the null set.

if $B^2 - 4AC > 0$, then S is a hyperbola or a pair of intersecting lines.

■ ■ ■ ■ ■ ■ ■

Example 3   What kind of conic section is the graph of the quadratic relation
$$2x^2 + 3xy - y^2 + 2x - 2y - 5 = 0?$$

Solution   Using $A = 2, B = 3,$ and $C = -1$, $B^2 - 4AC = 9 + 8 = 17$, which is greater than 0. Thus the conic is either a hyperbola or a pair of intersecting lines. To determine which, try to find some points on the graph. Begin with the given equation
$$2x^2 + 3xy - y^2 + 2x - 2y - 5 = 0.$$
If $x = 0$, then $-y^2 - 2y - 5 = 0$ or $y^2 + 2y + 5 = 0$, for which there is no real solution. This tells us that the graph does not cross the y-axis. This is impossible if the graph is two intersecting lines, so the graph must be a hyperbola.

762

If in Example 3 there had been points found on the graph, then we would have had to find a total of at least three points. The reasoning is as follows: If the graph is a pair of intersecting lines, then at least two of these points must be on one of the lines. If all three points are on the same line; that is, if part of the graph is a straight line, then the graph must be a pair of intersecting lines. If all three points are not on the same line, then try some other point (perhaps the midpoint) on each of the three lines connecting them. If the graph contains a straight line, then one of these lines must be one of the intersecting lines. If none of the midpoints satisfies the equation, then the graph does not contain any lines; the only possibility is that it is a hyperbola. Similar reasoning can be used to distinguish ellipses and parabolas from their degenerate forms.

Questions

Covering the Reading

In 1–3, **a.** rewrite the equation in standard form of a quadratic relation in two variables; **b.** give values for $A, B, C, D, E,$ and F in that form; and **c.** identify the conic section.

1. $\dfrac{x^2}{4} + \dfrac{y^2}{9} = 1$ 2. $(x-5)^2 + (y-2)^2 = 6$ 3. $xy = 1$
   See margin.   **See margin.**   **See margin.**

4. **a.** Show that the graph of $3x^2 + y^2 + 6x + 7y - 12 = 0$ is an ellipse, and
   b. find its center.   a) **See margin.** b) $\left(-1, \dfrac{7}{2}\right)$

In 5–8, name the conic section whose equation is given.

5. $3x^2 + 5y^2 - 6x + 2y - 9 = 0$ 6. $x^2 - 14x + y^2 + 2y + 50 = 0$
   ellipse   **a point**

7. $x + y + xy = 8$ **hyperbola**

8. $3x^2 - 6xy + 3y^2 - 7x + 2y + 5 = 0$ **parabola**

In 9 and 10, what is the magnitude of the rotation which will cause an equation for the image of the graph of the given equation to have no term in xy?

9. $x^2 + 3xy + y^2 = 8$ $\dfrac{\pi}{4}$

10. $-2x^2 - 4xy - 5y^2 + 4x - 3y + 27 = 0$ $\dfrac{1}{2}\tan^{-1}\left(\dfrac{4}{3}\right) \approx 0.464$

Applying the Mathematics

11. Examine the proof of the theorem on page 762. Do the algebraic manipulation to show that $(B')^2 - 4A'C' = B^2 - 4AC$.   **See margin.**

12. Consider the graph of $x^2 - xy = 6$.   **See below.**
    a. Use the $B^2 - 4AC$ criterion to determine the type of conic this is.
    b. Solve for y and graph this conic.
    c. At what angle(s) are the axes of this conic inclined to the x-axis?

13. Explain why $(2x + 5y - 7)(3x - 7y + 11) = 0$ is an equation for a degenerate conic.   **See margin.**

12a) $B^2 - 4AC = (-1)^2 - 4(1)(0) = 1$, so the conic is a hyperbola or a pair of intersecting lines. b) $y = \dfrac{x^2 - 6}{x} = x - \dfrac{6}{x}$ (see margin for graph); c) $-\dfrac{\pi}{8}$

NAME _____

LESSON **MASTER   12-5**
QUESTIONS ON **SPUR** OBJECTIVES

■SKILLS *Objective C (See pages 776– 778 for objectives.)*
In 1 and 2, rewrite the equation in the standard form of a quadratic relation in two variables and give values of $A, B, C, D,$ $E,$ and F for the equation.

1. $\dfrac{x^2}{25} + \dfrac{y^2}{9} = 1$ $9x^2 + 25y^2 - 225 = 0$;
   $A = 9, B = 0, C = 25, D = 0, E = 0, F = -225$

2. $(x-5)^2 + (y+2)^2 = 40$ $x^2 + y^2 - 10x + 4y - 11 = 0$;
   $A = 1, B = 0, C = 1, D = -10, E = 4, F = -11$

■PROPERTIES *Objective E*
In 3–5, name the degenerate conic section for the graph of $Ax^2 + Cy^2 + Dx + Ey + F = 0$ under the given condition.

3. $AC = 0$ **single line**
4. $AC > 0$ **point**
5. $AC < 0$ **pair of intersecting lines**

■REPRESENTATIONS *Objective H*
In 6 and 7, graph each equation.

6. $xy + 24 = 0$ 7. $y = \dfrac{x^2 - 1}{x}$

■REPRESENTATIONS *Objective I*
In 8–10, describe the graph of the relation represented by the given equation.

8. $9x^2 + 5y^2 - 6x + 4y + 10 = 0$ **the null set**
9. $10x + 10y - 5xy = 8$ **hyperbola**
10. $x^2 + 6xy + 6y^2 + 4x - 3y + 10 = 0$ **hyperbola**

124   *Functions, Statistics, and Trigonometry* © Scott, Foresman and Company

763

MORE PRACTICE
For more questions on SPUR Objectives, use *Lesson Master 12-5*, shown on page 763.

EXTENSION
Small Group Work You might separate the class into groups and have them work on the following questions: What is an equation for the conic whose graph is (3, 4)? $((x-3)^2 + (y-4)^2 = 0)$ What is an equation for the conic whose graph is the union of the intersecting lines $3x - 4y = 2$ and $x + y = 7$? $((3x - 4y - 2)(x + y - 7) = 0)$ Then ask for generalizations of both the questions and answers. For example, what is an equation for a relation whose graph is the union of two lines? (Sample: If the lines are $a_1 x + b_1 y = c_1$ and $a_2 x + b_2 y = c_2$, the equation of the relation is $(a_1 x + b_1 y - c_1)(a_2 x + b_2 y - c_2) = 0$.)

PROJECTS
Project 6 on pages 772 and 773 is related to the content of this lesson.

EVALUATION
Small Group Work
Groups of students should write their own version of a chapter test. Point out that the test items should cover the most important ideas of the chapter. It might be interesting to have the groups switch tests and actually take each others tests.

ADDITIONAL ANSWERS
16.a., 17.a. See margin, page 763.

22.a.

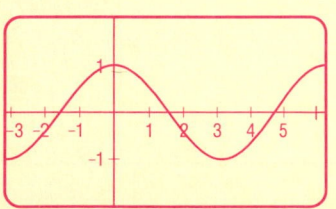

Review

14. Find an equation for the image of $x^2 - y^2 = 1$ under the rotation $R_{-45°}$. *(Lesson 12-4)* **$xy = -\frac{1}{2}$**

15. What makes a hyperbola a *rectangular hyperbola?* *(Lesson 12-4)* **See below.**

In 16 and 17, **a.** sketch the curve satisfying the conditions, and **b.** find an equation for the curve. *(Lessons 12-2, 12-3)*

16. ellipse with vertices $(\pm 5, 0)$ and minor axis of length 4 **a) See margin. b) $(x^2/25) + (y^2/4) = 1$**

17. hyperbola with vertices $(0, \pm 6)$ and asymptotes $y = \pm 4x$ **a) See margin. b) $(y^2/36) - (4x^2/9) = 1$**

In 18 and 19 simplify each expression. *(Lessons 5-7, 11-5)*

18. $\cos(\pi - \theta)$ **$-\cos\theta$** 19. $\cos\left(\frac{\pi}{2} + \theta\right)$ **$-\sin\theta$**

20. *True* or *false.* For all real numbers x, $\sin\left(\frac{\pi}{2} - x\right) = \sin\left(x - \frac{\pi}{2}\right)$. Justify your answer. *(Lessons 5-7, 11-5)* **See below.**

21. Given $z = 3 + 2i$ and $w = 5 - i$, find *(Lesson 9-6)*

 a. $z + 4w$ **$23 - 2i$** **b.** zw **$17 + 7i$** **c.** $\frac{z}{w}$. **$\frac{1}{2} + \frac{1}{2}i$**

22. **a.** Use an automatic grapher to graph $y = \sin x \cot x$. **See margin.**
 b. Find a simpler equation for the same graph.
 c. Use the definition of $\cot x$ to justify your answer. *(Lesson 6-7)*
 b) $y = \cos x$, $x \neq k\pi$ for all integers k; c) See below.

Exploration

23. Explore the graph of the equation $x^2 + 2xy + Cy^2 - 12 = 0$ for various values of C. How many different types of conics are possible? What values of C lead to degenerate conics? Are there any properties that all members of this family of quadratic relations satisfy?
 sample: Three types of conics are possible, one of them degenerate. If $C = 1$, the conic is a pair of parallel lines. If $C > 1$, the conic is an ellipse. If $C < 1$, the conic is a hyperbola.

15) A hyperbola is called a rectangular hyperbola when its asymptotes are perpendicular.

20) False; if $x = 0$, then $\sin\left(\frac{\pi}{2} - x\right) = \sin\left(\frac{\pi}{2}\right) = 1$, but $\sin\left(0 - \frac{\pi}{2}\right) = \sin\left(-\frac{\pi}{2}\right) = -1$.

22c) $\cot x = \frac{\cos x}{\sin x}$, so for all $x \neq k\pi$, where k is any integer,
$\sin x \cdot \cot x = \sin x \cdot \frac{\cos x}{\sin x} = \cos x$.

12-6

It's All Done with Mirrors

LESSON 12-6

RESOURCES
■ Teaching Aid 92: Diagrams for **Questions 1–4**

An object that hits a flat hard surface without spin will bounce off that surface in a path that is the reflection image over that surface of what its path would have been if the wall were not there. This reflection principle is often seen in handball, tennis, billiards, basketball, baseball, hockey, and many other sports.

In physics this principle is stated as *the angle of incidence equals the angle of reflection,* where the two angles in question are measured from the ray perpendicular to the surface at the point of contact. Furthermore, the two rays of the path of the object and the third perpendicular ray all lie in the same plane, so that the mathematics of this situation can be analyzed in a plane.

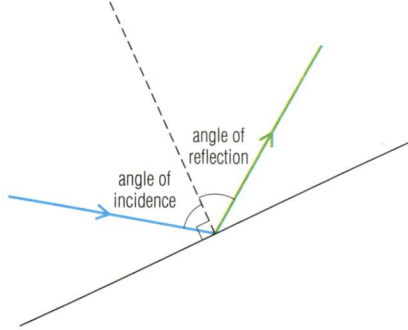

These reflection principles apply also to electromagnetic radiation (such as light waves and radio waves) and sound waves. When you look in a mirror and see the image of an object, that image is located in the position of the reflection image of the object. Acoustical engineers use reflection principles to design concert halls so that the sound waves from the stage are not distorted as they bounce off the walls, ceiling, seats, and floor.

The reflection principles apply even when surfaces are curved as long as the curvature of the surface is smooth, that is, as long as there is a plane tangent to the surface at the point of contact. Think of it this way: when the object hits such a surface, it does not know anything about the surface except at the exact point where it touches, so it bounces off the surface as if the surface is flat at that point. This is equivalent to saying that the object bounces off in a direction determined by the plane tangent to the surface at the point, as pictured below. The path and the perpendicular are coplanar, and they form equal angles with the perpendicular to the tangent plane at the point of contact, just as if the surface were flat there.

The conic sections possess reflection properties that are even more special. If an ellipse is rotated about its major axis, the surface formed is called an **ellipsoid**.

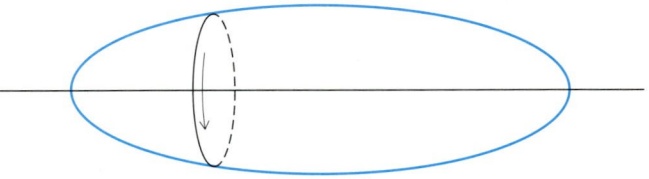

In a number of museums and other buildings (the U.S. Senate is one), one of the rooms is designed so that its walls (from head-level up) and ceiling make up half of an ellipsoid.

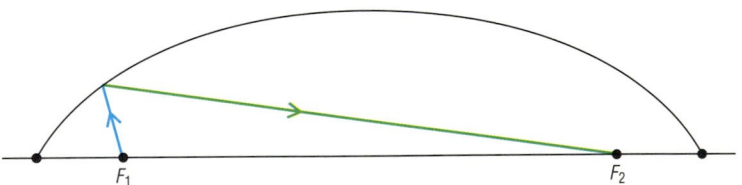

F_1 F_2

If a person is talking at one focus of the ellipsoidal room, the sound will bounce off the walls and ceiling of the room and travel to the other focus. The dispersed sound is so "focused" that even if the two foci are quite far apart, whispers from one focus can be easily heard at the other focus. For this reason, a room with this property is called a **whispering gallery**.

Below is a proof of this property of ellipses. The object of the proof is to show that if P is any point on an ellipse with foci F_1 and F_2, then $\overrightarrow{PF_1}$ and $\overrightarrow{PF_2}$ form equal angles with the line tangent to the ellipse at P. The proof requires one property of ellipses we have not discussed, namely that if k is the focal constant, then a point Q is outside the ellipse if and only if $QF_1 + QF_2 > k$. The tangent to the ellipse at P, because it only intersects the ellipse at one point, consists of such points Q that satisfy $QF_1 + QF_2 > k$ and also the single point P satisfying the defining property $PF_1 + PF_2 = k$.

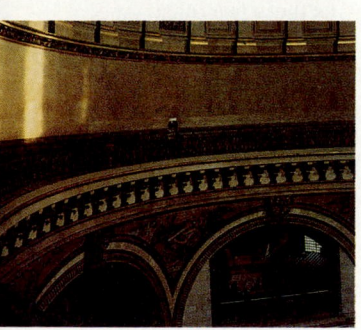

The Whispering Gallery, St. Paul's Cathedral, London

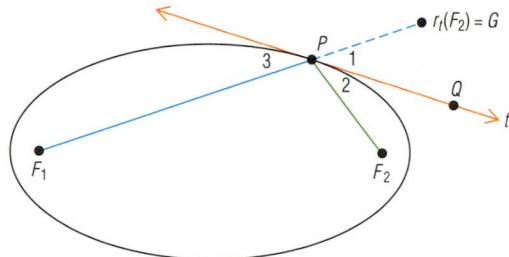

Proof Let line t be the tangent to the ellipse at P, and let $G = r_t\,(F_2)$. Now $PG = PF_2$, and $QG = QF_2$ for all Q on t, because reflections preserve distance. We wish to show that G, P, and F_1 are collinear. This will occur if $\overline{F_1G}$ intersects t at the point P. Suppose $\overline{F_1G}$ intersects t at some other point Q. Then,

$$F_1G = F_1Q + QG \qquad \text{betweenness}$$
$$= F_1Q + QF_2 \qquad \text{reflections preserve distance}$$
$$> k \qquad\qquad\quad \text{because } Q \text{ is outside the ellipse.}$$

But $F_1P + PF_2 = k$ from the definition of the ellipse, and thus $F_1P + PG = k$ because reflections preserve distance. So the shortest path from F_1 to G is through P, the point of tangency, and so P is on $\overline{F_1G}$. As a result, $\angle 1$ and $\angle 3$ are vertical angles, so $m\angle 1 = m\angle 3$. Because reflections preserve angle measure, $m\angle 1 = m\angle 2$. Consequently, $m\angle 2 = m\angle 3$ and so $\overrightarrow{PF_1}$ and $\overrightarrow{PF_2}$ make equal angles with the tangent to the ellipse.

A reflecting property for the parabola has more common uses than that for the ellipse. When a parabola is rotated about its axis, the three-dimensional surface formed is called a **paraboloid**. Any cross section of this paraboloid formed by a plane containing the axis is a parabola whose focus is the same as the focus of the original parabola; that point is also the focus of the paraboloid. If an object traveling parallel to the axis hits the "inside" of the paraboloid, then the object will be reflected through the focus of the paraboloid.

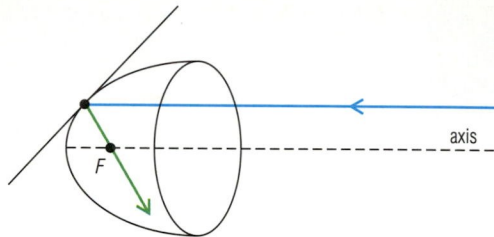

This property makes the paraboloid a particularly appropriate surface for collecting light or radio waves that come from a far distance; the waves are close to parallel, so after reflection they concentrate together at the focus where they can be collected. For this reason, satellite dishes to receive television or radio transmissions, and most telescope mirrors to collect light from stars, are parts of paraboloids.

Automobile headlights use this property in reverse. The light bulb is placed at the focus of a parabolic (actually, *paraboloidal*) mirror. All the light from the bulb that hits the surface of the mirror bounces off in a direction parallel to the axis of the paraboloid. Usually the only difference between "brights" and normal headlights in a car is that the parabolic mirror is turned so that its axis is parallel to the ground, and so the light travels without hitting the ground. A proof of this property of the parabola is not easy. You are asked only to apply the property in the questions.

Questions

Covering the Reading

In 1–3, a laser beam hits the mirror at point P. Trace the picture and draw the path of the beam after it hits the mirror.

1.

2.

3.

4.
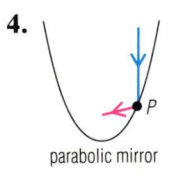

768

5. Find an equation for the tangent line to the circle $x^2 + y^2 = 25$ at the point $(3, 4)$. (Hint: The tangent is perpendicular to the radius at the point of tangency.) $y = -\frac{3}{4}x + \frac{25}{4}$

In 6 and 7, **a.** sketch an example and **b.** identify the three-dimensional figure obtained by each of the following.

6. rotating an ellipse 360° around the major axis
   a) See margin. b) ellipsoid

7. rotating a parabola around its axis of symmetry
   a) See margin. b) paraboloid

8. An equation for the tangent line t to a parent ellipse $\frac{x^2}{a^2} + \frac{y^2}{b^2} = 1$ at any point on it can be found using the following procedure.

Step 1: By a scale change, map the ellipse onto the unit circle, and let P' be the image of P under this scale change.

Step 2: Use the idea of Question 5 to find an equation for t', the tangent to the circle at P'.

Step 3: Use the inverse of the transformation used in Step 1 to find an equation for the original tangent.

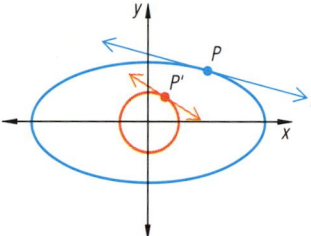

Use this procedure to find an equation for the tangent line t to the ellipse $\frac{x^2}{200} + \frac{x^2}{450} = 1$ at the point $(14, 3)$. $y = -\frac{21}{2}x + 150$

9. a. State a reflecting property of a paraboloid.   **See below.**
   b. Give a practical application of this property.   **sample: Satellite dishes use this property to collect incoming radio waves.**

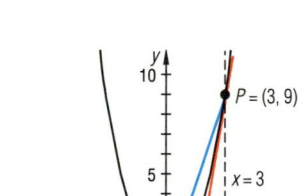

10. The focus F of the parabola $y = x^2$ is at the point $\left(0, \frac{1}{4}\right)$. It can be shown that the tangent to this parabola at the point $P = (3, 9)$, has equation $y = 6x - 9$. By drawing triangles and using trigonometry, show that $\overline{FP}$ and the vertical line $x = 3$ make equal angles with the tangent. (Hint: show that $m\angle TPQ = \frac{1}{2}m\angle FPQ$.)   **See margin.**

9a) If an object traveling parallel to the axis of a paraboloid hits the "inside" of the paraboloid, then the object will be reflected through the focus of the paraboloid.

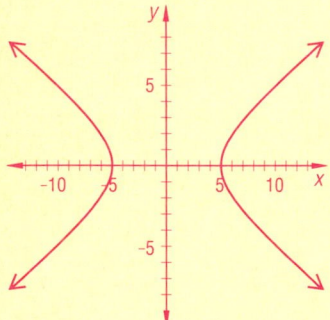
Review

In 11 and 12, name the conic section whose equation is given.   *(Lesson 12-5)*

11. $4x^2 + 5xy - 6y^2 + 12x + 36 = 0$ **hyperbola**

12. $x^2 + xy + y^2 + x + y - 5 = 0$ **ellipse**

13. Suppose the curve with equation $xy = -6$ is rotated $\frac{\pi}{4}$ about the origin. What is an equation for the image hyperbola?   *(Lesson 12-4)* $\frac{y^2}{2} - \frac{x^2}{2} = -6$

In 14 and 15, graph.   *(Lessons 12-2, 12-3)*

14. $\frac{x^2}{25} + \frac{y^2}{8} = 1$ **See margin.**   **15.** $\frac{x^2}{25} - \frac{y^2}{8} = 1$ **See margin.**

16. Name the conic section(s) with the following symmetry.
(Lessons 12-1, 12-3, Previous course)
   a. exactly one line of symmetry   **b.** at least two lines of symmetry
   c. 90° rotation symmetry   **a) parabola; b) ellipse, hyperbola c) circle**

17. The elliptical orbit of the moon has the center of the earth as one focus. If the closest distance of the moon to the center of the earth is 216 thousand miles and the largest distance to the center of the earth is 248 thousand miles, how far from the center of the earth is the second focus?
(Lesson 12-2)   **32,000 miles**

18. Find all complex solutions to $x^3 - 1000 = 0$.   *(Lesson 9-8)*
   $10, -5 + 5\sqrt{3}\,i, -5 - 5\sqrt{3}\,i$

In 19–22, *multiple choice*. Match each graph to its equation.   *(Lessons 5-6, 6-7)*
(a) $y = \tan x$ (b) $y = \cot x$ (c) $y = \sec x$ (d) $y = \csc x$

19.

(c)

20.

(d)

21.

(b)

22.

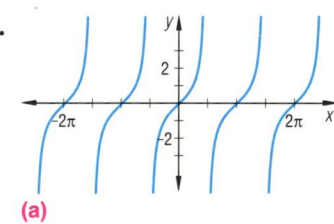

(a)

Exploration

23. There is a reflection property for the hyperbola. Find out what this is.
   If H is a hyperbola with foci F_1 and F_2, and P is any point on H, let ℓ be the tangent line to H through P, and let m be the line perpendicular to ℓ through P. Then, the angle between m and $\overline{PF_1}$ is equal to the angle between m and $\overline{PF_2}$.

770

Projects

1. The following methods always generate conic sections. For each of the following three methods, draw the curve, name the curve, and explain why the method always produces a certain conic section.
   a. Put a loop of string around two pins with some slack. With your pencil, make the loop taut and draw the curve of all points making the loop taut.
   b. Tie a pencil near the middle of a piece of string and pass the string around two pins pushed into a piece of paper. Using the pencil to keep the string taut, hold the two ends of the string together and draw them toward you around one of the pins, as shown below. Repeat by interchanging the roles of the two pins.

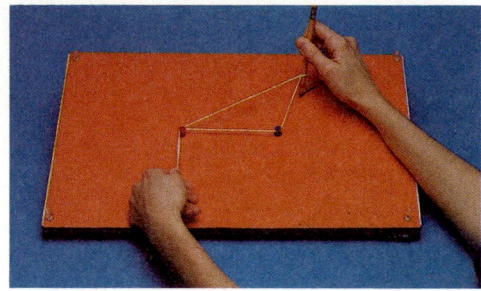

   c. Place a piece of string of length AB as shown below on a T-square. Place the T-square perpendicular to a line ℓ on the paper and place the ends of the string at B and at a point C on the paper. As the T-square slides on ℓ, draw the curve by keeping the pencil taut on the edge of the T-square.

2. Using wax paper, determine the curve that results from each of the following constructions.
   a. Draw a line ℓ and a point P not on the line. Fold and crease the paper so a point Q on the line coincides with P. Repeat at least a dozen times for different points Q. How does the distance between P and ℓ affect the resulting curve?
   b. Draw a circle and a point P in the interior of the circle (but not the center of it). Make different folds where a point on the circle coincides with P. How does the distance between P and the center of the circle affect the resulting curve?
   c. Draw a circle and a point P in the exterior of the circle. Make different folds where a point on the circle coincides with P. Discuss the effect of the distance between P and the center of the circle on the resulting curve.

3. Curves other than the conics can be described by *locus definitions,* as sets of all points which satisfy certain conditions. You may want tools such as a Spirograph™ for this project.
   a. The **cycloid** is the path of a point on a circle as the circle rolls along a line. A *prolate cycloid* is the path of a point on the interior of the circle as the circle rolls along the line, a *curtate cycloid* is the path of a point on the exterior of the circle as the circle rolls along the line. Draw examples of these three curves.
   b. Use the cycloid above to explain why, when a bicycle is in motion, the top of a bicycle wheel appears to be moving faster than the bottom.
   c. A **hypocycloid** is generated by a point of a circle rolling internally upon a given circle. Draw hypocycloids where the radius of the rotated circle is $\frac{1}{2}$ the given circle, $\frac{1}{3}$ the given circle (*deltoid*), and $\frac{1}{4}$ the given circle (*astroid*).

PROJECTS

The projects provide an opportunity for extended work on topics related to the content of the chapter. (See pages 66 and T37 for more information on the purposes of projects.) Some or all of these projects may be suitable for **small group work.**

Solutions to the projects are found at the end of the Additional Answers section at the back of this book.

Project 1 utilizes locus characterizations of the conic sections.

Project 2 determines *envelopes* for the conic sections, that is, sets of lines that are tangent to the particular conic being generated.

Project 5 should involve some manipulation of data.

The shape of the earth is sometimes described as an oblate spheroid, as mentioned in **Project 6.** That project might be appropriate for a group of students.

d. An **epicycloid** is generated by a point of a circle rolling externally upon a given circle. Draw epicycloids where the radius of the rotated circle is less than the given circle, equal to the given circle (*cardioid*), and greater than the given circle.

e. Experiment with curves generated by a point interior or exterior to a circle rolling internally or externally upon a given circle. Describe the procedure for the shape you like best.

4. As with the parabola, it is possible to define an ellipse and a hyperbola in terms of distances from a fixed point and a fixed line. If the curve has its major axis along the x-axis, then the distance from a point on the curve to the focus $(c, 0)$, divided by the distance from that point to the directrix line with equation $x = \dfrac{a^2}{c}$, is $\dfrac{c}{a}$.

a. Verify the previous statement for any point on the ellipse with equation $\dfrac{x^2}{a^2} + \dfrac{y^2}{b^2} = 1$, and any point on the hyperbola with equation $\dfrac{x^2}{a^2} - \dfrac{y^2}{b^2} = 1$.

b. The ratio $\dfrac{c}{a}$ is called the **eccentricity** of the conic and is denoted as e.
Multiple choice. Match the following equations with their curves.

i.	$0 \le e < 1$	(I)	ellipse
ii.	$e = 1$	(II)	hyperbola
iii.	$e > 1$	(III)	parabola

c. Both the ellipse and hyperbola have two foci and two *directrices*. Give an equation for both directrices.

d. Draw curves with increasing eccentricities such as .25, .5, .75, .9, 1.5, 2, 2.5, 3, 4, 5, and 10. Make some conjectures as to the relationship between the eccentricities and the shape of the curve.

e. What is the eccentricity of the most pleasing ellipse to your eye? Some people believe it to be the reciprocal of the golden ratio or the reciprocal of the number e. Do you agree?

5. a. Collect data on the elliptical orbits of the planets. Arrange the eccentricities (see Project 4) of the orbits of the planets from smallest to largest.

b. Find data on the elliptical orbits of comets or other celestial bodies. Does there appear to be any relation between the period of the comet (*i.e.*, how long it takes to complete one trip) and the eccentricity of its path? If so, what is it?

c. Can a heavenly body travel in a path that is parabolic? hyperbolic? Why or why not?

6. As the circle has three-dimensional analogues in the sphere and cylinder, the other conic sections have three-dimensional analogues as well. These are called quadric surfaces and can be expressed by the following general second degree equation in three variables,

$$Ax^2 + By^2 + Cz^2 + Dxy + Exz + Fyz + Gx + Hy + Iz + J = 0,$$

where the coefficients are real numbers.

a. For instance, an **ellipsoid** has as its general equation

$$\frac{x^2}{a^2} + \frac{y^2}{b^2} + \frac{z^2}{c^2} = 1.$$

If two of a, b, and c in the above equation are equal, then the surface is a **spheroid**. A *prolate spheroid*, somewhat like a football, is created by rotating an ellipse about its major axis. An *oblate spheroid*, somewhat like a doorknob, is created by rotating an ellipse about its minor axis. Draw some ellipsoids and spheroids. For instance, graph $\dfrac{x^2}{4} + \dfrac{y^2}{9} + z^2 = 1$, and graph both a prolate and an oblate spheroid from the ellipse $\dfrac{x^2}{4} + \dfrac{y^2}{25} = 1$. If you have access to computer software that does three-dimensional graphics, compare your plots to those produced by computer.

b. Other quadric surfaces have the following general equations.

hyperboloid of one sheet:

$$\frac{x^2}{a^2} + \frac{y^2}{b^2} - \frac{z^2}{c^2} = 1$$

hyperboloid of two sheets:

$$-\frac{x^2}{a^2} - \frac{y^2}{b^2} - \frac{z^2}{c^2} = 1$$

elliptic paraboloid:

$$z = \frac{x^2}{a^2} + \frac{y^2}{b^2}$$

hyperbolic paraboloid (the "saddle-shaped" surface):

$$z = \frac{y^2}{b^2} - \frac{x^2}{a^2}$$

Graph examples of each of these curves, for instance,

$$\frac{x^2}{9} + \frac{y^2}{4} - \frac{z^2}{16} = 1,$$

$$-\frac{x^2}{4} - y^2 + \frac{z^2}{9} = 1,$$

$$z = \frac{x^2}{4} + \frac{y^2}{25}, \text{ and}$$

$$z = \frac{y^2}{16} - \frac{x^2}{4}.$$

c. Many of the quadric surfaces can be approximated by string models.

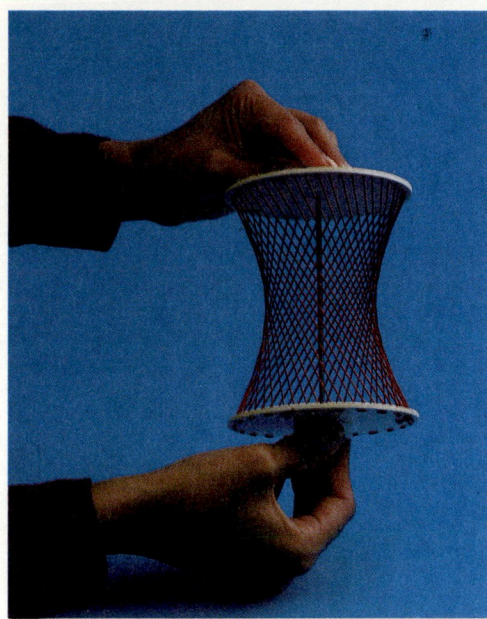

The hyperboloid of one sheet pictured above is a surface which is *doubly ruled,* meaning that through each point on the surface there are two lines which lie entirely on the surface! Use string to build the above model and one quadric surface not pictured here. Explain how the string model above illustrates this property.

Summary

When one of two intersecting lines is rotated about the other, an unbounded cone of two nappes is formed. The sections of this cone lead to the three primary conic sections: the parabola (which was studied earlier), the ellipse, and the hyperbola.

Both the ellipse and the hyperbola are defined by their foci, two fixed points in the plane. An ellipse is the set of all points where the sum of the distances to each of the points is the focal constant. A hyperbola is the set of all points where the absolute value of the difference of the distances to each of the points is the focal constant. Relating these definitions to the description of the conic sections as intersections of a plane with a cone is elegantly done by the use of Dandelin spheres.

Both the ellipse and hyperbola have two symmetry lines. These contain the major axis and the minor axis of the curves. If these symmetry lines are the x- and y-axes, then the ellipse can be described by the standard form equation $\frac{x^2}{a^2} + \frac{y^2}{b^2} = 1$, and the hyperbola can be described by the standard form equation $\frac{x^2}{a^2} - \frac{y^2}{b^2} = 1$ or $\frac{y^2}{a^2} - \frac{x^2}{b^2} = 1$. Translation images can be described by applying the Graph Translation Theorem.

When neither symmetry line is horizontal, the conics are rotation images of an ellipse or hyperbola in standard form. If the center of the rotation is the origin, then the Graph Rotation Theorem can be applied to find an equation for the conic.

The general quadratic equation

$$Ax^2 + Bxy + Cy^2 + Dx + Ey + F = 0$$

gives the equation for all conic sections as well as the degenerate conic sections which are the empty set, a point, a line, or two intersecting lines. The quantity $B^2 - 4AC$ tells you the possible shapes of the curve. By using the equations $\frac{B}{C-A} = \tan 2\theta$ or $B \cos 2\theta = 0$, the amount of rotation of the curve can be determined.

The reflection property of an ellipse—that a path through one focus will reflect off the ellipse through the other focus—is used in constructing whispering galleries. The reflection property of a parabola—that a path parallel to the axis will reflect off the parabola through the focus—is used in constructing satellite dishes and headlights. Systems of hyperbolas are used in locating and tracking objects that emit sound waves.

Vocabulary

For the starred (*) items you should be able to give a definition of the term.
For the other terms you should be able to give a general description and a specific example of each.

Lesson 12-1
* conic section
quadratic relation
axis, vertex, nappes of a cone
cross section, section
* ellipse
focus, foci of an ellipse
focal constant for an ellipse
astronomical units (a.u.)

Lesson 12-2
* standard form equation for an
  ellipse
major axis, minor axis
center of an ellipse
horizontal axis, vertical axis
eccentricity

Lesson 12-3
* hyperbola
foci, focal constant of a hyperbola
* standard form equation for a
  hyperbola
asymptotes
vertices, axes
center of a hyperbola

Lesson 12-4
Graph Rotation Theorem
rectangular hyperbola
isometry

Lesson 12-5
standard form of a quadratic relation
  in two variables
degenerate conic sections

Lesson 12-6
angles of incidence, reflection
ellipsoid
whispering gallery
paraboloid

F_1 F_2

Progress Self-Test

See margin for answers not shown below.

Take this test as you would take a test in class. You will need graph paper, a calculator, and an automatic grapher. Then check the test yourself using the solutions at the back of the book.

1. How is the ''cone'' which generates conic sections generated?

In 2–4, graph.

2. $\dfrac{x^2}{25} + \dfrac{y^2}{81} = 1$

3. $\dfrac{x^2}{9} - y^2 = 1$

4. $\dfrac{(x-2)^2}{36} + \dfrac{(y+1)^2}{20} = 1$

5. Find an equation for the hyperbola with foci $(7, 0)$ and $(-7, 0)$ and focal constant 10.

6. Draw an ellipse in which the distance between the foci is 8 cm and the focal constant is 10 cm.

7. Describe the symmetry lines of a hyperbola.

8. Give equations for the asymptotes of the hyperbola $\dfrac{x^2}{81} - \dfrac{y^2}{100} = 1$. $y = \frac{10}{9}x, y = -\frac{10}{9}x$

1) The cone is formed by rotating one of two intersecting, non-perpendicular lines about the other.

5) $\dfrac{x^2}{25} - \dfrac{y^2}{24} = 1$

7) The symmetry lines of a hyperbola are the axes of the hyperbola. One axis is the line through the foci. The other is the perpendicular bisector of the segment joining the foci.

9. Find an equation for the image of $y = x^2 - 3x - 4$ under $R_{\pi/3}$.

10. Tell whether or not $\dfrac{y^2}{30} - \dfrac{x^2}{4} = 1$ is a rectangular hyperbola. Explain your answer.

In 11 and 12, describe the graph of the relation represented by the given equation.

11. $2x^2 - 6xy + 18y^2 - 14x + 6y - 110 = 0$ **ellipse**

12. $9x^2 - 2xy + y^2 - 4x + 22 = 0$ **the null set**

13. Rewrite $\dfrac{x^2}{9} - \dfrac{y^2}{64} = 1$ in quadratic standard form and give values of $A, B, C, D, E,$ and F for that equation.

14. The elliptical orbit of the planet Jupiter has the center of the sun as one focus. If the closest distance to the center of the sun is 460 million miles and the farthest distance is 508 million miles, how far from the center of the sun is the second focus?   **48 million miles**

Voyager 1 photo of Jupiter

9) sample: $x^2 + 2\sqrt{3}xy + 3y^2 + 2(\sqrt{3}-3)x - 2(3\sqrt{3}+1)y = 16$

10) No; the asymptotes $y = \dfrac{\sqrt{30}}{2}x$ and $y = -\dfrac{\sqrt{30}}{2}x$ are not perpendicular.

13) sample: $64x^2 - 9y^2 - 576 = 0$;
$A = 64, B = 0, C = -9, D = 0, E = 0, F = -576$

The Progress Self-Test provides an opportunity for feedback and correction. The Chapter Review provides additional opportunities for practice.

USING THE PROGRESS SELF-TEST
Assign the Progress Self-Test as a one-night assignment. (See pages 69 and T35 for more information.)

ADDITIONAL ANSWERS
2.

3.

4.

6. See margin, page 774.

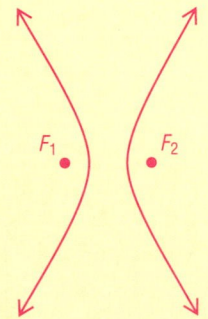
Chapter Review

Questions on **SPUR** Objectives

SPUR stands for **S**kills, **P**roperties, **U**ses, and **R**epresentations.
The Chapter Review questions are grouped according to the SPUR Objectives for this chapter.
See margin for answers not shown below.

SKILLS deal with the procedures used to get answers.

■ **Objective A:** *Use properties of ellipses and hyperbolas to write equations describing them.* (Lessons 12-2, 12-3)

1. Find an equation for the ellipse with foci $(0, 15)$ and $(0, -15)$ and focal constant 34.

2. Find an equation for the hyperbola with foci $(4, 0)$ and $(-4, 0)$ and focal constant 6.

3. Give an equation for the ellipse whose center is the origin, horizontal axis has length 18, and vertical axis has length 7.

4. Give equations for the asymptotes of the hyperbola $\dfrac{x^2}{16} - \dfrac{y^2}{4} = 1$. $y = \frac{1}{2}x, y = -\frac{1}{2}x$

■ **Objective B:** *Find equations for rotation images of figures.* (Lesson 12-4)

In 5–8, find an equation for the image of the given figure under the given rotation.

5. $\dfrac{x^2}{9} + \dfrac{y^2}{36} = 1; R_{45°}$ $5x^2 + 6xy + 5y^2 = 72$

6. $9x^2 - y^2 = 1; R_{\pi/3}$ $\frac{3}{2}x^2 - 5\sqrt{3}xy + \frac{13}{2}y^2 = 1$

7. $y = x^2 - 8; R_{\pi/3}$

8. $x^2 + y^2 = 20; R_{55°}$ $x^2 + y^2 = 20$

1) $\dfrac{x^2}{64} + \dfrac{y^2}{289} = 1$

2) $\dfrac{x^2}{9} - \dfrac{y^2}{7} = 1$

3) $\dfrac{x^2}{81} + \dfrac{4y^2}{49} = 1$

7) $-\frac{3}{4}x^2 - \frac{\sqrt{3}}{2}xy - \frac{1}{4}y^2 + \frac{\sqrt{3}}{2}y - \frac{1}{2}x + 8 = 0$

9. Suppose the hyperbola $xy = 120$ is rotated $\frac{\pi}{4}$ about the origin. What is an equation for the image hyperbola? $-\frac{1}{2}x^2 + \frac{1}{2}y^2 = 120$

10. Find an equation for the image of the parabola $y = (x + 4)^2 - 5$ when rotated 60° about the point $(-4, 5)$.

■ **Objective C:** *Rewrite equations of conic sections in standard form of a quadratic relation in two variables.* (Lesson 12-5)

In 11–14, rewrite the equation in the standard form of a quadratic relation in two variables and give values of $A, B, C, D, E,$ and F for the equation.

11. $\dfrac{x^2}{25} + \dfrac{y^2}{49} = 1$

12. $(x + 3)^2 + (y - 8)^2 = 35$

13. $7xy = 8$

14. $\dfrac{(x - 2)^2}{4} - y^2 = 1$

10) $-x^2 - 2\sqrt{3}xy - 3y^2 + (8\sqrt{3} - 8)x + (32 - 8\sqrt{3})y + 32\sqrt{3} - 101 = 0$

11) $49x^2 + 25y^2 - 1225 = 0$;
    $A = 49, B = 0, C = 25, D = 0, E = 0, F = -1225$

12) $x^2 + y^2 + 6x - 16y + 38 = 0$;
    $A = 1, B = 0, C = 1, D = 6, E = -16, F = 38$

13) $7xy - 8 = 0$;
    $A = 0, B = 7, C = 0, D = 0, E = 0, F = -8$

14) $x^2 - 4y^2 - 4x = 0$;
    $A = 1, B = 0, C = -4, D = -4, E = 0, F = 0$

PROPERTIES deal with the principles behind the mathematics.

■ **Objective D:** *State and apply properties of ellipses and hyperbolas to draw and describe them.* (Lessons 12-1, 12-3, 12-4)

15. Draw an ellipse in which the distance between the foci is 6 cm and the focal constant is 8 cm.

16. Draw a hyperbola in which the distance between the foci is 2 inches and the focal constant is 1 inch.

17. How many symmetry lines does an ellipse have?

18. Prove that the perpendicular bisector of the segment joining the foci of a hyperbola is a symmetry line for the hyperbola.

In 19 and 20, tell whether or not the hyperbola is a rectangular hyperbola.

19. $\dfrac{x^2}{6} - \dfrac{y^2}{12} = 1$ **No**

20. $5xy = 80$ **Yes**

17) at least 2

■ **Objective E:** *Describe the intersections of a plane and a cone of 2 nappes.* (Lessons 12-1, 12-5)

21. How is the "cone" in conic sections generated?

22. Draw a plane intersecting a cone where the section is a hyperbola.

23. Name the three degenerate conic sections.

24. *True* or *false*.   A plane must intersect a given cone.   **True**

21) The cone is formed by rotating one of two intersecting, non-perpendicular lines about the other.

23) a point, a line, two intersecting lines

USES deal with applications of mathematics in real situations.

■ **Objective F:** *Determine information about elliptical orbits.* (Lesson 12-2)

25. The elliptical orbit of the planet Mercury has the center of the sun as one focus. If the closest distance to the center of the sun is 28,440,000 miles and the farthest distance is 43,560,000 miles, how far from the center of the sun is the second focus?   **15,120,000 miles**

26. The closest distance the planet Mars gets to the center of the sun is 129 million miles and its largest distance is 154 million miles. Is the second focus of Mars's orbit nearer or farther from the center of the sun than the planet Mercury (refer to Question 25)?   **closer**

22.

27.

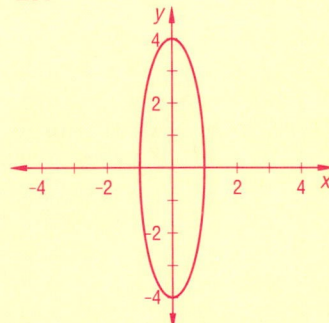

REPRESENTATIONS deal with pictures, graphs, or objects that illustrate concepts.

■ **Objective G:** *Graph or identify graphs of ellipses and hyperbolas with equations in their standard form.* (Lessons 12-2, 12-3)

In 27–32, graph.

27. $x^2 + \dfrac{y^2}{16} = 1$

28. $x^2 - \dfrac{y^2}{16} = 1$

29. $\dfrac{y^2}{40} - \dfrac{x^2}{9} = 1$

30. $\dfrac{x^2}{49} + \dfrac{y^2}{20} = 1$

31. $y^2 - x^2 = 1$

32. $\dfrac{x^2}{100} - \dfrac{y^2}{9} = 1$

28.

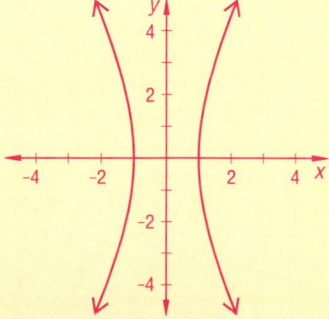

29–32. See margin, page 778.

In 33 and 34, give equations for the following ellipses.

33.

34.

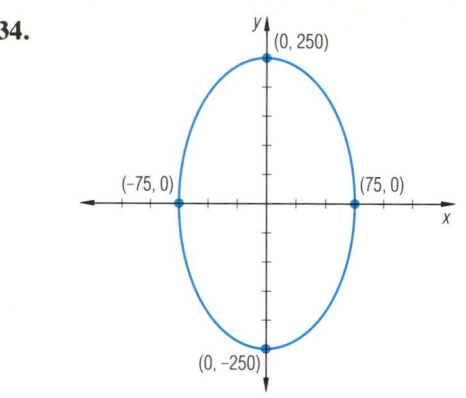

33) $\dfrac{x^2}{64} + \dfrac{y^2}{16} = 1$

34) $\dfrac{x^2}{5625} + \dfrac{y^2}{62,500} = 1$

■ **Objective H:** *Graph transformation images of ellipses and hyperbolas.* (Lessons 12-2, 12-3, 12-5)

In 35 and 36, graph.

35. $\dfrac{(x-4)^2}{25} + \dfrac{(y+2)^2}{4} = 1$

36. $\dfrac{(x+8)^2}{16} - (y-3)^2 = 1$

37. Graph $xy = 20$.

38. Graph $x^2 - xy = 8$ by solving for y.

■ **Objective I:** *Describe graphs of quadratic equations.* (Lesson 12-5)

In 39–42, describe the graph of the relation represented by the given equation.

39. $8x^2 + 3y^2 - 12x + 6y + 18 = 0$ **the null set**

40. $8x + 8y - 4xy = -12$ **hyperbola**

41. $x^2 + 4xy + 4y^2 + 8x + 2y + 9 = 0$ **parabola**

42. $3y^2 - 4y = 2x^2 + 9x - 3$ **hyperbola**

778

CHAPTER 13 ■ ADVANCED TOPICS IN TRIGONOMETRY

OBJECTIVES ■ CHAPTER 13

The objectives listed here are the same as in the Chapter 13 Review on pages 831–833 of the student text. The Progress Self-Test on page 830 and the tests in the Teacher's Resource File cover these objectives. For recommendations regarding the handling of this end-of-chapter material, see the notes in the margins on the corresponding pages of this Teacher's Edition.

OBJECTIVES FOR CHAPTER 13 (Organized into the SPUR categories— Skills, Properties, Uses, and Representations)	Progress Self-Test Questions	Chapter Review Questions	Teacher's Resource File	
			Lesson Masters*	Chapter Test Forms A & B
SKILLS				
A Perform operations with complex numbers.	12, 13	1–6	13-6	11
B Represent complex numbers in different forms.	8	7–12	13-5, 13-6	9, 10
PROPERTIES				
C Prove trigonometric identities.	3b, 14	13–16	13-1, 13-2	1, 3
D Describe singularities of functions.	1, 3a	17–20	13-2	2
E Find powers and roots of complex numbers.	9, 10	21–24	13-7	12
F Use an automatic grapher to test a proposed identity.	2	25–28	13-1, 13-2	7
G Given polar coordinates of a point, determine its rectangular coordinates, and vice versa.	7	29–34	13-3	4, 6
USES				
None				
REPRESENTATIONS				
H Plot points in the polar coordinate system.	4	35–38	13-3, 13-4	13
I Graph and interpret graphs of polar equations.	11, 15	39–44	13-4	8
J Graph complex numbers.	5, 6	45–48	13-5, 13-6	5

*The masters are numbered to match the lessons, and each is intended for use with that lesson.

OVERVIEW ■ CHAPTER 13

This chapter is necessary to provide a complete course in trigonometry. It covers the proofs of trigonometric identities, the use of trigonometry in polar coordinates and polar graphs, and the relationships between trigonometry and complex numbers, culminating in DeMoivre's Theorem and its applications. All of these topics are covered in detail in UCSMP *Precalculus and Discrete Mathematics*, so if students will be taking that course next, this chapter can be omitted.

In Lessons 13-1 and 13-2, students examine trigonometric and other identities. An automatic grapher is employed to determine if an equation appears to be an identity. If it appears to be an identity, then an algebraic proof is worth pursuing; if not, then students are instructed to find a counterexample. In Lesson 13-1, the domain of the identity or potential identity is given. In Lesson 13-2, students are expected to determine the domain.

The topic of Lessons 13-3 and 13-4 is the polar coordinate system. In the former, students apply their knowledge of trigonometry to plot points in the polar plane and convert between the rectangular and polar systems. In the latter, students graph simple polar equations and verify their shapes by rewriting the equations in rectangular form. A function grapher with a polar graphing option is a very useful tool in these lessons.

Lessons 13-5 through 13-7 present properties and graphs of complex numbers, and develop theorems to aid in calculations with complex numbers. In 13-5, complex numbers and the sum of two complex numbers are graphed. In Lesson 13-6, complex numbers are rewritten in polar form and in trigonometric form, and the product of two complex numbers can be expressed in either form. In Lesson 13-7, students generalize the product of two complex numbers to obtain DeMoivre's Theorem for calculating powers of a complex number. This theorem is then used to generalize a theorem for calculating roots of complex numbers.

Lesson 13-8 presents a beautiful and recent (since 1980) application of complex numbers and computers, in describing what is now known as the Mandelbrot Set. It combines algebra, geometry, and sequences in a fitting way to end this book.

PERSPECTIVES ■ CHAPTER 13

The Perspectives provide the rationales for the inclusion of topics or approaches, give some mathematical background, and make connections with other lessons and to other UCSMP texts.

13-1
PROVING TRIGONOMETRIC IDENTITIES
There is no special logic to finding or proving trigonometric identities. One objective in this lesson is to determine whether an equation of the form $f(x) = g(x)$, where f and g may involve trigonometric functions, is true for all x in the domains of f and g. The technique is to examine the graphs of $y = f(x)$ and $y = g(x)$ over a reasonably wide domain. If the graphs are identical, then one can proceed to attempt to prove that the equation is an identity.

13-2
RESTRICTIONS ON IDENTITIES
This lesson examines the domain for which an identity is true. Many identities are not true for all real numbers. Even $\frac{\sin x}{\cos x} = \tan x$ is not true for all x. Determining the values for which an identity $f(x) = g(x)$ is not true often is equivalent to finding the singularities of the functions f and g. The functions $f(x) = \frac{\sin x}{\cos x}$ and $g(x) = \tan x$ have singularities at each value of x such that $\cos x = 0$. Algebra needs to be used to find these singularities because they can be impossible to locate in graphs.

13-3
POLAR COORDINATES
We study polar coordinates to expand the horizons of students regarding coordinate systems, so that students will be able to consider other systems later. Also, they afford a very useful representation of complex numbers, and certain relations graphed in this system are enjoyable and pleasing to see.

In this lesson students graph points and convert from one coordinate system to the other. They also graph the basic polar equations $r = k$ for $k > 0$.

The use of a polar reference system is quite old. It is the basis of a compass used for orientation and map making. It is the basis of radar and satellite communication and detection systems, navigation procedures, and astronomical representations. In general, a polar reference system determines position in a plane based on distance from an origin at some angle to a fixed ray (the polar axis).

13-4
POLAR GRAPHS
The mechanics of graphing in polar coordinates are nearly the same as graphing in rectangular coordinates. The differences are that the variables are customarily r and θ rather than x and y and that different values of r and θ may represent the same point.

The differences in the shapes of the graphs are more substantial, however. This lesson provides only an introduction to such graphs, concentrating on the circles with equations of the form $r = a \cos \theta$ and $r = a \sin \theta$ and the rose curves of the form $r = a \sin b\theta$ and $r = a \cos b\theta$, though other curves are considered in the questions.

13-5
THE GEOMETRY OF COMPLEX NUMBERS
To graph a point (x, y) in rectangular coordinates, we think of "going over x and up y." This process is similar to applying the translation $T_{x,y}$ to the origin. To add two complex numbers (a, b) and (c, d), we can think of applying the translation $T_{c,d}$ to the first point. That is, we move the first point over c and up d. The result is the sum $(a + c, b + d)$. This addition is the same as vector addition of the vectors (a, b) and (c, d). The two addends, the sum, and the origin are vertices of a parallelogram.

13-6
TRIGONOMETRIC FORM OF COMPLEX NUMBERS
We often wish to both add and multiply complex numbers, and we want to do this without having to convert between polar form and rectangular form. The trigonometric form $r(\cos \theta + i \sin \theta) = [r, \theta]$ resolves this difficulty. It exhibits r and θ and so the polar form ideas can be used, but it is very much like $a + bi$ form so the rectangular coordinate ideas can be used.

This lesson begins by translating numbers from $a + bi$ form to trigonometric form and back. Then multiplication and division are pictured and described in this form.

13-7
DeMOIVRE'S THEOREM
In this lesson, DeMoivre's Theorem provides an easy way to calculate powers of complex numbers, but only if they are written in polar or trigonometric form. It is far more efficient than applying the Binomial Theorem to calculate powers of complex numbers. The theorem is presented by appealing to student's intuitions after analyzing the pattern of powers of one particular complex number. The proof of DeMoivre's Theorem requires mathematical induction, and is found in UCSMP *Precalculus and Discrete Mathematics*. DeMoivre's Theorem provides a formula for the nth power of any complex number. When the nth powers are graphed, they lie on a spiral.

The beauty of the geometry of complex numbers extends also to the nth roots of a number. If $[r, \theta]^n = z$, then so also does $\left[r, \theta + \dfrac{2\pi}{n}\right]^n = z$. Therefore, if $[r, \theta]$ is an nth root of z, another nth root is the image of the first under a rotation of $\dfrac{2\pi}{n}$ about the origin. This process can be continued with $n - 1$ successive rotations finding all n nth roots, and so the nth roots are vertices of a regular n-gon centered at the origin. It is a wonderful result to have at the close of the year!

13-8
THE MANDELBROT SET
The mathematics of dynamical systems was first studied early in this century by French mathematicians. But only in recent years has the subject taken off, because the computer has afforded the opportunity to analyze such systems in great detail. The Mandelbrot Set is one of the important sets in this field, and its popularity is probably due more to the intriguing visual nature of the set as to its algebraic properties. The purpose of this lesson is to explain how the Mandelbrot Set is constructed and to give pictures of this set in various windows in the complex plane. A computer program is given by which students can generate graphs themselves.

DAILY PACING CHART ■ CHAPTER 13

For optimal student performance, the Progress Self-Test and the Chapter Review should be covered. (See *General Teaching Suggestions: Mastery* on page T35 of this Teacher's Edition.) By following the pace of the Full Course given here, students can complete 12 or 13 chapters of the book by the end of the year. Students following the pace of the Minimal Course spend more time to complete lessons that are combined with quizzes and to consider the Chapter Review. When all lessons are covered in order (the recommendation of the authors), students in the Minimal Course can cover all but the last two chapters. For more information on pacing, see *General Teaching Suggestions: Pace* on page T34 of this Teacher's Edition.

DAY	MINIMAL COURSE	FULL COURSE
1	13-1	13-1
2	13-2	13-2
3	13-3	13-3
4	Quiz (TRF); Start 13-4	Quiz (TRF); 13-4
5	Finish 13-4	13-5
6	13-5	13-6
7	13-6	Quiz (TRF); 13-7
8	Quiz (TRF); Start 13-7	13-8
9	Finish 13-7	Progress Self-Test
10	13-8	Chapter Review
11	Progress Self-Test	Chapter Test (TRF)
12	Chapter Review	
13	Chapter Review	
14	Chapter Test (TRF)	

TESTING OPTIONS

■ Quiz for Lessons 13-1 through 13-3   ■ Chapter 13 Test, Form A   ■ Chapter 13 Test, Cumulative Form
■ Quiz for Lessons 13-4 through 13-6   ■ Chapter 13 Test, Form B   ■ Comprehensive Test, Chapters 1–13

A Quiz and Test Writer is available for generating additional questions, additional quizzes, and additional forms of the Chapter Test.

PROVIDING FOR INDIVIDUAL DIFFERENCES

The student text has been written for, and tested with, average students. It has also been used successfully with students who are better prepared as well as with students who are poorly prepared.

The Lesson Notes often include Error Analysis and Alternate Approach features to help you assist those students who need more help. Students of all abilities often learn from their peers and may benefit from small group work referenced as appropriate throughout the Notes. A blackline Lesson Master (in the Teacher's Resource File), keyed to the chapter objectives, is available for each lesson to provide more practice. (However, we advise you to use these masters sparingly; again refer to the suggestions on pace on page T34.)

Extension activities are provided in the Lesson Notes for use with students who need to be challenged even more than the requirements of completing the Full Course provide.

Further Work with Trigonometry

evening primrose

trillium

clematis

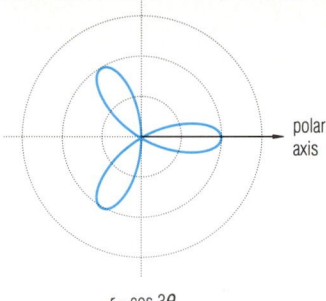

polar axis

$r = \cos 2\theta$

polar axis

$r = \cos 3\theta$

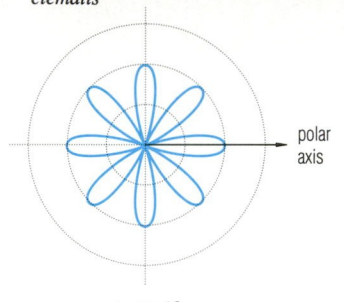

polar axis

$r = \cos 4\theta$

This chapter covers three topics in trigonometry. First are identities involving the six circular functions—sine, cosine, tangent, secant, cosecant, and cotangent.

Second is a coordinate system different from rectangular coordinates, the *polar coordinate system*. When plotted with polar coordinates, the graphs of trigonometric functions often are quite beautiful. Patterns like those above are examples.

Third is the representation of complex numbers in both the rectangular and polar coordinate systems. Operations with complex numbers can be expressed simply and elegantly in these systems, and the trigonometric functions play a major role.

CHAPTER 13

This chapter should take 8–10 days for the lessons and quizzes, 1 day for the Progress Self-Test, 1 or 2 days for the Chapter Review, and 1 day for a Chapter Test. (See Daily Pacing Chart on page 779D.)

USING PAGE 779
The three graphs on this page are examples of rose curves. Students will study them in Lesson 13-4. They strike many people as quite beautiful. There is much beautiful mathematics in this chapter, ranging from the relationships among the trigonometric functions to the graphs of the nth powers and nth roots of a complex number.

Alternative Assessment
You might ask students: If you had to pick a piece of mathematics that you have studied as being beautiful, what would you pick? Note that beauty does not have to be geometric beauty; it could be the beauty of a logical argument or the beauty in the way ideas fit together.

RESOURCES
■ Lesson Master 13-1

OBJECTIVES

C Prove trigonometric identities.

F Use an automatic grapher to test a proposed identity.

TEACHING NOTES

An important objective of the lesson is for students to realize that a graph does *not* prove or disprove the identity. It points out the reasonableness of constructing a proof in the former case, and indicates counterexamples in the latter. Bring differences in graphs due to round-off error to the attention of the class as evidence that algebraic proof is necessary.

Technology There are two distinct identical graphs at the top of page 780 because of the limitations of having to write on paper. Students using an automatic grapher should graph one function on top of the other, and see dynamically that the second function does not appear to get graphed at all. If any new points are added and the identity really is true, then the computer has displayed round-off error.

The proofs in **Examples 2– 4** illustrate three distinct approaches to proving identities. In **Example 2,** one side of the identity is manipulated until the other side is obtained. In **Example 3** we manipulate both sides of the identity until they are identical. In **Example 4,** a known identity, the Pythagorean Identity, is manipulated to generate the desired identity.

Proving Trigonometric Identities

In this lesson you will learn various ways to derive and prove trigonometric identities. Although the proof techniques will be illustrated using only trigonometric functions, they are applicable to any function.

You have seen many instances where graphs are used to decide whether or not a particular statement is true. In particular, if you want to see whether an equation in one variable is an identity, you can consider each side of the equation as a separate function of x, and graph the two functions. If the graphs coincide, the original equation is likely to be an identity.

As an example, consider the Pythagorean Identity,
$$\cos^2 x + \sin^2 x = 1.$$
If each side of this equation is considered as a separate function of x, specifically if $f(x) = \cos^2 x + \sin^2 x$ and $g(x) = 1$, the following graphs result.

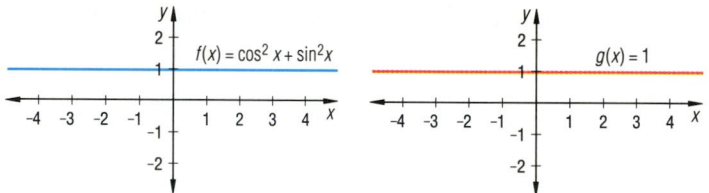

As expected, the graphs are identical. Had they been plotted on the same set of axes, the graphs would have coincided.

In contrast, $(\cos x - \sin x)^2 = \cos^2 x - \sin^2 x$ is not an identity. The graphs of $f(x) = (\cos x - \sin x)^2$ and $g(x) = \cos^2 x - \sin^2 x$ are shown below.

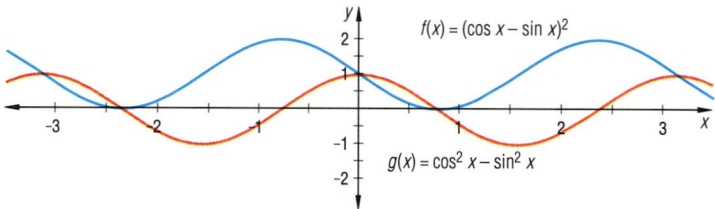

It is apparent that f and g are not the same function, so the given equation is not an identity.

Notice that the graph of g appears to be a cosine function with amplitude 1 and period π. This suggests that an expression identically equal to $\cos^2 x - \sin^2 x$ is $\cos 2x$. This agrees with what you proved in Lesson 11-6: $\cos 2x = \cos^2 x - \sin^2 x$ for all x.

780

Example 1 Find an expression, involving a single circular function, equal to $(\cos x - \sin x)^2$.

Solution Expand the expression.
$$(\cos x - \sin x)^2 = \cos^2 x - 2 \cos x \sin x + \sin^2 x$$
$$= \cos^2 x + \sin^2 x - 2 \cos x \sin x$$
Recall that $\cos^2 x + \sin^2 x = 1$, and $2 \cos x \sin x = \sin 2x$. So,
$$(\cos x - \sin x)^2 = 1 - \sin 2x.$$

Check Graph $f(x) = (\cos x - \sin x)^2$ and $g(x) = 1 - \sin 2x$. Check that the graphs coincide. As shown below, they do.

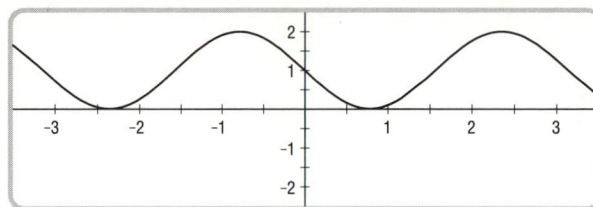

In Example 1, one expression was manipulated to equal a second expression. This technique can be employed to prove that an equation is an identity. Start with one side of a proposed identity, and rewrite it using definitions, known identities, or algebraic properties, until it equals the other side.

Example 2 Prove that $1 + \tan^2 x = \sec^2 x$, for all x in the domains of these functions.

Solution Substitute into the left expression and rewrite until you get the right expression. For all x with $\cos x \neq 0$,

$1 + \tan^2 x = 1 + \dfrac{\sin^2 x}{\cos^2 x}$	Definition of tangent
$= \dfrac{\cos^2 x}{\cos^2 x} + \dfrac{\sin^2 x}{\cos^2 x}$	Forming a common denominator
$= \dfrac{\cos^2 x + \sin^2 x}{\cos^2 x}$	Adding fractions with a common denominator
$= \dfrac{1}{\cos^2 x}$	Pythagorean Identity
$= \sec^2 x$	Definition of secant

So $1 + \tan^2 x = \sec^2 x$ for all x in the domain of $\tan x$ and $\sec x$.

Check Let $f(x) = 1 + \tan^2 x$ and $g(x) = \sec^2 x$. Below, each function is shown on a separate set of axes. In practice, you should graph both on the same set of axes.

Because the graphs appear to be identical, it checks.

LESSON 13-1 *Proving Trigonometric Identities* **781**

All of these methods are legitimate proof techniques and in our opinion, none of these methods should be discounted in favor of another.

In all the Examples, identities can be deduced without students worrying about singularities, covered in Lesson 13-2.

ADDITIONAL EXAMPLES
1. Find an expression involving a single circular function equal to tan x csc x for all values for which this expression is defined.
sec x

2. Prove that for all x,
$$\cos \left(x + \frac{\pi}{4} \right) = \frac{\sqrt{2}}{2} \, (\cos x - \sin x).$$

cos $\left(x + \dfrac{\pi}{4} \right)$ = cos x ·
cos $\dfrac{\pi}{4}$ – sin x · sin $\dfrac{\pi}{4}$ =
$\dfrac{\sqrt{2}}{2}$ cos x – $\dfrac{\sqrt{2}}{2}$ sin x =
$\dfrac{\sqrt{2}}{2}$ (cos x – sin x)

3. Prove that for all x, $\cos^2 x - \cos^4 x = \sin^2 x - \sin^4 x$.
cos² x – cos⁴ x =
cos² x(1 – cos² x) =
cos² x(sin² x); sin² x –
sin⁴ x = sin² x(1 – sin² x) =
sin² x(cos² x);
So cos² x – cos⁴ x =
sin² x – sin⁴ x.

4. Prove that $\dfrac{\sin 2x}{\sin x} = 2 \cos x$ for all x for which $\sin x \neq 0$.
Begin with sin 2x =
2 cos x sin x; divide both
sides by sin x to obtain
the desired result.

NOTES ON QUESTIONS
Question 5: A related identity is cot x · sec x · csc $x = \dfrac{1}{\sin^2 x}$.

Making Connections for Questions 7 and 8: Students are asked to explore more identities like these in **Question 25.** You might use this as a lead-in to the next lesson.

Question 9: Simplifying the right side to equal the left is probably easiest.

A second proof technique is to rewrite each side of a proposed identity independently until equal expressions are obtained. When using this technique, because you cannot be sure that the proposed identity is true until you have finished, you should not write an equal sign between the two sides until the end. We draw a vertical line between the two sides as a reminder of this restriction. Thus, a proof that A = B is an identity based on this technique has the following form.

$$
\begin{array}{c|c}
A & B \\
= \ldots & = \ldots \\
= \ldots & = \ldots \\
= E & = E
\end{array}
$$
$$\text{So } A = B.$$

Example 3 Prove that $\sin\left(x + \frac{\pi}{4}\right) = \cos\left(x - \frac{\pi}{4}\right)$ for all x.

Solution Use the Addition Identities for the sine and cosine.

$$
\begin{array}{c|c}
\sin\left(x + \frac{\pi}{4}\right) & \cos\left(x - \frac{\pi}{4}\right) \\
= \sin x \cos \frac{\pi}{4} + \cos x \sin \frac{\pi}{4} & = \cos x \cos \frac{\pi}{4} + \sin x \sin \frac{\pi}{4} \\
= \sin x \cdot \frac{\sqrt{2}}{2} + \cos x \cdot \frac{\sqrt{2}}{2} & = \cos x \cdot \frac{\sqrt{2}}{2} + \sin x \cdot \frac{\sqrt{2}}{2}
\end{array}
$$
$$\text{So } \sin\left(x + \frac{\pi}{4}\right) = \cos\left(x - \frac{\pi}{4}\right).$$

A third proof technique is to begin with a known identity and derive statements equivalent to it until the proposed identity appears.

Example 4 Prove that $1 + \cot^2 x = \csc^2 x$ for $x \ne n\pi$, n an integer.

Solution Begin with the Pythagorean Identity.
For all real numbers x,

$$\cos^2 x + \sin^2 x = 1.$$
$$\frac{\cos^2 x}{\sin^2 x} + \frac{\sin^2 x}{\sin^2 x} = \frac{1}{\sin^2 x} \qquad \text{Divide both sides by } \sin^2 x.$$
$$\text{(Note that } \sin^2 x \ne 0 \text{ when } x \ne n\pi.\text{)}$$
$$\cot^2 x + 1 = \csc^2 x \qquad \text{Definitions of } \cot x \text{ and } \csc x$$

Check Below is a graph of both $f(x) = 1 + \cot^2 x$ and $g(x) = \csc^2 x$. They appear to coincide.

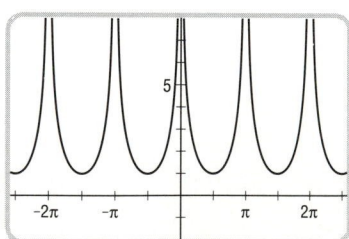

As shown in the solution to Example 3, the identity $1 + \cot^2 x = \csc^2 x$ can be derived quickly from the Pythagorean Identity. Hence it is sometimes called a *corollary* to the Pythagorean Identity. A **corollary** is a theorem that can be quickly deduced from another theorem. In the Questions you are asked to use this technique to show that $1 + \tan^2 x = \sec^2 x$ is also a corollary to the Pythagorean Identity.

Questions

Covering the Reading

1. *True* or *false*. $(\cos x - \sin x)^2 = \cos^2 x - \sin^2 x$ **False**

2. a. Is the sentence $(\cos x + \sin x)^2 = \cos^2 x + \sin^2 x$ an identity? Justify your answer with a graph. **No, see margin for graph.**
  b. Find an expression involving a singular circular function equal to $(\cos x + \sin x)^2$. **$(\cos x + \sin x)^2 = \sin 2x + 1$**

3. State three different techniques for proving that an equation is an identity. **See margin.**

4. Prove that $1 + \cot^2 x = \csc^2 x$, $x \ne n\pi$, using
  a. the first technique of this lesson
  b. the second technique of this lesson. **a,b) See margin.**

5. Prove: For all x for which all the functions are defined,
$$\sin x \cdot \cos x \cdot \tan x = \frac{1}{\csc^2 x}.$$ **See margin.**

6. Show that $1 + \tan^2 x = \sec^2 x$ is a corollary to the Pythagorean Identity. That is, prove the identity $1 + \tan^2 x = \sec^2 x$, for $x \ne \frac{\pi}{2} + n\pi$, using the third technique of this lesson. **See margin.**

Applying the Mathematics

In 7 and 8, the graph of a product of trigonometric functions is drawn.
a. What identity is suggested by the graph? (Hint: The answer involves one of the six parent trigonometric functions.) **b.** Prove your answer to part **a.**
c. Over what domain is the identity true?

7. $y = \cot x \sec x$ **See margin.**

8. $y = \sec x \tan x \cos x$ **See below.**

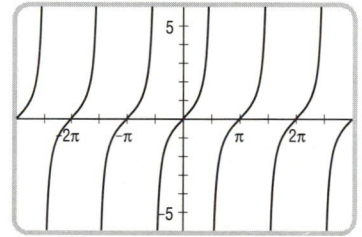

8a) $\sec x \tan x \cos x = \tan x$; b) $\sec x \tan x \cos x = \dfrac{1}{\cos x} \cdot \dfrac{\sin x}{\cos x} \cdot \cos x = \dfrac{\sin x}{\cos x} = \tan x$; c) $x \ne \dfrac{\pi}{2} + n\pi$, for any integer n

---

(Margin, right column)

5. $\sin x \cdot \cos x \cdot \tan x =$
$\sin x \cdot \cos x \cdot \dfrac{\sin x}{\cos x} =$
$\sin^2 x$; $\dfrac{1}{\csc^2 x} = \sin^2 x$; So
$\sin x \cdot \cos x \cdot \tan x =$
$\dfrac{1}{\csc^2 x}$ for $x \ne \dfrac{n\pi}{2}$, n an integer.

6. $\sin^2 x + \cos^2 x = 1 \Rightarrow$
$\dfrac{\sin^2 x}{\cos^2 x} + \dfrac{\cos^2 x}{\cos^2 x} = \dfrac{1}{\cos^2 x} \Rightarrow$
$\tan^2 x + 1 = \sec^2 x$ for
$x \ne \dfrac{\pi}{2} + n\pi$, n an integer

7.a. $\cot x \sec x = \csc x$

b. $\cot x \sec x =$
$\dfrac{\cos x}{\sin x} \cdot \dfrac{1}{\cos x} = \dfrac{1}{\sin x} =$
$\csc x$ for $x \ne \dfrac{\pi}{2} + n\pi$ and
$x \ne n\pi$, n an integer

c. $x \ne \dfrac{\pi}{2} + n\pi$ and $x \ne n\pi$,
n an integer

MORE PRACTICE
For more questions on SPUR Objectives, use *Lesson Master 13-1*, shown on page 783.

PROJECTS
The projects for Chapter 13 are described on page 828. Project 1 is related to the content of this lesson.

EXTENSION
Small Group Work
Groups of students can apply the following procedure to construct their own identities. Then they can prove each other's identities.
1. Start with a statement that is always true: 1 = 1.
2. Replace a part(s) of the beginning statement with an equivalent trigonometric expression: $\sin^2 \theta + \cos^2 \theta = 1$.
3. Repeat step 2 as many times as one wishes:

$$\sin^2 \theta + \cos^2 \theta = \frac{\sin \theta}{\sin \theta}$$

(if $\sin \theta \neq 0$).
4. Apply procedures for changing an equation to an equivalent equation:

$$\sin^2 \theta + 2 \sin \theta \cos \theta + \cos^2 \theta$$
$$= \frac{\sin \theta}{\sin \theta} + 2 \sin \theta \cos \theta;$$
$$(\sin \theta + \cos \theta)^2 =$$
$$\sin \theta\left(\frac{1}{\sin \theta} + 2 \cos \theta\right);$$
$$(\sin \theta + \cos \theta)^2 =$$
$$\sin \theta (\csc \theta + 2 \cos \theta)$$

To prove the above identity, one has to essentially perform the opposite operations of the steps above in reverse order.

ADDITIONAL ANSWERS

9.a.

$y = \cos x$
$y = -\cos (\pi - x)$

9.b., 10.a.,b., 11.a.,b., 12.a.,b., 13.a., 15. See Additional Answers in the back of this book.

In 9–12, **a.** use an automatic grapher to test whether the equation may be an identity. **b.** Prove the identity or give a counterexample.

9. $\cos x = -\cos (\pi - x)$
See margin.

10. $\sin 3x = 3 \sin x \cos x$
See margin.

11. $\dfrac{\csc x}{\sec x} = \cot x$
See margin.

12. $\tan^2 x \cos^2 x = 1 - \cos^2 x$
See margin.

13. a. Graph $f(x) = \sin x$ and $g(x) = x - \dfrac{x^3}{6} + \dfrac{x^5}{120}$ for $-\pi \leq x \leq \pi$.

 b. Calculate $f(1) - g(1)$. $\approx -.0001957$

 c. Is $\sin x = x - \dfrac{x^3}{6} + \dfrac{x^5}{120}$ an identity? **No**

 d. For what values of x are $f(x)$ and $g(x)$ approximately equal?
 a) See margin. d) sample: $-1.75 \leq x \leq 1.75$

14. Let θ be in the interval $270° < \theta < 360°$ and $\tan \theta = -\dfrac{3}{8}$. Use the Pythagorean Identity or one of its corollaries to find the following.
 a. $\cot \theta$ $-8/3 = -2.\overline{6}$ **b.** $\sec \theta$ **c.** $\sin \theta$
 $\sqrt{73}/8 \approx 1.068$ $-3/\sqrt{73} \approx -0.351$

15. Prove that for all x, $\cos^4 x + \sin^2 x = \sin^4 x + \cos^2 x$. See margin.

In 16 and 17, use the techniques in this lesson to test whether the statement is or is not an identity. Justify your answer.

16. $x^3 + x^4 = x^7$
Let $x = 1$. Then $x^3 + x^4 = 2$ and $x^7 = 1$, so $x^3 + x^4 = x^7$ is not an identity.

17. $\log x^3 + \log x^4 = \log x^7$
$\log x^3 + \log x^4 = \log x^3 x^4 = \log x^7$, so it is an identity.

18. a. Write the first six terms of the sequence *(Lesson 8-1)*
$$\begin{cases} a_1 = 1 \\ a_n = 2a_{n-1} + 1. \end{cases}$$ **1, 3, 7, 15, 31, 63**

 b. Tell whether the sequence is arithmetic, geometric, or neither.
 neither

In 19 and 20, let $\triangle ABC$ be a right triangle with right angle C. Express each function in terms of the sides a, b, and/or c. *(Lesson 6-7)*

19. $\csc A$ **c/a**

20. $\cot B$ **a/b**

In 21 and 22, evaluate in radians. *(Lesson 6-5)*

21. $\sin^{-1}(-1)$ $-\pi/2$

22. $\tan^{-1}(\sqrt{3})$ $\pi/3$

23. Refer to the circle of radius 4 drawn at the left. Give the coordinates of point A in terms of θ. *(Lesson 5-4)* **(4 cos θ, 4 sin θ)**

24. If a sector of a circle has a central angle of $\dfrac{\pi}{8}$ and an area of 5π square inches, what is the radius of the circle? *(Lesson 5-2)*
$r = \sqrt{80} \approx 8.94$ inches

25. Refer to Questions 7 and 8.
 a. Find another pair of parent functions whose product seems to be a parent function, and prove the identity.
 b. Find another set of three parent functions whose product seems to be another parent function, and prove the identity. **Answers will vary.**
 a) sample: $\sin x \sec x = \tan x$, for $x \neq \dfrac{\pi}{2} + n\pi$, n an integer.

 Proof: $\sin x \sec x = \sin x \cdot \dfrac{1}{\cos x} = \tan x$

784

13-2

Restrictions on Identities

Recall that an identity is a statement that is true for *all* values of a variable in a particular domain. In general, when trying to prove an identity you should consider the largest domain on which all the relevant functions are defined. In particular, when testing or proving trigonometric identities you should recall that only two of the parent functions—the sine and cosine—are defined for all real numbers. The others have the following restrictions on their domains.

function	not defined for	
tan x	$\frac{\pi}{2} + n\pi$	
cot x	$n\pi$	n an integer
sec x	$\frac{\pi}{2} + n\pi$	
csc x	$n\pi$	

Thus, in Lesson 13-1 we proved that $1 + \cot^2 x = \csc^2 x$ for all real numbers except where cot x and csc x are not defined. That is, $1 + \cot^2 x = \csc^2 x$ for all real numbers x except $x = n\pi$, where n is an integer.

An isolated value for which a function is undefined is called a **singularity**. The singularities of the parent tangent, cotangent, secant, and cosecant functions are signalled graphically by vertical asymptotes. For instance, when $x = n\pi$, n an integer, x is a singularity of $f(x) = \cot x$, and the lines with equations $x = n\pi$ are vertical asymptotes of the graph of cot x. Singularities of other functions may not be represented by asymptotes, and may not be obvious on a graph. Thus when using an automatic grapher to test a potential identity you should always consider the restrictions on the domain.

Example 1   Consider cos x tan x = sin x. Use an automatic grapher to test whether or not the equation seems to be an identity. If it seems to be, prove the identity over the largest possible domain.

Solution   The graphs of $f(x) = \cos x \tan x$ and $g(x) = \sin x$ as they appear on our automatic grapher are shown below. It appears that the graphs coincide, so an algebraic proof of the identity is worth pursuing.

Rewrite the left side. $\cos x \tan x = \cos x \cdot \dfrac{\sin x}{\cos x} = \sin x$

Notice that the tangent function is defined only when cos $x \neq 0$, that is, when $x \neq \frac{\pi}{2} + n\pi$, n an integer.

So $\cos x \tan x = \sin x$ for $x \neq \frac{\pi}{2} + n\pi$, n an integer.

LESSON 13-2

RESOURCES
■ Lesson Master 13-2
▣ Computer Master 31

OBJECTIVES

C Prove trigonometric identities.
D Describe singularities of functions.
F Use an automatic grapher to test a proposed identity.

TEACHING NOTES

This lesson gives more time to practice proving identities.

There are two kinds of singularities, *essential* and *removable*. An essential singularity is signaled by a vertical asymptote and can usually be determined from a graph.

This lesson is concerned more with removable singularities, those which are not represented by vertical asymptotes, but by holes in a graph. An automatic grapher cannot represent such holes. So a critical analysis of the symbolic form of an identity is the safest way to identify singularities.

In both **Examples 1 and 2,** the restrictions on the domain of the identity are the singularities of the trigonometric functions involved. The only other possible restrictions arise when a denominator is zero, as in **Question 11.**

Notice that although the graph in Example 1 suggested that $\cos x \tan x = \sin x$ is an identity for all x, it is an identity only when $\tan x$ is defined. Geometrically, you can think of the singularities of the function $f(x) = \cos x \tan x$ as "holes" in the graph of $g(x) = \sin x$ at the points where x is an odd multiple of $\frac{\pi}{2}$. A singularity of this type, where the break in the graph can be "removed" by adding a single point, is called a **removable singularity**. On some automatic graphers, if you plot $f(x) = \cos x \tan x$ and repeatedly zoom in around the point $\left(\frac{\pi}{2}, 1\right)$ you may eventually see the singularity represented as a small hole in the graph. Try this with your technology.

Example 2 also involves an identity on a restricted domain.

■ ■ ■ ■ ■ ■ ■ ■ ■

Example 2 Consider the statement $\sin 2\theta = \frac{2\cot\theta}{1+\cot^2\theta}$.

a. Determine any restrictions on the domain of θ.
b. Prove the statement over the domain in part **a**.

Solution
a. Singularities occur where $\cot\theta$ is undefined, so $\theta \neq \pi + n\pi$, for n an integer. Since $1 + \cot^2\theta > 0$ for all θ, there are no other restrictions.
b. Rewrite both sides of the equation until equivalent expressions remain. (In the Questions, you are asked to give justifications for each of the four numbered steps.)

	$\sin 2\theta$	$\frac{2\cot\theta}{1+\cot^2\theta}$	
1.	$= 2\sin\theta\cos\theta$	$= \frac{2\cot\theta}{\csc^2\theta}$	2.
		$= \frac{\frac{2\cos\theta}{\sin\theta}}{\frac{1}{\sin^2\theta}}$	3.
		$= 2\sin\theta\cos\theta$	4.

So $\sin 2\theta = \frac{2\cot\theta}{1+\cot^2\theta}$ for all $\theta \neq n\pi$.

Questions

1. What is a singularity of a function?
an isolated value for which a function is undefined

In 2 and 3, identify **any** singularities of the function.

2. $f(x) = \cot x$ ** $n\pi$, n an integer**

3. $g(x) = \cos x \tan x$
$\pi/2 + n\pi$, n an integer

4. *True* or *false*. All singularities of functions are signalled by asymptotes. **False**

5. Give reasons for the four numbered steps in Example 2.
(1) double angle formula; (2) Pythagorean Identity; (3) definitions of cot, csc; (4) operation of division

In 6–9, **a.** use an automatic grapher to test whether the equation may be an identity. **b.** If the equation is an identity, prove it and give its domain.

6. $\sin x \cot x = \cos x$
See margin.

7. $\sec x = \tan x \csc x$
See margin.

8. $\csc x = \dfrac{\tan x}{\sec x}$
a) See margin. b) not an identity

9. $\sin 2x = \dfrac{2 \tan x}{1 + \tan^2 x}$
See margin.

Applying the Mathematics

In 10 and 11, **a.** graph the function, **b.** identify its domain, **c.** propose an identity based on this graph, and **d.** prove the identity.

10. $f(x) = \dfrac{1 - \tan^2 x}{1 + \tan^2 x}$
a,d) See margin. b) $x \neq \frac{\pi}{2} + n\pi$, n an integer;
c) $\frac{1 - \tan^2 x}{1 + \tan^2 x} = \cos 2x$

11. $f(x) = \dfrac{1 + \cos 2x}{\sin 2x}$
a,d) See margin. b) $x \neq n\pi$, n an integer;
c) $\frac{1 + \cos 2x}{\sin 2x} = \cot x$

In 12 and 13, **a.** give any restrictions to the domain of the proposed identity. **b.** Prove or disprove the proposed identity over your domain in part **a.**

12. $\tan x + \cot x = \sec x \csc x$
a) $x \neq \pi/2 + n\pi$ and $x \neq n\pi$, n an integer;
b) See margin.

13. $\csc^2 x \sin x = \dfrac{\sec^2 x - \tan^2 x}{\sin x}$
a) $x \neq \frac{\pi}{2} + n\pi$ and $x \neq n\pi$, n an integer;
b) See margin.

14. a. Graph $y = \dfrac{x^2 - x - 12}{x - 4}$ and $y = x + 3$ on the same set of axes.

b. For what values of x do the two graphs appear to coincide?

c. Find one value of x for which $\dfrac{x^2 - x - 12}{x - 4} \neq x + 3$. $x = 4$

d. What is the domain for the identity $\dfrac{x^2 - x - 12}{x - 4} = x + 3$? $x \neq 4$

e. Prove the identity. a) See margin. b) all values
e) $\frac{x^2 - x - 12}{x - 4} = \frac{(x - 4)(x + 3)}{x - 4} = x + 3$ for $x \neq 4$.

Review

15. Let α be in the interval $\frac{\pi}{2} < \alpha < \pi$, and $\cot \alpha = -4$. Use trigonometric identities to evaluate each of the following.
a. $\csc \alpha$ $\quad \sqrt{17} \approx 4.123$ **b.** $\sin \alpha$ $\quad 1/\sqrt{17} \approx .243$ **c.** $\sin 2\alpha$
(Lessons 6-7, 11-6, 13-1) $\qquad\qquad\qquad\qquad\qquad\qquad\qquad\qquad -8/17 \approx -.471$

16. Graph, on the same set of axes,
a. $\dfrac{x^2}{16} - \dfrac{y^2}{9} = 1$ **b.** $\dfrac{x^2}{16} + \dfrac{y^2}{9} = 1$.
c. Show algebraically that the graphs in parts **a** and **b** intersect at only two points. *(Lessons 12-2, 12-3)* a–c) See margin.

17. *True* or *false*. The matrix $\begin{bmatrix} 1 & 0 \\ 0 & 1 \end{bmatrix}$ does not have an inverse.
(Lesson 11-7) **False**

18. Find the number z such that 75% of the area under the standard normal curve lies between $-z$ and z. *(Lesson 10-5)* $z \approx 1.15$

ADDITIONAL ANSWERS
6.a.

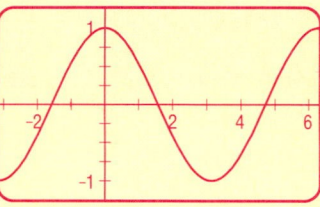

6.b., 7.a.,b., 8.a., 9.a.,b., 10.a.,d., 11.a.,d., 12.b., 13.b., 14.a., 16.a.,b.,c. See Additional Answers in the back of this book.

EXTENSION
Small Group Work Allow students to work together to prove the following identities.

a. $\dfrac{\cos^4 \theta - \sin^4 \theta}{1 - \tan^4 \theta} = \dfrac{1}{\sec^4 \theta}$

b. $\left(\dfrac{1 + \tan x}{1 - \tan x}\right)^2 = \dfrac{1 + \sin 2x}{1 - \sin 2x}$

c. $\sqrt{\dfrac{1 - \sin t}{1 + \sin t}} = |\sec t - \tan t|$

d. In $\triangle ABC$, $\tan A + \tan B + \tan C = (\tan A)(\tan B)(\tan C)$. Determine any restrictions on this identity.

ADDITIONAL ANSWERS
22.

24.

27.a.

x	$\sqrt{x^2 - 1}$	$x - \dfrac{1}{2x}$
1	0	.5
5	4.899	4.9
9	8.9443	8.9444
1	9.9499	9.95

b. all values of x larger than about 2.395

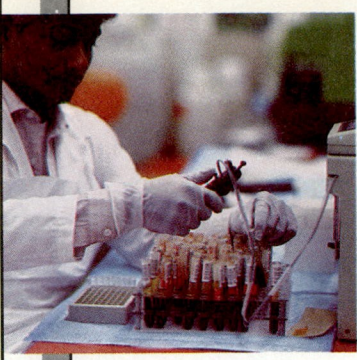

In 19 and 20, consider the ELISA test which was introduced in the mid-1980s to screen donated blood for the presence of antibodies to the AIDS virus. *(Lessons 10-1, 10-2)*

19. When presented with AIDS-contaminated blood, ELISA gives a positive response in about 98% of all cases. Suppose that among the blood that passes through a blood bank in a year there are 25 units containing AIDS antibodies.
  a. What is the probability that ELISA will detect all 25 of these units?
  b. What is the probability that more than 2 of the 25 contaminated units will escape detection?   ≈ **0.013**
  c. What is the mean number of units among the 25 that will be detected by ELISA?   **24.5**
  d. What is the standard deviation of the number detected?   **0.7**
  a) $(.98)^{25} ≈ .60$

20. ELISA gives a positive response (that is, AIDS antibodies are present) in uncontaminated blood about 7% of the time. Such a result is called a *false positive*. Suppose a blood bank contains 20,000 uncontaminated units of blood.
  a. What is the expected number of false positives among this group?
  b. What is the standard deviation of the number of false positives?
  a) **1400; b) ≈ 36**

21. Consider the row of Pascal's Triangle that begins with 1, 12, ... *(Lesson 8-7)*
  a. What is the next entry in the row?   **66**
  b. What is the sum of all the numbers in the row?   **4096**

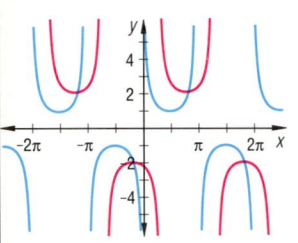

22. Copy the graph of $f(x) = \csc x$ at the left. On the same set of axes, sketch a graph of $g(x) = 2 \csc \left(x - \dfrac{\pi}{4}\right)$.   *(Lessons 6-3, 6-7)*   **See margin.**

23. Solve the equation $\tan \theta = 2.5$ on the given domain.   *(Lesson 6-6)*
  a. $-\dfrac{\pi}{2} \le \theta \le \dfrac{\pi}{2}$ **b.** $0 \le \theta \le 2\pi$ **c.** all real numbers
  $\theta ≈ 1.19$ $\theta ≈ 1.19$ or $\theta ≈ 4.33$ $\theta ≈ 1.19 + n\pi$ for any integer n

24. On the same set of axes, graph $y = \tan x$ and $y = \tan^{-1} x$.   *(Lesson 6-5)*
  See margin.

In 25 and 26, solve. Give your answer to the nearest tenth.   *(Lessons 4-8, 4-9)*

25. $2 = (1.07)^{y/100}$ **1024.5**   **26.** $100 = 12.2 \ln t + 4.8$ **2448.6**

Exploration

27. The sentence $\sqrt{x^2 - 1} = x - \dfrac{1}{2x}$ is not an identity for $x \ge 1$. In fact, the expressions on the two sides never have the same value.   **See margin.**
  a. Evaluate the two expressions in the sentence at $x = 1$, $x = 5$, $x = 9$, and $x = 10$ to support the preceding claims.
  b. Notice that the two expressions $\sqrt{x^2 - 1}$ and $x - \dfrac{1}{2x}$ have closer and closer values as x gets bigger. Use an automatic grapher to determine the values of x for which $\sqrt{x^2 - 1}$ is within .01 of $x - \dfrac{1}{2x}$.

788

Polar Coordinates

The rectangular coordinate system that you have been using for many years dates back to the early 1600s, when René Descartes and Pierre de Fermat worked to develop analytic geometry. Later in that same century other mathematicians, notably Isaac Newton and Jakob Bernoulli, introduced other coordinate systems. This lesson introduces you to the system of *polar coordinates*.

To construct a polar coordinate system, first select a point O as the **pole** of the system. Then select any line through O as the **polar axis**. Usually the polar axis is drawn horizontally as shown below. Coordinatize this line so O has coordinate 0. Any point P has **polar coordinates** $[r, \theta]$ if and only if P is the image, under a rotation θ about the pole O, of the point on the polar axis with coordinate r.

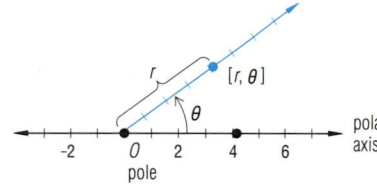

Recall that in a rectangular coordinate system, every point in the plane can be identified by a unique ordered pair of numbers (x, y) representing the point's distances and direction from two perpendicular axes, and every ordered pair of numbers (x, y) represents a unique point in the plane. In a **polar coordinate system**, a pair of numbers $[r, \theta]$ again represents a unique point. Square brackets [,] are used to distinguish polar coordinates from rectangular coordinates. Here r or $-r$ is a distance, but θ is a magnitude of rotation measured in degrees or radians.

Example 1 Plot each point $[r, \theta]$, where θ is in radians.

a. $[2, \frac{\pi}{3}]$ **b.** $[3, -\frac{\pi}{2}]$ **c.** $[-2, \frac{7\pi}{5}]$

Solution

a. Rotate the point on the polar axis with coordinate 2 counterclockwise by $\frac{\pi}{3}$ radians.

RESOURCES
- Lesson Master 13-3
- Quiz for Lessons 13-1 through 13-3
- Teaching Aid 93: Polar coordinate graph paper (1 grid)
- Teaching Aid 94: Polar coordinate graph paper (4 grids)
- Teaching Aid 95: Graphs of polar coordinates for **Example 1**

OBJECTIVES

G Given polar coordinates of a point, determine its rectangular coordinates, and vice versa.

H Plot points in the polar coordinate system.

TEACHING NOTES

Polar coordinates and rectangular coordinates are both ways of naming a point, just as decimals and fractions are both ways to name a rational number.

Because of their background using the unit circle, students should be able to plot points in the polar plane, as in **Example 1,** for which a teaching aid is provided.

Error Analysis The most common difficulties seem to occur when r is negative, and in accepting the fact that any point in the polar plane has an infinite number of polar coordinate pairs which describe it. **Example 2** and **Questions 1–7** are intended to help confront these difficulties, and can be used to guide class discussion.

b. Rotate the point on the polar axis with coordinate 3 clockwise by $\frac{\pi}{2}$ radians.

c. Rotate the point on the polar axis with coordinate -2 counterclockwise by $\frac{7\pi}{5}$ radians.

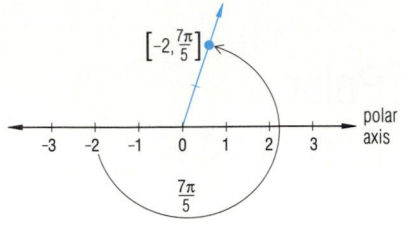

While any pair $[r, \theta]$ identifies a unique point in the polar plane, there are an infinite number of pairs which identify any given point. Example 2 illustrates how some of those pairs of coordinates can be found.

▪ ▪ ▪ ▪ ▪ ▪ ■ ■

Example 2   Give four different polar pairs for the point A graphed below.

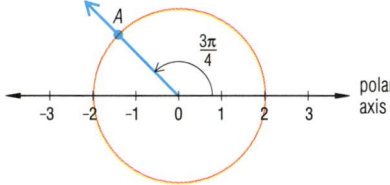

Solution   There are an infinite number of answers. One of them is $[2, \frac{3\pi}{4}]$ because A is the image of the point on the polar axis with coordinate 2 under a rotation of $\frac{3\pi}{4}$ radians. Rotating the same point another 2π yields $[2, \frac{11\pi}{4}]$. A can also be considered as the image of the point -2 on the polar axis under a rotation of either $-\frac{\pi}{4}$ or $\frac{7\pi}{4}$, so $[-2, -\frac{\pi}{4}]$ and $[-2, \frac{7\pi}{4}]$ also are polar coordinates for point A.

The following theorem summarizes ways to find possible polar coordinate for points in the plane.

Theorem

For any particular values of r and θ, the following polar coordinate representations name the same point.
a. $[r, \theta]$
b. $[r, \theta + 2\pi n]$, for all integers n
c. $[-r, \theta + (2n + 1)\pi]$, for all integers n

Polar coordinate graph paper, with grids like the *polar grid* pictured below, is very helpful for plotting points and sketching curves in the polar plane.

Each of the concentric circles in the grid represents a value of *r*, and each ray from the pole represents a value of θ. When plotting using polar coordinates, you should identify the positive polar axis with an arrow and put a scale on it to indicate values of *r*.

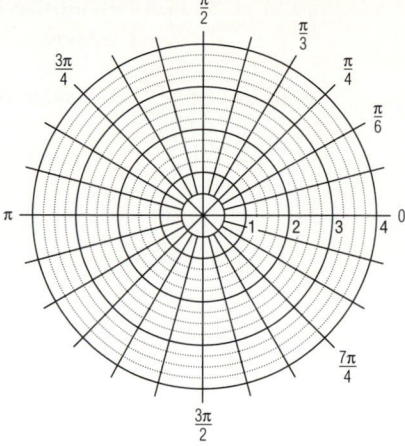

Example 3   **a.** Sketch all solutions [*r*, θ] to the equation *r* = 3.
b. Sketch all solutions [*r*, θ] to the equation $\theta = \frac{-\pi}{3}$.

Solution

a. The equation *r* = 3 describes all points 3 units from the pole. The graph is a circle of radius 3 centered at the pole. This circle is drawn in blue below.
b. $\theta = \frac{-\pi}{3}$ is the line obtained by rotating the polar axis by $\frac{-\pi}{3}$. This line is drawn in orange below.

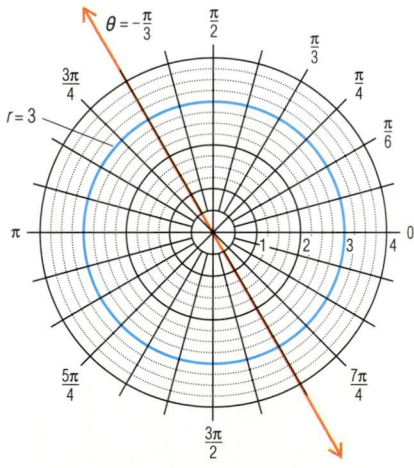

1. Plot the following points on a polar coordinate system.

a. [6, 0]   **b.** $\left[2.5, \frac{2\pi}{3}\right]$

c. [-1, -40°]

2. Give four additional pairs of polar coordinates for the point $\left[5, \frac{\pi}{3}\right]$.

$\left[5, \frac{7\pi}{3}\right]$, $\left[5, \frac{13\pi}{3}\right]$, **or in general any pair of the form**
$\left[5, \frac{\pi}{3} + 2n\pi\right]$, $\left[-5, \frac{4\pi}{3} + 2n\pi\right]$

3. Sketch all solutions [*r*, θ] to each equation.

a. $\theta = \pi$
the line containing the polar axis

b. $r = \pi$
a circle with center at the pole and radius π

4. Find rectangular coordinates for the point with polar coordinates $\left[1.3, \frac{\pi}{5}\right]$.

$\left(1.3 \cos \frac{\pi}{5}, 1.3 \sin \frac{\pi}{5}\right) \approx$
(1.05, .76)

5. Find a set of polar coordinates for the point (6, -1).

sample: $\left[\sqrt{37}, \tan^{-1}\left(\frac{-1}{6}\right)\right] \approx$
[6.1, -9.46°]

Often polar and rectangular coordinate systems are superimposed on the same plane. Then the polar axis coincides with the *x*-axis and the pole is the origin. When this is done, you can use trigonometry to find the unique rectangular coordinate representation for any point whose polar coordinate representation is known.

Consider the diagram below.

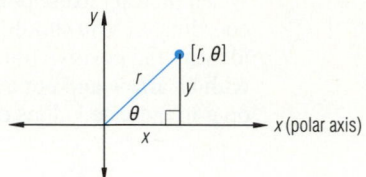

If $r \neq 0$, then $\cos\theta = \dfrac{x}{r}$ and $\sin\theta = \dfrac{y}{r}$. So $x = r\cos\theta$ and $y = r\sin\theta$. Note that if $r = 0$, these equations are still true. This proves the following theorem.

Theorem

If *P* has polar coordinates $[r, \theta]$, then the rectangular coordinates (x, y) of *P* are given by $x = r\cos\theta$ and $y = r\sin\theta$.

Example 4 Find the rectangular coordinates for the point with polar coordinates $[4, 300°]$.

Solution By the preceding theorem the coordinates are $x = 4\cos 300°$ and $y = 4\sin 300°$. So

$$(x, y) = \left(4 \cdot \tfrac{1}{2}, 4 \cdot \tfrac{-\sqrt{3}}{2}\right)$$
$$= (2, -2\sqrt{3}).$$

Check Use a graph. Both $[4, 300°]$ and $(2, -2\sqrt{3})$ are in the 4th quadrant, and the points agree.

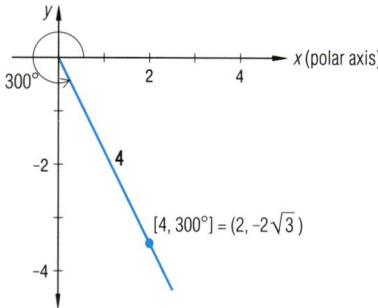

Converting from rectangular coordinates to polar coordinates also requires trigonometry. Suppose a point in the first quadrant has polar coordinates $[r, \theta]$ and rectangular coordinates (x, y). By the Pythagorean Theorem, $r^2 = x^2 + y^2$, so

$$|r| = \sqrt{x^2 + y^2},$$

or

$$r = \pm\sqrt{x^2 + y^2}.$$

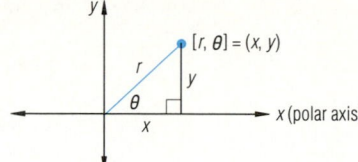

Also, $\tan \theta = \dfrac{y}{x}$.

Notice that r may be positive or negative and that there are infinitely many values of θ for which $\tan \theta = \dfrac{y}{x}$. So there are infinitely many polar coordinates for any given point.

▪ ▪ ▪ ▪ ▪ ▪ ▪ ▪ ▪

Example 5 Find two different sets of polar coordinates for the point whose rectangular coordinates are (-2, -5).

Solution Plot the point and draw a triangle.

$$r^2 = 5^2 + 2^2$$

So $r = \pm\sqrt{5^2 + 2^2}$
$$= \pm\sqrt{29}.$$

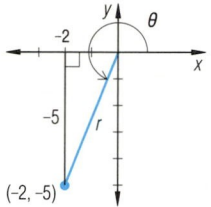

Also $\tan \theta = \dfrac{-5}{-2} = 2.5$. So one value of θ is $\tan^{-1} 2.5 \approx 68°$. Thus

$\theta \approx 68° \pm n \cdot 180°$. Now the choice of particular values of θ and r depend on matching the angle to the quadrant. If you use $r = \sqrt{29}$ then because (-2, -5) is in the third quadrant, a value of θ that can be used is $68° + 180° = 248°$. So one set of polar coordinates for (-2, -5) is about $[\sqrt{29}, 248°]$. If the value $-\sqrt{29}$ for r is used, the pair $[-\sqrt{29}, 68°]$ is another set of polar coordinates for this point.

Questions

Covering the Reading

In 1–4, use a protractor and ruler to estimate the polar coordinates $[r, \theta]$ of the point. Measure r in cm and θ in degrees.

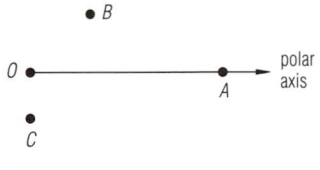

1. A
   ≈ **[2.8, 0°]**
2. B
   ≈ **[1.2, 43°]**
3. C
   ≈ **[0.7, 270°]**
4. D
   ≈ **[2.5, 215°]**

LESSON 13-3 Polar Coordinates **793**

5. Plot all the points on the same polar grid.
   $P = [1, \frac{\pi}{2}], Q = [-1, \frac{\pi}{4}], R = [3.25, -\frac{5\pi}{6}], S = [-4, -75°]$ **See margin.**

6. Suppose $P = [4, \frac{5\pi}{6}]$. Give another set of polar coordinates for P:

   a. with $r = 4, \theta \neq \frac{5\pi}{6}$,   **b.** with $r = -4$. **samples:** $\left[-4, \frac{11\pi}{6}\right], \left[-4, \frac{-\pi}{6}\right]$
      See below.

7. *Multiple choice.*   Which polar coordinates do not name the same point as $[6, -40]$?
   (a) $[6, 320°]$ (b) $[6, -400°]$
   (c) $[-6, 40°]$ (d) $[-6, 140°]$ **(c)**

8. On the same polar grid sketch all solutions to the following equations.

   a. $r = 5$ **b.** $\theta = \frac{5\pi}{6}$ **a,b) See margin.**

In 9 and 10, suppose (x, y) in the rectangular coordinate system names the same point as $[r, \theta]$ in the polar coordinate system. *True* or *false*

9. $x = r \cos \theta$ **True**   10. $|r| = \sqrt{x^2 + y^2}$ **True**

In 11 and 12, find the rectangular coordinates for the point P whose polar coordinates are given.

11. $[10, \frac{3\pi}{2}]$ **(0, -10)**   12. $[8, -60°]$ **(4, -4√3)**

In 13 and 14, give one pair of polar coordinates for the (x, y) pair.

13. $(5, 5\sqrt{3})$ **sample: [10, 60°]**   14. $(-6, -8)$
   sample: $[10, \tan^{-1}(\frac{4}{3})] \approx [10, -127°]$

Applying the Mathematics

15. The point $[r, \theta]$ has $r = 8$. On what geometric figure must this point lie?   **a circle with radius 8 and center at the pole**

16. The point $[r, \theta]$ has $\theta = \frac{3\pi}{4}$ radians. On what geometric figure must this point lie?   **See below.**

In 17 and 18, consider the point $P = [6, \frac{\pi}{3}]$. State one pair of polar coordinates for the image of P under the given transformation.

17. reflection over the polar axis   **samples:** $\left[-6, \frac{\pi}{3}\right], \left[6, \frac{4\pi}{3}\right]$

18. reflection over the line $\theta = \frac{\pi}{2}$ **sample:** $\left[6, \frac{2\pi}{3}\right]$

19. Airport runways are often numbered in a way that is related to polar coordinates. If you land from the north, you see a runway numbered 0. If you land from the west, you see a runway numbered 9. Each additional runway number unit corresponds to 10° counterclockwise from the previous runway number, the highest being 35. From what direction do you land on a runway with the given number?
   a. 18   **b.** 1   **c.** 22   **d.** 8
   a) south; b) from 10° W of N; c) from 50° S of E; d) from 80° W of N

6a) samples: $\left[4, \frac{17\pi}{6}\right], \left[4, \frac{29\pi}{6}\right], \left[4, -\frac{7\pi}{6}\right]$

16) the line that is the image of the polar axis under a rotation of $\frac{3\pi}{4}$ about the pole

20. a. Plot on polar graph paper.

r	0	$\frac{1}{2}(\sqrt{6}-\sqrt{2})$	1	$\sqrt{2}$	$\sqrt{3}$	$\frac{1}{2}(\sqrt{2}+\sqrt{6})$	2
$\theta°$	0	15	30	45	60	75	90

b. The points above all satisfy the equation $r = 2 \sin \theta$.
Let $\theta = 105°, 120°, \ldots$, and find six more points satisfying this equation. Plot these points.

c. Make a conjecture about the graph of all points $[r, \theta]$ satisfying $r = 2 \sin \theta$.

a,b) See margin. c) It is a circle with radius 1 and center (0, 1).

Review

In 21 and 22, if the equation is an identity, give a proof and state the domain on which it is true. If the equation is not an identity, provide a counter-example. *(Lessons 11-5, 13-2)*

21. $\sin\left(\frac{\pi}{2} + x\right) = \cos x$ **See below.**

22. $\sec x + \cot x \csc x = \sec x \csc x$ **See below.**

In 23 and 24, simplify using trigonometric identities. *(Lessons 5-7, 13-1)*

23. a. $\sin^2 \theta + \cos^2 \theta$ **1**   **b.** $25 \sin^2 \theta + 25 \cos^2 \theta$ **25**

24. a. $1 + \tan^2 \theta$ **sec² θ**   **b.** $r^2 + r^2 \tan^2 \theta$ **r² sec² θ**

In 25 and 26, graph on the same set of axes. *(Lessons 12-2, 6-1)*

25. a. $x^2 + y^2 = 1$ **b.** $x^2 + y^2 = 9$ **a,b) See margin.**

26. a. $y = \cos 2x$ **b.** $y = \cos 3x$ **a,b) See margin.**

Exploration

27. How might points be described in space using polar coordinates?
Sample: One way is by using cylindrical coordinates. A point (x, y, z) would be described by [r, θ, z] in cylindrical coordinates, where, as in polar coordinates, x = r cos θ and y = r sin θ.

---

21) $\sin\left(\frac{\pi}{2} + x\right) = \sin\frac{\pi}{2}\cos x + \cos\frac{\pi}{2}\sin x$
$= (1)\cos x + (0)\sin x$
$= \cos x$ for all x

22) To show it is not an identity, let $x = 30°$.
Then $\sec x + \cot x \csc x = \sec 30° + \cot 30° \cdot \csc 30°$
$= \frac{2}{\sqrt{3}} + \frac{\sqrt{3}}{1} \cdot \frac{2}{1} \approx 4.6$

and $\sec x \csc x = \sec 30° \cdot \csc 30° = \frac{2}{\sqrt{3}} \cdot \frac{2}{1} \approx 2.3$.

LESSON 13-3 Polar Coordinates   **795**

MORE PRACTICE
For more questions on SPUR Objectives, use *Lesson Master 13-3*, shown below.

EVALUATION
A quiz covering Lessons 13-1 through 13-3 is provided in the Teacher's Resource File.

25.a.,b.

26.a.,b.

NAME _____

LESSON **MASTER 13-3**
QUESTIONS ON **SPUR** OBJECTIVES

■**PROPERTIES** *Objective G (See pages 832–834 for objectives.)*
In 1–4, convert from polar coordinates to rectangular coordinates.

1. $\left[2, \frac{\pi}{4}\right]$ __$(\sqrt{2}, \sqrt{2})$__   2. $\left[-1, \frac{\pi}{3}\right]$ __$\left(\frac{1}{2}, -\frac{\sqrt{3}}{2}\right)$__

3. $[-3, 70°]$ __≈ (-1.03, -2.82)__   4. $[4, 150°]$ __$(-2\sqrt{3}, 2)$__

In 5–8, give one pair of polar coordinates for each (x, y) pair.

5. $(2, 3)$ __sample: $[\sqrt{13}, \approx 56°]$__   6. $(-5, 1)$ __sample: $[\sqrt{26}, \approx 169°]$__

7. $(-3, 0)$ __sample: $[3, 180°]$__   8. $(-4, -3)$ __sample: $[5, 217°]$__

■**REPRESENTATIONS** *Objective H*
In 9 and 10, choose the polar coordinate pair that does not name the given point.

9. $[3, 140°]$ __b__   10. $[-2, 230°]$ __c__

a. $[3, 500°]$ a. $[-2, 590°]$
b. $[3, 220°]$ b. $[-2, -130°]$
c. $[-3, 320°]$ c. $[2, 590°]$
d. $[-3, 680°]$ d. $[2, 410°]$

In 11 and 12, plot $[r, \theta]$ on the polar grid, where θ is in radians.

11. $\left[2, \frac{\pi}{4}\right]$ 12. $\left[-1, \frac{\pi}{6}\right]$

Functions, Statistics, and Trigonometry © Scott, Foresman and Company   **127**

OBJECTIVES

H Plot points in the polar coordinate system.
I Graph and interpret graphs of polar equations.

TEACHING NOTES

Technology The graphs of equations in the polar plane can be as simple as a circle, or as delightful as a rose. Unfortunately, these graphs have not been part of the secondary curriculum in large part because of the difficulty of drawing them with pencil and paper. An automatic grapher with the option to graph polar curves takes the tedium away and opens this beautiful bit of mathematics to students.

Even if you do not have access to a polar grapher, we have designed the questions so that students are asked to graph only equations whose graphs have relatively simple shapes—probably the most difficult are the cardioid and spiral in **Questions 14 and 15.** We do not expect students to be able to graph equations more complicated than the following:

$r = k$ (circle)
$\theta = k$ (line through origin)
$r = a \sin \theta$ (circle)
$r = a \cos \theta$ (circle)
$r = a \sin b\theta$ (rose curve)
$r = a \cos b\theta$ (rose curve)
$r \sin \theta = k$ (horizontal line)
$r \cos \theta = k$ (vertical line).

LESSON

13-4

Polar Graphs

Any equation which involves only the variables r and θ can be graphed using polar coordinates. For instance, in the previous lesson the equations $r = 3$ and $\theta = -\frac{\pi}{3}$ were graphed. Some automatic graphers allow you to graph polar relations. The relation is usually entered in the form $r = f(\theta)$ or $r = f(t)$.

You should be able to draw by hand graphs of relatively simple equations of the form $r = f(\theta)$. In particular, you should study the patterns that emerge when graphs of the parent circular functions are made.

■ ■ ■ ■ ■ ■ ■ ■ ■

Example 1 Sketch a graph of the polar equation $r = \cos \theta$.

Solution Make a table of values for $[r, \theta]$ for $0 \le \theta \le 2\pi$, and plot the points. A table for some values of θ between 0 and π, and the corresponding graph, are below. When making the table, it is common to consider θ to be the independent variable and r to be the dependent variable.

θ	0	$\frac{\pi}{6}$	$\frac{\pi}{4}$	$\frac{\pi}{3}$	$\frac{\pi}{2}$	$\frac{2\pi}{3}$	$\frac{3\pi}{4}$	$\frac{5\pi}{6}$	π
r	1	0.866	0.707	0.500	0	−0.500	−0.707	−0.866	−1

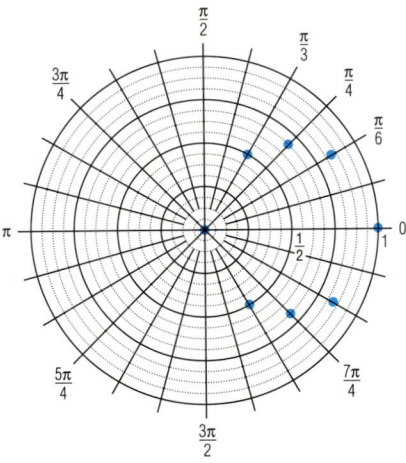

Notice that as θ increases from 0 to $\frac{\pi}{2}$, r decreases from 1 to 0, and points above the polar axis are generated. As θ increases from $\frac{\pi}{2}$ to π, r decreases from 0 to −1, which produces points below the polar axis.

796

When θ is between π and $\frac{3\pi}{2}$, $r = \cos \theta$ is negative. All such points are in the first quadrant, and coincide with the points generated when $0 \le \theta \le \frac{\pi}{2}$. For instance, if $\theta = \frac{5\pi}{4}$, $r = \cos \frac{5\pi}{4} \approx -0.707$. The point $[-0.707, \frac{5\pi}{4}]$ coincides with the point $[0.707, \frac{\pi}{4}]$, which has already been plotted.

Similarly, when $\frac{3\pi}{2} \le \theta \le 2\pi$, the points generated coincide with those generated when $\frac{\pi}{2} \le \theta \le \pi$. The complete graph is drawn at the right.
The graph appears to be a circle.

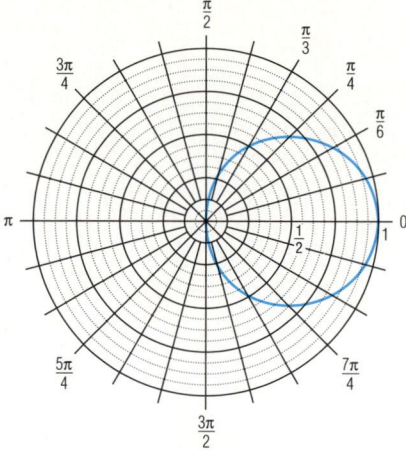

By converting from polar to rectangular coordinates, the graph of $r = \cos \theta$ can be proved to be a circle.

■ ■ ■ ■ ■ ■ ■ ■

Example 2   Prove that the graph of $r = \cos \theta$ is a circle.

Solution   Derive a rectangular equation for $r = \cos \theta$, and show that it is of the form $(x - h)^2 + (y - k)^2 = r^2$.

Since $x = r \cos \theta$, $\cos \theta = \frac{x}{r}$.
Now substitute $\frac{x}{r}$ for $\cos \theta$ in $r = \cos \theta$ and rewrite.

If $r = \cos \theta$,

then $r = \frac{x}{r}$.

So $r^2 = x$.

But $r^2 = x^2 + y^2$, and substituting for r^2 in the preceding equation gives
$$x^2 + y^2 = x.$$
Subtracting x from each side of the equation gives
$$x^2 - x + y^2 = 0.$$
Now complete the square in x.
$$\left(x^2 - x + \frac{1}{4}\right) + y^2 = \frac{1}{4}$$
$$\left(x - \frac{1}{2}\right)^2 + y^2 = \frac{1}{4}$$

Rose curves of the above forms have b petals if b is odd but $2b$ petals if b is even. (See **Question 10b.**)

Alternate Approach
There are two general ways to graph polar equations. One is to plot points, as in **Example 1.** (A teaching aid is provided with the table and graph.) The other is to utilize the corresponding graph in rectangular coordinates, as in **Example 3.** (A teaching aid is provided with all the graphs from this example.) The method of **Example 1** has the advantage of being more direct; the method of **Example 3** can help one see some of the larger features of the curve.

Point out that we can do **Example 2** because we know rectangular coordinate descriptions of a circle.

ADDITIONAL EXAMPLES
1. Sketch a graph of the polar equation $r = 4 \sin \theta$.

2. Prove that the graph in Additional Example 1 is a circle.
Substitute $\sqrt{x^2 + y^2}$ for r and $\frac{y}{r} = \frac{y}{\sqrt{x^2 + y^2}}$ for $\sin \theta$. This yields $\sqrt{x^2 + y^2} = \frac{y}{\sqrt{x^2 + y^2}}$, which is equivalent to $x^2 + y^2 = y$, or $x^2 + y^2 - y = 0$. From completing the square, or from the theorem about quadratic relations in Lesson 12-6, this is seen to be a circle.

3. Graph all $[r, \theta]$ with $r = 2 \cos 3\theta$.

Question 4: The answer to this question need not be in standard form, but it should involve only x and y, and not r and θ.

Question 14: This question is difficult if one does not do some algebra before considering points. Since $\sec \theta = \dfrac{1}{\cos \theta}$, the equation is equivalent to $r \cos \theta = 1$, that is, $x = 1$, which explains why the graph is a vertical line.

ADDITIONAL ANSWERS

1.a.

3.a. $[2, 0]$, $\left[\sqrt{3}, \dfrac{\pi}{6}\right]$, $\left[\sqrt{2}, \dfrac{\pi}{4}\right]$, $\left[1, \dfrac{\pi}{3}\right]$, $\left[0, \dfrac{\pi}{2}\right]$, $\left[-1, \dfrac{2\pi}{3}\right]$;

This equation, $\left(x - \dfrac{1}{2}\right)^2 + y^2 = \dfrac{1}{4}$, is an equation in the rectangular coordinate system for the circle with center at $\left(\dfrac{1}{2}, 0\right)$ and radius $\dfrac{1}{2}$. It is the same circle as graphed in Example 1 (though shown here smaller).

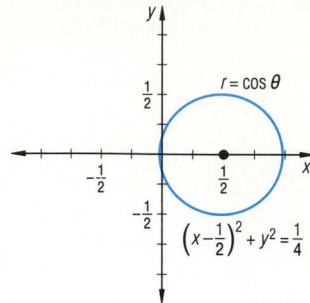

In the Questions you are asked to show that any equation of the form $r = a \cos \theta$, where $a \neq 0$, is a circle.

Recall that in the rectangular coordinate system, graphs of functions of the form $y = \cos b\theta$, where b is a positive integer, are sine waves with amplitude 1 and period $\dfrac{2\pi}{b}$. The graphs of polar equations in the form $r = \cos b\theta$, where b is a positive integer, are quite different and beautiful. Below are graphs for $b = 2, 3,$ and 4 made by an automatic grapher.

$r = \cos 2\theta$

$r = \cos 3\theta$

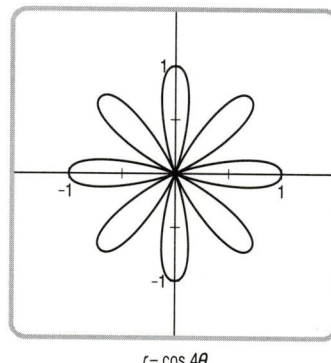

$r = \cos 4\theta$

These graphs are part of a family of polar graphs called *rose curves* or *petal curves*. The following example shows how to draw a rose curve by hand without having to plot too many points.

Example 3 Sketch a graph of all $[r, \theta]$ with $r = 3 \sin 2\theta$ in polar coordinates.

Solution First sketch a graph of $y = 3 \sin 2x$ for $0 \le x \le 2\pi$ in rectangular coordinates. As shown below it has amplitude 3 and period $\frac{2\pi}{2} = \pi$.

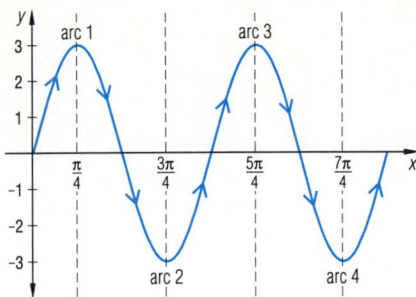

The possible y-values of the function on the rectangular coordinate graph are the possible values of r in the polar graph. In this case, $-3 \le r \le 3$.

The x-intercepts of the rectangular graph indicate when $r = 0$, that is, when the polar graph passes through the pole. Notice also that when $0 \le x \le 2\pi$, the rectangular coordinate graph has 4 congruent arcs, each symmetric to a dotted vertical line where x is an odd multiple of $\frac{\pi}{4}$. In the polar graph, this reflection symmetry gives rise to symmetry in the "petals."

To sketch the polar graph, begin with the point $[0, \theta]$. As θ increases from 0 to $\frac{\pi}{4}$, r increases from 0 to 3. This part of the graph of $r = 3 \sin 2\theta$ is pictured at the right.

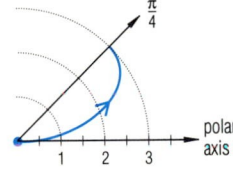

As θ continues to increase from $\frac{\pi}{4}$ to $\frac{\pi}{2}$, the value of r decreases from 3 to 0. The reflection symmetry in arc 1 above results in symmetry over the line $\theta = \frac{\pi}{4}$ for the corresponding arc in the polar graph. Thus, the loop shown at the right has been completed.

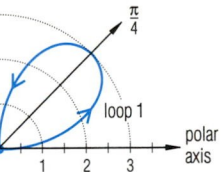

b. For $r \ne 0$, $r = 2 \cos \theta \Leftrightarrow r = 2\left(\frac{x}{r}\right) \Leftrightarrow r^2 = 2x \Leftrightarrow x^2 + y^2 = 2x \Leftrightarrow x^2 - 2x + y^2 = 0 \Leftrightarrow (x - 1)^2 + y^2 = 1$, **which is the equation of a circle with center (1, 0) and radius 1.**

4.a.

7.a.

c.,d. $[3, 0]$, $\left[1.5, \frac{\pi}{6}\right]$, $\left[3, \frac{\pi}{2}\right]$;

9.

10.b.

$r = \cos 5\theta$

$r = \cos 6\theta$

11.

12.a.

Similarly, as θ increases from $\frac{\pi}{2}$ to $\frac{3\pi}{4}$, r decreases from 0 to -1; and as θ goes from $\frac{3\pi}{4}$ to π, r increases from -1 to 0. These points are on loop 2, as noted on the sketch at the right.

A similar analysis of arcs 3 and 4 indicate that there are two more loops in the polar graph. A complete graph of $r = 3 \sin 2\theta$ is sketched below. It is a *4-petaled rose*.

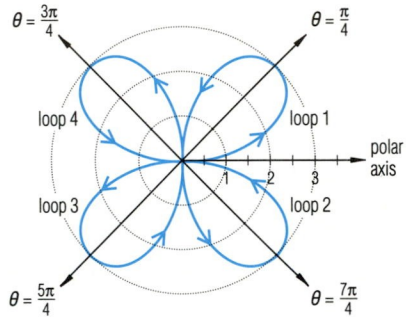

In general, polar graphs of trigonometric functions with equations of the form $r = c + a \sin b\theta$ or $r = c + a \cos b\theta$, where b is a positive integer and $a \neq 0$, are beautiful curves. You are asked to draw some of these in the Questions.

Questions

Covering the Reading

1. Consider the polar equation $r = 2$.
   a. Sketch a graph of the equation.   **See margin.**
   b. Give an equation in x and y for this equation. $x^2 + y^2 = 4$

2. Describe the graph of $r = a$, where a is a constant nonzero real number.
   It is a circle with center at the pole and radius a.

3. Consider the equation $r = 2 \cos \theta$ in the polar plane.
   a. Find and plot at least six points on its graph.
   b. Prove that the graph is a circle.   **a,b) See margin.**

4. Consider $r = \sin \theta$.
   a. Sketch a graph of this function in a polar coordinate system.
   b. Derive an equation for this curve in rectangular coordinates.
   a) See margin.

   b) For $r \neq 0$, $r = \sin \theta \Leftrightarrow r = \frac{y}{r} \Leftrightarrow r^2 = y \Leftrightarrow x^2 + y^2 = y \Leftrightarrow x^2 + y^2 - y = 0 \Leftrightarrow$
   $x^2 + (y-1)^2 = \frac{1}{4}$

800

In 5 and 6, refer to the rose curves on page 798. *True* or *false*.

5. The point $[4, \pi]$ is on the graph of $r = \cos 4\theta$.   **False**

6. The point $[0, 0]$ is on the graph of $r = \cos 2\theta$.   **True**

7. Consider the equation $r = 3 \cos 2\theta$.
  a. Sketch a graph of this equation in the rectangular plane.
  b. What are the maximum and minimum values of r?   **3, -3**
  c. Find r when $\theta = 0$, $\frac{\pi}{6}$, and $\frac{\pi}{2}$, and plot these points in the polar plane.
  d. Use the technique described in Example 3 to sketch a complete graph of this equation in polar coordinates.   **a, c, d) See margin.**

8. At the right is the graph of $r = \sin 2\theta$. Give the coordinates of four points on the graph.

  samples: $\left[1, \frac{\pi}{4}\right]$,

  $[0, \pi]$, $\left[-1, \frac{3\pi}{4}\right]$, $[0, 2\pi]$

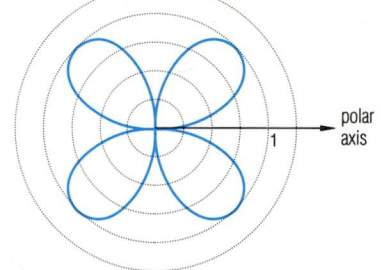

polar axis

9. Sketch a graph of $r = a \sin 2\theta$, where a is a positive real number, in polar coordinates.   **See margin.**

10. Refer to the graphs of $r = \cos 2\theta$, $r = \cos 3\theta$ and $r = \cos 4\theta$ in the reading.
  a. Make a conjecture regarding the number of petals on the polar curve $r = \cos n\theta$, where n is a positive integer.
  b. Test your conjecture by drawing graphs of $r = \cos 5\theta$ and $r = \cos 6\theta$ with an automatic grapher.   **a) If *n* is odd, there are *n* petals. If *n* is even, there are 2*n* petals. b) See margin.**

11. The graph of $r = \theta$, for $\theta > 0$, in the polar plane is called an *Archimedean spiral*, named after Archimedes. Graph this curve.   **See margin.**

12. The graph of $r = a(\cos \theta - 1)$ in the polar plane is called a *cardioid*.
  a. Graph this curve for $a = 1$.   **See margin.**
  b. Why is *cardioid* an appropriate name for this curve?
  It is shaped like a heart.

13. The graph of $r = a^\theta$, where $a > 0$, is called a *logarithmic spiral*.
  a. For $r = 2^\theta$, find the coordinates of the points in the interval $0 \le \theta < 2\pi$ where θ is a multiple of $\frac{\pi}{4}$.   **See below.**
  b. Graph the curve.   **See margin.**

14. Consider the equation $r = \sec \theta$.
  a. Graph the equation in polar coordinates.
  b. Give an equation of the graph in rectangular coordinates.
  a) See margin. b) x = 1

13a) $[1, 0]$, $\left[1.72, \frac{\pi}{4}\right]$, $\left[2.97, \frac{\pi}{2}\right]$, $\left[5.12, \frac{3\pi}{4}\right]$, $[8.82, \pi]$,

$\left[15.21, \frac{5\pi}{4}\right]$, $\left[26.22, \frac{3\pi}{2}\right]$, $\left[45.19, \frac{7\pi}{4}\right]$

13.b.

14.a.

801

MORE PRACTICE
For more questions on SPUR
Objectives, use *Lesson Mas-
ter 13-4*, shown on page 801.

PROJECTS
Project 2 on page 828 is
related to the content of this
lesson.

ADDITIONAL ANSWERS
16.a.

17.a.

$y = \tan 2x$ $y = 2 \tan x$

18. sample: $\begin{bmatrix} 1 & 2 \\ 1 & 2 \end{bmatrix} \cdot$

$\begin{bmatrix} 3 & 4 \\ 3 & 4 \end{bmatrix} = \begin{bmatrix} 9 & 12 \\ 9 & 12 \end{bmatrix}$

while $\begin{bmatrix} 3 & 4 \\ 3 & 4 \end{bmatrix} \cdot$

$\begin{bmatrix} 1 & 2 \\ 1 & 2 \end{bmatrix} = \begin{bmatrix} 7 & 14 \\ 7 & 14 \end{bmatrix}$

**Multiplication of 2 × 2
matrices is not
commutative.**

Review

15. Which polar pairs describe the same point?

(a) $[0.5, \frac{5\pi}{6}]$ (b) $[-\frac{1}{2}, -\frac{\pi}{6}]$

(c) $[\frac{1}{2}, -\frac{7\pi}{6}]$ (d) $[-0.5, -\frac{5\pi}{6}]$ *(Lesson 13-3)*   **(a), (b), and (c)**

In 16 and 17, **a.** use an automatic grapher to test whether the equation seems
to be an identity. **b.** Prove your conclusion in part **a.**   *(Lessons 13-1, 13-2)*

16. $\sin^2 x = \frac{1}{2}(1 - \cos 2x)$ **a) See margin. b) See below.**

17. $\tan 2x = 2 \tan x$ **a) See margin. b) See below.**

18. Show that multiplication of 2×2 matrices is not commutative.
(Lesson 11-1)   **See margin.**

19. Suppose $f(x) = x^3 - 729$. Find all complex numbers such that $f(x) = 0$.
(Lesson 9-8) $9, -\frac{9}{2} + \frac{9\sqrt{3}}{2}i, -\frac{9}{2} - \frac{9\sqrt{3}}{2}i$

20. Let $u = 3 - 4i$ and $v = 5 + i$. Write each of the following in $a + bi$
form.   *(Lesson 9-6)*
a. $u + v$ **8 − 3i**   **b.** uv **19 − 17i**   **c.** $\frac{u}{v}$ $\frac{11}{26} - \frac{23}{26}i$

Exploration

21. Consider equations of the form $r = a + b \sin \theta$, where $a > 0$ and $b > 0$.
Experiment with an automatic grapher using various values of a and b.
a. Find an equation whose graph looks like the one below.

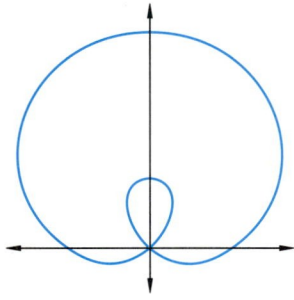

b. In general, what is true about a and b if the graph of $r = a + b \sin \theta$
has a loop as above?
a) sample: $r = 1 + 2 \sin \theta$; **b)** $b > a$

16b) $\sin^2 x \mid \frac{1}{2}(1 - \cos 2x)$

$= \frac{1}{2}(1 - (\cos^2 x - \sin^2 x))$

$= \frac{1}{2}(1 - \cos^2 x + \sin^2 x)$

$= \frac{1}{2}(2 \sin^2 x)$

$= \sin^2 x$

So $\sin^2 x = \frac{1}{2}(1 - \cos 2x)$ for all x.

17b) For $x = 30°$, $2 \tan x = 2 \cdot \frac{1}{\sqrt{3}} = \frac{2\sqrt{3}}{3}$ and $\tan 2x = \tan 60° = \sqrt{3}$.

So $\tan 2x = 2 \tan x$ is not an identity.

The Geometry of Complex Numbers

About 1800, Caspar Wessel, a Norwegian surveyor, and Jean Robert Argand, a Swiss mathematician, independently invented a geometric representation of complex numbers. Their diagrams, sometimes called *Argand diagrams,* involve graphs in a *complex plane*. In a complex plane, the horizontal axis is called the **real axis** and the vertical axis is called the **imaginary axis**. To graph the complex number $a + bi$ in the complex plane, first write it as the ordered pair (a, b). Then plot (a, b) as you normally would in a rectangular coordinate system.

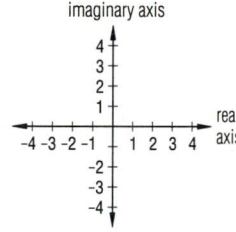

Notice that a real number is of the form $a + 0i$, so it equals $(a, 0)$ and is plotted on the real axis in the complex plane. Similarly, imaginary numbers are of the form $0 + bi$, and $(0, b)$ is plotted on the imaginary axis. The complex number $0 = 0 + 0i = (0, 0)$, and it is graphed at the origin.

∙ ∙ ∙ ∙ ∙ ∙ ∙ ∙ ∙ ∎ ∎ ∎

Example 1   Graph in the complex plane.
   a. $-6i$ **b.** $2 + 5i$ **c.** $-4 - i$

Solution
   a. $-6i$ can be rewritten as $0 - 6i$, which equals the ordered pair $(0, -6)$.
   b. $2 + 5i = (2, 5)$
   c. $-4 - i = (-4, -1)$

RESOURCES
- Lesson Master 13-5
- Teaching Aid 99: Graphs for Geometric Addition Theorem and **Example 2**

OBJECTIVES

B Represent complex numbers in different forms.
J Graph complex numbers.

TEACHING NOTES

Making Connections
There are at least four analogies that you can make in this lesson between real numbers and complex numbers. (1) Real numbers are identified with points on a line, called the *real number line*. Complex numbers are identified with points on a plane, the *complex plane*. (2) Real number addition can be done using arrows on a number line. The picture of complex number addition is identical, except it involves two dimensions. (A teaching aid is provided.) (3) When x is a real number, $|x|$ is the distance of x from 0, and when $z = a + bi = (a, b)$ is a complex number, $|z| = \sqrt{a^2 + b^2}$ is the distance of z from $(0, 0)$. (4) If x and y are real numbers, then $|x - y|$ is the distance on a number line from x to y. If z and w are complex numbers, then $|z - w|$ is the distance between them in the complex plane. (See **Question 15.**)

Make certain that students understand that complex numbers can be written in a variety of ways. For instance, from **Example 3,** note that $3 - 5i = (3, -5) \approx [\sqrt{34}, -59°]$. This is analogous to a real number $18\frac{1}{3} = \frac{55}{3} \approx 18.33$.

804

Although there are many polar coordinates [r, θ] for a given complex number, in practice the preferred value of r is positive, and θ is usually in the interval $0 \le \theta < 2\pi$. This is similar to the preference of having a fraction in lowest terms.

Error Analysis   A common error is for students to put an i in the second coordinate of the complex number. That is, they will write (a, bi) for $a + bi$. Explain that the order of the coordinates tells which is the real part and which is the imaginary part, so no i is needed.

ADDITIONAL EXAMPLES

1. At what point in the complex plane would each number be graphed?
a. $3 - 7i$ **b.** $4i$ **c.** 8.21
a. (3, -7); b. (0, 4);
c. (8.21, 0)

2. Let $z = 5 - 2i$ and $w = 1 + 6i$.

a. Graph z, w, $z + w$, $z - w$, and $w - z$.
z = (5, -2), w = (1, 6),
z + w = (6, 4), z - w = (4, -8),
and w - z = (-4, 8).

b. How are the graphs of these numbers related?
Sample: In the triangle with vertices $z + w$, $z - w$, and $w - z$, the midpoints of the three sides are z, w, and 0.

3. Find polar coordinates for $-7 + 7i$.

Sample: $\left[7\sqrt{2}, \dfrac{3\pi}{4}\right]$

Addition of complex numbers has a nice geometric interpretation in the complex plane. For instance, the sum of $3 + 7i$ and $2 - 4i$ is $5 + 3i$. As shown in the diagram below, the complex numbers (3, 7), (2, -4), their sum (5, 3), and the origin (0, 0) are the vertices of a parallelogram.

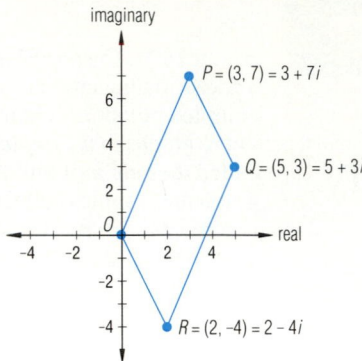

This is easy to verify. Because the slopes of the opposite sides of $OPQR$ are equal, the opposite sides are parallel and the quadrilateral is a parallelogram.

$$\text{slope of } \overline{OP} = \frac{7-0}{3-0} = \frac{7}{3}$$

$$\text{slope of } \overline{PQ} = \frac{3-7}{5-3} = \frac{-4}{2} = -2$$

$$\text{slope of } \overline{QR} = \frac{-4-3}{2-5} = \frac{-7}{-3} = \frac{7}{3}$$

$$\text{slope of } \overline{OR} = \frac{-4-0}{2-0} = \frac{-4}{2} = -2$$

This proves one instance of the following theorem. You are asked to prove the general case in the Questions.

Geometric Addition Theorem

Given two complex numbers $a + bi$ and $c + di$ that are not on the same line through the origin in the complex plane, the sum $(a + c) + (b + d)i$ is the fourth vertex of a parallelogram with consecutive vertices $a + bi$, 0, and $c + di$.

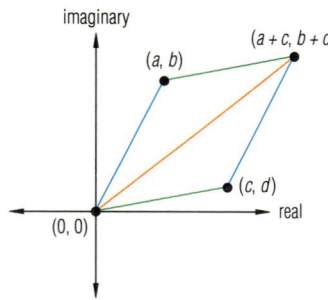

804

Example 2 Let $z = 8 + 6i$ and $w = 2 - 4i$. Represent geometrically the quantities $z + w$, $z - w$, $w - z$, and $-(z + w)$.

Solution First perform the operations:
$$z + w = 10 + 2i$$
$$z - w = 6 + 10i$$
$$w - z = -6 - 10i$$
and $-(z + w) = -10 - 2i$.

Notice that $z - w$ is the opposite of $w - z$ and $-(z + w)$ is the opposite of $z + w$, as you would expect. These complex numbers are graphed in the complex plane below.

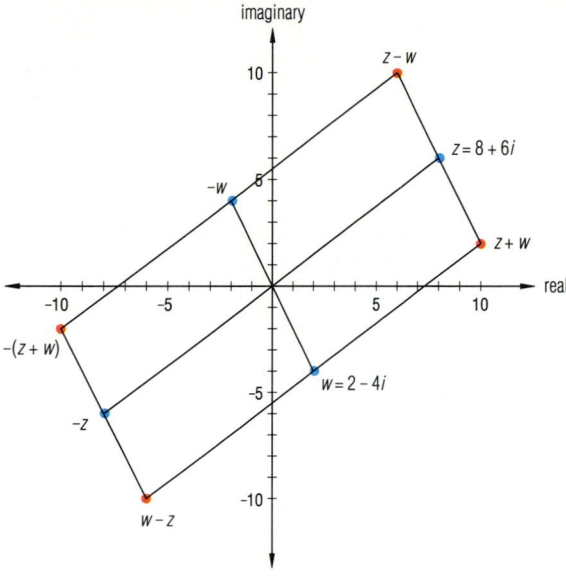

Complex numbers can also be represented with polar coordinates. Let $[r, \theta]$ with $r \geq 0$ be polar coordinates for (a, b). Then from Lesson 13-3 you know that $r = \sqrt{a^2 + b^2}$ (because r is not negative) and $\tan \theta = \frac{b}{a}$. There are usually two values of θ between 0 and 2π which satisfy $\tan \theta = \frac{b}{a}$. The correct value can be determined by examining the quadrant in which (a, b) is located.

Example 3 Find polar coordinates for the complex number $3 - 5i$.

Solution Let $[r, \theta]$ be polar coordinates for $3 - 5i = (3, -5)$. This is a point in the 4th quadrant.

$$r = \sqrt{3^2 + (-5)^2} = \sqrt{34}$$
$$\tan \theta = -\frac{5}{3}$$
So $\theta = \tan^{-1}\left(-\dfrac{5}{3}\right) \approx -59°$.

Thus one pair of polar coordinates for $(3, -5)$ is about $[\sqrt{34}, -59°]$.

For any complex number $z = [r, \theta]$ with $r \geq 0$, r is its distance from the origin. That distance is also called the **absolute value** or **modulus** of the complex number and is written $|z|$. If $z = a + bi$, then $|z| = |a + bi| = \sqrt{a^2 + b^2}$. The direction θ is called an **argument** of the complex number and can be found using trigonometry: $\tan \theta = \dfrac{b}{a}$. An argument may be measured in degrees or radians. Because of periodicity, more than one argument exists for each complex number. In Example 3, the modulus of the complex number is $\sqrt{34}$, and an argument is $-59°$. Other arguments are $-59° + 360°n$, where n is an integer.

The form $[r, \theta]$ for a complex number is called **polar form**. In the next lesson you will see that the polar form of complex numbers is very useful for describing the product or quotients of two complex numbers.

Questions

Covering the Reading

1. In a complex plane the horizontal axis is called the __?__ axis, and the vertical axis is called the __?__ axis.   **real, imaginary**

In 2 and 3, **a.** rewrite each complex number as an ordered pair (a, b), and **b.** graph the number in the complex plane.

2. $4 - 6i$ **a) (4, -6); b) See margin.**   3. $-5 + 4i$ **a) (-5, 4); b) See margin.**

4. Write each complex number pictured at the right in $a + bi$ form.
$A = 3 + 0i$, $B = -1 + 6i$, $C = -3 - 3i$, $D = 0 - 4i$

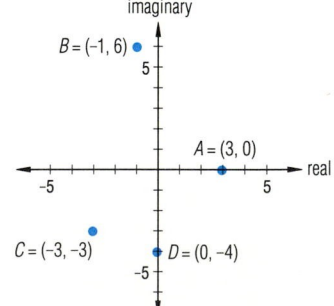

806

5. Let $U = 3 - 2i$, $V = -5 + 6i$, and $O = 0 + 0i$.
  a. Find $U + V$.   **-2 + 4i**
  b. Graph U, V, and $U + V$ in the same coordinate plane.
  c. Verify that U, O, V, and $U + V$ are vertices of a parallelogram.
     b,c) See margin.

6. Refer to Example 2.
  a. How many parallelograms are determined by the complex numbers pictured?   **9**
  b. Prove that the largest quadrilateral pictured is a parallelogram.
     See margin.

In 7 and 8, give polar coordinates $[r, \theta]$ for each complex number. Let $r \geq 0$ and $0 \leq \theta \leq 2\pi$.

7. $4 - 3i$ $\approx$ **[5, 5.64]**   **8.** $\frac{1}{2} + \frac{\sqrt{3}}{2}i$ $[1, \frac{\pi}{3}]$

9. How is the absolute value of the complex number $a + bi$ calculated?
   $\sqrt{a^2 + b^2}$

In 10 and 11, **a.** find the absolute value, and **b.** find the argument θ. Let $r \geq 0$ and $0° \leq \theta < 360°$.

10. $7i$ **a) 7; b) 90°**   **11.** $-3 + 6i$ **a)** $\sqrt{45} \approx$ **6.7;**
                                                            b) $\approx$ **116.6°**

Applying the Mathematics

12. Name and graph four complex numbers with a modulus of 1.
   samples: 1, i, .6 + .8i, -.8 + .6i
13. Prove the Geometric Addition Theorem.   **See margin.**

14. a. Draw the quadrilateral with vertices
       $P = 6 + i$, $Q = 6 - i$, $P + Q$, and $(0, 0)$.   **See margin.**
   b. What special type of parallelogram is this?   **rhombus**
   c. Determine the length of the longer diagonal of the quadrilateral.   **12**

15. On the real number line, the distance between points with coordinates u and v is $|u - v|$. Determine whether the distance from $u = a + bi$ to $v = c + di$ in the complex plane is $|u - v|$. Justify your answer.
   See margin.
16. Consider $u = -3 + 4i$ and $v = -12i$. Evaluate.
   a. $|u| + |v|$ **17**   **b.** $|u + v|$ $\sqrt{73}$
   c. Use the Triangle Inequality to explain why $|u + v| < |u| + |v|$ in general.   **See margin.**

Review

In 17–20, graph in the polar plane.   *(Lesson 13-4)*

17. $\theta = \frac{3\pi}{4}$ **See margin.**   **18.** $r = 3\theta$, when $0 \leq \theta \leq 4\pi$
                                                              See margin.
19. $r = 2 \sin \theta$ **See margin.**   **20.** $r = \sin 5\theta$ **See margin.**

21. Give an equation in the rectangular coordinate system for the polar equation in Question 19.   *(Lesson 13-3)* $x^2 + (y - 1)^2 = 1$

22. Find three other pairs of polar coordinates for the point $[-4, \frac{\pi}{6}]$.
   (Lesson 13-3)
       samples: $\left[4, \frac{7\pi}{6}\right]$, $\left[4, \frac{19\pi}{6}\right]$, $\left[4, \frac{-5\pi}{6}\right]$

16.c. |u + v| is the length of one side of the triangle with vertices (0, 0), u, and v. |u| and |v| are the lengths of the other two sides. By the Triangle Inequality, the length of one side of a triangle is less than the sum of the lengths of the other two sides. Therefore, |u + v| < |u| + |v|.

17.

18.

19.

20.

23. Seven pairs of 16 year old, black female twins took part in a study of the heights (in inches) and weights (in pounds) of identical twins. Here are the (height, weight) data for each twin in each pair.

Pair	(Height, Weight)	(height, weight)
1	A = (68, 148)	a = (67, 137)
2	B = (65, 124)	b = (67, 126)
3	C = (63, 118)	c = (63, 126)
4	D = (66, 131)	d = (64, 120)
5	E = (62, 123)	e = (65, 124)
6	F = (62, 119)	f = (63, 130)
7	G = (66, 114)	g = (66, 104)

A scatter-plot of the data is below.

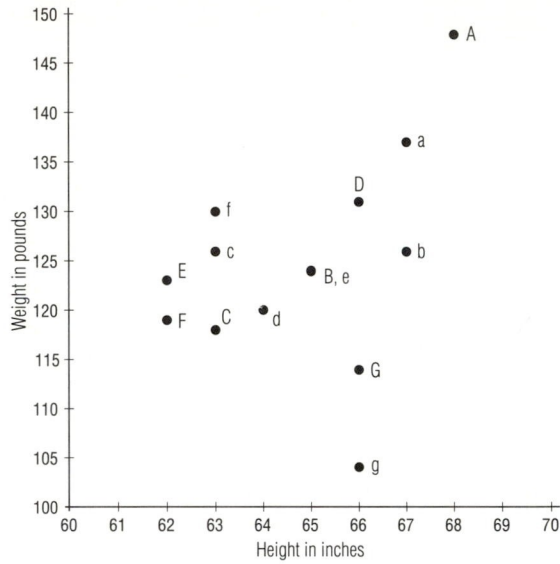

By calculating various statistics from the data, respond to the following questions.
a. Which pair do you think is most different on these variables?
b. Which pair do you think is most alike on these variables?
c. Defend your answers to parts **a** and **b**. *(Lessons 1-4, 1-8, 10-6)*

Sample: Calculate z-scores for the 14 heights and for the 14 weights. Then, pair 4 is the most different because the total difference of the z-scores of D and d for height and weight is greatest; pair 3 is most similar because the total difference of the z-scores for C and c is least.

24. Give the coordinates of the image of $(11, 6)$ under a rotation of θ.
(Lesson 11-4)   **$(11 \cos \theta - 6 \sin \theta, 11 \sin \theta + 6 \cos \theta)$**

25. Give the complex conjugate.   *(Lesson 9-6)*
  a. $-1 + 5i$ **$-1 - 5i$**   **b.** $5i$ **$-5i$**   **c.** -1 **-1**

26. Given $z_1 = 2 + i$ and $z_2 = 3 + 2i$, put in $a + bi$ form.   *(Lesson 9-6)*

  a. $z_1 z_2$ **$4 + 7i$**   **b.** $\dfrac{z_1}{z_2}$ **$\dfrac{8}{13} - \dfrac{1}{13}i$**

27. Two tracking stations are 30 miles apart. They measure the elevation angle of a weather balloon to be α and β, respectively. How high is a balloon sighted with $\alpha = 40°$ and $\beta = 70°$?   *(Lessons 5-3, 6-7)*

≈ 36 miles

Exploration

28. Let $w = 1 + i$.
  a. Write w^0, w^2, w^3, and w^4 in $a + bi$ form.
  b. Graph w^0, w, w^2, w^3, and w^4 in the complex plane.   **See margin.**
  c. Describe the pattern that emerges in the graph of $(1 + i)^n$ for positive integers n.   **The points lie on a spiral.**
  a) $w^0 = 1 + 0i$; $w^2 = 0 + 2i$; $w^3 = -2 + 2i$; $w^4 = -4 + 0i$

OBJECTIVES

A Perform operations with complex numbers.
B Represent complex numbers in different forms.
J Graph complex numbers.

TEACHING NOTES

Alternate Approach To graph a point $[r, \theta]$ in polar coordinates, we can think of applying the composite $R_\theta \circ S_r$ to $[1, 0]$. To multiply two complex numbers $[r, \theta]$ and $[s, \phi]$, think of applying the transformation $R_\phi \circ S_s$ to the first point. That is, rotate the first point ϕ and apply a size change of magnitude s. The result is the product $[rs, \theta + \phi]$. Thus multiplication of complex numbers in polar coordinates is analogous to adding complex numbers in rectangular coordinates.

After students have read this lesson, review the four ways for writing complex numbers: binomial form ($a + bi$ form):
$a + bi$
rectangular coordinate form:
(a, b)
polar form:
$[r, \theta]$
trigonometric form:
$r (\cos \theta + i \sin \theta)$.
Continue the analogies with real numbers. Generally, real number addition and subtraction are easier if the real numbers are represented as decimals rather than fractions. Real number multiplication and division, however,

---

LESSON

13-6

Trigonometric Form of Complex Numbers

Consider the complex number $z = a + bi$ with polar coordinates $[r, \theta]$, $r \geq 0$.

By the theorem of Lesson 13-3, $a = r \cos \theta$ and $b = r \sin \theta$. So
$$z = a + bi$$
$$= r \cos \theta + (r \sin \theta) i$$
$$= r(\cos \theta + i \sin \theta).$$

The expression $r(\cos \theta + i \sin \theta)$ is called the **trigonometric form of a complex number** because it uses the cosine and sine of the argument θ. Like polar form, the trigonometric form denotes a complex number in terms of r and θ, its absolute value and argument. But unlike polar form, a complex number in trigonometric form is still in the rectangular coordinate form $a + bi$. Because of this link between polar and rectangular coordinates, the trigonometric form of complex numbers is quite useful.

■ ■ ■ ■ ■ ■ ■ ■ ■ ■

Example 1 Write the complex number $-2 - 2\sqrt{3}\, i$ in trigonometric form. Use a value of θ in $0 \leq \theta < 2\pi$.

Solution The process is quite similar to that used in finding polar coordinates for a point. First sketch a graph.

$$r = \sqrt{(-2)^2 + (-2\sqrt{3}\,)^2}$$
$$= \sqrt{4 + 12}$$
$$= 4$$
$$\tan \theta = \frac{-2\sqrt{3}}{-2}$$
$$= \sqrt{3}$$

Since $-2 - 2\sqrt{3}i$ is in the third quadrant,

$$\theta = \pi + \frac{\pi}{3}$$
$$= \frac{4\pi}{3}.$$

Therefore, $-2 - 2\sqrt{3}\, i = 4(\cos\frac{4\pi}{3} + i \sin \frac{4\pi}{3})$.

810

Consider again the number of Example 1. Since the sine and cosine have period 2π, $-2 - 2\sqrt{3}\,i = 4\left(\cos\left(\frac{4\pi}{3} + 2n\pi\right) + i\sin\left(\frac{4\pi}{3} + 2n\pi\right)\right)$, where n is any integer. In general, every complex number has infinitely many trigonometric forms.

■ ■ ■ ■ ■ ■ ■ ■

Example 2 Write the complex number $5\left(\cos\frac{2\pi}{3} + i\sin\frac{2\pi}{3}\right)$ in $a + bi$ form.

> **Solution** Distribute the 5 and simplify.
> $$5\left(\cos\frac{2\pi}{3} + i\sin\frac{2\pi}{3}\right) = 5\cos\frac{2\pi}{3} + i\left(5\sin\frac{2\pi}{3}\right)$$
> $$= -\frac{5}{2} + \frac{5\sqrt{3}}{2}\,i$$

The trigonometric form of complex numbers exhibits a nice geometrical representation of complex number multiplication. An instance is given in Example 3, where three complex numbers are each multiplied by $2i$.

■ ■ ■ ■ ■ ■ ■ ■

Example 3 Consider $\triangle ABC$ in the complex plane with vertices $A = 3i$, $B = 4 - 3i$, and $C = 5$.

a. Multiply each vertex by $2i$ to get A', B', and C'.
b. Describe the transformation that maps $\triangle ABC$ to $\triangle A'B'C'$.

Solution
a. $A' = 2i(3i) = -6$
$B' = 2i(4 - 3i) = 6 + 8i$
$C' = 2i(5) = 10i$
b. The graphs of $\triangle ABC$ and $\triangle A'B'C'$ are shown at the right.

It appears that $\triangle A'B'C'$ is the image of $\triangle ABC$ under the composite of a size change of magnitude 2 and a rotation of 90°.

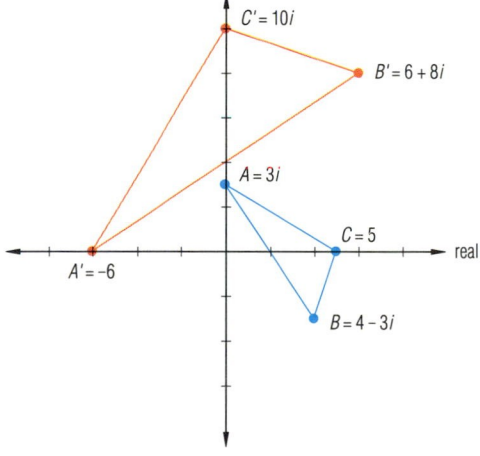

are often easier with fractions. Similarly, complex number addition and subtraction are usually easier in $a + bi$ form or rectangular coordinate form, while complex number multiplication and division are easier in polar or trigonometric form.

Examples 1 and 2 cover the translation in and out of trigonometric form. **Example 3** illustrates the geometry of complex number multiplication. A teaching aid is provided.

Regarding the Product of Complex Numbers Theorem, the most important objective is that students be able to apply the theorem as in **Example 4** and the Questions. The graph from **Example 4** is on a teaching aid.

The graphical representation of products of complex numbers is interesting, but time-consuming to draw. You might expect students to be able to explain the representation shown on page 813, but ask them to graph a product only once or twice (**Question 7c**). The fact that multiplication rotates a complex number will be applied in Lesson 13-7 when students examine the nth roots of a complex number.

Example 5 indicates how easy division is with numbers in trigonometric form. Note how long the check is compared to the solution.

ADDITIONAL EXAMPLES
1. Write $3 + 4i$ in trigonometric form.
$\approx$ **5(cos 53° + i sin 53°)**

2. Write $2\left(\cos\frac{\pi}{4} + i\sin\frac{\pi}{4}\right)$ in $a + bi$ form.
$\sqrt{2} + \sqrt{2}\,i$

3. Consider the $\triangle XYZ$ in the complex plane with $X = -3 - 6i$, $Y = 4i$, and $Z = 2 + i$.

a. Multiply each vertex by $1 - i$ to obtain X', Y', and Z'.
$X' = -9 - 3i$; $Y' = 4 + 4i$, and $Z' = 3 - i$

811

In trigonometric form, $2i = 2(\cos 90° + i \sin 90°)$. Thus the magnitudes of the transformations in Example 3 are the absolute value and argument of $2i$, respectively. That is, multiplication by $2i$ can be considered as the composite of a size change of magnitude 2 (its absolute value) and a rotation through 90° (its argument) around the origin.

In general, multiplying a complex number z_1 by the complex number $z_2 = [r_2, \theta_2] = r_2(\cos \theta_2 + i \sin \theta_2)$ applies to the graph of z_1 the composite of a size change of magnitude $|z_2|$ and a rotation θ_2 about the origin.

Product of Complex Numbers Theorem (Trigonometric Form)

If $\qquad z_1 = r_1(\cos \theta_1 + i \sin \theta_1)$ and
$\qquad\qquad z_2 = r_2(\cos \theta_2 + i \sin \theta_2),$
then $\qquad z_1z_2 = r_1r_2(\cos (\theta_1 + \theta_2) + i \sin (\theta_1 + \theta_2)).$

Proof The proof applies, perhaps surprisingly, the sum formulas for the cosine and sine.

$$z_1z_2 = (r_1(\cos \theta_1 + i \sin \theta_1)) (r_2(\cos \theta_2 + i \sin \theta_2))$$
$$= r_1r_2(\cos \theta_1 + i \sin \theta_1)(\cos \theta_2 + i \sin \theta_2)$$
$$= r_1r_2(\cos \theta_1 \cos \theta_2 + i \cos \theta_1 \sin \theta_2 + i \sin \theta_1 \cos \theta_2 + i^2 \sin \theta_1 \sin \theta_2)$$
$$= r_1r_2((\cos \theta_1 \cos \theta_2 + i^2 \sin \theta_1 \sin \theta_2) + i(\cos \theta_1 \sin \theta_2 + \sin \theta_1 \cos \theta_2))$$
$$= r_1r_2((\cos \theta_1 \cos \theta_2 - \sin \theta_1 \sin \theta_2) + i(\sin \theta_1 \cos \theta_2 + \cos \theta_1 \sin \theta_2))$$
$$= r_1r_2(\cos(\theta_1 + \theta_2) + i \sin(\theta_1 + \theta_2))$$

This is the trigonometric form for a complex number with absolute value r_1r_2 and argument $\theta_1 + \theta_2$.

In polar coordinates, the above theorem states that the product of the complex numbers $[r_1, \theta_1]$ and $[r_2, \theta_2]$ is $[r_1r_2, \theta_1 + \theta_2]$.

■ ■ ■ ■ ■ ■ ■ ■ ■

Example 4 If $z_1 = 10i$ and $z_2 = 3(\cos 75° + i \sin 75°)$, find the product z_1z_2 in trigonometric form.

Solution In trigonometric form,
$z_1 = 10(\cos 90° + i \sin 90°).$
Now use the Product of Complex Numbers Theorem.
$z_1z_2 = 10 \cdot 3 (\cos (90° + 75°) + i \sin (90° + 75°))$
$\qquad = 30(\cos 165° + i \sin 165°)$

Check   Multiplying by $z_2 = 3 (\cos 75° + i \sin 75°)$ should apply the composite of a size change of magnitude 3 and a rotation of 75° to the graph of the complex number $z_1 = 10 (\cos 90° + i \sin 90°)$. The graph below illustrates that this is so.

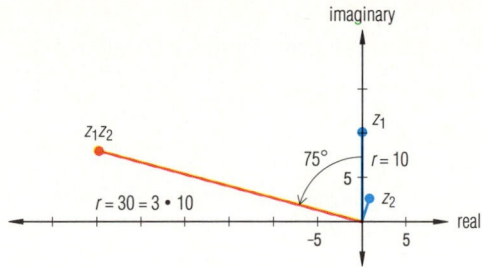

To perform division of a number z by a complex number in $a + bi$ form, it was useful to multiply numerator and denominator of $\dfrac{z}{a + bi}$ by the complex conjugate $a - bi$. In trigonometric form, the conjugate of $r (\cos \theta + i \sin \theta)$ is

$$r (\cos \theta - i \sin \theta) = r \cos \theta - i(r \sin \theta)$$
$$= r \cos (-\theta) + ir \sin (-\theta)$$
$$= r (\cos (-\theta) + i \sin (-\theta)).$$

This result is needed to prove the following theorem, which you are asked to do in the Questions.

Division of Complex Numbers Theorem (Trigonometric Form)

If $z_1 = r_1(\cos \theta_1 + i \sin \theta_1)$ and
$z_2 = r_2(\cos \theta_2 + i \sin \theta_2)$,

then $\dfrac{z_1}{z_2} = \dfrac{r_1}{r_2} (\cos (\theta_1 - \theta_2) + i \sin (\theta_1 - \theta_2)).$

Geometrically, division of the complex number $z_1 = [r_1, \theta_1]$ by the complex number $z_2 = [r_2, \theta_2]$ applies to z_1 the composite of a size change of magnitude $\dfrac{1}{|z_2|}$ and a rotation of $-\theta_2$ about the origin. So in polar form, the above theorem states that the quotient $\dfrac{[r_1, \theta_1]}{[r_2, \theta_2]}$ is $\left[\dfrac{r_1}{r_2}, \theta_1 - \theta_2\right]$.

2.a.

3.a.

4.a.

5.a.

6.a.

7.c.

14b. $\dfrac{z_1}{z_2} = \dfrac{\text{-}18 + 0i}{\dfrac{3\sqrt{3}}{2} + \dfrac{3}{2}i} =$

$\dfrac{\text{-}18\left(\dfrac{3\sqrt{3}}{2} - \dfrac{3}{2}i\right)}{\left(\dfrac{3\sqrt{3}}{2} + \dfrac{3}{2}i\right)\left(\dfrac{3\sqrt{3}}{2} - \dfrac{3}{2}i\right)} =$

$\dfrac{\text{-}27\sqrt{3} + 27i}{9} = \text{-}3\sqrt{3} + 3i$, **and**

$6\left(\cos\dfrac{5\pi}{6} + i\sin\dfrac{5\pi}{6}\right) =$

$6\left(\dfrac{\sqrt{3}}{2} + \dfrac{1}{2}i\right) = 3\sqrt{3} + 3i$

15b. $\dfrac{z_1}{z_2} = \dfrac{10 - 10\sqrt{3}i}{\dfrac{5}{2} + \dfrac{5\sqrt{3}}{2}i} =$

$\dfrac{10 - 10\sqrt{3}i}{\dfrac{5}{2}(1 + \sqrt{3}i)} = \dfrac{4 - 4\sqrt{3}i}{1 + \sqrt{3}i} =$

$\dfrac{(4 - 4\sqrt{3}i)(1 - \sqrt{3}i)}{(1 + \sqrt{3}i)(1 - \sqrt{3}i)} = \dfrac{\text{-}8 - 8\sqrt{3}i}{4}$
$= \text{-}2 - 2\sqrt{3}i$, **and** $[4, 240°] =$
$4(\cos 240° + i\sin 240°) =$
$4\left(\text{-}\dfrac{1}{2} - \dfrac{\sqrt{3}}{2}\right) = \text{-}2 - 2\sqrt{3}i.$

21.a.

Example 5 Given the complex numbers

$$z_1 = 10\left(\cos\tfrac{\pi}{2} + i\sin\tfrac{\pi}{2}\right) \text{ and } z_2 = 2\left(\cos\tfrac{\pi}{3} + i\sin\tfrac{\pi}{3}\right),$$

find $\dfrac{z_1}{z_2}$.

Solution From the Division of Complex Numbers Theorem,

$$\frac{z_1}{z_2} = \frac{10}{2}\left(\cos\left(\tfrac{\pi}{2} - \tfrac{\pi}{3}\right) + i\sin\left(\tfrac{\pi}{2} - \tfrac{\pi}{3}\right)\right)$$

$$= 5\left(\cos\tfrac{\pi}{6} + i\sin\tfrac{\pi}{6}\right).$$

Check Convert to $a + bi$ form.
$$z_1 = 10\,(0 + i(1)) = 10i$$
$$z_2 = 2\left(\tfrac{1}{2} + i\tfrac{\sqrt{3}}{2}\right) = 1 + \sqrt{3}i$$

Now compute the quotient by multiplying the numerator and denominator of $\dfrac{z_1}{z_2}$ by the conjugate of $1 + \sqrt{3}\,i$.

$$\frac{z_1}{z_2} = \frac{10i}{1 + \sqrt{3}i}$$

$$= \frac{10i(1 - \sqrt{3}\,i)}{(1 + \sqrt{3}\,i)(1 - \sqrt{3}\,i)}$$

$$= \frac{10i + 10\sqrt{3}}{4}$$

$$= \frac{5\sqrt{3} + 5i}{2}$$

$$= 5\left(\frac{\sqrt{3}}{2} + \frac{1}{2}i\right)$$

$$= 5\left(\cos\tfrac{\pi}{6} + i\sin\tfrac{\pi}{6}\right) \quad \text{It checks.}$$

Notice in Example 5 how using the trigonometric form greatly simplifies division of complex numbers. In the next lesson you will see how using the trigonometric form leads to beautifully simple ways to calculate powers and roots of complex numbers.

Questions

Covering the Reading

In 1–4, **a.** graph each number on the complex plane. **b.** Convert it to trigonometric form. Use an argument in $0° \leq \theta < 360°$.

1. $3 + 3i$ a) See margin.
b) $3\sqrt{2}\,(\cos 45° + i\sin 45°)$

2. $4\sqrt{3} - 4i$ a) See margin.
b) $8(\cos 330° + i\sin 330°)$

3. $\text{-}5$ a) See margin.
b) $5(\cos 180° + i\sin 180°)$

4. $\text{-}2 - 5i$ a) See margin.
b) $\approx \sqrt{29}\,(\cos 248° + i\sin 248°)$

In 5 and 6, **a.** graph each number on the complex plane.
b. Convert to $a + bi$ form.

5. $3\left(\cos\tfrac{3\pi}{2} + i\sin\tfrac{3\pi}{2}\right)$
a) See margin. b) $0 - 3i$

6. $2\left(\cos\left(\text{-}\tfrac{2\pi}{3}\right) + i\sin\left(\text{-}\tfrac{2\pi}{3}\right)\right)$
a) See margin. b) $\text{-}1 - \sqrt{3}i$

814

7. a. Multiply $z_1 = 2(\cos 65° + i \sin 65°)$ by $z_2 = 4(\cos 40° + i \sin 40°)$, and express the result in trigonometric form.   **8(cos 105° + _i_ sin 105°)**

b. The composite of which two transformations maps z_1 to $z_1 z_2$?

c. Illustrate the multiplication with a diagram showing the appropriate size transformation and rotation.   **See margin.**
  b) size change with magnitude 4 and rotation around the origin of 40°

In 8–10, **a.** find $z_1 z_2$, **b.** give its absolute value, and **c.** give an argument of the product. Use exact values if possible.

8. $z_1 = 3(\cos 150° + i \sin 150°)$, $z_2 = 2(\cos 60° + i \sin 60°)$
  a) 6(cos 210° + _i_ sin 210°); b) 6; c) sample: 210°

9. $z_1 = 10\left(\cos \frac{11\pi}{4} + i \sin \frac{11\pi}{4}\right)$, $z_2 = 5\left(\cos \frac{\pi}{2} + i \sin \frac{\pi}{2}\right)$
  a) 50(cos 5π/4 + _i_ sin 5π/4); b) 50; c) sample: 5π/4

10. $z_1 = 2 + 3i$, $z_2 = -4 + i$
  a) -11 − 10_i_; b) $\sqrt{221}$; c) sample: ≈ 222.3°

In 11 and 12, find the conjugate of the complex number.

11. $3\left(\cos \frac{\pi}{6} + i \sin \frac{\pi}{6}\right)$ **12.** $[5, 175°]$ **[5, 185°]**

  3(cos 11π/6 + _i_ sin 11π/6)

13. a. Divide $z_1 = 12(\cos 220° + i \sin 220°)$ by $z_2 = 5(\cos 100° + i \sin 100°)$ and express the result in trigonometric form.   **2.4 (cos 120° + _i_ sin 120°)**

b. The composite of which two transformations maps z_2 to $\frac{z_1}{z_2}$?

  a size change with magnitude 1/5 and a rotation about the origin of -100°

In 14 and 15, **a.** find the quotient $\frac{z_1}{z_2}$ and express the result in trigonometric form. **b.** Check your result by converting to $a + bi$ form.

14. $z_1 = 18(\cos \pi + i \sin \pi)$, $z_2 = 3\left(\cos \frac{\pi}{6} + i \sin \frac{\pi}{6}\right)$ **a) $6\left(\cos \frac{5\pi}{6} + i \sin \frac{5\pi}{6}\right)$**

15. $z_1 = [20, 300°]$, $z_2 = [5, 60°]$ **a) [4, 240°] = 4(cos 240° + _i_ sin 240°)**
  14b, 15b) See margin.

Applying the Mathematics

16. A complex number z has absolute value 7 and argument $\frac{2\pi}{3}$.
a. Express z in polar form.
b. Express z in $a + bi$ form.   **a) $\left[7, \frac{2\pi}{3}\right]$; b) $-\frac{7}{2} + \frac{7\sqrt{3}}{2}i$**

17. Prove the Division of Complex Numbers Theorem.   **See margin.**

18. The complex number $z = 3(\cos 40° + i \sin 40°)$ undergoes a transformation that multiplies its absolute value by 5 and rotates it 75° about the origin.
a. What is the image of z under the transformation?
b. Identify the mathematical operation and the complex number that will accomplish the transformation.
  a) 15(cos 115° + _i_ sin 115); b) multiplication by 5(cos 75 + _i_ sin 75°)

19. Given: $z = 2(\cos 15° + i \sin 15°)$.
a. Find z^2, z^3, z^4, and z^5.
[Hint: Use the fact that $z^2 = z \cdot z$, $z^3 = z^2 \cdot z$, and so on.]
b. Look for a pattern in the results of part **a**. Use the pattern to predict what z^{10} should be.   **$z^{10} = 2^{10}(\cos 150° + i \sin 150°)$; the pattern is $z^n = 2^n(\cos (n \cdot 15°) + i \sin (n \cdot 15°))$.**

**19a) $z^2 = 4(\cos 30° + i \sin 30°)$, $z^3 = 8(\cos 45° + i \sin 45°)$
$z^4 = 16(\cos 60° + i \sin 60°)$, $z^5 = 32(\cos 75° + i \sin 75°)$**

21.b.

22.a.

NAME _____

LESSON **MASTER   13-6**
QUESTIONS ON **SPUR** OBJECTIVES

■**SKILLS** _Objective A_ (See pages 832–834 for objectives.)
In 1 and 2, determine $z_1 z_2$. Leave your answer in the form of the original numbers.

1. $z_1 = 2(\cos 47° + i \sin 47°)$, $z_2 = 6(\cos 58° + i \sin 58°)$
  12(cos 105° + _i_ sin 105°)

2. $z_1 = \left[3, \frac{\pi}{2}\right]$, $z_2 = \left[5.5, \frac{\pi}{6}\right]$ _____ $\boxed{16.5, \frac{2\pi}{3}}$

In 3 and 4, given z_1 and $z_1 z_2$, determine z_2. Leave your answer in the form of the original numbers.

3. $z_1 = 3\left(\cos \frac{7\pi}{6} + i \sin \frac{7\pi}{6}\right)$, $z_1 z_2 = 24\left(\cos \frac{3\pi}{2} + i \sin \frac{3\pi}{2}\right)$
  $z_2 = 8\left(\cos \frac{\pi}{3} + i \sin \frac{\pi}{3}\right)$

4. $z_1 = \left[4, \frac{\pi}{4}\right]$, $z_1 z_2 = \left[12, \frac{7\pi}{6}\right]$ _____ $z_2 = \left[3, \frac{11\pi}{12}\right]$

In 5 and 6, give a. the absolute value, and b. the argument of the complex number.

5. $5 + 6i$ 6. $4\left(\cos \frac{\pi}{6} + i \sin \frac{\pi}{6}\right)$
a. $\sqrt{61} \approx 7.81$ a.   4
b. $\theta \approx 50.2°$ b. $\frac{\pi}{6}$

■**SKILLS** _Objective B_
In 7 and 8, write the complex number in $a + bi$ form.

7. $4(\cos 210° + i \sin 210°)$ $-2\sqrt{3} - 2i$

8. $\left[3, \frac{5\pi}{6}\right]$ $-\frac{3\sqrt{3}}{2} + \frac{3}{2}i$

■**REPRESENTATIONS** _Objective J_
In 9 and 10, graph each number on the complex plane.

9. $2\left(\cos \frac{\pi}{4} + i \sin \frac{\pi}{4}\right)$ 9.   10.

10. $3\cos \frac{\pi}{6} + 3i \sin \frac{\pi}{6}$

20. What is the length of the diagonal $\overline{OT}$ of the parallelogram formed in the complex plane by the origin, $F = 3 - i$, $R = \text{-}1 - 2i$, and T?   *(Lesson 13-6)*
$\sqrt{13}$

21. Graph $r = \csc \theta$
    a. in rectangular coordinates,
    b. in polar coordinates.   *(Lessons 6-7, 13-4)*   **a,b) See margin.**

22. a. Sketch a graph of $r = \text{-}\sin \theta$ in polar coordinates.
    b. Prove that the graph is a circle.   *(Lesson 13-4)*   **See margin.**
    See below.

23. a. Represent $\triangle ABC$ of this lesson as a 2×3 matrix. $\begin{bmatrix} 0 & 4 & 5 \\ 3 & \text{-}3 & 0 \end{bmatrix}$
    b. Verify by matrix multiplication that $\triangle A'B'C'$ is the image of $\triangle ABC$ under the composite of a rotation of 90° and a size change of 2.
    (Lesson 11-3)   **See margin.**

24. A fair coin is tossed 35 times. To the nearest .0001, what is the probability of getting
    a. exactly 25 heads;   **.0053**
    b. fewer than 5 heads?   *(Lesson 8-9)*   **0.0000017 ≈ 0.0000**

In 25 and 26, expand using the Binomial Theorem.   *(Lesson 8-8)*

25. $(a - b)^5$ 26. $(2p + 3)^4$ **See below.**
$(a - b)^5 = a^5 - 5a^4b + 10a^3b^2 - 10a^2b^3 + 5ab^4 - b^5$

27. The wooden part of a pencil is a right, regular hexagonal prism with radius 3.5 mm and length 175 mm. The "lead" is a solid cylinder with diameter 2 mm running the length of the wood. What is the volume of wood in the pencil?   *(Previous course, Lesson 5-3)*   **≈ 5019.8 mm³**

28. Give at least one way to prove that $\triangle ABC$ and $\triangle A'B'C'$ in Example 3 are similar triangles.   **sample: find the lengths of all 6 sides, and show that the ratios of corresponding sides are equal.**

22b) For $r \neq 0$, $r = \text{-}\sin \theta \Leftrightarrow r = -\dfrac{y}{r} \Leftrightarrow r^2 = \text{-}y \Leftrightarrow x^2 + y^2 = \text{-}y \Leftrightarrow x^2 + y^2 + y = 0 \Leftrightarrow x^2 + \left(y + \dfrac{1}{2}\right)^2 = \dfrac{1}{4}$; that is the equation of a circle with center $\left(0, -\dfrac{1}{2}\right)$ and radius $\dfrac{1}{2}$.

26) $(2p + 3)^4 = 16p^4 + 96p^3 + 216p^2 + 216p + 81$

816

DeMoivre's Theorem

Abraham DeMoivre

RESOURCES
■ Lesson Master 13-7
■ Teaching Aid 101: Graph for **Example 1**

OBJECTIVE

E Find powers and roots of complex numbers.

TEACHING NOTES

Example 1 shows that the various powers of a complex number lie on a logarithmic spiral when plotted in the complex plane. This fact has no analogy with the real numbers. A teaching aid is provided.

The application of DeMoivre's Theorem to finding roots of complex numbers, which is presented on pages 819–820, may take some time to discuss. You might wish to do either Additional Example 2 or **Question 4** as an example before discussing the general Roots of a Complex Number Theorem. Emphasize the equal spacing of roots around the origin in the complex plane; sketching roots can be a quick way to check the accuracy of calculations.

Point out that since every real number is a complex number, the nth roots of any nonzero real number are vertices of a regular polygon when $n > 2$.

The number of questions for this lesson may seem fewer than usual, but calculating roots takes some time.

Gauss (1777–1855) was the first person to call the numbers you have been studying in the last two lessons "complex." He applied complex numbers to the study of electricity; a unit of electromagnetism was named after him. But during the century before Gauss, several mathematicians explored complex numbers and discovered many remarkable properties of them. This lesson presents a theorem about powers of complex numbers which is named after Abraham DeMoivre (1667–1754; his name is pronounced *de mwav′*).

Consider expanding $(2 + 2\sqrt{3}\,i)^4$. One way to do it is to use the Binomial Theorem.

$$(2 + 2\sqrt{3}\,i)^4$$
$$= 2^4 + 4 \cdot 2^3(2\sqrt{3}\,i)^1 + 6 \cdot 2^2(2\sqrt{3}\,i)^2 + 4 \cdot 2^1(2\sqrt{3}\,i)^3 + (2\sqrt{3}\,i)^4$$
$$= 16 + 64\sqrt{3}\,i + 24(-12) + 8(24\sqrt{3}\,(-i)) + 16 \cdot 9$$
$$= -128 - 128\sqrt{3}\,i$$

Another approach is to rewrite the number $2 + 2\sqrt{3}\,i$ in trigonometric form and use the theorems of the previous lesson. You may use the formulas or geometric inspection to find r and θ. We give all angles in degrees.

$$z = 2 + 2\sqrt{3}\,i$$
$$r = \sqrt{2^2 + (2\sqrt{3}\,i)^2} = 4$$
$$\theta = \tan^{-1}\left(\frac{2\sqrt{3}}{2}\right) = 60°$$

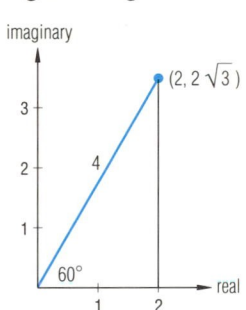

So $\quad z = 4(\cos 60° + i \sin 60°)$.

Then $\quad z^2 = 4(\cos 60° + i \sin 60°) \cdot 4(\cos 60° + i \sin 60°)$
$\qquad\quad = 4^2(\cos 120° + i \sin 120°)$,

$\qquad z^3 = 4^2(\cos 120° + i \sin 120°) \cdot 4(\cos 60° + i \sin 60°)$
$\qquad\quad = 4^3(\cos 180° + i \sin 180°)$,

and $\quad z^4 = 4^3(\cos 180° + i \sin 180°) \cdot 4(\cos 60° + i \sin 60°)$
$\qquad\quad = 4^4(\cos 240° + i \sin 240°)$
$\qquad\quad = 256\left(-\frac{1}{2} - i\,\frac{\sqrt{3}}{2}\right)$
$\qquad\quad = -128 - 128\sqrt{3}\,i.$

This second approach may seem tedious, but a simple pattern is developing. Note that for each $n = 1, 2, 3,$ or 4, $z^n = 4^n(\cos n \cdot 60° + i \sin n \cdot 60°)$.

This result, generalized, is called *DeMoivre's Theorem*.

DeMoivre's Theorem

If $\quad z = r(\cos\theta + i\sin\theta)\quad$ and n is an integer,

then $\quad z^n = r^n(\cos n\theta + i\sin n\theta)$.

In polar form, DeMoivre's theorem states that if $z = [r, \theta]$, then $z^n = [r^n, n\theta]$ for all integers n. According to some historians, DeMoivre only proved the theorem for $r = 1$, but it is true for any r. The proof is beyond the scope of this course.

Example 1 For $z = 2\left(\cos\frac{\pi}{3} + i\sin\frac{\pi}{3}\right)$,

a. find z^n for $n = 1, 2, 3, 4, 5, 6,$ and 7.

b. Plot the powers in the complex plane.

Solution

a. Use DeMoivre's Theorem and convert to polar form for easy graphing.

$z^1 = 2\left(\cos\frac{\pi}{3} + i\sin\frac{\pi}{3}\right)\quad = \left[2, \frac{\pi}{3}\right]$

$z^2 = 2^2\left(\cos\frac{2\pi}{3} + i\sin\frac{2\pi}{3}\right)\quad = \left[4, \frac{2\pi}{3}\right]$

$z^3 = 2^3(\cos\pi + i\sin\pi)\quad = [8, \pi]$

$z^4 = 2^4\left(\cos\frac{4\pi}{3} + i\sin\frac{4\pi}{3}\right)\quad = \left[16, \frac{4\pi}{3}\right]$

$z^5 = 2^5\left(\cos\frac{5\pi}{3} + i\sin\frac{5\pi}{3}\right)\quad = \left[32, \frac{5\pi}{3}\right]$

$z^6 = 2^6(\cos 2\pi + i\sin 2\pi) = [64, 2\pi]$

$z^7 = 2^7\left(\cos\frac{7\pi}{3} + i\sin\frac{7\pi}{3}\right)\quad = \left[128, \frac{7\pi}{3}\right]$

b. The points are plotted at the left below. The smooth curve connecting them, shown below at the right, is called a *logarithmic spiral*.

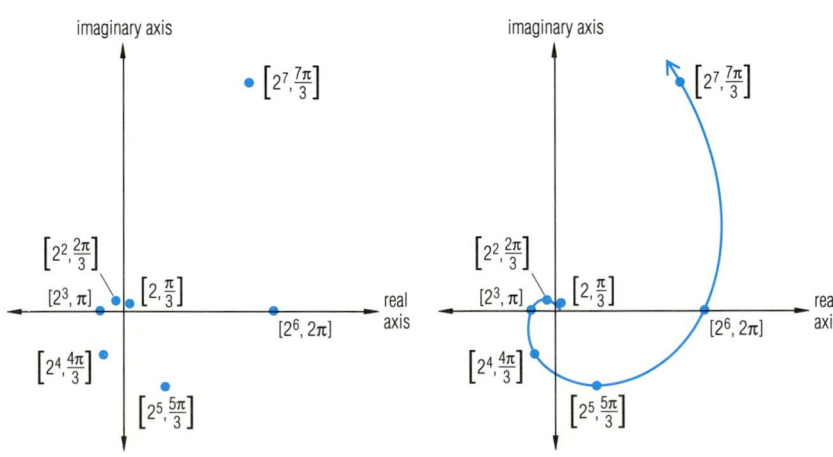

818

Two complex numbers in polar and trigonometric form are equal if and only if their absolute values are equal and their arguments differ by an integral multiple of 360°. With this knowledge and DeMoivre's Theorem you can find roots of complex numbers.

Consider, for instance, a cube root z of the complex number $8i$. By definition of cube root,
$$z^3 = 8i.$$

By examining the graph of $8i$ at the right, you can see that an argument of $8i$ is 90° and its absolute value is 8. So in trigonometric form, $8i = 8(\cos 90° + i \sin 90°)$. Substituting this and $z = r(\cos \theta + i \sin \theta)$ into $z^3 = 8i$ gives $(r(\cos \theta + i \sin \theta))^3 = 8(\cos 90° + i \sin 90°)$. Applying DeMoivre's Theorem, we get $r^3(\cos 3\theta + i \sin 3\theta) = 8(\cos 90° + i \sin 90°)$.

For these complex numbers to be equal,

$$r^3 = 8 \quad \text{and} \quad \cos 3\theta + i \sin 3\theta = \cos 90° + i \sin 90°.$$

So $\quad r = 2 \quad$ and $\quad\quad\quad\quad\quad 3\theta = 90° + 360n°.$

Thus $\quad\quad\quad\quad\quad\quad\quad\quad\quad\quad\quad \theta = 30° + 120n°$, where n is an integer.

Therefore, the cube roots of $8i$ are of the form
$$z = 2(\cos (30° + 120n°) + i \sin (30° + 120n°)).$$

This solution may seem complicated, but actually there are only three distinct roots. For $n = 0$, 1, and 2, the roots are

$$2(\cos 30° + i \sin 30°) \quad\quad = \sqrt{3} + i,$$
$$2(\cos 150° + i \sin 150°) = -\sqrt{3} + i,$$

and $\quad 2(\cos 270° + i \sin 270°) = -2i$.

For any integer $n > 2$, you will find that these values are repeated.

These three roots of $8i$ are plotted below. Because they all have the same absolute value 2 and are 120° apart, they lie equally spaced around a circle with center at the origin and radius 2.

Generalizing the technique used to find the cube roots of $8i$ leads to a proof of the following theorem.

Roots of a Complex Number Theorem

Let z be any nonzero complex number and n be any positive integer. Then there are n distinct nth roots of $z^n = r(\cos \theta + i \sin \theta)$. They are
$$z = \sqrt[n]{r}\left(\cos\left(\frac{\theta}{n} + k\,\frac{360°}{n}\right) + i \sin\left(\frac{\theta}{n} + k\,\frac{360°}{n}\right)\right)$$
where $k = 0, 1, 2, ..., n - 1$.

LESSON 13-7 DeMoivre's Theorem **819**

ADDITIONAL ANSWERS
3.b

4.a. z =
3 (cos 50° + i sin 50°),
3 (cos 170° + i sin 170°), or
3 (cos 290° + i sin 290°)

4.b.

10.a. [2, 15°], [2, 105°],
[2, 195°], [2, 285°]

b.

11.a. [2, 15°], [2, 75°],
[2, 135°], [2, 195°],
[2, 255°], [2, 315°]

11.b.

12.b.

13.a. $z = \sqrt{2} + \sqrt{2}i$, $\sqrt{2} - \sqrt{2}\,i$, $-\sqrt{2} + \sqrt{2}i$, $-\sqrt{2} - \sqrt{2}\,i$

b.

17.a.

Example 2 Find the 5th roots of $16 + 16\sqrt{3}\,i$ and graph them in the polar coordinate system.

Solution Calculate the absolute value and an argument of $16 + 16\sqrt{3}\,i$ or examine its graph to determine that $r = 32$ and $\theta = 60°$.

So in trigonometric form,
$16 + 16\sqrt{3}\,i = 32(\cos 60° + i \sin 60°)$.
Use the theorem above. The fifth roots of z are

$$z = \sqrt[5]{32}\left(\cos\left(\frac{60°}{5} + k\,\frac{360°}{5}\right) + i \sin\left(\frac{60°}{5} + k\,\frac{360°}{5}\right)\right)$$
$$= 2(\cos(12° + 72k°) + i \sin(12° + 72k°)),$$

where $k = 0, 1, 2, 3,$ and 4.
Thus the roots are $2(\cos 12° + i \sin 12°)$, $2(\cos 84° + i \sin 84°)$, $2(\cos 156° + i \sin 156°)$, $2(\cos 228° + i \sin 228°)$, and $2(\cos 300° + i \sin 300°)$. They are graphed below. Note that the 5 fifth roots of $16 + 16\sqrt{3}\,i$ are equally spaced on a circle centered at the origin with radius $32^{1/5} = 2$.

In general for $n > 2$, the graphs of the nth roots of any nonzero complex number are vertices of a regular n-gon centered at the origin!

Questions

Covering the Reading

In 1 and 2, use DeMoivre's Theorem to find each power. Write your answer in trigonometric form.   **See below.**

1. $\left(3\left(\cos\frac{\pi}{5} + i \sin\frac{\pi}{5}\right)\right)^4$

2. $\left(2\left(\cos\frac{4\pi}{7} + i \sin\frac{4\pi}{7}\right)\right)^3$

3. Consider $z = 2(\cos\frac{\pi}{6} + i \sin\frac{\pi}{6})$.
   a. Use DeMoivre's Theorem to calculate z^n for $n = 2$ to 6.   **See below.**
   b. Plot $z^1, z^2, z^3, z^4, z^5,$ and z^6 in the complex plane.   **See margin.**
   c. Describe the pattern in the graphs of the powers of z.
      They lie on a spiral.

4. **a.** Solve the equation $z^3 = 27(\cos 150° + i \sin 150°)$.   **See margin.**
   b. Plot the solutions in the complex plane.   **See margin.**
   c. Describe the graph in part **b**.
      They are the vertices of an equilateral triangle.

1) $81\left(\cos\frac{4\pi}{5} + i \sin\frac{4\pi}{5}\right)$; 2) $8\left(\cos\frac{12\pi}{7} + i \sin\frac{12\pi}{7}\right)$

3a) $z^2 = 4\left(\cos\frac{\pi}{3} + i \sin\frac{\pi}{3}\right)$; $z^3 = 8\left(\cos\frac{\pi}{2} + i \sin\frac{\pi}{2}\right)$; $z^4 = 16\left(\cos\frac{2\pi}{3} + i \sin\frac{2\pi}{3}\right)$;

$z^5 = 32\left(\cos\frac{5\pi}{6} + i \sin\frac{5\pi}{6}\right)$; $z^6 = 64(\cos\pi + i \sin\pi)$

820

In 5–7, find

5. the 3 cube roots of $125(\cos 30° + i \sin 30°)$;
$5(\cos 10° + i \sin 10°), 5(\cos 130° + i \sin 130°), 5(\cos 250° + i \sin 250°)$

6. the 4 fourth roots of $7\left(\cos \frac{4\pi}{5} + i \sin \frac{4\pi}{5}\right)$; **See below.**

7. the square roots of -1.   **$i, -i$**

6) $\sqrt[4]{7}(\cos \pi/5 + i \sin \pi/5)$; $\sqrt[4]{7}(\cos 7\pi/10 + i \sin 7\pi/10)$; $\sqrt[4]{7}(\cos 6\pi/5 + i \sin 6\pi/5)$;
$\sqrt[4]{7}(\cos 17\pi/10 + i \sin 17\pi/10)$

In 8 and 9, use DeMoivre's Theorem to find each power in $a + bi$ form.

8. $(3i)^4$
$(3(\cos 90° + i \sin 90°))^4 = 81$

9. $(-\sqrt{3} + i)^6$
$(2(\cos 150° + i \sin 150°))^6 = -64$

In 10 and 11, **a.** write the roots in polar form. **b.** Plot the roots in a complex plane.

10. the fourth roots of $8 + 8\sqrt{3}\,i$
See margin.

11. the sixth roots of $64i$
See margin.

In 12 and 13, **a.** solve each equation over the set of complex numbers.
b. Plot the solutions in a complex plane.

12. $z^3 = 8$
a) $z = 2, -1 + \sqrt{3}i, -1 - \sqrt{3}i$; b) **See margin.**

13. $z^4 = -16$ **a,b) See margin.**

14. A ninth root of z is $2(\cos 30 + i \sin 30°)$. Find z in polar form.
$[512, 270°]$

(Lesson 13-6)
15. Write $7(\cos 10° + i \sin 10°)$ in $a + bi$ form.   *(Lesson 13-6)*
$7 \cos 10° + (7 \sin 10°)i \approx 6.89 + 1.22\,i$

16. If $z_1 = 8(\cos 70° + i \sin 70°)$ and $z_2 = 5(\cos 155° + i \sin 155°)$,
find $z_1 \cdot z_2$.   *(Lesson 13-6)*   **40 $(\cos 225° + i \sin 225°)$**

17. a. Sketch a graph of $r = e^{\theta/3}$.   **See margin.**
b. Show that the points on the curve in part **a** satisfy the relationship $\ln r = k\theta$ for some constant k, and identify k.   *(Lessons 4-6, 13-4)*
See below.

18. Consider $\sin^2 \alpha = \frac{1}{2}(1 - \cos^2 \alpha)$.

a. Use a function grapher to test whether the equation may be an identity.
b. Verify your conclusion in part **a**.   *(Lesson 13-1)*
a,b) See margin.

19. Determine an equation for the hyperbola with major and minor axes of 4 and 1, centered around the point $(3, -2)$, and whose major axis is rotated counterclockwise by $30°$ from the horizontal.   *(Lesson 12-3)*   **See margin.**

20. Multiply. $\begin{bmatrix} -1 & 0 & 1 \\ 0 & 1 & -1 \\ 1 & -1 & 0 \end{bmatrix} \cdot \begin{bmatrix} 1 & 2 & 3 \\ 2 & 3 & 1 \\ 3 & 1 & 2 \end{bmatrix}$ **21.** Evaluate $\lim\limits_{n\to\infty} \dfrac{n+3}{4 - 2n}$.

(Lesson 11-1) $\begin{bmatrix} 2 & -1 & -1 \\ -1 & 2 & -1 \\ -1 & -1 & 2 \end{bmatrix}$ *(Lesson 8-2)* $-\dfrac{1}{2}$

22. a. Graph the first ten terms of each sequence $a_n = z^n$.   **a,b) See margin.**
i. $z = [2, 40°]$ **ii.** $z = [1, 40°]$ **iii.** $z = [.9, 40°]$
b. Compare and contrast the three sets of points, and predict what happens for higher powers.

17b) $r = e^{\theta/3} \Leftrightarrow \ln r = \ln e^{\theta/3} = \frac{\theta}{3} \ln e = \frac{\theta}{3}(1) = \frac{1}{3}\theta$, so $k = \frac{1}{3}$

Left margin labels:
Applying the Mathematics
Review
Exploration

LESSON 13-8

TEACHING NOTES

We close this book with this lesson for several reasons. The Mandelbrot Set is an example of an application of a topic that students often think has few applications. The set is an example of a recent development of mathematics, and hopefully students will appreciate that mathematics is a growing field of study—not a dead collection of numbers and algorithms to be memorized. The set is a result of the computer age—it could not have been studied so extensively without them. And students at this level know enough mathematics and computer science to generate the set for themselves. Point this out to your students—they have successfully learned enough mathematics to study a topic which is at the cutting edge of mathematics research.

You will need to explain how the Mandelbrot Set is generated. The question is: Is a particular complex number z in the set or not? For this, we set up a sequence $z_1 = z$, $z_2 = z_1^2 + z$, $z_3 = z_2^2 + z$, $z_4 = z_3^2 + z$, and so on. Then we ask: Are the terms of this sequence bounded in the sense that they stay within a particular distance of the origin? If so, then z is in the Mandelbrot Set. If not, then z is not in the set. Pictures of the set are given on two teaching aids.

822

LESSON

13-8

The Mandelbrot Set

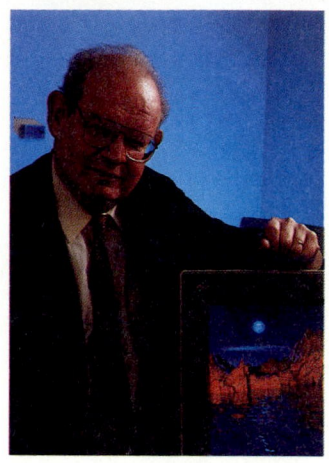

Benoit Mandelbrot

The last lesson ended with a geometrically beautiful result that has been known for about 200 years. This lesson discusses a geometric set that has been known only since 1980.

In recent years, a new field of mathematics has arisen, called *dynamical systems,* in which complex numbers play an important role. Some of the products of this field are beautiful computer-generated drawings which have won awards in art competitions. Among these are drawings involving the *Mandelbrot set* (named for Benoit Mandelbrot, a mathematician at IBM), which is a particular set of complex numbers plotted in the complex plane. Below is a graph of this set in part of the plane. Points colored black are in the Mandelbrot set. Points colored red or yellow are outside the set.

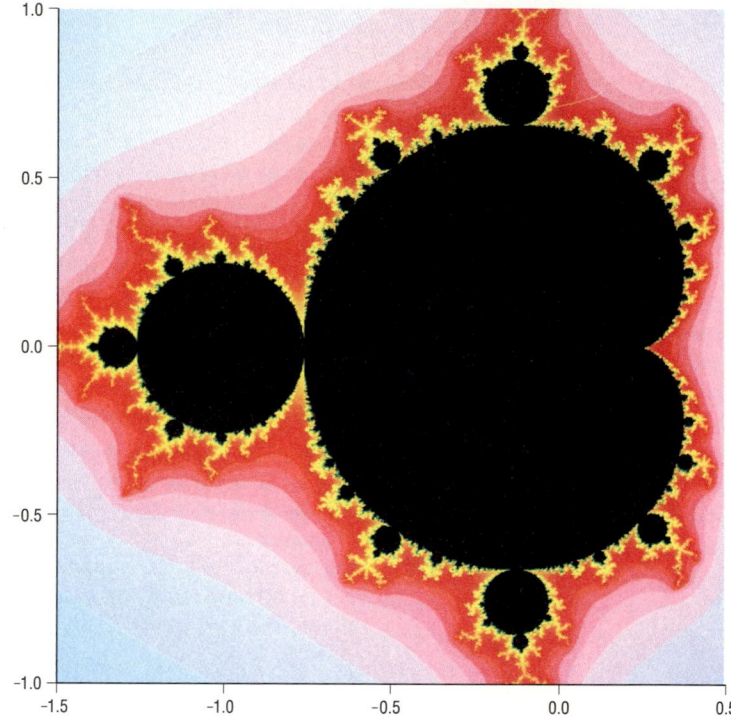

Points in the Mandelbrot set are defined recursively. For a complex variable z and a complex constant c (that is, c is a fixed complex number, and z_j depends on c and the subscript j),

$$\begin{cases} z_1 = c; \\ z_n = (z_{n-1})^2 + c \text{ for } n > 1. \end{cases}$$

The recursion squares the previous z value, adds c to the result, then repeats these two steps indefinitely. For many values of c, the limit of $|z_n|$ is infinity. Points in the Mandelbrot set are those values of c for which the limit of $|z_n|$ is not infinity.

For instance, for $z_1 = c = -1.5 - 1.0i$, the point at the lower left corner of the window, the recursion begins with the following values.

$z_1 = c = -1.5 - 1.0i$, so $|z_1| \approx 1.80$.
$z_2 = (-1.5 - 1.0i)^2 + (-1.5 - 1.0i) = -0.25 + 2i$, so $|z_2| \approx 2.02$.
$z_3 = (-0.25 + 2i)^2 + (-1.5 - 1.0i) = -5.4375 - 2i$, so $|z_3| \approx 5.79$.

It can be shown that $|z_n|$ will go to infinity if and only if at some stage of the iteration, $|z_n| > 2$. It also turns out that for most values of c which iterate to infinity, $|z_n|$ reaches 2 rather quickly. For these reasons, $c = -1.5 - 1.0i$ would be discarded as not being part of the Mandelbrot set. Note that the point $(-1.5, -1.0)$ is not colored black in the graph on page 822.

By varying the initial c and the size of the window, the Mandelbrot set can be studied to virtually any degree of detail. The three graphs below show the set as you zoom in on the region around the point $-0.55 + 0.62i$. Note how the "bump" on the original graph is very similar in shape to the original graph itself. For this reason the Mandelbrot set is sometimes called "self similar." In fact, however, most of the parts of the graph differ in some way.

Technology If you have access to a Macintosh computer, you might want to get a copy of a much faster and versatile graphics program called Super MANDELZOOM. It is available as a public domain item (a donation is suggested) from the developer Robert P. Munafo, 239 Clinton Road, Brookline, MA 02146. Send a two-sided (800K) disk with a self-addressed (two stamps) envelope. With Super MANDELZOOM, students have the ability to generate all the graphs that we have generated and we suggest that they do so. Students should not ignore the BASIC program or **Questions 1–8,** however, because the algorithm used in Super MANDELZOOM is the same. Making the graphs is one thing, understanding how they are created is where mathematics enters the picture.

Error Analysis If you wish to use MANDELZOOM on an Apple, please read the lesson before giving the program to your students—the graphs take a very long time to draw. And without careful choices of the parameters, the result could be nothing but a solid-colored square on your monitor.

If your library has back issues of *Scientific American*, it is worth the effort to find the article by A. K. Dewdney in the August 1985 issue. It has beautiful full-color pictures of the set (the color comes by assigning different hues to values of $|z_n|$ in different intervals). The August 1990 issue of *Scientific American* also includes color photos in an article on fractals by Hartmut Jürgens and others.

In the 1980s, John H. Hubbard, a mathematician at Cornell University, generated graphs of the Mandelbrot set using a computer. Hubbard's pioneer programs have inspired many programs for drawing the Mandelbrot set. The pictures in this lesson were created with *The Beauty of Fractals Lab*, written by Thomas Eberhardt and Marc Parmet. This program makes use of sophisticated computing techniques and languages in order to increase the speed of the calculations and drawings. You can draw a small version of the set, however, using the following BASIC program.

```
10    REM MANDELZOOM
20    INPUT "ENTER THE REAL AND IMAGINARY PARTS OF C"; ACORNER,
      BCORNER
30    INPUT "ENTER THE SIZE OF THE VIEWING WINDOW"; SIZE
40    CLS
50    P = 50
60    DIM PIC(P,P)
70    GAP = SIZE / P
80    FOR J = 1 TO P
90      FOR K = 1 TO P
100       AC = ACORNER + J*GAP
110       BC = BCORNER + K*GAP
120       AZ = 0:BZ = 0
130       FOR COUNT = 1 TO 100
135         OLDAZ = AZ
140         AZ = AZ*AZ – BZ*BZ + AC
150         BZ = 2*OLDAZ*BZ + BC
160         MAGZ = SQR(AZ*AZ + BZ*BZ)
170         IF MAGZ>2 THEN 200
180       NEXT COUNT
190       PIC(J,K) = 1
200     NEXT K
210   NEXT J
220   FOR J = 1 TO P
230     FOR K = 1 TO P
240       IF PIC(J,K) = 1 THEN PSET (J,P–K)
250     NEXT K
260   NEXT J
```

The program has four major steps. (You do not need to understand every detail of the program.)

1. The program asks for the initial value of c and the size of the window in the complex plane (lines 20 and 30).

2. It scales the information from step (1) into "pixel-units" and creates an array PIC(J, K) for each pixel in the graph (lines 50–70). In this program, the rectangle is 50 pixels on a side (P = 50).

3. For every pixel with coordinates (AC, BC) in the viewing window, the program calculates $z_n = z_{n-1}^2 + c$ until $|z_n|>2$ or $n = 100$, whichever comes first (lines 80–210). The bigger n is, the more accurate the graph, but the longer the time it takes to do the calculations. If $|z_n| \leq 2$ after 100 iterations of the squaring process, then the pixel at PIC(J, K) is set to 1 (line 190). Lines 140–160 apply the mathematics of complex numbers you learned in Lesson 13-6. For $z = $ AZ + BZi, the real part of z^2 is $(AZ)^2 - (BZ)^2$ and the imaginary part is 2(AZ)(BZ). The variable MAGZ is the absolute value of the new iteration of z.

824

4. Finally the graph is drawn (lines 220–260). If PIC(J, K) = 1, the point c is in the set and the pixel is turned on (line 240).

(Note: this program is written to work on Macintosh computers running MS-BASIC. For other computers and other versions of BASIC there are slight variations. Run on a Macintosh computer with $c = -2 - 1.25i$ and size $= 2.5$, MANDELZOOM took 45 minutes to produce the graph at the left below. When run with P = 200 (line 50) and a maximum count of 200 rather than 100 (line 130), it took 12 hours and produced the graph at the right below.)

Using a BASIC program is obviously not the best way to graph the Mandelbrot set, but it shows how you can use the computing and mathematics skills you have studied in this text. With better programs and faster computers, you can apply the mathematics you know to generate gorgeous views like the ones below, centered at about $-0.745 + 0.113i$.

The boundary of a Mandelbrot set is an example of a *fractal*. Fractals can be generated by special functions embodying randomness. Thus, in a way, the Mandelbrot set is related to all the themes of this book: functions, statistics, and trigonometry. We hope you agree that it is a fitting subject for the last lesson of this book.

Questions

Covering the Reading

1. **a.** Calculate z_1, z_2, and z_3 in the Mandelbrot recursion for $c = 0 + 0i$.
   b. Is $c = 0 + 0i$ in the Mandelbrot set? Why or why not?
   a) $z_1 = 0 + 0i$, $z_2 = 0 + 0i$, $z_3 = 0 + 0i$; b) Yes; $\lim_{n\to\infty}|z_n|$ is finite.

In 2–4, based on $|z_4|$, is the point in the Mandelbrot set?

2. $-1 + 0i$ $|z_4| = 0$; **Yes**

3. $0 + 0.5i$ $|z_4| \approx .407$; **Yes**

4. $1 + i$ $|z_4| = \sqrt{9410}$; **No**

5. Refer to the BASIC MANDELZOOM program in the lesson. Let ACORNER $= 0.2$, BCORNER $= 0$, and SIZE $= 0.1$.
   a. Give the values of AC and BC the first time the computer executes lines 100 and 110.   **AC = .202; BC = .002**
   b. Give the values of AZ, BZ, and MAGZ the first time the computer executes lines 140–160.   **AZ = .202; BZ = .002; MAGZ ≈ .202**
   c. Will the computer turn on PIC(1, 1)?   **Yes**

In 6–8, run the program MANDELZOOM for complex numbers $x + yi$ in the indicated viewing window.

6. Describe the graph a computer generates for $-0.5 \leq x \leq -0.49$, $0 \leq y \leq 0.01$.   **a square filling the entire viewing window**

7. Draw a rough sketch of the graph a computer generates for $-1.5 \leq x \leq -0.9$, $-0.3 \leq y \leq 0.3$, $c = -1.5 - 0.3i$.   **See margin.**

8. **a.** Use the close-up views around $c = -0.5507 + 0.6259i$ to hypothesize whether c is actually in the set.
   b. Test your hypothesis in part **a** by computing $|z_4|$ for this value of c.
   a,b) See below.

Review

In 9 and 10, find each power. Write your answers in $a + bi$ form.   *(Lesson 13-8)*

9. $\left(\dfrac{\sqrt{2}}{2} + \dfrac{\sqrt{2}}{2}i\right)^5$ $-\dfrac{\sqrt{2}}{2} - \dfrac{\sqrt{2}}{2}i$

10. $(1 + i\sqrt{3})^6$ **64 + 0i**

**8a) It is probably in the set because the part of the graph near that point is black.
b) $|z_4| \approx 1.111$ and $|z_1| = |c| \approx .834$, so it would seem that $|z_n|$ is close to 1 and that c is actually in the set.**

826

In 11 and 12, **a.** write each root in trigonometric form. **b.** Plot each root in a complex plane. *(Lesson 13-8)*

11. the 3 cube roots of $10(\cos 12° + i \sin 12°)$ **a)** $\sqrt[3]{10}(\cos 4° + i \sin 4°)$, $\sqrt[3]{10}(\cos 124° + i \sin 124°)$, $\sqrt[3]{10}(\cos 244° + i \sin 244°)$; **b) See margin.**

12. the 5 fifth roots of i $\cos 18° + i \sin 18°$, $\cos 90° + i \sin 90°$, $\cos 162° + i \sin 162°$, $\cos 234° + i \sin 234°$, $\cos 306° + i \sin 306°$; **b) See margin.**

In 13 and 14, **a.** express z_1 and z_2 in trigonometric form. **b.** Evaluate $z_1 \cdot z_2$ and $\dfrac{z_1}{z_2}$. *(Lesson 13-6)*

13. $z_1 = 3 + \sqrt{2}\,i$, $z_2 = -3 - \sqrt{2}\,i$ **a)** $z_1 \approx 3.32\,(\cos 25.2° + i \sin 25.2°)$, $z_2 \approx 3.32\,(\cos 205.2° + i \sin 205.2°)$; **b) See below.**

14. $z_1 = [0, \frac{\pi}{6}]$, $z_2 = [2, 0]$ **a)** $z_1 = 0\left(\cos \frac{\pi}{6} + i \sin \frac{\pi}{6}\right)$, $z_2 = 2(\cos 0 + i \sin 0)$; **b)** $z_1 z_2 = 0; \frac{z_1}{z_2} = 0$

15. Find $\left|z_1 z_2\right|$ when $z_1 = 112 - 15i$ and $z_2 = 0.01 + 4i$. *(Lesson 13-6)* ≈ 452

16. The graph of $r = 2 \sin \theta \cos^2 \theta$ is called a *bifolium*. *(Lesson 13-5)*
   a. Sketch the bifolium for $0 \le \theta \le \pi$. **See margin.**
   b. Graph $r = 2$ on the same axes as the graph in part **a**. Does $2 \sin \theta \cos^2 \theta = 2$ for any θ in $0 \le \theta \le \pi$? **No**

17. A point $[r, \theta]$ has $\theta = -180°$. *(Lesson 13-4)*
   a. Where must this point be? **on the polar axis**
   b. Give a rectangular equation for all such points. **$y = 0$**

In 18 and 19, **a.** use a function grapher to test whether the equation may be an identity. **b.** Prove your conclusion in part **a**. *(Lessons 13-2, 13-3)*

18. $\dfrac{1 + \cos \alpha}{\sin \alpha} = \cot \dfrac{\alpha}{2}$ [Hint: $\cos \alpha = \cos\left(2 \cdot \frac{\alpha}{2}\right)$ and $\sin \alpha = \sin\left(2 \cdot \frac{\alpha}{2}\right)$.]
   a) See margin. b) See below.

19. $\sec \theta = \sec (\pi - \theta)$ **a) See margin. b) See below.**

Exploration

20. Explore the Mandelbrot set on intervals of your own choosing. **Answers will vary.**

13b) $z_1 z_2 \approx 11(\cos 230.4° + i \sin 230.4°) \approx -7 - 8.48i$
$\dfrac{z_1}{z_2} = 1(\cos (-180°) + i \sin (-180°)) = \cos 180° + i \sin 180° = -1$

18b) $\dfrac{1 + \cos \alpha}{\sin \alpha} = \dfrac{1 + \cos 2\left(\frac{\alpha}{2}\right)}{\sin 2\left(\frac{\alpha}{2}\right)} = \dfrac{1 + \cos^2\left(\frac{\alpha}{2}\right) - \sin^2\left(\frac{\alpha}{2}\right)}{2 \sin \frac{\alpha}{2} \cos \frac{\alpha}{2}} = \dfrac{2 \cos^2\left(\frac{\alpha}{2}\right)}{2 \sin \frac{\alpha}{2} \cos \frac{\alpha}{2}} = \dfrac{\cos \frac{\alpha}{2}}{\sin \frac{\alpha}{2}} = \cot \frac{\alpha}{2}.$

So $\dfrac{1 + \cos \alpha}{\sin \alpha} = \cot \frac{\alpha}{2}$ is an identity for $\alpha \ne n\pi$ where n is an integer.

19b) Let $\theta = \frac{\pi}{6}$. Then $\sec \frac{\pi}{6} = \dfrac{1}{\cos \frac{\pi}{6}} = \dfrac{2}{\sqrt{3}}$, and $\sec\left(\pi - \frac{\pi}{6}\right) = \sec \frac{5\pi}{6} = \dfrac{1}{\cos \frac{5\pi}{6}} = -\dfrac{2}{\sqrt{3}}$.

So $\sec \theta = \sec (\pi - \theta)$ is not an identity.

FOLLOW-UP

PROJECTS
Project 3 on page 828 is related to the content of this lesson.

EXTENSION
If you have students that are interested in the program and have the patience to explore with it, suggest Project 3 to them. Students with computers at home may be encouraged to run the program there for larger values of P and COUNT. If they discover special regions of the set they could print the screen and show the results to the class.

18.a.

19.a.

$y = \sec \theta$

$y = \sec (\pi - \theta)$

Projects

1. Using calculus, it can be shown that if x is in radians,

$$\cos x = 1 - \frac{x^2}{2!} + \frac{x^4}{4!} - \frac{x^6}{6!} + \dots$$

and $\sin x = x - \frac{x^3}{3!} + \frac{x^5}{5!} - \frac{x^7}{7!} + \dots$.

These series are very likely used in your calculator to approximate values of sine and cosine.

   a. Approximate $\cos(0.2)$ using the first three terms of the appropriate series. Also, find the fourth term of the series. Show that the difference between your approximation and the calculator value of $\cos(0.2)$ is less than the absolute value of the fourth term.

   b. Repeat part **a** for $\sin(0.2)$.

   c. Give an explicit definition for the nth term of each series.

   d. Use the series expansion for $\sin x$ and $\cos x$ to find series expansions for $\sin 2x$ and $\cos 2x$. Check your answers with a calculator.

   e. In Lesson 4-6 you studied the following series expression for e^x.

$$e^x = 1 + x + \frac{x^2}{2!} + \frac{x^3}{3!} + \frac{x^4}{4!} + \dots$$

   It can be proved that this series converges even if the exponent is a complex number. Find a series expansion for e^{ix}.

   f. Use the result of part **e** to show that
$$e^{ix} = \cos x + i \sin x.$$

   g. Find a complex number in the form $a + bi$ for $e^{i\pi}$. (The answer to this is known as *Euler's Theorem*, and is one of the most extraordinary results in mathematics.)

   h. Use the result of part **f** to show that $i = e^{i \cdot \pi/2}$.

   i. Find i^i. (It may surprise you that i^i is a real number.)

2. Explore some of the following classic polar equations with an automatic grapher. If the grapher does not have an option for entering equations in the form $r^2 = f(\theta)$, graph $r = +\sqrt{f(\theta)}$ and $r = -\sqrt{f(\theta)}$ simultaneously. In all the equations, a may be any nonzero real number. Graph each equation a few times, with different values of a.

Describe the patterns that you find.
   a. Cardioid: $r = a(\cos \theta - 1)$
   b. Cissoid of Diocles: $r = a \sin \theta \tan \theta$
   c. Cochleoid (Ouija board curve): $r = \dfrac{a \sin \theta}{\theta}$
   d. Folium of Descartes: $r = \dfrac{3a \sin \theta \cos \theta}{\sin^3 \theta + \cos^3 \theta}$
   e. Strophoid: $r = a \cos 2\theta \sec \theta$
   f. Lemniscate of Bernoulli:
   $r^2 = a^2 \cos 2\theta$; $r^2 = a^2 \sin 2\theta$
   g. Lituus: $r^2 = \dfrac{a^2}{\theta}$

3. Use the BASIC MANDELZOOM program in Lesson 13-8. Experiment with the program. Leave P = 50 and try various values of c and SIZE. Remember to be patient—the program as written takes up to 45 minutes to run. You can speed it up by changing line 130 to 130 FOR COUNT = 1 TO 50, but this decreases detail and may give you a black square as the image.

   a. Reproduce the 50×50-pixel Mandelbrot graph in the lesson. Let $c = -2 - 1.25i$, SIZE = 2.5.

   b. The image below has $c = 0.26 + 0i$ and SIZE = 0.01, P = 100, and COUNT = 1 to 100. It took about $1\frac{1}{2}$ hours. Try a P = 50 version if you don't have time.

   c. Look around the "neck" of the set. This is the area near $-0.76 + 0i$. (Do not expect results like the fancy drawings in the lesson.)

828

Summary

Graphs can be used to test whether an equation is an identity: check if the graphs of the functions determined by the two sides of the proposed identity coincide. While the graphs cannot prove an identity, they can help in finding a counterexample. Some of the techniques of proving identities include: start with one side and rewrite it until it equals the other side; rewrite each side independently until equal expressions are obtained on both sides; begin with a known identity and derive equivalent statements until the proposed identity appears.

Many identities involving circular functions hold only on restricted domains, excluding points where one or more of the functions in the identity are not defined. A point where a function is undefined is called a singularity. Singularities of functions show up as vertical asymptotes or missing single points on the graphs.

In a polar coordinate system, a point is identified by $[r, \theta]$, where r is the distance of the point from the origin on a ray which is the image of a horizontal ray, called the polar axis, under a rotation of θ. Every point has infinitely many polar coordinate representations. The four relations

$x = r \cos \theta$, $y = r \sin \theta$, $r = \sqrt{x^2 + y^2}$, and $\tan \theta = \frac{y}{x}$ relate

polar coordinates $[r, \theta]$ and rectangular coordinates (x, y).

Graphs of sets of points which satisfy equations involving r and θ include familiar figures such as lines and circles and beautiful spirals, rose curves, and other curves that do not have simple descriptions in terms of rectangular coordinates.

The complex number $a + bi$ is represented in rectangular coordinates as the point (a, b). It can also be represented in polar coordinates by the point $[r, \theta]$, where $r = \sqrt{a^2 + b^2}$ and $\tan \theta = \frac{b}{a}$. The trigonometric form of this number is $r(\cos \theta + i \sin \theta)$.

If $z = a + bi$, the absolute value of z is $|z| = \sqrt{a^2 + b^2}$. Addition of complex numbers $A = a + bi$ and $B = c + di$ can be represented by a parallelogram in the complex plane with vertices at the origin, and points A, B, and $(a + c) + (b + d)i$. For $z_1 = r_1(\cos \theta_1 + i \sin \theta_1)$ and $z_2 = r_2(\cos \theta_2 + i \sin \theta_2)$, the product $z_1 z_2$ is equal to $|z_1 z_2|$ $((\theta_1 + \theta_1) + i \sin (\theta_1 + \theta_1))$. This product is represented graphically as the composite of a size change of z_1 by a magnitude of $|z_2|$ and a rotation of θ_2. Similarly, $\frac{z_1}{z_2} = \frac{|z_1|}{|z_2|}$ $(\cos (\theta_1 - \theta_2) + i \sin (\theta_1 - \theta_2))$, representing the composite of a size change of $\frac{1}{|z_2|}$ and a rotation of $^-\theta_2$ applied to z_1.

Repeated multiplications of a single complex number $z = r(\cos \theta_1 + i \sin \theta_1)$ lead to DeMoivre's Theorem: $z^n = r^n(\cos n\theta_1 + i \sin n\theta_1)$. Working backwards leads to a theorem for finding nth roots:
If $z^n = r(\cos \theta + i \sin \theta)$ with $r > 0$,

then $z = \sqrt[n]{r} \left(\cos \left(\frac{\theta}{n} + k \frac{360°}{n} \right) + i \sin \left(\frac{\theta}{n} + k \frac{360°}{n} \right) \right)$.

The Mandelbrot set is a recently-discovered, important, and intriguing graph of a set of complex numbers. Each number c in the set has the property that the sequence defined by $\begin{cases} z_1 = c \\ z_n = z_{(n-1)} + c, \ n > 1 \end{cases}$ is bounded.

SUMMARY

The Summary gives an overview of the entire chapter and provides an opportunity for students to consider the material as a whole. Thus, the Summary can be used to help students relate the various concepts presented in the chapter.

VOCABULARY

Terms, symbols, and properties are listed by lesson to provide a checklist of concepts a student must know.

Vocabulary

For the starred (*) items you should be able to give a definition of the term.
For the other terms you should be able to give a general description and a specific example of each.

Lesson 13-1
corollary

Lesson 13-2
singularity
removable singularity

Lesson 13-3
pole, polar axis
polar coordinates, $[r, \theta]$
polar coordinate system
polar grid

Lesson 13-4
rose curve, petal curve

Lesson 13-5
complex plane
real axis, imaginary axis
Geometric Addition Theorem
* absolute value, modulus of
   a complex number
* argument of a complex
   number
polar form of a complex
   number

Lesson 13-6
* trigonometric form of a
   complex number
Product of Complex Numbers
   Theorem
Division of Complex
   Numbers Theorem

Lesson 13-7
DeMoivre's Theorem
Roots of a Complex Number
   Theorem

Lesson 13-8
Mandelbrot set
MANDELZOOM

Progress Self-Test

ADDITIONAL ANSWERS
4.

[4, -185°]

5. imaginary

-4 + 3i

6. imaginary

$2\left(\cos\frac{\pi}{2} + i\sin\frac{\pi}{2}\right)$

**10. 3 (cos 65° + i sin 65°);
3 (cos 155° + i sin 155°);
3 (cos 245° + i sin 245°);
3 cos (335° + i sin 335°)**

**11.b., 12.a.,b., 13.b., 14.,
15. See Additional
Answers in the back of
this book.**

See margin for answers not shown below.

Take this test as you would take a test in class. You
will need graph paper, a calculator, and an automatic
grapher. Then check the test yourself using the
solutions at the back of the book.

1. Give the apparent singularities of the function
   whose graph is shown below.

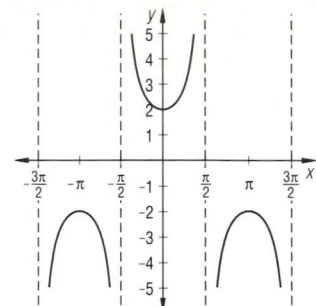

odd multiples of π/2

In 2 and 3, an automatic grapher shows the same
graph for $f(x) = 2\cos^2 x \tan x$ and $g(x) = \sin 2x$.

2. *True* or *false*.   Drawing this graph is a proof
   that the equation $2\cos^2 x \tan x = \sin 2x$ is an
   identity.   **False**

3. **a.** What values are not in the domain of the
   identity in Question 2?
   b. Prove that the equation in Question 2 is an
   identity.   **See below.**
   3a) π/2 + nπ for all integers n

In 4–6, plot on an appropriate coordinate system.

4. $[r, \theta] = [4, -185°]$ 5. $-4 + 3i$

6. $2\left(\cos\frac{\pi}{2} + i\sin\frac{\pi}{2}\right)$

7. The rectangular coordinates of a point are
   $(5, -5\sqrt{3})$. Find two pairs of polar coordinates
   $[r, \theta]$ that name this same point. Give θ in
   radians.   **sample: $\left[10, -\frac{\pi}{3}\right], \left[-10, \frac{2\pi}{3}\right]$**

8. Rewrite $-\frac{1}{2} + \frac{\sqrt{3}}{2}i$ in **a.** polar form,   **See below.**
   b. trigonometric form.   **cos 120° + i sin 120°**

9. Simplify $(3(\cos 20° + i \sin 20°))^4$. Write your
   answer in $a + bi$ form.   **≈ 14.07 + 79.77i**

10. Find the 4 fourth roots of
    $81(\cos 260° + i \sin 260°)$.

3b) $2\cos^2 x \tan x = 2\cos^2 x \cdot \frac{\sin x}{\cos x} = 2\cos x \sin x = \sin 2x$

8a) sample: [1, 120°]

---

11. **a.** Give the coordinates of two points on the
    graph of $r = \sin 3\theta$.
    b. Trace the graph below and label on your
    copy the points you gave in part **a.**
    a) samples: $\left[1, \frac{\pi}{6}\right], \left[-1, -\frac{\pi}{6}\right]$

In 12 and 13, given z_1 and z_2, **a.** perform the
indicated operation. **b.** Describe the effect of the
operation on the absolute value and argument of z_1.

12. $z_1 = -3i$, $z_2 = 5\left(\cos\frac{\pi}{12} + i\sin\frac{\pi}{12}\right)$; find $z_1 z_2$.

13. $z_1 = [6, \pi]$, $z_2 = [2, \pi]$; find $\frac{z_1}{z_2}$.   **a) [3, 0]**

14. Prove that $(1 - \sin^2\theta)(1 + \tan^2\theta) = 1$ is an
    identity.

15. Graph $r = 6\cos\theta$ in the polar coordinate
    system.

"We did the whole room over in fractals."

Chapter Review

Questions on **SPUR** Objectives

SPUR stands for **S**kills, **P**roperties, **U**ses, and **R**epresentations.
The Chapter Review questions are grouped according to the SPUR Objectives for this chapter.
See margin for answers not shown below.

SKILLS deal with the procedures used to get answers.

■ **Objective A:** *Perform operations with complex numbers in polar or trigonometric form. (Lesson 13-6)*

In 1 and 2, determine $z_1 z_2$. Leave your answer in the form of the original numbers.

1. $z_1 = 5(\cos 32° + i \sin 32°)$,
   $z_2 = 3(\cos 157° + i \sin 157°)$

2. $z_1 = [7, 0]$, $z_2 = [1.3, \frac{\pi}{4}]$ **[9.1, $\frac{\pi}{4}$]**

1) 15(cos 189° + i sin 189°)

In 3 and 4, given z_1 and $z_1 z_2$, determine z_2. Leave your answer in the form of the original numbers.

3. $z_1 = 9\left(\cos \frac{\pi}{5} + i \sin \frac{\pi}{5}\right)$,
   $z_1 z_2 = 18\left(\cos \frac{\pi}{2} + i \sin \frac{\pi}{2}\right)$ **$z_2 = 2\left(\cos \frac{3\pi}{10} + i \sin \frac{3\pi}{10}\right)$**

4. $z_1 = [2, \frac{\pi}{6}]$, $z_1 z_2 = [5, \frac{7\pi}{6}]$ **$z_2 = \left[\frac{5}{2}, \pi\right]$**

In 5 and 6, give **a.** the absolute value and **b.** an argument of the complex number.

5. $-10 + 12i$

6. $\cos \frac{\pi}{3} + i \sin \frac{\pi}{3}$ **a) 1; b) sample: $\frac{\pi}{3}$**

5a) $\sqrt{244} \approx 15.62$; b) sample: $\tan^{-1}(-1.2) \approx 129.8°$

■ **Objective B:** *Represent complex numbers in different forms. (Lessons 13-5, 13-6)*

In 7–9, write the complex number in polar form. Use an argument θ in $0° \leq \theta < 360°$.

7. $-\sqrt{3} + i$ **[2, 150°]**

8. $2 + 5i$ **$[\sqrt{29}, \tan^{-1} \frac{5}{2}] \approx [5.39, 68.2°]$**

9. $\frac{1}{8}\left(\cos \frac{\pi}{5} + i \sin \frac{\pi}{5}\right)$ **$\left[\frac{1}{8}, \frac{\pi}{5}\right]$**

In 10 and 11, write the complex number in $a + bi$ form.

10. $3(\cos(-120°) + i \sin(-120°))$ **$-\frac{3}{2} - \frac{3\sqrt{3}}{2}$**

11. $[4, \frac{2\pi}{5}]$ **$4\cos\frac{2\pi}{5} + \left(4\sin\frac{2\pi}{5}\right)i \approx 1.2 + 3.8i$**

12. Give two different representations of the complex number $[-2, 45°]$ in trigonometric form.
samples:
2(cos 225° + i sin 225°)
2(cos(-135°) + i sin(-135°))

PROPERTIES deal with the principles behind the mathematics.

■ **Objective C:** *Prove trigonometric identities. (Lessons 13-1, 13-2)*

13. Prove the identity $(1 - \cos^2 x)(1 + \cot^2 x) = 1$ by starting with the left hand side and rewriting it until the other side appears.

14. Prove that $\sin\left(\theta - \frac{\pi}{4}\right) = -\cos\left(\theta + \frac{\pi}{4}\right)$ for all θ by rewriting each side of the equation independently until equal expressions are obtained.

13) $(1 - \cos^2 x)(1 + \cot^2 x) = \sin^2 x\left(1 + \frac{\cos^2 x}{\sin^2 x}\right)$

$= \sin^2 x + \cos^2 x = 1$

15. Prove that $\csc^2 x - \cot^2 x = 1$ for all $x \neq n\pi$, where n is an integer, starting with the Pythagorean Identity.

16. Prove: For all θ for which the equation is defined, $\frac{\sin \theta}{\cos \theta \cdot \tan \theta} = 1$.

16) $\frac{\sin \theta}{\cos \theta \tan \theta} = \frac{\sin \theta}{\cos \theta \cdot \frac{\sin \theta}{\cos \theta}} = \frac{\sin \theta}{\sin \theta} = 1$.

So $\frac{\sin \theta}{\cos \theta \tan \theta} = 1$ is an identity for all $\theta \neq \frac{\pi}{2} + n\pi$

and $\theta \neq n\pi$ where n is an integer.

RESOURCES
■ Chapter 13 Test, Form A
■ Chapter 13 Test, Form B
■ Chapter 13 Test, Cumulative Form
■ Comprehensive Test, Chapters 1–13
■ Quiz and Test Writer

CHAPTER REVIEW

The main objectives for the chapter are organized here into sections corresponding to the types of understanding this book promotes: Skills, Properties, Uses, and Representations (called SPUR objectives).

USING THE CHAPTER REVIEW
Students should be able to answer questions like these with about 85% accuracy by the end of the chapter. (See pages 72 and T38–39 for more information.)

ADDITIONAL ANSWERS

14. $\sin\left(\theta - \frac{\pi}{4}\right) =$

$\sin \theta \cos \frac{\pi}{4} - \cos \theta \sin \frac{\pi}{4} =$

$\sin \theta \frac{1}{\sqrt{2}} - \cos \theta \frac{1}{\sqrt{2}} =$

$\frac{1}{\sqrt{2}}(\sin \theta - \cos \theta);$

$-\cos\left(\theta + \frac{\pi}{4}\right) = -$

$\left[\cos \theta \cos \frac{\pi}{4} - \sin \theta \sin \frac{\pi}{4}\right] =$

$-\left[\cos \theta \frac{1}{\sqrt{2}} - \sin \theta \frac{1}{\sqrt{2}}\right] =$

$-\left[\frac{1}{\sqrt{2}}(\cos \theta - \sin \theta)\right] =$

$\frac{1}{\sqrt{2}}\sin \theta - \cos \theta)$

$\frac{1}{\sqrt{2}}$

15. $\sin^2 x + \cos^2 x = 1;$
$1 - \cos^2 x = \sin^2 x;$
$\frac{1 - \cos^2 x}{\sin^2 x} = \frac{\sin^2 x}{\sin^2 x} = 1;$
$\frac{1}{\sin^2 x} - \frac{\cos^2 x}{\sin^2 x} = 1;$
$\csc^2 x - \cot^2 x = 1$

831

b. Yes; the graphs coincide.

27.a.

$y = \tan x$

$y = \dfrac{\csc x}{\sec x}$

b. No; the graphs do not coincide.

28.a.

b. Yes; the graphs seem to coincide.

35.,36.

41.

■ **Objective D:** *Describe singularities of functions.*
(Lesson 13-2)

17. Explain why the restriction $x \neq n\pi$ (n an integer) is necessary for the identity in Question 15.

18. Consider the identity given in Question 16.
   a. Determine all the singularities of the functions mentioned in the equation.
   b. Give the biggest domain on which the identity holds. $\{\theta: \theta \neq \pi/2 + n\pi \text{ and } \theta \neq n\pi\}$

19. a. Determine the singularities (if any) of the functions
   i. $y = \dfrac{x^3 - 27}{x - 3}$ $x = 3$
   ii. $\dfrac{x^3 - 27}{x^2 + 3x + 9}$.   **none**
   b. *True* or *false.*   The proposed identity is true for all real x. Explain your answer.
   i. $\dfrac{x^3 - 27}{x - 3} = x^2 + 3x + 9$
   ii. $\dfrac{x^3 - 27}{x^2 + 3x + 9} = x - 3$

20. *True* or *false.*   The functions
$f(x) = \dfrac{1}{1 - \cos x}$ and $g(x) = \dfrac{1}{1 - \cos^2 x}$
have the same singularities. Explain your answer.

■ **Objective E:** *Find powers and roots of complex numbers.* *(Lesson 13-7)*

In 21 and 22, find z^n for the given z and n.

21. $z = \sqrt{2} + \sqrt{2}\,i$, $n = 5$

22. $z = 3(\cos 240° + i \sin 240°)$, $n = 4$

In 23 and 24, find the indicated roots.

23. the 4 fourth roots of $256(\cos 12° + i \sin 12°)$

24. the 6 sixth roots of -2

17) The values must be excluded because csc x and cot x are undefined for $x = n\pi$.

18a) $\theta = \pi/2 + n\pi$ or $\theta = n\pi$

19b) i) False; it is not true for $x = 3$.
   ii) True; the denominator is not zero for any real value of x.

20) False; $\dfrac{1}{1 - \cos x}$ has a singularity for cos $x = 1$, or for x any even multiple of π. The function $\dfrac{1}{1 - \cos^2 x}$ has singularities for $\cos^2 x = 1$, which is for x any multiple of π.

21) $z^5 = 32(\cos 225° + i \sin 225°) = -16\sqrt{2} - 16\sqrt{2}i$

22) $z^4 = 81(\cos 240° + i \sin 240°) = -\dfrac{81}{2} - \dfrac{81\sqrt{3}}{2}i$

23) $4(\cos 3° + i \sin 3°)$, $4(\cos 93° + i \sin 93°)$, $4(\cos 183° + i \sin 183°)$, $4(\cos 273° + i \sin 273°)$

24) $\sqrt[6]{2}(\cos 30° + i \sin 30°)$, $\sqrt[6]{2}(\cos 90° + i \sin 90°)$, $\sqrt[6]{2}(\cos 150° + i \sin 150°)$, $\sqrt[6]{2}(\cos 210° + i \sin 210°)$, $\sqrt[6]{2}(\cos 270° + i \sin 270°)$, $\sqrt[6]{2}(\cos 330° + i \sin 330°)$

USES deal with applications of mathematics in real situations.

There are no objectives relating to uses in this chapter.

REPRESENTATIONS deal with pictures, graphs, or objects that illustrate concepts.

■ **Objective F:** *Use an automatic grapher to test a proposed identity.* *(Lessons 13-1, 13-2)*

25. *True* or *false.*   Showing that the graphs of the functions on the two sides of a given equality in a single variable coincide when created by an automatic grapher proves that the equation is an identity.   **False**

In 26–28, **a.** graph the functions related to the two sides of the proposed identity, and **b.** decide if a person should attempt to prove the identity. Explain your reasoning.

26. $\sin^2 x (1 + \tan^2 x) = \tan^2 x$

27. $\dfrac{\csc x}{\sec x} = \tan x$

28. $\dfrac{x^2 - 4x - 5}{x + 1} = x - 5$

832

Objective G: *Given polar coordinates of a point, determine its rectangular coordinates, and vice versa.* *(Lesson 13-3)*

In 29 and 30, convert from polar coordinates to rectangular coordinates.

29. $[4, \frac{3\pi}{2}]$ **(0, -4)**   **30.** $[-3, 85°]$
(3 cos 265°, 3 sin 265°) ≈ (-.26, -2.99)

In 31 and 32, give one pair of polar coordinates for the (x, y) pair.

31. $(5, 2) ≈$ **[5.4, 21.8°]**   **32.** $(-2, -3) ≈$ **[3.6, 236.3°]**

33. A point is located at $(-4\sqrt{3}, 4)$ in a rectangular coordinate system. Find three pairs of polar coordinates $[r, \theta]$ that name this same point. Assume θ is in radians.   **See below.**

34. When the coordinates of P are written in polar form, $\theta = \frac{\pi}{6}$. When the coordinates of P are written in rectangular form, $x = 5$. Find polar and rectangular coordinates for P.

$$P = \left[\frac{10}{\sqrt{3}}, \frac{\pi}{6}\right] = \left(5, \frac{5}{\sqrt{3}}\right)$$

Objective H: *Plot points in the polar coordinate system.* *(Lessons 13-3, 13-4)*

In 35 and 36, plot $[r, \theta]$, where θ is in radians.

35. $[3, \frac{\pi}{6}]$ **36.** $[-2, 0]$

37. *Multiple choice.* Which polar coordinate pair does not name the same point as $[4, 250°]$?
  (a) $[4, -110°]$ (b) $[-4, 110°]$
  (c) $[-4, 70°]$ (d) $[4, 610°]$ **(b)**

38. A point P has polar coordinates $[2, \frac{7\pi}{6}]$. Give two pairs of polar coordinates for P where $r = -2$.   **samples:** $\left[-2, \frac{\pi}{6}\right], \left[-2, -\frac{11\pi}{6}\right]$

33) samples: $\left[8, \frac{5\pi}{6}\right], \left[8, -\frac{7\pi}{6}\right], \left[-8, \frac{11\pi}{6}\right]$

Objective I: *Graph and interpret graphs of polar equations.* *(Lesson 13-4)*

39. Verify that $[2, \frac{2\pi}{3}]$ is on the graph of $r = 2\cos 6\theta$.

40. Give the coordinates of two points on the graph of $r = \csc \theta$.   **samples: [1, π/2], [√2, π/4]**
39) 2 cos 6θ = 2 cos (6(2π/3)) = 2 cos 4π = 2

In 41 and 42, graph the equation in the polar coordinate system.

41. $r = 4\sin \theta$ **42.** $r = 1 - \cos \theta$

In 43 and 44, **a.** use an automatic grapher to graph the given equation. **b.** Verify the shape of the graph in part **a** by finding a rectangular coordinate equation for the relation.

43. $r = 4\sec \theta$ **44.** $r = \frac{1}{3}\cos \theta$

Objective J: *Graph complex numbers.* *(Lessons 13-5, 13-6)*

In 45 and 46, graph in the complex plane.

45. $4 - 5i$

46. $8(\cos 130° + i\sin 130°)$

47. a. Graph the origin, $A = 3 + i$, $B = 1 - 3i$, and $A + B$ on one coordinate system.
  b. Prove that the figure with vertices $(0, 0)$, A, B, and $A + B$ is a parallelogram.

48. Consider $z = 2(\cos 72° + i\sin 72°)$.
  a. Graph z, z^2, z^3, z^4, and z^5 on one complex coordinate system.
  b. Verify that z is a solution to $z^5 = 32$.

EVALUATION

Three tests are provided for this chapter in the Teacher's Resource File. Chapter 13 Test, Forms A and B, cover just Chapter 13. The third test is Chapter 13 Test, Cumulative Form. About 50% of this test covers Chapter 13, 25% covers Chapter 12, and 25% covers earlier chapters. A Comprehensive Test covering chapters 1–13 is also included with this chapter. For information on grading, see *General Teaching Suggestions: Grading* on page T43 in this Teacher's Edition.

42.

43.a.

b. For $\theta \neq \frac{\pi}{2} + n\pi$, where n is an integer, $r = 4\sec \theta = \frac{4}{\cos \theta} \Leftrightarrow r\cos \theta = 4 \Leftrightarrow x = 4$

44.a.

44.b., 45., 46., 47.a.,b., 48.a.,b. See Additional Answers in the back of this book.

Parent Functions and Their Graphs

Type of Function	Parent Function f	Graph of f	Inverse Function f^{-1}	Graph of f^{-1}
linear—constant	$f(x) = k$		none	
linear	$f(x) = x$		$f^{-1}(x) = x$	
absolute value	$f(x) = |x|$		none	
greatest integer	$f(x) = \lfloor x \rfloor$		none	

Type of Function	Parent Function f	Graph of f	Inverse Function f^{-1}	Graph of f^{-1}
polynomial quadratic	$f(x) = x^2$		$f^{-1}(x) = \sqrt{x}$, $y \geq 0$	
polynomial cubic	$f(x) = x^3$		$f^{-1}(x) = \sqrt[3]{x}$	
polynomial of higher degree	$f(x) = x^n$, n an odd integer		$f^{-1}(x) = \sqrt[n]{x}$	
	$f(x) = x^n$, n an even integer		$f^{-1}(x) = \sqrt[n]{x}$ $y \geq 0$	

Type of Function	Parent Function f	Graph of f	Inverse Function f^{-1}	Graph of f^{-1}
inverse linear	$f(x) = \dfrac{1}{x}$		$f^{-1}(x) = \dfrac{1}{x}$	
inverse quadratic	$f(x) = \dfrac{1}{x^2}$		$f^{-1}(x) = \sqrt{\dfrac{1}{x}}$ $y > 0$	
exponential any base	$f(x) = b^x$ $b > 1$		$f^{-1}(x) = \log_b x$	
	$f(x) = b^x$ $0 < b < 1$		$f^{-1}(x) = \log_b x$	

Type of Function	Parent Function f	Graph of f	Inverse Function f^{-1}	Graph of f^{-1}
exponential natural base	$f(x) = e^x$		$f^{-1}(x) = \ln x$	
circular— sine	$f(x) = \sin x$		$f^{-1}(x) = \sin^{-1}x$ $-\dfrac{\pi}{2} \le y \le \dfrac{\pi}{2}$	
circular— cosine	$f(x) = \cos x$		$f^{-1}(x) = \cos^{-1}x$ $0 \le y \le \pi$	
circular— tangent	$f(x) = \tan x$		$f^{-1}(x) = \tan^{-1}x$ $-\dfrac{\pi}{2} \le y \le \dfrac{\pi}{2}$	

Parent Function *f*	Reciprocal Function *g*	Graph of Reciprocal Function *g*
$f(x) = \sin x$	$g(x) = \csc x$	
$f(x) = \cos x$	$g(x) = \sec x$	
$f(x) = \tan x$	$g(x) = \cot x$	

The BASIC Language

In the computer programming language BASIC (Beginner's All Purpose Symbolic Instruction Code), the arithmetic symbols are + (addition), − (subtraction), * (multiplication), / (division), and $^\wedge$ (powering). In some versions of BASIC, ↑ is used for powering.

Variables are represented by strings of letters and digits. The first character of a variable name must be a letter, but later characters may be digits. Examples of variable names are X, X2, SDEV, and WEIGHT. Variable names are generally chosen to describe the variable they represent. In some versions of BASIC, only the first two characters of a variable are read by the computer. (So ARC and AREA would both be read as AR, for example.) Check to see if this is true for your computer.

The computer evaluates expressions according to the usual order of operations. Parentheses ( ) may be used for clarity or to change the usual order of operations. The comparison symbols, $=$, $>$, $<$, are also used in the standard way, but BASIC uses $<=$ instead of $\leq$, $>=$ instead of $\geq$, and $<>$ instead of $\neq$.

Each time you enter information on the screen, whether it is a command, an input to a program, or a new program line, finish by pressing the key marked RETURN or ENTER.

Commands

The BASIC commands used in this course and examples of their uses are given below.

REM	This command allows remarks to be inserted in a program. These may describe what the program does and how it works. REM statements are especially important for complex programs, programs that others will use and programs you may not use for long periods.

        10 REM PYTHAGOREAN THEOREM   The statement has no effect when the program is run. It appears when LIST is used.

LET	A value is assigned to a given variable. Some versions of BASIC allow you to omit the word LET in the assignment statement.

        LET X = 5   The number 5 is stored in a memory location called X.

        LET NEXT = NEXT + 2   The value in the memory location called NEXT is increased by 2 and then restored in the location called NEXT.

INPUT	The computer prompts the user to enter a value and stores that value to the variable named. Using a message helps the user know what to type.

        INPUT X   The computer prompts you with a question mark and then stores the value you type in memory location X.

        INPUT "HOW OLD"; AGE   The computer prints HOW OLD? and stores your response in the memory location AGE.

GOTO The computer goes to whatever line of the program is indicated. GOTO statements should be avoided wherever possible, as they interrupt smooth program flow and make programs hard to interpret.

GOTO 70   The computer goes to line 70 and executes that command.

DEF FN With this command, the computer allows the user to define a function for later use. Whenever the function is referred to in the program, it is written as FN, followed by the name of the function and an argument in parentheses.

10 DEF FN C(X) = 5 * (X-32)/9   Line 10 defines a function C to convert temperatures in
20 INPUT "FAHRENHEIT";F   degrees Fahrenheit to degrees Celsius, according to the
30 PRINT F,FN C(F)   rule $C(X) = \dfrac{5(X-32)}{9}$. The function can have any variable or number for its argument.

IF... The computer performs the consequent (the THEN part) only if the antecedent (the IF part) is true. When
THEN   the antecedent is false, the computer *ignores* the consequent and goes directly to the next line of the program.

10 IF X>6 THEN END   If the X value is less than or equal to 6, the computer
20 PRINT X   ignores END, goes to the next line, and prints the value stored in X. If the X value is greater than 6, the computer stops and the value stored in X is not printed.

FOR ... The FOR command assigns a beginning and ending value to a variable using the word TO. The first time
NEXT   through the loop, the variable has the beginning value in the FOR command. When the computer reaches
STEP   the line reading NEXT, the value of the variable is increased by the amount indicated by STEP. If STEP is not written, the computer increases the variable by 1 each time through the loop. The commands between FOR and NEXT are then repeated. When the incremented value of the variable is larger than the ending value in the FOR command, the computer leaves the loop and executes the next line of the program.

10 FOR N = 3 TO 6 STEP 2   The computer assigns 3 to N and then prints the value of
20 PRINT N   N. On reaching NEXT, the computer increases N by 2
30 NEXT N   (the STEP amount), and the second time through the
40 END   loop prints 5. The next N would be 7 which is too large. So, the computer executes the command after NEXT, ending the program.

PRINT The computer prints on the screen what follows the PRINT command. If what follows is a constant or variable the computer prints the value of that constant or variable. If what follows is in quotes, the computer prints exactly what is in quotes. Commas or semicolons are used to space the output.

PRINT COST   The computer prints the number stored in memory location COST.

PRINT "X-VALUES"   The computer prints the phrase X-VALUES.

PRINT X,Y,Z   The computer prints the numbers stored in the memory locations X, Y and Z in that order with a tab space between each.

PRINT "MEAN IS ";XBAR   The computer prints the phrase MEAN IS, followed immediately by the number in memory location XBAR.

END The computer stops running the program. A program should not contain more than one END statement.

Functions

The following built-in functions are available in most versions of BASIC. These can be used directly, in a PRINT statement, or within a program.

ABS   The absolute value of the number or expression that follows is calculated.

         LET X = ABS(-8)   The computer calculates $|-8|$ and stores the value 8 in the memory location X.

ATN   The arctangent or inverse tangent of the number or expression that follows is calculated. The result is always in radians.

         LET P = ATN(1)   The computer calculates $\tan^{-1} 1 = \frac{\pi}{4}$ and stores the value 0.785398 in the memory location P.

EXP   The number e raised to the power of the number or expression that follows is calculated.

         LET K = EXP(2)   The computer calculates e^2 and stores the value 7.389056 in the memory location K.

LOG   The natural logarithm (that is the logarithm to the base e) of the number or expression that follows is calculated.

         LET J = LOG(6)   The computer calculates ln 6 and stores the value 1.791759 in the memory location J.

INT   The greatest integer less than or equal to the number or expression that follows is calculated.

         LET W = INT(-7.3)   The computer calculates $\lfloor -7.3 \rfloor$ and stores the value -8 in the memory location W.

RND(1)   A random number greater than 0 and less than 1 is generated. The argument of the RND function is always 1.

         LET D = 2 * RND(1)   The computer generates a random number between 0 and 1 and stores twice the value in the memory location D.

SIN   The sine, cosine or tangent of the number or expression that follows is calculated. The argument of these
COS   circular functions is always in radians. To convert degrees to radians, multiply by (3.14159/180).
TAN

         LET R = 5*SIN(0.7)   The computer finds 5 sin (0.7) and stores the value 3.221088 in the memory location R.

SQR   The square root of the number or expression that follows is calculated.

         LET C = SQR(A * A + B * B)   The computer calculates $\sqrt{A^2 + B^2}$ using the values stored in A and B and stores the result in C.

SGN   The signum function of the number or expression that follows is calculated. The function has value 1 for positive numbers, -1 for negative numbers and 0 for zero.

         PRINT SGN(-6)   The computer prints -1.

Programs

In many versions of BASIC, every line in a BASIC program must begin with a line number. It is common to start numbering at 10 and count by tens, so that intermediate lines can be added later if necessary. The computer reads and executes a BASIC program in the order of the line numbers. It will not go back to a previous line unless told to do so.

To run a new program, first type NEW to erase the previous one; then type in the lines of the program. Finally, type RUN.

To run a program already saved on a disk you must know the exact name of the program, including any spaces. To run a program called TABLE SOLVE, type RUN TABLE SOLVE, then press the RETURN or ENTER key.

If you type LIST, the program currently in the computer memory will be listed on the screen. A program line can be changed by re-typing the line, including the line number.

In most versions of BASIC, you can stop a program completely while it is running by typing Control-C (CTRL C). This is done by pressing the C key while holding down the CTRL key. A program may be stopped temporarily (for example, to see part of the output) by typing CTRL S and resumed by typing CTRL Q. Some versions of BASIC use different commands for these purposes.

10 PRINT "A DIVIDING SEQUENCE"	The computer prints A DIVIDING SEQUENCE.
20 INPUT X	You must give a value to store in the location X. Suppose you use 16. (Press the RETURN or ENTER key after entering 16.)
30 LET Y = 2	2 is stored in location Y.
40 FOR Z = -5 TO 2	Z is given the value $^-5$. Each time through the loop, the value of Z will be increased by 1.
50 IF Z = 0 THEN GO TO 70	When $Z = 0$ the computer goes directly to line 70. If $Z \neq 0$ the computer executes line 60.
60 PRINT (X * Y)/Z	On the first pass through the loop, the computer prints $^-6.4$ because $(16 \cdot 2)/(^-5)$ is $^-6.4$.
70 NEXT Z	The value in Z is increased by 1 and the computer goes back to line 50.
80 END	After going through the FOR ... NEXT loop with $Z = 2$, the computer stops.

$^-6.4$
$^-8$
$^-10.666667$
$^-16$
$^-32$
32
16

The output of this program is shown.

Table of Random Numbers

col. 1	2	3	4	5	6	7	8	9	10	11	12	13	14
row													
1 10480	15011	01536	02011	81647	91646	69719	14194	62590	36207	20969	99570	91291	90700
2 22368	46573	25595	85393	30995	89198	27982	53402	93965	34095	52666	19174	39615	99505
3 24130	48360	22527	97265	76393	64809	15179	24830	49340	32081	30680	19655	63348	58629
4 42167	93093	06423	61680	17856	16376	39440	53537	71341	57004	00849	74917	97758	16379
5 37570	39975	81837	16656	06121	91782	60468	81305	49684	60672	14110	06927	01263	54613
6 77921	06907	11008	42751	27756	53498	18602	70659	90655	15053	21916	81825	44394	42880
7 99562	72905	56420	69994	98872	31016	71194	18738	44013	48840	63213	21069	10634	12952
8 96301	91977	05463	07972	18876	20922	94595	56869	69014	60045	18425	84903	42508	32307
9 89579	14342	63661	10281	17453	18103	57740	84378	25331	12566	58678	44947	05585	56941
10 85475	36857	43342	53988	53060	59533	38867	62300	08158	17983	16439	11458	18593	64952
11 28918	69578	88231	33276	70997	79936	56865	05859	90106	31595	01547	85590	91610	78188
12 63553	40961	48235	03427	49626	69445	18663	72695	52180	20847	12234	90511	33703	90322
13 09429	93969	52636	92737	88974	33488	36320	17617	30015	08272	84115	27156	30613	74952
14 10365	61129	87529	85689	48237	52267	67689	93394	01511	26358	85104	20285	29975	89868
15 07119	97336	71048	08178	77233	13916	47564	81056	97735	85977	29372	74461	28551	90707
16 51085	12765	51821	51259	77452	16308	60756	92144	49442	53900	70960	63990	75601	40719
17 02368	21382	52404	60268	89368	19885	55322	44819	01188	65255	64835	44919	05944	55157
18 01011	54092	33362	94904	31272	04146	18594	29852	71585	85030	51132	01915	92747	64951
19 52162	53916	46369	58586	23216	14513	83149	98736	23495	64350	94738	17752	35156	35749
20 07056	97628	33787	09998	42698	06691	76988	13602	51851	46104	88916	19509	25625	58104
21 48663	91245	85828	14346	09172	30168	90229	04734	59193	22178	30421	61666	99904	32812
22 54164	58492	22421	74103	47070	25306	76468	26384	58151	06646	21524	15227	96909	44592
23 32639	32363	05597	24200	13363	38005	94342	28728	35806	06912	17012	64161	18296	22851
24 29334	27001	87637	87308	58731	00256	45834	15398	46557	41135	10367	07684	36188	18510
25 02488	33062	28834	07351	19731	92420	60952	61280	50001	67658	32586	86679	50720	94953
26 81525	72295	04839	96423	24878	82651	66566	14778	76797	14780	13300	87074	79666	95725
27 29676	20591	68086	26432	46901	20849	89768	81536	86645	12659	92259	57102	80428	25280
28 00742	57392	39064	66432	84673	40027	32832	61362	98947	96067	64760	64584	96096	98253
29 05366	04213	25669	26422	44407	44048	37937	63904	45766	66134	75470	66520	34693	90449
30 91921	26418	64117	94305	26766	25940	39972	22209	71500	64568	91402	42416	07844	69618
31 00582	04711	87917	77341	42206	35126	74087	99547	81817	42607	43808	76655	62028	76630
32 00725	69884	62797	56170	86324	88072	76222	36086	84637	93161	76038	65855	77919	88006
33 69011	65797	95876	55293	18988	27354	26575	08625	40801	59920	29841	80150	12777	48501
34 25976	57948	29888	88604	67917	48708	18912	82271	65424	69774	33611	54262	85963	03547
35 09763	83473	73577	12908	30883	18317	28290	35797	05998	41688	34952	37888	38917	88050
36 91567	42595	27958	30134	04024	86385	29880	99730	55536	84855	29080	09250	79656	73211
37 17955	56349	90999	49127	20044	59931	06115	20542	18059	02008	73708	83517	36103	42791
38 46503	18584	18845	49618	02304	51038	20655	58727	28168	15475	56942	53389	20562	87338
39 92157	89634	94824	78171	84610	82834	09922	25417	44137	48413	25555	21246	35509	20468
40 14577	62765	35605	81263	39667	47358	56873	56307	61607	49518	89656	20103	77490	18062

Standard Normal Distribution Table

This table gives the area under the standard normal curve to the left of a given positive number a.

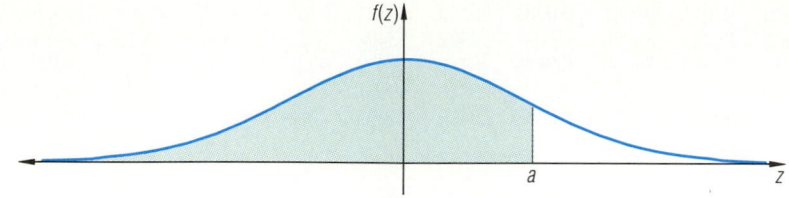

					$P(z < a)$ for $a \geq 0$					
a	**0**	**1**	**2**	**3**	**4**	**5**	**6**	**7**	**8**	**9**
0.0	.5000	.5040	.5080	.5120	.5160	.5199	.5239	.5279	.5319	.5359
0.1	.5398	.5438	.5478	.5517	.5557	.5596	.5636	.5675	.5714	.5753
0.2	.5793	.5832	.5871	.5910	.5948	.5987	.6026	.6064	.6103	.6141
0.3	.6179	.6217	.6255	.6293	.6331	.6368	.6406	.6443	.6480	.6517
0.4	.6554	.6591	.6628	.6664	.6700	.6736	.6772	.6808	.6844	.6879
0.5	.6915	.6950	.6985	.7019	.7054	.7088	.7123	.7157	.7190	.7224
0.6	.7257	.7291	.7324	.7357	.7389	.7422	.7454	.7486	.7517	.7549
0.7	.7580	.7611	.7642	.7673	.7704	.7734	.7764	.7794	.7823	.7852
0.8	.7881	.7910	.7939	.7967	.7995	.8023	.8051	.8078	.8106	.8133
0.9	.8159	.8186	.8212	.8238	.8264	.8289	.8315	.8340	.8365	.8389
1.0	.8413	.8438	.8461	.8485	.8508	.8531	.8554	.8577	.8599	.8621
1.1	.8643	.8665	.8686	.8708	.8729	.8749	.8770	.8790	.8810	.8830
1.2	.8849	.8869	.8888	.8907	.8925	.8944	.8962	.8980	.8997	.9015
1.3	.9032	.9049	.9066	.9082	.9099	.9115	.9131	.9147	.9162	.9177
1.4	.9192	.9207	.9222	.9236	.9251	.9265	.9279	.9292	.9306	.9319
1.5	.9332	.9345	.9357	.9370	.9382	.9394	.9406	.9418	.9429	.9441
1.6	.9452	.9463	.9474	.9484	.9495	.9505	.9515	.9525	.9535	.9545
1.7	.9554	.9564	.9573	.9582	.9591	.9599	.9608	.9616	.9625	.9633
1.8	.9641	.9649	.9656	.9664	.9671	.9678	.9686	.9693	.9699	.9706
1.9	.9713	.9719	.9726	.9732	.9738	.9744	.9750	.9756	.9761	.9767
2.0	.9772	.9778	.9783	.9788	.9793	.9798	.9803	.9808	.9812	.9817
2.1	.9821	.9826	.9830	.9834	.9838	.9842	.9846	.9850	.9854	.9857
2.2	.9861	.9864	.9868	.9871	.9875	.9878	.9881	.9884	.9887	.9890
2.3	.9893	.9896	.9898	.9901	.9904	.9906	.9909	.9911	.9913	.9916
2.4	.9918	.9920	.9922	.9925	.9927	.9929	.9931	.9932	.9934	.9936
2.5	.9938	.9940	.9941	.9943	.9945	.9946	.9948	.9949	.9951	.9952
2.6	.9953	.9955	.9956	.9957	.9959	.9960	.9961	.9962	.9963	.9964
2.7	.9965	.9966	.9967	.9968	.9969	.9970	.9971	.9971	.9973	.9974
2.8	.9974	.9975	.9976	.9977	.9977	.9978	.9979	.9979	.9980	.9981
2.9	.9981	.9982	.9982	.9983	.9984	.9984	.9985	.9985	.9986	.9986
3.0	.9987	.9987	.9987	.9988	.9988	.9989	.9989	.9989	.9990	.9990

For specific details on the use of this table, see page 620.

Professional Sourcebook For UCSMP

1 Overview of UCSMP

The Reasons for UCSMP

■ Recommendations for Change

The mathematics curriculum has undergone changes in every country of the world throughout this century, as a result of an increasing number of students staying in school longer, a greater number of technically competent workers and citizens being needed, and because of major advances in mathematics itself. In the last generation, these developments have been accelerated due to the widespread appearance of computers with their unprecedented abilities to handle and display information.

In the last 100 years, periodically there have been national groups examining the curriculum in light of these changes in society. (A study of these reports can be found in *A History of Mathematics Education in the United States and Canada,* the 30th Yearbook of the National Council of Teachers of Mathematics, 1970.) The most recent era of reports can be said to have begun in the years 1975–1980, with the publication of reports by various national mathematics organizations calling attention to serious problems in the education of our youth.

Beginning in 1980, these reports were joined by governmental and private reports on the state of American education with broad recommendations for school practice. Two of these are notable for their specific remarks about mathematics education.

1983:   National Commission on Excellence in Education. *A Nation At Risk.*

"The teaching of mathematics in high school should equip graduates to: (a) understand geometric and algebraic concepts; (b) understand elementary probability and statistics; (c) apply mathematics in everyday situations; and (d) estimate, approximate, measure, and test the accuracy of their calculations. In addition to the traditional sequence of studies available for college-bound students, new, equally demanding mathematics curricula need to be developed for those who do not plan to continue their formal education immediately." (p. 25)

1983:   College Board (Project EQuality). *Academic Preparation for College: What Students Need to Know and Be Able to Do.*

All students (college-bound or not) should have: "The ability to apply mathematical techniques in the solution of real-life problems and to recognize when to apply those techniques. Familiarity with the language, notation, and deductive nature of mathematics and the ability to express quantitative ideas with precision. The ability to use computers and calculators. Familiarity with the basic concepts of statistics and statistical reasoning. Knowledge in considerable depth and detail of algebra, geometry, and functions." (p. 20)

The specific remarks about school mathematics in these documents for the most part mirror what appeared in the earlier reports. Thus, **given what seemed to be a broad consensus on the problems and desirable changes in pre-college mathematics instruction, it was decided at the outset of UCSMP, that UCSMP would not attempt to form its own set of recommendations, but undertake the task of translating the existing recommendations into the reality of classrooms and schools.**

At the secondary (7–12) level, these reports respond to two generally perceived problems pursuant to mathematics education.

GENERAL PROBLEM 1: Students do not learn enough mathematics by the time they leave school.

Specifically:

(A) Many students lack the mathematics background necessary to succeed in college, on the job, or in daily affairs.

(B) Even those students who possess mathematical skills are not introduced to enough applications of the mathematics they know.

(C) Students do not get enough experience with problems and questions that require some thought before answering.

(D) Many students terminate their study of mathematics too soon, not realizing the importance mathematics has in later schooling and in the marketplace.

(E) Students do not read mathematics books and, as a result, do not learn to become independent learners capable of acquiring mathematics outside of school when the need arises.

These situations lead us to want to **upgrade students' achievement.**

GENERAL PROBLEM 2: The school mathematics curriculum has not kept up with changes in mathematics and the ways in which mathematics is used.

Specifically:

(A) Current mathematics curricula have not taken into account today's calculator and computer technology.

(B) Students who do succeed in secondary school mathematics are prepared for calculus, but are not equipped for the other mathematics they will encounter in college.

(C) Statistical ideas are found everywhere, from newspapers to research studies, but are not found in most secondary school mathematics curricula.

The UCSMP grades 7–12 can be considered to be the first mathematical sciences curriculum.

(D) The emergence of computer science has increased the importance of a background in discrete mathematics.

(E) Mathematics is now applied to areas outside the realm of the physical sciences, as much as within the field itself, but these applications are rarely taught and even more rarely tested.

(F) Estimation and approximation techniques are important in all of mathematics, from arithmetic on.

These existing situations lead us to a desire to **update the mathematics curriculum.**

Since the inception of UCSMP, reports from national groups of mathematics educators have reiterated the above problems, and research has confirmed their existence. Three reports are of special significance to UCSMP.

Universities have for many years had to recognize that mathematics encompasses far more than algebra, geometry, and analysis. The term **mathematical sciences** is an umbrella designation which includes traditional mathematics as well as a number of other disciplines. The largest of these other disciplines today are statistics, computer science, and applied mathematics. In 1983, the Conference Board of the Mathematical Sciences produced a report, *The Mathematical Sciences Curriculum: What Is Still Fundamental and What Is Not*. THE UCSMP GRADES 7–12 CAN BE CONSIDERED TO BE THE FIRST MATHEMATICAL SCIENCES CURRICULUM.

The Second International Mathematics Study (SIMS) was conducted in 1981–82 and involved 23 populations in 21 countries. At the eighth-grade level, virtually all students attend school in all those countries. At the 12th-grade level, the population tested consisted of those who are in the normal college preparatory courses, which, in the United States, include precalculus and calculus classes.

At the eighth-grade level, our students scored at or below the international average on all five subtests: arithmetic, measurement, algebra, geometry, and statistics. We are far below the top: Japan looked at the test and decided it was too easy for their 8th-graders, and so gave it at 7th grade. Still, the median Japanese 7th-grader performed at the 95th percentile of United States 8th-graders. These kinds of results have been confirmed in other studies, comparing students at lower-grade levels.

At the twelfth-grade level, about 13% of our population is enrolled in precalculus or calculus; the mean among developed countries is about 16%. Thus, the United States no longer keeps more students in mathematics than other developed countries, yet our advanced placement students do not perform well when compared to their peers in other countries. SIMS found:

1987:   Second International Mathematics Study (SIMS). *The Underachieving Curriculum*.

In the U.S., the achievement of the Calculus classes, the nation's **best** mathematics students, was at or near the average achievement of the advanced secondary school mathematics students in other countries. (In most countries, **all** advanced mathematics students take calculus. In the U.S., only about one-fifth do.) The achievement of the U.S. Precalculus students (the majority of twelfth grade college-preparatory students) was substantially below the international average. In some cases the U.S. ranked with the lower one-fourth of all countries in the Study, and was the lowest of the advanced industrialized countries. (*The Underachieving Curriculum*, p. vii.)

The situation is, of course, even worse for those who do not take precalculus mathematics in high school. Such students either have performed poorly in their last mathematics course, a situation which has caused them not to go on in mathematics, or they were performing poorly in junior high school and had to take remedial mathematics as 9th-graders. If these students go to college, they invariably take remedial mathematics, which is taught at a faster pace than in high school, and the failure rates in such courses often exceed 40%. If they do not go to college but join the job market, they lack the mathematics needed to understand today's technology. IT IS NO UNDERSTATEMENT TO SAY THAT UCSMP HAS RECEIVED ITS FUNDING FROM BUSINESS AND INDUSTRY BECAUSE THOSE WHO LEAVE SCHOOLING TO JOIN THE WORK FORCE ARE WOEFULLY WEAK IN THE MATHEMATICS THEY WILL NEED.

SIMS recommended steps to renew school mathematics in the United States. **The UCSMP secondary curriculum implements the curriculum recommendations of the Second International Mathematics Study.**

In 1986, the National Council of Teachers of Mathematics began an ambitious effort to detail the curriculum it would like to see in schools. The "NCTM Standards," as they have come to be called, involve both content and methodology. The *Standards* document is divided into four sections, K–4, 5–8, 9–12, and Evaluation. Space limits our discussion here to just a few quotes from the 5–8 and 9–12 standards.

1989:   National Council of Teachers of Mathematics. *Curriculum and Evaluation Standards for School Mathematics*

"The 5–8 curriculum should include the following features:

■ Problem situations that establish the need for new ideas and motivate students should serve as the context for mathematics in grades 5–8. Although a specific idea might be forgotten, the context in which it is learned can be remembered and the idea can be re-created. In developing the problem situations, teachers should emphasize the application to real-world problems as well as to other settings relevant to middle school students.

■ Communication with and about mathematics and mathematical reasoning should permeate the 5–8 curriculum.

■ A broad range of topics should be taught, including number concepts, computation, estimation, functions, algebra, statistics, probability, geometry, and measurement. Although each of these areas is valid mathematics in its own right, they should be taught together as an integrated whole, not as isolated topics; the connections between them should be a prominent feature of the curriculum.

It is no understatement to say that UCSMP has received its funding from business and industry because those who leave schooling to join the work force are woefully weak in the mathematics they will need.

■ Technology, including calculators, computers, and videos, should be used when appropriate. These devices and formats free students from tedious computations and allow them to concentrate on problem solving and other important content. They also give them new means to explore content. As paper-and-pencil computation becomes less important, the skills and understanding required to make proficient use of calculators and computers become more important.'' (pp. 66–67)

''The standards for grades 9–12 are based on the following assumptions:

■ Students entering grade 9 will have experienced mathematics in the context of the broad, rich curriculum outlined in the K–8 standards.

■ The level of [paper and pencil] computational proficiency suggested in the K–8 standards will be expected of all students; however, no student will be denied access to the study of mathematics in grades 9–12 because of a lack of computational facility.

■ Although arithmetic computation will not be a direct object of study in grades 9–12, conceptual and procedural understandings of number, numeration, and operations, and the ability to make estimations and approximations and to judge the reasonableness of results will be strengthened in the context of applications and problem solving, including those situations dealing with issues of scientific computation.

■ Scientific calculators with graphing capabilities will be available to all students at all times.

■ A computer will be available at all times in every classroom for demonstration purposes, and all students will have access to computers for individual and group work.

■ At least three years of mathematical study will be required of all secondary school students.

■ These three years of mathematical study will revolve around a core curriculum differentiated by the depth and breadth of the treatment of topics and by the nature of applications.

■ Four years of mathematical study will be required of all college-intending students.

The UCSMP secondary curriculum is the first full mathematics curriculum that is consistent with the recommendations of the NCTM Standards.

■ These four years of mathematical study will revolve around a broadened curriculum that includes extensions of the core topics and for which calculus is no longer viewed as *the* capstone experience.

■ All students will study appropriate mathematics during their senior year.'' (pp. 124–125)

THE UCSMP SECONDARY CURRICULUM IS THE FIRST FULL MATHEMATICS CURRICULUM THAT IS CONSISTENT WITH THE RECOMMENDATIONS OF THE NCTM STANDARDS.

In 1989, the Mathematical Science Education Board (MSEB), a committee of the National Research Council that coordinates efforts for improvement of mathematics education in the United States, came out with the report *Everybody Counts*, emphasizing the need for the mathematics curriculum to be appropriate for as many students as possible. This thrust reflects the UCSMP position that as many students as possible be accommodated with the curriculum taken by those who go to college.

The entire movement of the past 15 years is summarized in a second MSEB report, *Reshaping School Mathematics: A Philosophy and Framework for Curriculum*. Six changes are identified there as affecting the context of mathematics education:

■ Changes in the need for mathematics.
■ Changes in mathematics and how it is used.
■ Changes in the role of technology.
■ Changes in American society.
■ Changes in understanding of how students learn.
■ Changes in international competitiveness.

As the reader can plainly see, in the development of the UCSMP secondary curriculum we have attempted to respond to each of these changes.

■ Accomplishing the Goals

We at UCSMP believe that the goals of the various reform groups can be accomplished, but not without a substantial reworking of the curriculum. It is not enough to insert applications, a bit of statistics, and take students a few times to a computer. Currently the greatest amount of time in arithmetic is spent on calculation, in algebra on manipulating polynomials and rational expressions, in geometry on proof, in advanced algebra and later courses on functions. These topics are the most affected by technology.

It is also not enough to raise graduation requirements, although that is the simplest action to take. Increases in requirements characteristically lead to one of two situations. If the courses are kept the same, the result is typically a greater number of failures and even a greater number of dropouts. If the courses are eased, the result is lower performance for many students as they are brought through a weakened curriculum.

The fundamental problem, as SIMS noted, is the curriculum, and the fundamental problem in the curriculum is **time**. There is not enough time in the current 4-year algebra-geometry-algebra-precalculus curriculum to prepare students for calculus, and the recommendations are asking students to learn even more content.

Fortunately, there is time to be had, because the existing curriculum wastes time. It underestimates what students know when they enter the classroom and needlessly reviews what students have already learned. This needless review has been documented by Jim Flanders, a UCSMP staff member ("How Much of the Content in Mathematics Textbooks is New?" *Arithmetic Teacher,* September, 1987). Examining textbooks of the early 1980s, Flanders reports that at grade 2 there is little new. In grades 3–5, about half the pages have something new on them. But over half the pages in grades 6–8 are totally review.

And then in the 9th grade the axe falls. Flanders found that almost 90% of the pages of first-year algebra texts have content new to the student. The student, having sat for years in mathematics classes where little was new, is overwhelmed. Some people interpret the overwhelming as the student "not being ready" for algebra, but we interpret it as the student being swamped by the pace. When you have been in a classroom in which at most only 1 of 3 days is devoted to anything new, you are not ready for a new idea every day.

This amount of review in grades K–8, coupled with the magnitude of review in previous years, effectively decelerates students at least 1–2 years compared to students in other countries. It explains why almost all industrialized countries of the world, except the U.S. and Canada (and some French-speaking countries who do geometry before algebra), can begin concentrated study of algebra in the 7th or 8th grade.

Thus we believe that algebra should be taught one year earlier to most students than is currently the case.

However, we do not believe students should take calculus one year earlier than they do presently. It seems that most students who take four years of college preparatory mathematics successfully in high schools do not begin college with calculus. As an example, consider the data reported by Bert Waits and Frank Demana in the *Mathematics Teacher* (January, 1988). Of students entering Ohio State University with exactly four years of college preparatory high-school mathematics, only 8% placed into calculus on the Ohio State mathematics placement test. The majority placed into precalculus, with 31% requiring one semester and 42% requiring two semesters of work. The remaining 19% placed into remedial courses below precalculus.

Those students who take algebra in the 8th grade and are successful in calculus at the 12th grade are given quite a bit more than the normal four years of college preparatory mathematics in their "honors" or "advanced" courses. It is not stretching the point too much to say that they take five years of mathematics crammed into four years.

Thus, even with the current curriculum, four years are not enough to take a typical student from algebra to calculus. Given that the latest recommendations ask for students to learn more mathematics, WE BELIEVE FIVE YEARS OF COLLEGE PREPARATORY MATHEMATICS *BEGINNING WITH ALGEBRA* ARE NECESSARY TO PROVIDE THE TIME FOR STUDENTS TO LEARN THE MATHEMATICS THEY NEED FOR COLLEGE IN THE 1990s. The UCSMP secondary curriculum is designed with that in mind.

. . . we believe five years of college preparatory mathematics *beginning with algebra* are necessary to provide the time for students to learn the mathematics they need for college in the 1990's.

The UCSMP Secondary Curriculum

The UCSMP curriculum for grades 7–12 consists of these six courses:

Transition Mathematics

Algebra

Geometry

Advanced Algebra

Functions, Statistics, and Trigonometry

Precalculus and Discrete Mathematics

EACH COURSE IS MEANT TO STAND ALONE. Each course has also been tested alone. HOWEVER, TO TAKE BEST ADVANTAGE OF THESE MATERIALS, AND TO HAVE THEM APPROPRIATE FOR THE GREATEST NUMBER OF STUDENTS, IT IS PREFERABLE TO USE THEM IN SEQUENCE.

Each course is meant to stand alone. . . . However, to take best advantage of these materials, and to have them appropriate for the greatest number of students, it is preferable to use them in sequence.

Content Features

Transition Mathematics: This text weaves three themes—applied arithmetic, pre-algebra and pre-geometry—by focusing on arithmetic operations in mathematics and the real world. Variables are used as pattern generalizers, abbreviations in formulas, and unknowns in problems, and are represented on the number line and graphed in the coordinate plane. Basic arithmetic and algebraic skills are connected to corresponding geometry topics.

Algebra: This text has a scope far wider than most other algebra texts. It uses statistics and geometry as settings for work with linear expressions and sentences. Probability provides a context for algebraic fractions, functions, and set ideas. There is much work with graphing. Applications motivate all topics, and include exponential growth and compound interest.

Geometry: This text integrates coordinates and transformations throughout, and gives strong attention to measurement formulas and three-dimensional figures in the first two-thirds of the book. Work with proof-writing follows a carefully sequenced development of the logical and conceptual precursors to proof.

Advanced Algebra: This course emphasizes facility with algebraic expressions and forms, especially linear and quadratic forms, powers and roots, and functions based on these concepts. Students study logarithmic, trigonometric, polynomial, and other special functions both for their abstract properties and as tools for modeling real-world situations. A geometry course or its equivalent is a prerequisite, for geometric ideas are utilized throughout.

Functions, Statistics, and Trigonometry (FST): FST integrates statistical and algebraic concepts, and previews calculus in work with functions and intuitive notions of limits. Computers or sophisticated calculators are assumed available for student use in plotting functions, analyzing data, and simulating experiments. Enough trigonometry is available to constitute a standard precalculus course in trigonometry and circular functions.

Precalculus and Discrete Mathematics (PDM): PDM integrates the background students must have to be successful in calculus (advanced work with functions and trigonometry, an introduction to limits and other calculus ideas), with the discrete mathematics (number systems, combinatorics, recursion, graphs) helpful in computer science. Mathematical thinking, including specific attention to formal logic and proof, is a theme throughout. It is assumed that computers or graphing calculators are available to students throughout this course.

■ General Features

Wider Scope: Geometry and discrete mathematics are present in all courses. Substantial amounts of statistics are integrated into the study of algebra and functions. The history of concepts and recent developments in mathematics and its applications are included as part of the lessons themselves.

Reality Orientation: Each mathematical idea is studied in detail for its applications to the understanding of real-world situations, or to the solving of problems like those found in the real world. The reality orientation extends also to the approaches allowed the student in working out problems. Students are expected to use scientific calculators. Calculators are assumed throughout the series (and should be allowed on tests), because virtually all individuals who use mathematics today use calculators.

Problem Solving: Like skills, problem solving must be practiced. When practiced, problem solving becomes far less difficult. All lessons contain a variety of questions so that students do not blindly copy one question to do the next. Explorations are a feature of the first four years, and Projects are offered in the last two years. Some problem-solving techniques are so important that at times they (rather than the problems) are the focus of instruction.

Enhancing Performance: Each book's format is designed to maximize the acquisition of both skills and concepts, with lessons meant to take one day to cover. Within each lesson there is review of material from previous lessons from that chapter or from previous chapters. This gives the student more time to learn the material. The lessons themselves are sequenced into carefully constructed chapters. Progress Self-Test and Chapter Review questions, keyed to objectives in all the dimensions of understanding, are then used to solidify performance of skills and concepts from the chapter, so that they may be applied later with confidence. (See pages T34–T35 for more detail.)

Reading: Reading is emphasized throughout. Students can read; they must learn to read mathematics in order to become able to use mathematics outside of school. Every lesson has reading and contains questions covering that reading. The last lesson of each chapter is a special reading lesson. (See pages T36–T37 for more detail.)

Understanding: Four dimensions of understanding are emphasized: skill in carrying out various algorithms; developing and using mathematical properties and relationships; applying mathematics in realistic situations; and representing or picturing mathematical concepts. We call this the SPUR approach: **S**kills, **P**roperties, **U**ses, **R**epresentations. On occasion, a fifth dimension of understanding, the historical dimension, is discussed. (See pages T38–T39 for more detail.)

Technology: Scientific calculators are recommended because they use an order of operations closer to that found in algebra and have numerous keys that are helpful in understanding concepts at this level. Work with computers is carefully sequenced within each year and between the years, with gradual gain in sophistication until *FST*, where computers are an essential element. In all courses, integrated computer exercises show how the computer can be used as a helpful tool in doing mathematics. Students are expected to run and modify programs, but are not taught programming. (See pages T40–T42 for more detail.)

■ Target Populations

We believe that all high-school graduates should take courses through *Advanced Algebra,* that all students planning to go to college should take courses through *Functions*, *Statistics*, *and Trigonometry*, and that students planning majors in technical areas should take all six UCSMP courses.

The fundamental principle in placing students into the first of these courses is that entry should not be based on age, but on mathematical knowledge. Our studies indicate that about 10% of students nationally are ready for *Transition Mathematics* at 6th grade, about another 40% at 7th grade, another 20% at 8th grade, and another 10–15% at 9th grade. We caution that these percentages are national, not local percentages, and the variability in our nation is enormous. We have tested the materials in school districts where few students are at grade level, where *Transition Mathematics* is appropriate for no more than the upper half of 8th-graders. We have tested also in school districts where as many as 90% of the students have successfully used *Transition Mathematics* in 7th grade.

However, the percentages are not automatic. Students who do not reach 7th-grade competence until the 9th-grade level often do not possess the study habits necessary for successful completion of these courses. At the 9th-grade level, *Transition Mathematics* has been substituted successfully either for a traditional pre-algebra course or for the first year of an algebra course spread out over two years. It does not work as a substitute for a general mathematics course in which there is no expectation that students will take algebra the following year.

On page T27 is a description of this curriculum and the populations for which it is intended. The percentiles are national percentiles on a 7th-grade standardized mathematics test using 7th-grade norms, and apply to students entering the program with *Transition Mathematics*. Some school districts have felt that students should also be reading at least at a 7th-grade reading level as well. See page T28 for advice when starting with a later course.

Top 10%: The top 10% nationally reach the 7th-grade level of competence a year early. They are ready for *Transition Mathematics* in 6th grade and take it then. They proceed through the entire curriculum by 11th grade and can take calculus in 12th grade. We recommend that these students be expected to do the Extensions suggested in this Teacher's Edition. Teachers may also wish to enrich courses for these students further with problems from mathematics contests.

50th–90th percentile: These students should be expected to take mathematics at least through the 11th grade, by which time they will have the mathematics needed for all college majors except those in the hard sciences and engineering. For that they need 12th-grade mathematics.

30th–70th percentile: These students begin *Transition Mathematics* one year later, in 8th grade. The college-bound student in this curriculum is more likely to take four years of mathematics because the last course is hands-on with computers and provides the kind of mathematics needed for any major.

15th–50th percentile: Students who do not reach the 7th-grade level in mathematics until 9th grade or later should not be tracked into courses that put them further behind. Rather, they should be put into this curriculum and counseled on study skills. The logic is simple: mathematics is too important to be ignored. If one is behind in one's mathematical knowledge, the need is to work more at it, not less.

Even if a student begins with *Transition Mathematics* at 9th grade, that student can finish *Advanced Algebra* by the time of graduation from high school. That would be enough mathematics to enable the student to get into most colleges.

UCSMP Target Populations in Grades 6–12

Each course is meant to stand alone. However, to take best advantage of these materials, and have them appropriate for the greatest number of students, it is preferable to use them in sequence. Although it is suggested that students begin with *Transition Mathematics,* students may enter the UCSMP curriculum at any point. Below is a brief description of the UCSMP curriculum and the populations for which it is intended. All percents should be considered using national, not local, comparisons.

The top 10% of students are ready for *Transition Mathematics* at 6th grade. These students can proceed through the entire curriculum by 11th grade and take calculus in the 12th grade.

Students who reach 7th-grade level in mathematics by 7th grade **(in the 50th–90th percentile** on a standardized test) should take *Transition Mathematics* in 7th grade.

Students who do not reach the 7th-grade level in mathematics until the 8th grade **(in the 30th–70th percentile)** begin *Transition Mathematics* in 8th grade.

Students who don't reach the 7th-grade level in mathematics until the 9th grade **(in the 15th–50th percentile)** begin *Transition Mathematics* in the 9th grade.

Grade				
6	Transition Mathematics			
7	Algebra	Transition Mathematics		
8	Geometry	Algebra	Transition Mathematics	
9	Advanced Algebra	Geometry	Algebra	Transition Mathematics
10	Functions, Statistics, and Trigonometry	Advanced Algebra	Geometry	Algebra
11	Precalculus and Discrete Mathematics	Functions, Statistics, and Trigonometry	Advanced Algebra	Geometry
12	Calculus (Not part of UCSMP)	Precalculus and Discrete Mathematics	Functions, Statistics, and Trigonometry	Advanced Algebra

■ Starting in the Middle of the Series

From the beginning, every UCSMP course has been designed so that it could be used independently of other UCSMP courses. Accordingly, about half of the testing of UCSMP courses after *Transition Mathematics* has been with students who have not had any previous UCSMP courses. We have verified that any of the UCSMP courses can be taken successfully following the typical prerequisite courses in the standard curriculum.

ALGEBRA

No additional prerequisites other than those needed for success in any algebra course are needed for success in UCSMP *Algebra*. Students who have studied *Transition Mathematics* tend to cover more of UCSMP *Algebra* than other students because they tend to know more algebra, because they are accustomed to the style of the book, and because they have been introduced to more of the applications of algebra.

UCSMP *Algebra* prepares students for any standard geometry course.

GEOMETRY

No additional prerequisites other than those needed for success in any geometry course are needed for success in UCSMP *Geometry*. UCSMP *Geometry* can be used with faster, average, and slower students who have these prerequisites. Prior study of *Transition Mathematics* and UCSMP *Algebra* insures this background, but this content is also found in virtually all existing middle school or junior high school texts.

Classes of students who have studied UCSMP *Algebra* tend to cover more UCSMP *Geometry* than other classes because they know more geometry and are better at the algebra used in geometry.

Students who have studied UCSMP *Geometry* are ready for any second-year algebra text.

ADVANCED ALGEBRA

UCSMP *Advanced Algebra* should not be taken before a geometry course but can be used following any standard geometry text. Students who have studied UCSMP *Advanced Algebra* are prepared for courses commonly found at the senior level, including trigonometry or precalculus courses.

FUNCTIONS, STATISTICS, AND TRIGONOMETRY

FST assumes that students have completed a second-year algebra course. **No additional prerequisites other than those found in any second-year algebra text are needed for success in *FST*.**

PRECALCULUS AND DISCRETE MATHEMATICS

PDM can be taken successfully by students who have had *FST*, by students who have had typical senior level courses that include study of trigonometry and functions, and by top students who have successfully completed full advanced algebra and trigonometry courses.

PDM provides the background necessary for any typical calculus course, either at the high school or college level, including advanced placement calculus courses.

Development Cycle for UCSMP Texts

The development of each text has been in four stages. First, the overall goals for each course are created by UCSMP in consultation with a national advisory board of distinguished professors, and through discussion with classroom teachers, school administrators, and district and state mathematics supervisors.

The Advisory Board for the Secondary Component at the time of this planning consisted of Arthur F. Coxford, Jr., University of Michigan; David Duncan, University of Northern Iowa; James Fey, University of Maryland; Glenda Lappan, Michigan State University; Anthony Ralston, State University of New York at Buffalo; and James Schultz, Ohio State University.

As part of this stage, UCSMP devoted an annual School Conference, whose participants were mathematics supervisors and teachers, to discussion of major issues in a particular area of the curriculum. Past conferences have centered on the following issues:

1984   Changing the Curriculum in Grades 7 and 8

1985   Changing Standards in School Algebra

1986   Functions, Computers, and Statistics in Secondary Mathematics

1987   Pre-College Mathematics

1988   Mathematics Teacher Education for Grades 7–12

At the second stage, UCSMP selects authors who write first drafts of the courses. Half of all UCSMP authors currently teach mathematics in secondary schools, and all authors and editors for the first five courses have secondary school teaching experience. The textbook authors or their surrogates initially teach the first drafts of Secondary Component texts, so that revision may benefit from first-hand classroom experience.

After revision by the authors or editors, materials enter the third stage in the text development. Classes of teachers not connected with the project use the books, and independent evaluators closely study student achievement, attitudes, and issues related to implementation. For the first three years in the series, this stage involved a formative evaluation in six to ten schools, and all teachers who used the materials periodically met at the university to provide feedback to UCSMP staff for a second revision. For the last three years in the series, this stage has involved a second pilot.

The fourth stage consists of a wider comparative evaluation. For the first three books, this evaluation has involved approximately 40 classrooms and thousands of students per book in schools all over the country. For the last three books in the series, this stage has involved a careful formative evaluation. As a result of these studies, the books have been revised for commercial publication by Scott, Foresman and Company, into the edition you are now reading. (See pages T45–T49 for a summary of this research.)

2 Functions, Statistics, and Trigonometry

Problems *FUNCTIONS, STATISTICS, AND TRIGONOMETRY* is Trying to Address

This book is meant to follow UCSMP *Advanced Algebra* or an equivalent study of two years of algebra and one year of geometry. But it is quite different from most precalculus or high school analysis courses. The differences are due to its attempt to respond to six serious problems in mathematics education that cannot be treated by small differences in content or approach. Here are the problems that this book addresses and the UCSMP response to each.

PROBLEM 1: The traditional sequence of four years of high school mathematics does not prepare most students for a successful study of calculus.

As we pointed out on page T23, data from studies of mathematics placement testing show that only a small minority—less than 10% in one study—of the students who take the traditional four years of college preparatory mathematics in the United States place into calculus in college (Waits & Demana, 1988). Most students who have taken mathematics through precalculus in high school need to take one or two courses in order to qualify for college calculus. Furthermore, international comparisons show that the mean performance of U.S. precalculus students on tests of algebra, geometry, and functions ranks among the lowest quarter of industrialized nations (SIMS, 1987; see page T21).

The UCSMP response to this problem is to provide an extra year of mathematics for students. We offer a more challenging, less repetitive program (*Transition Mathematics*) for average students in grade 7 and recommend starting the

study of *Algebra* in grade 8—even for average college-intending students. This leaves the average student an additional year to prepare for college mathematics, and so UCSMP has created two courses to replace the traditional year of precalculus mathematics. In the first of these courses, *Functions, Statistics, and Trigonometry*, students study the various meanings of function (set of ordered pairs, mapping, equation, graph), and they study polynomial, exponential, logarithmic, and trigonometric functions in depth. They also extend the work of previous courses with quadratic relations and the concepts of inverse and composition of functions. In the second text, *Precalculus and Discrete Mathematics*, students continue the study of functions, trigonometry, sequences, limits, and continuity found in this book and study other precalculus topics such as rational functions, operations with functions such as addition and multiplication, vectors, parametric equations, and also are introduced to the derivative and the integral.

PROBLEM 2: Even when students are prepared for calculus, they are often not prepared for the other mathematics they will encounter in college. In particular, the traditional curriculum ignores fundamental ideas of discrete mathematics.

Statistics involves the gathering, display, and analysis of data; from the analysis we often wish to infer patterns in a larger population from what has been collected from a sample. These statistical ideas are found everywhere in our society, but they are not found in most high school mathematics curricula. In college, more majors require a statistics course than a calculus course. These majors, which typically include business, psychology, political science and other social sciences, and sometimes the biological sciences, often enroll a student who is not as mathematically inclined as the physical science or engineering student. This fact, along with inadequate preparation, is the reason why students often fear the study of statistics. This is unfortunate, since statistics are basic to the study of these fields.

The UCSMP response to this problem is to deal with data in all of its courses, but particularly in *Transition Mathematics* and *Algebra* (which are intended for almost all students) and in this course *Functions, Statistics, and Trigonometry*. This approach develops competence in descriptive statistics, in finding functions that model data sets, and connects this with probability and counting so as to provide an introduction to making statistical inferences. This course is not meant as a replacement for the study of statistics in college, but rather as a solid preparation for understanding the statistical ideas students will encounter in their careers and college courses. As a bonus, we have found that by connecting the study of statistics and functions, students' learning of each content area is enhanced. (See Schultz and Rubenstein (1990).)

The emergence of computer science as a field of study has increased the importance of a background in topics from discrete mathematics that are not usually taught in high schools. In *Functions, Statistics, and Trigonometry,* students study counting, probability, and simulation, particularly as they relate to statistics. They also extend the study of sequences and series previously begun. These ideas are continued in the next course.

PROBLEM 3: Many students do not realize the importance mathematics has in later schooling, on the job, or in daily affairs.

Perhaps one of the most telling indicators of the lack of a connection between the mathematics studied in traditional courses and the mathematics found outside of school is that at the end of a course students are less likely to see the value in what they took than before they studied the course. Many teachers tell students that the reason for studying a course beyond second year algebra in high school is to be successful in college mathematics, but do not tell the students why they are required to study mathematics in college.

As in all UCSMP texts, the *Functions, Statistics, and Trigonometry* response to this problem is to present reasons for all the mathematics being studied. Applications are used to motivate the study of skills and concepts. Students use mathematics to investigate facts about themselves and their environment. We make a serious effort to blur the distinctions between the world of the classroom and the real world. We believe it necessary for students to see real-world applications of what they learn, to use these applications to motivate, develop, reinforce, and expand their knowledge. We have attempted to pick contexts which are relevant to young adults. When possible, we use actual data and ask the kinds of questions that workers, concerned citizens, or curious observers would ask. Teachers who have used earlier versions of this and UCSMP texts frequently remark that students no longer ask "Why are we studying this?"

PROBLEM 4: Despite the nearly universal availability of calculators and widespread availability of computers, most contemporary textbooks do not take full advantage of the existence of this technology.

Unlike the other problems identified here, this problem has not been with us for a long time. Four-function hand-held calculators first appeared in 1971. The first personal computer appeared in 1977. The first easy-to-use calculator with the capability to graph functions appeared in 1985. Only in recent years has there been effective, easy-to-use computer software for graphing and for statistics. Many books do not accommodate contemporary technology because that technology was not available when the book was written.

From its inception, UCSMP has had a policy of using the latest in technology. Scott, Foresman has enhanced this policy by developing software so that appropriate computer technology is available with each of the UCSMP texts.

Scientific calculators and computers allow students to work with realistic numbers, to practice estimation skills, to avoid tedious or repetitive calculations, and to solve problems that would not be solvable otherwise. For these reasons all UCSMP secondary courses refer to calculators and computers in the lessons, and assume that students have access to scientific calculators for every lesson and on all tests.

In this course learning about both functions and statistics is enhanced if students and teachers have access to both function graphing and statistical analysis software. Such technology promotes students' ability to visualize functions, to explore relations between equations and their graphs, to simulate experiments, to generate or analyze data, and to develop the concept of limit.

Thus, in *Functions, Statistics, and Trigonometry* computers are assumed to be available to the teacher for demonstration and to students for laboratory work whenever they are needed. Any microcomputer, minicomputer, or mainframe computer with graphics capabilities can be used. The necessary software for the computer includes a function grapher, a statistics package, and the BASIC programming language. For most homework questions, a graphics calculator can be used instead of a computer.

More detailed information about technology is included on pages T40–T42.

PROBLEM 5: *Students do not read mathematics texts.*

Students tell us that they do not read because (1) the text is uninteresting, and (2) they do not need to read—their teacher explains everything for them. But students *must* learn to read for future success in mathematics. Our response to (1) is to include informative and interesting reading in every lesson. The text of the lesson develops concepts and principles, provides examples and non-examples, and shows relations between important ideas. Many of the questions themselves require reading to set the context of a problem. Also, the last lesson of every chapter is a special reading lesson that simulates the reading a student might do in non-mathematics courses where mathematics is used. Reading lessons cover a wide range of applications, theoretical discussions, and historical themes that students will find quite interesting.

Our response to (2) above is to encourage teachers not to explain everything *in advance* to students. Because students can read and understand the text, the teacher has the freedom to teach in a variety of ways and it is not necessary for the teacher to explain every day what the text says. The teacher can concentrate on helping students with difficult new symbols or vocabulary, and on developing further examples and explanations specifically tailored to his or her class. More detailed information about reading and reading lessons can be found on pages T36–T37.

PROBLEM 6: *Students do not get enough experience with problems and questions which require some thought before answering.*

In general, the evidence is that students are rather skillful at simple routine problems, but have a great deal of difficulty with complicated numbers, different wordings, or new contexts. Throughout this book, Applying the Mathematics questions are given with all sorts of numbers, a variety of wordings, and the many different contexts which arise naturally from applications. Exploration questions confront students with open-ended problems, problems requiring more sophisticated generalization, or with problems requiring the use of outside references.

Constructing mathematical models is a central theme of this course. Students collect data, use statistical and algebraic techniques to analyze relationships among the data, and search for functions which model those relationships. At times the study of a particular data set extends over several chapters as students learn different or more sophisticated methods for constructing models.

In preparation for the more sophisticated analysis and synthesis expected in college, the end of each chapter contains a set of Projects. These are activities requiring more elaborate data collection or more extensive analyses than the questions in the regular lessons. They provide a wonderful opportunity for students to develop reasoning and communication skills. It is expected that most projects will result in a written report with some accompanying visual models. Many teachers have found it useful to have students work on the projects in small groups and to present their projects orally. (See page T37 for more detail.)

We want students to view their study of mathematics as worthwhile, as full of interesting and entertaining information, as related to almost every endeavor.

Goals of *FUNCTIONS, STATISTICS, AND TRIGONOMETRY*

Obviously, among our goals is the removal of all the problems detailed above, and to do this while studying the content areas named in the title of the text. But we have been guided by other goals as well. We want to convey to the student, through historical references and references to recent

Section 2: Functions, Statistics, and Trigonometry

mathematical work that mathematics was developed by people and continues to develop. Through applications, WE WANT STUDENTS TO VIEW THEIR STUDY OF MATHEMATICS AS WORTHWHILE, AS FULL OF INTERESTING AND ENTERTAINING INFORMATION, AS RELATED TO ALMOST EVERY ENDEAVOR. Through explorations and projects, we want students to look for and recognize mathematics in places they haven't before, to use the library, to search through newspapers or almanacs, to get excited by knowledge.

As a new course, there was some latitude in selecting and organizing the content of this book. Guiding principles for including a topic in the curriculum were:

(1) it is important for daily living, career development, or future study of mathematics;

(2) it contributes to a balanced preparation for college mathematics courses in calculus or statistics;

(3) it is at a suitable level of difficulty for average 11th-grade students and is accessible to them with current technology;

Functions, Statistics, and Trigonometry provides materials for students to learn to display, describe, transform, and interpret numerical information in the form of data, graphs, or equations. Statistical concepts are integrated with graphical and algebraic ones through the contruction of mathematical models of relationships in data sets. Generating models is facilitated by using a computer statistics package, so students are exposed to a wide range of models including linear, quadratic, higher-order polynomial, exponential, logarithmic, and trigonometric functions. Statistical modeling extends through binomial and normal distributions, and gives students some exposure to hypothesis testing.

With the aid of an automatic grapher, students can accurately approximate relative extrema and zeros of virtually any function, not just ones easily manipulated using pencil and paper. Therefore, the formal algebraic manipulations usually covered in courses at this level need less attention in this course. In exchange, students are able to graph many more functions than they could by hand and to build their understanding of the properties and uses of functions. This course exploits technology to help students see the parallel effects of transformations on functions and on data.

The trigonometry in this book includes the right triangle and unit circle definitions of the trigonometric functions, their application to the finding of lengths and angle measures in any triangle, their graphs, trigonometric identities, and their applications to polar coordinates and to the finding of powers and roots of complex numbers. While the content includes all the development needed by someone who has never seen the trigonometric functions before, the pace of the course assumes that students have encountered the sine and cosine functions before, graphed them, and applied them to solving triangles.

Who Should Take *Functions, Statistics, and Trigonometry*?

Virtually all students who expect to go to college should take this course.

Upon entering the course, students should have a strong background in solving linear and quadratic equations, and linear systems. They should also have studied how to make and read coordinate graphs of linear, quadratic and power functions. And they need to have mastered finding an equation for a line given either the coordinates of two points or a point and the slope of that line. In addition, this course assumes that the student has previously studied exponents, logarithms, triangle trigonometry, and certain geometric transformations (reflections, rotations, translations, and size changes). Some familiarity with the BASIC computer language is also beneficial.

Students who do not have all the prerequisite skills can still be successful in this course. However, you may need to spend more time on certain chapters. We have made comments in the lesson Perspectives and in the Teaching Notes on topics where teachers using the prepublication editions of the book had to slow down because their students had little or no previous background. For example, if your students have not previously studied transformations, Chapter 3 will take longer than the recommended pace. If they have not previously studied logarithms, Chapter 4 on Exponential and Logarithmic Functions will take a bit more time to complete. If students have not had any trigonometry, then Chapter 5 will take longer.

3 General Teaching Suggestions

Functions, Statistics, and Trigonometry is a textbook which assists students in learning mathematics. It is not meant to substitute for a teacher, nor does it attempt to prescribe rigidly how to teach mathematics. We believe that it can be adapted to a variety of models of teaching—from direct instruction through cooperative learning. We also believe that our teaching suggestions for this book are consistent with the *Professional Standards for Teaching Mathematics* (NCTM, 1991). In the sections which follow, and in the detailed comments on each chapter and lesson elsewhere in these teacher's notes, we present ideas on specific teaching issues. They should not be construed as rigid: students, teachers, classes, schools, and school systems vary greatly. But they should not be ignored. These suggestions come from users, from our extensive discussions with teachers of earlier versions of these materials, and from test results. We encourage you to try them, and hope that you find them to be effective.

Optimizing Learning

■ Pace

As in other UCSMP texts, virtually all lessons in *Functions, Statistics, and Trigonometry* are intended to be read and discussed in one day. However, unlike earlier UCSMP courses, because the questions are more advanced and more complex, and because many of them involve the use of technology, their discussion often requires more time than is available in one day. Thus you must make decisions on what to discuss in class, and what to leave for students to do. You will find suggestions in the notes for each lesson regarding particular questions to highlight and ideas to cover.

Students adjust to the pace set by the teacher. There is a natural tendency, when using a new book, to go more slowly, to play it safe should you forget something. Teachers using these materials for the first time have almost invariably said that they would move more quickly the next year. Do not be afraid to move quickly. We know from our studies that a lesson-per-day pace produces the highest performance levels. Students need to be exposed to content in order to learn it. To set a tone of high expectations, it is especially important that Chapter 1 be taught at this pace. At the end of this and other chapters, spend a few days on the Progress Self-Test and Chapter Review to cinch the major ideas.

We recommend that a typical homework assignment be one of the following:

> 1. read Lesson n; write answers to all Questions in Lesson n; or
> 2. read Lesson n; write answers to the Covering the Reading questions in Lesson n, and Applying the Mathematics, Review, and Exploration questions in Lesson $n - 1$.

Thus, virtually every day students should be expected to do the equivalent of a complete set of questions from a lesson. The questions have been designed to cover the key skills, properties, uses and representations in the lesson. Questions were not written with an odd-even assignment plan in mind. Skipping questions may lead to gaps in student understanding. The Exploration questions may be assigned for all to do, or left as optional work for extra credit.

There are times when it will be difficult to maintain this pace. But be advised: a slow pace can make it too easy to lose perspective and the opportunity to relate ideas. You need to get to later content to realize why you were asked to learn earlier content! If you spend too much time in the lessons, you may find that your slowest students may have learned more by having gone through content slowly, but all the other students will have learned less. Thus, it is advisable to strike a balance by going quickly enough to keep things interesting but slowly enough to have time for explanations. David R. Johnson's booklets *Every Minute Counts* and *Making Minutes Count Even More* (see reference list) give excellent practical suggestions on making use of class time.

Section 3: General Teaching Suggestions

The Lesson Masters (included in the Teacher's Resource File) are particularly appropriate for students who wish more practice; for helping students individually with problems they have not seen; for helping students whose performance on a chapter test was too low; and for in-class work on days when a normal class cannot be conducted. However, they should be used sparingly. *The Lesson Masters should not be part of the normal routine* and should seldom be used when they would delay moving on to the next lesson.

Average classes who have studied from UCSMP *Transition Mathematics*, *Algebra*, *Geometry*, and *Advanced Algebra* should be able to cover about 12 chapters of FST. Classes with better than average students have been able to complete the entire text. Average classes with no prior UCSMP experience have been able to complete about 10 chapters of the Field Trial edition; and we expect such students to complete about 11 chapters of this edition.

If your students will take *Precalculus and Discrete Mathematics*, then they do not need to study Chapter 13.

■ Review

Every lesson includes Review questions. These questions serve a variety of purposes. First, they develop competence in a topic. Because we do not expect students to master a topic on the day they are introduced to it, these questions, coming *after* introduction of the topic, help to solidify the ideas. Second, they maintain competence from preceding chapters. This review is particularly effective with topics that have not been studied for some time.

At times, we are able to give harder questions in reviews than we could expect students to be able to do on the day they were introduced to the topic. Thus the reviews sometimes serve as questions which integrate ideas from previous lessons.

Finally, we occasionally review an idea that has not been discussed for some time, just before it is to surface again in a lesson. The Notes on Questions usually alert you to this circumstance.

Teachers in classes that perform the best assign all the review questions, give students the answers each day, and discuss them when needed. Those who do not assign all

reviews tend to get poorer performance; their students never get enough practice to solidify and master the ideas and, even when mastered, the ideas are forgotten. **The Review questions must be assigned to ensure optimum performance.**

■ Mastery

Some students master the contents of one lesson in one day. However, many do not. Why then do we suggest that you spend only one day per lesson? We do so because the combination of Review questions in each lesson and the end-of-chapter material has proved to be a powerful vehicle for achieving mastery, while allowing teachers and students to cover a substantial amount of material.

The components of the mastery strategy are: a Progress Self-Test (the "formative test" in the parlance of some master learning literature), solutions to that test in the student's textbook (the "feedback"), Review questions tied to the same objectives used to make up the Self-Test (the "correctives"), and finally a Chapter Test covering the same objectives.

Following the strategy means assigning the Progress Self-Test as a homework assignment to be done *under simulated test conditions*. The next day should be devoted to answering student questions about the problems and doing some problems from the Chapter Review if there is time. If a particular topic is causing a great deal of trouble, the corresponding Lesson Master (organized by objectives) may be of help.

For most classes, as a second night's assignment, we suggest the *even-numbered* Chapter Review questions on the SPUR Objectives. Neither solutions nor answers to these questions are in the student text, and students will have to work on their own without these aids. The next day, discuss these review questions in class.

Give the test on the third day. The odd-numbered Chapter Review questions, for which answers are given in the student text, can be useful for studying for that test. If at all possible, return the test to students so they see where they made errors and can better study for midterm and final exams. In some classes, a third day before the test may be needed. If so, either the odd-numbered Chapter Review questions or selected Lesson Masters can be used as sources of problems.

We strongly recommend that, except for classes of exceptionally talented students (where less review may be needed), teachers follow this strategy. THE EVIDENCE IS SUBSTANTIAL THAT USE OF THE CHAPTER END-MATTER MATERIALS PROMOTES HIGHER LEVELS OF PERFORMANCE.

■ Reading in UCSMP Texts

Both students and teachers who are first-time users of UCSMP materials are often surprised by the nature and extent of the text in each lesson. Some teachers are not quite sure how to get students to read the lessons; nor are they sure how to structure class time once the students have read the book.

A student in *Functions, Statistics, and Trigonometry* with prior UCSMP experience will generally be accustomed to reading mathematics. But students new to UCSMP courses may require some period of adjustment to a new style of text and new types of questions. Such students may never have been asked to read mathematics. As a result, it is common for students to ask why they have to read.

We tell them: You must read because you must learn to read for success in all future courses that use mathematics, not just in mathematics courses; because you must learn to read for success in life outside of school and on any job; because the reading will help you understand the uses of mathematics; because the reading contains interesting information; because the reading tells you how the material from one lesson is related to other material in the book; because there is not enough time in class to spend doing something that you can do in a study period or at home.

Students often do not know how to read a mathematics text. They read too quickly and they gloss over little words ("if", "but", and so on) that may be very important to the meaning of a statement. They may be unable to verbalize mathematical symbolism, such as

$$\sum_{i=1}^{5} 2^i$$

("the sum of 2 to the i, as i changes from i equals 1 to i equals 5"). They may not realize that text and graphics are often related and they should move back and forth from one to the other.

For classes that need some help getting used to reading, we suggest that during the first few days in class you have students read parts out loud for you, and provide feedback on their ability to read technical words and symbols correctly. You might have students answer the Covering the Reading questions orally, and point out how answers to such questions can be found literally in the reading or by mimicking worked examples. Other days you might give a brief summary of the key ideas in the reading, have students read silently in class, and then ask them to identify where in the exposition those main ideas are covered. Still other teachers find it useful at times to have students read difficult passages slowly out loud in class, and have one or two students summarize key ideas orally before the specific Covering the Reading questions are done. If students simply balk at reading, you might give 2–3 minute "reading quizzes" at the beginning of class. Doing so for 3 or 4 consecutive days early in the year will let students know you are serious about their reading.

The Covering the Reading questions in each lesson are meant to test comprehension of the material in the text. They can be used as oral exercises during or after oral reading of the text, or as part of a written assignment. Once students are comfortable with the format of the lessons, we suggest you begin to expect that reading be done outside of class on a regular basis.

It is not uncommon for students who are using UCSMP materials for the first time to resist your suggestions to read the text. However, in our trials of earlier versions of this text, teachers have reported that even the most recalcitrant students have learned to read the text by the end of the first marking period. Furthermore, teachers report that students who have studied from previous UCSMP texts where they were expected to read carry this expectation over into *Functions, Statistics, and Trigonometry*. That is, students'

The evidence is substantial that use of the chapter end-matter materials promotes higher levels of performance.

willingness to read regularly carries over into their study of mathematics in following years.

Although we believe that reading their texts is an important strategy by which students learn mathematics, we know that it is not the only way they learn. In particular, if you want to give a brief overview of the new lesson before students read it, please do. We do, however, wish to discourage the practice of *always* explaining how to do questions before the students have had the opportunity to learn on their own. Particularly counterproductive is to tell students that certain problems do not have to be tried ''because we have not yet done them in class.'' This only teaches students that they cannot learn on their own and that they must be dependent on you.

Students learn enormous amounts from discussing alternate strategies to problems with you and their classmates, from engaging in well-constructed computer laboratory activities, and from attacking open-ended Exploration questions in the text and the Extension activities in the Teacher's Edition. By teaching students to read outside of class, you are free to use class time more creatively and effectively.

■ Special Reading Lessons

The last lesson of each chapter is a special type of lesson which we call a ''reading lesson.'' These lessons give students practice in the reading that might appear in courses other than mathematics courses, or in the recreational reading of mathematics. Each reading lesson provides an opportunity to examine an application of mathematics in greater depth than in regular lessons, or extends the theory of the chapter to discuss a particular problem or set of related problems at length. It is not essential that students master the content of these lessons; there are no Applying the Mathematics questions; the content does not appear in the end-of-chapter Objectives; and the content does not appear on any of the tests. However, we do not consider these lessons optional. Our experience is not only that most students find these lessons interesting, but that they appreciate the opportunity to study content without the pressure of having it ''count.'' Furthermore, these lessons often integrate many ideas of the chapter, thereby helping students learn the content from the earlier lessons. And the Review questions in these lessons are essential for solidifying students' mastery of previous content.

We suggest that these lessons be assigned in a manner similar to other lessons. We discourage spending more than one day on them. If a student wishes to delve into the content more deeply, you may be able to find or suggest a project related to this content.

■ Projects

After the reading lesson of each chapter is a collection of projects, usually 4 to 8 in number. Each project is an extended set of activities, usually somewhat open-ended, meant to take from 3 to 10 hours to complete. Many of them provide an opportunity for students to engage in research and library work. Many teachers find that some projects work best when assigned to small groups of students. This illustrates that many problems in real life are not solved in isolation. Students profit from working together and sharing ideas—and as a result they spend less time on each project. When some projects are assigned to groups of 4 students, there will be a small enough number of projects completed to enable all of them to be presented during a class period. The projects serve many purposes:

(a) Students experience using real data in a mode comparable to that actually used by people in business, science, and many other careers.

(b) Students understand that a higher level of persistence than normal is expected. Too often in mathematics the greatest demand we make of students is to apply 5 to 15 minutes of effort on a single task. Longer-term projects demand more peristence and stretch a student's personal level of expectation.

(c) Projects, with allowances for student choice, provide a sense of ''ownership'' of a task.

(d) Projects provide a chance for students to share their learning publicly in a visual or oral presentation.

(e) Projects provide an opportunity for students to apply graphic, writing, and oral talents in mathematical situations.

(f) Projects provide an alternative medium for the evaluation of student achievement.

(g) The use of small groups teaches students that success or failure often is the result of a group effort.

T37

We strongly encourage you to allow students some latitude in the selection of projects, and to grade projects on effort as well as on execution. You can schedule work on them in a number of ways. Here are two suggestions: (1) Assign one project when you reach the middle of a chapter, to be due in the middle of the next chapter. (2) Assign one project per grading period from any of the chapters covered in that period. Notice that we do not recommend that the Projects for a particular chapter be completed during the course of the chapter in which they appear; they take too long and would disrupt the flow of the chapter.

We know of no class in which every project in the book was done by some student. We do know of classes in which teachers began the year thinking they would skip the projects, but when they talked to teachers who had done the projects, they became convinced of their value. Students often get very excited about these projects and do outstanding work. Thus, we strongly recommend that, however you use the projects, please do not avoid them. In fact, you may find students to be encouraged if they are given time to put together a first-rate presentation to be displayed on a bulletin board or a school display case. Teachers from one of the schools which used the pilot edition of this book remarked that by the end of the year graphs appeared in the school yearbook.

■ Understanding— The SPUR Approach

"Understanding" is an easy goal to have, for who can be against it? Yet understanding means different things to different people. In UCSMP texts an approach is taken that we call the SPUR approach. It involves four different aspects, or dimensions, of understanding.

Skills: For many people, understanding mathematics means simply knowing *how* to get an answer to a problem. This may be done by paper and pencil, or mentally, or with a calculator or computer. In UCSMP texts, all of these ways of getting answers to problems are aspects of the same kind of understanding, the understanding of algorithms (procedures) for obtaining a result. This is the S of SPUR, the Skills dimension, and it ranges from the rote application of simple algorithms to the invention of new algorithms.

Properties: It is obvious that some students have skills without knowing what it is they are doing. During the 1960s, understanding *why* emerged to become at least as important as understanding *how*. Mathematicians often view this kind of understanding as the ultimate goal. For instance, mathematics courses for prospective elementary school teachers assume these college students can do arithmetic and instead teach the properties and principles behind that arithmetic. This is the P of SPUR, the Properties dimension, and it ranges from the rote identification of properties to the discovery of new proofs.

Uses: To the person who applies mathematics, neither knowing how to get an answer nor knowing the mathematical reasons behind the process is as important as being able to *use* the answer. For example, a person does not possess full understanding of mathematical logic until the person can use it appropriately in real situations. This dimension ranges from the rote application of ideas (for instance, the finding of the median of a set of scores on a test) to the discovery of new applications or models for mathematical ideas. *Functions, Statistics, and Trigonometry* is notable for its attention to this dimension of understanding.

Representations: To some people, even having all three dimensions of understanding given above does not comprise full understanding. They require that students represent a concept and deal with the concept in that representation in some way. Ability to use concrete materials and models, or graphs and other symbolic representations, demonstrates this dimension of understanding. This is the R of SPUR, the Representations dimension, and it ranges from the various ways to display data discussed in the opening chapter to graphs in a polar coordinate system in the final chapter.

The four types of understanding have certain common qualities. For each there are people for whom that type of understanding is preeminent, and who believe that the other types do not convey the *real* understanding of mathematics. Each has aspects that can be memorized and that have the potential for the highest level of creative thinking. Each can be, and often is, learned in isolation from the others. We know there are students who have S and none of the others; in the 1960s some students learned P and none of the others; there are people on the street who have U and none of the others; and some people believe that children cannot really acquire any of the others without having R. There are continual arguments among educators as to which dimension should come first and which should be emphasized.

Because of this, IN UCSMP TEXTS WE HAVE ADOPTED THE VIEW THAT THE UNDERSTANDING OF MATHEMATICS IS A MULTI-DIMENSIONAL ENTITY. We believe each dimension is important, that each dimension has its easy aspects and its difficult ones. Some skills (for example, long division) take at least as long to learn as proofs of trigonometric identities; some uses are as easy as putting together beads.

For a specific example of what understanding means in these four dimensions, consider what would constitute evidence of understanding of the concept of end behavior of a rational function.

Skills understanding means knowing how to get an answer. (For example, find the standard deviation for a particular distribution.)

Properties understanding means knowing the mathematical underpinnings and connections of the idea. (Determine how the standard deviation is affected by multiplying each number in a data set by k.)

Uses understanding means knowing situations in which you could apply the idea. (Know that the standard deviation is a measure of spread and that in a normal distribution about 68% of the data lie within one standard deviation of the mean.)

Representations understanding means having a representation of the solving process or a graphical way of interpreting the solution. (Know that the points of inflection of a normal curve are one standard deviation from the mean.)

We believe there are students who prefer one of these dimensions over the others when learning mathematics. Some students prefer applications, some would rather do drawings, some prefer to know the theory, and still others like the representations best. Many students need to work with several dimensions of understanding before a concept makes sense to them. Thus the most effective teaching allows students opportunities in all these dimensions.

In this book, you see the SPUR categorization at the end of each chapter in a set of Objectives and corresponding Chapter Review questions. The Progress Self-Test for each chapter and the Lesson Masters (in the Teacher's Resource File) are keyed to these objectives. We never ask students (or teachers) to categorize tasks into the various kinds of understanding; that is not a suitable goal. The categorization is meant only to be a convenient and efficient way to ensure that the book provides the opportunity for teachers to teach and for students to gain a broader and deeper understanding of mathematics than is normally the case.

You may wonder why the Chapter Review and the Lesson Masters (in the Teacher's Resource File) are organized using the SPUR categorization. It is because we believe that practice is essential for mastery, but that blind practice, in which a student merely copies what was done in a previous problem to do a new one, does not help ultimate performance. The practice must be accompanied by understanding. The properties, uses, and representations enhance understanding and should be covered to obtain mastery, even of the skills.

. . . in UCSMP texts we have adopted the view that the understanding of mathematics is a multi-dimensional entity.

PROFESSIONAL SOURCEBOOK

Using Technology

We use calculators and computers in UCSMP because they make important mathematical ideas accessible to students at an early age; they relieve the drudgery of calculation and graphing, particularly with numbers and functions encountered in realistic contexts; and they facilitate exploration and open-ended problem solving by making multiple instances easy to examine. Furthermore, test results with earlier UCSMP books show that our use of technology has resulted in no loss of paper-and-pencil skill in arithmetic or algebra, and has freed up time in the curriculum to spend on other topics that lead to overall better performance by UCSMP students.

■ Scientific Calculators

In 1975, a national commission recommended that hand calculators be used on all tests starting in eighth grade, and in 1980 the National Council of Teachers of Mathematics recommended that calculators be used in all grades of school from kindergarten on. The College Board exams will be using scientific calculators on the achievement tests beginning in 1990–91, on the Advanced Placement tests beginning in 1992, and on the SATs beginning in 1992–93. Several standardized test batteries are being developed with calculators. Slowly but surely use of calculators is being expected on more and more licensing exams outside of school.

The business and mathematics education communities generally believe that paper-and-pencil algorithms are becoming obsolete. We should not be surprised. The long division algorithm we use was born only in the late 1400s; it can have a death as well. (Before that time, the abacus was used almost exclusively to get answers to problems. So mechanical means to do problems are really older than paper-and-pencil means.) Banks require that their tellers do all arithmetic using a calculator.

Inevitably, calculators will be considered as natural as pencils for doing mathematics. In the next century, people will be amazed when they learn that some students as recently as the 1990s went to schools where calculators were not used. Students of the future will no doubt consider it cruel and unusual punishment. Students of the present may already feel that way!

In all courses of the UCSMP secondary curriculum, we assume that all students have access to scientific calculators. Although there are times when we ask students not to use calculators (such as in giving values of trigonometric functions for arguments that are multiples of $30°$ or $\frac{\pi}{6}$), students in *Functions, Statistics, and Trigonometry* will be greatly hampered if they do not have such calculators available for all lessons. Such a calculator should have keys for π; powers and roots; logarithms (both base 10 and base e); sines, cosines, and tangents (with arguments in degrees or radians); inverse function capability; and factorials. The calculator should use algebraic logic and be able to represent large and small numbers in scientific notation. Virtually all scientific calculators today can do all these things. Some can do much more, and may be able to calculate some (but not all) of the kinds of statistics called for in this course.

In UCSMP courses, students learn that there are three ways to get answers to mathematics problems: by paper and pencil, mentally, or by using some automatic means (a table, a calculator, a computer, etc.). Some problems can be solved using any of the methods, while others require more than one method. However, through experience, students discover which methods are the most efficient to use—thus enabling them to become wise appliers of mathematics.

Good arithmeticians do a lot of calculations mentally, either because they are basic facts such as 3×5 or because they can be gotten by simple rules such as $2/3 \times 4/5$ or 100×4.72. They may not use a calculator on these because the likelihood of making an error entering or reading is greater than the likelihood of making a mental error. We do not often say, "Do not use calculators here." because we want students to learn for themselves when calculator use is appropriate and when it is not. However, you may feel the need to prod some students to avoid the calculator. An answer of 0.9999999 for $\sin^2 20° + \cos^2 20°$ should be strongly discouraged.

■ Graphics Calculators

In *Functions, Statistics, and Trigonometry*, students need automatic graphers in almost all lessons for graphing functions and/or statistical analysis. Computers can be used for these tasks, but graphing calculators have an advantage over computers because of their lower cost, easier access,

T40

and the fact that they can easily be taken home. However, there are certain things computers can do that even the most sophisticated calculators cannot. (Refer to the section "Computer Requirements" immediately below to identify the technology capabilities required for this course.)

Generally, graphing and statistical analysis calculators extend the capabilities of scientific calculators. One major difference, however, concerns input of functions. For example, the key sequence 30 $\boxed{\sin}$ has been standard on scientific calculators; but today's more sophisticated calculators display algebraic formulas and enter them in the order that they are written. They use the key sequence $\boxed{\sin}$ 30 to calculate sin 30°. We encourage students and teachers to read their owner's manuals and practice using the functions on arguments with known values.

■ ## Computer Requirements

While graphics calculators are indepensable to students, they are not by themselves sufficient for the needs of *Functions, Statistics and Trigonometry*. Computer use is necessary in some parts of the course. The computer is used as a *tool* for doing mathematics; not as an instructor. The materials have been prepared with these assumptions:

(a) The class can meet in a computer laboratory when necessary.
(b) At a minimum, there are enough computers available for each pair of students to have unrestricted access to a machine when necessary.
(c) A demonstration computer is available for use in the classroom when necessary.
(d) Students have access to computers including a printer for assignments when necessary.
(e) Teachers and students already are, or are willing to become, familiar with the hardware, operating system, and types of software needed for this course.

Functions, Statistics and Trigonometry was not written for any particular hardware or software. During 1987–90, the pilot and field study schools successfully used Apple II, Macintosh, and IBM or IBM-compatible computers for this book. Other standard microcomputers can also be used.

Three types of software are needed: (a) a statistics package, (b) a function grapher, and (c) BASIC programming.

The *statistics package* should be capable of displaying data in scatter plots, histograms, and box plots, and calculating descriptive statistics (mean, median, mode, standard deviation, variance, correlation and regression equations). Ideally, it can also transform data via translations, scale changes, and logarithmic transformations. *StatExplorer*, a statistics package specifically designed for this course, is available from Scott, Foresman. Separate versions have been developed for the Macintosh and IBM computers. This state-of-the-art software enables students to enter their own data sets, graph the data, and perform statistical analyses of the data. In particular, students can construct their own linear, quadratic, and exponential models and then compare them to corresponding computer generated models. The software also has simulation capabilities. Other suitable statistical packages are listed in the Bibliography on page T52.

The *function grapher* should graph in rectangular coordinates any polynomial, rational, exponential, logarithmic, or trigonometric functions input by the user. It should be capable of graphing two or more curves simultaneously and should have a "zoom" feature. Polar coordinate plotting and the ability to plot inverses of functions are beneficial, but not essential. Ideally, the software can also graph some curves which are not functions, such as vertical lines, circles, and ellipses.

GraphExplorer, a function grapher software package specifically designed for this course and for *Precalculus and Discrete Mathematics*, is available from Scott, Foresman. Separate versions have been developed for the Macintosh and IBM computers. The capabilities of this easy-to-use, but powerful function grapher, includes performing compositions of functions, performing transformations, and exploring families of functions. Other suitable function grapher software is listed in the Bibliography on page T52.

Copies of BASIC programs appear in the *Functions, Statistics, and Trigonometry* lessons. The programs are generally very short—few are longer than ten lines. Programming per se is not an objective of the course. However, some programming activities appear as a means to an end. For example, students study a program to add successive terms of a geometric sequence. They then use this program to find how many terms are needed to reach a certain value. In another lesson, user-defined BASIC functions with several arguments are presented to allow for easy computations. Appendix B in the student edition is provided for students unfamiliar with BASIC.

We re-emphasize that this course is not written with particular hardware, software, or combinations of these in mind. Any combinations meeting the above substantive requirements will suffice. Schools using the *Functions, Statistics, and Trigonometry* courses are expected to provide hardware and the above software in sufficient quantity to meet the educational needs of students, and not to violate the copyright laws.

■ Managing technology

We have tried to make the use of technology in the classroom as easy as possible. The first lesson requiring computer use, for example, is Lesson 1–7. It is the seventh lesson in part because it gives you and your students a week to get settled into a class routine. By far the simplest way to manage the technology is to arrange the allocation of rooms so that the FST class meets in or next to a computer lab every day. While computers are not required every day, the mathematics they are used for is not predictable. In some chapters they are not very helpful, but in others, students can use them every day and on homework assignments. In general, computer uses are embedded in the lesson, both in text and questions.

If you have never done much work with computers, do not be alarmed. This is a mathematics course, not a programming course. Computer language is explained in the text whenever needed.

There are two management suggestions regarding computer use which we cannot overemphasize. (1) Orient yourself by running programs in advance of class. This will better prepare you to anticipate student problems with hardware or software. The various versions of BASIC and Logo and each computer have slightly different characteristics, and our generic programs may need to be modified slightly for your system. Check with a computer expert in your school (perhaps a student) on how this might be done. This is *very* important and a nontrivial finding from the FST study. (2) When you use the computer to demonstrate statistical analysis, or if students are all going to be analyzing the same data set during class, enter the data into a file on a disk before class begins. We recommend finding student assistants who can help you with both these tasks.

In addition to the references in the text to BASIC, Logo, and automatic drawers, Computer Masters are provided in the Teacher's Resource File. These blackline masters are keyed to specific lessons in each of Chapters 1–13, with almost all requiring automatic graphing or statistics software. They are meant for use by a class in a laboratory setting, or by individuals for extra credit work. It is not necessary to do these masters to be successful in *Functions, Statistics, and Trigonometry*. Since each one stands alone, you may do as many or as few as you think appropriate for your class.

■ Grading

No problem seems more difficult than the question of grading. If a teacher has students who perform so well that they all deserve *A*s and the teacher gives them *A*s as a result, the teacher will probably not be given plaudits for being successful but will be accused of being too easy. This suggests that the grading scale ought to be based on a fixed level of performance, which is what we recommend. We recommend this because the performance that gives an *A* in one school or with one teacher may only rate a *C* in another, and we think it unfair.

Seldom in this book are there ten similar questions in a row. To teach students to be flexible, the wording of questions is varied, and principles are applied in many contexts. The problems are varied because that is the way problems come in later courses and in life outside of school. Learning to solve problems in a variety of contexts or to discern relationships between properties is more difficult than learning to perform a routine skill. Thus, a natural question that arises is "How should I grade students in UCSMP courses?"

First, we believe in a multifaceted approach to evaluation. Such areas as class participation and preparation, homework, presentation of Projects, along with performance on written tests and quizzes, should be taken into account when grades are determined. The Follow-Up section of the lesson notes in the Teacher's Edition often provides recommendations for Alternative Assessment.

Second we believe a student should be able to do each set of objectives at about the 85% level. An 85% score on a test deserves no less than a high *B*, and probably an *A*. In the past, our tests have often led us to the following grading scale: $85-100=A$, $72-84=B$, $60-71=C$, $50-59=D$, $0-49=F$. Such a scale may alarm some teachers, but STUDENTS IN UCSMP COURSES GENERALLY LEARN MORE MATHEMATICS OVERALL THAN STUDENTS IN COMPARISON CLASSES. We believe that the above grading policy rewards students fairly for work well done.

Some teachers are accustomed to the grading scale $90-100=A$, $80-90=B$, $70-79=C$, $60-69=D$, and $0-59=F$. In fact, in some districts, this is the required grading scale. If this is your situation, you may want to subtract points from, say, 110.

One January a teacher of *Transition Mathematics* remarked, "I've never had a class that learned so much, but my grades are lower." Later she said, "I have students who are failing. But I can't switch them to another class [using another book at the same level] because they know too much." Her problem is not unusual: percentages correct on tests of higher order thinking are generally lower than scores on tests of routine skills. We often make a basketball analogy. In many courses all of the shots students ever have are lay-ups (exercises) and an occasional free throw (easy problems). They shoot these over and over again, from the same spot ("Do the odd-numbered exercises from 1–49."). In *Functions, Statistics, and Trigonometry* almost every question is a different shot (a problem), some close in, some from middle distance, a few from half-court. To expect percentages of correct shots to be the same is unrealistic.

In basketball, there are times when a person wants to practice one specific shot to make it automatic, and so it is at times with the content of *Functions, Statistics, and Trigonometry*. We suggest focussing in on a few topics for quizzes. The Teacher's Resource File contains quizzes and tests for each chapter. If students perform well on quizzes and tests, it has a positive effect on interest and motivation. You should endeavor to use grading as a vehicle for breeding success as well as for evaluating students.

Students in UCSMP courses generally learn more mathematics overall than students in comparison classes.

In the light of all this, we strongly urge you to let students know what they need to know in order to get good grades. All research on the subject indicates that telling students what they are supposed to learn increases the amount of material covered and tends to increase performance. Also, let students know that you expect *all of them* to perform well. Do not arbitrarily consign some of your students to low grades. Let them know that it is possible for all of them to get *A*s if they learn the material.

Some teachers have found that because of the way that the Review questions maintain and improve performance, cumulative tests at the end of each marking period give students an opportunity to do well. For your convenience, the Teacher's Resource File has Chapter Tests in Cumulative Form (in addition to the regular Chapter Tests) for each chapter, beginning with Chapter 2. These Cumulative Tests allow students a chance to show what they have learned about topics studied in a previous chapter along with the topics of the current chapter. Beginning with Chapter 3, the questions in the Cumulative Tests are about 50% from the current chapter, 25% from the preceding chapter, and 25% from all other earlier chapters.

■ Standardized Tests

At our first conference with school teachers and administrators in the early days of UCSMP, we were told in the strongest terms, "Be bold, but remember that you will be judged by old standardized tests."

There are many types of standardized tests in high school mathematics: standardized tests for a particular course; general test batteries in which there may be a mathematics subtest; local school or state assessments; and college entrance examinations. There is strong evidence that

performance on any of these can be enhanced by practice on questions like those which will appear. Since these tests often have bearing on student placement you should, when possible, inform students of the kinds of questions they are likely to encounter.

Based on our testing and what teachers and schools have reported, here is what we conclude about these various kinds of tests. Students in *Functions, Statistics, and Trigonometry* will perform as well as (if not better than) students from other classes on traditional standardized tests. Yet, since UCSMP students outperform other students on applications, graphing, and statistics, and some other content found in *Functions, Statistics, and Trigonometry,* it is unfair to UCSMP students if they are judged only by a traditional standardized test.

From our studies, we believe that students studying UCSMP courses are likely to do better on the College Board SAT-M than other students. The SAT-M is a problem-solving test, and far more questions on the SAT-M cover content likely to be unfamiliar to other students than to UCSMP students. The ACT exams seem at least as favorable to our students as to students in other books. Furthermore, if UCSMP suggestions are followed and more students take *Geometry* in the 9th grade, then there is no question that students in the UCSMP curriculum will far outscore comparable students who are a course behind. Preliminary evidence with PSAT scores indicates that 10th graders who have had all UCSMP courses through *Advanced Algebra* score as well as non-UCSMP juniors of equivalent ability in second-year algebra.

Finally, a note about curriculum projects. Many people treat books from curriculum projects as "experimental." Few commercial textbooks, however, have been developed on the basis of large scale testing. Few have been produced as part of a coherent 7–12 curriculum design. This text has gone through extensive prepublication analysis and criticism. It has been revised on the basis of comments and evaluations of the many teachers who have been in our evaluation studies. It is the intention of UCSMP to provide materials which are successful as well as teachable. FUNCTIONS, STATISTICS, AND TRIGONOMETRY SHOULD BE VIEWED AS A NEW TEXT, NOT AS AN EXPERIMENTAL TEXT.

Functions, Statistics, and Trigonometry should be viewed as a new text, not as an experimental text.

4 Research and Development of Functions, Statistics, and Trigonometry

Planning and Selection of Authors

In the early days of UCSMP, the last two courses in the secondary curriculum were entitled simply "Pre-College Mathematics 1" and "Pre-College Mathematics 2." After some discussions with the advisory board to the project, it was decided that each of these courses should have clear themes, for otherwise teachers might favor familiar over newer content even if the newer content were more important for students. The development of functions over a multi-year period, felt to be essential for a thorough preparation for calculus, made that a natural first theme. Because statistics was judged to be important to any person wishing to go to college, statistics was made a second theme of the first course.

It was only after this selection that it was realized that the teaching and learning of these two topics, functions and statistics, were singularly affected by technology. Statistics has gained its importance due to the ability of computers to store and process data. The analysis of real-valued functions has been irrevocably changed since the advent of software capable of graphing them. The title of the first draft became "Functions and Statistics with Computers."

The task of weaving these themes together and connecting them with other books in the UCSMP series could only be given to individuals familiar with both this content and that of the previous books. Rheta Rubenstein, then the mathematics department chair at Renaissance H.S. in Detroit, Michigan, and James Schultz, Associate Professor of Mathematics at Ohio State University, were asked to be a two-person team to write the first draft of the materials. Dr. Rubenstein had been a member of the *Advanced Algebra* writing team and had initiated a course in statistics at her high school. Dr. Schultz was a member of the UCSMP secondary component advisory board whose special expertise included the use of technology and making connections within the broad range of content that such a course would involve.

First Pilot Study

Writing began in the spring and summer of 1986 and continued during the 1986–87 school year. The manuscript was edited at UCSMP by James Flanders and Barry Kissane, each of whom had experience with teaching this kind of content. The materials were loose-leaf and sent to schools one chapter at a time. The work of each author was assisted by two pilot classes near them using these materials. No formal testing was done at this stage.

In addition to the normal revisions one makes in improving a text from one version to the next, two broad ideas emerged. In writing the pilot, it was thought that the computer requirements of the course could be handled if one day per week were scheduled in a computer lab. (All teachers at a conference held the previous spring had indicated that in their schools this was feasible.) This turned out to be unrealistic; the day of the week too often came on a day with shortened periods or a holiday, and students needed to use computers more often. Thus the decision was made to recommend that classes using the book meet in or near a computer lab.

The second broad idea was that the two subject matters of statistics and functions were more closely related than their typical separation in school curricula would suggest. The use of functions to model data, not typically studied in college statistics courses but a key idea in applied mathematics, became a recurring theme. Transformations could be done of data, of graphs of functions, and of equations. Coordinate graphs themselves were natural in both settings.

Second Pilot Study

A major revision was made of the materials during the summer and fall of 1987, with an experienced UCSMP author, Margaret Hackworth assisting Rheta Rubenstein under the supervision first of Zalman Usiskin, then of Sharon Senk. The revised materials were used during the 1987–88 school year in schools in Michigan, Ohio, Indiana, and Illinois. These materials were again loose-leaf, sent to schools one chapter at a time.

Teachers kept detailed records on each lesson and sent these back to the project as the year progressed. All teachers were invited to visit the University of Chicago three times during the year for full-day meetings to discuss all aspects of teaching and learning from the materials.

This pilot showed a non-trivial result: although the materials were still uneven in difficulty, the use of technology seemed to make it possible for the materials to work well with a wide variety of students, from the slowest students at this level to the best-prepared. The richness of the problems made all the teachers enthusiastic about the course, though all also agreed that there were places in which revisions were needed to smooth out the rough spots. Length of the manuscript was seen as a problem, but there was also a concern that students be able to obtain a credit for a semester of trigonometry after completing this course. Consequently, the title of the course was changed to *Functions, Statistics, and Trigonometry with Computers*. We then started calling the course *FST*, for short.

Formative Evaluation and Test Results

In the summer of 1988, the materials were again revised. Because of the need to have more materials ready by the beginning of the school year, two additional experienced UCSMP authors, John McConnell, and Steven Viktora were asked to assist Rheta Rubenstein in the revisions. The materials were printed in three spiral-bound parts; students and teachers received the course one part at a time, with only the first part delivered by the beginning of the 1988–89 school year.

Nine teachers from eight schools participated in the formative evaluation, conducted by a team from the Evaluation Component of UCSMP headed by Catherine Sarther. The nine teachers included five of the six teachers from the second pilot. A sixth teacher was one of the authors. Three other teachers had no prior FST experience. No special training was given the teachers. During the school year, all nine teachers met in Chicago in November, February, and May to discuss how things were going in their classrooms and how they felt about the materials. Their classes were observed and the teachers and some students were interviewed by Ms. Sarther. The following summarizes the evaluation report.

At the beginning of the school year, 239 students completed a Student Information Sheet, 52% male and 48% female. Most (79%) were seniors; the rest (21%) were juniors. Almost all (95%) had used a scientific calculator before, and 11% had used a graphing calculator. Virtually all (96%) had used a computer before, 83% for games, 73% for word processing, and 52% for mathematics. About half had a computer at home capable of doing more than playing games. The racial/ethnic background of the students was 60% white, 30% black, 4% Hispanic or Latino, 4% Asian or Pacific Islander, 1% American Indian or Alaskan Native, and 1% other. All of these students planned to go to college; the most common intended majors were business and commerce (28%), engineering (13%), social sciences or history (11%), and health professions (9%).

Section 4: Research and Development

The teachers in this study implemented all facets of the book: the lessons, the use of technology, the Projects, and the end-of-chapter materials. Teachers generally had students read the text, moved at the recommended pace, and covered the Review questions in the lessons. On average, teachers completed 11 chapters by the end of the year. Novice computer users had difficulty at the beginning of the course but by the end they were using the computer as frequently as the other teachers in the study. According to the needs of the students, teachers chose the reading lessons, Projects, and Chapter Review material. Six of the nine teachers supplemented the materials, mostly to spend more time on trigonometry. Both teachers and students were enthusiastic about the Self-Tests at the ends of chapters. In only one class (taught by a first-year teacher) was implementation a problem.

No comparison classes were involved in this study, for it would be impossible to give a test whose content was fair to both UCSMP and comparison classes. Instead, students were compared with broad groups of similar students, as is described below.

During the first week of school, a 36-item test covering content of traditional second-year algebra texts and UCSMP *Advanced Algebra* was given to all students as a pretest. This test had been given previously to students in a UCSMP *Advanced Algebra* study. The range of *FST* class means was from 7.3 to 25.1, a very wide range indeed, but not dissimilar to the range of means in the *Advanced Algebra* study, 8.0 to 20.8. Thus FST students fell within the same range of ability as students who took part in the previous study.

No classes in this study used calculators; all used computers. Computer use varied among classes, and it varied among students in them. 10% of students reported using a computer several times a week on average, 40% once or twice a week, 33% once or twice a month, with the remainder once a month or less. Only 30% used the computer for homework outside of class, but 79% used the computer on tests, 89% for solving problems in class, 93% for studying functions, and 92% for analyzing data.

Teachers typically required one Project per marking period. Only two teachers used Projects in every chapter. Projects that involved collecting data or indirect measurement were the most popular. Teacher reaction to the Projects was mixed; some were enthusiastic because the Projects enabled students to be creative and ask their own questions. The biggest problem noted was the time it takes to discuss them.

The posttest consisted of two parts: a 38–question multiple-choice test taken by all students; and an open-ended part that had two forms, each taken by half the students. On the multiple-choice test, 12 items were repeated from the pretest. These items involved exponents and logarithms, matrices, trigonometry, and functions. The mean percent correct on these item was 32% on the pretest and 53% on the posttest; the gains on items ranged from 10% to 39%. It could be concluded that students learned quite a bit in the course.

All of the eight open-ended items (four on each test) required the use of technology; six required computers. These items were hard items, requiring thorough knowledge of ideas in order to be answered correctly. In general, students could use a function grapher: to approximate zeros of a function (95% could make some headway; 63% could finish the task accurately); to graph a function with a complicated formula (56% and 11%); to find the maximum value of a polynomial function (63% and 24%). Students could use a statistics package to find a line of best fit and could explain the limitations of extrapolation from data (70% and 53%). The evaluators concluded, ''students in *FST* were able to use a function grapher and a statistics package.''

Teacher attitudes towards the book were compared with attitudes of teachers in a study by Iris Weiss of the Research Triangle Institute for the 1985–86 National Survey of Science and Mathematics Education.

Percent of teachers giving favorable ratings about textbooks

Item	UCSMP n = 9	Weiss n = 517
explains concepts clearly	100	73
clear and well-organized	89	82
develops problem-solving skills	89	68
examples of applications of mathematics	89	40
good suggestions for computers	89	31
good suggestions for calculators	89	27
appropriate reading level	78	87
good suggestions for activities/assignments	78	55
interesting to students	78	43
high quality supplements	44	33
has exercises for practice and drill	33	60

Students liked the end-of-the-chapter tests the most and the practical applications. They were most bothered by the length of assignments and errors (in that edition). Generally, in 6 of 9 classes student attitudes were positive; in 1 class attitudes were neutral; in 2 classes attitudes were negative. Seventy percent of the students reported spending 30 minutes or less per night on homework. This agrees with the mean of 32 minutes for students in grades 10–12 found by Weiss in her study.

Continuing Research and Development

The studies of *Functions, Statistics, and Trigonometry with Computers* convinced us that the approach we have taken is fundamentally sound. Still, this Scott, Foresman edition differs from the edition used in the formative evaluation in many ways. Under the direction of Sharon Senk and with Dora Aksoy, many lessons and several chapters have been reworked to simplify the exposition and clarify connections between ideas. A chapter devoted to polynomial functions has been added to make the transition between the study of functions in UCSMP *Advanced Algebra* and the study of rational functions in *Precalculus and Discrete Mathematics*.

Perhaps the most major change is with the technology due to the increasing use of graphics calculators. Many teachers report that student ability to take these calculators with them greatly enhances their homework performance and, as a consequence, their understanding of the content. The increasing popularity of these calculators has been accommodated in the exposition. It also caused us to delete "with Computers" from the title.

The present version of *Functions, Statistics, and Trigonometry* also includes the appearance of four colors, of attractive pictures, an expanded Teacher's Edition, and many more supplements for teachers and students. These supplements include Lesson Masters for teachers who wish more questions on a particular lesson, Teaching Aids to help in explanations, and extra quizzes and tests. Because of these changes, we expect students to perform as well, if not better, with the Scott, Foresman version than with our previous versions.

Each November since 1985, we have held a conference at the University of Chicago at which users and prospective users of secondary component UCSMP materials can meet with each other and with authors. Beginning in August 1989, we have conducted a free full day in-service for any teachers or supervisors who will be using the materials. We encourage new users to attend one of these meetings.

We desire to know of any studies school districts conduct using these materials. UCSMP will be happy to assist school districts by supplying a copy of the noncommercial tests we have used in our evaluations (except for certain open-ended items) and other information as needed.

Both Scott, Foresman and UCSMP welcome comments on these books. Please address comments either to Mathematics Product Manager, Scott, Foresman, 1900 East Lake Avenue, Glenview, IL 60025, or to Zalman Usiskin, University of Chicago, 5835 S. Kimbark Avenue, Chicago, IL 60637.

5 Bibliography

■ REFERENCES for Sections 1–4 of Professional Sourcebook:

College Board. *Academic Preparation for College: What Students Need To Know and Be Able To Do.* New York: College Board, 1983.

Flanders, James. **"How Much of the Content in Mathematics Textbooks Is New?"** *Arithmetic Teacher*, September 1987, pp. 18–23.

Jones, Philip, and Coxford, Arthur F. *A History of Mathematics Education in the United States and Canada,* 30th Yearbook of the National Council of Teachers of Mathematics. Reston, VA: NCTM, 1970.

Mathematical Sciences Education Board. *Reshaping School Mathematics: A Philosophy and Framework for Curriculum.* Washington, DC: National Academy Press, 1990.

McNight, Curtis, et al. *The Underachieving Curriculum: Assessing U.S. School Mathematics from an International Perspective.* Champaign, IL: Stipes Publishing Company, 1987.

National Commission on Excellence in Education. *A Nation at Risk: The Imperative for Educational Reform.* Washington, D.C.: U.S. Department of Education, 1983.

National Council of Teachers of Mathematics. *Curriculum and Evaluation Standards for School Mathematics.* Reston, VA: NCTM, 1989.

National Council of Teachers of Mathematics. *Professional Standards for Teaching Mathematics.* Reston, VA: NCTM, 1991.

National Research Council. *Everybody Counts.* Washington, DC: National Academy Press, 1989.

Schultz, James E., and Rubenstein, Rheta N. **"Integrating Statistics into a Course on Functions."** *Mathematics Teacher,* November 1990, pp. 612–616.

Waits, Bert, and Demana, Franklin. **"Is Three Years Enough?"** *Mathematics Teacher*, January 1988, pp. 11–15.

■ Additional References on Statistics

Gnamadesikam, Mrudulla, Richard L. Scheaffer, and Jim Swift. *The Art and Techniques of Simulation.* (part of the *Quantitative Literacy Series*) Palo Alto, California: Dale Seymour Publications, 1987.

Ekeland, Ivar. *Mathematics and the Unexpected.* Chicago: The University of Chicago Press, 1988.

Hollander, Myles, and Frank Proschan. *The Statistical Exorcist: Dispelling Statistical Anxiety.* New York: Marcel Dekker, Inc., 1984.

Jaffe, A.J., and Herbert F. Spier. *Misused Statistics: Straight Talk for Twisted Numbers.* New York: Marcel Dekker, Inc., 1987.

Katz, Samuel, and Donna F. Stroup. *Educated Guessing: How to Cope in an Uncertain World.* New York: Marcel Dekker, Inc., 1983.

Landwehr, James M., and Ann E. Watkins. *Exploring Data.* (part of the *Quantitative Literacy Series*) Palo Alto, California: Dale Seymour Publications, 1986.

Landwehr, James M., Jim Swift, and Ann E. Watkins. *Exploring Surveys and Information from Samples.* (part of the *Quantitative Literacy Series*) Palo Alto, California: Dale Seymour Publications, 1987.

Moore, David S. *Statistics: Concepts and Controversies.* (Second Edition). New York: W.H. Freeman and Co., 1985.

Mosteller, Frederick, and William H. Kruskal, Richard F. Link, Richard S. Pieters, Gerald R. Rising, (editors). *Statistics by Example.* Reading, Massachusetts: Addison-Wesley Publishing Company, 1973.

Newman, Claire, M., Thomas E. Obremski, and Richard L. Sheaffer. *Exploring Probability.* (part of the *Quantitative Literacy Series*) Palo Alto, California: Dale Seymour Publications, 1987.

Shulte, Albert P., and James R. Smart. *Teaching Statistics and Probability,* the 1981 Yearbook of the National Council of Teachers of Mathematics. Reston, Virginia: The National Council of Teachers of Mathematics, 1981.

Tanur, Judith (ed.), and Frederick Mosteller, William H. Kruskal, Richard F. Link, Richard S. Pieters, Gerald R. Rising. *Statistics: A Guide to the Unknown.* Monterey, California: Wadsworth & Brooks/Cole Advanced Books & Software, 1985.

Tufte, Edward R. *The Visual Display of Quantitative Information.* Chesire, Connecticut: Graphics Press, 1983.

Wallis, W. Allen, and Harry V. Roberts. *The Nature of Statistics.* New York: The Free Press, A Division of the MacMillan Publishing Co., Inc., 1965.

Zeisel, Hans. *Say It with Figures.* Sixth Edition. New York: Harper & Row, Publishers, 1985.

■ ## SOURCES for Data or Additional Problems

Consumer Reports. Yonkers, NY: Consumers Union of U.S. Inc., monthly.

The Diagram Group. *Comparisons.* New York. St. Martin's Press, 1980.

Eves, Howard. *An Introduction to the History of Mathematics.* 5th ed. Philadelphia. Saunders College Publishing, 1983.

Garfunkel, Solomon, and Steen, Lynn A., editors. *For All Practical Purposes: An Introduction to Contemporary Mathematics.* New York: W.H. Freeman, 1988.

Hanson, Viggo P., and Zweng, Marilyn J, editors. *Computers in Mathematics Education,* 1984 Yearbook of the National Council of Teachers of Mathematics. Reston, VA: NCTM, 1984.

Hoffman, Mark, editor. *The World Almanac and Book of Facts.* New York: World Almanac, yearly.

Johnson, Otto, executive ed. *The Information Please Almanac.* Boston: Houghton Mifflin Company, yearly.

Joint Committee of the Mathematical Association of America and the National Council of Teachers of Mathematics, *A Sourcebook of Applications of School Mathematics.* Reston, VA: NCTM, 1980.

Kastner, Bernice. *Applications of Secondary School Mathematics.* Reston, Va.: National Council of Teachers of Mathematics, 1978.

Kastner, Bernice. *Space Mathematics.* Washington, D.C.: U.S. Government Printing Office, 1985.

Sharron, Sidney, and Reys, Robert E., editors. *Applications in School Mathematics,* 1979 Yearbook of the National Council of Teachers of Mathematics. Reston: Va.: NCTM, 1979.

USA Today. Arlington, VA: Gannett & Co., Inc., daily.

U.S. Bureau of the Census. *Statistical Abstract of the United States: 1990.* 110th ed. Washington D.C., 1990.

■ UCSMP SOFTWARE

GraphExplorer. Scott, Foresman and Company, 1900 East Lake Avenue, Glenview, IL 60025. (Macintosh, IBM)

StatExplorer. Scott, Foresman and Company, 1900 East Lake Avenue, Glenview, IL 60025. (Macintosh, IBM)

UCSMP Algebra/Advanced Algebra Graphing Software. Scott, Foresman and Company, 1900 East Lake Avenue, Glenview, IL 60025. (Apple II, IBM)

■ OTHER SOFTWARE

ANUGraph. The Australian Association of Mathematics Teachers, 18 Blue Hills Road, O'Halloran Hill, South Australia 5158. (Macintosh)

Avery, Chris, and Soler, Frank. *Interactive Statistics with Probability.* DeAnza College, Cupertino, CA 95014. (Apple II)

Calculus T/L. Brooks/Cole Publishing, 511 Forest Lodge Road, Pacific Grove, CA 93950. (Macintosh Plus)

Edwards, Lois, and Keogh, K.M. *Data Insights.* Sunburst Communications, Pleasantville, NY 10570. (Apple II)

Schwartz, Judah. *The Function Analyzer.* Sunburst Communications, Pleasantville, NY 10570. (IBM)

StatView™. Brain Power, Inc., 24009 Ventura Blvd., Suite 250, Calabasas, CA 91302. (Macintosh)

Statview SE & Graphics. Abacus Concepts, Inc., 1984 Bonita Avenue, Berkeley, CA 94704. (IBM)

Statworks™. Cricket Software, 3508 Market Street, Suite 206, Philadelphia, PA 19104. (Macintosh)

TrueSTAT. TrueBASIC, Inc., 39 South Main Street, Hanover, NH 03755. (IBM, Macintosh)

Visualizing Algebra: The Function Analyzer. Sunburst Communication, Inc., 39 Washington Avenue, P.O. Box 40, Pleasantville, NY 10570. (IBM)

Waits, Bert, and Demana, Frank. *Master Grapher.* Addison-Wesley Publishing Company. (Apple II, IBM, Macintosh)

SELECTED ANSWERS

Full-size versions of these pages appear in the Student Edition. Full-size versions of the Progress Self-Test charts are found on pages T96–T100 in this Teacher's Edition.

LESSON 1-1 (pp. 4–9)
5. a. all of the cookies in the batch b. the ten cookies whose raisins the inspector actually counts c. the number of raisins in one cookie 7. a. 1.5 million people from the U.S. and abroad; about 26,000 people from the U.S.; about 100,000 people from abroad b. the entire world population 9. a sample in which every member of the population under study has an equal chance of being selected 11. about 62 turtles 13. a. 75% b. about 1558 women 15. a. 228,834,146 people b. Sample: 1) Some people did not respond to the census. 2) Homeless people were not counted. 17. a. The student did not use a representative sample. b. 11 pieces 19. about 106,000,000 people 21. (a) 23. a. 90° b. 135°

LESSON 1-2 (pp. 10–16)
1. $424 3. True 5. about 5 times 7. a. about 75–80% b. 1988 c. Both bars for those having attended college ("4 years of high school and some college" and "4 years or more of college") are higher for 1988 than for 1980 or 1970. 9. a. about 49,000 b. about 6000 c. False d. about 6.7% 11. a. ≈31.6% b. ≈6.1% c. about 760,000 d. about 13,200,000 13. Sample: The more education a householder has, the more likely that householder will earn more money. The less education a householder has, the more likely his or her family will live in poverty. 15. a. See below. b. Sample: The number of polio cases would be zero or close to zero because the vaccine for polio has almost completely eliminated the disease since 1960. The number of measles cases in 2000 may be back down to the 1980 level because the increase of cases in 1990 will bring a greater awareness of the need for measles immunization. Since the number of syphilis cases has steadily been decreasing, a good prediction would be about 35,000 cases. The number of AIDS cases is expected to increase, unless a cure is found. The number of gonorrhea cases may decrease because of the threat of AIDS. 17. a. 27.8% b. 54° 19. Sample: Not everyone has an equal chance of being selected. For example, no one whose last name starts with A would be surveyed. 21. a. Sample: Those who called in owned a TV, were watching the ABC channel when the station asked the viewers to call in, and cared enough to take the time and to pay the expense of calling in to register their opinions. This group was not necessarily representative of the American public as a whole. Furthermore, in a survey of this type there is no way to prevent the same individuals from calling in numerous times. b. Sample: The random survey is far more likely to have been representative. 23. about 19,000 people per year

15. a.

LESSON 1-3 (pp. 17–23)
1. slope 3. sample: 1960 to 1970 5. a. minimum b. maximum c. range 7. Sample: The individual data values are not lost. 9. a. 8 home runs in a season by Maris b. 15 years c. Ruth hit 60; Maris hit 61. d. Maris's 61 11. (a) 13. a. $209 million b. $76 million c. 6 movies 15. a. the 20 AIDS patients to whom the drug is administered b. all AIDS patients c. number of infections each patient contracts 17. The various costs of owning and operating a car are listed in thousands of dollars. 19. (b)

LESSON 1-4 (pp. 24–30)
5. False; Sample counterexample: The mean of the set {1, 4, 4} is 3, the median is 4. 7. There are 2 supervisors with a salary of $25,000 and 4 sales representatives earning $21,000. 9. a. 104 b. 76 11. the mean 13. a. mean = 26.1 home runs; median = 24.5 home runs b. mean = about 43.9 home runs; median = 46 home runs c. Sample: Babe Ruth was a better home run hitter because his home run totals for typical years are much larger than Roger Maris's home run totals for a typical year. 15. The mean is likely to be $141,200 and the median $117,800. The mean is larger because of outliers—very expensive houses. 17. a. B b. A c. Sample: The numerical intervals for letter grades vary and are not uniformly spaced. d. If the two middle grades are equal, then the median can be found. If the two middle grades are not equal, then the median cannot be found because it does not make sense to average the two middle grades. 19. a. $\sum_{i=1}^{8} m_i$
b. $424,082,000,000 21. sample: (1, −2) and (4, 13) 23. a. between 1983 and 1984 b. about 160 heliports per year c. about 3360 heliports d. Sample: The demand for heliports might change drastically over the next 20 years.

LESSON 1-5 (pp. 31–37)
1. $3870 3. 40th percentile 5. 90th percentile 7. 75th percentile 9. a. maximum, minimum, and the 3 quartiles b. the median c. the first and third quartile d. the maximum and minimum 11. a. median = $31; lower quartile = $18.50; upper quartile = $47.50 b. The values $97 and $113 are outliers. c. See page 846. d. Sample: About half the values are clustered in the relatively small interval from $18.50 to $47.50. The spread between the extreme values is almost 4 times the spread between the first and third quartiles. Also the spread among the "big spenders" is less than that among those who spend less. 13. a. See page 846. b. False
c. Division C 15. sample: $\frac{1}{100}\sum_{i=1}^{100} a_i$ 17. a. mean = 6′$\frac{1}{2}$″; median = 6′ 1″ b. mean = 6′$\frac{3}{5}$″; median = 6′ 1″ c. mean = 6′ 3″; median = 6′ 2″ d. sample: median, because four of the five team members are shorter than 6′ 3″, the mean e. median
19. a. $\frac{1}{2}$ b. sample: $y = \frac{1}{2}x$

11. c.
Amount of money spent at grocery store
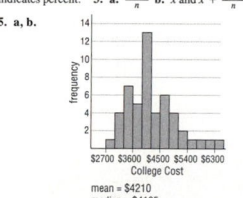

13. a.
Division A
Division B
Division C

LESSON 1-6 (pp. 38–45)
1. Sample: Frequency indicates count, and relative frequency indicates percent. 3. a. $\frac{y-x}{n}$ b. x and $x + \frac{y-x}{n}$
5. a, b.

$2700 $3600 $4500 $5400 $6300
College Cost
mean = $4210
median = $4105

7. a. cannot be determined from the histogram b. 25–44 9. a.

Number of Representatives	Frequency
1 to 5	22
6 to 10	17
11 to 15	3
16 to 20	2
21 to 25	3
26 to 30	1
31 to 35	1
36 to 40	0
41 to 45	1

b. See next column. 11. Sample: The intervals are not of uniform length; the first two intervals overlap at 10; the last two intervals miss 40. 13. x = mode, y = mean, z = median 15. a. 3 b. 13 c. 100 d. the mean number of pairs of shoes owned 17. a. It has steadily increased. b. 1975: ≈ 400 acres, 1985: ≈ 440 acres c. about 30,000 farms per year 19. about 119 cm 21. sample: $x + 6y = ^-43$

9. b.

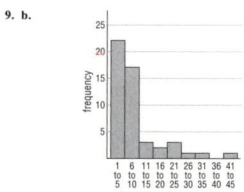

LESSON 1-7 (pp. 46–50)
5. False 13. b. sample: size = 16; sum = 882; mean = 55.125; median = 55; minimum = 42; maximum = 69; range = 27 c. Answers will vary depending on the statistics package used. 15. a. 10c b. 1.5c c. 2c 17. about 167 elk 19. about 9,860 21. about −2.3° per hour

LESSON 1-8 (pp. 51–58)
1. a. center b. spread c. spread d. center e. spread f. spread 3. a. 0 inches b. 0 inches c. about 1.37 inches 5. a. about 4.5 b. about 1.6 7. a. 430 b. 470 or 436 c. Jerry's 9. Variance is the sum of squared deviations divided by $n − 1$, so it must be nonnegative. 11. a. (i) Group Z, (ii) Group Y, (iii) Group X b. Group Y has the greatest standard deviation and Group Z has the smallest. c. s_X = 6.5, s_Y = 8.8, s_Z = 5.2 13. 360 to 600 15. sample: 70 seconds, 80 seconds, 140 seconds, 150 seconds; s = 40.8 seconds 17. a. 15 b. not possible c. about 60.7 d. ≈ 7.8 19. $\frac{(n − 1)a + b}{n}$ 21. a. sample: See below. b. The first shows more variability. 23. a. about 3300 women per year b. about 5700 men per year

21. a.

LESSON 1-9 (pp. 59–65)
1. to persuade the citizens of New York to ratify the Constitution 3. 50 of Madison's papers were studied. 5. about 16% 7. a. 9. cliometrics 11. a. mean = 4; standard deviation = 1.6 b. sample: 2, 3, 3, 3, 4, 5, 6, 6 13. a. lb b. lb c. lb² d. lb 15. a. Jakarta: mean = 6.1, standard deviation ≈ 3.3; Sydney: mean = 4, standard deviation ≈ 1.0 b. Sydney c. See page 847. 17. a. about 42 years old b. 46 years c. Age 80 is an outlier. d. Yes e. Sample: There is about a 10-year gap in the ages at which men and women win Oscars. The median age at which men win Oscars is about 42, while for women it is 33. In fact, 75% of the actresses who win Oscars are 38 or younger, while 75% of the actors are 38 or older.

19. a. $x + \frac{y-x}{n}, x + 2\frac{y-x}{n}, x + 3\frac{y-x}{n}$
b. $x + (n − 2)\left(\frac{y-x}{n}\right), x + (n − 1)\left(\frac{y-x}{n}\right)$
c. $\frac{(n − 1)x + y}{n}, \frac{(n − 2)x + 2y}{n}, \frac{(n − 3)x + 3y}{n}, \frac{2x + (n − 2)y}{n}, \frac{x + (n − 1)y}{n}$ 21. a. $y + 5 = 2(x + 4) \Leftrightarrow y + 5 = 2x + 8 \Leftrightarrow y = 2x + 3$ b. slope = 2, y-intercept = 3 c. See below.

15. c.
Jakarta

Inches of precipitation

Sydney
Inches of precipitation

21. c.

CHAPTER 1 PROGRESS SELF-TEST (pp. 69–71)
1. In 1980, the U.S. civilian noninstitutional population included a total of 15,900,000 females with activity limitations due to chronic medical conditions. 2. This is the second number in the "65 years and over column."; 24.5 million. 3. The largest number in the "45–64" years column in the second half of the chart is 22.8 across arthritis and rheumatism. So, arthritis and rheumatism most frequently limited activity by individuals 45–64 years of age in 1985. 4. We can estimate that $\frac{3}{17}$ of the geese in the flock are tagged. Let x be the number of geese in the flock, then $\frac{3}{17} = \frac{12}{x}$. So $x = 68$. There are about 68 geese in the flock.

5. Samples: The capture locations are representative of all possible places the members of this flock can be found. No unusual events occurred during the week. Tagging the geese did not affect their chance of being captured. 6. 2 P.M. is 14 hours after 12 midnight. The point on the graph located directly above 14 on the time axis shows a temperature of about 29° Celsius. 7. average rate of temperature change = $\frac{\text{total temperature change}}{\text{total time elapsed}}$ = $\frac{\text{(temp. at 4 P.M.)} − \text{(temp. at 6 A.M.)}}{\text{(4 P.M.)} − \text{(6 A.M.)}}$ = $\frac{\text{(temp. at time 16)} − \text{(temp. at time 6)}}{\text{(16 hours after midnight)} − \text{(6 hours)}}$ = $\frac{(31° \text{ Celsius}) − (21° \text{ Celsius})}{(16 \text{ hours}) − (6 \text{ hours})}$ = $\frac{10° \text{ Celsius}}{10 \text{ hours}}$ = 1° Celsius per hour 8. The apartment is coolest at about 4 A.M. to 5 A.M., warmest at about 5 P.M. 9. Samples: At 4 A.M. to 5 A.M., the apartment has had all night to cool off. The apartment heats up during the day. The very sharp fall at 5 P.M. probably corresponds to the cooling system being activated when the temperature gets too high. 10. The total number of observations is 2 + 2 + 3 + 1 + 3 + 4 + 3 + 1 + 1 = 20 observations. So, the median is the average of the 10th and 11th observations. The 10th observation = 4 heads, 11th observation = 4 heads, so the median = $\frac{4 \text{ heads} + 4 \text{ heads}}{2}$ = $\frac{8 \text{ heads}}{2}$ = 4 heads. 11. Percentile rank of 3 heads = $\frac{\text{\# of observations of 3 heads or fewer}}{\text{total \# of observations}}$ × 100 = $\frac{2 + 2 + 3 + 1}{20}$ × 100 = $\frac{8}{20}$ × 100 = $\frac{800}{20}$ = 40 = the 40th percentile 12. A frequency of 1 in 20 tosses means a relative frequency of $\frac{1}{20}$. A frequency of 2 means a relative frequency of $\frac{2}{20}$ or $\frac{1}{10}$. Just relabel the vertical axis as "relative frequency" and divide all the scale labels by 20. 13. (b). Graph (b) has lower frequencies near the mean and greater frequencies further away from the mean; therefore, it has the greater standard deviation. 14. $\sum_{i=1}^{9} x_i = x_1 + x_2 + x_3 + x_4 + x_5 + x_6 + x_7 + x_8 = 1 + 1 + 2 + 1 + 12 + 2 + 1 + 1 = 21$ 15. mean = (sum of data items) ÷ (number of data items) = 21 minutes ÷ 8 = 2.625 minutes, or 2 minutes and 37.5 seconds 16. Only one call exceeded the length of the mean because that one call was an isolated outlier, far from all the other data items, which pulled the mean up much higher than it would have been otherwise. 17. range = maximum value − minimum value = 72 − 40 = 32 years 18. 51 (location of the lower quartile) 19. (c). The median age for vice presidents is at 54, the 75th percentile is at 60.5. The median age for presidents, 55, falls between these two values. 20. Using the 1.5 × IQR (interquartile range) criterion:
presidents:
IQR = upper quartile − lower quartile = 58 − 51 = 7
1.5 × IQR = 1.5 (7) = 10.5
upper quartile + 1.5 × IQR = 58 + 10.5 = 68.5
lower quartile − 1.5 × IQR = 51 − 10.5 = 40.5.
Therefore any presidential ages greater than 68.5 or less than 40.5 would be considered outliers. There are no presidential ages less than 40.5 (the minimum is 42), but the maximum presidential age is 69, so there is at least one outlier among the presidential ages.
vice presidents:
IQR = upper quartile − lower quartile = 60.5 − 50 = 10.5

$1.5 \times \text{IQR} = 1.5 (10.5) = 15.75$
upper quartile $+ 1.5 \times \text{IQR} = 60.5 + 15.75 = 76.25$
lower quartile $- 1.5 \times \text{IQR} = 50 - 15.75 = 34.25$.
Therefore any vice presidential ages greater than 76.25 or less than 34.25 would be considered outliers. Since the maximum vice presidential age is 72 and the minimum is 40, there are no outliers among the vice presidential ages. **21.** Ordering the data gives: 64, 113, 146, 244, 329, 365, 402, 402, 475, 578, 661, 975, and 1,570. The median of an ordered set of 13 items is the 7th item: median = 402. The lower quartile is the median of the data items below the median. For this data there are 6 items below the median, so the lower quartile is the mean of the 3rd and 4th items: lower quartile $= \frac{146 + 244}{2} = \frac{390}{2} = 195$. The upper quartile is the median of the data items above the median. For this data there are 6 items above the median, so the upper quartile is the mean of the 10th and 11th items: upper quartile $= \frac{578 + 661}{2} = \frac{1239}{2} = 619.5$. **22.** 64 and 1570 appear to be outliers. **23.** The only repeated value is 402. So 402 is the mode with a frequency of 2. **24.** Using a statistics package, $s = 408$ millions of barrels.

25.

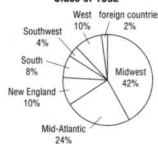

As the range of the data is a little over 1500, a good range for the histogram is 1750 (0–1750), giving intervals of 350. Then, there are five countries in the first interval, six in the second, one each in the third and fifth, and none in the fourth.

The chart below keys the **Progress Self-Test** questions to the objectives in the **Chapter 1 Review** on pages 72–78. This will enable you to locate those **Chapter 1 Review** questions that correspond to questions you missed on the **Progress Self-Test.** The lesson where the material is covered is also indicated in the chart.

Question	1–3	4–5	6–9	10–12	13	14	15	16	17–18	19	20
Objective	D	E	G	I	F	B	A	F	H	C	F
Lesson	1-2	1-1	1-3	1-6	1-8	1-4	1-4	1-4	1-5	1-5	1-5
Question	21–25										
Objective	J										
Lesson	1-7										

CHAPTER 1 REVIEW (pp. 72–78)

1. a. 8 **b.** 7.5 **2. a.** 7 **b.** 6 **c.** about 2.4 **3.** 15
4. a. 4.3 **b.** 10.0 **c.** 5.7 **5.** 8.7 **6. a.** 7.2 **b.** 5.6 **c.** 8.8
7. a. 145 **b.** 175 **8.** $77\frac{4}{9}$ **9.** $\sum_{i=1}^{14} g_i$ **10. a.** 95 **b.** 75
11. (b) **12.** (b) **13.** (a) **14.** (b) **15.** the mean
16. a. False **b.** sample data set: 1, 2, 3, 4, 5; mean = median = 3 **17.** Sample answer: (1) Studying the entire population may be prohibitively expensive. (2) Members of the population actually studied may be destroyed in the process. **18. a.** Sample answer: This is not random. Teens who are employed are not likely to be found at the arcade during working hours. **b.** Sample answer: This is not random. Teens who are employed are more likely to be found at the malls after 6 P.M. than those who are unemployed. **19. a.** about 75 foxes **b.** Sample answer: (1) The foxes captured in each sample are representative of the entire population of foxes. (2) When the samples have been chosen randomly, the percentage of tagged animals among the animals caught in the second sample is nearly the same as the percentage of tagged animals in the entire population. **20. a.** about 60 words per paragraph **b.** Sample answer: The sample was selected according to an arbitrary criterion, rather than randomly; paragraphs apart from those in

the Chapter 1 Summary had no chance to be included. The paragraphs in the Chapter 1 Summary are not necessarily representative of those in the rest of the book. Furthermore, the sample chosen was too small—just a few paragraphs—to be reliable. **21.** the numbers in the first column, from 14,531 to 11,055 **22.** Between 1987 and 1988, 7.4% of the 20–24 year old segment of the U.S. population moved to a different county within the same state. **23.** The numbers are expressed in units of 1,000, so each figure is a rounded figure. The total does not come out precisely correct because of cumulative rounding error. **24.** about 411,198 people **25.** about 678,900 people **26.** the 20–24 year old age group **27. a.** mean $\approx$ 79.2, standard deviation $\approx$ 28.6 **b.** 11, 42, 174, and 176 are the outliers. **c.** mean $\approx$ 76.5, standard deviation $\approx$ 10.2 **d.** Sample: The members of the troop had sales averaging about 75 to 80 boxes each; the mean, median, and mode figures were 79.2, 74, and 77, respectively. Most sales were fairly close to this average range, with the exception of two scouts who sold over 170 boxes each, one who sold only 42 boxes, and one who sold only 11 boxes. **28. a.** maximum = 265.7, minimum = 68.7, median = 106.9, upper quartile = 186.2, lower quartile = 76.2 **b.** Sample: The price of housing in the largest metropolitan areas in the U.S. averages in the range of about

$70,000 to $270,000. Prices vary more widely at the upper end of the scale than at the lower end. **29. a.** Minneapolis-St. Paul **b.** Minneapolis-St. Paul **c.** Minneapolis-St. Paul has the higher average temperature. Sample: The median of its temperature figures is 58°, whereas the median of the figures for Juneau is 47°. **d.** Minneapolis-St. Paul has the more variable temperatures. Sample: The IQR of the temperature figures for Minneapolis-St. Paul is 42.5°, whereas for Juneau it is only 23°. **30. a.** mean = 13; variance = 10; standard deviation = $\sqrt{10} \approx 3.2$ **b.** The fertilizer makes the plants tend to flower at nearly the same time and to flower sooner. **31. a.** for males: minimum = 70, lower quartile = 106, median = 119, upper quartile = 146.5, maximum = 188; for females: minimum = 90, lower quartile = 116.5, median = 138, upper quartile = 153, maximum = 200 **b.** In the male group, the scores 180 and 188 are outliers. The score of 200 among females is an outlier. **c.** The primary difference between the groups is that the females scored higher; their scores were 6.5 to 20 points higher than the males' scores at each point of the five-number summary. The primary similarity is in the spread; the range and IQR for the two groups are comparable (118 and 40.5 for males, 110 and 36.5 for females) with the males' scores being only slightly more widely dispersed. **32. a.** The 0–9 age group comprises about twice as great a percentage of the Guatemalan population as it does of the U.S. population. **b.** Sample: Wider availability and greater social acceptability of contraception might account for the fact that the 0–9 age group forms a smaller part of the population in the U.S. than it does in Guatemala. Wider access to better health care and better hygiene might account for the fact that the proportion of the population formed by each succeeding age group declines far less rapidly in the U.S. than it does in Guatemala. **33. a.** 27.5% **b.** in the $15,000–$19,999 level **34. a.** about $60,500 million **b.** about $60,500 million **35. a.** ≈ 1.5 hours **b.** the steepness of the slope up and the shallowness of the slope down **c.** about 11:30 P.M. to 4:30 A.M. **d.** slope = $-\frac{.15}{7} \approx -0.02$ percent blood-alcohol concentration per hour **36.** about $27,000 **37.** about $52,000 **38.** about $6,000 **39.** about $38,000 **40.** about $15,000 and about $49,000 **41.** Scores below the 10th percentile and above the 90th percentile are not very representative of the population. **42.** 11 students **43. a.** 7 students **b.** 8 students **c.** 5 students **44.** $33\frac{1}{3}\%$ **45.** Since the number of observations is essentially arbitrary, the actual frequency with which a particular result was observed is less meaningful than the relative frequency. **46.** about 17 million people **47.** about 57% **48.** False

49. sample:

University of Chicago Class of 1992

West 10% / foreign countries 2% / Southwest 4% / South 8% / New England 10% / Mid-Atlantic 24% / Midwest 42%

50. sample:

Class of 1992 SAT scores

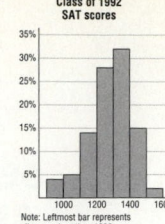

Note: Leftmost bar represents all scores less than 1000

51. sample:

Boxes sold

52. sample:

Median home prices ($1000s)

53. sample:

Daily Maximum Temperatures Juneau, Alaska

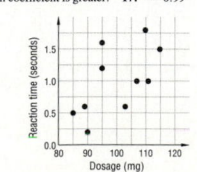
temperature (°F)

Daily Maximum Temperatures Minneapolis-St. Paul, Minnesota

temperature (°F)

Direct comparison of the histograms shows that Minneapolis-St. Paul has **(a)** higher summer temperatures and **(b)** lower winter temperatures. The temperatures in Minneapolis-St. Paul are spread fairly evenly over a wider range, whereas the temperatures in Juneau are clustered in a narrower range. Thus, Minneapolis-St. Paul has **(d)** greater variability in temperature. The mean and median cannot be read directly from the graph, but you can estimate that the median high temperature in Minneapolis-St. Paul is between 50° and 60°F while the median high temperature in Juneau is between 40°

and 50°F. So, Minneapolis-St. Paul has **(c)** the higher average temperature.

54. sample:

Test scores

females / males
60 80 100 120 140 160 180 200

LESSON 2-1 (pp. 80–85)

3. a. C is a function of n. **b.** $6.00
c. $C = \begin{cases} .75n & \text{if } n < 8 \\ .75n - 1.50 & \text{if } n \geq 8 \end{cases}$ **d.** See below.
5. a. 23 **b.** 19 **c.** 18 **7. a.** Yes **b.** No **9. a.** Yes **b.** No
11. a. See below. **b.** function; domain = range = the set of all real numbers **13. a.** See below. **b.** not a function
15. a. $P = 0.8d$ **b.** domain = range = the set of all nonnegative real numbers. **c.** sample: (10, 8), (50, 40), (100, 80) **d.** See below. **17. a.** domain = the set of all real numbers, range = $\{y : y \geq -7\}$ **b.** sample: (0, 9), (4, -7) **19.** sample: $y = -x$; domain is the set of all positive integers
21. No; $3 \cdot 2 + 4 = 100$ **23.** $a = \frac{3}{2}, b = \frac{19}{2}$

3. d. art: AA2,1a

11. a. art: AA2,1b

13. art: AA2,1d

15. d. art: AA2,1f

LESSON 2-2 (pp. 86–91)

3. a. about 9 laps **b.** $\frac{5}{3}$ **c.** the average time (in minutes) per lap **d.** 10 seconds **5. a.** interpolation **b.** extrapolation
7. (b)

9. a,b.

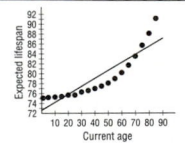
Expected lifespan
Current age

c. sample: $y = 0.16x + 72.6$ **d.** 72.6 years **e.** (iii)
f. 75 years **g.** 0.4 years **h.** 0 to 20 years, 70 to 85 years

11. a.

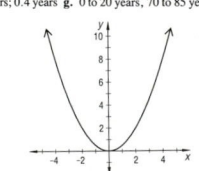

b. The domain is the set of all real numbers; the range is the set of all nonnegative real numbers.
13. $a = \frac{3}{5}, b = 7$ **15.** about 58.7%

LESSON 2-3 (pp. 92–98)

3. a. 1 **b.** 3 **c.** -2, 1.2, -1.5 **d.** 7.69 **5. a.** $y = 0.5x + 1$
b.

c. 6 **d.** The center of gravity is on this line, and the sum of the squares of the deviations is smaller than in Question 4.

7. a, d.

60 / 55 / 50 / 45 / 40 / 35 / 30 / 25 / 20 / 15
1955 1960 1965 1970 1975 1980 1985
Year
Area (millions of hectares)

b. Answers will vary. **c.** $y = 1.7320x - 3381$ **d.** The equation in part **c** is a better model, since it contains the center of gravity of the data and has a smaller total for the sum of the squares of the errors.

9. a.

96 / 94 / 92 / 90 / 88 / 86 / 84 / 82 / 80 / 78 / 76 / 74 / 72 / 70
10 20 30 40 50 60 70 80 90
Current age
Expected lifespan

b. $y = 0.165x + 72.3$ **c.** 74.8 years; 0.6 years **d.** This model produces the least amount of error. **11. a.** 3 **b.** $\{x : x \geq -7\}$
c. range = the set of all nonnegative real numbers

13. a.

b. domain = the set of all real numbers except zero; range = the set of all positive real numbers **15.** interpolation **17.** 16
19. (a), (b)

LESSON 2-4 (pp. 99–104)

1. 25 **3.** 0 **5.** all integers **7. a.** g **b.** $f(40) \approx 4.45$, $g(40) = 4$, $h(40) = 4.013$ **e.** f **9. a.** $f(n) = \left\lceil \frac{n}{40} \right\rceil$ **b.** Draw a graph of $f(n) = \left\lceil \frac{n}{40} \right\rceil$ and $g(n) = -\left\lceil \frac{-n}{40} \right\rceil$. **See above for graph.** The graphs are identical, so g also gives the number of buses needed. **11.** (c) **13. a.** $270 **b.** $B(M) = 135 \left\lceil \frac{M}{1000} \right\rceil$

15. a.

X	X + 0.5	INT(X + 0.5)
12.4	12.9	12
12.7	13.2	13
4.49	4.99	4
5.50	6.00	6

b. The output is X rounded to the nearest integer.
17. a. function **b.** function **c.** function **d.** not a function
19. a. sample: $y = -6.106x + 2232.55$ **b.** 1988

9. b.

$f(n)$

$g(n)$

LESSON 2-5 (pp. 105–111)

1. True **3.** False **5.** (a) **7.** (c) **9.** (d) **11.** ≈ 0.76
13. about 0.88 **15. a.** The correlation between team ERA and Proportion of Games Won should be negative since a lower ERA indicates stronger pitching, which should help a team win more games.; $r \approx -0.79$; The prediction is verified and the correlation is strong.; $r^2 \approx 0.63$ **b.** Earned run average, since the correlation coefficient is greater. **17.** ≈ -0.99

19. a.

Reaction time (seconds)
Dosage (mg)

b. Sample: 0.6; A linear model seems appropriate.

21. a. 100 mg **b.** 1.0 sec **c.** The point (100, 1) is the center of gravity for the (dosage, reaction time) data; it lies on the line of best fit found in **20 a**. The point (1, 100) is the center of gravity for the (reaction time, dosage) data; it lies on the line of best fit found in **20 b**. **23.** (b) **25.** $a = 3, b = \frac{51}{13}, c = \frac{14}{13}$

LESSON 2-6 (pp. 112–119)
1. (a) **3. a.** −4 **b.** $\frac{1 + \sqrt{33}}{4} \approx 1.69, \frac{1 - \sqrt{33}}{4} \approx -1.19$
c.

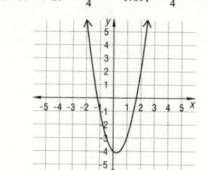

5. 32 ft/sec² **7. a.** $h = -16t^2 + 44t + 200$ **b.** 188 feet **c.** about 5.2 seconds after it is thrown **9.** about 115°F; sample: population growth, expanded area covered **11. a.** 1.185 cm/sec **b.** 0.519 cm/sec **c.** 8×10^{-4} cm from the axis of symmetry **d.** $0 \le r \le 8 \times 10^{-4}$
e.

13. a.

b. (33, 36) **c.** negative **15. a.** ≈ 0.95 **b.** ≈ 0.81 **c.** True **17. a.** $V = 40m + 45{,}000$ **b.** 49,200 liters **c.** 4:10 P.M. **19. a.** 1704 bass **b.** The bass in the original sample were representative of the population of bass in the lake; the locations from which the bass were taken were representative of the lake as a whole; marking the bass did not affect their chances of being recaptured.

LESSON 2-7 (pp. 120–125)
1. a. Sample: $y - 20 = \frac{2}{3}(x - 10)$ **b.** Because it was constructed using the same two points, and two points determine a line uniquely. **3.** $f(5) = -1.625(5)^2 + 11(5) + 5.625 = 20$ **5. a.** about −47% **b.** extrapolation **c.** It is unlikely that giving the pigs an unusually large dose of food supplement will make them lose half of their body weight. **7.** $f(x) = x^2 + 4x - 1$

9. a.

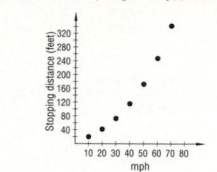

b. $y = 0.058x^2 + 0.4x + 9.25$ **c.** $y = 0.072x^2 - 0.38x + 15.6$ **d.** model b: $f(20) = 40.5; f(60) = 242.1$; model c: $f(20) = 36.8$; $f(60) = 252$; model b gives a smaller sum of squared errors. **e.** $f(x) = 0.072x^2 - 0.49x + 19.714; f(60) \approx 249.5$ **f.** $f(x) = 0.0005x^3 + 0.013x^2 + 1.55x + 1.75$ **g.** Yes
11. True **13.** True **15.** $n = \left\lceil \frac{m}{20{,}000} \right\rceil$ **17.** not a function
19. a. range = 8; mean = 1.6; median = 1; mode = 0
b, c.

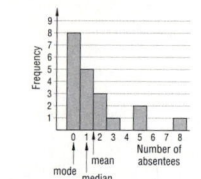

LESSON 2-8 (pp. 126–132)
1. 226.32 seconds **3.** about 254 yards **5.** in the year 2035 **7. a.** The increments of time being considered are small and the values of the independent variable are large. Even slight changes in the coefficients result in significant changes in the predicted time. **b.** 1960 3:48.2; 2020 3:27.2; 3000 − 2:15.8 **c.** The answers in **b** predict much faster running times. **9. a.** about 27.2 kph **b.** about 16.9 mph **11.** (b) **13.** (b) **15. a.** 204 billion **b.** 1976 **17. a.** independent: per capita cigarette consumption in 1930; dependent: lung cancer deaths per million males in 1950 **b.** Iceland, Norway, Sweden, Canada, and the U.S.A. **c.** Denmark, Holland, Switzerland, Finland, and Great Britain **d.** the U.S.A **e.** Sample: because the effects of cigarette smoking on lung cancer incidence are not immediately noticeable; someone who began smoking in 1930 would probably not die from lung cancer in the same year.

19. The domain is the set of all real numbers; the range is the set of all integers. **21.** r has no units.
23. a.

Copying machine downtime

b. 4 customers

CHAPTER 2 PROGRESS SELF-TEST (pp. 136–137)
1. a. Yes, since no x value is matched with more than one y value. **b.** The domain of the function seems to be the set of all real numbers; since both of the branches stay above the x-axis, the range is the set of all nonnegative real numbers or $\{y: y \ge 0\}$. **2. a.** The ordered pairs (4, 9) and (4, 7) both have the same first element, so the relation is not a function. **b.** The domain is the set of first elements: {0, 3, 4, 6}. **b.** The range is the set of second elements: {−1, 4, 7, 9, 12}. **3. a.** $k(-3) = (-3)^2 + 7 = 9 + 7 = 16$ **b.** The square of any real number is nonnegative. Then 7 is added to these values. So, the range is all real numbers greater than or equal to 7, $\{y: y \ge 7\}$. **4. a.** $6\pi \approx 18.85$. So $f(6\pi) = \lceil 6\pi \rceil + 1 = 19 + 1 = 20$ **b.** See below. **5. a.** $131 = 4 \cdot 30 + 11$, so there will be four full buses and a fifth one with eleven passengers. **b.** This is a ceiling function situation since having extra places on a bus is preferred to leaving passengers behind. Hence, divide the number of passengers by the individual bus capacity, and round functions up to the next integer. $n = \left\lceil \frac{p}{30} \right\rceil$ **6.** The correlation between x and y is given by the correlation coefficient, $r = \sqrt{r^2} = \sqrt{0.37} \approx \pm 0.61$. As the equation indicates a positive linear relation (slope is positive), $r = 0.61$.
4. b.

7. a.

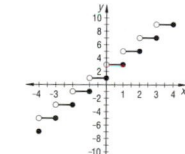

b. sample: Vehicle efficiency increased consistently between 1970 and 1986. **c.** sample: The rate of increase in vehicle efficiency (which corresponds to the slope of the linear model) is not constant. **8.** $y = 1990 - 1900 = 90$. So $m = 0.025$ (90)² − 3.67(90) + 145.84 = 18.04 mpg. **9.** The flat fee is $60. The quantity $1 - h$, if it is not zero, gives −1 times the remaining hours and fractions of hours. A floor function pulls this value to the nearest negative integer bigger than its value. Multiplying this integer by $45 gives the additional fee as a negative number. Hence subtracting this final amount from $60 gives her total charge, C. $C = 60 - 45 \lfloor 1 - h \rfloor$, or (d). **10.** Choice (d), since −0.82 is the value closest in absolute value to 1, which indicates perfect linear relation. **11. a.** We need to solve for a, b, and c in $h(t) = at^2 + bt + c$. Using (1, 299), (2, 311), and (3, 291), we get the following equations:
$$299 = a(1)^2 + b(1) + c = a + b + c$$
$$311 = a(2)^2 + b(2) + c = 4a + 2b + c$$
$$291 = a(3)^2 + b(3) + c = 9a + 3b + c$$
Subtracting the first from the second and the third from the second we get,
$$12 = 3a + b$$
$$20 = -5a - b$$
Adding the two gives,
$$32 = -2a, \text{ or } a = -16.$$
Then $b = 60$ and $c = 255$. So $h(t) = -16t^2 + 60t + 255$. **b.** $h(t) = 80 = -16t^2 + 60t + 255$, or $-16t^2 + 60 + 175 = 0$. Use the quadratic formula:
$$t = \frac{-60 \pm \sqrt{60^2 - 4(-16)(175)}}{-32} = \frac{-60 + \sqrt{14800}}{-32}$$
so $t \approx -1.93$ or $t \approx 5.68$. Choose the positive time, about 5.68 seconds after being thrown. **12. a.** Using a statistics package we get $y = -.202x + 503$, where x is the year of the Olympic Games. **b.** $x = 2000$, so $y = -0.202(2000) + 503 \approx 99$ seconds. **c.** 1 minute 40 seconds is 100 seconds, so $y = 100 = -0.202x + 503$, or $-0.202x + 403 = 0$. Then $x \approx 1995$. The nearest Olympic Games are in 1996. **d.** Using a statistics package, $r \approx -0.87$ **e.** The equation for the model gives $y = -0.202(1988) + 503 \approx 101.42$ seconds, or 1 minute 41.42 seconds. Thus the error would be about 2.03 seconds. **f.** The year, 1996, is beyond the range of the data; so the prediction is an extrapolation.

The chart below keys the **Progress Self-Test** questions to the objectives in the **Chapter 2 Review** on pages 138–142. This will enable you to locate those **Chapter 2 Review** questions that correspond to questions you missed on the **Progress Self-Test**. The lesson where the material is covered is also indicated in the chart.

Question	1	2	3	4a	4b	5	6	7	8	9	10
Objective	I	D	A	A	H	F	D	J	G	F	D
Lesson	2-1	2-1	2-1	2-4	2-4	2-4	2-5	2-6	2-6	2-4	2-5
Question	11a	11b	12a,b	12c	12d	12e	12f				
Objective	B	G	E	C	B	F	F				
Lesson	2-7	2-7	2-3	2-2	2-5	2-3	2-2				

CHAPTER 2 REVIEW (pp. 138–142)
1. a. 3 **b.** $\frac{1}{3}$ **2. a.** False; $f(2) + f(-2) = 9 + \frac{1}{9} = \frac{82}{9} \ne 0$ **b.** True; $f(2) \cdot f(3) = 3^2 \cdot 3^3 = 3^5 = f(5)$ **3. a.** 4 **b.** No; $g(4) - g(2) = 19 - 7 = 12 \ne 7$ **4. a.** False **b.** True **5. a.** 12 **b.** −5 **c.** 3 **d.** −4 **6. a.** −2 **b.** 9 **7.** For (0, 1): $2 \cdot 0^2 - 5 \cdot 0 + 1 = 1$; (2, −1): $2 \cdot 2^2 - 5 \cdot 2 + 1 = -1$; (3, 4): $2 \cdot 3^2 - 5 \cdot 3 + 1 = 4$ **8.** $y = -\frac{7}{12}x^2 + \frac{9}{4}x + \frac{16}{3}$ or $y = -.583x^2 + 2.25x + 5.33$ **9. a.** two **b.** $y = 0.27x + 0.33$ **10. a.** 3 **b.** sample: $-5 = 0a + 0b + c$; $9 = 4a + 2b + c$; $47 = 16a + 4b + c$ **c.** $a = 3, b = 1, c = -5$; $y = 3x^2 + x - 5$ **11. a.** the set of all real numbers **b.** the set of all integers **12. a.** the set of all nonnegative real numbers **b.** the set of all nonnegative real numbers **13. a.** the set of all real numbers except zero **b.** the set of all real numbers except zero **14. a.** the set of all real numbers **b.** $\{y: y \ge -18\}$ **15. a.** the set of all real numbers **b.** the set of all real numbers **16. a.** $\{-2, -1, 0, 2, 5\}$ **b.** $\{2\}$ **17. a.** x **b.** y **18. a.** t **b.** $f(t)$ **19.** at $x = 0$ **20.** at all integer values of x **21.** A correlation coefficient of 0.95 means that there is a strong linear relationship in the data. A scatterplot would reveal that the data lie very close to a straight line with positive slope. **22.** 0.73, −0.73 **23.** ≈ -0.89 **24.** False **25.** True **26.** False **27.** (b) **28.** −1.0

29. a.

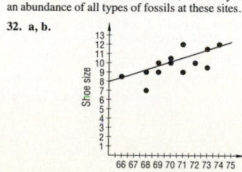

b. The production values for 1978 and 1980 appear to be outliers. **c.** 994 million tons **d.** 77 million tons **e.** 1985 **f.** To predict for the year 2010 would be to extrapolate far beyond the range of the data. Other sources of energy may be in use by 2010, affecting the amount of coal produced. **30. a.** $C = 35.99 + 0.18m$ **b.** 355.6 miles **31. a.** 0.95 **b.** The correlation suggests that a linear model is appropriate. **c.** Sample: The claim seems reasonable since high numbers of flintstones were generally found at sites with a large number of charred bones. However, there may have simply been an abundance of all types of fossils at these sites.

32. a, b.

32. c. sample: $y = 0.43x - 20$ **d.** Sample: The slope indicates the expected change in shoe size per inch of height. The y-intercept (−20) is meaningless since it is impossible to speak of negative shoe sizes. Clearly, the linear model holds only within a certain range of data. **e.** sample: **i.** 11 **ii.** 8 **33. a.** $y = 0.426x - 20.13$ **b.** $11\frac{1}{2}$ **c.** between about 62 and 78 inches **34.** (b) **35.** $p = \left\lfloor \frac{b}{6} \right\rfloor$ **36.** $b = 167.5 \left\lceil \frac{k}{1000} \right\rceil$ **37.** (b) **38.** $d = 0.33 + 0.22 \left\lceil t - 1 \right\rceil$ **39. a.** 460 feet above sea level **b.** about 9 seconds

40. a.

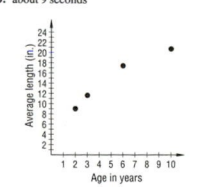

b. Sample: Like most other animals, the bass probably reach a mature stage and stop growing. **c.** sample: $y = -0.16x^2 + 3.38x + 2.9$ **d.** 19.7 inches **e.** 3.8 years of age **41. a.** Temperature and wind speed are independent variables; wind chill is dependent. **b.** See below. **c.** sample: $.025x^2 - 1.993x + 34.69$

41. b.

42.

43.

44.

45.

46. not a function **47.** function **48.** function **49.** not a function **50.** domain: $\{x: -2 \le x < 3\}$; range: $\{-2, -1, 0, 1, 2\}$ **51.** domain: the set of all real numbers; range: $\{y: y \le 3\}$ **52.** domain: the set of all real numbers; range: $\{y: y \le 2\}$ **53.** domain: $\{x: x \le -2 \text{ or } x \ge 2\}$; range: the set of all real numbers **54. a.** linear **b.** negative **55. a.** linear **b.** positive **56. a.** curvilinear **b.** approximately zero **57. a.** linear **b.** approximately zero **58.** 12 students **59. a.** 80 percent **b.** ≈ 72 percent **c.** ≈ 8 percent **60. a.** (6.75, 73.5) **b.** $1.887 \cdot 6.75 + 60.762 \approx 73.5$ It checks. **61.** 0.71 **62. a.** 80 percent **b.** approximately 78 percent **c.** approximately 2 percent **63.** Sample: approximately 9 hours; Studying too many hours may point to lack of adequate previous preparation, which cannot be helped much by cramming. Also, the fatigue caused by too many study hours may impede performance during exam. **64.** The quadratic model **65.** Sample: The quadratic model, since it fits the data points better. It also shows a decline in performance past a certain point of studying, which is likely.

LESSON 3-1 (pp. 144–150)
3. sample: $x \approx -40, x \approx 100$, and $x \approx 240$ **5. a.** See below. **b.** −4 and 6 **c.** $x = -4$ or $x = 6$ **d.** (1, −25) **7. a.** The graph of $y = \sqrt{x} - 5$ is the image of the graph of $y = \sqrt{x}$ under a translation 5 units to the right. **b.** See below.

5. a.

7. b.

9. a.

b. for $f(x)$, $x = 0$; for $g(x)$, $x = -6$; for $h(x)$, $x = 0$ **c.** The graph of $g(x)$ is the image of the graph of $f(x)$ under a translation 6 units to the left. The graph of $h(x)$ is the image of the graph of $f(x)$ under a translation 6 units up. **11. a.** sample: $-30 \le x \le 15, -20 \le y \le 5$

11. b.

13. a.

b. True **c.** The graph of $g(x)$ is higher than the graph of $f(x)$ for all x in the domains of f and g. Also, since x^2 is always positive when $x \neq 0$, $\frac{4}{x^2} > \frac{1}{x^2}$ when $x \neq 0$. **15. a.** 35 competitors
b. mean ≈ 2.23; median = 2; mode = 2 **c.** 1.8 **d. See below.**
17. a, b. See below. c. The figures are congruent. $P'Q'R'S'$ is the image of $PQRS$ under a translation 5 units to the right and 2 units down. **19. a.** 1970: about \$1250 per student; 1986: about \$4690 per student **b.** increase

15. d.

17. a, b.

LESSON 3-2 (pp. 151–155)
1. $(3, -1)$ **3.** $(r + a, v + b)$ **5.** $(p - 2, q + 5)$
7. a. $T(-2, 2) = (-8, 7)$; $T(-1, 1) = (-7, 6)$; $T(0, 0) = (-6, 5)$
b. $7 - 5 = |-8 + 6|$, $2 = 2$; $6 - 5 = |-7 + 6|$, $1 = 1$;
$5 - 5 = |-6 + 6|$, $0 = 0$ **9. a.** $g(x) = |x - 5| - 3$
b. See below. 11. a. $(x, y) \to (x - 3, y)$ **b. See below.**
13. $y = x^2 - 2$ **15.** a circle of radius 4 with center at $(-7, 6)$

17. (a) **19. a.** $\bar{x} = \dfrac{\sum\limits_{i=1}^{n} x_i}{n}$ $\sqrt{\dfrac{\sum\limits_{i=1}^{n}(x_i - \bar{x})^2}{n - 1}}$ **21.** (c)

9. b.

11. b.

LESSON 3-3 (pp. 156–161)
3. 20 **5.** 48 **7.** 25 **9.** Samples: variance, standard deviation, interquartile range, and range **11.** 26 **13. a. See page 857. b.** $x \to x + 11$ **c.** original scores: range = 16; mode = 10; mean = 13; median = 11
transformed scores: range = 16; mode = 21; mean = 24; median = 22

15. a. $\sum\limits_{i=1}^{n}(x_i + a) = [(x_1 + a) + (x_2 + a) + \ldots + (x_n + a)]$
$= [x_1 + x_2 + \ldots + x_n + a + a + \ldots + a]$
$= \left(\sum\limits_{i=1}^{n} x_i\right) + na$
b. from a, $\sum\limits_{i=1}^{n}(x_i - \bar{x}) = \sum\limits_{i=1}^{n}(x_i) - n\bar{x}$
$= \dfrac{n\sum\limits_{i=1}^{n} x_i}{n} - n\bar{x}$ because $\dfrac{n}{n} = 1$
$= n\bar{x} - n\bar{x}$ because $\dfrac{\sum\limits_{i=1}^{n} x_i}{n} = \bar{x}$
$= 0$
17. (b), (d), (e)

19. a, b.

c. $(x, y) \to (x + 7, y + 10)$ **21.** $y = |x - 9|$
23. a. x-intercepts: $-3 + \sqrt{5} \approx -0.8$, $-3 - \sqrt{5} \approx -5.2$; y-intercept: 4 **b. See below. 25. a.** $5x$ **b.** $36x$ **c.** $36x^2$

13. a.

● original scores
○ transformed scores

23. b.

LESSON 3-4 (pp. 162–166)
1. a. i, iii **b.** i, iv **3.** We must show that $q(-x) = -q(x)$ for all x. $q(-x) = (-x)^7 = [(-1)(x)]^7 = (-1)^7 x^7 = -x^7 = -q(x)$ Since $q(-x) = -q(x)$ for all x, q is an odd function.
5. a. neither **7.** sample: $f(2) = 2^3 - 2 = 6$, $f(-2) = (-2)^3 - 2 = -10$ **9. a. See below. b.** k and m are even; n is neither. **c.** k is the parent; m is a translation of k by 2 units down; n is a translation of k by 4 units to the left and 2 units down. **d.** The graph of g is a reflection image of the graph of f over the x-axis. **13.** 15 **15. a.** $x = 3$ or $x = -3$ **b.** $x = 3$ or $x = 0$ or $x = -3$ **c.** $x = \sqrt{5}$ or $x = 0$ or $x = -\sqrt{5}$ **17. a.** about \$8 billion **b.** about \$2 billion
19. a. $(-1, 8)$ **b.** $(-2, -6)$ **c.** $(1, 2)$

9. a.

LESSON 3-5 (pp. 167–173)
1. (ax, by) **3. a.** $S(-3, 9) = \left(-\frac{3}{2}, 27\right)$; $S(0, 0) = (0, 0)$;
$S\left(\frac{1}{2}, \frac{1}{4}\right) = \left(\frac{1}{4}, \frac{3}{4}\right)$ **b.** $\frac{27}{2} = 9 = (-3)^2 = \left(2 \cdot \frac{-3}{2}\right)^2$; $\frac{0}{3} = 0 = (0)^2 = (2 \cdot 0)^2$; $\frac{3/4}{3} = \frac{1}{4} = \left(\frac{1}{2}\right)^2 = \left(2 \cdot \frac{1}{4}\right)^2$; **5.** False

7. a. Yes **b.** No **c.** Yes **d.** No
9. a, b. See below. c. The image is the preimage reflected over the x-axis. **d.** $g(x) = -x^2$ **11. a.** x-intercepts of $y = x^2 - 16$ and of $\frac{y}{4} = x^2 - 16$: -4 and 4. **b. See below.**
13. $S: (x, y) \to (3x, y)$ **15. a. See below. b. i.** $(3, 0)$ and $(12, 0)$ **ii.** $\left(0, \frac{1}{2}\right)$ **iii.** $\left(-6, \frac{3}{2}\right)$ **iv.** $\left(6, -\frac{1}{2}\right)$
17. a. $\sqrt{24} = 2\sqrt{6}$ **b.** about 25.5 mph **c.** about 126 feet **d.** $S(s, d) = \left(s, \frac{5}{9}d\right)$ **19. a.** neither
21. a. 24 **b.** mean ≈ 4.7; median = 2; mode = 1 **c.** Sample: the mean; the mean is affected by the outlier score, 46. Only 4 students were absent more than 4 days. **d.** 9.8 days
23. a. $125y^3$ **b.** x^3y^3 **c.** x^py^p

9. a, b.

11. b.

15. a.

LESSON 3-6 (pp. 174–179)
1. A scale change of a data set is a transformation that maps each element x_i of the data set to ax_i, where a is a nonzero constant. **3.** about \$759 **5. a.** 50 **b.** $5k$ **c.** $\frac{5}{m}$
7. mean ≈ 145 grams, range ≈ 95 grams

---

856   857

---

858   859

---

9. a. $S: x_i \to \frac{3}{2}x_i$ **b.** original scores: range = 16, mode = 10, mean = 13, median = 11; scaled scores: range = 24, mode = 15, mean = 19.5, median = 16.5 **c. See below.** Sample: The distribution of scaled scores has a wider range and therefore appears to be more spread out. However, the shapes of the distributions are similar. **11. a. See below.**
b. $w = .342h - 18.43$ **c. See below.** The scatter plot is stretched by a factor of $\frac{1}{2.54}$ in the horizontal direction.
d. $w = .869h - 18.43$ **e. See below.** The scatter plot is stretched by a factor of $\frac{1}{2.54}$ in the horizontal direction and $\frac{1}{.454}$ in the vertical direction. **f.** $w = 1.914h - 40.59$ **13.** Let s be the standard deviation of the data set. Dividing each element of the data set by s is the same as multiplying each element of the data set by $\frac{1}{s}$. Therefore, the standard deviation of the scaled data set is $\left|\frac{1}{s}\right| \cdot s$. Since the standard deviation of any data set is nonnegative, $\left|\frac{1}{s}\right| \cdot s = \frac{1}{s} \cdot s = 1$. The variance of the scaled data set is $\left(\frac{1}{s}\right)^2 \cdot s^2 = 1$. **15.** (a) **17.** (d) **19.** functions g and h
21. Yes; each side of the second triangle has a length 2.5 times the corresponding side of the first triangle.
23. a. 30 **b.** $a^2 + 7a + \frac{7}{4}$ **c.** $a^2 - a - 12$

9. c.

original scores

scaled scores

11. a.

11. c.

11. e.

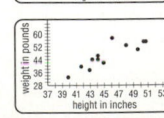

LESSON 3-7 (pp. 180–184)
1. maternal grandmother **3. a.** $f(g(0)) = f(0 - 5) = f(-5) = (-5)^2 + (-5) = 25 - 5 = 20$; $g(f(0)) = g((0)^2 + 0) = g(0) = 0 - 5 = -5$ **b.** $f(g(2.5)) = f(2.5 - 5) = f(-2.5) = (-2.5)^2 + (-2.5) = 6.25 - 2.5 = 3.75$; $g(f(2.5)) = g((2.5)^2 + 2.5) = g(8.75) = 8.75 - 5 = 3.75$ **5. a.** $\frac{4}{15}$ **b.** $(n \circ m)(x) = \frac{4}{x + 8}$
c. the set of all real numbers except -8 **7. a.** 9 **b.** 9
9. a. 2 **b.** $\sqrt{5}$ **c.** $g(f(x)) = \sqrt{x + 1}$ **d.** $\{x: x \geq -1\}$
11. a. $.945x$ dollars **b.** It doesn't matter; both have the same effect. **13.** No, for example, let $f(x) = 2x^2$ and $g(x) = x^2$. Then $f(g(x)) = f(x^2) = 2x^4$, which is not a quadratic function.
15. The mean and the standard deviation will be tripled.
17. neither; If f were odd, then $f(-x) = -f(x)$ for all x. If f were even, then $f(-x) = f(x)$ for all x. However, $f(-9) = 0$. Since $f(9) \neq 0$, f is neither even nor odd. **19.** $y = (x - r)^2 - s$
21. a. Yes **b.** No **23. a.** $y = \frac{12}{x}$ for $x \neq 0$
b. $y = \frac{10}{x} + 1$ for $x \neq 0$ **c.** $y = \frac{4}{x}$ for $x \neq 0$
d. $y = \frac{1}{x + 2}$ for $x \neq -2$ **25.** samples: $f(x) = x$, $f(x) = \frac{1}{x}$

LESSON 3-8 (pp. 185–189)
3. a. $\{(4, 5), (6, 6), (8, 7), (10, 8)\}$ **b. See below.**
c. reflection over $y = x$ **d.** reflection over $y = x$

3. b.

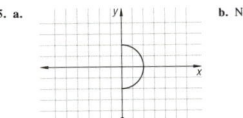

5. a. $f^{-1}(x) = \frac{x}{3} - 2$ **b.** Yes **7. a.** $f^{-1}(x) = \sqrt[3]{x}$ **b.** Yes
9. -1 **11. a.** $(g \circ f)(x) = g(2x + 1) = \frac{1}{2}(2x + 1) - 1 = x + \frac{1}{2} - 1 = x - \frac{1}{2}$; $(f \circ g)(x) = f\left(\frac{1}{2}x - 1\right) = 2\left(\frac{1}{2}x - 1\right) + 1 = x - 2 + 1 = x - 1$; they are not inverses.
b.

13. a. $M(x) = 2670x$; $U(x) = \frac{x}{2670}$ **b.** \$7.49 **c.** Yes

15. a.

b. No

17. Let $q = \frac{1}{p}$. To form the inverse, switch p and q. Then $p = \frac{1}{q}$.
Taking the reciprocal of both sides, we get $q = \frac{1}{p}$. So $h^{-1}(p) = \frac{1}{p}$
for $p \neq 0$. **19. a.** $f^{-1}(x) = \frac{x - b}{m}$ **b.** False; the inverse of the linear function $f(x) = 0$ for all x, is $x = 0$ which is not a function.
21. a. It stretches the graph horizontally by a factor of 3 and translates it 2 units down. **b. See below.**
23. a. $p(-t) = 5 - |-t| = 5 - |t| = p(t)$ **b. See below.**
25. a. $y = x^2 - 16x + 73$ **b.** $y = (x + 6)^2 - 39$

21. b.

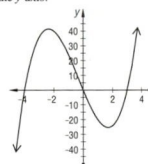

23. b.

LESSON 3-9 (pp. 190–195)
1. a. $(0, 0)$ **b.** $0 = 0^2$; $0 = \frac{1}{2}(0)^2$ **3. a.** $S(x, y) \to (25x, 25y)$ **b.** True **5. a.** $y = 2(x - 2)^2 + 5$ **b.** $(x, y) \to (x + 2, y + 5)$ **c.** $(x, y) \to \left(\frac{1}{2}x, \frac{1}{2}y\right)$ **7. a.** $y = \frac{b}{a^3}x^3$
b. $\frac{b}{a} = \frac{1}{5}$ **c.** sample: $a = \sqrt{5}$, $b = 1$ **9.** False **11.** 6
13. a. symmetric with respect to the line $x = 6$ **b.** False
15. (b) **17. a.** $(x, y) \to \left(\frac{2}{3}x, \frac{1}{2}y\right)$ **b.** $(x, y) \to \left(\frac{2}{3}x, 2y\right)$
19. -0.78 **21.** $2^5 = 32$ **b.** x^{15} **c.** $3^p x^{4p}$

CHAPTER 3 PROGRESS SELF-TEST (p. 196)
1. To apply $S(x, y) = (x - 1, y + 5)$ to $y = 3x^2$, replace x by $x + 1$ and y by $y - 5$.
$$y - 5 = 3(x + 1)^2$$
$$y - 5 = 3(x^2 + 2x + 1)$$
$$y - 5 = 3x^2 + 6x + 3$$
$$y = 3x^2 + 6x + 8$$
2. The vertex of the original parabola is $(0, 0)$. $S(0, 0) = (0 - 1, 0 + 5) = (-1, 5)$ **3. See below.** The scale change $S: (x, y) \to (-x, 2y)$ is a stretch by 2 in the vertical direction and a reflection over the y-axis.

4. An equation for the image of $y = 12x + x^2 - x^3$ under $S: (x, y) \to (-x, 2y)$ is found by replacing x by $-x$ and $\frac{y}{2}$ in the equation of the preimage.
$$\frac{y}{2} = 12(-x) + (x)^2 - (-x)^3$$
$$\frac{y}{2} = -12x + x^2 + x^3$$
$$y = -24x + 2x^2 + 2x^3$$
5. a. The median of the translated data is changed by the amount of the translation, so the median of the original data is $7.4 + 20 = 27.4$ g. **b.** Translations do not affect the range so the range of the original data is $11.8 - 3.4 = 8.4$ g. **c.** Translations do not affect the variance so the variance of the original data is $1.3^2 = 1.69$ g^2. **6.** Multiply each statistic by the scale factor .0353: mean = $(27.9)(.0353) \approx 0.98$ oz; standard deviation = $(1.3)(.0353) \approx 0.046$ oz.
7. $\sum\limits_{i=1}^{6}(g_i + 6) = (8 + 6) + (7 + 6) + (2 + 6) + (-4 + 6) + (-11 + 6) + (5 + 6) = 8 + 7 + 2 + -4 + -11 + 5(6) = 2 + 30 = 32$
8. $\sum\limits_{i=1}^{5} 4g_i = 4(8) + 4(7) + 4(2) + 4(-4) + 4(-11) = 4(8 + 7 + 2 + -4 + -11) = 4(2) = 8$
9. $(b \circ a)(2) = b(a(2)) = b(18 - 3(2)) = b(12) = 4(12)^2 = 4(144) = 576$ **10.** $a(b(x)) = a(4x^2) = 18 - 12x^2$ **11.** The domain of a is the set of all real numbers. The domain of b is the set of all real numbers and the range of b is in the domain of a. Thus the domain of $a \circ b$ is the set of all real numbers. **12.** If $f(x) = y = \frac{2}{x - 3}$, then the inverse is $x = \frac{2}{y - 3}$. So $\frac{y - 3}{1} = \frac{2}{x}$ $\Rightarrow$ $y - 3 = \frac{2}{x} \Rightarrow y = \frac{2}{x} + 3$. **13. See page 860.** **14.** Yes, $y = \frac{2}{x} + 3$ is a function because each value of x yields exactly one value of y.

15. See next column. **16.** From the graph, the object hits the ground ($h(t) = 0$) at $t = 4$. **17.** We need to show that $f(-x) = -f(x)$ for all x. $f(-x) = 4(-x)^3 - 2(-x) = -4x^3 + 2x = -(4x^3 - 2x) = -f(x)$ **18.** Sample: The point $(1, 3)$ is the image of $(1, 1)$, so the scale change is $(x, y) \to (x, 3y)$, and that transformation changes $y = |x|$ by $\frac{y}{3} = |x|$ or $g(x) = y = 3|x|$.
19. g is symmetric with respect to the y-axis. **20.** False; under a scale change of magnitude k, the variance of the scaled data is k^2 times the variance of the original data.
21. True; a translation preserves distances, so the image is congruent to the preimage.

13.

15.

The chart below keys the **Progress Self-Test** questions to the objectives in the **Chapter 3 Review** on pages 197–200. This will enable you to locate those **Chapter 3 Review** questions that correspond to questions you missed on the **Progress Self-Test**. The lesson where the material is covered is also indicated in the chart.

Question	1–2	3	4	5	6	7	8	9–10	11	12–14	15–16
Objective	C	J	C	F	I	E	E	A	H	B	K
Lesson	3-2	3-2	3-5	3-3	3-3	3-3	3-6	3-7	3-7	3-8	3-1
Question	17	18	19	20	21						
Objective	G	K	G	I	D						
Lesson	3-4	3-5	3-4	3-6	3-2						

CHAPTER 3 REVIEW (pp. 197–200)
1. a. -14 **b.** -72 **2. a.** $f(g(t)) = -6t^2 + 6t - 2$
b. $g(f(t)) = -36t^2 + 30t - 6$ **3.** 13 **4.** 1 **5. a.** -3 **b.** $4\frac{1}{2}$
6. a. $(m \circ n)(x) = \frac{3}{x+5}$ **b.** $(n \circ m)(x) = \frac{3}{x} + 5$ **b.** Yes
7. a. $\{(7, 3), (8, 4), (9, 5), (10, 6), (11, 7)\}$ **b.** Yes
8. a. $y = \frac{x-7}{2}$ **b.** Yes **9. a.** $x = |y|$ **b.** No **10. a.** $y = \frac{2}{x} - 1$
b. Yes **11.** The image is translated h units to the right and k units up. **12.** The image is stretched by a factor of a horizontally and by a factor of b vertically. **13.** (b) **14.** (c)
15. $y = x^2 - 14x + 46$ **16.** $y = \frac{x^2}{12}$ **17.** $y = 5|4x|$ **18.** $y = |x-1|$
19. $(x, y) \to \left(\frac{x}{10}, y\right)$ **20.** $(x, y) \to (x + 8, y + 9)$ **21.** True
22. True **23.** scale change **24.** $(x, y) \to (-x, y)$
25. a. 18 **b.** 10 **c.** 56 **d.** $8k$
26. True. $\sum_{i=1}^{n} (x_i - 7) = (x_1 - 7) + (x_2 - 7) + \dots +$
$(x_n - 7) = (x_1 + x_2 + \dots + x_n) - (n)(7) = \left(\sum_{i=1}^{n} x_i\right) - 7n$ **27.** $k = \frac{1}{2}$
28. $\sum_{i=1}^{n} (x_i - y_i) = (x_1 - y_1) + (x_2 - y_2) + \dots + (x_n - y_n)$
$= (x_1 + x_2 + \dots + x_n) + (-y_1 - y_2 - \dots - y_n)$
$= (x_1 + x_2 + \dots + x_n) - (y_1 + y_2 + \dots + y_n)$
$= \sum_{i=1}^{n} x_i - \sum_{i=1}^{n} y_i$

29. The mean goes up by 10. **30.** The standard deviation is unchanged. **31.** The median is multiplied by k. **32.** The variance is multiplied by k^2. **33.** The standard deviation is unchanged. **34.** The standard deviation is divided by 2.
35. True **36.** False **37. a.** even **b.** odd **c.** even
d. neither **38. a.** odd **b.** We want to show that $f(-x) = -f(x)$ for all x.
$f(-x) = 8(-x)^3 = -8x^3 = -f(x)$
39. a. even **b.** We want to show that $s(-t) = s(t)$ for all t.
$s(-t) = 5(-t)^2 - (-t)^4 = 5t^2 - t^4 = s(t)$
40. a. neither **b.** $g(1) = 1, g(-1) = 7$ **41.** $\{x : x \geq -10\}$
42. reflection over the line $y = x$. **43.** (d) **44.** (b) **45.** 109
46. a. $s \to 5s$ **b.** mode = 7; mean = 6; median = 7
c. mode = 35; mean = 30; median = 35 **d.** Multiplying each element of a data set by the factor a multiplies each of the mode, mean, and median by a factor of a.
47. mean = 8:29.64; standard deviation = 1.15 seconds
48. a. 44.70 mm **b.** 76.71 mm **c.** ≈ 181 mm^2
49. See page 861. **50. a.** See page 861. **b.** $x = -6, y = 5$
c. x-intercept: $\left(-\frac{31}{5}, 0\right)$; y-intercept: $\left(0, \frac{31}{6}\right)$ **d.** parent equation:
$y = \frac{1}{x}$; transformation: $(x, y) \to (x - 6, y + 5)$
51. a. See page 861. **b.** sample: $x = 0, x = 5, x = -10$
52. a. See page 861. **b.** False **53.** $y = (x + 2)^2 + 5$
54. $y = |x + 3| - 4$ **55.** $y = \sqrt{2x}$ **56.** $y = \frac{1}{(x-4)^3} - 2$

49.

$y = |x - 5| + 7$

$y = |x|$

50. a.

51. a.

52. a.

$g(x) = 2x^3$; $f(x) = x^3$

57.

58. a. $(2.2, 0), (-3.8, 0)$ **b.** $(0, 16)$ **c.** $(-0.8, 17)$
59. a.

b. domain: the set of all real numbers not equal to -5; range: the set of all nonzero real numbers **60. a.** See below. **b.** domain: $\{x : x 5\}$; range: $\{y : y \geq 0\}$ **61. a.** See below. **b.** domain: the set of all real numbers; range: $\{y : y \geq 10\}$
62. a. See below. **b.** Yes **c.** $y = \frac{3}{x}$ **63. a.** See below. **b.** No **c.** $x = y^2$ **64.** No **65.** Yes

60. a.

61. a.

62. a.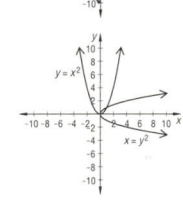

$h(x) = h^{-1}(x) = \frac{3}{x}$

63. a.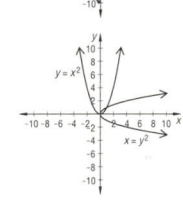

$y = x^2$; $x = y^2$

860

861

862

863

LESSON 4-1 (pp. 202–207)
1. $7^5 = 16,807$ **3. a.** 2 **b.** -8 and 8 **5.** 1 **7.** -6
9. False **11. a.** See below. **b.** The domains of f and g can be restricted to the set of all nonnegative real numbers.
13. True **15.** 1.260 units **17. a.** See below. **b.** $(x, y) \to (x - 2, y + 3)$ **c.** $(x, y) \to (x + 2, y - 3)$ **19.** They are not inverse functions. Sample: The point $(2, 1)$ is on the graph of $f(x)$, but the point $(1, 2)$ is not on the graph of $g(x)$. **21. a.** 3-bedroom homes in the city. **b.** 3-bedroom homes in a 30-block region on one side of the city **23. a.** $z = 4$ **b.** $y = 5$
25. mean = 22.7; standard deviation = 6.5; The mean and standard deviation are both higher than the national figures.

11. a.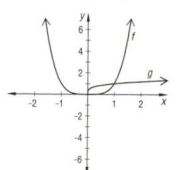

f ; g

17. a.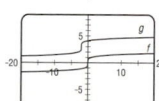

g ; f

LESSON 4-2 (pp. 208–214)
1. a. 4 **b.** 512 **c.** $\frac{1}{8}$ **d.** $\frac{1}{16}$ **3.** $k^{3/10}$ **5.** $(\sqrt[5]{a})^7$ or $\sqrt[5]{a^7}$ **7.** $\frac{1}{x^{1/2}}$
9. $0.1^{0.3}$ **11. a,b.** See below. **c.** $y = x^{2/3}$ **13. a.** $V = \left(\frac{A}{6}\right)^{3/2}$
b. 13.824 cm^3 **15. a.** -5 **b.** 5 **c.** No **17. a.** See above. **b.** False **19.** $3^4, 4^3, \frac{3}{4}, 3^{1/4}, 3^4$
21. a. $f^{-1}(x) = (x + 2)^3$ **b.** See above. **c.** the set of all real numbers **23. a.** In 1986, Tennessee farmers harvested 567 pounds of cotton per acre from about 335,000 acres, for which they received about 49.0 cents/lb. As a result, farms in Tennessee were valued at approximately $93,000,000.
b. 1986; the yield per acre was lower than in 1987 and 1988. The acreage harvested was much lower than 1987 and 1988. This caused farm value in 1986 to be much lower than in the other years.
11. a,b.

$y = x^{5/2}$; $y = x^{2/5}$

17. a.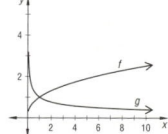

g ; f

21. b.

f ; f^{-1}

LESSON 4-3 (pp. 215–219)
1. about 113 **3. a.** Yes **b.** No **c.** No **d.** Yes
5. a. ≈ 0.54 **b.**
7. False **9.** Sample: Exponential functions with bases in the range $0 < b < 1$ are always decreasing. As x gets smaller, $f(x)$ grows without bound. As x gets larger, $f(x)$ remains positive but approaches zero. Exponential functions with $b > 1$ are always increasing. As x gets larger, $f(x)$ grows without bound. As x gets smaller, $f(x)$ remains positive but approaches zero.

5. b.

$y = 10^x$

11. a.

g ; f

b. $g(x) = 3^x$ **c.** $(x, y) \to (-x, y)$ **13. a.** 1.62 kg
b. $2(0.9)^t$ **15.** 8 **17.** 16
19. a. 900 **b.** 300 **c.** $P = \frac{2700}{3\sqrt{3}}$
21. a. $(f \circ g)(t) = 3t^2 + 6t + 13$ **b.** domain = the set of all real numbers; range = $\{x : x \geq 10\}$

LESSON 4-4 (pp. 220–224)
1. False **3. a.** 37 **b.** 1.32 **5. a.** 20 $= ab^3$; 40 $= ab^5$
b. $f(t) = \sqrt{50}(\sqrt{2})^t$ **c.** $\sqrt{50}$ **d.** $\sqrt{2}$ **7. a.** $A(t) = m(0.5)^{t/1620}$
b. about 0.54 kg **9.** about 89% **11.** 2.1%
13. a.

Minutes (y-axis) vs Number of people (x-axis)

b. sample: $y = 2x - 3$ (on graph) **c.** sample: (using $(4, 3.4)$ and $(7, 7.4)$); $f(x) = 1.2(1.3)^x$ **d.** the exponential model, because the errors are much smaller than the errors using the linear model
15. about $0.001A_0$ **17. a.** the set of all real numbers **b.** the set of all positive real numbers **c.** $y = 0$ **19.** $\frac{1}{16}$
21. a. **b.** $x = y^2 + 3$ **c.** No

$y = x^2 + 3$; $x = y^2 + 3$

LESSON 4-5 (pp. 225–229)
1. (d) **3.** $6^3 = 216$ **5.** 4 **7.** -1 **9.** 3 **11.** -2
13. ≈ 1321 **15. a.** the set of all positive real numbers **b.** True **c.** the y-axis **17.** about 7.1 **19.** (c) **21.** about 75.3 Haugh units **23.** 2 **25. a.** $P = 6,900,000(1.026)^y$
b. about 9,400,000 people **27.** $P(t) = 14.6(2)^{t/28}$ where $P(t)$ is in millions of people and t is years after 1980.
29. a. $d = \sqrt{\frac{k}{P}}$ **b.** $d = \left(\frac{k}{P}\right)^{1/2}$

LESSON 4-6 (pp. 230–235)
1. a. i. 2.716924 **ii.** 2.718281 **b.** True **3. a.** $1058.50
b. $1057.64 **5.** ≈ 0.45 **7.** ≈ -0.8 **9.** 0 **11.** 5
13. a. ln 17 **b.** log 17 ≈ 1.2, ln 17 ≈ 2.8 **15. a.** about 4,065,000 **b.** This prediction is slightly higher than the prediction in Lesson 4-3. The estimate in Lesson 4-3 does not assume continuous growth. **17. a.** 6.2 **b.** 8.1 **c.** ln $(e^a) = a$; $e^{\ln a} = a$ **d.** the part of the line $y = x$ in the first quadrant
19. about 604 days **21.** $g^{-1}(x) = 6^x$ **23.** 27.8576 **25.** $<$
27. mean = 49.1; standard deviaton = 10.3
29. a. See above.

29. a.

$y = e^x$; $y = 1 + x$

LESSON 4-7 (pp. 236–240)
1. 1 **3.** 3 **5.** 2.112 **7.** 0.8922 **9.** $a = \frac{b^3}{c^5}$
11. a. i. sample: overflow error **ii.** 2.48×10^{111}
iii. $1.64 \times e^{256}$ **b.** Method **ii** is best because the method in **i** gives an overflow error with most scientific calculators and the method in **iii** gives an answer that is more difficult to interpret.
c. 112 **13.** $\log_b \frac{1}{n} = \log_b 1 - \log_b n = 0 - \log_b n$. Therefore, $\log_b \frac{1}{n} = -\log_b n$. **15.** 10 dB **17.** The decibel level for $10P$ is 10 more than the decibel level for P. **19. a.** ii **b.** ii **c.** i
21. 5 **23. a.** $A = 5, b = 0.56$ **b.** ≈ 1.58 liters **25. a.** (b)

LESSON 4-8 (pp. 241–246)
1. a. $x \approx 2.465$; It checks. **b.** $x \approx 1.75$; $7^{1.75} \approx 30$
5. ≈ 1.67 **7.** 2017 **9.** about 9.9 years **11. a.** 5 **b.** $\frac{1}{5}$
c. They are reciprocals. **d.** $\log_6 7776 = \frac{\log 7776}{\log 6}$
$\log_{7776} 6 = \frac{\log 6}{\log 7776}$ **13. a.** In Example 4, $t_{0.5} = 5730$ and $r \approx -0.000121$.
Substituting $\frac{P}{A(t)}$ for $\frac{N_0}{N}$ gives
$\frac{t_{0.5}}{0.693} \ln\left(\frac{P}{Pe^{-0.000121t}}\right) = \frac{t_{0.5}}{0.693} \ln\left(\frac{1}{e^{-0.000121t}}\right)$
$= \frac{t_{0.5}}{0.693} \ln\left(e^{0.000121t}\right)$
$= \frac{t_{0.5}}{0.693} \cdot 0.000121t$
$= \frac{5730}{0.693} \cdot 0.000121t$
$= t$. It checks.
b. Use the Continuous Change Formula $A(t) = Pe^{rt}$.
From the half-life of the substance, $A(t_{0.5}) = \frac{1}{2}P$. So, $\frac{1}{2}P = Pe^{rt_{0.5}}$, and therefore $\frac{1}{2} = e^{rt_{0.5}}$. Taking logs, $\ln \frac{1}{2} = \ln (e^{rt_{0.5}})$ or $-0.693 = rt_{0.5}$. So, $r = \frac{-0.693}{t_{0.5}}$. Substituting into the original formula, $A(t) = Pe^{(-0.693/t_{0.5})t}$, and therefore $e^{(0.693/t_{0.5})t} = \frac{P}{A(t)}$.
Take the natural logarithm of both sides to get $\frac{0.693}{t_{0.5}} t = \ln\left(\frac{P}{A(t)}\right)$. Solving for t yields $t = \frac{t_{0.5}}{0.693} \cdot \ln\left(\frac{P}{A(t)}\right)$ as required.
15. 1 **17.** 0.03 **19.** (c) **21.** about 5 seconds
23. a. $y = 2.22x + 5.2$ **b.** For every increase of one year in age, there is an increase, on average, of 2.22 inches in length. **c.** about 31.8 inches **d.** The growth of the catfish might tend to slow as the fish ages.

LESSON 4-9 (pp. 247–252)

1. $\log k = 0.303$, $\log b = 0.121$ **3.** about 57 cm² **5. a.** True **b.** linear **7.** $C = 1995.3 \cdot 15.8^t$ **9.** about 45 sec **11. a.** $\log P = 0.01282y + 6.7255$ **b.** No, it predicts slightly below 250 million. **c.** sample: The Civil War and World War I, less immigration during this period **d.** $P = (5,314,960)10^{0.01282y}$ or $5,314,960(1.03)^y$ **13. a.** $P = 36.2 \log t + 30.1$ **b.** about 62 seconds **c.** It is the same answer. The answer doesn't seem to depend on whether common or natural logarithms are used. **15.** $t \approx 3.684$ Check: $5^{3.684} \approx 376$ **17.** $t = \frac{\ln 3}{.01r}$ **19.** False **21. a.** $y = 0.40625x^2 - 1.0625x + 3.45625$ **b.** (0, 3.45625); (8, 20.95625); (10, 33.45625) **c.** exponential

LESSON 4-10 (pp. 253–258)

1. $\frac{x^3}{5 \cdot 4 \cdot 3 \cdot 2 \cdot 1}$, $\frac{x^6}{6 \cdot 5 \cdot 4 \cdot 3 \cdot 2 \cdot 1}$ **3.** $\ln (0.9) = (-0.1) - \frac{(-0.1)^2}{2} + \frac{(-0.1)^3}{3} - \frac{(-0.1)^4}{4} + \frac{(-0.1)^5}{5} - \dots$ **5.** 4.12713 **7.** $11^{2.6} = e^{\ln 11^{2.6}} = e^{2.6 \ln 11} \approx 510.1$

9. a.

Current age (years) / Expected Lifespan (years)

b. $y = ab^t$ **c.** $y = 72.6(1.002)^t$ **d.** the model in part **c** **e.** Both models predict an expected lifespan of less than 90 years. Neither model is suitable for extrapolation. **11.** $x \approx 1.44$

13.

15. 2

17. a.

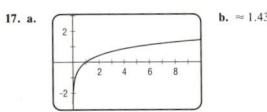

b. ≈ 1.43

19. a. $\frac{605}{18} \pi$ in.² or about 105.6 in.²

b. $\frac{55}{9} \pi$ in. or about 19.2 in.

CHAPTER 4 PROGRESS SELF-TEST (p. 262)

1. $64^{2/3} = (\sqrt[3]{64})^2 = 4^2 = 16$ **2.** $\sqrt[5]{x^{10}y^3} = (x^{10}y^3)^{1/5} = (x^{10})^{1/5}(y^3)^{1/5} = x^{10/5}y^{3/5} = x^2y$ **3.** $\left(\frac{9}{25}\right)^{-1/2} = \left(\frac{25}{9}\right)^{1/2} = \frac{25^{1/2}}{9^{1/2}} = \frac{\sqrt{25}}{\sqrt{9}} = \frac{5}{3}$ **4.** $\log 1 = 0$ because $10^0 = 1$. **5.** $\ln e^2 = 2$ $\ln e = 2 \cdot 1 = 2$. **6.** $\log_3 27 = 3$ because $3^3 = 27$. **7.** Since $x^{1/4}$ means an even root, the domain is all non-negative reals; $x^{1/4}$ is always a non-negative number, so the range is also all non-negative reals. **8.** $7^x = 2401 \Rightarrow \log 7^x = \log 2401 \Rightarrow x \log 7 = \log 2401 \Rightarrow x = \frac{\log 2401}{\log 7} = 4$ **9.** $5^x = 47 \Rightarrow \log 5^x = \log 47 \Rightarrow x \log 5 = \log 47 \Rightarrow x = \frac{\log 47}{\log 5} \approx 2.392$. **10.** Using the model $P(t) = ab^t$, where $P(t)$ is the population t years after 1960, a is the initial (1960) population, and b is the annual growth factor, $P(0) = ab^0 = a = 180 \times 10^6$. Then, in 1970, $P(10) = 180 \times 10^6 \cdot b^{10} = 209 \times 10^6$. Thus $b^{10} = \frac{209 \times 10^6}{180 \times 10^6}$. $1.161 \Rightarrow b = 1.161^{1/10} \approx 1.015$. Thus, $P(t) = (180 \times 10^6)(1.015)^t$. The population in 1990 is $P(30) = (180 \times 10^6)(1.015)^{30} \approx 281 \times 10^6$. **11.** Using the model $S(t) = ab^t$ where t is the time after initial inspection, a is the initial amount, and b is the rate of decay, $S(25) = 10b^{25} = 5$. Then $b^{25} = .5$ or $b \approx .973$. In 30 years, $S(30) = 10(.973)^{30} \approx$ 4.35 grams. **12.** $\log_5 16 = \frac{\log 16}{\log 5} \approx \frac{1.204}{.699} \approx 1.723$. **13. a.** $y = b^x$ and $\log_b x = y$ are inverse functions so $g^{-1}(x) = \log_3 x$ **b.** $g^{-1}(8) = \log_3 8 = \frac{\log 8}{\log 3} \approx 1.9$ **14.** $\frac{1}{2}\log a + \log b = \frac{1}{2}\log c \Rightarrow \log a^{1/2} + \log b = \log c^{1/2} \Rightarrow \log(a^{1/2}b) = \log c^{1/2} \Rightarrow a^{1/2}b = c^{1/2}$ **15.** $b^{\log_b x} = x$ for any positive b and x. So $10^{\log 4} = x$ and $e^{\ln x} = x$. True.

16. a, b.

17. $L = \frac{k}{R^2} \Rightarrow R^2 = \frac{k}{L} \Rightarrow R = \sqrt{\frac{k}{L}}$ **18.** (b), (d) **19.** (c) **20. a.** $\ln y = \ln 80 + t \ln(0.875) \Rightarrow e^{\ln y} = e^{(\ln 80 + t \ln(0.875))} = e^{\ln 80} \cdot e^{t \ln(0.875)} = e^{\ln 80} \cdot (e^{\ln 0.875})^t \Rightarrow y = 80 \cdot 0.875^t$ **b.** 27.5°

The chart below keys the **Progress Self-Test** questions to the objectives in the **Chapter 4 Review** on pages 263–266. This will enable you to locate those **Chapter 4 Review** questions that correspond to questions you missed on the **Progress Self-Test**. The lesson where the material is covered is also indicated in the chart.

Question	1	2	3	4	5	6	7	8	9	10	11
Objective	A	A	A	C	C	C	D	B	B	G	G
Lesson	4-2	4-1	4-2	4-5	4-6	4-5	4-3	4-8	4-8	4-3	4-3
Question	12	13	14	15	16	17	18	19	20		
Objective	C	D	E	E	I	F	J	J	H		
Lesson	4-6	4-5	4-7	4-8	4-5	4-1	4-5	4-9			

CHAPTER 4 REVIEW (pp. 263–266)

1. 5 **2.** 3125 **3.** $\frac{1}{25}$ **4.** –5 **5.** $\sqrt[5]{a^7}$ or $(\sqrt[5]{a})^7$ **6.** $\frac{1}{\sqrt[7]{7}}$ or $\frac{1}{(\sqrt[7]{7})}$ **7.** $s^{1/5} \cdot 7^{7/5}$ **8.** $a^{-1/3}b^{-1/2}$ **9.** about 60 **10.** about 0.33 **11.** < **12.** > **13.** ≈ 2.08 **14.** – 1.37 **15.** $m \geq 1.45$ **16.** ≈ 12.7 **17.** 4 **18.** – 3 **19.** – 1.73 **20.** 11 **21.** $\frac{4}{3}$ **22.** 5.3 **23.** $\log 5, \log_3 5, \ln 5, \log_2 5$ **24.** ≈ 0.461 **25.** ≈ 2.39 **26. a.** the set of all real numbers **b.** the set of all positive real numbers **27. a.** $b > 1$ **b.** $0 < b < 1$ **28.** $y = 0$ **29.** $e = \lim_{n \to \infty} \left(1 + \frac{1}{n}\right)^n$ **30.** (0, 1) **31.** k models exponential growth; m models exponential decay. **32.** (b) **33. a.** the set of all positive real numbers **b.** the set of all real numbers **34.** ≈ 1.969 **35.** ≈ 0.353 **36.** ≈ 0.239 **37.** $y = 17x^3$ **38.** $\log A = 3\log C - \log D$ **39.** $\ln y = \ln 8 + (\ln 5)x$ **40.** 189 **41.** True **42.** True **43.** $\log_a \frac{x}{y} + \log_a \frac{y}{z} + \log_a \frac{z}{x} = (\log_a x - \log_a y) + (\log_a y - \log_a z) + (\log_a z - \log_a x) = \log_a x - \log_a x + \log_a y - \log_a y + \log_a z - \log_a z = 0$ **44.** $N = k^3 T^3$ **45.** $d = \sqrt[3]{\frac{k}{W}}$ **46. a.** $T(4) = 0.76T_0$ **b.** $d \approx 53$ days **c.** 24 days **47. a.** $P = 4$ **b.** 6.4 **c.** 2025 **48.** $P = 3 \cdot 2^d$ **49.** $y \approx 9.99(1.45)^x$ **50. a.** $I = 100(.325)^d$ **b.** about 57 units **c.** about 5.5 meters **51.** (a) **52. a.** $P = (24.3)e^{0.035x}$, in millions **b.** about 38.3 million **c.** 1993 **53.** 4.0% **54.** 3.4% **55. a.** $L = A(.841)^b$ or $L = Ae^{.25\ln(0.5) \cdot b}$ **b.** about 30% **56.** about 21,450 **57.** about 74 pounds **58.** $b = 0.126(10)^{0.38c}$ **59.** $T = (.179)a^{3/2}$

60. a.

Week / Average score

b. (iii)

60. c.

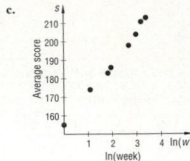
ln(week) / Average score

d. $s = 16.783 \ln w + 74.882$ **e.** about 140 **61. a.** (i) **b.** about 30 minutes

62. a.

b. h **c.** g

63. a.

b. $\sqrt{10} \approx 3.2$; $10^{1.6} \approx 39.8$ **64.** (a) and (c) **65.** They are reflections of each other over the line $y = x$. **66. See below. 67. a.** linear **b. See below. 68.** (1, 0) **69.** (a) **70.** (c)

66.

67. b.

LESSON 5-1 (pp. 268–273)

1. 120° **3.** 154.8° **5. a.** π units **b.** π radians **7. a. See below. b.** 60° **9.** $\frac{-5\pi}{4}$ **11.** 6.458 radians **13.** – 240.642° **15.** 57 **17.** 22° 22′ 30″ **19.** faster **21.** x radians $\cdot \frac{180°}{\pi \text{ radians}} = \frac{180x}{\pi}$ degrees **23.** $y = \frac{1}{2}x^2 + 1$ **25.** The area of circle 2 is not twice the area of circle 1. When measuring the amount of spaghetti needed for 2 portions with these instructions, the result will not be equal to twice the amount needed for one portion. **27. a.** 5π cm **b.** 25π cm²

7. a.

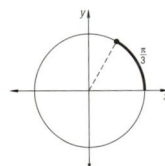

LESSON 5-2 (pp. 274–278)

1. $\frac{15\pi}{4}$ square units **3.** 6π cm **5.** $\frac{32}{15}\pi$ cm **7.** $\frac{15\pi}{2}$ ft² **9.** 218 cm **11.** $\frac{25\pi}{6}$ square units **13. a.** 22π inches **b.** 3300π inches **c.** about 9.8 miles per hour **15.** $\frac{\pi}{2}$ radians **17.** – 30° **19.** 48° **21.** 15a **23. a.** $\frac{11}{2}$ **b.** $\frac{11\sqrt{3}}{2}$

LESSON 5-3 (pp. 279–285)

1. $\frac{5}{7}$ **3.** $\frac{2\sqrt{6}}{5}$ **5.** 0.99 **7.** 0.30 **9.** $y = (\tan 30°)x + 4 \approx 0.577x + 4$ **11.** $\approx 46.4°$ **13. a.** $\sin M = \frac{1}{\sqrt{2}}$ or $\frac{\sqrt{2}}{2}$, $\cos M = \frac{1}{\sqrt{2}}$ or $\frac{\sqrt{2}}{2}$, $\tan M = 1$ **b.** $\sin M = \sin 45° = 0.707 \approx \frac{1}{\sqrt{2}}$, $\cos M = \cos 45° = 0.707 \approx \frac{1}{\sqrt{2}}$, $\tan M = \tan 45° = 1$ **15.** about 47.7 ft **17. a.** 13,800 feet **b.** maximum $\approx$ 16,600 feet; minimum $\approx$ 11,800 feet **19. a.** 1.3 radians **b.** about 74.5° **c.** $2\pi - 1.3$ radians **d.** ≈ 0.65 square units **21.** IV **23.** II **25.** $(-m, n)$

LESSON 5-4 (pp. 286–291)

1. – 0.985 **3.** – 4.381 **5.** 1 **7.** undefined **9.** B **11.** D **13.** III **15. a.** the set of all real numbers **b.** $\{y : -1 \leq y \leq 1\}$ **17.** sample: $\theta = 0, 2\pi, -8\pi$ **19.** 0.8 **21.** decreases **23.** about 48 miles east and 20 miles north of its port **25.** about 19 times **27.** $\ln 3125$ **29.** two angles whose sum is 180°

LESSON 5-5 (pp. 292–297)

1. a. $\sqrt{3}$ and 2 **b.** False **3.** $-\frac{\sqrt{2}}{2}$ **5.** $-\frac{1}{2}, -\frac{\sqrt{3}}{2}$ **9.** $\frac{1}{2}$ **11.** $150\sqrt{3}$ cm² **13. a.** $\theta = -\frac{7\pi}{4}, -\frac{3\pi}{4}, \frac{\pi}{4}, \frac{5\pi}{4}$ **b.** 1 **15.** $\sqrt{\frac{2}{3}}$

---

17. In the diagram, let P be the point (1, 0) and Q the intersection of $\overline{AB}$ and the y-axis. Then, $m\angle POA = \frac{\pi}{3}$, $m\angle POB = \frac{2\pi}{3}$, and $m\angle POQ = \frac{\pi}{2}$. So, $m\angle AOQ = m\angle POQ - m\angle POA = \frac{\pi}{2} - \frac{\pi}{3} = \frac{\pi}{6}$. Also, $m\angle QOB = m\angle AOB - m\angle AOQ = \frac{\pi}{3} - \frac{\pi}{6} = \frac{\pi}{6}$. Now A and B are the images of P under the rotation about the origin, therefore $OA = OB = OP = 1$. It follows that $\triangle AOQ \cong \triangle BOQ$ by SAS because $\overline{OA} \cong \overline{OB}$, $\angle AOQ \cong \angle BOQ$, and $\overline{OQ} \cong \overline{OQ}$. This implies that $BQ = AQ$ and $m\angle AQO = m\angle BQO$. But $m\angle AQO + m\angle BQO = m\angle AQB = \pi$, therefore $m\angle AQO = m\angle BQO = \frac{\pi}{2}$. So, the y-axis is the perpendicular bisector of $\overline{AB}$. **19. a.** if $\cos \theta \neq 0$, $\cos \theta \cdot \tan \theta = \cos \theta \cdot \frac{\sin \theta}{\cos \theta} = \frac{\cos \theta}{\cos \theta} \cdot \sin \theta = \sin \theta$ **b.** $\tan \theta$ would be undefined **21.** about 64.3 m **23.** about 1041 $\frac{\text{miles}}{\text{hour}}$ **25.** $y = \frac{c^n}{3}$ **27.** $\sqrt{170}$

LESSON 5-6 (pp. 298–303)

1. a.

b.

3. sample: $1.016 + 2\pi \approx 7.299$, $1.016 - 4\pi \approx -11.550$, $1.016 + 10\pi \approx 32.432$. **5.** sample: $\theta = 1.250$, $\theta = 1.250 + 2\pi$, $\theta = 1.250 - 4\pi$ **7.** sine, cosine **9.** sine **11.** cosine **13.** $\approx 1.249, \approx 4.391, \approx 7.532, \approx 10.674$ **15.** $(x, y) \rightarrow \left(x - \frac{\pi}{2}, y\right)$ **17. a, b. See page 867.** **c.** The graph of f has the same period as $y = \sin \theta$, but its range is $\{y : -2 \leq y \leq 2\}$. The graph of g has the same range as $y = \sin \theta$, but its period is π. **19.** $\left(-\frac{\sqrt{2}}{2}, -\frac{\sqrt{2}}{2}\right)$ **21.** $-\frac{1}{2}$ **23.** $\frac{\sqrt{3}}{3} - 1$

---

17. a, b.

LESSON 5-7 (pp. 304–309)

1. True

3.

$\cos (40°) = \cos (-40°)$, $(\cos (40°), \sin (40°))$, $(\cos (-40°), \sin (-40°))$

5. $-\frac{3}{7}$ **7.** $-\frac{9}{24}$ **9.** – 0.375 **11. a.** False **13. a.** True **b.** Opposites Theorem

15. a.

b. The two graphs are reflection images of each other over the y-axis and over the x-axis. **c.** $\tan(-\theta) = -\tan \theta$ (Opposites Theorem)

17. a.

(graph)

b. The two graphs are identical. **c.** $\cos\left(\frac{\pi}{2} - x\right) = \sin x$ (Complements Theorem) **19.** 0 **21.** sample: $\theta = 0.2013$ and $\theta = 2.9403$ **23.** $\frac{\sqrt{3}}{2}$ **25. a.** about $(-0.7314, -0.6820)$ **b.** 0.9325

LESSON 5-8 (pp. 310–314)

1. ≈ 105 ft **3. a.** $\tan C$ **b.** $\tan C$ **c.** A, D, and C are all points on a line through point C whose slope is $\tan C$. Since there is only one such line, the three points are collinear. **5.** $\approx 32.1°$ **7.** (a) **9.** (c) **11.** ≈ 16.2 **13. a.** Let C be the largest angle. By the Law of Cosines, $(15)^2 = (5)^2 + (9)^2 - 2(5)(9)\cos C$. This implies that $\cos C = -\frac{119}{90}$. But this is impossible; therefore there is no such triangle. **b.** By the Triangle Inequality, the sum of any two sides of a triangle must be greater than the third side. But $5 + 9 < 15$, so there is no such triangle. **15.** Periodicity Theorem **17.** 30°, 330° **19.** $\frac{1}{2}$ **21.** $\sqrt{3}$ **23.** $\frac{\pi}{3}, \frac{5\pi}{3}$ **25.** ≈ 10.34 square units

LESSON 5-9 (pp. 315–320)

3. ≈ 7.3 units **4.** ≈ 12.2 **5.** $a = 6\sqrt{2}$, $b = 6\sqrt{3} + 6 \approx 16.4$ **7. a.** $\approx 34.4°$ **b.** ≈ 565 ft **9.** False **11.** (a) **13.** (b) **15.** (c) **17.** 756 m **19.** ≈ 14.2 **21.** (c) **23. a.** 4 **b.** 4 **25.** $-\frac{1}{2}$

LESSON 5-10 (pp. 321–328)

1. (a) **3.** Greenwich meridian or prime meridian **5.** the equator **7.** ≈ 190 miles **9.** about 20 minutes **11. a.** ≈ 4796 miles **b.** ≈ 340 miles **d.** $\approx 7\%$ **13.** $\approx \pm .917$ **15.** –.4 **17.** $XZ \approx .59$, $YZ \approx .46$ **19.** ≈ 8.04 **21.** domain: the set of all real numbers; range: $\{y : -1 \leq y \leq 1\}$ **23. a.** $f(g(x)) = 2\pi x^2$ **b.** $g(f(x)) = 4\pi^2 x^2$ **25.** $(x, y) \rightarrow (x - 2, y - 5)$

CHAPTER 5 PROGRESS SELF-TEST (pp. 332–333)

1. The conversion factor is $\frac{180°}{\pi \text{ radians}}$. So 1.4 radians $\cdot \frac{180°}{\pi \text{ radians}} \approx$ 80.21°. **2.** $67° \cdot \frac{\pi \text{ radians}}{180°} = \frac{67\pi}{180}$ radians ≈ 1.17 **3.** The arc between Adelaide and the South Pole has a central angle of $90° - 34° 55' = 55° 05' = 55\frac{5}{12}° \cdot \frac{\pi \text{ radians}}{180°} \approx .961$ radians. So, $s = r\theta = 6400 \cdot .961 = 6150$ km. **4.** $r = 12$ and $\theta = \frac{5\pi}{4}$, so $A = \frac{1}{2}r^2\theta = \frac{1}{2}(12)^2 \cdot \frac{5\pi}{4} = 90\pi$ cm² ≈ 282.7 cm² **5.** $r = 20$ and $\theta = 40° = 40 \cdot \frac{\pi}{180} = \frac{2\pi}{9}$ radians, so $A = \frac{1}{2}r^2\theta = \frac{1}{2}(20)^2\left(\frac{2\pi}{9}\right) = \frac{400\pi}{9}$ m² ≈ 139.6 m²

6. Using a calculator, $\sin 47° \approx 0.731$. **7.** Using a calculator, $\tan(-3) \approx 0.143$. **8.** Using a calculator, $\cos 16°35' \approx 0.958$. **9. a.** the set of all real numbers **b.** $\{y: -1 \leq y \leq 1\}$ **10.** By definition, $P = (x, y) = \left(\cos\frac{4\pi}{5}, \sin\frac{4\pi}{5}\right) \approx (-.81, .59)$.

11. $\sin M = \frac{\text{opposite}}{\text{hypotenuse}} = \frac{15}{17}$ **12.** $\tan B = \frac{\text{opposite}}{\text{adjacent}} = \frac{8}{15}$ **13.** $m\angle B = \tan^{-1}\left(\frac{8}{15}\right) \approx 28.1°$ **14.** The sine function has a period of 2π, so $\sin(\theta - \pi) = -\sin\theta$ for all θ. Then $\sin(\theta - 2\pi) = 0.80$. **15.** By the Complements Theorem, $\cos\left(\frac{\pi}{2} - \theta\right) = \sin\theta = 0.80$. **16.** By the Opposites and Complements Theorems, $\sin(\theta - \pi) = -\sin(\pi - \theta) = -\sin\theta = -0.80$. **17.** Since $\sin^2\theta + \cos^2\theta = 1$, $\cos\theta = \sqrt{1 - \sin^2\theta} = \sqrt{1 - 0.80^2} = \sqrt{0.36} = 0.60$. **18.** $\theta = \sin^{-1}(0.80) \approx .93$ radian or $53°$ **19.** $\sin\frac{3\pi}{4} = \frac{\sqrt{2}}{2}$

20. $\tan\frac{\pi}{3} = \frac{\sin\frac{\pi}{3}}{\cos\frac{\pi}{3}} = \frac{\frac{\sqrt{3}}{2}}{\frac{1}{2}} = \sqrt{3}$ **21. a.** See right; The cosine function has a period of 2π, so $\cos(2\pi - \theta) = \cos\theta$ for all θ. Then the graph of $f(x) = \cos(2\pi - \theta)$ is identical to that of $\cos\theta$. **b.** 2π **c.** The cosine function

The chart below keys the **Progress Self-Test** questions to the objectives in the **Chapter 5 Review** on pages 334–338. This will enable you to locate those **Chapter 5 Review** questions that correspond to questions you missed on the **Progress Self-Test**. The lesson where the material is covered is also indicated in the chart.

Question	1	2	3	4	5	6	7	8	9	10	11
Objective	A	A	H	E	H	B	C	B	F	K	B
Lesson	5-1	5-1	5-2	5-2	5-2	5-3	5-4	5-3	5-6	5-4	5-3
Question	12	13	14	15	16	17	18	19	20	21	22
Objective	B	D	G	G	G	G	B	C	B	L	I
Lesson	5-3	5-3	5-6	5-7	5-7	5-7	5-5	5-5	5-6	5-3	
Question	23	24	25								
Objective	I	J	J								
Lesson	5-3	5-9	5-8								

CHAPTER 5 REVIEW (pp. 334–338)

1. a. $72°$ **b.** $\frac{2\pi}{5}$ **2. a.** $-240°$ **b.** $-\frac{4\pi}{3}$ **3.** $\frac{\pi}{6}$ **4.** $\frac{3\pi}{2}$ **5.** $105°$ **6.** $-22.5°$ **7.** -5.463 **8.** 0.838 **9.** $114.59°$ **10.** $-492.74°$ **11.** $\frac{3}{8}$ **12.** $2\frac{3}{4}$ **13.** $16.242°$ **14.** $103°25'12''$ **15. a.** $\frac{24}{25}$ **b.** $\frac{7}{25}$ **c.** $\frac{7}{24}$ **16.** $\frac{\sqrt{3}}{2}$ **17.** $\frac{\sqrt{2}}{2}$ **18.** $\frac{\sqrt{3}}{3}$ **19.** 1.965 **20.** $.287$ **21.** $.137$ **22.** $.33$ **23.** $.46$ **24.** $\frac{2}{3}$ **25.** $.14$ **26.** $-.49$ **27.** $.55$ **28.** $-.97$ **29.** $-\frac{\sqrt{2}}{2}$ **30.** $-\frac{1}{2}$ **31.** -1 **32.** 1 **33.** $(-.37, .93)$ **34.** ≈ 1.25, ≈ 4.39 **35.** $77.6°, 282.4°, 437.6°, 642.4°$ **36.** $-\frac{\pi}{6}, \frac{7\pi}{6}, \frac{11\pi}{6}$ **37.** ≈ 9.33 **38.** ≈ 39.44 **39.** $\approx 82.82°$ **40.** $\approx 34.75°$ **41.** $22.6°, 67.4°, 90°$ **42.** 8.0 units **43.** ≈ 20.67 cm^2 **44.** ≈ 179.76 ft^2 **45.** 6π in. **46.** 18 ft **47.** $\frac{25\pi}{3}$ m^2 **48.** $152.8°$ **49. a.** the set of all real numbers **b.** $\{y: -1 \leq y \leq 1\}$ **50.** odd multiples of $\frac{\pi}{2}$ **51.** True **52.** True **53.** (d) **54.** $\frac{\pi}{2} < \theta < \pi$ **55.** (a) **56.** False **57.** True **58.** False **59. a.** $\pm\frac{\sqrt{15}}{4}$ **b.** $\pm\sqrt{15}$ **60.** sample: $-2\pi + 2.45 \approx -3.83$; $-\pi + 2.45 \approx -.69$; $\pi + 2.45 \approx 5.59$ **61.** $\approx .469$ **62.** $\approx .469$ **63.** $\approx -.469$ **64.** $\approx .469$ **65.** $-\sin\left(\frac{\pi}{2} - \theta\right) = -\cos\theta$ (by the Complements Theorem) $= \cos(\pi - \theta)$ (by the Supplements Theorem)

868

---

22. $\cos\theta = \frac{\text{adjacent}}{\text{hypotenuse}} = \frac{2}{9}$, so $\theta = \cos^{-1}\left(\frac{2}{9}\right) \approx 77°$ **23.** $\tan 41° = \frac{h}{6} \Rightarrow h = 6\tan 41° \approx 5.2$ ft **24.** Using the Law of Sines, $\frac{\sin 15°}{x} = \frac{\sin 140°}{5}$. Then $x = \frac{5\cdot\sin 15°}{\sin 140°} \approx 2.0$. **25.** Using the Law of Cosines, $d^2 = 200^2 + 300^2 - 2(200)(300)\cos 55° \approx 61,171$. So, $d = \sqrt{61,171} \approx 247.3$ m.

21. a.

66. A formula for the area of a triangle, given two sides a and b and the included angle C, is $A = \frac{1}{2}ab\sin C$. Since $\sin 70° = \sin 110°$, the two triangles must have the same area. **67.** In the Law of Cosines, $c^2 = a^2 + b^2 - 2ab\cos C$. If $C = 90°$, then $\cos C = 0$ and $c^2 = a^2 + b^2$. **68.** Tom (This is the SsA condition.) **69.** ≈ 2094 m^2 **70.** $\frac{868\pi}{3}$ m^2 ≈ 909 m^2 **71.** 16 ft **72. a.** ≈ 13.8 mi **b.** ≈ 11 mph **c.** ≈ 4.56 hours after 1 P.M., at 5:33 and 36 seconds; around 5:30 P.M **73.** ≈ 9.5 m **74.** $\approx 56.4°$ **75. a.** ≈ 8.52 mi **b.** $.013$ hr ≈ 47 seconds **76.** ≈ 17.3 cm **77.** ≈ 105.2 m **78. a.** ≈ 10.03 mi **b.** ≈ 6.1 mi

84.

$(\cos\theta, \sin\theta)$
$(\cos(-\theta), \sin(-\theta))$

88. a

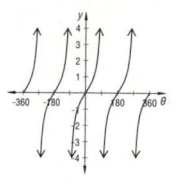

869

---

79. a. $122.1°$ **b.** ≈ 743.1 mm^2 **80.** (b) **81.** (g) **82.** (c) **83.** (h) **84.** See below left. **85.** $y = (\tan 20°)x + (3 + 2\tan 20°) \approx .364x + 3.728$ **86.** $18.4°$ **87. a.** See below. **b.** 2π **88. a.** See below left. **b.** $180°$ **c.** sample: $\theta = 90°$, $\theta = 270°$ **89. a.** $T(x, y) \rightarrow \left(x + \frac{\pi}{2}, y\right)$ **b.** $T(x, y) \rightarrow \left(x - \frac{\pi}{2}, y\right)$ **90. a.** See below. **b.** $y = \cos x$ and $y = \cos(-x)$ are identical, and each is the reflection image of $y = \cos(-x)$ over the x-axis. **c.** $\cos(-x) = \cos x$ represents the Opposites Theorem, and $\cos(\pi - x) = -\cos x$ represents the Supplements Theorem.

87. a.

90. a.

LESSON 6-1 (pp. 340–345)

1. a. $y = \frac{1}{3}\sin\frac{x}{4}$ **b.** See right. **c.** $\frac{1}{3}$ **d.** 8π **3. a.** 4 **b.** π **c. i.** 4 **ii.** 2 **5. a.** period $= \frac{2\pi}{5}$, amplitude $= \frac{1}{3}$ **b.** See right. **7.** (d) **9. a.** $g(x) = -\sin\left(\frac{x}{3}\right)$ **b.** period $= 6\pi$; amplitude $= 1$ **c.** See page 870. **11. a.** See page 870. **b.** $(x, y) \rightarrow \left(\frac{x}{2}, y\right)$ **c.** $(x, y) \rightarrow (x, 2y)$ **d.** f and h **e.** None of the functions have a maximum or minimum. **13. a.** domain $=$ the set of all real numbers; range $= \{y: 1 \leq y \leq 3\}$ **b.** maximum $= 3$; minimum $= 1$ **c.** 2 **15.** True

1. b.

5. b.

870

9. c.

11. a.

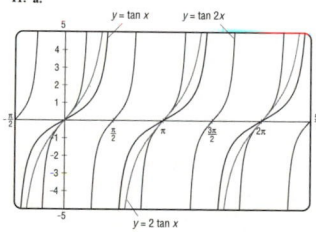
$y = \tan x$ $y = \tan 2x$
$y = 2\tan x$

LESSON 6-2 (pp. 346–351)

1. a. $-\frac{\pi}{2}$ **b.**

3.

5. (b) **7.** (d) **9.** sample: $y = \tan\left(x - \frac{\pi}{2}\right) + 1$ **11.** sample: $y = \sin x + 2$ **13.** samples: $y = -\sin x$; $y = -\sin(x - 2\pi)$; $y = \cos\left(x + \frac{\pi}{2}\right)$; $y = -\cos\left(x - \frac{\pi}{2}\right)$

15. a.

b. $\alpha = -\frac{\pi}{3}, \alpha = \frac{2\pi}{3}, \alpha = \frac{5\pi}{3}$

17. a.

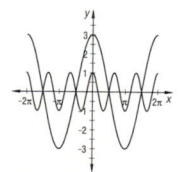

b. ≈ 1.57

19. $\frac{\sqrt{2}}{2}$ **21. a.** $9x^2 - 12xy + 4y^2$ **b.** $9x^2 - 12xy + 4y^2 - 114x + 76y + 361$ **c.** $9\sin^2 x - 12(\sin x)(\cos x) + 4\cos^2 x$

LESSON 6-3 (pp. 352–357)

1. scale changes and translations

3. a.

b. amplitude $= 1$; period $= 4\pi$; phase shift $= -\pi$. **c.** $(x, y) \rightarrow (2x, y)$ followed by $(x, y) \rightarrow (x - \pi, y)$ **5. a.** amplitude $= 2$; period $= 2$; phase shift $= 0$; vertical shift $= 1$ **b.** See page 871. **7.** ≈ 6.5 seconds **9.** $y = 4\sin 4(x - \pi)$ **11. a.** $(x, y) \rightarrow (3x + \pi, y)$ **b.** See page 871. **c.** period $= 3\pi$; phase shift $= \pi$ **13. a.** False **b.** $(S \circ T)(x, y) = S(x + h, y + k) = (a(x + h), b(y + k)); (T \circ S)(x, y) = T(ax, by) = (ax + h, by + k) \neq (a(x + h), b(y + k))$ **15. a.** See page 871. **b.** $0.45, 0.63, 2.24, 3.14$

---

17. a. $f(g(x)) = (x^{1/3} + 1)^3 - 1$ or $x + 3x^{2/3} + 3x^{1/3}$, $g(f(x)) = (x^3 - 1)^{1/3} + 1$ **b.** No **19. a.** See below. **b.** 392 feet **c.** after 6 seconds

5. b.

11. b.

15. a.

19. a.

LESSON 6-4 (pp. 358–363)

1. After substituting $x = 92$ (June 21 is the 92nd day after March 21) into the equation $\frac{y - 12.25}{3.75} = \sin\left(\frac{x}{365/2\pi}\right)$ the result is $y \approx 16.0$. After substituting $x = 172$ (June 21 is the 172nd day of the year) into the equation $\frac{y - 12.25}{3.75} = \sin\left(\frac{x - 80}{365/2\pi}\right)$, the result is $y \approx 16.0$. **3. a.** sample: $\frac{y - 12.25}{3.75} = \sin\left(\frac{x - 171.25}{365/2\pi}\right)$, where x represents the day of the year. **b.** about 15.9 hours **5.** ≈ 3180 km **7.** sample: $\frac{y}{4500} = \sin\left(x + \frac{15}{60\pi}\right)$ **9. a, b.** See next column. **c.** 12 **d.** sample : about 26 degrees **e.** (iii) **f.** sample: $\frac{y - 46}{-26} = \sin\left(\frac{x + 3}{6}\right)$

11. a. $\frac{1}{3}$ **b.** 10π **c.** 0 **d.** 0 **e.** See next column.

---

13. sample: $y = \tan\left(x + \frac{\pi}{3}\right)$ **15.** $0, \pi, 2\pi, 3\pi, 4\pi, 5\pi,$ or 6π **17.** 0.9 **19.** sample: See below.

9. a, b.

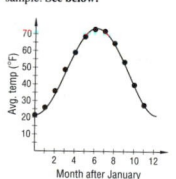
Avg. temp (°F)
Month after January

11. e.

19.

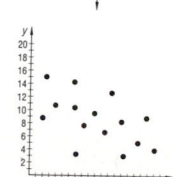

LESSON 6-5 (pp. 364–370)

1. Theta is the angle whose sine is k. **3. a.** $\frac{\pi}{6}$ **b.** $30°$ **5. a.** $\frac{-\pi}{4}$ **b.** $-45°$ **7.** ≈ 2.691 **9. a.** True **b.** False **11. a.** $\angle HSC = \sin^{-1}\left(\frac{r}{r + h}\right)$ **b.** $\approx 61.2°$ **c.** $\cos\alpha = \sin(\angle HSC)$ **13. a.** $y = \tan^{-1}\left(\frac{3}{x}\right)$ **b.** $y = \tan^{-1}\left(\frac{4}{x}\right) - \tan^{-1}\left(\frac{1}{x}\right)$ **15.** For x between -1 and 1, let $y = \sin^{-1} x$. By the definition of $\sin^{-1} x$, $\sin y = x$. Therefore, $\sin(\sin^{-1} x) = x$ for all x such that $-1 \leq x \leq 1$. **17. a.** domain $=$ the set of all real numbers except $\frac{\pi}{4} + n\pi$ where n is an integer **b.** range $=$ the set of all real numbers **c.** period $= \pi$ **d.** amplitude does not exist **19.** 4π in.^{2} **21.** $\frac{1}{2}, -1$

T59

23. a. about 101 **b.** February 1981 **c.** 313
d. Sample: The number of deaths by gas poisoning is probably about the same each winter (maximum values) and also about the same each summer (minimum values).
e. sample : amplitude ≈ 80; period ≈ 12 months **25.** 2

LESSON 6-6 (pp. 371–376)
1. a. ≈ 1.159 **b.** ≈ 1.159, ≈ 5.124 **c.** θ ≈ 1.159 + 2πn or θ ≈ 5.124 + 2πn for any integer n **3. a.** 70° **b.** 70°, 290° **c.** 430°, 650°, 790°, 1010° **5.** ≈ 1.772, ≈ 4.511
7. a. 0, $\frac{\pi}{3}$, $\frac{3\pi}{2}$ **b.** x = 2nπ or x = $\frac{\pi}{3}$ + 2πn or x = $\frac{5\pi}{3}$ + 2πn, for any integer n **9. a.** no solution **b. See below.** Sample: The graphs of y = 3 cos θ and y = 7 do not intersect, so there is no solution to 3 cos θ = 7. **11. a.** $\left|\frac{b}{a}\right| > 1$ **b.** $\left|\frac{b}{a}\right| = 1$
c. $\left|\frac{b}{a}\right| < 1$ **13.** $\frac{7\pi}{12}$ **15.** about 37° **17. a.** the set of all real numbers **b. See below.** **19.** sample: y = 2sin(8x)
21. a. sample: y = 17.5 cos $\left(\frac{\pi t}{60}\right)$ + 12.5 **b.** 10
c. i. (17.5 * (COS(3.14159265 * X/60)) + 12.5 ; 45
ii.

TIMBER	LENGTH
1	30.000
2	29.404
3	27.655
4	24.874
5	21.250
6	17.029
7	12.500
8	7.971
9	3.750
10	0.126

23. log₁₀8, log₆8, ln 8, log₂8

9. b.

17. b.

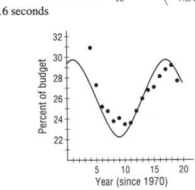

LESSON 6-7 (pp. 377–381)
1. a. $\sqrt{2}$ **b.** $\sqrt{2}$ **c.** 1 **3.** $\frac{-2\sqrt{3}}{3}$ **5.** $\frac{2\sqrt{3}}{3}$ **7.** −1
9. 14.9 cm **11. a.** $\frac{\pi}{2}$, $\frac{3\pi}{2}$ **b.** none **13. a.** $\frac{x}{y}$ **b.** $\frac{x}{x}$ **c.** $\frac{x}{y}$ **d.** $\frac{x}{y}$
15. about 31 feet tall
17. a.

b. π

19. a. $\frac{3\pi}{2}$ **b.** $\frac{3\pi}{2}$ + 2nπ for any integer n
21. a. $\frac{-16}{3}$, $\frac{-8}{3}$, $\frac{-4}{3}$, $\frac{4}{3}$, $\frac{8}{3}$, $\frac{16}{3}$
b.

23. 120° **25.** about 238 m² **27.** 55

LESSON 6-8 (pp. 382–387)
1. 2π **3.** 2π **5. a.** .7856 **b.** −.7847
c. .7939 **d.** −.7850 **7.** −2 **9. a.** even
b. sec(−x) = $\frac{1}{\cos(-x)}$ definition of secant
= $\frac{1}{\cos x}$ Opposites Theorem
= sec x   definition of secant
11. {y : y ≤ −1 or y ≥ 1} **13.** ≈ −2.78, ≈ −0.37, ≈ 0.59, ≈ 2.55 **15. a.** sample: y = 4sin(240πt), where t is time in minutes **b.** 0 cm **17. a. See below. b.** f(x): domain = the set of all real numbers x such that x ≠ nπ + $\frac{\pi}{2}$ for any integer n; range = the set of all real numbers; g(x): domain = the set of all real numbers; range = {y : −$\frac{\pi}{2}$ < y < $\frac{\pi}{2}$} **19. a.** d₁, d₃, d₂
b. d₃, d₁, d₂

17. a.

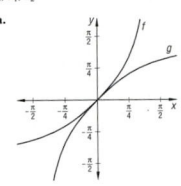

CHAPTER 6 PROGRESS SELF-TEST (p. 392)
1. a. the set of all real numbers **b.** Since the y values of the parent sine function are multiplied by $\frac{1}{2}$, the range is {y : −$\frac{1}{2}$ ≤ y ≤ $\frac{1}{2}$} **c.** The amplitude is cut in half, so it is $\frac{1}{2} \cdot 1 = \frac{1}{2}$ **d.** The factor of $\frac{1}{2}$ indicates a period twice as long as the parent function, or 4π. **2. a.** a phase shift of $\frac{\pi}{6}$ to the right; that is, (x, y) → $\left(x + \frac{\pi}{6}, y\right)$ **b.** $\frac{\pi}{6}$

3. a.

b. y = cos(3x + π) = cos$\left(3\left(x + \frac{\pi}{3}\right)\right)$, so the period is $\frac{1}{3}$ of the parent cosine function, or $\frac{2\pi}{3}$. **c.** Since the x in the parent cosine function is replaced by x + $\frac{\pi}{3}$, the phase shift is $\frac{\pi}{3}$ units to the left, or −$\frac{\pi}{3}$. **d.** Use an automatic grapher, or solve the equation cos(3x + π) = 0.5.
3x + π = cos⁻¹(0.5) = $\begin{cases} \frac{\pi}{3} + 2\pi n \text{ for } n \text{ an integer} \\ \frac{5\pi}{3} + 2\pi n \text{ for } n \text{ an integer} \end{cases}$
Take $\frac{5\pi}{3}$. Then 3x + π = $\frac{5\pi}{3}$ ⇒ x = $\frac{2\pi}{9}$ ≈ 0.70 radian.
4. y = cos 2x + 1 has a period of $\frac{2\pi}{2}$ = π and a vertical shift of 1 up from the graph of the parent cosine function. This is graph (c). **5.** cos⁻¹$\left(\frac{1}{2}\right)$ = 60° or $\frac{\pi}{3}$ radians **6.** Arctangent and tangent are inverse functions so Arctan$\left(\tan \frac{\pi}{4}\right)$ = $\frac{\pi}{4}$.
7. cot −$\frac{\pi}{6}$ = $\frac{\cos -\frac{\pi}{6}}{\sin -\frac{\pi}{6}}$ = $\frac{\cos \frac{\pi}{6}}{-\sin \frac{\pi}{6}}$ = $\frac{\sqrt{3}/2}{-1/2}$ = −$\sqrt{3}$.
8. sec 405° = $\frac{1}{\cos 405°}$ = $\frac{1}{1/\sqrt{2}}$ = $\sqrt{2}$ **9. a.** Using the Graph Standardization Theorem, h = −π and 2π|a| = $\frac{\pi}{2}$, so |a| = $\frac{1}{4}$.

Then, a possible function is y = cos$\left(\frac{x - (-\pi)}{\frac{1}{4}}\right)$ = cos(4(x + π)) = cos(4x + 4π). **b. See below.** **10.** Solve 12 = 14cos 5πt. cos 5πt = $\frac{12}{14}$ = $\frac{6}{7}$ ⇒ 5πt = cos⁻¹$\left(\frac{6}{7}\right)$ ≈ .541 ⇒ t ≈ 0.034 second. **11.** sin² θ − $\frac{1}{4}$ = 0 ⇒ sin θ = ±$\frac{1}{2}$ ⇒ θ = sin⁻¹$\left(\frac{1}{2}\right)$ or θ = sin⁻¹$\left(-\frac{1}{2}\right)$ ⇒ θ = $\frac{\pi}{6}$ or θ = −$\frac{\pi}{6}$ So a general solution is θ = $\frac{\pi}{6}$ + 2πn or θ = −$\frac{\pi}{6}$ + 2πn for any integer n. **12.** The period of the vibrations is $\frac{1}{60}$ second and the amplitude is 0.1. By the Graph Standardization Theorem, |b| = 0.1 and 2π|a| = $\frac{1}{60}$. Then $\frac{y}{0.1}$ = sin$\left(\frac{t}{1/120\pi}\right)$ or y = 0.1sin(120πt) **13.** False. sin⁻¹ θ is defined for −$\frac{\pi}{2}$ ≤ θ ≤ $\frac{\pi}{2}$. **14.** False. For example, tan 243.4° ≈ 2, but, by definition, tan⁻¹2 ≈ 63.4 ≠ 243.4.
15. a. See below. b. Since sec(x − π) = $\frac{1}{\cos(x - \pi)}$, the period is 2π. **c.** Sample: Asymptotes occur where cos(x − π) = 0. So x = $\frac{3\pi}{2}$ and x = −$\frac{\pi}{2}$ are asymptotes. In general, x = $\frac{\pi}{2}$ + πn is an asymptote for any integer n. **16. a.** amplitude = $\frac{\text{maximum } y - \text{ minimum } y}{2}$ = $\frac{71 - 22}{2}$ = 24.5° **b.** A full cycle for the temperature data is one year, so the period is 12 months. **c.** There is a horizontal shift of 6 and a vertical shift of (24.5 + 22) = 46.5. Using the Graph Standardization Theorem, k = 46.5, h = 6, 2π|a| = 12, and |b| = 24.5. Then $\frac{T - 46.5}{24.5}$ = cos$\left(\frac{n - 6}{6/\pi}\right)$ **d.** T = 24.5cos$\left(\frac{\pi}{6}n - \pi\right)$ + 46.5 **e.** We need to find T at n = 1. T = 24.5cos$\left(\frac{\pi}{6} \cdot 1 - \pi\right)$ + 46.5 ≈ 25°F.

9. b.
15. a.

The chart below keys the **Progress Self-Test** questions to the objectives in the **Chapter 6 Review** on pages 393–396. This will enable you to locate those **Chapter 6 Review** questions that correspond to questions you missed on the **Progress Self-Test**. The lesson where the material is covered is also indicated in the chart.

Question	1	2	3	4	5	6	7	8	9a	9b	10
Objective	D	D	D	I	A	B	B	I	H	G	
Lesson	6-1	6-2	6-3	6-3	6-5	6-5	6-7	6-7	6-3	6-3	6-1
Question	11	12	13	14	15a	15b	15c	16			
Objective	C	G	E	E	J	F	J	G			
Lesson	6-7	6-1	6-6	6-6	6-7	6-7	6-7	6-4			

CHAPTER 6 REVIEW (pp. 393–396)
1. 30° **2.** 45° **3.** 135° **4.** 1.5 **5.** 0.2 **6.** 2.6 **7.** $\frac{4}{3}$
8. $\frac{\pi}{3}$ **9.** 1 **10.** −1 **11.** $\frac{2\sqrt{3}}{3}$ **12.** 1.13 **13.** −1.89
14. 0.65 **15.** 2 **16.** 0 **17.** 6 **18.** θ ≈ 2.76 or 5.90
19. θ ≈ 0.72 or 5.56 **20.** θ ≈ 1.03 + nπ for any integer n
21. x ≈ 2.87 + nπ or x ≈ 0.84 + nπ for any integer n
22. θ ≈ 5.89 + 2nπ or θ ≈ 3.53 + 2nπ for any integer n
23. θ ≈ 1.32 + 2nπ or θ ≈ 4.97 + 2nπ for any integer n
24. a. 2 **b.** 1.12 **25. a.** 4π **b.** 5 **c.** 0
26. a. $\frac{2}{3}$ **b.** 2 **c.** 0 **27. a.** 2π **b.** 2 **c.** $\frac{\pi}{3}$
28. a. $\frac{\pi}{8}$ **b.** does not exist **c.** 0 **29. a.** sample : 3y + 3 = sin$\left(\frac{x - 1}{2}\right)$ **b.** amplitude = $\frac{1}{3}$; period = 4π; phase shift = 1; vertical shift = −1 **30. a.** sample: y = −2 cos (3x − 18) **b.** (x, y) → $\left(\frac{x}{3} + 6, -2y\right)$ **31. a.** 4 **b.** 6π **c.** $\frac{1}{6\pi}$
32. a. 15 **b.** 5 **33.** −$\frac{\pi}{3}$ ≤ θ ≤ $\frac{\pi}{3}$ **34.** (d)
35. a. {x : −1 ≤ x ≤ 1} **b.** {y : 0 ≤ y ≤ π} **36.** False
37. 2π **38.** for x = nπ for any integer n **39.** False
40. False **41. a.** maximum = 40 amps; minimum = −40 amps **b.** 30 **42. a.** after about 0.0031 second
b. t = $\frac{n}{30}$ seconds for any integer n **43.** y = 30 cos 40πt
44. a. $\frac{1}{2}$ radian **b.** $\frac{1}{\pi}$ **c.** 5π seconds, or about 15.7 seconds
45. a. 10 **b.** 15 **c.** sample: $\frac{d - 7}{10}$ = sin$\left(\frac{t - 1.851}{7.5/\pi}\right)$ **d.** after about 5.6 seconds
46.

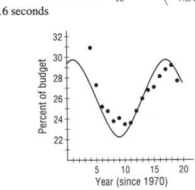

b. sample: about 4 **c.** sample: about 13
d. Answers may vary.

47.

48.

49.

50. a.

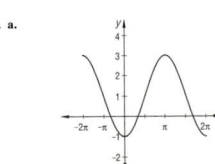

b. period = 2π; maximum = 3
51. a.

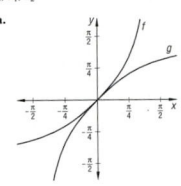

b. period = $\frac{\pi}{3}$; maximum = 7

52. a.

b. period = $\frac{\pi}{3}$; maximum = 11
53. a. y = sin $\left(x - \frac{\pi}{6}\right)$
b.

54. a. y = cos $\left(x + \frac{4\pi}{3}\right)$
b.

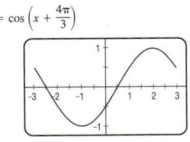

55. (b) **56.** (a) **57.** (d) **58.** (c)
59. sample: $\frac{y - 3}{4}$ = sin (2πx)
60. sample: y = tan$\left(x + \frac{\pi}{6}\right)$
61.

62.

63. False **64.** (d)

LESSON 7-1 (pp. 398–402)
1. About 27.7% **3.** 50 **5. a.** {(1, 1), (2, 2), (3, 3), (4, 4), (5, 5), (6, 6)} **b.** $\frac{1}{6}$ **7.** $\frac{3}{8}$ **9.** Sample: A probability of 0 means an event is impossible; a relative frequency of 0 means an event has never happened. **11.** 4% **13.** $\frac{15}{25}$ = .6
15. a. ≈ 0.318 **b.** .326 **17.** 5 to 3 **19.** 13.5 **21.** A coin tossed 4 times showed no heads.

LESSON 7-2 (pp. 403–408)
1. a. No **b.** 1 **c.** 8 **3.** $\frac{3}{5}$ **5.** $\frac{119}{350}$
7. a. 0.9 **b.** 0.5 **c.** 0.1 **9.** A and B are mutually exclusive events **11.** True **13.** sample: hockey **15. a.** 0.54 **b.** 0.83 **c.** 0.32 **d.** 0.44 **17. a.** ≈ 0.162 **b.** ≈ 0.630
19. ≈ 0.2 **21.** 10⁷ **23.** 50 ± $\sqrt{2521}$

LESSON 7-3 (pp. 409–413)
1. 18 **3. a.** 18 **b.** See below. **5.** $\frac{1}{128}$ ≈ .0078
7. 2¹⁰ · 5²⁰ ≈ 9.8 × 10¹⁶ **9.** 720 **11.** 10! = 3,628,800
13. n(n − 1)(n − 2) **15.** 5 **17.** 5³⁰ · 4²⁰ = $\frac{10^{50}}{2^{30}}$ · 2⁴⁰ =
2¹⁰ · 10³⁰ = 1,024 · 10³⁰ ≈ 10³ · 10³⁰ = 10³³ **19.** $\frac{4}{37}$ ≈ 0.11

21. $\frac{\pi}{4}$ **23.** $\frac{\pi}{2}$ **3. a.** domain = the set of all real numbers; range = {y : 0 ≤ y ≤ 2} **b.** maximum value = 2; minimum value = 0 **c.** 4

3. a.

LESSON 8-7 (pp. 494–498)

1. a. 7 **b.** 2 **c.** 3 **d.** 64 **e.** 2^6 **3. (a) 5.** $_nC_0 = \frac{n!}{0!(n-0)!} = \frac{n!}{(n-n)!} = {_n}C_n$ **7. a, b.** $1 + 8 + 28 + 56 + 70 + 56 + 28 + 8 + 1 = 256 = 2^8$ **9.** False **11.** 27, 1 **13.** See below. sample: The "hockey stick" pattern. In any shape as shown below, with any number of terms in the "handle" but only one in the "blade," the sum of the terms in the handle is equal to the term in the blade. For example, $1 + 3 + 6 + 10 = 20$. **15.** 705,432 **17.** 128 **19. a.** 8 **b.** 6 **21.** 96 units³ **b.** 24π units³ **c.** 8π units³

13.

LESSON 8-8 (pp. 499–502)

1. a. 3 **b.** The coefficient is the number of combinations of x-terms taken two at a time (or 3 y-terms taken one at a time). **3.** 15 **5.** $x^6 - 6x^5 + 15x^4 - 20x^3 + 15x^2 - 6x + 1$ **7.** $1000p^3 + 600p^2 + 120p + 8$ **9.** True **11.** True **13.** $\frac{11}{16} = 68.75\%$ **15.** $2^n = (1 + 1)^n = \sum_{r=0}^{n} {_n}C_r \cdot 1^{n-r} \cdot 1^r = \sum_{r=0}^{n} {_n}C_r$ **17.** 210 **19. a.** permutation **b.** 94,109,400 **21.** ≈ 0.000606 **23.** 128

LESSON 8-9 (pp. 503–507)

3. No, the trials are not independent and do not have the same probability of success. **5.** is a binomial experiment **7.** ≈ 0.92 **9.** ≈ 0.00086 **11. a.** 4 **b.** 3.5 **13. a.** $\frac{1}{3}$

b.

x	0	1	2	3	4
$P(x)$	.198	.395	.296	.099	.012

c. See below. **d.** 0.111 **e.** .802 **15. a.** $1 - a$ **b.** $35a^3(1-a)^4$ **17.** $a^6 + 12a^5b + 60a^4b^2 + 160a^3b^3 + 240a^2b^4 + 192ab^5 + 64b^6$ **19.** $\frac{1}{_{55}C_8} = \frac{1}{28,989,675} \approx 3.4 \times 10^{-8}$ **21. a.** Since "never" is less than once a month, it might be thought that the last two categories overlap. But since the percentages shown total 100%, it seems reasonable to suppose that "less than once a month" really means "sometimes, but less than once a month." Under this interpretation, all five categories are mutually exclusive. **b.** 56% **23.** True

13. c.

LESSON 8-10 (pp. 508–513)

1. No sample can ever have more than (the) one defective nail, so all boxes will be accepted. **3.** when there are 82 or more defective nails in the box **5.** True **7. a.** .03 **b.** ≈ .6108 **c.** ≈ .6333 **9.** $(x^3 - y^2)^6 = x^{18} - 6x^{15}y^2 + 15x^{12}y^4 - 20x^9y^6 + 15x^6y^8 - 6x^3y^{10} + y^{12}$ **11. a.** ≈ .2717 **b.** ≈ .408 **13.** 5 **15.** Yes; 100 **17.** 0.11

CHAPTER 8 PROGRESS SELF-TEST (pp. 517–518)

1. Since line 40 uses the previous G value to calculate the new one, the formula for G is recursive.

2. $\begin{cases} G_0 = 0; \\ G_n = G_{n-1} + n^2 \end{cases}$ so,

$G_1 = 0 + 1^2 = 1$, $G_2 = 1 + 2^2 = 5$, $G_3 = 5 + 3^2 = 14$, and $G_4 = 14 + 4^2 = 30$. **3.** $a_{15} = a_1 + (15 - 1)(d) = -12 + 14d$. **4. a.** There is a common ratio of $\frac{1}{3}$ between consecutive terms, so the sequence is geometric. **b.** $g_n = 9 \cdot \left(\frac{1}{3}\right)^{n-1}$,

so $g_7 = 9 \cdot \left(\frac{1}{3}\right)^6 = \frac{1}{81}$ **5. a.** The employee's salary is $(19,000)(1.05)$ during the second year, $(19,000)(1.05)^2$ during the third year, and $(19,000)(1.05)^3 \approx \$21,994.88$ during the fourth year. **b.** This is a geometric sequence, so the salary during the nth year is $S_n = \$19,000(1.05)^{n-1}$. **6. a.** See below; Some initial values are: $a_1 = 3$, $a_2 = 3 - \frac{1}{2}(2) = 2$, $a_3 = 2 - \frac{1}{2}(3)$, $a_4 = \frac{1}{2} - \frac{1}{2}(4) = -\frac{3}{2}, \ldots$ **b.** As n grows large, the a_ns become larger negative numbers. So the sequence does not converge. **7.** This is a geometric series with $g_1 = 75$ and $r = 0.70$. Then $S = \frac{g_1}{1-r} = \frac{75}{1-0.70} = 250$ cm. **8. a. i.** The new term in the series, $\frac{3}{l}$, needs to be added. Thus, $3/(l^*l^*l)$. **ii.** The new partial sum up to the ith term is to be printed. Thus, SUM. **b.** 3.606157 **9.** This is an arithmetic series with $a_1 = 100$, $d = 1$, so $S_n = \frac{n}{2}(a_1 + a_n)$. Then, $100 + 101 + \ldots + 200 = S_{101} = \frac{101}{2}(100 + 200) = 15150$. **10.** This is a geometric series with $g_1 = 7 \cdot \frac{3}{4} = \frac{21}{4}$, $r = \frac{3}{4}$, and $n = 10$. Then, $S_{10} = \frac{g_1(1 - r^{10})}{1 - r} = \frac{\frac{21}{4}\left(1 - \left(\frac{3}{4}\right)^{10}\right)}{1 - \frac{3}{4}} \approx 19.8174$. **11.** The distances Lillian runs each day are given by an arithmetic sequence with $a_1 = 0.5$ and $a_{12} = 3.25$. The twelfth partial sum, $S_{12} = \frac{12}{2}(0.5 + 3.25) = 22.5$ miles. **12.** $_nC_2 = \frac{n!}{(n-2)!2!} = \frac{n(n-1)(n-2)!}{(n-2)!2}$ $= \frac{n(n-1)}{2}$; $_nC_{n-2} = \frac{n!}{(n-n+2)!(n-2)!} = \frac{n!}{2!(n-2)!}$ $= \frac{n(n-1)}{2}$; So $_nC_2 = {_n}C_{n-2} = \frac{n(n-1)}{2}$. **13. a.** $(x + y)^3 = x^3 + 3x^2y + 3xy^2 + y^3$ **b.** The coefficients are the terms in row 3 of Pascal's Triangle. **14.** By the Binomial Theorem, the 4th term is $_{10}C_3(2c)^{10-3}(-b)^3 = \frac{10!}{3!7!}(2c)^7(-b)^3 = \frac{10 \cdot 9 \cdot 8}{3!}(2c)^7(-b)^3$, or choice (d).

6. a.

15. Three out of 50 are chosen without respect to order. So $_{50}C_3 = \frac{50!}{3!47!} = \frac{50 \cdot 49 \cdot 48}{3!}$, or choice (d). **16. a.** Because the range of RND(1) is $0 \le x < 1$, $x < 0.63$ represents 63% of the outcomes, or a 0.63 probability of success. **b.** The first number is less than 0.63, the next two are not. So, the first three lines of output would be: Yes No No **17.** This is a binomial experiment with $p = 0.86$ and $n = 5$. Then $P(\text{at least 4 successes}) = P(4 \text{ successes}) + P(5 \text{ successes}) = {_5}C_4(0.86)^4(0.14) + {_5}C_5(0.86)^5 \approx 0.85$. **18.** sample: In row 1: $_1C_0 = {_1}C_1 = 1$. In row 3: $_3C_0 + {_3}C_2 = 1 + 3 = 3 + 1 = {_3}C_1 + {_3}C_3$. In row 5: $_5C_0 + {_5}C_2 + {_5}C_4 = 1 + 10 + 5 = 5 + 10 + 1 = {_5}C_1 + {_5}C_3 + {_5}C_5$.

The chart below keys the **Progress Self-Test** questions to the objectives in the **Chapter 8 Review** on pages 519–522. This will enable you to locate those **Chapter 8 Review** questions that correspond to questions you missed on the **Progress Self-Test**. The lesson where the material is covered is also indicated in the chart.

Question	1	2	3	4a	4b	5	6	7	8	9	10
Objective	A	A	A	B	B	I	F	G	C		C
Lesson	8-1	8-1	8-1	8-1	8-1	8-1	8-2	8-5	8-5	8-3	8-4

Question	11	12	13a	13b	14	15	16	17	18
Objective	I	H	D	L	D	J	K	K	L
Lesson	8-5	8-6	8-8	8-7	8-8	8-6	8-9	8-9	8-7

CHAPTER 8 REVIEW (pp. 519–522)

1. 0, 2, 6, 12, 20 **2.** $-6, 0, 9, 21, 36$ **3.** 1; 1; 30 **30.** 0.00111111 **4.** $t_n = \left(\frac{1}{n}\right)^2$ **5.** $k_n = 22,000 \cdot (.8)^{n-1}$ **6.** $\begin{cases} A_1 = 27; \\ A_n = A_{n-1} - 5 \text{ for } n > 1 \end{cases}$ **7.** 400 **8.** 100 **18.933697** **9.** -749 **10.** $24(-3.5)^{11} = -23,171,797.77$ **11.** 2358 **12.** $143v - 13u$ **13.** 7,174,453 **14.** $\frac{x - x^5y^{21}}{1 - xy}$ **15.** 40 **16.** ≈ 24.999 **17.** $a^2 + 2ab + b^2$ **18.** $(x + y)^6 = x^6 + 6x^5y + 15x^4y^2 + 20x^3y^3 + 15x^2y^4 + 6xy^5 + y^6$ **19.** $(2x - 5)^4 = 16x^4 - 160x^3 + 600x^2 - 1000x + 625$ **20.** False **21.** False **22.** $10pq^9$ **23.** arithmetic **24.** neither **25.** arithmetic **26.** neither **27.** geometric **28.** neither **29.** arithmetic **30.** neither **31.** limit does not exist **32.** Yes; 4 **33.** Yes; 0 **34.** limit does not exist **35.** Yes; 0 **36.** limit does not exist **37. a.** limit does not exist **b.** 1.5 **38.** divergent **39.** not convergent **40.** convergent; 35 **41.** $\frac{5}{6}a$ **42.** $\approx .69$ **43.** -1.0369 **44. a.** $b^2 < 1$ **b.** $\frac{a^2b^2}{1 - b^2}$ **45.** $_nC_r = \frac{n!}{(n-r)!r!} = \frac{n!}{(n-r)!(n-(n-r))!} = {_n}C_{n-r}$ **46.** $_nC_1 = \frac{n!}{1!(n-1)!} = \frac{n!}{(n-1)!(n-(n-1))!} = {_n}C_{n-1}$ **47.** $_{n-1}C_r + {_{n-1}}C_{r-1} = \frac{(n-1)!}{r!(n-1-r)!} + \frac{(n-1)!}{(r-1)!(n-r)!} = \frac{(n-1)!}{(n-r)!r!} \cdot \frac{n-r}{(n-r)} + \frac{(n-1)!}{(r-1)!(n-r)!} \cdot \frac{r}{r} = \frac{(n-1)!(n-r)}{r!(n-r)!} + \frac{(n-1)!r}{r!(n-r)!} = \frac{(n-1)!(n-r+r)}{r!(n-r)!} = \frac{(n-1)!(n)}{r!(n-r)!} = \frac{n!}{r!(n-r)!} = {_n}C_r$ **48.** False **49. a.** $a_n = 500(500)^{n-1}$ **b.** $\begin{cases} a_1 = 500; \\ a_n = a_{n-1} \cdot 500 \text{ for } n > 1 \end{cases}$ **c.** $100(500)^5 = 3.125 \times 10^{15}$ **50.** 93 in. **51.** ≈ 224.7 million **52. a.** 8 **b.** 10 **53. a.** $\$364.33$ per month **b.** $\$48,057$ **54.** 7218π cm $\approx 22,676$ cm **55.** 286 **56.** 76,904,685 **57.** 190 **58. a.** 201,376 **b.** 4368 **59.** .32768 **60.** .67232 **61.** $\approx 30\%$ **62.** $\approx 32\%$ **63.** $\frac{1}{4^{10}} \approx 5.8 \times 10^{-11}$ **64.** True in every row; each is symmetric **65.** Every element off the border; each is the sum of the two elements above it. **66.** True in every row **67.** True in every row

880 | 881

882 | 883

LESSON 9-1 (pp. 524–529)

1. a. 6 **b.** -5 **c.** True **3. a.** Yes **b.** No **c.** Yes **d.** No **e.** Yes **f.** No **5.** $a_nx^n + a_{n-1}x^{n-1} + \ldots + a_1x + a_0$ **7.** $\$587.40$ **9.** $\$6,403.04$ **11.** 11 **13.** sample: $x^3yz + xyz$ **15. a.** c_n **b.** n **17.** True **19.** $t(n) = \frac{n(n+1)}{2} = \frac{1}{2}n^2 + \frac{1}{2}n$ **21.** yearbook **23.** Sample: In general, drama clubs have a higher ineligibility rate than other activities. Half of the schools declared at least 30% of drama club members ineligible, and at least one school declared all drama club members ineligible. On the other hand, yearbook enjoys the lowest median ineligibility rate, with baseball a close second. Band and track are in the middle with median ineligibility rates of about 22% and 25% respectively.
Yearbook displays the widest diversity among schools with three-fourths of the schools declaring from 0 to about 38 percent of the yearbook staff ineligible. In contrast, baseball has the least diversity among schools while having a median ineligibility rate only slightly higher than yearbook. Half of the schools declared from 10 to 20 percent of baseball players ineligible. **25.** $2x^2 + 4xy - 16y^2$

LESSON 9-2 (pp. 530–535)

1. a. $f(-1) = 3$, $f(0) = 5$, $f(1) = 3$, $f(2) = -3$, $f(3) = -13$, $f(4) = -27$, $f(5) = -45$, $f(6) = -67$, $f(7) = -93$ **b.** 1st differences $2 -2 -6 -10 -14 -18 -22 -26$; 2nd differences $-4 -4 -4 -4 -4 -4 -4$ **3.** 4 **5. a.** Yes **b.** 3 **7. a.** No **9. a.** f is a third degree polynomial because its third difference is constant. 1st difference: 5, 9, 19, 35, 57; 2nd difference: 4, 10, 16, 22; 3rd difference: 6, 6, 6 **b.** $f(x) = x^3 - x^2 + 5x$ **11. a.** 30, 55, 91 **b.** $f(x) = \frac{1}{3}x^3 + \frac{1}{2}x^2 + \frac{1}{6}x$ **c.** $f(12) = 650$ **13. a.** $f(1) = a + b + c + d$, $f(2) = 8a + 4b + 2c + d$, $f(3) = 27a + 9b + 3c + d$, $f(4) = 64a + 16b + 4c + d$, $f(5) = 125a + 25b + 5c + d$, $f(6) = 216a + 36b + 6c + d$ **b.** 1st differences: $7a + 3b + c$, $19a + 5b + c$, $37a + 7b + c$, $61a + 9b + c$, $91a + 11b + c$; 2nd differences: $12a + 2b$, $18a + 2b$, $24a + 2b$, $30a + 2b$; 3rd differences: $6a$, $6a$, $6a$ **15. a.** r **b.** $n - 1$ **17. a.** $(x + h)(y + h) = xy + xh + yh + h^2$ **b.** $xy + xh + yh + h^2$ **c.** See below. **19.** 6 **21.** $(x - 5)^2$ **23. a.** See below.

b.

n	t
1	3
2	6
3	10
4	15
5	21
6	28

c. $t(n) = \frac{1}{2}n^2 + \frac{3}{2}n + 1$

17. c.

23. a.

LESSON 9-3 (pp. 536–543)

1. True **3.** 2.41 **b.** 2.41 **5. a.** See below. **b.** -2 and -1, -1 and 0, 1 and 2 **7. a.** $V(x) = x(100 - 2x)(70 - 2x) = 4x^3 - 340x^2 + 7000x$ **b.** See below. **c.** $x < 0$ or $35 < x < 50$; x cannot have a value in these intervals because the volume must be a positive quantity. (Actually $0 < x < 35$ since each of the three dimensions must be positive.) **d.** $\approx (13.5, 42,400)$ **e.** $x \approx 13.5$ cm **f.** about 42,400 cm³ **9. a.** ≈ 8 sec and ≈ 23 sec **b.** about 3900 ft **c.** ≈ 31 sec **11. a.** See below. **b.** 1 minimum

5. a.

7. b.

11. a.

LESSON 9-4 (pp. 544–549)

1. $3x + 5$ **3.** $3x^2 - \frac{3}{2}x - 3$ **5.** dividend = (divisor)(quotient) + remainder **7. a.** -4
$\begin{array}{r} y^2 + 7y + 12 \\ \end{array}$
b. $y - 3)y^3 + 4y^2 - 9y - 40$
$\quad \underline{y^3 - 3y^2}$
$\qquad 7y^2 - 9y$
$\qquad \underline{7y^2 - 21y}$
$\qquad\qquad 12y - 40$
$\qquad\qquad \underline{12y - 36}$
$\qquad\qquad\qquad -4$
9. 0 **11. a.** $q(a) = a + 5$, $r(a) = 5a + 15$ **b.** degree of remainder is 1, degree of divisor is 2 **c.** $(a^2 - 5)(a + 5) + (5a + 15) = a^3 + 5a^2 - 5a - 25) + (5a + 15) = a^3 + 5a^2 - 10$ **13. a.** True **b.** See below. **c.** $x^3 - 12x^2 + 48x - 64 = (x^2 - 8x + 16)(x - 4)$ **15. a.** See below.

13. b.

15. a.

b. increasing on $-3.69 < x < -0.13$; $x > 1.57$; decreasing on $x < -3.69$; $-0.13 < x < 1.57$ **c.** positive $x < -5$, $-1 < x < 1$, $x > 2$; negative $-5 < x < -1$, $1 < x < 2$ **17. a.** $f(x) = x^3 + x^2 - x + 2$ **b.** -2 **c.** $(-1, 3.0)$, $(.3, 1.8)$ **19. a.** $x = -\frac{5}{2}$ or $x = \frac{1}{3}$ **b.** $6x^2 + 13x - 5 = (3x - 1)(2x + 5)$ **c.** $(x - a)$ is a factor $\Leftrightarrow f(a) = 0$

LESSON 9-5 (pp. 550–555)

1. For a polynomial $f(x)$, a number c is a solution to $f(x) = 0$ if and only if $(x - c)$ is a factor of f. **3.** Samples: $(x + 1)$ is a factor of $g(x)$; -1 is an x-intercept of the graph of $y = g(x)$; $g(-1) = 0$; the remainder when g is divided by $(x + 1)$ is zero. **5. a.** See below; zeros at $x = -2$, $x = 5$, $x = \frac{1}{7}$. **b.** $r(x) = (x + 2)(x - 5)(7x - 1)$ **7. a.** See below. **b.** Yes **9.** $(x + 2)$, $(x - 1)$ **11.** $-2, -5$ **13.** $f(x) = (x + 8)x(x - 5)(x - 10) = x^4 - 7x^3 - 70x^2 + 400x$ **15.** $m(2) = 30$ **17.** $x^4 + 3x^3 + 9x^2 + 27x + 81$ **19. a.** m, p, r **b.** $(n, f(n))$, $(q, f(q))$ **c.** $x < n$, $x > q$ **d.** $n < x < q$ **e.** $n < x < p$, $x > r$ **f.** $x < m$, $p < x < r$ **21. a.** .53 **b.** .83 **c.** .48 **d.** .0901 **e.** $.2444 \approx .33$ **e.** $\approx .67$

21. a.
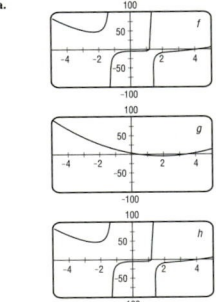

Graphs shown separately for clarity.

5. a.

21. a. See below. **b.** Sample: The graphs of f and h are identical. Both have vertical asymptotes at $x = \pm\sqrt{2}$. The graph of g is a parabola with line of symmetry $x = 2$. As x increases, points on the graph of f get closer and closer to the graph of g.

13. a. See below. **b.** 1 minimum **15. a.** $f(x) = 750x^2 + 750x + 750$ **b.** $\$2430.06$ **17. a.** $P(n) = 52.9(1.028)^n$ **b.** $P(12) = 73.7$ million **c.** in the year 2011 **19. a.** 0, 7 **b.** -9, $\frac{7}{3}$ **c.** 5, -8, -15

13. a.

LESSON 9-6 (pp. 556–561)

1. $5i$ **3.** -49 **5.** 1 **7.** $(i\sqrt{7})^2 = i^2 \cdot 7 = -1 \cdot 7 = -7$; $(-i\sqrt{7})^2 = (-i)^2 \cdot 7 = -1 \cdot 7 = -7$ **9.** $a + bi = c + di$ if and only if $a = c$ and $b = d$ **11. a.** $2 + 3i$ **b.** $-6i + 5$
13. $1 - i$ **15.** $29 + 2i$ **17.** $-\frac{1}{5} - \frac{13}{5}i$ **19. a.** $9z^2 + 18 = 9(z^2 + 2) = 9(z - i\sqrt{2}\sqrt{2} + i\sqrt{2})(z + i\sqrt{2})$ **b.** $9(z - i\sqrt{2})(z + i\sqrt{2}) = 9(z^2 + \sqrt{2}iz - \sqrt{2}iz - 2i^2) = 9(z^2 + 2) = 9z^2 + 18$
21. a. $\frac{2}{3} + 5i, \frac{2}{3} - 5i$ **b.** They are complex conjugates. **23.** $59 + 19i$ **25.** $(x - 1 - \sqrt{2}i)(x - 1 + \sqrt{2}i)$ **27.** $(x - b)$ is a factor of $g(x)$ **29.** All are of the form $p(x) = k(x)(x - 2)(x - 4)\left(x + \frac{6}{5}\right)$ for $k(x)$ a polynomial or nonzero constant function. **31. a.** $P(x_i) \geq 0$ for all x_i and $\sum_{i=1}^{8} P(x_i) = 1$. **See below. c.** $\frac{109}{24} \approx 4.542$ **d.** Over a long period of time, the average of the numbers the spinner lands on will approach $\frac{109}{24}$.

31. b.

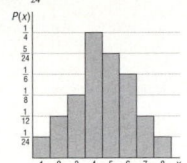

LESSON 9-7 (pp. 562–568)

5. True **7.** True **9.** True **11.** 0, ± 2, $\pm 2i$ **13. a.** $p(i) = i^4 + 2i^3 + 11i^2 + 2i + 10 = 1 + 2(-i) + 11(-1) + 2i + 10 = 1 - 2i - 11 + 2i + 10 = 0$ **b.** $-i, -1 + 3i, -1 - 3i$ (all of multiplicity one) **15.** $\pm 1, \pm 1i$ **17.** $p(x) = x - 3)(x - 1 + 2i)(x - 1 - 2i) = x^3 - 5x^2 + 11x - 15$ **19. a.** Yes; the graph will cross a horizontal line at most 2 times. **b.** Yes; the graph will cross a horizontal line at most 4 times. **c.** No; the graph crosses the x-axis 6 times, so the degree of the polynomial must be at least 6. **21. a.** $3 + 8i$ **b.** $6 + 0i$ **c.** $0 + 16i$ **d.** $73 + 0i$ **e.** $-\frac{55}{73} + \frac{48}{73}i$ **23.** $(2p + 3)(2p - 3)$ **25.** Multiply the quotient and divisor, then add the remainder. The result should equal the dividend. **27. a.** $V(r) = \frac{2}{3}\pi r^3 + 7\pi r^2$ **b.** 3 **c.** $\frac{2}{3}\pi r^3$ **d.** maximum $= \frac{75}{2}\pi \text{ m}^3 \approx 812 \text{ m}^3$, minimum $= 81 \pi \text{ m}^3 \approx 254 \text{ m}^3$ **29.** True

LESSON 9-8 (pp. 569–573)

1. $\left(-\frac{5}{2} - \frac{5\sqrt{3}}{2}i\right)^3 = \left(-\frac{5}{2}\right)^3 - 3\left(-\frac{5}{2}\right)^2\left(\frac{5\sqrt{3}}{2}i\right) + 3\left(-\frac{5}{2}\right)\left(\frac{5\sqrt{3}}{2}i\right)^2 - \left(\frac{5\sqrt{3}}{2}i\right)^3 = -\frac{125}{8} - \frac{3\sqrt{3}}{8} \cdot 125i + \frac{9}{8} \cdot 125 + \frac{3\sqrt{3}}{8} \cdot 125i = 125$ **3. a.** $x^3 - 64 = (x - 4)(x^2 + 4x + 16)$; real zero is $x = 4$ **b. See below. 5. a.** difference of squares **b.** $(a - 8x)(a + 8x)$ **7. a.** difference of cubes **b.** $(3 - 4a)(9 + 12a + 16a^2)$ **9. a.** $(x - 2)(x^4 + 2x^3 + 4x^2 + 8x + 16)$ **b.** $(x - 2)(x^4 + 2x^3 + 4x^2 + 8x + 16) = x^5 + 2x^4 + 4x^3 + 8x^2 + 16x - 2x^4 - 4x^3 - 8x^2 - 16x - 32 = x^5 - 32$

3. b.

11. False; if you divide $x^4 + y^4$ by $x + y$ the remainder is $2y^4$. **13.** $7x^2(2y + x)(2y - x)$ **15.** $(2xy + 7z)(4x^2y^2 - 14xyz + 49z^2)$ **17. a.** $x^5 - 5x^4 + 4x = x(x + 2)(x - 2)(x + 1)(x - 1)$, so the zeros are $x = 0, \pm 1, \pm 2$.

7. b.

19. a. $(t - 2)(t^8 + 2t^7 + 4t^6 + 8t^5 + 16t^4 + 32t^3 + 64t^2 + 128t + 256)$ **b.** $(t - 2)(t^8 + 2t^7 + 4t^6 + 8t^5 + 16t^4 + 32t^3 + 64t^2 + 128t + 256) = t^9 + 2t^8 + 4t^7 + 8t^6 + 16t^5 + 32t^4 + 64t^3 + 128t^2 + 256t - 2t^8 - 4t^7 - 8t^6 - 16t^5 - 32t^4 - 64t^3 - 128t^2 - 256t - 512 = t^9 - 512$ **21.** True **23. a.** True **b.** False **c.** True **d.** True **e.** True **25. a.** x-intercepts: 0, $3\sqrt{3}$, $-3\sqrt{3}$; y-intercept: 0 **b.** odd **c.** x-intercepts: 2, $2 + 3\sqrt{3}$, $2 - 3\sqrt{3}$; y-intercept: 46 **d.** neither

LESSON 9-9 (pp. 574–577)

1. $(x^3 - 9x) + (2x^2 - 18) = x(x^2 - 9) + 2(x^2 - 9) = (x + 2)(x^2 - 9) = (x + 2)(x - 3)(x + 3)$ **3.** $x^3 + 3x^2 - 4x - 12 = x^2(x + 3) - 4(x + 3) = (x^2 - 4)(x + 3) = (x - 2)(x + 2)(x + 3)$ **5.** $2x^3 - 5x^2 + 2x - 5 = x^2(2x - 5) + 1(2x - 5) = (2x - 5)(x^2 + 1)$ **7.** $2x^3 - 5x^2 + 6x - 15 = x^2(2x - 5) + 3(2x - 5) = (x^2 + 3)(2x - 5)$, so the zeros of f are $\pm i\sqrt{3}, \frac{5}{2}$ **b. See below. 9.** $(x - z)(2x + y)$ **11.** $xy - y^3 - x^3 + x^2y^2 = 0$ **13.** $(x - 2)(x^2 + 2x + 4)$ **15.** $(6x^3 - y^2)(6x^3 + y^2)$ **17.** $\sqrt[3]{10}, -\frac{\sqrt[3]{10}}{2} \pm \frac{(\sqrt{3}x}{2}\sqrt[3]{10}i$ **19. a.** $V(r) = \pi r^2(10 - 2r) + \frac{4}{3}\pi r^3$ **b. See below. c.** $\frac{500\pi}{3} \text{ ft}^3 \approx 524 \text{ ft}^3$ **21. a.** $\frac{8!}{4!} = 1680$ **b.** $7 \cdot 6 \cdot 5 = 210$ **23. a. See below. b.** $-1.5, 0, 0.7, 1.5$ **c.** $f(x) = x(2x + 3)(3x - 2)(2x - 3)$, so the zeros are $0, -\frac{3}{2}, \frac{2}{3}, \frac{3}{2}$

7. b.

19. b.

23. a.

LESSON 9-10 (pp. 578–582)

1. samples: $x^4 - 1 = 0$; $x^4 - x^3 + 4x^2 - 3x - 1 = 0$; $5^{x-1} = 1$; $\log(2x - 1) = 0$ **3.** $5 \pm \sqrt{15}i$ **5.** sample: $x^3 - 6x^2 + 11x - 6 = 0$ **7. a.** 0 **b.** -60 **9.** -1 (multiplicity 2), $\frac{1}{2} \pm \frac{\sqrt{3}}{2}i$ **11. a.** $2i$ **b.** $-24i$, 48 **c.** $3 \cdot 2^{100}$ **13.** One solution is $x = -\frac{4}{3}$. **15.** $p(x) = x^4 - 12x^3 + 5x - 61$ **17.** $-(.208)^4 \approx .2\%$ **19.** Answers will vary.

CHAPTER 9 PROGRESS SELF-TEST (p. 586)

1. Make a difference table for the y-values.

-4		-3		-2		4		17		36		61		92
	1		4		13		19		25		31				
		6		6		6		6		6		6			

Because the second differences are constant and non-zero, the polynomial has degree two. Let $y = ax^2 + bx + c$. Using the first three points, form a system of equations.

$$\begin{cases} a + b + c = -4 \\ 4a + 2b + c = -3 \\ 9a + 3b + c = 4 \end{cases} \Rightarrow \begin{cases} 3a + b = 1 \\ 5a + b = 7 \end{cases} \Rightarrow 2a = 6 \Rightarrow$$

$a = 3, b = -8$, and $c = 1$
Thus, $y = 3x^2 - 8x + 1$ **2.** Using an automatic grapher, $y = 0$ when $x \approx 1.42$.

3.
$$\begin{array}{r} 3x^3 - 13x^2 + 34x - 82 \\ x + 2\overline{)3x^4 - 7x^3 + 8x^2 - 14x - 10} \\ \underline{3x^4 + 6x^3} \\ -13x^3 + 8x^2 \\ \underline{-13x^3 - 26x^2} \\ 34x^2 - 14x \\ \underline{34x^2 + 68x} \\ -82x - 10 \\ \underline{-82x - 164} \\ 154 \end{array}$$
So, $q(x) = 3x^3 - 13x^2 + 34x - 82$; $r(x) = 154$
4. $9x^2 - 3x - 2 = (3x - 2)(3x + 1)$ **5.** Use the Sums and Differences of Cubes Theorem. $t^6 + 1000y^3 = (t^2)^3 + (10y)^3 = (t^2 + 10y)(t^4 - 10t^2y + 100y^2)$ **6.** By the Factor-Solution-Intercept Equivalence Theorem, $(x - 1)$ is a factor of $f(x)$, so the answer is (a). **7. a.** $2z + w = 2(2 + 5i) + (3 - i) = 4 + 10i + 3 - i = 7 + 9i$. **b.** $\frac{z}{w} = \frac{2 + 5i}{3 - i} = \frac{(2 + 5i)(3 + i)}{(3 - i)(3 + i)} = \frac{1 + 17i}{10} = \frac{1}{10} + \frac{17}{10}i$. **8.** The squared factor $2x + 3$ indicates that $x = -\frac{3}{2}$ is a zero of multiplicity two. **9.** If all the factors are multiplied, the highest power of x would be $3 + 2 + 1 = 6$, so g is of degree 6. **10.** Since there are three distinct linear factors, by the Factor-Solution-Intercept Equivalence Theorem, the graph of $g(x)$ has three x-intercepts. **11.** By the Number of Zeros of a Polynomial Theorem, a polynomial of degree seven has exactly seven complex zeros, counting multiplicities. Non-real zeros occur in conjugate pairs according to the Conjugate Zeros Theorem, so there are none, two, four, or six of them. This leaves at least one real zero. **12.** By the Conjugate Zeros Theorem, another solution is $-i$. So, $x^4 + 11x^2 + 10 = (x + i)(x - i) p(x) = (x^2 + 1)p(x)$. Polynomial long division can be used to find $p(x) = x^2 + 10$. Factoring gives $x^2 + 10 = (x + i\sqrt{10})(x - i\sqrt{10})$. Thus, the solutions for $x^4 + 11x^2 + 10x = 0$ are $i, -i\sqrt{10}i$, and $-\sqrt{10}i$. **13.** Let $p(x) = ax^2 + bx + c$ be the price of an x-inch pizza.

$$\begin{cases} p(10) = 100a + 10b + c = 5.95 \\ p(12) = 144a + 12b + c = 6.95 \\ p(14) = 196a + 14b + c = 8.50 \end{cases} \Rightarrow \begin{cases} 44a + 2b = 1.00 \\ 52a + 2b = 1.55 \end{cases}$$

$\Rightarrow 8a = .55 \Rightarrow a \approx .069, b \approx -1.013, c \approx 9.2$. So $p(x) = .069x^2 - 1.013x + 9.2$ and $p(16) \approx \$10.66$. **14.** Using the Pythagorean Theorem, $(x^3)^2 = (x^2)^2 + (x)^2$. Then $x^6 - x^4 - x^2 = 0$. Using an automatic grapher we get that $x \approx 1.27$. **15. a.** In the part shown, there seems to be a negative zero of multiplicity two and a positive zero. So, there are a total of at least three real zeros. **b.** $r(x)$ must have at least three linear factors, and thus must have a degree of at least three.

The chart below keys the **Progress Self-Test** questions to the objectives in the **Chapter 9 Review** on pages 587–590. This will enable you to locate those **Chapter 9 Review** questions that correspond to questions you missed on the **Progress Self-Test**. The lesson where the material is covered is also indicated in the chart.

Question	1	2	3	4	5	6	7	8	9	10	11
Objective	A	B	C	D	D	G	E	F	F	F	H
Lesson	9-2	9-3	9-4	9-9	9-8	9-5	9-6	9-3	9-1	9-3	9-7

Question	12	13	14	15	16
Objective	H	I	J	K	J
Lesson	9-7	9-2	9-2	9-3	9-2

---

16. $V_{\text{insulation}} = V_{\text{outer cylinder}} - V_{\text{inner cylinder}} = 25\pi r^2 - 25\pi(r - .35)^2 = 25\pi(r^2 - (r - .35)^2) = 25\pi(r^2 - r^2 + .70r - .1225) = 17.5\pi r - 3.0625\pi m^3$

CHAPTER 9 REVIEW (pp. 587–590)

1. a. 4 **b.** 5 **2. a.** Yes **b.** $3x^2 - 2x + 5$ **3. a.** No **4. a.** $t_n = \frac{n(n + 1)}{2} = \frac{1}{2}n^2 + \frac{1}{2}n$ **b.** 5050 **5.** -1.62 **6.** $(-.7, 4.5)$ **7.** By the Factor-Solution-Intercept Equivalence Theorem, $(t - 3)$ is a factor of $g(t)$ if and only if 3 is a t-intercept. **8.** 3.16 **9.** quotient: $2t^3 - 9t + 1$; remainder: $27t + 8$ **10.** quotient: $t^3 + 1$; remainder: 1 **11.** quotient: $x^3 - 2x^2 + x - 1$; remainder: 0 **12.** quotient: $(2 - w)y^3 + 3xy + (3 - w)y^2$; remainder: $(2 - w)y^3$ **13.** True **14.** $-2, 1, 2$ **15.** $(3t - w)(81t^4 + 27t^3u + 9t^2u^2 + 3tu^3 + u^4)$ **16.** $(x + \sqrt{2})(x - \sqrt{2})(x + 1)(x^2 - x + 1)$ **17.** $(z - 2)(z + 1 - \sqrt{3}i)(z + 1 + \sqrt{3}i)$ **18.** $(4x + 3)(2x + 1)$ **19.** $(10x + 1)(2x - 1)$ **20.** $(t - 2)(t + 2)(t^2 + 4)(t - 1)$ **21. a.** $8 - 2i$ **b.** $-2 + 6i$ **c.** $23 - 2i$ **d.** $\frac{3}{4} + \frac{1}{2}i$ **22. a.** $-1 + 3i$ **b.** $3 + 9i$ **c.** $16 - 15i$ **d.** $\frac{1}{4} + \frac{3}{2}i$ **23.** $\frac{14}{25} + \frac{27}{25}i$ **24.** $-\frac{10}{5} - \frac{22}{5}i$ **25.** 0 **26.** $(-2 + 2\sqrt{3}i)^3 = (-2 + 2\sqrt{3}i)(-2 + 2\sqrt{3}i)^2 = (-2 + 2\sqrt{3}i)(-8 - 8\sqrt{3}i) = 64$ **27. a.** 7 **b.** 3 **c.** 0 **d.** -127 **e.** 7 (counting multiplicities) **28.** sample: $4x^3 + 1$ **29.** False **30. a.** 3 **b.** -2 **c.** 1 **31.** False **32.** True **33.** True **34.** 3 **35. a.** For all a, $a = a + 0i$. If a is real, then its complex conjugate is $a - 0i$. But $a - 0i = a$. Thus every real number is equal to its complex conjugate. **b.** True; an odd degree polynomial has an odd number of zeros, counting multiplicities. Since complex zeros occur in conjugate pairs, the total number of non-real zeros must be even. Therefore, there must be at least one real zero. **36.** They are not real. **37.** $x = 3i, -3i, 2, -2$ **38.** True; $2 + i$ and $2 - i$ are zeros since zeros come in conjugate pairs. If $1 + i$ were a zero then $1 - i$ would also have to be a zero. But, since the polynomial is a cubic it only has three zeros. The third zero must be real. **39. a.** $100x^{12} + 40x^{11} + 40x^{10} + 40x^9 + 40x^8 + 40x^7 + 40x^6 + 40x^5 + 40x^4 + 40x^3 + 40x^2 + 40x + 40$ **b.** $\approx \$602$ **40. a.** $\ell = 510 - w$; $A(w) = w(510 - w) = 510w - w^2$ **b.** 41,000 ft² **i.** 65,000 ft² **iii.** 63,000 ft² **c.** $\ell = 255$ ft, $w = 255$ ft **41. a.** $R(c) = 2.50c + .001c^2 - .00002c^3$ **b.** ≈ 220 **c.** \$385 **42.** $V = \frac{1}{3}\pi r^2h + \frac{2}{3}\pi r^3$ **43. a.** $V(r) = 2\pi r^2 + \frac{2}{3}\pi r^3$ **b.** 3; volume is a 3-dimensional quantity. **c.** $\frac{2}{3}\pi$

44. a. $V(x) = \frac{1}{3}\pi(25 - x^2)(5 + x) = \frac{125}{3}\pi + \frac{25}{3}\pi x - \frac{5}{3}\pi x^2 - \frac{1}{3}\pi x^3$ **b.** 3 **c.** 4 **45.** $s^3 + 3s^2h + 3sh^2 + h^3$ **46.** 24 cm² **47. a. See below. b.** maximum (5, 8) **c.** 3, 7 **48. a. See below. b.** relative maximum $(-2.6, .4)$; relative minimum $(-1.4, -.4)$ **c.** $-3, -2, -1$ **49. a. See below. b.** relative maximum $(-.4, 84)$; relative minima $(1.5, -28.7)$, $(-7.4, -3566)$ **c.** $-9.9, -1.3, .9, 2$ **50. a. See below. b.** relative maximum $(-1, 0)$; relative minimum $(0, -1)$ **c.** $1, -1$ **51. a.** 3 **b.** f has two nonreal zeros. **52.** False; The equation has exactly three real solutions. **53.** Since f has only three real zeros, it does not cross the x-axis again. Thus there are no more maxima, minima, or intercepts.

47. a.

48. a.

49. a.

50. a.

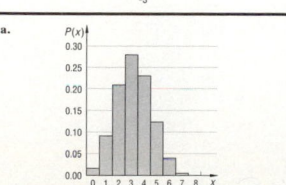

LESSON 10-1 (pp. 592–599)

1. a.

x	0	1	2	3	4
$P(x)$	.01680	.08958	.20902	.27869	.23224

x	5	6	7	8	
$P(x)$	.12386	.04129	.00786	.00066	**See right.**

b. about 5.0% **c.** Sample: The mode in part a is less than the mode in Example 1; this graph is a little more spread out. **3. a.** k **b.** n **c.** binomial **d.** p **5.** False

1. a.

9. a.

9. b. The mode increases; the graph approximates a bell-shaped curve; the distribution "spreads out" and "flattens."

11. a. .03125 **b.** Sample: It seems effective because the probability of all 5 people recovering without an effective drug is only about .03. **c.** about .022 **d.** Sample: Again, the drug seems to be effective. The second experiment is more conclusive because the result of 18 people out of 25 recovering is less likely to happen by chance (.022) than the result of 5 out of 5 people recovering (.031). **13.** Yes **15.** Sample:

```
10 N = 20: S = 0: M = 0: L = 0
20 FOR I = 1 TO N
30   P = RND(1)
40   IF P < .35 THEN S = S + 1
50   IF P >= .35 AND P < .85 THEN M = M + 1
60   IF P >= .85 THEN L = L + 1
70 NEXT I
80 PRINT S; "SMALL WATERMELONS"
90 PRINT M; "MEDIUM WATERMELONS"
100 PRINT L; "LARGE WATERMELONS"
110 END
```

17. $\approx 117.5°$

LESSON 10-2 (pp. 600–605)

1. a. 3 **b.** 2.10 **c.** -1.45 **3. a.** $66\frac{2}{3}$ **b.** $\frac{\sqrt{200}}{3} \approx 4.7$

5. a.

x	0	1	2	3	4
$P(x)$	.316	.422	.211	.047	.004

b. See below. c. 1 **d.** $\frac{\sqrt{3}}{2} \approx .87$ **7. a.** $40\% \cdot 200 = 80$ **b.** $\mu = np = 200 \cdot 0.4 = 80$ **9.** $n = 225, p = .2$ **11. a.** about .988 **b.** $\mu = 4, \sigma = \sqrt{3.2} \approx 1.8$ **c.** about 41% **13.** 0.5 **15. a.** $\approx .048$ **b.** $\approx .0075$ **c.** $P(\text{living near plant and having polluted water}) \approx .0075$; $P(\text{living near plant}) \cdot P(\text{having polluted water}) = (.048)(.03) = .00144$. Since $P(\text{living near plant and having polluted water}) \neq P(\text{living near plant}) \cdot P(\text{having polluted water})$, the events are not independent. **17.** $\pm 3\sqrt{110}$ **19. a.** all employees **b.** executives

5. b.

LESSON 10-3 (pp. 606–612)
1. a. .99 square units b. .99 3. 0 5. The area triples.
7. a. See below. b. $A = \frac{1}{2}bh = \frac{1}{2}(4)(.5) = 1$;
$A = \left|1\cdot\frac{1}{10}\right|\cdot\frac{1}{2}(4\cdot5) = \left|\frac{1}{10}\right|\cdot10 = 1$ c. ≈ .28 9. a. .875
b. On $\triangle O'M'N'$, (area of region to the left of $x = 3$) = (area of $\triangle O'M'N'$) − (area to the right of $x = 3$) = $1 - \frac{1}{2}(1)\left(\frac{1}{4}\right) = .875$ c. 0 15. sample: $(x, y) \to \left(x, \frac{2}{\pi}y\right)$ 17. a. 5
b. $\sqrt{3.75} \approx 1.94$ 19. $2a^3(1 + 243b^5c^{10})$ 21. Sample: For $x = 0$, $\tan 0 = 0 = \tan(-0)$
7. a.
23. a.
b.
c.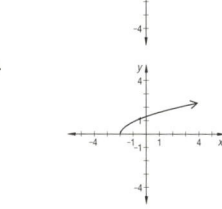

LESSON 10-4 (pp. 613–617)
1. True 3. a. $x = 0$ b. $y = 0$ 5. True
7. a. .50 b. .78 c. .28 9. a. reflection-symmetric about the y-axis b. $y = 0$ c. See below. d. (1, .61), (−1, .61)
11. $(x,y)\to(\sqrt2, y)$ 13. The area is about 2.13.
15. a. .7 b. $(.7)^5 \approx .168$ 17. $81x^4 - 540x^3 + 1350x^2 - 1500x + 625$ 19. a. all students in the nation who take the test b. students in the school who take the test
9. c.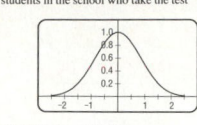

LESSON 10-5 (pp. 618–625)
1. a. See below. b. $f(z) = \frac{1}{\sqrt{2\pi}}e^{-\frac{z^2}{2}}$

3. a. .0548 b. .4452 c. .8904 d. .9452 5. .4721
7. .9772 9. True 11. $P(-2 < z < 2) = 2P(0 < z < 2) = 2(.9772 - .5) = 2(.4772) = .9544$ 13. Only a tiny percent of scores are beyond 3.09 standard deviations from the mean.
15. ≈ 1.96 17. ≈ .67 19. a. ≈ .954 b. ≈ .997
21. $n = 27$, $p = .7355$ 23. ≈ .7 inch 25. 792
27. a. $e^{5/3}$
b.

LESSON 10-6 (pp. 626–633)
1. normal curve 3. See below. 5. ≈ .37 7. ≈ 4 or 5
3.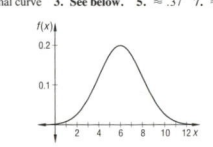

9. a. $z = \frac{t - 100}{4}$ b. mean = 0; s.d. = 1 11. np and nq are each ≥ 5 13. Calculations are simpler. 15. almost 0
17. a. $f(x) = \frac{1}{2\sqrt{2\pi}}e^{-\frac{1}{2}\left(\frac{x-8}{2}\right)^2}$ b, c. See below. d. $g(x) = \frac{1}{6\sqrt{2\pi}}e^{-\frac{1}{2}\left(\frac{x-6}{6}\right)^2}$ e. The maximum height of the first curve, ≈ .20, is about 3 times as great as the maximum height of the second curve, ≈ .066. f. The area under each is 1. 19. 2.58
21. $\frac{1}{\sqrt{\pi}}$ units^2 ≈ .56 units^2 23. $p = 300\cdot(2.4)^h$ 25. ≈ .998
27. a. 50π m² b. $\frac{125}{9}\pi \approx 43.6$ m² c. 50π m²
17. b, c.

LESSON 10-7 (pp. 634–638)
1. a. i. ≈ .042 ii. ≈ .9207 b. Yes 3. P(5 female) ≈ .016; using the .05 level, the evidence supports the contention that the office discriminates. 5. H_0: Gas mileage is not increased with the new oil. H_1: Gas mileage is increased with the new oil.
7. a. Yes b. Yes 9. P(11 or more hits) ≈ .027; using the 0.05 significance level, the batter is doing better this season.
11. about 95% 13. z ≈ 2.33 15. a. 1 dollar b. The sum of the probabilities is 1, and each probability is between 0 and 1.
c. negative 50 cents d. negative 5 dollars e. $3.5 million
17. False 19. False

LESSON 10-8 (pp. 639–646)
3. True 5. False 7. mean ≈ 63.7 in., s.d. ≈ 0.26 in.
9. ≈ .9938 11. (a) 13. No; the distribution for $n = 2$ is skewed. If the population were normal, the sample distributions would be closer to normal. 15. ≈ .28 17. a. H_0: The baby has no color preference between red and blue. H_1: The baby prefers red to blue. b. P(3 or 4 red balls) ≈ .26, so we cannot reject H_0. There is no evidence to support a claim of preference for red balls. 19. ≈ .0082 or about 0.8%
21. True 23. a, c. See below. b. sample: $y = -1.476x^2 + 20.81x + 4.29$ d. Sample: Weather is generally periodic; a sine curve is more appropriate.
23. a, c.

LESSON 10-9 (pp. 647–653)
1. a. 434 ≤ μ ≤ 446 b. ±6 3. 430 ≤ μ ≤ 450 5. True
7. lower the confidence level; increase the sample size
9. a. 25% to 31% b. No 11. a. 170 ≤ μ ≤ 190 b. No
13. False 15. a. normal b. They are approximately equal.
17. $1 - p$ 19. $1 - p$ 21. (b) 23. (c) 25. a. 49
b. arithmetic 27. a. 840 b. 360

CHAPTER 10 PROGRESS SELF-TEST (p. 657)
1. By the Binomial Probability Theorem, for each value of x, $P(x$ of 10 survive$) = {}_{10}C_x(0.89)^x\cdot(.11)^{10-x}$. Evaluating this expression for each possible value of x gives the following table:

x	$P(x$ of 10 survive)
0	2.6×10^{-10}
1	2.1×10^{-8}
2	7.6×10^{-7}
3	1.65×10^{-5}
4	2.3×10^{-4}
5	.0022
6	.015
7	.071
8	.214
9	.385
10	.312

These data are graphed in the histogram below.

2. domain would increase; range would "flatten" or decrease; shape would approach normal distribution 3. $\mu = np = .2\cdot60 = 12$; $\sigma = \sqrt{npq} = \sqrt{60\cdot.2\cdot.8} = \sqrt{9.6} \approx 3.1$
4. First note that, if s is a side of the square, then by the Pythagorean theorem $s^2 + s^2 = 8^2 \Rightarrow 2s^2 = 64 \Rightarrow s^2 = 32 \Rightarrow s = \sqrt{32} = 4\sqrt2$. Now,
P(point is in shaded region) $= \dfrac{\text{area of shaded region}}{\text{area of circle}}$
$= \dfrac{\text{area of circle} - \text{area of square}}{\text{area of circle}}$
$= \dfrac{\pi4^2 - 4\sqrt2\cdot4\sqrt2}{\pi4^2}$
$= \dfrac{16\pi - 32}{16\pi}$
$\approx .36$

5. $x = 0$ 6. False; the standard normal curve has equation $f(x) = \frac{1}{\sqrt{2\pi}}e^{-\frac{x^2}{2}}$. 7. Using the Standard Normal Distribution table we get $P(z < 1.3) = .9032$ 8. $P(-1.5 < z < 1.5) = 2\cdot P(z < 1.5) = 2[P(z < 1.5) - P(z < 0)]$. Using the Standard Normal Distribution table we see that this equals $2(.9332 - .5) = .8664$ 9. a. H_0: The mean salary of union members is not lower than the mean salary of the population. H_1: The mean salary of union members is lower than the mean salary of the population. b. Let $\bar{x}$ be the mean salary of 50 randomly selected construction workers. The Central Limit Theorem and its corollary imply that $\bar{x}$ is approximately normally distributed with a mean of 20,100 and a standard deviation of $\frac{\sigma}{\sqrt{n}} = \frac{2100}{\sqrt{50}} \approx 297$. Thus the variable $z = \frac{\bar{x} - 20100}{297}$ has a standard normal distribution. Since $P(\bar{x} < \$18{,}200) \approx P(z < -6.40) \approx 0$, the union members' salary is significantly lower than the national

level at the 0.05 level. 10. False; the theorem is also true when the population is normally distributed. 11. We must find a value $a > 0$ such that $P(-a < z < a) = .95$. Since $P(-a < z < a) = 2\cdot P(0 < z < a) = 2[P(z < a) - P(z < 0)] = 2[P(z < a) - .5]$, we get $2[P(z < a) - .5] = 2P(z < a) - 1 < .95 \Rightarrow 2P(z < a) < 1.95 \Rightarrow P(z < a) < \frac{1.95}{2} = .975$. Using the Standard Normal Distribution Table, the probability value .975 corresponds to the z-value $a = 1.96$. 12. standard normal 13. Let x be the number of miles a student travels to school. Then, because x is normally distributed with a mean of 2.15 miles and a standard deviation of 0.36, the variable $z = \frac{x - 2.15}{0.36}$ has a standard normal distribution. So $P(x > 3) = 1 - P(x < 3) = 1 - P\left(z < \frac{3 - 2.15}{0.36} \approx 2.36\right) \approx 1 - .9909 \approx .0091$.

The chart below keys the **Progress Self-Test** questions to the objectives in the **Chapter 10 Review** on pages 658–662. This will enable you to locate those **Chapter 10 Review** questions that correspond to questions you missed on the **Progress Self-Test**. The lesson where the material is covered is also indicated in the chart.

Question	1	2	3	4	5	6	7	8	9	10	11
Objective	H	C	A	I	D	D	B	B	G	E	D
Lesson	10-1	10-1	10-2	10-3	10-4	10-5	10-5	10-5	10-7	10-8	10-6

Question	12	13a	13b
Objective	D	J	F
Lesson	10-6	10-6	10-6

CHAPTER 10 REVIEW (pp. 658–662)
1. $\mu = 36, \sigma = \sqrt{14.4} \approx 3.8$ 2. $\mu = 30, \sigma = \sqrt{21} \approx 4.58$
3. $\mu = 10, \sigma = \sqrt5 \approx 2.24$ 4. $\mu = 5, \sigma = \sqrt{3.75} \approx 1.94$
5. 100 6. $n = 188, p = .842$ 7. a. .7994 units² b. .7994
8. .9772 9. .1359 10. .0096 11. $P(-3 < z < 3) = 2P(0 \le z < 3) = 2(.9987 - .5) = 2(.4987) = .9974 \approx 99.7\%$
12. a. $\frac{p}{5}$ $\frac{1-p}{2}$ c. $1 - p$ 13. False 14. True
15. True 16 a. They are reflection images of each other over the vertical line $k = 5$. b. Given n and p, the graphs of $B(k, n, p)$ and $B(k, n, 1 - p)$ are reflections of each other over the vertical line $k = \frac{n}{2}$ 17. a. $f(-x) = e^{-(-x)^2} = e^{-x^2} = f(x)$
b. reflection symmetry over the y-axis 18. a. True
19. False 20. True 21. ≈ 1.21 22. .40 23. a. $f(x) = \frac{1}{\sqrt{2\pi}}e^{-\frac{1}{2}(x - 2)^2}$ b. $\frac{1}{\sqrt{2\pi}}$ c. ≈ .4 24. μ 25. True 26. False
27. almost zero 28. It is normally distributed with mean 63 and standard deviation 1.5. 29. ≈ .35 30. in about 1% of the cases

31.
level	interval
highly extroverted	49 or above
extroverted	42–48
average	29–41
introverted	22–28
highly introverted	21 or lower

32. .52% 33. (c) 34. $P(\bar{x} \le 1.14) = .014 < .05$; the owner can reject the null hypothesis and conclude that the settings are truly off.

35. $P(\text{weight} > 27.8 \text{ kg}) \approx 0$; the officers cannot reject the null hypothesis that passengers are checking in luggage that does not weigh more than 20 kg. 36. $P(|\text{average speed} - 63| > 4.25)$ is almost 0. The police officer can claim the report is wrong.
37. The sum of the probabilities is not equal to 1.
38. a. $P(3) = \frac{1}{4}, P(4) = \frac{1}{16}$ b. See below.
38. a.
39. a.
x	$P(x)$
0	.296
1	.444
2	.222
3	.037

b.

40.

41. a. ≈ 10 b. ≈ 0.25 42. Samples: it is continuous and not discrete; $P(x) > 0$ for arbitrarily large values of x.
43. $\frac{1}{9}$ 44. True; the area of a point is zero, and ratios of areas represent probability values.

45. a. 12.5 units² b. i. .6 ii. $.6\overline{3}$ 46. $k = \sqrt{7.5} \approx 2.74$
47. False; each area is divided by 4, so the ratios of areas do not change. 48. a. .9544 b. .4772 c. .9772
49. a. ii b. iii c. i 50. See below. 51. Sample: Much less than 99.7% of the area lies within 3 standard deviations of the mean. 52. a, b. See below. c. True d. Sample: The curve in a is not a normal distribution; the area beneath it is not 1. The curve in b is a normal distribution.
50.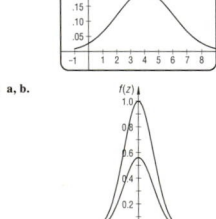
52. a, b.

LESSON 11-1 (pp. 664–669)
3. 23 5. The total number of members of the House of Representatives for each particular year. 7. 3 × 4
9. [74] 11. $\begin{bmatrix} ax + by \\ cx + dy \end{bmatrix}$ 13. sample: 3 × 2 and 3 × 4
15. a. $\begin{bmatrix} 2860 \\ 2918 \\ 2640 \end{bmatrix}$ b. $2860 17. $x = -2$
19. a. $\begin{bmatrix} 4 & 0 \\ -8 & 0 \end{bmatrix}$ b. $\begin{bmatrix} 0 & -8 \\ -8 & 0 \end{bmatrix}$ 21. a. 0, 1, 2, 3, or 4
b.
b	0	1	2	3	4
$P(b)$	.282	.470	.217	.031	.001

c.

23. $f(g(x)) = 9x^2 - 6x + 3$
25. a. (v) b. (i) c. (ii) d. (iv) e. (iii)

LESSON 11-2 (pp. 670–676)
1. (c) 3. a. (iii) b. (v) c. (ii) d. (i) e. (iv)
5. a. $T' = (-4.5, 3)$; $R' = (1.5, 3)$; $Z' = (3, -1.5)$; $D' = (-4.5, -1.5)$ b. It is a size change of 1.5.
7. $\begin{bmatrix} a & 0 \\ 0 & a \end{bmatrix}$ 9. $\begin{bmatrix} -1 & 0 \\ 0 & 1 \end{bmatrix}\begin{bmatrix} x \\ y \end{bmatrix} = \begin{bmatrix} -x \\ y \end{bmatrix}$ So, the matrix $\begin{bmatrix} -1 & 0 \\ 0 & 1 \end{bmatrix}$ maps each point (x, y) to $(-x, y)$, which is a reflection over the y-axis. 11. a. $\begin{bmatrix} -2 & -3 & 0 \\ 1 & -2 & -4 \end{bmatrix}$ b. It is a rotation of 180° around the origin. 13. a. $\begin{bmatrix} 2 & 0 \\ 0 & 2 \end{bmatrix}$ b. $\begin{bmatrix} 0.5 & 0 \\ 0 & 0.5 \end{bmatrix}$
c. sample: $\begin{bmatrix} 0.3 & 0 \\ 0 & 5 \end{bmatrix}$ 15. $\frac{\pi}{12}$ 17. $\begin{bmatrix} 2x + 3y \\ 4y \end{bmatrix}$
19. a. $\begin{bmatrix} 394 & 436 & 477 \\ 675 & 705 & 760 \end{bmatrix}$ b. The first row represents production costs in Oakland in thousands of dollars for each of the years 1988, 1989, and 1990. The second row represents the same numbers for Charleston. c. No. A 3 × 3 matrix cannot be multiplied by a 2 × 3 matrix, because the number of columns in the first matrix must correspond to the number of rows in the second. 21. ≈ .62% 23. all values except $\theta = \frac{k\pi}{2}$ for integers k

LESSON 11-3 (pp. 677–683)
1. r_x 3. $\begin{bmatrix} 0 & -1 \\ 1 & 0 \end{bmatrix}$ 5. a. R_{90} b. $\begin{bmatrix} 0 & -1 \\ 1 & 0 \end{bmatrix}$
c. $\begin{bmatrix} 0 & -1 \\ 1 & 0 \end{bmatrix}\begin{bmatrix} 4 & 1 \\ 6 & -2 \end{bmatrix} = \begin{bmatrix} -6 & 2 & -4 \\ 4 & 1 & 1 \end{bmatrix}$
7. a. $\begin{bmatrix} 0 & 0 & 2 \\ 0 & 6 & 6 \end{bmatrix}$ b. $\begin{bmatrix} 0 & 0 & 4 \\ 0 & 3.6 & 3.6 \end{bmatrix}$ c. $\begin{bmatrix} 0 & 2 \\ .6 & 0 \end{bmatrix}$

9. a. $(0,-1)$ **b.** $(-1, 0)$ **c.** $\begin{bmatrix} 0 & -1 \\ -1 & 0 \end{bmatrix}$

11. $\begin{bmatrix} -14 & 9 & -89 \\ -14 & 3 & -53 \end{bmatrix}$

13. $R_{90} \circ R_{270} = \begin{bmatrix} 0 & -1 \\ 1 & 0 \end{bmatrix} \begin{bmatrix} 0 & 1 \\ -1 & 0 \end{bmatrix} = \begin{bmatrix} 1 & 0 \\ 0 & 1 \end{bmatrix}$

15. a. $R_{90}(3,4) = (-4, 3)$; $R_{180}(3,4) = (-3,-4)$; $R_{270}(3,4) = (4,-3)$ **b. See below. c.** square **d.** center $(0, 0)$, radius 5 **17. a.** w **b.** x **c.** y **d.** z **19.** $x = -2$; $y = 7$

21. ≈ 4 feet **23. a, b. See below. c.** $(x, y) \to \left(\frac{x}{2}, \frac{y}{2}\right)$

25. $\frac{5\sqrt{3}}{4}$

15. b.

23. a, b.

$f(\theta) = 2\cos\theta$
$g(\theta) = \cos 2\theta$

LESSON 11-4 (pp. 684–689)

1. $\begin{bmatrix} \cos t & -\sin t \\ \sin t & \cos t \end{bmatrix}$ **3.** $\begin{bmatrix} 0.891 & -0.454 \\ 0.454 & 0.891 \end{bmatrix}$

5. $R_{90} = \begin{bmatrix} \cos 90° & -\sin 90° \\ \sin 90° & \cos 90° \end{bmatrix} = \begin{bmatrix} 0 & -1 \\ 1 & 0 \end{bmatrix}$

7. $R_{270} = \begin{bmatrix} \cos 270° & -\sin 270° \\ \sin 270° & \cos 270° \end{bmatrix} = \begin{bmatrix} 0 & 1 \\ -1 & 0 \end{bmatrix}$

9. $\begin{bmatrix} 4.8 & 7.0 \\ 2.5 & 3.6 \end{bmatrix}$ **11. a.** $\begin{bmatrix} 0 & -4\sqrt{3} & -2\sqrt{3} \\ 0 & -4 & -6 \end{bmatrix}$

11. b.

13. a. A clockwise rotation of θ is the same as a counterclockwise rotation of $(-\theta)$. So,
$R_{-\theta} = \begin{bmatrix} \cos(-\theta) & -\sin(-\theta) \\ \sin(-\theta) & \cos(-\theta) \end{bmatrix} = \begin{bmatrix} \cos\theta & \sin\theta \\ -\sin\theta & \cos\theta \end{bmatrix}$
by the Opposites Theorem.

b. $\begin{bmatrix} \cos 90° & \sin 90° \\ -\sin 90° & \cos 90° \end{bmatrix} = \begin{bmatrix} 0 & 1 \\ -1 & 0 \end{bmatrix}$

15. a. Yes **b.** No **c.** Yes **d.** No **17.** 1

19. a. $y = 3\sin\left(x - \frac{\pi}{4}\right)$ **b.** Sample: $y = 3\sin\left(x - \frac{\pi}{4}\right)$ is the image of the graph of $y = \sin x$ under $(x, y) \to \left(x + \frac{\pi}{4}, 3y\right)$. The transformation takes the point $\left(\frac{\pi}{4}, \frac{\sqrt{2}}{2}\right)$, which is on the curve $y = \sin x$, and maps it to $\left(\frac{\pi}{2}, \frac{3\sqrt{2}}{2}\right)$. **21.** False

23. a. $y = x^2 - 2x + 1$ **b.** sample: $y = x^3 + 4x^2 - 3x - 2$

25. (a) **27. a.** $c = .12n + 5$ **b.** $245

LESSON 11-5 (pp. 690–694)

1. We know $R_\beta(1,0) = (\cos\beta, \sin\beta)$. So, $R_\alpha \circ R_\beta(1,0) = R_\alpha(R_\beta(1,0)) = R_\alpha(\cos\beta, \sin\beta)$.

3. $\sin(75° + 15°) = \sin(90°) = 1$ **5.** $\frac{\sqrt{6} - \sqrt{2}}{4}$

7. $\cos(\alpha - \beta) = \cos\alpha\cos(-\beta) - \sin\alpha\sin(-\beta)$
$= \cos\alpha\cos\beta - \sin\alpha(-\sin\beta)$
$= \cos\alpha\cos\beta + \sin\alpha\sin\beta$

9. $\cos\left(\frac{\pi}{2} - x\right) = \cos\frac{\pi}{2}\cos x + \sin\frac{\pi}{2}\sin x = \sin x$

11. $\frac{-\sqrt{2} - \sqrt{6}}{4}$ **13.** $\begin{bmatrix} \frac{\sqrt{6} - \sqrt{2}}{4} & \frac{-\sqrt{6} - \sqrt{2}}{4} \\ \frac{\sqrt{6} + \sqrt{2}}{4} & \frac{\sqrt{6} - \sqrt{2}}{4} \end{bmatrix}$ **15. a.** f and h

b.

$y = g(x)$ · $y = f(x) = h(x)$

c. $f(x) = 2\sin\left(x + \frac{\pi}{6}\right) = 2\left(\sin x \cos\frac{\pi}{6} + \cos x \sin\frac{\pi}{6}\right)$
$= 2\left(\frac{\sqrt{3}}{2}\sin x + \frac{1}{2}\cos x\right) = \sqrt{3}\sin x + \cos x = h(x)$

17. $\sin\left(x - \frac{3\pi}{2}\right) = (\sin x)\cos\left(-\frac{3\pi}{2}\right) + \cos x \sin\left(-\frac{3\pi}{2}\right)$
$= (\sin x)(0) + (\cos x)(-1)$
$= -\cos x$

---

19. a. $\frac{1 - 2\sqrt{6}}{6}$ **b.** $\frac{2\sqrt{6} - 1}{6}$ **21. a.** $\begin{bmatrix} 2 & 12 & 16 & 6 \\ -1 & -1 & -4 & -4 \end{bmatrix}$ **b.** It is a scale change: $(x, y) \to (2x, -y)$. **23. a.** 15 **b.** Friday **c.** HR **d.** $53.80 **25. a.** 3 **b.** 4 **c.** True. Since g is at least of degree 4, and since the tails of the graph both point in the same direction, the degree of g must be an even number greater than 2. Since nonreal roots come in pairs, there must be an even number of nonreal roots and thus an even number of real roots. So, there cannot be only 3 real roots; there must be at least 4.

27. (c)

LESSON 11-6 (pp. 695–699)

1. ≈ 118 m **3.** $\cos 2\theta = \cos(\theta + \theta) = \cos\theta\cos\theta - \sin\theta\sin\theta = \cos^2\theta - \sin^2\theta = (1 - \sin^2\theta) - \sin^2\theta = 1 - 2\sin^2\theta$

5. a.

b. sample: $R(x) = \cos(2x)$ **7.** $\cos 50°$

9. $\cos\frac{3\pi}{4}$ **11.** $\sin 4\theta = 4\sin\theta\cos^3\theta - 4\sin^3\theta\cos\theta$

13. a. $\frac{3}{5}$ **b.** $\frac{24}{25}$ **15.** $r_L = \begin{bmatrix} \cos(2\cdot 0°) & \sin(2\cdot 0°) \\ \sin(2\cdot 0°) & -\cos(2\cdot 0°) \end{bmatrix} =$
$\begin{bmatrix} 1 & 0 \\ 0 & -1 \end{bmatrix} = r_x$ **17. a.** $r_L = \begin{bmatrix} \frac{1}{2} & \frac{\sqrt{3}}{2} \\ \frac{\sqrt{3}}{2} & -\frac{1}{2} \end{bmatrix}$ **b.** $r_L\begin{bmatrix} 1 \\ 0 \end{bmatrix} = \begin{bmatrix} \frac{1}{2} \\ \frac{\sqrt{3}}{2} \end{bmatrix}$

c.

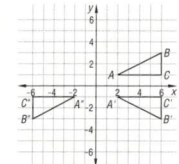

19. a.
$\begin{array}{rl} & -\cos 2\theta = \sin^2\theta - \cos^2\theta \\ + & 1 = \sin^2\theta + \cos^2\theta \\ \hline & 1 - \cos 2\theta = 2\sin^2\theta \end{array}$
$\frac{1 - \cos 2\theta}{2} = \sin^2\theta$
$\sin\theta = \pm\sqrt{\frac{1 - \cos 2\theta}{2}}$

b. $\sin\frac{\pi}{8} = \pm\sqrt{\frac{1 - \cos\frac{\pi}{4}}{2}} = \pm\sqrt{\frac{1 - \frac{\sqrt{2}}{2}}{2}} = \pm\sqrt{\frac{2 - \sqrt{2}}{4}} = \pm\frac{\sqrt{2 - \sqrt{2}}}{2}$. Since $\frac{\pi}{8}$ is the first quadrant, $\sin\frac{\pi}{8} > 0$. So $\sin\frac{\pi}{8} = \frac{\sqrt{2 - \sqrt{2}}}{2}$ **21.** $\frac{\sqrt{2} - \sqrt{6}}{4}$ **23.** (a)

25. $a = 9$; $b = 3$ **27. a.** $6 + 8i$ **b.** i

LESSON 11-7 (pp. 700–708)

1. $\begin{bmatrix} \cos\theta & -\sin\theta \\ \sin\theta & \cos\theta \end{bmatrix}\begin{bmatrix} \cos\theta & \sin\theta \\ -\sin\theta & \cos\theta \end{bmatrix} =$
$\begin{bmatrix} \cos^2\theta + \sin^2\theta & \cos\theta\sin\theta - \cos\theta\sin\theta \\ \cos\theta\sin\theta - \cos\theta\sin\theta & \sin^2\theta + \cos^2\theta \end{bmatrix}$
$= \begin{bmatrix} 1 & 0 \\ 0 & 1 \end{bmatrix}$ **3.** $S = \begin{bmatrix} 4 & 0 \\ 0 & 4 \end{bmatrix}$

b. a size change $T(x, y) = \left(\frac{1}{4}x, \frac{1}{4}y\right)$ **c.** $T = \begin{bmatrix} \frac{1}{4} & 0 \\ 0 & \frac{1}{4} \end{bmatrix}$

d. True, $ST = \begin{bmatrix} 1 & 0 \\ 0 & 1 \end{bmatrix} = TS$ **5.** The inverse does not exist.

7. 87 **9.** Its determinant is zero.

11. a. $\begin{bmatrix} 3 & 21 \\ 2 & 19 \end{bmatrix}$ **b.** $\begin{bmatrix} 2 & 1 \\ 3 & 2 \end{bmatrix}$ **c.** $\begin{bmatrix} 3 & 21 \\ 2 & 19 \end{bmatrix} = \begin{bmatrix} 8 & 61 \\ 13 & 101 \end{bmatrix}$

c. HIMW **13. a.** True **b.** $-\det\begin{bmatrix} Y & Z \\ W & X \end{bmatrix}$

15. a. $\begin{bmatrix} 13 & 16 \\ 11 & 14 \end{bmatrix}$ **b.** $\begin{bmatrix} \frac{7}{3} & -\frac{8}{3} \\ -\frac{11}{6} & \frac{13}{6} \end{bmatrix}$ **c.** $\begin{bmatrix} \frac{4}{3} & -\frac{5}{3} \\ -\frac{1}{2} & \frac{3}{2} \end{bmatrix}$ **d.** $\begin{bmatrix} 2 & 1 \\ -1.5 & -0.5 \end{bmatrix}$

e. $A^{-1}B^{-1} = \begin{bmatrix} \frac{31}{6} & \frac{13}{6} \\ -\frac{5}{2} & -2 \end{bmatrix}$, $B^{-1}A^{-1} = \begin{bmatrix} \frac{7}{3} & -\frac{8}{3} \\ -\frac{11}{6} & \frac{13}{6} \end{bmatrix}$

f. $(AB)^{-1} = B^{-1}A^{-1}$ **17.** $2\sin a\cos b$ **19.** $\cos\frac{\pi}{13}$

21. a. R_a **b.** r_x **23. a.** $\begin{bmatrix} 2 & 5 \\ -1 & -5 \end{bmatrix}$ **b. See below.**

c. $\begin{bmatrix} 1 & 1 & 5 \\ -2 & -5 & -5 \end{bmatrix}$ **d. See below. e.** $\begin{bmatrix} -1 & -1 & 5 \\ 2 & 5 & 5 \end{bmatrix}$

25. a. 729 **b.** 729.00

23. b.

23. d.

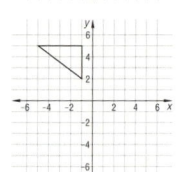

LESSON 11-8 (pp. 709–717)

1. 6×10^4 **3.** $(2, 0, -1)$ **5.** $\begin{bmatrix} -1 & 0 \\ 0 & -1 \end{bmatrix}$ **7.** $\approx (146.4, 127.7)$

9. Sample: The bottom row of a 4×4 matrix representing 3-D graphics has 3 zeros and a 1, so when such a 4×4 matrix is multiplied by a point in homogeneous coordinates, the last number in the result is a 1 and does not have to be calculated. So, there are only 3 vector multiplications that need to be made—one each for x', y', and z'. **11.** no more than about 16.7 frames per second **13. a.** $\begin{bmatrix} .985 & .174 \\ -.174 & .985 \end{bmatrix}$ **b.** $R_{10} = \begin{bmatrix} .985 & -.174 \\ .174 & .985 \end{bmatrix}$

15. If T_1 and T_2 are scale changes with T_1: $(x, y) \to \left(3x, \frac{1}{2}y\right)$ and T_2: $(x, y) \to \left(\frac{1}{3}x, 2y\right)$, then $T_1 \circ T_2 = I$, where I is the identity transformation. **17.** $f(x) = 5(2\cos^2 x - 1) = 5\cos(2x)$, which has amplitude 5. **19.** a counterclockwise rotation of 25°

21. $\begin{bmatrix} -22 & -28 \\ -49 & -64 \\ -76 & -100 \end{bmatrix}$ **23.** ≈ 0.67

CHAPTER 11 PROGRESS SELF-TEST (p. 722)

1. Multiply the number of voters in each region by the percentage of people expected to vote Republican in that region and then add the three results; $3.4(.29) + 0.6(.41) + 2.1(.58) = 2.45$ million. **2.** $VP = \begin{bmatrix} .50 & .47 & .29 \\ .29 & .41 & .58 \\ .21 & .12 & .13 \end{bmatrix} \cdot \begin{bmatrix} 3.4 \\ 0.6 \\ 2.1 \end{bmatrix} =$

$\begin{bmatrix} .50(3.4) + .47(0.6) + .29(2.1) \\ .29(3.4) + .41(0.6) + .58(2.1) \\ .21(3.4) + .12(0.6) + .13(2.1) \end{bmatrix} = \begin{bmatrix} 2.591 \\ 2.45 \\ 1.059 \end{bmatrix}$ **3.** To obtain the first row of VP, multiply the percentage of urban voters expected to vote Democrat by the total number of registered urban voters, add the percentage of rural voters expected to vote Democrat times the total number of rural voters, and then add the percentage of suburban voters expected to vote Democrat times the total number of suburban voters. Thus the entry in the first row is the total number of registered voters expected to vote Democrat. Similarly, the second and third rows represent the total number of voters expected to vote Republican and Independent, respectively. **4.** $\begin{bmatrix} 5 & -1 \\ 2 & 0 \end{bmatrix}\begin{bmatrix} 3 & 8 \\ -1 & 1 \end{bmatrix} =$
$\begin{bmatrix} 5 \cdot 3 + -1 \cdot -1 & 5 \cdot 8 + -1 \cdot 1 \\ 2 \cdot 3 + 0 \cdot -1 & 2 \cdot 8 + 0 \cdot 1 \end{bmatrix} = \begin{bmatrix} 16 & 39 \\ 6 & 16 \end{bmatrix}$ **5.** Since M is a 3×4 matrix and N is a $4 \times t$ matrix, MN is a $3 \times t$ matrix, so $t = 5$. **6. a.** $A = (0, 2)$, $B = (-3, 0)$, and $C = (4, -1)$ so the point matrix for $\triangle ABC$ is $\begin{bmatrix} 0 & -3 & 4 \\ 2 & 0 & -1 \end{bmatrix}$.

7. $XM = \begin{bmatrix} -1 & 0 \\ 0 & 2 \end{bmatrix}\begin{bmatrix} 0 & -3 & 4 \\ 2 & 0 & -1 \end{bmatrix}$
$= \begin{bmatrix} -1 \cdot 0 + 0 \cdot 2 & -1 \cdot -3 + 0 \cdot 0 & -1 \cdot 4 + 0 \cdot -1 \\ 0 \cdot 0 + 2 \cdot 2 & 0 \cdot -3 + 2 \cdot 0 & 0 \cdot 4 + 2 \cdot -1 \end{bmatrix}$
$= \begin{bmatrix} 0 & 3 & -4 \\ 4 & 0 & -2 \end{bmatrix}$

8. See right. 9. a. The reflection $r_{y=x}$: $(x, y) \to (y, x)$ maps $(1, 0)$ to $(0, 1)$ and $(0, 1)$ to $(1, 0)$. Therefore, the matrix representing $r_{y=x}$ is $\begin{bmatrix} 0 & 1 \\ 1 & 0 \end{bmatrix}$. **b.** $\begin{bmatrix} 0 & 1 \\ 1 & 0 \end{bmatrix}\begin{bmatrix} 0 & 1 \\ 1 & 0 \end{bmatrix} = \begin{bmatrix} 1 & 0 \\ 0 & 1 \end{bmatrix}$, which is the identity matrix.

10. $R_{30} = \begin{bmatrix} \cos 30° & -\sin 30° \\ \sin 30° & \cos 30° \end{bmatrix} = \begin{bmatrix} \frac{\sqrt{3}}{2} & -\frac{1}{2} \\ \frac{1}{2} & \frac{\sqrt{3}}{2} \end{bmatrix}$ which is matrix (b).

11. a. r_x **b.** The matrix representing r_x is $\begin{bmatrix} 1 & 0 \\ 0 & -1 \end{bmatrix}$, and the matrix representing R_{180} is $\begin{bmatrix} -1 & 0 \\ 0 & -1 \end{bmatrix}$ so the matrix representing the composite $R_{180} \circ r_x$ is $\begin{bmatrix} -1 & 0 \\ 0 & -1 \end{bmatrix}\begin{bmatrix} 1 & 0 \\ 0 & -1 \end{bmatrix} = \begin{bmatrix} -1 & 0 \\ 0 & 1 \end{bmatrix}$ **12.** $\triangle PQR$ is represented by the point matrix $\begin{bmatrix} 0 & 1 & 0 \\ 0 & 0 & 1 \end{bmatrix}$, and the rotation R_θ is represented by $\begin{bmatrix} \cos\theta & -\sin\theta \\ \sin\theta & \cos\theta \end{bmatrix}$, so the image $\triangle P'Q'R'$ of $\triangle PQR$ is represented by the point matrix
$\begin{bmatrix} \cos\theta & -\sin\theta \\ \sin\theta & \cos\theta \end{bmatrix}\begin{bmatrix} 0 & 1 & 0 \\ 0 & 0 & 1 \end{bmatrix} = \begin{bmatrix} 0 & \cos\theta & -\sin\theta \\ 0 & \sin\theta & \cos\theta \end{bmatrix}$. Thus, $P' = (0, 0)$, $Q' = (\cos\theta, \sin\theta)$, and $R' = (-\sin\theta, \cos\theta)$ **13.** $\sin 83°\cos 42° - \cos 83°\sin 42° = \sin(83° - 42°) = \sin 41°$. **14.** $\cos(\pi + \theta) = \cos\pi\cos\theta - \sin\pi\sin\theta =$
$-1 \cdot \cos\theta - 0 \cdot \sin\theta = -\cos\theta$ **15.** $\sin 2A = 2\sin A\cos A =$
$2\sin A(\sqrt{1 - \sin^2 A}) = 2 \cdot .40 \cdot \sqrt{1 - .40^2} = 2 \cdot .40 \cdot \sqrt{.84} \approx .73$

16. $C^{-1} = \begin{bmatrix} \frac{5}{1 \cdot 5 - 3 \cdot -1} & \frac{-(-1)}{1 \cdot 5 - 3 \cdot -1} \\ \frac{-3}{1 \cdot 5 - 3 \cdot -1} & \frac{1}{1 \cdot 5 - 3 \cdot -1} \end{bmatrix} = \begin{bmatrix} \frac{5}{8} & \frac{1}{8} \\ -\frac{3}{8} & \frac{1}{8} \end{bmatrix}$

17. $\det G = (3x)(10) - (5x)(6) = 30x - 30x = 0$. A matrix with a zero determinant has no inverse.

18. a. $\begin{bmatrix} 3 & -1 \\ 5 & -2 \end{bmatrix}\begin{bmatrix} x \\ y \end{bmatrix} = \begin{bmatrix} 6 \\ 11 \end{bmatrix}$ **b.** $\begin{bmatrix} 3 & -1 \\ 5 & -2 \end{bmatrix}^{-1}$
$\begin{bmatrix} \frac{-2}{3 \cdot -2 - 5 \cdot -1} & \frac{-(-1)}{3 \cdot -2 - 5 \cdot -1} \\ \frac{-5}{3 \cdot -2 - 5 \cdot -1} & \frac{3}{3 \cdot -2 - 5 \cdot -1} \end{bmatrix} = \begin{bmatrix} 2 & -1 \\ 5 & -3 \end{bmatrix}$. Multiplying both sides of the equation $\begin{bmatrix} 3 & -1 \\ 5 & -2 \end{bmatrix}\begin{bmatrix} x \\ y \end{bmatrix} = \begin{bmatrix} 6 \\ 11 \end{bmatrix}$ on the left by this inverse gives the solution $\begin{bmatrix} x \\ y \end{bmatrix} = \begin{bmatrix} 1 \\ -3 \end{bmatrix}$.

8.

The chart below keys the **Progress Self-Test** questions to the objectives in the **Chapter 11 Review** on pages 723–726. This will enable you to locate those **Chapter 11 Review** questions that correspond to questions you missed on the **Progress Self-Test**. The lesson where the material is covered is also indicated in the chart.

Question	1	2	3	4	5	6	7	8	9a	9b	10
Objective	F	A	F	A	C	I	I	I	G	D	G
Lesson	11-1	11-1	11-1	11-1	11-1	11-2	11-2	11-2	11-2	11-7	11-4
Question	11	12	13	14	15	16	17	18			
Objective	H	I	E	E	E	D	D	B			
Lesson	11-3	11-4	11-5	11-5	11-6	11-7	11-7	11-7			

CHAPTER 11 REVIEW (pp. 723–726)

1. (c) **2.** $\begin{bmatrix} -\frac{5}{2} & \frac{5}{2} \\ 6 & 25 & -6 \end{bmatrix}$ **3.** $[x - 21]$ **4.** $\begin{bmatrix} \frac{61}{4} & -\frac{3}{2} \\ -\frac{5}{2} & 16 \end{bmatrix}$

5. 5 **6. a.** $\begin{bmatrix} 1 & 3 \\ 2 & -1 \end{bmatrix}\begin{bmatrix} x \\ y \end{bmatrix} = \begin{bmatrix} 11 \\ 8 \end{bmatrix}$ **b.** $\begin{bmatrix} \frac{1}{7} & \frac{3}{7} \\ \frac{2}{7} & -\frac{1}{7} \end{bmatrix}$

c. $\begin{bmatrix} \frac{1}{7} & \frac{3}{7} \\ \frac{2}{7} & -\frac{1}{7} \end{bmatrix}\begin{bmatrix} 1 & 3 \\ 2 & -1 \end{bmatrix}\begin{bmatrix} x \\ y \end{bmatrix} = \begin{bmatrix} \frac{1}{7} & \frac{3}{7} \\ \frac{2}{7} & -\frac{1}{7} \end{bmatrix}\begin{bmatrix} 11 \\ 8 \end{bmatrix} \Rightarrow \begin{bmatrix} x \\ y \end{bmatrix} = \begin{bmatrix} 5 \\ 2 \end{bmatrix}$

d. $x = 5$ and $y = 2$ **7.** $x = \frac{3}{2}$ and $y = \frac{3}{2}$ **8.** $x = 2$ and $y = -2$ **9.** $x = 3$, $y = 2$, and $z = 1$ **10.** 2×5 **11.** 5×2 **12.** (d) **13.** Sample: a rotation about the origin, a reflection over a line containing the origin, and the identity transformation.

14. $\begin{bmatrix} \frac{4}{13} & -\frac{1}{13} \\ \frac{5}{13} & \frac{2}{13} \end{bmatrix}$ **15.** $\begin{bmatrix} \cos 40° & \sin 40° \\ -\sin 40° & \cos 40° \end{bmatrix}$ **16.** $\begin{bmatrix} 4 & -\frac{1}{2} \\ 3 & 1 \end{bmatrix}$

17. (a) and (e) **18.** $\frac{\sqrt{6} + \sqrt{2}}{4}$ **19.** $\frac{\sqrt{6} - \sqrt{2}}{4}$ **20.** $(.5)(.6) + (.8)\frac{\sqrt{3}}{2} \approx .99$ **21.** $2(.8)(.6) = .96$ **22.** $-\cos\theta$ **23.** $-\cos\theta$ **24.** $\cos 4\theta = 8\cos^4\theta - 8\cos^2\theta + 1$ **25.** $\sin(2A)$ **26.** $\cos\left(\frac{23\pi}{15}\right)$ **27.** $\cos 50°$ **28.** The exchange rates in terms of the number of U.S. dollars for each pound, Canadian dollar, yen, and U.S. dollar, respectively, in the row given to the left of the row. **29.** $6.80 **30.** $523.82 **31.** True

32. a. $\begin{bmatrix} 6,950 & 10,900 \\ 7,400 & 11,800 \\ 13,225 & 20,450 \\ 3,475 & 5,450 \end{bmatrix}$ **b.** It is the total cost of production (first column) and total cost to the consumer (second column) resulting from the produce from each farm (represented by the rows). **33. a.** $13,225 **b.** 3rd row, 1st column **34.** $10,900

35. $\begin{bmatrix} 1 & 0 \\ 0 & -1 \end{bmatrix}$ **36.** $\begin{bmatrix} 1 & 0 \\ 0 & \frac{1}{3} \end{bmatrix}$ **37.** A size change by a factor of 11.

38. $r_{y=x}$ **39.** $\begin{bmatrix} 0 & -1 \\ 1 & 0 \end{bmatrix}$ **40.** R_{180} **41.** $\begin{bmatrix} \cos 32° & -\sin 32° \\ \sin 32° & \cos 32° \end{bmatrix}$

42. $\begin{bmatrix} -\frac{1}{2} & -\frac{\sqrt{3}}{2} \\ \frac{\sqrt{3}}{2} & -\frac{1}{2} \end{bmatrix}$ **43. a.** $\begin{bmatrix} 5 & 0 \\ 0 & -1 \end{bmatrix}$ **b.** $(5, -1)$

44. a. $\begin{bmatrix} -1 & 0 \\ 0 & 1 \end{bmatrix}\begin{bmatrix} 0 & -1 \\ 1 & 0 \end{bmatrix}$ **b.** $\begin{bmatrix} 0 & 1 \\ 1 & 0 \end{bmatrix}$ **c.** $r_{y=x}$

45. A counterclockwise rotation of 90°, followed by a reflection over the line $y = x$, followed by a counterclockwise rotation of 32°. **46. a.** $\begin{bmatrix} 0 & 1 \\ -1 & 0 \end{bmatrix}$ **b.** A clockwise rotation of 90°.

47. $\begin{bmatrix} -1 & 0 \\ 0 & 1 \end{bmatrix}$ **b.** r_y **48. a.** $\begin{bmatrix} 1 & 10 & -5 \\ 3 & -6 & 0 \end{bmatrix}$ **b. See below.**

49. a. sample: $\begin{bmatrix} 0 & 0 & 7 \\ 0 & 1 & 0 \\ 0 & -1 & 0 \end{bmatrix}$ **b. See below.**

50. a. $\begin{bmatrix} 1 & 0 \\ 0 & -1 \end{bmatrix}\begin{bmatrix} 0 & -1 \\ -1 & 0 \end{bmatrix} = \begin{bmatrix} 7 & 0 & -1 \\ 0 & 2 & -1 \end{bmatrix}$ **b.** $\begin{bmatrix} 0 & -2 & 1 \\ -7 & 0 & 1 \end{bmatrix}$

51. $R_{-75} = \begin{bmatrix} \cos(-75°) & -\sin(-75°) \\ \sin(-75°) & \cos(-75°) \end{bmatrix} = \begin{bmatrix} \cos 75° & \sin 75° \\ -\sin 75° & \cos 75° \end{bmatrix}$

48. b.

49. b.

LESSON 12-1 (pp. 728–733)
1. The cone studied in geometry is the set of all points between the points on a given circle and a given point not in the plane of the circle. It is finite in extent in the sense that the distance from any point on the cone to the vertex is bounded. In contrast, the cone that gives rise to conic sections is formed by rotating one of two intersecting lines about the other. It is infinite in extent since it contains infinitely many lines through the vertex.
3. a. Sample: The center of the sphere is on the axis of the cone. The intersection of the cone and the sphere is a circle whose points are all the same distance from the vertex of the cone.
b.

5. a. The smaller of the spheres is between the plane and the vertex of the cone, tangent to the plane at one focus of the ellipse. The plane is between the vertex and the larger of the spheres; it is tangent to the sphere at the other focus. **b.** Each Dandelin sphere is tangent to the plane. The foci are the points of tangency. **7.** True **9.** The ellipse should have the shape of the one below.

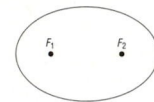

11. Let F_1 and F_2 be the foci of an ellipse with focal constant k. Let n be the perpendicular bisector of $\overline{F_1F_2}$. It must be shown that the ellipse coincides with its reflection image over n. First we show that the image of any point on the ellipse is on the ellipse. Let P be any point on the ellipse and $P' = r_n(P)$. Since P is on the ellipse, $PF_1 + PF_2 = k$. Also, by definition of r_n, $r_n(F_1) = F_2$ and $r_n(F_2) = F_1$. Therefore, $r_n(PF_1) = P'F_2$ and $r_n(PF_2) = P'F_1$. So, because reflections preserve distance, $P'F_2 + P'F_1 = PF_1 + PF_2 = k$. Thus P' is on the ellipse. Now if Q is any point on the ellipse, then $Q = r_n(r_n(Q))$. By the above argument, $r_n(Q)$ is on the ellipse; therefore, the entire ellipse is the image. **13. a.** There is no smallest focal constant, but it must be larger than 1 meter. **b.** As the focal constant gets larger, the ellipse looks more and more like a circle. **c.** There is no largest focal constant; it can be arbitrarily large as long as it is larger than 1 meter.
15. a. $\begin{bmatrix} 1 & 0 \\ 0 & -1 \end{bmatrix}$ **b.** $\begin{bmatrix} 2 & 4 & 5 \\ -1 & -3 & -1 \end{bmatrix}$

c.

17. 15,150 **19.** $4ab$ **21.** $\sqrt{(a-x)^2 + (b-y)^2}$

LESSON 12-2 (pp. 734–740)
1. $\sqrt{(x-11)^2 + (y-6)^2} + \sqrt{(x+2)^2 + (y-7)^2} = 15$
3. a.

b. $(3, 0)$ and $(-3, 0)$ **c.** 10 **5. a.** $(0, 0)$ **b.** $(a, 0)$ and $(-a, 0)$ **c.** $2a$ **7. a.** The graph is the translation image of the graph in Example 1a under $T(x, y) \rightarrow (x + 1, y - 7)$
b.

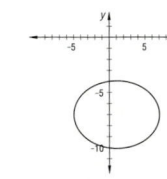

9. farther **11. a.** $\frac{x^2}{100} + \frac{y^2}{49} = 1$ **b.** $\frac{(x-3)^2}{100} + \frac{(y+6)^2}{49} = 1$
13. major axis = 10, minor axis = 10; Multiplying both sides of the equation by 25 gives the equation $x^2 + y^2 = 25$. This is an equation for a circle of radius 5. Both major and minor axes are diameters of the circle and so have length 10. **15. a.** 0 **b.** $0 \le$ eccentricity < 1

17. a.

b. sample: $\left(1, \frac{3\sqrt{3}}{2}\right), \left(1, -\frac{3\sqrt{3}}{2}\right), \left(-1, \frac{3\sqrt{3}}{2}\right), \left(-1, -\frac{3\sqrt{3}}{2}\right)$
19.

21. $\approx 125°$ **23. a.** about 3971 megabytes
b. 5000 megabytes

LESSON 12-3 (pp. 741–749)
1. True

3. a, b.

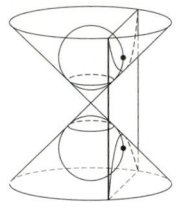

c. The spheres are tangent to the plane at the foci of the hyperbola.

5. a.

b. True

7. a.

b.

c. They are reflection images of each other over the line $y = x$.
9. Let F_1 and F_2 be the foci of a hyperbola with focal constant k. If P is any point on the hyperbola then $|PF_1 - PF_2| = k$. Let P' be the image of P under the reflection r over $\overline{F_1F_2}$. Since $r(F_1) = F_1$ and $r(F_2) = F_2$ and reflections preserve distance, $P'F_1 = PF_1$ and $P'F_2 = PF_2$. Therefore $|P'F_1 - P'F_2| = |PF_1 - PF_2| = k$ and so P' is on the hyperbola. This shows that the image of any point on the hyperbola is also on the hyperbola. To show that the image is the entire hyperbola, note that if Q is on the hyperbola, then $r(r(Q)) = Q$ and $r(Q)$ is on the hyperbola by the above argument. Thus any point on the hyperbola is the image of some point on the hyperbola.
11. Let $A = (5000, 0)$ and $B = (-5000, 0)$. An equation for the hyperbola is $\frac{x^2}{1,562,500} - \frac{y^2}{23,437,500} = 1$.
13. a. $(-1, -3), (-3, -3)$ **b.** $y = x - 1, y = -x - 5$
c.

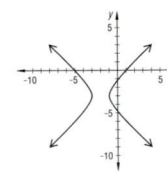

15. a. $\frac{x^2}{4} + \frac{y^2}{9} = 1$ **b.** $(0, \sqrt{5}), (0, -\sqrt{5})$ **17. a.** $\begin{bmatrix} 0 & 1 \\ -1 & 0 \end{bmatrix}$
b. $\begin{bmatrix} \cos\theta & -\sin\theta \\ \sin\theta & \cos\theta \end{bmatrix}$ **c.** $\begin{bmatrix} \cos\theta & \sin\theta \\ -\sin\theta & \cos\theta \end{bmatrix}$
19. a. i. $(1 + 1)^0 = {}_0C_0 \cdot 1^0 = 1$ **ii.** $(1 + 1)^1 = {}_1C_0 \cdot 1^1 + {}_1C_1 \cdot 1^1 = 1 + 1 = 2$ **iii.** $(1 + 1)^2 = {}_2C_0 \cdot 1^2 + 2C_1 \cdot 1^1 \cdot 1^1 + {}_2C_2 \cdot 1^2 = 1 + 2 + 1 = 4$ **iv.** $(1 + 1)^3 = {}_3C_0 \cdot 1^3 + {}_3C_1 \cdot 1^2 \cdot 1^1 + {}_3C_2 \cdot 1^1 1^2 + {}_3C_3 \cdot 1^3 = 1 + 3 + 3 + 1 = 8$ **b.** True; by definition, the elements of the nth row of Pascal's triangle are ${}_nC_r$ for $0 \le r \le n$. By the Binomial Theorem, the terms in the binomial expansion of $(1 + 1)^n$ are ${}_nC_r \cdot 1^r \cdot 1^{n-r} = {}_nC_r$. The terms are the same. **21.** 7

LESSON 12-4 (pp. 750–757)
1. the influence of heredity on height **3.** sample: $91x^2 - 18\sqrt{3}xy + 73y^2 = 1600$ **5. a.** rectangular **b.** rectangular **c.** rectangular **d.** not rectangular **7.** The equation of the image of the ellipse is found by substituting linear combinations of x and y for x and y. Since the original equation has degree 2, the resulting equation also has degree 2. **9. a.** sample: $139x^2 - 78\sqrt{3}xy + 217y^2 = 6400$ **b.** major axis: $y = \frac{\sqrt{3}}{3}x$; minor axis: $y = -\frac{\sqrt{3}}{3}x$ **11. a.** sample: $y^2 - x^2 = 2k$ **b.** True
13. a.

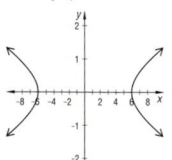

b. $y = \frac{1}{6}x, y = -\frac{1}{6}x$ **c.** $\frac{y^2}{36} - x^2 = 1$
15. an equation which is true for all values in the domain of the variables **17.** (a) **19.** (d)

LESSON 12-5 (pp. 758–764)
1. a. $9x^2 + 4y^2 - 36 = 0$ **b.** $A = 9, B = 0, C = 4, D = 0, E = 0, F = -36$ **c.** ellipse **3. a.** $xy - 1 = 0$ **b.** $A = 0, B = 1, C = 0, D = 0, E = 0, F = -1$ **c.** hyperbola
5. ellipse **7.** hyperbola **9.** $\frac{\pi}{4}$ **11.** Let $s = \sin\theta$ and $t = \cos\theta$. $B'^2 - 4A'C' = (2(A - C)ts + B(t^2 - s^2))^2 - 4(At^2 - Bts + Cs^2)(As^2 + Bts + Ct^2) = 4(A - C)^2t^2s^2 + 4(A - C)B(t^2 - s^2)ts + B^2(t^2 - s^2)^2 - 4A^2s^2t^2 - 4B^2s^2t^2 + 4C^2s^2t^2 + 4AB(st^3 - s^3t) + 4AC(s^4 + t^4) + 4BC(s^3t - st^3) = 4A^2t^2s^2 - 8ACt^2s^2 + 4AB(t^3s - ts^3) - 4BC(t^3s - ts^3) + B^2t^4 - 2B^2t^2s^2 + B^2s^4 - 4A^2s^2t^2 + 4B^2s^2t^2 - 4C^2s^2t^2 - 4AB(st^3 - s^3t) - 4AC(s^4 + t^4) - 4BC(s^3t - st^3) = B^2(t^2 + s^2)^2 - 4AC(t^2 + s^2)^2 = B^2 - 4AC$

13. $(2x + 5y - 7)(3x - 7y + 11) = 0$ if and only if $2x + 5y - 7 = 0$ or $3x - 7y + 11 = 0$. The graph of each of the latter two equations is a line, so the graph of the conic is a pair of lines. Since the slopes of the lines are not equal, the lines intersect. **15.** A hyperbola is called a rectangular hyperbola when its asymptotes are perpendicular.
17. a.

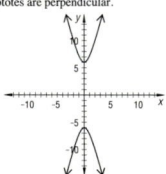

b. $\frac{y^2}{36} - \frac{4x^2}{9} = 1$ **19.** $-\sin\theta$ **21. a.** $23 - 2i$ **b.** $17 + 7i$
c. $\frac{1}{2} + \frac{1}{2}i$

LESSON 12-6 (pp. 765–770)
1. See below. **3.** See below. **5.** $y = -\frac{3}{4}x + \frac{25}{4}$ **7. a.** See below. **b.** paraboloid **9. a.** If an object traveling parallel to the axis of a paraboloid hits the "inside" of the paraboloid, then the object will be reflected through the focus of the paraboloid. **b.** sample: Satellite dishes use this property to collect incoming radio waves. **11.** hyperbola
13. $\frac{y^2}{2} - \frac{x^2}{2} = -6$
1.

3.

7.

axis of symmetry

15.

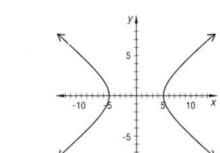

17. 32,000 miles **19.** (c) **21.** (b)

CHAPTER 12 PROGRESS SELF-TEST (p. 775)
1. The cone is formed by one of two intersecting, non-perpendicular lines rotated about the other.
2.

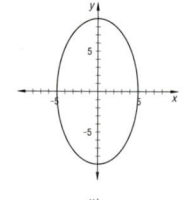

3.

4.

5. The equation for a hyperbola with foci $(c, 0)$ and $(-c, 0)$ and focal constant $2a$ has equation $\frac{x^2}{a^2} - \frac{y^2}{b^2} = 1$, where $b^2 = c^2 - a^2$. Here $c = 7$ and $2a = 10 \Rightarrow a = 5$. Thus, $b^2 = 7^2 - 5^2 = 24$, and the equation for the hyperbola is $\frac{x^2}{25} - \frac{y^2}{24} = 1$.
6.

The ellipse should have the shape of the ellipse below.

7. The symmetry lines of a hyperbola are the axes of the hyperbola. One axis is the line through the foci. The other is the perpendicular bisector of the segment joining the foci.
8. A hyperbola of the form $\frac{x^2}{a^2} - \frac{y^2}{b^2} = 1$ has asymptotes with equations $y = \frac{b}{a}x$ and $y = \frac{-b}{a}x$. Thus the hyperbola $\frac{x^2}{81} - \frac{y^2}{100} = 1$ has asymptotes $y = \frac{10}{9}x$ and $y = -\frac{10}{9}x$.
9. By the Graph Rotation Theorem we can find the image of $y = x^2 - 3x - 4$ under $R_{\pi/3}$ by replacing x by $x\cos\frac{\pi}{3} + y\sin\frac{\pi}{3} = \frac{1}{2}x + \frac{\sqrt{3}}{2}y$ and y by $-x\sin\frac{\pi}{3} + y\cos\frac{\pi}{3} = -\frac{\sqrt{3}}{2}x + \frac{1}{2}y$. This gives $-\frac{\sqrt{3}}{2}x + \frac{1}{2}y = \left(\frac{1}{2}x + \frac{\sqrt{3}}{2}y\right)^2 - 3\left(\frac{1}{2}x + \frac{\sqrt{3}}{2}y\right) - 4 = \frac{1}{4}x^2 + \frac{\sqrt{3}}{2}xy + \frac{3}{4}y^2 - \frac{3}{2}x - \frac{3\sqrt{3}}{2}y - 4$. Simplifying, we get $0 = \frac{1}{4}x^2 + \frac{\sqrt{3}}{2}xy + \frac{3}{4}y^2 + \frac{-3 + \sqrt{3}}{2}x + \frac{-1 - 3\sqrt{3}}{2}y - 4$. **10.** No; this hyperbola has asymptotes $y = \frac{\sqrt{30}}{2}x$ and $y = -\frac{\sqrt{30}}{2}x$. Since the slopes of these lines are not negative reciprocals, the lines are not perpendicular. Thus, the hyperbola is not rectangular. **11.** $B^2 - 4AC = (-6)^2 - 4 \cdot 2 \cdot 18 = 36 - 144 < 0$, so the graph of this equation is either an ellipse, a single point, or the null set. Substituting 0 for x we get $18y^2 + 6y - 110 = 0$. Solving using the quadratic formula gives $y = \frac{-1 \pm \sqrt{21}}{6}$. Since the points $\left(0, \frac{-1 + \sqrt{21}}{6}\right)$ and $\left(0, \frac{-1 - \sqrt{21}}{6}\right)$ are on the graph it cannot be a single point or the null set. Thus the graph is an ellipse. **12.** $B^2 - 4AC = (-2)^2 - 4 \cdot 9 \cdot 1 = 4 - 36 < 0$, so the graph of this equation is either an ellipse, a single point or the null set. Substitute a value, call it a, for x to get the quadratic equation $y^2 - 2ay + (9a^2 - 4a + 22) = 0$. This equation has solution $y = \frac{2a \pm \sqrt{4a^2 - 4(9a^2 - 4a + 22)}}{2} = a \pm \sqrt{a^2 - 9a^2 + 4a - 22} = a \pm \sqrt{-8a^2 + 4a - 22}$. The value under the radical sign is always negative (this can be seen by graphing the parabola $y = -8x^2 + 4x - 22$), thus $9x^2 - 2xy + y^2 - 4x + 22 = 0$ is not defined for any value of x so its graph must be the null set. **13.** $64x^2 - 9y^2 - 576 = 0; A = 64, B = 0, C = -9, D = 0, E = 0, F = -576$

14. The situation is illustrated at night. The center of the orbit is the origin and the sun is at $(c, 0)$ where the unit is one million miles. The maximum and minimum distances from the sun will be achieved when Jupiter intersects the x-axis. From the given information $a - c = 460$ and $c - (-a) = 508$. Adding these equations gives $2a = 968$, so $a = 484$. Thus $c = 24$. So the second focus is $(-24, 0)$ and is 48 million miles from the sun.

The chart below keys the **Progress Self-Test** questions to the objectives in the **Chapter 12 Review** on pages 776–778. This will enable you to locate those **Chapter 12 Review** questions that correspond to questions you missed on the **Progress Self-Test.** The lesson where the material is covered is also indicated in the chart.

Question	1	2	3	4	5	6	7	8	9	10	11
Objective	E	G	G	H	A	D	D	A	B	D	B
Lesson	12-1	12-2	12-3	12-2	12-3	12-1	12-3	12-3	12-4	12-4	12-4

Question	12	13	14	15
Objective	I	I	C	F
Lesson	12-5	12-5	12-5	12-2

CHAPTER 12 REVIEW (pp. 776–778)

1. $\frac{x^2}{64} + \frac{y^2}{289} = 1$ **2.** $\frac{x^2}{9} - \frac{y^2}{7} = 1$ **3.** $\frac{x^2}{81} + \frac{4y^2}{49} = 1$

4. $y = \frac{1}{2}x, y = -\frac{1}{2}x$ **5.** $5x^2 + 6xy + 5y^2 = 72$

6. $\frac{3}{2}x^2 - 5\sqrt{3}xy + \frac{13}{2}y^2 = 1$ **7.** $-\frac{3}{4}x^2 - \frac{\sqrt{3}}{2}xy - \frac{1}{4}y^2 -$ $\frac{1}{2}x + \frac{\sqrt{3}}{2}y + 8 = 0$ **8.** $x^2 + y^2 = 20$ **9.** $-\frac{1}{2}x^2 + \frac{1}{2}y^2 = 120$
10. $-x^2 - 2\sqrt{3}xy - 3y^2 + (8\sqrt{3} - 8)x + (32 - 8\sqrt{3})y + 32\sqrt{3} - 101 = 0$ **11.** $49x^2 + 25y^2 - 1225 = 0; A = 49,$ $B = 0, C = 25, D = 0, E = 0, F = -1225$ **12.** $x^2 + y^2 + 6x - 16y + 38 = 0; A = 1, B = 0, C = 1, D = 6,$ $E = -16, F = 38$ **13.** $7xy - 8 = 0; A = 0, B = 7, C = 0,$ $D = 0, E = 0, F = -8$ **14.** $x^2 - 4y^2 - 4x = 0; A = 1,$ $B = 0, C = -4, D = -4, E = 0, F = 0$
15. The ellipse should have the shape of the ellipse below. **16.** The hyperbola should have the shape of the hyperbola below.

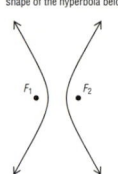

17. at least 2 **18.** Let F_1 and F_2 be the foci of a hyperbola with focal constant k. Let m be the perpendicular bisector of $\overline{F_1F_2}$. If P is any point on the hyperbola, then $|PF_1 - PF_2| = k$. Let $P' = r_m(P)$. Since $r_m(F_1) = F_2$ and $r_m(F_2) = F_1$ and reflections preserve distance $P'F_1 = PF_2$ and $P'F_2 = PF_1$. Therefore $|P'F_1 - P'F_2| = |PF_2 - PF_1| = k$ and so P' is on the hyperbola. This show that the image of any point on the hyperbola is on the hyperbola. To show that the image is the entire hyperbola, note that if Q is on the hyperbola, then $r_m(r_m(Q)) = Q$ and $r(Q)$ is on the hyperbola by the above argument. Thus any point on the hyperbola is the image of some point on the hyperbola.
19. No **20.** Yes **21.** The cone is formed by rotating one of two intersecting, non-perpendicular lines about the other.
22.

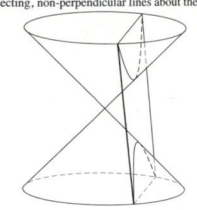

23. a point, a line, two intersecting lines **24.** True
25. 15,120,000 miles **26.** closer

27.

28.

29.

30.

31.

32.

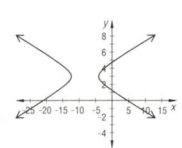

33. $\frac{x^2}{64} + \frac{y^2}{16} = 1$ **34.** $\frac{x^2}{5625} + \frac{y^2}{62,500} = 1$

35.

36.

37.

38.

39. the null set **40.** hyperbola **41.** parabola **42.** hyperbola

LESSON 13-1 (pp. 780–784)

1. False **3.** manipulate one expression to equal a second; rewrite each side independently until equal expressions are obtained; begin with a known identity and derive statements equivalent to it until the proposed identity appears

5. $\begin{array}{l|l} \sin x \cdot \cos x \cdot \tan x & \frac{1}{\csc^2 x} \\ = \sin x \cdot \cos x \cdot \frac{\sin x}{\cos x} & = \sin^2 x \\ = \sin^2 x \end{array}$

So $\sin x \cdot \cos x \cdot \tan x = \frac{1}{\csc^2 x}$ for $x \neq \frac{\pi}{2} + n\pi$ and $x \neq n\pi$, n an integer. **7. a.** $\cot x \cdot \sec x = \csc x$ **b.** $\cot x \cdot \sec x = \frac{\cos x}{\sin x} \cdot \frac{1}{\cos x} = \frac{1}{\sin x} = \csc x$ for $x \neq \frac{\pi}{2} + n\pi$ and $x \neq n\pi$, n an integer. **c.** $x \neq \frac{\pi}{2} + n\pi$ and $x \neq n\pi$, n an integer
9. a.

$y = \cos x$
$y = -\cos(\pi - x)$

b. $\begin{array}{l|l} \cos x & -\cos(\pi - x) \\ & = -(-\cos x) \\ & = \cos x \end{array}$

So, $\cos x = -\cos(\pi - x)$ for all x.
11. a.

$y = \frac{\csc x}{\sec x}$
$y = \cot x$

b. $\begin{array}{l|l} \frac{\csc x}{\sec x} & = \frac{\frac{1}{\sin x}}{\frac{1}{\cos x}} = \frac{1}{\sin x} \cdot \frac{\cos x}{1} = \frac{\cos x}{\sin x} = \cot x \text{ for } x \neq \frac{\pi}{2} + n\pi \end{array}$

and $x \neq n\pi$, n an integer. **13. a.** See below.
b. $\approx -.0001957$ **c.** No **d.** Sample: For $-1.75 \leq x \leq 1.75$, $|f(x) - g(x)| < .01$. **15.** Begin with the Pythagorean identity. For all x, $\cos^2 x + \sin^2 x = 1$ $(\cos^2 x + \sin^2 x)(\cos^2 x - \sin^2 x) = \cos^2 x - \sin^2 x$ $\cos^4 x - \sin^4 x = \cos^2 x - \sin^2 x$ $\cos^4 x + \sin^2 x = \sin^4 x + \cos^2 x$
17. $\log x^3 + \log x^4 = \log x^3 x^4 = \log x^7$, so it is an identity.
19. $\frac{c}{a}$ **21.** $-\frac{\pi}{2}$ **23.** $(4\cos\theta, 4\sin\theta)$
13. a.

$y = x - \frac{x^3}{6} + \frac{x^5}{120}$

$y = \sin x$

LESSON 13-2 (pp. 785–788)

3. $\frac{\pi}{2} + n\pi$, n an integer **5.** (1) double angle formula; (2) Pythagorean Identity; (3) definitions of cot, csc; (4) operation of division
7. a.

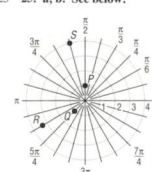

b. $\begin{array}{l|l} \sec x & \tan x \cdot \csc x \\ = \frac{1}{\cos x} & = \frac{\sin x}{\cos x} \cdot \frac{1}{\sin x} \\ & = \frac{1}{\cos x} \end{array}$

So $\sec x = \tan x \cdot \csc x$ for $x \neq \frac{\pi}{2} + n\pi$ and $x \neq n\pi$, n an integer.

9. a.

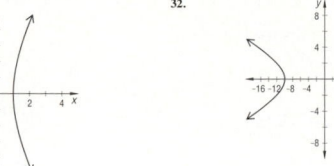

b. $\begin{array}{l|l} \sin 2x & \frac{2\tan x}{1 + \tan^2 x} \\ = 2\sin x \cos x & = \frac{2\tan x}{\sec^2 x} \\ & = 2\tan x \cos^2 x \\ & = 2\frac{\sin x}{\cos x} \cdot \cos^2 x \\ & = 2\sin x \cos x \end{array}$

So $\sin 2x = \frac{2\tan x}{1 + \tan^2 x}$ for $x \neq \frac{\pi}{2} + n\pi$, n an integer.
11. a.

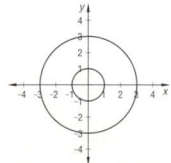

b. $x \neq n\pi$, n an integer **c.** $\frac{1 + \cos 2x}{\sin 2x} = \cot x$
d. $\begin{array}{l} \frac{1 + \cos 2x}{\sin 2x} = \frac{1 + \cos^2 x - \sin^2 x}{2\sin x \cos x} \\ = \frac{\cos^2 x + \cos^2 x}{2\sin x \cos x} \\ = \frac{2\cos^2 x}{2\sin x \cos x} \\ = \frac{\cos x}{\sin x} \\ = \cot x \end{array}$ for $x \neq n\pi$, n an integer
13. a. $x \neq \frac{\pi}{2} + n\pi$ and $x \neq n\pi$, n an integer
b. $\begin{array}{l|l} \csc^2 x \cdot \sin x & \frac{\sec^2 x - \tan^2 x}{\sin x} \\ = \frac{1}{\sin^2 x} \cdot \sin x & = \frac{1}{\sin x} \\ = \frac{1}{\sin x} \end{array}$

So $\csc^2 x \cdot \sin x = \frac{\sec^2 x - \tan^2 x}{\sin x}$ for $x \neq \frac{\pi}{2} + n\pi$ and $x \neq n\pi$, n an integer. **15. a.** $\sqrt{17} \approx 4.123$ **b.** $\frac{1}{\sqrt{17}} \approx .243$ **c.** $-\frac{8}{17} \approx -.471$ **17.** False **19. a.** $(.98)^{25} \approx .60$ **b.** ≈ 0.013 **c.** 24.5 **d.** 0.7 **21. a.** 66 **b.** 4096 **23. a.** $\theta \approx 1.19$ **b.** $\theta \approx 1.19$ or $\theta \approx 4.33$ **c.** $\theta \approx 1.19 + n\pi$ for any integer n **25.** 1024.5

LESSON 13-3 (pp. 789–795)

1. $\approx [2.8, 0°]$ **3.** $\approx [0.7, 270°]$ **5.** See below. **7.** (c)
9. True **11.** $(0, -10)$ **13.** sample: $[10, 60°]$ a circle with radius 8 and center at the pole **17.** samples: $[-6, \frac{\pi}{3}]$, $[6, \frac{4\pi}{3}]$ **19. a.** south **b.** from 10° W of N **c.** from 50° S of E **d.** from 80° W of N
21. $\sin\left(\frac{\pi}{2} + x\right) = \sin\frac{\pi}{2} \cdot \cos x + \cos\frac{\pi}{2} \cdot \sin x$ $= 1 \cdot \cos x + 0 \cdot \sin x$ $= \cos x$
So $\sin\left(\frac{\pi}{2} + x\right) = \cos x$ for all x.
23. a. 1 **b.** 25 **25. a, b.** See below.
5.

25. a, b.

LESSON 13-4 (pp. 796–802)

1. a.

b. $x^2 + y^2 = 4$ **3. a.** $[2, 0]$, $[\sqrt{3}, \frac{\pi}{6}]$, $[\sqrt{2}, \frac{\pi}{4}]$, $[1, \frac{\pi}{3}]$, $[0, \frac{\pi}{2}]$, $[-1, \frac{2\pi}{3}]$;

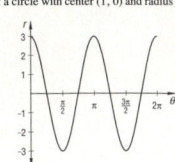

b. For $r \neq 0$, $r = 2\cos\theta \Leftrightarrow r = 2\frac{x}{r} \Leftrightarrow r^2 = 2x \Leftrightarrow x^2 + y^2 = 2x \Leftrightarrow x^2 - 2x + y^2 = 0 \Leftrightarrow (x - 1)^2 + y^2 = 1$, which is the equation of a circle with center $(1, 0)$ and radius 1. **5.** False

7. a.

b. $3, -3$ **c, d.** $[3, 0]$, $[1.5, \frac{\pi}{6}]$, $[-3, \frac{\pi}{2}]$;

$[1.5, \frac{\pi}{6}]$ $[3, 0]$ $[-3, \frac{\pi}{2}]$

9.
(polar graph)

11.

13. a. $[1, 0]$, $[1.72, \frac{\pi}{4}]$, $[2.97, \frac{\pi}{2}]$, $[5.12, \frac{3\pi}{4}]$, $[8.82, \pi]$, $[15.21, \frac{5\pi}{4}]$, $[26.22, \frac{3\pi}{2}]$, $[45.19, \frac{7\pi}{4}]$

b.

15. (a), (b), and (c)

17. a. See below. For $x = 30°$, $2\tan x = 2 \cdot \frac{1}{\sqrt{3}} = \frac{2\sqrt{3}}{3}$ and $\tan 2x = \tan 60° = \sqrt{3}$. So $\tan 2x \neq 2\tan x$ is not an identity. **19.** $9, -\frac{9}{2} + \frac{9\sqrt{3}}{2}i, -\frac{9}{2} - \frac{9\sqrt{3}}{2}i$

17. a.

$y = \tan 2x$ $y = 2\tan x$

LESSON 13-5 (pp. 803–809)

3. a. $(-5, 4)$

b.

imaginary $-5 + 4i$

5. a. $-2 + 4i$

b.

c. Let $W = U + V$. Then, the slope of $\overline{OU} = \frac{-2 - 0}{3 - 0} = -\frac{2}{3}$; the slope of $\overline{UW} = \frac{4 - -2}{-2 - 3} = -\frac{6}{5}$; the slope of $\overline{WV} = \frac{6 - 4}{-5 - -2} = -\frac{2}{3}$; and the slope of $\overline{OV} = \frac{6 - 0}{-5 - 0} = -\frac{6}{5}$. Because the slopes of the opposite sides of $OUWV$ are equal, the opposite sides are parallel and the quadrilateral is a parallelogram. **7.** $\approx [5, 5.64]$ **9.** $\sqrt{a^2 + b^2}$ **11. a.** $\sqrt{45} \approx 6.7$ **b.** $\approx 116.6°$ **13.** Given two complex numbers $a + bi$ and $c + di$ that are not on the same line through the origin, let $O = 0 + 0i$, $P = a + bi$, $Q = (a+c) + (b+d)i$, and $R = c + di$. The slope of $\overline{OP}$ is $\frac{b-0}{a-0} = \frac{b}{a}$; the slope of $\overline{PR}$ is $\frac{(b+d) - b}{(a+c) - a} = \frac{d}{c}$; the slope of $\overline{RQ}$ is $\frac{d - (b+d)}{c - (a+c)} = \frac{b}{a}$; and the slope of $\overline{OQ}$ is $\frac{d - 0}{c - 0} = \frac{d}{c}$. Because the slopes of the opposite sides of $OPRQ$ are equal, the opposite sides are parallel and the quadrilateral is a parallelogram. **15.** Yes, the distance from $u = a + bi$ to $v = c + di$ in the complex plane is $|u - v|$. $u - v = (a - c) + (b - d)i$, and $|u - v| = \sqrt{(a - c)^2 + (b - d)^2}$, which is the distance between the points (a, b) and (c, d). **17.** See below. **19.** See below. **21.** $x^2 + (y - 1)^2 = 1$ **23. a–c.** Sample: Calculate z-scores for the 14 heights and for the 14 weights. Then, pair 4 is the most different because the total difference of the z-scores of D and d for height and weight is greatest; pair 3 is most similar because the total difference of the z-scores for C and c is least. **25. a.** $-1 - 5i$ **b.** $-5i$ **c.** -1 **27.** ≈ 36 miles

17. $\theta = \frac{3\pi}{4}$ **19.**

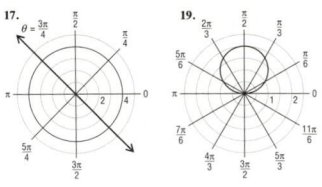

LESSON 13-6 (pp. 810–816)
1. a. See below. **b.** $3\sqrt{2}(\cos 45° + i\sin 45°)$ **3. a.** See below. **b.** $5(\cos 180° + i\sin 180°)$ **5. a.** See below. **b.** $0 - 3i$ **7. a.** $8(\cos 105° + i\sin 105°)$ **b.** size change with magnitude 4 and rotation around the origin of 40° **c.** See below.

1. a.

imaginary • $3 + 3i$

3. a.
(graph)

5. a.

$3\left(\cos\frac{3\pi}{2} + i\sin\frac{3\pi}{2}\right)$

7. c.

z_1z_2 $r = 8 = 2 \cdot 4$ z_1 $r = 2$ $40°$

9. a. $50\left(\cos\frac{5\pi}{4} + i\sin\frac{5\pi}{4}\right)$ **b.** 50 **c.** Sample: $\frac{5\pi}{4}$
11. $3\left(\cos\frac{11\pi}{6} + i\sin\frac{11\pi}{6}\right)$ **13. a.** $2.4(\cos 120° + i\sin 120°)$
b. a size change with magnitude $\frac{1}{5}$ and a rotation about the origin of $-100°$ **15. a.** $4(\cos 240° + i\sin 240°)$
b. $\frac{z_1}{z_2} = \frac{10 - 10\sqrt{3}i}{\frac{5}{2} + \frac{5\sqrt{3}}{2}i} = \frac{10 - 10\sqrt{3}i}{\frac{5}{2}(1 + \sqrt{3}i)} = \frac{4 - 4\sqrt{3}i}{1 + \sqrt{3}i} = \frac{(4 - 4\sqrt{3}i)(1 - \sqrt{3}i)}{(1 + \sqrt{3}i)(1 - \sqrt{3}i)} = \frac{-8 - 8\sqrt{3}i}{4} = -2 - 2\sqrt{3}i$, and $[4, 240°] = 4(\cos 240° + i\sin 240°) = 4(-\frac{1}{2} - \frac{\sqrt{3}}{2}i) = -2 - 2\sqrt{3}i$.
17. $\frac{z_1}{z_2} = \frac{r_1(\cos\theta_1 + i\sin\theta_1)}{r_2(\cos\theta_2 + i\sin\theta_2)} = \frac{r_1(\cos\theta_1 + i\sin\theta_1)(\cos\theta_2 - i\sin\theta_2)}{r_2(\cos\theta_2 + i\sin\theta_2)(\cos\theta_2 - i\sin\theta_2)} = \frac{r_1}{r_2}\cdot\frac{(\cos\theta_1 + i\sin\theta_1)(\cos(-\theta_2) + i\sin(-\theta_2))}{(\cos^2\theta_2 + \sin^2\theta_2)} = \frac{r_1}{r_2}(\cos(\theta_1 - \theta_2) + i\sin(\theta_1 - \theta_2))$
19. a. $z^2 = 4(\cos 30° + i\sin 30°)$, $z^3 = 8(\cos 45° + i\sin 45°)$, $z^4 = 16(\cos 60° + i\sin 60°)$, $z^5 = 32(\cos 75° + i\sin 75°)$
b. $z^{10} = 2^{10}(\cos 150° + i\sin 150°)$; the pattern is $z^n = 2^n(\cos(n \cdot 15°) + i\sin(n \cdot 15°))$
21. a. See below. **b.** See below.
23. a. $\begin{bmatrix} 0 & 4 & 5 \\ 3 & -3 & 0 \end{bmatrix}$ **b.** $R_{90} = \begin{bmatrix} 0 & -1 \\ 1 & 0 \end{bmatrix}$, $S_2 = \begin{bmatrix} 2 & 0 \\ 0 & 2 \end{bmatrix}$ and $\begin{bmatrix} 2 & 0 \\ 0 & 2 \end{bmatrix}\begin{bmatrix} 0 & -1 \\ 1 & 0 \end{bmatrix}\begin{bmatrix} 0 & 4 & 5 \\ 3 & -3 & 0 \end{bmatrix} = \begin{bmatrix} -6 & 6 & 0 \\ 0 & 8 & 10 \end{bmatrix}$, which is the matrix for $\triangle A'B'C'$. **25.** $(a - b)^5 = a^5 - 5a^4b + 10a^3b^2 - 10a^2b^3 + 5ab^4 - b^5$ **27.** ≈ 5019.8 mm³

21. a.
(graph)

b.
(polar graph)

LESSON 13-7 (pp. 817–821)
1. $81\left(\cos\frac{4\pi}{5} + i\sin\frac{4\pi}{5}\right)$
3. a. $z^2 = 4\left(\cos\frac{\pi}{3} + i\sin\frac{\pi}{3}\right)$
$z^3 = 8\left(\cos\frac{\pi}{2} + i\sin\frac{\pi}{2}\right)$
$z^4 = 16\left(\cos\frac{2\pi}{3} + i\sin\frac{2\pi}{3}\right)$
$z^5 = 32\left(\cos\frac{5\pi}{6} + i\sin\frac{5\pi}{6}\right)$
$z^6 = 64(\cos\pi + i\sin\pi)$

b.

imaginary

c. They lie on a spiral.
5. $5(\cos 10° + i\sin 10°)$, $5(\cos 130° + i\sin 130°)$, $5(\cos 250° + i\sin 250°)$ **7.** $i, -i$ **9.** $(-\sqrt{3} + i)^6 = (2(\cos 150° + i\sin 150°))^6 = 2^6(\cos 900° + i\sin 900°) = 64(\cos 180° + i\sin 180°) = -64$ **11. a.** $[2, 15°]$, $[2, 75°]$, $[2, 135°]$, $[2, 195°]$, $[2, 255°]$, $[2, 315°]$ **b.** See below. **13. a.** $z = \sqrt{2} + \sqrt{2}i$, $\sqrt{2} - \sqrt{2}i$, $-\sqrt{2} + \sqrt{2}i$, $-\sqrt{2} - \sqrt{2}i$ **b.** See below.

11. b.

imaginary

13. b.
(graph)

15. $7\cos 10° + (7\sin 10°)i \approx 6.89 + 1.22i$ **17. a.** See below. **b.** $r = e^{\pi\theta/3} \Leftrightarrow \ln r = \ln e^{\pi\theta/3} = \frac{\pi}{3}\ln e = \frac{\pi}{3}(1) = \frac{1}{3}\pi\theta$, so $k = \frac{1}{3}$ **19.** $-13x^2 + 34\sqrt{3}xy - 47y^2 + (68\sqrt{3} + 78)x + (-102\sqrt{3} - 188)y - 321 - 204\sqrt{3} = 0$ **21.** $-\frac{1}{2}$

17. a.

LESSON 13-8 (pp. 822–827)
1. a. $z_1 = 0 + 0i$; $z_2 = 0 + 0i$; $z_3 = 0 + 0i$ **b.** Yes; $\lim_{n\to\infty} |z_n|$ is finite. **3.** $|z_4| \approx .407$; Yes **5. a.** AC = .202; BC = .002 **b.** AZ = .202; BZ = .002; MAGZ $\approx$ .202 **c.** Yes

7.

9. $-\frac{\sqrt{2}}{2} - \frac{\sqrt{2}}{4}i$ **11. a.** $\sqrt[10]{10}(\cos 4° + i\sin 4°)$; $\sqrt[10]{10}(\cos 124° + i\sin 124°)$; $\sqrt[10]{10}(\cos 244° + i\sin 244°)$

b.

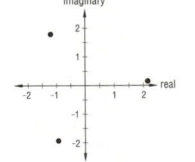

imaginary

13. a. $z_1 \approx 3.32(\cos 25.2° + i\sin 25.2°)$; $z_2 \approx 3.32(\cos 205.2° + i\sin 205.2°)$ **b.** $z_1z_2 \approx 11(\cos 230.4° + i\sin 230.4°) \approx -7 - 8.48i$; $\frac{z_1}{z_2} = \cos(-180°) + i\sin(-180°) = \cos 180° + i\sin 180° = -1$ **15.** ≈ 452
17. a. on the polar axis **b.** $y = 0$

19. a.

$y = \sec\theta$

$y = \sec(\pi - \theta)$

b. Let $\theta = \frac{\pi}{6}$. Then $\sec\frac{\pi}{6} = \frac{1}{\cos\frac{\pi}{6}} = \frac{2}{\sqrt{3}}$, and $\sec\left(\pi - \frac{\pi}{6}\right) = \sec\frac{5\pi}{6} = \frac{1}{\cos\frac{5\pi}{6}} = -\frac{2}{\sqrt{3}}$. So $\sec\theta = \sec(\pi - \theta)$ is not an identity.

CHAPTER 13 PROGRESS SELF-TEST (p. 830)
1. Singularities are signalled by vertical asymptotes which occur at $x = \frac{\pi}{2} + n\pi$, for n an integer. **2.** False, a graph can only show the functions on a finite domain; a proof must show that the equation is an identity for all values for which the functions are defined. Also, an automatic grapher may not indicate isolated singularities. **3. a.** Singularities occur where $\tan x$ is not defined, that is, at $x = \frac{\pi}{2} + n\pi$, for all integers n
b. $2\cos^2 x \cdot \tan x = 2\cos^2 x \cdot \frac{\sin x}{\cos x}$
$= 2\cos x \sin x$
$= \sin 2x$
4.

$[4, -185°]$ polar axis

5.

• $-4 + 3i$

6.

$2 \cdot 2\left(\cos\frac{\pi}{2} + i\sin\frac{\pi}{2}\right)$

7. $r^2 = x^2 + y^2 = 5^2 + (5\sqrt{3})^2 = 25 + 25 \cdot 3 = 25 + 75 = 100$, so $r = \pm 10$. $\tan\theta = \frac{y}{x} = \frac{-5\sqrt{3}}{5} = -\sqrt{3}$, so $\theta = -\frac{\pi}{3} \pm n\pi$. The choice of values of θ and r depend on matching the angle to the quadrant. If we choose $r = 10$ then, since $(5, -5\sqrt{3})$ is in the fourth quadrant, some possible values for θ are $\frac{2\pi}{3}$ or $\frac{5\pi}{3}$. If we choose $r = -10$, possible choices for θ are $\frac{2\pi}{3}$ or $-\frac{4\pi}{3}$. Thus, $[10, -\frac{\pi}{3}], [10, -\frac{5\pi}{3}], [-10, \frac{2\pi}{3}]$, and $[-10, \frac{-4\pi}{3}]$ are all polar coordinate representations of $(5, -5\sqrt{3})$. **8. a.** $r^2 = a^2 + b^2 = \left(-\frac{1}{2}\right)^2 + \left(\frac{\sqrt{3}}{2}\right)^2 = \frac{1}{4} + \frac{3}{4} = 1$, so $r = \pm 1$. $\tan\theta = \frac{b}{a} = \frac{\frac{\sqrt{3}}{2}}{-\frac{1}{2}} = -\sqrt{3}$. Since $\left(-\frac{1}{2}, \frac{\sqrt{3}}{2}\right)$ is in the second quadrant, we can choose $r = 1$ and $\theta = 120°$ giving the polar form $[1, 120°]$. **b.** $-\frac{1}{2} + \frac{\sqrt{3}}{2}i = r\cos\theta + ir\sin\theta = \cos 120° + i\sin 120°$. **9.** By DeMoivre's Theorem, $[3(\cos 20° + i\sin 20°)]^4 = 3^4(\cos 4\cdot20° + i\sin 4\cdot20°) = 81(\cos 80° + i\sin 80°) \approx 14.07 + 79.77i$. **10.** By the Roots of a Complex Number Theorem the fourth roots of $81(\cos 260° + i\sin 260°)$ are $\sqrt[4]{81}\left(\cos\left(\frac{260°}{4} + k\frac{360°}{4}\right) + i\sin\left(\frac{260°}{4} + k\frac{360°}{4}\right)\right) = \sqrt[4]{81}\left[\cos(65° + k\cdot90) + i\sin(65° + k\cdot90)\right]$ where $k = 0, 1, 2, 3$. Thus the four fourth roots are $3(\cos 65° + i\sin 65°)$; $3(\cos 155° + i\sin 155°)$; $3(\cos 245° + i\sin 245°)$; $3(\cos 335° + i\sin 335°)$.

11. a. Sample: When $\theta = \frac{\pi}{6}$, $r = \sin 3\left(\frac{\pi}{6}\right) = \sin\frac{\pi}{2} = 1$. When $\theta = -\frac{\pi}{6}$, $r = \sin 3\left(-\frac{\pi}{6}\right) = \sin-\frac{\pi}{2} = -1$, so two points on the graph of $r = 2\sin 3\theta$ are $[1, \frac{\pi}{6}], [-1, -\frac{\pi}{6}]$
b. sample:

$[-1, \frac{\pi}{6}]$ $[1, \frac{\pi}{6}]$

polar axis

12. a. Converting z_1 to trigonometric form gives $z_1 = -3i = 3\left(\cos\frac{3\pi}{2} + i\sin\frac{3\pi}{2}\right)$. By the Product of Complex Numbers Theorem $z_1z_2 = 3\cdot5\left(\cos\left(\frac{3\pi}{2} + \frac{\pi}{12}\right) + i\sin\left(\frac{3\pi}{2} + \frac{\pi}{12}\right)\right) = 15\left(\cos\frac{19\pi}{12} + i\sin\frac{19\pi}{12}\right)$. **b.** The absolute value of z_1 is multiplied by 5 which is a scale change of magnitude 5. $\frac{\pi}{12}$ is added to the argument which is a rotation around the origin by an angle of $\frac{\pi}{12}$. **13. a.** By the Division of Complex Numbers Theorem $\frac{z_1}{z_2} = [\frac{6}{2}, \pi - \pi] = [3, 0]$. **b.** The absolute value of z_1 is divided by 2 which is a size change of magnitude $\frac{1}{2}$. Then π is subtracted from the argument which is a rotation around the origin by an angle of $-\pi$. **14.** For $\theta \neq \frac{\pi}{2} + n\pi$ where n is an integer, $(1 - \sin^2\theta)(1 + \tan^2\theta) = (\cos^2\theta)\left(1 + \frac{\sin^2\theta}{\cos^2\theta}\right) = \cos^2\theta + \sin^2\theta = 1$.

15.

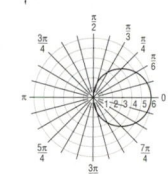

The chart below keys the **Progress Self-Test** questions to the objectives in the **Chapter 13 Review** on pages 831–833. This will enable you to locate those **Chapter 13 Review** questions that correspond to questions you missed on the **Progress Self-Test**. The lesson where the material is covered is also indicated in the chart.

Question	1	2	3a	3b	4	5	6	7	8	9	10
Objective	D	F	C	C	H	J	J	G	B	E	E .
Lesson	13-2	13-2	13-2	13-1	13-3	13-5	13-6	13-3	13-6	13-7	13-7
Question	11	12	13	14	15						
Objective	I	A	A	C	A						
Lesson	13-4	13-6	13-6	13-2	13-4						

CHAPTER 13 REVIEW (pp. 831–833)
1. $15(\cos 189° + i\sin 189°)$ **2.** $[9.1, \frac{\pi}{4}]$ **3.** $z_2 = 2\left(\cos\frac{3\pi}{10} + i\sin\frac{3\pi}{10}\right)$ **4.** $z_2 = [\frac{5}{2}, \pi]$ **5. a.** $\sqrt{244} \approx 15.62$
b. sample: $\tan^{-1}(-1.2) \approx 129.8°$ **6. a.** 1 **b.** sample: $\frac{\pi}{3}$
7. $[2, 150°]$ **8.** $[\sqrt{29}, \tan^{-1}\frac{5}{2}] \approx [5.39, 68.2°]$ **9.** $[\frac{1}{8}, \frac{\pi}{4}]$
10. $-\frac{3}{2} - \frac{3\sqrt{3}}{2}$ **11.** $4\cos\frac{2\pi}{5} + \left(4\sin\frac{2\pi}{5}\right)i \approx 1.2 + 3.8\,i$
12. samples: $2(\cos 225° + i\sin 225°)$, $2(\cos(-135°) + i\sin(-135°))$ **13.** $(1 - \cos^2 x)(1 + \cot^2 x) = \sin^2 x\left(1 + \frac{\cos^2 x}{\sin^2 x}\right) = \sin^2 x + \cos^2 x = 1$
14.

$$\sin\left(\theta - \frac{\pi}{4}\right) \;\Big|\; -\cos\left(\theta + \frac{\pi}{4}\right)$$
$$= \sin\theta \cdot \cos\frac{\pi}{4} - \cos\theta \cdot \sin\frac{\pi}{4} \;\Big|\; = -\left[\cos\theta \cdot \cos\frac{\pi}{4} - \sin\theta \cdot \sin\frac{\pi}{4}\right]$$
$$= (\sin\theta)\frac{1}{\sqrt{2}} - (\cos\theta)\frac{1}{\sqrt{2}} \;\Big|\; = -\left((\cos\theta)\frac{1}{\sqrt{2}} - (\sin\theta)\frac{1}{\sqrt{2}}\right)$$
$$= \frac{1}{\sqrt{2}}(\sin\theta - \cos\theta) \;\Big|\; = -\left(\frac{1}{\sqrt{2}}(\cos\theta - \sin\theta)\right)$$
$$\Big|\; = \frac{1}{\sqrt{2}}(\sin\theta - \cos\theta)$$

15. $\sin^2 x + \cos^2 x = 1$
$$1 - \cos^2 x = \sin^2 x$$
$$\frac{1 - \cos^2 x}{\sin^2 x} = \frac{\sin^2 x}{\sin^2 x}$$
$$\frac{1}{\sin^2 x} - \frac{\cos^2 x}{\sin^2 x} = 1$$
$$\csc^2 x - \cot^2 x = 1$$

16.
$$\frac{\sin\theta}{\cos\theta \cdot \tan\theta} \;\Big|\; 1$$
$$= \frac{\sin\theta}{\cos\theta \cdot \frac{\sin\theta}{\cos\theta}}$$
$$= \frac{\sin\theta}{\sin\theta}$$
$$= 1$$
So $\frac{\sin\theta}{\cos\theta \cdot \tan\theta} = 1$ is an identity for all $\theta \neq \frac{\pi}{2} + n\pi$ and $\theta \neq n\pi$, where n is an integer. **17.** The values must be excluded because $\csc x$ and $\cot x$ are undefined for $x = n\pi$, where n is an integer. **18. a.** The left-hand side of the equation has singularities at $\theta = \frac{\pi}{2} + n\pi$ or $\theta = n\pi$, where n is an integer.

b. $\{\theta: \theta \neq \frac{\pi}{2} + n\pi$ and $\theta \neq n\pi$, where n is an integer$\}$
19. a. i. $x = 3$ **ii.** none **b. i.** False; it is not true for $x = 3$. **ii.** True; the denominator is not zero for any real value of x.
20. False; $\frac{1}{1 - \cos x}$ has a singularity for $\cos x = 1$, or for $x = 2n\pi$, where n is an integer. The function $\frac{1}{1 - \cos x}$ has singularities for $\cos^2 x = 1$, or for $x = n\pi$, where n is an integer. **21.** $z^5 = 32(\cos 225° + i\sin 225°) = -16\sqrt{2} - 16\sqrt{2}i$
22. $z^4 = 81(\cos 240° + i\sin 240°) = -\frac{81}{2} - \frac{81\sqrt{3}}{2}i$
23. $4(\cos 3° + i\sin 3°)$, $4(\cos 93° + i\sin 93°)$, $4(\cos 183° + i\sin 183°)$, $4(\cos 273° + i\sin 273°)$
24. $\sqrt[4]{2}(\cos 30° + i\sin 30°)$, $\sqrt[4]{2}(\cos 90° + i\sin 90°)$, $\sqrt[4]{2}(\cos 150° + i\sin 150°)$, $\sqrt[4]{2}(\cos 210° + i\sin 210°)$, $\sqrt[4]{2}(\cos 270° + i\sin 270°)$, $\sqrt[4]{2}(\cos 330° + i\sin 330°)$
25. False **26. a.** See below. **b.** Yes; the graphs seem to coincide. **27. a.** See below. **b.** No; the graphs do not coincide.
26.

27.

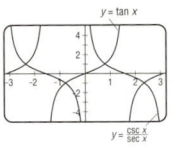

$y = \tan x$

$y = \frac{\csc x}{\sec x}$

---

28. a.

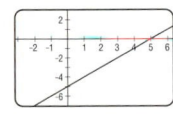

b. Yes; the graphs seem to coincide. **29.** $(0, -4)$
30. $(3\cos 265°, 3\sin 265°) \approx (-.26, -2.99)$ **31.** $\approx [5.4, 21.8°]$
32. $\approx [3.6, 236.3°]$ **33.** samples: $[8, \frac{5\pi}{6}], [8, -\frac{7\pi}{6}], [-8, \frac{11\pi}{6}]$
34. $P = [\frac{10}{\sqrt{3}}, \frac{\pi}{6}] = (5, \frac{5}{\sqrt{3}})$ **35, 36.** See below. **37.** (b)
38. samples: $[-2, \frac{\pi}{6}], [-2, \frac{-7\pi}{6}]$ **39.** $2\cos 6\left(\frac{2\pi}{3}\right) = 2\cos 4\pi = 2(1) = 2$, so $[2, \frac{2\pi}{3}]$ is on the graph of $r = 2\cos 6\theta$.
40. samples: $[1, \frac{\pi}{2}], [\sqrt{2}, \frac{\pi}{4}]$ **41.** See below. **42.** See below.

35, 36.

$[3, \frac{\pi}{6}]$
$[-2, 0]$

41.

42.

43. a. See below. **b.** For $\theta \neq \frac{\pi}{2} + n\pi$, where n is an integer, $r = 4\sec\theta = \frac{4}{\cos\theta} \Leftrightarrow r\cos\theta = 4 \Leftrightarrow x = 4$, which is a line.
44. a. See below. **b.** $r = \frac{1}{3}\cos\theta \Leftrightarrow r\frac{1}{3}\cdot\frac{x}{r} \Leftrightarrow r^2 = \frac{x}{3} \Leftrightarrow x^2 + y^2 = \frac{x}{3} \Leftrightarrow x^2 - \frac{x}{3} + y^2 = 0 \Leftrightarrow x^2 - \frac{x}{3} + \frac{1}{36} + y^2 = \frac{1}{36} \Leftrightarrow \left(x - \frac{1}{6}\right)^2 + y^2 = \frac{1}{36}$, which is a circle. **45.** See below.
46. See below.

43.

44. a.

45.

imaginary

• $4 - 5i$

46.

imaginary

• $8(\cos 130° + i\sin 130°)$

47. a.

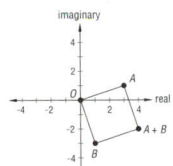

imaginary

A
O
real
$A + B$
B

b. Let $C = A + B$. The slope of $\overline{OA} = \frac{1 - 0}{3 - 0} = \frac{1}{3}$; the slope of $\overline{AC} = \frac{-2 - 1}{4 - 3} = -3$; the slope of $\overline{BC} = \frac{-2 - -3}{4 - 3} = \frac{1}{3}$; and the slope of $\overline{OB} = \frac{-3 - 0}{1 - 0} = -3$. Because the slopes of the opposite sides of $OACB$ are equal, the opposite sides are parallel and the quadrilateral is a parallelogram.

48. a.

imaginary

z z^5
real
z^4

b. $z^5 = (2(\cos 72° + i\sin 72°))^5 = (2^5(\cos 5\cdot72° + i\sin 5\cdot72°)) = 32(\cos 360° + i\sin 360°) = 32 + 0i = 32$

absolute value of a complex number The distance of the graph of the number from the origin or pole. Also called *modulus*. (806)

acceleration due to gravity The acceleration of a free-falling object toward a more massive object caused by gravitational forces, on the surface of the Earth approximately 32 feet (or 9.8 meters) per second per second. (113)

Addition Counting Principle If two finite sets A and B are mutually exclusive, then $N(A \cup B) = N(A) + N(B)$. (403) (General Form): For any finite sets A and B, $N(A \cup B) = N(A) + N(B) - N(A \cap B)$. (404)

Addition Identities for the Cosine and Sine
For all real numbers α and β,
$$\cos(\alpha + \beta) = \cos \alpha \cos \beta - \sin \alpha \sin \beta$$
$$\sin(\alpha + \beta) = \sin \alpha \cos \beta + \cos \alpha \sin \beta. \quad (690)$$

alternating current Current created by rotating a rectangular wire through the magnetic field created by two opposite poles of a magnet. (382)

alternative hypothesis, H_1 An alternative hypothesis to the null hypothesis. (635)

ambiguous case The situation in which two noncongruent triangles each satisfy a given SSA condition. (316)

amplitude One-half the difference between the maximum and minimum values of a sine wave. (340)

angle The union of two rays (its **sides**) with the same endpoint (its **vertex**). (268)

angle of incidence The angle at which an object hits a surface, measured from the ray perpendicular to the surface at the point of contact. (765)

angle of reflection The angle at which an object bounces off of a surface, measured from the ray perpendicular to the surface at the point of contact. (765)

Arccos function See *inverse cosine function*.

Archimedean spiral The graph of $r = k\theta$, for $\theta > 0$, in the polar plane. (801)

Arcsin function See *inverse sine function*.

Arctan function See *inverse tangent function*.

Area Scale Change Theorem If the scale change $S:(x, y) \rightarrow (ax, by)$ is applied to a region, then the area of the image is $|ab|$ times the area of the preimage. (608)

area under a curve The area between the curve and the x-axis. (606)

Argand diagram See *complex plane*.

argument of a complex number For the complex number $[r, \theta]$, θ. (806)

arithmetic sequence A sequence in which the difference between adjacent terms is constant. (457)

arithmetic series An indicated sum of the first n terms of an arithmetic sequence. (470)

astronomical units (a. u.) A unit of distance equal to the radius of the earth's orbit. (732)

asymptote A line that the graph of a function $y = f(x)$ approaches as the variable x approaches a fixed value or increases or decreases without bound. (147)

asymptotes of a hyperbola Two lines that the hyperbola approaches farther and farther away from its center. (744)

automatic grapher A calculator or computer software on which the graph of a relation can be displayed. (144)

average rate of change between two points The slope of the segment joining the points. (18)

axes of a hyperbola The symmetry lines of a hyperbola. (746)

axis of a cone The fixed line in the generation of a cone. (728)

axis of symmetry In a plane, a reflecting line ℓ over which a figure can be mapped onto itself; in space, a line around which a figure can be rotated onto itself. (164)

back-to-back stemplot A stemplot in which the stem is written in the center of the display, with one set of leaves to the right of the stem and another set of leaves to the left. (19)

bar graph A two-dimensional display of data in which one axis labels categories or variables and the other is a numerical scale typically with counts or percents. (12)

base The number r in the expression r^n. (203)

base of a logarithmic function The number b in the logarithmic function on $f(x) = \log_b x$. (225)

base of an exponential function The number b in the exponential function $f(x) = ab^x$. (216)

bearing An angle measured counterclockwise from due north. (284)

bell-shaped curve See *normal curve*.

biased sample A sample that is not random. (6)

binomial A polynomial with two terms. (526)

binomial coefficients The coefficients in the series expansion of $(x + y)^n$; the combinations $_nC_k$. (500)

binomial distribution function, $B(k, n, p)$ The function which maps the ordered triple (k, n, p) onto the probability of getting exactly k successes in n binomial trials, each of which has probability p of success. (594)

binomial experiment An experiment with a fixed number of trials in which each trial has only two possible outcomes (often called **success** or **failure**), the trials are independent, and each trial has the same probability p of success. (503)

BINOMIAL GENERATOR A particular BASIC program that gives values in a binomial probability distribution. (593)

binomial probability distribution The probability distribution generated from the probability of x successes in a binomial experiment. (505, 592)

Binomial Theorem When n is a nonnegative integer,
$(x + y)^n = {}_nC_0x^n + {}_nC_1x^{n-1}y + {}_nC_2x^{n-2}y^2 + \ldots$
$+ \; {}_nC_kx^{n-k}y^k + \ldots + {}_nC_ny^n = \sum_{k=0}^{n} {}_nC_kx^{n-k}y^k.$ (500)

bivariate data Data involving two variables. (79)

box plot A visual representation of the five-number summary of a data set in which a box represents the part of the data from the first to the third quartiles and two segments protruding from the box represent part or all of the rest of the data. Also called *box-and-whiskers plot*. (32)

capacitance Property of an alternating current circuit created when the current flow leads the voltage by $\frac{\pi}{2}$. (351)

capture-recapture method A method of approximating the number of animals in an area by capturing and tagging some animals, releasing them back into the area, then capturing a second group of animals and seeing how many of the tagged animals are caught again. (5)

cardioid The graph of $r = a\,(\cos\theta - 1)$ or $r = a\,(\sin\theta - 1)$ in the polar plane. (801)

ceiling function See *rounding-up function*.

center of a hyperbola The point of intersection of the axes of the hyperbola. (746)

center of an ellipse The intersection of the axes of the ellipse. (735)

center of gravity of a data set The point whose coordinates are the means of the corresponding coordinates of the points in the data set. (94)

center of symmetry for a figure The center of the rotation under which a point-symmetric figure is mapped onto itself. (164)

Central Limit Theorem Suppose random samples of size n are selected from any population with mean μ and standard deviation σ. Then, as n increases,
1. the mean M of the distribution of sample means approaches μ
2. the standard deviation s of the distribution of sample means approaches $\frac{\sigma}{\sqrt{n}}$; and
3. the distribution of sample means approaches a normal curve. (641)

Change of Base Theorem For all values of a, b, and c for which the logarithms exist: $\log_b a = \dfrac{\log_c a}{\log_c b}$. (242)

characteristics of a binomial distribution The numbers n and p of the binomial distribution function $B(k, n, p)$. (594)

circle graph A display of numerical data in which the total is represented by the entire circle and sectors of the circle represent parts of the total in proportion to their contribution to the total. Also called *pie chart*. (11)

Circular Arc Length Formula If s is the length of the arc of a central angle of θ radians in a circle of radius r, then $s = r\theta$. (274)

circular functions The trigonometric functions, when defined in terms of the unit circle. (289)

circular motion Movement of a point around a circle. (355)

Circular Sector Area Formula If A is the area of the sector formed by a central angle of θ radians in a circle of radius r, then $A = \frac{1}{2}r^2\theta$. (275)

cliometrics The discipline of applying mathematics to history. (61)

coding matrix A matrix that is used to generate a code and whose inverse decodes. (704)

coefficient The numbers $a_n, a_{n-1}, a_{n-2}, \ldots a_0$ of the polynomial $a_nx^n + a_{n-1}x^{n-1} + \ldots + a_1x + a_0$; more generally, a constant factor of a variable term. (525)

column matrix A matrix with only one column. (665)

combination A selection of objects in which the order of the objects does not matter. (490)

combination of n things taken r at a time A subset of r objects from a set of n objects. (490)

common logarithm A logarithm with base 10. (226)

complementary events Two events that are mutually exclusive and whose union is the entire sample space. (406)

Complements Theorem For every θ,
$\sin\left(\frac{\pi}{2} - \theta\right) = \cos\theta$, and $\cos\left(\frac{\pi}{2} - \theta\right) = \sin\theta$. (307)

complex conjugates   A pair of complex numbers of the form $a + bi$ and $a - bi$. (558)

complex number, *a + bi*   A number of the form $a + bi$, where a and b are real numbers and $i = \sqrt{-1}$. (557)

complex plane   A coordinate plane for representing complex numbers. Also called *Argand diagram*. (803)

composite   The function $f \circ g$ defined by $(f \circ g)(x) = f(g(x))$, whose domain is the set of values of x in the domain of g for which $g(x)$ is in the domain of f. (180)

composition of functions   The binary operation that maps two functions f and g onto their composite $f \circ g$. (180)

concave down   Technically, a property of a graph whose center of curvature is below the graph. (614)

concave up   Technically, a property of a graph whose center of curvature is above the graph. (614)

confidence interval   An interval within which a certain percentage of outcomes from an experiment would be expected to occur. (647)

conic section   The intersection of a two-napped cone with a plane: an ellipse, a parabola, or a hyperbola. (727)

Conjugate Zeros Theorem   Let $p(x) = a_n x^n + a_{n-1} x^{n-1} + \ldots + a_1 x + a_0$ where $a_n, a_{n-1}, \ldots a_1, a_0$ are all real numbers, and $a_n \neq 0$. If $z = a + bi$ is a zero of $p(x)$, then the complex conjugate of z, namely $a - bi$, is also a zero of $p(x)$. (565)

constant   A variable whose values do not change in the course of a problem. (previous course)

constant function   Any function with an equation $f(x) = k$, where k is a fixed value. (18, 437)

Continuous Change Formula   If an initial quantity P grows or decays continuously at an annual rate r, the amount $A(t)$ after t years is given by $A(t) = Pe^{rt}$. (232)

continuous function   Informally, a function whose graph can be drawn without lifting the pencil off the paper. (100)

convergent sequence   A sequence that has a finite limit. (464)

convergent series   A series for which the sequence of partial sums has a finite limit. (484)

corollary   A theorem that can be quickly deduced from another theorem. (783)

correlation coefficient, *r*   A measure of the strength of the linear relation between two variables. (105)

cosecant (csc) of a real number *x* $\dfrac{1}{\sin x}$, for $\sin x \neq 0$. (377)

cosine (cos) of an acute angle in a right triangle $\dfrac{\text{side adjacent to the angle}}{\text{hypotenuse}}$. (279)

cosine (cos) of a real number *x*   The first coordinate of the image of $(1,0)$ under a rotation of magnitude x about the origin. (286)

cosine function   A function that maps x onto $\cos x$ for all x in its domain. (289)

cotangent (cot) of a real number *x* $\dfrac{\cos x}{\sin x}$, for $\sin x \neq 0$. (377)

cross section of a figure   The intersection of a plane with the figure. Also called *section*. (728)

cubic model   A model for a data set of the form $g(x) = ax^3 + bx^2 + cx + d$. (123)

cumulative binomial probability table   A table that gives, for a fixed n and p, the probabilities for k or fewer successes. (599)

cycle   One period of a sine wave. (299)

cycloid   The path of a point on the interior (a **prolate cycloid**) or the exterior (a **curtate cycloid**) of a circle as the circle rolls along a line. (771)

cylinder graph   A three-dimensional display of numerical data in which the total is represented by an entire cylinder and slices of the cylinder represent parts of the total in proportion to their contribution to the total. (15)

data   The plural of *datum*, the Latin word for fact; a piece of information. (3)

decreasing   A function is decreasing on an interval if the segment connecting any two points on the graph of the function over that interval has negative slope. (18)

deductive reasoning   Reasoning adhering to strict principles of logic. (61)

default domain   A domain that is set into an automatic grapher until changed by the user. (144)

default range   A range that is set into an automatic grapher until changed by the user. (144)

default settings   Settings which are set automatically and stay that way until the user changes them. (47)

default window   A viewing window which is preset into an automatic grapher until changed by the user. (144)

degenerate conic sections   The intersection of a cone and a plane containing the vertex of the cone. (759)

degree   A unit for measuring angles or rotations. (268)

degree of a polynomial   In a polynomial with a single variable, the number n in the polynomial $a_n x^n + a_{n-1} x^{n-1} + a_{n-2} x^{n-2} + \ldots + a_1 x + a_0$; in a polynomial with more than one variable, the largest sum of the exponents of the variables in any term. (525)

DeMoivre's Theorem If $z = r(\cos \theta + i \sin \theta)$ and n is an integer, then $z^n = r^n(\cos n\theta + i \sin n\theta)$. (818)

demonstration file A data set prestored on a system as an example. (46)

dependent events Events A and B such that $P(A \cap B) \neq P(A) \cdot P(B)$. (415)

dependent variable The second variable in a relation. (80)

determinant, det A particular number associated with a matrix. For the 2×2 matrix $\begin{bmatrix} a & c \\ b & d \end{bmatrix}$, $ad - bc$. (704)

deviation See *error (in a prediction)*.

deviation from the mean The difference of a data value from the mean of a distribution. (51)

dimensions of a matrix $m \times n$, where m is the number of rows and n is the number of columns of a matrix. (664)

discontinuous function A function that is not continuous. (100)

disjoint sets Two sets with no elements in common. (403)

disk The union of a circle and its interior. (275)

divergent sequence A sequence that does not have a finite limit. (464)

divergent series A series for which the sequence of partial sums does not have a finite limit. (484)

dividend The number or expression a when a is divided by b. (544)

Division of Complex Numbers Theorem If $z_1 = r_1(\cos \theta_1 + i \sin \theta_1)$ and $z_2 = r_2(\cos \theta_2 + i \sin \theta_2)$, then $\frac{z_1}{z_2} = \frac{r_1}{r_2}(\cos (\theta_1 - \theta_2) + i \sin (\theta_1 - \theta_2))$. (813)

divisor The number or expression b when a is divided by b. (544)

domain The set of first elements of a function; more generally, the replacement set for a variable. (80)

dot frequency diagram A graph of a frequency distribution in which over each individual value is a number of dots equal to its frequency. (43)

Double Angle Identities For all real numbers θ,
$$\sin 2\theta = 2 \sin \theta \cos \theta; \text{ and}$$
$$\cos 2\theta = \cos^2\theta - \sin^2\theta$$
$$= 2 \cos^2\theta - 1$$
$$= 1 - 2 \sin^2\theta.$$
Also called *Double Argument Identities*. (695)

doubling time In an exponential growth situation, the time it takes a quantity to grow to double its original amount. (221)

dynamical system A finite set S and a function whose domain is S and whose range is a subset of S. (822)

e $\lim\limits_{n \to \infty} \left(1 + \frac{1}{n}\right)^n$; the base of natural logarithms. (231)

eccentricity of an ellipse $\frac{c}{a}$ for the ellipse with equation $\frac{x^2}{a^2} + \frac{y^2}{b^2} = 1$ where $c = \sqrt{|a^2 - b^2|}$. (739)

edit To change data values or lines in a program. (46)

element An object in a matrix. A member of a set. (664, previous course)

ellipse Given two points F_1 and F_2 (the **foci**) and a positive real number k (the **focal constant**) with $k > F_1F_2$, the set of all points P in the plane which satisfy $PF_1 + PF_2 = k$. (731)

ellipsoid The three-dimensional surface formed by rotating an ellipse about its major axis. (766)

epicycloid The curve generated by a point of a circle rolling externally upon a given circle. (772)

error (in a prediction) The difference between the observed and the expected value of a variable. Also called *deviation*. (92)

even function A function f such that for all x in its domain, $f(-x) = f(x)$. (162)

expected value See *mean of a probability distribution*.

experiment A situation to be studied. (398)

explicit formula A formula which determines the nth term of a sequence directly from n. (456)

exponent The number n in the expression r^n. (203)

exponential decay A situation that can be modeled by an exponential function $f(x) = ab^x$ with base $0 < b < 1$. (218)

exponential decay curve The graph of an exponential function $f(x) = ab^x$ with base $0 < b < 1$. (217)

exponential equation An equation with a variable exponent. (241)

exponential function A function with a formula of the form $f(x) = ab^x$, where a $\neq 0$, $b > 0$ and $b \neq 1$. (216)

exponential growth A situation that can be modeled by an exponential function $f(x) = ab^x$ with base $b > 1$. (217)

exponential growth curve The graph of an exponential function $f(x) = ab^x$ with base $b > 1$. (217)

exponential model An exponential function $f(x) = ab^x$ that is a model of a situation. (220)

extrapolation Estimating a value beyond known values of data. (89)

f^{-1}   The function which is the inverse of the function f. (186)

Factor-Solution-Intercept Equivalence Theorem   For any polynomial f, the following are equivalent: $f(c) = 0$; c is a zero of f; $(x - c)$ is a factor of f; c is an x-intercept of the graph of $y = f(x)$; the remainder when f is divided by $(x - c)$ is 0. (551)

Factor Theorem   For a polynomial $f(x)$, a number c is a solution to $f(x) = 0$ if and only if $(x - c)$ is a factor of f. (550)

fair experiment   An experiment with a sample space in which all outcomes are equally likely. Also called an *unbiased experiment*. (399)

file   A program or a set of data organized on a disk. (46)

file name   The name by which a file is accessed. (46)

finite sequence   A sequence whose terms can be put in one-to-one correspondence with the set of integers from 1 to n. (456)

finite series   An indicated sum of the terms of a finite sequence. (470)

first (lower) quartile   In a data set, the median of the numbers smaller than the median. (32)

five-number summary   The quartiles together with the minimum and maximum of the data set. (32)

floor function   See *greatest integer function*.

focal constant of a hyperbola   See *hyperbola*.

focal constant of an ellipse   See *ellipse*.

focus (foci) of a hyperbola   See *hyperbola*.

focus (foci) of an ellipse   See *ellipse*.

Formula for $_nP_r$ Theorem   The number of permutations of n objects taken r at a time is

$$_nP_r = n(n-1)(n-2) \cdot \ldots \cdot (n-r+1) = \frac{n!}{(n-r)!}. \quad (421)$$

frequency   The number of times an event occurs; the reciprocal of the period of a sine wave; the number of cycles per time period. (38, 341)

frequency distribution   A function mapping events onto their frequencies. (38)

frequency table   A table defining a frequency distribution. (38)

function   A set of ordered pairs in which each first element is paired with exactly one second element. A correspondence between two sets A and B in which each element of A corresponds to exactly one element of B. (80)

function notation   The notation $f(x)$ for the value of a function f when the value of the independent variable is x. (82)

fundamental function   The function which represents a pure sound tone. (383)

Fundamental Theorem of Algebra   If $p(x)$ is any polynomial of degree $n > 1$ with complex coefficients, then $p(x)$ has at least one complex zero. (563)

Galton board   See *quincunx*.

general solution to a trigonometric equation   The solution that arises from considering all real values of the variable. (371)

Geometric Addition Theorem   Given two complex numbers $a + bi$ and $c + di$ that are not on the same line through the origin in the complex plane, their sum $(a + c) + (b + d)i$ is the fourth vertex of a parallelogram with consecutive vertices $a + bi$, 0, and $c + di$. (804)

geometric sequence   A sequence in which the ratio of adjacent terms is constant. (458)

geometric series   An indicated sum of the terms of a geometric sequence. (476)

gradient (grad)   A unit for measuring angles. 100 grads = 90 degrees. (271)

Graph Rotation Theorem   In a relation described by a sentence in x and y, the following two processes yield the same graph:
 1. replacing x by $x \cos \theta + y \sin \theta$ and y by $-x \sin \theta + y \cos \theta$;
 2. applying the rotation of magnitude θ about the origin to the graph of the original equation. (752)

Graph Scale Change Theorem   In an equation for a relation, if x is replaced by $\frac{x}{a}$ and y by $\frac{y}{b}$, where $a \neq 0$ and $b \neq 0$, the graph of the resulting equation is the image of the original equation under a scale change mapping (x, y) to (ax, by). (168)

Graph Standardization Theorem   In a relation described by a sentence in x and y, the following processes yield the same graph:
 1. applying the scale change $(x, y) \rightarrow (ax, by)$ where $a \neq 0$ and $b \neq 0$, followed by applying the translation $(x, y) \rightarrow (x + h, y + k)$;
 2. applying the linear change $(x, y) \rightarrow (ax + h, by + k)$ to the graph of the original sentence;
 3. replacing x by $\frac{x-h}{a}$ and y by $\frac{y-k}{b}$. (352)

Graph Translation Theorem   In an equation for a relation, if x is replaced by $x - h$ and y by $y - k$, the graph of the resulting equation is the image of the graph of the original equation under the translation which maps (x, y) to $(x + h, y + k)$. (152)

great circle   A circle on a sphere that has the same center as the sphere. (321)

great circle route   A path between two points on the surface of the Earth along a great circle arc. (321)

greatest integer function, $\lfloor \; \rfloor$ The function f such that $f(x)$ is the greatest integer less than or equal to x. Also called *rounding-down function* or *floor function*. (100)

Greenwich meridian   See *prime meridian*.

grouping   A technique used to factor polynomials which contain groups of terms with common factors. (574)

growth factor   In an exponential growth situation, the base of the exponential function. (220)

half-life   In an exponential decay situation, the time it takes a quantity to decay to half its original amount. (221)

hard copy   A printed copy of text or displays from a screen or file. (46)

harmonics   Functions whose frequencies are multiples of the fundamental function for a pure sound tone. (383)

harmonic series   The sum of the terms of a sequence of reciprocals of the positive integers; more generally, the sum of the terms of a sequence of reciprocals of an arithmetic sequence. (486)

histogram   A bar graph in which the range of values of a numerical variable are broken into non-overlapping intervals (usually of equal width), and side-by-side bars display the number of values that fall into each interval. (38)

homogeneous form   The increasing of the dimension of a point or matrix in order to be treated: for the point (x, y) the homogeneous form is $(x, y, 1)$. For the 2×2 transformation matrix $\begin{bmatrix} a & b \\ c & d \end{bmatrix}$ the homogeneous form is $\begin{bmatrix} a & b & 0 \\ c & d & 0 \\ 0 & 0 & 1 \end{bmatrix}$. (710)

homogeneous population   A population in which members are very similar on some measure. (57)

horizontal scale change   A transformation that maps (x, y) to (ax, y) for all (x, y), where $a \neq 0$ is a constant. (167)

horizontal scale factor   The number a in the transformation that maps (x, y) to (ax, by). (168)

hyperbola   Given two points F_1 and F_2 (the **foci**) and a positive real number k (the **focal constant**) with $k < F_1F_2$, the set of all points in the plane which satisfy $|PF_1 - PF_2| = k$. (742)

hypocycloid   The curve generated by a point of a circle rolling internally upon a given circle. (771)

hypothesis   In statistics, a statement to be tested. (635)

hypothesis testing   The use of a statistical test to determine the likelihood that a statement is true. (635)

identity   An equation that is true for all values of the variable(s) for which the expressions are defined. (304)

identity function   A function that maps each element in its domain onto itself. (187)

image   The result of a transformation. (156)

imaginary axis   The vertical axis (axis of second coordinates) in a complex plane. (803)

imaginary numbers   The number $i = \sqrt{-1}$ and its nonzero, real-number multiples. (556)

imaginary part of a complex number   The real number b in the complex number $a + bi$. (557)

increasing   A function is increasing on an interval if the segment connecting any two points on the graph of the function over that interval has positive slope. (18)

independent events   Events A and B such that $P(A \cap B) = P(A) \cdot P(B)$. (414)

independent variable   The first variable in a relation. (80)

index   A variable indicating the position of a number in an ordered list or sequence. (26)

inductance   Property of an alternating current circuit created when the current flow lags behind the voltage. (348)

inferential reasoning   Reasoning based upon principles of probability. (61)

infinite loop   A sequence of steps in a computer program that will repeat again and again until the program is interrupted. (464)

infinite sequence   A sequence whose terms can be put in one-to-one correspondence with the set of all positive integers.

infinite series   An indicated sum of the terms of an infinite sequence. (470)

infinity, ∞ The limit of a sequence whose terms after a given point become larger than any fixed number one might choose; greater than any given number. (463)

inflection point   A point on the graph of a function where the graph changes from concave down to concave up or vice-versa. (614)

initial value   In a function f modeling a situation, the value $f(0)$. (220)

in-phase circuit   An alternating current circuit in which the voltage and current flow coincide. (348)

interest rate In an investment, the percent by which the principal is multiplied to obtain the interest paid to the investor. (230)

International Date Line The meridian which is $180°W$ (and $180°E$) of the prime meridian. (321)

interpolation Estimating a value between known values of data. (89)

interquartile range (IQR) The difference between the third quartile and the first quartile. (32)

intersection, $A \cap B$ The set of elements that are in both A and B. (405)

invariant Unchanged by a particular transformation. (158)

inverse circular functions The inverses of the sine, cosine, and tangent functions with appropriate restrictions on their domains. (365)

inverse cosine function, $\cos^{-1}$, Arccos The function that maps x onto the number or angle y whose cosine is x, for $0 \le y \ge \pi$. (365)

Inverse Function Theorem Given any two functions f and g, then f and g are inverse functions if and only if $f(g(x)) = x$ for all x in the domain of g and $g(f(x)) = x$ for all x in the domain of f. (187)

inverse matrices Two matrices whose product is the identity matrix for matrix multiplication. (700)

inverse of a function The relation formed by switching the ordered pairs of a given function. (185)

Inverse of a 2 x 2 Matrix Theorem If

$$ad - bc \ne 0, \text{ then } \begin{bmatrix} a & b \\ c & d \end{bmatrix}^{-1} = \begin{bmatrix} \dfrac{d}{ad-bc} & \dfrac{-b}{ad-bc} \\ \dfrac{-c}{ad-bc} & \dfrac{a}{ad-bc} \end{bmatrix}. \text{ (702)}$$

inverse sine function, $\sin^{-1}$, Arcsin The function that maps x onto the number or angle y whose sine is x, for $-\frac{\pi}{2} \le y \le \frac{\pi}{2}$. (365)

inverse tangent function, $\tan^{-1}$, Arctan The function that maps x onto the number or angle y whose tangent is x, for $-\frac{\pi}{2} < y < \frac{\pi}{2}$. (365)

inverse transformations Two transformations S and T such that $S \circ T$ and $T \circ S$ map each point onto itself. (700)

isometry A composite of translations, rotations, and reflections; a distance-preserving transformation. (187, 754)

latitude The approximate number of degrees along a meridian that a point is north or south of the equator. (322)

Law of Cosines In any $\triangle ABC$, $c^2 = a^2 + b^2 - 2ab \cos C$. (310)

Law of Sines In any triangle ABC, $\dfrac{\sin A}{a} = \dfrac{\sin B}{b} = \dfrac{\sin C}{c}$. (315)

leading coefficient of a polynomial The coefficient of the term of highest degree of the polynomial. (525)

least squares line See *line of best fit*.

limit of a sequence A number to which the terms of a sequence get closer so that beyond a certain term of the sequence all terms are as close as desired to that number. Also called *limiting value*. (463)

linear change A composite of scale changes and translations. (352)

linear function A function which can be described by an equation of the form $y = mx + b$, where m and b are constants. (86)

line graph The graph of a set of ordered pairs connected by line segments in order. (17)

line of best fit The line that fits a set of data points with the smallest value for the sum of the squares of the errors (vertical distances) from the data points to the line. Also called *regression line*, *least squares line*. (92)

line of symmetry See *axis of symmetry*.

locus definitions Definitions that describe curves as sets of all points satisfying certain conditions. (771)

logarithm function A function which maps x onto $\log_b x$ for a fixed b and all positive real numbers x. (226)

logarithmic spiral The polar graph of $r = ka^\theta$, where $a > 0$. (801)

logarithmic transformation A transformation under which a variable x is replaced by $\log x$. (248)

Logarithm of 1 Theorem For any base b, $\log_b 1 = 0$. (236)

Logarithm of a Power Theorem For any base b, any positive real number x and any real number p, $\log_b x^p = p \log_b x$. (237)

Logarithm of a Product Theorem For any base b and for any positive real numbers x and y, $\log_b(xy) = \log_b x + \log_b y$. (236)

Logarithm of a Quotient Theorem For any base b and for any positive real numbers x and y,
$\log_b \left(\dfrac{x}{y} \right) = \log_b x - \log_b y$. (237)

logarithm of x to the base b, $\log_b x$ The power to which b must be raised to equal x; that is, the number y such that by $b^y = x$. (225)

longitude The number of degrees that a meridian is E or W of the prime meridian, used as a coordinate of a location on earth. (321)

lower quartile See *first quartile*.

major axis of an ellipse The axis that contains the foci of the ellipse. (735)

Mandelbrot set The set of complex numbers c for which the limit of the sequence defined by

$$\begin{cases} z_1 = c \\ z_n = z_{n-1}^2 + c \text{ for } n > 1 \end{cases} \text{ is not infinity. (822)}$$

MANDELZOOM A computer program for generating graphs of the Mandelbrot set. (824)

MANDELZOOM A computer program for generating graphs of the Mandelbrot set. (824)

margin of error Half the length of a confidence interval. (647)

mathematical model A mathematical description of a real situation, often involving some simplification and assumptions about that situation. (79)

matrix An array of mn elements arranged in a rectangle with m rows and n columns. (664)

Matrix Basis Theorem Suppose t is a transformation represented by a 2×2 matrix. If $t(1, 0) = (x_1, y_1)$ and

$t(0, 1) = (x_2, y_2)$, then t has the matrix $\begin{bmatrix} x_1 & x_2 \\ y_1 & y_2 \end{bmatrix}$. (680)

matrix multiplication An operation on an $m \times n$ matrix A and an $n \times p$ matrix B whose result is the product matrix $A \cdot B$, an $m \times p$ matrix whose element in row i and column j is the sum of the products of elements in row i of A and corresponding elements in column j of B. (665)

matrix representing a transformation A matrix M such that if F is a matrix for a geometric figure, then $M \cdot F$ is a matrix for the image of F under a transformation. (671)

maximum The largest value in a set. (19)

mean The sum of the elements of a numerical data set divided by the number of items in the data set. Also called *average*. (24)

Mean of a Binomial Distribution Theorem The mean μ of a binomial distribution with n trials and probability p of success on each trial is given by $\mu = np$. (601)

mean of a probability distribution For the probability distribution $\{(x_1, P(x_1)), (x_2, P(x_2)), \ldots, (x_n, P(x_n))\}$, the number $\sum_{i=1}^{n}(x_i \cdot P(x_i))$. Also called *expected value*. (432)

measure of an angle A number that represents the size (and, sometimes, the direction of rotation) used to generate an angle. (268)

measure of center A statistic describing a typical value of a numerical data set; the *mean* and *median,* and sometimes the *mode*. Also called *measure of central tendency*. (24)

measure of central tendency See *measure of center*.

measure of spread A statistic that describes how far data are from a center of a distribution. (51)

median The middle value of a set of data placed in increasing order. (24)

meridian A semicircle of a great circle on the surface of the Earth from the north pole to the south pole. (321)

method of least squares The process of finding the line of best fit. (93)

middle quartile See *second quartile*.

minimum The smallest value in a set. (19)

minor axis of an ellipse The axis perpendicular to the major axis. (735)

minute A unit for measuring angles. 60 minutes $= 1$ degree. (269)

mode The most common item(s) in a data set. (25)

modulus See *absolute value of a complex number*.

monomial A polynomial with one term. (526)

Monte Carlo method The method of using random numbers and related probabilities to simulate events for the purpose of solving a problem. (443)

Multiplication Counting Principle Let A and B be any finite sets. The number of ways of choosing one element from A and then one element from B is $N(A) \cdot N(B)$. (409)

multiplicity of a zero For a zero r of a polynomial, the highest power of $(x - r)$ that appears as a factor of the polynomial. (563)

mutually exclusive events Two events with no outcomes in common. (403)

nappe The part of a cone on one side of the cone's vertex. (728)

natural logarithm, *ln* A logarithm to the base e. (233)

nautical mile A unit of distance frequently used in navigation, sometimes defined as the length of an arc AB on the circumference of the earth with a central angle of one minute. (285)

$_nC_r$ Calculation Theorem For all whole numbers n and r, with $r \le n$, $_nC_r = \dfrac{n!}{(n-r)! \ r!}$. (491)

Negative Exponent Theorem For all $x \ne 0$ and n for which x^n is defined, $x^{-n} = \dfrac{1}{x^n}$. (210)

negative relation   A situation in which there is a negative correlation between two variables. (106)

nested loops   A loop in a computer program that is inside another loop. (425)

n factorial, $n!$ For n a positive integer, the product of the positive integers up to and including n. In symbols, $n! = n \cdot (n-1) \cdot (n-2) \cdot (n-3) \cdot \ldots \cdot 3 \cdot 2 \cdot 1$. (411)

nonlinear models   A model for a data set that is not a linear function. (112)

normal curve   The graph of a normal distribution. Also called *bell-shaped curve*. (613)

normal distribution   The probability distribution function $f(z) = \frac{1}{\sqrt{2\pi}} e^{-\frac{z^2}{2}}$ or any of its offspring. (613)

not A   The set of outcomes that is in the sample space but not in the event A. (406)

nth partial sum   The sum of the first n terms of a sequence. (476)

nth root   A number whose nth power is a given number. (202)

nth root function   A function with an equation of the form $y = x^{1/n}$, where n is an integer with $n \geq 2$. (205)

nth roots of unity   The zeros of $x^n - 1$. (567)

null hypothesis, H_0   A hypothesis that a particular phenomenon does not occur. (635)

Number of Zeros of a Polynomial Theorem   A polynomial of degree $n \geq 1$ with complex coefficients has exactly n complex zeros, if multiplicities are counted. (564)

observed values   Data collected from sources such as experiments or surveys. (92)

odd function   A function f such that for all x in its domain, $f(-x) = -f(x)$. (163)

1.5 X IQR criterion   A criterion under which those elements of a data set greater than $1.5 \times$ IQR plus the third quartile or less than the first quartile minus $1.5 \times$ IQR are considered to be outliers. (34)

operating characteristic curve   A graph of the number of defective items in a batch of a fixed quantity versus the probability that the batch will be accepted. (510)

Opposites Theorem   For every θ, $\sin(-\theta) = -\sin\theta$, $\cos(-\theta) = \cos\theta$, and $\tan(-\theta) = -\tan(\theta)$. (305)

oscilloscope   An instrument for representing the oscillations of varying voltage or current on the fluorescent screen of a cathode-ray tube. (339)

outcome   A possible result of an experiment. (398)

outlier   An element of a set of numbers which is very different from most or all of the other elements. (19)

out-of-phase circuit   An alternating current circuit in which the current flow lags behind the voltage. (348)

paraboloid   The three-dimensional surface formed by rotating a parabola about its axis. (768)

parent (function)   A simple form or the simplest form of a class of functions, from which other members of the class can be derived by transformations. (145)

partial sum   See *nth partial sum*.

Pascal's Triangle   The values of $_nC_r$ arranged in an array in the form of a triangle; the $(r+1)$st term in row n of Pascal's Triangle is $_nC_r$. (494)

pentagonal numbers   Terms of the sequence $p(n) = \frac{3}{2}n^2 - \frac{1}{2}n$, that is, 1, 5, 12, 22, ... (555)

percentile   The pth percentile of a set of numbers is a value in the set such that p percent of the numbers are less than or equal to that value. (33)

percentile curve   A graph in which each ordered pair (x, y) represents the percent y of the data in the set that are less than or equal to x. (654)

perfect correlation   A correlation of 1 or -1; a situation in which all data points lie on the same line. (106)

periodic function   A function f for which there is a positive real number p such that $f(x + p) = f(x)$ for all x. The smallest such positive number is called the **period** of the function. (299)

Periodicity Theorem   For every θ, and for every integer n: $\sin(\theta + 2\pi n) = \sin\theta$, $\cos(\theta + 2\pi n) = \cos\theta$, and $\tan(\theta + \pi n) = \tan\theta$. In degrees, $\sin(\theta + n \cdot 360°) = \sin\theta$, $\cos(\theta + n \cdot 360°) = \cos\theta$, and $\tan(\theta + n \cdot 180°) = \tan\theta$. (301)

permutation   An arrangement of a set of objects. (420)

permutation of n objects taken r at a time   An arrangement of r objects from a set of n objects. (421)

Permutation Theorem   There are $n!$ permutations of n different elements. (420)

petal curve   See *rose curve*.

phase shift   The least positive or the greatest negative horizontal translation that maps the graph of a circular function onto a given sine wave. (346)

pie chart   See *circle graph*.

point matrix   The matrix $\begin{bmatrix} a \\ b \end{bmatrix}$ when it represents the point (a, b). (670)

point of discontinuity   A point at which a function is not continuous. (100)

point-symmetric figure   A figure that can be mapped onto itself by a rotation of 180°. (164)

polar axis   A ray, usually horizontal and drawn to the right, through the pole of a polar coordinate system, from which magnitudes of rotations are measured. (789)

polar coordinates, [r, θ]   Description of a point in a polar coordinate system. (789)

polar coordinate system   A system in which a point is identified by a pair of numbers [r, θ] where $|r|$ is the distance of the point from a fixed point (the **pole**), and θ is a magnitude of rotation from the polar axis. (789)

polar grid   A grid of rays and concentric circles for plotting points and sketching curves in polar coordinates. (791)

pole   See *polar coordinate system*.

Polynomial Difference Theorem $y = f(x)$ is a polynomial function of degree n if and only if, for any set of x-values that forms an arithmetic sequence, the nth differences of corresponding y-values are equal and non-zero. (531)

polynomial function   A function whose rule can be written as a polynomial. (526)

polynomial in x   An expression of the form $a_nx^n + a_{n-1}x^{n-1} + a_{n-2}x^{n-2} + \ldots + a_1x + a_0$, where n is a nonnegative integer and $a_n \neq 0$. (524)

polynomial model   A polynomial function used to estimate data in a set. (123)

population   The set of all individuals or objects to be studied. (4)

population standard deviation   See *standard deviation*.

population variance   See *variance*.

position of a term in a sequence   The domain value of a term in a sequence. (456)

positive relation   A situation in which there is a positive correlation between two variables. (106)

power   An expression of the form r^n. (203)

predicted values   Data predicted by a model. (92)

prime meridian   The meridian through Greenwich, England. Also called the *Greenwich meridian*. (321)

probabilities   Numbers which indicate the measure of certainty of an event. (397)

probability   The branch of mathematics that studies models of uncertainty. (398)

probability distribution   A function that maps each value of a random variable onto its probability. (431)

probability of an event   A number from 0 to 1 that measures the certainty or uncertainty of the event. In a finite sample space S in which each outcome in S is equally likely, the probability of the event E occurring,
$$P(E) = \frac{\text{number of outcomes in the event}}{\text{number of outcomes in the sample space}}. \quad (399)$$

Probability of a Union Theorem   If A and B are mutually exclusive events in the same finite sample space, then $P(A \cup B) = P(A) + P(B)$.
(General Form): If A and B are any events in the same finite sample space, then $P(A \text{ or } B) = P(A \cup B) = P(A) + P(B) - P(A \cap B)$. (403, 405)

Probability of Complements Theorem   If A is any event, then $P(\text{not } A) = 1 - P(A)$. (406)

Product of Complex Numbers Theorem   If $z_1 = r_1(\cos \theta_1 + i \sin \theta_1)$ and $z_2 = r_2(\cos \theta_2 + i \sin \theta_2)$, then $z_1z_2 = r_1r_2(\cos (\theta_1 + \theta_2) + i \sin (\theta_1 + \theta_2))$. (812)

product matrix   See *matrix multiplication*.

Properties of Probabilities   Let S be the sample space associated with an experiment, let E be any event in S, and let $P(E)$ be the probability of E, then:
(i) $0 \leq P(E) \leq 1$.
(ii) If $E = S$, then $P(E) = 1$.
(iii) If $E = \varnothing$, then $P(E) = 0$. (401)

pseudo-random numbers   Numbers, generated by computers or calculators, that simulate random numbers. (437)

pure tone   A tone in which air pressure varies sinusoidally with time. (339)

Pythagorean Identity   For every θ, $\cos^2 \theta + \sin^2 \theta = 1$. (304)

quadratic models   A quadratic function used to estimate data in a set. (112)

quadratic relation   The set of ordered pairs (x, y) satisfying an equation of the form $Ax^2 + Bxy + Cy^2 + Dx + Ey + F = 0$. (717)

quadric surface   A surface in three dimensions whose equation is of the second degree in x, y, and z; an ellipsoid, paraboloid, or hyperboloid. (772)

quartic model   A fourth degree polynomial function used to estimate data in a set. (123)

quartiles   The three values which divide an ordered set into four subsets of approximately equal size. See *first (lower) quartile, second (middle) quartile, third (upper) quartile*. (32)

quincunx   An apparatus designed by Sir Francis Galton to illustrate binomial experiments. Also called a *Galton board*. (654)

quotient The answer to a division problem. The polynomial $q(x)$ when $f(x)$ is divided by $d(x)$ where $f(x) = q(x) d(x) + r(x)$, and either $r(x) = 0$ or the degree of $r(x)$ is less than the degree of $d(x)$. (544)

radian A unit for measuring angles or rotations. 2π radians $= 360°$. (269)

radical The symbol $\sqrt{}$ used to denote square roots or nth roots. (203)

random numbers A set of numbers such that each number has the same probability of occurring, each pair of consecutive numbers has the same probability of occurring, each trio of consecutive numbers has the same probability of occurring, and so on. (437)

random outcomes Outcomes of an experiment that have the same probability. (4)

random sample A sample chosen in a way so that every member of the population had an equal chance of being chosen. (4)

random variable A variable whose values are numbers determined by the outcome of an experiment. (431)

range The difference between the largest and smallest values in a set. The set of second elements of a function. (80)

rank ordered Sequenced in order on some scale. (31)

rational exponent An exponent that is a rational number. (208)

rational power functions A function with an equation of the form $y = ax^{m/n}$, where m and n are nonzero integers. (211)

real axis The horizontal axis (axis of first coordinates) in a complex plane. (803)

real part of a complex number The real number a in the complex number $a + bi$. (557)

reciprocal trigonometric functions The secant, cosecant, and cotangent functions. (377)

rectangular hyperbola A hyperbola whose asymptotes are perpendicular. (754)

recursive formula A formula for a sequence in which the first term or first few terms are given, and the nth term is expressed in terms of the preceding term(s). (456)

reflection-symmetric figure A figure that can be mapped onto itself by a reflection over some line ℓ. (164)

regression line See *line of best fit*.

relation A set of ordered pairs. (80)

relative frequency The ratio of the number of times an event occurred to the number of times it could have occurred. (40)

relative frequency distribution A function mapping events onto their relative frequencies. (38)

remainder The polynomial $r(x)$ when $f(x)$ is divided by $d(x)$ and $f(x) = q(x) d(x) + r(x)$. Either $r(x) = 0$ or the degree of $r(x)$ is less than the degree of $d(x)$. (545)

Remainder Theorem If a polynomial $f(x)$ is divided by $x - c$, then the remainder is $f(c)$. (547)

removable singularity A point of discontinuity of a function which can be ''removed'' by adding a single point to the graph of the function. (786)

rescaling See *scaling*.

revolution A unit for measuring rotations. 1 revolution counterclockwise $= 360°$. (268)

RND(1), RND In BASIC, a distribution of pseudo-random numbers with decimal values between 0 and 1. (437)

Roots and Coefficients of Polynomials Theorem For the polynomial equation $a_n x^n + a_1 x^{n-1} + a_2 x^{n-2} + \ldots + a_{n-1} x + a_n = 0$, the sum of the roots is $-a_1$, the sum of the products of the roots two at a time is a_2, the sum of the products of the roots three at a time is $-a_3$, $\ldots$, and the product of all the roots is $\begin{cases} a_n \text{ if } n \text{ is even,} \\ -a_n, \text{ if } n \text{ is odd.} \end{cases}$ (580)

Roots of a Complex Number Theorem For any positive integer n, the n distinct roots of $z^n = r(\cos\theta + i\sin\theta)$, $r > 0$, are
$z = \sqrt[n]{r}\left[(\cos\frac{\theta}{n} + k\frac{360°}{n}) + i\sin\left(\frac{\theta}{n} + k\frac{360°}{n}\right)\right]$ where $k = 0, 1, 2, \ldots, n - 1$. (819)

rose curve The graph of the polar equation $r = c + a\sin b\theta$ or $r = c + a\cos b\theta$, where b is a positive integer and $a \neq 0$. Also called *petal curve*. (798)

Rotation Matrix Theorem The matrix for R_θ, the rotation of magnitude θ about the origin, is $\begin{bmatrix} \cos\theta & -\sin\theta \\ \sin\theta & \cos\theta \end{bmatrix}$. (684)

rotation-symmetric figure A figure that can be mapped to itself by a nonzero rotation. (164)

rounding-down function See *greatest integer function*.

rounding-up function, $\lceil \ \rceil$ The function which pairs each number x with the smallest integer greater than or equal to x. Also called *ceiling function*. (100)

row matrix A matrix with only one row. (665)

Rule of 72 The formula $t = \frac{72}{r}$, used to estimate the length of time t it takes to double an investment at an interest rate of $r\%$. (245)

sample   The subset of a population that is studied in an experiment. (4)

sample space   The set of all possible outcomes of an experiment. (398)

sample standard deviation   See *standard deviation*.

sample variance   See *variance*.

sampling distribution   A distribution of the means of samples from the same population. (640)

scale change (in the plane)   The transformation that maps (x, y) to (ax, by), where $a \neq 0$ and $b \neq 0$ are constants. (167)

scale change (of data)   A transformation that maps each data value x_i in a set of data $\{x_1, x_2, \ldots, x_n\}$ to ax_i, where $a \neq 0$ is a constant. (174)

scale factor   The non-zero constant by which each data value is multiplied in a scale change. (174)

scaling   Applying a scale change to a data set. Also called *rescaling*. (174)

scatter plot   A graph of a finite set of ordered pairs in the coordinate plane. (17)

secant (sec) of a real number x $\dfrac{1}{\cos x}$, for $\cos x \neq 0$. (377)

second   A unit for measuring angles. 60 seconds = 1 minute. (269)

second (middle) quartile   The median. (32)

section   See *cross section*.

sector of a circle   That part of the circle's disk that is on or in the interior of a given central angle. (275)

Selections without Replacement Theorem   Let S be a set with n elements. If a choice of an element from S is made n times without replacement, then there are $n!$ possible arrangements of elements. (411)

Selections with Replacement Theorem   Let S be a set with n elements. If a choice of an element from S is made k times with replacement, then there are n^k possible arrangements of elements. (410)

sequence   A function whose domain is a set of consecutive integers greater than or equal to a fixed integer k. (456)

sequence of partial sums   The sequence whose nth term is the sum of the first n terms of a given sequence. (482)

series   An indicated sum of terms of a sequence. (470)

series expansion of $(x + y)^n$ The rewriting of the power of the binomial as a series in which the $(k + 1)$st term is $_nC_k\, x^{n-k} y^k$. (499)

sexagesimal   Based on 60; the number system with base 60. (269)

sides of an angle   The two rays whose union is the angle. (268)

Σ (sigma)   The symbol for sum. (26)

sigma-notation   See *summation notation*.

significance level   The probability level (often .05 or .01) for accepting a hypothesis in an experiment. (635)

similar figures   Figures which are the images of each other under a composite of reflections, rotations, translations, and size changes. (187)

simulated data   Data for a real event that have been obtained without actual observation of the event. (438)

sine function   A function that maps x onto $\sin x$ for all x in its domain. (289)

sine (sin) of an acute angle in a right triangle $\dfrac{\text{side opposite the angle}}{\text{hypotenuse}}$. (279)

sine (sin) of a real number x   The second coordinate of the image of $(1, 0)$ under a rotation of magnitude x about the origin. (286)

sine wave   The image of the graph of the sine or cosine function under a composite of translations and scale changes. (340)

singularity   An isolated value for which a function is undefined. (785)

size change   A scale change in which the scale factors are equal; a transformation that maps (x, y) to (kx, ky), where k is a constant. (168)

slide rule   An instrument with a sliding rule and a logarithmic scale that allows calculations to be performed mechanically. (238)

slope   For the segment joining (x_1, y_1) and (x_2, y_2), the number $\dfrac{y_2 - y_1}{x_2 - x_1}$. (18)

Spherical Law of Cosines   If ABC is a spherical triangle with sides a, b, and c, then $\cos c = \cos a \cos b + \sin a \sin b \cos C$. (324)

Spherical Law of Sines   If ABC is a spherical triangle with sides a, b, and c, then $\dfrac{\sin a}{\sin A} = \dfrac{\sin b}{\sin B} = \dfrac{\sin c}{\sin C}$. (324)

spherical triangle   A triangle whose sides are arcs of great circles. (324)

square matrix   A matrix with the same number of rows as columns. (704)

square numbers   Terms of the sequence $s_n = n^2$, that is, $1, 4, 9, 16, 25, \ldots$ (528)

standard deviation   The square root of the sample variance (s) or population variance (σ). (51)

standard form equation for a hyperbola $\dfrac{x^2}{a^2} - \dfrac{y^2}{b^2} = 1$ (743)

standard form equation for an ellipse $\dfrac{x^2}{a^2} + \dfrac{y^2}{b^2} = 1$ (735)

standard form of a quadratic relation in two variables $Ax^2 + Bxy + Cy^2 + Dx + Ey + F = 0$, where at least one of A, B, and C is not zero. (758)

standard normal curve The graph of the standard normal distribution. (618)

standard normal distribution The probability distribution with equation $f(z) = \dfrac{1}{\sqrt{2\pi}} e^{-\frac{z^2}{2}}$. (618)

Standard Normal Distribution Table A table that gives the area under the standard normal curve to the left of a given positive number a. (620, 841)

standardizing a variable The process of getting z-values from an original data set with mean m and standard deviation s by applying the transformation $x \rightarrow \dfrac{x - m}{s}$. (627)

standard score For a variable x with a normal distribution with mean m and standard deviation s, the value of the transformed variable $z = \dfrac{x - m}{s}$. Also called a z-score (627)

statistical inference Judgments using probabilities derived from statistical tests. (634)

statistics The branch of mathematics dealing with the collection, organization, analysis, and interpretation of information, usually numerical information. (3)

statistics package A collection of programs for doing statistics. (46)

stem-and-leaf diagram A display of data in which certain digits are the stems while digits with smaller place values are placed side by side as the leaves. Also called *stemplot*. (19)

step function A function whose graph looks like a series of steps (99)

strong relation A relation for which the data in a data set falls close to a line or other prespecified curve. (106)

summation notation The use of the symbol Σ to represent sums. Also called *sigma-notation*, *Σ-notation*. (26)

sum of an infinite series, S_∞, $\displaystyle\sum_{i=1}^{\infty} a_i$ The limit of the sequence of partial sums S_n of the sequence a, provided the limit exists and is finite. In symbols, $S_\infty = \displaystyle\sum_{i=1}^{\infty} a_i = \lim_{n\to\infty} S_n = \lim_{n\to\infty} \sum_{i=1}^{n} a_i$. (484)

Sums and Differences of Cubes Theorem For all x and y, $x^3 + y^3 = (x + y)(x^2 - xy + y^2)$ and $x^3 - y^3 = (x - y)(x^2 + xy + y^2)$. (570)

Sums and Differences of Odd Powers Theorem For all x and y and for all odd positive integers n, $x^n + y^n = (x + y)(x^{n-1} - x^{n-2}y + x^{n-3}y^2 - \ldots - xy^{n-2} + y^{n-1})$ and $x^n - y^n = (x - y)(x^{n-1} + x^{n-2}y + x^{n-3}y^2 + \ldots + xy^{n-2} + y^{n-1})$. (570)

Supplements Theorem For every θ, $\sin(\pi - \theta) = \sin\theta$, $\cos(\pi - \theta) = -\cos\theta$, and $\tan(\pi - \theta) = -\tan\theta$. (306)

survey A gathering of facts or opinions through an interview or questionnaire; to gather such facts. (4)

symmetric to the origin A relation such that if (x, y) is on its graph, then so is $(-x, -y)$. (163)

symmetric with respect to the x-axis A relation such that, for each point (x, y) on its graph, the point $(x, -y)$ is on the graph. (162)

symmetric with respect to the y-axis A relation such that, for each point (x, y) on its graph, the point $(-x, y)$ is on the graph. (162)

table A display of data organized by rows and columns. (10)

TABULATE A particular BASIC program which prints out values of a function. (538)

tangent function A function that maps x onto $\tan x$, for all x in its domain. (289)

tangent (tan) of an acute angle in a right triangle $\dfrac{\text{side opposite the angle}}{\text{side adjacent to the angle}}$. (279)

tangent of a real number $\dfrac{\sin x}{\cos x}$, if $\cos x \neq 0$. (288)

Tangent-Slope Theorem Let m be the slope of a line that forms an acute angle θ with the positive x-axis. Then $m = \tan\theta$. (282)

term An element in the range of a sequence. (456)

tessellations Tilings of the plane with congruent objects. (514)

tetrahedral numbers Terms of the sequence $a_n = \dfrac{n(n+1)(n+2)}{6}$ that is, 1, 4, 10, 20, 35, ... (530)

The Federalist papers Essays written between 1787 and 1788 by James Madison, Alexander Hamilton, and John Jay under the pen name "Publius" to persuade the citizens of the State of New York to ratify the U.S. Constitution. (59)

third (upper) quartile In a data set, the median of the numbers greater than the median. (32)

time-series data A function with finite domain in which the domain coordinate is time. (17)

transformation   A one-to-one correspondence between sets of points. (143)

translation (in the plane)   The transformation that maps each point (x, y) to $(x + h, y + k)$, where h and k are constants. (151)

translation (of data)   A transformation that maps each x_i of a data set to $x_i + h$, where h is a constant. (156)

trial   One of the instances of an experiment. (398)

triangular numbers   Terms of the sequence $T_n = \frac{n(n+1)}{2}$, that is, 1, 3, 6, 10, 15, … (528)

trigonometric equation   An equation involving values of trigonometric functions. (371)

trigonometric form of a complex number   The form $r(\cos\theta + i\sin\theta)$ of the complex number $[r,\theta]$. (810)

trigonometric functions   The sine, cosine, tangent, cotangent, secant, and cosecant functions and their offspring. Also called *circular functions*. (289)

trigonometric ratios   Ratios of the lengths of sides of right triangles. (279)

trigonometry   The branch of mathematics that deals with the relations between the sides and angles of triangles, and with the circular functions. (267)

trinomial   A polynomial with three terms. (526)

2 × 2 Identity Matrix   The matrix $\begin{bmatrix} 1 & 0 \\ 0 & 1 \end{bmatrix}$. (672)

unbiased   See *fair*.

uniform distribution   A distribution that may be represented by a constant function. (437)

union, $A \cup B$ The set of elements that are either in A or in B. (403)

unit circle   A circle with radius 1 centered at the origin. (269)

univariate data   Data involving a single variable. (79)

upper quartile   See *third quartile*.

variable   (in statistics) A characteristic of a person or thing which can be classified, counted, ordered, or measured. (4)

Variance and Standard Deviation of a Binomial Distribution Theorem   In a binomial distribution with n trials, probability p of success and probability q of failure on each trial, the variance $\sigma^2 = npq$, and the standard deviation $\sigma = \sqrt{npq}$. (602)

variance of a probability distribution   For a probability distribution $\{(x_i, P(x_i)\}$ with n points and mean μ,

$$\sigma^2 = \left[\sum_{i=1}^{n} x_i^2 \cdot P(x_i) \right] - \mu^2. \text{ (601)}$$

variance, s^2, σ^2   In a data set, the sum of the squared deviations divided by one less than the number of elements in the set (sample variance s^2) or by the number of elements in the set (population variance σ^2). (51)

vector processors   Processors designed to multiply a row of a matrix by a column of a matrix in one step. (713)

Venn diagram   A display that pictures unions and intersections of sets. (404)

vertex of a cone   The point of intersection of the two lines used to generate the cone; the point of intersection of the two nappes. (728)

vertex of an angle   The common endpoint of its sides. (268)

vertical line test   A test to determine whether a set of ordered pairs in the coordinate plane is a function: if there exists a vertical line that intersects the set in more than one point, then the set is not a function. (82)

vertical scale change   A transformation that maps (x, y) to (x, by), where $b \neq 0$ is a constant. (167)

vertical scale factor   The number b in the transformation that maps (x,y) to (ax,by). (168)

vertices of a hyperbola   The points of intersection of the hyperbola with its axes. (745)

viewing window   The rectangular region that can be seen on the display of an automatic grapher. Also called *window* or *viewing rectangle*. (144)

weak relation   A relation for which, although a linear trend can be seen, many points are not very close to the line. (106)

whispering gallery   An ellipsoidal room where the whispers from one focus are heard at the other focus even though the two foci are quite far apart. (767)

window   See *viewing window*.

world coordinates   Coordinate systems which are used to orient 3-D objects. (712)

yield   The number $\left(1 + \frac{r}{n}\right)^n - 1$, when a principal is invested at an annual interest rate r compounded n times per year. (230)

Zero Exponent Theorem   If b is any nonzero real number, $b^0 = 1$. (210)

zoom   A feature that enables a window of an automatic grapher to represent a smaller rectangle. (146)

z-score   See *standard score*.

SYMBOLS

$\approx$	is approximately equal to	r_x	reflection over the x-axis
$\pm$	positive or negative	r_y	reflection over the y-axis
e	the base of the natural logarithms $\approx 2.71828\ldots$	$r_{y=x}$	reflection over the line $y = x$
		r_ℓ	reflection over line ℓ
π	pi	$\lim_{n\to\infty} a_n$	limit of sequence a
∞	infinity	a_n	nth term of sequence a
$!$	factorial		
$|x|$	absolute value of x	$\sum_{i=1}^{n} x_i$	summation notation; the sum $x_1 + x_2 + \ldots + x_n$
$\sqrt{x}$	positive square root of x		
$\sqrt[n]{x}$	nth root of x	S_∞	sum of the infinite series S
$a + bi$	complex number	$\log x$	common logarithm of x
(a, b)	rectangular coordinates; rectangular form of a complex number	$\log_b x$	logarithm of x to the base b
		$\ln x$	natural logarithm of x
$[r, \theta]$	polar coordinates; polar form of a complex number	$\lfloor x \rfloor$	greatest integer function of x, or floor function of x
$r(\cos\theta + i\sin\theta)$	trigonometric form of a complex number	$\lceil x \rceil$	ceiling function of x
		f^{-1}	inverse function of f
$\bar{z}$	complex conjugate of a complex number	$f \circ g$	composite of functions f and g
		$x\to\infty$	x approaches infinity
$|z|$	modulus of a complex number	$_nP_r$	number of permutations of n elements taken r at a time
i	imaginary unit, $\sqrt{-1}$		
$\begin{bmatrix} a & b \\ c & d \end{bmatrix}$	2×2 matrix	$_nC_r$ or $\binom{n}{r}$	number of combinations of n elements taken r at a time
$\overleftrightarrow{AB}$	line through A and B	$A \cup B$	union sets A and B
$\overrightarrow{AB}$	ray from A passing through B	$A \cap B$	intersection of sets A and B
$\overline{AB}$	segment with endpoints A and B	$N(A)$	number of elements of set A
AB	distance from A to B	$P(E)$	probability of event E
$\angle ABC$	angle ABC	$\bar{x}$	mean of a data set or of a sample
$m\angle ABC$	measure of angle ABC	s	standard deviation of a data set or of a sample
$\triangle ABC$	triangle with vertices A, B, and C		
$ABCD$	polygon with vertices A, B, C, and D	s^2	variance of a data set or of a sample
$\parallel$	is parallel to	μ	Greek letter mu, mean of a population, or expected value of a probability distribution
$\cong$	is congruent to		
$\sim$	is similar to	σ	Greek letter sigma, standard deviation of a population or of a probability distribution
$T_{h,k}$	translation of h units horizontally and k units vertically		
		σ^2	variance of a population or of a probability distribution
$S_{a,b}$	scale change with horizontal magnitude a and vertical magnitude b	M	mean of a sampling distribution
		$B(k, n, p)$	binomial distribution function
S_a	size change of magnitude a	H_0	null hypothesis
R_θ	rotation of magnitude θ	H_1	alternative hypothesis

T84

LESSON 1-2 (pages 10–16)
6.

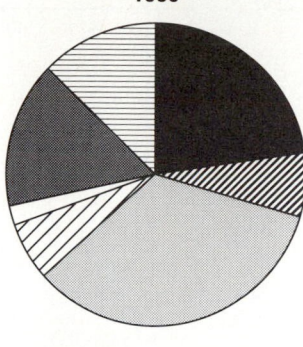

Estimated average cost of raising a child from birth to age 18 in a Midwestern rural area in 1986

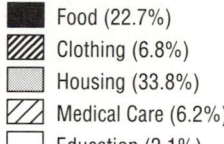

- Food (22.7%)
- Clothing (6.8%)
- Housing (33.8%)
- Medical Care (6.2%)
- Education (2.1%)
- Transportation (15.6%)
- All others (12.9%)

12.a. (continued)

Percent of families in which householder has completed 4 or more years of college by income level

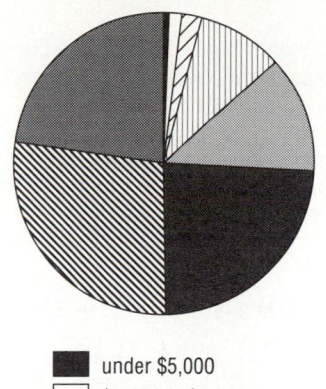

- under $5,000
- $5,000 to $9,999
- $10,000 to $14,999
- $15,000 to $24,999
- $25,000 to $34,999
- $35,000 to $49,999
- $50,000 to $74,999
- $75,000 and over

12.b.

Percent of families in which householder has completed 4 or more years of college by income level

12.a.

Percent of families in which householder has completed 1 to 3 years of high school by income level

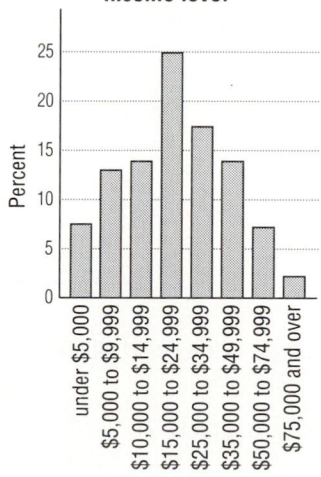

15.a.

Total Number of Reported Cases of Selected diseases between 1950 and 1990

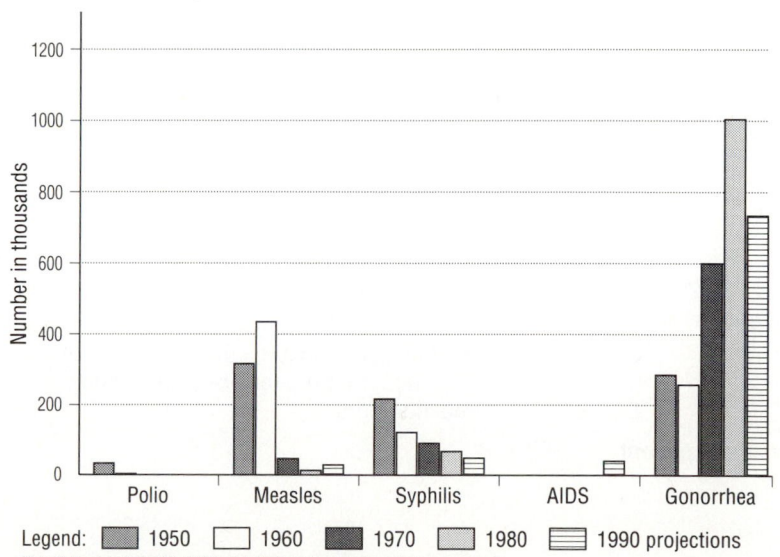

Legend: 1950, 1960, 1970, 1980, 1990 projections

Note: Polio data for 1970, 1980, and 1990 are too small to display graghically.

ADDITIONAL ANSWERS

12.b. (continued)

Percent of families in which householder has completed 1 to 3 years of high school by income level

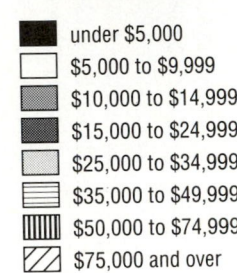

- ■ under $5,000
- □ $5,000 to $9,999
- ▨ $10,000 to $14,999
- ▨ $15,000 to $24,999
- ▨ $25,000 to $34,999
- ▤ $35,000 to $49,999
- ▥ $50,000 to $74,999
- ▧ $75,000 and over

22.a. sample: (0, -4); (1, -1); (5, 11); (-2, -10); (10, 26)

b.

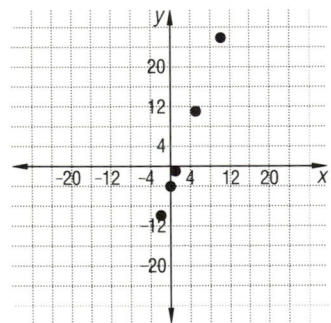

LESSON 1-3 (pages 17–23)

15.a. the 20 AIDS patients to whom the drug is administered

b. all AIDS patients

c. number of infections each patient contracts

18.a. For large cars, costs of repairs and maintenance went up sharply from 1970 to 1981 but then leveled off between 1981 and 1984. For compact cars, costs went up from 1976 to 1981 but went down from 1981 to 1984.

b. For both large and compact cars, the cost of gasoline went up until 1981 and went down after 1981.

20.a. Sample: The areas taken up for red, green, and orange do not match the percents reported for them.

LESSON 2-1 (pages 79–85)

11.a.

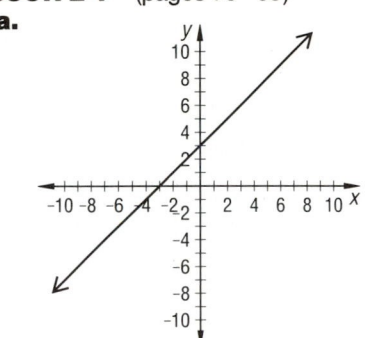

b. function; domain = range = the set of all real numbers

12.a.

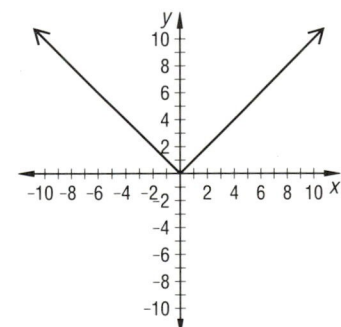

b. function; domain = the set of all real numbers; range = the set of all nonnegative real numbers

---

13.a.

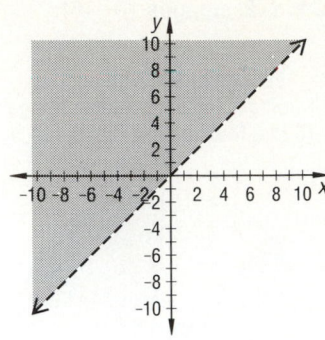

b. not a function

14.a.

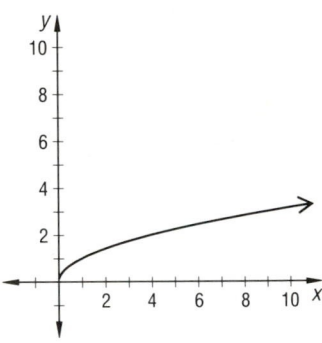

b. function; domain = range = the set of all nonnegative real numbers

15.c. sample: (10, 8), (50, 40), (100, 80)

d.

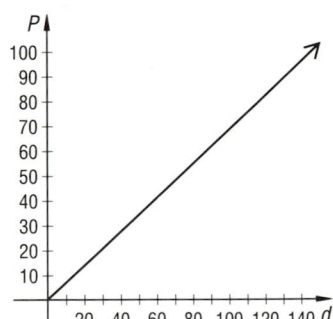

17.a. domain = the set of all real numbers, range = $\{x: x \geq -7\}$

b. sample: (0, 9), (4, -7)

20.a. domain = {2, 3, 4, 5, 6, 7, 8, 9, 10, 11, 12}

22.a. $y = mx + b$, where m is the slope and b is the y-intercept.

b. $y - y_1 = m(x - x_1)$, where m is the slope and (x_1, y_1) is a point on the line.

c. $y = -3x + 13$

LESSON 2-2 (pages 86–91)

1. A linear function is a function which can be described by an equation of the form $y = mx + b$, where m and b are constants.

4.a. the expected change in rating with a unit change in cost

6.d.

8.a.,b.

d. (i) ≈ 64.5% (ii) -29%

9.a.,b.

10.c. False; $f(1) = 2$, $f(6) = -13$, so $f(1) + f(6) = -11$ whereas $f(1 + 6) = f(7) = -16$. So $f(1 + 6) \neq f(1) + f(6)$.

11.a.

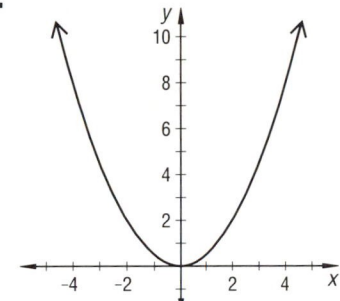

b. The domain is the set of all real numbers; the range is the set of all nonnegative real numbers.

12.b. $E(0) - M(0)$ is about 80˚ C; it is the difference between the temperatures of Earth and Mars at their respective equators.

d. Mars is, on the average, colder, since at all latitudes it has a lower temperature than Earth.

LESSON 2-3 (pages 92–98)

13.a.

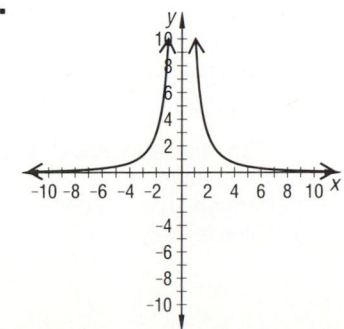

b. domain = the set of all real numbers except zero; range = the set of all positive real numbers

LESSON 2-4 (pages 99–104)

18.a.

21. In 1991, the rate for first-class mail was 29¢ for the first ounce and 23¢ for each additional ounce or fraction of an ounce. This can be modeled by the step function $R(w) = 29 + 23\lceil w - 1 \rceil$ where $R(w)$ is the rate in cents for mail which weighs w ounces.

LESSON 2-7 (pages 120–125)

6.a.
$$a + b + c = 13$$
$$9a + 3b + c = 24$$
$$36a + 6b + c = 16$$

b. $a = -\frac{49}{30} \approx -1.633$;

$b = \frac{361}{30} \approx 12.033$;

$c = \frac{13}{5} = 2.6$

c. The solutions differ, but not greatly.

d. The system in Example 2 was easier to solve.

8.b.

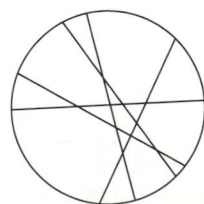

This model predicts 16 slices. It checks.

ADDITIONAL ANSWERS

9.a.

d. model b: $f(20) = 40.5$, $f(60) = 242.1$; model c: $f(20) = 36.8$, $f(60) = 252$; model b gives a smaller sum of squared errors.

18. The domain is the set of all real numbers; the range is the set of all positive real numbers.

19.b.,c.

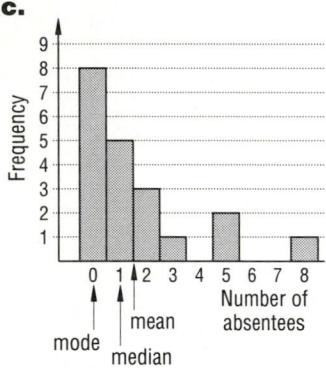

CHAPTER 2 REVIEW (pages 138–142)

29.a.

43.

44.

45.

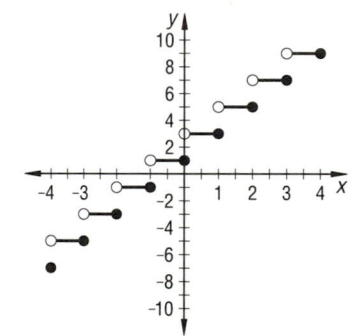

63. Sample: A student should study approximately 9 hours. Studying too many hours may point to lack of adequate previous preparation, which cannot be helped much by cramming; also, the fatigue caused by too many study hours may impede performance during exam.

65. Sample: The quadratic model, since it fits the data points better. It also shows a decline in performance past a certain point of studying, which is likely.

LESSON 3-1 (pages 144–150)

10.b.

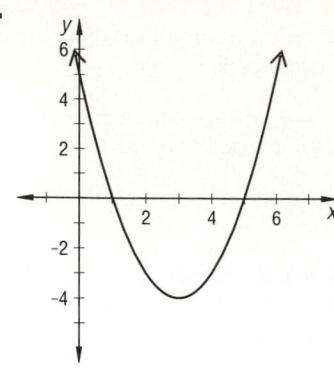

11.a. sample: $-30 \leq x \leq 15$, $-20 \leq y \leq 5$

b.

12.a.

i.

ii.

T88

iii.

iv.

13.a.

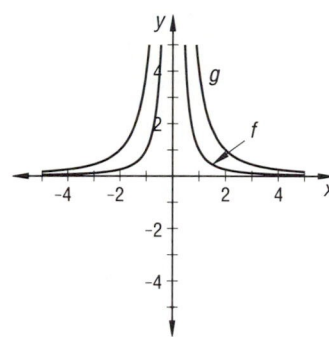

c. The graph of $g(x)$ is higher than the graph of $f(x)$ for all x in the domains of f and g. Also, since x^2 is always positive when $x \neq 0$, $\frac{4}{x^2} > \frac{1}{x^2}$ when $x \neq 0$.

14.a.

15.d.

16.a.,b.

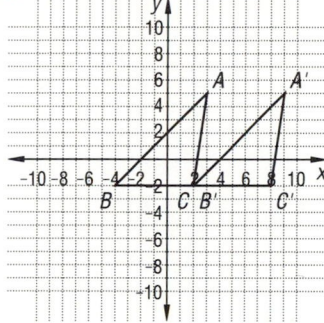

c. The triangles are congruent. $\triangle A'B'C'$ is the image of $\triangle ABC$ under a translation 6 units to the right.

17.a.,b.

c. The figures are congruent. $\square P'Q'R'S'$ is the image of $\square PQRS$ under a translation 5 units to the right and 2 units down.

LESSON 3-7 (pages 180–184)
8.a. $(g \circ h)(t) = (2t + 1)^2 = 4t^2 + 4t + 1$;
$(h \circ g)(t) = 2t^2 + 1$

b.

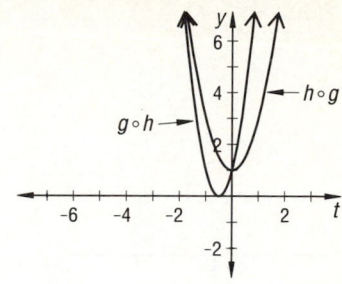

12.c. Let $f(x) = ax + b$ and let $g(x) = cx + d$. Then $f(g(x)) = f(cx + d) = a(cx + d) + b = acx + ad + b$ which is a linear function.

24.a.,b.

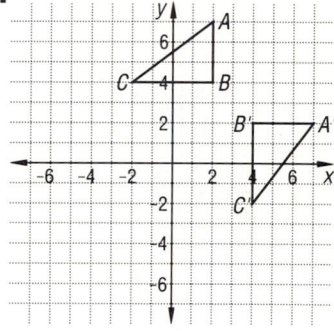

LESSON 3-8 (pages 185–189)
10.b.

11.a. $(g \circ f)(x) = g(2x + 1) = \frac{1}{2}(2x + 1) - 1 = x + \frac{1}{2} - 1 = x - \frac{1}{2}$; $(f \circ g)(x) = f(\frac{1}{2}x - 1) = 2(\frac{1}{2}x - 1) + 1 = x - 2 + 1 = x - 1$; they are not inverses.

b.

T89

ADDITIONAL ANSWERS

12.a. $(g \circ f)(x) = g\left(\frac{2}{x} - 5\right) =$

$\frac{2}{\left(\frac{2}{x} - 5\right) + 5} = \frac{2}{\frac{2}{x}} = x$;

$(f \circ g)(x) = f\left(\frac{2}{x+5}\right) = \frac{2}{\frac{2}{x+5}} - 5 =$

$x + 5 - 5 = x$; they are inverses.

b.

15.a.

b. No

16.a.

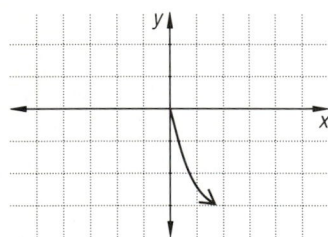

b. Yes

17. Let $q = \frac{1}{p}$. To form the inverse, switch p

and q. Then $p = \frac{1}{q}$. Taking the reciprocal of

both sides, we get $q = \frac{1}{p}$. So, $h^{-1}(p) = \frac{1}{p}$ for

$p \neq 0$.

19.b. False; the inverse of the linear function $f(x) = 0$ for all x, is $x = 0$ which is not a function.

20.c. $(g \circ f)(m) = 2\pi \sqrt{\dfrac{\frac{m}{10} - 30}{980}}$

e. $g \circ f$ is a function of m, the mass (in grams) of a pendulum, which gives t, the time (in seconds) required for the pendulum to swing across and back once.

21.a. It stretches the graph horizontally by a factor of 3 and translates it 2 units down.

b.

23.b.

24.a.

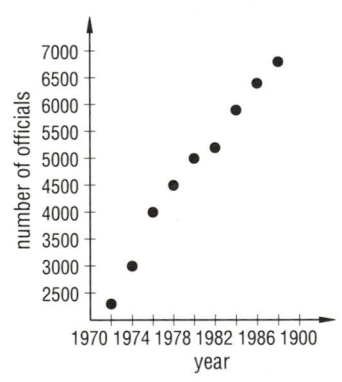

d. sample: a change in American attitude toward electing black officials

CHAPTER 3 REVIEW (pages 197–200)

52.a.

57.

59.a.

b. domain: the set of all real numbers not equal to -5; range: the set of all nonzero real numbers

60.a.

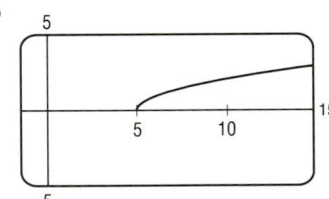

b. domain: $\{x: x \geq 5\}$; range: $\{y: y \geq 0\}$

61.a.

b. domain: the set of all real numbers; range: $\{\, y: y \geq 10 \,\}$

62.a.

b. Yes

c. $y = \dfrac{3}{x}$

63.a.

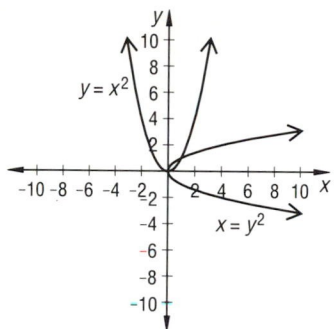

b. No

c. $x = y^2$

LESSON 6-1   (pages 340–345)

11.a.

LESSON 6-3   (pages 352–357)

20.a.,b.,c.

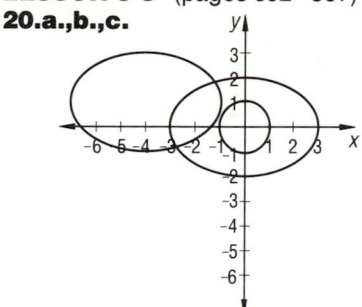

LESSON 6-7   (pages 377–381)

26.b.

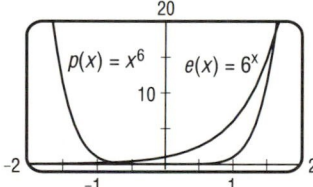

28. The graph of $\left(\dfrac{y-k}{b}\right) = \sec\left(\dfrac{x-b}{a}\right)$ is the image of the graph of $y = \sec x$ under the transformation $S \circ T$ where $T: (x, y) \rightarrow (x + h, y + k)$ $S: (x, y) \rightarrow (ax, by)$
Since $y = \sec x$ has asymptotes at $x = \dfrac{\pi}{2} + n\pi$ for all integers n, $\left(\dfrac{y-k}{b}\right) = \sec\left(\dfrac{x-h}{a}\right)$ has asymptotes at $x = a\left(\dfrac{\pi}{2} + n\pi\right) + h$.
Since $y = \sec x$ has relative extrema at $(2\pi n, 1)$ and $(\pi + 2\pi n, -1)$, $\left(\dfrac{y-k}{b}\right) = \sec\left(\dfrac{x-h}{a}\right)$ has relative extrema at $(2\pi na + h, b + k)$ and $((\pi + 2\pi n)a + h, -b + k)$.

CHAPTER 8 REVIEW   (pages 519–522)

47. $_{n-1}C_r + {}_{n-1}C_{r-1}$

$$= \frac{(n-1)!}{r!(n-1-r)!} + \frac{(n-1)!}{(n-r)!(r-1)!}$$

$$= \frac{(n-1)!}{r!(n-r-1)!} \cdot \frac{n-r}{n-r} + \frac{(n-1)!}{(n-r)!(r-1)!} \cdot \frac{r}{r}$$

$$= \frac{(n-1)!(n-r)}{r!(n-r)!} + \frac{(n-1)!r}{r!(n-r)!(n-r)}$$

$$= \frac{(n-1)!(n-r+r)}{r!(n-r)!} + \frac{(n-1)!n}{r!(n-r)!}$$

$$= \frac{(n)!}{r!(n-r)!} = {}_nC_r$$

LESSON 9-8   (pages 569–573)

1. $\left(-\dfrac{5}{2} - \dfrac{5\sqrt{3}}{2}\right)^3 = \left(-\dfrac{5}{2}\right)^3 - 3\left(-\dfrac{5}{2}\right)^2 - \left(\dfrac{5\sqrt{3}}{2}\,i\right) +$

$3\left(-\dfrac{5}{2}\right)\left(\dfrac{5\sqrt{3}}{2}\,i\right)^2 - \left(\dfrac{5\sqrt{3}}{2}\,i\right)^3 = -\dfrac{125}{8} - \dfrac{3\sqrt{3}}{8} \cdot 125i$

$+ \dfrac{9}{8} \cdot 125 + \dfrac{3\sqrt{3}}{8} \cdot 125i = 125$

4.a.

$$\require{enclose}
\begin{array}{r}
x^2 + 3x + 9 \\
x - 3 \enclose{longdiv}{x^3 + 0x^2 + 0x - 27} \\
\underline{x^3 - 3x^2} \\
3x^2 \\
\underline{3x^2 - 9x} \\
9x - 27 \\
\underline{9x - 27} \\
0
\end{array}$$

ADDITIONAL ANSWERS

LESSON 9-9 (pages 574–577)
19.b.

23.a.

LESSON 10-1 (pages 591–599)
9.a. (continued)

$B(k, 25, 0.3)$

$B(k, 100, 0.3)$

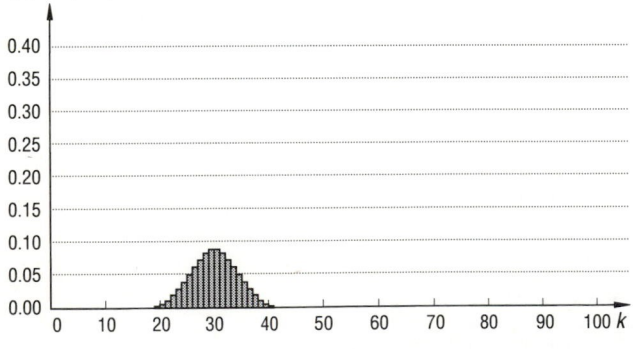

LESSON 11-2 (pages 670–676)
12. sample:

$$\begin{bmatrix} 0 & 1 & 2 & 3 & 4 & 5 & 5 & 4 & 1 & 0 & 0 \\ 4 & 6 & 4 & 4 & 6 & 4 & \text{-}3 & \text{-}4 & \text{-}4 & \text{-}3 & \text{-}4 \end{bmatrix}$$

LESSON 11-5 (pages 690–694)
7. $\cos(\alpha - \beta) = \cos\alpha\cos(\text{-}\beta) - \sin\alpha\sin(\text{-}\beta)$
   (Addition Identity for the Cosine)
$= \cos\alpha\cos\beta - \sin\alpha(\text{-}\sin\beta)$
   (Opposites Theorem)
$= \cos\alpha\cos\beta + \sin\alpha\sin\beta$

16. $\sin(x + 2\pi n) =$
$\sin x\cos(2\pi n) + \cos x\sin(2\pi n)$
   (Addition Identity for Sine)
$= (\sin x)(1) + (\cos x)(0)$ (Periodicity Theorem)
$= \sin x$

17. $\sin\left(x - \dfrac{\pi}{2}\right)$

$= \sin x\cos\left(\text{-}\dfrac{\pi}{2}\right) + \cos x\sin\left(\text{-}\dfrac{\pi}{2}\right)$
   (Addition Identity for Sine)
$= (\sin x)(0) + (\cos x)(\text{-}1) = \text{-}\cos x$

LESSON 11-6 (pages 695–699)
18.a.
$$\cos 2\theta = \cos^2\theta - \sin^2\theta$$
$$+ \quad 1 = \cos^2\theta + \sin^2\theta$$
$$\overline{1 + \cos 2\theta = 2\cos^2\theta}$$

$$\frac{1 + \cos 2\theta}{2} = \cos^2\theta$$

$$\cos\theta = \pm\sqrt{\frac{1 + \cos 2\theta}{2}}$$

b. $\cos\dfrac{\pi}{8} = \pm\sqrt{\dfrac{1 + \cos\dfrac{\pi}{4}}{2}}$

$= \pm\sqrt{\dfrac{1 + \dfrac{\sqrt{2}}{2}}{2}}$

$= \pm\sqrt{\dfrac{2 + \sqrt{2}}{4}} = \pm\dfrac{\sqrt{2 + \sqrt{2}}}{2}$.

Since $\dfrac{\pi}{8}$ is in the first quadrant,

$\cos\dfrac{\pi}{8} > 0$, so $\cos\dfrac{\pi}{8} = \dfrac{\sqrt{2 + \sqrt{2}}}{2}$.

19.a.
$$\text{-}\cos 2\theta = \sin^2\theta - \cos^2\theta$$
$$+ \quad 1 = \sin^2\theta + \cos^2\theta$$
$$\overline{1 - \cos 2\theta = 2\sin^2\theta}$$

$$\frac{1 - \cos 2\theta}{2} = \sin^2\theta$$

$$\sin\theta = \pm\sqrt{\frac{1 - \cos 2\theta}{2}}$$

b. $\sin\dfrac{\pi}{8} = \pm\sqrt{\dfrac{1 - \cos\dfrac{\pi}{4}}{2}}$

$= \pm\sqrt{\dfrac{1 - \dfrac{\sqrt{2}}{2}}{2}}$

$= \pm\sqrt{\dfrac{2 - \sqrt{2}}{4}} = \pm\dfrac{\sqrt{2 - \sqrt{2}}}{2}$.

Since $\dfrac{\pi}{8}$ is in the first quadrant, $\sin\dfrac{\pi}{8} > 0$. So

$\sin\dfrac{\pi}{8} = \dfrac{\sqrt{2 - \sqrt{2}}}{2}$.

LESSON 11-7 (pages 700–708)
1. $\begin{bmatrix} \cos\theta & \text{-}\sin\theta \\ \sin\theta & \cos\theta \end{bmatrix}\begin{bmatrix} \cos\theta & \sin\theta \\ \text{-}\sin\theta & \cos\theta \end{bmatrix} =$

$\begin{bmatrix} \cos^2\theta + \sin^2\theta & \cos\theta\sin\theta - \cos\theta\sin\theta \\ \cos\theta\sin\theta - \cos\theta\sin\theta & \sin^2\theta + \cos^2\theta \end{bmatrix}$

$= \begin{bmatrix} 1 & 0 \\ 0 & 1 \end{bmatrix}$

4.
$$\begin{bmatrix} \dfrac{d}{ad-bc} & \dfrac{-b}{ad-bc} \\ \dfrac{-c}{ad-bc} & \dfrac{a}{ad-bc} \end{bmatrix} \begin{bmatrix} a & b \\ c & d \end{bmatrix} =$$

$$\begin{bmatrix} \dfrac{ad}{ad-bc} - \dfrac{bc}{ad-bc} & \dfrac{bd}{ad-bc} - \dfrac{bd}{ad-bc} \\ \dfrac{-ac}{ad-bc} + \dfrac{ac}{ad-bc} & \dfrac{-bc}{ad-bc} + \dfrac{ad}{ad-bc} \end{bmatrix}$$

$$= \begin{bmatrix} \dfrac{ad-bc}{ad-bc} & 0 \\ 0 & \dfrac{ad-bc}{ad-bc} \end{bmatrix} = \begin{bmatrix} 1 & 0 \\ 0 & 1 \end{bmatrix}$$

16.c.
$$\begin{bmatrix} \frac{1}{7} & \frac{2}{7} \\ \frac{3}{7} & -\frac{1}{7} \end{bmatrix} \begin{bmatrix} 1 & 2 \\ 3 & -1 \end{bmatrix} \begin{bmatrix} x \\ y \end{bmatrix} =$$

$$\begin{bmatrix} \frac{1}{7} & \frac{2}{7} \\ \frac{3}{7} & -\frac{1}{7} \end{bmatrix} \begin{bmatrix} 11 \\ 12 \end{bmatrix}$$

$$\begin{bmatrix} 1 & 0 \\ 0 & 1 \end{bmatrix} \begin{bmatrix} x \\ y \end{bmatrix} =$$

$$\begin{bmatrix} 5 \\ 3 \end{bmatrix} \begin{bmatrix} x \\ y \end{bmatrix} = \begin{bmatrix} 5 \\ 3 \end{bmatrix}$$

LESSON 11-8   (pages 709–717)

10. A 3×3 matrix in homogeneous form

has the following form: $\begin{bmatrix} a_1 & a_2 & a_3 \\ a_4 & a_5 & a_6 \\ 0 & 0 & 1 \end{bmatrix}$,

where a_1, a_2, a_3, a_4, a_5, and a_6 are any real

numbers. Suppose $A = \begin{bmatrix} a_1 & a_2 & a_3 \\ a_4 & a_5 & a_6 \\ 0 & 0 & 1 \end{bmatrix}$ and

$B = \begin{bmatrix} b_1 & b_2 & b_3 \\ b_4 & b_5 & b_6 \\ 0 & 0 & 1 \end{bmatrix}$. Then $AB =$

$$\begin{bmatrix} a_1b_1 + a_2b_4 & a_1b_2 + a_1b_5 & a_1b_3 + a_2b_6 + a_3 \\ a_4b_1 + a_5b_4 & a_4b_2 + a_5b_5 & a_4b_3 + a_5b_6 + a_6 \\ 0 & 0 & 1 \end{bmatrix}$$

CHAPTER 12 REVIEW   (pages 776–778)

32.

35.

36.

37.

38.

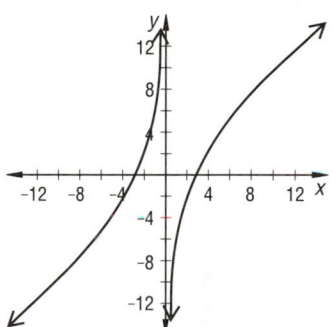

LESSON 12-5   (pages 758–764)

11. Let $s = \sin\theta$ and $t = \cos\theta$. $B'^2 - 4A'C' = (2(A - C)ts + B(t^2 - s^2))^2 - 4(At^2 - Bts + Cs^2)(As^2 + Bts + Ct^2)$

$= 4(A - C)^2t^2s^2 + 4(A - C)B(t^2 - s^2)ts + B^2(t^2 - s^2)^2 - 4(A^2s^2t^2 - B^2s^2t^2 + C^2s^2t^2 + AB(st^3 - s^3t) + AC(s^4 + t^4) + BC(s^3t - st^3))$

$= 4A^2t^2s^2 - 8ACt^2s^2 + 4AB(t^3s - ts^3) - 4BC(t^3s - ts^3) + B^2t^4 - 2B^2t^2s^2 + B^2s^4 - 4A^2s^2t^2 + 4B^2s^2t^2 - 4C^2s^2t^2 - 4AB(st^3 - s^3t) - 4AC(s^4 + t^4) - 4BC(s^3t - st^3)$

$= B^2(t^2 + s^2)^2 - 4AC(t^2 + s^2)^2$

$= B^2 - 4AC$

ADDITIONAL ANSWERS

LESSON 13-1 (pages 780–784)

9.b.

$$\cos x \quad \bigg| \quad -\cos(\pi - x)$$
$$= -(-\cos x)$$
$$= \cos x$$

So, $\cos x = -\cos(\pi - x)$ for all x.

10.a.

$y = \sin 3x$
$y = 3 \sin x \cos x$

b. Let $x = \dfrac{\pi}{6}$. Then $\sin 3x = \sin \dfrac{\pi}{2} = 1$, and

$3 \sin \dfrac{\pi}{6} \cos \dfrac{\pi}{6} = 3\left(\dfrac{1}{2}\right)\left(\dfrac{\sqrt{3}}{2}\right) = \dfrac{3\sqrt{3}}{4} \neq 1$. So

$\sin 3x \neq 3 \sin x \cos x$ for all x.

11.a.

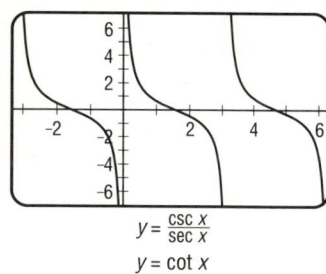

$y = \dfrac{\csc x}{\sec x}$

$y = \cot x$

b. $\dfrac{\csc x}{\sec x} = \dfrac{\dfrac{1}{\sin x}}{\dfrac{1}{\cos x}} = \dfrac{1}{\sin x} \cdot \dfrac{\cos x}{1} = \dfrac{\cos x}{\sin x} = $

$\cot x$, for $x \neq \dfrac{\pi}{2} + n\pi$ and $x \neq n\pi$, n an integer

12.a.

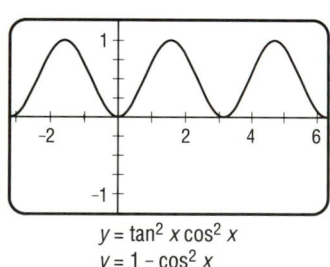

$y = \tan^2 x \cos^2 x$
$y = 1 - \cos^2 x$

b. $\tan^2 x \cos^2 x = \dfrac{\sin^2 x}{\cos^2 x} \cdot \cos^2 x = \sin^2 x = $

$1 - \cos^2 x$, for $x \neq \dfrac{\pi}{2} + n\pi$, n an integer

13.a.

$y = x - \dfrac{x^3}{6} + \dfrac{x^5}{120}$

$y = \sin x$

15. Begin with the Pythagorean Identity; for all x, $\cos^2 x + \sin^2 x = 1$. $(\cos^2 x + \sin^2 x) \cdot (\cos^2 x - \sin^2 x) = \cos^2 x - \sin^2 x$; $\cos^4 x - \sin^4 x = \cos^2 x - \sin^2 x$; $\cos^4 x + \sin^2 x = \cos^4 x + \cos^2 x$

LESSON 13-2 (pages 785–788)

6.b. $\sin x \cot x = \sin x \cdot \dfrac{\cos x}{\sin x} = \cos x$, for

$x \neq n\pi$, for n an integer

7.a.

b.
$$\sec x \quad \bigg| \quad \tan x \csc x$$
$$= \dfrac{1}{\cos x} \quad \bigg| \quad = \dfrac{\sin x}{\cos x} \cdot \dfrac{1}{\sin x}$$
$$\bigg| \quad = \dfrac{1}{\cos x}$$

So $\sec x = \tan x \csc x$ for $x \neq \dfrac{\pi}{2} + n\pi$ and

$x \neq n\pi$, n an integer.

8.a.

b. $\sin 2x = $
$2 \sin x \cdot \cos x$

$$\bigg| \quad \dfrac{2 \tan x}{1 + \tan^2 x} = \dfrac{2 \tan x}{\sec^2 x}$$
$$= 2 \tan x \cos^2 x$$
$$= 2\left(\dfrac{\sin x}{\cos x}\right) \cdot \cos^2 x$$
$$= 2 \sin x \cos x$$

So $\sin 2x = \dfrac{2 \tan x}{1 + \tan^2 x}$ for $x \neq \dfrac{\pi}{2} + n\pi$, n an integer.

10.a.

d. $\dfrac{1 - \tan^2 x}{1 + \tan^2 x} = \dfrac{1 - \tan^2 x}{\sec^2 x} = (1 - \tan^2 x) \cos^2 x$

$= \left(1 - \dfrac{\sin^2 x}{\cos^2 x}\right) \cos^2 x = \cos^2 x - \sin^2 x$

$= \cos 2x$

11.a.

d. $\dfrac{1 + \cos 2x}{\sin 2x} = \dfrac{1 + \cos^2 x - \sin^2 x}{2 \sin x \cos x} = $

$\dfrac{\cos^2 x + \cos^2 x}{2 \sin x \cos x} = \dfrac{2 \cos^2 x}{2 \sin x \cos x} = \dfrac{\cos x}{\sin x} = $

$\cot x$, for $x \neq n\pi$, n an integer

12.b.
$$\tan x + \cot x \quad \bigg| \quad \sec x \csc x$$
$$= \dfrac{\sin x}{\cos x} + \dfrac{\cos x}{\sin x} \quad \bigg| \quad = \dfrac{1}{\cos x} \cdot \dfrac{1}{\sin x}$$
$$= \dfrac{\sin^2 x + \cos^2 x}{\sin x \cos x} \quad \bigg| \quad = \dfrac{1}{\sin x \cos x}$$
$$= \dfrac{1}{\sin x \cos x}$$

So $\tan x + \cot x = \sec x \csc x$ for $x \neq \dfrac{\pi}{2} + n\pi$ and $x \neq n\pi$, n an integer.

13.b.
$$\csc^2 x \sin x \quad \bigg| \quad \dfrac{\sec^2 x - \tan^2 x}{\sin x}$$
$$= \dfrac{1}{\sin^2 x} \cdot \sin x \quad \bigg| \quad = \dfrac{1}{\sin x}$$
$$= \dfrac{1}{\sin x}$$

So $\csc^2 x \sin x = \dfrac{\sec^2 x - \tan^2 x}{\sin x}$ for $x \neq \dfrac{\pi}{2} + n\pi$

and $x \neq n\pi$, n an integer.

14.a.

16.a.,b.

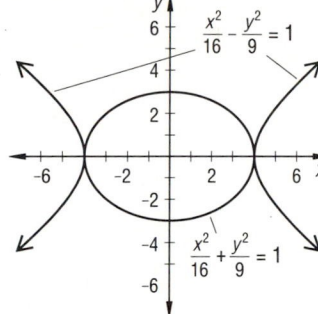

c. To find the points of intersection, find the solutions of the system of equations

$\frac{x^2}{16} - \frac{y^2}{9} = 1$ and $\frac{x^2}{16} + \frac{y^2}{9} = 1$.

Add the equations to get $\frac{2x^2}{16} = 2$, or $x^2 = 16$.

So $x = 4$ or $x = -4$. From the first equation,

$\frac{x^2}{16} - \frac{y^2}{9} = 1$. If $x^2 = 16$, then $y = 0$.

So the solutions are (4, 0) and (-4, 0).

LESSON 13-6 (pages 810–816)

17. $\frac{z_1}{z_2} = \frac{r_1(\cos \theta_1 + i \sin \theta_1)}{r_2(\cos \theta_2 + i \sin \theta_2)} =$

$\frac{r_1(\cos \theta_1 + i \sin \theta_1)(\cos \theta_2 - i \sin \theta_2)}{r_2(\cos \theta_2 + i \sin \theta_2)(\cos \theta_2 - i \sin \theta_2)} =$

$\frac{r_1}{r_2} \frac{(\cos \theta_1 + i \sin \theta_1)(\cos (-\theta_2) + i \sin (-\theta_2))}{(\cos^2 \theta_2 + \sin^2 \theta_2)} =$

$\frac{r_1}{r_2} \frac{(\cos \theta_1 \cos (-\theta_2) - \sin \theta_1 \sin (-\theta_2)) + i (\cos \theta_1 \sin (-\theta_2) + \sin \theta_1 \cos (-\theta_2))}{1}$

$= \frac{r_1}{r_2} (\cos (\theta_1 - \theta_2) + i \sin (\theta_1 - \theta_2))$

LESSON 13-7 (pages 817–821)
19. $-13x^2 + 34\sqrt{3}\,xy - 47y^2 + (68\sqrt{3} + 78)x + (-102\sqrt{3} - 188)y - 321 - 204\sqrt{3} = 0$

22.a. i.

ii.

iii.

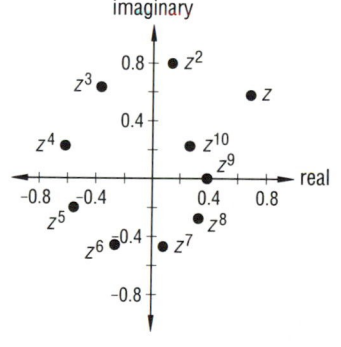

b. The points in part **i** spiral outward, the points in part **ii** fall on a circle of radius 1 with center at the origin, the points in part **iii** spiral inward. As the powers in part **i** increase, the absolute values of the terms increase so the points continue to spiral farther and farther from the origin. Since $1^n = 1$ for every n, the absolute values of the terms in part **ii** are all 1, so the points will continue to fall on the circle. As n increases the absolute values of the terms in part **iii** get smaller and smaller, so the points continue to spiral inward toward the origin.

PROGRESS SELF-TEST 13 (page 830)
11.b. sample:

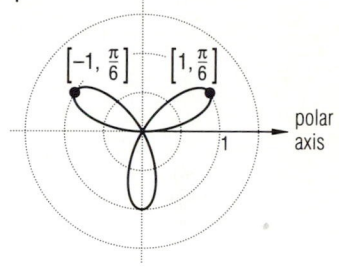

12.a. $15\left(\cos \frac{19\pi}{12} + i \sin \frac{19\pi}{12}\right)$

b. size change of magnitude 5 and a rotation around the origin by an angle of $\frac{\pi}{12}$

13.b. size change of magnitude $\frac{1}{2}$ and a rotation around the origin by an angle of $-\pi$

14. for $\theta \neq \frac{\pi}{2} + n\pi$, where n is an integer,
$(1 - \sin^2 \theta)(1 + \tan^2 \theta) =$
$(\cos^2 \theta)\left(1 + \frac{\sin^2 \theta}{\cos^2 \theta}\right) = \cos^2 \theta + \sin^2 \theta = 1$

15.

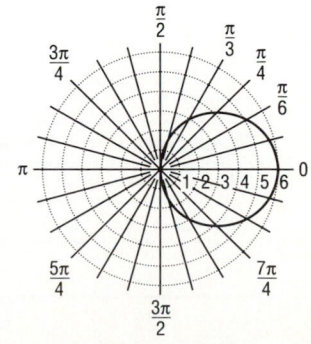

ADDITIONAL ANSWERS

CHAPTER 13 REVIEW (pages 831–833)

44.b. $r = \frac{1}{3}\cos\theta \Leftrightarrow r = \frac{1}{3}\cdot\frac{x}{r} \Leftrightarrow r^2 = \frac{x}{3} \Leftrightarrow$

$x^2 + y^2 = \frac{x}{3} \Leftrightarrow x^2 - \frac{x}{3} + y^2 = 0 \Leftrightarrow$

$x^2 - \frac{x}{3} + \frac{1}{36} + y^2 = \frac{1}{36} \Leftrightarrow \left(x - \frac{1}{6}\right)^2 + y^2 = \frac{1}{36}$

45.

46.

47.a.

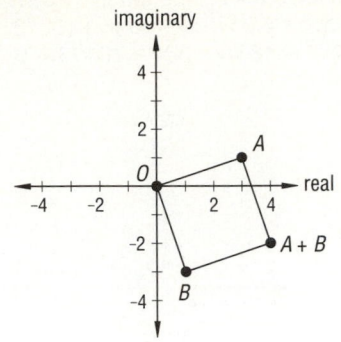

b. Let $C = A + B$. The slope of $\overline{OA} = \frac{1-0}{3-0} = \frac{1}{3}$; the slope of $\overline{AC} = \frac{-2-1}{4-3} = -3$; the slope of $\overline{BC} = \frac{-2--3}{4-3} = \frac{1}{3}$; and the slope of $\overline{OB} = \frac{-3-0}{1-0} = -3$. Because the slopes of the opposite sides of $OACB$ are equal, the opposite sides are parallel and $OACB$ is a parallelogram.

48.a.

b. $z^5 = [2(\cos 72° + i\sin 72°)]^5 =$
$[2^5(\cos 5\cdot 72° + i\sin 5\cdot 72°)] =$
$32(\cos 360° + i\sin 360°) = 32 + 0i = 32$

Progress Self-Test Charts

The chart below keys the **Progress Self-Test** questions to the objectives in the **Chapter 1 Review** on pages 72–78. This will enable you to locate those **Chapter 1 Review** questions that correspond to questions you missed on the **Progress Self-Test.** The lesson where the material is covered is also indicated in the chart.

Question	1–3	4–5	6–9	10–12	13	14	15	16	17–18	19	20
Objective	D	E	G	I	F	B	A	F	H	C	F
Lesson	1-2	1-1	1-3	1-6	1-8	1-4	1-4	1-4	1-5	1-5	1-5

Question	21–25
Objective	J
Lesson	1-7

The chart below keys the **Progress Self-Test** questions to the objectives in the **Chapter 2 Review** on pages 138–142. This will enable you to locate those **Chapter 2 Review** questions that correspond to questions you missed on the **Progress Self-Test.** The lesson where the material is covered is also indicated in the chart.

Question	1	2	3	4a	4b	5	6	7	8	9	10
Objective	I	D	A	A	H	F	D	J	G	F	D
Lesson	2-1	2-1	2-1	2-4	2-4	2-4	2-5	2-6	2-6	2-4	2-5

Question	11a	11b	12a,b	12c	12d	12e	12f
Objective	B	G	E	E	D	F	F
Lesson	2-7	2-7	2-3	2-2	2-5	2-3	2-2

The chart below keys the **Progress Self-Test** questions to the objectives in the **Chapter 3 Review** on pages 197–200. This will enable you to locate those **Chapter 3 Review** questions that correspond to questions you missed on the **Progress Self-Test.** The lesson where the material is covered is also indicated in the chart.

Question	1–2	3	4	5	6	7	8	9–10	11	12–14	15–16
Objective	C	J	C	F	I	E	E	A	H	B	K
Lesson	3-2	3-2	3-5	3-3	3-3	3-3	3-6	3-7	3-7	3-8	3-1

Question	17	18	19	20	21
Objective	G	K	G	I	D
Lesson	3-4	3-5	3-4	3-6	3-2

The chart below keys the **Progress Self-Test** questions to the objectives in the **Chapter 4 Review** on pages 263–266. This will enable you to locate those **Chapter 4 Review** questions that correspond to questions you missed on the **Progress Self-Test.** The lesson where the material is covered is also indicated in the chart.

Question	1	2	3	4	5	6	7	8	9	10	11
Objective	A	A	A	C	C	C	D	B	B	G	G
Lesson	4-2	4-1	4-2	4-5	4-6	4-5	4-3	4-8	4-8	4-3	4-3

Question	12	13	14	15	16	17	18	19	20
Objective	C	D	E	E	I	F	J	J	H
Lesson	4-6	4-5	4-7	4-8	4-5	4-1	4-5	4-6	4-9

PROGRESS SELF-TEST CHARTS

The chart below keys the **Progress Self-Test** questions to the objectives in the **Chapter 5 Review** on pages 334–338. This will enable you to locate those **Chapter 5 Review** questions that correspond to questions you missed on the **Progress Self-Test**. The lesson where the material is covered is also indicated in the chart.

Question	1	2	3	4	5	6	7	8	9	10	11
Objective	A	A	H	E	H	B	C	B	F	K	B
Lesson	5-1	5-1	5-2	5-2	5-2	5-3	5-4	5-3	5-6	5-4	5-3
Question	12	13	14	15	16	17	18	19	20	21	22
Objective	B	D	G	G	G	G	B	C	B	L	I
Lesson	5-3	5-3	5-6	5-7	5-7	5-7	5-3	5-5	5-5	5-6	5-3
Question	23	24	25								
Objective	I	J	J								
Lesson	5-3	5-9	5-8								

The chart below keys the **Progress Self-Test** questions to the objectives in the **Chapter 6 Review** on pages 393–396. This will enable you to locate those **Chapter 6 Review** questions that correspond to questions you missed on the **Progress Self-Test.** The lesson where the material is covered is also indicated in the chart.

Question	1	2	3	4	5	6	7	8	9a	9b	10
Objective	D	D	D	I	A	A	B	B	I	H	G
Lesson	6-1	6-2	6-3	6-3	6-5	6-5	6-7	6-7	6-3	6-3	6-1
Question	11	12	13	14	15a	15b	15c	16			
Objective	C	G	E	E	J	F	J	G			
Lesson	6-7	6-1	6-6	6-6	6-7	6-7	6-7	6-4			

The chart below keys the **Progress Self-Test** questions to the objectives in the **Chapter 7 Review** on pages 451–454. This will enable you to locate those **Chapter 7 Review** questions that correspond to questions you missed on the **Progress Self-Test.** The lesson where the material is covered is also indicated in the chart.

Question	1a	1b	2	3	4	5	6	7	8	9	10
Objective	A	B	B	B	C	D	D	E	F	G	H
Lesson	7-1	7-3	7-3	7-2	7-5	7-3	7-5	7-1	7-4	7-5	7-4
Question	11	12	13	14	15						
Objective	I	I	J	K	L						
Lesson	7-5	7-5	7-7	7-8	7-6						

The chart below keys the **Progress Self-Test** questions to the objectives in the **Chapter 8 Review** on pages 519–522. This will enable you to locate those **Chapter 8 Review** questions that correspond to questions you missed on the **Progress Self-Test.** The lesson where the material is covered is also indicated in the chart.

Question	1	2	3	4a	4b	5	6	7	8	9	10
Objective	A	A	B	E	B	I	F	I	G	C	C
Lesson	8-1	8-1	8-1	8-1	8-1	8-1	8-2	8-5	8-5	8-3	8-4

Question	11	12	13a	13b	14	15	16	17	18
Objective	I	H	D	L	D	J	K	K	L
Lesson	8-5	8-6	8-8	8-7	8-8	8-6	8-9	8-9	8-7

The chart below keys the **Progress Self-Test** questions to the objectives in the **Chapter 9 Review** on pages 587–590. This will enable you to locate those **Chapter 9 Review** questions that correspond to questions you missed on the **Progress Self-Test.** The lesson where the material is covered is also indicated in the chart.

Question	1	2	3	4	5	6	7	8	9	10	11
Objective	A	B	C	D	D	G	E	F	F	F	H
Lesson	9-2	9-3	9-4	9-9	9-8	9-5	9-6	9-3	9-1	9-3	9-7

Question	12	13	14	15	16
Objective	H	I	J	K	J
Lesson	9-7	9-2	9-2	9-3	9-2

The chart below keys the **Progress Self-Test** questions to the objectives in the **Chapter 10 Review** on pages 658–662. This will enable you to locate those **Chapter 10 Review** questions that correspond to questions you missed on the **Progress Self-Test.** The lesson where the material is covered is also indicated in the chart.

Question	1	2	3	4	5	6	7	8	9	10	11
Objective	H	C	A	I	D	D	B	B	G	E	D
Lesson	10-1	10-1	10-2	10-3	10-4	10-5	10-5	10-5	10-7	10-8	10-6

Question	12	13a	13b
Objective	D	J	F
Lesson	10-6	10-6	10-6

ADDITIONAL ANSWERS

PROGRESS SELF-TEST CHARTS

The chart below keys the **Progress Self-Test** questions to the objectives in the **Chapter 11 Review** on pages 723–726. This will enable you to locate those **Chapter 11 Review** questions that correspond to questions you missed on the **Progress Self-Test.** The lesson where the material is covered is also indicated in the chart.

Question	1	2	3	4	5	6	7	8	9a	9b	10
Objective	F	A	F	A	C	I	I	I	G	D	G
Lesson	11-1	11-1	11-1	11-1	11-1	11-2	11-2	11-2	11-2	11-7	11-4

Question	11	12	13	14	15	16	17	18
Objective	H	I	E	E	E	D	D	B
Lesson	11-3	11-4	11-5	11-5	11-6	11-7	11-7	11-7

The chart below keys the **Progress Self-Test** questions to the objectives in the **Chapter 12 Review** on pages 776–778. This will enable you to locate those **Chapter 12 Review** questions that correspond to questions you missed on the **Progress Self-Test.** The lesson where the material is covered is also indicated in the chart.

Question	1	2	3	4	5	6	7	8	9	10	11
Objective	E	G	G	H	A	D	D	A	B	D	B
Lesson	12-1	12-2	12-3	12-2	12-3	12-1	12-3	12-3	12-4	12-4	12-4

Question	12	13	14	15
Objective	I	I	C	F
Lesson	12-5	12-5	12-5	12-2

The chart below keys the **Progress Self-Test** questions to the objectives in the **Chapter 13 Review** on pages 831–833. This will enable you to locate those **Chapter 13 Review** questions that correspond to questions you missed on the **Progress Self-Test.** The lesson where the material is covered is also indicated in the chart.

Question	1	2	3a	3b	4	5	6	7	8	9	10
Objective	D	F	D	C	H	J	J	G	B	E	E
Lesson	13-2	13-2	13-2	13-1	13-3	13-5	13-6	13-3	13-6	13-7	13-7

Question	11	12	13	14	15
Objective	I	A	A	C	I
Lesson	13-4	13-6	13-6	13-2	13-4

CHAPTER 1 PROJECTS (pages 66–67)

1. Answers will vary.

2.a. sample:

car #	style	color	year	manufacturer
1	van	white	1990	Ford
2	sports car	red	1988	Toyota
3	truck	brown	1979	Chevrolet
...	...	...	...	...

Because it may be difficult to determine the year, the student might want to estimate the year, or classify the cars into decades.
b. samples: a table, histogram, or pie chart showing the number of occurrences of various styles, colors, years, or manufacturers; a histogram showing the styles of cars produced by different manufacturers
c. Answers will vary.
d. Sample: A parking lot at a senior citizen center might have more large cars. An executive office garage might have more sports cars. These samples would not be representative of the entire population.

3.a. i., ii., iii. Answers will vary.
b. i. $\bar{x} \approx 5$, $s \approx 1.5$;
ii. $\bar{x} \approx 3$, $s \approx 1.2$;
iii. $\bar{x} \approx 10.5$, $s \approx 3$
c. Sample: Histograms with smaller intervals are more precise because they display more information.

4.a.–d. Sample: Out of a coffee can filled with red beans, a student picks 30 beans. She replaces these red beans with 30 white beans and mixes the beans well. She then selects 32 beans from the can and finds that 4 of them are white. Since one-eighth of the beans in this second sample are white, she concludes that approximately one-eighth of the beans in the entire population are white. Since there are 30 white beans in the population, she estimates the population to be $30 \cdot 8 = 240$.
e. Sample: The student takes several more samples of varying sizes from the can of beans. For each sample, she counts the white beans and makes an estimate of the population. After analyzing each sample, she returns the beans to the can and mixes the beans well. The table below summarizes her results.

sample size	# of white beans in sample	estimate of population
12	1	360
20	3	200
25	4	188
29	3	290
38	5	228
51	6	255

g. sample: a sample scatter plot of the population estimate versus the sample size

h. In this experiment, there were 236 beans.
i. Sample: Answers will vary. Ideally, a larger sample will result in greater accuracy.

5. This answer should be a histogram with one bar for each sample of writing and with the bar height corresponding to the relative frequency of the word in the sample. This frequency graph really is not enough to conclude that the student's writing style is closer to one or the other selected styles. A full study would have to be done comparing many different words.

6.,7. Answers will vary.

8.b. This database should have 12 categories with an entry for each student under each category.
c. Answers will vary. Any display using any variable would be appropriate.
d. Answers will vary. All appropriate measures of center and spread should be mentioned.

CHAPTER 2 PROJECTS (pages 133–134)

1.a. The following data were taken from *Progression of World Best Performances and IAAF Approved World Records*. Because of the difficulty in verifying track and field records, a student might find slightly different data.

The History of the World Record for the Women's Mile

Time	Athlete	Country	Year
5:27.5	R. Christmas	G. Britain	1932
5:23.0	G. Lunn	G. Britain	1936
5:20.8	G. Lunn	G. Britain	1937
5:17.0	G. Lunn	G. Britain	1937
5:15.3	E. Forster	G. Britain	1939
5:11.0	A. Oliver	G. Britain	1952
5:09.8	E. Harding	G. Britain	1953
5:08.0	A. Oliver	G. Britain	1953
5:02.6	D. Leather	G. Britain	1953
5:00.3	E. Treybal	Romania	1953
5:00.2	D. Leather	G. Britain	1954
4:59.6	D. Leather	G. Britain	1954
4:50.8	D. Leather	G. Britain	1955
4:45.0	D. Leather	G. Britain	1955
4:41.4	M. Chamberlain	N. Zealand	1962
4:39.2	A. Smith	G. Britain	1967
4:37.0	A. Smith	G. Britain	1968
4:36.8	M. Gommers	Holland	1969
4:35.3	E. Tittel	W. Germany	1971
4:29.5	P. Cacchi-Pigni	Italy	1973
4:23.8	N. Marasescu	Romania	1977
4:22.1	N. Marasescu	Romania	1979
4:21.68	M. Decker	USA	1980
4:20.89	L. Veselkova	USSR	1981
4:18.08	M. Decker-Tabb	USA	1982
4:17.44	M. Puica	Romania	1982
4:16.21	M. Slaney-Decker	USA	1985
4:15.61	P. Ivan	Romania	1989

c. The data show a linear trend. The line of best fit for predicting world record time y from the year x is $y = 2988.056 - 1.377x$.
d. This model predicts that a woman will break the existing mens record in 2006. This is near enough in the future to have some faith in the extrapolation.
e. Sample: The women are improving their times at a much faster rate than the men. We can predict that the men and the women will both be running the mile in 3:38 in the year 2012.

2.a. i. It would be relatively linear, with the line of best fit roughly at a slope of 1.
ii. Answers will vary.
iii. An outlier is a person whose height and arm span are not roughly equal. After removing these data, the regression line should be close to the line $y = x$, where x represents arm span and y represents height.
b. Answers will vary. Some possibilities are height vs. age, arm span vs.gender (for example, let male = 0, female = 1), and hand span vs. foot length.
i. Answers will vary.
ii. Answers will vary. Note that the line of best fit for the switched variables is not necessarily the inverse of the original line of best fit.
c.,d. Answers will vary.

3. Answers will vary. Below is an example.
a. For this example, this chart shows the percent of U.S. males 20 years and older who were smokers in various years.

1965	1970	1974	1976	1977	1978
50.2	44.3	43.4	42.1	40.9	39.0

1979	1980	1983	1985	1987
38.4	38.5	35.5	33.2	31.5

b. Sample: For the above data, a line graph, bar graph, or scatter plot is appropriate.

c. Sample: A good model for the above data is $y = -0.819x + 103.333$, where x is the year after 1900 and y is the % of males who smoke. This model predicts that only 21.4% of U.S. males will smoke in the year 2000.
d. Sample: An argument might be made that because smoking is a health hazard, a smaller percent of males are smoking.

e. Sample: The tobacco industry might argue that fewer men are smoking because it is no longer "in style," rather than because it is unhealthy. They could also point out that while the percent of males who smoke has declined, the number who smoke has increased.

4. Answers will vary. Here we use the men's 1500-meter run.
a.

b. In this case, a linear model is appropriate.
c. on plot
d. The equation for the line of best fit is $y = -.371x + 240.443$, where x is the year after 1900 and y is time in seconds.
e. There are no outliers in this data.
f. samples: 1992—3:26.32; 1995—3:25.21; 2000—3:23.35
g. This model would predict that the time of 3:20 will be broken in the year 2009.

5.a.

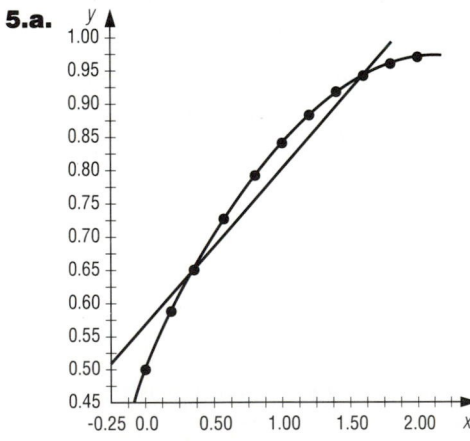

b. Sample: The line shown on the graph passes through (0.4, .6554) and (1.6, .9452) and has equation $y = 0.2415x + .5588$.

c. Sample: The parabola shown on the graph passes through (0.2, .5793), (1.0, .8413), and (1.8, .9641) and has equation $y = -0.10875x^2 + 0.458x + 0.49205$.
d. Sample: The linear model in part **a** predicts the points (0.3, 0.6321), (0.72, 0.7327), (1.55, 0.9331), and (1.812, 0.9964). The quadratic model in part **b** predicts the points (0.3, 0.6197), (0.72, 0.7654), (1.55, 0.9407), and (1.812, 0.9649).
e. Sample: The values for $(x, N(x))$ for the suggested x-values are (0.3, 0.6179), (0.72, 0.7642), (1.55, 0.9394), and (1.812, 0.9651). The quadratic model fits the table much better than the linear model.

6.a.

Place	Latitude
Acapulco, Mex.	16.85°
Bakersfield, CA	35.43°
Caribou, ME	46.87°
Charleston, SC	32.90°
Chicago, IL	41.98°
Dallas–Fort Worth, TX	32.90°
Denver, CO	39.77°
Duluth, MN	46.83°
Great Falls, MT	47.48°
Juneau, AK	58.30°
Kansas City, MO	39.32°
Los Angeles, CA	33.93°
Mexico City, Mex.	19.42°
Miami, FL	25.82°
New Orleans, LA	29.98°
New York, NY	40.78°
Ottawa, Can.	45.43°
Phoenix, AZ	33.43°
Quebec, Can.	46.80°
Salt Lake City, UT	40.78°
San Francisco, CA	37.62°
Seattle, WA	47.45°
Vancouver, Can.	49.30°
Washington, DC	38.85°

b.

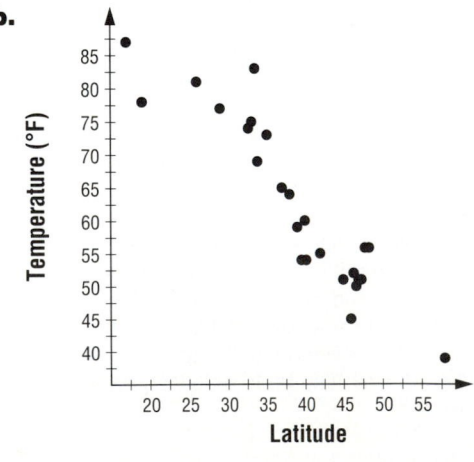

c. The line of best fit is $y = -1.214x + 110.247$, where x is the latitude in degrees and y is the temperature in °F.
d. The slope of the regression line is negative indicating a negative correlation between latitude and temperature for North American cities. That is, in general, the greater the latitude, the lower the temperature. The magnitude of the slope indicates that in general, for every change of one degree in latitude, there is a -1.214 degree change in temperature.
e. sample: about 10° to about 65° latitude
f. Detroit, MI: 59°; Tampa, FL: 76°
g. Detroit, MI: 25%; Tampa, FL: 6%
h. Sample: Phoenix, AZ, is an outlier. It is warmer than the model would predict because it is in the desert.
i. Sample: A linear model for the relation between latitude, or degrees from the equator, in South America and temperature should show that, in general, temperature decreases as latitude increases. Following are some sample data. We do not consider whether a city is north or south of the equator since the above conjecture assumes that, in general, temperature goes down as one moves away from the equator in either direction.

City	Latitude	Temp. (°F)
Arica, Chile	18.50°S	73°
Asuncion, Paraguay	25.25°S	82°
Barranquilla, Columbia	11.00°N	91°
Belem, Brazil	1.45°S	88°
Belo Horizonte, Brazil	19.90°S	81°
Buenos Aires, Argentina	34.67°S	72°
Caracas, Venezuela	10.58°N	82°
Cordoba, Argentina	31.42°S	77°
Cuzco, Peru	13.53°S	72°
Guayaquil, Ecuador	2.22°S	88°
La Paz, Bolivia	16.50°S	66°
Lima, Peru	12.10°S	75°
Manaus, Brazil	3.10°S	86°
Montevideo, Uruguay	34.92°S	72°
Porto Alegre, Brazil	30.05°S	77°
Punta Arenas, Chile	53.17°S	48°
Quito, Ecuador	0.23°S	70°
Recife, Brazil	8.10°S	86°
Rio de Janeiro, Brazil	22.88°S	81°
Salvador (Bahia), Brazil	12.97°S	82°
Santa Cruz, Bolivia	17.75°S	82°
Santiago, Chile	33.00°S	73°
Sao Paulo, Brazil	23.55°S	77°

The equation for the line of best fit is $y = -0.455x + 82.028$. This agrees with the conjecture. However, the slope of the line for North American cities is much steeper, indicating that the effect of latitude on temperature is weaker in South America than in North America.

j. Answers will vary.

CHAPTER 3 PROJECTS (page 194)
1.a. Sample: The functions graphed below are $y = x^2 + 8x + 5$, $y = x^2 + 5x + 5$, $y = x^2 - 6x + 5$, $y = x^2 - x + 5$, $y = x^2 + 3x + 5$, and $y = x^2 - 4x + 5$.

b. All parabolas of the form $y = x^2 + bx + 5$ pass through (0, 5) since $0^2 + b \cdot 0 + 5 = 5$.
c. The vertices of the curves drawn are $(-4, -11)$, $\left(-\frac{5}{2}, -\frac{5}{4}\right)$, $(3, -4)$, $\left(\frac{1}{2}, \frac{19}{4}\right)$, $\left(-\frac{3}{2}, \frac{11}{4}\right)$, and $(2, 1)$. These points appear to lie on a parabola.
d. The vertex of any parabola of the form $y = x^2 + bx + 5$ lies on the parabola $y = 5 - x^2$. To justify this, start with $y = x^2 + bx + 5$ and complete the square. $y = x^2 + bx + \frac{b^2}{4} - \frac{b^2}{4} + 5 \Rightarrow y = \left(x + \frac{b}{2}\right)^2 - \frac{b^2}{4} + 5 \Rightarrow y + \frac{b^2}{4} - 5 = \left(x + \frac{b}{2}\right)^2$. Thus, the parabola with equation $y = x^2 + bx + 5$ is the image of the graph of $y = x^2$ under a translation $T: (x, y) \rightarrow \left(x - \frac{b}{2}, y - \frac{b^2}{4} + 5\right)$. Since the vertex of the parabola represented by $y = x^2$ is the point (0, 0), the vertex of the parabola represented by $y = x^2 + bx + 5$ is $\left(-\frac{b}{2}, -\frac{b^2}{4} + 5\right)$. If we substitute x for $-\frac{b}{2}$, we get $(x, 5 - x^2)$. So each vertex must lie on the curve $y = 5 - x^2$.
e.(a) Sample: We let $a = -2$ and graph parabolas of the form, $y = -2x^2 + bx + 5$. The equations for the parabolas below are $y = -2x^2 + 8x + 5$, $y = -2x^2 + 5$, $y = -2x^2 - 9x + 5$, $y = -2x^2 + 3x + 5$, $y = -2x^2 - 5x + 5$, and $y = -2x^2 + x + 5$.

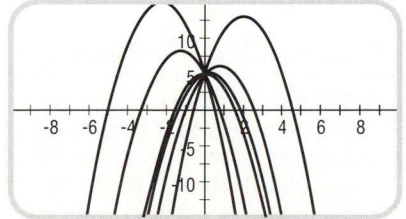

(b) The graphs all pass through (0, 5).
(c) The vertices of these parabolas are (2, 13), (0, 5), $\left(-\frac{9}{4}, \frac{121}{8}\right)$, $\left(\frac{3}{4}, \frac{49}{8}\right)$, $\left(-\frac{5}{4}, \frac{65}{8}\right)$, and $\left(\frac{1}{4}, \frac{41}{8}\right)$. These vertices appear to lie on a parabola.
(d) The vertices of the graphs all lie on the same parabola. We can verify this as in **d**.
$y = -2x^2 + bx + 5 \Rightarrow y = -2\left(x^2 - \frac{b}{2}x - \frac{5}{2}\right) \Rightarrow$
$y = -2\left(x^2 - \frac{b}{2}x + \frac{b^2}{16} - \frac{b^2}{16} - \frac{5}{2}\right) \Rightarrow y =$
$-2\left(\left(x + \frac{b}{4}\right)^2 - \frac{b^2}{16} - \frac{5}{2}\right) \Rightarrow y = -2\left(x + \frac{b}{4}\right)^2 + \frac{b^2}{8} + 5$
$\Rightarrow y - \frac{b^2}{8} - 5 = -2\left(x + \frac{b}{4}\right)^2$. So, the graph of $y = -2x^2 + bx + 5$ is the image of the graph of $y = -2x^2$ under the translation $T: (x, y) \rightarrow \left(x - \frac{b}{4}, y + \frac{b^2}{8} + 5\right)$. Since the vertex of the parabola $y = x^2$ is (0, 0), the vertex of the parabola $y = x^2 + bx + 5$ is $\left(-\frac{b}{4}, \frac{b^2}{8} + 5\right)$. Substituting x for $-\frac{b}{4}$, we get $(x, 2x^2 + 5)$. So each vertex must lie on the curve $y = 2x^2 + 5$.
f. Sample: Given an equation of the form $y = ax^2 + bx + c$, if we keep a and c constant and let b vary, all of the graphs pass through the point $(0, c)$ and all of the vertices of the graphs lie on the parabola $y = -ax^2 + c$.

2.a. Answers will vary.
b. Answers will vary. The comparison should confirm the results that, if each member of a data set is multiplied by a, then the mean, median, and mode are all multiplied by a, the standard deviation is multiplied by $|a|$, and the variance is multiplied by a^2.
c. Sample: The correlation and y-intercept are invariant under these transformations.

3.a. i. Let $S_1(x, y) = (ax, by)$ and $S_2(x, y) = (cx, dy)$. Then $(S_2 \circ S_1)(x, y) = S_2[S_1(x, y)] = S_2(ax, by) = (cax, dby)$. Because the real numbers are closed under multiplication, ca and db are real numbers. So $S_2 \circ S_1$ is a scale change.
iii. Let $S_1(x, y) = (ax, by)$, $S_2(x, y) = (cx, dy)$, and $S_3(x, y) = (ex, fy)$. Then, using **ii.**, $[(S_3 \circ S_2) \circ S_1](x, y) = [S_1 \circ (S_3 \circ S_2)](x, y) = S_1 \circ [S_3(S_2(x, y))] = S_1 \circ [S_3(cx, dy)] = S_1(ecx, fdy) = (ecax, fdby)$ and $[S_3 \circ (S_2 \circ S_1)](x, y) = S_3 \circ [S_2(S_1(x, y))] = S_3 \circ [S_2(ax, by)] = S_3(cax, dby) = (ecax, fdby)$. So $(S_3 \circ S_2) \circ S_1 = S_3 \circ (S_2 \circ S_1)$.
iv. Let $S(x, y) = (ax, by)$ and $I(x, y) = (x, y)$. Then $S \circ I = I \circ S$ by **ii.** Also, $(S \circ I)(x, y) = S(I(x, y)) = S(x, y) = (ax, by)$. So $S \circ I = I \circ S = S$.

v. Let $S(x, y) = (ax, by)$. For every scale change such that $a, b \neq 0$, we can let $S^{-1}(x, y) = \left(\frac{1}{a}x, \frac{1}{b}y\right)$. Now, $S \circ S^{-1} = S^{-1} \circ S$ by **ii**, and $(S \circ S^{-1})(x, y) = S(S^{-1}(x, y)) = S\left(\frac{1}{a}x, \frac{1}{b}y\right) = \left(a\frac{1}{a}x, b\frac{1}{b}y\right) = (x, y) = I$. So, $S \circ S^{-1} = S^{-1} \circ S = I$.

b. Yes, here are the first two parts of the proof and outline of the rest.

i. Let $T_1(x, y) = (x + h, y + k)$ and let $T_2(x, y) = (x + m, y + n)$, for real numbers m and n. Then $(T_2 \circ T_1)(x, y) = T_2[T_1(x, y)] = T_2(x + h, y + k) = (x + h + m, y + k + n)$. Since real numbers are closed under addition, $h + m$ and $k + n$ are real numbers and thus $T_2 \circ T_1$ is a transformation.

ii. Let $T_1(x, y) = (x + h, y + k)$ and let $T_2(x, y) = (x + m, y + n)$, for real numbers m and n. Then $(T_2 \circ T_1)(x, y) = T_2[T_1(x, y)] = T_2(x + h, y + k) = (x + h + m, y + k + n) = (x + m + h, y + n + k) = T_1(x + m, y + n) = T_1[T_2(x, y)] = (T_1 \circ T_2)(x, y)$. Thus $T_2 \circ T_1 = T_1 \circ T_2$.

iii. Let $T_1(x, y) = (x + h, y + k)$ and let $T_2(x, y) = (x + m, y + n)$, and $T_3(x, y) = (x + p, y + q)$, for real numbers m and n. Show $[T_1 \circ (T_2 \circ T_3)](x, y) = [(T_1 \circ T_2) \circ T_3](x, y)$.

iv. Show that $I(x, y) = (x, y)$ is the identity.

v. Show that if $T_1(x, y) = (x + h, y + k)$, then $T_1^{-1}(x, y) = (x + (-h), y + (-k))$ is the inverse of T_1.

4.a.,b. Answers will vary.

c. Sample: The Declaration of Independence was signed on Thursday, July 4, 1776.

d. Sample: $\left\lfloor \frac{y}{4} \right\rfloor$ adds one extra day every four years (leap year). Subtracting $\left\lfloor \frac{y}{100} \right\rfloor$ eliminates a leap day on all years divisible by 100, allowing for the fact that a year is not exactly 365.25 days long. However, this does not account for the length of a year exactly, so every 400 years, or at the beginning of every fourth century, we do have a leap year. Thus, we must add $\left\lfloor \frac{y}{400} \right\rfloor$.

CHAPTER 4 PROJECTS (pages 259–260)

1.a. f is mapped to g by a scale change of a factor of 3. This appears as a translation on the semi-log paper because, instead of graphing $y = 1.6^x$ and $y = 3(1.6)^x$, we are essentially graphing $y = \log[(1.6)^x]$ and $y = \log[3(1.6)^x] = \log 3 + \log[(1.6)^x]$.

b. The graph of any exponential function on semi-log paper is a straight line.

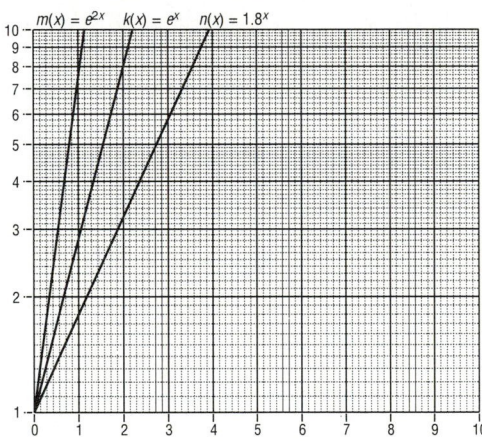

$m(x) = e^{2x}$ $k(x) = e^x$ $n(x) = 1.8^x$

c. Sample: The following are population data for the state of Illinois:

Year	Population	Year	Population
1870	2,540,000	1930	7,631,000
1880	3,078,000	1940	7,897,000
1890	3,826,000	1950	8,712,000
1900	4,822,000	1960	10,081,000
1910	5,639,000	1970	11,114,000
1920	6,485,000	1980	11,427,000

From the graph, we can predict the population to be about 12 million in 1990. The actual estimate for the 1990 population in Illinois is 11.6 million according to the *1990 Statistical Abstract of the United States*. There seem to be two sections of the graph that can be modeled by a linear model—one from 1870 to 1930 and one from 1930 to 1980, when population growth slowed. Lower immigration rates, the Great Depression, and world wars might have slowed population growth. A linear model of the log of population versus year may not be the most useful. In fact a linear model fits population vs. year much better than the log of population vs. year. See bottom of page T105 for graphs for **c.** and **d.**

d. The graph of $y = x^m$ is a straight line with slope m.

2.a.,b. The graphs appear to be the same.

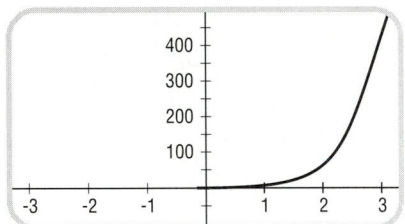

c. $y = 8^x \Rightarrow y = e^{\ln 8^x} \Rightarrow y = e^{x \ln 8}$. Because $8^x = e^{x \ln 8} \approx e^{2.07944x}$, the graphs of $f(x)$ and $g(x)$ are approximately equal.

d. $8 \approx 6^{1.16056}$, $8^x \approx (6^{1.16056})^x = 6^{1.16056x}$

e. $f(x) \approx 3^{1.8928x}$

f. $f(x) = ab^x = ae^{x \ln b}$, which is in the form ae^{kx}, where $k = \ln b$.

3.a. Malthus wrote this book largely in reaction to the liberal views of his father and other economists. Malthus argued that social happiness could never be achieved and that population would always expand to the limits of substinence and would stay there due to starvation, war, and sickness. Malthus said that "vice" (birth control), "misery," and "self-restraint" were the only things that could check population growth. Malthus's ideas were bitterly opposed by the "enlightened" economists of his day. Malthus's book was generally used to blame the problems of the poor on their lack of foresight. Although there was renewed interest in Malthus's ideas in the early 1900's, especially during the depression, today most of his ideas have been disproved or discarded.

b. Malthus did not anticipate agricultural advances that increase food production.

c. Sample: Population growth since 1950 has followed an exponential trend in Africa, Asia, and South America. Growth in the other parts of the world is better approximated by a linear model. Total world population has increased exponentially. The population of Africa can be described by the equation $y = 55.4(1.0274)^x$, where y is population in thousands and x is year after 1900. The population of Asia can be described by the equation $y = 484(1.0212)^x$. The population of South America can be described by the equation $y = 43.6(1.0274)^x$. The population of the world can be described by the equation $y = 969(1.0192)^x$. Following are some graphs to demonstrate the exponential growth of population in these areas.

Africa

Asia

4.a. Sample: Logarithms were used to simplify the process of evaluating certain expressions. These values are easily found in a table of logs. Logarithms were also used for the same problems for which we use them today.

b.,c. Answers will vary.

CHAPTER 5 PROJECTS (pages 329–330)

1. Answers will vary. This problem can be done like similar problems in the chapter. Many distances, such as heights of buildings, can be looked up in reference sources.

2.a. The athletes run counterclockwise. If the starting marks were not staggered, the athletes on the inside part of the track would have a shorter distance to travel than those on the outside part of the track. Because the radii of the semi-circles that form the inner lanes are shorter than the radii of the semi-circles of the outer lanes, the arc lengths of

the inner lanes are smaller than the arc lengths of the outer lanes.

b. Sample: All six runners must run 200 m. The track could have straight sections that are 100 m long, and curved sections such that the semi-circles for the innermost lane have an arc length of 100 m. The finish line should then be placed at the end of a straight section. The starting position of the various lanes must be staggered according to the following chart. Lane 1 is the innermost lane, and θ is the angle shown in the diagram.

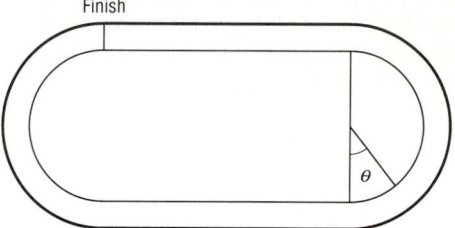

Finish

Lane	θ	Lane	θ
1	0.0°	4	28.5°
2	10.6°	5	36.2°
3	20.1°	6	43.0°

c. Sample: For lane 1, the start line and the finish line are the same. The starting positions for the other lanes must be staggered according to the chart below, where θ is the angle shown in the diagram.

Lane	θ	Lane	θ
1	0.0°	4	57.1°
2	21.3°	5	72.3°
3	40.2°	6	86.1°

1.c.

1.d.

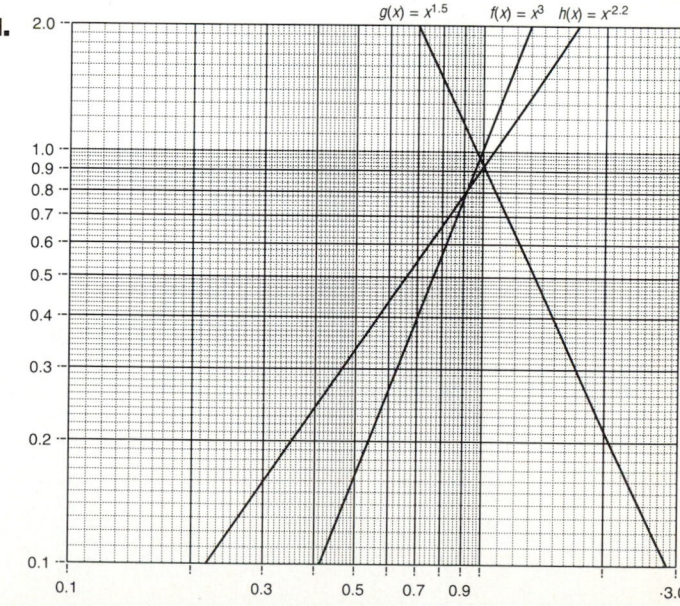

$g(x) = x^{1.5}$ $f(x) = x^3$ $h(x) = x^{2.2}$

Finish

3.a. A benchmark is a fixed point that lies at a known elevation. A benchmark may be an already existing object, or it may be a brass plate on top of a concrete post that has been specially placed by a surveyor. Surveyors can measure the elevation of a point by using a level telescope to determine the point's distance above an established benchmark. Once the elevation of the new point is determined, it may become a new benchmark.
b.,c. Answers will vary.

4.a. $\dfrac{\sin C}{c} = \dfrac{\sin A}{a}$ (Law of Sines) $\Rightarrow \dfrac{a}{c} =$

$\dfrac{\sin A}{\sin C} \Rightarrow \dfrac{a}{c} - 1 = \dfrac{\sin A}{\sin C} - 1 \Rightarrow \dfrac{a}{c} - \dfrac{c}{c} =$

$\dfrac{\sin A}{\sin C} - \dfrac{\sin C}{\sin C} \Rightarrow \dfrac{a - c}{c} = \dfrac{\sin A - \sin C}{\sin C}$

b. $\dfrac{\sin B}{b} = \dfrac{\sin A}{a}$ (Law of Sines) $\Rightarrow \dfrac{a}{b} = \dfrac{\sin A}{\sin B}$

c. $\dfrac{b - c}{c} = \dfrac{\sin B - \sin C}{\sin C}$ (from part **a**) $\Rightarrow$

$\dfrac{c}{b - c} = \dfrac{\sin C}{\sin B - \sin C}$, and $\dfrac{b - c}{b} =$

$\dfrac{\sin B - \sin C}{\sin B}$ (from part **a**) $\Rightarrow \dfrac{b}{b - c} =$

$\dfrac{\sin B}{\sin B - \sin C}$. So $\dfrac{b}{b - c} + \dfrac{c}{b - c} =$

$\dfrac{\sin B}{\sin B - \sin C} + \dfrac{\sin C}{\sin B - \sin C} \Rightarrow \dfrac{b + c}{b - c} =$

$\dfrac{\sin B + \sin C}{\sin B - \sin C}$.

d. $\angle BAC$, then $\angle BAD = \angle CAD$
Definition of angle bisector
$\sin \angle BAD = \sin \angle CAD$

$\dfrac{\sin \angle BAD}{x} = \dfrac{\sin \angle BDA}{c}$ Law of Sines

$\dfrac{\sin \angle BAD}{\sin \angle BDA} = \dfrac{x}{c}$

$\dfrac{\sin \angle CAD}{y} = \dfrac{\sin \angle CDA}{b}$ Law of Sines

$m\angle BDA + m\angle CDA = 180°$
$m\angle BDA = 180° - m\angle CDA$
$\sin \angle BDA = \sin \angle CDA$ Supplements Theorem

$\dfrac{\sin \angle BAD}{y} = \dfrac{\sin \angle BDA}{b}$ Substitution

$\dfrac{\sin \angle BAD}{\sin \angle BDA} = \dfrac{y}{b}$

$\dfrac{x}{c} = \dfrac{y}{b}$ Substitution

$\dfrac{x}{y} = \dfrac{c}{b}$,

e. $\dfrac{\sin A}{a} = \dfrac{\sin B}{b} = \dfrac{\sin C}{c}$, by the Law of Sines. Since ABC is a scalene triangle, one side is longer than the others. Suppose this side is a. Then $a > b$ and $a > c$, so $\dfrac{a}{b} > 1$, $\dfrac{a}{c} > 1$. Now $\dfrac{\sin A}{\sin B} = \dfrac{a}{b}$, $\dfrac{\sin A}{\sin C} = \dfrac{a}{c}$, so $\dfrac{\sin A}{\sin B} > 1$, $\dfrac{\sin A}{\sin C} > 1$. Thus, $\sin A > \sin B$, $\sin A > \sin C$. Since this triangle is acute, all the angles have measures between 0° and 90°. Since the sine function is strictly increasing over that interval, $A > B$ and $A > C$. So, A is the largest angle.

f. First, suppose that ABC is an equilateral triangle. Then $a = b = c$. The Law of Sines gives $\dfrac{\sin A}{a} = \dfrac{\sin B}{b} = \dfrac{\sin C}{c}$, substitution gives us $\dfrac{\sin A}{a} = \dfrac{\sin B}{a} = \dfrac{\sin C}{a}$. Multiplying each expression by a we get $\sin A = \sin B = \sin C$. Since all angles are between 0° and 180°, and since the sine function is one-to-one on this interval, $A = B = C$. That is, the triangle is equiangular.

Next, suppose that the triangle is equiangular. Then $A = B = C$, so $\sin A = \sin B = \sin C$. By the Law of Sines we get $\dfrac{\sin A}{a} = \dfrac{\sin B}{b} = \dfrac{\sin C}{c} \Rightarrow \dfrac{a}{\sin A} = \dfrac{b}{\sin B} = \dfrac{c}{\sin C}$. Substituting and then multiplying all the resulting expressions by $\sin A$, we get $\dfrac{a}{\sin A} = \dfrac{b}{\sin A} = \dfrac{c}{\sin A} \Rightarrow a = b = c$.

5.a. We used Figure A. See below.
b. Sample: For Figure A, the Pythagorean Identity does not apply. Consider the case $\theta = \dfrac{\pi}{4}$. Then side θ = coside $\theta = 1 \Rightarrow$ side^2 θ + coside^2 $\theta = 2$. However, there is an interesting result. If $k\pi - \dfrac{\pi}{4} \le \theta \le k\pi + \dfrac{\pi}{4}$ for all integers k, then side^2 θ + coside^2 θ = sec^2 θ. For all other θ, side^2 θ + coside^2 θ = csc^2 θ.
c. Answers will vary.

5.a.

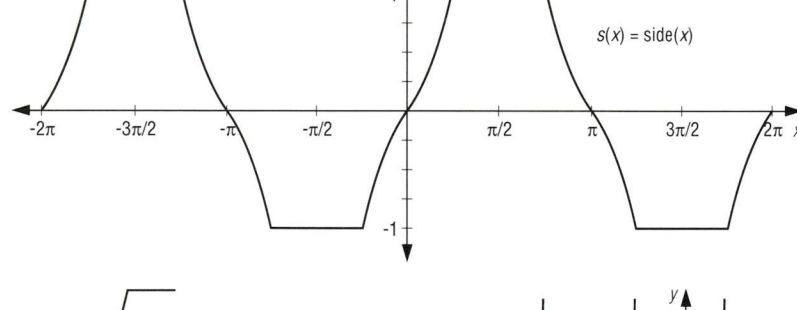
$s(x) = \text{side}(x)$

5.a. cont.

$c(x) = \cos(x)$

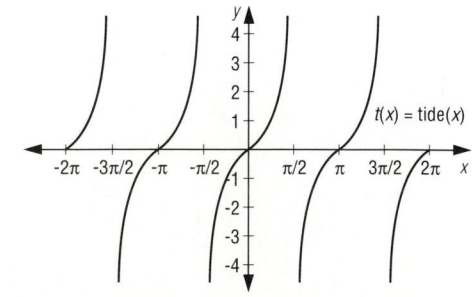
$t(x) = \text{tide}(x)$

1. Sample: We will use the sunrise and sunset data for Chicago, IL.

a. The sunrise data can be modeled by the equation $\frac{y - 6.433}{1.55} = \sin\left(\frac{x - 274}{58.09}\right)$. The sunset data can be modeled by the equation $\frac{y - 18.583}{1.59} = \sin\left(\frac{x - 74}{58.09}\right)$.

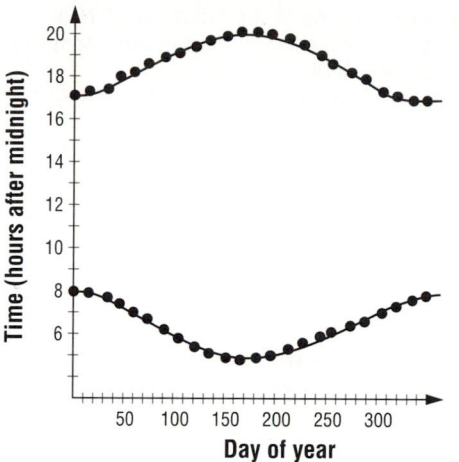

b. The graph would have two discontinuities—one where the graph would jump up 60 minutes and another where the graph would jump down 60 minutes.

c. If both equations were in the same units (e.g., minutes after midnight), the expression describing time of sunrise could be subtracted from the expression describing time of sunset.

d. In the Northern Hemisphere, as distance from the equator increases (i.e., as the absolute value of latitude increases), sunrise becomes earlier in the summer and later in the winter. Likewise, sunset becomes later in the summer and earlier in the winter. In the Southern Hemisphere, the effect is the opposite. There is a small effect on times of sunrise and sunset by longitude. This is caused by the location of the city within its time zone. These facts can be supported by comparing data of times of sunset and sunrise of cities throughout the world.

2.a. i. The curve is an ellipse. The length of the minor axis was determined by the diameter of the candle. The length of the major axis was determined by the slant of the cut.

ii. The curve is a sine wave. The period was determined by candle sizes. The amplitude was determined by the slant of the cut.

b. Answers will vary.

3.a.

$y = \sin 10x + \sin 12x$

$y = \sin 9x + \sin 8x$

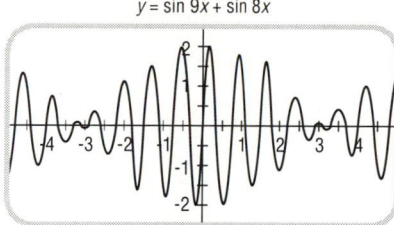

b. Beats are produced when two sound waves traveling in the same medium have frequencies that differ by a small amount. The amplitude of the resulting sound wave at a given point is equal to the sum of the amplitudes of the two waves at that point. The waves start in phase at a point. At first there is *constructive interference*. That is, the amplitude of the resulting wave is greater than the amplitude of either of the original waves. As they move from the point they become more and more out of phase. When the two waves are 180° out of phase, their amplitudes "cancel each other out," resulting in a wave of amplitude zero. Further away from the point the waves may again be in phase. The effect is a diminished loudness during destructive interference and increased loudness during constructive interference.

4.a. Each of the graphs has a period of 2π and an amplitude less than 2. As n gets larger, the graph of $S_n(x) = \sum_{i=1}^{n} \frac{\sin ix}{i}$ becomes more and more "N" shaped. In other words, there are segments of the graph that are almost vertical and there are diagonal segments connecting the top of one almost vertical segment to the bottom of the next.

$S_1(x) = \sin x$

$S_2(x) = \sin x + \frac{1}{2} \sin 2x$

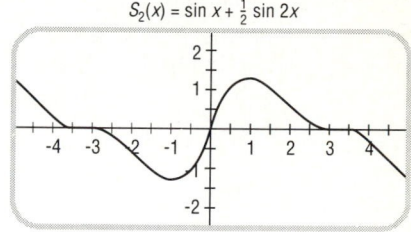

$S_3(x) = \sin x + \frac{1}{2} \sin 2x + \frac{1}{3} \sin 3x$

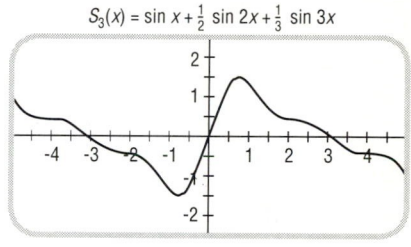

$S_{10}(x) = \sum_{i=1}^{10} \frac{(\sin ix)}{i}$

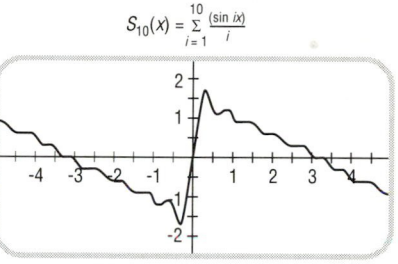

b. These functions have a period of 2π as do the functions in part **a**. Their amplitudes are all less than three. Their graphs are "U" shaped.

$S_{10}(x) = \sum_{i=1}^{10} \frac{(\cos ix)}{i}$

5.a. The graphs are all sine waves.

$f(x) = \sin x + \cos x$

$f(x) = \sqrt{3} \sin x + \cos x$

$f(x) = \sin x - \cos x$

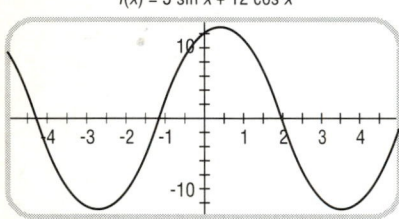
$f(x) = 5 \sin x + 12 \cos x$

b. The graph of $f(x) = 3 \sin x + 4 \cos x$ has amplitude 5. In general, the amplitude of the graph of $f(x) = a \sin x + b \cos x$ is $\sqrt{a^2 + b^2}$.

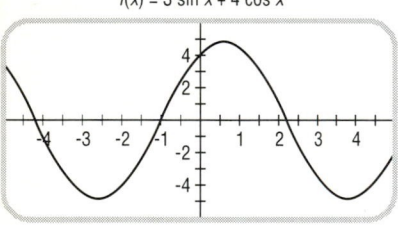
$f(x) = 3 \sin x + 4 \cos x$

c. The phase shift of $f(x) = a \sin x + b \cos x$ from $y = \cos x$ is $\tan^{-1} \frac{a}{b}$, if $\frac{a}{b}$ is positive, and $\pi + \tan^{-1} \frac{a}{b}$, if $\frac{a}{b}$ is negative. This formula would predict that the graph of $f(x) = \sin x + \sqrt{3} \cos x$ is the graph of $y = 2 \cos x$ after a phase shift of $\frac{\pi}{6}$. In other words, the graph of $f(x) = \sin x + \sqrt{3} \cos x$ is the graph of $y = 2 \cos x$ under a translation of $\frac{\pi}{6}$ units to the right.

$f(x) = \sin x + \sqrt{3} \cos x$

d. The amplitude is the length of the hypotenuse. The phase shift of the graph from the graph of $y = \cos x$ is the angle between the leg of length b and the hypotenuse.
e. If a and b are positive, the graph of $y = a \sin x - b \cos x$ is the graph of $y = a \sin x + b \cos x$ after a phase shift of $2 \tan^{-1} \frac{a}{b}$. If a or b is negative, the graph of $y = a \sin x - b \cos x$ is the graph of $y = a \sin x + b \cos x$ after a phase shift of $-2 \tan^{-1} \frac{a}{b}$.

6.a. The boarding point was $-\frac{10\pi}{15} = -\frac{2\pi}{3}$ radians from the top of the wheel.
b. The function has its next maximum at 40.

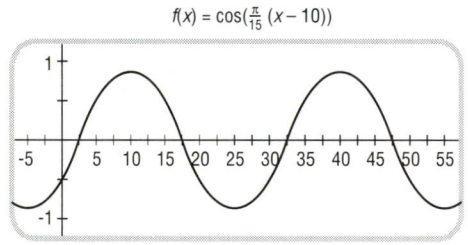
$f(x) = \cos(\frac{\pi}{15}(x - 10))$

c. $f(25) = 25 + 20 \cos \frac{\pi}{15}(25 - 10) =$
$25 + 20 \cos \pi = 5$
d. Answers will vary.

7. Students' reports may include some of the following main points:
1. mid-18th century Belgian composer François Fetiz;
2. composers Charles Ives, Bela Bartok, and Alexander Scriabin;
3. Austrian composers Arnold Schonberg, Alban Berg, and Anton Webern;
4. instruments used to play atonal music;
5. the 12-tone system.

8.a. Each term of the series is being defined in terms of the previous terms. For example, the third term, $\frac{x^5}{5}$, is $-x^2 \cdot \frac{5-2}{2}$, or $-\frac{3}{5} x^2$ times the previous term.

b. Sample: The program calculates arctan 1 to be between .782 and .788. The program calculates arctan -0.5 to be about -.464. This is the actual value to three decimal places. The program does not work for $|x| > 1$. This can be checked by testing several appropriate values of x. The series does not converge for any $|x| > 1$.
c. Sample: Change line 50 by increasing the maximum value for N. When the program reaches N = 4900 the value for SUM is 0.784. This is the value of arctan 1 to three decimal places.
d. The series gives a value of 0.5235988. This is the exact value of arctan $\frac{\sqrt{3}}{3}$ to seven decimal places. This series converged much more quickly than the series for arctan 1. The output was accurate to seven decimal places when the program reached N = 23.
e. Answers will vary.

CHAPTER 7 PROJECTS (page 448)
1.a. Answers will vary.
b. Answers will vary. The frequency distributions should be close to uniform. Each frequency should be about $\frac{1}{6}$.

c. Answers will vary. The frequency distribution should be close to the one discussed in the chapter for the sum of two dice. In other words, the frequency for a sum of 2 or 12 should be about $\frac{1}{36}$; the frequency for a sum of 3 or 11 should be about $\frac{2}{36}$; etc.

d. Answers will vary.
e. Answers will vary. For a large number of tosses, the relative frequencies should be close to the probabilities. If this is not the case, the dice might not be fair.

2.a. 6
b. Since rotation images are regarded as the same, we can seat a person and regard them as the "starting point." There are now $(n - 1)$ people that can be seated to the left of this person. After the second person is seated, there are $(n - 2)$ people remaining who can sit to the left of her, and so on. So the number of ways that n people can sit around a table is
$(n - 1)(n - 2)(n - 3) \cdot \ldots \cdot 1 = (n - 1)!$.

3.a. Answers will vary.
b. $f(x)$ will have real roots whenever $r^2 - 4s \geq 0$ or equivalently when $\frac{1}{4}r^2 \geq s$. We can graph $s = \frac{1}{4}r^2$ and count the integral points that satisfy the inequality.

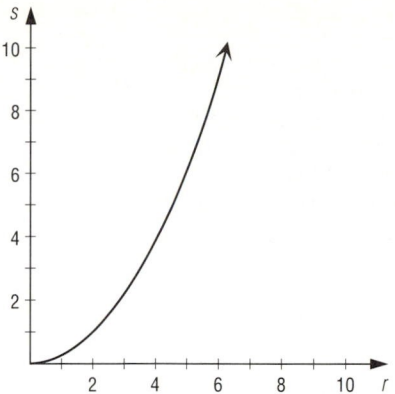

Any point below or on the graph will satisfy the inequality. Since we are interested only in the points where r and s are both integers between 0 and 9 we get (0, 0), (1, 0), (2, 0), (2, 1), (3, 0), (3, 1), (3, 2), (4, 0), (4, 1), (4, 2), (4, 3), (4, 4), (5, 0), (5, 1), (5, 2), (5, 3), (5, 4), (5, 5), (5, 6), and all points for which $r = 6, 7, 8,$ or 9 and s is between 0 and 9 satisfy the inequality. The total number of points is 59. Since there are 100 outcomes in the sample space, the probability that $f(x)$ has real roots is 0.59.
c. Here is the modified program:

```
10    REM REAL ROOTS AREA
20    RANDOMIZE
30    INPUT "HOW MANY POINTS"; N
40    C = 0
50    FOR P = 1 TO N
60    X = INT(RND*10): Y = INT(RND*10)
70    IF Y < X*X/4 THEN C = C + 1
80    NEXT P
90    AREA = C/N
100   PRINT "THE APPROXIMATE AREA
          IS", AREA
110   END
```

This program should give an answer of about 0.6.

4.a. To simplify the problem, assume that it is equally likely for someone to have a birthday on any of the 366 possible days of the year. For $n = 2$, the probability is $\frac{365}{366}$, or about 0.997. For $n = 3$, the probability is

$\frac{365}{366} \cdot \frac{364}{366}$, or about 0.992. Using the same method, the probability for $n = 4$ is about 0.984. The probability for $n = k$ is $\frac{365^{P_k}}{366^k}$.

b. The smallest value for which this probability is less than .5 is $n = 22$.
4.c., 5. Answers will vary.

CHAPTER 8 PROJECTS (pages 514–515)
1.a. Sample: We verify the first property for 4, 22, and 31. $2 \cdot 4 + 1 = 9$, $2 \cdot 22 + 1 = 45$ and $2 \cdot 31 + 1 = 63$ which are not prime numbers. We verify the second property for 5, 20, and 21. $2 \cdot 5 + 1 = 11$, $2 \cdot 20 + 1 = 41$, and $2 \cdot 21 + 1 = 43$ which are all prime numbers.
b. The values of d for the first five rows are 3, 5, 7, 9, and 11 in that order. In general, if n is the number of the row, the common difference is $d = 2n + 1$.
c. The kth term in the nth row is $2nk + n + k$. If N is the kth term in the nth row in the table, $2N + 1 = 2(2nk + n + k) + 1 = 4nk + 2n + 2k + 1 = (2n + 1)(2k + 1)$. Since $2n + 1$ and $2k + 1$ are integers greater than 1, $2N + 1$ is not a prime.
d. If $2N + 1$ is not a prime, then it is the product of two integers x and y such that $x > 1$ and $y > 1$. Since $2N + 1$ is odd, x and y must also be odd. Thus $x = 2a + 1$ and $y = 2b + 1$ for $a > 0$ and $b > 0$. So $2N + 1 = xy = (2a + 1)(2b + 1) = 4ab + 2a + 2b + 1$. That implies that $2N = 4ab + 2a + 2b$ and $N = 2ab + a + b$. So N is the bth term in the ath row of the table.
e. No, 2 is prime, but 2 cannot be written as $2N + 1$ for any integer N.

2. Answers will vary.

3.a. The first term is 1, the second term is $\frac{1}{2}$, the sum of the third and fourth terms is $\frac{1}{2}$, the sum of the fifth, sixth, seventh, and eighth terms is $\frac{1}{2}$, the sum of the ninth through sixteenth terms is $\frac{1}{2}$, and so on. So at any point in the series, we can add $\frac{1}{2}$ by grouping together more terms. This sequence cannot have a limit. If it did, when the sequence approached this limit, we could add $\frac{1}{2}$ again by grouping together more terms.

b. $1 = 1, \frac{1}{2} = \frac{1}{2}, \frac{1}{3} > \frac{1}{4}, \frac{1}{4} = \frac{1}{4}, \frac{1}{5} > \frac{1}{8}$, and so on. Since every term of the harmonic series is greater than the corresponding term of series A, any partial sum of the harmonic series is greater than the corresponding partial sum of series A. Since the sequence of partial sums for A does not have a limit neither does the sequence of partial sums for the harmonic series. Thus the harmonic series diverges.
c. $B \approx 0.693147$
d. The series converges to 11.015%.

4.a. If a square with sides of length 1 is considered the first stage of the dragon curve, then at the nth stage the dragon curve has perimeter $4 \cdot 2^{(n-1)}$ and area 1.
b.,c. Answers will vary.

5.a. $t_4 = \frac{38}{22} \approx 1.727$, $t_5 = \frac{104}{60} \approx 1.733$, $t_6 = \frac{284}{164} \approx 1.7317$, $t_7 = \frac{776}{448} \approx 1.7321$, $t_8 = \frac{2120}{1224} \approx 1.7320$, $t_9 = \frac{5792}{3344} \approx 1.7321$.
The limit appears to be about $1.7321 \approx \sqrt{3}$.
b. The sequence converges to 1.732051, which is approximately $\sqrt{3}$.
c. This sequence also converges to 1.732051, or $\sqrt{3}$.
d. This sequence converges to 2.236068, or approximately $\sqrt{5}$.

6. Answers will vary.

7. Newton stated, but never proved, that the binomial expansion held for fractional and negative n. To understand this, we must first rewrite the binomial expansion as $(x + y)^n = x^n + nx^{n-1}y + \frac{n(n-1)}{1 \cdot 2}x^{n-2}y^2 + \ldots + \frac{n(n-1)\ldots(n-r+1)}{1 \cdot 2 \cdot 3 \cdot \ldots \cdot r}x^{n-r}y^r + \ldots + y^n$.
Notice that, when n is an integer, these coefficients are $_nC_r = \binom{n}{r}$, and the expansion is equivalent to the binomial expansion in the text. Newton claimed that this expansion holds even when n is negative or a fraction. In these cases the expansion is an infinite series.

$(x+y)^{-3} =$
$\quad x^{-3} - 3x^{-4}y + 6x^{-5}y^2 + 10x^{-6}y^3 + \ldots;$
$(x+y)^{2/3} =$
$\quad x^{2/3} + \frac{2}{3}x^{-1/3}y - \frac{1}{9}x^{-4/3}y^2 + \frac{4}{81}x^{-7/3}y^3 + \ldots;$
$(x+y)^{-5/6} =$
$\quad x^{-5/6} - \frac{5}{6}x^{-11/6}y + \frac{55}{72}x^{-17/6}y^2 - \frac{935}{1296}x^{-23/6}y^2 + \ldots$

CHAPTER 9 PROJECTS (pages 583–584)

1.a. sample: $y = x^3 - 5x^2 + 11x - 6$
b. Sample: If the third difference is 1, we get $y = \frac{1}{6}x^3 + \frac{11}{6}x - 1$.
c. The second differences must be the same (i.e., they must both be 2). Working backwards we get $y = 16$.
d. Given n data points, there is a polynomial of degree $n - 1$ that fits the data. It's possible that a polynomial of lesser degree fits. Given three data points that all lie on the same line, a polynomial of degree 1 (a linear function) is enough to fit the data. There are an infinite number of polynomials of degree greater than $n - 1$ that fit n data points.

2.a. A point-symmetry can be observed.
b. If $f(x) = ax^3 + px$, then $f(-x) = a(-x)^3 + p(-x) = -ax^3 - px = -(ax^3 + px) = -f(x)$, so f is odd. Suppose a point $(b, f(b))$ is on the curve. Since f is odd $(-b, f(-b)) = (-b, -f(b))$ is on the curve. The line connecting these two points is $(y - f(b)) = \frac{f(b) - (-f(b))}{(b - (-b))}(x - b)$ or equivalently $y = \frac{f(b)}{b}x$. This line passes through the origin. Also the distance from the origin to $(b, f(b))$ is the same as the distance from the origin to $(-b, -f(b))$. Thus $(-b, -f(b))$ is the image of $(b, f(b))$ under a rotation of $180°$ about the origin. That is, $(0, 0)$ is a center of symmetry for $f(x) = ax^3 + px$.
c. The center of symmetry is $(0, q)$.
d. (h, q)
e. $a(x - h)^3 + p(x - h) + q =$
$a(x^3 - 3x^2h + 3xh^2 - h^3) + p(x - h) + q =$
$ax^3 - 3ax^2h + 3axh^2 - ah^3 + px - ph + q =$
$ax^3 - 3ahx^2 + (3ah^2 + p)x + (-ah^3 - ph + q)$.
So, if $f(x)$ is of the form $ax^3 + bx^2 + cx + d$,
$-3ah = b \Rightarrow h = -\frac{b}{3a}$. Also $c = 3ah^2 + p$. Substituting for h gives $c = 3a\left(-\frac{b}{3a}\right)^2 + p \Rightarrow p = c - \frac{b^2}{3a}$. Finally, we have $d = -ah^3 - ph + q$.

Substituting for h and p gives $d = -a\left(-\frac{b}{3a}\right)^3 -$
$\left(c - \frac{b^2}{3a}\right)\left(-\frac{b}{3a}\right) + q = \frac{b^3}{27a^2} + \frac{bc}{3a} - \frac{b^3}{9a^2} + q =$
$-\frac{2b^3}{27a^2} + \frac{bc}{3a} + q \Rightarrow q = d + \frac{2b^3}{27a^2} - \frac{bc}{3a}$. So the center of symmetry for the graph is $\left(-\frac{b}{3a}, d + \frac{2b^3}{27a^2} - \frac{bc}{3a}\right)$.

3. Move from any diagonal of Pascal's Triangle toward the side that is parallel to that diagonal. Each diagonal contains differences of the diagonal preceding it. Since all the elements in the side of the triangle are the same, you get a constant difference when you reach the side. We can let the elements in the original diagonal be the y-values corresponding to the x-values 1, 2, 3, . . . , and then find a polynomial to fit the data. In the first drawing below, the first differences are constant, and we can show that the polynomial $y = x$ fits the data. In the second triangle, the second differences are constant, so the elements on this diagonal are on a quadratic curve, namely $y = \frac{x(x + 1)}{2} = \frac{1}{2}x^2 + \frac{1}{2}x$.

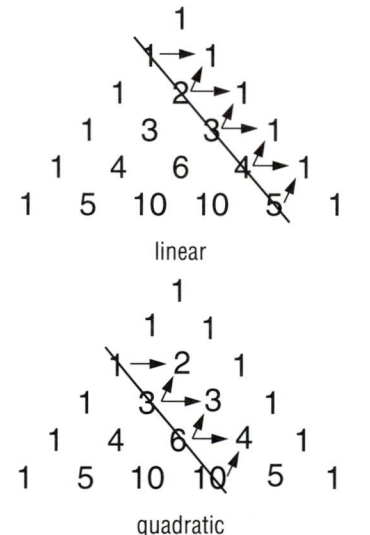

linear

quadratic

4. This topic is discussed in the *Precalculus and Discrete Mathematics* book of the UCSMP series.

5. Algebraic numbers are roots of polynomial functions with rational coefficients. π or e can never be the solution of a polynomial equation. All transcendental numbers are irrational (but not all irrationals are transcendental). Although there are infinitely many transcendental numbers, the two best

known, e and π, are used in exponential modeling and trigonometry. Georg Cantor first proved the existence of transcendentals in the early 1860's. In 1873, Charles Hermite proved that e is transcendental, and, in 1882, F. Lindemann proved that π is transcendental.

6. Students' reports may include some of the following main points:
1. Abel's birth in Norway (1802);
2. Abel's proof that general quintic cannot be solved by means of radicals;
3. *Abel's integral equation, Abel's Theorems, Abel's convergence test, Abelian* groups;
4. Abel's tragic death in 1829.

1. Galois' birth in Paris (1811);
2. difficulty in school;
3. important results submitted to the French academy at age 17;
4. Galois' theory of groups;
5. Galois' theory of equations;
6. death in a duel over a love affair (1832).

CH. 10 PROJECTS (pp. 654–655)

1.a. Observe a pattern of increasing and then decreasing columns of balls.
b. If there are n rows of pins, the binomial distribution has $2n$ trials and 0.5 probability of success in each trial. Success can be defined as a fall to the right of a pin and failure as a fall to the left.

2.–4. Answers will vary.

5.a. Using a random number generator, generate 30 numbers between 591 and 662 (page numbers of this chapter). For each number, turn to the appropriate page, count the number of sentences, and then generate another random number between 1 and the number of sentences. Use the result to select a sentence from that page. To obtain the percent of e occurences, count the total number of letters in each sentence.
b. sample: $10.4 \le$ % of e's ≤ 13.8
c.,d.,6. Answers will vary.

7.a. sample:

$P(s \le n)$

n	$P(s \le n)$	n	$P(s \le n)$
0	0	10	.62
1	.02	11	.70
2	.04	12	.72
3	.14	13	.86
4	.18	14	.96
5	.26	15	.98
6	.32	16	.98
7	.36	17	1.00
8	.42	18	1.00
9	.5		

b. sample:

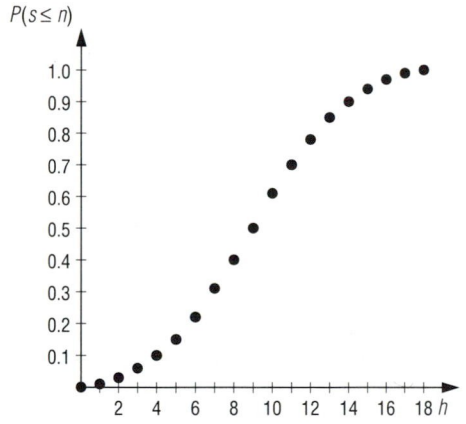

$P(s \le n)$

n	$P(s \le n)$	n	$P(s \le n)$
0	0	10	.605
1	.006	11	.698
2	.025	12	.778
3	.056	13	.846
4	.099	14	.901
5	.154	15	.944
6	.222	16	.975
7	.302	17	.994
8	.395	18	1.000
9	.500		

c. The graphs are very similar. In fact, if the graph in **a** did not depend on experiment, but was theoretical, the graphs would be the same.

CH. 11 PROJECTS (pp. 718–720)

1. The program should allow the user to input the vertices of a figure and the kind of transformation to be applied (i.e., translation, reflection, scale change, or rotation). It should contain a subroutine for each type of transformation. You may also want to have separate subroutines to multiply the matrices and to draw the figure. Here is an outline of the program. Notice that, for reasons discussed in Lesson 11-8, the matrices have three rows even though the figure is two-dimensional. The third row contains only 1s.

Main body of program
Input the number of vertices for the figure; N.
Input the vertices in a $3 \times N$ array VERT (i.e., $x_j =$ VERT(1,J), $y_j =$ VERT(2,J), $z_j = 1 =$ VERT(3,J)).
Draw the figure.
Input the type of transformation.
GOSUB to appropriate subroutine.
Draw the transformed figure.

Reflection subroutine
Input the horizontal and vertical translations; H, K.
Define a 3×3 array to represent the translation matrix $M = \begin{bmatrix} 1 & 0 & H \\ 0 & 1 & K \\ 0 & 0 & 1 \end{bmatrix}$.
Multiply $M \times$ VERT and store results in VERT. Return to main program.

Reflection subroutine
Input the axis of reflection (i.e., x-axis, y-axis, $x = y$).
Define a 3×3 array to represent the appropriate reflection matrix
(i.e., $M = \begin{bmatrix} 1 & 0 & 0 \\ 0 & -1 & 0 \\ 0 & 0 & 1 \end{bmatrix}$, $M = \begin{bmatrix} -1 & 0 & 0 \\ 0 & 1 & 0 \\ 0 & 0 & 1 \end{bmatrix}$,
or $M = \begin{bmatrix} 0 & 1 & 0 \\ 1 & 0 & 0 \\ 0 & 0 & 1 \end{bmatrix}$).
Multiply $M \times$ VERT and store results in VERT. Return to main program.

Scale change subroutine
Input the size of the scale change; S.
Define a 3×3 array to represent the scale change matrix $M = \begin{bmatrix} S & 0 & 0 \\ 0 & S & 0 \\ 0 & 0 & 1 \end{bmatrix}$.
Multiply $M \times$ VERT and store results in VERT. Return to main program.

Rotation Subroutine
Input the angle of rotation in radians; ANGLE.
Define a 3×3 array to represent the rotation matrix
$M = \begin{bmatrix} \cos(\text{ANGLE}) & -\sin(\text{ANGLE}) & 0 \\ \sin(\text{ANGLE}) & \cos(\text{ANGLE}) & 0 \\ 0 & 0 & 1 \end{bmatrix}$.
Multiply $M \times$ VERT and store results in VERT. Return to main program.

2. To rotate an object in 3 dimensions counterclockwise about one of the coordinate axes, find a matrix representation for that object and then multiply on the left by one of the following rotation matrices.

rotation about the x-axis $\begin{bmatrix} 1 & 0 & 0 \\ 0 & \cos\theta & -\sin\theta \\ 0 & \sin\theta & \cos\theta \end{bmatrix}$

rotation about the y-axis $\begin{bmatrix} \cos\theta & 0 & -\sin\theta \\ 0 & 1 & 0 \\ \sin\theta & 0 & \cos\theta \end{bmatrix}$

rotation about the z-axis $\begin{bmatrix} \cos\theta & -\sin\theta & 0 \\ \sin\theta & \cos\theta & 0 \\ 0 & 0 & 1 \end{bmatrix}$

Note that by counterclockwise we mean counterclockwise relative to the plane perpendicular to the axis of rotation.

You can also rotate geometric figures about a coordinate axis. For example, to rotate the pyramid with vertices (1, 0, 0), (0, 1, 0), (-1, 0, 0), (0, -1, 0), and (0, 0, 1), 90° counterclockwise about the x-axis, first represent the pyramid with a point matrix,
$\begin{bmatrix} 1 & 0 & -1 & 0 & 0 \\ 0 & 1 & 0 & -1 & 0 \\ 0 & 0 & 0 & 0 & 1 \end{bmatrix}$. Then compute the matrix for a 90° counterclockwise rotation about the x-axis,
$\begin{bmatrix} 1 & 0 & 0 \\ 0 & \cos 90° & -\sin 90° \\ 0 & \sin 90° & \cos 90° \end{bmatrix} = \begin{bmatrix} 1 & 0 & 0 \\ 0 & 0 & -1 \\ 0 & 1 & 0 \end{bmatrix}$.
Finally multiply the matrices. Thus the rotated pyramid has vertices (1, 0, 0), (0, 0, 1), (-1, 0, 0), (0, 0, -1), and (0, -1, 0).

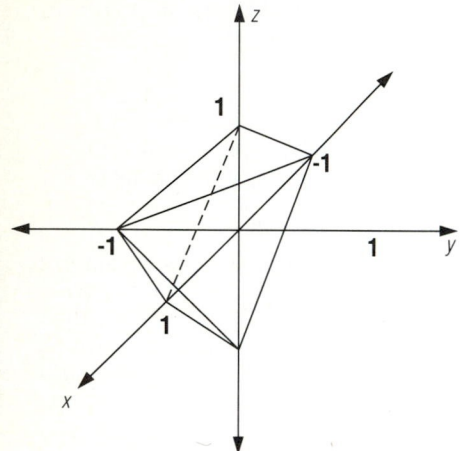

3.a.

θ	d	θ	d
20	14.95	40	19.01
25	16.47	45	18.97
30	17.71	50	18.54
35	18.57		

Using intervals of 0.5°, we find that the maximum distance is approximately 19.05 meters and occurs when $\theta = 42°$.
b. ≈ 18.87 meters
c. Answers will vary. The Olympic shot-putting record is $73'8\frac{3}{4}'' \approx 22.47$ m.

4.a. i. $\begin{bmatrix} 2 & -3 \\ 3 & 2 \end{bmatrix}$ **ii.** $\begin{bmatrix} -1 & 0 \\ 0 & -1 \end{bmatrix}$

iii. $\begin{bmatrix} -4 & -8 \\ 8 & -4 \end{bmatrix}$ **iv.** $\begin{bmatrix} 0 & -1 \\ 1 & 0 \end{bmatrix}$

b. i. $3 + 7i$ **ii.** $-4i$

c. i. $z = \begin{bmatrix} a & -b \\ b & a \end{bmatrix}$; $w = \begin{bmatrix} c & -d \\ d & c \end{bmatrix}$

ii. The matrix representation of $z \cdot w$ is
$$\begin{bmatrix} a & -b \\ b & a \end{bmatrix}\begin{bmatrix} c & -d \\ d & c \end{bmatrix} = $$
$$\begin{bmatrix} ac - bd & -(ad + bc) \\ ad + bc & ac - bc \end{bmatrix}.$$

iii. $z \cdot w = (a + bi)(c + di) =$
$(ac - bd) + (ad + bc)i$
iv. The answer for part **ii** is the matrix form of the result of part **iii**.
d. Let $z = a + bi$ and $w = c + di$. Adding the matrix representations for z and w we get
$$\begin{bmatrix} a & -b \\ b & a \end{bmatrix}\begin{bmatrix} c & -d \\ d & c \end{bmatrix} = \begin{bmatrix} a+c & -(b+d) \\ b+d & a+c \end{bmatrix}.$$
This is the matrix representation for $(a + c) + i(b + d) = z + w$. Thus matrix addition corresponds to addition of complex numbers.

e. $\det \begin{bmatrix} a & -b \\ b & a \end{bmatrix} = a^2 + b^2 = |a + bi|^2$. So the absolute value of a complex number is the square root of the determinant of its matrix representation.
f. matrices with determinant 1

5.a. $\begin{bmatrix} 2 & 1 & 1 \\ 1 & 4 & -3 \\ -1 & -3 & 3 \end{bmatrix} \cdot \begin{bmatrix} \frac{3}{7} & -\frac{6}{7} & -1 \\ 0 & 1 & 1 \\ \frac{1}{7} & \frac{5}{7} & 1 \end{bmatrix} = $

$\begin{bmatrix} 1 & 0 & 0 \\ 0 & 1 & 0 \\ 0 & 0 & 1 \end{bmatrix}$;

$\begin{bmatrix} \frac{3}{7} & -\frac{6}{7} & -1 \\ 0 & 1 & 1 \\ \frac{1}{7} & \frac{5}{7} & 1 \end{bmatrix} \cdot \begin{bmatrix} 2 & 1 & 1 \\ 1 & 4 & -3 \\ -1 & -3 & 3 \end{bmatrix} = $

$\begin{bmatrix} 1 & 0 & 0 \\ 0 & 1 & 0 \\ 0 & 0 & 1 \end{bmatrix}$

b. $\begin{bmatrix} 2 & 1 & 1 \\ 1 & 4 & -3 \\ -1 & -3 & 3 \end{bmatrix}\begin{bmatrix} x \\ y \\ z \end{bmatrix} = \begin{bmatrix} 0 \\ 5 \\ 7 \end{bmatrix} \Rightarrow$

$\begin{bmatrix} 2x + y + z \\ x + 4y - 3z \\ -x - 3y + 3z \end{bmatrix} = \begin{bmatrix} 0 \\ 5 \\ 7 \end{bmatrix} \Rightarrow 2x + y + z = 0,$

$x + 4y - 3z = 5$, and $x - 3y + 3z = 7$

c. $\begin{bmatrix} \frac{3}{7} & -\frac{6}{7} & -1 \\ 0 & 1 & 1 \\ \frac{1}{7} & \frac{5}{7} & 1 \end{bmatrix} \cdot \begin{bmatrix} 2 & 1 & 1 \\ 1 & 4 & -3 \\ -1 & -3 & 3 \end{bmatrix}\begin{bmatrix} x \\ y \\ z \end{bmatrix} = $

$\begin{bmatrix} \frac{3}{7} & -\frac{6}{7} & -1 \\ 0 & 1 & 1 \\ \frac{1}{7} & \frac{5}{7} & 1 \end{bmatrix}\begin{bmatrix} 0 \\ 5 \\ 7 \end{bmatrix};$

$\begin{bmatrix} 1 & 0 & 0 \\ 0 & 1 & 0 \\ 0 & 0 & 1 \end{bmatrix}\begin{bmatrix} x \\ y \\ z \end{bmatrix} = \begin{bmatrix} -\frac{79}{7} \\ 12 \\ \frac{74}{7} \end{bmatrix} \Rightarrow \begin{bmatrix} x \\ y \\ z \end{bmatrix} = $

$\begin{bmatrix} -\frac{79}{7} \\ 12 \\ \frac{74}{7} \end{bmatrix} \Rightarrow x = -\frac{79}{7}, y = 12, z = \frac{74}{7}$

d. INVERTER gives $\begin{bmatrix} 2 & 1 & 1 \\ 1 & 4 & -3 \\ -1 & -3 & 3 \end{bmatrix}^{-1} \approx$

$\begin{bmatrix} .42857 & -.85714 & -1 \\ 0 & 1 & 1 \\ .142857 & .714286 & 1 \end{bmatrix} \approx$

$\begin{bmatrix} \frac{3}{7} & -\frac{6}{7} & -1 \\ 0 & 1 & 1 \\ \frac{1}{7} & \frac{5}{7} & 1 \end{bmatrix}$, so the program is correct.

e. Using INVERTER, we get
$\begin{bmatrix} 3 & 1 & 1 \\ 2 & -2 & -1 \\ 3 & -1 & -1 \end{bmatrix}^{-1} \approx$

$\begin{bmatrix} .16667 & 0 & .16667 \\ -.16667 & -1 & .83333 \\ .66667 & 1 & -1.33333 \end{bmatrix} \approx \begin{bmatrix} \frac{1}{6} & 0 & \frac{1}{6} \\ -\frac{1}{6} & -1 & \frac{5}{6} \\ \frac{2}{3} & 1 & -\frac{4}{3} \end{bmatrix}.$

Since $\begin{bmatrix} \frac{1}{6} & 0 & \frac{1}{6} \\ -\frac{1}{6} & -1 & \frac{5}{6} \\ \frac{2}{3} & 1 & -\frac{4}{3} \end{bmatrix} \cdot \begin{bmatrix} 4 \\ 2 \\ 1 \end{bmatrix} = \begin{bmatrix} \frac{5}{6} \\ -\frac{11}{6} \\ \frac{10}{3} \end{bmatrix}$, the

solution to the system is $x = \frac{5}{6}$, $y = -\frac{11}{6}$, and $z = \frac{10}{3}$. Substituting these values into the left side of each original equation we get
$3\left(\frac{5}{6}\right) + -\frac{11}{6} + \frac{10}{3} = 4$, $2\left(\frac{5}{6}\right) - 2\left(-\frac{11}{6}\right) - \frac{10}{3} = 2$, and $3\left(\frac{5}{6}\right) - \left(-\frac{11}{6}\right) - \frac{10}{3} = 1$. So our solution checks.

CH. 12 PROJECTS (pp. 771–773)
1.a. The curve is an ellipse with the two pins as foci and the length of the string as the focal constant. As you move the pencil, the sum of the distances of the pencil point from the two pins remains constant.
b. The curve is a hyperbola. In the illustration below, we let k, m, and n be the lengths of the sides of the triangle formed by the string. Since the pencil is at the midpoint of the string, $k + n = m$. As you pull the string toward you m and n change, but k (the distance between the pins) is constant. Thus any point on the curve lies on a hyperbola with the pins as foci and with focal constant k.

T112

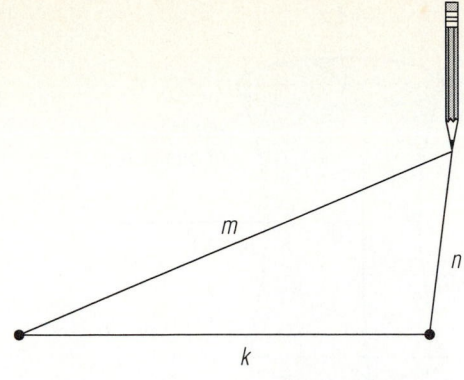

m

n

k

c. Look at the drawing below. If the distance from B to the pencil point is n, then the distance from the pencil point to A must be $AB - n$. Since the string has length AB, then the distance from C to the pencil point is also $AB - n$. Thus at any point on the resulting curve, the distance from C to the point is the same as the distance from the point to ℓ. So, by definition, the curve must be a parabola with focus C and directrix ℓ.

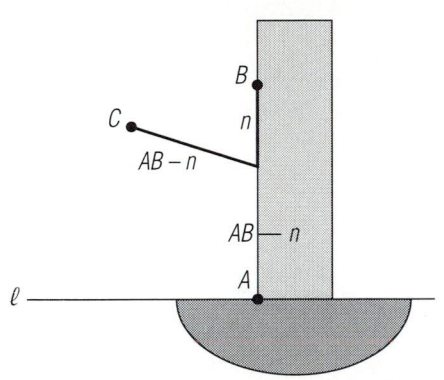

B

C

n

$AB - n$

$AB \mathbin{\rule[0.5ex]{1em}{0.4pt}} n$

A

ℓ

2.a. The curve is a parabola. When P is far from the line, the parabola is wide and flat. When P is close to the line, the parabola is more curved.
b. The curve is an ellipse. When P is far from the center, the ellipse is elongated. When P is close to the center, the ellipse is almost circular.
c. The curve is a branch of a hyperbola. When P is far from the circle, the hyperbola is wider and flatter. When P is close to the circle, the parabola is more curved.

3.a.

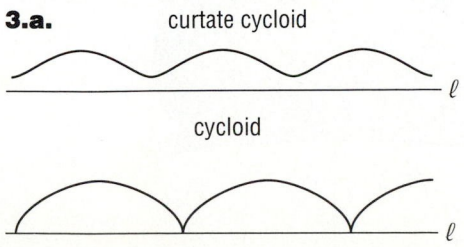

curtate cycloid

ℓ

cycloid

ℓ

prolate cycloid

ℓ

b. Start drawing the cycloid with your pencil point on the line. Examine the path after each quarter turn of the circle. Notice that the parts of the path the pencil draws during the 2nd and 3rd quarter turns are longer than the parts it draws during the 1st and 4th quarter turns. Notice also that the top of the cycloid is drawn during the 2nd and 3rd quarter turns, while the part of the cycloid closer to the line is drawn during the 1st and 4th quarter turns. So, if we assume that the wheel of a bicycle is turning at a constant speed, then a point on the wheel must "trace out" the top part of the cycloid in the same amount of time it takes to trace out the bottom part. Since the length of the top part is longer than the length of the bottom part, the point must move faster while it is on the top of the cycloid.

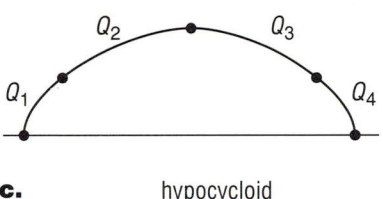

Q_2 Q_3

Q_1 Q_4

c.

hypocycloid
with radius$_{\text{inner}} = \frac{1}{2}$ radius$_{\text{outer}}$

deltoid

astroid

d.

radius$_{\text{external}}$ < radius$_{\text{given}}$

radius$_{\text{external}}$ = radius$_{\text{given}}$

radius$_{\text{external}}$ > radius$_{\text{given}}$

e. Answers will vary.

4.a. Let (x, y) be a point on the ellipse. Let d_1 be the distance from this point to the foci, and let d_2 be the distance from this point to the directrix line, $x = \frac{a^2}{c}$. We must show that $\frac{d_1}{d_2} = \frac{c}{a}$. Since $\frac{d_1}{d_2} = \frac{c}{a} \Rightarrow \frac{(d_1)^2}{(d_2)^2} = \frac{c^2}{a^2} \Rightarrow (d_1)^2 = \frac{c^2}{a^2}(d_2)^2$, we will show $(d_1)^2 = \frac{c^2}{a^2}(d_2)^2$. Now $(d_1)^2 = (x - c)^2 + (y - 0)^2 = x^2 - 2cx + c^2 + y^2$. From the equation for the ellipse, we know that $y^2 = b^2\left(1 - \frac{x^2}{a^2}\right) = b^2 - \frac{b^2 x^2}{a^2}$. We also know that for an ellipse, $c^2 = a^2 - b^2$. Substituting we get $(d_1)^2 = x^2 - 2cx + a^2 - b^2 + b^2 - \frac{b^2 x^2}{a^2} = x^2 - 2cx + a^2 - \frac{b^2 x^2}{a^2}$. Combining the x^2 terms we get $(d_1)^2 = \frac{(a^2 - b^2)}{a^2}x^2 - 2cx + a^2 = \frac{c^2}{a^2}x^2 - 2cx + a^2$. Now, since the directrix is a vertical line, $(d_2)^2 = \left(x - \frac{a}{c}\right)^2 = x^2 - \frac{2a^2}{c}x + \frac{a^4}{c^2}$. Multiplying through by $\frac{c^2}{a^2}$, we get $\frac{c^2}{a^2}(d_2)^2 = \frac{c^2}{a^2}\left(x^2 - \frac{2a^2}{c}x + \frac{a^4}{c^2}\right) = \frac{c^2}{a^2}x^2 - 2cx + a^2 = (d_2)^2$, which is what we wished to prove.

Let (x, y) be a point on the hyperbola. Let d_1 be the distance from this point to the foci, and let d_2 be the distance from this point to the directrix line, $x = \dfrac{a^2}{c}$. We must show that

$\dfrac{d_1}{d_2} = \dfrac{c}{a}$. As above, we will prove the equiva-

lent statement $(d_1)^2 = \dfrac{c^2}{a^2}(d_2)^2$. Now $(d_1)^2 =$

$(x - c)^2 + (y - 0)^2 = x^2 - 2cx + c^2 + y^2$.
From the equation for the hyperbola, we

know that $y^2 = b^2\left(\dfrac{x^2}{a^2} - 1\right) = \dfrac{b^2 x^2}{a^2} - b^2$. We

know that for a hyperbola $c^2 = a^2 + b^2$. Substituting we get $(d_1)^2 = x^2 - 2cx + a^2 + b^2 +$

$\dfrac{b^2 x^2}{a^2} - b^2 = x^2 - 2cx + a^2 + \dfrac{b^2 x^2}{a^2}$. Combining

the x^2 terms we get $(d_1)^2 = \dfrac{a^2 + b^2}{a^2} x^2 - $

$2cx + a^2 = \dfrac{c^2}{a^2} x^2 - 2cx + a^2$. The distance,

d_2, from (x, y) to the directrix satisfies $(d_2)^2 = $

$\left(x - \dfrac{a^2}{c}\right)^2 = x^2 - \dfrac{2a^2}{c} x + \dfrac{a^4}{c^2}$. Multiplying

through by $\dfrac{c^2}{a^2}$, we get $\dfrac{c^2}{a^2}(d_2)^2 = $

$\dfrac{c^2}{a^2}\left(x^2 - \dfrac{2a^2}{c} x + \dfrac{a^4}{c^2}\right) = \dfrac{c^2}{a^2} x^2 - 2cx + a^2 = $

$(d_1)^2$, which is what we wished to prove.
b. i. I   **ii.** III   **iii.** II

c. $x = \pm \dfrac{a^2}{c}$

d.

$e = .25$

$e = 1.5$

$e = .5$

$e = 2$

$e = 2.5$

$e = .75$

$e = .9$

$e = 4$

When the eccentricity of an ellipse is close
to 0, it is nearly circular. When the eccentricity is close to 1, the ellipse is long and narrow. When the eccentricity of a hyperbola is
close to 1, the branches of the hyperbola are
very curved. When the eccentricity is larger,
the branches become straighter.
e. Answers will vary.

5.a.

planet	eccentricity
Venus	.007
Neptune	.009
Earth	.017
Uranus	.047
Jupiter	.048
Saturn	.056
Mars	.093
Mercury	.206
Pluto	.254

b. Here are some data relating period to
eccentricity for several comets. Overall,
periods seem to get longer as eccentricity
increases. In particular, there is a large jump
in period length as the eccentricities become
larger than .900.

e	period (yr)	e	period (yr)
.379	7.35	.810	5.46
.463	5.98	.821	13.61
.553	6.42	.847	3.3
.559	6.8	.919	27.87
.600	6.82	.955	70.87
.604	7.02	.967	76.04
.761	10.99	.969	164.3
.774	17.97		

c. A heavenly body can travel in a path that
is hyperbolic or parabolic. In fact, a majority
of the newly discovered comets have parabolic orbits. If an object is traveling fast
enough, it can escape the gravitational pull
of the sun and follow a parabolic or
hyperbolic path to infinity.

6.a. sample: $\dfrac{x^2}{4} + \dfrac{y^2}{25} + z^2 = 1$

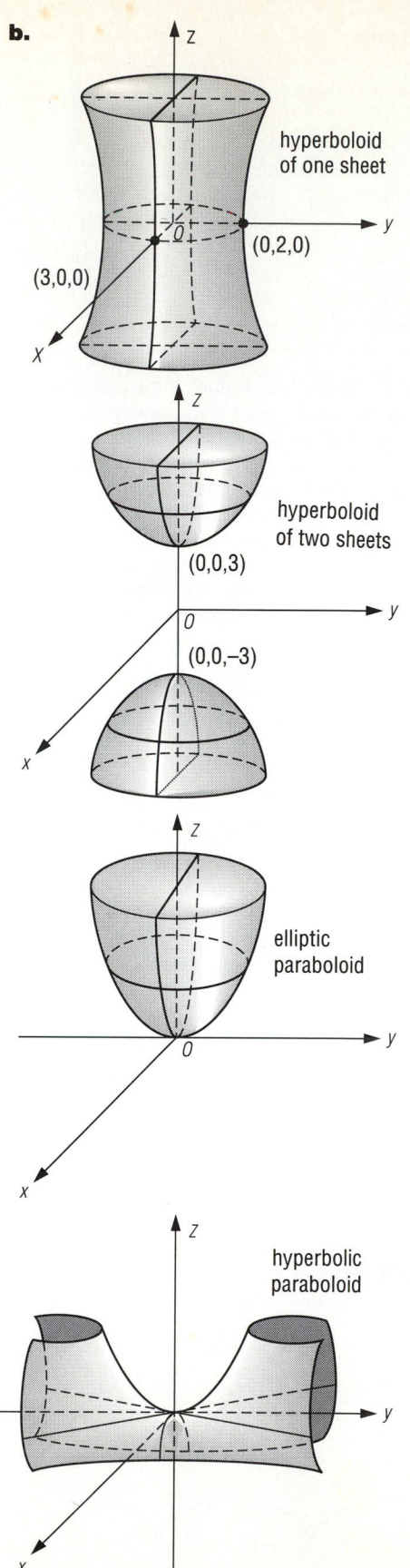

b.

hyperboloid
of one sheet

$(3,0,0)$ $(0,2,0)$

hyperboloid
of two sheets

$(0,0,3)$

$(0,0,-3)$

elliptic
paraboloid

hyperbolic
paraboloid

c. Each point where the strings intersect is the intersection of two lines (the strings) which lie entirely on the surface.

CH. 13 PROJECTS   (page 828)

1.a. The sum of the first three terms is $\approx .9800666667$. The value of the fourth term is $\approx -8.89 \times 10^{-8}$. Using a calculator we get $\cos (0.2) \approx .9800665778$. The difference between the estimate and the actual value is $\approx .9800666667 - .9800665778 = 8.88 \times 10^{-8}$, which is less than 8.89×10^{-8}, the absolute value of the fourth term.

b. The sum of the first three terms is $\approx .1986693333$. The value of the fourth term is $\approx -2.54 \times 10^{-9}$. Using a calculator we get $\sin (0.2) \approx .1986693308$. The difference between the estimate and the actual value is $\approx .198669333 - .198669331 = 2.50 \times 10^{-9}$ which is less than 2.54×10^{-9}, the absolute value of the fourth term.

c. The nth term of the series for $\cos x$ is $(-1)^{n-1}\dfrac{x^{2(n-1)}}{(2n-1)!}$. The formula for the nth term of the series for $\sin x$ is $(-1)^{n-1}\dfrac{x^{2n-1}}{(2n-1)!}$.

d. $\cos 2x = 1 - \dfrac{(2x)^2}{2!} + \dfrac{(2x)^4}{4!} - \dfrac{(2x)^6}{6!} + \ldots +$ $(-1)^{n-1}\dfrac{(2x)^{2(n-1)}}{(2n-2)!} + \ldots$; $\sin 2x = 2x - \dfrac{(2x)^3}{3!} +$ $\dfrac{(2x)^5}{5!} - \dfrac{(2x)^7}{7!} + \ldots + (-1)^{n-1}\dfrac{(2x)^{2n-1}}{(2n-1)!} + \ldots$

e. $e^{ix} = 1 + ix + \dfrac{(ix)^2}{2!} + \dfrac{(ix)^3}{3!} + \dfrac{(ix)^4}{4!} + \ldots =$ $1 + ix - \dfrac{x^2}{2!} - i\dfrac{(x)^3}{3!} + \dfrac{x^4}{4!}$

f. If we group the real and imaginary parts of the expansion of e^{ix}, we get
$e^{ix} = \left(1 - \dfrac{x^2}{2!} + \dfrac{x^4}{4!} - \dfrac{x^6}{6!} + \ldots\right) +$
$i\left(x - \dfrac{(x)^3}{3!} + \dfrac{x^5}{5!} - \dfrac{x^7}{7!} + \ldots\right) = \cos x + i \sin x.$

g. From part **f**, $e^{i\pi} = \cos \pi + i \sin \pi = -1.$

h. From part **f**, $e^{i\cdot\pi/2} = \cos \dfrac{\pi}{2} + i \sin \dfrac{\pi}{2} = i.$

i. From part **h**, $i^i = (e^{i\cdot\pi/2})\, i = e^{-\pi/2} \approx .208.$

2.a.
$r = a\,(\cos \theta - 1)$
$a > 0$

$a < 0$

b.
$r = a \sin \theta \tan \theta$
$a > 0$

$a < 0$
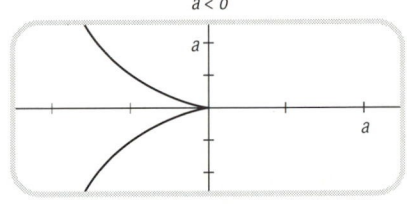

c.
$r = \dfrac{a \sin \theta}{\theta}$
$a > 0$

$a < 0$
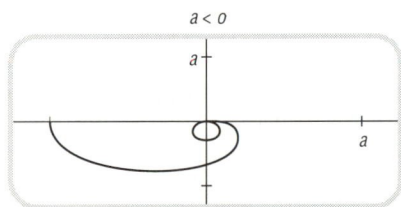

d.
$r = \dfrac{3a \sin \theta \cos \theta}{\sin^3 \theta + \cos^3 \theta}$
$a > 0$

$a < 0$

e.
$r = a \cos 2\theta \sec \theta$
$a > 0$

$a < 0$

f.
$r^2 = a^2 \cos 2\theta$
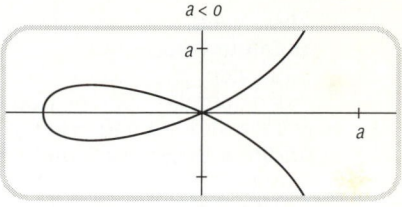

$r^2 = a^2 \sin 2\theta$

g.
$r^2 = \dfrac{a^2}{\theta}$
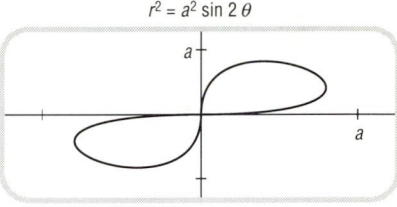

3.a. See the Mandelbrot graph on page 825.
b. See graph on bottom of page 828.
c. Answers will vary.

T115

ACKNOWLEDGMENTS

For permission to reproduce indicated information on the following pages, acknowledgment is made to: **3** Bar chart by Mark Seidler © *National Geographic Society.* **750** Chart: Regression towards mediocrity in hereditary statue. *Journal of the Anthropological Institute* 15:246–263.

Unless otherwise acknowledged, all photographs are the property of Scott, Foresman. Page positions are as follows: (T) top, (C) center, (B) bottom, (L) left, (R) right, (INS) inset.

5 Stephen J. Kraseman/DRK Photo **18T** Library of Congress **18B** John Coletti/Stock Boston **21** AP/Wide World **25** D. Luria/FPG **28** Focus on Sports **31** Deborah Davis/Photo Edit **34** M. Wyville/Bruce Coleman Inc. **44** Milt & Joan Mann/Cameramann International, Ltd. **49** The Bettmann Archive **57** Deborah Davis/Photo Edit **59L** James Monroe Memorial Library **59R** Copyright by the White House Historical Association/Photo: National Geographic Society **61** Rare Book Room/New York Public Library, Astor, Lenox and Tilden Foundations **63L** N. Myers/Bruce Coleman Inc. **63R** Milt & Joan Mann/Cameramann International, Ltd. **64ALL** The Kobal Collection/SuperStock International **73** Steve Kaufman/Peter Arnold, Inc. **79** Courtesy American Air Lines **90** Ms. Mukul Roy **91** NASA **97** Tony Duffy/ALLSPORT USA **98** Lettau/FPG **101T** Porterfield/Chickering/Photo Researchers **101B** Milt & Joan Mann/Cameramann International, Ltd. **103** Courtesy GTE Airfone and DELTA Airlines. Photo: Milt & Joan Mann/Cameramann International, Ltd. **109** Joe Patronite/ALLSPORT USA **113** The Granger Collection, New York **120** Milt & Joan Mann/Cameramann International, Ltd. **127** UPI/Bettmann **129** Gray Mortimore/ALLSPORT USA **157** Milt & Joan Mann/Cameramann International, Ltd. **174** Milt & Joan Mann/Cameramann International, Ltd. **178** Alan Pitcairn from Grant Heilman Photography **188** Sidney Harris **189** Brent Jones **201** AP/Wide World **202** Ed Cooper/Shostal/Superstock **215** Centers for Disease Control, Atlanta **220** Courtesy S. C. Johnson Company. Photo: Milt & Joan Mann/Cameramann International, Ltd. **225** Courtesy of IBM **229** Bob Daemmrich **234** Ted Kerasote/Photo Researchers **244** Photo courtesy of West Virginia University Photography Dept. **251** Liane Enkelis/Stock Boston **256** Milt & Joan Mann/Cameramann International, Ltd. **267T** The Bettmann Archive **267B** NASA **268** GEOPIC (TM)/Earth Satellite Corp. **276** Grant Heilman Photography **309** Bruce Wellman/Stock Boston **310** Milt & Joan Mann/Cameramann International, Ltd. **321** Porterfield/Chickering/Photo Researchers **332** Marie Veda/Tony Stone Worldwide **339ALL** Milt & Joan Mann/Cameramann International, Ltd. **340** Milt & Joan Mann/Cameramann International, Ltd. **357** Mark Antman/The Image Works **358** Jay Lurie/Bruce Coleman Inc. **361** NASA **375** Milt & Joan Mann/Cameramann International, Ltd. **367T** Joseph Nettis/Stock Boston **397** Milt & Joan Mann/Cameramann International, Ltd. **407** Brent Jones **410** Reprinted by permission of United Feature Syndicate, Inc. **418** Hans and Judy Beste/ANIMALS ANIMALS **420** AP/Wide World **422** Bob Daemmrich **423** Mickey Gibson/ANIMALS ANIMALS **436** Brent Jones **441** Eric Heyer from Grant Heilman Photography **451** National Museum of Natural History, Smithsonian Institution **453** M. Austerman/ANIMALS ANIMALS **455** Hank Morgan/Science Source/Photo Researchers **455INS** Custom Medical Stock Photo. All Rights Reserved. **459** Courtesy MAACO Auto Painting and Bodyworks **460** Head/ANIMALS ANIMALS **471** Walter Sanders, Courtesy Stadtische Museum, Brunswick, Germany **475** Milt & Joan Mann/Cameramann International, Ltd. **479** From THE GREAT ANCESTOR HUNT by Lila Perl, Drawings copyright © 1989 Ericka Weihs. Text copyright © 1989 by Lila Perl. Clarion Books, a Houghton Mifflin Company imprint. Used with permission. **486** The Image Works **488** Douglas Faulkner/Photo Researchers **489** Bob Daemmrich/The Image Works **491** Milt & Joan Mann/Cameramann International, Ltd. **498** Milt & Joan Mann/Cameramann International, Ltd. **502** Tony Freeman/Photo Edit **513** Robert Caputo **523** Courtesy Lear Inc. **525** Bob Daemmrich **527** Tony Freeman/Photo Edit **529** Tony Freeman/Photo Edit **543** Robert Frerck/Odyssey Productions, Chicago **544** Courtesy of The Newberry Library, Chicago **549** Phil Borden/Photo Edit **555** Focus On Sports **563** The Granger Collection, New York **568** Larry Lefever/Grant Heilman Photography **573** Milt & Joan Mann/Cameramann International, Ltd. **579** The Granger Collection, New York **582** Tony Freeman/Photo Edit **591** Focus On Sports **598** Grant Heilman Photography **604** David R. Frazier Photolibrary **606** Brent Jones **612** Jon Feingersh/The Stock Market **625** Milt & Joan Mann/Cameramann International, Ltd. **630** Brent Jones **637** Courtesy Zenith. Photo: Milt & Joan Mann/Cameramann International, Ltd. **646** Milt & Joan Mann/Cameramann International, Ltd. **648** Courtesy University of California, Berkeley **654** From THE HISTORY OF STATISTICS by Stephen M. Stigler. Copyright © 1986 by the President and Fellows of Harvard College, Belknap Press of Harvard University Press, Cambridge, MA and London, England 1986. **667** AP/Wide World **668** Dawn Murray **676** Walter Chandoha **690** Courtesy Computervision Division: Prime Computers, Inc. Photo: Milt & Joan Mann/Cameramann International, Ltd. **693** Bob Daemmrich **699** Custom Medical Stock Photo. All Rights Reserved. **709** Courtesy Evans and Sutherland **714** TM & (c) Lucasfilm Ltd. (LFL)1981. All Rights Reserved. COURTESY OF LUCASFILM LTD. **718** Focus On Sports **727** The Bettmann Archive **739** NASA **741** Zedenek Burian from LIFE BEFORE MAN **750** The Granger Collection, New York **765** Brent Jones **767** The Granger Collection, New York **775** NASA **779L** Walter Chandoha **779C** Walter Chandoha **779R** Walter Chandoha **788** David Pollack/The Stock Market **808** Aaron Haupt/David R. Frazier Photolibrary **822** Hank Morgan **830** Sidney Harris